WORLD *of* PHYSICS

ISSN 1531-0809

WORLD *of* PHYSICS

Kimberley A. McGrath, *Editor*

Volume 1

A-L

GALE GROUP

Detroit
New York
San Francisco
London
Boston
Woodbridge, CT

STAFF

Kimberley A. McGrath, *Senior Editor*

Christine Jeryan, *Contributing Managing Editor*

Stacey Blachford, *Coordinating Assistant Editor (Illustrations)*

Zoran Minderovic and Ellen Thackery, *Associate Editors*

Melissa C. McDade, *Assistant Editor*

Mark Springer, *Technical Specialist*

Margaret A. Chamberlain, *Permissions Specialist*
Shalice Shah-Caldwell, *Permissions Associate*

Mary Beth Trimper, *Manager, Composition and Electronic Prepress*
Evi Seoud, *Assistant Manager, Composition, Purchasing, and Electronic Prepress*
Dorothy Maki, *Manufacturing Manager*
Rhonda Williams, *Buyer*

Kenn Zorn, *Design Manager*
Michelle DiMercurio, *Art Director*
Michael Logusz, *Graphic Artist*

Barbara J. Yarrow, *Manager, Imaging and Multimedia Content*
Robyn V. Young, *Project Manager, Imaging and Multimedia Content*
Dean Dauphinais, *Senior Editor, Imaging and Multimedia Content*
Kelly A. Quin, *Editor, Imaging and Multimedia Content*
Leitha Etheridge-Sims, Mary K. Grimes, and David G. Oblender, *Image Catalogers*
Pam A. Reed, *Imaging Coordinator*
Randy Bassett, *Imaging Supervisor*
Robert Duncan, *Senior Imaging Specialist*
Dan Newell, *Imaging Specialist*
Christine O'Bryan, *Graphic Specialist*

Indexing provided by Synapse, the Knowledge Link Corporation.
Illustrations created by Argosy, West Newton, Massachusetts.

ISBN: 0-7876-3651-7 (set)
    0-7876-5052-8 (volume 1)
    0-7876-5053-6 (volume 2)

ISSN: 1531-0809

Printed in the United States of America
10 9 8 7 6 5 4 3 2 1

Contents

Welcome to the *World of Physics*, a volume devoted to the physical world around you. Physics is the study of the basic laws of nature, their application to the real world, and the relationships among these laws. The *World of Physics* is a collection of interesting and useful entries—more than 800 in number—both up-to-date and historical. They range from short definitions to concise but detailed articles, plus biographies of many of the most important physicists of the past and present, all in everyday language (the most important terms in the jargon used by professional physicists are carefully explained).

The articles in the book are meant to be understandable by anyone with curiosity about the laws of nature. Cross-references to related articles, definitions, and biographies in this collection are indicated by **bold-faced type**, and these cross-references will help explain and expand the individual entries.

The term physics originally meant natural science in general. Currently, physics means the group of sciences pertaining to matter and energy. Chemistry (which deals with different forms of matter), biology (having to do with living matter), engineering (applied physics, including mechanics and electromagnetism), astronomy (the science of extraterrestrial objects), and geophysics (the science of the earth) are taught in separate departments in most universities, yet these disciplines have large overlaps with physics itself. So does mathematics, often described as the language of physics, and computer sciences, of vital importance in determining the implications of physical laws. I like to define physics as the study of the laws of nature, including both the basic laws and those derived from them, as well as the logical relationships among these laws and—most important—the experimental underpinning and elaboration of these laws.

There are many ways to look at the subject of physics. In one sense it is beautiful, an elegant and even artistic way to look at nature. In a second sense, physics provides the tools to manipulate nature; it is the basis of engineering and chemistry and can help you cope with even very complex natural situations. Finally, physics can be considered the "blue print" or plan of the complicated ways in which natural phenomena act, providing a schematic diagram of a complex clockwork. I hope that as you read this book, you will keep all three of these perspectives in mind: Think of the laws of physics as beautiful, elegant, an organic part of the real world, concepts to be enjoyed and cherished. Also, look on physics as if you wanted to become a magician or sorcerer, with the ability to bend nature, or parts of it, to your will. Finally, regard physics from a mechanistic point of view, in which the physical laws of nature codify what you see around you.

Physicists are divided into experimentalists and theorists more than any other group of scientists. I am a theorist, for example, and my main interest is in the mathematical expression of physical laws, particularly the laws of spacetime, general relativity. Yet I have participated in experimental programs and have helped design and run our teaching laboratories here at The University of Texas. Although few physicists work professionally as both theorists and experimentalists, few physics departments would want to be divided into theoretical and experimental physics. Those of us who are theorists are motivated by the thought that even the most abstract mathematical laws help explain the real world. Physicists who investigate real world phenomena are motivated by the idea that behind their observations and tests lie elegant and universal truths. This set, the *World of Physics*, contains articles on both theory and experiment, with selected applications to chemistry, biology, engineering, astronomy, and geophysics. Here, too, are articles on the mathematical and computational techniques used in physics.

It is not necessary to have a background in higher mathematics to read this book. Certainly, mathematics is used by physicists everywhere—and I do mean everywhere. Geometry was the mathematics of choice in the sixteenth century, and calculus was developed in the seventeenth century in large part because of its usefulness in physics. The methods of variational calculus became important in the eighteenth and nineteenth centuries, when complex numbers and complex calculus were also found to be important for physics. The twentieth century saw even more abstract mathematics being used: differential geometry, knot theory, path integration, to name a few. I throw out the names of these mathematical areas not to impress and certainly not to frighten. Rather, you should realize that the logical way of thinking, which is entailed in mathematics—theorems proved from other theorems in a rigorous way—is important to the understanding of physics: Remember that physics involves the laws of nature and the relationships among these laws. As you read this book, please follow the many cross-references, for they will help you understand that physics, far from being a hodgepodge of observations and facts is a unified and carefully constructed discipline.

Physics has progressed over the last 400 years at a pace that has continually increased. The scientific revolution of the sixteenth and seventeenth centuries led to an understanding that the real world is comprehensible and that all of the universe obeys rules which may be tested by experiments. The eighteenth century, in physics, was dominated by Newtonian ideas and Newtonian gravity, yet inklings of what was to come were evident in the growing science of electricity. In the nineteenth century, electricity and magnetism came into their own, and were united into the field of electromagnetism. The physics of many-particle systems was also developed. The application of physical principles led to the telegraph and then the telephone, to advances in manufacturing, to revolutions in transportation. And again, the seeds of the enormous flowering of twentieth century physics were sown, when light was recognized as an electromagnetic phenomenon and x rays, the electron, and radioactivity were discovered.

No introduction of mine can possibly do justice to twentieth-century physics. In fact, most of this book is devoted to the tremendous strides that have been made in mankind's understanding of the physical world in the past 100 years. As the twenty-first century begins, look for these advances in physics: at the atomic level, experiments in the manipulation of single atoms and molecules and collections of a small number of particles will revolutionize both chemistry and timekeeping: at the sub-atomic level (which corresponds to the high energy found in the largest particle accelerators), the unification of strong and weak nuclear forces with electromagnetism and even gravity will become a reality, both in theory and in experimental tests. The physics of low temperatures and the solid state, which led to the transistor and to modern computing, will reach unprecedented levels of precision and will quickly lead to the science of nanostructures. The field of non-linear physics and chaos will become more and more the language of pattern formation in the real world. Relativity, that is general relativity, will lead to an almost complete understanding of how the matter in the universe evolved from a hot, almost unimaginable beginning to the structure now seen in stars and galaxies. Yet, these predictions are easy; even a decade from now they will seem naïve.

Of course, problems remain to be solved. The production of economical and safe electric power by fusion has remained elusive; perhaps the advances in fusion physics in the next decades will find a solution. The application of quantum theo-ry to relativity has never been achieved; perhaps string theory will show that twentieth century notions of spacetime structure must be superceded by a more fundamental theory, governed more by quantum ideas than by geometrical ones. Even the very basis of quantum theory itself, as one might apply it to the universe as a whole, remains a bone of contention. Again, these problems may or may not be solved, perhaps in a decade or two they will seem irrelevant because a new understanding of physical laws will show that they may be safely bypassed.

And there are other problems, whose roots lie in physics. Nuclear war and nuclear terrorism remain political and military problems, which one may hope will be solved in this new century. The growth of the availability of information, the erosion of privacy, the difficulty of teaching and learning science are due in great part to the successes of physics. Who knows what role physics will play in resolving these problems?

I close with a few personal remarks. When I was a lad, I served a term as a flutist in our high school orchestra (which played *H.M.S. Pinafore*, among other works). It was there that I, like many physics students, learned to appreciate the arts, music, and by extension, literature. It was in high school and in college where I was exposed to the beauty of mathematics; again, many physics students are as enamored as I was by mathematical puzzles and ideas. And yet, I wanted to understand the world around me, not just the achievements of man, not just the worlds created by the mathematicians and the artists, but the real world. Physics is for me the way to understand nature at the most fundamental level.

I think you will find, on reading this book, that the many sides of physics will help you, too, understand the world. Choose those topics that seem most intriguing or exciting, or perhaps choose the biography of a physicist whom you admire. Follow the cross-references. Soon you will see as I do that physics is the right way to view nature (the organic approach), that physics will help you control and use physical objects (the magical approach), and that much of the world becomes understandable in an elegant way (the mechanistic approach). And yet you will also find that much about the real world remains to be understood. Maybe, you will become a physicist, also!

Lawrence C. Shepley
Associate Professor, Retired
Physics Department
The University of Texas at Austin

How to Use the Book

This first edition of *World of Physics* has been designed with ready reference in mind:

- **Entries are arranged alphabetically**, rather than by chronology or scientific field.
- **Bold-faced terms** direct reader to related entries.
- **"See also" references** at the end of entries alert the reader to related entries not specifically mentioned in the body of the text.
- A **Sources Consulted** section lists the most worthwhile print material we encountered in the compilation of this set. It is there for the inspired reader who wants more

information on the people and discoveries covered in this edition.

- The **Historical Chronology** includes more than 300 important events in the biological sciences spanning the period from 900 B.C. through 2000.
- A **comprehensive General Index** guides the reader to all topics and persons mentioned in the book. Bolded page references refer the reader to the term's full entry.

Advisory Board

In compiling this edition, we have been fortunate in being able to call upon the following people—our panel of advisors—who contributed to the accuracy of the information in this premiere edition of *World of Physics*, and to them we would like to express sincere appreciation:

Russell J. Clark
Research Physicist
Carnegie Mellon University
Pittsburgh, Pennsylvania

Leonard Cohen
Professor Emeritus
Drexel University
Philadelphia, Pennsylvania

K. Lee Lerner
Professor Fellow
Science Research & Policy Institute
Washington, D.C./London, U.K. and
Advanced Physics & Chemistry
Shaw School
Mobile, Alabama

Steven Rea
Physics Department
Plymouth Canton High School
Canton, Michigan

Lawrence C. Shepley
Associate Professor, Retired
Physics Department
The University of Texas at Austin
Austin, Texas

The editor would like to add a special thank you to Adrienne and Lee Wilmouth Lerner for their research and to Brenda Wilmouth Lerner for her copyediting assistance. Thanks are due also, to advisor K. Lee Lerner, who wrote, revised, and copyedited multitudinous entries with competence and unflappable enthusiasm under an unyielding production schedule.

Cover

The image on the cover depicts an example of the de Broglie wavelength.

A

ABSOLUTE ZERO

Absolute zero is the lowest **temperature** theoretically possible. At absolute zero atoms have the minimum amount of vibration (i.e., **motion**) possible. Although this state cannot be achieved physically, the temperature has been extrapolated from experimental data beginning with observations correlating decreasing **pressure** and temperature in gases. This temperature is -459.69°F, or 0K on the Kelvin scale (-273.16°C, 0°R on the Rakine scale).

Heat is actually the motion of atoms. The hotter something seems to be, the faster its atoms are moving, vibrating back and forth. If the vibrations become frenzied enough, phase transitions can occur, such as when **boiling** water changes to steam. Conversely, when vibrations slow, cooling takes place, as when water freezes to ice. Just as the **velocity** of objects can approach but never reach the speed of **light**, however, neither can atoms be absolutely motionless. At this minimum vibratory motion, no work can be done by the atoms on their surrounding environment. Accordingly, atoms at this minimum vibratory level have zero heat and the temperature measuring this heat is at a minimum on whatever scale is used.

The Heisenberg Uncertainty Principle bars absolute zero from being attained. This inviolable law of **quantum mechanics** states that either the **momentum** or the position of particles can be known, but never both simultaneously. Once one of these values is established, it becomes impossible to establish the other. Any **atom** at absolute zero would violate this immutable law, as both its momentum and its position would be known.

Even in the **vacuum** of deep **space**, temperatures measure approximately 3K due to **background radiation** left over from the **big bang**. Scientists can do better in the laboratory, but even there absolute zero is an unobtainable goal. Scientists can use various methods to slow the motion of atoms and thus cool them to temperatures very near absolute zero. For example, in 1995, Eric Cornell led a group of researchers at the Joint Institute of Laboratory Astrophysics in Boulder, Colorado, in creating a Bose-Einstein condensate, a previously theoretical state of **matter** existing at a few billionths of a degree above absolute zero. In February, 1999, a group at Harvard University, utilizing a Bose-Einstein condensate and laser cooling techniques, achieved a temperature of 50 billionths of a degree above absolute zero and, in the process, slowed the **speed of light** from its speed in a vacuum (186,000 miles per second) to an astonishingly low 38 mph (61.15 km/h).

Although absolute zero remains out of reach, scientists continue to look for new ways to reduce the temperature of atoms. Continuing research into Bose-Einstein condensates as well as other esoteric low-temperature **states of matter** such as Fermi degenerate gases promises to lead to fresh discoveries whose applications will be as important for industry as for science. The booming field of superconductivity, for example, began with experiments conducted at temperatures near absolute zero, where certain materials acquired the ability to carry an **electric charge** with no loss of **energy**.

See also Heat; Temperature; Temperature and its measurement; Thermodynamics

ACCELERATED REFERENCE FRAMES

The coordinate system to which a particular observer refers his or her measurements is called a reference frame. In general, there are two kinds of **reference frames**: inertial and accelerating. Suppose a person is in a small closed room on Earth and drops a ball. The ball will fall to the floor due to the **force** of **gravity** pulling it down. This is considered an inertial **frame of reference** in a uniform **gravitational field**, meaning the coordinate system is stationary, or, at constant **velocity**. Now suppose this room is instead placed on a spaceship with no gravitational field, but the ship is uniformly accelerating so that an apparent force of one "g" is pushing the person to the floor. If a ball is dropped here in this accelerated reference frame, the ball will seem to fall in exactly the same manner as the inertial frame described on Earth. In reality, the ball is stationary, since there is no gravity and the floor rushes up to hit the ball. The floor's movement upward causes the apparent force that would be mistakenly called gravity.

Einstein's **general relativity** states that both situations are identical. In fact, general relativity claims you cannot tell the difference between the two reference frames; a uniformly accelerated reference frame is equivalent to an inertial reference frame in a uniform gravitational field. This is known as the equivalence principle.

Perhaps the most interesting example of an accelerated reference frame is an a stronaut in a "weightless" environment. If an astronaut drops a ball in this example, the ball will "float" in place. In reality, the astronaut and the ball are both falling towards the Earth and so the ball appears stationary. Being in orbit around Earth (or some other massive object with sufficient gravity) creates this **microgravity** environment. Other examples of accelerating reference frames and the effects associated with them are the **Coriolis force**, artificial gravity, and centripetal forces.

ACCELERATION

The term acceleration, used in physics, is a vector quantity. This means that acceleration contains both a number which is called its magnitude and a specific direction. An object is said to be accelerating if its rate of change of **velocity** is increasing or decreasing over a period of **time** and/or if its direction of **motion** is changing. The units for acceleration include a distance unit and two time units. Examples are m/s^2 and mi/hr/s. Sir **Isaac Newton** in his Second Law of Motion defined acceleration as the ratio of an unbalanced **force** acting on an object to the **mass** of the object.

The study of motion by **Galileo** Galilei in the late sixteenth and early seventeenth centuries and by Sir Isaac Newton in the mid-seventeenth century was one of the major cornerstones of modern Western experimental science. Over a period of 20 years, Galileo observed the motions of objects rolling down various inclines and attempted to time these events. He discovered that the distance an object traveled was proportional to the square of the time that it was in motion. From these experiments came the first correct concept of accelerated motion. Newton wanted to know why acceleration occurred. In order to produce a model that would help explain how the known **universe** of the seventeenth century worked, Newton had to give to science and physics the concept of a force which was mostly unknown at that time. With his Second Law of Motion, he clearly demonstrated that acceleration is caused by an unbalanced force (commonly called a push or a pull) acting on an object. What we call **gravity**, Newton showed was nothing more than a special type of acceleration. The interaction of the acceleration of gravity on the mass of our body produces the force which is called **weight**. A general definition of mass is that it refers to the quantity of **matter** in a body.

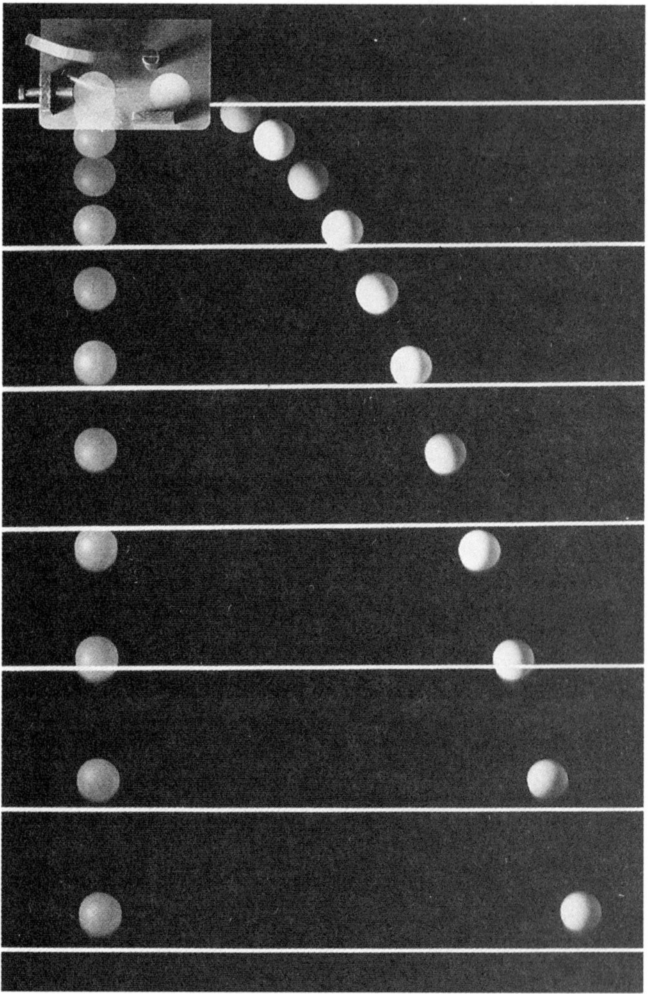

Acceleration of projected and dropped balls. *(© 1990 Richard Megna. Fundamental Photographs, New York. Reproduced by permission.)*

An object that is moving in a straight line is accelerating if its velocity (sometimes incorrectly referred to as speed) is increasing or decreasing during a given period of time. Acceleration (a) can be either positive or negative depending on whether the velocity is increasing (+a) or decreasing (-a). An automobile's motion can help explain linear acceleration. The speedometer measures the velocity. If the auto starts from rest and accelerates to 60 mph in 10 seconds, what is the acceleration? The auto's velocity changed 60 mph in 10 seconds. Therefore, its acceleration is 60 mph/10 s = + six mi/hr/s. That means its acceleration changed six miles per hour every second it was moving. Notice there are one distance unit and two time units in the answer. If the auto had started at 60 mph and then stopped in 10 seconds after the brakes were applied, the acceleration would be = -6 mi/hr/s. If this automobile changes direction while moving at this constant acceleration, it will have a different acceleration because the new vector will be different from the original vector. The mathematics of **vectors** are quite complex.

In circular motion, the velocity may remain constant but the direction of motion will change. If our automobile is going down the road at a constant 60 mph and it goes around a curve in the road, the auto undergoes acceleration because its direction is constantly changing while it is in the curve. Roller coasters and other amusement park rides produce rapid changes in acceleration (sometimes called centripetal acceleration) which will cause such effects as "g" forces, "weightlessness" and other real or imaginary forces to act on the body, causing dramatic experiences to occur. The astronauts experience as much as 7 "gs" during lift-off of the **space** shuttle but once in orbit it appears that they have lost all their weight. The concept of "weightlessness" in space is a highly misunderstood phenomena. It is not caused by the fact that the shuttle is so far from the Earth; it is produced because the space shuttle is in free fall under the influence of gravity. The shuttle is traveling 17,400 mph (28,014 kph) around the Earth and it is continually falling toward the Earth, but the Earth falls away from the shuttle at exactly the same rate.

Before the time of Sir Isaac Newton, the concept of force was unknown. Newton's Second Law was a simple equation and an insight that significantly affected physics in the seventeenth century as well as today. In the Second Law, given any object of mass (m), the acceleration (a) given to that object is directly proportional to the net force (F) acting on the object and inversely proportional to the mass of the object. Symbolically, this means a = F/m or in its more familiar form F = ma. In order for acceleration to occur, a net force must act on an object.

ACCELERATORS

The term accelerators most commonly refers to **particle accelerators**, devices for increasing the **velocity** of **subatomic particles** such as protons, electrons, and positrons. Particle accelerators were originally invented for the purpose of studying the basic structure of **matter**, although they later found a number of practical applications. Particle accelerators can be subdivided into two large sub-groups: linear and circular accelerators. Machines of the first type accelerate particles as they travel in a short line, sometimes over very great distances. Circular accelerators move particles along a circular or spiral path in machines that vary in size from less than a few feet to many miles in diameter.

The simplest particle accelerator was invented by Alabama-born physicist Robert Jemison Van de Graaff (1901-1967) in about 1929. The machine that now bears his name illustrates the fundamental principles on which all particle accelerators are based.

In the Van de Graaff accelerator, a silk conveyor belt collects positive charges from a high-voltage source at one end of the belt and transfers those charges to the outside of a hollow dome at the other end of the belt located at the top of the machine. The original Van de Graaff accelerator operated at a potential difference of 80,000 volts, although later improvements raised that value to 5,000,000 volts.

The Van de Graaff accelerator can be converted to a particle accelerator by attaching a source of positively charged **ions**, such as protons or He⁺ions, to the hollow dome. These ions feel an increasingly strong **force** of repulsion as positive charges accumulate on the dome. At some point, the ions are released from their source, and they travel away from the dome with high **energy** and at high velocities. If this beam of rapidly-moving particles is directed at a target, the ions of which it consists may collide with atoms in the target and break them apart. An analysis of ion-atom collisions such as these can provide a great deal of information about the structure of the target atoms, about the ion "bullets" and about the nature of matter in general.

Linear Accelerators

In a Van de Graaff generator, the velocity of an electrically charged particle is increased by exposing that particle to an electric field. The velocity of a **proton**, for example, may go from zero to 100,000 miles per second (160,000 km per second) as the particle feels a strong force of repulsion from the positive charge on the generator dome. Linear accelerators (linacs) operate on the same general principle except that a particle is exposed to a series of electrical fields, each of which increases the velocity of the particle.

A typical linac consists of a few hundred or a few thousand cylindrical metal tubes arranged one in front of another. The tubes are electrically charged so that each carries a charge opposite that of the tube on either side of it. Tubes 1, 3, 5, 7, 9, etc., might, for example, be charged positively, and tubes 2, 4, 6, 7, 10, etc., charged negatively.

Imagine that a negatively charged **electron** is introduced into a linac just in front of the first tube. In the circumstances described above, the electron is attracted by and accelerated toward the first tube. The electron passes toward and then into that tube. Once inside the tube, the electron no longer feels any force of attraction or repulsion and merely drifts through the tube until it reaches the opposite end. It is because of this behavior that the cylindrical tubes in a linac are generally referred to as drift tubes.

At the moment that the electron leaves the first drift tube, the charge on all drift tubes is reversed. Plates 1, 3, 5, 7, 9, etc. are now negatively charged, and plates 2, 4, 6, 8, 10, etc. are positively charged. The electron exiting the first tube now finds itself repelled by the tube it has just left and attracted to the second tube. These forces of attraction and repulsion provide a kind of "kick" that accelerates the electron in a forward direction. It passes through the **space** between tubes 1 and 2 and into tube 2. Once again, the electron drifts through this tube until it exits at the opposite end.

The electrical charge on all drift tubes reverses, and the electron is repelled by the second tube and attracted to the third tube. The added energy it receives is manifested in a greater velocity. As a result, the electron is moving faster in the third tube than in the second and can cover a greater distance in the same amount of **time**. To make sure that the electron exits a tube at just the right moment, the tubes must be of different lengths. Each one is slightly longer than the one before it.

The largest linac in the world is the Stanford Linear Accelerator, located at the Stanford Linear Accelerator Center (SLAC) in Stanford, California. An underground tunnel 2 mi (3 km) in length passes beneath U.S. highway 101 and holds 82,650 drift tubes along with the magnetic, electrical, and auxiliary equipment needed for the machine's operation. Electrons accelerated in the SLAC linac leave the end of the machine traveling at nearly the speed of **light** with a maximum energy of about 32 GeV (gigaelectron volts).

The term electron volt (ev) is the standard unit of energy measurement in accelerators. It is defined as the energy lost or gained by an electron as it passes through a potential difference of one volt. Most accelerators operate in the megaelectron volt (million electron volt; MeV), gigaelectron volt (billion electron volt; GeV), or teraelectron volt (trillion electron volt; TeV) range.

Circular Accelerators

The development of linear accelerators is limited by some obvious physical constraints. For example, the SLAC linac is so long that engineers had to take into consideration the Earth's curvature when they laid out the drift tube sequence. One way of avoiding the problems associated with the construction of a linac is to accelerate particles in a circle. Machines that operate on this principle are known, in general, as circular accelerators.

The earliest circular accelerator, the **cyclotron**, was invented by University of California professor of physics Ernest Orlando Lawrence in the early 1930s. Lawrence's cyclotron added to the design of the linac one new fundamental principle from physics: a charged particle that passes through a magnetic field travels in a curved path. The shape of the curved path depends on the velocity of the particle and the strength of the magnetic field.

The cyclotron consists of two hollow metal containers that look as if a tuna fish can had been cut in half vertically. Each half resembles a uppercase letter D, so the two parts of the cyclotron are known as dees. At any one time, one dee in the cyclotron is charged positively and the other negatively. But the dees are connected to a source of alternating current so that the signs on both dees change back and forth many times per second.

The second major component of a cyclotron is a large magnet that is situated above and below the dees. The presence of the magnet means that any charged particles moving within the dees will travel not in straight paths, but in curves.

Imagine that an electron is introduced into the narrow space between the two dees. The electron is accelerated into one of the dees, the one carrying a positive charge. As it moves, however, the electron travels toward the dee in a curved path.

After a fraction of a second, the current in the dees changes signs. The electron is then repelled by the dee toward which it first moved, reverses direction, and heads toward the opposite dee with an increased velocity. Again, the electron's return path is curved because of the magnetic field surrounding the dees.

Just as a particle in a linac passes through one drift tube after another, always gaining energy, so does a particle in a cyclotron. As the particle gains energy, it picks up speed and spirals outward from the center of the machine. Eventually, the particle reaches the outer circumference of the machine, passes out through a window, and strikes a target.

Lawrence's original cyclotron was a modest piece of equipment—only 4.5 in (11 cm) in diameter—capable of accelerating protons to an energy of 80,000 electron volts (80 kiloelectron volts). It was assembled from coffee cans, sealing wax, and leftover laboratory equipment. The largest accelerators of this design ever built were the 86 in (218 cm) and 87 in (225 cm) cyclotrons at the Oak Ridge National Laboratory and the Nobel Institute in Stockholm, Sweden, respectively.

Cyclotron Modifications

At first, improvements in cyclotron design were directed at the construction of larger machines that could accelerate particles to greater velocities. Soon, however, a new problem arose. Physical laws state that nothing can travel faster than the **speed of light**. Thus, adding more and more energy to a particle will not make that particle's speed increase indefinitely. Instead, as the particle's velocity approaches the speed of light, additional energy supplied to it appears in the form of increased **mass**. A particle whose mass is constantly increasing, however, begins to travel in a path different from that of a particle with constant mass. The practical significance of this fact is that, as the velocity of particles in a cyclotron begins to approach the speed of light, those particles start to fall "out of synch" with the current change that drives them back and forth between dees.

Two different modifications-or a combination of the two-can be made in the basic cyclotron design to deal with this problem. One approach is to gradually change the rate at which the electrical field alternates between the dees. The goal here is to have the sign change occur at exactly the moment that particles have reached a certain point within the dees. As the particles speed up and gain **weight**, the rate at which electrical current alternates between the two dees slows down to "catch up" with the particles.

In the 1950s, a number of machines containing this design element were built in various countries. Those machines were known as **frequency** modulated (FM) cyclotrons, synchrocyclotrons, or, in the Soviet Union, phasotrons. The maximum particle energy attained with machines of this design ranged from about 100 MeV to about 1 GeV.

A second solution for the mass increase problem is to alter the magnetic field of the machine in such a way as to maintain precise control over the particles' paths. This principle has been incorporated into the machines that are now the most powerful cyclotrons in the world, the **synchrotrons**.

A synchrotron consists essentially of a hollow circular tube (the ring) through which particles are accelerated. The particles are actually accelerated to velocities close to the speed of light in smaller machines before they are injected into the main ring. Once they are within the main ring, particles receive additional jolts of energy from accelerating chambers placed at various locations around the ring. At other locations around the ring, very strong magnets control the path followed by the particles. As particles pick up energy and tend to spiral outward, the magnetic fields are increased, pushing particles back into a circular path. The most powerful synchrotrons now in operation can produce particles with energies of at least 400 GeV.

In the 1970s, nuclear physicists proposed the design and construction of the most powerful synchrotron of all, the superconducting super collider (SSC). The SSC was expected to have an accelerating ring 82.9 kilometers (51.5 miles) in circumference with the ability to produce particles having an energy of 20 TeV. Estimated cost of the SSC was originally set at about $4 billion. Shortly after construction of the machine at Waxahachie, Texas began, however, the United States Congress decided to discontinue funding for the project.

Applications

By far the most common use of particle accelerators is basic research on the composition of matter. The quantities of energy released in such machines are unmatched anywhere on Earth. At these energy levels, new forms of matter are produced that do not exist under ordinary conditions. These forms of matter provide clues about the ultimate structure of matter.

Accelerators have also found some important applications in medical and industrial settings. As particles travel through an accelerator, they give off a form of **radiation** known as synchrotron radiation. This form of radiation is somewhat similar to **x rays** and has been used for similar purposes.

ACOUSTICS

Acoustics is the science that deals with the production, transmission, and reception of **sound**. Sound may be produced when a material body vibrates; it is transmitted only when there is some material body, called the medium, that can carry the vibrations away from the producing body; it is received when a third material body, attached to some indicating device, is set into vibratory **motion** by that intervening medium. However, the only vibrations that are considered sound (or sonic vibrations) are those in which the medium vibrates in the same direction as the sound travels, and for which the vibrations are very small. When the rate of vibration is below the range of human **hearing**, the sound is termed infrasonic; when it is above that range, it is called ultrasonic. (The term supersonic refers to bodies moving at speeds greater than the **speed of sound**, and is not normally involved in the study of acoustics.)

Production of Sound

There are many examples of vibrating bodies producing sounds. Some are as simple as a string in a violin or piano, or a column of air in an organ pipe or in a clarinet; some are as complex as the vocal chords of a human. Sound may also be caused by a large disturbance which causes parts of a body to vibrate, such as sounds caused by a falling tree.

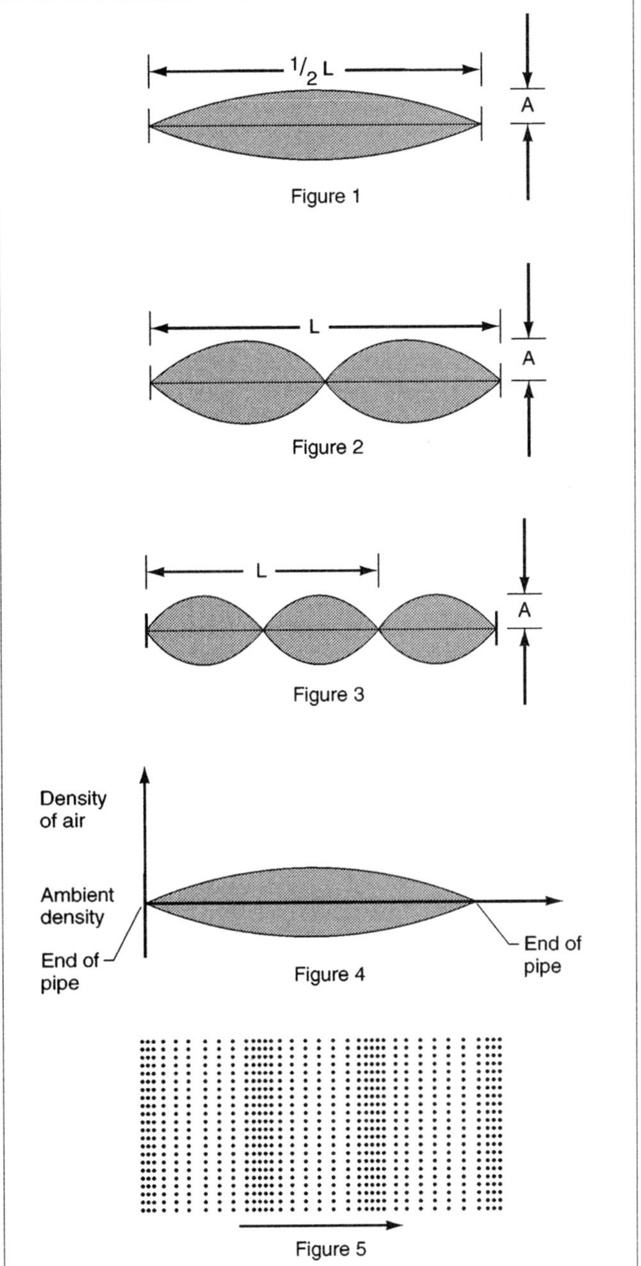

Acoustics, three figures illustrating acoustics.

Vibrations of a String

To understand some of the fundamentals of sound production and propagation it is instructive to first consider the small vibrations of a stretched string held at both ends under tension. While these vibrations are not an example of sound, they do illustrate many of the properties of importance in acoustics as well as in the production of sound. The string may vibrate in a variety of different ways, depending upon whether it is struck or rubbed to set it in motion, and where on the string the action took place. However, its motion can be analyzed into a combination of a large number of simple motions. The simplest, called the fundamental (or the first harmonic), appears in Figure 1, which shows the outermost extensions of the string carrying out this vibration.

The second harmonic is shown in Figure 2; the third harmonic in Figure 3; and so forth (the whole set of **harmonics** beyond the first are called the overtones). The rate at which these vibrations take place (number of times per second the motion is repeated) is called the **frequency**, denoted by f (the reciprocal of the frequency, which is the **time** for one cycle to be competed, is called the period). A single complete vibration is normally termed a cycle, so that the frequency is usually given in cycles per second, or the equivalent modern unit, the hertz (abbreviated Hz). It is characteristic of the stretched string that the second harmonic has a frequency twice that of the fundamental; the third harmonic has a frequency three times that of the fundamental; and so forth. This is true for only a few very simple systems, with most sound-producing systems having a far more complex relationship among the harmonics.

Those points on the string which do not move are called the **nodes**; the maximum extension of the string (from the horizontal in the Figures) is called the amplitude, and is denoted by A in Figures 1-3. The distance one must go along the string at any instant of time to reach a section having the identical motion is called the **wavelength**, and is denoted by L in Figures 1-3. It can be seen that the string only contains one-half wavelength of the fundamental, that is, the wavelength of the fundamental is twice the string length. The wavelength of the second harmonic is the length of the string. The string contains one-and-one-half (3/2) wavelengths of the third harmonic, so that its wavelength is two-thirds (2/3) of the length of the string. Similar relationships hold for all the other harmonics.

If the fundamental frequency of the string is called f_0, and the length of the string is l, it can be seen from the above that the product of the frequency and the wavelength of each harmonic is equal to $2f_0 l$. The dimension of this product is a **velocity** (e.g., feet per second or centimeters per second); detailed analysis of the motion of the stretched string shows that this is the velocity with which a small disturbance on the string would travel down the string.

Vibrations of an Air Column

When air is blown across the entrance to an organ pipe, it causes the air in the pipe to vibrate, so that there are alternate small increases and decreases of the **density** of the air (condensations and rarefactions). These alternate in **space**, with the distance between successive condensations (or rarefactions) being the wavelength; they alternate in time, with the frequency of the vibration. One major difference here is that the string vibrates transversely (perpendicular to the length of the string), while the air vibrates longitudinally (in the direction of the column of air). If the pipe is open at both ends, then the density of the air at the ends must be the same as that of the air outside the pipe, while the density inside the pipe can vary above or below that value. Again, as for the vibrations of the string, the density of the air in the pipe can be

analyzed into a fundamental and overtones. If the density of the air vibrating in the fundamental mode (of the open pipe) is plotted across the pipe length, the graph is as in Figure 4.

The "zero value" at the ends denotes the fact that the density at the ends of the pipe must be the same as outside the pipe (the ambient density), while inside the pipe the density varies above and below that value with the frequency of the fundamental, with a maximum (and minimum) at the center. The density plot for the fundamental looks just like that for the fundamental of the vibrating stretched string (Figure 1). In the same manner, plots of the density for the various overtones would look like those of the string overtones. The frequency of the fundamental can be calculated from the fact that the velocity, which is analogous to that found for vibrations of the string, is the velocity with which sound travels in the air, usually denoted by c. Since the wavelength of the fundamental is twice the pipe length, its frequency is $c/2\,l$, where l is the length of the organ pipe. (While the discussion here is in terms of the density variations in the air, these are accompanied by small variations in the air **pressure**, and small motions of the air itself. At places of increased density the pressure is increased; where the pressure is changing rapidly, the air motion is greatest.) When a musician blows into the mouthpiece of a clarinet, the air rushing past the reed causes it to vibrate which then causes the column of air in the clarinet to vibrate in a manner similar to, but more complicated than, the motion of the organ pipe. These vibrations (as for all vibrations) can also be analyzed into harmonics. By opening and closing the keyholes in the clarinet, different harmonics of the clarinet are made to grow louder or softer causing different tones to be heard.

Sound Production in General

Thus, the production of sound depends upon the vibration of a material body, with the vibration being transmitted to the medium that carries the sound away from the sound producer. The vibrating violin string, for example, causes the body of the violin to vibrate; the "back-and-forth" motion of the parts of the body of the violin causes the air in contact with it to vibrate. That is, small variations in the density of the air are produced by the motion of the violin body, and these are carried forth into the air surrounding the violin. As the sound is carried away, the small variations in air density are propagated in the direction of travel of the sound.

Sounds from humans, of course, are produced by forcing air across the vocal cords, which causes them to vibrate. The various overtones are enhanced or diminished by the size and shape of the various cavities in the head (the sinuses, for example), as well as the placement of the tongue and the shape of the mouth. These factors cause specific wavelengths, of all that are produced by the vocal cords, to be amplified differently so that different people have their own characteristic voice sounds. These sounds can be then controlled by changing the placement of the tongue and the shape of the mouth, producing speech. The frequencies usually involved in speech are from about 100 to 10,000 Hz. However, humans can hear sounds in the frequency range from about 20 to 18,000 Hz.

These outer limits vary from person to person, with age, and with the loudness of the sound. The density variations (and corresponding pressure variations) produced in ordinary speech are extremely small, with ordinary speech producing less than one-millionth the **power** of a 100 watt light bulb! In the sonic range of frequencies (those produced by humans), sounds are often produced by loudspeakers, devices using electronic and mechanical components to produce sounds. The sounds to be transmitted are first changed to electrical signals by a microphone (see Reception of sounds, below), for example, or from an audio tape or compact disc; the frequencies carried by the electrical signals are those to be produced as the sound signals. In the simplest case, the wires carrying the electrical signals are used to form an electromagnet which attracts and releases a metal diaphragm. This, in turn, causes the variations in the density in the air adjacent to the diaphragm. These variations in density will have the same frequencies as were in the original electrical signals.

Ultrasonic vibrations are of great importance in industry and medicine, as well as in investigations in pure science. They are usually produced by applying an alternating electric **voltage** across certain types of crystals (quartz is a typical one) that expand and contract slightly as the voltage varies; the frequency of the voltage then determines the frequency of the sounds produced.

Transmission of Sound

In order for sound to travel between the source and the receiver there must be some material between them that can vibrate in the direction of travel (called the propagation direction). (The fact that sound can only be transmitted by a material medium means that an explosion outside a spaceship would not be heard by its occupants!) The motion of the sound-producing body causes density variations in the medium (see Figure 5, which schematically shows the density variations associated with a sound wave), which move along in the direction of propagation. The transmission of sounds in the form of these density variations is termed a wave since these variations are carried forward without significant change, although eventually **friction** in the air itself causes the wave to dissipate. (This is analogous to a water wave in which the particles of water vibrate up and down, while the "wave" propagates forward.) Since the motion of the medium at any point is a small vibration back and forth in the direction in which the wave is proceeding, sound is termed a longitudinal wave. (The water wave, like the violin string, is an example of a transverse wave.) The most usual medium of sound transmission is air, but any substance that can be compressed can act as a medium for sound propagation. A fundamental characteristic of a wave is that it carries **energy** and **momentum** away from a source without transporting **matter** from the source.

Since the speed of sound in air is about 331 meters per second (about 1,088 feet per second), human speech involves wavelengths from about 3.3 meters to 3.3 centimeters (from about 11 feet to 1.3 inches). Thus, the wavelengths of speech are of the size of ordinary objects, unlike light, whose wavelengths are extremely small compared to items that are part of everyday life. Because of this, sound does not ordinarily cast "acoustic shadows" but, because its wavelengths are so large, can be transmitted around ordinary objects. For example, if a light is shining on a person, and a book is placed directly between them, the person will no longer be able to see the light (a shadow is cast by the book on the eyes of the observer). However, if one person is speaking to another, then placing a book between them will hardly affect the sounds heard at all; the sound **waves** are able to go around the book to the observer's ears. On the other hand, placing a high wall between a highway and houses can greatly decrease the sounds of the traffic noises if the dimensions of the wall (height and length) are large compared with the wavelength of the traffic sounds. Thus, sound waves (as for all waves) tend to "go around" (e.g., ignore the presence of) obstacles which are small compared with the wavelength of the wave; and are reflected by obstacles which are large compared with the wavelength. For obstacles of approximately the same size as the wavelength, waves exhibit a very complex behavior known as **diffraction**, in which there are enhanced and diminished values of the wave amplitude, but which is too complicated to be described here in detail.

The speed of sound in a gas is proportional to the square root of the pressure divided by the density. Thus, **helium**, which has a much lower density than air, transmits sound at a greater speed than air. If a person breathes some helium, the characteristic wavelengths are still determined by the shape of the mouth, but the greater sound speed causes the speech to be emitted at a higher frequency-thus the "Donald Duck" sounds from someone who speaks after taking a breath of helium from a balloon.

In general, the speed of sound in liquids is greater than in gases, and greater still in solids. In sea water, for example, the speed is about 1,447 meters per second (about 4,750 feet per second); as for a gas, the speed increases as the pressure increases, and as the density decreases. Typical speeds of sound in solids are 5,000 meters per second, but vary considerably from one solid to another.

Reception of Sound

By far the most important sound receiver in use is the human ear. While the details of the workings of the ear are too complicated to be described here, the process consists of the sound wave (the density variations carried through the air) striking the eardrum, causing it to vibrate with the same set of frequencies as had been carried in the wave. This vibration is transmitted through a set of bones in the ear to a liquid-filled chamber; small hairs in the liquid are set into motion and excite nerves which then transmit these frequencies to the brain.

In the sonic range of frequencies, the microphone, a device using electrical and mechanical components, is the common method of receiving sounds. One simple form is to have a diaphragm as one plate of an electrical condenser. When the diaphragm vibrates under the action of a sound wave, the current in the circuit varies due to the varying **capacitance** of the condenser. This varying current can then be used to activate a meter or oscilloscope or, after suitable processing, make an audio tape or some such permanent record.

Applications

The applications of acoustical devices are far too numerous to describe; one only has to look around our homes to see some of them: telephones, radios and television sets, compact disc players and tape recorders; even clocks that "speak" the time! Probably one of the most important from the human point of view is the hearing aid, a miniature microphone-amplifier-loudspeaker that is designed to enhance whatever range of frequencies a person finds difficulty hearing.

However, one of the first large-scale "industrial" uses of sound propagation was by the military in World War I, in the detection of enemy submarines by means of **sonar** (for *so*und *n*avigation *a*nd *r*anging). This was further developed during the period between then and World War II, and since then. The ship hunting submarine has a sound source and receiver projecting from the ship's hull that can be used for either listening or in an echo-ranging mode; the source and receiver are directional, so that they can send and receive an acoustic signal from only a small range of directions at one time. In the listening mode of operation, the operator tries to determine what are the sources of any noise that might be heard: the regular beat of an engine heard underwater can tell that an enemy might be in the vicinity. In the echo-ranging mode, a series of short bursts of sound is sent out, and the time for the echo to return is noted; that time interval multiplied by the speed of sound in water indicates (twice) the distance to the reflecting object. Since the sound source is directional, the direction in which the object lies is also known. This is now such a well developed method of finding underwater objects that commercial versions are available for fishermen to hunt for schools of fish.

Ultrasonic sources, utilizing **pulses** of frequencies in the many millions of cycles per second (and higher!), are now used for inspecting metals for flaws. The small wavelengths make the pulses liable to reflection from any imperfections in a piece of metal. Castings may have internal cracks which will weaken the structure; welds may be imperfect, possibly leading to failure of a metal-to-metal joint; metal fatigue may produce cracks in areas impossible to inspect byeye. The use of ultrasonic inspection techniques is increasingly important for failure prevention in bridges, aircraft, and pipelines, to name just a few.

The use of ultrasonics in medicine is also of growing importance. The detection of kidney stones or gallstones is routine, as is the imaging of fetuses to detect suspected birth defects, cardiac imaging, blood flow measurements, and so forth.

Thus, the field of acoustics covers a vast array of different areas of use, and they are constantly expanding. Acoustics in the communications industry, in various phases of the construction industries, in oil field exploration, in medicine, in the military, and in the entertainment industry, all

attest to the growth of this field and to its continuing importance in the future.

ADDITION OF VELOCITIES (EINSTEIN LAW) • See Velocity

AHARONOV, YAKIR (1932-   )
Israeli theoretical condensed matter physicist

Yakir Aharonov is a theoretical condensed **matter** physicist studying nonlocal and topological effects in **quantum mechanics**, relativistic quantum field theories, and interpretations of **quantum mechanics**. Aharonov received his B.S. in 1956 from Technion University in Haifa, Israel. He then studied at Bristol University in England and received a Ph.D. degree in 1960. Dr. Aharonov then held a postdoctoral position for one-year at Brandeis University in Massachusetts before returning to Israel. There, he took an Assistant Professorship at Yeshiva University, becoming an Associate Professor in 1964 and then Professor in 1967. At that time, he accepted and held a joint appointment with Yeshiva University and Tel Aviv University until 1973. In 1973, he began a joint position at Tel Aviv University and the University of South Carolina, which he still holds as of 2000.

In 1998, Aharonov was a co-recipient of the Wolf Prize recognizing the discovery of the Aharonov-Bohm (AB) effect proposed in 1959. The AB effect predicts that electrons passing through field-free regions surrounding a region of **magnetic flux** will acquire different phases depending on whether they pass to the left or to the right of the flux tube. The phase difference, which can be measured in an **interference** experiment, depends on the flux enclosed. The AB effect has since been observed, and has become an experimental tool in the domain of mesoscopic physics.

Aharonov has also been recognized for this work by the 1995 Hewlett-Packard Europhysics Prize. Other honors he has received include the Weizmann Prize and Rothschild Prize in 1984, the Israel National Prize in Physics in 1989, and the Elliot Cresson Medal in 1991. Aharonov has also received a Miller Research Professorship Award at Berkeley, 1988-89; the Alex Maguy-Glass Chair in Theoretical Physics, Tel Aviv University; and a Chair in Theoretical Physics, University of South Carolina. Honorary Doctorates have been awarded to Aharonov from the Israel Institute of Technology, the University of South Carolina, Bristol University, and the University of Buenos Aires. Professor Aharonov was elected a Fellow of the American Physical Society in 1981, and a Member of the National Academy of Sciences in Israel in 1990, followed by the United States in 1993. He also received The Distinguished Scientist Governor Award of South Carolina in 1993.

See also Mesoscopic systems

AL HAZAN, IBN AL-HAITHAM (965-CA. 1040)

Iraqi physicist and mathematician

As Europe was nearing the end of the dark ages, science flourished in the Arab world. From Baghdad to Cairo, scholars studied the science of the ancient Greeks and expanded on it with new investigations. Ibn al-Haitham Al Hazan, often referred to as Alhazen or Alhazan, was one of the great Arab experimentalists and a pioneer of optical science.

Al Hazan, who was born in 965, was commonly called al-Basrī , indicating that he was from Basra, Iraq. He also was known as al-Misrī , meaning "of Egypt." Although his earliest studies were of a religious nature, he soon turned to mathematics, physics, philosophy, and the works of **Aristotle**. The details of Al Hazan's life are unclear. According to one source, the caliph al-Hākim, who had founded the "House of Science" in Cairo, summoned Al Hazan to Fātimid, Egypt, after hearing of his plan to control the periodic flooding of the Nile River. When Al Hazan's plan failed, al-Hākim gave him a post in a government office. But fearing for his life, Al Hazan feigned insanity until the caliph's death in 1021. Other sources, however, contend that Al Hazan was a government minister in Basra, who feigned madness so as to be relieved of his duties. According to these accounts, Al Hazan did not go to Egypt until after al-Hākim's death. In either case, he spent the remaining years of his life in Cairo, teaching, writing, and earning his living by copying scientific books, until his death about 1040.

Al Hazan's most important work, *Optics*, consisted of seven volumes of experiments, mathematics, and inductive reasoning, without reliance on previous authorities. **Euclid**, **Ptolemy**, and other ancient Greek scientists had believed that vision resulted from **light** rays emitted by the eye. Al Hazan originated the theory that vision was the result of illuminated rays reaching the eye. He believed that light rays emanated in straight lines, in a spherical direction, from every point of a luminous object. He studied the properties of various types of **lenses**, mirrors, and magnifying glasses and conducted major studies on refraction, the angle at which light is bent when passing from one medium to another. Latin translations of *Optics* influenced European scientists, such as Roger Bacon, **Johannes Kepler**, Pierre de Fermat, and **René Descartes**, from the end of the twelfth century into the seventeenth century.

Al Hazan addressed the "moon illusion," the ancient question of why the **sun** and **moon** appear larger near the horizon. He suggested that objects on the horizon influence our optical perception of the moon. Although the illusion holds even at sea where there are no objects on the horizon, he was correct in considering it to be a problem of visual perception, wherein the brain is unable to accurately interpret optical information about size and distance.

Al Hazan wrote at least 92 works on mathematics, physics, and metaphysics, as well as treatises on logic, politics, religion, ethics, poetry, and music. He wrote summaries of the works of Aristotle, the Roman physician Galen, Ptolemy, and Euclid. About 20 of Al Hazan's mathematical works are extant. One of these became known as "Al Hazan's problem." This is the mathematical problem of finding the point, on a surface of a given shape, that will reflect the light from a point opposite the surface to a second point opposite the surface.

At least 20 of Al Hazan's surviving works deal with **astronomy** and he authored an important commentary on the *Almagest*, the astronomical text written by Ptolemy in the second century. Al Hazan's most famous astronomical work, *On the Configuration of the World*, was translated into Spanish in the thirteenth century, and from Spanish into Latin. It also was translated into Hebrew and then into Latin and influenced the astronomers of the early Renaissance.

ALPHA DECAY

Alpha decay occurs when an atomic **nucleus** disintegrates by emitting alpha particles. An **alpha particle** consists of two protons and two neutrons; it is the nucleus of a **helium atom**. Alpha decay is most often observed in high- as opposed to low-mass nuclear decay. Elements of high atomic **mass** are relatively less stable than objects of lower atomic mass, and are more likely to undergo the process of alpha decay.

These high-mass elements can be compared to a top-heavy pick-up truck. If the truck is filled beyond a stable capacity, it may throw off some of its contents. In the same way, alpha particles are released by radioactive material as a means of achieving greater stability. When radioactive material or other unstable nuclei break down, there is a loss in the total mass of the products of this reaction. A loss of mass results in a corresponding amount of **energy** produced; therefore, energy is released with the emission of alpha particles. Part of this energy is seen in the form of **alpha radiation**, which occurs in the process of alpha decay. When alpha particles are emitted as a spontaneous nuclear reaction, the radioactive material transmutates toward a more stable **isotope** or element. Eventually the **matter** is stable and no longer radioactive. The stages it goes through to reach stability are known as **half-life** periods.

The principles of alpha decay are used in **radioactive dating**, in which half-lives play an important part. The half-life of radioactive matter is the time before half of any given amount of nuclei will break down through alpha decay. A half-life is a measurable, stable period of time for any given element. Scientists can compare the number of breakdowns in any given time and the intensity of the **radioactivity** among elements and isotopes. The number of half-lives the radioactive material has gone through gives a good estimate of the number of years the object has been in existence.

Another practical use of alpha decay is in smoke detectors. A smoke detector consists of two metal plates with a small space between them. These plates have wires connected to a battery and current monitor. One of the plates has a slight amount of radioactive material on it giving off alpha particles. Through these alpha particles a circuit is completed and the monitor is dormant. When smoke goes between the two metal plates, the alpha particles are deflected and the circuit is not completed. The current monitor is then activated and turns on the buzzer.

ALPHA PARTICLE

The alpha particle is emitted by certain radioactive elements as they decay to a stable element. It consists of two protons and two neutrons; it is positively charged. The element that undergoes "alpha decay" changes into a new element whose atomic number is down two and atomic **mass** is down four from the original element. **Alpha decay** occurs when a **nucleus** has so many protons that the strong nuclear **force** is unable to counterbalance the strong repulsion of the electrical force between the protons. Because of its mass, the alpha particle travels relatively slowly (less than 10% the speed of **light**), and it can be stopped by a thin sheet of aluminum foil.

When Henri Becquerel first discovered the property of **radioactivity** in 1896, he did not know that the **radiation** consisted of particles as well as **energy. Ernest Rutherford** began experimenting to determine the nature of this radiation in 1898. One experiment demonstrated that the radiation actually consisted of three different types: a positive particle called "alpha," a negative particle, and a form of electromagnetic radiation that carried high energy. By 1902 Rutherford and his colleague Frederick Soddy were proposing that a different **chemical element** is formed whenever a radioactive element decays, a process known as transmutation. Rutherford was awarded the Nobel Prize in chemistry in 1908 for discovering these basic principles of radioactivity. In Rutherford's classic "gold foil" experiment to determine the structure of an **atom**, his assistants **Hans Geiger** and E. Marsden used positively charged high speed alpha particles that were emitted from radioactive polonium to bombard a gold foil. The results of this experiment showed that an atom consisted of mostly empty **space**, with essentially all its mass concentrated in a very small, dense, positively charged center called the nucleus. In fact, the alpha particle itself was identified as the nucleus of the **helium** atom! Soddy, working with William Ramsay, who in 1895 had discovered that the element helium was a component of earth minerals, verified that helium is produced when radium is allowed to decay in a closed tube.

The first production in the laboratory of radioisotopes— as opposed to the naturally occurring radioactive isotopes-was achieved by bombarding stable isotopes with alpha particles. In 1919, Rutherford succeeded in producing oxygen-17 by bombarding ordinary nitrogen-14 with alpha particles; a **proton** was set free in the reaction. The 1935 Nobel Prize in chemistry was awarded to Irène and Frédéric Joliet-Curie for their **work** in producing **radioisotopes** through the bombardment of stable elements with alpha particles from polonium decay. After bombarding aluminum with alpha particles, they found that when the nucleus absorbed an alpha particle, it changed into a previously unknown radioisotope of phosphorus; an unidentified neutral particle was set free in the reaction. **James Chadwick** repeated their experiment using a beryl-

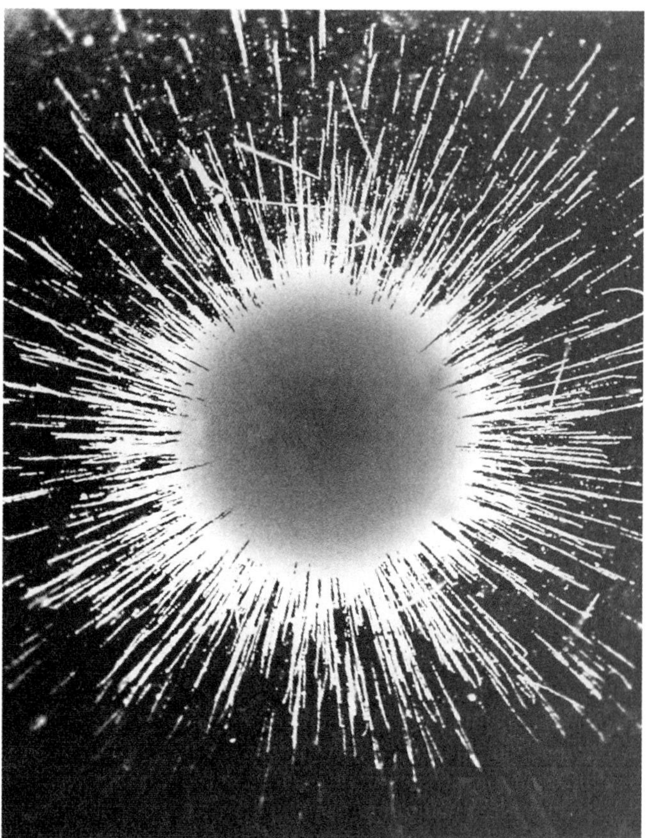

The radioactive emission of alpha particles from radium. Radium is the most powerful radioactive substance known. It emits radiation a million times more intensely than uranium. *(Photo by C. Powell, P. Fowler, and D. Perkins, Science Source/Photo Researchers. Reproduced by permission.)*

lium target, then captured the particle that was emitted and used principles of classical physics to identify its mass as same as that of the proton. This particle was named the **neutron**, and Chadwick received the 1935 Nobel Prize in physics for its discovery. Subsequently, scientists began accelerating alpha particles to very high energies in specially constructed devices like the **cyclotron** and linear accelerator before shooting them at the element they wanted to transmute.

A number of radioisotopes, both natural and man made, have been identified as alpha emitters. The decay chain of the most abundant uranium **isotope**, uranium-238, to stable lead-206 involves eight different alpha decay reactions. Uranium has a very long half life (4.5 billion years) and because it is present in the earth in easily measured amounts, the ratio of uranium to lead is used to estimate the age of our planet Earth. More recently, Guenther Lugmair at the University of California at San Diego introduced agedating using samarium-147, which undergoes alpha decay to produce neodymium-143.

Most of the transuranic radioactive elements undergo alpha decay. During the 1960s when an instrument was needed to analyze the lunar surface, Anthony Turkevich developed an alpha **scattering** instrument, using for the alpha source curium-242, a transuranic element produced by the alpha decay of americium-241. This instrument was used on the moon's surface by Surveyors 5, 6, and 7. Plutonium has been used to power more than twenty spacecraft since 1972; it is also being used to power cardiac pacemakers.

Ionization smoke detectors detect the presence of smoke using an ionization chamber and a source of **ionizing radiation** (americium-241). The alpha particles emitted by the **radioactive decay** of the americium 241 ionize the **oxygen** and nitrogen atoms present in the chamber, giving rise to free electrons and **ions**. When smoke is present in the chamber, the electrical current produced by the free electrons and ions (as they move toward positively and negatively charged plates in the detector) is neutralized by smoke particles; this results in a drop in electrical current and sets off an alarm.

Because of their low penetrability, alpha particles do not usually pose a threat to living organisms, unless they are ingested. This is the problem presented by the radioactive gas radon, which is formed through the natural radioactive decay of the uranium (present in rock, soil, and water throughout the world). As radon gas seeps up through the ground, it can become trapped in buildings, where it may build up to toxic levels. Radon can enter the body through the lungs, where it undergoes alpha decay to form polonium, a radioactive solid that remains in the lungs and continues to emit cancer-causing radiation.

ALPHA RADIATION

Alpha **radiation** is the stream of alpha particles emitted when radioactive materials disintegrate. When radioactive material breaks down, there is a loss in the total **mass** of the products. According to the law of conservation of mass-energy there must be a corresponding **energy** release, which is the energy associated with the emission of the **alpha particle**. An alpha particle is a heavy **nucleon**, which is actually the **nucleus** of a **helium atom**. They are most active in unstable elements of high atomic mass.

The stability of an atom relies on the balance of two forces, the **force** of repulsion among protons and the strong nuclear force. The number of protons in the atom influences the electromagnetic force by which the positively charged **proton** repels other (positively charged) protons. The strong nuclear force that holds the nucleus together is enhanced by the number of neutrons in the atom. There is a limit to the degree of influence that the number of neutrons has over the cohesion of the nucleus. When that limit is exceeded, the atom starts to lose particles. Because the alpha particle (the helium nucleus) with two protons and two neutrons is especially cohesive, the release of nuclear material is often in this form.

Alpha radiation is a result of **alpha decay** where a parent atom is profoundly changed through the release of alpha particles. With the release of each alpha particle, the atomic mass of the parent atom is reduced by 4 and the atomic number is reduced by 2. After a period of time, all atoms of a given amount of a radioactive element will reduce to a more stable **isotope** of that element or to a more stable element. This time period is measured by the time it takes for half of a given amount of atoms to be stabilized. One unit of this period is known as the **half-life** of the element. A half-life through alpha decay can be over five billion years (as with uranium) or as short as a fraction of a second (as with polonium).

The nature of alpha radiation was a mystery for a long time after its discovery. It is a fundamental part of the **radioactivity** first discovered by **Marie Curie**. Those studying radioactivity realized that the rays of radioactivity were not of the same composition. **Ernest Rutherford** was the one who first identified three types of rays as alpha, beta, and gamma. By introducing these rays to an **electromagnetic field**, Rutherford found that the beta rays were highly attracted to the cathodes and so must have been made up of negatively charged particles with low mass. Alpha rays were moderately deflected in a direction opposite to that of the beta rays. From these experiments, he was able to describe the alpha radiation as a stream of particles and not as a form of **light** as was previously believed. He discovered the alpha particles to be rapidly moving, positively charged particles with a high atomic mass. This mass was found to be nearly 7,300 times more massive than an **electron**.

One of the first practical uses of alpha radiation was as a nuclear probe. Long before artificial probes were available, subatomic researchers such as Rutherford relied on concentrated alpha radiation. Rutherford noted the rate of deflection of alpha particles shot at a gold foil. From this study, he developed his model of the atom as mostly composed of empty **space** but with a massive nucleus. Today, alpha radiation is used in smoke detectors. In these devices, alpha particles from radioactive material are allowed to complete an **electric current**.

ALPHA RAYS • See Alpha radiation

ALTERNATING CURRENT CIRCUITS • See
Electronic current

ALTERNATING CURRENT POWER TRANS-MISSION • See Electric current

ALTERNATING VOLTAGES AND CURRENTS
• See Electric current

ALVAREZ, LUIS (1911-1988)
Hispanic American physicist

Luis Alvarez's scientific contributions to the military during World War II included the development of a narrow beam **radar** system that allows airplanes to land in inclement **weather**. He was also involved in the **Manhattan Project** to develop the world's first **nuclear weapons**. One of Alvarez's more controversial theories involved the possibility of a massive colli-

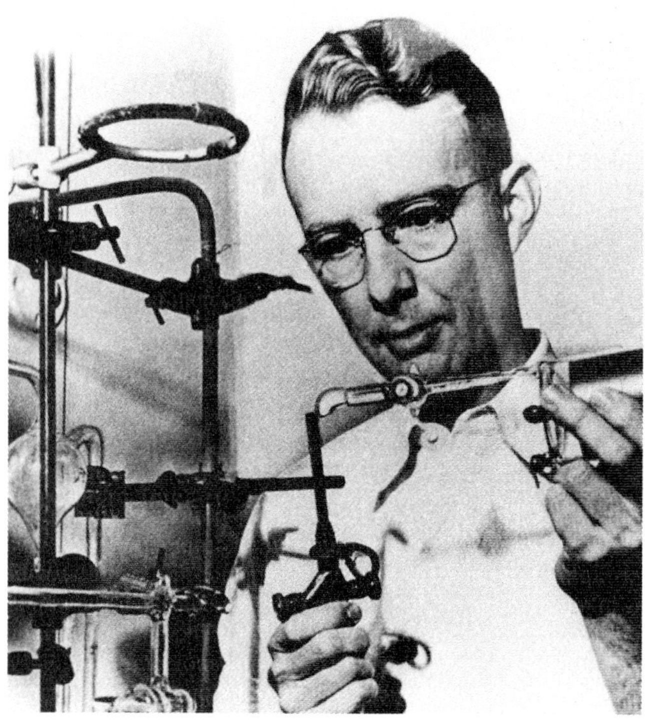

Alvarez, Luis W. (1911-1988).

sion of a meteorite with the Earth 65 million years ago, an event that Alvarez believed may account for the disappearance of the dinosaurs. Among the many honors that Alvarez received was the 1968 Nobel Prize for physics for his development of giant **bubble chambers** used to detect a variety of **subatomic particles**.

Luis Walter Alvarez was born in San Francisco, California, on June 13, 1911. His father, Dr. Walter Clement Alvarez, was a medical researcher at the University of California at San Francisco and also maintained a private practice. Luis' mother was the former Harriet Skidmore Smythe. His grandfather Alvarez was born in Spain but ran away to Cuba, and later made a fortune in Los Angeles real estate before moving to Hawaii and then to San Francisco. Luis' mother's family, originally from Ireland, established a missionary school in Foochow, China. Alvarez's parents met while studying at the University of California at Berkeley.

Alvarez attended grammar school in San Francisco and enrolled in the city's Polytechnic High School, where he avidly studied science. When his father accepted a position at the prestigious Mayo Clinic, the family moved to Rochester, Minnesota. Alvarez reported in his autobiography *Alvarez: Adventures of a Physicist,* that his science classes at Rochester High School were "adequately taught [but] not very interesting." Dr. Alvarez noticed his son's growing interest in physics and hired one of the Mayo Clinic's machinists to give Luis private lessons on weekends. Alvarez enrolled at the University of Chicago in 1928 and planned to major in chemistry. He was especially interested in chemistry, but soon came to despise

the mandatory chemistry laboratories. Alvarez "discovered" physics in his junior year and enrolled in a laboratory course, "Advanced Experimental Physics: Light" about which he later wrote in his autobiography: "It was love at first sight." He changed his major to physics and received his B.S. in 1932. Alvarez stayed at Chicago for his graduate work and his assigned advisor was Nobel Laureate **Arthur Compton**, whom Alvarez considered "the ideal graduate advisor for me" because he visited Alvarez's laboratory only once during his graduate career and "usually had no idea how I was spending my time."

Alvarez earned his bachelor's, master's, and doctoral degrees at the University of Chicago before joining the faculty at the University of California at Berkeley, where he remained until retiring in 1978. His doctoral dissertation concerned the **diffraction** of **light**, a topic considered relatively trivial, but his other graduate work proved to be more useful. In one series of experiments, for example, he and some colleagues discovered the "east-west effect" of **cosmic rays**, which explained that the number of cosmic rays reaching the Earth's atmosphere differed depending on the direction from which they came. The east-west effect was evidence that cosmic rays consist of some kind of positively charged particles. A few days after passing his oral examinations for the Ph.D. degree, Alvarez married Geraldine Smithwick, a senior at the University of Chicago, with whom he later had two children, Walter and Jean. Less than a month after their wedding the Alvarezes moved to Berkeley, California, where Luis became a research scientist with Nobel Prize-winning physicist Ernest Orlando Lawrence, and initiated an association with the University of California that was to continue for forty-two years.

Alvarez soon earned the title "prize wild idea man" from his colleagues because of his involvement in such a wide variety of research activities. Within his first year at Berkeley, he discovered the process of K-electron capture, in which some atomic nuclei decay by absorbing one of the electrons in its first orbital (part of the nuclear shell). Alvarez and a student, Jake Wiens, also developed a mercury vapor lamp consisting of the artificial **isotope** mercury–198. The **wavelength** of the light emitted by the lamp was adopted as an official standard of length by the U.S. Bureau of Standards. In his research with Nobel Prize-winning physicist Felix Bloch, Alvarez developed a method for producing a beam of slow moving neutrons, a method that was used to determine the **magnetic moment** of neutrons (the extent to which they affect a magnetic field). Just after the outbreak of World War II in Europe, Alvarez discovered tritium, a radioactive isotope (a variant **atom** containing a different number of protons) of hydrogen.

World War II interrupted Alvarez's work at Berkeley. In 1940 he began research for the military at Massachusetts Institute of Technology's (MIT's) **radiation** laboratory on radar (**radio** detecting and ranging) systems. Over the next three years, he was involved in the development of three new types of radar systems. The first made use of a very narrow radar beam to allow a ground-based controller to direct the "blind" landing of an airplane. The second system, code-named "Eagle," was a method for locating and bombing objects on the ground when they could not be seen by a pilot. The third invention became known as the microwave early-warning system, a mechanism for collecting images of aircraft movement in overcast skies.

In 1943, Alvarez left MIT to join the Manhattan Project research team working in Los Alamos, New Mexico. His primary accomplishment with the team was developing the detonating device used for the first plutonium bomb. Alvarez flew in the B–29 bomber that observed the first test of an atomic device at Alamogordo, south of Los Alamos. Three weeks later, Alvarez was aboard another B–29 following the bomber "Enola Gay" as it dropped the first **atomic bomb** on Hiroshima, Japan. Like most scientists associated with the Manhattan Project, Alvarez was stunned and horrified by the destructiveness of the weapon he had helped to create. Nonetheless, he never expressed any doubts or hesitation about the decision to use the bombs, since they brought a end to the war. Alvarez became one of a small number of scientists who felt strongly that the United States should continue its nuclear weapons development after the war and develop a fusion (hydrogen) bomb as soon as possible.

After the war, Alvarez returned to Berkeley where he had been promoted to full professor. Determining that the future of physics lay in high **energy** research, he focused his research on powerful particle accelerator—devices that accelerate electrons and protons to high **velocity**. His first project was to design and construct a linear accelerator for use with protons. Although his machine was similar in some ways to the **electron accelerators** that had been available for many years, the **proton** machine posed a number of new problems. By 1947, however, Alvarez had solved those problems and his forty-foot-long proton accelerator began operation.

Over the next decade, the science of **particle physics** (the study of atomic components) developed rapidly at Berkeley. An important factor in that progress was the construction of the 184-inch synchrocyclotron at the university's radiation laboratory. The synchrocyclotron was a modified circular particle accelerator capable of achieving much greater velocities than any other type of accelerator. The science of physics involves two fundamental problems: creation of particles to be studied in some type of accelerator and detection and identification of those particles. After 1950, Alvarez's interests shifted from the first to the second of these problems, particle detection, because of a chance meeting in 1953 with University of Michigan physicist **Donald Glaser**. Glaser had recently invented the bubble chamber, a device that detects particles as they pass through a container of superheated fluid. As the particles move through the liquid, they form **ions** that act as nuclei on which the superheated material can begin to boil, thereby forming a track of tiny bubbles that shows the path taken by the particles. In talking with Glaser, Alvarez realized that the bubble chamber could be refined and improved to track the dozens of new particles then being produced in Berkeley's giant synchrocyclotron. Among these particles were some with very short lifetimes known as resonance states.

Improving Glaser's original bubble chamber involved a number of changes. First, Alvarez decided that liquid hydrogen would be a more sensitive material to use than the diethyl **ether** employed by Glaser. In addition, he realized that sophisticated equipment would be needed to respond to and record the resonance states that often lasted no more than a billionth of a second. The equipment he developed included relay systems that transmitted messages at high speeds and computer programs that could sort out significant from insignificant events and then analyze the former. Finally, Alvarez aimed at constructing larger and larger bubble chambers to record a greater number of events. Over a period of about five years, Alvarez's chambers grew from a simple one-inch glass tube to his most ambitious instrument, a seventy-two inch chamber that was first put into use in 1959. With these devices, Alvarez eventually discovered dozens of new elementary particles, including the unusual resonance states.

The significance of Alvarez's work with bubble chambers was recognized in 1968 when he was awarded the Nobel Prize for physics. At the awards ceremony in Stockholm, the Swedish Academy of Science's Sten von Friesen told Alvarez that, because of his work with the bubble chamber, "entirely new possibilities for research into high- energy physics present themselves....Practically all the discoveries that have been made in this important field [of particle physics] have been possible only through the use of methods developed by Professor Alvarez." Alvarez attended the Nobel ceremonies with his second wife, Janet Landis, whom he married in 1958. Largely as a result of their war-related separation, Alvarez and his first wife had divorced. With Janet, Alvarez had two more children, Donald and Helen.

Advancing years failed to reduce Alvarez's curiosity on a wide range of topics. In 1965 he was in charge of a joint Egyptian-American expedition whose goal was to search for hidden chambers in the pyramid of King Kefren at Giza. The team aimed high-energy muons (subatomic particles produced by cosmic rays) at the pyramid to look for regions of low **density** that would indicate possible chambers. However, none were found. Alvarez shared the last major scientific achievement with his son Walter, who was then a professor of geology at Berkeley. In 1980, the Alvarezes accidentally discovered a band of sedimentary rock in Italy that contained an unusually high level of the rare metal iridium. Dating techniques set the age of the layer at about 65 million years. The Alvarezes hypothesized that the iridium came from an asteroid that struck the Earth, thereby sending huge volumes of smoke and dust (including the iridium) into the Earth's atmosphere. They suggested that the cloud produced by the asteroid's impact covered the planet for an extended period of time, blocked out sunlight, and caused the widespread death of plant life on Earth's surface. The loss of plant life in turn, they theorized, brought about the extinction of dinosaurs who fed on the plants. While the theory has found favor among many scientists and has been confirmed to some extent by additional findings, it is still the subject of debate.

Alvarez's hobbies included flying, golf, music, and inventing. He made his last flight in his Cessna 310 in 1984, almost exactly 50 years after he first learned to fly. In 1963 he assisted the Warren Commission in the investigation of President John F. Kennedy's assassination. Among his inventions were a system for **color** television and an electronic indoor golf-training device developed for President Eisenhower. In all, he held 22 patents for his inventions. Alvarez died of cancer in Berkeley, on September 1, 1988.

AMPÈRE, ANDRÈ MARIE (1775-1836)

French mathematician and physicist

André Ampère was born on January 22, 1775, in Lyon, France. He was the son of a well-to-do merchant. He mastered advanced mathematics by the age of twelve and learned Latin so he could read the works of the learned scientists.

Ampère was interested in a variety of subjects, including psychology, philosophy, astronomy, physics, and chemistry. He studied the nature of chlorine and iodine, but the credit went to Humphry Davy. Ampère analyzed **Boyle's law** and, in 1814, studied the molecular makeup of gases, independently coming up with what has become known as **Avogadro's number**. He spent most of his life teaching and giving lectures, primarily in mathematics and chemistry, at various institutions.

In 1819 a discovery occurred that had a profound effect on science in general, and Ampère in particular: Danish physicist Hans Christian Ørsted placed a magnetized needle next to a wire carrying an **electric current** and saw the needle deflect. This was the first indication that **electricity** generated a magnetic field, linking the two forces. Ørsted published the results of his experiment in 1820; **electromagnetism** subsequently became Ampère's life work.

Ampère and his countryman Dominique-Françios Arago became leaders in investigating the link. Within a week of learning of Ørsted's work, Ampère had devised the right-hand screw rule to show the direction in which a magnetic needle will be deflected. This introduced the concept of lines of **force**, which figured prominently in the work of **Michael Faraday**. After Ørsted's magnetic needle was deflected by an electric current, Ampère realized that it could be used to measure the flow of that current if placed next to a graduated scale.

The only problem with the right-hand screw rule was in determining the direction in which the current flowed. **Benjamin Franklin**, who pioneered the concept of positive and negative electricity, had made the assumption that current moved from the abundant pole (positive) to the deficient pole (negative). That seemed logical to Ampère, and he used it to determine the direction of current flow. Alas, Franklin's assumption was incorrect; fortunately, because of consistent reversal, this made no difference to Ampère's results. To differentiate between his flowing current and the static charge with which Franklin had worked, Ampère coined the term *electrostatics*.

Continuing to experiment with electricity, Ampère discovered that two wires, placed end to end, attracted each other when they were carrying current flowing in the same direction, and repelled each other if the current flowed in opposite directions. If one wire was allowed to rotate around the other, it would move through a semicircle and stop when its current matched the flow of the current in the other wire.

What would happen if the wire was circular, Ampère wondered. In theory, the electric current flowing through a wire helix, or coil, would set up a magnetic field and create an electro-magnet that would act like a bar magnet. Ampère dubbed the wire helix a solenoid, a device with which **Joseph Henry** experimented, leading to a revolution in communications.

In 1823 Ampère published a mathematical theory that suggested **magnetism** was electricity in **motion**. He showed that a magnet's properties could be explained if one assumed there were many tiny electric currents circling around within it. Numerous scientists refused to accept this concept, which was far ahead of its time; 60 years later the negatively charged **electron** was discovered.

In 1827 Ampère published his *Theory of Electrodynamic Phenomena*, in which he established precise mathematical formulations of electromagnetism. *Ampère's Law* relates the **electromotive force** that is created by the currents in two parallel conductors to the product of their currents and the distance between the conductors.

Ampère died in Marseille, France, on June 10, 1836. His memory was honored in 1883 when William Thomson (Lord Kelvin) proposed measuring the quantity of current passing a given point at a given time in amperes. It was very appropriate; Ampère had been the first to distinguish between the rate in which current passed and the force (measured in volts, after **Alessandro Volta**) that pushes it.

ANAXAGORAS OF CLAZOMENAE (c.500 B.C.-428 B.C.)

Greek philosopher

Anaxagoras was an early Greek philosopher and scientist who developed remarkable insights concerning the differentiation and structure of **matter**. Born around 500 B.C. in Clazomenae, now part of Turkey, Anaxagoras came from a noble and wealthy family. As legend has it, he was drawn early in life to philosophy and science and neglected his material possessions. He is noted as the person who introduced philosophy to Athens, where he moved when he was around the age of 20. According to **Plato**, Anaxagoras taught Pericles (c.495-429 B.C.), a Greek leader who helped develop democracy in Athens and make the city a political and cultural center of Greece. Anaxagoras may also have taught the Greek dramatist Euripides (c.484-406 B.C.) and, possibly, Socrates (c.470-399 B.C.).

Following in the footsteps of Pythagorus, Anaxagoras dealt with many questions in geometry, including the problem of squaring the circle. However, it was his theories on the creation of the solar system and matter that remain a stunning achievement in ancient Greek science and thought. Although only fragments of his only known published book still exist, these fragments and other reports reveal a keen mind that was not afraid to question widely held beliefs of the time. Anaxagoras held that an infinite number of elements made up matter, which contradicted the belief that four basic elements (earth, air, fire, and water) made up all things. His concept that "in all things there is a portion of everything..." and that the formation of different matter is caused by the union, separation, and arrangement of numerous elements anticipated the atomic theories of Democritus. Also central to his philosophy was the idea of *nous* (mind or reason) as the creator and driving force of the world.

Anaxagoras's insightful claim that the **Sun** was a hot stone instead of a god, and that the **Moon** reflected the Sun's **light**, caused him to be arrested and imprisoned. His troubles were compounded because of his friendship with Pericles, who had many political opponents. Pericles intervened and had Anaxagoras released from prison. However, Anaxagoras had to leave Athens. He returned to Asia Minor and is believed to have founded a school at Lampsacus. After his death, the citizens of Lampsacus erected an alter in his memory dedicated to mind and truth. The anniversary of his death became a holiday for school children.

See also Atomic theory

ANDERSON, CARL DAVID (1905-1991)

American physicist

Carl David Anderson discovered two of the elementary particles of matter—the **positron** ("positive electron") and the **meson**, also known as **muon** (identical to the negatively charged **electron** in almost every aspect except **mass**). His discoveries dramatically expanded physicists' understanding of the nature and structure of the **atom** and confirmed theoretical predictions about the existence of such **subatomic particles**. Anderson's experimental identification of the positron earned him a Nobel Prize in 1936.

Anderson was born on September 3, 1905, in New York City, New York, the only son of Swedish immigrants from farming families. His mother was Emma Adolfina (Ajaxson) Anderson. His father, also named Carl David Anderson, came to the United States in 1896. When Anderson was a boy, the family moved to Los Angeles, where the father was just able to support the family by managing small restaurants. Anderson attended the Los Angeles Polytechnic High School. He had ambitions to be a track star—a high jumper—but his future career interest was more accurately indicated by his membership in Polytech's science club. After high school graduation in 1924, Anderson could not afford to live away from home and pursue his studies, so he commuted to the California Institute of Technology (Caltech) in nearby Pasadena. Although financial necessity dictated his choice of school, it was a happy one, as Anderson remained at Caltech for his

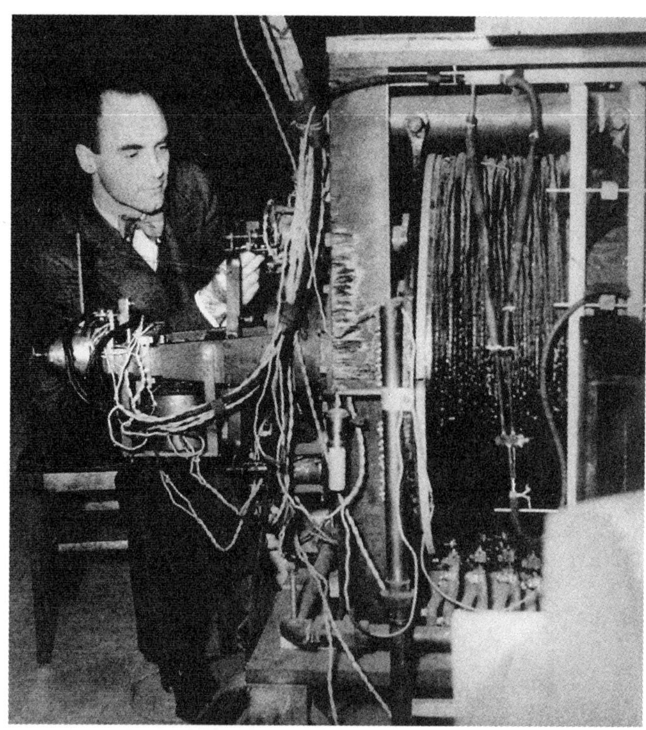

Anderson, Carl David (1905-1991).

entire career. Originally, he intended to study electrical engineering, but in his sophomore year he chose physics as his major. His outstanding undergraduate work brought him tuition grants and a prize of a trip to Europe. He earned his bachelor's degree in physics and engineering in 1927, after just three years of study, and became a member of the honor society Sigma Xi in recognition of original scientific work he had done.

Continuing his studies at Caltech in physics and mathematics, Anderson was Coffin research fellow from 1927 to 1928 and a teaching fellow in physics from 1928 to 1930. One of his graduate courses was physicist J. Robert Oppenheimer's first course in quantum **mechanics**, which the professor persuaded Anderson to stick with after everyone else had dropped out. Anderson agreed even though the course focused on theoretical physics, while he preferred experimental work. His graduate research concentrated on **x rays** and electrons; his doctoral thesis dealt with the **space** distribution of photoelectrons (electron exposed to **radiation**) scattered from various gases by x rays. After receiving his Ph.D. magna cum laude in 1930, Anderson stayed at Caltech as a research fellow in physics. In 1933 he joined Caltech's faculty as an assistant professor of physics; in 1937 he became an associate professor and in 1939, a full professor.

While Anderson was a graduate student, his research director was physicist Robert A. Millikan, whose scientific reputation (he had won the 1923 Nobel Prize in physics for successfully measuring the charge of an electron) and fundraising talent had helped to make Caltech a premier research institute. Millikan was at the time interested in measuring energies of **cosmic rays** (very penetrating radiation from outer space). Soon after Anderson received his doctorate, Millikan asked him to stay at Caltech and help him develop a type of **cloud chamber** that could make the cosmic ray measurements. Anderson agreed to Millikan's request and, as it turned out, this was the work that won Anderson his Nobel Prize.

The cloud chamber was developed by Scottish physicist C. T. R. Wilson in the early 1900s as a means of detecting charged particles. Particles with an **electric charge** pass through a cloud of moist air inside the chamber, ionizing (charging) gas molecules along their path. Water vapor condenses on the ionized molecules, and the hail of these water droplets can be photographed through a glass cover on one side of the chamber. Particle types can be identified by their characteristic trails. The cloud chamber is positioned between the poles of a electromagnet, which causes positively and negatively charged particles to curve in different directions within the chamber. In order to study cosmic rays, Anderson and Millikan designed a cloud chamber with an extremely powerful electromagnet capable of deflecting the high-energy cosmic particles. Millikan then left his assistant in charge of the project. Anderson's task was to photograph clouds hoping to capture shots of particles jolted into action by a cosmic ray. The particles' curvature trails would yield a measurement of their **energy**. As so often happens in science, the experiment yielded a unexpected result.

Anderson made and studied a thousand photographs, at 15-second intervals day and night, during 1930 and 1931. He noticed that some photographs showed paths of particles curving in opposite directions. The negatively charged particles were electrons; the particles curving away from them must be positively charged. At the time, only two elementary particles were known to exist—the electron and the positively charged **proton**. Continuing his experiments to prove that the mysterious particles were protons, Anderson instead found that the particles seemed to have a mass comparable to the electron, much less than that of the proton. His suggestion to Millikan that he had found either positively charged electrons or a new, previously unknown particle resulted in what Anderson described later as "frequent and at times somewhat heated discussions" between the two physicists, with Millikan holding out for the particles' identity as protons (*American Journal of Physics*). Anderson then inserted a lead plate across the middle of the cloud chamber—which would lessen the particles' energy and therefore clarify the direction of the particles' **motion**. This resulted on August 2, 1932, in a very clear photograph of an upwardly moving positively charged particle that measurements showed to have a mass consistent with an electron. Additional photographs confirmed the finding, a difficult feat as subsequent investigations showed that these particles have a average life of just a few billionths of a second. Anderson concluded, and Millikan concurred, that he had found a new particle, which he called the positron, or positive electron. At Millikan's urging, Anderson announced his discovery in a letter to *Science* on September 9, 1932.

The existence of the positron (and other antiparticles, with opposite charges from existing known particles) had been predicted by **Paul Dirac** in 1928. Still, other researchers had ignored the diverging tracks that caught Anderson's attention. And, as Anderson stated in the *American Journal of Physics,* "The discovery of the positron was wholly accidental," although, he added, if a researcher had taken "the Dirac theory at face value he could have discovered the positron in a single afternoon." (Anderson himself had not been familiar with Dirac's theory at the time of his experiments.) While the discovery confirmed the existence of **antimatter**, the positron's existence was not greeted enthusiastically by Anderson's fellow physicists, because it seemed to complicate rather than clarify attempts to understand **atomic structure**. As a reporter in the *New York Times* noted, "Dr. Anderson's positron was a bomb. When it exploded it burst all conceptions of the atom's constitution apart." By the spring of 1933, however, researchers at Great Britain's Cavendish Laboratory had confirmed the existence of positrons. Also in 1933, Anderson and Seth Neddermeyer, his student and then colleague, succeeded in producing positrons in the laboratory by directing powerful gamma rays from thorium C" (a radioactive substance) at atomic nuclei. Further investigations by Anderson and others of electron-positron interactions provided proof of Albert Einstein's famous formula relating **energy and mass**. Anderson and other researchers found that positrons and electrons annihilate one another, creating gamma rays (transforming mass into energy), and that gamma rays could combine to convert into an electron-positron pair (changing energy into mass). Anderson also made contributions to science concerning the energy spectrum of cosmic rays and the energy loss experienced by the rays as they pass through **matter**.

As Anderson and Neddermeyer continued their experiments with cosmic rays, using the same cloud chamber in which the positron was discovered, they began to suspect the existence of yet another new elementary particle. They painstakingly tracked it down over a period of act four years, taking thousands of cloud-chamber photographs in Pasadena, in Panama, and at the top of Pike's Peak in Colorado. The new particle was charged, so it had to be either an electron (negative), positron (positive), or a proton (positive). But it was far more penetrating than allowed by current **atomic theory**, and its mass seemed to be somewhere between that of the electron and the proton. Anderson hinted at the existence of the new particle in a paper he presented to the International Conference on Physics in London in 1934, and mentioned it again at the end of his Nobel Prize lecture in 1936. Finally, in 1938 he announced the existence of the mesotron (soon shortened to meson), which could carry either a positive or negative electric charge and has an average life span of two-millionths of a second.

As Anderson pointed out in the *American Journal of Physics,* this discovery—unlike that of the positron—was not accidental but rather was the result of a "series of careful, systematic investigations." As with the positron, however, the existence of the meson had been predicted by a theoretical physicist, in this case **Hideki Yukawa** of Japan, in 1935, although Yukawa's theory was little known. Also like the positron, the positive identification of the meson was not welcomed by physicists because it too complicated understanding of subatomic structure. On this point, Anderson wryly commented in a 1948 Sigma Xi address (reprinted in the Smithsonian Institution's annual report of 1949), "Apparently the Creator does not favor a world of too great simplicity." In 1947 physicists Cecil Frank Powell and Giuseppe P. S. Occhialini found a meson somewhat heavier than Anderson's, which was identified as the particle predicted by Yukawa. Anderson's meson, which he has called "an oddball particle," was renamed muon, an example cited by Anderson in his Sigma Xi address of the great difficulties faced by physicists who must name new discoveries or concepts before they have accumulated a full set of knowledge about them.

At the age of 31 in 1936, Anderson became one of the youngest Nobel laureates ever, recognized, as the Nobel presenter told him, for "utilizing ingenious devices" to find "one of the buildingstones of the universe" (*Physics* [Nobel Lectures, 1922–1941]). He shared the Nobel Prize in physics with Austrian physicist Victor Hess, who discovered cosmic radiation and thus provided the foundation for the investigations of Anderson's career. During World War II, Anderson worked on rocket research for the Office of Scientific Research and Development and the National Defense Research Committee. He was offered the directorship of the **Manhattan Project**, the government effort to develop the **atomic bomb**, but turned it down. In 1947, after the war, Anderson took his cloud chamber up in a Navy plane to photograph cosmic ray energy in the stratosphere and captured the disintegration of a meson on film.

At the time Anderson won his Nobel Prize, a journalist in the *New York Times* described him as "shy and retiring," while an interviewer for *Scientific American* concluded that Anderson's "very approachable manner brands him as one of the more human scientists." In 1946, Anderson married Lorraine Elvira Bergman; they had two sons, Marshall (who became a mathematician and computer analyst) and David (who became an engineer). Anderson served as chairman of Caltech's division of physics, mathematics, and astronomy from 1962 to 1970. He maintained his research and teaching activities, focusing on cosmic radiation and elementary particles, until his retirement in 1976 with the title professor emeritus. He published many of his findings in *Science* and *The Physical Review*. Anderson died at his home in San Marino, California, near Pasadena on January 11, 1991, at the age of 85 of undisclosed causes. He had earned many honorary degrees and awards, including medals from the American Institute of the City of New York, the Franklin Institute, and the American Society of Swedish Engineers. His memberships included the National Academy of Science and the American Academy of Arts and Sciences.

ANGULAR MOMENTUM

Angular **momentum** is the tendency of an object in rotational movement to continue to rotate in the absence of any interfering **force**. An object's momentum will not change unless it is acted upon by an outside force. When the object (or a system of objects) is spinning, other factors have to be considered in discussing the momentum of the rotational movement. This momentum is referred to as angular momentum. The principles of angular momentum applies equally to spinning objects on Earth or to orbiting objects in **space**.

To understand the factors of angular momentum, the various influences on rotational movement must be considered. Three of these influences are 1) the speed of points at different distances from the center of rotation (angular **velocity**), 2) the moment of force, a product of two **vectors** (also known as **torque**), and 3) a figure derived from the summation of the distribution of **mass** throughout the object and their distance from a given line (moment of **inertia**).

Angular velocity is defined as the rate of displacement a particular point undergoes in relation to another point over **time**. Using a wheel as an example, a point on the rim and a point on the axle will have completed a rotation in the same amount of time. However, the point on the rim will have gone a further distance than the point on the axle in completing the rotation. It is a property of rotational movement that speed is dependent on the distance from the center.

Torque, or rotation moment, is a product rather than a force itself. In the example of the wheel mentioned above, torque can cause the wheel to try to rotate on an axis passing through both the axle and the rim. The effect of the torque depends on where, between the center of rotation and the outermost point, the initial force is applied. The further away from the center of rotation that the torque is applied the more velocity it causes. Torque is used in reference to the product of forces that slow down or stop the object as well as those that start the object spinning.

The moment of inertia is important in the study of angular momentum because its interaction with the object's rotational velocity will never change. When the distribution of mass changes, the other factor has to compensate. When there is a reduction in the moment of inertia, then there will be a reciprocating increase in rotational velocity. The two factors always have to be balanced for the angular momentum to remain unchanged. One way the distribution of mass can changed without any loss of total mass is by a reduction in the average distance from the center of rotation. This is why a figure skater spins faster when she pulls her arms in close to her body.

Angular momentum, like all other forms of **motion**, follows the general principle stated in Newton's third law. In weak form, N3 says that the force of object A on object B is equal in magnitude and opposite in direction to the force of object B on A. In strong form, the forces are pictured as acting along the line joining A and B. When the force is due to an **electromagnetic field**, these assumptions break down; hence, electromagnetic **waves** can carry angular momentum.

An example of angular momentum and its conservation is given in Kepler's second law. This law states that for a planet orbiting the **sun**, a line from the planet to the sun sweeps across equal areas in equal times. In an elliptical orbit, as the planet moves further from the sun, it also moves slower in its orbit. The line from the Sun to the planet covers as much area when the planet moves a short distance between two points in the apogee (the most distant portion) of the ellipse as when the planet moves a greater distance at the perigee (the closest portion) of the ellipse.

Newton explained in mathematical formulations how the Sun's gravitational pull causes the planet to speed up as it gets closer to the sun and to slow down as it gets further away from the sun. The numbers associated with the speeding up and slowing down of the planet in its orbit follow the formula for the conservation of angular momentum. The statistic representing angular momentum in the formula for the Earth's orbit around the Sun does not change; neither does the mass of the object. What does change is the velocity and the radius of the ellipse. This is similar to a change in the moment of inertia for a spinning object being compensated by change in velocity, keeping angular momentum a constant.

Angular momentum is also important for scientists trying to understand subatomic physics. Since the beginning of the twentieth century, physicists have used the observed principles of angular momentum to understand the **motion of particles** relative to other **matter** in the **universe** (the Mach principle), the classification of certain **subatomic particles** by their spin, and the apparent peculiar behavior of **light** (sometimes it behaves like a particle, sometimes like a wave). The entire field of **quantum mechanics** owes much of its existence to the study of angular momentum.

ANODE

The term anode, from the Greek words *hodos*, meaning way, and *ana*, meaning up, was used by experimenters in the late ninteenth and early twentieth century to designate the conducting to which negatively charged particles, electrons, were drawn when they subjected partially evacuated tubes to high **voltage**. The tubes were made of glass and typically had a voltage source connected to two pieces of conducting materials sealed into the glass and separated by a few inches. The piece of conducting material from which the electrons emanated was called the cathode and the one to which they were attracted was called the anode.

In **chemistry** the most common use of the term anode occurs in electrochemistry. Electrochemical cells consist of one half-cell in which oxidation occurs and a second half-cell in which reduction occurs. Electrons flow though a conductor connecting the two half-cells. At the two ends of the conductor are the electrodes, which gather and disperse the electrons. The anode is the electrode at which oxidation takes place and the cathode is the electrode at which reduction takes place.

A voltaic cell, or a sources of **energy**, is an electrochemical cell is one in which an oxidation-reduction reaction takes place simultaneously. The everyday batteries we commonly use are voltaic cells. Because electrons originate at the anode in a voltaic cell, the anode has a (-) charge; electrons enter the cathode, which has a (+) charge. This polarity of the

terminals and the role of the electrodes are accurate for any cell that is operating spontaneously and producing **electricity**.

In an electrolytic cell, energy must be supplied in the form of **electric current** from an external source. Electrolytic cells are used to electroplate tableware and jewelry and to produce useful chemicals from naturally occurring materials. In an electrolytic cell, the **power** source pumps electrons into the electrode to which it is attached; that electrode is the cathode. Therefore, in electrolytic cells, the cathode is the negative terminal and the anode is the positive terminal.

The polarities of the anode and cathode are reversed in comparison to their assignments in voltaic cells, but the function of the anode, being the site of oxidation, and of the cathode, being the site of reduction, is the same.

The materials that are useful for anodes must be good conductors and must not corrode too easily under oxidizing conditions. If the anode itself is the substance that one wishes to oxide, it should be the most easily oxidized material in the half-cell. If another substance is to be oxidized, the anode material should not be readily oxidized. Some materials commonly used as unreactive anodes are platinum and graphite.

One of the most common materials using an anode is zinc metal. The zinc containers of carbon-zinc and alkaline dry cell batteries and mercury batteries serve as the anode and are oxidized as the battery is used. In NiCad batteries, the anode is composed of cadmium metal and cadmium hydroxide. As the battery is used, cadmium metal is oxidized to cadmium hydroxide and as it is recharged the process is reversed.

Electrochemical techniques are also widely used in chemical analysis. In electrochemical analysis by such methods as anodic stripping voltammetry, cyclic voltammetry, differential pulse polarography, and potentiometry, the anode always functions as the site of oxidation. In general, the anode is a highly conductive and relatively inert material such as platinum, graphite or mercury.

ANTHROPIC PRINCIPLE

The anthropic principle is a form of philosophical reasoning that, with regard to **cosmology** (the study of the structure and origin of the **Universe**), states that the present existence and state of life on Earth places limits on the evolution of the Universe. The term anthropic principle is actually misleading, for despite its derivation from the Greek *anthropos* meaning human, it is not concerned exclusively with human beings but with intelligent life in the universe.

Scientists, philosophers and theologians have long noted that the values of a multitude of universal constants such as **gravity**, **electromagnetism**, and the nuclear forces fall precisely within a narrow range required for the universe to have evolved as it has. Had a single value been other than it is, the resulting universe would be devoid of life as we know it. These values, which almost seem to have been fine-tuned for the emergence of intelligent life, are known as anthropic coincidences.

One of the most famous anthropic coincidences was discovered by the English physicist Sir **Fred Hoyle** in 1954. While researching stellar **nucleosynthesis**, the process by which heavy elements are formed within the nuclear furnaces of **stars**, Hoyle realized that carbon-12, an element essential to the existence of carbon-based life, could only be produced by a sequence of reactions taking place in precise order and within a narrow range of **energy** levels

Another anthropic coincidence involves the fact that neutrons outweigh protons. The decay of neutrons into protons is essential for the existence of the element hydrogen, whose **nucleus** is a single **proton**. If protons outweighed neutrons, it would be protons that decayed into neutrons rather than the other way around. Without isolated protons, hydrogen wouldn't exist. Neither would the **sun**. And, so anthropic reasoning proceeds, neither would human beings.

Anthropic principles attempt to account for what appear to be the special circumstances surrounding the existence of intelligent life in the universe. Scientists like Brandon Carter, John D. Barrow, and Frank J. Tipler have elaborated on various forms of the anthropic principle in an effort to make sense of these and other mind-boggling examples of what almost seems to be evidence of a cosmic fine-tuning for intelligent life.

The Weak Anthropic Principle (WAP) states that because intelligent life forms such as human beings exist in the universe, the universe, or portions thereof, possess the characteristics (i.e., anthropic coincidences) necessary for such life forms to exist. Hoyle used WAP in making his prediction about carbon-12. WAP can also account for the age of the universe; because human beings are present now, there must have been time enough for the conditions necessary for the evolution of intelligent life to occur.

The Strong Anthropic Principle (SAP) goes a lot further in attempting to explain anthropic coincidences. SAP contends that the fundamental constants and laws of nature are such that intelligent life must arise. SAP is highly controversial, as it seems to imply that the universe is a kind of machine or entity devoted to the creation of intelligent life. Another interpretation of SAP uses the many worlds interpretation of **quantum mechanics**: in a multiverse of infinite universes, there will inevitably be universes whose fundamental laws allow for, or require, intelligent life, and we are in one such universe.

The Participatory Anthropic Principle (PAP) derives from the Copenhagen interpretation of **quantum mechanics**, which stresses the indeterminacy of events at the quantum level as well as the influence of observers on those events. PAP turns SAP on its head and argues that without intelligent life, or observers, the universe could not exist, and that the universe exists because of the presence of intelligent observers.

The Final Anthropic Principle (FAP) goes even further, stating that once intelligence has emerged it will become immortal and god-like, and ultimately will be (or will have been) responsible for its own creation.

ANTIMATTER

Antimatter is **matter** composed of antiparticles, such as positrons, antineutrons and antiprotons. For each type of ordinary particle (e.g., electrons or protons), there is an **antiparticle**. Antiparticles have opposite charges and certain other quantum properties of opposite sign.

Collectively, antiparticles with **mass**, and atoms composed of them, are referred to as antimatter. Particles and antiparticles have the same mass and spin. When a particle and antiparticle are brought together they annihilate each other, converting all of their mass and **energy** into a burst of electromagnetic **radiation**, and sometimes into other particles. Some neutral particles such as photons are their own antiparticles, while others like neutrons have antiparticles with opposite magnetic moments.

In general, the distinction between matter and antimatter is somewhat arbitrary. With protons, neutrons, electrons, and their antiparticles, the matter-antimatter distinction is based on the fact that all of the ordinary objects we encounter are composed of what we call matter, and therefore it is the rarer particles are termed antiparticles. Antimatter is believed to respond to **gravity** in the same way as ordinary matter, and it is possible that **stars**, **planets**, and living organisms could exist made up of antimatter, though there is no astronomical evidence of this.

Paul Dirac predicted the existence of antimatter with his relativistic wave equation in 1928. The first experimental evidence of its existence came from **cloud chamber** tracks produced during a cosmic ray experiment performed by **Carl Anderson** in 1932. He detected an antielectron, also known as a **positron**, which produced tracks similar to an ordinary **electron**, but which moved in the opposite direction in the presence of an electric field, indicating an opposite charge. Heavier antiparticles required more energy to produce in the laboratory, and it was not until 1955 that the antiproton was discovered.

Antimatter is routinely created in high-energy physics laboratories though the process of pair production, where a **gamma ray** passing near an atomic **nucleus** is converted into a particle and an antiparticle, which are then separated by an electric field. A complete **atom** of antimatter was assembled for the first **time** at CERN in 1995 by combining an antiproton and an antielectron to make antihydrogen. It is also possible to produce short lived positronium, where an electron and positron orbit each other in a way similar to the ordinary atom, with its positive **proton** orbited by a negative electron, except that positronium quickly self-annihilates.

Certain **radioisotopes**, such as fluorine-18, emit antielectrons as they decay. This effect is used in medicine in the PET (Positron Emission Tomography) scanner. In a PET scan a patient is injected with or inhales a small amount of a substance containing radioisotopes which will be absorbed by the tissues to be studied. For example, because the more active regions of the brain use more glucose, radioactive glucose is used to evaluate activity levels in differing areas of the brain. The decaying atoms emit positrons, which encounter electrons and are annihilated, releasing pairs of gamma rays. Detectors and **computers** then analyze the gamma rays and assemble a picture of the activity inside the brain.

The fact that pair production always results in equal numbers of particles and antiparticles poses a cosmological question: why is antimatter so rare? Conservation laws suggest that the production of matter from radiation should be accompanied by the creation of an equal amount of antimatter, and yet no large bodies such as planets, stars, or galaxies are known to be made of antimatter. The asymmetry between the amounts of matter and antimatter in the **universe** remains unexplained.

See also Particle physics; Subatomic particles; Symmetry

ANTIPARTICLE

One of the seminal scientific discoveries of the twentieth century was **Louis Victor de Broglie's** theory of the wave nature of the **electron**. By integrating theories by **Albert Einstein** dealing with **mass** and **energy** and Max Planck's Quantum relationship between energy and **wavelength**, de Broglie posited that any material particle should also exhibit wave like properties, just as Einstein earlier had shown that **light**, which had been described as a wave phenomenon, also exhibited particle like behavior.

Building on de Broglie's theory, **Erwin Schrödinger** developed his famous wave equation for both free electrons and those bound to atoms. A limitation of Schrodinger's equation was that it did not include relativity. In the late 1920s, the English physicist **Paul Adrien Maurice Dirac** extended Schrodinger's theory by incorporating relativity into the wave equation for the electron. He showed that particles, such as. the electron should always exist in two energy states: one positive and one negative. When applied to the electron, the theory suggested that, in addition to the negatively-charged electron already known, there should also exist a positively-charged "twin." The twin would be identical to the electron in every respect except for its electrical charge.

Dirac's prediction, announced in 1930, was confirmed experimentally within two years. In his 1932 studies of cosmic ray interactions, **Carl David Anderson** found the positive electron that Dirac had anticipated. Anderson suggested the name **positron** for the new particle. Anderson discovered that the collision of an electron and a positron resulted in the annihilation of both particles, with their mass being converted into energy. Similarly, an electron-positron pair can be created when a high-energy **photon** interacts with **matter**. The mathematical equivalence of mass-energy calculated for these conversions provided dramatic confirmation of Einstein's mass-energy equation ($e = mc^2$) of 1905.

Dirac's theory applied not only to electrons, but to all other particles. Just as he predicted an antielectron, he also suggested the existence of an antiproton, a particle identical to a **proton** in every respect except for its electrical charge.

The negatively-charged proton proved to be much more elusive than had been the positron. With a mass more than 1,800 times greater than that of the positron, the antiproton requires far greater energy for its production. Such energies are available in **cosmic rays**, but the rate of antiproton production from this source is too low to have resulted in their discovery.

It was not until the invention of particle **accelerators** that a systematic search for the antiproton could be launched. Then, in 1955, the particle was discovered by a research team led by **Emilio Segrè** and Owen Chamberlain at the University of California. The antiproton was produced when protons from a **cyclotron** were used to bombard a copper target. As predicted by Dirac, the antiproton was similar to the proton in every respect except for its electrical charge.

Is there also an antineutron? Lacking an electrical charge, an exact analogue to the electron and proton for the **neutron** is obviously impossible. However, Dirac's original research had anticipated and solved this problem. Along with the existence of antiparticles, Dirac had predicted that all particles possess an intrinsic property known as *spin*. The spin of a particle is designated as either "up" or "down." An antineutron would, Dirac suggested, be identical to a neutron except for its spin. In the presence of an external magnetic field, an antineutron and a neutron would have equal but opposite intrinsic spins. The first antineutron was discovered at the University of California at Berkeley only a year after the discovery of the first antiproton.

One can imagine atoms that are made of antiparticles: antiprotons, antineutrons, and positrons. Substances composed of such atoms are known as **antimatter**. Thus far, the antideuteron, made of one antiproton and one antineutron, and an **isotope** of antihelium-3 have been made in the laboratory.

Theory suggests that, at the moment the **universe** was created, equal amounts of matter and antimatter were formed. Today, however, we are able to detect only matter in the universe. One of the great questions of modern physics, therefore, is what happened to the original antimatter that was created at the **time** of the **big bang**.

Modern theory suggests that all particles have antiparticles. For example, the **neutrino**, hypothesized to explain energy changes that occur during some **nuclear reactions**, was found in 1963 to have its own antiparticle, the antineutrino. In some cases, the theory produces somewhat bizarre predictions. For example, since the photon has no mass and no electrical charge, it is, in fact, its own antiparticle.

APPLETON, EDWARD (1892-1965)
English physicist

The existence and location of an ionized reflecting layer in the Earth's atmosphere was identified by Edward Appleton, whose research responded to the first long distance **radio** messages transmitted across the Atlantic Ocean. After **Guglielmo Marconi** made his successful transmission in 1901, two physicists, **Oliver Heaviside** and Arthur E. Kennelly postulated the existence of an ionized band inthe atmosphere. More than twenty years later, Appleton, then Wheatstone Professor of Physics at King's College at the University of London verified that hypothesis and later discovered two more layers, one of which is now named in his honor. For his work on atmospheric structure, Appleton was awarded the 1947 Nobel Prize in physics.

Edward Victor Appleton was born in Bradford, Yorkshire, England, on September 6, 1892. His parents were Peter and Mary (Wilcock) Appleton. Appleton attended Barkerend Elementary School from 1899 to 1903 and Hanson Secondary School from 1903 to 1911. He gave evidence of a brilliant future at an early age, passing the entrance examination at the University of London with first class honors at the minimum age permitted of 16. Two years later he won a scholarship at St. John's College, Cambridge, from which he graduated, again, with honors in physics in 1913. Upon graduation, Appleton continued his studies in **crystallography** under **William Henry Bragg**, but his work was interrupted by the outbreak of World War I. In August of 1914, he enlisted in the British infantry and eventually rose to the rank of captain in the Royal Engineers. After his discharge, Appleton returned to Cambridge and became first a fellow at St. John's College and later demonstrator in physics at the Cavendish Laboratories. His major area of interest was **vacuum** tubes, a topic he first learned about during the war. Some years later, he was to publish a monograph on this research, *Thermionic Vacuum Tubes.*

In 1924, at the age of 32, Appleton was appointed Wheatstone Professor of Physics at King's College at the University of London. The first major research topic to which he turned his attention was the propagation of radio signals. While serving his country, he had become aware of the problem that radio signals have—a tendency to fade out at various times of the day. That same year, assisted by graduate student Miles Barnett, he designed and carried out a series of experiments to elucidate the mechanism by which radio **waves** are transmitted in the atmosphere.

Appleton and Barnett were able to convince the British Broadcasting Corporation (BBC) to allow them to use the network's radio signals outside of regular broadcasting hours. They varied the **frequency** of the BBC signal on a regular basis, looking for points at which **interference** occurred between direct waves from the ground and reflected waves from the atmosphere. Their results indicated the existence of a reflecting layer in the atmosphere approximately 60 mi (100 km) above the Earth's surface. The layer corresponded to a region that had been predicted more than twenty years earlier by Britain's Oliver Heaviside and American physicist A. E. Kennelly. That region was later named the Heaviside-Kennelly, or "E," layer. It is now known to be the lower boundary of the ionosphere.

Appleton continued his research on the atmosphere throughout his life. In 1926, he discovered a second reflecting layer above the E-layer, established as the F- or Appleton-layer. He calculated the lower boundary of this region to be about 150 mi (230 km) above the Earth's surface. The discovery of these reflecting bands in the ionosphere made possible later developments in radio, shortwave, and **radar** technologies. Appleton's research also led him to discover a number of important characteristics of the reflecting layers. In 1927, for example, his observations of a solar eclipse convinced him

Sir Edward Victor Appleton.

that the existence of such layers are the consequence of solar **radiation** bombarding the Earth's atmosphere.

Investigative studies on the atmosphere brought Appleton a number of honors. He was elected a fellow of the Royal Society in 1927, was knighted in 1941, and received the Nobel Prize in physics in 1947 for "his work on the physical properties of the upper atmosphere, and especially for the discovery of the so-called Appleton layer." In addition, he was awarded many honorary doctorates, including those from the universities of Aberdeen, Glasgow, London, Oxford, and Cambridge.

In 1936, Appleton succeeded to the Jacksonian Chair of Natural Philosophy at Cambridge, left vacant two years earlier by the retirement of Charles Thomson Rees Wilson. He brought with him most of his research staff from King's College, began the construction of a new field laboratory for atmospheric research, and initiated an active program for the study of ionized gases at the Cavendish Laboratories at Cambridge.

Appleton remained at Cambridge for only three years, however, before accepting an appointment as secretary of the Department of Scientific and Industrial Research (DSIR). In that post, he rapidly converted the DSIR from a civilian research agency to one making preparations for the war then looming on the horizon in Europe. Included among the DSIR's later military activities was Great Britain's first research on

nuclear weapons. At the war's conclusion, Appleton returned to academia, becoming Principal and Vice-Chancellor of the University of Edinburgh, where he remained until his death on April 21, 1965.

Appleton married Jessie Longson in 1916, with whom he had two daughters, Marjery and Rosalind. Lady Appleton died in 1964 from a stroke suffered three years earlier. A month before he died, Appleton was married for a second time, to Mrs. Helen Allison, his secretary of more than 13 years. His biographer in *Biographical Memoirs of Fellows of the Royal Society,* J. A. Ratcliffe, says that Appleton was known for many admirable qualities, among which were "his wide humanity, his ability as a public speaker, and his continuing dedication to his scientific researches."

AQUINAS, ST. THOMAS (1225-1274)
Italian philosopher, theologian, and scientist

St. Thomas Aquinas was born Thomas d'Aquino near Naples in 1225 to a noble family. His parents, Lundulph, Count of Aquino, and Theodora, Countess of Teano, were related to the Emperors Henry VI and Frederick II, and to the Kings of Aragon, Castile, and France. Known as perhaps the supreme theologian of Catholicism, Aquinas was enrolled in the Benedictine monastery of Monte Cassino at the age of five, where he was noted early for his diligent study and philosophical leanings. Despite his calling to a religious life, Aquinas's mother and family had him kidnapped in May of 1244 after he had become a Dominican friar the month before. They kept him for more than year to persuade him to give up his membership in the Dominican order. Aquinas would not renounce his calling, and they allowed him to return to his order in 1245. He went on to attend the University of Naples and studied under the German philosopher Albertus Magnus (c.1200-1280), author of writings on the Aristotelian *corpus,* from 1245-1252. Ordained a priest in 1250, Aquinas began teaching at the University of Paris in 1252. Humble, taciturn, and heavy set, he was referred to as the "Dumb Ox." However, Magnus recognized Aquinas's keen intellect and proclaimed that "this ox will one day fill the world with his bellowing."

Although known primary as a theologian and philosopher, Aquinas was also a man of science and sought to bring philosophy and science together in a common field of understanding. Under the tutelage of his mentor, Magnus, Aquinas became a defender of Aristotle's approach to philosophical and scientific inquiry. He classified physics and mathematics as theoretical sciences that seek to understand the world. In his commentary on Aristotle's *Sense and Sensibilia,* Aquinas states that physics is the first step in the study of the natural world. Although Aquinas was not an experimental scientist, he tackled, on a philosophical basis, concepts of **matter** and form. Despite his strong religious faith, Aquinas also believed in the importance of scientific discovery and that the study of the humblest acts is related to the highest truths.

Aquinas's best-known work is *Summa Theologiae,* which he began in 1266. In it he discuses the existence of God,

a system of ethics, and Jesus Christ. In addition to teaching in Paris, Aquinas also taught in Naples and Rome. In 1272, the Catholic Church assigned him to Naples as a regent in theology. On December 6, 1273, Aquinas professed to having a heavenly revelation while celebrating mass in the chapel of St. Nicholas in Naples. Aquinas is reported to have said, "I can do no more... such secrets have been revealed to me that all I have written now appears to be of little value." While attending the Council of Lyon, Aquinas fell and injured himself. His health rapidly deteriorated, and he died on March 7, 1274, at the age of 49. A shrine was erected in 1628, where Aquinas's body remained until the shrine was destroyed during the French Revolution. He was then moved to the Church of St. Sernin, where the body is held in a sarcophagus of gold and silver. He remains the patron saint of all Catholic universities, academies, colleges, and schools.

See also Cosmology

ARCHIMEDES (287 B.C.-212 B.C.)
Greek mathematician

Archimedes is remembered as a mathematician, philosopher, and inventor. It seems, however, that he did not think as much of his numerous inventions—important and fundamental as they were—as he did of his work in the field of mathematics.

Archimedes was born around 287 B.C. in Syracuse, a town in the Greek colony of Sicily. His father was the astronomer Phidias, and he was related to King Hieron II (308 B.C.?-216 B.C.). Archimedes went to Alexandria about 250 B.C. to study under Conon and other mathematicians who were former students of **Euclid**. He later returned to Syracuse where he apparently stayed for the rest of his life. He was executed by a Roman soldier in 212 B.C.

Archimedes earned the honorary title "father of experimental science" because he not only discussed and explained many basic scientific principles, but he also tested them in a three-step process of trial and experimentation. The first of these three steps is the idea that principles continue to work even with large changes in size. The second step proposes that mechanical **power** can be transferred from "toys" and laboratory work to practical applications. The third step states that a rational, step-by-step logic is involved in solving mechanical problems and designing equipment.

Some of Archimedes's accomplishments were with mathematical principles such as his calculation of the value of pi to figure the areas and volumes of curved surfaces and circular forms. During this process, Archimedes used a method similar to calculus, which was not to be defined for almost another 2,000 years. He also created a system of exponential notation to allow him to prove that nothing exists that is too large to be measured.

Adherence to these principles led Archimedes to such inventions as block and tackle systems, devices for driving objects using axles and drums, and the water snail screw. He performed countless experiments on screws, **levers**, and pulleys.

Bust of Archimedes. *(Photo Researchers, Inc. Reproduced by permission.)*

The water snail (also known as the Archimedean screw) is still used in certain parts of the world to raise and move water. This idea of enclosing a screw inside a cylinder to create, in essence, the first water pump, is perhaps his most remembered invention. The Archimedean screw has been the basis for the creation of many other tools, such as the combine and auger drills.

Archimedes's work with levers and pulleys led to the inventions of compound pulley systems and cranes. His compound pulleys are highlighted in a story that reports that Archimedes moved a fully-loaded ship single-handedly while seated at a distance. His crane was reportedly used in warfare during the Roman siege of his home, Syracuse. Other wartime inventions attributed to Archimedes include rock-throwing catapults, grappling hooks, and **lenses** or mirrors that could allegedly reflect the Sun's rays and cause ships to burn.

Archimedes deemed the importance of one of his inventions so great, that he devoted an entire book to the subject. This invention was a self-moving celestial model representing the **Sun, Moon,** and constellations. The model was incredibly accurate and even showed eclipses in a time-lapse manner. This invention utilized a system of screws and pulleys that moved the globe in their various courses and speeds. His theories in the realm of "statics," particularly in the studies of **gravity,** balance, and **equilibrium,** were based on experiments with levers. He also formed bases for hydrostatics (the study of **fluids**) and discovered the principle of buoyancy and displacement.

ARCHIMEDES'S PRINCIPLE

Archimedes's principle states that an object immersed in a liquid will be supported by a **pressure** equal to the **weight** of the liquid displaced by the object. This is caused by an **equilibrium** of forces. The two forces seeking equilibrium are **gravity** and pressure from the surrounding liquid. Gravity pulls the object down through the liquid. The pressure of the surrounding liquid is resisting the free fall of the immersed object. The degree to which the weight of the object corresponds to the weight of the displaced liquid determines whether the object sinks quickly or slowly or floats. The intensity of the pressure is in direct relationship to the depth.

A cubic foot of water weighs 62.4 lb (28.3 kg). Therefore, if an object placed in water has a volume of a cubic foot and weighs more than 62.4 lb, it will sink. Rocks sink because most rocks have a greater **mass** than water (except pumice). A block of wood will likely float in water because most wood has a mass less than 62.4 lb per cubic foot. If an object has a mass exactly 62.4 lb per cubic foot, weighing exactly the same as the water it displaces, it will neither sink nor float but remain suspended where it is placed.

The story of the discovery of the principle is apocryphal; however it is a good example of an application of the principle. **Archimedes** lived in the Greek city-state of Syracuse on the island of Sicily about 200 B.C. and was friends with the king, Heron. As the story goes, Heron suspected that his crown was not pure gold but gilded silver. He asked Archimedes to determine if the crown was truly gold or mostly silver. Archimedes thought of the problem while he stepped into a full bath. As the water ran over the sides of the tub, he realized the solution. He immediately ran to Heron's house, through the streets of Syracuse, naked, yelling "Eureka", which means "I found it".

Archimedes realized that he could determine the volume of the crown by placing it in water and measuring the amount of displaced water. Since the mass of gold and the mass of silver were already known, he would need only to multiply the calculated volume of the crown by its weight. The result was then compared to the known mass of gold and silver. Another way the principle could be used to determine the substance of an object is to weigh the object before placing it in water and again while it is in the container of water. Determine the difference in the weight of the object outside of water and the weight of the object when submerged. The ratio of that difference to the weight outside the water should equal the ratio of the mass of water to the mass of the substance. Using the crown as an example again, the ratio of the mass of gold to the mass of water is 1/18. The submerged crown would be expected to weigh 17/18 as much as the crown outside of water, if it was indeed gold.

ARISTOTLE (384 B.C.-322 B.C.)
Greek philosopher and biologist

While he is highly regarded as a philosopher and father of logic and reasoning, Aristotle is also known for accomplishments in and contributions to other sciences. Throughout his life, he wrote several biological works which laid the foundations for comparative anatomy, taxonomy (classification), and embryology.

Aristotle was born in the northern Greek village of Stagira. His father was the court physician to the king of Macedonia, and it was at the Macedonian court where Aristotle spent much of his early boyhood. His father died before Aristotle was ten years old and the boy was raised by friends of the family.

At age seventeen, Aristotle was sent to the Academy of **Plato** in Athens where he plunged wholeheartedly into Plato's pursuit of truth and goodness, and soon became Plato's best pupil, earning the nickname "intelligence of the school." Twenty years after Aristotle's arrival, Plato died; Aristotle then left the Academy to travel. His journeys led him through the Greek empire and Asia Minor for twelve years during which he began his research into natural history and biology.

In 342 B.C., Philip II invited Aristotle to return to the Macedonian court and teach his son, Alexander. Aristotle's student later became known in history as Alexander the Great.

After the death of Philip and Alexander's rise to the throne, Aristotle left the court for a brief visit to his hometown; he soon returned to Athens to resume his scientific studies. In 335 B.C., he founded a university called the *Lyceum*. He had renounced some of Plato's theories and began his own style of brilliant teaching at the newly-established school. In the mornings, he would stroll through the Lyceum gardens, discussing problems and theories with his advanced students. Because he walked about while teaching, Athenians nicknamed his school the *Peripatetic*—the Greek term meaning "to walk about." Like their headmaster, Lyceum pupils performed research in nearly every existing field of knowledge. They dissected animals and studied the habits of insects, helping Aristotle to compile data for his classification system.

The school became the basic building block for the great library and museum in the area. Unfortunately, in about the year 323 B.C., the ruling emperor Alexander died, forcing Aristotle to leave Athens due to anti-Macedonian sentiment and accusations of impiety. He went to his mother's homeland of Chalcis where he died a year later.

He contributed much to the field of biology, especially through his early work on classification. In helping devise a classification system, he established the basic principles of dividing and subdividing plants and animals. At that time, only about a thousand species were known and he was able to group them into simple categories of animals with red blood (with backbones), and animals with no red blood (without backbones). Plants were divided into different categories that dealt more with size and appearance.

Aristotle's classification system remained intact for almost 2,000 years, until in the 1500s, scientists recognized that the growth of knowledge called for an expanded system. Modification came slowly, with great debate, and it revealed the complexity of the process. In the late 1700s, the Aristotelian classification system was finally replaced by a

much more comprehensive and systematic one developed by Swedish naturalist, Carl Linnaeus.

Aristotle's work on classification was not accidental. He was a painstaking observer, believing nature never created anything without a reason. He was particularly fascinated by sea creatures, often dissecting them and studying their natural habitats. This approach to anatomy led him to look for correlations between structure and function and to a belief that each biological part has its own special uses.

Ultimately, he established a teleological approach through constant inquiry about the ultimate purpose of things. This approach persisted in biological thinking well into the twentieth century.

Besides classifying animals and plants in nature, Aristotle also was the first to define and classify the various branches of knowledge. He sorted them into physics, metaphysics, rhetoric, poetics, and logic. In doing so, he laid the foundation of most of the sciences.

ARROW OF TIME

Arrow of **time** refers to the capability of distinguishing previous events from subsequent events by observing significant differences between the two events. The concept throughout history has represented a philosophical issue. However, it became a scientific issue with the advent of the theory of relativity and again with the discovery of phenomena at the subatomic level.

The second law of **thermodynamics**, or the law of **entropy**, states that the **universe** and all of its **energy** systems will increase in disorder as time moves forward. In physics, disorder means the breakdown of energy into useless **heat**, from which no **work** can be done.

The concept of entropy, the increase in disorder over time, can be expected to provide a **direction of time**. A movement into the future can be recognized by observing deterioration and decay of things in the past. The laws of entropy may mistakenly appear to have a challenge in an open system such as Earth. On Earth, forests can turn into wastelands over many years but wastelands can also grow into forests over many years. It should be noted that Earth, as an open system, has the **Sun** as an outside energy source. The seeming reverse of entropy on Earth is minute compared to the tremendous breakdown of atomic order necessary for the fusion within the Sun. If we could measure the entropy of the whole system (Sun included) we could use this total breakdown of order as an arrow of time.

The question of the reversibility of time has been an intriguing component of the theory of relativity. According to that theory, the perception of time and the effects of time are reduced under various conditions. This inconstancy of time leads to the theoretical question of the possible reversibility of time. A phenomenon illustrating reversibility of time is the barely perceptible lag of atomic clocks transported on jets traveling at high speeds. The people on such flights are a microproportion of time behind the people at the flight's origin and the people at the flight's destination. Is this discrepancy in the measure of time truly a lag or actually a slight reversal of time?

In studying **subatomic particles**, there are few means of detecting time as a systematic sequence of events. Particles continuously come together to form new particles only to separate to form particles identical to the original particles, which again interact to form new particles. The laws of physics allow that the incidence of new particles reverting back to the original particles may occur as easily as the incidence of original particles interacting to create new particles. Viewing the steps in this process, one has little clue which event came before which. A common standard for measuring time, observing entropy, is very difficult at the subatomic level.

In the late twentieth century, an arrow of time may have been discovered at the subatomic level. The important characteristic of the reversibility of time is the concept of charge-parity symmetry. This concept implies that there is no breakdown in the order of the subatomic particles at the different phases of interaction and separation. In 1998, the Fermilab in Illinois and the CERN (European Organization for Nuclear Research) in Switzerland reported independent results from work with subatomic particles. Both laboratories found significant differences associated with time in the activity of neutral particles called kaons. Therefore, as one becomes aware of what to look for, even at the subatomic level, one will find an arrow of time.

ASTRONOMY

Astronomy, the oldest of all the sciences, seeks to describe the structure, movements and processes of celestial bodies.

Some of the most ancient ruins of early civilization provide evidence that our most remote ancestors observed and attempted to understand the workings of the **cosmos**. Although not always fully understood, these ancient ruins demonstrate that early man attempted to mark the progression of the seasons as related to the changing of the apparent changing positions of the **Sun, stars, planets** and **Moon** on the celestial sphere. Archaeologists speculate that such observations made more reliable the determination of times for planting and harvest in developing agrarian communities and cultures.

The regularity of the heavens also profoundly affected the development of indigenous religious beliefs and cultural practices. For example, according to **Aristotle** (384-322 B.C.) the Earth occupied the center of the Cosmos, and the Sun and planets orbited the Earth in perfectly circular orbits at an unvarying rate of speed. The word astronomy itself is a Greek term meaning star arrangement. Although heliocentric (Sun centered) theories were also advanced among ancient Greek and Roman scientists, the embodiment of the **geocentric theory** conformed to prevailing religious beliefs and, in the form of the Ptolemaic model, were subsequently embraced by the growing Christian church that dominated Western thought until the rise of empirical science during the **Scientific Revolution** of the sixteenth and seventeenth centuries.

In the East, Chinese astronomers, carefully charted the night sky and noted the appearance of "guest stars" (**comets**, novae, etc.). As early as 240 B.C. the records of Chinese astronomers record the passage of one such guest star known now as Comet Halley. In A.D. 1054 the records indicate that one star became bright enough to be seen in daylight. Archaeoastronmers argue that this transient brightness was a supernova explosion, the remnants of which now constitute the Crab Nebula. The appearance of this supernova was also recorded in carvings made by the Anasazi Indians of the American Southwest.

Observations and astronomical recordkeeping were not, however, limited to spectacular celestial events. After decades of patient observation the Mayan peoples of Central America were able to accurately predict the movements of the Sun, Moon and stars and to devise a calendar that accurately predicted the length of a year, to what would now be measured to be within six seconds.

Early in the sixteenth century Polish astronomer Nicolaus Copernicus (1473-1543) reasserted the heliocentric theory abandoned by the Greeks and Romans. Although the Copernican model was deeply flawed by an insistence on circular orbits, it sparked a revoution n astronomy and science. Based upon the Coperican model, Danish astronomer Tycho Brahe's (1546-1601) precise observations of the celestial movements allowed German astronomer and mathematician **Johannes Kepler** (1571-1630) to fomulate his laws of planetary **motion** that correctly described the elliptical orbits of the planets.

Italian astronomer and physicist **Galileo** Galilei (1564-1642) was the first scientist to utilize a newly invented **telescope** to make recorded observations of celestial objects. In a prolific career, Galileo's discoveries, including phases of Venus and moons orbiting Jupiter dealt a death blow to geocentric theory.

In the seventeenth century English physicist and mathematician Sir Isaac Newton's (1642-1727) development of the laws of motion and gravitation marked the birth of **Newtonian physics** and modern **astrophysics**. In addition to developing calculus, Newton made tremendous advances the understanding of **light** and **optics** critical to the development of astronomy. Without question, Newton's seminal 1687 work, *Philosophiae Naturalis Principia Mathematica (Mathematical Principles of Natural Philosophy)*, dominated the Western intellectual landscape for more than two centuries and proved the direct impetus for rapid advancements in the study of celestial **dynamics**.

Theories surrounding celestial **mechanics** during the eighteenth century were profoundly shaped by important contributions by French mathematician Joseph-Louis Lagrange (1736-1813), French mathematician Pierre Simon de Laplace (1749-1827), and Swiss mathematician Leonhard Euler (1707-1783) that explained small discrepancies between Newton's predicted planetary orbits and the observed orbits of the planets. These explanations strengthened concepts of a clockwork-like mechanistic **universe** that operated according to knowable physical laws.

Just as primitive astronomy influenced early religious concepts, during the eighteenth century advancements in astronomy caused significant changes in Western scientific and theological concepts based upon an unchanging, immutable God who ruled over a static universe. During the course of the eighteenth century there developed a growing scientific disregard for understandings based upon divine revelation and a growing acceptance of an understanding of Nature based upon the application of scientific laws. Whether God intervened to operate the mechanisms of the universe through miracles or signs (such as comets) became a topic of lively philosophical and theological debate. Concepts of the divine became increasing identified with the assumed eternity or **infinity** of the Cosmos. Theologians argued that the assumed immutability of a static universe, a concept shaken by the discoveries of Copernicus, Kepler, Galileo and Newton, offered proof of the existence of God. In this regard, the clockwork universe could be viewed as confirmation of the existence of a God of infinite power who was the "prime mover" or creator of the universe. For many scientists and astronomers, however, the revelations of a mechanistic universe left no place for the influence of supernatural forces, and they discarded their religious views. These philosophical shifts sent sweeping changes across the political and social landscape of the Enlightenment.

Astronomers increasingly sought to explain "miracles" in terms of natural phenomena. Accordingly, by the nineteenth century the appearance of comets were no longer viewed as direct signs from God but rather a natural, explainable and predictable result of a deterministic universe. Explanations for catastrophic events (e.g., comet impacts, extinctions, etc.) increasingly came to be viewed as the inevitable results of time and statistical probability.

The need for greater accuracy and precision in astronomical measurements, particularly those used in navigation, spurred development of improved telescopes and pendulum driven clocks that greatly increased the pace of astronomical discovery. In 1781, improved mathematical techniques combined with technological improvements allowed English astronomer **William Herschel** to discover the planet Uranus.

Until the twentieth century, astronomy essentially remained concerned with the discovery and accurate description of planets and stars. Developments in electromagnetic theories of light and the formulation of quantum and relativity theories, however, allowed astronomers to probe the inner workings of the celestial objects. Influenced by German-American physicist Albert Einstein's theories of relativity and the emergence of **quantum theory**, Indian-born American astrophysicist **Subrahmanyan Chandrasekhar** first articulated the evolution of stars into supernova, **white dwarfs**, **neutron stars** and accurately predicting the conditions required for the formation of **black holes**. The articulation of the stellar evolutionary cycle allowed rapid advancements in cosmological theories regarding the creation of the universe. In particular, American astronomer Edwin Hubble's discovery of redshifted spectra from stars provided evidence of an expanding universe that, along with increased understanding of **stellar evolution**,

ultimately led to the abandonment of static models of the universe and the formulation of **big bang** based cosmological models.

In 1932, American engineer Karl Janskey (1905-1945) discovered existence of **radio waves** of emanating from beyond the Earth. Janskey's discovery led to the birth of **radio astronomy** that ultimately became one of the most productive means of astronomical observation and spurred continuing studies of the Cosmos across all regions of the **electromagnetic spectrum**.

Profound questions regarding the birth and death of stars led to the stunning realization that, in a very real sense, because the heavier atoms of which humans are comprised are derived from neucleosythesis in dying stars, mankind too was a product of stellar evolution. After millenniums of observing the Cosmos, by the dawn of the twenty-first century advances in astronomy allowed humans to gaze into the night sky and realize thatthey were looking at the light from stars distant in **space** and time, and that they themselves were made from the very dust of stars.

See also Chandrasekhar limit; Chandrasekhar mass; Cold stars; Comets; Cosmic background explorer (COBE); Cosmic background radiation; Cosmic inflation; Cosmic rays; Cosmological arrow of time; Cosmological constant; Cosmological heat death; Cosmological horizon; Cosmological principle; Cosmology; Dark matter; Doppler effect; Event horizon; Extraterrestrial life; Gravitational collapse; Gravity; Hawking radiation; Hubble constant; Hubble Space Telescope; Impacts; Kepler's laws of planetary motion; Kepler's second law and angular momentum; Microwave background radiation; Milky Way galaxy; Nebulae; Neptune; Newton's law of universal gravitation; Novae and super novae; Nuclear fusion; Nucleosynthesis; Olbers's paradox; Oort cloud; Primordial black holes; Pulsars; Quasars; Radiation; Radio astronomy; Red dwarf/Red giants; Red shift; Rydberg constant; Seeing (astronomy); Singularity; Solar-thermal energy; Space-time; Space-time geometry; Spectral lines; Spectroscopy; Speed of light; Standard model of cosmology; Stellar evolution; Stellar life cycle; Sun; Supernovae; Time; Universe; White dwarf; Worm holes; Zeeman effect

ASTROPHYSICS

Astrophysics is the field of science that combines astronomical observations with physical principles. It may be argued that the first true astrophysicist was Sir **Isaac Newton**, who explained the orbits of the **planets**, which had been known from antiquity, from his laws of **motion** and his universal law of **gravity**. As a field, astrophysics first began to mature in the latter half of the nineteenth century. This was due primarily to several key technical innovations that revolutionized the field of **astronomy**.

First, as the nineteenth century progressed, so did the sizes of the largest telescopes, and hence the furthest distances, as well as the smallest and dimmest features astronomers could observe. Whereas earlier astronomers usually worked in relative isolation using small telescopes, increased support from both governments and private individuals led to the founding of more observatories (like the U.S. Naval Observatory in 1844, the Lowell Observatory in 1894, the Yerkes Observatory in 1897, and the Mount Wilson Observatory in 1904) with improved equipment and larger telescopes.

Next was the invention of the photograph plate in the first half of the nineteenth century by Louis Daguerre. The first known astronomical application of the photographic plate was a Daguerrotype of the **Moon** taken by the American J. W. Draper in 1840. The impact of photographic plates on the field of astronomy was tremendous—images of the sky could be taken and stored for years, and accurate positions and brightnesses could be inferred from analyzing the plates at leisure afterwards.

Last was the invention of the **diffraction** grating for use in analyzing spectra from astronomical bodies. Seminal **work** on the diffraction grating was done by Joseph von Fraunhofer beginning in 1821, who demonstrated that the gratings could produce spectra of unprecedented quality. Although the grating was invented by an American astronomer, David Rittenhouse, in 1785, it was von Fraunhofer who first used it to study absorption lines in the spectrum of the **Sun**, and who derived equations governing their behavior. Scientists began to use the diffraction grating in their research after other instrument makers began to produce them in bulk after 1850. Accurate spectra of astronomical bodies would have an even greater impact on astrophysics than the photographic plate: by comparison with spectra obtained in gases in terrestrial laboratories, scientists could deduce the composition of vastly distant astronomical bodies by using the unique spectral "fingerprints" left by atomic and molecular species. Moreover, under the right conditions, the **density**, temperature, **velocity** (along the line of sight), and the magnetic field of the emitting gas, as well as similar properties for absorbing gas along the line of sight, could all be inferred from a spectral measurement. The impact which the advent of spectroscopy would have on astrophysics cannot be overemphasized: whereas previously **stars** and other astronomical bodies were simply lights in the sky, afterwards they were physical systems whose properties could be measured, modeled, and understood.

The advent of **quantum mechanics**, **relativity theory**, and **nuclear physics** during the first half of the twentieth century led to some of the most impressive strides in modern astrophysics. Although late nineteenth century astronomers had cataloged and archived volumes of data, they often lacked the understanding of basic physical principles to comprehend their observations. With a deeper understanding of fundamental physical principles, enormous strides were made in all branches of astrophysics during the first half of the 20th century. In **cosmology**, Einstein's theory of general relativity was able to explain Edwin Hubble's observation that galaxies seemed to recede from us in all directions: in Einstein's theory, the **universe** itself was dynamic, and expanding in **time**. Building on a century of work in stellar physics, **Hans Bethe** finally cracked the riddle of how stars like our Sun shine by drawing from terrestrial knowledge of **nuclear reactions**. The

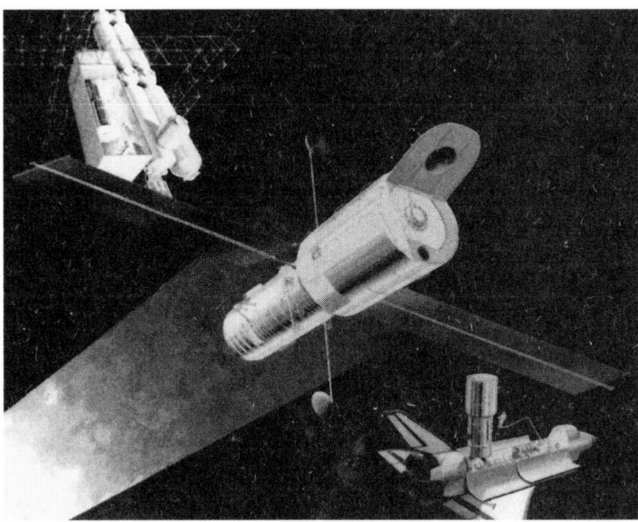

NASA's x-ray astrophysics facility.

list of such contributions of physical principles with astronomical observations goes on and on: the union of the two has been quite successful indeed.

The period from the 1950s to today has been one of unprecedented expansion in astrophysics, by any standard of measurement: volumes of data, numbers of papers published, and numbers of researchers in the field. Just as in the birth of modern astrophysics during the period after 1850, the reasons are primarily technological. The **space** age led to the launching of new satellites and missions to other planets and the Sun opening up entirely new frontiers in our knowledge of the universe. Also from space, astronomers can gaze through many bands of the **electromagnetic spectrum** not visible from the ground because the earth's atmosphere blocks them out. Indeed, only **radio waves** and optical **light** are easily transmitted through the earth's atmosphere. Finally, the invention of the modern computer, which is ideally suited to storing and manipulating enormous volumes of data, was absolutely crucial in assisting astronomers in their capacity to deal with the ever-increasing tide of information from their observations. Indeed, the modern computer has led to a third branch of astrophysics, alongside observation and theory: computational astrophysics, which uses the computer to simulate nature under the model conditions put forth by a scientist.

Today, astrophysicists delve deeper into the discoveries of previous generations, seeking to understand the myriad of new questions that surface with every discovery. Cosmologists are narrowing in on the lifetime, geometry, and future of the universe itself. Extragalactic astrophysicists puzzle over the nature of fantastically energetic bursts of **gamma radiation**, which are now known to sometimes originate in galaxies far away from our own. Scientists studying our own galaxy seek to understand how it formed, and whether it has a supermassive black hole at its center as do many other galaxies. Other scientists studying the physics of the interstellar medium (ISM) try to comprehend the vastly complicated processes by which the tenuous ISM eventually forms stars, and how those stars replenish the ISM with **energy** and new material from their nuclear forges through winds and explosions. Experts on very compact **white dwarfs, neutron stars**, and **black holes** study the enormously energetic and fascinatingly rich array of phenomena that these systems exhibit. Still others try to make sense of revolutionary discoveries of planets around other stars, which are quite unlike those in our own solar system. At the dawn of the new millennium, astrophysicists are learning how the universe works, and the end is still nowhere in sight.

ATMOSPHERIC REFRACTION

There are three common phenomena that are attributed to atmospheric refraction—the colors of the sky at sunset (and sunrise), mirages, and looming. **Light** traveling from one medium to another will undergo refraction, i.e. it will be bent from its original straight-line path. When traveling from a medium of low **density** to one of a higher density, the light is bent towards the normal. It is bent away from the normal if traveling from high to low density.

Sunlight traveling from **space** into the earth's atmosphere will undergo a large amount of refraction. The amount of refraction is dependent upon the angle of incidence and the **wavelength** of the light. When the **Sun** is overhead, the incident angle is very small and so the Sun appears white. When on the horizon, the sunlight hits at an angle that causes the colors to separate much like a prism separates light into a rainbow. The particles in the atmosphere enhance this separation of **color**. As the Sun dips below the horizon, the sky goes through color changes from red to indigo.

A mirage is caused by a small refraction of light near a hot surface. The hotter, denser air refracts the light coming from the horizon and the viewer sees an image of the sky near the horizon. Since the image is blue or gray, it is commonly mistaken for a pool of water. Here, colors are not separated, since the density difference in the mediums is not high.

Looming is caused by a small refraction of light off a cold surface. Light from a distant mountain is bent downward by the less dense air and the mountain appears to be looming over the observer. The mountain looks bigger and taller and thus closer to the observer on a cold day than on a warm day.

ATOM

Atoms are the elementary building blocks of material substances. Although the term atom, derived from the Greek word *atomos*, meaning indivisible, would seem inappropriate for an entity that, as science has established, is divisible, the word atom still makes sense, because, depending on the context, atoms can still be regarded as indivisible. Namely, once split, the atom loses its identity. For example, an atom of gold is the building block of gold. If we split a gold atom, there is no

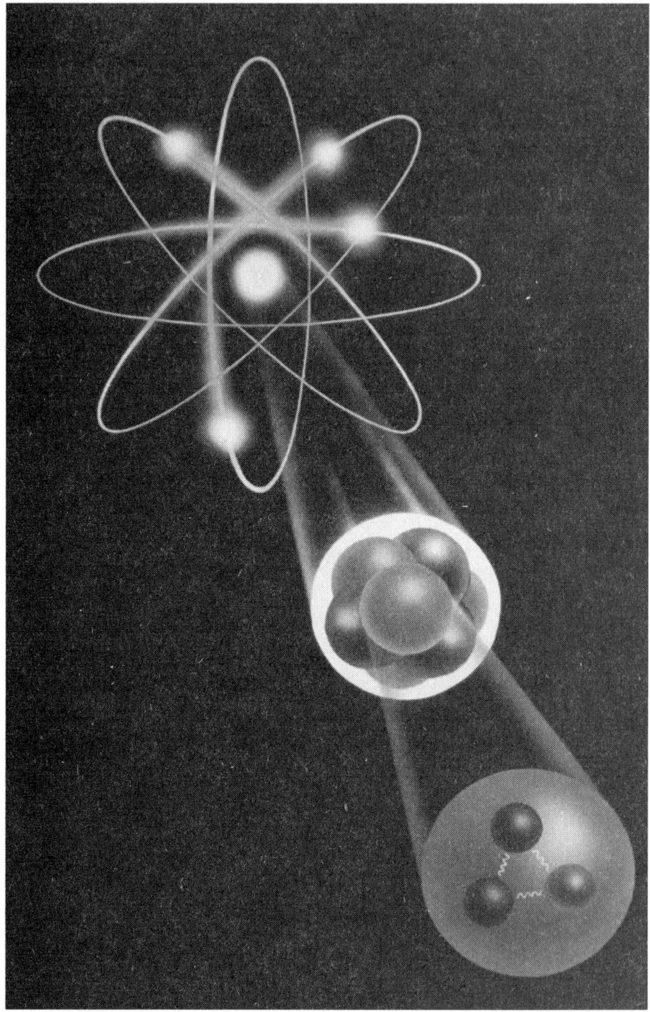

Atom, diagram of the structure of the atom. (Image by Michael Gilbert. Photo Researchers, Inc. Reproduced by permission.)

more gold. As Carl H. Snyder has written, an atom is "the smallest particle of an element that we can identify as that element."

Atoms share many characteristics with other material objects: they can be measured, and they also have **mass** and **weight**. For example, a gold atom, which is could be regarded as an atom of average size, measures approximately $3\times10^{10-8}$ in diameter, and its mass is about 3.3×10^{-22} g. Because traditional methods of measuring are difficult to use for atoms and **subatomic particles**, scientists have created a new unit, the atomic mass unit (amu), which is defined as the one-twelfth of the mass of the average **carbon** atom.

The principal subatomic particle are the **nucleus**, the **proton**, the **neutron**, and the **electron** (*elektron* is Greek for yellow amber; the ancient Greeks were among the first to observe static **electricity**, produced when a piece of amber is rubbed). A nucleus, the atom's core, consists of protons, which are positively charged particles, and neutrons, particles without any charge. Electrons are negatively charged particles with negligible mass which move around the nucleus. Compared to an atom's total mass, the subatomic particles are truly miniscule. For example, neutrons and protons have the mass of 1.67 $\times$ 10^{-24}, while electrons have the hardly detectable mass of 0.0009 $\times$ 10^{-24}. An electron's mass is so small that it is usually given a 0 value in atomic mass units, compared to the value of 1 assigned to neutrons and protons. In fact, as the nucleus represents more than 99% of an atom's mass, it is interesting to note that an atom is mostly space. For example, if a hydrogen atom's nucleus were enlarged to the size of a marble, the atom's diameter would be around 0.5 mi (800 m).

Scientists used to believe that electrons circled around the nucleus in planet-like orbits. Because all subatomic particles, including electrons, exhibit wave-like properties, it is makes no sense to conceptualize the movement of electrons as planetary rotation. Scientists therefore prefer terms like "electron cloud patterns," or "shells," indicating an electron's pattern of movement in relation to the nucleus. Thus, for example, hydrogen has one electron shell, containing one electron; lithium has two shells, with one electron in the inner shell, and two in the outer. Scientists have posited the existence of four shells, each containing one or more electrons, and each defined by a particular level of **energy**.

While subatomic particles may, in a certain way, be regarded as generic and interchangeable, they, in fact, determine an atom's identity. For example, we know that an atom with a nucleus consisting of one proton must be hydrogen (H). An atom with two protons is always a **helium** (He) atom. Thus, we see that the key to an atom's identity is to be found in the atom's inner structure. In addition, a **chemical element** is an instance of atomic **equilibrium**: for example, in a chemical element, the number of positively charged particles (protons) always equals the number of negatively charged particles (electrons). While elements are not always stable, sometimes appearing as different isotopes, the balance of forces within an atom is regarded as a fundamental feature.

Research into the atom's nucleus has uncovered a variety of subatomic particles, including **quarks** and gluons. Considered by some researchers the true building blocks of **matter**, quarks are the particles which form protons and neutrons. Gluons hold smaller clusters of quarks together.

Although the atom, particularly its mysterious inner world of quantum laws and wave-like particles, may seem far removed from the practical concerns of **chemistry**, and more relevant to theoretical physics than to anything else, there is a clear connection between **atomic structure** and the tangible, observable characteristics of substances, which is obviously the domain of chemistry. In particular, chemists study reactions, the behavior of elements in interaction, and reactions, such as those leading to the formation of chemical compounds, involve electrons. For example, the formation of sodium chloride, also known as table salt, would be impossible without specific changes at a subatomic level. The genesis of sodium chloride (NaCl) starts when a sodium (Na) atom, which has 11 electrons, loses an electron. With ten electrons, the atom now has one proton too many, thus becoming a positively charged ion, or cation. The electron lost by the sodium atom is gained by a chlorine atom, which has 17 electrons distributed in three shells. The newly acquired electron goes into the outer shell, also known as the valence shell, which contains seven electrons. The chlorine atom's behavior also exemplifies the general tendencies of atoms to enter reactions in order to attain eight electrons in the valence shell. But the chlorine atom not only reaches an octet balance; it also becomes a negatively charged ion, or anion, since it now has 18 electrons to its 17 protons. Crystals of table salt consist of equal numbers of sodium cations and chlorine anions, cation-anion pairs being held together by a **force** of attraction.

Interestingly, not long after scientists realized that an atom is divisible, transmutation, or the old alchemic dream of turning one substance into another, became a reality. Scientists even succeeded in creating gold by bombarding platinum-198 with neutrons to create platinum-199 which decays to gold-199. Although clearly demonstrating the reality of transmutation, this particular is by no means a cheap method of producing gold—quite the contrary, since platinum, particularly the platinum-199 **isotope** is more expensive than gold. However, the symbolic value of the experiment is immense, as it shows that the idea, developed by ancient alchemists and philosophers, of material transmutation does not essentially contradict our understanding of the atom.

ATOMIC BOMB

An atomic bomb is a weapon of mass destruction which uses **nuclear fission** to produce vast amounts of **energy**. Its basic principle is assembling a **critical mass** of a radioactive element with properties to spark a swift **chain reaction**. Its devastating effect is twofold: first, from the **heat** of the thermonuclear reaction, and second, from the **radioactive fallout** which continues after the bomb has been deployed.

The difference between an atomic bomb and a nuclear reactor is in the speed and control of the release of energy. It is precisely those features which are undesirable in a nuclear reactor that are vital in an atomic weapon. An atomic weapon must chain react swiftly enough that the entire reaction happens in a matter of minutes. Uranium-235 and plutonium are generally used for atomic bombs because of their ability to fission when hit by "slow" neutrons. That is, these substances can fission at all energy levels, not just at very high ones.

In order for an atomic bomb to only harm its targets and not its creators, the critical mass must remain separated until the bomb is supposed to go off. There were two original design solutions for this problem: the gun method and the implosion method. The gun method involves shooting a plug of radioactive matter at lower than critical mass into a cavity in a block of the same substance. The plug and block together then form critical mass. The implosion method packs explosive around the outside of two hemispheres. When the explosives go off, the hemispheres are pushed together into a sphere of critical mass.

In a typical uranium bomb, the critical mass of uranium is assembled inside a moderator. The moderator provides neu-

Atomic bomb blast (mushroom cloud, ground level view). *(Photo from National Archives and Records Administration. Reproduced by permission.)*

trons, which are captured by the uranium atoms. The atoms then undergo nuclear fission, producing daughter nuclei and more neutrons. The neutrons produced by the fission have to be going at the right speed to be captured by other uranium atoms, which continues the process.

The first atomic bomb was exploded on July 16, 1945, in Alamogordo, New Mexico. It was called the Trinity Test. The bomb had been designed by members of the **Manhattan Project**, the American wartime effort in **atomic physics**. In August of the same year, American Air **Force** troops dropped two bombs over the Japanese cities of Hiroshima and Nagasaki. Since the Second World War, atomic weapons have been developed by many countries but not used in warfare. The doctrine of Mutually Assured Destruction originally held that the **nuclear weapons** hold the world's superpowers in check because each side was aware of the devastating **power** of the atomic bomb. The absence of a second superpower has done nothing to decrease fear and awe of this work of science and technology. The creation of the atomic bomb did, however, prompt the formation of several professional groups of physicists opposed to nuclear warfare.

See also Alpha radiation; Binding energy; Chain reaction; Nuclear reactions; Nuclear reactors; Nuclear weapons; Radioactive fallout; Radioactive isotopes

ATOMIC CLOCK

The atomic clock is the most precise measurement of **time** that exists. The National Institute of Standards and Technology based in Boulder, Colorado houses a cesium atomic clock (so-called because it uses atoms of cesium) that is accurate to one second in six million years. Before the atomic clock, the rotational period of Earth was taken as the standard of time. The second was defined as 1/86,400 of a solar day. But the Earth's rotation is irregular, and the second needed to be redefined.

American physicist Isador Rabi suggested in the 1940s that a clock could be based on an atom's internal **oscillations**. Atoms are accurate because of their reliability: an atom's **motion** will never falter or deviate from a fixed period. In 1949, the U.S. National Bureau of Standards (now the NIST) built the first atomic clock using a molecule of ammonia. A

Atomic clock. *(Corbis-Bettmann. Reproduced by permission.)*

few years later, the National Bureau of Measures built a clock using cesium atoms. Since 1967, the official international length of the second has been defined by atomic standards.

In an **atom**, electrons orbit a **nucleus**. According to the laws of **quantum mechanics**, these electrons may exist only in certain states. In other words, they may orbit the nucleus in one of two ways. They may move between these states if they are exposed to electromagnetic **radiation** (**light** or **radio waves**) at a specific frequency. The correct frequency depends on the amount of **energy** it takes for the electrons to move from one state to another. Electrons "know" when they are exposed to radiation at the right frequency, and only then will they shift from one energy state to another. The atomic clock is based on the period of that shift.

In an atomic clock, atoms are sprayed through electromagnetic radiation (such as **microwaves**) that oscillate at the frequency needed to make the electrons jump from one energy state to another. If the frequency is correct, then a majority of the original atoms should make the transition from their original state. A counter in the clock tracks the time it takes for most of the atoms to make the shift. In a cesium clock, it has been determined that one second is equal to 9,192,631,770 oscillations or transitions.

Atomic clocks are useful for satellite navigation, which allow us to measure distances on Earth. They also helped con-

firm Einstein's prediction that time slows down for observers moving at great speeds.

Atomic force microscope

In recent years, tremendous advances have been made in the field of microscopy. In 1985, the atomic force microscope (AFM) was invented by **Gerd Binnig** (co-inventor of the scanning tunneling microscope), Christoph Gerber in Zurich, Switzerland, and Calvin Quate in California. The AFM represents the technological pinnacle of microscopy.

The AFM uses a tiny needle made of diamond, tungsten, or silicon, much like those used in the STM. While the STM relies upon a subject's ability to conduct **electricity** through its needle, the AFM scans its subjects by actually lightly touching them with the needle. Like that of a phonograph record, the AFM's needle reads the bumps on the subject's surface, rising as it hits the peaks and dipping as it traces the valleys. Of course, the topography read by the AFM varies by only a few molecules up or down, so a very sensitive device must be used to detect the needle's rising and falling. In the original model, Binnig and Gerber used an STM to sense these movements. Other AFM's use a fine-tuned laser.

The AFM has already been used to study the supermicroscopic structures of living cells, objects that could not be viewed with the STM. American physicist Paul Hansma and his colleagues at the University of California, Santa Barbara, are quickly becoming experts in AFM research. In 1989, this team succeeded in observing the blood-clotting process within blood cells. Hansma's team presented their findings in a 33-minute movie, assembled from AFM pictures taken every ten seconds.

Other scientists are utilizing the AFM's ability to remove samples of cells without harming the cell structure. By adding a bit more force to the scanning needle the AFM can scrape cells, making it the world's most delicate dissecting tool. Scientists hope to apply this method to the study of living cells, particularly floppy protein cells, whose fragility makes them nearly impossible to view without distortion.

Atomic orbitals • See Atomic structure

Atomic physics

Atomic physics is a branch of physics devoted to the study of the material composition and energetic processes of atoms and molecules.

Concepts of the **atom** and of **atomic structure** date to the ancient Greek philosopher **Democritus**, who proposed that all things are made of atoms and that, although they are too small to be seen in their ceaseless **motion**, they could collide and accumulate.

The contemporary study of physics (i.e., the study of **matter** and **energy** and the interaction between them) is rough-

　　　　　　　　　　　　　　　　　　　　WORLD OF PHYSICS

ly divided into two categories: classical physics and modern physics. Classical physics includes the traditional branches of physics (e.g. **mechanics**) developed before the beginning of the 20th century. Most of classical physics is concerned with behavior of energy and matter on the macroscopic scale of observation. On the other hand, atomic, particle, and **nuclear physics** comprise areas of modern physics. Unlike classical physics, modern physics is concerned with energy and matter under extreme conditions or on the microscopic scale. On this small-scale ordinary, commonly accepted aspects of motion, **space**, matter and energy are no longer applicable. Accordingly, atomic physics relies on the principles of **quantum mechanics** to describe the discrete nature of matter at the atomic and subatomic levels. Atomic physics involves specifically investigations into the electronic parts of the atom.

Near the end of the 19th century, English physicist J. J. Thomson's (1856-1940) discovery of the **electron** proved that the atom was not an indivisible particle. Subsequent to Thomson's discovery, various models of atomic structure evolved to describe atomic structure. Thomson's model of the atom was a solid sphere that had electrons embedded in a positive part thought to be a fluid. He thought that the positive part made up the bulk of the atom's **mass** and volume. In 1909-1911 English physicist **Ernest Rutherford** (1871-1937) developed an atomic model based upon the observation of the deflection of alpha particles bombarding thin foils. Rutherford's model of the atom contained negatively charged particles surrounding a **nucleus** of tightly packed positive charges. Importantly, Rutherford proposed that most of the atom was empty space.

Rutherford's proposal led to the birth of atomic physics. Subsequent studies of the atom were divided into investigations of the electronic parts of the atom, atomic physics, and investigations of the nucleus itself, nuclear physics. The division arose because of the huge difference in energy required to produce nuclear as opposed to electronic changes.

Although Rutherford's **atomic theory** had two major problems it was the basis for studies involving the atom. In 1913 Danish physicist **Niels Bohr** (1885-1962), who had studied under Rutherford, corrected the problems with Rutherford's atomic theory. Instead of existing in arbitrary **orbitals** Bohr proposed that electrons are found only in certain states. Using German physicist Maxwell Planck's (1858-1947) theoretical insights, **Planck's constant**, Bohr was able to create a formula used to find the energy levels of different atoms. Although insightful and useful as a conceptual model, in reality the Bohr model could only be applied to the simple **hydrogen atom**. Detailed models of more complex atoms proved elusive and mathematically complex.

In 1923 French physicist **Louis de Broglie** (1892-1987) proposed that electrons have a **particle-wave duality** wherein they have properties of both particles and **waves**. This idea supported Planck and German-American physicist **Albert Einstein's** theory that the fundamental entity of electromagnetic **radiation**, the **photon**, has both particle and wave properties. In 1924, using Einstein's special theory of relativity, de Broglie showed that particles have waves associated with

them and that electrons cannot be pictured as localized particles in space but rather should be thought of as clouds of negative charge spread out over the entire orbit. These clouds represent the regions around the nucleus where the probability of finding an electron is the largest. Electrons occupy different energy levels in atoms and have different wave properties. Because electrons behave both as particles and waves it became obvious that a specific type of partial differential equation should be used to describe the position and path of electrons.

In 1926, Austrian physicist **Erwin Schrödinger** (1887-1961) used partial differential equations and the Hamiltonian function to develop a powerful equation that relates the energy of the electron and the energy of the electric field in which it is situated. This equation, the Schrödinger equation, relates the energy of a system to its wave properties and allows one to predict the energy of the electron and its future behavior (i.e. the probability of finding the electron in a particular region around the nucleus). Shortly after the advancement of the Schrödinger equation German physicist **Max Born** postulated that the **wave function** could be used to determine the probability of finding a particle in a particular region at a specific **time**. In 1927 German physicist **Werner Heisenberg** developed a formula describing atoms in terms of the frequencies of their atomic **spectral lines** and set forth the Heisenberg uncertainty principle that states that one cannot know both the position and the **velocity** of a particle simultaneously.

Subject to the limitations of the Heisenberg uncertainty principle, the advancement of atomic physics and **quantum physics** allowed increasingly accurate descriptions of complex atoms. The delineation of atomic substructure and mechanisms of subatomic processes evolved into the modern study of **particle physics**.

See also Atom; Atomic orbitals; Bohr theory; Bohr atom; Quantum theory

ATOMIC SPECTROSCOPY • See Spectroscopy

ATOMIC STRUCTURE

Atomic theory, first put in a quantitative conceptual framework by **John Dalton**, and **quantum theory**, which emerged in the 1920s as a result of the work of **Werner Heisenberg, Erwin Schrödinger**, and **Max Planck**, are the cornerstones of our present-day view on atomic structure. Atomic theory holds that **matter** consists of vast numbers of small particles called atoms which combine together to form molecules existing in the three main **states of matter** as gases, liquids or solids. Atoms are very small, their diameter lies between 0.7 and 6 Ångstroms (1 Ångstrom = 10^{-10} m), as revealed by x-ray **crystallography**, the method of choice to determine the size of atoms from a pattern of **x rays** diffracted on a crystal. At the turn of the century, the work of **J.J. Thomson, Robert Millikan** and **Ernest Rutherford** showed that atoms are electrically neutral and consist of a central positively charged **nucleus** surrounded by a cloud of negatively charged electrons.

Thomson's experiments led him to propose a "plum-pudding" view of the **atom** in which a continuous distribution of positive **mass** extends over the size of the atom with negative "plums" of much smaller mass (i.e. electrons) inserted into it. This model was overthrown by a series of **alpha particle scattering** experiments carried out by Rutherford, **Hans Geiger** and Ernest Marden. They were able to observe back-scattering of alpha particles emitted by a piece of radium as they were being shot through a thin gold foil. The fact that alpha particles are positive and that some of them were scattered back could only be explained by proposing that the positive charge and mass in the gold atoms making up the foil could not be continuous and had to be concentrated in a very small region and that the negative region had to be large enough to let some alpha particles through. This led to a view of the atom in which the positive nucleus is central, very dense and significantly smaller than the size of the overall atom.

Modern **neutron** scattering experiments have shown that the radius of a nucleus is proportional to the cubic root of its mass number and that atomic radii, including **electron** clouds, are about twenty thousand times bigger with a spherical shape or elongated like a football.

Atomic nuclei are now considered to consist of two different types of particles of almost equal mass: protons, which have a positive charge and a mass of 1.6726×10^{-24} g and neutrons, which have no charge and a mass of 1.6750×10^{-24} g. The nucleus of the lightest element, hydrogen, consists of a single **proton**. Atoms with the same number of protons and neutrons are called nuclides and atoms with a unequal number of protons and neutrons are called isotopes.

The total number of protons and neutrons in a nucleus is its mass number, A. The mass of the nucleus, in atomic mass units (amu), is nearly equal to its mass number. One amu is one twelfth of the mass of one **isotope** (1 amu = 1.660244×10^{-24} g).

The mass of an electron is 9.1094×10^{-28} g. The electron has the same charge as a proton: 1.6022×10^{-19} coulombs, but is oppositely charged, i.e. negative. Overall, atoms are neutral and the number of protons in their nuclei equals the number of electrons. This number (Z) is called the atomic number. For example, the atomic number of Na is 11; this means that sodium has 11 electrons occupying concentric **energy** levels around the nucleus, which consists of 11 protons.

The view that electrons can be thought of as being arranged in successive shells of increasing energy around the nucleus is called the Bohr model of the atom and one of the most significant contributions of quantum theory has been to show that these energy levels are quantized. This led to the orbital model of the atom, in which the discrete energy levels accomodating the electrons are ordered by increasing energy. Each level can only have one s (spherical) orbital containing a maximum of 2 electrons. In every **energy level**, these **orbitals** are always filled first and the others (p, d, f, g,...) afterwards, following the rules of the aufbau principle.

When the lowest energy levels are thus successively occupied by the electrons of an atom, it is said to be in the ground state. The addition of energy, for example **heat** or **light**, can promote electrons to higher energy levels. The atom is then said to be in an excited state. The study of the interaction of electromagnetic energy and matter is the subject of spectroscopy which is used to gain insight into the properties of matter and atomic structure.

ATOMIC THEORY

Atomic theory is the description of atoms, the smallest units of elements. The scientific evidence for the existence of atoms and its even smaller constituents is so vast that most people now consider the existence of atoms to be a fact and not just a theory.

History

Beginning in about 600 B.C., many Greek philosophers struggled to understand the nature of **matter**. Some said everything was made of water, which comes in three forms (solid ice, liquid water, and gaseous steam). Others believed that matter was made entirely of fire in ever-changing forms. Still others believed that whatever comprised matter, it must be something that could not be destroyed but only recombined into new forms. If they could see small enough things, they would find that the same "building blocks" they started with were still there. One of these philosophers was named **Democritus**. He imagined starting with a large piece of matter and gradually cutting it into smaller and smaller pieces, finally reaching the smallest piece. This tiniest building block that could no longer be cut he named *a-tomos*, Greek for "no-cut." *Atomos* has been changed in modern times to "atom." The atoms Democritus envisioned differed only in shape and size. In his theory, different objects looked different because of the way the atoms were arranged. **Aristotle**, one of the most influential philosophers of that **time**, believed in some kind of "smallest part" of matter but not with Democritus's descriptions. Aristotle said there were only four elements (earth, air, fire, water) and that these had some smallest unit that made up all matter. Aristotle's teachings against the idea of Democritus's **atom** were so powerful that the idea of the atom fell out of philosophical fashion for the next 2,000 years.

Although atomic theory was abandoned for this long period, scientific experimentation, especially in **chemistry**, flourished. From the Middle Ages (A.D. 1100) onward, many chemical reactions were studied. By the seventeenth century, some of these chemists began thinking about the reactions they were seeing in terms of smallest parts. They even began using the word atom again. One of the most famous chemists of the end of the eighteenth century was Antoine Lavoisier. His chemical experiments involved very careful weighing of all the chemicals. He reacted various substances until they were in their simplest state. He found two important factors: (1) the simplest substances, which he called elements, could not be broken down any further, and (2) these elements always reacted with each other in the same proportions. The same more complex substances he called compounds. For example, two

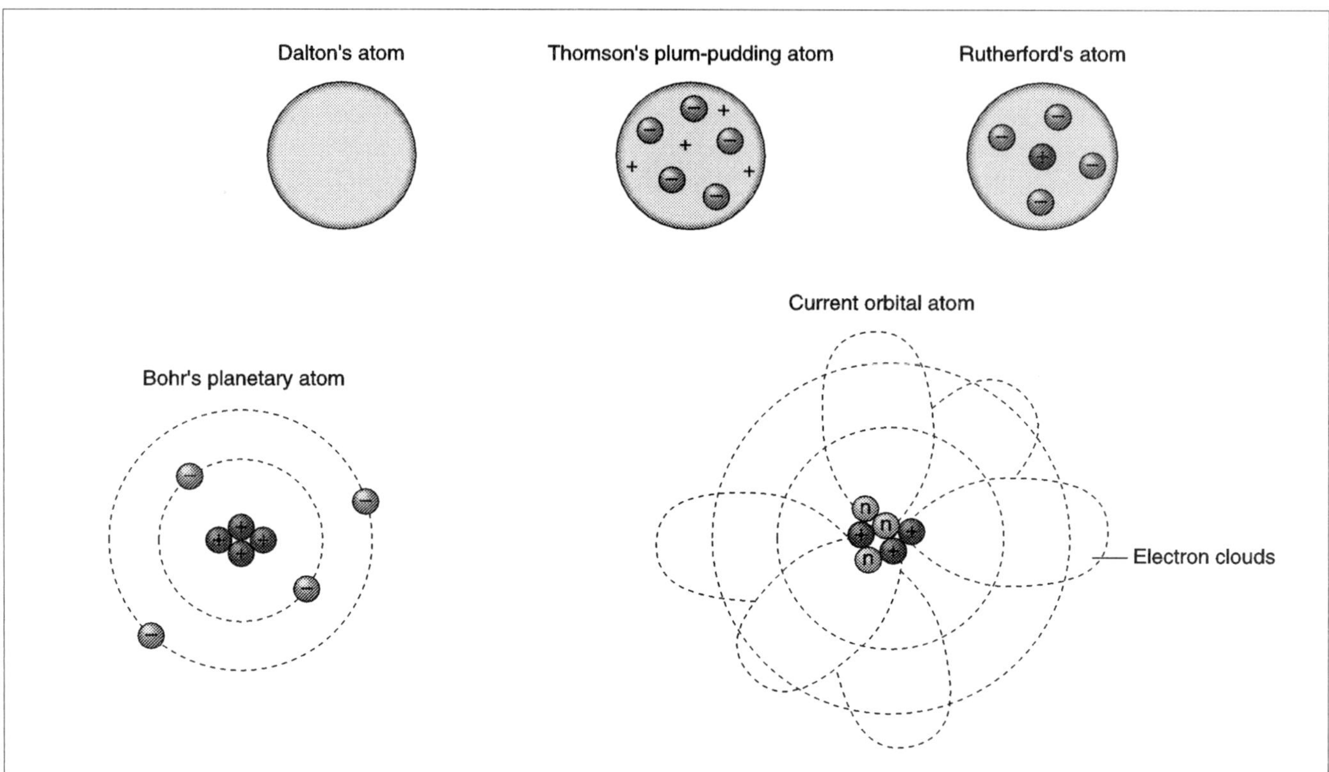

Atomic theory.

volumes of hydrogen reacted exactly with one volume of **oxygen** to produce water. Water could be broken down to always give exactly two volumes of hydrogen and one volume of oxygen. Lavoisier had no explanation for these amazingly consistent results. However, his numerous and careful measurements provided the clue to another chemist named **John Dalton**.

Dalton realized that if elements were made up of atoms, a different atom for each different element, atomic theory could explain Lavoisier's results. If two atoms of hydrogen always combined with one atom of oxygen, the resulting combination of atoms, called a molecule, would be water. Dalton published his explanation in 1803. This year is considered the beginning of modern atomic theory. Scientific experiments that followed Dalton were attempts to characterize how many elements there were, what the atoms of each element were like, how the atoms of each element were the same and how they differed, and, ultimately, whether there was anything smaller than an atom.

Describing Characteristics of Atoms

One of the first attributes of atoms to be described was relative atomic **weight**. Although a single atom was too small to weigh, atoms could be compared to each other. The chemist Jons Berzelius assumed that equal volumes of gases at the same temperature and **pressure** contained equal numbers of atoms. He used this idea to compare the weights of reacting gases. He was able to determine that, for example, oxygen atoms were 16 times heavier than hydrogen atoms. He made a

list of these relative atomic weights for as many elements as he knew. He devised symbols for the elements by using the first letter or first two letters of their Latin names, a system still in use today. The symbol for hydrogen is H, for oxygen is O, for sodium (natrium, in Latin) is Na, and so on. The symbols also proved useful in describing how many atoms combine to form a molecule of a particular compound. For example, to show that water is made of two atoms of hydrogen and one atom of oxygen, the symbol for water is H_2O. One oxygen atom can even combine with one other oxygen atom to produce a molecule of oxygen with the symbol O_2.

As more and more elements continued to be discovered, it became convenient to begin listing them in symbol form in a chart. In 1869, Dmitri Mendeleev listed the elements in order of increasing atomic weight and grouped elements that seemed to have similar chemical reactions. For example, lithium (Li), sodium (Na), and potassium (K) are all metallic elements that burst into flame if they get wet. Similar elements were placed in the same column of his chart. Mendeleev began to see a pattern among the elements, where every eighth element on the atomic weight listing would belong to the same column. Because of this periodicity or repeating pattern, Mendeleev's chart is called the **Periodic Table** of the Elements. The Table was so regular, in fact, that when there was a "hole" in the table, Mendeleev predicted that an element would eventually be discovered to fill the place. For instance, there was a **space** for an element with an atomic weight of about 72 (72 times heavier than hydrogen) but no known element. In 1886, 15 years after its prediction, the element Germanium (Ge) was isolated and found to have an atomic weight of 72.3. Many more elements continued to be predicted and found in this way. However, as more elements were added to the Periodic Table, it was found that if some elements were placed in the correct column because of similar reactions, they did not follow the right order of increasing atomic weight. Some other atomic characteristic was needed to order the elements properly. Many years passed before the correct property was found.

As chemistry experiments were searching for and characterizing more elements, other branches of science were making discoveries about **electricity** and **light** that were to contribute to the development of atomic theory. **Michael Faraday** had done much **work** to characterize electricity; **James Clerk Maxwell** characterized light. In the 1870s, **William Crookes** built an apparatus, now called a Crookes tube, to examine "rays" being given off by metals. He wanted to determine whether the rays were light or electricity based on Faraday's and Maxwell's descriptions of both. Crookes's tube consisted of a glass bulb, from which most of the air had been removed, encasing two metal plates called electrodes. One electrode was called the **anode** and the other was called the cathode. The plates each had a wire leading outside the bulb to a source of electricity. When electricity was applied to the electrodes, rays appeared to come from the cathode. Crookes determined that these cathode rays were particles with a negative electrical charge that were being given off by the metal of the cathode plate. In 1897, **J. J. Thomson** discovered that these negatively charged particles were coming out of the atoms and must have been present in the metal atoms to begin with. He called these negatively charged **subatomic particles** "electrons." Since the electrons were negatively charged, the rest of the atom had to be positively charged. Thomson believed that the electrons were scattered in the atom like raisins in a positively-charged bread dough, or like plums in a pudding. Although Thomson's "plum-pudding" model was not correct, it was the first attempt to show that atoms were more complex than just homogeneous spheres.

At the same time, scientists were examining other kinds of mysterious rays that were coming from the Crookes tube that did not originate at its cathode. In 1895, Wilhelm Roentgen noticed that photographic plates held near a Crookes tube would become fogged by some invisible, unknown rays. Roentgen called these rays "x rays," using "x" for unknown as in mathematics. Roentgen also established the use of photographic plates as a way to take pictures of mysterious rays. He found that by blocking the **x rays** with his hand, for instance, bones would block the x rays but skin and tissue would not. Doctors still use Roentgen's x rays for imaging the human body.

Photographic plates became standard equipment for scientists of Roentgen's time. One of these scientists, Henri Becquerel, left some photographic plates in a drawer with uranium, a new element he was studying. When he removed the plates, he found that they had become fogged. Since there was nothing else in the drawer, he concluded that the uranium must have been giving off some kind of rays. Becquerel showed that this **radiation** was not as penetrating as x rays since it could be

blocked by paper. The element itself was actively producing radiation, a property referred to as **radioactivity**. Largely through the work of Pierre and **Marie Curie**, more radioactive elements were found. The attempts to characterize the different types of radioactivity led to the next great chapter in the development of atomic theory.

In 1896, **Ernest Rutherford**, a student of J. J. Thomson, began studying radioactivity. By testing various elements and determining what kinds of materials could block the radiation from reaching a photographic plate, Rutherford concluded that there were two types of radioactivity coming from elements. He named them using the first two letters of the Greek alphabet, alpha and beta. **Alpha radiation** was made of positively charged particles about four times as heavy as a **hydrogen atom**. **Beta radiation** was made of negatively charged particles that seemed to be just like electrons. Rutherford decided to try an experiment using the alpha particles. He set up a piece of thin gold foil with photographic plates encircling it. He then allowed alpha particles to hit the gold. Most of the alpha particles went right through the gold foil. But a few of them did not. A few alpha particles were deflected from their straight course. A few even came straight backward. Rutherford wrote that it was as surprising as if one had fired a bullet at a piece of tissue paper only to have it bounce back. Rutherford concluded that since most of the alpha particles went through, the atoms of the gold must be mostly empty space, not Thomson's space-filling plum-pudding. Since a few of the alpha particles were deflected, there must be a densely packed positive region in each atom that he called the **nucleus**. With all the positive charge in the nucleus, the next question was the arrangement of the electrons in the atom.

In 1900, physicist **Max Planck** had been studying processes of light and **heat**, specifically trying to understand the light radiation given off by a "black-body," an ideal cavity made by perfectly reflecting walls. This cavity was imagined as containing objects called oscillators which absorbed and emitted light and heat. Given enough time, the radiation from such a black-body would produce a colored-light distribution called a spectrum that depended only on the temperature of the black-body and not on what it was made of. Many scientists attempted to find a mathematical relationship that would predict how the oscillators of a black-body could produce a particular spectral distribution. Max Planck found that correct mathematical relationship. He assumed that the **energy** absorbed or emitted by the oscillators was always a multiple of some fundamental "packet of energy" he called a quantum. Objects that emit or absorb energy do it in discrete amounts, called quantums.

At this same time, there was a physicist working with Thomson and Rutherford named **Niels Bohr**. Bohr realized that the idea of a quantum of energy could explain how the electrons in the atom are arranged. He described the electrons as being "in orbit" around the nucleus like **planets** around the **Sun**. Like oscillators in a black-body could not have just any energy, electrons in the atom could not have just any orbit. There were only certain distances that were allowed by the energy that an **electron** had. If an electron of a particular atom absorbed the precisely right quantum of energy, it could move farther away from the nucleus. If an electron farther from the nucleus emitted the precisely right quantum of energy, it could move closer to the nucleus. What the precisely right values were differed for every element. These values could be determined by a process called atomic spectroscopy, an experimental technique that looked at the light spectrum produced by atoms. An atom was heated so that all of its electrons were moved far away from the nucleus. As they moved closer to the nucleus, the electrons would begin emitting their **quanta** of energy as light. The spectrum of light produced could be examined using a prism. The spectrum produced in this way did not show every possible **color**, but only those few that matched the energies corresponding to the electron orbit differences. Although later refined, Bohr's "planetary model" of the atom explained atomicspectroscopy data well enough that scientists turned their attention back to the nucleus of the atom.

Rutherford, along with Frederick Soddy, continued work with radioactive elements. Soddy, in particular, noticed that as alpha and beta particles were emitted from atoms, the atoms changed in one of two ways: 1) the element became a totally different element with completely new chemical reactions, or 2) the element maintained the same chemical reactions and the same atomic spectrum but only changed in atomic weight.

He called atoms of the second group isotopes, atoms of the same element with different atomic weights. In any natural sample of an element, there may be several types of isotopes. As a result, the atomic weight of an element that was calculated by Berzelius was actually an average of all the **isotope** weights for that element. This was the reason that some elements did not fall into the right order on Mendeleev's Periodic Table—the average atomic weight depended on how much of each kind of isotope was present. Soddy suggested placing the elements in the Periodic Table by similarity of chemical reactions and then numbering them in order. The number assigned to each element in this way is called the atomic number. The atomic numbers were convenient ways to refer to elements.

Meanwhile, Thomson had continued his work with the Crookes tube. He found that, not only were cathode rays of electrons produced, but so were positive particles. After much painstaking work, he was able to separate the many different kinds of positive particles by weight. Based on these measurements, he was able to determine a fundamental particle, the smallest positive particle produced, called a **proton**. Since these were being produced by the atoms of the cathode and since Rutherford showed that the nucleus of the atom was positive, Thomson realized that the nucleus of an atom must contain protons. A young scientist named **Henry Moseley** experimented with bombarding atoms of different elements with x rays. Just as in atomic spectroscopy, where heat gives electrons more energy, x rays give protons in the nucleus more energy. And just as electrons give out light of specific energies when they cool, the nucleus emits x rays of a specific energy when it "de-excites." Moseley discovered that the energy of the emitted x rays for every element followed a simple math-

ematical relationship. The energy depended on the atomic number for that element, and the atomic number corresponded to the number of positive charges in the nucleus. So the correct ordering of the Periodic Table is by increasing number of protons in the atomic nucleus. The number of protons equals the number of electrons in a neutral atom. The electrons are responsible for the chemical reactions. Elements in the same column of the Periodic Table have similar arrangements of electrons with the highest energies, and this is why their reactions are similar.

Only one problem remained. Electrons had very little weight, 1/1,836 the weight of a proton. Yet the protons did not account for all of the atomic weight of an atom. It was not until 1932 that **James Chadwick** discovered the existence of a particle in the nucleus with no electrical charge but with a weight slightly greater than a proton. He named this particle the **neutron**. Neutrons are responsible for the existence of isotopes. Two atoms of the same element will have the same number of protons and electrons but they might have different numbers of neutrons and therefore different atomic weights. Isotopes are named by stating the name of the element and then the number of protons plus neutrons in the nucleus. The sum of the protons and neutrons is called the **mass** number. For example, uranium-235 has 235 protons and neutrons. We can look on a Periodic Table to find uranium's atomic number (92) which tells us the number of protons. Then by subtracting, we know that this isotope has 143 neutrons. There is another isotope of uranium, U-238, with 92 protons and 146 neutrons. Some combinations of protons and neutrons are less stable than others. Picture trying to hold 10 bowling balls in your arms. There will be some arrangement where you might be able to manage it. Now try holding 11 or only nine. There might not be a stable arrangement and you would drop the bowling balls. The same thing happens with protons and neutrons. Unstable arrangements spontaneously fall apart, emitting particles, until a stable structure is reached. This is how radioactivity like alpha particles is produced. Alpha particles are made of two protons and two neutrons tumbling out of an unstable nucleus.

Hydrogen has three kinds of isotopes, hydrogen-1, hydrogen-2 and hydrogen-3.

The atomic weights of the other elements were originally compared to hydrogen without specifying which isotope. It is also difficult to get single atoms of hydrogen because it usually reacts with other atoms to form molecules like H_2 or H_2O. So a different element's isotope was chosen for comparison. The atomic weights are now based on carbon-12. This isotope has six protons and six neutrons in its nucleus. Carbon-12 was defined to be 12 atomic mass units. (Atomic mass units, abbreviated amu, are units used to compare the relative weights of atoms. One amu is less than 200 sextillionths of a gram.) Every other isotope of every other element is compared to this. Then the weights of a given element's isotopes are averaged to give the atomic weights found on the Periodic Table.

Until this point in the story of the atom, all of the particles comprising the atom were thought of as hard, uniform spheres. Beginning in 1920 with the work of **Louis de Broglie**, this image changed. De Broglie showed that particles like electrons could sometimes have properties of **waves**. For instance, if water waves are produced by two sources, like dropping two pebbles into a pond, the waves can interfere with each other. This means that high spots add to make even higher spots. Low spots add to make even lower regions. When electrons were made to travel through a double slit, with some electrons going through one slit and some through the other, they effectively created two sources. The electrons showed this same kind of **interference**, producing a pattern on a collection plate. The ability of electrons and other particles to sometimes show properties of particles and sometimes of waves is called wave-particle duality. This complication to the nature of the electron meant that Bohr's idea of a planetary atom was not quite right. The electrons do have different discrete energies, but they do not follow circular orbits. In 1925, **Werner Heisenberg** stated that the precise speed and location of an electron cannot both be known at the same time. This "Heisenberg Uncertainty Principle" inspired **Erwin Schrödinger** to devise an equation to calculate how an electron with a certain energy moves. Schrödinger's equation describes regions in an atom where an electron with a certain energy is *likely* to be but not exactly where it is. This region of probability is called an orbital. Electrons move so fast within these **orbitals** that we can think of them as blurring into an **electron cloud**. Electrons move from one orbital into another by absorbing or emitting a quantum of energy, just as Bohr explained.

Applications of Atomic Theory

Early studies of radioactivity revealed that certain atomic nuclei were naturally radioactive. Some scientists wondered that if particles could come out of the nucleus, would it also be possible to **force** particles into the nucleus? In 1932, Cockcroft and Walton succeeded in building a particle accelerator, a device that could make streams of charged particles move faster and faster. These fast particles, protons for example, were then aimed at a thin plate of a lighter element like lithium (Li). If a lithium atom nucleus "captures" a proton, the nucleus becomes unstable and breaks apart into two alpha particles. This technique of inducing radioactivity by bombardment with accelerated particles is still the most used method of studying nuclear structure and subatomic particles. Today, **accelerators** race the particles in straight lines or, to save land space, in ringed paths several miles in diameter.

The spontaneous rearrangement of the atomic nucleus always results in a release of energy in the form of kinetic **motion** in fast-moving neutrons. When a large nucleus falls apart to form smaller atoms, the process is called fission. When lighter atoms are forced together to produce a heavier atom, the process is called fusion. In either case, fast neutrons are released. These can transfer their **kinetic energy** to the surroundings, heating it. This heat can be used to boil water, producing steam to run a turbine that turns an electric generator. Fusion is the process occurring in the center of the Sun and other **stars**. So much energy can be released quickly that the process has also been used for the hydrogen bomb. However, fusion is not yet controlled enough for running a **power** plant. Research continues to find a controlled method of using fusion energy.

On the other hand, fission reactions have also been used for very powerful weapons. The first **atomic bomb** was detonated in 1945. Since then, however, fission energy has also been controlled enough to operate the many **nuclear power** plants around the world.

While an atom is the smallest part of an element which still is that element, atoms are not the smallest particles that exist. Even the protons and neutrons in the atomic nucleus are believed to made of even smaller particles called **quarks**. Current research in **atomic physics** focuses on describing the internal structure of atoms. By using **particle accelerators**, scientists are trying to characterize quarks which may combine in a number of ways to produce other types of subatomic particles.

No one has ever seen a single atom even with the best optical microscopes. Special types of microscopes called **scanning tunneling microscopes** and atomic force microscopes make use of the forces produced by the electrons to obtain images of the electron clouds. These clouds indicate how atoms are arranged but we cannot "see" through the cloud to the nucleus. Because of the limitations of size, we will never see an atom with our own eyes. Everything we know about atoms must be deduced from larger-scale experiments. As a result, the description of atoms is still called a theory. However, this theory explains atomic experiments so well that we usually think of the existence of atoms as a fact.

ATOMIC WEIGHT/MASS

The atomic **mass** of an element is the mass of the number of particles in its **nucleus** given in atomic mass units. Although it is sometimes referred to as atomic **weight**, this is inaccurate, since weight is a measurement that includes the effect of **gravity**. One atomic mass unit (amu) is 1.66×10^{-24} grams. This is equal to one-twelfth the mass of an **atom** of carbon-12. This standard was adopted internationally in 1961, replacing an arbitrarily assigned value of 16.000 amu for the atomic mass of an atom of **oxygen**.

The instrument most often used to measure atomic mass is the mass spectrometer (spectrograph). In this instrument, atoms are vaporized and then changed to positively charged particles by knocking off electrons. These charged particles are passed through a magnetic field that causes them to be deflected different amounts, depending on the size of the charge and mass. The particles are eventually deposited on a detector plate where the amount of deflection can be measured and compared with the charge. Very accurate relative masses are determined in this way.

In the early years of the study of the behavior of **matter**, it became generally accepted that mass was conserved in physical and chemical change. By the end of the eighteenth century, studies of compounds suggested that when elements combine, the compounds that form follow the law of constant (or definite) proportions. This law was formally stated by the French chemist Joseph Proust (1754-1826) in 1799: the ratio of the masses of the elements in a compound will always be the same, independent of the method of preparation or the proportions that are mixed in its preparation.

Early in the nineteenth century, **John Dalton** (1766-1844), the English chemist who is sometimes called the father of modern **chemistry**, introduced the concept of atomic mass when he formulated a model of the atom as a round, solid, indivisible particle that represents the elementary unit involved in chemical change. He postulated that all the atoms of the same element have the same atomic mass, while the atoms of a different element have a different atomic mass. He determined that compounds always have the same percentage composition by mass. His explanation was that compounds have a fixed ratio based on the number of atoms of each element in the compound. He expanded on the law of constant proportions, which deals with compounds made of any two single elements, adding to it the law of multiple proportions, to explain situations where more than one compound is formed from the same two elements. The law of multiple proportions states that the ratio of the element to itself in multiple compounds will be a small whole number, like carbon monoxide and carbon dioxide.

In 1815, the English chemist William Prout (1785-1850) observed that the atomic weights of several gases were whole number multiples of the atomic weight of hydrogen. His conclusion based on this experimental result was that every atom is made up of a number of hydrogen atoms that are somehow held together.

In 1864, the English chemist J.A.R. Newlands (1837-1898) and the German chemist Lothar Meyer (1830-1895) recognized independently that there was a connection between the atomic masses of the elements and the periodic repetition of chemical and physical properties. In 1869, the Russian chemist Dmitry Mendeleev (1834-1907) tried to arrange the 63 known elements in a pattern that would reflect similar chemical behavior. He found that the arrangement was essentially one of increasing atomic mass with three exceptions where he had to reverse the order of atomic mass in order to maintain the pattern of similar chemical behavior.

After the turn of the century, changes in the model of the atom led to an awareness of a relationship between the atomic mass and the number of particles in the nucleus. At first, the nuclear particles were thought to be only those that carry the positive charges of the atom, but it was soon obvious that, except for hydrogen, the weight of an atom was around twice the weight of the number of protons it contained. Ernest Rutherford (1871-1937) suggested in 1920 that the atom must have neutral particles (neutrons) whose mass is similar to that of the **proton**. This particle was identified by the English physicist **James Chadwick** (1891-1974) in 1932.

Avogadro, Amedeo (1776-1856)
Italian chemist

Avogadro was born in Turin, Italy, on August 9, 1776, the son of Count Filippo Avogadro and Anna Maria Vercellone. Count Avogadro was a lawyer, civil servant, and senator for the state

of Piedmont. Amedeo followed his father into the law and received his doctorate in ecclesiastical law in 1796.

He practiced law for only a few years, however, before his interests shifted to the sciences. He studied mathematics and physics on his own and, in 1806, was appointed demonstrator in physics at the Academy of Turin. In 1809, he was appointed professor of natural philosophy at the College of Vercelli. A decade later he became the first professor of mathematics at Turin although he lost his chair in the revolution of 1822. He was reappointed in 1835 and remained at Turin until he retired in 1850. He died in Turin on July 9, 1856.

Avogadro's name is best known in connection with his hypothesis about the composition of gases. In 1809, the French chemist and physicist **Joseph Gay-Lussac** had published reports on his research on the combining volumes of gases. Gay-Lussac reported that, under conditions of equal temperature and **pressure**, gases combine with each other in simple whole number ratios. For example, when hydrogen and **oxygen** combine to form water, they do so in the proportions of two volumes of hydrogen to one volume of oxygen, producing two volumes of water vapor in the process.

The conclusion suggested by these results was obvious to Avogadro. Equal volumes of all gases, he said, must contain equal numbers of particles. This statement is known as Avogadro's hypothesis.

The same thought had occurred to **John Dalton**, originator of the **atomic theory**. But Dalton had decided to reject this hypothesis, as well as Gay-Lussac's experimental results. Dalton believed that gases were made of atoms that were in contact with each other. Since he thought that atoms of different elements were of different sizes, then equal volumes could not contain equal numbers of particles. He suggested that Gay-Lussac's research was faulty and carried out experiments of his own that produced results more consistent with his own ideas.

It was Dalton, however, who was in error this time. Gay-Lussac and others were able to demonstrate that the particles in a gas are not in contact with each other and actually make up only a very small fraction of the gas itself. Most of the gas is empty **space**. So the size and shape of atoms in the gas are irrelevant.

Avogadro made one further suggestion to clarify Gay-Lussac's results. Consider the case with water again. Gay-Lussac showed that: two volumes of hydrogen + one volume of oxygen = two volumes of water.

If we assume the simplest case, in which each volume consists of only one particle, the same result can be expressed as: two particles of hydrogen (H H) + one particle of oxygen (O) = two particles of water (H-O H-O). (At the time, water was thought to consist of one **atom** of hydrogen and one atom of oxygen.)

Obviously, this situation is impossible. One cannot begin with one particle of oxygen and end with two. The solution Avogadro suggested is that the smallest particle of an element can sometimes consist of *two*atoms joined together in a molecule, a term he invented for the new particle. Thus, if the smallest particle of oxygen is a *molecule* of oxygen, we can explain Gay-Lussac's results as follows: two particles of hydrodgen (H H) + one particle of oxygen (O-O) = two particles of water (H-O H-O).

Amedeo Avagadro.

We now know the smallest particle of hydrogen also consists of two atoms together. If we draw the reaction with modern forms of the components and products, Avogadro's solution works as follows: H-H H-H + O-O = H-O-H H-O-H.

Avogadro's hypothesis can also be applied to non-gasses. Equal atomic weights of any solid or liquid have the same number of particles. For example, an atom of **carbon** has an atomic **weight** of 12, while a molecule of water is composed of an atom of oxygen with an atomic weight of 16 plus two atoms of hydrogen with atomic weights of one each for a total of 18. Therefore, twelve grams of carbon will have the same number of particles as eighteen grams of water. This number of particles turns out to be 6.026×10^{23} and is called **Avogadro's number**.

Avogadro's suggestions expanded and improved on Dalton's atomic theory. For a number of reasons, however, his ideas were largely ignored for half a century. Not until Stanislao Cannizarro began to spread Avogadro's ideas in the 1850s did chemists finally understand and adopt them.

Avogadro's law

Avogadro's law (sometimes called Avogadro's principle) states that equal volumes of different gases at the same tem-perature and **pressure** contain equal numbers of molecules. Amedeo Avogadro (1776-1856), an Italian physi-cist, formulated this law in 1811 as a direct consequence of Gay- Lussac's law which was artic ulated by the French chemist **Joseph-Louis Gay-Lussac** (1778- 1850) in 1802.

Gay-Lussac's law states that gases combine in simple whole number ratios (all volumes measured at the same tem-perature and pressure). Avogadro's law explains Gay- Lussac's law without violating John Dalton's idea that atoms are indi-visible. Avogadro's law also led to the concepts of **Avogadro's number** and the mole, and to the **ideal gas law**. Avogadro's ideas were largely ignored for half a century until Stanislao Cannizarro (1826-1910) began to promote them in the 1850s.

Avogadro's number

Historically, Avogadro's number (or the Avogadro constant) is the number of particles, atoms, formula units, or molecules, in one mole of a given substance. The metric system now more precisely defines it as the number of atoms in exactly 0.024 lb (12g) of ^{12}C. In equations, Avogadro's number is given the symbol L; numerically it is equal to 6.023×10^{23}.

Avogadro's number is the number of particles present when the amount of material is the same as the atomic **weight** (or relative molecular **mass**, or relative atomic mass or weight) expressed in grams. This is one mole of the substance. For example, with water, H_2O, the relative molecular mass is 18 (16 for **oxygen** and 1 for each of the two hydrogens), so in 0.036 lb (18 g) of water there are 6.023×10^{23} HOH molecules. With a gas there is a slight difference because the gas may be encountered in the diatomic state in the atmosphere. For exam-ple, oxygen and nitrogen in the atmosphere are generally encountered as O_2 and N_2 respectively. With these diatomic molecules, there is an Avogadro's number of diatomic mole-cules in the amount of gas that is equivalent to the relative molecular mass. There may be an ambiguity if correct termi-nology is not applied. One mole of oxygen could refer to one mole of oxygen atoms (6.023×10^{23} atoms) or it could refer to one mole of diatomic, gaseous oxygen, one mole of oxygen molecules (12.046×10^{23} oxygen atoms). The first case is ele-mental oxygen and the latter is molecular oxygen. A mole of anything contains the Avogadro number of those objects, whether it be a mole of atoms, molecules, or **ions**.

Amedeo Avogadro was an Italian physicist who lived from 1776-1856, and his main contribution to **chemistry** is Avogadro's hypothesis which gave us **Avogadro's Law**, which in turn lead to Avogadro's number. These were worked out by looking at volumes of gas reacting together and Avogadro noticed that they always reacted in a constant ratio of whole numbers of volumes of gas. In other words, one volume of gas A would always react completely with one volume of gas B or two volumes of gas C. Avogadro realized there was a constant relationship, and by further observation and hypothesizing, he eventually came up with what we now know as Avogadro's law and number.

A number as large as Avogadro's number is difficult to comprehend, but the following example is often quoted to give an indication of exactly how large this number is. When standing on a beach with an uninterrupted view to the horizon both left and right, the number of sand grains that are present are still not enough to make one mole of sand grains. Another way of looking at this would be to consider an Avogadro's number of soft drink cans: if they were stacked together, they would cover the entire surface of the earth to a depth of 200 mi (322 km).

The numerical value of Avogadro's number has been confirmed by several different experimental techniques, including **Brownian motion**, electronic charge, and the counting of alpha particles.

Avogadro's hypothesis states that equal volumes of gases at the same temperature and **pressure** contain equal numbers of molecules. Avogadro's law states that a gas at constant temperature and pressure has a volume directly proportional to the number of moles of gas. From these two statements, and from what has been previously said, it can be seen that the volume of a gas is directly proportional to the number of molecules present. Also, one mole or an Avogadro's number of molecules of any gas would occupy the same volume as one mole or one Avogadro's number of any other gas, at constant temperature and pressure, irrespective of the size of the molecules considered. Avogadro's number can be used in working out the amount of substance required to completely take part in a **chemical reaction**. Obviously dealing routinely with numbers of the size of Avogadro's number is unwieldy. To overcome the problems associated with this, the mole is used.

Avogadro's number is not just true for gases; it holds true whatever the state of **matter** under consideration. When a substance is dissolved in a solvent, the strength of solution can be discussed in terms of Avogadro's number. When an Avogadro's number of particles is dissolved into 0.264 gal (1 L) of solvent the strength of the solution is one molar. This is the molarity of the solution. Molality is a similar concept to molarity. Instead of dealing with the volume of the solution or solvent, molality is concerned with the mass of solution or solvent. As such a one molal solution has one mole dissolved in 2.205 lb (1 kg) of solution. A one molal solution has an Avogadro's number of particles of the solute.

Avogadro's number is a conversion factor between the number of moles present and the actual number of physical particles. For example, the number of particles present in 0.1 lb (50 g) of carbon dioxide can easily be calculated. The **molecular weight** of carbon dioxide is 12 for the carbon plus 16 for each of the two oxygens, which totals a molecular weight of $12 + 16 + 16 = 44$. One mole of carbon dioxide would weigh 0.088 lb (44 g). Moles are calculated by dividing the weight by the weight of one mole, this is 50 divided by 44, which equals 1.14. In this example, the result is 1.14 moles of carbon dioxide in 0.1 lb (50 g) of carbon dioxide. The number of molecules of carbon dioxide is the number of moles multiplied by Avogadro's number, or $1.14 \times 6.023 \times 10^{23}$ which gives us an answer of 6.866×10^{23}. In other words, in 50 g or 1.14 moles of carbon dioxide there are 6.866×10^{23} molecules of carbon dioxide. The calculations listed here will work in any direction relating moles to particles to weights of materials, simple rearrangement of the equations will provide the appropriate answer.

Avogadro's number is a powerful and useful concept. It illustrates how much of a material is actually participating in a reaction at a basic level. The numbers that it generates are too large to comprehend with any validity so the mole concept is used. One mole of substance contains an Avogadro's number of particles. It is easier to visualize one mole simply because the number one is a more comfortable number to work with than 6.023×10^{23}, even though, by definition, the quantities involved are exactly the same. Conversion is easy between Avogadro's number of particles, moles, weights and also the figures used in chemical equations. Such figures help to optimize a reaction so reactants are not overused.

B

BACKGROUND RADIATION

Background **radiation** is the low intensity radiation from the small amounts of **radioisotopes** in the environment to which we are all exposed.

Background radiation is a result of naturally occurring **radioactive decay** from a number of elements found in the air, rocks, soil, and in all living things, as well as a component of bombardment by **cosmic rays**. There is also an amount that has been added by humans. This latter radiation comes from all uses that **radioactivity** is put to, from nuclear bombs to power stations, x-ray machines, and irradiation of food. The greatest component of background radiation is naturally occurring radon gas. This accounts for 55% of all background radiation. The next largest component at 11% is medical x-rays, which is the same level of contribution as radioisotopes naturally occurring in the human body. The level of exposure to background radiation for a person for one year is approximately 360 mrem. A typical dental x-ray would be the equivalent of 0.5 mrem. There are no detectable clinical effects for a short term exposure to **ionizing radiation** below 25 rem. Since the 1940s when nuclear material started to be used by humans the level of background radiation has increased by a tiny amount. This increase has to be taken into account when calculating dates using the carbon-14 dating technique, but, other than such techniques that rely on measuring minute quantities of radioactive material, the increase in background radiation has had no noticeable affect.

Background radiation is mostly natural levels of radiation which have been around throughout most of the history of the planet.

BALLISTICS

Ballistics is the study of projectile **motion**. A projectile is an object that has been launched, shot, hurled, thrown or by other means projected and which continues in motion due to its own **inertia**. The path of the projectile is determined by its initial **velocity** (direction and speed) and the forces of **gravity** and air resistance. For objects projected close to Earth and with negligible air resistance, the flight path is a parabola. When air resistance is significant, however, the shape and rotation of the object are important and determining the flight path is more complicated. Ballistics influences many fields of study ranging from analyzing a curve ball to developing missile guidance systems.

In order to understand projectile motion it is first necessary to understand the motion of free-falling bodies, that is objects which are simply dropped from a certain height above the Earth. For the simplest case, when air resistance is negligible and when objects are close to the Earth's surface, **Galileo** Galilei was able to show that two objects fall the same distance in the same amount of time, regardless of their weights. It is also true that the speed of a falling object will increase by equal increments in equal time periods. For example, a ball dropped from the top of a building will start from rest and increase to a speed of 32 ft (9.8 m) per second after one second, to a speed of 64 ft (19.5 m) per second after two seconds, to a speed of 96 ft (29.4 m) per second after three seconds, and so on. Thus, the *change* in speed for each one second time interval is always 32 ft per second. The change in speed per time interval is known as the **acceleration** and is constant. This acceleration is equal to 1 g which stands for the acceleration due to the **force** of gravity. By comparison, a pilot in a supersonic jet pulling out of a nose dive may experience an acceleration as high as 9 g (of course, a jet is not in free fall but is being accelerated by its engines and gravity).

The acceleration of gravity, g, becomes smaller as the distance from the Earth increases. However, for most Earth bound applications, the value of g can be considered constant (it only changes by 0.5% due to a 10 mi [16 km] altitude change). Air resistance, on the other hand, can vary greatly depending on altitude, wind, and properties and velocity of the projectile itself. It is well know that sky divers may change their altitude relative to other sky divers by simply changing the shape of their body. Also, it is obvious that a rock will fall more quickly than a feather. Therefore, when treating problems in ballistics, it is necessary to separate the effects due to gravity, which are fairly simple, and the effects due to air resistance, which are more complicated.

The motion of projectiles, without air resistance, can be separated into two components. Motion in the vertical direction where the force of gravity is present, and horizontal motion where the force of gravity is zero. As **Isaac Newton** (1642-1727) proposed, an object in motion will remain in motion unless acted upon by an external force. Therefore, a projectile in motion will remain with the same horizontal velocity throughout its flight, since no force exists in the horizontal direction, but its velocity will change in the vertical direction due to the force of gravity. For example, a cannon ball is fired in the horizontal direction. The velocity of the cannon ball will remain constant, in the horizontal direction, but the ball will accelerate toward the Earth, in the vertical direction, with an acceleration of 1 g. The combination of these two effects produces a path which describes a parabola. Since the vertical motion is determined by the same acceleration which describes the motion of objects in free fall, a second cannon ball which is dropped, at precisely the same instant as the first cannon ball is fired, will reach the ground at precisely the same instant. Therefore, the motion in the horizontal direction does not affect the motion in the vertical direction. This fact can be confirmed by knocking one coin off the edge of the desk with a good horizontal whack, while a second coin is simultaneously knocked off the desk with a gentle nudge. Both coins will reach the ground at the same time.

By increasing the amount of gun powder behind the cannon ball, one could increase the horizontal velocity of the cannon ball as it leaves the cannon and cause the cannon ball to land at a greater distance. If it were possible to increase the horizontal velocity to very high values, there would come a point at which the cannon ball would continue in its path without ever touching the ground, similar to an orbiting satellite. To attain this orbiting situation close to the Earth's surface, the cannon ball would have to be fired with a speed of 17,700 mph (28,500 Km/h)! In most instances, projectiles, like cannon balls, are fired at some upward angle to the Earth's surface. As before, the flight paths are described by parabolas. (The maximum range is achieved by aiming the projectile at a 45° angle above the horizontal.) Note that angles which are equally greater or less than 45° will produce flight paths with the same range (for example 30° and 60°).

If projectiles were only launched from the surface of the **moon** where there is no atmosphere, then the effects of gravity, as described in the previous section, would be sufficient to determine the flight path. On Earth, however, the atmosphere will influence the motion of projectiles. As opposed to the situation due to purely gravitational effects, projectile motion with air resistance will be dependent on the **weight** and shape of the object. As one would suspect, lighter objects are more strongly affected by air resistance. In many cases, air resistance will produce a drag force which is proportional to the

velocity squared. The effects of increased air drag on an object such as a cannon ball will cause it to fall short of its normal range without air resistance. This effect may be significant. In World War I, it was realized that cannon balls would travel farther distances if aimed at higher elevations, due to the decreased air **density** and decreased drag.

More subtle effects of air resistance on projectile motion are related to the shape and rotation of the object. Clearly, the shape of an object can have an effect on its projectile motion, as anyone has experienced by wadding up a piece of paper before tossing it into the waste can. The rotation of an object is important also. For example, a good quarterback always puts a spin on a football when making a pass. By contrast, to produce an erratic flight, a knuckle ball pitcher in baseball puts little or no spin on the ball. The physical property which tends to keep spinning objects spinning is the conservation of **angular momentum**. Not only do spinning objects tend to keep spinning but the orientation of the spin axis tends to remain constant. This property is utilized in the design of rifle barrels which have spiral grooves to put a spin on the bullet. The spinning of the bullet around its long axis will keep the bullet from tumbling and will increase the accuracy of the rifle. This property is also utilized in designing guidance systems for missiles. These guidance systems consist of a small spinning device called a gyroscope which keeps a constant axis orientation and thus helps to orient the missile. Small deviations of the missile with respect to the orientation of the gyroscope can be measured and corrections in the flight path can be made.

BARDEEN, JOHN (1908-1991)

American physicist

John Bardeen has the unique distinction of being the first—and so far only—person to have won two Nobel Prizes in physics. The first, shared with Walter Brattain and William Shockley, came in 1956 for his role in the discovery of the transistor. Only a few months after receiving this award, Bardeen completed another research project that would lead to his second Nobel Prize in 1972, this one for his part in the development of the BCS theory of superconductivity, named for its three inventors, Bardeen, Leon Cooper, and J. Robert Schrieffer, who shared the Nobel Prize. The BCS theory accounts for the tendency of certain materials to lose all **electrical resistance** when they are cooled to temperatures close to **absolute zero**. The theory not only accounted for all known superconductivity phenomena, but also suggested an extensive agenda of new research on the topic.

John Bardeen was born in Madison, Wisconsin, on May 23, 1908. His father was Charles Russell Bardeen, a professor of anatomy and Dean of the Medical School at the University of Wisconsin. His mother was Althea Harmer Bardeen, an artist and teacher. John had two younger brothers, Thomas and William, and a sister, Helen. When John was twelve, his mother died; his father was later remarried to Ruth Hames and had another daughter with her. John's parents encouraged his intellectual development while he was still young. They introduced him to logical word problems in mathematics, for example, before he even entered school.

Bardeen attended Madison public schools through the third grade before transferring to the University High School, skipping three grades and entering as a seventh grader. It was there that he had his first course in general science and discovered his interest in the subject. Nonetheless, mathematics was his first love. He was allowed to take algebra when he was only ten years old, and at the end of each school year, he was named outstanding student in the class. Bardeen later transferred to Madison Central High School, from which he graduated in 1923.

When he enrolled at the University of Wisconsin, Bardeen decided to major in electrical engineering with minors in mathematics and physics. He was awarded his bachelor's degree in 1928 and stayed on to complete his master of science degree in electrical engineering a year later. For his master's degree, he carried out research on the **radiation** emitted by antennas and on problems of applied geophysics.

Upon graduation from Wisconsin, Bardeen accepted a research job at the Gulf Research and Development Corporation in Pittsburgh. His work there put him in the forefront of a new and growing field of physics, geophysics. He worked with Leo J. Peters on the development of new methods for locating deposits of petroleum. Although the work was interesting and well paid ($3,000 per year in the midst of the Depression), Bardeen concluded that he wanted a different kind of life. He had heard about the establishment of the new Institute for Advanced Studies at Princeton University and hoped to have an opportunity to work with some of the great scientists associated with the Institute. He decided, therefore, to apply for the doctoral program in mathematics at Princeton, where the Institute was to be housed. He was accepted and was assigned the great Eugene Paul Wigner as his advisor. That assignment was propitious, as Wigner was one of the world's leading authorities on the application of **quantum mechanics** to solid-state physics. It was not long before Bardeen, too, became deeply involved in this subject. For his doctoral research, Bardeen chose to study the forces that act on electrons in metals. In the midst of that research, he left Princeton to accept an appointment as a Junior Fellow at Harvard University in 1935, an appointment that lasted officially until 1938. At Harvard, Bardeen worked with **Percy Bridgman**, the world's leading authority on high **pressure** physics, and with **John Van Vleck**, who was to win the 1977 Nobel Prize for his research on the electronic structure of magnetic and disordered systems. Bardeen eventually completed his dissertation at Princeton and was awarded a Ph.D. in mathematics and physics in 1936.

Bardeen's first academic appointment was as assistant professor of physics at the University of Minnesota in 1938. While at Minnesota, he continued to study the behavior of electrons in metals; in 1939, he published a paper on this topic with his former advisor, Van Vleck. When World War II began, Bardeen left Minnesota to take a job as physicist at the Naval Ordnance Laboratory in Washington, D.C. His work there

John Bardeen. *(AP/Wide World Photos, Inc. Reproduced by permission.)*

dealt with the magnetic fields of ships as they move through salt waters.

When World War II ended, Bardeen took a job at the Bell Telephone Laboratories at Murray Hill, New Jersey. One reason for this move was that Bardeen felt that he could no longer support his family on the salary of an academic. He had married Jane Maxwell on July 18, 1938, and by 1945 they had three children, James Maxwell, William Allen, and Elizabeth Ann.

The move to Bell Labs was an extremely fortunate one for Bardeen. He was assigned to work there with Brattain and Shockley, who had been studying the properties of **semiconductors** for many years. The impetus for this research was the general dissatisfaction among physicists with the fundamental problems posed by existing electronic systems. Those systems depended on **vacuum** tubes, which were bulky, fragile, and huge consumers of electrical **power**. Brattain and Shockley, among others, were convinced that the solution to this problem lay in the use of semiconductors, materials that conduct an electrical current better than do insulators like rubber and plastics, but not nearly as well as conductors such as copper and silver.

Shockley had developed an idea for using semiconductors as amplifiers as early as 1939, but his work had been interrupted by the war. When Bardeen arrived at Bell Labs in 1945, Shockley had had no success in producing a model of his idea that actually worked. As Bardeen began to think about this problem, it occurred to him that the atoms on the surface of the semiconductor might be acting as a barrier to incoming electrical signals, preventing the material from behaving as Shockley had expected.

This idea led to a new approach to Shockley's work, one in which he, Bardeen, and Brattain began to study the fundamental properties of semiconductors. By 1948, enormous progress had been made, and Bardeen and Brattain had constructed the first working model of Shockley's concept, the transistor. In the transistor (short for "transfer resistor"), a semiconductor is placed between two wires, which serve as electrical contacts for the current. For their work, the three Bell scientists were jointly awarded the 1956 Nobel Prize in physics. Shockley's part of the prize was given for his fundamental theoretical work, while Bardeen and Brattain were honored for their accomplishment in producing a working model of Shockley's theory.

The invention of the transistor revolutionized **electronics**. It also brought honors and awards to Bardeen, as well as to Brattain and Shockley. Among these were the Stuart and Ballantine Medal of the Franklin Institute (1952), the John Scott Medal of the city of Philadelphia (1955), the Oliver E. Buckley Solid-State Physics Prize of the American Physical Society (1954), and the Fritz London Award (1962), as well as election to the National Academy of Sciences in 1954.

The collaboration among Bardeen, Brattain, and Shockley came to an end in 1951, when Bardeen accepted an appointment as Professor of Electrical Engineering and Physics at the University of Illinois. Once settled at Illinois, Bardeen turned to the problem of superconductivity, which he had been interested in as a graduate student at Princeton but had set aside because of World War II. The phenomenon of superconductivity had first been described by the Dutch physicist **Heike Kamerlingh Onnes** in 1911. Kamerlingh Onnes had reported that some metals lose all resistance to the flow of electrical current, becoming "superconductive," as they are cooled to temperatures close to absolute zero. The heightened conductivity derives from a free flow of electrons in the metal. Scientists were intrigued by this phenomenon for both theoretical and practical reasons. An understanding of superconductivity could potentially revolutionize the electronics industry. Every device that operates by means of an **electric current** wastes a huge amount of **energy** in overcoming electrical resistance. If superconducting materials could be used in those devices, virtually no energy would be lost to resistance. The practical problem posed by Kamerlingh Onnes's discovery was the low temperature required for superconductivity. In the intervening half century, scientists had been singularly unsuccessful in improving on his work. By the 1950s, the highest temperature at which superconductivity had been observed was still only about 20° Kelvin, 20° above absolute zero.

As his coworkers on the problem of superconductivity Bardeen selected a postdoctoral associate at Illinois, Leon Cooper, and a graduate student, J. Robert Schrieffer. One of their first breakthroughs on the problem came in 1956, when

Cooper showed that an **electron** moving through a crystal causes a slight deformation of the crystal and, in the process, attracts a second electron to itself. Eventually, electrons throughout the crystal become joined to each other in pairs (now known as "Cooper pairs"). The difficulty lay in showing how Cooper pairs could be used to explain the free flow of *all* valence electrons in a superconductor. That problem required a sophisticated mathematical analysis, which Schrieffer set out to perform. While Bardeen was at Stockholm to receive his first Nobel Prize, Schrieffer found an answer to the problem. The theory that eventually evolved from Cooper pairs and Schrieffer's mathematical analysis is now known as the BCS theory of superconductivity, in honor of its three creators, Bardeen, Cooper, and Schrieffer. Often described as one of the most important developments in theoretical physics since **quantum theory**, it brought the 1972 Nobel Prize for physics to its three co-discoverers.

Superconductivity promises to revolutionize technology in much the way **transistors** did. For example, research is now being conducted on the development of maglev railway trains. The term *maglev* is shorthand for *mag*netic *lev*itation, a phenomenon that occurs when one object (such as a rail car) is suspended above another object (such as a track) by the repulsive **force** of magnetic fields. Long a dream of inventors, maglev trains have only recently become a realistic possibility with the development of powerful, low-cost magnets made from superconducting materials.

In 1959, Bardeen moved to the Center for Advanced Study at the University of Illinois, where he continued to carry out research on solid-state physics and low-temperature phenomena. He retired officially in 1975 and was designated professor of physics emeritus. Bardeen spent his leisure time in golfing, swimming, and traveling. He died in Boston on January 30, 1991, as the result of heart failure following surgery that had revealed the presence of lung cancer.

BARYONS AND BARYONIC MATTER

Baryons, taken from the Greek word for heavy, is a term that refers to a class of **subatomic particles** of high **mass**. With the discovery of the **neutron** in 1932, the neutron and **proton** were grouped together as the baryons, in contrast to the considerably lighter **electron**. With the discovery of subparticles, new categories had to be introduced. A particle called the **photon** was assigned its own category, while the leptons category includes the electron and the **neutrino**.

The baryon classification took on new members, including particles more massive than the proton called hyperons. These new baryons included the lambda particle, the sigma particles and the negatively charged omega-minus. The lambda particle is a neutrally charged hyperon that can replace particles in the **nucleus** of the **atom**. The sigma particles can be positive, negative, or neutral. The omega-minus particle is another hyperon classified as a baryon. An important characteristic of the baryons is that they contribute to the intensity of the strong nuclear **force**.

Murray Gell-Mann introduced a new classification system in which baryons became a subclassification of a new category called hadrons. Hadrons were on the same categorical level as photons and leptons, except hadrons are not elementary particles as photons and leptons are. Hadrons are made up of elementary particles called **quarks**. Gell-Mann distinguished baryons from mesons, the other **hadron** subclassification, by the number of quarks constituting their make-up. Baryons have three quarks while mesons have two. Gell-Mann found that by classifying known hadrons, he could discover patterns that reveal the probability of the existence of previously unknown particles. Using a method of mathematically grouping known hyperons, he found an empty **space** in the pattern. He predicted this position in the pattern would be filled through the discovery of another hyperon. He named this empty space omega-minus because it was the place for the last member of the pattern and the particle to fill it was predictably negatively charged. Later the hyperon fulfilling this prediction was discovered and had characteristics Gell-Mann described to the detail.

Baryons are a necessary part of a classification system that represents algebraically the reactions that are known to be possible. With the discovery of subparticles, theoretical speculation held that a proton could break down into a **positron** and multitudinous photons, although this was never observed. With the many years of research on the proton there was no evidence of the proton breaking down into a positron or any other subparticle. Such speculation was contrary to what was known about the proton's stability and abundance, but there was no clear explanation of the improbability of such an event. The explanation of this phenomenon was developed based on the theoretical assumption that the number of baryons must be conserved.

The law of the conservation of baryon number states that the total number of baryons must be the same before and after any subatomic event. In describing **nuclear reactions** algebraically all baryons are assigned a value of +1; the antibaryons are assigned a value of -1; and **light** particles, including electrons, photons, and neutrinos, are given a 0. The baryonic number, which is equivalent to the atomic mass number, has to remain constant for a reaction. Baryons may change within their classification but a baryon will not change into a **lepton** nor into an antibaryon. This law has been found to hold true with nuclear reactions involving antiparticles by assigning the antiparticles the opposite number of their counterparticle.

BASOV, NIKOLAI (1922-   )

Russian physicist

Nikolai Basov is one of the inventors of the laser and its technical predecessor, the maser. **Lasers** have become one of the most widely used of all twentieth-century inventions. They find diverse applications from delicate surgery to the cutting of steel, astronomical research, and even popular entertainment. Working with theoretical concepts first developed by **Albert Einstein** four decades earlier, Basov found ways of amplifying a beam of incoming electromagnetic **radiation** until it becomes a discrete, intense, monochromatic and amplified version of itself, a source of high-intensity radiation. For this discovery, Basov shared the 1964 Nobel Prize for physics with his teacher Aleksandr Prokhorov and **Charles Townes**, an American who made the same discovery independently.

Nikolai Gennadiyevich Basov was born on December 14, 1922, in the small village of Usman, outside the city of Voronezh, Russia. His mother was the former Zinaida Andreyevna Molchanova, and his father, Gennady Fedorovich Basov, was a professor at the Voronezh Forest Institute. Basov attended local schools in Voronezh and then in 1941 was drafted into the Soviet army. He was trained as a medical assistant and served on the Western front until his discharge in December of 1945.

As a civilian, Basov entered the Moscow Engineering and Physics Institute (also referred to as the Moscow Institute of Physical Engineers or the Moscow Institute of Engineering Physics), where he studied theoretical and experimental physics. Five years later, he was awarded his candidate's degree, comparable to a master's degree in Western universities. While still at the Moscow Institute, he became a laboratory assistant at the P. N. Lebedev Physical Institute of the Academy of Sciences of the U.S.S.R. In 1956, he was awarded his doctorate in physico-mathematical sciences. After receiving his degree, Basov remained at the Lebedev Institute, eventually becoming deputy director and director of the laboratory. In 1963, Basov also became professor of physics at his alma mater, the Moscow Engineering and Physics Institute, a post he still holds. In that same year, he founded the Laboratory of Quantum Radio Physics at the Lebedev Institute and became its director. Since 1967, he has also been editor of the prestigious Soviet science magazine *Priroda* (*Nature*) and the more specialized *Soviet Journal of Quantum Electronics*.

During the early 1950s, a number of physicists had begun to consider the possibility of developing a device that amplifies a given electromagnetic wave, that is, a device that increases the strength of an incoming wave while preserving its original phase. Such a device had been made possible by theoretical research carried out by Albert Einstein around 1917. According to classical physics, electrons are capable of absorbing discrete amounts of **energy** that cause them to jump from lower energy levels to higher energy levels. When they do so, they remain in the higher **energy level** only instantaneously before re-emitting the absorbed energy and returning to their ground state.

Einstein re-studied this problem from the standpoint of **quantum mechanics** and made an intriguing discovery. He found that in the presence of specific types of radiation, an **electron** already present in a higher energy level could jump to a lower energy level, emitting energy as it did so. Since this change is contrary to classical laws of physics, it was designated as a "stimulated emission" of energy.

Research in the 1950s was aimed at translating this theoretical concept into a practical device. The most difficult challenge was to find a way of promoting electrons from lower to higher energy levels where they could be stimulated by an external source of radiation. That problem was solved independently and almost simultaneously by Charles H. Townes in the United States and by Basov and Prokhorov in Russia. In 1952, the two Russian scientists read a paper before the All-Union (Soviet Union) Conference on Radio Spectroscopy in which they described a "molecular generator." The "molecular generator" was identical with an amplification device that became better known as a maser (for *m*icrowave *a*mplification by *s*timulated *e*mission of *r*adiation).

Over the next three decades, Basov continued and expanded his research on the amplification of electromagnetic radiation. In 1955, he and Prokhorov suggested a more elegant method of promoting electrons, called the three-level technique, since electrons are maintained at three different energy levels within an **atom**. Basov also examined the possibility of using **semiconductors** in the manufacture of masers and lasers and, in 1968, was able to initiate a thermonuclear fusion reaction by means of an especially powerful laser. Basov's accomplishments have earned him a number of honors, most notably the Nobel Prize for physics in 1964, the Lenin Prize in 1959 and 1964, and gold medals from the Czechoslovakian Academy of Sciences, the Italian Physical Society, and the Slovak Academy of Sciences. He was married to Kseniya Tikhonova Nasarova on July 18, 1950, with whom he has two sons, Gennadiy and Dmitriy. He reports that his favorite leisure time activities are skiing and **photography**.

BECQUEREL, ANTOINE-HENRI (1852-1908)

French physicist

Antoine-Henri Becquerel's landmark research on **X rays** and his discovery of **radiation** laid the foundation for many scientific advances of the early twentieth century. X rays were discovered in 1895 by the German physicist Wilhelm Conrad Röntgen, and in one of the most serendipitous events in science history, Becquerel discovered that the uranium he was studying gave off radiation similar to X rays. Becquerel's student, **Marie Curie**, later named this phenomenon **radioactivity**. His later research on radioactive materials found that at least some of the radiation produced by unstable materials consisted of electrons. For these discoveries, Becquerel shared the 1903 Nobel Prize in physics with Marie and **Pierre Curie**. Becquerel's other notable research included the effects of **magnetism** on **light** and the properties of **luminescence**.

Becquerel was born in Paris on December 15, 1852. His grandfather, Antoine-César Becquerel, had fought at the Battle of Waterloo in 1815 and later earned a considerable reputation as a physicist. He made important contributions to the study of electrochemistry, **meteorology**, and agriculture. Antoine-Henri's father was Alexandre-Edmond Becquerel, who also made a name for himself in science. His research included studies on **photography**, **heat**, the conductivity of hot gases, and luminescence.

Becquerel's early education took place at the Lycée Louis-le-Grand from which he graduated in 1872. He then enrolled at the Ecole Polytechnique, and two years later he moved on to the Ecole des Ponts et Chaussées. It appears that there was never any question about the direction of Becquerel's career, as he concentrated on scientific subjects throughout his schooling. In 1877 he was awarded his engineering degree and accepted an appointment as an *ingénieur* with the National Administration of Bridges and Highways.

During his years at the Ecole des Ponts et Chaussées, Becquerel became particularly interested in English physicist **Michael Faraday's** research on the effects of magnetism on light. Faraday had discovered in 1845 that a plane-polarized beam of light (one that contains light **waves** that vibrate to a specific pattern) experiences a rotation of planes when it passes through a magnetic field; this phenomenon was called the Faraday effect. Becquerel developed a formula to explain the relationship between this rotation and the refraction the beam of light undergoes when it passes through a substance. He published this result in his first scientific paper in 1875, although he later discovered that his initial results were incorrect in some respects.

Although the Faraday effect had been observed in solids and liquids, Becquerel attempted to replicate the Faraday effect in gases. He found that gases (except for **oxygen**) also have the same ability to rotate a beam of polarized light as do solids and liquids. Becquerel remained interested in problems of magneto-optics for years, and returned to the field with renewed enthusiasm in 1897 after Dutch physicist **Pieter Zeeman's** discovery of the **Zeeman effect**, whereby **spectral lines** exposed to strong magnetic fields split, provided new impetus for research.

In 1874 Becquerel had married Lucie-Zoé-Marie Jamin, daughter of J.-C. Jamin, a professor of physics at the University of Paris. She died four years later in March of 1878, shortly after the birth of their only child, Jean. Jean later became a physicist himself, inheriting the chair of physics held by his father, grandfather, and great-grandfather before him. Two months prior to Lucie's death, Becquerel's grandfather died. At that point, his son and grandson each moved up one step, Alexandre-Edmond to professor of physics at the Musée d'Histoire Naturelle, and Antoine-Henri to his assistant. From that point on, Becquerel's professional life was associated with the Musée, the Polytechnique, and the Ponts et Chaussées.

In the period between receiving his engineering degree and discovering radioactivity, Becquerel pursued a variety of research interests. In following up his work on Faraday's magneto-optics, for example, he became interested in the effect of the earth's magnetic field on the atmosphere. His research determined how the earth's magnetic field affected **carbon** disulfide. He proposed to the International Congress on Electric Units that his results be used as the standard of electrical current strength. Becquerel also studied the magnetic properties of a number of materials and published detailed information on nickel, cobalt, and ozone in 1879. He also reported the surprising discovery that nickel-plated iron becomes magnetic when heated to redness.

In the early 1880s Becquerel began research on a topic his father had been working on for many years—luminescence, or the emission of light from unheated substances. In particular, he made a detailed study of the spectra produced by luminescent materials and examined the way in which light is absorbed by various crystals. Becquerel was especially interested in the effect that polarization had on luminescence. For this work Becquerel was awarded his doctoral degree by the University of Paris in 1888, and he was once again seen as an active researcher after years of increasing administrative responsibility.

When his father died in 1891, Becquerel was appointed to succeed him as professor of physics at the museum and at the conservatory. The same year he was asked to replace the ailing Alfred Potier at the Ecole Polytechnique. Finally, in 1894 he was appointed chief engineer at the Ecole des Ponts et Chaussées. Becquerel married his second wife, Louise-Désirée Lorieux, the daughter of a mine inspector, in 1890; the couple had no children.

The period of quiescence in Becquerel's research career came to an end in 1895 with the announcement of Röntgen's discovery of x rays. The aspect of the discovery that caught Becquerel's attention was that x rays appeared to be associated with a luminescent spot on the side of the cathode-ray tube used in Röntgen's experiment. Given his own background and interest in luminescence, Becquerel wondered whether the production of x rays might always be associated with luminescence.

To test this hypothesis Becquerel wrapped photographic plates in thick layers of black paper and placed a known luminescent material, potassium uranyl sulfate, on top of them. When this assemblage was then placed in sunlight, Becquerel found that the photographic plates were exposed. He concluded that sunlight had caused the uranium salt to luminesce, thereby giving off x rays. The x rays then penetrated the black paper and exposed the photographic plate. He announced these results at meeting of the Academy of Sciences on February 24, 1896.

Through an unusual set of circumstances the following week, Becquerel discovered radioactivity. He began work on February 26th as usual by wrapping his photographic plates in black paper and taping a piece of potassium uranyl sulfate to the packet. Since it was not sunny enough to conduct his experiment, however, Becquerel set his materials aside in a dark drawer. He repeated the procedure the next day as well, and again a lack of sunshine prompted him to store his materials in the same drawer. On March 1st Becquerel decided to develop the photographic plates that he had been prepared and set aside. It is not clear why he did this since, according to his hypothesis, little or no exposure would be expected. Lack of sunlight had meant that no luminescence could have occurred; hence, no x rays could have been emitted.

Surprisingly, Becquerel found that the plates had been exposed as completely as if they had been set in the **sun**. Some form of radiation—but clearly not x rays—had been emitted from the uranium salt and exposed the plates. A day later, according to Oliver Lodge in the *Journal of the Chemical Society,* Becquerel reported his findings to the academy, pointing out: "It thus appears that the phenomenon cannot be attrib-

Antoine-Henri Becquerel.

uted to luminous radiation emitted by reason of **phosphorescence**, since, at the end of one-hundredth of a second, phosphorescence becomes so feeble as to become imperceptible."

With the discovery of this new radiation Becquerel's research gained a new focus. His advances prompted his graduate student Marie Curie to undertake an intensive study of radiation for her own doctoral thesis. Curie later suggested the name radioactivity for Becquerel's discovery, a phenomenon that had until that time been referred to as Becquerel's rays.

Becquerel's own research continued to produce useful results. In May of 1896, for example, he found uranium metal to be many times more radioactive than the compounds of uranium he had been using and began to use it as a source of radioactivity. In 1900 he also found that at least part of the radiation emitted by uranium consists of electrons, particles that were discovered only three years earlier by Joseph John Thomson. For his part in the discovery of radioactivity, Becquerel shared the 1903 Nobel Prize in physics with Curie and her husband Pierre.

Honors continued to come to Becquerel in the last decade of his life. On December 31, 1906, he was elected vice president of the French Academy of Sciences, and two years later he become president of the organization. On June 19, 1908, he was elected one of the two permanent secretaries of the academy, a post he held for less than two months before his death on August 25, 1908, at Le Croisic, in Brittany. Among his other honors and awards were the Rumford Medal of the Royal Society in 1900, the Helmholtz Medal of the Royal Academy of Sciences of Berlin in 1901, and the Barnard Medal of the U.S. National Academy of Sciences in 1905.

BEDNORZ, J. GEORG (1950-   )
German physicist

German-born physicist J. Georg Bednorz, now a Swiss resident, has a particular genius for finding solutions to problems of equipment and technique in physics research. He is the discoverer, with Swiss physicist K. Alex Müller, of ceramic substances that become superconducting, or allow electrical current to flow without resistance, at temperatures higher than possible with any previous superconductors. The discovery promises to lead to a host of new superconduction applications. Bednorz was only thirty-seven and had had his Ph.D. only four years when he shared the 1987 Nobel Prize in Physics with Müller for the research on superconductivity. He has also won a number of other prizes for his work.

Johannes Georg Bednorz was born on May 16, 1950, in Nevenkirchen in what was then West Germany. He graduated from the University of Münster in 1976 and received his doctorate from the Swiss Federal Institute of Technology in Zurich in 1982. That year he went to work at the IBM Zurich Research Laboratory and met Müller, who invited him to participate in his study of superconductors—generally metals and alloys, whose **electrical resistance** becomes virtually nonexistent at temperatures close to **absolute zero**. Bednorz's own interest in the applications of superconductivity, especially in connection with the design of high-speed trains, made the project particularly appealing to him. Superconducting levitation would reduce **power** requirements in rail transportation: if a magnet is placed on a superconducting surface, it will float above it when the temperature of the surface is lowered.

Scientists have long sought materials that were superconductive at as high a temperature as possible to provide for the most effective electrical conduction for ordinary and practical use. After World War II, superconductors were put to many practical uses. Among these was the operation of large electromagnets, but they only worked at very low temperatures. Finding new materials would open up huge possibilities in reducing power requirements and preventing power losses in overhead power transmissions.

All of the metallic elements and many alloys had been studied, but superconductivity could not be found at a temperature higher than 23° Kelvin (23° above absolute zero). Superconductivity was only possible when materials were cooled by expensive liquid **helium**. If materials could be made superconductive at the temperature of liquid nitrogen, 77° Kelvin, superconductivity would be more practical, since liquid nitrogen is relatively inexpensive, safe, and easy to make. In 1983, Bednorz and Müller began testing ceramic substances made of mixtures of metallic oxides in the hope that they would serve as superconductors at higher temperatures. Bednorz was largely responsible for making and testing the

oxides. On January 27, 1986, the researchers found that a barium-lanthanum-copper oxide became superconductive at 35° Kelvin (–238° Celsius), more than 10° Kelvin above the highest temperature for previous superconductors.

Many scientists were skeptical about the results until teams from the University of Tokyo, the University of Houston, and Bell Laboratories confirmed them. In 1987 researchers at the University of Houston, working with similar ceramics, announced their discovery of superconductors effective at 90° Kelvin, above the **boiling** point of liquid nitrogen. The same year President Ronald Reagan announced an initiative to develop superconductivity. Scientists around the world rushed to study ceramics and superconductivity in the liquid nitrogen temperature range.

The Nobel Prize committee took note of the excitement when they awarded Bednorz and Müller the physics prize, stating that the discovery had generated "an explosive development in which hundreds of laboratories the world over" were taking part. Still, the speed with which the award was given—only two years after the discovery was made—generated some surprise. Prize winners often wait much longer to be recognized.

Despite Bednorz's and Müller's discovery, no theoretical explanation for the behavior of the ceramic substances has yet been made. Nor did their ceramic materials immediately lend themselves to technological applications, since they are not easily made into wire. The Japanese have gone farthest in the field: the Nippon Steel Company developed a "melt-processing" technique to produce a superconducting wire out of the ceramics. Room-temperature superconductivity has yet to be achieved.

BEGINNING OF TIME • See Big bang

BELL, JOHN STEWART (1928-1990)
British theoretical physicist

John Stewart Bell was a humorous, peaceful quantum physicist of the twentieth century. His work encompassed many of the important issues in the development of **quantum physics**. Although he is remembered as a theorist, he demonstrated that he was skilled in the practical side of physics, as well. His doctoral dissertation addressed the issue of the reversibility of **matter** at the subatomic level. At other times in his career, he worked to improve the **proton** synchrotron and the particle accelerator. He is best known for the theorem that bears his name.

Bell's theorem seeks to explain the relationship between two particles traveling in opposite directions at equal speeds. This explanation relies on instant communication across distances and presents an equation for calculating the distant interaction of these particles. Although his logic is solid, there is little room for direct testing of his assumptions. Bell warned that all of the factors contributing to the communication between particles are currently unknown and unobservable.

Still, his equation has made other experiments possible and has opened up a previously closed direction for research. An interesting interpretation of this theorem is the possibility that communication between particles may occur faster than **light**; however, Bell never upheld this conjecture.

Bell was born in 1928 to working-class parents in Belfast, Northern Ireland. He attended Belfast's Queens College. Because of his parents' financial condition, Bell worked while attending college as an assistant in the physics laboratory. He found it to be good preparation for his career and, after graduation, went to work at the Atomic **Energy** Research Establishment at Harwell.

When he was twenty-five, he took a year's leave to work on his doctoral dissertation at Birmingham University where he was remembered by Rudulf Peierls as being experienced and imaginative. Peierls especially remembers Bell as having a talent for readily offering an everyday example to illustrate even the most complex concept. After earning his doctorate, he married a physicist, and, in 1960, the Bells moved to Switzerland where he worked at the Centre European de Recherches Nucleaires (CERN). He was still working at CERN thirty years later when he died of a stroke at age 62.

BELL'S THEOREM

Bell's theorem is a logical argument in support of the completeness of **quantum theory**. The theorem argues against the existence of any hidden or unknown variables that might deterministically explain otherwise seemingly random events predicted by quantum **mechanics**.

Bell's theorem was devised in 1964 by British physicist **John Stewart Bell**. Sometimes known as Bell's inequalities, in **quantum mechanics** the theorem is an analysis of a paradox first advanced by physicists **Albert Einstein**, Boris Podolsky and Nathan Rosen (EPR) in a 1935 *Physical Review* article titled "Can Quantum Mechanical Description of Physical Reality be Considered Complete?" As a consequence, Bell's theorem is used to argue against any incompleteness or hidden variables in the quantum mechanical description of nature.

Bell's theorem examines the expected results of EPR-type experiments when it is assumed that particle properties such as **momentum** and position have real values prior to measurement. This assumption is made by the hidden-variable theories that have been advanced as alternatives to quantum mechanics and the uncertainty principle. The theorem demonstrates that hidden-variable theories give results that are consistent with quantum mechanics in special cases, but that in more general experimental situations, hidden-variable theories predict results that are inconsistent with quantum mechanics.

For example, examining the correlations between measurements of the two particles in an EPR-type experiment gives statistical predictions of the outcome. By comparing the statistical predictions of a quantum mechanical model of the experiment with the predictions of alternative models that assume variables (e.g., momentum and position) have values

prior to measurement, Bell's theorem shows that the predictions of the two models differ by a significant amount unless it is assumed that measurements on one particle can affect those on the other particle instantaneously, no **matter** how far apart they are separated. Since this condition would involve faster-than-light travel, it would violate special relativity, and thus it suggests that the criticisms of quantum mechanics in the EPR paradox do not seriously threaten the validity of the uncertainty principle.

Experiments by physicist Alain Aspect and others have tended to support Bell's theorem and strengthen arguments particles cannot have simultaneous values for complementary properties such as position and momentum unless physics is non-local and permits particles to interact instantaneously regardless of how far apart they are. David Bohm (1917-1992) and others have constructed such non-local hidden-variable theories, but they have not been widely accepted. Because the quantum mechanical description of nature is so different from everyday experience its interpretation continues to be a subject of debate, and Bell's theorem raises deep questions about the physical world that have yet to be completely answered.

See also Complementarity principle; Determinism; Particle-wave duality; Quantum theory

BELOUSOV-ZHABOTINSKI REACTION

The Belousov-Zhabotinsky is an oscillating **chemical reaction** used to demonstrate aspects of chaos theory that, in turn, is used to model the overall behavior of complex reactions and systems. The Belousov-Zhabotinski (B-Z) reaction is a nonlinear chemical reaction which maintains **oscillations** and propagating **pulses** of color.

It was long argued that true oscillating chemical reactions would violate the **Second Law of Thermodynamics**. Some scientists contended, however, that the second law was simply being misapplied (or oversimplified) for calculations involving non-equilibrium reactions. In 1951 Soviet physicist Boris P. Belousov's work with candidate chemical oscillators associated with the Krebs cycle, resulted in the discovery of a reaction involving citric acid, acidified bromate, and a ceric salt that oscillated periodically between yellow and clear. Russian physicist, Anatol M. Zhabotinsky continued Belousov's work and replaced citric acid with malonic acid. When Zhabotinsky left a thin layer of the solution undisturbed, colored concentric circles and Archemedian spirals propagated across the medium. Ultimately known as the Belousov-Zhabotinski reaction, the B-Z reaction produces a visible wave pattern from color changing (red to blue, then red again) spirals. These **waves** are on the order of 0.04 inches (one millimeter) in length with a period of 1 Hz and a **velocity** of 5.4710^{-5} ft/s (1.67×10^{-5} m/s).

The Belousov-Zhabotinsky reaction remained a curiosity until the underlying reaction mechanism was examined in the early 1970s by University of Oregon scientists, Richard M. Noyes, Richard J. Field, and Endre Koros. The researchers eventually articulated a reaction mechanism consisting of twenty-one distinct chemical species and eighteen distinct reaction steps.

The B-Z initial reactants (in moles/liter) are 0.2 M/L of malonic acid, 0.3 M/L of sodium bromate, 0.3 M/L of sulfuric acid and 0.005 M/L of ferroin. The solution undergoes a series of cyclical reactions which are color producing. The cycle begin when the Ferroin, a red substance, is oxidized to create Ferriin, a blue substance. The oxidation continues until a maximum amount of Ferriin is produced and the blue color is prominent. The Ferriin then begins to break down, due to organic reactions, and slowly the concentration of Ferriin decreases while Ferroin increases. This ultimately leads to a return to red as the dominant color. The cycle repeats, propagating out from the center in a spiral pattern.

Being a nonlinear system, the B-Z reaction is considered a complex system. Complexity is the study of how complicated systems can generate simple behavior, such as the sustained spiral pattern seen here. More precisely, the B-Z reaction is a phenomena of spatio-temporal chaos. Spatio-temporal chaotic phenomena, with additional degrees of freedom than other phenomena, change in both **space** and **time**.

The study of B-Z reaction mechanisms advances an understanding of non-linear phenomena and the development of chaos theories that are now an important tools in the study of complex dynamical physical, chemical and biological phenomena ranging from the mechanisms of explosions to predicting population trends.

See also Chaos and order; Entropy; Non-linear dynamics; Oscillations; Thermodynamics

BERNOULLI, DANIEL (1700-1782)

Dutch Swiss mathematician and physicist

Bernoulli's work on **fluids** pioneered the sciences of hydrodynamics and aerodynamics. Born in the Netherlands and spending most of his life in Switzerland, Bernoulli was one of a large family of brilliant scientists and mathematicians that included his father, Jean Bernoulli, and uncle, Jacques Bernoulli.

Ignoring his family's pleas to enter the world of business, Bernoulli pursued a degree in medicine and then, after graduation, a career as a professor of mathematics. He began teaching in 1725 at a college in St. Petersburg, Russia, eventually returning to Switzerland in 1732. While a professor at the University of Basel, he became the first scientist outside of Great Britain to fully accept **Newtonian physics**. It was also here that Bernoulli performed the research on fluid behavior that would make him famous.

The 1738 publication *Hydrodynamica* developed the prominent theories of hydrodynamics, or the movement of water. Paramount among these was the fact that, as the **velocity** of a fluid increases, the **pressure** surrounding it will decrease. Called **Bernoulli's principle**, this pressure drop was also shown to occur in moving air, and it is the reason boats and planes experience lift as water or air passes around them.

Daniel Bernoulli. *(Corbis-Bettmann. Reproduced by permission.)*

This effect is easily shown by blowing between two pieces of paper; the drop in pressure will cause the papers to bend toward each other. Bernoulli 's research marked the first attempt to explain the connection pressure and temperature have with the behavior of gas and fluids.

Bernoulli's experiments with fluids caused him to devise a series of hypotheses about the nature of gases. He was certainly one of the first to formulate principles dealing with gases as groups of particles, which later became the basis for atomic theories. As groundbreaking as this work was, it was paid little attention by his peers, and subsequently it was nearly a century before the **atomic theory** rose again.

BERNOULLI'S PRINCIPLE

Bernoulli's principle states that flowing **fluids** like air and water press less than still fluids and that **pressure** decreases quadratically with speed; i.e., with speed-squared.

One quarter of a millennium ago, **Daniel Bernoulli** pioneered use of kinetic theory that molecules moved and bumped things. He also knew that flowing fluids pressed less, but he did not connect these ideas logically. In *Hydrodynamica*, Daniel's logic that flow reduced pressure was obscure, and his formula was awkward. Daniel's father Johann, amid controversy, improved his son's insight and presentation in

Hydraulica. This research was centered in St. Petersburg where Leonard Euler, a colleague of Daniel and a student of Johann, generalized a rate-of-change dependence of pressure and **density** on speed of flow. Bernoulli's Principle for liquids was then formulated in modern form for the first time.

In this same group of scientists was D'Alembert, who found paradoxically that fluids stopped ahead of obstacles, so frictionless flow did not push.

Progress then seems to have halted for about a century and a half until Ludwig Prandtl or one of his students solved Euler's equation for smooth streams of air in order to have a mathematical model of flowing air for designing wings. Here, speed lowers pressure more than it lowers density because expanding air cools, and the ratio of density times degrees-kelvin divided by pressure is constant for an ideal-gas.

More turbulent flow, as in atmospheric winds, requires an alternative solution of Euler's equation because mixing keeps air-temperature fixed.

Bernoulli's principle is regarded by many as a paradox because currents and winds upset things, but standing a stick in a stream of water helps to clarify the enigma. You will see calm, smooth, level water ahead of the stick and a cavity of reduced pressure behind it. Calm water pushes the stick, as lower pressure downstream fails to balance the upsetting **force**.

Bernoulli's principle never acts alone; it also comes with molecular entrainment. Molecules in the lower pressure of faster flow aspirate and whisk away molecules from the higher pressure of slower flow. Solid obstacles such as airfoils carry a very thin stagnant layer of air with them. A swift low-pressure airstream takes some molecules from this boundary layer and reduces molecular impacts on that surface of the wing across which the airstream moves faster.

For Bernoulli's principle to dominate a dynamic situation, **friction** must be less dominant. Elastic molecular impacts are frictionless-no heating. Molecules of dry air, even more than those of water, collide elastically; so Bernoulli's principle with its molecular-entrainment agent is in fact, to use a vernacular cliche, the only game in town for windy air.

BETA DECAY

Beta decay describes a number of processes involving the transformation of atomic nuclei and the resulting release of **radiation** in the form of beta particles. Beta decay is one of three principal forms of **radioactive decay**: alpha, beta, and gamma. During the process of radioactive decay, unstable nuclei are converted into more stable forms as they release electromagnetic radiation, **subatomic particles**, or both.

There are various mechanisms and forms of beta decay (e.g., negative beta decay, **neutron** beta decay, double beta decay, inverse beta decay, etc.). During beta decay the **nucleon** (protons and neutrons) composition of the atomic **nucleus** changes, either by the transformation of a neutron into a **proton** or by the transformation of a proton into a neutron. Beta decay, like other radioactive decay processes, is a spontaneous event among unstable atoms. The processes of beta decay are very fast, usually occurring in a fraction of a billionth of a second.

Beta decay results in the radiation of electrically charged subatomic particles, termed beta particles. Beta particles are approximately 1/1837 the **mass** of a positively charged proton and can be negatively or positively charged. Negatively charged beta particles (sometimes called negatrons) are identical to electrons that orbit nuclei. Positively charged beta particles, the **antiparticle** of electrons, are called positrons.

Atomic nuclei with an unstable surplus of neutrons generally undergo negative beta decay (also termed neutron beta decay). During this process a neutron composed of one up quark and two down **quarks** (udd) undergoes a quark transformation to become a proton with two up quarks and one down quark (uud). During this process (mediated by a virtual W-particle) the nucleus expels a negative beta particle (**electron** or negatron) as radiation. As a result of the nucleon transformations, a new proton is left behind in the nucleus and the atomic number of the **atom** is raised by one.

If the nucleus of the an atom has too few neutrons to be stable, however, **positron** emission occurs. Positron emission results from the transformation of a proton into a neutron. During this process the atomic number of the atom is reduced by one.

During beta decay, **energy** is conserved by the simultaneous emission of a **neutrino** or antineutrino along with the negatron or positron.

Beta radiation is a common process. The carbon-14 used in **radioactive dating**, for example, decays into nitrogen through the process of negative or neutron beta decay. Large doses of beta radiation, however, can cause skin burns, and long-term exposure to lower levels of beta radiation (e.g., resulting from the beta decay of Radon, a colorless, odorless, gas sometimes found in the foundations of homes) has been linked to lung cancers. Before beta particles were discovered, the products of beta decay were termed beta rays.

See also Antiparticle; Particle physics; Radioactivity; Subatomic particles

BETA RADIATION

Beta **radiation** is the emission of an **electron** from the **nucleus** of a radioactive **isotope**. This electron comes from one of the neutrons in an unstable nucleus. The weak nuclear **force** is involved, and the **neutron** is converted into a **proton** when the beta particle is emitted. This produces an isotope of the next element in the **periodic table**, a process known as transmutation. The emitted beta particle travels through air at close to the speed of **light**. However, it can be stopped by a sheet of aluminum foil greater than 0.12 in (3 mm) thick.

When French physicist Henri Becquerel (1852-1908) first discovered the property of **radioactivity** in 1896, he did not know that radiation consists of particles as well as **energy**. Beginning in 1898, **Ernest Rutherford** conducted experiments to determine the nature of this radiation. One experiment demonstrated that the radiation actu ally consisted of three different types: a positive particle, a negative particle, and a form of electromagnetic radiation that carried high energy. Further studies on the mass/charge ratio of the particles supported the idea that the negative radiation, which he had labeled "beta", had the same charge and **mass** as the particle identified by **J.J. Thomson** as the electron. By 1902 Rutherford and his colleague Frederick Soddy (1877-1956) proposed that a different **chemical element** is formed whenever a radioactive element decays. Rutherford was awarded the 1908 Nobel Prize in chemistry for his work in explaining these processes.

Although they had discovered a new type of reaction, physicists strongly believed that the conservation laws of classical physics would still apply. This meant that the decay reaction should exhibit **conservation of energy**, conservation of **linear momentum**, conservation of **angular momentum**, conservation of **electric charge**, and conservation of the number of particles in the nucleus. Three of the quantities were conserved when the beta particle was emitted; however, energy and angular momentum appeared to be "missing." To satisfy these last two conservation laws, **Wolfgang Pauli** suggested in 1934 that **beta decay** must involve a third particle that is neutral, has negligible rest mass and a spin of one-half, and carries away from the reaction the energy that appears to be missing. **Enrico Fermi** named this particle the **neutrino**, Italian for "little neutral one"—its existence was accepted without evidence. It was not until 1953 that **Frederick Reines** devised an experiment that successfully detected the neutrino; he received the 1995 Nobel Prize in Physics for that work.

During the 1930s, when scientists were experimenting with reactions that were produced by neutrons, Enrico Fermi and his associates discovered that a heavier isotope of uranium than is found in nature, uranium-239, will spontaneously give off a beta particle and change into a new element of one higher atomic number, 93; this element will also undergo beta decay to change into an additional new element, 94. The new elements were named neptunium and plutonium, respectively. This is the first record of the production of synthetic elements, known as transuranic elements. Early in 1999 synthesis of the element with atomic number 114 was reported.

More detailed explanations of beta decay were developed in the late twentieth century, following theories of the existence of unique particles that transmit the weak nuclear force. These particles, identified as W and Z, were finally discovered in 1983 by **Carlo Rubbia**, who was awarded the Nobel Prize in physics the following year (1984).

The decay chain of the most abundant uranium isotope uranium-238 to stable lead-206 involves six different beta decay reactions. In addition, a number of **radioisotopes**, both natural and man-made, have been identified as beta emitters. Rubidium-87, one of the relatively abundant minerals in the Earth's crust, is a beta emitter; its **half-life** is 49 billion years. The ratio of the amount of the element produced (strontium-87) to rubidium remaining in the sample is one of the methods that is being used by geologists to determine the age of the rocks of the Earth. Some of these isotopes have been put to

practical use in industry and medicine. Often their application depends on the half-life of the element. Tritium, the radioactive isotope of hydrogen that is produced in the atmosphere, is also a beta emitter. Its half-life of 12.33 years is the right range for age dating fine wines and other materials that contain a high percentage of water. Cobalt-60 is an example of a beta emitter of shorter half-life (5.27 years); it is being used in medical applications. Iodine-131 (half-life 8.04 days) is also used in medical applications.

BETA RAYS • See Beta radiation

BETHE, HANS (1906-   )
German American physicist

Hans Bethe is one of the premier physicists of the twentieth century, and one of only a very small number whose understanding and contributions span nearly all of the subfields of physics. His achievements range widely through science, his most well-known achievement being a theory that accounts for the **energy** production of **stars**. He also contributed to theories regarding **quantum mechanics**, fundamental particles, **nuclear reactors**, **nuclear weapons**, and **astrophysics**, and has been active for many years in the scientific community's efforts to limit the spread of nuclear weapons.

Hans Albrecht Bethe was born in Strasbourg (then part of Germany) on July 2, 1906. He was the only child of Albrecht Theodore Julius Bethe, a *privatdocent* (unsalaried lecturer) in physiology at the University of Strasbourg, and Anna Kuhn, the daughter of a professor. Bethe showed an early precocity in mathematics. He recalls walking with his mother at age five and noting to her that it was strange that a zero at the end of a number means a lot, while one at the beginning means nothing. At seven he learned about exponents and filled a book with the powers of two and three. The family moved to Kiel in 1912 when Bethe's father became chair of the physiology department at the University of Kiel. In 1915 Albrecht Bethe was invited to start a department of physiology at the University of Frankfurt, and the family moved again.

The period between the two world wars was a time of severe inflation for the German economy. While in Frankfurt, Bethe's father was paid twice a week to keep up with the continuing devaluation of the German mark. Bethe's excellent grasp of numbers earned him the responsibility of collecting his father's salary in the morning and spending it on food for the family before the new value of the mark was calculated in the afternoon and prices increased dramatically. This early preoccupation with money undoubtedly contributed to Bethe's lifelong amateur fascination with economics. Due to his frail health, Bethe did not attend school until the family went to Frankfurt. In Kiel he went to a tutor several times a week with several other children. In Frankfurt he entered the Goethe Gymnasium. His father recognized his son's talent in mathematics but did not want him to get too far ahead of his peers. In fact, when Bethe wanted to borrow his father's trigonometry and calculus books, he had to do so secretly. He excelled in his academic subjects, but did poorly at physical education. In his last few years in the Gymnasium, he took some elective physics courses and found the subject stimulating.

Bethe entered the University of Frankfurt in 1924, continuing to live at home during that time. In 1926 he went to the University of Munich, where he spent two and a half years. He earned his Ph.D. in theoretical physics under **Arnold Sommerfeld** in 1928. While a graduate student, Bethe became interested in **quantum mechanics**, the new theory of interactions between **matter** and **radiation**. In 1927 he wrote a paper on **electron diffraction** by crystals, using quantum mechanics to explain how electrons behave like **waves**. After his graduation Bethe returned to the University of Frankfurt as an instructor. One year later he went to the Technical College of Stuttgart to work with Paul Ewald. Bethe was welcomed into the Ewald family, and was often invited over for dinner and Sunday walks. In 1929 Bethe was appointed a *privatdocent* at the University of Munich but spent much of the next three years outside Germany. A Rockefeller Foundation fellowship gave him the opportunity to go to Cambridge University. He also spent some time working in Rome with **Enrico Fermi**, a physicist who had earned worldwide respect for his scientific contributions and who already, at age thirty, was a full professor.

Bethe went to the University of Tübingen in 1932 as an assistant professor but lost that job one year later when Hitler came to power. Like others of Jewish ancestry, Bethe was summarily dismissed. During his time at Tübingen he had many fruitful discussions with **Hans Geiger**, a professor of theoretical physics and the inventor of the Geiger counter. Geiger regularly attended Bethe's classes on quantum mechanics and in turn explained his own work to Bethe. To Bethe's sorrow, however, once he was dismissed, Geiger withdrew his friendship. In contrast, Arnold Sommerfeld spent much of the summer of 1933 searching for jobs for Bethe and other Jewish academics.

Bethe spent the years 1933–35 at the Universities of Manchester and Bristol in England as a postdoctoral refugee scientist. Then in 1935 he was invited to Cornell University in Ithaca, New York, as an assistant professor. He recounts in Jeremy Bernstein's *Hans Bethe, Prophet of Energy,* "I met a physicist who had been there. He said, 'Don't go to that place. It is a terrible place. It is so straitlaced that you have to go to church every Sunday. You won't like it at all.' I accepted anyway, and was offered a salary of $3,000 a year. I considered myself immensely rich." Bethe became a full professor at Cornell in 1937, and with the exception of a hiatus during World War II, has remained there. At Cornell Bethe found his colleagues eager but not very knowledgeable. He devoted much time to explaining theoretical physics to the physics faculty who were mainly experimentalists. In 1936 he worked on a series of articles with American physicists Robert F. Bacher and M. Stanely Livingston, and his old mentor, Arnold Sommerfeld, which summarized the current knowledge in **nuclear physics**. These articles were so seminal in their description and presentation that they were considered essen-

tial reading for two generations of graduate students in physics.

In 1938 Bethe attended a meeting on astrophysics in Washington, D.C., convened by **Edward Teller** and **George Gamow**. It was here that Bethe began to consider the question of the mechanism whereby the **Sun** and other stars get their energy. Many had already concluded that the necessary reactions must be thermonuclear, but no one had been able to determine the reactions that could account for the observed data. In characteristic fashion, Bethe solved the problem in six weeks. He proposed two different mechanisms by which different types of stars released their massive amounts of energy. One of these, which had already been suggested independently by Carl Friedrich von Weizsäcker of Germany, begins with the fusion of two protons and proceeds through a series of steps to ultimately result in a **helium nucleus** and two excess protons that are released to repeat the process (a proton-proton reaction). The other process for more massive stars, derived solely by Bethe, involves a series of six steps that begins with carbon-12 and finishes with one carbon-12 **atom** and one atom of helium-4. Energy is released as a consequence of each of the six steps in the process. Generally considered to be Bethe's greatest achievement, the end product was published as one page in *Physical Review* in 1939. Polish-born scientist and author Jacob Bronowski noted in *The Ascent of Man*: "Hans Bethe's explanation is as vivid to me as my own wedding day, and the subsequent steps that followed as the birth of my own children. Because what was revealed in the years that followed (and finally sealed in what I suppose to be the definitive analysis in 1957) is that in all the stars there are going on processes which build up the atoms one by one into more and more complex structures."

Bethe is also well known throughout the world of science for another famous paper he coauthored with Ralph Alpher and George Gamow. Bethe was added at the suggestion of Gamow to make the sequence of authors' names resemble alpha, beta, and gamma, the first three letters of the Greek alphabet. This paper dealt with the early history of the **universe** and presented a mathematical case for what is now termed the **big bang**. As **Stephen Hawking**, who occupies Newton's chair at Cambridge University, explains in *A Brief History of Time,* "In this paper they made the remarkable prediction that radiation (in the form of photons) from the very hot early stages of the universe should still be around today, but with its temperature reduced to only a few degrees above **absolute zero** (−273°C). It was this radiation that [American physicist Arno] Penzias and [American physicist Robert] Wilson found in 1965."

On September 14, 1939, Bethe married Rose Ewald, the daughter of his former mentor Paul Ewald, in New Rochelle, New York. She had emigrated to the United States and was a student at Smith College. That same year his mother emigrated to the United States, as life became increasingly more difficult for Jews in Nazi Germany.

Like many others Bethe was convinced of the inevitability of the participation of the United States in World War II. He wished to help the war effort, but prior to his naturalization in

1941 was not permitted to do any classified research. He did work for a time with **Edward Teller** on **shock waves** that would later prove particularly useful on the **atomic bomb** project. After receiving a security clearance in 1941, Bethe worked on the development of **radar** at the Massachusetts Institute of Technology (MIT). While working on this project, he developed a device used to measure an increase in electromagnetic waves, now known as a Bethe coupler.

Bethe renewed his acquaintance with **Robert Oppenheimer**, who later became the director of the **Manhattan Project**, at an American Physical Society meeting in Seattle in 1941. The two had met briefly in Germany in 1929. In 1942 Bethe had become aware of the work of **Enrico Fermi** and Edward Teller in Chicago on a graphite nuclear reactor. At that point Teller was already looking past the atomic bomb to the more powerful hydrogen bomb. Despite his wife Rose's reservations, Bethe accepted Oppenheimer's invitation to join the Manhattan Project at its Los Alamos, New Mexico, facilities as director of the theoretical physics division. Rose was placed in charge of housing for all those working on the project. Bethe believed the threat of the Germans producing such a device was real and had to be countered by an American device. The project required him to pull together his knowledge of nuclear physics, shock waves, and electromagnetic theory to explain in advance how the atomic bomb would work and with what effect. This extended to mathematically tackling serious questions such as whether the entire atmosphere would explode when the bomb was detonated.

It was a relatively young group of scientists who gathered at Los Alamos. Bethe was thirty-seven at the time and Oppenheimer was thirty-nine. In fact the average age of those working on the project was twenty-seven. One outgrowth of the work there was Bethe's collaboration with **Richard P. Feynman**, a young physicist whom Bethe considered the brightest individual with whom he had ever worked, to devise a formula to calculate the efficiency of a nuclear weapon. Known as the Bethe-Feynman formula, it is still used today. Richard Feynman, recalling those days in his autobiography, *Surely You're Joking, Mr. Feynman!,* wrote of Bethe's mathematical abilities: "I had a lot of fun trying to do arithmetic fast, by tricks, with Hans. It was very rare that I'd see something he didn't see and beat him to the answer, and he'd laugh his hearty laugh when I'd get one. He was nearly always able to get the answer [mentally] to any problem within a percent. It was easy for him—every number was near something he knew."

Bethe left Los Alamos in 1946 to return to Cornell over the objections of Teller. Teller for some time had considered the Russians as great a threat as the Germans and wanted the Manhattan Project to continue to work on the hydrogen bomb. Bethe, however, longed to return to teaching physics and dealing with theoretical questions in physics that did not have military applications.

After the war Bethe was active in the disarmament movement. He advocated civilian control of atomic energy in this country and international control of atomic energy worldwide. Along with other scientists, such as Oppenheimer and **Leo Szilard**, he tried to educate the public about the severe destruction caused by nuclear weaponry. Bethe had a chance to return to Los Alamos in 1949, but was deterred by his wife's arguments and Los Alamos scientist and author Victor F. Weisskopf's description of the destructive capabilities of the H-bomb. He was active at the Geneva test ban discussion in 1958. From 1956 to 1959 he served on the President's Science Advisory Committee.

Not surprisingly, Bethe's research after the war demonstrated his breadth of knowledge. In 1947 Willis E. Lamb, Jr. and R. C. Retherford measured a tiny shift in the energy levels of the electron in a **hydrogen atom**, now known as the **Lamb shift**. Previous theoretical methods were unable to account for the shift, but in 1947 Bethe was able to produce the first theoretical calculation of the Lamb shift. During the late 1940s and 1950s Bethe served as a consultant to several laboratories that were developing nuclear reactors. He was also involved in work with **lasers, rockets**, and astrophysics.

Bethe was awarded the Nobel Prize in physics in 1967 with commendation "for his contributions to the theory of **nuclear reactions**, especially his discoveries concerning the energy production in stars." In his presentation speech **Oskar Klein** of the Royal Swedish Academy of Sciences remarked that several of Bethe's discoveries were individually worthy of the Nobel Prize. Bethe's other awards include the Presidential Medal of Merit (1946), the Henry Draper Medal of the National Academy of Sciences (1947), the Max Planck Medal of the German Physical Society (1955), the Enrico Fermi Award (1961), the National Medal of Science (1976), and the Vannevar Bush Award (1985).

Retired from Cornell as an emeritus professor since 1975, Bethe still maintains an office there and remains busy lecturing, consulting, and keeping up with correspondence. In earlier years he enjoyed skiing and mountain climbing, and remains interested in economics as he has been for much of his life. The Bethes have two children, a son Henry and a daughter Monica.

BIG BANG

The big bang is the cosmic event that is theorized to have marked the origin of the **universe**. At that instant all **matter** and **energy** in the entire physical universe, and the four dimensions of **time** and **space**, were created from a state of enormous **density, temperature**, and **pressure**. Big bang theory is the widely held set of scientific explanations relating to the primordial big bang. **Cosmology** is the study of the creation, evolution, present structure, and ultimate fate of the universe; the big bang theory is, at present, the central, unifying model that guides cosmology.

Three subdivisions of cosmology (quantum, particle, and standard cosmology) deal with the big bang according to the length of time after the big bang event.

Quantum cosmology ranges from about 10^{-43} seconds to about 10^{-11} seconds after the big bang. Because no agreed-upon physical theory exists among scholars for this time period (called the Planck epoch), the process and events contained therein remain highly speculative. In addition, the uncertainty principle of **quantum mechanics** prevents speculation on times shorter than 10^{-43} seconds after the big bang.

Particle cosmology deals with the period from about 10^{-11} seconds to one hundredth of a second after the big bang. This area of cosmology is less speculative; involving some estimated equations and unverified assumptions.

Standard cosmology studies the period from about one hundredth of a second after the big bang to the present day. This subdivision has withstood many tests and provides a reliable basis for the bulk of information related to cosmological development and evolution.

Big bang theory is essentially based on two major theoretical assumptions. The first assumption, derived from German-American physicist **Albert Einstein's** general theory of relativity, correctly describes the gravitational interaction of all matter and **radiation**. The second assumption, called the **cosmological principle**, states that at sufficiently large scales, the average properties of different regions of the universe are the same. As a result, even though average densities of matter and radiation vary wildly within small regions, such as the **Milky Way galaxy**, over vast volumes of space the average densities and types of matter are the same from one region to another. These two premises allow physicists to make determinations regarding the history of the universe (from the Planck epoch) using field equations that were developed by Russian physicist Alexander (Aleksandr Aleksandrovich) Friedmann (also spelled as Fridman, 1888-1925). Field equations are solutions mathematically derived from Einstein's general theory of relativity. In addition, in 1929 American astronomer **Edwin Hubble** provided supporting evidence for the theory with his discovery that the **light** of galaxies is universally redshifted. His studies showed that galaxies are generally moving away from each other and that the universe is uniformly expanding. During the 1940s Russian-born American cosmologist and nuclear physicist **George Gamow** developed the modern version of the big bang model that fit with Friedmann's solutions.

Super-dense theory, another name for the big bang theory, proposes that approximately 10-20 billion years ago the universe was composed of a very high concentration of matter and radiation mixed in an extremely dense conglomerate at very high temperatures. This initial **singularity**, or "primordial atom", then expanded rapidly in its first 10^{-32} second, to about 10^{50} times its original size. The universe then slowed its expansion and cooled, allowing the single **force** that existed at the beginning of the universe to separate into the forces of **gravity, electromagnetism**, the strong nuclear force, and the weak nuclear force. Temperatures quickly dropped below 10^{12} K as the production of protons and heavy particles, and then electrons and light particles, rapidly decreased. At about one second into the life of the universe, the majority of exist-

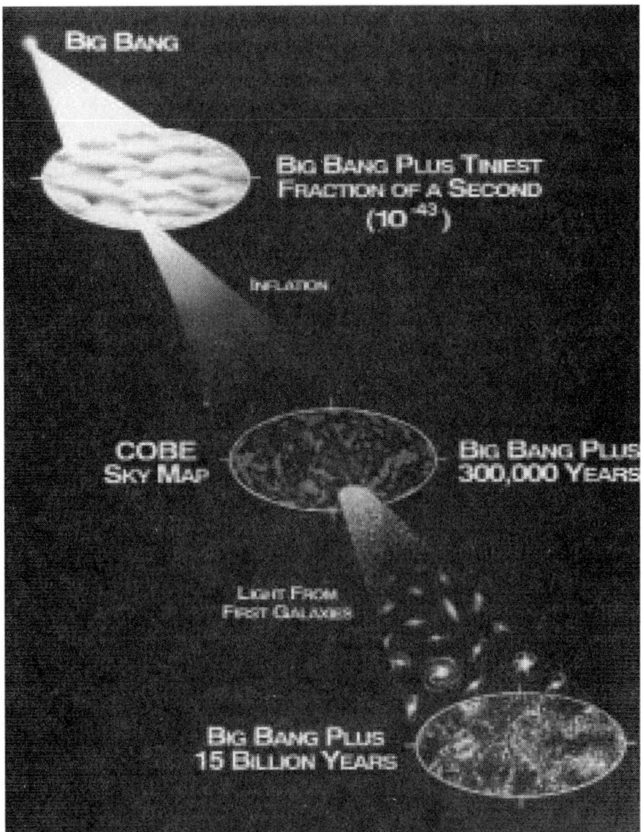

Big bang theory as conceptualized by NASA, 1992.

ing particles and antiparticles had been destroyed to produce photons of electromagnetic radiation. The domination of photons during this time was called the radiation era. During this expansion radiation quickly lost energy and energy density, causing matter to overtake radiation in total energy density after an elapse of approximately 10,000 years and at a temperature of about 10,000K. This change of domination from energy to matter began the matter era. The matter was composed primarily of an ionized gas of electrons, protons, and **helium** nuclei. At a temperature of about 3,000K electrons combined with protons to form neutral hydrogen. At around 100,000 years after the big bang this formation of neutral hydrogen brought about a cooling of radiation that has continued to the present at the cosmic microwave **background radiation** observed at 2.7K. The expansion of the universe also caused red-shifting of traces of early radiation. With matter no longer coupled with radiation, existing matter was able to form **planets**, **stars**, galaxies, and other large-scale structures observed today. The heavier elements were created later within the interiors of stars and spread widely in supernova explosions.

A widely accepted version of the big bang theory, one that solves a number of problems within cosmology, involves an inflationary model that predicts the universe can be either open or closed. The theory was advanced in the 1980s by Alan Harvey Guth and was elaborated upon by Paul Steinhardt,

Andrei Linde, and Andreas Albrecht. If the universe is open (with the amount of matter less than a critical density), it will expand forever ultimately leading to both entropic death and a universe filled with **subatomic particles** too far apart to form larger atomic structures. If the universe is closed (with the amount of matter more than a critical density), the expansion of the universe will eventually stop and then reverse to begin a contraction leading to an eventual collapse termed the big crunch. The inflationary model currently predicts that the universe is on the boundary between being open and closed, with scientists differing upon whether there is sufficient matter and **mass** in the universe to meet the requirements of **critical mass**.

The 1965 observation of the 2.7K microwave background radiation by **Arno Penzias** and Robert Wilson from Bell Laboratories was a critical confirmation big bang based cosmological models. In cosmological analysis, because the laws of physics break down as one regressively approaches the time of the big bang, there is no satisfactory explanation for the cause the big bang itself.

See also Inflationary universe; Microwave background radiation; Relativity, general

BIG CRUNCH • See Big bang

BINARY PULSAR

A binary pulsar, a rotating **neutron** star that generates regular **pulses** of **radiation**, exists in a binary system in which the pulsar has a companion stellar object (e.g., another neutron star, white dwarf or a main-sequence star). In binary systems both stellar bodies orbit around a common **center of mass**.

Most solitary **pulsars** have periods of around one-half second. Approximately two to seven percent of pulsars, however, have shorter periods on the order of milliseconds (millisecond pulsars). The period of normal pulsars increases with age, resulting in a corresponding decrease in the magnetic field, such that after 10^7 years the **radio** emission is undetectable. The slowdown rates of millisecond pulsars are extremely small, indicating a lifetime of 10^9 years. Such pulsars are clearly a population separate from normal pulsars. That pulsars are part of a binary system is indicated by Doppler shifts in the pulse period and the change in arrival **time** of the pulse; both a result of the orbital **motion** of the pulsar.

The increase in the rotation rate of the pulsar is a result of **mass** transfer between the pulsar and the companion star. At late stages in **stellar evolution** a massive star will expand, filling what is known as the Roche surface, beyond which the gravitational pull of the neutron star will exceed that of the companion star. Stellar **matter** will then fall either directly onto the surface of the neutron star, or first form an accretion disk before falling onto the surface. This infalling mat-

ter results in a transfer of **angular momentum** from the orbital motion of the companion star to the rotation of the neutron star. The spot where such infalling matter impacts the surface of the neutron star, perhaps one square kilometer by 100 meters thick, is extremely hot, on the order of 10^7 K, and thermally emits x-rays. Such binary systems are called x-ray pulsars.

Not all millisecond pulsars remain in a binary system. In these cases, it is argued that a companion star can be evaporated by the impacts of highly energetic charged particles from the pulsar. Evidence for such a process is seen in the pulsar PSR 1957+20. The companion star of this system is a white dwarf roughly the size of the earth. However, based on the length of time this star occults the neutron star during its orbit, it's disk is over 100 times larger. Changes in the shape of the radio pulse over time immediately before and after occultation indicate that the pulse has passed through ionized gas. Within a million years, the white dwarf will be completely evaporated.

Of the first 450 pulsars discovered, only seven were definitely in a binary system, despite the fact that most O-stars, the precursors of pulsars, are in multiple star systems. Models have shown however, that during the supernovae explosion which creates the neutron star, if more than half of the mass of the supernovae is expelled, the binary system will be disrupted. If the system is not disrupted, the companion star will evolve into a white dwarf, if its original mass is equal or less than the mass of the **sun**. If the companion is more massive, it will also go through a supernova explosion, and the system will be composed of two **neutron stars**. The pulsar PSR 1913+16, which was the first binary pulsar to be discovered, is probably such a system.

PSR 1913+16 is composed of two bodies, both of roughly 1.4 solar masses, in an highly eccentric orbit (e = 0.617), with a semi-major axis of 70,000 km. Because the high orbital **velocity** of the two bodes (about 0.1% of the speed of **light**) the effects of general relativity can be directly detected.) The wavelengths of the radio **waves** emitted by the pulsar are lengthened as a result of the effects of the **gravitational field** of the companion star (gravitational redshift). In addition, the duration of the transmission pulse is increased as a result of the orbital motion (transverse Doppler shift). Moreover, the arrival of the pulse is delayed by 50 microseconds, when the path of the radio waves travels near the Sun. The periastron of the system, which is the point in the orbit at which the **stars** are closest, precesses by 4.2 degrees per year (in comparison, the orbit of Mercury precesses by .75 seconds of arc per century) and the orbital period of the system lengthens by about one part in 10^{12}. A coupling between the angular **momentum** of the orbit and the angular momentum of the rotating pulsar causes the axis of rotation of the pulsar itself to precess.

See also Binary stars; Hertzsprung-Russell diagrams; Stellar evolution; Stellar life cycle

BINARY STARS

Binary **stars**, also called double stars, are composed of two stars that orbit around a common **center of mass**. Binary stars are quite common. Astronomers estimate that 85% of the stars in the galaxy are actually systems consisting of two or more stars.

Many double star systems can be detected with a small **telescope** and a few can even be seen as two stars by the unaided eye. English astronomer **William Herschel** first determined in 1801 that binary stars actually orbit each other, and are not two unrelated stars that happen to be in the same direction. Double stars that can be observed separately by telescope observation are called visual binary stars. These stars are generally separated by distances of several astronomical units or more. The orbital periods of visual binary stars can be several centuries, so many observations are needed to confirm that they are actually binary stars. Because the members of the system are so widely separated, there is very little transfer of **matter** between them and they have very little effect on each other.

A second type of binary star is detected by observing the spectrum of the system. These spectroscopic binary stars are generally much closer to each other than visual binaries, as close as a few hundredths of an astronomical unit. As they orbit each other, their spectrum as viewed from earth changes. As the stars orbit their **center of gravity**, the stars have a different **velocity** relative to the earth. This causes a periodic Doppler shift of the spectrum of one or both members of the pair. The amount and period of the shift is used to calculate the orbital velocity, size of the stars, and their distance from one another. The orbital periods of spectroscopic binaries are as short as a few days.

A third type of double stars involves a pair of stars revolving about their common center of **mass** in an orbit whose plane passes through or very near the Earth. An observer on the Earth sees one star pass in front of the other. As the stars alternately eclipse one another, there is a periodic decrease in their observed brightness. This type of system is called an eclipsing binary, or an eclipsing variable star. The star Algol was the first recognized as an eclipsing binary by Dutch-born astronomer John Goodricke (1764-1786) in 1782. Several thousand are now known. By comparing the observed duration of the eclipse to the period of the orbit, as determined spectroscopically, astronomers can find the diameters of the stars relative to the size of their orbits. From this they can calculate the densities of the stars and their temperatures. Most of the thousands of eclipsing binary stars that are known have orbital periods of a few days and are less than one astronomical unit apart.

Most stars are actually binary or multiple stars. Many binary stars are themselves part of larger multiple star systems. For example, Castor, in the constellation Gemini, is a visual binary with an orbital period of 467 years. Each of the components of the binary is itself a spectroscopic binary, with a period of several days.

Several mechanisms have been proposed to explain the formation of binary stars. Widely separated binaries, such as visual binary stars, may have formed by tidal capture. Two clouds of matter lose **energy** to tidal effects as they approach each other and then each cloud evolves into a star. An alternative is conucleation, in which the two stars form from the material of a single cloud after two fragments begin orbiting each other. Theories to account for the formation of closely separated binaries include fragmentation, in which a collapsing cloud breaks apart during star formation to produce two or more stars, and fission, in which the newly formed star splits into two pieces, each of which becomes a star.

Binary stars that are close together transfer mass between them. This transfer changes the **angular momentum** of the stars, and changes their separation and orbital periods. As matter builds in the **space** between the stars, the increase in energy can cause matter to be ejected from the system as a nova or supernova. One of the stars then becomes a **neutron** star or a black hole. Close binary systems that include a small, very dense star can become extremely energetic as the transferred mass is heated, creating **pulses** of **x rays**. These systems are known as x-ray **pulsars**. Binary pulsars include a neutron star or a white dwarf star. As their companions transfer mass, the rotation rate of the systems increase, shortening the orbital period. Some binary pulsars have orbital periods of only a few milliseconds.

BINDING ENERGY

Binding **energy** is the term used to specify the amount of energy necessary to hold components of a system together, usually in describing the interactions within constituents of molecular, atomic or nuclear systems. However, it can also be used in describing astrophysical systems. Through binding energy, the system is relatively stable and allowed to act as a unit. The measure of the energy necessary to hold a system together is essentially equal to amount of energy necessary to break the system apart. Therefore "binding energy" is used in reference to both processes.

At the astrophysical level, the **force** of **gravity** provides much of the binding energy necessary to keep solar systems as units. Gravity holds the components of a system together because no matter how small the system's "particle" (Newton's term) is, it will have an attraction to another object proportional to the **mass** of the two objects. This attraction of the two objects is inversely limited by the distance between the two objects. As massive as Earth and the **Sun** are, they will exert a strong attraction on each other. Because Earth and the other **planets** rotate around the Sun, there are forces such as **angular momentum** and centrifugal force that prevent the planets from leaving their orbits and crashing into the Sun.

Binding energy among atomic particles prevents the **atom** from breaking down into a number of hydrogen atoms with extra neutrons. Francis Aston (1877-1945) investigated the phenomenon of atoms weighing less than the sum of their particles. He calculated the total **weight** of the hydrogen atoms and neutrons that would be produced by the atom breaking apart minus the weight of the original atom. His relating the mass of the constituent parts of an atom to the mass of the whole atom identified a slight difference. Aston termed this difference the packing fraction. He concluded that in bringing the constituent particles together to form the atom some matter was transformed into energy. An equivalent amount of energy would be necessary to split the atom apart. **Albert Einstein** calculated this energy as equaling the mass defect (the difference between the sum of the mass of the atomic particles and the mass of the atom as a whole) times a constant equaling the square of the speed of **light**.

At the particle level binding energy is measured as the difference of the mass of the whole **nucleus** and the sum of the constituent parts. When the binding energy of the nucleus is determined, oftentimes this difference is divided by the number of nucleons for a convenient average per **nucleon**. A nucleus with low binding energy and low mass may reconfigure with other structures of low stability to form a more massive and stable structure through the process of **nuclear fusion**. A nucleus with high mass and low stability may intercept a **neutron** and begin a process known as fission. During fission, a nucleus splits into two nuclei of less mass with greater stability. The more binding energy the structure has, the less likely it is to reconfigure or breakdown into another structure. Binding energy is increasingly low for atoms with an atomic mass greater than 100. Fission is more likely to occur in these atoms since all matter tends to progress toward a more stable state of high binding energy. Among the elements, the standard of stability is set by iron and nickel with atomic masses of 50 and 60, respectively. Elements with atomic mass less than that of iron or nickel have extremely low binding energy. Among these elements, fusion is more likely to occur.

BINNIG, GERD (1947-   )

German physicist

Along with his research colleague **Heinrich Rohrer**, Gerd Binnig invented the first **microscope** that opened the individual **atom** to view. The Royal Swedish Academy of Sciences found this **scanning tunneling microscope** (STM) so important that it awarded the device's inventors half of the 1986 Nobel Prize in physics just five years after the first successful test of the STM. The academy declared that even though development of the STM was in its infancy, it was already clear that "entirely new fields are opening up for the study of the structure of matter." Binnig was only thirty-nine years old when he received the honor.

Binnig was born in Frankfurt am Main, then West Germany, on July 20, 1947, the son of Ruth Bracke Binnig, a drafter, and Karl Franz Binnig, a machine engineer. Binnig pursued interests in both physics and music—classical and rock—and earned both a diploma and a Ph.D. in physics from

Johann Wolfgang Goethe University in Frankfurt. Immediately after receiving his doctorate for work on superconductivity in 1978, Binnig joined the staff of the research laboratory operated by International Business Machines (IBM) in Zurich, Switzerland, and began his collaboration with Rohrer on the development of the STM.

Rohrer had been at the IBM lab since 1963 and also had a background in superconductivity. Together Binnig and Rohrer became interested in exploring the characteristics of the surface of materials. This was a challenging proposition: the **atomic structure** of the surface of a solid differs from the atomic structure of the solid's interior in that atoms on the surface can interact only with atoms at, on, and immediately below the surface, making surface structures frustratingly complex. In the words of physicist **Wolfgang Pauli**, quoted by Binnig and Rohrer in the introduction to their description of the STM in *Scientific American* (August 1985), "The surface was invented by the devil."

To accomplish their goal, Binnig and Rohrer turned to a phenomenon of quantum **mechanics** known as **tunneling**. **Quantum mechanics** had earlier revealed that the wavelike nature of electrons permits them to escape the surface boundary of a solid—they "smear out" beyond the surface and form an **electron cloud** around the solid. Electrons can "tunnel" through touching and overlapping clouds between two surfaces. Ivar Giaever of General Electric verified this experimentally in 1960. Binnig had investigated tunneling in superconductors during his graduate studies. Now he and Rohrer decided to make electrons tunnel through a **vacuum** from a sample solid surface to a sharp, needlelike probe. This proved surprisingly easy to accomplish. As the needle tip approaches within a nanometer (one billionth of a meter) of the sample, their **electron** clouds touch and a tunneling current starts to flow. The probe's tip follows this current at a constant height above the surface atoms, producing a three-dimensional map of the solid's surface, atom by atom.

In order to insulate their microscope against the serious problem of distorting vibration and noise, Binnig and Rohrer made a series of technical advances that included the creation of a probe tip consisting of a single atom. The colleagues and their research team soon demonstrated practical uses of the STM, revealing the surface structure of crystals, observing chemical interactions, and scanning the surface of DNA (deoxyribonucleic acid) chains. Using the STM, Binnig became the first person to observe a virus escape from a living cell. The tremendous importance of the STM lies in its many applications—for basic research in chemistry, physics, and biology and for applied research in semiconductor physics, microelectronics, metallurgy, and bioengineering. On winning the Nobel Prize for his development of the STM, Binnig was quoted in the *New York Times* as having mixed emotions: "It was beautiful and terrible at the same time"—beautiful because it signaled a great success but terrible because it concluded "an exciting story of discovery."

However, Binnig had continued the "exciting story" while he was on leave at Stanford University in California in 1985. Freed from his constant work on the STM, Binnig had time to contemplate using the atomic **force** between atoms, rather than tunneling current, to move the scanning tip over a solid's surface. Binnig shared his ideas with Christoph Gerber of IBM Zurich and Calvin Quate of Stanford, and soon they had produced a prototype of a new type of scanner, the **atomic force microscope** (AFM), which started a new field of microscopy. The AFM made it possible for the first time to image materials that are not electrically conductive.

Binnig became group leader at IBM's Zurich lab in 1984, and also an IBM Fellow. He was visiting professor at Stanford from 1986 to 1988. In 1969 he married Lore Wagler, a fellow student, who became a psychologist. They had a daughter in 1984 and a son, born in the United States, in 1986. Binnig is interested in a number of outdoor pursuits including sailing, golf, tennis, soccer, and skiing, and he has maintained his involvement in music as a composer, player of various instruments, and singer. He was the author of a popular German book on human creativity and chaos titled *Aus dem Nichts* (*Out of Nothing*), which argued that creativity arises from disordered thoughts. He and Rohrer shared a number of prestigious international awards for their pioneering research in microscopy.

BINOCULARS

Binoculars are an optical devices constructed to allow the magnification of objects without sacrifice of depth perception. As their name implies, binoculars are optical devices intended for use with two eyes in order to preserve human stereoscopic vision. In reality, binoculars are simply a pair of telescopes strapped side by side. The physics of the **telescope** must be understood before one can appreciate the clever engineering in binoculars.

Good binoculars require the following properties: an appropriate level of magnification; a bright image; a good depth of view; and reasonable portability. If we consider these four properties in terms of the telescope we will start to appreciate the limitations forced on us by the physics of the instrument.

First consider magnification. Binoculars use two **lenses** (objective and eyepiece). The magnification of an object by the use of two lenses totally depends on the focal lengths of the objective and eyepiece lenses. In fact, it is the ratio of the lenses' focal lengths which gives us the size of the magnification. The magnification is simply the focal length of objective lens divided by the focal length of eyepiece.

$$\text{Magnification} = \frac{\text{Focal length of objective lens}}{\text{Focal length of eyepiece}}$$

When a lens is used to focus the sun's rays onto a piece of paper the distance of the paper from the lens is called the focal length. Typically thin lenses have a relatively long focal length and thicker lenses a shorter focal length. Binoculars usually allow between five to eight times magnification. Anything larger than that will present problems with steadi-

Man with binoculars. *(Photo by Robert J. Huffman. Reproduced by permission.)*

ness as the binoculars will magnify natural hand tremor. This means that if the eyepiece has a focal length of 1.5 in (3.8 cm) then for eight times magnification the objective focal length must be 12 in (30.5 cm) long. This means that the binocular would need to be about 13.5 in (34.3 cm) long which decreases their portability.

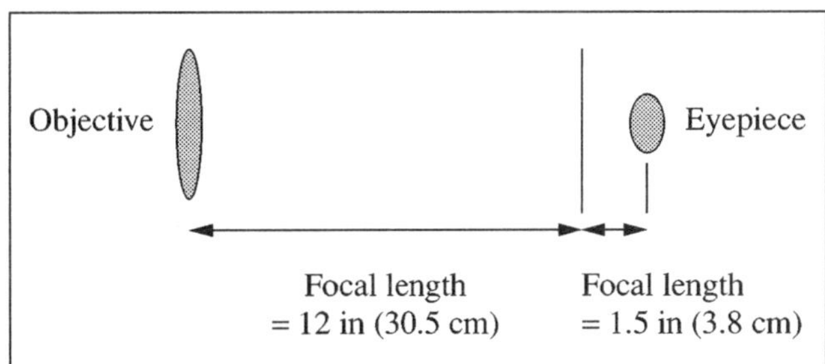

There is an additional problem with using the two lens design. The image, as seen by the eye, will be inverted (upside down). Also the left appears to be on the right, and vice versa. Obviously these are an extremely undesirable features.

The second property, a bright image, simply requires that the surface of the **light** collecting lens (the objective lens) must be large enough. The larger the objective lens, the greater the **photon** capture, and the brighter the image. Having two objective lenses, one for each eye, doubles the light collecting capability of the instrument.

The third property, an appropriate depth of field, is dependent on the diameter (or aperture) of the eyepiece lens. The bigger the diameter, the larger the depth of field. Depth of field is an important issue because a shallow depth of field means that everything, but the object focused on, is fuzzy and out of focus. This means that the user has constantly to refocus the lenses when moving from one close object to another. Accordingly, a large depth of field is a very desirable feature in binocular design.

The fourth property, portability, means that all the issues mentioned above—length, lens diameters and inverted images—must be dealt with in a small lightweight design.

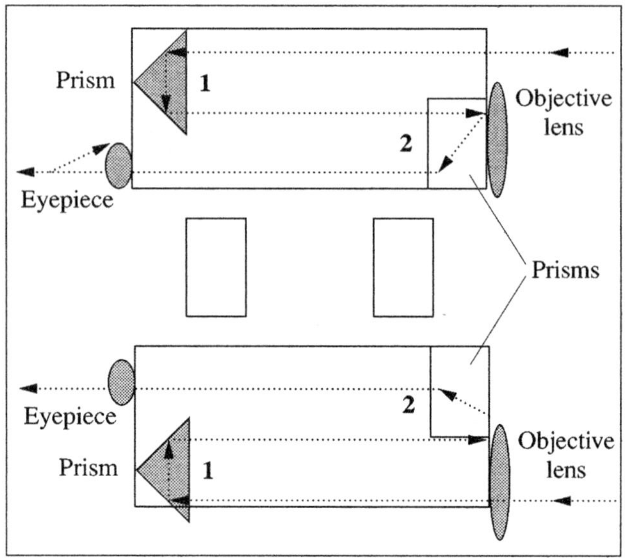

As shown, the key to binocular design is the use of four prisms. The prisms immediately solve the magnification problem because of the availability of long focal length lenses in which then the distance between then is effectively quartered. A four-inch long binocular is equivalent to a sixteen-inch long telescope. Also prism 1 in both arms of the binoculars corrects the left to right reversal problem. Prism 2 in both arms of the binoculars is set at right angles to prism I and this corrects the inverted image problem.

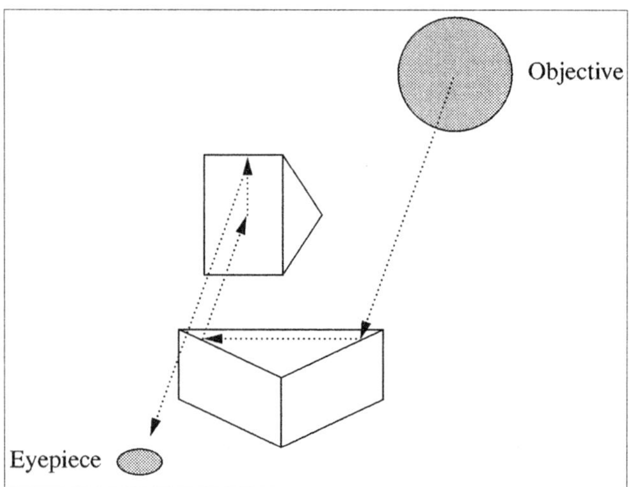

The remaining properties of field of view and brightness of image are comparatively simple issues. A large aperture eyepiece will increase the filed of view and a large diameter, well coated objective lens will enhance brightness issue. Binoculars are specified by both their magnification and objective lens diameter. The specification 6 x 30 means the binoculars have a magnification power of six times and the diameter of the objective lens is 1.2 in (30 mm). For low light conditions and maximum magnification a pair of binoculars with the specification 8 x 50 is usually suitable.

BIOPHYSICS

Biophysics applies the laws and methods of physics to the study of biological systems. It has only recently emerged as a subdivison of physics and as such, it represents one of the most specialized area of study in physics.

Its range of applicability is broad and seeks to gain understanding of every aspect of the biological world. Major areas of current interest are very diversified and include: bioenergetics, biophysical theory and modeling, cell biophysics, membrane, muscle and contractility, nucleic acid research, photobiophysics, imaging and supramolecular assemblies.

Today, it is becoming increasingly difficult to distinguish among biochemistry, genetics, biophysics, **chemistry**, and all the other sciences investigating biological phenomena since these almost always involve an overlap of approaches and conceptualizations. What is then the contribution of physics to biophysics? First, biophysics is responsible for the development of new methods of investigation to address the mechanisms of biological processes at every level. For example, the contribution of x-ray **crystallography** to understanding the structure and function of macromolecular assemblies as large as viruses or ribosomes is now acknowledged. Spectroscopy is also used to investigate structure-function relationships. **Nuclear magnetic resonance** (NMR) is a complementary approach that provides not only structural knowledge about biomolecules, but also provides insights in intramolecular **dynamics**. **Electron** microscopy and image processing also allows visualization of biological **molecular structure**. And this arsenal of physical methods is constantly supplemented by newer techniques

Second, physics, more than any other science, provides a conceptual framework of choice for dealing with biological phenomena. This is due to the universality of its laws on **time** and **space** time-scales compatible with the processes under investigation. Another contribution of physics lies in its ability to reduce complex systems to models that can be treated theoretically and compared with experiment. An example is provided by molecular biophysics that reduces biological phenomena to a set of algorithms forming a consistent whole into which a biomolecular assembly can fit. In this sense, it is similar to **thermodynamics** which also provides a logical framework for the occurrence of all physical phenomena described by a few laws.

Biophysics can then be considered as the interface between two apparently different views of nature, those of physics and those of biology. One of the interacting surfaces is turned towards biological investigations and the other towards the concepts, models and methods of physics. The biophysical interface then serves the purpose of providing quantitative interpretations for biological phenomena using the laws of physics, as shown by the following examples.

In the physics of the inanimate world, the **first law of thermodynamics** is concerned with the bulk behavior of **matter**. In biophysics, thermodynamics is used to predict the position of **equilibrium** in a system and the **energy** changes involves in biological processes such as enzyme catalyzed reactions, the mechanical **work** performed by muscles, the electrical work required to charge nerve membranes, the chemical work involved in the synthesis of biomolecules or in the emission of **light** by phosphorescent marine organisms. Another example is provided by the field of protein dynamics. As in the physical world, **motion** is of crucial importance at the molecular level of biology. Detailed studies of the atomic motions of proteins and nucleic acids are of relatively recent origin and it is now recognized that an appreciation of molecular flexibility and dynamics is essential to the understanding of the activity of biomolecules and to the design of new molecules with specific uses, such as drugs or vaccines.

A variety of biophysical methods are available to study the dynamics of biomolecules, one being *molecular dynamics simulation*, in which the classical equations of motion for the atoms in the sytem are solved by numerical techniques for time intervals ranging from the picosecond to the nanosecond range. The second method is *normal mode analysis*, in which the motion is described as a superposition of harmonic vibrations whose characteristics depend on the shape of the **potential energy** surface near an energy minimum. These theoretical approaches have the advantage that the models can be experimentally verified using vibrational *spectroscopy* techniques.

Another active area of biophysical research concerns the study of the interactions involving charged, polar and polarizable groups of atoms in proteins. This is important, because they are believed to be responsible for biomolecular recognition and assembly processes, as well as fundamental biochemical processes such as electron and energy transfer. Recent advances in understanding electric field effects lead to viewing proteins as dynamic assemblies that can control and reorganize their charge distribution, provided by their amino acid constituents. Since proteins are also dynamic systems undergoing constant fluctuations, the electric fields generated by a protein are also changing and it has been proposed that this is how proteins adopt the conformations required for their biological function.

It can be said that biophysics is now firmly at the forefront of the studies presently investigating the most important problems in biology, such as the protein folding problem (how do proteins fold their long amino acid sequence into their characteristic and functionally important structure?), the molecular recognition problem (how can an inhibitor bind to the HIV-1 virus?), the long-range electron-transfer problem (how are electrons transferred by protein active centers over long distances?), to name but a few.

BLACK HOLES

The modern theoretical prediction of black holes came as a spectacular consequence of German-American physicist **Albert Einstein's** general theory of relativity. Einstein's theory predicted that massive **stars** ultimately collapse into black holes with gravitational fields so intense that not even **light** can escape. In 1969, American physicist **John A. Wheeler** popularized the term *black hole*. During the later half of the twentieth century the discovery of, and physics associated with, black holes became one of the preeminent quests of modern **astronomy**.

In the 1930s Indian-born American astrophysicist **Subrahmanyan Chandrasekhar** mathematically proved that black holes were the remains of massive stars and fully articulated the evolution of stars into supernova, **white dwarfs**, **neutron stars** or black holes. Before the intervention of WWII, American physicist **J. Robert Oppenheimer**, who ultimately supervised project Trinity (the making of the first atomic bombs), made detailed calculations reconciling Chandrasekhar's predictions with general **relativity theory**. In the late 1960s English mathematician and physicist **Stephen Hawking** along with English mathematician **Roger Penrose** drew from both quantum and relativity theory to show that within a black hole there must exist a **singularity** (a geometric point without **space**) of infinite **density**. Penrose advanced a scientific and philosophic concept known as the Law of Cosmic Censorship, an assertion that regions around the singularity (i.e., black holes) are regions of space cut off from direct human observation. In 1963 New Zealand astrophysicist, Roy Patrick Kerr (1934-) used Einstein's field equations to predict the existence of rotating black holes.

Essentially, throughout the life of a star a tug-of-war exists between the compressing **force** of the star's own **gravity** and the expanding pressures generated by **nuclear reactions** at its core. After cycles of swelling and contraction associated with the burning of progressively heavier nuclear fuels, the star eventually runs out of useable nuclear fuel. The spent star then contracts under the pull of it own gravity. Modern understandings of **astrophysics** allow three possible fates for such a collapsing star.

The particular fate for any star determined by the **mass** of the star left after blowing away its outer layers during its paroxysmal death spasms. A star less than 1.44 times the mass of the **Sun** (termed the **Chandrasekhar limit**) collapses until the **pressure** in the increasing compacted **electron** clouds exerts enough pressure to balance the collapsing **gravitational force**. Such stars become white dwarfs contracted to the a radius of only a few thousand miles (roughly the size of a planet). Because most stars in the visible **universe** are low-mass stars, this is the fate of most stars. If the star retains between 1.4 and roughly three times the mass of the Sun, the pressure of the electron clouds is insufficient to stop the gravitational collapse. In such stars contraction continues to produce a **neutron** star. Only a few miles in radius, within a neutron star the nuclear forces and the repulsion of the compressed atomic nuclei balance the crushing force of gravity. With more massive stars, however, there is no known force in the universe that can withstand the gravitational collapse. Such extraordinary stars will continue their collapse to form a singularity—a star collapsed to a point of infinite density. As such a star collapses its **gravitational field** warps **space-time** so intensely that not even light can escape and a black hole forms around the singularity.

The **event horizon** is the boundary of a black hole. The size of the event horizon is termed the Schwarzschild radius (named after the German astronomer Karl Schwarzschild). Inside the event horizon the gravitational attraction of the sin-

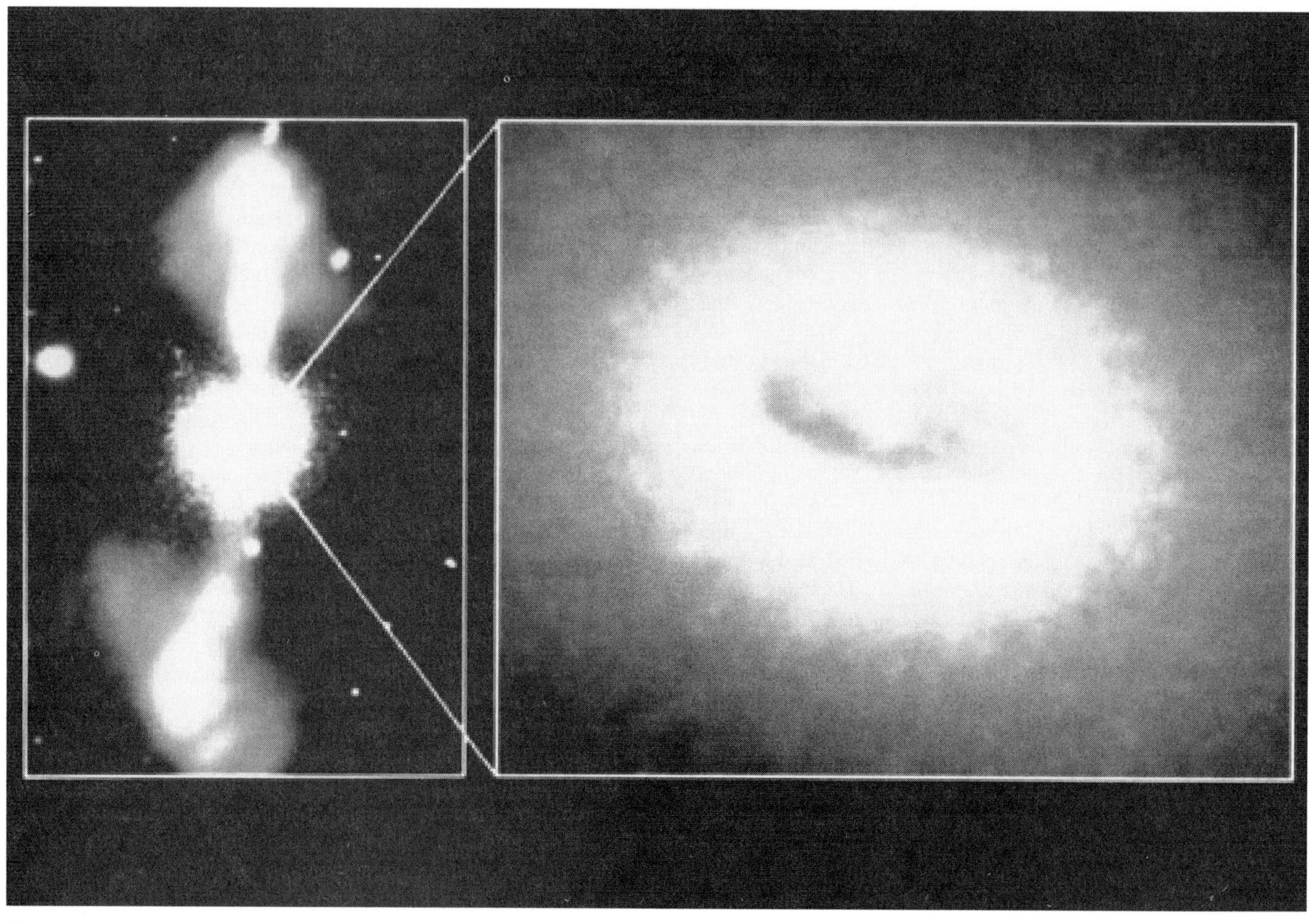

Black Hole, core of Galaxy NGC 4261, view from Hubble Space Telescope. *(Image courtesy of NASA).*

gularity is so strong that the required escape **velocity** is greater than the **speed of light**. As a consequence, because no object can exceed the speed of light, not even light itself can escape from the region of space within the event horizon.

Because no information generated within the black hole can escape to us, the event horizon is an important observational boundary. Essentially all we can know with any certainty regarding the processes of the singularity are the external gravitational effects exerted by its tremendous mass. Astronomers consider systems good candidates to contain a black hole wherever a star orbits around an unseen companion and there is strong electromagnetic **radiation** from an unidentified source near the center of rotation. In binary star systems (two stars orbiting each other), for example, although the black hole can not be directly observed its gravitational strength by the orbit its observable companion star. In the last two decades of the twentieth century, astronomers have found a half-dozen or so binary star systems where black holes may exist.

In 1994 the **Hubble Space Telescope** provided arguably conclusive evidence for the existence of a supermassive black hole located at the center of the M87 galaxy. Similar evidence indicates that a black hole also lies at the center of our **Milky Way galaxy**.

An accretion disk forms as **matter** accelerates toward the event horizon of the black hole. As the matter in the accretion disk spirals toward the black hole it is heated to very high temperatures and emits strong highly energetic electromagnetic radiation (e.g., **x rays** and gamma rays). Some astronomers assert that such a mechanism, working on a galactic scale may account for the phenomena associated with **quasars**.

Although the events that occur inside a black hole remain an enigma, some physicist have attempted to speculate about the nature of time dilations and contractions near the event horizon. **Hawking radiation**, involving massless **virtual particles** and particle-antiparticle pairs, for example, may explain mass and radiation leakage from blackholes. Astronomers also suggest that in the early Universe, black holes could have formed as a consequence of the collection and collapse of large volumes of **interstellar gas**.

The existence of black-holes opens the possibility for the existence of another stunning concept known as worm-holes (also termed Einstein-Rosen bridges). A wormhole is a theoretical opening in **space-time** that connects two black holes.

See also Quasars; Stellar evolution; Stellar life cycle; Worm holes

BLACKBODY RADIATION

The term blackbody **radiation** refers to electromagnetic radiation emitted by a completely opaque object. Such an object is referred to as a blackbody since it absorbs all of the radiation that falls on it and thus appears to be colorless, or black. According to Kirchoff's law, any object that qualifies as a blackbody must also be a perfect emitter of radiation.

In fact, no real object fits the definition of a blackbody since even the most opaque of materials reflects some small fraction of the **light** incident upon it. Soot, **carbon** black, platinum black, and carborundum are among the materials that come closest to a blackbody in the real world. The concept of an idealized blackbody is still of major importance in physics. It serves as a standard for **heat** and **temperature** measurements just as the **wavelength** of light emitted by krypton-86 atoms is a standard for measurements of length.

For research purposes, physicists replicate the principle of a blackbody with a device known as a cavity radiator. A cavity radiator is a hollow sphere with a small hole through which radiation can enter and leave. Radiation that enters the hole is reflected continuously within the sphere until it is completely absorbed (as would be the case with a blackbody). It follows, then, that any radiation emitted by the cavity radiator corresponds to the definition of blackbody radiation.

The study of blackbody radiation was of considerable interest to physicists in the late 1800s. Experiments showed that for any given temperature, the intensity (brightness) of blackbody radiation is a maximum for a relatively narrow range of wavelengths, dropping off sharply at shorter and longer wavelengths. A number of attempts were made to use classical electromagnetic theory to derive a mathematical formula that would describe the intensity/wavelength relationship, but all failed for one or another part of the curve. Finally, in 1900, the German physicist **Max Planck** solved the problem. By assuming that radiation travels not in continuous **waves**, but in discrete "packages" (called **quanta**), Planck was able to derive a formula for the blackbody radiation curve. That formula is:

$$I = (\hbar/\lambda)[\exp(\hbar c/\lambda kT - 1]^{-1}$$

where $\hbar$ is **Planck's constant**, k is Boltzmann's constant, λ is the wavelength of the radiation, and T is the absolute temperature.

BOHR, NIELS (1885-1962)

Danish physicist

Niles Bohr received the Nobel Prize in physics in 1922 for the quantum mechanical model of the **atom** that he had developed a decade earlier, the most significant step forward in scientific understanding of **atomic structure** since English physicist **John Dalton** first proposed the modern **atomic theory** in 1803.

Bohr founded the Institute for Theoretical Physics at the University of Copenhagen in 1920, an Institute later renamed for him. For well over half a century, the Institute was a powerful **force** in the shaping of atomic theory. It was an essential stopover for all young physicists who made the tour of Europe's center of theoretical physics in the mid-twentieth century. Also during the 1920s, Bohr thought and wrote about some of the fundamental issues raised by modern **quantum theory**. He developed two basic concepts—the principles of complementarity and correspondence—that he said must direct all future work in physics. In the 1930s, Bohr became interested in problems of the atomic **nucleus** and contributed to the development of the liquid-drop model of the nucleus, a model used in the explanation of **nuclear fission**.

Niels Henrik David Bohr was born on October 7, 1885, in Copenhagen, Denmark. He was the second of three children born to Christian and Ellen Adler Bohr. Bohr's early upbringing was enriched by a nurturing and supportive home atmosphere. His mother had come from a wealthy Jewish family involved in banking, government, and public service. Her father, D. B. Adler, had founded the Commercial Bank of Copenhagen and the Jutland Provincial Credit Association. Bohr's father was a professor of physiology at the University of Copenhagen. His closest friends met every Friday night to discuss events, so that, as a young boy, Bohr "learned much from listening to these conversations," according to J. Rud Nielsen in *Physics Today.*

Bohr became interested in science at an early age. His biographer, Ruth Moore, has written in her book *Niels Bohr: The Man, His Science, and the World They Changed* that as a child he "was already fixing the family clocks and anything else that needed repair." Bohr received his primary and secondary education at the Gammelholm School in Copenhagen. He did well in his studies, although he was apparently overshadowed by the work of his younger brother Harald, who later became a mathematician. Both brothers were also excellent soccer players.

On his graduation from high school in 1903, Bohr entered the University of Copenhagen, where he majored in physics. He soon distinguished himself with a brilliant research project on the **surface tension** of water as evidenced in a vibrating jet stream. For this work he was awarded a gold medal by the Royal Danish Academy of Science in 1907. In the same year, he was awarded his B.S. degree, to be followed two years later by his M.S. Bohr then stayed on at Copenhagen to work on his doctorate, which he gained in 1911. He thesis dealt with the **electron** theory of metals and confirmed the fact that classical physical principles were sufficiently accurate to describe the qualitative properties of metals but failed when applied to quantitative properties. Probably the main result of this research was to convince Bohr that classical **electromagnetism** could not satisfactorily describe atomic phenomena. The stage had been set for Bohr's attack on the most fundamental questions of atomic theory.

Bohr decided that the logical place to continue his research was at the Cavendish Laboratory at Cambridge University. The director of the laboratory at the time was

Niels Bohr

English physicist **J. J. Thomson**, discoverer of the electron. Only a few months after arriving in England in 1911, however, Bohr discovered that Thomson had moved on to other topics and was not especially interested in Bohr's thesis or ideas. Fortunately, however, Bohr met English physicist **Ernest Rutherford**, then at the University of Manchester, and received a much more enthusiastic response. As a result, he moved to Manchester in 1912 and spent the remaining three months of his time in England working on Rutherford's nuclear model of the atom.

On July 24, 1912, Bohr boarded ship for his return to Copenhagen and a job as assistant professor of physics at the University of Copenhagen. Also waiting for him was his bride-to-be Margrethe Nørlund, whom he married on August 1. The couple later had four sons: Hans, Erik, Aage, and Ernest. Two other sons, one named Christian, died young. Aage Bohr earned a share of the 1975 Nobel Prize for his work on the structure of the atomic nucleus.

The field of **atomic physics** was going through a difficult phase in 1912. Rutherford had only recently discovered the atomic nucleus, which had created a profound problem for theorists. The existence of the nucleus meant that electrons must be circling it in orbits somewhat similar to those traveled by **planets** in their **motion** around the **Sun**. According to classical laws of **electrodynamics**, however, an electrically charged particle would continuously radiate **energy** as it trav-

eled in such an orbit around the nucleus. Over time, the electron would spiral ever closer to the nucleus and eventually collide with it. Although electrons clearly *must* be orbiting the nucleus, they could *not* be doing so according to classical laws.

Bohr arrived at a solution to this dilemma in a somewhat roundabout fashion. He began by considering the question of atomic spectra. For more than a century, scientists had known that the heating of an element produces a characteristic line spectrum; that is, the specific pattern of lines produced is unique for each specific element. Although a great deal of research had been done on **spectral lines**, no one had thought very deeply about what their relationship might be with atoms, the building blocks of elements.

When Bohr began to attack this question, he decided to pursue a line of research begun by the German physicist Johann Balmer in the 1880s. Balmer had found that the lines in the hydrogen spectrum could be represented by a relatively simple mathematical formula relating the **frequency** of a particular line to two integers whose significance Balmer could not explain. It was clear that the formula gave very precise values for line frequencies that corresponded well with those observed in experiments.

When Bohr's attention was first attracted to this formula, he realized at once that he had the solution to the problem of electron orbits. The solution that Bohr worked out was both simple and elegant. In a brash display of hypothesizing, Bohr declared that certain orbits existed within an atom in which an electron could travel *without* radiating energy; that is, classical laws of physics were suspended within these orbits. The two integers in the Balmer formula, Bohr said, referred to orbit numbers of the "permitted" orbits, and the frequency of spectral lines corresponded to the energy released when an electron moved from one orbit to another.

Bohr's hypothesis was brash because he had essentially no theoretical basis for predicting the existence of "allowed" orbits. To be sure, German physicist **Max Planck**'s quantum hypothesis of a decade earlier had provided some hint that Bohr's "quantification of space" might make sense, but the fundamental argument for accepting the hypothesis was simply that it worked. When his model was used to calculate a variety of atomic characteristics, it did so correctly. Although the hypothesis failed when applied to detailed features of atomic spectra, it worked well enough to earn the praise of many colleagues.

Bohr published his theory of the "planetary atom" in 1913. That paper included a section that provided an interesting and decisive addendum to his basic hypothesis. One of the apparent failures of the Bohr hypothesis was its seeming inability to predict a set of spectral lines known as the Pickering series, lines for which the two integers in the Balmer formula required half-integral values. According to Bohr, of course, no "half-orbits" could exist that would explain these values. Bohr's solution to this problem was to suggest that the Pickering series did not apply to hydrogen at all, but to **helium** atoms that had lost an electron. He rewrote the Balmer formula to reflect this condition.

Within a short period of time spectroscopists in England had studied samples of helium carefully purged of hydrogen and found Bohr's hypothesis to be correct. Although a number of physicists were still debating Bohr's theory, at least one—Rutherford—was convinced that the young Danish physicist was a highly promising researcher. He offered Bohr a post as lecturer in physics at Manchester, a job that Bohr eagerly accepted and held from 1914 to 1916. He then returned to the University of Copenhagen, where a chair of theoretical physics had been created specifically for him. Within a few years he was to become involved in the planning for and construction of the University of Copenhagen's new Institute for Theoretical Physics, of which he was to serve as director for the next four decades.

In many ways, Bohr's atomic theory marked a sharp break between classical physics and a revolutionary new approach to natural phenomena made necessary by quantum theory and relativity. He was very much concerned about how scientists could and should now view the physical world, particularly in view of the conflicts that arose between classical and modern laws and principles. During the 1920s and 1930s, Bohr wrote extensively about this issue, proposing along the way two concepts that he considered to be fundamental to the "new physics." The first was the principle of complementarity that says, in effect, that there may be more than one true and accurate way to view natural phenomena. The best example of this situation is the wave-particle duality discovered in the 1930s, when particles were found to have wavelike characteristics and **waves** to have particle-like properties. Bohr argued that the two parts of a duality may appear to be inconsistent or even in conflict and that one can use only one viewpoint at a time, but he pointed out that both are necessary to obtain a complete view of particles and waves.

The second principle, the **correspondence principle**, was intended to show how the laws of classical physics could be preserved in light of the new **quantum physics**. We may know that **quantum mechanics** and relativity are essential to an understanding of phenomena on the atomic scale, Bohr said, but any conclusion drawn from these principles must not conflict with observations of the real world that can be made on a macroscopic scale. That is, the conclusions drawn from theoretical studies must correspond to the world described by the laws of classical physics.

In the decade following the publication of his atomic theory, Bohr continued to work on the application of that theory to atoms with more than one electron. The original theory had dealt only with the simplest of all atoms, hydrogen, but it was clearly of some interest to see how that theory could be extended to higher elements. In March, 1922, Bohr published a summary of his conclusions in a paper entitled "The Structure of the Atoms and the Physical and Chemical Properties of the Elements." Eight months later, Bohr learned that he had been awarded the Nobel Prize in physics for his theory of atomic structure, by that time universally accepted among physicists.

During the 1930s, Bohr turned to a new but related, topic: the composition of the atomic nucleus. By 1934, scientists had found that the nucleus consists of two kinds of particles, protons and neutrons, but they had relatively little idea how those particles are arranged within the nucleus and what its general shape was. Bohr theorized that the nucleus could be compared to a liquid drop. The forces that operate between protons and neutrons could be compared in some ways, he said, to the forces that operate between the molecules that make up a drop of liquid. In this respect, the nucleus is no more static than a droplet of water. Instead, Bohr suggested, the nucleus should be considered to be constantly oscillating and changing shape in response to its internal forces. The greatest success of the Bohr liquid-drop model was its later ability to explain the process of nuclear fission discovered by German chemist **Otto Hahn**, German chemist Fritz Strassmann, and Austrian physicist **Lise Meitner** in 1938. It is somewhat ironic that the Nobel Prize won by Bohr's son Aage in 1975 was given for the latter's elucidation of his father's nuclear model (to which many other scientists had also contributed).

Bohr continued to work at his Institute during the early years of World War II, devoting considerable effort to helping his colleagues escape from the dangers of Nazi Germany. When he received word in September, 1943, that his own life was in danger, Bohr decided that he and his family would have to leave Denmark. The Bohrs were smuggled out of the country to Sweden aboard a fishing boat and then, a month later, flown to England in the empty bomb bay of a bomber. The Bohrs then made their way to the United States, where both Bohr and his son became engaged in work on the **Manhattan Project** to build the world's first atomic bombs.

After the War, Bohr, like many other Manhattan Project researchers, became active in efforts to keep control of atomic weapons out of the hands of the military and under close civilian supervision. For his long-term efforts on behalf of the peaceful uses of atomic energy, Bohr received the first Atoms for Peace Award given by the Ford Foundation in 1957. Meanwhile, Bohr had returned to his Institute for Theoretical Physics and become involved in the creation of the **European Center for Nuclear Research (CERN)**. He also took part in the founding of the Nordic Institute for Theoretical Atomic Physics (Nordita) in Copenhagen. Nordita was formed to further cooperation among and provide support for physicists from Norway, Sweden, Finland, Denmark, and Iceland.

Bohr reached the mandatory retirement age of seventy in 1955 and was required to leave his position as professor of physics at the University of Copenhagen. He continued to serve as director of the Institute for Theoretical Physics until his death in Copenhagen on November 18, 1962.

Bohr was held in enormous respect and esteem by his colleagues in the scientific community. American physicist **Albert Einstein**, for example, credited him with having a "rare blend of boldness and caution; seldom has anyone possessed such an intuitive grasp of hidden things combined with such a strong critical sense." Among the many awards Bohr received were the Max Planck Medal of the German Physical Society in 1930, the Hughes (1921) and Copley (1938) medals of the Royal Society, the Franklin Medal of the Franklin Institute in

1926, and the Faraday Medal of the Chemical Society of London in 1930. He was elected to more than twenty scientific academies around the world and was awarded honorary doctorates by a dozen universities, including Cambridge, Oxford, Manchester, Edinburgh, the Sorbonne, Harvard, and Princeton.

BOHR ATOM

In 1913, Danish physicist **Niels Bohr** developed the first model of the **atom** that used the concepts of **quantum theory** to explain atomic spectra. The Bohr atom depicted electrons "in orbit" around a heavy **nucleus**, similar to the **motion** of **planets** around the **sun**. The model was based on Ernest Rutherford's 1911 demonstration that the atom consists of a central nucleus surrounded by a cloud of electrons. The electrons, Rutherford postulated, would have to be in constant motion, since the attraction of opposite charges would otherwise pull them into the nucleus. Because electrons are charged particles, they should radiate electromagnetic **energy** during orbits, eventually losing **momentum** and spiraling into the nucleus. The radiated energy would be expected to change continuously across the **electromagnetic spectrum** as the momentum of the moving particle changed.

This was not the observed behavior of atoms. Under normal conditions, they do not radiate energy at all. When an atom is exposed to an environment in which energy can be transferred to it (a very strong electric field or very high **temperature**), it does emit **light**. This light is observed only at particular wavelengths instead of a continuous spectrum. These wavelengths are referred to as the **spectral lines** of the substance. Each element, when it is excited, gives off certain colors or wavelengths, a color "fingerprint" that identifies the particular atoms emitting the light. These fingerprints are sometimes referred to as spectral lines. The light coming from these atoms does not take the shape of lines but, when separated using an array of **lenses** and slits, the light produces an image in the shape of a series of lines, each corresponding to a particular **wavelength**. Every type of atom has a characteristic set of spectral lines at discrete wavelengths.

To explain this phenomenon, Bohr applied the idea of quantized energy previously developed by Planck. His first assumption was that, in contrast to classical **mechanics**, where an infinite number of orbits are possible, an **electron** could be in only one of a discrete set of orbits, which he termed stationary states. His proposed atom restricted electrons to orbits of certain "allowed" radii. An electron orbiting in one of these "allowed" orbits has a defined energy state, does not radiate energy, and does not spiral into the nucleus. Even though the electron is in constant motion, it does not emit electromagnetic **radiation** from a stationary state. Bohr concluded that energy changes in an atom occur only when electrons move from one allowed orbit to another. Energy must be absorbed for an electron to move to a higher state (further from the nucleus) and emitted when the electron moves to an orbit of lower energy (closer to the nucleus).

The overall change in energy associated with this "orbit jumping" is the difference in energy levels between the initial and final orbits. The difference in energy levels corresponds to a particular wavelength of light multiplied by **Planck's constant**. The spectrum for each atom is defined by these wavelengths. The line spectrum of an atom—even a simple **hydrogen atom** with only one electron—can be very complex. Although the electron can only move from one **energy level** to another with no stopping in between, the energy differences between levels are not all the same. This means that a move from the third level to the second will emit a quantum of light energy (now known as a **photon**) with a different wavelength than that of a move between the second level and the first. To further complicate the spectrum, the electron can move several levels at once, emitting yet another wavelength. Each stable orbit has room for a limited number of electrons. Once it is full, no more electrons can move to that particular energy. A final rule of electron movement states that the electron cannot move beyond the highest energy level into the nucleus.

The Bohr atom model successfully accounted for most known spectral lines for hydrogen and predicted new spectral wavelengths that were later confirmed. The theory also explained the observation that more spectral lines were observed for hydrogen in **interstellar space**. These lines are the result of transitions from very high energy levels. Electrons can only remain at these levels in the high **vacuum** of interstellar **space**, where collisions between atoms are much less frequent. Although the theory provided a good explanation for the spectrum of light from an atom, it did have some major deficiencies: it did not account for intensities of the various lines in the spectrum; it did not explain chemical bonding; and it did not work for atoms with more than one electron. It was only useful for predicting the behavior of atoms with a single electron (H, He^+, and Li^{2+} **ions**).

Even so, Bohr's theory was a significant advance in understanding **atomic structure**. It provided several very important features that have survived in the current **quantum mechanics** models: notably, the existence of stationary, nonradiating states and the relationship of radiation **frequency** to the energy difference between the initial and final states in a quantum transition. After its introduction by Bohr in 1913, the model was expanded and modified by Bohr and other scientists, most notably **Arnold Sommerfeld**. In 1922, Niels Bohr was awarded the Nobel Prize in physics. Bohr's model of the atom was important because it introduced quantized energy states for the electrons. Although the model itself was eventually replaced, the concept of quantized energy states remained as a key element of all later models.

BOHR THEORY

The Bohr theory or, more properly, the Bohr model of **atomic structure** was developed by Danish physicist and Nobel laureate **Niels Bohr**. Published in 1913, Bohr's model improved the classical atomic models of physicists Sir **Joseph John Thomson** and **Ernest Rutherford** by incorporating **quantum** *theory*. While working on his doctoral dissertation at Copenhagen University, Bohr studied German physicist **Max Planck**'s quantum theory of **radiation**. After graduation Bohr worked in England with Thomson and subsequently with Rutherford. During this **time** Bohr developed his model of atomic structure.

Before Bohr, the classical model of the **atom** was similar to the Copernican model of the solar system, where, just as the **planets** orbit the **Sun**, electrically-negative electrons moved in orbits around a relatively massive, positively-charged **nucleus**.

The classical model of the atom allowed electrons to orbit at any distance from the nucleus. This predicted that when, for example, a **hydrogen atom** was heated it should produce a continuous spectrum of colors as it cooled because its **electron**, moved away from the nucleus by the **heat energy**, would gradually give up that energy as it spiraled back closer to the nucleus. Spectroscopic experiments, however, showed that hydrogen atoms produced only narrow bands of **color** when heated. In addition, Scottish physicist **James Clerk Maxwell**'s influential studies on electromagetic radiation (**light**) predicted that an electron orbiting around the nucleus according to Newton's laws would continuously lose energy and eventually fall into the nucleus. To account for the observed properties of hydrogen, Bohr proposed that electrons existed only in certain orbits and that, instead of traveling between orbits, electrons made instantaneous quantum leaps or jumps between allowed orbits.

In the Bohr model the most stable, lowest **energy level** is found in the innermost orbit. This first orbital forms a shell around the nucleus and is assigned a principal quantum number (n) of n=1. Additional orbital shells are assigned values n=2, n=3, n=4, etc. The orbital shells are not spaced at equal distances from the nucleus and the radius of each shell increases rapidly as the square of n. Increasing numbers of electrons can fit into these orbital shells according to the formula $2n^2$. The first shell can hold up to two electrons, the second shell (n=2) up to eight electrons, the third shell (n=3) up to 18 electrons. Subshells or suborbitals (designated *s, p, d,* and *f*) with differing shapes and orientations allow each element a unique electron configuration.

As electrons move farther away from the nucleus they gain **potential energy** and become less stable. Atoms with electrons in their lowest energy orbits are in a "ground" state, those with electrons jumped to higher energy orbits are in an "excited" state. Atoms may acquire energy that excites electrons by random thermal collisions, collisions with **subatomic particles**, or by absorbing a **photon**. Of all the photons (quantum packets of light energy) that an atom can absorb, only those having an energy equal to the energy difference between allowed electron orbits will be absorbed. Atoms give up excess internal energy by giving off photons as electrons return to lower energy (inner) orbits.

The electron quantum leaps between orbits proposed by the Bohr model accounted for Plank's observations that atoms emit or absorb electromagnetic radiation only in certain units called **quanta**. Bohr's model also explained many important

properties of the **photoelectric effect** described by physicist **Albert Einstein**.

According to the Bohr model, when an electron is excited by energy it jumps from its ground state to an excited state (i.e., a higher energy orbital). The excited atom can then emit energy only in certain (quantized) amounts as its electrons jump back to lower energy orbits located closer to the nucleus. This excess energy is emitted in quanta of electromagnetic radiation (photons of light) that have exactly same energy as the difference in energy between the orbits jumped by the electron. For hydrogen, when an electron returns to the second orbital (n=2) it emits a photon with energy that corresponds to a particular color or spectral line found in the Balmer series of lines located in the visible portion of the electromagnetic (light) spectrum. The particular color in the series depends on the higher orbital from which the electron jumped. When the electron returns all the way to the innermost orbital (n=1) the photon emitted has more energy and forms a line in the Lyman series found in the higher energy, ultraviolet portion of the spectrum. When the electron returns to the third quantum shell (n=3), it retains more energy and therefore the photon emitted is correspondingly lower in energy and forms a line in the Paschen series found in the lower energy, infrared portion of the spectrum.

Because electrons are moving charged particles, they also generate a magnetic field. Just as an ampere is a unit of **electric current**, a magneton is a unit of magnetic dipole moment. The orbital **magnetic moment** for the hydrogen atom is called the Bohr magneton.

Bohr's work earned a Nobel Prize in 1922. Subsequently, more mathematically complex models based on the work of physicists **Louis Victor de Broglie** and **Erwin Schrödinger** that depicted the particle and wave nature of electrons proved more useful to describe atoms with more than one electron. The standard model incorporating quark particles further refines the Bohr model. Regardless, Bohr's model remains fundamental to the study of **chemistry**, especially the valence shell concept used to predict an element's reactive properties.

The Bohr model remains a landmark in scientific thought that poses profound questions for scientists and philosophers. The concept that electrons make quantum leaps from one orbit to another, as opposed to simply moving between orbits, seems counter-intuitive, that is, outside the human experience with nature. Bohr said, "Anyone who is not shocked by quantum theory has not understood it."

BOILING

Boiling is a physical process in which heating converts a liquid into a gas or vapor. The boiling point or boiling **temperature** is that point where the vapor **pressure** of a substance is equal to the atmospheric pressure. For a given atmospheric pressure the boiling point for pure substances is well defined and impurities in a substance broaden the boiling point range.

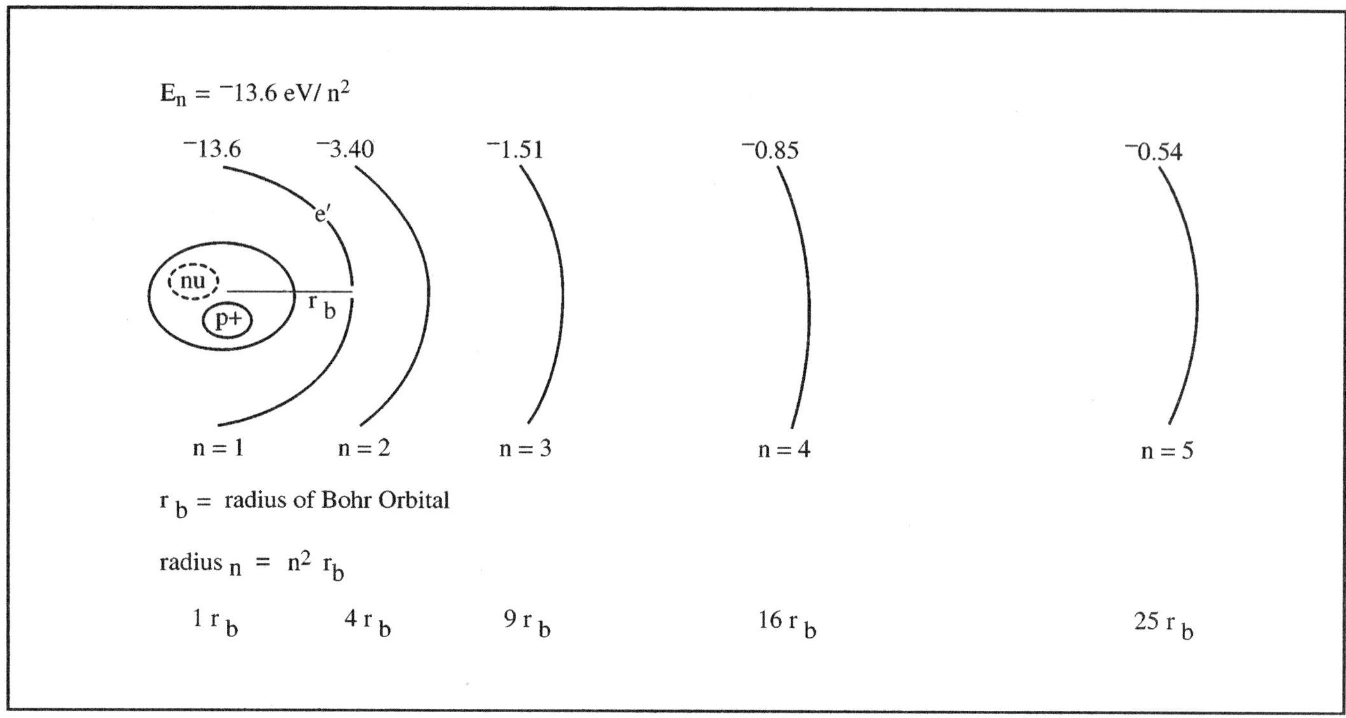

Bohr theory.

All **matter** generally exists in one of three physical phase states commonly described as solid, liquid, or gas. When matter undergoes a phase change, for example from solid to liquid or liquid to gas, there is a significant change in the amount of atomic or molecular **motion** (i.e., **heat**). For example, as the motion of molecules in liquid water increases (i.e. the water heats) the collisions of water molecules at the liquid/gas boundary increase in intensity due to the increased kinetic **energy** of the molecules. As a result, as water heats, more molecules gain sufficient energy to escape into the vapor phase (i.e. the vapor pressure of the substance increases). Because temperature is a measure of heat, the boiling temperature is that temperature where vapor pressure of the water equals the atmospheric pressure.

The exact temperature at which a liquid begins to boil at normal atmospheric pressure is called the normal boiling point and is a unique physical characteristic of individual elements and compounds. For example, the normal boiling point of water is 212°F (100°C) while the normal boiling point of ethanol, is 172.9°F (78.3°C).

The boiling point of liquids changes in response to changes in pressure. As pressure lessens, the boiling point of liquids decreases. For example, water boils at lower temperatures as pressure decreases with increasing elevation. This is the reason that food cooked at mountain elevations often seems undercooked. Similarly, a glass of water can be made to boil at room temperature if it is placed inside of a **vacuum** chamber. At the other extreme, because the boiling point temperature of water will increase as pressure increases, pressure cookers cook food much faster than normal cooking methods.

Boiling of liquids is also influenced by the presence of dissolved substances, or solutes. In general, when liquids contain dissolved substances, their boiling point temperatures are elevated. For example, when salt is added to water, it boils at a higher temperature. Likewise, if sugar is added to water, boils at a higher temperature. This characteristic is used in automobile engine coolants that must be altered to boil at temperatures much higher than the normal boiling point of water.

BOLTZMANN, LUDWIG EDUARD (1844-1906)

Austrian physicist

Ludwig Eduard Boltzmann, a brilliant physicist, was called "The Father of Statistical Mechanics," the science which became the foundation of **quantum mechanics**. Boltzmann applied the laws of mechanics to describe the random **motion** of molecules (atoms) and its effects on the physical properties of **matter**. His statistical analysis included the kinetic-molecular theory of gases and **thermodynamics**, the nature of **heat** and its conversion into other forms of **energy**. He also described mathematically the principle of blackbody, an ideal substance that absorbs all radiant energy that strikes it and reflects none. So radical and complicated were his hypotheses that many other scientists violently disputed them. Suffering serious illness, a depressive disorder, and perhaps feeling hopeless that his scientific endeavors would never be recognized or understood, he committed suicide just before his theories were proven.

Boltzmann, whose father was a tax official, was born on February 20, 1844 in Vienna, Austria. He received his doctorate from the University of Vienna in 1866, choosing for his thesis the kinetic theory of gases. This theory explains the physical action of gases by assuming that the extremely minute particles (molecules) of which any gas is composed, are also extremely far apart in relation to their size. The only time they affect each other (exert a **force**) is during a "rare, perfectly elastic collision." The application of the laws of mechanics applied to the behavior of gas in this "ideal state," establishes "gas laws." **Pressure** is then the collision of large numbers of molecules against the walls of the container in which the gas is held.

After graduating, Boltzmann first became assistant to his instructor, Josef Stefan (1869-1893), then taught at the University of Graz before moving first to Heidelberg and then Berlin. In 1869 he became chair of theoretical physics at Graz, retaining the post for four years before becoming chair of mathematics at Vienna. Once again, he returned to Graz as chair of experimental physics only to return in 1894 to Vienna as chair of theoretical physics.

A year later, **Ernest Mach** became chair of history and philosophy of science at Vienna. The two clashed badly, not just differing in scientific philosophy, but personally as well. Boltzmann moved to Leipzig in 1900 where he began working with Wilhelm Oswald (1853-1932), one of his most adamant opponents. On good terms personally, their scientific arguments apparently led Boltzmann to an unsuccessful suicide attempt. When Mach retired from Vienna because of ill health in 1901, Boltzmann returned to Vienna in 1902 to assume the still vacant position of chair of theoretical physics. Here, he taught **mathematical physics**, and was also given the philosophy class of his old opponent, Mach. His lectures became so popular that his audience grew too large for the lecture hall.

Boltzmann's work continued to receive severe opposition from many contemporaries, however. The difference in their analytical processes has been likened to a "scientific war between the atomists (those who based their calculations on the movement of atoms) and the energists (who declared all physical science was based on energy only)." Boltzmann was one of the first European physicists to understand the importance of a theory of electromagnetics proposed by England's James Maxwell. He also derived a hypothesis known as the Maxwell-Boltzmann kinetic theory of gases, which simplistically states that the average amount of energy used for each different direction in which an **atom** moves is exactly the same. It involved the statistical probability theory, that **temperature** and heat involved only molecular movement and that molecules at high temperature have only a high *probability* of moving toward those at low temperature. This flew in the face of the firmly-held concept of certainty, heat flowing from hot to cold.

Boltzmann published a series of papers in 1870 in which he declared that the laws of mechanics and the probability theory, which explains and predicts how the properties of atoms

(**mass**, charge, and structure), could be successfully applied to determine the visible properties of matter (**viscosity**, thermal conductivity, and diffusion).

In 1904, Boltzmann travelled to the United States to attend the World's Fair in St. Louis. He lectured on applied mathematics and visited Berkeley and Stanford where he saw new discoveries being made in relation to **radiation**. Tragically, he did not understand these discoveries would ultimately prove his theories to a doubting scientific community. He continued on, however, attempting to explain in *Populäre Schiriften* in 1905 how the physical world could be adequately described by equations yet fail to describe the underlying **atomic structure**. He wrote, "May I be excused for saying with banality that the forest hides the trees for those who think that they disengage themselves from atomistics by the consideration of differential equations."

Suffering throughout life from severe mood swings, Boltzmann's depression had a negative impact upon his professional and personal relationships. Continuing to undergo attacks upon his work, and suffering from depression and failing health, Boltzmann hanged himself on October 5, 1906 while on vacation with his wife and daughter at the Bay of Duino near Trieste. Shortly thereafter, his mathematical interpretation of the atomic world would be confirmed and the foundation laid for **quantum mechanics**.

BOLTZMANN'S CONSTANT K

The fundamental physical constant, often denoted as k and equal to 1.38×10^{-23} J/K, is called Boltzmann's constant. This constant plays an important role in **statistical mechanics**. In the late nineteenth century Austrian physicist Ludwig Eduard Boltzmann first recognized the important connection between **entropy** and probability. He began studies investigating the relationship between the **temperature** and the **energy** distribution of molecules in a gas. During these studies he laid the foundations of statistical **mechanics** by applying the laws of mechanics and the theory of probability to the motions of atoms.

The kinetic theory of gases pictures a gas as composed of a very large number of molecules whose size is small compared with the average distance between molecules. The molecules move rapidly and freely through **space**. From 1870 to 1910 a controversy raged between two groups of scientists, the Energetics and the Atomists regarding the kinetic theory of gases. Boltzmann led the Atomists who held that atoms and molecules were in fact real entities. The Energetics denied the existence of atoms and molecules and argued that the kinetic theory of gases was at best a mechanical model that imitated the properties of gases but did not correspond to the true structure of **matter**. Eventually the Atomists view of the kinetic theory of gases was proved to be correct by **Albert Einstein** and **Jean-Baptiste Perrin** around the turn of the century. Boltzmann's work with the kinetic theory of gases led him to develop the Boltzmann distribution law which states that the number of molecules in a state decreases with increasing energy of the state and is written mathematically as N2/N1

= exp(-De/kT) where N1 and N2 are the numbers of molecules in states 1 and 2 respectively, De is the difference in energy of the two states, k is Boltzmann's constant, and T is temperature.

The main conclusion from Boltzmann's work with gasses is that entropy is a measure of the probability of a state; that is, apart from an additive constant, the entropy is proportional to the log of the probability of the thermodynamic state. In this particular system, the absolute temperature of an ideal gas is directly proportional to the average **kinetic energy** of the molecules. **Boyle's law** in conjunction with the Charles-Gay-Lussac law form the basis for an important relationship called the **ideal gas law**, PV = nRT, where P is **pressure**, V volume, n moles of gas, R the universal gas constant that can be determined from experiments, and T temperature. The ideal gas law is often expressed in terms of the total number of molecules, N. By substituting N/NA, where NA is **Avogadro's number**, the above equation can be rewritten as PV =(N/NA)RT. Boltzmann's constant, often referred to as the gas constant per molecule, can be substituted for R/NA which yields PV = NkT, where P is pressure, V volume, N the number of molecules, k Boltzmann's constant, and T temperature. As well as being referred to as the gas constant per molecule Boltzmann's constant is also often called the universal conversion factor and is incorporated into many thermodynamic equations relating environmental conditions to the energy and state of systems. This constant also plays an important role in the kinetic theory of gases and the Boltzmann distribution law.

See also Gas laws

Born, Max (1882-1970)
German English physicist

Max Born's early scientific research involved a study of the **dynamics** of crystal lattices, a topic to which he returned from time to time throughout his life. The theories he developed eventually came to form the basis of modern solid-state physics. Born is perhaps better known for his role in establishing and clarifying a number of fundamental concepts in modern **quantum theory**. In the early 1920s, he collaborated with two of his students, German physicist **Werner Heisenberg** and Pascual Jordan, in the development of matrix **mechanics**, one of the two fundamental mathematical systems for working with the new concepts of quantum theory. He later attacked the problem of finding a physical interpretation for the mathematical formulations that had been developed to describe the wave-particle character of **matter**. It was primarily for these contributions that Born shared the 1954 Nobel Prize in physics with German physicist **Walther Bothe**.

Born was born on December 11, 1882, in Breslau, then capital of the Prussian province of Silesia, and later re-named Wroclaw when ceded to Poland after World War I. His father was Gustav Born, an eminent embryologist and professor of anatomy at the University of Breslau, and his mother was the former Margarete Kaufmann, daughter of a successful textile merchant. Born's mother died when he was four years old; in

1890, his father married Bertha Lipstein. Born's home life was a strong factor in his early educational development. His mother was particularly fond of music, which may explain Born's own lifelong love of music. Born's father introduced him to a number of important scientific figures of the time, including German bacteriologist Paul Ehrlich and German physician Albert Neisser.

Born attended the Kaiser Wilhelm Gymnasium in Breslau and then, in 1901, enrolled at the University of Breslau and spent the summer semesters of 1902 and 1903 studying in Heidelberg and Zurich. He studied a wide range of subjects, including chemistry, physics, zoology, philosophy, logic, mathematics, and astronomy, finding the last two of particular interest. In 1904, he enrolled at the University of Göttingen, which housed three of the most brilliant German mathematicians of the day, **David Hilbert**, Felix Klein, and **Hermann Minkowski**. Born quickly impressed his teachers with his own brilliance and was invited to become Hilbert's special assistant. Born also seemed to have a knack for frustrating his prestigious instructors. On one occasion, Klein based the topic for a university competition on a paper Born had written on the **elasticity** of wires and tapes. Born at first refused to enter because he was more interested at the time in learning about Minkowski's work on relativity, but he eventually relented and his paper won him both the competition and, in 1907, a Ph.D. in physics.

After receiving his doctorate, Born began his year of compulsory military service with the German army, but was discharged early because of a severe asthmatic condition. His army experience apparently engendered in Born an anti-military feeling that was to remain with him for the rest of his life. He then left for Cambridge University, where he spent six months studying with English physicist **J. J. Thomson** and English mathematician Joseph Larmor in order to learn about electrons.

Returning to Breslau in 1908, Born first became familiar with American physicist Albert Einstein's papers on the special theory of relativity, which had been published three years earlier. Combining Minkowski's mathematical approach to relativity with the theory put forth by Einstein, Born developed a more accurate method of calculating an electron's electromagnetic **mass**. Intrigued by Born's work, Minkowski invited him back to Göttingen to work on relativity. That collaboration lasted only a few weeks, however, as a result of Minkowski's unexpected death following an operation for appendicitis.

Born remained at Göttingen and was eventually appointed Privatdozent under Woldemar Voigt. During this period, Born carried out a study of the **heat** capacity of solids, particularly crystals, with Hungarian aerodynamicist Theodore von Kármán, resulting in the Born-Kármán theory of specific heats. During his tenure at Göttingen, Born met Hedwig Ehrenberg, the daughter of a professor of law at the University. They married on August 2, 1913, and had three children, Irene, Margaret, and Gustav, who would later become head of the department of pharmacology at Cambridge University.

Two events at the outset of World War I almost simultaneously changed Born's life. In 1915, he was offered a post as extraordinary (assistant) professor at the University of Berlin so that he might relieve German physicist **Max Planck** of his lecturing duties. At the same time, Born was inducted into the Germany army. Through a stroke of good fortune, he was able to get an assignment in Berlin, allowing him to form and maintain a connection with the University while carrying out his military obligations in the army's ordinance testing division. While not particularly productive for Born, his years in Berlin seemed especially enjoyable for him. Part of Berlin's appeal at the time had been the opportunity it gave Born to work with such notable scientists as Planck and **Albert Einstein**, with whom he would form a lifelong friendship. Born, a talented pianist, also shared with Einstein a love of music and was known to have accompanied Einstein, a violinist, in chamber music.

In 1919, Born moved once again, this time to the University of Frankfurt-on-the-Main. He had been asked by German physicist Max von Laue, then professor of physics at Frankfurt, to exchange teaching positions so that von Laue could work with Planck in Berlin. At Frankfurt, Born carried out a series of experiments on the mean free path of gas molecules with his assistant, Elisabeth Bormann.

Born's reputation at Göttingen had not dimmed during his absence and when Dutch American chemical physicist Peter Debye left as director of the University's Physical Institute in 1921, the post was offered to Born. He accepted and spent perhaps the most productive period of his life during his twelve years at Göttingen. Born had initially planned to continue his earlier research on the **thermodynamics** of crystal lattices but soon became interested in a new topic, one that had been suggested by Danish physicist **Niels Bohr**, architect of modern **atomic theory**. Bohr had elucidated in 1922 his "correspondence principle," suggesting that a new quantum theory could be developed by finding its transitionary relationship to classical physics. Challenged by Bohr's ideas, Born set out to derive a new formulation of physical laws using, according to his own terminology, a "quantum mechanics."

The real breakthrough in this effort came in 1925, when one of Born's assistants, Heisenberg, showed him a paper he had written on this general topic. Heisenberg had begun with the premise that only observable quantities should be used in the development of theoretical principles. With this assumption, Heisenberg had derived a new mathematical approach for dealing with such principles. When Born read Heisenberg's paper, he realized that the mathematical system described there was essentially the same as that of matrix algebra. Since Heisenberg had gone on vacation after giving Born his paper, Born asked another of his students, Jordan, to assist him in developing and refining Heisenberg's paper. Ultimately, a series of three papers—the original by Heisenberg, a collaborative effort by Born and Jordan building on the Heisenberg paper, and a final paper coauthored by all three—was published. These three papers outlined the fundamentals of what was later to become known as matrix mechanics.

Only a year later, an entirely different approach to the treatment of **quantum physics** was proposed by Austrian physicist **Erwin Schrödinger**, an approach that became known as wave mechanics. Eventually, Schrödinger was able to

Max Born.

demonstrate that these two mathematical formulations—matrix mechanics and wave mechanics—are entirely consistent with each other. The most familiar portion of the Heisenberg-Born-Jordan matrix theory is the equation relating the position (q) and **momentum** (p) of a particle. Born himself regarded this accomplishment as the most significant of his career and had the equation, $pq - qp = h/2\pi i$, engraved on his tombstone.

For most physicists, Schrödinger's wave mechanics were easier to use and understand than matrix mechanics. Wave mechanics rather quickly gained acceptance, therefore, and became the tool used to interpret particle phenomena. The one troubling problem, however, was that Schrödinger's wave mechanics was an entirely mathematical analysis of particle behavior based on a few fundamental properties of matter. The wave equation produced by this analysis was an elegant and obviously useful way of analyzing particle properties, but it had no apparent physical meaning. Physicists wondered, in particular, how they were to interpret the **wave function**, ψ, that constituted the heart of wave mechanics.

In the summer of 1926, Born proposed a solution for this dilemma. He suggested that the square of the wave function, ψ^2, should be understood as the probability of finding a particle at any particular point along the wave. Thus, if one thinks of the particle as being carried along on the wave, the square of the amplitude of the wave at any particular point is equal to the likelihood of finding the particle at that location. For this work, as well as for his earlier studies on **quantum mechanics**, Born shared the 1954 Nobel Prize in physics with Bothe.

One of the early laws passed by German leader Adolf Hitler's regime was the "Gesetz zür Wiederherstellung des Berufsbeamtentums," requiring the removal of Jews from civil service. As a result, Born was fired from his post at Göttingen on April 25, 1933, after which he traveled to the Italian Tyrol. In June, he was invited by English physicist Patrick Maynard Stuart Blackett to take up a Stokes Lectureship at Cambridge University, where he worked on problems of nonlinear **electrodynamics** and wrote two immensely popular books, *Atomic Physics* and *The Restless Universe*. When his fellowship at Cambridge expired in 1936, Born traveled to Bangalore, India, where he visited physicist and Nobel laureate C. V. Raman. He then returned to Great Britain and became Tait Professor of Natural Philosophy at the University of Edinburgh. Born's tenure at Edinburgh was spent primarily in the supervision of graduate students, and he produced relatively little new research of his own.

Born retired in 1953 at the age of 70; he returned to Germany and built a home at Bad Pyrmont, a small town near Göttingen. The German government returned the personal property that had been confiscated from him in 1933 and restored the pension that had been denied him by the Nazis. In the last decade of his life, Born became especially active in writing and speaking about the social responsibilities of scientists. In July 1955, for example, he was one of the signers of the Mainau Declaration, condemning the development of atomic weapons. In addition to the Nobel Prize, Born was awarded the Max Planck Medal of the German Physical Society in 1948, and was a member of many scientific academies throughout the world. Born was elected a member of the Royal Society in 1939 and received its Hughes Medal in 1950. He was also awarded the Macdougall-Brisbane Prize of the Royal Society of Edinburgh in 1945. Born died in Göttingen on January 5, 1970.

BOSE, SATYENDRANATH (1894-1974)

Indian physicist

Satyendranath Bose was a physicist whose derivation of **Max Planck**'s black body **radiation** law—a fundamental law of **quantum theory** that is concerned with the concept of **energy** transfer associated with **light**, **x rays**, and other radiations— verified **Albert Einstein**'s concept of photons and played a major role in the development of quantum statistics and theoretical physics. Einstein used his influence to have Bose's groundbreaking paper published while building upon it to develop a system of quantum **statistical mechanics** known as Bose-Einstein statistics. As a result of Bose's contribution to the field, bosons—certain **subatomic particles** of finite **mass** studied in **quantum physics** using the Bose-Einstein statistical approach—were named after him.

Although Bose published a relatively low number of original papers throughout his scientific career, his interests in

physics and mathematics were wide ranging and included statistical **mechanics**, electromagnetics, x-ray **crystallography**, and the unified field theory. Still, Bose failed to develop the worldwide reputation enjoyed by Einstein, who readily acknowledged Bose's contribution as the major step in the development of quantum statistical mechanics. Within India, Bose was highly regarded for both his scientific and political accomplishments, being among a group of pioneering nationalist scientists and intellectuals who helped spur colonial India's independence movement from Great Britain.

Born in Calcutta, India, on January 1, 1894, Bose was the son of Surendranath Bose, an accountant who would eventually found the East India Chemical and Pharmaceutical Works, and Amodini Raichaudhuri. His education began in the English-language schools set up by the British during their colonial reign over India, but he was soon transferred by his father to a Bengali-language school during the resurgence of Bengali nationalism in 1907 (Bengal is a region in the northeast section of the Indian peninsula). After graduation from secondary school, Bose attended Presidency College in Calcutta, where he studied under noted Indian physicist Jagadischandra Bose (no relation). In 1915 Bose continued his postgraduate studies at the university and finished first in his class with an M.Sc. degree in mathematics. Two years later he became a lecturer in the physics department at the University of Calcutta's college of science, which was established in 1914 and was the first Indian college to offer advanced science studies.

Bose's interest and natural gifts in mathematics soon became evident with the publication of two papers he coauthored with Meghnad Saha on the equation of state. In 1919 they coedited one of the first anthologies in English of Einstein's scientific papers on relativity. Then, in the next year, Bose published his first paper on quantum statistics in the *Philosophical Magazine*.

In 1921, Bose accepted a position as a reader, or professor, at the University of Dacca, a newly established university in East Bengal. At Dacca, Bose focused his attention on the statistics of photons, a quantum of electromagnetic energy that has no charge or mass but carries energy such as light and x rays in both wave and particle form. In an interview with American physicist William A. Blanpied for the *American Journal of Physics,* Bose commented that he "spent many sleepless nights" contemplating Planck's law. Finally, according to Bose, a new theory for quantum mathematics dawned on him while he was lecturing in class one day. By 1923, he had written and submitted a paper on his new theory to *Philosophical Magazine,* but it was rejected.

Bose was persistent, however, and sent the article to Einstein along with a letter requesting his assistance. Einstein, who in 1905 had first tried to prove that electromagnetic radiation had an **atomic structure** made up of a measurable, or quantum, amount of electromagnetic energy, was impressed with Bose's theory. Essentially, Bose had succeeded in substantiating Einstein's proposal where Einstein himself and others had failed. With Einstein's endorsement and translation into German, Bose's paper, "Planck's Law and the Hypothesis of Light Quanta," was published in *Zeitschrift Für Physik* in 1924.

Bose's paper focused on how to derive Planck's black body radiation law, in which the black body is a theoretical ideal body that absorbs all radiation and reflects none. Planck's equation describes the spectral energy distribution from such a body. Einstein's 1905 paper questioned Planck's assumption that his law could be applied *ad hoc* to unquestionable classical **electrodynamics** laws (electrodynamics is a discipline within physics concerned with the inter-relatedness of electric and other currents and magnets). In Einstein's thermodynamic approach, the quantum structure of electromagnetic radiation could be viewed the same as an ordinary gas and its atomic structure. Without reference to classical electrodynamics, Bose used a phase-space approach that treated radiation as an ideal gas to show that Einstein's model and Planck's law were consistent. Bose's paper both vindicated Einstein's theory and pointed the way for future developments in electrodynamics.

Unfortunately, Bose concentrated on the mathematical implications of his **work** and was not as quick to see the far-reaching ramifications it had in the field of physics, especially in the areas of electrodynamics and quantum gases. As a result, he failed to gain the international renown of such famous physicists as Planck and Einstein, who, building on Bose's work, developed the foundation for the Bose-Einstein statistics. This statistical approach to quantum physics was the first of two approaches to determine the distribution of certain subatomic particles among the various possible energy values. Depending on the approach used, the particles that adhere to these mathematical laws are known as bosons—named after Bose—or **fermions**, named after the developer of the second approach, **Enrico Fermi**.

Although Einstein's endorsement and further support led to Bose leaving India for two years to study in Germany and France, Bose's dream of working with Einstein never materialized; in fact, the two scientists had only one brief personal meeting. In 1926, Bose returned to Dacca and became a professor of physics. Dedicated to his teaching duties, Bose's production of scientific papers was, by many standards, meager, amounting to only twenty-six original papers focusing on mathematical statistics, electromagnetic properties of the ionosphere, x-ray crystallography, thermoluminescence, and the unified field theory.

Bose was appointed in 1945 as the Khaira Professor of Physics at Calcutta University. In 1956, he became vice-chancellor of Visva-Bharati University, which was established by Rabindranath Tagore, who had won the Nobel Prize in poetry. He was then appointed a national professor by the Indian government in 1959.

Bose married Ushabala Ghosh in 1914; the couple had two sons and five daughters. Devotedly nationalistic during the days of British rule, Bose abhorred the fact that some people thought he was German. He remained dedicated to his country and particularly to the Bengali cultural renaissance. He was among a core group of Indian intellectuals who assisted with India's political emergence, primarily by instilling pride in the intellectual capabilities of the populace. The founder of the Science Association of Bengali in 1948, which helped popularize science in his native language, Bose also served in the Indian parliament from 1952 to 1958 and as president of the National Institute of Sciences of India. He received India's Padma Vibhushan award from the government in 1954, and, in 1958, was elected to the Royal Society. Fluent in Bengali, Sanskrit, English, and French, Bose loved poetry.

Bose died on February 4, 1974, in Calcutta. Jagadish Sharma, who studied physics under Bose, commented in Bose's obituary in *Physics Today* that the scientist's remarkable rise to "the highest echelons of science" must be placed in the context of an India that was ruled by Great Britain. Such an environment was not conducive to fostering the native Indian intellect, Sharma noted, evidenced by the "very few Indian scientists of international reputation." Sharma also added that Bose's kindness was endearing to all and that he "was liberal with his pocketbook whenever the situation demanded."

BOSE-EINSTEIN CONDENSATION

Although predicted by physicists **Satyendranath Bose** (1894-1974) and **Albert Einstein** (1879-1955) in the early 1920s, the unusual state of **matter** known as a Bose-Einstein condensate was not created until 1995, when a team led by Eric Cornell and Carl Weiman at the Joint Institute for Laboratory Astrophysics in Boulder, Colorado, brought the first known example of this non-naturally occurring substance into existence by cooling rubidium atoms to less than 170 billionths of a degree above **absolute zero**.

Bose wrote equations describing the behavior of photons under the laws of **quantum mechanics**; these equations were elaborated by Einstein to include fundamental particles and the atoms made of those particles. The resulting Bose-Einstein equations describe the behavior of certain types of particles. According to the Standard Model developed in the later half of the twentieth century, there are two classes of particles in the **universe**: bosons, named in honor of Bose, and **fermions**, named after the physicist **Enrico Fermi**. The difference between the two is of profound importance.

Fermions are the building blocks of matter. They come in two varieties, elementary and composite. There are two types of elementary fermions (so-called because they cannot be divided into smaller particles): leptons (e.g., electrons, the negatively charged particles surrounding the **nucleus** of an **atom**) and **quarks**, the particles that form protons and neutrons. Composite fermions are particles made of an odd number of elementary fermions; they behave according to the same rules that govern elementary fermions even though they may also contain bosons. Particles made of an even number of elementary fermions, however, behave according to the Bose-Einstein equations and are therefore regarded as bosons even though they include fermions.

Bosons, such as photons, are the basic units of matter and **energy**. As with fermions, there are two types: elementary and composite. Elementary bosons are mediators of the four fundamental forces (electromagnetic, strong, weak, and gravitational) that govern the behavior of matter and energy. The bosons projected to mediate the **gravitational force** have not yet been observed. Composite bosons, or mesons, help to hold atoms together.

Fermions obey the **Pauli exclusion principle**. This principle, named after its discoverer, physicist **Wolfgang Pauli**, states that two identical fermions cannot occupy the same **space** at the same **time**. This is why electrons must arrange themselves in various **orbitals** about the nucleus of atoms.

Bosons are not bound by the exclusion principle. Unlike fermions, identical bosons can occupy the same space. It is this property that makes Bose-Einstein condensation possible. Bosons and fermions also differ in their respective spins, a measurement of the rotation and **angular momentum** of a particle. Bosons have spins that are whole numbers (e.g. 0,1, 2 etc.), while the spins of fermions are odd multiples of one-half (e.g., 1/2, 3/2, 5/2, etc.)

According to the Bose-Einstein equations, rubidium atoms should demonstrate extraordinarily interesting properties at temperatures close to absolute zero. Implicit in the equations is the prediction that at near absolute zero temperatures these bosonic atoms will fall to the same low **energy level**. In quantum mechanical terms, their wave functions will overlap, causing the atoms to become locked in phase, thereby rendering them identical in every measurable respect. Instead of a dense collection of super-cooled atoms, there will be a single superatom.

Not until 1995 was anyone able to produce temperatures cold enough to verify this hypothesis. Cornell and Weisman devised an ingenious two-step procedure to achieve their historic result. The first step, known as laser cooling, involved bombarding rubidium atoms with laser beams to slow and, with the help of a magnetic field, contain them. The **temperature** of the trapped atoms dropped to less than one-ten thousandth of a degree above absolute zero. This was cold, but not cold enough. To further reduce the temperature, a second step utilized evaporative cooling. Just as the escape of hot molecules from a cup of coffee in the form of steam results in the cooling of the liquid left behind, the hottest rubidium atoms were ejected from the magnetic containment field, allowing the rest to grow colder and colder, until Bose-Einstein condensation occurred.

Bose-Einstein condensates give physicists the opportunity to study quantum phenomena on a macroscopic level. Many of the properties of this fascinating form of matter remain to be discovered, but scientists expect the applications to be vast. For example, Bose-Einstein condensates could be useful in the creation of atom **lasers** that utilize beams of coherent matter rather than **light**.

See also Low-temperature physics; Standard model of particle physics

BOYLE, ROBERT (1627-1691)

English physicist and chemist

Boyle is considered by many as the father of modern chemisty. He offered the first accurate definitions of elements and chemical reactions, is credited with pioneering modern **scientific method**, and also formulated the law that bears his name which describes the relationship between **pressure** and volume in gases.

For centuries people believed that everything was made of just three or four substances, which were mistakenly called elements. It was Boyle who first set science on the right track and asserted the true nature of elements and compounds.

Boyle's father was an Englishman who made his fortune in Ireland and became a successful landowner there. Boyle was his seventh son and the youngest of fourteen children. By the time Boyle was born, his father had become an earl and was one of the wealthiest men in the country. Like his father, Boyle was an industrious worker; before he entered the prestigious Eton school at the age of eight, he was already speaking Greek and Latin. His passion for reading and learning continued to grow, and Boyle proved to be a gifted student with an excellent memory.

At an early age, Boyle and his brother went to Europe with a tutor to study French, mathematics, and many other subjects. For six years, they lived in Switzerland and traveled extensively through France and Italy, where Boyle learned of Galileo's experiments on the effects of **gravity** and other physical laws. When Boyle was 13, he witnessed a sudden, violent thunderstorm that changed his whole outlook on life. But in Boyle's view, his religious faith did not contradict scientific beliefs but reinforced his admiration for the creator of such a complex **universe**. From then on, he was a devout Christian, and he learned several ancient languages, including Hebrew and Aramaic, so that he could read the Bible in its original texts.

A civil war between England and Ireland demanded Boyle's return from continental Europe, and he reached home in 1645. By this time, Boyle had become interested in performing experiments in order to understand the way things work. Previously, people believed in making up a theory and then judging how well the facts fit the theory. Boyle, however, agreed with the philosopher Francis Bacon (1561-1626) that facts should be observed first, and then a theory should be developed to explain them. Boyle and other scholars interested in experimentation began meeting regularly in London, England, to discuss these new ideas. At first, they called themselves the "Invisible College," but when King Charles II (1630-1685) was restored to the throne in 1663, he granted a charter to the group of scientists; the group thus became known as the Royal Society.

When Boyle moved to Oxford in 1654, he met many more scientists and became interested in chemistry, because of its relation to medicine. During his 14 years at Oxford, Boyle contributed greatly to scientific philosophy in the fields of physics and chemistry. He set up an elaborate research laboratory and hired skilled assistants to conduct experiments.

Unlike most scientists of his day, Boyle believed in meticulously recording his experiments and publishing them so that others could repeat his tests and confirm the data. He is credited with pioneering modern scientific method and today this practice is universal in the research world.

In 1661 Boyle published his most famous work, *The Sceptical Chymist*, which revolutionized scientific thought and formed the basis of modern chemistry. In this work, Boyle defined an element as the simplest form of **matter**, one that cannot be broken down into any simpler form or changed into a different substance. Boyle's ideas contradicted beliefs held ever since the ancient Greeks proposed that all things are made of only four elements—air, earth, fire, and water—which could be changed, or transmuted, into other substances. In another version of this idea, only three substances existed in nature (salt, sulfur, and mercury). But according to Boyle, none of these substances were true elements. Boyle argued that elements could be identified only by scientific experimentation. He also pointed out that a compound will usually have chemical properties that are very different from its parent elements.

Boyle's concept of an element arose from his experiments with gases, and he was the first scientist to succeed in collecting hydrogen in a device now called a pneumatic trough. In 1660 Boyle discovered a fundamental law of physics that helps explain the behavior of gases. When a gas is pressurized, Boyle found that the amount of **space** it takes up is related to the amount of pressure being exerted on it, as long as the gas's temperature doesn't change. For example, if the pressure on a given quantity of gas is doubled, the gas's volume is cut in half; if pressure is tripled, volume is reduced to one-third. (This relationship is called inversely proportional.) **Boyle's law**, along with a similar law that explains the effects of temperature, allows chemists today to calculate the volume of gases under any pressure or temperature conditions. Boyle also realized that if air could be compressed, it must be composed of tiny particles separated by space. It was this conclusion that led Boyle to envision a universe composed of numerous tiny particles, and, in doing so, he anticipated the modern concept of **atomic theory**.

Vacuums were poorly understood but of much scientific interest to Boyle and his assistant, Robert Hook. Airpumps were used in early laboratories to create a **vacuum** inside cylinders. Robert Hook built an improved airpump based on German engineer, Otto von Guerricke's airpump design. Together, Boyle and Hook developed a better vaccum. This new vacuum had better placement of the pumps valves and a preferable method of cranking its piston and supporting the air pump's cylinder. Boyle also proved for the first time that all objects, no matter how **light** or heavy, fall through a vacuum at the same speed. This showed, as **Galileo** had predicted, that the **force** of gravity is uniform. In another experiment, Boyle demonstrated that the **sound** of a clock ticking could not be heard in a vacuum, proving that sound **waves** depend on air for their transmission. Boyle showed, however, that electrical attraction could be felt through a vacuum.

Boyle was also interested in the nature of **color**, and he accurately described how the absorption and reflection of light

Robert Boyle.

produces the appearance of black and white, studying the changes in color that occur in certain plant extracts, such as *litmus*. He discovered that these substances, now called *indicators*, can be used to distinguish acids from bases. Boyle went on to develop tests for identifying other substances, such as copper, silver, and sulfur, via **chemical reaction**. He not only coined the term *analysis* in its modern sense, but also encouraged generations of chemists to determine the composition of substances through meticulous experimentation. In the late 1660s Boyle became the first scientist to study the phenomenon of bioluminescence, showing that certain bacteria and other organisms will glow in the dark if supplied with air. Boyle also found that water begins to expand just before it freezes. Throughout this period, Boyle and his staff published immense amounts of information for use by other scholars and scientists.

In 1668 Boyle returned to London to live with his favorite sister. In 1680 he invented the first match by coating a piece of course paper with phosphorus. He produced a flame by drawing a sulfur-tipped wooden splint through a fold in the paper. Also in 1680, he was elected president of the Royal Society, but he declined the honor, believing that the oath of office would conflictwith his strict religious beliefs, and he continued to refuse all titles and other honorary positions. In his later years, Boyle wrote about medicine and diseases and devoted greater effort to promoting Christian ideals; in his will he left money for a series of lectures to defend Christianity from atheists and other "infidels."

BOYLE'S LAW

Boyle's law is one of the **gas laws**. It states that at constant **temperature**, the volume of a fixed **mass** of gas is inversely proportional to the **pressure**. The other way of expressing this law is pV = constant, where p is the pressure and V is the volume of the gas. It is named after British physicist and chemist **Robert Boyle** (1627-1691).

Boyle's law is sometimes known as the constant temperature law. It can be combined with **Charles' law** and the pressure law to give the **ideal gas law** (also known as the universal gas law), pV = nRT, where p is pressure, V is volume, n is the number of moles of gas, R is the universal gas constant, and T is temperature.

Boyle's law is an approximation and works perfectly only for an ideal(theoretical) gas. In practice it works best at low pressures but once the pressures become high the predictions become less accurate. This inaccuracy is due to the size of the gas molecules and weak **intermolecular forces**, such as van der Waals forces. At high pressures the molecules are forced together whereas at low pressures the molecules are free to move with very little interaction from neighboring molecules.

Boyle's law is a good indicator of how a gas will react if the temperature is kept constant and the pressure and volume are altered.

BRAGG, WILLIAM HENRY (1862-1942)

English physicist and mathematician

Sir William Henry Bragg was a noted English physicist, mathematician, and teacher whose reputation rests on his pioneering work on the determination of crystal structure by the use of x-ray **diffraction**. For this achievement, he was awarded the 1915 Nobel Prize for physics with his eldest son, **William Lawrence Bragg**. Bragg's work advanced understanding of the way atoms bond together to form molecules, and resulted in practical repercussions throughout industry. He is also noted for his efforts to win physics a popular audience, particularly through his acclaimed Christmas lectures for the Royal Society.

Bragg was born in Cumberland, England, to Robert John Bragg, a former Merchant Marines officer, and Mary Wood, a vicar's daughter. After the death of his mother when he was only seven years of age, Bragg went to live with his uncle, a pharmacist, and attended school at Market Harborough in Leicestershire. After six years there, Bragg completed his early schooling at King William's College on the Isle of Man. After winning a partial scholarship to Trinity College, Cambridge, Bragg studied pure maths under Dr. E. J. Routh. In his last year, he also attended lectures at the Cavendish Laboratory given by its chair, **J. J. Thomson**, and Lord Rayleigh, the esteemed English physicist.

In 1885, Bragg accepted a professorship in mathematics and physics at the University of Adelaide, Australia. He arrived in 1886 and immediately adapted to the country's easy-going pace and outdoor life. He wrote that "going to

Australia, to a new work and an assured position, the people I met there, the sunshine, the fruit and flowers, was a marvelous change for me. I know that I had been lucky enough in England, but I am not ungrateful when I say that going to Australia was like sunshine and fresh invigorating air." In 1889 Bragg married Gwendoline Todd, the daughter of Charles Todd, government astronomer and postmaster general. The Braggs had three children, two boys and a girl. Their eldest son, Lawrence, was to prove as gifted and pioneering a physicist as his father.

Bragg's lifelong interest in education was sparked in Adelaide, where he found his students to be of a low caliber. This set him wondering what could be wrong with the educational system that it had produced such a poor quality student and how it could be ameliorated. He wrote what would be his first of many dissertations on the subject in 1888. Perhaps surprisingly, Bragg published no important research studies while in Adelaide. For one thing, his laboratory was simply not equipped for the task. In fact, he was driven to apprentice himself to a company of equipment manufacturers in the town to learn how to make his own laboratory supplies. The other explanation was Bragg's lack of technical training and his isolation from the throes of the exciting new developments unfolding in the field at that time. He wrote in 1927, looking back: "Perhaps it may seem wrong that a professor of mathematics and physics should be so long content with the ordinary round of teaching and management, but there may be some explanation in the fact that I had no laboratory training nor had I come into contact with research. I had indeed never studied physics as now understood."

Nonetheless, Bragg was far from idle. He closely followed developments in England and continental Europe and carried out what basic experiments his inadequate equipment would permit. A trip home to England via Egypt and Italy in 1897 allowed Bragg to meet with his peers and catch up on all the exciting advances that had been made, especially at Cambridge, in his absence.

Upon his return to Adelaide, he continued his usual schedule of lecturing, both to his students and to the public. His speeches on subjects such as radium, the **electron**, and **x rays** did much to increase the public understanding of the new scientific and technological developments.

It was through his work as a teacher and administrator at the University of Adelaide that Bragg chanced upon the subject that was to ignite his interest in experimental physics and start him on the path of fame. Asked to deliver the presidential address to the mathematical and physical section of the Australasian Association for the Advancement of Science, Bragg decided to review the latest developments in physics, concentrating largely on **radioactivity**. As he prepared for his speech, he reviewed the recent work of Madame **Marie Curie** in France, and was suddenly seized by an explanation for her results that had hitherto not been proposed. Curie had shown that when a radium **atom** breaks into two parts, the smaller part, which is an atom of **helium**, otherwise known as an **alpha particle**, is released into the air as **alpha radiation**. Bragg was intrigued as to how the alpha particles managed to travel straight through the air when clearly their way was blocked by atoms of air. He concluded that they must pass through the air atoms, which meant that at the moment the alpha particles met the air atoms they occupied the same **space**. The alpha particles would collide with the air atoms and scatter only if the protons of the two atoms met: a chance of about one in a million. Bragg confirmed his suspicions in an experiment in which he passed alpha particles straight through an inch of air. This proved that the atom is not a solid body and that two atoms can occupy the same space.

Between 1904 and 1908, Bragg continued his experiments on the passage through **matter** of beta, and gamma, and especially alpha rays emitted from radioactive substances. He discerned that alpha rays have a definite range, **velocity**, and ability to produce ionization in gases, that is, to change their atomic configuration by the addition or subtraction of one or more electrons. It was this ionizing effect that, Bragg deduced, caused the loss of **energy** of an alpha ray as it sped through the air. He concluded that at a certain distance from the point of the alpha ray's departure, the ionizing effect would stop. Exactly where would depend on the speed at which the alpha ray was traveling and the nature of the substance through which it passed, a distance Bragg referred to as its range. He discovered that alpha rays with different velocities, and thus different ranges, exist, each corresponding to a different kind of radioactivity in its source. As a result of these findings, it became possible to identify different radioactive substances by the measurement of their alpha rays. Bragg was able to show, for instance, that helium atoms of four different ranges were expelled from a preparation of radium, confirming the existence of the four different radioactive substances, identified by English physicist **Ernest Rutherford**, who won the Nobel Prize in chemistry in 1908. Bragg lost no time in conveying his results to Rutherford himself, who encouraged him to publish them in one of the British journals. As a result, in 1907, his papers appeared in *The Philosophical Magazine* and *The Philosophical Transactions of the Royal Society.*

Bragg's important findings were the spur he needed to begin the research that would occupy him the rest of his life. He began publishing papers every few months, was elected a fellow of the Royal Society of London, and in 1908 was appointed to the Cavendish Professorship of Physics at the University of Leeds. The position represented a major change from the free and easy life in Adelaide and it took Bragg some years to fit in. During his first years in Leeds, the research that had brought him there ground to a halt and Bragg was more miserable than ever before in his life. Only his friendship with Rutherford helped to sustain him during this difficult period. One particular source of irritation to Bragg was an ongoing feud with Professor Charles Glover Barkla concerning the nature of x rays. Many angry letters were exchanged between them in the pages of *Nature,* with Barkla propounding his wave theory of x rays and Bragg clinging to the corpuscular, or particle, theory that he had formed in Adelaide. Actually, both men were right: x rays are both **waves** and particles. They had formed their different views, however, by studying different aspects of x rays.

Bragg's spirits were raised in 1911 and 1912 with the discovery of x-ray diffraction by crystals, or the bending of the rays as they pass through the crystal. He immediately grasped its significance and realized that he could no longer deny the wave-like properties of x rays. He understood that a theory was necessary that could account for both its wave and corpuscular properties.

During the next two years, Bragg worked with his son Lawrence on the study of crystal structure by means of x rays, using the x-ray spectrometer he had designed in 1913. Together, they worked out a method to determine the **atomic structure** of a number of crystals using x rays and were thus able to account for their behavior and properties.

Their findings were jointly published in 1915 in *X-rays and Crystal Structure,* a study which led to much greater understanding of the nature of solid bodies. As Bragg said in one of his Christmas lectures, popular talks at the Royal Institution given in simple language on physics: "We are now able to look ten thousand times deeper into the structure of the matter that makes up our **universe** than when we had to depend on the **microscope** alone." For their findings, Bragg and his son shared the Nobel Prize in physics in 1915. The mathematical equation derived from their discovery, called Bragg's law, is now used to study the **molecular structure** of complex substances.

Bragg's research using x rays to gain a greater understanding of crystal structure took a back seat during the First World War as Bragg served his government through his work for the Admiralty. During this time, he invented the hydrophone, a device to detect submarines. In 1916, he was made resident director of research at the Naval Experiment Station at Hawkcraig, with a staff of two physicists and a mechanic. For his work for the Royal Admiralty, Bragg was knighted in 1920.

By 1919, Bragg was finally able to take up the Quain Professorship of Physics at University College, London, that he had been offered in 1915, and began building the college's reputation as a research establishment. In 1923, Bragg became director of the Royal Institute of Great Britain, and moved most of his research team there to its Davy-Faraday lab. As a result of Bragg's leadership, the lab gained a worldwide reputation for excellence, especially for its work surrounding crystal structure. There, Bragg and his son continued their analyses of organic crystals, work that has been of fundamental importance in the development of molecular biology. At the Royal Institute, Bragg was also responsible for the spread of his methodology to other fields, such as work on the structures of organic molecules.

During his lifetime, Bragg was honored with countless awards and a total of sixteen honorary doctorates by universities in Britain and overseas. In 1916, he was presented with the Rumford Medal by the Royal Society of London, and with the Copley Medal in 1930. Bragg was named president of the Royal Society in 1935.

Bragg is remembered not only as a great physicist but also as a gifted teacher who was able to communicate the complexities of physics in a clear and dynamic manner. His annual Christmas lectures at the Royal Institute, originally geared toward children, made Bragg a household name and succeeded in communicating the complexities of physics to the general public. He also published no fewer than 237 books and research papers. He believed that science should not offer solutions to the conundrum of our existence, but gather knowledge to increase our understanding of the physical world. Bragg was a family man, devoted to the relationships in his life as well as to his work. It is not surprising, then, that after his death from old age on March 12, 1942, at the age of eighty, he was remembered in the obituary notices of the Fellows of the Royal Society as "a gentle man."

BRAGG, WILLIAM LAWRENCE (1890-1971)

English physicist

William Lawrence Bragg shared a remarkable two-year collaboration with his father and fellow physicist, **William Henry Bragg**, during which they founded the new science of x-ray **crystallography**. The methods developed by this father-son team made it possible to explore the **atomic structure** of **matter** very precisely and in great detail. The Braggs shared the 1915 Nobel Prize in physics for their work.

Bragg was born in Adelaide, Australia. His father was professor of physics and mathematics at the University of Adelaide; his mother, Gwendoline Todd Bragg, was the daughter of Sir Charles Todd, South Australia's postmaster general and government astronomer. Bragg had a brother one year younger than he, Robert, who was killed at Gallipoli in Turkey during World War I, and a sister, Gwendolen, seventeen years his junior. The children's parents were a contrast; Gwendolen later wrote in the biography *William Henry Bragg* that their father wanted his children to be absolutely free and avoided advising them on what they should do, whereas their mother "always knew exactly what one ought to do, and said so." Foreshadowing the field of the Braggs' future work, Bragg's father built a primitive x-ray machine within weeks after Wilhelm Conrad Röntgen's 1896 announcement of his discovery of the rays. Soon thereafter, five-year-old W. L. fell from his tricycle and broke his elbow. Professor Bragg used his device to reveal the location and extent of the injury. This was the first recorded use of **x rays** for medical diagnosis in Australia.

As a child, Bragg was a loner, being academically ahead of boys his age and poor at sports because, as he explained it, he wasn't aggressive or self-confident enough. His sister quoted him as telling her, "You and I find *things* easier than people." He expressed his interest in natural sciences by collecting shells; a new cuttlefish species he discovered, *Sepia braggii,* was named after him. With his brother, he enjoyed creating mechanical devices out of discarded scraps, and a chemistry master aroused his interest in scientific experimentation.

A gifted scholar, Bragg attended St. Peter's College (a secondary school) in Adelaide and entered the University of Adelaide in 1905 at the age of fifteen. His father was beginning to experiment with x rays at the university at that time,

and W. H. often talked to his son about his results. "I lived in an inspiring scientific atmosphere," Bragg wrote of those years in a chapter he contributed to *Fifty Years of X-Ray Diffraction* entitled "Personal Reminiscences." He graduated in 1908, after just three years, with first-class honors in mathematics. In January 1909 the family left Australia for England, where W. H. Bragg had accepted a professorship at the University of Leeds. W. L. enrolled at Trinity College, Cambridge, where he began by studying mathematics. At the end of his first year, taking an exam while sick in bed with pneumonia, he won a major scholarship. Then, following his father's suggestion, he switched his concentration to physics. Again he graduated with first-class honors, this time in natural sciences, in 1912. He stayed at Cambridge, doing research under J. J. Thomson at the Cavendish Laboratory, and began his very fruitful collaboration with his father in the fall of 1912.

Earlier that year, **Max Laue** had discovered the **diffraction** of x rays in crystals: x rays passing through crystals are bent, producing distinct patterns. At the time, physicists were engaged in a lively debate about the nature of x rays: Were they particles or **waves**? Bragg's father had a deep interest in x rays and favored the particle theory. He was excited by Laue's discovery, even though the diffraction patterns could be explained only if x rays were waves, and he thought further investigation of the new phenomenon might provide the missing evidence to support either the wave or the particle theory, or even a new theory incorporating both wave and particle properties. W. L. Bragg, too, was excited by Laue's discovery, especially as it related to the structure of crystals.

During the summer holiday of 1912, father and son talked often about Laue's discovery, and when W. L. returned to Cambridge that fall, he launched into a series of experiments using x-ray diffraction. He concluded that Laue's wave interpretation was correct, but that his explanation of diffraction was unnecessarily complex. W. L. suggested that since atoms are arranged in a regular way in crystals, the diffraction patterns might be caused by x rays reflecting off the planes of atoms within the crystals. He developed an equation relating angles of rays, **wavelength**, and perpendicular distance between atomic planes, which became known as Bragg's law, with the glancing angle of the x rays called Bragg's angle. Both Braggs saw the great implications of W. L.'s idea: the reflected patterns of the x rays would reveal the previously hidden arrangement of the atoms within the crystals. The Braggs plunged into investigations of crystals, with W. L. first analyzing sodium chloride (table salt) and potassium chloride (a crystalline salt used as a fertilizer), though they were hampered by inadequate equipment. Bragg's father solved this problem in 1913 by his invention of the x-ray spectrometer, a device that measured x-ray angles and intensities precisely and allowed for analysis of complex crystals. The elder Bragg used the spectrometer to continue his studies of **radiation**, while W. L. used it to pursue his interest in analyzing the structure of crystals.

W. L. Bragg published his first suggestion about the wave nature of x rays as shown by Laue's discovery in November 1912. W. L. and his father published their first joint paper early in 1913 in the *Proceedings of the Royal Society,* which established the basic principles of the new science of crystal analysis by x-ray diffraction. W. L. followed this a few months later with his own paper on the structure of sodium chloride, which he showed was not made up of molecules, as had been thought, but rather of **ions** (charged atoms) of sodium and chloride. This finding became very important in the analysis of solutions, as it was the first distinction made between compounds that consist of ions and those made up of molecules. In July 1913 the Braggs published a joint paper on the structure of diamond, and W. L. published another paper in November describing more crystal structure discoveries. By 1914 the Braggs had set the standards for x-ray crystallography. They had also transformed crystallography from its role as a secondary, if interesting, branch of science to its new position as a fundamental branch of modern physics with applications to many other sciences. In "Personal Reminiscences" W. L. said of this period of collaboration with his father, "We had a thrilling time together in an intense exploitation of the new fields of research."

The Braggs were recognized for their ground-breaking achievements in exploring the arrangements of atoms by the 1915 Nobel Prize in physics, awarded—in the Nobel Committee's words—"for their contributions to the study of crystal structure by means of x rays." W. L. was only twenty-five years old, the youngest Nobel laureate ever. But the "thrilling time" of collaboration had ended abruptly with the outbreak of World War I in August 1914. Bragg volunteered for service and found himself assigned to a horse artillery battery among "hunting men." After a year, he was sent to France to adapt the new method of locating enemy guns by **sound**, called sound ranging, for use by the British forces. He later recalled that it was while setting up a sound-ranging base in Belgium that he learned of his Nobel Prize. The parish priest in whose house he was lodged broke out a bottle of wine to celebrate. After the war, Bragg returned to Cambridge as a lecturer.

In 1919 Bragg was named to succeed **Ernest Rutherford** as professor of physics at the University of Manchester. The first years were difficult, as Bragg was inexperienced and he had to handle classes full of war veterans. Gradually, however, he built up a fine research facility, concentrating on improving methods of using x rays to determine crystal structure. He published a list of atomic sizes and measured absolute intensities of x-ray reflections. He then turned to the silicate family of minerals, whose complex structure had proved elusive. This work, completed around 1930, was of fundamental importance to the science of mineralogy. In 1927 Bragg spent four months at the Massachusetts Institute of Technology as a visiting professor, lecturing on x-ray diffraction and crystal structure. During the 1930s Bragg oversaw and encouraged investigations into metals and metal alloys, which provided new basic knowledge about the chemistry of metals, and he encouraged the application of these x-ray diffraction techniques to industrial firms in northern England, where the university was located.

In 1938 Bragg left Manchester to became director of the National Physical Laboratory. But a year later Rutherford died, and Bragg was invited to take his place as professor of physics at the Cavendish Laboratory at Cambridge. During World War II, Bragg remained at Cambridge (except for eight months in Canada as scientific liaison), advising the British navy on methods of underwater detection of submarines and assisting with further development of sound ranging. After the war had ended, Bragg organized an international meeting of crystallographers in London, which resulted in the formation of the International Union of Crystallography; he was named its first president in 1948. He also secured funds from British industries to help establish *Acta Crystallographica,* a scientific journal first published in 1948 and devoted to the new crystallography.

Although in his later years Bragg was more involved with administration than with hands-on, day-to-day scientific research, he continued to have a powerful influence on the development of x-ray analysis. He organized and found resources to support the very challenging work of Max Perutz, later joined by John C. Kendrew, in investigating globular proteins, molecular structures that contain many thousands of atoms. This work dramatically culminated in the structural analysis of hemoglobin by Perutz and Kendrew and of DNA (deoxyribonucleic acid, the molecule that carries the genetic formula) by Francis Crick, James Watson, and Maurice Wilkins.

In 1954 Bragg became director of the Davy-Faraday Research Laboratory as well as professor of chemistry at London's Royal Institution. He revitalized the laboratory with much-needed infusions of research funds from sources such as the Rockefeller Foundation, the National Institutes of Health, and industrial firms, and by pooling resources with the Cavendish Laboratory. He was actively interested in the Royal Institution's long-standing tradition of sponsoring public lectures on science to nonscientific audiences, and he took particular pleasure in developing a program to bring the excitement of science to schoolchildren. Bragg himself was a popular lecturer owing to his ability to explain complex scientific concepts in clear, simple terms, and he gave a series of scientific talks on television. In the 1960s, at the government's request, he created a series of elementary science lectures to provide civil servants with basic scientific knowledge.

After his retirement in 1966, Bragg continued lecturing, especially to young people, and writing. He completed a definitive book on the history of his field, *The Development of X-Ray Analysis,* only a week before he died on July 1, 1971. In 1921 Bragg had married Alice Hopkinson, the daughter of a doctor. Mrs. Bragg was mayor of Cambridge in 1945, and served on a number of public bodies. The couple had four children, two sons and two daughters. Bragg was knighted in 1941 and was named a Companion of Honour in 1967. He received many awards and honorary doctorates and was a member of numerous scientific academies around the world. His contributions to science were immense, touching as they did on a number of fields in addition to physics, including chemistry, metallurgy, mineralogy, and molecular biology. These contributions came both from Bragg's own direct research and from his notably energetic, effective, and committed organization and leadership of the research efforts of others. In *William Henry Bragg,* Bragg's sister summed up the lives of both W. H. and W. L. Bragg in these words: "To each, science was an art, research an adventure, and life an experimental journey which they lived with enthusiasm."

BRAHE, TYCHO (1546-1601)

Danish-Swedish astronomer and philosopher

Counted among the greatest pre-telescopic astronomers, Tycho Brahe was renowned for his many achievements, including revealing irregularities in the Moon's orbit and developing the wall quadrant and other instruments leading to improved stellar instrumentation. **Johannes Kepler** worked for Brahe and used many of his mentor's observations to construct his laws of planetary movement.

The son of Otto Brahe and Beatte Bille, both from high nobility in Denmark, Brahe was born on December 14, 1546, in Knudstrup, which is now in Sweden. His uncle, Jörgen Brahe, a vice admiral, legally adopted Brahe. By the time he was seven, Brahe was studying Latin to prepare for a career in law. His family sent him to study in Copenhagen, where, at the age of 13, he witnessed a partial eclipse of the **Sun**. Brahe was profoundly impressed that the event was accurately predicted and occurred precisely on schedule. Brahe continued on to Leipzig to specialize in law but became obsessed with **astronomy**. While studying further at the universities of Wittenberg, Rostock, and Basel, Brahe bought books and astronomical instruments with the money he inherited from his uncle, who died from pneumonia after rescuing King Frederick II of Denmark from drowning. Brahe also had a duel with another student in Wittenberg in 1566. Said to be an argument over who was the better mathematician, the duel cost Brahe part of his nose, and he used a metal insert of gold and silver to cover it for the rest of his life.

Brahe was convinced that accurate measurements were fundamental for the future of astronomy. Brahe was especially noted for building new, complex, and exceedingly large astronomical instruments. King Frederick II of Denmark gave Brahe the island of Hveen and funded the building of an observatory, called Uraniburg. From this observatory, Brahe fully embarked on a career that made many contributions to astronomy and physics. He made accurate and continuous measurements of the heavens for 20 years, mapping 777 **stars** and measuring the length of the year to within one second. Brahe is the first astronomer credited with making measurements that accounted for **atmospheric refraction** (a phenomenon that changes the direction of **light** rays as they pass from **space** into the atmosphere and causes celestial object to appear in a location other than their actual ones). Brahe also discovered a variation of the Moon's orbital speed, now known to be associated with the Sun's gravitational pull.

Although convinced that the Polemic/Aristotelian geocentric model (in which the Earth was the center of all celestial motions) was false, Brahe also did not embrace the Copernican heliocentric system (in which the Sun is the center of **motion**). Brahe created his own unique "Tychonic" system in which the Earth remained the center of the **universe**. Essentially, Brahe proposed that the Sun and the **Moon** revolved around the Earth, but that the other **planets** revolved around the Sun. Brahe eventually employed the German astronomer **Johannes Kepler**, who used some of Brahe's work

Tycho Brahe.

to refute the geocentric model and show that the planets move in elliptical orbits around the Sun.

Brahe lost royal support and was forced to leave his island in 1597. Afterward, Brahe moved to Prague and became imperial mathematician to the Holy Roman Emperor Rudolf II. Fortunately, Brahe had built all his instruments so that they could be dismantled and transported. His famed observatory on the island burned within a few years of Brahe's departure. Brahe died on October 14, 1601, in Prague. Reportedly, Kepler wrote down his mentor's last words, *"Ne frustra vixisse videar"* ("May I not seem to have lived in vain").

See also Atmospheric refraction; Cosmology

BRIDGMAN, PERCY WILLIAMS (1882-1961)

American physicist

Percy Williams Bridgman was an experimental physicist whose principal focus was on developing apparatus for producing high pressures and on measuring the effects of high pressures on materials. His work in high-pressure physics won him the Nobel Prize in 1946. The results of his work continued to have implications for such diverse fields as solid-state physics, geophysics, and **cosmology**.

As an academic concerned with doing experiments rather than constructing theories, Bridgman made a number of discoveries that had profitable applications. His work on the electrical and thermal conductivity of metals became important to industrial metallurgists. The refrigeration industry was intrigued when Bridgman's high-pressure experiments produced "hot ice," a form of water that was solid at eighty degrees centigrade (the normal melting point of water is zero degrees centigrade). Hot ice turned out to be of no practical value, but other entrepreneurs speculated that a similar process might produce artificial diamonds. Bridgman felt that such a process was possible but not practical and did not pursue it in earnest. Later, though, in 1955, much of his work on high-pressure physics became the basis of the development of synthetic diamonds by General Electric.

Percy Bridgman was born on April 21, 1882, in Cambridge, Massachusetts. His father, Raymond Landon Bridgman, was a writer and journalist as well as a devout Protestant. His mother was Mary Ann Maria Williams. Both came from established New England families. Bridgman entered Harvard in 1900. He graduated in 1904 and remained at Harvard to earn his M.A. in 1905 and his Ph.D. in 1908. He joined the Harvard Physics department as a research fellow in 1908 and became an instructor in 1910, a full professor in 1919, Hollis Professor in 1926, and Higgins Professor in 1950. He retired from Harvard in 1954.

Bridgman was known to be shy but tenaciously principled. He had absorbed his father's Protestant values of honesty and hard work to such an extent that one biographer, Maila Walter, called him "a scientific puritan." However, in spite of his upbringing, from an early age he steadfastly rejected religion in any form, although later he did not object to his wife's religious practices or his children's participation in Sunday school. He viewed religion as another form of absolutism and as such an impediment to human progress. So strong were his convictions about religion that in 1948 he declined to participate in a conference on science, philosophy, and religion because the word "religion" appeared in its title.

Bridgman conducted his work by collecting data through measurement and then progressing toward theories. But most of his effort went toward the former. Bridgman took great pride in making things work. He paid scrupulous attention to the details of his experimental apparatus. For example, when nothing of the quality he needed was commercially available, he bored his own pipes for one of his high-pressure experiments. Bridgman did not avoid theorizing, but he preferred the laboratory, where his work was guided more by motor and visual operations than by abstract concepts. Although Bridgman spent his life in academia, he did not consider himself a good lecturer. He worked hard at his presentations, but he often expressed frustration over his apparent inability to communicate well with his students. As a result he much preferred the laboratory to the lecture hall. Nevertheless, Bridgman had considerable impact on many of his students, who included John C. Slater and **J. Robert Oppenheimer.**

Bridgman is best known for his experiments to fashion apparatus strong enough to produce higher and higher pressures. In his early work he attained pressures of up to 6,000 kg/cm^2, which was twice the limit set by nineteenth-century French experimenters. In 1909 he pushed his equipment to the point of collapse at 7,000 kg/cm^2, twice that attained by any other experimenter at the time. Through a combination of meticulous craftsmanship and innovative metallurgical techniques Bridgman soon was able to reach pressures of up to 20,000 kg/cm^2. In his forty years of research in high-pressure physics, Bridgman eventually attained pressures of up to 500,000 kg/cm^2.

Bridgman entered his field of study, high-pressure physics, by accident. In the course of working on his doctorate at Harvard, he stumbled on a form of packing that allowed him to construct an assembly capable of pressures higher than ever attained before. The pressures he could create were limited only by the strength of the metal parts of his equipment. Years later, discussing this discovery in his textbook *The Physics of High Pressure,* he recalled, "The whole high-pressure field opened up at once before me like a vision of the promised land."

As Bridgman developed equipment capable of producing higher pressures, he needed gauges to measure them. Bridgman knew, in fact, that these new pressures were of no use to science unless they could be measured accurately. Bridgman had worked with **pressure** gauges since his student days at Harvard, where they were the subject of his Ph.D. dissertation, "Mercury Resistance as a Pressure Gauge." From this dissertation Bridgman in 1909 wrote three papers, the second of which described the construction of a scale that correlated the **electrical resistance** of mercury with pressure. The gauges he developed could measure pressures up to 7,000 kg/cm^2, at which point the steel used in their construction began to weaken. In 1911 Bridgman developed another pressure gauge, using manganin, an alloy of copper, manganese, and nickel. This gauge could measure pressures up to 13,000 kg/cm^2 and was more accurate than the earlier mercury-based gauge because it had a more linear (direct) response between pressure and electrical resistance.

Bridgman felt he was too busy doing experiments and writing up results for publication to enter the discussions over relativity, which **Albert Einstein** had introduced at the time. But in 1914 the death of a colleague at Harvard forced Bridgman to teach a course in advanced **electrodynamics**, a field that had just been revolutionized by Einstein's work. As a classically trained physicist, Bridgman lacked the background to teach this new material, so he was forced to study it himself. What he learned disturbed him and altered his view of physics.

For Bridgman and most other physicists of his generation, Einstein's relativity rendered obsolete their Newtonian beliefs in absolutes like time and **space**. Bridgman was so troubled by this development that in 1927 he wrote *The Logic of Modern Physics,* in which he confessed his despair over the disarray in the foundations of physics that resulted from relativity. His struggles to reconcile the classical physics of his own training with the revolutionary physics that emerged during his career were protracted and arduous.

Bridgman was able to regain some measure of equanimity when he extracted from Einstein's work a way for physicists to interpret the world by maintaining contact with experience and not by searching for ideals like absolute time. Bridgman called this approach to physics operationalism. It was a method that scientists could use to distinguish what is physically real from what is not. Operationalism required scientists to define things with respect to how their physical qualities are measured. For Bridgman operationalism implied that knowledge was not possible without some human presence.

No sooner had Bridgman regained his balance after the assault by relativity than he was tripped up by another revolutionary idea, **quantum mechanics**. Bridgman felt in particular the effects of Werner Heisenberg's uncertainty principle, which states that it is impossible to know both the exact position and the exact **momentum** of a particle. This principle forced Bridgman to concede that beyond a certain point in nature science has no place. Operationalism did nothing to resolve this dilemma, but as it turned out, the discussions of **quantum mechanics** were indebted to operationalism for their vocabulary.

Even though Bridgman felt he had made peace with relativity, many aspects of it bothered him until his death. In fact, he wrote *A Sophisticate's Primer of Relativity* as a critical examination of special relativity late in his life, by which time the theory had gained general acceptance. In his *Primer* Bridgman spelled out his criticisms of Einstein's theory. For example, he felt that **light** was no more than the illumination of things, not, as relativity said, something that traveled. Bridgman doubted the constancy of the **speed of light**, and he searched without success for ways to verify through experiments Einstein's notions about light.

Bridgman was fiercely individualistic. He valued individual freedom not only in the context of science but also in the broadest areas of life. He believed that society existed for the sake of the individual, not the individual for society. On that basis Bridgman espoused the principle that society should interfere as little as possible with the activities of the individual, a practice that makes the fewest assumptions and creates the fewest expectations. On the other hand, Bridgman felt that the individual must not take more from society than he gives, and, in fact, he should give a bit more just for good measure.

Bridgman believed that science could liberate people and protect their freedom because it is a practice based on intellectual integrity. This practice can help humankind to ignore false gods and subdue base urges and emotions. Science can do this, Bridgman believed, because it rests not on authority but on fact.

As an individualist, Bridgman bristled under any show of authority. During World War I he refused to get caught up in the intense patriotism that emerged in the United States. Maila Walter reports that when he found out that a paper of his was going to be edited, he demanded that the byline read "amplified and edited from the manuscript of P. W. Bridgman."

Bridgman objected most to religious authority. He felt that in yielding to the ethical system of some divine will, peo-

ple avoid responsibility for themselves. Bridgman believed that ethics ought to come not from some absolute authority but from an analysis of all the outcomes of various actions. In an article for *Harper's,* Bridgman wrote that he welcomed the reported decline of religious belief because he saw it as an indication of greater intellectual honesty brought about by the popularization of science.

In the spring of 1961 Bridgman began to suffer from what he thought was a muscular rheumatism. By June he felt worse, and in July he was told that he had a nonoperable cancer in his pelvis. Bridgman believed that his condition would worsen quickly, and he lamented that he would not have enough time to finish the preparation of his complete scientific papers, which Harvard University Press would eventually publish posthumously in 1964. In spite of his pain and despair, Bridgman continued to work until August, finishing *A Sophisticate's Primer of Relativity* and submitting a book review for *Science*. On August 20, 1961, Percy Bridgman died by his own hand at his home in Randolph, New Hampshire. He was survived by his wife, Olive Ware Bridgman, and two children, Jane and Robert.

BROWNIAN MOTION

Brownian **motion** is the incessant random motion exhibited by microscopic particles immersed in a fluid. The effect is named after its discoverer, Robert Brown, a Scottish botanist who first noticed the effect in a suspension of pollen grains in 1827. It was determined quickly by Brown and others that the effect is exhibited by small particles of any material when they are suspended in a liquid or gas, and is not unique to pollen grains or other once-living matter.

We now understand Brownian motion as the result of the many millions of collisions with molecules of the fluid experienced by the particle being observed each second. While the average **velocity** of the molecules in a stationary fluid is zero, the average velocity of the small number of molecules which will collide with the particle at any instant will fluctuate about zero, producing a net **force** continuously changing in magnitude and direction. As a result, each particle executes a highly random motion. Since the path of each particle executing Brownian motion will be different, a theory of Brownian motion must deal with the statistical properties of the paths of a large number of particles. It is found by observation that the average square of the displacement of Brownian particles over any **time** interval is directly proportional to the elapsed time. The proportionality constant is a characteristic of the fluid, the particle size, and the **temperature**.

Albert Einstein, who chose the topic for his doctoral dissertation at the University of Zurich, provided the first detailed explanation of the effect. He also submitted a paper on the subject for publication in 1905, the same year as the publication of his famous theory of relativity. An important result of his theory was a method of measuring the number of molecules in a volume of liquid, by observing the motion of the much larger Brownian particles.

Einstein's paper, and its subsequent validation by the French physicist **Jean-Baptiste Perrin** (1870-1942), for which the latter was awarded the Nobel prize in 1926, is regarded by some as the final proof of the existence of atoms and molecules. By the end of the Nineteenth Century, as a result of **Maxwell**'s kinetic theory of gases and **Avogadro**'s molecular interpretation of **Gay-Lussac's law** of combining volumes for reacting gases, it had been generally accepted that gases consisted of molecules moving rapidly through **space**. Scientists were less certain that the structure of solids and liquids was also atomic in character, since it was conceivable that atoms only formed as a liquid or solid evaporated. Some eminent scientists, including the Austrian physicist and philosopher, **Ernst Mach** (1838-1916) opposed basing physical theory on the existence of unobservable entities. Einstein's work showed that not only was it possible to demonstrate the continuing existence of molecules in the liquid, but that the Brownian particles could be viewed themselves as gigantic molecules in thermal **equilibrium** with the other molecules in solution. In accordance with the fundamental principles of **statistical mechanics**, this means that the average **kinetic energy**, equal to one-half the **mass** times the square of the velocity, for each Brownian particle is the same as the average **kinetic energy** for the molecules of the solution. Because the mass of the Brownian particles is so much greater than that of the liquid molecules, the velocity is smaller, but the difference is only quantitative, not qualitative. The diffusive motion of Brownian particles is of the same character as that of the molecules of the liquid.

The paths of particles exhibiting Brownian motion is of some mathematical interest in its own right. A motion picture of a Brownian particle's path will look pretty much the same if it is viewed at double speed but magnified four times or at triple speed and magnified nine times. The motion is thus *self-similar* and has no single characteristic length scale. Further the path taken by the Brownian particle, on all but the fastest time scales is highly irregular. The trajectory of the particle is constantly changing directions. It is, in fact, one of the objects characterized by mathematician Benoit B. Mandelbrot as *fractal*, that is an object of fractional dimension. This is most easily seen if one thinks of drawing line segments connecting the locations of the particle at equal time intervals. For a particle moving along a smooth curve, the total length of the line segments over a given total time will begin to approach a clear limit as the time interval becomes smaller and smaller. For a Brownian particle, this distance appears to increase without limit, as long as the time intervals are long enough for several atomic collisions to occur. In this sense fractal geometry assigns to the Brownian paths a dimensionality greater than one but less than two.

Brownian motion is very important from the standpoint of theory but it also has practical application in the study of colloidal materials. One of the simplest ways of demonstrating that an apparent solution is actually a suspension of microscopic particles is through **light scattering**. A true solution will scatter light only weakly while a colloidal suspension will display a pronounced Tyndall effect, scattering a concentrated beam of light to the side with an efficiency that is dependent on both the number and volume of the colloidal particles. Because the colloidal particles will also be undergoing Brownian motion, the scattered light will also exhibit a Doppler shift towards highe r or lower **frequency** depending on the distribution of particle velocities. Measuring the amount of light scattering and the range of Doppler shifts together permits accurate determination of particle size. Such information is important in developing better, more stabilized colloidal suspensions for use in the paint, ink, and dairy industries. Noncolloidal suspensions having larger particle sizes, such as fine sand grains dispersed in water, tend to settle out with time.

BRUNO, GIORDANO (1548-1600)
Italian philosopher

Giordano Bruno was a philospher/theologian and an outspoken critic of many of the religious, philosophical, and scientific beliefs of his time. Although he was not a scientist and had limited understanding of disciplines like **astronomy**, he was a strong defender of the heliocentric theory of our solar system as presented by **Nicholas Copernicus**. Passionate and confrontational, Bruno was eventually executed as a heretic.

Bruno was born "Fillipo" Bruno in Nola, near Naples, in 1548 to Biovanni Bruno, a soldier, and Fraulissa Savolino. He took the name of Giordano when he entered the Catholic Dominican order in 1563. He was ordained a priest in 1572. His original views and criticisms of accepted theological doctrines soon garnered him an official accusation of heresy in 1576. He first fled to Rome and then France. At Lyons he completed his work *Clavis Magna* or "Great Key," which described his art of memory-training. Bruno's expertise in memory-training brought him to the attention of many patrons around the world. In 1583 he went to England where, for a short period, he won the favor of Queen Elizabeth. Living in the house of the French ambassador in London, he completed his book *Cena de le Cenert* ("The Ash Wednesday Supper"), in which he unreservedly accepted the Copernican heliocentric model of our solar system, in which the **Sun** is the center of the solar system and not the earth, as was widely believed at the time. He also completed *De L'Infinito, Universo e Mondi* ("On the Infinite **Universe** and Worlds"), in which he stated that the universe is infinite, with an infinite number worlds, and that all the worlds are inhabited by intelligent beings.

Bruno invariably outraged both his patrons and the institutions that supported him. His travels took him to Wittenberg, Prague, Helmstedt, and Frankfurt, where he taught, wrote extensively on philosophy and theology, and lived off the largesse of others. In 1591 he accepted an invitation from a patron to live in Venice. However, he was arrested by the Catholic Inquisition and tried as a heretic. He initially recanted and was sent to Rome in 1592, where he was imprisoned for the next several years. When he refused to recant his beliefs permanently, he was burned at the stake in 1600.

See also Nicholas Copernicus

Statue of Giordano Bruno.

BUBBLE CHAMBERS

The bubble chamber is a device that allows physicists to see particle tracks directly in nuclear collision experiments. It was most widely used from the mid-1950s to the 1970s but is still useful because it helps scientists to visualize individual particle interactions.

In 1952, **Donald Glaser** invented the bubble chamber; he won the Nobel Prize for Physics for his invention in 1960. His prototype used xenon in its chamber, while later models used liquid hydrogen, propane, and freons (such as CF_2Cl_2 and CF_3Br) instead.

In an active bubble chamber, these liquids are superheated: heated to beyond their usual **boiling** point under high pressures, then quickly reduced to lower pressures. This leaves them extremely unstable, and bubbles will form anywhere a positive ion occurs. Researchers then shoot particle beams into the liquid chamber and take photographs, using the bubbles to track the paths of positive **ions**. Some bubble chambers can measure as large as 3.7 m in size and contain up to 30 m³ of liquid.

The bubble chamber has several advantages in collision research. It has excellent resolution, is truly a three-dimensional detector, and can be used for long periods of **time**. Magnetic fields can easily be set up to run through the chamber, allowing physicists to consider electromagnetic properties of the particles. Experiments also can be repeated for thousands or millions of photographic images with very little **energy** and cost in resetting the observation chamber.

The bubble chamber does not enjoy the popularity it did in the 1970s, mostly because of one limitation: the liquid in the chamber is both the target and the observation device. Therefore, experiments are limited to targets which can clearly show bubble trails. Within this limitation, however, the bubble chamber is one of the most versatile tools of **nuclear physics**.

BUCKMINSTERFULLERENE

Buckminsterfullerene (also known as buckyballs) are one of a class of hollow, aromatic **carbon** compounds constructed of carbon atoms arranged in 12 pentagonal and variable numbers of hexagonal faces.

The non-metallic element carbon, atomic number 6, is a highly versatile element that can exist in many solid forms, called allotropes. Graphite, for example, is a relatively soft allotrope of pure carbon that is used in lead pencils. Diamond is another allotrope of pure carbon. Diamond, the hardest substance known, is a colorless crystalline form of solid carbon that is a famed, highly valuable, gem widely used in industry and jewelry. A third, allotrope of solid carbon is comprised of a class called fullerenes. Fullerenes are cage-like, hollow configurations of chemically bonded carbon atoms. Shaped into hollow spheres, the carbon atoms of fullerenes are arranged into alternating, conjoined hexagons and pentagons that create faceted, globular surfaces resembling soccer balls. Fullerenes containing from 32 to 600 carbon atoms have been detected. The most stable fullerene is a sphere containing 60 carbon atoms termed Buckminsterfullerene

Buckminsterfullerene are fullerenes specifically composed of 60 carbon atoms. They are found in the soot that results from burning carbon-containing organic materials in the presence of too little **oxygen**. Buckminsterfullerenes or buckyballs, were first discovered in 1985 by the Nobel Prize winning scientists Richard E. Smalley, Robert F. Curl, and Harold W. Kroto. Buckminsterfullerenes were later isolated in pure form in 1990. The structure of this allotrope of carbon resembles the structure of geodesic domes and so was named after the architect who invented the domes, Buckminster Fuller. Currently, chemists are developing practical applications for the durable and highly stable buckminsterfullerenes. Potential uses include use in superconductor materials and in drug-delivery systems, where an active drug can be attached to the inert and stable buckyball for delivery to a target tissue.

See also Carbon; Molecular structure

BUOYANCY, PRINCIPLE OF

The principle of buoyancy is called Archimedes's Principle, since it was discovered by this Greek mathematician in the third century B.C. The principle states that the buoyant **force** acting on an object placed in a fluid is equal to the **weight** of the fluid displaced by the object. An object completely

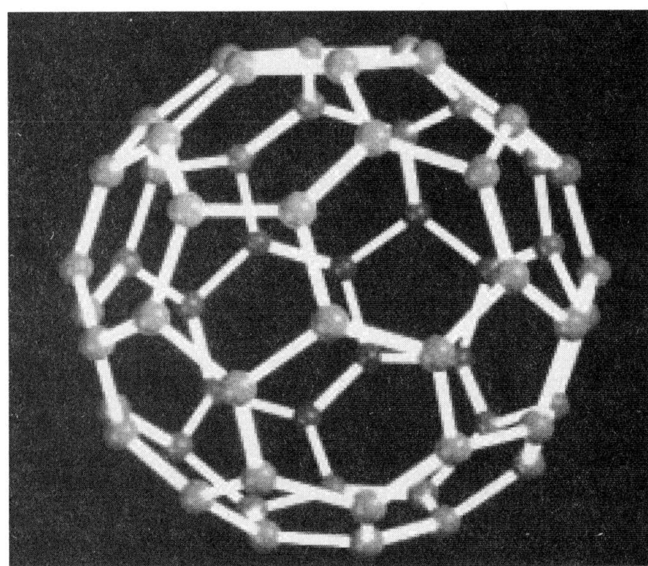

Computer simulation of the atomic structure of Buckminsterfullerene. (Photo by J. Bernholc et al, North Carolina State University. Photo Researchers, Inc. Reproduced by permission.)

immersed in a fluid (liquid or gas) displaces a volume of fluid exactly equal to the volume of the object. The weight of that volume of displaced fluid is the buoyant force acting on the object.

Fluids such as water or air exert **pressure** in all directions and the amount of pressure depends on the depth of the fluid. The pressure on the bottom of an object immersed in a fluid will be greater than the pressure on the top of the object. The imbalance of pressure acting on the object creates an upward force called the buoyant force. If the buoyant force is greater than the weight of the object, the object will float. If the buoyant force is less than the weight of the object, the object will sink in the fluid.

The **density** of a fluid is its weight per unit of volume. Liquids and gases exhibit widely different densities. The buoyant force, or the weight of the volume of displaced fluid, will depend on the density of the fluid as well as the displaced volume. Fresh water has a density of 62.4 pounds per cubic foot (pcf), salt water density is, on average, 64 pcf. Air at sea level has a density of 0.08 pcf and at 10,000 ft (3,050 m), 0.06 pcf. Salt water is denser than freshwater because of its salt content, and, as a result, a swimmer is more buoyant in the ocean than in a freshwater lake. The density of salt water depends on its salinity and varies around the world. The **molecular structure** of water expands when it freezes, therefore, ice is less dense than liquid water. As a result, ice cubes float and lakes freeze from the top down rather than the bottom up.

Steel has a density of 487 pcf, about eight times that of water. Steel boats float, however, because they are hollow and shaped to displace a volume of water that weighs more than the boat's weight. Ships are often rated by their displacement. Displacement is measured in units called tons which are the weight of the water displaced by the ship. As a ship is loaded with cargo, it settles deeper into the water. This displaces an additional volume of water and produces the greater buoyant force required to support the added load. Plimsoll marks are painted onto the hull of cargo ships to indicate the depth to which the ship could be loaded. The different marks refer to fresh and salt water and to the various seasons where **temperature** also effect water density.

Fish can alter their buoyancy by changing the volume of their internal swim bladder. Scuba divers can inflate their external buoyancy compensator vest to change its volume. Both of these changes alter the amount of displaced water and, thus, the buoyant force acting on the body. With this control, divers and fish can ascend or descend at will as they observe each other at play.

The principle of buoyancy applies to all fluids, including gases. A blimp is filled with very light **helium** gas with a density of 0.01 pcf. As a result, the weight of the blimp is less than the weight of the air that it displaces and the blimp will float in air. By dropping ballast or venting helium, the blimp can control its buoyancy and, thus, its altitude. A hot air balloon gets it buoyancy because hot air is less dense than cold air. The density of air at 200°F (93°C) at sea level is 0.06 pcf and serves the same function as the light helium gas in the blimp. Although highly flammable, hydrogen gas is much less dense than helium and was used for lift in dirigibles up until 1937, when the German airship *Hindenberg* burned and crashed.

C

CALORIC THEORY

In the seventeenth century, scientists held a clear association between **heat** and **motion** of constituent particles. Heat became recognized as a fluid that flowed from hot objects to cold ones. During **Galileo**'s time, this heat fluid was known as phlogiston and was considered the soul of **matter**. Phlogiston had **mass** and was released by or absorbed by an object when burning. In the late eighteenth century Antoine Lavoisier refined the view that heat was a liquid, overthrew the current phlogiston theory, and developed the caloric theory of heat. In 1787 Lavoisier coined the term caloric to represent the heat fluid. Caloric was thought to be massless, colorless and conserved in total in the **universe**. The individual particles making up the fluid were elastic and repelled each other but were attracted by particles of other substances, the magnitude of the attraction being different for different substances. It was thought that caloric could be "sensible" in that it diffused among the particles of the material it was acting upon thereby surrounding each particle with an atmosphere of caloric. It could also combine with the particles of the material in a manner similar to chemical combinations and be "latent." By the beginning of the nineteenth century most scientists accepted the caloric theory as the correct theory of heat. The theory offered many plausible explanations regarding **heat transfer** where other theories had failed. It was a simple theory and had many successful applications. These facts made the theory widely accepted but also made it extremely difficult to overthrow.

Because the caloric theory was so powerful, it took some 50 years to overthrow. A dispute rumbled concerning whether caloric had **weight**. Benjamin Thompson showed that cooling and heating of a substance had no detectable effect on its weight. He studied the heat produced by **friction** in boring of cannons in 1798 and showed that just as much heat was produced when a blunt boring tool was used and no metal was cut as when a sharp instrument was employed. It appeared that the heat produced by friction was inexhaustible, and, therefore, not a conserved quantity as required by caloric theory. Thompson continued his attacks on the caloric theory into the early nineteenth century. Eventually, in the 1840s, James Joule explained the source of heating in Thompson's experiments and recognized that heat is another form of **energy**, resulting from the motion or **kinetic energy** of atoms and molecules. This led to the downfall of the caloric theory and formation of the principle of **conservation of energy**, and the kinetic theory of heat.

CALORIMETRY

Calorimetry is the experimental process of studying the conservation of **heat energy** in closed systems. According to the laws of **thermodynamics**, when **matter** is transformed through chemical or physical reactions, energy is released as heat or **work**. Calorimetry provides a measurement of the released heat.

For example, imagine the steam from a kettle operating a small paddlewheel. Some of the heat is retained by the water in the kettle and the escaping steam, some is absorbed by the surrounding air from the steam, and some is translated into the work necessary to move the paddlewheel. If you could measure the heat in the water and the heat absorbed by the air, you would find the sum of all the heat equaled the energy invested (supposedly by the stove) plus the energy transformed into work. Still, how could you measure the energy from all those sources? In the natural world, energy not retained by the transformed matter is absorbed by the environment too readily to be measured precisely. Through calorimetry, the environment of an experiment is isolated and controlled enough to be able to identify all sources of input (specifically, energy) and to measure precisely all necessary output, such as change in heat, transfer of energy, effects of heat, etc. The controlled environment through which the transfer of heat is measured is called the calorimeter.

Most calorimetric studies involve a calorimeter that consists of a container that, when closed, is completely insulated from outside heat. This container has enough room inside for a beaker or other vessel to hold the experimental material. The calorimeter also includes a thermometer that can be read from

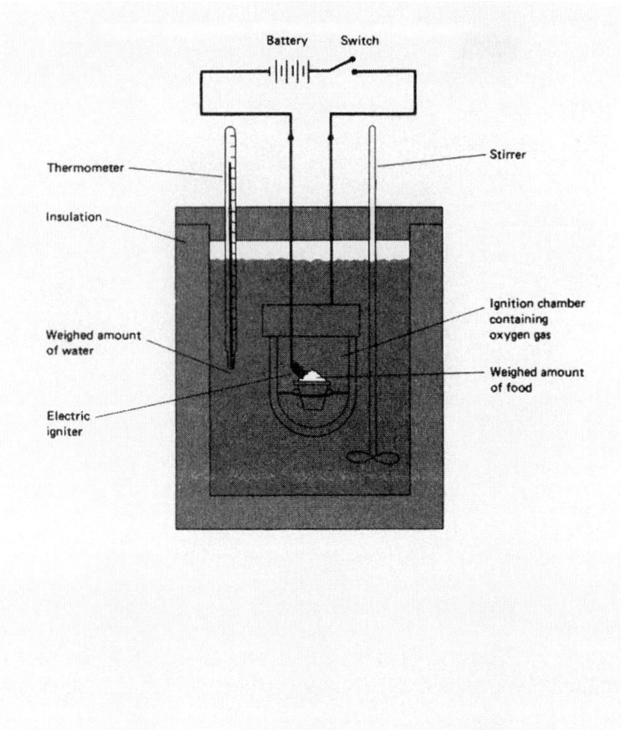

Diagram of a calorimeter. *(Image by Robert L. Wolke. Reproduced by permission.)*

the outside. The calorimeter serves as the controlled environment for the experiment. The goal of the controlled environment is to assure that conditions essential to the study of heat and **heat transfer** are measured and regulated. The **temperature** of the environment and equipment is recorded at the outset of the experiment, and the **pressure** of the environment is kept at a controlled level. The experimenter must also monitor the input and output energy. The only input energy should be that producing the heat as an independent variable. The output energy could be measured as heat or mechanical work. Whereas most calorimeters are designed to maintain a constant environmental temperature, the Bunsen ice calorimeter uses an encasement of ice for a low starting temperature. The substances placed into the calorimeter then heat up the ice. The volume of the melted ice is included as a measure of the heat transfer.

CAPACITANCE

Capacitance is an electrical effect that opposes change in **voltage** between conducting surfaces separated by an insulator. Capacitance stores electrical **energy** when electrons are attracted to nearby but separate surfaces. The voltage across an unchanging capacitance value will stay constant unless the quantity of charge stored is changed.

The unit of capacitance is the Farad, in honor of **Michael Faraday**'s work with electrostatics. When a 1-Farad capacitance stores 1 Coulomb the result will be 1 volt. The Coulomb

is the basic unit of electrical charge, equal to 6.2422×10^{18} charges the size carried by an **electron** or by a **proton**.

An electrical component that introduces capacitance is called a capacitor. Practical capacitors may have as small a value as a few trillionths of a Farad or as large as several Farads.

Work is performed to accumulate charge in a capacitor. Each additional electron stored must overcome the repelling **force** caused by the charge previously stored. Energy storage increases as the square of the voltage across a capacitor. This often considerable energy can be used later.

Capacitors used as energy reservoirs can deliver powerful **pulses** of energy. A capacitor can discharge quickly then slowly recharge until the next **power** demand. The power source needs only to be large enough to supply the average energy. Inexpensive audio amplifiers often use large capacitors to provide high power peaks required by occasional loud sounds. Quiet intervals allow the capacitor to recharge before the next power burst.

A capacitor effectively conducts alternating current even though electrons do not cross from one plate to other plate. Alternating current that appears to pass through a capacitor is actually, the charge and discharge current resulting from the constantly-changing voltage across the capacitor.

An uncharged capacitor always appears as a short circuit because its voltage must equal zero when its stored charge is zero. A capacitor carrying an alternating current continually charges and discharges, spending much of the time in a near-zero charge state. The resulting low voltage across its terminals means that it is often less significant in limiting circuit than other components in the circuit.

A capacitor's opposition to alternating current is called reactance. Higher capacitance introduces less reactance and higher frequencies result in lower reactance.

In a direct-current circuit a series capacitor will permit only a single pulse of charging current when the circuit voltage is changed. The charging current in quickly falls to almost zero as a capacitor charges from a constant-voltage source. Capacitors are sometimes used in circuits to oppose direct current. They may block direct current while simultaneously passing a superimposed alternating currents. A blocking capacitor is commonly used to separate alternating and direct current components.

Dielectrics are the insulating materials used between the conducting plates of capacitors. Dielectrics increase capacitance or provide better insulation between the plates. Dielectrics materials exhibit very little ability to conduct **electric charge**. Mylar, paper, mica, and ceramics are commonly-used dielectrics. When extremely-high capacitance is required, a thin film of aluminum oxide on etched aluminum plates is used as a dielectric.

Dielectrics have a property called polarizability. A dielectric placed within an electric field appears to have electric charge on its surfaces even though the insulator remains electrically neutral. Each of the dielectric's molecules is stretched when the electric field causes its negative charges to be pulled toward the positive-charged capacitor plate and the molecule's positive charges are pulled toward the negative

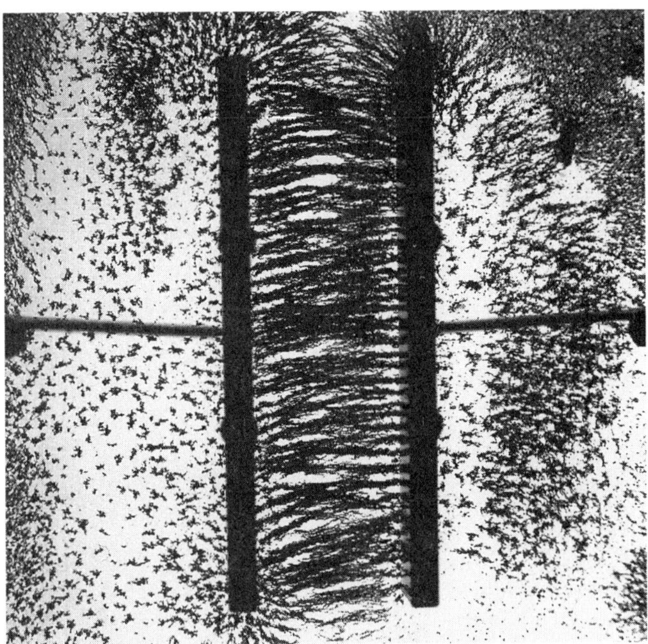

Electric field lines between plates of capacitor. *(Image by Yoav Levy. Phototake, New York. Reproduced by permission.)*

plate. This polarization strain causes each dielectric molecule to act as a voltage source. These voltages add in series aiding as do the voltage from several cells making up the battery in a flashlight. A phantom charge appears on each surface of the dielectric canceling much of the electric field produced by the real charges. The greater the polarization developed by a dielectric the larger the quantity of real charge the capacitor must store to develop a given voltage. The capacitance appears to increase as a result of dielectric polarization.

The capacitance multiplier for any dielectric is called its dielectric constant. The dielectric constant of a perfect **vacuum** is defined as exactly 1. Common dielectrics have dielectrics constants in the range of 2-4. Using a higher quality dielectric increases the capacitance by a factor equal to the dielectric constant.

Dielectric strength is the measure of a dielectric's ability to resist electric stress without losing its insulating capabilities. A high dielectric constant does not always correspond to high dielectric strength. Distilled water has a fairly high dielectric constant but it has poor dielectric strength. Water, therefore, is not a useful dielectric for capacitors because it breaks down too easily. Some ceramics have dielectric constants as high as 10,000. These materials would be extremely valuable if they had better dielectric strength.

Working voltage

If the voltage across a capacitor is increased until charges jump from one plate to the other, the capacitor will probably fail, either momentarily or permanently. Capacitors are rated to specify the maximum continuous voltage that can be applied across the dielectric before the capacitor will fail.

Failed capacitors are a common cause of electronic-equipment breakdowns. When a capacitor's dielectric is destroyed the resulting short circuit may cause other components to fail. Capacitors also develop open circuits, causing the loss of the capacitance.

Electrolytic capacitors are generally less reliable than other types, a tradeoff made to secure very-high capacitance in a small package. They tend to fail if stored without a voltage across their terminals. The electrolytic paste may dry in time, causing a loss of capacitance. Experienced electronic technicians consider electrolytic capacitor failures as a likely cause of an equipment fault that is not otherwise immediately obvious.

Capacitance, inductance, and resistance are the passive electrical properties affecting electrical circuits. Understanding capacitance is an essential part of the study of **electricity** and **electronics**.

CARBON

Carbon, is a element that forms diamonds, graphite, coal, and is the essential element around which all organic molecules are constructed. Carbon atoms are a product of solar **helium** fusion and are expelled into **space** during supoernova explosions. Carbon atoms contain six protons (i.e., an atomic number of 6). The element carbon is designated by the elemental symbol (C). Although the average atomic **weight** of all carbon isotopes is 12.01115 amu, the prevalent **isotope** of carbon, carbon-12 (i.e, carbon atoms containing 6 protons and 6 neutrons) is, by definition, 12.0000 amu and is the standard by which all other atomic masses and weights are determined.

Although carbon is not abundant in the crust of the Earth it is found dissolved in limestone and in water. Ironically, carbon comprises more than 17% of the human body by weight, second only to **oxygen** atoms. Combined with oxygen, carbon form oxides which are abundant in the atmosphere as carbon dioxide and carbon monoxide.

Carbon is the least metallic element in Group 14 of the **periodic table** and is considered a non-metal. Because of the number and configuration of its outer (valence) electrons, carbon exhibits great versatility in forming chemical bonds. Hybridized **electron orbitals** determine the particular bonding geometry and properties of carbon compounds. Among a variety of shapes, carbon atoms can form four single bonds in a tetrahedral structures, double bonds (2 covalent bonds) with other carbon atoms to create molecules of fixed orientation that have important stereochemical properties, or triple bonds. Carbon atoms can also bond with other carbon atoms in rings or arrange into hollow multifaceted spheres with pentagonal and hexagonal faces (buckyballs).

Major allotropes of carbon include graphite and diamonds. Graphite is an important industrial lubricant and is also used in pencils. Diamond, the crystalline form of solid carbon, is the hardest mineral and is a highly valuable gem widely used in industry and jewelry.

Combined exclusively with hydrogen, carbon forms hydrocarbons. Important hydrocarbon fuels include natural gas, petroleum, and coal. Carbon is also found in coke, which is important making steel from iron. Additionally, carbon is present in very important inorganic molecules like carbon dioxide, a product of respiration, and in poisonous carbon monoxide, an important industrial reducing agent that is also found in automobile exhaust and cigarette smoke.

The importance and diversity of carbon-containing organic compounds cannot be overstated. Biological sugars, DNA, and amino acids are just a few examples of organic, carbon-containing molecules.

See also Buckminsterfullerene; Carbon-nitrogen-oxygen process; Nucleosynthesis, Stellar evolution; Stellar life cycle

Carnot, Nicolas Léonard Sadi
(1796-1832)
French physicist and engineer

Nicholas "Sadi" Carnot studied the process by which **heat** converts into other forms of **energy**. His unique mathematical theory pertaining to the conversion relationship between heat and work (energy) preceded and predicted the first and second laws of **thermodynamics**. A major interest to Carnot was industrial development and, in particular, the steam engine. He focused much of his research attempting to improve its effectiveness and efficiency, studying the relationship between **pressure**, **temperature**, and energy output.

Carnot was born in Palais du Petit-Luxembourg, the eldest son of Lazare Carnot, a French revolutionary and Napoleon's minister of war. In 1807, Lazare resigned his post to devote his time to educating his two sons, Sadi (named after a medieval poet and philosopher Sa'di of Shiraz) and Hippolyte. Carnot was an exceptional student and entered the Ecole Polytechnique at the age of 16, the youngest age for entry into the elite institute. Carnot interrupted his studies when he and several other students volunteered to join Napoleon in the ultimately unsuccessful battle to defend Vincennes. He graduated in 1814, ranked sixth in his class, and embarked on a two-year course at the Ecole du Génie at Metz in military engineering. When Napoleon was again defeated after returning from exile to assume what became known as his "Hundred Days Rule," Carnot's father was exiled and lived in Germany for the remainder of his life.

Remaining in France as a second lieutenant in the Metz engineering regiment, Carnot was transferred by the military from place to place on different assignments, which seriously interfered with his intellectual scientific activities. Tiring of the humdrum assignments, he sought relief when, in 1819 he, sat for and passed a difficult exam that enabled him to join the General Staff Corps in Paris. Immediately, he took leave with half pay, moved into his father's empty apartment in Paris, and began taking courses at the Sorbonne, the College de France, and other prestigious institutions. He developed a keen interest in industry and, in particular, the steam engine, which ultimately led him to the study of gases.

Nicolas-Leonard Sadi Carnot.

In 1821, Carnot took a trip to visit his brother and exiled father in Magdeburg where, three years earlier, the town's first steam engine became operational. Upon his return to Paris, his excitement for steam **power** ultimately led to a manuscript, *Réflexions*, the only work published in his lifetime. This 118-page essay gave a succinct report of the "industrial, political, and economic importance of the steam engine." In it, he also introduced his concept of the ideal engine, which he called the "Carnot engine," the idea of reversibility, "that motive power can be used to produce the temperature difference within the engine." He also addressed two significant areas which remained elusive in the advancement of the steam engine: 1) Is there a limit to the motive power of heat, and 2) Do agents exist which are more capable of producing motive power than steam? He perceived his entire essay as a "deliberate examination of these questions."

To answer these questions, researchers had been using (unsuccessfully) two methods: 1) Empirical observation of the relationship between fuel input and energy output of individual engines; and 2) Application of mathematical analysis of gases to abstract engine functions. Carnot's methodology was radically different and became the focal point of what would ultimately become his important contribution to the development of thermodynamics. In *Réflexions*, he set out three premises: 1) The impossibility of perpetual **motion** in relationship to **mechanics**; 2) His "caloric" theory of heat, which proposed that heat (*calorique*) must be viewed as a "weightless fluid that could neither be created nor destroyed in any process ... and that the quantity of heat absorbed or released by a body depends only on the initial and final states of the body;" and 3) Whenever a temperature difference exists, motive power can be produced ... by the movement of caloric from a warm body to a cold body.

Réflexions was released by the foremost scientific publishing company in Paris in June 1824 and received positive acclaim upon its presentation at the Académie des Sciences in July. However, a critique published sometime later in the *Revue encyclopédique* focused only the application of Carnot's methodology in relation to **steam engines**; it would be years before the originality of his mathematical reasonings in heat-to-energy transference would be recognized and appreciated.

In 1831, Carnot turned his primary focus to studying physical properties of gases and vapors in relation to temperature and pressure. His work was tragically cut short when, in June 1832 he developed scarlet fever. Because of his weakened state, he died on August 24, 1832, just 24 hours after contracting cholera. Due to the practice of burning all personal effects of those suffering with the disease, most of Carnot's research notes were lost. His only existing writings include *Réflexions*, some other manuscript notes (which years later became noted as a major work), fragments of mathematics and physics lecture notes, and "Recherche." This latter piece was a 21-page manuscript compiled around 1823, the earliest of his written works, in which he was seeking a mathematical expression for motive power as well as a general solution to encompass the heat/energy **dynamics** for every type of steam engine. Although never published, its careful preparation in comparison to his notes indicated he intended it to be so. This manuscript remained undiscovered until 1966.

Following his death, Carnot's work received brief attention in 1834 then virtually none until William Thomson (1824-1907) who, in 1850, working directly from *Réflexions*, confirmed some of Carnot's theories. Extensions of these theories by Thomson and others ultimately led to the development of the **second law of thermodynamics**.

Twenty-three sheets of rough and disjointed manuscript notes, not discovered nor published until 1878, contained details of a new theory for **kinetic energy**. Carnot was obviously aware that this theory refuted some theories developed in his *Réflexions*; however, undaunted, he calculated a "conversion coefficient for heat and work and went on to assert that the total quantity of motive power in the **universe** was constant." These hypotheses hinted at almost all the basic theories that would ultimately become the **first law of thermodynamics**.

CATHODE AND CATHODE RAYS

A cathode is the negative electrode of a battery or electrolytic capacitor. A cathode is also the reduction electrode in an electrochemical cell. Cathode rays are beams of electrons that emanate from a cathode and are most commonly produced by heating a filament in a **vacuum** tube or by bombarding a target with positive **ions**.

Cathode ray tube.

Metals are excellent conductors of **electricity** because the outer orbiting electrons (negative charges) in the metal **atom** are very loosely bound to the **nucleus**. A traditional and still useful conceptual model, depicts a metal as containing a cloud of electrons with the ability to migrate throughout the metal. Accordingly, any electrical **force** (**voltage**) applied to a metal will cause this cloud of electrons to move in response to the force. In contrast, insulators are materials (e.g., rubber and plastic) in which there are few free electrons. Without these responsive electrons, electrical force has little or no effect. Electrons, usually do not actually leave the metal. If the metal is heated, however, some of the electrons become energetic enough to be expelled from the metal. The hotter the metal the greater the number of electrons exuded from the surface. If the metal is placed in a vacuum the stream of electrons meets little resistance (e.g., from air molecules) and the stream or cloud of exuded electrons is free to move in response to electrical and gravitational fields. Because electrical forces are vastly stronger than gravitational forces, gravitational effects can usually be ignored and the drift of the **electron cloud** or direction of movement of the **electron** stream is controlled solely by the application of electromagnetic fields.

The cloud of electrons causes a green glow to appear on the glass tube wall near the cathode. The glass fluoresces because of the bombardment by the electrons. When first discovered, researchers thought that the bombarding rays must have been emitted from the cathode, and termed these cathode rays. Within a year or two, researchers managed to pull the electrons in a constant stream from the cathode and across a **space** in a glass vacuum tube. The cloud of electrons can be induced to move away from the hot electron exuding metal by placing a positively charged object (**anode**) nearby. The electrons will cross the vacuum with **velocity** directed toward the anode with a speed directly proportional to the strength of the positive charge on the anode.

The first to realize that the cathode rays were, indeed, streams of charged negative particles (electrons) was the English physicist **William Crookes** (1832-1919) who demonstrated that cathode rays could be deflected by a magnet. In 1897, English physicist **J. J. Thomson** (1856-1940) demonstrated that the cathode rays were attracted to positively charged bodies. These two experiments proved beyond doubt that cathode rays were negatively charged particles, and in 1906 Thomson received the Nobel Prize for establishing the existence of the electron.

Cathode rays (electron streams) are utilized in many electrical devices including television screens, computer monitors and x-ray machines. Before the advent of **transistors** and integrated circuits (chips) electrical devices relied upon glass vacuum bottles called **vacuum tubes**.

See also Electric charge; Electric field and forces; Vacuum

CAUSATION

Often two events are thought of as causally linked. A set of conditions labeled A causes an effect B to occur. From a physical point of view, Newton's second law, F = ma, contains the idea that forces create accelerations. The change in the state of an object results as a consequence of those forces that act upon the object. Something that is not influenced by such forces remains unaffected, but once the object realizes the set of conditions A, it necessarily follows that the action B must occur. In other words, whenever A is satisfied, then the event B cannot fail to occur. An example illustrates the point. Consider a cue ball striking a stationary billiard ball. The cue ball imparts some of its **momentum** to the billiard ball, which was initially at rest, and the second ball starts to roll. The collision of the cue ball is the cause; the **motion** of the billiard ball is the effect. The cue ball cannot collide with the billiard ball and fail to cause it to move. Likewise, a billiard ball at rest cannot move until an external **force** acts to put it in motion.

In classical **mechanics**, the notion of causality is closely related to the idea of **determinism**, which argues that the initial conditions and the forces acting within a system evolve it to its final state. In **quantum mechanics**, this is no longer the case because the uncertainty principle prohibits the exact knowledge of the initial state. A sensible physical theory cannot have events that precede their causes. The maximum speed at which information can be transmitted—the speed of light—constrains how forces can act.

The relation between causes and effects is a significant point of contact between physical science and philosophy.

Philosophers argue that it is our experience that helps us determine how causes and effects correlate. The first ingredient is the concept of universality. Every effect must have a cause. Secondly, the future resembles the past. This notion of uniformity informs us that the laws that describe a physical process tomorrow are the same laws that described a similar physical process yesterday. Our past knowledge of causes and effects serves to predict what effects will follow from future causes. Thirdly, experience tells us which states are impossible (i.e., which events cannot be causally connected). These premises allow us to argue that the set of conditions A is necessary for the event B to take place.

The philosopher David Hume rejects the idea of necessity. He argues that when we see the cue ball hit the billiard ball and then see the billiard ball move, we observe a pair of isolated events. The notion of causality is something extra that we impose from without. The set of conditions A may be "constantly conjoined" with the event B, but it cannot be said that A caused B. Other empiricists adopted the weaker position that causality is necessary only insofar as dictated by the laws of motion. **Immanuel Kant** refutes Hume's ideas and champions a necessary correlation between causes and effects because causality is the organizing principle by which the human mind makes sense of the world. Causality is one of a class of "synthetic" *a priori* propositions.

Many modern descriptions of causality, such as the one presented by Richard Taylor, return to the Aristotelian notion of a cause producing an effect through its efficacy or potency. This line of reasoning permits the differentiation of cause and effect even when the two are simultaneous by selecting an active agent that imparts a change in the state of a passive object whenever certain conditions are satisfied. Each of those conditions is necessary, and together they are sufficient to ensure that the effect occurs. For example, a hand that moves a piece of chalk across a blackboard is an active agent. Provided that certain conditions are true (e.g., the blackboard is not wet), the motion of the chalk causes a change in the state of the blackboard by leaving writing behind.

CAVENDISH, HENRY (1731-1810)

English physicist and chemist

Today, Henry Cavendish is recognized as one of the most brilliant scientists of his day, but his contemporaries knew him to be an extremely unusual person. Throughout his life, Cavendish avoided women entirely; he even went so far as to order his female servants to stay out of his way, or they would be fired. He was so shy he almost never left his house, except to attend the weekly dinners and meetings of the scientific Royal Society. Sometimes, he would hang around outside the meeting room, waiting to slip in when no one would notice him. If a stranger came near him, he quickly walked away. The only picture we now have of Cavendish is a watercolor sketch that was painted without his knowledge. He never changed his style of dress; he wore old-fashioned clothes, which he replaced with an identical outfit according to a regular schedule.

Whatever he lacked in social graces, Cavendish, a Cambridge scholar, made up for in scientific excellence. His laboratories and experiments were all that he cared about, and he managed to take part in the scientific activities of his day even though he preferred to work alone. After leaving Cambridge without getting a degree, which was not unusual then, Cavendish moved in with his father in London, England, where he set up a laboratory and workshop. Cavendish also maintained a large library in a separate building in London, and he had a country house equipped with another laboratory.

In 1766 Cavendish presented a paper to the Royal Society telling how he discovered hydrogen gas, which he called *inflammable air*. Cavendish produced hydrogen by combining metals with acids. For example, he found that hydrogen is produced when zinc, iron, or tin is dropped into hydrochloric acid or a diluted sulfuric acid. Cavendish's inflammable air was given its modern name of hydrogen in 1791 by Antoine-Laurent Lavoisier. Although a few chemists had collected hydrogen previously, Cavendish was the first to distinguish it from ordinary air and to investigate its specific properties. He found that hydrogen was extremely **light** in comparison to other gases. Eventually, this discovery was used to develop hydrogen-filled balloons, which could travel higher and farther than balloons filled with hot air. In some of his most important experiments, done during 1784 and 1785, Cavendish discovered that when hydrogen is combined with air, the mixture explodes and creates water. Cavendish even figured out approximately how much hydrogen and air combine to produce water, which anticipated the development of water's modern chemical formula (H_2O). At that time, many people still believed that water itself was a basic **chemical element**, along with fire, air, and earth.

Cavendish also discovered the inert gas argon while experimenting with nitrogen. Cavendish isolated nitrogen, although he is not credited with its discovery because his work remained unpublished, and then combined it electrically with **oxygen**. When finishing the experiment, he found that about one percent of the original gas was unaccounted for in the reaction. In 1894 this mysterious gas was identified as argon by Sir William Ramsey. Cavendish also discovered the composition of nitric acid. By dissolving the nitrogen-oxygen gas in water, he created nitric acid, proving that it is composed of nitrogen, oxygen, and hydrogen.

In a series of about 400 experiments, Cavendish established that the composition of air is the same regardless of the air's geographic origin. He determined that air contains about twenty-one percent oxygen, and he studied the properties of **carbon** dioxide, comparing its specific properties with those of air. During the same period Cavendish studied the freezing points of mercury and various acids.

Cavendish also performed many important physical and electrical experiments. He was the first scientist to calculate the Earth's mass—and he was right, to within about 10% of the best modern estimates. In this spectacular experiment, Cavendish made the most important advance in the understanding of **gravity** since **Isaac Newton** proposed the *law of gravity* in 1665. Newton's law lacked one key ingre-

Henry Cavendish.

dient—a number called the gravitational constant that represents gravity's **force** of attraction between objects. Without this number scientists could not calculate gravity 's effect on the orbits of the **moon** and **planets**, the ebb and flow of the ocean's tides, or the path of falling objects and projectiles. Cavendish measured gravity's force of attraction, using a method suggested earlier by John Michell (1724-1793). The gravitational attraction between objects that are small enough to work with in the lab is very weak, so Cavendish's equipment had to be very sensitive to detect the force. His equipment consisted of a lightweight rod suspended by a wire; small balls were placed at both ends of the rod. When Cavendish placed larger balls near the small ones, the gravitational attraction between each pair of balls caused the rod to twist. From this he calculated the strength of attraction, which enabled him to solve Newton's equation for the gravitational constant. Once that number was known, Cavendish was then able to calculate the **mass** of the Earth as well as its average **density**. Cavendish had made the law of gravity complete for the first time.

In Cavendish's day, electrical experimentation had become quite fashionable in scientific and popular circles throughout England and continental Europe. Cavendish anticipated many basic concepts that were later developed by **Michael Faraday**, **Charles-Augustin Coulomb**, and others. He explained how charged objects attract or repel each other, and he introduced the idea of **voltage**, or electrical potential. He also showed that different materials conduct different amounts of **electric current** and studied the capacity of those materials to store an electrical charge.

Although Cavendish published two papers on **electricity** during the 1770s, his work remained unknown until his notebooks and manuscripts were discovered and published by **James Clerk Maxwell** in 1879. Much of Cavendish's other scientific research in **optics**, geology, **magnetism**, and pure mathematics was not revealed during his lifetime, but Cavendish did not care whether his research was appreciated or whether he got credit for his work. He studied science to satisfy his own curiosity. When Cavendish died, his relatives left much of his fortune to Cambridge University, which established its Cavendish Professorship and the world-renowned Cavendish Laboratory.

CAVENDISH, DUCHESS OF NEWCASTLE, MARGARET (1623-1673)

English natural philosopher

Margaret Cavendish was one of the first prolific female science writers. As the author of approximately 14 scientific or quasi-scientific books, she helped to popularize some of the most important ideas of the **scientific revolution**, including the competing vitalistic and mechanistic natural philosophies and atomism. A flamboyant and eccentric woman, Cavendish was the most visible of the "scientific ladies" of the seventeenth century.

Margaret Lucas was born into a life of luxury near Colchester, England, in 1623, the youngest of eight children of Sir Thomas Lucas. She was educated informally at home. At the age of eighteen, she left her sheltered life to become Maid of Honor to Queen Henrietta Maria, wife of Charles I, accompanying the queen into exile in France following the defeat of the royalists in the civil war. There she fell in love with and married William Cavendish, the Duke of Newcastle, a 52 year-old widower, who had been commander of the royalist forces in the north of England. Joining other exiled royalists in Antwerp, the couple rented the mansion of the artist Rubens. Margaret Cavendish was first exposed to science in their informal salon society, "The Newcastle Circle," which included the philosophers Thomas Hobbes, René Descartes and Pierre Gassendi. She visited England in 1651–52 to try to collect revenues from the Newcastle estate to satisfy their foreign creditors. It was at this time that Cavendish first gained her reputation for extravagant dress and manners, as well as for her beauty and her bizarre poetry.

Cavendish prided herself on her originality and boasted that her ideas were the products of her own imagination, not derived from the writings of others. Cavendish's first anthology, *Poems, and Fancies,* included the earliest version of her natural philosophy. Although English **atomic theory** in the seventeenth century attempted to explain all natural phenomena as **matter** in **motion**, in Cavendish's philosophy all atoms contained the same amount of matter but differed in size and shape; thus, earth atoms were square, water particles were round, atoms of air were long, and fire atoms were sharp. This led to her

Margaret Cavendish. *(Corbis-Bettmann. Reproduced by permission.)*

humoral theory of disease, wherein illness was due to fighting between atoms or an overabundance of one atomic shape. However in her second volume, *Philosophical Fancies,* published later in the same year, Cavendish already had disavowed her own atomic theory. By 1663, when she published *Philosophical and Physical Opinions,* she had decided that if atoms were "Animated Matter," then they would have "Free-will and Liberty" and thus would always be at war with one another and unable to cooperate in the creation of complex organisms and minerals. Nevertheless, Cavendish continued to view all matter as composed of one material, animate and intelligent, in contrast to the Cartesian view of a mechanistic **universe**.

Cavendish and her husband returned to England with the restoration of the monarchy in 1660 and, for the first time, she began to study the works of other scientists. Finding herself in disagreement with most of them, she wrote *Philosophical Letters: or, Modest Reflections upon some Opinions in Natural Philosophy, maintained by several Famous and Learned Authors of this Age, Expressed by way of Letters* in 1664. Cavendish sent copies of this work, along with *Philosophical and Physical Opinions,* by special messenger to the most famous scientists and celebrities of the day. In 1666 andd again in 1668, she published *Observations upon Experimental Philosophy,* a response to Robert Hooke's *Micrographia,* in which she attacked the use of recently-developed microscopes and telescopes as leading to false observations and interpretations of the natural world. Included in the same volume with *Observations* was *The Blazing World* was a semi-scientific utopian romance, in which Cavendish declared herself "Margaret the First."

More than anything else, Cavendish yearned for the recognition of the scientific community. She presented the universities of Oxford and Cambridge with each of her publications and she ordered a Latin index to accompany the writings she presented to the University of Leyden, hoping thereby that her work would be utilized by European scholars.

After much debate among the membership of the Royal Society of London, Cavendish became the first woman invited to visit the prestigious institution, although the controversy had more to due with her notoriety than with her sex. On May 30, 1667, Cavendish arrived with a large retinue of attendants and watched as **Robert Boyle** and **Robert Hooke** weighed air, dissolved mutton in sulfuric acid and conducted various other experiments. It was a major advance for the scientific lady and a personal triumph for Cavendish.

Cavendish published the final revision of her *Philosophical and Physical Opinions*, entitled *Grounds of Natural Philosophy*, in 1668. Significantly more modest than her previous works, in this volume Cavendish presented her views somewhat tentatively and retracted some of her earlier, more extravagant claims. Cavendish acted as her own physician, and her self-inflicted prescriptions, purgings and bleedings resulted in the rapid deterioration of her health. She died in 1673 and was buried in Westminster Abbey.

Although her writings remained well outside the mainstream of seventeenth-century science, Cavendish's efforts were of major significance. She help to popularize many of the ideas of the scientific revolution and she was one of the first natural philosophers to argue that theology was outside the parameters of scientific inquiry. Furthermore, her work and her prominence as England's first recognized woman scientist argued strongly for the education of women and for their involvement in scientific pursuits. In addition to her scientific writings, Cavendish published a book of speeches, a volume of poetry, and a large number of plays. Several of the latter, particularly *The Female Academy*, included learned women and arguments in favor of female education. Her most enduring work, a biography of her husband, included as an appendix to her 24 page memoir, was first published in 1656 as a part of *Nature's Pictures*. This memoir is regarded as the first major secular autobiography written by a woman.

CELESTIAL MECHANICS

Celestial **mechanics** is the branch of **astronomy** that deals with the theory and application of the laws of **dynamics** and gravitation to the motions of bodies in **space**. The solution of problems involving these motions is the primary goal of celestial mechanics.

The heliocentric theory of Polish astronomer Nicolaus Copernicus, the telescopic observations of Italian astronomer **Galileo** (Galileo Galilei), the positional observations of Danish astronomer **Tycho Brahe**, and the formulation of the laws of planetary **motion** by German mathematician and astronomer **Johannes Kepler** provided the foundation for English physicist Sir Isaac Newton's advancement of the laws of gravitation in his important and influential 1687 **work**, *Philosophiae Naturalis Principia Mathematica (Mathematical Principles of Natural Philosophy)*. In *Principia*, Newton provided the first comprehensive and mathematically consistent explanation of celestial mechanics.

Included in *Principia* were Newton's three laws of motion and a law of universal gravitation. Together, these laws permitted the statement of a problem in celestial mechanics (sometimes called gravitational astronomy) in the form of ordinary differential equations of the second order. These basic equations form the framework of classical celestial mechanics. The actual calculations of celestial motions are often complicated because many independent forces are acting continuously over very long intervals of time, and the bodies under observation are continuously moving.

Newton was the first to analytically solve the two-body problem (i.e., the motion of two bodies under their mutual gravitational attractions and disregarding all other forces). He was able to derive Kepler's laws and thus explain the motions of Jupiter's satellites, the motion of the **Moon** about the Earth, and the motion of the Earth about the **Sun**. In the classical theory of celestial mechanics, only the problem of two isolated moving bodies mutually attracted by gravitation can be solved exactly. The solution to the two-body problem calculates the relative position and velocities of two objects at successive points in time, and yields the **mass** of each object. However, the two-body problem is an idealized situation that does not exist in nature because there are always other gravitational forces due to other material bodies. On the other hand, in many cases (such as binary star systems) the two-body solution adequately describes the motion. When a two-body approach is not applicable (for example, a globular cluster of **stars**) the task of calculating the individual motions of the celestial bodies involved can become extremely complex. In fact, the n-body problem is often said to have no solution because it is not possible to write an equation that will (analytically) describe the orbit of a body in such a situation. In reality, many n-body problems are solved to high precision with the use of elaborate numerical calculations that are very accurate in predicting the actual motions within the system.

A simpler case of two-body motion can be examined: consider an artificial satellite in a circular orbit around the Earth. For a typical communications satellite, its mass is on the order of 10^{21} times smaller than the Earth's. The gravitational effect upon the Earth by the satellite is therefore insignificant, and any **acceleration** of the Earth by the satellite can be neglected. Since the satellite is in a circular orbit about the Earth, its inward acceleration is v^2/r, where v is the satellite's speed and r is the radius of its orbit. (Newton was the first to derive this centripetal acceleration that holds true for any object traveling in a circular path). Equating the centripetal **force** on the satellite to the **gravitational force** producing it yields the following equation: $m_s v^2/r = -GM_E m_s/r^2$, where M_E and m_s are the mass of the Earth and the satellite, respectively, and G is the gravitational constant. The right hand term is the force that the satellite experiences due to its gravitational attraction to the Earth found by applying **Newton's law of universal gravitation**. The negative signs on both sides of the equation show that the forces involved are inward (i.e., attractive). Solving this equation for v we have: $v = \sqrt{(GM_E/r)}$. This simple equation gives the speed of any particle in any circular orbit about a much more massive body as a function solely of the orbit's radius. As noted previously, however, when applied to actual physical problems the previous equation is only approximately true because of the effects of other material objects and forces. For an actual artificial satellite that is orbiting the Earth in a circular orbit, its actual speed will deviate slightly from that calculated with the equation above. This deviation has a number of causes, including (but not limited to) the non-spherical shape and mass distribution of the Earth, gravitational forces due to the Moon and Sun, solar **radiation pressure**, and atmospheric drag (if the orbit is close to the Earth). The effects caused by these sources are called non-central forces and are taken into account by adding perturbation terms to the two-body equations of motion. In other words, in actual satellite operations the two-body model is retained, but modified to obtain an accurate prediction of the satellite's future motion.

It is now recognized that the Newtonian laws of motion and the law of universal gravitation supply solutions that are approximations to the actual motions of celestial bodies. For example, classical theory was unable to fully explain the advance of the planet Mercury's perigee in its orbit about the Sun. Einstein's general theory of relativity, when applied to this problem, predicted a motion for Mercury much closer to observed values. Despite fundamental differences, general relativity, for the most part, reduces in a first approximation to Newtonian mechanics. Most dynamical problems in celestial mechanics, however, can be solved satisfactorily using only classical mechanics. A good example of the continued utility of classical celestial mechanics is in its use with rocketry and artificial satellites. The equations used to predict the motions of these objects rely almost exclusively upon the methods of classical mechanics.

See also Kepler's laws of planetary motion; Newton's laws of motion; Relativity, general

CELSIUS, ANDERS (1701-1744)

Swedish astronomer

Celsius is a familiar name to much of the world since it represents the most widely accepted scale of **temperature**. It is ironic that its inventor, Anders Celsius, the inventor of the Celsius scale, was primarily an astronomer and did not conceive of his temperature scale until shortly before his death.

The son of an astronomy professor and grandson of a mathematician, Celsius chose a life within academia. He studied at the University of Uppsala where his father taught, and in 1730 he, too, was given a professorship there. His earliest research concerned the aurora borealis (northern lights), and he was the first to suggest a connection between these lights and changes in the earth's magnetic field.

Celsius traveled for several years, including an expedition into Lapland with French astronomer **Pierre-Louis Maupertuis** (1698-1759) to measure a degree of longitude. Upon his return he was appointed steward to Uppsala's new observatory. He began a series of observations using colored glass plates to record the magnitude of certain **stars**. This constituted the first attempt to measure the intensity of starlight with a tool other than the human eye.

The work for which Celsius is best known is his creation of a hundred-point scale for temperature, although he was not the first to have done so since several hundred-point scales existed at that time. Celsius' unique and lasting contribution was the modification of assigning the freezing and **boiling** points of water as the constant temperatures at either end of the scale. When the Celsius scale debuted in 1747 it was the reverse of today's scale, with zero degrees being the boiling point of water and one hundred degrees being the freezing point. A year later the two constants were exchanged, creating the temperature scale we use today. Celsius originally called his scale centigrade (from the Latin for "hundred steps"), and for years it was simply referred to as the Swedish thermometer. In 1948 most of the world adopted the hundred-point scale, calling it the Celsius scale.

CENTER OF GRAVITY

In the study of **mechanics** the center of **gravity** refers to the point in an object that moves in a **gravitational field** as though the entire **mass** of the object were concentrated at that point. For other than extremely large objects approaching planetary dimensions, when considering application of forces the object moves as if the **gravitational force** acting on the object was applied at the **center of mass**.

Every object is a collection of an incredibly large number of atoms. Each **atom** is attracted to the center of the Earth by the pull of gravity. The **weight** of any object is the total pull on all the atoms that make up the object. It is usually impossible, however, to analyze the countless forces acting on different atoms in the object. It is much more convenient to think of all the forces acting as though they were applied to one point within the body. The resultant of the application of such forces is the same as if all of the mass of the body were located at the center of mass (i.e., the point of **force** application).

Often, the balancing point of an object is determined by its center of mass and the center of gravity of an object, therefore, is that point in the object through which the total weight (force of gravity) appears to act.

In most cases the center of gravity is considered to be the point in a object about which the weight of the object body is uniformly distributed. It is possible to obtain a reasonable estimate of the position of the center of gravity of an object by experiment. Whenever an object is suspended at a point it will turn about the center of gravity until the center of gravity lies beneath the suspension point. The object can be suspended from various points and at each position a line drawn vertically downward. The center of gravity will be located at the intersection of the lines.

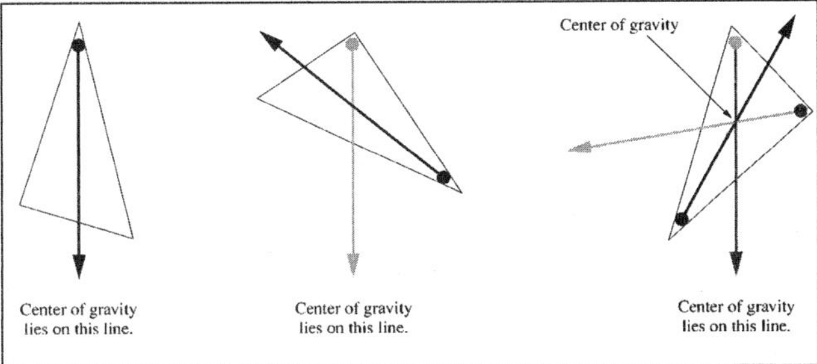

CENTER OF MASS

The center of **mass** of an object is a point where, for purposes of measurements, all of the object's mass is imagined to be concentrated. It is also the point at which an object will balance on a single support or spin around if thrown into the air.

Knowing where the center of mass is located becomes particularly useful when calculating the **motion** of a body as it is acted upon by external forces. A perfectly symmetrical ball thrown through the air will travel in an arc, or parabola. This motion can be easily tracked with the naked eye or mathematically. But, for instance, if we throw a hammer into the air, it will wobble and spin, and appear not to travel in a parabolic motion. If we track the end of the hammer on its journey, it will be mathematically very complicated. There is, however, a location (the center of mass) that follows a smooth parabolic path.

In a perfectly symmetrical object, such as a barbell with equal weights on each end, the center of mass is easy to find: it is in the middle of the bar, at the exact midpoint between the two weights.

If we replace the **weight** on one side with one twice as heavy, the center of mass shifts towards the more massive end of the barbell. The new center of mass is 2/3 of the way from the **light** end to the heavy end. This is the point at which the barbell would balance on a sawhorse. To calculate the center of mass in the lopsided barbell, we find the sum of all the segments of mass of the object multiplied by the position of each segment of mass, divided by the entire mass of the object. The center of mass may not even reside inside the object: It is between the legs of a chair and in the center of a hoop.

CENTRIFUGAL AND CENTRIPETAL FORCE

Centripetal **force** is the central force felt by an object moving in a circle. It is the force that is necessary to keep the object moving in this circle, otherwise, the object would follow a straight line path, in accordance with Newton's Laws.

Assuming this circle is of radius, r, and the object moves with constant **velocity**, v, then the centripetal **acceleration** is:

$$a = \frac{v^2}{r}$$

The centripetal force, Fc, which is directed towards the center of the circle, is then:

$$F_c = ma = \frac{mv^2}{r}$$

Centrifugal force is not a true force at all; it is an effect seen when the observer is assuming a rotating reference frame is an inertial frame. This is done quite often with systems such as the earth, since it is only a factor when the rotation is high in comparison to the size of the system.

To see how this force behaves, consider a two liter soda being carried home in the car. On the way home, you round a sharp left hand curve and the soda bottle slides to the right and crashes into the door. To you, it seems that the bottle has been moved by some invisible force towards the door. Hence we have the fictitious centrifugal force. To an outside observer, the bottle has just obeyed Newton's Law; an object in **motion** will continue to move in a straight line unless acted upon by an outside force. This is merely an application of **inertia**. Thus, since you neglected to restrain the soda, it continued to move in a straight line while you turned. It was the force of the door pressing against it that caused it to move, in the end.

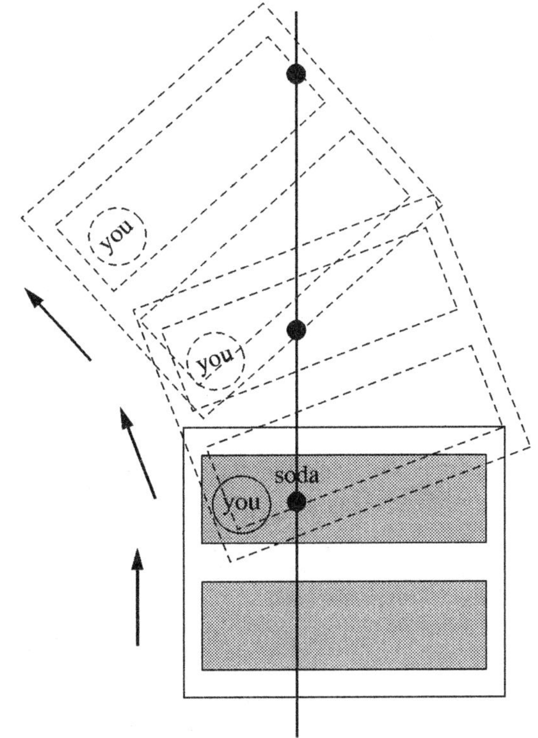

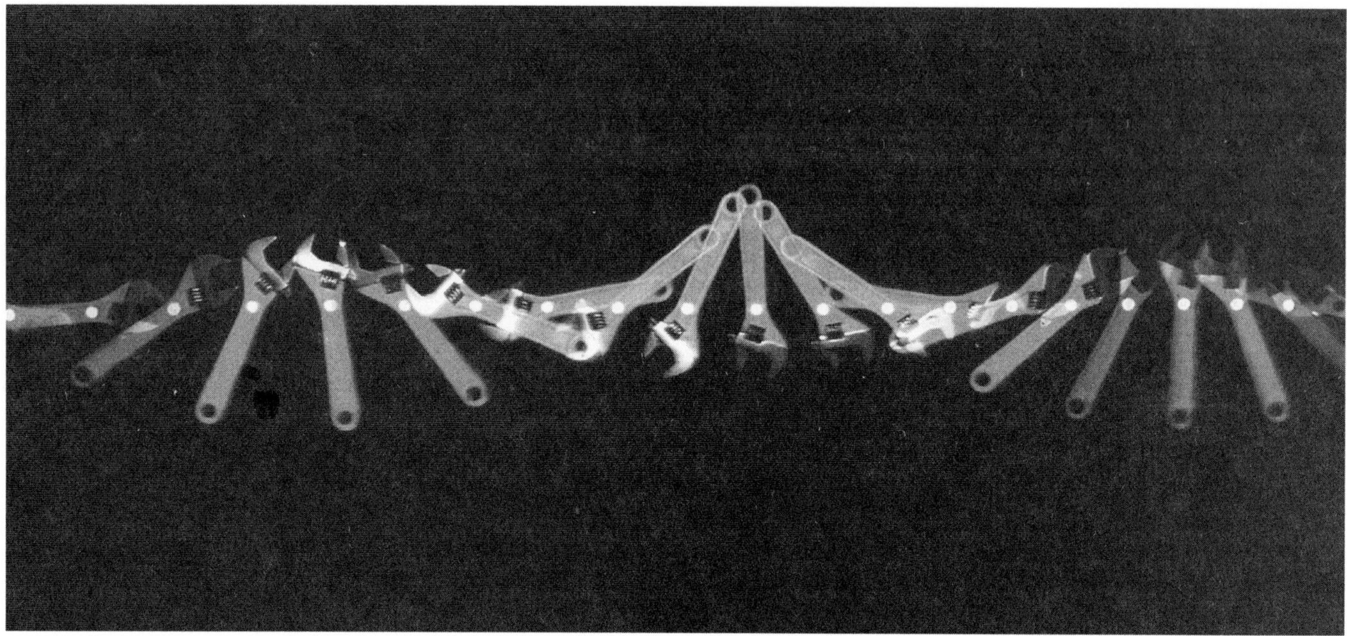

Rotation around center of mass. *(Photo by © 1990 Richard Megna. Fundamental Photographs, New York. Reproduced by permission.)*

Even though centrifugal force isn't a "real" force, it is a real effect and can be applied to our advantage. Consider the Centrifuge apparatus; its use is wide spread. It uses the centrifugal force, which is really an application of inertia, to separate **fluids** into layers such as cream and milk or plasma and serum. The liquid is a solution made up of particles with different masses. Each particle has an inherent amount of inertia proportional to its **mass**. As the liquid is spun, the amount of inertia, or the amount of resistance to force, causes the particles to resist being moved in a circle. Thus, more massive particles, being more resistive to movement off their straight line path end up at the bottom of the vessel while less massive particles end up at top.

CHADWICK, JAMES (1891-1974)
English physicist

James Chadwick, who is remembered principally for his work in **nuclear physics**, received the Nobel Prize in 1935 for his discovery of the **neutron**. That discovery gave rise to new approaches and techniques in the physical sciences, new developments in the biological sciences, and a new form of warfare. Chadwick himself was actively involved with the development of the **atomic bomb**. Although he was shy and did not have the public visibility of other physicists of his time, his work in science, diplomacy, and administration left permanent marks in twentieth-century history. Knighted in 1945, Chadwick received the Medal of Merit from the United States government in 1946, and the Copley Medal of the Royal Society in 1950.

Chadwick was born in Bollington, not far from Manchester, England, on October 20, 1891, to John Joseph Chadwick and Ann Mary Knowles. Chadwick senior owned a laundry business in Manchester. At the age of sixteen, Chadwick won a scholarship to the University of Manchester, where he had intended to study mathematics. However, because he was mistakenly interviewed for admittance to the physics program and was too shy to explain the error, he decided to stay in physics.

Initially Chadwick was disappointed in the physics classes, finding them too large and noisy. But in his second year, he heard a lecture by experimental physicist **Ernest Rutherford** about his early New Zealand experiments. Chadwick established a close working relationship with Rutherford and graduated in 1911 with first honors. Chadwick stayed at Manchester to work on his master's degree. During this time he made the acquaintance of others in the physics department, including **Hans Geiger** and **Niels Bohr**. Chadwick completed his M.Sc. in 1913 and won a scholarship that required him to do his research away from the institution that granted his degree. At this time Geiger returned to Germany, and Chadwick decided to follow him.

Chadwick had not been in Germany long when World War I broke out. Unsure of his plans, he postponed leaving the country. Soon he was arrested and sat in a Berlin jail for ten days until Geiger's laboratory interceded for his release. Eventually Chadwick was interned for the duration of the war, as were all other Englishmen in Germany. Chadwick spent the war years confined at a race track, where he shared with five other men a stable intended for two horses. His four years there were quiet, cold, and hungry. To keep up his morale, he participated in a scientific society formed by a group of internees. He managed to maintain correspondence with Geiger, and he even persuaded his German captors to supply some basic scientific equipment. One such piece he obtained was a bunsen burner. However, it lacked a bellows, so Chadwick enlisted one of his fellow captives to blow air through a tube into the burner. Although the work he did under such harsh conditions was not very fruitful, Chadwick later believed that it helped greatly to keep up his spirits at the time. He also felt that the experience of internment contributed to his maturity. Moreover, when Chadwick returned to England, he found that no one else had made much progress in nuclear physics during his time away.

In 1919 Ernest Rutherford was appointed director of Cambridge University's Cavendish Laboratory, the leading research center for nuclear physics at the time. Rutherford brought Chadwick with him, and the two began a partnership that lasted sixteen years. Chadwick earned his Ph.D. in physics from Cambridge in 1921 and was made a fellow at Gonville and Caius College, enabling him to continue working with Rutherford at the Cavendish lab. Rutherford had been working on **atomic structure** and had discovered many of the essential properties of the **proton**. One of Chadwick's first tasks was to help Rutherford establish a unit of measurement for **radioactivity**, to aid in experiments with the **radiation** of atomic nuclei. Chadwick then developed a method to measure radioactivity that required the observation of flashes, called scintillations, in zinc sulfide crystals under a **microscope** and in complete darkness. Although this method was difficult and prone to error, later experiments using newer detection techniques confirmed Chadwick's radiation measurements. In 1922 Chadwick became assistant director of research under Rutherford. In this post, Chadwick supervised all the research at the laboratory. Chadwick and Rutherford spent much time experimenting with the transmutation of elements, attempting to break up the **nucleus** of one element so that different elements would be formed. These experiments involved bombarding the nuclei of nitrogen and other elements with alpha particles (**helium** nuclei). This work eventually led to other experiments to gauge the size and map the structure of the atomic nucleus.

Throughout the years of work on the transmutation of elements, Chadwick and Rutherford struggled with an inconsistency. They saw that almost every element had an atomic number that was less than its atomic **mass**. In other words, an **atom** of any given element seemed to have more mass than could be accounted for by the number of protons in its nucleus. Rutherford initially thought that the extra mass was made up of proton-electron pairs, which would add mass but no additional charge to the nucleus. However, the nucleus did not seem to have room enough for such pairs. Rutherford then suggested the possibility of a particle with the mass of a proton and a neutral charge, but for a long time his and Chadwick's attempts to find such a particle were in vain.

For twelve years, Chadwick looked intermittently and unsuccessfully for the neutrally-charged particle that Rutherford proposed. In 1930 two German physicists, **Walther Bothe** and Hans Becker, found an unexpectedly penetrating radiation, thought to be gamma rays, when some elements were bombarded with alpha particles. However, the element beryllium showed an emission pattern that the gamma-ray hypothesis could not account for. Chadwick suspected that neutral particles were responsible for the emissions. Work done in France in 1922 by physicists **Frédéric Joliot-Curie** and **Irène Joliot-Curie** supplied the answer. Studying the hypothetical gamma-ray emissions from beryllium, they found that radiation increased when the emissions passed through the absorbing material paraffin. Although the Joliot-Curie team concluded that gamma rays emitted by beryllium knocked hydrogen protons out of the paraffin, Chadwick immediately saw that their experiments would confirm the presence of the neutron, since it would take a neutral particle of such mass to move a proton. He first set to work demonstrating that the gamma-ray hypothesis could not account for the observed phenomena, because gamma rays would not have enough **energy** to eject protons so rapidly. Then he showed that the beryllium nucleus, when combined with an **alpha particle**, could be transmuted to a **carbon** nucleus, releasing a particle with a mass comparable to that of a proton but with a neutral charge. The neutron had finally been tracked down. Other experiments showed that a boron nucleus plus an alpha particle results in a nitrogen nucleus plus a neutron. Chadwick's first public announcement of the discovery was in an article in the journal *Nature* with a title characteristic of his unassuming personality, "Possible Existence of a Neutron." In 1935 Chadwick was awarded the Nobel Prize for Physics for this discovery. That same year, Chadwick took a position at the University of Liverpool to establish a new research center in nuclear physics and to build a particle accelerator.

Chadwick's discovery of the neutron made possible more precise examinations of the nucleus. It also led to speculations about uranium fission. Physicists found that bombarding uranium nuclei with neutrons caused the nuclei to split into two almost equal pieces and to release energy in the very large amounts predicted by Einstein's formula $E = mc^2$. This phenomenon, known as **nuclear fission**, was discovered and publicized on the eve of World War II, and many scientists immediately began to speculate about its application to warfare. Britain quickly assembled a group of scientists under the Ministry of Aircraft Production, called the Maud Committee, to pursue the practicality of an atomic bomb. Chadwick was put in charge of coordinating all the experimental efforts of the universities of Birmingham, Cambridge, Liverpool, London, and Oxford. Initially Chadwick's responsibilities were limited to the very difficult and purely experimental aspects of the research project. Gradually, he became more involved with other duties in the organization, particularly as spokesperson.

Chadwick's work in evaluating and presenting evidence convinced British government and military leaders to move ahead with the project. Chadwick's involvement was broad and deep, forcing him to deal with scientific details of uranium supplies and radiation effects as well as broader issues of scientific organization and policy. His correspondence during this time referred to issues ranging from Britain's relationship with the United States to the effects of cobalt on the health of sheep.

As the pressures of war became greater, the British realized that even with their theoretical advances, they did not have the practical resources to develop a working atomic bomb. In 1943 Britain and the United States signed the Quebec

Agreement, which created a partnership between the two countries for the development of an atomic bomb. Chadwick became the leader of the British contingent involved in the **Manhattan Project** in the United States. Although he was shy and used to the isolation of the laboratory, Chadwick became known for his tireless efforts at collaboration and his keen sense of diplomacy. He maintained friendly Anglo-American relations despite a great variety of scientific challenges, political struggles, and conflicting personalities. On July 16, 1945, he witnessed the first atomic test in the New Mexico Desert.

After the war Chadwick's work continued to focus on **nuclear weapons**. He was an advisor for the British representatives to the United Nations regarding the control of atomic energy around the world. He believed that the British had to have their own atomic weapons, and he pushed hard for Britain's atomic independence. This issue was particularly evident to him when the United States and Britain argued over an earlier agreement for dividing up the uranium supplies they had jointly accumulated during the development of the atomic bomb. Chadwick used his diplomacy to effect a compromise that enabled the British to reclaim the uranium needed for their own atomic projects. Chadwick's postwar involvement with nuclear energy was not limited to weapons. He also was interested in medical applications of radioactive materials, and he worked to develop ways of regulating radioactive substances.

Chadwick was a dedicated and tireless scientist who balanced his commitments to science with a commitment to his family. He and his wife, Aileen Stewart-Brown, whom he married in 1925, had twin daughters. Though he was shy and serious, he did not lack a sense of humor: to celebrate the operation of a new **cyclotron** (particle accelerator) at Liverpool, Chadwick passed around laboratory beakers full of champagne.

Chadwick had an exacting sense of discipline and a tireless attention to detail. When he was at the Cavendish laboratory, all papers that went out for publication passed under his critical gaze. In an article in the *Bulletin of the Atomic Scientists,* Mark Oliphant, who had been a research student under Chadwick, remarked that "he was a severe critic of English usage and seemed to carry in his head the whole of Roget's *Thesaurus of English Words and Phrases.*'‘

Much of Chadwick's early career went toward searching for the neutron. When he found it, the physics community quickly seized on the prospect of developing from it the atomic bomb. Much of the rest of Chadwick's career was spent in assisting that

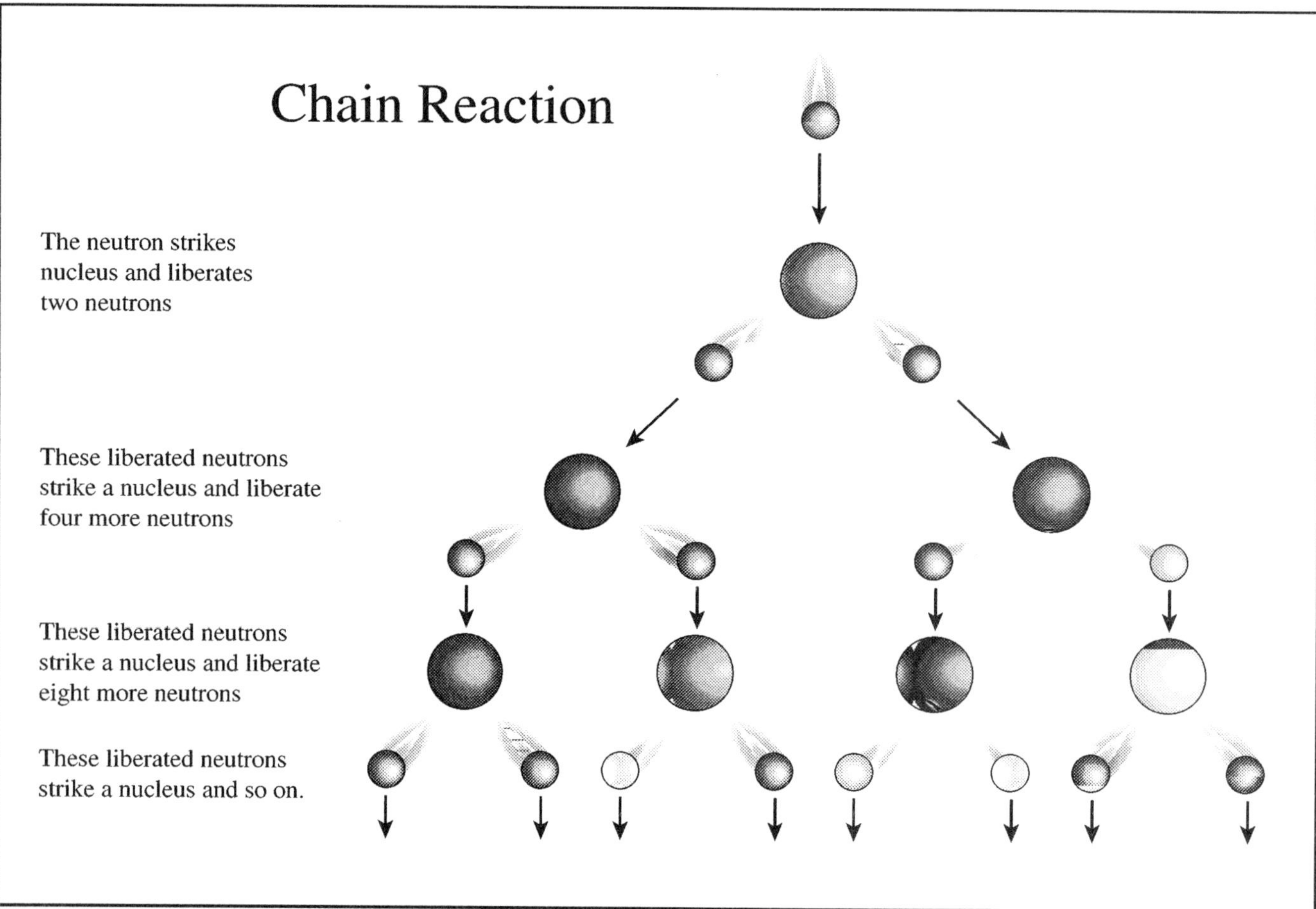

Neutron chain reaction.

development. Although he pushed for atomic policy issues as much as he pushed for scientific solutions, Chadwick eventually saw the uselessness of the atomic bomb. Margaret Gowing, in her article, "James Chadwick and the Atomic Bomb," quoted Chadwick as saying of the bomb, "Its effect in causing suffering is out of all proportion to its military effect." James Chadwick died in Cambridge, England, on July 24, 1974.

CHAIN REACTION

A chain reaction is a situation in which one action causes or initiates a similar action. In a nuclear chain reaction, for example, a **neutron** strikes a uranium-235 **nucleus**, causing the nucleus to undergo fission, which in turn produces a variety of products. Among these products is one or more neutrons. Thus, the particle needed to initiate this reaction (the neutron) is itself produced as a result of the reaction. Once begun, the reaction continues as long as uranium-235 nuclei are available. Nuclear chain reactions are important sources of fission and fusion **energy**.

CHANDRASEKHAR, SUBRAHMANYAN (1910-1995)

Indian American astrophysicist and applied mathematician

Subrahmanyan Chandrasekhar was an Indian-born American astrophysicist and applied mathematician whose work on the origins, structure, and **dynamics** of **stars** has secured him a prominent place in the annals of science. His most celebrated work concerns the **radiation** of **energy** from stars, particularly white dwarf stars, which are the dying fragments of stars. Chandrasekhar demonstrated that the radius of a white dwarf star is related to its **mass**: the greater its mass, the smaller its radius. Chandrasekhar has made numerous other contributions to **astrophysics**. His expansive research and published papers and books include topics such as the system of energy transfer within stars, **stellar evolution**, stellar structure, and theories of planetary and stellar atmospheres. For nearly twenty years, he served as the editor-in-chief of the *Astrophysical Journal,* the leading publication of its kind in the world. For his immense contribution to science, Chandrasekhar has received numerous awards and distinctions, most notably, the 1983 Nobel Prize in physics for his research into the depths of aged stars.

Chandrasekhar, better known as Chandra, was born on October 19, 1910, in Lahore, India (now part of Pakistan), the first son of C. Subrahmanyan Ayyar and Sitalakshmi (Divan Bahadur) Balakrishnan. Chandra came from a large family: he had nine siblings. As the firstborn son, Chandra inherited his paternal grandfather's name, Chandrasekhar. His uncle was the Nobel Prize-winning Indian physicist, Sir C. V. Raman.

Chandra received his early education at home, beginning when he was five. From his mother he learned Tamil, from his father, English and arithmetic. He set his sights upon becoming a scientist at an early age, and to this end, undertook at his own initiative some independent study of calculus and physics. The family moved north to Lucknow in Uttar Pradesh when Chandra was six. In 1918, the family moved again, this time south to Madras. Chandrasekhar studied with private tutors until 1921, when he enrolled in the Hindu High School in Triplicane. With typical drive and motivation, he studied on his own and steamed ahead of the class, completing school by the age of fifteen.

After high school, Chandra attended Presidency College in Madras. For the first two years, he studied physics, chemistry,, English, and Sanskrit. For his B.A. honors degree he wished to take pure mathematics, but his father insisted that he take physics. Chandra resolved this conflict by registering as an honors physics student but attending mathematics lectures. Recognizing his brilliance, his lecturers went out of their way to accommodate Chandra. Chandra also took part in sporting activities and joined the debating team. A highlight of his college years was the publication of his paper, "The Compton **Scattering** and the New Statistics." These and other early successes while he was still an eighteen-year-old undergraduate only strengthened Chandra's resolve to pursue a career in scientific research, despite his father's wish that he join the Indian civil service. A meeting the following year with the German physicist **Werner Heisenberg**, whom Chandra, as the secretary of the student science association, had the honor of showing around Madras, and Chandra's attendance at the Indian Science Congress Association Meeting in early 1930, where his work was hailed, doubled his determination.

Upon graduating with a M.A. in 1930, Chandra set off for Trinity College, Cambridge, as a research student, courtesy of an Indian government scholarship created especially for him (with the stipulation that upon his return to India, he would serve for five years in the Madras government service). At Cambridge, Chandra turned to astrophysics, inspired by a theory of stellar evolution that had occurred to him as he made the long boat journey from India to Cambridge. It would preoccupy him for the next ten years. He also worked on other aspects of astrophysics and published many papers.

In the summer of 1931, he worked with physicist **Max Born** at the Institut für Theoretische Physik at Göttingen in Germany. There, he studied **group theory** and quantum **mechanics** (the mathematical theory that relates **matter** and radiation) and produced work on the theory of stellar atmospheres. During this period, Chandra was often tempted to leave astrophysics for pure mathematics, his first love, or at least for physics. He was worried, though, that with less than a year to go before his thesis exam, a change might cost him his degree. Other factors influenced his decision to stay with astrophysics, most importantly, the encouragement shown him by astrophysicist Edward Arthur Milne. In August 1932, Chandra left Cambridge to continue his studies in Denmark under physicist **Niels Bohr**. In Copenhagen, he was able to devote more of his energies to pure physics. A series of Chandra's lectures on astrophysics given at the University of Liège, in Belgium, in February 1933 received a warm reception. Before returning to Cambridge in May 1933 to sit for his doctoral exams, he went back to Copenhagen to work on his thesis.

Subrahmanyan Chandrasekhar, with King Carl Gustaf of Sweden, receiving the Nobel Prize 1983. *(AP/Wide World Photos. Reproduced by permission.)*

Chandrasekhar's uncertainty about his future was assuaged when he was awarded a fellowship at Trinity College, Cambridge. During a four-week trip to Russia in 1934, where he met physicists Lev Davidovich Landau, B. P. Geraismovic, and Viktor Ambartsumian, he returned to the work that had led him into astrophysics to begin with, **white dwarfs**. Upon returning to Cambridge, he took up research of white dwarfs again in earnest.

As a member of the Royal Astronomical Society since 1932, Chandra was entitled to present papers at its twice monthly meetings. It was at one of these that Chandra, in 1935, announced the results of the work that would later make his name. As stars evolve, he told the assembled audience, they emit energy generated by their conversion of hydrogen into **helium** and even heavier elements. As they reach the end of their life, stars have progressively less hydrogen left to convert and emit less energy in the form of radiation. They eventually reach a stage when they are no longer able to generate the **pressure** needed to sustain their size against their own gravitational pull and they begin to contract. As their **density** increases during the contraction process, stars build up sufficient internal energy to collapse their **atomic structure** into a degenerate state. They begin to collapse into themselves. Their electrons become so tightly packed that their normal activity is suppressed and they become white dwarfs, tiny objects of enormous density. The greater the mass of a white dwarf, the smaller its radius, according to Chandrasekhar.

However, not all stars end their lives as stable white dwarfs. If the mass of evolving stars increases beyond a certain limit, eventually named the *Chandrasekhar limit,* and calculated as 1.4 times the mass of the **Sun**, evolving stars cannot become stable white dwarfs. A star with a mass above the limit has to either lose mass to become a white dwarf or take an alternative evolutionary path and become a supernova, which releases its excess energy in the form of an explosion. What mass remains after this spectacular event may become a white dwarf but more likely will form a **neutron** star. The neutron star has even greater density than a white dwarf and an average radius of about.15 km. It has since been independently proven that all white dwarf stars fall within Chandrasekhar's predicted limit, which has been revised to equal 1.2 solar masses.

Unfortunately, although his theory would later be vindicated, Chandra's ideas were unexpectedly undermined and ridiculed by no less a scientific figure than astronomer and physicist Sir **Arthur Stanley Eddington**, who dismissed as absurd Chandra's notion that stars can evolve into anything other than white dwarfs. Eddington's status and authority in the community of astronomers carried the day, and Chandra, as the junior, was not given the benefit of the doubt. Twenty years passed before his theory gained general acceptance among astrophysicists, although it was quickly recognized as valid by physicists as noteworthy as **Wolfgang Pauli**, Niels Bohr, Ralph H. Fowler, and **Paul Dirac**. Rather than continue sparring with Eddington at scientific meeting after meeting,

Chandra collected his thoughts on the matter into his first book, *An Introduction to the Study of Stellar Structure* and departed the fray to take up new research around stellar dynamics. An unfortunate result of the scientific quarrel, however, was to postpone the discovery of **black holes** and **neutron stars** by at least twenty years and Chandra's receipt of a Nobel Prize for his white dwarf work by fifty years. Surprisingly, despite their scientific differences, he retained a close personal relationship with Eddington.

Chandra spent from December 1935 until March 1936 at Harvard University as a visiting lecturer in cosmic physics. While in the United States, he was offered a research associate position at Yerkes Observatory at Williams Bay, Wisconsin, staring in January 1937. Before taking up this post, Chandra returned home to India to marry the woman who had waited for him patiently for six years. He had known Lalitha Doraiswamy, daughter of Captain and Mrs. Savitri Doraiswamy, since they had been students together at Madras University. After graduation, she had undertaken a masters degree. At the time of their marriage, she was a headmistress. Although their marriage of love was unusual, as both came from fairly progressive families and were both of the Brahman caste, neither of their families had any real objections. After a whirlwind courtship and wedding, the young bride and groom set out for the United States. They intended to stay no more than a few years, but, as luck would have it, it became their permanent home.

At the Yerkes Observatory, Chandra was charged with developing a graduate program in astronomy and astrophysics and with teaching some of the courses. His reputation as a teacher soon attracted top students to the observatory's graduate school. He also continued researching stellar evolution, stellar structure, and the transfer of energy within stars. In 1938, he was promoted to assistant professor of astrophysics. During this time Chandra revealed his conclusions regarding the life paths of stars.

During the Second World War, Chandra was employed at the Aberdeen Proving Grounds in Maryland, working on ballistic tests, the theory of shock **waves**, the Mach effect, and transport problems related to neutron diffusion. In 1942, he was promoted to associate professor of astrophysics at the University of Chicago and in 1943, to professor. Around 1944, he switched his research from stellar dynamics to radiative transfer. Of all his research, the latter gave him, he recalled later, more fulfillment. That year, he also achieved a lifelong ambition when he was elected to the Royal Society of London. In 1946, he was elevated to Distinguished Service Professor. In 1952, he became Morton D. Hull Distinguished Service Professor of Astrophysics in the departments of astronomy and physics, as well as at the Institute for **Nuclear Physics** at the University of Chicago's Yerkes Observatory. Later the same year, he was appointed managing editor of the *Astrophysical Journal,* a position he held until 1971. He transformed the journal from a private publication of the University of Chicago to the national journal of the American Astronomical Society. The price he paid for his editorial impartiality, however, was isolation from the astrophysical community.

Chandra became a United States citizen in 1953. Despite receiving numerous offers from other universities, in the United States and overseas, Chandra never left the University of Chicago, although, owing to a disagreement with Bengt Strömgren, the head of Yerkes, he stopped teaching astrophysics and astronomy and began lecturing in **mathematical physics** at the University of Chicago campus. Chandra voluntarily retired from the University of Chicago in 1980, although he remained on as a post-retirement researcher. In 1983, he published a classic work on the mathematical theory of black holes. He later studied colliding waves and the Newtonian two-center problem in the framework of the general theory of relativity. His semi-retirement also left him with more time to pursue his hobbies and interests: literature and music, particularly orchestral, chamber, and South Indian.

During his long career, Chandrasekhar received many awards. In 1947, Cambridge University awarded him its Adams Prize. In 1952, he received the Bruce Medal of the Astronomical Society of the Pacific, and the following year, the Gold Medal of the Royal Astronomical Society. In 1955, Chandrasekhar became a Member of the National Academy of Sciences. The Royal Society of London bestowed upon him its Royal Medal seven years later. In 1962, he was also presented with the Srinivasa Ramanujan Medal of the Indian National Science Academy. The National Medal Science of the United States was conferred upon Chandra in 1966; and the Padma Vibhushan Medal of India in 1968. Chandra received the Henry Draper Medal of the National Academy of Sciences in 1971 and the Smoluchowski Medal of the Polish Physical Society in 1973. The American Physical Society gave him its Dannie Heineman Prize in 1974. The crowning glory of his carer came nine years later when the Royal Swedish Academy awarded Chandrasekhar the Nobel Prize for Physics. ETH of Zurich gave the Indian astrophysicist its Dr. Tomalla Prize in 1984, while the Royal Society of London presented him with its Copley Prize later that year. Chandra also received the R. D. Birla Memorial Award of the Indian Physics Association in 1984. In 1985, the Vainu Bappu Memorial Award of the Indian National Science Academy was conferred upon Chandrasekhar. In May 1993, Chandra received the state of Illinois's highest honor, Lincoln Academy Award, for his outstanding contributions to science.

While his contribution to astrophysics has been immense, Chandra always preferred to remain outside the mainstream of research. He described himself to his biographer, Kameshar C. Wali, as "a lonely wanderer in the byways of science." Throughout his life, Chandra tried to acquire knowledge and understanding, according to an autobiographical essay published with his Nobel lecture, motivated "principally by a quest after perspectives." He died in 1995.

CHANDRASEKHAR LIMIT

By combining special relativity with quantum **mechanics, Subrahmanyan Chandrasekhar** showed that if the **mass** of a star were greater than a critical value, then that star could not evolve into a white dwarf. It would instead collapse into a **neutron** star or a black hole. The critical value of mass, about 1.4 times the mass of the **Sun**, is called the Chandrasekhar limit.

Once a main sequence star, like our Sun, expends the amount of fuel available for **nuclear fusion** in its core, it begins to collapse. The star is supported against its self-gravity by a quantum mechanical effect called electron-degeneracy **pressure**. This degeneracy pressure is a consequence of Wolfgang Pauli's exclusion principle, which prohibits any pair of electrons from having the same set of **quantum numbers**, and Werner Heisenberg's uncertainty principle, which presents a fundamental uncertainty in the precision to which the electron's position and **momentum** can be simultaneously known.

By extending earlier investigations of Ralph Fowler and **Arthur Eddington**, Chandrasekhar realized that the **density** at the center of a white dwarf is many times greater than its average density. Assigning different momentum states to each **electron**, one eventually reaches a point where the momentum becomes comparable to the electron's rest **energy**. At that point, the relativistic mass of the electron needs to be taken into account.

Non-relativistic considerations alone would suggest that the radius of a white dwarf scales as $M^{-1/3}$, where M is the mass of the star. Given two white dwarf **stars** of different radii, the more massive star is the smaller one. The radius of a white dwarf reaches zero in the limit of infinite mass. Because a relativistic electron gas provides less pressure support at a given density than a non-relativistic electron gas, introducing the relativistic corrections to the analysis means that the radius of the white dwarf hits zero at a finite value of M. This critical value, which can be expressed in terms of various fundamental constants, is the Chandrasekhar limit. Stars that are more massive than the critical value cannot be supported by the degeneracy pressure of electrons. Rather, they evolve into something else, **neutron stars** or **black holes**.

CHANDRASEKHAR MASS • See Chandrasekhar limit

CHANGES OF PHASE • See Phase changes and phase equilibrium

CHAOS AND ORDER

Chaos and order, as used in chaos theory, are terms used to describe conditions of complex systems in which, out of seemingly random, disordered (e.g. aperiodic) processes there arise processes that are deterministic and predictable. Accordingly, despite its name, chaos theory attempts to identify and quantify order in apparently unpredictable systems.

Along with quantum and relativity theories, chaos theory—with its inclusive concepts of chaos and order—is widely regarded as one of the great intellectual leaps of the twentieth century. The modern physical concepts of chaos and order, however, actually trace their roots to classical mechanical con-

cepts introduced in English physicist Sir Isaac Newton's 1686 work, *Philosophy Naturalis Principia Mathematica (Mathematical Principles of Natural Philosophy)*. It was Newton, one of the inventors of the calculus, who revolutionized **astronomy** and physics by showing that the behavior of all bodies, celestial and terrestrial, was governed by the same laws of **motion**, which could be expressed as differential equations. These differential equations relate the rates of change of physical quantities to the values of those quantities themselves. Such calculated predictability of physical phenomena led to the concept of a mechanistic, clockwork **universe** that operated according to deterministic laws. The idea that the universe operated in strict accord with physical laws was profoundly influential on science, philosophy and theology.

Most physical models are devoted to the understanding of simple systems (e.g., kinetic molecular theories often rely on concepts related to a ball bouncing in a box). From fundamental laws, using easily quantifiable behavior of such simple systems, theorists often attempt to project the behavior of more complex systems (e.g., the collision and **dynamics** of hundreds of balls bouncing in a box). It was long thought by physicists that, with regard to these types of models, the complexity of a system simply veiled an underlying fundamental simplicity.

For example, according to classical deterministic concepts, the accurate analysis and prediction of complex systems (e.g., the determination of the **momentum** of a particular ball among hundreds of other balls bouncing and colliding in a box) could be calculated if the initial or starting conditions were accurately known. The fact that is usually impossible to predict the exact condition or behavior of a system (especially considering that such interactions or measurements of a systems must also alter the system itself) is explained away as the result of a lack of knowledge regarding starting conditions or a lack of calculating vigor (e.g., inadequate computing power).

Complex phenomena are those generally regarded as having too many variables (or too many possible conditions or states) to yield to conventional quantitative analysis. The motion of molecules in swirling smoke or the turbulent hydraulics of a river current, for example, are systems that exhibit such chaotic complexity.

Used together, the terms chaos and order fundamentally describe conditions related to the thermodynamic concept of **entropy**. Entropy is a measure of thermodynamic **equilibrium** used to explain irreversibility in physical and chemical processes. The second law of **thermodynamics** specifies that in an isolated system, increasing entropy corresponds to changes in the system over time and that entropy tends toward (a statistical mechanical concept) maximization. The **second law of thermodynamics** dictates that in natural processes, without work being done on a system, there is a movement from order to disorder.

According to the laws of thermodynamics, in all natural processes there is—when considering both system and surroundings—tendency a movement from the ordered state to a more the chaotic (disordered) state. Conversely, according to some chaos theory models, chaotic, unpredictable, and **irreversible processes** may, evolve into or produce ordered states.

Because entropy in natural processes increases over time, even very straightforward linear-type relationships must eventually take on a degree of irregularity (i.e., of seemingly disordered complexity). In accord with the second law of thermodynamics, apparently chaotic phenomenon arise from initially ordered (i.e., lower entropy) systems. This dual tendency toward increasing entropy and chaos from an initially stable state can take place spontaneously. Small perturbations in initial conditions intensify these tendencies. Chaos theorists describe such departures as the *Butterfly Effect*.

The study of such mathematical irregularities involving chaos and order remained a relatively unnoticed corner of advanced mathematics until the advent of the digital computer. In 1956, Edward Lorenz, a professor of **meteorology** at the Massachusetts Institute of Technology was studying the numerical solution to a set of three differential equations in three unknowns, a highly simplified version of the type of equations meteorologists were then using to describe atmospheric phenomena. Lorenz came to the conclusion that his set of differential equations displayed a sensitive dependence on initial conditions, a sensitivity of the same type that French mathematician **Jules-Henri Poincaré** had discovered for the Newtonian equations when those equations were applied to celestial dynamics. Lorenz, however, gave this phenomenon a new and highly appealing name, the butterfly effect, suggesting that, in the extreme, the flapping of a butterfly's wings in Kansas might be responsible for a monsoon in India a month later.

Precise definitions of chaos and order are hotly debated among physicists. The terms themselves presuppose definitions for entropy, complexity, and irreversibility that are anything but well-settled in the modern physics community. Regardless, physicists generaly agree that the discovery of underlying order—instead of complete randomness—in seemingly chaotic systems is perfectly consistent with the proper application of the laws of thermodynamics.

See also Belousov-Zhabotinski reaction; Causation; Determinism; Reversible processes

CHARGES IN ELECTRIC FIELDS

An **electric charge** in a uniform external electric field will feel a **force** due to that field given by, F = qE, where F is the force, q is the charge and E is the external electric field (i.e. the charge is not affected by its own electric field.). This force points in the direction of the field if the charge is positive and in the opposite direction if negative.

One of the most famous experiments using this force is the Millikan oil-drop experiment. A microscopic drop of oil was placed in an electric field between two plates. From the **motion** of the drop, the value of q was determined. Millikan discovered this value was always some integer times a constant, e. E was later determined to be the elementary charge of an **electron**. This experiment proved that charge was quantized and Millikan won the Nobel Prize in 1923 for his work.

Two common applications of charges in electric fields are inkjet printing and a television set. Both work on the same principle; charges are accelerated through and electric field. The field bends the charges off of the straight line path according to its strength. The charges (electrons in a TV or ink droplets in a printer) are then deposited at a specific location.

When a dipole is placed in a uniform external electric field, it will experience a net **torque**:

$$\tau = p \times E$$

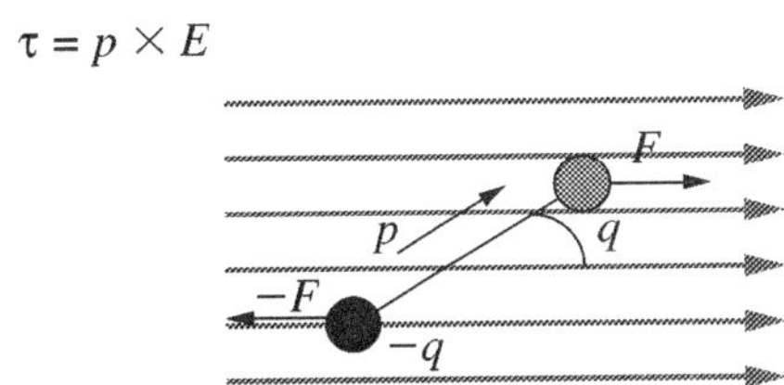

where p is the **magnetic moment** and E is the electric field. One end of the dipole is positive while the other is negative. Thus, the positive end will feel a force in the direction of E, while the negative end will feel a force in the direction of $-E$. A dipole in an electric field can easily be considered to be oscillatory; this being said, a **potential energy** dependent on the angle between the magnetic moment, p and the electric field E can be defined (U = –p•E). When the angle, θ, is zero, we have minimum potential **energy**; when the angle is 90, we have maximum.

Microwave ovens use this torque, and the energy it can produce, to heat items. Specifically, **microwaves** heat the water in the items by causing the water molecules to rotate. Due to the orientation of the hydrogen and **oxygen** atoms in a molecule of water, each molecule is a dipole. These dipoles have a tendency to align themselves positive to negative, much like north and south poles of a magnet. The microwaves in the oven produce an electric field that rotates the molecules in a group, breaking them free. Since they naturally want to be in groups, they will release the energy they just gained as **thermal energy** and reform groups that can be broken again, creating a cyclical event until the microwaves stop.

CHARLES, JACQUES-ALEXANDRE-CÉSAR (1746-1823)
French experimental physicist

Jacques-Alexandre-César Charles is famous for his contribution to ballooning, (the science of aerostation). He was the first to use hydrogen ("inflammable air") instead of hot air in an aeronautical balloon, and designed and developed almost all the essential components of a balloon so perfectly that few changes or additions have been made in the ensuing 200 years of ballooning.

Charles was born in Beaugency on November 12, 1746. The only information surviving about his childhood is that he received a liberal education with no scientific focus. When he moved to Paris, he worked at the bureau of finances and in 1799 became interested in nonmathematical, experimental physics. By 1781, he began giving public lectures and experimental demonstrations, attracting a large number of attendees, among them people who had already achieved considerable notoriety. He was ultimately appointed a resident member of the Académie de Sciences in 1795. He also became professor of experimental physics and librarian at the Conservatoire de Arts et Métiers and, in 1816, president of the Class of Experimental Physics at the academy. In 1804, he married Julie-Françoise Bouchard des Hérettes, who died in 1817 after suffering from a long illness.

Charles' association with ballooning began when, in 1782, Joseph Montgolfier (1740-1810) discovered that hot air entering a paper bag held over an open fire made the bag rise to the ceiling. He and his brother, Étienne (1745-1799), experimented with larger "envelopes" and, on April 25, 1783, launched a balloon large enough that it was considered capable of carrying a person. On June 4 they successfully launched, from Annonay in France, a manned balloon made of linen covered with paper. Their creation, called *montgolfières*, consisted of an open fire hanging in a cage under the basket. The fire had to be fed by the pilot for the balloon to remain afloat.

News of the Montgolfier's experiment quickly reached Paris and the French Academy of Science assigned Charles the task of studying the invention. He had already learned of a recent scientific discovery by British scientist **Henry Cavendish** of a gas 14 times lighter than air and which was being called "flammable air." (It would ultimately be named hydrogen.)

Hydrogen was probably first produced by alchemists in the fifteenth century. Although the element itself was unknown, alchemists were aware of metal/acid reactions that, according to Phillippus Aureolus Paracelsus caused "air to arise and break forth like the wind." French chemist Antoine-Laurent Lavoisier, continuing Cavendish's experiments, became the first to produced both hydrogen and **oxygen** in a laboratory by dissolving metals in acid. He named them "flammable air" and "life-sustaining air" respectively.)

Charles either assumed Montgolfier's balloon utilized this new gas, or simply decided to use it in his own ballooning experiments. Regardless, on August 27, 1783, he launched his first balloon from the Champs de Mars in Paris (the site on which the Eiffel Tower was ultimately constructed), creating the gas by pouring sulfuric acid over scrap iron. Among the spectators at this launch was the American ambassador to France, **Benjamin Franklin**. The balloon descended after about 45 minutes, landing in a field where terrified farmers attacked the "monster" with pick axes and shovels, "inspired by the beast's behavior to sigh and groan and emit a horrible smell."

Charles' creation included almost all the essential components still used in ballooning. He constructed the balloon sack of silk, coating it with a solution of rubber dissolved in turpentine to prevent the gas from escaping. At the top of the sack was an apex, open tube through which expanded gas could escape to prevent the balloon from exploding. A network of ropes covering the sack attached to a wooden hoop holding a wicker basket suspended below the sack. This basket would ultimately hold pilot and passengers. Charles also invented a valve line, allowing the pilot to release gas to initiate descent.

In December 1783, Charles, along with Nicholas Marie-Noel Robert (1761-1826), one of two brothers who helped in the design of the balloon sack, entered the basket and launched his creation called *Charlière*. This was the first manned gas balloon flight. Leaving Tuileries, the two aeronauts travelled for about two hours, landed in the village of Nesles-la- Vallée, and Charles then ascended alone to soar for another half hour. For some reason, this was the only flight Charles made, even though he lived until 1823.

Because of Charles' success with hydrogen, hot air balloons quickly disappeared, and it was not until recent times that they made a resurgence in popularity.

CHARLES' LAW

One of the **gas laws**, Charles' law states that at a constant **pressure** the volume of a fixed **mass** of gas is directly proportional to the absolute **temperature**. This law also can be expressed by the equation V/T = constant where V is the volume of the gas and T is the temperature in Kelvin.

Charles' law also is known as **Gay-Lussac's law** and the constant pressure law. This law was discovered independently by French physicist Jacques Charles (1746-1823) in 1787 (who did not publish his findings) and French chemist Joseph Gay-Lussac (1778-1850) in 1802. Although Gay-Lussac was the first scientist to announce this law, it is commonly known as Charles' law for its earlier discoverer. Charles' law can be combined with **Boyle's law** and the pressure law to give the **ideal gas law** (also known as the universal gas law), pV = nRT, where p is pressure, V is volume, n is the number of moles of gas, R is the universal gas constant, and T is temperature.

Charles' law is one of the ways in which **absolute zero** is calculated. This is done by extrapolating a graph of volume against temperature. At the point on the graph where the volume of the gas is zero (gas cannot have a volume less than zero), the theoretical temperature is absolute zero, the coldest temperature possible. This point has never actually been achieved, since gases liquefy before they reach this temperature and they are no longer subject to the gas laws.

Charles' law is a good indicator of what happens to a gas, since the volume and temperature are altered while the pressure remains constant.

CHARPAK, GEORGES (1924-　)
Polish French physicist

Georges Charpak received the Nobel Prize in physics in 1992 for his invention and development of **particle detectors**, most notably the multiwire proportional chamber. A number of his colleagues, who received the Nobel Prize before him, had used his invention to make important discoveries in physics. Charpak is credited with creating instrumentation that is used by thousands of other scientists at CERN, the European laboratory for **particle physics** located in Geneva, Switzerland, as well as by researchers in other prominent laboratories involved in the study of the nature of **matter**.

Born on August 1, 1924, in Dabrovica, Poland, to Maurice Charpak and Anna Szapiro, Charpak moved to France with his family in 1929. In 1943 the French Vichy government accused the young Charpak of being a terrorist and sentenced him to the concentration camp in Dachau, West Germany (now Germany). Charpak remained in the camp until its liberation in 1945. Upon his return to France, he completed a degree in civil engineering from the Ecole des Mines in Paris, and in 1946 he became a French citizen.

Two years later, as a graduate student in nuclear physicsat the Collège de France in Paris, Charpak went to work in the laboratory of physicist **Frédéric Joliot-Curie**. It was in this laboratory that Charpak began building the equipment he needed to perform his experiments (he constructed his own equipment out of necessity, as the laboratory had none). Charpak contends that he was not good at invention but had to learn in order to perform his experiments. In 1955 he received his Ph.D. from the Collège de France.

In awarding the Nobel Prize to Charpak, the Swedish Academy of Sciences traced the history of the development of detector devices in physics. The **cloud chamber** and the bubble chamber were two earlier inventions that had received recognition from the academy. Both relied on photographic techniques to capture particle events. In 1958 Charpak was invited by experimental physicist **Leon Lederman**, who had heard Charpak lecture in Padua, to come to CERN to work on sparking devices to detect particles. Though these devices were an improvement over existing techniques, they, too, relied on photographic recording, which was slow and cumbersome to analyze. Charpak turned to the problem of a **spark chamber** reading without photographic film, and built his first multiwire proportional tracking chamber in 1968. The multiwire proportional chamber extended the technology of the Geiger-Muller tube, bubble chamber, and chamber in two ways. The multiwire proportional chamber replaced the single positively charged wire of the Geiger-Muller tube, which attracts electrons in a chamber of ionized gas, with a multiwire device. It also replaced photographic analysis of a trail of bubbles with computerized electronic analysis of current produced in the wires as they attract electrons. Charpak credited his background in **nuclear physics** with the success of his invention.

Since liberating physics from dependence on film readings, Charpak has turned his interests to medicine and aerospace problems. Work he has done in the latter area makes it possible to produce an x-ray radiograph of turbine blades as they spin. In the field of medicine, his chamber is able to analyze the structure of a protein with **x rays** a thousand times faster than was previously possible. He also is working on imaging problems to identify receptors in the brain.

Charpak's particle work mimics the state of the **universe** as it was a fraction of a second after the **big bang**. It is believed that some of the particles have not existed in nature since that time, and the ability of physicists to study them will reveal and increase the understanding of the relationships among the forces of nature. Whereas Charpak built on the work of his predecessors with the bubble and spark chambers, others have used his invention to make their own contributions to the field of physics. A group led by **Samuel Ting** discovered the first manifestation of charmed **quarks** at Brookhaven in 1974, and **Carlo Rubbia** led a group to discover the W and Z particles. (Charmed quarks and W and Z particles are subatomic particles.)

Charpak has worked on behalf of other scientists imprisoned by repressive governments. He is the founder of the SOS committee at CERN. This association worked diligently on the part of Soviet dissidents, such as Andrei Sakharov, Yuri Orlov, and Anatoly Sharansky, when they were deprived of their civil rights under the former Soviet Union.

Charpak married Dominique Vidal in 1953; they have two sons and one daughter. Some of Charpak's leisure interests include skiing, music, and hiking. He has been a member of the French Academy of Sciences since 1985. In addition to his continued association with CERN, Charpak is also the Joliot-Curie Professor at the Ecole Supérieure de Physique et Chimie in Paris, a position he has held since 1984. In 1989 he received the High **Energy** and Particle Physics Prize from the European Physical Society. He has published numerous papers in scientific journals.

CHEMICAL ELEMENT

A chemical element is a substance that cannot be changed, broken down, or decomposed by ordinary chemical or physical methods into a simpler form. A chemical element is composed of atoms, identical and unique for every element, that vary only in their mixture of protons, neutrons, and electrons.

The number of elements known to scientists constantly changes as advances in technology allow the synthesis of heavier elements. As of 2000, the International Union of Pure and Applied Chemistry (IUPAC), an international scholarly scientific organization with the responsibility for naming elements, had named elements as heavy as the element Meitnerium (atomic number 109). Observations of heavier elements (up through atomic number 118) awaited confirmation. Of the recognized elements, 89 occur in a free or combined state in the Earth's crust. The remaining elements are prepared or synthesized in laboratories. Most of the natural elements are found in various combinations with each other as chemical compounds. Only a few of the elements, such as, gold, mercury, lead, silver, sulfur, and copper can be found in their pure state in nature.

The properties associated with the chemical elements differ enough that they are classified as two general classes, metals and nonmetals. Metallic elements are malleable, ductile, and good conductors of **heat** and **electricity**. Their luster allows them to reflect heat and **light**. Some examples of metals include gold, silver, copper, potassium, and zinc. At standard **temperature** and **pressure**, mercury is a liquid metal. Sulfur, iodine, and **carbon** are examples of solid nonmetallic chemical elements. These elements are poor conductors of heat and electricity, and are also brittle and dull in appearance. At standard temperature and pressure, bromine is a liquid nonmetal and nitrogen and **oxygen** are gaseous members of this class. Some elements are considered borderline because they display some characteristics of metals (such as, luster) and other properties associated with nonmetals (brittle). These elements are referred to as metalloids and include arsenic and germanium. Silicon, a metalloid, does not conduct heat and electricity as well as a metal, but is a better conductor than a nonmetal. These properties make it a good semiconductor for use in the computer industry.

The Swedish chemist, Jons Jakkob Berzelius (1779-1848) developed a system for representing each chemical element by a one or two letter symbol. These abbreviations are divided into three different types. The first type of symbol is derived from the first letter of the common name of the chemical element. Accordingly, the symbol for hydrogen is H, C represents carbon, and O represents oxygen. To provide for increasing numbers of elements, Berzelius suggested the use of a second letter representing the second letter of the chemical element or a letter whose **sound** is conspicuous when the name of the element is pronounced. Thus, calcium, represented by the symbol Ca and chromium, represented by the symbol Cr because of the predominant "r" in the pronunciation, represent a second class of element symbols. The final type of chemical symbol is derived from the Latin name for the element. The abbreviation for lead is Pb because the Latin name for lead is *plumbum*. The symbols for iron, Fe, and sodium, Na, come from the Latin *ferrum* and *natrium*. The first letter of an element's symbol is always capitalized and the second letter is normal or lower case. For example, the symbol for cobalt is Co, while CO represents carbon monoxide, a molecule composed of the elements carbon (C) and oxygen (O).

See also Atom; Atomic structure; Periodic table

CHEMICAL REACTION

Chemical reactions describe the changes between reactants (the initial substances that enter into the reaction) and products (the final substances that are present at the end of the reaction). Chemical equations are notations that used to concisely summarize and convey information regarding chemical reactions.

In a balanced chemical reaction all of the **matter** (i.e., atoms or molecules) that enter into a reaction must be accounted for in the products of a reaction. Accordingly, associated with the symbols for the reactants and products are numbers (stoichiometry coefficients) that represent the number of molecules, formula units, or moles of a particular reactant or product. Reactants and products are separated by addition symbols (addition signs). The addition signs represent the interaction of the reactants and are used to separate and list the products formed. The chemical equations for some reactions may have a lone reactant or a single product. The subscript numbers associated with the chemical formula designating individual reactants and products represent the number of atoms of each element that are in each molecule (for covalently bonded substances) or formula unit (for ironically associated substances) of reactants or products.

For a chemical reaction to be balanced, all of the atoms present in molecules or formula units or moles of reactants to the left of the equation arrow must be present in the molecules, formula units and moles of product to the right of the equation arrow. The combinations of the atoms may change (indeed, this is what chemical reactions do) but the number of atoms present in reactants must equal the number of atoms present in products.

Charge is also conserved in balanced chemical reactions and therefore there is a conservation of electrical charge between reactants and products.

Although chemical equations are usually concerned only with reactants and products chemical reactions may proceed through multiple intermediate steps. In such multi-step reactions the products of one reaction become the reactants (intermediary products) for the next step in the reaction sequence.

Reaction catalysts are chemical species that alter the **energy** requirements of reactions and thereby alter the speed at which reactions run (i.e., control the rate of formation of products).

Combustion reactions are those where **oxygen** combines with another compound to form water and **carbon** dioxide. The equations for these reactions usually designate that the reaction is exothermic (**heat** producing). Synthesis reactions occur when two or more simple compounds combine to form a more complicated compound. Decomposition reactions reflect the reversal of synthesis reactions (e.g., reactions where complex molecules are broken down into simpler molecules). The **electrolysis** of water to make oxygen and hydrogen is an excellent example of a decomposition reaction.

Equations for single displacement reactions, double displacement, and acid-base reactions reflect the appropriate reallocation of atoms in the products.

In accord with the laws of **thermodynamics**, all chemical reactions change the energy state of the reactants. The change in energy results from changes in the in the number and strengths of chemical bonds as the reaction proceeds. The heat of reaction is defined as the quantity of heat evolved or absorbed during a chemical reaction. A reaction is called exothermic if heat is released or given off during a chemical transformation. Alternatively, in an endothermic reaction, heat is absorbed in transforming reactants to products. In endothermic reactions, heat energy must be supplied to the system in order for a reaction to occur and the heat content of the products is larger than that of the reactants. For example, if a mixture of gaseous hydrogen and oxygen is ignited, water is formed and heat energy is given off. The chemical reaction is an exothermic reaction and the heat content of the product(s) is lower than that for the reactants. The study of energy utilization in chemical reactions is called chemical **kinetics** and is important in understanding chemical transformations.

See also Atomic theory; Avogadro's law; Avogadro's number; Conservation of energy; Conservation of heat; Conservation of mass; Energy conservation; Irreversible processes; Molecules and molecular forces

Sodium bicarbonate (an ingredient in Alka-Seltzer) producing a fizzing reaction due to reactons with water molecules. *(Photo by Robert J. Huffman. Fieldmark Publications. Reproduced by permission.)*

CHEMISTRY

Chemistry is the science that studies the properties and composition of **matter**, the changes in composition and structure that occur, and the **energy** phenomena accompanying these changes. The term chemistry is a shortened form of the word alchemy. Alchemy describes the medieval art that searched for the philosopher's stone to change basic metals into gold and heal all diseases. In this quest, those skilled in alchemy started the use of experimentation to test various hypotheses and standardized various common laboratory techniques, such as how to **heat**, distill, and combine substances.

The beginning of the modern science of chemistry is attributed to Antoine Lavoisier (1743-1794). In 1774, this French scientist demonstrated that **oxygen** is a critical component of air needed for combustion. This observation led to a better understanding of the changes in composition and structure of matter. Lavoisier later published the first list of elemental substances that eventually evolved into the **Periodic Table** of the Elements. Other contributions important to early chemistry were made by **John Dalton** with the concept of the **atom**, Amadeo Avogadro with his theory that molecules are made up of atoms, and Sir Edward Frankland (1825-1899), who developed the understanding of the **chemical reaction**.

The field of chemistry is divided into four traditional disciplines: organic, inorganic, analytical, and physical chemistry. Each discipline studies a different facet of the structure and composition of materials and their changes in composition and energy. As molecules and scientific problems become more complex, the traditional areas of chemical investigation begin to overlap with other physical sciences. For example, the fields of biochemistry, medicinal chemistry, electrochemistry and nuclear chemistry have emerged as important non-traditional areas in chemistry.

Organic chemistry is the study of compounds that contain **carbon** atoms. The term organic was first introduced by the Swedish scientist Jons Berzelius (1779-1848) to refer to substances isolated from living systems. Inorganic compounds were those isolated from nonliving sources. At the time, it was believed that a "vital force" present only in living systems was necessary for the preparation of organic compounds. In 1828, Friedrich Wöhler (1800-1882) first synthesized urea, an organic compound isolated from urine, by evaporating a water solution of the inorganic compound ammonium cyanate. Eventually, the "vital force" theories died and organic chemistry became the investigation of the more than seven million carbon-containing compounds. Today, organic chemists work primarily to synthesize new molecules to be used in pharmaceuticals, surfactants, paints, and coatings. They are also involved in scaling reactions from grams to tons in industrial research laboratories.

Inorganic compounds, at the time of the "vital force" theories, were those materials isolated from nonliving sources. Now, inorganic chemistry is the chemistry of all the elements except for carbon. This includes: the transition metals which coordinate with organic ligands and make up hemoglobin; the very reactive alkali metals used to make organometallic compounds in the manufacture of pharmaceutical materials; and the semi-metallic elements that have unusual electronic properties used in solar cells for the conversion of **light** into **electricity**. Inorganic chemists find employment in the production of glass, ceramics, semi-conductors, and advanced synthetic catalysts.

In 1909, Wilhelm Ostwald (1853-1932) was awarded the Nobel Prize in chemistry for his work with catalysis, a very useful technique in industrial manufacturing. This German scientist is referred to as the father of physical chemistry. This branch of chemistry is the investigation of the underlying physical processes that are responsible for chemical properties and phenomena. Physical chemistry describes the influence of **temperature**, **pressure**, concentration, and catalyst used in organic and inorganic reactions. These data give important insight into the mechanisms of the chemical change and predict the best experimental methodology for a specific manufacturing process. Physical chemists are employed in industrial, academic, and governmental laboratories to study and calculate the fundamental properties of elements and molecular compounds. This information is used to develop more efficient devices, new applications of chemicals, and better methods for measuring chemical phenomena.

Analytical chemistry is the branch of chemistry involved with the measurement and characterization of materials. Chemical analysis is divided into classical and instrumental methods. Classical, or wet, chemical analysis is the oldest form of analytical chemistry and involves the use of chemical reactions utilizing gravimetric and volumetric methodology to analyze material compositions. The use of instrumental methods for analytical analysis provides comprehensive information about chemical structure. The instrumental technique includes methods for measuring molecular spectroscopy, such as **infrared spectroscopy** (IR), **nuclear magnetic resonance** spectroscopy (NMR), **mass** spectroscopy (MS) and x-ray **crystallography**. Gas chromatography, liquid chromatography, and **electrophoresis** are examples of separation methods that utilize instrumental methods. There is a need for analytical chemists in governmental, industrial, and academic research organizations to characterize new materials and determine chemical compositions.

Aside from these traditional areas, chemistry is important in the study of polymers, biotechnology, environmental health and safety, and many non-traditional jobs such as law, marketing, and sales.

CHERENKOV, PAVEL A. (1904-1990)
Russian physicist

Pavel A. Cherenkov was a physicist whose work helped build the foundation for modern **nuclear physics**. He is best known for his discovery in 1934 of Cherenkov **radiation**, which he first noticed as a faint blue **light** in water that was absorbing radiation. In 1937, Ilya Frank and Igor Tamm established that this light was caused by radioactive particles that move faster through water than the **speed of light** does, and for their work the three men shared the 1958 Nobel Prize in physics. Cherenkov radiation has been used to design and build a sensitive tool for measuring high-energy particles, known as the Cherenkov counter, which has important uses in experimental physics.

Cherenkov was born to peasant parents in the Voronezh region of Russia on July 28, 1904. He went to Voronezh University, where he graduated in 1928 with a degree in physics and mathematics. From there he taught in a high school, then moved to Leningrad in 1930, where he began graduate studies in physics. He studied at the Institute of

Pavel A. Cherenkov.

Physics and Mathematics under Sergei I. Vavilov ; soon after he began his affiliation there, the institute was transferred to Moscow and renamed the P. N. Lebedev Institute of Physics. Cherenkov would remain there for the rest of his life.

It had long been observed that irradiated liquids produced certain kinds of **luminescence**, and Cherenkov was still a graduate student when he began researching the causes of this phenomenon. He was quickly able to establish that in most cases this light was the result of atomic and molecular activity in the **fluids** themselves, a phenomenon known as **fluorescence**. But he discovered that high-energy gamma rays, when passed through liquid, produced a light that was not so easily explained. It was Cherenkov's assumption that this light was caused by the radiation itself and not the liquid, and he set out to prove this theory in a series of taxing experiments. He used little more than a weak source of **gamma radiation** and a variety of liquids and measured the light with only his eyes. In an obituary for Cherenkov in *Physics Today*, Alexander E. Chudakov writes: "I imagine a young and enthusiastic fellow who for several years started his working day by spending an hour in a totally dark room to prepare his eyes to observe faint light, and who scrupulously repeated the observations again and again, varying the liquids and the geometry of the experiment, trying to find the clue to the nature of the puzzling radiation that now bears his name."

Cherenkov was first able to show that the gamma rays produced the same light in a number of different liquids, thus establishing that the light was not affected by the medium through which the high-energy particles passed. After verifying that the effect was not the result of the medium, he proved that this light lacked a characteristic which fluorescence had: The light was not dimmed or "quenched" by the introduction of additives. The light produced by gamma rays was also not polarized in the same way fluorescent light was, and the results of this last experiment established definitively that what Cherenkov had been observing was a new and distinct phenomenon.

Cherenkov was only able to explain the radiation he had discovered with the theoretical assistance of Ilya Frank and Igor Tamm. Frank and Tamm developed a mathematical theory that explained the phenomenon as the result of high-energy particles moving faster than the speed of light within the liquid. Light travels more slowly in a transparent medium than it does in a **vacuum**, and gamma rays have enough **energy** to pass it, creating a visible cone or wave of light, analogous to a supersonic boom when an airplane exceeds the speed of **sound**. Cherenkov, Frank, and Tamm were able to prove these theories after an additional series of experiments. Though the majority of their scientific colleagues initially showed little interest in their work, the growth of nuclear physics during and after World War II made the importance of their discovery increasingly apparent, and they were awarded the State Prize from the Soviet Government in 1946 and the Nobel Prize in 1958.

Cherenkov was always modest about his role in the identification and explanation of Cherenkov radiation. As Chudakov notes in his obituary: "Limiting his own contribution to the period of the 1930s, Cherenkov always emphasized the crucial role of Sergei Vavilov, Frank and Tamm in the discovery." Tamm went on to work with Andrei Sakharov and others on the development of the first Soviet hydrogen bomb in 1953; he also helped formulate the theoretical basis for controlled thermonuclear fusion—the use of thermonuclear **power** for generating **electricity**. The Cherenkov effect has been used by others to develop counters to identify the **velocity** of high-energy particles. Cherenkov detectors are used in instruments sent up in satellites or balloons to study primary **cosmic rays**, exploring the effect of some particles that reach Earth's atmosphere from further out in the galaxy. The detectors have also been used in studies of Cherenkov radiation under the Antarctic ice cap.

Cherenkov earned his doctorate in 1940. He became a professor of experimental physics in 1959, then a senior scientific officer at the institute. Cherenkov was elected a member of the U.S.S.R. Academy of Sciences in 1970, and he headed the department of high-energy physics of the Lebedev Institute until he died January 6, 1990.

CHERENKOV EFFECT

The water that surrounds the core of a nuclear reactor often emits a blue glow. That glow is produced by the Cherenkov effect. This phenomenon was discovered in the 1930s by the Russian physicist Pavel Alexeyevich Cherenkov.

Cherenkov was born in the Voronezh region of Russia on July 15, 1904. He graduated from the Voronezh State University in 1928 and, in 1930, was appointed a senior scientific officer at the Lebedev Physical Institute in Moscow. It was at this institute that Cherenkov carried out the elaborate experiments that revealed the cause of the **radiation** that was eventually given his name.

The Cherenkov phenomenon can be observed when high-energy radiation passes through a transparent material, such as water. It had long been considered a form of **luminescence**, somewhat similar to the **light** emitted by certain dyes when they are exposed to light. By means of an ingenious series of experiments, Cherenkov was able to show that such is not the case.

He demonstrated, for example, that "quenching" does not occur with this form of radiation. The term *quenching* refers to the gradual decrease in light intensity coming from a luminescent object. Cherenkov radiation does not display a quenching effect.

Cherenkov concluded that the radiation was produced when high-speed charged particles pass through a material at a speed that is greater than the **speed of light** in the material. The effect is similar to that of the sonic boom caused by an airplane when it travels through air faster than does the **sound** wave it produces.

For all his experimental accomplishments, Cherenkov was unable to provide a theoretical explanation for his discovery. That theory was developed by two Russian colleagues, Ilya Mikhailovich Frank and Igor Evgenevich Tamm, in 1937. Frank was born in St. Petersburg on October 23, 1908. He graduated from Moscow University in 1930 and later worked at the State Optical Institute and at the Physics Institute of the U.S.S.R. In 1944, he became professor at Moscow University.

Tamm was also born in Vladivostok, on July 8, 1895. He graduated from Moscow University in 1918. After teaching at several universities, he accepted a position at the Physics Institute of the U.S.S.R. in 1934. He died in Moscow on April 12, 1971. For their work on the Cherenkov effect, Cherenkov, Frank, and Tamm received the Nobel Prize for physics in 1958.

An important application of this discovery is the Cherenkov counter. Particles with very short lives (less than 10^{-10}) seconds) can be detected by these counters. By constructing counters from suitable materials, they can also be used to determine the **velocity** of these particles.

CHLADNI, ERNST FLORENZ FRIEDRICH (1756-1827)
German physicist

As a musician, Ernst Chladni was fascinated by **acoustics** and vibrations. Travelling throughout Europe, he demonstrated the patterns of vibrations that were produced when fine powder or sand was spread on glass plates of various shapes and sizes and a violin bow was run over the edges of the plates. Visible lines, called **nodes**, appeared at the motionless points where the powder collected, forming patterns with bilateral symmetry. The opposite sides of each line corresponded to opposing vibratory motions.

Chladni was born in Wittenberg, Germany, in 1756, the son of Ernst Martin and Johanna Sophia Chladni. His father was a jurist who forced his son to study for the law and Chladni earned his degree from the University of Leipzig in 1782. However, as soon as his father died, Chladni turned his attention to scientific investigations. Earlier scientists had explained the vibrations of strings, but vibrations on solid plates had never been examined. By classifying the shapes and identifying the corresponding pitch, Chladni demonstrated how the patterns and sounds of the plates corresponded to the **harmonics** of strings.

On his tours, Chladni performed on the keyboard instruments he had invented, types of glass harmonicas called the "euphonium" and the "clavicylinder." When not travelling, Chladni resided in Wittenberg, where he conducted experiments and wrote treatises on acoustics. His early work on vibrating plates was published in 1787. Chladni also studied the vibrations of various types of rods and measured **sound velocity** in solid rods. To measure sound velocity in gases, Chladni compared the pitch of wind instruments filled with air or with another gas. He demonstrated the existence of longitudinal **waves** that were discordant with common transverse waves and he showed how a chime should be supported and struck, so as not to excite the discordant waves. He also studied meteorites, suggesting that they came from outer **space**. Chladni died in Breslau, Germany, in 1827.

Chladni's vibrating figures inspired major research efforts into **elasticity** and acoustics. After his demonstration before Napoleon and an audience of scientists at the Paris Academy in 1808, the academy offered a prize for the best mathematical analysis of the plates. In 1816, the French mathematician Sophie Germain was awarded the prize for her differential equation that could predict the patterns formed on Chladni's plates.

See also Savant, Félix

CLAUSIUS, RUDOLF JULIUS EMMANUEL (1822-1888)

German physicist

Clausius was responsible for significant advances in the field of **thermodynamics**, and he is generally regarded as the author of the **second law of thermodynamics**, better known as **entropy**.

He was born in Köslin, Prussia (now Koszalin, Poland). His father was a schoolmaster who taught Clausius during his early school years. In 1840 he enrolled at the University of Berlin, where he studied history and science. After graduation Clausius pursued his postgraduate degree at the University of Halle, receiving his Ph.D. in 1847. Clausius went on to serve

Rudolf Julius Emmanuel Clausius.

as professor at the Zurich Polytechnic as well as the Universities of Würtzburg and Bonn.

Clausius's strength was in theoretical physics—the examination, explanation and application of theories previously introduced by other scientists. Although he rarely conducted experiments of his own, preferring to dissect and interpret the work of others, he gave new understanding to previously neglected ideas.

In 1850, just three years after receiving his doctorate, Clausius published his most influential work, an essay discussing the degradation of **energy** that had first been proposed by **Nicolas-Léonard Sadi Carnot** and later advanced by **William Thomson (Lord Kelvin)**. These earlier scientists had suggested that all energy will eventually be transformed into unusable **heat** and dissipate into the environment.

Clausius proposed that this phenomenon could be expressed as the ratio of heat content to absolute **temperature**. As the heat content within a system rose, so did the ratio, provided that the system was closed, and, thus, allowing no energy to enter or escape. In such a system heat could be created but never removed, causing the ratio to increase until the entire system reached a static state. Since scientists could not create a closed system in the laboratory, Clausius's concept remained purely theoretical; however, his explanation did serve as proof of the inevitable degradation of energy.

Fifteen years after this significant finding Clausius labeled this concept *entropy*, a word derived from the Greek for "transformation." It is now known as the Second Law of Thermodynamics, second in importance only to the knowledge that heat cannot move from a cool substance to a hot substance. In 1850, however, it was not universally welcomed by the scientific community. Only with the support of Scottish physicist **James Clerk Maxwell** were Clausius's theories eventually accepted.

One of the most controversial ramifications of the concept of entropy is its application to the future of the **universe**. It is generally assumed that the universe is the only truly closed system. If this is so, and if entropy does increase inexorably in such a system, then it must be assumed that at some point all of the energy in the universe will be converted to unusable heat. At this time the universe would exist at a single, very low temperature, deprived of the energy that keeps it going. The universe would, in effect, die.

This theory is known as the heat-death of the universe, and, although it stems from verified thermodynamic concepts, it is not widely subscribed to by cosmologists. They claim that, though entropy works on Earth, such a theory is invalid because it is possible that none of the known laws of physics may apply elsewhere in the universe.

In addition to his explanation of entropy, Clausius provided new interpretations of Hermann von Helmholtz's law of **conservation of energy**, as well as new insight into the relationship between the temperature and the **pressure** of gases. Clausius began much of the work for the kinetic theory of gases, later completed by Maxwell and **Ludwig Boltzmann**. In 1857 Clausius introduced the idea of dissociation of certain elements by an **electric current**; this concept, known as **electrolysis**, was furthered many years later by Svante August Arrhenius.

In 1870 Clausius was wounded in the Franco-Prussian war, and in 1875 his wife died, leaving him with six children. These two events combined to effectively halt his scientific career. The work he did in his later life never approached the magnitude of his first publication.

CLAUSIUS-CLAPEYRON EQUATION

According to the second law of **thermodynamics**, all Carnot engines operated between two given temperatures will have the same efficiency. A Carnot cycle, then, is any reversible cycle having two adiabatic (no **heat** added) transitions and two isothermal (no **temperature** change) transitions. If we assume that an engine works between two temperatures that differ greatly and that our working substance can undergo a phase change, then the slope of the **equilibrium** lines in a pressure-temperature diagram is the Clausius-Clapeyron equation.

Examining the Carnot cycle, we can get a clearer understanding of the above assertion. The engine contains a liquid and vapor (such as water and steam) in equilibrium at temperature (T) and **pressure** (p). The specific volume of the liquid is V_l and the specific volume of the vapor is V_v. The first stage

(from a to b in the figure below) is to undergo an isothermal expansion at temperature (T) and pressure (p) until a **mass** (m) has vaporized. The heat, Q, absorbed by the system in this step is the product of the mass converted and the latent **heat of vaporization** of the mass: Q = mL

In the next stage (from b to c), the system expands adiabatically, i.e., it absorbs no heat, until the cylinder is at a new temperature, T_2 and new pressure p_2. An isothermal compression is carried out in the third stage, maintaining T_2 and p_2. The final stage in another adiabatic process to bring the system back to point a at temperature (T) and pressure (p).

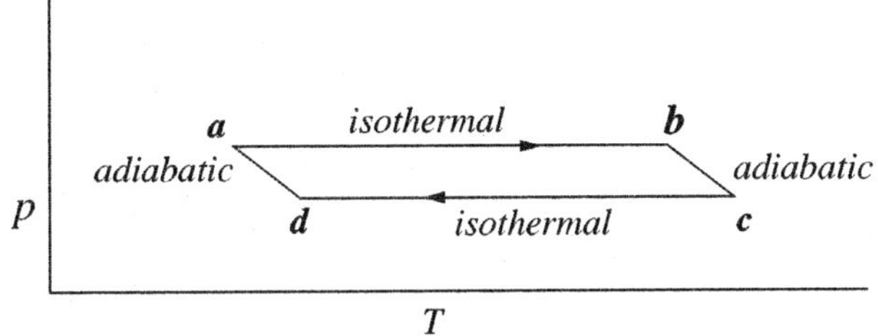

Knowing that the efficiency of such an engine is:

$$\frac{dW}{Q} = \frac{dT}{T_2}$$

and assuming the volume changes during the adiabatic processes are small (and so can be neglected), then

$$dW = m(V_v - V_\ell)dp$$

Substituting for work and solving for the ratio of the change in pressure and temp, we find:

$$\frac{dp}{dT} = \frac{L}{T(V_v - V_\ell)}$$

Now, if make the assumption that the volume on the vapor is much greater that the volume of the liquid, and that it behaves as an ideal gas, this reduces to:

$$\frac{dp}{dT} = \frac{L}{TV_v}$$

Knowing that L = ΔH and

$$V_v = \frac{nRT}{p}$$

the equation becomes:

$$\frac{dp}{dT} = \frac{p\Delta H}{nRT^2}$$

An integration over p gives us

$$\ln\left(\frac{p_2}{p_1}\right) = -\frac{\Delta H}{nR}\left(\frac{1}{T_2} - \frac{1}{T_1}\right)$$

which is the Clausius-Clapeyron equation. Comparing this equation to y=mx+b, the equation for a straight line, we see that y corresponds to ln (p) and x corresponds to 1/T. Thus, the term $\Delta H/nR$ must be the slope of the line. Using these facts, we can glean a variety of data from the equation; we can calculate ΔH just given a pressure and two temperatures and so forth.

CLOUD CHAMBER

A cloud chamber allows the observation of the tracks of high-speed particles as they pass through a saturated vapor that condenses to form drop around **ions** left in the particle's wake. Cloud chambers were an early type particle detector that provided the prototype for many modern **particle detectors**. The cloud chamber was designed and first constructed by British physicist **Charles T. R. Wilson** (1869-1959) in 1896. Wilson's scientific curiosity and ability to see beyond the immediate experiment gave scientists their first look at the world of **particle physics** as represented in a cloud chamber.

The vapor in a cloud chamber cools via adiabatic expansion. Accordingly, cloud chamber trails left by particles look somewhat like the vapor trails of jet aircraft. Just as these atmospheric vapor trails can remain in the sky for several minutes after the plane has passed and thereby provide a record of the plane's movements, the trails left in a cloud chamber provide evidence for the passage of **subatomic particles.**

In 1894, Wilson built a glass chamber which was fitted with a piston and filled with water vapor. When Wilson withdrew the piston quickly, the sudden expansion cooled the gas so that a mist formed in the cold damp atmosphere of the chamber. The water vapor had been supercooled, but instead of falling into a solution it remained a dense gas in the chamber. Wilson discovered that if he made repeated small expansions of the chamber without allowing in fresh air, the mist would disappear due to lack of ions forming on dust particles. Wilson was astounded when, despite clearing the dust and making very large expansions, he still produced a thin mist, no **matter** how many times he expanded the chamber. Wilson surmised that the drops were condensing on the ions (electrically charged particles) known to be present in the atmosphere.

Wilson set his chamber next to an X-ray source and observed a tremendous increase of condensation. At this point Wilson put his cloud chamber aside until 1910, when he passed alpha and **beta radiation** through the chamber and saw their tracks as the charged particles produced ions. For this discovery Wilson was awarded the Nobel prize in 1927.

English physicist Patrick Blackett (1897-1974), a cohort of Wilson's, designed an improved chamber. Blackett received the Nobel prize in physics in 1948 for his interpretation of data from the improved apparatus. The cloud chamber subsequently served as the prototype for **bubble chambers,**

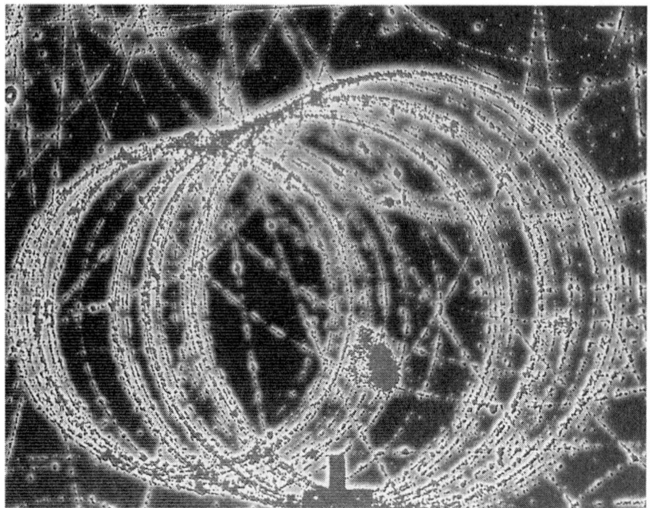

Cloud chamber. (*Photo by Lawrence Berkeley. National Audubon Society Collection/Photo Researchers, Inc. Reproduced by permission.*)

spark chambers, streamer chambers, proportional chambers, and a vast array of computer assisted devices useful in the study of modern particle physics.

See also Spark chamber

CNO CYCLE

The CNO cycle (**Carbon** Nitrogen **Oxygen** cycle) is a process carried out in the cores of **stars** in which four hydrogen nuclei are fused into one **helium nucleus** with the production of nitrogen and oxygen as intermediate products of the reaction. ^{12}C acts as an initiating catalyst for the reaction. Within a star, the CNO cycle becomes the dominant **energy** source once the **temperature** of the core exceeds about 16 million degrees Kelvin and is the main source of energy for stars with masses greater than about 1.1 solar masses where such temperatures are common. Stars the size of the **Sun** or smaller are unable to generate the core temperatures required to drive the CNO cycle and at these lower temperatures the proton-proton (p-p) fusion chain predominates. In the Sun, about 10% of the energy comes from the CNO cycle. In both processes 0.7% of the total **mass** is converted into energy, with a yield of 4.1 x 10^{-12} ergs per each helium nucleus created. The CNO cycle produces slightly less energy that can be used by the star than the p-p chain, since it produces more neutrinos. Neutrinos interact only extremely weakly with **matter**, and essentially all escape the stellar interior without causing any heating.

The rate at which the CNO cycle progresses is extremely temperature dependent. A temperature increase of 1% in the Sun's core would increase the p-p chain rate by 46%, but the CNO cycle rate would increase at 350%. This temperature sensitivity greatly changes the structure of the star's interior and affects the star's later evolution. In massive stars, 50% of energy production takes place within the innermost (by mass) 2% of the star, as opposed to solar mass stars, in which 50% of energy is produced in the innermost 10% of the star. As a result of this huge energy production within a relatively small volume, the energy is not able to escape by radiating away, and convection currents begin throughout the stellar interior. As a result, material in the core is mixed with the outer regions of the star, and a degenerate helium core never forms. In late **stellar evolution** helium fusion begins without the helium flash seen in low mass stars. Mixing also results in core material enriched by fusion with carbon, nitrogen, and oxygen being brought to the surface of the star. In medium mass stars this material is expelled as a planetary nebula where the elemental abundances can be measured. Thus, CNO enrichment serves as a benchmark for theories of stellar evolution.

There are six fundamental steps in the CNO cycle. In the initiating step in the CNO cycle, a ^{12}C captures a **proton** and becomes a radioactive ^{13}N nucleus, with the emission of gamma rays. The second step in the cycle takes place, with the ^{13}N **atom** emission of a **positron** and a **neutrino** during a transformation to the ^{13}C **isotope** of carbon. The ^{13}C atom then captures a proton, becoming an ^{14}N atom. The fourth step is the bottleneck of the CNO cycle and ultimately determines the rate of the reaction process. The likelihood of an ^{14}N atom for capturing a proton (termed the cross section) is rather low in solar mass stars. As a result, the CNO causes significant enrichment in ^{14}N in the star. Once the proton capture occurs, however, the resulting ^{15}O atom quickly decays into ^{15}N. In the final step, ^{15}N captures a proton, and decays into ^{12}C and an **alpha particle** (a helium nucleus). The net effect is to combine four protons into a helium nucleus, with carbon acting as a catalyst. In all of the CNO reaction steps there is a release of **radiation** and/or position and neutrino emissions

In the last step, one time in 2500 the ^{15}N captures another proton before decaying to become ^{16}O. Eventually, however, a ^{14}N nucleus and an alpha particle are produced. In this branch of the cycle, the net effect is to turn six protons and a ^{12}C nucleus into an alpha particle and an ^{14}N nucleus, with an enrichment in fluorine also a consequence.

See also Nuclear fusion; Nuclear reactions; Stellar evolution; Stellar life cycle; Sun

COCKCROFT, SIR JOHN DOUGLAS (1897-1967)

English physicist

Sir John Douglas Cockcroft made substantial contributions to the field of **nuclear physics**, designing the first particle accelerator and discovering the transmutations of atomic nuclei induced by artificially accelerated particles. Along with his collaborator, Irish physicist Ernest T. S. Walton (1903-1995), Cockroft received the Nobel Prize in physics in 1951 for research that profoundly influenced the subsequent course of nuclear physics.

Cockcroft was born in Todmorden, England, on May 27, 1897, where his family manufactured cotton products for several generations. He studied mathematics at Manchester University and served in World War I in the Royal Field Artillery. After the war, Cockroft returned to study mathematics and electrical engineering at St. John's College, Cambridge. Graduating in 1924, Cockroft began working under English physicist **Ernest Rutherford** (1871-1937) at the Cavendish Laboratory at Cambridge and collaborated with Russian physicist Pyotr Kapitza (1894-1984) on the production of intense magnetic fields and low temperatures. Cockcroft then turned to nuclear physics, focusing on how to accelerate protons with high voltages. In 1929, Cockcroft and Walton developed the **voltage** multiplier, which became known as the Cockcroft-Walton generator. It disintegrated lithium atoms by bombarding them with protons and, in the process, produced alpha particles. For the first time, a nuclear transmutation was produced artificially by means entirely under human control. Cockcroft and Walton continued working on the "splitting" of other atoms and, in the process, pioneered the use of **accelerators** as tools for nuclear research.

As World War II dawned, Cockcroft was appointed assistant director of scientific research in the British Ministry of Supply, where he began developing a new **radar** defense system. In 1944, he took charge of the Canadian Atomic Energy Project and in 1946, became director of the British Atomic Energy Research Establishment in Harwell, England. Cockcroft, who married Eunice Elizabeth Crabtree in 1925, had four daughters and a son. In addition to the Nobel Prize, Cockcroft received numerous awards and recognition, including a knighthood for his accomplishments in 1948. Author of *Problems of Disarmament*, a treatise on nuclear bomb proliferation and disarmament, he was awarded the Atoms for Peace Award in 1961. He died in Cambridge, England, at the age of 70.

COHEN-TANNOUDJI, CLAUDE (1933-   )

French physicist

It took a Nobel Prize for Claude Cohen-Tannoudji to receive popular attention, though he had been long recognized for his accomplished research by his peers. The French physicist was one of the winners of the Nobel Prize in 1997 (with Americans Steven Chuand William Phillips), but even before then had a distinguished career. While Cohen-Tannoudji's research focused on the general areas of atomic and molecular physics as well as quantum **optics**, his award-winning work narrowly concerned cooling atoms to near **absolute zero** (or ultracold) temperatures (such a measurement describes the rate of movement of atoms within **matter**) and trapping them, via the **light** of a laser. By slowing (i.e., cooling) the atoms, scientists can study them in more detail.

Cohen-Tannoudji was born April 1, 1933, in Constantine, Algeria, to Abraham Cohen-Tannoudji and Sarah (nee Sebba) Cohen-Tannoudji. He studied at Ecole Normale Superieure from 1953 until 1957, when he received his *agrégation*. He married Jacqueline Veyrat on November 24, 1958, and with her had three children, Alain, Joelle, and Michel. He earned his D.Sc. (Ph.D.) from the University of Paris in physics in 1962. While a graduate student, he studied under

1966 Nobel Prize winner Alfred Kastler. His thesis advisor was Jean Brossel. Following in both of his mentors' theoretical footsteps, Cohen-Tannoudji has arguably become their most distinguished student. Also, while still a student as well as shortly after graduation, from 1960-64, Cohen-Tannoudji worked as a researcher at the Centre National La Recherche Scientifique in Paris. He then served as a professor at the University of Paris from 1964-73.

In his early research, Cohen-Tannoudji studied the phenomenon that formed the foundation of his later work. Even at this stage of his career, Cohen-Tannoudji worked methodically, predicting theoretically, then demonstrating his theories through experiments. His investigations concerned atoms and their **energy** levels. He found that energy levels shifted a small amount when, in a field intensely radiated with light, they absorbed and emitted photons of light. (Photons are the smallest pieces of electromagnetic energy.) This is Cohen-Tannoudji's so-called light shift discovery.

In 1973, in addition to publishing his seminal book on quantum **mechanics**, Cohen-Tannoudji began working at the atomic and molecular physics laboratory, and as chair of the atomic and molecular physics department at the College of France. The laboratory has become recognized worldwide as a leader in research; many of the best physicists have done research there.

It was in the 1970s that Cohen-Tannoudji came up with another theory that eventually was applied specifically to **lasers**. His theory explained the interactions between high-intensity electromagnetic fields and the atoms within them. When an **atom** is surrounded by photons, it is constantly absorbing and re-emitting them. He described the atom as being "dressed" by the photons, so to speak. This observation became important when it was applied to lasers in the late 1970s and beyond.

In recognition of the continuing importance of his work, Cohen-Tannoudji became member of the Academy of Sciences in 1981. He also held memberships in other scholarly societies, among them the American Academy of Arts and Sciences, and the National Academy of Sciences.

By the 1980s, Cohen-Tannoudji theorized specifically about the light (of a laser) and its effect on the **motion** of an atom. His experiments in laser cooling were a response to other scientists' work with laser beams, cooling atoms, and trapping atoms. Cohen-Tannoudji, leading his own team of experts, focused on the modification of atomic motion when it interacted with photons. He had his first breakthrough in 1985, when he saw that an atom in motion slows in a standing light wave. As the laser beam intensifies, the atom slows increasingly. Because a material's **temperature** measures how fast its atoms are moving, this effect came to be known as "Sisyphus cooling," named for the Greek mythological personage who had to roll a stone up a hill for eternity. However, this method was limited in its cooling effects.

By 1988 Cohen-Tannoudji brought together his "dressed" atom ideas with the problems raised by the cooling and trapping atoms theories in order to try to bring the temperature of atoms closer to absolute zero. A few years later,

Cohen-Tannoudji had an unexpected breakthrough when studying an atom trapped in a standing light wave. He found that it was dependent on optical pumping, an advanced laser technology in which emits light energy in **pulses**. This approach keeps the atoms from rebounding, in effect "trapping" them. This type of Sisyphus cooling brought the temperature to 0.18 millionths of degree above absolute zero. The phenomenon had been anticipated by **Albert Einstein** and others.

In the 1990s, Cohen received important recognition for his work. In 1992 he won the Julius Edgar Lilienfeld Prize from the American Physics Society. In 1993 he won the Charles Hard Townes medal from the Optical Society of America. In 1996 he won the Harvey prize in science and technology from Technion, Israel. In that same year, Cohen-Tannoudji also won the Médaille d'Or of the National Center for Scientific Research, the highest honor a scientist can receive in France. (He was President of the organization for four years at one point.) He won the Médaille specifically for his research in atoms and related cooling methods. In 1997 Cohen-Tannoudji shared the Nobel Prize in physics for his explanation of the phenomenon and his "trapping" discovery.

Cohen-Tannoudji's work has opened up long range research possibilities, but affects basic physics as well. It can be used to study **atomic structure**, as well as help bridge the frontier between classical and **quantum physics**. Cohen-Tannoudji's theories also have proven useful for practical technology now and in the future. Ultracold atoms can be used to create more precise atomic clocks. (Atomic clocks are devices that use the constant **frequency** associated with some atomic and subatomic phenomena to define an accurate and reproducible time scale.) His theories are also influential in the development of measuring instruments of extreme precision. Indeed, an atomic laser might be possible. (An atomic laser would be like an ordinary one, save atoms would take the place of photons used in ordinary lasers.)

In 1994 Cohen-Tannoudji's 30 plus years of papers were published in one book, *Atoms in Electromagnetic Fields*. By putting them in one book, with some comments from Cohen-Tannoudji, the spectrum and development of his ideas over the period is clearly shown. Certainly, this does not mean he is finished with this area of research. Cohen-Tannoudji continued to try to cool atoms even further. As Cohen-Tannoudji told *Science* magazine, "It's exciting to see how far one can go."

COHERENCE AND INCOHERENCE

Coherence is the degree to which the difference in phase between two or more **waves** is constant over some region of time or **space**.

Beams of wave **radiation** from two sources are said to be mutually coherent when the difference between the phases of the two waves is constant. Coherence is necessary for the observation of **interference**, as in Thomas Young's slit experiment. In an interference experiment, the signal **power** (square of the wave amplitude) will be constant provided the phase difference between interfering waves does not change.

Perfect self-coherence describes the case when the phase difference of all waves in some beam of radiation is constant over all time and space. Maximally incoherent radiation is the superposition of waves such that the phase difference between interfering waves is random. Then the phase of the entire signal, the superposition of waves, is ill-defined. Most **natural radiation** has intermediate coherence.

Very coherent and very incoherent signals are commonly obtained in the laboratory. Sources of thermal noise commonly produce very incoherent radiation. For example, blackbodies emit incoherent **light**; similarly, a source of acoustic white noise radiates incoherent **sound** waves, and the Johnson noise in a resistor produces incoherent electrical signals. On the other hand, slitted screens and beam splitters are used to produce multiple beams that are mutually coherent. Highly coherent (i.e. self-coherent) light is most often obtained from lasing devices (**lasers**, masers, etc.). Stimulated emission, the process by which lasers amplify a given light **frequency**, duplicates the phase and frequency of an incoming **photon** in two outgoing photons. All photons produced then share a common phase, together comprising a very coherent laser beam.

Coherence length and *coherence time* are quantities which describe the interval of length or time for which the coherence of a beam of waves is acceptably high for a given purpose. The coherence time of a beam with a bandwidth of frequencies δf is approximately equal to $1/\delta f$.

COLD FUSION

Cold fusion is the term proposed to describe controlled **nuclear fusion** reactions occurring at or near room **temperature**. The attainment of cold fusion, with a net release of **energy**, was announced at a press conference in Salt Lake City, Utah, on March 23, 1989, by B. Stanley Pons of the University of Utah and Martin Fleishman of Southampton U niversity in England. The following day, a paper claiming the achievement of cold fusion at a much lower rate was mailed to the British journal Nature by Steven E. Jones, a professor of physics at Brigham Young University, and a group of his colleagues. The announcement by Pons and Fleishman received worldwide media coverage, but also a great deal of criticism as it did not provide adequate technical details for other scientists to independently verify what was claimed. As the details of the experiment became known, attempts to reproduce it in a number of other laboratories gave inconsistent results. In addition, Pons and Fleishman found it necessary to retract some of their initial claims. To deal with the controversy, the Energy Research Advisory Board of the United States Department of Energy appointed a fact-finding panel. By the end of the year, this panel concluded that there was "no convincing evidence that useful sources of energy would result from" the cold fusion **work**. Within a few years, the majority of chemists and phys icists reached consensus that the Fleishman and Pons experiment had failed to demonstrate the release of fusion energy on a significant scale. While some scientists believe that Jones and others following his approach had demonstrated an interesting

new effect, there is currently only limited interest in funding research pertaining to cold fusion as a **potential energy** source.

Nuclear fusion is the process responsible for the production of **heat** energy in the **sun** and other **stars**. It can be understood using a few basic ideas from **nuclear physics**. The structure of the **nucleus** is the result of two different fundamental forces: the long range electrical repulsion which exists between protons, and the short range nuclear strong **force** which is attractive and of roughly equal strength between protons and protons, protons and **neutron** and neutrons and neutrons. Just as electrons outside the **atom**, the protons and neutrons separately occupy set of energy levels within the nucleus, with the number in each level limited by the **Pauli exclusion principle**. A third fundamental force, the nuclear weak force, allows for neutrons to decompose into a **proton** plus an **electron** plus a massless particle called an antineutrino, or a proton, if it has enough energy available, to split into a neutron and a **positron** plus a massless **neutrino**. These weak-force induced properties account for the roughly equal number of protons and neutrons in the common elements, since an excess of protons or neutrons requires that the excess particles be in higher energy levels and can spontaneously change into the other type of particle.

In the sun and other stars, hydrogen is fused into **helium** in a cyclic reaction in which two protons first combine to form a deuteron (heavy hydrogen nucleus), a third proton is added to form a nucleus of helium-3, and then two helium-3 nuclei combine to form helium-4 and release two protons. This reaction occurs only in the core of the sun at a temperature of about ten million degrees Kelvin. The high kinetic energies of the particles in this case are sufficient to allow the particles to approach closely enough to allow the short-range strong force to act. Each day approximately 1.5×10^{19} kJ of energy reaches the earth's surface as a result of the fusion reactions taking place in the interior of the sun. The direct fusion of deuterons is also possible, with three possible outcomes: the formation of a helium-three nucleus and release of a neutron, the formation of a tritium (hydrogen-3) nucleus with the release of a proton, and the direct formation of a helium-4 nucleus with the emission of a **gamma ray**. Both theoretical and experimental studies indicate that the third possibility occurs with the least **frequency** and the first two with nearly equal frequency.

The harnessing of fusion energy to replace fossil and nuclear (fission) fuels has been a goal of energy researchers since the 1960s. The main obstacle to controlled thermonuclear fusion had been being able to confine plasma, a mixture of **ions** and electrons, at the extremely high temperature required for thermonuclear fusion—temperatures hot enough to melt the walls of any container. The cold fusion researchers had in mind another possibility. The tunnel effect of quantum **mechanics** allows the spontaneous fusion of the two nuclei in a diatomic deuterium molecule, although at an astronomically slow rate. If one of the bonding electrons is replaced by a **muon**, an elementary particle with the same charge as the electron and 200 times the **mass**, the bond is much shorter and **tunneling** becomes far more probable. Indeed, Steven E. Jones, was a recognized researcher in muon-catalyzed fusion.

The premise behind the cold fusion experiments was that one could achieve a comparable increase in the rate of deuterium fusion by electrochemically driving the nuclei into one of the transition metals known to have a large capacity to absorb hydrogen gas. Palladium, for example, is able to absorb six hydrogen atoms for every ten palladium atoms. The experiment would require energy input to establish the required current in an electrochemical cell, but the energy output could be determined by **calorimetry**. The initial report by Pons and Fleishman claimed a fourfold return on the energy input and a net production of helium. The experiment did not however, show a significant production of neutrons, as would be expected from the helium-3 producing process, or the level of gamma ray production to be expected with helium-4, and the original claims of heat and helium production were also brought into doubt by subsequent investigation. Many other investigators have elaborated the more careful experiment of Jones, with generally lower estimates obtained for the possible fusion rate. These studies, however, leave open the possibility that some real but as of yet unexplained phenomenon may be at the root of the reported results. It may be several years, perhaps even decades, before all of the experimental research pertaining to cold fusion is explained to everyone's satisfaction. In the meantime small pockets of research are still being conducted at several prominent research laboratories both in the United States and abroad.

COLD STARS

Stars are often classified by **temperature** and cold stars describe a variety of stars and protostars with low stellar temperatures. Cold stars have surface temperatures around as low as 1,000K to 3,000K. In contrast, hot stars have temperatures around 40,000K. Cold stars encompass lower temperature brown dwarfs (protostars in which fusion has yet to commence), and also to compact objects such as old **neutron stars**, in which fusion has ceased.

Brown dwarfs are failed protostars of less than 0.084 solar masses. Below this **critical mass**, protostars are not massive enough to develop the high core temperatures (approximately 3 million degrees) needed to fuse hydrogen into **helium**. As a result, the **energy** emitted from these stars is limited to the **heat** generated through gravitational contraction. The spectrum of these stars may be identified by the prescience of lithium, which in normal stars is destroyed during the fusion process, and also by the presence of molecules such as methane, which is also destroyed at relatively low temperatures. It has been suggested that a large fraction of the **dark matter** in the **universe** may be contained in brown dwarfs. Brown dwarfs are difficult to detect, with no reliable estimate of their numbers feasible at this time. Current theories of star formation differ greatly on the number of brown dwarfs that might exist, with estimates ranging from numbers greater than M-class stars (a majority of the main-sequence stars that comprise over 90% of observed stars), to numbers that would make brown dwarfs a scarce cosmological species. Because of their extreme dimness, these stars were

not discovered until the 1990s. The star Gliese 229B in the constellation of the Pleiades is thought to be a brown dwarf.

All protostars form within dense interstellar clouds. Gravitational perturbations arising from a variety of sources create regions of slightly higher **gravity** in which matter within the interior of the cloud begins to collapse, eventually forming a hot rotating disk within the cloud. The disk continues condensing into a protostar. A strong stellar wind begins, possibly as a result of compression of the magnetic field, which emerges along the axis of the disk's rotation. These bipolar flows, or jets, dispel the cloud, revealing the new star. The protostars which emerge are known as T Tauri stars, and exhibit a strong infrared spectrum, a result of the obscuring dust in the interstellar birth cloud. Protostars with sufficient **mass** continue to condense, begin hydrogen fusion and take their place among the hydrogen fusing main-sequence stars found on Hertzsprung-Russell diagrams.

At the other extreme of **stellar evolution**, cold stars may encompass old cold neutron stars that are the massive core remnants of stars which have undergone a series of supernovae explosions. These stars, more massive than the **Sun**, are composed mainly of neutrons and are on the order of only ten kilometers across. The **Hubble Space telescope** has successfully imaged a non-pulsar dim neutron star that will dim and cool over millions of years.

Cooler stars are redder than hotter stars. A blue-visible **color** index (B-V index) quantifies the spectra of stars. Using two different filters the blue (B) filter allows measurement of only narrow range blue-range wavelengths while the visible (V) filter only allows measurement of green-yellow wavelengths. The B-V index for stars ranges from cool stars at 2.0 to hot stars at 0 (or negative). Other methods for determining stellar temperature involve the use of Wein's law. The peak of the continuous spectrum for cool stars is found in the longer (redder) wavelengths. As stellar temperatures increase, the continuous spectrum peak shifts to shorter (bluer) wavelengths. It is also possible to relate stellar temperatures to the strength of different absorption lines in a star's spectrum.

See also Absorption spectrum; Astrophysics; Measurement of thermal energy; Stars; Stellar life cycle

COLOR

Color is a complex and fascinating subject. Several fields of science are involved in explaining the phenomenon of color. The physics of **light**, the **chemistry** of colorants, the psychology and physiology of human emotion are all related to color. Since the beginning of history, people of all cultures have tried to explain why there is light and why we see colors. Some people have regarded color with the same mixture of fear, reverence, and curiosity with which they viewed other natural phenomena. In recent years scientists, artists, and other scholars have offered interpretations of the **sun**, light, and color differences.

Color perception plays an important role in our lives and enhances the quality of life. This influences what we eat and wear. Colors help us to understand and appreciate the beauty of sunrise and sunset, the artistry of paintings, and the beauty of a bird's plumage. Because the sun rises above the horizon in the East, it gives light and color to our world. As the sun sets to the West, it gradually allows darkness to set in. Without the sun we would not be able to distinguish colors. The **energy** of sunlight warms the earth, making life possible; otherwise, there would be no plants to provide food. We are fortunate to have light as an essential part of our planet.

Colors are dependent on light, the primary source of which is sunlight. It is difficult to know what light really is, but we can observe its effects. An object appears colored because of the way it interacts with light. A thin line of light is called a ray; a beam is made up of many rays of light. Light is a form of energy that travels in **waves**. Light travels silently over long distances at a speed of 190,000 mi (300,000 km) a second. It takes about eight minutes for light to travel from the sun to the earth. This great speed explains why light from shorter distances seems to reach us immediately.

When we talk about light, we usually mean white light. When white light passes through a prism (a triangular transparent object) something very exciting happens. The colors that make up white light disperse into seven bands of color. These bands of color are called a spectrum (from the Latin word for image). When a second prism is placed in just the right position in front of the bands of this spectrum, they merge to form invisible white light again. **Isaac Newton** was a well known scientist who conducted research on the sun, light, and color. Through his experiments with prisms, he was the first to demonstrate that white light is composed of the colors of the spectrum.

Seven colors constitute white light: red, orange, yellow, green, blue, indigo, and violet. Students in school often memorize acronyms like ROY G BIV, to remember the seven colors of the spectrum and their order. Sometimes blue and indigo are treated as one color. In any spectrum the bands of color are always organized in this order from left to right. There are also wavelengths outside the visible spectrum, such as ultraviolet.

A rainbow is nature's way of producing a spectrum. One can usually see rainbows after summer showers, early in the morning or late in the afternoon, when the sun is low. Rain drops act as tiny prisms and disperse the white sunlight into the form of a large beautiful arch composed of visible colors. To see a rainbow one must be located between the sun and raindrops forming an arc in the sky. When sunlight enters the raindrops at the proper angle, it is refracted by the raindrops, then reflected back at an angle. This creates a rainbow. Artificial rainbows can be produced by spiraling small droplets of water through a garden hose, with one's back to the sun. Or, indoors, diamond-shaped glass objects, mirrors, or other transparent items can be used.

Refraction is the bending of a light ray as it passes at an angle from one transparent medium to another. As a beam of light enters glass at an angle, it is refracted or bent. The part of the light beam that strikes the glass is slowed down, causing the entire beam to bend. The more sharply the beam bends, the more it is slowed down.

Each color has a different **wavelength**, and it bends differently from all other colors. Short wavelengths are slowed more sharply upon entering glass from air than are long wavelengths. Red light has the longest wavelength and is bent the least. Violet light has the shortest wavelength and is bent the most. Thus violet light travels more slowly through glass than does any other color.

Like all other **wave phenomena**, the **speed of light** depends on the medium through which it travels. As an analogy, think of a wagon that is rolled off a sidewalk onto a lawn at an oblique angle. When the first wheel hits the lawn, it slows down, pulling the wagon toward the grass. The wagon changes direction when one of its wheels rolls off the pavement onto the grass. Similarly, when light passes from a substance of high **density** into one of low density, its speed increases, and it bends away from its original path. In another example, one finds that the speed and direction of a car will change when it comes upon an uneven surface like a bridge.

Sometimes while driving on a hot sunny day, we see pools of water on the road ahead of us, which vanish mysteriously as we come closer. As the car moves towards it the pool appears to move further away. This is a mirage, an optical illusion. Light travels faster through hot air than it does through cold air. As light travels from one transparent material to another it bends with a different refraction. The road's hot surface both warms the air directly above it and interacts with the light waves reaching it to form a mirage. In a mirage, reflections of trees and buildings may appear upside down. The bending or refraction of light as it travels through layers of air of different temperatures creates a mirage.

Similar colors can be seen in a thin film of oil, in broken glass and on the vivid wings of butterflies and other insects. Scientists explain this process by the terms, **diffraction** and **interference**. Diffraction and refraction both refer to the bending of light. Diffraction is the slight bending of light away from its straight line of travel when it encounters the edge of an object in its path. This bending is so slight that it is scarcely noticeable. The effects of diffraction become noticeable only when light passes through a narrow slit. When light waves pass through a small opening or around a small object, they are bent. They merge from the opening as almost circular, and they bend around the small object and continue as if the object were not there at all. Diffraction is the sorting out of bands of different wavelengths of a beam of light.

When a beam of light passes through a small slit or pin hole, it spreads out to produce an image larger than the size of the hole. The longer waves spread out more than the shorter waves. The rays break up into dark and light bands or into colors of the spectrum. When a ray is diffracted at the edge of an opaque object, or passes through a narrow slit, it can also create interference of one part of a beam with another.

Interference occurs when two light waves from the same source interact with each other. Interference is the reciprocal action of light waves. When two light waves meet, they may reinforce or cancel each other. The phenomenon called diffraction is basically an interference effect. There is no essential difference between the phenomena of interference and diffraction.

Light is a mixture of all colors. One cannot look across a light beam and see light waves, but when light waves are projected on a white screen, one can see light. The idea that different colors interfere at different angles implies that the wavelength of light is associated with its colors. A spectrum can often be seen on the edges of an aquarium, glass, mirrors, chandeliers or other glass ornaments. These colored edges suggest that different colors are deflected at different angles in the interference pattern.

The color effects of interference also occur when two or more beams originating from the same source interact with each other. When the light waves are in phase, color intensities are reinforced; when they are out of phase, color intensities are reduced.

When light waves passing through two slits are in phase there is constructive interference, and bright light will result. If the waves arrive at a point on the screen out of phase, the interference will be destructive, and a dark line will result. This explains why bubbles of a nearly colorless soap solution develop brilliant colors before they break. When seen in white light, a soap bubble presents the entire visible range of light, from red to violet. Since the wavelengths differ, the film of soap cannot cancel or reinforce all the colors at once. The colors are reinforced, and they remain visible as the soap film becomes thinner. A rainbow, a drop of oil on water, and soap bubbles are phenomena of light caused by diffraction, refraction, and interference. Colors found in birds such as the blue jay are formed by small air bubbles in its feathers. Bundles of white rays are scattered by suspended particles into their components colors. Interference colors seen in soap bubbles and oil on water are visible in the peacock feathers. Colors of the mallard duck are interference colors and are iridescent changing in hue when seen from different angles, beetles, dragonflies, and butterflies are as varied as the rainbow and are produced in a number of ways, which are both physical and chemical. Here, the spectrum of colors are separated by thin films and flash and change when seen from different angles.

Light diffraction has the most lustrous colors of mother of pearl. Light is scattered for the blue of the sky which breaks up blue rays of light more readily than red rays.

Materials like air, water, and clear glass are called transparent. When light encounters transparent materials, almost all of it passes directly through them. Glass, for example, is transparent to all visible light. The color of a transparent object depends on the color of light it transmits. If green light passes through a transparent object, the emerging light is green; similarly if red light passes through a transparent object, the emerging light is red.

Materials like frosted glass and some plastics are called translucent. When light strikes translucent materials, only some of the light passes through them. The light does not pass directly through the materials. It changes direction many times and is scattered as it passes through. Therefore, we cannot see clearly through them; objects on the other side of a translucent object appear fuzzy and unclear. Because translucent objects are semi-transparent, some ultraviolet rays can go through

them. This is why a person behind a translucent object can get a sunburn on a sunny day.

Most materials are opaque. When light strikes an opaque object none of it passes through. Most of the light is either reflected by the object or absorbed and converted to **heat**. Materials such as wood, stone, and metals are opaque to visible light.

We do not actually see colors. What we see as color is the effect of light shining on an object. When white light shines on an object it may be reflected, absorbed, or transmitted. Glass transmits most of the light that comes into contact with it, thus it appears colorless. Snow reflects all of the light and appears white. A black cloth absorbs all light, and so appears black. A red piece of paper reflects red light better than it reflects other colors. Most objects appear colored because their chemical structure absorbs certain wavelengths of light and reflects others.

The sensation of white light is produced through a mixture of all visible colored light. While the entire spectrum is present, the eye deceives us into believing that only white light is present. White light results from the combination of only red, green, and blue. When equal brightnesses of these are combined and projected on a screen, we see white. The screen appears yellow when red and green light alone overlap. The combination of red and blue light produces the bluish-red color of magenta. Green and blue produce the greenish-blue color called cyan. Almost any color can be made by overlapping light in three colors and adjusting the brightness of each color.

Scientists today are not sure how we understand and see color. What we call color depends on the effects of light waves on receptors in the eye's retina. The currently accepted scientific theory is that there are three types of cones in the eye. One of these is sensitive to the short blue light waves; it responds to blue light more than to light of any other color. A second type of cone responds to light from the green part of the spectrum; it is sensitive to medium wavelengths. The third type of light sensitive cone responds to the longer red light waves. If all three types of cone are stimulated equally our brain interprets the light as white. If blue and red wavelengths enter the eye simultaneously we see magenta. Recent scientific research indicates that the brain is capable of comparing the long wavelengths it receives with the shorter wavelengths. The brain interprets electric signals that it receives from the eyes like a computer.

Nearly 1,000 years ago, Alhazen, an Arab scholar recognized that vision is caused by the reflection of light from objects into our eyes. He stated that this reflected light forms optical images in the eyes. Alhazen believed that the colors we see in objects depend on both the light striking these objects and on some property of the objects themselves.

Some people are unable to see some colors. This is due to an inherited condition known as color blindness. **John Dalton**, a British chemist and physicist, was the first to discover color blindness in 1794. He was color blind and could not distinguish red from green. Many color blind people do not realize that they do not distinguish colors accurately. This is potentially dangerous, particularly if they cannot distinguish between the colors of traffic lights or other safety signals.

Those people who perceive red as green and green as red are known as red-green color blind. Others are completely color blind; they only see black, gray, and white. It is estimated that seven percent of men and one percent of women on earth are born color blind.

We often wonder why the sky is blue, the water in the sea or swimming pools is blue or green, and why the sun in the twilight sky looks red. When light advances in a straight line from the sun to the earth, the light is refracted, and its colors are dispersed. The light of the dispersed colors depends on their wavelengths. Generally the sky looks blue because the short blue waves are scattered more than the longer waves of red light. The short waves of violet light (the shortest of all the light waves) disperse more than those of blue light. Yet the eye is less sensitive to violet than to blue. The sky looks red near the horizon because of the specific angle at which the long red wavelengths travel through the atmosphere. Impurities in the air may also make a difference in the colors that we see.

There are three main characteristics for understanding variations in color. These are hue, saturation, and intensity or brightness. Hue represents the observable visual difference between two wavelengths of color. Saturation refers to the richness or strength of color. When a beam of red light is projected from the spectrum onto a white screen, the color is seen as saturated. All of the light that comes to the eye from the screen is capable of exciting the sensation of red. If a beam of white light is then projected onto the same spot as the red, the red looks diluted. By varying the intensities of the white and red beams, one can achieve any degree of saturation. In handling pigments, adding white or gray to a hue is equivalent to adding white light. The result is a decrease in saturation.

A brightly colored object is one that reflects or transmits a large portion of the light falling on it, so that it appears brilliant or luminous. The brightness of the resulting color will vary according to the reflecting quality of the object. The greatest amount of light is reflected on a white screen, while a black screen would not reflect any light.

Color fills our world with beauty. We delight in the golden yellow leaves of Autumn and the beauty of Spring flowers. Color can serve as a means of communication, to indicate different teams in sports, or, as in traffic lights, to instruct drivers when to stop and go. Manufacturers, artists, and painters use different methods to produce colors in various objects and materials. The process of mixing of colorants, paints, pigments and dyes is entirely different from the mixing of colored light.

Colorants are chemical substances that give color to such materials as ink, paint, crayons and chalk. Most colorants consist of fine powders that are mixed with liquids, wax or other substances that facilitate their application to objects. Dyes dissolve in water. Pigments do not dissolve in water, but they spread through liquids. They are made up of tiny, solid particles, and they do not absorb or reflect specific parts of the spectrum. Pigments reflect a mixture of colors.

When two different colorants are mixed, a third color is produced. When paint with a blue pigment is mixed with paint that has yellow pigments the resulting paint appears green.

When light strikes the surface of this paint, it penetrates the paint layer and hits pigment particles. The blue pigment absorbs most of the light. The same color looks different against different background colors. Each pigment subtracts different wavelengths.

All color is derived from two types of light mixture, an additive and a subtractive process. Both additive and subtractive mixtures are equally important to color design and perception. The additive and subtractive elements are related but different. In a subtractive process, blended colors subtract from white light the colors that they cannot reflect. Subtractive light mixtures occur when there is a mixture of colored light caused by the transmittance of white light. The additive light mixture appears more than the white light.

In a subtractive light mixture, a great many colors can be created by mixing a variety of colors. Colors apply only to additive light mixtures. Mixing colored light produces new colors different from the way colorants are mixed. Mixing colorants results in new colors because each colorant subtracts wavelengths of light. But mixing colored lights produces new colors by adding light of different wavelengths.

Both additive and subtractive mixtures of hues adjacent to each other in the spectrum produce intermediate hues. The additive mixture is slightly saturated or mixed with light, the subtractive mixture is slightly darkened. The complimentary pairs mixed subtractively do not give white pigments with additive mixtures.

Additive light mixtures can be used in a number of slide projectors and color filters in order to place different colored light beams on a white screen. In this case, the colored areas of light are added to one another. Subtractive color mixing takes place when a beam of light passes through a colored filter. The filter can be a piece of colored glass or plastic or a liquid that is colored by a dye. The filter absorbs and changes part of the light. Filters and paints absorb certain colors and reflect others.

Subtractive color mixing is the basis of color printing. The color printer applies the colors one at a time. Usually the printer uses three colors: blue, yellow, and red, in addition to black for shading and emphasis. It is quite difficult for a painter to match colors. To produce the resulting colors, trial and error, as well as experimenting with the colors, is essential. It is difficult to know in advance what the resulting color will be. It is interesting to watch a painter trying to match colors.

People who use conventional methods to print books and magazines make colored illustrations by mixing together printing inks. The color is made of a large number of tiny dots of several different colors. The dots are so tiny that the human eye does not see them individually. It sees only the combined effects of all of them taken together. Thus, if half of the tiny dots are blue and half are yellow, the resulting color will appear green.

Dying fabrics is a prehistoric craft. In the past, most dyes were provided solely from plant and animal sources. In antiquity, the color purple or indigo, derived from plants, was a symbol of aristocracy. In ancient Rome, only the emperor was privileged to wear a purple robe.

Today, many dyes and pigments are made of synthetic material. Thousands of dyes and pigments have since been created from natural coal tar, from natural compounds, and from artificially produced organic chemicals. Modern chemistry is now able to combine various arrangements and thus produce a large variety of color.

The color of a dye is caused by the absorption of light of specific wavelengths. Dyes are made of either natural or synthetic material. Chemical dyes tend to be brighter than natural dyes. In 1991 a metamorphic color system was created in a new line of clothes. The color of these clothes changes with the wearer's body **temperature** and environment. This color system could be used in a number of fabrics and designs.

Sometimes white clothes turn yellow from repeated washing. The clothes look yellow because they reflect more blue light than they did before. To improve the color of the faded clothes a blue dye is added to the wash water, thus helping them reflect all colors of the spectrum more evenly in order to appear white.

Most paints and crayons use a mixture of several colors. When two colors are mixed, the mixture will reflect the colors that both had in common. For example, yellow came from the sap of a tree, from the bark of the birch tree, and onion skins. There were a variety of colors obtained from leaves, fruits, and flowers.

Three colorants that can be mixed in different combinations to produce several other colors are the primary colorants. In mixing red, green, and blue paint the result will be a muddy dark brown. Red and green paint do not combine to form yellow as do red and green light. The mixing of paints and dyes is entirely different from the mixing of colored light.

By 1730, a German engraver named J. C. LeBlon discovered the primary colors red, yellow, and blue are primary in the mixture of pigments. Their combinations produce orange, green, and violet. Many different three colored combinations can produce the sensation of white light when they are superimposed. When two primary colors such as red and green are combined, they produce a secondary color. A color wheel is used to show the relationship between primary and secondary colors. The colors in this primary and secondary pair are called complimentary. Each primary color on the wheel is opposite the secondary color formed by the mixture of the two primary colors. And each secondary color produced by mixing two primary colors lies half-way between them on a color wheel. The complimentary colors produce white light when they are combined.

Color influences many of our daily decisions, consciously or unconsciously from what we eat and what we wear. Color enhances the quality of our lives, it helps us to fully appreciate the beauty of colors. Colors are also an important function of the psychology and physiology of human sensation. Even before the ancient civilizations in prehistoric times, color symbolism was already in use.

Different colors have different meanings which are universal. Colors can express blue moods. On the other hand, these could be moods of tranquility or moods of conflict, sorrow or pleasure, warm and cold, boring or stimulating. In several parts of the world, people have specific meanings for different colors. An example is how Eskimos indicate the different numbers of snow conditions. They have seventeen separate words for white snow. In the west the bride wears white, in China and in the Middle East area white is worn for mourning. Of all colors, that most conspicuous and universal is red.

The color red can be violent, aggressive, and exciting. The expression "seeing red" indicates one's anger in most parts of the world. Red appears in more national and international colors and red cars are more often used than any other color. Homes can be decorated to suit the personalities of the people living in them. Warm shades are often used in living rooms because it suggests sociability. Cool shades have a quieting effect, suitable for study areas. Hospitals use appropriate colors depending on those that appeal to patients in recovery, in surgery, or very sick patients. Children in schools are provided bright colored rooms. Safety and certain color codes are essential. The color red is for fire protection, green is for first aid, and red and green colors are for traffic lights.

Human beings owe their survival to plants. The function of color in the flowering plants is to attract bees and other insects-to promote pollination. The color of fruits attract birds and other animals which help the distribution of seeds. The utmost relationship between humans and animals and plants are the chlorophylls. The green coloring substance of leaves and the yellowish-green chlorophyll is associated with the production of carbohydrates by photosynthesis in plants. Life and the quality of the earth's atmosphere depends on photosynthesis.

COMETS

Comets are small astronomical bodies that are composed chiefly of dust and ice crystals that orbit the **Sun** along highly elliptical orbital paths. As comets approach the Sun a portion of the comet is vaporized to form a characteristic head and tail.

Observations of comets date back to the fifth century B.C., when the Greeks mistook comets for clusters of **stars**. **Aristotle** (384-322 B.C.) argued that comets arose from the Earth. During the 18th century French mathematician Joseph-Louis Lagrange (1736-1813) postulated that comets were expelled from **planets**. Significant theories of comets include the hypothesis of British astronomer R. A. Lyttleton in 1948, that a comet was actually a "flying gravel-bank" of dust particles. In 1950, American astronomer Fred L. Whipple countered with the "dirty iceball" theory. Whipple believed the **nucleus** of a comet was a conglomerate of rocky fragments and "ices" made from frozen methane, ammonia, **carbon** dioxide, water, and other substances. Even before the international efforts to observe Halley's Comet in 1985 and 1986, evidence indicted that Whipple's dirty iceball model of the comet was correct, partly because of observations made of Kohoutek's Comet in 1974.

Until the twentieth century, most new comets were discovered with the naked eye or with a **telescope**. In contemporary **astronomy**, the majority of comets are seen on photographic plates or **electron** detectors.

The overall **mass** of a comet is very small in comparison with a planet and spectroscopic analysis is utilized to determine the actual composition of comets. In 1864, Italian astronomer Giovanni Battista Donati was first astronomer to

Comet West (falling to Earth), 1976. *(Photo courtesy of NASA.)*

study the spectrum of a comet contained within the reflected solar **light** returned from a comet.

While some cometary processes are still not completely understood, the typical comet appears to contain a nucleus that may measure a few miles in diameter in larger comets. The nucleus is surrounded by a coma composed of small particles of material such as iron, nickel and magnesium in an icy mixture with frozen water, ammonia and methane. The coma contains **ions**, molecules that have lost at least one electron. These give off emissions after being excited by solar **radiation**.

A tail extends from the coma, but not every comet will have an observable tail. The tail is formed when icy particles evaporate or when small particles are driven out of the nucleus. The tail is composed of gaseous ions and/or dust. Comet tails are actually formed by different processes. When gases from a comet mix with solar wind, ion tails are created. In almost every case, the tail will point away from the sun. Most astronomers believe the direction of a comet's tail is influenced by a form of solar wind that consists of electrified low-energy particles streaming from Sun. However, the details of this process remain a subject of astronomical investigation.

Ultraviolet observations have revealed that some comets are surrounded by large clouds of hydrogen. Comet clouds have been seen on **space** probes, first in 1968 by Orbiting Astronomical Observatory No. 2, which observed the Tago-Sato-Kosaka Comet to have a hydrogen cloud 1 million miles in diameter.

In particular, Halley's Comet has long fascinated both scientists and the general public. There are records of the comet's appearance dating back to 240 B.C.. Until the 17th century the comet's appearance was often interpreted as a omen or sign from God. The name of the comet recognizes that in 1682 English astronomer **Edmund Halley** predicted the return of the comet after reading records and recognizing the periodic return of the same comet. Halley also demonstrated that the orbit of the comet was in accord with English physicist **Sir Isaac Newton**'s (1642-1727) law of gravitational attraction. The return of the comet in 1910 spurred some hysteria over whether gases from the comet might prove dangerous as the Earth passed through the path of the comet tail. In fact, although some of the gases in a comet can be toxic in dense form, they are many millions of times more rarefied than air, and thus what few molecules might survive entry into Earth's atmosphere would be harmless. As part of a concerted international study, in 1986 six spacecraft made specific studies of Halley's Comet.

See also Astronomy; Spectroscopy

COMPLEMENTARITY PRINCIPLE

The complementarity principle states that physical phenomena on the atomic scale exhibit a wave particle-duality. The principle was first advanced by Danish physicist **Niels Bohr** (1885-1962), one of the founders of **quantum theory**, in 1928.

The complementarity principle was intended to be a general framework through which the new and unfamiliar features of quantum **mechanics** could be understood. Unlike in classical physics, in **quantum mechanics** there are limits on the types of properties that can be simultaneously determined. For example, a particle such as an **electron** cannot have both its **momentum** and position determined exactly at the same time. This effect is one of the unique aspects of **quantum physics**, and is described by the Heisenberg **uncertainty principle**. Bohr's view of such physical systems was that they could exhibit features that seemed contradictory or mutually exclusive, and that this should be viewed as a guiding principle in the understanding of nature, replacing the classical concept of nature which taught that all of the physical variables in a system could theoretically be exactly determined.

The counterintuitive and paradoxical wave-particle duality of quantum particles is the central example of the complementarity principle, and it demonstrates the apparent impossibility of completely describing a particle such as an electron as being either strictly a wave or a particle. Further, the complimentarity principle, asserts that it is impossible to observe both the wave and particle characteristics at the same time. The wave-particle duality of phenomena means that elements of both particle and wave descriptions of phenomena must be synthesized to give a more accurate description of phenomena (e.g., an electron). Moreover, the principle also implies that description of natural phenomena often depends upon the type of experiment performed.

Rather than attempt, as German-born American physicist **Albert Einstein**, to eliminate these paradoxical features of quantum mechanics, Bohr chose to embrace them as the basis for a new world-view.

COMPLETE CIRCUIT • See Circuits

COMPOUND MICROSCOPE

Microscopes have been in use in various forms for more than 3,000 years. The first types were extremely simple magnifiers made of globes of water-filled glass or chips of transparent

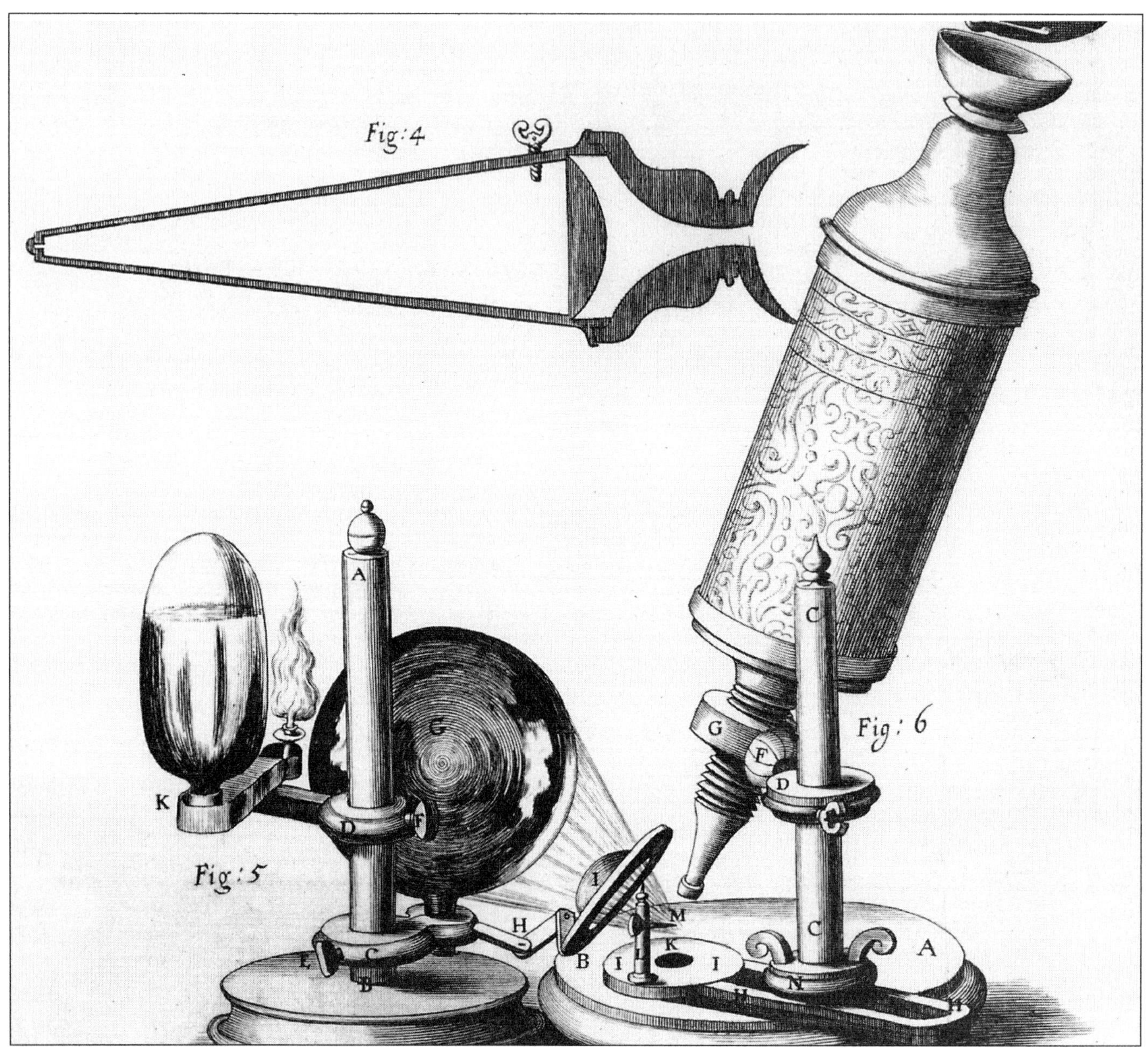

Robert Hook's compound microscope (1665).

crystal. Ancient Romans were known to use solid, bead-like glass magnifiers; Emperor Nero (A.D. 37-68) often used a bit of cut emerald to augment his poor vision.

The first **lenses**, which were used in primitive eyeglasses, were manufactured in Europe and China in the late thirteenth century. By this time, lenscrafters realized that most clear glass or crystal could be ground into a certain shape (generally with the edges thinner than the center) to produce a magnifying effect. All of these single-lens magnifiers are called simple microscopes.

Until the turn of the seventeenth century, most simple microscopes could provide a magnification of 10 (magnifying a specimen to ten times its diameter). About this time, the Dutch draper and amateur optician Antoni van Leeuwenhoek (1632-1723) began constructing magnifying lenses of his own. Though still relying upon single lenses, Leeuwenhoek's unparalleled grinding skill produced microscopes of very high power, with magnifications ranging to 500 power.

In order to achieve such results, Leeuwenhoek manufactured extremely small lenses, some as tiny as the head of a pin. Because of the very short focal length of these lenses, the **microscope** had to be held a fraction of an inch away from both the observed specimen and the observer's eye. Through his minute lenses Leeuwenhoek observed tiny "animalcules"—what we now know as bacteria and protozoa—for the first time. His findings earned him international acclaim, and the simple microscopes he designed are still among the best-crafted. However, the limitations of the single-lens magnifier were apparent to scientists, who labored to develop a practical system to increase microscope magnification.

The next breakthrough in microscopy was the invention of the compound microscope; however, the origin of this device, as well as the identity of its inventor, is the subject of some debate. Generally, credit for the invention of the compound microscope has been given to another Dutchman, the optician Zacharias Janssen (1580-c.1638). Around 1590, Janssen reportedly stumbled upon an idea for a multiple-lens microscope design, which he then constructed. Though he affirmed its ability, no record exists of Janssen actually using his invention. Currently, it is believed that Janssen's son fabricated the story. Meanwhile, yet another Dutch-born scientist, Cornelius Drebbel, claimed that he had constructed the first compound microscope in 1619. **Galileo** also reported using a two-lens microscope to examine and describe the eye of an insect.

Regardless of its inventor, the design of the original compound microscope is very similar to those used today. Two or more lenses are housed in a long tube. Individually, none of the lenses are particularly powerful; however, the image produced by the first lens is further magnified by the second (and the third and fourth, in a multiple-lens system), producing a greatly enlarged image. In addition, multiple lenses allow for a much longer focal length, permitting both the specimen and the eye a greater distance from the lenses.

The first scientist to further improve the compound design was the Englishman **Robert Hooke** in the 1670s. Hooke was the first to use a microscope to observe the structure of plants, consisting of tiny walled "chambers" that he called cells. After Hooke, little advancement occurred in microscopy until the work of Carl Zeiss (1816-1888) and Ernst Abbe in the mid-1800s. Abbe, generally recognized as the first optical engineer, took over the design duties at the Zeiss Optical Works in 1876. The scientific instruments that resulted from Zeiss and Abbe's collaboration set new standards for optical equipment. Among their inventions were lenses that corrected blurring and **color** aberrations.

By the twentieth century, the essential design and shape of the compound microscope had evolved into those we know today. Microscopes used in schools and small laboratories can achieve magnification of up to 400 power. More advanced microscopes used in research laboratories can magnify a specimen to almost 1000 power. These research microscopes often have binocular eyepieces, relying upon a series of prisms to split the image so that it may be viewed with both eyes. Even trinocular microscopes—creating a third image for a camera to view—have been designed.

At its most powerful, the practical limit for any compound microscope is 2,500 power. This limited magnification capability frustrated scientists, who, in the early twentieth century, were anxious to view the world on submicroscopic and subatomic levels. In 1931, Ernst Ruska constructed the **electron microscope** to permit such investigations.

Designed much like a compound microscope, the electron microscope uses a beam of electrons focused through magnetic lenses. Since electrons have much smaller wavelengths than does visible **light**, the electron microscope can provide much higher magnification than light-based instruments. Through electron microscopes, scientists first viewed strands of DNA. Since Ruska's invention, instruments like the **scanning tunneling microscope** and the field ion microscope have been developed with the ability to observe the activities and structures of individual atoms.

COMPTON, ARTHUR HOLLY (1892-1962)
American physicist

Arthur Holly Compton's research on the **scattering** of x-rays, which he explained by assuming that the rays consist of tiny,discrete particles now called photons, earned him a share of the 1927 Nobel Prize for physics. Compton also was a member of the **Manhattan Project** team of scientists convened during World War II, serving as director of the Metallurgical Laboratory at the University of Chicago.

Compton was born on September 10, 1892, in Wooster, Ohio. His father, Elias Compton, was a Presbyterian minister and professor of philosophy and dean at the University (later the College) of Wooster. Arthur's oldest brother, Karl, was to become professor of physics and later president of the Massachusetts Institute of Technology. His mother was the former Otelia Catherine Augspurger, who received a honorary LL.D. degree by Western College in Oxford, Ohio; in all, the six members of the Compton family eventually received more than seventy earned and honorary degrees.

Compton completed his early education at the Wooster Elementary and Grammar School, and at the Wooster Preparatory School. He entered the College of Wooster in the fall of 1909, and, under the advice of his father and brother Karl, chose science over religion as his major. He graduated from Wooster with a B.S. degree in 1913. Compton's biographers have commented on the strong religious influences exerted on him by family and his early education. Robert S. Shankland, in the *Dictionary of Scientific Biography,* for example, claims that Compton's years at the College of Wooster "had a decisive influence on [him] throughout his career. His attitudes towards life, science, and the world in general were almost completely determined as he grew to manhood and received his basic education there."

For his graduate studies, Compton chose to follow his older brothers, Karl and Wilson, to Princeton University. He had originally planned on majoring in engineering, since that field held more promise as a way of solving human problems. But he became fascinated with the challenges of pure physics and abandoned his plans for an engineering career. Working first under Nobel laureate Owen W. Richardson and later under H. L. Cooke, Compton earned his M.A. degree at Princeton in 1914 and his Ph.D. two years later. Shortly after graduation Compton married a former Wooster classmate, Betty Charity McCloskey, on June 28, 1916. They had two sons, Arthur Alan, who later served as an officer in the U.S. State

Department, and John Joseph, who became professor and head of the philosophy department at Vanderbilt University.

The three years following Compton's graduation from Princeton were marked by some uncertainty about the future direction of his career. He taught physics for one year at the University of Minnesota before accepting a job as research engineer at the Westinghouse Lamp Company in East Pittsburgh, Pennsylvania. At Westinghouse, he was active in the early development of sodium vapor and fluorescent lamps and, during the war years of 1917 and 1918, worked on aircraft instruments for the U.S. Signal Corps.

The work at Westinghouse convinced Compton that his interests lay with questions of basic research rather than with applied research or development. Consequently, he applied for and received a National Research Fellowship for 1919 to 1920 which allowed him to spend a year working at the Cavendish Laboratories at Cambridge University. There, at one of the busiest and most exciting research centers in the world at the time, Compton was able to work closely with **J. J. Thomson** and **Ernest Rutherford** in their pioneering research on the structure of atoms.

Upon his return to the United States in 1920, Compton was appointed professor of physics and head of the department at Washington University in St. Louis, Missouri. He focused his full attention on the scattering of x-rays, making use of techniques originally developed by the father-and-son team of **William Henry Bragg** and **William Lawrence Bragg**. Compton studied the changes in wavelengths of x-rays by **light** elements, especially **carbon** (in the form of graphite). He found, contrary to the laws of classical physics, that some of the scattered x-rays had wavelengths greater than those of the incident **radiation**. In order to explain this phenomenon—later named the **Compton effect** in his honor—Compton adopted the principles of **quantum theory**. Suppose, he argued, that one conceives of an x-ray as a beam of discrete particles, similar to Max Planck's "quanta." When one of these particles collides with an **electron** in a carbon **atom**, it yields some of its **energy** to the electron; the resulting lower levels of energy in the particle reflect a lower **frequency** of radiation (longer **wavelength**). Going one step further, Compton was able to derive an equation that could be used to predict the wavelength of a scattered wave under various conditions. Later research confirmed the accuracy of that equation. These results were especially important because they provided strong support for the argument that electromagnetic **waves** have a dual character. Apart from validating **Albert Einstein**'s light quantum theory proposed in 1905, Compton's discovery of wave-particle signatures inspired the development of quantum theory during the years of 1924 to 1927.

The timing of Compton's finding was also significant because it came at nearly the same moment in history when Prince **Louis Victor de Broglie** demonstrated the wave character of electrons. By the 1930s, therefore, it had become apparent that both **matter** and energy can be interpreted in terms of particles or as **wave phenomena**. In recognition of his discovery and explanation of the Compton effect, Compton was

awarded a share (with **C. T. R. Wilson**) of the 1927 Nobel Prize for physics.

Shortly after presenting conclusive proof of the wave-particle behavior of **X rays**, Compton accepted an appointment as professor of physics at the University of Chicago. During his first decade at Chicago, he extended his research on the scattering of x-rays while carrying a full teaching load. He was well liked and widely respected by both colleagues and students. Biographer Shankland says that "his contagious enthusiasm, friendliness, and great mental powers made his classes and laboratory meetings memorable experiences for all who were privileged to attend them."

Early in the 1930s, Compton became interested in a new research topic, the nature of cosmic radiation.Cosmic rays strike all parts of the earth 's atmosphere and appear to originate from deep **space**. Though first discovered by Victor Hess in 1912, their true nature remained ambiguous even by the 1930s. Some scientists argued that they were a highly energetic form of electromagnetic radiation, while others believed them to be rapidly moving charged particles.

One approach to the resolution of this debate was to study the distribution of **cosmic rays** at various points on the earth's surface. If cosmic rays are a form of energy, they will not be deflected by the earth's magnetic field but detectable at equal intensities anywhere on the earth. If they are charged particles, they will be deflected by the earth's magnetic field and will be found in greater concentrations in some latitudes than in others.

Throughout the mid–1930s, Compton recorded measurements of cosmic ray intensities at many points on the earth's surface. Fortunately, Compton and his wife loved to travel, and he was in great demand as a speaker and consultant. By 1938, Compton had collected enough data to show without question that the intensity of cosmic rays is greatest at the poles, decreasing to a minimum at the equator. It was clear that at least some cosmic radiation consists of charged particles. A similar conclusion had been obtained ten years earlier by the Dutch physicist Jacob Clay, although his results were not well known to other scientists, including Compton.

At the conclusion of his cosmic ray research, Compton began to feel the need to withdraw from active research. He had become involved in the administration of higher education and of science and had told a friend and colleague, Samuel K. Allison, that he was ready to turn over his experimental work "to younger men who could do it better." But the onset of World War II justified Compton's decision to remain in the laboratory.

As news of the discovery of **nuclear fission** reached the United States in the early 1940s, physicists began to lobby politicians and government officials about its potential application for weapons development. In 1941, President Franklin D. Roosevelt appointed a committee to explore the possibility of developing a fission bomb. The president asked Compton to serve as chairman of that committee.

The decision reached by the committee culminated in the birth of the Manhattan Project, through which the world's first atomic bombs were designed and built. As part of that effort, Compton created a facility at the University of Chicago known as the Metallurgical Laboratory and became its director. It was at the Metallurgical Laboratory that the world's first nuclear reactor was constructed and put into operation in December 1942.

At the war's conclusion, Compton was offered the post of chancellor of Washington University. He served in that position from 1945 to 1954 when he retired, but continued at Washington as Distinguished Service Professor of Natural Philosophy until resigning in 1961. For the last year of his life, Compton divided his time among St. Louis, the College of Wooster, and the University of California at Berkeley. He died of a cerebral hemorrhage in Berkeley on March 15, 1962.

COMPTON EFFECT

The Compton effect (sometimes called Compton **scattering**) occurs when an x ray collides with an **electron**. In 1923, Arthur H. Compton did experiments bouncing **x rays** off the electrons in graphite atoms. Compton found the x rays that scattered off the electrons had a lower **frequency** (and longer **wavelength**) than they had before striking the electrons. The amount the frequency changes depends on the scattering angle, the angle that the x ray is deflected from its original path. Why?

Imagine playing pool. Only the cue ball and 8 ball are left on the table. When the cue ball strikes the 8 ball, which was initially at rest, the cue ball is scattered at some angle. It also loses some of its **momentum** and kinetic **energy** to the 8 ball as the 8 ball begins to move. The x-ray **photon** scattering off an electron behaves similarly. The x ray loses energy and momentum to the electron as the electron begins to move. The energy and frequency of **light** and other electromagnetic **radiation** are related so that a lower frequency x-ray photon has a lower energy. The frequency of the x ray decreases as it loses energy to the electron.

In 1905, **Albert Einstein** explained the **photoelectric effect**, the effect that causes solar cells to produce **electricity**, by assuming light can occur in discrete particles, photons. This photon model for light still needed further experimental confirmation. Compton's x-ray scattering experiments provided additional confirmation that light can exhibit particle-like behavior. Compton received the 1927 Nobel Prize in physics for his work. Additional experiments show that light can also exhibit wave-like behavior and has a wave particle duality.

COMPUTERS

What makes computers so powerful is that they can process very quickly. This computing power is very useful in physics and science.

Traditionally there are theoretical physicists and experimental physicists. Theorists do paper and pencil calculations to derive formulae and make predictions from their formulae. Experimentalists can design experiments to test the predictions and the theories. Experimentalists may also discover unexplained phenomena in their experiments. Theorists then try to devise some theory to explain the phenomena. The problem is that formulae can not be compared directly with experiment results. It is usually difficult to get numerical results from an equation to compare with experiments. The formulae are often very complicated differential equations. Many are so complicated that they can't be solved by hand at all.

After the invention of computers, people found a new way to solve equations numerically using computers. Consequently, a new branch of physics emerged, called computational physics. Computational physicists use computers to get much more accurate results from equations and compare them with experiments. More generally, computational science can be used in engineering design where good numerical results are often crucial. Computation is also very useful to experimentalists. Changing experimental settings is usually expensive. Physicists can use computers to test different experimental settings to determine the best setting quickly and easily before actually doing the experiment. Theorists can take advantage of computers' capacity for computation as well. They can save a lot of time by using computer software to do analytical calculations such as solving equations in symbols.

Most of the important physics equations are partial differential equations. There are mainly two classes of methods to solve them numerically. The first class includes the finite difference method and the finite element method. The basic idea of the finite difference method is to set up a mesh on the region we want to study, approximating the differentiation by taking the difference of the values at mesh points, for example:

$$\frac{dy}{dt} \approx \frac{y_{i+1} - y_i}{t_{i+1} - t_i}$$

The resulting linear equations can then be solved by computers. The mesh can be rectangular, triangular, or any shape that is convenient. The finite element method divides the whole region into many small regions, often into small triangles. Then the differential equation can be approximated by some linear equations. In both methods, the linear equations are solved to get approximate numerical results for the original equations. The finite difference method is often more straightforward to implement, resulting in simpler equations to solve by computers. The finite element method has no mesh, so it can be used for any irregular region such as the surface of an airplane. The denser the mesh or the smaller the region, the more accurate the result.

Another class of methods is called the Monte Carlo method. This method is named after the famous gambling city in Europe because it uses coin tosses to do calculations. Suppose we want to get the area of a disc on a square. We throw darts randomly into the square. The probability of the dart falling into the disc is proportional to its area. We can assume that in a total of N darts, n of them fall into the disc. The area of the square is easy to determine. If we assume it is a unit square with an area of one unit, then Monte Carlo would claim the area of the disc is approximately n/N. The larger N is, the more accurate the result is.

The Monte Carlo method is easy to implement. When we solve a problem with no more than four independent variables, Monte Carlo method is usually slower than the finite difference and finite element methods. When the problem gets more complicated with more variables, Monte Carlo becomes superior compared to the other two. These methods are the main tools for computation. We have to choose the most suitable for each specific problem.

Computers always have limited computational power. On the problem side, people are always searching for better approximations and better algorithms. On the computer side, there are two approaches. The first approach is to design faster and larger computers called supercomputers. For example, the National Center for Supercomputing Applications (NCSA) has some of the largest supercomputers in the world. One of them has 512 CPUs. These supercomputers can compute with lightning speed and are called high performance computers (HPC). The biggest problem with HPC is the high cost. A less expensive approach to faster computing is available thanks to network technology. Many small computers are connected together by fast network, with many programs running in cooperation in the cluster. They usually can not do calculations as quickly, but they can do a large amount of **work** during a sustained period. This is called high through-put computing (HTC). This system has a much lower cost compared to HPC.

CONDENSED MATTER PHYSICS

Condensed **matter** physics is the study of crystalline solids, liquids, and irregularly structured materials. It is also known as **solid state physics**.

Many new technologies have resulted from condensed matter physics investigations concerning how systems of many particles are organized and how they behave. The size of the systems studied range from quantum mechanical electrons and atoms to classical macromolecular systems and solid bodies. Glasses, ceramics, organics, polymers and composite material studies are all the subject of condensed matter physics research.

Advances in condensed matter physics have broadly enhanced the quality of life. Beginning in 1910 scientists used x-ray **diffraction** to study crystalline materials. Using x-ray diffraction scientists are able to see the atoms of materials in a crystalline material and are able to formulate atomic maps and models from the x-ray diffraction data. Many complex systems, such as proteins, have been characterized using this technique.

In 1911 Dutch physicist Heike Kammerlingh-Onnes (1853-1926) found that mercury loses it **electrical resistance** at temperatures near **absolute zero**. This was the first observation of superconductivity, a phenomenon whereby conductors lose all electrical resistance. In the 1920s scientists began extensive studies on superconductor systems and the study of **semiconductors** and new conductor materials remains on the cutting edge of scientific discovery.

In 1938 a Soviet physicist studying **helium** near absolute zero observed **superfluidity** and, following WWII the pace of technological advance quickened the pace of discoveries in condensed matter physics. In 1947 the first transistor, the fore-

runner of integrated circuits and memory chips, was developed. In 1949 the first practical hardware allowing computer memory was invented by American engineer Jay Forrester (1918-). The next year x-ray diffraction studies yielded the structure of penicillin. In 1951 the first commercial computer was built and used to predicted the outcome of the 1952 United States Presidential election. In 1952 the first x-ray studies of DNA were carried out, the results of which would lead to the eventual determination of its structure. In 1954 Bell Telephone Labs developed the photovoltaic cell, a device which utilizes **light energy** to generate an electrical current.

In 1955 a Dutch physicist greatly improved the transmission of light by a glass fiber after coating the glass fiber with a material with a high refractive index. This lead to the development of efficient fiber **optics**. American physicists **John Bardeen**, Leon Cooper (1930-) and J. Robert Schrieffer (1931-) advanced a microscopic theory of superconductivity in 1957 that was awarded a Nobel Prize (1972). Their theory describes superconductivity as a quantum mechanical phenomenon wherein the conduction electrons move in pairs and hence showing no electrical resistance. Later, in 1962, British physicist **Brian Josephson** proposed the existence of **oscillations** in the electrical current flowing through two superconductors separated by a thin insulating layer in a field. The same year semiconductor **lasers** were invented by a group of scientists studying gallium arsenide. During the 1990s billions of these devices are made and sold for use in telecommunications and CD players.

In 1970 Intel made silicon chips into memory devices for **computers** and realized $9 million in sales the first year. In 1979 French physicist Pierre-Gilles de Gennes (1932-) presented his contributions to the theory of liquid crystals and polymers. After developing new methods of manufacturing these materials both types of materials now appear in a variety of applications varying from computer displays to automobile bodies. In 1980 a German physicist discovered a quantized electrical effect in semiconductors. This result is currently employed in new, highly accurate electrical standards.

In 1981 a device was developed that maps the outlines of individual atoms on surfaces by measuring the electrical current between the surface and a fine metallic tip. The next year CDs and CD players were introduced and revolutionize the music industry.

In 1995 a new state of matter was achieved when American physicists trapped 2,000 cooled metallic atoms to produce a Bose-Einstein condensate. Two years later this phenomenon was employed to build the first atomic laser.

The band theory of solids is one of the most important ideas of condensed matter physics to date. The **atomic theory** says that electrons bound to an isolated **atom** can exist only in a distinct set of atomic energy levels as determined by its **quantum numbers**. However, in a solid where atoms are bound together in a regular array, all of the energy levels are split into bands of allowed energies and separated by forbidden energies. Accordingly, each atom contributes valance electrons to fill the allowed bands.

At absolute zero all bands are filled and no **work** can be done by a body. At higher temperatures electrons are thermally excited and will move to higher energy levels. The Fermi level is the dividing line above which energy levels are usually empty and below, are full. This level helps define the conduction properties of materials. If the Fermi level is in the middle of an allowed band then the material is a conductor because electrons can move freely when the material is exposed to an electrical field. If the Fermi level is near the top of the allowed band and there is a relatively large gap separating this band and the next higher allowed band then the electrons can not move from band to band easily and the material acts as an insulator. If the Fermi level is near the top of an allowed band and there is a small energy gap between that level and the next allowed level then the material is said to be a semiconductor.

Superconducting materials have lost almost all electrical resistance but such materials usually only exist in the superconducting state at low temperatures. Recent advances in warmer **temperature** superconductors have opened new possibilities for applications of these materials.

See also Bose-Einstein condensation; Conductors and insulators; Superconductors and superconductivity; Superfluidity

CONDUCTION (HEAT)

Heat or thermal conduction is a transport phenomenon. wherein **energy transformations** result in **work** done by systems or bodies that results in a lowering of the **thermal energy** level of the originating system and a concordant increase in the thermal **energy** level of the destination system or body.

Simply put, thermal energy lowers in a body of hotter gas, liquid or solid, and increases in regions with initially lower temperatures (an indirect measure of heat). The direction of the transportation of energy always takes place in the direction in which the **temperature** decreases. The means by which this occurs, however, will vary depending on whether the body is a gas, liquid or a solid.

The simplest model of thermal "transport" or heat "flow" takes place in a gas. In the kinetic molecular model concepts of heat and temperature directly reflect molecular velocities and rates of collisions. In an ideal gas, **intermolecular forces** are neglected and so the temperature of the gas is really just a measure of the average speed of the gas molecules. A high temperature means a faster average speed for the gas. If the gas is hot in one region of its volume and colder in another region then there is a gradient of molecular velocities. In the hot region, molecules on average move faster and in the colder region the molecules will move, on average, correspondingly slower. The faster molecules, however, will give the neighboring molecules increased speed through multiple collisions. Over a period of time the faster gas molecules will slow down as their **kinetic energy** is transferred to the slower molecules. Eventually the average speed of the molecules in all regions of the gas will become the same. In our (macro) world we would say that the heat has gone from the hot region to warm up the colder region. In reality, through multiple collisions, the faster molecules have speeded up the slower molecules and the slower molecules have slowed down the faster molecules until a point of **equilibrium** is reached.

Liquid transfer operates differently because the liquid molecules, by definition, are much more tightly packed than gas molecules and therefore experience significant intermolecular forces. The molecules in a liquid are loosely tied together so that, if one molecule is displaced, neighboring molecules will immediately respond and move also. Because of the coupling that occurs between the molecules, they tend to vibrate or flow in layers rather than move freely as in a gas. Regardless, the speed of the molecules still determines the temperature of the liquid, just as they do in a gas. The more energetic vibrations are perceived as higher temperature because the molecules are moving faster. Obviously, a region of highly energetic vibration (a hot region) will directly induce movement in less energetic (colder) portions of the liquid until an equilibrium is reached. This will again be perceived as heat being "transported" from the hot region to the colder region.

A solid operates in much the same way as a liquid. Compared to liquids, the molecules that make up solids are bound into stronger lattices with regular patterns so the vibration of one molecule has an immediate effect on its neighbors. The mechanism for transferring vibrating energy is identical to the liquid example explained above. However, metallic solids and liquids, besides having their molecules bound together, also have clouds of free flowing electrons drifting through the lattices of the metal. In many important ways these electrons will behave very similarly to the molecules of gas described above. The vibrating molecules can shake their neighbors and transport kinetic energy throughout the region but the electrons can freely move and collide with each other and the molecules in the lattice. This provides a separate and additional mechanism for transporting kinetic energy from high energy to low energy regions. Because of this double mechanism, metals tend to transport heat energy very quickly compared to nonmetals. Metals are said to be good conductors of heat.

Different substances have different rates for conducting the kinetic energy (or heat) of their molecules throughout a medium. It is interesting to compare a few conductivity rates. Assuming air as a baseline, pure hydrogen is about six times faster in conducting heat. Asbestos (considered a poor conductor) is about three times faster than air. And copper (considered an excellent conductor) is about fifteen thousand times faster than air.

CONDUCTORS AND INSULATORS

Conductors and insulators are materials involved in the transfer of **sound, heat, light** or electrical charge. In general, conductors provide a medium of transfer while insulators offer a barrier or resistance to transfer.

A conductor, in the context of electrical charge, is a material through which electrical charges can easily move under the influence of electrical forces. These materials typically have an **energy** band that is only partially filled with electrons, a conduction band. The energy gap between the conduction band and the valance band, **orbitals** that contain the outer most electrons in atoms and are involved in chemical bonding, is typically small enough that an **electron** can easily move from the valence band to the conduction band thereby allowing transfer of electrical charge.

An insulator, in the context of electrical charge, is a material in which electrical charges are not allowed to easily move. In these materials the energy gap separating the conduction band and the valance band, which is typically filled in these materials, is usually large thereby effectively prohibiting electrons to make the transition between the two bands. There are materials with conductivity properties somewhat in-between conductors and insulators known as **semiconductors**. Although the valance band in these materials is filled, the energy gap between the conduction and valance bands is somewhat small and thus allows some electrons to make the transition between bands.

A conductor can be charge by a process known as induction. When a conductor is connected to the Earth by means of a conducting device it is said to be grounded since the Earth can be considered an infinite reservoir for electrons (it can accept or supply an unlimited number of electrons. In the process of induction a charged object is brought near a conductor that is insulated so that there is no conducting path to ground. The charges on the charged object repel those charges of like sign that are on the conductor, forcing them to the far side of the conductor and leaving an excess of the opposite charge on the conductor near the charged object. If a ground is then connected to the conductor some of the repelled charges leave the conductor and travel to the Earth. If the **grounding** wire is then removed, it leaves the conductor with an excess of one type of charge, either positive or negative. The charged object can then removed from the vicinity of the conductor but the induced charge is still present on the conductor. The induced charge remains on the conductor and is evenly distributed over the entire material. Charging an object by induction does not require contact between the object inducing the charge and the conductor. An insulator can be charged by rubbing, but only the rubbed area becomes charged and there is no tendency for the charge to move into other regions of the material.

A good conductor is a material that contains electrons that are not really bound to any one **atom** but are free to move around in the material but cannot easily leave. The state in which there is no net **motion** of charge within a conductor is called electrostatic **equilibrium**. When a conductor is in this state there are some important characteristics associated with the electric field in and around the material. The electric field of a conductor in this state is zero everywhere inside of the material and so no **work** is required to move a charge between any two points inside the material. Any excess charge, such as that from induction, resides entirely

Semiconductor wafers. *(Photo by John Carter, Photo Researchers, Inc. Reproduced by permission.)*

on the surface of the conductor and the **electric potential** is constant at every point on the surface of the material. If the conductor is irregularly shaped the charge tends to accumulate at the locations where the curvature is the largest (e.g., at sharp points). The electric field surrounding the conductor in electrostatic equilibrium is perpendicular to the surface of the material.

Electrical conduction is a transport activity in which electrical charges move through a system. The electrical current I is defined as the rate of flow of charge through the conducting material: $I = dQ/dt$, where dQ is the change of charge or the charge that passes through a cross section of the conductor and dt is the time it takes for the charge to move. The **electric current density**, measured in amperes, is the electrical current per unit cross-sectional area of conductor. A charge in a conductor moves because it experiences an electric **force** due to an electrical field. The higher the conductivity, the greater the current density that flows for a given applied electric field. The reciprocal of the conductivity is the resistivity, usually expressed in units of ohms, and is an important quantity in electrical applications. The resistance and current in a conductor are related via the **voltage** by $R = VI$, where R is the resistance, V is the voltage and I is the current.

See also Electric conductor; Electric current; Electric field and forces; Electrical conductivity; Electrical resistance; Electricity; Electronics

CONSERVATION OF ENERGY

Energy is defined as the capacity to do **work**. The English physicist and physician **Thomas Young** (1773-1829) was the first person to use the term energy in this sense. The concept of energy unites almost all branches of science through its various manifestations (**light**, **heat**, atomic and subatomic behavior, etc.). Energy can be converted from one form to another and the total energy in any closed system remains constant. In classical physics, this principle was known as conservation of energy; in modern physics, it is termed the **conservation of mass** and energy.

In classical physics, there are two types of energy: kinetic and potential. (Modern physics recognizes a third type, rest-mass energy.) When two moving objects collide, the first object may bounce off the second in a new direction but with no change in speed, or it may slow down, or it may be blown to pieces. The work performed by a moving object in coming to rest is known as **kinetic energy**. The larger an object is and the faster it moves, the greater will be its kinetic energy. Mathematically, the kinetic energy of a moving object is equal to one-half the product of its **mass** times the square of its **velocity**.

An object at rest near the edge of a cliff has **potential energy**. If the object falls toward the bottom of the cliff, the object's potential energy is converted into kinetic energy. The magnitude of the object's kinetic energy on impact at the bottom of the cliff will be equal to whatever work was required to raise it to its position at the top of the cliff from the ground.

The action of a pendulum provides an excellent example of the interconversion of potential and kinetic energy. The potential energy achieves maximum value when the pendulum is at either extreme of its swinging **motion**; it decreases to zero when the pendulum is vertical to the ground. The kinetic energy is zero when the pendulum is at its extreme position, and maximum when the pendulum is in the vertical position.

Another example is provided by the toss of a ball into the air. The ball's kinetic energy gradually decreases as it rises, and eventually reaches zero (having been completely transformed to potential energy) at the point that the ball stops ascending and begins descending. The fact that the ball regains its kinetic energy while descending can be taken as evidence that potential energy can be converted back to kinetic energy.

In the **Sun**, the process of thermonuclear fusion converts atoms of hydrogen into **helium** atoms, producing radiant energy. Some of this radiant energy reaches the Earth. Some of this energy reaches plants, which may eventually form coal or be used by other organisms dependent on plants for food. Part of the solar energy contributes to the evaporation of ocean water, which returns to Earth in the form of rain. And this rainwater may be converted into **hydroelectric energy** by generating plants.

In mechanical systems, there is always a certain amount of energy lost as heat due to frictional processes and inelastic collisions between moving parts. The notion that heat is a form of energy derives in part from observations in an eighteenth century arsenal that the mechanical energy expended in boring a cannon is roughly equivalent to the heat produced in the process. The discovery that heat is a form of energy led to the formulation of the first law of **thermodynamics**, i.e., that energy can neither be created nor destroyed, only converted into another form, which is a restatement of the law of conservation of energy.

As science evolved, other forms of energy were recognized. When **James Clerk Maxwell** succeeded in formulating the laws of **electromagnetism**, it became possible to recognize electrical energy as yet another manifestation of energy.

The discovery of the **neutrino** was predicted on the basis of the principle of conservation of energy. When the energy changes accompanying the decay of certain radioactive nuclei were not found to properly balance, physicists speculated that an unseen particle must be responsible for the discrepancy. It was 30 years before scientists were able to verify the existence of this particle, but eventually it was found.

CONSERVATION OF HEAT

Heat is a form of **energy (thermal energy)** derived from the **temperature** difference between a body and its surrounding system. Accordingly, the principle of the conservation of heat is implied by the **conservation of energy** contained in the first law of **thermodynamics** that states that energy cannot be created or destroyed, merely transformed from one form to another.

Thermodynamics is the study of the transformation of energy. Thermodynamics deals with quantities of energy and **entropy** known as state functions. The **first law of thermodynamics** states that energy is conserved; it can neither be created nor destroyed. The **second law of thermodynamics** states that, in a isolated system, entropy, a measure of amount of energy in a system unavailable to do **work**, must increase as time passes. The third law of thermodynamics states that the entropy of a perfect crystal is zero when the temperature of the crystal is equal to **absolute zero** (0K). There is a fourth law called the zeroth law that simply states that if substance A is in thermal **equilibrium** with substance C and substance B is in thermal equilibrium with substance C, then substances A and B must also be in thermal equilibrium. If A = C and B = C then A = B. The first law explains the relationship between energy, heat and work. The change in internal energy(ΔU) of the system of an engine is equal to the heat (Q) added to the system minus the work (W) done by the system.

The second law puts forth that the change in entropy (ΔS) of a system is equal or greater than zero. These laws lead to several conclusions that are important to the application of the laws of thermodynamics and to considerations involving thermal energy changes (heat flow).

The conservation of heat and principle of heat exchange describe the heat flow in a closed, isolated system. In natural processes, whenever two substances at different temperatures are allowed to mix (e.g., hot coffee with cold cream) the hotter body or system must lose thermal energy while the cooler system or body must gain thermal energy. This change in thermal energy is often commonly referred to as heat flow. Accordingly, the flow of heat in natural processes will always be from the hotter substance to the colder one, never from the colder to the hotter body or system.

The conservation of heat dictates that the amount of heat energy lost by the warmer body or system must equal the gain in heat energy by the cooler body or system. The natural process continues until thermal equilibrium is reached and the system settles at a temperature between the original temperatures of the two substances. To reverse the heat flow requires work be done on the system (at the expense of energy of the source of the work). Accordingly, in natural processes we never find hot coffee getting hotter and the cold cream freezing.

The efficiency of an engine is determined by thermodynamic limits on efficiency. No engine can be perfectly efficient, i.e. no device perfectly transfers its heat energy completely into work.

The most efficient engine cycle possible is described by the Carnot cycle. Carnot's reversible engine uses isothermal and adiabatic processes between two heat reservoirs at temperatures T_h and T_c. An isothermal process is one in which the temperature is held constant, but in which the volume and **pressure** change. An adiabatic process is one in which the volume is held constant but the temperature and pressure change. Accordingly, the conservation of heat is an important theoretical underpinning to the concept of the Carnot engine. Looking more closely at the ideal (as opposed to a real) Carnot engine, we can see how these processes work. A pressure-volume (P-V) diagram shows us what is happening. The cycle occurs as follows: from point a to point b, there is an isothermal expansion; from point b to c, an adiabatic process; from point c to d, an isothermal compression; from d to a, another adiabatic process.

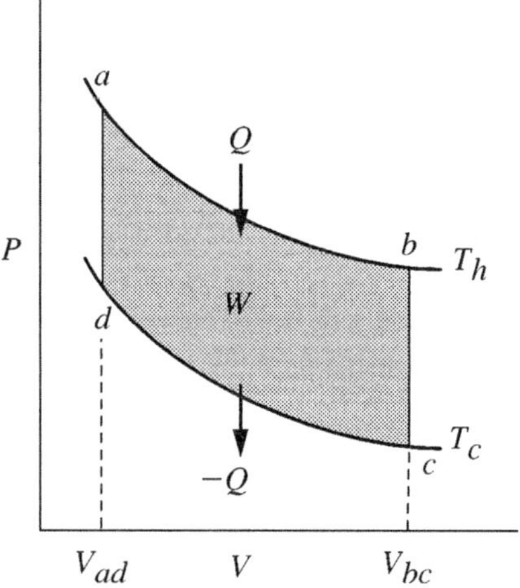

A more detailed description of the cycle would be that following the curve a-b: The gas in the cylinder is allowed to expand. Heat is added to maintain T_h and the gas does W amount of work on the piston. Along the curve b-c the volume is held constant while the gas cools. Along the curve c-d there is a reversal of the process followed along curve a-b but at T_c. An amount -W amount of work is done on the gas by the pis-

ton. Along the curve d-a there the volume is held constant while the pressure increases. This results in an increased temperature.

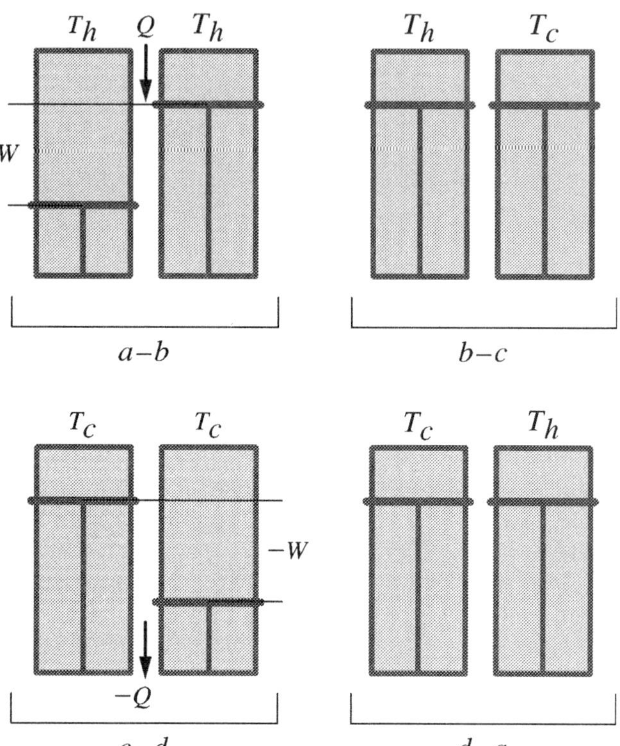

Although the conservation of heat dictates there is no net loss of heat, In practice, a perfect engine, such as the ideal Carnot engine, cannot be built. According to Carnot's Principle, an irreversible heat engine operating between T_h and T_c cannot have an efficiency greater than that of a reversible heat engine operating between the same two temperatures. A truly reversible engine is an ideal engine, but in the real world, some heat is always transferred (lost) to the outside system and so the process is irreversible. A fuel must be used to make up for the lost heat. Accordingly the efficient an engine the less fuel it requires to replace the heat lost.

A corollary to this principle is that two ideal engines operating between the same two temperatures have the same efficiency. This stems from the fact that the work done by the gas is path independent; i.e. the work done is related only to the difference in temperature, not to the path taken to reach these temperatures. Since an ideal engine will return to its original state after each cycle, the change in internal energy is zero. The change in internal energy is also equal to the heat minus the work done. Thus,

$$Q - W = 0$$
$$(Q_h - Q_c) - W = 0$$
$$(Q_h - Q_c) = W$$

The efficiency of the engine, e, is, by definition, the ratio of the work done to the heat energy utilized, Q_h.

$$\varepsilon = \frac{W}{Q_h}$$

[Substituting for work and rearranging, we see that the efficiency for an ideal engine is only temperature dependent:

$$\varepsilon = \frac{\left(Q_h - Q_c\right)}{Q_h}$$

$$\varepsilon = 1 - \frac{Q_c}{Q_h} = 1 - \frac{T_c}{T_h}$$

Thus, two engines with the same hot and cold temperatures will have the same efficiency. A real engine based on a model ideal engine will always have an efficiency that is less than that of the ideal.

A refrigerator, in contrast to an engine, uses work to transfer heat from a cold reservoir to a hot one. A cold reservoir will always have some thermal energy to be extracted unless it is at $T_c = 0$K (absolute zero), which according to the third law of thermodynamics can not be reached. happen (see the third law of thermodynamics). Refrigerators have an efficiency, but it is commonly referred to as the coefficient of performance, K. It is also exclusively temperature dependent.

$$K = \frac{Q_c}{Q_h - Q_c} = \frac{T_c}{T_h - T_c}$$

As there are no perfect engines, there are also no perfect refrigerators. Mechanical refrigeration without work being done to reverse the natural heat flow would violate the second law of thermodynamics because the entropy of the system would spontaneously decrease (have a negative sign).

See also Energy and work; Entropy; Heat; Heat engines; Heat transfer

CONSERVATION OF MASS

Although Flemish chemist Jan van Helmont (1579-1644) and British chemist **Robert Boyle** stated versions of a law of indestructibility of **matter**, it is French chemist Antoine Lavoisier (1743-1794) who is usually given the credit for first clearly expressing and proving the law of conservation of **mass**. He performed a series of careful experiments and quantitative measurements that showed clearly that there is no change in total mass in a **chemical reaction**. English chemist **John Dalton** (1766-1844) explained this observation by proposing (in 1803) that chemical compounds are made up of combinations of elements and that these elements consist of indivisible, indestructible atoms. In a chemical reaction, atoms are not destroyed nor are they changed into atoms of other elements. Atoms are simply rearranged in the product compounds. Since the products contain the same atoms as the reactants, mass is neither destroyed nor created in a chemical process.

With the discovery of the **electron** by **J. J. Thomson** in 1897 and of the **nucleus** by **Ernest Rutherford** in 1911, it became evident that atoms are, in fact, not indivisible. Atoms are made up of smaller particles: electrons, protons, and neutrons. In 1905 **Albert Einstein** (1879-1955) proposed that in very high-energy processes the mass of these **subatomic particles** may be converted into **energy**. Subsequent experimental evidence proved his proposal to be correct. In these processes, the law of conservation of mass is not obeyed. Since, however, these high-energy transformations do not occur in ordinary chemical reactions, chemists may assume that the law of conservation of mass is followed in the processes with which they are normally involved.

COORDINATE TRANSFORMATIONS

Physics is an experimental science, and therefore science of measurement. Of all possible measurements, the most fundamental is position. To quantitatively locate an object in **space** (or **space-time**) requires a coordinate system (CS). The choosing of a coordinate system is nothing more than the assignment of a label to each point in space. (In general, such a label is an *n*-tuplet of numbers, where *n* is the dimension of the space. Thus it requires three numbers to label a point in space, and four numbers to label an event in space-time.) In order to compare measurements between coordinate systems a set of equations relating the two different labels assigned to the same physical point is required. These relations are known as "coordinate transformations."

The simplest CS is the well-known Cartesian system, with three orthogonal (mutually-perpendicular) axes, usually labeled *x*, *y*, and *z*. It is not always the best choice, however. It is usually most convenient (and calculatingly most simple) to choose a CS that possess the same symmetries as the situation being modeled. Hence, for problems possessing spherical symmetry, spherical coordinates, are chosen. A point is still labeled by three numbers, but they are now the distance from the origin *r* and two angular coordinates. The coordinate transformations relate the two systems.

Two coordinate systems can differ only in their choice of origin; one is then translated with respect to the other. If the origins coincide, the axis of one system can be rotated with respect to the corresponding axis of the other. Again, a simple set of equations relates the labels of a point in one CS to the labels of the same point in the other.

It is of very deep significance that the laws of physics are invariant (have the same mathematical form) under certain coordinate transformations. No matter what CS is chosen, physical law must be invariant under coordinate translations; that symmetry leads to **momentum conservation**. Invariance under time translations leads to **energy** conservation, while the invariance of the laws of **mechanics** under rotations is at the heart of **angular momentum** conservation.

Coordinate transformations between different CSs may involve one observer, or two. One observer may analyze a problem using spherical coordinates, because the mathematics will then be simpler, but then express the result in terms of

Cartesian coordinates. Coordinate transformations between the two systems allow him to do this. In addition, two observers, each looking at the same situation, may each employ a different CS. This may occur for any of a number of reasons. Perhaps they are not at the same place (the origins of their CSs differ), or one observer must look in a different direction" than the other (his CS is rotated with respect to his colleague's). There is no physical significance attached to the use of these different systems. Each physicist will (if he analyzes the situation correctly) predict that the same thing will happen at the same place—although the coordinates used to label that place will differ. The coordinate transformations will show that the two sets of labels refer to the same point in space.

Another possibility is that the two physicists are in **motion** with respect to each other. Suppose one physicist moves with a constant **velocity** with respect to another. The space and time coordinates of the two physicist's CSs are related by a set of transformations named after the Dutch physicist **Hendrik Lorentz**. In 1905 German-American physicist **Albert Einstein** showed that the laws of physics must be invariant under the **Lorentz transformations**, which led directly to the Special theory of relativity. The requirement that physical law be invariant under general (arbitrary) coordinate transformations led Einstein to the General theory of relativity. In **relativity theory** the terms "Special" and "General" refer to the type of coordinate transformations under which the theories are invariant.

See also Accelerated reference frames; Gauge symmetry; Relativistic velocity transformations; Relativity theory; Relativity, general; Relativity, special

COPERNICUS, NICHOLAS (1473-1543)

Polish mathematician and astronomer

Nicholas Copernicus was born into a well-to-do family on February 19, 1473. His father, a copper merchant, died when Copernicus was ten, and the boy was taken in by an uncle who was a prince and bishop.

Copernicus was well able to afford a good education. He entered the University of Cracow in 1491 and studied mathematics and painting. In 1496 he went to Italy for ten years where he studied medicine and religious law. Two things happened in the year 1500 that influenced Copernicus: he attended a conference in Rome dealing with calendar reform and, on November 6, 1500, witnessed a lunar eclipse.

The tables of planetary positions that were in use at the time were very complex and inaccurate. Predicting the positions of the **planets** over long periods of time was haphazard at best, and the seasons were out of step with the position of the **Sun**. Copernicus realized that tables of planetary positions could be calculated much more easily, and accurately, if he made the assumption that the Sun, not the Earth, was the center of the solar system and that the planets, including the Earth, orbited the Sun. He first proposed this theory in 1507.

Copernicus was *not* the first person to introduce such a radical concept. Aristarchus had come up with the idea in ancient Greece long before, but the teachings of **Ptolemy** had

Nicholas Copernicus.

been dominant for 1,300 years. Ptolemy claimed the Earth was at the center of the **universe**, and all the planets (including the Sun and **Moon**) were attached to invisible celestial spheres that rotated around the Earth.

Copernicus not only wished to refute Ptolemy's universe, he claimed that the Earth itself was very small and unimportant compared to the vast vault of the **stars**. This marked the beginning of the end of the influence of the ancient Greek scientists.

Copernicus was not an especially good astronomical observer. It is said he never saw the planet Mercury (which never gets very far from the Sun), and he made an incorrect assumption about planetary orbits—he decided they were perfectly circular. Because of this, he found it necessary to use some of Ptolemy's cumbersome **epicycles** (smaller orbits centered on the larger ones) to reduce the discrepancy between his predicted orbits and those observed. It wasn't until **Johannes Kepler**'s time that this was corrected and the true nature of planetary orbits was understood.

Even so, the heliocentric model developed by Copernicus fit the observed data better than the ancient Greek concept. For example, the periodic "backward" **motion** in the sky of the planets Mars, Jupiter, and Saturn and the lack of such motion for Mercury and Venus was more readily explained by the fact that the former planets' orbits were *outside* the Earth's. Thus, the Earth "overtook" them as it circled the Sun. Planetary positions could also be predicted much more accurately using Copernicus's model.

Copernicus was very reluctant to make his ideas public. He fully realized his theory not only contradicted the Greek scientists, it went against the teachings of the Church, the consequences of which could be severe. In 1530 he allowed a summary of his ideas to circulate among scholars, who received it with great enthusiasm, but it was not until just shortly before his death in 1543 that his entire book was published. It took the efforts of the mathematician Rheticus to convince Copernicus to grant him permission to print it. Unfortunately, Rheticus had fallen afoul of official doctrine himself, and found it wise to leave town. Overseeing the publication for Copernicus's book was transferred to the hands of a Lutheran minister named Andreas Osiander (1498-1552).

Osiander now found he was in a tight spot; Martin Luther (1483-1546) had come out firmly against Copernicus' new theory, and Osiander was obligated to follow him. "This Fool wants to turn the whole Art of Astronomy upside down," Luther had said. Copernicus had dedicated his book to Pope Paul III, perhaps to gain favor, but Osiander went one step further; he wrote a preface in which he stated the heliocentric theory was not being presented as actual fact, but just as a concept to allow for better calculations of planetary positions. He did not sign his name to the preface, making it appear that Copernicus had written it and was debunking his own theory. Copernicus, suffering from a stroke and close to death, could do nothing to defend himself. It is said he died only hours after seeing the first copy of the book. (Kepler discovered the truth about the preface in 1609 and exonerated Copernicus.)

The immediate reaction to the book, *De Revolutionibus Orbium Coelestium* (*Revolution of the Heavenly Spheres*), was subdued. This was primarily due to Osiander's preface which weakened Copernicus' reputation. In addition, only a limited number of books were printed, they were very expensive, and, consequently, had limited circulation. The book did achieve a number of converts, but one had to be a mathematician to fully understand the theories. Still, it was placed on the Roman Catholic Church's list of prohibited books where it remained until 1835.

Almost as significant as proving the heliocentric solar system was possible, was Copernicus's questioning of the ancient Greek scientists. Ptolemy had bent the facts to fit his preconceived theory and his teachings had been accepted, without question, for centuries. Copernicus, on the other hand, did his best to develop his theory to match observed facts. It was the dawning of modern scientific methods.

CORIOLIS, GASPARD-GUSTAVE DE (1792-1843)

French mathematician and engineer

Gaspard-Gustave de Coriolis was a French engineer and mathematician best known for his discovery of the Coriolis effect, or **force**, which has great significance in **astrophysics**, stellar **dynamics**, and the earth sciences, such as **meteorology** and

Gustave-Gaspard de Coriolis. *(Culver Pictures. Reproduced by permission.)*

oceanography. Born in Paris in 1792 to an aristocratic family who fled Paris during the French Revolution, Coriolis and his family settled in Nancy, where his father became an industrialist. He attended the Napoleonic École Polytechnique to train as a civil servant and continued his studies at the École des Ponts et Chaussées. After several years in the corps of engineers working in the Vosges mountains, his poor health coupled with financial needs led him to accept a post as a tutor in mathematical analysis and **mechanics** at the École Polytechnique in 1816.

Coriolis published his first major work, "On the Calculation of Mechanical Action," in 1829. In it, Coriolis introduced the terms "work" and "kinetic energy" in their modern scientific meanings. Six years later he published his most famous work, "On the Equations of Relative **Motion** of a System of Bodies." Although what has become known as the Coriolis effect has figured prominently in the study of atmosphere dynamics, the scientific paper grew of Coriolis's research into industrial rotating machines, like water wheels, which were an integral part of nineteenth-century industry and engineering. The Coriolis effect is a theory of relative motion in a rotating **frame of reference**, such as an object moving longitudinally along the Earth's atmosphere. In essence, in addition to the ordinary effects of a body in motion, Coriolis said that another inertial force was acting on the body at right angles to its direction of motion. By adding this extra inertial force to the equations of motion, many atmospheric and oceanographic phenomenon could be explained, such as the curved paths of falling bodies in the Earth's atmosphere and of winds across the Earth. Contrary to popular belief, the Coriolis effect does not explain why water in a sink or toilet flows in different directions depending on whether one is in the Northern or Southern Hemisphere.

Coriolis was known for his ability to unify theory and application. He went on to write "Mathematical Theory of the Game of Billiards," published in 1835. "Treatise on the Mechanics of Solid Bodies" was published posthumously in 1844, one year after Coriolis died in Paris.

CORIOLIS FORCE

Coriolis **force** is not a true force at all; instead it is an apparent force. It manifests itself when the observer is in a rotating reference frame, such as the Earth, but the frame is assumed to be at rest, as we often assume with our environment on Earth. For example, consider two children on a merry-go-round. The children are throwing a ball to each other. To us, outside the rotating reference frame, the ball is moving in a straight line, but the children are rotating under the ball. To the children, the ball is being pushed off the straight line path by some unseen force. This fictitious force is the Coriolis force.

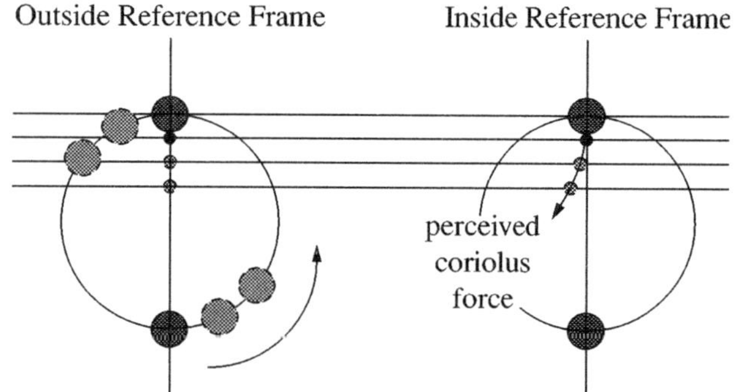

Perhaps the most prevalent application of this force is seen in nature. Since the Earth is a rotating reference frame that we often assume is not, the atmosphere experiences the Coriolis force, according to observers here on Earth. Due to the spin of the Earth, the clouds in the atmosphere experience a counter-clockwise force in the Northern Hemisphere while the clouds in the Southern Hemisphere experience a clockwise force. Hence, hurricanes and typhoons above the equator spin counter-clockwise while those below the equator spin opposite. A common misuse of the Coriolis force it to explain the swirling of the water as it goes down the drain in your sink. The misconception explains that if fill your sink with water and then remove the plug, the water will swirl down the drain counter-clockwise if you are above the equator; if you are below, it will swirl clock-wise. However, the Coriolis force, on this scale, is so small that it is negligible. It is more likely that the sink is not level or the surface is not smooth and so the water swirls for other fluid mechanical reasons.

Another atmospheric display of the Coriolis force is the curvature of a distant rain shower. If you have ever been in the position to watch an on-coming rain without any objects blocking your view, you can see how the rain appears to be curved when coming down out of the cloud. Some of this, of course, might be due to winds, but if the winds are calm, the effect you are seeing is one of the Coriolis; the rain is falling straight, but we are turning beneath it. Since we assume that we are not moving, it appears to us that a force to bending the path of the rain.

There are two main reasons for the appearance of the Coriolis force: one, the Earth rotates eastward; and two, the tangential **velocity** of a point on the Earth is a function of latitude (meaning an object's velocity is essentially zero at the poles and is a maximum at the Equator). Thus, if a cannon were fired northward from a point on the Equator, the cannonball would land to the east of its original northern target. This is due to the fact that the cannonball originally had a large eastward velocity, but then landed at a latitude that had less eastward velocity. Likewise, if the cannon were fired toward the south from the Equator, the projectile would again land to the west of its true path. The Coriolis force is therefore related to the **motion** of the object, the motion of the Earth, and the latitude. For this reason, the magnitude of the effect is given by $2v\,\omega\sin\theta$ where v is the velocity of the object, ω is the angular velocity of the Earth and θ is the angle of latitude.

CORRESPONDENCE PRINCIPLE

The correspondence principle, which **Niels Bohr** proposed in 1923, asserts that a new theory of physics should also account for the predictions of an older theory when the older theory is consistent with observations. In particular, if **Planck's constant** (h), which sets the characteristic scale for quantum mechanical phenomena is sent to zero, then in this limit, quantum **mechanics** recovers the classical **dynamics** of **Isaac Newton** and the classical **electrodynamics** of **James Clerk Maxwell**. This dictum formed a part of the Copenhagen Interpretation of **quantum theory**.

An example illustrates how to apply the idea. **Quantum mechanics** holds that all particles exhibit wave-like properties with a **wavelength** given by the relation $\lambda = 2\pi h/p$ where p is the particle's **momentum**. Since momentum is the product of **mass** and **velocity** ($p = mv$), this wavelength decreases as the mass increases. Classical mechanics, on the other hand, holds that a particle has no wavelength. Sending h to zero in the quantum theory yields this result. The classical limit is the appropriate limit to consider when studying the trajectory of a baseball, which is massive enough that quantum considerations are negligible. The limit is not appropriate to the study of an **electron**.

Bohr introduced the correspondence principle while quantum mechanics was still in its infancy. Bohr's theory of the hydrogen **atom** was consistent with the classical picture in the limit of large **quantum numbers**, but it did not explain the properties of systems more complicated than a single atom.

Newer formulations of quantum theory that sought to do more had also to explain the predictions of Bohr's model. The correspondence principle requires as much. Erwin Schrödinger's wave mechanics, which was equivalent to Werner Heisenberg's matrix mechanics, does this.

The correspondence principle serves as a useful philosophical tool in choosing a theory because it requires a model to explain more than just one particular set of phenomena. This idea also applies to theories other than quantum mechanics. For example, in the limit where speeds are much smaller than that of **light**, the predictions of the special theory of relativity match the predictions of Newtonian mechanics. The new theory, special relativity, applies to a regime where the previous theory, Newtonian mechanics, does not, and in addition, it reproduces the result of the older theory. In this way, the correspondence principle acts as a unifying thread in the development of physical theories.

Cosmic Background Explorer (COBE)

Launched on November 18, 1989, Cosmic Background Explorer (COBE) was a satellite sent into orbit to observe the cosmic background. Created through the effort of over one thousand people, the satellite carried three separate instruments for observation: DMR (Differential Microwave Radiometer), DIRBE (Diffuse Infrared Background Experiment), and FIRAS (Far Infrared Absolute Spectrophotometer). COBE is most famous for the data produced by FIRAS and DMR regarding the cosmic microwave background.

The primary goal of FIRAS was to compare the spectrum of the cosmic microwave background to that of an ideal blackbody. The instrument collected data by comparing the intensity of the cosmic microwave background to the intensity of a calibrated blackbody for a given **wavelength** of **radiation**. FIRAS covered the range of the spectrum from 0.1 mm to 10 mm in wavelength. After ten months of operation, results showed that the spectrum of the cosmic microwave background was nearly identical to that of an ideal blackbody of **temperature** 2.728K.

DMR was designed to measure the uniformity of the cosmic microwave background and to search for fluctuations in intensity across the sky. This instrument searched for intensity differences at frequencies of 31.5 GHz, 53 GHz, and 90 GHz, frequencies at which little galactic emission is assumed to interfere with results. A full sky survey conducted over four years provided conclusive evidence that the cosmic microwave background was uniform at 2.728K to better than one part in a thousand. Results also demonstrated nonuniformities that appeared as splotches across the sky. These splotches differ from the otherwise uniform background by only one part in 100,000. The experiment was designed to measure to a precision of seven angular degrees (the **moon** is half a degree in diameter). So when compared to a clear part of the sky, the typical fluctuation had a size of seven angular degrees and an intensity of 35 microkelvin. There are of

Cosmic Background Explorer (COBE), designed to study the big bang theory (1989). *(Image courtesy of NASA.)*

course smaller ripples in the fabric of the cosmic microwave background that are currently under study with ground-based and balloon-based instruments.

The results of the COBE experiment support several predictions of the **big bang** theory. First of all, theorists had predicted that a big bang implied a cosmic microwave background that cooled with expansion exactly as a blackbody. The FIRAS results proved that the spectrum of the cosmic **background radiation** is nearly identical to that of a 2.728K blackbody. Secondly, theory predicted that the **universe** should expand evenly in all directions, therefore the cosmic microwave background should appear fairly uniform across the sky. The first result of the DMR experiment was to show that the cosmic microwave background is indeed constant in all directions. Finally, and possibly most important, the DMR experiment provided an explanation for large scale structure. The splotches that appear in the cosmic microwave background provide information about the distribution of **matter** in the early universe. The radiation that comprises the cosmic microwave background last scattered from matter when the universe was only a few hundred thousand years old (the present age of the universe is thought to be near 15 billion years). The ripples observed by the DMR experiment result from primordial pockets of matter that later evolved into galactic clusters, superclusters and possibly larger structures. In fact, many of the anisotropies in the cosmic microwave background are larger than any structures known today.

See also Blackbody radiation; Cosmology

COSMIC BACKGROUND RADIATION

The 1965 discovery of cosmic **background radiation** by Arno A. Penzias and Robert W. Wilson provided important evidence in support of **big bang** based cosmological models first proposed after American astronomer Edwin Hubble's (1889-1953) discovery that the **universe** was expanding. Big bang models assert that the universe started as an extremely small and highly energetic plasma of all kinds of particles. As this **matter** expanded, the form of the matter changed as it underwent many transitions. Cosmic background **radiation** is a relic of one of these cosmological transitions, and it remains the best evidence to verify the big bang theory. In addition, precise measurements of cosmic background radiation allow physicists to study the formation of structure in the early universe.

The early stages of the universe were radiation dominated, because all particle energies were so high that they could be considered without **mass**, as in photons. **Scattering** reactions between charged particles were constantly taking place and photons were continuously absorbed and emitted. As the mean free path of a **photon** was extremely short, the universe was essentially opaque. As the universe expanded, it cooled, and particle energies decreased. As energies decreased, certain scattering reactions could no longer take place, and in other cases, bound states of particles formed.

At a **temperature** of about 3,500K, the energies of free nuclei and electrons became low enough that they began to form atoms. Once the atoms had formed, interactions between **light** and matter became increasingly infrequent, and the universe became transparent to light. The radiation, mostly in the visible and nearby **frequency** range, could suddenly propagate over long distances without being absorbed. Because the radiation was emitted long ago, and the universe has undergone much expansion, the light has been redshifted to a lower frequency. The redshift is similar to the Doppler shift; because every point of the universe is expanding away from us, the frequency of the light emitted is decreased. The redshift also can tell us when radiation was emitted by what measure it has been shifted. In the case of the background radiation, the redshift is significant enough that the visible frequencies are shifted down to the microwave region, indicating that the atoms formed when the universe was about a million to ten million years old (the universe is approximately 10 to 15 billion years years old.)

The spectrum of radiation matches the spectrum of a blackbody at about 3K. It is extremely uniform; the temperature difference between any two points in the sky is found to be less than 0.01%. This temperature uniformity is one of the hallmarks of the big bang theory. The atoms forming at the time the background radiation was emitted began to clump together due to **gravity**, eventually leading to galaxies. Some differences in the spectrum at different points in the sky, therefore, can be expected due to structure formation. Satellites observing the cosmic background radiation continue to provide data to be used in new theories of how these structures formed.

See also Blackbody radiation; Cosmology

COSMIC INFLATION • See Inflationary universe

COSMIC RAYS

Cosmic rays consist of **subatomic particles** which have high **energy** (those not from our own **sun**) due to their high velocities. Victor Franz Hess (1883-1964) showed in 1911 that the ionization of the Earth's atmosphere increases with altitude. Hess made ten ascents in balloons with detectors over 1911-12. He reached altitudes of 5000 meters. He came to the conclusion that a powerful **radiation** entered Earth's atmosphere and then diminished as it progressed toward the ground. This radiation he called "cosmic rays".

Bruno Rossi (1905-) demonstrated how penetrating cosmic rays were. At first thought to be **gamma radiation** it was proved in the 1920's that cosmic rays are actually high energy particles and are a constant "cosmic rain" falling onto the Earth. It was also discovered that the intensity of the radiation was dependent on latitude. This finding indicated that the particles which make up cosmic rays are electrically charged and are deflected by the Earth's magnetic field.

Each cosmic ray particle has an **electric charge**, a rest **mass**, and energy which is derived from its rest mass and **velocity**. As the particles hurl through the atmosphere they collide with the nuclei of atoms in the upper atmosphere producing showers of secondary particles such as protons, neutrons, **light** nuclei, and charged and neutral **pions**. The pions shortly decay into **gamma ray** photons which go on to produce electrons and positrons. Charged pions can transmutate into penetrating muons and neutrinos which can travel deeply underground.

About 87% of primary cosmic rays are protons and 12% are alpha particles (**helium** nuclei) but heavier elements are also present. They are classified as light (lithium, beryllium, and boron), medium (**carbon, oxygen**, nitrogen, and fluorine), and heavy (the remainder of the elements—even up to uranium).

The origin of cosmic rays remains somewhat enigmatic. The sun emits low energy cosmic rays during solar flares. But the heavier and highly energetic particles for the most part are believed to come from supernovas, giant exploding **stars**. There is disagreement about whether the supernovas involved are only from the Milky Way or are also from distant galaxies. It is considered as of the year 2000 that the majority of cosmic rays arise within our own galaxy. This conclusion is based on fragmentation of particles which should be much different if greater distances were involved—there should be far fewer large particles. Also, with the great distances that would have to be traveled a large number of particle—photon collisions would occur by the cosmic ray electrons, greatly decreasing the energy from what is actually recorded. Finally, satellite mapping has shown a strong concentration of high-energy gamma rays toward our Milky Way galactic equator. A more widely spread distribution would be expected if the primary 700MeV (million electron volts) protons which collide with interstellar hydrogen to produce neutral pions which then decay into high-energy gamma ray photons were extragalactic.

There is also some disagreement as to whether Type I or Type II supernovas are the primary producers. Type I supernovas are thought to occur in the late stages of binary star system evolution. One of the stars is a white dwarf while its partner is a normal star with an extended atmosphere allowing material to collect at the surface of the high **gravity** white dwarf. Once the mass of the white dwarf reaches a certain limit a massive explosion occurs blasting particles into **space**.

The Type II supernovas are more abundant and probably are the final stages in the evolution of a star similar to our sun. These stars eventually become as small as 6.2 mi (10 km) across but with a magnetic field 10 the strength of the Earth's. Rapid rotation and strong magnetic field combine to hurl cosmic ray particles into space. Recent data using the VLA (very large array) radiotelescopes in New Mexico indicate that indeed the Type II supernovas are the source of the great majority of cosmic rays, if not the only source. And again most of those sources are considered to be intragalactic.

The paths of cosmic rays are incredibly tangled. The particles are subject to the interstellar magnetic field since the particles are charged. It can be deduced that a typical carbon or oxygen particle has traveled a distance of almost thirty times the Milky Way galactic diameter between its origin and its entry into Earth's atmosphere. Obviously, there is interstellar **matter** with which the particles may react thus transmutating, or undergoing multiple glancing collisions.

There are cosmic rays with little or no mass. Essentially this category includes electrons and neutrinos. All photons are not included but are used as diagnostics of distant cosmic rays.

Electrons are produced along with protons and heavier nuclei in supernova explosions. To these electrons are added others which result from interstellar collisions of nuclear particles. Most of these collisions are cosmic ray protons colliding with hydrogen nuclei (also a **proton**). Approximately equal numbers of positively and negatively charged pions are produced which decay almost instantaneously into muons, which, after about 2 microseconds, decay into electrons and positrons. About 20% of these primary cosmic rays are positrons. As these particles collide with Earth's atmosphere a cascade of photons occurs.

Right now you are being pierced again and again by neutrinos, you do not feel them and they have no effect on you. But they do represent our third and final type of cosmic ray. Along with the heavier nuclear particles, and the nearly massless **electron** (or **positron**), the **neutrino** is a true cosmic ray. But it has neither mass or electric charge. It does carry energy and **momentum** and therefore can interact with nuclei and electrons but rarely does so. In fact while passing through the

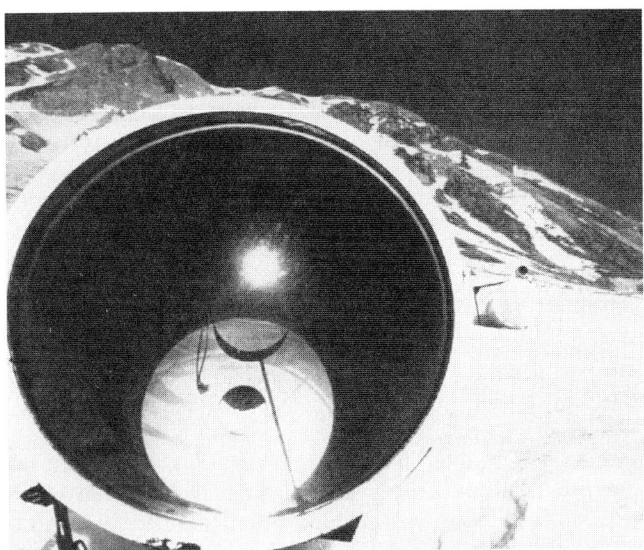

Cosmic rays research. *(Photo by Tommaso Guicciardini. Photo Researchers, Inc. Reproduced by permission.)*

entire Earth a neutrino has only a few chances in a trillion of having a collision.

Neutrinos can be emitted in the decay of radioactive nuclei or come from mesons. When a pion decays to a **muon** a neutrino is produced, and when the muon decays to an electron 2 neutrinos are produced. Neutrinos are generated from supernova explosions and in our sun's core. Our galaxy is in a hailstorm of neutrinos.

But neutrinos are difficult to study, Since they have no charge we must wait for them to be picked up by special detectors such as deep underground chambers filled with ultrapure water or extremely pure perchlorethylene (a cleaning fluid), and even then the results are confusing. Particle physicists know that there are 3 types of neutrinos: the electron-neutrino, the muon-neutrino, and the tau-neutrino. And as of the year 2000 they are manipulated at will by scientists at high-energy physics facilities. But they remain fundamentally elusive. Some theorists have arrived at a mass for the neutrino, usually no more than 1/50,000 of an electron mass. Some have said that neutrinos may change our ideas regarding our sun, and supernovas. But speculation begs for proof, and we can only await the next intragalactic supernova to gain more information. Such data may change many of our present cosmological ideas and theories.

COSMOLOGICAL ARROW OF TIME • See Arrow of time

COSMOLOGICAL CONSTANT

The cosmological constant is a mathematical term (symbolized by the Greek letter lambda, λ) that German-born physicist

Albert Einstein placed in his general relativity equations in 1917. The theory of general relativity predicted that the **universe** must either expand or contract, but Einstein did not believe this was possible. Einstein attempted to fix his equations by adding the cosmological constant, which served to halt the expansion. It can therefore be thought of as a positive **energy** attributed to **space** that, if exactly balanced against the **force** of **gravity**, will make the universe static.

In the decade following the formulation of general relativity, scientists proved that the universe is not static, contrary to Einstein's belief. In the early 1920s, Russian meteorologist Alexander Friedmann proposed an expanding universe model, now the generally accepted the **big bang** theory. A few years later, astronomer **Edwin Hubble** calculated that the majority of the galaxies he observed recede from the Milky Way in speeds proportional to their distance from Earth, indicating that the universe expands. Einstein conceded that he had made a mistake and declared the cosmological constant to be "the biggest blunder of my life."

Many scientists continue to believe that the cosmological constant exists. Although the universe is expanding, and Einstein's original reason for inventing the constant is obsolete, there may be room for it elsewhere in cosmological theory.

Einstein made up the cosmological constant because he was attempting to reconcile a theory with the established view of what the universe actually looks like. When current scientists attempt to do the same thing, they experience problems. A cosmologist who thinks about the fate of the universe may ask, Will the universe continue to expand forever? Will the force of gravity slow it down enough so that it eventually collapses under its own **weight**? Will the universe slow down in another way? Knowing the **density** of the universe and its rate of expansion can help answer these questions.

Cosmologists and astronomers have predicted the age of the universe based on estimates of its density and observations of **stars** in distant galaxies. If the **matter** in the universe is very dense, as many cosmologists assert, then the universe is quite young. The predicted age of the universe (between 8 and 14 billion years) is younger than the calculated age of some stars. As this prediction appears impossible, it is attractive to scientists that there may exist a cosmological constant.

One explanation is that the universe is slowing in its expansion due to gravity. The rate that it slows depends upon its density and, as mentioned above, many scientists believe the density is such that the universe is younger than expected. The existence of a cosmological constant, a positive energy attributed to space, could allow the universe to slow down irregularly. In other words, a non-zero cosmological constant (i.e., one that exists) means the expansion of the universe happens at different rates. If this is true, cosmologists can justifiably add more time to the age of the universe. A non-zero cosmological constant could therefore reconcile theory and observation.

Some scientists believe that this kind of reconciliation is oversimplified. It is strange, they argue, that a number should have the exact value needed to explain the universe as it exists in current theories.

See also Relativity, general

COSMOLOGICAL HEAT DEATH

Cosmological **heat** death is the name given to the fate of the **universe** if it continues to expand forever. In standard cosmological models, the universe is finite, so it has a finite amount of **energy**. Because the universe is expanding, the energy is being spread throughout a larger and larger volume; less energy in a given volume means that the **temperature** in that volume is decreasing. Eventually, the energy **density** becomes so low that the temperature approaches **absolute zero**. At this point, heat death sets in and, for all practical purposes, all activity ceases from that **time** onward.

Soon after the discovery of general relativity by **Albert Einstein** in 1915, the theory was applied to the entire universe by Friedmann, Robertson, and Walker (FRW). The solution of Einstein's equations in this case predicted that the universe was expanding, prompting Einstein to introduce his flawed **cosmological constant** to counteract the expansion. **Edwin Hubble**'s 1929 discovery that the universe was indeed expanding embarrassed Einstein, but opened up a new era in cosmological predictions.

The most important parameter in the FRW model is the density of the universe today. A critical density determines whether the universe will eventually contract back to a point (the big crunch) or whether it will continue expanding forever. If the density is greater than the critical density, the universe will at some point begin to contract. If the density is less than the critical density, then the universe will expand forever at an accelerating rate. If the density is equal to the critical density, then the expansion will continue forever but will approach zero after a long time.

When the density is less than or equal to the critical density, the expansion never stops, leading to cosmological heat death. Based on current observations, this heat death should occur when the universe is about 10^{98} years old; at this time, galactic **black holes** will evaporate, leaving no observable structure in the universe. The universe is about 10^{10} years old now, so it is highly unlikely that humans will exist to observe the occurrence of the heat death.

COSMOLOGICAL HORIZON

Observations of the **universe** are limited by the physical properties of **light**. Terrestrial observations are limited by the fact that light propagates in straight lines, while the surface of Earth is curved. The distance one can see can be increased by observing from a higher vantage point, but there is a limit: an observer on an infinitely tall platform could only see one quarter of the way around Earth's circumference. Astronomical observations are limited by the fact that light propagates with a finite **velocity**, c. The only objects that can be seen are those whose emitted light has had enough time, since the beginning of the Universe, to reach Earth today. In both cases, the physical limit to vision is called the horizon.

The mathematics of the terrestrial horizon is straightforward; knowing the observer's height above the surface, and the radius of Earth, simple trigonometry will suffice to calculate the horizon distance. The mathematics of the cosmological horizon is also straightforward, but far from simple. Exactly how far into the Universe can be seen depends upon a large number of cosmological parameters. The rate at which the Universe is expanding, the rate at which that expansion is slowing down (due to the mutual gravitational attraction between **matter**), the curvature of the Universe (whether it is negative, zero, or positive), and the matter content of the Universe (whether it is dominated by massive or massless particles), all enter into the calculation. Unlike the terrestrial horizon distance, the cosmological horizon distance is a dynamical quantity; it evolves in time, and does not expand at the same rate as the Universe expands. The ratio of the horizon volume to the Universe volume (that fraction of the Universe that can be in causal contact) changes as the Universe evolves, with important cosmological implications.

The calculation of the horizon distance simplifies if the Universe is flat, and its matter content is dominated by massive particles (matter-dominated). A universe whose content is primarily **radiation**, massless photons, is said to be radiation-dominated. Certainly the luminous matter, what can be seen with telescopes, is composed of massive particles. This may constitute as little as ten percent of the entire **mass** content, however, and it remains an open question as to whether the **dark matter** is massive or massless. In this case, the horizon distance at any time, $d_H(t)$, is given by $d_H(t)=3ct$, where t is the time since the **big bang**.

In the hot big bang model, a flat, matter dominated universe grows slower than the rate at which the horizon distance grows. If the big bang is considered as happening backwards, the horizon distance shrinks faster than does the Universe. What constitutes the observable Universe today was, in the past, composed of many regions which were causally disconnected. There had not been enough time since the bang for light signals to propagate between these regions. This problem with the standard big bang model helped lead to the theory of cosmic inflation

COSMOLOGICAL PRINCIPLE

A principle used by astronomers and cosmologists that asserts that, disregarding identifiable local irregularities, the **universe** must looks essentially the same to all observers. The cosmological principle may be found in three subtly varying statements: the weak cosmological principle, the ordinary cosmological principle, and the perfect cosmological principle.

Often, to make reasonable progress towards solving difficult theoretical problems, astrophysicists must make simplifying assumptions. In this spirit, the cosmological principle provides a fundamental theoretical basis for the study of modern **cosmology**. The perfect cosmological principle, as it is presently phrased, was invoked in the early 1900s by cosmologists including **Albert Einstein**, Aleksander Friedmann (1888-1925), and others who sought to simplify the solution of relativistic equations in order to reconcile their cosmological theories with the concept of a static (non-expanding) universe.

The weak cosmological principle asserts that the universe is homogeneous (uniform throughout, in structure or composition). All observers in the universe, no matter their location, see the same state of the universe around them. In other words, there is no preferred position in the universe.

The ordinary cosmological principle states that the universe is homogeneous and isotropic (identical in every direction). No observer in the universe can ascertain, from observations of large-scale structure, where they are located. Observers have no way to determine the any absolute direction with regard to where they are looking. This is tantamount to saying that there is no preferred position or preferred direction in the universe (e.g. there is no center).

The perfect cosmological principle augments the weak and ordinary principles; it qualifies that the universe appears to be the same at all times. Thus, there is no preferred position, direction, or epoch for the universe.

At the dawn of the twenty-first century, the ordinary cosmological principle is accepted by cosmologists as fundamental to the theoretical framework of modern **astrophysics**. The perfect cosmological principle was useful for supporting now generally discredited steady-state based cosmological theories. By convention, however, these steady-state theories proposed that the universe had no beginning or will it have an end. So, given that the most widely accepted cosmological theory, the **big bang** theory, strongly asserts that the universe did have a beginning, the ordinary cosmological principle is more consistent with modern cosmological theory and has replaced the perfect cosmological principle in most cosmological models.

Fundamentally, the ordinary cosmological principle asserts that the universe, and the laws that govern it, are not arbitrary. The same local physical laws which describe phenomena in the Solar System must apply to stellar systems millions of light-years away. In addition, as we observe the expansion of the universe via receding motions of galaxies, so too should other observers at every location in the universe see the same quality of expansion.

Modern observational evidence supports cosmological principle. Since the 1930s, it has been clear from both the distribution of galaxies and the Hubble law of expansion that the universe is nearly isotropic. Similarly, in 1989, NASA's **Cosmic Background Explorer (COBE)** satellite verified that the cosmic **background radiation** that permeates the universe is isotropic and homogeneous to one part in 100,000.

The application of the cosmological principle is most successful and apparent only over scales of hundreds of millions of light-years (the approximate dimension of galaxy super-clusters). The cosmological principle is not necessarily valid over local matter distributions. After all, it is evidentthat the matter distribution on Earth or in the Solar System is homogeneous and isotropic.

When Einstein contemplated the cosmological principle in 1917, the universe was thought to comprise the **Milky Way galaxy**, home to our Solar System, and a presumed void of

space beyond. Inarguably, the matter distribution in our galaxy is not homogeneous nor isotropic; astronomers clearly detect increased star number **density** in the spiral arms and have located the center of the Milky Way in the direction of the constellation Sagittarius. One must therefore accept that, with the information he had at the time, Einstein had no solid physical grounds for postulating the present form of the cosmological principle. That the principle seems valid today is testament to Einstein's extraordinary prescience (acumen) as a theorist.

See also Relativity theory

COSMOLOGY

Cosmology is the study of the origin, structure and evolution of the **universe**.

The origins of cosmology predate the human written record. The earliest civilizations constructed elaborate myths and folk tales to explain the wanderings of the **Sun, Moon** and **stars** through the heavens. Ancient Egyptians tied their religious beliefs to celestial objects and Ancient Greek and Roman philosophers debated the composition and shape of the Earth and the **cosmos**. For more than 13 centuries, until the **Scientific Revolution** of the sixteenth and seventeenth centuries the Greek astronomer **Ptolemy**'s model of an Earth-centered Cosmos composed of concentric crystalline spheres dominated the Western intellectual tradition.

Polish astronomer **Nicolaus Copernicus**'s reassertion of the once discarded heliocentric (sun-centered) theory sparked a revival of cosmological thought and work among the astronomers of the time. The advances in empiricism during the early part of the Scientific Revolution, embraced and embodied in the careful observations of Danish astronomer **Tycho Brahe**, found full expression in the mathematical genius of the German astronomer **Johannes Kepler** whose laws of planetary **motion** swept away the need for the errant but practically useful Ptolemaic models. Finally, the patient observations of the Italian astronomer and physicist **Galileo**, in particular his observations of moons circling Jupiter and of the phases of Venus, empirically laid to rest cosmologies that placed the Earth at the center of the Cosmos.

English physicist and mathematician Sir **Isaac Newton**'s, important *Philosophiae Naturalis Principia Mathematica (Mathematical Principles of Natural Philosophy)* quantified the laws of motion and **gravity** and thereby enabled cosmologists to envision a clockwork-like universe governed by knowable and testable natural laws. Within a century of the publication of Newton's *Principia,* the rise of the concept of a mechanistic universe led to the quantification of celestial **dynamics** that, in turn, led to a dramatic increase in the observation, cataloging and quantification of celestial phenomena. In accord with the development of natural theology, scientists and philosophers argued conflicting cosmologies that argued the existence and need for a supernatural God who acted as "prime mover" and guiding **force** behind a clockwork uni-

verse. In particular, French mathematician, **Pierre Simon de Laplace** argued for a completely deterministic universe, without a need for the intervention of God. Most importantly to the development of modern cosmology, Laplace asserted explanations for celestial phenomena as the inevitable result of time and statistical probability.

By the dawn of the twentieth century advances in mathematics allowed the development of increasingly sophisticated cosmological models. Many of these advances pointed toward a universe not necessarily limited to three dimensions and not necessarily absolute in time. These intriguing ideas found expression in the intricacies of relativity and theory that, for the first time, allowed cosmologists a theoretical framework upon which they could attempt to explain the innermost workings and structure of the universe both on the scale of the subatomic world and on the grandest of galactic scales.

As a direct consequence of German-American physicist **Albert Einstein**'s (1879-1955) **relativity theory**, cosmologists advanced the concept that **space-time** was a creation of the universe itself. This insight set the stage for the development of modern cosmological theory and provided insight into the evolutionary stages of stars (e.g., **neutron stars, pulsars, black holes**, etc.) that carried with it an understanding of nucleosynthesis (the formation of elements) that forever linked the physical composition of **matter** on the Earth to the lives of the stars.

Twentieth-century progress in cosmology has been marked by corresponding and mutually beneficial advances in technology and theory. American astronomer **Edwin Hubble**'s discovery that the universe was expanding, Arno A. Penzias and Robert W. Wilson's observation of cosmic **background radiation**, and the detection of the elementary particles that populated the very early universe all proved important confirmations of the **big bang** theory. This theory asserts that all matter and **energy** in the universe and the four dimensions of time and **space** were created from the primordial explosion of a **singularity** of enormous **density, temperature**, and **pressure**.

During the 1940s Russian-born American cosmologist and nuclear physicist **George Gamow** (1904-1968) developed the modern version of the big bang model based upon earlier concepts advanced by Russian physicist Alexander (Aleksandr Aleksandrovich) Friedmann (also spelled as Fridman, 1888-1925) and Belgian astrophysicist and cosmologist **Abbé Georges Lemaître**. Big bang based models replaced static models of the universe that described a homogeneous universe that was the same in all directions (when averaged over a large span of space) and at all times. Big bang and static cosmological models competed with each other for scientific and philosophical favor. Although many astrophysicists rejected the steady state model because it would violate the law of mass-energy conservation, the model had many eloquent and capable defenders. Moreover, the steady model was interpreted by many to be more compatible with many philosophical, social and religious concepts centered on the concept of an unchanging universe. The discovery of **quasars** and of a permeating **cosmic background radiation** eventually tilted the cosmological argument in favor of big bang-based models.

Technology continues to expand the frontiers of cosmology. The **Hubble Space Telescope** has revealed gas clouds in the cosmic voids and images of fledgling galaxies formed when the universe was less than a billion years old. Analysis of these pictures and advances in the understanding of the fundamental constituents of nature continue to keep cosmology a dynamic discipline of physics and the ultimate fusion of human scientific knowledge and philosophy.

See also Astronomy; Astrophysics; Chandrasekhar limit; Chandrasekhar mass; Cold stars; Comets; Cosmic background explorer (COBE); Cosmic inflation; Cosmic rays; Cosmological arrow of time; Cosmological constant; Cosmological heat death; Cosmological horizon; Cosmological principle; Dark matter; Doppler effect; Electromagnetic spectrum; Event horizon; Extraterrestrial life; Geocentric theory; Gravitational collapse; Hawking radiation; Hubble constant; Impacts; Kepler's laws of planetary motion; Microwave background radiation; Milky Way galaxy; Nebulae; Neptune; Newton's law of universal gravitation; Newtonian physics; Novae and super novae; Nuclear fusion; Nucleosynthesis; Olbers's paradox; Quantum theory; Radio astronomy; Red dwarf/Red giants; Red shift; Rydberg constant; Space-time geometry; Spectral lines; Spectroscopy; Speed of light; Stellar evolution; Stellar life cycle; Supernovae; Telescope; White dwarf; Worm holes; Zeeman effect

COSMOS

A synonym for the **universe**, the cosmos is the entirety of **matter** and **energy**. The word cosmos derives from the Greek word *kosmos* meaning the structure and harmony of the entire world.

The entire physical universe, all matter and energy, comprise the cosmos. Matter is any physical substance that occupies **space** and has **mass**. The smallest and most basic units of matter are **subatomic particles**, like protons, **quarks**, positrons, neutrinos, and electrons. The identity of a singular unifying subatomic unit that comprises all matter in the cosmos remains elusive. Energy, in contrast, is an ambiguous concept, a function of state, that can most usefully be described as the capacity to do **work**. Energy exists in many forms, including potential and **kinetic energy**.

Occasionally, the term cosmos is improperly used to denote the universe apart from Earth. In its truest sense, however, the cosmos includes the Earth as well. Early theories regarding the structure of the cosmos reflected indigenous religious philosophies. Although ancient Greek philosophers and scientists offered proof that the Earth was spherical and that it rotated about the **Sun**, such proofs were rejected upon religious grounds. The Ptolemaic model, formalized in the second century, held that the Earth was the center of the Cosmos, with the **stars** and **planets** rotating about it. In Western science, because the Ptolemaic view was embraced by the Christian church, it was only with considerable contro-

versy that this view discarded in the 16th century in favor of the Copernican theory which held that the Sun was the center of the cosmos. Subsequent observations by astronomers led to a realization that the Sun and its system of planets are but one small system located on a spiral arm of in a vast galaxy termed the Milky Way that is comprised of billions of stars. Moreover, the **Milky Way galaxy** is but one among billions of other galaxies in the observable universe.

The known cosmos contains gigantic masses of ionized gases called **nebulae** (singular nebula), rapidly rotating **neutron stars** called **pulsars** that emit oscillating bursts of electromagnetic **radiation**, **black holes** consisting of immeasurably dense matter from which **light** can not escape, and many other objects of interest to astronomers. The cosmos is dynamic. Supernova explosions resulting from the gravitational collapse of stars propel matter in all directions to aid in the formation of new nebulae in which new stars are created by gravitational forces. The cosmos is expanding, and some current estimates predict the width of the cosmos to be least 10-15 billion light-years in diameter.

The study of the structure, components, origins, and evolution of the universe is called **cosmology**. A cornerstone of modern cosmology is the **big bang** theory, which accounts for the creation of the material universe and the evolution of physical laws from an explosion that occurred approximately 10-20 billion years ago. The explosion is postulated to have occurred from an immeasurably dense **singularity**, or dimensionless point that contained all of the matter and energy of the current cosmos. After the big bang, the universe was extremely hot and began expanding, creating space as it expanded. As the cosmos cooled it allowed the formation of subatomic particles, atoms which in turn evolved the material universe.

COULOMB, CHARLES AUGUSTIN (1736-1806)

French physicist

Charles-Augustin Coulomb was born at Angoulême, Charente, France, on June 14, 1736. He was educated as a military engineer, and graduated from the Ecole du Génie in 1761. He served at various posts, and ended up at Paris in 1776.

An interest in **magnetism** induced Coulomb to enter a competition sponsored by the Paris Academy of Science in 1777. He shared first prize for his paper on magnetic compasses and, in 1781, he received a double first prize for his work on **friction**. These honors helped get him elected to the Academy.

While Coulomb had been experimenting with compasses, he noticed slight errors were introduced by friction acting on the pivot that held the magnetized needle. He invented a compass that had its needle suspended by a fine thread. The thread was an improvement, but it twisted. Coulomb realized that the amount of twisting was related to the amount of **force** the Earth's magnetic field applied to the needle. This realization led him to the invention of the torsion balance in 1777.

Charles August de Coulomb.

(English geologist John Michell had actually anticipated Coulomb, devising a torsion balance around the year 1750.)

The torsion balance allowed Coulomb to measure extremely small weights. With it, he was able to measure the force acting upon two electrically charged balls, discovering that the force changed as the distance between the balls varied. In 1785 Coulomb issued what would become known as **Coulomb's Law**, which states that the repulsion and attraction of two electrically charged objects depends on the product of their charges and the inverse square of the distance between them. This "inverse-square" law was very similar to the law of **gravity** discovered by **Isaac Newton**. Once again, Coulomb had been anticipated; **Henry Cavendish** had independently discovered "Coulomb's Law," but neglected to publish his work. Cavendish's work didn't come to light until fifty years after his death.

Coulomb investigated static electrical charges and discovered they adhered to the same inverse-square law as magnetic forces. This established a relationship between electric and magnetic forces that led **André Ampère** to discover the link between magnetism and **electric current**. Coulomb also found that the electric charges were located on the surface of a charged body and not within its center.

The French Revolution erupted in 1789 and Coulomb left Paris. Two years later, Coulomb decided it would be wise to resign his military commission. After the Revolution he was restored to his former posts and returned to Paris, where he died on August 23, 1806. In the years between 1781 and 1806 he had published twenty-five papers dealing with magnetism, **electricity**, and the torsion balance. In his honor the unit of **electric charge** is named the coulomb.

COULOMB'S LAW

Coulomb's law states that the magnitude of the **force** that one charged particle exerts on a second charged particle is proportional to the magnitude of the charges and inversely proportional to the square of the distance between them. It is written as $F = k(Q_1 Q_2/r^2)$, where F is the force, Q_1 is the charge on the first particle, Q^2 is the charge on the second particle, r is the distance between the two particles, and k is a proportionality constant. The direction of the force is along the line joining the two particles. Coulombic force is one of the principal forces involved in atomic reactions.

The proportionality constant k in Coulomb's law is associated with the environment of the charged particles. In a **vacuum** the proportionality constant is written as 1/4pe0 where e0 is the permittivity of vacuum and is obtained experimentally. In SI units $k = 8.99 \times 10^9$ (Nm^2/C^2). In a liquid Coulomb's law is altered by the dielectric constant of the liquid and k is written as 1/4pe0er where er is the dielectric constant of the liquid. This means that the force exerted by Q_1 on Q_2 is reduced by 1/er when compared to the force between the particles in vacuum. Also since **intermolecular forces** are electrical, the dielectric constant of the solvent influences **equilibrium** constants and reaction rate constants for charges in that solvent.

In 1777 **Charles Coulomb**, a French physicist, invented the torsion balance for measuring the force of magnetic and electrical attraction. He attempted to investigate the law of electrical repulsion as stated by Joseph Priestley using this device. It was later, in 1785, that he was able to formulate the principle now known as Coulomb's law from examining data obtained using the torsion balance. This law basically describes the principles governing the interaction between electrical charges. Although this law was not published until 1785 it was known, from experiments performed by John Robinson in 1769, that electrical repulsion went as the square of the distance between the particles involved. Coulomb's **work** on electric charges also carried over and led him to establish the inverse square law of repulsion and attraction of like and unlike **magnetic poles**.

It is known that an electrical charge Q_1 produces an electric field in the **space** surrounding the charge. The electric force is operative between charges down to distances of at least 3.33×10^{-17} ft (10^{-16} m) or about one-tenth of the diameter of atomic nuclei. If there are two charges in the vicinity of one another then they can either attract or repel each other depending upon their respective charges. Coulomb's law describes the force that one of these particles exerts on the other particle. The law applies exactly only when the charged particles are much smaller than the distance separating them and therefore can be treated approximately as point charges. Coulomb's law, in combination with principles of **quantum physics**, helps to describe the forces that bind electrons to the **nucleus** in atoms, that bind atoms together to form molecules, and that hold the molecules in liquids and solids together. Coulomb's law for electrical charges is very similar to the force governing gravitational fields of one body acting on another.

COWAN, CLYDE LORRAIN (1919-1974)

American physicist

Clyde Lorrain Cowan was a physicist and educator who is best remembered for his work in uncovering the properties and proving the existence of neutrinos, a fundamental nuclear particle that is electrically neutral. After serving as a captain in the United States Army Air Forces in World War II, Cowan received his doctorate from Washington University in St. Louis in 1949. Cowan then went to work for the Los Alamos Scientific Laboratory in the **nuclear weapons** test division.

With no charge and negligible **mass**, the **neutrino** is extremely difficult to detect. In 1951, Cowan began a productive collaboration with American physicist **Frederick Reines** (1918-1998) at Los Alamos, initially focusing on detecting neutrinos emitted from a nuclear explosion. Cowan and Reines soon realized, however, that **nuclear reactors** could provide a large neutrino flux of 10^3 neutrinos per square centimeter per second. In 1953, they decided to use the Hanford nuclear reactor, moving operations to the Savannah River nuclear reactor in 1955, primarily for safety considerations. The two scientists' success in detecting the rare neutrino capture event in a nuclear reaction created the exciting field of experimental neutrino physics.

Cowan joined the faculty at the Catholic University of America in 1958 as professor of physics. His work included pioneering techniques of particle detection in elementary **particle physics**, monitoring of low-level **radioactivity**, and the medical uses of radioactive isotopes. He was a fellow of the American Physical Society and a Guggenheim fellow, and he served as a consultant to the United States Atomic Energy Commission and the Smithsonian Institution. In 1995, Cowan's partner, Reines, was awarded the Nobel Prize in Physics for their work. Because the award is not given posthumously, Cowan, who died in 1974, was not included.

CRITICAL MASS

A **chain reaction** is a self-propagating reaction that continues without additional outside assistance once it is started. Like a series of dominoes falling in regular succession after the first is toppled, chain reactions proceed by themselves once started. Nuclear chain reactions advance in a somewhat similar way, where the effects of one reaction induce the start of the next. With regard to nuclear chain reactions, however, critical **mass** must first be achieved.

Critical mass is the amount of material necessary to sustain a nuclear chain reaction at a constant rate. The amount is critical because if the threshold mass is not met, the chain reaction will not occur. Controlled nuclear chain reactions are important in nuclear **power** plants. Nuclear **energy** is harnessed in these plants to produce **electricity**. The nuclear chain reaction requires a critical mass of nuclear material. Similarly, atomic explosions require a critical mass of fissionable or fusionable material in order for the nuclear chain reaction to occur. In **nuclear fission** explosions, where atoms are split apart to release immense quantities of energy, critical mass is needed to slow the emitted neutrons of the nuclear reaction. When atoms are split, they eject neutrons at very high speeds. The critical mass provides the minimum material to ensure that an emitted **neutron** will adequately collide with another **atom** causing it in turn to split, initiating the chain reaction. In fusion reactions, samples of less than critical mass are forced together to create one portion with supercritical mass that then proceeds as a nuclear chain reaction.

CROOKES, SIR WILLIAM (1832-1919)

English chemist and physicist

The English chemist and physicist Sir William Crookes discovered the element thallium and invented the radiometer, the spinthariscope, and the Crookes tube.

William Crookes was born in London on June 17, 1832. His education was limited, and despite his father's wish that he become an architect, he chose industrial **chemistry** as a career. He entered the Royal College of Chemistry in London, where he began his researches in chemistry. In 1859, he founded the *Chemical News,* which made him widely known, and remained its editor and owner all his life.

Most notable among Crookes's chemical studies is that one which led to his 1861 discovery of thallium. Using spectrographic methods, he had observed a green line in the spectrum of selenium, and he was thus led to announce the existence of a new element, thallium. While determining the atomic **weight** of thallium, using a delicate **vacuum** balance, he noticed several irregularities in weighing, which he attributed to the method. His investigation of this phenomenon led to the construction in 1875 of an instrument that he named the radiometer.

In 1869, J. W. Hittorf first studied the phenomena associated with electrical discharges in **vacuum tubes**. Not knowing of this, Crookes, 10 years later, made a parallel but more extensive investigation. In his 1878 report, he pointed out the significant properties of electrons in a vacuum, including the fact that a magnetic field causes a deflection of the emission. He suggested that the tube was filled with **matter** in what he called the "fourth state;" that is, the mean free path of the molecules is so large that collisions between them can be ignored. Tubes such as this are still called "Crookes tubes," and his work was honored by naming the **space** near the cathode in low **pressure** "Crookes dark space."

Sir William Crookes.

Crookes also made useful contributions to the study of **radioactivity** in 1903 by developing the spinthariscope, a device for studying alpha particles. He foresaw the urgent need for nitrogenous fertilizers, which would be used to cultivate crops to meet the demands of a rapidly expanding population. Crookes did much to popularize phenol (carbolic acid) as an antiseptic; in fact, he became an expert on sanitation. Mention should also be made of the serious and active interest he took in psychic phenomena, to which he devoted most of four years.

Crookes was knighted in 1897. His marriage lasted from 1856 until the death of his wife in 1917; they had 10 children. He died in London on April 4, 1919.

CRYOGENICS

Cryogenics is the science of producing and studying low-temperature environments. The word cryogenics comes from the Greek word "cryos," meaning cold; combined with a shortened form of the English verb "to generate," it has come to mean the generation of temperatures well below those of normal human experience.

More specifically, a low-temperature environment is termed a cryogenic environment when the **temperature** range is below the point at which permanent gases begin to liquefy.

Permanent gases are elements that normally exist in the gaseous state and were once believed impossible to liquefy. Among others, they include **oxygen**, nitrogen, hydrogen, and **helium**. The origin of cryogenics as a scientific discipline coincided with the discovery by nineteenth century scientists, that the permanent gases can be liquefied at exceedingly low temperatures. Consequently, the term cryogenic applies to temperatures from approximately -148°F (-100°C) down to **absolute zero**.

The temperature of a sample, whether it be a gas, liquid, or solid, is a measure of the **energy** it contains, energy that is present in the form of vibrating atoms and moving molecules. Absolute zero represents the lowest attainable temperature and is associated with the complete absence of atomic and molecular **motion**. The existence of absolute zero was first pointed out in 1848 by **William Thompson** (later to become **Lord Kelvin**), and is now known to be -459°F (-273°C). It is the basis of an absolute temperature scale, called the Kelvin scale, whose unit, called a kelvin rather than a degree, is the same size as the Celsius degree. Thus, -459°F corresponds to -273°C corresponds to 0K (note that by convention the degree symbol is omitted, so that 0K is read "zero kelvin"). Cryogenics, then, deals with producing and maintaining environments at temperatures below about 173K (-148°F [-100°C]).

In addition to studying methods for producing and maintaining cold environments, the field of cryogenics has also come to include studying the properties of materials at cryogenic temperatures. The mechanical and electrical properties of many materials change very dramatically when cooled to 100K or lower. For example, rubber, most plastics, and some metals become exceedingly brittle, and nearly all materials contract. In addition, many metals and ceramics lose all resistance to the flow of **electricity**, a phenomenon called superconductivity, and very near absolute zero (2.2K) liquid helium undergoes a transition to a state of **superfluidity**, in which it can flow through exceedingly narrow passages with no **friction**.

The development of cryogenics as a low temperature science is a direct result of attempts by nineteenth century scientists to liquefy the permanent gases. One of these scientists, **Michael Faraday**, had succeeded, by 1845, in liquefying most of the gases then known to exist. His procedure consisted of cooling the gas by immersion in a bath of **ether** and dry ice and then pressurizing the gas until it liquefied. Six gases, however, resisted every attempt at liquefaction and were known at the time as permanent gases. They were oxygen, hydrogen, nitrogen, **carbon** monoxide, methane, and nitric oxide. The noble gases, helium, neon, argon, krypton, and xenon, were yet to be discovered.

Of the known permanent gases, oxygen and nitrogen, the primary constituents of air, received the most attention. For many years investigators labored to liquefy air. Finally, in 1877, Louis Cailletet in France and Raoul Pictet in Switzerland, succeeded in producing the first droplets of liquid air, and in 1883 the first measurable quantity of liquid oxygen was produced by S.F. von Wroblewski at the University of Cracow. Oxygen was found to liquefy at 90K (-297°F [-183°C]), and nitrogen at 77K (-320°F [-196°C]).

Following the liquefaction of air, a race to liquefy hydrogen ensued. James Dewar, a Scottish chemist, succeeded in 1898. He found the **boiling** point of hydrogen to be a frosty 20K (-423°F [-253°C]). In the same year, Dewar succeeded in freezing hydrogen, thus reaching the lowest temperature achieved to that time, 14K (-434°F [-259°C]). Along the way, argon was discovered (1894) as an impurity in liquid nitrogen, and krypton and xenon were discovered (1898) during the fractional distillation of liquid argon. (Fractional distillation is accomplished by liquefying a mixture of gases each of which has a different boiling point. When the mixture is evaporated, the gas with the highest boiling point evaporates first, followed by the gas with the second highest boiling point, and so on.) Each of the newly discovered gases condensed at temperatures higher than the boiling point of hydrogen, but lower than 173K (-148°F [-100°C]).

The last element to be liquefied was helium gas. First discovered in 1868 in the spectrum of the **Sun**, and later on Earth (1885), helium has the lowest boiling point of any known substance. In 1908, the Dutch physicist **Heike Kamerlingh Onnes** finally succeeded in liquefying helium at a temperature of -452°F (4.2K).

Methods of Producing Cryogenic Temperatures

There are essentially only four physical processes that are used to produce cryogenic temperatures and cryogenic environments: heat conduction, evaporative cooling, cooling by rapid expansion (the Joule–Thompson effect), and adiabatic demagnetization. The first two are well known in terms of everyday experience. The third is less well known but is commonly used in ordinary refrigeration and air conditioning units, as well as cryogenic applications. The fourth process is used primarily in cryogenic applications and provides a means of approaching absolute zero.

Heat conduction is familiar to everyone. When two bodies are in contact, heat flows from the higher termperature body to a lower termperature body. Conduction can occur between any and all forms of matter, whether gas, liquid, or solid, and is essential in the production of cryogenic termperatures and environments. For example, samples may be cooled to cryogenic termperatures by immersing them directly in a cryogenic liquid or by placing them in an atmosphere cooled by cryogenic refrigeration. In either case, the sample cools by conduction of heat to its colder surroundings.

The second physical process with cryogenic applications is evaporative cooling, which occurs because atoms or molecules have less energy when they are in the liquid state than when they are in the vapor, or gaseous state. When a liquid evaporates, atoms or molecules at the surface acquire enough energy from the surrounding liquid to enter the gaseous state. The remaining liquid has relatively less energy, so its temperature drops. Thus, the temperature of a liquid can be lowered by encouraging the process of evaporation. The process is used in cryogenics to reduce the temperature of liquids by continuously pumping away the atoms or molecules as they leave the liquid, allowing the evaporation process to cool the remaining liquid to the desired temperature. Once the desired temperature is reached, pumping continues at a reduced level in order to maintain the lower temperature. This method can be used to reduce the temperature of any liquid. For example, it can be used to reduce the temperature of liquid nitrogen to its freezing point, or to lower the temperature of liquid helium to approximately -458°F (1 K).

The third process makes use of the Joule-Thompson effect, and provides a method for cooling gases. The Joule-Thompson effect involves cooling a pressurized gas by rapidly expanding its volume, or, equivalently, creating a sudden drop in **pressure**. The effect was discovered in 1852 by **James P. Joule** and William Thompson, and was crucial to the successful liquefaction of hydrogen and helium.

A valve with a small orifice (called a Joule-Thompson valve) is often used to produce the effect. High pressure gas on one side of the valvedrops very suddenly, to a much lower pressure and temperature, as it passes through the orifice. In practice, the Joule-Thompson effect is used in conjunction with the process of **heat** conduction. For example, when Kamerlingh Onnes first liquefied helium he did so by cooling the gas, through conduction, to successively lower temperatures, bringing it into contact with three successively colder liquids: oxygen, nitrogen, and hydrogen. Finally, he used a Joule-Thompson valve to expand the cold gas, and produce a mixture of gas and liquid droplets.

Today, the two effects together comprise the common refrigeration process. First, a gas is pressurized and cooled to an intermediate temperature by contact with a colder gas or liquid. Then, the gas is expanded, and its temperaturedrops still further. Ordinary household refrigerators and air conditioners work on this principle, using freon, which has a relatively high boiling point. Cryogenic refrigerators work on the same principle but use cryogenic gases such as helium, and repeat the process in stages, each stage having a successively colder gas until the desired temperature is reached.

The fourth process, that of adiabatic demagnetization, involves the use of paramagnetic salts to absorb heat. This phenomenon has been used to reduce the temperature of liquid helium to less than a thousandth of a degree above absolute zero in the following way. A paramagnetic salt is much like an enormous collection of very tiny magnets called magnetic moments. Normally, these tiny magnets are randomly aligned so the collection as a whole is not magnetic. However, when the salt is placed in a magnetic field, say by turning on a nearby electromagnet, the north poles of each **magnetic moment** are repelled by the north pole of the applied magnetic field, so many of the moments align the same way, that is, opposite to the applied field. This process decreases the **entropy** of the system.

Entropy is a measure of randomness in a collection; high entropy is associated with randomness, zero entropy is associated with perfect alignment. In this case, randomness in the alignment of magnetic moments has been reduced, resulting in a decrease in entropy. In the branch of physics called **thermodynamics**, it is shown that every collection will naturally tend to increase in entropy if left alone. Thus, when the electromagnet is switched off, the magnetic moments of the salt will tend to return to more random orientations. This requires energy, though, which the salt absorbs from the surrounding liquid, leaving the liquid at a lower temperature. Scientists know that it is not possible to achieve a temperature of absolute zero, however, in their attempts to get ever closer, a similar process called nuclear demagnetization has been used to reach temperatures just one millionth of a degree above absolute zero.

Scientists have demonstrated another method of cooling via the reduction of energy, using **lasers** instead of electromagnets. Laser cooling operates on the principle that temperature is really a measure of the energy of the atoms in a material. In laser cooling, the **force** applied by a laser beam is used to slow and nearly halt the motion of atoms. Slowing the atoms reduces their energy, which in turn reduces their temperature.

In a laser cooling setup, multiple lasers are aimed from all directions at the material to be cooled. Photons of **light**, which carry **momentum**, bombard the atoms from all directions, slowing the atoms a bit at a time. One scientist has likened the process to running through a hailstorm—the hail hits harder when you're running, no **matter** which direction you run, until finally you just give up and stop. Although laser-cooled atoms do not completely stop, they are slowed tremendously—normal atoms move at about 1000 miles per hour; laser-cooled atoms travel at about three feet per hour.

Using laser cooling and magnetic cooling techniques, scientists have cooled rubidium atoms to 20 billionths of a degree above absolute zero, creating a new state of matter called a Bose-Einstein condensate, in which the individual atoms condense into a superatom that acts as a single entity. Predicted many years before by **Albert Einstein** and **Satyendra Nath Bose**, Bose-Einstein condensate has completely different properties from any other kind of matter, and does not naturally exist in the **Universe**.

The researchers first used laser cooling to bring the temperature of the rubidium atoms down to about 10 millionths of a degree above absolute zero, which was still too warm to produce Bose-Einstein condensate. The atoms held in the laser light trap were then subjected to a strong magnetic field that held them in place. Called a time-averaged orbiting potential trap, the magnetic trap had a special twist that allowed the scientists to remove the most energetic (hottest) atoms, leaving only the very cold atoms behind.

Since the initial demonstration in 1995, scientists have continued to work with Bose-Einstein condensate, using it to slow light to less than 40 miles per hour, and even to produce an **atom** laser which produces bursts of atoms with laser-like properties. Multiple studies are underway to help researchers understand the properties of this baffling material.

Following his successful liquefaction of helium in 1908, Kamerlingh Onnes turned his attention almost immediately to studying the properties of other materials at cryogenic temperatures. The first property he investigated was the **electrical resistance** of metals, which was known to decrease with decreasing temperature. It was presumed that the resistance would completely disappear at absolute zero. Onnes discovered, however, that for some metals the resistance dropped to zero very suddenly at temperatures above absolute zero. The effect is called superconductivity and has some very important applications in

today's world. For example, superconductors are used to make magnets for particle **accelerators** and for **magnetic resonance imaging** (MRI) systems used in many hospitals.

The discovery of superconductivity led other scientists to study a variety of material properties at cryogenic temperatures. Today, physicists, chemists, material scientists, and biologists study the properties of metals, as well as the properties of insulators, **semiconductors**, plastics, composites, and living tissue. In order to chill their samples they must bring them into contact with something cold. This is done by placing the sample in an insulated container, called a dewar, and cooling the inner **space**, either by filling it with a cryogenic liquid, or by cooling it with a cryogenic refrigerator.

Over the years, this research has resulted in the identification of a number of useful properties. One such property common to most materials that are subjected to extremely low temperatures is brittleness. The recycling industry takes advantage of this by immersing recyclables in liquid nitrogen, after which they are easily pulverized and separated for reprocessing. Still another cryogenic material property that is sometimes useful is that of thermal contraction. Materials shrink when cooled. To a point (about the temperature of liquid nitrogen), the colder a material gets the more it shrinks. An example is the use of liquid nitrogen in the assembly of some automobile engines. In order to get extremely tight fits when installing valve seats, the seats are cooled to liquid nitrogen temperatures, whereupon they contract and are easily inserted in the engine head. When they warm up, a perfect fit results.

Cryogenic liquids are also used in the space program. For example, cryogens are used to propel **rockets** into space. A tank of liquid hydrogen provides the fuel to be burned and a second tank of liquid oxygen is provided for combustion. A more exotic application is the use of liquid helium to cool orbiting infrared telescopes. Any object warmer than absolute zero radiates heat in the form of infrared light. The infrared sensors that make up a telescope's "lens" must be cooled to temperatures that are lower than the equivalent temperature of the light they are intended to sense, otherwise the **telescope** will literally be blinded by its own light. Since temperatures of interest are as low as 3K (-454°F [-270°C]), liquid helium at 1.8K (-456°F [-271°C]) is used to cool the sensors.

Finally, the production of liquefied gases has itself become an important cryogenic application. Cryogenic liquids and gases such as oxygen, nitrogen, hydrogen, helium, and argon all have important applications. Shipping them as gases is highly inefficient because of their low densities. This is true even at extremely high pressures. Instead, liquefying cryogenic gases greatly increases the **weight** of cryogen that can be transported by a single tanker.

CRYPTOLOGY

Cryptography is a division of applied mathematics concerned with developing methods to enhance the privacy of communications and protect data from tampering through the use of coded information. The application of quantum principles to information processing and cryptology fuses mathematical theory and physical reality. Quantum cryptology utilizes the Heisenberg uncertainty principle to describe a system in which a measurement of the system disturbs the system and in so doing yields valuable information about the state of the system.

Cryptography allows its users to maintain confidentiality in their communications and, ideally, cryptographic schemes should only be able to be decoded and understood by authorized recipients (i.e., to be crack proof) who have a specific key (i.e., procedures to decode) that can be used to reverse the encoding procedure. Cryptography is also essential to the development and use of non-reputable transactions in which parties can not later deny involvement in making or authorizing certain transactions (e.g., entering into contracts). Courts around the world are increasingly faced with disputes regarding electronic communications and the development of a global electronic economy is dependent on the development of verifiable (non-reputable) transactions that carry the legal weight of traditional paper contracts.

The art of cryptography is ancient, the word cryptography originally derives from the Greek word *Kryptos* (to hide). Throughout history, wars and diplomatic negotiations have often turned on the ability of military and political leaders to read the messages of their adversaries. During World War II, for example, the Allied Forces enjoyed a tremendous strategic and tactical advantage from being able to intercept and read the secret messages of Nazi Germany. Although the Germans had encoded their communications with a cipher machine called Enigma, the Allies were able to crack the German code. Both direct physical examination of the encoding machines and mathematical analysis of encrypted communications were essential to the Allied success in cracking the German code. The United States also gained a decided advantage over Imperial Japanese forces through the development of operation MAGIC—a program that cracked Japan's diplomatic codes.

Although earlier cryptographic algorithms were developed for use by government intelligence agencies, in 1977 Ronald Rivest, Adi Shamir, and Leonard Adleman published what would become known as the RSA algorithm based upon the factoring of very large composite numbers. The RSA algorithm became an essential component in the first cryptologic systems widely available to the public. By the end of the twentieth century, the RSA algorithm became the most commonly used encryption and authentication algorithm in the world as it was incorporated into security programs used by Internet browsers, spreadsheets, email, and word processing programs.

Encryption systems can be simple (e.g., involving only the replacement of letters with numbers) or they can involve the use of highly secure one-time pads (also known as Vernam ciphers). In some cryptologic schemes, encryption is accomplished by calculating the products of certain prime numbers as basis for further mathematical operations. In addition to developing such mathematical keys, the data itself is divided into blocks of specific and limited length to thwart mathematical analysis based on statistical methods. Decryption is usually accomplished by following a reconstruction procedure that itself often involves unique or complex mathematical opera-

tions. In other cases, decryption is accomplished by performing the inverse of mathematical operations performed during encryption.

In some cryptographic systems, encoding and decoding procedures are accomplished using two separate keys (mathematical procedures). One key is used to code or lock a message, the other key is used to decode or unlock the message. In a two-key cryptologic systems such as a public key system, the public key is distributed so that a sender can encode a message and send it to the holder of a related private key. Accordingly, although the sender uses the recipient's public key to encode the message, the message can only be decoded with the recipient's private key. Beginning in 1991 the two-key public key method was used by cryptographers and computer technologists to enhance Internet security through a freely distributed package called Pretty Good Privacy (PGP).

Modern cryptographic systems rely on functions associated with advanced mathematics, including a specialized branch of mathematics termed number theory that explores the properties of numbers and the relationships between numbers. Specialized mathematical derivations and equations dealing with elliptical curves are also making an increasing impact on cryptology. Although, in general, larger and more elaborate cryptographic keys provide increased security, applications of number theory and elliptical curves to cryptological algorithms allow the use smaller keys or shorter procedures without a loss of security.

Cryptologic applications increased with the incorporation of security protocol in transistor-based electronic technology originally developed for use with **RADAR**. Although subsequent microchip technology made possible the production of higher-speed **computers**, they also rendered cryptologic systems increasingly vulnerable to mathematical analysis. Fortunately, despite advances in number theory and computing power, it remains easier to generate the large composite numbers that underlie most modern cryptologic systems than it is to factor those numbers. Recent advances in number theory, however, now allow factoring of large numbers that, if factored by traditional methods or hand calculations, might have previously taken, at a minimum, millions of man-hours to factor. Accordingly, the use of advanced computers adds a powerful tool to both encryption and decryption programs. Further advances in number theory may lead to the discovery of polynomial time-factoring algorithms that further reduce required computing time

Application of the principles derived from **quantum physics** to quantum communications systems would make it impossible to interrupt or intercept such a system without alerting the users of said **interference**.

The National Institute of Standards and Technology (NIST), oversees the development of cryptography standards. As of 2000, cryptographic algorithms were classified as munitions by the United States government. As such, they remained subject to severe export control and restrictions inhibiting their distribution and use.

See also Heisenberg uncertainty principle; Quantum theory

CRYSTALLOGRAPHY

Crystallography is the study of materials in which the atoms stack in a three-dimensionally ordered geometric arrangement. In a single crystal a single pattern extends throughout the entire material. Polycrystalline materials have discontinuities in the periodicity of the material. In amorphous materials, there is still less periodicity within the material; the amount of non-ordered atoms in amorphous materials is at least comparable with that exhibiting periodicity.

Detailed analyses of crystal structures are carried out by x-ray **diffraction**. In 1912, Max von Laue predicted that the spacing of crystal layers is small enough to diffract **light** of the appropriate **wavelength**. **William Henry Bragg** and his son, **William Lawrence Bragg**, were awarded the Nobel Prize in **chemistry** (1915) for their development of crystal structure analysis using x-ray diffraction. The Braggs' found that when x-ray **radiation** was scattered by a crystalline material both constructive and destructive **interference** occurred; this interference occurs because the wavelength of the **x rays** is of similar magnitude to the spaces between the atoms of the crystal. Analysis of the resulting diffraction pattern, the position of the lines of the scattered radiation along with their relative intensities, is the basis of the x-ray diffraction technique; this analysis allows the determination of the precise location of the atoms in the crystal.

The x-ray diffraction technique has been one of the most important structural methods throughout the twentieth century. It has expanded our knowledge by providing detailed structures of vitamins, proteins (enzymes, bacterial membranes, liquid crystals, polymers, organic compounds and inorganic compounds.

The idea that a crystal is composed of identical structural subunits was first proposed in 1784 based on observations of the cleavage of calcite. Subsequent investigations have shown that the structures of these subunits can be inferred from a crystal's symmetry. Even casual observation suggests that the symmetry of a crystal as a whole is related to some smaller subunit within it. The subunit is called a unit cell and it contains all of the essential information, such as the symmetry and elemental composition, of the crystal. Repeated translation along the edges of the unit cell can be used to derive the entire crystal lattice; in other words, the crystal lattice is the unit cell repeated many times in a periodic fashion.

The unit cells are often categorized in terms of **space** lattices called Bravais lattices; in such lattices, imaginary points called lattice points replace all of the atoms of the crystals. There are only fourteen distinct types of Bravais lattices, and these are associated with seven crystal systems. The crystal systems are all parallelepipeds whose shapes are completely defined by the lengths of the three sides and by the three angles characterizing the parallelepiped.

The most important symmetry elements in the consideration of crystal structures are axes of rotation and mirror planes. An n-fold rotation axis brings the crystal into self-coincidence after rotation by 360° a mirror plane occurs when the crystal can be bisected in such a way that one half is the reflection of the other. The seven crystal systems, distinguished by their axes of symmetry are as follows: 1) triclinic no symmetry elements; all unit cell lengths different, no 90° unit cell angles. 2) monoclinic one 2-fold axis or one plane; all unit cell lengths different, two 900 unit cell angles and one non-900 unit cell angle. 3) orthorhombic three mutually perpendicular axes, or two planes intersecting in a 2-fold axis; all unit cell lengths different, all unit cell angles = 900. 4) tetragonal one 4-fold axis or a 4-fold inversion axis; two unit cell lengths are identical and one is different, all unit cell angles = 900. 5) trigonal one 3-fold axis; all unit cell lengths are identical, all unit cell angles are identical but are not = 900. 6) hexagonal one 6-fold axis; two unit cell lengths identical and one different, two unit cell angles = 900 and the third = 1200. 7) cubic four 3-fold axes; all bond lengths identical, all bond angles = 900. The cubic form is the most symmetric, and the triclinic form the least. Generally speaking, crystals of lower symmetry are more common in nature than those of higher symmetry.

Another issue in crystallography is the physical basis for the crystal packing. How and why do crystals form? In what patterns do different types of materials crystallize? Crystals form since the ordered states of most solids are energetically more favorable (lower in **energy**) than disordered or irregularly packed ones. Hardness is one indication of the packing efficiency of atoms in a material. Materials with small atoms, packed closely together with strong covalent bonds throughout tend to be the hardest materials. The softest materials may contain metallic bonds or weaker van der Waals interactions.

The placement of atoms in a crystal can be described in terms of their layering. Within each layer, the most efficient packing occurs when the particles are staggered with respect to one another, leaving small triangular spaces between the particles. The second layer is placed on top of the first, in the depressions between the particles of the first layer. In one packing arrangement, the third layer lies in the depressions of the second and directly over depressions of the first layer. These so-called *face-centered cubic* close-packed structures are common in metals such as aluminum, rhodium, iridium, copper, silver, and gold. If the third layer particles are directly over atoms of the first, one has *hexagonal close packing*. This packing arrangement also is observed for many metals, including rubidium, osmium, cobalt, zinc, and cadmium.

Although ionic solids follow similar patterns as described above for metals, the detailed arrangements are more complicated, because the positioning of two different types of **ions**, cations and anions, must be considered. In general, it is the larger ion (usually, the anion) that determines the overall packing and layering, while the smaller ion fits in the holes (spaces) that occur throughout the layers.

The energy associated with crystal formation, the lattice energy, can be calculated by consideration of the different types of bonds within the solid: van der Waals bonds, ionic bonds, hydrogen bonds, covalent bonds, and/or metallic bonds. In the case of van der Waals solids (e.g., Ne, CO_2), the lattice energy can be calculated by summing up the pair potentials of interacting atoms using a secondary bond potential for atomic interactions. For ionic solids (e.g., NaCl,

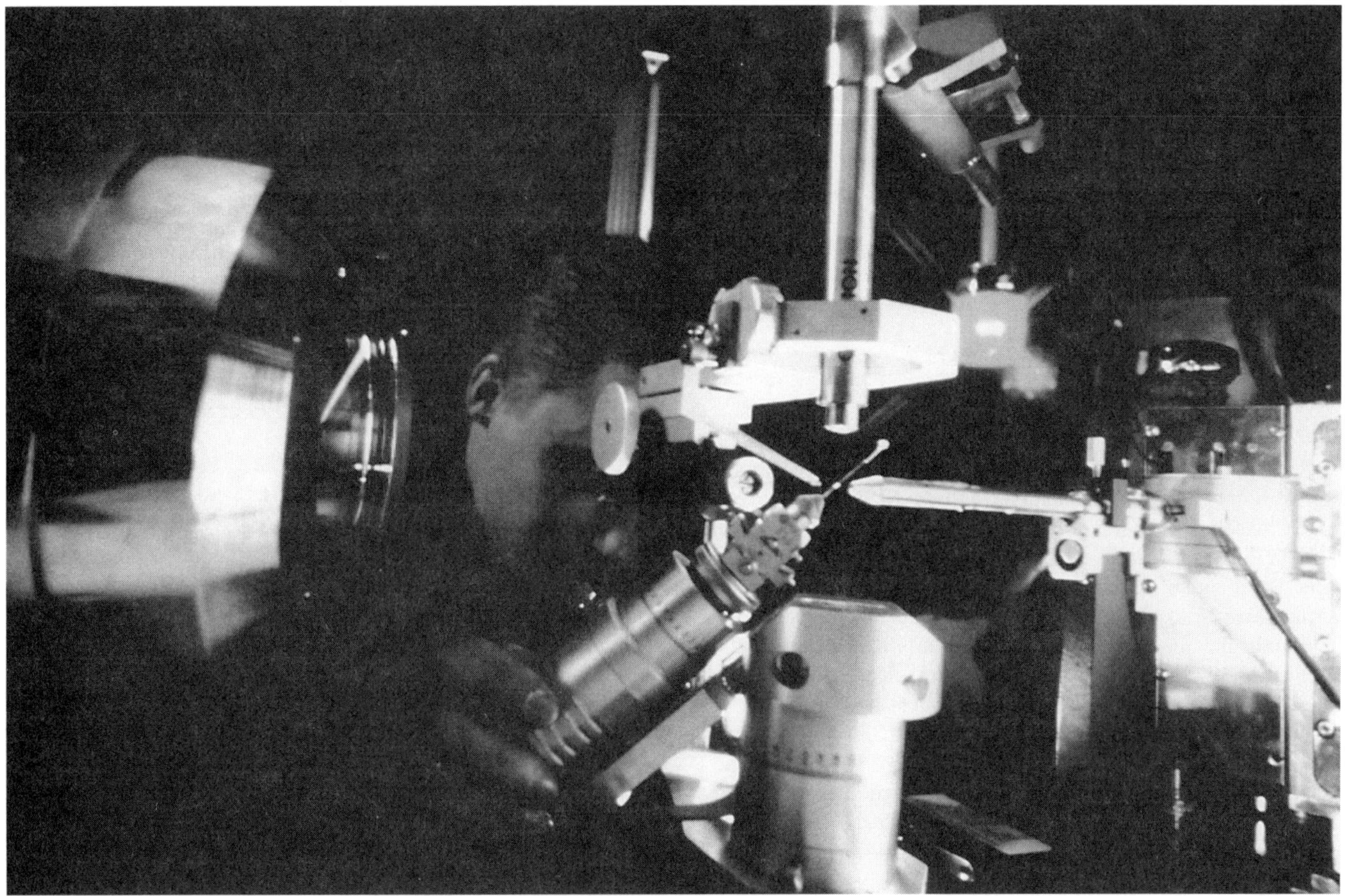

X-ray diffraction crystallography. *(Photo by James King-Holmes. Reproduced by permission.)*

ZnS), the Coulomb interaction, supplemented with a strongly repulsive **force**, is used in place of the secondary bond potential. In covalent (e.g., diamond, graphite) and hydrogen-bonded (e.g., H_2O) materials, the calculation of the lattice energy is much more complicated, and the lattice energy cannot simply be calculated as a sum over pair potentials acting between atoms.

The growth and size of any crystal depends on the conditions of its formation. Temperature, **pressure**, the presence of impurities, etc., will affect the size and perfection of the crystal. As a crystal grows, different imperfections may occur which can be classified as either point defects, line defects (or dislocations), or plane defects. Point defects include missing atoms or substituted atoms; line defects are defects that extend along straight or curved lines in a crystal; plane defects extend along true planes or curved surfaces within crystals.

Sometimes, imperfections are introduced to crystals intentionally. For example, the conductivity of **semiconductors** such as silicon and germanium can be modulated by the intentional addition of arsenic or antimony impurities. This procedure is called "doping." In this case, the additional electrons provided by arsenic or antimony impurities result in increased conductivity of the semiconductors.

Experiments in decreased **gravity** conditions aboard the space shuttles and in Spacelab I demonstrated that proteins formed crystals rapidly, and with fewer imperfections, than is possible under normal gravitational conditions. This is important because macromolecules are difficult to crystallize, and usually will form only crystallites whose structures are difficult to analyze. Protein analysis is important because many diseases (including Acquired Immune Deficiency Syndrome, AIDS) involve enzymes, which are the highly specialized protein catalysts of chemical reactions in living organism. The analysis of other biomolecules may also benefit from these experiments. It is interesting that similar advantages in crystal growth and degree of perfection have also been noted with crystals grown under high gravity conditions.

There is continued scientific interest in learning new ways to grow crystals due to the promise of new or improved crystalline materials with valuable properties. Although methods of synthesizing larger diamonds are expensive, polycrystalline diamond films can be made cheaply by a method called chemical vapor deposition (CVD). The technique involves methane and hydrogen gases, a surface on which the film can deposit, and a microwave oven. Energy from **microwaves** breaks the bonds in the gases, and, after a series of reactions, **carbon** films in the form of diamond are produced. The method holds much promise for a) the tool and cutting industry since diamond is the hardest known substance; b) **electronics** applications since diamond is a conductor of **heat**, but not **electricity**; and c) medical applications since it is tissue-compatible and tough. Thus it is suitable for use in items such as joint replacements and heart valves.

CURIE, MARIE (1867-1934)

French physicist and radiation chemist

Marie Curie was the first woman to win a Nobel Prize, and one of very few scientists ever to win that award twice. In collaboration with her physicist-husband **Pierre Curie**, Marie Curie developed and introduced the concept of **radioactivity** to the world. Working in very primitive laboratory conditions, Curie investigated the nature of high **energy** rays spontaneously produced by certain elements, and isolated two new radioactive elements, polonium and radium. Her scientific efforts also included the application of **x rays** and radioactivity to medical treatments.

Curie was born to two schoolteachers on November 7, 1867, in Warsaw, Poland. Christened Maria Sklodowska, she was the fourth daughter and fifth child in the family. By the age of five, she had already begun to suffer deprivation. Her mother Bronislawa had contracted tuberculosis and assiduously avoided kissing or even touching her children. By the time Curie was eleven, both her mother and her eldest sister Zosia had passed away, leaving her an avowed atheist. Curie was also an avowed nationalist (like the other members in her family), and when she completed her elementary schooling, she entered Warsaw's "Floating University," an underground, revolutionary Polish school that prepared young Polish students to become teachers.

Curie left Warsaw at the age of seventeen, not for her own sake but for that of her older sister Bronya. Both sisters desired to acquire additional education abroad, but the family could not afford to send either of them, so Marie took a job as a governess to fund her sister's medical education in Paris. At first, she accepted a post near her home in Warsaw, then signed on with the Zorawskis, a family who lived some distance from Warsaw. Curie supplemented her formal teaching duties there with the organization of a free school for the local peasant children. Casimir Zorawski, the family's eldest son, eventually fell in love with Curie and she agreed to marry him, but his parents objected vehemently. Marie was a fine governess, they argued, but Casimir could marry much richer. Stunned by her employers' rejection, Curie finished her term with the Zorawskis and sought another position. She spent a year in a third governess job before her sister Bronya finished medical school and summoned her to Paris.

In 1891, at the age of twenty-four, Curie enrolled at the Sorbonne and became one of the few women in attendance at the university. Although Bronya and her family back home were helping Curie pay for her studies, living in Paris was quite expensive. Too proud to ask for additional assistance, she subsisted on a diet of buttered bread and tea, which she augmented sometimes with fruit or an egg. Because she often went without heat, she would study at a nearby library until it closed. Not surprisingly, on this regimen she became anemic and on at least one occasion fainted during class.

In 1893, Curie received a degree in physics, finishing first in her class. The following year, she received a master's degree, this time graduating second in her class. Shortly thereafter, she discovered she had received the Alexandrovitch Scholarship, which enabled her to continue her education free of monetary worries. Many years later, Curie became the first recipient ever to pay back the prize. She reasoned that with that money, yet another student might be given the same opportunities she had.

Friends introduced Marie to Pierre Curie in 1894. The son and grandson of doctors, Pierre had studied physics at the Sorbonne; at the time he met Marie, he was the director of the École Municipale de Physique et Chimie Industrielles. The two became friends, and eventually she accepted Pierre's proposal of marriage. Their Paris home was scantily furnished, as neither had much interest in housekeeping. Rather, they concentrated on their work. Pierre Curie accepted a job at the School of Industrial Physics and Chemistry of the City of Paris, known as the EPCI. Given lab space there, Marie Curie spent eight hours a day on her investigations into the magnetic qualities of steel until she became pregnant with her first child, Irene, who was born in 1897.

Curie then began work in earnest on her doctorate. Like many scientists, she was fascinated by French physicist **Antoine-Henri Becquerel**'s discovery that the element uranium emitted rays that contained vast amounts of energy. Unlike **Wilhelm Röntgen**'s x rays, which resulted from the excitation of atoms from an outside energy source, the "Becquerel rays" seemed to be a naturally occurring part of the uranium ore. Using the piezoelectric quartz electrometer developed by Pierre and his brother Jacques, Marie tested all the elements then known to see if any of them, like uranium, caused the nearby air to conduct **electricity**. In the first year of her research, Curie coined the term "radioactivity" to describe this mysterious **force**. She later concluded that only thorium and uranium and their compounds were radioactive.

While other scientists had also investigated the radioactive properties of uranium and thorium, Curie noted that the minerals pitchblende and chalcolite emitted more rays than could be accounted for by either element. Curie concluded that some other radioactive element must be causing the greater radioactivity. To separate this element, however, would require a great deal of effort, progressively separating pitchblende by chemical analysis and then measuring the radioactivity of the separate components. In July, 1898, she and Pierre successfully extracted an element from this ore that was even more radioactive than uranium; they called it polonium in honor of Marie's homeland. Six months later, the pair discovered another radioactive substance—radium—embedded in the pitchblende.

Marie Curie, working in lab with husband Pierre Curie. *(AP/Wide World Photos. Reproduced by permission.)*

Although the Curies had speculated that these elements existed, to prove their existence they still needed to describe them fully and calculate their atomic **weight**. In order to do so, Curie needed an abundant supply of pitchblende and a better laboratory. She arranged to get hundreds of kilograms of waste scraps from a pitchblende mining firm in her native Poland, and Pierre Curie's EPCI supervisor offered the couple the use of a laboratory space. The couple worked together, with Marie performing the physically arduous job of chemically separating the pitchblende and Pierre analyzing the physical properties of the substances that Marie's separations produced. In 1902 the Curies announced that they had succeeded in preparing a decigram of pure radium chloride and had made an initial determination of radium's atomic weight. They had proven the chemical individuality of radium.

Pierre Curie's father had moved in with the family and assumed the care of their daughter, Irene, so the couple could devote more than eight hours a day to their beloved work. Pierre Curie's salary, however, was not enough to support the family, so Marie took a position as a lecturer in physics at the École Normal Supérieure; she was the first woman to teach there. In the years between 1900 and 1903, Curie published more than she had or would in any other three-year period, with much of this work being coauthored by Pierre Curie. In 1903 Curie became the first woman to complete her doctorate in France, summa cum laude.

The year Curie received her doctorate was also the year she and her husband began to achieve international recognition for their research. In November the couple received England's prestigious Humphry Davy Medal, and the following month Marie and Pierre Curie—along with Becquerel—received the Nobel Prize in physics for their efforts in expanding scientific knowledge about radioactivity. Although Curie was the first woman ever to receive the prize, she and Pierre declined to attend the award ceremonies, pleading they were too tired to travel to Stockholm. The prize money from the Nobel, combined with that of the Daniel Osiris Prize—which she received soon after—allowed the couple to expand their research efforts. In addition, the Nobel bestowed upon the couple an international reputation that furthered their academic success. The year after he received the Nobel, Pierre Curie was named professor of physics of the Faculty of Sciences at the Sorbonne. Along with his post came funds for three paid workers, two laboratory assistants and a laboratory chief, stipulated to be Marie. This was Marie's first paid research job.

In December, 1904, Marie gave birth to another daughter, Eve Denise, having miscarried a few years earlier. Despite the fact that both Pierre and Marie frequently suffered adverse effects from the radioactive materials with which they were in constant contact, Eve Denise was born healthy. The Curies continued their work regimen, taking sporadic vacations in the French countryside with their two children. They had just returned from one such vacation when on April 19, 1906, tragedy struck; while walking in the congested street traffic of Paris, Pierre was run over by a heavy wagon and killed.

A month after the accident, the University of Paris invited Curie to take over her husband's teaching position. Upon acceptance she became the first woman to ever receive a post in higher education in France, although she was not named to a full professorship for two more years. During this time, Curie came to accept the theory of English physicists **Ernest Rutherford** and Frederick Soddy that radioactivity was caused by atomic nuclei losing particles, and that these disintegrations caused the transmutation of an atomic **nucleus** into a different element. It was Curie, in fact, who coined the terms disintegration and transmutation.

In 1909, Curie received an academic reward that she had greatly desired: the University of Paris drew up plans for an Institut du Radium that would consist of two branches, a laboratory to study radioactivity—which Curie would run—and a laboratory for biological research on radium therapy, to be overseen by a physician. It took five years for the plans to come to fruition. In 1910, however, with her assistant André Debierne, Curie finally achieved the isolation of pure radium metal, and later prepared the first international standard of that element.

Curie was awarded the Nobel Prize again in 1911, this time "for her services to the advancement of chemistry by the discovery of the elements radium and polonium," according to the award committee. The first scientist to win the Nobel twice, Curie devoted most of the money to her scientific studies. During World War I, Curie volunteered at the National Aid Society, then brought her technology to the war front and instructed army medical personnel in the practical applications of radiology. With the installation of radiological equipment in ambulances, for instance, wounded soldiers would not have to be transported far to be x-rayed. When the war ended, Curie returned to research and devoted much of her time to her work.

By the 1920s, Curie was an international figure; the Curie Foundation had been established in 1920 to accept private donations for research, and two years later the scientist was invited to participate on the League of Nations International Commission for Intellectual Cooperation. Her health was failing, however, and she was troubled by fatigue and cataracts. Despite her discomfort, Curie made a highly publicized tour of the United States in 1921. The previous year, she had met Missy Meloney, editor of the *Delineator,* a woman's magazine. Horrified at the conditions in which Curie lived and worked (the Curies had made no money from their process for producing radium, having refused to patent it), Meloney proposed that a national subscription be held to finance a gram of radium for the institute to use in research. The tour proved grueling for Curie; by the end of her stay in New York, she had her right arm in a sling, the result of too many too strong handshakes. However, with Meloney's assistance, Curie left America with a valuable gram of radium.

Curie continued her work in the laboratory throughout the decade, joined by her daughter, **Irene Joliot-Curie**, who was pursuing a doctoral degree just as her mother had done. In 1925, Irene successfully defended her doctoral thesis on alpha rays of polonium, although Curie did not attend the defense lest her presence detract from her daughter's performance. Meanwhile, Curie's health still continued to fail and she was forced to spend more time away from her work in the laboratory. The result of prolonged exposure to radium, Curie contracted leukemia and died on July 4, 1934, in a nursing home in the French Alps. She was buried next to Pierre Curie in Sceaux, France.

CURIE, PIERRE (1859-1906)
French physicist

Pierre Curie was a noted physicist who became famous for his collaboration with his wife **Marie Curie** in the study of **radioactivity**. Before joining his wife in her research, Pierre Curie was already widely known and respected in the world of physics. He discovered (with his brother Jacques) the phenomenon of piezoelectricity—in which a crystal can become electrically polarized—and invented the quartz balance. His papers on crystal symmetry, and his findings on the relation between **magnetism** and **temperature** also earned praise in the scientific community. Curie died in a street accident in 1906, a physicist acclaimed the world over but who had never had a decent laboratory in which to work.

Pierre Curie was born in Paris on May 15, 1859, the son of Sophie-Claire Depouilly, daughter of a formerly prominent manufacturer, and Eugène Curie, a free-thinking physician who was also a physician's son. Dr. Curie supported the family with his modest medical practice while pursuing his love for the natural sciences on the side. He was also an idealist and an ardent republican who set up a hospital for the wounded during the Commune of 1871. Pierre was a dreamer whose style of learning was not well adapted to formal schooling. He received his pre-university education entirely at home, taught first by his mother and then by his father as well as his older brother, Jacques. He especially enjoyed excursions into the countryside to observe and study plants and animals, developing a love of nature that endured throughout his life and that provided his only recreation and relief from work during his later scientific career. At the age of 14, Curie studied with a mathematics professor who helped him develop his gift in the subject, especially spatial concepts. Curie's knowledge of physics and mathematics earned him his bachelor of science degree in 1875 at the age of sixteen. He then enrolled in the Faculty of Sciences at the Sorbonne in Paris and earned his *licence* (the equivalent of a master's degree) in physical sciences in 1877.

Curie became a laboratory assistant to Paul Desains at the Sorbonne in 1878, in charge of the physics students' lab work. His brother Jacques was working in the mineralogy laboratory at the Sorbonne at that time, and the two began a productive five-year scientific collaboration. They investigated pyroelectricity, the acquisition of electric charges by different faces of certain types of crystals when heated. Led by their knowledge of symmetry in crystals, the brothers experimentally discovered the previously unknown phenomenon of piezoelectricity, an electric polarization caused by **force** applied to the crystal. In 1880 the Curies published the first in a series of papers about their discovery. They then studied the opposite effect—the compression of a piezoelectric crystal by an electric field. In order to measure the very small amounts of **electricity** involved, the brothers invented a new laboratory instrument: a piezoelectric quartz electrometer, or balance. This device became very useful for electrical researchers and would prove highly valuable to Marie Curie in her studies of radioactivity. Much later, piezoelectricity had important practical applications. Paul Langevin, a student of Pierre Curie's, found that inverse piezoelectricity causes piezoelectric quartz in alternating fields to emit high-frequency **sound waves**, which were used to detect submarines and explore the ocean's floor. Piezoelectric crystals were also used in **radio** broadcasting and stereo equipment.

In 1882 Pierre Curie was appointed head of the laboratory at Paris' new Municipal School of Industrial Physics and Chemistry, a poorly paid position; he remained at the school for 22 years, until 1904. In 1883 Jacques Curie left Paris to become a lecturer in mineralogy at the University of Montpelier, and the brothers' collaboration ended. After Jacques's departure, Pierre delved into theoretical and experimental research on crystal symmetry, although the time available to him for such work was limited by the demands of organizing the school's laboratory from scratch and directing the laboratory work of up to 30 students, with only one assistant. He began publishing works on crystal symmetry in 1884, including in 1885 a theory on the formation of crystals and in 1894 an enunciation of the general principle of symmetry. Curie's writings on symmetry were of fundamental importance to later crystallographers, and, as Marie Curie later

wrote in *Pierre Curie,* "he always retained a passionate interest in the physics of crystals" even though he turned his attention to other areas.

From 1890 to 1895 Pierre Curie performed a series of investigations that formed the basis of his doctoral thesis: a study of the magnetic properties of substances at different temperatures. He was, as always, hampered in his work by his obligations to his students, by the lack of funds to support his experiments, and by the lack of a laboratory or even a room for his own personal use. His magnetism research was conducted mostly in a corridor. In spite of these limitations, Curie's work on magnetism, like his papers on symmetry, was of fundamental importance. His expression of the results of his findings about the relation between temperature and magnetization became known as Curie's law, and the temperature above which magnetic properties disappear is called the **Curie point**. Curie successfully defended his thesis before the Faculty of Sciences at the University of Paris (the Sorbonne) in March 1895, thus earning his doctorate. Also during this period, he constructed a periodic precision balance, with direct reading, that was a great advance over older balance systems and was especially valuable for chemical analysis. Curie was now becoming well-known among physicists; he attracted the attention and esteem of, among others, the noted Scottish mathematician and physicist William Thomson (Lord Kelvin). It was partly due to Kelvin's influence that Curie was named to a newly created chair of physics at the School of Physics and Chemistry, which improved his status somewhat but still did not bring him a laboratory.

In the spring of 1894, at the age of 35, Curie met Maria (later Marie) Sklodowska, a poor young Polish student who had received her *licence* in physics from the Sorbonne and was then studying for her *licence* in mathematics. They immediately formed a rapport, and Curie soon proposed marriage. Sklodowska returned to Poland that summer, not certain that she would be willing to separate herself permanently from her family and her country. Curie's persuasive correspondence convinced her to return to Paris that autumn, and the couple married in July, 1895, in a simple civil ceremony. Marie used a cash wedding gift to purchase two bicycles, which took the newlyweds on their honeymoon in the French countryside and provided their main source of recreation for years to come. Their daughter Irene was born in 1897, and a few days later Pierre's mother died; Dr. Curie then came to live with the young couple and helped care for his granddaughter.

The Curies' attention was caught by Henri Becquerel's discovery in 1896 that uranium compounds emit rays. Marie decided to make a study of this phenomenon the subject of her doctor's thesis, and Pierre secured the use of a ground-floor storeroom/machine shop at the School for her laboratory work. Using the Curie brothers' piezoelectric quartz electrometer, Marie tested all the elements then known to see if any of them, like uranium, emitted "Becquerel rays," which she christened "radioactivity". Only thorium and uranium and their compounds, she found, were radioactive. She was startled to discover that the ores pitchblende and chalcolite had much greater levels of radioactivity than the amounts of uranium and thorium they contained could account for. She guessed that a new, highly radioactive element must be responsible and, as she wrote in *Pierre Curie,* was seized with "a passionate desire to verify this hypothesis as rapidly as possible."

Pierre Curie too saw the significance of his wife's findings and set aside his much-loved work on crystals (only for the time being, he thought) to join Marie in the search for the new element. They devised a new method of chemical research, progressively separating pitchblende by chemical analysis and then measuring the radioactivity of the separate constituents. In July 1898, in a joint paper, they announced their discovery of a new element they named polonium, in honor of Marie Curie's native country. In December 1898, they announced, in a paper issued with their collaborator G. Bémont, the discovery of another new element, radium. Both elements were much more radioactive than uranium or thorium.

The Curies had discovered radium and polonium, but in order to prove the existence of these new substances chemically, they had to isolate the elements so the atomic **weight** of each could be determined. This was a daunting task, as they would have to process two tons of pitchblende ore to obtain a few centigrams of pure radium. Their laboratory facilities were woefully inadequate: an abandoned wooden shed in the School's yard, with no hoods to carry off the poisonous gases their work produced. They found the pitchblende at a reasonable price in the form of waste from a uranium mine run by the Austrian government. The Curies now divided their labor. Marie acted as the chemist, performing the physically arduous job of chemically separating the pitchblende; the bulkiest part of this work she did in the yard adjoining the shed/laboratory. Pierre was the physicist, analyzing the physical properties of the substances that Marie's separations produced. In 1902 the Curies announced that they had succeeded in preparing a decigram of pure radium chloride and had made an initial determination of radium's atomic weight. They had proven the chemical individuality of radium.

The Curies' research also yielded a wealth of information about radioactivity, which they shared with the world in a series of papers published between 1898 and 1904. They announced their discovery of induced radioactivity in 1899. They wrote about the luminous and chemical effects of radioactive rays and their **electric charge**. Pierre studied the action of a magnetic field on radium rays, he investigated the persistence of induced radioactivity, and he developed a standard for measuring time on the basis of radioactivity, an important basis for geologic and archaeological dating techniques. Pierre Curie also used himself as a human guinea pig, deliberately exposing his arm to radium for several hours and recording the progressive, slowly healing burn that resulted. He collaborated with physicians in animal experiments that led to the use of radium therapy—often called "Curie-therapie" then—to treat cancer and lupus. In

1904 he published a paper on the liberation of **heat** by radium salts.

Through all this intensive research, the Curies struggled to keep up with their teaching, household, and financial obligations. Pierre Curie was a kind, gentle, and reserved man, entirely devoted to his work—science conducted purely for the sake of science. He rejected honorary distinctions; in 1903 he declined the prestigious decoration of the Legion of Honor. He also, with his wife's agreement, refused to patent their radium-preparation process, which formed the basis of the lucrative radium industry; instead, they shared all their information about the process with whoever asked for it. Curie found it almost impossible to advance professionally within the French university system; seeking a position was an "ugly necessity" and "demoralizing" for him (*Pierre Curie*), so posts he might have been considered for went instead to others. He was turned down for the Chair of Physical Chemistry at the Sorbonne in 1898; instead, he was appointed assistant professor at the Polytechnic School in March 1900, a much inferior position.

Appreciated outside France, Curie received an excellent offer of a professorship at the University of Geneva in the spring of 1900, but he turned it down so as not to interrupt his research on radium. Shortly afterward, Curie was appointed to a physics chair at the Sorbonne, thanks to the efforts of Jules Henri Poincaré. Still, he did not have a laboratory, and his teaching load was now doubled, as he still held his post at the School of Physics and Chemistry. He began to suffer from extreme fatigue and sharp pains through his body, which he and his wife attributed to overwork, although the symptoms were almost certainly a sign of **radiation** poisoning, an unrecognized illness at that time. In 1902, Curie's candidacy for election to the French Academy of Sciences failed, and in 1903 his application for the chair of mineralogy at the Sorbonne was rejected, both of which added to his bitterness toward the French academic establishment.

Recognition at home finally came for Curie because of international awards. In 1903 London's Royal Society conferred the Davy medal on the Curies, and shortly thereafter they were awarded the 1903 Nobel Prize in physics—along with Becquerel—for their work on radioactivity. Curie presciently concluded his Nobel lecture (delivered in 1905 because the Curies had been too ill to attend the 1903 award ceremony) by wondering whether the knowledge of radium and radioactivity would be harmful for humanity; he added that he himself felt that more good than harm would result from the new discoveries. The Nobel award shattered the Curies' reclusive work-absorbed life. They were inundated by journalists, photographers, curiosity-seekers, eminent and little-known visitors, correspondence, and requests for articles and lectures. Still, the cash from the award was a godsend, and the award's prestige finally prompted the French parliament to create a new professorship for Curie at the Sorbonne in 1904. Curie declared he would remain at the School of Physics unless the new chair included a fully funded laboratory, complete with assistants. His demand was met, and Marie was named his laboratory chief. Late in 1904 the Curies' second daughter, Eve, was born. By early 1906,

Pierre Curie was poised to begin work—at last and for the first time—in an adequate laboratory, although he was increasingly ill and tired. On April 19, 1906, leaving a lunchtime meeting in Paris with colleagues from the Sorbonne, Curie slipped in front of a horse-drawn cart while crossing a rain-slicked rue Dauphine. He was killed instantly when the rear wheel of the cart crushed his skull. The world mourned the untimely loss of this great physicist. True to the way he had conducted his life, he was interred in a small suburban cemetery in a simple, private ceremony attended only by his family and a few close friends. In his memory, the Faculty of Sciences at the Sorbonne appointed Curie's widow Marie to his chair.

CURIE POINT

Substances with positive magnetic susceptibilites are said to be paramagnetic, which implies that they will have induced magnetic moments in the direction of an applied magnetic field. A ferromagnet, on the other hand, has a spontaneous **magnetic moment**, i.e., a magnetic moment even in the absence of an applied magnetic field.

In ferromagnetic materials, a transition occurs between the magnetized state (for example, of iron or nickel) at low temperatures and an unmagnetized state at high temperatures. The magnetization exhibits decreases gradually to zero rather than changing abruptly, but there does exist a **temperature** (called the Curie point after **Pierre Curie** who investigated it) at which the drop in magnetization becomes especially pronounced. The Curie point is thus considered to be the temperature where the spontaneous magnetization vanishes. It separates the disordered paramagnetic phase (at higher temperatures) from the ordered ferromagnetic phase (at lower temperatures).

The magnetic susceptibility of a paramagnetic material is inversely proportional to absolute temperature. The modified law proposed by Curie and French physicist Pierre Weiss (1865-1940) for ferromagnetic materials above their Curie points, which states the susceptibility is proportional to the reciprocal of the difference between the absolute temperature and the Curie temperature, is often obeyed.

CURRENT AND ELECTRICITY • See Electric current

CYCLOTRON

The cyclotron is a type of particle accelerator invented by physicist **Ernest O. Lawrence** (1901-1958) in 1930. The first working cyclotron was built by Lawrence and an assistant, M. Stanley Livingston, at the University of California, Berkeley, in 1931. Lawrence was awarded the 1939 Nobel Prize in physics for this pioneering achievement, which gave scientists direct experimental access to the atomic and subatomic world.

Particle **accelerators** produce a beam of fast-moving, electrically charged atomic or **subatomic particles** for use in fundamental research into the structure of nuclei, the nature of the nuclear forces, the interactions of elementary subatomic particles, and the creation and study of nuclei not found in nature. Of special interest are the **nucleon** rich nuclei (i.e., atomic nuclei with large numbers of protons and neutrons) found in the transuranic elements, so called because they are heavier (i.e. have a greater atomic **mass**) than uranium, the heaviest naturally occurring element. The transuranic element lawrencium, was discovered in 1961 and named in honor of the inventor of the cyclotron. The most famous, or infamous, transuranic element is plutonium, the deadliest substance known in terms of leathality per dose. A fissionable **isotope** of plutonium, plutonium-239, is used in **nuclear weapons** and reactors.

Following the 1919 discovery by physicist **Ernest Rutherford** of a reaction between a nitrogen **nucleus** and an **alpha particle**, scientists had searched for a way to artificially accelerate **ions** to higher energies than those found in nature. In 1928, a German physicist, Rolf Wideröe, devised a linear accelerator which, while demonstrating the possibility of artificial **acceleration**, did not provide sufficient **velocity** to the accelerated particles to initiate a nuclear reaction in the target nucleus.

Lawrence chanced to read an account of the linear accelerator and realized immediately that the solution to the velocity problem lay in accelerating the particles along a spiral path shaped by application of a magnetic field. Thus was born the magnetic resonance accelerator, or cyclotron.

Cyclotrons modeled on Lawrence's invention are called classical cyclotrons. They consist of two hollow semicircular electrodes (called "dees" because of their shape) mounted back to back and separated by a narrow gap in a **vacuum** chamber between the poles of a magnet. An electric field, alternating in polarity, is created in the gap by a radio-frequency oscillator. The particles to be accelerated are introduced into the gap and propelled by the electric field into one of the dees. There the magnetic field takes over and guides them in a semicircular path back toward the gap. By the time the particles reach the gap again, however, the electric field has oscillated, or reversed, accelerating them into the other dee. The speed of each particle and the radius of its orbit increase each time it crosses the gap, until finally it smashes into the target nucleus (hence the name **atom** smashers) with sufficient velocity to break the nucleus apart.

The velocity of accelerated particles in a classical cyclotron is limited by the laws of relativistic physics. The faster an object goes, the more its mass increases. As accelerated particles approach the speed of **light**, their increasing mass causes their orbital **frequency** to decrease until they cross the gap between the dees out of phase with the **oscillations** of the electric field. Instead of kicking them up to a faster speed, the field begins to decelerate them.

Cyclotron chamber.

Physicist **Edwin Mattison McMillan** (1907-1991) solved this problem in 1945, when he found that by altering the frequency of the accelerating **voltage**, he could keep the particles in phase and thus avoid the limitations of relativistic mass increase. McMillan called his improved cyclotron a synchrocyclotron. Subsequent improvements in the design and functionality of **particle accelerators**, along with technological advances, have yielded still more powerful accelerators such as betatrons, **synchrotrons**, and storage rings. Lawrence's first cyclotron, whose diameter was a mere 4.5 inches and was capable of accelerating particles to only 80,000 **electron** volts (80 keV) of **energy**, is dwarfed in size and **power** by particle accelerators such as the Cosmotron at Brookhaven National Laboratory in New York, the Tevatron at Fermilab in Illinois, and the Large Electron-Positron Collider (LEP) at the European Organization for Nuclear Research (CERN) outside Geneva, Switzerland.

Cyclotrons and their technological offspring have important applications not only in **nuclear physics** but in archeology (radiocarbon dating), medicine (cancer diagnosis and treatment), and industry (polymerization of plastics).

See also Accelerators; Atomic weight/mass; Nucleon; Nucleus; Particle accelerators; Particle physics; Synchrotrons

D

DALTON, JOHN (1766-1844)
English chemist

Although he dabbled in **meteorology** and studied color blindness, the contribution for which Dalton is most famous, without question, is his **atomic theory** and the list of atomic weights for the known elements that he created.

Dalton was born on approximately September 5, 1766, in Eaglesfield, England. Since his Quaker parents did not register his birth officially, no one knows his exact birth date.

Strongly influenced by his Quaker faith, Dalton grew up in modest surroundings and continued to live simply throughout his life. He attended the local school at Eaglesfield until the age of 11. A year later, he returned as a teacher at the same school, confronted with the task of instructing a number of students his own age or older. Then, as throughout his life, Dalton continued studying a number of topics on his own.

At the age of 15, Dalton became an assistant at the Quaker school in Kendal. About four years later, he was appointed principal of the school, a post he held for eight more years. While at Kendal, Dalton also offered a series of public lectures on natural philosophy. The series was unsuccessful, however, as Dalton was shy and a less than fascinating speaker.

In 1793, Dalton accepted an assignment as tutor in mathematics and natural philosophy at the New College in Manchester. Before long, however, he decided that the post took far more time than he cared to give and, in 1799, he resigned his post at New College. For the rest of his life, he supported himself as a private tutor in mathematics and natural philosophy.

Dalton's earliest scientific interests were in the field of meteorology. For 57 years he kept a daily record of the **temperature**, barometric **pressure**, rainfall, dew point, and other **weather** conditions. In his lifetime, he accumulated more than 200,000 individual observations.

Another topic of interest was color blindness, a condition that afflicted both Dalton and his brother. His first scientific paper, read to the Literary and Philosophical Society of Manchester on October 31, 1794, dealt with this topic. As a consequence of his interests, the condition of color blindness was given the name (and is still sometimes referred to as) Daltonism.

Without question, the contribution for which Dalton is most famous is his atomic theory. Dalton's interest in weather caused him to think about the nature and composition of air. He eventually concluded—as had a few scholars before him—that air consists of tiny, individual particles. But Dalton went beyond the musings of his predecessors and hypothesized that *all* forms of matter—solid, liquid, and gas—consisted of these tiny particles.

He first developed a formal theory about these particles between 1803 and 1805. The ideas he presented were by no means new ones. The Greek philosopher **Democritus** of Abdera had proposed such a theory of **matter** in the fourth century B.C. He had called the tiny particles *atomos*, Greek for "indivisible."

Democritus's ideas were quite different from those of Dalton. He used the term *atomos* to describe an abstract, theoretical concept, not the concrete, marble-like particles Dalton had in mind. The Greek's ideas were not widely accepted by his contemporaries and, over the centuries, they became less fashionable than competing theories of matter proposed by **Aristotle**.

Still, Democritus's ideas never really died out. They filtered in and out of the writings of natural philosophers and scientists for 22 centuries until Dalton expressed them in modern form.

The *atoms* that Dalton had in mind (for he retained the ancient Greek term) were tiny, indivisible particles that constituted all chemical elements. If you could imagine looking at a piece of gold, for example, with the very highest magnification imaginable, what you would see is a collection of gold atoms. Similarly, an examination of a piece of copper under maximum magnification would reveal a collection of copper atoms.

John Dalton, engraving.

The atoms of any one element, Dalton said, were all exactly alike. A single gold **atom** would look like any other gold atom anywhere in the world. But atoms of different elements were different from each other. A gold atom looked different from a copper atom.

Dalton's theory also dealt with the composition of compounds. The smallest particle of any compound, he said, was a compound atom. Thus, he taught that water was composed of *compound water atoms*. Two decades later, **Amedeo Avogadro** was to clarify the difference between atoms and compound atoms, which he proposed calling *molecules*.

Dalton is called the father of modern atomic theory partly because of his clear statement of that theory and partly because of his emphasis on atomic weights. No proponent of the concept of atoms had previously made clear the fact that atoms must have weights that can be determined experimentally. Dalton did. He said that finding the weights of atoms was a relatively straightforward task that any chemist could accomplish.

For example, he said, suppose that we assume that the elements **oxygen** and hydrogen are both composed of atoms. When oxygen gas combines with hydrogen gas to form water, then, what happens on the simplest level is that one atom of oxygen combines with one atom of hydrogen to form one compound atom of water, or: H + O = HO (Dalton had no way of knowing the actual combining ratio of hydrogen and oxy-gen in water [2:1], so he assumed the simplest possible ratio [1:1]).

Next, Dalton said, it is easy to determine in the laboratory what the *weight ratio* of hydrogen to oxygen is. A fixed **weight** of hydrogen will consume eight times its weight in oxygen to form water. It followed that each atom of oxygen must weigh eight times as much as each atom of hydrogen.

Dalton's discussion of atomic weights gave chemists a concrete plan for exploring at least this aspect of the atomic theory. Within a few years, Dalton had prepared a list of atomic weights for the known elements.

Dalton's theory was quickly accepted by the vast majority of chemists. One reason for its rapid success was that it explained certain experimental results that had recently been announced, most notably Joseph Proust's law of constant composition which stated that a chemical compound always contains the same constituents in the same proportions.

In his later years, Dalton was flooded with honors and awards. He died quietly at his home in Manchester on July 27, 1844.

DARK MATTER

Dark **matter** is defined to be any form of matter whose existence can be determined only through its gravitational interaction with visible matter. Dark matter is not visible because, by definition, it emits no detectable electromagnetic **radiation** by which to observe it. Studies of celestial **mechanics** involving the **motion** of galaxies imply that perhaps as much as 90% of the matter in a typical galaxy is dark. Astronomers speculate that dark matter may also exist in intergalactic **space**, and the resolution of this question may provide insight into questions regarding the ultimate fate of the **universe**.

Dark matter presumably reveals itself through its large-scale effects in the universe, especially in the formation of galaxies or galaxy clusters. Objects smaller than stellar clusters may not have detectable amounts of dark matter associated with them.

According to the laws of **gravity**, to exert a **gravitational force** on other celestial objects dark matter must have **mass**. Beyond this, the composition of dark matter is undetermined. Candidate dark matter may be separated into two primary classes: normal baryonic matter, composed of baryons (protons, neutrons, and electrons); or some form of non-baryonic matter, such as massive neutrinos or **weakly interacting massive particles** (WIMPs). As indicated from astronomical observations of **galaxies and galaxy clusters**, as well as theoretical **work** in **cosmology** concerning the **big bang** theory and inflationary universe theory, dark matter, baryonic or non-baryonic, seems to comprise a substantial portion (some estimates run as high as 90%) of all the matter in the universe.

The possible candidates for baryonic dark matter are numerous. Compact objects that do not emit much **light** per unit mass could retain a large fraction of the missing mass if such objects reside in large numbers in the halos of galaxies. Examples of compact objects include large Jupiter-mass **plan-**

ets, cool brown dwarf **stars**, white dwarf stars, **neutron stars**, and **black holes**. Alternatively, baryonic dark matter may be a more diffuse, highly energized, intergalactic hydrogen gas that does not readily interact with light.

Two possible forms of non-baryonic dark matter have been proposed. Fundamentally, non-baryonic dark matter may be distinguished by the speed at which it travels. Particles moving with velocities near the **speed of light** are designated as hot dark matter. Moving at such velocities would make it difficult, however, for hot dark matter to be constrained by the gravitational **binding energy** of galaxy halos. Thus, such hot dark matter should be found not only within galaxies but between them as well. By far, the most attractive hot dark matter candidate is the **neutrino**.

In 1998 an international research group based in Japan conducted the Super-Kamiokande experiment to determine whether neutrinos have mass. Studies of neutrino **oscillations** indicated a small mass for neutrinos, approximately one billionth the mass of a **proton**. Because the results of the Super-Kamiokande experiment are not in conformity with predictions made by the standard model, as of 2000 scientists continue to evaluate both the results of the Super-Kamiokande experiment and the standard model.

A cousin of hot dark matter, cold dark matter, is much less energetic. Moving sluggishly, cold dark matter may be gravitationally bound to galaxy halos and clusters. Cold dark matter presents an exotic solution to the dark matter problem given that the existence of cold particles like WIMPs have not yet been experimentally verified. They are merely theoretical entities that may someday be discovered in high-energy particle accelerator experiments.

The determination of the existence and amount of dark matter has important consequences for cosmology. If there is sufficient mass in the universe the mutual gravitational attraction of all particles will eventually slow the expansion of the universe and the universe will be fated to undergo cyclic contractions and expansions. If there is insufficient matter in the universe the expansion of space will continue and the fate of the universe is entropic death.

See also Baryons and baryonic matter; Cosmic inflation; Interstellar space

DAVISSON, CLINTON (1881-1958)

American physicist

A major focus of Clinton Davisson's research at Bell Telephone Laboratories (formerly Western Electric Company) was the **scattering** of electrons by crystals. In 1927, while working with a colleague named **Lester Germer**, Davisson accidentally discovered evidence to support the earlier hypothesis of Prince **Louis Victor de Broglie** that a beam of electrons has wave properties. For this discovery, Davisson was awarded a share of the 1937 Nobel Prize in physics. He remained at Bell until his retirement in 1946.

Clinton Joseph Davisson was born in Bloomington, Illinois, on October 22, 1881. His father, Joseph Davisson, was a veteran of the Union Army who worked as a contract painter; his mother was the former Mary Calvert, a schoolteacher before her marriage. The Davissons also had one daughter, Carrie. Young Clinton attended local schools and graduated from Bloomington High School in 1902. He entered the University of Chicago on a scholarship, as a physics major.

Davisson's intellectual promise soon became apparent. At the end of his sophomore year, one of his professors, Nobel laureate **Robert A. Millikan**, recommended Davisson as a replacement for a physics instructor at nearby Purdue University who had recently died. Davisson took that post and spent the last half of the 1903–04 academic year at Purdue. He returned to Chicago in the fall of 1904, but was there only a year when he was offered a position as instructor in physics at Princeton University the following autumn. At Princeton he also worked as a research assistant under another Nobel winner, **Owen W. Richardson**. Over the next six years, Davisson worked on both his B.S. at Chicago by attending summer sessions and his Ph.D. at Princeton during the regular academic year. He received the former in 1908 and the latter in 1911.

Shortly after graduation from Princeton, on August 4, 1911, Davisson was married to the former Charlotte Sara Richardson, his mentor's sister. Davisson had met Richardson while she was staying with her brother at Princeton on a visit from England. The Davissons had four children, Clinton Owen Calvert, James Willans, Richard Joseph, and Elizabeth Mary. A few weeks after his marriage, Davisson began his new job as assistant professor of physics at Carnegie Institute of Technology in Pittsburgh.

Davisson's experience at Carnegie over the next six years was largely unsatisfactory. His teaching load was so heavy that he had virtually no time to spend on research. Indeed, as Mervin J. Kelly writes in his obituary of Davisson in the National Academy of Sciences' *Biographical Memoirs,* "he had been able to carry only one research to the point where, with his lifelong high standards, he would publish it." As a result, Davisson was not at all disappointed in May 1917 to receive a leave of absence from Carnegie to take on war-related research in the Engineering Department of the Western Electric Company (later, the Bell Telephone Laboratories). When the war ended, Davisson chose to continue on at Western Electric rather than to return to academic life. He remained with the company for another 29 years, retiring in 1946 at the age of 65.

Laboratory Accident Leads to Discovery of Electron Diffraction

Davisson's first research assignment at Western Electric was carried out in connection with a patent suit in which the company was involved regarding rights on **vacuum** tubes. That research concerned a study of **electron** emission patterns produced when a metal filament is bombarded by positive **ions**. When that research was completed, Davisson moved on to a variation of that topic, the characteristics of electron emission when metals are bombarded by electrons rather than by positive ions. For about six years, Davisson and a colleague, C. H. Kunsman, bombarded a variety of metals with electrons and measured the angles at which secondary electrons were emitted from the metal surface. Throughout this period, however, they were unable to develop a theoretical model that explained their results.

Finally, in April 1925, a laboratory accident provided some key new information. A vacuum tube being used by Davisson and Lester Germer (who had replaced Kunsman) exploded, resulting in the rapid oxidation of the nickel target inside the tube. Intending to restore the target for re-use in another vacuum tube, Davisson was forced to re-heat the nickel for an extended period of time. When new experiments were begun using **vacuum tubes** containing the rejuvenated nickel target, some remarkable results were obtained. Specifically, electrons were not just reflected from the nickel target, but diffracted as well. Later studies revealed that the heating process had converted many tiny nickel crystals in the original sample to a few much larger crystals in the restored sample.

The new findings were significant because **diffraction** is a characteristic of **waves**. Although the Davisson-Germer results suggested that electron beams reflected from the nickel target behaved as if they were traveling in waves, Davisson did not fully appreciate the meaning of his results until the summer of 1926. While attending a meeting of the British Association for the Advancement of Science, he learned for the first time about Prince Louis de Broglie's recently announced hypothesis of the wave nature of electrons. Using de Broglie's theory, Davisson was able to calculate the **wavelength** of the electrons diffracted in his experiment. His results matched de Broglie's predictions so closely that they could be regarded as confirmation of the French physicist's theory.

Davisson continued his research on electron scattering over the next two decades of his tenure, first at Western Electric, and then at Bell Laboratories. After his retirement in 1946 he accepted an appointment as visiting professor of physics at the University of Virginia. He resigned from that post in 1954 and died in Charlottesville, Virginia, on February 1, 1958. In addition to the 1937 Nobel Prize, Davisson received a number of other honors and awards, including the Comstock Prize of the National Academy of Sciences in 1928, the Elliot Cresson Medal of the Franklin Institute in 1935, the Hughes Medal of the Royal Society in 1935, and the University of Chicago Alumni Medal in 1941.

DE BROGLIE, LOUIS VICTOR (1892-1987)

French theoretical physicist

Louis Victor de Broglie, a theoretical physicist and member of the French nobility, is best known as the father of wave **mechanics**, a far-reaching achievement that significantly changed modern physics. Wave mechanics describes the behavior of **matter**, including **subatomic particles** such as electrons, with respect to their wave characteristics. For this

groundbreaking work, de Broglie was awarded the 1929 Nobel Prize for physics.

Louis Victor Pierre Raymond de Broglie was born on August 15, 1892, in Dieppe, France, to Duc Victor and Pauline d'Armaille Broglie. His father's family was of noble Piedmontese origin and had served French monarchs for centuries, for which it was awarded the hereditary title *Duc* from King Louis XIV in 1740, a title that could be held only by the head of the family. A later de Broglie assisted the Austrian side during the Seven Years War and was awarded the title *Prinz* for his contribution. This title was subsequently borne by all members of the family. Another of de Broglie's famous ancestors was his great-great-grandmother, the writer Madame de Stael.

The youngest of five children, de Broglie inherited a familial distinction for formidable scholarship. His early education was obtained at home, as befitted a great French family of the time. After the death of his father when de Broglie was 14 his eldest brother Maurice arranged for him to obtain his secondary education at the Lycée Janson de Sailly in Paris.

After graduating from the Sorbonne in 1909 with baccalaureates in philosophy and mathematics, de Broglie entered the University of Paris. He studied ancient history, paleography, and law before finding his niche in science, influenced by the writings of French theoretical physicist Jules Henri Poincaré. The work of his brother Maurice, who was then engaged in important, independent experimental research in **x rays** and **radioactivity**, also helped to spark de Broglie's interest in theoretical physics, particularly in basic **atomic theory**. In 1913, he obtained his Licencié ès Sciences from the University of Paris's Faculté des Sciences.

De Broglie's studies were interrupted by the outbreak of World War I, during which he served in the French army. Yet even the war did not take the young scientist away from the country where he would spend his entire life; for its duration, de Broglie served with the French Engineers at the wireless station under the Eiffel Tower. In 1919, after what he considered to be six wasted years in uniform, de Broglie returned to his scientific studies at his brother's laboratory. Here he began his investigations into the nature of matter, inspired by a conundrum that had long been troubling the scientific community: the apparent physical irreconcilability of the experimentally proven dual nature of **light**. Radiant **energy** or light had been demonstrated to exhibit properties associated with particles as well as their well-documented wave-like characteristics. De Broglie was inspired to consider whether matter might not also exhibit dual properties. In his brother's laboratory, where the study of very high frequency **radiation** using spectroscopes was underway, de Broglie was able to bring the problem into sharper focus. In 1924, de Broglie, with over two dozen research papers on electrons, **atomic structure**, and x rays already to his credit, presented his conclusions in his doctoral thesis at the Sorbonne. Entitled "Investigations into the Quantum Theory," it consolidated three shorter papers he had published the previous year.

In his thesis, de Broglie postulated that all matter— including electrons, the negatively charged particles that orbit an atom's nucleus—behaves as both a particle and a wave. Wave characteristics, however, are detectable only at the atomic level, whereas the classical, ballistic properties of matter are apparent at larger scales. Therefore, rather than the wave and particle characteristics of light and matter being at odds with one another, de Broglie postulated that they were essentially the same behavior observed from different perspectives. Wave mechanics could then explain the behavior of all matter, even at the atomic scale, whereas classical Newtonian mechanics, which continued to accurately account for the behavior of observable matter, merely described a special, general case. Although, according to de Broglie, all objects have "matter waves," these **waves** are so small in relation to large objects that their effects are not observable and no departure from classical physics is detected. At the atomic level, however, **matter waves** are relatively larger and their effects become more obvious. De Broglie devised a mathematical formula, the matter wave relation, to summarize his findings.

American physicist **Albert Einstein** appreciated the significance of de Broglie's theory; de Broglie sent Einstein a copy of his thesis on the advice of his professors at the Sorbonne, who believed themselves not fully qualified to judge it. Einstein immediately pronounced that de Broglie had illuminated one of the secrets of the **universe**. Austrian physicist **Erwin Schrödinger** also grasped the implications of de Broglie's work and used it to develop his own theory of wave mechanics, which has since become the foundation of modern physics. Still, many physicists could not make the intellectual leap required to understand what de Broglie was describing.

De Broglie's wave matter theory remained unproven until two separate experiments conclusively demonstrated the wave properties of electrons—their ability to diffract or bend, for example. American physicists **Clinton Davisson** and **Lester Germer** and English physicist George Paget Thomson all proved that de Broglie had been correct. Later experiments would demonstrate that de Broglie's theory also explained the behavior of protons, atoms, and even molecules. These properties later found practical applications in the development of magnetic **lenses**, the basis for the **electron microscope**.

De Broglie devoted the rest of his career to teaching and to developing his theory of wave mechanics. In 1927, he attended the seventh Solvay Conference, a gathering of the most eminent minds in physics, where wave mechanics was further debated. Theorists such as German physicist Werner Karl Heisenberg, Danish physicist **Niels Bohr**, and English physicist **Max Born** favored the uncertainty or probabilistic interpretation, which proposed that the wave associated with a particle of matter provides merely statistical information on the position of that particle and does not describe its exact position. This interpretation was too radical for Schrödinger, Einstein, and de Broglie; the latter postulated the "double solution," claiming that particles of matter are transported and guided by continuous "pilot waves" and that their movement is essentially deterministic. De Broglie could not reconcile his pilot wave theory with some basic objections raised at the conference, however, and he abandoned it.

Louis Victor Pierre Raymond de Broglie.

The disagreement about the manner in which matter behaves described two profoundly different ways of looking at the world. Part of the reason that de Broglie, Einstein, and others did not concur with the probabilistic view was that they could not philosophically accept that matter, and thus the world, behaves in a random way. De Broglie wished to believe in a deterministic **atomic physics**, where matter behaves according to certain identifiable patterns. Nonetheless, he reluctantly accepted that his pilot wave theory was flawed and throughout his teaching career instructed his students in probabilistic theory, though he never quite abandoned his belief that "God does not play dice," as Einstein had suggested.

In 1928, de Broglie was appointed professor of theoretical physics at the University of Paris's Faculty of Science. De Broglie was a thorough lecturer who addressed all aspects of wave mechanics. Perhaps because he was not inclined to encourage an interactive atmosphere in his lectures, he had no noted record of guiding young research students.

In 1929, at the age of 37, de Broglie was awarded the Nobel Prize in physics in recognition of his contribution to wave mechanics. In 1933, he accepted the specially created chair of theoretical physics at the Henri Poincaré Institute—a position he would hold for the next 29 years—where he established a center for the study of modern physical theories. That same year, he was elected to the Académie des Sciences, becoming its life secretary in 1942; he used his influence to urge the Académie to consider the harmful effects of nuclear explosions as well as to explore the philosophical implications of his and other modern theories.

In 1943, anxious to forge stronger links between industry and science and to put modern physics, especially **quantum mechanics**, to practical use, de Broglie established a center within the Henri Poincaré Institute dedicated to applied mechanics. He was elected to the prestigious Academie Francaise in 1944 and, in the following year, was appointed a counsellor to the French High Commission of Atomic Energy with his brother Maurice in recognition of their work promoting the peaceful development of nuclear energy and their efforts to bridge the gap between science and industry. Three years later, de Broglie was elected to the National Academy of the United States as a foreign member.

During his long career, de Broglie published over 20 books and numerous research papers. His preoccupation with the practical side of physics is demonstrated in his works dealing with cybernetics, atomic energy, particle **accelerators**, and wave-guides. His writings also include works on x rays, gamma rays, atomic particles, **optics**, and a history of the development of contemporary physics. He served as honorary president of the French Association of Science Writers and, in 1952, was awarded first prize for excellence in science writing by the Kalinga Foundation. In 1953, Broglie was elected to London's Royal Society as a foreign member and, in 1958, to the French Academy of Arts and Sciences in recognition of his formidable output. With the death of his older brother Maurice two years later, de Broglie inherited the joint titles of French duke and German prince. De Broglie died of natural causes on March 19, 1987, at the age of 95, having never fully resolved the controversy surrounding his theories of wave mechanics.

DE BROGLIE WAVELENGTH

In 1923, Louis Victor Pierre Raymond de Broglie postulated that since **light** particles had also been shown to exhibit wave characteristics, perhaps other particles, such as the **electron**, might as well. From relativity, he knew that the **momentum** of a **photon** was: $E = pc$. Substituting the **energy** in one equation with the energy of the other and solving for **wavelength** (λ), he found: $\lambda = h/p = h/mv$. However, from the **photoelectric effect**, he knew: $E = hc/\lambda$.

This postulate was verified later by the Davisson-Germer experiment. This used the concept of constructive **interference** resulting from the path length difference that incident beams travel when hitting a lattice structure. If the electrons did not behave as **waves**, no constructive interference would be seen and thus no maximum (or minimum) intensities would be measured. However, if the electrons did behave as waves, there would be a maximum intensity at an angle related to the distance between the lattice layers of the **scattering** material.

The electrons were accelerated by a potential of 54 V and were incident on a nickel plate. The intensity of scattered electrons was determined by measuring the current produced. A plot of current versus detection angle was then plotted. The electrons were scattered through a wide range of angles, but a maximum intensity was seen at 50°. Since the distance between the layers of the nickel lattice was 2.15Å, a wavelength based on the experiment of 1.65Å was calculated. Using the de Broglie wavelength equation, a wavelength of 1.67Å was predicted, thus proving that indeed, electrons behave as waves: $\lambda = 2.15\text{Å} \sin(50°) = 1.65\text{Å}$. Thus the electron was shown to behave as a wave as well as a particle.

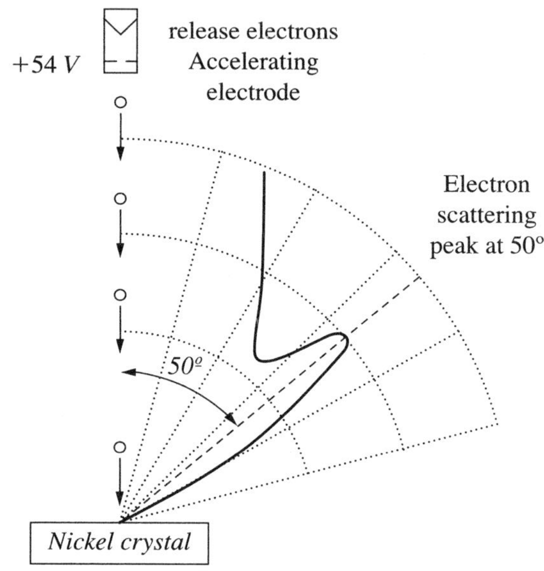

See also Davisson, Clinton Joseph; Germer, Lester Halbert; Relativity theory

DE FOREST, LEE (1873-1961)

American engineer and inventor

Lee de Forest was one of several scientists who contributed to the development of **radio**. A controversial and litigious man who seemed nearly as much concerned with fame as with science, de Forest is generally credited with the invention of the **triode**, or three-electrode **vacuum tube**, which he called the Audion. This tube made possible the amplification of electrical **energy** by introducing a third element into the existing two-electrode vacuum tube, allowing control over the flow of electrons, or **ions**, through ionized gas in the tube. De Forest's invention paved the way for radio signals to be received through the airwaves, without wires.

De Forest was born August 26, 1873, the son of the Reverend Henry Swift De Forest (he retained the capital D in his name, while his son preferred the lowercase) and the former Anna Robbins. His father, president of Talladega College in Alabama, had hoped that his son would also become a minister, but early in life Lee had shown an intense interest in inventing. As preparation for what his father hoped would be a religious career, de Forest was enrolled at Mount Hermon school in Massachusetts. The school's emphasis on hard work (students did most of the chores in order to defray the expenses) left de Forest little time for inventing, and he was unhappy there. Nevertheless, after graduation he was able to convince his parents to allow him to attend the Sheffield Scientific School at Yale University.

An unpopular and unsociable youth (he was voted the "homeliest boy in school"), de Forest was deeply concerned, perhaps even obsessed, with receiving recognition from his peers. Unlike many of his classmates, de Forest had to work to supplement his scholarship money, and still was often in debt. In an attempt to raise his status, de Forest sought money and fame by inventing several devices that he hoped to sell to companies or enter in contests with large prizes. Although none of his inventions were accepted, de Forest was certain of future success; he wrote in his diary that "I must be brilliant, win fame, show the greatness of genius and to no small degree," as Tom Lewis related in *Empire of the Air.*

De Forest received his bachelor's degree in 1896 and went on to receive a Ph.D. in physics from Yale in 1899; his dissertation, "Reflection of Hertzian Waves from the Ends of Parallel Wires," was a pioneering study of a phenomenon of radio waves. By 1902 he had founded the De Forest Wireless Telegraph Company, based on his invention of an improved wireless receiver. The company was the first of a series of businesses that were to fail, largely because of questionable practices by de Forest's business partners.

De Forest's Breakthrough: The Audion

Despite his creativity, de Forest seemed to be weak in theoretical understanding; he frequently did not see the potential in his inventions and often misunderstood why they worked. De Forest's triode was based on the Fleming valve, which was itself a variation on inventor Thomas Edison's incandescent lamp. Edison had noticed that the insides of his incandescent lamps tended to blacken, as if particles were being emitted from the **carbon** filament. When he inserted a metal plate into the lamp and connected it to the "positive" side of the filament, he found that current somehow flowed to the positively charged plate. At this time nothing was known of the existence of electrons, the negatively charged particles of **electricity**, and Edison did not pursue this line of research, although he did receive a patent for the two-element lamp in 1883.

The emission of particles to the plate, known as the "Edison effect," became the focus of experiments by **John Ambrose Fleming**, an engineer who recognized that current flowed in only one direction, from the filament to the plate, and could therefore convert an input of alternating current (AC), or current continuously switching polarity, to an output of direct current (DC). The Fleming valve, invented in 1904, was physically no different from Edison's two-element lamp, but unlike Edison, Fleming understood both its operation and its potential.

In the meantime, de Forest was working on an improved detector for wireless telegraph signals. Existing detectors used a gas flame to ionize gas; by employing a filament in a par-

tially evacuated vacuum tube instead of an open flame, he found that he could create a more stable detector. His tube differed from Fleming's chiefly in that it retained some gas in the tube, on the theory that this residual gas was necessary for ionization. This tube, the first of two to which he gave the name Audion, became the subject of litigation by Fleming, although de Forest claimed that his source was the gas-flame detector, not Fleming's valve.

De Forest's further modification of this tube, however, was his own work, and it was to be his most significant development; this was the triode, created by the addition of a third element, the grid, between the two existing components of the vacuum tube, the filament, and the plate. When current was applied to this grid, the current received by the plate was strengthened, or amplified, allowing weak telegraph signals—and, after engineer Ernst F. W. Alexanderson's development of Reginald Fessenden's high-frequency alternator, radio signals—to be heard more easily. De Forest did not fully understand how this tube worked; he believed, mistakenly, that the operation of the triode was dependent on the presence of gas (ions) within the tube. He failed to realize that the presence of gas limited the tube's life, weakened its signal, and generated noise, making it only marginally useful as an amplifier. It was not until 1912 that other scientists showed that the total evacuation of the tube produced a stronger, less noisy tube that had potential not only as an amplifier of weak radio signals but as an oscillator—a necessary component in the construction of a radio transmitter. By this time it was too late for de Forest to take full advantage of his invention; others with a clearer understanding of the technology had left him behind.

A Pioneer in the Broadcasting Industry

De Forest clearly had foresight, however, in his understanding of the potential of radio broadcasting as a public service—an opportunity to bring educational, cultural, and informational material to a wide spectrum of people. Once it was demonstrated (by 1906) that high-frequency radio signals could carry sound—of music and of the human voice—the possibilities of radio as a tool for communication unfolded. De Forest's vision for the medium frequently exceeded its existing technical capabilities, but he created a compelling popular image of what radio could achieve. On January 13, 1910, he attempted the first live broadcast from the stage of the Metropolitan Opera House in New York City, a performance of Mascagni's *Cavalleria rusticana* featuring Emmy Destinn and Enrico Caruso. The medium was still primitive—employing an antenna strung across the roof of the opera house, an arc generator to provide the radio-frequency signal, and a telephone transmitter to send the signal out over the airwaves—and the broadcast could be heard only at a few locations in the city, its **sound** quality compromised by heavy background noise. Nevertheless, it represented an attempt to use radio to raise the cultural level of Americans. De Forest also made what may have been the first news broadcast, an inaccurate announcement of the outcome of the 1916 presidential election.

De Forest continued producing inventions related to radio, but none had the importance of the triode. In 1912 he developed a method for transmitting two signals over a single line—a diplex system. That same year he developed a cascade amplifier circuit, which fed the output of one triode into the input of another. It increased the amplification by a factor of three; a series of three tubes thus produced 27 times the original signal.

A more significant outcome of this research was the feedback, or regenerative, circuit. In this circuit, the output of one triode was fed back to become its own input. This not only increased the output even further, but it allowed the tube to produce **oscillations**, that is, alternating current (AC). This enabled the tube to be used as a transmitter. But de Forest did not realize the true potential of this discovery; by the time he applied for a patent on the feedback circuit in 1915, the device had been patented by inventor E. Howard Armstrong. Litigation over the patent dragged on until 1934; although the courts found in de Forest's favor, the industry did not; it viewed the feedback circuit as Armstrong's invention and credited de Forest only with the invention of the triode.

Even the Audion triode, de Forest's most successful invention, did not produce the financial rewards he sought. Although American Telephone and Telegraph showed interest in the tube, the firm feared legal action by the giant Marconi companies, which owned the rights to the Fleming valve, of which the Audion was still seen as a possible patent infringement. De Forest did eventually sell the rights to the Audion to AT&T, but for a low sum; he had been advised to do so by one of his partners who, unknown to de Forest, was associated with AT&T.

Expanding Horizons: Sound Recording, The Phonofilm, and Television

During the early 1920s, de Forest turned his attention to sound recording, intending to develop a method for recording sound using electricity—based on his Audion—as opposed to the purely mechanical methods then in use. He succeeded in recording sound on magnetic wire, which had been done by engineer Valdemar Poulsen in 1898, but without the advantage of de Forest's Audion. He later succeeded in recording sound on film using what were, in effect, photographed sound waves read by a photoelectric cell; this, too, was derivative of an earlier inventor's work—Eugen Lauste had made recordings using **light** as early as 1903—but again, the Audion provided the amplification lacking in the earlier device.

It was a small conceptual step from this idea to that of recording sound directly on **motion** picture film—avoiding the synchronization problems of the existing sound films using disk recordings—a goal he achieved by 1921, which he called "Phonofilm." A series of sound films followed, and theaters began to be equipped with Phonofilm equipment by 1924, a full three years before Warner's Vitaphone sound-on-disk method produced *The Jazz Singer,* the first full-length film using sound and the one usually credited with launching the sound film era. Inadequate interest by Hollywood studios, de Forest's lack of expertise as a producer, his decision to make only short films, and legal troubles with the inventor of the photoelectric cell used in the process all contributed to the

demise of the De Forest Phonofilm Corporation by 1925. Ironically, this light-on-film soundtrack method was eventually adopted by the motion picture industry.

After 1925 de Forest developed a television system. This was a mechanical (rather than purely electronic) device for transmitting and receiving pictures over the airwaves. Later, in 1946, he developed a **color** television system; this was also a mechanical system, and it could not compete with the all-electronic color television developed by RCA in the early 1950s. As with radio, de Forest believed that television had the potential to raise the level of civilization. He became severely critical of the radio and televison programming of his time, however, feeling that it perpetuated the lowest forms of entertainment.

De Forest's life was punctuated with business reverses and long-running litigation that gave him as much notoriety in the radio industry as fame. In his private life as well, de Forest encountered considerable difficulty. There was a series of failed marriages. The first, to Lucile Sheardown, lasted barely a month before they separated; they were divorced in 1907. His second wife was Nora Stanton Blatch, a civil engineer at Cornell University who was the granddaughter of Elizabeth Cady Stanton and daughter of suffragist Harriot Stanton Blatch. Nora had a strong interest in electrical engineering—she had studied under the inventor Michael Pupin—and she worked alongside her husband in his laboratory. This did not conform with de Forest's idea of a woman's place, however, and relations between them became strained. A daughter, Harriot Stanton de Forest, was born in 1909, but by this time her parents had separated, their divorce becoming final in 1911. De Forest married his third wife, singer Mary Mayo, in December 1912. Apparently depressed about giving up her stage career, and suffering from severe rheumatism, she became an alcoholic. After three difficult pregnancies—two daughters survived, but the third child, the son for whom de Forest had long hoped, died after two days—Mayo left her husband; they were divorced in 1929. His fourth and final marriage, to Marie Mosquini, a motion picture actress, took place in 1930.

Throughout his life, de Forest held to the belief that he was the "father of radio"—in fact, this was the title of his autobiography, published in 1950. He could not understand why he did not receive this recognition, and he saw his life as a series of hurdles placed in his path by business partners and other scientists in an effort to rob him of the glory he felt was his due. Others in the industry, however, found engineer **Guglielmo Marconi** or Fleming as more deserving of the appellation "father of radio," pointing out that de Forest's inventions were adaptations of others' and, in any case, that he did not completely understand how they worked. Radio was not the invention of a single person; rather, it evolved from the separate, sometimes overlapping, inventions of several people, including Edison's incandescent lamp, Marconi's wireless telegraph, Fleming's valve, and Fessenden's electrolytic detector. Lee de Forest died of heart failure on June 30, 1961, and is remembered not as the "father of radio," as he intended, but as one of several contributors to its development.

DEGRADATION OF ENERGY • See Entropy

DEMOCRITUS (460-370 B.C.)

Greek philosopher and naturalist

Democritus's background is uncertain. Many stories of his life are tradition or later inventions. Among the few undisputed facts are his birthplace (Abdera, Thrace) and his teacher (Leucippus). Only fragments of his writings survived, and many of them concerned ethics. Most of the knowledge of Democritus's scientific ideas came from his students, especially Nausiphanes, the teacher of Epicurus.

According to Democritus, the physical **universe** was composed of atoms and void. An infinite number of atoms were in perpetual **motion** throughout the void. An **atom** was an eternal, unchangeable, solid body without pores or voids, so small it could not be seen or divided (the word *atomon* means indivisible). Atoms only differed from each other in shape, size, arrangement, and position, which explained the properties of different substances. For example, water atoms were smooth and round so water flowed easily and had no permanent shape. Contrarily, atoms of fire, which were thorny, caused pain, and atoms of earth, which were jagged and rough, held together in a definite form. A substance's feel and taste depended upon the observer's sense organs and convention while **color** was a result of the position of the atoms of compounds. A substance's change of nature, such as water turning into steam, was the result of atoms separating and rejoining in a new pattern.

Democritus was one of the earliest mechanists who believed that the eternal, unbreakable laws of nature, not the whims of gods or demons, determined the creation and operation of the universe, and even the human mind and soul.

DENSITY

The density of an object is defined simply as the **mass** of the object divided by the volume of the object. For a concrete example, imagine you have two identical boxes. You are told that one is filled with feathers and the other is filled with cement. You can tell when you pick up the boxes, without looking inside, which is the box filled with cement and which is the box filled with feathers. The box filled with cement will be heavier. It would take a very large box of feathers to equal the **weight** of a small box of cement because the box of cement will always have a higher density.

Density is a property of the material that does not depend on how much of the material there is. One pound of cement has the same density as one ton of cement. Both the mass and the volume are properties that depend on how much of the material an object has. Dividing the mass by the volume has the effect of canceling the amount of material. If you are buying a piece of gold jewelry, you can tell if the piece is solid gold or gold plated steel by measuring the mass and volume of the piece and computing its density. Does it have the density of gold? The mass is usually measured in kilograms or grams and the volume is usually measured in cubic meters or cubic centimeters, so the density is measured in either kilograms per cubic meter or in grams per cubic centimeter.

The density of a material is also often compared to the density of water to give the material's specific **gravity**. Typical rocks near the surface of Earth will have specific gravities of 2 to 3, meaning they have densities of two to three times the density of water. The entire Earth has a density of about five times the density of water. Therefore the center of Earth must be a high density material such as nickel or iron. The density provides an important clue to the interior composition of objects, such as Earth and **planets**, that we can't take apart or look inside.

DERHAM, WILLIAM (1657-1735)

English philosopher, scientist, and minister

A scholar with many interests, from botany and **meteorology** to astronomy and physics, William Derham made the most accurate measurements of the speed of **sound** up to his time. English physicist Sir **Isaac Newton** accepted these measurements and used them in his landmark publication *Principia*. Derham was born in Stoughton, England, in 1657. Little is known about his family other than his father's name, Thomas Derham, and that the family was poor. Derham attended Oxford University, Trinity College, where he received his B.A. and M.A. degrees. Ordained a priest in the Anglican Church in 1682, he served as the Vicar of Wargrave in Berkshire and the Vicar of Upminster in Essex. Derham also fulfilled the role of village physician.

With a wide range of interests, Derham wrote on philosophy, theology, and the sciences. His first publication was a treatise on clocks, titled *The Artificial Clockmaker* (1696). His extensive writings included papers on wildlife, the behavior of mercury barometers, telescopes, and astronomy. His best known works were *Physico-Theology* (1713) and *Astro-Theology* (1714). In *Physico-Theology*, Derham discussed atmosphere, **light**, **gravity**, and biology. While studying the **speed of sound**, Derham instructed friends to fire shotguns on a set time with synchronized pocket watches from distant locations (often the towers of churches). Derham then observed the interval between the flash and the arrival of sound, using telescopes and a half-second pendulum. Derham also invented an instrument for finding the meridian.

Derham was married to Anne Scott, daughter of George Scott, who, like Derham, was a fellow of the Royal Society. Derham had several children, including an eldest son, William, who became president of St. John's College, Oxford. Derham died in 1735.

DESCARTES, RENÉ (1596-1650)
French mathematician and philosopher

René Descartes is often called the father of modern philosophy for his break with the Scholastic tradition that had previously dominated Western thought. Unlike the Scholastic philosophers, who respected the authority of **Aristotle**, Descartes believed that he could attain the greatest possible degree of certitude in both his philosophic and scientific investigations by relying on reason. Descartes's emphasis on objectivity and reason became fundamental to the methodology of later philosophers and scientists. In the field of mathematics, Descartes's most important contribution was his discovery of the principles that ultimately developed into the field of analytic geometry.

Descartes was born in the town of La Haye (now La Haye-Descartes), France. He received his early education at a Jesuit school, where he displayed a precocious facility in mathematics, and studied law at the University of Poitiers, from which he graduated in 1616. Two years later he went to the Netherlands and joined the army of Maurice of Nassau. This venture, like his subsequent service in Bavarian and Hungarian armies, was undertaken by Descartes because it afforded him the opportunity to travel, observe, and experience what he described as "the book of the world." During this active period in Descartes's life he formed a friendship with the Dutch philosopher and scientist Isaac Beeckman, who familiarized Descartes with contemporary developments in mathematics.

In 1628, Descartes settled in the Netherlands, where he lived most of his remaining years. It was during this time that he wrote almost all of his philosophical and mathematical treatises. His most influential work, *Discours de la méthode* (*Discourse on Method*) originally appeared as the preface of a 1637 volume devoted to mathematics and physical science. The *Discourse* and accompanying studies reflect Descartes's principal ambition as a thinker: to compose a method for analyzing and explaining all phenomena of the physical world.

Descartes's concern with attaining a comprehensive knowledge of the workings of the **universe** involved an investigation into the nature of knowledge itself. In the *Discourse* he initially doubts that anything can be known to exist. Ultimately he concludes that he can be certain of the existence of his own doubt, and from this primal certainty infers the existence of a thinking being who is the source of this doubt. Declaring "Cogito, ergo sum" ("I think, therefore I am"), Descartes believed he had found an indubitable fact that could serve as the basis for further knowledge.

A devout Catholic, Descartes maintained that religious truth cannot be attained through scientific reasoning. Descartes made a radical distinction between mind and **matter**, viewing the latter, whether organic or inorganic, as strictly mechanical in its nature and functions. Accordingly, he regarded the human body as subject only to physical laws that had no connection with an individual's mental or spiritual life. This disjunction between the material and nonmaterial realms is often referred to as Cartesian dualism. Descartes neverthe-

René Descartes.

less believed that God enables human beings to exert mental control over their bodies. Descartes's name is also used in the branch of mathematics known as analytic geometry, which employs Cartesian coordinates to plot the location of objects on a two-dimensional plane or in a three-dimensional **space**.

In September 1649, Descartes accepted the invitation of Queen Christina of Sweden to join her court and act as her instructor in philosophy. The following winter he contracted pneumonia. He died on February 11, 1650.

DETERMINISM

Determinism maintains that precise information about the initial conditions of a system provides an exact knowledge of how that system evolves in **time**. Newtonian **mechanics**, **electrodynamics**, and relativity are all deterministic theories. If we knew the properties (**mass**, charge, position, **momentum**, etc.) of every particle within a classical system to arbitrary precision today, then the equations of **motion** of the system that tell us how the particles interact can in principle predict the future state of the system tomorrow. It is only our incomplete knowledge of the state of the system today and the difficulty in performing the necessary calculations that prevents us from divining the future. Some philosophers hold that in such a framework, there is no such thing as free will because actions are determined exclusively by the initial conditions. The **universe** evolves as the equations tell it to.

Chaos theory, a branch of mathematics with a wide range of physical applications, holds that tiny differences between a pair of initial conditions can produce profound differences in how a system evolves. Even for simple mechanical systems such as the double pendulum, if the initial conditions (which include environmental factors) are even slightly different, the final state and the path the system traverses to get there can be drastically modified. Therefore, any imperfections in the knowledge of the initial conditions render an attempt to deduce the future state futile. Because the initial conditions need to be measured and the devices used to make those measurements do not provide arbitrarily precise data, information about the initial conditions is always imperfect.

Quantum mechanics is not a deterministic theory. It holds that the initial conditions of a system cannot be measured to arbitrary precision even if the measuring devices are capable of providing that level of precision. The uncertainty principle states that $\delta x \cdot \delta p \geq h/2$. That is to say, the product of the uncertainty in a measurement of the position of a particle and the uncertainty in a measurement of its momentum is always greater than a dimensionful constant. Because of this intrinsic property of **matter**, we cannot obtain a perfect knowledge of the initial conditions, and the evolution of any system is not deterministic.

DIELECTRICS

Dielectrics are non-conductors. Their **electric charge** will not flow when an electric field is applied, but there will be a displacement of electric charges. This displacement of charges can occur at both the atomic level and molecular level.

At the atomic level, electrons are all confined in atoms in dielectrics. Electrons and nuclei carry negative and positive charges respectively. In the absence of an electric field, electrons move symmetrically around the **nucleus** inside the **atom**. The atom appears as a neutral particle. When an electric field is applied, electrons will be pulled toward the higher **voltage** end, while nuclei will be pulled toward the lower voltage end. Total charge of the atom is still zero, but there is a nonzero distance between the average position of electrons and nucleus. This is called a dipole. Under the influence of the electric field, each atom becomes a dipole lined up in the direction of the field. The sum of them is a nonzero dipole. The stronger the field is, the larger the displacement and the dipole are.

At the molecular level, some materials' molecules have an uneven charge distribution. As a result, each molecule is already a dipole. These dipoles point to random directions in **space** in the absence of an external field, summing up to zero on average. When an external field is applied, they become more or less aligned with the field direction, summing up to a nonzero dipole pointing in the direction of the field. The stronger the field is, the more aligned the molecules are, thus the larger the dipole.

The dipole produced at the atomic level is usually much weaker than that at the molecular level. When a material's molecules have dipoles at the molecular level, the atomic level effect can be ignored. Such a material is called a polar compound. Otherwise, the material is called a nonpolar compound. Water is a polar compound while methane is nonpolar. To quantify how easily a dipole can be produced when an electric field is applied, permittivity or relative dielectric constant (ε) is defined. The larger the displacement is, the larger the permittivity becomes. Permittivity is generally a complicated complex tensor. It is approximated by a real number in many cases.

Any material has a certain permittivity, including the conductor. Permittivity of a dielectric can often be approximated as a real number while that of a conductor is always a complex number with a large imaginary part. The imaginary part is called conductivity. Permittivity is a macroscopic quantity. It is only meaningful in a large region as an average property. In today's semiconductor technology, where several layers of atoms can be sandwiched in a bulk material, what the suitable value to use as the permittivity in that thin layer remains a research problem. We can always do the calculation on an atom-by-atom basis, but that is usually complicated.

DIFFRACTION

Diffraction is the deviation from a straight path that occurs when a wave such as **light** or **sound** passes around an obstacle or through an opening. The importance of diffraction in any particular situation depends on the relative size of the obstacle or opening and the **wavelength** of the wave that strikes it. The diffraction grating is an important device that makes use of the diffraction of light to produce spectra. Diffraction is also fundamental in other applications such as x-ray diffraction studies of crystals and **holography**.

All **waves** are subject to diffraction when they encounter an obstacle in their path. Consider the shadow of a flagpole cast by the **Sun** on the ground. From a distance the darkened zone of the shadow gives the impression that light traveling in a straight line from the Sun was blocked by the pole. But careful observation of the shadow's edge will reveal that the change from dark to light is not abrupt. Instead, there is a gray area along the edge that was created by light that was "bent" or diffracted at the side of the pole.

When a source of waves, such as a light bulb, sends a beam through an opening or aperture, a diffraction pattern will appear on a screen placed behind the aperture. The diffraction pattern will look something like the aperture (a slit, circle, square) but it will be surrounded by some diffracted waves that give it a "fuzzy" appearance.

If both the source and the screen are far from the aperture the amount of "fuzziness" is determined by the wavelength of the source and the size of the aperture. With a large aperture most of the beam will pass straight through, with only the edges of the aperture causing diffraction, and there will be less "fuzziness." But if the size of the aperture is comparable to the wavelength, the diffraction pattern will widen. For

Laser light sample of diffraction. *(Photograph by Hank Morgan, Photo Researchers Inc. Reproduced by permission.)*

example, an open window can cause sound waves to be diffracted through large angles.

Fresnel diffraction refers to the case when either the source or the screen are close to the aperture. When both source and screen are far from the aperture, the term Fraunhofer diffraction is used. As an example of the latter, consider starlight entering a **telescope**. The diffraction pattern of the telescope's circular mirror or lens is known as Airy's disk, which is seen as a bright central disk in the middle of a number of fainter rings. This indicates that the image of a star will always be widened by diffraction. When **optical instruments** such as telescopes have no defects, the greatest detail they can observe is said to be diffraction limited.

Applications

Diffraction gratings

The diffraction of light has been cleverly taken advantage of to produce one of science's most important tools—the diffraction grating. Instead of just one aperture, a large number of thin slits or grooves—as many as 25,000 per inch—are etched into a material. In making these sensitive devices it is important that the grooves are parallel, equally spaced, and have equal widths.

The diffraction grating transforms an incident beam of light into a spectrum. This happens because each groove of the grating diffracts the beam, but because all the grooves are parallel, equally spaced and have the same width, the diffracted waves mix or interfere constructively so that the different components can be viewed separately. Spectra produced by diffraction gratings are extremely useful in applications from studying the structure of atoms and molecules to investigating the composition of **stars**.

X-ray diffraction

X rays are light waves that have very short wavelengths. When they irradiate a solid, crystal material they are diffracted by the atoms in the crystal. But since it is a characteristic of crystals to be made up of equally spaced atoms, it is possible to use the diffraction patterns that are produced to determine the locations and distances between atoms. Simple crystals made up of equally spaced planes of atoms diffract x rays according to Bragg's law. Current research using x-ray diffraction utilizes an instrument called a diffractometer to produce diffraction patterns that can be compared with those of known crystals to determine the structure of new materials.

Holography

When two laser beams mix at an angle on the surface of a photographic plate or other recording material, they produce an **interference** pattern of alternating dark and bright lines. Because the lines are perfectly parallel, equally spaced, and of equal width, this process is used to manufacture holographic diffraction gratings of high quality. In fact, any hologram (holos-whole:gram-message) can be thought of as a complicated diffraction grating. The recording of a hologram involves the mixing of a laser beam and the unfocused diffraction pattern of some object. In order to reconstruct an image of the object (holography is also known as wavefront reconstruction) an illuminating beam is diffracted by plane surfaces within the hologram, following Bragg's law, such that an observer can view the image with all of its three-dimensional detail.

DIODES

A diode is an electrical circuit component that allows current to flow in only one direction. A diode consists of two electrodes sealed in a **vacuum**. One electrode, the cathode, is attached to a negative **electric potential**, the other electrode, the **anode**, is attached to a positive electrical potential. The word diode is derived from the term di-electrode (i.e., two electrodes).

The diode was first developed by English scientist Sir **John Ambrose Fleming**. Historically the original diode was an evacuated glass bottle or tube with two metal plates (electrodes) inside, each connected to opposite electrical charges. The negatively charged plate (cathode) was heated to a glowing red **temperature** and the positively charged electrode (anode) was left cold. Thermionic diode tubes (originally thermionic valves) were the mainstay of the **electronics** industry until the late 1950s when **transistors** and, later, integrated circuit (IC) chips became the norm. Today, thermionic diode tubes still have use, particularly when large electrical currents are needed.

The principle of the thermionic diode is based on the fact that in a vacuum a hot conductor will expel free electrons from its surface. If the **electron** emitter is connected to a negative potential, the free-floating electrons will experience an electrical **force** directed away from the negative cathode. If an additional electrode attached to a relatively positive electrical potential (i.e., the anode) is placed some distance away from the cathode, the electrons will experience an attractive force toward the positive anode. Accordingly, as the charged particles migrate, a current of **electricity** is said to flow across the **space** between the cathode and the anode.

If the electrical charge on the two electrodes is now reversed the flow of electrons stops. This is because the hot cathode is still expelling electrons but the electron cannot cross to the original anode because of the repulsive negative charge. They are also pulled back to the original cathode, which is when it becomes positively charged. In other words, in a thermionic diode, current can flow only one way across the space between the cathode and anode, and that is when the hot cathode is negatively charged with respect to the cold relatively positive anode.

Most electrical current in electronic equipment is alternating in two opposite directions. Circuit designers often need to rectify this current from an alternating current (AC) to a direct current (DC). The diode achieves this effect. When an alternating potential is applied across a conductor, an alternating current results; when it is placed across a diode, current flows in one direction only.

The transistor eventually replaced the thermionic diode with what became known as the p-n junction diode. This diode had no glass bottle, no heated cathode, and no vacuum, instead it used two substances termed p-type and n-type **semiconductors** that have opposite charge flow characteristics. One substance, n-type germanium, allows its electrons to move through itself relatively easily. The other semiconductor, p-type germanium, tends to allow its positive charges or, more accurately, positive holes (vacancies formerly occupied by electrons) to move much more efficiently one direction than in the other.

In a p-n junction diode, these two substances are fused together. Their junction becomes a place where the drifting positive charges or holes from the p-type germanium and the drifting electrons from the n-type germanium can combine (in the process forming new positive holes elsewhere in the p-type germanium and new free electrons elsewhere in the n-type germanium). This drifting of the positive holes and the negative electrons towards each other at the junction is similar to an **electric current** through the bonded materials. The current

is accentuated by applying a positive charge to the p-type germanium thus repelling the positive holes more strongly towards the junction. A corresponding negative charge applied to the n-type germanium strongly repels the electrons towards the junction and a strong electrical current appears to flow through the junction.

Applying the opposite charges to the materials has the same effect as in the thermionic diode, basically there will be no electric current. This is because the reversed forces on the positive holes and the electrons will pull them both away from the junction where they can combine. Pulling the holes and electrons apart effectively stops the current. Accordingly, the thermionic diode and the p-n junction diode produce the same effect. Applying an electric potential one way across them will result in current flow. Reversing that potential stops current flow and allows rectification of alternating current.

DIRAC, PAUL (1902-1984)
English physicist

Paul Adrien Maurice Dirac was one of the twentieth century's leading theoretical physicists. Instrumental in developing quantum **mechanics**, the theoretical study of **atomic structure** and properties, and **quantum electrodynamics**, the study of the electrical interactions between atomic particles, Dirac also postulated the existence of the **positron**, a positive-charge **electron**, which led to later discoveries concerning **antimatter**. For his work on the development of **quantum mechanics**, Dirac shared the 1933 Nobel Prize in physics with Austrian physicist **Erwin Schrödinger**.

Dirac was born in Bristol, England, on August 8, 1902. His father, Charles Adrien Ladislas Dirac, was a Swiss immigrant, and his mother, Florence Hannah (Holten) Dirac, was British. Charles Dirac took a position as a teacher of French at the Merchant Venturer's Technical College, where young Dirac enrolled for his early schooling. After graduating in 1918, Dirac entered Bristol University, where he majored in electrical engineering and received his bachelor's degree in 1921. His hopes for employment after graduation were postponed, however, by Great Britain's postwar depression, and he decided instead to return to school, accepting a two-year scholarship from the department of mathematics at Bristol.

When Dirac's scholarship at Bristol came to an end, he was accepted at St. John's College, Cambridge, as a research student in mathematics. He soon became familiar with the latest developments in **atomic theory**, partly through his class work and partly by reading the works of Danish physicist **Niels Henrik David Bohr**, German physicist **Max Born**, German physicist **Arnold Sommerfeld**, English astronomer **Arthur Stanley Eddington**, and other leaders in the field. In addition, he had the opportunity to hear lectures by Bohr, German physicist **Werner Karl Heisenberg**, and others during their visits to Cambridge. Dirac's first research papers were published in 1925, while he was still a student. He received his Ph.D. in physics from Cambridge in 1926. His thesis involved an elab-

Paul Dirac.

oration of quantum mechanical concepts that had originally been developed by Heisenberg.

In the fall of 1926, Dirac traveled to Copenhagen, where he spent much time talking with Bohr. In February 1927 he moved on to Göttingen, where he came into contact with Born, American physicist **J. Robert Oppenheimer**, American physicist **James Franck**, and Russian physicist Igor E. Tamm, among others. In late 1927, Dirac returned to England where he was elected a fellow at St. John's College, Cambridge. It was there during the winter of 1927–28 that he made an improvement on Schrödinger's wave equation.

Develops Wave Equation

Schrödinger's wave equation sought to explain the behavior of an electron in an **atom**, but Schrödinger chose to ignore relativistic effects in his calculations. Dirac's approach was to begin with only the simplest information known about the electron—its **mass** and charge—and devise a mathematical theory that would describe the electron's properties more fully than Schrödinger's equation did. He was successful in this effort, producing equations that accurately predicted the electron's spin, magnetic charge, and other properties, as measured in experimental work. In recognition of their work on the wave equation, Dirac and Schrödinger shared the 1933 Nobel Prize in physics.

One consequence of Dirac's mathematical analysis led to his hypothesis of the positive electron. In working with his equations for the electron, Dirac discovered that two solutions were possible, suggesting that the electron could either have positive or negative kinetic **energy**. Dirac's theory also suggested experiments a researcher might use to look for the positive electron, examining situations in which positively charged electrons would be produced (always in connection with negatively charged electrons and always in such a way that the two would annihilate each other). Exactly these conditions were noted in 1932 when American physicist **Carl David Anderson** first observed the positive electron or, as it was later named, the positron.

By extension, Dirac's theory also suggested the existence of other forms of antimatter, including the antiproton and antineutron. Those predictions took much longer to be fulfilled, but were eventually confirmed. Today, considerable speculation exists as to the possibility of a whole **universe** of **matter** made out of antiprotons, positrons, and antineutrons, as well as to the ultimate consequences of the collision of such a universe with our own.

Appointments at Cambridge and Worldwide Travel after 1929

In 1929, Dirac was appointed to the post of university lecturer and praelector in **mathematical physics** at St. John's. The appointment carried with it very few specific duties, allowing him to spend as much time as he liked on research, writing, and travel, the latter of which had become a great passion for him. In fact, he spent five months abroad in 1929, during which time he taught briefly at the University of Michigan and the University of Wisconsin. At the end of this trip, he returned to England in 1930 by way of Japan and Siberia. Upon his return, he was elected to the Royal Society and, two years later, was chosen for the post of Lucasian Professor of Mathematics at Cambridge, a position once held by the seventeenth-century English physicist and mathematician **Isaac Newton**.

In 1934, Dirac returned to the United States, spending most of the 1934–35 academic year at the Institute for Advanced Studies at Princeton University. While there, he met his future wife, Margit ("Manci") Wigner, sister of the physicist **Eugene Wigner**. Dirac and Margit returned to Cambridge and were eventually married in London in January 1937. They had two daughters, Mary Elizabeth, born in 1940, and Florence Monica, born in 1942. Dirac remained at Cambridge during World War II, but became involved in a number of government projects related to the development of atomic energy. He was especially interested in techniques for the separation of uranium isotopes, although none of his suggestions was specifically incorporated into later weapons development programs.

During the last half century of his life, Dirac became increasingly interested in problems of **cosmology**, a branch of **astronomy** that is concerned with the origins of the universe. One of the first topics in this field to capture his attention was that of "large numbers." During the 1920s, Eddington had become intrigued by the realization that certain fundamental physical constants had relationships to each other that fall in the range of 10^{39-40}. For example, the ratio of the gravitational attraction between an electron and **proton** to their electrostatic attraction is about $10^{40}:1$, as is the ratio of the radius of the universe to the radius of the electron. Dirac argued that this recurring ratio was variable over time, a hypothesis not yet proven by experiment.

In 1969, Dirac retired from his post at Cambridge and moved to the United States. He stayed briefly at the Center for Theoretical Studies of the University of Miami before accepting an appointment as professor of physics at Florida State University in Tallahassee in 1972. He continued to travel, write, and speak during the next decade. After 1982, however, his health began to deteriorate, and he died in Tallahassee on October 20, 1984.

DIRECTION OF TIME

The directionality of **time** is a fundamental and unexplained characteristic of the **universe**. Unlike the three dimensions of **space**, which have no obvious preferred direction, time is unequivocally separated into the past, from which we may receive information, the present, and the future, which is influenced by the present. There exists an **arrow of time** that separates cause from effect.

Curiously, reversal of time has no effect on nearly all microscopic physical processes. The laws of **electromagnetism**, for example, are invariant (do not change form) when time is reversed: every possible electromagnetic process, if observed to "run backwards," obeys the same laws. Thus, no electromagnetic phenomenon can define a direction of time. For electromagnetic systems the distinction between forward and reverse **motion** is not well-defined. Similarly, classical and quantum **mechanics**, and Einstein's theory of gravitation are all time-reversal invariant. Yet some physical phenomena must violate the laws of physics when time is simply reversed—otherwise the experimentally obvious distinction between the past and future would not exist.

Strong empirical evidence does exist that the distinction between past and future is not simply an artifact of human perception. The second law of **thermodynamics** states that **entropy** (disorder) of the universe will not decrease with the (forward) advance of time for all physical processes. That is, the universe moves toward increasing disorder. Thus the "arrow of time" can be defined as the direction of increasing entropy. The unyielding increase of entropy as expressed in the **second law of thermodynamics**, having been observed in countless systems, can be explained by the use of probability theory. Probability inherently requires a notion of past and future, however, and a satisfactory explanation of the connection between fundamental, microscopic physical laws and the source of the universal arrow of time has not been found.

Violation of time-reversal invariance (also called time-reversal symmetry breaking) is exhibited by nuclear processes, namely those that involve the weak nuclear **force** (weak interactions) as opposed to electromagnetic interactions. In 1964 J. W. Cronin and V. L. Fitch indirectly observed time-reversal symmetry breaking in the decay of neutral **K mesons**. In 1997 investigators reported observations of violation of time-reversal invariance in the behavior of tunnel junctions in a superconductor.

DISPERSION

Dispersion refers to the observed dispersion of microwave **radiation** when it encounters an obstruction. Perhaps the most common form of dispersion is that of **light**, when it encounters an object and is reflected back in wavelengths that the human eye perceives as colors. Dispersion theory also has many uses in nonrelativistic **quantum mechanics**, in **wave motion**, and in **sound** mechanics.

The ratio of light's **velocity** in a substance to its velocity in a **vacuum** is called the index of refraction. It is represented by the equation $n = c/v$, where n is the index of refraction, c is the absolute **speed of light** in a vacuum, and v is the light's observed velocity while passing through a substance. In almost every case the index of refraction is equal to or greater than one. The refraction of light depends on passage either through a medium with varying index of refraction or from one medium to another.

This effect can be seen whenever white light, which is made up of all wavelengths of light, is passed through a prism to produce a spectrum. The light is separated into its constituent colors by dispersion. The incident ray is bent away from the normal through an angle, δ. According to Snell's law of refraction, light bends at an angle wherever two media form a border. This angle is dependent on the **wavelength** of the light. Because the human eye perceives light of different wavelengths as different colors, we say that blue light has a higher δ than red.

Another common display of dispersion is a rainbow. Contrary to popular belief, raindrops are spherical, not tear shaped. Light enters the sphere of water and is reflected off the back of the raindrop. Depending on where it enters, it is refracted varying amounts. There is one angle, for each wavelength, which is an extremal (maximum or minimum depending on the number of internal reflections). At this extremal, the most brightness is seen. The extremal angle depends on wavelength. Red is bent the least, so it is seen at the top of the rainbow while violet is bent the most, so it appears at the bottom.

The rainbow is seen in the sky at an angle between 40° and 42° relative to the observer, which describes an arc. Thus, the rainbow is curved. If you could view the entire sky, you would see the rainbow as a complete circle. You might also see a second rainbow, with the colors reversed. This is due to light being reflected in the raindrop twice.

Another common way of producing a spectrum is through a **diffraction** grating. A grating is a series of very fine slits placed very close together, perhaps 600 or more lines per millimeter. Diffraction gratings, like prisms, disperse white light into individual colors. What is projected by a diffraction grating is called a diffraction pattern consisting of a bright central maximum with lesser maxima to either side. These lesser (higher order) maxima get increasingly dimmer as they get farther away from the central maximum. How many maxima are seen depends in part on the condition of the grating.

If the grating spacing, d, is known and careful measurements are made of the angles at which light of a particular **color** has a maximum, the wavelength of the light can be measured. This is particularly useful when doing **spectroscopy**. Gases, when excited, emit only certain wavelengths of light due to the excitation and decay of **electron** orbits. This creates an emission spectra. Each element, in its gaseous state, will emit an emission spectrum. Using this spectral fingerprint, elements can be identified in any spectrum. For example, an emission spectrum is gathered for a particular star and compared to emission spectra of known elements. Thus, astronomers can classify the star by what elements are found in the spectrum. Once classified, a great deal of information about the star can be known.

DOPPLER, CHRISTIAN JOHANN (1803-1853)

Austrian physicist

Known primarily for his theories on the behavior of **sound** waves, Doppler was one of the first major Austrian physicists, and his research on the **Doppler effect** led to tremendous advances in scientific thinking, especially in the field of **astronomy**.

Born in Salzburg, Doppler intended to follow his father into the stonemason's trade. However, chronic health problems eliminated the possibility of so physically taxing a career, and he soon redirected his efforts toward the business field. His natural talent for mathematics placed him at the head of his classes and, at the advice of one instructor, Doppler enrolled at the Polytechnic Institute in Vienna. Unsatisfied with the curriculum, he returned to Salzburg to complete his education privately. After graduation he worked as a tutor and mathematical assistant, but he longed desperately for a professorship—so much, in fact, that he was on the brink of emigrating to America, where opportunities were less scarce, when he received an offer from the State Secondary School in Prague.

While teaching in Prague, Doppler became fascinated with the nature of sound and particularly with finding an explanation for a rather common but perplexing phenomenon. It had been observed for years that the pitch of a sound varies as the source moves toward or away from the listener. The most familiar example of this phenomenon is the train whistle: when the train is approaching, the whistle is high pitched; when it recedes, the whistle sounds deeper. Doppler knew of the existence of sound **waves**, and he also knew that the pitch of a sound is dependent on how far apart those waves are. If a source is moving toward the listener, he supposed, the waves in front of the source would be squeezed together, creating a higher **frequency**, while the waves behind would be stretched out, creating a lower frequency. He soon worked out the mathematical formula that governed this shift, presenting his theories to the scientific community in 1842.

In 1845, to substantiate his theory, Doppler arranged for his colleague C. H. D. Buys Ballot (1817-1890) to conduct what would be one of the most unusual experiments in history. Buys Ballot arranged for a locomotive to pull an open car full of trumpeters while, to the track's side, a group of musicians with perfect pitch listened. For two days the trumpeters were towed back and forth while the listeners recorded the changing pitches of their instruments. Fortunately for the reputations of Doppler and Buys Ballot, the findings provided absolute confirmation of the theory.

The Doppler effect has had two important applications in modern science. The distance and bearing of a target can be found by measuring the acoustical shift of a bounced signal. This effect forms the basis for both **sonar** and **radar**, as well as for speed-guns used by traffic police. Although his theory was never proven during his lifetime, Doppler also predicted that it should apply to **light** waves. In 1901, the first experiment was completed showing a Doppler shift in starlight; light that was receding from the observer was shifted toward the red end of the visible spectrum, while light approaching the observer was shifted toward the blue end. This shift, which showed that many **stars** (including our own) were moving away from some central point in the **universe**, became the foundation for the **big bang** theory of creation.

DOPPLER EFFECT

The Doppler effect was named after **Christian Johann Doppler**. This Austrian physicist observed and explained the changes in pitch and **frequency** of **sound** and **light** waves, as well as all other types of **waves**, caused by the **motion** of moving bodies. The general rule of the Doppler effect is that the wave frequencies of moving bodies rise as they travel toward an observer and fall as they recede from the point of observation.

While Doppler, in 1842, demonstrated the phenomenon named after him in the area of sound waves, in the same year he also predicted that light waves could be shown to exhibit the same response to the movement of bodies similar to those of sound waves.

The response of sound waves to moving bodies is illustrated in the example of the sounding of the locomotive whistle of a moving train. When the train blows its whistle while it is at rest in the station, stationary listeners who are either ahead of the engine or behind it will hear the same pitch made by the whistle, but as the train advances, those who are ahead will hear the sound of the whistle at a higher pitch. Listeners behind the train, as it pulls further away from them, hear the pitch of the whistle begin to fall.

The faster the train moves the greater will be the effect of the rising and falling of the pitch. Also, if the train remains at rest but the listeners either move toward the sounding train whistle or away from it, the effect will be the same. Those who move toward the train will hear a higher pitch, while those who travel away from the train will hear a lower pitch.

When the train is at rest it is the center of the sound waves it generates in circles around itself. As it moves for-

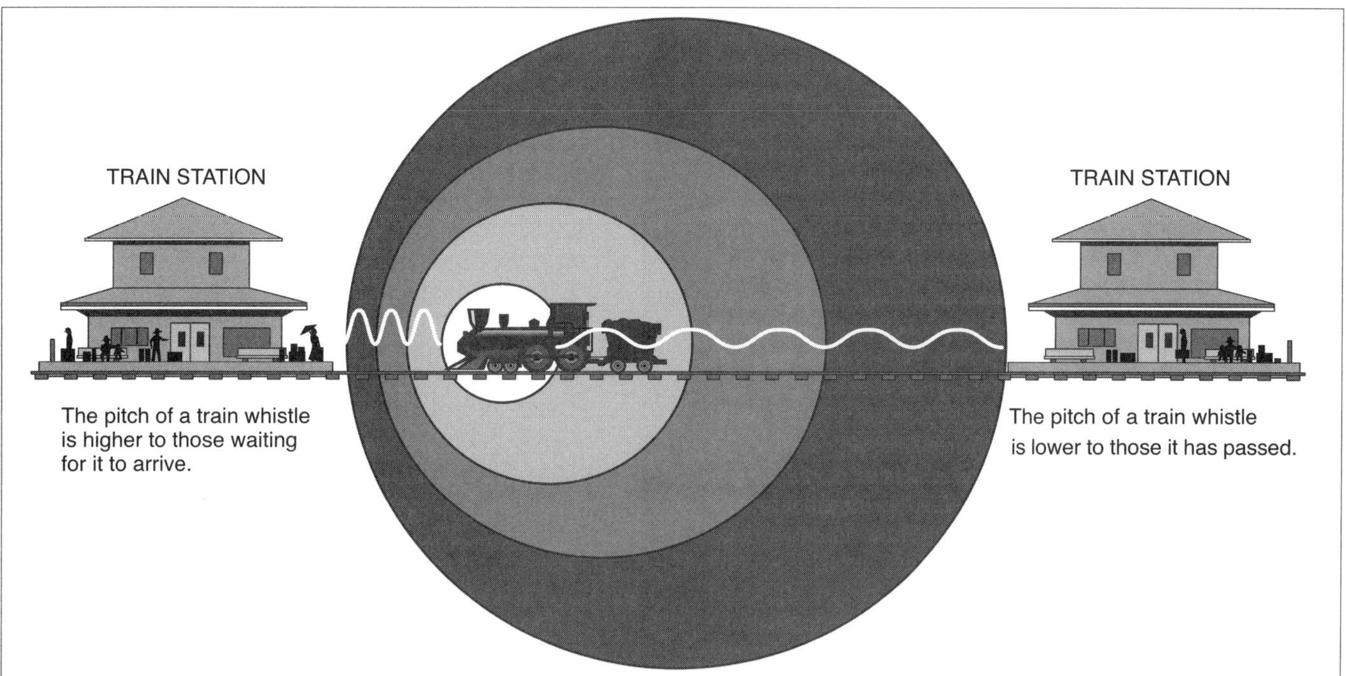

Demonstration of Doppler effect.

ward, it ceases to be the center of the sound waves it produces. The sound waves move in the same direction of the train's motion. The train is chasing or crowding its waves up front, compressing them, so that the listener in front of the direction of its movement hears more waves per second, thus producing the effect of a higher frequency. The listener standing behind the train hears a lower pitch because the waves have spread out behind the forward motion of the train. Thus, there are fewer waves per second. The listener is now **hearing** a lower frequency than is actually being produced by the whistle.

In 1845, the Doppler effect received further confirmation in an elaborate experiment devised by a Dutch meteorologist, Christoph Hendrick Buys Ballot. He placed a band of trumpet players on an open railroad flatcar and had it ride by listeners with perfect pitch who recorded their impressions of the notes produced by the whistle. Their written recordings of the pitches clearly demonstrated the Doppler wave effect.

The Doppler effect in light waves can be observed by the spectral analysis of light emitted by luminous objects.

The light from a stationary distant object whose chemical composition is known is refracted at a specific band of light on a spectroscope. That band is known as its index of refraction. If the light, instead, appears at another frequency band in the spectroscope, it can be inferred from the Doppler effect that the body is in motion. When the light appears at a higher frequency band, then the body is no longer stationary but moving toward the observer. The Doppler effected light wave is displaced toward the higher frequency band, which is the blue end of the spectroscope. If the known body's light waves appear at a lower frequency band of the spectroscope, towards the red end, then the body is now in motion away from the observer.

With the use of the spectroscope, astronomers have been able to deduce the chemical composition of the **stars**. The Doppler effect enables them to determine their movements. In our own galaxy, all stars will be shifted either to the blue or red end because of a slight Doppler effect, indicating either a small movement toward or away from Earth. In 1923, however, **Edwin Hubble**, an American astronomer, found that the light from all the galaxies outside our own were shifted so much toward the red as to suggest that they were all speeding away from our own at very great velocities. At the same time he saw that the recession of galaxies nearer to us was much less than those further away.

In 1929, Hubble and Milton Humason established a mathematical relationship that enables astronomers to determine the distance of galaxies by determining the amount of the galaxy's red shifts. This mathematical relationship is known as Hubble's law or Hubble's constant. Hubble's law shows that the greater the **velocity** of recession, the further away from Earth the galaxy is.

The concept of the expanding **universe** along with the corollary idea of the **big bang**, that is, the instant creation of the universe from a compressed state of **matter**, owes much of its existence to Hubble's work, which in turn is an important development of the Doppler effect in light waves. While some recent research challenges the **red shift** phenomenon for galaxies, most astronomers continue to accept Hubble's findings.

In addition to its uses in science, the Doppler effect has many practical applications. In maritime navigation, **radio** waves are bounced off orbiting satellites to measure shifts that indicate changes in location. In highway traffic speeding detection, **radar** employs the Doppler effect to determine automobile speeds. There are also a number of medical applications of the Doppler effect found in ultrasonography, echocardiography, and radiology, all of which employ ultrasonic waves.

See also Hubble constant; Spectroscopy

DOUBLE-SLIT INTERFERENCE EXPERIMENT

Throughout the eighteenth century, two theories about the nature of **light** were in strong competition for acceptance. One theory, proposed by Dutch physicist and astronomer **Christiaan Huygens** (also spelled as Huyghens) during the seventeenth century asserted that light traveled as small **waves** (wave theory). Although the wave theory successfully accounted for many of the observed properties of light (e.g., **interference**), a discernible **ether** or medium of propagation thought necessary for the propagation of such waves continued to elude physicists. The second theory, propounded by English physicist Sir **Isaac Newton** in his 1687 work, *Philosophiae Naturalis Principia Mathematica (Mathematical Principles of Natural Philosophy)* was that light consisted of a series of tiny particles or "corpuscles." The particle theory explained why light traveled in straight lines and why it did not need a medium to travel through. But several problems remained with this particle theory, including trying to explain why light was able to cross paths without being deflected. The two theories vied for acceptability for more than two centuries and many prominent scientists took opposing views and argued their position with great vehemence.

In 1801, English physicist **Thomas Young** performed an experiment that swung opinion away from Newton's particle theory of light and towards Huygens's wave theory. Young forced light to travel through two narrow slits and to land on a screen placed behind the slits. The diagram below illustrates Young's experimental layout.

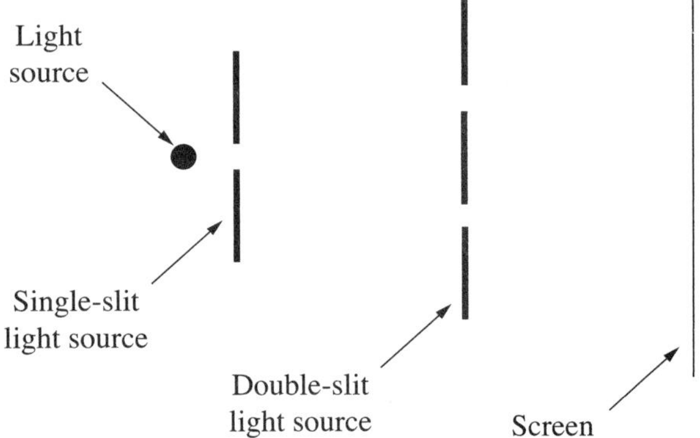

The genius of Young's experiment lay in the fact that the pattern projected on the screen by the light coming from the two slits would, if it showed interference patterns, confirm that the light had at least a partial wave nature. On the other hand, if light traveled like particles, Young expected to see two bright lines or bands appear on the screen directly opposite the slits.

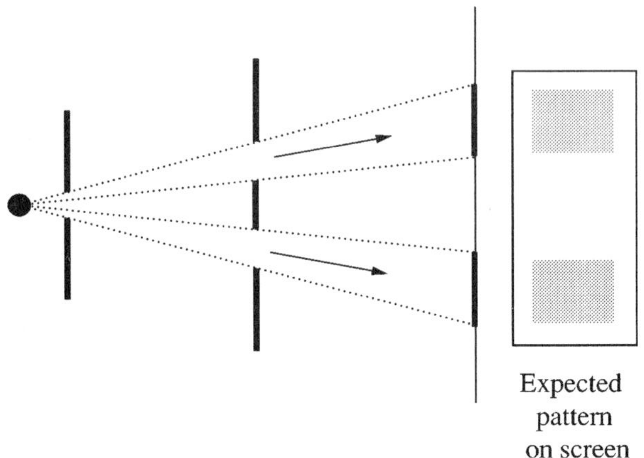
Expected result from Newton's "Particle" theory

If Huygens's wave theory were correct, however, one would predict a different result. If the light from the two slits overlapped in patterns characteristic of waves, Young expected to see a brighter band or line at the center. If light travels in waves then the light would bend a little at the exit of the two slits. Also, where the path of the light from the two slits overlaps, the waves would interfere with each other. For example, when two water waves intersect, a new wave pattern is set up. Two wave peaks can coincide and form an accentuated or amplified peak. Likewise, two wave troughs can coincide to form a deeper trough in a relationship known as constructive interference. Where a peak and trough coincide, however, they cancel each other out in a phenomenon known as destructive interference. On Young's screen, constructive interference would be evidenced by bright bands or lines, destructive interference would be evidenced by dark lines or bands.

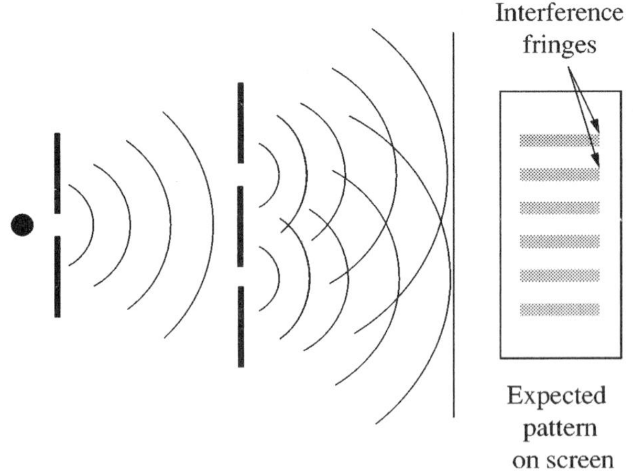
Expected results from Huygens's wave theory

Young's double-slit experiment clearly showed constructive and destructive interference patterns on the screen. In addition, this experiment enabled Young to calculate the wave length of the light used (approximately 0.00008 centimeter). In addition, Young was able to show that different colors of light had different wavelengths. This was a tremendous development in the understanding of light and the double-slit experiment stands as one of the key experiments in the **history of physics.**

The results from the double-slit experiment did much to convince physicists that light was, indeed, wave-like in nature and for a brief time Newton's particle theory of light fell into disuse. Early experiments with nuclear **radiation** at the beginning of the twentieth century revitalized the particle theory of light and led to the development of a dual wave-particle model of light (electromagnetic radiation). This apparent dual nature of light, which had been the source of much mental agony for physicists over many decades, finally gave rise to the **quantum theory** of **matter** that attempts to integrate these two characteristics of light.

When electrons are passed through experimental apparatus similar in design to Young's experiment, these elementary "particles" also show wave patterns of constructive and destructive interference. Although counterintuitive, experimental results also show that even a single **electron** shows evidence of simultaneously passing through both slits and of producing wave-like interference patterns.

See also Light; Particle-wave duality; Wave interference; Wave motion; Wave phenomena; Wave superposition; Wavelength

DYNAMICS

Dynamics is the subdivision of *mechanics* that describes the forces responsible for **motion** in classical physics. Since it deals with causes of motion, the study of dynamics leads to concepts of *mass* (defined as an object's resistance to being moved), *work* (**energy** required to move an object a certain distance), *inertia* (the tendency of a body to resist any attempt to change its state of motion), *energy* (the capacity to do **work**), and *power* (the amount of work done in a given **time** interval). The study of dynamics is also important for the modern understanding of **quantum mechanics** and for describing the motion of **waves.**

The original equations of dynamics are *Newton's laws*, which describe how bodies respond to an external **force.** Newton's first law states that a body at rest stays at rest, and that a body in motion stays in motion with a constant **velocity**, if there is no external force between the body and its surroundings. Newton's second law states that the **acceleration** of a body is proportional to the force required to move it and is inversely proportional to the body's own **mass.** Newton's third law holds that for every action, there is an equal and opposite reaction. Applications of these laws include the description of the behavior of fluid and elastic solids and the motions of planetary bodies and projectiles.

Before the contributions of **Galileo**—credited with the invention of mechanics—and **Isaac Newton**, dynamics was mostly concerned with **ballistics** applications. The first real perception of dynamics as a specific field of importance had to await **Newton's laws of motions** and his law of **gravity** as well as

his specific vision of dynamics as the science of the rate of change of the variables associated with motion. After Newton, the main sphere of application of dynamics remained the development of models to account for planetary motion. New contributions were made by **Henri Poincaré** (1854-1912) who, realizing that the solar system was not predictable, laid the foundations of what would later develop into the *topology* of nonlinear systems. The field of **mechanics** was further refined by the contributions of mathematicians Sir William Hamilton (1805–1865), Joseph-Louis Lagrange, and Pierre Simon, Marquis de Laplace, who developed formulas that described motion in dynamic systems through time.

By the start of the twentieth century, dynamics as a field of physics was increasingly relegated to the calculation of orbits and other related astronomical phenomena. The spectacular developments of the new **quantum theory** also increasingly mobilized the efforts and interest of the physics community. World War II with its urgent requirements for weapon systems and the beginnings of **space** exploration in the 1950s signaled, however, a renewed interest for the field. Combined with the development of (mostly mathematical) chaotic theories attempting to describe the unpredictable motions of systems that are sensitive to their initial conditions and the need to model nonlinear dynamical processes, the field was finally able to reach maturity.

But it was really the advent of **computers** that allowed dynamics to explode to the forefront of physics by providing physicists with the means to model change, be it linear or not. The first successful computer modeling of a nonlinear dynamical system was achieved in the 1960s by an Massachusetts Institute of Technology meteorologist, Edward Lorenz. This fueled interest for chaotic dynamics and *fractal* modeling, first introduced by the mathematician Benoit Mandelbrot (1924-) as somewhat esoteric geometric representations with sponge-like properties until it was understood that they are created and transformed as a result of dynamic processes.

See also Conservation of energy; Energy and work; Fractals; Harmonic motion; Kinetic energy; Meteorology; Nonlinear dynamics; Perturbation theory; Potential energy; Weight

DYSON, FREEMAN J. (1923-)

English American physicist

Freeman J. Dyson developed a general theory of **quantum electrodynamics** that integrates a number of specific concepts previously developed by **Richard P. Feynman, Julian Schwinger,** and **Sin-Itiro Tomonaga,** among others. As a result of his work with **Edward Teller** on nuclear **power** plants and the fusion bomb, he became active in the debate over the nuclear test ban treaty, arguing first one side of the issue and then, at a later date, the opposite side. Since the late 1950s, Dyson has also been interested in **space** travel and in research on the possible existence of intelligent life elsewhere in the **universe.**

Freeman John Dyson was born on December 15, 1923, in Crowthorne, a village in the south of England. His father,

George Dyson, was a music teacher at Winchester College, and later became director of the Royal College of Music in London. His mother, Mildred Atkey Dyson, was a lawyer. He has one sister, Alice. From an early age, Dyson had a passionate interest in mathematics, so much so that his mother expressed concern about his becoming too asocial if he followed that career. He was not deterred by her warnings, however, and, on one occasion when he was 15, taught himself calculus over the Christmas holidays from a mail-order textbook. After completing primary school, he enrolled at Winchester College and then, in 1941, entered Cambridge University to major in mathematics. His schooling was interrupted by World War II, however, when he was assigned to work in the operational research section of the Royal Air Force Bomber Command. His experience in trying to make bombing runs more effective revealed to him the horrible loss of human life that was taking place and left a lifelong impression on him. He was particularly upset by his own role in promoting the war effort and later said that the only difference he could see between his own wartime work and that of Nazis who were tried and convicted at Nuremberg was that "they were sent to jail or hanged as war criminals and I went free."

At the war's completion, Dyson returned to Cambridge and received his bachelor of arts degree in mathematics in 1945. He then stayed on at Trinity College, Cambridge, for two years before winning a Commonwealth Fund Fellowship that allowed him to study physics at Cornell University. The next three years were especially significant in his life since he had nearly daily contact at Cornell with **Hans Albrecht Bethe**, Feynman, and Schwinger, three physicists who were at the forefront of research on quantum electrodynamics.

Quantum electrodynamics (QED) is a field of physics that attempts to understand the interaction between electromagnetic fields and atoms. The origins of QED go back to the research of **Paul Dirac**, **Werner Karl Heisenberg**, **Wolfgang Pauli**, and other physicists during the early 1930s. As successful as early theories of QED were, many unsolved questions and problems remained nearly two decades later. By the late 1940s, however, many of these problems had begun to yield to the analysis of Bethe, Feynman, Schwinger, and Tomonaga.

Dyson's contribution to this effort was primarily that of a synthesizer. He showed how the independent theories of his colleagues could be brought together into a single unified theory of QED. The solution that he discovered occurred to him on a bus ride from California to the East Coast in the summer of 1948. He has explained that after working on the problems for many months, he had decided to take a vacation and to stop thinking about QED. But then, returning from his vacation, the solution for which he had been looking "suddenly became, somehow, transcendentally clear. It was one of those moments of illumination...which every scientist hopes for. It only happens once in a lifetime, but, anyway, it did happen."

In 1951, Dyson was invited to become professor of physics at Cornell, but he remained there for only two years. He had become more interested in the **philosophy of physics** and spent some time with **J. Robert Oppenheimer** discussing this topic. When he had the opportunity in 1953 to move to the Institute for Advanced Studies at Princeton University, where Oppenheimer was director, he did so eagerly. He has remained at the institute ever since; he became a naturalized U.S. citizen in 1957.

During the 1950s, Dyson became particularly interested in the possibilities of **nuclear power** as a commercial source of **energy**. He began spending his summers at the General Atomic Company, a division of General Dynamics. This work brought him into contact with, among others, Edward Teller, father of the hydrogen bomb and another fervent advocate of the use of nuclear power. Together, the two developed a nuclear reactor called the High Temperature Graphite Reactor that they regarded as "safe even in the hands of an idiot." The model was never put into use in the United States, however, because the nuclear power industry judged the initial construction costs to be too high.

The launch of the first artificial satellite by the Soviet Union in 1957 brought an immediate and concerned response from U.S. political and scientific leaders, Dyson among them. To help keep the United States from falling behind the Soviets in space technology, Dyson took a leave of absence from Princeton to join a research and development effort, the Orion Project in La Jolla, California, to construct a nuclear-powered satellite. The project experienced some exciting successes early on, and Dyson later referred to his first year at La Jolla as "the most exciting and in many ways the happiest of my scientific life." The U.S. government eventually decided not to use nuclear power for its satellite systems, however, and in 1965, the Orion Project was officially terminated.

One of the programs with which Dyson's name is often connected is the search for the existence of intelligent life elsewhere in the universe. He points out that, as a young boy, the books of Jules Verne and H. G. Wells were among his favorite reading material. He even tells of finding among his mother's papers, after her death, an incomplete science fiction story about travel to the **Moon** that he had started while still a teenager, and then forgotten about. His imaginative writings about possible extraterrestrial contact include the book *Disturbing the Universe,* published in 1979. Dyson has advocated the exploration and colonization of deep space and remains a fervent backer of attempts to communicate with alien civilizations.

Dyson was elected to the National Academy of Sciences in 1964 and continues to accumulate honors and prizes, including the Lorentz Medal of the Royal Netherlands Society (1966), the Hughes Medal of the Royal Society (1968), the Max Planck Medal of the German Physical Society (1969), the Harvey Prize of the Israeli Institute of Technology (1977), and Israel's Wolf Prize (1981). Dyson married the former Verene Haefeli-Huber in 1950, and the couple had two children, Esther and George. They were divorced in 1958. Dyson then married Imme Jung, with whom he had four more children, Dorothy, Emily, Miriam, and Rebecca. Although he never earned a degree higher than the B.A., Dyson has been awarded honorary doctorates by a number of institutions, including Yeshiva University, Princeton University, and the University of Glasgow.

EARTH (PLANET)

Earth is the third planet from the **Sun** in the solar system. Earth's average distance from the Sun is 92,900,000 mi (149,476,000 km), and defines one astronomical unit. Earth is an oblate spheroid with a surface area of 196,938,800 square miles (510,485,000 square kilometers). The elliptical orbit of Earth around the Sun takes approximately 365.26 days with respect to the background **stars**, and defines one sidereal year. Radiometric dating has led to a widely accepted date for Earth's age of about 4.6 billion years.

The bulk of scientific knowledge about Earth has accumulated only within the past few centuries and has evolved into the disciplines of geophysics and geology. Geophysics is the study of Earth as a planet and the **space** surrounding it, and the interactions between Earth and extraterrestrial forces. Geophysics contains three major divisions of study: the solid-earth, the atmosphere and hydrosphere, and the magnetosphere.

Solid-earth studies are concerned with Earth's surface and interior. The surface of Earth is very young compared to its age. In a period of about 500 million years erosion and tectonic processes destroy and re-create most of the surface features. Earth's surface is unique from the other **planets** in the solar system because it has liquid water in large quantities.

The knowledge learned about the interior of the solid-earth is mostly based on evidence and deductions made from the propagation of seismic **waves** through Earth. Studies of seismic waves propagating through the planet show that the interior consists of three main layers: the crust, mantle, and core.

The "crust," immediately below the surface, is of variable thickness. In some areas the crust is only a few miles or kilometers thick—in other areas it may measure 25 mi (40 km) in thickness. It has an average **density** of about three times that of water. The crust consists primarily of quartz and other silicates in the form of sedimentary rocks resting on a base of igneous rocks. It is divided into several separate solid tectonic plates that float independently on top of the mantle.

The mantle generally extends from a depth of 25-1,802 mi (40-2,900 km), with a density reaching 5.5 times that of water. Its composition is thought to include a high proportion of silicon, magnesium, and **oxygen**, with some iron, calcium, and aluminum.

Earth's core extends from a depth of 1,802-3,964 mi (2,900-6,378 km), with an increase in density from 10 times that of water at its mantle boundary to 13 times that of water at the center. Its composition may be similar to nickel-iron meteorites although it is possible that lighter elements also may exist. The outer portion of Earth's core is known to be in a liquid state and it is the rotation of the ion filled outer core around the inner core that is thought responsible for Earth's magnetic field. Temperatures at the core of Earth may range as high as 7,500K (hotter than the Sun's surface).

The material that comprises Earth's interior has an average strength similar to steel, but maintains a fluidic consistency. As a result of Earth's consistency and **angular momentum**, materials tend to migrate to the equator causing the equatorial radius 3,964.2 mi (6,378.4 km) to be 13.2 mi (21.3 km) larger than the polar radius, and thereby give Earth its oblate spheroid shape.

The atmosphere of Earth is a mixture of gases, with dry air composed of 78% nitrogen, 21% oxygen, 1% argon, and trace percentages of **carbon** dioxide and other gases. The majority of these gases exist at the lowest layer of the atmosphere, the troposphere, which extends an average thickness of 4-8 mi (7-13 km). The stratosphere, mesosphere, ionosphere, and the outermost layer, the exosphere, lie in progressive layers above the troposphere.

Two-thirds of the planet Earth is covered by water (hydrosphere) at an average ocean depth of 12,795 ft (3,900 m). The water area is 1.394×10^8 mi^2 (3.617×10^8 km^2).

The magnetosphere of Earth extends from the upper atmosphere to thousands of miles into space. The magnetic field of Earth dominates this region, trapping various charged particles. Similar to a bar magnet, lines of **force** stretch between the **magnetic poles** of Earth, whose positions shift over time. The rotation of Earth causes the magnetic field to be aligned approximately with the rotation axis. In 2000 the position of the north magnetic pole was 77.3 degrees north latitude and longitude 101.8 degrees west, and located on the Bathurst Island in northern Canada. The overall magnetic field strength of Earth is fairly weak, averaging about one-half gauss at Earth's surface.

The sidereal rotation of Earth about its axis takes approximately 23 hours, 56 minutes, 4.099 seconds. The sidereal rotation is variable because it changes slightly as Earth's rate of spin varies.

Earth has one natural satellite, the **Moon**. Earth's average distance from the Moon is 238,857 mi (384,321 km).

The **mass** of Earth is approximately 5.977×10^{24} kg. Earth's volume is 260.3 billion mi^3 (1,084.2 billion km^3).

See also Geocentric theory; Geothermal energy; Greenhouse effect

EDDINGTON, ARTHUR STANLEY (1882-1944)

English astronomer

Arthur Stanley Eddington was born in Kendal, Westmoreland, England, on December 28, 1882. He was the son of a school headmaster and distinguished himself at a very early age with his grasp of mathematics. He was appointed professor of astronomy at Cambridge University in 1913 and became director of the Cambridge observatory one year later.

Eddington's theoretical investigations into the structure of **stars** helped solve a question that had been plaguing scientists: if the **Sun** was only a giant ball of gas, what kept it from collapsing under its own **gravity** to become a white dwarf star? In the 1920s, Eddington proposed that the stellar **matter** at the core of the star was under extreme **pressure**, causing the production of tremendous **heat**. The **temperature** would reach millions of degrees, and the outward pressure of this **radiation** would balance the inward pull of gravity. This high temperature would be crucial to establishing the nuclear process as worked out later by **Hans Bethe**.

Eddington's theory about the interior of stars also disproved a hypothesis regarding the formation of the solar system. Some scientists held that the solar system had been formed when another star passed by the Sun and the star's gravitational tug pulled material from the core of the Sun that coalesced into the **planets**. Eddington, however, postulated that any material pulled from a star's core would expand violently into a thin gas once released from the pressure that contained it.

In 1924 Eddington proposed his mass-luminosity law: a more massive star would have greater interior pressure, higher temperature, and be more luminous. A star with 50 times the **mass** of the Sun would be blown apart by the **force** of its radiation. Another theory Eddington formulated concerns Cepheid variable stars; they pulsate regularly, changing both their size and brightness, because they are at the edge of stability.

In addition to his theories about stars, Eddington was one of the first to understand the importance of **Albert Einstein**'s theories of relativity.

Eddington was an extraordinarily gifted writer. For the student of physics, he wrote texts that made the notorious difficulties of the subject clear and intuitive. He also wrote numerous books that introduced these difficult concepts to the public, becoming one of the early astronomy "popularizers." He was knighted in 1938 and died on November 22, 1944.

EDISON, THOMAS ALVA (1847-1931)

American inventor

Thomas Alva Edison's nickname, "The Wizard of Menlo Park," reflects his amazing inventive talent. Over his lifetime, more than 1,300 patents were issued in his name, far more than have been credited to any other individual in American history. Among the best known of his inventions are a stock-ticker machine, the incandescent **light** bulb, an automatic telegraphy machine, the phonograph, and the **motion** picture machine. Edison's one major accomplishment in scientific research was the discovery of the emission of electrons from a heated cathode, a phenomenon now known as the Edison effect.

Thomas Alva Edison was born in Milan, Ohio, on February 11, 1847. He was the seventh and youngest child of Samuel Ogden Edison Jr. and the former Nancy Elliott. Edison's parents had met in Canada, where the family had moved after the Revolutionary War. Sam Edison, in turn, had fled Canada in 1837 when an insurrection against the government in which he was involved failed. He established a successful grain and lumber business in Milan that allowed him to bring his wife and four children to the United States in 1839. Eight years later, Thomas Alva was born.

Disaster struck Milan in 1854 when a new railroad line bypassed the city and isolated it from commercial traffic. Before long, eight percent of the city's population had left the area, including the Edison family. Samuel Edison relocated his family to Port Huron, Michigan, began again, and established a grain and lumber business that was soon very successful. In Port Huron, young Thomas began his education in a one-room school taught by the Reverend G. B. Engle and his wife. That schooling was to last only a few months, however, because young Tom heard Mrs. Engle refer to him as "addled." She had apparently become convinced that Tom's impatience with formal schooling was a sign of mental inferiority. Furious at her son's report of this remark, Nancy Edison withdrew him from school and from that point on, provided for his education at home.

Carries Out Chemical Experiments at Home

Nancy Edison introduced her son to natural philosophy, a mixture of physics, chemistry, and other sciences, among other subjects. Before long, Tom became fascinated with chemistry in particular and built a chemical laboratory in a corner of the cellar. By the age of 10, he was conducting various original experiments in his home.

Edison's entrepreneurial spirit became apparent while he was still quite young. He obtained permission to ride the train that ran between Port Huron and Detroit, selling newspapers and magazines, candy, apples, sandwiches, molasses, peanuts, tobacco, and other materials along the way. During the layover in Detroit, Edison spent his time at the city's public library. Writing about these experiences later, he said, "I didn't read a few books. I read the library." Over time, Edison expanded his activities on the train, receiving permission to use an empty part of the baggage car to set up first a small chemistry laboratory and later a printing press on which he printed a small newspaper.

Begins to Lose His Hearing

It was on one of the early train runs that a famous event occurred that affected Edison's future life. While Edison was running to catch the train, one of the conductors reached down, grabbed him by the ears, and pulled him aboard. Edison later reported that he felt something snap inside his head and that shortly thereafter he began to lose his hearing until he was almost completely deaf. Some have theorized, however, that Edison's deafness may have had an organic basis, since his son later also became deaf, or that it may have been caused by a childhood bout with scarlet fever.

His loss of hearing marked a critical turning point in Edison's life. Matthew Josephson in *Edison: A Biography* pointed out that his deafness not only caused him to become "more solitary and shy," but also changed his viewpoint on the reading and study to which he had devoted so much time. "He had been only 'playing,' hitherto, with his books and his 'experiments,'" Josephson writes. "Now he put forth tremendous efforts at self-education, for he had absolutely to learn everything for himself." One of the lifelong habits he developed during this period was to work long hours with only short naps to keep him going. Later in life, it was not unusual for Edison to spend 20 out of every 24 hours at work.

Learns about Telegraphy

One of the subjects in which Edison became interested during his railroad days was telegraphy, a means of communicating over a great distance by using coded signals transmitted by wire. In 1862, J. U. MacKenzie, whose three-year-old son Edison had saved from being run over by a rail car, offered Edison the chance to learn telegraphy. Within a year, Edison had become skilled enough as a telegrapher to hold jobs in a variety of cities and towns in the Midwest, including Adrian, Michigan; Fort Wayne, Indiana; Cincinnati, Ohio; Nashville, Tennessee; and Louisville, Kentucky. At each location, he not only performed his assigned duties but also worked on ways of improving the telegraphy apparatus itself.

Edison's Midwest travels finally ended in the fall of 1867 when he returned to Port Huron. He found his family home in a shambles, his mother ill, and his father distraught and remote. Thomas decided to strike out on his own again

Thomas Alva Edison, working in lab.

and headed to Boston, where he found a job with the Western Union Telegraph Company. Although the job itself was boring, Edison found enough time to continue his practice of studying and inventing on his own. An important breakthrough came when he came across a copy of **Michael Faraday**'s *Experimental Researches in Electricity,* read the two-volume work through in one sitting, and concluded that he could repeat such experiments himself. Edison was about to begin a new era in his life.

Receives First Patent in 1869

The first invention that resulted from his experimentation was a device for electronically recording the voice votes taken in a legislative assembly. For that invention, Edison was awarded patent number 90,646 on June 1, 1869, the first patent of his career. Unfortunately, he discovered that legislators were not interested in speeding up the process of vote recording and thus concluded that in the future, he would investigate the necessity of a product before inventing it.

June 1869 may have been a milestone in Edison's career as an inventor, but he was still desperately poor, trying to survive day-to-day on very low wages. He decided that nothing was to be lost by making yet another move, this time down the coast to New York City. He arrived in the city without a cent in his pockets and was saved by a remarkable stroke of good fortune. While Edison was waiting to interview for a job in the offices of the Gold Indicator Company, the central transmitting machine broke down. Edison quickly found the problem and fixed the machine. The next day, the owner of the company, Dr. S. S. Laws, offered Edison a job as general manager of the firm, with the princely salary of $300 a month.

The job Laws offered Edison allowed him to think more seriously about a career in invention. Only four months after accepting the job, therefore, Edison joined with two other men, Franklin L. Pope and James N. Ashland, to form a new company. Pope, Edison and Company was to be, they announced, a firm of "electrical engineers" and "constructors of various types of electric devices and apparatus." They were prepared to "design telegraphic instruments, test materials, build telegraph lines, design fire alarms, construct experimental apparatus, and render other services." Within a year, the company was bought by the Gold and Stock Telegraphic Company, who also paid Edison $40,000 for a device that kept stock tickers working in unison. This was quite a fortune for a 23-year-old man from Port Huron with essentially no formal education.

Opens First "Invention Factory"

The profit from the Gold and Stock deal allowed Edison to put into practice a scheme about which he had been thinking for some time. He opened a firm consisting of about 50 consulting engineers in Newark, New Jersey, which was to be, as he described it, an "invention factory." The Newark plant operated for six years, turning out a variety of inventions primarily related to improvements in stock tickers and telegraphy equipment, but also dozens of minor products. In all, Edison was granted about 200 new patents for work completed in the Newark laboratories. Those laboratories are generally regarded as the first formally organized non-academic research center in the United States.

In 1876, Edison had outgrown his Newark facilities and arranged for the construction of a new plant at Menlo Park, New Jersey. His most productive work was accomplished at this location over the next decade. Three inventions in particular stand out among the dozens produced by Edison and his coworkers: a telephone, the phonograph, and the incandescent light bulb. Although Alexander Graham Bell invented the telephone, Edison developed a carbon-resistant transmitter that greatly improved the quality of Bell's invention that same year.

Invents the Phonograph and Improves the Light Bulb

In 1877, Edison produced what he himself named his favorite invention: the phonograph. The first model, in which **sound** vibrations were transferred by means of a steel stylus to a cylinder wrapped in tin foil, produced recognizable sounds, but of poor quality. Edison originally saw no application for the phonograph and in a newspaper interview called it "a mere toy, which has no commercial value." He then set it aside and did not return to work on it again for a decade. Josephson has described this decision to abandon research on the phono-

graph as "one of the inventor's greatest blunders, one that in the end was to cost him dearly."

A year later, in 1878, Edison began work on yet another invention, one that was not original with him but that he made commercially feasible: the incandescent light bulb. As early as 1848, the English physicist Sir Joseph Swan had investigated the possibility of converting an electrical current into light. The concept was simple enough: when an electrical current passes through a thin wire, it encounters resistance that causes the wire to become warmer. At some point, the wire becomes hot enough to glow, that is, to reach incandescence.

The problem that Swan encountered was that, as wire gets hot, it tends to oxidize, or burn up. In theory, the solution to this problem is to encase the wire in a **vacuum**. In Swan's time, however, it was not possible to make vacuums good enough to prevent wires from oxidizing. As a result, a wire might be brought to incandescence and to produce light for a short time, but it quickly burned up and the light went out.

Swan continued to work on the incandescent light bulb for the next two decades and finally solved the problem at about the same time that Edison did. An important key to Edison's success was that much better vacuums were available in 1878 than had been obtainable in 1848. In addition, Edison discovered an ideal material for use as the "wire" in a light bulb, a charred length of cotton thread. On October 21, 1879, Edison first demonstrated in public an incandescent light bulb, made with his charred cotton thread, that burned continuously for 40 hours.

The incandescent light bulb was, of course, a remarkable success, and Edison spent the next few years adapting it for large-scale use. He found it necessary, for example, to invent the generating, switching, and transmitting devices needed to supply **electricity** to a large number of light bulbs at the same time. Within three years he had solved many of these problems and was operating the world's first **power** station on Pearl Street in New York City. When the plant began operation on September 4, 1882, it supplied power to 400 incandescent light bulbs owned by 85 customers.

During his work on the incandescent light bulb, Edison made his one and only important scientific discovery. In attempting to modify the construction of the bulb, he introduced a wire into the bulb adjacent to the filament. When the lamp was turned on, Edison observed that electrical current flowed from the filament to the wire. He saw no practical application of this discovery and so did no further work on it. The Edison effect, as the discovery is now called, later had important applications as a way of rectifying direct current.

Invents Early Motion Picture Machine

By 1887, the laboratories at Menlo Park had become too small for Edison's many products, and he moved to a new site in West Orange, New Jersey. There he constructed a new building 10 times larger than the Menlo Park facility. At that point, the Edison laboratories were practically an invention mill where, at peak production, Edison was receiving an average of one new patent every five days.

Probably the best-known invention to come out of this period was the kinetograph, a primitive form of the moving picture. Edison developed a method for arranging a series of photographs on a strip of celluloid film and then running the film through a projector. He used this technique to produce *The Great Train Robbery* in 1903, one of the first moving pictures, and then lost interest in the technology.

A tendency to go from subject to subject was characteristic of Edison's personality. In some cases he stayed with a project long enough to see it to commercial production. In other cases, he spent time developing the early stages of an idea and then moved on to something new. Among the inventions to which he made at least some contribution were the lead storage battery, the mimeograph machine, the dictaphone, and the fluoroscope. He also developed an interest in iron mining and processing and in cement production. As he went from project to project and interest to interest, he made and lost a fortune many times over.

Edison was married twice, the first time on Christmas day, 1871, to Mary Stilwell, whom he had met in Newark. They had three children, Marion Estell, Thomas Alva, and William Leslie. Mary Edison died on August 8, 1884, and two years later, on February 24, 1886, Edison was married a second time, to Mina Miller. Three children, Charles, Madeleine, and Theodore, were born to this marriage. Edison died in West Orange, New Jersey, on October 18, 1931. In recognition of his accomplishments he had been elected to the National Academy of Sciences in 1927, and he was elected to the Hall of Fame for Great Americans in 1960. In addition he was awarded the Albert Medal of the British Society of Arts in 1892, the John Fritz Medal of the American Engineering Societies in 1908, and the Rumford Medal of the American Academy of Arts and Sciences. In 1920 the Department of the Navy awarded Edison the Distinguished Service Medal for his wartime service. The inventor declined the honor, however, because of his disappointment with the way in which he thought he had been treated by the military. Eight years later Edison did accept a special gold medal awarded to him by the U.S. Congress in recognition of his life's work.

EIGENFUNCTION

As used in quantum **mechanics** to describe solutions for Schrödinger equations, the term Eigenfunction derives from the Germanic *Eigenfunktion*, meaning characteristic or proper function. Eigenfunctions correspond to particular eigenvalues (derived from the Germanic *Eigenwert* to mean the characteristic or proper values). In **quantum physics** eigenfunctions yield defined and discrete eigenvalues related to quantized values used to describe various attributes of a body or system (e.g., **quantum numbers**).

The postulates of **quantum mechanics** state that the physical state of a system can be fully described by specific types of wave functions that can be obtained by solving **Schrödinger's equation**. Such functions that are possible solutions of this equation are called eigenfunctions or characteris-

tic functions. They exist only for specific eigenvalues or characteristic values of **energy**. In quantum mechanical calculations, every dynamic variable that correlates with a physically observable property (such as position, **linear momentum**, **angular momentum**, **time**, **kinetic energy**, **potential energy**, etc.) is represented by a linear operator. Operators are derived from classical expressions for these properties and, as their name implies, they operate on a function to yield another function, an eigenfunction, that only differs from the first by a constant factor (an eigenvalue).

For example, in quantum mechanical calculations to obtain specific values for energy, the wavefunction is operated with the Hamiltonian operator associated with energy. The solutions of the Schrödinger equation yield only quantized values of energy (i.e., eigenvalues of energy).

See also Quantum theory

EINSTEIN, ALBERT (1879-1955)
American physicist

Albert Einstein ranks as one of the most remarkable theoreticians in the history of science. During a single year, 1905, he produced three papers that are among the most important in twentieth-century physics, and perhaps in all of the recorded history of science, for they revolutionized the way scientists looked at the nature of **space**, **time**, and **matter**. These papers dealt with the nature of particle movement known as **Brownian motion**, the quantum nature of electromagnetic **radiation** as demonstrated by the **photoelectric effect**, and the special theory of relativity. Although Einstein is probably best known for the last of these works, it was for his quantum explanation of the photoelectric effect that he was awarded the 1921 Nobel Prize in physics. In 1915, Einstein extended his special theory of relativity to include certain cases of accelerated **motion**, resulting in the more general theory of relativity.

Einstein was born in Ulm, Germany, on March 14, 1879, the only son of Hermann and Pauline Koch Einstein. Both sides of his family had long-established roots in southern Germany, and, at the time of Einstein's birth, his father and uncle Jakob owned a small electrical equipment plant. When that business failed around 1880, Hermann Einstein moved his family to Munich to make a new beginning. A year after their arrival in Munich, Einstein's only sister, Maja, was born.

Although his family was Jewish, Einstein was sent to a Catholic elementary school from 1884 to 1889. He was then enrolled at the Luitpold Gymnasium in Munich. During these years, Einstein began to develop some of his earliest interests in science and mathematics, but he gave little outward indication of any special aptitude in these fields. Indeed, he did not begin to talk until the age of three and, by the age of nine, was still not fluent in his native language. His parents were actually concerned that he might be somewhat mentally retarded.

In 1894, Hermann Einstein's business failed again, and the family moved once more, this time to Pavia, near Milan, Italy. Einstein was left behind in Munich to allow him to fin-

Albert Einstein.

ish school. Such was not to be the case, however, since he left the *gymnasium* after only six more months. Einstein's biographer, Philipp Frank, explains that Einstein so thoroughly despised formal schooling that he devised a scheme by which he received a medical excuse from school on the basis of a potential nervous breakdown. He then convinced a mathematics teacher to certify that he was adequately prepared to begin his college studies without a high school diploma. Other biographies, however, say that Einstein was expelled from the *gymnasium* on the grounds that he was a disruptive influence at the school.

In any case, Einstein then rejoined his family in Italy. One of his first acts upon reaching Pavia was to give up his German citizenship. He was so unhappy with his native land that he wanted to sever all formal connections with it; in addition, by renouncing his citizenship, he could later return to Germany without being arrested as a draft dodger. As a result, Einstein remained without an official citizenship until he became a Swiss citizen at the age of 21. For most of his first year in Italy, Einstein spent his time traveling, relaxing, and teaching himself calculus and higher mathematics. In 1895, he thought himself ready to take the entrance examination for the Eidgenössiche Technische Hochschule (the ETH, Swiss Federal Polytechnic School, or Swiss Federal Institute of Technology), where he planned to major in electrical engineering. When he failed that examination, Einstein enrolled at a Swiss cantonal

high school in Aarau. He found the more democratic style of instruction at Aarau much more enjoyable than his experience in Munich and soon began to make rapid progress. He took the entrance examination for the ETH a second time in 1896, passed, and was admitted to the school. (In *Einstein,* however, Jeremy Bernstein writes that Einstein was admitted without examination on the basis of his diploma from Aarau.)

The program at ETH had nearly as little appeal for Einstein as had his schooling in Munich, however. He apparently hated studying for examinations and was not especially interested in attending classes on a regular basis. He devoted much of this time to reading on his own, specializing in the works of **Gustav Kirchhoff**, **Heinrich Hertz**, **James Clerk Maxwell**, **Ernst Mach**, and other classical physicists. When Einstein graduated with a teaching degree in 1900, he was unable to find a regular teaching job. Instead, he supported himself as a tutor in a private school in Schaffhausen. In 1901, Einstein also published his first scientific paper, "Consequences of Capillary Phenomena."

In February 1902 Einstein moved to Bern and applied for a job with the Swiss Patent Office. He was given a probationary appointment to begin in June of that year and was promoted to the position of technical expert, third class, a few months later. The seven years Einstein spent at the Patent Office were the most productive years of his life. The demands of his work were relatively modest and he was able to devote a great deal of time to his own research.

The promise of a steady income at the Patent Office also made it possible for Einstein to marry. Mileva Marić (also given as Maritsch) was a fellow student in physics at ETH, and Einstein had fallen in love with her even though his parents strongly objected to the match. Marić had originally come from Hungary and was of Serbian and Greek Orthodox heritage. The couple married on January 6, 1903, and later had two sons, Hans Albert and Edward. A previous child, Liserl, was born in 1902 at the home of Marić's parents in Hungary, but there is no further mention or trace of her after 1903 since she was given up for adoption.

In 1905, Einstein published a series of papers, any one of which would have assured his fame in history. One, "On the Movement of Small Particles Suspended in a Stationary Liquid Demanded by the Molecular-Kinetic Theory of Heat," dealt with a phenomenon first observed by the Scottish botanist Robert Brown in 1827. Brown had reported that tiny particles, such as dust particles, move about with a rapid and random zigzag motion when suspended in a liquid.

Einstein hypothesized that the visible **motion of particles** was caused by the random movement of molecules that make up the liquid. He derived a mathematical formula that predicted the distance traveled by particles and their relative speed. This formula was confirmed experimentally by the French physicist **Jean-Baptiste Perrin** in 1908. Einstein's work on the Brownian movement is generally regarded as the first direct experimental evidence of the existence of molecules.

A second paper, "On a Heuristic Viewpoint Concerning the Production and Transformation of Light," dealt with another puzzle in physics, the photoelectric effect. First observed by

Heinrich Hertz in 1888, the photoelectric effect involves the release of electrons from a metal that occurs when **light** is shined on the metal. The puzzling aspect of the photoelectric effect was that the number of electrons released is not a function of the light's intensity, but of the **color** (that is, the **wavelength**) of the light.

To solve this problem, Einstein made use of a concept developed only a few years before, in 1900, by the German physicist **Max Planck**, the quantum hypothesis. Einstein assumed that light travels in tiny discrete bundles, or "quanta," of **energy**. The energy of any given light quantum (later renamed the **photon**), Einstein said, is a function of its wavelength. Thus, when light falls on a metal, electrons in the metal absorb specific **quanta** of energy, giving them enough energy to escape from the surface of the metal. But the number of electrons released will be determined not by the number of quanta (that is, the intensity) of the light, but by its energy (that is, its wavelength). Einstein's hypothesis was confirmed by several experiments and laid the foundation for the fields of quantitative photoelectric chemistry and quantum **mechanics**. As recognition for this work, Einstein was awarded the 1921 Nobel Prize in physics.

A third 1905 paper by Einstein, almost certainly the one for which he became best known, details his special theory of relativity. In essence, "On the Electrodynamics of Moving Bodies" discusses the relationship between measurements made by observers in two separate systems moving at constant **velocity** with respect to each other.

Einstein's work on relativity was by no means the first in the field. The French physicist **Henri Poincaré**, the Irish physicist George Francis FitzGerald, and the Dutch physicist **Hendrik Lorentz** had already analyzed in some detail the problem attacked by Einstein in his 1905 paper. Each had developed mathematical formulas that described the effect of motion on various types of measurement. Indeed, the record of pre-Einsteinian thought on relativity is so extensive that one historian of science once wrote a two-volume work on the subject that devoted only a single sentence to Einstein's work. Still, there is little question that Einstein provided the most complete analysis of this subject. He began by making two assumptions. First, he said that the laws of physics are the same in all frames of reference. Second, he declared that the velocity of light is always the same, regardless of the conditions under which it is measured.

Using only these two assumptions, Einstein proceeded to uncover an unexpectedly extensive description of the properties of bodies that are in uniform motion. For example, he showed that the length and **mass** of an object are dependent upon their movement relative to an observer. He derived a mathematical relationship between the length of an object and its velocity that had previously been suggested by both FitzGerald and Lorentz. Einstein's theory was revolutionary, for previously scientists had believed that basic quantities of measurement such as time, mass, and length were absolute and unchanging. Einstein's work established the opposite—that these measurements could change, depending on the relative motion of the observer.

In addition to his masterpieces on the photoelectric effect, Brownian movement, and relativity, Einstein wrote two more papers in 1905. One, "Does the Inertia of a Body Depend on Its Energy Content?," dealt with an extension of his earlier work on relativity. He came to the conclusion in this paper that the **energy and mass** of a body are closely interrelated. Two years later he specifically stated that relationship in a formula, $E=mc^2$ (energy equals mass times the **speed of light** squared), that is now familiar to both scientists and non-scientists alike. His final paper, the most modest of the five, was "A New Determination of Molecular Dimensions." It was this paper that Einstein submitted as his doctoral dissertation, for which the University of Zurich awarded him a Ph.D. in 1905.

Fame did not come to Einstein immediately as a result of his five 1905 papers. Indeed, he submitted his paper on relativity to the University of Bern in support of his application to become a *privatdozent,* or unsalaried instructor, but the paper and application were rejected. His work was too important to be long ignored, however, and a second application three years later was accepted. Einstein spent only a year at Bern, however, before taking a job as professor of physics at the University of Zurich in 1909. He then went on to the German University of Prague for a year and a half before returning to Zurich and a position at ETH in 1912. A year later Einstein was made director of scientific research at the Kaiser Wilhelm Institute for Physics in Berlin, a post he held from 1914 to 1933.

In recent years, the role of Mileva Einstein-Marić in her husband's early work has been the subject of some controversy. The more traditional view among Einstein's biographers is that of A. P. French in his "Condensed Biography" in *Einstein: A Centenary Volume.* French argues that although "little is recorded about his [Einstein's] domestic life, it certainly did not inhibit his scientific activity." In perhaps the most substantial of all Einstein biographies, Philipp Frank writes that "For Einstein life with her was not always a source of peace and happiness. When he wanted to discuss with her his ideas, which came to him in great abundance, her response was so slight that he was often unable to decide whether or not she was interested."

A quite different view of the relationship between Einstein and Marić is presented in a 1990 paper by Senta Troemel-Ploetz in *Women's Studies International Forum.* Based on a biography of Marić originally published in Yugoslavia, Troemel-Ploetz argues that Marić gave to her husband "her companionship, her diligence, her endurance, her mathematical genius, and her mathematical devotion." Indeed, Troemel-Ploetz builds a case that it was Marić who did a significant portion of the mathematical calculations involved in much of Einstein's early work. She begins by repeating a famous remark by Einstein himself to the effect that "My wife solves all my mathematical problems." In addition, Troemel-Ploetz cites many of Einstein's own letters of 1900 and 1901 (available in *Collected Papers*) that allude to Marić's role in the development of "our papers," including one letter to Marić in which Einstein noted: "How happy and proud I will be when both of us together will have brought our work on relative motion to a successful end." The author also points out the somewhat unexpected fact that Einstein gave the money he received from the 1921 Nobel Prize to Marić, although the two had been divorced two years earlier. Nevertheless, Einstein never publicly acknowledged any contributions by his wife to his work.

Any mathematical efforts Mileva Einstein-Marić may have contributed to Einstein's work greatly decreased after the birth of their second son in 1910. Einstein was increasingly occupied with his career and his wife with managing their household; upon moving to Berlin in 1914, the couple grew even more distant. With the outbreak of World War I, Einstein's wife and two children returned to Zurich. The two were never reconciled; in 1919, they were formally divorced. Towards the end of the war, Einstein became very ill and was nursed back to health by his cousin Elsa. Not long after Einstein's divorce from Marić, he was married to Elsa, a widow. The two had no children of their own, although Elsa brought two daughters, Ilse and Margot, to the marriage.

The war years also marked the culmination of Einstein's attempt to extend his 1905 theory of relativity to a broader context, specifically to systems with nonzero **acceleration**. Under the general theory of relativity, motions no longer had to be uniform and relative velocities no longer constant. Einstein was able to write mathematical expressions that describe the relationships between measurements made in *any* two systems in motion relative to each other, even if the motion is accelerated in one or both. One of the fundamental features of the general theory is the concept of a **space-time** continuum in which space is curved. That concept means that a body affects the shape of the space that surrounds it so that a second body moving near the first body will travel in a curved path.

Einstein's new theory was too radical to be immediately accepted, for not only were the mathematics behind it extremely complex, it replaced Newton's theory of gravitation that had been accepted for two centuries. So, Einstein offered three proofs for his theory that could be tested: first, that relativity would cause Mercury's perihelion, or point of orbit closest to the **Sun**, to advance slightly more than was predicted by Newton's laws. Second, Einstein predicted that light from a star will be bent as it passes close to a massive body, such as the Sun. Last, the physicist suggested that relativity would also affect light by changing its wavelength, a phenomenon known as the red-shift effect. Observations of the planet Mercury bore out Einstein's hypothesis and calculations, but astronomers and physicists had yet to test the other two proofs.

Einstein had calculated that the amount of light bent by the Sun would amount to 1.7 seconds of an arc, a small but detectable effect. In 1919, during an eclipse of the Sun, English astronomer **Arthur Eddington** measured the deflection of starlight and found it to be 1.61 seconds of an arc, well within experimental error. The publication of this proof made Einstein an instant celebrity and made "relativity" a household word, although it was not until 1924 that Eddington proved the final hypothesis concerning **red shift** with a spectral analy-

sis of the star Sirius B. This phenomenon, that light would be shifted to a longer wavelength in the presence of a strong **gravitational field**, became known as the "Einstein."

Einstein's publication of his general theory in 1916, the *Foundation of the General Theory of Relativity,* essentially brought to a close the revolutionary period of his scientific career. In many ways, Einstein had begun to fall out of phase with the rapid changes taking place in physics during the 1920s. Even though Einstein's own work on the photoelectric effect helped set the stage for the development of **quantum theory**, he was never able to accept some of its concepts, particularly the **uncertainty principle**. In one of the most-quoted comments in the history of science, he claimed that **quantum mechanics**, which could only calculate the probabilities of physical events, could not be correct because "God does not play dice." Instead, Einstein devoted his efforts for the remaining years of his life to the search for a unified field theory, a single theory that would encompass all physical fields, particularly gravitation and **electromagnetism**.

Since the outbreak of World War I, Einstein had been opposed to war, and used his notoriety to lecture against it during the 1920s and 1930s. With the rise of National Socialism in Germany in the early 1930s, Einstein's position became difficult. Although he had renewed his German citizenship, he was suspect as both a Jew and a pacifist. In addition, his writings about relativity were in conflict with the absolutist teachings of German leader Adolf Hitler's party. Fortunately, by 1930, Einstein had become internationally famous and had traveled widely throughout the world. A number of institutions were eager to add his name to their faculties.

In early 1933, Einstein made a decision. He was out of Germany when Hitler rose to power, and he decided not to return. Instead he accepted an appointment at the Institute for Advanced Studies in Princeton, New Jersey, where he spent the rest of his life. In addition to his continued work on unified field theory, Einstein was in demand as a speaker and wrote extensively on many topics, especially peace. The growing fascism and anti-Semitism of Hitler's regime, however, convinced him in 1939 to sign his name to a letter written by American physicist **Leo Szilard** informing President Franklin D. Roosevelt of the possibility of an **atomic bomb**. This letter led to the formation of the **Manhattan Project** for the construction of the world's first **nuclear weapons**. Although Einstein's work on relativity, particularly his formulation of the equation $E=mc^2$, was essential to the development of the atomic bomb, Einstein himself did not participate in the project. He was considered a security risk, although he had renounced his German citizenship and become a U.S. citizen in 1940.

After World War II and the bombing of Japan, Einstein became an ardent supporter of nuclear disarmament. He also lent his support to the efforts to establish a world government and to the Zionist movement to establish a Jewish state. In 1952, after the death of Israel's first president, Chaim Weizmann, Einstein was invited to succeed him as president; he declined the offer. Among the many other honors given to Einstein were the Barnard Medal of Columbia University in 1920, the Copley Medal of the Royal Society in 1925, the

Gold Medal of the Royal Astronomical Society in 1926, the Max Planck Medal of the German Physical Society in 1929, and the Franklin Medal of the Franklin Institute in 1935. Einstein died at his home in Princeton on April 18, 1955, after suffering an aortic aneurysm. At the time of his death, he was the world's most widely admired scientist and his name was synonymous with genius. Yet Einstein declined to become enamored of the admiration of others. He wrote in his book *The World as I See It:* "Let every man be respected as an individual and no man idolized. It is an irony of fate that I myself have been the recipient of excessive admiration and respect from my fellows through no fault, and no merit, of my own. The cause of this may well be the desire, unattainable for many, to understand the one or two ideas to which I have with my feeble powers attained through ceaseless struggle."

See also Electrodynamics; Inertia; Relativity theory; Relativity, general; Relativity, special

EINSTEIN'S LAW OF ADDED VELOCITIES •

See Relativity, special

ELASTICITY

Elasticity is the ability of a material to return to its original shape and size after being stretched, compressed, twisted or bent. Elastic deformation (change of shape or size) lasts only as long as a deforming force is applied to the object, and disappears once the force is removed. Greater forces may cause permanent changes of shape or size, called plastic deformation.

In ordinary language, a substance is said to be "elastic" if it stretches easily. Therefore, rubber is considered a very elastic substance, and rubber bands are even called "elastics" by some people. Actually, however, most substances are somewhat elastic, including steel, glass, and other familiar materials.

The simplest description of elasticity is Hooke's Law, which states, "The stress is proportional to the strain." This relation was first expressed by the British scientist, **Robert Hooke**. He arrived at it through studies in which he placed weights on metal springs and measured how far the springs stretched in response. Hooke noted that the added length was always proportional to the **weight**; that is, doubling the weight doubled the added length.

In the modern statement of Hooke's law, the terms "stress" and "strain" have precise mathematical definitions. Stress is the applied force divided by the area the force acts on. Strain is the added length divided by the original length.

To understand why these special definitions are needed, first consider two bars of the same length, made of the same material. One bar is twice as thick as the other. Experiments have shown that both bars can be stretched to the same additional length only if twice as much weight is placed on the bar that is twice as thick. Thus, they both carry the same stress, as defined above.

The special definition of strain is required because, when an object is stretched, the stretch occurs along its entire length, not just at the end to which the weight is applied. The same stress applied to a long rod and a short rod will cause a greater extension of the long rod. The strain, however, will be the same on both rods.

The amount of stress required to produce a given amount of strain also depends on the material being stretched. Therefore, the ratio of stress to strain is a unique property of materials, different for each substance. It is called the elastic modulus (plural: moduli). It is also known as Young's modulus, after **Thomas Young** who first described it. It has been measured for thousands of materials. The greater the elastic modulus, the stiffer the material is. For example, the elastic modulus of rubber is about 600 psi (pounds per square inch). That of steel is about 30 million psi.

All deformations, no matter how complicated, can be described as the result of combinations of three basic types of stress. One is tension, which stretches an object along one direction only. Thus far, our discussion of elasticity has been entirely in terms of tension. Compression is the same type of stress, but acting in the opposite direction.

The second basic type of stress is shear stress. This results when two forces push on opposite ends of an object in opposite directions. Shear stress changes the object's shape. The shear modulus is the amount of shear stress divided by the angle through which the shape is strained.

Hydrostatic stress, the third basic stress, squeezes an object with equal force from all directions. A familiar example is the **pressure** on objects under water due to the weight of the water above them. Pure hydrostatic stress changes the volume only, not the shape of the object. Its modulus is called the bulk modulus.

The greatest stress a material can undergo and still return to its original dimensions is called the elastic limit. When stressed beyond the elastic limit, some materials fracture, or break. Others undergo plastic deformation, taking on a new permanent shape. An example is a nail bent by excessive shear stress of a hammer blow.

The elastic modulus and elastic limit reveal much about the strength of the bonds between the smallest particles of a substance, the atoms or molecules it is composed of. However, to understand elastic behavior on the level of atoms requires first distinguishing between materials that are crystalline and those that are not.

Metals are examples of crystalline materials. Solid pieces of metal contain millions of microscopically small crystals stuck together, often in random orientations. Within a single crystal, atoms are arranged in orderly rows. They are held by attractive forces on all sides. Scientists model the attractive force as a sort of a spring. When a spring is stretched, a restoring force tries to return it to its original length. When a metal rod is stretched in tension, its atoms are pulled apart slightly. The attractive force between the atoms tries to restore the original distance. The stronger the attraction, the more force must be applied to pull the atoms apart. Thus, stronger atomic forces result in larger elastic modulus.

Stresses greater than the elastic limit overcome the forces holding atoms in place. The atoms move to new positions. If they can form new bonds there, the material deforms plastically; that is, it remains in one piece but assumes a new shape. If new bonds cannot form, the material fractures.

The ball and spring model also explains why metals and other crystalline materials soften at higher temperatures. **Heat** energy causes atoms to vibrate. Their vibrations move them back and forth, stretching and compressing the spring. The higher the **temperature**, the larger the vibrations, and the greater the average distance between atoms. Less applied force is needed to separate the atoms because some of the stretching **energy** has been provided by the heat. The result is that the elastic modulus of metals decreases as temperature increases.

To explain the elastic behavior of materials like rubber requires a different model. Rubber consists of molecules, which are clusters of atoms joined by chemical bonds. Rubber molecules are very long and thin. They are polymers, long chain-like molecules built up by repeating small units. Rubber polymers consist of hundreds or thousands of atoms joined in a line. Many of the bonds are flexible, and can rotate. The result is a fine structure of kinks along the length of the molecule. The molecule itself is so long that it tends to bend and coil randomly, like a rope dropped on the ground. A piece of rubber, such as a rubber band, is made of vast numbers of such kinked, twisting, rope-like molecules.

When rubber is pulled, the first thing that happens is that the loops and coils of the "ropes" straighten out. The rubber extends as its molecules are pulled out to their full length. Still more stress causes the kinks to straighten out. Releasing the stress allows the kinks, coils, and loops to form again, and the rubber returns to its original dimensions. Materials made of long, tangled molecules stretch very easily. Their elastic modulus is very small. They are called elastomers because they are very "elastic" polymers.

The "kink" model explains a very unusual property of rubber. A stretched rubber band, when heated, will suddenly contract. It is thought that the added heat provides enough energy for the bonds to start rotating again. The kinks that had been stretched out of the material return to it, causing the length to contract.

Elasticity is involved whenever atoms vibrate. An example is the movement of sound **waves**. A **sound** wave consists of energy that pushes atoms closer together momentarily. The energy moves through the atoms, causing the region of compression to move forward. Behind it, the atoms spring further apart, as a result of the restoring force.

The speed with which sound travels through a substance depends in part on the strength of the forces between atoms of the substance. Strongly bound atoms readily affect one another, transferring the "push" due to the sound wave from each **atom** to its neighbor. Therefore, the stronger the bonding force, the faster sound travels through an object. This explains why it is possible to hear an approaching railroad train by putting one's ear to the track, long before it can be heard through the air. The sound wave travels more rapidly through the steel of the track than through the air, because the elastic modulus of steel is a million times greater than the bulk modulus of air.

The most direct way to determine the elastic modulus of a material is by placing a sample under increasing stresses, and measuring the resulting strains. The results are plotted as a graph, with strain along the horizontal axis and stress along the vertical axis. As long as the strain is small, the data form a straight line for most materials. This straight line is the "elastic region." The slope of the straight line equals the elastic modulus of the material. Alternatively, the elastic modulus can be calculated from measurements of the **speed of sound** through a sample of the material.

ELECTRIC CHARGE

Rub a balloon or styrofoam drinking cup against a wool sweater. It will then stick to a wall (at least on a dry day) or pick up small bits of paper. Why? The answer leads to the concept of electric charge.

Electromagnetic forces are one of the four fundamental forces in nature. The other three are gravitational, strong nuclear, and weak nuclear forces. The electromagnetic **force** unifies both electrical and magnetic forces. The magnetic forces occur whether the charges are moving or at rest. Electric charge is our way of measuring how much electric force an object can exert or feel.

Electric charge plays the same role in electric forces as **mass** plays in gravitational forces. The force between two electric charges is proportional to the product of the two charges divided by the distance between them squared, just as the force between two masses is proportional to the product of the two masses divided by the distance between them squared. These two force laws for electric and gravitational forces have exactly the same mathematical form.

There are, however, differences between the electric and gravitational forces. One difference is the electrical force is much stronger than the **gravitational force**. That is why the styrofoam cup mentioned above can stick to a wall. The electrical force pulling it to the wall is stronger than the gravitational force pulling it down.

The second major difference is that the gravitational force is always attractive. The electrical force can be either attractive or repulsive. There is only one type of mass, but there are two types of electric charge. Like charges will repel each other and unlike charges will attract. Most **matter** is made up of equal amounts of both types of charges, so electrical forces cancel out over long distances. The two types of charge are called positive and negative, the names given by **Benjamin Franklin**, the first American physicist. Contrary to what many people think, the terms positive and negative don't really describe properties of the charges. The names are completely arbitrary.

An important property of electric charges that was discovered by Benjamin Franklin is that charge is conserved. The total amount of both positive and negative charges must remain the same. Charge conservation is part of the reason the balloon and cup mentioned above stick to the wall. Rubbing causes electrons to be transferred from one object to another, so one has a positive charge and the other has exactly the same negative charge. No charges are created or destroyed; they are just transferred. The objects then have a net charge and electrical forces come into play.

Electric charge forms the basis of the electrical and magnetic forces that are so important in our modern electrical and electronic luxuries.

ELECTRIC CIRCUIT

An electric circuit is a system of conducting elements designed to control the path of **electric current** for a particular purpose. Circuits consist of sources of electric **energy**, like generators and batteries; elements that transform, dissipate, or store this energy, such as resistors, capacitors and inductors; and connecting wires. Circuits often include a fuse or circuit breaker to prevent a **power** overload.

Devices that are connected to a circuit are connected to it in one of two ways: in series or in parallel. A **series circuit** forms a single pathway for the flow of current, while a parallel circuit forms separate paths or branches for the flow of current. **Parallel circuits** have an important advantage over series circuits. If a device connected to a series circuit malfunctions or is switched off, the circuit is broken, and other devices on the circuit cannot draw power. The separate pathways of a parallel circuit allow devices to operate independently of each other, maintaining the circuit even if one or more devices are switched off.

The first electric circuit was invented by **Alessandro Volta** in 1800. He discovered he could produce a steady flow of **electricity** using bowls of salt solution connected by metal strips. Later, he used alternating discs of copper, zinc, and cardboard that had been soaked in a salt solution to create his voltaic pile (an early battery). By attaching a wire running from the top to the bottom, he caused an electric current to flow through his circuit. The first practical use of the circuit was in **electrolysis**, which led to the discovery of several new chemical elements. **Georg Ohm** discovered some conductors had more resistance than others, which affects their efficiency in a circuit. His famous law states that the **voltage** across a conductor divided by the current equals the resistance, measured in *ohms*. Resistance causes **heat** in an electrical circuit, which is often not wanted.

ELECTRIC CONDUCTOR • See Conductors and

insulators; Semiconductors; and Transistors

ELECTRIC CURRENT

Electric current is the result of the relative **motion** of net **electric charge**. In metals, the charges in motion are electrons. The magnitude of an electric current depends upon the quantity of charge that passes a chosen reference point during a specified time interval. Electric current is measured in amperes, with one ampere equal to a charge-flow of one coulomb per second.

A current as small as a picoampere (one-trillionth of an ampere) can be significant. Likewise, artificial currents in the millions of amperes can be created for special purposes. Currents between a few milliamperes to a few amperes are common in **radio** and television circuits. An automobile starter motor may require several hundred amperes.

The total charge transferred by an unvarying electrical current equals the product of current in amperes and the time in seconds that the current flows. If one ampere flows for one second, one coulomb will have moved in the conductor. If a changing current is graphed against time, the area between the graph's curve and the time axis will be proportional to the total charge transferred.

Electrical currents move through wires at a speed only slightly less than the **speed of light**. The electrons, however, move from **atom** to atom more slowly. Their motion is more aptly described as a drift. Extra electrons added at one end of a wire will cause extra electrons to appear at the other end of the wire almost instantly. Individual electrons will not have moved along the length of the wire but the electric field that pushes the charge against charge along the conductor will be felt at the distant end almost immediately. To visualize this, imagine a cardboard mailing tube filled with Ping-Pong balls. When you insert an extra ball in one end of the tube, an identical ball will emerge from the distant end almost immediately. The original ball will not have traveled the length of the tube, but since all the balls are identical it will seem as if this has happened. This mechanical analogy suggests the way that charge seems to travel through a wire very quickly.

Heat results when current flows through an ordinary electrical conductor. Common materials exhibit an electrical property called resistance. **Electrical resistance** is analogous to **friction** in a mechanical system. Resistance results from imperfections in the conductor. When the moving electrons collide with these imperfections, they transfer **kinetic energy**, resulting in heat. The quantity of heat **energy** produced increases as the square of the current passing through the conductor.

A magnetic field is created in **space** whenever a current flows through a conductor. This magnetic field will exert a **force** on the magnetic field of other nearby current-carrying conductors. This is the principle behind the design of an electric motor.

An electrical generator operates on a principle similar to an electric motor. In a generator, mechanical energy forces a conductor to move through a magnetic field. The magnetic field forces the electrons in the conductor to move, which causes an electric current.

A current in one direction only is called a direct current, or DC. A steady current is called pure DC. If DC varies with time it is called pulsating DC.

If a current changes direction repeatedly it is called an alternating current, or AC. Commercial electrical **power** is transported using alternating current because AC makes it possible to change the ratio of **voltage** to current with **transformers**. Using a higher voltage to transport electrical power across country means that the same power can be transferred using less current. For example, if transformers step up the voltage by a factor of 100, the current will be lower by a factor of 1/100. The higher voltage in this example would reduce the energy loss caused by the resistance of the wires to 0.01% of what it would be without the use of AC and transformers.

When alternating current flows in a circuit the charge drifts back and forth repeatedly. There is a transfer of energy with each current pulse. Simple electric motors deliver their mechanical energy in **pulses** related to the power line **frequency**.

Power lines in North America are based on AC having a frequency of 60 Hertz (Hz). In much of the rest of the world the power line frequency is 50 Hz. Alternating current generated aboard aircraft often has a frequency of 400 Hz because motors and **generators** can work efficiently with less iron, and therefore less **weight**, when this frequency is used.

Alternating current may also be the result of a combination of signals with many frequencies. The AC powering a loudspeaker playing music consists of a combination of many superimposed alternating currents with different frequencies and amplitudes.

We cannot directly observe the electrically charged particles that produce current. It is usually not important to know whether the current results from the motion of positive or negative charges. Early scientists made an unfortunate choice when they assigned a positive polarity to the charge that moves through ordinary wires. It seemed logical that current was the result of positive charge in motion. Later it was confirmed that it is the negatively charged **electron** that moves within wires.

The action of some devices can be explained more easily when the motion of electrons is assumed. When it is simpler to describe an action in terms of the motion of electrons, the charge motion is called electron flow. Current flow, conventional current, or Franklin convention current are terms used when the moving charge is assumed to be positive.

Conventional current flow is used in science almost exclusively. In **electronics**, either conventional current or electron flow is used, depending on which flow is most convenient to explain the operation of a particular electronic component. The need for competing conduction models could have been avoided had the original charge-polarity assignment been reversed.

ELECTRIC FIELDS AND FORCES

An electric field describes the region of **space** extending from a charged particle in which another charged particle experiences a **force** related to its charge. Electric fields surround all charged particles.

Surrounding every charged particle there is a region of space in which there is a potential electrostatic force. The strength of the electric field surrounding a point charge, as measured by the electrostatic force, is directly proportional to the charge and inversely proportional to the square of the distance from the charge. The electrostatic force requires no medium to act (i.e., it can act in a **vacuum**).

Just as all particles with **mass** exert a gravitational influence on all other particles, all charged particles exert an electrostatic force on all other charged particles. Because the strength of the force is the inverse of the square of the distance between the charged particles, however, there are some practical limitations regarding the measurability of the an electrical field or force. In practical terms an electrical or electrostatic force arises whenever a charged object is sufficiently near another charged particle so that the resulting force is measurable. The strength of the electric field, E, at the position of a charged object in that field is defined to be the magnitude of the electric force F acting on the object divided by the magnitude of the charge Q of that object, $E = F/{-}Q{-}$.

The electric field is a vector quantity that is expressed in terms of force per charge (usually Newtons per Coulomb. The direction of this vector at any point is defined to be the direction of the electric force that would be exerted on a positive charge placed at that point. Accordingly, if the origin of the electric field is a negatively charged particle the force on a positively charged particle is directed toward the negative charge (opposite charges attract). If the origin of the electric field is a positive charge the force on another positive charge is directed away from the field originating charge (like charges repel).

Coulomb's law states that the electric force between two stationary charged particles q_1 and q_2 separated by a distance r is given by the equation $F = k{-}q_1{-}q_2{-}/r^2$, where k (9×10^9 N m^2 C^{-2}) is the Coulomb constant.

Electric forces can be added vectorially. To determine the electric field at a point P when there are two charged particles in the vicinity of this point, the electric fields at P due to each charged particle must be determined separately. The respective vector components of each field can be vectorially added to give the overall field strength at point P.

When a battery or some other device possessing a **electromotive force** is connected to two conducting plates held parallel to one another, an electric field is set up. The magnitude of the electric field in this case is given by the potential difference divided by the plate separation. A positive charge will gain electrical **potential energy** when it is moved in a direction opposite to the electric field. A negative charge will, however, lose electrical potential **energy** when it is moved in the direction opposite to the electric field. To move a particle from one place in an electric field to another place in the electric field takes **work**. The energy required to do this work is the **electric potential** between these two points.

The physical description of electric fields and forces can be traced back to 300 B.C. when Greek philosopher Theophrastus asserted that amber acquired a **power** to attract **light** objects when rubbed. The first published study of electrical phenomena appeared in 1600 A.D. when English physician and scientist **William Gilbert** first published his studies on **magnetism** and electrical attractions. Gilbert was the first to use the term electric, which he derived from the Greek *elektron* meaning amber, when describing a force that such substances exert after being rubbed. Gilbert differentiated between magnetic and electric action as well.

In 1672 German physicist **Otto von Guericke** (1602-1686) described the first machine for producing an **electric charge**. In the eighteenth century the empirical foundations for intensive studies in electromagnetic phenomena were kindled by the work of American scientist **Benjamin Franklin**. Franklin spent much of his life researching electrical phenomena and developed a theory that all **matter** consisted of a fluid that was believed to be **electricity**. He explained the forces associated with electricity to be either an excess or lack of this fluid. In 1766 English scientist Joseph Priestley (1733-1804) developed a law relating the force between electric charges and the distance between them. He also discovered that no electric field of force exists within a metal sphere whose surface is charged. In the 1780s, French physicist **Charles-Augustin de Coulomb** invented the torsion balance and used it to accurately measure the force exerted by electrical charges.

In the nineteenth century there were consistent advancements in the understanding of electric fields and forces. In 1800, Italian physicist **Alessandro Volta** demonstrated the first electric battery. The work of German mathematician **Carl Friedrich Gauss**, French scientist **André-Marie Ampère**, Danish scientist **Hans Christian Ørsted**, and English physicist **Michael Faraday** resulted in the unification of electrical theory with magnetic theory in Scottish physicist **James Clerk Maxwell**'s set of equations describing **electromagnetism**.

See also Electric current; Electromagnetic field; Electromagnetic induction; Electromagnetism; Electrostatics and electrostatic devices; Faraday's laws; Maxwell's equations

ELECTRIC POTENTIAL

Electric potential is amount of **work** required to move a unit charge from a reference point or surface to a specified point in an electric field. Accordingly, the **energy** possessed by a charged particle is due to its position with respect to the electric field. Electric potential is a scalar quantity and can result from the introduction of a particle into an electric field produced by a source of potential difference such as a battery, or by another nearby charged particle.

Electric potential can be compared to the gravitational **potential energy** of an object in a **gravitational field**. Just as the gravitational field results in a **force** on an object with **mass** or **momentum**, an electrical field results in a force upon a charge. The work done by the electric field depends only on the initial and final positions of the object, not the path followed by the object between the two points. In addition, just as a gravitational field results in a gravitational potential energy for an object relative to a specified surface (e.g., the surface of Earth), an electric field results in charged particle's electric potential energy. The magnitude of the electric potential is dependent upon the particle's charge and the electric field strength. If the particle has no charge then an electric field has no effect on it.

Relative to a particle, A, when an electric field exists due to another charged particle, B, the electric potential of particle A is dependent upon the charges of the particles and the distance between the two particles. The point of zero electric potential is taken to be when particle B is at infinite distance from particle A. If the two particles are of the same sign then the electric potential is positive because like charges repel one another.

The electric potential at a distance r from an isolated point charge Q, or from the center of a equipotential sphere, is directly proportional to the charge and inversely proportional to the distance r. The electric potential for an isolated point charge is $V=kQ/r$, where V is the electric potential in volts, Q is the charge of the particle, r is the distance from the point charge, and k (9×10^9 N m^2 C^{-2}) is a constant derived from **Coulomb's law**. The electric potential of a system of two or more charges can be obtained by applying the superposition principle. The superposition principle states that the total potential at some point P due to several point charges is the algebraic sum of the potentials due to the individual charges. So for a system of two charges the potential at point P is $V_P=k(q_1/r_1+q_2/r_2)$, where q_1 is the charge of particle 1 that is a distance r_1 from point P, and q_2 is the charge of particle 2 that is a distance r_2 from point P.

The difference in electric potential between two points is referred to as the **voltage** or **electromotive force**. It is not actually a force but a difference in potential that is capable of doing work. The actual force is dependent upon the charge and distance involved.

See also Electric charge; Electric field and forces; Electromagnetic field; Work and potential energy

ELECTRIC POWER AND ENERGY

Electrical **energy** is a function of state associated with all charged particles due to their position in an electric field. Electric **power** is a measure of the electrical energy utilization (or change) per unit time. In more general terms, electrical energy is the ability of electrical charges to do **work**. In quantitative terms, electrical power is defined as the current flow through an object multiplied by the **voltage** across it.

A battery is any device that is capable of storing and converting chemical, thermal, solar, or nuclear energy into electrical energy. In essence, batteries supply electrical power to a conductive circuit, and this power is used by devices that offer resistance to current flow. For example, in a car battery, the battery transforms chemical **potential energy** into electrical potential energy that can be used to establish a current in a conductive circuit connected to the battery.

In an **electric circuit** with a resistor, it takes work to move a charge through the resistor. Electric power then becomes a measure of the rate at which the charge is moved through the resistor. Electric power is also dissipated as electrical energy is converted to **thermal energy** (i.e., the heating of the resistor, or Joule heating). Although the conversion of elec-

Interior of electrical power plant, converting hydroelectric power to electric power. *(JLM Visuals. Reproduced by permission.)*

trical power to **heat** is often considered wasted electrical energy, it is in some devices, such as toasters and electrical heaters, the intended result.

Electrical power is expressed as energy per unit time. In the International System of Units (SI), electrical power is expressed in watts (one watt is equivalent to one joule per second). A watt represents the generation or use of electrical energy of one joule (the work done when a charge of one coulomb travels across a potential difference of one volt) per second. Accordingly, when a charge of one coulomb travels across a potential difference of one volt in one second, a watt of power is used. Another common unit of power is electrical horsepower. The term horsepower was originally developed by Scottish inventor **James Watt** to describe the work done by a steam engine (later the Watt engine) in terms of work done by a horse. One horsepower equals the amount of power required to lift 33,000 pounds a distance of one foot in one minute. With regard to electrical power, one horsepower is approximately 746 watts.

A common misconception related to electrical power is that utility or power companies bill customers for electrical power. Ironically power companies actually bill customers for electrical energy, using units of kilowatt-hours.

See also Electric charge; Electric field and forces; Electric motors and meters; Electric potential; Electricity; Electromotive force

ELECTRICAL CONDUCTIVITY

Electrical conductivity is the ability of different types of **matter** to conduct an **electric current**. The electrical conductivity of a material is defined as the ratio of the current per unit cross-sectional area to the electric field producing the current. Electrical conductivity is an intrinsic property of a substance, dependent on the **temperature** and chemical composition, but not on the amount or shape. The unit of electrical conductivity in the International System of Units (SI) is the *siemens per meter*, where the siemens is the reciprocal of the ohm, the unit of **electrical resistance**, represented by the Greek capital letter omega. An older name for the siemens is the mho, which, of course, is ohm spelled backwards (which was written as an inverted Greek omega). Electrical conductivity is due to the presence of electrons or **ions** able to move through the material of interest.

Electrical conductivity is the inverse quantity to electrical resistivity. For any object conducting **electricity**, one can define the resistance in ohms as the ratio of the electrical potential difference applied to the object to current passing through it in amperes. For a cylindrical sample of known length and cross-sectional area, the resistivity is obtained by dividing the measured resistance by the length and then multiplying by the area. The conductivity is then found by taking the reciprocal of the resistivity.

Metals generally have very high electrical conductivity. The electrical conductivity of copper at room temperature, for instance, is over 70 million siemens per meter. On an atomic level this high conductivity reflects the unique character of the metallic bond in which pairs of electrons are shared not between pairs of atoms, but among all the atoms in the metal, and are thus free to move over large distances. Many metals undergo a transition at low temperatures to a superconducting state, in which the resistance disappears entirely and the conductivity becomes infinite. The superconduction process involves a coupling of **electron** motion with the vibration of the atomic nuclei and inner-shell electrons, to allow net current flow without **energy** loss.

Electrical conductivity in the liquid state is generally due to the presence of ions. Substances that give rise to ionic conduction when dissolved are called electrolytes. The conductivity of one molar electrolyte is of the order of 0.01 siemens per meter, far less than that of a metal, but still very much larger than that of typical insulators. Sodium chloride (common table salt), composed of sodium ions and chloride ions, is a very poor conductor in the solid state. If it is dissolved in water, however, it becomes a good ionic conductor. Likewise, if it is melted, it becomes a good conductor. Substances such as hydrogen chloride or acetic acid are nonconductors in the pure state but give rise to ions and thus electrical conductivity when dissolved in water. In modern electrochemistry, substances of the sodium chloride type, which are actually composed of ions, are termed true electrolytes, while those that require a solvent for ion formation, like hydrogen chloride, are termed potential electrolytes.

As might be expected, the conductivity of a dissolved electrolyte depends on its concentration and so physical chemists find it useful to define molar conductivity of an electrolyte as the electrical conductivity of the solution divided by the concentration of the electrolyte in moles per liter. Because different electrolytes can give rise to ions of different charge, for example, Mg^{2+} ions instead of Na^+ ions, an even more useful measure of conductivity is given by the equivalent conductivity, obtained by dividing the molar conductivity by the number of moles of positive charge released by one mole of the electrolyte. Because the ions of an electrolyte solution interact with each other, the equivalent conductivity is found to decrease gradually as the concentration increases. For dilute solutions, the equivalent conductivity can be expressed as a constant, the equivalent conductivity at infinite dilution minus a term proportional to the square root of the concentration. This important relation was discovered by the German physicist Friedrich Wilhelm Georg Kohlrausch (1840-1910). It is the limiting conductivities at infinite dilution that are additive for the ions involved. Thus, there are tabulations in handbooks of these conductivities for different ions (Na^+, K^+, Mg^+, Cl^-, Br^-, etc.) from which the corresponding conductivities of different electrolytes ($NaCl$, KCl, KBr, etc.) can be calculated. The equivalent conductivities of the individual ions are proportional to their electrical mobilities (ratio of particle **velocity** to electrical field strength).

Ionic crystals do exhibit a limited electrical conductivity even in the crystalline state. This arises from point defects present at any temperature above **absolute zero**. For compounds like sodium chloride, composed of ions of roughly equal size, there will be a certain number of cation and anion vacancies, which allows a jumping motion of the corresponding cations and anions. For compounds like silver chloride, in which one of the ions, the cation, is much larger than the other ion, the anion, vacancies and interstitials (out of place ions) of the smaller ion will allow an ionic conductivity. In addition, compounds like silver chloride tolerate a range of stoichiometry, that is a deviation of the Ag to Cl ratio of one to one. An excess of silver is accommodated by a delocalization of the extra electrons to allow an electronic conductivity in addition to the conductivity of silver ions. A deficit of silver introduces mobile electron vacancies, or holes, into the electronic structure of the material, adding a component to the conductivity, which behaves as if positively charged.

Semiconductors are covalently bonded materials for which **heat** energy is sufficient to promote electrons from bonding energy levels to delocalized energy levels. The conductivity of semiconductors is a sensitive function of impurity content (the substitution of semiconductor atoms by atoms with one fewer or more valence electrons) introducing positive (hole) or negative charge carriers into the material.

ELECTRICAL MOTORS AND METERS

Electrical motors are devices that convert electrical **energy** into mechanical energy. Electric meters use electric motors and counting devices to translate current flow into a measured use of electrical energy. Electrical motors perform the inverse transformations performed by electrical **generators** (i.e., the transformation of mechanical energy into electrical energy). To a limited extent, all electrical motors perform these inverse transformations as they operate to produce a counter or back **electromotive force**.

The modern form of the electrical generator evolved from the research and engineering work of nineteenth-century Danish scientist **Hans Christian Ørsted** and English physicist **Michael Faraday** who developed a rotator (later to be known as an electric motor) containing a conductive (i.e., electrical current carrying) wire rotating around a magnet. Working independently, American inventor **Joseph Henry** also developed an early model of the electric motor. Although it was another half century before widespread production, the first practical electric motors were all designed according to principles first documented by Faraday and Henry. Moreover, Faraday's work with **electromagnetic induction** in 1831 formed the basis of a collection of papers eventually published as *Experimental Researches in Electricity*, a publication that became the standard authoritative reference on **electricity** and **magnetism** for nineteenth-century scientists and inventors. During the last decades of the nineteenth century, to serve the needs of the European and American industrial revolutions, electric motors drove an increasing number of time and labor saving

Cross section of a direct current electric motor. *(Photo by Bruce Iverson. Science Photo Library, National Audubon Society Collection/Photo Researchers Inc. Reproduced by permission.)*

machines. Safer and more productive than steam or combustion engines, electrical motors took on an important role in the creation of modern technological society.

Electrical motors contain a set of magnets upon which the forces created by an electrical current act. An electrical motor contains a spinning electromagnetic armature that interacts with a set of stationary field magnets. All of the magnets contain north and south poles and these poles are mutually repulsive to one another (i.e., like poles repel and opposite poles attract). As electrical current flows through a conducting coil with the proper orientation to the magnets, it also produces a magnetic field that likewise contains north and south poles. The interaction of the armature's north and south poles with the north and south poles of the field magnets and the magnetic field induced by the electrical current produces a torque **force** on the armature.

As a result of the application of **torque** to the armature, the armature spins. DC motors (drawing DC or direct current) utilize split-ring commutator devices containing conducting brushes that allow current to flow to the armature to regulate the reversal of **magnetic poles**. As the armature rotates, its reversing magnetic poles interact with the poles of the permanent field magnets to produce continuing torque. The armature can be connected to a shaft that, as armature rotates, is capable of doing **work** (e.g., to drive such labor-saving devices as **power** tools, fans, etc.).

When electrical motors first start, they require a greater quantity of electrical current to spool up to operating speed than they do to maintain their operating speed. For example, when large motors are turned on they often dim lights hooked to the same power source as they require this extra current. The extra current requires greater electromotive force (emf). In fact, because all electrical motors have current-carrying coils that turn in an electric field, all electrical motors also generate emf. In electrical motors this emf is termed *back emf*, and it acts against the applied emf driving the motor. Because of increased **friction** during motor start-up, the back emf can significantly reduce the efficiency of the applied emf. This requires the device to draw significant amounts of extra current until the device reaches efficient operating speed.

Electric motors can range in size and power from those able to drive large ships to microscopic devices incorporated into computer hardware.

Most electrical meters utilize electric motor-driven shafts and a series of gears to measure the use of electrical energy on an analog counter consisting of a series of rotating disks. A known and regulated fraction of the total current flowing through a circuit is utilized to operate the meter, and the speed of the motor is directly proportional to the current flow. Accordingly, as the current flows faster, the motor spins faster and the counter reflects the increased usage. The gear-driven rotating disks are usually marked in increasing powers of kilowatt-hours (1,000 watt-hours).

Induction-type meters measure alternating current circuit usage, mercury-based and commutator-type meters are used to measure usage in DC circuits.

ELECTRICAL RESISTANCE

The electrical resistance of a wire or circuit is a way of measuring the resistance to the flow of an electrical current. A good electrical conductor, such as a copper wire, will have a very low resistance. Good insulators, such as rubber or glass insulators, have a very high resistance. The resistance is measured in ohms, and is related to the current in the circuit and **voltage** across the circuit by **Ohm's law**. For a given voltage, a wire with a lower resistance will have a higher current.

The resistance of a given piece of wire depends on three factors: the length of the wire, the cross-sectional area of the wire, and the resistivity of the material composing the wire. To understand how this works, think of water flowing through a hose. The amount of water flowing through the hose is analogous to the current in the wire. Just as more water can pass through a fat fire hose than a skinny garden hose, a fat wire can carry more current than a skinny wire. For a wire, the larger the cross-sectional area, the lower the resistance; the smaller the cross-sectional area, the higher the resistance. Now consider the length. It is harder for water to flow through a very long hose simply because it has to travel farther. Analogously, it is harder for current to travel through a longer wire. A longer wire will have a greater resistance. The resistivity is a property of the material in the wire that depends on the chemical composition of the material but not on the amount of material or the shape (length, cross-sectional area) of the material. Copper has a low resistivity, but the resistance of a given copper wire depends on the length and area of that wire. Replacing a copper wire with a wire of the same length and area but a higher resistivity will produce a higher resistance. In the hose analogy, it is like filling the hose with sand. Less water will flow through the hose filled with sand than through an identical unobstructed hose. The sand in effect has a higher resistivity to water flow. The total resistance of a wire is then the resistivity of the material composing the wire times the length of the wire, divided by the cross-sectional area of the wire.

ELECTRICITY

The word electricity comes from the Greek word, **electron**, meaning amber. Amber is a translucent, yellowish mineral made of fossilized resin. When rubbed with a cloth, amber becomes charged with static electricity. We now know that such electrostatic effects arise whenever a body has an excess or deficiency of electrons or protons. These effects usually accompany the transfer of electrons, such as occurs when two dissimilar materials are rubbed together. A fundamental rule of electrostatics is that bodies of like charge repel each other, and bodies of opposite charge attract each other. The Greeks described these attractive and repulsive electrostatic forces as electric **force**.

At the time of the American Revolution, little more was known about electricity than at the time of the Greeks. **Benjamin Franklin** was one of the first persons to experiment with electricity. In 1821, **Hans Christian Ørsted** demonstrated that electricity and **magnetism** were interrelated. Later, **James Clerk Maxwell** synthesized all that was known about electricity and magnetism into a set of equations that explain why a moving **electric charge** gives rise to a magnetic field, and why a changing magnetic field produces an electric field.

In the classical picture of the **atom**, an **electron cloud** surrounds a small **nucleus** (composed of neutrons and protons). The **proton** and electron are considered the fundamental particles of electricity, the two particles having equal and opposite charges (convention assigns a positive charge to the proton, and a negative one to the electron). Under normal conditions, the number of electrons equals the number of protons, so the atom is electrically neutral.

Since its discovery at the end of the nineteenth century, the electron has been considered a fundamental building block of nature that cannot be decomposed into more basic particles. Electrons are intrinsically stable, and possess **mass**, charge, and spin (i.e., intrinsic **angular momentum**). In recent years, despite extensive experimental and theoretical research in elementary **particle physics**, many physicists have concluded that the electron is indeed an ultimate, indivisible constituent of **matter** that will not be found to have more fundamental parts. The electron is thus considered a basic unit of electricity and matter.

When electrons in a conducting material drift under the influence of an electric field, they produce an **electric current**. Electric currents can also be produced by the **motion** of positive nuclei, as well as by **ions** moving in a liquid or gaseous electrical discharge.

In order for electric current to flow, there must exist an excess of electrons and a conductor to carry the current. As electrons enter the conductor, they displace the electrons in the conductor's atoms. In turn, these displaced electrons displace other electrons. Thus, the electron that exits the conductor will not be the same one that entered it. When a current flows in a conductor, the charges are impeded by collisions with the atoms that make up the material. The charges give up **energy** to the atoms, often as **heat**.

The electrons in the atoms of the conductor have fixed orbits. The maximum number of electrons in any orbit is fixed.

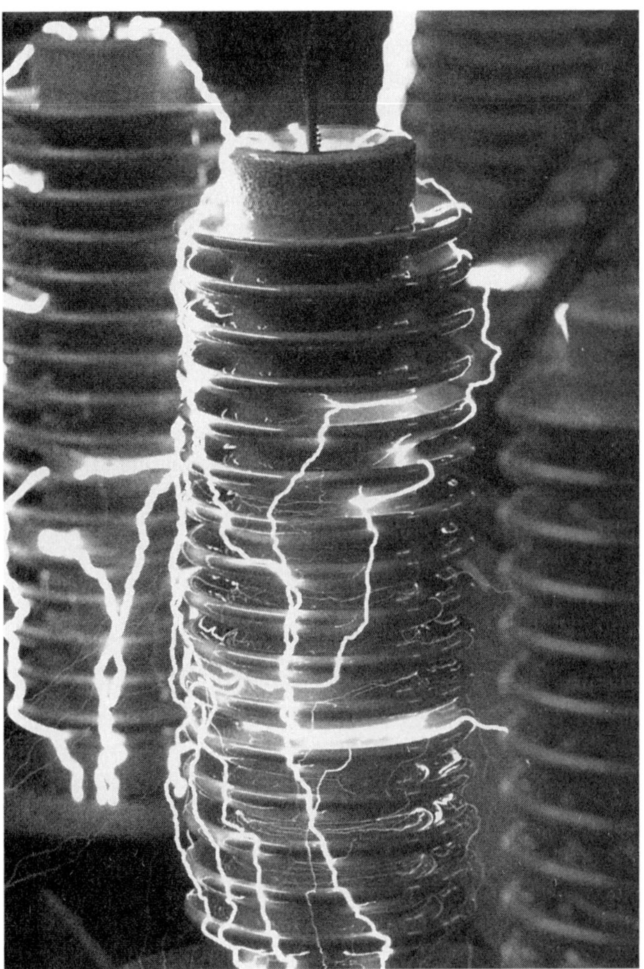

Ceramic insulators covered in web of electricity. *(Photo by Robert Essel/Bikderberg. Stock Market. Reproduced by permission.)*

For all orbits except the first one, this maximum number is eight. The number of valence electrons in an atom is the number of electrons that would be required to fill the outer orbit. If an electron orbit is full and unable to accept electrons, there must be eight electrons in that orbit. When an outside electron enters the conductor, it joins the partially empty orbit of one of the conductor's atoms, giving the atom a net negative charge. In the presence of an electric field, the displaced electron moves on to another atom, and the process repeats itself.

It is the electrons in the outer orbit that are available for electron flow. These electrons determine whether a particular element is classified as a conductor, semiconductor, or insulator. Freeing one of the electrons in the inner orbits of the atom requires considerably more energy than required to displace one in the outer orbit, and usually does not occur. Metals are excellent conductors, especially silver, copper, and aluminum. Metals are sometimes represented as a gas of free electrons. In the alkali metals, the outer electrons are only loosely bound to the nucleus, so are free to move from ion to ion in the solid state. To describe the conducting properties of more complex metals, the interactions of the various atoms and the periodic potential in which the electrons move must be taken into account. In pure **semiconductors**, there are no free electrons at **absolute zero** temperature. As the **temperature** is increased, some of the electrons are excited to current-carrying states (the conducting band); the valence states that are left unoccupied (called holes) are also available to carry electric current. Insulators may be thought of as semiconductors with very large gaps between the valence and conduction bands. In some insulating solids such as the alkali halides, ions carry current by hopping between vacant atomic sites in the crystal.

Electricity as a natural phenomenon shows up in many forms, including piezoelectricity, static electricity, atmospheric effects, and **cosmic rays**. When a piezoelectric material such as Rochelle salt, quartz, or tourmaline is squeezed, mechanical energy is converted into electrical energy. Earth resembles a large battery, with electrons continuously being lost to the atmosphere. Thunderstorms return the electrons that have been conducted away from Earth's surface back to Earth. When cosmic rays strike the upper atmosphere, they release electrons, giving rise to such phenomena as the northern lights.

Although electricity has been known to man since ancient times, and scientists have identified many of its consistent and predictable characteristics, it is still not completely understood.

ELECTRODYNAMICS

Electrodynamics is the study of the relationship between electric, magnetic, and mechanical phenomena.

A stationary charged particle has an electric field associated with it. When the charged particle is put in **motion** a magnetic field is established. These fields have forces that can do **work** or cause effects on other objects nearby. Accordingly, the study of electrodynamics is concerned with charges in motion.

The flow of electric charges constitutes an **electric current**. In order to move charges in a conductor, an electromotive driving **force** is required. This **electromotive force**, also known as an electrical potential difference, with a conductor connected across the device's two terminals constitutes a circuit. There are two types of current: direct current and alternating current. Direct current is a flow of charge and therefore current in one direction at a constant rate. Alternating current is a flow of current that increases in magnitude from zero to a maximum, decreases back to zero, increases to a maximum in the opposite direction, and then decreases back to zero. Most household currents are alternating currents.

The interaction of an electric field with a charged particle is actually the result of the interaction of the electric field with the quantum field of the charged particle. **Quantum theory** establishes that elementary particles must have corresponding quantum fields. **Quantum electrodynamics (QED)** is the **quantum field theory** of electrons and photons. It is the theory that describes how charged particles and electric fields interact. The theory predicts that the atomic **energy** levels of ele-

mentary particles must change in discrete steps, that is, they are quantized. QED was developed to reconcile classical electrodynamic theory with quantum theory. QED initially posed formidable obstacles because of the presence of "infinities" (infinite values) in the mathematical calculations. QED was ultimately rendered reliable by a process termed **renormalization** independently developed by American physicist **Richard Feynman**, American physicist **Julian Schwinger**, and Japanese physicist **Sin-Itiro Tomonaga**. The renormalization of QED allowed positive infinities to cancel out negative infinities and thus allowed measured values of **mass** and charge to be used in QED calculations. Because QED is compatible with special **relativity theory**, and special relativity equations are part of QED equations, QED is termed a relativistic theory. QED is also termed a gauge-invariant theory, meaning that it makes accurate predictions regardless of where applied in **space** or time. The theory of quantum electrodynamics has been one of the most successful theories of modern physics in predicting values that agree remarkably well with experiments.

See also Electric field and forces; Electromagnetism; Relativistic electromagnetism

ELECTROLYSIS

Electrolysis is the process of causing a **chemical reaction** to occur by passing an **electric current** through a substance or mixture of substances, most often in liquid form. Electrolysis frequently results in the decomposition of a compound into its elements. To carry out an electrolysis, two electrodes, a positive electrode (**anode**) and a negative electrode (cathode), are immersed into the material to be electrolyzed and connected to a source of direct (DC) electric current.

The apparatus in which electrolysis is carried out is called an electrolytic cell. The roots *-lys* and *-lyt* come from the Greek *lysis* and *lytos*, meaning to cut or decompose; electrolysis in an electrolytic cell is a process that can decompose a substance.

The substance being electrolyzed must be an electrolyte, a liquid that contains positive and negative **ions** and therefore is able to conduct **electricity**. There are two kinds of electrolytes. One kind is an ion compound solution of any compound that produces ions when it dissolves in water, such as an inorganic acid, base, or salt. The other kind is a liquefied ionic compound such as a molten salt.

In either kind of electrolyte, the liquid conducts electricity because its positive and negative ions are free to move toward the electrodes of opposite charge—the positive ions toward the cathode and the negative ions toward the anode. This transfer of positive charge in one direction and negative charge in the opposite direction constitutes an electric current, because an electric current is, after all, only a flow of charge, and it doesn't matter whether the carriers of the charge are ions or electrons. In an ionic solid such as sodium chloride, for example, the normally fixed-in-place ions become free to move as soon as the solid is dissolved in water or as soon as it is melted.

During electrolysis, the ions move toward the electrodes of opposite charge. When they reach their respective electrodes, they undergo chemical oxidation-reduction reactions. At the cathode, which is pumping electrons into the electrolyte, chemical reduction takes place—a taking-on of electrons by the positive ions. At the anode, which is removing electrons from the electrolyte, chemical oxidation takes place—a loss of electrons by the negative ions.

In electrolysis, there is a direct relationship between the amount of electricity that flows through the cell and the amount of chemical reaction that takes place. The more electrons are pumped through the electrolyte by the battery, the more ions will be forced to give up or take on electrons, thereby being oxidized or reduced. To produce one mole's worth of chemical reaction, at least one mole of electrons must pass through the cell. A mole of electrons, that is, 6.02×10^{23} of them, is called a faraday. The unit is named after **Michael Faraday**, the English chemist and physicist who discovered this relationship between electricity and chemical change. He is also credited with inventing the words anode, cathode, electrode, electrolyte, and electrolysis.

Various kinds of electrolytic cells can be devised to accomplish specific chemical objectives.

Perhaps the best-known example of electrolysis is the electrolytic decomposition of water to produce hydrogen and **oxygen**:

$$2H_2O + energy \rightarrow 2H_2 + O_2$$

Because water is such a stable compound, we can only make this reaction go by pumping energy into it—in this case, in the form of an electric current. Pure water, which doesn't conduct electricity very well, must first be made into an electrolyte by dissolving an acid, base, or salt in it. Then an anode and a cathode, usually made of graphite or some non-reacting metal such as platinum, can be inserted and connected to a battery or other source of direct current.

At the cathode, where electrons are being pumped into the water by the battery, they are taken up by water molecules to form hydrogen gas:

$$4H_2O + 4e^- \rightarrow 2H_2 + 4OH^-$$

At the anode, electrons are being removed from water molecules:

$$2H_2O - 4e^- \rightarrow O_2 + 4H^+$$

The net result of these two electrode reactions added together is

$$2H_2O \rightarrow 2H_2 + O_2$$

(Note that when these two equations are added together, the four H^+ ions and four OH^- ions on the right side are combined to form four H_2O molecules, which then cancel four of the H_2O molecules on the left side.) Thus, two molecules of water have been decomposed into two molecules of hydrogen and one molecule of oxygen.

The acid, base, or salt that makes the water into an electrolyte is chosen so that its particular ions cannot be oxidized or reduced (at least at the **voltage** of the battery), so they don't react chemically and serve only to conduct the current through the water. Sulfuric acid, H_2SO_4, is commonly used.

By electrolysis, common salt, sodium chloride, NaCl, can be broken down into its elements, sodium and chlorine. This is an important method for the production of sodium; it is used also for producing other alkali metals and alkaline earth metals from their salts.

To obtain sodium by electrolysis, we'll first melt some sodium chloride by heating it above its melting point of 1,474°F (801°C). Then we'll insert two inert (non-reacting) electrodes into the melted salt. The sodium chloride must be molten in order to permit the Na^+ and Cl^- ions to move freely between the electrodes; in solid sodium chloride, the ions are frozen in place. Finally, we'll pass a direct electric current through the molten salt.

The negative electrode (the cathode) will attract Na^+ ions and the positive electrode (the anode) will attract Cl^- ions, whereupon the following chemical reactions take place.

At the cathode, where electrons are being pumped in, they are being grabbed by the positive sodium ions:

$$Na^+ + e^- \rightarrow Na$$

At the anode, where electrons are being pumped out, they are being ripped off the chloride ions:

$$Cl^- - e^- \rightarrow Cl$$

(The chlorine atoms immediately combine into diatomic molecules, Cl_2.) The result is that common salt has been broken down into its elements by electricity.

Another important use of electrolysis is in the production of magnesium from seawater. Seawater is a major source of that metal, since it contains more ions of magnesium than of any other metal except sodium. First, magnesium chloride, $MgCl_2$, is obtained by precipitating magnesium hydroxide from seawater and dissolving it in hydrochloric acid. The magnesium chloride is then melted and electrolyzed. Similar to the production of sodium from molten sodium chloride, above, the molten magnesium is deposited at the cathode, while the chlorine gas is released at the anode. The overall reaction is $MgCl_2 \rightarrow Mg + Cl_2$.

Sodium hydroxide, NaOH, also known as lye and caustic soda, is one of the most important of all industrial chemicals. It is produced at the rate of 25 billion pounds a year in the United States alone. The major method for producing it is the electrolysis of brine or "salt water," a solution of common salt, sodium chloride, in water. Chlorine and hydrogen gases are produced as valuable by-products.

When an electric current is passed through salt water, the negative chloride ions, Cl^-, migrate to the positive anode and lose their electrons to become chlorine gas.

$$Cl^- + e^- \rightarrow Cl$$

(The chlorine atoms then pair up to form Cl_2 molecules.) Meanwhile, sodium ions, Na^+, are drawn to the negative cathode. But they don't pick up electrons to become sodium metal atoms as they do in molten salt, because in a water solution the water molecules themselves pick up electrons more easily than sodium ions do. What happens at the cathode, then, is

$$2H_2O + 2e^- \rightarrow H_2 + OH^-$$

The hydroxide ions, together with the sodium ions that are already in the solution, constitute sodium hydroxide, which can be recovered by evaporation.

This so-called chloralkali process is the basis of an industry that has existed for well over a hundred years. By electricity, it converts cheap salt into valuable chlorine, hydrogen, and sodium hydroxide. Among other uses, the chlorine is used in the purification of water, the hydrogen is used in the hydrogenation of oils, and the lye is used in making soap and paper.

The production of aluminum by the Hall process was one of the earliest applications of electrolysis on a large scale, and is still the major method for obtaining that very useful metal. The process was discovered in 1886 by Charles M. Hall, a 21-year-old student at Oberlin College in Ohio, who had been searching for a way to reduce aluminum oxide to the metal. Aluminum was a rare and expensive luxury at that time, because the metal is very reactive and therefore difficult to reduce from its compounds by chemical means. On the other hand, electrolysis of a molten aluminum salt or oxide is difficult because the salts are hard to obtain in anhydrous (dry) form and the oxide, Al_2O_3, doesn't melt until 3,762°F (2,072°C).

Hall discovered that Al_2O_3, in the form of the mineral bauxite, dissolves in another aluminum mineral called cryolite, Na_3AlF_6, and that the resulting mixture could be melted fairly easily. When an electric current is passed through this molten mixture, the aluminum ions migrate to the cathode, where they are reduced to metal:

$$Al^{3+} + 3e^- \rightarrow Al$$

At the anode, oxide ions are oxidized to oxygen gas:

$$2O^{2-} - 2e^- \rightarrow O_2$$

The molten aluminum metal sinks to the bottom of the cell and can be drawn off.

The production of aluminum by the Hall process consumes huge amounts of fuel. The recycling of beverage cans and other aluminum objects has become an important energy conservation measure.

Unlike aluminum, copper metal is fairly easy to obtain chemically from its ores. But by electrolysis, it can be refined and made very pure—up to 99.999%. Pure copper is important in making electrical wire, because copper's electrical conductivity is reduced by impurities. These impurities include such valuable metals as silver, gold, and platinum; when they are removed by electrolysis and recovered, they go a long way toward paying the electricity bill.

In the electrolytic refining of copper, the impure copper is made from the anode in an electrolyte bath of copper sulfate, $CuSO_4$, and sulfuric acid, H_2SO_4. The cathode is a sheet of very pure copper. As current is passed through the solution, positive copper ions, Cu^{2+}, in the solution are attracted to the negative cathode, where they take on electrons and deposit themselves as neutral copper atoms, thereby building up more and more pure copper on the cathode. Meanwhile, copper atoms in the positive anode give up electrons and dissolve into the electrolyte solution as copper ions. But the impurities in the anode do not go into solution because silver, gold, and platinum atoms are not as easily oxidized (converted into positive ions) as copper is. So

the silver, gold, and platinum simply fall from the anode to the bottom of the tank, where they can be scraped up.

Another important use of electrolytic cells is in the electroplating of silver, gold, chromium, and nickel. Electroplating produces a very thin coating of these expensive metals on the surfaces of cheaper metals, to give them the appearance and the chemical resistance of the expensive ones.

In silver plating, the object to be plated (let's say a spoon) serves as the cathode of an electrolytic cell. The anode is a bar of silver metal, and the electrolyte (the liquid in between the electrodes) is a solution of silver cyanide, AgCN, in water. When a direct current is passed through the cell, positive silver ions (Ag^+) from the silver cyanide migrate to the negative anode (the spoon), where they are neutralized by electrons and stick to the spoon as silver metal:

$$2H_2O + energy \rightarrow 2H_2 + O_2$$

Meanwhile, the silver anode bar gives up electrons to become silver ions:

$$Ag - e^- \rightarrow Ag^+$$

Thus, the anode bar gradually dissolves to replenish the silver ions in the solution. The net result is that silver metal has been transferred from the anode to the cathode, in this case the spoon. This process continues until the desired coating thickness is built up on the spoon—usually only a few thousandths of an inch—or until the silver bar has completely dissolved.

In electroplating with silver, silver cyanide is used in the electrolyte rather than other compounds of silver such as silver nitrate, $AgNO_3$, because the cyanide ion, CN^-, reacts with silver ion, Ag^+, to form the complex ion $Ag(CN)_2^-$. This limits the supply of free Ag^+ ions in the solution, so they can deposit themselves only very gradually onto the cathode. This produces a shinier and more adherent silver plating. Gold plating is done in much the same way, using a gold anode and an electrolyte containing gold cyanide, AuCN.

ELECTROMAGNETIC FIELD

An electromagnetic field is an area in which electric and magnetic forces are interacting. It arises from electric charges in **motion**. Electromagnetic fields are directly related to the strength and direction of the **force** that a charged particle, called the "test" charge, would be subject to under the electromagnetic force caused by another charged particle or group of particles, called the source.

An electromagnetic field is best understood as a mathematical function or property of **space-time**, but may be represented as a group of **vectors**, arrows with specific length and direction. For a static electric field, meaning there is no motion of source charges, the force $\vec{F}$ on a test charge is $\vec{F} = q\vec{E}$, where q is the value of the test charge and $\vec{E}$ is the vector electric field. For a static magnetic field (caused by a moving charge inside an overall neutral group of charges, or a bar magnet, for example) the force is given by $\vec{F} = q\vec{v} \times \vec{B}$, where $\vec{v}$ is the charge **velocity**, $\vec{B}$ is the vector magnetic field, and the indicates a cross-product of vectors.

A stationary charge produces an electric field, while a moving charge additionally produces a magnetic field. Since velocity is a relative concept dependent on one's choice of reference frame, **magnetism** and **electricity** are not independent, but linked together, hence the term **electromagnetism**.

Since all charges, moving or not, have fields associated with them, we must have a way to describe the total field due to all randomly distributed charges that would be felt by a positive charge, such as **proton** or **positron**, at any position and time. The total field is the sum of the fields produced by the individual particles. This idea is called the principle of superposition.

Let us look at arrow or vector representations of the fields from some particular charge distributions. There is an infinite number of possibilities, but we will consider only a few simple cases.

According to **Coulomb's law**, the strength of the electric field from a nonmoving point charge depends directly on the charge value q and is inversely proportional to the distance from the charge. That is, farther from the source charge will be subject to the same strength of force regardless of whether it is above, below, or to the side of the source, as long as the distance is the same. A surface of the same radius all around the source will have the same field strength. This is called a surface of equipotential. For a point source charge, the surface of equipotential is a sphere, and the force F will push a positive charge radially outward. A test charge of **mass** m and positive charge q will feel a push away from the positive source charge with an **acceleration** a directly proportional to (æ) the field strength and inversely proportional to the mass of the test charge. The equation is written: $F = ma$ æ q/d^2, so that a æ $q/(md^2)$ where $F = ma$ is given by Newton's second law, which tells us that force causes acceleration. The letter d represents the distance from the source charge to the test charge.

If a charge does not move because it is acted upon by the electromagnetic force equally from all directions, it is in a position of stable **equilibrium**.

Now let us consider the field from two charges, one positive and one negative, a distance d apart. We call this combination of charges a dipole. Remember, opposite charges attract, so this is not an unusual situation. A hydrogen **atom**, for example, consisting of an **electron** (negative charge) and a proton (positive charge) is a very small dipole, as these particles do not sit right on top of each other. According to the superposition principle mentioned above, we can just add the fields from each individual charge and get a rather complicated field. If we only consider the field at a position very far from the dipole, we can simplify the field equation so that the field is proportional to the product of the charge value and the separation of the two charges. There is also dependence on the distance along the dipole axis as well as radial distance from the axis.

Next, we consider the field due to a group of positive charges evenly distributed along an infinite straight line, defined to be infinite because we want to neglect the effect of the endpoints as an unnecessary complication here. Just as the field of a point charge is directed radially outward in a sphere, the field of a line of charge is directed radially outward, but at any specific radius the surface of equipotential will be a cylinder.

Recall that the force of a stationary charge is $F\rightarrow = qE\rightarrow$, but if the charge is moving the force is $F\rightarrow = qE\rightarrow + qv\rightarrow \times B\rightarrow$. A steady (unchanging in time) current in a wire generates a magnetic field. **Electric current** is essentially charges in motion. In an electrical conductor like copper wire, electrons move, while positive charges remain steady. The positive charge cancels the **electric charge** so the overall charge looks like zero when viewed from outside the wire, so no electric field will exist outside the wire, but the moving charges create a magnetic field from $F\rightarrow = qv\rightarrow \times B\rightarrow$ where $B\rightarrow$ is the magnetic field vector. The cross product results in magnetic field lines circling the wire. Because of this effect, solenoids (current-carrying coils of wire that act as magnets) can be made by wrapping wire in a tight spiral around a metallic tube, so that the magnetic field inside the tube is linear in direction.

Relating this idea to Newton's first law of motion, which states that for every action there is an equal and opposite reaction, we see that an external magnetic field (from a bar magnet, for example) can exert a force on a current-carrying wire, which will be the sum of the forces on all the individual moving charges in the wire.

A simple example of a combination of electric and magnetic fields is the field from a single point charge, say a proton, traveling through **space** at a constant speed in a straight line. In this case, the field vectors pointing radially outward would have to be added to the spiral magnetic field lines (circles extend into spirals because an individual charge is moving) to get the total field caused by the charge.

A description of the field from a current that changes in time is much more complicated, but is calculable owing to **James Clerk Maxwell**. His equations, which have unified the laws of electricity and magnetism, are called **Maxwell's equations**. They are differential equations that completely describe the combined effects of electricity and magnetism, and are considered to be one of the crowning achievements of the nineteenth century. Maxwell's formulation of the **theory of** electromagnetic **radiation** allows us to understand the entire **electromagnetic spectrum**, from **radio** waves through visible **light** to **gamma rays**.

ELECTROMAGNETIC INDUCTION

Electromagnetic induction is the production of an **electromotive force** that results from the movement of a conductor through a magnetic field. An electromotive **force** also results from the changing of **magnetic flux** in a closed loop circuit.

Magnetic fields and electric fields are both components of an electromagnetic wave. In such a wave the electrical and magnetic fields potentiate each other.

Within most electrical circuits, an electromotive force, or EMF, is produced by a source of potential difference within the circuit. The source of potential difference is usually some form of battery or capacitor. Electromotive force drives current flow between areas of differing voltages in a conducting material. The movement of charges or current flow is

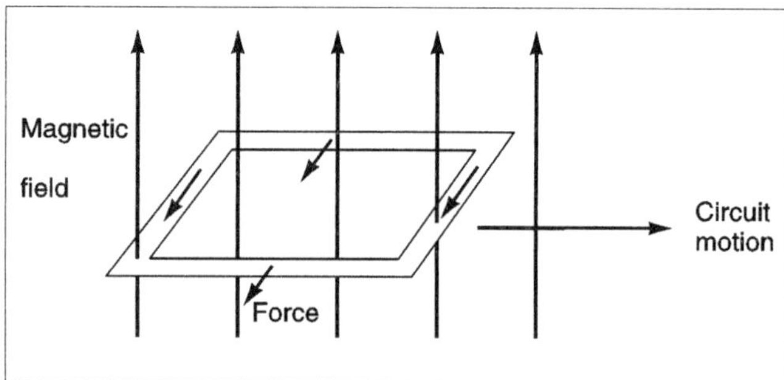

Electromagnetic induction.

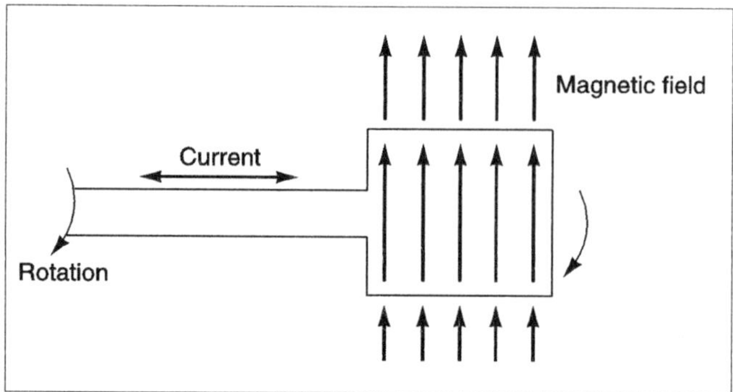

Electromagnetic induction.

directly proportional to the EMF and inversely related to the resistance of the conducting material.

An EMF that causes current within a circuit can, however, be created from an outside source as well. In its most usual form, electromagnetic induction refers to the induction of an electromotive force within a conductor by an external magnetic field. A current that is produced in a circuit (for example) without a battery or other source of electrical potential is called an induced current.

The electromagnetic induction of current by a magnetic field occurs only when the magnetic field is changing. Shifting magnitude or direction of a magnetic field is called magnetic flux change. A current is induced within a conductor existing inside of a magnetic field when magnetic flux changes with time. This principle, that the induced electromotive force through a circuit is equal to the rate of change of magnetic flux through it, is known as Faraday's law of induction, first demonstrated by English physicist and chemist **Michael Faraday**.

To illustrate, consider a copper wire looped to form a coil with both ends connected to a galvanometer, a device that measures electrical current. The resulting system then forms a circuit. Without a source of EMF within the circuit (such as a battery), no current is measured by the galvanometer. Suppose, then, that a strong bar magnet is placed near the copper wire coil. If the magnet is held motionless, no current registers. As soon as the magnet is moved, however, current is indicated on the galvanometer. Electromagnetic induction of a current in the circuit occurs only as the magnet flux changes. The current that is produced within the conductor is proportional to the force applied to the magnet and the strength of the magnetic field. Interestingly, the very same phenomena is observed if the magnetic field is held stationary and the conductor is moved through the magnetic field. Here, magnetic flux still changes relative to the conductor, thus Faraday's law of induction still applies and an induced current results.

Electromagnetic induction is the key phenomenon behind **energy** converting devices such as electric motors, **generators**, and **transformers**. Electrical motors are devices designed to convert electrical energy into mechanical energy. Electric motors utilize electromagnetic induction principles to induce movement from current flowing within a magnetic field. Electric motors are widely used in tools and appliances.

Electromagnetic induction is also used to produce **electricity** in electric generators. Here, an outside force is used to create current by moving a magnet, thus supplying a changing magnetic flux. For example, hydroelectric generators use the force of moving water to create large amounts of electromagnetically induced current. Similarly, steam generators use the force provided by steam to rotate turbines that provide changing magnetic flux to produce induced current. The steam is produced by the combustion of fossil fuels like gas or coal, or by controlled **nuclear reactions** in nuclear **power** plants.

See also Electric field and forces; Electric motors and meters; Electric power and energy; Electricity; Electrodynamics; Electromagnetic field; Faraday's laws; Induction motor; Inductors and inductance; Magnetic fields and forces

ELECTROMAGNETIC SPECTRUM

The electromagnetic spectrum encompasses a continuous range of frequencies or wavelengths of electromagnetic **radiation**, ranging from long **wavelength**, low energy radio **waves** to short wavelength, high **frequency**, high **energy** gamma rays. The electromagnetic spectrum is traditionally divided into regions of **radio** waves, **microwaves**, infrared radiation, visible **light**, ultraviolet rays, **x rays**, and **gamma rays**.

Scottish physicist **James Clerk Maxwell**'s development of a set of equations that accurately described electromagnetic phenomena allowed the mathematical and theoretical unification of electrical and magnetic phenomena. When Maxwell's calculated **speed of light** fit well with experimental determinations of the speed of light, Maxwell and other physicists realized that visible light should be a part of a broader electromagnetic spectrum containing forms of electromagnetic radiation that varied from visible light only in terms of wavelength and wave frequency. Frequency is defined as the number of wave cycles that pass a particular point per unit time; it is commonly measured in Hertz (cycles per second). Wavelength defines the distance between adjacent points of the electromagnetic wave that are in equal phase (e.g., wavecrests).

Exploration of the electromagnetic spectrum quickly resulted in practical advances. German physicist **Heinrich Rudolf Hertz** regarded **Maxwell's equations** as a path to a

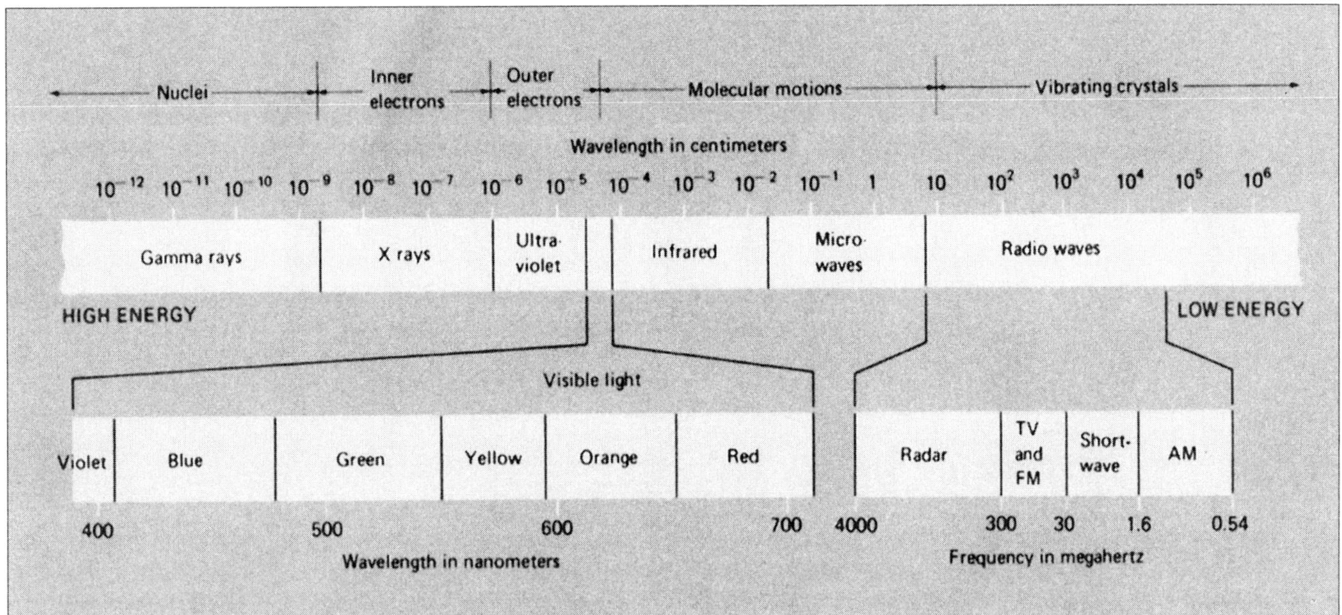

Spectrum of electromagnetic radiation (chart). *(Image by Robert L. Wolke. Reproduced by permission.)*

"kingdom" or "great domain" of electromagnetic waves. Based on this insight, in 1888 Hertz demonstrated the existence of radio waves. A decade later **Wilhelm Röntgen**'s discovery of high energy electromagnetic radiation in the form of x rays quickly found practical medical use.

At the beginning of the twentieth century, German physicist **Max Planck** proposed that atoms absorb or emit electromagnetic radiation only in certain bundles, termed **quanta**. In his work on the **photoelectric effect**, German-born American physicist **Albert Einstein** used the term **photon** to describe this electromagnetic quanta. Planck determined that energy of light was proportional to its frequency (i.e., as the frequency of light increases, so does the energy of the light). **Planck's constant** ($h = 6.626 \times 10^{-34}$ joule-second in the meter-kilogram-second system) relates the energy of a photon to the frequency of the electromagnetic wave and allows a precise calculation of the energy of electromagnetic radiation in all portions of the electromagnetic spectrum.

Although electromagnetic radiation is now understood as having both photon (particle) and wave-like properties, descriptions of the electromagnetic spectrum generally utilize traditional wave related terminology (i.e., frequency and wavelength).

Electromagnetic fields and photons exert forces that can excite electrons. As electrons transition between allowed **orbitals**, energy must be conserved. This conservation is achieved by the emission of photons when an **electron** moves from a higher potential orbital energy to a lower potential orbital energy. Accordingly, light is emitted only at certain frequencies characteristic of every **atom** and molecule. Correspondingly, atoms and molecules absorb only a limited range of frequencies and wavelengths of the electromagnetic spectrum, and reflect all the other frequencies and wavelengths of light. These reflected frequencies and wavelengths are often the actual observed light or colors associated with an object.

The region of the electromagnetic spectrum that contains light at frequencies and wavelengths that stimulate the rod and cones in the human eye is termed the visible region of the electromagnetic spectrum. **Color** is the association the eye makes with selected portions of that visible region (i.e., particular colors are associated with specific wavelengths of visible light). A nanometer (10^{-9} m) is the most common unit used for characterizing the wavelength of visible light. Using this unit, the visible portion of the electromagnetic spectrum is located between 380nm-750nm and the component color regions of the visible spectrum are Red (670-770 nm), Orange (592-620 nm), Yellow (578-592 nm), Green (500-578 nm), Blue (464-500 nm), Indigo (444-464 nm), and Violet (400-446 nm). The color regions of the visible spectrum can easily be remembered by use of the acronym ROYGBIV. Because the energy of electromagnetic radiation (i.e., the photon) is inversely proportional to the wavelength, red light (longest in wavelength) is the lowest in energy. As wavelengths contract toward the blue end of the visible region of the electromagnetic spectrum, the frequencies and energies of colors steadily increase.

Like colors in the visible spectrum, other regions in the electromagnetic spectrum have distinct and important components. Radio waves, with wavelengths that range from hundreds of meters to less than a centimeter, transmit radio and television signals. Within the radio band FM radio waves have a shorter wavelength and higher frequency than AM radio waves. Still higher frequency radio waves with wavelengths of a few centimeters can be utilized for **radar** imaging.

Microwaves range from approximately a foot in length to the thickness of a piece of paper. The atoms in food placed

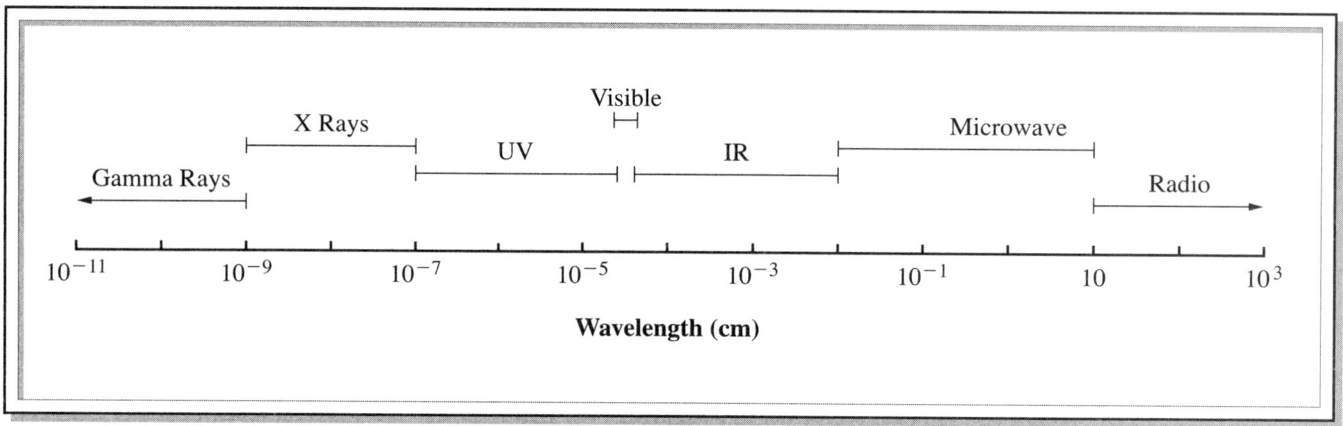

Another representation of electromagnetic radiation. *(Image by Robert L. Wolke. Reproduced by permission.)*

in a microwave oven become agitated (heated) by exposure to microwave radiation. Infrared radiation comprises the region of the electromagnetic spectrum where the wavelength of light is measured region from one millimeter (in wavelength) down to 400 nm. Infrared waves are discernible to humans as **thermal radiation** (**heat**). Just above the visible spectrum in terms of higher energy, higher frequency, and shorter wavelengths is the ultraviolet region of the spectrum with light ranging in wavelength from 400 to 10 billionths of a meter. Ultraviolet radiation is a common cause of sunburn even when visible light is obscured or blocked by clouds. X rays are a highly energetic region of electromagnetic radiation with wavelengths ranging from about 10 billionths of a meter to 10 trillionths of a meter. The ability of x rays to penetrate skin and substances makes them useful in both medical and industrial radiography. Gamma rays, the most energetic form of electromagnetic radiation, are comprised of light with wavelengths of less than about 10 trillionths of a meter and include waves with wavelengths smaller than the radius of an atomic **nucleus** (10^{-15} m). Gamma rays are generated by **nuclear reactions** (e.g., **radioactive decay** and nuclear explosions).

Cosmic rays are not a part of the electromagnetic spectrum. Cosmic rays are not a form of electromagnetic radiation but are actually high-energy charged particles with energies similar to, or higher than, observed gamma electromagnetic radiation energies.

See also Bohr theory; Electromagnetic radiation; Spectroscopy

ELECTROMAGNETISM

Electromagnetism is a branch of physical science that involves all the phenomena in which **electricity** and **magnetism** interact. This field is especially important to **electronics** because a magnetic field is created by an **electric current**. The rules of electromagnetism are responsible for the way charged particles of atoms interact.

Some of the rules of electrostatics, the study of electric charges at rest, were first noted by the ancient Romans who observed the way a brushed comb would attract particles. It is now known that electric charges occur in two different types, called positive and negative. Like types repel each other, and differing types attract.

The **force** that attract positive charges to negative charges weakens with distance, but is intrinsically very strong. The fact that unlike types attract means that most of this force is normally neutralized and not seen in full strength. The negative charge is generally carried by the atom's electrons, while the positive resides with the protons inside the atomic **nucleus**. There are other less well known particles that can also carry charge. When the electrons of a material are not tightly bound to the atom's nucleus, they can move from **atom** to atom and the substance, called a conductor, can conduct electricity. On the contrary, when the **electron** binding is strong, the material is called an insulator.

When electrons are weakly bound to the atomic nucleus, the result is a **semiconductor**, often used in the electronics industry. It was not initially known if the electric current carriers were positive or negative, and this initial ignorance gave rise to the convention that current flows from the positive terminal to the negative. In reality we now know that the electrons actually run from the negative to the positive.

Electromagnetism is the theory of a unified expression of an underlying force, the so-called electromagnetic force. This is seen in the movement of **electric charge**, which gives rise to magnetism (the electric current in a wire being found to deflect a compass needle), and it was a Scotsman, **James Clerk Maxwell**, who in 1865 published the theory unifying electricity and magnetism. The theory arose from former specialized work by **Carl Friedrich Gauss, Charles-Augustin de Coulomb, André-Marie Ampère, Michael Faraday, Benjamin Franklin,** and **Georg Simon Ohm**. However, one factor that did not contradict the experiments was added to the equations by Maxwell so as to ensure the conservation of charge. This was done on the theoretical grounds that charge should be a conserved quantity, and this addition led to the prediction of a **wave phenomena** with a certain anticipated **velocity**. **Light**, which has the expect-

Region	Wavelength (Angstroms)	Wavelength (centimeters)	Frequency (Hz)	Energy (eV)
Radio	$>10^9$	>10	$< 3 \times 10^9$	$<10^{-5}$
Microwave	$10^9 - 10^6$	$10 - 0.01$	$3 \times 10^9 - 3 \times 10^{12}$	$10^{-5} - 0.01$
Infrared	$10^6 - 7000$	$0.01 - 7 \times 10^{-5}$	$3 \times 10^{12} - 4.3 \times 10^{14}$	$0.01 - 2$
Visible	$7000 - 4000$	$7 \times 10^{-5} - 4 \times 10^{-5}$	$4.3 \times 10^{14} - 7.5 \times 10^{14}$	$2 - 3$
Ultraviolet	$4000 - 10$	$4 \times 10^{-5} - 10^{-7}$	$7.5 \times 10^{14} - 3 \times 10^{17}$	$3 - 10^3$
X-Rays	$10 - 0.1$	$10^{-7} - 10^{-9}$	$3 \times 10^{17} - 3 \times 10^{19}$	$10^3 - 10^5$
Gamma Rays	< 0.1	$< 10^{-9}$	$>3 \times 10^{19}$	$>10^5$

Further break-down of the electromagnetic spectrum. *(Image by Robert L. Wolke. Reproduced by permission.)*

ed velocity, was found to be an example of this electromagnetic **radiation**.

Light had formerly been thought of as consisting of particles (photons) by **Isaac Newton**, but the theory of light as particles was unable to explain the wave nature of light (**diffraction** and the alike). In reality, light displays both wave *and* particle properties. The resolution to this duality lies in **quantum theory**, where light is neither particles or wave, but both. It propagates as a wave without the need of a media and interacts in the manner of a particle. This is the basic nature of quantum theory.

Classical electromagnetism, useful as it is, contains contradictions (acausality) that make it incomplete and drive one to consider its extension to the area of **quantum physics**, where electromagnetism, of all the fundamental forces of nature, is perhaps the best understood.

There is much symmetry between electricity and magnetism. It is possible for electricity to give rise to magnetism, and symmetrically for magnetism to give rise to electricity (as in the exchanges within an electric transformer). It is an exchange of just this kind that constitutes electromagnetic **waves**. These waves, although they don't need a medium of propagation, are slowed when traveling through a transparent substance.

Electromagnetic waves differ from each other only in amplitude, **frequency,** and orientation (polarization). Laser beams are particular in being very coherent, that is, the radiation is of one frequency, and the waves coordinated in **motion** and direction. This permits a highly concentrated beam that is used not only for its cutting abilities, but also in electronic data storage, such as in CD-ROMs.

The differing frequency forms are given a variety of names, from **radio** waves at very low frequencies through light itself, to the high frequency x and **gamma rays**.

Many a miracle depends upon the broad span of the **electromagnetic spectrum**. The ability to communicate across long distances despite intervening obstacles, such as the walls of buildings, is possible using the radio and television frequencies. **X rays** can see into the human body without opening it. These things, which would once have been labeled magic, are now ordinary ways we use the electromagnetic spectrum.

The unification of electricity and magnetism has led to a deeper understanding of physical science, and much effort has been put into further unifying the four forces of nature. The remaining known forces are the so-called weak, strong, and gravitational forces. The weak force has now been unified with electromagnetism, called the **electroweak force**. There are proposals to include the strong force in a **grand unified theory**, but the inclusion of **gravity** remains an open problem.

Maxwell's theory is in fact in contradiction with Newtonian **mechanics**, and in trying to find the resolution to this conflict, **Albert Einstein** was led to his theory of special relativity. **Maxwell's equations** withstood the conflict, but it was Newtonian mechanics that were corrected by relativistic mechanics. These corrections are most necessary at velocities close to the **speed of light**. The many strange predictions about **space** and **time** that follow from special relativity are found to be a part of the real world.

Paradoxically, magnetism is a counterexample to the frequent claims that relativistic effects are not noticeable for low velocities. The moving charges that compose an electric current in a wire might typically only be traveling at several feet per second (walking speed), and the resulting Lorentz contraction of special relativity is indeed minute. However, the electrostatic forces at balance in the wire are of such great magnitude that this small contraction of the moving (negative) charges exposes a residue force of real world magnitude, namely the magnetic force. It is in exactly this way that the magnetic force derives from the electric. Special relativity is indeed hidden in Maxwell's equations, which were known before special relativity was understood or separately formulated by Einstein.

Before the advent of technology, electromagnetism was perhaps most strongly experienced in the form of lightning, and electromagnetic radiation in the form of light. Ancient man kindled fires that he thought were kept alive in trees struck by lightning.

Much of the magic of nature has been put to work by man, but not always for his betterment or that of his surroundings. Electricity at high voltages can carry **energy** across extended distances with little loss. Magnetism derived from that electricity can then **power** vast motors. But electromagnetism can also be employed in a more delicate fashion as a means of communication, either with wires (as in the telephone), or without them (as in radio communication). It also drives our electronics devices (as in **computers**).

Magnetism has long been employed for navigation in the compass. This works because Earth is itself a huge magnet, thought to have arisen from the great **heat**-driven convection currents of molten iron in its center. In fact, it is known that Earth's **magnetic poles** have exchanged positions in the past.

See also Relativity, special

ELECTROMOTIVE FORCE

In an **electric circuit**, electromotive **force** is the **work** done by a source on an electrical charge. Because it is not really a force, the term is actually a misnomer; it is more commonly referred to by the initials EMF. EMF is another term for electrical potential, or the difference in charge across a battery or **voltage** source. For a circuit with no current flowing, the potential difference is called EMF.

Electrical sources that convert **energy** from another form are called seats of EMF. In the case of a complete circuit, such a source performs work on electrical charges, pushing them around the circuit. At the seat of EMF, charges are moved from low electrical potential to higher electrical potential.

Water flowing downhill in a flume is a good analogy for charges in an electric circuit. The water starts at the top of the hill with a certain amount of **potential energy**, just as charges in a circuit start with high electrical potential at the battery. As the water begins to flow downhill, its potential energy drops, just as the electrical potential of charges drops as they travel through the circuit. At the bottom of the hill, the potential energy is minimum, and work must be performed to pump it to the top of the hill to travel through the flume again.

Similarly, in an electrical circuit, the seat of EMF performs work on the charges to bring them to a higher potential after their trip through the circuit.

ELECTRON

An electron is one of the three **subatomic particles** that make up an **atom**. The other two subatomic particles are protons and neutrons, which are found in the **nucleus** (center) of the atom. Electrons are found in the area outside the nucleus of an atom. They have a negative electrical charge (as opposed to protons, which are positively charged, and neutrons, which have no charge) and are quite small. An electron is only about 1/2,000 the size of a **proton** or a **neutron**. Protons and neutrons have masses of approximately 1 atomic **mass** unit (amu) each, whereas electrons only have a mass of .0006 amu (9.11 x 10^{-28}g). The atomic number of an element is equal to the number of protons in one atom of the element. In a neutral atom, the number of electrons is equal to the number of protons, therefore, the atomic number also indicates the number of electrons in an atom. An atom can gain or lose electrons at which point it gains an **electric charge** and is called an ion.

Exactly where the electrons in an atom are located has been the topic of much research and debate. Until the early 1800s the idea that elements were made up of smaller particles called atoms was unknown. English chemist **John Dalton** performed various experiments that led him to develop his **atomic theory**. This theory stated that all elements are made up of atoms. According to Dalton, atoms were the smallest particles possible and could not be divided into smaller particles. Atoms of the same element were exactly alike, and atoms of different atoms were different. Atoms of different elements combined to form compounds. Protons, neutrons, and electrons were still undiscovered.

In 1897, English physicist **Joseph John Thomson** was experimenting with passing an **electric current** through a gas. The gas was made of uncharged atoms, but when an electric current passed through it, negatively charged particles in the form of rays were given off. Thomson was faced with the problem of explaining where these negatively charged particles originated. He reasoned that the only place these particles could have come from was the individual atoms in the gas. Therefore, the atom must be made up of even smaller particles. He used the term *corpuscles* to describe the negatively charged particles that we now call electrons.

Now that Thomson had identified electrons, the problem he faced was determining the placement of these particles inside the atom. Thomson developed a model of electrons being scattered throughout a positively charged material, much like plums would be scattered throughout plum pudding (this model of the atom is often referred to as the "plum pudding" model). In 1911, British physicist **Ernest Rutherford** proposed that there are positively charged particles in an atom called protons. The protons are located in the nucleus of an atom, and the electrons are scattered outside of the nucleus around the edge of the atom. The negatively charged electrons are held in place by the **force** that results from the attraction toward the positively charged protons. This force is called the electromagnetic force. This model explained further the idea of subatomic particles, but still could not describe the exact location of the electrons in an atom.

A few years later, in 1913, Danish physicist **Niels Bohr** improved upon the Rutherford model. He developed the idea of **energy** levels in an electron. According to this new atomic model, the electrons were placed in definite orbits—called energy levels—around the nucleus. Each **energy level** is a certain distance away from the nucleus. This model likens electrons orbiting the nucleus to **planets** orbiting the **Sun**. Today, scientists realize that this is not exactly correct. Electrons do not move in a definite orbit, always a certain distance from the nucleus. In fact, the exact location of the electrons in an atom cannot be determined. Only the approximate location where an electron is likely to be found can be predicted.

The electrons in an atom take up a particular amount of **space** as they move around the nucleus at a rate of billions of time per second. This space is referred to as the **electron cloud**. The electron cloud is not a definite shape or size, rather it is simply the space where electrons are likely to be located. Each electron appears to stay in a certain area within the cloud. The area of the cloud in which any particular electron is likely to be found depends upon how much energy the electron has. This area is an energy level within the electron cloud. Electrons with low energies are in the lowest energy level, closest to the nucleus. Electrons with higher energies are in higher energy levels, increasing in distance from the nucleus as energy increases. Each energy level can hold a certain number of electrons. The lowest energy level can hold only two electrons; the second level, eight; and the third level, 18.

How the energy levels are filled in an atom determines the properties of the different elements. This electron arrangement is important in predicting the chemical properties of an element. One of the most important properties determined by electron arrangement is the bonding characteristics of an element. Some elements bond readily to other elements and others hardly bond at all. This is determined by the number of electrons—called valence electrons—present in the outermost energy shell. An atom is most stable when its valence electrons satisfy what is called the octet rule. This means that an atom with eight electrons in its outermost energy level is very stable. An atom will bond with another in order to satisfy the octet rule. When two atoms combine, they can gain or lose electrons. When one atom transfers electrons to another, it is called ionic bonding. When electrons are shared between two atoms, it is called covalent bonding.

The number of valence electrons of an atom can also be used to group elements together into families on the **periodic table**. Elements within a family share many properties. Metals have one, two, three, or four valence electrons. These electrons are in the outer energy level quite weakly, so they are lost easily. This is the reason metals react readily with water or other atmospheric elements in a reaction known as corrosion. Elements that are nonmetals have five, six, seven, or eight valence electrons. Nonmetals tend to gain electrons during chemical reactions. Nonmetals with eight valence electrons are chemically unreactive.

The first family on the periodic table is the alkali metals. Atoms of these elements have only one valence electron, so they tend to bond easily. In fact, they are so reactive that they are rarely found in nature not combined with other elements. Family 2 on the periodic table is the alkaline earth metals with two valence electrons. They lose these electrons easily. Since they have two electrons that they must lose, they are not quite as reactive as the alkali metals. The transition elements have one or two valence electrons, which they lose when they react. They can also lose an electron from the energy level that is next to the highest. Atoms of these elements can also share electrons when they react with other elements. Family 13, the boron family, has three valence electrons. Family 14 is known as **carbon** family, and atoms of these elements have four valence electrons. Family 15, the nitrogen family, has five valence electrons. Atoms of these elements tend to gain or acquire electrons when they react with other atoms. Family 16 is called the **oxygen** family. Atoms of these elements have six valence electrons and tend to gain or acquire electrons in chemical reactions. Members of the halogen family, family 17, have seven valence electrons. A halogen atom needs only to gain one electron to fill its outer electron shell. This property makes elements in family 17 the most active nonmetals. These elements are also rarely found uncombined in nature. Halogens tend to react with the alkali metals rather easily. The last family, family 18, contains the noble gases. These are the elements with eight valence electrons, or complete energy levels. These elements are unreactive.

Electrons not only determine an element's reactivity, they also are the cause of various phenomena, for example, **beta radiation**. Some electrons can be formed inside of the nucleus of an atom when a neutron breaks apart. When this occurs, the electron shoots out of the atom and is called a beta particle, a type of **radioactivity**. Another example of an electron-caused phenomenon is static **electricity**. Sometimes the electrons in the outermost energy levels are easily lost from an atom. When an atom loses an electron, it becomes positively charged. When it gains an electron, it becomes negatively charged. An object can also become charged if the atoms of which it is composed gain or lose electrons. An example of an entire object becoming charged is when a balloon is rubbed with a piece of cloth. The cloth loses electrons that are transferred to the balloon, and the balloon acquires a negative charge. When the balloon is held up to a wall, the negative charge causes the electrons in the wall to move away from the area. That area of the wall then becomes positively charged. The negative balloon is attracted to the positive wall, and will be held on the wall by the electromagnetic force.

This example illustrates the phenomenon of induction—an electrical charge built up due to the rearrangement of atoms. The wall became charged because of induction. When electrons flow from one object to another it is called conduction. Certain materials—called conductors—allow for this electron flow better than others. Metals are good conductors because the electrons in these atoms are loosely held and free to move.

Materials that do not allow for electrons to flow are called insulators. Insulators do not conduct electricity well because the electrons are tightly held and are not free to move.

Another well-known phenomenon that occurs because of electrons is lightning. Clouds can build up a negative charge, and if they pass near the surface of Earth, Earth can become positively charged because of induction. Electrons then are attracted to the positively charged Earth and jump from the cloud. As they move through the air, they produce a great quantity of **light** and **heat**. The light is the lightning bolt we see. The heat causes the air around it to expand quickly, which is the thunder we hear.

Many electronic devices and equipment are made possible because of electrons. Televisions, computer monitors, and video games all use a device called a cathode-ray tube. The cathode-ray tube is a **vacuum tube** that emits a beam of electrons onto a screen. As the electrons hit the screen, which is coated with a fluorescent material, they cause the screen to glow, producing an image. Photocells in cameras are based on a phenomenon called the **photoelectric effect**. When light shines on a metal, the loosely held electrons are shot off of the surface of the metal. For any particular metal, a particular **frequency** of light must be used for the photoelectric effect to take place.

ELECTRON CLOUD

The term **electron** cloud is used to describe the area around an atomic **nucleus** where an electron will probably be. It is also described as the "fuzzy" orbit of an atomic electron.

An electron bound to the nucleus of an **atom** is often thought of as orbiting the nucleus in much the same manner that a planet orbits a **sun**, but this is not a valid visualization. An electron is not bound by **gravity**, but by the Coulomb **force**, whose direction depends on the sign of the particles' charge. (Remember, opposites attract, so the negative electron is attracted to the positive **proton** in the nucleus.) Although both the Coulomb force and the **gravitational force** depend inversely on the square of the distance between the objects of interest, and both are central forces, there are important differences. In the classical picture, an accelerating charged particle, like the electron (a circling body changes direction, so it is always accelerating) should radiate and lose **energy**, and therefore spiral in towards the nucleus of an atom...but it does not.

Since we are discussing a very small (microscopic) system, an electron must be described using quantum mechanical rules rather than the classical rules that govern planetary **motion**. According to quantum **mechanics**, an electron can be a **wave** or a particle, depending on what kind of measurement one makes. Because of its wave nature, one can never predict where in its orbit around the nucleus an electron will be found. One can only calculate whether there is a high probability that it will be located at certain points when a measurement is made.

The electron is therefore described in terms of its probability distribution or probability **density**. This probability distribution doesn't have definite cutoff points; its edges are somewhat fuzzy. Hence the term "electron cloud." This

cloudy probability distribution takes on different shapes, depending on the state of the atom. At room **temperature**, most atoms exist in their lowest energy state or "ground" state. If energy is added—by shooting a laser at it, for example—the outer electrons can "jump" to a higher state (think larger orbit, if it helps). According to quantum mechanical rules, there are only certain specific states to which an electron can jump. These discrete states are labeled by *quantum numbers*. The letters designating the basic **quantum numbers** are n, l, and m, where n is the principal or energy quantum number, l relates to the orbital **angular momentum** of the electron, and m is a magnetic quantum number. The principal quantum number n can take integer values from 1 to **infinity**. For the same electron, l can be any integer from 0 to (n - 1), and m can have any integer value from $-l$ to l. For example, if $n = 3$, we can have states with $l = 2$, 1, or 0. For the state with $n = 3$ and $l = 2$, we could have $m = -2$, -1, 0, 1, or 2.

Each set of n, l, m quantum numbers describes a different probability distribution for the electron. A larger n means the electron is most likely to be found farther from the nucleus. For $n = 1$, l and m must be 0, and the electron cloud is spherical about the nucleus. For $n = 2$, $l = 0$, there are two concentric spherical shells of probability about the nucleus. For $n = 2$, $l = 1$, the cloud is more barbell-shaped. We can even have a daisy shape when $l = 3$. The distributions can become quite complicated.

Experiment has verified these distributions for one-electron atoms, but the **wave function** computations can be very difficult for atoms with more than one electron in their outer shell. In fact, when the motion of more than one electron is taken into account, it can take days for the largest computer to output probability distributions for even a low-lying state, and simplifying approximations must often be made.

Overall, however, the quantum mechanical wave equation, as developed by **Erwin Schrödinger** in 1926, gives an excellent description of how the microscopic world is observed to behave, and we must admit that while **quantum mechanics** may not be precise, it is accurate.

ELECTRON DIFFRACTION

When a beam of electrons passes through a material, an electron-sensitive film placed beyond the sample will show, when developed, a pattern of concentric circles around the central intense **electron** beam. This phenomenon is due to the **diffraction** of the electron beam by the material through which it passes. **Clinton Davisson**, **Lester Halbert Germer**, and George P. Thomson (1892-1975) used this method to prove the hypothesis of **Louis de Broglie** that electrons have wave characteristics. Further research on electron diffraction showed that the angles of the diffracted electrons and the intensity of the diffracted electron beams can be used to determine the structure of the molecules of the gas.

Diffraction is a phenomenon that is characteristic of all types of **waves**. Perhaps the simplest case to understand is the diffraction of monochromatic **light**. Whenever an advancing wavefront of light encounters a barrier in which there is a hole, this hole acts as if it were a point source of light, with **radiation** of the same **wavelength** moving beyond the barrier away from the hole. When the barrier contains a series of holes, then each hole acts as a point source. The radiation observed beyond the barrier will be the sum of the radiation from all of these individual point sources. If the holes are uniformly placed at distances similar to the wavelength of the light, the waves from all these individual point sources will add together to produce beams of light only in certain definite directions. This is caused by constructive **interference** and results from the circumstance that the waves from all the point sources are in phase only in these directions. In other directions, destructive interference occurs among the waves, and they tend to cancel each other out. This phenomenon is called diffraction.

If, instead of holes, a series of parallel narrow slits are cut in the barrier at uniform distances similar in magnitude to the wavelength of the impending wave, the result is a diffraction grating. A photographic film placed a short distance beyond the grating will show a series of lines when exposed. The direction of each line on the film from the diffraction grating is given by an equation that relates the sine of the angle of the diffracted line with an integral multiple of the wavelength divided by the distance between the slits. **Joseph von Fraunhofer** invented the diffraction grating. The results of his experiments on the diffraction of light played a large part in convincing the scientific community that light was made up of waves.

In 1859, Julius Plücker (1801-1868) discovered that when a high **voltage** difference is established between two electrodes in a container of dilute gas or in a **vacuum**, an electrical discharge occurs between the electrodes and cathode rays are produced. In 1897, **Joseph John Thomson** showed that cathode rays are made up of electrons. In subsequent experiments, the **mass** of the electron was shown to be 9.1×10^{-28} gram with an electrical charge of 1.6×10^{-19} coulomb.

Max von Laue recognized that the regularly spaced planes of atoms and molecules in crystals resemble a diffraction grating. In 1912, he demonstrated that **x rays**, with wavelengths in the range of 10^{-8} cm, comparable to the magnitude of crystal spacings, was diffracted when passed through crystalline substances.

In 1923, de Broglie proposed that moving particles possess some of the properties commonly associated with waves. He predicted that the wavelength of such a particle wave would be equal to **Planck's constant** (6.6×10^{-27} erg-second) divided by the product of the particle's mass and the **velocity** with which the particle is moving (the particle's **momentum**). For a "particle" with a mass as large as a baseball, the wavelength would be very small indeed, and could not be detected. For a particle whose mass is of similar magnitude to Planck's constant, such as an electron, the wavelength should be large enough for the effects of its wave nature to be observed.

The wavelength that de Broglie predicted for electrons is determined by their velocity and can be adjusted to the same order of magnitude as atomic separations in molecules and crystals by changing the voltage of the charged electrodes

between which the electrons move. Davisson, Germer, and Thomson employed this approach when they proved the wave nature of electrons. They used von Laue's observation that the regular spaced atoms in crystals act as a diffraction grating for wavelengths of the order of magnitude of 10^{-8} cm. Diffraction of a beam of electrons by a crystal of nickel was observed when the voltage of the electrodes was adjusted to produce electron beams with the appropriate wavelength.

Von Laue showed that x rays are diffracted in specific directions determined by the distances of separation between the parallel planes of molecules in crystals. X-ray crystallographers subsequently demonstrated that the intensities of these diffracted beams were determined by the arrangement of the atoms within the molecules themselves.

A similar phenomenon is observed when electron beams pass through gases. A diffraction pattern of concentric circles is obtained from an exposed electron-sensitive film placed beyond the diffracting gas. A detailed analysis of this **scattering**, begun by R. Wierl in his pioneering research and expanded by other researchers, led to an equation (known as the Wierl equation) that relates the intensity of the scattered beam at a particular angle to the interatomic distances within the molecules. Refinements developed by Linus Pauling, L. O. Brockway, and others have resulted in techniques that have been useful in determining molecular structures with accurate values of interatomic distances and angles.

Further advances in theory, technique, and instrumentation have led to the development of electron microscopes that are used to examine the fine structure of biological samples.

ELECTRON MICROSCOPE

Described by the Nobel Society as "one of the most important inventions of the century," the electron microscope is a valuable and versatile research tool. The first working models were constructed by German engineers Ernst Ruska and Max Knoll in 1932, and since that time, the electron microscope has found numerous applications in **chemistry**, engineering, and medicine.

At the turn of the twentieth century, the science of microscopy had reached an impasse: because all microscopes relied upon visible **light**, even the most powerful could not detect an image smaller than one **wavelength** of light. This was tremendously frustrating for physicists, who were anxious to study the structure of **matter** on an atomic level. Around this time, French scientist **Louis de Broglie** theorized that **subatomic particles** sometimes act like **waves**, but with much shorter wavelengths. Ruska, then a student at the University of Berlin, wondered why a microscope couldn't be designed that was similar in function to a normal **microscope** but used a beam of electrons instead of a beam of light. Such a microscope could resolve images thousands of times smaller than a wavelength.

There was one major obstacle to Ruska's plan, however. In a **compound microscope**, a series of **lenses** are used to focus, magnify, and refocus the image. In order for an electron-based instrument to perform as a microscope, some device was required to focus the electron beam. Ruska knew that electrons could be manipulated within a magnetic field, and in the late 1920s, he designed a magnetic coil that acted as an electron lens. With this breakthrough, Ruska and Knoll constructed their first electron microscope. Though the prototype model was capable of magnification of only a few hundred power (about that of an average laboratory microscope), it proved that electrons could indeed be used in microscopy.

The microscope built by Ruska and Knoll is very similar in principle to a compound microscope. A beam of electrons is directed at a specimen sliced thin enough to allow the beam to pass through. As they travel through, the electrons are deflected according to the **atomic structure** of the specimen. The beam is then focused by the magnetic coil onto a photographic plate; when developed, the image on the plate shows the specimen at very high magnification.

Scientists worldwide immediately embraced Ruska's invention as a breakthrough in optical research, and they directed their own efforts toward improving upon its precision and flexibility. A Canadian American physicist, James Hillier, constructed a microscope from Ruska's design that was nearly 20 times more powerful. In 1939, modifications made by Vladimir Kosma Zworykin enabled the electron microscope to be used for studying viruses and protein molecules. Eventually, electron microscopy was greatly improved, with microscopes able to magnify an image 2,000,000 times. One particularly interesting outcome of such research was the invention of **holography** and the hologram by Hungarian-born engineer **Dennis Gabor** in 1947. Gabor's work with this three-dimensional **photography** found numerous applications upon development of the laser in 1960.

There are now two distinct types of electron microscopes: the transmission variety (such as Ruska's), and the scanning variety. Scanning electron microscopes, instead of being focused by the scanner to peer through the specimen, are used to observe electrons that are scattered from the surface of the specimen as the beam contacts it. The beam is moved along the surface, scanning for any irregularities. The scanning electron microscope yields an extremely detailed three-dimensional image of a specimen but can only be used at low resolution; used in tandem, the scanning and transmission electron microscopes are powerful research tools.

Today, electron microscopes can be found in most hospital and medical research laboratories. One of the more interesting applications for the invention is that by petroleum companies, who use electron microscopy to study the molecular links in petrochemicals. The results of their studies help in the search for petroleum substitutes and synthetic fuels, especially those derived from coal.

The advances made by Ruska, Knoll, and Hillier have contributed directly to the development of the field ion microscope (invented by Erwin Wilhelm Muller) and the **scanning tunneling microscope** (invented by **Heinrich Rohrer** and **Gerd Binnig**), now considered the most powerful optical tools in the world. For his work, Ruska shared the 1986 Nobel Prize for physics with Binnig and Rohrer.

ELECTRONICS

Electronics is a field of engineering and applied physics that grew out of the study and application of **electricity**. Electricity concerns the generation and transmission of **power** and uses metal conductors. Electronics manipulates the flow of electrons in a variety of ways and accomplishes this by using gases, materials such as silicon and germanium that are **semiconductors**, and other devices like solar cells, **light-emitting diodes** (LEDs), masers, **lasers**, and microwave tubes. Electronics applications include **radio**, **radar**, television, communications systems and satellites, navigation aids and systems, control systems, **space** exploration vehicles, microdevices like watches, many appliances, and **computers**.

The history of electronics is a story of the twentieth century and three key components: the **vacuum** tube, the transistor, and the integrated circuit. In 1883, **Thomas Alva Edison** discovered that electrons will flow from one metal conductor to another through a vacuum. This discovery of conduction became known as the Edison effect. In 1904, **John Fleming** applied the Edison effect in inventing a two-element **electron** tube called a **diode**, and **Lee de Forest** followed in 1906 with the three-element tube, the **triode**. These **vacuum tubes** were the devices that made manipulation of electrical **energy** possible so it could be amplified and transmitted.

The first applications of electron tubes were in radio communications. **Guglielmo Marconi** pioneered the development of the wireless telegraph in 1896 and long-distance radio communication in 1901. Early radio consisted of either radio telegraphy (the transmission of Morse code signals) or radio telephony (voice messages). Both relied on the triode and made rapid advances thanks to armed forces communications during World War I. Early radio transmitters, telephones, and telegraph used high-voltage sparks to make **waves** and **sound**. Vacuum tubes strengthened weak audio signals and allowed these signals to be superimposed on radio waves. In 1918, Edwin Armstrong invented the "superheterodyne receiver" that could select among radio signals or stations and could receive distant signals. Radio broadcasting grew astronomically in the 1920s as a direct result. Armstrong also invented wide-band **frequency** modulation (FM) in 1935; only AM or amplitude modulation had been used from 1920 to 1935.

Communications technology was able to make huge advances before World War II as more specialized tubes were made for many applications. Radio as the primary form of education and entertainment was soon challenged by television, which was invented in the 1920s but didn't become widely available until 1947. Bell Laboratories publicly unveiled the television in 1927, and its first forms were electromechanical. When an electronic system was proved superior, Bell Labs engineers introduced the cathode-ray picture tube and **color** television. But Vladimir Zworykin, an engineer with the Radio Corporation of America (RCA), is considered the "father of the television" because of his inventions, the picture tube and the iconoscope camera tube.

Development of the television as an electronic device benefited from many improvements made to radar during World War II. Radar was the product of studies by a number of scientists in Britain of the reflection of radio waves. An acronym for *RA*dio *D*etection *A*nd *R*anging, radar measures the distance and direction to an object using echoes of radio **microwaves**. It is used for aircraft and ship detection, control of weapons firing, navigation, and other forms of surveillance. Circuitry, video, pulse technology, and microwave transmission improved in the wartime effort and were adopted immediately by the television industry. By the mid-1950s, television had surpassed radio for home use and entertainment.

After the war, electron tubes were used to develop the first computers, but they were impractical because of the sizes of the electronic components. In 1947, the transistor was invented by a team of engineers from Bell Laboratories. **John Bardeen**, Walter Brattain, and William Shockley received a Nobel Prize for their creation; but few could envision how quickly and dramatically the transistor would change the world. The transistor functions like the vacuum tube, but it is tiny by comparison, weighs less, consumes less power, is much more reliable, and is cheaper to manufacture with its combination of metal contacts and semiconductor materials.

The concept of the integrated circuit was proposed in 1952 by Geoffrey W. A. Dummer, a British electronics expert with the Royal Radar Establishment. Throughout the 1950s, **transistors** were mass produced on single wafers and cut apart. The total semiconductor circuit was a simple step away from this; it combined transistors and diodes (active devices) and capacitors and resistors (passive devices) on a planar unit or chip. The semiconductor industry and the silicon integrated circuit evolved simultaneously at Texas Instruments and Fairchild Semiconductor Company. By 1961, integrated circuits (ICs) were in full production at a number of firms, and designs of equipment changed rapidly and in several directions to adapt to the technology. Bipolar transistors and digital integrated circuits were made first, but analog ICs, large-scale integration (LSI), and very-large-scale integration (VLSI) followed by the mid-1970s. VLSI consists of thousands of circuits with on-and-off switches or gates between them on a single chip. Microcomputers, medical equipment, video cameras, and communication satellites are only a few examples of devices made possible by integrated circuits.

Electronic components

Integrated circuits are sets of electronic components that are interconnected. Active components supply energy and include vacuum tubes and transistors. Passive components absorb energy and include resistors, capacitors, and inductors.

Vacuum tubes or electron tubes are glass or ceramic enclosures that contain metal electrodes for producing, controlling, or collecting beams of electrons. A diode has two elements, a cathode and an **anode**. The application of energy to the cathode frees electrons that migrate to the anode. Electrons flow only during one half-cycle of an alternating current (AC).

A grid inserted between the cathode and anode can be used to control the flow and amplify it. A small **voltage** can cause large flows of electrons that can be passed through circuitry at the anode end.

Special purpose tubes use photoelectric emission and secondary emission, as in the television camera tube that emits and then collects and amplifies return beams to provide its output signal. Small amounts of argon, hydrogen, mercury, or neon vapors in the tubes change its current capacity, regulate voltage, or control large currents. The finely focused beam from a cathode-ray tube illuminates the coating on the inside of the television picture tube to reproduce images.

Transistors are made of silicon or germanium containing foreign elements that produce many electrons or few. N-type semiconductors produce a lot of electrons, and p-type semiconductors do not. Combining the materials creates a diode, and when energy is applied, the flow can be directed or stopped depending on direction. A triple layer with either n-p-n or p-n-p creates a triode, which, again, can be used to amplify signals. The field-effect transistor or FET superimposes an electric field and uses that field to attract or repel charges. The field can amplify the current much like the grid does in the vacuum tube. FETs are very efficient because only a small field controls a large signal. A controlling terminal or gate is called a JFET or junction FET. Addition of metal semiconductors, metal oxides, or insulated gates produce other varieties of transistor that enhance different signal-transmitting aspects.

An integrated circuit consists of tens of thousands of transistors and other circuit elements that are fabricated in a substrate of inert material. That material can be ceramic or glass for a film-integrated circuit or silicon or gallium-arsenide for a semiconductor integrated circuit. These circuits are small pieces or chips that may be 0.08-0.15 sq in long (2-4 sq mm long). Designers are able to place these thousands of components on a chip by using photolithography to place the components and minute conducting paths in the proper patterns for the purpose of each type of circuit. Many chips are made simultaneously on a 4-in-sq (10-cm-sq) wafer.

Several methods are used to introduce impurities into the silicon in the planar process. A mask with some regions isolated is placed over the surface or plane of the wafer, and the surface of the silicon is altered or treated to modify its electrical character. Crystals of silicon are grown on the substrate in a process called epitaxy; another method, thermal oxidation, grows a film of silicon dioxide on the surface that acts as a gate insulator. During solid-state diffusion, impurities diffused as a gas or spread in a beam of **ions** are distributed or redistributed in regions of the semiconductor. The number of impurities diffused into the crystal can be carefully controlled so the movement of electrons through the chip will also be specific. Coatings can also be added by chemical vapor deposition, evaporation, and a method called sputtering used to deposit tungsten on the substrate; the results of all these methods are coatings on the substrate or disturbed surfaces of the substrate that are only atoms thick. Etching and other forms of lithography (using electron beams or **x rays**) are also used to pattern the wafer surface for the interconnection of the surface elements.

Resistance to the flow of current can be controlled by the conductivity of the material, dimensions over which current flows, and the applied voltage. In electronic circuits, metal films, mixtures containing **carbon**, and resistance wire are used to make resistors. Capacitors have the ability to retain charge and voltage and to act as conductors, especially when currents change in flow. Inductors regulate rapid changes in signals and current intensity.

Sensors are specialized electronic devices that detect changes in quantities such as **temperature**, electrical power levels, chemical concentrations, physical position, fluid flow, or mechanical properties like **velocity** or **acceleration**. When a sensor responds to change, it usually requires a transducer to convert the quantity the sensor has measured into electrical signals that are translated into printouts, electronic readouts, recordings, or information that is returned to the device to control the change measured. Specialized resistors and capacitors are sometimes used as combined sensors and transducers. Variable resistors respond to mechanical motions by changing them to electrical signals. The thermistor varies its resistance with temperature; a thermocouple also measures temperature changes in the form of small voltages as temperatures are measured at two different junctions on the thermocouple. Usually, sensors produce weak electronic signals, and added circuits amplify these. But sensors can be operated from a distance and in conditions such as extreme **heat** or cold or contaminated environments where working conditions are unpleasant or hazardous to humans.

Amplifiers are electronic devices that boost current, voltage, or power. Audio amplifiers are used in radios, televisions, cassette recorders, sound systems, and citizens band radios. They receive sound as electrical signals, amplify these, and convert them to sound in speakers. Video amplifiers increase the strength of the visual information seen on the television screen by regulating the brightness of the image-forming **light**. Radio frequency amplifiers are used to amplify the signals of communication systems for radio or television broadcasting and operate in the frequency range from 100 kHz to 1 GHz and sometimes into the microwave-frequency range. Video amplifiers increase all frequencies equally up to 6 MHz; audio amplifiers, in contrast, usually operate below 20 kHz. But both audio and video amplifiers are linear amplifiers that proportion the output signal to the input received; that is, they do not distort signals. Other forms of amplifiers are nonlinear and do distort signals usually to some cutoff level. Nonlinear amplifiers boost electronic signals for modulators, mixers, oscillators, and other electronic instruments.

Oscillators

Oscillators are amplifiers that receive an incoming signal and their own output as feedback (that is, also as input). They produce radio and audio signals for precision signal-

ing, such as warning systems, telephone electronics between individual telephones and central telephone stations, computers, alarm clocks, high-frequency communications equipment, and the high-frequency transmissions of broadcasting stations.

Electronic equipment usually operates on direct current (DC) power supplies because these are more easily regulated. Power supplies in electrical outlets, however, are alternating currents (AC), so electronic equipment must be able to convert AC to DC. A team of devices is used for this conversion. The piece of equipment has an internal transformer that adjusts the voltage it receives from the outlet up or down to suit operation of the equipment. The transformer is also a ground, a type of insulation that reduces the possibility of electrical shock. A rectifier converts AC to DC, and a capacitor filters the converted voltage to level out any fluctuations. A voltage regulator may take the place of the capacitor, especially in more sophisticated equipment; modern voltage regulators are manufactured as integrated circuits.

Microwaves are the frequencies of choice for many forms of communications especially telephone and television signals that are transmitted long distances through overland methods, broadcast stations, and satellites. Microwave electronics are also used for radar.

Microwaves are within the frequency of 3 GHz to about 300 GHz; because of their high-frequency spectrum, microwaves can carry large numbers of channels. They also have short wavelengths from 10 cm to 0.1 cm; **wavelength** dictates the size of antenna that can be used to transmit that particular wavelength, so the small antennae for microwave communications are very practical. They do require repeater stations to make long-distance links.

Electronic devices like capacitors, inductors, oscillators, and amplifiers were not usable with microwaves because their high frequency and the speeds of electrons are not compatible. This complication of component size was studied in detail in the 1930s. Finally, it was found that the velocity of the electrons could be modulated to the advantage of microwave applications. The modulating device, the klystron, was a tube that amplified the microwave signal in a resonating cavity. The klystron could amplify only a narrow range of microwave frequencies, but the traveling-wire tube (invented in 1934)—a similar velocity modulator—could amplify a wider frequency band using a wire helix instead of a resonating cavity

High-powered and high-pulsed microwave use especially for radar required another device, the magnetron. The magnetron was perfected in 1939 and was a tube with multiple resonating cavities. While these devices were successful for their specialized uses, they were expensive and bulky (like other vacuum tubes); they have been replaced completely by semiconductors and integrated circuits with equally sophisticated and specialized solutions for handling the high frequencies of microwaves that fit much smaller spaces and can be mass produced economically.

Microwave electronics have also required adaptations of other parts of transmission systems. Conventional wires can't carry microwaves because of the energy they give off; instead, coaxial cables can carry microwaves up to 5 GHz in frequency because their self-shielding conductors prevent radiating energy. Waveguides are used for higher-frequency microwave transmission; waveguides are hollow metal tubes with a refractive interface that reflects energy back. Microstrips are an alternative to waveguides that connect microwave components and work by separating two conductors with dielectric material. Microstrips (also called striplines) can be manufactured using integrated circuit technology and are compatible with the small size of ICs.

Masers were first developed in 1954. The maser (*M*icrowave *A*mplification by *S*timulated *E*mission of *R*adiation) can be used for amplifying and oscillating microwaves in signals from satellites, atomic clocks, spacecraft, and radio. Masers focus molecules in an excited energy state into a resonant microwave cavity that then emits them as stimulated emission of **radiation** through the microwave output.

Optical electronics involve combined applications of optical (light) signals and electronic signals. Optoelectronics have a number of uses, but the three general classifications of these uses are to detect light, to convert solar energy to electric energy, and to convert electric energy to light. Like radio waves and microwaves, light is also a form of electromagnetic radiation except that its wavelength is very short. Photodetectors allow light to irradiate a semiconductor that absorbs the light as photons and converts these to electric signals. Light meters, burglar alarms, and many industrial uses feature photodetectors.

Solar cells convert light from the **Sun** to electric energy. They use single-crystal doped silicon to reduce internal resistance and metal contacts to convert over 14% of the solar energy that strikes their surfaces to electrical output voltage. Cheaper, polycrystalline silicon sheets and other **lenses** are being developed to reduce the cost and improve the effectiveness of solar cells.

Light-emitting diodes (LEDs) direct incoming voltage to gallium-arsenide semiconductors that, when agitated, emit photons of light. The wavelength of the emitted light depends on the material used to construct the semiconductor. The LED is used in many applications where illuminated displays are needed on instruments and household appliances. Liquid crystal displays (LCDs) use very low power levels to produce reflected or scattered light; they cannot be seen in the dark like LED displays because they don't produce light. Conductive patterns of electrodes overlie parallel-plate capacitors that hold large molecules of the liquid crystal material that works as the dielectric.

Like microwaves, optical electronics use waveguides to reflect, confine, and direct light. The most familiar form of optical waveguide is the optic fiber. These fine, highly specialized glass fibers are made of silica that has been doped with germanium dioxide.

Digital electronics are the electronics that transformed our lives beginning in the 1970s. The personal computer is one of the best examples of this transformation because it has simplified tasks that were difficult or impossible for individuals to accomplish. Digital devices use simple "true-false" or "on-off" statements to represent information and to make decisions. In contrast, analog devices use a continuous system of values. Because digital devices recognize only one of two permissible signals, they are more tolerant to noise (unwanted electronic signals) and a range of components than analog devices. Digital systems are built from a collection of components that process, store, and transmit or communicate information. The basis of these components is the logic circuit that makes the true-false decision from what may be many true-false signals. The logic circuit is an integrated circuit from any one of a number of families of digital logic devices that use switches, transducers, and timing circuits to function. Digital logic gates are the most elementary inputs and outputs in a logic device. A logic gate is based on a simple operation in Boolean algebra (a form of mathematics that uses logic variables to express thought processes). For example, a logic gate may perform an "or," "and," or "not" function; to make it capable of a "nor" function, an "or" gate is followed by an inverter. By linking combinations of these gates, any decision is possible.

The most popular form of logic circuit is probably the transistor-transistor logic (TTL) circuit. High-speed systems use emitter coupled logic (ELC), and the complementary metal oxide semiconductor (CMOS) logic uses lower speeds to also lower power levels. Logic gates are also combined to make static-memory cells. These are combined in a rectangular array to form the random-access memory (RAM) familiar to home computer users. The binary digits that make up this memory are called "bits," and typical large-scale integrated (LSI) circuit memory chips have over 16,000 bits of static memory. Dynamic memory cells use capacitors to send memory to a selected cell or to "write" to that cell. Very-large-scale chips with 256,000 bits per chip were made beginning in the 1980s, and dynamic memory made these possible because of its high **density**.

Microprocessors have replaced combinations of switching and timing circuits. They are programmed to perform sets of tasks and a wider variety of logic functions. Electronic games and digital watches are examples of microprocessor systems. Digital methods have revolutionized music, library storage, medical electronics, and high definition television, among thousands of other tools that influence our lives daily. Future changes to so-called computer architecture are directed at greater speed; ultra-high-speed computers may operate by using superconducting circuits that operate at extremely cold temperatures, and integrated circuits that house hundreds of thousands of electronic components on one chip may be commonplace on our desktops.

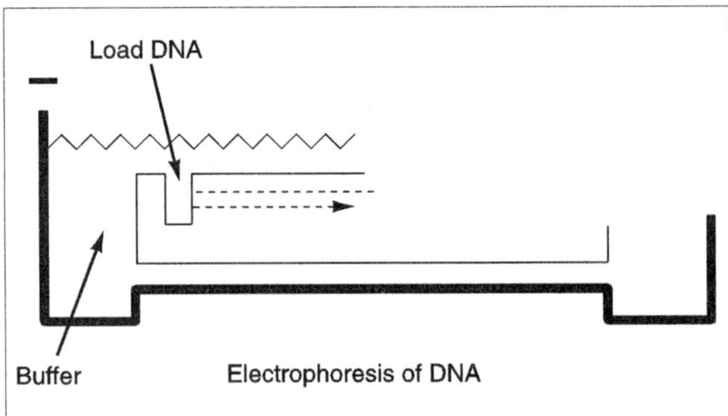

Electrophoresis of DNA.

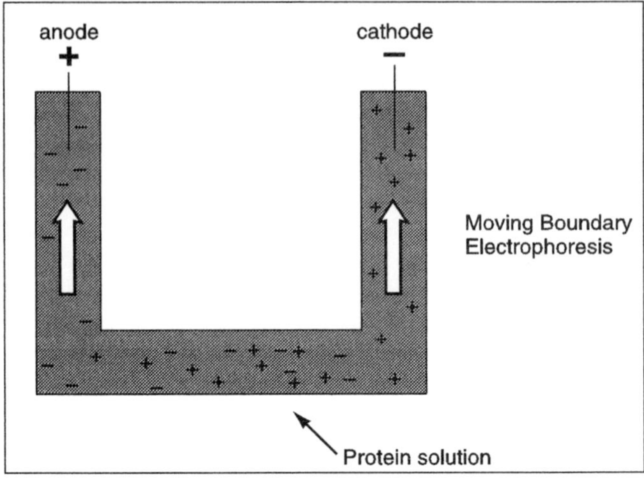

Electrophoresis, moving boundary electrophoresis.

ELECTROPHORESIS

Electrophoresis (also termed cataphoresis) is the physical movement and separation of charged molecules in a solution medium under the influence of an electric field. The rate and distance of molecular movement depends upon the size, shape, and charge of the molecules subjected to the electric field, the field strength, and the conditions of the medium (e.g., electrolyte concentration, pH, ionic strength, **viscosity**, **temperature**) or gel in which the molecules move. Advances in electrophoresis have revolutionized methods of protein analysis and have found wide-ranging applications.

Swedish biochemist Arne Tiselius pioneered research in electrophoretic analysis. Tiselius was able to separate serum proteins in a tube (the Tiselius tube) subjected to an electric field.

Modern electrophoresis allows the controlled differentiation of molecules into bands or regions in differing mediums (e.g., agarose, polyacrylamide gel, paper). In addition to electrical forces, some mediums or support matrices are porous gels that can also act as a physical sieve for macromolecules.

There are many variations on electrophoresis and many of these variations have found wide-ranging applications. One specialized technique is termed blotting. Specific blots are named according to the technique used and the molecule(s) under study. Regardless, modern electrophoresis techniques have developed into an increasingly important application of **biophysics** and a critical component in genetics research (e.g., gene expression and the diagnosis of heritable diseases). In particular, electrophoretic analysis allows the identification of biological species—including bacterial and viral strains—and is finding increasing acceptance as a powerful forensic tool in criminal investigations.

In general, the medium is infused with buffers that allow transport of the **electric charge** applied to the medium/buffer matrix. The matrix is placed in a tray and samples of molecules to be separated are loaded into wells at one end of the matrix. As electrical current is applied to the tray and the matrix takes on this charge and develops positively and negatively charged ends. Charged molecules are thus attracted toward the oppositely charged end and, simultaneously, repulsed by the fields created by the like-charged end of the matrix. If the electric field at a particular point is known, the **force** a charge q experiences when it is placed at that point is given by: $F = qE$ where electric field E is a vector. Accordingly, the net force on the molecule is the vector sum of the force exerted by each end of the matrix.

Because molecules have differing shapes, sizes, and charges, they are pulled through the matrix at different rates and this, in turn, causes a separation of the molecules. Generally the smaller and more charged a molecule, the faster the molecule moves through the matrix.

A wide range of molecules may be separated by electrophoresis, including, but not limited to DNA, RNA, and protein molecules. Because nucleic acids always carry a negative charge, differential separation of nucleic acids occurs strictly by molecular size.

Proteins carry net charges determined by their charged groups of amino acids and therefore proteins can be amphoteric (a compound that can take on a negative or positive charge). An amphoteric protein might carry a positive charge in a particular medium and thus migrate toward the negative end of the matrix. In another matrix the same protein might carry a negative charge and migrate toward the positive end of the matrix.

Under some circumstances, the suspending medium or other liquids in the matrix respond to the electric fields more than the particles (termed electroosmosis).

In the 1960s, sodium dodecyl sulfate (SDS) polyacrylamide gel electrophoresis was used as a means of protein fractionation. Unfortunately, the protein bands did not always separate sharply and therefore only small concentrations of molecules could be separated. Development in the 1970s of a two-dimensional electrophoresis technique (actually two separate separations) allowed higher concentrations of molecules to be physically separated. In two-dimensional electrophoresis the first separation is achieved by isoelectric focusing that separates protein polypeptide chains according to amino acid com-

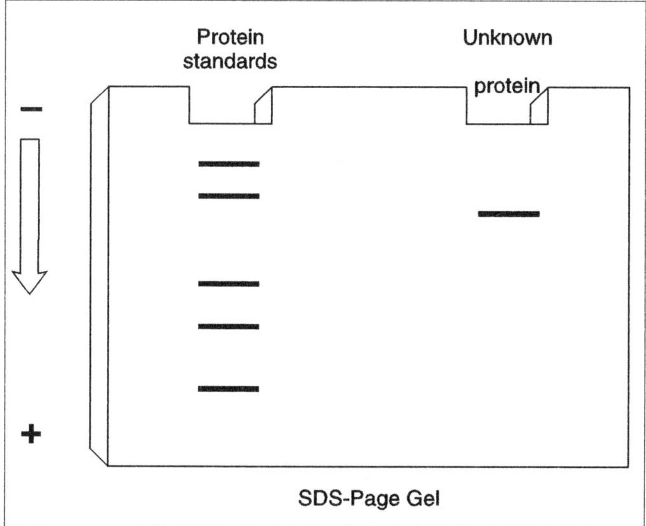

Electrophoresis, SDS-Page gel.

position. The proteins separate after being exposed to a pH gradient that forces them to move to their isoelectric point. The second separation is completed using SDS slab gel electrophoresis to separate molecules by molecular size. Instead of a hazy mixture of bands, a resulting two-dimensional pattern of spots (each with a unique protein or protein fragment) allows additional analysis. In addition, some techniques involve applications of radioactive labels to test proteins in order to study the characteristic movement of labels under specific conditions.

ELECTROSTATIC DEVICES

Electrostatics is the study of the behavior of electric charges that are at rest. The phenomenon of static **electricity** has been known for well over 2,000 years, and a variety of electrostatic devices have been created over the centuries.

The ancient Greek philosopher Thales (624-546 B.C.) discovered that when a piece of amber was rubbed, it could pick up light objects, a process known as triboelectrification. The Greek name for amber, *elektron*, gave rise to many of the words we use in connection with electricity. It was also noted that lodestone had the natural ability to pick up iron objects, although the early Greeks did not know that electricity and **magnetism** were linked.

In the late sixteenth century, **William Gilbert** began experimenting with static electricity, pointing out the difference between static electric attraction and magnetic attraction. Later, in the mid-1600s, **Otto von Guericke** built the first electrostatic machine. His device consisted of a sulfur globe that was rotated by a crank and stroked by hand. It released a considerable static **electric charge** with a large spark.

A similar device was invented by Francis Hawkesbee in 1706. In his design, an iron chain contacted a spinning globe and conducted the electric charge to a suspended gun barrel; at the other end of the barrel another chain conducted the charge.

In 1745, the first electrostatic storage device was invented nearly simultaneously by two scientists working independently. Pieter van Musschenbroek, a professor at the University of Leyden, and Ewald von Kleist of the Cathedral of Camin, Germany, devised a water-filled glass jar with two electrodes. A Leyden student who had been using a Hawkesbee machine to electrify the water touched the chain to remove it and nearly died from the electric shock. This device, known as the Leyden jar, could accumulate a considerable electric charge, and audiences willingly received electric shocks in public displays. One of these displays aroused the curiosity of **Benjamin Franklin**, who obtained a Leyden jar for study. He determined that it was not the water that held the electric charge but the glass insulator. This is the principle behind the electrical condenser (capacitor), one of the most important electrical components in use today.

Charles F. Du Fay (1698-1739) discovered that suspended bits of cork, electrified with a statically charged glass rod, repelled each other. Du Fay concluded that any two objects that had the same charge repelled each other, while unlike charges attracted. The science of electrostatics, so named by **André-Marie Ampère**, is based on this fact.

French physicist **Charles Coulomb** became interested in the work of Joseph Priestley (1733-1804), who had built an electrostatic generator in 1769, and studied electrical repulsion. Coulomb used his torsion balance to make precise measurements of the **force** of attraction between two electrically charged spheres and found they obeyed an inverse square law. The mathematical relationship between the forces is known as **Coulomb's law**, and the unit of electric charge is named the coulomb in his honor.

Alessandro Volta invented a device in 1775 that could create and store an electrostatic charge. Called an electrophorus, it used two plates to accumulate a strong positive charge. The device replaced the Leyden jar, and the two-plate principle is behind the electrical condensers in use today.

Several other electrostatic machines have been devised. In 1765 John Reid, an instrument maker in London, built a portable static electric generating machine to treat medical problems. In 1783, John Cuthbertson built a huge device that could produce electrical discharges 2 ft (61 cm) in length. The gold leaf electroscope, invented in 1787, consists of two leaves that repel each other when they receive an electric charge. In 1881, British engineer James Wimshurst invented his Wimshurst machine, two glass discs with metal segments spinning opposite each other. Brushes touching the metal segments removed the charge created and conducted it to a pair of Leyden jars where it was stored for later use.

The most famous of all the electrostatic devices is the Van de Graaff generator. Invented in 1929 by Robert J. Van de Graaff, it uses a conveyor belt to carry an electric charge from a high-voltage supply to a hollow ball. It had various applica-

Children with Van de Graaff generator. *(Photo by Lester V. Bergman, Corbis. Reproduced by permission.)*

tions. For his experiments on properties of atoms, Van de Graaff needed to accelerate **subatomic particles** to very high **velocity,** and he knew that storing an electrostatic charge could result in a high potential. Another generator was modified to produce **x rays** for use in the treatment of internal tumors. It was installed in a hospital in Boston in 1937. Van de Graaff's first generator operated at 80,000 volts, but was eventually improved to five million volts. It remains one of the most widely used experimental exhibits in schools and museums today.

ELECTROWEAK FORCE

Electroweak **force** is the name given to the unified electromagnetic and weak interactions. The two interactions are viewed as different combinations of one underlying force, hence the term unified. The electroweak force acts differently between particles depending on the direction of their spins. Along with the familiar **photon,** three more particles, called the Z, W^+, and W^-, mediate the electroweak force.

Electromagnetic interactions involve **electric charge** and electric currents, as well as electric and magnetic fields. Weak interactions are mainly involved in the decay of unstable particles, such as nuclear **beta decay** or **muon** decay. At first glance, weak interactions may seem different and unrelated. However, in much the same way that **electricity** and **magnetism** were found to be difference aspects of the same interaction by **James Clerk Maxwell**, a correct description of

the weak force cannot be done without also including the electromagnetic force in the theory.

Progress on finding the correct theory of the weak force was slow. In the 1950s, **quantum electrodynamics (QED)** was spectacularly successful and was verified by many experiments. A generalization of QED, called non-Abelian gauge theory, was discovered in the 1950s, and many scientists thought it would easily find use as the theory of the weak force. There were complications involved, however, that were yet to be discovered.

Probably the largest leap in knowledge of the weak force came when it was realized that it violated **parity.** Parity is a discrete **space-time** symmetry; under a parity transformation all position **vectors** are reversed. QED was invariant under parity, and it was believed at the time that all theories must also be invariant under parity. But decays of two **strange particles** (theta and tau) with identical masses and lifetimes into different final states convinced physicists that the weak force must violate parity.

The consequence of parity violation is that particles of different spin interact differently. A common definition of spin in **particle physics** is to say that a particle is left-handed or right-handed. A left-handed particle has spin pointing opposite to its **momentum,** and a right-handed particle has spin pointing along its momentum. Only massless particles can be completely left- or right-handed. Left-handed particles have an internal SU(2) isospin symmetry, which is a mathematical way to say that electrons can be traded for neutrinos, or down **quarks** for up quarks without changing the appearance of the theory.

As a result, in electroweak interactions right-handed particles can interact only via photons and **Z bosons**, but left-handed particles can interact through all the **electroweak particles**. Another result is that there are no right-handed neutrinos, because their existence would cause disagreement between the theory and the experimental evidence.

The final, useful form of the **electroweak theory** was constructed by physicists **Sheldon Glashow**, **Abdus Salam**, and **Steven Weinberg** in the early 1970s, for which they eventually recieved the Nobel Prize. This theory contained one extra ingredient that overcame an important difficulty. As stated above, only massless particles can be completely left- or right-handed, but the electroweak force makes a distinction between left and right, so particles cannot be mixtures of left- and right-handed states. Also, the W and Z bosons were not discovered experimentally until long after the theory was finished, meaning that they also had to be very massive. Explicit **mass** terms cannot be included into the theory without spoiling its good points. Another solution, therefore, had to be found.

The elegant solution was termed the Higgs mechanism. A new field was postulated, the Higgs field, which interacts with all particles. The **quanta** of the field are known as **Higgs bosons**. The Higgs field has a nonzero **vacuum** expectation value, meaning that even if there are no Higgs bosons around, other particles still interact with the Higgs field. This constant interaction yields an effective mass term for every particle that does not spoil the successes of the electroweak theory.

As of 2000, the Higgs boson is the last undiscovered ingredient of the electroweak theory (the discovery of the tau **neutrino** was announced in July 2000). There is, however, much indirect evidence that it exists. The success of the theory in explaining the results of **scattering** experiments, and the agreement of the masses of the W and Z bosons with the theoretical predictions, would be difficult (perhaps impossible) to reproduce without the existence of Higgs bosons.

Together with quantum chromodynamics, the theory of the strong force, the electroweak theory makes up the **standard model of particle physics**. The success of unifying the electromagnetic and weak interactions has led physicists to attempt to further unify the strong and electroweak forces. So far, their attempts have not been verified, but have led to further features that may be discovered in future experiments.

See also Electromagnetism; Weak force and interactions

ELECTROWEAK PARTICLES

The **electroweak theory** of **particle physics** is a unification of the electromagnetic and weak interactions. Because it is a gauge field theory, the **force** is transmitted between **matter** particles by other particles known as gauge bosons. The electroweak particles are the **photon**, the W boson, and the **Z boson**. In addition, there is the **Higgs boson**. All matter particles (**quarks** and **leptons**) interact through the electroweak interactions.

The photon is the familiar force-carrier of **electromagnetism**. It is massless, electrically neutral, and interacts with all particles that carry **electric charge**. The only matter particles with which the photon does not interact are the three **neutrinos**, which are electrically neutral. **Albert Einstein** is usually credited with the discovery of the photon because he used its existence to explain the **photoelectric effect** in 1905, although **Isaac Newton** was the first to guess that **light** had a particle-like character.

The Z boson is one of the force carriers of the weak force. The Z boson is massive, with a **mass** of 91 GeV/c². It is similar to the photon in its interactions because it is electrically neutral, with the exception that it also interacts with neutrinos. The Z boson was discovered at the Large Electron Positron Collider (LEP) at the **European Center for Nuclear Research (CERN)** in Geneva, Switzerland, in 1983.

The W bosons are also carriers of the weak force. The two W bosons, W+ and W-, are antiparticles of each other. Each W particle has a mass of approximately 80 GeV/c². The W+ has electric charge +e and the W- has charge -e. W bosons interact with all particles, and are frequently produced as intermediate particles in the decay of other charged particles. The W boson was discovered along with the Z boson at LEP in 1983. In subsequent experimental runs at LEP, pairs of W bosons were also produced.

The Higgs bosons do not carry the **electroweak force**, as do the gauge bosons, but have the equally important purpose of generating masses for all of the particles. Higgs bosons are **quanta** of the Higgs field, a mathematical function that has a value at each point in **space-time**. Unlike other fields, which have the value zero when there are no particles present, the Higgs field has a constant nonzero component, called its **vacuum** expectation value, or VEV. Interactions of other particles with the constant Higgs component results in mass terms in the mathematical expression for the particle's **energy**. Without this Higgs interaction, the other particles would be required to have no mass by the **gauge symmetry** principle. However, the Higgs boson has not yet been discovered, and its discovery is the highest priority for experimental particle physicists. There are high hopes that the Higgs boson will be discovered soon at either LEP, the newly refurbished Tevatron at Fermilab in Chicago, or the new **Large Hadron Collider** at CERN.

See also Quantum field theory; Standard model of particle physics; W mesons

ELECTROWEAK THEORY

In the early 1970s, the theories describing the electromagnetic and weak forces were unified into a single electroweak theory. This was the first unification of forces to occur since the late nineteenth century, when the theories of **electricity** and **magnetism** were unified into electromagnetic theory. Before the unification, the weak interaction was only poorly understood; there was a collection of ad hoc rules, but no fundamental theoretical unification existed. This unification of two seemingly unrelated interactions is one of the most important accomplishments in theoretical physics in the twentieth century, and it sparked a whole industry of physicists trying to further unify the remaining fundamental forces.

The fundamental objects of the electroweak theory are the particles of **matter** called **quarks** and **leptons**, which are represented by quantum fields, mathematical functions with a value at each point in **space-time**. The fields consist of complex numbers, which have a real and an imaginary component. There are also the gauge bosons, or force-carrier particles, which are required by a symmetry called **gauge symmetry**. In a nutshell, if a theory respects gauge symmetry, then the point of view an observer takes at one point in space-time cannot affect the laws of physics at a different point in space-time.

The **force** carrier of **electromagnetism** is the **photon**, and the force carriers of the weak force are the massive W and **Z bosons**.

Before the electroweak theory was developed, the **quantum theory** of electromagnetism was well developed. It only involved, however, the interactions of photons with charged particles. When **neutrinos** were included in the picture, it was predicted that **photons** should also have interacted with neutrinos, which did not occur. The proposed solution to this problem is more than one photon-like field interacting with the matter particles. The photon is actually a combination of two other fields and does not interact with neutrinos, but the "perpendicular" combination of these fields does interact with neutrinos. This perpendicular combination is called the Z boson. Therefore, in order to make the theory of electromagnetism self-consistent, the theory of weak interactions is treated simultaneously. It is in this sense that the two interactions are unified.

The other weak interactions, which describe the decay of charged particles, are explained by the exchange of the charged W bosons. Because of the gauge symmetry, it is found that the coupling strength of the W bosons is related to the coupling strength of the Z boson and the photon. This coupling strength relation has been measured and verified in collider experiments.

See also Electroweak force; Electroweak particles; Quantum field theory; Standard model of particle physics; Symmetry and symmetry principles

EMPEDOCLES OF ACRAGAS (C.490-C.430 B.C.)

Greek philosopher

A philosopher, poet, politician, and visionary, Empedocles of Acragas developed radical new ideas about the nature of the **universe**. His philosophy of the four elements in the universe and the definition of **matter** as the various ratios of these elements foreshadowed later developments in **atomic theory** by philosophers such as **Democritus** of Abdera.

Empedocles was born in Acragas, Sicily. His father, Meto, was wealthy, and his grandfather, also named Empedocles, was renowned for winning a horse race in the Olympia. Empedocles is believed to have traveled to Thourioi shortly after it was established approximately 444 B.C.

Empedocles, 490-430 B.C.

Empedocles's keen intellect enabled him to combine talents in philosophy, natural history, poetry, and politics, and to achieve superstar status in his day. According to the Greek philosopher **Aristotle**, Empedocles was the inventor of rhetoric, a talent Empedocles often utilized as a statesman. He became popular among his fellow citizens through his support of democracy.

Empedocles's scientific inquiries usually included mysticism. However, his philosophies contained early insight into basic laws of physics, including atomic theory. Although sometimes labeled a Pythagorean, Empedocles followed the Greek philosopher Parmenides (c.515-c.445 B.C.) in the belief that matter (or, "what is") is indestructible. Empedocles claimed that matter was the only principle of all things and that four elements in the universe—air, fire, earth, and water—made up all things according to various ratios of these elements. Empedocles further stated that two forces that he called love and hate, or eros and strife, controlled how the four elements come together or move apart. In addition to creating a philosophy that closely resembles modern atomic theory, Empedocles also studied the nature of change in the universe. Empedocles asserted that the cyclical nature of the universe introduces the possibility of reincarnation because nothing that comes into being can be destroyed but only transformed. Empedocles later wrote a poetic treatise *On Nature* (*Peri Physeos*) containing the ideas of evolution, the circulation of the blood, and atmospheric pressure. He stated that the **Moon** shone by reflected **light** and estimated that the Moon was one-third the distance from Earth to the **Sun**.

The object of admiration, Empedocles, according to Aristotle, was offered a kingship but refused to be considered king. Nevertheless, some scholars claim that Empedocles assumed royal status and went so far as to claim himself a deity. Accordingly, legends grew of various miracles he performed, such as healing the sick and even raising the dead. Viewed by some as a demi-god and by others as a charlatan, Empedocles made important contributions to the philosophy of science in his day. Galen (c.130-c.200), the physician to several Roman emperors, also credits Empedocles with founding the Italian school of medicine. In addition, Empedocles was an accomplished poet. However, little remains of his writings except for segments of his poems *On Nature* and *Purifications* (*Katharmoi*). That Empedocles had a flair for self-promotion and public relations is evident in scholarly writings. Empedocles was reported to have leapt into the crater of the Mt. Etna volcano so he would have a death befitting a god. The English poet, Matthew Arnold, wrote a poem about the episode entitled *Empedocles on Etna*. Some scholars dispute the story of Empedocles's firey death. According to the writings of Aristotle, Empedocles died at the age of 60.

ENERGY

Energy is defined as the capacity to perform **work** or to produce **heat**. The Système International d'Unités (SI) unit for energy is the joule (J). Lifting a medium-sized potato a distance of one meter against **gravity** would require approximately one joule of energy. Energy is often expressed as the calorie (cal), which is the amount of heat needed to raise the **temperature** of one gram of water by 1°C. One calorie is equal to 4.184 joules. The Calorie (Cal), which is how the energy in food is expressed, is 1,000 calories. Energy is not only involved in chemical reactions and technology, but also in our basic bodily functions. It is because of the energy of the reactions inside of our body that we maintain a constant body temperature, our heart and lungs can function, and our muscles can move. Energy can take many forms, all of which can be classified into three categories. These categories are radiant energy, **kinetic energy**, and **potential energy**.

An example of radiant energy is sunlight. Kinetic energy is the energy of an object in **motion**. Examples of kinetic energy are mechanical energy (caused by motion of parts) and **thermal energy** (caused by the random **motion of particles** of **matter**). Potential energy is the energy of position of objects. It is also known as stored energy. An example of potential energy is water in an elevated tank. If the water is allowed to fall on a wheel and the wheel turns, the wheel can be used to produce **electricity**. The water in the tank has the potential to perform work once it falls, therefore it has gravitational potential energy (gravity is what converted the potential energy of the water as it fell into kinetic energy). Other forms of potential energy include chemical potential energy, the energy stored in fuels such as kerosene or in the food you eat, and electrical

potential energy, the energy when objects with different electrical charges are separated, such as in batteries.

Two forms of energy that are most familiar are **light** and heat. Light is a form of radiant energy. Scientists used to believe that light was a beam of energy that took the form of **waves**. In the 1900s, however, experiments were performed that showed that light behaved like very small, fast-moving particles. Scientists today realize that light has properties of both waves and particles and describe it accordingly.

Heat is a form of thermal energy that is transferred from one object to another when there is a temperature difference between them. More specifically, it is the transfer of kinetic energy from something of greater temperature to something of lesser temperature. Heat energy can cause physical or chemical changes. A physical change that heat can cause is the melting of ice into water, or the evaporation of water into vapor. This water vapor can undergo a chemical change if enough additional heat is added, when it decomposes into **oxygen** gas and hydrogen gas.

An English scientist named **James Joule** performed an experiment in 1845 that demonstrated a relationship between energy and heat. The experiment was not complicated—he placed a paddle wheel in a tank of water and measured the temperature of the water. He then cranked the wheel in the water for a period of time, and read the temperature of the water again. He found that the temperature of the water rose as he cranked the paddle wheel. He quantified this observation and discovered that an equal amount of energy was always required to raise the temperature of water by one degree. He also discovered that it did not have to be mechanical energy, it could be energy in any form. He obtained the same results with electrical or magnetic energy as he did with mechanical energy. Joule's experiments showed that different forms of energy are equivalent and can be converted from one form to another. In the process of converting energy, no energy is lost and no energy is created.

These observations led to what is now called the "Law of conservation of energy." The law of **conservation of energy** states that any time energy is transferred between two objects, or transformed from one form to another, no energy is created, and none is destroyed. The total amount of energy involved in the process remains the same. This law applies to any process except those that occur in the **Sun** and in **nuclear reactors**, which involve nuclear energy, for a combination of **mass** and energy are conserved but mass and energy are interconverted.

Energy is not only involved in macroscopic processes, but in microscopic processes as well. Most chemical reactions involve changes in energy. A **chemical reaction** is simply breaking bonds between atoms and making new ones. When bonds are broken in a chemical reaction, the reactants involved are colliding with enough kinetic energy to break the atoms apart. During the collision, the kinetic energy is transformed into potential energy as the atoms are rearranged and new bonds are formed. In 1888, Svante Arrhenius (1859-1927), a Swedish chemist, theorized that the particles involved in a chemical reaction must have a certain minimum amount of kinetic energy in order to react. This minimum amount of energy is called the activation energy of a reaction. Bond breaking requires energy, while bond making releases energy. Most chemical reactions, therefore, involve the overall absorption or release of energy. The energy flow that results is heat.

Energy is involved in physical as well as chemical changes. As the temperature of a substance rises, the individual particles of the substance tend to gain kinetic energy. As the particles gain kinetic energy, they obtain different physical properties, and this results in a phase change. A phase change is the conversion of a substance from one of the three **states of matter** (solid, liquid, or gas) to another. A phase change always involves a change of energy. The particles that make up a solid have very low kinetic energies, so they stay fixed together and immobile. As the temperature increases, they gain kinetic energy, and they can move around more easily. When the particles in a substance have enough kinetic energy so that the substance can flow, it is in the liquid phase. As the temperature increases, and the kinetic energy of the particles increase, the particles that make up the liquid begin to move in a speedy, random fashion. At this point, the liquid has become a gas, the phase with the highest kinetic energy. As the temperature of a gas increases, the particles begin to move faster and faster, and their collisions with themselves and their container become more forceful. As this occurs, gas **pressure** increases.

There are many practical applications to the study of energy. One is the development of fuels. A fuel is a chemical compound that burns easily and releases a large amount of heat. When gasoline is burned in an automobile engine, for example, it releases 48 kilojoules of heat per gram. This heat then undergoes several **energy transformations** to **power** the automobile. Two other familiar fuels are oil and coal. Both of these fuels release a large amount of energy when burned. Oil releases 20,900 British thermal units (Btu) per pound, and coal releases 9,600 Btu per pound (one Btu is the amount of heat needed to raise the temperature of one pound of water 1°F). Plastic products could also release similar quantities of energy. The plastic in polyethylene terephthalate (PET) soda bottles can release 10,900 Btu per pound, and the plastic used in high **density** polyethylene (HDPE) milk jugs releases 18,700 Btu per pound. The potential of re-using these plastic containers by burning them in a "waste-to-energy incinerator" is currently under investigation.

The use of alternative forms of fuel, such as the plastic containers described above, is being investigated in order to conserve energy. The law of conservation of energy states that energy can neither be created nor destroyed. Why, then, is there concern about "running out" of energy? The energy scientists are concerned about is usable energy. The majority of our energy comes from coal and petroleum, two of the three fossil fuels (the third fossil fuel is natural gas). Approximately 90% of our energy comes from fossil fuels. Coal is burned in power plants to generate electricity and petroleum is converted into the fuel that is used for transportation. When the potential energy stored in these fossil fuels is released, it is converted into other forms of energy, such as heat and **sound**, which cannot be recaptured and reused. As Earth's coal and petroleum are used up, we will

need to find different sources of energy. Alternatives that are currently under investigation are wind, water, **geothermal energy**, and solar energy.

The many forms of energy are an integral part of daily life. The amount of energy in the **universe** remains constant, but the amount of usable energy is decreasing. The study of energy and its different forms will become increasingly important as the need for alternative forms of energy rises.

ENERGY AND MASS

In Newtonian **mechanics**, **mass** and **energy** are two separate but related quantities. In a closed system, however, the total sum of mass and energy remain constant. To give the concept of energy a solid footing, it is best to consider mass first.

Often, mass is given a rather imprecise definition as the "amount of stuff" in an object. **Galileo** laid the foundation for a more rigorous idea of mass with his principle of **inertia**. The principle states that an object moving at a constant **velocity** in a straight line will continue moving at a constant velocity in a straight line as long as it is not disturbed, and an object sitting still will remain sitting still.

In his laws of **motion**, **Isaac Newton** linked inertia to mass. Newton's first law is essentially Galileo's principle of inertia, but with an important addition. The first law reveals what will "disturb" a moving object: a **force**. The second law defines what a force is. Forces cause masses to accelerate, $F = ma$. **Acceleration** is a change in velocity with **time**. The second law says that a force, F, on an object produces an acceleration, a. The force is directly proportional to the acceleration by the mass, m. Hence, mass is that property of an object that resists changes to its velocity when a force is applied. Newton's laws coincide with the intuitive notion of mass as an "amount of stuff" because common experience demonstrates that the more stuff an object is made out of, the harder it is to get moving.

In addition to resisting changes in velocity, mass generates gravitational attraction. According to **Newton's law of universal gravitation**, $F = Gm_1m_2/r^2$, the force of **gravity** between two objects is directly proportional to the product of their masses. Mass acts as a "gravitational charge" similar to the **electric charge** in **Coulomb's law**. However, there is no *a priori* reason for the equivalence of inertial and gravitational masses. They are the same thing, and their equivalence has been verified experimentally to one part in 10^{12}.

Mass is popularly confused with **weight**. The distinction between them is actually quite simple. Weight is the force needed to prevent an object from moving under the influence of gravity. For example, it is the normal force needed to prevent a chair from falling through the floor. The mass of the chair and the mass of Earth together produce the **gravitational force** holding the chair on the floor. Mass is a property intrinsic to the chair, but its weight will decrease if it is moved to the **Moon** because the Moon has one-sixth the surface gravity of Earth.

Energy is the capacity to do **work**. The work done on an object, W, is defined as the force, F, applied to the object times the distance, d, over which the force is applied, $W = Fd$. Two types of energy in classical mechanics can do work, **kinetic energy** and **potential energy**. The kinetic energy, K, in an object is proportional to its mass, m, and the square of its velocity, v: $K = mv^2/2$. The amount of work that can be done by an object with potential energy, K, is W.

Some types of potential energy, such as gravitational potential energy, are also proportional to mass. However, not all potential energy is proportional to mass. The potential energy of a mass on a spring, for example, depends on the amount the spring is stretched from its **equilibrium** length.

Finally, perhaps the most famous equation in all of physics is **Albert Einstein**'s mass/energy relation, $E = mc^2$. Einstein's equation says that mass has an energy associated with it, even if the mass is not moving and has no potential energy. Atomic bombs exploit Einstein's equation by converting some of the mass in uranium atoms into energy. Fortunately, the conversion is difficult enough that ordinary massive objects do not convert to energy in our daily lives.

See also Conservation of energy; Conservation of mass; Relativistic mass and energy

ENERGY AND WORK

Energy and **work** are two very central, closely related concepts in the physics of **motion**. In fact, energy can itself be defined as the capacity to do work. Work, in turn, is defined as the **force** required to move an object some distance. Therefore, a transfer of energy occurs with work as one body moves another against an opposing force, such as **gravity** or **friction**.

There are many forms of energy. **Heat** energy, **light** energy, and **potential energy** are examples. With regard to work, however, **kinetic energy** is central. Kinetic energy is the energy of motion. Because all moving **matter** has **mass** and **velocity**, the kinetic energy of moving objects can be expressed in terms of these. Kinetic energy, then, is defined as one-half the product of the mass of an object and the square of its velocity. Therefore, the greater the mass of a moving object, the greater is its kinetic energy. Similarly, the greater the velocity or **momentum** of a moving object, the greater is its energy of motion. For instance, a car moving at 20 mph (32 kph) has far more (kinetic) energy than a bicycle moving at the same speed. Likewise, an object completely at rest has negligible kinetic energy.

The relationship between kinetic energy and work is important. The work that is performed on an object is equal to the change in its kinetic energy as the result of an applied force. The greater the change in kinetic energy, or movement, of an object, the greater is the amount of work done. If, for example, a car traveling at 60 mph (96 kph) is slowed to 40 mph (64 kph), the work done by the brakes over the distance that they were applied equals the change in kinetic energy of the moving vehicle as it is slowed. Similarly, the work done by your leg muscles as you accelerate from walking to running is equal to the change in kinetic energy of your body between walking and running. Interestingly, work is not done by mus-

cles as a textbook is held motionless outward at arm's length because there is no movement of the book, and thus no change in kinetic energy of the text.

An important principle regarding energy and work is the concept of **conservation of energy**. A fundamental law of nature, conservation of energy states that in any closed system, the total amount of energy is constant. Energy is not gained or lost but merely changes form or location as it is transferred. For example, the total energy is not changed when **electricity** is used to create heat in an electric oven. The total energy that starts as electricity is equal to the heat energy liberated plus the (lessened) electricity leaving the heating elements to complete the circuit. Electric energy is not lost, but converted into heat energy such that the total energy remains constant. Considering work, then, if the change in kinetic energy is negative, the energy of motion is not actually lost, but converted into another form, such as heat. For instance, when a baseball strikes the catcher's glove and its movement is suddenly changed, the kinetic energy of the ball is converted into heat as it comes abruptly to a halt. Similarly, with positive work, where kinetic energy is increased, energy is converted from another form into motion within a closed system.

ENERGY BANDS IN SOLIDS

In **solid state physics**, band theory accounts for the structure and properties of solid materials. Band theory proposes that the available **energy** states for electrons in solids are not discrete, as in the case of free atoms, but are merged into continuous, or wide, bands. This model is very successful for understanding the difference between conductors, insulators, and **semiconductors**.

The continuous energy bands are referred to as energy bands, and they are visualized as consisting of overlapping molecular **orbitals**. For example, sodium has a total of 11 electrons distributed as follows by increasing order of orbital energy: two electrons each in the $1s$ and $2s$ orbitals, six electrons in the $2p$ orbitals, and one **electron** in the $3s$ orbital. The next orbital ($3p$) with 3 suborbitals is empty. To make a sodium crystalline solid, sodium atoms are packed together in a configuration, in which the atoms are equally spaced, and **space** between the atoms is minimized. Although the interactions between one Na **atom** and an adjacent Na atom are weak, each sodium atom in the metallic solid is associated with many neighboring atoms. The bonding interactions sum to form a bonded metallic solid.

When the $3s$ orbital of one sodium atom overlaps the $3s$ orbital of its immediate neighbor in a crystal lattice, two molecular orbitals are formed: a bonding orbital of lower energy and an antibonding orbital of higher energy. When a third atom overlaps the first two, a third molecular orbital is formed, a nonbonding orbital encased by the lower energy bonding orbital and the higher energy antibonding orbital of the other two sodium atoms. Similarly, when N atoms come together in a single line, N molecular orbitals are formed with N distinct energy levels. These N molecular orbitals are very closely spaced in energy and form an essentially continuous band covering a range of discrete energies. The wide 3*s* band contains the freely moving valence electrons and is referred to as the valence band. The 3*p* band, empty of electrons, is called the conducting band. The extent of the gap between the valence and conducting bands of various materials determines their electrical conducting properties. If the gap is small, electrons will be able to jump from the valence to the conducting band and will be good conductors; if the gap is too large, electrons will remain in the valence band and the material will be a good insulator. The best conductors are metals, because in these materials, the valence and conduction bands overlap. Semiconductors have a small gap between their valence and conduction bands so that thermal or other excitations can bridge the gap by allowing electrons to jump to the conducting band. Because the gap in semiconductors is small, the presence of a small percentage of a doping material can also significantly increase conductivity.

Most solid substances are insulators, which means that in most solid substances, a large, forbidding gap exists between the energy of the valence electrons and the energy at which the electrons can move freely through the conduction band. A good example is provided by glass. Glass is an insulating material, but the fact that it is transparent to visible **light** allows its **color** to be changed. Visible light photons do not have enough energy to bridge the band gap of the glassy material and promote the electrons up to the conduction band. The addition of a very small percentage of impurity atoms to the glass can provide specific available energy levels to absorb certain colors of visible light, but the band gap is too large to transform glass into a good conductor by doping.

On the other hand, a semiconductor's conductivity can be significantly altered by doping. A semiconductor's band gap is small, and, in addition to the **energy level** of its valence and conduction bands, it is also characterized by a flexible Fermi level. This level, located between the valence and conduction bands, represents the maximum energy level that can be occupied by an electron at **absolute zero**. The Fermi level can be shifted by the addition of impurity atoms, which can drastically modify the conducting properties of semiconductors. For example, the application of band theory to n-type semiconductors shows that the addition of impurities shifts the Fermi level near the top of the band gap so that electrons can easily jump into the conduction band. In p-type semiconductors, the Fermi level is shifted close to the valence band, which results in extra holes in the band gap allowing excitation of valence band electrons, leaving mobile holes in the valence band.

See also Atomic structure; Crystallography; Molecular orbital theory

ENERGY CONSERVATION • See Conservation of energy

ENERGY LEVEL

Energy level is a particular value of the total energy that an atomic particle or group of particles can acquire, according to **quantum theory**. In this theory the energy levels for **matter** and **radiation** must fall into a discrete, quantized set.

Based on observations of absorption and emission spectra, a number of scientists pointed out in the latter half of the nineteenth century that the elements each have a set of characteristic wavelengths of electromagnetic radiation that they absorb when cool and emit when hot. An element only absorbs or emits radiation of its characteristic wavelengths. Any useful model of the **atom** must account for this observation.

Joseph John Thomson proved the existence of the **electron** in 1897. In 1911 **Ernest Rutherford** proposed the nuclear model of the atom with a heavy **nucleus** at the center and much lighter electrons moving in orbits around the nucleus, analogous to the solar system. In 1913, **Niels Bohr** combined Rutherford's nuclear model of the atom with the quantum ideas of **Max Planck** to propose a model of the **hydrogen atom** that introduced the concept of quantized energy levels and explained the characteristic absorption and emission spectra of hydrogen.

Bohr assumed that the energy that an electron in an atom may have is quantized, i.e., it can possess only certain amounts of energy. If an electron is to change energy, it can do so only by moving from one of the allowed quantized energy levels to another allowed level. It absorbs energy to move from a permitted level of lower energy to one of higher energy. It emits energy when it moves from a higher energy level to a lower energy level. Each element would be expected to have its own characteristic set of allowed quantized energy levels, based on the attractive **force** between its nucleus and electrons. This explained the absorption and emission spectra of an element; only frequencies of radiation whose energies correspond exactly with the energy gaps between allowed electron energy levels could be absorbed or emitted. Bohr's model predicted the frequencies that the quantized hydrogen atom could absorb or emit; the predicted values matched the experimentally observed values. Unfortunately, efforts to extend Bohr's theory to atoms with more than one electron failed.

In 1923, **Louis de Broglie** proposed that particles possess wave properties, and this was confirmed in 1927 by **Clinton Davisson**, **Lester Halbert Germer**, and George Thomson (1892-1975). In 1926, based on de Broglie's proposal, **Erwin Schrödinger** proposed a theory, known as wave or **quantum mechanics**, that is based on a quantized wave equation and could potentially explain the activity and properties of particles such as electrons in atoms. The solutions of this equation for hydrogen match exactly the predictions of the **Bohr theory**. Although Schrödinger's wave equation is not soluble exactly for systems of more electrons than hydrogen, appropriate approximations led to solutions for the allowed energies of electrons in other atoms that match observed absorption and emission spectra. Predictions made by wave mechanics for other properties also agree with experiment. Thus the working model of the atom has been revolutionized, and the concept of quantized energy levels has become firmly established.

Each solution of **Schrödinger's equation** for atomic systems is called an orbital. Physically, the orbital describes the **space** in which the electron can move. Each orbital has a set of four **quantum numbers** associated with it. The principal quantum number n may have any positive, nonzero integer value (1, 2, 3...). It is the primary determinant of the allowed energy levels of electrons. The **angular momentum** quantum number l may have only integer values from 0 to (n-1). The energy of an orbital depends slightly on the value of l; **orbitals** with higher values of l are higher in energy. Different values of l correspond to different orbital shapes that are designated s, p, d, and f. (The letters are derived from spectroscopic observations and refer to "sharp," "principal," "diffuse," and "fine.") The s orbital is spherical while the p orbital is dumbbell shaped.

The magnetic quantum number m_l may have any of the integer values -1 to +1 including 0 (...-1, 0, +1...). The spin quantum number s may have only the values +1/2 and -1/2. The energy is not affected by the value of the magnetic and spin quantum numbers. Each orbital has a different set of these four quantum numbers.

In 1925, **Wolfgang Pauli** showed that no two electrons in an atom may occupy the same orbital. Since the spin quantum number may have only the values +1/2 and -1/2 and since the value of spin does not affect the energy nor any other part of the **wave function**, often we speak of two electrons being in each orbital, one with spin of +1/2 and the other with spin of -1/2.

The quantum numbers corresponding to the possible energy levels of an atom are as follows (with energy increasing as the value of n and l increase):

There is only one orbital with n = 1, because when n = 1, l = 0 and m = 0. This is called the 1s orbital, and two electrons may occupy it (one with spin +1 and one with spin -1).

There are five orbitals with n = 2: a 2s orbital with n = 2, l = 0, m = 0 containing two electrons; three orbitals, called 2p orbitals, of slightly higher energy for which n = 2, l = 1, and m has the values -1, 0, and +1; each of these p orbitals may contain two electrons, and six electrons may thus be accommodated in these p orbitals.

There are nine orbitals with n = 3: a 3s orbital that may contain two electrons; three 3p orbitals of slightly higher energy, which may contain as many as six electrons; and five orbitals with m values of -2, -1, 0, +1, and +2 that are called 3d orbitals and that have slightly higher energy than the 3p orbitals. The 3d orbitals may contain a total of 10 electrons. Similar orbital schemes may be constructed for larger values of the principal quantum number.

The ground state of an atom is the state of lowest energy. It exists when the electrons of the atom are occupying the orbitals of lowest energy. An excited state is one in which one or more of the electrons occupy orbitals of higher energy than in the ground state.

The lone electron of the hydrogen atom will occupy the 1s orbital in the ground state. It could absorb energy and move into a 2s, 2p, 3s, or higher energy level orbital. It would then be in an excited state. The **frequency** of radiation absorbed in the transition from the ground state to one of the other orbitals corresponds to the difference in energy between the ground state and the excited state.

A chlorine atom has 17 electrons. In the ground state, its 17 electrons will be located in the orbitals of lowest possible energy: 2 electrons in the 1s orbital, 2 in 2s, 6 in 2p, 2 in 3s, and 5 in 3p. The following notation is used to described chlorine's electron configuration: $1s^2\ 2s^2\ 2p^6\ 3s^2\ 3p^5$. A bromine atom has 35 electrons, distributed in atomic orbitals as follows: 2 in 1s, 2 in 2s, 6 in 2p, 2 in 3s, 6 in 3p, 10 in 3d, 2 in 4s, and 5 in 4p ($1s^2\ 2s^2\ 2p^6\ 3s^2\ 3p^6\ 3d^{10}\ 4s^2\ 4p^5$). The absorption spectra of these atoms can be correlated with the change of energy of electrons in the transition between the ground state configuration to an excited state configuration: e.g., an electron in a 5p orbital of bromine excited to a 6s orbital.

The energy levels of electrons in a molecule can be understood in a similar way. Generally only the occupied orbital of the highest energy, called the valence shell, in the bonded atoms needs to be considered. The electrons in orbitals of energy below the valence shell remain associated with their own atom and their energy is negligibly affected by the other atoms of the molecule. The wave theory treatment of the shared electrons of a molecule results in a set of molecular orbitals that spread over two or more atomic centers. A molecular orbital is a bonding orbital if its energy is lower than that of the isolated atomic orbitals. Each of the electrons in the valence shell that are shared by the bonding atoms belongs to a molecular orbital. The ground state of the molecule is that in which the bonding electrons are in the molecular orbitals of lowest energy. An excited state occurs when an electron in a low energy molecular orbital moves to an orbital of higher energy. The absorption and emission spectra of molecules can be correlated with such transitions of electrons between allowed quantized energy levels.

The **wavelengths** of electromagnetic radiation that correspond to transitions between energy levels in atomic and molecular systems lie in the visible and ultraviolet regions of the **electromagnetic spectrum**.

The energy of the chemical bond is the energy required to break the atoms apart with the shared electrons in the molecular orbitals reverting to atomic orbitals on the individual atoms. The bond energy is less for a molecule in an excited state than for one in the ground state.

Chemical species possess energy levels other than those of their electrons. There are also translational, rotational, vibrational, and nuclear energies. The energy states of the nucleus of an atom are of little consequence in **chemistry** and may be ignored in this discussion.

The quantum mechanical treatment of molecular rotational energy yields a set of quantized energy levels with allowed energies determined by the product J(J+1), where J is a quantum number with integer values, and reciprocally related to the moment of **inertia** of the molecule. The moment of inertia is related to the **mass** of the molecule's atoms and to the bond distance. The result is a set of quantized energy levels corresponding to the various integer values of J, the lower values of J having lower energy. The molecule will be in its ground state when J = 0, which corresponds to no rotation at all. An electromagnetic absorbtion can occur when the energy of the radiation corresponds to the energy gap between an occupied wave function and an excited state wave function. An absorption spectrum results. The wavelengths of electromagnetic radiation that rotating molecules are capable of absorbing lies in the **microwave** region of the spectrum.

In addition to translation and rotation, a molecule vibrates about an **equilibrium** position. The quantum mechanical treatment of the vibration of molecules yields a set of wave functions with associated energies. The energies of the allowed quantized vibrational states are related to **Planck's constant**, a quantum number v, the strength of the bond, and reciprocally to the masses of the atoms involved. The stronger the bond, the higher the energy required to make it vibrate. The molecule absorbs radiation whose frequency corresponds to the difference in energy between vibrational energy levels. Molecular absorption spectra are observed in the infrared and microwave portion of the electromagnetic radiation spectrum. The strength of the molecular bond can be determined from data obtainable from the vibrational absorption spectrum.

ENERGY TRANSFORMATIONS

Energy is the capacity to do **work** or to produce **heat**. There are many forms of energy, each of which can be classified into three categories: radiant energy, **kinetic energy**, and **potential energy**. Energy can be changed, or transformed, from one form into another. Energy transformation is also called energy conversion. The Système International d'Unités (SI) unit for energy is the joule (J), after **James Joule**, who demonstrated that work can be converted into heat. Lifting a medium-sized potato a distance of 1 m (3.28 ft) would require approximately one joule of energy. Energy is often expressed as the calorie (cal), which is the amount of heat needed to raise the **temperature** of one gram of water by 1°C. One calorie is equal to 4.184 joules. The calorie (Cal), which is used to express the energy in food, is 1,000 calories.

Kinetic energy is the energy of an object in **motion**. An object in motion can cause another object to do work by colliding with it, causing it to move a particular distance. The colliding objects can be a hammer swinging down on a nail, or two atoms colliding in a **chemical reaction**. Examples of kinetic energy are mechanical energy (caused by motion of parts) and **thermal energy** (caused by the random **motion of particles** of **matter**). An object that has potential energy is not moving or doing work. At some point, that object had work performed on it that resulted in energy storage. One example of performing work on an object to give it potential energy is the action of winding a watch. As it is being tightened, the spring is gaining potential energy. Winding a watch, therefore, transforms mechanical into potential energy. Another example of potential energy is water in an elevated tank. If water is allowed to fall on a wheel and the wheel turns, the wheel can be used to

produce **electricity**. The water in the tank has the potential work once it falls; therefore, it has gravitational potential energy (**gravity** is what converted the potential energy of the falling water into kinetic energy). The potential energy of the water was transformed into mechanical energy of the wheel, which was further transformed into electrical energy. Conversion between potential and kinetic energies is quite common. As the potential energy of an object increases, its kinetic energy tends to decrease. Likewise, if the kinetic energy of an object increases, its potential energy decreases. For example, a rubber ball held out of a third-story window has potential energy. If it is released, gravity causes it to fall. As it falls, it loses potential energy because its height decreases. It gains kinetic energy because its **velocity** is increasing. Eventually, when the ball reaches the ground, it has lost all of its potential energy (because it cannot fall any farther) and at the same time reached a maximum kinetic energy.

In 1845, Joule performed an experiment that demonstrated energy transformation both qualitatively and quantitatively. The experiment was not complicated—he placed a paddle wheel in a tank of water and measured the temperature of the water. He then cranked the wheel in the water for a period of time, and read the temperatutre again. He found that the temperature of the water rose as he cranked the paddle wheel. He quantified this observation and discovered that an equal amount of energy was always required to raise the temperature of the water by one degree. He also discovered that it did not have to be mechanical energy; it could be energy in any form. He obtained the same results with electrical or magnetic energy as he did with mechanical energy. Joule's experiments showed that different forms of energy are equivalent and can be converted from one form to another.

These observations led to what is now called the "Law of **conservation of energy**." This law states that any time energy is transferred between two objects, or converted from one form into another, no energy is created and none is destroyed. The total amount of energy involved in the process remains the same. It should be noted, however, that this does not apply to processes, such solar explosions and the generation of energy in **nuclear reactors**, which involve nuclear energy.

Most chemical reactions involve transformations in energy. A chemical reaction is simply the process whereby bonds are broken between atoms and new ones are made. When bonds are broken in a chemical reaction, the reactants involved are colliding with sufficient kinetic energy to break the atomic bonds apart. During the collision, the kinetic energy is transformed into potential energy as the atoms are rearranged and new bonds are formed. Most chemical reactions involve the overall absorption or release of energy. The energy flow resulting from transforming kinetic energy into potential energy is heat.

There are many everyday examples of energy transformation that can be cited. One simple example is the act of lighting a match, which transforms chemical energy stored in the match head into heat and **light**, two other forms of energy. Another example of energy transformation is the stored chemical energy in flashlight batteries. When the flashlight is turned on, the chemical energy is first transformed into electrical energy and ultimately into light energy. Compressing a spring gives it potential energy; releasing the spring converts the potential energy into kinetic energy. Similarly, stretching a rubber band between your fingers gives the rubber band potential energy, and when it is released, the rubber band is given the kinetic energy to fly away. Electromagnetic energy in an electric motor is converted into mechanical energy. In a heat engine, the chemical energy in the fuel is converted into heat energy as it is burned. The heat energy is then converted into mechanical energy. Some energy transformations seem to go "full circle." In a microphone-loudspeaker system, for example, **sound** (which is a form of mechanical energy) is converted by the microphone into electromagnetic energy (electricity). The electricity reaches the loudspeaker, which converts it back into sound.

Quite often, a whole series of energy transformations is needed to complete a particular job. For example, many transformations occur in the operation of an electric hair dryer. The dryer needs to use electromagnetic energy (electricity), which does not simply originate in the outlet on the wall. A fuel source, such as coal, is burned in a **power** plant, releasing the chemical energy of fuel. The chemical energy is converted to heat energy as it is burned. This heat energy is used to change water into steam, and the rising steam is used to turn a turbine. In this process, heat energy has been transformed into mechanical energy. The mechanical energy performs work on a generator, which converts the mechanical energy into electromagnetic energy. This energy reaches the hair dryer in the form of electricity. The electricity is changed to mechanical energy inside the dryer, where it is transformed again into heat and sound.

The ultimate example of energy transformation is that of the radiant energy of the **Sun**. All of the energy on Earth originated from the Sun. The Sun's radiant energy is converted by plants into chemical energy through the process of photosynthesis. This chemical energy is stored in the form of sugars and starches. When these plants are eaten by animals, this chemical energy is either transformed into another form of chemical energy (fats or muscle) or used for mechanical or thermal energy. Whenever an animal eats another animal, a similar process occurs. Petroleum originated from ancient plants and animals buried under many layers of sediment. Ultimately, then, even the fuels we use come from the Sun's energy. The energy we use every day is the result of the transformation of solar energy over millions of years.

Energy transformations occur every time a chemical reaction takes place, a light switch is turned on, or a person takes a breath. Joule's work enabled scientists to understand energy transformations that in turn made many technological advancements possible. Understanding the nature of energy transformation is one step toward understanding how the world works.

ENERGY-MOMENTUM FOUR-VECTOR

In relativistic equations, **momentum** and **energy** conservation are unified. Because momentum and energy are combined

there must be conservation of all four components of the energy-momentum four-vector. Because momentum (p) and energy (E) form a four-vector it is possible to construct an invariant, frame independent, quadratic analysis.

The fundamental change brought to the concept of **space-time** by relativity required a revision of the classical laws of **mechanics** that held only for non-relativistic velocities based on the observation that **time** passes differently for moving observers than it does for stationary ones. This has usually been addressed by discussion of separate moving objects and separate inertial frames of reference, including **Lorentz transformations** between them.

Another approach has been to define two time variables when describing the **motion** of a single object with respect to a single inertial frame. These time variables are the map or coordinate-time, and the moving object or proper-time. This results in considering not one, but two velocities, namely the coordinate velocity and the proper velocity. The coordinate **velocity** measures map-distance traveled per unit map time, and the proper velocity measures map-distance traveled per unit moving object-time. Each of these velocities can be calculated from the other by knowing the velocity-dependence of the moving object's speed of map-time.

The concepts of energy and momentum are closely related, as **space** and time are, and this close relationship can be demonstrated using a 4-vector, or covariant, representation starting with a 4-vector displacement that, for two events in space-time, can be written in terms of the position and time coordinates of the two events in question.

$$\Delta X = \begin{bmatrix} c\Delta t \\ \Delta x \\ \Delta y \\ \Delta z \end{bmatrix}$$

The dot product of this 4-vector is the square of the frame-invariant proper time interval between the two events: $\Delta X \bullet \Delta X = (c\Delta\tau)^2 = (c\Delta t)^2 - (\Delta x^2 + \Delta y^2 + \Delta z^2)$. The dot product may be positive or negative and this implies that the proper time intervals can be either real (e.g., time) or imaginary (e.g., space). The expression can be rearranged to yield:

$$\gamma = dt / d\tau = 1 / \sqrt{1 - (dx / dt)^2}$$

Using γ and the components of proper velocity w = dx/dτ, the momentum-energy 4-vector can then be written as:

$$P = m \begin{bmatrix} c\gamma \\ w_x \\ w_y \\ w_z \end{bmatrix} = \begin{bmatrix} E / c \\ p_x \\ p_y \\ p_z \end{bmatrix}$$

The dot product of this vector yields the well-known relativistic relation between the total energy E, the momentum p and the frame-invariant rest **mass** energy, equal to mc^2:

$$c^2 P \bullet P = (mc^2)^2 = E^2 - (cp)^2$$

The force-power 4-vector can also be obtained since it is defined as the proper time derivative of the momentum-energy 4-vector.

$$F = dP / d\tau = mA = \begin{bmatrix} c(dy / d\tau) \\ dw_x / d\tau \\ dw_y / d\tau \\ dw_z / d\tau \end{bmatrix} = \begin{bmatrix} 1 / c(dE / d\tau) \\ dp_x / d\tau \\ dp_y / d\tau \\ dp_z / d\tau \end{bmatrix}$$

The dot product of this vector is always negative and it can then be used to define the frame-invariant proper **acceleration**.

See also Event horizon; Minkowski, Herman; Momentum conservation; Relativistic interval; Relativistic mass and energy; Relativistic velocity transformations; Relativity, special; Relativity theory

ENTROPY

Entropy is a measure of the disorder present in a system. Entropy is disorder or randomness.

Entropy is a thermodynamic quantity, its value is equal to the amount of **heat** absorbed or emitted divided by the thermodynamic **temperature** (the absolute temperature, in Kelvin). The units of entropy are joules per kelvin per mole. In chemical equations entropy is normally given the symbol S. The higher the entropy value the more disordered a system is, so for example, of the three states of **matter** a gas, with the greatest degree of disorder, has the highest entropy value. Next are liquids and finally, of the three **states of matter** the most ordered form are the solids and they have the lowest degree of entropy. Any change that increases entropy is a positive entropy change. Most spontaneous thermodynamic processes are accompanied by an increase in entropy.

Entropy is a measure of disorder, it is **energy** in a system that is unavailable for **work**. All compounds have an absolute entropy level that is an indication of the level of randomness of the molecules and the amount of energy that cannot be freed for use. For example, at 68°F (20°C), liquid water has an absolute entropy level of 70 J· mol[-1] and water in the vapor phase has an absolute entropy value of 189 J· mol[-1]. The values of absolute entropy for some other compounds are **oxygen** gas 205, **carbon** monoxide gas 198, carbon dioxide gas 214, solid magnesium 33, and solid magnesium oxide 27. All values are in J· mol[1] at 68°F (20°C). Absolute entropy values are different at different temperatures.

The third law of **thermodynamics** states that for a perfect crystal at a temperature of **absolute zero** on the Kelvin scale the entropy value is zero. This agrees with the work of **Albert Einstein** who predicted that the specific heat of a sub-

stance would approach zero at 0K. Subsequent to Einstein, **Max Planck** said that the entropy value of pure solids and liquids approaches zero as they approach 0K. Obviously if a substance is not pure then the entropy value will be different.

When a reaction occurs spontaneously the entropy of the system and surroundings increases as the change takes place. As time progresses the reaction starts to slow down and it eventually stops. The system is in **equilibrium** and the entropy no longer changes. A diamond is made of carbon in a very ordered manner. As this ordered form becomes more random the diamond reverts back to carbon. In other words, as the entropy of the system increases the diamond reverts to a more disordered form, an incredibly slow process.

A gas will expand spontaneously to fill a given volume, ice will melt spontaneously at a temperature above its melting point, and other similar reactions occur. With these examples the products are in a more random state than the reactants, meaning an increase in entropy. In ice the water molecules are held rigidly in place in a crystal lattice. As the water melts, the water molecules are now free to move about. In solid ice the water molecules are rigidly held in place, it can be reasonably shown where an individual molecule is at any given time. Once that ice has melted, the water molecules are in constant **motion** with respect to each other, it is now much harder to predict where the molecules are, the randomness and entropy of the system have increased. Many other systems show this change, for example, changes from solid to liquid to gas and many substances dissolving in a solvent.

The total change of entropy in a system is dependent only upon the start and final points of the system, the pathway taken from start to finish is irrelevant. A measure of something that does not depend upon the pathway taken is known as a state function, consequently entropy is a state function. A positive entropy value indicates an increase in disorder, a negative entropy value shows that there is a move towards a more ordered state.

The **second law of thermodynamics** is normally stated as being that heat cannot of itself pass from a colder to a warmer body. The second law can be rewritten in the following way (taking into account entropy): any system of its own accord will always undergo change in such a way as to increase entropy. When applying the second law of thermodynamics it is important to take into account the surroundings as well as the system under consideration. In a reversible process, the total change in entropy in the system and the total change in entropy in the surroundings is zero. In an irreversible process, the total change in entropy in the system and in the surroundings is greater then zero, it is a positive value. The effect of this is that in the **universe** as a whole, entropy increases. Unlike energy, entropy is not conserved and the universal level of entropy is constantly increasing.

Cleaning a house can be used as an example of increasing entropy. The house starts in a random, untidy form. A person comes along and cleans so the house is now neat. The house (the system) is now more ordered, it has had a decrease in entropy. The person doing the work, however, has had to metabolize food, this has given out heat (energy) into the surroundings. The second law of thermodynamics tells us that the increase in entropy in the surroundings must be greater than the decrease in entropy in the system, so the overall entropy level in the universe has been increased by the act of tidying up a house, even though at the level of a human observation it might appear that the entropy has decreased as the house is now more ordered.

Humans add order to the environment around them. They extract metals, build things, and generally produce apparent order out of chaos. However, the overall level of disorder is increasing. For each of the events listed, energy is required, which will add heat to the surroundings and cause entropy to increase in the universe. As solid food is ingested some of the waste and by-products are released as liquids and gases, states with higher levels of entropy. In a **chemical reaction,** when two substances react together, chemical bonds are formed. These chemicals now have the constituent components joined together, reducing their freedom to move independently, so the molecules concerned now have less entropy than before. The overall entropy of the universe will still increase in some other way, for example, heat may be liberated by the reaction.

Ludwig Boltzmann realized that the disorder of a system is related to the number of possible arrangements of the molecules. Boltzmann devised the equation $S = k \ln W$, where S is entropy, k is the Boltzmann constant (1.38×10^{-23}J K^1) and ln W is the natural log of the number of possible arrangements in the system. For example, a pure crystal at absolute zero has only one possible arrangement.

Entropy is a measure of randomness in a system. The universe tends towards disorder as a result of constantly increasing entropy.

EPICYCLES

Epicycles were mathematical modifications added to the **geocentric theory** to ensure that the model was consistent with astronomical observations. Geocentrism proposed that Earth was a stationary object at the center of the **universe** and that the **planets**, the **Moon**, the **Sun**, and the sphere of the fixed **stars** revolved about Earth in a series of concentric orbits. This model implies that the planets always traverse the night sky in the same direction, from west to east with respect to the stars. Three planets visible to the naked eye (Mars, Jupiter, and Saturn) do not behave in this manner. Rather, their orbits exhibit the property of retrograde **motion** in which the planet appears to slow down in its eastward motion, stop, and briefly change its direction, traveling west with respect to the stars. It again stops and resumes its eastward journey. To explain this phenomenon, the astronomer **Ptolemy** proposed the notion of epicycles.

Epicycles are smaller circles pinned to a circle around Earth. The planet revolves around the center of the epicycle, which is called its deferent. The deferent is what revolves around Earth. Suppose that the deferent is rotating clockwise around Earth and that the planet is also rotating clockwise around the deferent. When the planet is closest to Earth, it travels in the direction opposite that of the deferent. Projecting its motion against the distant stars, it appears to have changed direction. In this way retrograde motion is explained.

A further emendation is required in order that the planets move with uniform angular **velocity**. Earth is not positioned at the center of the deferent's orbit, but it is instead displaced to one side by a distance called the eccentricity. A planet travels with uniform angular velocity with respect to another point, which is also displaced from the center by the eccentricity in the opposite direction as Earth. Such a circle, where Earth is not at the center, is called an equant.

Combining epicycles, deferents, and equants, and sometimes invoking epicycles nested upon epicycles, it becomes possible to fit the observed data regarding a planet's orbit to the geocentric model. Ptolemy's theory is therefore not a predictive one. Newer and more accurate measurements invariably necessitate further fine-tuning, and with a sufficient amount of tinkering, the model can reproduce the observations. The model does not tell us how or why the planet moves as it does. Also, the refinements introduced by Ptolemy are inconsistent with the philosophical ideas of **Aristotle**, who believed that planets traveled on crystalline spheres. With epicycles, the planets would crash into those spheres. In a heliocentric model with planets orbiting the Sun in ellipses, the ad hoc complications of geocentrism disappear.

EPR (EINSTEIN, PODOLSKY, ROSEN) PARADOX

The EPR paradox was proposed by physicists **Albert Einstein**, Boris Podolsky, and Nathan Rosen (EPR) in a 1935 *Physical Review* article titled "Can Quantum Mechanical Description of Physical Reality Be Considered Complete?" The EPR paradox was intended to demonstrate the inadequacy of the prevailing interpretation of quantum **mechanics**, specifically in regard to the relationships between distant particles predicted by quantum nonlocality. The paradox argues for incompleteness or hidden variables in the quantum mechanical description of nature.

One of the foundations of **quantum theory** is the **uncertainty principle** that places limits on the accuracy of measurements on the atomic scale. According to the uncertainty principle, it is impossible to measure two complementary variables such as the **momentum** and the position of a particle with exact precision. Likewise, the components of a particle's spin cannot be all known simultaneously.

If, for example, two particles have some correlated property, such as momentum, spin, or polarization state, such that they have opposite values of that property (in the case of spin this could result from a spin 0 particle decaying into two spin 1/2 particles (one with +1/2 spin the other with -1/2 spin). The particles, designated A and B, are then allowed to travel a long distance from each other in order to rule out the possibility of an exchange or signal traveling between them during the course of later measurements. A measurement performed on A to determine its z component spin, would fix A's z component spin at the expense of uncertainty in other quantum properties. According to EPR, if particle B were separated from A by **light** years of distance, according to special **relativity theory** (wherein faster-than-light travel is not possible), the measurement of A cannot instantly affect the measurement of B.

The paradox is established because the particles are also bound by the law of conservation of **angular momentum**. Thus if the measurement on A finds its z component spin to be +1/2, then the z component of B's spin must, at that instant be -1/2 (i.e., if they together originally had zero spin, then their spins must still add up to zero). We could now choose to measure some other component of the spin of particle B. If a measurement of the x component is performed, then since the z component is already known through the correlation of the spins, the result is that we know values for both the x and z components of the spin simultaneously. This extra information would contradict the limitations of the uncertainty principle.

One of the motivations of the EPR experiment was the desire of the proponents to demonstrate the existence of "objective reality"; that is, that a physical system contains more information than is specified by the Schrödinger equation, which gives probabilities for finding a particle in a given state but does not simultaneously define all of the physical parameters of the system. According to the EPR proponents, a system has at all times real values for momentum, position, spin, and all other parameters regardless of whether they are measured or not, a belief Einstein summed up in his remark that "God does not play dice." Their explanation of the fact that the EPR experiment provides more information about a system than is strictly allowed by the uncertainty principle is that the information was there all along, and that all of the components of spin were defined from the beginning of the experiment. Theories of this sort are sometimes referred to as hidden-variable theories, because they suggest that there may be parameters that completely define the state of a particle or system, but that are not necessarily accessible to experiment due to the fact that measurement disturbs the system. Such theories were countered in 1964 by **Bell's theorem**, which placed restrictions on the types of hidden-variable theories that could reproduce the results of **quantum mechanics**.

Subsequent tests, such as the French physicist Alain Aspect experiment in 1982, have verified Bell's theorem and unfavorably resolved the EPR paradox against its proponents. By studying quantum properties of paired photons shot away from each other, Aspect demonstrated that the only way a hidden-variable theory can be valid is if the principle of locality is abandoned and it is assumed that the two particles, A and B, can interact instantaneously no matter how far separated they are. As this would violate the speed-of-light **velocity** limit of special relativity, it casts severe doubt on the validity of the concept of objective reality at the atomic level.

See also Bell's theorem; Complementarity principle; Quantum theory; Schrödinger's equation

EQUILIBRIUM

Equilibrium describes a state or process in that is in balance. When dealing with two or more opposing forces, a state of equilibrium specifies that the opposing forces balance. With regard to **mechanics**, at equilibrium there is zero net **force** on a body (i.e., the body must have constant **velocity**). In **chemistry**, equilibrium specifies a state where processes occur at equal rates, and result in zero net change in reactants and products. Static, rotational, and translational equilibrium all refer to the **motion** of objects. Thermal equilibrium is used when referring to **temperature**. Physical equilibrium deals with changes between gas, liquid, and solid states.

Static equilibrium is a state in which all the forces acting on an object are balanced so that the body experiences constant velocity (including, with regard to an appropriate reference frame, a state of rest). There are two physical conditions that must be met for an equilibrium state to be achieved. First, there must be zero net external forces acting on a body in equilibrium. This condition assumes that all the forces affecting an object are concurrent (i.e., the forces that act along lines that pass through a common point). Forces acting in this manner do not cause rotation of the object. An object with a zero sum of concurrent forces might still rotate, however, unless there is also zero net angular force upon the object. The rotational effect of a force depends on its distance from an anchor or pivot point, as well as its direction and magnitude. **Torque** is the quantity used to measure the turning or rotational effect. For an object to be at equilibrium, there must be zero net torque on an object at any and all pivot points.

A static equilibrium can have three different defined states: stable, unstable, and neutral equilibrium. A stable equilibrium is a situation in which any small change or shift in the equilibrium will result in the object moving back into the original equilibrium position. For example, a brick sitting on the ground is at static equilibrium. If you lift the brick up a small amount, the brick will fall to the ground and back into its original equilibrium position. The original forces (**gravity**) acting on the object pull it back into the position it had before it was disturbed.

An example of an unstable static equilibrium would be a wine bottle balanced upside down on its opening. If the bottle is moved slightly, it will topple and fall. An unstable equilibrium is a situation in which any shift from the equilibrium position causes the object to move further away from the initial equilibrium state.

Neutral static equilibrium describes the case where the forces acting on an object have no effect in maintaining or shifting the equilibrium position of the body. A ball sitting on the floor is in equilibrium with regard to the **gravitational field**. If you rolled it a few inches, the ball will have moved relative to the floor but not relative to the gravitational field. The forces acting to roll the ball had no effect on its gravitational equilibrium.

Chemical equilibrium is a state of balance in which the rates of opposing chemical reactions are exactly equal. This dynamic condition is one in which the rate for consuming reactants to make a product is identical to the speed at which the product breaks down to re-form the reactants. At equilibrium, the concentration of all the substances in the **chemical reaction** does not change with time. A reaction system at equilibrium shows the same tendency to proceed forward to make the products as it does to move in the reverse direction to convert the product back to its starting reactants.

An example of a physical equilibrium is a saturated solution of sugar. When household sugar is added to a glass of water, a point is eventually reached when the solid sugar will no longer dissolve in the water solution. At this saturation point, the rate of sugar molecules in the liquid phase combining to reform the solid equals the rate at which the solid sugar crystals dissolve into molecules that go into solution.

At thermal equilibrium, bodies are, by definition, at the same temperature and are incapable of **heat** transfer. In natural processes, **heat transfer** allows systems to reach thermal equilibrium by lowering the average molecular kinetic **energy** (i.e., temperature) of the warmer body and raising the average molecular **kinetic energy** of the relatively initially cooler body.

See also Force, mass, and acceleration; Moving objects and velocity; Newton's laws of motion; Oscillations; Perturbation theory; Phase changes and phase equilibrium; Phase diagram

EQUIVALENCE OF MASS AND ENERGY

The equivalence of **mass** and **energy**, also called the conservation of mass-energy, states that in any closed system of interacting bodies the total mass (m) plus the total energy (E) is constant. The mass-energy equivalence is expressed by the equation $E=mc^2$ with the constant c^2 being the square of the **speed of light**.

The law of conservation of mass-energy is a generalization of the classical laws of **conservation of energy** and **conservation of mass**. Under the classical (or Newtonian) laws of physics, mass and energy are two separate entities, both with separate conservation laws. The conservation law of energy stated by Prussian physicist **Hermann von Helmholtz** (1821-1894) asserts that in a closed system energy is conserved; that is, the total energy of the system remains constant over time. Energy could be transformed from one form to another (for example, from **kinetic energy** to **heat**, or chemical energy to electrical energy), but the total energy of the closed system must remain unchanged. Similarly, the conservation law of mass, first advanced by French chemist Antoine-Laurent Lavoisier (1743-1794), asserts that the total mass in a closed system must remain constant. The mass was thought to be an intrinsic property of **matter**, unchanged by its speed or position in **space**.

The equivalence of mass and energy is suggested by, and formulated with **Albert Einstein**'s theory of special relativity. Einstein himself originally thought that it might be impossible to find a physical process that could convert mass to energy in usable quantities. Early in the twentieth century, Ernest Thomas Sinton Walton provided the first experimental

confirmation for the equivalence of mass and energy. His experiments demonstrated that atomic nuclei contain enormous energies, and that energy can be transferred between forms of **energy and mass**.

In Einstein's special theory of relativity, mass and energy are intricately connected, so that in a closed system the total "mass-energy" of the system is conserved. With the advent of **particle accelerators**, electrons and positrons can be observed annihilating each other completely, and yielding energy in precisely the amounts predicted by Einstein's theory. A direct consequence of this view about mass and energy is that mass and energy are not only intertwined, but actually interchangeable. Stated another way, mass is simply another form of energy.

To better understand the view of mass as a form of energy, a comparison of the concept of kinetic energy from both the classical and relativistic viewpoints is useful. The kinetic energy (KE) of a particle is due to its **motion**. In the classical view, the kinetic energy of a particle is $1/2mv^2$, where m is the particle's mass and v its **velocity**. If an object is at rest, its kinetic energy is zero. Any change to the kinetic energy of a particle is due entirely to a change in its velocity v because the mass of the particle does not change with time and it is unaffected by the particle's position or speed. Mass is said to be invariant; that is, the mass is constant.

In contrast to the Newtonian view, **relativity theory** states that mass is not invariant, but increases as the speed of the particle increases. The Lorentz-FitzGerald contraction is an equation for calculating the increase in mass of a particle as a function of its speed. By applying the binomial theorem, an approximation to the Lorentz-FitzGerald contraction becomes $m_1=m_0(1 + v^2/2c^2)$, where m_1 is the mass in motion and m_0 is the mass at rest. At ordinary velocities much smaller than the speed of light, $v^2/2c^2$ is very small, resulting in m_1 equaling m_0 to a high degree of accuracy. In other words, at small velocities, there is very little noticeable change in the mass of an object. But when speed v of a particle approaches the speed of light c, the increase in mass $(M=m_1 - m_0)$ becomes $M=1/2m_0v^2/2c^2$. Because kinetic energy is equal to $1/2m_0v^2$, the equation reduces to $M=KE/c^2$, or the more familiar $E=mc^2$.

When waveforms are considered, even the energy for massless particles (e.g., photons) can be calculated as a product of their **frequency** and **Planck's constant**.

An important example of the mass-energy relationship is **nuclear fusion**. In the **Sun**'s interior (and the interiors of many other **stars**) hydrogen atoms are fused under enormously high pressures and temperatures into **helium** atoms. The process heats up the core of the Sun and maintains the fusion reaction. Excess heat from this process migrates from the core and eventually finds its way to the surface where some of the Sun's energy is radiated into space in the form of electromagnetic **radiation** (including visible **light**). Energy for these processes is derived from the hydrogen-to-helium conversion process because a tiny fraction of the total mass involved has been converted into radiation. In order for the Sun to maintain its radiation output, it must convert approximately 4,600,000 metric tons (4,173 million kilograms) of mass into energy each second. This is a vast quantity by human standards, but is insignificant to the Sun. At this rate the Sun can continue to radiate in an essentially unchanged fashion for approximately another five billion years.

See also Energy transformations; Length contraction; Lorentz transformations; Newton's laws of motion; Relativistic mass and energy; Relativity, special

EQUIVALENCE PRINCIPLE

In modern physics the equivalence principle states that an observer obtains the same measurement for the **mass** of an object whether such calculation of mass is derived from the **inertia** of the object or is measured by the **force** of **gravity** on the object.

The equivalence principle is applied to two related concepts of gravitational theory. The weak equivalence principle relates to **Galileo**'s famous experiment that demonstrated, by dropping objects of various composition or rolling them down inclined planes to determine if they fall at the same rate, that all bodies fall at an equal rate regardless of their mass or chemical nature. **Isaac Newton**'s theory of gravity advanced the weak equivalence principle in a slightly different form in its assumption that inertial mass and gravitational mass are equal in a given body. This equivalence, which Newton also tested by comparing the **motion** of pendula of various materials, simplifies the mathematical formulation of gravity. Experimentally, the assumption implies that bodies of different masses will fall at the same rate, because although gravity exerts a force proportional to a body's mass, the body also resists that force by an amount proportional to that mass, resulting in an **acceleration** that does not depend on the mass of the body.

Prior to the acceptance of the weak equivalence principle it was believed that heavier objects fell faster than lighter ones, a theory that can be traced to **Aristotle**, and that was accepted until it was finally put to experimental test by Galileo.

The strong equivalence principle, also called Einstein's equivalence principle, relates acceleration and gravity, and led him to conceive of his general theory of relativity. **Albert Einstein**'s equivalence principle involving the relationship between the force of gravity and accelerated bodies was, as he described in his writings, "the happiest thought of his life." In his thought experiment Einstein imagined falling off a roof and realized that, while falling, a person experiences no force of gravity, and that any other objects dropped while in free-fall would appear to be motionless, since they would be falling at exactly the same rate as the observer. He concluded that such an observer is in the same condition as if the observer were unaccelerated and far from a **gravitational field**, and that physics should describe both states in the same way.

Likewise, in a closed laboratory aboard a uniformly accelerating spacecraft, the falling of objects and other phenomena will be identical to phenomena in an earthbound laboratory experiencing an equivalent acceleration from gravity. In the spacecraft, the floor "rushes up" to meet falling objects because of the ship's acceleration, and on Earth the objects accelerate to the floor. In both cases the apparent motion of the objects is identical. Strictly speaking, this similarity is not exact in most real situations because of **tidal forces**. Gravitating bodies such as **stars** and **planets**, being spherical, attract objects to their centers, and so falling objects tend to converge rather than fall parallel to each other. Because of this, the strong equivalence principle is only exact for infinitesimally small regions, or in gravitational fields that lack tidal effects, which are unlikely to occur in nature. Nevertheless, the strong equivalence principle was a key influence in guiding Einstein towards the development of the highly successful general theory of relativity. In free-fall there are no discernible effects due to gravity and special relativity provides an accurate description of motion identical to the inertial state.

See also Frame of reference; Relativistic velocity transformations; Relativity, general; Relativity, special

ERATOSTHENES OF CYRENE (C.276-C.195 B.C.)

Greek mathematician and geographer

Eratosthenes was one of the most versatile scholars of the ancient Greek world. As a mathematician and chief librarian at the Museum of Alexandria, in Egypt, he worked at the center of Greek science. He was the founder of mathematically based geography and was famous for his accurate measurement of Earth's circumference. Eratosthenes, along with his friend **Archimedes**, **Euclid**, and other third-century B.C. mathematicians, raised Greek science to unprecedented heights.

Eratosthenes was born in Cyrene, a prosperous Greek city-state in North Africa (modern-day Libya), about 276 B.C., although his birthdate is sometimes placed as early as 285 B.C.. The son of Aglaos, he studied with the grammarian Lysanias in Cyrene and perhaps with the poet Callimachus in Alexandria and Eratosthenes achieved early renown as a poet. He then spent several years in Athens, studying astronomy and geometry with Arcesilaus, and philosophy with Bion the Cynic, Ariston of Chios, and Apelles, at the Athenian Academy. Subsequently, Eratosthenes was invited by **Ptolemy** III (Euergetes I) to return to Alexandria as tutor to the royal family. About 235 B.C., he became head of the Alexandrian Library.

Prior to Eratosthenes, geography was based primarily on the stories written down by the poet Homer. Eratosthenes argued that poetry was for enjoyment and should not be used as the basis for scientific geography. So he set about mapping the inhabited world, from the Atlantic Ocean to the Ganges River. With the knowledge of the Great Library at his disposal, he began by dividing the known world into northern and southern divisions with an east-west line that passed through the Straits of Gibraltar and the island of Rhodes, reaching the Himalayan Mountains in the east. At right angles to this line, he placed his fundamental north-south meridian, passing through Alexandria. Parallel lines were drawn through major landmarks such as Sicily, the Pillars of Hercules, the Euphrates River, and the tip of India. He pointed out that it would be possible to sail due west from Spain across the Atlantic Ocean and reach India. In the following century, Eratosthenes's work was extended by his vocal critic, Hipparchus of Nicaea, who used astronomical observations to accurately define longitude and latitude.

Eratosthenes's most famous accomplishment was his accurate determination of the circumference of Earth. He had heard from travelers that there was a well at Syene, a Greek city in Egypt near modern Aswan, where, at noon on the summer solstice, the **sun** did not cast a shadow, indicating that it was directly overhead. From his map, Eratosthenes thought that Syene was due south of Alexandria, on the same meridian. Therefore, he measured the shadow of a rod in Alexandria at noon on June 21 and, using simple geometry, calculated that the sun was 7°12' from overhead. Since this was one-fiftieth of a full circle, he deduced that Earth's circumference was 50 times the distance from Alexandria to Syene. He determined this distance based on how far a camel traveled in one day, and the number of days it took to make the journey. Although his calculation for the angle of the sun was extremely accurate, his camel calculations were less so. Nevertheless, his was the most accurate value for Earth's circumference available until modern times.

Geographica, which dealt with mathematical and physical geography as well as ethnogeography, was Eratosthenes's most influential work. His terrestrial globe divided Earth into frigid circular zones at each pole (the Arctic and Antarctic), two temperate zones, and the tropics or torrid zone bisected by the equator. The Greeks had long been puzzled by the reverse currents found in straits, but Eratosthenes analyzed them accurately and explained that they were tidal, like the ocean.

Eratosthenes discussed many different mathematical problems in his *Platonicus*, including proportion, progression, and the theory of musical scales. The Greeks solved problems in geometry by using proportions and Eratosthenes attacked the Delian problem, or the duplication of the cube. This was the question of how to find an infinite number of mean proportionals between two straight lines. He designed an instrument called a mesolabe to demonstrate his mathematical method for solving this problem. The sieve of Eratosthenes, his simple method for finding prime numbers, was a major contribution to number theory.

In addition to his mathematical and geographical treatises, Eratosthenes wrote works on grammar and philosophy. *Ancient Comedy* was his most important work of literary criticism. He also compiled a historical chronology of Greek literature and politics, *Chronography*, with a surprisingly accurate dating of events. To achieve this, he developed a system of dating based on the timing of the Olympiads, the original Olympic Games.

An influential member of the royal court, Eratosthenes remained at the Alexandrian Library until his death at about the age of 80. After going blind, he committed suicide by starvation in *c.*195 B.C. Only fragments of his writings have survived.

ETHER

Ether, sometimes spelled *aether*, is a hypothetical substance once thought to fill all of empty **space** and act as a medium for the propagation of electromagnetic **waves**. Ether was also known as the luminiferous ether because of its assumed association with **light** transmission.

Until advancement of the equations of **electromagnetism** put forth by Scottish physicist **James Clerk Maxwell** in the nineteenth century it was assumed that light required a medium of propagation just as **sound** pressure waves require air, water, or solid **matter** for propagation. Accordingly, the ether was thought to be the medium that allowed light to move through the supposed empty space between the **Sun** and Earth. In **Newtonian physics** the ether was also considered to be the basis for determining absolute rest, with all **motion** being absolutely defined in relation to such an ether.

The quest to identify an ether did not end abruptly with the publication of **Maxwell's equations**. There was such a strong sentiment that such an ether was at least necessary to define absolute rest that the quest to identify it continued until the start of the twentieth century. In 1887 the experiments of American scientists **Albert Michelson** and **Edward Morley** demonstrated the constancy of the **speed of light** regardless of the direction the light traveled relative to Earth's revolution about the Sun. The findings were ultimately seized upon by German American physicist **Albert Einstein** to develop the special theory of relativity. Einstein realized that the Michelson-Morley experiments demonstrated that there was no discernible ether or propagation medium to alter the measured **velocity** of light.

Michelson and Morley devised an experiment that used light **interference** to measure effects that might result from Earth's motion through the ether. In particular, the Michelson-Morley experiments of 1887 indicated that no ether-related effects were detected. Irish physicist George Francis FitzGerald (1851-1901) and Dutch physicist **Hendrik Antoon Lorentz** attempted to explain this negative result by postulating that the lengths of objects were changed slightly by their motion through the ether; the problem, however, was not finally solved until 1905, when Einstein published his special theory of relativity. Einstein's theory abandoned the notion of the ether entirely while retaining the Lorentz-FitzGerald contraction effect in a different form.

Postulating that the speed of light is constant for all observers, and that all motion of constant velocity is relative, there was no longer a need for an ether to provide a standard to measure all motion against because the concept of absolute motion was discarded after the emergence of **relativity theory**.

See also Cosmology; Electromagnetic waves; Relativity, special

EUCLID (C.325-C.275 B.C.)
Greek mathematician

Virtually nothing is known about the personal life of Euclid. Where he was born is a mystery. Some stories describe Euclid as having attended **Plato**'s Academy in Athens, Greece. If that is true, he probably attended during the generation after Plato's

Euclid.

death, which occurred around 340 B.C. This circumstance, and the record that Euclid set up a school of mathematics in Alexandria around 300 B.C., places a probable date for the birth of Euclid at around 325 B.C. Euclid's disciple, Apollonius of Perga, is reported to have been the director of Euclid's Alexandrian school around 250 B.C., so a probable date for Euclid's death lies somewhere between 300 and 250 B.C.

Euclid studied and analyzed the entire body of mathematical knowledge that had accumulated in the ancient world up to his time and compiled a book called *The Elements*. The work is divided into 13 smaller books on the subjects of plane geometry, number theory, irrational numbers, and solid geometry. Most of the treatise is not original research, but a brilliant synthesis of the results of centuries of mathematical work. The theorems and axioms of *The Elements* are clearly stated and arrived at through rigorous logic devoid of any superfluous reasoning. The concise style of *The Elements* became the world standard for scientific discourse over the next 2,000 years.

Perhaps best known is Euclid's reputation as the principal codifier of plane, or flat, geometry. Indeed, this branch of mathematics is known as Euclidean geometry. The principles of Euclidean geometry were so powerful in their presentation that for thousands of years people assumed that the workings of the entire **universe** could be described according to them. It was not until the nineteenth century, when the work of Nikolai

Lobachevski, János Bolyai, and Georg Riemann demonstrated the usefulness of **non-Euclidean geometry**, that mathematicians began to seriously consider examining the foundations of Euclid's thought in a critical manner. **Albert Einstein**'s revelation of the non-Euclidean nature of real **space** finally awakened scientists to the limitations of Euclidean geometry. This discovery did not diminish Euclid's work, but showed it to be only one expression of the many subtle ways in which the universe demonstrates relationships between objects in **space** and events in **time**.

From the modern point of view, Euclid's work has certain flaws. Improvements were made on Euclid's postulates in 1899 by **David Hilbert** and others with the intention of making Euclidean principles the basis from which the entire structure and language of mathematics could be derived using logic alone. The revolutionary discovery by Kurt Gödel in the 1930s that no self-consistent mathematical system could be derived in this way put an end to attempts to extend Euclid's work.

EUCLIDEAN SPACE • See Space-time

EUROPEAN CENTER FOR NUCLEAR RESEARCH (CERN)

The European Center for Nuclear Research (CERN) houses the largest magnet in the world. During its course of operation, many of the world's foremost particle physicists have worked or conducted research at the CERN facility. Medical imaging and the World Wide Web were both derived from CERN research. The World Wide Web portion of the Internet actually began as a method for computer scientists at CERN to track high-energy collaborations from physicists all over the world.

CERN was completed in 1954 and is located near Geneva, Switzerland. The facility was one of the first joint scientific ventures built in Europe. Membership has grown from 12 original countries to more than 20 participating countries. About 3,000 people work at CERN, along with visiting scientists who customarily remain on the roster of their home university or research institute. It is not uncommon for researchers at CERN to represent more than 500 universities and more than 80 nationalities.

CERN houses the world's largest particle accelerator, the Large Electron Positron (LEP) collider, built in 1989. **Power**-generated by electromagnets, the LEP sends **subatomic particles** around a circular tunnel at the rate of 11,200 times per second. The tunnel has four points where the magnets can push **electrons** and their **antimatter** positrons into collision. Banks of **computers** record the reactions.

The LEP has helped scientists learn about the conditions in the early **universe** as it evolved following the **big bang**, about 15 billion years ago. In October 2000, the LEP was scheduled for shut down to make way for a new experiment, the **Large Hadron Collider** (LHC), which is scheduled to be operational by 2005. The LHC will be able to make particles collide with greater **velocity** and at much higher energies than the LEP.

When a particle and an **antiparticle** collide, they annihilate and produce a burst of electromagnetic **radiation**. A landmark experiment performed at CERN in September 1995 used xenon gas to force particles and antiparticles together. Most of the protons and antiprotons were destroyed, but a few of the collisions resulted in a positron by-product. From these **positrons**, an even smaller number combined with antiprotons to form antihydrogen. Over the three-week course of the experiment, nine antihydrogen atoms were observed. Once researchers proved that antihydrogen exists, they started looking for other methods of creating it. Scientists at CERN continue to experiment with a technique that traps and cools positrons and antiprotons with a system of magnets, electrical fields, and **lasers**.

Plans for international physics research did an about-face in 1993 when the U.S. Congress canceled an $11 billion superconducting supercollider project, after first spending $1.9 billion to dig a tunnel for the supercollider. The Congressional cancellation of the supercollider project in the United States put pressure on CERN to upgrade. When the $2.15 billion construction of the Large Hadron Converter is completed at CERN in 2005, protons will be able to collide at **energy** levels never attempted before (e.g., about 7,000 billion electron volts or more than 140 times more energy than CERN collisions currently produce).

Other research at CERN involves the use of a super **proton** synchrotron. When LHC is ready, two of the main experiments will involve compact **muon** solenoid (CMS), working with unstable leptons; and ATLAS, doughnut-shaped apparatus specifically designed to work with LHC. Both experiments will register high-momentum transfer events. Other experiments planned for LHC include ALICE, a large ion collision experiment to track collisions between heavy **ions**; and LHC-B, an experiment to study charge conjugation and **parity** inversion.

See also Accelerators; Particle physics; Synchrotrons

EVAPORATION AND CONDENSATION

Evaporation and condensation are the same process going in different directions: evaporation is the process of a liquid turning into a gas, and condensation is the process of a gas turning into a liquid. The term evaporation refers to the process when it is unforced: evaporation occurs well below the **boiling** point, as can be seen in daily life. These processes are studied as a part of the larger field of **thermodynamics**. Each substance will have its own energies of condensation and evaporation, due to the size and bond structure of the molecules.

Evaporation occurs with those molecules of a liquid that have greater kinetic **energy**. These molecules can have enough energy to escape the **surface tension** of the liquid and the attractive molecular forces within the liquid. The escape of the more energetic particles will make the average **temperature** of the

remaining particles less, since temperature is an expression of **kinetic energy**. This is why wiping one's forehead with a cool cloth will cause a decrease in temperature. The evaporation process is the same process that rules the escape of particles into **space** from Earth's atmosphere; it explains why there is not much free hydrogen or **helium** in the air on this planet.

Condensation is the opposite process: it involves gaseous molecules cooling down to the liquid stage. This process requires the **mass** of molecules to give up a certain amount of **heat**, which was equal to the heat they would have had to gain to be vaporized or evaporated. The energy goes into the molecular organization and strength of bonds present in a liquid but not present in the equivalent gas. The heat given up is the same as the **work** going into the molecular attraction when the molecules are in liquid form.

Evaporation and condensation can be seen in simple things, from the water beads on a glass on a hot summer's day to the steam on the mirror after a hot shower. They are widely used in heating and cooling technologies, including refrigeration and some forms of climate control. In some parts of the country, buildings are heated and cooled on the same steam system, which will run in condensing reverse in the summer from its winter evaporative steam heating procedure. Evaporating and recondensing are also used in purification systems: the system may be heated to the point where much of the water (or other desired liquid) evaporates off. The air containing the evaporated molecules is then shunted into a different region, where it is cooled to condensation in a purer form. This process is used industrially on waters and alcohols, among other substances.

EVENT HORIZON

The boundary of a **black hole** is termed the event horizon. The size of the event horizon surrounding a black hole is termed the **Schwarzschild radius**, named after the German astronomer Karl Schwarzschild (1873-1916), who studied the properties of geometric **space** around a **singularity** when warped according to general **relativity theory**. **Light** is bent by **gravitational fields**, and within a black hole the **gravity** is so strong that light is actually bent back upon itself. Another explanation in accord with the wave-particle duality of light is that inside the event horizon the gravitational attraction of the singularity is so strong that the required escape **velocity** is greater than the **speed of light**. As a consequence, because no object can exceed the speed of light, not even light itself can escape from the region of space within the event horizon.

The event horizon is an observational boundary on the cosmic scale because no information generated within the black hole can escape. Processes occurring at or near the event horizon, however, are observable and offer important insights into the physics associated with black holes.

An accretion disk surrounding a black hole forms as **matter** accelerates toward the event horizon. The accelerating matter heats and emits strong, highly energetic electromagnetic **radiation**, including **x rays** and **gamma rays**, that may be associated with some of the properties exhibited by **quasars**.

Although light and matter cannot escape a black hole by crossing the event horizon, English physicist **Stephen Hawking** formed an important concept later named **Hawking radiation** that offers a possible explanation to possible matter and **energy** leakage from black holes. Relying on quantum theories regarding **virtual particles** (particle-antiparticle pairs that exist so briefly only their effects [not their masses] can be measured), Hawking radiation occurs when a virtual particle crosses the event horizon and its partner particle cannot be annihilated and thus becomes a real particle with **mass** and energy. With Hawking radiation, mass can thus leave the black hole in the form of new particles created just outside the event horizon.

In general usage, an event horizon is a limiting boundary beyond which nothing can be observed or measured with regard to a particular phenomena (e.g., the cosmic event horizon beyond which nothing can be observed from Earth).

EVOLUTION AND ENTROPY

In physics and science the term "evolution" simply refers to change over time. The second law of **thermodynamics** states that, in an isolated system, **entropy**, a measure of the amount of **energy** in a system unavailable to do **work**, must increase as time passes.

Thermodynamics is the study of the transformation of energy (a state function of a body or system) and changes in entropy. In accord with the laws of thermodynamics, physical systems evolve or undergo changes so that the system seeks a state reflecting the lowest possible energy (unless it is totally isolated, in which case the total amount of energy will remain constant), and a state of maximum thermodynamic disorder, that is, maximum entropy. Considering both the system and surroundings, in accord with the **second law of thermodynamics** dealing with entropy, all real and natural processes are irreversible. In this regard, entropy provides time and reaction directionality to all processes in the **cosmos**.

Entropy may be treated in two different ways. In statistical thermodynamics, entropy is the number of states that a system can assume. The number of possible configurations (i.e., the number of states available to the system) is a measure of the amount of disorder, or entropy.

In classical thermodynamics, the change in entropy is defined as $dS = Q/T$ where Q is amount of **heat** flowing into, or out of, the system, and T is the absolute **temperature** of the system. Accordingly, entropy is a quantitative basis for the fact that naturally occurring processes have a direction, i.e., in the direction of increasing entropy.

The classical example idealizes that two bodies of different temperatures are brought into contact with each other so that the change in entropy at any instant is then $dS = Q/T_1 - Q/T_2$, where Q reflects the loss of heat in the hotter body and the corresponding increase of heat in the cooler body. When $T_1 = T_2$, then $dS = 0$ and entropy reaches a maximum, and changes in heat (commonly referred to as "heat flow") between the two bodies stops.

The proper application of entropy concepts requires consideration of both system and surroundings. Key to understanding the effect of entropy on any evolutionary process is the fact that the drive toward maximum entropy occurs in the absence of work done to maintain or increase the ordered state of a system. Such increased order is achieved at the expense of increased entropy in the system's surroundings. It would be an errant application of entropy theory, for example, to suggest that molecular organization and other evolutionary processes could not occur on Earth because such organization, reflecting a more ordered state, would violate the thermodynamic law concerning entropy. Solar **radiation**, cosmological radiation, gravitational heating, and nuclear processes provide highly energetic sources of work available to evolutionary processes that result in more ordered states within Earth's localized system.

See also Absolute zero; Cosmology; Reversible processes

EVOLUTION OF THE UNIVERSE • See

Cosmology

EXPANSION, THERMAL

Thermal expansion is the change in size of an object as the **temperature** changes. Normally, as the temperature increases, the size of an object also increases. Conversely, the object will shrink as the temperature drops. Look at electrical **power** lines on a hot summer day. They will sag more than on a cold winter day. If you look at many bridges, they have interlocking metal fingers forming an expansion joint, where the bridge joins the road. This is to accommodate thermal expansion. (One notable exception to this general rule is water. It usually undergoes a normal thermal expansion, but as it approaches its freezing point, it will expand instead of shrink.) As an object expands or contracts with a temperature change, its change in length depends on three quantities: the original length, the temperature change, and the thermal properties of the material composing the object. Think about a metal rod that is 3 ft (1 m) long and has its temperature increased to the point where it expands by one hundredth of a millimeter. Now consider two such identical rods placed end to end. They are the equivalent of a single rod 6 ft (2 m) long that will expand by two hundredths of a millimeter for the same temperature increase. Hence a larger object will have a greater change in length with temperature than a smaller object, simply because the larger object has more material to expand. Secondly, the change in length also depends on the temperature change. An object will expand twice as much for a 20° temperature change as for a 10° temperature change. In a third variable, different materials also expand at different rates. For example, with a given temperature change, a block of wood will not expand as much as a similar block of metal. The coefficient of thermal expansion, found by experimentation, is a property of the material that accounts for the different expansion rates for different materials. Using these three factors, the change in length of an object is equal to the coefficient of thermal expansion multiplied by the original length multiplied by the temperature change.

The length of an object is not the only dimension that expands or contracts. The height and width also change. As all the dimensions of the object change in size, the volume of the object changes. The change in volume is equal to the coefficient of volume expansion times the original volume times the temperature change. Notice that the change in volume is similar to the change in length, but the coefficient of volume expansion is roughly three times the coefficient of expansion used for the length because an object expands in three dimensions.

When designing bridges, power lines, or anything else that might be subjected to wide temperature variations, engineers must take into account the effects of thermal expansion. Depending on the temperature range in a particular area, they may choose building materials that expand with **heat** or not, different types of joint structures, and what other kind of reinforcements might be needed.

EXPONENTIAL DECAY • See Half Life

EXTRATERRESTRIAL LIFE

Since the dawn of recorded history, Man has pondered the possibility of life beyond Earth. Our ideas about extraterrestrial life have become an integral part of our culture, permeating everything from cinema to psychiatry. We have sent messages aboard deep-space probes, we scan the **electromagnetic spectrum**, and we slice apart ancient meteorites in an attempt to answer ages-old questions concerning the possibility of extraterrestrial life.

Initially, a distinction must be drawn between extraterrestrial life and extraterrestrial intelligence. Extraterrestrial life can encompass anything from microscopic amoebas to intelligent animal forms. To seek extraterrestrial intelligence, however, is to embark upon a search for extraterrestrial culture and communicative abilities. We assume that intelligent extraterrestrial civilizations would be the easiest for us to discover, either through direct contact (which seems unlikely, considering the vast interstellar distances that would presumably have to be crossed) or through an exchange of communications.

Thus far, exploitation of the electromagnetic spectrum holds the greatest promise for successful communication. Not long after it was discovered that we on Earth were constantly leaking our **radio** transmissions into **space**, astronomer Frank Drake (1930-) suggested that alien cultures might be doing the same thing. In 1959, Drake and others from Cornell University began sweeping the sky with radio telescopes, scanning for signals that stood out among the billions of frequencies of background noise. Because radio **waves** cross the vastness of space at no faster than the speed of **light**, it was understood from the outset that any signals received would most likely be thousands or millions of years old. Regardless, radio detection remains the only practical way to search for signs of intelligence outside of our solar system.

By the 1970s, National Aeronautics and Space Administration (NASA) had taken over Drake's work, and named the project SETI (Search for ExtraTerrestrial Intelligence). SETI scientists had at their disposal the large arrays of radio telescopes and powerful supercomputers. Despite the enormous effort, however, technicians were only able to monitor a very small portion of the both the sky and the electromagnetic spectrum in an effort to identify any non-naturally occurring radio source. There were frequent false alarms, but confirmation of an extraterrestrial radio signal never came. In the early 1990s, SETI's funding was cut by Congress, and SETI was forced to continue privately and in a much scaled-back form.

In the summer of 1996, NASA scientists announced the possible detection of artifacts associated with fossils within an asteroid known to have originated from the planet Mars. The story made headlines around the world and remained a focal point of news attention for months. Ultimately, however, most scientists came to the conclusion that based upon the evidence, what at first had appeared to be artifacts associated with fossilized microscopic lifeforms were actually rather common inert crystals.

Although scientists failed to confirm the historical existence of extraterrestrial life on Mars, the generally calm and scholarly debate surrounding the potential evidence of existence of extraterrestrial life largely put to rest fears expressed by some sociologists and theologians that discovery of life beyond Earth might engender mass hysteria.

See also Astronomy; Radio astronomy; Speed of light

EYES AND EYESIGHT

Human eyes work much like a camera. They allow images of a person's surroundings to be perceived by the brain and permit him or her to perceive different shapes, sizes, and colors. The eye is the classic subject of study of **geometric optics**, an important subclassification of modern physics. From studies of the eye, optical physicists have derived not only modern cameras and telescopes, but **lenses** to correct abnormal vision. They have also used the science of **optics** to understand the nature and **motion** of **light**.

The eye sits in a small, hollow area in the skull called the eye socket. Eyelids protect the eye from excessive light and foreign objects. The iris is the central, colored, part of the eye. It has muscles that control how much light enters the pupil, that black circular hole in the center of the iris. The pupils dilate and contract depending on the amount of light— they get bigger in dim light and they shrink in size as the light gets brighter. Constricting the pupil prevents the eyes from being blinded by too much light.

The white, outer part of the eyeball is called the sclera, and it surrounds the cornea. The cornea is a clear dome-shaped surface that covers the front of the eye. A clear fluid, called the aqueous humor, helps the cornea keep its rounded shape. Good vision requires that the cornea is free of any cloudy or opaque areas. The lens is suspended by a bundle of fibers attached to the ciliary muscle. The ciliary muscle squeezes and changes the shape of the lens, depending on the distance of the object being viewed, so that the image of the object can be focused on the retina. When a person looks at something up close, the lens thickens. When something farther away is viewed, the lens becomes thinner. The lens also must remain clear and colorless for proper vision.

Once light hits the lens, it is projected as an inverted image onto the retina. The retina is a tiny postage-stamp-sized tissue full of millions of photoreceptor cells. These cells, called rods and cones, change light into nerve signals that the brain can understand. The optic nerve sends the picture to the brain and the person "sees" the image.

Rod cells are long and thick. They help decipher shape and movement. Rod cells also contain large amounts of visual pigment for vision in low-light conditions. Cone cells sense **color**. The human eye can perceive only a tiny amount of the **electromagnetic spectrum**. This small section is called "visi-

ble light." Cones are sensitive to different wavelengths of light and allow humans to see red, green, and blue.

Sometimes the cornea and lens do not focus images precisely at the retina due to the shape of the eye itself. If the eye is stretched from its classical spherical shape so that it is longer from the front to the back of the skull than it is from the top to the bottom of the skull, the cornea and lens focus light rays in front of the retina. This condition is called nearsightedness (myopia). Objects close-up can be seen clearly, but those at a distance are blurred. If the eye is stretched so that it is taller than it is wide, light is focused behind the retina. This condition is called far-sightedness (hyperopia). In both cases, glasses or contact lenses, using models of geometrical optics, can help focus the image on the retina.

If a doctor describes a person's vision as 20/20, it means that he or she can see clearly at 20 ft (6 m) those things that should normally be seen at that distance, i.e., it is simply an indication of how sharp or clear a person's vision is at a distance. If someone has 20/100 vision, he or she must be as close as 20 ft to see what a person with normal vision can see at 100 ft (30 m).

F

FAHRENHEIT, DANIEL GABRIEL (1686-1736)

Dutch physicist

Daniel Gabriel Fahrenheit invented the first truly accurate thermometer using mercury instead of alcohol and water mixtures. In the laboratory, he used his invention to develop the first **temperature** scale precise enough to become a worldwide standard.

The eldest of five children born to a wealthy merchant, Fahrenheit was born in Danzig (Gdansk), Poland. When he was 15, his parents died suddenly and he was sent to Amsterdam to study business. Instead of pursuing this trade, Fahrenheit became interested in the growing field of scientific instruments and their construction. Sometime around 1707 he began to wander the European countryside, visiting instrument makers in Germany, Denmark, and elsewhere, learning their skills. He began constructing his own **thermometers** in 1714, and it was in these that he used mercury for the first time.

Previous thermometers, such as those constructed by **Galileo** and Guillaume Amontons, used combinations of alcohol and water; as the temperature rose, the alcohol would expand and the level within the thermometer would increase. These thermometers were not particularly accurate, however, since they were too easily thrown off by changing air **pressure**. The key to Fahrenheit's thermometer was a new method for cleaning mercury that enabled it to rise and fall within the tube without sticking to the sides. Mercury was an ideal substance for reading temperatures since it expanded at a more constant rate than alcohol and is able to be read at much higher and lower temperatures.

The next important step in the development of a standard temperature scale was the choosing of fixed high and low points. It was common in the early eighteenth century to choose as the high point the temperature of the body, and as the low point the freezing temperature of an ice-and-salt mixture—then believed to be the coldest temperature achievable in the laboratory. These were the points chosen by Claus Roemer, a German scientist whom Fahrenheit visited in 1701. Roemer's scale placed blood temperature at 22.5°F and the freezing point of pure water at 7.5°F. When Fahrenheit graduated his own scale he emulated Roemer's fixed points; however, with the improved accuracy of a mercury thermometer, he was able to split each degree into four, making the freezing point of water 30°F and the temperature of the human body 90°F. In 1717 he moved his points to 32°F and 96°F in order to eliminate fractions.

These points remained fixed for several years, during which time Fahrenheit performed extensive research on the freezing and **boiling** points of water. He found that the boiling point was constant, but that it could be changed as atmospheric pressure was decreased (such as by increasing elevation to many thousand feet above sea level). He placed the boiling point of water at 212°F, a figure that was actually several degrees too low. After Fahrenheit's death, scientists chose to adopt this temperature as the boiling point of water and to shift the scale slightly to accommodate the change. With 212°F as the boiling point of water and 32°F as the freezing point, the new normal temperature for the human body became 98.6°F.

In 1742 Fahrenheit was admitted to the British Royal Society despite having had no formal scientific training and having published just one collection of research papers.

FAIRBANK, WILLIAM M. (1917-1989)

American physicist

William Fairbank was a pioneer and advocate of **gravity** research, **space** research, and condensed **matter** physics. In the late 1950s Fairbank accepted a position at Stanford University to assist in the establishment of a low **temperature** laboratory. Fairbank served as a professor of physics at Stanford from 1959-89.

William M. Fairbank, 1965. *(UPI/Corbis-Bettmann. Reproduced by permission.)*

In 1961 Fairbank was one of the discoverers of flux quantization. Fairbank was also instrumental in the development of an early type of superconducting accelerator that laid the engineering foundations for later **accelerators.**

In 1963 Fairbank agreed to serve on National Aeronautics and Space Administration's (NASA) first Physics Committee. In an effort to provide experimental evidence regarding the general theory of relativity, Fairbank and others worked on the Gravity Probe B Project that was designed to develop the procedures and technology that would enable researchers to make very sensitive measurements of gravity. The project, utilizing the principles of gyroscopic precession, became an important component in NASA's gravitational physics research program and a further confirmation of general **relativity theory.**

In the 1970s Fairbank and other scientists began to study the effects of gravity on the phase transition related critical point phenomena (e.g., changes in such properties as **viscosity** at or near a phase transition critical point). Of major interest were the potential effects of **microgravity** on such properties and transition. Microgravity describes the reduced gravity found in space. Although there is no region of space where **gravitational force** equals zero, as the distance from Earth increases, Earth's gravitational **force** decreases as an inverse square of the distance. Fairbank contributed his expertise derived from research in the normal fluid-superfluid

critical point transition of **helium** where microchanges in gravity such as those found between the top and bottom of a flask created **pressure** differences that affected phase-sensitive transitions. Fairbank realized that for a sample to be able undergo more simultaneous phase transitions, a microgravity environment with a more uniform pressure was needed.

Along with other scientists Fairbank helped to design experiments related to critical phenomena experiments that were incorporated into NASA's Physics and Chemistry Experiments research program. In the 1980s Fairbank's work was incorporated with other research projects to form NASA's Microgravity Research Program. In particular, Fairbank designed the Lambda Point Experiment to explore the properties of microgravity. The experiment, one of the NASA microgravity program's first fundamental physics experiments, was performed aboard a 1992 flight of the space shuttle.

FARADAY, MICHAEL (1791-1867)
English physicist and chemist

Michael Faraday's early life had a remarkable resemblance to that of **Benjamin Franklin**. Faraday was born on September 22, 1791, in Newington, Surrey, England. Like Franklin, Michael Faraday was part of a large family. His father was a blacksmith who lacked the resources to obtain a formal education for Michael. Franklin had been apprenticed to a printer; young Faraday went a similar route, becoming apprenticed to a bookbinder. In each case, this led to a voracious love of books. Michael was especially interested in chemistry and electricity. He studied the articles about electricity in the *Encyclopaedia Britannica*. His employer not only allowed him to read all that he wanted, he encouraged the boy to attend scientific lectures.

A turning point in Faraday's life occurred in 1812 when he obtained tickets to hear the lectures of Humphry Davy at the Royal Institution. Faraday took careful notes extending to 386 pages, which he had bound in leather. He sent a copy to Sir Joseph Banks (1743-1820), president of the Royal Society of London, who wielded great influence over European scientific investigation. Faraday hoped to make a favorable impression, but if Banks ever looked through the book with its carefully drawn and colored diagrams, Faraday never knew it. Determined not to be ignored, Faraday sent a copy of his notes to Davy and included an application for a job as Davy's assistant. Davy was extremely impressed, but did not offer Faraday work since he already had an assistant. However, after firing the assistant in 1813, Davy contacted Faraday. The job description was not quite what Faraday had in mind. A trustee of the Royal Institution had said, "Let him wash bottles. If he is any good, he will accept the work; if he refuses, he is not good for anything." Faraday accepted, even though it meant he would be paid less than what he was making as a bookbinder.

Shortly thereafter, Davy resigned his post at the Royal Institution, married a wealthy widow, and decided to travel through Europe. Faraday accompanied the couple and met such illustrious men as Italian physicist **Alessandro Volta** and French chemist Louis-Nicolas Vauquelin.

The next major event in Faraday's life occurred in 1820 when Danish physicist **Hans Christian Ørsted** amazed scientists with the discovery that **electric current** produced a magnetic field. Faraday had a greater goal in sight: Ørsted had converted electric current into a magnetic **force**; Faraday intended to reverse the process and create electricity from **magnetism**. Within a year Faraday, now back in England, constructed a device that essentially consisted of a hinged wire, a magnet, and a chemical battery. When the current was turned on, a magnetic field was set up in the wire, and it began to spin around the magnet. Faraday had just invented the electric motor. Faraday's motor was certainly an interesting device, but it was treated as a toy. At this point, Davy, realizing that Faraday had the potential to eclipse him, jealously claimed that Faraday had taken his own idea for the experiment.

Faraday's first major contribution to chemistry came a few years later. In 1823 he unknowingly became the "father" of **cryogenics** by producing laboratory temperatures that were below freezing. He discovered how to liquify **carbon** dioxide, chlorine, hydrogen sulfide, and hydrogen bromide gases by placing them under **pressure**. But, once again Davy took credit for Faraday's work. Two years later Faraday discovered the compound benzene, which became his greatest contribution in chemistry. In 1865 German chemist Friedrich Kekulé added to Faraday's discovery by determining the structure of benzene, which led to a greater understanding of molecular structures in general.

Electricity and magnetism were still the main interests of Faraday, so he elaborated on Davy's pioneering work in electrochemistry. Davy had passed an electric current through a variety of molten metals and created new metals in the process. Faraday named this process **electrolysis**. He also bestowed names that were suggested by the British scholar Whewell and are still in use today: electrolyte, for the compound or solution that conducts electricity; electrode, for the metal rod inserted in the object; and **anode** and cathode for the positively and negatively charged electrodes, respectively.

In 1832 he devised what became known as **Faraday's laws** of electrolysis, which holds that the **mass** of the substance liberated at an electrode during electrolysis is proportional to the amount of electricity going through the solution; and the mass liberated by a given amount of electricity is proportional to the **atomic weight** of the element liberated and inversely proportional to the "combining power" of the element liberated. The two laws showed there was a connection between electricity and chemistry. They also supported the suggestion that Franklin had made nearly 100 years earlier when he claimed electricity was composed of particles, a theory that would be another 50 years in the making.

In another experiment Faraday sprinkled iron filings on a paper that was held over a magnet and noticed the filings had arranged themselves along what he called "lines of force." The connections along the lines showed where the strength of the field was equal. With the magnetic field now "visible," scientists began to wonder if **space** itself was filled with interacting fields of various types and this helped establish a new way of thinking about the **universe**. Up to this point most scientists

Michael Faraday.

had believed in the mechanical nature of the universe as established by **Galileo** and **Isaac Newton**. Taking the concept of his lines of force one step further, Faraday realized that when an electric current began to flow it caused lines of force to expand outward. When the current stopped, the lines collapsed. If the lines expanded and collapsed across an intervening wire, an electric current would be induced to flow through it, first forward then in reverse.

By this time Faraday was giving public lectures, which were very popular, at the Royal Institute, just as Davy had. Faraday reasoned that if electricity could induce a magnetic field, then it should be possible for the reverse to be true. Taking an iron ring during one demonstration, Faraday wrapped half of it with a coil of wire that was attached to a battery and switch. **André Ampère** had shown that electricity would set up a magnetic field in the coil. The other half of the ring was wrapped with a wire that led to a galvanometer. In theory, the first coil would set up a magnetic field that the second coil would intercept and convert back to electric current which the galvanometer would register. Faraday threw the switch: the experiment worked. He had just invented the **transformer**. However, the result was not exactly what he expected. Instead of registering a continuous current, the galvanometer moved only when the circuit was opened or closed. Ampère had observed the same effect a decade earlier but ignored it because it did not fit his theories. Deciding to make the theory fit the observation, instead of the other way around, Faraday concluded that when the current was turned on or off, it caused magnetic "lines of force" from the first coil to expand or contract across the second coil, inducing a momentary flow of current in the second coil. With this observation Faraday had defined electrical induction. Meanwhile, in the United States, physicist **Joseph Henry** had independently made the same discovery.

Faraday's affiliation with Davy had been suffering because Davy was extremely jealous of his former assistant, who was now overshadowing him. The situation escalated following Faraday's invention of the transformer; Davy claimed the idea for the experiment had been his. When Faraday was nominated to become a member of the Royal Society in 1824, Davy cast the only negative vote.

In 1839, at the age of 48, Faraday had a nervous breakdown. Like Newton, who had suffered a similar fate, Faraday never completely recovered. It is also possible that he was afflicted with a low-grade chemical poisoning. This was a common ailment affecting chemists at the time; Davy had suffered from it as well. In any event, Faraday's failing memory forced him to leave the laboratory. He published a book describing his lines of force in 1844, but because he lacked a formal education it was written without mathematical equations and it was therefore not taken seriously. When **James Clerk Maxwell** investigated the subject, he essentially came to the same conclusion as Faraday had, but used mathematics to prove his theory.

Although out of the laboratory, Faraday was by no means inactive. He investigated the effect of weak magnetic fields on nonmetallic substances and coined the terms *paramagnetic* and *diamagnetic* to differentiate between the force of attraction and repulsion. Development of a theory to explain the two opposing forces, however, had to wait more than 50 years for the work of Paul Langevin. In the 1850s, during the Crimean War, the British government sought Faraday's opinion on the feasibility of using poisonous chemical weapons and asked him to oversee their development. Faraday immediately said the project was very feasible, but he refused to have anything to do with its initiation.

Faraday died on August 25, 1867, at Hampton Court, Middlesex, England. The word "farad," which is a unit of **capacitance**, was named in his honor.

FARADAY'S LAWS

English chemist and physicist **Michael Faraday** developed the first scientific understanding of fundamental relations in electrochemical processes. He discovered that the amount of a substance produced or consumed in an electrochemical process depends quantitatively on the amount of **electricity** that flows as the reaction takes place. For example, the thickness of the silver plate that is deposited electrochemically on a spoon is greater if the electrical current is allowed to flow longer.

Just as the amount of water flowing through a pipe is equal to the rate of flow, or the current, multiplied by the length of time that the water flows, the amount of electricity that passes through a wire is equal to the electrical current multiplied by the time. The unit for current flow is the ampere. One ampere or amp is equivalent to one coulomb per second; a coulomb is the measure of charge in the International System of Units (SI) and is equal to 6.6242×10^{18} electrons.

Faraday discovered several relationships that apply to the quantitative measure of electricity. His laws can be summarized using the modern term for quantities of electricity named after him, the faraday. One faraday (1 F) of electricity is equal to one mole of electrons, which is equal to 96,487 coulombs of electricity. If one faraday of electricity flows through an external circuit, 6.022×10^{23} electrons have passed through the wire. Faraday's laws may be condensed into one statement: during an electrochemical process, the passage of one faraday through a circuit results in the transfer of one mole of electrons at the cathode and the transfer of one mole of electrons at the **anode**. Because we know the relationship between numbers of moles and the **mass** of chemical substances involved in a **chemical reaction**, we can determine the mass of a material that is oxidized at the anode or reduced at the cathode in an electrochemical cell if we know how many electrons are transferred for each **atom** of the substance that undergoes the chemical reaction.

Faraday's laws allow us to determine the amount of electricity (the number of coulombs) required to obtain the electrochemical conversion of a known quantity of a substance. Given a particular current flow in coulombs per second, we can also determine the time required to produce a certain amount of a substance electrochemically. The important quantities we must use to carry out these calculations are the molecular weights of the substances, the number of electrons required to oxidize or reduce each molecule of the substance, and, in the case of determining the time required for an electrochemical conversion, the rate of current flow.

FERMI, ENRICO (1901-1954)

Italian American physicist

Enrico Fermi's fame rests on accomplishments in the fields of both theoretical and experimental physics. At the age of 25, he developed a statistical method for describing the behavior of a cloud of electrons that later came to be known as Fermi-Dirac statistics. In combination with another system of mathematics, Bose-Einstein statistics, it provides a method for analyzing any system of discrete particles, such as photons, electrons, or neutrons. Fermi also devised an explanation for the process of **beta decay**, an event in which an atomic **nucleus** emits an **electron** and changes into a new nucleus. Fermi's major experimental contributions involved the study of nuclear changes brought about as a result of **neutron** bombardment of nuclei. This field of research eventually led to Fermi's participation in the **Manhattan Project**, during which the first controlled fission

Enrico Fermi.

reactions were carried out. For his work on neutron bombardment, Fermi was awarded the 1938 Nobel Prize for physics.

Fermi was born in Rome on September 29, 1901. His father, Alberto Fermi, was employed by the Italian state railway system, while his mother, Ida de Gattis, had been a schoolteacher before her marriage. The Fermis had two other children, a son, Giulio, and a daughter, Maria. Giulio died when Enrico was 14 with apparently traumatic effects on the younger brother. Fortunately, Enrico soon made the acquaintance of a young man, Enrico Persico, who was to become his best friend and a close professional colleague for the rest of his life.

Obtains His Early Education in Italy and Northern Europe

Fermi exhibited unusual intellectual talents at an early age. He did well in school but also read a great deal on his own. A friend of his father, Adolfo Amidei, assumed some responsibility for Fermi's intellectual development, arranging to have books on mathematics and physics given to him in the proper sequence. According to his friend and colleague, **Emilio Segrè**, Fermi knew as much classical physics by the time he graduated from high school as did the typical university graduate student. That knowledge had come not only from books, but also from experiments designed and carried out by Fermi and Persico.

In November 1918 Fermi applied for a scholarship at the Reale Scuola Normale in Pisa. His entrance essay dealt with the mathematics and physics of vibrating reeds and convinced the examiners that he was a candidate of unusual promise. Four years later, Fermi was granted his doctorate in physics, magna cum laude, from Pisa. His thesis dealt with experiments he had conducted using **x rays**.

Since the scientific community was not then well developed in Italy, Fermi set out for northern Europe to continue his studies. He spent seven months with **Max Born** at the University of Göttingen and then returned to Rome. He taught mathematics for one year in Rome before traveling north again, this time to the University of Leiden. There he studied with the great Dutch physicist Paul Ehrenfest and became close friends with **Samuel A. Goudsmit** and George Uhlenbeck, both of whom later emigrated to the United States.

Fermi returned to Italy in 1925, hoping to be appointed to a chair in **mathematical physics**. When this opportunity was offered to another man instead, Fermi accepted a teaching position at the University of Florence. During his first year at Florence, Fermi wrote a paper that was to establish his name among physicists almost immediately. The paper dealt with an application of **Wolfgang Pauli**'s exclusion principles, discovered earlier the same year, to the atoms in a gas. Pauli's exclusion principles restrict the possible location of electrons. Fermi postulated that the same rules developed by Pauli for electrons might also be applied to the atoms in a gas. The mathematical system he invented, later developed independently by **Paul Dirac**, has become known as Fermi-Dirac statistics.

This accomplishment came about at a propitious time for Fermi. In Rome, Orso Mario Corbino had just begun a campaign to revitalize Italian science, which had been in a long period of decline. Corbino, as both chairman of the physics department at the University of Rome and a powerful figure in the Italian government, obtained authorization to create a new chair of theoretical physics at Rome, a position that he immediately offered to Fermi. In 1926, Fermi left Florence to assume his new position in Corbino's department. Over the next half dozen years, Corbino's dreams were realized as a number of first-class physicists and young students were attracted to Rome.

Great Discoveries Made in Theoretical and Experimental Physics

During that period, Fermi and his colleagues concentrated on the latest developments in **atomic theory**. Along with many other chemists and physicists, they were working out the implications of quantum and wave **mechanics**, relativity, and uncertainty to atomic theory. By the turn of the decade, however, many of those problems had been solved, and Fermi began to look elsewhere for challenges. He, like many others, settled on the atomic nucleus as a new focus of research.

One of the first problems to which he turned was beta decay. Beta decay is the process by which a neutron in an atomic nucleus breaks apart into a **proton** and electron. The electron is then emitted from the nucleus as a beta particle. The mechanics of beta decay appeared to violate known physical principles and were, therefore, the subject of intense study in the early 1930s. Wolfgang Pauli had suggested, for example, that the apparent violation of conservation laws observed during beta decay could be explained by assuming the existence of a tiny, essentially massless particle, also released during the event. Fermi was later to name that particle the **neutrino** ("little neutron"). In 1933, Fermi proposed a theory for beta decay. He said that the event occurs because a neutron moves from a state of higher **energy** to one of lower energy as it undergoes conversion to a proton and electron. To explain

the process by which this occurs, Fermi postulated the existence of a new kind of **force**, a force now known as the **weak force**.

Fermi's interest in the nucleus also prompted him to return to the laboratory and to design a number of experiments in this field. An important factor motivating this research was the recent discovery of artificial **radioactivity** by **Irène Joliot-Curie** and her husband, **Frédéric Joliot-Curie**. The Joliot-Curies had found that bombarding stable isotopes with alpha particles would convert the original isotopes into unstable, radioactive forms. Fermi reasoned that this type of experiment might be even more effective if neutrons, rather than alpha particles, were used as the "bullets." Neutrons have no electrical charge and are, therefore, not repelled by either the negatively charged electrons or the positively charged nuclei in an **atom**.

Beginning in 1934, Fermi and his colleagues systematically submitted one element after another to bombardment by neutrons. Progress was slow until a key discovery was made. In contrast to previous expectations, it turned out that slow neutrons—neutrons that have been slowed by passing through substances containing hydrogen—are more effective in bringing about nuclear changes than are fast neutrons. After revising his experimental procedure based on this discovery, Fermi had greater success. Over a period of months, he was able to show that 37 of the 63 elements he studied could be converted to radioactive forms by neutron bombardment.

The element among all others that especially intrigued Fermi was uranium. He knew that one reaction that could reasonably be expected as the result of neutron bombardment of uranium was the production of a new element with an atomic number one greater than that of uranium. But no such element exists naturally. So, the successful bombardment of uranium with neutrons might well result in the formation of the first synthetic element.

When this experiment was actually performed, the results were ambiguous. Fermi did not feel that he had enough evidence to announce the preparation of a new element. His sponsor, Corbino, had no such reluctance, however. Motivated by a desire to exalt the "new Italian science," Corbino reported that Fermi had indeed discovered element #93 and suggested that it be named italium.

In fact, Fermi did not recognize the revolutionary nature of his experimental results. Bombardment of uranium with neutrons had resulted not in a simple nuclear transformation, but in **nuclear fission**, a process during which the uranium atom is split, unleashing tremendous amounts of energy. The true nature of this reaction was later elucidated by **Lise Meitner** and Otto Robert Frisch.

As the 1930s came to a close, Fermi's future became more uncertain. Fascist influences were growing not only in Hitler's Germany, but also in Mussolini's Italy. These political changes troubled Fermi in a number of ways. For one thing, they threatened the nature of his scientific work since political factors now determined what was "correct" research and what was not. In addition, Fermi's family felt threatened since his wife was Jewish and subject, therefore, to the dangers of state-enforced anti-Semitism.

Defection to the West Marks the Beginning of New Life

Fortunately, a solution for his dilemma presented itself in November 1938 when Fermi was informed that he had won the Nobel Prize for physics. The award had been given for his research on the bombardment of elements by slow neutrons. The Fermis and a few friends decided that their trip to Stockholm to accept the prize would be a one-way passage. After the ceremony, he would defect to the West. On January 2, 1939, he arrived in the United States and assumed a new post as professor of physics at Columbia University.

The timing of Fermi's arrival in the United States was indeed fortuitous. At almost the same time, **Niels Bohr** was delivering to American physicists news of the discovery of nuclear fission by **Otto Hahn** and **Fritz Strassmann** in Germany. The significance of this discovery was immediately apparent to Fermi and his colleagues. A movement was soon underway to inform the U.S. government of the political and military implications of the Hahn-Strassmann discovery. That movement culminated in the famous August 2, 1939, letter to President Franklin D. Roosevelt, written by **Leo Szilard** and signed by **Albert Einstein**, that discussed the importance of nuclear research.

Fermi's immediate future was now laid out for him. He became involved in the Manhattan Project to determine whether a controlled nuclear **chain reaction** was possible and, if so, how it could be used in the construction of a nuclear weapon. Although his initial work was carried out at Columbia, Fermi eventually moved to the University of Chicago. He began work there in April 1942 with a team of the nation's finest physicists attempting to produce the world's first sustained nuclear fission reaction.

That effort came to a successful conclusion under the squash courts at the university on December 2, 1942. At 3:21 P.M., instruments indicated that a self-sustaining chain reaction was taking place in the world's first atomic "pile," a primitive nuclear reactor. The message sent to Washington confirming this event stated, "The Italian navigator has landed in the New World." In response to that message, James Conant, director of the project, asked, "Are the natives friendly?," indicating that the team was ready to go ahead with further research. The reply from Fermi's team was, "Yes, very friendly," and the race to build a bomb was on its way.

For the next two years, Fermi continued his research on nuclear fission at the Argonne National Laboratory outside of Chicago. Towards the end of that work, on July 11, 1944, Fermi and his wife became naturalized citizens of the United States. A few weeks later, Fermi left for Hanford, Washington, where he briefly assisted in the design and construction of a new plutonium processing plant. In September, he moved on to Los Alamos, New Mexico, where final assembly of the first **nuclear weapons** was to occur. There Fermi was placed in charge of his own special division (the F division) whose job it was to solve special problems as they arose during bomb construction.

After the first successful tests at Los Alamos, on July 16, 1945, and after the bombs were dropped on Japan a month later, Fermi concluded his work with the Manhattan Project. He returned to the University of Chicago, where he became Charles H. Swift Distinguished Service Professor of Physics and a member of the newly created Institute for Nuclear Studies. He remained at Chicago for the rest of his life, a period during which he received many honorary degrees and other awards. Included among the former were honorary doctorates from Washington University, Yale, and Harvard. He also received the Civilian Medal of Merit in 1946 for his work on the Manhattan Project, the Franklin Medal from the Franklin Institute in 1947, and the Transenter Medal from the University of Liége in Belgium in 1947.

Fermi's health began to fail in 1954, and exploratory surgery showed that he had stomach cancer. He refused to give up his work, however, and continued his research almost until he died, in his sleep, on November 30, 1954. He was buried in Chicago. Shortly after his death, he was posthumously awarded the first Enrico Fermi Award, given by the U.S. Atomic Energy Commission. Since his death, Fermi has received additional honors, including the naming of element #100, *fermium*, after him and the choice of the unit *fermimeter* for nuclear dimensions.

In addition to his great mental genius, Fermi was a highly respected and well-loved person. His long-time colleague, Segrè, has written of Fermi's special pleasure in working with younger colleagues, a tendency that helped maintain his own youthful spirit.

See also Pauli exclusion principle

FERMI NATIONAL ACCELERATOR LABORATORY (FERMILAB)

The Fermi National Accelerator Laboratory, Fermilab, is one of the world's largest research facilities, with more than 2,000 workers. Over the years, more than 5,000 scientists have come from around the world to work at Fermilab near Chicago, Illinois. The lab houses equipment used to study **subatomic particles**, the particles in the **nucleus** of an **atom**. By smashing atoms, scientists hope to learn more about nature and the behavior of **matter**.

The U.S. Congress funded construction of the lab with $250 million in 1969. The facility was originally called the National Accelerator Laboratory until 1974, when it was dedicated to the memory of **Enrico Fermi**, the nuclear physicist who helped develop the first nuclear reaction at the University of Chicago. The Universities Research Association Inc. (URA), a consortium that now includes 77 universities in the United States and Canada, operates Fermilab for the U. S. Department of Energy.

Fermilab contains one of the world's largest proton **synchrotrons**, a type of oversized **microscope** for studying subatomic particles. The main particle accelerator is contained within the Main Ring, a 4-mi-long (6.4-km-long) circular tunnel that is used for examining particles one ten-millionth of one-billionth of a centimeter in size. Two kinds of particles—

Fermi National Accelerator Laboratory, Batavia, Illinois. *(Photo by Michael Yamashita. Corbis-Bettmann. Reproduced by permission.)*

quarks and **leptons**—are believed to be the building blocks of matter. **Protons** and neutrons contain **quarks**. The accelerator accelerates protons in a way through equipment that tests them and notes their behavior under different conditions.

Fermilab runs 24 hours a day, 7 days a week, with the capacity to perform up to 15 experiments simultaneously. In March 1995, a new particle was discovered at the facility: the top quark, which is the sixth. In July 2000, a team of 54 internationally renowned physicists announced the discovery of the tau **neutrino**, one of two remaining major undetected particles in the standard model of theoretical physics. Work continues at Fermilab, searching for the remaining unseen particle, the **Higgs boson**.

For a cybertour of Fermilab, please see www.fnal.gov.

FERMIONS

Fermions and **bosons** are the two classes of elementary particles. Fermions are the particles that actually compose **matter** whereas bosons are associated with mediating the fundamental forces of nature between fermions. Fermions are particles whose quantum mechanical spin number is a half-integral (1/2, 3/2, 5/2, etc.) and obeys Fermi-Dirac statistics. In 1926 **Paul Dirac**

and **Werner Heisenberg** applied the Pauli principle, which states that the complete **wave function**, including both spatial and spin coordinates, of a system of identical fermions must be antisymmetric with respect to interchange of all of the coordinates of any two particles. This means that in a system of identical fermions each spin-orbit can hold no more than one fermion.

There are two classes of fermions, elementary fermions, such as leptons and **quarks**, and composite fermions, such as protons and neutrons. The leptons subclass is broken down into three groups or generations that each consist of one charged **lepton** and one **neutrino**. The first generation leptons compose everyday matter and consist of the **electron**, a particle with a charge of -1, and the electron neutrino, a neutral particle. The second and third generation leptons are created only during high-energy collisions between particles and exist on Earth during collisions of **cosmic rays** with atoms in the atmosphere and in the laboratory using **particle accelerators**. In outer **space** they are created in stellar explosions such as supernovas. The second generation leptons are the **muon**, with a charge of -1, and the muon neutrino, a neutral particle. The third generation leptons are the tau, with a charge of -1, and the tau neutrino, a neutral particle.

Quarks, the other subclass of elementary fermions, are particles with a fraction of the charge on an electron; they are never found alone but exist only in pairs or triplets. This subclass also is broken down into three generations. The first generation composes ordinary matter and consists of up-type quarks, +2/3, and down-type quarks, -1/3. These particles combine to create protons and neutrons. Again, the second and third generations are created only during high-energy collisions between particles. The second generation consists of charm, up-type, and strange, down-type. The third generation consists of top, up-type, and bottom, down-type.

Because protons consist of two up-type quarks this seems to present a conflict with the **Pauli exclusion principle** until the characteristic of **color** charge is introduced. Color charge is similar to the concept of **electric charge** but important to the strong **force** instead of the electromagnetic force. There are three values of color charge: red, blue, and green. As long as the quarks contained in the composite fermion have different color charges they do not violate the exclusion principle.

Composite fermions are particles that are composed of odd numbers of quarks and so follow the same rules of behavior as elementary fermions. (If the particles contain even numbers of fermions then they are bosons.) Although there is no limit to the size of a composite fermion, the simplest one contains three quarks and is known as a **baryon**. Protons and neutrons are baryons and consist of three quarks held together by gluon (a boson). Gluon transfers color charge from one quark to another and so the constant change in color charge binds the quarks together. Protons consist of two up-type quarks and one down-type quark and have a net charge of +1. On the other hand, neutrons are composed of one up-type quark and two down-type quarks and have no net charge. Many other baryons exist but are unstable. These baryons, such as lambda and **sigma particles**, contain second and third generation quarks such as charm, strange, and bottom.

In 1897 Sir **Joseph John Thomson** isolated and measured the electron and so began the research on fermions. Fermion research has continued since then. For example, in 1925 **Wolfgang Pauli** developed the Pauli exclusion principle that states that two fermions cannot occupy the same quantum mechanical state. The next year **Enrico Fermi** employed the exclusion principle to develop a set of equations that explain the behavior of particles—fermions—that obey the exclusion principle and so these particles were named in his honor. Particle accelerators developed in the 1950s allowed scientists to discover several new particles that had been unobserved before then. In 1964 two American physicists published the first theory of quarks and classified them as fermions. Finally, in 1995 the last elementary fermion, top quark, was discovered at the Stanford Linear Accelerator Center in California.

See also Lambda particles; Tau particles

FERROMAGNETISM

Ferromagnetism is the effect that creates permanently magnetized materials, such as ordinary horseshoe magnets. Although the effect is most common in pure iron, it can also be reproduced in iron-like substances, including compounds like barium ferrite, and heterogeneous compounds that unite iron with aluminum, nickel, cobalt, and tin.

Ferromagnetism occurs when magnetic moments of the atoms inside a material interact, leading to the overall **magnetism** of the object. Ferromagnetic substances will maintain their magnetism until the atoms are forced out of alignment. The easiest way to do this is to **heat** the substance. Above a certain **temperature**, the heat is enough to break the alignment and that subsequent cooling is not enough to allow realignment to take place. This temperature is known as the Curie temperature, T_c, named after **Marie Curie**'s husband, **Pierre Curie**.

Magnetic moments of atoms are created by the contribution of individual magnetic moments of the constituent particles: the **proton**, the **neutron**, and the **electron**. The electron's **magnetic moment**, however, is approximately 1,000 times stronger than either the proton or the neutron, so electrons essentially account for all the magnetic moment of the individual **atom**. The magnetic moment comes from the spin of the electron.

A particle's spin is an inherent quantum mechanical property, not a result of the particle's actual angular **motion** around another particle. Instead, spin comes from particles behaving as if they were charge distributions spinning around an axis. Thus, each particle is like a tiny magnetic dipole. This causes each particle to have a tiny magnetic moment, contributing to the overall magnetic moment of the atom. There is evidence that an electron orbiting a **nucleus** does, in fact, create a magnetic moment component for the atom, but the effect is so small that it is negligible when compared to the electon's spin contribution.

It is not the just the magnetic moment of the atoms in the substance that creates a magnet; it is the interaction of these magnetic moments. There is an interaction **energy** between adjacent magnetic moments that causes them to correlate between themselves; this is called the exchange energy. The strength of the interaction and the alignment of the magnetic moments depends on the distance between atoms and the atoms' individual **quantum states**.

In a substance, there are three ways these magnetic moments can align: all in the same direction, alternating pairs of directions, or in random directions (which usually gives a small net magnetism to the substance). Those substances where all the atoms' moments are aligned in the same direction are ferromagnetic substances. Those substances with small net magnetism from an overall random effect are weak magnets. Finally, substances that have alternating pairs of directions are considered anti-ferromagnetic.

FEYNMAN, RICHARD PHILLIPS (1918-1988)

American physicist

Richard Feynman was born in New York City on May 11, 1918. He was educated in the public schools of New York City and rapidly impressed his teachers with his remarkable grasp of difficult concepts. It is reported that his high school physics teacher allowed Feynman to sit in the back of the room and solve assigned problems using calculus while other students struggled along with simple algebra.

Feynman attended the Massachusetts Institute of Technology, from which he received a bachelor's degree in 1939. His doctoral work was completed at Princeton University in 1942. Along with nearly all physicists of the time, Feynman spent World War II working in the **Manhattan Project** on the development of the first **nuclear weapons**. After the war, he accepted a teaching position at Cornell University. In 1950, Feynman became professor of theoretical physics at the California Institute of Technology, where he remained until his death.

Feynman's special field of interest was **quantum electrodynamics**. That term refers to the study of the interactions among electrons, **positron**s, and photons. During the 1920s, a great deal of effort was expended on the application of modern **quantum mechanics** to classical electromagnetic theory. A number of leading physicists, including **Paul Adrien Maurice Dirac**, **Werner Heisenberg**, **Wolfgang Pauli**, and **Enrico Fermi**, all looked for ways to salvage **James Clerk Maxwell**'s theory of **electromagnetism** by incorporating quantum principles.

Many of these efforts were at least partially successful. However, they too often failed to account for detailed phenomena, such as the specific **energy** levels observed for electrons in the hydrogen **atom**. There was some concern that either classical theory or its modern modifications might have to be totally abandoned.

While at Princeton University, Feynman attacked this problem. He developed a program for "renormalizing" some of the fundamental terms in equations that had been giving troublesome results. This program eliminated many of the nonsensical answers obtained from the equations. It became apparent that existing theories could be modified and successfully applied. For his work on quantum electrodynamics, Feynman shared the 1965 Nobel Prize for physics with **Julian Schwinger** and **Sin-Itiro Tomonaga**. Schwinger and Tomonaga had independently developed methods of analysis similar to those of Feynman's.

Feynman was widely known and admired outside the field of science. His autobiography, *Surely You're Joking, Mr. Feynman*, became a best-seller. On another level, his *Feynman Lectures on Physics* is widely regarded as a lucid explanation of some fundamental ideas in modern physics.

Feynman died in 1988 after a long battle with cancer.

FEYNMAN DIAGRAMS

American physicist Richard Feynman's work and writings were fundamental to the development of quantum electrodynamic theory (QED theory). With regard to QED theory, Feynman is perhaps best remembered for his invention of what are now known as Feynman diagrams, which portray the complex interactions of atomic particles. Moreover, Feynman diagrams allow visual representation and calculation of the ways in which particles can interact through the exchange of

virtual photons and thereby provide a tangible picture of processes outside the human capacity for observation. Because Feynman diagrams allow physicists to depict subatomic processes and develop theories regarding particle interactions, the diagrams have become an indispensable and widely used tool in **particle physics**.

Feynman diagrams derive from QED theory. **Quantum electrodynamics (QED)**, is a fundamental scientific theory that is also known as the **quantum theory** of **light**. QED describes the quantum properties (properties that are conserved and that occur in discrete amounts called **quanta**) and **mechanics** associated with the interaction of electromagnetic **radiation** (of which visible light is but one part of an **electromagnetic spectrum**) with **matter**. The practical value of QED rests upon its ability, as set of equations, to allow calculations related to the absorption and emission of light by atoms and thereby allow scientists to make very accurate predictions regarding the result of the interactions between photons and charged atomic particles (e.g., electrons).

Feynman diagrams are a form of shorthand representation outlining the calculations necessary to depict electromagnetic and weak interaction particle processes. QED, as **quantum field theory**, asserts that the electromagnetic **force** results from the quantum behavior of the **photon**, the fundamental particle responsible for the transmission of electromagnetic radiation. According to QED theory, particle vacuums actually consist of electron-positron fields and electron-positron pairs (positrons are the positively charged **antiparticle** to electrons) are created when photons interact with these fields. QED accounts for the subsequent interactions of these electrons, positrons, and photons. Photons, unlike the particles of everyday experience, are **virtual particles** constantly exchanged between charged particles. As virtual particles, photons cannot be observed because they would violate the laws regarding the **conservation of energy** and **momentum**. QED theory therefore specifies that the electromagnetic force results from the constant exchange of virtual photons between charged particles that causes the charged particles to constantly change their **velocity** (speed and/or direction of travel) as they absorb or emit virtual photons. Accordingly, only in their veiled or hidden state do photons act as mediators of force between particles and only under special circumstances do photons become observable as light.

In Feynman time-ordered diagrams, **time** is represented on the x axis and a depicted process begins on the left side of the diagram and ends on the right side of the diagram. All of the lines comprising the diagram represent particular particles. Photons, for example, are represented by wavy lines. Electrons are denoted by straight lines with arrows oriented to the right. Positrons are depicted by a straight line with the arrow oriented to the left. Vertical y axis displacement in Feynman diagrams represents particle **motion**. The representation regarding motion is highly schematic and does not usually reflect the velocity of a particle.

Feynman diagrams depict electromagnetic interactions as intersections (vertices) of three lines and are able to describe the six possible reactions of the three fundamental

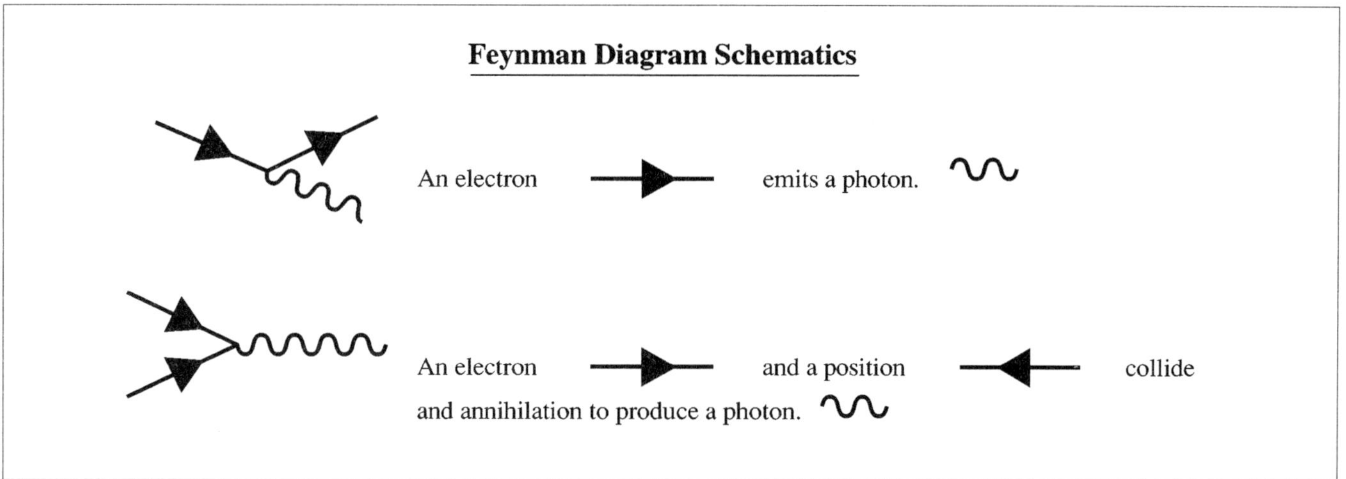

QED particles (i.e., the **electron**, **positron**, and the photon). Accordingly, the diagram can depict the emission and absorption of photons by either electrons or positrons. In addition, it is possible to depict the photon production of an electron-positron pair. Lastly, the diagrams can depict the collisions of electrons with positrons that results in their mutual annihilation and the production of a photon.

Feyman diagrams presuppose that **energy** and momentum are conserved during every interaction and hence at every vertex. Although lines on the diagram can represent virtual particles, all lines entering or leaving a Feynman diagram must represent real particles with observable values of energy, momentum, and **mass** (which may be zero). By definition, virtual particles acting as intermediate particles on the diagrams do not have observable values for energy, momentum, or mass.

Although QED theory allows for an infinite number of processes (i.e., an infinite number of interactions) the theory also dictates that interactions of increasing numbers of particles become increasingly rare as the number of interacting particles increases. Correspondingly, although Feynman diagrams can accommodate or depict any number of particles, the mathematical complexities increase as the diagrams become more complex.

Feynman developed a set of rules for constructing his diagrams (appropriately named Feynman rules) that allow QED theorists to make very accurate calculations that closely match experimental findings. The Feynman rules are very simple, but as greater accuracy is demanded the diagrams become more complex. Feynman diagrams derive from the Feynman path integral formulation of **quantum mechanics**. Using asymptotic expansions of the integrals that describe the interactions, physicists are able to calculate the interactions of particles with great (but not unlimited) accuracy. The mathematical formulae associated with the diagrams are added to arrive at what is termed a Feynman amplitude, a value that is subsequently used to calculate various properties and processes (e.g., decay rates).

The development of QED theory and the use of Feynman diagrams allowed scientists to predict how subatomic processes create and destroy particles. Over the last half-century of research in particle physics, QED has become, arguably, the best-tested theory in science history. Most atomic interactions are electromagnetic in nature and, no matter how accurate the equipment yet devised, the predictions made by modern QED theory hold true. Some tests of QED, predictions of the mass of some **subatomic particles**, for example, offer results accurate to six significant figures or more. Although some predictions can be made using one Feynman diagram and a few calculations, others may take hundreds of Feynman diagrams and require the use of high-speed **computers** to complete the necessary calculations.

FIRST LAW OF THERMODYNAMICS

Thermodynamics is the study of **heat**, a discipline that has its roots in the early 1800s. The first law of thermodynamics is the application of the law of **conservation of energy** to thermal systems. A system is any group of "things" under defined conditions. For example, a system may be the steam in a steam engine, the components of the human body, or all of the plants in a garden. Anything outside of the set boundaries defined by the system is said to be a part of the surroundings. A thermal system is any system that involves the transfer of heat.

Heat was originally thought of as a fluid that flowed from hot to cold objects. This fluid was called caloric. This theory of heat, called the **caloric theory**, was the basis for the study of thermal systems in the eighteenth century. Early physicists realized that the amount of caloric in thermal systems was always conserved. This discovery led to the development of the law of conservation of **matter**, which states that matter can neither be created nor destroyed. In any interaction, matter can be transferred from one substance to another, but the total amount of matter in a system remains constant. Therefore, the amount of caloric that flowed out of a hot object must equal the amount of caloric that flowed into a corresponding cold object.

In the 1840s, the caloric theory was disproved. Heat was now described not as a fluid, but as a form of **energy**. Energy, like matter, can neither be created nor destroyed. This realization led to the first law of thermodynamics, which states that whenever heat is added to a system, it can be transformed into an equal amount of another form of energy. The total energy of the system remains the same. The amount of **thermal energy** in a system can be increased either by adding heat or by doing **work** on the system.

Work is performed whenever a **force** moves an object. The amount of work done can be calculated by multiplying the magnitude of the force acting on an object by the distance the object moves. Using the English system of measurement, the force is measured in pounds (lb) and the distance is measured in feet (ft). The product of the two, or the work done, is then measured as foot-pounds (ft-lb). If a machine performs work using rotation, the work performed is the product of **torque** and the angle of rotation. With the metric, or International System of Units (SI), system, the unit of force is the Newton (N) and the unit of distance is the meter (m). The resulting unit for work is the Newton-meter (N-m), otherwise known as the joule (J). One joule is equal to 0.7376 ft-lb. The unit for work, the joule, is the same as the unit for energy. When work is performed, energy is transferred from one object to another through mechanical means. The first law of thermodynamics, therefore, states that the total increase of the thermal energy of a system is the sum of the work done on it and the amount of heat added to it (change in heat = work + change in energy).

Doing work on the system can also increase the energy of a system, even if there is no change in heat. If the change in heat is equal to 0 J, then the above equation reads "0 = work + change in energy." This equation can be rearranged to read "-work = change in energy." In other words, if work is done on the system, the energy of the system will increase, raising the **temperature** without adding heat. For example, if you rub your hands together, you will begin to feel a warm sensation. You are doing work on your skin (through **friction**), increasing your skin's temperature without directly adding heat. In this case, mechanical energy was transformed into thermal energy. Another example of converting mechanical energy into thermal energy is when you use a hand pump to inflate your bicycle tires. As you pump on the handle, the pump and the tire become hot. As you perform mechanical work on the system, the energy of the system, thus its temperature, is increased.

It is quite easy to transform mechanical energy into thermal energy; however, transforming thermal energy into mechanical energy is not as easy. Adding heat energy to a system does one of two things to the system. If the heat energy remains in the system, it increases the energy of the system. If an airtight container were placed on a hot stove, for example, all of the heat from the stove would be used to increase the energy, or the temperature, of the air in the container. Nothing would be moved, so no work would be performed. If the heat energy leaves the system, it allows the system to perform work. An example of the transfer of thermal energy into mechanical energy is the steam engine. If heat is supplied to a steam engine, some of the heat energy increases the energy (or

temperature) of the engine, some increases the energy (or temperature) of the steam, and some is transformed into mechanical work. Not all of the thermal energy is transformed into mechanical energy. However, the sum of the increased energy of the engine and steam and the work performed is equal to the amount of thermal energy put into the system.

Other forms of energy besides mechanical energy can also be transformed into heat. **Light**, **sound**, and electrical energy can all be converted into thermal energy according to the first law of thermodynamics. This law simply states that whatever energy is placed into a system is transformed into new forms of energy within the system, and that the sum of the new energy must equal the amount of energy added.

FLAME ANALYSIS

Flame analysis is a qualitative test used to identify the components of a substance or mixture. Flame analysis, also known as atomic emission spectroscopy (AES), is based on the physical and chemical principle that atoms—after being heated by flame—return to their normal **energy** state via the emission of electromagnetic **radiation** in the form of photons. The frequencies of the photons (i.e., the frequencies of the **light**) given off are characteristic for each element.

Flame tests are useful means of determining the composition of substances. In general, an unknown substance to be tested by flame analysis is either sprayed into a flame or placed on a thin wire that is then put into a flame. The colors produced by the flame test are compared to known standards to confirm the presence of individual elements in the sample.

Working together, nineteenth-century Prussian physicist **Gustav Kirchhoff** and German chemist Robert Bunsen (1811-1899) developed methods to analyze **matter** via flame analysis. In particular, Bunsen's invention of what is now known as the Bunsen burner allowed development of one of the fundamental techniques of **spectroscopy**. When air is admitted at the base of a Bunsen burner it mixes with the gas to produce a very hot flame at approximately 3,272°F (1,800°C) that is sufficient to cause the emission of light from certain elements. Bunsen used a simple apparatus that consisted of a prism, slits, and a magnifying glass or photo-sensitive film to examine the colors of light emitted and spectra of substances (including known elements) subjected to such intense flame. Bunsen determined that the **color** of the flame and its component colors or spectra was unique for each element, that is, the individual spectra of elements that emitted light contained only specific portions of the total colors in the spectrum. Using flame analysis techniques, Bunsen discovered the elements cesium and rubidium.

Flame analysis is often used to determine the presence of metal elements in water. The water is vaporized and the emission spectrum of the vaporized metals analyzed.

In some cases, special techniques must be used to properly interpret the results of a flame analysis test. The colors produced by a potassium flame (pale violet), for example, can usually be observed only with the assistance of glass that can filter out interfering colors. Some colors are similar enough that their line spectrum must be carefully examined to make a complete and accurate identification.

The flame test does not work on all elements. Those that produce a measurable spectrum when subjected to flame include, but are not limited to, lithium, sodium, potassium, rubidium, cesium, magnesium, calcium, strontium, barium, zinc, and cadmium. Other elements may need hotter flames to produce measurable spectra. In addition, when subjected to flame, very volatile elements (such as chlorides) produce intense colors. The yellow color of sodium, for example, can be so intense that it overwhelms other colors. To prevent this the wire to be coated with the unknown sample is usually dipped in hydrochloric acid and subjected to flame to remove sodium and volatile impurities.

Using the spectroscopic principles pioneered by Bunsen during his development of flame analysis, scientists have been able to determine the chemical composition of **stars**. Bunsen's observation that flamed elements that emit light only at specific **wavelengths** and that every element produces a characteristic spectra provided a theoretical base for the subsequent development of **quantum theory** by **Max Planck**, **Niels Bohr**, and others.

FLAT UNIVERSE

Flat universe is the term used to describe a **universe** that is expanding and will continue to expand forever, albeit at a rate that will eventually approach zero. The solution of Einstein's general relativistic equations for the entire universe predicts one of three possibilities: the universe can start out expanding, but eventually stop expanding and begin to contract; it can be flat; or it can expand forever at an ever-increasing rate. The case of a flat universe is special, because unless the universe is very close to being flat, its lifetime is calculated to be much shorter than the observed age of the universe.

There is a critical **energy** density required for the universe to be flat; the ratio of the actual energy **density** to the critical density is a parameter called Omega (Ω). For a flat universe, Ω is exactly one. Until recently, it was believed that the universe was flat, but recent evidence shows that the universal expansion may actually be accelerating, leading us to believe that the universe is not completely flat.

See also Cosmology; Relativity, general

FLEMING, JOHN AMBROSE (1849-1945)
English electrical engineer

John Ambrose Fleming was a pioneering engineer who made numerous contributions both to the theoretical aspects and practical applications of **electricity**. Fleming played an important role in the development of lighting, heating, and **radio,** and, as a consultant in private industry, was a proponent of their widespread conventional use. Fleming's most wide-ranging practical contribution to the field of electrical engineering

was the development of the thermionic (or radio) valve, which acts as a rectifier for high **frequency** currency, permitting the current to flow in only one direction. Also known as the Fleming valve or **diode**, this precursor to the **transistor** revolutionized the infant field of radio as it became an essential part of nearly every radio transmitter and receiver for more than three decades; it was an important component of early televisions and **computers** as well. Fleming was also an accomplished educator who helped train a trailblazing cadre of electrical engineers.

Born in Lancaster, England, on November 29, 1849, Fleming was the eldest of seven children born to James Fleming, a minister, and the daughter of John Bazley White, a pioneer in the manufacturing of portland cement. In 1854, Fleming's family moved to London, where his father became minister of the Kentish Town Congregational Chapel. Fleming's aptitude for science and **mechanics** was evident early in his life as he excelled in geometrical drawing at a private school for boys. When he was 11, he had already set up his own workshop, where he spent his time building model engines and ships rather than playing with other children. In 1863, he entered the University College School in London. Despite his lack of aptitude for the required Latin courses (he was at the bottom of his class), Fleming excelled in mathematics and passed the London matriculation at the age of 17.

In 1867, Fleming entered University College to study physics and mathematics. Unfortunately, as one of seven children, he received little financial support from his family; within a year, he was forced to leave school and go to work. He first took a position with a Dublin shipbuilding firm, believing that he could gain some practical experience, but left after only a few months when he quickly grew bored with tracing drawings. He next worked as a clerk for a firm on the London Stock Exchange. Although far from his area of interest, this job, which he held for two years, provided ample free time during the evenings to study. He received his bachelor of science degree in 1870 from University College, graduating as one of the two top students in his class.

After a year and a half as the science master at Rossall School, Fleming had saved enough money to continue his scientific education. He returned to London to attend the Royal School of Mines in South Kensington, where he studied advanced chemistry under Edward Frankland and Frederick Guthrie. By 1874, Fleming had assumed teaching duties in the school's Advanced Chemical Laboratories. Although his career seemed to be heading in the direction of chemistry, his attention turned to physics as he worked with Guthrie on experiments in electrically charged bodies, such as an iron ball. This early work in the ability of certain elements to store or discharge negative and positive charges laid some of the groundwork for Fleming's later work on his radio valve. Guthrie also presented Fleming with the opportunity to present the first scientific paper to the newly formed Physical Society of London. Fleming's discussion, published in the *Proceedings of the Society of London,* focused on the contact theory of the galvanic cell.

But Fleming's education was to be interrupted once more by the need for money; in 1884, he became science master at Cheltenham College, where he incorporated hands-on experimentation by the students into his teaching approach. By 1877 Fleming was able to return to his studies, enrolling at St. John's College in Cambridge. (He received extra financial support for winning the school's entrance exhibition.) Before returning to college, Fleming had become interested in the work of **James Clerk Maxwell**, who had written the groundbreaking two-volume book *Electricity and Magnetism*. After spending the first six months at St. John's in an intensive study of mathematics, Fleming began his study of electricity and mathematics under Maxwell at the school's Cavendish Laboratory. Devoted to his studies, Fleming spent little time socializing, even going so far as to move off the college's campus where he felt he could spend more time studying without being interrupted by the social aspects of college life.

During his studies at the Cavendish Laboratory, Fleming worked on improving the Carey-Foster Bridge, a method for measuring the difference between two nearly equal resistances in electrical conduction. Fleming's improvement made the measuring device faster and more accurate and was called Fleming's banjo by Maxwell because of the measuring wire's circular shape.

After receiving his doctor of science degree in 1880, Fleming became a lecturer in applied mechanics in St. John's new engineering laboratories and was elected a fellow of the college in 1883. By this time, Fleming had begun acting as a consultant for private industry, including the Edison Telephone and Electric Light Companies, the Swan Lamp Factory, and the London National Company. In his capacity as an adviser to these burgeoning electrical companies, Fleming was a primary contributor to the development of electrical generator stations and distribution networks. With both the Swan and Edison companies, Fleming lent his expertise to photometry (the measurement of electric **light** intensity) and was largely responsible for the large-bulb incandescent lamp with an aged filament as the light source. Fleming was also asked by a number of towns in England to be an adviser for the development of municipal electric lighting systems. In 1899, he was hired by the Marconi Wireless Telegraph Company, where he worked as a scientific adviser and helped design the Poldhu Power Station in Cornwall, England. This station, the largest wireless station in England, was the source of the first transatlantic wireless telegraph transmission. Although the message was simple (the Morse code equivalent of the letter S, which is three dots), it marked the beginning of intercontinental wireless communication. Fleming maintained his association with the Marconi company for 30 years, making many marked improvements in early radio devices. Overall, his consulting work resulted in many new methods and instruments for measuring high-frequency currents and for new designs of **transformers**.

Fleming's most important contribution to electrical engineering was in the field of the radio telegraph and the telephone. Although he developed the 2-electrode, or Fleming valve in 1904, he laid the foundation for this discovery many years earlier when, in 1884, **Thomas Alva Edison** announced his discovery of the Edison effect, which described the escape of electrons or **ions** from a heated solid or liquid. Fleming repeated Edison's experiments in 1899 using both direct and alternating currents, but he found little practical use for the discovery. But in 1904, while searching for a more efficient and reliable detector of weak electrical currents, he had what he described as a "sudden very happy thought," as noted in a biography by J. T. MacGregor-Morris in *Notes and Records of the Royal Society*. Fleming's inspiration was to make a new lamp, or tube, that would have a hot filament and an insulated plate sealed inside a high **vacuum tube**. Further experimentation showed that the tube's sensitivity to extremely weak currents alternating at very high frequencies made it a major advance in radio technology.

Fleming dubbed his radio tube a thermionic valve because it acted much like a check valve that allows **fluids** to pass in only one direction. Essentially, the valve worked by utilizing a cathode that was kept hot, thus causing electrons to evaporate into the vacuum tube. The second electrode, called the **anode**, remained cool to prevent the evaporation of electrons from it. When an alternating current was applied, the electrical currents flowed in one direction. This discovery revolutionized radio science by providing the first truly reliable method to measure high-frequency radio **waves**. Two years later, **Lee de Forest** took Fleming's diode valve and added a third electrode, thus effectively separating the high-frequency circuit from that of the filament, making amplification possible.

Although the Fleming valve was eventually replaced by the transistor, it remained an important component of radios for nearly three decades and was also used in the early days of computers and television. Fleming himself helped support the infant technology of television by becoming president of the Television Society of London, which published a monthly journal and sponsored readings of new papers on the subject.

Fleming's long and productive scientific career spanned several decades. Despite increasing deafness, which began in middle age, he continued to work in his field, using an assistant to take notes at the lectures he attended to keep up-to-date on the latest advances in the field of electrical engineering. He gave his last presentation to the Physical Society in 1939, 65 years after presenting the society's inaugural address in 1874. His awards were numerous and included the Hughes Medal in 1910, the Faraday Medal of the Institution of Electrical Engineers in 1928, and the Gold Medal of the Institute of Radio Engineers in 1933. He was knighted in 1929.

In addition to his scientific research, Fleming was also an accomplished educator and, as professor of electrical technology at University College for 41 years, he helped train a new group of engineers who would help advance electrical engineering to new heights. He retired in 1936 and was appointed an emeritus professor.

Despite his teaching, research, and consultation duties, Fleming maintained many outside interests, including mountain climbing and painting in watercolors. He was also an avid amateur photographer, having made his own first camera in his workshop when he was only a teenager. Fleming's first wife, Clara Ripley, died in 1917. He married Olive Franks in 1933. Fleming died on April 18, 1945, at the age of 95.

FLUID DYNAMICS

Fluid **dynamics** is the study of the flow of liquids and gases, usually in and around solid surfaces. The flow patterns depend on the characteristics of the fluid, the speed of flow, and the shape of the solid surface. Scientists try to understand the principles and mechanisms of fluid dynamics by studying flow patterns experimentally in laboratories and also mathematically, with the aid of powerful **computers**. The two **fluids** studied most often are air and water. Aerodynamics is used mostly to look at air flow around planes and automobiles with the aim of reducing drag and increasing the efficiency of **motion**. Hydrodynamics deals with the flow of water in various situations such as in pipes, around ships, and underground. Apart from the more familiar cases, the principles of fluid dynamics can be used to understand an almost unimaginable variety of phenomena such as the flow of blood in blood vessels, the flight of geese in V-formation, and the behavior of underwater plants and animals.

The **viscosity**, **density**, and compressibility of a fluid are the properties that determine how the liquid or gas will flow. Viscosity measures the internal **friction** or resistance to flow. Water, for instance, is less viscous than honey and so flows more easily. All gases are compressible whereas liquids are practically incompressible and cannot be squeezed into smaller volumes. Flow patterns in compressible fluids are more complicated and difficult to study than those in incompressible ones. Fortunately for automobile designers, air, at speeds less than 220 mph or one-third the **speed of sound**, can be treated as incompressible for all practical purposes. Also, for incompressible fluids, the effects of **temperature** changes can be neglected.

The speed of flow is another factor that determines the nature of flow. The speed of flow is either that of a liquid or gas moving across a solid surface or, alternatively, the speed of a solid object moving through a fluid. The flow patterns in either case are exactly the same. That is why airplane designs can be tested in wind tunnels where air is made to flow over stationary test models to simulate the flight of actual planes moving through the air.

The speed of flow is related to the viscosity by virtue of the fact that a faster moving fluid behaves in a less viscous manner than a slower one. Therefore, it is useful to take viscosity and speed of flow into account at the same time. This is done through the Reynolds number named after the English scientist Osborne Reynolds (1842-1912). This number characterizes the flow. It is greater for faster flows and more dense fluids and smaller for more viscous fluids. The Reynolds number also depends on the size of the solid object. The water flowing around a large fish has a higher Reynolds number than water flowing around a smaller fish of the same shape.

As long as the shape of the solid surface remains the same, different fluids with the same Reynolds number flow in exactly the same way. This very useful fact is known as the principle of similarity or similitude. Similitude allows smaller-scale models of planes and cars to be tested in wind tunnels where the Reynolds number is kept the same by increasing the speed of air flow or by changing some other property of the fluid. The Ford Motor Company has taken advantage of the principle of similarity and conducted flow tests on cars under water. Water flowing at 2 mph was used to simulate air flowing at 30 mph.

Flow patterns can be characterized as laminar or turbulent. In laminar or streamlined flow, the fluid glides along in layers that do not mix so that the flow takes place in smooth continuous lines called streamlines. This is what we see when we open a water faucet just a little so that the flow is clear and regular. If we continue turning the faucet, the flow gradually becomes cloudy and irregular. This is known as turbulent flow. This change to **turbulence** takes place at high Reynolds numbers. The critical Reynolds number at which this change occurs differs from case to case.

An important idea in fluid flow is that of the **conservation of mass**. This implies that the amount of fluid that goes in through one end of a pipe is the same as the amount of fluid that comes out through the other end. Thus the fluid has to flow faster in narrower sections or constrictions in the pipe. Another important idea, expressed by **Bernoulli's principle**, is that of the **conservation of energy**.

Daniel Bernoulli was the first person to study fluid flow mathematically. He imagined a completely nonviscous and incompressible or "ideal" fluid in order to simplify the mathematics. Bernoulli's principle for an ideal fluid essentially says that the total amount of **energy** in a laminar flow is always the same. This energy has three components—potential energy due to **gravity**, **potential energy** due to **pressure** in the fluid, and **kinetic energy** due to speed of flow. Since the total energy is constant, increasing one component will decrease another. For instance, in a horizontal pipe in which **gravitational energy** stays the same, the fluid will move faster through a constriction and will, therefore, exert less pressure on the walls. In recent years, powerful computers have made it possible for scientists to attack the full mathematical complexity of the equations that describe the flow of real, viscous, and compressible fluids. Bernoulli's principle, however, remains surprisingly relevant in a variety of situations and is probably the single most important principle in fluid dynamics.

Even though Bernoulli's principle works extremely well in many cases, neglecting viscosity altogether often gives incorrect results. This is because even in fluids with very low viscosity, the fluid right next to the solid boundary sticks to the surface. This is known as the no-slip condition. Thus, however fast or easily the fluid away from the boundary may be moving, the fluid near the boundary has to slow down gradually and come to a complete stop exactly at the boundary. This is what causes drag on automobiles and airplanes in spite of the low viscosity of air.

The treatment of such flows was considerably simplified by the boundary layer concept introduced by Ludwig Prandtl (1875-1953) in 1904. According to Prandtl, the fluid slows down only in a thin layer next to the surface. This boundary layer starts forming at the beginning of the flow and slowly increases in thickness. It is laminar in the beginning but becomes turbulent after a point determined by the Reynolds number. Since the effect of viscosity is confined to the boundary layer, the fluid away from the boundary may be treated as ideal.

Moving automobiles and airplanes experience a resistance or drag due to the viscous **force** of air sticking to the surface. Another source of resistance is pressure drag that is due to a phenomenon known as flow separation. This happens when there is an abrupt change in the shape of the moving object, and the fluid is unable to make a sudden change in flow direction and stays with the boundary. In this case, the boundary layer gets detached from the body, and a region of low pressure turbulence or wake is formed below it. This creates a drag on the vehicle due to the higher pressure in the front. That is why aerodynamically designed cars are shaped so that the boundary layer remains attached to the body longer, creating a smaller wake and, therefore, less drag. There are many examples in nature of shape modification for drag control. The sea anemone, for instance, continuously adjusts its form to the ocean currents in order to avoid being swept away while gathering food.

FLUIDS

A fluid is a substance that can flow. A liquid is a fluid, but smoke and air can also be considered fluids. Fluids also have the tendency to assume the shape of the container housing them. There are many physical phenomena associated with fluids and there is an entire branch of physics concerned with studying fluids and the motions of such systems called fluid **dynamics**. Gases with a large number of highly charged particles (plasmas) are also considered fluids, and the study of plasmas is an important subcategory of **fluid dynamics**.

When a fluid is in **motion** it can move in several different ways. At one extreme the flow of the fluid can be laminar or streamlined; at the other, the motion can be turbulent. In a laminar flow every point that passes a particular point moves along the exact path followed by the particles that preceded it. When the flow is laminar each particle of the fluid moves along a smooth path called a streamline. There can be many parallel streamlines in a laminar flow but they can never cross each other under steady-flow conditions. The streamline at any point in the flow coincides with the direction of fluid **velocity** at that point.

Sometimes a fluid's flow becomes turbulent or irregular above a certain velocity or near a boundary housing the flow. These interruptions cause changes in velocity and irregular motions of the fluid called eddy currents. The flow of a fluid is often dependent upon its **viscosity**. Viscosity characterizes the degree of internal **friction** or resistance of two adjacent layers of fluid to move relative to each other. A fluid such as smoke has a lower viscosity than olive oil or syrup. The vis-

cosity of a fluid is dependent upon the **temperature** of the fluid. Usually as temperature rises, viscosity decreases.

Physics classifies an ideal fluid as one that has certain properties. An ideal fluid is nonviscous, which means that it flows without dissipating its **energy**. This removes the possibility that the velocity or **density** or **pressure** at any point is changing in **time**. Also, ideal fluids flow without **turbulence**; hence, no eddy currents exist.

The study of fluid dynamics has several interesting concepts, including the equation of continuity, which tells about the speed of a fluid with respect to the cross-sectional area of a container carrying its flow. In the early 1800s Jean-Léonard-Marie Poiseuille (1797-1869), a French scientist, derived a relationship between the rate of flow, the pressure difference from one side of the flow to the other, the dimensions of the container that carries the flow, and the viscosity of the fluid. This relationship is known as Poiseuille's law.

Bernoulli's equation relates pressure, **kinetic energy**, and **potential energy** of a fluid's movement. It usually takes the form $(V^2/2) + (p/\rho) + U =$ const. (a constant). V and U are both components of fluid velocity, ρ is fluid density, and p is pressure. The Bernoulli effect is a common demonstration of the physical principle of this relationship, and items such as chimneys, perfume atomizers, and airbrushes use the effect in everyday life.

Another concept related to turbulent flow is the Reynolds number. It is a dimensionless factor that determines the onset of turbulence of flow in a tube. The number is derived from a relationship between the density of the fluid, the average velocity of the fluid, the diameter of the container housing the flow, and the viscosity of the fluid. If the Reynolds number is above 3,000 turbulence occurs; if the number is below 2,000 the flow is laminar; between 2,000 and 3,000 the flow is unstable and can move in a combination of laminar and turbulent flow.

Typically when a fluid flows through a tube it is because of a pressure difference across the ends of the tube. There is another method of fluid transportation called diffusion or osmosis. This mechanism depends on a concentration difference between two points in the fluid. If the concentration of molecules is higher at one point than another the molecules will move from the position where the concentration is higher to a position where the concentration is lower. The rate of the diffusion process is proportional to the change in concentration per unit distance, the concentration gradient, and the cross-sectional area. Osmosis is similar to diffusion in that it is a movement of fluid from a point of higher concentration to a point of lower concentration but it also involves transport of the fluid across a selectively permeable membrane. With both of these transport mechanisms they will continue until the concentrations on the two sides of the membrane or in the entire system are equal. Osmosis is considered to play a vital role in living cells and the transport of necessary nutrients to and from them.

FLUORESCENCE

Fluorescence is the process by which a substance absorbs electromagnetic **radiation** (visible or invisible **light**) from another source, then re-emits the radiation with a **wavelength** that is longer than the wavelength of the illuminating radiation. It can be observed in gases at low **pressure** and in certain liquids and solids, such as the ruby gemstone. Fluorescence is the principle that is the basis of the common fluorescent lamp used for lighting; it is also a useful laboratory diagnostic tool.

Matter interacts with electromagnetic radiation (such as ultraviolet and visible light) through the processes of absorption and emission. The internal structure of atoms and molecules is such that absorption and emission of electromagnetic radiation can occur only between distinct **energy** levels. If the **atom** is in its lowest **energy level** or ground state, it must absorb the exact amount of energy required to reach one of its higher energy levels, called excited states. Likewise an atom that is in an excited state can only emit radiation whose energy is exactly equal to the difference in energies of the initial and final states. The energy of electromagnetic radiation is related to its wavelength as follows: shorter wavelengths correspond to greater energies, longer wavelengths correspond to lower energies.

In addition to emitting radiation, atoms and molecules that are in excited states can give up energy in other ways. In a gas, they can transfer energy to their neighbors through collisions that generate **heat**. In liquids and solids, where they are attached to their neighbors to some extent, they can give up energy through vibrations. The observation of fluorescence in gases at low pressure comes about because there are too few neighboring atoms or molecules to take away energy by collisions before the radiation can be emitted. Similarly, the structure of certain liquids and solids permits them to exhibit strong fluorescence.

If the wavelength of the radiation that was absorbed by the fluorescent material is equal to that of its emitted radiation, the process is called resonance fluorescence. Usually though, the atom or molecule loses some of its energy to its surroundings, so that the emitted radiation will have a longer wavelength than the absorbed radiation. In this process, simply called fluorescence, Stoke's law says that the emitted wavelength will be longer than the absorbed wavelength. There is a short delay between absorption and emission in fluorescence that can be a millionth of a second or less. There are some solids, however, that continue to emit radiation for seconds or more after the incident radiation is turned off. In this case, the phenomenon is called phosphorescence.

Ruby is a crystalline solid composed of aluminum, **oxygen**, and a small amount of chromium that is the atom responsible for its reddish **color**. If blue light strikes a ruby in its ground state, it is absorbed, raising the ruby to an excited state. After losing some of this energy to internal vibrations the ruby will settle into a *metastable state*—one in which it can remain longer than for most excited states (a few thousandths of a second). Then the ruby will spontaneously drop to its ground state emitting red radiation whose wavelength (longer than the blue radiation) measures 6,943 Angstroms. The fluorescent efficiency of the ruby—the ratio of the intensity of fluorescent radiation to the intensity of the absorbed radiation—is very high. For this reason the ruby was the material used in building the first **laser**.

The most well-known application of fluorescence is the fluorescent lamp, which consists of a glass tube filled with a gas and lined with a fluorescent material. **Electricity** is made to flow through the gas, causing it to radiate. Often mercury vapor, which radiates in the violet and ultraviolet, is used. This radiation strikes the coating, causing it to fluoresce visible light. Because the fluorescence process is used, the fluorescent lamp is more efficient and generates less heat than an incandescent bulb.

Resonance fluorescence can be used as a laboratory technique for analyzing different phenomena such as the gas flow in a wind tunnel. Art forgeries can be detected by observing the fluorescence of a painting illuminated with ultraviolet light. Painting medium will fluoresce when first applied, then diminish as time passes. In this way paintings that are apparently old, but are really recent forgeries, can be discovered.

FORCE

Force is the term used for an outside influence exerted by one body on another that produces a change in state of **motion** or state of configuration. This limited meaning in science compared to our everyday usage is most important because of the specific results of this outside influence.

Force producing a change in state of motion gives a body **acceleration**. If forces acting on a body produce no acceleration, the body will experience some change in configuration: a change of size (longer or shorter), a change of shape (twisted or bent), or a positional change (relative to other masses, charges, or magnets). Changes of size or shape involve elastic properties of materials.

Forces are given various names to indicate some specific character. For example, a wagon can be made to go forward by pushing from behind or pulling from the front, so push or pull is more descriptive. Electrical and magnetic forces can result in attraction (the tendency to come together) or repulsion (the tendency to move apart) but **gravitational force** results only in attraction of masses. The gravitational force exerted by Earth on a body is called **weight**. A body moving, or attempting to move, over another body experiences a force opposing the motion called **friction**. When wires, cables, or ropes are stretched, they then in turn exert a force that is called tension. Specific names give information about the nature of the force, what it does, and direction of action.

FORCE AND THE LAW OF MOTION • See

Newton's laws of motion

FORCE, MASS, AND ACCELERATION

Today most people will have very little difficulty with the idea of **motion** continuing indefinitely. Since the advent of the **space** age we have grown accustomed to the idea that space probes can leave the solar system, traveling endlessly onward through empty space. We also are familiar with the notion that

Electric drill balanced in a piece of wood due to opposing forces of their weight. (Photo by Charles D. Winters, National Audubon Society Collection/Photo Researchers Inc. Reproduced by permission.)

atoms, molecules, and subatomic particles are all in a continuous state of ceaseless motion. We are comfortable with the concept of the **planets**, **stars**, and galaxies all spinning, rotating, and forever moving through empty space. A citizen of earlier centuries, if told these facts, would have immediately asked, "What is driving these bodies? What is keeping them moving?" In earlier times, all moving bodies seemed to eventually stop. Continuous motion seemed against nature. Early ideas about the movements of planets had mythical horses pulling them through the sky. The concept of **perpetual motion** seemed unattainable, whereas, ironically, we now know that perpetual motion is the normal state of things.

It took the genius of **Isaac Newton** to see that if bodies are left alone they will continue to move at a constant **velocity** in a straight line forever. It has taken the rest of the world nearly 400 years to get comfortable with this idea.

Although constant velocity in a straight line is the normal state of a totally isolated body, it is obvious that most bodies do not exist in that state. Planets travel in elliptical paths (not straight lines), moving bodies often do stop, and velocities do change. Speeding up and slowing down is an extremely common occurrence. So what is going on?

From a physicist's perspective, when a body speeds up, slows down, or changes direction it is demonstrating that a **force** is acting upon it. A force is simply an agent of motion change. When an object like a car accelerates, it is because there is a force pushing it to greater speeds. When a speeding

bullet drops to the earth, it is because the **friction** of the air is supplying a retarding force opposing the motion of the bullet thus slowing down. Throwing a ball vertically into the air is a very good example of what a force can do. At first a force (**gravity**) pulls against the ball reducing the speed of the ball until, for a moment it stops at the height of its climb. Then, since the force of gravity never ceases, the ball will move from rest at the top of its climb and start to get faster and faster as it falls to the earth. To re-emphasize, when acted upon by a force, objects will change speed and direction.

When several forces act on a body they can be added together as **vectors**. The change in velocity experienced by the body reflects the effect of all the forces acting on it. This explains a lot of the phenomena we see around us. It might be argued that, based on what was said earlier, an airplane flying at 500 miles per hour should constantly get faster while the jet engines continue to push it. After all, we said that a force makes bodies accelerate. The fact that the plane does not get consistently faster is because it is being acted upon by two forces not one. The force of the engines is one force, the resisting force of the surrounding air is another. The force of the engines and the force of the air are acting in opposition to each. In fact, the airplane has a total of zero force acting on it because the push of the engines and the push of the wind exactly cancel each other out. And because the airplane has no resulting force acting on it, it will continue at a constant velocity in a straight line.

Comparing an airplane surrounded by air and a rocket in space is useful for clarifying this point. An airplane needs the push of an engine to equal the opposite force of the wind in order to remain at a constant velocity. A rocket moving through space, well away from suns, planets, and moons, needs no engines to keep at a constant velocity. There are no forces acting on it so constant velocity is assured. Unlike an airplane, when a rocket engine is firing in space the rocket will not cruise at a constant speed, on the contrary it will continue to accelerate until it is switched off. In this case the engine is the only force acting on the rocket, thus we get **acceleration**.

Inertial Mass

Two road vehicles line up to race a one-mile stretch of highway. One vehicle is a small sports car, the other a large 24-wheel truck. To make the race fair, the truck has had its engine removed and replaced with an engine identical to the sports car. In this way the two vehicles are to be accelerated by exactly the same force. The race is not fair though because intuitively we know that the truck needs more force to accelerate it than the sports car. Our intuition is right. Physicists say that the truck has a greater inertial **mass** than the sports car. How quickly an object reacts to a force is a measure of its inertial mass. If two different objects have identical forces applied to them, one might move off with a slower acceleration. If that is the case, the one that is slower to build up speed is said to have the greater inertial mass. It is important to understand that this phenomenon has nothing to do with **weight** and gravity. Take the truck and the sports car way out to the far reaches of space, well away from any gravity, attach identical rocket motors to each of them and the result is the same. The truck will not accelerate as quickly as the sports car. Acceleration, force, and inertial mass are inextricably tied together. Greater inertial mass means slower acceleration for the same force.

Gravitational Mass

Two objects floating close together far out in intergalactic space will be attracted to each other and if left alone will slowly drift towards each other until they collide and stick together. The attractive force exerted by each body on the other will not be very strong but it will increase the closer the two bodies are to each other. This is the mysterious force of gravity. The strange thing about this force is that its strength is also dependent upon the mass of the object. From childbirth each of us experiences the force exerted by Earth on our bodies. If we want relief from this force we would need to live on the **Moon** where the attractive force is one-sixth of the force on Earth simply because the mass of the Moon is one-sixth the mass of Earth. We now have two absolutely different ways of regarding mass. One way is to observe its acceleration when pushed or pulled. This measurement is called its inertial mass. The other way is to measure how much **gravitational force** the object will exert on another body. This we call its gravitational mass.

Gravitational Mass versus Inertial Mass

Gravitational mass and inertial mass are two different things. Gravitational mass decides how strong its gravitational force on other bodies will be. Inertial mass decides how fast the body will accelerate when acted upon by a force. The two masses are measured in totally different ways. Physicists have not, however, been able to find any difference in their measurements. If object A accelerates at one-tenth the acceleration of object B given the same force, we can say that object A has an inertial mass 10 times bigger than object B. We then find that object A exerts exactly 10 times more force on an object than object B showing that A's gravitational mass is also exactly 10 times bigger than B's. Experiments of incredible accuracy have tried to detect a difference in the two masses but without success. **Albert Einstein** finally reconciled these two concepts in his general theory of relativity.

Units of Force, Mass, and Acceleration

Mass units are not derived from any other units. They are standard units. In the United States we commonly use a standard pound as our mass measure. Physicists use the kilogram unit, which is a block of platinum-iridium stored in Sèvres, France.

Acceleration is the rate of change of velocity. If an individual goes from 30 mph (48.3 kph) to 50 mph (80.5 kph) the velocity change is 20 mph (32.2 kph). This shows nothing about the acceleration until it is determined how long it took for this velocity change to take place. If it took one minute then the acceleration is 20 mph in one minute (written 20m/h/min). Clearly we have **time** units a little mixed up here so it is more usual to write velocity measurement as feet per

second, and acceleration measure as velocity change per second. This reads as feet per second each second or ft/sec/sec. In metric measure it is usually meters per second each second or m/sec/sec.

The units for force are derived from the units of mass and acceleration. Physicists use the kilogram of mass, the meter of length, and the second of time to derive their unit of force. Taking Newton's concepts they define the unit of force as that amount of force that would accelerate a unit mass with a unit amount of acceleration. In other words, if a one-kilogram mass is having its velocity changed (accelerated) at one meter per second each second, then physicists say that there is a unit force acting on the object. The name they give to that unit is not surprisingly, the newton.

See also Relativity, general

FOUCAULT, JEAN-BERNARD-LÉON (1819-1868)
French physicist

Jean-Bernard-Léon Foucault performed two of the most important experiments in nineteenth-century physics: the accurate determination of the **velocity** of **light** on Earth and the demonstration of Earth's rotation, using his famous pendulum. He also made important improvements to the **telescope**, carried out significant experiments in **optics**, and invented several widely used instruments. Persistently in poor health, Foucault lived a quiet life devoted to scientific experimentation and his numerous discoveries and inventions brought about major advances in several scientific fields.

Born in Paris in 1819, Foucault was the son of a bookseller and publisher. A frail child, Foucault was educated at home. Planning to become a surgeon, he entered the École de Médecine. However, Foucault couldn't stand the blood and abandoned his studies. Instead, he became an assistant in clinical microscopy to Alfred Donné and coauthored a textbook with him. In 1845, Foucault succeeded Donné as the science reporter for the *Journal des débats* and for many years he wrote a science column for the general reader. Between 1844 and 1846, Foucault also published arithmetic, geometry, and chemistry textbooks.

Throughout the 1840s, Foucault carried out experiments in his home laboratory. With his collaborator Armand Fizeau, Foucault made significant studies in optics They also began experimenting with daguerreotype (early **photography**) and made several improvements to the process. They took the first picture of the surface of the **Sun** in 1845. Since this required that the telescope remain pointed at the Sun for long exposure times, Foucault designed a drive with a pendulum to move the instrument as the position of the Sun changed. This device suggested to Foucault a means for demonstrating the rotation of Earth. He constructed a pendulum and demonstrated that the plane of its swing moved with Earth's rotation. In 1851, using a pendulum that was 12 yd (11 m) long, he demonstrated that the plane of the pendulum's swing was inversely proportional

Jean-Bernard-Léon Foucault. (UPI/Corbis-Bettmann. Reproduced by permission.)

to the sine of the latitude. This proved that Earth's rotation resulted in a deflecting **force**. Foucault then built a massive pendulum and held a public demonstration at the Panthéon. The following year, he invented the gyroscope to better illustrate these principles. For this work, Foucault was awarded the Cross of the Legion of Honor.

Foucault examined conductivity in liquids and he demonstrated the conversion of mechanical work into **heat**. He made improvements to the arc lamp and invented a number of different mechanical regulators, first used in the heliostat that kept his telescope pointed at the Sun. He also built a siderostat to keep his telescope pointed at a star. Later, his regulators were used in **steam engines**. Foucault was the first to use glass in reflecting telescopes and developed new methods for correcting mirrors and **lenses**. With Jules Regnault, he showed that different images seen separately by each eye are combined into a single image by the human brain. In 1862 Foucault joined the Bureau des Longitudes, where he made improvements to surveying instruments.

In 1850, Foucault was able to announce that light travels faster in air than it does in water. This lent support to the theory that light was a **wave** rather than a particle. He also demonstrated that radiant heat was a wave identical to light. After earning his doctorate in physics in 1853 for this work, Napoleon III appointed Foucault as physicist at the Paris

Observatory. In 1862, he was able to accurately measure the **speed of light** in air.

Foucault was awarded the Copley Medal of the Royal Society of London in 1855 and became a foreign member of that society in 1864. He also was a foreign member of the scientific academies of Berlin and St. Petersburg. However, he was not elected to the French Académie des Sciences until 1865. He died of a brain disease in Paris in 1868, at the age of 48.

FOUR-DIMENSIONAL SPACE TIME • See Space-time

FOURIER, JEAN-BAPTISTE-JOSEPH (1768-1830)
French physicist and mathematician

Jean-Baptiste-Joseph Fourier was a French mathematician and physicist. Fourier was active in the French Revolution and later worked for Napoleon and the French government as an administrator and Egyptologist. Best known for his contribution to the mathematical analysis of **heat** flow, Fourier laid the groundwork for later efforts on trigonometric series and the theory of functions of a real variable.

Born in Auxerre, France, Fourier was the son of a tailor. At a young age he showed great promise in various areas of study, including Latin and literature. He was drawn towards mathematics and had completed a comprehensive study of Bezout's *Cours de Mathematique* by the time he was 14 years of age. Although he briefly studied for the priesthood in a Benedictine abbey, Fourier continued his interest in mathematics and decided to leave the monastery. Fourier went on to teach at the College de France and the École Polytechnique. Fourier then joined Napoleon on his invasion of Egypt in 1798 as a scientific adviser. Upon his return to France in 1801, Napoleon appointed Fourier prefect of the Department of Isère in Grenoble, where he spent much time working on his *Description of Egypt* and on the mathematics for his theory of heat.

In his 1807 memoir, *On the Propagation of Heat in Solid Bodies*, Fourier laid the groundwork for his mathematical theory of heat, which was culminated with the 1822 publication of *Analytical Theory of Heat*. In his theory, Fourier describes the partial differential equation governing heat diffusion and its solution by using infinite series of trigonometric functions. The method of mathematical analysis is now known as the Fourier series. Although other mathematicians had used the series before Fourier, he investigated them in much greater detail. He was initially criticized for his work's lack of vigor but later was proven correct. Through Fourier's work, it became accepted that nearly any function of a real variable can be represented by a series involving the sines and cosines of integral multiples of variables.

Fourier's work in mathematics represented a high point in the history of pure and applied mathematics in the eighteenth century. Fourier, who was elected to the French Academy of Sciences, died at the age of 62.

Jean-Baptiste-Joseph Fourier.

FOWLER, WILLIAM A. (1911-1995)

American physicist

William A. Fowler was noted for his theories dealing with how **stars** produce **heat** and **light** based on his explanations of the synthesis of chemical elements in the **universe**. Fowler received the 1983 Nobel Prize in physics in recognition of "his theoretical and experimental studies of the **nuclear reactions** of important atoms in the formation of the chemical elements in the universe." His contributions have been of benefit to the fields of **astronomy**, astrophysics, **cosmology**, and geophysics in addition to **nuclear physics**.

Fowler was born on August 9, 1911, in Pittsburgh, Pennsylvania, to John MacLeod, an accountant, and Jennie Summers (Watson) Fowler. The Fowlers had two other children, Arthur Watson, born in 1913, and Nelda, born in 1919. When William was two years old, his family moved to Lima, Ohio, where he attended Horace Mann Grade School and Central High School. At Central, Fowler was president of the senior class, a varsity football and baseball player, and valedictorian of the graduating class of 1929.

Fowler received his bachelor's degree from Ohio State University in 1933 and his Ph.D. in nuclear physics at the Kellogg Radiation Laboratory at the California Institute of Technology (Cal Tech) in 1936. Immediately, he was offered a job as research fellow at Kellogg and then, over the next 45 years, was promoted from assistant to full professor. He retired from the California Institute of Technology in 1982 and was named emeritus professor of physics.

Fowler entered Ohio State University in the fall of 1929 intending to major in ceramic engineering. Two years later, however, he switched to engineering physics, a field in which he would earn his bachelor of science. Although he had to work throughout his college years in order to support himself, Fowler was able to record the highest grade point average in his graduating class.

Begins Nuclear Research at Cal Tech

Upon graduation from Ohio State, Fowler decided to enter the California Institute of Technology for his graduate work. There, he was assigned to assist the director of the W. K. Kellogg Radiation Laboratory, C. C. Lauritsen, whom Fowler credited as being the greatest influence in his life, according to an article in *Physics Today*. For his doctoral dissertation, Fowler studied the production of radioactive **isotopes** as the result of bombarding light elements with **protons** and deuterons. He was granted his Ph.D. in physics summa cum laude in 1936.

Lauritsen was well satisfied with the work of the young Fowler and asked him to stay on as a research fellow in nuclear physics. Three years later, Fowler began his climb up the academic ladder with an appointment as assistant professor at Cal Tech and then, in 1942, with a promotion to associate professor. At this point, World War II interrupted the normal research taking place at Cal Tech. Lauritsen and Fowler were assigned to work in Washington, D.C., on the development of proximity fuses (fuses for projectiles that use **radar** to detect the presence of a target within effective ranges) for bombs, shells, and **rockets**. Later in the war, Fowler became involved in the development of the **atomic bomb**. For his wartime contributions, Fowler was given the U.S. Government Medal for Merit in 1948.

Explains the Origins of the Elements

Scientists have long been intrigued by the question of how the chemical elements are formed in the universe. A major revelation took place in 1939 when physicist **Hans Albrecht Bethe** at Cornell University and Carl F. Von Weizsäcker at the University of Berlin proposed a mechanism by which hydrogen is converted into **helium** in a star. The CN cycle (for the **carbon** and nitrogen involved in the process) not only explained the conversion of hydrogen to helium, but also showed how **energy** is released in the process.

The question remained, however, as to how elements heavier than helium can burn, and thus be formed in a star. At one point, **George Gamow** had suggested a simple and reasonable explanation. The capture of a **neutron** by one **atom** could result in the formation of a new atom one atomic number greater than the original. But the problem with Gamow's hypothesis was that it could not be confirmed experimentally. Researchers at Kellogg had demonstrated that no stable **mass** of 5 or 8 could exist. With these gaps, Gamow's theory became undefendable.

By the early 1950s, Fowler had become convinced that the production of heavier elements can take place through the fusion of helium atoms. By 1954, the details of that process were becoming clear. Fowler spent the 1954–55 academic year at Cambridge University working with the eminent astrophysicist **Fred Hoyle** and the husband and wife team of Geoffrey Burbidge and Margaret Burbidge. Together the four researchers identified a process by which helium can be converted to carbon, carbon to iron, and eventually iron to the heavier elements (by neutron capture).

In 1975, Fowler, Hoyle, and the Burbridges published one of the classic papers of modern science, "Synthesis of the Elements in Stars," which often referred to the authors' initials as B_2FH. The ideas presented in the paper were the basis for the Nobel Prize committee's decision to award a share of the 1983 physics prize (along with **Subrahmanyan Chandrasekhar**) to Fowler.

The mechanisms by which elements are formed in the universe continued to dominate Fowler's research agenda for another two decades. Working often with Hoyle, he developed hypotheses about the formation of elements in bodies other than stars, such as the recently discovered radiogalaxies. He also became increasingly interested in the study of neutrinos and other astronomical phenomena.

Fowler was in great demand as a lecturer and visiting scholar. He was a Fulbright lecturer and Guggenheim Fellow at the University of Cambridge twice, in 1954–55 and again in 1961–62, and also at St. John's College in the latter visit, and lectured at the University of Washington, the Massachusetts Institute of Technology, and the Institute of Theoretical Astronomy at Cambridge. He received honorary doctorates from the University of Chicago, Ohio State University, Denison University, Arizona State University, the University of Liège, and the Observatorie de Paris.

Fowler was married to the former Ardiane Foy Olmsted on August 24, 1940. They had two daughters, Mary Emily and Martha Summers. Mary Fowler became his second wife after Ardiane's death in 1988. Fowler's colleague Bethe, writing in *Science,* described him as "full of humor and cheerfulness," and noted that Fowler's most outstanding characteristic was that "he loved people." Fowler died of kidney failure in Pasadena, California, at the age of 83.

FRACTALS

Fractals are an important part of chaos theory, which is crucial to understanding physics at many, perhaps all, levels. Chaos theory is often one of the more misunderstood portions of modern physics. A chaotic function or system is not the same as a random one-thus, chaos theory does not explain the randomness in **quantum physics**. Rather, a chaotic system is one in which the resulting action of the system is widely divergent from a very small divergence in initial conditions. That is, dis-

placement by a tiny fraction of a percent of the measurement used will cause results that diverge by orders of magnitude.

Chaotic behavior can occur in both classical and quantum mechanical systems. The physical world has many examples of chaotic behavior, from magnetic pendula to **planets** orbiting double star systems. Each system follows known equations of **motion**, but each system is heavily reliant on the position of the moving object at the beginning of the observation period. Thus, the behavior is predictable from its initial conditions only so much as they are precise: if the researcher is off by a tiny amount, or if the researcher's computer can't hold all of the relevant decimal places, the physics predictions and the physical results will appear to be nonsense.

Fractals are self-similar curves: they display the same behavior on any scale. That is, if an observer zooms in on a fractal, it will have the same shape characteristics on the smaller and larger scales. Physically, stretched and folded objects will end up resembling fractals, with the location of any point on the object heavily dependent on where it began. One of the properties of fractal objects or objects of fractal dimension is that their measurements will depend on the size of the measuring object used. Natural formations such as clouds, fjords, rivers, and trees usually have some sort of fractal behavior in this regard.

The concept of fractal dimension is useful when determining how chaotic a system is. Fractals can have fractional dimension, unlike the normal dimensions people are used to dealing with, which only have integer values. (That is, one normally hears of three-dimensional objects but never of 3.6-dimensional objects—unless one is talking about chaos theory.)

Fractals have been widely used in problems ranging from measurement to polymers to nuclear reactor safety. They, and other aspects of chaos theory, have proven essential for any study of physics in the modern age. Phenomena such as **turbulence**, which were previously regarded as extremely "messy" fields to study, can be made somewhat more manageable with the application of fractal dimensions. Some subjects, both classical and quantum mechanical, which were previously thought to be closed or insoluble, have now become much more vital parts of physics research.

FRAME OF REFERENCE

To measure the position and **time** of an event, we rely upon frames of reference. The frame of reference is an imaginary three-dimensional coordinate grid that allows us to specify where an object is in time and **space** as compared to other objects in different **reference frames**.

In the early seventeenth century, Italian astronomer **Galileo** reasoned that everything in **motion** can only be measured in relation to something else. **Isaac Newton** later elaborated on this idea, using frames of reference to determine that measurements of speed depend on who measures them. We observe such inconsistencies every day. To a person standing on the side of the road, a cup sitting on the dashboard of a car moves at the speed of the car. To the driver of the car, the cup is stationary. Newton concluded that any physics experiment, such as measuring the **force** of **gravity**, conducted either by the person in the moving car or the person on the street would yield the same results. Therefore, it is just as correct to say that the cup is at rest as it is to say that the cup is moving at the speed of the car. Newton called the two viewpoints inertial frames of reference, because there is no difference in the perspective of an observer moving at a constant speed and one who is stationary.

Albert Einstein borrowed Newton's idea that the laws of **mechanics** are valid in all inertial frames of reference for his special theory of relativity published in 1905. A fundamental concept of that theory is that no one viewpoint from which measurements can be made is privileged over any other, if the viewpoints are either stationary or moving at a constant **velocity**. To Einstein, there exists no absolute frame against which we can measure distance, time, or velocity. The only absolute, from any frame of reference, is the **speed of light**.

Though Einstein used old ideas as the basis of his theory, his conclusion was revolutionary for physics. It meant that time and space are not absolute: two people may measure the same object with entirely different results.

FRANCK, JAMES (1882-1964)

German American physicist

James Franck was a physicist whose experimental work with atoms and electrons proved **Niels Bohr**'s theory that atoms are quantized—that they transmit and absorb **energy** in discrete quantities or packages. Along with collaborator **Gustav Hertz**, he was awarded the 1925 Nobel Prize in physics. Franck was also known for his outspoken opposition to the use of the **atomic bomb**, which he helped develop during World War II.

Franck was born in Hamburg, Germany, on August 26, 1882, to Jacob Franck, a German Jewish banker, and Rebecka Nachum Drucker. Although Jacob Franck was deeply religious—he observed Jewish holidays with fasting and chanting—his spiritual devotion did not, on the whole, pass on to James, who would later declare science and nature as his true love and religion. He attended school at the Wilhelm Gymnasium in Hamburg before enrolling at the University of Heidelberg. Franck's father wanted him to study law and economics with the hope that his son would take over the family business. Out of a sense of duty, Franck complied, but after attending law lectures for a short time, he determined to follow his own path and enrolled in the faculty of chemistry.

Heidelberg was where Franck met **Max Born**, the German physicist with whom he formed his closest friendship. After two terms studying chemistry, Franck enrolled in the doctoral program at the University of Berlin. Under the influence of its physics professor, Emil Warburg, he became interested in physics and switched fields. He began a study to determine the mobility of **ions** using a method invented by Cambridge University physicist **Ernest Rutherford**.

After graduating with a D.Phil. in 1906, Franck continued to pursue the same lines of research, exploring the forces between electrons and atoms at the physics faculty of the University of Frankfurt-on-Main. He returned to Berlin in 1908 to become an assistant to Professor Heinrich Rubens. There, Franck began collaborating with the German physicist Gustav Hertz on a series of experiments that would provide direct proof of Bohr's theoretical model of **atomic structure**, demonstrate the quantized energy transfer from kinetic, or moving, energy to **light** energy, and establish both of their reputations.

Bohr had postulated that an atom's **nucleus**, or core, is surrounded by "orbits" of negatively charged electrons. Bohr theorized that these orbits revolve around the nucleus at set distances known as shells. The number of electrons and, thus, the number of shells, vary according to the type of **atom**. Atoms ranking high on the **periodic table** of elements contain more electrons than simple elements such as hydrogen, which has just one **proton** and one **electron**. These extra electrons are contained in extra shells, according to a definite pattern. The first shell contains two electrons; the second, eight; the third, 18; the fourth, 32; and fifth, 50, and so on. As soon as the first shell is full, electrons begin to fill up the second shell, then the third, up to the last shell.

In their natural, unexcited state, the electrons try to stay as close to the nucleus as possible, that is, in an inner shell. Bohr suggested that electrons would jump from one shell to another if energy were applied to them. The distance they would jump would depend on the amount of energy supplied; when the energy source was withdrawn, they would fall back to their original position. The energy emitted by electrons falling back in toward the nucleus would be exactly equivalent to that absorbed by them when jumping to an outer shell. Most importantly, atoms receiving energy could not absorb just any amount but only the specific amount they would need to make a leap. Thus, Bohr spoke of the atom as being "quantized."

Experimentally Proves Bohr's Theory of the Quantum Atom

Franck and Hertz did not set out to prove Bohr's theory. In fact, they were not even familiar with his work at the time they were carrying out their experiments. Rather, they were interested in measuring the energy needed to ionize atoms of mercury. To this end, they bombarded atoms of mercury vapor with electrons moving at controlled speeds. Below a certain speed, the electrons would bounce off the atoms with perfect **elasticity**, indicating that the electrons did not possess sufficient energy to ionize the mercury atom, that is, to transfer enough energy to the mercury to enable *its* electrons to jump from one atomic shell into another. Above a certain speed, Franck and Hertz discovered that resonance occurred. At this point, energy was transferred from the electrons to the atoms, causing the mercury gas to glow. They found that energy had been transferred from the electrons to the atoms in discrete amounts. The energy value of the light emitted from the ionized atoms was equivalent to the energy given to them by the electrons. This experiment proved that the quantized energy had changed from the **kinetic energy** of the moving electrons to the electromagnetic energy given off by the glowing mercury. It also provided direct experimental evidence for Bohr's theory of the quantized atom, a crucial step in the development of twentieth-century physics.

This experiment was also significant because it led to the realization that the light spectrum of an atom holds the key to its atomic structure. The discontinuous bands of light in an atomic spectrum, each representing a particular **energy level**, correspond to the range of possible jumps that an excited electron could make as it drops from the outer shells, where the absorption of energy had sent it, back to its original inner shell.

Decorated with Iron Cross for War Service

Franck's work was unexpectedly interrupted with the outbreak of World War I. He signed up and became an officer. He served through 1918, working with a group of physicists who prepared and later directed chemical warfare. Franck received the Iron Cross for his valor; he also received a serious leg injury, which almost claimed his life. Returning to academia in 1918, he was named as the head of the physics division at the Kaiser Wilhelm Institute for Physical Chemistry, later renamed the Max Planck Institute. There, Franck pursued his work on electron impact measurements. It was also at the institute that he met Niels Bohr, with whom he developed a lasting friendship. Franck always regarded Bohr as a physicist second to none and consulted him regularly. "I never felt...such hero worship as [I did] to[ward] Bohr," he said in an interview excerpted in *Redirecting Science: Niels Bohr, Philanthropy, and the Rise of Nuclear Phhysics.*

In 1920, with the influence of Born, Franck was appointed professor and director of the Second Physical Institute of the University of Göttingen. The friendship between Franck and Born blossomed into a close working relationship, with Franck the experimenter complementing Born the theorist. During their 12 years at Göttingen, the pair used one another as sounding boards for their ideas, discoveries, and publications, although they collaborated on only a few joint papers. The only contention between them was Franck's habit of holding frequent consultations with Bohr, a practice that tended to slow down their work. More than 60 letters between Franck and Born have survived from the 1920s.

In the spring of 1921, at Bohr's invitation, Franck paid a visit to Copenhagen in time for the March opening of Bohr's Institute of Theoretical Physics. By now, his reputation preceded him and his visit made front-page news in Denmark. His meeting with Swedish physicist **Oskar Klein** and Norwegian Svein Rosseland convinced him to continue his experimental work on Bohr's theories.

Back at the University of Göttingen a couple of months later, Franck concentrated on building a research facility of international repute. He afforded his students considerable academic freedom. Scientific discussions between teacher and pupils would occur as often during a walk or bicycle ride as in the laboratory. The standards for admission to his school were extremely high but once accepted, a student was assured of his

unwavering support and friendship, both professionally and personally.

Franck continued to investigate collisions between atoms, the formation and disassociation of molecules, **fluorescence**, and chemical processes. In 1925, building on three previously unconnected theories, he published a paper dealing with the elementary processes of photochemical reactions. In it he set out the connection between electron transition and the **motion** of nuclei, and described a general rule for vibrational energy distribution. This rule was later expressed by American physicist Edward U. Condon in terms of quantum **mechanics** (a mathematical interpretation of particle structures and interactions) and became known as the Franck-Condon principle, which is applied to a large number of chemical and spectroscopic phenomena. In 1926 Franck published a book summarizing his work in this area.

Also in 1926, Franck traveled to Sweden to accept the 1925 Nobel Prize in physics, awarded jointly to him and Hertz for their experiments proving Bohr's **atomic theory**. He returned to Göttingen to begin his next project, the study of photosynthesis, but had no sooner begun his experiments when Adolf Hitler's arrival on the German political stage changed his life.

Resigns from Göttingen to Protest Anti-Semitism

When Hitler's anti-Semitic Nazi regime took control of Germany, a new law was declared that barred Jews from the civil service, excepting those who had served in World War I. Although Franck's position was secure, he could not in good conscience continue to work for a regime dedicated to racism, so on April 17, 1933, he sent letters to the minister of education and to the rector of the university, announcing his resignation and decrying the government's discriminatory policy. Hoping to remain in Germany, Franck searched for another position. Two possibilities presented themselves, one being the chair of physics at the University of Berlin, which would shortly be open. Though it was a position Franck would have coveted under other circumstances, it would have meant working for the government. The other possibility was the directorship of the Kaiser Wilhelm Institute for Physical Chemistry, a position that retiring director Fritz Haber hoped Franck would accept. Internal problems in the institute, however, prevented Franck from assuming this post as well. Franck decided to accept a visiting lectureship at the Johns Hopkins University in America. After the three-month period of that position he returned to Göttingen to contemplate his uncertain future. Tentative offers were made from universities in the United States, but they did not promise the permanency Franck was seeking. He decided to accept an offer from Bohr for a year's work at his Institute of Theoretical Physics.

Franck arrived in Copenhagen in April 1934, and, with his assistant Hilde Levi, set to studying the fluorescence of green plants, an extension of his previous work studying energy exchanges in complex molecular systems. Under Bohr's direction, he also began administering experimental nuclear research at the institute. He was frustrated by poor facilities and slow coworkers and, as stated in *Redirecting Science,* wrote of this period: "My **nuclear physics** exhausts itself at present in work which is just about to be completed when someone else publishes it in *Nature.*" Working with a master theorist such as Bohr also proved difficult for Franck. "Bohr's genius was so superior. And one cannot help that one would get so strong inferiority complexes in the presence of such a genius that one becomes sterile," he later said in an interview quoted in *Redirecting Science.* After being used to having his own laboratory and students, it was hard for Franck to get used to working in Bohr's shadow.

Emigrates to United States

The combination of numerous frustrations spurred Franck to accept an offer to settle in the United States. In late 1935, he became a professor at Johns Hopkins University, where he spent three years before moving to the University of Chicago to fill its chair of physical chemistry. With the help of the Samuel Fels Fund, a laboratory dedicated to research into photosynthesis was built, which Franck directed until his retirement in 1949, though he continued to work there for many years subsequently. He became an American citizen in the early 1940s.

When World War II broke out, Franck played a leading role in the **Manhattan Project**, the American government-sponsored atomic bomb project. Like the other German scientists on the team, he was driven by a desire to beat Hitler to the production of a nuclear weapon. But he firmly believed that the bomb should be used as a mode of deterrence, not as a means of aggression. When the United States finally developed the bomb and subsequently deployed it against the Japanese, Franck was a harsh critic.

In 1942, a crisis struck in Franck's private life with the death of his wife, Ingrid Josephson, who had been sick for many years. He coped with the loss by immersing himself in his work. He chaired a committee of scientists charged with exploring the social and political implications of detonating an atom bomb. That committee's findings, titled the Franck Report, was submitted to the U.S. Secretary of War, Henry Stimson, in 1945, and warned the U.S. government against the use of the bomb as a military weapon. The report also speculated on the dangers of embarking upon an arms race and also urged the United States to restrict nuclear testing to areas where human life would not be endangered. The Franck Report has been seen as a testament to Franck's integrity, conviction, and sense of scientific responsibility.

With the end of the war, Franck returned to his post at the University of Chicago where he continued his work with photosynthesis. He was particularly curious as to how plants are able to transform visible light into a form of energy that they use for sustenance and growth. He began experiments on the emanation of electromagnetic **radiation** of chlorophyll, a key ingredient in the photosynthesis process. Happy to be back at work, Franck experienced joy in his personal life as well. In 1946, he married Hertha Sponer, a professor of physics at Duke University in North Carolina, whom Franck knew from

Göttingen and Berlin. They had two daughters, Dagmar and Elizabeth.

Franck was honored with numerous awards during his long career. In addition to the Nobel Prize, he was awarded the highest honor of the German Physical Society, the Max Planck Medal, in 1953. Two years later, he received the Rumford Medal of the American Academy of Arts and Sciences. He became a foreign member of the Royal Society of London in 1964 and a member of the U.S. National Academy of Sciences.

During a visit in 1964 to Göttingen, the city where he had spent his most productive years and that had made him an honorary citizen in 1953, Franck died suddenly on May 21. He was 81. He was remembered by his colleagues as a brilliant experimentalist, a dedicated scientist, and a kind and generous man.

FRANKLIN, BENJAMIN (1706-1790)
American statesman and scientist

Born on January 17, 1706, in the British colony of Boston, Massachusetts, Franklin was the fifteenth of 17 children. His father was an impoverished candlemaker, unable to afford to send young Benjamin to school. As a result, he received only two years of formal education. Franklin was working in his father's shop at the age of 10, and later was apprenticed to his brother, a printer, where he developed a love for books. In 1724 he went to London where he became skilled at printing, returning to Philadelphia two years later. In Philadelphia he made a name for himself, as well as a small fortune, publishing the *Pennsylvania Gazette* and *Poor Richard's Almanack.*

Franklin's first major invention, around 1740, was the Pennsylvania fireplace, which eventually became known as the Franklin stove. Improving on an existing design, the Franklin stove had a flue around which room air could circulate. The flue acted like a radiator, increasing heating efficiency. Franklin claimed it made a room twice as warm, with one-quarter of the wood.

In addition to his pursuit of printing, Franklin became interested in the study of **electricity** in 1746. During this period, scientists around the globe, many of whom had advanced degrees, were investigating the phenomena of static electricity. A less confident man might have felt inadequate to compete, but Franklin, who was essentially self-educated, obtained a Leyden jar and began his own research.

The Leyden jar, invented by Pieter van Musschenbroek, was a water-filled bottle with a stopper in the end. Through the stopper was a metal rod that extended into the water. A machine was used to create a static **electric charge**, which could be stored in the jar. A person who touched the end of the charged rod received an electrical jolt. Public demonstrations, in which many people joined hands and received a simultaneous shock, were very popular. Franklin saw such a demonstration, and that initiated his interest in electricity.

It was Franklin's originality and tenacity that earned him the reputation as a leading scientist. He was the first per-

Benjamin Franklin, modeling bifocal lenses.

son to wonder how the Leyden jar actually worked, and performed a series of experiments to find the answer. He poured the "charged" water out of the jar into another bottle, and discovered the water had lost its charge. This indicated that it was the glass itself, the material that insulated the conductor, that produced the shock. To verify that, he took a window pane and placed a sheet of lead on each side. He "electrified" the lead, removed each sheet one at a time, and tested for a charge. Neither sheet gave so much as a single spark, but the window pane had been charged. Franklin had unknowingly invented the electrical condenser. The condenser, also known as a capacitor, was destined to be one of the most important elements in electric circuits. Today the condenser, which received its name from **Alessandro Volta**, is used in radios, televisions, telephones, **radar** systems, and many other devices.

Drawing a parallel between the sparking and crackling of the charged Leyden jar and lightning and thunder, Franklin wondered if there was an electrical charge in the sky. He planned to erect a long metal rod atop Christ Church in Philadelphia to conduct electricity to a sentry box in which a man, standing on an insulated platform, would be able to collect an electric charge. Because he was a proponent in the free exchange of ideas, Franklin had written a book outlining his theories that received wide circulation in Europe. A French scientist named D'Alibard stole Franklin's idea and performed the experiment himself on May 10, 1752, charging a Leyden jar with lightning. Franklin generously gave D'Alibard credit for being the first to "draw lightning from the skies." If nothing else, Franklin did receive credit for the invention of the lightning rod.

While waiting for the rod to be installed atop Christ Church, Franklin had come up with an idea of a faster way to get a conductor into the sky. He tied a large silk handkerchief to two crossed wooden sticks, attached a long silken thread with a metal key at the end, and waited for a thunderstorm. The rain made the thread an excellent conductor, and the static charge traveled down to the key. When Franklin brought his knuckle to the key, a spark jumped from the key to his hand, proving the existence of electricity in the sky.

Franklin had been wise enough to connect a ground wire to his key; two other scientists, attempting to duplicate the experiment but neglecting the ground wire, were killed when they were actually struck by lightning. Still, Franklin was lucky he was not hit by lightning himself. Franklin invented the lightning rod from his work with electricity. The lightning rod became indispensable for protecting buildings from the destructive **force** of lightning. Because he had discovered he could get the Leyden jar to spark over a greater distance with a sharply pointed rod, Franklin's lightning rods had very sharp points. (In 1776, after the unpleasantness between the colonies and King George III had broken out, the king ordered that lightning rods with *blunt* ends be installed on his palace.) By 1782 there were 400 lightning rods in Philadelphia.

His discovery of sky-borne electricity led Franklin to speculate on the nature of the aurora borealis, the "northern lights" that illuminate the sky. Franklin thought they might be electrical in nature, and suggested that conditions in the upper atmosphere might be responsible.

His work on electricity led to a plethora of new words (battery, condenser, conductor, armature, charge, and discharge, to name a few) and concepts. He suggested that electrical charge was due to the abundance or lack of "something" that resulted in attraction and repulsion, and he established the concept of positive and negative charges, believing (incorrectly) that electrical flow went from positive to negative. In fact, the opposite is true.

Continuing his observations of the **weather**, he noticed there was a prevailing pattern as it moved from west to east and suggested the circulation of air masses was responsible, establishing the concept of high and low **pressure**. He went on to show that the **boiling** point of water was affected by air pressure; as he created a **vacuum** in a sealed water bottle, the **temperature** needed to boil the water dropped. He also charted the flow on the Gulf Stream in the Atlantic Ocean.

Franklin was caught off-base on occasion. When Cotton Mather proposed inoculating people to avert smallpox, Franklin was one of those who disagreed. Ironically Franklin's own son became a casualty of the disease, after which Franklin became a proponent of inoculation.

Volumes have been written about Franklin's life as a statesman. He founded service organizations, became postmaster of Philadelphia, and established a college that eventually became the University of Pennsylvania. He returned to London in 1757 as an agent of the Pennsylvania Assembly and remained there until 1775. After warning that the "Stamp Tax" was not a good way to obtain revenue from the American colonies, he returned and joined the committee drafting the Declaration of Independence.

During Franklin's long life he developed many inventions (such as bifocal **lenses** and the Franklin stove), received numerous honors, and achieved an international reputation, becoming the only American of colonial days to do so. He died on April 17, 1790, in Philadelphia, at the age of 84.

FRANKLIN, ROSALIND ELSIE (1920-1958)
English molecular biologist

The story of a great scientific discovery usually involves a combination of inspiration, hard work, and serendipity. While all these ingredients play a part in the discovery of DNA, the relationships between the four individuals who pieced together the double-helix model of the master molecule provides a subplot tainted by controversy. At the center of this quartet stands British geneticist Rosalind Franklin, who made key contributions to studies of the structures of colds and viruses, in addition to providing the scientific evidence upon which James Watson and Francis Crick based their double-helix model. Compounding the irony that Franklin died four years before Watson, Crick, and Maurice Wilkins shared the Nobel Prize for this discovery (the Nobel Committee honors only living scientists), is James Watson's characterization of Franklin in his personal chronicle of the search for the double-helix as a competitive, stubborn, unfeminine scientist. Despite his account, Franklin has been depicted elsewhere as a devoted, hard-working scientist who suffered from her colleagues' reluctance to treat her with respect.

Franklin was born in London on July 25, 1920, to a family with long-standing Jewish roots. Her parents, who were both under the age of 25 when she was born, were avowed socialists. Ellis Franklin devoted his life to fulfilling his socialist ideals by teaching at the Working Men's College, while his wife, Muriel Waley Franklin, cared for their family, in which Rosalind was the second of five children and the first daughter. From an early age, Franklin excelled at science. She attended St. Paul's Girls' School, one of the few educational institutions that offered physics and **chemistry** to female students. A foundation scholar at the school, Rosalind decided at the age of 15 to pursue a career in science, despite her father's exhortations to consider social work. In 1938, Franklin enrolled at Newnham College, Cambridge, the second-youngest student in her class.

She graduated from Cambridge in 1941 with a high second degree and accepted a research scholarship at Newnham to study gas-phase chromatography with future Nobel Prize winner Ronald G. W. Norrish. Finding Norrish difficult to work with, she quit graduate school the following year to accept a job as assistant research officer with the British Coal Utilization Research Association (CURA). At CURA, she applied the physical chemistry experience she had garnered at Cambridge to studies concerning the microstructures of coals, using **helium** as a measurement unit. From 1942 to 1946, she wrote five papers, three of them as sole author, and submitted her thesis to Cambridge. Franklin moved to Paris in 1947 to take a job with the Laboratoire Central des Services

Chimiques de l'Etat. There she became fluent in French and, under the tutelage of Jacques Mering, learned the technique known as x-ray **diffraction**. Using this technique, Franklin was able to describe in exacting detail the structure of **carbon** and the changes that occur when carbon is heated to form graphite. In 1951, she left Paris for an opportunity to try her new skills on biological substances. As a member of Sir John T. Randall's Medical Research Council at King's College, London, Franklin was charged with the task of setting up an x-ray diffraction unit in the laboratory to produce diffraction pictures of DNA.

Begins Work on Structure of DNA

Eager to apply Franklin's x-ray diffraction skills to the problem of DNA structure, Randall had lured her to his lab with a Turner Newall Research Fellowship and the promise that she would be working on one of the more pressing research problems of the era—puzzling out the structure and function of DNA. When she arrived in Randall's research unit, she started working with a student, Raymond Gosling, who had been attempting to capture pictures of the elusive DNA. No stranger to the sexism rife in science at that time, Franklin made no apologies for the fact she was a woman. Maurice Wilkins, already well ensconced in the lab and working on the same problem as Franklin, took a disliking to her the first time they met. Franklin's biographers have difficulty ascertaining exactly why Wilkins and Franklin did not find common ground. Anne Sayre has suggested that the discomfort might have stemmed from the fact that Wilkins, only four years older than Franklin, may have misinterpreted her presence in his lab as a subordinate, whereas she considered herself an equal. Their mutual dislike of one another was not helped by the fact that the staff dining room was open only to the male faculty. In addition, she was the only Jew on staff. But the animosity between the two did not detract Franklin from her work, and shortly after arriving at King's, she started x-raying DNA fibers that Wilkins had obtained from a Swiss investigator.

Within a few months of joining Randall's team, Franklin gave a talk describing preliminary pictures she had obtained of the DNA as it transformed from a crystalline form, or A pattern, to a wet form, or B pattern, through an increase in relative humidity. The pictures showed, she suggested, that phosphate groups might lie outside the molecule. In the audience that November day sat James Watson, a 24-year-old American who was also working on unraveling the **molecular structure** of DNA. Working with Francis Crick at Cambridge, Watson was even more disinclined than Wilkins to like and respect Franklin. Compounding his dislike for her was Franklin's refusal to set aside hard crystallographic data in favor of model building. Perhaps for that reason, Franklin remained publicly scornful of the notion gradually gaining adherents that perhaps the DNA molecule had a helical structure. In her unpublished reports, however, she suggested the probability that the B form of DNA exhibited such a structure, as did, perhaps, the A form. Throughout the spring of 1952, she continued studying the A form, which seemed to produce more readable x-ray photographs. This presumed legibility proved deceptive, however, because the A form does not show the double-helical structure as clearly as the B form.

Research Provides Evidence of DNA's Double-Helical Structure

In the late spring of 1952, Franklin traveled to Yugoslavia for a month, where she visited coal research labs. When she returned, she and Gosling continued to investigate the A form, to no avail. In January 1953, she started model building, but could think of no structure that would accommodate all of the evidence she had gleaned from her diffraction pictures. She ruled out single and multi-stranded helices in favor of a figure-eight configuration. Meanwhile, Watson and Crick were engaged in their own model building, hastened by the fear that the American scientist Linus Pauling was nearing a discovery of his own. Although Watson had not befriended Franklin in the past two years, he had grown quite close to Wilkins. In *The Double-Helix,* Watson recalls how Wilkins showed him the DNA diffraction pictures Franklin had amassed (without her permission), and immediately he saw the evidence he needed to prove the helical structure of DNA. Watson returned to Crick in Cambridge and the two began writing what would become one of the best-known scientific papers of the century: "A Structure for Deoxyribose Nucleic Acid." Franklin and Gosling, who had been working on a paper of their own, quickly revised it so that it could appear along with the Watson and Crick paper. Although it is unclear how close Franklin was to a similar discovery—in part because of the misleading A form—unpublished drafts of her paper reveal that she had deduced the sugar-phosphate backbone of the helix before Watson and Crick's model was made public.

On April 25, 1953, Watson and Crick published their article in the British science journal *Nature,* along with a corroborative article by Franklin and Gosling providing essential evidence for the double-helix theory. In July 1953, she and Gosling published another paper in *Nature* that offered "evidence for a 2-chain helix in the crystalline structure of sodium deoxyribonucleate." But Franklin's interest in the world of DNA research had already begun to wane by the spring of 1953. Despite all the excitement surrounding the double-helical structure, she had decided to move on to a lab that she hoped would offer a more congenial working environment. When she informed Randall of her intention to leave King's College for J. D. Bernal's unit at Birkbeck College, he made it clear that the DNA project was to stay in his lab. Although Gosling had been warned against further associating with Franklin, they continued to meet in private and finish their DNA work. She also continued to work on coal, but devoted the bulk of her efforts to applying crystallographic techniques to uncover the structure of the tobacco mosaic virus (TMV).

Franklin did, in fact, find the Birkbeck lab more to her liking, even though she complained to some of her friends that Bernal, a strong Marxist, attempted to foist his political views on anyone who would listen. In comparison to the situation at King's College, however, she found this bearable. She did not even complain about her lab situation. At Birkbeck she worked in a small lab on the fifth floor while her x-ray equipment sat in the basement. Because there was no elevator in the building, she made frequent treks up and down the stairs. The roof leaked, and she had to set up pots and pans to catch the water. But Franklin didn't mind adversity. In fact, she told friends she preferred the challenge it presented, whether at work or even while traveling. She loved to travel and once journeyed to Israel in the steerage of a slow boat sheerly for the adventure of it. She said she preferred to travel with little money "because then you need your wits," an attitude that stood her in good stead in 1955 when the Birkbeck lost its backing from the Agricultural Research Council, in part, Franklin thought, because they did not approve of a project headed by a woman. Franklin successfully sought funding from another government source—the U.S. Public Health Service. The year after Franklin began at Birkbeck, the South African scientist Aaron Klug joined the laboratory. By 1956, Franklin had obtained some of the best pictures of the crystallographic structure of the TMV and, along with her colleagues, disproved the then-standard notion that TMV was a solid cylinder with RNA in the middle and protein subunits on the outside. While Franklin confirmed that the protein units did lie on the outside, she also showed that the cylinder was hollow, and that the RNA lay embedded among the protein units. Later, she initiated work that would support her hypothesis that the RNA in the TMV was single-stranded.

Franklin spent the summer of 1956 in California with two American scientists, learning from them techniques by which to grow viruses. Upon returning to England, Franklin fell ill, and friends began to suspect she was in pain a great deal of the time. That fall, she was operated on for cancer and, the following year, she had a second operation, neither of which stopped either her work or the disease. Franklin knew she was dying, but did not let that impede her progress. She began working on the polio virus, even though people warned her it was dangerous and highly contagious. She died of cancer at the age of 37 on April 16, 1958. Four years later, Watson, Crick, and Wilkins won the Nobel Prize in medicine or physiology, and Watson penned his potboiler account of the discovery of DNA. Although he vilifies her throughout his account, he tones down his earlier depiction of her as the mad, feminist scientist in an epilogue: "Since my initial impressions of her, both scientific and personal (as recorded in the early pages of this book), were often wrong, I want to say something here about her achievements." He continues that he and Crick "both came to appreciate greatly her personal honesty and generosity, realizing years too late the struggles that the intelligent woman faces to be accepted by a scientific world which often regards women as mere diversions from serious thinking."

FRAUNHOFER, JOSEPH VON (1787-1826)
German physicist and optician

From humble beginnings as the son of a poor glass grinder, Joseph von Fraunhofer pursued a successful business career in **optics** and, in the process, made important contributions to physics and the theory of **light** through his study of the dark lines in the solar spectrum. Fraunhofer was born in Straubing, Bavaria (now Germany). At the age of 12, after experiencing the death of both of his parents, Fraunhofer became an apprentice to a lens and mirror maker. Fraunhofer received little formal education during his childhood, and he possibly would have remained a glassmaker if not for an accident that nearly took his life.

In 1801, the house in which Fraunhofer worked collapsed. Buried but virtually unharmed, Fraunhofer came to the notice of Joseph Utzschneider, an entrepreneur, and Maximilian Joseph, who would later became King Maximilian I of Bavaria. Both men offered Fraunhofer financial and fatherly support. Scholars suggest Utzschneider gave Fraunhofer books and an informal education in the field of optics and physics. Through Utzschneider's influence, Fraunhofer became an assistant in 1806 at a mathematical and technical institute started by Utzschneider and his associates. Fraunhofer then became a partner and manager in Utzschneider's optical institute at Benediktbeuern near Munich. In 1814, Fraunhofer and Utzschneider formed a new optical firm together.

Fraunhofer initiated the field of spectrum analysis with his study of optics, along with his interest in designing achromatic objective **lenses** for telescopes. During his work, he found numerous dark lines in a solar light spectrum. English chemist and physicist William H. Wollaston (1766-1828) had noticed several of these lines earlier in 1802, but paid them little attention. Fraunhofer, however, was interested in the **spectral lines** because they could be used as **wavelength** standards to determine the index of refraction of optical glasses. Fraunhofer accurately plotted hundreds of the spectral lines. By measuring the wavelength of more than 300 of the lines he saw in his solar spectrum, Fraunhofer discovered that the lines' relative positions in the spectra of elements are constant. Fraunhofer assigned the most prominent lines with letters from A to Z, a nomenclature still in use today. Other physicists, however, quickly recognized that the **Fraunhofer lines**, as they became known, could help discover properties of the solar atmosphere. Building on Fraunhofer's discovery, Prussian-born physicist **Gustav Kirchhoff** and others developed the science of **spectroscopy**, which revolutionized solar physics. In fact, most data obtained about the **Sun** and **stars** is still obtained through spectroscopy. Fraunhofer also built the first **diffraction** grating, which he used to measure wavelengths of specific colors and dark lines in the solar spectrum. Fraunhofer did not taint his work with deductive interpretations. Rather, he confined himself to scientific empirical observations. The **optical instruments** he developed allowed him to conduct fundamental research in the fields of light and optics.

A master craftsman who was also successful as an inventor, entrepreneur, and researcher, Fraunhofer was awarded an honorary doctorate degree from the University of Erlangen in 1822. He was appointed director of the Physics Museum in Munich and taught at the University of Bavaria. He was made a Knight of the Danish order of Danebrog in 1824. Fraunhofer died two years later at age 39 from tuberculosis.

Joseph von Fraunhofer *(Image from Science Photo Library/Photo Researchers Inc. Reproduced by permission.)*

FRAUNHOFER LINES

Fraunhofer lines are dark absorption lines in the solar spectrum that can be seen when sunlight is passed through a prism to separate it into the colors of the rainbow. They occur because cooler gas, which is higher in the Sun's atmosphere, absorbs some colors of the **light** emitted by hotter gas lower in the Sun's atmosphere. Sir **Isaac Newton** discovered that if white light is passed through a prism, it separates into a rainbow, which is called a spectrum. While studying the spectrum that sunlight made, **Joseph von Fraunhofer** discovered some dark lines scattered among the colors. These dark lines were segments of colors missing from the complete spectrum. Fraunhofer counted 574 of these lines, which we now call Fraunhofer lines. Today, using much more sophisticated techniques, astronomers have discovered tens of thousands of Fraunhofer lines. Why doesn't the **Sun** emit these missing colors? Or, if the Sun does emit these colors, what happens to the light before it reaches Earth? The answer lies at the surface of the Sun.

When we look at a picture of the Sun, the surface that we see is called the photosphere. The photosphere is a region, several hundred kilometers thick, in which the Sun changes from opaque to transparent. It is not actually the outermost surface: the Sun extends for thousands of kilometers beyond the photosphere, but it is not usually visible from Earth. The photosphere is interesting because within this thin layer of the Sun (thin compared to the whole Sun, of course), sunlight is created, and some of the colors are lost almost immediately. The lower region of the photosphere has a **temperature** of about 10,000°F (about 5,500°C) and glows white-hot. Any object that glows due to a high temperature gives off a complete spectrum, that is, it has all the colors of the rainbow. As this light proceeds upwards in the Sun into a higher region of the photosphere, the temperature drops several thousand degrees. Although most of the light passes right through, some of the light is absorbed by the cooler gas. Only certain colors are removed because the chemical elements in the photosphere can absorb only certain wavelengths of light, and different wavelengths correspond to different colors. For example, sodium absorbs some yellow light at a **wavelength** of about 5.89×10^{-7}m. These absorbed colors cause the Fraunhofer lines. By measuring precisely the wavelengths of the missing colors, that is, the Fraunhofer lines, and how much light is actually absorbed, astronomers have learned much about the temperature inside the Sun and its chemical composition.

We can also learn about other **stars** in the sky by looking at the absorption lines in their spectra. By studying the similarities and differences that they have with the Fraunhofer lines, we can learn a lot about the similarities and differences that other stars have with our Sun.

FREQUENCY

Any process that is repetitive or periodic has an associated frequency. The frequency is the number of repetitions, or cycles, during a given time interval. The inverse of the frequency is called the period of the process.

Pendulums, as in a grandfather clock, also have a frequency of a certain number of swings per minute. A complete oscillation for a pendulum requires the pendulum bob to start and finish at the same location. Counting the number of these **oscillations** during one minute will determine the frequency of the pendulum (in units of oscillations/minute). This frequency is proportional to the square root of the **acceleration** due to **gravity** divided by the pendulum's length. If either of these are changed, the frequency of the pendulum will change accordingly. This is why you adjust the length of the pendulum on your grandfather clock to change the frequency, which changes the period, which allows the clock to run faster or slower.

Vibrating strings also have an associated frequency. Pianos, guitars, violins, harps, and any other stringed instrument require a particular range of vibrational frequencies to generate musical notes. By changing the frequency, generally by changing the length of the string, you change the pitch of the note you hear.

In any type of wave, the frequency of the wave is the number of wave crests (or troughs) passing a fixed measuring position in a given time interval; and, is also equal to the wave's speed divided by the **wavelength**. As a wave passes by a fixed measurement point, a specific number of wave crests (or troughs) pass a fixed point in a given amount of time. In the case of **waves**, the frequency is also equal to the speed of the wave divided by the wavelength of the wave.

Light also exhibits the characteristics of waves; so, it too has a frequency. By changing this frequency, you also change the associated **color** of the light wave.

FRESNEL, AUGUSTIN-JEAN (1788-1827)
French physicist

At the onset of the nineteenth century, the most widely accepted belief among physicists was that **light** was a particle that traveled through **ether**, an invisible substance that made up the heavens. Soon after the turn of the century, however, came the revival of an old belief; that light was actually a wave that needed no medium for travel. The debate that arose among scientists was furious, and there were few more successful spokesmen for the wave theory than Augustin Fresnel.

The son of an architect, Fresnel was born in Normandy in 1788. His future as a scientist was by no means obvious—in fact, Fresnel was eight before he learned to read. He entered the École Polytechique in Paris in 1804, with plans to pursue a career in engineering. After three more years, he became a civil engineer, working for the government for most of his adult life.

Fresnel's interest in **optics** began under unusual circumstances. In 1814, while he was working on the imperial highway, France's political stability was threatened by Napoleon's return from exile. Seeing this as an attack on civilization, Fresnel abandoned his position with the government to assist the opposing Royalist forces under Louis XVIII. Subsequently, he was suspended and put under police surveillance. During this time, Fresnel furthered his budding theories concerning light, duplicating several experiments **Thomas Young** had developed a decade earlier. Soon after, Napoleon was defeated at Waterloo and Fresnel was welcomed back, but his brief diversion had planted within him a firm belief in the wave nature of light and an obsession with expanding his own ideas into a comprehensive mathematical theory.

Until 1824, Fresnel's studies advanced almost without interruption, and he was aided by several government appointments to Paris, where he conducted the bulk of his research during this time. The steadfastness with which he attacked his subject was founded upon a few maxims: Fresnel believed, foremost, that for a theory to be true in nature it must be simple and in agreement with both experience and experiment. Nature itself, he believed, aims at producing varied effects by simple causes. It was this belief that led him to doubt the particle theory of light, with its imponderable ether and its multiple **fluids**. However, it is unlikely that Fresnel was familiar with the earlier light-wave theories authored by **Christiaan Huygens** and Leonhard Euler (1707-1783), so he was forced to start from scratch.

Fresnel's initial experiments dealt with the reaction of light when diffracted and led to his discovery of light as a transverse wave, rather than a longitudinal wave as theorized by Huygens in the seventeenth century. The transverse wave theory was validated by Danish astronomer Thomas Bartholin's (1616-1680) experiments, in which it was shown that two rays of light emerging from a piece of Iceland spar differed in properties—a discovery that was completely inconsistent with either particle or longitudinal theories. Fresnel furthered these experiments, advancing many of the first hypotheses dealing with polarized light.

Fresnel's increasing notoriety once again attracted the attention of the government, which, hoping to direct his talents toward more profitable ends, transferred him to the Lighthouse Commission in 1824. This new appointment put great demands upon his time and energy. Though he worked to develop a new echelon lens for lighthouses, his research in light theory dwindled. Added to this new pressure was his losing battle with tuberculosis, an affliction that had plagued him throughout his career and ultimately ended his life.

Many consider Fresnel's work the first major challenge to the theories of particles and imponderables, and his findings a basis for the development of nineteenth-century energetics. Just a half year before his death, he was awarded the Rumford Medal from the Royal Society.

FRICTION

Friction is the **force** that resists **motion** when the surface of one object slides over the surface of another. Frictional forces are always parallel to the surfaces in contact, and they oppose any motion or attempted motion. No movement will occur unless a force equal to or greater than the frictional force is applied to the body or bodies that can move.

While friction is often regarded as a nuisance because it reduces the efficiency of machines, it is, nevertheless, an essential force for such items as nails, screws, pliers, bolts, forceps, and matches. Without friction we could not walk, play a violin, or pick up a glass of water.

Gravity and friction are the two most common forces affecting our lives, and while we know a good deal about **gravitational forces**, we know relatively little about friction. Frictional forces are believed to arise from the adhesive forces between the molecules in two surfaces that are pushed together by **pressure**. The surface of a material may feel smooth, but at the atomic level it is filled with valleys and hills a hundred or more atoms high. Pressure squeezes the hills and valleys in the two surfaces together and the molecules adhere to one another. The actual contact area, from a microscopic perspective, is much less than the apparent area of contact as viewed macroscopically. As the **weight** of an object resting on a surface increases, it squeezes the two surfaces together and the actual area of contact increases. The actual contact area is believed to be proportional to the weight pushing the bodies together.

In addition to the adhesive forces between molecules, there are other factors that affect friction. They include the force needed to raise one surface over the high places of another; the fact that a rough region along a hard surface may "plough" a groove in a softer material; and electrical forces of

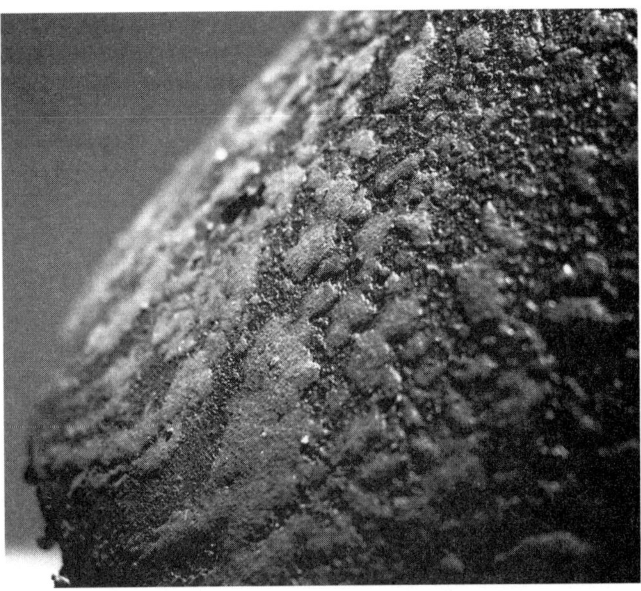

Close up of worn-out tire from NASCAR race car. Heat caused by friction has melted the rubber and worn away all tread marks. (Photo by Dennis Winn.)

attraction required to separate oppositely charged regions of the surfaces.

There are three laws that apply to friction: (1) The force of friction between an object and the surface on which it rests is proportional to the weight of the object. (The magnitude of the frictional force depends on the nature of the two surfaces.); (2) The force of friction between an object and the surface on which it rests is independent of the surface area of the object. (Remember, the actual contact area depends on the weight. If the weight remains constant so will the actual area of contact regardless of what the apparent area may be.); (3) The force of friction between an object and the surface on which it rests is independent of the speed at which the object moves as long as the speed is not zero.

The third law applies only to moving objects. Static friction, the force required to make an object at rest begin to move, is always greater than kinetic friction, which is the resistance to motion of an object moving across a surface. The reduction of friction that arises with motion is the result of fewer areas of contact once a body is in motion; the molecules are not in contact long enough to form firm bonds. Rolling friction for an object mounted on wheels or rollers is far less than kinetic friction. The reason that rolling friction is so small is probably the result of minimal contact area between wheel and surface, particularly if both are very hard, and because any molecular adhesions are pulled apart by vertical shearing rather than horizontal tearing.

In breaking molecular adhesions, molecular vibrations increase, causing a rise in **temperature**. You can easily verify this by simply rubbing your hands together. In machines, the adhesion and tearing of molecular bonds between surfaces causes wear. To reduce wear, we add a lubricant (often oil). The oil decreases the actual area of contact between surfaces.

As a result, it reduces the wear associated with tearing surface molecules apart and thereby keeps **heat** and wear at lower levels than would otherwise be the case.

FRIEDMAN, JEROME (1930-   )

American physicist

Jerome Friedman shared the 1990 Nobel Prize for physics with physicists **Henry Kendall** and Richard Taylor for the trio's research leading to the discovery of **quarks**—supercharged **subatomic particles** found in protons and neutrons. Friedman had originally become interested in atomic research while at Stanford University, where he worked with physicist **Robert Hofstadter** on his studies of the atomic **nucleus**. This research went on until 1960, when Friedman left Stanford to join the physics faculty of the Massachusetts Institute of Technology (MIT); Kendall and Taylor joined Friedman at MIT in the early 1960s. Originally, the three scientists intended to merely extend Hofstadter's original findings; instead, they made a discovery once described by a colleague as "one of the pivotal contributions to physics in this century."

Friedman was born in Chicago on March 28, 1930, the younger of two sons of Selig and Lillian Warsaw Friedman. Selig Friedman had immigrated to the United States in 1913; he later became the owner of a sewing machine repair business. Friedman's early interests revolved around art and music, but he redirected his energies after a chance encounter with **Albert Einstein**'s book *Relativity*. Even though he was offered a scholarship to the Art Institute of Chicago, Friedman decided to pursue this new interest in science by studying physics at the University of Chicago in the late 1940s. While at the university, he had the opportunity to work on projects with physicists **Enrico Fermi** and Valentine Telegdi. Remaining at the University of Chicago for the bulk of his studies, Friedman received his bachelor's degree in physics in 1950, his master's degree in 1953, and his doctorate in 1956.

The focus of Friedman's research—nuclear physics—was determined early in his academic career. While still at Chicago, Friedman carried out studies on the "weak force," which explains changes such as beta, or organic particle, decay. He chose to continue in this field when he left school in late 1956 to accept an appointment as research associate at Stanford University's High Energy Physics Laboratory. Robert Hofstadter—winner of 1961 Nobel Prize in physics for his research on the structure of protons and neutrons—was director of the laboratory. Hofstadter's work during the 1950s constituted the latest in physicists' efforts to define the smallest building blocks of **matter**. When scientist **John Dalton** first announced his **atomic theory** in 1808, he indicated that atoms constituted these basic components. Near the end of the nineteenth century, Sir **Joseph John Thomson** showed Dalton's hypothesis was incorrect by identifying electrons. Two decades later, physicist **Ernest Rutherford** discovered the atomic nucleus and a second subatomic particle, the **proton**. In 1932, Rutherford's student **James Chadwick** discovered a third particle, the **neutron**.

For some time, the belief that atoms consisted of three fundamental particles—proton, neutron, and electron—appeared to explain much observed phenomena. The invention of **particle accelerators** in the 1930s, however, soon invalidated this view. First dozens, and later hundreds, of new elementary particles were discovered as the by-products of these "atom smashers," which sped subatomic fragments toward atoms at high velocities. Upon being struck, the atoms would break and their fragmentary contents could be studied. In 1964, physicist **Murray Gell-Mann** postulated that some of these tiny fragments—which he named quarks—were the smallest components of all "fundamental" particles. The problem was that no one had ever observed a quark (indeed, many physicists thought that the concept was no more than a mathematical construction for bringing order to **particle physics**). Hofstadter's research provided the first hint that quarks might actually have measureable physical properties. He bombarded atomic nuclei with the high-powered **electron** beams of the Stanford Linear Accelerator (SLA) and found evidence for structure *within* protons and neutrons. From the way electron beams were scattered by nucleons (protons and neutrons), Hofstadter concluded that these particles were not discrete points, but "fuzzy little balls" that probably had a more detailed structure as yet undetectable in his research.

It was into this research setting that Friedman came in 1957. He disagreed with researchers who felt that Hofstadter's work was essentially complete. Friedman's stance was supported by two factors. First, improvements in the SLA vastly increased the ability of its electron beam to penetrate the atomic nucleus, allowing researchers a clearer look at protons and neutrons. Secondly, a suggestion by theoretical physicist James Bjorken prompted Friedman, Henry Kendall, and Richard Taylor to look more closely at the nature of inelastic collisions (in which nucleons are actually blown apart) rather than at elastic collisions (in which electrons are scattered by nucleons, which then remain intact). In following up on Bjorken's suggestion, Friedman, Kendall, and Taylor found that their data strongly suggested the existence of tiny points of matter within nucleons that could be identified by their tendency to deflect electrons during inelastic **scattering**. This confirmation of the existence of quarks and gluons (the matter that holds quarks together) not only gave scientists a more complete picture of **atomic structure**, it also validated the theories advanced by Gell-Mann and Hofstadter.

Over the course of his career in physics, Friedman has held a number of academic posts. He rose from researcher to professor at MIT in 1960; he also served as director of MIT's Laboratory of Nuclear Science from 1980 to 1983 and as head of its physics department from 1983 to 1988. In addition to his academic research, Friedman has been involved in many administrative and political endeavors in the scientific community (he has, for example, served as a member of the Department of Energy's High Energy Advisory Panel and as chairman of the Science Policy Committee for the construction of the Superconducting Super Collider). Along with the Nobel Prize, Friedman and his associates also received the 1989 W. K. H. Panofsky Prize from the American Physical Society. In

1956, Friedman married the former Tania Letesky-Baranovsky, with whom he has four children: Ellena, Joel, Martin, and Sandra.

FUNDAMENTAL CONSTANT

Physicists use fundamental constants to help determine the scale at which a theory is valid. In order for a quantity to be a fundamental constant, it must appear as a constant in a physical law, and it must take the same value in every law in which it appears. Three of the most well-known constants are Newton's gravitational constant (G), the **speed of light** in a **vacuum** (c), and **Planck's constant** (h).

English physicist **Isaac Newton** discovered the first fundamental constant with his law of universal gravitation, $F = Gm_1m_2/r^2$. The law of universal gravitation states that the **force**, F, between two masses, m_1 and m_2, is directly proportional to the product of the masses, and inversely proportional to the square of the distance between them, r^2. The proportionality constant, G, sets the scale of the gravitational interaction. G is very small ($G=6.67 \times 10^{-11}$ Nm/kg^2), which makes **gravity** the weakest of the four fundamental forces. In fact, gravity is so weak that the repulsive electrical force between two protons is far more significant than the attractive **gravitational force** between them. Accordingly, the gravitational attraction of two cups sitting on a table is essentially negligible. Significant **mass**, however, produces significant gravitational force. The immense mass of Earth, for example, produces a gravitational force strong enough to prevent everything on its surface from floating off into **space**.

Albert Einstein proposed that the speed of light is constant for all observers who are not accelerating. In other words, an observer traveling in a car at a constant speed who observes a light ray coming towards him will measure the ray to have the same speed that an observer standing still on the side of the road would measure. Einstein was forced to this non-intuitive conclusion because all experiments intended to demonstrate its falsity had failed, and he based his special theory of relativity on it. The consequences of the special theory of relativity are stranger still. Since the observer in the car measures the same speed for the light ray as the observer on the ground, their measurements of **time** and distance do not agree. Because light travels very quickly ($c = 3.00 \times 10^8$ m/s), the car must be moving very fast for relativistic corrections to become significant, which is why the effects of special relativity are not observed in day-to-day life.

A third fundamental constant was proposed by German physicist **Max Planck** in his explanation of why objects radiate light when hot. Planck's constant, commonly written h, or the reduced quantity $h = h/2\pi$, sets the scale at which classical (Newtonian) **mechanics** no longer applies. At that scale the world becomes quantum mechanical, and objects are treated as wavefunctions rather than as either **waves** or particles. This leads to effects such as quantum mechanical **tunneling**. Because h is very small ($h=6.626 \times 10^{-34}$ Js), **quantum mechanics** applies at small distance scales. It is useful for describing physics at atomic distances.

The three constants G, c, and h may be combined to form mass, length, and time scales that characterize a theory that combines gravitation, special relativity, and quantum mechanics. The distance scale, called the Planck length, is approximately 10^{-33} cm. **String theory** models the **universe** at these distances.

The list of fundamental constants presented here is not all-inclusive. Other fundamental constants include the charge of the **electron**, Boltzmann's constant, and the Hubble expansion rate. The term fundamental constant is often applied loosely to quantities like the mass of the **proton** or the Bohr radius of the hydrogen **atom**. These are derived from the fundamental constants of a more precise theory that describes physics at higher energies. For example, the proton is not an indivisible particle. Rather, it is composed of **quarks** and gluons that interact through the electromagnetic, strong, and weak forces. The **energy** (mass) of the proton is a consequence of the properties of its constituent quarks and gluons. A theory where the mass of the proton is treated as a fundamental constant is only an effective description of more complicated physics taking place at shorter distance scales. Accordingly, the history of science suggests that the universe exhibits finer structure when it is examined in greater detail.

See also Newton's law of universal gravitation; Relativity, special

G

GABOR, DENNIS (1900-1979)

Hungarian English physicist

Dennis Gabor is best known for inventing and developing **holography**, which earned him the Nobel Prize in physics in 1971. Early in his career, Gabor built a crude prototype of the **electron microscope**, but did not pursue its development, and his regret over not continuing his work on this idea lasted throughout his lifetime. The discovery of holography was Gabor's triumph, though he could not have realized at the time the widespread application it has today. Among Gabor's other achievements was the quartz mercury lamp.

Dennis Gabor, the oldest of three sons, was born on June 5, 1900, in Budapest, Hungary. His mother, Ady Jacobvits, was a former actress. His father, Berthold, grandson of Russian Jewish immigrants, ultimately became the director of the Hungarian General Coal Mines. Although Gabor's family became Lutherans in 1918, religion appeared to play a minor role in his life. He maintained his church affiliation through his adult years but characterized himself as a "benevolent agnostic."

Gabor showed an early interest in science. By age 15 he and his younger brother George were replicating experiments they read about in science journals in their homemade laboratory. After being called up for military service in the Austro-Hungarian army in 1918, Dennis Gabor joined the Officers Training Corps and trained in artillery and horsemanship. He served briefly in Italy before World War I ended. Following the war he entered the Budapest Technical University in 1918 for a four-year course in mechanical engineering. His third year was interrupted, however, when he was once again called up for military service. Since he was opposed to serving in the military for what he felt was a reactionary monarchy, he left the country and went to Berlin.

Begins Professional Scientific Investigations

Gabor studied at the Technische Hochschule in Berlin, where he earned his diploma in 1924; he received his doctor-ate in engineering from the same institution in 1927. His doctoral dissertation involved the measure of lightning-induced fast surges in high-voltage **power** lines. To measure these surges, Gabor developed a cathode-ray oscilloscope with a fast response. (An oscilloscope temporarily displays in a visible wave form the variations in a fluctuating electrical quantity.) Rather than using a coil of wire in the form of a long cylinder to carry the current (a selenoid) common at the time, Gabor chose a short, iron-encased coil. This was done to confine the magnetic field within the shorter area of the coil and prevent it from being influenced by stray magnetic fields. In this device, Gabor had the crude forerunner for the magnetic lens of an **electron microscope**. Gabor chose not to pursue this work, however, a choice he came to regret. He later told physicist **Leo Szilard** that, although he felt he had the expertise necessary at the time to build the electron microscope, he lost interest in electrons after finishing his doctoral work.

Later, Gabor would describe his years in Berlin as some of the happiest of his life. He frequently went to the University of Berlin to hear lectures by the renowned scientists of the day, including **Max Planck**, Walther Nernst, and **Max von Laue**. He attended a seminar conducted by **Albert Einstein** and enjoyed the company of many other future expatriate Hungarians who would flee Hitler's Germany, including Szilard, who was instrumental in convincing the U.S. government to develop the **atomic bomb**. When Gabor completed his doctorate in 1927, he secured a position in the physics lab of Siemens and Halske in Siemensstadt, Germany. His most notable achievement there was the invention of the quartz mercury lamp. His contract was terminated within weeks of Adolf Hitler's ascent to power in 1933.

Gabor then returned to Hungary, where he worked on the development of the plasma lamp, a new kind of fluorescent lamp. He was unable to sell the patent there but was able to work out an inventor's agreement with British Thomson-Houston Company (BTH) to work on the lamp in England. Gabor traveled to Rugby, England, and began his long association with BTH in 1934. He worked on his lamp until 1937, and was appointed to the permanent staff that year despite the fact that his efforts on the plasma lamp proved unsuccessful. From 1937 to 1948 he worked in the area of electron **optics**, which involves directing and focusing beams of electrons. On August 8, 1936, Gabor married Marjorie Louise Butler, a fellow employee of BTH. The Gabors had no children.

As a non-British citizen, Gabor initially was not permitted to have anything to do with the war effort during World War II. Ultimately, he was granted special status, which meant that a special research hut was built for him outside the high security area at BTH. He was not privy to any of the classified research being conducted there during the war. Indeed the only journal he was able to receive regularly was *Nature*. Because Gabor was unaware of the development of **radar** by fellow British scientists and its successful employment in the war against the Axis powers, he spent much of his wartime working on a system to detect airplanes by the **heat** of their engines. The war years also brought personal sadness to Gabor. His parents had visited him in 1938, but he could not persuade them to stay in England. They returned to Hungary shortly before Hitler invaded Poland. Then in 1942 his father, who had greatly influenced him, passed away.

Discovers Holography

After the war Gabor again turned his attention to the electron microscope. Bothered by his earlier failure to develop the instrument, he was determined to make a comeback in the field. His goal was to be able to "see" individual atoms. During this period the resolution power of the electron microscope was limited by a technical barrier caused by the electron lens then in use. At a certain level of magnification, the lens distorted the image, and some information was lost.

Gabor's idea, which came to him while he was sitting on a bench at a tennis club in London in 1947, was to take an electron picture, distorted by problems with the lens, but one that could be corrected by optical means utilizing **light**. His theory was expressed in papers published that year in which he first coined the term "hologram" (Greek for "completely written"). In practical terms, Gabor's idea found only limited use as an electron-optical technique. Its implications for beams of light, however, awaited only the invention of the **laser**.

During these postwar years Gabor seriously considered accepting one of the many offers he received to come to the United States, feeling he lacked support at BTH for his work. His naturalization as a British citizen in 1946 was one of several factors that guaranteed he would remain in England. Gabor finally did leave BTH in 1949, when he accepted a position as a reader in electron physics at the Imperial College of Science and Technology of the University of London (a reader is the equivalent of an associate professor in an American university). Although for a time Gabor continued to entertain the possibility of moving to the United States, a number of highly skilled postgraduate students at Imperial College allowed him to work on many of his ideas. He and his students built a number of devices, including a flat television tube, a Wilson **cloud chamber**, and an analog computer. In 1958 he was appointed professor of applied electron physics.

During the same time period, Associated Electrical Industries, the parent company of BTH, had begun work in 1950 on a holographic electron microscope. While not directly involved in the day-to-day operations of the project, Gabor was closely involved as a consultant. By 1953 a system had been built, but the images were no better than those from other methods. In addition to random disturbances in the pictures, there were two images instead of one. No further work was done on the project, and in 1955 it was permanently shut down. Gabor was bitterly disappointed.

Nobel Prize Brings Recognition

The invention of the laser in 1960 sparked renewed interest in holography (the process of making or using holograms), since a constant narrow light source where all **waves** were in phase was now available to experimenters. In that same year, Emmett N. Leith and Juris Upatnieks from the Willow Run Laboratory at the University of Michigan replicated Gabor's earlier experiments and were able to eliminate the second image that had so consistently plagued Gabor as he worked on the holographic electron microscope. The first laser holograms in 1962 ensured Gabor's reputation.

In 1967, Gabor retired from Imperial College, but remained a professor emeritus and research fellow while serving as a part-time consultant for CBS Laboratories in Stamford, Connecticut. He also found time to continue his own research. He was able to show the application of holography to computer data processing, where it has been particularly useful in data compression. Additionally, he was in great demand as a speaker on the subject of holography. He received his highest honor in December 1971, when he was awarded the Nobel Prize in physics for his work in holography. His Nobel lecture was illustrative of the issues that dominated his concerns in later years, namely the role of science and technology in society.

In the 1960s Gabor saw much that contributed to his belief in the irrationality of human behavior. He opposed both the Vietnam War and the **space** program. His first book on the subject, *Inventing the Future,* published in 1963, outlined his belief in the need for scientists and engineers to develop socially useful inventions. This book and its sequel, *The Mature Society,* published in 1972, are surprisingly optimistic in their outlook, given Gabor's own pessimism. He was highly influenced by novelist and critic Aldous Huxley and the British eugenicists, believing that scientific progress and enlightenment would enable humanity to move toward a socially engineered society free of the failures that have plagued human societies throughout history.

After Gabor was awarded the Nobel Prize, Lawrence Bartell of the University of Michigan set out to develop a holographic electron microscope by forming images of electron-clouds in gas-phase atoms. He shared his work with Gabor in April 1974, and Gabor immediately began designing his own holographic electron microscope. In the summer of that year, however, Gabor suffered a stroke that left him unable to read or write. Despite this he was able to maintain contact with his colleagues. He was even able to visit the Museum of Holography in New York City when it opened in 1977. He died in a London nursing home in February 1979.

Gabor was the author of numerous works, with translations in many European languages. He was the recipient of many awards and an honored member of scientific societies around the world. The legacy of his pioneering efforts in holography is evident in science and everyday life—from medicine to computing, mapmaking, **photography**, and supermarket checkouts. Futurists predict that three-dimensional holographic images will play an even larger role in conveying visual information and providing entertainment.

GALAXIES AND GALAXY CLUSTERS

By definition, galaxies are gravitationally bound aggregates of many (one million [10^6] to one trillion [10^{12}]) **stars**, gas, dust, and **dark matter**. Astronomers estimate that there are at least 100 billion galaxies in the known **universe**. What is truly fascinating about galaxies, however, is the huge range of **mass**, size, composition, and brightness that they span. For example, the mass of a galaxy may be 10 million to a trillion times the mass of the **Sun**. In addition, with diameters ranging from 1,500-300,000 light-years, a galaxy may comprise a range of stars and interstellar material; old stars and little interstellar material to young stars and copious gas and dust. Finally, galaxies may have luminosity a million to a trillion times that of the Sun.

Every galaxy is associated with one of three distinct classes. These categories are: spiral (S), elliptical (E), and irregular (Irr). Even though the division of galaxy classes is largely based upon appearance, galaxy types differ in terms of internal motions of stars, chemical abundances, type and number of stars, as well as distribution and **color** of **light** emitted.

The spiral class is further divided into two subclasses: barred (SB) and normal (S). Both normal and barred spiral galaxies exhibit highly flattened stellar systems resembling pancakes. The disk material, mostly stars and gas, which rotates about an axis perpendicular to the plane of the system, is also observed to have a bright nuclear bulge at its center. Also within the spiral galaxy's disk lie the whirlpool-like spiral arms, which are delineated by very young stars. Old stars with low chemical abundances are found in the galactic **nucleus** and in a spherical stellar halo surrounding the disk. Finally, the entire multicomponent system is contained within a spherical halo of dark matter, whose speculative existence is discerned through gravitational effects.

Elliptical (E) galaxies are characterized by an oblate spheroid shape, ranging from nearly round to highly elongated, with a bright nuclear region. These galaxies exhibit no axis of rotation as individual star orbits are found to be randomly oriented. In addition, the stellar populations in elliptical systems are dominated by old low-mass stars grouped in star clusters called globular clusters. Finally, relatively small quantities of **interstellar gas** and dust are found in elliptical galaxies.

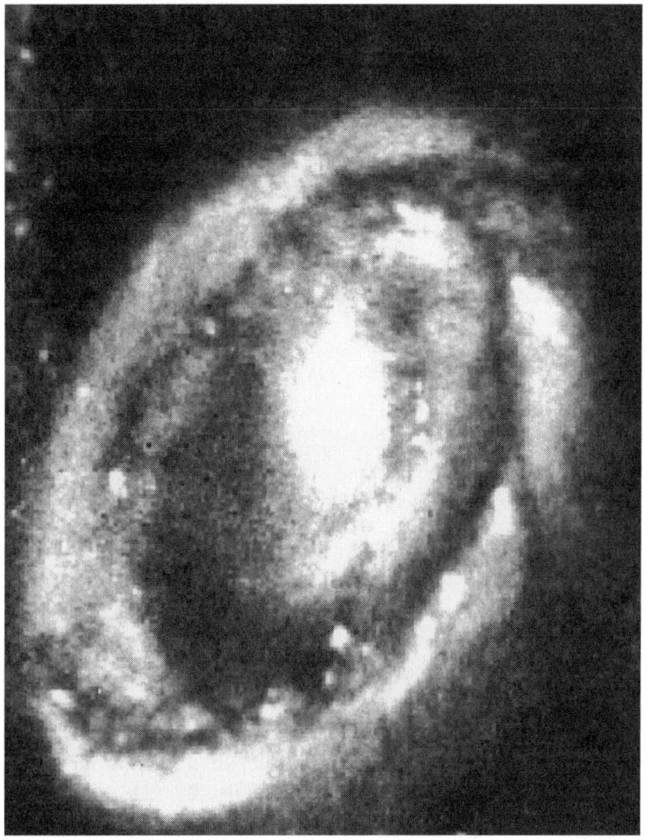

Cartwheel galaxy. (*NASA. Reproduced by permission.*)

Irregular galaxies (Irr) exhibit rather chaotic, asymmetrical shapes. The stellar properties of irregular galaxies are very similar to those found in spiral galaxies. Two types of irregular galaxies dominate the subclass. Irr (I) galaxies lack symmetry or defined spiral arms, yet may show the presence of several bright knots. Irr (II) galaxies also lack symmetry, but are very smooth with no bright knots. Furthermore, large amounts of interstellar dust may be found in Irr (II) systems.

Many galaxies look almost like one of the Hubble classifications described above, but with some unusual features. For example, imagine an elliptical galaxy that looks as if someone sliced it through the center, pulled it apart a little bit, and displaced each half sideways. These galaxies are called peculiar and the designation, pec, is added to the classification. Whatever causes a galaxy to look as if it were ripped apart as described would require large amounts of **energy**. Peculiar galaxies are therefore interesting because they often tend to be the active galaxies that emit large amounts of energy.

Active galaxies emit far more energy than normal galaxies. A galaxy is considered active if it emits more than 100 times the energy of the **Milky Way galaxy**. Active galaxies often have a very compact central source of energy, much of which is emitted as **radio** waves rather than optical light. These radio **waves** are emitted by electrons moving in a helical path in a strong magnetic field at speeds near that of light. Active galaxies also often have a peculiar photographic appearance, which can include jets of material streaming out from the nucleus or the appearance of either explosions or implosions. They tend to vary erratically in brightness on rapid **time** scales. There are a number of varieties of active galaxies, including: compact radio galaxies, extended radio galaxies, Seyfert galaxies, BL Lac objects, and **quasars**.

Compact radio galaxies appear photographically as ordinary giant elliptical galaxies. Radio telescopes, however, reveal a very energetic compact nucleus at the center. This radio nucleus is the source of most of the energy emitted by the galaxy. Perhaps the best-known compact radio galaxy is M87. This giant elliptical galaxy has both a very compact energetic radio source in the nucleus and a jet consisting of globs of material shooting out from the nucleus. Recent observations from the **Hubble Space Telescope** provide strong evidence that this core contains a supermassive **black hole**.

Extended radio galaxies consist of two giant lobes emitting radio waves. These lobes are on either side of a peculiar elliptical galaxy. The lobes can appear straight or curved as if the galaxy is moving through **space**. These lobes are the largest known galaxies and can stretch for millions of light years.

Seyfert galaxies look like spiral galaxies with a hyperactive nucleus. The spiral arms appear normal photographically, but they surround an abnormally bright nucleus. Seyfert galaxies also have evidence for hot turbulent interstellar gas.

BL Lacertae objects look like stars. In reality they are most likely to be very active nuclei of elliptical galaxies. However, BL Lacertae objects have sufficiently unusual behavior, including extremely rapid and erratic variations in observed properties, that their exact nature is not known for certain.

Quasars also look like stars, but they are perhaps the most distant and energetic objects in the universe known so far. Most astronomers consider them the very active nuclei of distant galaxies in the early stages of evolution. As for the other types of active galaxies they produce large amounts of energy in a very small volume. Most astronomers currently think that the energy source is a supermassive black hole.

The road to a twenty-first century understanding of galaxies was certainly long and interesting. Surprisingly, it was not definitively known until 1923 that galaxies exist outside of the Milky Way galaxy. The prevailing view until then was that the universe consisted of the Milky Way and a presumed void of space beyond.

As of 1610, when **Galileo** used a **telescope** to discover that the Milky Way comprises individual stars not visible to the naked eye, scientists have known that the Sun resides in a disk-like distribution of many stars. During the mid-eighteenth century, philosopher **Immanuel Kant** reasoned that **gravity** should act between stars in the Milky Way in the same way that gravity is responsible for the motions of the **planets** in the solar system. Moreover, he postulated that, just like in the solar system, the mutual inward gravitational pull of the stars should be balanced by rotational **motion** of the stars about a common **center of mass** resulting in a disk structure with a rotation axis perpendicular to the plane of the disk.

In a rather bold step, Kant ultimately suggested that the Milky Way may not be alone in the universe. He proposed that the faint fuzzy patches discovered by astronomers, called **nebulae**, might actually be "island universes." Great improvements in telescope technology during the late eighteenth century eventually allowed astronomers to discern the conspicuous spiral pattern in several of the 103 nebulae cataloged in 1780 by French astronomer Charles Messier (1730-1817). It is now known that these "spiral" nebulae are external galaxies similar to the Milky Way.

Astronomers of the nineteenth and early twentieth centuries divided into two camps with respect to the nature of the spiral nebulae: those who believed spiral nebulae lie inside the Milky Way and those who believed them to lie outside. Heber D. Curtis (1872-1942) was the champion for the latter while Harlow Shapely was representative of the former. Both men met on April 26, 1920, at the National Academy of Sciences in Washington, D.C., to argue their opposing views in a public forum that has since become known as the "Great Debate." The argument was finally put to rest in the early 1920s by **Edwin P. Hubble**, Vesto M. Slipher (1875-1969), Milton Humason (1891-1972), and a host of other astronomers working on the problem. In 1923, Hubble found that the Andromeda nebula clearly lay outside of the Milky Way; it has since come to be known as the Andromeda galaxy.

Galaxies cluster in agglomerations called groups, clusters, or superclusters. The smallest of the agglomerations, the group, is characterized by less than 50 member galaxies. By all accounts, a diameter of one million light-years is typical for galaxy groups. Galaxy clusters comprise between 50 and 10,000 member galaxies and may extend 20 million light-years on the diameter of their roughly spherical shape. Clusters are usually divided into two subclasses: sparse clusters, with approximately 100 member galaxies and a mixture of spiral and elliptical galaxies; and rich clusters, with mostly elliptical galaxies and thousands of cluster members. Finally, the behemoth superclusters may contain thousands of member clusters, each containing thousands of galaxies, and may span 330 million light-years in length.

On the whole, galaxy clustering environments create dynamic conditions where many phenomena can occur. New stars may form, material may be lost from tidal interactions of galaxies, and galaxies may merge to form new galaxies.

See also Gravitational forces; Interstellar space

GALILEO (1564-1642)

Italian mathematician and astronomer

Galileo Galilei is credited with establishing the modern experimental method. Before Galileo, knowledge of the physical world that was advanced by scientists and thinkers was for the most part a matter of hypothesis and conjecture. In contrast, Galileo introduced the practice of proving or disproving a scientific theory by conducting tests and observing the results. His desire to increase the precision of his observations led him to develop a number of inventions and discovery, particularly in the fields of physics and **astronomy**.

The son of Vincenzo Galilei (c.1520-1591), an eminent composer and music theorist, Galileo was born in Pisa. He received his early education at a monastery near Florence, and in 1581 entered the University of Pisa to study medicine. While a student he observed a hanging lamp that was swinging back and forth, and noted that the amount of time it took the lamp to complete an oscillation remained constant, even as the arc of the swing steadily decreased. He later experimented with other suspended objects and discovered that they behaved in the same way, suggesting to him the principle of the pendulum. From this discovery he was able to invent an instrument that measured time, which doctors found to be useful for measuring a patient's pulse rate, and **Christiaan Huygens** later adapted the principle of a swinging pendulum to build a pendulum clock.

While at the University of Pisa, Galileo listened in on a geometry lesson and afterward abandoned his medical studies to devote himself to mathematics. However, he was unable to complete a degree at the university due to lack of funds. He returned to Florence in 1585, having studied the works of **Euclid** and **Archimedes**. He expanded on Archimedes's work in hydrostatics by creating a hydrostatic balance, a device designed to measure the **density** of objects. The following year, he published an essay describing his new invention, which determined the specific **gravity** of objects by weighing them in water. With the hydrostatic balance, Galileo gained a scientific reputation throughout Italy.

In 1592 Galileo was appointed professor of mathematics at Padua University in Pisa, where he conducted experiments with falling objects. **Aristotle** had stated that a heavier object should fall faster than a lighter one. It is said that Galileo tested Aristotle's assertion by climbing the leaning tower of Pisa, dropping objects of various weights, and proving conclusively that all objects, regardless of **weight**, fall at the same rate.

Some of Galileo's experiments did not turn out as expected. He tried to determine the **speed of light** by stationing an assistant on a hill while he stood on another hill and timed the flash of a lantern between the hills. He failed because the hilltops were much too close together to make a measurement.

In 1593 he invented one of the first measuring devices to be used in science: the thermometer. Galileo's thermometer employed a bulb of air that expanded or contracted as **temperature** changed and in so doing caused the level of a column of water to rise or fall. Though this device was inaccurate because it did not account for changes in air **pressure**, it was the forerunner of improved instruments.

From 1602 to 1609 Galileo studied the **motion** of pendulums and other objects along arcs and inclines. Using inclined planes that he built, he concluded that falling objects accelerate at a constant rate. This law of uniform **acceleration** later helped **Isaac Newton** derive the law of gravity.

Galileo did not make his first contribution to astronomy until 1604, when a supernova abruptly exploded into view.

Galileo postulated that this object was farther away than the **planets** and pointed out that this meant that Aristotle's "perfect and unchanging heavens" were not unchanging after all. Ironically, Galileo's best-known invention, the **telescope**, was *not* his creation after all. The telescope was actually invented in 1608 by **Hans Lippershey**, a Danish spectacle maker. When Galileo learned of the invention in mid-1609, he quickly built one himself and made several improvements. His altered telescope could magnify objects at nine **power**, three times the magnification of Lippershey's model. Galileo's telescope proved to be very valuable for maritime applications, and Galileo was rewarded with a lifetime appointment to the University of Venice.

He continued his work, and by the end of the year he had built a telescope that could magnify at 30 power. The discoveries he made with this instrument revolutionized astronomy. Galileo saw jagged edges on the **Moon**, which he realized were the tops of mountains. He assumed that the Moon's large dark areas were bodies of water, which he called maria-maria (though we now know there is no water on the Moon). When he observed the **Milky Way**, Galileo was amazed to discover Jupiter, which resulted in his discovery of its four moons; he later called them "satellites," a term suggested by the German astronomer **Johannes Kepler**. Galileo named the moons of Jupiter, Sidera Medicea ("Medicean stars") in honor of Cosimo de Medici, the Grand Duke of Tuscany, whom Galileo served as "first philosopher and mathematician" after leaving the University of Pisa in 1610. Also, with repeated observation, he was able to watch the moons as they were being eclipsed by Jupiter and from this he was able to correctly estimate the period of rotation of each of the moons.

In 1610 he outlined planetary discoveries in a small book called *Siderus Nuncius* ("The Sidereal Messenger"). Venus, seen through the telescope, exhibited phases like the Moon, and for the same reasons: Venus did not produce its own light but was illuminated by the **Sun**.

Saturn was a mystery: Galileo's 30-power telescope was at the limit of its ability to resolve Saturn, and the planet appeared to have three indistinct parts. When Galileo looked at the Sun, he saw dark spots on its disc. The position of the spots changed from day to day, allowing Galileo to determine the rotational rate of the Sun.

In 1613 Galileo published a book in which for the first time he presented evidence for and openly defended the model of the solar system earlier proposed by the Polish astronomer **Nicholas Copernicus**, who argued that Earth, rather than being positioned at the center of the **universe**, as in the Ptolemaic design, was only one of several galactic bodies that orbited the Sun. While there was some support even among ecclesiastical authorities for Galileo's proof of the Copernican theory, the Roman Catholic hierarchy ultimately determined that a revision of the long-held astronomical doctrines of the church was unnecessary. Thus, in 1616 a decree was issued by the church declaring the Copernican system "false and erroneous," and Galileo was ordered not to support this system.

Following this run-in with the Catholic church and the Inquisition that forced his adherence to the Copernican theory of the solar system, Galileo focused on the problem of determining longitude at sea, which required a reliable clock. Galileo thought it possible to measure time by observing eclipses of Jupiter's moons. Unfortunately, this idea was not practical for eclipses could not be predicted with enough accuracy and observing celestial bodies from a rocking ship was nearly impossible.

Galileo wanted to have the edict against the Copernican theory revoked, and in 1624 traveled to Rome to make his appeal to the newly elected pope, Urban VIII. The pope would not revoke the edict but did give Galileo permission to write about the Copernican system, with the provision that it would not be given preference to the church-sanctioned Ptolemaic model of the universe.

With Urban's imprimatur, Galileo wrote his *Dialogue Concerning the Two Chief World Systems—Ptolemaic and Copernican*, which was published in 1632. Despite his agreement not to favor the Copernican view, the objections to it in the *Dialogue* are made to sound unconvincing and even ridiculous. Summoned to Rome to stand before the Inquisition, Galileo was accused of violating the original proscription of 1616 forbidding him to promote the Copernican theory. Put on trial for heresy, he was found guilty and ordered to recant his errors. At some point during this ordeal Galileo is supposed to have made his famous statement: "And yet it moves," referring to the Copernican doctrine of Earth's rotation on its axis.

While the judgment against Galileo included a term of imprisonment, the pope commuted this sentence to house arrest at Galileo's home near Florence. Although he was forbidden to publish any further works, he devoted himself to his work on motion and parabolic trajectories, arriving at theories that were later refined by others and made an important impact on gunnery. He died blind at the age of 78 in 1642.

GALVANI, LUIGI (1737-1798)

Italian anatomist

Luigi Galvani was not especially interested in **electricity**. At least, not at first. Born on September 9, 1737, at Bologna, Italy, Galvani was initially a student of theology, but he switched to the study of medicine, receiving his degree in 1762. In 1775 he was appointed professor of anatomy and gynecology at the University of Bologna, the school from which he graduated.

In 1771, Galvani happened to observe the dissection of a frog and noticed the dead frog's legs would twitch when a spark touched them. This was no big surprise; it was known that live muscle tissue twitched when touched with a spark, so dead tissue ought to react the same way.

At first Galvani had no further interest in the incident, but later he recalled that a generation earlier **Benjamin Franklin** had shown lightning was an electrical phenomena. Galvani reasoned that if Franklin was right, lightning should

Luigi Galvani. *(The Bettmann Archive. Reproduced by permission.)*

have the same effect on a frog's legs as the spark. He set out to confirm Franklin's findings.

Galvani found that twitching did indeed occur during a thunderstorm. In fact, muscles could be made to twitch every time they came into contact with two different types of metal. Electricity was obviously involved, but was it coming from the muscle or the metal?

Keep in mind that Galvani was an anatomist. He decided that an "animal electricity" was responsible, and he published his findings in 1791. His paper, suggesting a different type of electric phenomena, received general acceptance; however, there were a few exceptions. **Alessandro Volta** was among those who disagreed with Galvani's conclusion. A spirited controversy erupted between the two. Volta ultimately prevailed, showing that the twitching was caused by the combination of the two different metals and the fluid within the muscle tissue.

Galvani was extremely disappointed. Worse, when he refused to swear allegiance to Napoleon Bonaparte's government in 1797, he lost his position at the university. He died in poverty on December 4, 1798.

Galvani is nonetheless remembered in a number of ways. A person who abruptly jumps into action is said to be galvanized. Iron that has had a layer of zinc crystals applied with **electric current** is known as galvanized iron. The steady flow of elecricity that is created by the contact of two metals is known as galvanic electricity. The nicest tribute to Galvani's memory came from **André Ampère**, who invented a device to measure current in 1820, the galvanometer.

GAMMA DECAY

Gamma decay is one of the three **radioactive decay** modes available to atomic nuclei, along with alpha and **beta decay**. In alpha and beta decay, the atomic number of the **nucleus** changes, but in gamma decay the atomic number does not change. In gamma decay, a nucleus in an excited **energy** state decays to a lower-energy state by emitting a high-energy **photon**. The photons produced in this decay are consequently known as **gamma rays** and have a **wavelength** with an order of magnitude of about 1,000 femtometers, or 10^{-12} meters. The decay process is very similar to the absorption and emission of **light** by atoms in the ultraviolet, visible, and infrared spectrums.

Gamma decay often follows either alpha or beta decay in a decay process. Following an alpha or beta decay, the number of protons and neutrons in the nucleus has changed, and the resulting nucleus may not be in its lowest energy state (called the ground state). As a result, the nucleus will decay to the ground state by emitting one or more gamma-ray photons.

The lifetime of a nucleus that can undergo gamma decay is short, and direct measurements are difficult. A method called resonance **fluorescence** can be used to measure the lifetime of the excited state. In this method, an experimenter shines gamma rays on a nuclear target to excite it to a higher **energy level**. The probability of absorbing the gamma rays will be highest when they have the same energy as an energy level difference in the nucleus, called a resonance energy. The nuclei then re-emit the gamma rays, which are collected using a detector. Gamma rays with energy close to a resonance energy, however, may also be absorbed and emitted, leading to a peak with nonzero width in the measured spectrum. The Heisenberg **uncertainty principle** can be applied to this problem and by measuring the width of the spectrum one can find the lifetime of the nuclear energy state.

Although this process is straightforward, extra steps must be taken when actually performing the experiment. Absorbing and emitting a photon results in recoil energy due to conservation of **momentum**, and the nuclear recoil energy is much larger than the width of a typical gamma-ray spectrum peak. Resonance fluorescence, therefore, cannot be easily observed in many nuclear systems. By a special technique called Mössbauer **spectroscopy**, the effects of nuclear recoil are removed by cooling the entire experiment, allowing a precise measurement of the nuclear lifetime.

See also Alpha decay; Mössbauer effect; Nuclear reactions; Radiation; Resonance

GAMMA RADIATION

Gamma **radiation** is the emission of photons, the quantized units of **light**. The particles generally known as **gamma rays** are at the highest **energy** end of the **electromagnetic spectrum** and have the shortest **wavelength** and highest **frequency**. However, any electromagnetic wave emitted from a radiation reaction can be called **gamma decay**. Gamma radiation occurs when the **nucleus** of an **atom** decays from a higher energy state to a lower energy state. It often follows alpha or **beta decay**. Usually, the lifetime for gamma radiation to occur is quite short, but occasionally there will be a long-lived stable state called an isomer. Some common atoms that emit gamma radiation are isotopes of cobalt and cesium.

High-energy electromagnetic radiation was discovered by Wilhelm Röntgen, and he won the Nobel Prize in physics for it. Later exploration and classification of the types of radiation occurred in the laboratories of **Antoine-Henri Becquerel** and **Pierre** and **Marie Curie**.

Gamma radiation is the least damaging type of radiation when it comes in contact with human cells, but it also has the greatest capacity to penetrate vital organs, since it has a longer stopping distance than any of the other forms of radiation. It is commonly used for identifying isotopes from their decay patterns, also known as nuclear spectroscopy. Each **isotope** that shows gamma decay will emit gamma rays of specific energies and specific half-lives, with no two patterns of energy and **half-life** alike. For example, it became clear that the element of **atomic weight** 118 had been synthesized when a "machine-gun-like barrage" of gamma rays came out of the resultant atom as it decayed in energy.

Since its discovery, gamma radiation has opened the door to greater human understanding of physics on all scales, from the atom to the entire **universe**. It provides insight into the past as well as being a solid basis for current experiments. While it must be used very carefully to avoid damage to human tissue, gamma radiation is used and explored in almost every branch of physics.

GAMMA RAY

Gamma rays are a kind of electromagnetic **radiation**, or light **energy**. Just as visible **light**, infrared **waves**, **microwaves**, **radio** waves, ultraviolet rays, and **x rays** are forms of electromagnetic radiation, so too are gamma rays. Gamma rays are highly energetic waves with very short wavelengths. Gamma ray wavelengths range from 10^{-10}-10^{-13} meters long. Because of their extremely short length, gamma rays have very high frequencies. Typically, gamma rays have frequencies of greater than 1,018 hertz, or cycles per second. Each cycle is the complete passage of one full **wavelength** past a given point.

Gamma rays, also called **gamma radiation**, are photons that are emitted from very unstable atomic nuclei. Some gamma rays reach Earth's atmosphere from outer **space**. Of the kinds of radiation, gamma radiation is the most energetic and the most penetrating. Gamma rays can be thought of as energized x rays. While other forms of radiation can be blocked with paper and thin sheets of metal, glass, or plastic, gamma rays can easily pass through these. Shielding from gamma rays requires thick layers of concrete or lead. Passing unhindered through living tissue, gamma radiation is very damaging to cells. Intense exposure to gamma rays can cause immediate cell death by damaging protein and DNA. Mutated

DNA as the result of chronic low-level exposure, as with any form of radiation, can cause cancer. Gamma rays are often accompanied by alpha and **beta radiation** as well.

Like x rays, gamma rays are used in industry to inspect metal castings and welds. Another use of gamma rays is in gamma-ray **astronomy**. This branch of astronomy analyzes distant objects in space such as **stars**, **pulsars**, and supernovas by examining the gamma rays they emit. In 1992, the National Aeronautics and Space Administration (NASA) launched the Compton Gamma Ray Observatory into space. This large orbiting space **telescope** detects gamma-ray emissions from very distant sources for analysis back on Earth. Because they are of such high energy, gamma rays are some of the longest-reaching electromagnetic waves, illuminating the far reaches of space to our earthbound eyes.

GAMOW, GEORGE (1904-1968)

Russian American physicist

George Gamow was a celebrated physicist best known for his work dealing with the origins of the **universe**. Gamow also made important contributions to the unraveling of the genetic code in molecular biology. In addition to his research, Gamow was a great popularizer of physics and **astronomy**, particularly in his books dealing with a fictional character by the name of "Mr. Tompkins." Born in Russia, Gamow was trained in physics at the University of Leningrad shortly after the Russian Revolution, and his earliest important work on the theory of **alpha decay** of radioactive nuclei was carried out at the University of Göttingen in the late 1920s. As the political situation in his homeland gradually became more difficult, he decided to defect to the United States, and in 1933 became a professor at George Washington University.

Educates Himself in Mathematics and Science

George Gamow was born in Odessa, Russia, on March 4, 1904, the son of Anthony M. Gamow and Alexandra (Lebedinzeva) Gamow. Both of his parents were teachers, his father an instructor of Russian language and literature. Gamow's grandfather had been a general in the Russian army. Gamow was a precocious boy who taught himself mathematics and science. While his classmates were working their way through conventional school subjects, Gamow was teaching himself differential equations and the theory of relativity. Formal schooling was difficult at the time because of revolutionary battles, but Gamow eventually graduated from Odessa Normal in 1920.

As one of his earliest research projects, Gamow attempted to find out if communion bread from the local church was really different from ordinary bread by examining it under a **microscope**. He became fascinated with **optics** and astronomy in 1917 when his father gave him a **telescope** for his thirteenth birthday. In 1922, Gamow entered Novorossysky University in Odessa, planning to major in mathematics. He felt dissatisfied with his instruction there, however, at least partly because he was better educated than

were some of his teachers. A year later, therefore, Gamow transferred from Novorossysky to the University of Petrograd (later, the University of Leningrad).

Gamow's tenure at Petrograd also seems not to have been very satisfying. He has been quoted by George Greenstein in *American Scholar* as having later stated, "I was around the university, I was not in the university." Part of the problem was that he was passing courses, but generally not going to classes. His adviser assigned him a research problem in optics, but Gamow did not seem to be very interested in the work. In 1928, therefore, when given the opportunity to spend a summer at the University of Göttingen, he eagerly accepted. Most of his biographers say that Gamow was awarded his Ph.D. by Petrograd in 1928. But Greenstein, a colleague who interviewed him later in life, tells a different story. "He lost interest in his Ph.D. thesis research," Greenstein writes, "and sent the University of Leningrad his work on **radioactivity** instead. It seems that they never got around to awarding him a degree."

Quantum Mechanics Is Applied to Radioactivity

The "work on radioactivity" mentioned by Greenstein was Gamow's first major scientific accomplishment. While at Göttingen, he read about **Ernest Rutherford**'s failed efforts to bombard an atom's **nucleus** with alpha, or positively charged, particles. Gamow decided to see if **quantum theory**, about which he had only recently learned, would explain Rutherford's results. He eventually developed a theory that involved the process of "tunneling." **Tunneling** is a mechanism by which particles are able to break through (actually, to "tunnel under") a **potential energy** barrier that would, under classical laws of physics, appear to prevent their passage. The American physicists Edward Uhler Condon and Ronald Gurney were reaching the same conclusion at almost the same time.

News of Gamow's work soon spread throughout the physics community, and he was invited by **Niels Bohr** to become a Carlsberg fellow at his Institute for Theoretical Physics in Copenhagen from 1928 to 1929. There Gamow continued his research on nuclear phenomena and also began to study the physics of stars' interiors. After a brief trip home in the summer of 1929, Gamow was on the road again, this time to Cambridge as a Rockefeller fellow. At the Cavendish Laboratories at Cambridge, he worked with Rutherford, the preeminent nuclear scholar in the world, for a year. There Gamow's primary achievement was the calculation of the **energy** needed to split an atomic nucleus using protons, the positively charged element of the nucleus. Within a few months, **John Douglas Cockcroft** and **Ernest Walton** had begun the design of a machine—the world's first **particle accelerator**—to put Gamow's results into practical use.

At the conclusion of his year at the Cavendish Laboratories, Gamow was invited by Bohr to return to Copenhagen for a second year. His time there was to be devoted primarily to research on a quantum theory of the nucleus, a subject on which he had been asked to give a paper for the 1931 International Congress on Nuclear Physics in Rome. Gamow's return to the Soviet Union in 1931 was significant for two reasons. First, he married Loubov Wochminzewa. Second, he found a dramatically altered political climate, one in which foreign travel was much more closely supervised than it had ever been in his lifetime. In fact, the government told Gamow that he would not be permitted to attend the Rome conference. That decision convinced Gamow that he did not want to remain in the Soviet Union. Although he accepted an appointment as professor of physics at the University of Leningrad, he began devising a scheme to escape from the Soviet Union. The secret plans made by Gamow and his wife over the next two years read, according to Greenstein, "like a paperback thriller written by someone in an ironical frame of mind." The one actual attempt they made to flee to Finland was thwarted, however, by **weather** conditions, and they were forced to return to Leningrad.

Solvay Conference Provides Opportunity for Escape

As it turned out, the Soviet government solved Gamow's problem for him when, in the fall of 1933, it ordered him to attend the International Solvay Congress in Brussels and provided him and his wife with the visas and tickets needed to make the trip. When the Gamows left Leningrad in later 1933 for Brussels, they knew they would not be returning. At the conclusion of the Solvay conference, therefore, the Gamows set their sights on the West. Gamow visited the Pierre Curie Institute in Paris, the Cavendish Laboratories, and Bohr's institute for two months each before accepting a summer (1934) teaching assignment at the University of Michigan. The following year, he became professor of physics at George Washington University in Washington, D.C., a post he held until 1956, and he became a U.S. citizen in 1940. His earliest work at George Washington was carried out in conjunction with **Edward Teller**, a Hungarian physicist who had also defected to the United States. Together, the two worked out a theoretical explanation for the process by which a radioactive nucleus emits a beta particle, or high-speed **electron**. The theory became known as the Gamow-Teller selection rule. During World War II, Gamow was assigned to the U.S. Navy Bureau of Ordnance where he worked in the division of high explosives. After the war, he worked on the theory of war games for the U.S. Army and eventually became involved in research on the hydrogen bomb with Teller.

Gamow also began to focus on a new field, **astrophysics**. He decided to investigate the physical, chemical, and nuclear changes that might have occurred during the first few moments of the universe's creation. Gamow developed a theory that described the formation of the **light** elements, hydrogen and **helium**, but he did not see how heavier elements could have been produced as a result of the universe's beginnings. Eventually he concluded that these elements must have been formed much later in the history of the universe, in the center of **stars**. Over the years, Gamow developed mathematical formulations that described how this process of heavy element formation might have taken place. One of his most important papers on the subject of astrophysics was the now famous Alpha-Beta-Gamma paper that appeared in the April 1, 1948,

George Gamow (1904-1968).

issue of *Physical Review.* The paper's name is derived from the last names of its three authors, Ralph Alpher, **Hans Bethe**, and Gamow. It summarizes the hypothesized changes that took place during the first few moments following the **big bang.**

After 1953, Gamow began work on an entirely new subject, the young field of molecular biology. The impetus for this change of direction was the discovery by James Watson and Francis Crick of the structure of the deoxyribonucleic acid (DNA) molecule. An inherent question posed by that discovery was how the genetic information stored in DNA molecules could be used by cells in the manufacture of proteins. Gamow hypothesized the existence of a genetic "code" consisting of the four nitrogen bases found in a DNA molecule. A combination of three successive bases, Gamow suggested, might code for a specific amino acid. The arrangement of bases in a DNA molecule, therefore, could carry instructions for a sequence of amino acids, that is, for a protein molecule. Although the details of Gamow's analysis were not correct, his hypothesis of the genetic code was.

Ends Associations with George Washington University and Wife

In 1956, Gamow ended two long-term relationships, one with George Washington University and the other with his wife of 25 years, Loubov. He moved to Boulder, Colorado, where he became professor of physics at the University of Colorado. Two years later, he was married to Barbara Perkins. Gamow had long been interested in writing about science for the layperson, and his first effort along these lines described a tiny universe only five miles wide. No American publisher was interested in the story, but it was eventually accepted by C. P. Snow, editor of the British science magazine *Discovery.* The "toy universe" stories were later collected in Gamow's popular book, *Mr. Tompkins in Wonderland,* and then extended in a second book, *Mr. Tompkins Explores the Atom.*

The "Mr. Tompkins" of these stories is a bank teller who tends to fall asleep while listening to his father-in-law's lectures on physics. In his dreams, Mr. Tompkins is able to reinterpret those lectures in a more understandable setting, as readers of the books might. In other works, Gamow presented somewhat more serious descriptions of scientific ideas that were still easily accessible to most readers. Among these were *The Birth and Death of the Sun,* in 1940, *Biography of the Earth,* in 1941, and *The Creation of the Universe,* in 1952. Gamow's popular texts earned him the Kalinga Prize from the United Nations Educational, Scientific, and Cultural Organization (UNESCO) in 1956.

Gamow was tall, blond-haired, and blue-eyed, and was renowned for his sense of humor. Greenstein has described him as "exuberant, jovial, rough, and friendly," with qualities of "playfulness and inventiveness, and a resolute refusal to be trapped within a ponderous consistency." He held a special passion for magic and loved to regale his friends with the tricks he had mastered. Gamow had a lifelong habit of consuming large quantities of alcohol, and though he rarely—if ever—lost control because of his drinking, alcohol may ultimately have led to his early death at the age of 64. During the last few years of his life, he was constantly ill and finally died in Boulder on August 19, 1968.

GAS LAWS

A substance in the gaseous state of **matter** is one that has no definite shape or volume. The particles in a gas are moving with high energies and speeds. There are many familiar properties of gases. For example, a gas will take on the shape of any container in which it is placed. Likewise, a gas will occupy any volume that is made available to it. A gas can also be easily compressed when **pressure** is exerted on it. All gases show the same physical properties. They all have **mass**, they can move through each other, they exert pressure, and the pressure of all gases depends on their **temperature**.

The physical properties of gases are explained by a model called the kinetic-molecular theory of gases. This theory is composed of several postulates. First of all, a gas is made up of tiny particles. These tiny particles each have a particular mass. Second, the distance between the particles in gases is quite large in comparison with the distance between particles in liquids and solids. Third, the particles of a gas are in a constant **motion**. They move rapidly and randomly. Fourth, as the gas particles collide with each other or with the wall of a con-

tainer, their collisions are perfectly elastic. That means no **energy** is lost by a particle as it makes a collision. Fifth, the only factor that can affect the **kinetic energy** of the gas particles is the temperature of the gas. As the temperature of the gas increases, so does the kinetic energy of the particles. Finally, a gas particle does not exert a **force** on another gas particle.

The postulates of the kinetic-molecular model built the foundation for several theories that are now called the gas laws. The gas laws were formed by several scientists and collectively summarize the current understanding of the gaseous state of matter. Each of the gas laws assumes that the gas is under specific conditions, called standard temperature and pressure (STP). The standard temperature is zero degree Celsius or 273 Kelvin (K). The standard pressure is one atmosphere (atm), which is equivalent to 760 mm mercury (Hg) or 101.325 Pascal (Pa). These laws give a mathematical model for the observed properties of pressure, volume, temperature, and quantity of a gas.

The first of the gas laws is known as **Boyle's law.** Boyle's law describes the relationship between the pressure and the volume of a gas. Gases, as stated above, can be easily compressed. This is because the particles of a gas have a great distance between them. It is this property of gases that make them useful as cushioning devices such as the air bags in an automobile. An English chemist and physicist named **Robert Boyle** was the first scientist to describe this property of gases. In fact, he wrote about "the spring of air" in more than one publication.

Boyle performed many experiments that explored the "spring of air." In one such experiment, he trapped a certain volume of air and measured the new volume after changing the pressure. To do this, he used a J-shaped tube. The short end of the "J" was closed off, and the long end was left open. Boyle trapped some air in the end of the J-tube by adding mercury in the longer end. By changing the height of the mercury in the long end of the tube, Boyle could alter the pressure on the air in the short end. He measured the volume of the air in the short end of the tube at several different pressures, keeping the temperature constant throughout the experiment. He found that as he increased the pressure on the air, the volume of the air would decrease.

Boyle used these results to formulate his law. Boyle's law maintains that if the temperature of a gas is constant, the product of the pressure times the volume is constant. The mathematical formula that describes this law is $PV = k_1$. P is the pressure of the gas, V is the volume, and k_1 is the constant. The value of k_1 depends on the units of pressure and volume used, the amount of gas present, and the temperature of the gas. Using this equation, it can be stated that the volume of a gas and its pressure are inversely proportional to each other. This law can be useful if you measure the pressure and volume of a gas in two trials at constant temperature. If you rearrange the above equation accounting for the two trials, you will come up with $P_1V_1 = P_2V_2$, where P_1 and V_1 are the pressure and volume of the gas in the first trial, and P_2 and V_2 are the pressure and volume of the gas in the second trial.

The next of the gas laws is **Charles's law. Jacques Charles** formed this law in order to describe the relationship between the temperature of a gas and its volume. Charles also conducted several experiments in order to develop his theory. In one experiment, he trapped a gas sample in a cylinder with a moveable piston. The temperature of the gas sample was changed by placing the cylinder in water baths at different temperatures. As the gas's temperature changes, so does its volume, which is observed by the piston moving up or down. The volume of the gas can be measured by the position of the piston. Charles found, through these experiments, that the volume of a gas is directly proportional to its temperature. As the temperature rises, so does the volume of a gas. Anyone who has bought a **helium** balloon during cold **weather** has observed this phenomenon. When the balloon is removed from the heated store and walked through the cold parking lot, it collapses. As the temperature of the helium decreases, so does its volume. If the temperature is measured using the **Kelvin temperature scale** and the pressure is kept constant, the relationship can be described mathematically by the equation $V = k_2 T$. V is the volume of the sample, T is the temperature in Kelvin, and k_2 is called the Charles's law constant of proportionality. Again, this equation can be rearranged for an experiment with two trials to give the relationship $V_1 T_1 = V_2 T_2$.

The next of the gas laws is known as **Avogadro's law**. This law describes the relationship between the amount of a gas and its volume. It was proposed by the Italian chemist **Amedeo Avogadro** and states that two samples of gases of equal volume at the same temperature and pressure have the same number of particles in each. This relationship can be expressed by the equation $V = k_3 n$. V is the volume of the sample, n is the number of moles of the gas, and k_3 is called Avogadro's law constant.

The fourth gas law is called Dalton's law of partial pressures. **John Dalton**, an English chemist, explored the nature of gases in mixtures. Dalton determined that each gas present in a mixture exerts the same pressure as it would if it were not in a mixture at the same temperature. In a mixture, each gas exerts a pressure that is called the partial pressure of that gas. The total pressure of the mixture of gases is equal to the sum of all of the partial pressures of the individual gases present in the mixture.

Each of the above laws can be combined to give an equation called the ideal gas equation, $PV = nRT$. P is the pressure of the gas, V is the volume, n is the number of moles of the gas, T is the temperature, and R is called the gas constant (or, sometimes, the universal gas constant). At STP, R = 0.0821 atm-L/mol-K. This equation summarizes the physical properties of an "ideal" gas. An ideal gas is one that follows each of the postulates given by the kinetic-molecular theory of gases. In reality, no gas completely adheres to the kinetic-molecular theory, so there is no such thing as an ideal gas. Scientists still use the ideal gas equation to describe "real" gases. Real gases generally adhere to the kinetic-molecular postulates, it is only at high pressures and low temperatures that they deviate significantly.

The gas laws described above are a culmination of the works of many scientists. These laws are very useful in describing and predicting the behavior of gases. Even though no ideal gases exist, real gases, for the most part, adhere to the principles outlined above.

GAS PRESSURE

According to the kinetic theory of gases, gas **pressure** is the **force** exerted by a gas on a unit area. In a gaseous system the gas molecules are in continual **motion** and the hotter the gas the faster its molecules are moving. When these moving molecules (or atoms) strike an object they exert a force upon the object. Accordingly, gas pressure is an indirect measure of the kinetic **energy** of gas molecules and is a direct measure of the sum of the forces of collision over a defined unit area.

The speed with which a gas molecule moves is directly related to its **temperature**. The greater the rate of movement (molecular **velocity**) the greater number of molecular collisions with a given surface area. In addition, greater molecular velocity imparts a greater **momentum** at impact and means that the molecules carry a increased **kinetic energy** into collisions. All of these factors contribute to an increased gas pressure.

Just as pressure and temperature are connected, so are pressure and volume. Reducing the volume of a trapped gas means a shorter distance for the gas molecules to travel before colliding with the walls of the container. Trapping gas inside a container and then reducing the volume of the container will mean an increase in gas pressure. In fact, reducing the volume by one-half will double the number of collisions and thus double the pressure. This is the basis of **Boyle's law** formulated by English physicist **Robert Boyle** in 1662. Boyle's law states that, for an ideal gas at constant temperature, the pressure (p) of a gas varies inversely with volume (v). In equation form pv = k, a constant. Boyle's law is broadly applicable to real gases at lower pressure but fails to provide accurate results at high pressures or temperatures.

See also Gas laws; Gases, behavior and properties of; Kinetic molecular theory

GASES, BEHAVIOR AND PROPERTIES OF

Gases are the most diffuse form of **matter**. In a gas the molecules (the smallest particles of which the gas is made—for some gases these equate to atoms) are highly energetic and they can be widely separated from each other. The behavior of gases and their properties derive from these facts.

Sometimes the term vapor is used to describe a gas. Strictly speaking a gas is a substance at a **temperature** above its **boiling** point. A vapor is the gaseous phase of a substance that, under ordinary conditions, exists as a liquid or solid.

One of the more obvious characteristics of gases is their ability to expand and fill any volume they are placed in. This contrasts with the behavior exhibited by solids (fixed shape

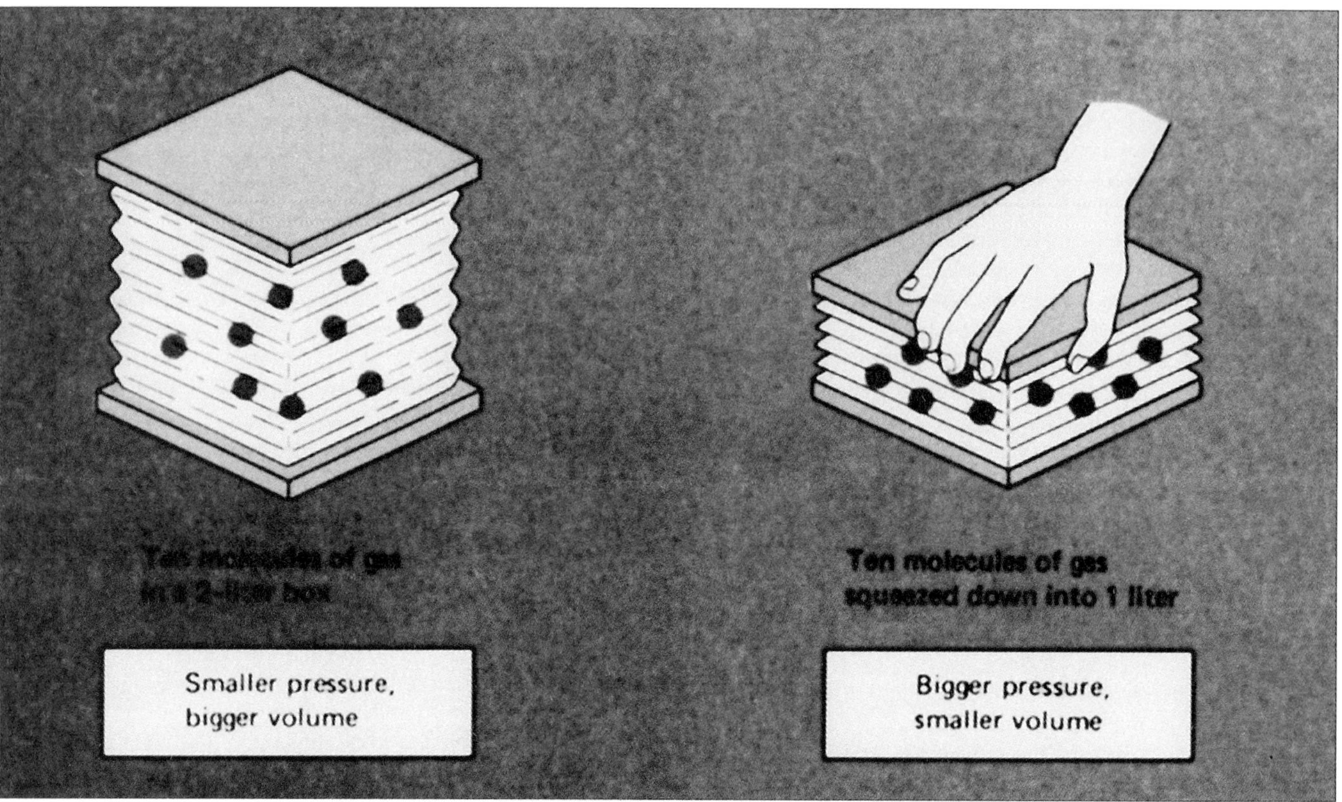

The effect of pressure change on the volume of a gas, with temperature held constant. *(Image by Robert L. Wolke. Reproduced by permission.)*

and volume) and liquids (fixed volume but indeterminate shape). Gases have an indeterminate volume and shape. This is due to the freedom of movement exhibited by the molecules comprising the gas. A gas can be compressed to a smaller volume or it can expand to fill a larger volume. Gases will form homogenous mixtures with each other. Providing no reaction occurs between the gases, the gases will disperse throughout the volume of the container they are kept in. The **pressure** inside the container is the sum of the pressures that each gas would exert if it were present alone. This is summarized in Dalton's law of partial pressures.

Gases are mostly **space**. For example, in the air that we breathe the actual molecules only account for 0.1% of the volume—the rest is empty space. The practical effect of this is that a molecule in a gas behaves as if it is in isolation, i.e., it does not react with the other molecules in the gas. This is why all gases have similar properties, irrespective of the type of gas.

Many substances are referred to as gases, since this is the state in which they are normally encountered. In reality many substances may exist only in the gaseous phase over a small range of conditions. Consequently when we refer to something as being a gas and we do not specify the conditions it is assumed that the condition is standard temperature and pressure (STP). Standard temperature and pressure is defined as a temperature of 32°F (0°C) and a pressure of 1 atmosphere.

The behavior of gases and their properties is explained by the **kinetic molecular theory**, which, in its current form, was first published by **Rudolf Clausius** in 1857. The kinetic molecular theory of gases states that gases are comprised of a large number of molecules that are in constant, random **motion**. The volume of all of the molecules of a gas is very small compared to the volume of the gas. Intermolecular forces are largely irrelevant within a gas due to the large intermolecular distances. While **energy** can be transferred between molecules during collisions, the average **kinetic energy** of molecules does not change (providing the temperature remains constant). The average kinetic temperature of the molecules of a gas is proportional to the absolute temperature (the temperature in Kelvin) and at any given temperature the molecules of all gases have the same kinetic energy.

The kinetic molecular theory allows us to understand the behavior of gases. The pressure exerted by a gas is a measure of how frequently (and how hard) the molecules of a gas strike the walls of the container in which it is being stored. The absolute temperature of a gas is a measure of the average kinetic energy of the molecules of the gas. Different gases at the same temperature have the same average kinetic energy. If the temperature of a gas is altered then the kinetic energy of the molecules is altered accordingly. If the absolute temperature of a gas is halved then the average kinetic energy of the molecules is also halved. The average kinetic energy is the same between all molecules of a gas, but some individual molecules in a gas are moving faster and some slower than the other molecules. The distribution of energies within a body of gas follows a normal distribution, if the number of molecules is plotted against the energy of the molecules.

Effusion (the movement of gas molecules through a pinhole in a container) and diffusion (the uniform spread of gas molecules throughout a mixture of gases) are both related to the energy levels of the gas. The higher the energy levels the more rapidly the molecules are moving and the more rapidly the processes of effusion and diffusion take place. This is due to the fact that the molecules are more likely to hit the pinhole the more rapidly they are moving, and the more rapidly they are moving the quicker they will spread into an unoccupied space. The law governing the rates of effusion of two gases is called Graham's law. Both effusion and diffusion are faster for lighter gases (diffusion will also occur with solids but the rate of diffusion is infinitesimal when compared to gases). Related to the rate of diffusion is the mean free path of a molecule. This is the distance traveled between collisions. The greater the distance traveled between collisions the more rapid will be diffusion, since collisions make gas molecules move from one place to another in a staggered manner. The length of the mean free path is governed by the **density** of the gas—the more the molecules are packed together the greater is the likelihood of a collision between molecules. At sea level the mean free path distance for a gas molecule in air is about 100 nm. At an altitude of 62 mi (100 km) the air density is much lower and the mean free path is some million times longer than at sea level. The mean free path in air is about 6 in (16 cm) at this altitude.

The universal gas law can be used to describe the behavior of a gas. Derived from **Boyle's law, Charles's law**, and the constant volume law, it can be written as pV/T = a constant, where p is the pressure, V is the volume, and T is the absolute temperature. Another way of writing this law is $p_1V_1/T_1 = p_2V_2/T_2$ where the subscript 1 represents the start conditions and the subscript 2 represents the end conditions. A third way of writing the universal gas law is as $pV = nRT$ where n represents the number of moles of gas being dealt with and R represents the universal gas constant. The universal gas constant is 0.08204 liter atmosphere per degree, or 8.314 $Jmol^1K^{-1}$.

Avogadro's hypothesis states that at equal pressure and temperature, equal volumes of gases contain the same number of molecules. This hypothesis led to **Avogadro's law**, which states that the volume of a gas maintained at constant temperature and pressure is directly proportional to the number of moles of the gas. So if the temperature and pressure of the gas remain constant and the number of moles of gas present is doubled then the volume will also double.

GAUGE SYMMETRY

If a field theory possesses a gauge symmetry, then the theory is invariant under certain local transformations of the fields. The description of the system has a built-in redundancy corresponding to the action of a gauge transformation, which is typically thought of as an element of an algebraic group. By choosing a particular gauge, the redundancy can be eliminat-

ed. Measurements that are made of any physical quantities are independent of the precise choice of gauge.

An example serves to illustrate the point. The **quantum field theory** that describes the interactions of charged point particles and **radiation** is **quantum electrodynamics (QED)**. This theory encodes the quantum mechanical properties of electrons and photons in a manner that is consistent with the theories of special relativity and classical **electrodynamics**. The electrons and photons are thought of as fields, an extension of the notion of wave functions from **quantum mechanics**. QED possesses a U(1) gauge symmetry. The group U(1) can be represented as the unit circle in the complex plane, or the collection of complex numbers $z = x + iy$ whose modulus—z—equals 1. Such numbers are called phases. The opportunity to redefine the fields in QED by a local phase without affecting the theory in any way is possible. Multiplying the **electron** field by a phase is compensated for by a corresponding shift in the **photon** field. The Lagrangian of the theory, which encodes all the **dynamics** of QED, remains invariant under this combined operation. Because it is impossible to add a **mass** term for the photon to the QED Lagrangian that respects the U(1) gauge symmetry, the photon must be massless. A continuous symmetry, such as a gauge symmetry, leads to a conserved current as a consequence of **Amalie Noether**'s theorem.

The theory of the particles observed in nature and their interactions (excluding **gravity**) is a gauge theory with gauge group SU(3) x SU(2) x U(1). This is called the standard model. Yang-Mills theory is the general name given to a model with a gauge group that is non-Abelian (the product of elements in the group does not commute).

GAUSS, CARL FRIEDRICH (1777-1855)
German mathematician

Carl Friedrich Gauss was as eccentric as he was brilliant. Known primarily for his work in mathematics, he also made contributions in physics, **astronomy**, and **magnetism**.

Born at Brunswick, Germany, on April 30, 1777, Gauss was a bona fide genius, able to make calculations before he was able to talk. At the age of three he was making corrections to his father's additions. He taught himself to read and regularly kept lists of numerical data, including such useless tables as how many days various famous men had lived. Gauss not only lived for numbers, he was consumed by them.

Gauss's father, recognizing the boy's talent with mathematics, encouraged his son. At the age of 14, Gauss was able to obtain a stipend from Duke Ferdinand of Brunswick to devote himself to science. Gauss eventually went on to the University of Göttingen in 1795 and obtained his doctor's degree in 1799.

Gauss made numerous mathematical discoveries while still in his teens. He advanced the work of Adrien-Marie Legendre by working out the method of least squares. His work in geometry went beyond the work of the ancient Greeks. In 1796 he showed how to construct an equilateral polygon with 17 sides using just a ruler and compass. In addition, he showed that only polygons with a certain number of sides could be constructed with those two tools. The inverse of this logic meant that there were certain geometric constructions that were impossible, and proving the impossible in mathematics later became extremely important in the work of Kurt Gödel.

At the age of only 24, Gauss calculated an orbit for *Ceres*, the first asteroid that Giuseppe Piazzi (1746-1826) had discovered in 1801 and then lost when it passed behind the **Sun**. The asteroid was recovered just where Gauss predicted it would be, and this remarkable calculation made his reputation. He also calculated theories of perturbations between the larger **planets** that helped Jean Urbain Leverrier (1811-1877) and John Couch Adams in their independent discovery of **Neptune**.

Eventually, Gauss came to be recognized as one of the greatest mathematicians of all times. He went on to work with Pierre de Fermat's theory of numbers and also worked out a **non-Euclidian geometry** that was based on axioms different from those of **Euclid**. Unfortunately, Gauss often tended to keep some of the results of his work secret for awhile, and two other mathematicians received the credit for the discovery by publishing first. In 1799 Gauss proved that every algebraic equation has a root of the form $a + bi$, which is one of algebra's fundamental theorems.

In 1807 Gauss was appointed director of the Göttingen Observatory, a position that he retained for the rest of his life. There, in 1821, he invented a device called a heliotrope, which afforded a more accurate way to survey locations on Earth. He became interested in Earth's magnetic field and established the first observatory to specialize in its study. Making geomagnetic observations with **Wilhelm Weber**, who would become famous for his work with **electricity**, Gauss was able to make a calculation that accurately located Earth's magnetic poles. They established the standard unit of measurement of magnetic influence, which was later named the *gauss*. Gauss and Weber became interested in **electromagnetism** and, making use of **Michael Faraday**'s 1831 discovery of magnetic induction, invented a telegraph. Their version differed from that of American physicist **Joseph Henry**, who was in the process of inventing his own telegraph at the same time. Henry's receiver used a metal arm that clicked up and down; Gauss's version consisted of a large coil over a magnet to create the current that deflected a magnetic needle at the receiver. In 1833 Gauss and Weber ran a wire across Göttingen from the physics laboratory to the observatory, but abandoned it when they were unable to gain support for additional development. Ironically, in 1837 Charles Wheatstone patented his telegraph in England, and in 1840 Samuel Morse did the same in the United States, much to the chagrin of Henry.

On February 23, 1855, Gauss died at Göttingen. His name has been honored twice: first with the standard unit of magnetic influence, which the metric system has since replaced with the *tesla* in honor of another eccentric genius, **Nikola Tesla**, and second with asteroid number 1,001, which is known as Gaussia.

GAY-LUSSAC, JOSEPH-LOUIS (1778-1850)
French chemist

Born about a decade before the French Revolution, Joseph Gay-Lussac was only 14 when his comfortable family life was disrupted by his father's arrest for suspected royalist sympathies. Although Gay-Lussac's tutor fled the country, the boy's education resumed at a private school in Paris, and he successfully competed for entrance to the new École Polytechnique. The hallmark of his scientific career was the analysis of gases, which he studied not only in the laboratory, using ingenious new techniques, but also on daring balloon flights that reached record altitudes.

Gay-Lussac's career began with a major revelation. In 1802, when he was just 24, he published a new fundamental law of physics that helps explain the behavior of gases. Although everyone knew that **heat** would make gases expand in volume, Gay-Lussac's careful experiments proved that different gases would all expand by the same amount with the same rise in **temperature**. Now known as **Gay-Lussac's law** (sometimes called **Charles's law**), this idea was crucial to **Amedeo Avogadro**'s discovery a few years later that equal volumes of gases at the same temperature and **pressure** contain the same number of molecules.

Not content to remain in the laboratory, Gay-Lussac took to the sky in a balloon, armed with various instruments to test Earth's magnetic field, the atmosphere's **electricity**, and the composition of high-altitude gases. On his second trip, in 1804, Gay-Lussac flew higher than the tallest mountain in the Alps. This record of more than 4 miles (6.4 km) above sea level was not reached again for another 50 years. The balloon tests showed that the composition of the air and the magnetic **force** of Earth were the same at high altitudes as they were on the ground. After these flights, Gay-Lussac spent two years studying Earth's magnetic intensity at ground level with his collaborator Alexander von Humboldt.

By then, however, word had spread of Sir Humphry Davy's astonishing results in discovering new elements by using electricity to split compounds. In the name of French nationalism, Napoleon Bonaparte (1769-1821) funded construction of a bigger, more powerful battery for Gay-Lussac and a coworker, Louis Thénard, to perform similar experiments. But after all, the battery wasn't needed; the two scientists were able to make potassium and sodium by purely chemical means, without electricity. What's more, they produced these elements in greater quantities than Davy had, making the metals available to others at lower cost. Pursuing this technique further, Gay-Lussac and Thenard discovered the element boron in 1808, just nine days before Davy did.

Around the same time, Gay-Lussac was also trying to determine the proportions in which various gases combined with each other. Collaborating with Humboldt, Gay-Lussac found that when gases form compounds, they always combine in simple proportions—for example, two parts of hydrogen by volume would react with one part **oxygen** to form water, or two parts **carbon** monoxide with one part oxygen to form carbon dioxide. Some scientists believe that this discovery, called the law of combining volumes, was even more important than Gay-Lussac's earlier gas law.

After some dispute with Davy over who had discovered chlorine, Gay-Lussac went on to study iodine, which had first been isolated by Bernard Courtois (1777-1838). Again, Gay-Lussac's research overlapped Davy's, but it was Gay-Lussac who gave the element its name, after the Greek word for "violet-colored." Gay-Lussac then performed a series of experiments on cyanides, proving conclusively that prussic acid contains no oxygen. At the time, chemists supposed that oxygen was present in all acids (in fact, hydrogen is). During these analyses, Gay-Lussac produced a carbon-nitrogen gas that he named cyanogen, from the Greek for "dark blue." He carefully examined several cyanide compounds, as well as a popular dye called Prussian blue.

Another basic idea in chemistry was formulated by Gay-Lussac in 1814—the concept of isomers, which are different compounds composed of the same elements in identical quantities, but differently arranged. Gay-Lussac, an ingenious technician, also created or perfected many of the techniques still practiced in analytical chemistry. In introducing a volumetric method for estimating the purity of silver—which was much more accurate than the primitive method then in use—Gay-Lussac popularized the use of titration (which involves the careful addition of tiny, exact volumes). He was also the first to use the terms pipette and burette to describe glass apparatus that is found in every chemistry lab today.

Although Gay-Lussac is better known for his achievements in inorganic chemistry, he and Thenard also studied organic compounds. They divided vegetable substances into three classes, one of which is now called carbohydrates, and they were the first to determine the elementary composition of sugar.

With another French chemist, Michel-Eugène Chevreul, Gay-Lussac also developed a new type of candle that greatly improved on the candles of his day. Made of certain fatty acids isolated by Chevreul, the new candle smelled better, gave more **light**, and burned more easily than tallow candles. It was also harder, meaning that it would soften less quickly during hot **weather**.

In a way, Gay-Lussac may also have been one of the first environmental scientists. At the time, noxious gases (nitrogen oxides) were released into the atmosphere during the manufacture of sulfuric acid. Gay-Lussac devised a way to capture these polluting gases in an absorption tower. His method was later adopted in a process to recover gases.

Most of Gay-Lussac's important work was done while he was relatively young. In his later years, he had to support a large family and perhaps became too overworked to produce the brilliant results of his youth.

GAY-LUSSAC'S LAW

By the late 1600s, scientists had learned that **heat** would make gases expand in volume, and British chemist **Robert Boyle** had begun to explain the relationship between the volume, **pres-**

sure, and **temperature** of gases. He summarized his ideas in an expression, known as **Boyle's law**, which states that the volume of a given amount of a gas varies inversely with pressure. This principle is important since chemistry often involves the study of gases that are hard to measure because they expand to fill whatever container they are in. The value of Boyle's law lies in its ability to calculate a gas's volume at a standard pressure without actually having to measure the volume at that pressure.

It was more than a century later, however, until French chemist **Joseph Gay-Lussac** discovered, with a series of careful experiments, that different gases all expand in volume by the same amount with the same rise in temperature. Although he was the first to announce this principle, in 1802, today it is more commonly known as **Charles's law** after French chemist **Jacques Charles**, who actually had the same idea earlier but did not publish his findings before Gay-Lussac.

Further, Gay-Lussac found that gases combine with each other in simple proportions—for example, two parts of hydrogen by volume react with one part **oxygen** to form water. This important elaboration came to be known as *Gay-Lussac's law of combining volumes*. These fundamental concepts were crucial to **Amedeo Avogadro**'s discovery a few years later that equal volumes of gases at the same temperature and pressure contain the same number of molecules. Although Gay-Lussac's laws strictly apply only to certain theoretical gases (called "ideal"), many simple substances that are gases at room temperature and normal air pressure do behave as if they were ideal gases. Thus, the laws have proved extremely useful in explaining the behavior of gases, and they are considered fundamental concepts in chemistry and physics.

GEIGER, HANS (1882-1945)

German experimental physicist

Hans Geiger was a German nuclear physicist best known for his invention of the Geiger counter, a device used for counting atomic particles, and for his pioneering work in **nuclear physics** with **Ernest Rutherford**.

Johannes Wilhelm Geiger was born in Neustadt an-der-Haardt (now Neustadt an-der-Weinstrasse), Rhineland-Palatinate, Germany, on September 30, 1882. His father, Wilhelm Ludwig Geiger, was a professor of philology at the University of Erlangen from 1891 to 1920. The eldest of five children, two boys and three girls, Geiger was educated initially at Erlangen Gymnasium, from which he graduated in 1901. After completing his compulsory military service, he studied physics at the University of Munich, and at the University of Erlangen where his tutor was Professor Eilhard Wiedemann. He received a doctorate from the latter institution in 1906 for his thesis on electrical discharges through gases.

That same year, Geiger moved to Manchester University in England to join its esteemed physics department. At first he was an assistant to its head, Arthur Schuster, an expert on gas ionization. When Schuster departed in 1907, Geiger continued his research with Schuster's successor,

Ernest Rutherford, and the young physicist Ernest Marsden. Rutherford was to have a profound influence on young Geiger, sparking his interest in nuclear physics. Their relationship, which began as partners on some of Geiger's most important experiments, was lifelong and is documented in a series of letters between them.

In addition to supervising the research students working at the lab, Geiger began a series of experiments with Rutherford on radioactive emissions, based on Rutherford's detection of the emission of alpha particles from radioactive substances. Together they began researching these alpha particles, discovering among other things that two alpha particles appeared to be released when uranium disintegrated. Since alpha particles can penetrate through thin walls of solids, Rutherford and Geiger presumed that they could move straight through atoms. Geiger designed the apparatus that they used to shoot streams of alpha particles through gold foil and onto a screen where they were observed as scintillations, or tiny flashes of **light**.

Manually counting the thousands of scintillations produced per minute was a laborious task. Geiger was reputedly something of a workaholic, who put in long hours recording the light flashes. David Wilson noted in *Rutherford: Simple Genius* that in a 1908 letter to his friend Henry A. Bumstead, Rutherford remarked, "Geiger is a good man and work[s] like a slave.... [He] is a demon at the work and could count at intervals for a whole night without disturbing his equanimity. I damned vigorously after two minutes and retired from the conflict." Geiger was challenged by the haphazardness of their methodology to invent a more precise technique. His solution was a primitive version of the "Geiger counter," the machine with which his name is most often associated. This prototype was essentially a highly sensitive electrical device designed to count **alpha particle** emissions.

Geiger's simple but ingenious measuring device enabled him and Rutherford to discern that alpha particles are, in fact, doubly charged nuclear particles, identical to the **nucleus** of **helium** atoms traveling at high **velocity**. The pair also established the basic unit of electrical charge when it is involved in electrical activity, which is equivalent to that carried by a single hydrogen **atom**. These results were published in two joint papers in 1908 entitled "An Electrical Method of Counting the Number of Alpha Particles" and "The Charge and Nature of the Alpha Particle."

In bombarding the gold with the alpha particles Geiger and Rutherford observed that the majority of the particles went straight through. However, they unexpectedly found that a few of the particles were deflected or scattered upon contact with the atoms in the gold, indicating that they had come into contact with a very powerful electrical field. Rutherford's description of the event as recorded by Wilson revealed its importance: "It was as though you had fired a fifteen-inch shell at a piece of tissue paper and it had bounced back and hit you." These observations were jointly published by Geiger and Marsden in an article entitled "On a Diffuse Reflection of the Alpha-Particles" for the *Proceedings of the Royal Society* in June 1909.

Thirty years later Geiger recollected, "At first we could not understand this at all," Wilson noted. Geiger continued to study the **scattering** effect, publishing two more papers about it that year. The first, with Rutherford, was entitled "The Probability Variations in the Distribution of Alpha-Particles." The second, referring to his work with Marsden, dealt with "The Scattering of Alpha-Particles by Matter." Geiger's work with Rutherford and Marsden finally inspired Rutherford in 1910 to conclude that the atoms contained a positively charged core or nucleus that repelled the alpha particles. Wilson noted Geiger's recollection that "One day Rutherford, obviously in the best of spirits, came into my [laboratory] and told me that he now knew what the atom looked like and how to explain the large deflections of the alpha particles. On the very same day, I began an experiment to test the relation expected by Rutherford between the number of scattered particles and the angle of scattering."

Geiger's results were accurate enough to persuade Rutherford to go public with his discovery in 1910. Nonetheless, Geiger and Marsden continued their experiments to test the theory for another year, completing them in June 1912. Their results were published in German in Vienna in 1912 and in English in the *Philosophical Magazine* in April 1913. Wilson noted that Dr. T. J. Trenn, a modern physics scholar, characterized Geiger and Marsden's work of this period: "It was not the Geiger-Marsden scattering evidence, as such, that provided massive support for Rutherford's model of the atom. It was, rather, the constellation of evidence available gradually from the spring of 1913 and this, in turn, coupled with a growing conviction, tended to increase the significance or extrinsic value assigned to the Geiger-Marsden results beyond that which they intrinsically possessed in July 1912."

In 1912 Geiger gave his name to the Geiger-Nuttal law, which states that radioactive atoms with short half-lives emit alpha particles at high speed. He later revised it, and in 1928, a new theory by **George Gamow** and other physicists made it redundant. Also in 1912 Geiger returned to Germany to take up a post as director of the new Laboratory for Radioactivity at the Physikalisch-Technische Reichsanstalt in Berlin, where he invented an instrument for measuring not only alpha particles but **beta rays** and other types of **radiation** as well.

Geiger's research was broadened the following year with the arrival at the laboratory of **James Chadwick** and Walther Bothe, two distinguished nuclear physicists. With the latter, Geiger formed what would be a long and fruitful professional association, investigating various aspects of radioactive particles together. However, their work was interrupted by the outbreak of World War I. Enlisted with the German troops, Geiger fought as an artillery officer opposite many of his old colleagues from Manchester including Marsden and **Henry G. J. Moseley** from 1914 to 1918. The years spent crouching in trenches on the front lines left Geiger with painful rheumatism. With the war over, Geiger resumed his post at the Reichsanstalt, where he continued his work with Bothe. In 1920, Geiger married Elisabeth Heffter, with whom he had three sons.

Geiger moved from the Reichsanstalt in 1925 to become professor of physics at the University of Kiel. His responsibilities included teaching students and guiding a sizable research team. He also found time to develop, with Walther Müller, the instrument with which his name is most often associated: the Geiger-Müller counter, commonly referred to as the Geiger counter. Electrically detecting and counting alpha particles, the counter can locate a speeding particle within about one centimeter in **space** and to within a hundred-millionth second in time. It consists of a small metal container with an electrically insulated wire at its heart to which a potential of about 1,000 volts is applied. In 1925, Geiger used his counter to confirm the **Compton effect**, that is, the scattering of **x rays**, which settled the existence of light quantum, or packets of **energy**.

Geiger left Kiel for the University of Tubingen in October 1929 to serve as professor of physics and director of research at its physics institute. Installed at the institute, Geiger worked tirelessly to increase the Geiger counter's speed and sensitivity. As a result of his efforts, he was able to discover simultaneous bursts of radiation called cosmic-ray showers, and concentrated on their study for the remainder of his career.

Geiger returned to Berlin in 1936 upon being offered the chair of physics at the Technische Hochschule. His upgrading of the counter and his work on **cosmic rays** continued. He was also busy leading a team of nuclear physicists researching artificial radioactivity and the by-products of **nuclear fission** (the splitting of the atom's nucleus). Also in 1936 Geiger took over editorship of the journal *Zeitschrift fur Physik,* a post he maintained until his death. It was at this time that Geiger also made a rare excursion into politics, prompted by the rise to power in Germany of Adolf Hitler's National Socialist Party. The Nazis sought to harness physics to their ends and engage the country's scientists in work that would benefit the Third Reich. Geiger and many other prominent physicists were appalled by the specter of political interference in their work by the Nazis. Together with **Werner Karl Heisenberg** and Max Wien, Geiger composed a position paper representing the views of most physicists, whether theoretical, experimental, or technical. As these men were politically conservative, their decision to oppose the National Socialists was taken seriously, and 75 of Germany's most notable physicists put their names to the Heisenberg-Wien-Geiger Memorandum. It was presented to the Reich Education Ministry in late 1936.

The document lamented the state of physics in Germany, claiming that there were too few up-and-coming physicists and that students were shying away from the subject because of attacks on theoretical physics in the newspapers by National Socialists. Theoretical and experimental physics went hand in hand, it continued, and attacks on either branch should cease. The memorandum seemed to put a stop to attacks on theoretical physics, in the short term at least. It also illustrated how seriously Geiger and his associates took the threat to their work from the Nazis.

Geiger continued working at the Technische Hochschule through the war, although toward the latter part he was increasingly absent, confined to bed with rheumatism. In 1938 Geiger was awarded the Hughes Medal from the Royal Academy of Science and the Dudell Medal from the London Physics Society. He had only just started to show signs of improvement in his health when his home near Babelsberg was occupied in June 1945. Suffering badly, Geiger was forced to flee and seek refuge in Potsdam, where he died on September 24, 1945.

GELL-MANN, MURRAY (1929-   )
American physicist

Murray Gell-Mann, a particle physicist, helped to develop the Stanford model, which describes the behavior of **subatomic particles** and their forces. He contributed some unique ways of categorizing and naming elementary particles including strangeness, the eight-fold way, and the quark. Though a theoretical physicist, Gell-Mann has made many important suggestions to experimental physicists for detecting the particles that he predicted in theory. For his work in classifying these particles and their interactions, Gell-Mann was awarded the Nobel Prize for physics in 1969.

Murray Gell-Mann was born on September 15, 1929, in New York City. His father, Arthur Gell-Mann, was the proprietor of a language school and very learned in the areas of **astronomy**, archeology, and mathematics. Like his father Murray Gell-Mann had many varied interests including archeology, linguistics, and ornithology. In an interview for *Omni* magazine he said that when he asked his father about studying one of them for a career, his father replied, "You'll starve." So he accepted his father's alternative: physics. Gell-Mann's first experiences with the subject were not positive. He studied physics in a high school for the intellectually gifted, but found it dull. It was, he said in the *Omni* interview "the only course I'd ever done badly in." He found language, mathematics, and history more to his liking. Nevertheless, he entered Yale University at the age of 15 and graduated in 1948 with a B.S. in physics. He then entered the Massachusetts Institute of Technology, and at 21 earned his Ph.D., again in physics.

Taming the Zoo

When Gell-Mann entered academia in 1951, **particle accelerators** were beginning to produce a variety of new subatomic particles. The number of particles grew to over 200, and came to be known as a "zoo." Nuclear physicists were troubled, because the sheer number of particles seemed to contradict the order and simplicity with which they preferred to view the **universe**. Moreover, these particles exhibited strange behavior, and did not decay at the rates expected. In 1953, while he was associate professor at the University of Chicago's Institute for Nuclear Studies, Gell-Mann published a paper explaining the unexpected behavior of these particles. Gell-Mann's principle, called *strangeness,* predicted numerous events in the decay of these particles; it also predicted the existence of a new particle called Xi zero. In 1959, scientists at the University of California found evidence for the exis-

Murray Gell-Mann, 1969. *(AP/Wide World Photos. Reproduced by permission.)*

tence of this particle in photographs of an **accelerator** experiment.

Gell-Mann's principle of strangeness described the electrical charge and spin of particles that resulted from collisions in large accelerators or **atom** smashers, particles that existed for only a few millionths of a second or less. With this principle, Gell-Mann classified particles by assigning appropriate strangeness numbers; he then formulated equations that represented the relationships between these particles and their interactions. The results of the equations were then compared to experimental results. With this system, Gell-Mann was able to predict which particles could be found experimentally and which could not, giving scientists a means of reducing the great variety of particles that confronted them.

Following the Eight-fold Way

In 1955 Gell-Mann became an associate professor of physics at the California Institute of Technology, and in the following year became full professor. During this time, Gell-Mann began to develop his notion of strangeness into a scheme for the orderly arrangement of newly discovered sub-atomic particles. He assigned to the particles eight **quantum numbers**, representing their various characteristics. Gell-Mann facetiously named his system the "eight-fold way" after the Buddhist list of eight virtues for achieving enlightenment.

With this system Gell-Mann formalized the patterns he saw among the particles' **mass**, charge, and spin.

Gell-Mann arranged the eight-fold way much like Dmitry Mendeleev had the **periodic table**, charting the known particles and leaving spaces for those he thought would be discovered later. In so doing, Gell-Mann predicted particles and their characteristics long before they were found through experimentation. Gell-Mann's subsequent recommendations for constructing larger, more powerful accelerators in order to find these predicted particles led to the discovery of the Eta particle in 1962, the Phi particle in 1963, and the Omega-minus particle in 1964. The Omega-minus particle turned out to have a mass of almost exactly the value that had been predicted for it in theory.

"Three Quarks for Muster Mark"

As Gell-Mann continued to examine subatomic particles, he began to see patterns in their characteristics and behavior; for example, he found that particles exhibited properties that were based on the number 3 and arrangements that formed triangles. Eventually, Gell-Mann suggested that the number 3 kept appearing because these particles were themselves composed of three sub-particles, which he proposed were the ultimate building blocks of **matter**.

Because of his interest in linguistics, Gell-Mann frequently played with words. In naming the new sub-particles he initially had thought that the **sound** *kwork* seemed suitable. He decided on the spelling *quark* while reading the phrase in James Joyce's novel *Finnegans Wake,* "three **quarks** for muster mark." Gell-Mann's playful naming of serious scientific concepts, however, did have a practical intent as well, for he felt that to continue the tradition of naming particles after Greek letters and roots was pretentious and that often these new terms quickly outgrew their validity. (**Proton**, from the Greek meaning "first," for example, is no longer considered the first particle in any sense, and atom, from the Greek "indivisible," has been subdivided several times since it was named.)

In keeping with his fondness for odd words, Gell-Mann named his three quarks *up, down,* and *strange.* With these curiously named entities he was able to explain the makeup of the over 200 subatomic particles confronting physicists at the time. Then, in two separate 1974 **antimatter** experiments, a particle emerged that the three-quark theory could not account for. Scientists suggested the existence of a fourth quark. Eventually, physicists discovered evidence of two additional quarks. Consistent with Gell-Mann's whimsical scheme for naming particles, the three new quarks were dubbed *charm, beauty,* and *truth.* All six became known as the "flavors" of quarks.

Gell-Mann's research on quarks was wholly theoretical, entirely based on mathematical formulations; however, physicists at various particle accelerators observed experimental results that they could account for only if Gell-Mann's quark theory was valid. No quarks have been isolated from a nucleus—and in fact Gell-Mann himself said in an interview for *Omni* magazine, "I believe that the confinement is absolute"—

but physicists in the 1970s found indirect evidence in the laboratory for the existence of quarks.

A Physicist's Mind at Work

Gell-Mann considered himself a theoretical physicist, and as such wrote equations based on measurable quantities that described particles and their interactions. Often he compared the results of his equations with experimental results in an attempt to predict new experimental figures. Gell-Mann felt that his brand of physics was fundamental to all other sciences in that their laws were ultimately based on the laws that elementary particles obey. In Theodore Berland's *The Scientific Life,* Gell-Mann said, "It's such a marvelous thing to contemplate a universe made up of all these absolutely identical— wherever you find them—particles."

Gell-Mann was interested in almost everything that confronted him, and his mind always seemed to have something simmering on the back burner. While he was in the middle of explaining another scientist's paper to a group in Princeton, he discovered an error in his work on the chalkboard; he stepped back from the board to sit down, and in that moment his theory of strangeness came to him. At first he felt that his strangeness principle was too silly to share with his colleagues; in 1953, however, he heard other physicists talking about a similar notion, which he felt was wrong, so he put his strangeness theory on paper and published it.

Gell-Mann had such faith in his **scientific method** that he felt it could apply to social contexts as well. For example, when he heard about the activities of the politically conservative John Birch Society, he researched the group; after learning of its viewpoints, Gell-Mann believed he could predict how this group would act in future circumstances, in much the same way that he predicted the future behavior of elementary particles.

A Man of Worldly Interests

Gell-Mann rose to prominence in the field of **particle physics** during the Cold War, drawing the U.S. military and the defense industries to seek out his expertise. Gell-Mann did some work for the Rand Corporation on antisubmarine and anti-ICBM projects, and served as a consultant to the Institute for Defense Analysis for issues like the detection of nuclear test detonations. In addition to his 1969 Nobel Prize, Gell-Mann has been honored with the American Physical Society's Heineman Prize, the U.S. Atomic Energy Commission's Ernest Orlando Lawrence Memorial Award for Physics, the Franklin Institute's Franklin Medal, and the National Academy of Sciences' John J. Carty Medal; Yale bestowed upon him an honorary degree in 1959.

As someone who is interested in almost everything, Gell-Mann has developed strong opinions about almost everything, as well. Regarding education, for example, with which he has had much contact, he feels that the traditional lecture approach is ineffective. He believes that the education system should model itself after the old practice of apprenticeship, where a young person would pair up with an older, experienced mentor. Gell-Mann feels that this approach to education would allow the teacher more time to discuss problems, answer questions, and inspire students.

When he isn't scratching equations on a chalkboard, Gell-Mann travels in search of adventure, seeking out the diversity of the world wherever he goes. Underneath this love of diversity is Gell-Mann's respect for the individual and his concern for nature; moreover, he feels that the lot of both man and nature could improve if, as he said in Berland's *The Scientific Life,* "there were fewer people and more wild animals."

GENERATORS

An electric generator is a device that is constructed to transform mechanical **energy** into electrical energy. Accordingly, a generator performs the opposite **energy transformations** as does an electrical motor (i.e., electrical to mechanical energy transformations). In some cases the same device can act, under different operating conditions, as either a generator or as an electric motor.

The most common generator design is based upon the principles of **electromagnetic induction**, a process by which a **voltage** (electromagnetic **force** or emf) is generated or induced in a circuit by changing **magnetic flux**. The quantitative relationship between the generated or induced voltage and the magnetic flux is described by **Faraday's law**.

In the early part of the nineteenth century Danish scientist **Hans Christian Ørsted** demonstrated a relationship between **electricity** and **magnetism** by placing a wire near a magnetic compass. When Ørsted applied an **electric current** to the wire, the magnetic compass needle moved. This movement reflected a change in the magnetic field. Advancing Ørsted's **work**, English physicist and chemist **Michael Faraday** developed a rotator (later to be known as an electric motor). Although Faraday's initial apparatus design contained a conductive (i.e., electrical current carrying) wire rotating around a magnet, Faraday found that he could also generate electrical current by rotating magnets about an electrically conductive wire. Accordingly, Faraday's law states that changing magnetic flux induces an emf in a coil of wire that is electrically conductive.

Whenever a magnetic field changes it creates an electric field that exerts a force upon the charge-carrying electrons in a system. A mere presence of a magnetic field is not enough to cause the generation of current. An induced voltage occurs only when there is a change in magnetic flux. Changes in the magnetic flux are usually caused by changes in the magnetic field, changes in coil orientation (spinning a coil), or changes in the shape of a coil. Rapid rates of change in magnetic flux result in the production of higher voltages.

Alternating current (AC) generators produce AC electricity through induction of an oscillating emf (**electromotive force**) that is produced by maintaining a constant rate of spin for a coil placed in a magnetic field. In the United States, standard AC electricity (e.g., that found in most electrical plug receptacles has a **frequency** of 60 Hz. Generators can also produce direct (DC) electrical current. Unlike its AC generator counterpart, the polarity of the voltage produced by a DC generator remains constant. DC generators supply an electrical output with unidirectional voltage and current utilizing the essentially the same principles of operation as synchronous generators. A voltage induced in coils by the changing magnetic field results in a current flow. DC generators are constructed with multiple coil loops in order to supply a steady voltage.

In general, all generator designs vary in the composition of coils (i.e., whether they are made of ferromagnetic materials), the number of coils, the total area or number of turns of each coil, the distance between coils, and the magnetic field strength. The orientation of the coils relative to one another is also an important design consideration because the emf induced in a coil is directly proportional to the production of current and magnetic fields in preceding coils. Through the principle of mutual inductance, as current is produced or passes through a coil a magnetic field is produced and the magnetic flux will (depending on the distance between the coils) affect the next coil.

Based on Faraday's law, the general design of modern electrical generators reflects simple but fundamental physical principles. For example, at hydroelectric plants the most common method of generating electricity is to keep the coil stationary and to spin permanent magnets (providing the magnetic field and flux) around the coil. The energy of falling water is used to spin permanent magnets around a fixed loop, producing AC **power**.

Thermoelectric generators (TEG) are designed to produce power by converting thermal (**heat**) energy into electricity. TEGs use semiconductor thermoelectric elements and **temperature** differences in thermopiles (i.e., hot and cold sides) to generate electricity. The principles of thermoelectric generation are used in some nuclear-fueled generators to produce electrical power by allowing the **radiation** from radioisotope decay to heat the hot side of a thermopile.

Other types of generators (e.g., motor generators) are designed to convert alternating current to direct current.

GEOCENTRIC THEORY

The geocentric theory of the **universe** describes the universe with Earth at its center and puts the other celestial bodies in circular orbits around it. The usual ordering placed the **Moon**, Mercury, Venus, the **Sun**, Mars, Jupiter, Saturn, and the fixed **stars** on a series of concentric spheres further and further away from the stationary Earth.

The heavens revolved around Earth once every 24 hours. While an outer sphere carried the Sun around Earth every day, it rotated on an inner sphere about an axis attached to the outer one. This second effect accounted for the Sun's yearly transit along the ecliptic, a plane held at a 23.5° angle with respect to the celestial equator (taking the North Star as the North Pole). Because the Sun's tilted orbit placed it at different angles relative to the equator at various times of the year, this model provided a natural explanation for the origin of seasons. Three more spheres are required to make the Sun's **motion** consistent with the dates on which the solstices and equinoxes fall. The motion of the Moon and the **planets** are treated in a similar fashion, and there were 27 nested spheres in all. All the celestial bodies traveled around Earth with uniform **velocity**, and the spheres were all transparent since the stars could be seen on the outermost one. This geocentric model, although shaped by the philosophical precepts of **Plato**, was largely designed by Eudoxus of Cnidus (c.400-353 B.C.).

Aristotle, another pupil of Plato, devised his own geocentric theory with 55 crystalline spheres. (The additional spheres were necessary to refine Eudoxus's model so that it better matched the observational data.) He also introduced an additional sphere belonging to the Prime Mover, which dwelt beyond the sphere of the fixed stars. The rotation of this sphere imparted motion to the other spheres.

Centuries later, **Ptolemy** dispensed with the crystalline spheres of Aristotle and introduced such modifications as **epicycles** and deferents to explain the retrograde motion of the outer planets. His classic compendium of **astronomy**, *Almagest* (*The Greatest*), served as the authoritative textbook on celestial **mechanics** for over a thousand years. Though there were in fact several geocentric theories, when historians or philosophers write about geocentrism, it is usually Ptolemy's system they infer.

Nicholas Copernicus supplanted the geocentric theory with a heliocentric model, which was much like one proposed by Aristarchus of Samos (310-c.230 B.C.). This new model, coupled with the elliptic orbits postulated by **Johannes Kepler**, explained the observations of the night sky without resorting to more and more ad hoc hypotheses.

GEOMETRIC OPTICS

Geometric **optics** is the study of the path of **light** rays through systems of **lenses**, mirrors, and other media. English physicist and mathematician Sir **Isaac Newto**n's *Opticks* published in 1704 formalized the description of image-formation in lenses and mirrors, providing the first thorough study of geometric optics.

The law of reflection states that the angle of an incident beam is equal to its angle of reflection. A medium that allows the passage of light without a change in **velocity** of light is said to have index of refraction (n) where the velocity of light propagation is determined as the velocity of light in a **vacuum** (c) divided by the index of refraction (n).

A ray of light that propagates across the interface (boundary) of two media (materials) obeys Snell's law of refraction. The media of incoming and outgoing rays, having indices of refraction n_1 and n_2, respectively, refract (bend) light rays at their interface according to Snell's law ($n_1 \sin \theta_1 = n_2 \sin \theta_2$), where θ_1 and θ_2 are the incoming and outgoing angles that rays of light make with lines perpendicular (normal) to the interface.

The principle of (time) reversibility is the observation, generally true in optics and for all electromagnetic fields, that the laws for propagation of light rays are the same when the **direction of time** is reversed. For the laws of refraction and reflection, time reversal simply means that light rays travel in the opposite direction; the form of each law remains the same.

The laws of **image formation** state that the reconvergence of light rays determines the location of an image of the point source. After reconverging, the rays diverge again from the image point, producing an apparent ray source. The image of an extended object is a set of the images of the object's constituent point sources. The location, size, shape, and orientation of images produced by lenses and mirrors is given by the laws of image formation. The classical tools for image formation are mirrors and lenses.

An object at a very large distance forms an image at the focal plane of a lens or mirror. The focal length of a human eye is approximately 1 in (2.5 cm), where the focal plane is located on or in front of the retina. The focal length of the objective (far lens) of a 10 ft (30 m) long **telescope** is 10 feet. Compound lens and mirror systems allow the manipulation of images. A second lens can produce a secondary image of an image produced by a first lens.

A lens or mirror with focal length (f) placed a distance (s) from an object will form an image at distance s' according to the Gaussian lens formula (1/f=1/s+1/s'). Given a medium with a known index of refraction (n), the lensmaker's formula, which can be derived geometrically from Snell's law, allows the design of thin lenses of arbitrary focal length:

$$\frac{1}{f} = (n-1)\left(\frac{1}{R_1} - \frac{1}{R_2}\right)$$

where R_1 and R_2 are the radii of curvature of the lenses spherical surfaces.

Geometrical aberrations are the characteristics of a lens that prevent an ideal image from being formed. The total geometric aberration of a lens may be approximately decomposed into quantitative measurements of five primary geometric aberrations, described by Ludwig van Seidel in the 1850s. Astigmatism occurs when a lens's shape is such that the focal plane is different for perpendicular lines. Distortion causes the magnification of either the center or edges of an image, producing a pincushion or barrel-shaped bending of lines in the image. Curvature of field gives the deviation of the focal surface from a plane (where a true focal plane has zero curvature of field). Coma is the formation of a comet-shaped image for every point on the object. Spherical aberration, which is observed in spherical lenses, comes about when the focal lengths from different regions of a lens are different, preventing the formation of a well-defined focal plane.

In a wide variety of circumstances, the rules of geometric optics give a close approximation to the macroscopic behavior of light. They predict the characteristics of eye lenses and allow the design of telescopes, microscopes, cameras, eyeglasses, and countless other inventions. The principles of geometric optics have been extended to other fields, such as the design of magnetic lenses in **particle accelerator** beam lines that allow experimentalists to focus beams of high-energy particles. Analysis of multiple images of galaxies produced by light refraction by **gravity** (gravitational lensing) in **space** requires consideration of principles of geometric optics, as does the modeling of the propagation of acoustic disturbances in the earth. Geometric optics does not account for the wave nature of light, which causes **interference** and **diffraction**, nor does it explain light's quantum mechanical properties.

See also Optical instruments; Optics

GEOTHERMAL ENERGY

During the seventh century the discovery that the **temperature** of Earth in deep mines exceeded the surface temperature, implied the existence of a source of **thermal energy** deep within Earth. Within the continental crusts, the temperature differential gradient averages about one micro calorie per square centimeter (equivalent to an increase of about 95°F per mile or 33°C per kilometer of increasing depth).

In some areas geothermal **energy** is a viable economic alternative to conventional energy generation. Commercially viable geothermal fields have the same basic structure. The source of **heat** is generally a magmatic intrusion into Earth's crust. The magma intrusion generally measures 1,112-1,652°F (600-900°C), at a depth of 4.3-9.3 m (7-15 km). The bedrock containing the intrusion conducts heat to overlying aquifers (i.e., layers of porous rock such as sandstone that contain significant amounts of water) covered by a dome-shaped layer of impermeable rock such as shale or by an overlying fault thrust that contains the heated water and/or steam. A productive geothermal field generally produces about 20 tons of steam, or several hundred tons of hot water, per hour. Historically, some heavily exploited geothermal fields have had decreasing yields due to a lack of replenishing water in the aquifer, rather than to cooling of the bedrock.

There are three general types of geothermal fields, hot water, wet steam, and dry steam. Hot water fields contain reservoirs of water with temperatures between 140-212°F (60-100°C), and are most suitable for space heating and agricultural applications. For hot water fields to be commercially viable, they must contain a large amount of water with a temperature of at least 60°C and lie within 2,000 meters of the surface.

Wet steam fields contain water under **pressure** and usually measure 100°C. These are the most common commercially exploitable fields. When the water is brought to the surface, some of the water flashes into steam, and the steam may drive turbines that can produce electrical **power**.

Dry steam fields are geologically similar to wet steam fields, except that superheated steam is extracted from the aquifer. Dry steam fields are relatively uncommon.

Because superheated water explosively transforms into steam when exposed to the atmosphere, it is much safer and generally more economical to use geothermal energy to gener-

Geothermal geysers in California, blowing off steam. *(Photo by UPI/Corbis-Bettmann. Reproduced by permission.)*

ate **electricity**, which is much more easily transported. Because of the relatively low temperature of the steam/water, geothermal energy may be converted into electricity with an efficiency of 10-15%, as opposed to 20-25% for coal or oil fired generated electricity.

To be commercially viable, geothermal electrical generation plants must be located near a large source of easily accessible geothermal energy. A further complication in the practical utilization of geothermal energy derives from the corrosive properties of most groundwater and steam. In fact, prior to 1950, metallurgy was not advanced enough to enable the manufacture of steam turbine blades resistant to corrosion. Geothermal energy sources for space heating and agriculture has been used extensively in Iceland, and to some degree Japan, New Zealand, and the former Soviet Union. Other applications include paper manufacturing and water desalination.

While geothermal energy is generally presented as a non-polluting energy source, water from geothermal fields often contains large amounts of hydrogen sulfide and dissolved metals, making its disposal difficult.

See also Boiling; Electric power and energy; Heat of vaporization; Heat transfer; Static pressure and depth

GERMER, LESTER HALBERT (1896-1971)

American physicist

Lester Germer, along with his colleague American physicist **Clinton Davisson**, conducted an experiment in 1927 that first demonstrated the **wave** properties of an **electron**. Germer's experiment confirmed an earlier hypothesis suggested by French physicist **Louis Victor de Broglie**. Germer's work came a critical time during the development of **quantum theory**. The experiments of Germer and Davisson helped validate the mathematically complex atomic models proposed by de Broglie and Austrian physicist **Erwin Schrödinger** that predicted a particle and wave duality for electrons (i.e., that electrons should show properties of both particles and electromagnetic waves).

Born in Chicago, Illinois, Germer exhibited precocious mathematical ability as a child, and later became a graduate student under the tutelage of Davisson at Columbia University. While working at the Bell Laboratories in New York City in 1927, Germer and Davisson experimented with directing a beam of electrons of known **energy** onto the surface of polycrystalline nickel. Germer and Davisson measured the various angles at which electrons bounced off of the surface. When measured, the scattered electrons showed peaks precisely where predicted by wave theory (i.e., an intense reflected beam was observed in accord with the Bragg condition for constructive **interference**). In a stroke of serendipity the experiment was altered when air accidentally entered the tube that contained target nickel. An oxide film formed on the surface of the nickel crystals and resulted in the production of crystalline structure consistent with wavelike electron defraction patterns. Germer and Davisson correctly asserted that their experiments argued for a **particle-wave duality** in electrons, a fundamental postulate of modern **quantum mechanics**.

Scientists use the wavelike properties of electrons discovered by Germer in a number of applications, including the **electron microscope**.

See also Bohr model; de Broglie wavelength; Diffraction

GIBBS, JOSIAH WILLARD (1839-1903)

American physicist

In the mid-1800s, while European scientists enjoyed recognition for their remarkable discoveries in chemistry, physics, and biology, American scientists remained almost completely unknown. However, beginning in 1876, the British physicist **James Clerk Maxwell** began heralding the work of a heretofore anonymous American, Willard Gibbs. Though acceptance of his theories came slowly, Gibbs would eventually be considered one of the greatest theoretical physicists ever.

Gibbs was born in New Haven, Connecticut, to a family of well-known academics. His father was a professor of sacred literature at Yale University, and it was a foregone conclusion that he would attend that school. Gibbs received his bachelor's degree in 1858, and in 1863 he became the first American to receive a Ph.D. in engineering. Gibbs realized that the opportunities for scientific advancement in the United States were slim, and after three years as a tutor at Yale, he traveled to Europe to continue his education.

It was in the lecture halls of the Sorbonne, the University of Berlin, and the University of Heidelberg that the foundation for Gibbs's future work was built. He studied under **Gustav Kirchhoff**, **Hermann von Helmholtz**, and Joseph Liouville (1809-1882), becoming more and more fascinated with the field of theoretical physics. Upon his return to the United States in 1871 he was offered a professorship at Yale—without salary. Still, he accepted the position, moved back into his childhood home, lived off a generous inheritance, and never again left the United States.

During the years from 1876 to 1878, Gibbs contributed groundbreaking essays regularly to the *Transactions of the Connecticut Academy of Sciences*. He discussed the principles of **thermodynamics**, applying them to the complex processes involved in chemical reactions. His discovered the concept of chemical potential, or the "fuel" that makes chemical reactions work.

Also in these essays were the beginnings of Gibbs's theories of phases of **matter**. He considered each state of matter a *phase*, and each substance a *component*. Thus, a system of ice and water would be one component in two phases, while solid and dissolved sugar in water would be two components in two phases. Gibbs took all of the variables involved in a chemical reaction—temperature, **pressure**, **energy**, volume, and entropy—and included them in one simple equation. By plugging some of the variables into the equation the rest could be easily deduced. This equation has come to be known as the Phase rule.

Over the course of the next 50 years, Gibbs's work initiated unprecedented insight into the nature of chemical reactions. Scientists would write more than 11,000 pages of discussion of the Phase rule, though Gibbs wrote only four pages to introduce it. Before 1950, four Nobel Prizes would be awarded for research stemming directly from Gibbs's work.

However, the monumental importance of Gibbs's findings was not immediately recognized. First, the *Transactions of the Connecticut Academy of Sciences* in which his essays were published was not well read even in America, let alone Europe. Second, when his publications *were* read, they were considered too mathematically complex for most chemists and too involved in chemistry for many mathematicians. It was not until Maxwell advocated Gibbs's work that European scientists began to understand its ramifications, and even then it was not until the 1890s that his ideas gained general acceptance. His memoirs were translated into German by Friedrich Wilhelm Ostwald and into French by Henry-Louis Le Châtelier. Before his death in 1903 he was awarded the Copley Medal of the British Royal Society, the most prestigious international science award.

GILBERT, WALTER (1932-   )

American molecular biologist

Walter Gilbert is a molecular biologist who shared the 1980 Nobel Prize in chemistry for his discovery of how to sequence, or chemically describe, deoxyribonucleic acid (DNA) molecules. Gilbert also identified repressor molecules, which modify or repress the activity of certain genes, and collaborated with Noble laureate biologist James Watson in his efforts to isolate messenger ribonucleic acid (RNA). Later in his career, Gilbert helped form and was chief executive officer of the biotechnology firm Biogen, and became a moving force in the medical research project known as the human genome project.

Gilbert was born in Cambridge, Massachusetts, on March 21, 1932. His father, Richard V. Gilbert, was an economist at Harvard University, and his mother, Emma Cohen Gilbert, was a child psychologist who provided her children's early education at home. In 1939 the family moved to Washington, D.C., where Gilbert initially performed poorly in school. He did, however, show a great deal of interest in science. He was fascinated by **astronomy**, and at age 12 he ground his own glass for telescopes; he also "nearly blew himself up brewing hydrogen in the pantry," writes Anthony Liversidge in *Omni*. As a senior at Sidwell Friends High School, he would go to the Library of Congress to expand his knowledge of **nuclear physics**.

In 1949, Gilbert entered Harvard University, where he majored in chemistry and physics, earning his B.A. summa cum laude in 1953. Gilbert remained at Harvard for a master's degree in physics, which he received in 1954. He went to Cambridge University for doctoral work in theoretical physics, studying under the physicist **Abdus Salam**. His doctoral dissertation focused on mathematical formulae that could predict the behavior of elementary particles in so-called **scattering** experiments. He received his Ph.D. in mathematics in 1957. Gilbert returned to Harvard as a National Science postdoctoral fellow in physics, and he gained an appointment as assistant professor in 1959.

Moves from Physics to Molecular Biology

While at Cambridge, Gilbert had met biologists James Watson and Francis Crick; just a few years before, these two men had established the structure of DNA and constructed a three-dimensional model of it. Their work had launched a new field of science called molecular biology, and when Watson moved to Harvard in 1960 he and Gilbert met again. Watson discussed with Gilbert his interest in isolating messenger RNA. This is the substance believed to be responsible for transmitting information from DNA to ribosomes, which are the cellular structures in which protein synthesis takes place. At Watson's invitation, Gilbert joined him and his colleagues to work on this project. This collaboration with Watson began Gilbert's move into molecular biology. He became convinced that his future lay in this field, and he made up for a lack of formal training in biochemistry through hard work. Within a few years he was publishing articles on molecular biology. He was made a tenured associate professor of **biophysics** at Harvard in 1964, and he became a full professor of biochemistry in 1968. In 1972, Harvard named him the American Cancer Society Professor of Molecular Biology.

In the middle of the 1960s, Gilbert began research on how genes are activated within cells. This was a problem that had been introduced by the French geneticists François Jacob and Jacques-Lucien Monod; DNA should, in theory, encode proteins continually, and they wondered what kept the genes from being activated. If cells contain some sort of element that in effect represses some genes, this would explain in part how cells perform different functions even though each cell contains the same complement of genetic instructions. Gilbert set out to determine whether actual "repressor" substances exist within each cell.

Repressor Molecules and Sequencing DNA

Working with the *Escherichia coli* bacterium, Gilbert attempted to find what he called the lac repressor. *E. coli* manufactures the enzyme betagalactosidase when the milk sugar lactose is present, and he hypothesized that the gene responsible for producing the enzyme was repressed by a substance that would only detach itself from the DNA molecule in the presence of lactose. If this hypothesis could be proven, it would confirm the existence of repressors. In 1966, Gilbert and his colleagues added radioactive lactose-like molecules to

a concentration of the bacteria as a means of tracing any potential lac repressor activity. As he had hoped, the lac repressor bonded with the radioactive material. By 1970, he was able to determine the precise region of DNA (called the lac operator) to which the repressor bonds in the absence of lactose.

The next phase of Gilbert's research focused on sequencing DNA. His aim was to identify and describe chemically any strand of DNA. Working with graduate student Allan Maxam, he began to sequence parts of the DNA strand. It was known that the molecules could be "broken" at specific chemical junctures by using certain chemical substances, and a colleague introduced Gilbert and Maxam to a procedure that broke DNA molecules into fragments that were easier to describe. After breaking radioactively labeled DNA into fragments, Gilbert and Maxam worked to separate them. They used a technique known as gel **electrophoresis**, in which an **electric current** causes the fragments to pass through a gel substance. Upon exposure to x-ray film, the fragments can be read and the chemical code of DNA can be identified. Working independently, the British scientist Frederick Sanger developed a similar sequencing technique. In recognition of both their contributions, Gilbert and Sanger were awarded half the 1980 Nobel Prize in chemistry. The other half went to the American biochemist Paul Berg for his work with recombinant DNA, more commonly referred to as "gene splicing."

From Researcher to Business Executive

With the breakthroughs that were being made by Gilbert, Sanger, Berg, and others, the concept of applying technological principles to biology came of age in the 1970s. The idea of being able to alter the genetic composition of a cell, for example, opened up such possibilities as curing or even eradicating many diseases. The possibilities in biotechnology intrigued not only scientists but the business community. Here was a concept that could be both revolutionary and lucrative—a company that held the patent to a definitive cure for cancer, for example, could become quite wealthy and powerful. Business leaders began to approach scientists, Gilbert among them. Most scientists were skeptical at first, but in 1978 Gilbert met with a group of venture capitalists who wanted to start a biotechnology firm. After receiving assurances that they would have considerable control over research and development, he and other scientists formed Biogen N.V., with Gilbert as the chairman of the scientific board of directors. Gilbert was so convinced of Biogen's potential that he left Harvard in 1981 to become the company's chief executive officer.

Despite widespread belief in the company's potential, Biogen had some difficult years in the beginning. After four years, it was still unprofitable and Gilbert had become increasingly disillusioned with business. He found the differences between the business world and the scientific community difficult to reconcile. The science of creating new products is vastly different from the business of bringing them to market. Scientists need to be patient because their breakthroughs might take years to obtain, but sound business practice dictates cutting one's losses when a project fails to produce after a reasonable time, and these differences led to conflicts with others at Biogen. Gilbert also found it time-consuming and expensive to run a company (although he personally profited from the venture), and in late 1984 he resigned his position at Biogen, while maintaining some involvement with the firm. In 1985, he was named H. H. Timken Professor of Science in Harvard's cellular and developmental biology department. In 1987 he became chairman of the department and Carl M. Loeb University Professor.

The Human Genome Project

Gilbert resumed his research, unhindered by the responsibilities of running a business. But he soon became interested in an undertaking that was bigger than Biogen: the human genome project. This project, wrote Robert Kanigel in the *New York Times Magazine,* "would reveal the precise biochemical makeup of the entire genetic material, or genome, of a human being.... It would grant insight into human biology previously held only by God." The plan was to create a map or library of human DNA. Such information would help researchers not only find cures for diseases but also identify potentially harmful gene mutations. Gilbert spoke out enthusiastically in favor of the genome project, and along with many others he encouraged Congress to support it with federal funds. Frustrated by the political process and believing that the project would be damaged by the bureaucracy federal participation would impose, he tried a different approach. In 1987 he announced plans to create his own company, which would sequence DNA, copyright the information, and sell it. Although he failed to get adequate backing for that project, he did win a two million dollar annual grant from the U.S. government to conduct his work at Harvard under the auspices of the National Institutes of Health.

In an interview with *Omni,* Gilbert has explained what he sees as some of the benefits of a complete genetic map of the human being: "The differences between people are what the genetic map [is] about. That knowledge will yield medicine tailored to the individual. One will first identify obvious genetic defects like cystic fibrosis. The next round of genetic mapping will show us clusters of genes for common diseases from arthritis to schizophrenia. We will be able to predict the side effects of those drugs and tailor the right dose to each person." Gilbert estimated that the project would take at least 10 to 20 years to complete.

Gilbert has also expressed concern as a researcher, arguing in favor of what he calls a "paradigm shift in biology." As scientific techniques are perfected, he contends, new scientists should be able to concentrate on new research, not repeating old research. Writing in *Nature,* he has noted that "in 1970, each of my graduate students had to make restriction enzymes in order to work with DNA molecules; by 1976 the enzymes were all purchased and today no graduate student knows how to make them." While it is important for scientists to understand what they are doing and why, he continues, "this is not the meaning of their education. Their doctorates should be testimonials that they [have] solved a novel problem, and in so doing [have] learned the general ability to find whatever new or old techniques were needed." As more and more of the technological problems of molecular biology are solved, he believes that biological research will begin with theoretical rather than experimental work.

In addition to the Nobel Prize, Gilbert shared with Sanger the Albert Lasker Basic Medical Research Award in 1979. He also won the Louisa Horwitz Gross Prize from Columbia University in 1979, and the Herbert A. Sober Memorial Award of the American Society of Biological Chemists in 1980. His memberships include the American Academy of Arts and Sciences, the National Academy of Sciences, and the British Royal Society.

Gilbert has been married to Celia Stone since 1953; the couple have a son and a daughter. Although his career has sometimes seemed almost as complex as the substances he studies, he is widely respected as a scientist. Younger scientists have called him "intimidating," but despite his successes he remains a committed teacher and researcher.

GLASER, DONALD (1926-   )
American physicist

The work for which Donald Glaser is best known, his bubble chamber invention for tracking the movement of high-energy particles, is said to have begun over a glass of beer. In the early 1950s, while teaching physics at the University of Michigan, Glaser followed a hunch that bubbles rising from a glass of beer might provide a clue for detecting high-energy **radiation**. Although his first attempt to prove this hypothesis, using beer, soda water, and ginger ale, failed, he kept working. In 1953 he created a small bubble chamber filled with superheated **ether** that was successful in capturing the trail of bubbles left by nuclear particles as they passed through the liquid. The bubble chamber invention won Glaser the 1960 Nobel Prize in physics and was a vital step in understanding atomic function. It also enabled the discovery of new atomic particles, such as the rho and omega minus particles, at the same time advancing visualization of charged-particle interactions, and furthering the study of particle **mass**, lifetime, and decay modes.

Donald Arthur Glaser was born in Cleveland, Ohio, on September 21, 1926, to William and Lena Glaser. Glaser's parents had come to the United States from Russia. His father operated a wholesale sundries business in Cleveland and Glaser attended elementary and secondary schools there. As a child he was given violin and viola lessons. Later he studied composition at the Cleveland Institute of Music. An accomplished musician, he became a member of a local symphony orchestra at age 16. Glaser remained in Cleveland for his undergraduate education, entering Case Institute of Technology (now Case Western Reserve University) to study mathematics and physics. He completed his graduate course work at the California Institute of Technology and received his Ph.D. in 1950, a year after he accepted a position as instructor at the University of Michigan. He remained at Michigan until 1959, and was made full professor there in 1957 at the age of 31. He left Michigan to accept a visiting professorship at the

University of California at Berkeley. That position was made permanent, and Glaser was to remain at Berkeley for the rest of his career, except for brief periods away on fellowships.

Bubble Chamber Fills a Gap

Other physicists before Glaser had attempted to make nuclear particles visible. The 1927 Nobel Prize was given to **C. T. R. Wilson**, a British scientist, for his **cloud chamber** method. In 1950 **Cecil F. Powell** received that honor for an emulsion method. But both these methods, while effective for elementary particle study, became inadequate with the advent of high-energy **particle acceleration** machines, such as that built at Berkeley in the late 1950s. These **accelerators** had capacities 1,000 times greater than those used for the cloud chamber and emulsion techniques.

Continuing to refine the bubble chamber method he had developed at the University of Michigan, Glaser next tried superheated liquids of higher **density**, such as liquid hydrogen and xenon gas. Experiments with these liquids provided glimpses of subatomic functions never before seen, and made possible the tracking of neutral as well as charged particles. As Glaser continued to refine his bubble chamber methods, information gathering increased a thousandfold. For example, in his first two years at Berkeley, Glaser was able to collect almost half a million tracking photographs. The machine that made this possible was Berkeley's new, two-million-dollar bubble chamber. Built by **Luis W. Alvarez**, the size of this bubble chamber was 6 ft (1.8 m) long, quite a jump from the 0.5-1.0 in (1.3-2.5 cm) **bubble chambers** Glaser had used earlier. The Berkeley bubble chamber was capable of measuring particle tracks at a rate of every 14 seconds, shooting three photographs of each instance. *Science* magazine reported that the Nobel Prize committee credited Glaser's bubble chamber invention with filling "the wide gap in range" left by the cloud chamber and the emulsion methods. Kai M. Siegbahn, speaking for the Royal Swedish Academy of Sciences, said: "Several other scientists also left important contributions to the practical shaping of different types of bubble chambers, but Glaser is the one who made the really fundamental contribution."

In the years after receiving the Nobel Prize, Glaser extended his knowledge of physics to the field of molecular biology. He studied microbiology at the University of Copenhagen in 1961, then returned to Berkeley, where he did research on bacterial evolution, regulation of cell growth, and the causes of cancer and genetic mutation. Using photoanalyzing equipment developed for the bubble chamber, Glaser was able to identify bacterial species through computer scanning. He was made professor of physics and molecular biology at Berkeley in 1964. Glaser retained his lifelong love of music, often playing with local chamber music groups. His other pastimes have included mountain climbing, tennis, and sailing. During his career, he has held membership in the American Physical Society and the National Academy of Sciences. In addition to the Nobel Prize in physics, he was awarded the Henry Russel Award of the University of Michigan in 1955, the Charles Vernon Boys Prize of the

Physical Society in London in 1958, the American Physical Society Prize in 1959, the Gold Medal of the Case Institute of Technology in 1967, and the Alumni Distinguished Service Award from the California Institute of Technology, also in 1967. He married the former Ruth Louise Thompson in 1960. They had two children, a son, William, and a daughter, Louise.

GLASHOW, SHELDON LEE (1932-)
American physicist

Sheldon Lee Glashow contributed to the independent work of **Steven Weinberg** and **Abdus Salam** to develop the **electroweak theory**, which shows how two fundamental forces—the weak and electromagnetic forces—can be viewed as separate manifestations of a single, more basic **force**, termed an **electroweak force**. Among his contributions to the **Weinberg-Salam theory** was his invention of the property of charm for elementary particles. For his research, he shared the 1979 Nobel Prize in physics with Weinberg and Salam.

Glashow was born in New York City on December 5, 1932. His parents, Lewis Gluchovski and Bella (Rubin) Gluchovski, had immigrated to New York from Bobruisk, Russia, to avoid anti-Semitic oppression by the Czarist government. Upon his arrival in the United States, the senior Gluchovski changed his name to Glashow and opened a plumbing business that would become very successful. The Glashows had two other sons, 14 and 18 years older than Sheldon; one became a dentist and the other had a career as a doctor.

Embarks on Career in Particle Physics

Glashow attended one of the nation's most prestigious high schools, the Bronx High School of Science. There he claims to have learned as much about physics from his classmates as he did from his instructors. Among those classmates were future fellow Nobel laureate Weinberg and later Columbia physicist Gerald Feinberg. By the time Glashow had graduated from Bronx High in 1950, he had decided his career: he wanted to be a particle physicist.

In order to pursue this goal, Glashow enrolled at Cornell University in the fall of 1950, choosing it in preference to the Massachusetts Institute of Technology and Princeton. Harvard had denied his application. He later told biographer Arthur Fisher in an interview for the book *A Passion to Know* that he was not very impressed with the faculty at Cornell. "I spent a good deal of time in the poolroom," he explained. "Most successful people in physics made it by going off by themselves and learning what they wanted to."

After receiving his bachelor's degree from Cornell in 1954, Glashow entered Harvard to do his graduate work. There, he studied under Nobel laureate **Julian Schwinger**, who was to become a major influence in Glashow's life. His doctoral thesis, "The Vector Meson in Elementary Particle Decay," was a preliminary effort to combine two of the basic forces of nature, the weak and the electromagnetic forces.

Sheldon L. Glashow. *(Photo courtesy of UPI/Corbis-Bettmann. Reproduced by permission.)*

Awarded Noble Prize for Electroweak Theory

Efforts to find ways of unifying the four basic forces of nature—the strong, weak, electromagnetic, and gravitational forces—go back to the turn of the twentieth century, especially to the work of **Albert Einstein**. These efforts are grounded in the belief among most physicists that these four fundamental forces are not actually distinct from each other, but are somehow four different manifestations of a single basic force.

One problem with this assumption is that the unification of forces is thought to be observable only at energies far greater than those encountered in everyday life. The belief is, for example, that the weak and electromagnetic forces actually merge into a single force only at the very high energies produced within **particle accelerators**. Higher unification may occur only at energies that once existed at the creation of the **universe**.

Development of unification theories has been, therefore, a highly complex, intricate theoretical exercise that may be testable initially only by checks of internal consistency and only later by experimental studies. In the 1960s, Weinberg and Salam devised such a theory, an explanation of the way in which the weak nuclear force and the electromagnetic force could be conceived of as manifestations of a single unified force, the electroweak force. That original theory, although very attractive, dealt only with one class of particles, the **leptons** (**electrons** and **neutrinos**).

Shortly after Weinberg and Salam announced their results, Glashow found a method for extending their theory to other elementary particles, such as **mesons** and **baryons**. In order to do so, he found it necessary to invent a new property for particles, a property he designated as "charm." In the decade following the formulation of the electroweak theory, experimental evidence supporting the theory gradually began to appear. In 1973, for example, researchers for the first time detected a previously unknown phenomenon known as "neutral currents" that had been predicted by the Salam-Weinberg theory. By 1979, support for the theory had become solid enough to justify the Nobel Prize committee's awarding the physics prize that year to the three researchers.

Since winning the Nobel Prize, Glashow has continued to work on unification theories. Now his goal is to find ways of incorporating the strong force into the electroweak force. He carries out that work at Harvard, where he has been on the faculty since 1966, and at Texas A & M University, where he accepted a joint appointment in 1983. Glashow was married to the former Joan Shirley Alexander in 1972. They have three sons, Jason David, Jordan, and Brian Lewis, and one daughter, Rebecca Lee. In addition to the Nobel Prize, Glashow was awarded the J. Robert Oppenheimer Memorial Medal in 1977, the George Ledlie Prize in 1978, and the Castiglione di Silica Prize in 1983.

GLOBAL WARMING

Global warming, as used in the popular context, is a scientifically controversial phenomenon that attributes an increase in the average annual surface **temperature** of Earth to increased atmospheric concentrations of **carbon** dioxide and other gases. Global warming describes only one of several components involved in climate change and specifically refers to a warming of Earth's surface outside of the range of normal fluctuations that have occurred throughout Earth's history.

Climate describes the long-term meteorological conditions or average **weather** for a region. Throughout Earth's history there have been dramatic and cyclic changes in climatic weather patterns corresponding to cycles of glacial advance and retreat that occur on the scale of 100,000 years. Within these larger cycles are shorter duration warming and cooling trends that last from 20,000 to 40,000 years. Scientists estimate that approximately 10,000 years have elapsed since the end of the last ice age and examination of physical and biological processes establishes that since the end of the last ice age there have been fluctuating periods of global warming and cooling.

Measurements made of weather and climate trends during the last decades of the twentieth century raised concern that global temperatures are rising not in response to natural cyclic fluctuations but rather in response to increasing concentrations of atmospheric gases that are critical to the natural and life-enabling **greenhouse effect** (infrared reradiation, mostly from water vapor and clouds, that warms Earth's surface).

Observations collected over the last century indicate that the average land surface temperature increased by 0.8-1°F (0.45-0.6°C). The effects of temperature increase, however, cannot be fully isolated and many meteorological models suggest that such increases temperatures also result in increased precipitation and rising sea levels.

Measurements and estimates of global precipitation indicate that precipitation over the world's landmasses has increased by approximately 1% during twentieth century. Further, as predicted by many global warming models, the increases in precipitation were not uniform. High latitude regions tended to experience greater increases in precipitation while precipitation declined in tropical areas.

Measurements and estimates of sea level show increases of 6-8 in (15-20 cm) during the twentieth century. Geologists and meteorologists estimate that approximately 25% of the sea-level rise resulted from the melting of mountain glaciers. The remainder of the rise can be accounted for by the expansion of ocean water in response to higher atmospheric temperatures.

Many scientists express concern that the measured increases in global temperature are not natural cyclic fluctuations but rather reflect human alteration of the natural phenomena known as the greenhouse effect by increasing concentrations of greenhouse-related atmospheric gases. Estimates of atmospheric greenhouse gases prior to the nineteenth century (extrapolated from measurements involving ice cores) indicate that of the last few million years the concentration of greenhouse gases, remained relatively unchanged prior to the European and American industrial revolutions. During the last two centuries, however, increased emissions from internal combustion engines and the use of certain chemicals have measurably increased concentrations of greenhouse gases that might result in an abnormal amount of global warming.

Although most greenhouse gases occur naturally, the evolution of an industrial civilization has significantly increased levels of these naturally occurring gases. In addition, new gases have been put into the atmosphere that potentiate (i.e., increase) the greenhouse effect. Important greenhouse gases in the modern Earth atmosphere include water vapor and carbon dioxide, methane, nitrous oxides, ozone, halogens (bromine, chlorine, and fluorine), halocarbons, and other trace gases.

The sources of the greenhouse gases are both natural and manmade. For example, ozone is a naturally occurring greenhouse gas found in the atmosphere. Ozone is constantly produced and broken down in natural atmospheric processes. In contrast, halocarbons enter the atmosphere primarily as the result of human use of products such as chlorofluorocarbons (CFCs). Water vapor and carbon dioxide are natural components of respiration, transpiration, evaporation, and decay processes. Carbon dioxide is also a by-product of combustion. Although occurring at lower levels than water vapor or carbon dioxide, methane is also a potent greenhouse gas. Nitrous oxides, enhanced by the use of nitrogen fertilizers, nylon pro-

duction, and the combustion of organic material, including fossil fuels, have also been identified as contributing to strong greenhouse effects.

Alterations in the concentrations of greenhouse gases results in a disruption of **equilibrium** processes. Both increased formation and retardation of destruction cause compensatory mechanisms to fail and result in an increased or potentiated greenhouse effect. For example, the amount of water vapor released through evaporation increases directly with increases in the surface temperature of Earth. Within normal limits, increased levels of water vapor are usually controlled by increased warming and precipitation. Likewise, within normal limits, concentrations of carbon dioxide and methane are usually maintained with specified limits by a variety of physical and chemical processes.

Measurement made late in the twentieth century showed that since 1800 methane concentrations have doubled and carbon dioxide concentrations measured at the highest values estimated to have existed during the last 160,000 years. In fact, increases in carbon dioxide over the last 200 years were exponential up until 1973 (the rate of increase has since slowed).

Although the effects of these increases in global greenhouse gases are debated among scientists, the correlation of the increased levels of greenhouse gases with a measured increase in global temperature during the twentieth century, have strengthened the arguments of models that predict pronounced global warming over the next few centuries. In the alternative, some scientists remain skeptical because Earth has not actually warmed to the same extent as predicted by these models. For example, where many models based upon the rate of change of greenhouse gases predicted a global warming of .8°F to 2.5°F (0.44°C to 1.39°C) over the last century, the actual measured increase is significantly less with a mean increase generally measured at .9°F (.5°C) and that this amount of global warming is within the natural variation of global temperatures.

One problem in reaching a scientific consensus regarding global warming is that the data used in many models is neither global nor a result of high-reliance systematic scientific measurement (i.e., that it generally neglects oceans and vast uninhabited areas). Other problems involve forming an accurate articulation of the interplay of global surface warming phenomena that include thermal conduction, greenhouse **radiation**, convective currents. Most scientists agree, however, that an enhanced greenhouse effect will result in some degree of global warming.

GLUONS • See Quarks and gluons

GOEPPERT-MAYER, MARIA (1906-1972)
American physicist

Maria Goeppert-Mayer was one of the inner circle of nuclear physicists who developed the atomic fission bomb at the secret laboratory at Los Alamos, New Mexico, during World War II. Through her theoretical research with nuclear physicists

Enrico Fermi and **Edward Teller**, Goeppert-Mayer developed a model for the structure of atomic nuclei. In 1963, for her work on nuclear structure, she became the first woman awarded the Nobel Prize for theoretical physics, sharing the prize with J. Hans D. Jensen, a German physicist. The two scientists, who had reached the same conclusions independently, later collaborated on a book explaining their model.

An only child, Goeppert-Mayer was born Maria Göppert on July 28, 1906, in the German city of Kattowitz in Upper Silesia (now Katowice, Poland). When she was four, her father, Dr. Friedrich Göppert, was appointed professor of pediatrics at the University at Göttingen, Germany. Situated in an old medieval town, the university had historically been respected for its mathematics department, but was on its way to becoming the European center for yet another discipline—theoretical physics. Maria's mother, Maria Wolff Göppert, was a former teacher of piano and French who delighted in entertaining faculty members with lavish dinner parties and providing a home filled with flowers and music for her only daughter.

Dr. Göppert was a most progressive pediatrician for the times, as he started a well-baby clinic and believed that all children, male or female, should be adventuresome risk-takers. His philosophy on child rearing had a profound effect on his daughter, who idolized her father and treasured her long country walks with him, collecting fossils and learning the names of plants. Because the Göpperts came from several generations of university professors, it was unstated but expected that Maria would continue the family tradition.

When Maria was just eight, World War I interrupted the family's rather idyllic university life with harsh wartime deprivation. After the war, life was still hard because of postwar inflation and food shortages. Maria Göppert attended a small private school run by female suffragists to ready young girls for university studies. The school went bankrupt when Göppert had completed only two of the customary three years of preparatory school. Nonetheless, she took and passed her university entrance exam.

From Quantum Mechanics to the Bomb

The University of Göttingen that Göppert entered in 1924 was in the process of becoming a center for the study of **quantum mechanics**—the mathematical study of the behavior of atomic particles. Many well-known physicists visited Göttingen, including **Niels Bohr**, a Danish physicist who developed a model of the **atom**. Noted physicist **Max Born** joined the Göttingen faculty and became a close friend of Göppert's family. Göppert, now enrolled as a student, began attending Born's physics seminars and decided to study physics instead of mathematics, with an eye toward teaching. Her prospects of being taken seriously were slim: there was only one female professor at Göttingen, and she taught for "love," receiving no salary.

In 1927 Göppert's father died. She continued her study, determined to finish her doctorate in physics. She spent a semester in Cambridge, England, where she learned English and met **Ernest Rutherford**, the discoverer of the **electron**. Upon her return to Göttingen, her mother began taking student

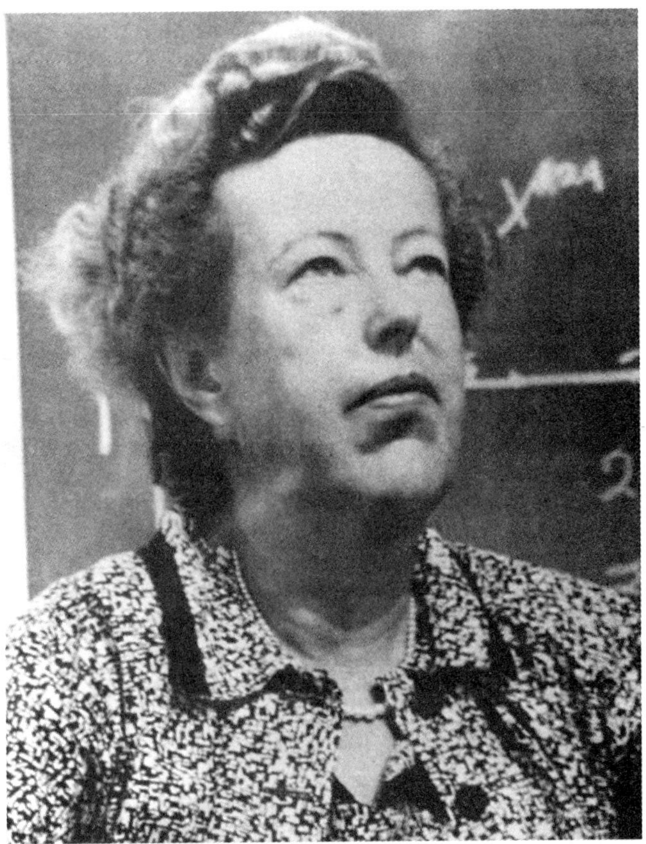

Maria Goeppert-Mayer.

boarders into their grand house. One was an American physical chemistry student from California, Joseph E. Mayer, studying in Göttingen on a grant. Over the next several years, Maria and Joe became close, going hiking, skiing, swimming, and playing tennis. When they married, in 1930, Maria adopted the hyphenated form of their names. (When they later moved to the United States, the spelling of her family name was anglicized to "Goeppert.") Soon after her marriage she completed her doctorate with a thesis entitled "On Elemental Processes with Two Quantum Jumps."

After Joseph Mayer finished his studies, the young scientists moved to the United States, where Mayer had been offered a job at Johns Hopkins University in Baltimore, Maryland. Goeppert-Mayer found it difficult to adjust. She was not considered eligible for an appointment at the same university as her husband, but rather was considered a volunteer associate, what her biographer Joan Dash calls a "fringe benefit" wife. She had a tiny office, little pay, and no significant official responsibilities. Nonetheless, her position did allow her to conduct research on **energy** transfer on solid surfaces with physicist Karl Herzfeld, and she collaborated with him and her husband on several papers. Later, she turned her attention to the quantum mechanical electronic levels of benzene and of some dyes. During summers she returned to Göttingen, where she wrote several papers with Max Born on beta ray decay—the emissions of high-speed electrons that are given off by radioactive nuclei.

These summers of physics research were cut off as Germany was again preparing for war. Born left Germany for the safety of England. Returning to the states, Goeppert-Mayer applied for her American citizenship and she and Joe started a family. They would have two children, Marianne and Peter. Soon she became friends with Edward Teller, a Hungarian refugee who would play a key role in the development of the hydrogen bomb.

When Joe unexpectedly lost his position at Johns Hopkins, he and Goeppert-Mayer left for Columbia University in New York. There they wrote a book together, *Statistical Mechanics,* which became a classic in the field. As Goeppert-Mayer had no teaching credentials to place on the title page, their friend Harold Urey, a Nobel Prize-winning chemist, arranged for her to give some lectures so that she could be listed as "lecturer in chemistry at Columbia."

In New York, Goeppert-Mayer made the acquaintance of Enrico Fermi, winner of the Nobel Prize for physics for his work on **radioactivity**. Fermi had recently emigrated from Italy and was at Columbia on a grant researching **nuclear fission**. Nuclear fission—splitting an atom in a way that released energy—had been discovered by German scientists **Otto Hahn**, **Fritz Strassmann**, and **Lise Meitner**. The German scientists had bombarded uranium nuclei with neutrons, resulting in the release of energy. Because Germany was building its arsenal for war, Fermi had joined other scientists in convincing the U.S. government that it must institute a nuclear program of its own so as not to be at Hitler's mercy should Germany develop a nuclear weapon. Goeppert-Mayer joined Fermi's team of researchers, although once again the arrangement was informal and without pay.

In 1941, the United States formally entered World War II. Goeppert-Mayer was offered her first real teaching job, a half-time position at Sarah Lawrence College in Bronxville, New York. A few months later she was invited by Harold Urey to join a research group he was assembling at Columbia University to separate uranium–235, which is capable of nuclear fission, from the more abundant **isotope** uranium–238, which is not. The group, which worked in secret, was given the code name SAM—Substitute Alloy Metals. The uranium was to be the fuel for a nuclear fission bomb.

Like many scientists, Goeppert-Mayer had mixed feelings about working on the development of an **atomic bomb**. (Her friend Max Born, for instance, had refused to work on the project.) She had to keep her work a secret from her husband, even though he himself was working on defense-related work, often in the Pacific. Moreover, while she loved her adopted country, she had many friends and relatives in Germany. To her relief, the war in Europe was over early in 1945, before the bomb was ready. However, at Los Alamos Laboratory in New Mexico the bomb was still being developed. At Edward Teller's request, Goeppert-Mayer made several visits to Los Alamos to meet with other physicists, including Niels Bohr and Enrico Fermi, who were working on uranium fission. In

August 1945 atomic bombs were dropped on the Japanese cities of Hiroshima and Nagasaki with a destructive ferocity never before seen. According to biographer Joan Dash, by this time Goeppert-Mayer's ambivalence about the **nuclear weapons** program had turned to distaste, and she was glad she had played a relatively small part in the development of such a deadly weapon.

The "Madonna of the Onion" Wins the Nobel Prize

After the war, Goeppert-Mayer returned to teach at Sarah Lawrence. Then, in 1946, her husband was offered a full professorship at the University of Chicago's newly established Institute of Nuclear Studies, where Fermi, Teller, and Urey were also working. Goeppert-Mayer was offered an unpaid position as voluntary associate professor; the university had a rule, common at the time, against hiring both a husband and wife as professors. However, soon afterwards, Goeppert-Mayer was asked to become a senior physicist at the Argonne National Laboratory, where a nuclear reactor was under construction. It was the first time she had been offered a position and salary that put her on an even footing with her colleagues.

Again her association with Edward Teller was valuable. He asked her to work on his theory about the origin of the elements. They found that some elements, such as tin and lead, were more abundant than could be predicted by current theories. The same elements were also unusually stable. When Goeppert-Mayer charted the number of protons and neutrons in the nuclei of these elements, she noticed that the same few numbers recurred over and over again. Eventually she began to call these her "magic numbers." When Teller began focusing his attention on nuclear weapons and lost interest in the project, Goeppert-Mayer began discussing her ideas with Enrico Fermi.

Goeppert-Mayer had identified seven magic numbers: 2, 8, 20, 28, 50, 82, and 126. Any element that had one of these numbers of protons or neutrons was very stable, and she wondered why. She began to think of a shell model for the **nucleus**, similar to the orbital model of electrons spinning around the nucleus. Perhaps the nucleus of an atom was something like an onion, with layers of protons and neutrons revolving around each other. Her magic numbers would represent the points at which the various layers, or "shells," would be complete. Goeppert-Mayer's likening of the nucleus to an onion led fellow physicist **Wolfgang Pauli** to dub her the "Madonna of the Onion." Further calculations suggested the presence of "spin-orbit coupling": the particles in the nucleus, she hypothesized, were both spinning on their axes and orbiting a central point—like spinning dancers, in her analogy, some moving clockwise and others counterclockwise.

Goeppert-Mayer published her hypothesis in *Physical Review* in 1949. A month before her work appeared, a similar paper was published by J. Hans D. Jensen of Heidelberg, Germany. Goeppert-Mayer and Jensen began corresponding and eventually decided to write a book together. During the four years that it took to complete the book, Jensen stayed with the Goeppert-Mayers in Chicago. *Elementary Theory of*

Nuclear Shell Structure gained widespread acceptance on both sides of the Atlantic for the theory they had discovered independently.

In 1959, Goeppert-Mayer and her husband were both offered positions at the University of California's new San Diego campus. Unfortunately, soon after settling into a new home in La Jolla, California, Goeppert-Mayer suffered a stroke that left an arm paralyzed. Some years earlier she had also lost the **hearing** in one ear. Slowed but not defeated, Goeppert-Mayer continued her work.

In November 1963 Goeppert-Mayer received word that she and Jensen were to share the Nobel Prize for physics with **Eugene Paul Wigner**, a colleague studying **quantum mechanics** who had once been skeptical of her magic numbers. Goeppert-Mayer had finally been accepted as a serious scientist. According to biographer Olga Opfell, she would later comment that the work itself had been more exciting than winning the prize.

Goeppert-Mayer continued to teach and do research in San Diego, as well as grow orchids and give parties at her house in La Jolla. She enjoyed visits with her granddaughter, whose parents were daughter Marianne, an astronomer, and son-in-law Donat Wentzel, an astrophysicist. Her son Peter was now an assistant professor of economics, keeping up Goeppert-Mayer's family tradition of university teaching.

Goeppert-Mayer was made a member of the National Academy of Sciences and received several honorary doctorates. Her health, however, began to fail. A lifelong smoker debilitated by her stroke, she began to have heart problems. She had a pacemaker inserted in 1968. Late in 1971, Goeppert-Mayer suffered a heart attack that left her in a coma. She died on February 20, 1972.

GOUDSMIT, SAMUEL A. (1902-1978)

Dutch American physicist

A prominent figure in American physics, Samuel A. Goudsmit was an authority on atomic **energy** and nuclear research, and was the co-discoverer of the **electron** spin. Goudsmit was educated in the Netherlands, where he received his Ph.D. in physics from the University of Leiden in 1927. While still a graduate student, Goudsmit, in collaboration with fellow physics student George Uhlenbeck, made the discovery for which he is most famous: electron spin. That discovery explained some important theoretical predictions by physicists **Wolfgang Pauli** and **Paul Dirac** as well as a number of anomalies in the existing **atomic theory**. Nobel Prize winner and physics professor **Isidore I. Rabi** was quoted by Daniel Lang in *New Yorker* as observing that the discovery "was a tremendous feat. Why those two men never received a Nobel Prize for it will always be a mystery to me." During World War II, Goudsmit led a group of physicists and military personnel in a search through war-torn Europe, in an attempt to locate German scientist **Werner Heisenberg** and determine what the Germans had accomplished in terms of their **atomic bomb**

research. For this mission he received the Medal of Freedom and the Order of the British Empire.

Samuel Abraham Goudsmit was born in the Hague, Netherlands, on July 11, 1902. His father, Isaac Goudsmit, was a prosperous dealer in bathroom fixtures, and his mother, Marianne Gompers Goudsmit, was the owner of a fashionable hat store called Au Louvre. Young Samuel developed a passionate interest in the millinery business early in life. He was excited by the tales his mother told of life in Paris, and he loved the challenge of trying to predict six months in advance what kind of hats the women of the Hague would be wearing.

Goudsmit's first introduction to the sciences came when he was 11 years old and off-handedly picked up his older sister's physics textbook. He was intrigued by a discussion on spectroscopic phenomena—the theory that the elements of Earth and the **stars** are identical. Although interested, Goudsmit developed no particular ongoing curiosity about science until, after graduation from high school in 1919, he was influenced by a physics teacher. At the University of Leiden, Goudsmit decided to major in physics—he had earned his best grades in science and mathematics—and found himself in a class taught by Paul Ehrenfest. Ehrenfest recognized in Goudsmit an inquiring intellect combined with an infallible intuition, and took a particular interest in helping his young student to develop those qualities. Goudsmit's moderate interest in physics soon became a passionate pursuit under the guidance of his new teacher. As a result of Ehrenfest's encouragement, Goudsmit published his first scientific paper in 1921 on the fine structure of atomic spectra.

That topic was one of immense importance in the 1920s. The quantum model of **atomic structure** proposed by **Niels Bohr** in 1913 had been an extraordinary breakthrough and solved many problems in the field of atomic theory. But a number of important questions remained. One of these concerned the nature of atomic spectra produced when atoms are placed within a magnetic field. In particular, the spectra produced in such instances always contain twice as many lines as were predicted by Bohr's theory. The answer to that puzzle came in 1925, shortly after Goudsmit met George Uhlenbeck, another of Ehrenfest's students. The two were assigned to spend the summer working together on the problem of double **spectral lines**. The match of the two young students turned out to be nearly ideal. According to Goudsmit's biographer, Stanley Goldberg, in *Dictionary of Scientific Biography,* "Goudsmit supplied the intuitions necessary to recognize and summarize regularities not immediately obvious in the data...[while] Uhlenbeck was more analytically oriented, more readily able to make connections between formal synthesis and traditional physical concepts."

As a result of this collaboration, Goudsmit and Uhlenbeck realized that the problem of double lines could be solved by assuming that the electron spins on its axis as it travels around the atomic **nucleus**. The two possible directions of spin—clockwise and counterclockwise—could then explain two different orientations of an electron in a magnetic field and, hence, two spectral lines that are very close together in all other respects. Scientists immediately recognized the significance of this bold hypothesis. Only a few months earlier, Pauli had proposed his "exclusion principle," according to which no two electrons in an **atom** can have exactly the same set of **quantum numbers**. A condition of that theory, however, was that a fourth quantum number was necessary to describe any given electron, a quantum number that could have the values of plus or minus one-half. Pauli made no guess as to what this quantum number might represent physically, but the Goudsmit-Uhlenbeck hypothesis immediately answered that question. The values of plus or minus one-half corresponded, they pointed out, to the two possible directions of electron spin. The discovery led to a fundamental change in the mathematical structure of **quantum mechanics**, as scientists recognized that spin is an integral property not only of electrons but also of protons and neutrons.

Goudsmit and Uhlenbeck no sooner announced their theory of electron spin before they were both offered appointments at the University of Michigan. For Goudsmit, the decision to leave the Netherlands was a difficult one. He hated leaving behind family, friends, Ehrenfest, and his other European colleagues. But he accepted the Michigan offer nonetheless. A few years later, Goudsmit referred to himself during an interview as a "has-been," he was quoted as saying by Daniel Lang in *New Yorker*. His remarks suggested that his greatest accomplishment was behind him, and he could look forward only to a rather mundane career in the future. "As a physicist's career goes...a scientist can do useful work all his life, but if he is to carry learning one big step forward, he usually does so before he is thirty. Youth has the quality of being radical, in the literal sense of the word—of going to the root.... Obviously, if one hits on something through this approach, it may well be outstanding. After a scientist passes his creative peak, it seems to me the most useful thing he can do is teach the status quo to youngsters." While he continued to remain active in research throughout the rest of his life, Goudsmit continued to teach—with one abrupt, but eventful interruption—until his death in 1978.

The interruption was World War II. At the outbreak of war, Goudsmit left the University of Michigan to conduct a secret research project at the Massachusetts Institute of Technology to test the theory of **radar**. In 1944 General Leslie R. Groves asked the physicist to serve as part of a secret intelligence mission. The project—code-name "Alsos"—was an effort to find out what German scientists had been able to discover about **nuclear weapons** and atomic-bomb research during the war. Goudsmit was placed in charge of a group of about 100 men, six of whom were scientists, sent to Europe to track down Werner Heisenberg, head of the German atomic weapons project. Goudsmit's team successfully found Heisenberg near Munich at the end of the war in Europe. Later, Goudsmit wrote a popular book titled *Alsos,* describing his experience.

Goudsmit's participation with the Alsos project engendered an important change in his outlook on the future. In a 1951 interview with *New Yorker* writer Daniel Lang, Goudsmit explained that he felt he could not simply go back to the routine of university life again, but needed to "take an active part in scientific developments in order to—yes, at the time I perhaps meant it literally—to help save the world." As a consequence, after a brief stay at Northwestern University (1946 to 1948), he accepted an appointment at the Brookhaven National Laboratory, where he was promoted to chair of the physics department in 1950. During his years at Brookhaven, he became very active in the political aspects of scientific research, including the development of scientific policy, the funding of research, and the defense of science against the attacks of McCarthyism—the extreme governmental opposition to communism that swept the United States in the 1940s and 1950s.

In the two decades between 1952 and 1974, Goudsmit took on another important responsibility with the editorship of the American Physical Society's (APS) *Physical Review.* During his long term of office, Goudsmit oversaw the expansion of the journal from a single publication of about 5,000 pages per year, to a group of five related journals with a combined size of more than 25,000 pages per year. He also recommended the creation of—and then put into production—an important new journal, *Physical Review Letters,* in 1958. Goudsmit retired from his editorial work at APS in 1974 and accepted a position as distinguished visiting professor at the University of Nevada at Reno. His only teaching assignment there was a large general education course in "physics appreciation." He was found dead of a heart attack in Reno on December 4, 1978.

GRAND UNIFIED THEORY

A long sought-after goal of modern physics is the formulation of a complete description of all interactions mediated by the four fundamental forces by a grand unified theory that would account for all phenomena within a single theoretical framework.

There are four fundamental forces that occur in nature: the strong, the electromagnetic, the weak, and the **gravitational force**. The strong **force** is aptly named because it represents the strongest force and is responsible for binding nuclear particles together. The standard model describes all of the fundamental forces except **gravity**, the weakest of the fundamental forces. Gravity is described by the general theory of relativity.

Although weak in comparison to the other fundamental forces, gravity acts on all particles that have **energy** and **mass** and it is a long-range force. As of 2000, there is no complete **quantum theory** of gravity. The postulated force carrier for gravity is the, as of yet, unobserved **graviton**. Because gravitational interactions depend on the mass of the gravitating system, and since it is the weakest force, it effectively only applies to the macroscopic **universe**.

The strong (nuclear) force is the force that holds **subatomic particles** together, it acts on particles such as **hadrons** and its effects are transmitted by gluons at a very short range, i.e., 10-15 m. The much weaker electromagnetic force describes electric and magnetic interactions between atoms and molecules. Accordingly the electromagnetic force, also a long-range force, acts on charged particles and it is carried by photons. The weak (nuclear) force acts in the **nucleus** on all **leptons** and hadrons and is associated with radioactive phenomena such as **beta decay**. Its particles of exchange are the weak **bosons**, W and Z, and its range of action is very short, i.e., ü10^{-17} m.

The first step towards developing a unified theoretical framework was the realization, mainly through the experiments of English physicist and chemist **Michael Faraday**, that **electricity** and **magnetism** were both manifestations of the same electromagnetic force. This work led to the formulation of the electromagnetic equations and theory by Scottish physicist **James Clerk Maxwell** and French scientist **André-Marie Ampère**. There were, however, experimental observations that classical electromagnetic theory could not account for, including the **photoelectric effect** and **blackbody radiation**. These phenomena were explained by the theoretical advancements of German Physicist **Max Planck**, French physicist **Louis-Victor de Broglie**, and German American physicist **Albert Einstein**, who were able to explain these phenomena by introducing the quantum mechanical concepts of the dual nature of **light** and the quantization of energy and photons.

Classical electromagnetic theory evolved into **quantum electrodynamics (QED)**, also known as the **quantum field theory** of the electromagnetic force. After first incorporating the contributions of English physicist **Paul Dirac**, advances developed independently by American physicists **Richard Feynman**, **Julian Schwinger**, and Japanese physicist **Sin-Itiro Tomonaga** allowed QED theory to accurately describe the interactions of **photons** and to provide a complete conceptual framework for interactions with fields of charged particles.

Other fundamental steps toward a still elusive unified theory included the incorporation of quantum **electrodynamics** into a field theory for the weak force known as **electroweak theory**. In this formulation, the electromagnetic and weak forces are considered different manifestations of the same force, the **electroweak force**. In parallel to these developments, another quantum chromodynamic theory (QCD) was refined during the 1970s to account for the manifestations of the strong force. QCD derives its name from a property of **quarks** and gluons termed "color."

A linkage of the electroweak and strong forces was achieved by the Yang-Mills generalization of the classical equations of **electromagnetism** in 1954 and the subsequent work of **Gerardus 'tHooft** and **Martinus Veltman**. In the 1970s these contributions resulted in the standard model that specifically accounted for interactions involving elementary particles such as electrons, neutrinos, **tau particles,** and force carriers such as photons, gluons, and W and **Z bosons**. In addition to the fact that the standard model does not yet incorporate gravity, physicists continue to strive to resolve several problems, including the identification of the **Higgs boson** and the inability to theoretically derive fractional charges for quarks. In addition, there are attempts to determine the nature of leptons and quarks as they relate to the multiplicity of elementary particles.

The need to refine the standard model has provided impetus for the development of several grand unified theories (GUTs), theoretical frameworks striving to link the electroweak interactions of quarks and leptons to the **strong interactions** of quarks. For the most part, they are based on the assumption that at sufficiently high energies, the strong, electromagnetic, and weak forces may be of the same magnitude and related to each other by a **gauge symmetry** that could occur in a high-energy regime. As energy is decreased, the symmetry breaks down and splits the three forces. A major difficulty is that the high energies required to experimentally validate the theory are of the order of 10^{16} GeV, which is well beyond the range of present-day **particle accelerators**. Another approach, first aimed at lowering the number of elementary particles, resulted in what has become known as supersymmetry. Although supersymmetry theory failed to lower the number of elementary particles, it succeeded in predicting the existence of partner particles that would have the same mass in a supersymmetry regime and with breakdown energies that, even if they are still too high for the currently available accelerators, may be studied at energy levels within realistic attainment in the very near future.

Another theoretical development of the 1970s was **string theory,** which describes elementary particles not as points in **space** but as extended straight, curved, or looped strings. Several variants of superstring theory were refined in the 1990s that were actually different manifestations of the same theory.

GRAVITATIONAL COLLAPSE • See Black holes

GRAVITATIONAL ENERGY

Gravitational **energy** is energy contained in the **gravitational field**. The relationship governing gravitational energy can be simply derived by noting that **gravity** is a conservative **force**, and the **work** done by a body in a gravitational field is equal to the vector product of the force and the displacement. **Newton's law of universal gravitation** states that the **gravitational force** is proportional to the product of the masses of any two bodies, and inversely proportional to the square of their separation, $F = G\, m_1 m_2 / r^2$, where m_1 and m_2 are the masses of the first and second bodies, respectively, and r is the magnitude of the vector displacement between the two bodies. It is straightforward to show, by integrating this force law against an infinitesimal outward radial vector displacement, and assuming that the total system energy is zero at infinite displacement, that the gravitational **potential energy** of the two bodies is $U = -Gm_1 m_2 / r$. Note that the minus sign arises because the vector gravitational force is attractive, and points oppositely to the outward radial vector displacement. Physically, this means that gravity acts to bind **matter** together, and hence makes a negative contribution to the total energy of the system.

It is similarly possible to compute the total energy necessary to break up a body and remove all of its pieces, layer-by-layer, to infinite distance. This quantity is known as the gravitational **binding energy** of the body; its size determines how strongly the body is bound together. Although one can calculate the binding energy of any body knowing its **mass** distribution, for the purposes of this discussion, we only need to note that for a body of mass M, the binding energy will always be of order GM^2/R, where M and R are the mass and radius of the body, respectively. Another way of putting this is that the gravitational binding energy of the body, *per unit mass* is of order GM/R. Note that the gravitational binding energy of the body per unit mass is roughly the square of the escape speed of the body, as one would expect from dimensionless grounds.

Gravitational energy is the ultimate source of energy for some of the most fascinating phenomena in the **universe**; gravity is extremely effective in accelerating matter onto more massive bodies. For instance, mass can flow onto **white dwarfs**, **neutron stars**, or **black holes**; when it reaches the surface of those bodies (or the surface of surrounding material, in the case of a black hole), the gas will be rapidly brought to a halt through a shock front. The shock will then convert the bulk **kinetic energy** into **heat**, and the accreted matter will radiate away virtually all of that energy.

GRAVITATIONAL FIELD

The gravitational field is a measure of the **gravitational force** present in a region of **space** due to material objects. A quantitative description of the gravitational field is developed in the following way: take the **force** felt by a test body and divide it by its own **mass**. The test body may be moved from place to place to map out the size of the field. This equation is **Newton's law of universal gravitation** extended to a many-particle (or n-particle) system. Although Newton's law of gravitation is applicable in many instances, it is not applicable when relativity (special or general) must be taken into account.

The gravitational field, represented by the symbol g, is defined as the gravitational force per unit mass. For any object whose mass is constant, Newton's second law of **motion** states that the net external force F on a body is the product of the body's mass m and **acceleration** a, or F=ma. Dividing both sides of this equation by the mass m yields the equation a=F/m. This equation shows that the gravitational field g is itself a type of acceleration. Specifically, it is the acceleration of the test mass m_0 due to **gravity**. Near Earth's surface the value of g is approximately 9.8 m/s^2 (32 ft/s^2).

Within the past half-century many discoveries have been made concerning our solar system's **planets**, planetary moons, asteroids, and other bodies. For some of these celestial bodies gravity maps have been constructed. The gravity map of a celestial body is the primary method for determining the mass distribution within that body. For example, the Magellan probe sent to the planet Venus was used not only to map its surface via **radar**, but to create a gravity map of the planet as well. By recording the **velocity** and position of the Magellan

probe as it orbited Venus, information about the gravitational force exerted on Magellan due to Venus was calculated. Dividing the gravitational force F by Magellan's mass m_0 at numerous points yielded the gravitational field just above Venus's atmosphere.

The gravitational field is useful in addressing the so-called action-at-a-distance problem. In fact, this was a criticism leveled at Newton's law of universal gravitation concerning the mechanism through which material bodies attract one another. A spring or lever through which forces are applied to various objects can be touched and held in one's hands. But the notion of all objects in the **universe** attracting one another through an invisible, permeating force seemed to some people during Newton's lifetime as being on the level of black magic. **Isaac Newton** was aware of this problem, but was unable to satisfactorily resolve it. The development of the gravitational field concept allowed people to think of the field itself as acting upon the masses present in a region of space. The gravitational field lines can be thought of as permeating a region of space so that when an object is brought into this space it is affected by, and interacts with, these lines. However, the action-at-a-distance problem was not completely remedied until the development of **relativity theory** in the twentieth century.

See also Force, mass, and acceleration; Many-body problem

GRAVITATIONAL FORCE

The gravitational **force** is one of fundamental forces in nature discovered to govern the interaction of material objects. The other three fundamental forces are the electromagnetic force, the strong nuclear force, and the weak nuclear force (which is now usually combined with the electromatic force as the **electroweak force**). The two nuclear forces are effective only at the subatomic scale. In contrast, the gravitational and electromagnetic forces act on objects from the atomic to extragalactic scales. Progress is also being made to treat gravitation as a side effect of string forces.

Johannes Kepler discovered (circa 1609) that the **planets** describe ellipses in their circuits about the **Sun**, with the Sun occupying a focal point of the ellipse. From this discovery scientists like **Edmond Halley**, **Robert Hooke**, and others, speculated that the force exerted by the Sun on the planets varied as the inverse square of the distance between the two bodies. In his *Principia* Sir **Isaac Newton** confirmed this suspicion with his law of universal gravitation as described in the following equation: $F = GMm/r^2$, where F is the magnitude of the gravitational forces between two bodies M and m separated by distance r. G is Newton's universal gravitational constant, defined as $6.67 \times 10^{-11} Nm^2 kg^{-2}$.

Encapsulated in this equation are the following important features: (1) the gravitational force between two particles is a "central" force: that is, the force acts along the (straight) line drawn from one particle to the other; (2) the gravitational force is a force of attraction, unlike the electrostatic force, which can be either attractive or repulsive; (3) the size of the force is proportional to the masses of each of the two particles. **Mass** must not be confused with **weight**. In Newton's system, the mass of a particle is an intrinsic property of that particle that remains unaltered when moved from point to point in **space**. By definition, weight is the gravitational force that one object applies to another. The weight of the same object on different planets, for instance, varies, but its mass is the same at each locale; (4) the gravitational force between two particles varies as the inverse square of the distance between them. Many phenomena in nature exhibit this "$1/r^2$" dependence. The gravitational force decreases with distance in the same way that the intensity of **light** decreases from an emitting object; and (5) the gravitational constant G is used to convert the term on the right side of the equation into units of force. The first person to determine a relatively accurate value of G was **Henry Cavendish**, when in 1798 he used a very sensitive torsion balance for that purpose. In addition to the previous points it must be noted that the total force on a particle due to many masses (the so-called n-body problem) is just the summation of the forces exerted by each mass upon the particle according to the equation above.

Newton's law of universal gravitation stood the test of time for over 200 years, until the appearance of the general theory of relativity in 1915 (constructed by **Albert Einstein**). In Newton's theory only particles possessing mass were influenced by gravitation. But in the general theory of relativity, both mass and **energy** are influenced by **gravity**. Through numerous experimental confirmations, the general theory of relativity has superseded Newton's laws of **motion** as being the best theoretical model for explaining the interactions of **matter** and energy. However, for most problems (such as stellar and planetary motion, or the ballistic flight of **rockets** and projectiles) Newton's laws, including his law of universal gravitation, are still adequate.

See also Relativity, general

GRAVITATIONAL RADIATION AND WAVES

The general theory of relativity predicts the existence of gravitational **radiation** or gravitational **waves**. Just as an accelerated charge produces electromagnetic radiation in the form of electric and magnetic fields, an accelerated **mass** produces gravitational radiation in the form of vibrations in the curvature of **space-time** analogous to the ripples in a pond.

The classical problem, which **Albert Einstein** solved in 1916, was to consider the radiation pattern from a thin rod rotating about an axis perpendicular to its center. The corresponding situation in **electromagnetism** produces a radiation pattern whose leading term is given by the electric-dipole moment. The mass-dipole moment is the **time** rate of change of the **momentum** of the spinning rod. Because this quantity vanishes by **momentum conservation**, the **power** of the radiation scales as the quadrupole moment, and its details are considerably more subtle.

Gravity wave research. *(Photo by Tommaso Guicciardini. Photo Researchers, Inc. Reproduced by permission.)*

The power of the radiation distribution makes the detection of gravitational waves very difficult. Consider a rod of length L and mass M. The moment of **inertia** through an axis perpendicular to the center is $I = 4ML^2/3$. If the rod is spinning with **angular momentum** Ω, the power is given by the relation: $P = 32GI^2\Omega^6/5\ c^5 = 1.73 \times 10^{-52}I^2\Omega^6$ where G is the gravitational constant and c is the **speed of light**. Assuming that the rod has a mass of 10^5 kg, a length of 10 m, and an angular **velocity** of 10 rad/s, the power dissipated in gravitational radiation is roughly 10^{-32} Watts. This is too small to be detected in the laboratory.

Joseph Weber devised an experiment in which a large cylinder of aluminum is supercooled under a **vacuum**. When gravitational waves strike the bar, it vibrates at its fundamental **frequency**. By considering a pair of such bars separated by hundreds of miles, Earth-based vibrations are subtracted out, and the apparatus records **gravity** waves incident from **space**. Weber claimed to have detected gravity waves in 1969, but independent measurements have not corroborated this result.

The existence of gravity waves has been inferred by examining the steady loss of orbital **energy** in **binary pulsar** systems such as PSR 1913+16. This work contributed to the 1993 Nobel Prize in physics, which was awarded to **Russell Hulse** and **Joseph Taylor**. The **Laser Interferometer Gravitational Observatory** (LIGO) is a present-day attempt to detect gravitational radiation originating in such energetic events as supernovae or black-hole collisions.

Just as there is a **quantum theory** of electromagnetism (called **quantum electrodynamics** or QED), there are hopes for a quantum theory of gravity. Though the technical issue of renormalizability complicates the issue, some generic comments can be made about the **graviton**, the gravitational analogue of the **photon**.

The fundamental forces of nature other than gravity (electromagnetism, the strong **force**, and the weak force) are gauge interactions that fit into a theory particle physicists call the standard model. Particle interactions within the standard model arise through the exchange of **vector bosons**. The photon is responsible for electromagnetism; the gluon, which carries **color**, is responsible for the strong force; and the intermediate vector bosons W+, W-, and Z are responsible for the weak interaction. Since these gauge particles each carry spin-1, the potential experienced between two particles that carry the same charge is repulsive. For example, in the case of electromagnetism, a pair of electrons repel, while oppositely charged particles like the **electron** and **positron** attract. This is the content of **Coulomb's law**.

Unlike the three gauge interactions, gravity is always attractive. **Newton's law of universal gravitation** states that any pair of objects experience a force of attraction proportional to the product of their masses and inversely proportional to the square of their separation. The general theory of relativity says precisely the same thing except that it argues that the force originates from the curvature of space in the presence of **matter** or energy. The universally attractive character of the potential means that gravity cannot be mediated by a vector (spin-1) boson. Rather, the particle must have spin-2. This tensor, the quantum of the **gravitational field**, which encodes the curvature of **space-time**, is called the **graviton**. According to **string theory**, the graviton is the lowest energy vibrational mode of the closed string.

See also Standard model of particle physics

GRAVITON

Gravitons are proposed quantum particles that form gravitational fields. Just as electromagnetic fields produce **photons**, which can be seen with image intensifiers, physicists believe that gravitational fields produce gravitons. However, because the gravitational equations proposed by **Albert Einstein** are nonlinear, unlike the linear electromagnetic particle equations, gravitons cannot (according to current theory) be observed directly. But scientists are convinced that gravitons exist, because their presence leaves perturbations that are observable, even if the particles themselves are not. Gravitons are supposed to have a rest **mass** and charge of zero, and a spin of 2.

Fundamental particles possess intrinsic **angular momentum**, or "spin." Spin is quantized, coming only in either integer or half-integer multiples of Planck's reduced constant ($\hbar$). Particles of half-integer spin are "**fermions**," while those of integer spin are "**bosons**." It can be shown that in order to produce a **force**, an exchange particle must be a boson. A photon is a spin-1 boson. And finally, electromagnetic fields do not interact with themselves, so a photon has no charge—photons do not couple to photons.

In an analogous manner, in a **quantum field theory** of **gravity** two massive particles interact gravitationally via the exchange of a quantum of the gravitational field—a virtual graviton. Gravitons couple to the source of the **gravitational field**, mass. A graviton is emitted by one massive particle, and absorbed by the other, resulting in the familiar **gravitational force** between the particles.

Graviton ride. *(Photo by Eunice Harris. Photo Researchers, Inc. Reproduced by permission.)*

A graviton's properties are such that its exchange results in the gravitational force. It, too, is massless, as gravity's range is infinite. It is a spin-2 boson; if it were of spin-1, gravity, like the electromagnetic force, could be either attractive or repulsive.

But here a crucial difference between gravitons and photons emerges. Einstein's mass-energy equivalence implies that gravitational potential energy—the **energy** stored in the gravitational field—must *itself* act as a source of that field. Thus, although they possess no mass, gravitons *do* couple to gravitons. It is this self-coupling (among other things) that makes the gravitational field so difficult to quantize. In fact, no fully consistent **quantum theory** of gravity yet exists.

GRAVITY

Gravity is the **force** of attraction between any two objects in the **universe** with **mass**. Although it is much weaker than any of the other fundamental forces (**electromagnetism**, the weak force, and the strong force) and is neglected in the interactions of elementary particles, which occur at short distances, the effects of gravity manifest themselves at larger length scales, most notably at astronomical distances.

Historically, gravity is the first of the fundamental forces of nature whose behavior was extensively studied and elucidated. **Aristotle** maintained the rate at which objects fell depended upon their masses. If two objects are dropped, the heavier one falls faster. **Galileo** showed that Aristotle's conclusion was false. Any two objects fall towards the center of Earth with the same rate of **acceleration**. This observation is an empirical one. It does not provide an explanation for why gravity acts as it does.

Isaac Newton offered an explanation through his law of universal gravitation, $F = Gm_1m_2/r^2$, which holds that the force of gravity between any pair of masses is directly proportional to the product of their masses and inversely proportional to the square of their separation. The factor G in this relation is a **fundamental constant** called the universal gravitational constant or Newton's constant. It is the same irrespective of whether one is talking about an apple that is dropped or a star that orbits the center of a galaxy.

An apple and a cannonball fall at the same rate because Newton's law tells us that they have the same acceleration, $g = GM/R^2$ where M is the mass and R is the radius of Earth. The **Moon** and Earth rotate about a common **center of mass**. Because Earth is roughly a hundred times heavier than the Moon, the center of mass of the Earth-Moon system resides within Earth. The force of attraction between the two bodies is centripetal. This fact, together with Kepler's third law and a knowledge of the distance between Earth and the Moon, the period of the Moon's orbit, and the radius of Earth, is sufficient to deduce that the ratio of Earth's and the Moon's masses $M_{Earth}/M_{Moon} = 81.3$ and $g = 9.8$ m/s^2. Similarly, it can be shown that Earth revolves around the **Sun**, which is roughly a million times more massive. Universal gravitation therefore vindicates the heliocentric models of Aristarchus (310-c.230 B.C.) and **Nicholas Copernicus**.

The constant G is typically determined using a torsion balance. **Henry Cavendish** performed the first such experiment by using a pair of solid lead spheres 5 cm in diameter attached to the ends of a light rod suspended by a thin fiber. When heavier 30 cm lead spheres were brought in proximity to this system, the gravitational attraction of the small balls to the larger ones produced a **torque** that twisted the fiber. Measuring this force allows G to be determined as 6.67×10^{-11} m^3/s^2-kg. A similar experiment also allowed Cavendish to measure the mass of Earth as 5.98×10^{24} kg.

The French astronomer Jean Leverrier (1811-1877) noticed that Mercury's perihelion, the radial point in its orbit nearest the Sun, precessed too rapidly by about 40 seconds of arc per century. This phenomenon suggested that Newton's theory of gravitation was wrong. A better model for gravity was provided by **Albert Einstein** in his general theory of relativity.

Newton's law of universal gravitation assumes that force acts at distance (i.e., without a medium that transmits the force) and does **work** instantaneously in that every object in the universe knows how far away and how massive any other other object in the universe is and immediately responds accordingly. **Newtonian physics** also invokes the existence of privileged inertial **reference frames**. Einstein's special theory of relativity contradicts these long-held beliefs by positing that the laws of physics are the same in every inertial reference frame and that the maximum speed at which the information about a particle's properties can travel is the **speed of light**. The general theory of relativity takes this argument one step further. Its starting point is the assumption that on a local scale, the physical consequences of an accelerated reference frame and the force of gravity cannot be distinguished. This is a restatement of the principle of equivalence, which maintains that the inertial mass, which determines how a body resists changes to its **inertia**, and the gravitational mass, which is used in the law of universal gravitation, are the same. In Einstein's theory, the **gravitational force** results from the curvature of **space-time** in the presence of **matter** (or **energy**). The measured deflection of starlight by the Sun, the delay of **radar** echoes when Venus passes behind the Sun, and the precession of Mercury's perihelion all provide spectacular evidence to support the general theory of relativity.

A technical problem remains. Unlike the other three fundamental forces, gravity cannot be treated as a renormalizable field theory. That is to say, it is inconsistent with **quantum mechanics**, and the short-distance physics of gravity leads to infinities that do not cancel. In order to treat all of the fundamental forces on an equal footing, it becomes necessary to spread the gravitational interaction out in **space**. **String theory** accomplishes this by treating the **graviton**, the particle that mediates gravity, as a loop of string. Such a theory is only consistent in 10 space-time dimensions. The details of string theory are still in the process of being worked out.

GRAY, STEPHEN (1666-1736)
English scientist

Not a great deal is known about Stephen Gray. He was born in 1666 and may have been educated, in part, by John Flamsteed (1646-1719), England's first Astronomer Royal. Gray died on February 25, 1736.

Gray became interested in the study of **electricity** and, in 1729, discovered that rubbing a glass tube caused an electric build-up that extended to a cork that was attached to the end. The cork had not been directly touched, yet the electric "fluid" had gone into it. Furthermore, a pine stick, inserted into the cork, transmitted the electricity to its end.

Experimenting with other substances, he found he could transmit a charge through a thread hanging vertically, but not horizontally unless the thread was suspended by silk. In other words, when he kept the horizontal thread from touching "ground," he could transmit the charge over a good distance.

Gray published his experiments in the *Philosophical Transactions*, which influenced the work of Charles Du Fay (1698-1739) and John Theophilus Desaguliers (1683-1744). Du Fay theorized the existence of "vitreous" and "resinous"

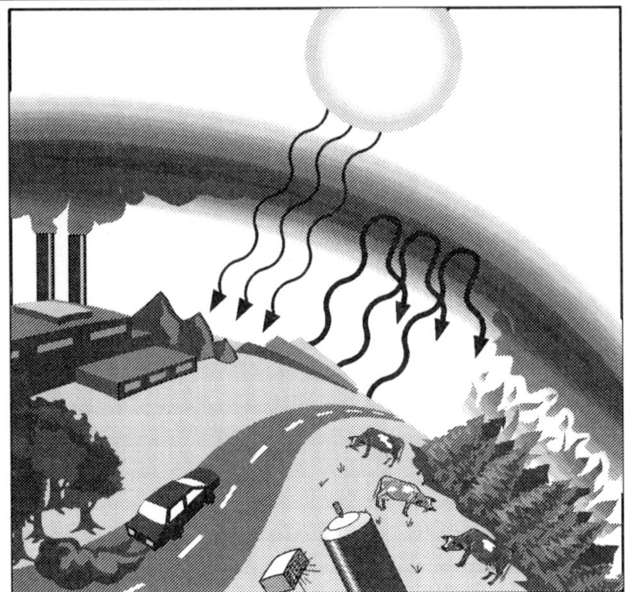

Greenhouse illustrations, under normal conditions and with greenhouse effect.

electricity, terms that **Benjamin Franklin** popularized as "positive" and "negative." Desaguliers first coined the word "conductor" to describe an object that transmitted electricity. An object that did not conduct was called an "insulator," based on the Latin word for "island," because an insulator could enclose an **electric current** just as the ocean could enclose an island.

GREENHOUSE EFFECT

In the greenhouse effect, Earth's gaseous atmosphere traps **heat** energy from sunlight and the planet's **temperature** increases. Scientists first theorized about Earth's greenhouse effect in the 1820s, but no one saw it as a problem until the 1950s, when a few researchers speculated that technology might be accelerating the rate at which Earth traps heat. Now there is international concern that the greenhouse effect leads to **global warming** and climate change. These unprecedented changes have the potential to affect rainfall, plant life, the level of the seas, and ultimately the life cycles of all plant and animal species.

Water vapor is one of the major gases that helps generate the greenhouse effect. Other greenhouse gases include **carbon** dioxide, methane, nitrous oxide, and chlorofluorocarbons (CFCs). The concentration of these gases has been observed to be increasing, and scientists believe that this increase of gases is a direct result of mankind's impact on the environment. It is thought that the increase of these gases is enhancing global warming.

Some scientists believe that, because the greenhouse effect boosts the amount of carbon dioxide in the atmosphere, the growth of green plants is affected. However, this acceler-

ated growth often leads to poor nutritional value, with unknown consequences for plant eaters from insects to mammals. There is already evidence that some leaf-eating insects are showing malnutrition and increased death rates. It is very difficult to set up controlled experiments with wildlife, so the effects of increased carbon dioxide are still unknown. Other scientists believe that the greenhouse effect, caused by various vapors, in turn increases the concentration of the vapors that cause it, leading to a run-away situation.

It is also difficult to measure the greenhouse effect globally. Instances of unusual **weather** such as flooding in Bangladesh, dry summers in the American Midwest, and lack of snow in Scandinavia, cannot be definitively tied to the greenhouse effect because there are so many interdependent factors. Many weather variations occur naturally. Detailed measurements of land temperatures have been kept since the 1850s and these correlate with increases in the greenhouse gases, but these statistics are not enough to prove cause and effect. There are ways of going back further to examine records. Scientists have looked at tree rings, at 160,000-year-old ice core data, and at changes in summer temperatures over the past 1,000 years. However, there is no agreement on how these records can be compared with current data, so efforts to correlate have been inconclusive.

Scientists do not agree on the significance of global observations. The long-range forecast is that Earth will continue to warm, simply as a function of increased population. There is some evidence that Earth's temperature has warmed one-quarter to one-half degree Centigrade over the past 100 years—an increase that, although small, can nonetheless have a significant impact on polar ice caps and cause a measurable rise in sea level. Some scientists believe urban pollution may be disguising the effects of global warming. If this is so, then pollution control methods may actually make the effects of global warming more prominent. In addition, there is no way of predicting the impact of genetically engineered organisms on the greenhouse effect.

The first review of the greenhouse effect and global warming was undertaken by three international groups: the International Council of Scientific Unions (ICSU), the United Nations Energy Program (UNEP), and the World Meteorological Organization (WMO). The Montreal Protocol on Substances that Deplete the Ozone Layer was signed by 24 countries in 1987 after scientific evidence showed that the stratospheric ozone layer is being damaged by halocarbons. Scientists working with the Intergovernmental Panel on Climate Change (IPCC) estimate that the greenhouse effect will be a gradual process over the next two to three centuries. In response, many industrialized countries have pledged to limit carbon emissions, with the exception of the United States.

The United States generates about 23% of greenhouse gas emissions worldwide and many countries are looking to the United States for leadership before taking a stand on greenhouse gases and climate change. Representatives from the G8 countries—the United States, Canada, Japan, the United Kingdom, Germany, France, Italy, and Russia—met in Otsu, Japan, in April 2000 to work out an agreement to reduce greenhouse gases. In November 2000, there will be a meeting at the Hague, Netherlands, to sign off on the details for the Kyoto Protocol, but before that meeting, representatives from more than 80 countries must work out their differences.

See also Ozone layer

GROUNDING

Grounding is the method by which electric circuits and other equipment are electrically connected to the earth. Grounding serves two purposes: to maintain an approximately constant **voltage** reference (usually designated as zero voltage) and to provide a charge reservoir (capacitor) for diverting dangerous or unwanted current and removing undesired static electric charges. Ground (or earth) refers to any point in a circuit that is grounded.

Electric instruments and equipment are frequently housed in a grounded conducting (metal) case, both to shield circuits from external electromagnetic **interference** and to reduce the risk of causing electric shocks, fires, and other accidents. Selected points in internal circuits are then grounded by connecting them via conductors to the grounded housing.

Modern jacks in most household **power** supplies provide three connections: two terminals that supply the alternating current, and a grounding wire. An alternating voltage is supplied between the power terminals, which in most households worldwide is typically 110 or 220 volts. The ground terminal (the third pin) is grounded locally and is used to ground the metal housing of many appliances, to help to prevent electrical accidents.

Lightning rods are a dramatic example of grounding: through a conductor between a lightning rod and the ground, current from a lightning strike on the rod can be safely diverted to the earth to prevent property damage or injury.

The term "virtual ground" refers to a point in the circuit that is held to the local ground voltage by an active electronic component, such as an operational amplifier. Unlike an ordinary grounded terminal, which is usually expected to behave as a voltage source, a terminal held to virtual ground may offer very high input impedance (i.e., accept very little current).

Complicated grounding techniques can be required to avoid ground-related electrical problems such as external interference and ground loops. Interference from nearby power supplies or from other strong electromagnetic signals emitted by nearby equipment and natural electromagnetic phenomena can produce local variations in ground voltages. These ground fluctuations can cause undesirable currents between different parts of a circuit that have been grounded at different locations, and thereby interfere with the circuit's intended function. Any unintended loop of current that passes through ground wires or the earth itself is known as a ground loop.

See also Electricity

GROUP THEORY

Group theory describes the rules of operation of sets of numbers or objects, including finite, discrete, topological, and Lie groups. Group theory became one of the central tools in elementary **particle physics** in the 1960s. Although **quantum electrodynamics (QED)** operated as the fundamental theory of **electricity** and **magnetism**, physicists tried to adapt it to other interactions by making minor changes to it. **Quantum electrodynamics** (QED) was invariant under transformations that are contained in an Abelian group (i.e., a group on which the defined binary operation is cummutative). Physicists developed a theory similar to QED, except it is invariant under transformations in non-Abelian groups. Eventually, this theory was applied to the strong and weak interactions and incorporated into the **standard model of particle physics**.

A group is a set of objects, plus an operation that obeys a set of four mathematical rules. The objects may be numbers or something more complicated. The operation may be addition, multiplication, or something more abstract. For the sake of simplicity, the operation between two different elements A and B of a group will be written as AB. The first rule of a group is that if A and B are elements of a group, then AB is also an element of the group. The second rule states that the operation is associative, that is, $(AB)C = A(BC)$. The third rule states that there is an identity element E of the group such that $EA = AE = A$. The fourth and final rule requires that in order to be a group, for any element A in the group, there is an inverse element A^{-1} such that $AA^{-1} = A^{-1}A = E$.

Groups are useful because the objects can be related to transformations of a system. For example, consider rotations in a plane. It can be shown that the set of rotations by any angle around a single axis in the plane form a group. First, if the plane is rotated by an angle (theta)$_1$ and then rotated again by another angle (theta)$_2$, the result is an element of the group corresponding to a single rotation by an angle ((theta)$_1$ + (theta)$_2$). The second property, associativity, is clearly satisfied, because the angles add in rotations, and addition is associative. For the third rule, rotation by the angle 0 degrees is the identity element, because it leaves the system unchanged. The final rule is satisfied, because if a rotation is performed by angle (theta), another rotation can be performed by angle -(theta), to return to the original configuration.

A commutative group, as previously illustrated, simply means that the order of operations does not matter, that is, AB = BA. If the rotation group is enlarged, however, to include three-dimensional rotations, this is no longer the case; the order of rotations in different planes becomes important. Commutative groups are called Abelian while non-commutative groups are called non-Abelian.

See also Rotation in three dimensions; Standard model of particle physics; Symmetry and symmetry principles; Vectors

GUERICKE, OTTO VON (1602-1686)
German physicist

Otto von Guericke was born on November 20, 1602, in Magdeburg, Germany. While he studied mathematics, law, and engineering, Guericke would become famous for his experiments with a **vacuum** and air **pressure**. He died at the age of 83, in Hamburg, Germany, on May 11, 1686.

Following travels to England and France, Guericke returned to Magdeburg in 1627 and became a politician. Unfortunately, this was during the Thirty Years' War; Guericke and his family had to flee the city in 1631. After the war, he returned and helped rebuild the city, becoming mayor in 1646. Twenty years later he became a noble and added "von" to his name. During Guericke's time, scientists were involved in an argument about whether a vacuum could exist. Guericke, who believed in the Copernican theory of the solar system, was extremely interested in understanding the nature of **space**. He wondered whether empty space could exist. Many scientists held on to the Aristotelian theory that a vacuum was impossible. **Aristotle** rightly suggested that as air became less dense an object would move faster. However, he went astray by claiming that if there wasn't any air, an object could move infinitely fast. Since he doubted the idea that infinitely increasing speed could exist, he concluded that a vacuum could not exist either.

Guericke was a rare breed of scientist who refused to accept previously held scientific facts; he believed in experimentation. He decided to try to successfully create a vacuum. In 1647 his first attempt failed, but in 1650 he built another air pump by putting a piston inside a cylinder that had two flap valves (a model later improved upon by **Robert Boyle**). He filled a cask with water to remove the air, sealed it, and used his pump to remove the water.

Unfortunately, the cask leaked air. He made a second attempt, placing the cask within another containing water. He hoped the water-filled cask would prevent air from entering the evacuated cask. He was right; air did not leak in, but water did.

Guericke decided to take another tack. He took a hollow copper sphere with a valve built in the bottom and used his pump to remove the air. The sphere promptly crumpled. This showed that when the sphere was empty, the external air pressure was strong enough to crush the sphere.

René Descartes held that space and **matter** were equivalent, so a vacuum could not exist. Guericke did not agree with Descartes's assertion. Guericke tried again with a more substantial sphere. This time he succeeded in creating his vacuum. He then undertook a series of grandstanding experiments to demonstrate the **power** of air pressure. In one of the most famous, in 1657, he placed two copper hemispheres together and removed the air. Sixteen horses were unable to pull the two halves apart. Obviously, the external air had substantial pressure; so much, in fact, that the hemispheres held together when the internal air was removed.

It is interesting to note that Guericke placed the valve at the bottom of the hemispheres because he believed that air, like water, would seek the lowest level. Later he found that air was distributed evenly, since he could create a vacuum regardless of the location of the valve. This led him to think about the **density** of air decreasing as one's altitude increased. He studied variations in air pressure and, in 1660, invented a barometer, which he used to make predictions about the **weather**.

Another of Guericke's experiments pitted 50 men against his piston. A rope was attached to a piston in a cylinder and the men were told to pull. Guericke created a vacuum on the opposite side of the piston and the men were unable to keep the external air pressure from pushing the pistons into the cylinder. The **force** of air pressure became very important later in the development of the steam engine.

In a different vein, Guericke also invented the first mechanical static **electricity** generator, similar to Robert Van de Graaff's generator. Guericke rotated a sulfur sphere on a shaft. When it was rubbed, it built up an electrical charge that emitted sizable sparks. Guericke did not give any special consideration to this electrical phenomenon, but during the next century others would continue to experiment with static electricity.

GUTH, ALAN (1947-   )
American physicist

In late 1979, while on sabbatical from Cornell University, Alan Guth made a theoretical discovery that changed the field of **astrophysics**. That discovery was the concept of an inflationary **universe**, a hypothesis that describes what happened in the first few moments after the creation of the universe. The inflationary hypothesis solved some of the critical problems that had long troubled cosmologists and suggested some exciting new directions for theoretical and experimental research. Guth's work was so impressive that he soon had a number of job offers to consider, one of which was from the

Massachusetts Institute of Technology (MIT). Guth accepted the position in 1980 and has since become the institution's Jerrold Zacharias Professor of Physics.

Alan Harvey Guth was born in New Brunswick, New Jersey, on February 27, 1947, to Hyman and Elaine Cheiten Guth. He grew up in nearby Highland Park in a middle-class family. He has described his childhood as "uneventful," as cited in *Omni,* but one dominated by an interest in mathematics. Coupled with this intense interest in math was a penchant for drawing highly detailed structures of rocket ships, which he hoped to someday construct.

After graduating from high school, Guth entered MIT, where he earned both a bachelor's and master's degree in physics in 1969. He then stayed on at MIT to earn a doctorate in 1972. A year before leaving MIT, he also married Susan Tisch, with whom he had grown up in New Jersey. The Guths have two children, Lawrence David and Jennifer Lynn.

Guth's first academic appointment was at Princeton, where he was instructor of physics from 1971 to 1974. He then received an appointment as postdoctoral research associate at Columbia University, from 1974 to 1977, and then a similar position at Cornell, from 1977 to 1979. Nearing the end of his Cornell appointment, Guth found himself in a difficult position. Most Ph.D.s would expect to have received a regular appointment within a few years of having earned their degree. Guth was now in his seventh postdoctoral year and knew, as he later said, that he was "nearing the end of the postdoctorate trail," as quoted in *Omni*. Fortunately, he was able to obtain an extension on his Cornell appointment by taking a one-year sabbatical at Stanford University's Linear Accelerator Center (SLAC). There he intended to carry out research on elementary particles while continuing to look for more permanent employment.

It was at Stanford on December 6, 1979, that Guth's life changed radically. He was working on a problem that had long intrigued him: how the basic theories of **particle physics** could be applied to an understanding of the origins of the universe. One of the ironies of modern physics is the close association that exists between these two extreme levels of physics: the smallest and most fundamental units of which all objects con-

sist, and the largest and most profound changes that have ever occurred (the big bang, the theory that the universe was created billions of years ago in an explosion of a single point of **energy** density).

Guth had been attempting to find a way to apply **Grand Unified Theory** to the earliest events of **cosmology**. Grand Unified Theories (GUTs) are theoretical efforts to show how three of the four fundamental forces of nature—strong nuclear **force**, weak nuclear force, and electromagnetic force—are associated with each other. GUTs assume that at some point in the earliest history of the universe, these three forces were all equivalent to and indistinguishable from each other. Guth had come to a study of this problem more from his own background in particle physics than from much experience in cosmology, a subject about which, he claims, he knew relatively little.

In any case, the evening of December 6, 1979, was characterized by a sudden breakthrough. As Guth explained to interviewer and fellow physicist Gregory Benford in *Omni,* he worked on this solution for a time on the evening of the sixth, then raced to his SLAC office on his bicycle the next morning to complete the solution. He later told Benford that "I broke my personal [bicycling] record dashing to SLAC to whip out my notebook and continue these calculations."

The result of those calculations is a hypothesis now known as the inflationary theory. According to this theory, the universe began to expand much more rapidly than had previously been imagined during the first moments after the big bang, doubling in size about once every 10^{-35} second. The theory was an almost instantaneous success because it accounted for a number of features of the universe—such as its present even distribution of background cosmic radiation—that had thus far remained unexplained.

As news of the inflationary theory spread among scientists, the job offers that Guth had been seeking began to appear. He decided to accept one from his alma mater, MIT, where he rose from visiting associate professor in 1980, to Jerrold Zacharias Professor of Physics—a post he currently holds—in 1989. Between 1984 and 1990, Guth was also on the staff at the Harvard Smithsonian Center for Astrophysics.

HADRON

Hadrons are **subatomic particles** that are affected by the strong **force**, the force that binds the **nucleus** together. Two hadrons—the **proton** and the neutron—are found in the atomic nucleus. All others are created by high-energy collisions, from **cosmic rays** and **particle accelerators**. Hadrons are combinations of basic particles, **quarks** and gluons, held together by the strong force. Quarks contribute **mass** and determine the hadrons properties, including electronic charge and spin. Gluons carry the strong force and hold the quarks together. All known subatomic particles except **bosons** (such as **photons**) and **leptons** (such as **electrons** and **neutrinos**) are hadrons. They are divided into two categories: baryons, made of three quarks, and mesons, made of a quark and an anti-quark.

All mesons and all the **baryons**, except the proton, are unstable and decay into other particles. Most of them exist for less than a millionth of a second, and some can survive only for the amount of time that it takes for a **light** wave to pass from one side to the other. Their existence is known by observation of the products and process of particle decay. Although a free **neutron** will decay in an average of about 15 minutes, neutrons bound within an atomic nucleus are stable. Some theories indicate that the proton may have some instability, but its **half-life** is very much longer than the age of the **universe**.

Particle physicists have discovered a large number of hadrons. The *Review of Particle Physics*, published every other year, contains more than 300 pages, listing particle types and their observed decay processes.

HAHN, OTTO (1879-1968)

German chemist

Otto Hahn is noted for his work on radioactive materials, which in 1938 led to his discovery, with physicist **Lise Meitner** and chemist **Fritz Strassmann**, of the process of **nuclear fission**. In recognition of their work, Hahn and Strassmann received the 1944 Nobel Prize in chemistry, and Hahn, Strassmann, and Meitner received the Fermi Award in 1966. Hahn was born in Frankfurt-am-Main on March 8, 1879, to Heinrich Hahn, a glazier, and Charlotte Giese Stutzmann Hahn. The Hahns' early years in Frankfurt were marked by poverty: according to R. Spence, writing in the *Biographical Memoirs of Fellows of the Royal Society,* the four Hahn boys—Otto, his two brothers, and his half-brother from Charlotte's first marriage—"slept in an unheated attic bedroom and took their weekly bath in a tub on the landing." Gradually, Heinrich's business became more successful, and the family attained "middle-class respectability." Otto, who attended the Klinger Realschule, demonstrated some early interest in science, carrying out simple chemical experiments in the family laundry house. But other subjects seemed more important to him, and the honors he received upon graduation were for gymnastics, religious studies, and music. Hahn's parents had hoped that he would pursue a career in architecture, but when he entered the University of Marburg in 1897, it was a course in chemistry that he decided to pursue. He interrupted his studies at Marburg to spend one year at the University of Munich, but then returned to Marburg, where he was awarded his doctorate in organic chemistry in 1901.

Biographers—and Hahn himself—mention Hahn's devotion to non-academic pursuits in college, especially cigar smoking and beer drinking. He felt obligated to join one of the student societies that were ubiquitous in German universities and on one occasion even challenged another student to a duel. Hahn's membership in the Nibelungia Society apparently brought him considerable happiness until he resigned in the 1930s in opposition to Nazi policies adopted by the group.

After graduation, Hahn enlisted in the infantry for one year and then returned as an instructor at Marburg. Soon thereafter, hoping to better his chances at a job in the German chemical industry, Hahn decided to spend a year in an English-speaking institution where he could polish his language skills while advancing his knowledge in chemistry. Through the efforts of a former teacher, Hahn was offered a research post

Otto Hahn.

with Sir William Ramsay in his laboratory at University College, London. Hahn left for England in September 1904.

Ramsay's fame rests primarily on his discovery of the inert gases argon, neon, krypton, and xenon. In the early 1900s, however, he had become interested in a new topic, **radioactivity**. When Hahn arrived in London, it was a problem in radioactivity, therefore, to which Ramsay assigned him. In some ways that decision was a peculiar one, since nothing in Hahn's background in organic chemistry had prepared him for research on radioactive materials. Ramsay seemed to believe that Hahn's lack of background in radioactivity might be an advantage, since he could proceed with no preconceived notions as to what to expect. The problem he assigned the young German chemist was to extract radium from a 100-gram sample of barium carbonate. After familiarizing himself with this new field of research, Hahn completed Ramsay's assignment, obtaining a few milligrams of radium. He discovered, however, that the radioactivity of the sample was greater than expected and eventually isolated a second radioactive material from the impure radium. He called the new substance "radiothorium," later identified as an **isotope** of thorium that decays into thorium-x.

At the conclusion of this research, both Ramsay and Hahn were convinced that the latter's future lay in radioactivity, not organic chemistry. Thus, Ramsay obtained for Hahn an appointment at Emil Fischer's Chemical Institute at the University of Berlin, where he could pursue this new interest.

In preparation for the Berlin post, however, Hahn decided to spend another year of study, this time with the world's foremost authority on radioactivity, **Ernest Rutherford**, at McGill University. During Hahn's year at McGill (1905–06) he discovered a second radioactive substance, radioactinium, now known to be an isotope of thorium that decays into actinium-x.

Hahn began work at the Chemical Institute in Berlin in the fall of 1907, was appointed a Privat dozent (university lecturer) a year later, and was promoted to professor of chemistry in 1910. One of the most significant events of this period was the arrival of Lise Meitner as a student at the Chemical Institute. Prejudice against women in the sciences was very strong at the time, and Meitner was not allowed to work in the same laboratories as male students. Fischer did, however, allow her to share a tiny makeshift laboratory with Hahn in a converted workshop. Hahn and Meitner worked well together, with the former's chemical approach to problems complementing Meitner's outlook as a physicist. Thus a collaboration began that was to last for three decades. During their work together at the Chemical Institute, Hahn and Meitner concentrated on a study of beta emitters and, in the process, discovered more new radioactive isotopes.

In 1912, Kaiser Wilhelm authorized the establishment of a new research institute to consist of several separate departments. Invited to head up the section on radioactivity in the new Institute for Chemistry at Berlin-Dahlem, Hahn asked Meitner to join him there. One of the advantages of the new setting—in addition to much more space—was that radioactive materials had never been used in the rooms before, so that Hahn and Meitner were able to detect far lower levels of **radiation** than they could in their former laboratories. This allowed them to discover weakly radioactive isotopes of potassium and rubidium that had not yet been observed.

During this period, Hahn met Edith Junghans, whose father was a member of the Stettin City Council. The two became engaged in November 1912. They were married in Stettin in March 1913, and eventually had one child, a son, Hanno, born in April 1922.

Hahn's personal and professional life were soon to be disrupted by the onset of World War I. In July 1914 he was called into the army. After serving with distinction in the infantry at the battlefront, he was ordered back to Berlin, where he was assigned to work with a poison-gas research unit headed by Fritz Haber. For the next three years, Hahn shuttled back and forth between Berlin and the battlefronts, developing and testing new gases. He was horrified by the results he saw when gases were used in battle, but he had become convinced that such atrocities might bring the war to an early end and save lives, the same argument eventually made for the use of nuclear bombs—built on the principle of nuclear fission discovered by Hahn—at the end of World War II.

During his stays in Berlin, Hahn was able to spend some time in his own laboratory at the Institute for Chemistry. One of the projects that he and Meitner pursued during this period was a study of a new radioactive element that had previously been announced by Kasimir Fajans and Oswald Göhring in 1913, an element they had named "brevium." Hahn and

Meitner found the element in the residues of pitchblende ore and showed its relationship to parent and daughter isotopes. The name they suggested for the element, "protoactinium" (now protactinium), eventually became preferred to that recommended by Fajans and Göhring.

By the 1930s, Hahn's fame had begun to spread. In 1933, he was invited to spend a year as visiting professor at Cornell University and to deliver the prestigious Baker Lectures there. His visit to the United States was cut short, however, by news of the Nazi uprisings taking place in Germany. He decided that it was best for him to return to Berlin, which he did in the summer of 1933.

One of the great ironies of the 1930s was that while political turmoil was sweeping through the world's greatest scientific nation, Germany, momentous scientific discoveries with profound historical significance were also being made there. Hahn returned to a Germany where many of the world's finest scientists were either being expelled from their university posts or were fleeing Hitler's wrath for the United Kingdom, the United States, or other destinations.

In the midst of the political chaos around him, Hahn turned his attention to an exciting new field of research: the **neutron** bombardment of uranium and thorium. Two important recent discoveries convinced Hahn that he needed to go beyond the now nearly exhausted field of radioactivity. The first of these was the discovery by **Irène Joliot-Curie** and her husband **Frédéric Joliot-Curie** that an element can be made radioactive by bombarding it with **alpha particles**. The second was the discovery of the neutron by **James Chadwick** in 1932. These two discoveries had been utilized in the early 1930s by **Enrico Fermi**, who saw that neutrons would be far more effective "bullets" for bombarding atomic nuclei than were alpha or beta particles. In a short time, Fermi had used this technique to convert dozens of stable elements to radioactive forms. He was especially interested in trying this technique on uranium, since the predicted result of that reaction would be an element with an atomic number one greater than uranium. Since that element had never been observed in nature, a successful result of this experiment would be the formation of the world's first synthetic element. When Fermi carried out this experiment, the results he obtained were ambiguous, and he could draw no firm conclusion as to what had happened as a result of bombarding uranium with neutrons.

It was this problem to which Hahn turned in 1934. To work with him and Meitner on the problem, Hahn selected another radiochemist who had joined the institute in 1929, Fritz Strassmann. Over the next four years, Hahn, Meitner, and Strassmann analyzed the complex mixture of isotopes formed when uranium is bombarded with neutrons. At first, they worked on the assumption that Fermi's hypothesis was correct and that isotopes with atomic numbers from about 90 to about 94 would predominate in the mixture. After all, the only **nuclear reactions** with which scientists were then familiar were those in which the atomic numbers of products differed from those of the original material by only one or two. By 1938, however, Hahn was convinced that something very different was taking place in this reaction. The evidence from his

chemical analyses had become overwhelming. The product that had originally seemed to be an isotope of radium was actually an isotope of barium, whose atomic number is 36 less than that of uranium. Hahn and Strassmann were not completely willing to accept the apparent meaning of these results, however. In their January 6, 1939, paper announcing their results, they noted that, as chemists, they felt certain that barium was one of the products of the reaction, but they admitted that as nuclear physicists they were not yet willing to accept the "big step" that this conclusion suggested.

The paper carried the names of Hahn and Strassmann only because their colleague Meitner had been forced to flee Germany as a result of the Nazi purge of Jewish scientists. From her new home in Sweden, however, Meitner stayed in contact with her colleagues in Berlin. When details of the forthcoming Hahn-Strassmann paper reached her, she continued to think about the problem. The solution came during a Christmas outing with her nephew Otto Robert Frisch. Meitner and Frisch finally realized that the Hahn-Strassmann results could be explained only by accepting that the uranium **nucleus** had been split into two large, roughly equal parts, a reaction to which Meitner gave the name *nuclear fission*. For his part in the discovery of nuclear fission, Hahn was awarded a share of the 1944 Nobel Prize in chemistry.

During World War II, Hahn remained in Germany. Although he had no love for the Nazi party, he felt a loyalty to his homeland. He was able to avoid working on the German **atomic bomb** project, however, and instead carried out research on fission fragments during the war. That research came to an end in March 1944 after the Institute for Chemistry was destroyed in a series of bombing raids. Hahn and his wife soon moved to the southern German town of Tailfinger, where they were captured by an advance team of U.S. intelligence officers in April 1945. The Hahns were sent to England, where they were held for almost a year.

On January 3, 1946, Hahn was permitted to return to Germany, where he became president of the Kaiser Wilhelm Society, then re-named the Max Planck Society. He devoted the next 15 years to the effort to rebuild the scientific community of his native land. In his later years, he was showered with honors and awards, including election to scientific societies in Berlin, Göttingen, Munich, Halle, Stockholm, Vienna, Madrid, Helsinki, Lisbon, Mainz, Rome, Copenhagen, and Boston. He also became active in the international movement to control **nuclear weapons** and helped draft the 1955 Mainau Declaration by Nobel laureates, warning of the dangers of misusing nuclear **energy**.

The last years of Hahn's life were filled with personal tragedy. He was shot in the back in 1951 by a disgruntled inventor and had scarcely recovered before his wife suffered a nervous breakdown. In 1960, his son Hanno and Hanno's wife were killed in an automobile accident in France, leaving their young son to Hahn's care. Distraught at the accident, Hahn's wife never recovered her health. In the spring of 1968, Hahn was seriously injured while getting out of a car. His health slowly deteriorated until he died on July 28. His wife also died two weeks later. He was buried in Stadtfriedhof, Göttingen, West Germany.

HALE, GEORGE ELLERY (1868-1938)

American astrophysicist

George Ellery Hale was an American astronomer and astrophysicist who pioneered spectroscopic research and the development and use of large telescopes. He is widely considered to be the father of modern solar observational **astronomy**, and is best known for his discovery in 1908 of the presence of magnetic fields in sunspots. Hale invented new instruments for studying solar and stellar spectra, including the spectroheliograph and the spectrohelioscope, and also founded three of the United States's leading observatories, at Yerkes, Mt. Wilson, and Palomar; the latter two were renamed the Hale Observatories in 1969.

Hale was born on June 29, 1868, in Chicago, Illinois. His father, William Ellery Hale, was a wealthy businessman who manufactured hydraulic elevators; his mother, Mary Scranton Browne, was the daughter of a Congregational minister who later became a doctor. Hale inherited from his mother an intense love of literature. As a child he read *Don Quixote,* the *Iliad,* Jules Verne's *From the Moon to the Earth,* and the poetry of Shelley and Keats. Hale was a sickly and overwrought child, subject to frequent bouts of illness; his delicate constitution troubled him throughout his life. As an adult he was overanxious and unable to relax, and suffered three nervous breakdowns.

Hale began his education at Oakland Public School, and later attended Adam Academy and the Chicago Manual Training School, where he was instructed in shopwork. His inventiveness manifested itself early, and was encouraged by his parents. As a boy, he tinkered in his self-constructed tool shop and laboratory, conducting simple experiments and building himself a **telescope** and other instruments. A neighbor and amateur astronomer, Sherburne W. Burnham, sparked his interest in astronomy, and particularly in **spectroscopy**, the study of solar and other spectra using a spectroscope. Burnham interested Hale in a secondhand Clark telescope that had come on the market, and Hale persuaded his father to buy it. The views it afforded of the **Moon** and other **planets** further excited Hale's interest in astronomy and strengthened his resolve to make it his life's work. In 1884, at the age of 16, he succeeded in photographing the solar spectrum using a prism spectroscope purchased for him by his father. In 1886, Hale entered the Massachusetts Institute of Technology (MIT) to study physics, chemistry, and mathematics. He found college less stimulating than his own experiments and found an outlet for his unsatiated intellectual energy in the independent study of spectroscopy and astronomy. Hale also volunteered as an assistant at the Harvard College Observatory, and, with money given by his father, set about building his own solar observatory at Kenwood, behind his house.

Hale made his first scientific breakthrough in 1889, when he hit on an idea for an invention that would permit him to photograph solar prominences in full daylight. Solar prominences are flames or masses of incandescent vapor, composed usually of **helium** and hydrogen, that extend beyond the **Sun**'s disk. They are usually of enormous height and brilliant red in color. Although they vary in appearance, they are usually manifested in an arch or bridge shape, and are found near sunspots and flares. They can remain inactive for months before entering a dynamic stage. During this period, they change shape, sometimes obtaining heights equivalent to the diameter of the Sun. They are visible during a solar eclipse or with the aid of a spectroscope. Unsuccessful attempts had previously been made to photograph the prominences; Hale's idea was to use an instrument of his own invention that he called a spectroheliograph. It relied on the **light** of one spectral line, selected by a slit, to photograph by continuously scanning either the whole or part of the Sun. In his senior thesis, he outlined the results he obtained at the Harvard Observatory using this apparatus. Hale also published his first paper, entitled "The New Astronomy," in 1889. It appeared in the July issue of the *Beacon,* a **photography** journal.

Hale and his fiancée, Evelina Conklin of Brooklyn, New York, married on June 5, 1890, the day after Hale graduated from MIT. They had two children, Margaret and Bill. During their honeymoon, the newlyweds visited the Lick Observatory on Mount Hamilton, near San Jose, California, where Hale viewed the 36-inch Lick telescope, then the largest in the world. Shortly afterward, Hale became the director of the Kenwood Observatory and persuaded his father to buy him a new 12-inch refracting telescope. He used it to continue his experiments with the spectroheliograph, and succeeded in photographing the bright calcium clouds (flocculi) and the prominences around the Sun's limbs for the first time.

The following year, Hale visited Europe and met many of that continent's leading astronomers and astrophysicists. They received him warmly. Upon his return, he founded the journal *Astronomy and Astrophysics* with William Payne, editor of the *Sidereal Messenger.* Hale became associate professor of **astrophysics** at the newly established University of Chicago in 1892. Later that year, hearing of the availability of two 40-inch **lenses**, he persuaded the traction magnate Charles T. Yerkes to sponsor a new observatory that would house a refracting telescope revolutionary in its focal length and light-gathering **power**. The vision and organizational ability demonstrated by Hale in this and later instances did much to advance astronomical research.

Hale traveled to Germany in 1893 while plans for the new observatory were being fine-tuned. At the University of Berlin, he worked with some of Germany's finest physicists, including **Max Planck**, **Hermann von Helmholtz**, and Heinrich Rubens. Although Hale's intention was to work towards a doctorate, this plan was later abandoned and Hale never received his Ph.D. On route back to the United States, he tried but failed to photograph the solar corona from Mt. Etna.

In 1895, with James Keeler as coeditor, Hale founded a second publication, the *Astrophysical Journal,* whose editorial board was composed of the world's leading astrophysicists. It quickly established itself as the leading journal in its field. The following year, Hale persuaded his father to provide the disk for a 60-inch reflecting telescope that would enable him to photograph the spectra of **stars** on so large a scale that their chemical makeup, the **temperature** and **pressure** in their atmospheres, and their motions could be investigated.

The Yerkes observatory opened in 1897. Hale staffed it with a small but highly talented and dedicated team of scientists, many of whom went on to become famous in their own right. They carried out spectroscopic, photographic, bolometric, and other types of optical experiments. Hale devoted himself to continued observation of sunspot spectra and the design of the Rumford spectroheliograph that he planned to attach to the 40-inch telescope. Hale used these tools to attain insight into the complete circulatory processes of the Sun. He also investigated the stars, especially the low-temperature, red stars known as Secchi's fourth type, which he was the first to photograph.

Besides being a tireless researcher, Hale devoted much of his energy to institutional work and strengthening ties between astrophysicists around the world. In 1899, he helped found and became the first vice president of the American Astronomical and Astrophysical Society (the name was changed to the American Astronomical Society in 1914). Upon election to the National Academy of Sciences in 1902, he threw himself into reforming that body. In 1904, Hale helped to establish a committee to organize the International Union for Cooperation in Solar Research, under the auspices of the academy, and went on to become its president.

With the Yerkes observatory up and running, Hale was anxious to apply his organizational powers to the establishment of a second observatory that would house his hoped-for 60-inch telescope. He obtained $150,000 from the Carnegie Institute of Washington as seed money to found the Mt. Wilson Solar Observatory in Pasadena, California, in 1904. Hale was its director from 1904 to 1923. It was from Mt. Wilson that Hale, using a Snow telescope of his own design, took the first photographs of a sunspot spectrum in 1905. Sunspots are visible spots on the Sun's disk, presumed to be of the nature of cyclonic storms in the Sun's atmosphere. On analysis, Hale's spectroscopic results revealed that sunspots are colder than other regions of the solar disk, rather than hotter as was previously assumed.

In 1906, Hale became a trustee of the little-known Throop Polytechnic School in Pasadena. He used his position to transform it into a first-rate technological college. Hale was passionately interested in Pasadena's future and wished to see the town invigorated by the establishment there of a world-class research center. His ambition was realized when the Los Angeles transport giant Henry Huntington agreed to finance such a center. For his work on behalf of Pasadena, Hale was given the city's highest accolade, the Noble Medal.

In 1908, Hale began construction of a 60-foot tower telescope with a 30-foot spectrograph in an underground dugout at Mt. Wilson. Using photographic plates sensitive to red light, he was able to detect hydrogen flocculi near sunspots. He was alerted to the possibility of there being powerful magnetic fields within sunspots and shortly afterwards was able to prove his hypothesis. Hale found that the magnetic fields in sunspots indirectly act as thermal shields and maintain sunspots' temperatures below that of the surrounding photospheres. Shortly afterwards, Hale achieved another breakthrough when he announced his fundamental polarity law. This states that 22-23-year intervals occur between the successive appearances in high latitudes of sunspots of the same magnetic polarity.

Having finally obtained his 60-inch reflecting telescope, then the world's largest, Hale hoped to achieve an understanding of the nature of **stellar evolution**. He also turned his attention to the question of whether the Sun itself might be a magnet. By 1912, work was completed on the construction of a 150-foot solar tower telescope with a 75-foot vertical spectrograph. Hale used it to measure the Sun's magnetic field. He found that the Sun has a dipole magnetic field of about 20 gausses. A gauss is equal to the intensity produced by a magnetic pole of unit strength at a distance of one centimeter. Later measurements taken by other astrophysicists found the field to be about four gausses. None of the results were conclusive, however. In fact, it was not until 1952 that the first accurate readings of the Sun's **magnetic field** were obtained by two American astrophysicists. They gave it as a polar field of about two gausses, with a polarity opposite to that of Earth. The astrophysicist Robert Howard said during a speech at the Hale Centennial Symposium at Dallas, Texas, in 1968, "It is clear to us now that magnetic fields hold the key to the phenomenon called solar activity, and it is a tribute to the genius of Hale that he recognized at such an early stage the great importance of these elusive magnetic fields."

During World War I, Hale was an emissary from the National Academy to President Woodrow Wilson, offering its services. The result was the establishment of the National Research Council in 1916, with Hale as its chairman. Its purpose was to stimulate research in the sciences and in the application of the sciences to socially useful fields, such as medicine, agriculture, and engineering, for both peaceful and defensive purposes. In 1918 Hale, inspired by the success of the National Research Council in promoting cooperation between scientists and other professionals nationally, proposed establishing a similar body on an international scale. Thus was born the International Research Council, which eventually replaced the International Association of Academies in 1919. Until the war ended, though, only scientists from allied countries were allowed membership. By 1931, when it was renamed the International Council of Scientific Unions, 40 countries had joined the council, and in 1932, Hale became its president.

Meanwhile, in his scientific work, by 1917 Hale was ready to begin using a new 100-inch telescope, which had been built with the support of the Carnegie Foundation and a Los Angeles businessman named John D. Hooker. It was used by two members of Hale's team to measure the diameter of the huge red star Betelgeuse, which was found to be 300 million miles. In 1923, **Edwin Hubble** used it to identify a Cepheid variable in a spiral nebula and prove that **nebulae** are external galaxies of dimensions similar to Earth. Without the hundred-inch, such a breakthrough would have been impossible.

In 1923, troubled by continued ill health, Hale retired his directorship of Mt. Wilson, and turned his attention to the building of another observatory, the Hale Solar Laboratory, at Pasadena. There, he invented and built the world's first spectrohelioscope, a spectroscope with an oscillating slit, for use in the study of solar phenomena. The following year, he had the satisfaction of seeing his long-held dream of reinvigorating the National Academy of Sciences come to fruition. With a grant of five million dollars from the Carnegie Corporation, a permanent scientific headquarters for the academy was opened in Washington. Hale's work on behalf of the academy went a long way towards revitalizing that institution.

In retirement, Hale was as busy as he had been at any time in his career. In 1928, he launched an ambitious plan to build a 200-inch telescope. He was once again able to persuade wealthy investors to contribute to his visionary plans. The International Educational Board of the Rockefeller Foundation donated six million dollars to the California Institute of Technology to build the telescope, in cooperation with the Mt. Wilson observatory. The completed instrument was set up atop Palomar Mountain in southern California. Unfortunately, Hale did not live to see it completed, although it was named the Hale telescope in his honor. He died on February 21, 1938, in Pasadena. Sir **James Jeans** remembered him thus: "Nature had not only endowed him with those qualities that make for success in science—a powerful and acute intellect, a reflective mind, imagination, patience, and perseverance—but also in ample measure with qualities that make for successes in other walks of life—a capacity for forming rapid judgments of men, of situations, and of plans of action; a habit of looking to the future, and thinking always in terms of improvements and extensions; a driving power which was given no rest until it had brought his plans and schemes to fruition; eagerness, enthusiasm, and above all a sympathetic personality of great charm."

HALF-LIFE

Half-life, in general terms, is the time required for half of something to undergo a process. The term has several uses but in all of those uses it means essentially the same thing. In dealing with radioactive isotopes, the term refers to the time required for half of the atoms of a radioactive substance to disintegrate. It can also be used in conjunction with ecological systems such as drug or radioactive tracer dissipation in the human body. In these circumstances it is the time required for half of the drug or tracer to be removed from the body. Half-life is also commonly used in the study of chemical **kinetics**. Here it is defined to be the time required for the reactant concentration to decay to one-half its initial value.

The half-life of radioactive isotopes is as varied as the number of isotopes known. Astatine, the heaviest known halogen and a radioactive element, has a half-life of 7.21 hours. It is often used as a radioactive tracer because it collects in the thyroid gland. Americium, a radioactive element discovered by Glenn T. Seaborg (1912-1999) in 1944 during bombardment of plutonium, is part of the actinide series. Many isotopes of this element have half-lives that range from 1.3 hours to over 7,000 years and are used in smoke detectors. It is clear from the use of this element that although some elements are radioactive they can still be effectively used in everyday functions for relatively long periods. The most sta-

ble **isotope** of actinium, a radioactive element discovered in 1899 by André Debierne in uranium residues from pitchblende and the first member of the actinide series, has a half-life of 21.6 years. So it is evident that there is a wide range of half-lives associated with **radioactive decay**. It should also be noted that the half-life of the shortest living elementary particles can be about the time for **light** to travel one nuclear diameter (these particles are better called resonances).

Half-life is also used in conjunction with chemical kinetics. In this situation the term usually refers to the time it takes for half of a substance to undergo a **chemical reaction** and be converted into a product, that is the time required for the reactant concentration to decay to one-half its initial value. In the mathematical analysis of such reactions the half-life is an important factor. If the reaction is first order, meaning simply that a reactant A goes to product B, the half-life is constant and depends only on the rate constant, k, that describes the reaction. So for a first order reaction, $t1/2 = ln2/k = 0.693/k$, where k is the rate constant for the reaction. This illustrates that the half-life of the reaction is independent of concentration for a first-order reaction. For a zero order reaction the half-life is found to be equal to the initial concentration of reactant divided by two times the rate constant for that reaction. For a second order reaction, one in which two molecules collide and form a product, the half-life is known to be equal to one divided by the initial concentration of reactant molecules times the rate constant for the reaction. It can be seen from this that the half-life of a chemical reaction depends upon the type of reaction being described.

HALL, EDWIN HERBERT (1855-1938)
American physicist

Edwin Herbert Hall was born on November 7, 1855, in Great Falls (now named North Gorham), Maine. After graduating from Bowdoin College in 1875, he became a graduate student at Johns Hopkins University, where he studied under experimental physicist **Henry Augustus Rowland**. Rowland, who was only seven years older than Hall, was very interested in **electricity**. One of his own experiments showed that a moving electrostatic charge had the same magnetic effect as an **electric current**.

Rowland stimulated Hall's interest in electricity and encouraged Hall to question a statement that **James Clerk Maxwell** had made in his 1873 *Treatise on Electricity and Magnetism*. Maxwell had said that the **force** acting on an electric conductor in a magnetic field acts on the conductor directly and not on the electric current.

That an electric current could produce a magnetic field was nothing new; **Hans Christian Ørsted** had discovered the principle in 1819. Shortly thereafter, **Michael Faraday** and **Joseph Henry**, working independently, showed that a magnetic field could induce an electric current. **Maxwell's equations** defined electricity and **magnetism** in terms of mathematics, unequivocally showing how one force affected the other.

Hall went to work, experimenting with one type of conductor after another. His persistence paid off when he discovered, in 1879, that an electric current flowing through a gold conductor in a magnetic field produced an **electric potential** that was perpendicular to both the current and the field. The strength of the potential was directly related to the strength of the magnetic field and created a transverse current in the conductor. **William Thomson Kelvin** said Hall's discovery was as great as any of Faraday's. It encouraged other scientists to begin their own studies, which eventually led to the discovery of three other transverse effects.

It was assumed that the "**Hall effect**" was due to the interaction between the current and the field. That was an excellent assumption, but an incorrect one. Other metal conductors produced an effect that was opposite to the predicted direction.

Hall received his Ph.D. for his research in 1880. The following year he went to Europe where he made measurements of the Hall effect in the laboratory of **Hermann von Helmholtz**, who was one of the greatest scientists of that time. When Hall returned to the United States, he took a job as an instructor at Harvard and became professor emeritus in 1921. Starting in 1911, he experimented with determining accurate values of the four transverse effects, a project in which he was still involved at the time of his death on November 20, 1938.

In 1966 it was discovered that the Hall effect influenced the electrons located at the interface between a semiconductor and an electric insulator. The investigation of objects at the subatomic level is the realm of **quantum theory**; at this level the interaction is known as the quantum Hall effect. There were some anomalies that had been blamed on "impurities," but physicist **Klaus von Klitzing**, using *metal oxide semiconductor field-effect transistors* (MOSFETs), showed this was not the case. Impurities were responsible for specific plateaus in the Hall resistance, but did not cause the resistance.

Von Klitzing's discovery, which earned him the Nobel Prize for physics in 1985, sparked a number of innovations. For example, the accuracy in the measurement of the quantum Hall effect led to its being adopted as the international standard for resistance; the MOSFET was found to be nearly identical to other components that might be important in future generations of **computers**; and the physical mechanisms of the "fractional" quantum Hall effect are very similar to those of high-temperature **semiconductors**.

HALL EFFECT

A current-carrying body placed in a magnetic field with the current direction unaligned with the field experiences a **force** leading to a transient sidewise drift of the charge carriers of the current. This drift continues until the force is balanced by an electric field produced by the charge accumulating at points on the body's surface in the direction of the drift. At points on the body's surface opposite the direction of the drift, there will clearly be an equal depletion of charge, which is equivalent to an accumulation of charge of opposite sign. The electric field created by this transient behavior is called the Hall field and results in a potential difference between corresponding points on the two oppositely charged surfaces. Which of the two surfaces is at the higher potential is determined by the sign of the charge carriers. If the carriers are positive, the surface in the direction of their drift will be at the higher potential; if the carriers are negative, the surface in the direction of their drift will be at the lower potential. The phenomenon thus described is called the Hall effect after **Edwin H. Hall**, who discovered it in 1879. A little over a century later, it was discovered by **Klaus von Klitzing** that the Hall potential in a semiconducting material experiences quantum jumps as the magnetic field is increased when subjected to temperatures far below room **temperature**. This remarkable discovery has made it possible to measure an important constant of physics, called the fine structure constant, to a heretofore unattainable accuracy. Also, it provides scientists with a readily achieved standard for making accurate determinations of conductivity. For this discovery, von Klitzing was awarded the Nobel prize in 1985.

Of monumental importance to today's technology is a class of materials whose ability to conduct **electric current** increases with temperature and whose charge carriers can be either positive or negative, depending on the impurity introduced into them. These materials are called **semiconductors**, prime examples of which are the elements silicon and germanium. When these elements are given traces of the appropriate impurity element, they can be made into either p-type (containing positive carriers called holes) or n-type (containing negative carriers called electrons). The Hall effect is then used to confirm which type of material one is dealing with. Furthermore, by measuring the Hall potential, the current, the magnetic field, and the sample geometry, it is easy to calculate the number of charge carriers per unit volume in the material tested. In the 1940s, it was found that junctions could be formed with these two different types of semiconductors across which current could flow only in one direction. Devices of this kind are called **rectifiers** or **diodes** and are vital for converting alternating current to direct current, adding or removing audio and video signals from their carrier **waves**, and many other applications. It was also found that more than two junctions could be formed in one device, and these were called **transistors**. These devices were capable of being employed in amplifier and oscillator circuits in radios and TVs. Previously, rectifiers, amplifiers, and oscillators used **vacuum tubes** as their essential components, which were generally bulky, used lots of **power**, and burned out frequently. The new semiconductor devices had none of these problems. In the 1950s and 1960s, it was learned how to create many transistor circuits on a small chip using integrated circuitry. Without this new technology, the powerful **computers** that were used in our **space** program and are now found universally in the form of compact personal computers would not have been possible. Even more importantly, without the discovery of the Hall effect and its use in the scientific investigation of semiconducting materials, this sequence of developments could not have even begun. Finally, with the discovery of its large-scale quantum behavior, the future role of the Hall effect in the advancement of science and technology may eventually prove to be even greater than its past role.

HALLEY, EDMOND (1656-1742)

English astronomer

The son of a wealthy merchant, Halley was attracted to **astronomy** after seeing two **comets** as a child. By the age of 18 he had found errors in authoritative tables on the positions of Jupiter and Saturn and by 19 had published a paper on the laws of **Johannes Kepler**. In 1676 Halley left England for St. Helena, an island west of Africa, to map the southern constellations, a task never before undertaken. Although the climate of St. Helena proved less than ideal for Halley's purposes, he was able to catalog 341 **stars** before returning to England. His pioneering work on the island assured his place in England's scientific community and Halley was awarded a master's degree from Oxford as well as election to the Royal Society.

In 1684 Halley entered into a conversation with biologist **Robert Hooke** and architect Christopher Wren (1632-1723) that concerned the **force** that drove the movement of the **planets**. Unable to reach a satisfactory conclusion, Halley turned to his friend **Isaac Newton**. Discovering that Newton had already answered the question using his law of **gravity**, Halley convinced his reticent friend to publish his findings. Using funds bequeathed to him by his father, Halley financed the publication of *Mathematical Principles of Natural Philosophy*, now considered one of the classic texts of modern scientific thought.

At the age of 39 Halley turned his attention to comets, which, as they streaked unexpectedly through the sky, appeared ungoverned by Newton's law. Halley, however, believed that gravity did indeed dictate their path and that the rarity of their appearances was due to the vast length of their orbit, which was elliptical. With the help of Newton, Halley compared the paths of past comets that had appeared in 1531, 1607, and 1682. From this data he was able to determine that these seemingly separate comets were indeed the same comet and accurately predicted its reappearance in 1758. In 1705 Halley published his findings in *A Synopsis of the Astronomy of Comets*. Eventually, the comet that he predicted was named for him.

In addition to his findings concerning comets, Halley undertook a lengthy study of solar eclipses and discovered that the so-called fixed stars actually moved with respect to each other. He also wrote in favor of the theory that the **universe** is limitless and has no center. Halley's scientific interests, however, extended beyond astronomy. He played a major role in transforming the Royal Society from a social club into a well-respected clearinghouse for scientific ideas. He devised the first **weather** map and calculated the amount of salt deposited by rivers into seawater over millions of years, which allowed him to draw conclusions about the age of **Earth**. He also invented, developed, and tested one of the first practical diving bells. He served as chief science adviser to Peter the Great when the Russian tzar came to England in an attempt to integrate Western advances into his country's society. From 1698 to 1700, Halley commanded the *Paramour*, a Royal Navy ship, for a scientific expedition that studied the effects of

Edmond Halley.

Earth's magnetic field on magnetic needle compasses. He became Astronomer Royal in 1720 and continued to make astronomical observations and attend scientific meetings until shortly before his death. His full life was a testimony to the value of the pursuit of knowledge.

HARMONIC MOTION

Harmonic **motion** describes periodic motion wherein the displacement path is comprised of one or more symmetric vibratory motions about an **equilibrium** position. A particular form of harmonic motion is called simple harmonic motion.

For objects undergoing simple harmonic motion, the **velocity** of the displaced object constantly changes, oscillating just as does the magnitude of displacement (as measured from equilibrium). When the object's displacement is maximum, the velocity is zero and when there is zero displacement the object's velocity is maximum. With simple harmonic motion, **acceleration** also oscillates.

A **mass** hanging from a coiled spring will exhibit simple harmonic motion. When disturbed from equilibrium, it will it will cyclically move upwards and downwards in a sinusoidal, oscillating motion. **Robert Hooke** first formulated the law used to describe such simple harmonic motion. Hooke's law states that the **force** a spring exerts on a body that is suspended is directly proportional to the displacement (extension) of the spring. This is expressed as: F = -kx, where F is the force exerted by the spring, x is the displacement, and k is the force constant that characterizes different springs. The value of the spring constant k will depend on the composition and dimensions of the spring that determine its elastic properties.

From Hooke's law and from the relation F = ma, it can be inferred that acceleration is opposite to displacement in direction, and proportional to displacement in magnitude. This explains why the further an object is displaced, the faster it accelerates back toward equilibrium.

The example of a mass suspended by a spring is also known as a simple harmonic oscillator, which represents the most basic oscillating system possible. Harmonic oscillation is possible whenever a system has the following two quantities: **elasticity** and **inertia**. When it is displaced from its equilibrium position, the elasticity of the system provides the restoring force required to bring it back to its equilibrium position. This restoring force is equivalent to the force constant in Hooke's law. Inertia (a property of a body, due to its mass, that indicates resistance to a change in motion) and **momentum** allow the system to cyclically swing past the equilibrium position.

The forces driving harmonic motion (e.g., spring force) depend on elasticity or a capacity of materials to be restored to their original shape after undergoing reversible deformation.

Objects undergoing simple harmonic motion are said to exhibit periodic behavior. One example is sinusoidal motion, mathematically expressed as sin(x). Simple harmonic motion is sinusoidal motion, because position, velocity, and acceleration behave as sine or cosine functions when plotted versus **time**.

The harmonic oscillator model is sometimes used to describe the vibrations of atoms in molecules. In this model, the atoms are considered as masses connected by springlike bonds. The force constant (k) becomes a characteristic property of the vibrating molecular system and it can be evaluated by vibrational **spectroscopy** methods (which measure the **frequency** of vibrations) since the frequency of harmonic oscillation is independent of amplitude.

The movement of a pendulum also represents simple harmonic motion. If displaced from rest (i.e., its equilibrium position), it executes simple harmonic motion as it swings back and forth, accelerated by the force of **gravity** and its momentum. Because a pendulum exhibits simple harmonic motion, frequency is independent of the mass of the pendulum.

Many timekeeping devices utilize harmonic motion, from **Galileo**'s simple pendulum to the modern quartz crystals whose harmonic **oscillations** are transformed into electrical signals in watches and clocks. Quartz crystals are much more precise than pendulums because they oscillate more frequently than a pendulum.

See also Force, mass, and acceleration; Newton's laws of motion; Waves

HARMONICS

What makes a note from a musical instrument **sound** rich? The volume of the sound is determined by the amplitude of the **oscillations** in a sound wave, the distance individual molecules oscillate. A larger amplitude produces a louder sound and transmits more **energy**. The pitch of a note is the **frequency** or number of oscillations per second. A higher frequency produces a higher pitched note. The richness or quality of a sound is produced by the harmonics.

A pure note consisting entirely of one frequency will sound boring. A musical instrument that only produced such pure notes would not sound pleasing. The harmonics are missing. The harmonics are integer multiples of the fundamental frequency. The first harmonic is the fundamental frequency, 264 cycles per second for middle C. The second harmonic will be twice this frequency, 528 cycles per second, which is an octave higher. The third harmonic will be three times the fundamental frequency, 792 cycles per second, and so on. These harmonics are also called overtones—the second harmonic is the first overtone, the third harmonic the second overtone, and so on.

The violin, piano, and guitar all produce sounds by vibrating strings. Playing the same note, say middle C, will produce a tone with a fundamental frequency of 264 cycles per second. Yet all three instruments sound different. Why? They have different harmonics. The amount of each harmonic present is what gives each musical instrument its own unique sound. A well-made instrument will sound richer than a poorly made one because it will have better harmonics. An instrument with no harmonics will sound like a tuning fork with only one fundamental frequency present.

For reasons that we do not understand, sounds composed of harmonics whose frequencies are integer multiples of each other sound pleasing to the human ear. They are music. On the other hand, sounds composed of frequencies that are not integer multiples of each other are dissonant noise to our ears. So turn on the stereo to your favorite music and enjoy the harmonics.

HAWKING, STEPHEN (1942-   )
English theoretical physicist

Stephen Hawking has been called the most brilliant theoretical physicist since **Albert Einstein**. His work concentrates on the puzzling cosmic bodies called **black holes** and extends to such specialized fields as **particle physics**, supersymmetry, and **quantum gravity**. The origin and fate of the **universe** are a central concern of Hawking's work. Though few people are able to understand these abstruse subjects, Hawking has gained a worldwide following, not only among other scientists, but among a great many laypeople. As an author and lecturer, he has become so famous as to approach rock-star celebrity.

Stephen William Hawking was born on January 8, 1942, in Oxford, England. He often refers to the fact that his birth

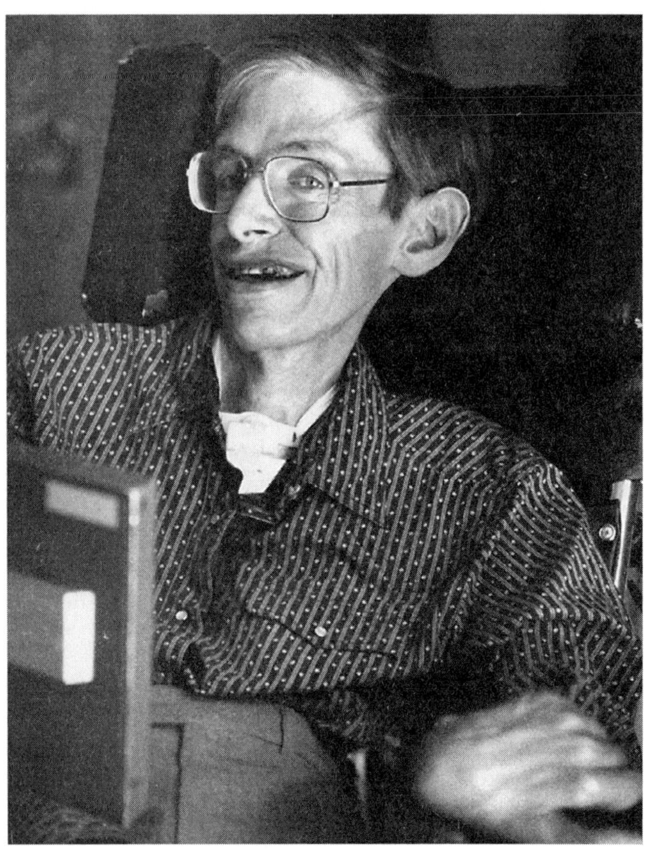

Stephen Hawking, 1992. *(AP/Wide World Photos. Reproduced by permission.)*

date coincided with the 300th anniversary of **Galileo**'s death. Hawking was the eldest child of an intellectual and accomplished family. His father, Frank Hawking, was a physician and research biologist who specialized in tropical diseases; his mother, Isobel, the daughter of a Glasgow physician and a well-read, lively woman, was active for many years in Britain's Liberal Party. After Stephen's birth, the Hawkings had two daughters, Mary and Philippa, and adopted a boy, Edward.

Stephen Hawking's earliest years were spent in Highgate, a London suburb. In 1950, when he was eight, the family moved to St. Albans, a cathedral town some 20 miles northwest of London. Two years later, his family enrolled him in St. Albans School, a private institution affiliated with the cathedral. As Michael White and John Gribbin describe the young schoolboy in *Stephen Hawking: A Life in Science,* "He was eccentric and awkward, skinny and puny. His school uniform always looked a mess and, according to friends, he jabbered rather than talked clearly, having inherited a slight lisp from his father." Young Hawking's abilities made little impact on his teachers or fellow students. But he already knew he wanted to be a scientist, and by the time he reached his middle teens, he had decided to pursue physics or mathematics.

Awarded Oxford Scholarship

Gangly and unathletic, Hawking formed close friendships with a small group of other precocious boys at school. Intrigued by subjects that focused on measurable quantities and objective reasoning, Hawking began to show increasing skill at mathematics, and soon he was outdistancing his peers with high grades while spending very little time on homework. In 1958 Hawking and his pals built a primitive computer that actually worked. In the spring of 1959, Hawking won an open scholarship in natural sciences to University College, Oxford—his father's old college—and in October he enrolled there. It was at Oxford that his unusual abilities began to become more obvious. Hawking's ease at handling difficult problems made it seem to others that he didn't need to study. In *Stephen Hawking's Universe,* John Boslough wrote, "He took an independent and freewheeling approach to studies although his tutor, Dr. Robert Berman, recalls that he and other dons were aware that Hawking had a first-rate mind, completely different from his contemporaries."

Pioneers Studies in Black Holes

In 1962, after receiving a first-class honors degree from Oxford, Hawking set off for Cambridge University to begin studying for a Ph.D. in **cosmology**. Now he was beginning to deal with some of the themes that would preoccupy him throughout his life. One of these was the poorly understood question of black holes. As scientists were later to realize, a black hole is a cosmic body that by its very nature can never be seen. One type of black hole is thought to be the remnant of a collapsed star, which possesses such intense **gravity** that nothing can escape from it, not even **light**. Hawking was also intrigued by "**space-time** singularities," those phenomena in the physical universe or moments in its history where physics seems to break down. In attempting to understand a black hole and the space-time **singularity** at its center, Hawking made pioneering studies, using formulas developed more than half a century earlier by Einstein.

Hawking received his Ph.D. in 1965 and obtained a fellowship in theoretical physics at Gonville and Caius College, Cambridge. He continued his work on black holes, frequently collaborating with **Roger Penrose**, a mathematician a decade his senior, who like Hawking was deeply interested in theories of space-time. Though still in his twenties, Hawking was beginning to acquire a reputation, and he would often attend conferences where he shocked people by questioning the findings of eminent scientists much older than himself.

In 1968, Hawking joined the staff of the Institute of Astronomy in Cambridge. He and Penrose began using complex mathematics to apply the laws of **thermodynamics** to black holes. He continued to travel to America, the Soviet Union, and other countries, and in 1973, he published a highly technical book, *The Large Scale Structure of Space-Time,* written with G. F. R. Ellis. Not long afterward, Hawking made a startling discovery: whereas virtually all previous thinking assumed that black holes could not emit anything, Hawking theorized that under certain conditions they could emit **subatomic particles**. These particles became known as **Hawking radiation**.

Receives Albert Einstein Award

Early in 1974, at the unusually young age of 32, Hawking was named a fellow of the Royal Society. Soon afterward, he spent a year as Fairchild Distinguished Scholar at the California Institute of Technology in Pasadena. On returning to England, he continued to work toward a theory of the origin of the universe. In this endeavor, he made progress toward linking the theory of relativity, which deals with gravity, with **quantum mechanics**, which deals with minuscule events inside the **atom**. Such a theoretical linkage, long sought by researchers, is called the **Grand Unified Theory**, or GUT. In 1978, Hawking received the Albert Einstein Award of the Lewis and Rose Strauss Memorial Fund, the most prestigious award in theoretical physics. The following year he coedited a book with Werner Israel, called *General Relativity: An Einstein Centenary Survey.* In 1979, Hawking was named Lucasian Professor of Mathematics at Cambridge—a position held three centuries earlier by Sir **Isaac Newton**. In the 1980s, his work was beginning to lead him to question the **big bang** theory, which most other scientists were accepting as the probable origin of the universe. Hawking now asked whether there really had ever been a beginning to space-time (a big bang), or whether one state of affairs (one universe, to put it loosely) simply gave birth to another without beginning or end. Hawking suggested that new universes might be born frequently through little-understood anomalies in space-time. He also investigated **string theory** and exploding black holes, and showed mathematically that numerous miniature black holes may have formed early in the history of our universe.

Publishes Popular Book on the Cosmos

In 1988, Hawking's *A Brief History of Time: From the Big Bang to Black Holes* was published. Intended for a general audience, it leapt onto best-seller lists in both America and Britain and remained there for several years. In that book, Hawking explained in simple language the evolution of his own thinking about the **cosmos**. Major articles followed in *Time, Popular Science,* and other magazines; films and television programs featured Hawking. He received honorary degrees from many institutions, including the University of Chicago, Princeton University, and the University of Notre Dame. His numerous awards included the Eddington Medal of the Royal Astronomical Society, in 1975; the Pius XI Medal, in 1975; the Maxwell Medal of the Institute of Physics, in 1976; the Franklin Medal of the Franklin Institute, in 1981; the Medal of the Royal Society, in 1985; the Paul Dirac Medal and Prize, in 1987; and the Britannica Award, in 1989.

In 1965, Hawking married Jane Wilde, and they had two sons and a daughter. The couple separated in 1990. Hawking suffers from amyotrophic lateral sclerosis, also called Lou Gehrig's disease, which confines him to a wheelchair and requires him to use a computer to speak.

HAWKING RADIATION

Hawking **radiation** is a quantum effect of general relativity. English theoretical physicist **Stephen Hawking** predicted that a black hole ought to emit **blackbody radiation**. He eventually formulated a quantum-mechanical mechanism of what is now termed Hawking radiation.

In 1958, David Finkelstein, a young postdoctoral fellow at the Stevens Institute of Technology discovered a new **reference frame** in which to investigate the Schwarzschild **space-time** geometry surrounding a collapsing star. Expressed in Finkelstein coordinates, the Schwarzschild horizon could be easily seen for what it actually was, a non-singular and well-behaved spherical surface in space-time that acted precisely as a semipermeable membrane. Anything could fall in through the horizon, but nothing could emerge. Inside the horizon the trajectory of any particle, including a **photon**, traveling initially in any direction, including radially outward, led inexorably towards and always ended at the **singularity** of infinite curvature.

In 1967, armed with the new understanding of gravitational collapse provided, in part, by Finkelstein's coordinate system, American physicist **John Wheeler** introduced the now-familiar name for the object left behind after the collapse of a star. Because no **light** emitted within the horizon could emerge, and any light incident on the horizon fell through without being reflected, he called the object a black hole.

In February 1974, the first Oxford University symposium on **quantum gravity** was held, and the conference proved historic. It was there that Cambridge physicist Stephen Hawking first announced that **black holes** were not black at all. He showed that when quantum-mechanical effects were included in the description (as they must be), black holes radiate. His calculation was lengthy and complicated, and his result was initially met with much skepticism, since no complete theory of **quantum gravity** existed (nor does one exist yet). Still, Hawking's result soon gained wide acceptance among physicists.

Hawking showed that the hole emits particles of all types (e.g., photons, gravitons, neutrinos), and that the **energy** distribution (each emitted particle has a certain energy; the distribution is simply the number of particles emitted at each such energy value) was that of a perfect black body.

Blackbody radiation depends only upon the emitter's **temperature**. In particular, the rate at which it radiates is proportional to the temperature. The only information about a black hole that can escape the horizon is its **mass**; mass is the only parameter appearing in the Schwarzschild description of the exterior **space-time geometry**. Hawking showed that the role of temperature in the Planckian energy distribution was played by the inverse of the hole's mass. Thus, as the hole radiates, its mass decreases (the process is known as black hole evaporation), and its gravitational temperature increases, increasing the rate at which it radiates. This is an unstable, runaway process that ultimately results in a final explosion, and the disappearance of the hole. Black holes formed by stellar collapse have an extremely low temperature. Such a hole would require more time than the present age of the **universe** to completely evaporate.

The secret of how the emitted radiation escapes the one-way horizon resides in the **vacuum** fluctuations of the particle quantum fields. According to **quantum field theory**, virtual particle-antiparticle pairs are constantly being formed at all points in **space**, leaving behind, for the moment, a region of space with a negative energy **density**. The particles separate only a small distance before they rejoin and annihilate one another. The energy released in the annihilation removes the negative energy density in the formation region.

But suppose such a pair of photons is born just outside the hole's horizon. The photons move apart, but now they are subject to the hole's gravity. Black holes produce large tidal gravitational effects, which pull the particles violently apart, feeding tremendous energy into the pair. One photon travels outward, the other falls through the horizon, to be forever lost. The gravitational **acceleration** feeds enough energy into the system to both convert them into real particles, able to separate an arbitrary distance, and to remove the negative energy density left in the formation region. The outgoing particle, now real, is free to fly off. The process converts the hole's gravitational **potential energy** into the energy of the outgoing photon, and, through Einstein's famous mass-energy relation, reduces the hole's mass. The outgoing particles thus produced constitute the Hawking radiation of the black hole.

It is possible to compute the energy loss associated with a black hole (i.e., the energy transformed into photons streaming away from the black hole). Assuming energy is conserved in quantum gravity, it is logical to conclude that black holes can eventually evaporate.

See also Schwarzschild radius; Worm holes

HEARING

Hearing is the ability to detect and interpret **sound**. Ears collect and process sound **waves** and transmit the signals to the brain. The ear has three parts: the outer ear, the middle ear, and the inner ear. The outer ear, or auricle, focuses sound waves into the channel of the middle ear. The middle ear is a small cavity containing the eardrum. A narrow passageway called the Eustachian tube connects the middle ear to the throat and helps keep air **pressure** equalized both inside and outside the ear, preventing the rupture of the inner ear's membrane, called the eardrum. Three tiny little bones in the inner ear, called the malleus, incus, and stapes, help transmit the vibrations further into the ear. These bones are also known as the hammer, anvil, and stirrup, respectively, because of the objects they resemble.

The inner ear houses the most important parts of the hearing mechanism: the cochlea and the vestibular labyrinth.

The cochlea is a tiny snail shell-like structure that is lined with tiny hairs. The cochlea converts vibrations into nerve impulses that are carried by the otic nerve to the brain for interpretation.

The vestibular labyrinth consists of three elaborately formed fluid-filled canals that help us keep our balance. In addition to hearing, the ear might also be considered as the organ of **equilibrium**. Sometimes certain movements, such as spinning, can cause the fluid inside these canals to brush up against sensory hairs the wrong way. If the fluid continues to stimulate these hairs after the movement has stopped, dizziness occurs.

The ear has a number of mechanisms to protect the organ from damage. Tiny hairs inside the ear serve to move sound vibrations along, to protect the ear from drastic **temperature** changes, and to help trap particles that enter the ear canal. Wax glands also help protect the middle ear and sensory hair cells from loud noises. However, very loud noises can kill the cells, and these cells do not regenerate. The result is irreversible hearing damage.

The ear is a very sensitive instrument and can detect sounds as high in **frequency** as 20,000 cycles per second (hertz) and as low in intensity as 10^{-12} watts per square meter. The threshhold of human hearing is used as the reference point for measuring relative intensity of sound. One bel, the basic unit of sound intensity, is 10 times the lowest threshold of human hearing (10^{-11} watts/m^2). Repeated exposure to sound levels exceeding 7 bels (70 decibels) can be harmful. Sounds exceeding 120 decibels—slightly louder than a car horn heard at a distance of 1 meter—can be physically painful.

HEAT

The concept of heat has always been an important consideration in scientific thought. Some things are warmer than others. When warm objects are placed in contact with cold ones, the warmer object cools down as the colder one heats up. Apparently something causes an object to be warm and flows from warm objects to cold ones. This "something" is called heat.

The controversy over whether heat is a material substance or is due to the kinetic **motion** of particles continued into the early twentieth century. **Isaac Newton, Christiaan Huygens, Robert Hooke, Henry Cavendish,** Humphry Davy, and **Benjamin Thompson Rumford** supported the kinetic interpretation. Hermann Boerhaave, Antoine-Laurent Lavoisier and **Pierre-Simon de Laplace** favored the material interpretation.

Joseph Black and Johann Wilcke (1732-1779) developed the concept of specific **heat capacity**, defined as the amount of heat that raises the **temperature** of a substance by one degree. The empirical definition of heat is based on this concept: the amount of heat involved in a process is equal to the product of the specific heat of the substance multiplied by the change in temperature that occurs when heat is added to or taken away from the substance. The unit given to the amount of heat is the calorie, defined as the amount of heat that raises the temperature of 1 g of water by 1°C.

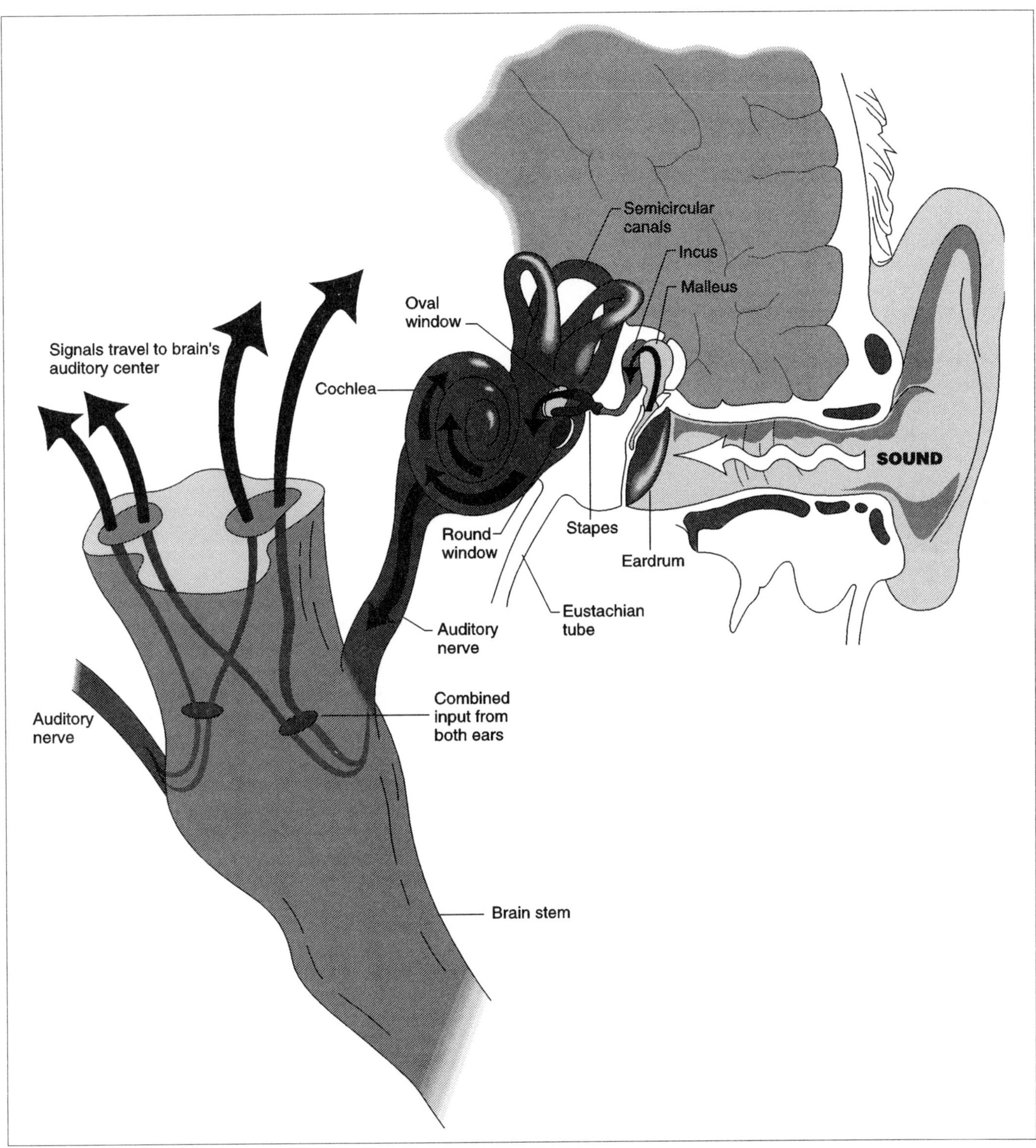

Diagram of a human ear.

James Joule showed that, unlike material substances, heat can be created and destroyed by changing it into **work**. One calorie of heat is equivalent to 4.2 joules of work. Subsequently, the kinetic theory proposed by A. K. Krönig (1822-1879) and **Rudolf Clausius** explained heat as the result of translational, rotational, and vibrational motions of molecules: the faster they move, the higher the temperature. The work of **Albert Einstein** finally clenched the argument in favor of the kinetic interpretation of heat.

HEAT CAPACITY

Heat capacity is the quantity of heat **energy** required to raise the **temperature** of a substance by 1°C. It is a measure of how well a substance stores heat. Numerically it is represented by the **mass** of a substance multiplied by that substance's specific heat. Historically heat energy has been defined in units of calories. One calorie is the quantity of heat energy required to raise the temperature of 1g of water from 14.5 to 15.5°C. This means that the specific heat of water is 1.0 cal g^{-1}°C^{-1}.

According to the **second law of thermodynamics**, heat flows from a warm body, one with a higher relative temperature, to a cooler body, one with a lower temperature. Heat flows between the two bodies until they reach the same temperature. The quantity of heat energy required to change the temperature of a substance is dependent upon how much the temperature is to be changed, the mass of substance undergoing the change and the type of substance it is composed of. Heat capacity, being the quantity of heat energy required to raise the temperature of a substance by 1°C, is clearly dependent upon the amount of substance undergoing the temperature change. The specific heat required to calculate the heat capacity is independent of the amount of substance by definition.

Another way to think about heat capacity is that it is equal to the heat energy added to the body, divided by the change in temperature. It is easier to think of heat capacity in this way when considering a substance's ability to retain heat. Substances with large heat capacities, like water, hold heat very efficiently. Other substances, such as lead, do not store heat well and therefore their temperature will rise significantly for a given amount of heat energy added.

An additional idea connected with heat capacity is that of the law of **conservation of energy** that states that in interactions among objects of substances, the total energy remains constant. This means that energy lost by one object must be gained by another object. This law plays an important role in determining the heat capacity of a substance. Since the total heat energy lost by one substance must be equal to that gained by another, one can easily determine the specific heats of substances experimentally. Once the specific heats are known, these values can be multiplied by each substance's mass to obtain the individual heat capacities.

HEAT ENGINES

A **heat** engine is defined as any engine that uses heat to perform **work**. As such, the physics underlying the operation of a heat engine is a direct consequence of the **first law of thermodynamics**. Heat engines take heat from a higher **temperature** heat source and transform it into work and a release of heat in the lower temperature surroundings. Heat engines can also convert internal **energy** (heat) into mechanical energy.

A classical description of a heat engine is provided by the Carnot cycle, named after its inventor **Nicolas-Léonard-Sadi Carnot**, who first designed it in 1824. The Carnot heat engine is only used as a standard of heat engine performance. It operates as a reversible process and consists of a gas confined by a piston moving back and forth in a cylinder. The process has four steps. First, the gas is heated and it expands, moving the piston up. This is called isothermal expansion to the higher temperature. The second step is adiabatic expansion of the gas to lower its temperature, followed by the third step, isothermal contraction, cooling the piston so that it will move back down. Finally, adiabatic contraction of the gas with warming to the higher temperature maintains a cycle of heating and cooling to displace the piston up and down. It is the **motion** of the piston that allows useful work to be done and the efficiency of a heat engine describes how efficient the engine is at turning heat into work. Two temperatures are then required to run a heat engine; at one temperature the system is heated, at the other temperature it is cooled.

The refrigerator and the air conditioner are examples of heat engines that run in reverse because they take heat from a lower temperature source through work performed by a compressor, and transfer it to higher temperature surroundings.

To be efficient, a heat engine must operate using a reversible process, that is, a process after which the thermodynamic system and its surroundings are returned to the state they were in before the process started. Theoretically, the most efficient heat engine is the Carnot cycle, although the Carnot cycle is not used for practical applications because of engineering difficulties that cannot be overcome.

An example of a thermodynamic cycle used to design practical heat engines is the Otto cycle (isentropic compression, reversible constant volume heating, isentropic expansion, reversible constant volume cooling) of which the gas engine is the most well-known application. Other examples of thermodynamic cycles used in heat engine design include the Diesel cycle (isentropic compression, reversible constant **pressure** heating, isentropic expansion, reversible constant volume cooling), the Brayton cycle (isentropic compression, isobaric heat supply, isentropic expansion, isobaric heat output) used for closed-cycle gas turbines in many industrial applications, the Stirling cycle (isochoric heating, isothermal expansion, isochoric cooling, isothermal compression), and the Clausius-Rankine cycle (isentropic expansion, isobaric heat rejection, isentropic compression, isobaric heat supply), which defines the operating principle of **steam engines** and turbines. Aside from their different operating principles and applications, these cycles also differ in their relative efficiency.

By dealing with thermodynamic quantities such as heat and work, the heat engine is also used to formulate the **second law of thermodynamics** in various ways. For example, the second law states that heat flows naturally from regions of higher temperature to regions of lower temperature and not the other way. Another form of the second law derived from the heat engine states it is impossible to build a heat engine that can use all the heat from a higher temperature surrounding it and turn it entirely into work. Accordingly, it is impossible to make a heat engine with a 100% efficiency.

See also Energy and work; Entropy; First law of thermodynamics; Heat transfer; Irreversible processes; Reversible processes; Thermodynamics

HEAT OF FUSION

Matter can exist in at least three physical states: solid, liquid, and gas. The physical state, or phase, of a substance is affected by **temperature**, or the amount of **heat energy** present, and by other factors such as **pressure**. Just as ice melts into water then boils into steam as heat is applied, many substances will pass through solid, liquid, and gaseous phases with the application of heat energy. The amount of heat energy required to melt a given **mass** of a solid at its melting temperature at a given pressure is called the heat of fusion of the substance. It is the amount of heat energy that is absorbed by the substance as it converts from solid to liquid without change in temperature. However, the heat of fusion may vary within a substance group because of small changes in the substance's composition.

Because a change in temperature does not occur during the phase shift, the heat of fusion is also sometimes referred to as **latent heat**. For example, when one gram of ice is steadily heated, its temperature will increase until it reaches a point where it begins to turn into liquid water. At this temperature, its melting point, the ice continues to melt into liquid water but the temperature remains the same even though heat is being applied. The temperature of the ice will remain the same until all is converted into water, after which the temperature will again begin to increase as heat is supplied. The latent heat energy required to change the gram of ice into a gram of water at the melting point, without a change in temperature, is the heat of fusion for water. Some gardeners use the principle of heat of fusion by putting buckets of water near plants when the temperature is expected to drop a bit below freezing to protect them from frost damage.

Many solids have characteristic heats of fusion and melting points, which are definitive physical characteristics of the substances. For example, ice requires about 80 calories of heat energy to melt one gram of ice into water at its melting point. Furthermore, the amount of heat that must be removed from a sample for it to freeze into a solid is equivalent to the heat of fusion required for melting.

HEAT OF VAPORIZATION

When a substance changes phase from liquid to gas, **heat energy** is required. The heat energy that is used to transform a liquid into a gas, or vapor, is called the heat of vaporization. As **thermal energy** is applied to a volume of a given liquid at a constant given **pressure**, the **temperature** of the liquid rises. With the continued application of heat, the temperature steadily increases until, at some characteristic point (called the **boiling** point) the liquid begins to shift from liquid to gas. While boiling, the temperature does not change even though heat energy is being added. The heat, instead of causing a rise in temperature, is absorbed by the molecules as they are liberated into gaseous form. The heat that is required in this process is the heat of vaporization for the liquid. However, the amount of heat required is always dependent on the external pressure exerted on the liquid and the actual composition of the liquid itself.

Because heat energy is consumed without a concomitant increase in temperature, it is also known as **latent heat** of transformation. Heat of vaporization for substances is dependent on a number of different factors, including pressure and external temperature. For example, the human body exploits the vaporization of water to cool itself through sweat—and it does so at temperatures far below water's boiling point. Substances like rubbing alcohol have vaporization points that are even lower than their boiling points. Heat of vaporization is often expressed as the amount of heat (in joules) that is required to change 1 gram of liquid into gas. For example, the heat of vaporization of water is 2,260 joules of heat per gram. At sea level, this translates into about 540 calories per gram of water vapor.

This relatively high value is an important characteristic of water because it allows water to absorb large amounts of heat as it changes phase. This quality explains how perspiration cools body temperature as it evaporates from skin, and why burn injuries from steam are more serious than from water at the same (boiling) temperature.

HEAT TRANSFER

Heat transfer is the net passage of **energy** as a result of **temperature** differences. This energy is transferred in the direction of decreasing temperature until thermal **equilibrium** (equality of temperatures) is achieved. The basic mechanisms involved in this process include **radiation** (the transfer of energy in the form of **electromagnetic waves**) and conduction (the transfer of **kinetic energy**). Heat transfer in **fluids** can occur at a faster rate, because large masses of a fluid can be displaced and can mix with other fluid masses of different temperatures. This process is considered a distinct mechanism called convection. In many heat transfer processes, radiation and convection or conduction work together, although one is often dominant.

Every object emits electromagnetic radiation in a **wave** spectrum related to its own temperature. An object cooler than its surroundings will absorb more energy in the form of radiation than it emits. This radiation can pass through both free **space** and transparent media. Heat transfer by radiation helps sustain life on Earth—energy received from the **Sun** is an example of this process.

The electromagnetic radiation associated with heat transfer is sometimes referred to as **blackbody radiation** (where "blackbody" is an ideal emitter and absorber) or as **thermal radiation**. Thermal radiation is often associated with infrared radiation, although more **thermal energy** is received from the visible potion of the Sun's spectrum than from the infrared portion.

The molecules of a hotter material move faster and therefore have higher kinetic energy than the molecules of a cooler material. When molecules collide with slower neighboring molecules, kinetic energy is transferred from one molecule to another. The rate of heat transfer is high for metals (which, therefore, are said to have higher conductivity) and quite low for gases like air.

The process of convection occurs when groups of molecules are displaced to the vicinity of slower or faster molecules and mix with them. Forced convection occurs when hotter or cooler parts of a fluid are moved by way of forces other than **gravity**, such as a pump. Natural or free convection occurs when fluids are heated from below (like a pot on a kitchen stove) or cooled from above (like a drink with ice cubes on top). Hotter portions of the fluid expand, become lighter, and move upwards, while cooler, heavier portions descend. Convection can be many times faster than conduction alone. Vertical and horizontal convection plays a major role in the distribution of heat on Earth through the movements of atmospheric and oceanic masses.

HEAVY WATER

Ordinary water consists of two **hydrogen atoms** and one **oxygen** atom. Heavy water consists of two deuterium **atoms** and one oxygen atom. Deuterium is an **isotope** of hydrogen; both have the same number of protons, and therefore have the same atomic number. However, deuterium has one **neutron** in its **nucleus** while hydrogen has none and so deuterium is heavier than hydrogen. Thus a molecule of heavy water has a **mass** of 20 atomic units while an ordinary water molecule has a mass of 18 units. Pure heavy water is 10% more dense than ordinary water; it freezes at 38.8°F (3.8°C) and boils at 214.5°F (101.4°C). It is not radioactive and occurs naturally in nature in limited quantities.

Heavy water, which can be harvested from natural water sources or created from electrolyzed water, is used as a moderator in nuclear **power** plants. The uranium-235 in the reactor can capture neutrons only if they are moving slowly enough; thus a moderator is needed to slow the particles. High-speed neutrons hit the deuterium and lose some of their **momentum**, slowing the neutron. Ironically, slowing down the neutrons will increase the **nuclear fission** rate. Ordinary water is used as a moderator as well, but absorbs most neutrons instead of just slowing them through collisions; these reactors must use enriched uranium to be able to sustain nuclear fission. Other uses for deuterium include quantum effect experiments, extensive use in **molecular structure** studies, superconductivity experiments, chemical **kinetics**, **nuclear magnetic resonance** shift studies, dipole moment studies, geochemical experiments, and experiments dealing with growth and morphology of plants and animals.

HEISENBERG, WERNER KARL (1901-1976)

German physicist

Werner Karl Heisenberg is best known for his discovery of the **uncertainty principle**, for which he was awarded the 1932 Nobel Prize in physics. The uncertainty principle states that it is impossible to specify precisely both the position and momentum of a particle at the same time. The enunciation of that principle contributed to the understanding of a number of problems in **atomic physics**. It was also to have a profound effect on a far more general level, in that it essentially invalidated the long-held and fundamental scientific thesis of cause and effect. One of Heisenberg's first great accomplishments was the development of a new approach for solving problems of **atomic structure**, matrix **mechanics**.

Heisenberg was born in Würzburg, Germany, on December 5, 1901. His father, August Heisenberg, taught Greek philology at the University of Würzburg and ancient languages at the Altes Gymnasium (secondary school) of the same city. His mother, Annie Wecklein, was the daughter of the headmaster of Münich's Maximilian Gymnasium. Werner entered elementary school at Würzburg, but at the age of nine he moved with his parents and his elder brother to Münich, where his father had accepted an appointment as professor of Greek philology at the university.

In Münich, Heisenberg enrolled in the Maximilian Gymnasium, where his grandfather was headmaster. His education was interrupted, however, in the summer of 1918, when he was called to assist the faltering wartime effort by harvesting crops on a Bavarian farm. Shortly thereafter, he returned to Münich and served as a messenger for the democratic socialist forces that managed after bitter street fighting to oust a communist-oriented government that had briefly taken control of the Bavarian state. He also became involved with a number of youth groups that had organized in an effort to build a new society out of the wreckage left in the wake of World War I. Eventually he was elected leader of a group associated with the Bund Deutscher Neupfadfinder (New Boy Scouts), which, according to David C. Cassidy writing in the *Dictionary of Scientific Biography,* "strove for a renewal of supposedly decadent German personal and social life through the direct experience of nature and the uplifting beauties of Romantic poetry, music, and thought." Already a gifted pianist, Heisenberg would maintain his interest in music throughout his life; later, after his marriage, the whole Heisenberg family often assembled for an evening of chamber music.

After the war, Heisenberg completed his studies at the gymnasium and, in 1920, decided to enter the University of Münich. He hoped to pursue a degree in mathematics, a subject in which he had long been interested and for which he had shown great talent; while still a student at the gymnasium he had taught himself calculus and had written a paper on number theory. Heisenberg changed his mind, however, when Münich's professor of mathematics, Ferdinand von Lindemann, refused to accept him as a student in an advanced seminar. On his father's advice, Heisenberg then applied for his second choice, physics. He requested an interview with **Arnold Sommerfeld**, then a professor of theoretical physics at Münich, who agreed to admit Heisenberg to his seminar.

As Sommerfeld's student, Heisenberg not only became rapidly involved in the most fundamental problems in theoretical physics, but also became acquainted with a group of scientists—including the physicists **Max Born**, **Niels Bohr**, **Wolfgang Pauli**, and **Enrico Fermi**—whose work would dominate the field of atomic physics for years to come. At Münich, Heisenberg immediately began investigating some of the problems related to refinements of the Bohr model of the **atom**. In 1913, Bohr had employed the new concept of **quantum physics** to develop a model of the atom that explained much empirical data with amazing accuracy. However, a number of problems remained with the Bohr model. For one thing, no theoretical basis existed for the concept of quantized **orbitals** professed by Bohr. Also, inadequacies in the model's predictive ability suggested that some essential points were still missing from the **Bohr theory**. One of the earliest improvements on the Bohr model had been suggested by Sommerfeld himself when, in 1916, he hypothesized the existence of elliptical orbitals for electrons.

The topic to which Heisenberg first addressed his attention was the **Zeeman effect**, the splitting of **spectral lines**, a phenomenon for which no explanation had yet been suggested. In a series of papers, two written with Sommerfeld, Heisenberg developed an explanation for the Zeeman effect. His approach required an abandonment of some fundamental principles in both classical physics and **quantum mechanics**. But the hypothesis was attractive, nonetheless, and, as Cassidy points out, was "the first indication of the radical changes required for solving the quantum riddle." Heisenberg obtained his Ph.D. in 1923, although not without some difficulty. As he was primarily interested in theoretical physics, he had spent little time on laboratory experimentation. During his oral examinations for the doctoral degree, he was unable to answer questions put to him on instrumental matters by **Wilhelm Wien**, a professor of experimental physics, who consequently recommended that Heisenberg not be passed. Largely as a result of Sommerfeld's negotiation, however, Wien was overruled and Heisenberg received his degree.

The Zeeman papers brought Heisenberg's name to the attention of physicists throughout Europe. Thus, when Sommerfeld spent a year as visiting professor at the University of Wisconsin in 1922–23, he made arrangements for Heisenberg to spend that year studying under Born at Göttingen. Born was so impressed with his new student's work that he recommended Heisenberg for a teaching position. This decision was quite remarkable since Heisenberg was only 22 at the time and had just taken his Ph.D. Heisenberg remained in Göttingen until 1926, with the exception of a semester back in Munich and another spent working under Bohr in Copenhagen in the fall of 1924.

One of the life-long medical problems with which Heisenberg had to deal was hay fever. He eventually became accustomed to leaving the inhospitable climate of Bavaria whenever a serious attack occurred and traveling to the less pollen-dense island of Heligoland. One such trip in June 1925 was to have profound significance for theoretical physics. With weeks of free time on his hands, Heisenberg devised a new approach to the problem of the Zeeman effect in particular and of **atomic theory** in general. Physicists had devoted too much effort, he said, to devising pictorial models that might have some physical reality. A better approach, he suggested,

Werner Karl Heisenberg.

would be to work strictly with experimental data and to determine the mathematical implications of that data.

In order to pursue this approach, Heisenberg devised a system that came to be known as matrix mechanics. Although, as he later learned, the mathematics of this technique was already familiar to professional mathematicians, its application to the problems of atomic physics was entirely new. After his return from Heligoland, he collaborated with Born and Born's assistant, Pascual Jordan, to publish a refined version of the theory of matrix mechanics. The paper marked a turning point in physics. As biographers Nevil Mott and Rudolf Peierls wrote in the *Biographical Memoirs of Fellows of the Royal Society,* "It is fair to regard this paper as the start of a new era in atomic physics, since any look at the physics literature of the next two or three years clearly shows the intensity and success of the work stimulated by this paper."

Matrix mechanics did prove to be very fruitful in the development of atomic theory. Still, many physicists were somewhat uneasy with the mathematical abstraction of the approach because it didn't provide them with physical models to which to relate. When Austrian physicist **Erwin Schrödinger** devised his wave mechanics about a year later, many physicists switched their allegiance to his more physically based approach. The conflict between the two theories was resolved fairly soon, however, when Schrödinger showed that matrix mechanics and wave mechanics are mathematically identical.

The Heisenberg-Born-Jordan paper was published in November 1925 and six months later Heisenberg left Göttingen to become Bohr's assistant in Copenhagen. The almost daily contact between the two great physicists was to lead to one of the most productive periods in Heisenberg's life, including his enunciation of the principle for which he is best known, the uncertainty principle.

Heisenberg turned his attention to uncertainty in February 1927. He was involved in a study of the fundamental quantum properties of an **electron** in its orbit around the **nucleus**. At one point in that analysis, he imagined using a gamma ray **microscope** to study an electron's **motion**. It occurred to him that the very act of measuring an electron's properties by shining **gamma rays** on it would disturb the electron in its orbit; the act of observing the electron, therefore, altered its behavior, and objectivity was lost. Out of this realization, he developed a fundamental principle of physics that is now familiar to most high school students of the subject. According to the uncertainty principle, one can measure the position of a particle *or* its momentum with as much precision as desired. However, the more precise one of these measurements becomes, the less precise the other will be, such that the product of their inaccuracies (to use Heisenberg's original term) must always be less than **Planck's constant** (the unvarying ratio of the **frequency** of **radiation** to its **quanta** of **energy**). The principle was quickly and widely accepted by most, though not all, physicists, who soon acknowledged the concept as a fundamental law of nature. For the development of the uncertainty principle, as well as his other contributions to theoretical physics, Heisenberg was awarded the Nobel Prize for physics in 1932.

In 1927, Heisenberg was offered chairs at the universities of Leipzig and Zürich. He chose the former and, upon assuming the post of professor theoretical physics in October of that year, became the youngest full professor in Germany. His duties at Leipzig, unlike those of modern professors, were extensive, including a full schedule of teaching and administrative responsibilities. Not surprisingly, his scientific output diminished somewhat as a result of these competing demands for his time and energy. Nonetheless, he continued to pursue research interests in a number of directions, including **ferromagnetism**, **quantum electrodynamics**, cosmic radiation, and **nuclear physics**. Perhaps his best-known contribution in this period occurred after English physicist **James Chadwick** discovered the **neutron** in 1932. Heisenberg proposed a model of the nucleus that consisted of protons and neutrons rather than protons and electrons, as favored by some physicists. This theory did not originate with Heisenberg, however.

The last half of Heisenberg's life was complicated by the political turmoil taking place in Germany and around the world in the years surrounding World War II. With the rise of National Socialism in the early 1930s, the vast majority of German scientists fled the country to take up residence in the United Kingdom, the United States, and other free nations. Heisenberg, like his colleague **Otto Hahn**, was one of the few world-class scientists who determined to remain behind and protect, as well as he could, Germany's scientific traditions and institutions.

At first, Heisenberg and his colleagues resisted German leader Adolf Hitler's efforts to "purify" and nationalize German science, but the forces they faced were too great. Soon Nazi students were in command of universities, including that of Leipzig, ensuring that non-Aryan and "politically unacceptable" faculty members were weeded out. Heisenberg's position was especially tenuous since the Nazis regarded theoretical physics (as opposed to experimental physics) as a suspect, "Jewish" science. In 1935, an invitation from Münich for Heisenberg to succeed his former teacher Sommerfeld was met with violent opposition from German political leaders and professional colleagues, one of whom branded Heisenberg a "white Jew." Any hope that Heisenberg may have had of accepting the Münich offer was dashed and, for a time, even his personal safety was at risk.

At the beginning of World War II, the Nazi government recognized Heisenberg's value and made him director of the German **atomic bomb** project. For the next five and a half years, all of his efforts were devoted to this project. Questions have since been raised as to the morality of Heisenberg's decision. The fact remains that he chose to stay in his native country and devote his abilities to its war effort.

When the war came to an end in 1945, Heisenberg and his bomb research team were captured in a remote region of southern Germany, where they had been sent to be safe from Allied bomb attacks. After a six-month internment in England, Heisenberg was allowed to return to Göttingen, where he reestablished the Kaiser Wilhelm Institute for Physics and renamed it the Max Planck Institute in honor of his colleague and friend, the originator of **quantum theory**.

In the postwar years, Heisenberg worked diligently to restore German science to its former high standing. He was largely responsible for the creation of the Deutscher Forschungsrat (German Research Council), of which he became president. In this role, he came to represent West Germany in many international organizations and at many international meetings. His continuing administrative and political responsibilities did not deter Heisenberg from his research interests, however. In the postwar years, he became particularly interested in the search for a unified field theory, which would link all physical fields, such as gravitation and **electromagnetism**. He also pursued his interest in more general topics, such as the relationship between physical theory and philosophy.

Heisenberg retired in 1970, although he continued to write on a variety of topics. His health began to fail in 1973, and two years later he became seriously ill. He died on February 1, 1976, in Münich. He was survived by his wife, the former Elisabeth Schumacher, whom he had married in 1937. He also left seven children, four daughters and three sons.

HELIUM

Helium, the second element in the **periodic table**, is the second-most abundant element in the **universe** after hydrogen. It is a noble gas that is unreactive (because of its filled outer electronic shell), colorless, and odorless. Helium is found in great abundance in many **stars**, such as the **Sun**, and is an important component in both the proton-proton reaction and the **carbon** cycle, which account for the **energy** of the Sun and stars. The energy released as hydrogen is fused into helium and is the source of the **power** contained in a hydrogen bomb. Under 25 atmospheres of **pressure**, helium has a **boiling** point of -268.9°C (-52.1°F or 4.22K) and a melting point of -272.2°C (-458°F or 0.95K). At this **temperature**, slightly above **absolute zero**, helium is transformed into helium II, superfluid helium, a liquid with unique properties.

French astronomer Pierre Janssen (1824-1907) first obtained evidence for the existence of helium during the solar eclipse of 1868 in India. During this eclipse he detected a new yellow line in the solar spectrum very close to that of the sodium D-line. Shortly thereafter it was recognized as an element not then known on **Earth**, and the British scientists, Sir Joseph N. Lockyer (1836-1920) and Sir Edward Frankland (1825-1899), suggested the name helium, from the Greek *helios* meaning sun, for the substance. Later, in 1895, British chemist Sir William Ramsay (1852-1916) isolated helium from cleveite, a uranium-bearing mineral, while simultaneously Swedish chemist Per Teodor Cleve (1840-1905) and Langlet did the same thing. In 1907 **Ernest Rutherford** and T. Royds demonstrated that alpha particles are helium nuclei. Two of the naturally occurring isotopes of helium on Earth are helium-3, now thought to be a product of the **radioactive decay** of the tritium **isotope** of hydrogen, and helium-4, the most common form, thought to come from the radioactive alpha emitters in rocks.

Superfluid helium has unique physical properties in that it has no exact freezing point and its **viscosity** is apparently zero. It passes readily through cracks and through minute pores. This form of helium was first discovered in 1908, but nearly 25 years passed before any explanations concerning its properties were published. It is one of the few phenomena supposed to show quantum effects on a large scale. The properties of superfluid helium are attributed to it behaving like a **boson** gas in that all of the atoms occupy the same **energy level** and can be described in terms of a single **wave function** and as such is known as a Bose-Einstein condensate.

All of the helium on Earth at this time is produced through radioactive decay. Natural gas is the major commercial source of helium. Helium is produced from radioactive alpha emitters in rocks and then seeps into natural gas where it is found at about 0.4%. Most of the helium produced commercially is extracted from natural gas found in the United States. At the current rate of use the United States will exhaust its helium supply by the year 2016.

Helium has a number of uses in various scientific fields, in manufacturing, and in other forms of business. Because helium's boiling point is close to absolute zero it is used in cryogenic research as well as studies concerning superconductivity. It is used in balloons since it is safer than hydrogen and will not burn. Helium's use as an inert gas shield for arc welding and growing silicon and germanium crystals as well as in titanium and zirconium production is widely known. In its liquid form it is used to cool **nuclear reactors**. Helium mixed with **oxygen** is used as an artificial atmosphere for divers and

Weather balloon being filled with helium. *(Archive Photos, Inc. Reproduced by permission.)*

others working under pressure because it reduces the chances of suffering from the bends.

HELMHOLTZ, HERMANN VON (1821-1894)

German physiologist and physicist

Hermann von Helmholtz was one of the few scientists to master two disciplines: medicine and physics. He conducted breakthrough research on the nervous system, as well as the functions of the eye and ear. In physics, he is recognized (along with two other scientists) as the author of the concept of **conservation of energy**.

Helmholtz was born into a poor but scholarly family; his father was an instructor of philosophy and literature at a gymnasium in his hometown of Potsdam, Germany. At home, his father taught him Latin, Greek, French, Italian, Hebrew, and Arabic, as well as the philosophical ideas of **Immanuel Kant** and J. G. Fichte (who was a friend of the family). With this background, Helmholtz entered school with a wide perspective. Though he expressed an interest in the sciences, his father could not afford to send him to a university; instead, he was persuaded to study medicine, an area that would provide him with government aid. In return, Helmholtz was expected to use his medical skills for the good of the government—particularly in army hospitals.

Helmholtz entered the Friedrich Wilhelm Institute in Berlin in 1898, receiving his M.D. four years later. Upon graduation he was immediately assigned to military duty, practicing as a surgeon for the Prussian army. After several years of active duty he was discharged, free to pursue a career in the academia. In 1848 he secured a position as lecturer at the Berlin Academy of Arts. Just a year later he was offered a professorship at the University of Konigsberg, teaching physiology. Over the next 22 years he moved to the universities at Bonn and Heidelberg, and it was during this time that he conducted his major works in the field of medicine.

Helmholtz began to study the human eye, a task that was all the more difficult for the lack of precise medical equipment. In order to better understand the function of the eye he invented the ophthalmoscope, a device used to observe the retina. Invented in 1851, the ophthalmoscope—in a slightly modified form—is still used by modern eye specialists. Helmholtz also designed a device used to measure the curvature of the eye called an ophthalmometer. Using these devices he advanced the theory of three-color vision first proposed by **Thomas Young**. This theory, now called the Young-Helmholtz theory, helps ophthalmologists to understand the nature of **color** blindness and other afflictions.

Intrigued by the inner workings of the sense organs, Helmholtz went on to study the human ear. Being an expert pianist, he was particularly concerned with the way the ear distinguished pitch and tone. He suggested that the inner ear is structured in such a way as to cause resonations at certain frequencies. This allowed the ear to discern similar tones, overtones, and timbres, such as an identical note played by two different instruments.

In 1852 Helmholtz conducted what was probably his most important work as a physician: the measurement of the speed of a nerve impulse. It had been assumed that such a measurement could never be obtained by science, since the speed was far too great for instruments to catch. Some physicians even used this as proof that living organisms were powered by an innate "vital force" rather than **energy**. Helmholtz disproved this by stimulating a frog's nerve first near a muscle and then farther away; when the stimulus was farther from the muscle it contracted just a little slower. After a few simple calculations Helmholtz announced the impulse **velocity** within the nervous system to be about one-tenth the **speed of sound**.

After completing much of the work on sensory physiology that had interested him, Helmholtz found himself bored with medicine. In 1868 he decided to return to his first love—physical science. However, it was not until 1870 that he was offered the physics chair at the University of Berlin and only after it had been turned down by **Gustav Kirchhoff**. By that time, Helmholtz had already completed his groundbreaking research on energetics.

The concept of conservation of energy was introduced by **Julius Mayer** in 1842, but Helmholtz was unaware of Mayer's work. Helmholtz conducted his own research on energy, basing his theories upon his previous experience with muscles. It could be observed that animal **heat** was generated by muscle action, as well as chemical reactions within a working muscle. Helmholtz believed that this energy was derived from food and that food got its energy from the **Sun**. He proposed that energy could not be created spontaneously, nor could it vanish—it was either used or released as heat. This explanation was much clearer and more detailed than the one offered by Mayer, and Helmholtz is often considered the true originator of the concept of conservation of energy.

While this was undoubtedly Helmholtz's greatest legacy, he also began several projects that were later completed by other scientists. He advanced a number of hypotheses on electromagnetic **radiation**, speculating that it lay far into the invisible ranges of the spectrum. This line of research was later resumed, very successfully, by one of Helmholtz's students, **Heinrich Rudolf Hertz**, the discoverer of **radio waves**. Helmholtz's theories on **electrolysis** were also the basis for future work conducted by Svante August Arrhenius.

Helmholtz had been a sickly child; even throughout his adult life he was plagued by migraine headaches and dizzy spells. In 1894, shortly after a lecture tour of the United States, he fainted and fell, suffering a concussion. He never completely recovered, dying of complications several months later.

HENRY, JOSEPH (1797-1878)
American physicist

Several parallels between the life and work of Joseph Henry and English physicist **Michael Faraday** exist. Like Faraday, Henry was born into a poor family. Both received little formal education and both were apprenticed at an early age, Faraday to a bookbinder, Henry to a watchmaker at the age of 13. Both, finally, made lasting contributions to the field of electrical research.

Henry's interest in science was sparked by an odd coincidence. He had chased his pet rabbit underneath a church. Noticing that some floorboards were missing, 16-year-old Henry climbed into the church and found a shelf of books. He began looking through *Lectures on Experimental Philosophy* and was instantly hooked on science. He entered the Albany Academy and later began to teach at country schools to earn an income. He graduated from the academy and was leaning toward studying medicine when a surveying job turned up and that steered him toward engineering. In 1826 he was back at the Albany Academy, but this time as a teacher of mathematics and science.

In 1820, Danish physicist **Hans Christian Ørsted** had discovered that the flow of an **electric current** produced a **magnetic field** in a wire. This amazed scientists and many, including Henry and Faraday, began to experiment with **magnetism**. In 1829 Henry learned that William Sturgeon (1783-1850) had built an electromagnet that could lift 9 pounds (4 kg). This was remarkable, but Henry believed he could create a magnet that was much stronger. The secret was to wrap more wire around the iron core, overlapping the levels. The problem with the wire in that era was that it was not insulated; wrapping one level over another caused a short circuit. Henry got around the problem by the laborious process of insulating copper by hand, using strips of his wife's silk petticoats.

Now that he had insulated wire, Henry proceeded to experiment. By 1831 he had created an electromagnet that could lift 750 pounds (340 kg). Later that same year he gave a demonstration at Yale University and lifted more than 2,000 pounds (900 kg). For this feat he received an appointment as professor.

In addition to his large electromagnets, Henry built small ones. In 1831 he ran a wire more than one and one-half kilometers (one mile) and attached a device consisting of an electromagnet, a movable iron arm, and a spring. At the other end of the wire was a battery and key switch. When Henry pressed the key, activating the current, the distant electromagnet engaged, attracting a metal arm with a click. Releasing the key cut the electric flow, and the spring forced the arm back to its rest position.

There was a practical limit to the length of the wire that could be used; the longer the wire, the greater the resistance, resulting in less current. **Georg Ohm** had devised a law by which resistance could be calculated. Henry found an easy way around the resistance problem in 1835 by inventing the electric relay.

Near the close of the 1830s, Henry had a chance encounter with a man who had very little electrical knowledge, but who was extremely interested in electromagnets and relays. Henry believed that the discoveries of science should be for the good of all mankind, so he never patented any of his devices. The man, Samuel F. B. Morse, received a great deal of advice and information from Henry. In 1840 Morse took out a patent on the electric telegraph and became a very rich man.

This was not the first time Henry had been cheated. In 1830, teaching at the Albany Academy monopolized Henry's time to such an extent, that he had only the month of August in which to conduct research and experiment. In August 1830 Henry discovered the principle of electric induction, the process in which an electric current in one coil of wire can set up a current in another coil. If the flow of **electricity** can produce a magnetic field, he reasoned, a magnetic field should be able to induce an electric current.

When the end of August arrived, Henry had not completed his investigation. He decided to set the work aside and return to it the following August. It was with considerable shock that Henry read Michael Faraday's announcement in 1831 of the discovery of electric induction.

Rushing back to his experiments, Henry published a report of his own discovery, but he was too late; Faraday received credit for the discovery. To his own credit, Henry never argued about Faraday's priority.

Henry, however, did get credit for including in his paper a discovery that Faraday had neglected to write about—the discovery of self-induction. A coil carrying electric current not only induces a flow in another coil, it can induce a current in itself. In 1834 Faraday discovered self-induction independently, but this time Henry got the credit. Estonian physicist

Joseph Henry. *(Photo courtesy of Corbis-Bettmann. Reproduced by permission.)*

Heinrich Lenz also made the discovery independently and took it further than both Henry and Faraday. Self-inductance became an important part in the design of electric circuits.

In 1831 Henry published a paper in which he described the workings of an electric motor. Ten years earlier Faraday had built a motor, but it was little more than a toy. Henry's motor was more practical, but until Faraday developed the electric generator, there was no way to adequately **power** the motor. It is Henry's design that is used for motors in electrical appliances today.

In 1842 Henry anticipated a discovery that has been credited to **Heinrich Rudolf Hertz**. Henry discovered that he could magnetize needles in a basement with an electric spark from two floors above, correctly ascribing it to electromagnetic wave propagation. In addition to describing the mechanism of an electric motor, Henry was involved in many other endeavors.

Henry always resented the thunder that had been stolen from him, but he kept his irritation to himself and continued to live for science. In 1846 he became the first secretary of the new Smithsonian Institution and encouraged the communication of scientific knowledge around the globe. Two years later he projected the image of the **Sun** on a screen, made careful measurements of **temperature**, and discovered that the mysterious sunspots were relatively cooler than the rest of the sun.

He used the telegraph to set up a system of obtaining **weather** reports from across the country, initiating a system that led to the founding of the U.S. Weather Bureau. During the Civil War he recommended the construction of ironclad warships, advice that was eventually followed. On May 13, 1878, Henry died in Washington, D.C., at the age of 80. He was finally honored in 1893 when the International Electrical Congress agreed to name the unit of inductance the henry.

HERSCHEL, WILLIAM (1738-1822)
German astronomer

William Herschel was born in Hanover, Germany, at a time when the city belonged to England under the rule of George II. As Herschel's father was a musician in the Hanoverian army, Herschel himself was trained in music in order to enter the same profession. The Seven Years War, however, made military life an unattractive option, and in 1757 Herschel arrived in England, where he began working as an organist and music teacher. Herschel learned of **astronomy** through his interest in the theory of music and the scientific basis for musical sounds, which led him to mathematics and then **optics**.

Isaac Newton's treatise on optics inspired Herschel with his desire to study the **stars**. Unable to find a **telescope** of a high enough resolution, he decided to grind his own **lenses** and to design his own instruments; he was helped by his sister Caroline, who came to England in 1772. As Herschel's first telescope, a six-foot (183-cm) *Gregorian reflector*, was one of the best of its kind, he decided that its first application would be to conduct a systematic survey of the stars and **planets**. Throughout his life he built numerous telescopes, each one more sophisticated and more powerful than the last.

Herschel's first major discovery occurred in 1781 during his second survey of the sky when he announced the existence of a new planet to be found in the constellation of Taurus. Herschel's name for the new planet was *Georgium Sidus*, George's star, in honor of King George III, but it eventually came to be known as Uranus, after the mythical father of Saturn. The discovery of Uranus, which effectively doubled the previously accepted size of the solar system, caused a popular and scientific sensation, and George III appointed Herschel to the position of King's Astronomer while providing him with a small annuity that allowed him to pursue astronomy full time.

Herschel's most significant achievements were in the area of sidereal astronomy, to which he contributed the first systematic body of evidence on the order and nature of the stars and the planets. Whereas plenty of theories had been put forward by prominent philosophers of the time on the systems that might govern the **universe**, none was supported by any scientific gathering of data. In 1783, Herschel began to search for **nebulae** in the sky, and raised their known total from little more than 100 to 2,500. Much of eighteenth-century astronomy set out to determine the distances between stars; trigonometrical calculations based on their apparent annual movement, however, had failed. **Galileo** had proposed the use of double stars, pairs of stars very close together, to calculate stellar distance, where the fainter member of the pair was so far away as to represent a fixed point from which the annual movement of its brighter companion could be measured. In Herschel's second survey, he searched for double stars, producing three catalogs over the next 40 years and listing 848 examples. It was later discovered by another astronomer who had seen Herschel's work that these double stars were in fact companions in **space** held together by gravitational forces and therefore equidistant from Earth; Herschel had assumed that companions in space would have been of equal brightness, and had therefore discounted this possibility. Nonetheless, much of Herschel's work was concerned with producing evidence for the powers of attraction between stars. In three of his papers delivered between 1784 and 1789, he proposed a cosmogony for the universe in which stars, initially randomly scattered throughout the universe, clustered together over time around the regions from which they originally developed.

Herschel was the first to embark upon a scientific study of the **Milky Way** and half of his work, though less influential, focuses upon the solar system. He studied the **Sun**, observing that what we see is not the Sun itself but the clouds of gases that cover its surface, and examined the nature of the infrared section of the spectrum by which some of the Sun's **heat** is transmitted. Besides calculating the height of lunar mountains, Herschel devoted most of his attention to the other known planets, Venus, Mars, Jupiter, and Saturn, determining their rotation period and checking the inclination of their axes, their shape, and the nature of their atmospheres. Herschel devoted most of his attention to examining Saturn and its rings, arguing at one point that the rings were solid, but later conceding that they were in fact composed of floating particles.

Herschel's work on nebulae had led him to conclude that they might well be other solar systems seen only as a luminous cluster of stars around a brighter one. As a result, he saw the Milky Way and Earth as only one rather insignificant part of the universe. In this sense, he changed the status of the solar system within the universe in much the same way as **Nicholas Copernicus** had Earth when he showed that the planets revolved around the Sun rather than Earth.

HERTZ, GUSTAV (1887-1975)
German physicist

Gustav Hertz's greatest fame came early in his career as the result of his collaboration with **James Franck**, a colleague at the University of Berlin. Hertz and Franck studied the **energy** changes that take place when an **electron** strikes an **atom**. Their results provided important confirmation of the **Bohr theory** of the atom, a theory announced only shortly before Hertz and Franck conducted their experiments. In recognition of their work, the Nobel Prize in physics was awarded jointly to Hertz and Franck in 1925.

Hertz was born in Hamburg, Germany, on July 22, 1887. His mother was Auguste Arning, and his father was Gustav Hertz, an attorney. Hertz was the nephew of the famous

Heinrich Rudolf Hertz, who carried out a number of important studies on **electromagnetic waves** in the 1880s, and for whom the unit of **frequency** is now named.

Hertz received his secondary education at the Johanneum Realgymnasium in Hamburg. Between 1906 and 1911, he studied mathematics and physics at the universities of Göttingen, Munich, and Berlin. He eventually decided to concentrate on a career in experimental physics. In 1911, he earned a Ph.D. from the University of Berlin for his study of the infrared absorption spectrum of **carbon** dioxide.

A position as research assistant at the physical institute of the University of Berlin was offered to Hertz in 1913. When he began work at the institute, Hertz met Franck, and the two decided to collaborate on their research. The first project they undertook involved a study of the emission of electrons from a metal surface when bombarded by a stream of electrons. This type of research can be traced to studies of the **photoelectric effect**, the emission of electrons from a metal surface that has been exposed to **light** energy. In 1902, German physicist **Philipp E. A. Lenard** carried out some experiments on this effect. Three years later, **Albert Einstein** provided a theoretical explanation of Lenard's results and of the photoelectric effect in general. Hertz and Franck attempted to determine the properties of electrons emitted from a metal surface bombarded by electrons rather than by light.

In their experiment, Hertz and Franck accelerated electrons from a hot tungsten wire by means of a positively charged metal gauze placed a few centimeters from the wire. The electrons were forced to pass through an atmosphere of mercury vapor. A second positively charged wire gauze was then arranged so as to detect electrons that had collided with mercury atoms and lost energy. The experiment consisted of gradually increasing the charge on the metal gauze and tracking the loss of energy for electrons reaching the detector screen. Hertz and Franck found that this loss of energy was negligible until the potential difference reached 4.9 volts. At that point, the electron current reaching the detector dropped nearly to zero.

For some months, Hertz and Franck were not able to interpret their results. They thought that 4.9 volts might represent the ionization potential of mercury, the energy needed to remove a single electron from a mercury atom. In fact, the correct explanation for their results was already available in **Niels Bohr**'s recently announced quantum model of the atom. Hertz and Franck eventually realized that the 4.9-volt result they observed corresponded to the transition between the first two electron energy levels (K to L) in the mercury atom. In fact, that numerical value precisely matched the energy difference that Bohr had predicted in his theory. As such, the Hertz-Franck results provided one of the first pieces of experimental confirmation for Bohr's revolutionary new theory. In recognition of this achievement, Hertz and Franck were jointly awarded the 1925 Nobel Prize in physics.

Less than a year after completing his momentous experiments, Hertz was inducted into the German army. He was seriously wounded at the battlefront in 1915 and spent more than a year in recuperation. After the war, he returned to Berlin, where he worked for three years as an unsalaried lecturer (privatdozent). He married Ellen Dihlmann in 1919, with whom he had two sons, Hellmuth and Johannes. In 1943, two years after Dihlmann died, Hertz was married a second time, to Charlotte Jollasse.

In 1920, Hertz accepted an offer of employment at the Philips Incandescent Lamp Works in Eindhoven, Netherlands. Philips was one of the first corporations to maintain a basic research laboratory. After five years at Philips, Hertz returned to Germany in October 1925 as professor of physics and director of the physical institute at the University of Halle. He remained there for three years before returning once more to Berlin. There, he became professor of physics at the Charlottenburg Technical University.

With the rise of National Socialism (the Nazi party), Hertz faced yet another career change. Unwilling to sign an oath of loyalty to the new government, he was forced to resign his post at the Technical University. Surprisingly, he was offered and accepted a job with the Siemens and Halske Company in Berlin in 1934. He remained in that job until the end of World War II, at which time he moved to the Soviet Union. He told friends that he hoped to be able to make a contribution to Soviet physics. That hope was not realized, however, since he and his German colleagues were assigned to a segregated community in Sukhumi on the Black Sea, where they worked on atomic energy, **radar**, and supersonics projects in isolation from Soviet scientists. At the conclusion of his 10-year commitment to the Soviet government, Hertz returned to East Germany, where he became director of the physics institute at Karl Marx University in Leipzig. He retired from that position in 1961 and returned to Berlin, where he died on October 30, 1975.

HERTZ, HEINRICH RUDOLF (1857-1894)

German physicist

Heinrich Hertz, best known for his work with **electromagnetism** and electromagnetic **radiation**, was born in Hamburg, Germany, on February 22, 1857. He initially studied engineering, but abandoned that subject to take up physics at the University of Berlin. At Berlin he studied under physicists **Gustav Kirchhoff** and **Hermann Helmholtz**. After Hertz received a Ph.D. in 1880, he became Helmholtz's assistant and began what became a lifelong friendship.

In 1883, while working at the University of Kiel, Hertz became especially interested in the electromagnetic equations that had been established by **James Clerk Maxwell**. The Berlin Academy of Science had established a prize for work on a specific problem, and Helmholtz encouraged Hertz to submit an entry. Hertz was interested, but lacked the facilities to carry out the experiments. Fortunately, he was offered a professorship in physics at Karlsruhe in 1885, a post that included a very well-equipped laboratory. Hertz began his work and ultimately took the prize, but what he discovered in the process went well beyond his expectations.

Heinrich Hertz.

To solve the Berlin Academy problem, Hertz had to delve into the matter of oscillating electric currents. He bent a copper wire into a loop and left a small gap between the two ends. He connected this to a circuit, discharged an induction coil into it and was rewarded with a spark jumping the gap. Attaching spheres to the ends of the gap, he passed an **electric current** backward and forward, causing each sphere alternately to become charged. When the charge reached a specific level, the spark jumped.

Realizing that Maxwell had predicted that electromagnetic radiation with an extremely long **wavelength** should be emitted by such an oscillating charge, Hertz wondered if these invisible **waves** were being generated by his spark. To see if this was so, he devised a simple test. He reasoned that in the same way an **oscillation** current in his loop created radiation, the radiation produced might set up an oscillating current in another loop. He bent a wire into another loop to serve as a "detector"—and he was able not only to detect the radiation, but to determine the shape and intensity of the invisible waves by moving the receiving loop around the room. He found the wavelength of the "Hertzian waves" was 2.2 feet (66 cm) which was a million times greater than the wavelength of visible **light**. Hertz had detected a new type of radiation—the **radio** wave—and shown that it behaved in accordance with Maxwell's theory. In England, physicist Sir Oliver Lodge (1851-1940) confirmed Hertz's conclusions.

Hertz was also the first to notice that under certain conditions, electrical flow could be affected by light. He had observed that shining an ultraviolet light on the negatively charged side of the gap in his loop made it easier for a spark to jump the gap, but there was no body of theory Hertz could call upon to make sense of this. It was not until the age of **quantum physics** that **Albert Einstein** explained this phenomenon, called the **photoelectric effect**.

Hertz's discovery paved the way for the age of radio, television, and satellite communication. Within 40 years scientists such as Aleksandr Popov (1859-1905), Lodge, **Guglielmo Marconi**, and Vladimir Zworykin (1889-1982) had learned how to carry Morse code, **sound,** and even moving pictures on radio waves.

Unfortunately Hertz did not live to see even radio become a major factor in communication. Plagued with ill-health for most of his life, he died of a chronic blood disease on New Year's Day, 1894, at the age of 36. The term "hertz," designating one single vibration or cycle per second, is named in his honor.

HERTZSPRUNG, EJNAR (1873-1967)

Danish astronomer

Born in Frederiksberg, Denmark, on October 8, 1873, Hertzsprung was trained as a chemical engineer, but he was also an amateur astronomer and so impressed Karl Schwarzschild (1873-1916) with his knowledge that Schwarzschild offered him a professorship in Göttingen, Germany, in 1909.

Hertzsprung thought there should be a standard by which the brightness of **stars** could be compared. A given star might be very luminous but appear dim to observers on Earth because of its distance. Likewise, a dim star might appear bright because it is closer. If all stars could be placed at exactly the same distance, their relative luminosity would be readily apparent.

In 1905, Hertzsprung set up a standard that he called the "absolute magnitude" of a star, defining the quantity as what a star's brightness would be at a distance of 10 parsecs from Earth (one parsec is equivalent to 3.26 light-years). Once he had determined the intrinsic brightness of a number of stars, it became apparent to him that there was a relationship between a star's **color** and its luminosity. He found that red stars, generally, are not as bright as blue stars, while yellow stars fall somewhere between the two. Furthermore, since the color of a star is related to its **temperature**, Hertzsprung concluded that a star's temperature and brightness were correlated. But he published his findings in a journal of **photography** and timidly kept his diagrams to himself.

Nearly 10 years passed before Hertzsprung's important contribution was finally recognized, when American astronomer **Henry Norris Russell** independently came to the same conclusions. Today the graph depicting the relationship between luminosity and temperature is called the **Hertzsprung-**

Russell diagram. The diagram has played an important role in providing clues to astronomers about **stellar evolution**.

In 1913 Hertzsprung began another important line of research. With the help of **Henrietta Leavitt**, he found a method to determine the distances to Cepheid variable stars. Their exploration of this Cepheid "yardstick" led Hertzsprung to make the first credible estimate for the distance of an extra-galactic object—the Small Magellanic Cloud. Their method, in turn, was used by **Harlow Shapley** to determine the size and dimensions of our galaxy, as well as to conclude that spiral **nebulae** are actually distant galactic systems.

Hertzsprung became director of the observatory at Leiden, Germany, in 1935 and retired 10 years later. He died in Denmark on October 21, 1967, at the age of 94.

HERTZSPRUNG-RUSSELL DIAGRAMS

Between 1911-13, Danish astronomer **Ejnar Hertzsprung** and American astronomer **Henry Norris Russell** independently developed what is now known as the Hertzsprung-Russell diagram. Used by astronomers, a Hertzsprung-Russell diagram is a two-dimensional graph or plot used to depict various types of **stars** and stages of the **stellar life cycle**. Absolute magnitude (or intrinsic luminosity on a logarithmic scale) on the vertical axis is plotted against stellar temperatures recorded on the horizontal axis.

Because the spectral appearance of stars depends on **temperature** of their photosphere, spectral classification is a grading of stars according to their surface temperature (i.e., on the horizontal axis of a Hertzsprung-Russell diagram) and does not directly reflect the star's chemical composition. Spectral classes group stars according to similarities in violet, blue, and green portions of visible spectrum. Stars are subdivided into seven temperature groups called spectral classes (designated as O, B, A, F, G, K, and M) on basis of the appearance of their absorption spectrums. Near the end of the twentieth century, some astronomers proposed the creation of a new class, L, to account for the discovery of dim brown dwarf stars barely capable of sustaining hydrogen fusion. Except for spectral class O, each class is further subdivided into 10 spectral types (e.g., A0 through A9). Spectral class O is subdivided into O4 through O9. Temperatures among the stellar classes range from O class stars with temperatures of 40,000K to M class stars at 3,800K. Stars are further characterized as early-type stars (spectral classes O, B, and A) or late-type stars (spectral classes G, K, and M).

Luminosity, plotted on the vertical axis of a Hertzsprung-Russell diagram, increases with distance from the origin. There are several distinct regions of a Hertzsprung-Russell diagram, including regions for main sequence stars, red giant stars, white dwarf stars, blue giant stars, and red supergiants. Apparent gaps in the diagram (i.e., regions where there are few stars) indirectly reflect fundamental changes in stars as they progress through the stellar life cycle, such as the differences in the **nuclear fission** fuels used in main-sequence and giant stars.

On the Hertzsprung-Russell diagram of all stars there is a conspicuous band or sequence of star types that curves its way from the upper left corner to the lower right corner. Star located in this band are termed main-sequence stars and this band contains most of stars in the **universe** (i.e., most of the stars that could be plotted). Stars in the upper left corner are extremely bright, hot stars. At the other extreme, in the lower right corner are faint, cool stars.

Main-sequence stars constitute a single class of stars. The distinguishing characteristic for all main-sequence stars is that they all utilize hydrogen to sustain fusion reactions in their core (i.e., they use hydrogen as their principal nuclear fuel). Although all main-sequence stars are luminosity-class five (V) stars, main-sequence stars range in luminosity from extremely luminous spectral-class O stars to very faint M dwarfs. Moving from left to right along the Hertzsprung-Russell diagram horizontal axis there is a decrease in temperature. Accordingly, temperatures for main-sequence stars range from 50,000K O spectral-class stars down to 3,000K M spectral-class stars. Radius and **mass** for main-sequence stars also decrease as one moves right along the horizontal axis (i.e., from O to M class stars). Luminosity of main-sequence stars is proportional to approximately the fourth **power** of their mass. Approximately 90% of the stars plotted on a Hertzsprung-Russell diagram are main-sequence stars and the majority of main-sequence stars observed in the **Milky Way galaxy** are class M stars (red dwarf stars).

The **Sun**, a G2 main-sequence star with a surface temperature of about 5,800K (the core temperature, not used for plotting, is 1.5×10^7K), plots near the middle of the Hertzsprung-Russell diagram in a region characterized by yellow dwarf stars.

The red giant region is comprised of stars with spectral classes ranging from F to M and bright but cool stars. Red giants are luminous stars that on a Hertzsprung-Russell diagram lie above the main-sequence band in a region that tends toward bright but cool stars in upper right corner. Although all red giants belong to the same luminosity class, they can vary by a factor of 100 or more in luminosity. Red giants are, on average, 100 times more luminous than the Sun and their surface temperatures range from 7,000K down to 3,000K. On a Hertzsprung-Russell diagram, within the red giant region there is no established relationship between mass-luminosity (i.e., mass and position on the diagram). The radii of red giant stars does, however, increase toward the upper right corner of the diagram. Red giants comprise less than .5% of the stars plotted on a Hertzsprung-Russell diagram.

The white dwarf region is comprised of stars with spectral classes ranging from B to F that are faint but hot cool stars. **White dwarfs** account for approximately 9% of the stars plotted on a Hertzsprung-Russell diagram. On the diagram, white dwarfs lie below the main sequence and are typically only a few thousandths as luminous as the Sun even though their surface temperatures may be greater than the Sun. Although white dwarfs are approximately the size of **Earth**, white dwarf masses range upward to 1.4 times the mass of the Sun. Accordingly, the densities of white dwarf stars is very high.

Blue supergiants and red supergiants are rare. The blue supergiant region is comprised of early-type stars with spectral classes ranging from O to A, with very bright, hot stars. The red supergiant region is comprised of late-type stars with spectral classes ranging from G to M that are very bright, hot stars. Both types of supergiants can be thousands of times more luminous than the Sun.

See also Stellar evolution

HERZBERG, GERHARD (1904-1999)
German Canadian physical chemist

Gerhard Herzberg is known as the founding father of molecular **spectroscopy**, the science that observes the interaction of **energy** with **matter** to obtain information on the identity and structure of molecules. For his "contributions to the knowledge of the electronic structure and geometry of molecules, especially free radicals," in 1971 Herzberg became the first Canadian to be honored by the Nobel Prize in chemistry. Herzberg also did pioneering work in other scientific fields, including **astrophysics**, and in association with the National Research Council of Canada, he founded the nation's premier molecular spectroscopy laboratory in 1948.

Herzberg was born December 25, 1904, to Albin and Ella (Biber) Herzberg. Raised and schooled in Hamburg, Germany, Herzberg graduated from the Darmstadt Institute of Technology with a B.S. in engineering in 1927. In 1928 he completed his Ph.D. with a thesis on the interaction of electromagnetic **radiation** with matter. Herzberg also studied at the University of Göttingen and the University of Bristol in England, before returning to Darmstadt in 1930 as an instructor.

Move to Canada

In 1935 Herzberg was forced to flee Germany as a result of Hitler's anti-Jewish policies. He subsequently obtained a position as Carnegie guest professor at the University of Saskatchewan, in Canada. Although the school lacked the resources and equipment Herzberg needed for his studies, the atmosphere was welcoming and supportive, and he accomplished a substantial amount of research while also publishing two books. While at Saskatchewan, Herzberg helped establish a graduate laboratory specializing in spectroscopy studies and obtained funding to improve the school's research facilities. Beginning in 1945, Herzberg spent three years as a professor at the Yerkes Observatory of the University of Chicago and then moved back to Canada to set up a spectroscopic research laboratory in Ottawa for the National Research Council. This laboratory was commended by the Swedish Royal Academy of Science, the Nobel Prize awarding institution, as "the foremost center for molecular spectroscopy in the world." Herzberg remained with the Canadian National Research Council for the remainder of his career, becoming the first Distinguished Research Scientist of that organization. He became a naturalized Canadian citizen in 1945.

Herzberg's Contributions to Molecular Spectroscopy

Although Herzberg considered himself a physicist, his research in molecular spectroscopy had special significance to chemistry. His research was of great importance as it developed a method of analyzing **molecular structure** by measuring **light** transmitted and absorbed by molecules. To do this, Herzberg used the spectroscope, a tool that enabled him to separate a molecule's radiant energy into different parts in the same way light is separated when passed through a prism. When molecules or atoms were passed through a spectroscope, the energy radiating from them separated into distinct lines or spectra, allowing an accurate analysis of their structure. At the time Herzberg began his experiments, spectroscopy, or the study of molecular structure, was a primitive science. Very little was known about atomic spectra when he began his analysis, and Herzberg made major contributions to the field by analyzing complicated spectra obtained from molecules, particularly diatomic molecules such as nitrogen, **oxygen**, and hydrogen. While still in Germany, for example, he and a colleague, Werner Heitler, showed that the nitrogen molecule's complex spectrum was much more significant than contemporary scientific belief acknowledged; the particles then believed to comprise nitrogen's **nucleus** could not account for the intensity of some of the spectral bands. The neutrons responsible for the inconsistency were discovered by English physicist **James Chadwick** some time later.

Herzberg also discovered several molecular species in outer **space** and the upper atmosphere while at the Yerkes Observatory. He found elemental hydrogen in some planetary atmospheres. He also discovered new bands, now called Herzberg bands, in the oxygen spectrum. Spectroscopy, however, allowed him to do much more throughout his career than just identify molecules. He accurately measured the **electron** energy levels in several species, using a combination of quantum mechanical theory and spectroscopy. By developing new experimental procedures, he was able to study short-lived molecules, or free radicals, that appear only briefly during chemical reactions. These chemical intermediates are difficult to study because they do not last long enough to apply other types of tests. Two important examples of his success in this area are the spectra of the methylene radical, CH_2, and the methyl radical, CH_3.

An Avid Researcher and Promoter of Science

Herzberg was also a voluminous writer throughout his career, publishing several hundred papers and many definitive books in the area of molecular and atomic spectra. He wrote in the introduction to his 1971 book, *The Spectra and Structure of Simple Free Radicals: An Introduction to Molecular Spectroscopy,* "My original plan, forty years ago, was to write a small book on [molecular spectroscopy] of no more than 200 pages. I was unable to prevent this original plan from leading to a three-volume work of over 2000 pages." Herzberg made a clear the distinction between scientific research and technological research. He believed that the former concentrated on finding out more about nature while the latter focused on the applications of science to society. In Herzberg's opinion, it was scientific research that was the true vocation of the scientist, and not technological research, which he believed to be the concern of politicians. He also championed the right of the scientist to work freely, without political or bureaucratic restrictions. Herzberg was awarded many honors in addition to the 1971 Nobel Prize in chemistry, including the Willard Gibbs Medal and the Linus Pauling Medal from the American Chemical Society, the Gold Medal of the Canadian Association of Physicists, and the Royal Medal of the Royal Society of London.

Herzberg married Luise H. Oettinger in 1929; the couple had two children, Paul and Agnes, both of whom chose teaching as their profession. Luise died in 1971 and Herzberg married Monika Tenthoff a year later. Herzberg's hobbies included music and mountain climbing. He died in 1999.

HIGGS BOSON

In the current standard model of **particle physics** (which does not account for the **gravitational force**), only one ingredient is yet to be directly discovered (the discovery of the tau neutrino—the other missing particle—was announced in July 2000). This elusive ingredient is known as the Higgs boson. The existence of the Higgs boson is indirectly predicted because most of the elementary particles have **mass**, which they could not have otherwise. The inclusion of the Higgs field in the standard model is an elegant solution of this problem.

The **standard model of particle physics** is a gauge field theory, meaning it is invariant under a certain set of symmetry transformations called gauge transformations. In effect, the gauge transformations allow the definitions of the fundamental objects of the theory (the **matter** fields) to be redefined differently at different points in **space**, without changing the observable physical effects. The upshot of this purely mathematical concept is that this invariance suggests there is an interaction between the basic objects (particles), carried by particles known as gauge bosons.

One caveat of the gauge principle is that it works only if the particles are initially assumed to be massless. If the particles have mass, then the mathematical terms associated with the particles' mass-energy are not invariant under gauge transformations, and would spoil features of the theory. It initially seemed to scientists, therefore, that gauge theories could not possibly be correct, since most matter particles have mass. In addition, the electroweak W and **Z bosons** were presumed not to be massless, or else they probably would have been discovered much earlier along with the **photon**.

Enter the Higgs boson to make order out of the confusion. A new object, called the Higgs field, was introduced. The Higgs field has a special property known as a nonzero **vacuum** expectation value. Most fields have the value zero when no particles of that field type are around; but the Higgs field has a nonzero constant component that exists even when no Higgs particles are in the vicinity. This constant component yields terms in the matter particles' **energy** formulas that looks just like mass energy. Therefore, the Higgs field gives mass to the particles. These mass terms are invariant under the **gauge symmetry**, because there are other Higgs field terms that cancel changes made under gauge transformations.

Because the Higgs boson has not yet been discovered, the exact form of the Higgs theory is not known. Many candidates exist, all with slightly different consequences that have not yet been measured. In the simplest version, the standard model Higgs, there are two Higgs particles, as well as their antiparticles. One of the particles is the neutral Higgs boson, but the other three are consumed by the W and Z bosons when these bosons acquire mass. In more complicated models, there are four Higgs particles and their antiparticles. These theories are called two Higgs doublet models. A special case of the two Higgs doublet models is supersymmetry, which has many other features. Each of these theories match the measured results of the standard model but have other consequences that differ. The exact form of the theory will not be known until Higgs bosons have been detected directly and these consequences have been studied experimentally.

HILBERT, DAVID (1862-1943)
German mathematician

David Hilbert was one of a group of nineteenth-century mathematicians like Nikolai Ivanovich Lobachevski (1792-1856), János Bolyai (1802-1860), and Georg Riemann (1826-1866) who for many decades had been reexamining the geometry of **Euclid**. The hoped-for goal of much of this work was to open up new possibilities for analysis of seemingly insoluble problems in conventional geometry and analytic geometry. But for some, like Hilbert, the redefining of Euclid's postulates promised to create a logical foundation upon which the entire structure of mathematics could be built.

Although it may seem obvious that mathematics is founded on logical principles, when properties of numbers are analyzed in depth, the reasons for their behavior are not immediately apparent. The logical source of seemingly simple and self-evident principles of arithmetic, such as the addition of numbers, is difficult to demonstrate when the only proof for its existence is based solely upon human intuition—what "feels right" or seems that it must be so. Intuition was the source that Euclid invoked in his axiomatic presentation of geometry in the *Elements*. But mathematicians like Hilbert wanted a more secure foundation for mathematics to rest upon than human intuition. The goal of discovering a consistent, logical basis of mathematics that did not rely on intuition was, in fact, a philosophical quest as much as a scientific pursuit, for many felt that if the **universe** had any objective reality at all, it should be definable in terms that do not require a human interpretation to give it form. It was this quest for the logical foundation of mathematics, indeed of all reality, to which Hilbert devoted much of his professional life.

Hilbert was born in East Prussia and later attended Königsberg University, an ancient and prestigious center of learning. After receiving his Ph.D. in 1885, Hilbert traveled to Leipzig, Germany, and Paris, France, before eventually settling into a lifelong professorial position at Göttingen University. He died at Göttingen in 1943. During his long life, Hilbert expended much effort to develop a logical basis for mathematics, as did such mathematicians as Alfred North Whitehead and Bertrand Russell. For these philosophers of mathematics, the disaster for this approach came in 1931, with Kurt Gödel's monumental proof that a collection of axioms could never lead to a complete and self-consistent mathematical system. This ended Hilbert's, and all others', attempts to discover an axiomatic "foundation" of mathematics.

HISTORY OF PHYSICS

Physics and **astronomy**, from which all other sciences derive their foundation, are attempts to provide a rational explanation for the structure and workings of the **cosmos**. The creation of the earliest civilizations and of mankind's religious beliefs were profoundly influenced by the movements of the **Sun, Moon,** and **stars** across the sky. As our most ancient ancestors instinctively sought to fashion tools through which they gained mechanical advantage beyond the strength of their limbs, they also sought to understand the mechanisms and patterns of the natural world. From this quest for understanding evolved the science of physics. Although these ancient civilizations were not mathematically sophisticated by contemporary standards, their early attempts at physics set mankind on the road toward the quantification of nature.

In ancient Greece, in a natural world largely explained by the whim of gods, the earliest scientists and philosophers of record dared to offer explanations of the natural world based on their observations and reasoning. **Pythagoras** argued about the nature of numbers, Leucippus (c. 440 B.C.), **Democritus**, and Epicurus (341-270 B.C.) asserted **matter** was composed of extremely small particles called **atoms**.

Many of the most cherished arguments of ancient science ultimately proved erroneous. For example, in **Aristotle**'s physics, a moving body of any **mass** had to be in contact with a "mover," and for all things there had to be a "prime mover." Errant models of the **universe** made by **Ptolemy** were destined to dominate the Western intellectual tradition for more than a millennium. Midst these misguided concepts, however, were brilliant insights into natural phenomena. More than 1,700 years before the Copernican revolution, Aristarchus of Samos (310-230 B.C.) proposed that **Earth** rotated around the Sun and **Eratosthenes of Cyrene**, while working at the great library at Alexandria, deduced a reasonable estimate of the circumference of Earth.

Until the collapse of the Western Roman civilization there were constant refinements to physical concepts of matter and form. Yet, for all its glory and technological achievements, the science of ancient Greece and Rome was essentially nothing more than a branch of philosophy. Experimentation would wait almost another 2,000 years for injecting its vigor into science. During the Dark and Medieval Ages in Europe science slumbered. In other parts of the world, however, Arab scientists preserved the classical arguments as they developed accurate astronomical instruments and compiled new works on mathematics and **optics**.

At the start of the Renaissance in Western Europe, the invention of the printing press and a rediscovery of classical mathematics provided a foundation for the rise of empiricism during the subsequent **Scientific Revolution**. Early in the sixteenth century Polish astronomer **Nicholas Copernicus**'s reassertion of heliocentric theory sparked an intense interest in broad quantification of nature that eventually allowed German astronomer and mathematician **Johannes Kepler** to develop laws of planetary **motion**. In addition to his fundamental astronomical discoveries, Italian astronomer and physicist **Galileo** made concerted studies of the motion of bodies that subsequently inspired seventeenth-century English physicist and mathematician Sir **Isaac Newton**'s development of the laws of motion and gravitation in his influential 1687 work, *Philosophiae Naturalis Principia Mathematica (Mathematical Principles of Natural Philosophy)*.

Following *Principia*, scientists embraced empiricism during an Age of Enlightenment. Practical advances spurred by the beginning of an industrial revolution resulted in technological advances and increasingly sophisticated instrumentation that allowed scientists to make exquisite and delicate calculations regarding physical phenomena. Concurrent advances in mathematics allowed development of sophisticated and quantifiable models of nature. More tantalizingly for physicists, many of these mathematical insights ultimately pointed toward a physical reality not necessarily limited to three dimensions and not necessarily absolute in **time** and **space**.

Nineteenth-century experimentation culminated in the formulation of Scottish physicist **James Clerk Maxwell**'s unification of concepts regarding **electricity**, **magnetism**, and **light** in his four famous equations describing **electromagnetic waves**.

During the first half of the twentieth century, these insights found full expression in the advancement of quantum and **relativity theory**. Scientists, mathematicians, and philosophers united to examine and explain the innermost workings of the universe—both on the scale of the very small subatomic world and on the grandest of cosmic scales.

By the dawn of the twentieth century more than two centuries had elapsed since Newton's *Principia* set forth the foundations of classical physics. In 1905, in one grand and sweeping theory of special relativity, German American physicist **Albert Einstein** provided an explanation for seemingly conflicting and counterintuitive experimental determinations of the constancy of the **speed of light**, **length contraction**, **time dilation**, and mass enlargements. A scant decade later, Einstein once again revolutionized concepts of space, time, and **gravity** with his general theory of relativity.

Prior to Einstein's revelations, German physicist **Max Planck** proposed that atoms absorb or emit electromagnetic **radiation** in discrete units of **energy** termed **quanta**. Although Plank's quantum concept seemed counterintuitive to well-established **Newtonian physics**, **quantum mechanics** accurate-

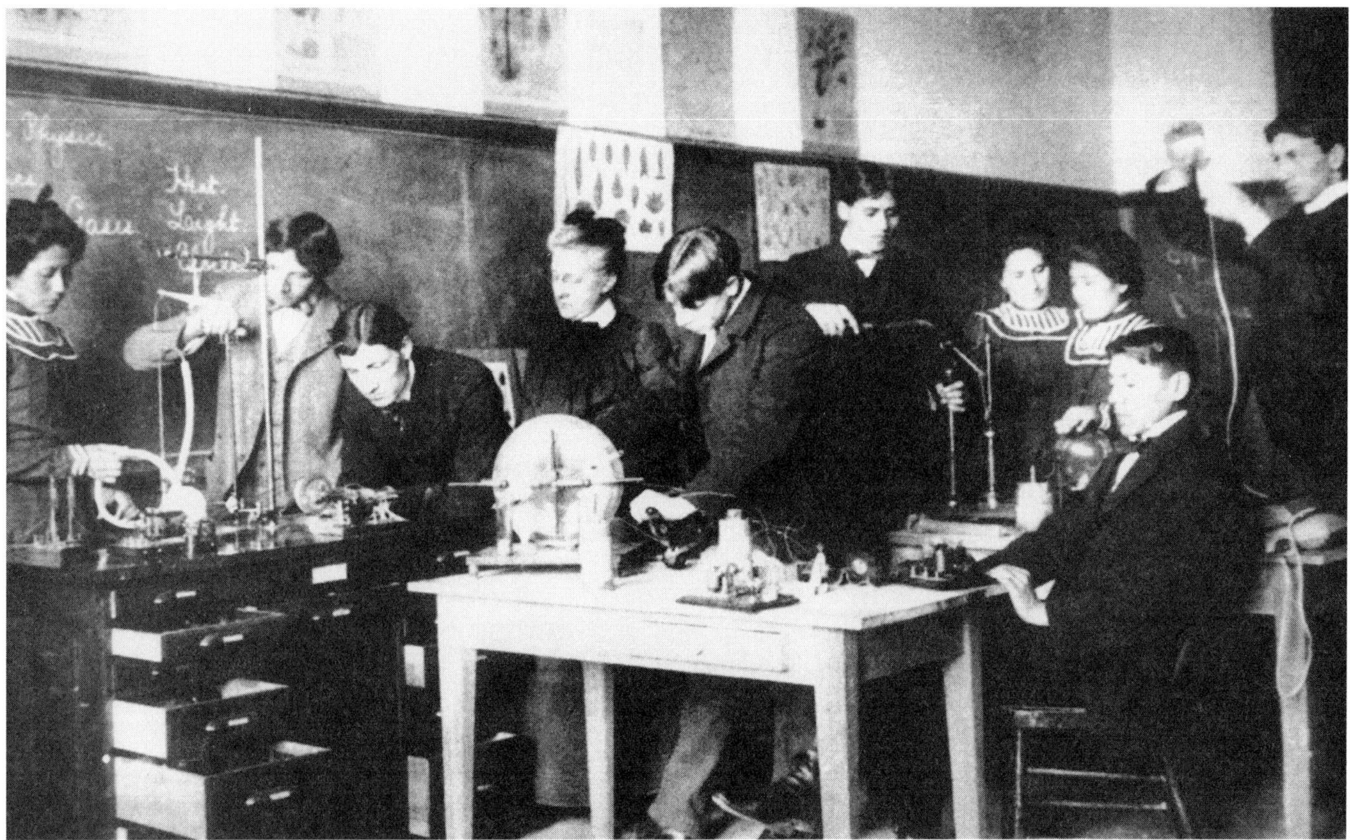

Students conducting physics experiments at Carlisle Indian School, Pennsylvania, in 1915.

ly described the relationships between energy and matter on atomic and subatomic scale and provided a unifying basis to explain the properties of the elements.

Concepts regarding the stability of matter also proved ripe for revolution. Far from the initial assumption of the indivisibility of atoms, advancements in the discovery and understanding of **radioactivity** culminated in renewed quests to find the most elemental and fundamental particles of nature. In 1913, Danish physicist **Niels Bohr** published a model of the hydrogen atom that, by incorporating **quantum theory**, dramatically improved existing classical Copernican-like atomic models. The quantum leaps of electrons between orbits proposed by the Bohr model accounted for Planck's observations and also explained many important properties of the **photoelectric effect** described by Einstein.

More mathematically complex atomic models were to follow based on the work of French physicist **Louis Victor de Broglie**, Austrian physicist **Erwin Schrödinger,** German physicist **Max Born**, and English physicist **Paul Dirac**. More than simple refinements of the Bohr model, however, these scientists made fundamental advances in defining the properties of matter—especially the **wave** nature of **subatomic particles**. By 1950 the articulation of the elementary constituents of atoms grew dramatically in numbers and complexity and matter itself was ultimately to be understood as a synthesis of wave and particle properties.

Against a maddeningly complex backdrop of politics and fanaticism that resulted in two world wars within the first half of the twentieth century, science knowledge and skill became more than a strategic advantage. The deliberate misuse of science scattered poisonous gases across World War I battlefields at the same time that advances in physical science (e.g., x-ray diagnostics) provided new ways to save lives. The dark abyss of World War II gave birth to the atomic age. In one blinding flash, the **Manhattan Project** created the most terrifying of weapons that could—in an instant—forever change course of history for all peoples of Earth.

The insights of relativity theory and quantum theory also stretched the methodology of science. No longer would science be mainly an exercise in inductively applying the results of experimental data. Experimentation, instead of being only a genesis for theory, became a testing ground to test the apparent truths unveiled by increasingly mathematical models of the universe. Moreover, with the formulation of quantum **mechanics**, physical phenomena could no longer be explained in terms of deterministic causality, that is, as a result of at least a theoretically measurable chain of causes and effects. Instead, physical phenomena were described as the result of fundamentally statistical, unreadable, indeterminist (unpredictable) processes.

The development of quantum theory, especially the delineation of **Planck's constant** and the articulation of the Heisenberg **uncertainty principle** carried profound philosophical implications regarding limits on knowledge. Modern cosmological theory (i.e., theories regarding the nature and formation of the universe) provided insight into the evolutionary stages of stars (e.g., **neutron stars, pulsars, black holes**) that carried with it an understanding of nucleosynthesis (the formation of elements) that forever linked mankind to the lives of the very stars that had once sparked the intellectual journey towards an understanding of nature based upon physical laws.

See also Astrophysics; Cosmology; Relativity, general; Relativity, special

HOFSTADTER, ROBERT (1915-1990)

American physicist

A noted physicist and researcher, Robert Hofstadter was best known for his research on the **nucleus** of the **atom**. During his several years of experimentation, Hofstadter discovered that protons and neutrons were complex components of the atom, and not as straightforward in design as previously conjectured. Hofstadter went on to expand on this discovery, and over the years he provided increasingly precise measurements of the components of an atom. For his contributions to the study of the atom, Hofstadter was awarded a share of the 1961 Nobel Prize in physics.

Born in New York City on February 5, 1915, Hofstadter was the third of four children born to Louis Hofstadter, a salesman, and Henrietta Koenigsberg. After attending public schools in New York City, Hofstadter entered the City College of New York (now the City University of New York), where he majored in physics, graduating with a B.S. in 1935. His degree was awarded magna cum laude and was accompanied by the Kenyon Prize, given for exceptional achievement in physics and mathematics. Hofstadter went on to receive both his M.A. and Ph.D. degrees from Princeton University in 1938, where he stayed on for postdoctoral work. Supported by a Proctor fellowship, Hofstadter studied photoconductivity in crystals. At the conclusion of his postdoctoral work, he took a position as instructor in physics at the University of Pennsylvania in 1939. He became Harrison Research Fellow at Pennsylvania before moving to the City College of New York, where he also became an instructor in physics.

Invents the Sodium Iodide-Thallium Scintillator

With the onset of World War II, Hofstadter took a job as research physicist at the National Bureau of Standards (NBS) in Washington, D.C. There, he worked with American physicist **James Van Allen** on the development of the proximity fuse, a device used to detonate a bomb when it has approached but not yet struck its target. After a year at NBS, Hofstadter moved to Norden Laboratories in New York City, where the famous Norden bombsight had been developed.

At the war's conclusion, however, Hofstadter returned to Princeton as assistant professor of physics. There, he became interested in problems of **radiation** detection. As studies of **radioactivity** for both theoretical and practical purposes

began to expand, the need for detection devices increased. In 1948, Hofstadter invented a device with a wide range of detection applications, the sodium iodide-thallium (NaI-Tl) scintillator. When radiation passes through the NaI-Tl scintillator, it causes the emission of **light**. The intensity of the light emitted provides a measure of the **energy** carried by the original radiation. Hofstadter's device is still used in **particle accelerators** today.

Hofstadter left Princeton in 1950 to accept an appointment as associate professor of physics at Stanford University. Four years later, he was promoted to full professor and then, in 1971, he was appointed Max H. Stein Professor of Physics. At his retirement in 1985, he was made emeritus professor of physics.

Begins Studies of the Atomic Nucleus

It was while working at Stanford University that Hofstadter became interested in studies of the atomic nucleus. One way that scientists learn about **matter** is simply to look at it. The light **waves** reflected off matter provide information about the gross structure of a material. But the use of light rays to observe matter is limited by the **wavelength** of visible light. Objects that are smaller than the wavelengths of visible light (such as an atom) do not reflect light, making it impossible to "see" them in the traditional sense.

Radiation with wavelengths smaller than that of light can, however, be used to see very small objects. **X rays**, **gamma rays**, and high-speed electrons are other forms of radiation that can be used to probe the submicroscopic structure of matter. When Hofstadter arrived at Stanford, he found that the university's linear accelerator (linac) was an excellent source of high-energy electrons that could be used to study the nucleus and its nucleons, the particles of which the nucleus is composed. He devoted a major part of his 35 years at Stanford to this kind of research.

In his earliest studies with the linac, Hofstadter bombarded a variety of atomic nuclei with **electron** beams and found that they all had similar structures. They differed in size, but all had a nearly uniform **density**. In the mid–1950s, Hofstadter turned to a study of the individual protons and neutrons of which nuclei are composed. He made the remarkable discovery that these particles are not solid, indivisible particles, but have detailed structure. By directing beams of very-high-energy electrons at gold, lead, tantalum, and beryllium targets, Hofstadter was able to obtain information about the structure of the nucleus (and later about the structure of protons and neutrons) from the way in which the electron beams were diffracted.

Hofstadter found that each **proton** or **neutron** consists of a dense, positively charged core surrounded by two shells of mesonic material. In the proton, the outer shell is also positively charged, making the particle as a whole, positive. In the neutron, on the other hand, one of the outer shells is negatively charged, making the particle as a whole, neutral. In recognition of this discovery, Hofstadter received the 1961 Nobel Prize in physics, as well as a number of other honors and awards, including City College of New York's Townsend

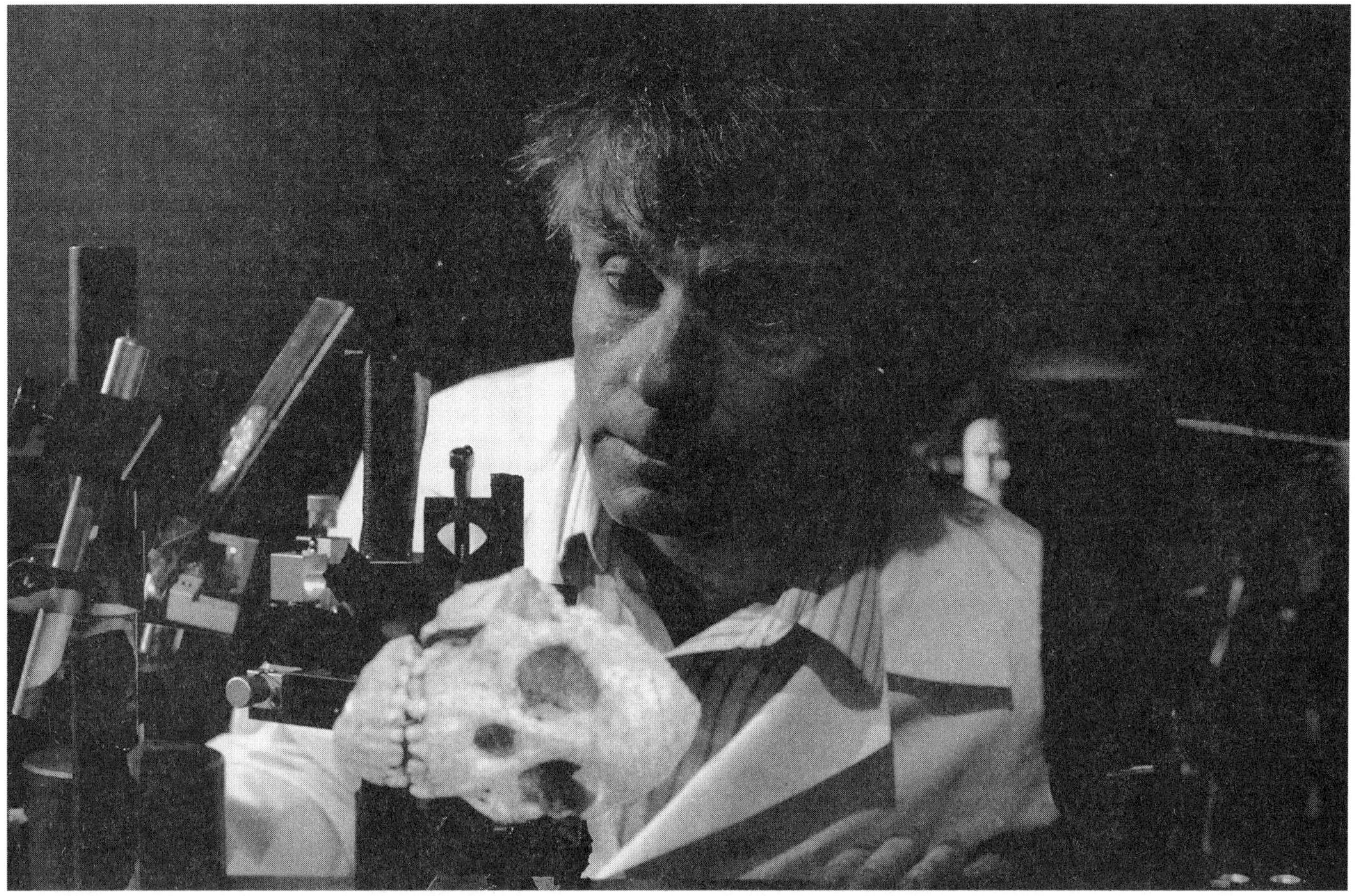

Man creating a hologram of a skull (close-up). *(Photo by Jonathon Blair. Corbis-Bettman. Reproduced by permission.)*

Harris Medal in 1962, the Röntgen Medal in 1985, Italy's Cultural Foundation Prize in 1986, the National Science Medal in 1986, and honorary doctorates from institutions, including the University of Padua, Carleton University, and the University of Clermont-Ferrand.

Hofstadter married Nancy Givan on May 9, 1942; they had three children, Douglas Richard, Laura James, and Mary Hinda. Hofstadter died of heart disease on November 17, 1990, in Stanford, California.

HOLOGRAPHY

Holography is the art of recording images with coherent **light**. A hologram, at its simplest, is a photographic recording of an **interference** pattern. The term "holography" was coined by Dr. George Stoke of Polaroid's research facilities, one of the initial leaders in the field of laser images.

Most holograms work with one **wavelength** of laser light. The image to be made into a hologram is illuminated with laser light at the chosen **frequency**. Each point blocks and reflects light at one point or another during the phase of the wave. The interference from these reflected light **waves** is recorded on a photographic plate. Then, shining light on the image from the same type of light source allows the person viewing the image to see the reflections at the same phase difference as was in the original. Coherent light—that is, light that is consistently emitted with the same phase—is extremely important to this process, because it is based entirely on the difference in phase at different points of the image. The referent laser light is the only thing that allows the points on the holographic image to form interference patterns with the incoming wave in a way that will reconstruct the original image.

Stephen Benton pioneered "white light" or "rainbow" holography, the type of holography most commercial holograms use. The desired image is formed with a laser as in normal holography. Then a second image is formed using Fourier **optics** to filter the light frequencies and make it visible in normal sunlight or room lighting.

Holograms have several useful properties. First, they are outstanding for their three-dimensionally realistic appearance. If the viewer moves his or her head around, he or she sees a complete, three-dimensional version of the object represented. Additionally, the hologram differs from normal **photography** in that the positive and negative forms of the hologram form the same image, rather than its color-opposite as a photographic negative would. The process of parallax works, as does lens action: looking through the image of a lens in a hologram will provide the same view as looking through the original lens itself. Finally, an entire holographic image can be reconstructed from a small fragment of the original. A partial hologram will be reduced in brightness and in some of its three-dimensional realism, but will otherwise be the same as the full-scale version. "White light" holograms share these features, except for a loss of scale perspective in one of the dimensions.

While holography was quite novel in the 1950s, it has now largely become taken for granted. Holograms have appeared on magazine covers, credit cards, jewelry, and anything else manufacturers can think of. Its imaging capabilities have both strengths and weaknesses, but are still one of the major triumphs of modern **laser** optics.

HOOKE, ROBERT (1635-1703)
English physicist

One of the preeminent scientists of the seventeenth century, Robert Hooke is perhaps best remembered for the wide variety of fields to which he contributed, including physics, **astronomy**, microscopy, biology, and architecture, among others. Although Hooke introduced many concepts previously unimagined or unexamined, his ability to formulate these ideas usually did not match his intuition, and the credit for many scientific breakthroughs inspired by Hooke's ideas is often given to such scientists as **Isaac Newton** and **Christiaan Huygens**, who brought the work to its fruition. Still, Hooke remains an important pioneer of science.

Born on Britain's Isle of Wight, Hooke was a sickly child who was not expected to see his fifth birthday. As a youth, his perpetual ill health made it impossible for him to attend classes regularly, and he was unable to enter the ministry as his father, a minister, had wished. Once freed from his father's agenda, Hooke was allowed to pursue his interest in **mechanics**, which he first demonstrated as a small child by constructing wonderfully elaborate toys. He attended Westminster School and later Oxford, where he became the laboratory assistant to **Robert Boyle**. It was in Boyle's lab that Hooke's talent for designing scientific instruments was noticed, as he constructed the improved air pump used to establish Boyle's **gas laws**. In fact, it has been speculated that Hooke himself may have been the author of **Boyle's law**, since, customarily, any findings from research done in the lab would have been credited to the professor.

Along with some of his colleagues from Oxford and the surrounding area, Hooke helped to establish what would soon become the Royal Society, to which he was appointed curator of experiments. This position gave Hooke the forum and the finances to conduct years of wide-ranging research. Around this same time many European inventors were vying to develop the first accurate device to determine longitude on a sailing ship. Already in use, the chronometer, essentially a modified clock, was unreliable since the pendulum used to regulate its

motion was thrown off by the ship's rocking. Sometime near 1660 Hooke introduced a chronometer design based upon a spring rather than a pendulum. Although his design was sound, he was unable to find investors to back him, and it was not until 1674 that Christiaan Huygens patented his own spring-driven chronometer. Hooke immediately claimed that Huygens's invention was a derivative of his own, beginning a dispute that remains unresolved to this day. During his time as curator he had many other successes attributed to him, such as the **compound microscope**, an improved barometer, the reflecting **telescope**, and the universal joint.

Although Hooke was not the first to experiment using a **microscope**, he was the first to dedicate a major intensive volume to microscopy. His 1665 publication *Micrographia* describes the structures of insects, fossils, and plants in unprecedented detail. While studying the porous structure of cork, Hooke noted the presence of tiny rectangular holes that he called cells—a word that has been adopted as the cornerstone of microbiology. *Micrographia* also contains illustrations in Hooke's own hand that remain among the best renderings of microscopic views.

A difficult man, Hooke was described as quarrelsome, miserly, and a dedicated hypochondriac. One of the most notorious tales about him concerns rivalry with Isaac Newton, which began in 1672. In that year Newton, then an unknown student, sent a paper to the Royal Society detailing his theory of colors, which contrasted greatly with that of Hooke. After Hooke quickly dismissed the young scientist's theory as irrelevant, Newton sent a scathing response to Hooke in a letter later published in the Royal Society's periodical. Newton published a second paper on **light** in 1675, introducing a theory of light as an undulatory wave; Hooke's reply was that Newton had stolen this wave theory outright from his own earlier publication, *Micrographia*. Hooke later made a similar claim to Newton's theory of gravitation. The feud between Hooke and Newton escalated, though the latter's increasing reputation in the British scientific community forced them to act cordially—at least in public. The two scientists' dislike for one another culminated in 1686 when Newton introduced to the world his theory of universal gravitation, a theory very similar to one Hooke had outlined in a letter seven years earlier. Though Hooke's theory was fundamentally flawed and was not supported mathematically, he insisted that Newton had plagiarized his idea. Modern scientists now credit Hooke with the ideas that inspired Newton to develop his own theory of gravitation. The verbal battles between these two scientists were very bitter, several times driving Newton to a nervous breakdown.

One theory that Hooke clearly can take full credit for concerns **elasticity**. While experimenting with systems of springs he found that the amount of **weight** added is proportional to the distance the spring stretches—that is, a four-pound weight will stretch a spring twice as far as a two-pound weight. Strictly speaking, Hooke's law states that, in an elastic system, the stress is proportional to the strain.

In the years following the great London fire of 1666, Hooke became a surveyor and, eventually, an architect, constructing numerous beautiful and famous buildings. Because his architectural interests took much time away from his scientific work, he was ultimately forced to retire as curator of experiments for the Royal Society in favor of his new vocation.

HOYLE, FRED (1915-   )
English astronomer

Fred Hoyle was born at Bingley, Yorkshire, England, in 1915. He attended Bingley Grammar School and went on to Emmanuel College at Cambridge. He eventually became a teacher of mathematics at Cambridge, a professor of astronomy and experimental philosophy, and a professor-at-large at Cornell University.

Hoyle delved deeply into the theory of **nuclear reactions** in **stars**, elaborating on gravitational, electrical, and nuclear fields, and on how various heavy elements are created within stars. In addition to writing technical books, he simplified complex theories for general readers and even wrote science fiction stories.

In 1948 Hoyle became involved with the **steady-state theory** of **cosmology**. First advanced by Thomas Gold and Hermann Bondi, steady-state cosmology proposed that as galaxies receded from one another over the eons, new **matter** is created in the empty **space** left behind and evolves into new galaxies. Hoyle, along with many other astronomers, became an advocate of steady-state cosmology because he felt its assertion that the **universe** would continue to expand indefinitely gave it a simplicity and symmetry lacking in **big bang** theory.

Hoyle, an extremely gifted writer, published several popular astronomy books that included elucidations of the new steady-state theory. He ran afoul of **George Gamow**, an equally talented theorist and supporter of the big bang theory of creation.

The greatest objection to the steady-state theory centered around the issue of the continuous creation of matter out of nothing throughout time. Hoyle felt it was easier to accept continuous creation over the theory that all the matter in the universe was created instantly from nothing by a mysterious big bang.

The debate between the two sides remained unresolved until 1963 when Maarten Schmidt discovered **quasars**. These unique objects did not fit into steady-state cosmology, which required that the universe contain similar objects everywhere. This tipped the scales toward the big bang theory, and the discovery of background microwave **radiation** by **Arno Penzias** and Robert W. Wilson in 1964 dealt a fatal blow to steady-state cosmology. Hoyle's steady-state theory is no longer widely accepted.

HUBBLE, EDWIN (1889-1953)
American astronomer

Edwin Hubble was an American astronomer whose impact on science has been compared to pioneering scientists such as

Fred Hoyle. *(Archive Photos. Reproduced by permission.)*

English physicist **Isaac Newton** and Italian astronomer **Galileo**. Hubble helped to change our perception of the **universe** in two very important ways. In an era when the **Milky Way** was perceived as the extent of the entire universe, Hubble confirmed the existence of other galaxies through his observations from the Mount Wilson Observatory in Pasadena, California. Furthermore, with the help of other astronomers of his time, Hubble showed that this newly discovered universe was expanding and developed a mathematical concept to quantify this expansion now known as Hubble's law.

Edwin Powell Hubble was born on November 20, 1889, in Marshfield, Missouri, to John P. Hubble, an agent in a fire insurance firm, and Virginia Lee James Hubble, a descendant of the American colonist Miles Standish. The third of seven children, Hubble spent his early childhood in Missouri, entering grade school in 1895. In 1898, John Hubble transferred to the Chicago office of his firm, and the Hubble family moved first to Evanston and then to Wheaton, both Chicago suburbs.

Hubble attended Wheaton High School, excelling in both sports and academics. He graduated in 1906 at the age of 16, two years earlier than most students. For his efforts, he received an academic scholarship to the University of Chicago, where he studied mathematics, physics, chemistry, and astronomy. In the summer, Hubble tutored and worked to earn money for his college expenses. In his junior year he received a scholarship in physics, and by his senior year he was working as a laboratory assistant to physicist **Robert A. Millikan**. Hubble graduated in 1910 with a B.S. in mathematics and astronomy. In addition to his academic career, the six-foot, two-inch Hubble was an amateur heavyweight boxer. According to one unconfirmed story, sports promoters urged him to become a professional boxer and fight against heavyweight champion Jack Johnson, an offer Hubble declined.

In 1910, Hubble was awarded a Rhodes Scholarship, following which he went to attend Queen's College at Oxford University in England. There he studied jurisprudence, completing the two-year course in 1912. He began working on a bachelor's degree in law during his third year, but renounced it for Spanish instead. He also continued his athletic endeavors, excelling in the high jump, broad jump, shot put, and running. In 1913, Hubble returned to the United States and began practicing law in Louisville, Kentucky, where his family was now living. Bored with his law career within a year, Hubble returned to the University of Chicago in 1914 to work towards his doctorate in astronomy.

The Turn to Astronomy

At the time Hubble attended the University of Chicago, the Yerkes Observatory was a waning institution that did not actually offer formal courses in astronomy. However, working under the supervision of Edwin B. Frost, the observatory's director, Hubble made regular observations on Yerkes' **telescope** and studied on his own. It is believed that Hubble's work at this time was influenced by a lecture he attended at Northwestern University. At the presentation, Lowell Observatory astronomer Vesto M. Slipher presented evidence that spiral **nebulae** (in that era, the term "nebulae" was used to describe anything not obviously identifiable as a star) had high radial velocities—the velocities with which objects appear to be moving toward or away from us in the direct line of sight. Slipher found spiral nebulae that were moving at much higher velocities than **stars** generally moved—evidence that the nebulae might not be part of the Milky Way.

During his term at Yerkes, Hubble also met astronomer **George Ellery Hale**, founder of the Yerkes Observatory and then the director of the Mount Wilson Observatory in Pasadena, California. Hale had heard of Hubble, and in 1916 invited him to join the Mount Wilson staff once he received his doctorate. However, Hubble's acceptance of this offer was delayed by World War I, which he joined in 1917. Hubble attained the rank of major, and after his discharge in 1919, he finally began work at Mount Wilson. The observatory had two telescopes, a 60-inch reflector and a newly operational 100-inch telescope, the largest in the world at that time. It was here that Hubble began the major portion of his life's work.

Discoveries at Mount Wilson

Hubble's first notable achievement at Mount Wilson was the confirmation of the existence of galaxies outside the Milky Way. From observations made in October 1923, Hubble was able to identify a type of variable star known as a Cepheid in the Andromeda nebula (known today as the Andromeda galaxy). By using information about the relationship between brightness, luminosity (how much **light** a star radiates), and the distances of Cepheid stars in our galaxy, Hubble was able to estimate the distance to the Cepheid in the Andromeda nebula to be about one million light years. Hubble also discovered other Cepheids, as well as other objects, and calculated the distances to them. Since scientists knew that the maximum diameter of the Milky Way was only 100,000 light years, Hubble's figures established the existence of galaxies outside our own. Eventually, he determined the distances to nine galaxies. Consistent with scientific terminology of his time, Hubble called these "extragalactic nebulae." The results of Hubble's work were publicly announced at the December 1924 meeting of the American Astronomical Society, settling one of the great scientific debates of that era.

Also in 1924, Hubble married Grace Burke Leib. His personal interests included dry-fly fishing (his favorite fishing haunts were in the Rocky Mountains and in England) and collecting antique books about the history of science. He served as a member of the Board of Trustees of the Huntington Library in San Marino, California, from 1938 until he died in 1953.

Hubble's work at Mount Wilson was interrupted during World War II, when he served as chief of exterior **ballistics** and director of the supersonic wind tunnel at the Ballistics Research Laboratory at Aberdeen Proving Ground in Maryland. He worked at Aberdeen from 1942 to 1946 and received a Medal of Merit for his efforts.

Returning to Mount Wilson after the war, Hubble continued his observations of galaxies. In 1925 he introduced a system for classifying them at a meeting of the International Astronomical Union; according to this system, galaxies were either "regular" or "irregular." In addition, regular galaxies were either spiral or elliptical, and each of these classes could be further subdivided. The system used to classify galaxies today is still based on Hubble's structure.

In 1927 Hubble was elected a member of the National Academy of Sciences, but another great achievement was yet to come. By combining his own work on the distances of galaxies with the work of American astronomers Vesto M. Slipher and Milton L. Humason, Hubble proposed a relationship between the high radial velocities of galaxies and distance. He systematically looked at a number of galaxies and found that except for a few nearby, all of the others were moving away from us at high speed. He discovered a correlation between this **velocity** and distance, and the result was a mathematical concept now known as Hubble's law. Simply put, Hubble's law states that the more distant a galaxy is from us, the faster it's moving away from us. Although Hubble didn't actually discover that the universe is expanding, he put the theory together in a coherent way. Today, the expanding universe is part of the big bang theory of the creation of the universe.

Besides his pioneering work in astronomy, Hubble helped develop the Mount Palomar Observatory in California, as well as the Hale 200-inch telescope. Hubble has been described as an accomplished speaker; in fact, many of his lectures, including several honorary lecture series at Yale and Oxford universities, were published as books, such as *The Realm of the Nebulae*. For his contributions to astronomy, Hubble received many awards, including that of honorary fellow of Queen's College, Oxford. The **Hubble Space Telescope**, launched by the National Aeronautics and Space Administration in 1990, is named after him. Hubble's final contribution to astronomy was a photographic guide to the classification of galaxies, which he was in the process of working on when he died. *The Hubble Atlas of Galaxies* was finished by Allan R. Sandage, who had worked with him. Hubble suffered from heart disease in the last years of his life, but continued to work at Mount Wilson and Palomar. He died of a cerebral thrombosis (a type of stroke) on September 28, 1953.

HUBBLE CONSTANT

The Hubble constant is used in **astrophysics** to characterize the expansion of the **universe**. The constant specifically relates the distance to a galaxy to that galaxy's **velocity** of recession. Ironically, because **gravitational forces** depend on the distance between objects, the Hubble constant is actually a function of the expansion of the universe (i.e., it changes as the universe expands).

In 1929, American astronomers **Edwin Hubble** and Milton Humason (1891-1972), expanding upon the work of Vesto M. Slipher (1875-1969), discovered that **light** from distant galaxies is systematically increased in **wavelength**. This increase in wavelength, called a **red shift**, is denoted (z). Hubble interpreted observed red shifts as a consequence of the **Doppler effect**, which implied that the galaxies in his sample were all receding from the **Sun** with a velocity (v) determined by the equation, $v = cz$, where (c) is the **speed of light**. Of profound significance to **cosmology**, and in contrast to prevailing theories describing a static universe, Hubble and Humason discovered proof of an expanding universe.

Hubble also realized that there was a linear relationship between the spatial velocity away from the Sun and the distance to each galaxy. The farther one looks into **space**, the faster galaxies are moving uniformly away from the Sun. Incidentally, this does not imply that the Sun is at the center of the universe. In accordance with the **cosmological principle**, there is no center and every location in the universe should see the same nature of the expansion of the universe.

Hubble quantified the linear relationship between expansion velocity (v) and distance (d) via the equation, $v = Hd$, known as Hubble's law. The constant term (H), determined from the slope of the linear relation between velocity and distance, has come to be known as the Hubble constant.

In an expanding universe, the Hubble constant provides the expansion rate of the universe in units of inverse time. Thus, the reciprocal of the Hubble constant, called the Hubble time, gives a rough estimate of the age of the universe. This is a rough estimate because it relies on the assumption that the expansion rate of the universe has been constant with time. In all likelihood, the expansion rate has slowed due to the mutu-

al gravitational attraction of the constituent material of the universe.

As of 2000, astronomers remain unable to determine a precise value of the Hubble constant better than to say it lies between 50 and 100 inverse time units of kilometers per second per Megaparsec. Such obscure units are used because it makes the analysis easier as red shifts are often measured in kilometers per second and galaxy distances are measured in Megaparsecs (the prefix Mega denotes one million and one parsec equals 3.26 light-years). Because of the uncertainty placed on the Hubble constant, astronomers have been loosely divided into two camps; those who subscribe, such as Alan Sandage and Gustav Tammann, to a value near 50; and those, like the late Gérard de Vaucouleurs, who argue for a value near 100. Finally, the modern accepted range of values for the Hubble constant place the age of the universe between 10 and 20 billion years.

The inability to pin down the Hubble constant is almost entirely due to the difficulty associated with measuring accurate distances to galaxies. Although obtaining accurate red shifts for galaxies is relatively easy with high-tech spectroscopic instruments, accurate distance determinations, especially to galaxies that are over 100 million light-years away, is highly prone to error. For example, Edwin Hubble's first calculation of the Hubble constant yielded a whopping value of 530 kilometers per second per Megaparsec because of a severe underestimate in the distances to galaxies resulting from a misidentification of the types of **stars** used to calibrate distance. Hubble's errors in distance determination propagated through his subsequent calculations and led to the extraordinarily high magnitude for the Hubble constant.

The manner in which a recessional velocity for a galaxy associates with a measured red shift is also highly dependent upon the cosmological model used. A direct correlation between Doppler velocity and red shift, as Hubble assumed and as outlined above, is valid only for a flat Euclidean space. The red shift that astronomers measure due to the expansion of the universe is known as a cosmological red shift. Most importantly, galaxies are not racing away from each other through space, but rather with space. As a result the entire universe is expanding. An important consequence of this expansion to astronomers is that the wavelengths of **radiation** from distant galaxies become stretched as a result their propagation through expanding space.

See also Big bang; Cosmological constant; Steady-state theory

HUBBLE SPACE TELESCOPE

The Hubble Space Telescope (HST) is a large Earth-orbiting astronomical telescope designed by the U.S. National Aeronautics and Space Administration (NASA) and the European Space Agency (ESA). Hubble observes the heavens from 380 miles above **Earth**, relaying pictures and data captured above the distortions of Earth's atmosphere. The HST is named after American astronomer **Edwin Hubble**, who early in

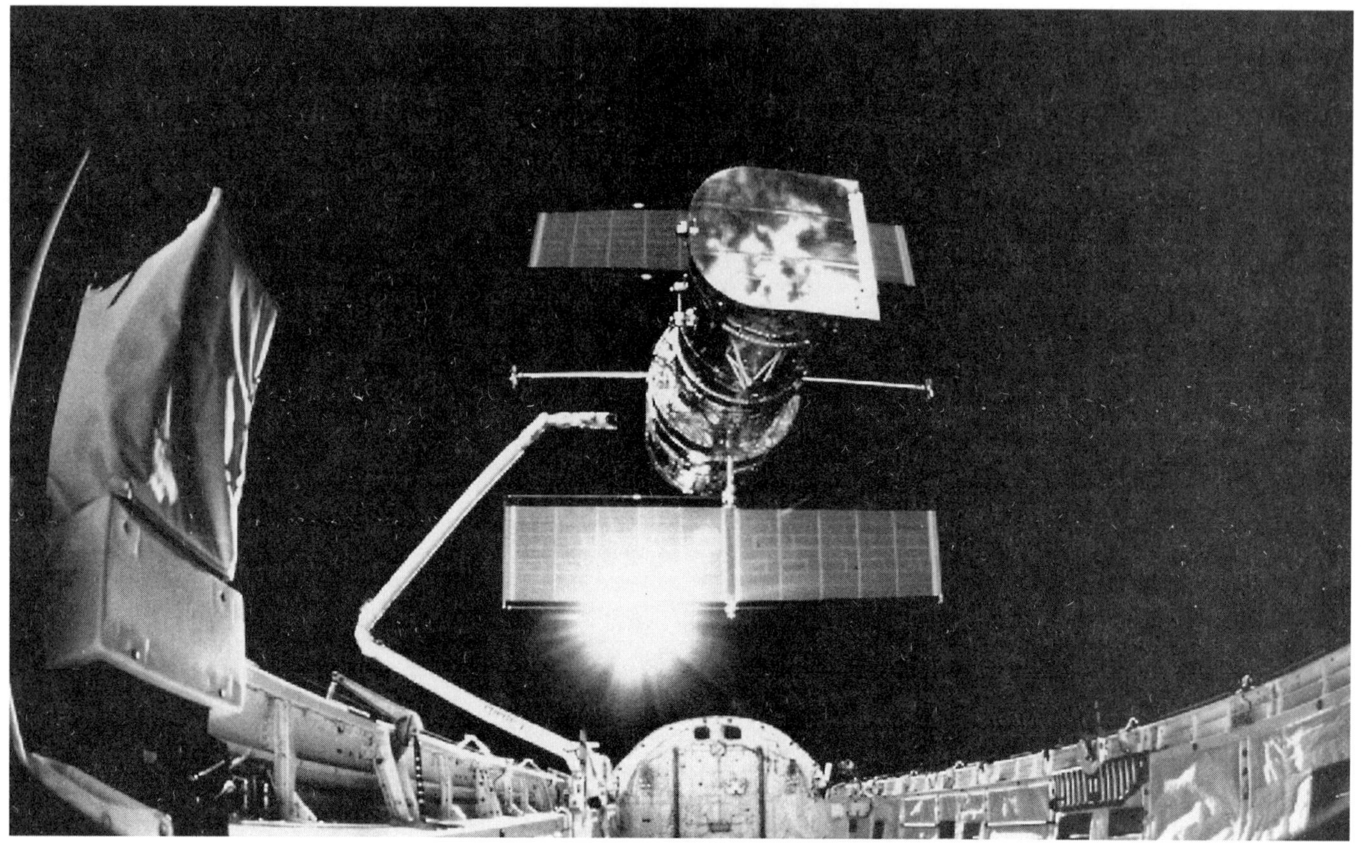

Hubble Space Telescope, seen moments after release from space shuttle *Discovery*, April 25, 1990. *(Photo courtesy of NASA. Reproduced by permission.)*

the twentieth century provided evidence of an expanding **universe** consisting of many galaxies beyond our **Milky Way galaxy**. The HST has provided scientists with the clearest views yet obtained of the universe. Moreover, stunning images and spectrographic data sent from the HST provide scientists with critical data relevant to studies regarding the birth of galaxies, the existence of **black holes**, and the workings of planetary systems around **stars**.

Deployed from the space shuttle *Discovery* on April 25, 1990, the Hubble Space Telescope was the culmination of a 20-year scientific effort to construct one of the largest and most complex satellites ever built. Astronomers first proposed the idea of building an orbiting observatory in the 1940s. The $1.5 billion project to build the Hubble Space Telescope began in earnest in 1977 after the U.S. Congress passed a resolution granting approval for the HST construction. By 1985, the HST was completed and ready for launch. The explosion of the space shuttle *Challenger* and loss of its crew in January 1986 delayed the Hubble's launch four years. As NASA officials re-evaluated the space shuttle program, the HST was relegated to storage at a maintenance cost of up to one million dollars a month.

The HST is roughly the size of a school bus, and is modular in design to facilitate in-orbit servicing. Like any reflecting **telescope**, the Hubble uses a system of mirrors to magnify and focus **light**. The primary mirror is concave, and a smaller convex secondary mirror is placed in front of the primary mirror to boost the telescope's total effective focal length. The telescope receives its main **power** from a pair of flexible, lightweight solar arrays. Each array is a large (40 foot by 8 foot) rectangle of light-collecting solar cells. Exterior thermal blanketing protects the HST from the extreme **temperature** changes encountered during each 95-minute orbit of Earth.

Shortly after the 1990 launch of the HST, scientists found the telescope was unable to adequately focus light to provide desired resolutions. Fuzzy halos appeared around objects observed by the HST. The culprit was found to be a defect in the primary mirror. As a result of an incorrect adjustment to a testing device, the mirror was precisely, but inaccurately, ground to a curvature that was too flat at its edge. Although the error measured less than a micron (one ten-thousandth of an inch), the defect caused a spherical aberration when light reflected by the mirror focused across a wider area than necessary for a sharp image. The problem was corrected in December 1993 when, following an orbital rendezvous between the space shuttle *Endeavour* and the HST, the crew of *Endeavour* completed the first Hubble servicing mission. During the 11-day operation, the Corrective Optics Space Telescope Axial Replacement (COSTAR) was installed. COSTAR corrected the spherical aberration of the HST primary mirror with a series of mirrors designed to act as corrective "eyeglasses" able to focus the blurred uncorrected image.

The Hubble Space Telescope carries a variety of onboard, scientific instruments designed to collect and send data to awaiting scientists. As needed, instruments are replaced or added during Hubble servicing missions. In 1977, two spectrographs were replaced with the Near-Infrared Camera and Multi-Object Spectrometer (NICMOS), and the Space Telescope Imaging Spectrograph (STIS). NICMOS allows the telescope to see objects in near-infrared wavelengths and these observations are important in the study of ancient light emitted from distant objects and in the study of the visible-light-obscuring gas and dust nebular clouds where stars are born. The STIS collects light from hundreds of points across a target and spreads it out into a spectrum, creating an image from which scientists can study individual wavelengths of **radiation** from a distant source. STIS is especially helpful to scientists studying regions of space where black holes are presumed to exist. In 1993, the HST's original Wide Field Planetary Camera was replaced with an updated version complete with relay mirrors spherically aberrated to correct for the spherical aberration on the Hubble's primary mirror. In 1999, the HST received a new high-speed computer.

Once the Hubble gathers data and pictures from celestial objects, its **computers** send the digitized information to Earth as **radio** signals. The HST signal is passed through a series of satellite relays, then to the Goddard Space Flight Center in Maryland before reaching the Space Telescope Science Institute at Johns Hopkins University. Here, the signal is converted back into pictures and data. Scientists at these institutions are responsible for the daily programming and operations of the HST.

Scheduled to serve until the year 2010, the Hubble Space Telescope continues to provide dramatic observations that stretch the boundaries of the known universe. Among its accomplishments so far, the HST has provided evidence of the existence of massive black holes at the centers of galaxies, captured the first detailed image of the surface of Pluto, detected protogalaxies (structures currently thought to have existed close to the time of the origin of the universe), and captured spectacular images of the comet Shoemaker-Levy as its parts collided with Jupiter.

In order to provide continuous and broader astronomical observations, NASA is expected to launch the Hubble's successor (tentatively named the Next Generation Space Telescope) more fully equipped with cameras and spectrographs sensitive to multiple regions of the **electromagnetic spectrum** prior to the end of the HST's expected service life.

See also Astronomy; Cosmology

Huygens, Christiaan (1629-1695)

Dutch astronomer, physicist, and mathematician

Though one of the most brilliant scientists in history, Christiaan Huygens enjoyed relatively little fame during his lifetime—primarily because he worked during the period directly after the death of **Galileo** and just before the ascent of **Isaac Newton**. Although his work went unregarded for many years after his death, he is today held as one of the chief contributors to the modern sciences of **mechanics**, physics, and **astronomy**.

Huygens was born in the Hague, Netherlands, in 1629. The environment in which he was raised was ideal for the nurturing of young minds: his father, Constantijn Huygens, was a diplomat and poet who understood the need for classical training, and he planned for his son a private education in mathematics, languages, literature, and music. Young Christiaan was also influenced by mathematician-philosopher **René Descartes**, a friend of the Huygens family and frequent visitor to their home. Huygens learned about the "mechanistic" philosophy of nature from Descartes and came to believe that all natural phenomena would one day be explained by science.

Huygens left his home in 1645 to study law and mathematics at the University of Leiden and, in 1647, he entered the College of Orange in Breda. He was dissatisfied with the university's approach to learning, however, and in 1649 he returned to the Hague. There he remained until 1666, living off an allowance from his wealthy father. Given the financial freedom to study as he pleased, Huygens began to perform some of his most important research. During the latter half of the seventeenth century, it was not unusual for a scientist to work in several different disciplines, attaining success in each. Such was the case with Huygens, who worked simultaneously in the fields of astronomy, physics, and applied mathematics.

In the early 1650s, Huygens spent much of his time learning to grind **telescope** lenses; though tutored by Dutch philosopher Baruch Spinoza (1632-1677), he was essentially self-taught in this art. By 1655 he had developed a new grinding technique that yielded **lenses** of unsurpassed clarity. Using his lenses, Huygens almost immediately discovered a large **moon** circling Saturn, which he named Titan. At this time there were six-known **planets** and six-known moons, and for a short time Huygens believed that, since this was such a convenient arrangement, no other heavenly bodies would be discovered.

In 1656 Huygens incorporated his lenses into telescopes of extreme length—some up to 23 feet (7 m)—whose long focal length allowed for even greater magnification. Using telescopes of such design Huygens was able to chart the surface features of the planet Mars, as well as discover the Great Orion nebula (a multicolored cloud of hot gas in the Orion constellation).

In that same year, Huygens discovered the truth about the "triple" planet Saturn—so named because it appeared in conventional telescopes to possess two smaller planets to either side. Using his vastly superior telescope Huygens found that Saturn was encircled within a thin ring. Huygens reported his discovery in a secret code, protecting it from other scientists while he continued his observations of Saturn.

To facilitate these observations, Huygens designed several new pieces of astronomical equipment. Chief among these were his lenses, which provided far greater resolution and magnification than any before. After several years he perfected an achromatic lens that corrected the "false color" fringes often associated with inferior lens systems; this lens, called the Huygensian eyepiece, is still used in many telescopes today.

In order to better view the sky, Huygens also modified the design of his telescopes. It was commonly known that longer telescopes, with their longer focal lengths, allowed for greater magnification. Huygens took this principle to its extremes, constructing telescopes up to 23 feet (7 m) long. While these remarkable instruments gave a magnificent view of the planets, Huygens remained unsatisfied: he believed that the conventional telescope design was too limiting, for it relied upon metal tubes that would bend if they were too long. His answer was a tubeless telescope called an aerial telescope. Because there was no tube to connect them, the large objective lens and the smaller eyepiece could be as far apart as practical lens construction would allow. The largest of these aerial telescopes was more than 100 feet (30 m) in length. In addition to optical aids, Huygens also invented a micrometer in 1658 that enabled him to measure the angular separations of objects (such as the apparent distance between Saturn and its moon) with a precision of a few seconds of arc.

While Huygens observed the heavens, the European scientific community was working toward the development of a reliable timekeeping device; such a device was in great demand by the trading industry, which required an accurate clock for use in the navigation of its sailing vessels. Previously, clocks were regulated by a slowly falling **weight** that would turn the device's gears. Unfortunately, the pace of the weight's descent was irregular, and the clocks were wildly inaccurate.

Years earlier, Galileo had noted that a pendulum would swing with a precise **motion**, taking the same amount of time to move in one direction as it did to return. He termed this effect *isochronicity* ("equal time") and suggested that it might be useful as a means for regulating timepieces; however, he was never successful in designing a working model.

In 1656 Huygens found that a swinging pendulum was not truly isochronic unless the arc it made was not quite circular. Using this knowledge he devised a system combining the pendulum with a weight-driven clock: the pendulum would swing exactly once each second, precisely regulating the motion of the clock's hands; the falling weight would drive the gears, as well as give the pendulum just enough **energy** to overcome the slowing forces of air resistance and **friction**. Before the invention of Huygens's pendulum clock there was no reliable means of measuring time. Within months of the introduction of the "grandfather clock" design, towns across the Netherlands (and, soon after, all of Europe) had large clock towers regulated by swinging pendulums.

As he experimented with pendulums, Huygens also noticed certain relationships between the **mass** of the weight and the **velocity** of its swing. From his observations he formulated a theory of **momentum** that was included in his 1673 publication *Horologux Oscillatoriux*; it is now better known as the law of conservation of momentum, wherein it is stated that the energy of a moving object remains constant until the object is stopped or its direction is changed. While the law of the conservation of momentum had been proposed earlier by John

Wallis, Huygens's theory was expressed more clearly and completely. This law was eventually expanded into **Hermann von Helmholtz**'s law of the **conservation of energy**, which was later expanded by **Albert Einstein** to include the special theory of relativity. Huygens also proposed hypotheses on centrifugal **force**, as well as the value for the **acceleration** of a falling body (g).

Meanwhile, a debate was raging through the European scientific community concerning the nature of **light**: was it a wave, like **sound**, or was it, as Isaac Newton had suggested, composed of particles? In his *Principia Mathematica*, Newton attempted to prove his theory by using a prism to split light, spreading the particles away from one another; white light, he concluded, must be a composite of all the colored particles traveling together through a **vacuum**.

This was a difficult concept for Huygens to envision because his earlier conversations with Descartes had taught him that nature is mechanical—that all phenomena are caused by **matter** influencing other matter. Thus, it seemed absolutely ridiculous that a group of particles could be created spontaneously and independently transfer themselves through a vacuum. He introduced his own wave theory of light in 1690, three years after Newton's *Principia*, that supported Descartes's view of the **universe**.

Huygens's wave theory of light was similar to that of the wave theory of sound. According to his theory, light was a series of pulsations similar to **shock waves**. These pulsations would travel longitudinally (with an in-and-out motion) through an invisible medium called the *ether*; each pulse would disturb the **ether**, pushing it in the direction of the light wave.

At the time of its publication, Huygens's wave theory of light was given little consideration, chiefly because scientists were reluctant to accept the notion of the invisible ether. As Newton's reputation grew, his particle theory became more and more popular, and the work of Huygens was all but forgotten. It was not until the nineteenth century that **Thomas Young** reintroduced the wave theory of light—this time without the dubious hypothesis of the ether.

It is not surprising that Huygens achieved little fame during his lifetime, for he published his work slowly and infrequently. He was also a recluse who chose not to take students. Still, he was instrumental in the foundation of the French Academy of Sciences and was a charter member of the British Royal Society. In addition to his clocks and his astronomical instruments, Huygens invented the manometer (a device used to measure the **pressure** of liquids and gases), as well as a prototype internal combustion engine using pistons.

HYDROELECTRIC ENERGY

Hydroelectric **energy** simply means producing electrical energy from water energy. Water is a heavy substance and as we watch a white-water river or, better yet, observe the Niagara Falls, we are left in no doubt as to the tremendous forces unleashed by cascading water.

Producing **electricity** is usually a matter of rotating a dynamo. Water passing over the blades of a turbine wheel is

Grand Coulee Dam on the Columbia River, Washington. *(Photo courtesy of Corbis-Bettmann. Reproduced by permission.)*

one method of doing this. In order to accomplish this a dam is usually constructed to block the flow of a river and to build up a reservoir of water for the turbines. The water is then piped from the bottom of the dam where its great steady **pressure** spins the turbines at a constant speed. The turbines are attached to a dynamo that then generates electrical energy.

Occasionally, a nation is lucky enough to have natural lakes high up in mountainous areas. Boring holes in the rock walls containing the lake and allowing the freed water to pour over turbines obviously achieves the same ends so long as there is a supply of water sufficient to replenish the lake. In certain cases even this is not essential. In a pumped storage plant there are two reservoirs, one at a higher elevation than the other. When electrical energy is required, water pours out of the upper reservoir or lake into the lower one, passing over the turbine blades on the way. Clearly, if this goes on for too long the upper lake will empty and the lower lake will flood. But at periods of low energy demand, such as weekends or at night, the process is reversed. The water in the lower lake is pumped back up into the upper lake to get ready for the next peak demand period.

The advantages of a hydroelectric energy plant are obvious. It is using a renewable energy source, it produces no air pollution and, once built, is comparatively cheap to run since it needs no fuel. In addition it can provide a new lake with recreational uses and promote new aquatic and bird life. There are, however, some negative aspects to hydroelectric energy plants. The plants can restrict the flow of the water below the dams altering the fish and animal life. The dams do block upwardly migrating fish such as salmon. On occasions large fish kills have resulted from salmon overcrowding the waters below dams as they try to move farther upstream. Also fish can be sucked from the upper reservoirs into the turbines and be killed.

Most of these disadvantages can be overcome with care and planning. Fish ladders can be constructed to allow passage of migrating fish and filters can be placed over the entrances to the turbines. Environmental damage during the construction phase can also be minimized with forethought.

Mountainous countries with heavy rainfall, such as Norway and Canada, rely extensively on hydroelectric energy production. Hydroelectric production accounts for approximately 20% of worldwide energy requirements. The United States relies less on this mode of production with about 10% of its energy requirements being produced this way.

HYDROGEN ATOM

Hydrogen, the simplest of all elements, is a Group 1A element possessing one **proton** and one **electron**. **Henry Cavendish** recognized it as a distinct substance in 1766 although it was produced on a regular basis much before that time. Antoine Lavoisier (1743-1794) named the element hydrogen which is based on the Greek *hydro*, meaning water and *genes*, meaning forming. Its atomic symbol is H and it has an atomic number of 1 with an **atomic weight** of 1.0079. It is estimated that 90% of all atoms in the **universe** are hydrogen, making it the most abundant element. Hydrogen has three isotopes, two of which are stable. The most common **isotope**, hydrogen-1, is called protium and is the most abundant of the three isotopes. In 1932, Harold Urey (1893-1981) prepared hydrogen-2, deuterium, and it is now known to exist as about one in 6,000 normal hydrogen atoms. Two years later tritium, hydrogen-3, the unstable isotope of hydrogen was discovered and determined to have a **half-life** of about 12.5 years.

The hydrogen **atom** is the simplest atom of all the elements. It is a system consisting of a proton **nucleus** circled by one electron in an s orbital interacting according to **Coulomb's law**. **Schrödinger's equation**, an equation predicting the **energy** levels in an atom, can be solved exactly for only the hydrogen atom because it has only one electron. In January 1926, **Erwin Schrödinger** formulated the Schrödinger equation and solved the time-independent version of the equation for the hydrogen atom obtaining the energy levels in agreement with the observed hydrogen spectrum. This was published in his first paper on **quantum mechanics** the same year. Subsequent ideas developed in treatment of the hydrogen atom provided a basis for dealing with other atoms possessing several electrons.

Hydrogen is a colorless, tasteless, odorless, nonpoisonous gas under normal atmospheric conditions. It is the lightest known substance, 14.5 times lighter than air, and 1 cu ft (0.036 cu m) of the gas has a lifting **power** of about 0.076 lb (0.035 kg) under normal conditions. It can be liquefied under extremely low temperatures, -259.14°C, and because of this low melting point it is important in cryogenic studies as well as superconductivity experiments. Hydrogen is not a very chemically active element and although it does not support combustion it does burn in the presence of **oxygen**. It plays an important role in proton-proton reactions as well as the carbon-nitrogen cycle that are responsible for the energy of the **Sun** and **stars**. It is used as a rocket fuel, for welding, filling balloons, production of hydrochloric acid, and reduction of metallic ores. Tritium is produced in **nuclear reactors** and is the isotope that is used in the production of hydrogen bombs. This isotope is also used as a tracer element and as an agent in luminous paints.

Hydrogen gas under ordinary atmospheric conditions is a mixture of two types of molecules, *ortho-* and *para*-hydrogen, which differ from one another by the spin of the electrons and nuclei. At room **temperature** 25% of the molecules exist in the *para*-state and the remaining 75% exist in the *ortho*-state. The *ortho*-state cannot be prepared in the pure state. Because these two types of hydrogen are different in energy they possess different physical properties. The melting and **boiling** points of *para*-hydrogen are about 0.1°C lower than those of hydrogen in its naturally existing state.

Hydrogen is produced by several means for commercial use. The most widely used procedures for producing hydrogen include **electrolysis** of water, displacement from acids or even water by certain metals, the action of steam on hot **carbon** and the decomposition of certain hydrocarbons. One of the most popular demonstrations in a basic **chemistry** course is the reaction of a piece of sodium metal in water. The sodium catalyzes the decomposition of water and thus releases hydrogen. It is possible to ignite the hydrogen gas liberated during this reaction on the surface of the water. It makes for quite a spectacular demonstration.

IDEAL GAS LAW

The relationship between three state variables, absolute **pressure** (P), volume (V), and absolute **temperature** (T), that is deduced from kinetic theory and characterizes an ideal gas is called the ideal gas law, PV = nRT, where P is absolute pressure, V is volume, n is the number of moles of a gas, R is the universal gas constant, and T is temperature. Emil Clapeyron first wrote down the ideal gas law in 1834 although it is a combination of the three empirical **gas laws** by **Robert Boyle**, **Jacques Charles** and **Amedeo Avogadro** that were known some time before that.

An ideal gas is one in which all collisions between molecules or atoms are perfectly elastic and in which there are no intermolecular attractive or repulsive forces. An ideal gas also obeys **Boyle's law**, **Charles's law,** and Avogadro's (or Gay-Lussac's) law. Although no real gas is actually ideal and so can only approach the behavior implied by the ideal gas equation, they do act as if they were ideal gases at and below atmospheric pressure and so the concept is very useful in reality.

Robert Boyle made the first quantitative measurements of gases in a systematic manner. Using a manometer, which measures differences in pressure, and a barometer, which measures the total pressure of the atmosphere, he developed what is now known as Boyle's law. This law states that at any constant temperature, the product of the pressure and the volume of any size sample of any gas is a constant (Vμ1/P). Several years later, after developing a quantitative temperature scale, French chemist Jacques Charles formulated a general law known as Charles's law. This law states that at any constant pressure, the volume of any sample of any gas is directly proportional to the temperature (VμT). Amedeo Avogadro formulated the last of the empirical laws required to derive the ideal gas law in 1811. This law is an interpretation of **Gay-Lussac's law**, which has to do with combining volumes of gases. **Avogadro's law** states that equal volumes of gases under the same conditions of temperature and pressure contain equal numbers of molecules (Vμn). This law is equivalent to the statement that volume is directly proportional to the number of atoms or molecules.

These three concepts in conjunction produce the ideal gas law, PV = nRT where R is the universal gas constant. The gas constant arises from a combination of the proportionality constants in the three empirical gas laws and has a value that depends only upon the units in which the pressure and volume are measured. Two of the more common values employed are 0.08206 L atm/mol K and 8.3145 J/mol K.

IDENTICAL PARTICLES

The concept of identical particles becomes important in the study of quantum **mechanics**. In a classical description from atomic physics—for instance, the one proposed by **Ernest Rutherford**—an **atom** resembles a miniature solar system. Protons and neutrons cluster together in the center like suns; electrons orbit around them like **planets**. In classical mechanics, each of the **subatomic particles** occupies a certain place in **space** and **time**, and (theoretically, at least) those places can be exactly defined. In **quantum mechanics**, however, the position of a given particle in space and time cannot be known exactly; it can only be predicted through probability. Under quantum mechanics, all particles of a given type become identical when they have the same probability of being in any particular space at any given time.

Quantum physicists divide all known particles into two basic types: **fermions**, which include all known particles of matter—**quarks** and **leptons**—and **bosons**, which include everything else: **photons, gravitons**, gluons, and others. There are some cases where collections of fermions can act like bosons because of their combined spin; the helium-4 **isotope** is an example. With respect to identical particles, bosons and fermions differ in a very important respect. According to the **Pauli exclusion principle**, no two fermions can occupy the same quantum state at the same time. According to Bose-Einstein statistics, however, any number of bosons can exist simultaneously in the same state. Under the system of investigation called wave mechanics, particles such as electrons behave in ways that can be predicted by a wave, a fermion must occupy a single point on the wave, but bosons can exist in quantity all along the wave form.

The theory of identical particles has had a great impact, not only on the study of **atomic structure**, but also in practical applications. The predicted behavior of fermions such as electrons has given rise to **electron** microscopy, new types of **lasers**, and solid-state **computers**.

IMAGE FORMATION

Image formation is the result of the **light** of a luminous object being refracted through a double convex lens to form an image, or identical representation, of the object. Refraction is a wave phenomenon light exhibits because it is an electromagnetic wave and it occurs whenever light enters a medium that can modify its speed. Another factor that impacts image formation is the index of refraction, which leads to chromatic **dispersion**.

Refraction is always accompanied by a change in the direction of travel, and the immediate consequence of light penetrating a lens is that it is bent to an extent described by Snell's law, which is derived from the Fresnel equations, and expressed as follows: $n_1/n_2 = \sin\theta_1/\sin\theta_2$, where n is a dimensionless unit called the index of refraction and is defined as the **speed of light** in **vacuum** divided by the speed of light in the medium, and where θ is the angle between the wave and the normal to the boundary between the two media, i.e., air and the lens. Small variations of the index of refraction with the **wavelength** of the light can occur and this phenomenon is called chromatic dispersion. The dispersion of a lens will affect the image produced because it will change the extent to which the light bends as it goes through the lens. As a general rule, the index of refraction decreases with increasing wavelength, because blue light (of shorter wavelength) travels more slowly in a lens than red light (of longer wavelength). The dispersion of a lens is measured by a standard parameter known as Abbe's number, or V number.

After penetrating the double convex lens, the light wavefront, consisting of several light rays, is bent in such a way that the upper fraction bends downward and the lower fraction upward. The resulting rays emerge and pass through a point located at a given distance from the lens called the focal point. The light rays travel beyond the focal point and produce an inverted real image. The upper features are due to the rays that had penetrated the lower part of the lens and were refracted upward and the lower features are due to the downward refracted light rays. The distance of the image from the lens can be determined using the lens equation expressed as follows:

1/object distance from lens + 1/image distance from lens = 1/focal length

For a thin double convex lens, all parallel rays are focused on a point called the principal focal point and the distance from the lens to that point is accordingly termed the principal focal length (f) of the lens. For a double concave lens, the principal focal length is the distance at which the back-projected rays come together and it always has a negative sign. The principal focal length is determined using the lensmaker's formula for thin **lenses**. The magnification, or enlargement, of the image is a function of the ratio of the image and object distances.

Images are either real or virtual. If the object is located at a distance farther away than the focal length of a convex lens, it then forms a real image on the opposite side of the lens. The position of the image can be determined from the lens equation or by using a ray diagram. If a concave lens is used, the lens equation yields negative values, and the image produced is a virtual image, and is located on the same side of the lens as the object. Virtual images are obtained using a diverging (or negative) lens instead of the converging (or positive) lens used to produce real images. The lens equation predicts the formation of both real and virtual images and applies to both converging and diverging lenses. It is also called the thin lens equation because it does not apply to thick lenses, which use Gullstrand's equation instead. With this equation, the distances are measured from hypothetical principal planes, which are two planes in a lens system at which all the refraction is considered to occur. For a given set of lenses and separations, the principal planes are fixed and do not depend upon the object position.

Image formation is an important phenomenon as it represents the operating principle of a wide variety of **optical instruments** such as the simple magnifier, the telephoto lens, the astronomical **telescope,** and the **compound microscope**. The most important manifestation of image formation, however, is the eye, which uses the cornea and its crystalline lens to form images on the retina that are perceived by the rods and cones. The same refraction phenomenon that makes image formation possible by optical instruments is used by the eye, in which 80% of the refraction occurs in the cornea and the remaining 20% occurs in the inner crystalline lens.

INDUCTION MOTOR • See Motors

INDUCTORS AND INDUCTANCE

Electrical inductance (L) is a property of conductors and it is defined in terms of the **electromotive force** (emf), or **voltage**, generated in a conductor to oppose a given change in current (I). This is expressed as: $emf = -L\,\Delta I/\Delta t$.

Inductance is illustrated by the behavior of a coil resisting any change in **electric current** going through it. If the current inside the coil increases, then a voltage opposing that increase will be created by the **magnetic field** of the coil. Inductance is a consequence of Faraday's law of induction,

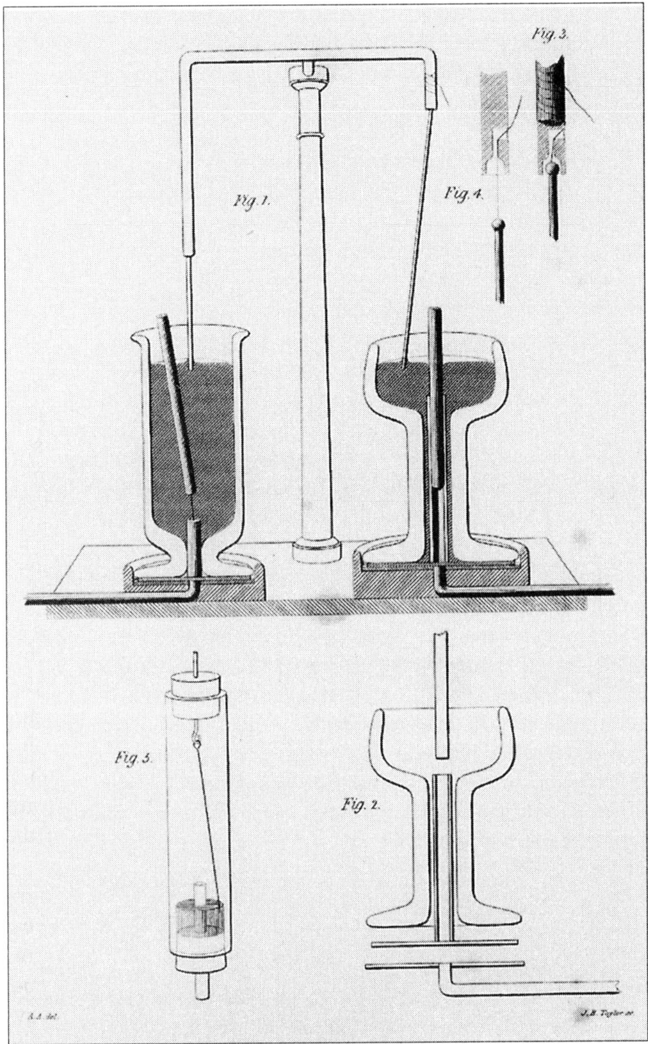

Diagram of an induction current.

where μ is the magnetic permeability of the coil, N is the number of turns of wire in the coil, and A is the surface area of the coil.

Inductors may be coupled. When a steady current flows in one coil, a constant magnetic field will be induced in the other coil. If this magnetic field is not changing, there will be no induced voltage in the secondary coil. But if the current is altered, there will be a change in the magnetic field and a voltage will be induced causing a current flow in the secondary coil that will then try to maintain the original magnetic field. Once the current is resumed, an induced current in the opposite direction will oppose the increase in the magnetic field. This generation of voltages to oppose changes in the magnetic field represents the operating principle behind **transformers**. It also describes a property known as mutual inductance, where a change in the current of one coil affects the current and voltage in the second coil.

Inductance is also an important concept in the design of active current (AC) motors, which operate by rotating a coil in a magnetic field and generating a **torque** on it. Since the current produced is alternating, such synchronous motors run smoothly only at the **frequency** of the sine wave. If the frequency is disrupted, the current will slip out of sync thus disrupting the output to the motor. The most common design is provided by the small AC induction motor, in which an electric current is not directly supplied, but is induced in the closed loop low-resistance rotating coils.

INERTIA

Isaac Newton's first law of **motion** is inertia. Inertia holds that objects at rest remain at rest and objects moving at a constant **velocity** continue to move at that velocity unless they are acted upon by an external **force**. Inertia is directly related to **mass**. The more massive an object, the more it will resist changes to its state of motion. Pushing a real car, for example, is harder than pushing a toy car.

Yet, behavior occurs that seems to contradict the law of inertia every day. When a hockey puck is given a push so that it slides across the floor, it stops after traveling only a short distance. No one has thrown a paper airplane that did not eventually slow down and crash. There may appear to be no forces acting on boxes and airplanes after they are pushed or thrown, but there is one, **friction**. The puck on an air hockey table continues moving for a long distance because the air jets in the table reduce the friction between the puck and the table, and a paper airplane thrown in outer **space** would continue moving in a straight line forever.

Though the law of inertia is cataloged among Newton's three laws of motion, Italian astronomer and physicist **Galileo** was the first to discover it. Galileo discovered that objects that are moving tend to continue moving forever. To Galileo's observation, Newton added the notion that a force is necessary to change an object's motion. Newton's first law is, therefore, an extension of Galileo's discovery. Newton's second law defines what is meant by a force and explains exactly how a

which states that an emf is induced in a conductor when the magnetic field around it changes. Thus, inductance results in opposition to an increase or decrease in current. This is consistent with **Lenz's law**, which states that when an emf is generated by a change in **magnetic flux** following Faraday's law, the polarity of the induced emf is such that it produces a current whose magnetic field opposes the change that produces it. This is why the induced magnetic field inside any loop of wire always acts to keep the magnetic flux in the loop constant.

The voltage induced in a conductor is also proportional to the rate of change of the magnetic flux and the faster the magnetic field is changing, the larger the induced voltage. This represents the operating principle of electric motors and **generators** and also explains the behavior of magnets and conductors when one is moving relative to the other. For example, when a conductor is moving across a magnetic field, a current is produced in the conductor inducing a magnetic field that acts to stop the conductor from moving. The inductance of a coil is expressed as follows: $L = \mu N^2 A / I$

force will change an object's state of motion. Some would argue that the first and second laws are not distinct, but two statements about the same thing. The first law, however, enables the definition of the inertial reference frame in which the second law is valid.

Galileo proposed that the laws of physics must be the same in every reference frame undergoing uniform motion. Such a reference frame is today called an inertial reference frame because it moves like a mass with no forces acting on it. Imagine two spaceships floating past each other with constant velocities. Observers on each spaceship performing the same experiment will obtain the same result. If one observer looks out the window to see the experiment in the passing spaceship, he will see the same distances and times in the other experiment as he sees in his own, and he will measure a velocity that is the sum of the velocity of his spaceship and the velocity in the experiment. This collection of ideas is known as Galilean relativity.

Albert Einstein showed that times, distances, and velocities do not transform between inertial **reference frames** in such a simple way. To show this, Einstein added the postulate, grounded in experiment, that observers in all inertial reference frames measure the same **speed of light**, regardless of their state of motion. The precise way in which this postulate affects transformations between inertial reference frames is worked out in the special theory of relativity.

The mass in **Newton's laws of motion** is called the inertial mass. **Newton's law of universal gravitation** also refers to mass. It states that the force of attraction between two objects is proportional to the product of their masses. There is no a priori reason why the gravitational mass should be the same as the inertial mass, but it is. In fact, the equivalence of gravitational and inertial masses has been verified to greater accuracy than most properties of classical physics. Gravitational and inertial masses are equivalent to one part in 10^{12}.

See also Relativiy, special

INERTIA AND ROTATIONAL KINETIC ENERGY

The kinetic **energy** of an object moving from one point to another is found by using the equation $1/2mv^2$. To determine the **kinetic energy** of an object rotating about an axis, the object is broken down into small-mass objects that are treated as point masses. The sum of the contributions of each small-mass object provides the kinetic energy of the rotating object. As this is done, a new quantity called the moment of **inertia** is introduced, which acts similar to **mass**.

First, consider a point mass m that is orbiting in a circular path of radius r around some fixed point. Its kinetic energy is given by $1/2mv^2$, where v is the magnitude of its **velocity** vector. Because the point mass repeats its **motion** each time it makes one complete orbit, its energy can be expressed in terms of an angular **frequency**, (omega) = v/r. The expression for the point mass's kinetic energy is then given by

$1/2mr^2$(omega)^2. In this expression, the quantity mr^2 is called the moment of inertia (I) of the point mass. Generally, for an extended object, the kinetic energy can be written as $1/2I$(omega)^2. The moment of inertia is the quantity that multiplies $1/2$(omega)^2 in the expression for kinetic energy.

In order to find an expression for the moment of inertia of an extended object, the object is broken down into small point masses, the moment of inertia is calculated for each point mass, and then added to find the total moment of inertia. For simple objects with constant **density**, this is an elementary math problem with simple formulas for the moment of inertia. For more complicated objects, such as **Earth**, the operation would have to be done using a computer. The main principle involved is that objects with greater mass farther away from the rotation axis have higher moments of inertia. For example, a hula hoop has a higher moment of inertia than a plastic disk of the same mass and radius, because all of the hula hoop's mass is all concentrated at its outer radius.

The moment of inertia acts much the way a mass does in linear motion. Essentially, mass opposes **force**, because for a constant force, heavier masses accelerate less. Similarly, for a constant **torque**, the rotational analogue of a force, objects with higher moments of inertia will have lower angular accelerations than objects of lower moments of inertia. In fact, the rotational kinematic equations are exact analogues of the linear kinematic equations; angle replaces position, angular frequency replaces velocity, angular **acceleration** replaces acceleration, torque replaces force, and **angular momentum** replaces **momentum**.

INFINITY

The concept of "infinity" is perhaps best approached through the concept of "the infinite." Intuitively, what is infinite has no end. If the **universe** is infinite—and you could travel forever in one direction, never reaching the edge of space—then the idea of "forever" requires that **time**, too, must be infinite.

These intuitions are of little use to physics, however, without some further mathematical precision. Mathematically, the idea of infinity is, in essence, the idea that there is no largest number (integer). For any number n that you advance as the largest number, I can counter with n + 1.

Calculus is built on the idea of infinity. Differential calculus finds the slope of a curve at an infinitely small point; integral calculus finds the area under a curve by adding up infinitely thin slivers (of the **space** bounded by the curve). In both cases, the properties of the infinitely small are defined by the tendencies of the very small. That is, the value of a function at infinity is defined as the limit of the function as it approaches infinity. Thus the limit of the function f(x) = 1/x as x approaches infinity is zero. As x gets very big, 1/x gets very small.

Other intuitions of infinity with mathematically interesting consequences emerge from Zeno's paradoxes. One example: to walk across a room, you must first walk across half the room. You must then cross half of the remaining half, and half of the next half, and half of the half after that, etc., ad infinitum. Though the distances to be traversed are smaller each time, they are still infinite in number, so, says Zeno, you can never reach the other side of the room.

The paradox relies on the intuition that an infinite series cannot have a finite sum; i.e., that $1/2 + 1/4 + 1/8 +...+ 1/2^n$ etc. must add up to infinity. In fact, that intuition is wrong. As the room-crossing example might suggest (the fractions being, after all, fractions of the length of the room), the sum of the $1/2 + 1/4 + 1/8 +...+ 1/2^n$ series is one (1). In other words, the sum or an infinite series can be a finite number.

Series that sum to a constant are said to converge, while those that sum to infinity are said to diverge—and telling them apart can be very difficult. The sums of infinite series that converge are often very useful constants. Fifty years ago, for example, cosmologist **Fred Hoyle** thought he had discovered such a constant and used it to defend a steady-state model of the universe (which is opposed to the currently accepted **big bang** model). A young graduate student named **Stephen Hawking** put himself on the physics map when he countered Hoyle's claim, telling Hoyle that the quantity he referred to diverges. "Of course it doesn't," replied Hoyle. "It does. I worked it out," said Hawking. Hawking was right; the series did diverge. Cosmologists' view of the universe is thus dependent upon the solution of infinite series problems.

In the first half of the twentieth century, infinity was a popular topic for philosophers. German mathematician Georg Cantor distinguished between different orders of infinity. The counting numbers, for instance, extend infinitely, but there are also infinitely many numbers (including an infinity of irrationals) between any two of them. Cantor described as "countable" any infinite set that could be placed in a one-to-one correspondence with the counting numbers, and his "diagonal proof" of the countability of the rationals was considered, by many, a great breakthrough. But these philosophical distinctions have in general proven to have little use for physicists.

INFRARED SPECTROSCOPY

Infrared spectroscopy is the study of the interaction between **matter** and electromagnetic **radiation** in the infrared region. This type of **spectroscopy** is used to study the vibrations of molecules; typically, the **energy** of molecular vibrations is the same as that of infrared **light**. Because of this, infrared spectroscopy can be used to determine molecular characteristics, such as bond length, bond angles, vibrational frequencies, and, indirectly, bond strengths. In addition, infrared spectroscopy is tremendously useful in determining the presence of certain characteristic groups in large molecules.

The **electromagnetic spectrum** consists of radiation having very high energy and **frequency** and short **wavelength** (such as gamma and **x rays**) to radiation having very low energy and frequency and long wavelength (such as **radio waves**). Visible light falls approximately in the middle of this spectrum. Infrared radiation has an energy that is lower than that of red light in the visible spectrum. Thus, infrared light lies in the

energy range of about 0.5 kcal (2 kJ)-9.5 kcal (40 kJ). The energy of most molecular vibrations also lies in this range.

Molecular spectroscopy is very complicated because molecules contain two or more atoms. In fact, the more atoms there are in the molecule, the more complicated the interpretation of the spectrum can be. However, it is possible to simplify the interpretation of a molecular spectrum based on the types of movement of molecules and the energies required for these movements.

All molecules in the gas phase are capable of three basic types of movement. The first is translation, or movement of the entire molecule in a random path in **space**. The second is rotation, or movement of the molecule around an axis (usually a bond) of the molecule. Rotational energies are usually in the same range as microwave or far infrared (longer wavelength) energy. The third type of **motion** is vibration, or movement between the atoms of a molecule, and this applies to molecules that are solid, liquid, or gaseous. In this type of movement, the atoms can be compared to balls attached to springs. If a **force** is applied to one of the balls (or atoms) in a particular direction, the other balls (or atoms) are affected. For example, in a bent molecule containing three atoms (such as water), there are three types of vibrations possible. The first is a symmetric stretch, in which the two end atoms are pulled down along their axes, and the center **atom** moves up. The second is an asymmetric stretch in which one end atom is pulled away from the center atom and the other end atom is pushed towards the center atom. The third is a bending motion in which the two end atoms move towards one another, causing the central atom to move upwards, much like the motion caused by waving one's arms up and down if the head is considered as the central atom. As molecules become bigger, more types of vibrations are possible and this complicates the interpretation of data.

Classically, the movement of two spheres connected by a spring can be described by a harmonic oscillator (the two spheres bounce back and forth with smaller and smaller regular **oscillations** until they finally come to rest). In classical **mechanics**, the vibrations of a harmonic oscillator can assume any energy value (just as it is possible to stop anywhere on a ramp). The movement of atoms in a molecule, however, is described by **quantum mechanics**. The equation that describes the energy of the vibrations limits the values that the energy can assume (just as it is possible to stand on one step or another, but not in between steps when climbing stairs). In the quantum mechanical model, the energy of the ground state (or the lowest **energy level** possible in the molecule) is not zero, but a finite quantity that is a function of **Planck's constant** and the vibrational quantum number. This minimum energy of the ground state is called the zero-point energy. Most molecules normally exist in the ground state. The energy required to excite a molecule from the ground state to the first excited state corresponds to the energy of radiation in the infrared region of the electromagnetic spectrum. Thus, if a molecule is exposed to infrared radiation and a certain wavelength of light is absorbed by the molecule, the molecule can move from the ground vibrational state to the first excited vibrational state, which in turn yields information about the molecule itself. This is how infrared absorption spectra are obtained.

Experimentally, infrared absorption spectra are obtained using infrared spectrometers. In general, a spectrometer consists of a source of radiation, a means of separating radiation into its distinct frequencies, and a method to detect the radiation that passes through a sample. Simply put, if all frequencies of infrared radiation are passed through a sample, and some of these frequencies are absorbed by the sample in going from the ground state to the first excited vibrational state, then the "spectrum" will show a plot of the frequencies that are absorbed by the sample. That is, an infrared spectrum is a plot of the intensity of light absorbed as a function of the frequency of vibration. Reading the frequencies where light is absorbed gives the vibrational frequencies of the molecules in the sample. This method works for solid, liquid (or solution), and gaseous samples.

As mentioned earlier, the infrared spectra of small molecules can provide information about particular stretches and bends in the molecule. As molecules increase in size, their spectra become more complex. One advantage, however, is that even in a large molecule, certain stretches and/or bends occur at characteristic frequencies. These characteristic frequencies remain unchanged in spite of the differences in the geometry, the size, the **mass**, or other features of the molecule. For example, a bond between **oxygen** and hydrogen (O-H) in a molecule shows a vibration at 3,600 wave numbers (a wave number is the reciprocal of the wavelength in the unit of centimeters), while a carbon-oxygen double bond (carbonyl bond) has a distinctive band at 1,700 wave numbers. These characteristic vibrations serve as molecular fingerprints. If the spectrum of an unknown compound is analyzed, it is possible to determine if there are certain groups in the compound based on the presence of certain absorption bands.

Infrared radiation is associated with the production of **heat**. The element on a stove is "hot" long before one can see the red **color** that is produced only when the element becomes very hot; this heat is due to infrared radiation that the eye cannot detect. A typical infrared spectrometer consists of a source of radiation, usually a ceramic element that is heated to relatively high temperatures. This radiation is passed through the sample and the frequencies that are absorbed by the sample are measured by a relatively complex apparatus called a Fourier transform infrared spectrometer (FTIR), details of which are beyond the scope of this article. In 1999, the apparatus was relatively inexpensive as scientific equipment goes (about $15,000) and it was completely computer controlled. In 1960, it took about 30 minutes to obtain an infrared spectrum. In 1999, approximately 30 infrared spectra could be obtained in one second!

One unusual aspect of infrared spectroscopy is that common materials do not transmit infrared radiation in the regions of interest where molecular vibrations occur. Simple materials such as glass are completely opaque in infrared regions of interest. Glass is mainly composed of silicon dioxide. This molecule strongly absorbs almost all infrared radiation and is thus inappropriate for studies of this type.

Therefore, in order to obtain a spectrum, the sample must be contained in a material that is transparent to this radiation—usually a substance that itself does not have vibrations in the infrared region. Interestingly, table salt (sodium chloride) does not absorb infrared radiation until the frequency is very low, and this is one of the most common materials now used to contain samples that are used in infrared spectroscopy. Today, there is a host of materials that can be used, and some of them are quite expensive and exotic (for example, cadmium telluride).

At the start of the twenty-first century, infrared spectroscopy is one of the standard workhorses of the manufacturing world, whether in **semiconductor** production, process control in chemical engineering, or environmental monitoring. It is relatively inexpensive, rapid, very sensitive, and can provide quick and efficient information on **chemical reaction** conditions through its ability to detect characteristic vibrations of molecules of interest in the process under investigation. Few techniques are so versatile at such a low cost, and it can easily be used as a remote sensor under extremely corrosive or dangerous reaction conditions, conditions that make many other potential in situ methods unsuitable.

INFRASOUND

Infrasound is the class of sonic vibrations that are lower than the human ear can hear. That is, they have lower frequencies and longer **wavelengths** than the sound **waves** perceived by human beings. The lower boundary on **sound** perceptible with the human ear is usually around 20 Hertz, or 20 vibrations per second. These waves still travel at the **speed of sound** (about 330 m/s in air), so their wavelengths can be extremely long.

Infrasound is often produced by large vibrating objects but can also be produced by living things. Elephants, for example, can emit infrasonic waves. Large bridges can be the source of infrasonic **pulses** when they sway, and nuclear explosions have been known to give off infrasonic waves at frequencies as low as one every hundred seconds (0.01 Hz).

One of the major uses of infrasound in modern science is in the detection of catastrophic storms such as tornadoes. The eye of a tornado generates infrasonic pulses that can be detected from as far away as 100 mi (60 km), allowing for better storm warning systems. While sound, including infrasound, has been studied for many years, new applications are still being explored.

INTERFERENCE

Interference is the interaction of two or more **waves**. Waves move along their direction of propagation characterized by crests and troughs. Wherever two or more waves, either from one source by different paths or from different sources, reach the same point in **space** at the same time, interference occurs.

When the waves arrive in-phase (the crests arrive together), constructive interference occurs. The combined

Interference patterns, overlapping rings on water surface. *(Photo by Martin Dohm. National Audubon Society Collection/Photo Researchers, Inc. Reproduced by permission.)*

crest is an enhanced version of the one from the individual wave. When they arrive out-of-phase (the crest from one wave and a trough from another), destructive interference cancels the wave **motion**. The **energy** of the wave is not lost; it moves to areas of constructive interference.

Interference occurs in **sound** waves, **light** waves, **shock waves**, **radio**, and **x rays**. Waves display crests and troughs like the wiggles along the length of a vibrating jump rope. We see interference when ripples from one part of the pond reach ripples from another part. In some places the combination makes a large wave; in other places the waves cancel each other and the water appears calm. Radio, visible light, x rays, and **gamma rays** are waves with crests and troughs in the alternating **electromagnetic field**. Interference occurs in all of these waves. Interference of sound waves causes some regions of a concert hall to have special behavior. Where the multiple reflections of the concert sound interfere destructively, the sound is muffled and appears "dead." Where the reflections are enhanced by adding constructively, the sound appears brighter, or "live." Switching the polarity of the wires on a stereo speaker also can result in the sound appearing flat because of interference effects.

The most striking examples of interference occur in visible light. Interference of two or more light waves appears as bright and dark bands called "fringes." Interference of light waves was first described in 1801 by **Thomas Young** when he presented information supporting the wave theory of light.

White light is a mixture of colors, each with a unique **wavelength**. When white light from the **Sun** reflects off a surface covered with an oil film, such as that found in a parking lot, the thickness of the film causes a delay in the reflected beam. Light of some colors will travel a path through the film where it is delayed enough to get exactly out of phase with the light reflected off the surface of the film. These colors destructively interfere and disappear. Other colors reflecting off the surface exactly catch up to the light traveling through the film. They constructively interfere, appearing as attractive **color** swirls on the film. The various colors on soap bubbles as they

float through the air are another example of thin film interference.

Modern technology makes use of interference in many ways. Active automobile mufflers electronically sense the sound wave in the exhaust system and artificially produce another wave out-of-phase that destructively interferes with the exhaust sound and cancels the noise. The oil-film phenomenon is used for filtering light. Precise coatings on optical **lenses** in **binoculars** or cameras, astronauts' visors, or even sunglasses cause destructive interference and the elimination of certain unwanted colors or stray reflections.

INTERFEROMETRY

Interferometry uses the principles of **interference** to determine properties about **waves**, their sources, or the wave propagation medium. Acoustic interferometry has been applied to study the **velocity** of **sound** in a fluid. **Radio** astronomers use interferometry to get accurate measurements of the position and properties of stellar radio sources. Optical interferometry is widely used to observe things without touching or otherwise disturbing them. **Light** beams are sent through various paths, and they combine at one observation region where interference fringes occur. Interpreting the fringes reveals information about optical surfaces, the precise distance between the source and the observer, spectral properties of light, or the visualization of processes such as crystal growth, combustion, diffusion, and shock wave **motion**.

The observation and explanation of interference fringes dates back to **Robert Hooke** and **Isaac Newton**, but the invention of interferometry is generally attributed to American physicist **Albert Michelson**. The **Michelson interferometer** consists of two perpendicular arms with a half-silvered mirror (a beam splitter) at the intersection.

Each arm has a mirror at one end. Light from the source enters the interferometer along one arm, and is equally split at the beam splitter. Half the light travels to mirror #1 and reflects back toward the beam splitter. It passes through the beam splitter and continues to the detector, which can be like a motion picture screen. The other half of the light first travels straight through the beam splitter, reflects off mirror #2, and then returns to the beam splitter, where it is reflected and sent to the detector. When the two beams combine at the detector, they interfere and produce a pattern of fringes that depends upon the path they traveled and the time it took them to travel this path.

In 1887, Michelson, along with physicist **Edward W. Morley**, set out to determine if the **speed of light** was dependent on the speed of the observer. According to the accepted theory of the time, light had to propagate in a medium called the **ether**. The motion of Earth traveling through the ether would affect the fringe pattern on Michelson's interferometer, because it would take the light longer to travel over one arm than the other arm. When the interferometer was rotated through 90°, the fringes would shift if the speed of light was not constant. In this most celebrated null experiment in the his-

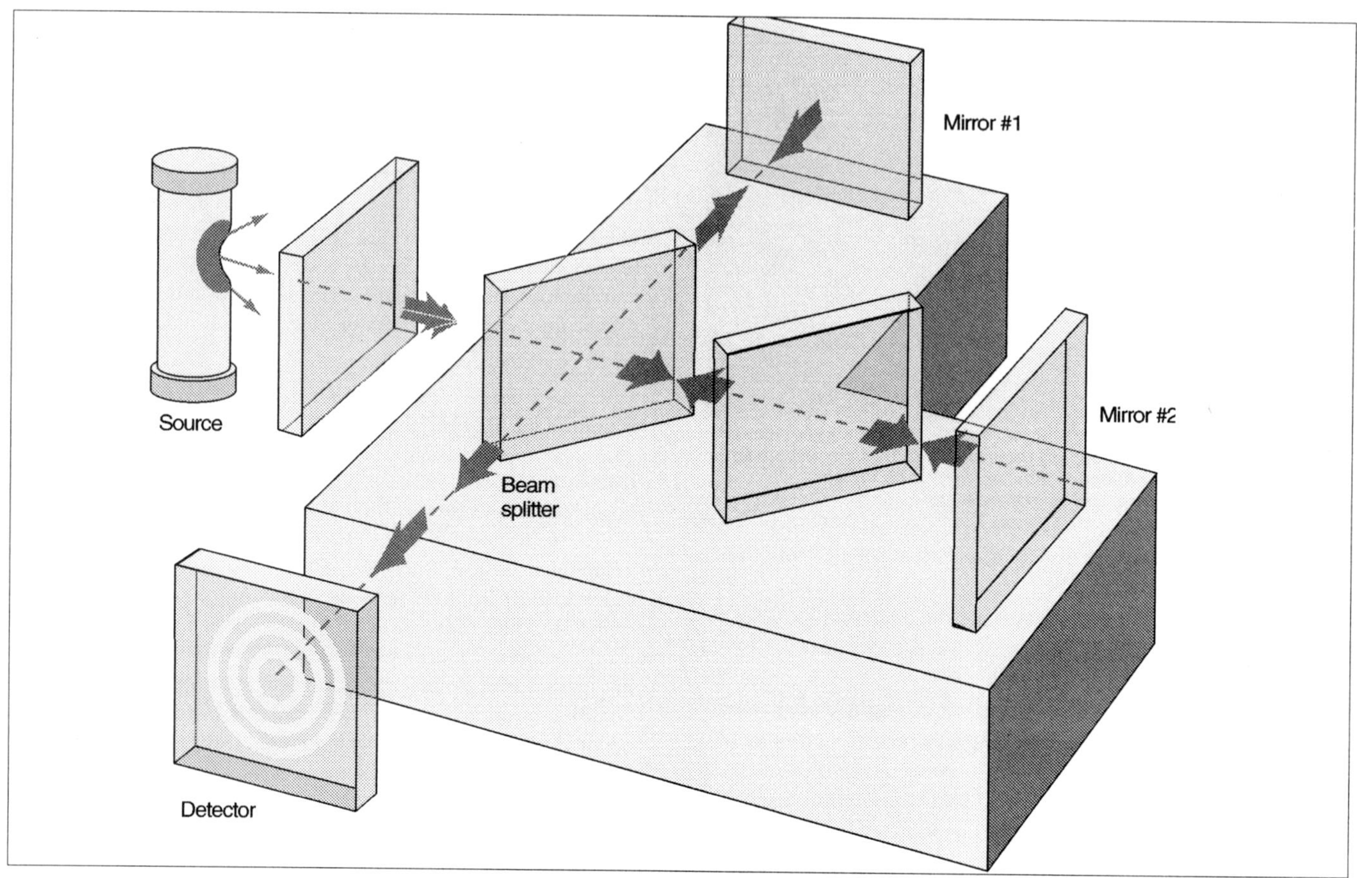

Figure 1. The Michelson interferometer.

tory of science, Michelson and Morley observed no changes in the fringes after many repetitions. The speed of light appears constant, regardless of the speed of its source, which was later explained by Einstein's theory of relativity.

Interferometers such as Michelson's can be used for many types of measurements. As a spectrometer, they can be used to accurately measure the mirror #1 mounted on a motor, so it can be moved along the direction of the beam at a constant speed. As the mirror moves, a fringe at a point in the interference plane will appear as an alternating bright and dark band. The **frequency** of the changing fringes is related to the **wavelength**, and can be calculated to better than several parts in 100 billion.

Modifications to the Michelson interferometer were introduced in 1916 by electrical engineer Frank Twyman (1876-1959) and A. Green for the purpose of testing **optical instruments**. If the element transmits light, like a prism or a lens, it is inserted between the beam splitter and mirror #1 in such a manner that the fringes that are observed become a measure of the element's optical quality. If the element to be tested is a mirror, and reflects light, it is substituted for mirror #1 altogether. Once again, the Twyman-Green interferometer fringes act as a map of irregularities of the optical element.

Another type of interferometer was introduced by L. Mach and L. Zehnder in 1891. Light leaves the sources, and is divided by beam splitter #1. One half travels toward mirror #1, and is reflected. The other half is reflected from mirror #2. The beams are combined by beam splitter #2, and propagate to the detector, where interference is observed. By virtue of the fact that the beams are separated, the objects to be tested can be quite large. The Mach-Zehnder two-beam interferometer is used for observing gas flows and **shock waves**, and for optical testing. It has also been used to obtain interference fringes of electrons that exhibit wavelike behavior.

Interference can occur if a beam is split, and one half travels in a clockwise path around the interferometer, while the other travels counterclockwise. Figure 3 shows the top view of a cyclic interferometer, named a Sagnac interferometer for its inventor, physicist Georges Sagnac (1869-1928). One half of the beam reflects off the beam splitter and travels from mirror #1, to mirror #2, to mirror #3, and again reflects off the beam splitter. The two halves interfere at the detector. The Sagnac interferometer is the basis for laser gyroscopes that were first demonstrated in 1963. They sense the interference pattern to determine the direction and the speed of rotation in a moving vehicle like an airplane or a spacecraft.

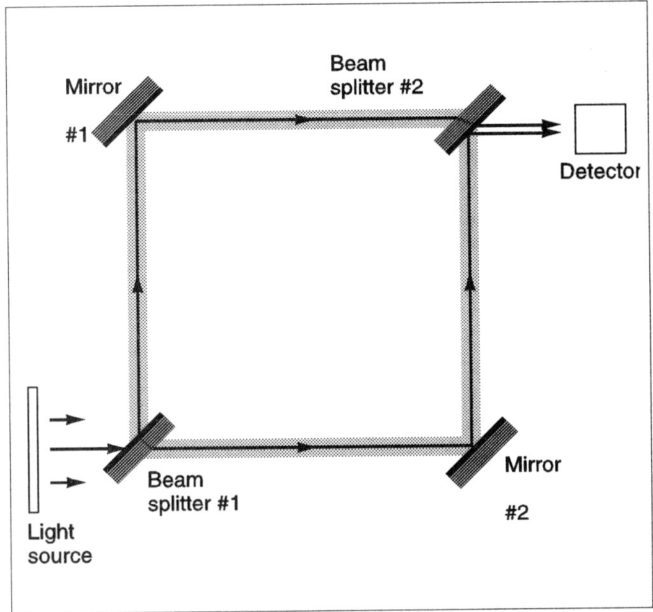

Figure 2. Top view of the interferometer introduced by Mach and Zehnder.

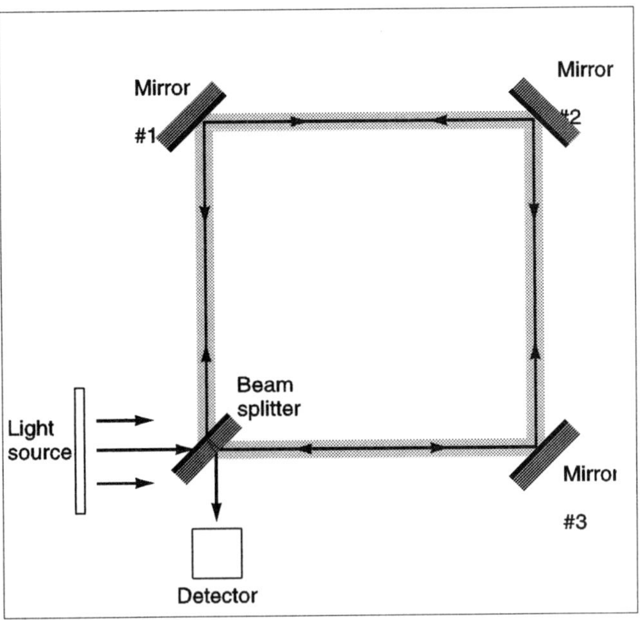

Figure 3. Top view of the Sagnac cyclic interferometer.

In 1899, physicists Charles Fabry (1867-1945) and Alfred Perot (1863-1925) introduced an interferometer designed to produce circular interference fringes when light passes through a pair of parallel half-silvered mirrors. Figure 4 shows the Fabry-Perot interferometer used with a broad light source. When the plates are separated by a fixed spacer, the interferometer is called a Fabry-Perot etalon. The diameter of the fringes from the etalon is related to the wavelength, so it can be used as a spectrometer. If two beams of slightly different wavelength enter the etalon, the position of the overlap in fringes can be used to determine the wavelength to better than one part in 100,000. If the two plates of the Fabry-Perot interferometer are aligned parallel to each other, the device becomes the device becomes the basic laser resonant cavity, since only certain wavelengths will add constructively as they propagate between the mirrors.

Rather than splitting the amplitude of the beam by beam splitters, one part of the beam can be made to interfere with another in the manner of Lloyd's mirror. One half of the beam from the source propagates directly to the detector. The other half reflects off the mirror and interferes with the direct beam at the detector. Information about the surface of the mirror is contained in the fringes. The surface of crystals can be studied with fringes from **x rays**. The surface of a lake or Earth's ionosphere can be studied by using interference from radio waves.

A variation of the Mach-Zehnder interferometer, introduced by W. J. Bates in 1947, made it possible to measure the wavefront (phase) of a beam without an error-free reference wave. By rotating one beam splitter in the Mach-Zehnder configuration, an incoming beam is split into two, and one half is shifted (sheared). Overlapping these beams results in an inter-

ference pattern that is a measure of the slope, or tilt, of the wavefront. Shearing interferometers are used in optical testing and in **astronomy** for measuring the distortions of the atmosphere.

The basic interferometer improves over the years as new technology appears. High-speed cameras and **electronics**, precise **optics**, and **computers** are brought together to make possible accurate interpretation of the fringes, as well as the extraction of new and exciting information.

Even though we cannot directly photograph and resolve the image of two **stars** close together, we can use interferometry to measure their separation. First proposed by physicist Armand Fizeau (1819-1896) in 1868, the method was first applied by Michelson and American astronomer Francis Pease (1881-1938) in 1920, and is commonly called Michelson stellar interferometry.

Light from two sources is collected by two telescopes that are a known distance apart. The light is filtered to restrict the wavelength, and then brought together. Each star exhibits a fringe pattern. The fringes will line up if the patterns of the two stars overlap. When the separation of the two telescopes is small, the fringes are visible. When the separation of the telescopes is exactly equal to the wavelength divided by twice the angle between the two sources, the fringes will disappear. By varying the separation of the telescopes, observing when the fringes disappear, the separation of the sources is calculated. In a similar way, a single remote star can be thought of as two halves that appear as point sources close together. By using stellar interferometry, the size of a star can be measured.

By placing an opaque screen with holes over the aperture of a **telescope**, each pair of holes will cause interference fringes. Stellar interferometry over a number of simultaneous separations is called aperture plane interferometry.

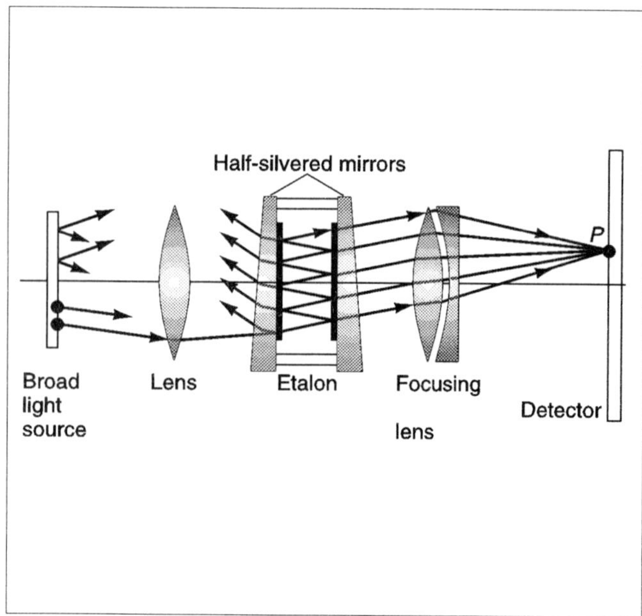

Figure 4. Fabry-Perot interferometer.

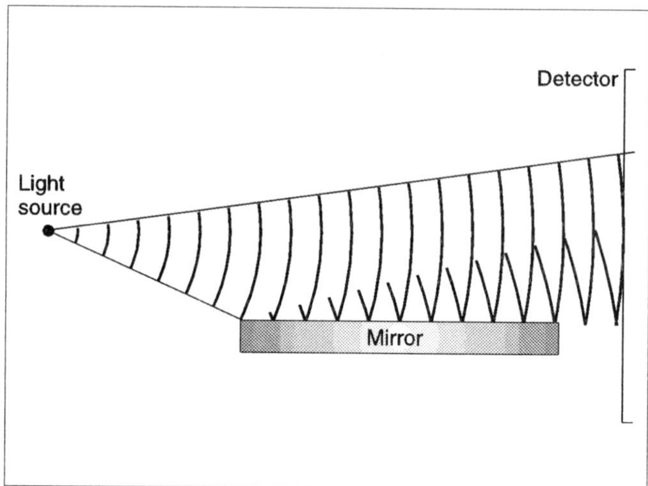

Figure 5. Wavefront splitting interferometer.

Radio astronomers R. Hanbury Brown and R. Q. Twiss, the first to use stellar interferometry in the radio region, measured the size of the star Sirius. Today, "Very Long Baseline Interferometry" links radio telescopes around the world to create interference fringes that can be used to measure stellar sizes in fractions of an arcsecond.

Invented by Antoine Labeyrie in 1970, speckle interferometry provides a method for large telescopes to see objects without being limited by the **turbulence** of the atmosphere. A star exposure for less than one-hundredth of a second appears speckled, because light from all points of the telescope interferes with each other. The speckle is similar to the speckled pattern of red light that reflects off the glass of a supermarket price scanner.

Averaging many short exposures (or taking a long exposure) smears out the speckles because the atmosphere is constantly moving around. Information in the image, smeared out by the blur, is lost. Because individual speckles themselves contain information about the object, the speckle interferometer gathers many short exposures, and a computer processes them to extract the information. Many measurements have been made in the last quarter century. The size and surface features of asteroids and the planet Pluto have been determined by speckle interferometry. The size of the nearby star Betelgeuse has also been measured. The angular separation of **binary stars**, measured by this technique, leads astronomers to calculate the star masses and develop theories about the evolution of the **universe**.

Holography was invented by **Dennis Gabor** in 1948. A hologram is recorded by splitting a light beam and letting half the beam scatter from an object while the other half travels undisturbed. The two beams combine on photographic film, where a complicated fringe pattern is formed. When light shines on the hologram, some of it will pass through the bright fringes, and some will be absorbed by the dark fringes. By observing the light from the hologram one reveals a three-dimensional replica of the original object.

Holographic interferometry is used to view small changes in an object. When two holograms, taken at different times, of the same object are superimposed, fringes will reveal the difference between the two objects. It is possible to see slowly varying changes of a growing plant, or rapidly varying changes of a vibrating object such as the face of a violin.

INTERMOLECULAR FORCES

The attractions of molecules for each other are called intermolecular interactions to distinguish them from covalent and ionic bonding, forms of intramolecular interactions. For very large molecules, such as proteins, nucleic acids, and synthetic polymers, noncovalent interactions of groups that make up the molecules (functional groups) have the same characteristics as intermolecular interactions even though they may actually take place between groups within the same molecule. The interactions that lead to specific shapes of enzymes, referred to as protein-folding interactions, is an example. Intermolecular interactions are most significant in liquid and solid phases where molecules are very close together. In fact, even in liquids and solids intermolecular interactions are strong only for molecules that are next to each other. The interactions of molecules in the liquid and solid states have significant consequences that are readily observable. The strength of intermolecular interactions affects numerous properties, including **boiling** points, miscibility, and solubility.

All neighboring molecules in liquids and solids attract each other. The nature and strength of these interactions depend on the types of groups of atoms or functional groups that comprise the molecules. All molecules interact with other molecules through London **dispersion** forces, also called van der Waals interactions. London dispersion forces are the attractive forces of one transient dipole for another. A transient dipole is a temporary imbalance of positive and negative charge. At particular instants, even atoms that are spherical on average, such as those of the noble gases, will have greater **electron** density on one side of the **atom** than another. At that instant, the atom will possess a temporary dipole with a negative charge concentration on the side of the atom with greater electron **density**. If this happens in the case of an argon atom in liquid argon, for example, the argon atoms next to the one with temporary dipole would feel the effect of the dipole. An atom near the negative end of the dipole would have its own electrons slightly repelled from the negative concentration of charge, developing a dipole with its positive end near the negative charge of the original atom. An argon atom on the other side of the original temporary dipole would feel its electrons attracted to the positive end of the dipole, developing a dipole with the opposite orientation. In this way, temporary dipoles are propagated through a liquid or solid. The **motion** of the molecules in the liquid or solid soon disrupts the pattern, but similar events take place continually. The larger the size of atoms and the more electrons they possess, the greater the probability of forming substantial transient dipole interactions. Molecules that are nonpolar and nonpolar functional groups of molecules only experience London dispersion or van der Waals interactions with other molecules or functional groups.

Some molecules and functional groups have permanent dipoles that result from a nonsymmetric geometric arrangement of atoms of different electronegativity. The portion of a molecule with a permanent dipole will be attracted to the portion of a neighboring molecule with its own permanent dipole. A polar-polar interaction is stronger than a van der Waals interaction.

The hydrogen bond is a specific and very important type of intermolecular interaction. A hydrogen bond occurs when the electrons forming the covalent bond between the **hydrogen atom** and the **oxygen**, nitrogen, or fluorine atom are pulled toward the electronegative atom. Only the first shell is occupied in the hydrogen atom, so when the electrons in the bond are pulled away, the **nucleus** is exposed. The extent of the exposure of positive charge on the hydrogen is the fundamental difference between it and other atoms that have an inner shell or several inner shells that continue to shield the nucleus even though the bonding electrons are pulled away. The **energy** of the hydrogen-bonding interaction can be as much as 100 kJ mol^{-1} for HF, but it is typically in the range of 20-40 kJ mol^{-1}. As such, it is the strongest of the intermolecular interactive forces. The specific geometric requirements for hydrogen bonding are profoundly important in biochemistry, notably in the operation of the genetic code.

The interactions of permanent dipoles with each other are stronger than London dispersion forces, and hydrogen-bonding interactions are stronger yet. Intermolecular interactions are additive, so that both the number of interactions between two molecules and the strength of each interaction determines the energy needed to separate molecules from each other. Molecules with many atoms and molecules that are

highly polar and/or contain hydrogen bonds are especially difficult to melt or vaporize. Many phenomena familiar in everyday life arise because of forces among molecules.

The effects of the various types of intermolecular interactions can be seen from the boiling points of the compounds of elements formed with hydrogen (the hydrides) in the periodic groups containing **carbon**, nitrogen, oxygen, and fluorine. For all the hydrides except water, ammonia, and hydrogen fluoride, the boiling points of the hydrides of each periodic group (for example, CH_4, SiH_4, GeH_4, and SnH_4) increase as the atomic number of the element increases. This is the normal effect due to London forces—the greater the polarizability of the **electron cloud**, the more the condensed phase is stabilized by transient dipoles.

The hydrides of the halogens (HCl, HBr, and HI) all have permanent dipoles. They boil somewhat higher than the hydrides of the elements in the corresponding row of the preceding group of the **periodic table** and also show the trend of increasing boiling point as the polarizability of their **electron clouds** increases. The hydrides of oxygen, nitrogen, and fluorine, however, require the input of more energy to boil than would be predicted from either London forces or dipole-dipole interactions. Water, ammonia, and hydrogen fluoride all form hydrogen bonds. Each water molecule can form four hydrogen bonds with other water molecules—two involving its two hydrogen atoms, and two involving the unshared pairs of electrons on its oxygen—so the effect of hydrogen bonding on its boiling point is especially great.

Molecules forming hydrogen bonds are stabilized when the molecules align properly. The **kinetic energy** of molecules always competes against attractive forces to determine physical properties because molecular motion opposes proper alignment. More **heat** is required to disrupt the stable associations in a solid containing hydrogen-bonded molecules to form a liquid, in which the molecules are more mobile and not as well aligned. Similarly, compared with nonhydrogen-bonding substances, liquids in which the molecules can hydrogen bond require more heat to form a gas, in which there is very little interaction between molecules. The stronger the intermolecular attractive forces, the more heat energy is required to cause changes of phase. Each hydrogen bond between molecules is much weaker than the covalent bonds holding the atoms together within the molecule. It may require a large amount of heat to convert liquid water into steam, but it requires significantly more energy to break a molecule of water apart into hydrogen and oxygen atoms.

To understand the differences in properties of larger molecules, the additivity of intermolecular interactions becomes important. In effect, the interaction of each group of atoms of a molecule with a group of atoms of a neighboring molecule can be considered to be independent of the interactions of other groups of atoms of the molecules. The total energy required to move two molecules apart is then the sum of all the energies of the individual interactions. The more groups and the stronger each individual interaction, the greater the sum of energy of interactions. Among the nonpolar linear alkanes, the boiling point for a molecule with many -CH$_2$

groups such as octane, $CH_3(CH_2)_6CH_3$, is higher than that of propane, $CH_3CH_2CH_3$, with octane being liquid at room **temperature** and propane a gas, because of the greater number of London dispersion interactions between the octane molecules. If there are 10 hydrogen-bonding interactions per molecule, as there would be for a large molecule such as a sugar with many -OH groups, the total energy of interaction because of hydrogen bonds is about 300kJ mol^{-1}, which is on the order of a strong chemical bond. Such molecules are typically solids at room temperature, while molecules containing only nonpolar groups of the same molecular **mass** may be liquid.

Life on Earth may exist because of the hydrogen bond. The physical properties of water are in large part due to its extensive network of hydrogen bonds. Our external environment is aqueous—two-thirds of Earth's surface is water—and two-thirds of our mass is water as well. Hydrogen bonding and intermolecular interactions are the very basis of the genetic code and the unique structures and shapes of the nonaqueous components of life: DNA, RNA, proteins, and other biomolecules making up living systems all owe their form and function to hydrogen bonds. The double helix of DNA is held together by hydrogen bonds. These are strong but still at least four times weaker than covalent bonds. Hydrogen bonds are the perfect strength to hold DNA together under most situations, but are weak enough to form and break readily to enable DNA to untwine for replication.

Stabilization through hydrogen bonding can also determine the ways in which molecules arrange themselves. The hydrogen bond is a directional bond, which means there is a specific architectural relationship among molecules hydrogen bonding with each other. An important illustration of the effect of the specific geometry of hydrogen-bonded molecules is the decrease in the density of water when it freezes to form ice. Most liquids increase in density as they solidify and, therefore, solid pieces of a substance typically sink when added to the liquid substance, whereas ice floats on water. For aquatic life and indeed for all life on Earth, this anomaly of water is important. As ice forms on the surface of a body of water, it eventually forms a sheet that tends to insulate the body of water below, helping to prevent the entire body of water from freezing solid. Aquatic life that would otherwise be killed by a freeze survives.

Many molecules are very complex, they have some polar parts, some nonpolar parts and some functional groups that can hydrogen bond. As is true for the London forces in alkanes and as we have already indicated in the discussion of hydrogen bonding, polar interactions are also additive. The boiling points of propane, $CH_3CH_2CH_3$, is -43.6°F (-42°C), while that of butane, $CH_3(CH_2)_2CH_3$, is 32°F (0°C). Adding one CH_2 group increases the boiling point by 75.6°F (42°C). When the nonpolar -CH_2- unit of propane is replaced with a polar -C = O group (this gives us acetone $CH_3C = OCH_3$), the boiling point increases by 180°F (100°C), from 43.6°F (-42°C) to 136.4°F (58°C). If additional groups are added, the boiling point increases accordingly.

Intermolecular interactions also affect the properties of molecular solids and some types of covalent network solids.

Molecular solids are composed of small molecules held together by London forces, dipole-dipole, and hydrogen-bonding interactions. Sugar is a molecular solid, composed of glucose molecules held in crystalline array by dipole-dipole interactions. Covalent network solids are composed of very large covalently bound molecules. Graphite is a covalent network. It consists of sheets of covalently bound carbon atoms held together by **van der Waals forces**. A diamond is a special kind of covalent network solid, because it consists entirely of covalently bonded carbon atoms. Each diamond is one molecule.

Some important observable consequences of intermolecular interactions are miscibility, solubility, the heat of mixing that occurs when different liquids are combined, and capillary action.

Some liquids mix and some do not. Familiar examples are the miscibility of alcohol and water and the immiscibility of oil and water, which can be explained by intermolecular interactions. A portion of the alcohol molecule (-OH) hydrogen bonds with water molecules and the interaction of the two molecules is energetically favorable. The pentane molecule, however, has neither hydrogen bonding nor polar groups to be attracted to water molecules. The water molecules tend to stick together and remain separate from the pentane. Because pentane is less dense than water, it floats above it. Substances that interact strongly with water, mixing well with it or dissolving in it, are called hydrophilic (water loving). Substances that interact poorly with water and tend to remain separate are called hydrophobic (water fearing).

The effect of polarity (charge asymmetry) is also evident in the solubility of compounds. For example, a bent, polar molecule like sulfur dioxide, SO_2, is much more soluble in polar solvents than is linear, nonpolar carbon dioxide, $CO\ SO_2$. This effect is also additive when molecules are complex. Alcohols with very few nonpolar groups are much more soluble in water than are alcohols with many nonpolar groups. Also, compounds with many polar groups, such as the sugars and alcohols like glycerol with three -OH groups, dissolve better in water and low **molecular weight** alcohols than alcohols with a lower ratio of polar to nonpolar groups. This tendency is frequently referred to as "like dissolves like"; e.g., polar solvents like water dissolve polar solutes like sugar, whereas nonpolar solutes like carbon tetrachloride dissolve nonpolar solutes like oils and other hydrocarbons.

A readily observed phenomenon associated with hydrogen bonding is the release of heat when two hydrogen-bonding solvents are mixed. If the average intermolecular interaction in the combined system is stronger than in the two separate liquids, heat is released. There can be a noticeable temperature rise upon mixing whenever intermolecular interactions between different molecules are stronger than the interactions between like molecules, whether these associations are due to London forces, dipole-dipole interactions or hydrogen bonding. It is typically easiest to observe a heat of mixing when hydrogen bonding occurs, because this is the strongest class of intermolecular interaction.

When a capillary, a narrow tube, touches the surface of a liquid, fluid rises into the tube. The extent to which a liquid

Prefix	Symbol	Meaning
exa-	E	10^{18}
peta-	P	10^{15}
tera-	T	10^{12}
giga-	G	10^{9}
mega-	M	10^{6}
kilo-	k	10^{3}
hecto-	h	10^{2}
deka-	da	10^{1}
deci-	d	10^{1}
centi-	c	10^{2}
milli-	m	10^{3}
micro-	μ	10^{6}
nano-	n	10^{9}
pico-	p	10^{12}
femto-	f	10^{15}

Figure 1. Common prefixes. *(Illustration by Electronic Illustrators Group.)*

rises is different for different liquids. When a narrow tube is inserted into water, the water rises in the tube. This occurs because the surface of glass is quite polar. As water molecules rise along the inside surface of the capillary, they pull up other water molecules to which they have formed hydrogen bonds. The balance of **gravity** and the attraction of the water for the glass surface determine the height to which the water rises. Other polar liquids besides water also rise in capillaries, but some nonpolar liquids show the opposite effect, the height of the liquid inside the capillary is less than outside. The molecules of these liquids are attracted to each other more than they are to the surface of the glass. Liquid mercury shows an especially large effect because the mercury atoms are attracted to each other much more strongly than they are attracted to the glass surface. Capillary action is also responsible for absorption of liquids into paper, such as paper towels.

INTERNATIONAL SYSTEM OF UNITS

Measurement is one of the hallmarks of civilization. How far is it? How much does it weigh? What time is it? These are all questions that we hear every day. Indeed, asking the time is probably the number one question that we ask, especially when we are doing something we don't like!

Simple units, such as the cubit, which is the distance between a man's elbow and the tip of his middle finger, or the span, the distance from the tip of the thumb to the tip of the little finger of an outstretched hand, were used for much of recorded history. After all, the human body is always available for measuring distance. But it does suffer from one inherent problem. Not all men are created equal! Some are taller than others so that not all forearms are the same length.

Of course, this led to disputes and difficulties. Solving such problems was probably the basis for the origin of civil law. How does one stop people from cheating by using a favorable measurement? The answer is to standardize the system of weights and measures. Records indicate that the Egyptians started this practice as early as 3000 B.C., probably as a result of the annual flooding of all of the arable land along the Nile. This necessitated accurate surveying every year and resulted in the early Egyptians developing an excellent understanding of geometry and surveying.

By the late eighteenth century, scientists and engineers were able to construct very precise measuring instruments but the standard units for measurement were poorly defined. To develop a systematic and precise set of measurement standards, a commission, including scientists **Pierre-Simon de Laplace**, Joseph-Louis Lagrange (1736-1813), and Antoine-Laurent Lavoisier, was established. They developed a system based on objects or phenomena that could be measured reproducibly; these natural measurements were independent of man. For instance, the meter (Greek for "to measure"), the basic unit of length, was calculated to be 1/10,000,000 of the distance between the North Pole and the equator on a line running through Paris. Other units were worked out to interconnect with the meter. For example, the liter (a derived unit of volume) is the volume occupied by one cubic decimeter, while the kilogram (the basic unit of **mass**) is the mass of a liter of water at 4°C. Smaller and larger units were available by multiplying and dividing by powers of 10.

This so-called metric system was by far the most logical and scientifically based system of measurement devised. Metric units are related decimally (i.e., by powers of 10). There have been many meetings and conferences to establish exact interpretations for the metric system. The original meter was defined as the distance between two marks on a platinum/iridium bar kept in a locked, air-conditioned vault maintained by the International Bureau of Weights and Measures in Paris. In 1960, the General Conference of Weights and Measurements redefined the meter as 1,650,763.73 wavelengths of one of the spectroscopic lines of an **isotope** of the noble gas, krypton. In 1983, it was further refined to be 1/299,792,458 of the distance that **light** travels in a second.

The General Conference of Weights and Measurements in 1960 established a revised metric system as the "International System of Units." The French name is "Le Système International d'Unités" or the "Système International," for short, and it is this latter term that leads to the abbreviation "SI," Scientists all over the world have agreed to SI units as the basis for measurement. Indeed, with the exception of the United States, all of the countries in the world use SI units for commerce as well. The SI system is also called MKSA (meter, kilogram, second, ampere). The seven base units of the SI system, each of which is used for a particular physical quantity, are shown in figure 1.

Modification of the units is accomplished decimally using the prefixes. For example, 5 teraseconds = 5 Ts = 5 x 10^{12}s, while 5 milliseconds = 5 ms = 5 x 10^{-3}s, and 5 femtoseconds = 5 fs = 5 x 10^{-15}s. While the base SI unit for **weight** is the kilogram (kg), for many purposes in **chemistry** the smaller weight unit, gram (g), is more convenient; 1 kg = 1,000 g.

All other physical quantities can be derived from appropriate combinations of the seven basic units. Examples of these derived units include: 1) The SI unit of **force** is defined as the newton (N) which is a kilogram times the **acceleration** in meters per second squared or N = kg m/s^2 · 2) The SI unit of **energy** is the joule (J); this unit is defined as the amount of energy required to apply a newton of force over the distance of one meter or J = N m · 3) The unit of charge, the coulomb (C), is the amount of **electricity** in 1 ampere during 1 second of time (C = A s) · 4) The SI unit of area is the square meter or m^2 · 5) The SI unit of volume is the cubic meter or m^3 · 6) The unit of **density** (d) is the kilogram per cubic meter or d = kg/m^3.

The United States has been reluctant to completely adopt the SI units and still favors older units in some cases. Traditional units for distance such as the mile (= 1.609 km) and the yard (= 0.9144 m) are commonly employed. Other commonly used non-SI units are the pound (= 0.45359 kg) for weight and the quart (= 0.9463 L = 0.9463 dm^3) for volume. Moreover, there has not been complete adoption of the SI units in the scientific community and in some instances traditional metric units, called CGS units, continue to find use. This reluctance to change completely to SI units is likely mostly due to comfort and tradition. Nevertheless, the older units are slowly being supplanted by the SI system.

Constructing derived units is relatively straightforward in the SI system, which is, in part, the reason that scientists have adopted it. However, it also has one other distinct advantage over some older systems of measurements and that is scalability. A meter is the basic unit of length but shorter and longer units are obtained by multiplication by factors of 10. For example, a centimeter is 1/100 of a meter, while a kilometer is 1,000 meters in length. Similarly, a centigram is 1/100 of a gram, while a kilogram is 1,000 grams. This commonality of prefixes for shifting powers of 10 makes scientific measurement very much easier, particularly where scientists are investigating the very small and the very large. For example, the **nucleus** of an **atom** is about 1 fm or 1 x 10^{-15} m in diameter and our galaxy is about 10 exameters or 1 x 10^{19} m in diameter.

INTERSTELLAR GAS

For many years, scientists believed that outer **space** was an empty **vacuum**, devoid of **matter**. This misconception has been changed with the discovery that the space between **stars** and other stellar bodies is not empty. Rather, seemingly empty regions of space actually do contain matter in very tiny quantities. Interstellar gas is the most prevalent form of matter that occupies the space between celestial bodies.

Interstellar gas makes up an essential part of the formation and destruction of larger structures, like stars. About 98% of interstellar matter is gaseous, and it is estimated that approximately 5% of the total matter in the **universe** is in the form of interstellar gas. Although this is a large amount of matter, the distribution of interstellar gas can be very diffuse. In some regions, it is far less dense than the best laboratory vacuum chambers can produce.

Interstellar gas, however, is not uniformly distributed between stars. Instead, clouds of gas form that are called giant molecular clouds (GMCs). Over time, these huge clouds of gas become so massive that they begin to coalesce. As a cloud becomes more and more dense, gravitational attraction increases so that the central portion collapses and begins to release **heat**. As the **density** of the interstellar gas continues to increase in the central core, **nuclear reactions** begin, eventually forming a new star. Over time, some stars completely collapse and explode in cataclysmic reactions that release new interstellar gas into space.

Most interstellar gas is composed of the elements hydrogen and **helium**. Very recently, however, the presence of gaseous water also has been detected in a kind of interstellar gas cloud, called a nebula. A nebula is a detectable (or visible) cloud of interstellar gas. **Nebulae** appear either as dark or bright clouds of gas. Bright nebulae, or emission nebulae, glow in a manner similar to neon lights, where excited atoms of hydrogen or helium emit characteristic wavelengths (or colors) of **light**. Over 300 bright nebulae have been described, including the famous Orion and Crab nebulae. Dark nebulae, in contrast, are made visible as silhouettes against a very bright

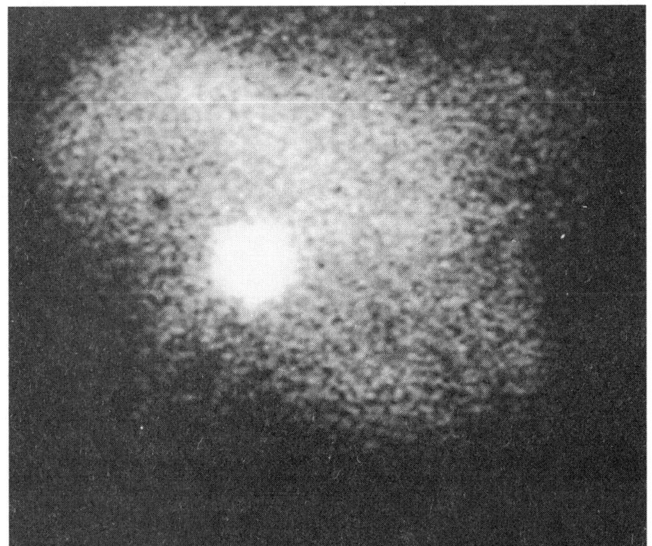

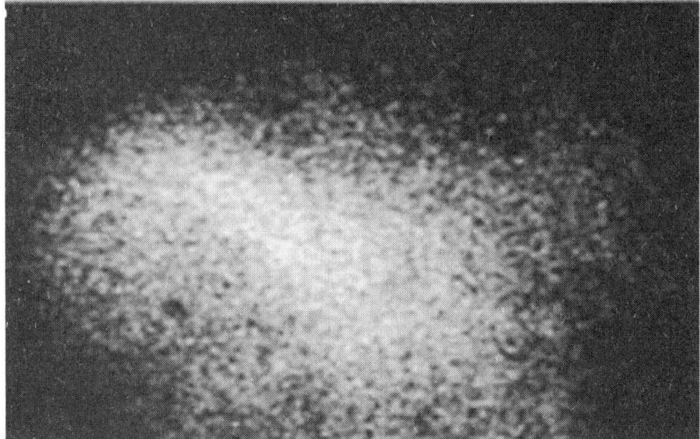

Crab Nebula, concentrated spot of gas near center of image. *(Image courtesy of NASA.)*

background. These dark clouds of interstellar gas block light from stars, revealing their shape to the observer. Some diffuse nebulae can reach 100 light years in diameter. A very well-known dark nebula is the Horsehead nebula, whose shape, as its name implies, resembles the head and neck of a horse.

INTERSTELLAR SPACE

Interstellar **space** (or the interstellar medium, or ISM, as it is sometimes more appropriately called) is the vast volume of space between the **stars** in our galaxy. Although so incredibly sparse that even its densest portions would constitute an excellent **vacuum** on **Earth**, it is a remarkably complex physical system. It is filled with an extremely dilute soup of **matter** (atoms, **ions**, electrons, molecules, and microscopic dust grains), **radiation** (in all bands of the **electromagnetic spectrum**, from long-wavelength Radio **waves** to very short **wavelength** gamma rays), and high **energy** particles (charged nuclei and electrons that have been accelerated to tremendously high energies). Moreover, as one immediately gathers by viewing

the sky, the ISM is highly inhomogeneous, clumpier and denser in some regions than others, and hotter and more turbulent in others.

The properties of the interstellar medium can be deduced by several primary methods. The first is to study stars with known distances along different lines of sight. By carefully examining their **color**, it is possible to determine that those further away tend to be redder than a star with similar physical characteristics would have nearby. The reason is simple: gas and dust along the line of sight to the star absorb and scatter radiation from the star, with shorter (i.e., bluer) wavelengths more preferentially affected, leaving redder wavelengths in the beam along the line of sight. This same phenomenon is observed in terrestrial sunsets: the **Sun** appears redder because the blue **light** is more easily scattered out of the beam along the line of sight. By measuring the reddening properties of many stars along many lines of sight, one can infer the mean column **density** (i.e., **mass** per square area) along the lines of sight, and hence build up a map of interstellar space.

It is also possible to directly map the ISM by observing the gas in emission. Hotter regions of the sky are seen in shorter wavelengths; colder regions in longer wavelengths. For instance, in the visible band, we can detect the so-called forbidden radiation from hydrogen atoms, which has a characteristically orange reddish color. In longer wavelengths—the **radio**, microwave, and infrared bands—we can map out the emission of extremely cold molecules. At the opposite end of the spectrum, at shorter wavelengths—the ultraviolet, x ray, and **gamma ray** bands—we see gas heated to extremely high temperatures through stellar radiation, stellar winds, and **novae and supernovae** explosions.

Lastly, it is possible to see the sky through "diffuse" radiation, which includes radiation from many sources that have been scattered into our line of sight. For instance, when viewing nearby stars in star-forming regions, we often detect a bluish diffuse halo surrounding them, which is actually their radiation scattered into our line of sight. (Such scattered radiation appears blue for the opposite reason that light undergoing absorption appears red, as described above.)

Because of its extremely complex nature, the ISM has proven to be an extraordinary challenge to the many generations of astronomers and astrophysicists who have sought to understand it. The presence of both very hot gas, with temperatures exceeding 10^6K, and blue, massive stars, tells us that the ISM within our galaxy is still a very active place. This is because such hot gas cools on timescales much shorter than the age of the galaxy, and similarly, such massive stars also live for a time much shorter than the age of the galaxy before ending their lives in supernovae explosions. Hence, there is a continual process by which gas cools from the ISM, condenses to form stars, with those stars driving winds and explosions into the ISM, from which other stars may be born. The most important details of that process—how stars form, and how the most massive stars explode—are themselves very fascinating subjects, and remain active topics of intense research by astronomers and astrophysicists today.

IONIZING RADIATION

Ionizing **radiation** is any **energy** that causes the ionization of the substance through which it passes. As the radiation is emitted from a source, it detaches a charged particle from an **atom** or molecule, leaving the atom or molecule with an excess charge. This charged particle is called an **ion**.

To remove an **electron** from an atom or molecule, the ionizing particles must have a **kinetic energy** exceeding the **binding energy** of the target species, typically a few electron volts. (An electron volt is a unit of energy defined as the **work** it takes one electron to move across a **voltage** difference of one volt.) Radiation of sufficient energy for this process to occur is commonly produced in nature.

Some common ionizing charged species are electrons, **positrons**, **protons**, and a-particles (**helium** nuclei). Electrons, positrons, and a-particles are emitted by radioactive elements. **Photons**, the uncharged particles of **light**, can also be emitted by radioactive nuclides, but can also be generated by x-ray devices. All the charged species, as well as neutrons, are currently produced at manmade **particle accelerators**, and **lasers** now generate photons of sufficient energy to exceed the binding energy of many atoms and molecules.

Most elements formed during the very early expansion of the **universe** were radioactive in the past, but over time became stable. Some, such as uranium, thorium, radium, and radon are still unstable, and spontaneously emit ionizing radiation. Here on **Earth** many rocks and minerals emit radon gas, a radioactive gas formed by the decay of radium. Other radioactive elements (^{3}H tritium and ^{14}C) can be produced by atmospheric interactions with **cosmic rays** (energetic particles continuously entering Earth's atmosphere from outer **space**).

Ionizing radiation is more damaging to human tissue than non-ionizing thermal-type radiation, as it is more likely to be localized and have a higher intensity (energy deposited per area per second). The damage is initialized by the ionizing particle when it knocks an electron off an atom or molecule in a living system, leaving an unpaired electron behind. The target atom or molecule is then a free radical, a highly reactive type that can spawn many more free radicals in the body. The induced chemical changes have been shown to cause cancer and genetic damage.

A unit called the rem (röntgen equivalent man) is used to measure the absorbed dose of ionizing radiation in living systems. An absorption of 0.5 rem annually is considered safe for a human being. By comparison, about 0.1 to 0.2 rem per year is contributed by natural sources, about 0.002 rem comes from dental x rays, and about 0.05 from a chest x ray.

IONS

An ion is a molecule of **atom** that carries a net positive or negative **electric charge**, that is it is not electrically neutral. The word *ion* is Greek, meaning the ones that move. A neutral atom has equal numbers of protons and electrons, as does a neutral molecule. If an atom or molecule has a deficit

Ion source machine at Los Alamos. *(Photo by Roger Ressmeyer/Corbis. Reproduced by permission.)*

or excess of electrons then it becomes electrically charged. Positive ions, cations, are formed when neutral molecules or atoms lose valence electrons and negative ions, anions, are those that have gained electrons. The process by which atoms or molecules become electrically charged is called ionization and can be induced by **radiation**, such as **x rays** or **light** with sufficient **energy** to rip electrons off of the valence shell, or by fast-moving particles, like those emitted by radioactive materials.

Typical levels in the atmosphere in the year 2000 are 3,000 cations and 4,000 anions per cubic centimeter. Indoors there are very few ions and there are more cations than anions. This is because negative ions exist for a short time relatively speaking and are destroyed by air pollution. Hot electrical discharges such as sparking, electric motors, and furnaces generate an excess of positive ions. Ions produced in air or water can carry electric currents. Although air is a pretty good insulator, when ions are present an electric charges can be carried along. This phenomenon was used in 1900 to detect radioactive emissions and measure their intensity.

There are several atoms that commonly lose an **electron**, like sodium, and become cations, Na$^+$, in an ionic solid such as sodium chloride, NaCl. Sodium chloride is composed of cations, Na$^+$, and anions, Cl$^-$. Although chloride ions have captured the outer electrons of the sodium atoms, the crystal as a whole has equal numbers of protons and electrons and so is neutral. When the crystal is dissolved in water the ions are separated from each other and surrounded by the polar water molecules that are oriented about the ions in a solvation sphere. The water molecules have a positive end and orient it towards the Cl$^-$ ions and a negative end that is directed towards the Na$^+$ ions. Ions can also consist of multiple atoms, such as sulfate ions (SO$_3^-$) and ammonium ions (NH$_4^-$), and exist in a similar way to that described for sodium chloride.

Radioactivity was first discovered in 1895 when heavy elements such as uranium were observed to emit rays, which could ionize air. **Ernest Rutherford** noted that these rays seemed to be composed of electrically charged components or opposite signs, positive alpha rays and negative beta rays. He was awarded the Nobel Prize for this work in 1908. Later these rays were identified as electrons, beta rays, and completely ionized **helium** nuclei He^{+2}, alpha particles. The **Sun** gets its energy by a process in which protons are combined to eventually form helium nuclei deep in the Sun's core. Since the helium **nucleus** is an unusually stable combination of particles, energy is released during this process. Today, ions are regularly produced and used to study the structure of **matter** and to produce a variety of new particles.

IRREVERSIBLE PROCESSES

On a small enough scale, virtually all of the laws of physics seem to be symmetrical in **time**. That is, when dealing with small enough objects—for instance fundamental particles—there is no preferential direction in which time flows. (Note that this is not absolutely true: recent experiments have shown that time asymmetry can be broken at the fundamental particle level, but only extremely rarely.)

On the macroscopic scale, however, we clearly observe a marked time asymmetry in the laws of nature; beyond the very smallest scale interactions, time always seems to flow in one direction. There are countless examples of the directionality of the flow of time. We observe traces of past occurrences all around us: footprints, photographs, memories, last night's dirty dishes, but we never stumble across the debris left behind by future events.

It seems a bit puzzling that laws of nature that have no built-in time asymmetry give rise to a world in which time clearly flows in one direction only, from past to future. The source of the asymmetry lies in the second law of **thermodynamics**, which states that any closed system tends to evolve so that its **entropy**, or disorder, increases. Thus, even though the laws that govern the behavior of individual bodies—for instance **Maxwell's equations**, Newton's laws, and the **Schrödinger's equation**—do not define a preferred direction in time, a system made up of a large number of bodies will evolve irreversibly in a way that leads to increased disorder or entropy.

One notable example of an irreversible process is the spontaneous flow of **heat**. The **second law of thermodynamics** requires that heat always flow from warmer to colder bodies, while forbidding it from flowing in the other direction. This is one of the "one-way" processes that define an "**arrow of time**" for the **universe**.

Other common irreversible processes are "mixing" processes. A drop of red dye will spontaneously diffuse throughout a cup of water, but the process will never occur in reverse, with all of the pre-mixed red dye spontaneously coalescing in a single drop suspended in the water. Again, this is due to the fact that the entropy or disorder of the well-mixed system is much greater than that of the drop of dye suspended in clear water.

ISOTOPE

Forms of an element that have the same number of protons but differ in their atomic weights are called *isotopes* of that element. The atomic number of an **atom** is equal to the number of protons found within its **nucleus**. The **atomic weight** of an atom includes the **mass** of protons and neutrons found in the nucleus. Therefore, because isotopes have the same number of protons, or the same atomic number, but different masses, they have different numbers of neutrons. Because they have the same number of protons, isotopes of the same element generally have identical chemical characteristics even though their masses differ. For example, the element **carbon**, atomic number 6, normally has an atomic weight of 12 atomic mass units. Its isotope, carbon-14, has the same atomic number, but an atomic mass of 14. Atoms of this isotope of carbon are heavier than normal because each atom has two additional neutrons in its nucleus. Approximately 275 isotopes exist for the 81 stable elements in the **periodic table** of elements. All elements have known isotopes.

Because of the particle imbalance, isotope nuclei having excess neutrons are unstable. Isotopes with unstable nuclei spontaneously convert to more stable forms over time. This process, called nuclear decay, releases **subatomic particles** and **energy** called **radiation**. Accordingly, such isotopes are called **radioisotopes**. Over 800 known radioisotopes exist. Radioisotopes are important elements used in biological and medical research. Radioisotopes that emit very tiny, but detectable, levels of radiation are used as labels, or molecular tags, to trace the paths of chemicals through tissues. Some radioisotopes, for example, are used to detect disease states or blockage in the human gastrointestinal tract. When an isotope solution is swallowed by the patient under investigation, x-ray film detects the presence of the radioisotope inside of the patient, providing valuable internal information without surgery. Similarly, radioisotopes such as technetium-99 can be used to detect bone cancer using sophisticated imaging techniques.

Other radioisotopes are highly radioactive and can actually cause harm. **Heavy water**, for example, is a dangerously radioactive by-product of nuclear **power** plants. Heavy water consists of molecular water that contains isotopes of hydrogen and/or **oxygen**. This "heavy" water (due to excess neutrons) is water that was used to cool uranium rods in the controlled **chain reaction** of **nuclear power** plants.

J

JOLIOT-CURIE, FRÉDÉRIC (1900-1958)

French nuclear physicist

Frédéric Joliot-Curie was a French nuclear physicist who, together with his wife, **Irène Joliot-Curie**, discovered artificial **radioactivity**, for which they received the 1935 Nobel Prize in chemistry. Their efforts made **nuclear fission** and the subsequent development of both nuclear **energy** and the **atomic bomb** feasible. Joliot met his wife while working as an assistant at the Radium Institute at the University of Paris. Irène Curie was the daughter of Marie and **Pierre Curie**, the Nobel Prize laureates who discovered radium and founded the Radium Institute. Irène became Frédéric's lifelong research collaborator and they usually published their findings under the combined form of their last names, Joliot-Curie. After World War II, Frédéric Joliot-Curie brought France into the atomic age as director of France's atomic energy commission.

Jean-Frédéric Joliot was born on March 19, 1900, in Paris to Henri Joliot and Emilie Roederer. According to tradition, all the Joliot men were named Jean in honor of Jean Hus, a champion for spiritual freedom who had been burned at the stake in the fifteenth century. Henri Joliot came from a long line of liberal thinkers and had been part of the French Communard movement at the end of the Franco-Prussian War. He became a Parisian dry goods merchant and the family was settled into a middle-class life, yet Henri remained passionate about his leftist political concerns. He also had a great love of the outdoors and of music, combining the two by composing a number of calls for the hunting horn. His son Frédéric would someday be an avid outdoorsman despite his busy scientific career. Frédéric's mother Emilie was also from a liberal family and Frédéric was exposed to progressive social ideas at a young age. The social and political leanings of his parents had a profound influence on young Frédéric and he was an atheist and political leftist his entire life.

Joliot was educated at the Lycée Lakanal in a suburb of Paris, then at the École Primaire Supérieure Lavoisier in Paris.

In 1920 he was admitted to the École Supérieure de Physique et de Chimie Industrielle of Paris, a preparatory school that turned out most of France's industrial engineers at the time. The director of the school, a brilliant physicist named Paul Langevin, recognized Joliot's interest and aptitude for scientific research and became Joliot's lifelong mentor.

After graduating at the head of his class, Joliot worked in industry for a short time. Following Langevin's advice, however, he took a position as research assistant at the Radium Institute in Paris under the guidance of Madame **Marie Curie**, a Nobel laureate in **nuclear physics**. Joliot held Marie and Pierre Curie in high esteem, going so far as to have pictures of them hanging on the wall of his makeshift home laboratory. Working at the institute, he was instructed by Madame Curie to be initiated into the rudiments of studying radioactivity by Irène Curie, the couple's elder daughter. For as affable and charming as Frédéric was, Irène was serious and aloof. Yet the two shared their love of research as well as political and social leanings and were married in 1926. They both adopted their combined last names to preserve the scientific Curie lineage.

Joliot-Curie began his research improving the Wilson chamber, a **cloud chamber** in which the charged particles of an **atom** can be detected as they leave a trail of water droplets in their wake. His engineering background made him a master at instrumentation and he redesigned the chamber so that the **pressure** could be lowered, making the tracks longer and more easily discernible. He also supervised the making of a camera that photographed what went on in the chamber. Further improvements allowed the energy of emissions to be measured using the amount of curvature shown by the tracks when placed in a magnetic field. Conducting experiments within the chamber had an elegance that Frédéric considered to be the most beautiful experience in the world.

In 1930, Joliot-Curie completed his doctoral thesis entitled *A Study of the Electrochemistry of Radioactive Elements,* thus beginning intensive research in the area of radioactivity. Later that same year, he and Irène became interested in the experiments of German physicists Walther Bothe and Hans

Frédéric Joliot-Curie.

The Joliot-Curies' discovery of artificial radioactivity was announced to the Academy of Science in 1934 and won them the Nobel Prize for chemistry in 1935. (The prize was awarded in chemistry rather than physics because of their synthesis of new radioactive elements.) Sadly, though she had predicted the success and recognition of her daughter and son-in-law, Marie Curie died of leukemia the year before the award, a casualty of lifelong exposure to radiation.

In the latter part of the 1930s, experiments by Joliot-Curie showed that a nucleus of a radioactive element such as uranium could be divided into two smaller nuclei of comparable size with a release of energy emissions. It was predicted that this phenomenon, known as nuclear fission, could occur very quickly in a **chain reaction** and therefore needed to be controlled or moderated. This conclusion led other researchers such as **Otto Hahn** and **Enrico Fermi** to work on nuclear fission experiments that eventually led to the creation of the atomic bomb as well as atomic energy.

The Joliot-Curies enjoyed these productive years in the company of their two children, Helene (born in 1927), and Pierre (born in 1932). Because of Irène's often fragile health, the family spent as much time away from work in Paris as possible, swimming or sailing at the beach or hiking or skiing in the mountains. It was said that rather than carry around pictures of his family in his wallet, Frédéric Joliot-Curie carried instead a picture of a giant pike he had once caught on a fishing trip. He was a very sociable man, telling not only fish stories, but entertaining both colleagues and family with his wit and charm.

By the time the Germans occupied Paris during World War II, Joliot-Curie, who was staunchly part of the French Resistance, continued to do research but kept a low profile. He wanted to prevent the Germans from gaining information about nuclear fission. In fact, he had a stockpile of **heavy water**, which was used as a moderator in nuclear fission experiments, sent out of the country so that the Germans could not gain possession of it. Inspired in part by the arrest of his friend and mentor, Paul Langevin, Joliot-Curie became more politically active. In 1942, when Langevin's son-in-law, Jacques Solomon, was tortured and shot by the Nazis for turning out a Resistance newspaper, Joliot-Curie joined the Communist Party. Arrested twice and fearing for his safety and that of his family, Joliot-Curie went underground in Paris and sent Irène and their two children to Switzerland.

After World War II, Joliot-Curie was instrumental in convincing the government of Charles de Gaulle to establish an Atomic Energy Commission in France, modeled after the similar agency in the United States. After being appointed its commissioner, Joliot-Curie oversaw the installation of a major nuclear research center. With the advent of the Cold War and tensions between the Soviet Union and the West, Joliot-Curie's membership in the Communist Party and his radical activism drew governmental concern over his reliability. He was relieved of his position as high commissioner of Atomic Energy.

Losing his position was a blow to Joliot-Curie both professionally and personally. He and his wife also began to suf-

Becker. The German pair had discovered that very strong **radiation** was emitted from some of the lighter elements when they were bombarded with alpha rays. An alpha ray is an energetic particle that resembles the **nucleus** of a **helium** atom and contains two positive charges. Such rays had only been discovered at the turn of the twentieth century as had the basic particles of the atom. Because of the readily available source of alpha rays—radium—stockpiled at the Radium Institute, Frédéric and Irène began doing research in which they bombarded nuclei of various elements with alpha particles.

In the course of experimentation, the Joliot-Curies had come very close to discovering two other **subatomic particles**, the **neutron** and the **positron**. But instead of investigating these anomalies, they focused their efforts on their alpha-particle research. As Joliot-Curie and his wife worked with radioactive polonium as a source of alpha rays, they found that when an element such as aluminum is bombarded, some alpha particles are absorbed by the nuclei, transmuting the atoms to phosphorus. Aluminum usually has 13 protons in its nucleus, but when bombarded with alpha particles, which contain two positive charges each, the protons were added to the nucleus, forming a nucleus of phosphorus, the element that normally contains 15 protons. The phosphorus produced was different from naturally occurring phosphorus because it gave off a strong radiation, as predicted by Bothe and Becker. It was a radioactive **isotope** produced artificially.

fer ill health from their lifelong exposure to radiation. Still dedicated to his leftist politics, he became president of the World Organization of the Partisans of Peace. In March 1956, Irène finally succumbed to leukemia. Frédéric succeeded her as head of the Radium Institute and continued her efforts to build a new physics laboratory south of Paris. At 58, his liver badly damaged by exposure to radiation, Frédéric Joliot-Curie died following an operation made necessary by an internal hemorrhage.

JOLIOT-CURIE, IRÈNE (1897-1956)
French chemist and physicist

Irène Joliot-Curie, elder daughter of famed scientists Marie and **Pierre Curie**, won a Nobel Prize in chemistry in 1935 for the discovery, with her husband **Frédéric Joliot-Curie**, of artificial **radioactivity**. She began her scientific career as a research assistant at the Radium Institute in Paris, an institute founded by her parents, and soon succeeded her mother as its research director. It was at the institute where she met her husband and lifelong collaborator, Frédéric Joliot. They usually published their findings under the combined form of their last names, Joliot-Curie.

Born on September 12, 1897, in Paris to Nobel laureates Marie and Pierre Curie, Irène Curie had a rather extraordinary childhood, growing up in the company of brilliant scientists. Her mother, the former Marie Sklodowska and her father, Pierre Curie, had been married in 1895 and had become dedicated physicists, experimenting with radioactivity in their laboratory. **Marie Curie** was on the threshold of discovering radium when little Irène, or "my little Queen" as her mother called her, was only a few months old. As Irène grew into a precocious, yet shy child, she was very possessive of her mother who was often preoccupied with her experiments. If, after a long day at the laboratory, the little Queen greeted her exhausted mother with demands for fruit, Marie Curie would turn right around and walk to the market to get her daughter fruit. Upon her father Pierre Curie's untimely accidental death in 1908, Irène was then more influenced by her paternal grandfather, Eugene Curie. It was her grandfather who taught young Irène botany and natural history as they spent summers in the country. The elder Curie was also somewhat of a political radical and atheist, and it was he who helped shape Irène's leftist sentiment and disdain for organized religion.

Curie's education was quite remarkable. Marie Curie made sure Irène and her younger sister, Eve Denise (born in 1904), did their physical as well as mental exercises each day. The girls had a governess for a time, but because Madame Curie was not satisfied with the available schools, she organized a teaching cooperative in which children of the professors from Paris's famed Sorbonne came to the laboratory for their lessons. Madame Curie taught physics, and other of her famous colleagues taught math, chemistry, language, and sculpture. Soon Irène became the star pupil as she excelled in physics and chemistry. After only two years, however, when Irène was 14, the cooperative folded and Irène enrolled in a private school, the College Sevigne, and soon earned her degree. Summers were spent at the beach or in the mountains, sometimes in the company of such notables as **Albert Einstein** and his son. Irène then enrolled at the Sorbonne to study for a diploma in nursing.

During World War I, Madame Curie went to the front where she used new x-ray equipment to treat soldiers. Irène soon trained to use the same equipment and worked with her mother and later on her own. Irène, who was shy and rather antisocial in nature, grew to be calm and steadfast in the face of danger. At age 21, she became her mother's assistant at the Radium Institute. She also became quite adept at using the Wilson **cloud chamber**, a device that makes otherwise invisible atomic particles visible by the trails of water droplets left in their wake.

In the early 1920s, after a jubilant tour of the United States with her mother and sister, Irène Curie began to make her mark in the laboratory. Working with Fernand Holweck, chief of staff at the institute, she performed several experiments on radium resulting in her first paper in 1921. By 1925 she completed her doctoral thesis on the emission of alpha rays from polonium, an element that her parents had discovered. Many colleagues in the lab, including her future husband, thought her to be much like her father in her almost instinctive ability to use laboratory instruments. Frédéric was several years younger than Irène and untrained in the use of the equipment. When she was called upon to teach him about radioactivity, Irène started out in a rather brusque manner, but soon the two began taking long country walks. They married in 1926 and decided to use the combined name Joliot-Curie to honor her notable scientific heritage.

After their marriage, Irène and Frédéric Joliot-Curie began doing their research together, signing all their scientific papers jointly even after Irène was named chief of the laboratory in 1932. After reading about the experiments of German scientists Walther Bothe and Hans Becker, their attention focused on **nuclear physics**, a field yet in its infancy. Only at the turn of the twentieth century had scientists discovered that atoms contain a central core or **nucleus** made up of positively charged particles called protons. Outside the nucleus are negatively charged particles called electrons. Irène's parents had done their work on radioactivity, a phenomenon that occurs when the nuclei of certain elements release particles or emit **energy**. Some emissions are called alpha particles, which are relatively large particles resembling the nucleus of a **helium atom** and thus contain two positive charges. In their Nobel Prize-winning work, the elder Curies had discovered that some elements, the radioactive elements, emit particles on a regular, predictable basis.

Irène Joliot-Curie had in her laboratory one of the largest supplies of radioactive materials in the world, namely polonium, a radioactive element discovered by her parents. The polonium emitted alpha particles that Irène and Frédéric used to bombard different elements. In 1933 they used alpha particles to bombard aluminum nuclei. What they produced was radioactive phosphorus. Aluminum usually has 13 protons in its nuclei, but when bombarded with alpha particles that contain two positive charges each, the protons were added to the nucleus, forming a nucleus of phosphorus, the element with 15 protons. The phosphorus produced is different from naturally occurring phosphorus because it is radioactive and is known as a radioactive **isotope**.

The two researchers used their alpha bombardment technique on other elements, finding that when a nucleus of a particular element combined with an **alpha particle**, it would transform that element into another, radioactive element with a higher number of protons in its nucleus. What Irène and Frédéric Joliot-Curie had done was to create artificial radioactivity. They announced this breakthrough to the Academy of Sciences in January 1934.

The Joliot-Curies' discovery was of great significance not only for its pure science, but for its many applications. Since the 1930s many more radioactive isotopes have been produced and used as radioactive trace elements in medical diagnoses as well as in countless experiments. The success of the technique encouraged other scientists to experiment with the releasing the **power** of the nucleus.

It was a bittersweet time for Irène Joliot-Curie. An overjoyed but ailing Marie Curie knew that her daughter was headed for great recognition but died in July of that year from leukemia caused by the many years of **radiation** exposure. Several months later the Joliot-Curies were informed of the Nobel Prize. Although they were nuclear physicists, the pair received an award in chemistry because of their discovery's impact in that area.

After winning the Nobel Prize, Irène and Frédéric were the recipients of many honorary degrees and named officers of the Legion of Honor. But all these accolades made little impact on Irène who preferred spending her free time reading poetry or swimming, sailing, skiing, or hiking. As her children Helene and Pierre grew, she became more interested in social movements and politics. An atheist and political leftist, Irène also took up the cause of woman's suffrage. She served as undersecretary of state in Leon Blum's Popular Front government in 1936 and then was elected professor at the Sorbonne in 1937.

Continuing her work in physics during the late 1930s, Irène Joliot-Curie experimented with bombarding uranium nuclei with neutrons. With her collaborator Pavle Savitch, she showed that uranium could be broken down into other radioactive elements. Her seminal experiment paved the way for another physicist, **Otto Hahn**, to prove that uranium bombarded with neutrons can be made to split into two atoms of comparable **mass**. This phenomenon, named fission, is the foundation for the practical applications of nuclear energy—the generation of **nuclear power** and the atom bomb.

During the early part of World War II, Irène continued her research in Paris although her husband Frédéric had gone underground. They were both part of the French Resistance movement and by 1944, Irène and her children fled France for Switzerland. After the war she was appointed director of the Radium Institute and was also a commissioner for the French atomic energy project. She put in long days in the laboratory and continued to lecture and present papers on radioactivity

although her health was slowly deteriorating. Her husband Frédéric, a member of the Communist Party since 1942, was removed from his post as head of the French Atomic Energy Commission in 1950. After that time, the two became outspoken on the use of nuclear energy for the cause of peace. Irène was a member of the World Peace Council and made several trips to the Soviet Union. It was the height of the Cold War and because of her politics, Irène was shunned by the American Chemical Society when she applied for membership in 1954. Her final contribution to physics came as she helped plan a large particle accelerator and laboratory at Orsay, south of Paris in 1955. Her health worsened and on March 17, 1956, Irène Joliot-Curie died as her mother had before her, of leukemia resulting from a lifetime of exposure to radiation.

JOSEPHSON, BRIAN D. (1940-)
English physicist

While still a graduate student, Brian D. Josephson made a discovery that was to earn him a share of the 1973 Nobel Prize for physics. That discovery concerned the flow of electrons across two superconducting materials separated by a thin nonconducting barrier. Josephson found that his own calculations predicted a far greater flow than would have been expected on the basis of traditional **quantum mechanics**. His predictions were experimentally confirmed shortly after his announcement of the effect.

Josephson was born in Cardiff, Wales, on January 4, 1940. His parents were Mimi and Abraham Josephson. After graduating from Cardiff High School, Josephson entered Trinity College, Cambridge. He earned his B.A. degree there in 1960 and his M.A. and Ph.D. in 1964. Between 1962 and 1969, he also held an appointment as junior research fellow at Trinity.

Studies Mössbauer and Tunneling Effects

Josephson's first research interest as a graduate student was the **Mössbauer effect**, the recoilless emission of **gamma rays** from a crystal. He found that very small **temperature** differences in emitter and detector in experiments using the Mössbauer effect could have significant effects on the outcome of those experiments. Researchers had not appreciated this point until Josephson published his calculations, and many had to repeat their work in order to take this effect into account.

In 1962, while still a graduate student, Josephson shifted his attention to the phenomenon of **tunneling** in superconducting materials. The model with which he worked consisted of a pair of metals maintained at temperatures close to **absolute zero** and separated by a thin layer of nonconducting material. According to traditional **quantum theory**, a small number of electrons should be able to "tunnel through" (pass across) the potential barrier represented by the nonconducting layer between the two conductors. To his great surprise, Josephson found that his own calculations predicted a much larger flow of current. In addition, he discovered that that current should be susceptible to an external magnetic field (i.e., very small changes in the magnetic field should result in large changes in **electron** flow).

Josephson was not able to detect these effects himself, but two colleagues, Philip Warren Anderson and J. M. Rowell, soon did so for direct-current currents, as did S. Shapiro for alternating-current currents. A decade later, Josephson was awarded a share of the Nobel Prize in physics (with Leo Esaki and Ivar Giaever) for this discovery, which has had enormous practical significance. The currents Josephson studied are highly sensitive to very small magnetic fields, and they have found application in a variety of detecting devices as well as in basic research.

Except for a year at the University of Illinois as a visiting research professor (1965–66), Josephson has spent his entire academic career at Cambridge. He has held appointments there as assistant director of research (1967–72), reader in physics (1972–74), and professor of physics (1974-). After receiving his Ph.D., Josephson continued his research on superconductivity and critical phenomena. Then, in the late 1960s, he began to drift away from conventional physics. The problem, he told an *Omni* interviewer, was that "conventional physics didn't offer much opportunity to achieve [the] sort of breakthrough" in which he had become interested. That breakthrough involved a study of intelligence, language, and higher states of consciousness and the paranormal.

Interest Shifts to Studies of the Mind

Josephson was powerfully influenced by a theorem developed by theoretical physicist **John Bell** in 1965. According to **Bell's theorem**, any two objects obtained from a single large source will always exert an influence on each other, no matter how far apart they may become in the **universe**. To Josephson, this type of phenomenon created the possibility of interactions with which conventional physical laws are incapable of dealing.

An important turning point in Josephson's life occurred in 1971, when he heard a **radio** announcement for a lecture on transcendental meditation (TM). He attended the lecture, became an adherent of TM, and has since become a student also of the Maharishi Mahesh Yogi. As part of his daily routine (which may include walking, ice skating, **photography**, and astronomical studies), Josephson also meditates for about two hours.

Josephson married Carol Anne Oliver in 1976. They have one daughter. Among his honors and awards are the 1969 New Scientist Award, the 1969 Research Corporation Award, the 1970 Fritz London Award, and the 1972 Hughes Medal of the Royal Society.

JOSEPHSON JUNCTION

The Josephson junction is named after the 1973 physics Nobel Prize winner **Brian Josephson** who carried out extensive theoretical work on its properties. The junction itself consists of two superconducting metals separated by a thin insulating oxide layer usually only 10-20 angstroms, or equivalently about 30 atomic layers, in width. The remarkable properties of the junction are due to a superconducting current that can flow through the oxide layer even at zero **voltage** difference between the outer plates. The origin of this current depends crucially on the quantum mechanical principles underlying superconductivity.

In a superconducting metal, electrons exert forces on the crystal lattice of the metal. This **force** displaces positive **ions** in the lattice and sets up a polarization that can in turn exert attractive forces on the original electrons. This net force is ordinarily much weaker than the usual Coulomb repulsion between electrons, but this Coulomb force is also suppressed by screening effects of positive ions in the lattice. At very low temperatures, in the absence of thermal fluctuations, the attractive force overcomes the repulsive force and electrons can actually bind with each other and form pairs called Cooper pairs.

In these Cooper pairs, the intrinsic spin of each **electron** cancels the other's spin and so the Cooper pairs all have total spin zero. The **Pauli exclusion principle** that applies to each individual electron then no longer applies to the Cooper pair as a whole, and so at very low temperatures all the Cooper pairs settle down to the same macroscopic quantum state. So in a Josephson junction, at low temperatures, on each side of the oxide layer there are two macroscopic **quantum states** each described by a wavefunction. However, the wavefunction need not be continuous across the junction. In particular the magnitude of the wavefunction can be the same, signifying equal charge densities of electrons on either side and thus no voltage difference across the junction, but the phase of the wavefunction may be different. Upon solving **Schrödinger's equation**, which governs time dependence in **quantum mechanics**, one finds that a current flows across the junction whose magnitude is proportional to the sine of the phase difference. This current is the famous Josephson effect and it requires no voltage difference to maintain itself.

The Josephson effect forms the basis for several extremely successful low **temperature** technologies that are in use today. One of the first applications was as extremely sensitive voltmeters. When a voltage is applied to a Josephson junction, the phase difference of the wavefunctions across the junction changes over time and this leads to an alternating-current as opposed to direct-current Josephson current. When an alternating voltage is applied, for example using microwave **radiation**, this alternating current rises in steps as the voltage is increased and the length of each step in voltage is proportional to the **frequency** of the Joesphson current. Since the frequency of the Josephson current can be measured extremely accurately, one can measure the voltages extremely accurately as well, often to the level of picovolts.

Josephson junctions are also crucial in the operation of superconducting quantum **interference** devices (**SQUIDs**), which are basically superconducting loops containing one or more Joesphson junctions inside. When these SQUIDs are coupled to a superconducting flux transformer, they can measure small spatial fluctuations in the magnetic field piercing the current loop. One of the far-ranging applications of this phenomenon includes the study of biomagnetism, especially the measurement of magnetic fields in the human brain. SQUIDs are also used in geophysical surveying, nondestructive testing, and **nuclear magnetic resonance.**

Perhaps the most promising applications of Josephson junctions are as switching elements in digital circuits. At low temperatures, the zero voltage Josephson current can be increased up to a certain point until the junction suddenly switches to a voltage carrying state. This switching occurs on the order of picoseconds, much faster than in conventional semiconductor circuit switching elements, and moreover this switching occurs at very little **power** loss. If Josephson junctions could be used as logic and memory gates then these gates could be placed much closer to each other than usual because of the negligible **heat** dissipation involved in switching states. Allowing gates to be closer to each other minimizes the amount of time it takes electromagnetic signals to propagate between them and results in a much faster chip. However, progress in this direction requires the development of high temperature superconductivity. Currently such gates have been operated at 77K and efforts are underway to improve upon this result. Clearly, the story of the full technological promise of Josephson junctions has yet to be told.

JOULE, JAMES PRESCOTT (1818-1889)
English physicist

James Prescott Joule is credited with the fundamental discovery that opened a new paradigm of the nature of **heat,** which allowed many new discoveries and inventions. A rather unassuming man in appearance, he had difficulty for a long time in convincing his contemporaries of the importance of his discovery. However, once he received the endorsement of an accepted scientist **William Thomson Kelvin**, he received nearly immediate acclaim. Though he continued his work throughout his life, all of the discoveries for which he is now famous were accomplished before he was 34 years old. One truly noteworthy condition of Joule's career was that he was self-reliant in his education and his research. He was self-educated and, coming from a wealthy family, he funded his own research. This condition, as respectable as it was, proved to be a great hindrance. The scientific community would not believe that Joule could discover something so important without the support, academically and financially, of an academic institution. Still, he knew that he had the answer to a long-puzzling scientific question and he persevered.

Joule is noted in the history of science as one of the most determined researchers. He even produced research while on his honeymoon. Joule and his bride were visiting Switzerland and Joule took the occasion to measure the **temperature** of a large waterfall as part of his research. By chance, Kelvin met him on the mountain that day and long after admired his determination. Joule was a great influence on Kelvin and many others for his breakthrough, his methods, and his persistence.

Joule was born in 1818 in Salford, England. He grew up in a well-established wealthy family, which had owned a brewery for generations. As a child, Joule was not very healthy and was educated at home through tutors and his own readings. When he was 16, he went to study for two years with **John Dalton**. Dalton first stimulated Joule's interested in chemistry and experimentation. After Dalton could no longer teach because of his poor health, Joule built his own laboratory at his family's home. He conducted experiments and began writing research papers when he was 20. In the autumn before he turned 23, Joule presented a paper at the meeting of the Manchester Literary and Philosophical Society. He was well received and was soon elected as a member of the society. He was an active member of the society for many years, culminating in his acceptance of the society presidency in 1860.

Joule never attended college. Instead, he read incessantly on whatever interested him and what did not interest him he avoided. He abhorred some of the topics usually considered part of a liberal college education. Still, he was fascinated with the study of physics, in particular **electromagnetic fields**. In 1838, he described in a paper what he called an electromagnetic engine. He believed that these engines could replace steam but later had to admit their inefficiency. In 1840, he wrote a paper describing what is now called the Joule effect. His premise was that the amount of heat produced by electrical current is a function of the amount of resistance and the square of the current.

From 1843 through 1878, he regularly attended the meetings of the British Association for Advancement of Science (BAAS). At his first meeting, he presented his seminal theory on the nature of heat. This theory stated that heat is derived from **work** and that heat and work are interchangeable. He worked to improve this theory and every year returned to the BAAS to present a paper reporting his progress. At the 1845 meeting, he brought a working model demonstrating the work-to-heat principle. This model consisted of water in a container that was agitated very fast by a paddle. The main problem with the conference attendees accepting his theory was that the experiments supporting the theory required highly precise measurements of heat. They did not appreciate that Joule, an amateur, had taken this preciseness into consideration.

After tolerating Joule for five years, the conference administrators requested that he give a summary of his findings and not a full presentation. This was the historic 1847 meeting where Kelvin rose to the defense of Joule's theory. After the short presentation, Kelvin asked so many questions that Joule had to go into his full presentation. Kelvin's fascination with the theory stirred up new interest among the scientists. Kelvin announced to the conference that he found Joule's methods and findings to be well developed and accurate. This was a turning point in Joule's career.

Only days after that turning point, Joule took a turning point in his life. He married Amelia Grimes who apparently was good-natured about his constantly measuring heat. When he saw the high waterfall on his honeymoon, he wanted to test his theory and she willingly helped him. He deduced that the water at the bottom of the fall should be hotter than at the top due to the work of the water running down the fall. As noted above, he met Kelvin again that day. Joule and Kelvin began collaboration from 1853 through 1862. During that time, they established the Joule-Thomson effect.

Joule's heat theory was the cornerstone of his research career. One of the embellishments involved the calculation of the **velocity** of a gas molecule in 1848. He developed relatively little more in his life but it was enough to eventually lead Kelvin and others to discover the second principle of **thermodynamics**. After 1862 when he was no longer doing research with Kelvin, Joule revised his theory as an expression of the **conservation of energy**. He died in 1889 at age 71.

Joule's heat-to-work theory was well founded in his experiments. He had done an exhaustive amount of work to determine that the relationship between heat and work was a constant. He even calculated the amount of work necessary to produce heat. Once his views were "approved" by Kelvin's endorsement, many researchers excitedly investigated his theory. They found that the number Joule used as his work-to-heat constant was too large. To find a more practical unit of measurement, they divided Joule's number by 10,000,000 and named the new unit after him. The measure of heat equivalent to work is now called the joule.

Working with Kelvin, Joule investigated different aspects of his theory. One of these was the difference in heat during expansion of a gas. They found that when an amount of gas was allowed to expand in a **vacuum** it tended to lose heat. They explained it in terms of heat-to-work, but it is now recognized as a property of the molecules separating one from another. By investigating this principle, industrialists founded the principles necessary to begin the refrigeration industry by the turn of the twentieth century.

Though it did not receive much recognition at the time, Joule's study of the velocity of gas molecules did much to encourage the kinetic theory of gases. His interest had been developed by the tutor from his teens, John Dalton. By this time, Joule was well recognized for his detailed thoughts and precise measurements. To have such a distinguished scientist present findings supporting the theory encouraged other researchers.

Joule received many awards in his lifetime. After presenting his research to the Royal Society in 1849, he was elected a fellow. As a member of the society, he earned the Royal

James Prescott Joule.

medal and the honored Copley medal for the same research. It was well noted that this was a very rare achievement, to win both for the same research. Another honored event occurred when his family's industry went broke. The queen herself offered him a pension. However, the most treasured award he received was perhaps the first, when he was reunited with Dalton at his first meeting of the Literary and Philosophical society. Dalton recognized how impressed the audience was and uncharacteristically asked the audience to applaud the young man. This applause, though faded for many years, eventually included the nation.

K

K MESONS

The K **meson**, or by its more preferred name, the kaon, is one of the family of **mesons** (a class of particles heavier than **leptons** but lighter than baryons). The k meson has zero spin, a nonzero strangeness quantum number, and a **mass** of approximately 494 MeV. The k meson is the lightest **hadron** that contains a strange quark. The mean lifetime of the k meson is 1.24×10^{-8}s.

In the 1930s physicists were trying to describe the strong nuclear **force**. Because the strong force acts only within the range of the **nucleus**, it acts only over a very short range of distance. Because the strong force acts on both protons and neutrons in the same way, the strong force must be independent of **electric charge**. In 1935 Japanese physicist **Hideki Yukawa** proposed that the nuclear force between protons was mediated by a massive particle, now called the pi-meson or pion. In 1947, British physicists George Rochester and Clifford Butler using a **cloud chamber** saw something unusual. Two tracks appeared from a single point beneath a lead plate, as if from nowhere. Rochester and Butler concluded a neutral particle had entered the chamber (leaving no track) and then decayed into two charged particles. At first this particle was called a V particle due to its V-shaped track upon decaying.

Further study revealed two kinds of V particles. One form decayed only to products always containing a **proton**, and which was called the hyperon. The other whose decay products consisted only of mesons was named the K meson or kaon.

The hyperons and kaons soon became known as **strange particles** because of their anomalous decay behavior. They were produced by the strong nuclear force, and so a decay time of about 10^{-23} seconds was expected. Instead their decay time was about 10^{-10} seconds, a decay time indicating decay via what is now termed the weak nuclear force!

In 1953 scientists at the Brookhaven Accelerator Test Facility were able to produce strange particles. In the same year American physicist **Murray Gell-Mann** and Japanese physicist Kazuhiko Nishijima proposed a new conservation law, strangeness, which applies only to the strong interaction, to explain the production and decay mechanism of hyperons and kaons. Each particle is assigned a quantum number of strangeness in addition to spin, intrinsic **parity**, and isospin. All the particles in a strong interaction must preserve strangeness. The decay of strange particles into nonstrange particles cannot proceed by the strong force, however, and proceeds by the weak force, which does preserve strangeness.

There are two charged strange kaons, K-plus and K-minus, and a neutral kaon. All have a mass of 494 MeV and are therefore about three times as massive as the **pions**. Like the pions, the kaons have zero spin and odd intrinsic parity. But the kaons do not have isospin $I = 1$ as do the pions. Consequently the neutral kaons must come in two versions with opposite strangeness. The two versions are the neutral kaon and the neutral antikaon.

As it turns out neutral kaons also come in two types, "short" (which has an average lifetime of less than 10^{-10} seconds), and "long" (which has a lifetime of slightly over 5×10^{-8} seconds). There is an equal probability that a neutral kaon will behave as a long or short neutral kaon.

The neutral kaon is produced by the strong force that respects strangeness (S). But after 10^{-10} seconds (the production took only 10^{-24} seconds) the weak force will be predominant.

When first discovered it was thought that the long and short kaons were two different particles (called the tau and theta particles, respectively). But it was soon shown that they are two species of the same particle. Kaons, like all mesons, are constructed of two **quarks**, an "up" quark and a "down" quark, always in a quark-antiquark doublet held together by the exchange of gluons. Although confusing and difficult to study, kaons have enabled physicists to unlock some secrets of the weak force and its interactions, and have served as useful probes of other quantum mechanical phenomena.

See also Nucleon; Particle accelerators; Particle detectors; Particle physics; Proton decay; Quantum field theory; Strong interactions; Weakly interacting massive particles (WIMPs)

KAMERLINGH ONNES, HEIKE (1853-1926)

Dutch physicist

Heike Kamerlingh Onnes was a Dutch experimental physicist distinguished for his work in the field of **low-temperature physics**. He was the first scientist to succeed in liquefying **helium**, a breakthrough that yielded a previously unattainable degree of cold. This accomplishment won him the 1913 Nobel Prize for physics, in addition to numerous other awards. He is also credited with the discovery of superconductivity—that is, the complete disappearance of **electrical resistance** in various metals at temperatures approaching **absolute zero**.

Kamerlingh Onnes was born in Groningen, the Netherlands, on September 21, 1853. His father owned a tile factory, and both his parents were strict, imbuing Kamerlingh Onnes and his brothers with an understanding of the value of hard work and perseverance. He was initially educated at Groningen High School under J. M. van Bemmelan, and in 1870 he enrolled in the physics program at the University of Groningen. The following year he submitted an essay on vapor **density** and won first prize in a contest sponsored by the University of Utrecht. In October 1871, Kamerlingh Onnes transferred to the University of Heidelberg in Germany, where he was taught by the eminent German chemist Robert Wilhelm Bunsen. He was one of only two students allowed to work in the private laboratory of German physicist **Gustav Robert Kirchhoff**. In April 1873, he returned to the University of Groningen, where he spent the next five years studying for his doctorate.

In 1878, Kamerlingh Onnes moved to Delft Polytechnic where he became an assistant to the professor of physics there. In 1879 he traveled to Groningen to defend his thesis, entitled "New Proof of the Earth's Rotation." He was awarded his physics doctorate magna cum laude. At Delft Polytechnic, Kamerlingh Onnes composed a paper on the general theory of **fluids** from the perspective of kinetic theory. He soon realized, though, that such a general theory of the nature of fluids required accurate measurements of volume, **pressure**, and **temperature** over as wide a range of values as possible. To this end, he turned his attention to the problem of attaining and maintaining very low temperatures.

Becomes Head of Cryogenic Laboratory

In 1882, at the age of 29, Kamerlingh Onnes accepted Holland's first chair in experimental physics at Leiden University. He also became the director of the laboratory there, where he was able to pursue his interest in low-temperature physics, also known as **cryogenics**. A dedicated experimentalist, Kamerlingh Onnes declared in his inaugural address: "I should like to write 'through measuring is know-ing' as a motto above each physics laboratory." He would spend the rest of his career at Leiden. During the next 42 years, he established it as the undisputed world headquarters of low-temperature research.

When Kamerlingh Onnes began his pioneering work, cryogenic physics was a relatively unknown science. Before him, the liquefaction of gases at very low temperatures was considered an end in itself, but Kamerlingh Onnes was interested in low-temperature physics in order to gather experimental evidence about the atomic nature of **matter**. When he set out to cool gases such as **oxygen**, hydrogen, and helium to extremely low temperatures, there were three means at his disposal. A cooling effect due to the rapid evaporation of a liquid had been discovered in 1877 by Swiss physicist R. P. Pictet. That same year, French physicist L. P. Cailletet had achieved low temperatures when he was able to cool oxygen by the application of intense pressure. The final method was based on the 1850 discovery by **James P. Joule** and **William Thomson Kelvin** that when a gas under pressure is released through very small openings, its temperature is lowered by an amount that depends on the nature of the gas. In Munich in 1895, Carl Linde constructed an apparatus that made use of the so-called Joule-Thomson effect; gas was put under pressure and repeatedly forced into a coil of tubes that also acted as a **heat** exchanger. This was known as the regenerative process. The amount of liquid gas produced by all of these means was, however, negligible.

In trying to achieve very low temperatures, Kamerlingh Onnes employed a combination of Pictet's and Linde's methods. His first objective was to liquefy oxygen—the creation of a bath of liquid oxygen being necessary for the liquefaction of other gases, particularly hydrogen. Kamerlingh Onnes vaporized oxygen, then liquefied it, and then forced it under pressure into a closed, circulating system. The system was bathed in gases that had achieved progressively lower temperatures than the circulating oxygen. This methodology proved successful and Kamerlingh Onnes was able to produce about 3.7 gal (14 l) of liquid air an hour.

The production of liquid helium and liquid hydrogen proved more difficult than the production of liquid air. Kamerlingh Onnes theorized that if he could begin from a point of normal pressure and liquefy oxygen by the application of immense pressure, then it would be possible to liquefy hydrogen, starting with the temperature of liquid oxygen. In 1892, he was midway through this painstaking process when Scottish chemist and physicist James Dewar succeeded in liquefying oxygen using a modified form of Pictet's cascade method. The process yielded about a pint of liquid oxygen.

Succeeds in Liquefying Helium

One practical advantage of Dewar's achievement was that Kamerlingh Onnes now had a source of cold with which to attempt to liquefy helium. He believed that if he started out from the freezing point of hydrogen—the lowest temperature to which it was possible to cool it—he could thereby liquefy helium using Linde's regenerative process. Kamerlingh Onnes constructed a system with a jacket of liquid hydrogen; the liquid hydrogen evaporated, which cooled the helium, and then the helium was forced under pressure through a small aperture that cooled it further, liquefying some of it. He then compressed the helium in a refrigerator, where it passed through an elaborate circuit surrounded by circuits of liquid hydrogen, which were themselves surrounded by liquid air, which was in turn surrounded by a flask in which warmed alcohol circulated.

In 1908, Kamerlingh Onnes finally succeeded in the long-elusive goal of liquefying helium. At first, he and his colleagues did not even notice what they had achieved: the liquid helium was colorless, and it was not until the circuit was almost full that they realized what had finally appeared before them. The accomplishment meant that a previously unattainable degree of cold was now at their disposal. Liquid helium was found to have a temperature of –451.8°F (–268.8°C), only about four degrees above absolute zero—absolute zero being a hypothetical temperature characterized by a complete absence of heat and equivalent to about –273.15°C or –459.67°F. Kamerlingh Onnes now set out to solidify the liquid helium in order to reach even lower temperatures, and in 1910, by **boiling** liquid helium under reduced pressure, he reached just over one degree above absolute zero.

Discovers Superconductivity at Low Temperatures

Kamerlingh Onnes used these temperatures to extend the range of his research into the properties of substances at low temperatures, and the results of these investigations were published regularly in English as "Communications from the Physical Laboratory at Leiden." In 1911, Kamerlingh Onnes made yet another breakthrough when he discovered the complete disappearance of electrical resistance in various metals at temperatures approaching absolute zero. The phenomenon of superconductivity is still not fully understood, although its implications are thought to be far reaching. Kamerlingh Onnes also discovered that the superconductor effect can be negated without changing the temperature by the application of a magnetic field.

Kamerlingh Onnes and his team remained preoccupied with the challenge of crystallizing helium. Their experiments yielded some intriguing if baffling results. In 1911, they found that the density of liquid helium peaked at a temperature of 2.2°K. When the various physical properties of liquid helium were measured, Kamerlingh Onnes discovered strange behavior in the helium in and around this temperature. Above 2.2K it was violently agitated, but at or below this temperature it seemed to lose its dynamic qualities. Kamerlingh Onnes was unable to explain this phenomenon. This was because he was attempting to understand it in terms of the classical laws of physics, but the behavior of liquid helium at these temperatures, unbeknownst to him, obeys the laws of **quantum mechanics**. Kamerlingh Onnes simply did not have the tools at his disposal to account for his findings. In the end, he and his colleagues put them down to some fault in their methodology and published only "definite and reliable" values for tem-

peratures above 2.2K, the position of the inexplicable maximum of the density of liquid helium.

Although he had yet to achieve absolute zero, in 1913 Kamerlingh Onnes was awarded the Nobel Prize in physics for his investigations into the properties of matter at low temperatures leading to the discovery of liquid helium. By 1921, Kamerlingh Onnes came within a degree of reaching absolute zero, and for the next three years he relentlessly pursued his quest. In 1926, he came across further peculiarities in the behavior of helium at 2.2K. This time Kamerlingh Onnes did not dismiss his findings as due to technical faults but began to seriously consider the possibility that some kind of fundamental change in helium occurred at this temperature. Unfortunately, although he was on the right track, Kamerlingh Onnes did not live long enough to resolve the mystery. His successor at Leiden, W. H. Keesom, came to the conclusion that helium above and below 2.2K is in fact two separate liquids, differing in fundamental ways. Keesom also completed another aspect of Kamerlingh Onnes's work: he succeeded in obtaining solid helium by cooling the liquid to about –457.6°F (–272°C) under pressure.

Kamerlingh Onnes was widely recognized for his work in low-temperature physics, and he received a number of awards in addition to the Nobel Prize. In 1904, he received the first of many distinctions when he was created Chevalier of the Order of the Netherlands Lion. In that same year, which was the twenty-fifth anniversary of his doctorate, his students and colleagues at Leiden issued a *Gedenkboek,* a survey of the work carried out at the laboratory from 1882 to 1904. A second *Gedenkboek* was issued in 1922, commemorating Kamerlingh Onnes's 40-year tenure as professor of experimental physics at Leiden. In 1912, he was awarded the Royal Society of London's Rumford medal, and four years later the society made him a foreign member. In 1923, he was elevated from chevalier to commander of the Netherlands Lion. Despite being known by his friends as "the gentleman of absolute zero," Kamerlingh Onnes died at Leiden on February 21, 1926, without ever having achieved it (German chemist Walther Nernst proved it was impossible to reach absolute zero in an experimental setting when he articulated the third law of **thermodynamics** in 1905). That same year, he was posthumously elected a corresponding member of the Prussian Academy of Sciences in Berlin.

Kant, Immanuel (1724-1804)

German philosopher

Considered one of the most important European philosophers of modern times, Immanuel Kant was not only a significant figure in the history of metaphysics, but also applied his keen intellect to fundamental issues in mathematics and physics. Born in Konigsberg, East Prussia (now Russia), Kant thrived in school at a young age and enrolled at the University of Konigsberg in 1740 as a theological student. Additionally, he was drawn to a wide range of central fields in philosophy and to the sciences, including mathematics, physics, and physical

Immanuel Kant.

geography. After graduation Kant tutored in private homes for several years. Later, Kant became a lecturer at Konigsberg, where he also received a master's degree in 1775.

During his lifetime, Kant received wide renown for his teachings and writings. One of Kant's first and most important scientific publications was a view of the **universe** called *General History of Nature and Theory of the Heavens* (1755). In it Kant hypothesized that the universe formed as the result of a spinning nebula, an idea that was later developed by French physicist and mathematician **Pierre-Simon de Laplace** and, eventually, disproved. However, Kant was on the mark with his proposal of the existence of many "island universes," such as our own **Milky Way**, which he described as a lens-shaped collection of **stars**. In the *Metaphysical Foundations of Natural Science* (1787), Kant stated that the study of nature must be grounded in mathematics. Although Kant is remembered for his efforts in metaphysical philosophy, he also contributed greatly to the philosophy of science. In his celebrated philosophical tome, *Critique of Pure Reason* (1781), which took 10 years to complete, Kant outlined his "Antinomies of Reason," or limits to understanding through scientific thought, which included Kant's arguments concerning philosophical issues such as the existence of God, free will, and immortality.

A small, stooped man who suffered from ill health throughout his life, Kant never traveled and spent his entire life in East Prussia. Nevertheless, his writings profoundly challenged many of the prevailing philosophical and scientific principles of his time. Kant's insights were applied to the physical sciences and to philosophical and religious movements such as transcendentalism. Kant died at the age of 79 and was buried in Konigsberg. His tombstone is inscribed "The starry heavens above me and the moral law within me," representing the two major focuses of his life.

See also Nebulae; Philosophy of physics

KAPITSA, PYOTR (1894-1984)
Russian physicist

Pyotr Kapitsa was a Russian physicist who directed the Mond Laboratory in England and the Physical Problems Institute in the Soviet Union during the 1930s and 1940s. He performed important research on high-density magnetic fields and high-energy plasma, but he is best known for his work on **low-temperature physics**, for which he shared the 1978 Nobel Prize in physics with **Arno Penzias** and Robert W. Wilson. Though many scientists believed that Kapitsa led the Soviet effort to develop an **atomic bomb** during World War II, Kapitsa, a longtime opponent of the Communist government, always denied the claim, and he is now generally not regarded as one of the bomb's designers.

Pyotr Leonidovich Kapitsa was born on July 8, 1894, in Kronstadt, an island near St. Petersburg, Russia. His father, Leonid Petrovich Kapitsa, was a lieutenant general in the engineers corps. His mother, Olga Ieronimovna Stebnitskaya, was an educator and a well-known folklorist. Kapitsa received his preparatory education in Kronstadt and then enrolled at the Polytechnic Institute of Petrograd. He married Nadeshda Tschernosvitova in 1916 and graduated in 1918 with a degree in electrical engineering. He became a lecturer at the institute, while developing a method of measuring **magnetism** with another student, Nikolai N. Semenov (who would one day win the Nobel Prize in chemistry). Their groundbreaking work was refined by **Otto Stern** in 1921, who himself won the Nobel Prize in physics in 1943.

Leaves Russia to Study in England

Kapitsa's research on magnetism at the Polytechnic Institute occurred in the midst of the collapse of czarist Russia and the turmoil of the Bolshevik Revolution. The country suffered widespread famine as a result of the civil unrest, and both his wife and their young child died. Kapitsa was repeatedly urged to study abroad, but his requests to do so were consistently refused by the new Communist government. Maxim Gorky, the Russian writer, interceded on his behalf, and in 1921 Kapitsa was allowed to leave for England where he worked with Nobel laureate **Ernest Rutherford** at Cambridge University's Cavendish Laboratory.

Rutherford initially balked at accepting Kapitsa, claiming that he had no room for another student. But Kapitsa was able to change Rutherford's mind with his innovative approach to research and his engineering background. His first experiments at the Cavendish Laboratory were on magnetic deflection of alpha and beta particles emitted by radioactive nuclei. He subsequently became involved in a study of high-density, temporary **magnetic fields**. Kapitsa developed an electric battery that discharged into a small copper magnetic coil, producing magnetic fields seven times stronger than any other previous mechanism had produced. These strong magnetic fields were the result of the short duration of the electric charges, that only lasted about 0.01 second each. This led Kapitsa to state that he was the world's highest paid physicist if his work were measured by the hour.

In 1923 Kapitsa received his doctorate from Cambridge University and won the James Clerk Maxwell fellowship, the first of his many awards. The following year he became the assistant director of magnetic research at Cavendish and shortly thereafter was made a fellow at Trinity College. Kapitsa seemed to feel at home in Cambridge and adopted many English habits—he rode a motor cycle, smoked a pipe, and wore tweeds.

In his work on high-density magnetic fields, Kapitsa devised ways of measuring how these fields affected the temperatures and properties of metals. It was this line of research that first interested him in the effects of **temperature**, and he began to work in low-temperature physics. Finding that the available supply of frigid liquefied gases was small, he invented his own apparatus for liquefying **helium**. The existing methods for liquefying helium, developed several decades earlier, produced only small quantities. Kapitsa developed a method that produced two liters an hour. His apparatus allowed the helium to expand rapidly, and the speed of the expansion caused it to cool before **heat** could be transferred into the helium chamber from the outside.

Kapitsa was elected a fellow of the Royal Society, the first foreigner to receive the honor in over 200 years, and in 1930 he became Messel Research Professor of the Royal Society. Rutherford, who considered Kapitsa his protégé, persuaded the society to build him a laboratory in which to study low-temperature physics. Endowed by German-born chemist and industrialist Ludwig Mond, the Mond Laboratory was custom-built for Kapitsa's experiments and opened in 1933. Kapitsa proved to be an industrious inventor and the new laboratory contained even more powerful magnets than he had developed earlier, as well as a **cryogenics** lab for low-temperature studies.

Remains in Russia after Detainment

By the time the Mond Laboratory opened, the Soviet government had repeatedly asked Kapitsa to return to his native country. He suggested he might if he was granted certain conditions, including the freedom to travel, but this was not acceptable to the government. In 1927 Kapitsa had married Anna Alekseevna Krylova, the daughter of a Soviet mathematician, and the couple frequently returned to the Soviet Union for personal and professional reasons. In 1934, however, at the end of his first year at the Mond Laboratory, Kapitsa and his wife left England for Russia, but once inside the country they were denied their exit visas. A series of confrontations

between Kapitsa and authorities in Russia followed, while Rutherford and others in England were unable to contact him and their inquiries to the government went unanswered. His wife was finally allowed to return to England for their two sons, only if Kapitsa remained, and by 1936 he had come to an agreement with them about his future.

In 1935 the Soviet government announced the creation of a new research facility, the S.I. Vavilov Institute for Physical Problems. For nearly a year Kapitsa refused the directorship of the institute because the government would not meet his conditions. By 1936 the government had relented and he accepted the position—though he still was not allowed to travel. Rutherford was disappointed by this turn of events; he had expected Kapitsa to succeed him as head of the Cavendish Laboratory. But Kapitsa's new position made him the leading force in Russian physics, and Rutherford arranged for the sale of the cryogenics equipment from the Mond Laboratory to the Soviet government. Thus, Kapitsa was able to continue his low-temperature research in the new laboratory in Russia.

The fact that Kapitsa had resisted Soviet leader Joseph Stalin's offers for a year placed him in a strong bargaining position, and his abilities at negotiation served him well under a totalitarian government. He not only developed one of the best physics staffs in the world, but he was able to acquire amenities for himself, such as custom-built living quarters that resembled a comfortable English cottage. He saw to it that the atmosphere of the new Physical Problems Institute was more Western and informal, devoid of the usual Soviet bureaucracy.

Low-Temperature Studies Lead to Superfluidity

In 1937 after the equipment dismantled from the Mond Laboratory was reassembled at the Physical Problems Institute, Kapitsa went back to low-temperature research. He focused on the properties of helium II, liquid helium that exists at temperatures hovering right above **absolute zero** (0Kelvin, or K). He measured the **viscosity** (thickness) of helium II when cooled below 2.17K. He coined the term "superfluid" to describe the behavior of helium II, which conducts heat extremely rapidly. His papers on **superfluidity**, which were published in 1938 and again in 1942, are considered his most important contributions.

Kapitsa's research was continued by **Lev Davidovich Landau**, who was able to describe helium II in a way that was consistent with **quantum mechanics**. Landau, who was head of the theoretical section at the institute, would win the Nobel Prize for this work in 1962. During the 1930s, the sharp-tongued Landau was arrested and charged with anti-Soviet activity and was even accused of being a Nazi spy despite the fact that he was Jewish. Kapitsa went to the Kremlin and told the authorities he would resign if Landau were not released. With this bold move, Kapitsa again demonstrated his strength of conviction and character in the face of authoritarian rule. Landau was released.

During World War II, many Western scientists believed Kapitsa was working on Russian development of the atomic bomb. However, it is known that he refused to cooperate with Lavrentii Beria—who was not only head of the Russian effort to develop an atomic bomb but also head of the secret police. The fact that Kapitsa was stripped of his directorship and placed under house arrest toward the end of the war corroborates his claim that he refused to participate in the making of the bomb. According to *Annual Obituaries,* "It was long rumored in Soviet scientific circles that Beria had wanted Kapitsa executed, but that Stalin refused." For the next decade he conducted experiments out of an improvised laboratory in his home. Kapitsa returned to the institute as director in 1955, two years after the death of Stalin. Kapitsa reportedly refused to have any KGB officers working at the institute and he regained the right to choose his own workers. As director, Kapitsa continued to conduct his own research as well as direct the research of others. In those postwar years he worked on the hydrodynamics of thin layers of fluid, the nature of ball lightning, and the standing **waves** produced by plasma.

The main focus of Kapitsa's research in the 1950s was high-power microwave **radiation** and **plasma physics**. Plasma, sometimes referred to as the fourth state of **matter**, is gas heated to temperatures so high that the molecules are stripped of their electrons and thus become electrified **ions**. Kapitsa found that a microwave generator can heat gas to an extremely high temperature—2,000,000K—over a very short space. His staff sought a way to use microwave radiation to heat plasma for thermonuclear fusion reactions, which require such high temperatures.

Kapitsa's skill and dedication as a teacher, as well as his administrative abilities, were almost as important to physics as the contributions of his research itself. He continued to improve the facilities at the Physical Problems Institute and he worked with most of the leading physicists in the Soviet Union. He remained an outspoken critic of Soviet bureaucracy, as well as Soviet science. He never joined the Communist Party and complained how the Soviet press was quick to cover scientific theories that agreed with Soviet dogma rather than those that demonstrated scientific validity. His English-style home, situated near the institute, was a gathering place for Soviet dissidents. He was friends with writers Aleksandr Solzhenitsyn and Andrei Sakharov, yet the government never placed him in the same category. He protested the pollution of Lake Baikal from nearby pulp mills, called for the abolition of the death penalty in Soviet Russia, and denounced the detainment of Soviet scientists.

In 1965 Kapitsa left the USSR for the first time in 31 years to receive the Niels Bohr International Gold Medal of the Danish Society and to lecture on high-energy physics. In 1966 he went back to England and toured his old laboratory, and several years later he and his wife visited the United States. In 1978, decades after his seminal work on helium II, he received the Nobel Prize in physics for his work in low-temperature physics. He shared the award with Americans Penzias and Wilson, who discovered cosmic microwave **background radiation**. In his later years Kapitsa was revered as a leader of Russian physics. He played chess, collected antique clocks, and watched his two sons become scientists. He died in Moscow at age 89 on April 8, 1984.

KARLE, ISABELLA (1921-   )
American chemist and physicist

Isabella Karle is a renowned chemist and physicist who has worked at the Naval Research Laboratory in Washington, D.C., since 1946 and heads the X-Ray Diffraction Group of that facility. In her research, she applied **electron** and x-ray **diffraction** to **molecular structure** problems in chemistry and biology. Along with her husband Jerome Karle, she developed procedures for gathering information about the structure of molecules from diffraction data. For her work, she has received numerous awards such as the Annual Achievement Award of the Society of Women Engineers in 1968, the Federal Woman's Award in 1973, and the Lifetime Achievement Award from Women in Science and Engineering in 1986. Her work has been described in the book *Women and Success.*

Isabella Lugoski Karle was born on December 2, 1921, in Detroit, Michigan. Her parents were Zygmunt A. Lugoski, a housepainter, and Elizabeth Graczyk, who was a seamstress. Both her parents were immigrants from Poland, and Karle spoke no English until she went to school. While still in high school, she decided upon a career in chemistry, even though her mother wanted her to be a lawyer or a teacher. She received her B.S. and M.S. degrees in physical chemistry from the University of Michigan in 1941 and 1942. Determined to continue her studies, Karle ran into serious financial problems since teaching assistant positions at the University of Michigan were reserved exclusively for male doctoral students. She managed to stay in school on an American Association of University Women fellowship, however, and in 1943 also became a Rackham fellow. She received her Ph.D. in physical chemistry from the University of Michigan in 1944, at the age of 22.

After receiving her doctorate, Karle worked at the University of Chicago on the **Manhattan Project** (the code name for the construction of the **atomic bomb**), synthesizing plutonium compounds. She then returned to the University of Michigan as a chemistry instructor for two years. In 1942 she had married Jerome Karle, then a chemistry student. In 1946 she and her husband joined the Naval Research Laboratory, where she worked as a physicist from 1946 to 1959. In 1959 she became head of the x-ray analysis section, a position she maintained through the 1990s.

Investigates Structure of Crystals

When Karle began her work at the Naval Research Laboratory, information about the structure of crystals was limited. Scientists had determined that crystals were solid units, in which atoms, **ions**, or molecules are arranged sometimes in repeating, sometimes in random, patterns. These patterns or networks of fixed points in **space** have measurable distances between them. Although chemists had been able to investigate the structure of gas molecules by studying the diffraction of electron or x-ray beams by the gas molecules, it was believed that information about the occurrences of the patterns—or phases—was lost when crystalline substances scattered an x-ray beam. The Karles, working as a team, gathered phase information using a heavy-atom or salt derivative. The position of a heavy **atom** in the crystal could be located by scattered x-ray reflections, even though light atoms posed more serious problems. When a heavy atom could not be introduced into a crystal, its structure remained a mystery. In 1950 Jerome Karle, in collaboration with chemist Herbert A. Hauptman, formulated a set of mathematical equations that would theoretically solve the problem of phases in light-atom crystals. It was Isabella Karle who solved the practical problems and designed and built the diffraction machine that photographed the diffracted images of crystalline structures.

While investigating structural formulas and the make-up of crystal structures using electron and x-ray diffraction, Karle made an important discovery. She found that only a few of the phase values—no more than three to five—are sufficient to evaluate the remaining values. She could then use symbols to represent these initial values and also numerical evaluations. This process for determining the location of atoms in a crystal was amenable to processing in high-speed **computers**. Eventually, it became possible to analyze complex biological molecules in a day or two that previously would have taken years to analyze. The rapid and direct method for solving crystal structure resounded through chemistry, biochemistry, biology, and medicine, and Karle herself has been active in resolving applications in a range of fields.

In addition to describing the structure of crystals and moleules, Karle also investigated the conformation of natural products and biologically active materials. After a crystallographer determines the chemical composition of rare and expensive chemicals, scientists can synthesize inexpensive substitutes that serve the same purpose. Karle headed a team that determined the structure of a chemical that repels worms, termites, and other pests and occurs naturally in a rare Panamanian wood. The team was then able to produce a synthetic chemical that mimics the natural chemical and is equally effective as a pest repellent. In another application, Karle studied frog venom. Using extremely minute quantities of purified potent toxins from tropical American frogs, the team headed by Karle established three-dimensional models, called stereoconfigurations, of many of the toxins and showed the chemical linkages of each of these poisons. The inexpensive substitutes of the toxins were of great importance in medicine. The venom has the effect of blocking nerve impulses and is useful to medical scientists studying nerve transmissions. Karle has also researched the effect of **radiation** on deoxyribonucleic acid (DNA), the carrier of genetic information. She demonstrated how the structural formulas of the configurations of amino acids and nucleic acids in DNA may be changed when exposed to radiation. Her research into structural analysis also established the arrangement of peptide bonds, or combinations of amino acids.

Karle has held several concurrent positions, such as member of the National Committee on Crystallography of the National Academy of Science and the National Research Council (1974–77). She has long been a member of the American Crystallographic Society and served as its president

in 1976. She was elected to the National Academy of Sciences in 1978. From 1982 to 1990 she worked with the Massachusetts Institute of Technology, and she has been a civilian consultant to the Atomic Energy Commission.

Karle has received numerous awards including the Superior Civilian Service Award of the Navy Department in 1965, the Hildebrand Award in 1970, and the Garvan Award of the American Chemical Society in 1976. She has received several honorary doctorates. Her most recent awards have been the Gregori Aminoff Prize from the Swedish Academy of Sciences in 1988 and the Bijvoet Medal from the University of Utrecht, the Netherlands, in 1990. She has written over 250 scientific articles.

Isabella and Jerome Karle have three daughters, Louise Isabella, Jean Marianne, and Madeline Diane. All three have become scientists like their parents. Jerome Karle, who is chief scientist at the Laboratory for Structure and Matter of the U.S. Naval Laboratory, received the Nobel Prize in chemistry in 1985 for developing a mathematical method for determining the three-dimensional structure of molecules.

See also Crystallography

KELVIN, SIR WILLIAM THOMSON (1824-1907)
Irish professor of natural philosophy

William Thomson Kelvin was born on June 26, 1824, in northern Ireland. He died on December 17, 1907, in Scotland. He was also known as Sir William Thomson, Baron William Thomson Kelvin of Largs, and Lord Kelvin. He was the son of James Thomson, a professor of engineering and mathematics at Belfast. Thomson was a major influence in the fields of physics and engineering during his time. His work on **thermodynamics**, **electricity**, and **magnetism** helped form the basis of modern physics.

Thomson's accomplishments were probably due, in part, to his personality. He was an active man, seeking answers and excitement and finding no rest until his curiosity was satisfied. He enjoyed traveling and rowing. He even risked his own life more than once while assisting in the laying of the first transatlantic cable. Thomson began his education at Glasgow, where he received numerous university awards. The professor of natural philosophy at the time, William Meickleham, had a profound influence on Thomson's career as a physicist. It was at Glasgow where Thomson discovered a love of mathematics and a respect for physical science. He was trained to question theories constantly and to work continuously to solve problems. Meickleham lent Thomson a copy of **Jean-Baptiste-Joseph Fourier**'s book *The Analytical Theory of Heat*, introducing the young student to abstract mathematics and the controversial theory of **heat** flow. Fourier's ideas became a driving force in Thomson's research career. At the age of 15, he received a gold medal for his article entitled *An Essay on the Figure of the Earth*. In writing this essay, Thomson demonstrated mathematical skills well beyond his

William Thomson Kelvin.

years. He referred to this essay quite often during his lifetime and used it as a basis for many of his theories.

After he matriculated in 1834, Thomson attended Cambridge. While at Cambridge, he published his first two articles, both defending Fourier's theories of heat flow, which were not widely accepted at the time. Shortly before graduation, he was given another book, George Green's (1793-1841) *An Essay on the Application of Mathematical Analysis to the Theories of Electricity and Magnetism.* This book, along with Fourier's, shaped the way Thomson regarded natural science, and as a result, his life's work. After graduating from Cambridge with high honors he traveled to Paris where he studied electricity under the direction of Henri-Victor Regnault (1810-1878). Working with Regnault, Thomson gained practical knowledge to complement the theoretical knowledge he obtained at Cambridge. After several years of study, he returned to Scotland at the age of 22 to replace his mentor, Meickleham, as the professor of natural philosophy. Thomson held this position for the rest of his career, until he retired at the age of 75.

Thomson's main assertion was that each of the numerous theories describing both **matter** and **energy** could be unified and described by one all-encompassing theory. He did not know if this theory would ever be articulated; however, all of his research was driven toward the goal of unification. One main tenet in Thomson's work was the demonstration that

energy could be converted into different forms. **James Prescott Joule** performed experiments showing that mechanical energy could be converted into heat energy, the basis for thermodynamics. Joule's work strongly influenced Thomson, who believed that electricity, magnetism, **light**, and energy were all interrelated. Thomson supported Joule's theory when he published his work, *On the Dynamical Theory of Heat.* In this essay, Thomson described his version of the **second law of thermodynamics**, bringing the unification of theories closer to reality.

Thomson himself influenced scientist **James Clerk Maxwell**. Thomson was Maxwell's mentor, and Maxwell used Thomson's ideas as the basis for his research. Maxwell developed his electromagnetic theory of light, demonstrating the interrelationship between electricity, magnetism, and light, based on Thomson's early research at Cambridge. Thomson also furthered the early theories of Fourier, Joule, and **Michael Faraday**. He was able to see the connections between individual experiments performed by these scientists in order to make generalizations about the **dynamics** of energy. Thomson was the first to show the similarity between heat flow in solid objects and electrical flow in conductors. He was also one of the first scientists to develop an absolute **temperature** scale, one in which the value of one degree is independent of temperature. His contributions to the fields of thermodynamics and electricity were critical to the development of the unifying theory he first imagined. This unifying theory, later developed as the laws of **conservation of energy** and of matter, was used to develop the absolute temperature scale used by scientists today, known as the Kelvin scale in Thomson's honor.

Thomson's work went well beyond that of theoretical electricity and thermodynamics. He also provided practical use of his theories, for example, when he worked on the first transatlantic cable project. He was first consulted to provide information regarding the rate of **electric current** passing through this long (3,000 mi [4,800 km]) cable. A delay was noted in the passing of current through the cable, and Thomson was asked to help solve this problem. Thomson demonstrated mathematically how this delay could be solved. The chief electrician for the project, E. O. W. Whitehouse, rejected Thomson's explanation, stating that his own practical experience outweighed the theoretical musings of Thomson. Whitehouse's plans were accompanied by numerous problems, and eventually Thomson's ideas were used. Thomson patented his first telegraph receiver, a mirror galvanometer, in 1858, which was eventually used on the transatlantic cable (only after Whitehouse was fired). Queen Victoria knighted Thomson in 1866 for his work on this project.

Thomson influenced the study of geology as well. He was the first to hypothesize that the **Sun** was slowly cooling down. He imagined early **Earth** underneath a sun with a much higher temperature than we know today. He stated that the temperature of Earth must also have been much greater, and that the planet was characterized by severe storms and completely different plant life. These theories opposed much of the theory of evolution proposed by Charles Darwin (1809-1882) and upset his scientific followers. Thomson had incorrectly estimated the age of Earth and the Sun, but had demonstrated that geologic theory followed the established theories of physics. Thomson also owned numerous patents on everything from an analog computer used to measure tides to electrical devices used to make measurements. He published a textbook as well, titled *Treatise on Natural Philosophy*. This textbook was used by and influenced modern physicists. Thomson was elected as a fellow of the Royal Society in 1851. He was the president of the Royal Society from 1890 to 1895. He published hundreds of papers in his lifetime, influencing physicists and engineers for years after his death.

KELVIN TEMPERATURE SCALE

The Kelvin **temperature** scale is based upon the kelvin, a unit of thermodynamic temperature, that is defined as the fraction 1/273.16 of the thermodynamic temperature of the triple point of water. The triple point is the temperature at which all three phases of water, solid, liquid, and gas, can coexist. The kelvin temperature scale is named after Scottish physicist **William Thomson Kelvin** who first described an absolute temperature scale based upon the thermodynamic defintions of **absolute zero**. The kelvin and the degree Celsius are both units of the International Temperature Scale of 1990 (ITS-90) adopted by the International Committee for Weights and Measures and the General Conference on Weights and Measures (CGPM, Conférence Générale des Poids et Mesures). When referring to the Kelvin temperature scale, the kelvin (symbol K) should be used instead of the term "degree Kelvin" (°K).

The Kelvin temperature scale is an absolute temperature scale used in scientific research. It is the official unit of measurement for temperature according to the Système Internationale of units, or SI units. The Kelvin scale is also known as the thermodynamic scale. According to the Kelvin temperature scale the temperature 0K represents the lowest possible temperature, called absolute zero. Accordingly, the Kevin scale is an absolute temperature scale.

By using the thermodynamic temperature of the triple point of water and assigning it the temperature of 273.16K, the difference from this temperature becomes the Celsius temperature of a body or substance. Accordingly, the thermal unit on the Kelvin temperature scale is identical to the thermal value of a unit on the Celsius temperature scale. Although the Kelvin temperature scale is one of four major temperature scales (the other major scales include the Celsius, Rakin, and Fahrenheit scales), the Kelvin scale is the most widely used and internationally accepted scale utilized by physicists.

Temperature is not direct measure of the **heat** energy of a material. Temperature is independent of the amount of material present, whereas the heat **energy** of an object does depend on the **mass** of the object.

Although temperature is not a measurement of the amount of heat energy in a substance, it is related to the **kinetic energy** of the substance. Temperature is proportional to the average kinetic energy of the molecules in an object. The heating of an object reflects a gain in molecular kinetic energy. As

their **motion** increases, so does the number of collisions the molecules make with each other. These collisions cause the kinetic energy to be transformed into heat energy. The more energetic the molecules, the more energetic their collisions become, and the greater the temperature of the object.

As the kinetic energy of the molecules in a substance decreases, the temperature of the substance decreases as well. As the molecules slow down they do not completely stop. As their kinetic energy approaches zero, the temperature of the substance also approaches a lower limit. This limit is called absolute zero, or 0K (-273.15°C and -459.67°F). This small amount of kinetic energy is called the zero-point energy.

Although it is theoretically impossible to experimentally reach absolute zero, scientists have come increasingly close to this value in laboratory experiments. Such extremely cold temperatures can be reached using techniques involving large **magnetic fields**. Researchers at the University of Colorado, the Massachusetts Institute of Technology, and Stanford University have lowered the temperature of atoms to within a millionth of a kelvin of absolute zero.

See also Temperature and its measurement; Thermodynamics

KENDALL, HENRY W. (1926-   )
American physicist

In awarding the 1990 Nobel Prize in physics to Henry W. Kendall, Richard Taylor, and **Jerome Friedman**, the Royal Swedish Academy of Sciences recognized the importance of the discovery of **quarks**. Building upon Nobel laureate **Ernest Rutherford**'s research into the structure of the **atom**, Kendall and his colleagues utilized the newly invented **particle accelerator** to explore the interior of protons and neutrons. Their results proposed that these components of atoms are composed of even smaller particles called quarks and that quarks are bound together by massless particles known as gluons.

Henry Way Kendall was born on December 9, 1926, in Boston, Massachusetts, the oldest of three children. His parents were Henry P. Kendall, a businessman, and Evelyn Way Kendall. Kendall completed his secondary education at the Deerfield Academy in western Massachusetts in 1945 and immediately entered the U.S. Merchant Marine Academy. He stayed at the academy one year before resigning in order to enroll at Amherst College in Amherst, Massachusetts.

Follows in Hofstadter's Footsteps at Stanford

Although he was interested in a wide variety of subjects, Kendall decided to major in mathematics and earned his bachelor of arts in 1950. He conducted his graduate work at the Massachusetts Institute of Technology (MIT) and was awarded his Ph.D. in nuclear and **atomic physics** in 1954. After two years as a postdoctoral fellow at the Brookhaven National Laboratory on Long Island, Kendall was appointed a research associate in a group headed by **Robert Hofstadter** at Stanford University's High Energy Laboratory.

At Stanford Kendall met two other researchers, Richard Taylor and Jerome Friedman, with whom he collaborated over the next decade. Their first project was to follow up on research begun in the early 1950s by Hofstadter—research that had won Hofstadter the 1961 Nobel Prize in physics. Hofstadter's research involved the bombardment of various atomic nuclei with high energy **electrons**. By studying the way electrons were reflected, Hofstadter was able to determine the structure of the nuclei. He found that nuclei consist of a central core surrounded by two outer shells of particles called **mesons**.

The work begun by Kendall and his colleagues in 1956 was not expected to yield anything especially remarkable. They assumed that they might get a better look at the **nucleus**, but they would not obtain new qualitative results. This assumption proved true in their earliest experiments, but soon two new factors changed the situation. Kendall, Taylor, and Friedman had gained access to a new particle accelerator—the two-mile long, 20 billion electron volt linear accelerator (linac)—at Stanford. The Stanford linac gave the researchers a far more powerful electron-beam probe than Hofstadter had ever had.

Another factor that influenced their research was a suggestion by a Stanford colleague, James Bjorken, who proposed that the team study inelastic collisions as well as the elastic collisions on which the team had been concentrating. In an elastic collision, an incoming electron strikes an atomic nucleus and bounces off with no loss of energy. By focusing on this type of reaction, Kendall and his colleagues had improved the precision of Hofstadter's findings, but had made no new breakthroughs. In an inelastic collision, an incoming electron strikes a nucleus with enough **force** to blow it apart. Bjorken argued that inelastic collisions were really the best way to get a close look at the protons and neutrons that make up the nucleus.

Bjorken's Suggestion Leads to the Discovery of Quarks

Kendall and his associates redesigned their equipment to detect inelastic collisions and found that protons and neutrons are composed of tiny, apparently solid subparticles. These subparticles appeared to be the quarks that had been postulated a few years earlier by physicist **Murray Gell-Mann** as the fundamental particles of which **matter** is composed. The results of their studies were so revealing that the team was even able to identify other particles—later called gluons—that appear to hold quarks together within protons and neutrons.

When it was announced that the 1990 Nobel Prize in physics had been awarded to Kendall, Taylor, and Friedman, the decision was met with wide approval in the scientific community. **Burton Richter**, director of the Stanford Linear Accelerator Center, was quoted in *Science* as saying "the only question in my mind is, why did it take so long?"

Kendall was appointed assistant professor at MIT in 1961 and became a full professor of physics in 1967. During this decade, he commuted from coast to coast so that he could teach and conduct research at MIT as well as continue his long-term work on quarks at Stanford. Kendall has long been involved in efforts to engage scientists in the debate over how their discoveries and inventions should be used in society. He was a founding member of the Union for Concerned Scientists in 1969 and served as the organization's chairperson for many years. He has also been an active member and officer of the Arms Control Association and a member of the board of directors of *The Bulletin of the Atomic Scientists.*

KEPLER, JOHANNES (1571-1630)
German astronomer

Johannes Kepler, born in Weil in Würtemberg (now southwestern Germany) in 1571, seemed destined for a life in the church. He obtained a B.A. degree in theology in 1588 and entered the University of Tübingen, one of the great centers of Protestant learning, in 1589. He graduated with an M.A. in 1591.

While at Tübingen, Kepler had learned mathematics, and when the high school in Graz, Austria, needed a mathematics teacher, he got the job. He turned out to be very poor at teaching; during his first year only a handful of students attended his lectures, and the following year he had no students at all. That gave him considerable free time to further his interest in astronomy and produce annual almanacs, which were often heavy on astrological content. He also raised additional income by casting horoscopes.

Kepler, always the mystic, published a book in 1596 in which he devised a relationship involving the distances of the **planets** from the **Sun** with geometric solid objects such as cubes and spheres. His *Mysterium Cosmographicum* (Mystery of the **universe**) showed considerable knowledge of astronomy and brought him to the attention of **Tycho Brahe** and **Galileo**.

Beginning in 1597, life for Protestants in the Catholic-dominated area in which Kepler lived became unpleasant due to religious persecution, so Kepler moved to Prague (Czech Republic) in 1600, where he began working with flamboyant Tycho Brahe.

Brahe was the antithesis of Kepler: Brahe had excellent eyesight that he used to compile the most detailed observational data in history; Kepler's eyesight was poor and he suffered from ill health all his life. Brahe was very well-off financially, and his extravagance drove Kepler, who always seemed to be in reduced circumstances, to distraction. Their working relationship was not the best, but Brahe died within 18 months of Kepler's arrival.

Brahe had spent years making accurate observations of Mars with the naked eye and assigned Kepler the task of devising a theory of planetary **motion** using his observational data. Kepler, the mathematician, was superbly suited for this task, which would end up occupying the majority of his time for the next 20 years.

In 1604, before he had made much progress with Mars, a supernova blazed into view, and Kepler wrote two pamphlets about it. The supernova, like Brahe's Star in 1572, rivaled Venus in brightness and has since come to be known as

Kepler's Star. He also wrote about applications of **optics** in astronomy and proposed a design for a **telescope**. After Galileo discovered the moons of Jupiter, Kepler used a telescope to prove to himself that they did, indeed, exist. He dubbed them satellites, a name that stuck.

The task with Mars was proving to be extremely difficult. The circular orbit Kepler calculated did not agree exactly with Brahe's observations. Kepler's creative imagination came up with theory after theory to account for the discrepancy; at one point, after three years of work, he had a geometrical scheme that disagreed with one observation by only 8 minutes of arc. (The full **Moon** is 31 minutes of arc in diameter; two objects 4 minutes of arc apart are barely noticeable to a person with average eyesight.) Kepler could not believe that Brahe's observations were anything less than perfect. He threw his scheme out and started again.

Kepler gave up on circles and **epicycles** and, out of desperation, tried working with an ellipse (oval). The results matched Brahe's data perfectly. Then he had to devise a law governing the variation of the speed of Mars as it moved along the ellipse. Here he got bogged down and lost his way, but he eventually formulated a simple law that matched observations. Finally, in 1609, Kepler published his first two laws of planetary motion: a planet orbits the Sun in an ellipse, not a circle as **Nicholas Copernicus** had believed; a planet moves faster when near the Sun, and slower when farther away. Kepler thought, incorrectly, that **magnetism** in the Sun was responsible for the variation.

The third law was a long time in coming. In 1619 Kepler determined that the square of the time it takes a planet to orbit the Sun is equal to the cube of its average distance. In other words, once you know how long it takes a planet to complete an orbit, you can calculate its relative distance from the Sun. However, you still have to have a definite measured distance for one planet to act as a yardstick to determine the distance of the others.

It seems likely that he happened on this formula, not by mathematical calculation, but by accident. He was constantly looking for mystical relations of numerical sequences to explain the "harmony of the heavens." In the same book in which the third law appears, Kepler devotes space to the "music of the spheres," assigning individual musical notes that each planet "sings."

Also in 1619, he published a book on **comets** in which he supports Brahe's contention that comets were celestial objects and not manifestations of Earth's atmosphere. Kepler incorrectly believed them to be objects that moved in straight lines, but he had a remarkably accurate explanation of the Sun's part in producing a comet's tail.

Once again, religious persecution of Protestants caused Kepler to move, and he relocated to Ulm (Germany) in 1626. One year later he published his final great work: the *Rudolphine Tables*, a tabular collection of the motions of planets, dedicated to his former patron Emperor Rudolph II. They were used as the standard astronomical tables for the next century. Kepler died after a short illness on November 15, 1630, at the age of 59.

Johannes Kepler.

KEPLER'S LAWS OF PLANETARY MOTION

Kepler's three laws are geometric relationships that describe the motions of the **planets** in the solar system. German astronomer **Johannes Kepler** derived them in the early 1600s, with the help of more than two decades' worth of detailed observations by Danish astronomer **Tycho Brahe**. Kepler worked as Tycho's assistant for the last 18 months of Tycho's life. Much of Tycho's work described the position of Mars in its orbit around the **Sun**.

Kepler was one of the first people to attempt to explain why planets move the way they do around the Sun. When Kepler joined Tycho's crew of assistants, he used Tycho's data to confirm his intuitions about the **motion** of Earth—that our planet is not privileged, but rather, moves about the Sun much like all the other planets. While working with Tycho's stacks of records on Mars, Kepler devised his laws of planetary motion. The first two were published in 1609, five years after he originally conceived them. Kepler published the third law in 1618. Together, these laws laid the groundwork for Sir **Isaac Newton**, whose laws of motion and universal gravitation were the foundation of modern physics. **Newton's law of universal gravitation** was mathematical proof of Kepler's laws.

The three laws of planetary motion are: 1) Planets revolve around the Sun in elliptical orbits. The Sun sits at one focus of the ellipse, rather than the center of a circular orbit as once thought. 2) Each planet moves so that a line connecting the planet and the Sun sweeps out equal areas of **space** in equal periods of time. In other words, a planet moves more quickly in its orbit when it is closer to the Sun. Kepler came upon this law after painstakingly calculating Mars' distance from the Sun at every degree of its orbit. He used Tycho's meticulously recorded observations for this work. From his calculations, Kepler deduced that Mars couldn't possibly revolve around the Sun in a circular orbit. Instead, the planet must revolve around the Sun in an orbit shaped like an ellipse. Thus, the first law of motion was born. (Kepler himself never numbered the laws.) 3) The square of a planet's period of orbit around the Sun is directly proportional to the cube of a planet's average distance from the Sun. That is, d^3/T^2 is the same for all planets.

Kepler's laws are true for any satellite that orbits a body—including artificial satellites scientists launch from Earth. As such, the laws are useful in predicting the motion of one object around another.

KEPLER'S SECOND LAW AND ANGULAR MOMENTUM

German astronomer **Johannes Kepler** is credited with demonstrating that abandoning an Earth-centered, or Ptolemaic, view of planetary **motion** for a Sun-centered, or Copernican, model implied that the motion of the **planets** had to be clearly elliptical. He reached this conclusion by analyzing the careful observations of his mentor, **Tycho Brahe**, on the motions of the heavenly bodies as well as his own observations of the orbit of Mars, which showed that planetary motion did not follow the path of a circle. From this insight, Kepler generated his three empirical laws of planetary motion. The first, the law of elliptical orbits, states that the orbits of the planets are ellipses with the **Sun** at one focus. The second, the law of equal areas, states that a line from the planet to the Sun sweeps over equal areas in equal intervals of time, and the third, the law of harmonies, states that the square of the period of revolution of a planet about the Sun is proportional to the cube of the semi-major axis of the planet's elliptical orbit. Kepler's laws imply a nonuniform speed of revolution for planets as they orbit around the sun; they also imply that their **velocity** also changes throughout the planet's year, being faster when the planet is close to the Sun and slower when the planet is far away. The theoretical framework of Kepler's laws was finally derived some 30 years after his death when **Isaac Newton** formulated his law of universal gravitation.

Kepler's second law is often taken to be equivalent to the statement of conservation of **angular momentum**, which states that the total angular **momentum** of a closed system remains constant. The law of equal areas describes the constantly varying speed at which any given planet moves while orbiting the Sun under the influence of **gravity**. This can be visualized by drawing an ellipse to describe the orbit of the planet around the Sun. If we draw a line from the center of the planet to the Sun, and another one after it has moved along the elliptical orbit for a given period, we obtain a triangle with a given surface area. Similarly, another such triangle can be drawn after the planet has moved an equal amount of time along the orbit. But the triangles will have different shapes, because the orbit is not circular, thus, the distance from the planet to the Sun will vary and so will the length of the sides of the triangles. With this approach, the areas of the triangles formed when Earth is close to the Sun yield short but wide triangles and the areas formed when Earth is far from the Sun are represented by long, narrow triangles. If the areas of all such possible triangles are calculated, Kepler's second law states that they are all the same size. Since the bases of these triangles are shorter when Earth is farthest from the Sun, Earth would have to be moving more slowly in order for this imaginary area to be the same size as when Earth is closest to the Sun.

Kepler's second law is related to the conservation of angular momentum as follows: If we describe a planet as experiencing rotational motion under the influence of gravity, we can use a vector quantity called **torque** (τ), which is produced by any **force** applied to a body that causes the body to rotate. Mathematically, this can be expressed as: $\tau = r \times F$, which is the cross product of **r**, the vector starting from the axis of rotation to where the force is applied and of the force applied, in this case, gravity. In this equation, F is given by Newton's universal law of gravitation, which states that whenever there are two objects that have a **mass**, they will exert a **gravitational force** on each other proportional to the product of their masses, and inversely proportional to the square of the distance between the two centers of mass. This is expressed as:

$$F = -(Gm_1m_2/r^2)r$$

in which F is the gravitational force, m_1 and m_2 are the respective masses of the two bodies separated by a distance r, G is the gravitational constant and r is a unit vector directed from the first mass to the second. Substituting in the expression for torque yields:

$$\tau = (rr)\ F = (rr) \times -(Gm_1m_2/r^2)r$$

Rearranging, $\tau = (r \times r) \times (Gm_1m_2/r)$. Since the cross product of a vector with itself is zero, the torque expression also reduces to zero, which is equivalent to stating that the sun does not exert a torque on an orbiting planet. The torque of a planet can also be defined as the instantaneous rate of change of its angular momentum (l) or: $\tau = dl/dt$. If the Sun does not exert a torque on a planet, it follows that dl/dt must also be a zero vector, which means there is no change in angular momentum; angular momentum remains constant, i.e., it is conserved.

Kepler derived these three laws strictly from observation as calculus had not yet been invented in his time. It is also interesting to note he was able to draw his conclusions because he happened to be observing Mars, the only planet visible to the naked eye with an eccentricity large enough so as to allow one to distinguish its elliptical orbit from a circular one with the rudimentary observation instruments available at the time.

KINETIC ENERGY

Energy is one of the central themes in science. There are many forms of energy, each with its own specific definition. Any form of energy can be converted into one of the other forms. This is the law of **conservation of energy**, which states that energy is conserved in any system. A general definition for energy is the ability to do **work**. This definition is oversimplified, for not all forms of energy can actually do work. It does, however, demonstrate the important relationship between **energy and work**.

Work is performed whenever a **force** moves an object. The amount of work done can be calculated by multiplying the magnitude of the force acting on an object by the distance the object moves. Using the English system of measurement, the force is measured in pounds (lb) and the distance is measured in feet (ft). The product of the two, or the work done, is then measured as foot-pounds (ft-lb). With the metric, or SI (Système Internationale) system, the unit of force is the newton (N) and the unit of distance is the meter (m). The resulting unit for work is the newton-meter (N-m), otherwise known as the joule (J). One joule is equal to 0.7376 ft-lb. The unit for work, the joule, is the same as the unit for energy. If work is performed using rotation, the work performed is the product of **torque** and the angle of rotation. Work and energy are scalar quantities.

An object in **motion** can do work on another object by exerting a force on the second object, moving it through a distance. Because an object in motion has the ability to do work, it has energy. This energy of motion is called kinetic energy, from the Greek word *kinetikos*, meaning motion. Mathematically, kinetic energy is defined as one-half of the product of the **mass** and the square of the **velocity** of a moving object. The kinetic energy of an object is therefore dependent on the speed of the object as well as the mass. The faster an object is moving, the more kinetic energy it possesses. The more massive an object, the more kinetic energy it will have. For example, if two bowling balls were rolled across the floor at different rates, the ball with the higher rate would have the most kinetic energy. If a bowling ball and a Ping-Pong ball are rolled along the floor at the same speed, the bowling ball will have more kinetic energy than the Ping-Pong ball because it has more mass. A group of objects has a total kinetic energy equal to the sum of the individual kinetic energies.

When an object with mass m (in kilograms) is moving in a straight line at a velocity v (in m/s), the kinetic energy (KE) of the object can be calculated using the following equation: $KE = 1/2mv^2$.

This form of kinetic energy, that of an object moving in a straight line, is called translational kinetic energy. For example, suppose a soccer player kicks the soccer ball across the midfield, and the ball is traveling at 5 m/s. If the mass of the soccer ball is 0.450 kg, then the kinetic energy is $(1/2)(.450$ kg$)(5$ m/s$)^2$, which equals 5.625 J.

When a moving object collides with another object, and no **heat** is produced in the collision, the kinetic energy is conserved. The total kinetic energy of the two objects is the same before and after the collision. If one object loses kinetic energy in the collision, the other object gains kinetic energy. A collision in which the total kinetic energy is conserved is called an elastic collision. In an elastic collision, the two objects return to their original shapes and move off separately after the collision. An example of an elastic collision is that between two billiard balls.

When a moving object collides with another object, producing heat, the kinetic energy is not conserved. These collisions are called inelastic collisions. The kinetic energy lost is converted into other forms of energy, usually heat and **sound**. The total energy of the system is conserved. In an inelastic collision, the objects change shape and sometimes stick together after the collision. An example of an inelastic collision is that between two automobiles.

When a moving object collides with another object, causing the second object to move, it does work on the object. The net work (W_{net}) done on an object in a collision is equal to the change in its kinetic energy (ΔKE). In equation form, $W_{net} = \Delta KE$. This is known as the work-energy theorem. If work is done on an object, the object's kinetic energy increases. This increase in kinetic energy is equal to the work done. For example, if a golf club collides with a golf ball, sending the golf ball down the fairway, the club has done work on the ball. At the same time, the ball has gained kinetic energy. The kinetic energy gained by the ball is equal to the work done on it by the club. If an object does work on another object, the object doing the work loses kinetic energy. This decrease in kinetic energy is also equal to the work done. The golf club slows down after it collides with the ball, so it loses kinetic energy. The amount of kinetic energy lost is equal to the work done on the ball. The work-energy theorem, then, shows that work is a method of energy transfer.

There are several forms of kinetic energy besides translational kinetic energy. Rotational, thermal, relativistic, and orbital energies are all forms of kinetic energy. Rotational kinetic energy is the energy possessed by an object rotating about an axis. The equation for rotational kinetic energy is $KE = 1/2I\omega^2$, where I is the moment of **inertia** of the object and ω angular velocity. Rotational kinetic energy is changed when a torque is applied to the object. **Temperature** is a measure of the average kinetic energy of the molecules of a substance, or the thermal kinetic energy. As the kinetic energy of molecules increases, the temperature increases. In other words, the faster the motion of the molecules of a substance, the higher the temperature will be. Relativistic kinetic energy describes the kinetic energy of objects at extremely high speeds and is described using Einstein's equation $E = mc^2$, where E is the energy of an object, m is the mass of the object, and c is the speed of **light**. An object in orbit, such as a satellite around a planet or a planet around the **Sun**, has kinetic energy that depends on the size of the orbit. For elliptical orbits, this kinetic energy will not remain constant throughout the orbit but the sum of kinetic and **potential energy**, called the mechanical energy, will remain constant.

KINETIC MOLECULAR THEORY

Several properties of a gaseous sample can be explained readily by the features of a model of gas behavior called the kinetic molecular theory or simply kinetic theory. This theory views the molecules as very small, very hard spheres that travel constantly and randomly at high speed.

Although scientists now think of kinetic molecular theory as a very simple model of the behavior of atoms and molecules in the gas phase, it took over 100 years for it to be accepted by the scientific community. **Daniel Bernoulli** published an article describing a theory on this subject in 1743. However, it wasn't until the latter half of the next century that a statistical view developed by **James Clerk Maxwell** and **Ludwig Boltzmann** convinced physicists and chemists that individual molecules in a gas could be viewed as independent, rapidly moving bodies.

The kinetic molecular theory is important to chemists and other scientists who need a molecular viewpoint in their work—biologists, material scientists, geologists, chemical engineers, medical scientists, environmental scientists, and others. They do not use this theory to perform calculations or predict properties of gases but apply it constantly in their consideration of how gaseous substances behave because it provides the most fundamental picture of **matter** at the molecular level. If we think of molecules as particles in constant agitation—bumping into each other at high **velocity** and exchanging energy—we develop a view that enables us to correlate a vast array of observations: the expandibility and compressibility of gases; their ability to take the shape of and fill whatever container they occupy; their miscibility; their rates of diffusion through membranes. In addition, all the **gas laws** can be derived from mathematical considerations based on the model in the kinetic theory.

The following statements are the basis of a model of gases. As such, the statements do not necessarily describe what a gas really is but rather suggest how we can think of a gas to explain its behavior. An underlying premise is that an ideal gas obeys the gas laws perfectly, and its behavior may be adequately described by kinetic theory.

Real gases sometimes behave ideally. The model for ideal gases has been improved to account for properties of non-ideal gases by comparing the properties observed for real gases with those predicted from the kinetic molecular model of an ideal gas.

The kinetic molecular model of an ideal gas consists of five statements:

1) Molecules are very small and far apart. Most of a container of gas is empty space. This picture is reasonable for gases because they are highly compressible. A common example of gas compression occurs when a bicycle tire is inflated with a hand pump. In fact, an entire industry has been built around compressed gases. To transport gaseous substances such as propane or acetylene economically, they must be compressed greatly to fit into containers of a reasonable size.

In an ideal gas, the total volume occupied by the molecules themselves is considered to be so small compared to the volume of the gas that it is negligible.

2) Molecules are very hard spheres that bounce off each other without losing **energy** in encounters called elastic collisions.

Very hard balls, such as billiard balls or steel ball bearings, undergo nearly elastic collisions. In elastic collisions, one molecule hits another and energy may be transferred, but the total energy of motion of the molecules remains the same. If one could construct a frictionless table (like an air hockey table) with rigid, hard sides, a group of very hard balls set in **motion** would keep moving—bouncing off each other and the sides of the table—for a long time. When this idea is applied to gases, it means the gas molecules in an insulated container with hard walls will bounce around continuously without losing any energy.

3) The motion of molecules is random, and properties that depend on the motion of the molecules will be the same in all directions.

Collisions of gas molecules with the walls cause the **pressure** that a gas exerts on its container. If a box contains a gas sample, a pressure gauge placed on any of the sides gives the same reading. (This assumes we use a laboratory-scale box so that the effect of **gravity** is negligible.)

4) Because molecules in gases are far apart, they act independently and neither attract nor repel each other.

Several kinds of attractive forces are known: gravitational, electrical, and magnetic. All of them decrease rapidly as two objects move farther apart. The **gravitational force** between objects as light as molecules is small and generally ignored. Because molecules in the gas phase are considered to be far apart in the kinetic molecular theory, these forces are ignored and the molecules are viewed as totally independent entities.

5) Properties of gases we observe, such as **temperature** and pressure, are due to average motions of a very large number molecules.

A great triumph of Boltzmann, Maxwell, and others was to convince physicists and chemists to view gases statistically. From this viewpoint, a sample of a gas is a collection of molecules with a distribution of speeds. At one given temperature, some molecules move much faster than others and some much slower. In the statistical model, a particular temperature is characterized by a certain average speed of the molecules. Ensembles of gas molecules that have the same average **kinetic energy** ($KE_{avg,1} = KE_{avg,2}$) have the same temperature ($T_1 = T_2$). Knowing the temperature or pressure of a gas tells us nothing about one particular molecule but does give us information about the entire sample.

If the simple view of the kinetic molecular theory governed all gases under all conditions, it would be very easy to predict their properties. But reality is more varied and interesting than this simple model. Points 1 and 4 in kinetic theory may be modified when the behavior of real gases is considered.

If molecules were really infinitely small, gases would be infinitely compressible, but this is not the case. As real gases are compressed, all become liquid or solid, and the liquid or solid is very difficult to compress further. At low pressures, the relationship of pressure and volume is accurately described by **Boyle's law**: the volume is inversely proportional to the pressure and (P)(V) is constant. The gases behave ideally. As the pressure increases and the volume becomes small, the straight-line relationship no longer holds; (P)(V) is no longer constant, and real behavior deviates from the ideal.

At high pressure, gases occupy smaller volumes than are predicted by ideal theory. For any real gas there is an abrupt change in the slope of the line of a plot of pressure versus volume that occurs as the gas condenses to form a liquid. This abrupt change in slope that marks the deviation from ideality occurs at different values of pressure for each gas.

If molecules in a gas acted independently, the pressure of a gas would depend directly on the number of molecules hitting the wall (the concentration) as long as the temperature is kept constant. This feature of the kinetic molecular theory also corresponds to Boyle's law. However, a direct proportionality is not observed over the entire range of pressure. If Boyle's law were valid for all gases under all conditions, there would be only one line for a plot of pressure times volume (PV) versus volume and that line would have a constant value of 22.414 L atm at all pressures. This is not the actual case.

The simple kinetic molecular theory outlined above and the **ideal gas law** are valid only when the pressure is low and the temperature high. At low pressures, molecules in a gas are far apart. Under these conditions the space taken up by the molecules is negligible compared to the total volume of the gas. This corresponds to the first statement in the kinetic theory. As the pressure increases, however, the volume occupied by the gas becomes smaller and the space taken up by the molecules is no longer negligible in comparison. At high pressures the volume of a real gas differs from ideal because the actual space taken up by the molecules becomes significant. For real gases, the model must be modified to account for the actual volume the molecules of a gas occupy.

When pressure is low, molecules are too far apart for any attractive forces to influence their behavior. Because both the effect of molecular size and attractive forces are insignificant at low pressures, gases behave ideally. At high pressures, molecules are forced closer together and attractive forces become significant. Furthermore, when the temperature is high, gas molecules move so rapidly that any attractive forces between them are overcome. Correspondingly, as temperature decreases, molecules move more slowly and attractive forces become significant. At low temperatures, gas volumes are smaller than the ideal volumes predicted by theory because the molecules move more slowly and the effects of intermolecular attractive forces have a chance to manifest themselves. This view requires modification of the simple model to include intermolecular attractive forces that become significant at high pressures and low temperatures.

Note these two effects work in opposite directions, for one increases and the other decreases volume. Real gases tend to behave ideally under moderate conditions where neither of the two effects is significant. Under conditions of low temper-

ature and/or high pressure, the observed deviation depends upon which effect is dominant.

A relatively early and quite successful attempt to modify kinetic molecular theory to provide a better approximation to the behavior of real gases was accomplished by **Johannes van der Waals**. He added terms to modify the pressure and volume in the ideal gas equation, adding an n2a/V2 term, where n is the number of mols, V is the observed volume and a is a constant with a different value of different gases, to the observed pressure, increasing the value of the pressure over the ideally predicted quantity, to correct for the loss due to the influence of attractive forces between the molecules and an nb term, where n is the number of mols and b is a constant with different values for different gases, to decreases the total volume occupied by the gas, thereby accounting for the portion of the volume taken up by the molecules themselves.

KINETICS

Chemical kinetics is the study of the rates of reactions and of the factors that determine the speed of a reaction. This information is used to help scientists understand reactions such as those involved in biochemical processes such as metabolism and photosynthesis; reactions that take place in the atmosphere, such as ozone formation and depletion, geochemical processes such as mineral formation and chemical changes that take place in the ocean and many other natural systems. Information from kinetics studies are also of great practical value. Industrial chemists use their knowledge of reaction rates and the factors that influence them to produce materials in a reasonable amount of time with the least loss of valuable reagents. Analytical chemists, including those who design medical diagnostic kits, use kinetics information to design detection reactions that can be carried out rapidly under moderate conditions and with a high degree of reproducibility. Besides the practical value of knowing how fast a reaction proceeds and what factors influence the rate of a particular reaction, studies of kinetics also provide chemists with a way to determine how reactions occur at the molecular level, often allowing a step-by-step analysis.

The study of chemical kinetics is concerned about what is going on during a reaction, not just about the identity or amounts of substances produced in a reaction. Stoichiometry and **thermodynamics** focus on reactants and products; what actually happens to the reacting species during a chemical change is not the focus of these approaches. Kinetics studies focus not only on how fast a reaction occurs under a particular set of conditions, but on trying to understand the steps that take place at the molecular level.

Some reactions, such as explosions and acid-base neutralizations, occur almost instantaneously. Others, such as the corrosion of the Statue of Liberty or the decomposition of marble statues by acid rain, are very slow. Some reaction occur when two molecules collide, while others are only complete after a complicated series of processes. As a result, the study of kinetics is complex and can involve many different techniques to allow the study of reactions that take place over a wide range of times and conditions.

Some reactions occur in a single step. For example, an **oxygen** molecule binds to hemoglobin in a single step when it collides with an iron **atom** inside the protein. However, many reactions are actually accomplished by a series of single-step reactions. For example, the formation of sugars from **carbon** dioxide and water by photosynthesis requires dozens of reactions. The sequence of steps by which a reaction proceeds is called the mechanism. An important objective of kinetics studies is to deduce reaction mechanisms. If a correct reaction mechanism can be found, we can understand what the most important factors are that determine the rate of a reaction and, in many cases, can learn how to adjust the rate of the reaction to suit our purposes. It may also be desirable to increase the rate of a reaction used or to decrease the rate of a reaction that causes a detrimental process like corrosion.

To alter the rate of a reaction, the kineticist typically determines the step of a reaction that is the slowest, the rate-determining step. Once the rate-determining step is identified, reaction conditions can often be modified to make it proceed faster and thereby speed up the entire reaction.

One method of speeding up reactions is to use a catalyst, a species that increases the rate of a **chemical reaction** but is not changed by the process. Ideally, the catalyst can be recovered in its original form when the reaction is over. The catalyst increases the rate of a step by providing an alternate route, allowing the overall reaction to be accomplished without proceeding through the slow, rate-determining step.

A scientist who studies reaction rates is called a kineticist. How does a kineticist evaluate reactions? The first step is to focus on one particular reaction. The kineticist may be interested in large problems like air pollution or the production of a pharmaceutical or may wonder how a particular enzyme works or how a new class of molecules that has just been prepared reacts. But the kineticist focuses on a particular reaction that seems to be most crucial to the process in question. This may be the one that limits the rate of the entire process or one that produces or destroys a particular substance of interest, for example, the destruction of ozone by a chlorofluorocarbon in the study of atmospheric reactions. The kineticist then chooses conditions that are relevant to the reaction in question, for example, conditions like those in the upper atmosphere where ozone destruction occurs.

Next the kineticist must choose appropriate experimental methods for studying the reaction. There must be a means of accurately determining when the concentrations of substances involved in the reaction change, and there must be a way to positively identify the products of the reaction. The sensitivity of the method chosen to study the reaction determines the quantities of reactants that can be used in the study. The experimental methods must also be capable of measuring the quantities of substances involved in the reaction under the actual conditions of the reaction, which may occur in the gas, liquid, or solid phase and at very high, moderate, or low **temperature**. The kineticist must also consider how reaction conditions can be varied to test the possible ways a reaction

occurs at the molecular level. The methods must allow different concentrations of substances to be used and the temperature at which the reaction takes place to be varied. Once the data are collected, the kineticist must be prepared to change the original experimental plan if unexpected results occur.

An important reason to carefully study reaction rates is to be able to predict the rate of a reaction under specific conditions, such as reactant concentrations, pH, temperature, and so forth. A rate law is a mathematical relationship that defines how a reaction depends on concentrations. For example, when a radioactive **isotope** decays, the rate at which the original isotope decreases and the new species is formed is directly proportional to the amount of the radioactive isotope originally present. This can also be stated: rate = k[isotope]. The isotope carbon-14 is used to date the time that ancient plants or animals died because this isotope is present at a very nearly constant concentration in the atmosphere but decreases in concentration when a plant or animal dies because it is no longer exchanging its carbon compounds with the atmosphere and other living things. The decrease in the amount of carbon-14 is given by the equation: rate = k[C-14].

This equation is the rate law for the reaction. It expresses the change in concentration of a particular species as a function time in terms of the concentration, or concentrations, of species involved in the reaction. The rate law is usually written in terms of concentrations that can be determined experimentally. The term "k" is the rate constant. The rate constant is typically represented by a small letter k, large K being used for **equilibrium** constants. The units of the rate constant are determined by the concentration terms of the rate law and the time units used. For carbon-14 dating, the units of the rate constant, k, can be expressed as per mole per liter per hour or $1/(mol/l \ hr)$.

To characterize the affect that changing a particular reactant has on the rate of a reaction, kineticists use the term "reaction order." When the rate of a reaction is directly related to the concentration of a substance, it is said to be "first order" in that substance. This is the case for radioactive decomposition.

The reaction of chlorine atoms and ozone, which has the rate law rate = $k[Cl][O_3]$ is first order in chlorine atoms and first order in ozone. The order of the entire rate law, called the reaction order, is the sum of all the exponents of the concentrations in the rate law. For **radioactive decay**, the order is 1, and the reaction is a first order reaction. For the reaction of chlorine atoms and ozone, the order is 2. This is a second order reaction. The order of the reaction is determined by the rate law, which is always determined experimentally. Only when a reaction is known to be an elementary reaction can we state the order of its rate law from its chemical equation.

Another important term used to characterize chemical reactions kinetics is "half-life." The term **half-life** is very commonly used in reference to radioactive decay reactions, but it is a useful concept for other reactions as well. The half-life of a reaction is the time required to convert half the initial concentration of a reagent to product. The half-life is given the symbol $t_{1/2}$. For a first order reaction the half-life is easily calculated: it is equal to the natural logarithm of 2 ($\ln 2 = 0.693$)

divided by the rate constant. For second or higher order reactions the calculation is considerably more complex.

Even though kinetics is concerned with the rates of reactions, there is an important relationship between rates of a reaction and the equilibrium constant for the reaction. Chemical changes are dynamic. When a reaction reaches equilibrium, there is no longer a net change in the concentrations of the reactants and products, but reactions still occur. The key feature of a system at equilibrium is that the net rate of formation of each species is equal to the rate at which it is changed into another species. The equilibrium constant for the reaction is equal to the rate constant of the forward reaction divided by the rate constant for the reverse reaction. This is a general result that applies to all reactions under equilibrium conditions.

The rate of a reaction can be changed not only by changing the concentrations of reactants, but also by changing temperature. Energy is required to distort molecules in such a way that chemical bonds will be broken and different bonds formed. The **energy** that is required to activate molecules for a chemical reaction is the activation energy of the reaction. There is a "hill" between the reactants and the products regardless of whether the energy of the products is lower or higher than that of the reactants.

Activation energy theory states that molecules must have a certain amount of energy available to initiate a reaction. This available energy comes from the **motion** of the molecules and is directly related to the temperature; the higher the temperature, the faster the molecules move. Consequently, if the temperature of a solution in which a reaction takes place is increased, a greater fraction of the molecules will have an energy equal to or above the activation energy and the reaction proceeds faster. By measuring reaction rates at more than one temperature, kineticists obtain the information they need to be able to predict reaction rates at other temperatures.

KIRCHHOFF, GUSTAV ROBERT (1824-1887)

German physicist

Gustav Robert Kirchhoff was reported to have a fastidious and dull personality. However, examining the results of his research career shows constant evidence of imagination, sincerity, and wit. He was able to do what was proclaimed impossible; that is, identify elements making up the **Sun**. Kirchhoff's collaboration with Robert Bunsen was due to a sincere desire to help a friend solve interesting puzzles. A story told about him shows that he did have wit. He told a banker friend that he was investing his money into his research identifying the elements of the Sun. The banker countered with "Why do I care if there is gold on the sun? I cannot bring it to earth." Later, Kirchhoff won an award for his work and was paid in British gold sovereigns. Kirchhoff showed the gold coins to his banker friend and said they had come from the Sun.

Kirchhoff was born in 1824 in the city of Königsberg, Prussia. The son of a law councilor, he grew up in a comfort-

Gustav Robert Kirchhoff.

ably affluent family. Because he attended the University of Königsberg, he did not have to leave his hometown until he was 23. While Kirchhoff was in college, he developed the laws of electrical circuits that bear his name. This discovery, published in 1845, was followed up two years later with a detailed application of these laws to circuit networks. He graduated in 1847. A third paper covering solid conductors was published the year after he graduated. These papers ensured that his reputation and influence in scientific research circles was beginning to rise.

After a short, unpaid period of teaching at the University of Berlin, Kirchhoff received an honored position of professor at Breslau in 1850. There, he began a long-term friendship with Robert Bunsen. When Bunsen received a position at Heidelberg, Kirchhoff followed him there. In 1854, he accepted a professorship that was not as high a position as the one that he held at Breslau. Kirchhoff remained at Heidelberg for 21 years. During that time, he achieved great success with both his collaborative projects with Bunsen and his independent research.

Kirchhoff and Bunsen were very busy in the years 1854 through 1860, working on developing and improving their spectroscope methodology. While they made many advancements, the major breakthrough came in 1857 with the invention of the Bunsen burner. The Bunsen burner could **heat** experimental substances with very little **light** of its own. Kirchhoff's major contribution to **spectroscopy** was recognizing that a prism could produce a measurable spectrum. During this time of initial spectroscope experimentation, they learned that they could identify the presence of an element in a substance through the spectrum of light given off when the substance was heated to the point of glowing.

Between 1859 and 1861, Kirchhoff and Bunsen began to publish their findings, including the "how-to" of spectroscopy. Kirchhoff presented his study of Fraunhofer **spectral lines** in 1859. In this study, he established the foundation for Kirchhoff's law of emissive to absorptive powers. This law stated that the ratio of emissive to absorptive powers was a function of **temperature** and **wavelength**. That is, all substances will have the same ratios at a specific temperature and wavelength. In addition, he announced that he had found a way to identify the elements present on the Sun. The report of this study gave some hints about the effectiveness of the new methodology, spectroscopy. The next year, Kirchhoff and Bunsen presented their seminal work on spectroscopy announcing the discovery of two previously unknown elements, cesium and rubidium.

Kirchhoff progressed with research independent of Bunsen and modified the spectroscope by adding prisms. In 1861, he announced the identification of even more terrestrial elements found on the Sun. At age 51, Kirchhoff left Heidelberg to accept the position of chair of **mathematical physics**. His later work was in the area of theoretical physics. He died in 1887 in Berlin at the age of 63.

In discussing the products (discoveries and inventions) of the collaboration between Kirchhoff and Bunsen, it should be noted that the collaboration was mutually beneficial. What brought the two together was a mutual respect for the talent of the other. Bunsen had the background in photochemistry, an exciting new field; Kirchhoff had the mathematical and Newtonian interests. Bunsen presented the problem: how to measure the light produced and absorbed in chemical reactions. Kirchhoff was the one who recognized the challenge as a measurement of the absence or presence of components of light. Since light passing through a prism is divided into a spectrum, he suggested integrating a prism into the structure of the light-measuring instrument. In this way, a spectrum developed through the light emitted in a **chemical reaction** could be compared with a "standard" spectrum, thus identifying the components that were present or not.

This insight was the foundation for their invention of the spectroscope. The methodology was completed with the invention of the Bunsen burner. This invention was needed because it was difficult to use other burners in measuring light emitted from heated elements because those other burners gave off their own light. An extra source of light interfered with an accurate measurement of the target spectrum. The Bunsen burner produced so little light as to be insignificant to the measurements.

With this new methodology, Kirchhoff and Bunsen began recording the spectral lines of all known elements. One day, they found a mineral that emitted previously unrecognized spectral lines. This could only mean that it was a newly discovered element. They named it after its strongest spectral line, which was blue. The name cesium is derived from the Latin word for blue. Less than a year later, they found another element with a strong red line. They named it rubidium, derived from the Latin word for red.

Working on his own, Kirchhoff recognized that there was an inverse relationship between the spectrum from sodium and the spectrum from solar light. The dark line of the solar spectrum recorded by Fraunhofer was in the same position as a bright line of the sodium spectrum. Kirchhoff reasoned that, if he shined them together through the spectroscope, the sodium bright line would cancel out the Sun's dark line. Instead, the dark line became darker. Investigating this phenomenon through many experiments, Kirchhoff discovered that, when light passes through a gas, the expected spectrum of that gas is filtered out. If light passes through sodium, it loses dominant components of the sodium spectrum. The only explanation, evident to Kirchhoff, for the Sun's spectral phenomenon was that the Sun's light must pass through sodium in the Sun's atmosphere. Many replicated his findings and a new means to study what was previously thought impossible was made real.

Most of Kirchhoff's work with Bunsen was based on the spectroscope methodology. Although the invention of the spectroscope was a joint project, it was based on the original research need of Bunsen and on Bunsen's concept of how to resolve that need. Kirchhoff influenced Bunsen as a colleague and Bunsen influenced a later generation of researchers as mentor. Kirchhoff could have shown the banker the financial benefit of science through the discovery of cesium. Kirchhoff had helped Bunsen with the work that led to that discovery. A student of Bunsen discovered that cesium could increase the illumination of gas lamps while reducing costs. This recommendation saved the previously declining gas light industry.

Kirchhoff did offer direct influence on future researchers. His contributions to the collaboration with Bunsen were essential and the important discoveries they made together. Furthermore, Kirchhoff made many independent contributions to physics. The field of chemistry (as well as other fields of science) is always affected by developments in physics.

See also Fraunhofer lines

KLEIN, OSKAR (1894-1977)
Swedish physicist

Oskar Klein, one of the most prominent of Swedish physicists, was born in Stockholm in 1894. At the age of 16, he started working in the laboratory of Svante Arrhenius, and published his first scientific paper the next year. Throughout his life, he worked to investigate the ways in which scientific theories are displayed in physical experience.

Using, the concept of a fourth **space** dimension, in addition to the three familiar space dimensions and the dimension of **time**, Klein formulated a theory, in 1927, that merged Maxwell's theories of **electromagnetism** and Einstein's theory of gravitation. This theory, known as the Kaluza-Klein Theory (it was developed independently by Theodor Kaluza), developed the concept of multiple undetected spatial dimensions. This concept is the foundation for string theories. Klein developed several other concepts key to modern physics. The Klein-Gordon equation is a relativistic counterpart to **Schrödinger's equation**. The Klein-Nishina formula shows the probability of different **scattering** angles after collisions between photons and electrons.

In 1954, Klein proposed the metalgalactic model of the **universe**, an alternative to the **big bang** theory. Klein suggested that the part of the universe that we can observe is not necessarily representative of the universe as a whole. The metagalaxy, our part of the universe, was formed from a thin gas cloud that, much more than 10 billion years ago, contracted and then reversed its **motion** due to **radiation pressure**. It is now in the process of expanding. Later, Klein proposed that the origin of the metagalaxy was a matter-antimatter annihilation.

Oskar Klein began working as an assistant to **Niels Bohr** in Copenhagen in 1918, remaining at Bohr's institute until 1931, receiving a position as reader in theoretical physics in 1928. During the two-year period from 1923-25, he left Copenhagen for an appointment as an instructor and then assistant professor at the University of Michigan. In 1930, Klein was appointed as a full professor of physics at Stockholm University where he remained until his retirement in 1962.

L

LAMB SHIFT

The Lamb shift describes a small but measurable difference in **electron** energy levels within the hydrogen **atom** that is accounted for by quantum electrodynamic theory. Specifically, the difference involves a shift between the $2s_{1/2}$ and the $2p_{1/2}$ states of the **hydrogen atom**. The $2s_{1/2}$ state has a **frequency** value 1,057.8 megahertz larger than the $2p_{1/2}$ state.

In early quantum mechanical theory of the hydrogen atom, these two atomic states for hydrogen were predicted to have the same **energy**, even when relativistic effects and coupling between the electron's spin and orbital **angular momentum** were taken into account. However, prior to World War II techniques were unavailable to verify the truth of this prediction regarding identical energy states. Work on microwave techniques during the war finally allowed for precision tests on hydrogen energy levels. In 1947 American physicists, Willis Lamb and Robert Retherford, first measured a frequency difference between the two energy levels of about 1,060 megahertz. Soon afterward, **Hans Bethe** calculated the energy splitting nonrelativistically to be 1,040 megahertz, and a later relativistic treatment by Victor F. Weisskopf (1908-) resulted in a theoretical prediction of 1,057.7 megahertz, in extremely good agreement with the experimental value.

In **quantum electrodynamics (QED)**, one can write down the self-energy of an electron, that is, the energy coming from an electron acting on itself. This self-energy leads to an **energy level** shift for atomic wavefunctions that are nonzero at the center of the atom; this explains why the $2s_{1/2}$ and $2p_{1/2}$ states are split, because the s-state wavefunctions are nonzero at the center of the atom while the p-state wavefunctions are zero there.

A problem with the first predictions of the level shift was that it was infinite. This problem was overcome in one of the first applications of **renormalization** theory. Following renormalization infinite values cancel each other out to leave a finite level shift. The accuracy of the theoretical prediction of the Lamb shift verifies that renormalization theory is a valid way to treat infinities in **quantum theory**.

See also Quantum states

LAMBDA PARTICLES

In the early 1950s, physicists were confronted by two varieties of **strange particles**. Although produced by the strong **force**, these particles lived longer than expected and seemed to decay by the electromagnetic or the weak force. Japanese physicist Kazuhiko Nishijima (1926-) proposed that these particles were produced by the strong force in pairs. The particles in question were the kaon, and the much heavier lambda particle that was discovered in 1947. They were termed strange particles because of their different mode of production and decay. Nishijima believed the particles to be produced in pairs and to be subject to the strong force only as long as they remained in pairs. When a pair of strange particles separated from one another, the strong force could no longer act, and the resulting decays were prolonged since they were by the weak force or the electromagnetic force.

The lambda particle is produced along with another strange particle, such as the kaon. The property of strangeness is preserved in the production of the pair. K^0 and Λ^0 particles have a neutral charge, and do not ionize the gas in the detection chamber. They then decay by the weak force, in a process where strangeness is not conserved, although **electric charge** is conserved. It is accepted that the electric charge is conserved under all circumstances.

In many ways, the lambda particle behaves as a heavy **neutron** with strangeness. It can be incorporated into a **nucleus** and join protons attracted electrons to form a strange **atom**. Because the lambda's lifetime is short, however, such atoms explode when the lambda decays. The lambda nuclei show that the lambda is an additional, albeit unexpected, member of the family containing neutrons and protons. Strange **matter** is unstable and studies are conducted on data derived from life-times measures in millionths of a second.

In terms of its place in the standard model, the lambda is a hyperon carrying zero charge, a strangeness of -1, with a **mass** of 1,156 MeV, about 20% more than that of the **proton**, making it the lightest strange **baryon**. It has a lifetime of 2.63 x 10 seconds (meaning decay via the weak interaction). It is composed of one up quark, one down quark, and one strange quark, in essence a neutron where one of the down **quarks** has been replaced by a strange quark. When the lambda decays, the strange quark converts into an up quark.

See also Quantum electrodynamics (QED); Standard model of particle physics; Strong interactions; Weak force and interactions

LAND, EDWIN H. (1909-1991)

American inventor

Edwin H. Land was the driving force behind the Polaroid Corporation's engineering and marketing successes. He was the first to figure out how to manufacture practical and useful polarized screens during the 1930s, and he produced revolutionary **optics** for the military during World War II. But it was the development of the instant camera that made his company famous, and he was able to dominate the instant-photography market with cameras that first produced pictures in sepia tones, then in black and white, and finally in **color**. One who routinely discarded conventional wisdom, Land believed that market research was not necessary; he claimed that any invention would sell if people believed it was something they could not live without.

An only child, Edwin Land was born to Martha F. and Harry Land on May 7, 1909, in Bridgeport, Connecticut. His father ran a salvage and scrap metal business; the family was well-off and Land had a comfortable upbringing. In his youth he dreamed of being an inventor and idolized **Michael Faraday**, **Thomas Alva Edison**, and Alexander Graham Bell. Even as a boy, Land was very interested in polarized **light**. He entered Harvard at age 17 in 1926. While walking along Broadway in New York City that same year, he was overwhelmed by the glare from headlights and store signs that shone in his eyes. Land perceived safety hazards in all that glare, and he determined that polarized lights could reduce it. He left Harvard at the end of the school year to pursue this idea and did not return for three years.

Land's parents provided an allowance that enabled him to stay in New York and work on this idea. He studied at the New York Public Library and even found a laboratory at Columbia University whose window was habitually unlocked. He would climb in at night and conduct various experiments. During this period Land met Helen Maislen, a graduate of Smith College who began assisting him in his research. They were married in 1929, and Land returned to Harvard that same year. This time the university provided him with a laboratory to conduct his research.

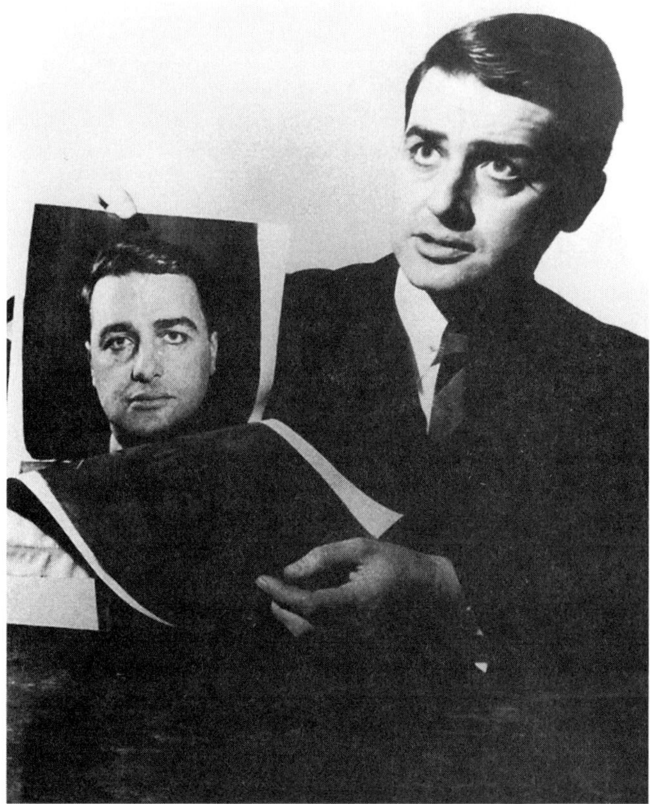

Edwin H. Land, sitting with his demonstration of the Land Polaroid.

Develops a Commercial Method to Polarize Light

It had been known since the eighteenth century that certain kinds of crystals could affect the direction of light **waves**. In his effort to develop a method for polarizing light, Land was searching for a crystal that could not only reduce glare but was stable and economical enough to be produced commercially. He conceived of the idea of two plates that would absorb the light waves that were not wanted and transmit those that were. He then succeeded in aligning millions of microscopic iodine crystals in one direction, thus creating the first polarizer. As Mark Olshaker wrote in the *Instant Image:* "Land's singular achievement was in discovering a way to synthesize a sheet material that could align light waves in the desirable planes of vibration. The invention was a combined achievement of chemistry and optics." Land presented a paper on his discovery at a physics colloquium at Harvard in February 1932. In June he left the university, one semester short of a degree, and never returned.

With a Harvard graduate student named George Wheelwright, who had been one of his teachers, Land formed Land-Wheelwright Laboratories, Inc. in June 1932. The two men worked on developing methods of manufacturing polarized sheets made of crystals trapped in nitrocellulose. On November 30, 1934, Eastman Kodak gave Land-Wheelwright an order for 10,000 dollars worth of polarizing filters. Kodak wanted a polarizer laminated between two sheets of optical glass, but neither Land nor Wheelwright had any idea how to manufacture such an item. Nonetheless, they accepted the order—a decision typical of the way Land would work in the future. Their persistence paid off, and Land-Wheelwright Laboratories invented what they dubbed "Polaroid," with which they fulfilled their contract with Eastman Kodak.

Land had a flair for the dramatic, which he put to good use in marketing his inventions. For example, when he was trying to sell his polarizers for use as sunglasses, he rented a room at a hotel and invited executives from the American Optical Company to meet him there. The late afternoon **sun** produced a glare on the windowsill; Land put a fishbowl there and the glare rendered the goldfish inside it invisible. When the executives arrived, Land handed them each a sheet of polarizer and they were able to see the fish instantly. Land told them that from now on their sunglasses should be made with polarized glass, and the company bought the idea.

Land gave his first press conference on polarization on January 30, 1936, at the Waldorf Astoria Hotel. He repeated it for the National Academy of Sciences and the New York Museum of Science soon after that. The press coverage Land expected from this last presentation was overshadowed by the abdication of King Edward VIII in England. But his sales ability once again came through. On August 10, 1937, a group of investors, impressed with Land-Wheelwright's accomplishments, put up 375,000 dollars to fund the Polaroid Corporation. Furthermore, they gave Land controlling interest in the company.

With the money Land purchased some competing patents on polarization and decided that the 1939 New York World's Fair would be an excellent way to demonstrate to automakers and the American public the virtues of polarized headlights. Chrysler rented Polaroid space in one of its booths, and Land played a three-dimensional movie he had invented that graphically illustrated how much improved polarized headlights were. The 12-minute film was well-received by the public; 150,000 people saw it.

LANDAU, LEV DAVIDOVICH (1908-1968)
Russian theoretical physicist

Lev Davidovich Landau was one of the twentieth century's finest theoretical physicists. Known as the last of the "Universalists," Landau was most remarkable for the breadth of his erudition and for his ability to move with ease between the various branches of physics. His *Collected Papers* record the scope of his interests—which included everything from **low-temperature physics** to the symmetry of space—and the exactitude with which he approached every challenge. Landau was a teacher no less than he was a theoretician; the towering standards that he set for himself were conveyed to his students at the School of Landau, many of whom later achieved recognition in their own right. Landau's work was widely and repeatedly recognized as evidenced by the copious honors

bestowed upon him, most notably, the 1962 Nobel Prize in physics "for his pioneering theories concerning condensed **matter**, especially liquid **helium**." Landau was born on January 22, 1908, in Baku on the Caspian Sea. Now the capital of Azerbaijan, Landau's hometown was, at that time, part of the Soviet Union. His parents were educated and well-to-do Russians; his father was a petroleum engineer at one of Baku's oil fields, his mother, independent and educated far beyond the standards for women of that time. She became, first, a midwife, before going on to become a physician, working during World War I as a field doctor and later as a teacher. One of Landau's biographers, Anna Livanova, writes in *Landau: A Great Physicist and Teacher* that "it may have been her example that gave Landau the foundation of his own versatility, his calling as both scientist and teacher." Lev also had a sister, Sofia, some years his senior, who became a chemical engineer.

Although he would later deny that he had been a child prodigy, he demonstrated that he was advanced at an early age when he finished school at age 13. He thought to study maths and physics, subjects that he loved, but his parents decided to send him, with his sister, for a year to an economics college since he was still very young. In 1922, he enrolled in the University of Baku in the faculties of math and physics and of chemistry.

In 1924, Landau transferred to the University of Leningrad to enroll in its department of physics. There, he attended classes twice weekly and spent the rest of the time pursuing his own lines of research and "dreaming formulae." While still an undergraduate, in 1926, Landau also became a supernumerary graduate student at Leningrad's Physicotechnical Institute. That same year, his first scientific paper, "On the Theory of the Spectra of Diatomic Molecules," was published to considerable interest.

Upon graduating from the university the following year, he joined the Physicotechnical Institute as a fully fledged graduate student. This was where he first came into contact with a group of other theoreticians and with "big physics," as the new quantum **mechanics** was known. He avidly read scientific papers written by the leading lights of **nuclear physics**, which enabled him to keep abreast of developments in Western Europe, the heartland of the new physics. Characteristic of him then and throughout his life was his extraordinary critical sense and his complete independence of thought. A paper published by him in 1927, when he was still only 19, on "The Damping Problem in Wave Mechanics," amply demonstrated that he was far from run of the mill. It was the first time that anyone had described the quantum state of systems with the aid of the **density** matrices. It was during this period that Landau was given the nickname by which he would always be known, becoming "Dau" to his teachers, fellow students, pupils, and to himself.

Meets Cream of Nuclear Physicists during European Trip

In 1929, Landau made his first trip abroad. Over a period of 18 months, he visited Switzerland, Germany, Denmark, England, Belgium, and Holland, meeting all the great physicists of the day with the exception of the one that he, perhaps, admired the most and was most often compared to: **Enrico Fermi**. Landau called Fermi "the second last of the Universalists"; after the Italian's death, Landau became, in his own words, "the last of the universal physicists," according to Livanova in her biography of the scientist.

The bulk of his time was spent in Copenhagen at Niels Bohr's Institute of Theoretical Physics, regarded by many as the mecca of physicists. Surrounded by the top names in the field, including **Wolfgang Pauli**, **George Gamow**, **Werner Karl Heisenberg**, Felix Bloch, and Paul Ehrenfest, Landau was far from daunted and proved himself a match of the greatest minds. Indeed, it was noted of him that he was mentally more agile than almost any of his contemporaries, being the first to solve a problem, often in the most unconventional way. In this, he was aided by his powerful grasp of mathematics, which he considered the first tool of the theoretician. The only regret he was heard to voice during this period was that he had not been born sooner. "All the nice girls have been snapped up and married, and all the nice problems have been solved. I don't really like any of those that are left," he remarked during a meeting in Berlin at the close of 1929, according to Livanova. By 1930, when Landau came to Denmark, the basics of **quantum mechanics** had already been figured out. Nonetheless, he subsequently managed to find enough unsolved problems to keep him busy throughout his life.

Bohr influenced Landau more than any other physicist. The Russian often said of his Danish mentor that he was his only teacher. Their relationship was mutually respectful and lifelong. "The success of the school of theoretical physics which Landau subsequently founded in the Soviet Union undoubtedly owed much to Bohr's example," said Russian scientist **Pyotr Kapitsa**, according to Livanova. Dau and Kapitsa would later work together.

In Cambridge, Landau worked with New Zealand-born Cavendish professor, **Ernest Rutherford**. Here, he met for the first time his fellow Russian, Kapitsa, and developed his theory of the diamagnetism of metals. This predicted the occurrence of unusual magnetic properties of free electrons in metals.

In Zurich, he worked with German physicist Wolfgang Pauli, to whom he was frequently compared for the sharp tongues and critical faculties they had in common. In Zurich, too, he wrote two important and well-received papers with German physicist Rudolf Peirels, one of which pertained to relativistic **quantum theory**.

Begins Teaching at Kharkov

Returning to Leningrad in 1931, he determined to begin teaching, as well as to continue his own work. First, he worked closely with Matvey Bronstein at the Physicotechnical Institute, before moving to Ukraine in 1932 to head the Theoretical Institute at its then capital, Kharkov. Here, finally, Landau was able to realize his wish to teach. Kharkov was where Landau devised his first "theoretical minimum" program in physics for members of the institute staff and began work on his magnum opus, the *Course of Theoretical Physics,* which remains, long after Landau's death, the bible of theoretical physics. These volumes, covering the spectra of the field, testify, perhaps more

than anything else Landau produced, to his genius. Of them, one of his pupils at Kharkov remarked, "It is Landau's true memorial. He alone could have created it; no other person. Books more striking than these will never be written, no matter what the subject," wrote Livanova. Another commented, "If there is no answer there, it will not be found anywhere." The course was co-written with one of Landau's favorite pupils, Evgeny Lifshitz, for Landau could not bear to write.

The theoretical minimum was an examination, with nine component parts, covering all of theoretical physics, including the necessary maths. It was Landau's conception of the minimum a theoretician should know before he would accept him as a pupil. Just how unusual it was can be gauged by the fact that the traditional approach is to view physics as a series of specialities. Not Landau. For him, physics was an indivisible whole, and anyone serious about studying under him had to prove that they, like him, were capable of moving with ease between the various branches. It was a challenge that few were up to: between 1933-61, just 43 people passed the examination, which was open to anyone who cared to tackle it.

Landau's unusual approach was seen, too, in the theoretical seminars that he introduced at Kharkov. These, too, covered the gamut of theoretical physics, an approach unheard of then or since. Landau was an active participant at these seminars; invariably, he knew more than anyone else there. "Landau knew everything because he was interested in everything," recalled Alexander Kompaneets, one of Landau's pupils, in Livanova's biography. Kompaneets, who incidentally placed first in the list of physicists who had passed the theoretical minimum, also noted: "There will not soon be another theoretical physicist with such erudition."

Investigates Low-Temperature Physics

In 1935, he accepted the chair of General Physics at the University of Kharkov. In addition to his administrative and teaching duties, he took a close interest in the groundbreaking work on low-temperature physics that was being carried out at the university. Landau's angle was the behavior of matter at low temperatures. He published four papers on the topic during this period (out of a total of 17 on a wide range of subjects). Even with this heavy workload, Landau found time for leisure pursuits: tennis, reading, movies, and fraternizing with his wide circle of friends. He was preoccupied, also, by his recent acquaintance with Kora Drobantseva, a food engineer at a chocolate factory. They married in 1937 and had one son, Igor, who went on to become an experimental physicist.

Landau and Kora moved to Moscow in 1937, where Kapitsa had invited him to head the theoretical division at the Institute of Physical Problems. Here, Landau felt completely at home. He continued working on a range of theoretical problems, being interrupted only briefly during World War II when the institute was evacuated to Kazan. There, Landau devised theories and made calculations of the processes governing the efficiency of armaments and published three papers on the detonation of explosives. For his war-related work, he was made a Hero of Socialist Labour in 1945.

When the war ended, Landau resumed his work on **superfluidity**. He succeeded in explaining various properties of liquid helium–4 mathematically; for instance, he explained why the element flows without **friction** below a **temperature** of 2K and has a thermal conductivity 800 times greater than copper at room temperature. This work laid the groundwork for later research into superconductivity, that is, the complete disappearance of **electrical resistance** in certain metals at temperatures near **absolute zero**. He forecast that **sound** would travel at two speeds in liquid helium; at that of a familiar **pressure** wave and at that of a temperature wave. This theory was experimentally verified in 1944 by Russian scientist Vasily Peshkov. Landau also continued to work for the government's **nuclear weapons** program.

In 1945, Landau investigated shock **waves** for the engineering company of the Soviet Army. The following year, he turned his attention to the **oscillation** of plasmas. He predicted that the **isotope** helium–3 would show unique properties at a temperature near absolute zero on the Kelvin scale, including a wave propagation called "zero sound" and sudden spinning.

Landau received many honors during his lifetime, including the State Prize three times and the Lenin Prize once. He was twice awarded the Order of Lenin, was elected an honorary member of the British Institute of Physics and Physical Society, a foreign member of the Royal Society of London, a member of the U.S. National Academy of Sciences, and a member of the Danish and Netherlands Royal Academies of Sciences. He received the Max Planck Medal, the Fritz London Prize, and the Nobel Prize in physics.

Landau's career was brought to a tragic and abrupt halt on January 7, 1962, when he was involved in a car crash. The base of his skull, ribs, and pelvic bones were broken, and for six weeks the doctors struggled to save his life as he slipped in and out of a coma. He survived, but was so badly injured that he could never work again.

Bedridden, Landau Wins 1962 Nobel Prize

Landau was still confined to his bed when it was announced that he had won the 1962 Nobel Prize in physics for his groundbreaking research into theories of condensed matter, especially liquid helium. As he was not well enough to travel to Stockholm, his wife and son accepted the award on his behalf. He spent the remaining years of his life battling his injuries until he died on April 1, 1968, after surgery to repair an intestinal blockage. He was 60 years old.

LAPLACE, PIERRE-SIMON DE (1749-1827)

French mathematician and astronomer

Because Pierre-Simon de Laplace was secretive about his background, little is known of his early life; he may have been ashamed of his past and kept the details to himself. He was born on March 23, 1749, at Beaumont-en-Auge, Normandy, France. It has been suggested that he came from a poor family and affluent neighbors helped him get an education. Other

Pierre-Simon de Laplace.

evidence, however, supports the contention that he came from a middle-class family. At any rate, Laplace was extremely intelligent; when only 18 years old he went to Paris and obtained a professorship in mathematics, based on a paper he had written on **mechanics**.

Laplace, working with Antoine-Laurent Lavoisier, conducted research on the heats of various substances. The two gave birth to the science of thermochemistry when they demonstrated in 1780 that the amount of **heat** necessary to decompose an object was equal to the heat evolved when the object had been originally formed. This concept would later resurface as the **conservation of energy**.

But the topic that interested Laplace most was the stability of the solar system and how to account for observed variations in the orbits of its members. Joseph-Louis Lagrange had been studying this subject as well. One of Laplace's greatest discoveries concerning orbits was his observation that the **Moon** was accelerating a bit faster than it was supposed to be according to existing calculations. He suggested that certain changes in Earth's orbit, caused by the gravitational tug of the other **planets** upon Earth, were generating an increase in the Moon's **motion**. Expanding on Lagrange's work, he suggested that gravitational tugging was the cause of variation in the orbits of Jupiter and Saturn as well.

Laplace realized that to maintain a constant **equilibrium** in the solar system, any change in one member had to produce a change in another. If one planet's eccentricity increased, there had to be a decrease in that of another. The solar system would remain the same indefinitely, so long as there was no change in the nature of the **Sun**. This supposition extended the work of **Isaac Newton**, earning Laplace the nickname of the "French Newton."

Laplace's gravitational theory was summed up in *Celestial Mechanics*, a five-volume book that appeared over the years from 1799 through 1825. An egotistic and boastful individual, Laplace was reluctant to give credit to the contributions of others such as Lagrange. To his credit, Lagrange did not protest his exclusion.

In 1704 Newton had suggested that short-range forces between particles could explain most chemical and physical phenomena. Laplace adopted this concept in 1796. He and his colleagues applied Newton's calculus to forces acting between particles of ordinary **matter**, **light**, heat, and **electricity**. By doing so, they determined equations for the refraction of light, the conduction of heat, capillary action, the **elasticity** of solids and the static distribution of electricity on conductors. There were a number of critics to the "Laplacian school of thought." It began to lose influence after 1810 and fell out of favor after 1820. The criticism was justified; by 1853 **Jean-Bernard-Léon Foucault** showed that light moved as a wave form and did not behave like a stream of particles.

Laplace also developed a hypothesis regarding the formation of the planets. All the planets orbit the Sun in the same direction and in about the same plane. (At least that was true for the seven Laplace knew about; Uranus had been discovered by **William Herschel** in 1781.) Laplace suggested that the Sun had formed from a rotating nebula, or cloud of gas. As the nebula contracted, rings of gas thrown off by centrifugal **force** were left behind. These rings condensed into the planets, which continued to orbit in the same direction as the original cloud.

This nebular hypothesis became very popular with some astronomers of the day, even though Laplace himself did not take it all that seriously. It had actually been suggested earlier by German philosopher **Immanuel Kant**, but that may have been unknown to Laplace. The hypothesis eventually fell by the wayside. Later, geologist Thomas Chrowder Chamberlin (1843-1928) and astronomer Forest Ray Moulton (1872-1952) put forward an alternative hypothesis, suggesting that the planets had formed by accretion of particles. The nebular hypothesis was resurrected by Carl von Weizäcker in the 1940s, and it became more popular than ever.

Laplace died in Paris on March 5, 1827. His last words were "That which we know is mere trifle, that we are ignorant of is immense."

LARGE HADRON COLLIDER

The Large Hadron Collider, or LHC, is a facility at the **European Center for Nuclear Research (CERN)** on the French-Swiss border. It is designed to take particles at high energies and smash them together, to see what results are obtained. The

energies of interest are above 1 TeV (tera-electron volt), or one million million **electron** volts.

The LHC was designed for versatility, allowing experiments in high-energy proton-proton collision, proton-electron collision, and collisions of heavy **ions** such as lead. The latter category of experiments could reach energies of thousands of TeV. In addition, the beam intensity is higher than other colliders of the time, allowing for more possible collision events. With each of these types of experiment, the LHC is designed to test the standard model of **quantum mechanics**. The heavy-ion collisions can provide particular insight into the nature of the strong nuclear **force**.

The LHC design includes a 16.7-mi-long (27 km) beam pipe through which the particles will travel to collide. Like most **accelerators**, it has a ring design, allowing for magnetized **acceleration** around the ring. The particles will remain in the system for fairly long times, raising concerns about synchrotron **radiation** potential chaotic effects. Synchrotron radiation—the tendency of charged particles that are being accelerated to radiate energy—is inevitable and must be compensated for in the beam itself. The magnetic guides are the most difficult part of damping the chaotic effects. Even small flaws in the magnetic system could be disproportionately deleterious to the experimental results.

The magnets used will be superconducting and will use the previously existing LEP (Large Electron Positron collider) tunnel for their accelerator chamber. The magnetic effects required are almost 100,000 times the strength of Earth's **magnetic field**. These superconducting magnets will be made of coils of copper-clad niobium-titanium cables, through which a strong current will pass. These magnets must be precisely coiled not just to prevent chaotic effects but also to prevent quenching breakdown, which costs the accelerator precious hours of beamtime. In order for these strong magnets to work, they must sit in 1.9K baths of superfluid **helium**. The **cryogenics** of this system have to be optimized for **temperature** and for fluid-tightness: spills of superfluid helium could be costly in both time and money.

The LHC is a highly ambitious project, lying before the cutting edge of **particle physics** at the time of the writing of this article. It is expected, among other things, to verify the existence of quark-gluon plasmas and to test other fundamental parts of current particle theory. The scheduled starting date of the LHC is 2005.

See also Hadrons; Positrons; Quarks and gluons

LASER INTERFEROMETER GRAVITATIONAL OBSERVATORY

The Laser Interferometer Gravitational Observatory (LIGO) is an ambitious experimental program designed to directly detect **gravitational radiation**. The LIGO project began construction in 1998 and was expected to begin taking data in 2000. LIGO will eventually consist of a number of sites worldwide, but as of 2000, two sites neared operational readiness, one near Richland,

Washington, and the other near Baton Rouge, Louisiana. Other sites will likely be built in Europe and Asia. As well as attempting to directly verify the existence of gravitational **waves**, the LIGO observatories will open new ways to observe and measure the composition and evolution of the **universe**.

Einstein's theory of general relativity relates the curvature of **space-time** at any point to the **mass** and **energy** that exists nearby. General relativity predicts a number of phenomena, such as bending of **light** by **stars** and the precession of the planet Mercury, which have been well verified by direct observations. General relativity also predicts the existence of gravitational waves, ripples in space-time curvature that travel at the **speed of light**. The existence of these ripples, also known as **gravitons**, was indirectly verified by Princeton University astronomers Joseph Taylor and Russell Hulse in 1974, by observing a pair of **neutron stars** in the **Milky Way galaxy**. One of the stars is a pulsar, a regular source of **radio** wave emissions that can be detected on Earth. After measuring the **frequency** of these **pulses** over a long time, it was found that the pulse frequency became shifted due to a loss of energy from the binary star system. This was in agreement with general relativity, which predicted loss of energy from the system associated with the emission of gravitational radiation.

However, no matter how well an indirect measurement supports a theory, it is possible to come up with an alternative explanation that will also reproduce that measurement. Accordingly, because direct measurements are always the most desirable, LIGO is designed to directly detect the gravitational radiation.

The LIGO apparatus consists of an L-shaped tunnel with two arms. At the end of each arm is a mirror, and where the two arms meet there is a laser beam splitter. Each arm has the same length. When a laser beam is shot into the tunnel, it goes through the beam splitter, so both beams have the same phase initially. The beams are reflected from the mirrors and recombine at the original entry point, where the intensity of the beam is measured by a photodetector. As long as both arms have the same length, the beams interfere constructively and the maximum intensity will be measured. If a gravitational wave passes through the apparatus, the length of the tunnel arms will change; one arm will become slightly shorter, and the other will become slightly longer. The laser beams will then be out of phase when they recombine, because they traveled different distances, and there will be some measurable destructive **interference**.

The challenge in making these measurements is to minimize vibrations caused by seismic activity or human activity that could disrupt the experiment and lead to false signals. To minimize these effects, sophisticated automatic mirror control systems have been developed to keep the mirrors from moving due to vibrations. In addition, new **vacuum** technology has been developed to minimize **scattering** from gas atoms, and high-precision **lasers** with minimal frequency variation are incorporated to avoid false interference patterns. However, even with these precautions, false signals can appear. This is the reason for having multiple LIGO sites; a seismic vibration in Washington will not appear at the same time in Louisiana,

so signals at one site can be compared to signals from other sites to weed out false signals. When a third site becomes operational, LIGO will be able to triangulate the signals and find their point of origin in the sky. In this way, LIGO will be a huge gravity-wave **telescope** capable of yielding new data about the composition and behavior of the universe.

See also Interferometry; Length contraction; Pulsars; Relativity, general

LASERS

A laser is a device that produces a very intense beam of **light** with properties that make it an essential piece of equipment across the whole spectrum of science.

Normal "white" light is a mixture of different wavelengths, each **wavelength** associated with a specific **frequency** and **energy**. This is why white light separates into a "rainbow" when it is beamed through a prism. Each **color** in the rainbow is a different wavelength of light, with short wavelength blue light at the energetic end of the spectrum and long wavelength red light at the low-energy end. Lasers are different because they produce light of only certain wavelengths, all of it traveling in the same direction. The light is also *coherent*, which means all the peaks and troughs of the different waveforms match up peak to peak and trough to trough. They are like soldiers that are not only marching shoulder to shoulder and at the same speed, but are in step too.

Lasers need three things to make them work. The first is a material called the *active medium*, which can be charged up with energy and then made to give up this energy in a controlled way. The second is a way of charging up the active medium with the necessary energy and the third is an optical cavity to amplify the **radiation**.

The word laser is an acronym for *l*ight *a*mplification by *s*timulated *e*mission of *r*adiation and suggests how lasers work. In a molecule, each **electron** can be assigned a specific energy. One way to understand this is to imagine the electrons to be whizzing round the nuclei in an "orbital" that is rather like the orbits of the **planets** revolving around the **Sun**. Each orbital and therefore each electron within it has a certain energy associated with it. Unlike the planets, however, an electron is given extra energy so it can jump from one orbital to one with higher energy.

In most materials in their normal state, there will be far more electrons in the unexcited lower **energy level** orbitals then there are in the more energetic excited ones. If energy that matches an energy gap between these different levels is put in, the electrons are excited from low-energy state **orbitals** to more energetic excited orbitals. If these excited oribitals are *metastable*, this means the electrons stay in these orbitals for a short time before decaying back to their original orbitals. Excitation of electrons to metastable orbitals therefore causes a *population inversion*, because there will be more molecules with electrons populating the higher energy levels than the lower.

Man at laser. *(Photo courtesy of Digital Stock. Reproduced by permission.)*

The electrons in active mediums give up this extra energy and decay back to their ground states in two ways. The first is called *spontaneous emission*. This involves an electron spontaneously jumping down from an excited orbital, giving off the energy as a particle of light called a **photon**. This photon is equal in energy to the gap between the energy levels, but is given off in a completely random direction. The second way is called *stimulated*, which occurs when a photon with energy that matches the gap in energy between the two levels hits an excited molecule. The photon then effectively "knocks" the electron down from its excited state by causing it to emit a photon. The important thing here is that the emitted photon not only travels in the same direction and with the same energy as the one that caused it to be emitted, but that it is also coherent with it. These identical photons can now travel through the active medium, knocking down more electrons from their excited states. This causes a snowball effect, releasing a whole cascade of coherent photons.

If a mirror is placed at either end of the active medium, an optical cavity is formed, where the light radiation bounces back and forth between the mirrors. As long as the active medium is charged up and there is "room" in the lower energy levels for the excited molecules to decay to, the light will sweep back and forth through the active medium getting *amplified* with each pass. Any photons produced in the cavity that do not travel in line with the mirrors are not reflected back into the active medium and lose energy and fade away. If the mirror at one end of the optical cavity is a partial mirror, it

reflects only part of the radiation back into the cavity, letting the rest out. The radiation emitted through this mirror is called a laser beam.

One factor determining the output of a laser is the length of the optical cavity. It is only possible for certain wavelengths—the so-called longitudinal modes—to exist in the cavity as all others will interfere and cancel each other. A particular active medium will also be capable of supporting only some of the longitudinal modes, and it is the combination of these two factors that shapes the output.

There are several different types of active medium, that the type used gives the laser its name. Gas lasers use gases such as argon, **carbon** dioxide, neon, and nitrogen. Solid-state lasers can use crystals (ruby for example), glass (Nd^{3+} in silicate glass, for example) or **semiconductors**. Another type uses a dye as the active medium and hence is called a dye laser. These are important in **chemistry** because it is very easy to change the frequency of the laser output and select a desired frequency for a particular experiment.

The nature of the active medium also determines how the energy used to cause the population inversion is supplied. An electrical discharge supplies the energy to a gas laser as **electricity**. Optical pumping, where a light is shone on the medium, is used in crystal and glass lasers. Chemical lasers work by reacting two species (such as hydrogen and chlorine gases) together to form products with an inverted population. Hybrid electrical discharges work in the same way, but an electric discharge excites the products first, causing them to react.

Lasers operate in two modes. The first is called continuous wave and occurs because the molecules remain in the upper, metastable excited levels longer than in the lower. This means there is always room in the lower levels where the molecules can decay. These lasers can then operate continuously. The second mode of operation is pulsed operation. In this mode the upper level is pumped up before the lower level has had time to empty and, consequently, there is no room in the lower levels for the metastable states to decay to. The radiation from these lasers are therefore produced in short **pulses**. With both modes of operation, there are advantages and disadvantages that dictate where these different types of lasers are used.

In many applications, pulses of an extremely short duration are required. There are several ways of achieving this. For a lower limit of nanoseconds (a one-thousand-millionth or 10-9 of a second), Q switching is used. This uses an optical switch, which "spoils" the optical cavity by blocking one of the mirrors. This allows the active medium to charge up, but there is nowhere for the energy to go. When a very short electric pulse is applied to the switch, the cavity is momentarily restored and the radiation rushes out in one big pulse.

For even shorter pulses, mode-locked lasers are used. One type of mode-locked laser contains a dye cell in the optical cavity, which absorbs light at the operational frequency of the laser. Normally the dye absorbs all of the energy of the laser and nothing happens. If by chance, however, the active medium releases a big pulse, it forces its way through the dye by saturating it with so much light that it cannot absorb any more and allows the rest of the pulse through. This pulse is then free to bounce back and forth in the cavity, soaking up all the energy from the active medium. When the big pulse is not traveling through it, the dye absorbs all other radiation, and so the only output of the laser is the big pulse. The frequency of the pulses in such a laser is fixed by the time it takes the pulse to complete the passage from one end of the cavity to the other. These pulses can be so short that quantum effects such as the **uncertainty principle** come into play and mean that instead of just one specific frequency, there are many. This can be solved to an extent by beaming in radiation from another laser that singles out just one of the possible frequencies present. These pulses can be even shorter than femtoseconds (a femtosecond is a thousand-million-millionth or 10-15 of a second!).

Lasers are extremely useful in spectroscopy, because they combine a focusable beam with a precise frequency and high energy. In absorption spectroscopy, a beam is split in two, one part being sent through the sample, the other going straight to a detector. By comparing the intensity of the two beams it is possible to see whether the one that passed through the sample was absorbed by it to any extent. By doing this at a number of different frequencies it is possible to get a very accurate absorption spectrum of even the most dilute samples.

Lasers with these extremely short pulses have given rise to hyperfast **spectroscopy**, in which it is possible to study processes such as bonds breaking step by step. This has huge implications for the study of the incredibly short-lived transition states that are characteristic of so many chemical reactions.

Lasers have also revolutionized Raman spectroscopy. Raman spectroscopy relies on the distortion—called the change in polarisability—of the **electron cloud** in a molecule when it is exposed to light radiation. Low-intensity normal light produces Raman spectra, but they are very faint. The intense radiation from lasers produces very clear Raman spectra.

Coherent anti-stokes Raman spectroscopy (CARS) has been developed on the back of developments in lasers using the nonlinear effects on the polarizability of molecules when placed in an even more powerful field than that used for normal Raman spectroscopy. The fact that the probe in Raman spectroscopy is light means that CARS can be used to look at the chemistry that occurs in hostile and otherwise inaccessible environments such as jet engine exhausts.

Lasers have also revolutionized the separation of chemical isotopes, which had otherwise been extremely difficult, if not impossible. Laser radiation of a very specific frequency picks out a molecule with **isotope** specific features in the spectrum and can then excite just the isotope leaving the rest of the molecules unaffected. The selectively excited isotope can then be captured either by ionizing it with more laser radiation or using the enhanced reactivity of an excited molecule to make it undergo a **chemical reaction**. This technique has facilitated the separation of isotopes of uranium, which has had a big impact on both nuclear **power** and **nuclear weapons**.

The quantum nature of light, in which it can be described as both a wave and a particle, has made it possible to use lasers to trap and cool individual atoms. The **momentum** of laser beams can be used to repeatedly "punch" atoms to slow them

down. The wave nature of light means that beams can interfere with each other, causing regions in **space** with high and low laser intensity. The combination of these two techniques have made it possible to trap individual atoms. First a laser is used to slow the atoms down, and then they are caught in an "optical cage" formed by the **interference** pattern that arises when many laser beams are crossed at a point in space.

LATENT HEAT

Usually when **heat** is added or extracted from a system the **temperature** changes. There are certain situations when the addition or subtraction of heat from a system does not result in a temperature change. In these situations the heat that is added or extracted is being converted into **energy** to effectuate a phase change from one physical state to another. This heat, the heat required to change a substance from one form to another, is known as the latent heat. It is the enthalpy change that accompanies a phase change at constant temperature and **pressure**. It is called latent because it is somewhat hidden from detection except for the physical change in a substance's form. The value of latent heat is dependent upon the exact nature of the phase change as well as on the specific properties of the substance. Latent heat is usually expressed in terms of calories per gram.

There are several forms of latent heat that reflect the exact phase change taking place. The latent **heat of fusion** refers to the heat used when the phase change is from a solid to a liquid. The latent heat of water is 79.7cal/g, a particularly high value. The latent **heat of vaporization** is the heat used when a liquid changes into a vapor. The latent heat of vaporization of water is 540cal/g. The latent heat of **sublimation** refers to the heat used when a material undergoes a phase change from solid to vapor in a single step. There is also a latent heat of transition that is the enthalpy change accompanying a solid-solid transition, such as a change in ice from one crystal structure to another. The latent heat of a physical transformation from one phase to another can be thought of as the amount of energy required to rearrange the molecules of a substance. When a solid is transformed into a liquid, the amplitude of the vibrations of the atoms about their **equilibrium** positions becomes large, large enough to overcome the attractive forces that bind the atoms together into a solid form. The latent heat is the energy required to break these bonds and transform the material from the ordered solid state to the disordered liquid state. Just as energy is required to break the bonds binding the atoms of a solid together, energy is also required to weaken the attractive forces between the molecules in a liquid in order for it to become a vapor. In a liquid the molecules are closer together than in a vapor phase and so the forces between them are stronger than in a gas. In order to separate the molecules these attractive forces must be broken. Since the average distance between molecules in a vapor are larger than either the liquid or the solid states, it is obvious that more **work** is required to separate the molecules to form a vapor. This explains why the latent heat of vaporization is much higher than the latent heat of fusion for a given substance.

LAUE, MAX VON (1879-1960)

German physicist

Max von Laue distinguished himself early in his career with the discovery of x-ray **diffraction**, a process that is now used to examine the structure of crystals and has been widely used in the research of proteins and other organic substances. The experiment also proved the **wavelength** of **x rays**. **Albert Einstein** viewed Laue's work as one of the greatest discoveries in physics. Laue was awarded the Nobel Prize in physics in 1914, at the age of 35. He subsequently enjoyed a long career, distinguished not only by his contributions to physics but also by his willingness to oppose the Nazi regime in his native Germany.

Laue was born on October 9, 1879, in Pfaffendorf, Germany, near Koblenz, to Julius Laue, a civil official of the Prussian military court system, and the former Minna Zerrenner. The family name was changed to von Laue when his father was raised to hereditary nobility in 1913. Due to his father's profession, Laue's family moved often during his childhood, requiring him to attend many different schools. He began to show a marked interest in physics by the age of 12. His mother arranged visits for him to Urania, a science society in Berlin that exhibited working models of various types of scientific machinery accompanied by explanatory demonstrations. Laue graduated from the Protestant gymnasium in Strasbourg in 1898. He then studied at the University of Strasbourg while simultaneously completing his one year of compulsory military service.

In the fall of 1899 Laue transferred to the University of Göttingen and ultimately to the University of Berlin, where he studied under Nobel laureate **Max Planck**. Laue graduated with highest honors in July 1903, earning a Ph.D. in physics. For his doctoral dissertation he studied the theory of **light interference** among parallel plates located in the same plane. This included a complicated discussion of the interaction between intersecting light **waves** that destroy or reinforce each other, depending on their relative phases. Laue spent the next two years at the University of Göttingen as a postdoctoral student. At the end of that time he passed the examination to teach in the gymnasiums.

In 1905, Laue was invited to become Planck's assistant at the Institute for Theoretical Physics in Berlin. While working together, he and Planck established a friendship that was to last a lifetime. Laue remained at the institute until 1909, pursuing Planck's idea of applying the concept of **entropy** (the second law of **thermodynamics** that states that all closed systems become increasingly disordered) to **radiation** fields. In 1906 he also began work at the University of Berlin as a *privatdocent*, an unsalaried lecturer whose sole income is derived from students who chose to attend his lectures.

Laue left Berlin in 1909 for the University of Munich, where he joined the physics faculty under **Arnold Sommerfeld**. There he lectured on **optics** and thermodynamics as a privatdocent. He was one of the early proponents of Einstein's theory of relativity, at a time when many in the field of physics remained skeptical. As early as 1907 Laue was working on a

mathematical proof for relativity drawn from his work in optics; when coupled with the famous **Michelson-Morley experiment**, this proof was instrumental in convincing physicists of the theory's validity. Laue published one of the first monographs on relativeity in 1911.

Discovers X-Ray Diffraction

It was while in Munich that Laue made his most significant discovery. Since **Wilhelm Conrad Röntgen** had discovered x rays in 1895, no one had been able to conclusively determine if they were very short electromagnetic waves. This was widely supposed but unproven, and in 1912 Laue took up this question. At the same time, he was writing a chapter on wave optics for an encyclopedia of mathematics in which he had to express mathematically how diffraction affected light waves. While he was thinking about the length of x rays and the physics of diffraction, Laue began a series of discussions with a colleague regarding the behavior of light waves in a crystal, whose internal physical structure was also a matter of some dispute. The discussion started Laue thinking about the possibility of using a pure crystal that could be exposed to x rays. The resulting diffraction of light **energy** through the crystal could provide a means to measure the wavelength of x rays.

On April 21, 1912, Walter Friedrich and Paul Knipping began an experiment in Sommerfeld's laboratory, exposing a copper sulfate crystal to x rays. This produced a pattern of dark points on a photographic plate behind the crystal—which is now known as a Laue diagram in honor of Laue's insightful speculations. The difficulty then was to deduce how to calculate the wavelengths of the x rays. As Laue was walking home after seeing the photographs, it suddenly struck him that his understanding of the mathematics of diffraction in optical gratings, which he had concurrently been examining, could be applied with some variations in this new context. Laue then worked out the calculations that related the diagram to the positions of atoms within the crystal and to x-ray wavelength. Another, much clearer set of photographs a few weeks later proved to the group's satisfaction that Laue's new equations were exactly correct. It was now possible to use x rays to determine indirectly but precisely the structure of crystals and to use crystals to determine the wavelengths of x rays.

For his discovery of x-ray diffraction, Laue was awarded the Nobel Prize for physics in 1914. In his presentation speech, quoted in *Reality and Scientific Truth*, G. D. Granqvist of the Royal Swedish Academy of Sciences said, "proof was thus established that these light waves are of very small wavelengths. However, this discovery also resulted in the most important discoveries in the field of **crystallography**.... It is now possible to determine the position of atoms in crystals, and much important knowledge has been gained in this connection." Of the discovery, Jacob Bronowski wrote in *The Ascent of Man* that it "was a double stroke of ingenuity, for it was the first proof that atoms are real, and also the first proof that x rays are electromagnetic waves."

Laue was a theoretical physicist and as such did not concern himself with studying the structure of individual substances. That he left to others, such as **William Henry Bragg** and

William Lawrence Bragg who went on to make substantial contributions to crystallography by building upon Laue's work. x-ray diffraction and x-ray spectroscopy have since revealed the molecular structures of penicillin, amino acids, and many other crystalline structures as nature yielded many secrets to these new techniques of the laboratory. Laue, however, continued to work on the theoretical aspects of x-ray diffraction and spectroscopy.

Laue was appointed an associate professor at the University of Zurich in 1912. Two years later he became a full professor at Frankfurt. He spent much of World War I working with **Wilhelm Wien** at the University of Würzburg on the development of **vacuum** tubes to improve telephone and wireless communication. In 1919, Laue returned to the University of Berlin as a professor of physics, where he once again was able to work with Planck. Laue was awarded the Max Planck Medal by the German Physical Society in 1932, perhaps the highest scientific honor in Germany at the time. During the 1930s, he contributed to Walther Meissner's theoretical work on superconductivity, a phenomenon where **electrical resistance** within materials completely disappears at **absolute zero**.

The rise to **power** of dictator Adolf Hitler's Nazi regime in 1933 brought significant changes to the administration of German science. Laue was one of only a few scientists who publicly opposed the Nazis' dismissal of Einstein, who was Jewish, as director of the Kaiser Wilhelm Institute for Physics. At a meeting of the German Physical Society on September 18, 1933, Laue likened the Nazi rejection of relativity, based on their ideological objection to its origination from a Jew, to the rejection of **Galileo**'s ideas in the seventeenth century by the Catholic Church. Although Einstein was not reinstated, Laue achieved some measure of success in his opposition of the Nazis. In December 1933, he led the move to block the admission of **Johannes Stark**, a Hitler lackey, to the Prussian Academy. Even though Laue met with some government censure during World War II, he was allowed to continue his work. During the war, he became the spokesperson for the physicists after the appointment of Stark to the presidency of the German Research Association. He continued to write to Einstein at the Institute for Advanced Study at Princeton University throughout the war, doing so at great personal risk and with a rather cavalier attitude. When asked why he persisted despite the risks he replied, as quoted in *Reality and Scientific Truth:* "They are so stupid, these people who check and censor my mail. I just write 'Professor Albert, Fine Hall, Princeton, USA,' and it arrives." He later said, "The regime seems prepared to provide some entertainment for me."

Even though Laue had not worked in the German uranium project he was nonetheless arrested with the other German scientists after the war and taken to England. They were kept at Huntingdon near London for six months, and not even their guards were permitted to know their names. Laue was allowed to return to Germany in 1946, and he became instrumental in the rebuilding of German science. It always saddened him that many German scientists did not return to Germany after the war. In 1950, at the age of 71, he became the director of the Fritz Haber Institute of Physical Chemistry in Berlin-Dahlem.

He held this post until 1959 when he was named director emeritus.

Laue had married Magdalene Degen on October 6, 1910. They had two children, a son and a daughter. He enjoyed mountain climbing, sailing, listening to classical music, and fast driving. On April 8, 1960, as he was hurrying to yet another scientific meeting at age 80, Laue's automobile collided with a motorcycle. On April 23, he died from injuries sustained in the accident; he was buried in Göttingen near Planck and Walther Nernst. Many throughout the world mourned the man who had contributed so much to theoretical physics and a lived a life of high moral principles. A younger colleague who had studied under Laue remarked, as quoted in *Reality and Scientific Truth:* "I know his ethical ideas were on a very high level. His consciousness that whatever he did was just and right, and his judgment of right or wrong, were very definite."

LAWRENCE, ERNEST ORLANDO (1901-1958)
American physicist

Known as the youngest physics professor at the University of California at Berkeley, Ernest Orlando Lawrence played a major role in the development of **nuclear physics**. From 1936 until his death he was director of the Radiation Laboratory at Berkeley, where he invented the **cyclotron**, a device that accelerates the speed of nuclear particles. The invention of the cyclotron inaugurated a new era in the study of nuclear physics, and for this achievement he was awarded the 1939 Nobel Prize in physics. At Lawrence's laboratory several radioactive **isotopes** were discovered; one of these was plutonium, an element essential to the development of the **atomic bomb**. Lawrence played an active role in the research on **nuclear weapons** both during and after World War II, and he remained a strong advocate of increasing America's nuclear arsenal.

Lawrence was born in Canton, South Dakota, on August 8, 1901, the eldest son of Carl Gustav Lawrence and Gunda Jacobson Lawrence. Both parents were of Norwegian ancestry; his mother was a teacher, and his father was a state superintendent of public education who then became president of a teacher's college. Lawrence completed public school in South Dakota at age 16; he began college at Saint Olaf's in Minnesota and then transferred to the University of South Dakota, from which he graduated with a B.S. in 1922. His gift for physics had become apparent, and he went on to earn a master's degree in 1923 at the University of Minnesota. He then followed his physics professor, W. F. G. Swann, first to the University of Chicago and then to Yale, where in 1925 Lawrence completed his doctorate. His dissertation was on the **photoelectric effect**, a phenomenon that occurs when certain electrons give off **light** energy as they change their position around the **nucleus** of an **atom**. By this time he had met such notable physicists as **Albert Michelson** and **Niels Bohr** and had gained a reputation for being an innovative experimenter.

Lawrence left Yale in 1928 for a job at the University of California at Berkeley, which was then a little-known state

Ernest Lawrence, with an atom smasher. *(Corbis-Bettman. Reproduced by permission.)*

school. But the seeds for growth in the field of nuclear physics were sown with the hiring of Lawrence and his contemporary, **J. Robert Oppenheimer**. Berkeley was soon to be a center for nuclear research equal to any other in the world. In 1930, a mere two years later, Lawrence became a full professor at just 29 years of age. While at Yale, Lawrence had met his future wife, Mary Kimberly Blumer, who was the daughter of the dean of Yale's medical faculty. Molly, as she was called, was just 16 at the time. They were married in 1932, and she joined him in California. They would have two sons and four daughters.

Develops the Cyclotron

At Berkeley, Lawrence became interested in the nucleus, or center, of atoms. In particular, he was curious about what kinds of reactions would occur if protons, the positively charged particles in the nucleus, were given high **energy** levels. He had read about a Norwegian engineer who had built a device that accelerated nuclei in a straight line, but it had failed to impart enough **power** to cause **nuclear reactions**. With graduate students Niels E. Edlefsen and M. Stanley Livingston, Lawrence developed a circular, rather than linear, accelerator. The particle accelerator had the shape of a flat circular can cut in two. When particles such as protons were placed in the center of the can, a **magnetic field** forced them to

travel in a circular path. Each time a particle went by, the magnetic field pushed it again. All of these small pushes increased the speed of the particle as it spiraled toward the rim of the can. When it approached the rim, the particle was made to hit a target—a nucleus. The result would be to disintegrate or "smash" the nucleus. Lawrence called the device a cyclotron and he and his team began to improve on it. They were fortunate enough to convince the Federal Telegraph Company to donate a huge 80-ton magnet that was no longer in use. With this magnet, and a bigger cyclotron chamber, Lawrence could impart a great deal of energy onto particles, measured in millions of **electron** volts.

The cyclotron ushered in a new era of experimentation in high-energy physics. In 1932, Lawrence and his team were able to disintegrate a lithium nucleus. As they smashed other nuclei, they were able to measure how much energy is needed to bind the nucleus of an atom together. In other experiments they bombarded other elements to obtain radioactive isotopes of **carbon**, iodine, and uranium. With his younger brother John Lawrence, who was also at Berkeley, Lawrence looked for medical uses for **radioisotopes**; one development was using them as tracers to study metabolism, an application that is still important today. In 1936, at age 35, Lawrence became the director of the Radiation Laboratory at Berkeley. In 1939, with the importance of the cyclotron proven, he received the Nobel Prize. He had already been the recipient of the Comstock Prize of the National Academy of Sciences and the Royal Society's Hughes Medal.

Participates in the Development of the Atomic Bomb

During World War II, the Radiation Laboratory, with Lawrence at the helm, played an important role in developing the atomic bomb. Urged on by German refugees who feared that the Nazis might produce an atomic weapon first, his laboratory worked to obtain the most useful materials for **nuclear fission**. When nuclear fission occurs, the nucleus of some radioactive elements disintegrates, releasing a great deal of energy. Research was done at Berkeley to find a way to separate uranium, and the technique was then used on a larger scale at the laboratory at Oak Ridge, Tennessee, which provided the uranium 235 for the atomic bomb that was eventually dropped on Hiroshima. While Lawrence remained at the Radiation Laboratory, J. Robert Oppenheimer went to direct the laboratory at Los Alamos, New Mexico, where the first bomb was made.

After the dropping of two fission bombs on Hiroshima and Nagasaki, resulting in destruction and loss of life never seen before in human history, there was a division in the ranks of the world's nuclear physicists. In the postwar era, Oppenheimer and others became convinced that the development of further nuclear weapons should cease in the interest of humanity. But Lawrence believed in continuing this work. He argued that it was in the interest of the United States to build even more powerful weapons, such as the hydrogen or thermonuclear bomb. The rift between these two men came to a head when Oppenheimer appeared before a Congressional hearing and subsequently lost his security clearance. The opinion of the government and the public at the time supported Lawrence's advocacy of nuclear buildup.

Lawrence and Hungarian-born physicist **Edward Teller** founded a second radiation laboratory in Livermore, California, for research on nuclear weapons. Bigger and better **accelerators** were created that could boost nuclear particles with energy measured in billions of electron volts; thus was born a powerful accelerator known as the bevatron. Lawrence's visions for physics laboratories on a grand scale set the stage for the **particle accelerators** built in the suburbs of New York and Chicago—Brookhaven National Laboratory on Long Island and the **Fermi National Accelerator Laboratory** (**Fermilab**) in Geneva, Illinois. He also influenced the creation of an accelerator at the **European Center for Nuclear Research** (**CERN**) in Switzerland. Each laboratory was set up with an international team of physicists, in much the same way the Radiation Laboratory had been.

During the 1950s, Lawrence oversaw operations in the nuclear research laboratories he had helped to fashion. His original cyclotron had been superseded by newer and faster ones designed by other scientists, but he continued to show his inventiveness, both in nuclear physics and in applied physics. He even designed a novel type of **color** television picture tube. In 1957 he was awarded the Fermi Award, the highest honor in physics in the United States. Lawrence was also busy as a government consultant on nuclear-energy issues for the Eisenhower administration. In the summer of 1958 he attended a conference in Geneva on the possibility of detecting violations of nuclear test agreements, but, ill and weary from recurrent ulcerative colitis, he flew home for an operation. He did not survive and died on August 27, 1958.

In 1961, the transuranium element 103 was discovered at Berkeley and was named lawrencium in his honor. The Lawrence Hall of Science at the university is a museum and research center dedicated to improving science education. The Radiation Laboratory is now known as the Lawrence Berkeley Laboratory, and the facilities in Livermore, California, are now called the Lawrence Livermore Laboratory.

LEAVITT, HENRIETTA (1868-1921)

American astronomer

Henrietta Leavitt's most famous discovery was the "period-luminosity" relation for variable **stars** (those changing in brightness), an important method of obtaining distances to far-off galaxies. She also identified 2,400 new variable stars and established brightness scales that helped other astronomers with their own observations.

Henrietta Swan Leavitt was born in Lancaster, Massachusetts, on July 4, 1868, to Henrietta Swan Kendrick and George Roswell Leavitt, a Congregationalist minister who had a parish in Cambridge. After attending public school in Cambridge, Leavitt moved with her family to Cleveland, Ohio, where she attended Oberlin College from 1885 to 1888. She switched, however, to the Society for the Collegiate

Instruction of Women (now Radcliffe College) in 1892. During her senior year, she took an **astronomy** course, which fired her interest in the subject. After receiving an A.B. from Radcliffe in 1892, Leavitt took another astronomy course, then spent a number of years home because of an illness that left her severely deaf. After some traveling, Leavitt volunteered as a research assistant at Harvard College Observatory in 1895, and was appointed to the permanent staff in 1902 by the astronomer Edward Pickering at a salary of 30 cents an hour. Leavitt worked at Harvard from 1902 until her death. While Pickering gave Leavitt little chance to do theoretical work, she became chief of the photographic photometry department at the observatory.

Establishes Scale of Star Brightnesses

In 1907 Pickering asked Leavitt to establish a "north polar sequence" of star brightnesses to serve as a standard for the entire sky. This standard was desirable because the photographic process in astronomy was complex—the brightness of star images on film was not proportional to their actual brightness, and each **telescope** gave different results for different **color** stars. Once determined, brightnesses could be estimated by comparing one star with another, rather than referring to photographic images. Leavitt used 299 plates from 13 telescopes, and compared stars ranging from the fourth to the twenty-first magnitude in brightness (each increasing unit of magnitude corresponds to a reduction in brightness by a factor of 2.512 on a logarithmic scale). The results were published in the *Annals of Harvard College Observatory*. In 1913 Leavitt's system was adopted by the International Committee on Photographic Magnitudes. She made this work a lifelong project, and established brightness sequences for 108 areas in the sky. When Pickering developed 48 "Harvard standard regions" in the sky, Leavitt derived secondary brightness standards for them. These were used as international standards until superseded by improved methods.

Discovers Period-Luminosity Relation

Leavitt discovered 2,400 new variable stars, about half those known at the time. Most notably, she studied photographs of the Magellanic Clouds (the Milky Way's two companion galaxies) taken at Harvard's observatory in Arequipo, Peru. Of the 1,800 variable stars Leavitt detected on the Magellanic Cloud pictures, some were Cepheid variables, whose change in brightness is extremely regular. (Cepheids were named after the first star of this type to be discovered, Delta Cephei.) In 1908 Leavitt found that the brighter the Cepheid was overall, the longer it took to change its magnitude. Leavitt reasoned that since the Cepheids in the Magellanic Clouds were nearly all the same distance from Earth, their periods were related to their **light** output: the longer the period of pulsation, the brighter the star. By 1912 Leavitt had proven that Cepheids' apparent brightness increased linearly with the logarithm of their periods. That year, she published a table of the length of 25 Cepheid periods—ranging from 1.253 days to 127 days, with an average of 5 days—and their apparent brightness.

Before this period-luminosity relation had been established, cosmic distances could be determined only out to about 100 light-years. With it, however, distances of out to 10 million light-years could be calculated. The intrinsic brightness of a Cepheid could now be gotten directly from a measure of its period, and this allowed the distance to be calculated. Astronomer **Ejnar Hertzsprung** adapted the period-luminosity relation so it could determine the actual distance of stars from Earth. Hertzsprung and astronomer **Harlow Shapley** found that the Magellanic Clouds were in the range of 100,000 light-years from Earth—an astonishing and unexpectedly high value. Leavitt was not aware as she worked that the Magellanic Clouds lay outside our own galaxy.

Leavitt died of cancer December 21, 1921, in Cambridge. During her lifetime she had little opportunity to give free rein to her intellect, but her talents did not go unnoticed. According to *Women of Science,* her colleague Margaret Harwood described her as possessing the best mind at the observatory, and contemporary astronomer Cecilia Payne-Gaposchkin said Leavitt was the most brilliant woman at Harvard.

LEDERMAN, LEON MAX (1922-)
American experimental physicist

Leon Max Lederman, physicist, educator, and advocate for basic research, has spent his scientific career designing experiments and devices for unlocking the structure of the **atom**. During the heyday of physics research at Columbia University in New York City, Lederman participated in pioneering experiments that proved the existence of a new kind of **neutrino**—a subatomic particle not observed previously. These experiments also established a new technique for probing physical phenomena involving the "weak" nuclear **force**, one of the three fundamental forces that rule the atom. For this work, Lederman shared in the 1988 Nobel Prize for physics.

Lederman, son of Russian immigrants Morris Lederman and Minna Rosenberg, was born on July 15, 1922, in New York City, where his father operated a laundry. Lederman pursued his undergraduate education at the City College of New York, earning a baccalaureate degree in chemistry in 1943. After serving three years in the U.S. Army, Lederman enrolled at Columbia University in New York City to study physics. He began his doctoral research in 1948, designing a special instrument to study **subatomic particles** produced by a new atom-smasher (the synchrocyclotron accelerator) then under construction at Columbia's Nevis Laboratories in Irvington, New York. In 1950, while still a graduate student, Lederman and physicist John Tinlot made special modifications to the Nevis accelerator, transforming the machine into a significantly more powerful research tool. In his doctoral research, Lederman used the new Columbia accelerator to measure the lifetimes of two different kinds of **pions**, a type of **meson** or fundamental particle that the

machine produced in abundance. Lederman received his Ph.D. in 1951 and held various teaching and research positions at Columbia University until 1989.

Conducts Subatomic Particle Research

A handful of experiments Lederman participated in during the 1950s proved especially significant. In 1956, he helped discover a new particle, the "long-lived neutral kaon." From 1959 to 1962 Lederman and his colleagues **Melvin Schwartz** and **Jack Steinberger** conducted research on subatomic particles for which they were later awarded the Nobel Prize in physics. The scientists used the accelerator at Brookhaven National Laboratories in Upton, New York, to find a previously unobserved particle—the **muon** neutrino. Notoriously elusive creatures, neutrinos are difficult to study because they have no **electric charge** and no detectable **mass**.

In the experiments, the scientists fired a beam of protons at a block of beryllium metal, producing a blast of subatomic debris, including neutrinos. A 40-ft-thick (12-m) barrier of steel plates culled from a battleship in the process of demolition absorbed all particles but the neutrinos. After eight months of shooting protons, the scientists found that some 56 neutrinos had betrayed their existence in the accelerator's 10-ton particle detector, the result of two different kinds of interactions involving the **weak force**, a fundamental physical force responsible for particle decay processes in **radioactivity**. One of the particles, the **electron** neutrino, had been known to physicists already. But the scientists also found the elusive muon neutrino, produced only during high-**energy** collisions.

This result provided a key theoretical insight, laying the cornerstone for the standard model of the atom: that there are two kinds of neutrinos, one intimately associated with electrons and the other with muons. But just as important, the celebrated "two-neutrino experiment" marked the first time anyone had artificially created a beam of neutrinos in the laboratory. The neutrino-beam technique, adopted at accelerator laboratories around the world, made possible subsequent experiments probing the weak nuclear force, which operates at short range and governs certain kinds of **radioactive decay**. Importantly, neutrino collisions produce only interactions involving the weak force.

A leader in the particle-physics community, Lederman served as an administrator at research institutions organized around the giant **particle accelerators** that came to dominate high-energy physics in the postwar era. From 1962 to 1979, he headed Columbia's Nevis Laboratories. During these years he was involved with original research much of his time. From 1976 to 1977, for instance, he participated in the discovery of a new, massive particle—the upsilon—that belonged to an entirely new grouping of subatomic particles. For this, Lederman shared the 1982 Wolf Prize, awarded by Israel, with colleague **Martin Perl**.

Then, from 1979 to 1989, Lederman directed the **Fermi National Accelerator Laboratory (Fermilab)** in Batavia, Illinois. Throughout his career, Lederman's research has remained centered on experiments and advanced equipment. It is not surprising, then, that in 1982 Lederman challenged his colleagues in the physics world to commit themselves to constructing the most powerful accelerator ever conceived—later christened the Superconducting Super Collider (SSC). Lederman and others argue that the giant SSC, designed to smash particles into each other at high energies, was needed to further explore the fundamental structure of **matter**.

Lederman played a significant role in advancing his field far enough that a large, next-generation accelerator like the SSC has become necessary to both confirm and extend existing theories. Lederman's work, done in collaboration with other researchers, helped to unveil a number of new subatomic particles. These discoveries enabled theorists to construct a "standard model" of the atom, listing the hierarchy of **quarks** and **leptons** thought to make up all matter in the **universe**.

After the Nobel Prize

Recognition for Lederman's research at Columbia University came in 1988, when he and former colleagues Schwartz and Steinberger were awarded the Nobel Prize in physics for the research on neutrinos they conducted from 1959 to 1962. After receiving the prize, Lederman ended his tenure as director of Fermilab, retaining the title of emeritus director. Concerned with the level of understanding and appreciation of science in the United States, Lederman returned to teaching in 1989, accepting a professorship at the University of Chicago. There he taught an undergraduate course popularly known as "physics for poets." Following his 1992 retirement from the University of Chicago, Lederman became a professor at the Illinois Institute of Technology in Chicago, teaching first-year physics.

In 1945, Lederman married Florence Gordon, with whom he had three children, Rena, Jesse, and Heidi. He later married Ellen Carr in 1981. A member of the National Academy of Sciences, Lederman served as president of the American Association for the Advancement of Science from 1990 to 1991. He has received numerous honorary degrees and awards, including the National Medal of Science in 1965. He has coauthored over 200 scientific papers.

LEE, TSUNG-DAO (1926-)

American physicist

Tsung-Dao Lee and his colleague, physicist Chen Ning Yang, developed the revolutionary theory that the unusual behavior of the **K meson** (a subatomic particle) is a result of its violating a supposedly inviolable law of nature, conservation of **parity**, which defines the basic symmetry of nature. A few months after their theory had been announced, fellow physicist **Chien-Shiung Wu** obtained experimental confirmation of their remarkable discovery. For their work, Lee and Yang were awarded the 1957 Nobel Prize in physics.

Lee was born in Shanghai, China, on November 24, 1926. He was the third of six children born to Tsing-Kong Lee, a businessman, and Ming-Chang Chang. Lee attended the Kiangsi Middle School in Kanchow and, after graduation,

entered the National Chekian University in Kweichow. After the invasion of Japanese troops in 1945, Lee fled to the south, where he continued his studies at the National Southwest Associated University in Kunming.

Immigrates to the United States

In 1946, Lee was presented with an unusual opportunity. One of his teachers at Kunming was the theoretical physicist Ta-You Wu. When Wu decided to return to the United States (where he had worked toward his Ph.D. degree), he invited Lee to accompany him. Lee accepted the offer, but found himself in a somewhat peculiar position. He had not yet received his bachelor's degree and found that only one American university would accept him for graduate study without a degree. He therefore decided to enroll in that institution, the University of Chicago.

At Chicago, Lee selected a topic in **astrophysics** for his doctoral research. Working under physicist **Enrico Fermi**, he completed that research and was awarded his Ph.D. in 1950 for his dissertation, on the hydrogen content of white dwarf **stars**. While at Chicago, Lee also renewed his friendship with physicist Chen Ning Yang. Lee and Yang had been acquaintances at Kunming, but they became very close friends after both reached the United States. They were separated in 1950 when Lee went to the Yerkes Astronomical Observatory at Lake Geneva, Wisconsin, and Yang went to the Institute for Advanced Studies at Princeton University. Lee then spent the next year (1950–51) as a research associate at the University of California at Berkeley. The two friends were reunited in 1951, however, when Lee accepted an appointment at the Institute for Advanced Studies.

Discussions with Yang Lead to Revolutionary Theory

Lee's departure from Princeton in 1953 for a post as assistant professor of physics at Columbia University seems to have had little effect on his collaboration with Yang. The two worked out a schedule that allowed them to continue meeting once a week, either in New York City or in Princeton. By the spring of 1956, these regular meetings had begun to focus on a particularly interesting subject, a subatomic particle known as the K meson. Discovered only a few years earlier, the K meson puzzled physicists because it appeared to be a single particle that decayed in two different ways. The decay schemes were so different that physicists had become convinced that two distinct forms of the K meson existed, forms they called the tau **meson** and theta meson.

The single difference between these two mesons was that one form had even parity and the other form had odd parity. The term "parity" refers to the theory that the laws of nature are not biased in any particular direction. That is, if one has two sets of interactions that are mirror images of each other, the physical laws describing those interactions are identical. This concept is known as the conservation of parity, a concept long held by physicists.

The problem that Lee and Yang attacked was that vast amounts of experimental evidence suggested that the theta and tau mesons were one and the same particle. The only contrary evidence was that the two mesons had opposite parity and, therefore, supposedly could not be identical. During an intense three-week period of work in the spring of 1956, Lee and Yang solved this puzzle. Their solution was to suggest, simply enough, that in some types of reactions, parity is not conserved. The **beta decay** of the (one and only) K meson was such a reaction. They then devised a series of experiments by which their theory could be tested. The fundamental elements in the Lee-Yang theory were announced in a paper sent to the *Physical Review* on June 22, 1956, and later given the title, "Question of Parity Conservation in Weak Interactions."

About six months later, the experiments suggested by Lee and Yang were carried out by one of their colleagues, Chien-Shiung Wu, first at Columbia and then at the National Bureau of Standards. The experiments confirmed the Lee-Yang prediction in every respect. Less than a year later, the two theorists were awarded the 1957 Nobel Prize in physics for their work.

After promotions to associate professor (1955) and professor (1956) at Columbia, Lee returned to the Institute for Advanced Studies in 1960 for three years. He then was appointed Enrico Fermi Professor of Physics at Columbia in 1963. In 1984, he was made University Professor at Columbia. Beginning in 1981, Lee held appointments as honorary professor at a number of Chinese universities, including the University of Science and Technology (1981), Jinan University (1982), Fudan University (1982), Quinghua University (1984), Peking University (1985), Nanjing University (1985), and Zhejiang University (1988). He married Hui-Chung Chin (also known as Jeanette) on June 3, 1950, while they were both students at Chicago. The Lees have two sons, James and Stephen.

LEIBNIZ, GOTTFRIED WILHELM (1646-1716)

German mathematician and philosopher

Gottfried Wilhelm Leibniz began life as a child prodigy, yet his brilliance and enormous talents remained in evidence throughout his entire lifetime. Liebniz's father, a professor of philosophy, died when he was only six years old. By the age of eight, Leibniz had taught himself Latin, and by 14 he had already mastered the Greek language. Educated at home, Leibniz often spent his spare time in his father's library, reading historical or classical literature. Leibniz entered the University of Leipzig when he was 15. At 17 he had completed his studies, but stayed on to earn his doctorate in law in 1665 at age 20.

During the next six years, Leibniz worked in the legal and diplomatic professions. While on a diplomatic mission to Paris in 1672, Leibniz became fascinated with mathematics and began to study the subject fervently. It was Leibniz's ambition to reform all of science by the use of a universal sci-

Gottfried Wilhelm Leibniz.

entific language and a calculus of reasoning. These thoughts, concepts that were years ahead of their time, were always at the forefront of his studies.

Leibniz particularly wished to reduce the tedium of **astronomy** and navigational calculations and believed that a mechanical calculating device could be the solution. After studying **Blaise Pascal**'s calculator, he realized that he could improve upon its design to perform multiplication and division—functions that Pascal's device could not do. In 1671 he introduced his calculator, which became quite a success. Multiplication and division were accomplished by reversing the addition and subtraction functions on the machine. Also quite innovative was Leibniz's use of multiplier wheels and a delay carry mechanism—used in calculators for the next 300 years. His machine is preserved in the state library at Hanover, Germany.

In addition to his calculator, Leibniz achieved great success in mathematics and science. As early as 1666, he published his first paper on mathematics. Entitled *De arte combinatorica*, this paper represented Leibniz's belief that the logic of thought could be reduced to mathematical calculations—concepts put to use years later in development of Boolean algebra and digital computer design.

In the years following his studies in symbolic logic, Leibniz focused his genius toward developing a system of differential and integral calculus. In 1675 he published his fundamental theorem of calculus. Reduced to its simplest form, it states that the process of summation of integration was the same as reversing the operation of differentiation. Although **Isaac Newton** had already done similar work some years earlier, there is no evidence that either was aware of the other's work. Leibniz's notation was considered superior and came to be preferred over that of Newton, causing a bitter controversy between the two. In the years following, Leibniz lost much of his favor among royal society.

In 1676 Leibniz left Paris and moved to Hanover where he served as a librarian and historian. His work with arithmetic, however, was far from over. In 1679 he began studying the use of binary numbers, a concept he traced back to the Chinese of the eleventh century. Leibniz was the first to recognize the importance of binary 0 and 1. The significance of this discovery was not realized until more than 250 years later when scientists began programming the digital computer.

In his later years, Leibniz's interests widened to include geology, psychology, history, philosophy, and genealogy. He published many significant papers and continued to hold several diplomatic positions. In one such paper, *Protogaea*, published posthumously in 1749, Leibniz explained his theory of the passage of Earth from vapor to a molten globular state into its present form. In addition, he accounted for geological irregularities in Earth's crust, expressing views that would be substantiated by geologists many years later.

Wishing to keep up to date with London and Paris scientists, he persuaded Prince Frederick of Prussia (1657-1713; crowned King Frederick I in 1701) to create the Berlin Academy of Science in 1700. Leibniz served as its first president. Unfortunately, by the time of his death in 1716, many in the scientific and mathematical communities had already forgotten him; the only mourner at his funeral was his secretary. An eyewitness wrote, "He was buried more like a robber than what he really was, the ornament of his country." While his contemporaries may have overlooked his innovative studies, future centuries have not neglected the genius of Leibniz's work and the many invaluable contributions he made.

LEMAÎTRE, GEORGES HENRI (1894-1966)
Belgian astrophysicist

Abbé Georges Lemaître was a man of varied talents. Born on July 17, 1894, in Charleroi, he attended the University of Louvain and studied civil engineering. He became an artillery officer when World War I started in 1914. After the war he received his Ph.D. and, in 1922, was ordained a priest. Then he went on to study **astrophysics** at Cambridge University, received a Ph.D. from the Massachusetts Institute of Technology in 1927, and returned to Louvain, Belgium, to teach. Lemaître became very interested in **Albert Einstein**'s equations involving **gravity**. When he solved the equations, he discovered they described a **universe** that should be expanding, not static as was commonly believed. Unknown to Lemaître, a Soviet meteorologist, Alexander A. Friedmann (1888-1925), had come to a similar conclusion in 1922. There

was no observational evidence to substantiate that idea, and Einstein would not support the theory until he had studied Lemaître's calculations. Within two years American astronomer **Edwin Powell Hubble** made observations that showed the galaxies were traveling away from Earth at incredible speeds. He concluded, unaware of Lemaître's theory, that the universe was expanding outward. Now direct observational evidence existed that supported Lemaître's theory. Einstein agreed that Lemaître's solutions to the equations of gravity were the best. Lemaître went one step further, reasoning it should be possible to reverse the concept of expansion and hypothesize backward in time. He suggested that all the **matter** in the universe originally existed as a compressed "cosmic egg" that had exploded, throwing matter outward. Based on Hubble's calculations, Lemaître's **big bang** would have occurred two billion years ago. That was impossible; scientists were very certain that Earth's crust was older than that. But in 1952, astronomer Walter Baade revised Hubble's figures and a new, much larger scale emerged. It pushed the origin of the big bang further into the past. The figure accepted today is 15 billion years. Lemaître also predicted that some background "noise," left over from the ancient big bang, should still be detectable. This too was proven by **Arno Penzias** and Robert Wilson (1936-) in 1965, the year before Lemaître's death.

LENARD, PHILIPP E. A. (1862-1947)
German physicist

Philipp E. A. Lenard was a brilliant experimental physicist who made important contributions to the study of the **photoelectric effect**, the characterization of cathode rays, the nature of **phosphorescence**, ionization potentials, and other phenomena. He was awarded the 1905 Nobel Prize in physics primarily for his work on cathode rays. He taught and carried on research at a number of German universities, including Heidelberg, Bonn, Breslau, and Kiel, as well as at the Technische Hochschule in Aachen. In spite of his gifts as an experimentalist, Lenard is also remembered as an anti-Semite and enthusiastic proponent of National Socialism. His unwillingness to embrace many of the new developments in science taking place in the 1920s and 1930s is thought to be at least partially due to his political feelings that these ideas were false pronouncements from "Jewish science." By the end of his life, Lenard had lost the personal respect and professional esteem of the great majority of his colleagues in the scientific community.

Philipp Eduard Anton Lenard was born on June 7, 1862, in Pressburg, Hungary (now Bratislava, Slovakia). His father was Philipp Lenard, a prosperous wine merchant, and his mother, the former Antonia Baumann. Young Philipp was expected to take over his father's business, but was not enthusiastic about that prospect. He attended technical universities in Budapest and Hungary, and spent a year working for his father before a trip to Germany rekindled his desire to study science. After attending the lectures of German chemist Robert Bunsen, Lenard decided to study physics at the University of Heidelberg in Germany, in the winter of 1883. He remained there for four semesters, spend-

  •

ing two more semesters at the University of Berlin before returning to Heidelberg, where he received a doctorate with high honors in 1886.

Takes Posts at Heidelberg, Bonn, Breslau, Aachen, and Kiel

Lenard was asked to stay on at Heidelberg as assistant to his former teacher, Georg Quincke. During his three years in that position, Lenard began research on phosphorescence, a topic that was to occupy him off and on for the next four decades. He was later to discover that phosphorescence occurs as a result of impurities in a material and that it attains some maximum intensity for a given concentration of impurity. After leaving Heidelberg in 1890, Lenard visited England where he worked at the electromagnetic and engineering laboratories of the City and Guilds of the London Central Institution. He soon came to dislike the English, however, and left after six months to take a post at the University of Breslau. After only one semester there, he accepted an appointment as an assistant to **Heinrich Hertz** at the University of Bonn, where he began to experiment with cathode rays.

When Hertz died unexpectedly in 1894 at the age of 36, Lenard became responsible for publishing Hertz's mammoth three-volume work *Gesammelte Werke,* interrupting his own research. He continued this task when he moved back to the University of Breslau in 1894 to become associate professor of physics. Once more, he remained at Breslau only briefly before taking consecutive posts at the Technische Hochschule in Aachen (1895–96) and the University of Heidelberg (1896–98). Finally, in 1898, he was appointed professor of physics and director of the physics laboratory at the University of Kiel, where he would remain for nine years.

Makes Important Discoveries about Cathode Rays and the Photoelectric Effect

In the years prior to 1914, Lenard made his most important scientific discoveries, particularly those relating to his studies of cathode rays and the photoelectric effect. Research on the former had been inspired by **William Crookes**'s 1879 discovery of cathode rays, beams of charged particles produced during the discharge of **electricity** in **vacuum tubes**. Lenard was interested (as were other researchers) in finding a way to study cathode rays outside of vacuum tubes. Based on Hertz's discovery that cathode rays could permeate thin metal sheets, Lenard found, in 1892, that cathode rays can be made to pass out of a vacuum tube through a thin aluminum window. Using this arrangement, Lenard was able to make a number of observations regarding the properties of cathode rays.

As a result of these studies, Lenard came to the conclusion that matter—and, therefore, the atoms of which **matter** consists—is largely empty **space**. He suggested that an **atom** consists of neutral pairs of dynamids, one positively charged and the other negatively charged. Though incorrect, this concept presaged the nuclear model of the atom developed by **Ernest Rutherford** a decade later. It was for his work on cath-

ode rays that Lenard was awarded the 1905 Nobel Prize in physics.

During the same period, Lenard also studied the photoelectric effect, the phenomenon that occurs when a beam of **light** strikes a metal plate, resulting in the emission of electrons from the plate. Lenard made the interesting discovery that the **velocity** of electrons emitted during this process was affected by the **wavelength** of the incident light, and that an increase in that light's intensity increased the number of electrons emitted, but not their speed. He was unable to explain this result, an anomaly unpredictable with the laws of classical physics. In fact, it was not until **Albert Einstein** applied the principles of **quantum theory** to this phenomenon that Lenard's results were correctly interpreted, an accomplishment that won for Einstein the 1921 Nobel Prize in physics.

Lenard was an enormously complex individual about whom a great deal has been written. He experienced a number of personal disillusionments, as when he failed to discover **x rays** only a short time before his countryman, **Wilhelm Conrad Röntgen**, did so. This disappointment was especially hard on Lenard since he had made substantial contributions to the theoretical and experimental aspects of Röntgen's discovery, contributions that Röntgen never acknowledged.

As Lenard grew older, he began to develop profound anti-Semitic feelings. In his unpublished autobiography, for example, quoted in Alan D. Beyerchen's *Scientists under Hitler: Politics and the Physics Community in the Third Reich,* Lenard describes the theory of relativity as "a Jewish fraud, which one could have suspected from the first with more racial knowledge than was then disseminated, since its originator Einstein was a Jew." Ultimately, Lenard rejected the traditional notion that science is an international activity that transcends national boundaries. Instead, he wrote in his textbook on experimental physics, *Deutsche Physik* (German Physics), "In reality, science, like everything man produces, is racially determined, determined by blood."

Before long, Lenard became one of the leaders of Aryan physics, an attempt to keep science pure of Jewish and other foreign influences, and he enthusiastically supported Adolf Hitler and his political agenda. When Hitler was named chancellor in 1933, Lenard wrote to him, offering his skills as a scientific adviser while pointing out how badly the nation's university system needed cleansing of Jewish influences.

In the decade after Hitler's ascent to power, Lenard was constantly involved in power struggles aimed at gaining control of the nation's scientific establishment, though, in his early seventies by then, he was probably too old to make many substantive contributions to the Nazi regime. In addition, he was defeated at nearly every turn by less-able, lesser-known scientific figures with more important contacts to inner circles of Nazi politics. Instead, he was given a number of awards and honorary appointments, such as the Eagle Shield of the Reich and election to the Executive Committee of the German Research Association. He survived the war, but was expelled from his post at the University of Heidelberg by Allied officials in 1945. After this disgrace, he settled in the village of Messelhausen, close to Heidelberg, where he died on May 20,

1947. Lenard had been married in 1897 to Katherine Schlehner. The one child of whom we know, Werner, died in February 1922, due to malnutrition brought on by wartime conditions in World War I.

LENGTH CONTRACTION

Length contraction is one of the primary phenomena predicted by **Albert Einstein**'s theory of special relativity published in 1905. The theory states that the measured distance between two points in **space** depends upon the observer's **frame of reference**. In brief, an object will always appear shorter from the frame of reference of an observer moving with respect to the object. Hence, the object appears to be contracted.

An observer who is stationary with respect to an object will measure the object's *proper length* L_0. The object can never be measured to be longer than the proper length. However, a second observer, moving with a constant **velocity** v with respect to the object and in a direction parallel to the object's length, will measure the length of the object to be shorter. The second observer will measure the length of the object to be

$$L = L_0 \sqrt{1 - \frac{v^2}{c^2}}$$

where c is the speed of **light** in a **vacuum**, measured to be 186,000 mi/s (300,000 km/s). This contraction takes place only along the direction of the object's **motion**. Therefore, the dimensions of the object perpendicular to the object's direction of motion appear the same as when the object is stationary.

Together with the prediction of **time dilation**, length contraction has been used to explain a wide variety of unusual phenomena dating back to the beginning of the twentieth century. One of the most striking observations of length contraction is in the phenomenon of **muon** decay. A *muon* is an unstable particle that looks like an **electron** with 207 times more **mass** and that lives for only 0.0000022 seconds before decaying into smaller particles. Muons can be produced from the collision of cosmic **radiation** with atoms and molecules in Earth's upper atmosphere, roughly 6.3 mi (10 km) above sea level. When produced, these muons are traveling at roughly 99.8% of the **speed of light**, but even at these speeds we expect the muon to decay before it reaches the ground. That is, the muon does not live long enough to travel such a long distance before decaying.

However, from the muon's frame of reference the height of the atmosphere is length contracted. To the muon, the atmosphere and surface of Earth are traveling toward the muon at 99.8% of the speed of light. Hence, the muon observes the total distance it must travel to be shorter by a factor of sqrt(1 − (0.998)²) = 0.063, or simply that it must travel only roughly 2,100 ft (640 m). From the muon's reference frame, the total distance traveled is roughly 2,200 ft (660 m),

and it is for this reason that we observe muons at the surface of Earth.

LENSES

A lens is a piece of transparent material shaped such that it bends parallel rays of **light**. The shape of a lens is a segment of a sphere, plane, or a cone (parabola, hyperbola, or ellipse). The bending of light causes the rays to cross, forming an image. This image can be larger or smaller than the actual object being viewed through the lens. The bending of light as it passes through a medium such as glass or plastic is called refraction. The early Greeks used lenses in magnifying glasses long before the ideas of refraction were understood. Today, lenses are used in many **optical instruments**, such as cameras, eyeglasses, and movie projectors.

Refraction is the bending of light that occurs when light travels through one medium into another. The amount a light ray is bent depends on the angle the ray makes with the interface between the two materials and the difference in the indices of refraction for the two materials. The index of refraction for a material is related to the **velocity** of light in that material. Light travels faster through air than through glass or plastic. Refraction is a result of this change in velocity of light.

Lenses can occur in one of two shapes. A lens that is thicker in the middle than at the edges is called a convex or converging lens, and a lens that is thinner in the middle is called a concave or diverging lens. A converging lens causes parallel light rays that pass through it to converge, or come together to a point, after it exits the lens. A diverging lens causes the light to diverge, or separate, after it exits.

Parallel light rays entering a converging lens will be bent to arrive at a point called the focal point. The optical axis of a lens is an imaginary line that passes through the two centers of curvature for the opposite surfaces. If entering light rays are parallel to this axis, the focal point will lie on this axis. If the entering light rays are parallel to each other but at an angle to the optical axis, they will arrive at a point in a plane that is perpendicular to the optical axis. This plane is called the focal plane. A lens has two focal points and two focal planes, each on either side of the lens, since light can travel through the lens in either direction.

Because a diverging lens does not focus light to a point, its focal point has a different definition. A diverging lens creates an image that appears to have originated from a point on the opposite side of the lens. This perceived point is the focal point. The focal length of a diverging lens is the same as that of a converging lens, and is defined as the distance between the center of the lens and its focal point.

A converging lens can produce both a virtual and a real image. A virtual image is one in which the light seems to diverge from the image point, when in fact it does not. An example is the image seen in a mirror. A virtual image occurs in a converging lens when the object is between the focal point and the lens. An example of a converging lens producing a virtual image is when a magnifying glass produces a larger

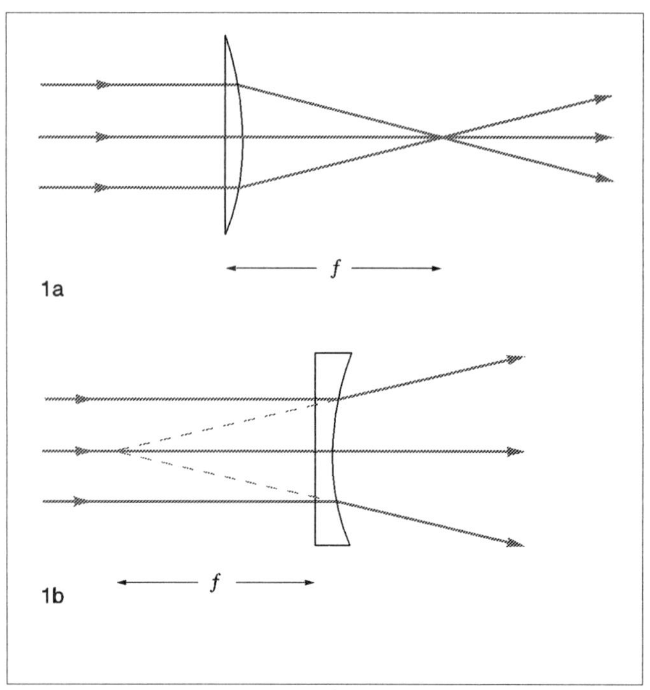

Lenses showing light dispersion.

image. The object to be magnified, for example, small print in a book, is placed between the focal point and the magnifying glass. The print appears larger as you look at it through the lens. A virtual image produced by a converging lens is always right-side-up and magnified. If an object is placed far enough away from a converging lens to be past the focal point, the light passing through the lens does converge to a point on the other side of the lens producing a real image. A real image is one in which the light passes through the image point. The real image produced by a converging lens is seen inverted, or upside-down. The size of the image depends on how far away the object is from the focal point. If the object is close to the focal point, the image will appear larger (it will focus farther away) than if the object is far from the focal point. The object viewed is always on the opposite side of the lens as its image. An example of a converging lens producing a real image is when a camera lens focuses an image on film.

A diverging lens can only produce a virtual image, because the light passing through a diverging lens never converges to a point. The virtual image produced by a diverging lens is always right-side-up and smaller than the original object. The image and the object viewed are always on the same side of the lens. Diverging lenses are used as viewfinders in cameras.

The image size and distance from the lens can be calculated using the following equation, called the thin-lens equation: $1/p + 1/q = 1/f$, where p is the distance from the object to the lens, q is the distance from the image to the lens, and f is the focal length for the lens. Certain sign conventions apply to the thin lens equation. The image distance, q, has a positive value for real images, and a negative value for virtual images.

Likewise, the object distance, p, is positive for real objects and negative for virtual objects.

The magnification of a lens can also be calculated using this equation: $M = h'/h = -q/p$, where M is the magnification, h' is the image height, and h is the object height. The same sign conventions for the thin-lens equation apply to the magnification equation. If the magnification is less than one, the image is smaller than the object. If the magnification is greater than one, the image is larger than the object. If the magnification is equal to one, the image is the same size as the object. A negative value for magnification means the image is real and inverted, and a positive value means the image is upright and virtual.

Lenses are used in many optical instruments such as magnifying glasses, cameras, **telescopes**, **microscopes**, and movie projectors. Lenses can be combined in compound optical instruments, such as the **compound microscope**. In this situation, the image of the first lens acts as the object for the second lens. This arrangement produces a better image because the second lens produces parallel rays that then enter the eye.

See also Speed of light

LENZ, HEINRICH (1804-1865)
Russian physicist

Heinrich Friedrich Emil Lenz, also called Emil Khristianovich, was a Russian physicist best known for his contributions to electromagnetic research. During his career as professor and dean at the University of St. Petersburg, Lenz published the two-volume *Handbook of Physics* along with 50 monographs and papers, most of which appeared in German periodicals. Early in his career, Lenz developed **Lenz's law** on the direction of induced current in an electromagnetic demonstration.

Very little is known of Lenz, partly because he worked far from the intellectual centers of Europe, and partly because Lenz, in his preoccupation with his scientific work, kept few personal records. He was born in Dorpat, Estonia, and studied theology before switching to science. He received a doctor of philosophy degree, which was common for scientists at the time, and traveled around the world at the age of 20, writing as a naturalist. Gradually his interests turned to physics, and Lenz was elected to the Imperial Academy of Sciences at St. Petersburg, where he had presented scientific papers.

Lenz's seventh paper, "On the Laws which Govern the Action of a Magnet upon a Spiral,"sealed his reputation as a physicist. In this paper, he spelled out Lenz's law, in which he observed that induced currents never support and always oppose the changes by which they are induced. Lenz's law is still included in standard physics texts. However, the full impact of Lenz's original work may have been obscured by the opening of the paper, in which he verified **Michael Faraday**'s findings. Lenz tried to establish quantitative or measurable relationships between different magnetic effects, and this set him apart from the electromagnetic research of **Joseph Henry** and Michael Faraday, whose work was more qualitative, or descriptive.

After his first electromagnetic studies, Lenz understood that he must map out the characteristics of the "electromotive spiral," as he called it. To carry out this research, Lenz designed a galvanometer, a soft iron cylinder that was wound with 72 turns of very thin wire. Lenz had read **Georg Simon Ohm**'s laws of electrical circuitry and incorporated these newly discovered findings into his own quantitative research. Lenz's notes suggest that he may have arrived at these laws on his own before reading Ohm. Lenz also discovered that **electrical resistance** depends on **temperature**.

LENZ'S LAW

In 1834, Russian physicist **Heinrich Lenz** discovered the directional relationships between induced **magnetic fields**, **voltage**, and current when a conductor is passed within the lines of a magnetic field. Lenz's law states:

"An induced electromotive force generates a current that induces a counter magnetic field that opposes the magnetic field generating the current."

Physically, Lenz's law merely tells us what direction an induced current flows in a conductor given a change in the magnetic field through the conductor. The induced current circulates in a sense such that the magnetic field it produces opposes the change in the field that induced the current to begin with. This opposition can be compared to the opposing **force** two magnets experience when pushed together with like poles facing each other. Mathematically, Lenz's law is used to keep track of the negative sign in Faraday's law, which gives the relationship between a changing magnetic field and the voltage produced by the change in the field.

Of significance to Lenz's law is its relationship to the **conservation of energy**. Lenz's law states the fact that **energy** is conserved for the case of induced currents and the magnetic fields they produce. This can be seen by assuming that if the induced current circulated in a sense that reinforced the change in magnetic field instead of opposing it, this would lead to a greater induced current, which would lead to greater reinforcing of the field. This cycle would continue ad infinitum, thus breaking the law of conservation of energy.

See also Faraday's Laws

LEONARDO DA VINCI (1452-1519)
Italian scientist and inventor

A true Renaissance man, Leonardo da Vinci was a painter, inventor, scientist, architect, engineer, mathematician, astronomer, and philosopher. Although centuries after his death Leonardo remains known primarily as the artist who painted the Last Supper and the Mona Lisa, Leonardo placed a stronger emphasis on his scientific rather than his artistic

Leonardo da Vinci.

endeavors. His investigations into almost every field of known science in his time resulted in plans for everything from airplanes to air conditioning systems. Leonardo was also prolific in the field of mathematics and physics, including squaring the circle and calculating the **velocity** of a falling object.

Born in Vinci, near Florence, Italy, Leonardo was the illegitimate son of Ser Piero da Vinci, a notary, and a peasant woman. Leonardo's father recognized his genius early and ensured that Leonardo received a proper education in reading, writing, and arithmetic at his home. Leonardo never attended a university. Rather, at the age of 15, Leonardo was sent to Florence, where he became an apprentice painter under Italian sculptor and painter Andrea del Verrocchio (1436-1488). It was during this apprenticeship that Leonardo became absorbed in science, and his interest in technical and mechanical skills was already leading him to sketch various machines. In 1482, Leonardo entered the service of the Duke of Milan as the court painter and adviser on architecture and military issues. According to one report, after studying **Euclid**, Leonardo became so interested in geometry that he neglected his duties as court painter.

Leonardo's interest in mathematics soon led him to provide several approaches to squaring the circle (constructing a square with the same area as a given circle) using mechanical methods. In his notebooks, Leonardo described and drew plans for both a **telescope** and a mechanical calculator. Leonardo also formulated several accurate astronomical theories, including one that stated that Earth rotates around the **Sun**, and another stating the **Moon** shines because of the Sun's reflected **light**. Leonardo postulated that the shadowing image of the full Moon that appears cradled between the horns of the crescent moon each month is illuminated by light reflected from Earth, a conclusion that was reached by German astronomer **Johannes Kepler** a century later. Through experimentation, Leonardo concluded that the velocity of a falling object is proportional to the time of its fall, predating Sir **Isaac Newton**'s mathematical theory of **force** and **gravity**. Leonardo's greatest contribution to science and physics, however, may have been his belief that much of nature could be explained scientifically through a strict adherence to mathematical laws, a fundamental tenet of the **philosophy of physics**.

The Duke of Milan was defeated by the French armies in 1499, and the following years were nomadic for Leonardo as he traveled to Mantua and then Venice, where he consulted on architecture and military engineering (Leonardo's notebook included plans for a triple-tier machine gun). Leonardo then returned to Florence briefly and in 1506, returned to Milan, where he worked on various engineering projects. Leonardo spent from 1513 to 1516 in Rome, then moved to France, where King Francis I employed him as a painter, architect, and mechanic. By this time, Leonardo worked little on painting and devoted himself primarily to his scientific studies. Leonardo's thousands of sketches and notes focusing on both practical matters of his day and visions of future scientific accomplishments remain as a testament to Leonardo's prolific genius.

See also Philosophy of physics

LEPTON

The leptons are a group of six elementary particles and their six antiparticle counterparts that do not interact with the strong **force**. They do, however, feel the weak force and the **gravitational force**, and electrically charged leptons interact by the electromagnetic force. The most familiar lepton is the **electron**. The electron is the most stable particle in the **universe** and it does not decay into other particles.

Leptons can either carry one unit of **electric charge** or be neutral. The charged leptons are the electrons, the muons, and the taus. Each of these types has a negative charge and a distinct **mass**. Electrons are the lightest charged leptons with a mass less than 1/2,000th that of the **proton**.

Muons are more than 200 times as massive as electrons and taus are about 3,700 times as massive as electrons. Each charged lepton has an associated neutral partner, called a neutrino—the electron-neutrino, the muon-neutrino, and the tau-neutrino.

Neutrinos have no electric charge and experiments indicate that their maximum possible mass is less than 1/30,000 the mass of an electron. Each of these six particles has an equivalent antiparticle. The mass of the antileptons is identical to that of the leptons, but all of the other properties, including charge, are reversed. Unlike the charged leptons, the neutral neutrinos do not come under the influence of the electromagnetic force. They experience only the weakest two of nature's forces, the weak nuclear force and **gravity**. For this reason, neutrinos react extremely weakly with **matter**. They can, for example, pass through Earth without interacting, which makes it difficult to detect neutrinos and to measure their properties.

Although electrically neutral, the neutrinos seem to carry an identifying property, which has been named "flavor," that associates them specifically with one type of charged lepton. All leptons have a positive flavor and antileptons have a negative flavor. The law of lepton flavor conservation, which applies to every known reaction of leptons, states that the total lepton flavor is the same before and after the reaction. For example, a **muon** decays, after an average lifetime of 2.2 microseconds, to produce an electron, an electron-antineutrino and a muon **neutrino**. The flavors of the electron and the antineutrino are opposite and cancel each other, leaving the muon flavor of the neutrino, which has the same flavor value as the original muon. A lepton can be destroyed only by interaction with its **antiparticle** and it can only change into a combination of particles that includes its neutrino. Neutrinos can interact with matter to form particles and release **energy**. In these interactions with matter, electron-neutrinos can only produce electrons, not muons. Muon-neutrinos produce muons but not electrons.

The tau particle is very large and can decay into many combinations of particles, but one of the products is always a tau-neutrino. For example, the tau can decay into a muon, plus a tau-neutrino and a muon-antineutrino; or it can decay into an electron, plus a tau-neutrino and an electron-antineutrino. Because the tau is heavy, it can also decay into particles containing **quarks**, always accompanied by production of a tau-neutrino. The leptons are considered to be elementary particles, since they apparently are not made up of smaller units of matter. They are the smallest known particles and interact with other matter as if they were points.

LEVERS

A lever is a rigid bar or rod that can pivot about a fixed position. This fixed position is called a fulcrum. Examples of levers include a seesaw, a crowbar, and a bottle opener. A lever is a type of simple machine.

Simple machines are the basic components of complex machines, such as a bicycle or an automobile. A machine is used to ease a load. Simple machines include the lever, wheel and axle, pulley, inclined plane, wedge, and screw. These machines can change the strength or direction of a **force** in order to do **work**. When a person does work on a machine, he or she is transferring **energy** to the machine. The machine then uses that energy to perform work on another object. The machine, therefore, transfers its energy to the object. Work is performed whenever a force moves an object. The amount of work done can be calculated by multiplying the magnitude of the force acting on an object by the distance the object moves. The work a person does on a machine is called the input work and the work done by the machine is called the output work. The output work can never be greater than the input work. The machine does not add energy to the system; it simply transfers energy in an efficient way.

The force the person exerts on the lever in order to perform work is called the effort force. The force that the lever exerts on an object, performing work, is called the resistance force. For example, if a screwdriver is used to pry the lid off a paint can, the force exerted by the person on the screwdriver is the effort force and the force the screwdriver exerts on the can is the resistance force.

When an effort force is applied to a lever, the lever overcomes the resistance force by pivoting on the fulcrum. In the above example, a person pushes down on the screwdriver, and the screwdriver pivots on its fulcrum, pushing the paint can lid up. In other words, the lever changes the direction of the force. The distance the screwdriver moves down (the effort distance) is greater than the distance the lid moves up (the resistance distance). Ideally, the work done by the screwdriver on the lid is equal to the work done by the person on the screwdriver. Since work is force times distance, and the amount of work is conserved, if the effort distance is greater than the resistance distance, the effort force must be smaller than the resistance force. In other words, the lever multiplies the effort force, easing a load.

Levers can be divided into three classes, based on the location of the fulcrum. First-class levers are those where the fulcrum is between the effort force and the resistance force. Examples of first-class levers include a crowbar or a seesaw. Second-class levers are those where the fulcrum is at the end of the bar or rod. These levers also multiply the effort force applied by decreasing distance, however, the direction of the force is not changed. Examples of second-class levers include a wheelbarrow or a bottle opener. Third-class levers are those where the fulcrum is at the same end as the effort force. In these levers, the resistance distance is actually greater than the effort distance. Third-class levers reduce effort force by increasing the resistance distance, so even though they do not multiply force, they still ease a load. Examples of third-class levers include hammers, tweezers, and fishing poles. Regardless of type, all levers perform work on an object in order to ease a load.

LIGHT

Light can be narrowly defined as the visible portion of the **electromagnetic spectrum**. A broader definition would include infrared, ultraviolet, and x-ray wavelengths, which are not visible to the eye. The nature of light has been the subject of controversy for thousands of years. Even today, while scientists know how light behaves, they do not always know why light behaves as it does.

The Greeks were the first to theorize about the nature of light. Led by scientists **Euclid** and Hero (first century A.D.), they came to recognize that light traveled in a straight line. However, they believed that vision worked by intromission— that is, that light rays originated at the eye and traveled to the object being seen. Despite this erroneous hypothesis, the Greeks were able to successfully study the phenomena of reflection and refraction and derive the laws governing them. In reflection, they learned that the angles of incidence and reflection were approximately equal; in refraction, they saw that a beam of light would bend as it entered a denser medium (such as water or glass) and bend back the same amount as it exited.

The next contributor to the embryonic science of **optics** was Arab mathematician and physicist **Alhazen**, who is sometimes called the greatest scientist of the Middle Ages. Experimenting around the year 1000, he showed that light comes from a source (the **Sun**) and reflects from an object to the eyes, thus allowing the object to be seen. He also studied mirrors and **lenses** and further refined the laws of reflection and refraction.

By the twelfth century, scientists felt they had solved the riddles of light and **color**. English philosopher Francis Bacon (1561-1626) contended that light was a disturbance in an invisible medium that could be detected by the eye; subsequently, color was caused by objects "staining" the light as it passed. More productive research into the behavior of light was sparked by the new class of realistic painters, who strove to better understand perspective and shading by studying light and its properties.

In the early 1600s, the refracting **telescope** was perfected by **Galileo** and **Johannes Kepler**, providing a reliable example of the laws of refraction. These laws were further refined by **Willebrord Snel**, whose name is most often associated with the equations for determining the refraction of light. By the mid-1600s, enough was known about the behavior of light to allow for the formulation of a wide range of theories.

Renowned English physicist and mathematician **Isaac Newton** was intrigued by the so-called phenomenon of colors—the ability of a prism to produce colors from white light. It had been generally accepted that white was a single color, and that a prism could somehow combine white light with others to form a multicolored mixture. Newton, however, doubted this assumption. He used a second prism to recombine the rainbow spectrum back into a beam of white light; this showed that white light must be a combination of colors, not the other way around.

Newton performed his experiments in 1666 and announced them shortly thereafter, subscribing to the corpuscular (or particulate) theory of light. According to this theory, light travels as a stream of particles that originate from a bright source and are absorbed by the eye. Aided by Newton's reputation, the corpuscular theory soon became accepted throughout Great Britain and in other parts of Europe.

In the European scientific community, many scientists believed that light, like **sound**, traveled in **waves**. This group of scientists was most successfully represented by Dutch physicist **Christiaan Huygens**, who challenged Newton's corpuscular theory. He argued that a wave theory could best explain the appearance of a spectrum as well as the phenomena of reflection and refraction.

Newton immediately attacked the wave theory. Using some complex calculations, he showed that particles, too, would obey the laws of reflection and refraction. He also pointed out that, if truly a wave form, light should be able to bend around corners, just as sound does; instead it cast a sharp shadow, further supporting the corpuscular theory.

In 1660, however, Francesco Grimaldi examined a beam of light passing through a narrow slit. As it exited and was projected upon a screen, faint fringes could be seen near the edge. This seemed to indicate that light did bend slightly around corners; the effect, called **diffraction**, was adopted by Huygens and other theorists as further proof of the wave nature of light.

One piece of the wave theory remained unexplained. At that time, all known waves moved through some kind of medium—for example, sound waves moved through air and kinetic waves moved through water. Huygens and his allies had not been able to show just what medium light waves moved through; instead, they contended that an invisible substance called **ether** filled the **universe** and allowed the passage of light. This unproven explanation did not earn further support for the wave theory, and the Newtonian view of light prevailed for more than a century.

The first real challenge to Newton's corpuscular theory came in 1801, when English physicist **Thomas Young** discovered **interference** in light. He passed a beam of light through two closely spaced pinholes and onto a screen. If light were truly particulate, Young argued, the holes would emit two distinct streams that would appear on the screen as two bright points. What was projected on the screen instead was a series of bright and dark lines—an interference pattern typical of how waves would behave under similar conditions.

If light is a wave, then every point on that wave is potentially a new wave source. As the light passes through the pinholes it exits as two new wave fronts, which spread out as they travel. Because the holes are placed close together, the two waves interact. In some places the two waves combine (constructive interference), whereas in others they cancel each other out (destructive interference), thus producing the pattern of bright and dark lines. Such interference had previously been observed in both water waves and sound waves and seemed to indicate that light, too, moved in waves.

The corpuscular view did not die easily. Many scientists had allied themselves with the Newtonian theory and were unwilling to risk their reputations to support an antiquated wave theory. Also, English scientists were not pleased to see one of their countrymen challenge the theories of Newton; Young, therefore, earned little favor in his homeland.

Throughout Europe, however, support for the wave nature of light continued to grow. In France, Étienne-Louis Malus (1775-1812) and **Augustin Jean Fresnel** experimented with polarized light, an effect that could only occur if light acted as a transverse wave (a wave that oscillated at right angles to its path of travel). In Germany, **Joseph von Fraunhofer** was constructing instruments to better examine the phenomenon of diffraction and succeeded in identifying within the Sun's spectrum 574 dark lines corresponding to different wavelengths.

In 1850 two French scientists, **Jean-Bernard-Léon Foucault** and Armand Fizeau, independently conducted an experiment that would strike a serious blow to the corpuscular theory of light. An instructor of theirs, Dominique-François Arago, had suggested that they attempt to measure the **speed of light** as it traveled through both air and water. If light were particulate it should move faster in water; if, on the other hand, it were a wave it should move faster in air. The two scientists performed their experiments, and each came to the same conclusion: light traveled more quickly through air and was slowed by water.

Even as more and more scientists subscribed to the wave theory, one question remained unanswered: through what medium did light travel? The existence of ether had never been proven—in fact, the very idea of it seemed ridiculous to most scientists. In 1872, **James Clerk Maxwell** suggested that waves composed of electric and magnetic fields could propagate in a **vacuum**, independent of any medium. This hypothesis was later proven by **Heinrich Rudolf Hertz**, who showed that such waves would also obey all the laws of reflection, refraction, and diffraction. It became generally accepted that light acted as an electromagnetic wave.

Hertz, however, had also discovered the **photoelectric effect**, by which certain metals would produce an electrical potential when exposed to light. As scientists studied the photoelectric effect, it became clear that a wave theory could not account for this behavior; in fact, the effect seemed to indicate the presence of particles. For the first time in more than a century there was new support for Newton's corpuscular theory of light.

The photoelectric effect was explained by **Albert Einstein** in 1905 using the principles of **quantum physics** developed by **Max Planck**. Einstein claimed that light was quantized—that is, it appeared in "bundles" of **energy**. While these bundles traveled in waves, certain reactions (like the photoelectric effect) revealed their particulate nature. This theory was further supported in 1923 by **Arthur Holly Compton**, who showed that the bundles of light—which he called photons—would sometimes strike electrons during **scattering**, causing their wavelengths to change.

By employing the quantum theories of Planck and Einstein, Compton was able to describe light as both a particle and a wave, depending upon the way it was tested. While this may seem paradoxical, it remains an acceptable model for explaining the phenomena associated with light and is the dominant theory of our time.

LIGHT-EMITTING DIODE

A light-emitting diode, commonly referred to as an LED, is a simple semiconductor device that uses little **energy** to release **light** when a current passes through it. The **wavelength** of light emitted depends on what material is used to make up the semiconductor crystals of the LED. The visible light emitting LED is most often used in alphanumeric displays where low-energy use is essential. The infrared emitting LED is most often used in remote control mechanisms.

The basic technology involved in the LED is the same as that which applies to all **diodes**. A diode is a simple rectifier that allows **electricity** to flow in one direction only. The first diodes were **vacuum tubes** in which electrons freed from the cathode would be attracted by the **anode** through a metal plate only when the anode is positively charged. The semiconductor diode consisting of two terminals and a simple positive-to-negative (p-n) junction achieves the same end. The semiconductor diode uses crystals made up of crystalline matter mixed with a small amount of nonconductive matter as part of the circuit to control the flow of current under various conditions. Crystalline matter usually consists of silicon, germanium, or gallium arsenide; nonconductive matter usually consists of boron or phosphorus.

LEDs constitute one kind of junction-type diodes that contain a junction of two different kinds of semiconductor material. In a process known as doping, traces of noncrystalline elements are mixed in with the crystalline material (from which crystals will be grown). These noncrystalline elements, known as donors or acceptors, produce regions, which are specifically p-type or n-type. A p-type region touching an n-type region forms a junction. Current will pass only one way through a junction. The purpose of a diode, rectifying the current, is thus fulfilled.

The model explaining the behavior of **semiconductors** holds that the doping material replaces atoms of the foundation material. In the n-type region of the junction, this produces "holes" of free electrons. When external electrons enter into the material the like charges dispel them toward the junction. At the junction, they are attracted to the holes in the p-type region. This migration of the electrons continues the current. If the external electrons enter the p-type region, the holes of the p-type region move toward the incoming electrons. Thus, they move away from the junction and there is a break in the continuity of the **electron** stream.

The remote-control capacity of LEDs derives from photoelectric and photovoltaic phenomena. A **photon** striking an electron can set it in **motion** to interact with other electrons. In the photovoltaic effect, the photons dislodge enough electrons near the junction of the semiconductor to bridge the gap and allow the electron flow to pass. In remote-control systems, the light from the LED is sufficient to create this photovoltaic effect in the receptor system.

See also Cathode and cathode rays; Electromagnetic spectrum; Photoelectric effect

Magnified image of LED display. *(Photo by Alfred Pasieka. Science Photo Library, National Audubon Society Collection/Photo Researchers, Inc. Reproduced by permission.)*

LINEAR MOMENTUM

In physics, Newton's first law of **motion** states that a moving body will remain in motion unless acted upon by an outside **force**. Similarly, a body at rest (no motion) will remain at rest unless acted upon by an outside force. This concept, called **inertia**, is a quality that is dependent upon the **mass** of the object under examination. The greater the mass of an object, the greater the force must be to oppose its inertia. A related concept is linear **momentum**, whose value is also determined by mass. Linear momentum, or simply momentum, is a vector quantity (having both magnitude and direction) whose value is equal to the product of the mass and **velocity** of an object moving in a linear (not angular) fashion. Therefore, if either (or both) the mass or velocity of an object is increased, its momentum is increased. For example, very tiny particles can have great momentum if they have immense velocity. A bullet shot from a revolver, for instance, has a large momentum. Likewise, very massive objects, such as railroad trains, have great momentum even when moving very slowly. Like inertia, the force required to oppose a body's momentum is proportional to its mass.

The central concept regarding momentum is its conservation. Conservation of momentum is the characteristic that within a system, total momentum is always the same, or constant. For example, consider a rocket being propelled

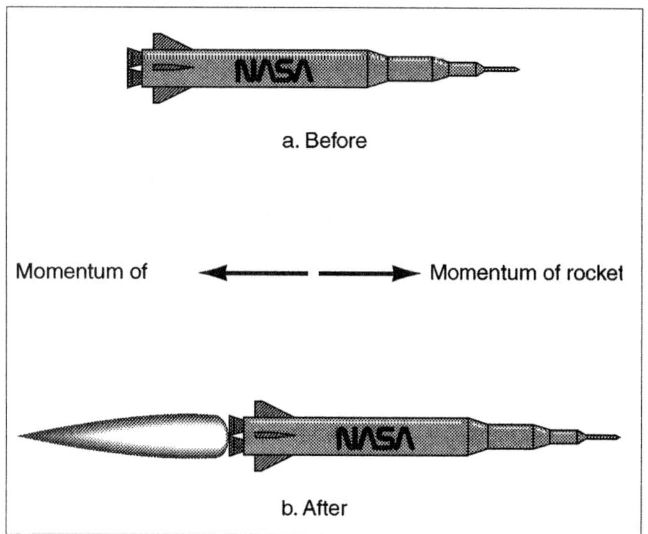

Conservation laws, linear momentum.

upward into **space** by the combustion of rocket fuel. Because the fuel is consumed and ejected at high velocity in the process of propulsion, the mass of the fuel carries momentum downward, in order that momentum be conserved, the rocket acquires an equal upward momentum, which propels it into space. Further, since the mass of the rocket is decreasing due to the expulsion of fuel, the velocity will increase so that the upward momentum will stay in balance with the downward momentum of the fuel.

This concept is also important when considering collisions. In any system, total momentum equals the sum of its components. For example, in billiards, the total momentum of the cue ball and the eight ball is equal to the momentum of the cue ball plus the momentum of the eight ball. If, during the final shot of the billiard game, the cue ball collides with the eight ball to propel it toward the corner pocket, conservation of momentum ensures that the total momentum does not change. Therefore, if the eight ball was not moving, some of the momentum of the cue ball will be transferred to the eight ball so that the total momentum remains constant. Conservation of linear momentum ensures that total momentum remains constant whether the collision is elastic or inelastic. Elastic collisions involve the complete conservation of both momentum and kinetic **energy** (the energy of motion). The example of billiard balls striking one another is an example of nearly perfectly elastic collisions. Inelastic collisions are collisions in which conservation of total momentum still occurs, but total **kinetic energy** is not conserved. For instance, if two balls of clay collide with one another, momentum is conserved, but some energy of movement is lost as **heat** and as **work** involved in deforming the clay.

Conservation of linear momentum acts on the **center of mass** of an object. The center of mass of an object is the singular point at which all of the mass of the object is acted upon by external forces. It is as if all of the mass of the body is con-

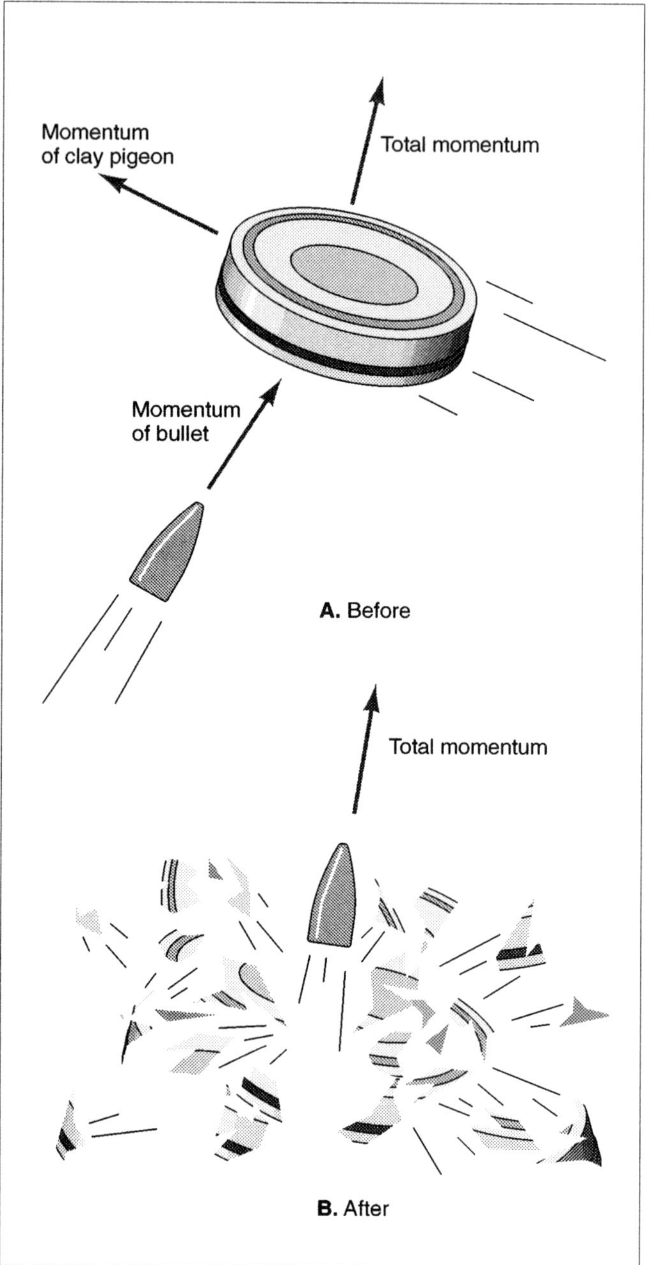

Conservation laws, linear momentum.

tained in one single point. Therefore, the total momentum of a system depends upon the velocity of its center of mass.

LINEAR SYSTEMATICS

The basic notions of classical physics concern particles moving in straight lines, or linear **motion**. The concepts of linear motion are usually grouped into two categories, kinematics and **dynamics**. Kinematics deals with predicting the position of a particle if its position, **velocity**, and **acceleration** at some reference time (usually set to be t = 0) is known. Velocity is the rate of change of position with respect to time, and acceleration is the rate of change of velocity with respect to time. The two kinematical equations are, for constant acceleration: $v(t) = v_0 + at$, and $x(t) = x_0 + v_0t + 1/2\ at^2$, where x is the position, v is the velocity (the subscripted v_0 indicates the velocity at the start or origin), a is the acceleration, and t is understood as the change in time.

Dynamics deals with how kinematical quantities such as velocity and acceleration react to forces. The basic dynamic rule is Newton's second law, which says that the acceleration of a particle is equal to the sum of all external forces on the particle, divided by the **mass** of the particle: $a = F_{ext}/m$. Equally important rules are the **conservation of energy**, and **linear momentum**. Conservation of **energy** states that as long no external forces are acting, the energy of a system of particles is constant, although it may change forms. Conservation of linear **momentum** states that if the momenta of all particles in a system is added, it is a constant, although the momentum may become differently distributed among the particles. Again, this principle works only if there are no external forces acting on the system. These rules allow us to analyze particles under the influence of external forces and particles colliding into each other.

Often particles are not moving in straight lines. In this case, the particle's motion can be separated into two components, which are straight lines. These components can be analyzed individually. For example, when a cannon is shot into the air at an angle, it has motion parallel to the ground and also motion perpendicular to the ground. The projectile motion of the cannonball can thus be analyzed in terms of these two sets of variables, both of which exhibit linear motion.

Because linear motion is well understood, other sorts of motion are often treated by approximating the motion as linear for a short time. An example is rotational motion, where particles travel in circles around a central point. The kinematical and dynamic equations for rotational motion are translated and transformed from the linear systematics.

See also Ballistics; Force; Rotation in three dimensions

LIPPERSHEY, HANS (1570-1619)

German Dutch lens maker

Although there is some debate, most historians agree that Hans Lippershey was the first inventor of the **telescope**. Born in Wesel (Germany), Lippershey migrated to Zeeland in the Netherlands. Little is known about Lippershey except that he married in 1594 and officially became a citizen of his adopted country in 1602.

According to a popular story, Lippershey, an eyeglass maker, came up with the idea for the telescope after watching two children in his eyeglass shop play with his **lenses**. The legend holds that children grasping two lenses and, looking through both simultaneously, noticed that a **weather** vane on a nearby church steeple seemed to appear larger and clearer than visible by normal sight. Lippershey realized the potential of this accidental discovery and proceeded to make a telescope by attaching lenses at the two ends of a tube.

Lippershey applied for a patent in 1608, but the device could not be kept a secret and the patent was eventually denied. Others came forth claiming to have invented the new device, including Jacob Metius and Sacharias Janssen, both of the Netherlands. Still, Lippershey's application for a patent remains the earliest record of a telescope, which he called a *kijker*, or "looker." Lippershey also benefited financially from his invention through the Dutch government, which paid him to construct several telescopes for military use.

LIPPMANN, GABRIEL (1845-1921)

French physicist

Gabriel Lippmann had a distinguished career as an inventor, theoretician and academic. At the Faculty of Sciences, a laboratory in Paris, France, he became professor of **mathematical physics** in 1883, a professor of experimental physics in 1886, and, later, director of the laboratory. He stayed active in this position, overseeing the laboratory's incorporation into the Sorbonne, until he died at sea on July 13, 1921. A member of the French Academy of Sciences and the Bureau des Longitudes, he was elected in 1908 as a Foreign Member of the Royal Society of London. Even before he had finished his doctorate in 1875, he embarked on a lifetime of publishing papers and the creation of measuring instruments to accompany research and observation in physics, **astronomy**, and seismology. Lippmann is most often remembered, however, for developing an early process of color **photography**. It was for this achievement that he received the 1908 Nobel Prize in physics.

Born in Hollerich, Luxemburg, on August 16, 1845, to French parents, Gabriel Jonas Lippmann began his education at home. When he was 13, his family settled in Paris, where he entered the Lycée Napoléon. After 10 years he attended the École Normale. During this time, he assisted with the publication of the *Annales de Chimie et de Physique* by summarizing German articles. In this manner he learned of recent discoveries in **electricity**. In 1873 he traveled, as part of a scientific mission, to Germany, where, at Heidelberg, he worked in physicist **Gustav Kirchhoff**'s laboratory. There, Wilhelm Kühne, a professor of physiology, showed him an experiment in which a drop of mercury, covered with diluted sulfuric acid, contracted when touched with a piece of iron wire—only to recover its original shape when the wire was removed. Lippmann theorized that the wire had somehow changed the tiny electrical current between the acid and the mercury, causing it to ball up. He obtained permission to systemically confirm his supposition with experiments in Kirchhoff's laboratory.

Creates the Capillary Electrometer and Pursues Its Theoretical Implications

Lippmann's investigations resulted in an 1873 publication that theoretically described the mercury phenomenon that Kühne had shown him as well as the device he developed from it, the capillary electrometer. This instrument, which came into wide use before the advent of solid-state **electronics**, could measure electrical currents as small as 1/1,000 of a volt. Its elegant design consisted of a narrow tube (or capillary), pitched at a slight horizontal angle, containing mercury covered with diluted acid. Any change in the **electric charge** between the two liquids caused a ripple at their interface to move up the tube. As a result of Lippmann's work on the capillary electrometer, the Sorbonne awarded him a doctorate in 1875.

In 1876 Lippmann published a paper showing that it was possible to reverse the electromagnetic phenomenon he had investigated in 1873. Returning to the experiment of mercury covered with acid, he demonstrated that altering the shape of the mercury by mechanical means, somehow squeezing it together, had an impact on the electrical field between the mercury and the acid. To demonstrate definitively the reversibility of the two processes, Lippmann devised an engine based on the principles of his capillary electrometer. This engine turned when electrified and produced electricity when turned mechanically. Lippmann built upon the earlier work of French engineer **Nicolas-Léonard-Sadi Carnot**. In 1824 Carnot demonstrated, with a reversible **heat** engine, the thermodynamic principle that there exists an inverse (or opposing) and measurable relationship between heat and **force**. Following this reasoning, Lippmann established a more general theorem that he published in 1881. It states that given any phenomenon, the reverse phenomenon also exists and that one can calculate its degree of change.

Advances Observational Methods in Astronomy and Seismology

Lippmann made important innovations in observation for physics, such as introducing high-speed photographs to record the behavior of pendulums. Not content to confine his efforts to one field, he also modernized observational instruments in astronomy and seismology. His most notable upgrade was a device called the coelostat. Using a mirror attached to a machine that reproduced the axis and rotation of Earth, the coelostat ensured that whole regions of the sky, rather than a single star, could be photographed without registering any **motion**. By increasing the area to be recorded, it improved on an earlier device, the siderostat. Lippmann created another instrument related to the coelostat, called the uranograph. This device produced a photographic map of the sky with its longitudes automatically imprinted on it. Lippmann also devised a method for measuring longitudinal differences between observatories through **radio** and photography. In his contributions to the field of seismology, Lippmann proposed using telegraphic signals for the early detection of earthquakes as well as for measuring how quickly they traveled. In addition, he suggested a seismograph that would record seismic **waves** while taking into account Earth's **acceleration**.

Wins 1908 Nobel Prize in Physics for Color Photography

The achievement that won Lippmann his widest recognition, and the 1908 Nobel Prize in physics, was his perfection of a photographic process with relatively permanent **color**. As early as the beginning of the nineteenth century, it was widely known that moist silver chloride could reproduce the colors of the spectrum. In 1848, Edmond Becquerel managed to reproduce colored objects on a silver plate covered with a layer of silver chloride. Unfortunately, Becquerel had no way of fixing the colors, which faded rapidly. In 1890, Otto Weiner confirmed, through an experiment, that Becquerel's phenomenon was the result of "interference"—**light** waves trapped at different levels in the layer of silver chloride.

In 1891 Lippmann published a method where a transparent plate with a layer of silver nitrate, gelatin, and potassium bromide in emulsion was placed, emulsion-side-down, in a holder with mercury in it. During exposure, which in the initial experiments lasted 15 minutes, light waves became fixed in the emulsion after repeatedly bouncing off the mercury, faithfully reproducing the colors in nature. After the plate was developed, the colors, seen in reflected light, were permanent. Although it presented a significant leap forward at the time, Lippmann's method proved impractical. There was no way to create copies and the exposure time, although later reduced to a minute, was still too long to suit the needs of mass production.

Lippmann married Mademoiselle Cherbuliez in 1888, their union producing no children. There is little public information regarding his personal life. What he is known for is his body of work, which is considered ingenious and progressive. Rather than the antiquated photographic process for which he received the Nobel Prize, however, many scientists believe Lippmann's real contributions to science lay in his work with the capillary electrometer and his theoretical papers.

LORENTZ, HENDRIK ANTOON (1853-1928)

Dutch physicist

Hendrik Antoon Lorentz was widely regarded as the world's leading theoretical physicist at the end of the nineteenth century and the beginning of the twentieth. His earliest work dealt with optical phenomena, but by 1892 he had begun to develop one of the concepts for which he is most famous, the **electron** theory. Based on the presumed existence of tiny charged particles in **matter**, this theory explained a number of well-known physical phenomena. In 1902 Lorentz received the Nobel Prize in physics for his work on the interaction of **radiation** and matter; he shared the award with **Pieter Zeeman**. In the first decade of the twentieth century, Lorentz worked on the effect of **motion** on the properties of a particle, foreshadowing some of the fundamental concepts that were later to become part of **Albert Einstein**'s theory of relativity.

Hendrik Antoon Lorentz was born in Arnhem in the Netherlands, on July 18, 1853. His father was Gerrit Frederik Lorentz, owner of a nursery; his mother, Geertruida van Ginkel Lorentz, died when Lorentz was four years old, and five years later his father was married for a second time, to Luberta Hupkes. Lorentz attended primary and secondary schools in Arnhem, and while he showed a special interest in the sciences, he did well in all subjects. In 1870, Lorentz entered the University of Leiden, where he studied physics and mathematics. He became especially interested in the lectures of Frederick Kaiser, a professor of astronomy, and Pieter Leonhard Rijke, the university's only professor of physics. Lorentz passed his candidate's examination (approximately equivalent to a bachelor's degree) in November 1871, only 18 months after entering the university.

Lorentz returned to Arnhem to prepare for his doctoral examination. Over the next two years he supported himself by teaching in the local high school; he passed the examination summa cum laude in 1873. He then began his doctoral research on the theory of reflection and refraction, an aspect of electromagnetic theory that **James Clerk Maxwell** had left unsolved in his earlier studies of the subject. Lorentz was awarded his Ph.D. in 1875 for his thesis, "The Theory of Reflection and Refraction of Light."

Lorentz remained in Arnhem for four years after receiving his doctorate, continuing to teach high school and trying to decide the future direction of his career. He was uncertain as to whether he should focus on mathematics or physics. That choice was made even more difficult in 1877 when he was offered a chair in mathematics at the University of Utrecht and a chair in theoretical physics at the University of Leiden. The position at Leiden was the first in theoretical physics created in the Netherlands and one of the first in the world. Lorentz chose physics, not only establishing his own future but that of the field of theoretical physics over the next half century.

During his first few years at Leiden, Lorentz concentrated on the study of optical phenomena. One of his accomplishments during this period was the discovery of a formula relating the **density** of a substance to its index of refraction. That formula is now known as the Lorentz-Lorenz equation in honor of his work and that of Danish physicist Ludwig Lorenz, who made the same discovery almost simultaneously.

Develops Theory of Electrons and Lorentz Transformations

In the early 1890s, Lorentz began work on one of his most important theoretical contributions, commonly known as his theory of electrons. Electrons had not yet been discovered, so the term is somewhat misleading; Lorentz posited the existence of tiny, charged particles to explain a number of physical phenomena, and these particles had properties very similar to those of the electron that was discovered by **Joseph John Thomson** in 1897. Lorentz argued, for example, that electromagnetic radiation is produced by the vibration of the tiny charged particles, and this has turned out to be true of electrons.

An important consequence of Lorentz's electron theory was the prediction that **spectral lines** would be split if the source of **light** from which they came was placed within a strong magnetic field. Spectral lines are an optical phenomenon, and they occur when the light from a gas flame is split into different colors. Lorentz hypothesized that the presence of the **magnetic field** would alter the motion of the charged particles that produced the spectral lines, causing them to shift

Hendrik Antoon Lorentz.

positions. Lorentz's theory was confirmed by a series of experiments conducted in 1896 by Lorentz's younger colleague at Leiden, Pieter Zeeman. For their discovery and explanation of the splitting of spectral lines, Lorentz and Zeeman were jointly awarded the 1902 Nobel Prize in physics.

In the last decade of the nineteenth century, Lorentz began to do more work in **electrodynamics**. One of the foundations of electrodynamic theory at the time was the posited existence of a substance called aether (or **ether**). The existence of aether was thought necessary to explain a number of physical phenomena, such as light **waves**, and it was believed to fill all **space**. In 1887, **Albert Michelson** and **Edward W. Morley** performed a classic experiment that failed to find any evidence for the existence of aether. In his own studies of electromagnetic phenomena, Lorentz had found it necessary to postulate the existence of an aether, so he was troubled by the results of the **Michelson-Morley experiment**.

Lorentz decided that one way to explain the Michelson-Morley results was to assume that objects traveling through the aether (in which he continued to believe) underwent a shortening in length in the direction in which they were moving. He derived a formula that predicted an amount of shortening that would exactly account for the failure of Michelson and Morley to find any evidence of the motion of light through the aether. At almost the same time, Irish physicist George Francis FitzGerald (1851-1901) postulated a formula identical to that of Lorentz's.

According to the Lorentz-FitzGerald transformation, the length of a body contracts by a factor of the square root of $1 - v^2/c^2$ as it moves through the aether (v equals **velocity** and c is the **speed of light**). Lorentz also calculated that the **mass** of a particle would increase by the same factor. The Lorentz-FitzGerald (1851-1901) transformation was later found by Albert Einstein to be a necessary consequence of the theory of relativity.

During his long tenure at Leiden, Lorentz had an enormous influence on the development of modern physics. That influence resulted not only from his own research and writing, but also from the students he trained and the professional organizations with which he became involved. In 1912, however, he resigned from his post at Leiden to become director of the Teyler's Stichting Museum in Haarlem, Netherlands. In his new job at Haarlem, Lorentz had his own, well-equipped laboratory, a facility that he had been promised for decades—though never received—at Leiden. During his 11-year tenure in Haarlem, Lorentz continued to teach at Leiden once a week and in 1919 became involved in the Dutch effort to reclaim land from the Zuider Zee. He was also very interested in science education in the Netherlands and served on the government board of education from 1919 until 1926.

Lorentz was held in high esteem by his professional colleagues and was offered a number of important positions in professional organizations. From 1911 to 1927, for example, he was chair of the International Solvay Congress in Physics. He also served as president of the physics section of the Royal Netherlands Academy of Sciences and Letters from 1909 to 1921. In 1923 he was elected as one of the seven members of the International Commission on Intellectual Cooperation of the League of Nations, becoming president of the group two years later. Among his many awards and honors were the 1908 Rumford Medal and the 1918 Copley Medal from the Royal Society. Lorentz was married in 1881 to Aletta Catherina Kaiser, a niece of his astronomy professor at Leiden. The couple had three children. Lorentz was still active in physics when he died in Haarlem on February 4, 1928.

LORENTZ TRANSFORMATIONS

The Lorentz transformations are a group of mathematical functions that are central to the special theory of relativity. They express the constancy of the speed of **light** in all frames of reference and the dependence of **space** and **time** measurements on the relative **motion** of the observer and the system observed.

The transformations describe how the coordinates of an event in one **frame of reference** compare with the coordinates as measured in another reference frame that is in motion at constant **velocity** relative to the first frame. Such transformations can be used to test the invariance of physical laws, showing whether the laws apply universally, regardless of the state of motion of the observer. In the Galilean transformations, which predate the Lorentz transformations, it is implicitly assumed that measurements of lengths and time intervals are not affected by the relative motion of the observers, and that the **speed of light** may have different values depending on the

observer's motion. In the equations below, the primed coordinates are those of one system, and the unprimed are in the other system. For simplicity, we assume that the origins of the coordinates overlap, and the motion is only along the x axis:

$$x' = x-vt$$
$$y' = y$$
$$z' = z$$
$$t' = t$$

Toward the end of the nineteenth century, electromagnetic theory began to suggest that some modification of the Galilean transformations was necessary. Following the **Michelson-Morley experiment** of 1887, there were many attempts to explain why it failed to detect changes in the velocity of light as a result of Earth's motion through the **ether**, as was expected. George Francis FitzGerald (1851-1901) was the first to suggest that motion through the ether caused a compression of the measuring apparatus that negated any **interference** effects in the light beams used in the experiment. A contraction formula was later advanced independently by **Hendrik Lorentz**. The length change, called Lorentz-FitzGerald **length contraction**, occurs only along the direction of motion, and causes a dramatic shortening of objects at speeds near the speed of light. Sir Joseph Larmor (1857-1942) and Lorentz independently developed what became known as the Lorentz transformations as part of their efforts to replace the Galilean transformation with equations consistent with electromagnetic theory. It was shown that the Lorentz-FitzGerald contraction could be derived from the Lorentz transformation equations. Using the same simplifications employed with the Galilean transformations, the Lorentz transformations are given by:

$$x' = \gamma(x-vt)$$
$$y' = y$$
$$z' = z$$
$$t' = \gamma(t-vx/c^2)$$
$$m' = \gamma m$$
$$\gamma = (1-v^2/c^2)^{-1/2}$$

The factor γ introduces a velocity dependence that becomes significant near c, the speed of light. Unlike the Galilean transformations, the time coordinate is different in the two **reference frames**.

According to the Lorentz transformations, as a body's relative motion approaches the speed of light, its length in the direction of motion will decrease, its **mass** will increase, and **time dilation** will occur. The formulas prohibit travel at the speed of light. Simultaneity is redefined by the transformations such that two events that are simultaneous in one frame of reference may be observed to occur at different times in another reference frame. The **twin paradox**, in which one of a pair of twins travels at high speed in a spacecraft and returns to Earth younger than the twin remaining behind, is a consequence of the Lorentz transformations.

In 1905 **Albert Einstein** derived the Lorentz transformations in an entirely new way with his special theory of relativity. Previously, the results of the Lorentz transformations were seen as effects of the ether on bodies, and were essentially an arbitrary device used to preserve the role of the ether in classical **mechanics** by explaining the anomalous results of the

Michelson-Morley experiment. Einstein dismissed ether theory, and instead made two simple assumptions about nature and determined the logical consequences. He postulated that the speed of light should be constant for all observers, and that physical laws are the same in all inertial reference frames, that is, reference frames that are unaccelerated. From these principles he deduced the Lorentz transformations, along with a new form of the law of addition of velocities that accommodated the constancy of the speed of light. He interpreted the transformations to mean that there is no ether or standard of absolute rest with which to define the "real" motion of bodies; only relative motion exists. Measurements of length, time, and mass are not absolute, but depend on the relative motion of the observer and the system being observed. For example, if two observers have identical measuring rods, masses, and synchronized clocks when at rest with respect to each other, when moving apart at a given speed they will each see the other's masses as greater, clocks as running slower, and measuring rods as shortened in the direction of travel. What distinguished Einstein's use of the Lorentz transformations from the work of contemporaries such as Lorentz and **Henri Poincaré** was Einstein's abandonment of absolute space, absolute time, and the ether, as well as the fact that he built his theory using only two basic principles and a system of light signals between observers.

The validity of the Lorentz transformations has been successfully tested by a variety of means. Unstable **subatomic particles** traveling at high speeds in **accelerators** or as **cosmic rays** are known to have much longer life spans than the same particles moving at low speeds, indicating that we observe time running more slowly for the particles when they are moving at high relative speeds. Time dilation has also been tested by comparing a ground-based **atomic clock** with one that has been traveling at high speed in an aircraft, and the predicted effect was confirmed to a high degree of accuracy.

See also Relativity, special

LOW-TEMPERATURE PHYSICS

Low-temperature physics studies the behavior of **matter** at temperatures approaching **absolute zero** (-273.15°C) with the use of experimental cryogenic methods. At such extremely cold temperatures, the optical, thermal, electric, and magnetic properties of materials undergo significant changes compared to those properties at warmer temperatures. Two classic examples of these changes are superconductivity and **superfluidity**. Major advances in low-temperature physics were made during the early twentieth century. Superconductivity was discovered in 1911 by Dutch physicist H. K. Onnes, who spent his scientific career developing cryogenic methods and investigating the properties of very cold materials. In 1908, he succeeded in liquefying **helium** by cooling it to -269°C (4K), thus providing experimentalists with the possibility to work at previously unreachable temperatures. By immersing samples in liquid helium with the use of cryostats or refrigerating units, physicists were able to cool other materials to temperatures

approaching absolute zero (0K), the coldest **temperature** possible and the temperature at which the **energy** of material becomes as small as possible. In 1911, Onnes began to study the electrical properties of metals at liquid helium temperatures. It was known that the **electrical resistance** of metals decreased with temperature, but the limiting value of their resistance was unknown, especially as the temperature approached zero kelvin. **William Thomson Kelvin** held that the **electron** flow would completely stop at such temperatures. Onnes believed that the resistance would dissipate, i.e., that there would be an incremental and gradual decrease in resistance and a corresponding increase in the conductivity of the cooled material. He passed a current through a mercury wire of high purity and measured its resistance as he gradually lowered the temperature to 4.2K, at which point the resistance suddenly disappeared. Superconductivity had been discovered and in 1972, the Nobel Prize for physics was conferred to **John Bardeen**, L. N. Cooper, and J. R. Schrieffer for formulating a fundamental theory of superconductivity, now called BCS theory in their honor.

The discovery of superfluidity in helium represents the other success story of low-temperature physics. Superfluidity is a consequence of the **atomic structure** of helium, which naturally occurs as two **isotopes**, 4-helium and 3-helium. With an even number of four particles in its **nucleus**, 4-helium is a **boson** and can thus undergo **Bose-Einstein condensation,** which brings it to a state of lowest possible energy where **friction** vanishes, allowing the helium to behave as a superfluid with the ability to flow through any opening, no matter how small. 4-helium superfluidity was discovered in 1938 by the Russian physicist **Pyotr Kapitsa** and 3-helium superfluidity was first observed by D. M. Lee, D. D. Osheroff and R. C. Richardson in the 1970s. Unlike 4-helium, 3-helium is a fermion because its nucleus has an odd number of particles and theoretically it should not be able to undergo Bose-Einstein condensation. However, BCS theory proposed a likely mechanism by which the 3-helium **fermions** could combine to form pairs and thus behave as bosons that could become superfluid at very low temperatures.

Present-day low-temperature physics focuses its efforts on investigating the properties of insulators, **semiconductors**, conductors, and **superconductors** in a dynamic search for improved materials and more fundamental insights into superconducting theory. For example, direct **tunneling** spectroscopy of the single-particle band gap in bulk Si:B was recently reported and the interesting question of how Coulomb interactions evolve across the metal-insulator transition could be examined. With advances in cryogenic technology, cryostats are now capable of cooling materials to temperatures of a few millikelvin. This is achieved by insulating the actual sample compartment from the surrounding room temperature within a concentric chamber, itself cooled with liquid nitrogen (77K). In other applications, low-temperature **scanning tunneling microscopes** have been developed to study the topographic and electronic structure of surfaces on the atomic scale. Such studies combine helium temperature measurements with the application of external magnetic fields reaching up to 7 tesla to investigate semiconducting materials, high superconducting crystals, spin glasses, nanotubes, and gold clusters. Techniques combining ultrahigh spatial and temporal resolution, based on near-field optical **spectroscopy** and femtosecond laser systems are also being used to study phenomena that are at the forefront of **quantum physics**, such as quantum wells, wires, and dots on the ultrafast time scales required, i.e., ranging from some hundred picoseconds to few femtoseconds. Another incentive for using low-temperature experimental methods in physics is that, whereas not all materials exhibit interesting properties that they do not have at normal temperatures, such as superconductivity, all matter experiences the loss of thermally induced vibrations at cryogenic temperatures. Low-temperature spectroscopy takes advantage of this "freezing-out" of vibrations in cooled materials, whose optical properties are improved in that their **spectral lines** become sharper, or more resolved, since the spectral broadening caused by atomic or molecular vibrations, which increases with temperature, is eliminated or significantly reduced.

See also Conduction; Conductors and insulators; Energy bands in solids; Entropy; Fluids; Heat capacity; Landau, Lev Davidovich; Lasers; Resistance, reactance, and impedance; Viscosity

LUMINESCENCE

Light generation by a process other than by heating is luminescence. For example, an incandescent light bulb, in which the filament is heated until it is literally white-hot, is not luminescent; a fluorescent light tube (which is cool to the touch) is luminescent. Luminescence is generated as part of a process in which atoms or molecules with electrons excited into higher **energy** states shed energy by emitting visible light.

People have observed luminescence in nature for centuries. In the early twentieth century, **Marie Curie**, in her doctoral thesis, mentioned that calcium fluoride glows when exposed to the radioactive material, radium. In the past 50 years, the use of luminescent devices, such as fluorescent lights and television screens, have become widespread.

Luminescence can be divided into categories by duration (**fluorescence** or **phosphorescence**) or by the mechanism that creates the light. By definition, fluorescent things stop emitting light very soon (about 10 ns) after the exciting energy is cut off. Phosphorescence continues for longer than fluorescence. Glow-in-the-dark stickers and watch hands that glow are examples of phosphorescence. A less obvious but more exact definition of the difference is that the amount of time phosphorescence continues after the material has been excited may change with **temperature**, but in fluorescence, this decay time does not change. Also, phosphorescence tends to occur at longer **wavelengths** than fluorescence.

Fluorescent dyes are included in many clothing detergents to make the clothes appear brighter. Because most organic materials fluoresce when excited by ultraviolet light, fluorescent **spectroscopy** is used to study organic molecules and atoms

by the "fingerprint" of their light emissions: the wavelengths, lifetime, polarization, and brightness of their fluorescence.

A common uses of phosphors (phosphorescent materials) is in televisions and computer monitors: small dots of red, green, and blue phosphors are grouped together on the inner surface of a cathode-ray tube. When electrons generated in the back of the tube hit the phosphors, they absorb the energy and then emit light.

Other types of luminescence are defined by the source of the energy that causes the light emission. These include chemiluminescence, bioluminescence, electroluminescence, **sonoluminescence**, triboluminescence, and thermoluminescence.

Chemical reactions provide the energy to generate photons in chemiluminescence. These chemical reactions often involve **oxygen**.

Cyalume sticks are chemiluminescent: when you bend the flexible tube enough to break the barrier that separates two substances, the tube glows for several hours until the chemical reactions are completed. A method called enhanced chemiluminescent detection, developed by researchers in Paris, offers a nonradioactive way of keeping track of genes and is being used in the international human genome mapping project.

Bioluminescence is a subset of chemiluminescence in which the chemical change occurs in living things, such as fireflies. Such reactions are very efficient. They occur when a substance called a luciferin is oxidized with the aid of a catalyst called a luciferase.

Electroluminescent devices glow when a current is applied, although not because of chemical changes. In neon lamps, current causes electroluminescence of the gas in the tube. Another fun example of this is the pickle trick: if electrical contacts are connected to either end of a cucumber pickled in brine, then current passing through the ionized pickling salts glows. (It also smells very bad.) Many flat-panel displays, such as in laptop **computers**, are made of electroluminescent materials (although the most common type uses a fluorescent light to backlight a liquid-crystal mask). Luminescence because of **electron** bombardment or an electrical field is related to electroluminescence. In fluorescent lamps, current ionizes the gas in the tube, and the **ions** activate a fluorescent coating on the inside of the lamps. The television example mentioned above is a case of cathodoluminescence, in which the phosphor is activated by a stream of electrons.

In sonoluminescence, the light is produced from energy provided by sound **waves**. This mechanism is fairly unusual. Recent research into sonoluminescence suggests that in some situations, **sound** may be concentrated to produce extremely high energies in small areas. This energy is dispelled as light, but it may be possible to harness that energy for other uses.

In triboluminescence, **friction** is responsible for the light. A famous example of this is the flash of light sometimes produced by crunching wintergreen flavor Life-Savers. (Don't confuse this with sparks given off by hitting flint and steel together. Those sparks, which can ignite a fire, are incandescent bits of metal.)

Thermoluminescence uses **heat** to release already-excited ions in a solid. When subjected to ultraviolet light, **x rays**, or

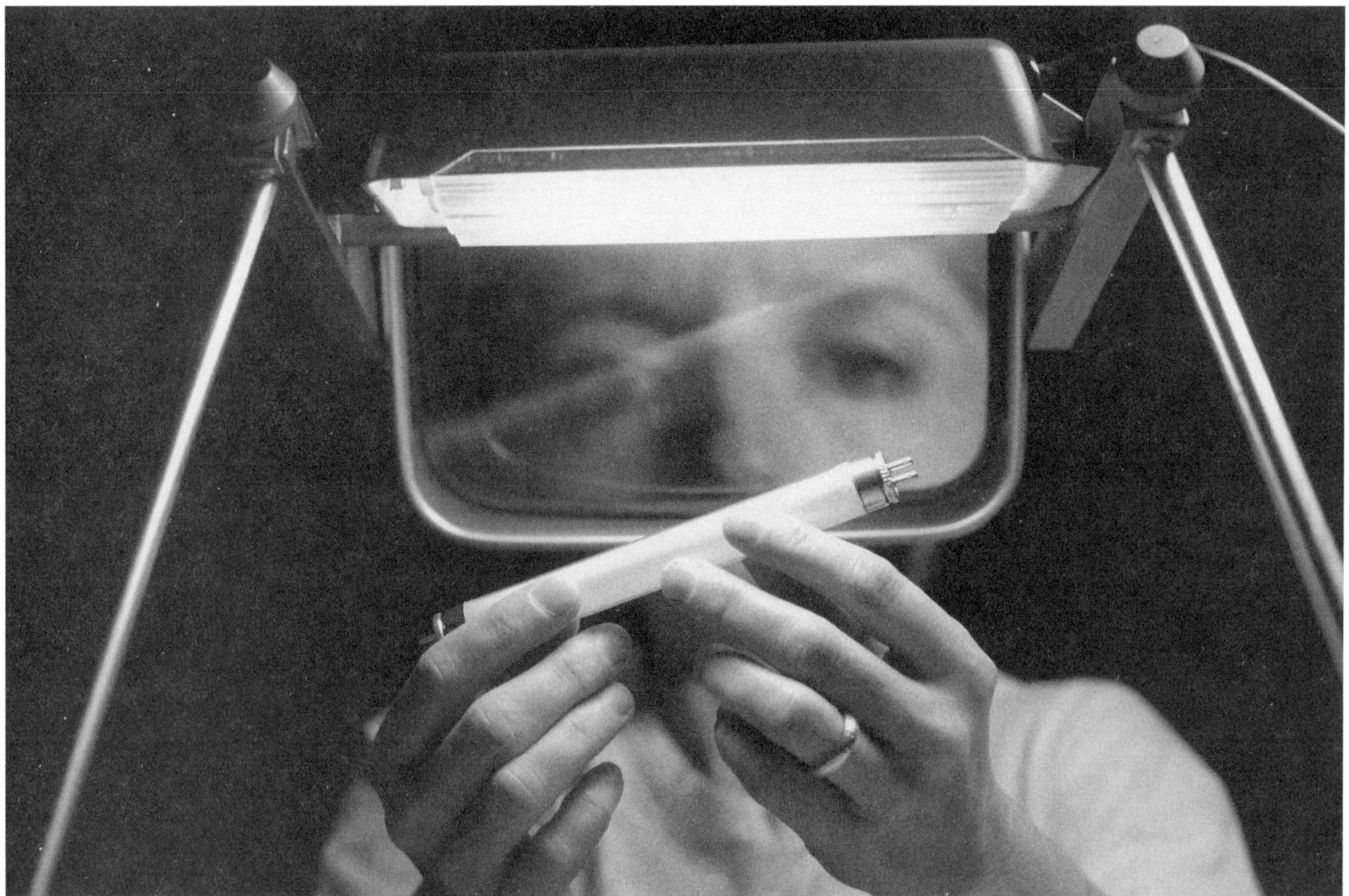

Woman examining small fluorescent light bulb. *(Photo by Robert J. Huffman. Field Mark Publications. Reproduced by permission.)*

gamma rays (all of which are energetic enough to separate electrons from **atoms**, thus forming ions), some electrons or ions become trapped in excited states. They are prevented from decaying back to a ground state because **quantum mechanics** forbids the transition. Heating allows the ion to rise to a higher state that can drop back to the ground state by emitting light.

Thermoluminescence can be used to measure how much **radiation** a material has been subjected to. It is used for dosimeters by people working around x rays or **radioactivity** who need to know how much **ionizing radiation** they have been exposed to. Thermoluminescence is also used for **radioactive dating** of pottery shards and for finding radioactive minerals.

In the early 1990s, L. T. Canham at the Royal Signal and Radar Establishment in England reported luminescence from porous silicon. This generated great interest, because if the luminescence could be controlled, then light-emitting devices could be integrated with silicon microelectronics. Although silicon photodetectors had been made, the material had not been known to emit light before. This could lead the way to relatively inexpensive optical computers, signal-processing devices, and optical communications devices.

Debate about what process creates the light is divided: those who believe the light is emitted as part of a quantum confinement effect and those who believe that the light is the product of a **chemical reaction** between the silicon and oxygen. If the quantum confinement theory proves correct and the effect can be prolonged, then useful light-emitting devices could be made from porous silicon. If the chemical theory is correct, then the luminescent period is probably inherently short-lived and the material would not make good reusable devices. Although no one has published reports that definitely debunk one theory, the chemical theory is most popular at the moment because of the instability of porous silicon devices.

WORLD *of* PHYSICS

ISSN 1531-0809

WORLD *of* PHYSICS

Kimberley A. McGrath, *Editor*

Volume 2

M-Z

General Index

GALE GROUP™

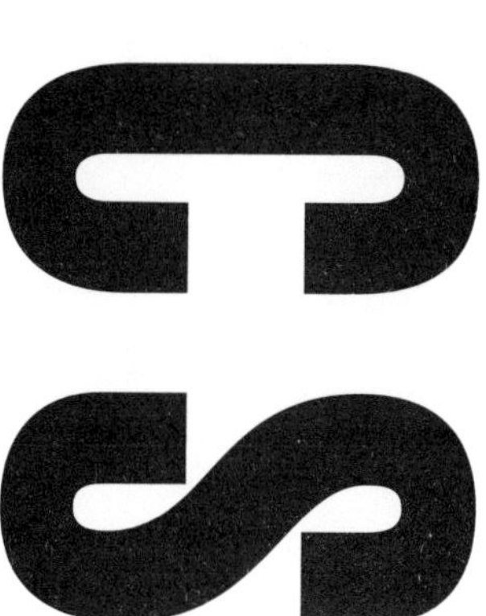

Detroit
New York
San Francisco
London
Boston
Woodbridge, CT

M

M-THEORY • See String theory

MACH, ERNST (1838-1916)
Austrian physicist

Through the efforts of such eminent scientific theorists of the mid-1800s, there seemed to be nothing in the **universe** that could not be explained through scientific hypotheses. However, there existed a small group of European scientists who would not accept these theoretical explanations. Calling themselves scientific positivists, they strove to understand nature through experimentation, rather than assumption and intuition. Mach, an Austrian physicist and philosopher, was a leading figure of the scientific positivists.

Mach's name is a familiar one to most people, who usually associate it with the velocities of supersonic aircraft. The bulk of Mach's experimental work was in the field of aerodynamics; he was the first to note the changes in the movement of air around an object as it approached the speed of **sound**. Consequently, the **speed of sound** through air of a given **temperature** is often called Mach 1, with twice that speed being Mach 2, and so on. This system is known as the Mach number.

Mach's aerodynamic findings came only after extensive research and repetition of experimentation, as was the creed of the positivist. Generally, he believed that the laws of nature were convenient fabrications that were used to tie together many different observations that contained no inherent truth. The only scientific studies that could be trusted were those with experiments founded in empirical evidence—that which can be shown and repeated. For this reason, most of the emerging theories of the time were dubious to Mach and his colleagues. Particularly infuriating to Mach was the near-universal acceptance of the **atomic theory**; since nobody had ever been able to actually see an **atom**, Mach claimed that it was ludicrous to believe in its existence.

The commitment to scientific positivism displayed by Mach and others helped to bring about a new trend in science, one in which experimentation and verification held as much importance as hypothesis. One of the scientists to benefit from this was **Albert Einstein**, whose theory of relativity was based highly upon the flaws Mach had found in **Newtonian physics**. Despite the aid he had given Einstein, Mach himself never accepted the theory of relativity and was in the process of writing a book illuminating its flaws when he died.

MACHINES • See Simple machines

MAGNETIC CONFINEMENT FUSION

Fusion is a nuclear process in which two atoms are combined to create a different **atom** (and some by-products). Magnetic confinement fusion uses a **magnetic field** to compress and **heat** gases into the plasma state so that fusion can take place. Plasma is a fourth state of **matter** not usually observed as a natural state on Earth; it exists as a natural state in **stars**. It is a state where the atoms of the gas have been stripped of their electrons so that the electrons, protons, and neutrons are free to interact. Very high temperatures, on the order of 150×10^6K, are needed to overcome nuclear forces, which hold individual atoms together, and atomic forces, which keep two atoms apart, and for fusion to take place. The magnetic field containing the plasma can be manipulated in order to increase the **pressure**, thereby increasing the **temperature**.

Ernst Mach.

The most common confinement shape is that of a toroid, or doughnut. This design is called the Tokamak class of fusion reactor and is originally of Russian design. Magnetic fields run along the doughnut as well as around it. The charged plasma then follows a helical pattern inside the toroid.

The plasma is made of a superheated gas, such as a mix of deuterium and tritium, which are **isotopes** of hydrogen. It is pumped into an evacuated chamber and heated in **pulses**. Eventually, the temperature is high enough for the fusion process. At this point, the plasma gives off **energy** and cools. Another pulse is then applied to the plasma and fusion occurs again. The cycle of heating and discharge continues until the fuel is exhausted. This is a rather inefficient form of fusion, since almost as much energy is put in as is given off by the plasma. Eventually, scientists wish to find a self-heating reaction that will sustain the fusion process, such as the nuclear ignition seen in the core of stars.

MAGNETIC FIELDS AND FORCES

Magnetic fields are produced by **magnetism**, the flow of **electric charge**, or fluctuating electric fields. In the first case, the magnetic field can be described as a region around a magnetic object, such as a magnet:

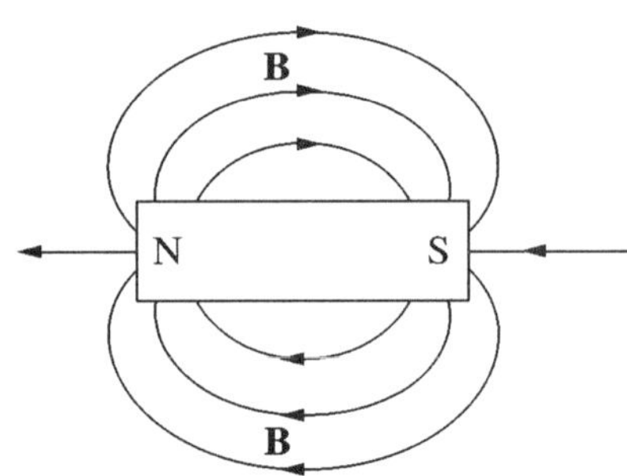

Inside the magnet, the magnetic field B is directed from the south (S) to the north pole (N). Outside the magnet, the direction of the field is represented by imaginary **force** lines called **magnetic flux** lines, which always flow in a closed loop from the north pole to the south pole. The **density** of these lines define the strength of the magnetic field, i.e., the more lines, the stronger the magnetic field. Magnetic field sources are always dipolar in nature, having north and south **magnetic poles**. But unlike electric monopoles—or charges—magnetic monopoles do not exist, i.e., an electric dipole consists of positive and negative monopoles and can be separated into isolated positive and negative charges. But separating a magnet will not yield a south and a north pole, it will yield two smaller magnets, each with a south and north pole.

A conductor through which a current is flowing will also produce a magnetic field, just as a magnet does. In this case, the magnetic field strength, H, (in units of A/m) is expressed as:

$$\vec{H} = \vec{B}/\mu_0$$

where μ_0 is the magnetic permeability of free space ($4\pi \times 10^{-7}$ N/A²). Electric and magnetic fields were first described as closely related by **James Clerk Maxwell** who formulated the basic principles of **electromagnetic induction**, which can be summarized by the following statements: Any electric field that changes over time will induce a magnetic field in the space surrounding it; any magnetic field that changes over time will also induce an electric field in the space surrounding it and the induced fields tend to have a loop pattern. These statements represent the fundamentals of **electromagnetism** and are mathematically expressed as **Maxwell's equations**. The second equation, also called Gauss's law for magnetic fields, and is expressed as:

$$div\vec{B} = 0$$

It states that the magnetic flux normal to a closed surface is zero. This is equivalent to a statement to the effect that there are no magnetic monopoles. For a magnetic dipole, the magnetic flux directed inward toward the south pole will equal the flux outward from the north pole and the net flux will always be zero for dipole sources. The third Maxwell equation, also called Faraday's law of induction, states that the line integral of the electric field E around a closed loop is equal to

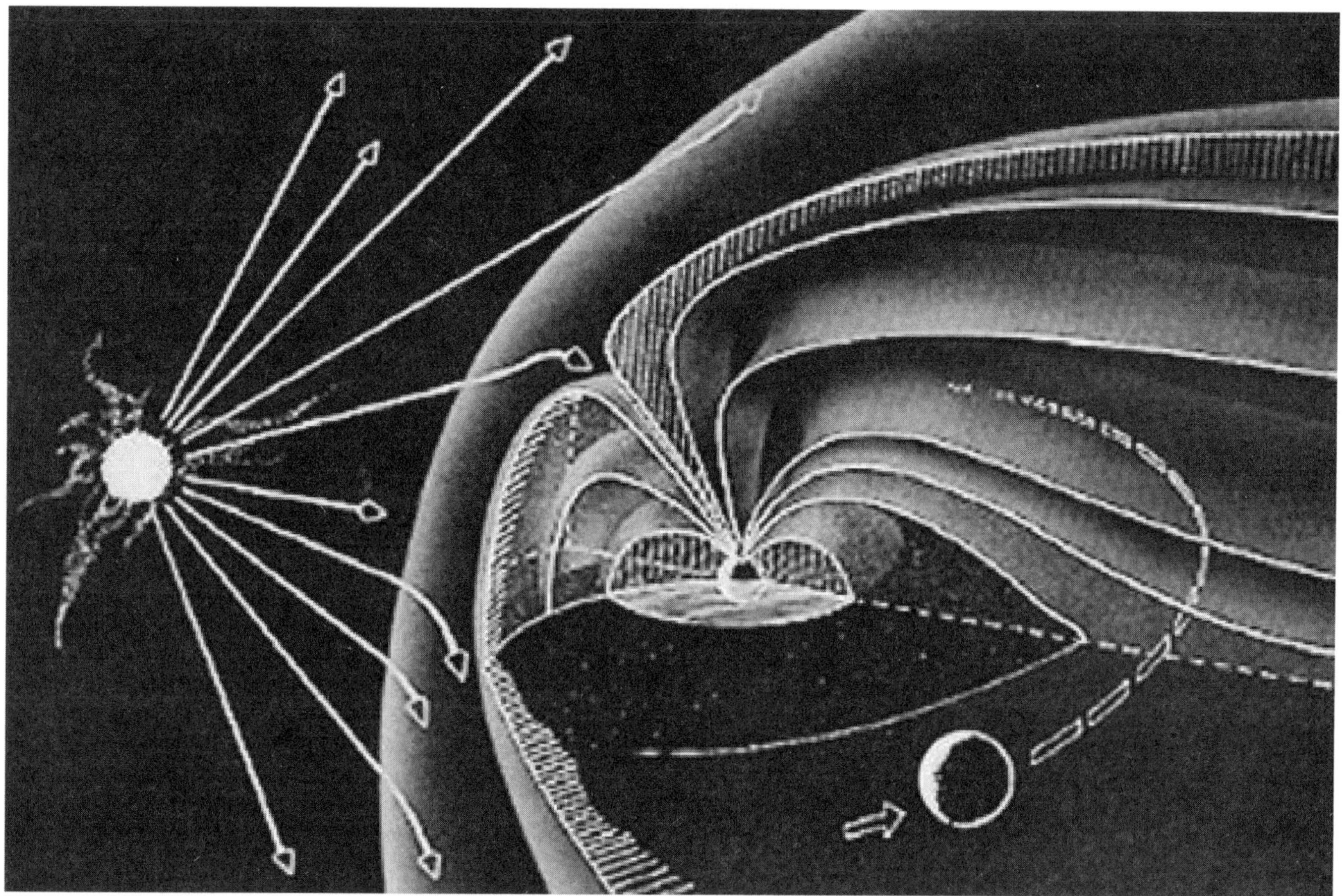

Van Allen belts, Earth's magnetic field. (Image courtesy of *NASA*.)

the negative of the rate of change of the magnetic field B through the area enclosed by the loop:

$$\nabla \times \vec{E} = -\partial \vec{B}/\partial t$$

The interaction of charge and magnetic field described by the above equation can be illustrated by a coil of wire moving in a magnetic field and generating an **electromotive force** (emf, or **voltage**). Faraday's law then states that an emf is induced in a coil when the magnetic field around it changes. It is expressed as: emf = -N $\delta\phi/\delta t$ where N is the number of turns in the coil and ϕ is the magnetic flux. The negative sign indicates that the emf is opposing the change that is producing it.

The fourth Maxwell equation is equivalent to Ampère's law and it quantitatively describes the relation of a magnetic field to the **electric current** or field inducing it. It states that for any closed loop path, the sum of the length elements (δl) times the magnetic field in the direction of the length element ($B_\Uparrow$)is equal to the magnetic permeability of free space (μ_0) times the electric current (I) enclosed in the loop.

$$\Sigma B_\Uparrow \Delta l = \mu_0 I$$

In the above formulation, the law can be used, for example, to calculate the magnetic field arising from the current flowing in a long, straight wire. The relationship between electric and magnetic forces is best expressed by the Lorentz force (F) which sums both forces. Mathematically, it is expressed as:

$$\vec{F} = q\vec{E} + q\vec{v} \times \vec{B}$$

where qE, the product of the charge q and of the electric field E, is the electric force that acts in the direction of the electric field when q is positive and where the second term of the equation is the magnetic force, in which v is the **velocity** of the charge and B the magnetic field. The direction of the magnetic force is given by a right-hand rule, such as commonly used for right-handed coordinate systems to determine which way the cross product of two vector quantities is directed.

Just as **Coulomb's law** relates electric fields to the point charges that are their sources, the Biot-Savart law relates magnetic fields to the currents that induce them and it is used to calculate such induced fields for any type of circuit. The law describes the magnetic field of a current element dB and is expressed as follows:

$$d\vec{B} = \frac{\mu_0 Id\vec{L} \times \vec{r}}{4\pi \vec{r}^2}$$

where I is the current, dL the infinitesimal length of the conductor—or wire—carrying the current and r is the unit vector that specifies the direction of the distance vector r from the current to the field point. This law is commonly used to calculate the magnetic field resulting from an electric current distribution in straight wires or in the center or on the axis of a current loop.

Some important practical applications of electromagnetic induction are the generator and the transformer. A generator basically consists of a permanent magnet that rotates within a coil of wire. The magnet is turned by a crankshaft and as it spins, the magnetic field surrounding it changes. This induces an electric field that in turn generates a current that flows around the coil. The induced current can then be used as a **power** source. An example of transformer is that of the type used to convert to lower voltages and higher currents to supply electrical power to households. The coil on the input side of such **transformers** induces a magnetic field. Since the current is alternating current, the magnetic field around the input coil is also alternating and this induces an electric field that drives the current to the output coil.

MAGNETIC FLUX

Magnetic flux describes the lines of a **magnetic field** emanating from, and in the region of space surrounding, a magnetized object. Quantitatively, magnetic flux describes the magnetic field lines of **force** that traverse a given cross-sectional area.

Magnetic fields exert forces on moving charged electrical particles. In this regard magnetic field lines are lines of force along which moving charged particles will be deflected. Magnetic fields are represented by continuous magnetic flux lines that are depicted to emerge from north-seeking **magnetic poles** and enter south-seeking magnetic poles.

Magnetic fields are characterized by magnetic flux **density** with the density of magnetic flux lines directly related to the magnitude of the magnetic field. The greater the magnitude of the field (field potential) the more dense the magnetic flux lines. At magnetic poles the magnitude of the magnetic field strength is at a maximum and the magnetic field lines are dense. Accordingly, as distance from the pole increases, the field lines diverge and become less dense. The quantitative unit of magnetic flux is the **weber (Wb)**. Because magnetic flux can arise due to the **motion** of electrons (current), at the atomic level magnetic flux can be shown to be quantized (i.e, related to **Planck's constant** and the quantum electrical charge).

Magnetic flux density is a vector (designated B) used to characterize magnetic field strength. The magnetic flux density vector measures the total magnetic field, including the fields created by other magnetic materials. The force on a charged particle moving through a magnetic field (i.e., magnetic flux lines) is related to the charge on the particle, the charged particle's **velocity** and the magnetic flux density: F = qv × B.

Because both magnetic flux and electric flux measure flow in terms of quantity per unit area per unit time, magnetic flux is defined in a very similar manner to the definition of flux through an electric field. There are, however, fundamental differences between magnetic and electrical fields. In the magnetic field there are no magnetic charges or poles to serve as the points of origin or termination for the magnetic field lines. In fact, there is no point of origin or termination for any magnetic field line. Magnetic field lines are continuous loops and the number of magnetic field lines leaving a surface must, therefore, equal the number of magnetic field lines entering a surface.

Because magnetic field lines form closed loops it is impossible to isolate magnetic poles (i.e., it is impossible to divide a magnet, one simply forms two new magnets). A surface that encloses one pole of a magnet has zero magnetic flux because each magnetic field line leaving the surface must reenter the surface.

Surrounding permanent magnets or conductive wires in which there is an established, steady, and unidirectional current flow, there is a magnetostatic field (i.e., a stationary magnetic field). The magnetostatic field is uniform in all directions. In such a uniform field the magnetic field lines are equally spaced and parallel. The magnetic field surrounding a conducting wire carrying an alternating current (AC) or variable direct current (DC), however, is not magnetostatic and the magnetic field changes in response to variations in current flow.

A changing magnetic flux induces an **electromotive force** in a loop of conducting coil. In the process of mutual inductance, emf is induced in one coil by altering current in another coil. For example, as current flows through the first in a series of coils it produces a magnetic field. Accordingly, magnetic flux induces an emf in the next coil in the series. The emf magnitude depends on the magnitude of magnetic flux and upon the distance, orientation, and composition of the coil. As current flows in the second coil, the induction process is repeated for a third coil, and all subsequent coils, in the series. The emf induced in the induced conductive coil is directly proportional to the change in magnetic flux. In turn, the magnetic flux is directly proportional to the change in current in the inducing coil.

Magnetic induction resulting from magnetic flux is used in **transformers** and ammeters designed for indirect measurement of current.

MAGNETIC MOMENT

Any current circulating in a planar loop produces a magnetic moment whose magnitude is equal to the product of the current and the area of the loop. When any charged particle is rotating, it behaves like a current loop with a magnetic moment. For a system of charges, the magnetic moment is determined by summing the individual contributions of each charge-mass-radius component.

It turns out that, both classically and quantum mechanically, there is a close connection between a particle's magnetic moment and its **angular momentum**. (The angular **momentum** of a particle moving in a circle is equal to the product of the particle's **linear momentum** and its perpendicular distance from the axis of revolution.) Classically, for the charged particle moving in a circle, the size of the magnetic moment is proportional to the magnitude of the particle's angular momentum. In **quantum mechanics**, angular momentum is quantized, i.e., it assumes only discrete values instead of the continuous range of possible values predicted by classical theory. A consequence of **quantum theory** and the quantization of angular momentum is that the magnetic moment also must be quantized.

The **magnetic properties of matter** can be thought of as arising from microscopic atomic currents that produce magnetic moments in that **matter**. In paramagnetic materials, permanent magnetic moments arise from the intrinsic angular momentum (spin) of individual electrons. In the absence of an applied **magnetic field**, these magnetic moments are randomly arranged, and there is no net magnetic moment. An applied magnetic field, however, aligns the moments in the field direction. In diamagnetic materials, all intrinsic magnetic moments are canceled out by the pairing of electrons. Any remaining magnetic effects in diamagnetic materials are produced by the orbiting electrons.

MAGNETIC POLES

Magnetic poles are the positive and negative sides of a magnetic system. Unlike electric charges, which can be separated into positive charges that have no corresponding negative charge and vice versa, magnetic poles do not exist in isolation. In other words, there is no such thing as a magnetic monopole. The magnetic component of the **electromagnetic force** is such that everything that is magnetized has positive and negative, or north and south, poles. This phenomenon can be observed with a handful of kitchen magnets: there will always be one side of a magnet that attracts and one side that repels, a given other magnet side. One of the basic principles of magnetics is that like poles repel and unlike poles attract.

The most common arrangement of magnetic poles is the dipole. This is a simple, linear polar arrangement, with a north pole on one side and a south pole on the other. Many natural objects, from kitchen magnets to molecules down to electrons, possess some dipole behavior. The macroscopic magnets we use in our daily life are magnetic because the molecules in them all have magnetic dipole moments that have aligned. In materials like iron, the alignment can be made permanent; in other materials, it can be temporarily induced. The magnetic polarization of a material is a measurement of how much the magnetic dipoles within it are aligned.

While dipole arrangements are the most basic and the most common, there are situations in which magnetic poles are said to be "of higher order." That is, the magnetic effects of the situation cannot be expressed in simply one line of **magnetism** with north and south poles. Instead, these situations require multipole expansions, with quadrupole and octopole arrangements the next higher level up. These arrangements are used to separate out different components of a magnetic field and can be useful in separating particles of different energies, as in accelerator experiments.

While classical physics does not allow for the existence of magnetic monopoles, there is nothing in **quantum theory** to prohibit them from existing. However, none have been observed in nature. Therefore, it is one of the major tasks for physicists in the twenty-first century to either observe magnetic monopoles or form a quantum theory that coherently explains why they do not exist. If they did exist, they would be extremely rare and would require changes in the equations that are thought to govern electromagnetic behavior.

MAGNETIC PROPERTIES OF MATTER

Magnetism is a property of **matter** and it occurs in different forms and degrees in various **conductors and insulators**. For example, at low temperatures, metallic systems exhibit either superconducting or magnetic order. The degree of magnetism of a substance is due to the intrinsic magnetic dipole moment of its electrons. The degree of magnetism is also called magnetization and it is defined as the net magnetic dipole moment of the substance per unit volume. The magnitude of the magnetic dipole moment of an **electron** is given by the Bohr magneton, $m_B = 9.27 \times 10^{-24} Am^2$. Magnetization is further defined by measuring a quantity called the magnetic susceptibility.

In the nineteenth century, **Michael Faraday** was the first to start classifying substances according to their magnetic properties. Faraday classified them as either diamagnetic or paramagnetic and he based his classification on the **force** exerted on the materials when placed in an inhomogeneous **magnetic field**.

Diamagnetic substances have a negative magnetic susceptibility, (i.e., they are materials in which the magnetization and magnetic field are opposite). The electrons in the atoms of diamagnetic materials are all paired and there is no intrinsic **magnetic moment**. When a material is placed into a magnetic field, its atoms acquire an induced magnetic moment pointing in a direction opposite to that of the external field, and the material becomes magnetic. The diamagnetic field produced by the material opposes the external field, although this diamagnetic field is very weak (except in superconductors). If the atoms of a material have no magnetic moment of their own, then diamagnetism is the only magnetic property of the material and the material is called diamagnetic. Copper exhibits such diamagnetism.

Paramagnetic substances have a weak positive magnetic susceptibility and their atoms usually have unpaired electrons of the same spin. Some metals, rare earth, and actinides are paramagnetic. All the magnetic moments of the electrons in their atoms do not completely cancel out, and each **atom** has a magnetic moment. Such materials thus have a permanent magnetic moment and they can interact with a magnetic field. An external magnetic field tends to align the magnetic moments in the direction of the applied field, but thermal **motion** tends to randomize the directions. If only a relatively small fraction of the atoms are aligned with the field, then the magnetization obeys Curie's law. Curie's law states that if the applied magnetic field is increased, the magnetization of the material also increases. This is because a stronger magnetic field will align a greater quantity of dipoles. Curie's law also states that the magnetization decreases with increasing **temperature**. The magnetic field produced by the aligned magnetic moments of paramagnetic materials strengthens the external field, but at standard temperatures it averages no more than 10 times stronger than a diamagnetic field and is, therefore, still very weak.

Ferromagnetic materials have the highest magnetic susceptibilities. In these materials, the spins of neighboring atoms do align even in the absence of an externally applied field through a quantum effect known as exchange coupling. Besides iron, examples of ferromagnetic materials are nickel, cobalt, and alnico, an aluminum-nickel-cobalt alloy. In these materials, all metals, the electrons give rise to permanent dipole moments that can align with those of their neighbors, creating magnetic domains that produce a magnetic field. Above a certain temperature, called the Curie temperature, a ferromagnetic material ceases to be ferromagnetic because the addition of **thermal energy** increases the motion of the atoms, thus destroying the alignment of the dipole moments. The material then becomes paramagnetic with weak magnetic susceptibility. The magnetic domains of ferromagnetic materials allow them to be turned into permanent magnets. If a ferromagnetic material is placed in a strong magnetic field, its magnetic domains converge into large domains aligned with the externally applied field. Upon removal of the external field, the electrons maintain the alignment and the magnetism remains.

The magnetic properties of matter are used in a wide variety of applications. For example, magnetization is the acting principle of magnetic memory and it is used to make audio and video tapes, as well as magnetic disk storage devices for **computers**. In such applications, the recording head of a tape recorder, or the write head of a disk drive, applies a field that magnetizes a small segment of the tape or disk, which remains magnetized until another magnetic field changes it. A magnetic tape reader can then take the information from the magnetic tape and convert it into an electrical signal that, in the case of a cassette player, is sent to a speaker and converted into **sound**. When a magnetic tape is passed under the reading device, a magnetic field is induced through the plane of the wire loops. Different orientations of the atoms in the tape material means that the magnetic field is changing, which subsequently induces a current. This electrical signal is then passed on to another device for further processing.

Diamagnets also exhibit useful properties in the presence of externally applied magnetic fields. In these materials, eddy currents consisting of moving electrons are induced and the magnetic effects cancel part of the applied external field. An example of a device incorporating a diamagnet is the metal detector. In this instrument, the magnetic field is generated by an electromagnet, which then forms eddy currents. The magnetic fields from the induced currents are in turn picked up by the metal detector in the form of small currents.

See also Atomic structure; Avogadro's number; Electric charge; Electric field and forces; Electricity; Electromagnetic field; Electromagnetism; Ferromagnetism; Hall effect; Magnetic poles; Semiconductors; Superconductors and superconductivity

MAGNETIC RESONANCE IMAGING

Magnetic resonance imaging (MRI) is the application of **nuclear magnetic resonance** to produce images. MRI is now a common, though relatively expensive, diagnostic technique. MRI images can be very detailed and informative and the technique is noninvasive and it does not use highly energetic and potentially dangerous **ionizing radiation**. Because of the association many people make of the term "nuclear" with dangerous **radiation**, the term "nuclear" was eliminated from the original term for the technique (nuclear magnetic resonance imaging).

Nuclear magnetic resonance was developed in the 1950s by physicists as a means of probing the properties of the atomic **nucleus**. Theoreticians predicted that nuclei with particular ratios of protons and neutrons, giving rise to nonzero nuclear spin numbers, would show an asymmetry, or anisotropy, in a magnetic field that would result in a greater number of nuclei aligned in one direction relative to the north-south orientation of the **magnetic field** than in the opposite direction. The **energy** necessary to cause a nucleus aligned with the field to lose its orientation was found to be in the range of **radio** frequencies for the magnetic field developed by strong electromagnets.

Very soon it was found that the chemical environment of the **atom** in which an atomic nucleus resides affects the energy the absorbed by the nucleus to change its orientation in a particular magnetic field. For example, the nuclei of the hydrogen atoms bound to the **carbon** atom of methyl alcohol, CH_3OH, absorb a radio **frequency** different from the **hydrogen atom** bound to the **oxygen** atom. Chemists quickly adopted nuclear magnetic resonance as a means of determining **molecular structure**. Because the absolute energy absorbed by the nucleus and the relative energies of nuclei in different chemical environments increase in proportion to the size of the magnetic field, chemists have continually worked with physicists and engineers to develop instruments with higher and higher magnetic fields. Modern instruments using superconducting magnets have magnetic fields over 10 times as great as those of the permanent- and electromagnet-based instruments used in the 1960s when high resolution nuclear magnetic resonance (NMR) **spectroscopy** was developed as a common technique for organic structure determination. Much larger, more com-

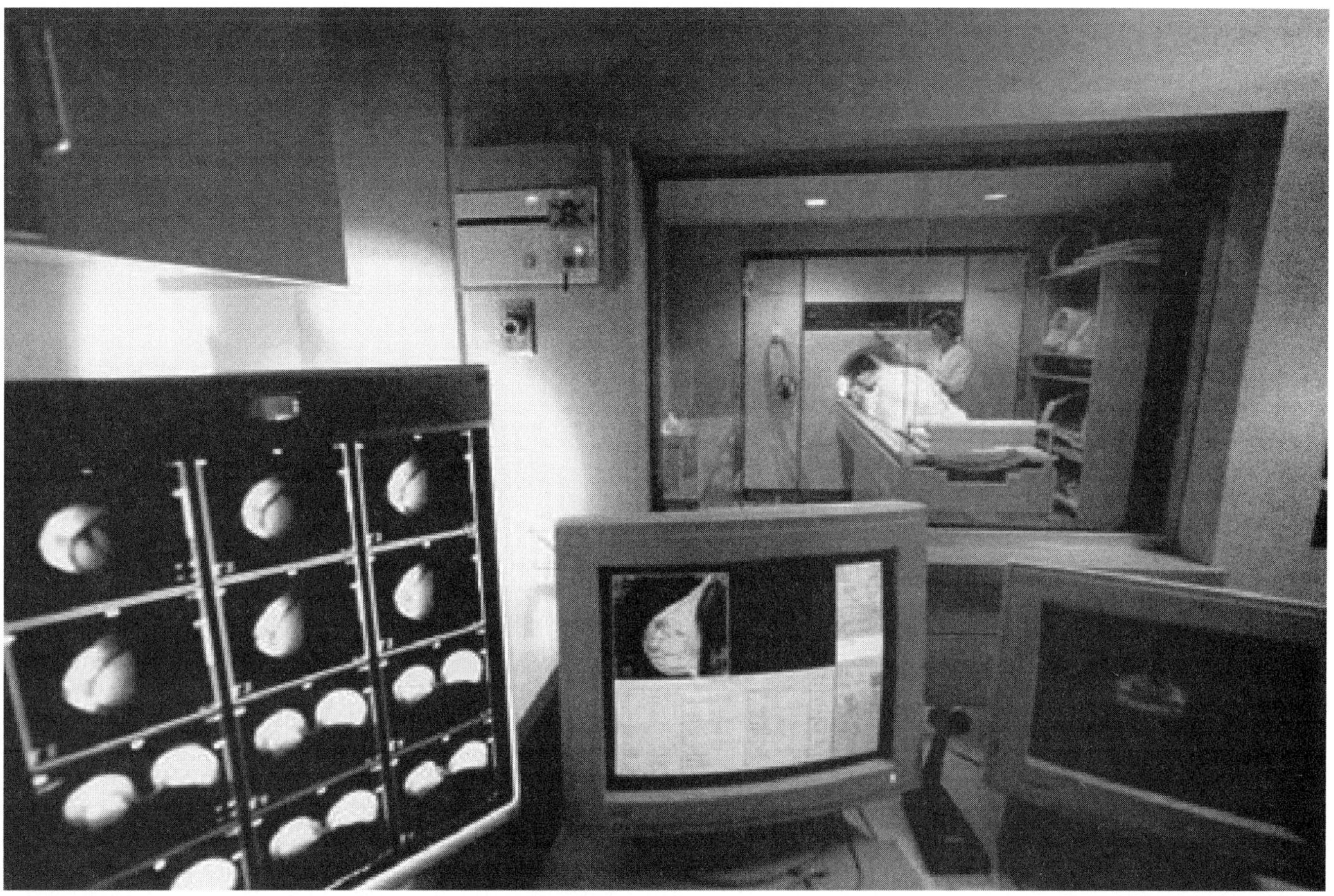

Magnetic resonance imaging system, room with equipment and viewing machine. *(Photo by Mason Morfit. FPG International Corp. Reproduced by permission.)*

plex molecules, including important biomolecules, can now be studied by this method.

Along with the effect of chemical environments on producing differences in energies, scientists found that the time it takes for nuclei that are disoriented in the magnetic field by applied the radio frequency to recover their **equilibrium** distribution orientation (the relaxation time) depends on their environment. If the radio transmitter is turned off and then turned back on again before the relaxation processes of the nuclei in the sample are complete, less energy will be absorbed from the second pulse of radio frequency energy. The relaxation times of nuclei are determined by using a sequence of carefully timed **pulses** and observing the amount of energy absorbed by each pulse.

Further experimentation revealed that the effect of the environment of a nucleus was not limited just to the atoms bonded to or even part of the same molecule that contains the particular nucleus, but to other molecules that are near it, such as the solvent, and the ability of the molecule in which the nucleus is contained to physically reorient itself in the magnetic field (its mobility). As instruments were developed that allowed scientists to use larger samples, and developments in high resolution NMR spectroscopy encouraged scientists to investigate more complex materials, experiments on biologi-

cal tissues indicated that the hydrogen nuclei in water contained within different tissues has different relaxation times— the basis of MRI.

A key to MRI is accurate location of the water whose relaxation time is measured by radio frequency pulse sequences. This is accomplished by the use of a magnetic field that varies in strength, referred to as a field gradient. The energy of radio frequency absorption by the hydrogen nuclei of the water molecules is great at the end of the field that has the greatest magnetic field strength and decreases proportionately as the field strength decreases, providing a means of locating the position of the water. The patient having an MRI scan performed is moved slowly through the magnet so that a series of images corresponding to field gradient "slices" can be obtained. The energy/time data are processed, enhanced, and rendered into photograph-like images by powerful computer programs. Image enhancement for MRI and other diagnostic scanning techniques is an active area of scientific research.

The high resolution and high contrast of different tissues afforded by MRI, coupled with its noninvasive safety, have made it the diagnostic method of choice in many situations despite its high cost. MRI is a practical, life-saving tool that came about only by the collaborative interplay of fundamental theory, technological advances, and experimental science.

MAGNETISM

Magnetism has intrigued humankind for millennia. Because magnetic and magnetically susceptible objects affect each other at a distance, without being in contact with each other, the effect appears to be magical. An understanding of magnetic effects that are much more subtle than the orientation of a compass needle by Earth's magnetic field has contributed greatly to our current concepts of chemical bonding and **molecular structure**.

Naturally magnetized pieces of magnetite, Fe_3O_4, have been known for thousands of years and for the past several hundred years sailors and explorers have used the effect of Earth's magnetic field on an iron needle to help them maintain their direction. The great advance in the understanding of magnetism that has been exploited in **chemistry** occurred only in the last century when physicists **James Clerk Maxwell** and others determined relationships that link **electricity** with magnetism. When Gilbert Newton Lewis convincingly demonstrated the importance of electrons in chemical bonding and **Erwin Schrödinger** developed a mathematical model for the interactions of electrons in atoms that could be expanded to include a description of the magnetic properties of electrons in molecules, the stage was set for using experimentally measured magnetic properties to deduce information about the chemical structure of materials.

Moving **electric charge** creates a magnetic field. If the distribution of moving electric charge is not symmetric, the magnetic field will also be unsymmetric. In some materials, such as the natural magnet lodestone, the unsymmetric magnetic fields of individual atoms are aligned in the same direction within the bulk material. If this condition persists even in the absence of an external magnetic field, the material is a permanent magnet. Magnetite that does not act as a magnet may not have been exposed for a sufficient time or under other appropriate conditions to become magnetized by Earth's magnetic field. Placing it close enough to a magnet or placing it in a strong **electromagnetic field** can magnetize nonmagnetized magnetite. Similarly, metallic iron and many other metals, alloys, oxides, sulfides, and other compounds containing iron and a number of other transition metals can be magnetized. Materials that can become magnetized are composed of atoms whose electrons can be predominately oriented asymmetrically and remain oriented asymmetrically once the external magnetic field is removed. The asymmetric **electron** distribution in such materials reinforces itself. Such materials are called ferro magnetic (like iron).

Most materials are not ferromagnetic, but there are a variety of magnetic effects that other materials exhibit. In most materials, the electrons of the atoms are all paired with each other. The molecules of most substances in noncrystalline bulk materials are randomly oriented in liquids and solids. A bulk sample of a randomly oriented, spin-paired material will not be attracted into a magnetic field. In fact, such materials are very slightly (compared to the magnitude of ferromagnetic effects) repelled by a magnetic field. These materials are said to be diamagnetic. Although diamagnetic

Iron recycling with an electromagnet. *(Photo by David R Frazier. Photo Researchers, Inc. Reproduced by permission.)*

effects are usually very small, one class of materials shows strikingly large diamagnetism: **superconductors**. Superconducting materials are so strongly repelled by magnetic fields that they are levitated by them. There have been proposals to utilize this property of superconductors (the Meissner effect) to construct levitated trains that would experience nearly zero **friction** by traveling above magnetized tracks.

Some materials, particularly compounds of the transition metals but also some compounds of nitrogen and other elements with an odd number of valence electrons, have unpaired electrons. The magnetic moments resulting from the unpaired electrons often align themselves with an external **magnetic field**, drawing the materials into the magnetic field. This effect can be so great for compounds with atoms that have several unpaired electrons, such as those of manganese (II) with five unpaired electrons per manganese (II) **atom**, that a sample can weigh 10% more in the field of a laboratory magnet than outside the field. Compounds that are attracted into a magnetic field are paramagnetic. The gain in **weight** at a particular magnetic field strength (the paramagnetic susceptibiliy) usually decreased inversely with the **temperature** on the absolute (Kelvin) scale, a phenomenon known as Curie's law.

By observing such factors as the degree to which a material becomes magnetized, the weight gained or lost in a magnetic field, the temperature dependence of such changes, and the dependence of such changes on the magnitude of the external magnetic field, chemists have determined that there are a number of different patterns that can be correlated with the interactions of electrons of one atom with neighboring atoms in a molecular or ionic structure or with the overall repeating structure of atoms in a crystalline structure. For example, in some structures, the magnetic moments of entire molecules within a crystal are aligned in opposite directions, a condition termed antiferromagnetism. If the magnetic moments of atoms within the molecules are aligned in opposite directions, the condition is antiferrimagnetism. Sometimes the **magnetic moment** in one direction is larger than that in the other and the material shows some degree of magnetization at high enough magnetic fields. These materials are called parasitic ferromagnets or parasitic ferrimagnets, respectively. The magnetic properties of some crystalline substances change drastically when they are subjected to **pressure** along one direction of the crystal. These substances are called piezomagnets, analogous to the terminology for piezoelectric crystals whose electrical conductivity changes when they are subjected to pressure.

The behavior of substances in a magnetic field may be used to deduce the number of unpaired electrons present in atoms of the substance, the ways that unpaired electrons on one atom interact with those on a nearby atom, and to deduce short- and long-range structural effects in crystalline compounds. Our understanding of magnetism has progressed from the mysterious to fundamental science useful in solving modern chemical questions about the structure of **matter**.

Majorana, Ettore (1906-1938)
Italian physicist

Ettore Majorana played an influential role in the development of **atomic physics** in the 1930s. First a student, then coworker with noted Italian physicist **Enrico Fermi**, in 1937 Majorana elaborated a symmetrical theory of weak interaction, where the **neutrino**, an elementary particle emitted during the decay of other particles, is identical to the antineutrino.

Born in Sicily in 1906, Majorana was the youngest son of physicist Fabio Majorana. Young Ettore showed early genius in mathematics and physics, but was also noted to have an eccentric personality. Majorana moved with his family to Rome and later entered the University of Rome to study engineering. While at the university, Majorana changed his focus to physics, earning his degree in 1930. As a member of a team of young Italian physicists, Majorana became known, along with Italian-born American physicist Enrico Fermi and others,

as one of "the boys of *via Panisperna*" involved in fundamental research on artificially induced **radioactivity**. Building upon these fundamental works, Fermi went on to win a Nobel Prize in 1938.

Majorana was named professor of theoretical physics at the University of Naples in 1938. During a boat trip from Palermo to Naples in March of the same year, Majorana suddenly disappeared. Although two notes were found, one declining the university position and the other expressing intent to end his life, investigations found no proof of Majorana's actions or whereabouts. At the time, Majorana was 31 years old.

Fermi, before his death in 1954, said of his former colleague, "There are many categories of scientists.... There are also people of first class, who make great discoveries, fundamental for the development of science. But then there are the geniuses, like [**Galileo**] Galilei and Newton. Well, Ettore Majorana was one of them."

Today, the Ettore Majorana International Center for Scientific Culture in Erice, Sicily, promotes international collaboration in science and technology between researchers throughout the world. Founded by the Sicilian government in 1963, the center is also home to the World Federation of Scientists.

Manhattan Project

The Manhattan Project was an epic, secret, wartime effort to design and build the world's first nuclear weapon. Commanding the efforts of the world's greatest physicists and mathematicians during World War II, the 20-billion-dollar project resulted in the production of the first uranium and plutonium bombs. The American quest for nuclear explosives was driven by the fear that Hitler's Germany would invent them first and thereby gain a decisive military advantage. The monumental project took less than four years, and encompassed construction of vast facilities in Oak Ridge, Tennessee, and Hanford, Washington, that were used for the purpose of obtaining sufficient quantities of the isotopes uranium-235 and plutonium-239, necessary to produce the fission **chain reaction** which released the bombs' destructive **energy**. After a successful test in Alamogordo, New Mexico, the United States exploded a nuclear bomb over the Japanese city of Hiroshima on August 6, 1945. Three days later another bomb dropped over the Japanese city of Nagasaki spurred the Japanese surrender that ended World War II.

In the 1930s and early 1940s, fundamental discoveries regarding the **neutron** and **atomic physics** allowed for the possibility of induced nuclear chain reactions. Danish physicist **Niels Bohr**'s compound **nucleus** theory, for example, laid the foundation for the theoretical exploration of fission, the process whereby the central part of an **atom**, the nucleus, absorbs a neutron, then breaks into two equal fragments. In certain elements, such as plutonium-239, the fragments release other neutrons that quickly break up more atoms, creating a chain reaction that releases large amounts of **heat** and **radiation**.

Hungarian physicist **Leo Szilard** conceived the idea of the nuclear chain reaction in 1933, and immediately became concerned that, if practical, nuclear energy could be used to make weapons of war. Szilard, who fled Nazi persecution first in his native Hungary, then again in Germany, conveyed his concerns to his friend and contemporary, noted physicist **Albert Einstein**. In 1939 the two scientists drafted a letter (addressed from Einstein) warning U.S. President Franklin D. Roosevelt of the plausibility of **nuclear weapons**, and of German experimentation with uranium and fission. In December 1941 after the Japanese attack at Pearl Harbor and the United States's entry into the war, Roosevelt ordered a secret U.S. project to investigate the potential development of atomic weapons. The Army Corps of Engineers took over and in 1942 consolidated various atomic research projects under the intentionally misnamed Manhattan Engineering District (now commonly known as the Manhattan Project) placed under the command of Brigadier General Leslie Richard Groves.

Groves recruited American physicist **J. Robert Oppenheimer** to became the scientific director for the Manhattan Project. Security concerns required the development of a central laboratory for physics weapon research in Los Alamos, New Mexico. Oppenheimer's leadership attracted many top young scientists, including American physicist **Richard Feynman**, who joined the Manhattan Project while still graduate student. Feynman and his mentor **Hans Bethe** calculated the **critical mass** fissionable material necessary to begin a chain reaction.

Fuel for the nuclear reaction was a primary concern. At the outset, the only materials seemingly satisfactory for sustaining an explosive chain reaction were either U-235 (derived from U-238) or Pu-239 (an **isotope** of the yet unsynthesized element plutonium). Additional requirements included an abundant supply of **heavy water** (e.g., deuterium and tritium). At Oak Ridge, the process of gaseous diffusion was used to extract the U-235 isotope from the 140 times more abundant U-235 found in uranium ore. At Hanford, production of P-239 was eventually made possible by leaving plutonium-238 in a nuclear reactor for an extended period.

In 1942 Italian physicist **Enrico Fermi** supervised the first controlled sustained chain reaction at the University of Chicago. Underneath the university football stadium, in modified squash courts, Fermi and his team assembled a lattice of 57 layers of uranium metal and uranium oxide embedded in graphite blocks to create the first reactor pile.

The Manhattan Project eventually produced four bombs. Little Boy, the code name for the uranium bomb, utilized explosives to crash pieces of uranium together to begin an explosive chain reaction. Fat Man, the code name for the plutonium bomb, was more difficult to design. It required a neutron-emitting source to initiate a chain reaction within a series of concentric nested spheres. The outermost shell was an explosive lens system surrounding a pusher/neutron absorber shell designed to reduce the effect of Taylor **waves**, the rapid drop in **pressure** that occurs behind a detonation front that could interfere with an implosion. The next nested sphere was a uranium tamper/reflector shell containing a plutonium pit and beryllium neutron initiator. The spheres were designed to implode, causing the plutonium to fuse, reach critical **mass**, then start the reaction.

The simple design of the uranium bomb left scientists confident of its success, but the complicated implosion trigger required by the plutonium bomb raised engineering concerns about reliability. On July 16, 1945, a plutonium test bomb code named Gadget was detonated in a remote area near Alamogordo, New Mexico. Observed by scientists wearing welder's glasses and suntan lotion for protection, the test blast (code named Trinity) was more powerful than originally thought, roughly equivalent to 20,000 tons of TNT and caused total destruction up to one mile from the blast center.

On August 6, 1945, an American B-29 "Flying Fortress," the *Enola Gay*, dropped the uranium bomb over Hiroshima. Sixty thousand people were killed instantly, and another 200,000 subsequently died as a result of burn and radiation injuries. Three days later, a plutonium bomb was dropped over Nagasaki. Although it missed its actual target by over a mile, the more powerful plutonium bomb killed or injured more than 65,000 people and destroyed half of the city. Ironically, ground zero, the point under the bomb explosion, turned out to be the Mitsubishi Arms Manufacturing Plant, at one time the major military target in Nagasaki. The fourth bomb remained unused.

Many Manhattan Project scientists eventually became advocates of the peaceful use of this nuclear **power** and advocates for nuclear weapons control.

MANY-BODY PROBLEM

The many-body problem involves describing a large number of interacting particles in a detailed way and, specifically, in a way that predicts their future behavior. In all of physics, there are few issues stated so simply, but the many-body problem is responsible for as many headaches as any other. In fact, the problem has not been completely solved, and it may never be. As with relations between people, interactions between particles, or bodies, become increasingly complex when more than one get involved. The many-body problem is not concerned with the bodies in and of themselves, for these can be successfully described; the central issue and the real difficulty lie in the effects that these particles have on one another. One body moving through **space** is easily understood and its future is predictable. When many bodies are involved, however, keeping track of their interactions quickly becomes nightmarish.

A simple many-body example appears in certain lotteries where a tumbler of numbered Ping-Pong balls determines winning numbers. Say there are 100 such balls in a large, rotatable drum. Even if the exact position of each ball is know in the beginning, it is virtually impossible to predict which number will sit on top after the drum rotates a few dozen times. The Ping-Pong balls are obviously interacting. When they collide, they bounce from one another, changing directions. It is these interactions that make for such a wild, umpredictable scene inside the drum and leave the lottery with a practically random result.

Fundamental particles (such as electrons) make for a much more challenging instance of the many-body problem because their interactions are more varied and complicated than those of Ping-Pong balls. Physicists can take on quite a few Ping-Pong balls before running into trouble, but they have their hands full when they consider only two fundamental particles.

Many-body literature has turned increasingly to the use of **Feynman diagrams** in the last 30 years. These diagrams, introduced by **Richard P. Feynman**, give physicists an elegant pictorial shorthand for complicated mathematical expressions. They involve a simple set of symbols (dots, arrows, various lines) and a short but rigid list of rules. Many-body physicists can carry on entire conversations and arguments using nothing more than these quick sketches.

MARCONI, GUGLIELMO (1874-1937)
Italian physicist and engineer

Before he was 30, Guglielmo Marconi entered the limited ranks of scientists whose reputations have spread beyond the confines of their community. An amateur, Marconi accomplished the first transatlantic transmission of **radio waves** without ever having received any academic degrees in physics. Although unquestionably gifted in that field, Marconi expressed his true genius in assembling the backing, the information, and the engineers he needed to pursue his vision of linking the globe in a network of radio waves. With the advantages of ready money, social access, and entrepreneurial drive, Marconi realized his dream. He received many honors and awards from governments, academic institutions, and scientific associations, but undoubtedly the principal one was the 1909 Nobel Prize in physics, which he shared with Karl Ferdinand Braun for their advancement of radio technology. It would not be accurate to the scientific record to call Marconi the father of radio. However, he greatly accelerated the establishment of radio in all its current applications—military, industrial, and commercial.

Marconi's father, Giuseppe, was a successful landowner who split his time between Bologna, Italy, and his estate, the Villa Grifone, 11 mi (17.7 km) outside of Bologna in a town called Pontecchio. When already a widower with one son, Luigi, Giuseppe met and fell in love with Annie Jameson, a young Irishwoman who had come to Italy to study operatic singing. Her father, a whiskey distiller from Dublin, and her mother did not approve of the relationship. Against the wishes of her parents, they were married in 1864. Marconi was born in Bologna on April 25, 1874. He had a brother nine years his senior called Alfonso. Life was very strict in his father's household, where he was educated in his early years. At age 12, he went to the Istituto Cavallero in Florence. The following year Marconi enrolled in the Technical Institute in Leghorn. His schooling there was augmented with private tutoring from Professor Vincenzo Rosa

Guglielmo Marconi.

of the Liceo Niccolini. Professor Rosa was an important early influence on the future scientist.

Accomplishes the First Successful Radio Transmission

In 1894 Marconi became acquainted with the work of **Heinrich Hertz**, who had died that year. He was so fascinated with Hertz's radio waves that he sought as much information as he could by auditing the courses of physicist and professor Augusto Righi, the expert on radio waves at the University of Bologna. At Villa Grifone, in 1895, Marconi built his first crude transmitter and receiver. The transmitter consisted of Hertz's oscillator, in which an electrical circuit generated radio waves as a spark jumped back and forth across a gap between two metal balls. In order to work up the **voltage** to **power** such a spark, Marconi used induction coils, two sets of wires wrapped around a soft iron core, to increase the current from a low-voltage battery. In the receiver, Marconi used a device introduced by French physicist Édouard-Eugéne Branly and

English physicist Oliver Joseph Lodge called a coherer—a glass tube full of metal filings connected to a battery—to pick up the radio signals. Through painstaking trial and error, he introduced improvements of his own, such as insulated wire antennae and **grounding** of both the transmitter and the receiver. In the fall of 1895, Marconi succeeded in sending radio signals across his father's estate, a distance of about a mile.

Giuseppe Marconi, who until that point had been skeptical of the practicality of his son's experiments, realized that Marconi's discoveries should become public domain. The family petitioned the Italian Minister of Post and Telegraph to underwrite the development of Marconi's invention. The minister showed no interest. Marconi's Irish mother suggested asking the English government, because surely England, as the greatest naval power at the time, would realize the value of wireless ship-to-shore communication. In February 1896, Marconi and his mother traveled to England. Once they arrived in London, his cousin Henry Jameson-Davis, as promised in correspondence prior to the trip, helped to guide Marconi through the English bureaucracy.

In 1896, after a successful radio transmission on England's Salisbury plain across a distance of nine miles, Marconi gained some notoriety, getting mail from all over the world and an offer to buy the rights to his system. On July 2, 1897, Patent Number 12,039 was granted for Marconi's invention of the "wireless," as radio was called then. Instead of selling his rights, however, Marconi formed the Wireless Telegraph and Signal Company, which was renamed Marconi's Wireless Telegraph Company in 1900. The English government, very impressed with the wireless, invited Marconi to become an English citizen in order to evade his three years of required service in the Italian military.

Marconi, a lifelong patriot of Italy, declined and through the auspices of Italy's ambassador in England became a cadet in training in the Italian navy with the understanding that he could pursue his work without interruption. Returning to Italy, Marconi performed a ship-to-shore demonstration of his system for the Italian navy at Spezia, where he reached a limit of 12 miles for good reception. In 1898 he returned to England for another demonstration, increasing the range to 18 miles. English physicist and mathematician **William Thomson Kelvin**, who had earlier disparaged radio communication, made the first paid wireless transmission in 1899 and thereafter became an ally. The same year, at Queen Victoria's invitation, Marconi set up a system so that the royal family could get news of the Kingstown regatta in Ireland. Although this feat had no scientific value, he shrewdly calculated that it would give him enormous publicity.

Permanent radio stations were beginning to be put into place. In March 1899, with the cooperation of the respective governments, radio communications began between Chelmsford in England and Wimereux in France, a new record of 85 miles. Marconi believed, unlike many other scientists of his day, that radio waves would follow Earth's curve rather than shoot straight into **space**. Therefore he decided to prove his point by transmitting across the Atlantic Ocean.

Demonstrates Radio's Potential for Transatlantic Communication

With the company he had created to finance him, he began to build stations at Poldhu, in Cornwall, England, and at Cape Cod, Massachusetts. Unfortunately, before Marconi was ready to test his theory, the antennas at both sites were destroyed in storms. The mast at Poldhu's antenna was repaired. On the other side of the Atlantic, Marconi, impatient to get on with his tests, improvised an antenna in St. John's, Newfoundland, Canada, by lofting a kite 400 ft (12.2 m) into the air with a wire attached to it. The first signal coming out of Poldhu, the Morse code for the letter *s,* was heard in St. John's on December 12, 1901, at a distance this time of 1,800 hundred miles. Marconi had proved to his own satisfaction that Earth's curvature was no obstacle.

The Anglo-American telegraph company, sensing, as did other telegraph companies, that its time was passing, fought back by bringing a suit against Marconi for trespassing on a monopoly it had been granted in Newfoundland. If the business community was convinced of Marconi's success, the scientific community was not. Many scientists expressed the opinion that the transatlantic transmission had been a hoax, or perhaps a fluke. No one knew then, as scientists later determined, that the type of radio waves Marconi generated are reflected by certain layers in Earth's atmosphere and thus can travel great distances despite Earth's curvature. Skepticism faded when in February 1902 Marconi, on the boat *Philadelphia* in front of technically reliable witnesses, received clear signals from Poldhu at 1,550 mi (2,494.5 km) and intermittent signals at 2,100 mi (3,379.6 km). The Canadian government decided that Marconi could be licensed to operate in Newfoundland but that he could charge no more than 10 cents a word, a rate far lower than the cable companies charged.

In 1902, not yet 30 years old, Marconi had reached the pinnacle of his career. His company was operating on two continents. The transmission across the Atlantic had won him the awe and admiration of his fellow scientists. He had overcome the limitations of the coherer in his receiver by designing and patenting a superior radio wave detector that operated with magnets and wires. The helpers that he had assembled were of the highest caliber: George S. Kemp, his assistant; Andrew Gray, chief engineer; and R. N. Vyvyan, engineer and expert finagler of bureaucracies.

Marconi returned to Italy in 1902, and the Italian navy offered him a ship, the *Carlo Alberto,* as his first floating laboratory. Although he successfully transmitted messages between U.S. President Theodore Roosevelt and England's King Edward VII on January 19, 1903, Marconi was having a hard time convincing anyone that transatlantic communication could become more reliable and profitable. The Marconi Company began making money that year, but not much. Most of the income came from ship-to-shore communications, although negotiations were underway with the London *Times* and the British Post Office. Unhappy with the English domination of radiotechnology, Germany was looking for ways to get around the Marconi patents. The Marconi Company's monopoly on radio broadcasting was beginning to slip.

The next year, 1904, was a year of mixed blessings. Marconi's father died and he met his future first wife, Beatrice O'Brien, the daughter of the thirteenth Baron Inchiquin. On March 16, 1905, they were married. They had three daughters—Lucia, who died as an infant, Degna, who later wrote a biography of her father, and Gioia Jolanda—and one son, Giulio. Marconi proved a difficult husband. Beatrice tried to weather his long absences and frequent philandering but, after falling in love with another man, asked Marconi for a divorce in 1923. Their marriage was annulled in 1927.

Shares the 1909 Nobel Prize in Physics with Braun

In December 1909 Marconi went to Stockholm to accept his Nobel Prize in physics. He was puzzled, as were many others, as to why he was splitting his award for radio with Karl Braun. In fact, Braun's modifications on Marconi's designs had increased radio's range fivefold. Something very close to Braun's improvements had allowed Marconi to transmit across the Atlantic. In spite of the awkward circumstances, when the two men met they became friends. Marconi was the first Italian in his field to receive the Nobel Prize.

With the onset of World War I, Marconi became a technical consultant to the Italian military, which was badly in need of modernization. He was also attached to the Italian War Mission to the United States. During his service in Italy, a new benchmark for radio range was reached in 1918 when England made its first transmission to Australia. In 1919 the king of Italy sent Marconi to the Peace Conference in Paris as a delegate. He signed the treaties with Austria and Bulgaria for the Italian government. After the war Marconi bought a yacht he christened the *Elettra,* which became his laboratory and home.

Establishes Worldwide Short-Wave Radio Network

In 1923 Marconi's experiments with short-wave radio began to pay off. Marconi had been plagued by the greatly reduced performance of long-wave radio during daylight but did not understand the reasons for it—physics had not advanced to that stage. Yet in October of that year he found that waves 32.8 yd (30 m) long—a tiny fraction of his original waves' length—could be transmitted across vast distances without **interference** from the Sun's **radiation**. In July 1924 the Marconi Company contracted with the British government to supply short-wave radio relay stations throughout the empire. By 1927 the world was completely encircled with a short-wave radio network.

After the annulment of his first marriage in the spring, Marconi married Cristina Bezzi-Scali, a daughter of the papal nobility, on June 12, 1927. They had a daughter, Elettra. After complex negotiations beginning in 1925, the Marconi Company merged with the telegraph companies in 1928, making Marconi a wealthy man. By then, he had received the hereditary title of *Marchese* or Marquis from the Italian king. His last 10 years, however, were marred by a decline in his

health and estrangement from his ex-wife and the children of his first marriage. Marconi's involvement in politics also ended on a sour note. His public embrace of fascist dictator Benito Mussolini alienated much of the European and American public after Italy invaded Abyssinia in 1935. When Marconi sought to explain Italy's conduct to his adopted country, England, the British Broadcasting Corporation would not cooperate.

Yet Marconi persevered. In 1932 he proved with tests from aboard the *Elettra* that **microwaves**, waves less than 1 centimeter in length, could be received well beyond the horizon-line limit that the theory of the day mandated. It was not until after the technological advances spawned by World War II that this discovery received commercial application. He even, in 1934, demonstrated a radio beacon, a primitive form of **radar**, with which he piloted the *Elettra* blind into a difficult harbor. After 1936 Marconi's heart condition confined him to his native Italy, where he died on July 20, 1937.

Musically gifted, involved in Italian and international politics, always in the society pages, Marconi was anything but one-sided. A man of enormous energy, he managed not only to invent long-wave and short-wave radio transmission but to run a company besides. With his business skill, scientific creativity, technical know-how, and determination, Marconi resembled another famous scientist-engineer, America's **Thomas Alva Edison**. Edison, in fact, greatly admired Marconi for his successful transatlantic radio transmission. In 1903, the old inventor invited the young Marconi to his home in Orange, New Jersey, where they spent a pleasant afternoon.

MASS

Mass is a measure of the amount of **matter** present in an object. Matter is anything that has mass and volume. Therefore, by definition, all matter has mass. The mass of a particular object is constant. The only way the mass of an object can change is if matter is added to or taken away from it. The more matter an object contains, the more massive the object is. Mass and volume are directly proportional to each other. If an object's mass is doubled, so is its volume. The Système International d'Unités (SI) unit for mass is the kilogram (kg). The kilogram is a rather large quantity, however, so for smaller objects the gram (one-thousandth of a kilogram) or the milligram (one-thousandth of a gram) is used.

People often confuse an object's mass with its **weight**. The weight of an object is often expressed, incorrectly, in grams. The mass of an object is determined by comparing that object with a standard mass, or a balance. The weight of an object is a measurement of how much gravitational pull is exerted on the object. A weight measurement is taken using a spring scale and is expressed in newtons (N) or pounds (lb). Mass and weight are related, however. A more massive object will weigh more than a less massive object. The weight of an object is actually the product of its mass and the **acceleration** due to **gravity**.

Mass is a measurement of all matter. An elephant has mass, as does an **electron** spinning around the **nucleus** of an **atom**. As can be imagined, the masses of atoms are very small. If the mass of an atom were expressed in grams, the number would be extremely tiny. For example, an atom of oxygen-16 has a mass of 0.657×10^{-23} g. Chemical calculations and measurements of such small numbers are very cumbersome. It is much more convenient in these cases to use relative atomic masses. Scientists created a relative scale of atomic mass by arbitrarily choosing the carbon-12 atom as a standard and expressing the mass of all other atoms by comparing them to this standard. The mass of the carbon-12 nuclide was assigned a mass of 12 atomic mass units (amu). One amu is exactly one-twelfth the mass of a carbon-12 atom. One amu is equal to 1.660540×10^{-27} kg. Every other atom is expressed in terms of the carbon-12 atom. For example, a hydrogen-1 atom has the mass of 1/12 a carbon-12 atom, or 1 amu. An oxygen-16 atom has about 16/12 the mass of a carbon-12 atom, or 16 amu. The atomic masses on the **periodic table** are expressed in amu. Looking at the periodic table, one would notice that the atomic masses are not whole numbers, as would be suggested above. This is because there are different **isotopes** for each atom, i.e., atoms with different atomic masses, and the atomic mass expressed on the periodic table is a weighted average of all of these isotopes. The masses of **subatomic particles** (electrons, protons, and neutrons) are also expressed using amu. An electron has a mass of .00005486 amu, a **proton** has a mass of 1.007276 amu, and the mass of a **neutron** is 1.008665 amu. These would be very small numbers if expressed in kilograms.

Two terms related to mass that are frequently encountered in **chemistry** are the mass number of an atom and the molar mass of a substance. The mass number of an atom is simply the sum of the protons and neutrons in the atom. Since the protons and neutrons have a mass of approximately 1 amu each, and the mass of an electron is so small, the mass number is a good approximation of the atomic mass of an element. The molar mass of a substance describes the relationship between moles and mass. The molar mass is equal to the mass that one mole of a substance has. It is expressed in units of grams per mole.

Chemists often refer to a theory called the law of **conservation of mass**, especially when balancing equations for chemical reactions. The law of conservation of mass states that matter can neither be created nor destroyed. This means that when two or more compounds or elements combine in a **chemical reaction**, the total mass of the reactants has to equal the total mass of the products. The law of conservation of mass applies to the individual atoms involved in a reaction as well. Because each atom has a particular mass, and mass is conserved, this means that the number of atoms of each element involved in a reaction must be the same before the reaction occurs and after the reaction is completed. A chemical reaction is the rearrangement, not the creation or destruction, of atoms. The law of conservation of mass is symbolized by a balanced equation for a chemical reaction, i.e., the same number of atoms of each element are found on the reactant side and the product side.

Another definition of mass is a measure of the **inertia** of an object. An object's inertia is its resistance to changes in its **motion**. It is more difficult to change the motion of a more massive object. For example, if you need to push a couch across the floor, you need to exert a **force** on the couch in order to make it move. As you move the couch, it resists your pushing or pulling on it. This resistance is the inertia of the couch. Once an object is moving, it requires a force to stop it. For example, a speeding car needs the force of **friction** applied from the brakes to slow it down to a stop. As you press on the brake pedal of the car, you should notice that the car is resisting. This is due to the car's inertia. The more mass an object has, the more inertia it has. The car has a greater inertia than the couch. Therefore, a larger force must be used on the car to overcome its inertia.

Another property dealing with the mass of moving objects is **momentum**. All moving objects have momentum. An object with a greater momentum will be harder to stop. The momentum of an object depends on two things, the mass of the object and its **velocity**. Therefore, two identical cars traveling at different speeds will have different momentum. The car that is traveling at a greater velocity will have the greater momentum. Likewise, a two-ton truck traveling at the same velocity as a sports car will have more momentum because it has more mass.

The mass of an object is also related to force. The force of an object is the product of its mass and its acceleration. If either the mass or the acceleration of an object is large, the object will have a large force. The force of gravity between two objects depends on mass. The size of the force of gravity depends on two things, the masses of the objects and the distance separating them. As the masses of the objects increase, so does the force of gravity. This is why we notice the force of gravity between us and Earth much more than the force of gravity between us and the person standing next to us. Gravitational forces of attraction always exist between two objects of any mass, but it takes an object as large as a planet for this force to become noticeable.

Mass is also directly related to **energy**. The **kinetic energy** of an object depends on the mass of the object and the velocity. A large mass or a high velocity will increase the kinetic energy of an object. **Albert Einstein** developed the theory of relativity in 1905 that related mass and energy. According to this theory, even objects with very small masses can have a very large amount of energy. Einstein showed that mass and energy can be converted into each other. His famous equation, $E = mc^2$, further explained the law of **conservation of energy**. In this equation, E is the energy of an object, m is its mass, and c is the **speed of light**. Einstein showed that matter can, in fact, be destroyed. When matter is destroyed, it is converted into energy. The same holds true for energy. When energy is destroyed, it is converted into matter. Einstein's theory explained why **nuclear reactions** did not seem to abide by the law of conservation of mass or the law of conservation of energy. Nuclear fusion is the process that changes mass to

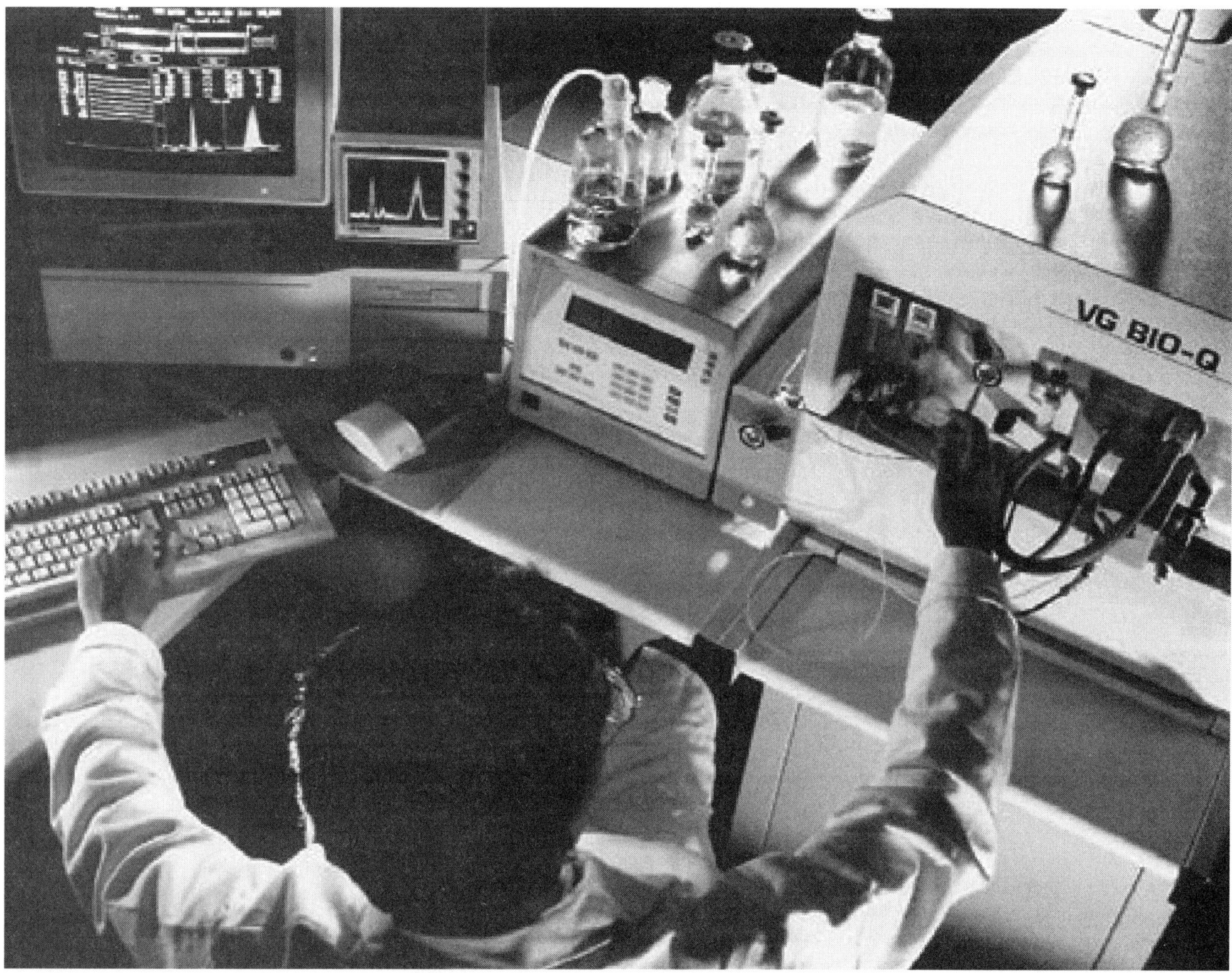

Mass spectrometer. (Photo by Geoff Tompkinson. Science Photo Library, National Audubon Society Collection/Photo Researchers, Inc. Reproduced by permission.)

energy. In fact, during this process, a very tiny mass produces a huge amount of energy.

The mass of an object is more than simply how much matter is present. It determines factors such as inertia, momentum, and force. The mass of an object is also directly related to energy. Mass is a property of an object that can account for many other behaviors of the object, whether the object is a jumbo jet or a proton.

MASS SPECTROMETRY

Mass spectrometry is an instrumental method of obtaining structure and mass information about either molecules or atoms by generating ionized particles and then accelerating them in a curved path through a **magnetic field**. Heavier particles are more difficult for the magnetic field to deflect around the curve, and thus travel in a straighter path than lighter par-

ticles. Consequently, by the time the particles reach the detector, a mixture of **ions** will have separated into groups by mass (or more specifically the mass-to-charge ratio of the individually weighted ions). The ions are produced from neutral molecules and atoms by stressing them with some form of **energy** to knock off electrons. In the case of molecules, fragmentation as well as ionization usually occurs. Each type of molecule breaks up in a characteristic manner, so a skilled observer can interpret a mass spectrum much like an archaeologist can reconstruct an entire skeleton from bone fragments. A mass spectrum can help establish values for ionization energy (the amount of energy it takes to remove an **electron** from a neutral **atom** or molecule) and molecular or atomic mass for unknown substances. The extremely high sensitivity of mass spectrometry makes it indispensable for analyzing trace quantities of substances, so it is widely used in environmental, pharmaceutical, forensic, flavor, and fragrance analysis. The petroleum industry has used mass spectrometry for decades to analyze hydrocarbons.

The basic principle underlying mass spectrometry was formulated by **Joseph John Thomson** (the discoverer of the electron) early in the century. Working with cathode-ray tubes, he was able to separate two types of particles, each with a slightly different mass, from a beam of neon ions, thereby proving the existence of **isotopes**. (Isotopes are atoms of the same element that have slightly different atomic masses due to the presence of differing numbers of neutrons in the nucleus.) The first mass spectrometers were built in 1919 by F. W. Aston and A. J. Dempster.

There are five major parts to a mass spectrometer: the inlet, the ionization chamber, the mass analyzer, the detector, and the electronic readout device. The sample to be analyzed enters the instrument through the inlet, usually as a gas, although a solid can be analyzed if it is sufficiently volatile to give off at least some gaseous molecules. In the ionization chamber, the sample is ionized and fragmented. This can be accomplished in many ways—electron bombardment, chemical ionization, laser ionization, electric field ionization—and the choice is usually based on how much the analyst wants the molecule to fragment. A milder ionization (lower electric field strength, less vigorous **chemical reaction**) will leave many more molecules intact, whereas a stronger ionization will produce more fragments. In the mass analyzer, the particles are separated into groups by mass, and then the detector measures the mass-to-charge ratio for each group of fragments. Finally, a readout device—usually a computer—records the data.

Mass spectrometers are often used in combination with other instruments. Since a mass spectrometer is an identification instrument, it is often paired with a separation instrument like a chromatograph. Sometimes two mass spectrometers are paired, so that a mild ionization method can be followed by a more vigorous ionization of the individual fragments.

MATHEMATICAL PHYSICS

Before English physicist and mathematician Sir **Isaac Newton**'s time, physicists experimented using simple mathematical calculations. Physics and related sciences actually began to develop after Newton and **Gottfried Wilhelm Leibniz** devised calculus.

Mathematical physics uses the laws of physics and the differential method of calculus to set up equations for the system being studied. The resulting differential equations can then be solved or analyzed. The tools of mathematics are calculus and other advanced mathematics, such as complex analysis. Many scientists have contributed to mathematical physics in the past several hundred years. Equations in mathematical physics were often named after those scientists who discovered them. The central topics of mathematical physics are the differential equations, or the partial differential equations (PDE). There are three classes of partial differential equations.

The first class is called the wave equation,

$$\frac{d^2 u}{dx^2} = \frac{1}{c^2}\frac{d^2 u}{dt^2} - f$$

where c is the speed of the wave and f is the source of wave. If a stone is thrown into a lake, the perturbation caused by the stone is the source of wave. The solutions for this wave equation are **waves** traveling in two opposite directions. There is also a wave equation whose solution has the wave traveling in one direction only. This unidirectional wave equation has only first-order differentiation. The quantum wave in **quantum physics**, however, obeys another equation called the Schrödinger equation, which is a complex partial differential equation. The wave equation is also called a hyperbolic equation.

The second class of partial differential equations used in mathematical physics are the transport equations, or diffusion equations,

$$\frac{d^2 u}{dx^2} = k\frac{du}{dt} + f$$

where f is the source. If a bomb explodes, the explosion provides the source. In thermal physics, the flow of **heat** between objects of different temperatures obeys the transport equation. The transport equation is also called a parabolic equation.

The last class of equations is the Poisson equation,

$$\frac{d^2 u}{dx^2} = f$$

f being the source. Electrical potential can be described by this equation. In this case, the charge is the source. If there is no source ($f = 0$), the resulting equation is called the Laplace equation. There is no time derivative in the Poisson equation, so it is for **equilibrium** cases only. In the case of an electrical field, the time to reach equilibrium is the **speed of light**. If we want to consider time, we should use the wave equation for electromagnetic waves. The Poisson equation is an elliptical equation. Sometimes f is proportional to the unknown. The resulting equation,

$$\frac{d^2 u}{dx^2} = -\lambda u$$

is called the Helmholtz equation.

The actual problem at first glance may not resemble any of these equations, but they can always be transformed into one of these. A differential equation usually has an infinite number of solutions called a solution family. In real problems there are boundary conditions related to each partial differential equation to guarantee a unique solution. There are three possible types of boundary conditions. The unknown function's value is given at the boundary (Dirichlet boundary condition). The derivative of the unknown function in normal direction is given at the boundary (Neumann boundary condition). A linear relation between the unknown function value and its derivative in the normal direction is given at the boundary (mixed boundary condition). As a rule of thumb, adding up the highest order of derivative of each independent unknown variable yields a result of the number of boundary conditions needed to obtain an unique solution. Any type of partial differential equation may be combined with any type of boundary condition, depending on the actual problem.

Advanced mathematical tools like complex analysis are often used in solving equations. One frequently used tool is series expansion. By expanding the possible solutions and source term into a series of simpler basis functions, the partial differential equation can be transformed into a series of algebraic equations. The solution to this equation is often an infinite series. The series solutions to some simple equations are already known. They are called special functions. Some of the well-known special functions are the Legendre polynomial family and the Bessel function family. They can be used as the basis functions in series expansion for more complicated equations. The source term is often complicated. When the source is a point source, the solution is called Green's function. Green's functions can be used to obtain solutions for an arbitrarily complicated source. In real problems, the source can be very complicated. For instance, an explosion is a very complicated source. **Computers** are used to determine solutions numerically. The most difficult equation to solve numerically is the wave equation. If a small error exists at the beginning, the error can travel with the wave and possibly become larger. The transport equation poses less challenge since a small error at the beginning will diffuse with time.

See also Schrödinger's equation

MATTER

In ancient Greece, some philosophers, most notably Heraclitus, believed that everything in the world was in a state of fluctuation. Others argued that there must be some permanence, otherwise it would not be possible to see anything as being real.

The fifth-century Greeks were apparently the first to attribute structure to matter. They postulated that matter consisted of very small particles that were firmly bound together in the solid state, but that could change position to accommodate compression and deformation. At higher temperatures, these particles were thought to slide past each other and to eventually separate, as the material object underwent melting and evaporation. The Greeks called these particles atoms, Greek for indivisible. This philosophical position offered a reconciliation between the views of a world in fluctuation and one that is permanent.

In 1687, **Isaac Newton** published his *Principia*, in which he described his laws of **motion**. As had his Greek predecessors, Newton viewed matter as passive and inert, and as consisting of "solid, massy, impenetrable, movable particles." Thus, for Newton and his contemporaries, there was little dis-tinction between the properties of matter in the material world and the building blocks of which it was composed.

In 1785, French chemist Antoine Lavoisier proposed his law of conservation of matter, which states that matter can neither be created nor destroyed, only changed into different forms. This law has since been superseded by the law of **conservation of mass** and **energy**, which takes into account **Albert Einstein**'s observation that **mass** and energy are interchangeable under certain conditions.

Today, scientists believe that matter is made up of molecules; that molecules consist of atoms bound together; that each **atom** contains a **nucleus** surrounded by electrons; that the nucleus contains protons and neutrons, bound together; and that protons and neutrons consist of **quarks** bound together by gluons.

The modern viewpoint is consistent with the definition of matter as anything that occupies space and has mass. Mass is the material property that shows up as **weight** when the object is acted upon by **gravity**. Mass can be measured from an object's tendency to resist moving, i.e., its **inertia**. The ratio of an object's mass to the volume it takes up is known as its **density**. The basic unit of mass in **chemistry** is the gram (1 kilogram = 1,000 grams).

Matter may exist in the solid, liquid, or gaseous state. Solids have definite sizes and shapes; liquids have definite volumes, but no fixed shapes; and gases possess neither shapes nor fixed volumes.

Matter that is made up of only one type of atom is called an element. Loosely, if a substance contains two or more kinds of atoms joined together and grouped in a definite way, that substance is called a compound; mixtures have a composition that is not uniform, but heterogeneous.

The physical properties of matter are those that can be observed when no change is taking place in the **atomic structure**. Examples include **color**, density, melting point, and hardness. Chemical properties have to do with a substance's tendency to react with other substances. An example is iron's tendency to rust in moist air.

In the case of ordinary chemical transformations, the mass of the products always equals the mass of the reactants.

MATTER WAVES

Classical physics predicts that **light** behaves like a wave; the **diffraction** and **interference** effects of **optics** come from the wave nature of light. However, by 1905, **Max Planck** and **Albert Einstein** had established that light also acts like a particle; by specifying that light **energy** comes in packets called **photons**, the classical inability to explain effects such as **blackbody radiation** spectra and the **photoelectric effect** had been resolved. Therefore, the nature of photons is somewhat ambiguous; depending on how you observe them, they sometimes behave as particles, and sometimes as **waves**.

In his 1924 doctoral dissertation, **Louis de Broglie** speculated that **matter** might also have this dual particle-wave nature. For the **frequency** and **wavelength** of the matter waves, he chose the same relations that hold for light: f = E/h and (lambda) = h/p, where h is **Planck's constant**. This idea was pure speculation at first, but in 1927 **Clinton Davisson** and **Lester Germer** verified the wave nature of electrons by **scattering** them from a nickel crystal. The electrons were only reflected at a few specific angles, which can be explained by assuming that the electrons behave like waves and are affected by interference. (The same effect is observed in x-ray scattering, where it is known as Bragg scattering.)

As a result of their wave nature, particles are actually .not located at points in **space-time**, but are "smeared out." According to **Werner Karl Heisenberg**'s famous **uncertainty principle**, the amount of smearing depends on how well we know the **momentum** of a particle. If we know the particle's momentum exactly, then the particle is smeared throughout all of **space**; if we don't know the particle's momentum at all, then the particle is not smeared out at all.

In 1925, **Erwin Schrödinger** introduced his famous wave equation, which allowed physicists to find the **wave functions** for electrons in the same way they can find the wave functions (electric and magnetic fields) for **photons**. The wave functions are the mathematical representation of the matter waves, allowing one to make rigorous predictions. In the same paper, Schrödinger solved the problem of the hydrogen **atom** using his equation and showed that it correctly predicted the spectrum of the **hydrogen atom**, which was the first verification of his new theory.

However, the interpretation of the wave functions was not immediately obvious. It took the insight of **Max Born** to realize that the wave functions make up a spatial probability **density** for a particle. The probability density function has maximum values where the particle is most likely to be found, and it may have zero points where the **electron** may never be found! This is just one of the bizarre predictions made by **quantum mechanics**, all based on the once-speculative hypothesis that matter can behave as both particles and waves.

See also de Broglie wavelength; Probability wave; Schrödinger's equation

MAUPERTUIS, PIERRE-LOUIS MOREAU DE (1698-1759)

French mathematician, biologist, and astronomer

A mathematician, biologist, and astronomer, Pierre-Louis Moreau de Maupertuis was a strong proponent of Sir **Isaac Newton**'s theory of gravitation, helped confirm Newton's theory on the exact shape of Earth, and formulated the principle of least action in physics. Born in Saint Malo, France, Maupertuis had a wide range of scientific interests. As a biologist, he wrote *Système de la Nature*, in which he provided the first accurate scientific record of a dominant hereditary trait transmitted among humans. He also introduced the theory of the survival of the fittest in his *Essai de Cosmologie*, a theory that Charles Darwin later expounded to wide acceptance.

Maupertuis may be best known for his formulation in 1744 of the principle of least action, also known as the minimum principle or Maupertuis's principle. Essentially, the principle states that any change that occurs in the **universe** and nature, such as a moving body or **light** rays, changes in the most economical path possible. For example, bubbles form in a shape that presents the smallest surface for a given volume of air. In *Essai de Cosmologie*, Maupertuis presented his theory as something that might help prove the existence of God by unifying the laws of the universe. In 1736, Maupertuis led a famous expedition to Lapland near the North Pole that proved Newton's theory that Earth is an oblate sphere (flattened at the poles). The proof was accomplished by measuring the length of degree along a meridian and comparing the findings with the findings of another expedition near the equator in Peru performing similar measurements.

Despite his many accomplishments, Maupertuis was considered arrogant by many of his fellow countrymen. Eventually, Maupertuis became a target of German mathematician Samuel Koenig, who accused him of plagiarism, and of French author Voltaire (1694-1778), whose satirical writings about Maupertuis were so savage that Maupertuis eventually left France. Maupertuis died in virtual exile in Basel, Switzerland, in the home of Swiss mathematician Johann Bernoulli (1667-1748).

See also Newton's law of universal gravitation; Newtonian physics

MAXWELL, JAMES CLERK (1831-1879)

Scottish mathematician and physicist

James Clerk Maxwell was a researcher who opened a new paradigm with his electromagnetic theory, influencing generations of researchers. He was without a doubt a child prodigy. At an early age, he was solving geometric problems and writing explanations that intrigued academics. Just as he considered how charged particles interact with their surrounding area, one might consider the interaction of the conditions of his inherent nature and the environment of his early childhood. Maxwell's life could make a good case study for the strength of the influences of heredity vis-à-vis environment as he had strong influences from both sources.

Maxwell was born into a distinguished family with notable accomplishments among his ancestors. He had an almost ideal childhood, living in the country in close warm association with his parents and relatives. Like **William Thomson Kelvin**, he experienced the death of his mother during his childhood, an experience that drove him closer to his father. Nevertheless, Maxwell did not have the personality of a high achiever. In fact, many reported him as being "eccentric," mostly due to his shyness and his rustic ways. Fortunately, with his eccentric personality came many admirable personality traits. He constantly demonstrated a great imagination, planning experiments to study what others considered impossible. He was a clear communicator, explaining complex topics to learned audiences.

James Clerk Maxwell was a descendent of the Clerk and the Maxwell families, both with a distinguished heritage. His father inherited a house in Edinburgh and land in the countryside. Maxwell was born in 1831 in Edinburgh while his parents were waiting for their country house to be built. They moved shortly after he was born. His father was a lawyer but was not very aggressive in pursuing new business. John Clerk Maxwell enjoyed studying science and building mechanical devices. As young as three years old, Jamesy was following his father insisting to know how everything worked. He was very close to his father all of his life. Maxwell's mother died suddenly when he was eight years old. For two years after his mother's death, he was educated by a series of tutors, but none were found suitable for Maxwell and his unique way of learning. His father and his aunt arranged for him to begin studies at the Edinburgh Academy. His first year at the academy was difficult as the children teased him for the way he talked and dressed. They considered Maxwell stupid and nicknamed him "Dafty." From the second year at the academy, he started to show his true capabilities and his classmates were less cruel.

About this time, Maxwell became interested in geometry. He did a study of ovals and double-foci ellipses. Just after he turned 15, he wrote a paper on what he had learned from his drawing and measuring of ovals. His father showed the paper to Professor Forbes who read it to the Edinburgh Royal Society. Those who were impressed with this work recognized that most of the conclusions were already known. However, they were amazed that the eloquence and intricacy in the description of the issues came from so young a person with so limited formal education. In 1847, at age 16, Maxwell began his college studies at the University of Edinburgh. He spent three years there and during this time, he contributed two more papers to the Edinburgh Royal Society. When he finished his studies at Edinburgh, his father sent him to Peterhouse, but shortly after beginning there, he transferred to Trinity where he believed he had a better chance for a fellowship. Maxwell studied at Trinity from early 1851 until he graduated in 1854. After graduation, he was awarded the fellowship. Maxwell then applied for a position at Marischal College to be close to his ailing father. However, his father did not live much longer. After his father's death in April 1855, he accepted the position at Marischal.

In 1858, he married the well-educated Katherine Dewar. Two years later, he had to leave Marischal, the victim of an institutional merger. He was immediately invited to teach at King's College, London. It was in London that he did his most prominent work. He remained there until he resigned his post (probably due to exhaustion) in the spring of 1865. He spent most of the next five years at his country home writing a book on his theory. He considered himself retired.

To stay involved in academia, he did consulting work for Cambridge. His encouraging Cambridge to offer courses on **heat** and **electromagnetism** directly influenced the foundation of the Cavendish Laboratory. It was only natural that the first Cavendish professorship should be offered to him and he

James Clerk Maxwell.

accepted. During his eight years as Cavendish professor, he worked to prepare for publication the experiment papers **Henry Cavendish** had written. It is well accepted that this self-imposed responsibility was influential in bringing due respect to Cavendish's work. In May 1879, as the school year wound down, it was obvious to many that Maxwell's health was beginning to fail. He tried to return to Cambridge in the autumn, but he could scarcely walk. On November 5, 1879, Maxwell died of abdominal cancer at the age of 48.

Maxwell's work leading to his kinetic theory of gases and his theory of electromagnetic fields was a logical advance from **James Prescott Joule**'s work. Both researchers measured the **velocity** of gas molecules and both recognized that heat was not the fluid that it once was thought to be. The importance of Maxwell's work was the direction that it gave to new understanding. Joule showed only the scientific community what was possible to measure and what might be proven. Maxwell went forward with detailed mathematical models that left no holes unfilled, with one important exception. Maxwell used statistics to show the high probability that proposed laws would direct the behavior of **matter**. Discussing the probability of natural law took science away from **determinism**. This opened the door for the modern study of physics. **Albert Einstein**'s theory of relativity and the recently nurtured chaos theory could not have been developed except for this new philosophical direction.

Maxwell began measuring the average velocity of a gas molecule with the objective to investigate whether the perceived random order of its movement could be predicted with some degree of accuracy. What he found was that the greater the velocity of the molecules, the greater the heat generated. There was a direct relationship between the amount of movement among the molecules and the amount of heat in a gas. In this experimental demonstration, heat was shown undeniably to be a property of particle movement and not a fluid flowing from one object to another. Furthermore, Maxwell's findings showed that the movement of particles could be controlled through increasing or reducing heat.

Maxwell understood **Michael Faraday**'s theory of electric and **magnetic fields**. He worked to demonstrate what Faraday could not explain himself through complex calculations. Assuming that the space surrounding a charged particle contained a field of **force**, Maxwell created a mathematical model demonstrating all the possible phenomena of electric and magnetic fields. Through this model, Maxwell demonstrated that the electric and magnetic fields worked together. He coined the term "electromagnetic" to name this new breakthrough.

This discovery is important for chemistry because it ultimately led to the discovery of the **electron**. **Joseph John Thomson** discovered the electron when he was investigating the effects of the **electromagnetic field** on gases, applying the principles that Maxwell had established. Research on the effects of **light** on elements was furthered by Maxwell's work. His subsequent work on the velocity of the **oscillation** of electromagnetic fields demonstrated that light should be considered a form of electromagnetic **radiation**. This consideration affected light effect theories.

Maxwell was a brilliant man who appeared at times to live in another world. Considering that he recognized deeply the order and potential in some things that others take for granted (such as light and heat), perhaps he did live in another world. However, it was evident by his appreciation of a good laugh with friends that he was not so distant. Furthermore, he could not ignore another brilliant mind that was not yet recognized. When he felt that **Josiah Willard Gibbs** did not receive the respect that he deserved, he built a three-dimensional model of the phenomenon Gibbs was describing. Maxwell named the model after Gibbs. He did this in his dying days; two weeks after he finished it, James Clerk Maxwell died.

Maxwell's equations

Advances in experimental techniques allowed nineteenth-century physicists to develop increasingly refined concepts of **electromagnetism**. The mathematical formulations of electromagnetism culminated in Scottish physicist **James Clerk Maxwell**'s development of a set of equations that accurately described electromagnetic phenomena. By the dawn of the twentieth century, Maxwell's equations allowed an increasing exploitation of the **electromagnetic spectrum**. **Radio waves**

were to spark a twentieth-century communications revolution and **x rays** quickly became the forerunner of many important medical diagnostic tools based on electromagnetic properties. Moreover, Maxwell's formulations profoundly shaped the development of modern relativity and quantum theories by providing a precise set of formulas with enormous predictive **power**. Twentieth-century physicists such as German physicist **Max Planck**, German American physicist **Albert Einstein**, and Danish physicist **Niels Bohr** all credited Maxwell with laying the essential foundations for modern physics.

Maxwell collected and first published **electromagnetic field** equations in 1864. In particular, Maxwell built his mathematical formulations upon the theoretical developments advanced by French scientist **André-Marie Ampère**. Maxwell's equations were, however, more than refined mathematical interpretations, they also contained an insightful fusion of experimentation and abstraction. The equations mathematically expressed the work and observations of many other important scientists and mathematicians, especially German mathematician **Carl Friedrich Gauss**, Danish scientist **Hans Christian Ørsted**, and English physicist and chemist **Michael Faraday**.

In four famous equations contained in his 1873 publication, *Electricity and Magnetism*, Maxwell formally united concepts regarding **electricity** and **magnetism** into elegant mathematical statements of the properties of electromagnetic waves (including visible **light**).

Because they are mathematically elegant (concise), a full understanding of Maxwell's equations requires an advanced level of mathematical sophistication (above the level of beginning calculus). In summary, Maxwell's first equation, also known as Gauss's law, relates the electric field (E) that passes through a particular surface area (A) (e.g., a sphere) to a charge (Q) within that surface. Maxwell's second equation asserts that there are no "magnetic charges" and that, therefore, magnetic lines of **force** must always form closed loops. Maxwell's third equation, also known as Ampère's law, states that a **magnetic field** (B) can be induced by a circular loop by charges moving through a wire (i.e., an **electric current**) (I) or by a changing electric flux. Maxwell's fourth equation, also known as Faraday's law, states that **voltage** is generated in a conductor as it passes through a magnetic field or as it cuts through magnetic lines of force.

Maxwell's equations established that the **electric charge** is a source of an electric field and that electric lines of force begin and end on electric charges (though this is not necessarily true in a changing magnetic field). Essentially, decades before the discovery of the **electron**, Maxwell defined the charge properties eventually associated with electrons.

Maxwell propositions allowed the unification of known electrical and magnetic phenomena into the electromagnetic spectrum. When Maxwell's calculated speed of electromagnetic propagation fit well with experimental determinations of the **speed of light**, Maxwell and other scientists realized that visible light was simply a part of a broader electromagnetic spectrum. Based on Maxwell's equations, all electromagnetic **radiation** could be understood to be manifestations of a common electromagnetic **wave phenomena** that varied only in terms of **wavelength** and **frequency**. Consequently, an electromagnetic spectrum implied that visible light was but one portion of the spectrum and that other portions of the spectrum remained to be explored. Based on Maxwell's equations, in 1888 German physicist **Heinrich Rudolf Hertz**, demonstrated the existence of radio waves at frequencies and energies lower than visible light. Subsequently, German physicist **Wilhelm Conrad Röntgen** (also spelled Roentgen) discovered a higher frequency and higher **energy** electromagnetic radiation in the form of x rays.

The demonstration of the existence of a transmission medium was of great interest to scientists. Prior to the publication of Maxwell's equations it was thought that all waves required a medium of propagation. An elusive **ether** through which light could pass was thought necessary to explain the wavelike behavior of light. The need for an ether seemed obvious when one considered propagation of **pressure** waves through water or **sound** through air. Maxwell's equations, however, established for scientists that, no matter how counterintuitive (outside of common experience), electromagnetic waves do not require such a medium. Further confirmation of Maxwell's equations subsequently came from the ingenious experiments of American scientists **Albert Michelson** and **Edward Morley** who demonstrated the constancy of the speed of the propagation of light regardless of the direction of travel (i.e., that there was no discernible ether or propagation medium that should have altered the measured **velocity** of light).

Maxwell's mathematical formulations, especially his predictions of the nature of propagation of electromagnetic waves, laid the foundation for the development of relativity and **quantum theory**. In the twentieth century, Maxwell's equations were used to describe the wavelike properties of the **photon** articulated by Planck and Einstein who separately proposed that atoms absorb or emit electromagnetic radiation in bundles of energy termed **quanta**. Electromagnetic radiation is now understood as having both particle and wavelike properties.

Maxwell's equation's were also used by twentieth-century scientists including French physicist **Louis-Victor de Broglie** and Austrian physicist **Erwin Schrödinger** to articulate the wavelike nature of the electron. Maxwell's equation's predicted that a particle-like electron, essentially in orbit around a **nucleus** according to Newton's laws, would continuously lose energy and eventually fall into the nucleus. Maxwell's equation's provided a foundation for the standing wave model of the electron advanced by Schrödinger's wave equations.

Maxwell's equations remain a powerful tool used by scientists to understand and predict the behavior of electromagnetic fields and waves in many engineering applications, including the design of electrical transmission lines, electromagnetic antenna (e.g., radio, television, microwave), radio telescopes, and other instruments used to measure portions of the electromagnetic spectrum. Faraday's **electromagnetic induction** (e.g., the production of electric current with magnets) embodied in Maxwell's fourth equation is a method still used by modern power **generators**.

MAYER, JULIUS ROBERT VON (1814-1878)

German physicist and physiologist

Julius Robert von Mayer was a brilliant but unlucky man. He was the originator of the concept of **conservation of energy**, now considered the first law of **thermodynamics**, as well as the first person to determine the mechanical equivalent of **heat**. He was a physician by training and did not have a reputation as a scholar in the scientific community. Thus, it was up to later scientists to "rediscover" the landmark research Mayer had performed years earlier.

Mayer was born in Heilbronn, Germany, in 1814. He was the youngest son of an apothecary, and the only son who chose not to enter his father's profession. He enrolled at the University of Tübingen and, despite being expelled in 1837 for his participation in a secret society, received his medical degree in 1838. In 1840 he signed on as a ship's doctor for a vessel sailing to the East Indies.

It was on this ship, near Java, that Mayer began to notice the remarkable redness of the blood drawn from the men aboard. Mayer supposed that, because of the intense tropical heat, the men's bodies required less **oxygen** to maintain their **temperature**; thus, more oxygen remained in the bloodstream, unused. This fascinated Mayer, for it seemed to indicate that oxygen was the true and single source of animal heat. He further proposed that food **energy** could be the only source of oxidation within a living organism. Thus, energy within the body could not be created from nothing, nor could it be reduced to nothing—it could only be used or stored.

What Mayer had conceived was the concept of conservation of energy. He planned to publish his ideas immediately but found he lacked the schooling in physics necessary to explain them. He began to study the nature of **work** and energy, conducting a number of experiments on the subject. One such experiment used a horse-powered device to stir a container full of paper pulp. By measuring the increase in temperature of the pulp in relation to the amount of work performed by the horse, Mayer came up with a figure for the mechanical value of heat (the amount of heat required to raise the temperature of a certain amount of water by 1°C). This figure was remarkably close to that of the present day, skewed only by the inaccurate values for specific heat used at that time.

Mayer announced both his theory of conservation of energy and his figure for the mechanical equivalent of heat in a paper published in 1842. It received very little attention, but Mayer became determined to explore the subjects in even greater detail. In 1845 he published another paper, this one applying energy conservation to the **motion** of the **planets**, the tides, the **Sun**, **magnetism**, **electricity**, and the living world. In many of these cases he identified solar energy as the source of most of the world's energy. In this way, Mayer explained the greater ramifications of the **first law of thermodynamics** to an extent no other scientist could match; still, he would never be given the credit he deserved.

In 1847 **James Joule** announced his own independent research into the mechanical equivalent of heat, and because of his international reputation, his work was given priority. That same year **Hermann von Helmholtz** published his work on the conservation of energy; he was also given credit over Mayer, whose research was still relatively unknown.

Two of Mayer's children died in 1848. This, coupled with the lack of respect paid him by the scientific community, drove him to despair. He attempted suicide in 1849, jumping from a third-story window. Mayer succeeded only in breaking his legs, laming himself. He was committed to a mental hospital in 1851 and remained there for nearly a decade.

While Mayer languished in the asylum, his work was finally beginning to gain status. Justus Liebig, who had originally published Mayer's papers, spread his theories during lectures. He also mistakenly announced, in 1858, that Mayer was dead. By the time he was released from the hospital's care, he had become a well-respected theoretical physicist. He was awarded the Copley Medal in 1871 from the Royal Society and was granted the privilege of adding "von" to his name—a very high German honor. He died of tuberculosis in 1878.

McMillan, Edwin M. (1907-1991)
American physicist

Edwin M. McMillan's first important discovery was made in 1940 when he, Philip Abelson, and Glenn T. Seaborg (1912-1999) produced and identified samples of transuranium elements, later named neptunium and plutonium. After World War II, McMillan became involved in the development of **particle accelerators**. Simultaneously with but independent of the Russian physicist V. I. Veksler, McMillan found a way to compensate for the relativistic **mass** increase that occurs in high **energy** accelerators. He won a share of the 1951 Nobel Prize for chemistry (with Seaborg) for his discovery of neptunium and a share of the 1963 Atoms for Peace Award (with Veksler) for his work on **accelerators**.

McMillan was born in Redondo Beach, California, on September 18, 1907. His father was Edwin Harbaugh McMillan, a physician, and his mother was Anna Marie Mattison. The McMillan family moved to Pasadena when Edwin was a year old. He attended local primary and secondary schools, graduating from Pasadena High School in 1924.

After completing his B.S. and M.S. degrees at the California Institute of Technology in 1928 and 1929, McMillan enrolled at Princeton University for his graduate study. In 1932, he was awarded a Ph.D. degree for his thesis, entitled "Electric Field Giving Uniform Deflecting Force on a Molecular Beam." McMillan then received a National Research fellowship that allowed him to begin his postdoctoral studies at the University of California at Berkeley.

In 1934, **Ernest Orlando Lawrence**, inventor of the **cyclotron**, established the Berkeley Radiation Laboratory on the University of California campus and invited McMillan to join its staff. McMillan maintained his relationship with the laboratory (later renamed the Lawrence Radiation Laboratory) for the next 40 years. He was made associate director of the

Edwin M. McMillan.

laboratory in 1954 and director in 1958, a post he held until his retirement in 1973.

In the late 1930s, McMillan turned his attention to one of the most exciting topics in scientific research at the time: the bombardment of atomic nuclei by neutrons. This type of research had been inspired by a series of experiments conducted by **Enrico Fermi** in the mid–1930s. Fermi had found that nuclei will often capture a **neutron** and undergo a nuclear transformation in which they are converted to a new element one place higher in the atomic table than the original element. Over a period of time, Fermi used this technique to transform more than 60 different elements.

The one element in which Fermi was most interested, however, was uranium. It was obvious that neutron capture by a uranium **nucleus** would result in the formation of the next heavier element, an element that does not exist naturally on Earth. When Fermi actually conducted this experiment, however, he obtained confusing results that could not be interpreted as the formation of a new element. A few years later, **Otto Hahn**, **Fritz Strassmann**, and **Lise Meitner** correctly interpreted Fermi's experiments, showing that the bombardment of uranium with neutrons had resulted in **nuclear fission**.

Working first with Abelson and later with Seaborg, McMillan repeated Fermi's original experiments. They bombarded uranium with neutrons and found that, while fission did occur, a small fraction of the uranium nuclei did undergo the kind of transformation to a heavier element that Fermi had anticipated. Later studies found that the uranium-235 **isotope** undergoes fission, while the uranium-238 isotope, under the proper conditions, is transformed to a heavier element. McMillan and Abelson suggested the name neptunium for the element after the planet **Neptune**, located one planet beyond Uranus, the namesake of uranium. A year later, McMillan and Seaborg found that the **radioactive decay** of neptunium produces yet another element, heavier than itself, an element they called plutonium. Again, the element's name was taken from that of a planet, Pluto, the farthest from the **Sun**. The two researchers were later to share the 1951 Nobel Prize in chemistry for their research on the transuranium elements.

With the onset of World War II, McMillan left Berkeley to conduct military research at the Massachusetts Institute of Technology, at the U.S. Navy Radio and Sound Laboratory in San Diego, and finally at the **Manhattan Project** laboratories at Los Alamos. At the war's conclusion, McMillan returned to Berkeley and the radiation laboratory, and immediately became immersed in problems of accelerator design. The cyclotron, invented by Lawrence in the early 1930s, had served the research community well for more than a decade, but the limits of the machine's usefulness were now becoming apparent. The most serious problem with traditional cyclotrons was relativistic mass increase.

Relativistic mass increase refers to the fact that as particles gain **velocity** in an accelerator, they also gain mass; the mass gain subsequently causes them to lose velocity. Before long, high-energy particles begin to fall out of phase with the electrical fields that are used to accelerate them, and they become lost inside the machine. McMillan's solution for this problem was simple and elegant. He modified the electrical field in the accelerator so that, as particles speed up and gain mass, the rate at which the electrical field changes direction slows down. In that way, the electrical field can be kept in phase with the particles even while they gain **energy and mass**.

For many years thereafter, this concept was employed in the development of more advanced circular accelerators, such as the **synchrotron** and synchrocyclotron. In 1963, McMillan was awarded a share of the Atoms for Peace Award with Veksler, who had developed the same concept at about the same time.

McMillan married Elsie Walford Blumer on June 7, 1941. She was the daughter of the former dean of the Yale University school of medicine and the sister of Mrs. Ernest Lawrence. The McMillans had three children, Ann, David, and Stephen. McMillan died of complications from diabetes on September 7, 1991, in El Cerrito, California.

MEASUREMENT OF THERMAL ENERGY

The measurement of **thermal energy** involves indirect measurement of the molecular kinetic energies of a substance. Rather than providing an absolute measure of molecular kinet-

ic **energy**, thermal measurements are designed to determine differences that result from **work** done on, or by, a substance (e.g., **heat** added to, or removed from, a substance). **Temperature** differences correspond to changes in thermal energy states and there are several analytical methods used to measure differences in thermal energy via measurement of temperature.

When dealing with the terminology associated with the measurement of thermal energy, one must be mindful that there is no actual substance termed "energy" and no actual substance termed "heat." Accordingly, when speaking of energy "transfer" or heat "flow," one is actually referring to changes in functions of state that can be raised or lowered only within a body or system. Neither energy or heat can really be "transferred" or "flow."

In **thermodynamics**, temperature is directly related to the average **kinetic energy** of a system due to the thermal agitation of its constituent particles. In accord with entropic law, temperature—as designated by a value on a scale (e.g., Celsius)—is commonly said to determine the direction of heat "flow" between systems (e.g., heat will always "flow" from a body of higher temperature to a body of lower temperature). More properly, thermal energy will decrease in the initially warmer body and increase in the initially cooler body.

In practical terms, temperature measures heat and heat measures the thermal energy of a system. In meteorological systems, for example, temperature, as an indirect measure of heat energy, reflects the level of sensible thermal energy of the atmosphere. Such measurements utilize **thermometers** and are expressed on a given temperature scale, usually Fahrenheit or Celsius. The common glass thermometer containing either mercury or alcohol uses the property of **thermal expansion** of the respective fluid as an indirect measure of the increase or decrease in the thermal energy of a body or system.

Other types of thermometers utilize properties such as resistivity, magnetic susceptibility, or **light** emission to measure temperature.

Electrical thermometers (e.g., thermoprobes, thermistor, thermocouples) that relate changes in electrical properties (e.g., resistivity) to changes in temperature are extensively used in scientific research and industrial engineering.

A technique known as differential thermal analysis measures the difference in temperature between a sample and a known reference as heat is applied to or lost from a system. Differential thermal analysis is sensitive to endothermic and exothermic processes and is a useful method to characterize various process (e.g., phase transitions).

Thermogravimetry measurements relate changes in the **mass** of a sample to temperature differences. These types of thermal measurements are often useful in determining the purity of a substance. Thermogravimetry measurements are sensitive indicators of water, carbonate, and organic content in a sample. Thermogravimetry studies are also useful in the study and characterization of various decomposition reactions.

A technique known as differential scanning **calorimetry** measures the rate of thermal energy change (i.e., heat flow) in a sample and relates that change to a standard at the same temperature (i.e., in thermal **equilibrium**). Data derived from differential scanning calorimetry are useful in measuring the heat capacities of various substances.

Because energy is commonly defined as the ability to do work, the thermal energy of a system is directly related to a system's ability to translate heat energy into work. Correspondingly, the measurement of the thermal energy of a system must be interpreted as the measurement of the changes in the ability of a system or body to do work. **Absolute zero** Kelvin (-459.69°F, -273.16°C, 0°R on the Rakine scale) is the lowest temperature theoretically possible. At absolute zero there is a minimum of vibratory **motion** and, by definition, no work can be done by a system on its surrounding environment. In this regard, such a system (although not motionless) would be said to have zero thermal energy.

See also Electrical resistance; Energy transformations; Entropy; Heat transfer; Kinetic molecular theory; Resistance, reactance, and impedance; Specific heat and statistical mechanics; Temperature and its measurement; Thermal radiation

MEASUREMENT THEORY

Measurement theory describes the methodology associated with the relation of numbers and mathematical operations to physical objects and processes. In a very practical sense, measurement theory deals with mathematical measurement and the analysis of mathematical measurement. A key component of mathematical analysis is error analysis that allows the estimations of measurement accuracy and precision. The development of measurement theory also involves justification of measurement and estimation of uniqueness of the measurement of physical phenomena. Measurements are considered unique if there is only one possible way to mathematically characterize the phenomena.

Measurement theory is as ancient as formalized mathematics. Aspects of measurement theory were included in **Euclid**'s elements and since that time measurement theory has evolved critical axioms that form the basis of modern data analysis. Measurement theory itself is broadly described as the set of rules related to the quantitative measurement of properties of physical entities (e.g., length) or the measurement of the duration between phenomena (e.g., time). Geometric measurements represent the spatial and dimensional properties of objects.

Well-ordered measurements are those in which the ultimate numerical assignments of a measurement closely correspond to the real or original order obtained in the actual observation or measurement process.

Measurement theory extends to a number of types of measurement. Extended measurements are those measurements that are created by the combination of measured attributes or that are linked together in a series or chain of meas-

urements to describe multiple objects. Comparative measurement relates the properties of a body or system to a known standard (e.g., **weight** and other standardized unit measures). Differential measurements usually deal with measurement of differences between systems or in the measurement of the intervals between phenomena. Conjointed measurements describe those attributes of a system or body that cannot be measured directly but can only be assigned a numerical quantity when related to changes in measurable phenomena (e.g., the measurement of intelligence by test scores).

Historically, the estimation of error in measurement evolved from direct estimation of numerical accuracy to a complex array of statistical techniques. Powerful computer-based data-analysis techniques referred to by mathematicians and statisticians as bootstrap statistics now allow the accurate determination of the reliability of data based upon only a small and representative sample of measurements. The techniques, invented in 1977 by Stanford University mathematician Bradley Efron, have found wide use in almost all fields of scholarship, including subjects as diverse as politics, economics, biology, and **astrophysics**.

Most scientific measurements are specified to contain some range of accuracy or degree of error. Errors are usually described as directed or random, and encompass deviations in mathematical measurements from the actual or true state. Errors can result from a variety of sources ranging from random human error to systematic errors that propagate through a series of mathematical equations. Modern error analysis still relies on the fundamental techniques advanced by French-Italian astronomer Joseph-Louis Lagrange (born Giuseppe Luigi Lagrandgia, 1736-1813) and French mathematician and astronomer **Pierre-Simon de Laplace**.

Prior to the formulation of **quantum theory**, measurement theory treated measurement error as a problem resolvable by refinement of technology and statistical techniques. The advent of quantum theory, however, provided limitations on the resolvability of error and made the actual measurement of phenomena an integral factor in scientific models. Electrons, for example, are be found to be in one of two possible spin states. Prior to any determination of this spin state the **electron** is described by a state vector that is a linear superposition of those two states. Accordingly, electron spin is not an intrinsic property of the electron unless, and until, it is measured. According to the Copenhagen interpretation of quantum theory, the act of measurement collapses the state vector and results in an electron with a defined spin.

See also Mathematical physics; Philosophy of physics; Quantum physics; Uncertainty principle; Units and standards

MECHANICS

In physics, mechanics is concerned with the **motion** of objects when acted upon by forces. For historical reasons, classical mechanics is usually limited to descriptions based upon **Newtonian physics** (i.e., **Newton's laws of motion**). In contrast, quantum mechanics utilizes **quantum theory** to describe the behavior of **subatomic particles**. Classical mechanics also differs from statistical or relativistic mechanics. **Statistical mechanics** is concerned with large numbers of objects. Relativistic mechanics studies objects moving at velocities close to the **speed of light**, or whose kinetic energies are on the scale of the product of their masses and the square of the **velocity** of light (mc^2). The subdivisions of mechanics include kinematics, which studies the movement of objects without explicit consideration of the forces responsible for the motion, and **dynamics**, which is concerned with the motion of moving objects and the forces and physical properties that affect them (e.g., **mass**, **momentum**, and **energy**). The special case in which any number of forces balance each other out and cause a body to remain at unaccelerated velocity (as in **equilibrium**), is covered by statistics.

The foundations of modern mechanics were laid in the early seventeenth century by Italian astronomer and physicist **Galileo** who studied the motion of simple objects. Galileo realized that, when not subjected to external forces, objects in motion continued to move and that the application of a **force** was required to change their motion. Besides this principle of **inertia**, Galileo discovered that free-falling objects move with constant **acceleration**. These concepts led English physicist Sir **Isaac Newton** to formulate his laws of motion, which represent the basic principles of mechanics, and later develop his law of universal gravitation.

In mechanics, motion is described using the concepts of velocity and acceleration. In turn, forces causing motion are studied by observing the resultant motion of bodies. The first law states that an object will remain at constant velocity (move at the same speed and in the same direction) unless acted upon by an external force. The law also implies the concepts of inertia and frames of reference. The second law defines force quantitatively as the product of a body's mass and acceleration ($F = ma$) and is a formulation of the fundamental cause of motion that applies to a wide range of physical phenomena. The third law, which states that there are no isolated forces and that forces come in pairs, implies a profound physical principle important to the understanding of forces, such as **friction** and thrust. Mechanics is also concerned with the study of the mechanical properties of a system interacting with its environment. These properties, such as energy, momentum, and **angular momentum**, are said to be conserved and the conservation laws that describe their behavior represent the most fundamental principles of mechanics. Mechanics is also concerned with **thermodynamics** quantities such as **work** (the application of force in the direction of the displacement of a body), energy (the capacity to perform work), and **power** (the rate at which work is performed).

Mechanics has a significant impact on day-to-day life, simply because almost everything moves in the physical world. Applications of mechanics include the study of rectilinear motion; displacement of motor-driven machines; motion under the influence of **gravity** (including projectile motion and **ballistics**); periodic motion; clock and time-keep-

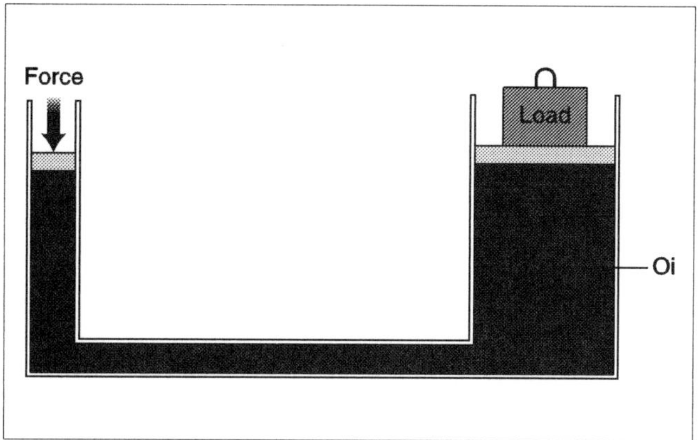

Fluid mechanics.

ing devices; circular motion; rotational motion; and motions of bodies following collisions.

See also Celestial mechanics; Momentum and Newton's second law; Momentum conservation; Motion of particles; Moving objects and velocity; Newton's law of universal gravitation; Relativity, special; Rockets; Rotation in three dimensions; Torque; Two-dimensional motion

MEITNER, LISE (1878-1968)

Austrian physicist

The prototypical female scientist of the early twentieth century was a woman devoted to her work, sacrificing family and personal relationships in favor of science; modestly brilliant; generous; and underrecognized. In many ways Austrian-born physicist Lise Meitner embodies that image. In 1938, along with her nephew Otto Robert Frisch, Meitner developed the theory behind **nuclear fission** that would eventually make possible the creation of the **atomic bomb**. She and lifelong collaborator **Otto Hahn** made several other key contributions to the field of **nuclear physics**. Although Hahn received the Nobel Prize for chemistry in 1944, Meitner did not share the honor—one of the more frequently cited examples of the sexism rife in the scientific community in the first half of the twentieth century.

Elise Meitner was born November 7, 1878, to an affluent Vienna family. Her father Philipp was a lawyer and her mother Hedwig traveled in the same Vienna intellectual circles as Sigmund Freud. From the early years of her life, Meitner gained experience that would later be invaluable in combatting—or overlooking—the slights she received as a woman in a field dominated by men. The third of eight children, she expressed interest in pursuing a scientific career, but her practical father made her attend the Elevated High School for Girls in Vienna to earn a diploma that would enable her to teach French—a much more sensible career for a woman. After completing this program, Meitner's desire to become a scientist was greater than ever. In 1899, she began studying with a local tutor who prepped students for the difficult university entrance exam. She worked so hard that she successfully prepared for the test in two years rather than the average four. Shortly before she turned 23, Meitner became one of the few women students at the University of Vienna.

At the beginning of her university career in 1901, Meitner could not decide between physics or mathematics; later, inspired by her physics teacher **Ludwig Boltzmann**, she opted for the latter. In 1906, after becoming the second woman ever to earn a Ph.D. in physics from the University of Vienna, she decided to stay on in Boltzmann's laboratory as an assistant to his assistant. This was hardly a typical career path for a recent doctorate, but Meitner had no other offers, as universities at the time did not hire women faculty. Less than a year after Meitner entered the professor's lab, Boltzmann committed suicide, leaving the future of the research team uncertain. In an effort to recruit the noted physicist **Max Planck** to take Boltzmann's place, the university invited him to come visit the lab. Although Planck refused the offer, he met Meitner during the visit and talked with her about **quantum physics** and **radiation** research. Inspired by this conversation, Meitner left Vienna in the winter of 1907 to go to the Institute for Experimental Physics in Berlin to study with Planck.

Soon after her arrival in Berlin, Meitner met a young chemist named **Otto Hahn** at one of the weekly symposia. Hahn worked at Berlin's Chemical Institute under the supervision of Emil Fischer, surrounded by organic chemists—none of whom shared his research interests in radiochemistry. Four months older than Hahn, Meitner was not only intrigued by the same research problems but had the training in physics that Hahn lacked. Unfortunately, Hahn's supervisor balked at the idea of allowing a woman researcher to enter the all-male Chemical Institute. Finally, Fischer allowed Meitner and Hahn to set up a laboratory in a converted woodworking shop in the institute's basement, as long as Meitner agreed never to enter the higher floors of the building.

This incident was neither the first nor the last experience of sexism that Meitner encountered in her career. According to one famous anecdote, she was solicited to write an article by an encyclopedia editor who had read an article she wrote on the physical aspects of **radioactivity**. When she answered the letter addressed to Herr Meitner and explained she was a woman, the editor wrote back to retract his request, saying he would never publish the work of a woman. Even in her collaboration with Hahn, Meitner at times conformed to gender roles. When British physicist Sir **Ernest Rutherford** visited their Berlin laboratory on his way back from the Nobel ceremonies in 1908, Meitner spent the day shopping with his wife Mary while the two men talked about their work.

Within her first year at the institute, the school opened its classes to women, and Meitner was allowed to roam the building. For the most part, however, the early days of the collaboration between Hahn and Meitner were filled with their investigations into the behavior of beta rays as they passed through aluminum. By today's standards, the laboratory in which they worked would be appalling. Hahn and Meitner frequently suffered from headaches brought on by their adverse

working conditions. In 1912 when the Kaiser Wilhelm Institute was built in the nearby suburb of Dahlem, Hahn received an appointment in the small radioactivity department there and invited Meitner to join him in his laboratory. Soon thereafter, Planck asked Meitner to lecture as an assistant professor at the Institute for Theoretical Physics. The first woman in Germany to hold such a position, Meitner drew several members of the news media to her opening lecture.

When World War I started in 1914, Meitner interrupted her laboratory work to volunteer as an x-ray technician in the Austrian army. Hahn entered the German military. The two scientists arranged their leaves to coincide and throughout the war returned periodically to Dahlem where they continued trying to discover the precursor of the element actinium. By the end of the war, they announced that they had found this elusive element and named it protactinium, the missing link on the **periodic table** between thorium (previously number 90) and uranium (number 91). A few years later Meitner received the Leibniz Medal from the Berlin Academy of Science and the Leibniz Prize from the Austrian Academy of Science for this work. Shortly after she helped discover protactinium in 1917, Meitner accepted the job of establishing a radioactive physics department at the Kaiser Wilhelm Institute. Hahn remained in the chemistry department, and the two ceased working together to concentrate on research more suited to their individual training. For Meitner, this constituted a return to **beta radiation** studies.

Throughout the 1920s, Meitner continued her work in beta radiation, winning several prizes. In 1928, the Association to Aid Women in Science upgraded its Ellen Richards Prize— billing it as a Nobel Prize for women—and named Meitner and chemist Pauline Ramart-Lucas of the University of Paris its first recipients. In addition to the awards she received, Meitner acquired a reputation in physics circles for some of her personal quirks as well. Years later, her nephew Otto Frisch, also a physicist, would recall that she drank large quantities of strong coffee, embarked on 10-mi (16-km) walks whenever she had free time, and would sometimes indulge in piano duets with him. By middle age, Meitner had also adopted some of the mannerisms stereotypically associated with her male colleagues. Not the least of these, Hahn later recalled, was absent-mindedness. On one occasion, a student approached her at a lecture, saying they had met earlier. Knowing she had never met the student, Meitner responded earnestly, "You probably mistake me for Professor Hahn."

Meitner and Hahn resumed their collaboration in 1934, after **Enrico Fermi** published his seminal article on "transuranic" uranium. The Italian physicist announced that when he bombarded uranium with neutrons, he produced two new elements—numbers 93 and 94, in a mixture of lighter elements. Meitner and Hahn joined with a young German chemist named **Fritz Strassmann** to draw up a list of all the substances, the heaviest natural elements produced when bombarded with neutrons. In three years, the three confirmed Fermi's result and expanded the list to include about 10 additional substances that resulted from bombarding these elements with neutrons. Meanwhile, physicists **Irène Joliot-Curie** and Pavle Savitch announced that they had created a new radioactive substance by bombarding uranium by neutrons. The French team speculated that this new mysterious substance might be thorium, but Meitner, Hahn, and Strassmann could not confirm this finding. No matter how many times they bombarded uranium with neutrons, no thorium resulted. Hahn and Meitner sent a private letter to the French physicists suggesting that perhaps they had erred. Although Joliet-Curie did not reply directly, a few months later she published a paper retracting her earlier assertions and said the substance she had noted was not thorium.

Current events soon took Meitner's mind off these professional squabbles. Although her father, a proponent of cultural assimilation, had all his children baptized, Meitner was Jewish by birth. Because she continued to maintain her Austrian citizenship, she was at first relatively impervious to the political turmoil in Weimar Germany. In the mid–1930s she had been asked to stop lecturing at the university but she continued her research. When Germany annexed Austria in 1938, Meitner became a German citizen and began to look for a research position in an environment hospitable to Jews. Her tentative plans grew urgent in the spring of 1938, when Germany announced that academics could no longer leave the country. Colleagues devised an elaborate scheme to smuggle her out of Germany to Stockholm where she had made temporary arrangements to work at the Institute of the Academy of Sciences under the sponsorship of a Nobel grant. By late fall, however, Meitner's position in Sweden looked dubious: her grant provided no money for equipment and assistance, and the administration at the Stockholm Institute would offer her no help. Christmas found her depressed and vacationing in a town in the west of Sweden.

Back in Germany, Hahn and Strassmann had not let their colleague's departure slow their research efforts. The two read and reread the paper Joliet-Curie had published detailing her research techniques. Looking it over, they thought they had found an explanation for Joliet-Curie's confusion: perhaps instead of finding one new substance after bombarding uranium, as she had thought, she had actually found two new substances! They repeated her experiments and indeed found two substances in the final mixture, one of which was barium. This result seemed to suggest that bombarding uranium with neutrons led it to split up into a number of smaller elements. Hahn immediately wrote to Meitner to share this perplexing development with her. Meitner received his letter on her vacation in the village of Kungalv, as she awaited the arrival of her nephew, Frisch, who was currently working in Copenhagen under the direction of physicist **Niels Bohr**. Frisch hoped to discuss a problem in his own work with Meitner, but it was clear soon after they met that the only thing on her mind was Hahn and Strassmann's observation. Meitner and Frisch set off for a walk in the snowy woods—Frisch on skis, with his aunt trotting along—continuing to puzzle out how uranium could possibly yield barium. When they paused for a rest on a log, Meitner began to posit a theory, sketching diagrams in the snow.

If, as Bohr had previously suggested, the **nucleus** behaved like a liquid drop, Meitner reasoned that when this drop of a nucleus was bombarded by neutrons, it might elongate and divide itself into two smaller droplets. The forces of electrical repulsion would act to prevent it from maintaining its circular shape by forming the nucleus into a dumbbell shape that would—as the bombarding forces grew stronger—sever at the middle to yield two droplets—two completely different nuclei. But one problem still remained. When Meitner added together the weights of the resultant products, she found that the sum did not equal the **weight** of the original uranium. The only place the missing **mass** could be lost was in **energy** expended during the reaction.

Frisch rushed back to Copenhagen, eager to test the revelations from their walk in the woods on his mentor and boss, Bohr. He caught Bohr just as the scientist was leaving for an American tour, but as Bohr listened to what Frisch was urgently telling him, he responded: "Oh, what idiots we have been. We could have foreseen it all! This is just as it must be!" Buoyed by Bohr's obvious admiration, Frisch and Meitner spent hours on a long-distance telephone writing the paper that would publicize their theory. At the suggestion of a biologist friend, Frisch coined the word "fission" to describe the splitting of the nucleus in a process that seemed to him analogous to cell division.

The paper "On the Products of the Fission of Uranium and Thorium" appeared in *Nature* on February 11, 1939. Although it would be another six and a half years before the American military would successfully explode an **atom** bomb over Hiroshima, many physicists consider Meitner and Frisch's paper akin to opening a Pandora's box of atomic weapons. Physicists were not the only ones to view Meitner as an important participant in the harnessing of nuclear energy. After the bomb was dropped in 1945, a **radio** station asked First Lady Eleanor Roosevelt to conduct a transatlantic interview with Meitner. In this interview, the two women talked extensively about the implications and future of nuclear energy. After the war, Hahn found himself in one of the more enviable positions for a scientist—the winner of the 1944 Nobel prize in chemistry—although, because of the war, Hahn did not accept his prize until two years later. Although she attended the ceremony, Meitner did not share in the honor.

But Meitner's life after the war was not without its plaudits and pleasures. In the early part of 1946, she traveled to America to visit her sister—working in the United States as a chemist—for the first time in decades. While there, Meitner delivered a lecture series at Catholic University in Washington, D.C. In the following years, she won the Max Planck Medal and was awarded numerous honorary degrees from both American and European universities. In 1966 she, Hahn, and Strassmann split the $50,000 Enrico Fermi Award given by the Atomic Energy Commission (AEC). Unfortunately, by this time Meitner had become too ill to travel, so the chairman of the AEC delivered it to her in Cambridge, England, where she had retired a few years earlier. Meitner died just a few weeks before her ninetieth birthday on October 27, 1968.

MERSENNE, MARIN (1588-1648)

French mathematician, philosopher, and scientist

Marin Mersenne was a French monk, mathematician, scientist, and philosopher known for his development of the Mersenne primes, which have played a significant role in number theory for several centuries. Born into a family of laborers in Oize in Maine, France, Mersenne attended the College of Mans for his grammar studies and then enrolled at the Jesuit college at La Fleche from 1604 until 1609. He then studied theology at the Sorbonne in Paris until 1611, when he joined the Minims, an order of Catholic friars dedicated to prayer and scholarship. Mersenne continued his studies at Minims' monasteries at Nigeon and Meaux and also taught philosophy at a monastery in Nevers. In 1619, Mersenne went to live at a monastery in Paris, where he spent the rest of his life.

A prolific writer, Mersenne's works in mathematics and physics were bolstered greatly by correspondence with scientists throughout Europe. As scientific journals or means of quick and easy communication were unavailable in Mersenne's time, scientists rarely had the opportunity to collaborate intellectually with their peers in other countries. Mersenne's prolific correspondence helped to bring scientists and scientific thought together. Many of the most famous scientists of the day corresponded with Mersenne or would visit him in his cell at the monastery, which is also where French philosopher **René Descartes** and mathematician **Blaise Pascal** first met.

The Mersenne prime developed from Mersenne's correspondence with French mathematician Pierre de Fermat (1601-1665) and his efforts to find a formula that would represent all prime numbers. Although Mersenne's formula ($2^P - 1$, in which P is the prime) works only for some primes, it has fostered many advances in number theory, especially in the search for new primes. Mersenne also furthered scientific thought by defending the ideas of Descartes and **Galileo**, and expressing opposition to magic, astrology, and other pseudo-sciences. Mersenne proposed that the pendulum would be a good timing device to **Christiaan Huygens**, the Dutch mathematician who is credited with its invention, and strongly supported the value of rigid experimentation in the study of nature. Mersenne's voluminous publications included *Cognita Physico-Mathematica*, in which he set forth his Mersenne prime formula; *Les Méchanique de Galilée*, about the work of Galileo; and *Traité d'Harmonie Universelle*, in which he discussed music and **acoustics**, establishing that the intensity of a **sound** is in inverse proportion to the distance from its source. Although a strong proponent of science, Mersenne equally defended his orthodox faith against atheism and skepticism.

See also Philosophy of physics

MESON

Mesons are particles composed of an even number of **quarks** and antiquarks, usually one quark and one antiquark, bound

Marin Mersenne. *(The Granger Collection. Reproduced by permission.)*

together by the exchange of gluons. Since they interact using the strong nuclear **force**, mesons are classified as hadrons, along with the **proton** and **neutron**. About 100 different mesons are known, varying from one-fifth the **mass** of a proton to about 10 times the proton's mass.

Mesons were first discovered in **cosmic rays**, as the products of the collisions of high-energy particles, mostly fast moving protons, with the nucleons of atoms in the atmosphere. Japanese physicist **Hideki Yukawa** predicted their existence in 1935 as the carriers of the strong force that binds the nucleons of an **atom** together. Since very few mesons reach the surface of Earth before interacting with particles in the atmosphere, they were not observed until 1946. These first mesons to be identified were called pi mesons, or **pions**. Although pions and other mesons are involved in building the atomic **nucleus**, their role is more complex than the original theory and is not yet completely understood.

Most current meson research is performed in high-energy **accelerators** by studying the sequence of particles produced by high-energy **hadron** collisions. One group of mesons, the **K mesons**, or kaons, is especially important to the study of basic particles because kaon decay modes have provided a better understanding of **parity** (the property of an elementary particle or physical system that indicates whether or not its mirror image occurs in nature) and the nonconservation of parity.

All mesons are unstable particles, most of them decaying in about 10-24 seconds. This is not long enough to observe the particle directly, so its existence is known only by studying the types and motions of particles formed when these mesons decay. There are, however, some types of mesons that exist longer, up to a few billionths of a second—a very long lifetime in subatomic terms. This is long enough to observe these mesons directly with **particle detectors**. The existence of quarks as the basic constituents of strongly interacting particles was discovered by observing the motions of mesons and the particles that are formed as they break apart. Most of the properties of quarks are determined by the interactions of mesons. For example, the decay rate of the pi meson into two **photons** was used to support the hypothesis that quarks have a characteristic property, called color (not related to **color** as we sense it) that exists in three different states. This property is the foundation of quantum chromodynamics.

The study of mesons also leads to identification of new quarks. The J/psi particle, discovered in 1974, is made of a previously unknown kind of quark, different from the three types that had been previously named—up, down, and strange. This meson was determined to be composed of a pair of quarks, labeled the charmed quark and the charmed antiquark. Charm was a new quantum number and its existence implies that quarks are related in pairs. The discovery of another heavy meson, called upsilon, revealed the existence of the bottom quark and its corresponding antiquark. After the discovery of the bottom quark, theoretical physicists predicted the existence a sixth quark, the top quark. Another meson, composed of a top quark and a top antiquark, was discovered in 1995, confirming some of the basic assumptions of the current quark model.

MESOSCOPIC SYSTEMS

The prefix "meso" means "in between" or "intermediate." Mesoscopic systems are those that are larger than atoms and yet very much smaller than the large-scale everyday objects that we can see and touch. They are 1,000-100,000 times smaller than the diameter of a human hair and they range in size from several hundred nanometers or billionths of a meter. That is why materials composed of mesoscopic parts are also known as nanostructures, and the technology based on these systems is referred to as **nanotechnology**. Since one nanometer is less than the width of 10 atoms in a row, mesoscopic systems are made up of less than a thousand atoms. Mesoscopic or nanoscale systems behave very differently from large-scale objects and they often have unusual physical and chemical properties. This makes them extremely interesting to scientists and engineers who hope that, by manipulating systems on the nanoscale, they can make **computers**, sensors, and other devices that have exactly the properties they want.

There are many examples of mesoscopic systems in nature. The molecules in our bodies that break down the foods in our stomachs and intestines, and the molecules that carry **oxygen** from the lungs to other parts of the body, are nothing but nanoscale machines. Artificial nanostructures, however,

have been studied and fabricated only in the last few decades. Multilayered nanostructures, made up of thin mesoscopic layers of different materials, have been investigated since the 1970s and have been used in devices such as high-efficiency **lasers** and **light-emitting diodes** (LEDs). More recently, researchers have begun studying and synthesizing **atom** clusters that are balls and tubes made up of 10-10,000 atoms. These clusters are sometimes put together to create new materials with novel properties. The properties depend on the size of the cluster. Copper, assembled from nanoscale clusters 5-7 nanometers in diameter, is five times harder than ordinary copper. Brittle ceramics become ductile when they are synthesized from clusters with sizes less than 15 nanometers. Cadmium selenide clusters of different sizes appear to have different colors.

Mesoscopic systems, particularly multilayered materials, are created and studied by a variety of well-established methods. One way of producing thin layers is to deposit atoms, from a chemical vapor or a beam of molecules, on to a base of some other material in an airless evacuated chamber. This is somewhat like spraying a thin layer of paint on a wall. Clusters of atoms are produced by chemical methods. An important tool in making and studying mesoscopic systems is the **scanning tunneling microscope** or STM. The STM creates an image of a surface at the atomic level by measuring the current that flows as the needle of the STM touches each atom. The STM can actually pick up individual atoms and move them around. In 1990, a group of researchers created a stir by writing "IBM" with 35 precisely placed xenon atoms on the surface of a nickel crystal.

The study of mesoscopic systems provides scientists with a picture of how the behavior of a material changes as it grows from a few atoms to large visible and tangible objects. This information will be useful in fabricating minuscule machines when nanotechnology finally becomes a reality in the coming decades. The first steps are already being taken. Researchers have used the STM to create a "switch," that could be used in computers, with a single atom that moves back and forth rapidly between two positions. Biologists are trying to manipulate large organic molecules such as proteins that could be used in the future for tasks such as the repair of damaged organs. Including wide-ranging applications in the computer industry, the possibilities of nanotechnology seem endless. Independent researcher Eric Drexler predicted that sometime in the future we may have minute machines that will repair clogged arteries, libraries that will fit into our pockets, and clothing that will change shape, **color**, and texture according to our needs.

METEOROLOGY

Meteorology is the science that investigates the physical processes occurring in Earth's atmosphere related to **weather** and climate. The knowledge obtained from the study of atmospheric conditions, such as **pressure, temperature**, wind, humidity, precipitation, clouds and solar **radiation** form the basis of meteorological research and weather forecasting.

Meteorological data and research are essential to the investigation of the changes affecting the Earth's climate, they are also instrumental in establishing the patterns of weather and climatic phenomena and for the determination of the **dispersion** of atmospheric pollutants. Meteorological data are critical to long term studies involving acid rain patterns or **ozone layer** depletion.

Although they describe fundamentally similar processes and atmospheric attributes (e.g., temperature, humidity, etc.), weather relates to short term meteorological phenomena that occurs over days or weeks. In contrast, climate relates to long term conditions and processes take place over months, seasons, years, etc.

Weather conditions guide daily activities and they can affect community health. On a larger scale, weather conditions can greatly impact economic activities, including resource industries (e.g., agriculture, fisheries and forestry) and transportation and travel sector industries. Military strategists expend great effort to maintain accurate weather maps and to train personnel to accurately interpret meteorological data.

Solar radiation provides the driving **force** for the circulation systems of the atmosphere and the oceans. Most of the physical processes taking place in the atmosphere result from self-regulating attempts to equalize the major differences that arise from inequalities in the distribution of atmospheric **heat**, moisture and pressure.

One principle of primary concern to meteorologists are the **dynamics** of differential pressure systems. In low pressure areas surrounding air masses converge toward the center of low pressure. Accordingly, such air masses are termed converging air masses and the area itself is also referred to as an area of convergence. In contrast, in high pressure areas air masses diverge outward from the center of high pressure (diverging air masses). In both converging and diverging areas there is a Coriolis deflection of the air masses to the right in the Northern Hemisphere and to the left in the Southern Hemisphere that imparts respective clockwise and counter-clockwise rotations to pressure systems. In addition, areas of convergence and divergence in the upper atmosphere are the determining factors in the formation of high and low pressure areas that will develop on the surface. In North America, the lateral driving force of upper level systems is the jet stream, an upper air current of strong **velocity** flowing from west to east, but oscillating on a north-south pattern. The moving air masses are also modified by thermodynamic and dynamic changes. For example, during the ascent of an air **mass** over a mountain range and its descent on the leeside there are adiabatic changes of temperature that can result in hot dry winds on the lee side of mountain ranges that can, in turn, lead to warm and dry downslope winds that can contribute to the formation of deserts or desert-like climates.

A major step towards understanding weather fluctuations was made when it was realized that the changes in weather seemed related to the formation and movement of boundaries between different air masses. A front defines the area of this conflict between opposing air masses. Front models and frontal depression wave models feature a warm front and a cold front separations. The two regions have distinct cloud cover, type of precipitation, barometric pressure and temperature. Frontal models provide the basis for synoptic weather maps and were a major step toward the understanding and forecasting of weather.

During the twentieth century, scientists used increasingly complex mathematical and physical models to characterize meteorological systems. For example, in 1922, L. F. Richardson proposed the use of numerical forecasting, a method which attempts to predict the physical processes in the atmosphere by means of Newton's equations of **motion**, the basic principle linking the rise or fall of surface pressure to mass convergence or divergence in the overlying air column. But with the tedious calculations of pre-computer days, this model was not considered particularly useful.

Advances in computer technology and their growing application for scientific research in the second part of the twentieth century made it possible to build statistical models capable of predicting weather based on historical trends. The first computer weather forecast was issued in April 1950 by a Princeton research team. The statistical model was replaced by the dynamic model that consisted of a succession of frames, each one slightly altered with respect to the previous one. In such frames, the first frame numerically represents the actual weather conditions (e.g., temperatures, pressures, humidity, visibility and other observations). These initial conditions are then entered on a grid superimposed on the map of the forecasting region. The conditions at each grid point are then recalculated by using equations which describe the dynamics of air and heat. The results of these calculations form the next frame. The computer simulation then proceeds frame-by-frame, moving each frame a few minutes closer to the specific forecast time.

However, in 1961, research findings by Edward Lorenz at MIT showed that a slight change of the initial conditions could lead to a significant change of the simulation results. Furthermore, Lorenz realized that the real atmosphere is characterized by chaotic behavior and that, in reality, weather conditions never precisely repeat themselves. The introduction of chaos has placed a limit on how far in the future prediction by computer modeling is possible. After an analysis of mathematical variables, Lorenz estimated this time limit to be of about two weeks. In addition, because initial conditions can not usually be determined with complete accuracy, chaotic effects often produce errant results if the time frame is extended too far.

The key to modern weather forecasting is the gathering of weather data from around the globe. The rapid collection, the assembling and processing of such data are crucial elements, because such data is required to prepare and distribute reasonably accurate forecasts. Historically, developments in data acquisition technologies have had immediate and profound impacts on meteorology. For example, in 1844, Samuel Morse's invention of wireless telegraphy made it possible, for the first time, to collect weather observations over long distances. In the 1920s meteorologists began exploring the upper atmosphere with kites, balloons and airplanes and by 1940 weather stations around the globe were making upper air observations. Contemporary meteorological studies, although they still rely on extensive direct monitoring by weather balloons and other modes of atmospheric sampling, are increasingly reliant upon satellite imaging, including infra-red thermal imaging and ultraviolet studies. Microwave studies provide information about wind conditions aloft and automatic readings from commercial airliners also provide data about temperature, pressure and wind at high altitudes.

Satellites, taking pictures of whole regions, transmit images of weather systems and have been instrumental in the prediction of the path and evolution of severe storms. On ground, **radar** provides a detailed picture of approaching precipitation systems and the Doppler radar system can supply additional information concerning the wind patterns occurring within thunderstorms and tropical storm systems.

While technological progress has greatly expanded the horizons of meteorological research and forecasting, Improved communication systems, involving **radio**, telephone, all-weather CABLE TELEVISION channels and the proliferation of numerous Internet sites have also dramatically increased access to weather information. Indeed, society's basic need for accurate weather information continues to be one of the motivating forces driving advancement in the meteorological sciences.

METRIC SYSTEM • See Units and Standards

MICHELSON, ALBERT (1852-1931)
Prussian American physicist

Albert Michelson devoted his life almost exclusively to the study of one subject: **light**. For his accomplishments in the study of **optics**, he was awarded the Nobel Prize for physics in 1907, the first American ever to receive that prestigious award. In experiments carried out over nearly half a century, Michelson repeatedly refined his attempts to determine the **speed of light**, a topic on which he became the world's foremost authority, and attained an accuracy of about three parts in a million for the value of this constant. In addition to his work on the speed of light, Michelson became an expert on the use of optical techniques to make a wide variety of physical measurements.

Albert Abraham Michelson was born on December 19, 1852, in Strelno, Prussia, not far from the Polish border. At the age of two, he came to the United States with his parents, Samuel and Rosalie Przlubska Michelson. The Michelsons landed in New York City, but shortly thereafter left for San Francisco. Samuel planned to take his family to the gold fields of California and Nevada where he could open a general store.

After a short stay in San Francisco, young Albert traveled with his parents to their new home in Murphys Camp, in California's Calaveras County. He was soon sent back to San Francisco, however, to complete his education. There he fell under the influence of Theodore Bradley, headmaster of the

Albert Michelson.

Boys' High School, where Michelson was enrolled. Bradley apparently recognized Michelson's talents and placed him in charge of the school's scientific equipment. At the age of 16, Michelson returned to live with his parents, whose home was now in Virginia City, Nevada. A brother, Charles, was born the following year and a sister, Miriam, two years later.

Perseverance Wins an Appointment to the Naval Academy

In 1869, Michelson decided to take the entrance examination for the U.S. Naval Academy in Annapolis. He passed the exam, but lost the appointment to another boy who had tied with him on the test. Undeterred, Michelson decided to travel to Washington and ask President Ulysses S. Grant personally for an appointment to the Academy. The Academy traditionally saved ten "open" appointments for special cases such as Michelson's, and he hoped to earn one of them. He got his appointment with Grant, but the President explained that all ten open appointments had already been made. He encouraged Michelson, however, to take his case directly to the Commandant in Annapolis. When Michelson did so, he was surprised to hear that the Commandant was willing to offer him a "special eleventh appointment." In later years, Michelson was to look back on this stroke of good fortune and explain that he owed his career in science to that early "illegal act."

Michelson graduated from the Naval Academy in 1873 and then served his required two years as an ensign on a series of cruises. At the end of this period, he accepted an appointment as an instructor of **chemistry** and physics at the Academy, a position he held until 1879. On April 10, 1877, Michelson married Margaret McLean Heminway, daughter of New York City banker and lawyer Albert Gallatin Heminway. The Michelsons' marriage lasted for twenty years and resulted in three children, Albert Heminway, Truman, and Elsa.

Michelson carried out his earliest scientific research during his years of teaching at the Naval Academy. The topic that captured his attention, and was to hold it during the rest of his life, was the speed of light. Prior to his own work, several other scientists, including physicist Jean Bernard Leon Foucault, had done significant research on finding the speed of light, all with varying degrees of success. Working on the problem at the same time was a colleague of Michelson's at the Academy, Simon Newcomb. Michelson and Newcomb both adopted the approach used by Foucault, and the extensive correspondence between the two Annapolis instructors about their respective experiments is now a classic part of the literature of American science.

By 1879, Michelson had devised a modification of Foucault's technique that gave a value for the speed of light of 186,500 miles per second, a result that was accurate to one part in 10,000. His research results were published in the April, 1879, issue of the *American Journal of Science*. In the same year, Michelson accepted a job at the Nautical Almanac office in Washington. He stayed only briefly, however, as he wanted to learn more about current research on the speed of light in particular and on optics in general, and that research was taking place largely in Europe. Michelson left for Europe in 1880 with his wife and two young children, Elsa and Truman. For the next two years, he visited a number of universities, including those in Berlin and Heidelberg, as well as the College de France and the Ecole Polytechnique in Paris.

His Research Fails to Detect the Luminiferous Ether

Michelson's first stop in Europe was at the laboratories of Hermann Helmholtz in Berlin. It was here that Michelson began his first studies of a question that was to earn him international fame: the movement of light through the "luminiferous ether." At the time, scientists believed that light travels as a wave, somewhat similar to water **waves**. But the existence of waves implies the presence of some material through which the waves can move. In the case of water waves, that "something" is water. In order to account for the existence of light waves, scientists had invented the concept of a "luminiferous ether." They envisioned this **ether** as a very thin substance that flowed much like water and permeated the whole **universe**. Light waves could then be explained as undulations within this ether. Efforts to detect the presence of this ether had been entirely unsuccessful, however. It was to this question that Michelson addressed himself during his stay at Helmholtz's laboratory in 1880–1881.

Michelson's investigation led him to create an experiment in which a beam of lightwas split into two parts and then projected at right angles to each other. The experiment was meant to create one beam of light that traveled parallel to the flow of the ether and a second beam that traveled perpendicular to it. As a simple analogy, think of a pair of boats traveling on a river, one moving downstream with the river's **motion**, and the other moving across the stream, at right angles to the river's flow. If both boats are traveling at the same speed, their actual motions will be somewhat different because of the additional **velocity** imparted by the motion of the river.

Michelson expected that in his experiment, one beam of light would be slowed down as it moved across the ether, while the other beam would be speeded up as it traveled with the ether. When the two beams were brought together by mirrors, then, they would no longer coincide, but would be slightly out of phase with each other. The device that Michelson developed to look for this change was an interferometer, a device that he was to use over and over again in many other applications in later years.

When Michelson actually carried out this experiment, however, he was able to detect no effect caused by an invisible ether. After being split, sent in two directions at right angles to each other, and then recombined, the light beams were exactly in phase with each other. There was no evidence that their travel had been at all affected by the ether. The result was especially troubling in Michelson's opinion because his equipment had been designed to detect very small changes in the speed of the two beams of light. Within a short time, flaws in the experiment's design were pointed out by Dutch physicist **Hendrik Lorentz**. These flaws, together with his inability to find a difference in speed between the two light beams, led Michelson to conclude that his effort was a failure. He began to think of modifications that would correct the problems in his experiment.

Meets Morley at Case

In the midst of his European tour, Michelson decided to resign his commission in the Navy. When he returned to the United States in 1882, he accepted an appointment in the physics department of the newly created Case School of Applied Science in Cleveland. While there, he continued his research on the speed of light. He also made the acquaintance of **Edward W. Morley** of Western Reserve College, also in Cleveland. Beginning in 1886, Michelson and Morley returned to the nagging problem of the ether, looking for ways to improve on Michelson's earlier "failed" experiment.

Over a five-day period in July of 1887, Michelson and Morley once more conducted a test for the differential velocity of light in the ether, this time using an interferometer sensitive to about one part in four billion. Again, the experiment produced no observable effect due to the ether. Somewhat disappointed about their results, Michelson and Morley abandoned their plans to repeat the experiment and went on to other fields of research. In particular, they were eager to explore other ways of using the very precise instruments they had built for the ether experiment.

The null results obtained by Michelson and Morley were not ignored by others, however. Indeed, a number of theorists began to ask what it meant that the speed of light appeared always to be the same. Several physicists devised equations to explain how this could happen in an ether. Irish physicist George Francis FitzGerald even argued that the lengths of objects shrink as their velocity increases, thus material measuring devices themselves are never completely accurate since they are affected by such things as the velocity of the Earth moving through **space**. (This theory became known as the Lorentz-FitzGerald contraction.) The Michelson-Morley results also had profound effects on **Albert Einstein**, then working on his theory of general relativity. One key assumption of Einstein's theory, that the speed of light is always a constant, can be traced directly to Michelson's "failed" experiment.

Honors and Awards Mark His Years at Clark and Chicago

In 1889, Michelson accepted an invitation to join the faculty at the newly created Clark University in Worcester, Massachusetts. While at Clark, he continued to collaborate with Morley on applications of their interferometer. Among the new uses to which the instrument was put during this period was a more precise measurement of the length of the meter. This was accomplished by measuring the Paris meter bar then used as an international standard and rendering that measurement in terms of red cadmium light waves (these lightwaves being unaffected by conditions that might render a material measuring device inaccurate). This experiment allowed the length of an object to be determined through spectroscopic study.

After four years at Clark, Michelson changed institutions once more, this time accepting an offer to become chairman of the department of physics at the University of Chicago. He remained at Chicago for the rest of his academic career, eventually assuming the chair of Distinguished Service Professor of Physics in 1925. During his first decade at Chicago, Michelson's international fame reached its zenith. He served for two years as president of the American Physical Society (1901–1903) and later as president of the American Association for the Advancement of Science (1910–1911). He also received a number of national and international awards, including the Copley Medal of the Royal Society (1907), the Elliott Cresson Medal of the Franklin Institute (1912), the Draper Medal of the National Academy of Sciences (1916), and the Nobel Prize for Physics in 1907.

Academic honors also came his way. Although he never earned a graduate degree himself, he was awarded honorary doctorates by a number of institutions including Cambridge (1889), Yale (1901), Pennsylvania (1906), Leipzig (1909), Clark (1909), and Gottingen (1911). While at Chicago, his personal life also underwent some major changes as he divorced his first wife of 20 years and married Edna Stanton of Lake Forest, Illinois, on December 23, 1899. With his new wife, Michelson had three more daughters, Madelaine, Beatrice, and Dorothy. Michelson's tenure at Chicago was interrupted by

World War I, during which he served as a naval officer. In the Navy, he used his expertise to develop an optical range finder and to improve optical glasses used by the military.

In the years following the war, Michelson turned his attention to a problem that had long interested him: measurements of astronomical objects and phenomena. He adapted his interferometer to determine the size of objects such as the moons of Jupiter and the star Betelgeuse. The latter accomplishment was recorded in daily newspapers across the nation.

Late in life, Michelson turned once more to the subject that never lost its fascination for him: the speed of light. At the invitation of astronomer George Ellery Hale, Michelson designed an experiment to measure the speed of a light beam transmitted over the 22-mile distance between Mount Wilson and Mount San Antonio near Pasadena, California. Work actually began on the project in 1930 and continued through the next year. Michelson had become so ill, however, that he could not carry out the actual measurements himself, but left that task to assistants Fred Pearson and Francis Pease. In the midst of the experiment that was ultimately to conclude in 1933, Michelson died of a cerebral hemorrhage on May 9, 1931.

Michelson was interested in a number of leisure time activities including billiards, sailing, and tennis. He also had some talent as a musician (he played the violin) and as an artist. At one point, in fact, he was persuaded to give an exhibition of his water colors at the University of Chicago. Michelson has been described as a quiet, withdrawn man with a "simplicity of character." In *Popular Astronomy,* F. R. Moulton described Michelson as "unhurried and unfretful.... He pursued his modest serene way along the frontiers of science, entering new pathways, and ascending to unattained heights as leisurely and as easily as though he were taking an evening stroll."

MICHELSON INTERFEROMETER

The most famous experiment designed to detect small changes in the speed of **light** was performed in 1887 by **A. A. Michelson** and **E. W. Morley**. It should be noted from the outset that the result of the experiment was negative, thus contradicting the then popular belief that Earth moved through an **ether** which served as the propagating medium for light. Prior to the **Michelson-Morley experiment** it was widely believed that light propagated through an hypothetical ether much like water **waves** propagate through water. The experiment was designed to determine the **velocity** of Earth with respect to this hypothetical ether. To test this hypothesis, Michelson devised the famed Michelson interferometer.

Based on the wave nature of light, the Michelson interferometer was designed with two mirrors, a light source, and a detector, all placed at right angles to each other with a third half-reflecting mirror positioned at an angle between the light source and one of the mirrors. The light source is pointed at the third mirror, and as light enters the third angled mirror, it is reflected to one mirror and transmitted to the other mirror. The light then reflects from both the first and second mirror heading back to the third angle mirror where the light rays combine and head towards the observer who sees an **interference** pattern (hence the name interferometer). An interference pattern is simply a series of light and dark bands signifying where the light wave crests interfere constructively (light band) and where they interfere destructively (dark band). By measuring changes in these bands one can determine very precisely differences in the distance between the first and third mirror and the distance between the second and third mirror.

Were there actually an ether through which Earth moved, then as Earth moved through this ether there would be an ether wind which, when light was propagating in the same direction as Earth relative to the wind, would slow the light, and which, when light was propagating in a direction opposite the direction of Earth in the ether, would speed the light. Thus, using the Michelson interferometer, changes in the **speed of light** could be detected based on the orientation of the arms of the interferometer. For any two differing orientations of the interferometer, the interference pattern would be different, signifying that there was indeed an ether wind. However, as stated above, no change was noticed in the interference pattern for different orientations of the arms.

This negative outcome was totally unexpected. The experiment was designed and originally performed by Michelson who was American. However, upon obtaining this astounding result, the European scientific community sent Morley (a British scientist) over to America (which at the time was still considered to be rather immature in the sciences) to validate the experiment. After many, many unsuccessful attempts to observe the expected change in the speed of light it was concluded that the ether did not indeed exist. This overturning of the hypothesis of the ether laid and important part of the groundwork necessary for Einstein's theory of relativity to be introduced and eventually accepted.

See also Speed of Light; Wave Interference

MICHELSON-MORLEY EXPERIMENT

In 1887 two American scientists, physicist **Albert Michelson** and physical chemist **Edward Morley**, performed an experiment that was designed to detect the **motion** of the Earththrough a hypothetical medium known as the luminiferous **ether** which was thought to be present throughout **space**. They made their measurements with a very sensitive optical instrument now called a **Michelson interferometer**. Their observations showed no indication of movement through the predicted ether. This outcome was unexpected and has become one of the fundamental experimental results in support of the theory of special relativity, developed by **Albert Einstein** in 1905.

During the 1800s scientists had become convinced that **light** was composed of **waves**, as opposed to a theory that light was made up of particles proposed more than a century earlier by **Isaac Newton**. They based their belief on experiments that demonstrated phenomena such as interference—the

change in intensity caused by mixing two or more beams of light; and diffraction—the fact that beams of light do not always travel in straight lines.

But if light was a wave, what medium did it travel through? Earthquakes produce seismic waves that are transmitted by the Earth's crust and a clanging bell makes **sound** waves carried by air. Scientists were certain that light had to be transmitted by something, so it was hypothesized that there existed a luminiferous ether. The term luminiferous means light-bearing, but the word ether was not so specific. No substance could be associated with the ether, especially so in space where sunlight and starlight travel in what otherwise appears to be a **vacuum**. The ether was predicted, but had not been observed.

Scientists thought that the ether should be everywhere and that it must be stationary, at rest with respect to absolute space which, following Newton, was believed to exist independently of the objects in it. It was thought that by measuring the motion of the Earth relative to the ether it would be possible to observe the latter.

Designing an experiment that would detect the Earth's movement through the ether was a formidable task, requiring the comparison of the **speed of light**, which was already known to be about 186,300 mi/sec (300,000 km/sec) and the speed of the Earth (almost 18 mi/sec or 30 km/sec). Michelson, who excelled in the art and science of measurement, built an instrument to do the job.

He made use of the **interference** that occurs between light waves. Light waves are transverse waves, which means that they vibrate perpendicularly to the direction in which they travel. If two waves of light of a single **color** (monochromatic light) arrive at a screen with their crests and troughs aligned, they will interfere constructively, adding up to make higher crests and lower troughs. If, on the other hand, the waves arrive so that crests coincide with troughs, they will cancel with each other, leaving the screen in darkness.

In Michelson's apparatus monochromatic light from a source was sent toward a beam splitter—a partially silvered mirror—where half of the beam continued on to mirror #2 while the other half was reflected along a perpendicular path toward mirror #1. A compensating plate placed in path #1 assured that both beams passed through equal thicknesses of glass. Following reflections at the mirrors the beams returned to the beam splitter where they joined and travelled to the **telescope**. Because the two rays are not exactly parallel and the wavefronts are not exactly plane the observer would not see all light or all dark, but rather a set of interference fringes-alternating dark and light parallel lines.

With his interferometer Michelson would have been able to measure movement through the ether by noting the change in the position of the fringes as the apparatus was rotated. To understand this first think of yourself to be at rest with the **interferometry**. From the instrument's point of view, it is the ether that moves, creating an ether wind which would push against the light beams. If the Earth moves in the direction of path #2, then the ether wind will be felt in the opposite direction. Beam #2 will act like a sailboat sailing first against the wind and then with it. It will travel slower when it opposes the ether wind but faster when the wind is at its back. In contrast, Beam #1 travels perpendicular to the ether wind on both parts of its trip. Because the ether wind affects each beam by different amounts there is a difference in the times it takes the beams to travel along their respective paths. That difference shows up as a fringe pattern.

The sole presence of the fringe pattern, however, does not allow measurement of the Earth's motion. That is accomplished by rotating the entire instrument. As the two beams change their orientation with respect to the ether wind, their travel times change. That causes the fringes to move or shift from their initial position. By measuring the fringe shift as the interferometer rotated, it should have been possible to measure the Earth's **velocity** through a stationary luminiferous ether, or, from the laboratory perspective, the velocity of an ether wind across a stationary interferometer.

Michelson performed the experiment for the first time in Germany in 1881. Contrary to his expectations no fringe shift could be observed. He repeated the experiment in 1887 in the United States, this time in collaboration with Morley. They placed their optical elements on a granite slab, and the slab on a vat full of liquid mercury. They lengthened the path each beam had to travel, and took good care to control the **temperature** in their laboratory to avoid thermal distortions.

According to their calculations the Michelson interferometer should have registered a fringe shift of about four-tenths (0.4) of a fringe. Instead, no fringe shift was observed. They were forced to conclude that their experiment had shown that the hypothesis of a stationary, luminiferous ether was not correct.

The Michelson-Morley experiment is a perfect example of a null experiment, one in which something that was expected to happen is not observed. The consequences of their observations for the development of physics were profound. Having proven that there could be no stationary ether, physicists tried to advance new theories that would save the ether concept. Michelson himself suggested that the ether might move, at least near the Earth. Others studied the possibility that rigid objects might actually contract as they travelled. But it was Einstein's theory of special relativity that finally explained their results.

The significance of the Michelson-Morley experiment was not assimilated by the scientific community until after Einstein presented his theory. In fact, when Michelson was awarded the Nobel Prize in physics in 1907, the first American to receive that honor, it was for his measurements of the standard meter using his interferometer. The ether wind experiment was not mentioned.

There has also been some controversy as to how the experiment affected the development of special relativity. **Albert Einstein** commented that the experiment had only a negligible effect on the formulation of his theory. Clearly it was not a starting point for him. Yet the experiment has been repeated by others over many years, upholding the original results in every case. Even if special relativity did not spring directly from its results, the Michelson-Morley experiment has

convinced many scientists of the accuracy of Einstein's theory and has remained one of the foundations upon which relativity stands.

MICROGRAVITY

The term microgravity means extremely low or zero **gravity** conditions. Ever since mankind evolved on this planet our gravity has been a condition of existence. The only individuals who ever experienced microgravity, albeit briefly, were those who fell from high places. Since the advent of **space** travel, however, microgravity has become an increasingly important area of research for many reasons.

Microgravity conditions can be achieved in several ways. As already mentioned, the condition of free fall will briefly give a zero gravity condition for the duration of the fall. Scientists use drop towers and shafts of varying heights to produce these conditions. At the Lewis Research Center in Cleveland, Ohio there are two facilities (a tower 75 ft tall and a shaft 400 ft deep) which allow experiments of 2.2 and 5.2 seconds respectively.

Another method of achieving microgravity is to push an aircraft through a parabolic arc. The aircraft climbs up through the arc turning into a dive at the top. If a parabolic path is followed, scientists inside the aircraft can experience microgravity for 20-25 seconds. If a longer period is required, a rocket can be used in a similar fashion to the aircraft, firing through a larger arc and falling back to earth. This can achieve several minutes of zero gravity for scientists to complete their experiments. Up to 17 days of microgravity conditions can be achieved in a space shuttle and experiments of several months duration can be done in a space station.

Research which needs microgravity conditions can come from the fields of biotechnology, materials science, fluid physics, combustion science and even fundamental physics research. When the **force** of the Earth's **gravitational field** is removed or masked, **matter** often behaves in new and unexpected ways. Crystals that grow in a solution will grow much larger in microgravity conditions. In the important field of biotechnology very pure and sizable crystals of proteins allow scientists a much better view of the protein's structure. Understanding the structure leads to better understanding of the protein's function in the body. Growing tissue in zero gravity conditions also yields much better results and cancer research has greatly benefited from the ability to grow cancerous cells more perfectly in microgravity conditions.

Once freed from the force of gravity, materials merge and combine in different ways. Alloys with completely new characteristics have already been formed during experiments in space. New electronic, composite, glass, ceramic, and polymer materials have been made under these conditions. Most importantly, insights have been gained into how to produce some of these results back on Earth. **Fluids** also behave differently. Areas of fluid physics research include uncovering how fluids interact with each other and with solid surfaces, how fluids behave when electrical or **heat energy** is applied, and how granular materials such as sand or soil behave under extreme **pressure**. Even flames behave differently in microgravity. Without the gravity induced convection currents of the air, flames burn spherically and produce no soot. They even burn with a different **color**. More understanding of combustion is essential if we are to address the pollution and **global warming** problems of this century.

Finally, fundamental questions about the nature of matter still need to be answered with ever increasing accuracy. Microgravity allows measurements of **mass** and **gravitational force** to be made with an accuracy many magnitudes greater than Earth-bound results. In addition, fundamental questions about the properties of atoms and molecules under weightless and low **temperature** conditions are being addressed. Superconductivity and laser cooling technology are two areas which have benefited from microgravity research.

MICROSCOPE

A microscope magnifies and resolves the image of an object that otherwise would be invisible to the naked eye. These objects include such items as human skin, the eye of a fly, cells of a living organism, viruses, individual molecules, and atoms.

The first optical microscopes, those that produce images through the use of visible **light**, used drops of water captured in a small hole to function as a magnifying lens. Later, magnifying glasses consisting of a single biconvex lens were developed and were capable of enlarging an object up to 20 times its original size as seen by the naked eye. (Scientists describe the **power** of magnification by writing a number followed by a times sign (x), so this would be expressed as 20x.)

Using several **lenses** in conjunction with each other is the secret of a microscope's success, however. Dutch spectacle maker, Zacharias Janssen, devised the first **compound microscope** in 1590. Today's basic optical compound microscopes consist of a two-lens system with an objective and an ocular lens, and can magnify up to 2,000x. The specimen sits close to the objective lens, which magnifies the images as an ordinary magnifying glass would. The lens forms an enlarged real image of the specimen near the ocular or eyepiece lens. The eyepiece works in conjunction with the objectives to correct for aberration and further magnify the image, creating the virtual image seen in a microscope. Light reflected from a mirror and concentrated by the condenser lenses located under the specimen stage, illuminates the object.

Various types of optical microscopes

The familiar monocular compound optical microscopes are being replaced in many laboratories with binocular styles. These microscopes have a single objective lens, but two ocular ones, each in its own eyepiece. Light coming through the objective lens is split into two beams by a prism. Each eye sees the exact same image, so there is no three dimensional effect.

For a three-dimensional view, scientists use a stereoscopic binocular microscope. This instrument consists of two separate sets of objective lenses as well as two separate ocular lenses. Prisms alter the angle of light coming through each pair of lenses, so each eye sees a slightly different image.

Living or stained specimens often yield poor images when viewed in bright-field illumination. To help with this, scientists developed a phase-contrast microscope that alters the phase differences in light **waves** as they pass through the specimen. This makes some parts of the object brighter and others darker than normal, allowing for a better view of the structural details of the object. Closely related to this type is the **interference** microscope that superimposes one field of view over a second to improve contrast.

Particularly useful for biological studies, a dark-field microscope uses a specialized illumination technique that capitalizes on indirect illumination to enhance contrast in specimens. The stop, an opaque disk set in a condenser under the stage, prevents illuminating light from shining directly on the specimen. Instead, illuminating light passes around the stop and is reflected off the condenser's walls.

To observe **color** in cells, scientists use polarizing microscopes. The microscope aligns the vibrations of a light wave by directing it through a specially cut prism. If two beams of polarized light are transmitted through the cell, as in a differential polarization microscope, researchers can make quantitative measurements.

Electron microscope

Prior to 1930, all microscopes were optical. In 1931, German physicist Ernst Ruska developed the **electron microscope** that used a beam of moving **electronics** to illuminate an object instead of light. Magnetic lenses or electric coils produce magnetic fields to deflect the electrons in the same manner that glass lenses bend light rays. The specimen has to be in a **vacuum**, however, because electrons cannot travel through air. **Electron** microscopes give point-to-point resolutions of less than 0.2 nanometers. This high resolution permits the direct visualization of many molecules and some atoms.

The transmission electron microscope (TEM) images specimens a fraction of a micrometer or less in thickness. In a TEM, the beam passes through the specimen so that some of the electrons are absorbed and some scattered. The remaining electrons are focused onto a fluorescent screen or special photographic plates via the use of magnetic lenses. The resulting image is in black and white.

In the scanning electron microscope (SEM), a narrowly focused electron beam is scanned over the surface of a solid object and used to build up an image of the details of the surface structure through reflection. Researchers use this type to study minute details on a surface of an object. These microscopes created 3D images that are magnified up to 50,000x.

Although there are several other special types of electron microscope, perhaps the most valuable is the electron-probe microanalyzer, which allows a researcher to make a chemical analysis of the composition of materials. This type of microscope uses the incident electron beam to excite the emis-

A scanning electron microscope (SEM). The images on the screen are fracture faces of engine components subjected to tension strength tests. *(Photo by Phillip Plailly. Eurelious/Science Photo Library, National Audubon Society Collection/Photo Researchers Inc. Reproduced by permission.)*

sion of characteristic x radiation by the various elements composing the specimen. Spectrometers built into the instrument detect and analyze the **x rays**. Viewing the resulting image, the researcher can easily correlate the structure and composition of the material.

Other types of microscopes

Scanning **tunneling** microscopes do not look like conventional microscopes at all. These are used to resolve individual atoms, identifying details down to one-tenth of an angstrom in height and less than two angstroms in width. Instead of lenses or mirrors, this microscope sports a tungsten rod with a tip is made up of a pyramid of atoms and three pieces of piezoelectric crystal, which compress and stretch in response to changes in the **voltage** of an **electric charge**. Electrons "tunnel" or flow through the vacuum or water from the tungsten tip to the atoms on an object's surface, creating a current that reacts with the crystal. Its inventors, **Gerd Binnig** of West Germany, and **Heinrich Rohrer** of Switzerland, won the Nobel Prize in 1986 for this development. They shared the prize with Ernst Ruska.

In 1986, the **atomic force microscope** (AFM) debuted. The AFM produces three dimensional images of surfaces both in air and under liquids at a resolution of nanometers, or billionths of a meter. In its contact mode, the AFM lightly touches a tip at the end of a 50-300 micrometer long leaf spring (the cantilever) to the sample. As the tip is scanned over the sample, a detector measures the vertical deflection of the cantilever, yielding the precise height of the sample at local points. The deflections of the cantilever are monitored by a laser beam that is reflected off the cantilever and into a position-sensitive detector. **Force** sensitivities and position accuracy as small as 10-15 Newtons/Hz1/2 and 0.01 nanometer, respectively, can be measured using microfabricated cantilevers. If the tip and sample are coated with two types of molecules, an AFM can measure force of attraction or repulsion between them, potentially at the level of a single hydrogen bond. Since its invention, the atomic force microscope has permitted high-resolution imaging at the subnanometer level. More recently, scientists introduced the microscope to a liquid environment and the resolution improved to the atomic level.

MICROWAVE BACKGROUND RADIATION

In 1964, a pair of **radio** astronomers, American physicist Robert Wilson (1936-) and German-born physicist **Arno Penzias**, working for the Bell Telephone Laboratories in Holmdel, New Jersey, stumbled upon the best evidence in existence to support the **big bang** model of cosmogenesis. Penzias and Wilson were recalibrating a radio antenna originally built to bounce signals off of Echo, the first telecommunications satellite. The same antenna was to both transmit and receive microwave signals from the second such satellite, Telstar. The astronomers-turned-radio engineers were measuring the antenna's gain (its power to amplify faint signals), and were looking out of the plane of the Milky Way. No matter where the antenna was pointed, whether at a radio source or at empty **space**, they detected a very faint, very uniform signal. The intensity of the **radiation** they found was identical to that emitted by a body at a **temperature** of about 3K. Radio astronomers characterize signals by comparing them with a black body distribution. Any body, at any temperature above **absolute zero**, emits a spectrum of radiation that is independent of the composition of the body, and depends only on its temperature. Wilson and Penzias were perplexed; the signal was clearly of cosmic origin, but no known cosmic source broadcast at the measured **wavelength**, 7.35 cm, and none known was so energetically uniform.

Meanwhile, a few miles away (and unbeknownst to the Bell labs duo) P. J. E. Peebles, a Princeton University astrophysicist, argued that if the **universe** had originated in the hot fireball of the big bang, its glow, though cooled by fifteen billion years of subsequent expansion, should be detectable. Peebles had independently rediscovered what had first been predicted in 1949 by cosmologists Ralph Alpher and Robert Hermann, who calculated that the present temperature of the big bang fireball should be about 5K. Inexplicably, their prediction was never acted upon, and by the 1960s was seemingly forgotten.

Wilson and Penzias discovered that relic glow, now known as the cosmic microwave **background radiation** (CMBR), for which they received the 1978 Nobel prize in physics.

For the first few hundred thousand years after the big bang it was too hot for electrically neutral atoms to exist. The Universe was filled with a soup of particles, electrons, protons, neutrons, neutrinos, and photons. The photons were constantly interacting with the charged particles, the electrons and protons. This **scattering**, the photon's transformations of **energy** related to interactions with charged particles, provided the mechanism by which the Universe attained thermal **equilibrium** (uniform temperature). The **photon** energies obeyed a black body distribution, reflecting the temperature of the Universe.

The Universe, however, was expanding, and thus cooling down. At some point (anywhere from 300,000 to 700,000 years after the bang, depending upon the details of the model), neutral atoms formed, which do not interact with the now free-streaming photons. The universe became transparent. The CMBR photons have not interacted with matter since that time, and thus provide a picture of the early Universe. The expansion of the Universe, the stretching of **space-time**, stretches the wavelengths of the CMBR photons and the black body photon distribution is red shifted as the Universe cools to yield its present temperature (which has been measured at all wavelengths and to great accuracy) of 2.73K.

The CMBR is, however, not perfectly uniform. One region of the sky is slightly hotter, while the diametrically opposed point is cooler. This dipole anisotropy reflects Earth's **motion** through the CMBR (in the general direction of the constellation Hydra) with a speed of about 550 km/s. High precision experiments performed by Earth-orbiting satellites and balloon-borne instruments have detected very tiny random fluctuations in temperature, which mirror **density** inhomogeneities in the early Universe. These regions of higher

than average density seem to be about the size needed to seed structure, galaxy and galactic cluster formation, resulting in the clumpy universe observed today.

See also Evolution of the Universe; Radio astronomy

MICROWAVES

Microwaves are electromagnetic **waves** that have lower frequencies (and longer wavelengths) than infrared **light**. They are higher on the **frequency** scale of the **electromagnetic spectrum** than **radio** waves. Their wavelengths range between 1 mm and 30 cm. The existence of microwaves was predicted by **James Clerk Maxwell** from his electromagnetic wave equations but only experimentally verified by Heinrich Hertz. While we do not usually think of microwaves as a form of light, they follow the same laws and possess the same properties as other forms of electromagnetic **radiation**, including visible light. Microwaves can show refraction, reflection, and **diffraction** patterns just as visible light does.

Microwaves are seen all over the universe-literally. The cosmic **background radiation**, which seems to be roughly isotropic, or the same in every direction, is composed of microwaves left over from the **energy** of the **big bang**. These microwaves have interacted with **matter** very little, if at all, since the Big Bang. Their existence must be explained in any theory that purports to explain the origins and age of the **universe**.

Microwave technology is widely used in **radar** detection systems. Microwave **pulses** are bounced off distant objects, and from the time it takes the reflected pulse to return, radar experts can determine the distance of the object. The **interference** pattern provided also gives a rough estimate of the shape of the object. Microwaves are also used for communications. Because of their shorter **wavelength**, they have a higher bandwidth than radio waves. They are used to transmit many different types of data including phone, cable television and internet traffic. It is often possible to spot microwave dish antennas at the tops of towers or tall buildings.

Some researchers hope microwaves could be used as a future energy source. They would plan to put a solar energy collector out in **space**, where the atmosphere would have less of an attenuating effect on the light intensity. The collector would then convert the collected energy into microwave energy and beam the microwaves down to **power** reception stations on Earth.

While most people use microwaves in their daily lives, their existence is usually taken for granted. They are commonly used to **heat** food in microwave ovens, taking advantage of the fact that water molecules excite easily in the microwave range. Since most foods have high water content, microwaves in microwave ovens take advantage of the water to heat the foods thoroughly. In addition, radiation at those wavelengths kills many types of bacteria that might otherwise be damaging to human beings. Microwave technology has thus become a widely used facet of electromagnetics.

Milky Way. *(Photo courtesy of UPI/Corbis-Bettmann. Reproduced by permission.)*

MILKY WAY GALAXY

The Milky Way Galaxy is the home galaxy of our solar system. It is a large spiral galaxy containing an aggregation of gas, dust, an estimated 400 billion **stars**, and thousands of globular clusters and **nebulae**. The Galaxy's **mass** is probably between 750 billion and one trillion solar masses, and its diameter is about 100,000 light-years (one light year is equal to 5,865,696 million miles (9,460,800 million kilometers)). All the objects composing our Galaxy orbit about their collective **center of mass**, called the galactic center. The Galaxy is bound together by the gravitational attraction between its parts, and its rotational **motion** prevents it from collapsing on itself.

The Milky Way is a term used to described how the stars of our galaxy appear to the naked eye from a position on the earth, away from the light pollution caused by modern cities. For thousands of years, people looked at the sky and saw a smear of light stretching across the sky that looked as if someone had spilled milk on the dome of the sky. In 1610 **Galileo** Galilei discovered (with a **telescope**) that the Milky Way was actually the integrated light of countless stars, individually too faint to see with the unaided eye. Nearly 200 years later, **William Herschel** used a more powerful telescope to count faint stars in different directions of the Galaxy. From this **work**

Herschel believed that the stars forming our Galaxy were grouped in a flattened distribution, with the **Sun** near the center and the concentration of stars marking the directions of the Galaxy's greatest extent. One hundred years later, in 1917, **Harlow Shapley** discovered that globular star clusters are distributed over a roughly spherical volume that was surrounding the center of the Galaxy. It was also discovered that the Sun was not near the center of the galaxy, but on the fringes of one of the great galactic arms.

The Milky Way Galaxy has three major components: a **nucleus**, a thin disk containing spiral arms, and an extended halo. The first major component of the Galaxy is the relatively massive "nucleus." The center of the galactic nucleus is obscured by interstellar dust particles that absorb both the visible and the ultraviolet light radiated by its components. However, scientists have been able to record and study emissions from the region at **radio**, infrared, X-ray, and gamma-ray wavelengths. The strong emissions of infrared **radiation** and X-rays, in particular, seem to indicate the presence of rapidly moving clouds of ionized clouds. These gas clouds are thought to be circling a supermassive object, possibly a black hole. The nucleus is a flattened spheroid of dimension 3,500 by 20,000 light-years. It is composed of closely packed old stars (about 10 billion years old) and hundreds of globular star clusters.

The thin and flat "disk" of the Milky Way Galaxy, its second component, contains five spiral arms (Cynus, Centarus, Sagittarius, Orion, and Perseus) that consist of diffuse nebulae, interstellar **matter**, and young and intermediate age stars (estimated at ages between a million and ten billion years old). These spiral arms are regions of active star formation. The disk is about 100,000 light-years in diameter and on the average 10,000 light-years thick (increasing up to 30,000 light-years near the nucleus).

Our solar system is situated within the outer regions of the Galaxy, well within the disk and only about 20 light-years above the equatorial symmetry plane. The solar system is about three-fifths of the way from the center of the Galaxy (about 26,000 to 36,000 light year from the center), located near the inner edge of the Orion arm. In its nearly circular orbit, the solar system takes about 220 million years to revolve once around the Galaxy.

The third component of the Galaxy is a diffuse, spherical, dark "halo" of star clusters that encompasses the Milky Way. It extends out in all directions from the edge of the visible portion of the Galaxy about 130,000 light-years. The halo contains a low **density** of old stars mainly in globular clusters. The stars are mostly **dark matter** (non-luminous), with their existence inferred from the gravitational pull on the visible matter within the Galaxy. The particular constituents, shape, and extent of its materials are still not known.

The Milky Way Galaxy belongs to a group of galaxies known as the Local Group. The group contains three large and over 30 small galaxies. The Milky Way is the second largest galaxy after the Andromeda Galaxy (also called M31), followed by the Triangulum Galaxy. M31 is about 2.9 million light-years away, the closest large galaxy to the Milky Way Galaxy. A number of faint galaxies are much closer. In fact, many of the dwarf local group members are satellites or companions of the Milky Way. SagDEG is a small galaxy that is currently in a close encounter with the Milky Way, and thus our closest known intergalactic neighbor at a distance away of 80,000 light-years. In turn, the Local Group belongs to a much larger group of galaxies known as the Virgo Supercluster.

See also Galaxies and galactic clusters

MILLIKAN, ROBERT A. (1868-1953)

American physicist

Robert A. Millikan vaulted from obscurity to international fame on the strength of his classic experiment designed to measure the charge on the **electron**. His "oil-drop" method for determining the electron charge earned him the 1923 Nobel Prize for physics. He solidified his role as a leader in American science by presiding over the rise of the California Institute of Technology into a world-famous center of scientific research.

Millikan was born March 22, 1868, the second of six children, to Silas Franklin Millikan and Mary Jane Andrews Millikan in Morrison, Illinois. His mother, a graduate of

Oberlin College in Ohio, served as dean of women at Olivet College in Michigan before moving to Illinois, and his father later earned a degree at the Oberlin Theological Seminary and became a Congregational preacher. The family moved to Maquoketa, Iowa in 1875.

Millikan graduated with high marks from Maquoketa High School in 1886. Following in his parents' footsteps, he then entered the Oberlin college preparatory program, and then the college itself the next year. He followed their classical course of study, taking classes in higher mathematics and basic physics, along with Latin and Greek. In 1889, he was asked to teach an introductory course in physics; the physics program at that time contained a large number of Greek terms, and he seemed to have been asked simply because he had done so well in Greek. His interest in physics began with his efforts to prepare himself for teaching this course. Millikan earned a B.A. in 1891, but stayed two more years as a physics tutor, taking additional science courses. He earned an M.A. in 1893 for his analysis of Silvanus P. Thomson's 1884 book, *Dynamic Electric Machinery.* On the strength of this achievement, Millikan earned a fellowship to Columbia University in New York as its sole graduate student in physics.

At Columbia, Millikan gravitated to Michael I. Pupin of the electrical engineering department. Pupin schooled Millikan in mathematical precision in experimentation. In the summer of 1894, Millikan, with the aid of Columbia professor of physics Ogden Rood, enrolled at the Ryerson Laboratory at the University of Chicago. There, Millikan first met **Albert Michelson**, the noted scientist who had measured the speed of **light** in 1879–80. Michelson's emphasis on precise measurement and rigorous attention to detail, as well as his faith in the evolutionary pace of scientific progress, appealed strongly to Millikan.

Millikan received his Ph.D. in 1895 for his research on the polarization of incandescent light. With financial assistance from Pupin, Millikan went to Europe for postgraduate study. This was a common practice for American scientists at the time, and he studied under such luminaries as Jules Henri Poincaré, **Max Planck**, and **Walther Nernst**. He was in Europe when the revolutionary discoveries of **x rays** and **radioactivity** were made. Millikan returned to the United States in 1896 to take a position as a physics instructor at the University of Chicago, where Michelson was still head of the department of physics and director of the Ryerson Laboratory. Millikan had a heavy teaching load at first; he was responsible for establishing the physics curriculum and preparing textbooks. In 1906, he published his *First Course in Physics,* which was widely used.

Despite his teaching success, Millikan remained anxious to pursue original research in physics. He was aware of the rapid progress in Europe on **atomic theory**, and he understood that advancement in his profession depended upon research results. Initial work in **thermodynamics** met with little success. In April of 1902 he took time off from research to marry Greta Irvin Blanchard, the daughter of a wealthy local businessman, and to travel in Europe. The couple later had three sons. After his honeymoon, Millikan focused his attention on radioactivity, and then on the behavior of electrons in metals. In particular, he studied the **photoelectric effect**, the ability of certain electromagnetic **waves** to detach electrons from metal surfaces.

In 1908, Millikan, now an assistant professor, began to focus on a problem involving precise measurement: the charge on the electron. The electron was playing an increasingly important role in the atomic theories then being assembled, and the precise measurement of its charge had become a pressing problem. The standard technique for measuring the charge had been developed by British physicist H. A. Wilson. First, water droplets would be given an **electric charge**, and the rate at which they fell would be measured. Next, an electric field would be introduced to attract the charged droplets upward and oppose the **force** of **gravity**. By determining the strength of the electric field and the **mass** of the water droplets, one could calculate the charge on the droplets and on the electrons.

Millikan tried Wilson's technique but found that the rapid evaporation of the water droplets made measurement difficult and the results erratic. He decided to measure the charge on a single drop of water by balancing it between the electric field and gravity. He observed in 1909 that the charge on any drop was an integral multiple of a fundamental value. His findings, however, were still complicated by evaporation of the droplets. In 1909 Millikan hit upon the idea of using slow-evaporating oil drops instead of water. He could now observe charged drops for several hours, instead of only seconds. By balancing charged oil drops between an electric field and gravity, Millikan proved that the electron was a fundamental particle with a fundamental charge. In 1913, he published his value for the charge, which remained the accepted value for decades. The determination of this value, and the ingenuity of the experiment, earned Millikan international recognition and the 1923 Nobel Prize in physics.

After determining the charge on the electron, Millikan returned to his research on the photoelectric effect. In 1905, **Albert Einstein** had explained the phenomenon by suggesting that small packages or **quanta** of light—photons—were responsible for knocking the electrons off the metal surfaces. Millikan, like many other physicists of the day, resisted this particle view of light, preferring instead to consider light as waves. He set out to test the validity of Albert Einstein's hypothesis by observing the liberation of electrons from metals by ultraviolet light. By careful observation and elimination of the errors of previous methods, in 1916 Millikan was able to verify the validity of Einstein's calculations, although Millikan himself refused to abandon the wave theory of light.

Europe was engulfed in World War I by this time. As it became more likely that the United States would enter the war, George Ellery Hale, the astronomer and observatory builder, successfully lobbied for the National Academy of Sciences to establish a National Research Council, dedicated to research for the war effort. Millikan assisted Hale in his plans and in 1917 took a leave of absence from the University of Chicago to serve as head of a committee on submarine detection. Soon after the United States entered the war, Millikan joined the U.S. Army Signal Corps and became director of the Corps

Robert A. Millikan.

Division of Science and Research, rising to the rank of lieutenant colonel. His work for the National Research Council in antisubmarine warfare helped to establish the council as a permanent body.

Millikan returned to the University of Chicago after the war. But in 1921, Hale and chemist Arthur A. Noyes persuaded him to chair the Executive Council of the California Institute of Technology in Pasadena. Millikan was also named director of its Norman Bridge Laboratory. Despite his initial misgivings, Millikan quickly turned the Institute into an internationally respected center of research. He lured talented students to the school by inviting such prominent scientists as Einstein, Michelson, **Arnold Sommerfeld**, Paul Ehrenfest, and C. V. Raman to take faculty assignments. By the end of the 1920s, the California Institute of Technology was the leading research institution in the United States, boasting of such faculty giants as Thomas Hunt Morgan, Theodore Kármán, and Linus Pauling.

Many Americans of the 1920s considered Millikan to be the foremost American scientist of the day. He continued to be active in the National Research Council, and from 1922 to 1932 he was the American representative to the Committee on International Cooperation of the League of Nations. He served as president of the American Association for the Advancement of Science in 1929. It was also during this period that he earned the Hughes Medal of the Royal Society of London and the Faraday Medal of the British Chemical Society. He would eventually hold honorary doctorates from twenty-five universities and belong to twenty-one foreign scientific academies.

Millikan still found time outside his administrative duties to conduct original research. With graduate student Carl F. Eyring, Millikan studied the ability of strong electric fields to draw electrons out of cold metals. His classical explanation for the phenomenon was eventually replaced by a quantum mechanical approach. He also studied the ultraviolet spectra of light elements with graduate student Ira S. Bowen. Their research supported Sommerfeld's relativistic explanation of spectra, rather than the atomic model of **Niels Bohr**; the conflict was resolved in 1925 by the hypothesis of electron spin.

Most of Millikan's research at the Institute, however, focused on **cosmic rays**, the penetrating rays that had been discovered by Austrian physicist Victor Hess in 1912. Millikan's initial objective was to determine whether the rays came from **space** or from radioactive elements in the earth. He launched a large number of sounding balloons, even mounting an expedition to the top of Pike's Peak in 1923, to detect and measure ionization of atmospheric gases by the rays. These experiments failed to settle the issue, and in 1925 Millikan decided to approach the problem by measuring the variation in ionization from Lake Arrowhead (5,000 ft [1,524 m] above sea level) to Muir Lake (12,000 ft [3,657.6 m] above sea level) in California. The atmosphere between the two lakes had the absorbing power of 6 ft (1.8 m) of water, and a cosmic source for the rays could be assumed if the intensity of ionization at Lake Arrowhead matched the intensity of ionization 6 ft (1.8 m) lower at Muir Lake. Millikan discovered this to be the case. Moreover, he observed that the ionization effects continued day and night, and that the rays were about eighteen times more energetic than the most energetic gamma rays then known. It was Millikan who dubbed these ionizing rays "cosmic."

Millikan became embroiled in controversy, however, when he attempted to build upon his cosmic ray work. He failed to convince other scientists that the cosmic rays were photons emitted by the spontaneous fusing of hydrogen atoms into heavier elements; he abandoned the theory himself by 1935. In 1932, **Arthur Holly Compton** detected that the intensity of the rays varied with latitude, suggesting that the rays were charged particles deflected by the earth's magnetic field. Millikan fiercely denied Compton's assertion, and he supported his argument by conducting his own experiments that failed to detect this latitude effect. In 1933, however, after determining that a local irregularity in cosmic ray intensity had interfered with his observations, Millikan ceded that some cosmic rays were charged particles.

Millikan remained a determined conservative throughout his career, both in politics and in science. True to his scientific training under Michelson and Pupin, he preferred to think of scientific progress as evolutionary and not revolutionary. Though he encouraged original research, he never quite accepted the revolutionary ideas of the new **quantum mechanics**. In his later years, he even fought back attempts to reexamine his determination of the charge on the electron. In

politics he was a staunch Republican; he was strongly opposed to the New Deal and, after the war, to the formation of the National Science Foundation. Despite his conservative inclinations, however, he was a religious modernist, promoting the compatibility of science and religion even as the 1925 Scopes trial pitted science against fundamentalist Christianity.

Millikan retired from his professorship and presidency of the California Institute in 1946. He then wrote his autobiography, which was published in 1950. He died in Pasadena on December 19, 1953.

MILLIKAN'S OIL DROP EXPERIMENT

The discovery of the **electron** in 1897 by English physicist **J. J. Thomson** lead to the beginning of knowledge of **atomic structure**. Thomson showed that charge came in discrete **quanta**, and indirectly measured the charge-to-mass ratio of the electron. However, Thomson's method did not allow him to measure the charge or **mass** ratio directly. Thomson attempted an experiment where he observed the behavior of falling charged water drops. This experiment, however, was inaccurate because the water drops evaporated during the experiment, and Thomson could only estimate how fast they evaporated in order to interpret the results. In 1909, American physicist **Robert Millikan** performed an experiment based on an Thomson's earlier attempt, but used charged oil drops instead of charged water drops. This eliminated the evaporation problem and considerably reduced the uncertainty of the experiment. Because of its success (and extreme accuracy, for the time) the experiment is referred to as Millikan's oil drop experiment.

In the experiment, oil drops were sprayed between the plates of a parallel-plate capacitor. A **microscope** installed in the capacitor observed the oil drops. The drops quickly attained terminal **velocity** downwards due to the balance of air resistance and their **weight**. The drops became charged in the spraying process, so an electric field provided by the capacitor was used to make the drops rise rather than fall. The mass of the drops was determined by measuring their radii, because their **density** was known. The major experimental uncertainty lied in determining the drag coefficient for the air resistance **force**. Millikan used a successive approximation procedure to estimate the drag coefficient, but was unable to pinpoint it exactly. By measuring the time necessary for a drop to rise or fall a certain distance, the charge on the drop was determined.

Millikan was able to observe a single drop for several hours. In that time, he observed when a drop gained or lost charges because its velocity changed. He then divided the difference in charge by different integers until he obtained the same result for every change in charge. He performed this experiment on thousands of drops of different composition and conductivity. This allowed him to find the value of the fundamental charge, which he reported as $e = 1.591 \times 10^{-19}$ coulombs. Compared to the present accepted value, $e = 1.602 \times 10^{-19}$ coulombs, Millikan's results are remarkably accurate.

MINKOWSKI, HERMANN (1864-1909)

Russian German physicist and mathematician

In spite of a relatively short career, Hermann Minkowski played an important role in the development of modern mathematics. His work formed the basis for modern functional analysis, and he did much to expand the knowledge of quadratic forms. He also developed the mathematical theory known as the geometry of numbers and laid the mathematical foundation for Albert Einstein's theory of relativity.

Minkowski was born in Alexotas, Russia on June 22, 1864, of German parents. The family returned to their native Germany in 1872, to the city of Königsberg, where Minkowski spent the rest of his childhood and also attended university.

Even as a student at the University of Königsberg, Minkowski demonstrated a rare mathematical talent. In 1881, the Paris Academy of Sciences offered a prize, the Grand Prix des Sciences Mathématiques, for a proof describing the number of representations of an integer as a sum of five squares of integers—a proof that, unbeknownst to the Paris Academy, the British mathematician J. H. Smith had in fact already outlined at the time. Minkowski produced the proof independently while Smith sent in his own work. In 1883 both Smith and Minkowski received the prize. At that time, the nineteen-year-old Minkowski was two years away from receiving his doctorate from the University of Königsberg. The work contained in the 140-page manuscript he submitted to the Academy was, in fact, considered a better formulation than Smith's because the young Minkowski used more natural and more general definitions in arriving at his proof.

After receiving his doctorate from the University of Königsberg, Minkowski taught at the University of Bonn until 1894. Returning to teach at the University of Königsberg for two years, he then taught until 1902 at the University of Zurich. One of his closest colleagues at Zurich was a former teacher, A. Hurwitz, who is best known for his theorem on the composition of quadratic forms.

After working on the arithmetic of quadratic forms for several years and making contributions particularly to work in *n* variables, Minkowski extended his work to what is most commonly known as the geometry of numbers. In 1889, he introduced what has been characterized as his most original achievement, when he included volume in his work with ternary quadratic forms. With this extension it became possible to give mathematical descriptions of the properties of, for example, convex bodies in both two and three dimensions. Common examples of convex regions are those bounded by circles, ellipses, and parallelograms. Using the two- and three-dimensional versions of Minkowski's convex body theorem, mathematicians are able to prove some fundamental facts about algebraic number theory and can derive new proofs of some theorems from elementary number theory. Minkowski extended these ideas to investigations of the geometrical properties of convex sets in *n*-dimensional **space**, and his observations ultimately formed the basis for modern functional analysis.

At the urging of a former classmate, the great mathematician **David Hilbert**, the University of Göttingen created a new professorship for Minkowski in 1902. It was during his tenure at Göttingen that Minkowski turned his attention to **relativity theory**. **Albert Einstein** had been one of Minkowski's pupils, and Minkowski was very interested in the special theory of relativity formulated by Einstein, which at the time competed with the more widely accepted **electron** theory of the Dutch physicist **Hendrik Lorentz** as an explanation of subatomic phenomena. Minkowski was the first to recognize the consequences of the relativity theory in consideration of time and space. "From now on," he said, "space by itself and time by itself are mere shadows and only a blend of the two exists in its own right." By 1907, Minkowski had placed a formal geometric interpretation upon relativity. He believed that in the **universe**, time and space exist as a fused "time-space." In his book *Raum und Zeit* (translated into English as *Time and Space*), he demonstrated that relativity made it necessary mathematically to take time into account as a fourth dimension besides the spatial dimensions of length, width, and depth. Einstein used Minkowski's ideas to develop his general theory of relativity, published nine years later, several years after Minkowski's death in Göttingen, Germany, on January 12, 1909.

MITCHELL, MARIA (1818-1889)

American astronomer

Maria Mitchell came by her love of learning naturally. Since her mother worked in libraries, she became an avid reader; her father enjoyed astronomy immensely, providing Maria early exposure to that field. In addition, she grew up in Nantucket, where people knew and appreciated natural phenomena. Although she completed her formal schooling at age sixteen, Mitchell studied further on her own, reading French authors, books on mathematics, and Bodwitch's *Practical Navigator*.

At seventeen she opened her own school, the beginning of a lifelong interest in educating others. She was unconventional in her approach to teaching, sometimes starting the school day before dawn to allow students to see certain birds, or extending class time into the night to encourage astronomical observations.

In 1836 she became a librarian, a post which gave her time to make an unusual astronomical discovery. Since the library was only open afternoons and Saturday evenings, she spent a great deal of time studying and observing the sky. On October 1, 1847, Mitchell discovered a new comet. Although a year passed before she was credited with the finding because of other claims, she eventually received a gold medal from the king of Denmark. In 1848 she was elected the first woman of the American Academy of Arts and Sciences and became the subject of numerous magazine articles. She also became close friends with **Joseph Henry**, a physicist and director of the Smithsonian Institution.

In 1849 she became an analyst for the *American Ephemeris and Nautical Almanac* and began to work for the United States Coast survey, helping calculate time, latitude, and longitude more accurately. As the chaperon for the daughter of a wealthy Chicago businessman, Mitchell traveled to Europe, where she visited observatories and met such famous individuals as author Nathaniel Hawthorne (1804-1864), astronomers William Herschel (1738-1822) and Mary Somerville (1780-1872), and Alexander von Humboldt.

In 1865 Mitchell became professor of astronomy and director of the observatory at Vassar College, a newly founded women's college in New York. She became a leading advocate of women's rights, arguing that women were suited for mathematics and other sciences. As she had done during her early years as an educator, Mitchell ignored the usual lecture method of instruction, stressing small classes and individual attention.

She toured Europe again in 1873 and that same year helped found the Association for the Advancement of Women, a moderate feminist group. She taught at Vassar for twenty-three years, retiring in 1888. Although she was offered a permanent residence at the school's observatory, Mitchell returned to her family's home in Lynn, Massachusetts, where she died in 1889.

In assessing her impact, it's fair to say she was more of an observer and teacher rather than a theoretical astronomer. As observer, she was especially interested in the **sun**, witnessing several total eclipses, and watching sunspots to try to understand their origin and mutations. She believed they were rotating, gaseous storms on the solar surface. She viewed Jupiter's clouds very much as modern scientists do: she felt they were not just atmospheric phenomenon, but were elements of the body itself. These clouds were seething upward and moving at different rates. As Mitchell did, astronomers today believe Jupiter is a planet without a solid interior. When she observed Saturn, she claimed the rings and the globe must be of different composition, again presaging modern thought. She also spent time observing and speculating about nebula and **binary stars**.

MOLAR QUANTITIES

A quantity of particles, atoms, or molecules of any substance that is equal to **Avogadro's number** is called one mole of that substance. Avogadro's number, a constant, is an immense number equal to 602,000,000,000,000,000,000,000, or 6.02×10^{23} particles of any material. Therefore, one mole of a substance is equal to 6.02×10^{23} particles. Molar quantities, then, are amounts of chemical substances expressed as moles. For example, one mole of salt is Avogadro's number of salt molecules, two moles of hydrogen **ions** equals two times Avogadro's number of protons.

Molar quantities, by definition have known numbers of particles. Such quantities, however, may have very different weights. For example, one mole of water and one mole of **carbon** dioxide each have the same number of molecules, but the mole of water weighs 18 g, while the mole of carbon dioxide weighs about 48 g. The difference in **mass** occurs because one mole of any substance is equal to its gram-molecular **weight**.

Gram-molecular weight is the **molecular weight** of a molecule expressed as grams. The molecular weight is, in turn, the sum of the atomic weights of the constituent atoms within the molecule. For example, the gram molecular weight of water, two hydrogen atoms bonded to one **oxygen atom**, is 18 grams (hydrogen has anatomic weight of about 1, and oxygen has an atomic weight of 16). Therefore, one mole of water contains 6.02×10^{23} molecules and weighs 18 g.

Molar quantities are used to express concentrations of substances dissolved in solution. Molarity is a molar quantity of concentration equal to the number of moles of a substance that are dissolved in 1 liter of liquid. Molarity, expressed as moles per liter, takes into account the total number of particles in solution rather than expressing concentration simply as a percentage. For example, a molar salt solution of 2.0 (mol/liter) is made by dissolving 2 moles of salt into 1 liter of water. In this solution, the exact number of salt molecules is known because it is a molar quantity. Also, the exact weight of salt needed is also known since two moles of salt equal two gram-molecular weights of sodium chloride, or about 116 g of salt.

Some molar quantities have exact volumes, called molar volumes, as well. For example, one mole of any gas always occupies approximately 22.4 liters of **space** at standard **temperature** and **pressure**. Also, one mole of liquid water always occupies 18 ml (cm^3) of space at standard temperature and pressure.

MOLECULAR ORBITAL THEORY

Atoms are the fundamental building blocks of all **matter**. This is the basis of **chemistry**. But if they are the bricks, then what is the mortar? What holds the atoms together to form molecules? The answer is "bonds" which, of course, doesn't really explain anything because it raises the next question—what is a bond? It is the nature of science that the answer to one question invariably leads to another question.

A simple thought experiment can be used to determine the basic nature of bonds. Electrostatics tells us that oppositely charged particles attract and that similarly charged particles repel. This means that nuclei will repel each other, as will electrons. Simply trying to push two nuclei together will not allow for the formation of a molecule. But attractions do exist between electrons and nuclei and this tends to favor molecule formation. Here's what happens when two hydrogen atoms approach to within bonding distance. Each **atom** has a single **electron** that is attracted to its own **nucleus** but also is attracted to the nucleus of the other atom. There are four such interactions—each of the two electrons are attracted to two nuclei. These four interactions overcome the electron-electron and nuclear-nuclear repulsions as they develop, allowing a molecule to form. In this case, a bond is a pair of shared electrons.

While this is qualitatively the picture of bonding, it is by no means complete. For example, if pure electrostatic interaction was all that was necessary then the more electrons and protons an atom had, the stronger it would bond.Metals, such as gold, would have so many bonding interactions that they would never melt or be malleable. The simplistic view of bonding outlined above is represented by Lewis structures, for example, and a great deal of chemistry can be done without ever pursuing the exact nature of the chemical bond any further. Most synthetic chemists don't need to know the details.

Various models have been put forward to explain molecular interactions. The two most dominant are the "valence bond approach" and "molecular orbital theory." They are very similar in the way that they describe molecules and come to roughly the same conclusion but from slightly different angles. In grossly simplified terms, the valence bond approach fills the atoms with electrons and then interacts the **orbitals**, whereas molecular orbital theory interacts the orbitals and then fills them with electrons.

Orbitals are a mathematical construct used to explain the location of electrons around atoms and their physical properties. No one has ever seen an orbital. But they are completely consistent with all that we know about atoms and the way that electrons behave in atoms. This may **sound** like metaphysics—and it is—but it is important to remember that orbitals are a consequence of the quantum **mechanics** and the equations that underlay it. This is important because representing orbitals by equations means that adding together equations should give new combinations or orbitals. This is exactly what molecular orbital theory does. By treating the molecule as a single unit and all of the atoms in the molecule as part of that unit, a new set of orbitals for the whole molecule are constructed out of linear combinations of atomic orbitals (LCAO).

Linear combinations of atomic orbitals have the property that orbitals are conserved. Thus, if two orbitals are combined, then the product is two orbitals. If three orbitals are combined, then three new orbitals are created. The two new orbitals created from the original orbitals, designated Ψ_1 and Ψ_2, are $(\Psi_1+\Psi_2)$ and $(\Psi_1-\Psi_2)$, i.e., the sum and difference of the original orbitals. This is what is meant by the term "linear combination".

Orbitals, though, are really electron **density** distributions and the symbol Ψ represents the solution to the **wave function**. The combination $(\Psi_1+\Psi_2)$ is a "piling up" of electron density. This is a simplistic view, but useful as this combination is bonding. It results in an increase in the electron density between the atomic nuclei. But if $(\Psi_1+\Psi_2)$ is an increase in electron density, then $(\Psi_1-\Psi_2)$ must represent the removal of electron density from between the nuclei. And if an increase in electron density is bonding, then what is a decrease? The answer is that $(\Psi_1-\Psi_2)$ represents an anti-bonding orbital—a combination that negates the bonding interaction. The **energy** of a bonding orbital is decreased relative to the atomic orbital and the anti-bonding orbital is increased. The energy differences are equal and opposite in sign so that energy is conserved in the system. The energy of the two atomic orbitals is the same as the combination of the bonding and anti-bonding orbitals.

Consider the interaction of two hydrogen atoms. The molecular combination has a bonding and anti-bonding pair of orbitals generated from the 1s atomic orbital on each atom (Figure 1). Each atom also has a single electron. In forming H_2, these electrons fill the molecular orbitals by the Aufbau principle—from the lowest energy orbital up. The result in H_2 is that both electrons reside in the bonding combination. The resulting molecule is stabilized relative to the individual atoms. On the other hand, consider trying to form the dihelium molecule, He_2. In this case, the 1s orbital also would form a bonding and anti-bonding pair. But with each atom contributing two electrons, the result would be the filling of both the bonding and anti-bonding combinations. The energy required to put the electrons into the anti-bonding orbital is slightly more (due the overlap integral) than the energy that is obtained from the electrons in the bonding orbital. This is the reason that He_2 doesn't form. It is energetically unfavorable relative to the separate atoms. Indeed, this is the reason that none of the noble gases occurs as a diatomic molecule. It would cost energy and there is no profit in it.

Now, consider the hydrogen molecule again. What are the consequences of combining a **hydrogen atom** with a hydrogen ion—H with H^+? In this case, the bonding orbital would only have one electron in it, but this enough to cause the molecule to form. The result is H_2^+, which is a real, experimentally observable molecule containing a half bond between the hydrogens. The energy of this bond is about one half that of the hydrogen-hydrogen interaction and the bond length is 106 pm versus the 74.2 pm of H_2, as would be expected for a weaker interaction. Similarly, it is possible to combine H^- with H to give H_2^-. Again, there is a half bond between the atoms, but in this case it is composed of two electrons in the bonding orbital and one in the anti-bonding. The bond order is defined as one half of the difference between the number of bonding and anti-bonding electrons.

Consider a slightly more complicated molecule, dilithium. It does exist, although it is unlikely to ever be used to **power** a starship, such as Star Trek's *Enterprise*. In fact, it is a rather ordinary molecule. In this case, each lithium atom contributes two orbitals to the molecule, a 1s and a 2s orbital. And much as you would expect, these result in bonding and anti-bonding combinations. The 1s orbitals overlap with each other and the 2s orbitals do likewise. The 1s orbital on one atom doesn't interact with the 2s on the other due to the relative energy differences. The 2s orbital is much higher in energy and inaccessible to the 1s orbital. Dilithium, therefore, has two bonding combinations and two anti- bonding combinations and with three electrons contributed from each atom, the net result is a single bond (i.e. one half of four minus two).

So far, as described, molecular orbital theory hasn't really differentiated itself from the Lewis model. Bonds are still the result of electron pairs residing in bonding molecular orbitals. It explains why simply adding more electrons and protons doesn't result in massive bonding. The anti-bonding orbitals insure that the bond order never gets too high. But this can be explained in a Lewis approach using lone pairs. What does molecular orbital theory tell us that makes chemists believe that it is a good model for bonds? Consider the p-orbitals. Unlike the s- orbitals, which are spherically symmetric, the p- orbitals have direction. They orient along the x, y, and z axes of a Cartesian coordinate system. If a line joining the two nuclei is defined as the z- axis, then it is fairly easy to see that the P_z orbitals will be pointing at each other. The result is a bonding and anti- bonding combination, usually designated σ_p and σ_p^*, respectively, to distinguish them from the σ_s and σ_s^* orbitals that are generated from the s- orbitals. The σ means that the orbital is spherically symmetric around the bond axis, while the asterisk is used to designate anti-bonding orbitals.

But what about the interaction of the P_x and P_y orbitals? Their interaction is much weaker because of the distance separating them, but they do overlap with their opposite partner to give a bonding and anti-bonding combination. These combinations are designated π_p and π_p^* where the symbol π means that they have a nodal plane that includes the bond axis. Viewed down the bond axis, σ-bonds look like s- orbitals and π-bonds look like p-orbitals.

Since P_x and P_y are equivalent, they have the same energy, but they don't interact with each other because of their spatial orientation. The result is that there are two sets of π_p and π_p^* orbitals created with the same energy. Furthermore, since the p-orbital overlap isn't as strong as the s-orbital overlap, the resulting difference in energy between the bonding and antibonding orbital is not as large. The result is that π_p is filled after σ_p, but π_p^* is filled before σ_p^*. That is, the filling order for a molecule such as **oxygen** is: σ_{1s}, σ_{1s}^*, σ_{2s}, σ_{2s}^*, σ_{2p}, $2(\pi_p)$, $2(pi_p^*)$, σ_{2p}^*. With each oxygen contributing eight electrons, the highest occupied molecular orbital (HOMO) is the antibonding π interaction and since there are two degenerate orbitals, each orbital gets one electron. This results in oxygen being a "diradical" and paramagnetic, both of which properties can be confirmed experimentally. This result is explained only by molecular orbital theory.

The HOMOs of oxygen could also be designated as SOMOs or singly occupied molecular orbitals. The next orbital up in energy, which is left unoccupied, is the LUMO or lowest unoccupied molecular orbital. Collectively, these are termed the "frontier orbitals", because it is at the frontier of the molecular orbitals that chemistry happens. For example, it is these orbitals that are involved when oxygen binds to hemoglobin or when it reacts with combustible materials.

Molecular orbital theory is the best explanation of molecular bonding that chemists have. In its simplest form, it provides qualitative explanations for the formation and reactivity of molecules. It can also be used with rigorous calculations to provide the **energy level** diagrams for molecules that qualitatively explain the results of photoelectron spectroscopy. In addition, it can also lead to daunting calculations as each atom contributes its orbitals to the whole molecule. While this explains such phenomena as the hyperfine and superhyperfine coupling observed for the magnetic resonance spectroscopies, it does destroy the intuitive picture of one bond/two electrons that was originally used to describe bonding. In the end, though, the advantages of molecular orbital theory vastly outweigh the disadvantages.

MOLECULAR STRUCTURE

The notion that molecules might have specific geometric structures arose in the second half of the nineteenth century. Prior to that time, much debate existed among chemists as to whether molecules had definite shapes that could be determined experimentally. For example, French chemist Charles Gerhardt (1816-1856) emphasized his belief that they did not by writing different formulas for the same compound. He wrote BaO, SO_3, BaS, O_4, and BaO_2, SO_2, for barium sulfate.

Friedrick Kekulé was one of the first chemists to attack the problem of molecular structure. In 1857, he suggested that the **carbon atom** was tetravalent, that is, it could bond with four other atoms. He developed the tool of structural formulas to illustrate this concept, although the formulas created by Archibald Couper (1831-1892) at about the same time were far simpler and more efficient to use than were those of Kekulé. Neither Kekulé nor Couper carried their analysis of molecular architecture much beyond the two-dimensional diagrams that could be drawn on paper. Couper, for example, showed the bonds on a carbon atom directed at the four corners of a square.

In 1874, Jacobus Van't Hoff and Joseph Le Bel (1847-1930) independently developed a three-dimensional concept of the carbon atom. They proposed that the four carbon bonds were directed towards the corners of a tetrahedron. When they constructed models of organic compounds using tetrahedral atoms, they were able to explain a number of phenomena, including the existence of optical isomer s, first discovered by Louis Pasteur in 1848. Other chemists were unimpressed by the ideas of Van't Hoff and Le Bel. Adolph Kolbe, for example, doubted the reality of atoms, molecules, and chemical bonds, and warned that thinking of them in concrete, structural terms was carrying theory too far.

The great German chemist Hermann Helmholtz expressed similar concerns. The work of Van't Hoff and Le Bel was justified, however, because of its successes in explaining many physical phenomena. Other chemists began to explore other consequences of molecular structure. The technique soon had spectacular success in the field of organic **chemistry**, where many puzzling experimental results were eventually explained. Attempts to describe the molecular structure of inorganic compounds were less successful. In fact, there was limited progress in this field in the twenty years following the work of Van't Hoff and Le Bel.

At first, structures for simple compounds, such as binary salts, could be drawn. But a number of more complex compounds escaped explanation. Then, in 1893, German chemist Alfred Werner successfully applied structural theory to inorganic compounds. Waking early one morning, Werner had arrived at a fully formed solution in his sleep. He began writing what turned out to be his most important scientific paper on the spot and had finished it by the next afternoon. His theory of *coordination compounds* showed that the structure of molecules could be understood in terms of geometry, rather than simply in terms of chemical bonds between atoms. In working with organometallic compounds, for example, he placed the metal atom at the center of a geometric figure (such as a cube) and surrounded it with other atoms, **ions**, and groups of atom at the corners of the figure. He suggested that the metallic atom was attached to the surrounding groups by "secondary valences." Werner's suggestion was enormously fruitful. He used the theory, for example, to predict the existence of geometric isomers, two forms of a compound that differ from each other only in the position of a single atom, ion, or group of atoms. In 1907, he synthesized a pair of geometric isomers that confirmed his predictions.

The theory of molecular structure reached its highest development in the work of Linus Pauling in the 1920s and 1930s. Pauling applied the theory of quantum **mechanics** to electrons and showed how the formation of chemical bonds resulted in more stable configurations than existed in free atoms. Pauling's work not only provided a clearer understanding of the chemical bond, but also explained the structures of molecules. He demonstrated that the most stable configuration for some molecules was some intermediary structure between two other structures. The bonds in a benzene molecule, for example, are neither pure single nor pure double bonds, but some kind of "compromise" that results from the shifting of electrons back and forth between the two possibilities. This theory of *resonance* explained a number of phenomena that had previously been a mystery.

In the 1940s, Pauling applied these ideas to the complex molecules that make up living systems, especially proteins and nucleic acids. One approach he found highly productive was model-building. Beginning with tinker-toy-type equipment, he constructed molecular structures that seemed reasonable based on available experimental evidence. Then he refined the models until he could produce a structure that accounted for all existing data. In the early 1950s, he used this technique to demonstrate how polypeptides could adopt a helical structure. He very nearly achieved similar success in unraveling the structure of nucleic acids. This work—using the physical structure of chemical molecules to explain biological phenomena—led to the development of the science now known as molecular biology.

Computer technology has improved to the point where it can be used to construct graphic models of molecular structures. This has led to the development of a new science known as computational chemistry. In this area of chemistry, scientists input data about compounds such as molecular composition or chemical behavior, and the computer displays a picture of the molecular structure. This technology will be an important part of the development of synthetic drugs and new catalysts.

MOLECULAR WEIGHT

The molecular **weight** of a molecule indicates how heavy that molecule is relative to an **atom** of **carbon** (with six protons and six neutrons). Saying that the molecular weight of water is 18 means that the water molecule is 18/12 as heavy as a carbon-12 atom, or 18/16 as heavy as an oxygen-16 atom. In general the molecular weight of a molecule is the sum of the atomic weights of its constituent atoms.

The gram molecular weight (GMW) of a substance is defined as the weight in grams of one mole (6×10^{23} molecules, known as Avogadro's number). The molecular weight of water is 18.02; its gram molecular weight is 18.02 g. Given the gram molecular weight of a substance, the weight of an individual molecule can be calculated by dividing that weight by **Avogadro's number**.

The first direct approach to determining molecular weight was proposed by two French scientists in 1819, Pierre Louis Dulong (1785-1838) and Alexis-Thérèse Petit (1791-1820). They suggested that the amount of **heat** required to raise the **temperature** of an atom of a solid material by a given amount should be independent of the type of atom, with the result that the gram atomic weight should be inversely proportional to the material's specific heat. Although the law of Dulong and Petit proved a fair approximation for many elements, it was far from exact for many others, and it was not at all helpful in determining the atomic weights of gaseous elements.

In 1811, the Italian physicist Lorenzo Romano Amadeo Carlo Avogadro di Quaregua e di Cerreto, known to posterity as **Avogadro**, concluded that equal volumes of all gases at the same temperature and **pressure** contain the same number of molecules. Unfortunately, Avogadro's ideas had little influence on the work of his contemporaries, and it was not until 1860 that another Italian scientist, Stanislao Cannizzaro, pointed out that Avogadro's hypothesis could be used as a basis for determining atomic and molecular weights.

The classical method of determining the molecular weight of a gas is to use **density** data. Experimentally, a few milliliters of a volatile liquid are placed in a stoppered flask containing a small orifice. The flask is then heated to a temperature above the **boiling** point of the liquid. As the liquid evaporates, its vapor replaces the air in the flask. The flask is then allowed to cool, and the vapor condenses as air re-enters the flask. The **mass** of the vapor is then calculated, and used in conjunction with the **ideal gas law** to determine the molecular weight of the liquid. This method works quite well for many gases and volatile liquids, but it cannot be used for substances that decompose on heating, such as urea.

The molecular weights of such substances have been determined by measuring those physical properties of solutions that depend primarily on the concentrations of solute particles and their molecular weight. Such properties include the lowering of vapor pressure, the elevation of boiling point, and the lowering of freezing point and osmotic pressure. Freezing point depressions have often been used because they are comparatively easy effects to observe. Osmotic pressure measurements have been used for solutes of high molecular weight.

Modern methods for measuring molecular weight include chromatographic methods and **mass spectrometry**.

MOLECULES AND MOLECULAR FORCES

Matter consists of atoms and molecules. Atoms are the smallest units of matter that retain the physical and chemical prop-

Two models that represent the arrangement of atoms in a chemical molecule. At the center is a ball and stick model with the balls representing the atoms and the sticks representing the bonds that hold them together. The cluster models are similar representations except that the chemical bonds are not represented. *(Photo courtesty of Tek Image, Science Source/Photo Researchers. Reproduced by permission.)*

erties of individual elements. Molecules, however, consist of two or more atoms chemically bonded together. Molecules generally display characteristics that are distinct from their constituent atoms. For instance, water, a compound whose molecules consist of two hydrogen atoms bonded to one **oxygen atom**, has drastically different physical and chemical properties than either hydrogen or oxygen alone. A molecule, then, is the smallest unit of a compound that retains the properties of the compound.

Ionic bonds are bonds between **ions** that attracted to each other because of their opposite electrical charges. The **electron** imbalance in ions can create negative or positive (e.g., an excess of electrons over protons creates a negative ion). Opposite ions attract one another by electrical **force** to create an ionic bond. Table salt, or sodium chloride, is a familiar example of ionic bonding. A second and stronger type of intramoleclar force is covalent bonding. Covalent bonds

between atoms within a molecule involve the sharing of electrons among nuclei. Covalent bonds are strong intramoleular atomic forces that bond atoms together to form molecules.

Intermolecular forces are those of attraction between separate molecules. Generally, intermolecular forces are weaker than ionic and covalent bonds, but nonetheless are important forces influencing compounds. Hydrogen bonds are intermolecular forces between molecules containing hydrogen, and molecules containing highly electronegative elements such as oxygen, sulfur, or nitrogen. The hydrogen of one molecule, itself bonded to an electronegative element, is attracted to an unshared pair of electrons of another electronegative element of a second molecule. Individually weak, but in numbers quite strong, hydrogen bonds are intermolecular forces that have important influences. For instance, hydrogen bonding between hydrogen and oxygen components of water molecules affects the melting and **boiling** points of the compound, as well as determining its cohesive and adhesive properties. The hydrogen bonding of water makes life as it is presently known to exist possible by influencing the properties of water. Also, hydrogen bonding is an essential molecular force in the structure of DNA. Hydrogen bonding holds the single strands together to form the characteristic double helix of DNA.

Dipole-dipole interactions are a second type of intermolecular force. Dipole-dipole interactions are less intense versions of hydrogen bonding. Some molecules have regions that are slightly more or slightly less negatively charged than other regions. Such partial charge differences are called dipoles. Opposite charged dipoles are attracted to one another. Thus dipole-dipole interactions are weak electrical forces between molecules. Because they are weak attractions, dipole-dipole intermolecular forces are significant only in closely associated molecules, such as cellular protein receptors and their ligands. One result of polarity in molecules is that found in hydrophobic and hydrophillic interactions. For example, fatty, hydrocarbon molecules that are insoluble in water are said to be hydrophobic. In contrast, charged molecules displaying dipoles are readily soluble in water, and are thus hydrophilic. When hydrophobic molecules, or molecules with hydrophobic regions, are placed in aqueous solution, the attractive forces among charged particles actively exclude the non-charged hydrophobic molecules. As a result, the fatty molecules are pushed together. The resulting intermolecular force pushing like-molecules together is called hydrophobic interaction. When oil and water are mixed, for example, the oil eventually separates out into a layer. The water-insoluble oil molecules are pushed together by hydrophobic interactions. Such intermolecular forces are important in the structure of complex biological molecules such as membranes, proteins, and the configuration of DNA. Combinations of intermolecular forces often determine the overall behavior of molecules.

Momentum

The momentum of an object is the **mass** of the object multiplied by the **velocity** of the object. The mass will often be measured in kilograms (kg) and the velocity, in meters per second (m/s), so the momentum will be measured in kilogram meters per second (kg m/s). Because velocity is a vector quantity, meaning that the direction is part of the quantity, momentum is also a vector. Just like the velocity, to completely specify the momentum of an object one must also give the direction.

A **force** multiplied by the length of time that the force acts is called the impulse. According to the impulse momentum theorem, the impulse acting on an object is equal to the change in the object's momentum. Notice the word change. The impulse is not equal to the object's momentum, but the amount the momentum changes. (This impulse momentum theorem is basically a disguised form of Newton's second law.) The force used to figure out the impulse here is the total

sum of all the external forces acting on an object. Internal forces acting within an object don't count.

The consequences of this impulse momentum theorem are rather profound. If there are no external forces acting on an object, then the impulse (force times time) is zero. The change in momentum is also zero because it is equal to the force. Hence, if an object has no external forces acting on it, the momentum of the object can never change. This law is the law of conservation of momentum. There are no known exceptions to this fundamental law of physics. Like other conservation laws (such as **conservation of energy**), the law of conservation of momentum is a very powerful tool for understanding the **universe** around us.

Rockets provide a dramatic application of **momentum conservation**. Before launch, the rocket is sitting on the launch pad, so its momentum is zero. When the engines fire, the spent fuel is hurled out the back, by an internal, not external, force. The total momentum of the rocket and fuel must remain zero. The fuel has a negative momentum because it is spit out backwards. The rocket must therefore have an equal positive momentum, so the total will be zero. (Remember momentum is a vector; direction matters.) So, the rocket moves forward. The greater the momentum of the fuel spit out backwards, the greater the forward thrust of the rocket. You can test this yourself. Many toy stores sell plastic water rockets that use air **pressure** to push the water out the back. Try one of these rockets with water and with air only. The dramatic difference is because the water has more mass and therefore more momentum.

Momentum and Newton's second law

English physicist and mathematician Sir **Isaac Newton**'s second law, usually written as F = ma where F is the **force**, m is the **mass** of the object and a is its **acceleration**, states that the acceleration of a body is proportional to the force applied to it and inversely proportional to its mass. Although the second law is usually written in this form it is not the form that Newton originally used in his *Philosophiae Naturalis Principia Mathematica (Mathematical Principles of Natural Philosophy)* nor can it be used in situations where the mass of the object may be changing. The original form was written in terms of **momentum** and is currently expressed as F = dp/dt where F is the force, dp is the change in momentum and dt is the time interval. Although Newton expressed his second law of **motion** in terms of momentum rather than acceleration there was no mention of the rate of change of momentum. This is because of a deep confusion that existed in the 17th century over the distinction between impulse force and continuous force. As Newton's second law was written it refers to an impulsive force. Newton he eventually used it for continuous forces applied over a finite length of time and the application of the second law evolved into its modern form.

The original formulation of Newton's second law came about in a series of steps. First Newton defined the quantity of

matter, mass, as an irreducible term arising from **density** and volume. The mass is the density multiplied by the volume of the object. Then Newton defined the quantity of motion, momentum, as an irreducible term arising from a measure of **velocity** and quantity of matter. Momentum is defined as the product of velocity and mass. In its original formulation, as Newton stated in *Principia*: the change of motion is proportional to the motive force impressed; and is made in the direction of the right line in which the force is impressed. When time is included, for continuous forces, the law states that the net force is equal to the rate of change of **linear momentum** with time. This is actually a more general form of Newton's second law which is true even in relativistic situations and in situations where mass may be changing.

The form of Newton's second law as it is usually written today is usually difficult for students to understand as the mass is usually taken as the object's **weight**. But the weight is actually caused by the mass of the object and **gravity**, which is in itself a force. Austrian physicist **Ernst Mach** explained that Newton's laws can be thought of as a single law: "When two compact objects act on each other, they accelerate in opposite directions, and the ratio of their accelerations is always the same." Stated in this way force and mass are removed and it is thought of only as acceleration of the two bodies.

Newton's second law, in conjunction with his first and third laws, is used to both qualitatively and quantitatively describe motion. It can be used with extended objects to describe the motion of the **center of mass** and with two body problems by employing their reduced masses. The formulation of the second law that employs momentum is more broadly useful since it takes into account the possibility of change in the object's mass.

See also Force and the law of motion; Force, mass, and acceleration; Newton's laws of motion; Newtonian physics

MOMENTUM CONSERVATION

Conservation of **momentum** is a fundamental conservation law of physics. It states that the momentum of a system is constant if there are no external forces acting on the system.

In physics, conservation laws state that, in a closed system, the total quantity of momentum of the system remains constant; that is, its rate of change is zero. Mathematically, this is expressed as $dX/dt = 0$ for a scalar quantity X, and as div F = 0 for a vector quantity F. The most well-known conservation law is the law of **conservation of energy**, which states that the total **energy** of a closed system remains constant during a transformation. This is the first law of **thermodynamics**, and can also be stated as the total energy of a closed system is equal to the sum of its potential and kinetic energies. Another conservation law, for which no exception has yet been recorded, is that the total **electric charge** of a closed system remains constant. Another example is the law of conservation of mass-energy expressed by German-American physicist **Albert Einstein** as $E = mc^2$ in his special **relativity theory**,

which states that the total energy of a particle (E) has a **mass** (m) equal to E/c^2, or the total mass of a particle has an energy equal to mc^2.

Conservation of momentum is implicit in English physicist and mathematician Sir **Isaac Newton**'s first law of **motion**, the law of **inertia**, which holds that a body at constant **velocity** (including a velocity of zero relative to a particular reference frame) will remain at constant velocity unless acted upon by an external **force**. Conservation of momentum applies to both linear and **angular momentum**. For translational or straight-line motion, the **linear momentum** vector is defined as the product of the mass of a particle and its velocity: $p = mv$. For rotational motion, the angular momentum vector is defined as the product of the moment of inertia of the particle and its angular velocity: $L = I\omega$. Conservation of linear momentum states that the total linear momentum (p) of a closed system is always conserved when there is no net acting force and conservation of angular momentum similarly states that the total angular momentum (L) of a closed system is always conserved when there is no net **torque**. This is also implicit in Newton's third law, or force-pair law, which states that whenever a body exerts a force on another body, the latter exerts a force of equal magnitude and opposite direction on the former. Therefore, the rate of change of momentum of the first body is equal and opposite to that of the second body, and the total rate of change of momentum in the system is zero, which is equivalent to stating that the momentum of the system is conserved.

Conservation of angular momentum is the underlying principle of the gyroscope, an instrument consisting of a rapidly spinning wheel set in a framework allowing the wheel to tilt in any direction which prevents the occurrence of a net torque. Thus, the axis of rotation points in a fixed direction, and the angular velocity and momentum does not change. No matter how the gyroscope is turned, it will always point in the same direction because no torque is exerted to change it.

See also Conservation of heat; Conservation of mass; Dynamics; Energy-momentum, four-vector; Force and the laws of motion; Inertia and rotational kinetic energy; Kepler's second law and angular momentum; Kinetic energy; Momentum and Newton's second law; Motion of particles; Newton's laws of motion; Newtonian physics; Potential energy; Scalar fields; Relativity, special; Vectors

MOMENTUM MEASUREMENT

Momentum is a property of **motion** that is classical physics is a vector (directional) quantity that in closed systems is conserved during collisions. In **Newtonian physics** momentum is measured as the product of the **mass** and component **velocity** of a body. For massless particles (e.g., photons) moving at the speed of **light** ($v = c$) the momentum (p) is equal to **Planck's constant** divided by the **wavelength**.

The first formal definitions and measurement of momentum date to the writing of French philosopher **René Descartes** (1596-1650). Descartes intended momentum to a quantifiable and measurable concept related to what he termed the "amount of motion".

Measurement of momentum often concentrates on rates of change in the momentum of bodies. In accord with the law of **inertia**, a body with no net **force** acting upon it experiences no change in momentum and therefore measurement of momentum reflect that momentum is conserved. Whenever a net force is applied to a body the change in momentum is proportional to the force applied but the conservation of momentum dictates that the momentum of the agent applying force to the body must correspondingly decrease so that the measured momentum of the combined systems remains unchanged.

Modern device used to measure momentum of **subatomic particles** often employ tracking devices located in strong magnetic fields. The paths of particles moving through these fields reveals their charge and momentum. The direction of deflection reveals the particles change and the momenta of particles can be calculated from the fact that the paths of particles with greater momentum deviate less than those of lesser momentum (i.e., those particles with higher momentum tend to travel along straighter or less bent paths).

Quantum theory dictates that the measurement of certain pairs of properties of particles, including position and momentum are limited by the Heisenberg uncertainty principle first advanced by German physicist **Werner Heisenberg**. In essence, although it is possible to measure either position or momentum they pair can not be measured simultaneously. The more exact the determination of position, the more uncertain becomes the measurement of momentum.

Although the uncertainty principle isn't relevant to the measurement of momentum of large objects, it places severe constraints on measurements of momentum of subatomic particles. Accordingly, quantum theory places a limitation on the experimental measurement of momentum. The more accuracy required in the determination of position, the less the accuracy possible with regard to the determination of momentum. For example, in attempting to make an accurate determination of the position of an **electron** it is necessary to bombard the electron with photons. In doing so the collisions between the photons and the electron alter the momentum of the electron and therefore introduce uncertainty in the measurement of the momentum of the electron.

Moreover, there are important philosophical ramifications to the measurement of momentum, In the Copenhagen interpretation of **quantum mechanics**, reality is dependent upon the observer's measurement. Essentially, the Copenhagen interpretation dictates that in the measurement of momentum or position of a two particle system, the measurement of momentum or position of one particle gives reality (a quantifiable value) to the momentum or position of the second particle. In this theoretical interpretation conflicting realities result when there is an attempt to measure the momentum of one particle and the position of the other. Because time-ordering of the measurements is dependent upon on the inertial frame varying **reference frames** yield differing realities and give rise to a problem in nonlocality problem related to the instantaneous propagation of information related to the measurement across and real **space**.

See also Momentum and Newton's second law; Momentum conservation

MOON

The Moon is the only natural satellite of the Earth. It has a diameter of 2,158.9 mi (3,475.9 km), lies at a mean distance from the Earth of 238,760 mi (384,404 km), and completes one orbit around the Earth in 27.32166 days. The Moon maintains approximately the same side (near side) toward the Earth by synchronously rotating on its axis with respect to the **stars** in the same average period that it revolves about the Earth. Still, the Moon's orbit (often called the lunar orbit) is constantly changing because of perturbations, with the largest perturbation caused by the gravitational attraction of the **Sun**. In some ways, the Sun has as much of an influence on the Moon as the Earth does, and because of this some scientists refer to the Earth and Moon as a double-planet system rather than a planet-satellite system.

The **mass** of the Moon is about 81.3 times less than the Earth's mass, with a value of 5.04×10^{21} slug (7.35×10^{22} kg). The Moon's average **density** is about three times that of water, similar to the Earth's crust. The volume of the Moon is about one forty-ninth that of the Earth's volume. Its surface **gravity** is one sixth that of the surface gravity on Earth. The lunar interior structure is currently believed to be composed of the major layers: crust, mantle, and core. The rocky "crust" averages 42 mi (68 km) in thickness, from a surface depth of 0-66 mi (0-107 km). A current model shows a partially molten "mantle" ranging from 66-868 mi (107-1,398 km) deep. Although not known for sure, lunar geologists believe that the Moon may have a small iron "core" ranging from a depth of 868-1,080 mi (1,398-1,738 km).

Earth-based observations have revealed much lunar information in the past, but more recently satellites and spacecraft have extensively photographed, measured, and sampled the Moon. Data from the Apollo program indicate that the Moon was formed about 4.6 billion years and was strongly heated during its first few hundred million years, causing melting and formation of igneous rock in its outer layers.

The Moon has an almost nonexistent atmosphere with only traces of the gases **helium**, neon, argon, and radon. Evidence from the spacecraft *Clementine* in 1994 suggest that frozen water might occur in deep craters near the Moon's south pole. The location of enhanced areas of hydrogen was found at both lunar poles by a **neutron** spectrometer on board the *Lunar Prospector* probe in 1999. It was inferred that significant amounts of water ice are trapped in permanently shadowed craters. The lunar magnetic field is less than 10^{-5} gauss (essentially zero). With this lack of atmosphere and the near absence of a global magnetic field, the Moon is exposed to extreme temperatures from +88°F to -140°F (+100°C to

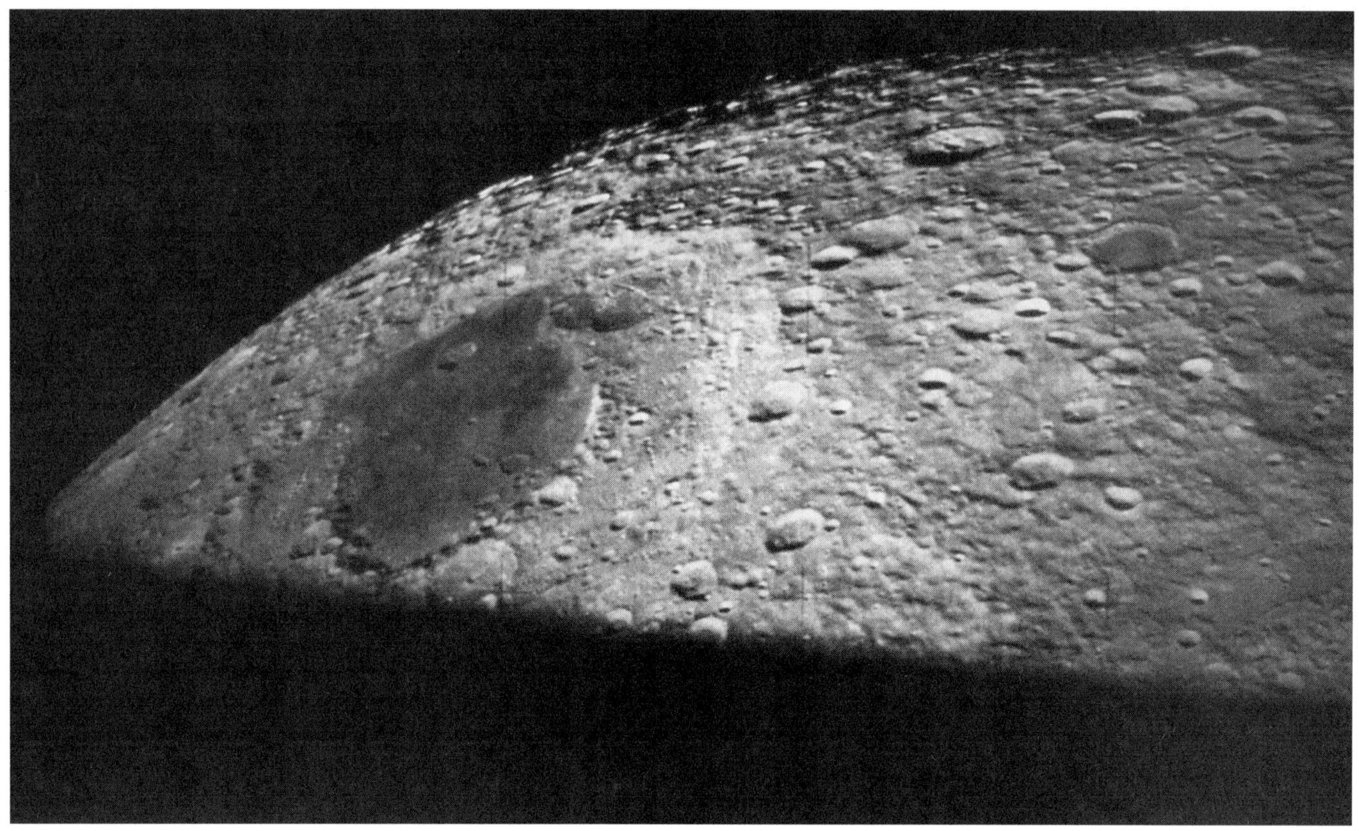

Surface of moon. *(Photo courtesy of NASA. Reproduced by permission.)*

-173°C), and the solar wind, **cosmic rays**, meteorites, and micrometeorites easily bombard its surface.

Because of the Moon's lack of an atmosphere, many features on its surface have remained unchanged for billions of years, showing a clear picture of its origin and primordial development. Most of the lunar surface consists of old, gray colored, heavily cratered highlands—craters caused by extralunar objects are the Moon's primary form of "weathering." They are primarily covered with regolith, a mixture of fine dust and rocky debris produced by foreign impacts. The craters, probably created from small asteroids and **comets** that hit the Moon, range in size from microscopic pits to the largest with a diameter of around 600 mi (1,000 km). Large impact basins and other depressions are believed to have been flooded about 3.8-3.2 billion years ago by basaltic lavas generated beneath the surface. This action formed the dark, relatively smooth and fairly young, maria plains (or lowlands), that comprises 16% of the surface, primarily on the near side. Few geological processes occur currently on the Moon, as there is virtually no atmosphere or water to permit erosive processes. The rate and magnitude of meteorite impact on the Moon is low today, but heavy cratering indicates that the rate was high in the first half-billion years of its history.

Prior to the study of Apollo samples there was no consensus about the origin of the Moon. Popular theories included: (1) the "co-accretion theory" where the Moon and the Earth formed at the same time from the solar nebula, (2) the "capture theory" where the Moon formed elsewhere and was eventually captured by the Earth, and (3) the "impact theory" where the Earth collided with a large object—possibly a comet, an asteroid, or some other large astronomical object—and the Moon was formed from the ejected material. Information learned from the Moon rocks returned by the Apollo missions, along with more recent computer simulations, has led to a wide acceptance of the "impact theory" as the current explanation of the Moon's formation.

See also Earth (planet)

MORLEY, EDWARD WILLIAMS (1838-1923)

American chemist and physicist

Originally trained for the ministry, Edward Williams Morley decided instead in 1868 to pursue a career in science, the other great love of his life. At first he devoted himself primarily to teaching, but gradually became engaged in more original research. His work can be divided into three major categories: the first two involved the determination of the **oxygen** content of the atmosphere and efforts to evaluate Prout's hypothesis; his third field of research involving experiments on the **veloc-**

ity of **light** with **Albert Michelson**, are those for which his name will always be most famous.

Morley was born in Newark, New Jersey, on January 29, 1838. His mother was the former Anna Clarissa Treat, a schoolteacher, and his father was Sardis Brewster Morley, a Congregational minister. According to a biographical sketch of Morley in the December 1987 issue of the *Physics Teacher,* the Morley family had come to the United States in early Colonial days and "was noted for its deep patriotism and religious devotion."

Morley's early education took place entirely at home, and he first entered a formal classroom at the age of nineteen when he was admitted to Williams College in Williamston, Massachusetts, as a sophomore. His plans were to study for the ministry and to follow his father in a religious vocation. But he also took a variety of courses in science and mathematics, including astronomy, chemistry, calculus, and **optics**. His courses at Williams were a continuation of an interest in science that he had developed at home as a young boy.

Morley graduated from Williams as valedictorian of his class with a bachelor of arts degree in 1860. He then stayed on for a year to do astronomical research with Albert Hopkins. Morley's biographers allude to the "careful" and "precise" calculations required in this work as typical of the kind of research Morley most enjoyed doing. Certainly the later research for which he is best known was also characterized by this high level of precision and accuracy.

After completing his work with Hopkins, Morley entered the Andover Theological Seminary to complete his preparation for the ministry, while concurrently earning his master's degree from Williams. Morley graduated from Andover in 1864 but rather than finding a church he took a job at the Sanitary Commission at Fortress Monroe in Virginia. There he worked with Union soldiers wounded in the Civil War.

His work completed at Fortress Monroe, Morley returned to Andover for a year and then, failing to find a ministerial position, took a job teaching science at the South Berkshire Academy in Marlboro, Massachusetts. It was at Marlboro that Morley met his future wife, Isabella Ashley Birdsall. The couple was married on December 24, 1868. They had no children.

Morley had finally received an offer in September, 1868, to become minister at the Congregational church in Twinsburg, Ohio. He accepted the offer but, according to biographers David D. Skwire and Laurence J. Badar in the *Physics Teacher,* became disenchanted with "the low salary and rustic atmosphere" at Twinsburg and quickly made a crucial decision: he would leave the ministry and devote his life to science.

The opportunity to make such a change had presented itself shortly after Morley arrived in Twinsburg when he was offered a job teaching chemistry, botany, geology, and mineralogy at Western Reserve College in Hudson, Ohio. Morley accepted the job and when Western Reserve was moved to Cleveland in 1882, Morley went also. While still in Hudson, Morley was assigned to a full teaching load, but still managed to carry out his first major research project. That project involved a test of the so-called Loomis hypothesis that during periods of high atmospheric **pressure** air is carried from upper parts of the atmosphere to the Earth's surface. Morley made precise measurements of the oxygen content in air for 110 consecutive days, and his results appeared to confirm the theory.

In Cleveland, Morley became involved in two important research studies almost simultaneously. The first was an effort at obtaining a precise value for the atomic **weight** of oxygen, in order to evaluate a well-known hypothesis proposed by the English chemist William Prout in 1815. Prout had suggested that all atoms are constructed of various combinations of hydrogen atoms.

Morley (as well as many other scientists) reasoned that should this hypothesis by true, the atomic weight of oxygen (and other elements) must be some integral multiple of that of hydrogen. For more than a decade, Morley carried out very precise measurement of the ratios in which oxygen and hydrogen combine and of the densities of the two gases. He reported in 1895 that the atomic weight of oxygen was 15.897, a result that he believed invalidated Prout's hypothesis.

Even better known than his oxygen research, however, was a line of study carried out by Morley in collaboration with Albert A. Michelson, professor of physics at the Case School of Applied Science, adjacent to Western Reserve's new campus in Cleveland. Morley and Michelson designed and carried out a series of experiments on the velocity of light. The most famous of those experiments were designed to test the hypothesis that light travels with different velocities depending on the direction in which it moves, a hypothesis required by current theories regarding the way light is transmitted through **space**. A positive result for that experiment was expected and would have confirmed existing beliefs that the transmission of light is made possible by an invisible "ether" that permeates all of space.

In 1886 Michelson and Morley published their report of what has become known as the most famous of all negative experiments. They found no difference in the velocity with which light travels, no matter what direction the observation is made. That result caused a dramatic and fundamental rethinking of many basic concepts in physics and provided a critical piece of data for **Albert Einstein**'s theory of relativity.

The research on oxygen and the velocity of light were the high points of Morley's scientific career. After many years of intense research, Morley's health began to deteriorate. To recover, he took a leave of absence from Western Reserve for a year in 1895 and traveled to Europe with his wife. When he returned to Cleveland, he found that his laboratory had been dismantled and some of his equipment had been destroyed. Although he remained at Western Reserve for another decade, he never again regained the enthusiasm for research that he had had before his vacation.

Morley died on February 24, 1923, in West Hartford, Connecticut, where he and Isabella had moved after his retirement in 1906; she predeceased him by only three weeks. Morley had been nominated for the Nobel Prize in chemistry in 1902 and received a number of other honors including the

Davy Medal of the Royal Society in 1907, the Elliot Cresson Medal of the Franklin Institute in 1912, and the Willard Gibbs Medal of the Chicago section of the American Chemical Society in 1917. He served as president of the American Association for the Advancement of Science in 1895 and of the American Chemical Society in 1899.

MOSELEY, HENRY GWYN JEFFREYS (1887-1915)

English physicist

Henry Gwyn Jeffreys Moseley's great contribution to science was his ordering of the elements by examining their x-ray spectra, thus creating the **periodic table**. His work, which ranks among the most profound discoveries of the twentieth century, was cut short when the scientist was killed during World War I. Although he never received a major award for his research, Moseley's accomplishments laid the groundwork for a number of later scientific developments.

Henry Gwyn Jeffreys Moseley (widely known as "Harry") was born on November 23, 1887, at Weymouth, England. He was the only son of Henry Nottidge Moseley, a professor of human and comparative anatomy at the University of Oxford. He had also served as a naturalist on the famous voyages of the *Challenger* in that vessel's studies of the world's oceans. Both sides of the family included a number of eminent scientists, including a paternal grandfather who was the first professor of natural philosophy at King's College, London, and a great grandfather who was an expert on tropical diseases.

Moseley's father died when Harry was four years old. The family then moved to the small town of Chilworth, in Surrey, where Harry and his two sisters began their education. Moseley's interest in science surfaced at a young age. He was fortunate in having the constant encouragement of his mother and friends in this interest which was to lead not only to a professional career in science, but also to a lifelong fascination with natural history.

In 1896, at the age of nine, Moseley was enrolled at the Summer Fields school, an institution that specialized in preparing boys for Eton. Five years later, he won a King's Scholarship that allowed him to enroll at Eton. At Eton, Moseley studied with T. C. Porter, one of the first scientists in England to work with **x rays**. Although Moseley studied a number of other subjects, x rays became the topic in which he was interested during his career in science.

After leaving Eton in 1906, Moseley was awarded a scholarship to continue his education at Trinity College, Oxford. Though he earned only second class honors in science, Moseley was able to get letters of recommendation that allowed him to take a position at the University of Manchester, where he worked with **Ernest Rutherford**. His assignment at Manchester involved a full teaching load, but he still found time to carry out an ambitious program of original research.

At the conclusion of his first year at Manchester, Moseley was relieved of his teaching responsibilities and allowed to devote all his time to research. The topic he selected for that research was the **diffraction** of x rays, a phenomenon that had just been discovered by the German physicist **Max Laue**. For a period of some months, Moseley collaborated with C. G. Darwin, a mathematical physicist, on a study of the general characteristics of diffracted x rays. In the fall of 1913, however, the two men went their separate ways, and Moseley began to focus on the nature of the spectra produced by scattered x rays.

Moseley saw that when x rays are beamed at certain crystalline materials, they are diffracted by atoms within the crystals, forming a continuous spectrum on which is superimposed a series of bright lines. The number and location of these lines are characteristic of the element or elements being studied. Much of the basic research on x-ray spectroscopy had been done in England by **William Lawrence Bragg**, with whom Moseley studied for a short period of time at the University of Leeds.

In his own research, Moseley devised a system that allowed him to study the x-ray diffraction pattern produced by one element after another in an orderly and efficient arrangement. Very quickly, he found that the frequencies of one set of **spectral lines**, the "K" lines, differed from element to element in a very consistent and orderly way. That is, when the elements were arranged in ascending order according to their atomic masses, the **frequency** of their K spectral lines differed from each other by a factor of one. To Moseley, the meaning of these results was clear. Some property inherent in the structure of atoms was responsible for the regular, integral change he observed. That property, he decided, was the charge on the **nucleus**. When the elements are arranged in ascending order to their atomic masses, he pointed out, they are also arranged in ascending order according to their nuclear charge. The main difference is that the variation in atomic masses between adjacent elements is never consistent, whereas the variation in nuclear charge is always precisely one. This property is such a clear defining characteristic of atoms and elements that it was given the special name of atomic number. Of all the properties of an **atom**, atomic number has come to be the single most important characteristic by which an atom can be recognized.

The implications of Moseley's discovery were manifold and profound. In the first place, the concept of a unique, identifying and characteristic number of an element—its nuclear charge—provided a new basis for the periodic table. The Russian chemist Dmitri Ivanovich Mendeleev's original proposal for arranging the elements on the basis of their atomic masses had worked extraordinarily well, but it did have its flaws. Moseley found, however, that the discrepancies found in Mendeleev's proposal disappeared when the elements were arranged according to their atomic numbers. It was obvious, therefore, that atomic number was an even more fundamental property of atoms and elements than was atomic **mass**.

Moseley's research also allowed him to predict the number and location of elements still missing from the periodic table. Since each element could be assigned its own unique atomic number (hydrogen=1, helium=2, lithium=3, beryllium=4, and so on), any missing numbers must mean that an element remains to be discovered for this particular position in the periodic table. Based on that logic, Moseley was able to predict with confidence the existence of an as-yet-undiscovered element #43, between molybdenum (#42) and ruthenium (#44) and another between neodymium (#60) and samarium (#62). Furthermore, he was able to predict the spectral pattern to be expected for each missing element and thereby provide a valuable tool in the search for those elements.

As his initial work on X-ray spectra was being concluded, Moseley decided to resign his position at Manchester and return to Oxford in the hope of securing a position as professor of experimental physics. At Oxford, Moseley continued to work on X-ray spectra, concentrating on the study of the relatively unknown group of elements known as the lanthanides.

In June of 1914, Moseley left for a meeting of the British Association for the Advancement of Science scheduled to be held in Australia. He received word while still aboard ship of the outbreak of World War I and decided to return to England immediately. He enlisted in the Royal Engineers and, after completing an eight-month training period, was sent to the Turkish battlefront at Gallipoli. There, during the battle of Sulva Bay on June 15, 1915, he was killed by a sniper's bullet. Moseley was never married and, in his short lifetime, received no great honors. His contributions to science, however, were enormous, in that many later advances in physics and chemistry were based on his work.

MÖSSBAUER EFFECT

Mössbauer effect is the recoil-free emission and resonant absorption of gamma (γ) rays from the nuclei of certain radioactive isotopes such as ^{57}Fe. These γ-rays, as emitted, do not lose **energy** from the recoil of the nuclei because the recoil **momentum** due to their emission or absorption is taken up by the whole bulky sample rather than by the emitting or absorbing **nucleus** itself. Under such conditions, the nuclei are bound rigidly in the lattice of a crystal, and the energy of recoil is negligible. Spectra based on Mössbauer effect have found numerous applications in studying the valence of an element in a chemical compound, the state of the irons and their electronic structure, as well as structural and magnetic properties of magnetic bulk and thin-film materials.

The Mössbauer effect was first reported by Rudolph Mössbauer in 1958. Three years later, he won the Nobel Prize with his discovery. Since then, it is believed that nuclear γ-ray emission and absorption process can take place in recoil-free fashion. In reality, of course we have both recoil and recoil-free events. Mössbauer also utilized the Doppler (**velocity**) shift to modulate the γ-ray energy so that Mössbauer effect could be developed into a **spectroscope** for material characterization. The emission of γ-rays with natural or nearly natural line width allows for observing in the γ-ray spectra the interaction between the nucleus and its **atom** in solids and viscous liquids.

A radioactive parent or Co, Ta, serves as the primary source of the excited nuclear state of the iron isotope $^{57}Fe*$. $^{57}Fe*$ decays rapidly and emits a γ-ray that strikes another ^{57}Fe or of the absorber or of the sample to be studied) in its ground state. If the energy of the γ-ray matches the ^{57}Fe-$^{57}Fe*$ energy separation, it will be absorbed via a resonant absorption process and ^{57}Fe is excited. Since we need to match the γ-ray **frequency** to the absorber energy levels, the energy of emitted γ-ray is modulated via the **Doppler effect** by oscillating the radioactive source. Therefore, absorption is observed as a function of relative velocities. The γ-radiation of interest is detected by a counter, and the data are processed by a multichannel analyzer in which each channel corresponds to a different velocity and a mechanism for synchronization is also installed. Certain elements and compounds have to be examined at cryogenic **temperature** such as -452°F (-269°C) to see the Mössbauer effect. The final results, or Mössbauer spectra, are usually plotted as curves of transmission (T) in % versus relative velocity and the transmitted intensity is a minimum at the velocity of exact resonance. These spectra consist of sharp lines due to recoil-free emission or absorption and of a featureless background due to recoil events and irrelevant **radiation**.

The transitions involved in the Mössbauer effect and the basic principle of Mössbauer spectroscopy, where the area enclosed by a dash box may be kept at cryogenic temperature.

In Mössbauer spectra, the resonance is shifted relative to the source frequency, generally called "isomer shift". Such a chemical shift is due to a difference in the s-electron **density** or the shielding of s-electrons between two cases. Therefore, a measure in the isomer shift provides us with the information on s-orbitals involved in the chemical bonding. In addition to the isomer shift, there are the second-order Doppler shift caused by temperature effects and the gravitational **red shift** caused by differences in the **potential energy** between the source and the absorber. Magnetic effects due to a net **electron** spin or an unpaired electron density, on the other hand, make the spectra often to be split into several lines. For a quadrupolar nucleus, under certain conditions or a change in the symmetry of the electron distribution, the energy of the nucleus exhibits a dependence on the nuclear orientation. This will then lead to the electric quadrupole splitting. When we look at Mössbauer spectra, the effect due to magnetic field and quadrupole splitting can be easily identified.

Based on the aforementioned mechanisms, Mössbauer spectroscopy has been applied successfully in all fields of natural science. For instance, Mössbauer effect study provides local information on both **magnetism** via the hyperfine interaction and structure via the electric quadrupole interaction in ferromagnetic films. It evaluates electromagnetic moments of excited states. That the differences in chemical bonding due to the change in valence state lead to variations of the occupation of the conduction band and the magnetic exchange coupling also manifests in the Mössbauer spectra. When a small amount of impurities is introduced in order to improve the magnetic properties and chemical stability of the material, Mössbauer spectroscopy can help find at which crystallographic sites of the matrix material the impurity atoms are located. Since iron occurs in a variety of biologically important compounds or heme proteins, Mössbauer spectroscopy also has found use in molecular biology and clinical diagnosis.

MÖSSBAUER, RUDOLF LUDWIG (1929-   )
German physicist

As a German graduate student, Rudolf L. Mössbauer discovered an unusual effect concerning how gamma rays are absorbed in certain materials that would have far reaching applications in solid-state physics, chemistry, and biology. Born in Munich, Germany, Mössbauer was the son of Ludwig and Erma Mössbauer. After graduating from secondary school in 1948, he worked for a year in industrial laboratories. Mössbauer studied physics at the Technical University in Munich and completed his thesis at the university's laboratory for applied physics in 1954.

While working on his doctorate from 1955 to 1957 at the Institute for Physics of the Max Planck Institute for Medical Research in Heidelberg, Mössbauer conducted a series of experiments on emissions of gamma rays from nuclear transitions. These investigations led him to be the first to observe recoilless nuclear resonance absorption, which became known as the **Mössbauer effect**. In essence, the Mössbauer effect demonstrated that nuclear resonant absorption increases at low temperatures, which was contrary to expectations.

Although initially doubted by many of his colleagues, Mössbauer proved his initial observations of recoilless nuclear resonance absorption through a series of experiments. His discovery and subsequent proofs were soon recognized as being of fundamental importance in numerous realms of physics, especially in conducting precise examinations of many important phenomena formerly beyond the limit of attainable accurate measurements. For example, Mössbauer's discovery has led to Mössbauer spectroscopy, and the Mössbauer effect has been used to predict gravitational red shifts in the laboratory. In 1961, Mössbauer received the Nobel Prize in Physics "for his researches concerning the resonance absorption of **gamma radiation** and his discovery in this connection of the effect which bears his name."

After receiving his doctorate, Mössbauer became scientific assistant at the Technical University in Munich and, the following year, went to the California Institute of Technology in Pasadena, California, to continue his studies of gamma **radiation**. Until 1972 he continued his work on the Mössbauer effect and then branched out into other areas of research in nuclear and solid-state physics. Mössbauer also studied the domain of **neutrino** physics, particularly neutrino oscillation experiments and solar neutrino experiments.

Mössbauer married Elisabeth Pritz and has two children, Peter and Regine. In addition to the Nobel Prize, he has received numerous other awards, including the Roentgen Prize of the University of Giessen and the Elliot Cresson Medal from the Franklin Institute. A member of the National Academy of Sciences in the United States, Mössbauer served as Director of the Institute Laue-Langevin and the associated French-German-British high flux reactor in Grenoble, France.

See also Gamma rays

MOTION

The process by which something moves from one position to another is referred to as motion; that is, a changing position involving **time**, **velocity** and **acceleration**. Motions can be classified as linear or translational (motion along a straight line), rotational (motion about some axis), or curvilinear (a combination of linear and rotational). A detailed description of all aspects of motion is called kinematics and is a fundamental part of **mechanics**.

The kinematical description of motion really began with **Galileo**. From observations Galileo introduced two concepts: velocity as the time rate of change of position and acceleration as the time rate of change of velocity. With velocity, acceleration, time and distance traveled (change of position) the complete kinematical description of motion was possible. Four algebraic equations resulted, each involving three variables and an initial position or velocity.

The position of an object must be given, or implied, relative to a **frame of reference** and its motion is then described relative to this frame. Within this frame, position, change of position, velocity, and acceleration require a magnitude (how much) and a direction, both being equally important for a complete description. Physical concepts having this nature are called **vectors** in contrast to scalar concepts which require only a magnitude for their description (for example, time and **mass**). Saying the mall is a 5 mi (8 km) drive may be true but doesn't guarantee one will find the mall. However, specifying 5 mi (8 km) north would give the mall's precise location. Magnitude and direction are equally important.

In circular motion velocity is always parallel to the direction of motion and perpendicular to the radius of motion. The acceleration required to change the velocity's direction, called centripetal acceleration, is always perpendicular to the velocity and toward the center of motion. To change the velocity's magnitude an acceleration is required in the direction of the velocity. Hence, acceleration is required to change both magnitude and direction of velocity and are in different directions. This is applicable to curvilinear motion in general.

MOTION OF PARTICLES

Mechanics is the study of the physical laws that describe the **motion** of objects. When studying moving objects, it is often useful to simplify their motion by treating them as single particles. In this context, a particle is defined as a small object whose microscopic structure is irrelevant to its motion. The fundamental concepts of mechanics are thus all introduced using this simplification, from kinematics, the branch of mechanics that quantitatively describes how a particle moves, to **dynamics**, which studies motion and its causes.

The simplest case of motion is treated by considering the rectilinear or translational motion of a particle along an axis. This type of motion is described using the following quantities: displacement, speed, **velocity** and **acceleration**. The position of a particle in **space** is a vector quantity because it has both magnitude and direction since it is specified according to a reference point. When a particle moves, its position changes as a function of **time** by a given distance, d. A change of position is also a vector quantity and it is called displacement, d. If the particle moves from x_1 at time t_1 to x_2 at time t_2, then its average velocity, or the rate at which it changes its position in the time interval $\Delta t = t_2-t_1$, is equal to its displacement divided by the time interval or $v = d/\Delta t$ and its average speed is the distance traveled in the same time interval. The instantaneous velocity is the velocity of the particle at an instant in time, $v = dr/dt$ where dr is the displacement during the time interval Δt. The instantaneous speed is the magnitude component of the instantaneous velocity vector. The acceleration of a particle is defined as the rate at which the particle changes its instantaneous velocity. If at time t_1 the particle has a velocity v_1, and at time t_2 it has a velocity v_2, then the change in instantaneous velocity is $\Delta v = v_2-v_1$ in the time interval $\Delta t = t_2-t_1$ and the average acceleration (a) during that time interval is $a = \Delta v/\Delta t$. The instantaneous acceleration is $a = dv/dt$ or the change in instantaneous velocity in the time interval Δt.

The motion of a particle in one dimension is easily represented using diagrams which plot the relevant quantities. For example, if a particle is moving with a constant, rightward (positive) velocity, the position vs. time diagram will show a straight line of constant and positive slope. If the particle is moving with a rightward (positive) and changing velocity, the position vs. time diagram will result in an upward curving line. This allows one to distinguish between constant velocity and accelerated motion. The rectilinear motion of a particle with constant acceleration is also described using kinematic equations showing how its velocity and displacement change. Thus, for one-dimensional constant acceleration, the velocity varies with time according to $v_x = v_{0x} + a_x t$, the displacement varies with time as $x = x_0 + v_{0x}t + \frac{1}{2}a_x t^2$ and the velocity varies with position according to $v_x^2 = v_{0x}^2 + 2a_x(x - x_0)$.

The three-dimensional motion of a particle with constant acceleration is described by including all three components, x, y and z, of the vector quantities. If acceleration is constant, then: $a = a_x\hat{\imath} + a_y j + a_z k$. The components a_x, a_y, and a_z are also constant and the average acceleration equals the instantaneous acceleration. If at $t = 0$, the particle is at position $r = x_0\hat{\imath} + y_0 j + z_0 k$ and has velocity $v = v_{0x}\hat{\imath} + v_{0y}j + v_{0z}k$, then, at time t, its velocity has changed by an amount $\Delta v = a\Delta t = at$ or, in terms of the vector equation incorporating the three components: $\Delta v_x\hat{\imath} + \Delta v_y j + \Delta v_z k = a_x t\hat{\imath} + a_y tj + a_z tk$, and yields the following set describing each component for each dimension: $\Delta v_x = a_x t$, $\Delta v_y = a_y t$, and $\Delta v_z = a_z t$. Then, at time t, $v_x = v_{0x} + \Delta v_x = v_{0x} + a_x t$, $v_y = v_{0y} + a_y t$, $v_z = v_{0z} + a_z t$, or $v = v_0 + at$.

See also Center of gravity; Center of mass; Centrifugal and centripetal forces; Coordinate transformation; Force and the law of motion; Force, mass and acceleration; Galileo; Inertia; Inertia and rotational kinetic energy; Moving objects and velocity; Newton, Isaac; Newton's laws of motion; Newtonian physics; Rigid bodies; Torque; Two-dimensional motion; Vectors

MOVING OBJECTS AND VELOCITY

When an object moves, its position changes as a function of **time**, which is the definition of displacement. The distance (d) of the displacement is a scalar quantity and the displacement (d) of the object is a vector quantity, i.e. it has both a magnitude and a direction. If at time t_1 the object is at position P_1 and if it is displaced to position P_2 at a time t_2, the displacement, d in the time interval $\Delta t = t_2- t_1$, is equal to P_2-P_1. The average **velocity** during the displacement is equal to $v = d/\Delta t$, another vector quantity. The distance traveled during the displacement divided by the time interval of the displacement is the average speed of the object, $v = d/\Delta t$, a scalar quantity. More useful when studying moving objects however, are the concepts of instantaneous velocity and speed, that is the velocity or speed at a given instant in time. The instantaneous velocity is defined as $v = dr/dt$ or the displacement (dr) in the time interval dt. The instantaneous speed is the magnitude of the instantaneous velocity. When the instantaneous velocity of the object is changing, then the object is accelerating. The change in velocity is equal to $\Delta v = v_2-v_1$ in the time interval $\Delta t = t_2-t_1$ and the average **acceleration** in that time interval is $a = \Delta v/\Delta t$. The instantaneous acceleration is $a = dv/dt$, where dv is the change in velocity in the time interval dt. These concepts and definitions are the most elementary ones in that branch of physics which concerns itself with understanding the **motion** of bodies under the action of forces, i.e. **mechanics**.

There are many types of motion and a basic distinction is between motion that does not change the shape of the object undergoing the motion (rigid body motion) and motion which changes the shape of the object. Another distinction is between center-of-mass motion and rotation. The motion of any rigid body can thus be motion of its **center of mass**, i.e. motion through **space**, rotational motion about an axis, or a combination of both types of motion.

Galileo Galilei is credited with fundamental contributions to the modern study of motion as shown by his mathematical descriptions of the behavior of freely falling bodies and parabolic trajectories. His insights about motion are summarized by his principle of **inertia** which states that no **force** is required to maintain motion with constant velocity in a straight line and that absolute motion does not cause any observable physical effects. In his systematic work on freely falling objects, i.e. objects not supported by anything and not acted on by any force except **gravity**, Galileo not only speculated that in addition to the force which pulls objects downward there is also a force exerted upward on falling bodies. He proved this experimentally by measuring the displacement of objects in water (so as to slow them down), thus showing that lighter objects took a longer time to reach the bottom. His experiments with balls rolling down inclined planes, to reduce the acceleration along the plane and thus reduce the rate of descent of the balls, also allowed him to show quantitatively that the velocity vs. time graph was linear. However, it was **Isaac Newton** who discovered the relationship between force and motion. Simply, he realized that there is only one cause for a change in motion, i.e. a force. Forces may be different, but they all produce changes in motion in conformity with the same laws. His first law of motion, the law of inertia, states that a moving object continues in its state of constant velocity unless it is acted upon by an external force. If the moving object is initially at rest, it will stay at rest; if it is moving with a given speed in a certain direction, it will keep on moving with the same velocity in the same direction. The first law of motion is a powerful restatement of Galileo's principle of inertia and the converse of the first law is also true: if an object is moving with constant velocity along a straight line, then the total force acting on it must be zero. Newton's second law applies to moving objects on which the total acting force is not zero, i.e. on accelerating objects, which experience changes in instantaneous velocity. It states that the acceleration of such an object is directly proportional to the net force acting on it, and inversely proportional to its **mass**. In other words, unbalanced forces cause acceleration and the net force is the vector sum of all forces acting on the object.

See also Dynamics; Force and the law of motion; Newton's laws of motion; Reference frames; Rigid bodies; Rotation in three dimensions; Vectors; Velocity

MULLIKEN, ROBERT S. (1896-1986)

American chemical physicist

Robert S. Mulliken's early interests and education were in the field of chemistry, especially in the structure of molecules. He eventually discovered that quantum mechanical theory offered an effective tool for the study of the topic and switched his allegiance to the field of physics, a subject that he taught for more than three decades at various institutions. His work ultimately contributed to the establishment of a new interdisciplinary subject, combining these two great fields of science in what is known as chemical physics. Mulliken's research has led to a new understanding of the way in which atoms are held together in molecules and the ways in which molecules interact with each other. In recognition of these accomplishments, he was awarded the 1966 Nobel Prize in chemistry.

Robert Sanderson Mulliken was born in Newburyport, Massachusetts, on June 7, 1896. His father was Samuel Parsons Mulliken, a professor of organic chemistry at the Massachusetts Institute of Technology (MIT). His mother was Katherine W. Mulliken, which was her maiden name as well as her married name. Robert's father was the first Mulliken in many generations not to choose a seafaring career. In fact, his grandfather (Samuel Parson's father) had chosen to become a

ship's captain rather than attend college, as his parents had wanted.

Mulliken's interest in chemistry developed early, largely as a result of his father's work. Through his father, Robert Mulliken made an important professional contact early in his life. Arthur A. Noyes, later to become one of the nation's and world's leading physical chemists, was also a resident of Newburyport and faculty member at MIT. While still in high school, the young Robert proofread galleys of his father's textbooks, books that were later to become standards in the field and make the elder Mulliken famous. Robert Mulliken wrote in "Molecular Scientists and Molecular Science: Some Reminisces" that the proofreading experience helped him to become "well acquainted with the rather formidable names of organic compounds."

In a 1975 interview with the *Journal of Chemical Education,* Mulliken praised the education he had at Newburyport High School. While there, he studied a full academic program, including Latin, French, German, physics, biology, and mathematics. The salutatory address he gave at his high school graduation made it clear that his scientific interests were fixed even at that early age. The topic of his address was "The Electron: What It Is and What It Does."

There seemed little doubt that Mulliken would follow his father to MIT and major in chemistry. In 1913, he was awarded a Wheelwright Scholarship, given to young men from Newburyport that allowed him to enroll at MIT. Once settled into MIT, Mulliken began to consider the possibility of concentrating in chemical engineering rather than chemistry. During one summer, he took part in a Chemical Engineering Practice School that required him to spend a week at each of a half dozen chemical plants in Massachusetts and Maine. He found the experience highly rewarding.

Mulliken eventually returned to chemistry, however, and graduated with a bachelor of science in that field in 1917, just as the United States was entering World War I. Instead of going on to graduate school, therefore, he took a job in war-related research at American University in Washington, D.C. His job was to work on the development of poison gases, but he was not very good at the work. On one occasion, he spilled mustard gas on the floor, much to the horror of his superior, James B. Conant. He later came down with a severe case of the flu and was still recovering in the hospital when the war ended. After the war, Mulliken briefly held a job at the New Jersey Zinc Company, where he studied the effects of zinc oxide and **carbon** black on the compounding of rubber. He apparently realized rather quickly that this was not the kind of work he wanted to do, and he enrolled in a doctoral program in chemistry at the University of Chicago.

He chose Chicago because he wanted especially to work with Professor W. D. Harkins on the study of atomic nuclei. The work to which he was assigned involved finding ways to separate the isotopes of mercury from each other. Mulliken experienced extraordinary success in this line of research and, in 1921, he received his Ph.D. for this work. He went on to improve the system by which mercury isotopes can be separated from each other and later called the construction of this "isotope factory" one of his proudest accomplishments.

One reason for his sense of pride was that it confirmed that he had finally become a "proper experimentalist." In his earlier years, he had not been very proficient in the laboratory (witness the mustard gas event), leading Noyes to express some doubt that he could become a successful researcher. Although Mulliken's practical skills obviously had improved at Chicago, it was ultimately as a theorist, and not an experimentalist, that he was to gain his greatest fame.

In 1923, Mulliken applied to have his National Research Council Fellowship extended for two more years. He was told, however, that he would have to go to a different institution and study a different field of chemistry. His request to work with British physicist **Ernest Rutherford** on **radioactivity** was denied by the board, so he chose instead to go to Harvard. At Harvard, he became interested in studying the **spectral lines** of diatomic molecules, a subject he began with the compound boron nitride (BN). His hypothesis was that when analyzed scientists would find isotopes (such as boron–10 and boron–11) present in the molecule. When he actually carried out this research, his hypothesis was confirmed. Faint bands reflecting the presence of less abundant isotopes—bands that no one had noticed to that point—were apparent.

A turning point in Mulliken's career was marked by two visits to Europe in 1925 and 1927. During these visits, he made an effort to visit all of the major researchers who were working on molecular spectra. But, inevitably, he also came into contact with physicists who were developing the **quantum theory**.

The 1920s were a decade of revolution in physics as researchers were developing, elaborating, and refining the new view of physics arising out of the work of **Max Planck**, Louis Victor Broglie, **Albert Einstein, Erwin Schrödinger**, and others. A very few chemists—Linus Pauling, perhaps most obviously—were also looking for ways in which quantum theory could be used to explain chemical phenomena.

Mulliken had already been introduced to quantum theory at MIT and Chicago. But it was his trips to Europe that really focused his attention on the problem that had now become foremost in his mind: how quantum theory could be used to explain the structure of molecules. He explained later in "Molecular Scientists and Molecular Science: Some Reminisces" that his earlier studies of spectra had "led naturally to attempts also to understand molecular electronic states as more or less like those of atoms."

Perhaps the most important single event on his visits to Europe was his meeting with the German chemist, Friedrich Hund. Hund had been working on many of the same problems as had Mulliken with one important exception. While Mulliken was using a deterministic form of **quantum mechanics**, stressing a more narrow range of likely outcomes for experiments in the field, Hund had employed the uncertainty principle developed by **Werner Karl Heisenberg** in his study of molecular electronic structure. The uncertainty principle states that it is impossible to specify precisely both the position and **momentum** of a particle at the same time. Applying such a concept to the study

Robert S. Mulliken.

of quantum mechanics opened up possibilities for Hund's work. Mulliken began to adopt Hund's approach and soon achieved success. In 1928, he published a classic paper that was crucial in his being awarded the Nobel Prize in 1966. In that paper, Mulliken proposed an entirely new model for **molecular structure**. Previously, chemists had assumed that when two or more atoms combine to form a molecule, the atoms tend to retain their independent characteristics. The molecule was thought to be an assemblage of essentially independent, recognizable atoms "tied together" in some fashion.

Mulliken argued that individual atoms lose their identify when they combine to form a molecule. Electrons that once "belonged" to one or another **atom** in the molecule now became part of an overall molecular structure. The bonding electrons involved in the molecular structure now had **energy** levels whose quantum mechanical descriptions were determined by *molecular* characteristics, not *atomic* characteristics.

In succeeding decades, Mulliken derived a number of specific applications from this general theory. For example, he found ways to measure the relative ionic character of a chemical bond and the properties of conjugated double bonds using this model.

Mulliken's first academic appointment was as assistant professor of physics at the Washington Square College of New York University in 1926. The appointment was significant in that Mulliken, trained as a chemist, was now recognized as a physicist also. He held the New York position until 1928 when he was invited to join the faculty at the University of Chicago. Again, this appointment was in the department of physics, as associate professor. In 1931, he was promoted to full professor.

Mulliken continued to hold an appointment at Chicago for the rest of his academic career, a period of more than five decades. From 1956 to 1961, he was Ernest de Witt Burton Distinguished Service Professor of Physics and Chemistry and, at the end of that period, he was given a joint appointment as professor of physics and chemistry. Beginning in 1965, Mulliken also spent part of each year at Florida State University in Tallahassee, where he was a distinguished research professor of chemical physics at the Institute of Molecular **Biophysics**.

Mulliken was married on December 24, 1929, to the former Mary Helen von Noé, daughter of a colleague in the geology department at the University of Chicago. The Mullikens had two daughters, Lucia Maria and Valerie Noé. Mrs. Mulliken accompanied her husband on many of his overseas trips, one of which was particularly noteworthy. During their 1932 trip to Europe, Mrs. Mulliken came down with an infection of the appendix, a condition that grew increasingly more severe as they moved from city to city on the continent. Eventually an operation in Berlin solved the problem. Their stay in Berlin was further troubled, however, by the growing aggressiveness of Nazi supporters. In "Molecular Scientists and Molecular Science: Some Reminisces" Mulliken described how upset his German colleagues had become about "coming events" and how he himself could "see and hear the Nazi storm troopers marching up and down in the street" below his hotel room early every morning.

During World War II, Mulliken served as director of editorial work and information at the University of Chicago's Plutonium Project, a division of the **Manhattan Project**, set up to develop the **atomic bomb**. He also served briefly as Scientific Attaché at the American Embassy in London in 1955. As Mulliken approached formal retirement age in 1961, he was offered a number of prestigious teaching and lecturing assignments. He was the Baker Lecturer at Cornell University in 1960 and the Silliman Lecturer at Yale in 1965. In the latter year, he also served as John van Geuns Visiting Professor at the University of Amsterdam.

Mulliken continued working on problems of molecular structure, molecular spectra, and molecular interactions throughout his life. In 1952, for example, he applied quantum mechanical theory to an analysis of the interaction between Lewis acid and base molecules. In the 1960s, his studies of molecular structure and spectra ranged from the simplest of cases (hydrogen and diatomic **helium**, for example) to complex molecular aggregates. During this decade, Mulliken also received most of the major awards available to chemists in the United States, including the Gilbert Newton Lewis, Theodore William Richards, John G. Kirkwood, and Willard Gibbs Medals and the Peter Debye Award. Mulliken died in Arlington, Virginia, on October 31, 1986.

MUON

The muon is a subatomic particle that is very similar to the **electron** but more massive. The **mass** of the muon is 207 times that of the electron. Muons were first detected in **cosmic rays** by American physicist Carl D. Anderson in 1936. The muon's mass led physicists to believe it was the particle predicted by the Japanese physicist Yukawa Hideki to explain the strong interaction that binds protons and neutrons together in atomic nuclei. That particle (later determined to be a pion) would interact via the strong **force**. Further investigation showed that the muon is a member of the **lepton** family of particles, which are not affected by the strong force. Muons are one of the main components of cosmic **radiation** at the earth's surface, produced by the decay of **ions** high in the atmosphere. Although they are unstable particles, decaying with an average life of 2.2 microseconds, muons penetrate deep into the earth, traveling at velocities close to the speed of **light**, and reacting weakly with **matter**.

Since muons appeared to be identical to electrons, except for their higher mass, they were expected to decay into an electron and a **gamma ray**. The observed decay products, however, were an electron and two neutrinos. The suggestion that the muon has some internal property that distinguishes it from the electron led to the development of the concept of flavor. Each charged lepton, the muon, the electron and the tau, which was discovered later, is associated with a particular type of **neutrino**. Each neutrino shares the property of flavor with its associated charged lepton. All reaction of leptons conserve this property. Due to this conservation, any decay or reaction of a muon, except the reaction with its anti particle, the positive muon, will produce a muon-neutrino as one of its products.

NANOTECHNOLOGY

Nanotechnology is an relatively new area of science in which extremely tiny machines, ranging from a few to a few hundred nanometers in size, are used to manipulate systems. This manipulation may take place at the cellular level or even the atomic level. A suggested name for these nanomachines is assemblers. Each assembler could function as a nanocomputer with a robotic arm. The computer could be programmed with instructions for placing atoms or molecules. The assemblers could even have the ability to replicate themselves, making production costs relatively cheap.

An example of a naturally occurring assembler is the protein-manufacturing ribosome. Ribosomes manufacture all the proteins existing in living cells and are relatively small, on the same scale as the suggested assembler. They are capable of building almost any protein. Each Ribosome does this by stringing together amino acids in a precise linear sequence. This means that it has some mechanism, such as the robotic arm of a possible assembler, to grasp a specific transfer RNA molecule and then chemically bond it to a specific amino acid by using a specific enzyme. It also uses this mechanism to manipulate the growing polypeptide, and to cause the specific amino acid to react with (and be added to) the end of the polypeptide chain creating the protein. The computer program that the ribosome follows is provided by messenger RNA, a polymer formed from four components, adenine, cytosine, guanine, and uracil. Each specific protein consists of a sequence of several hundred to a few thousand components. The ribosome receives the program from the messenger RNA and acts in a sequential fashion to build the required protein.

In the manufacturing industry, nanomachines (nanobots) might provide cheap, expendable machinery and labor. The process of manufacturing items would become more precise and efficient. Assemblers would be able to construct items **atom** by atom or molecule by molecule. They would be given the substances they need and would be able to use them with

little waste or environmental impact. Steel could be constructed without smelting, water could be sanitized without languishing pools of sewage. Even food could be manufactured instead of grown.

Nanomachines could also have many applications in the medical field. Topical creams could carry nanobots that remove dead or decaying tissue. Saline solutions could be injected into the body with cancer-cell hunting nanobots, or hormone producing assemblers. Dental hygene could become as simple as using a nanobot-rich oral rinse used like mouth wash.

The characteristic that makes nanomachines so flexible and gives them so much potential—their size—also makes them difficult to develop and use. Communicating with nanobots inside the body could be a little tricky; although ultrasound could be used to contact nanobots, any acoustic signal it would transmit would be attenuated within a few microns of flesh. Some suggest that the nanobots should first be instructed to build a large scale transmitter at a specific location just under the skin, which the doctor could then monitor. The nanobots would have inter-nanobot communication capabilities and then could relay messages to the transmitter. In this way, the tasks of the nanobots could be changed in a sequential manner as needed.

NATURAL RADIATION

Natural **radiation** describes a class of physical phenomena that share a spontaneous loss of **energy** in the radiating body or system that results in a propagation of particles and/or **waves** that carry the ability to do **work** and thus change the energy state of an impacted body or system. In general terms, natural radiation usually refers to the spontaneous emission and transmission of **subatomic particles** through **space** from a radioactive source. Although the imprecise term "radiation" can refer to a varied group of physical phenomena including electromagnetic, particle, and acoustic radiation, most often it refers to the particles given-off by naturally radioactive elements, such as uranium. Accordingly, natural radiation usually is a term applied to the spontaneous energetic particle emission from natural elements as they decay from unstable nuclides to more stable elemental species.

Many radioactive elements have been artificially fabricated. Such elements, not found naturally on Earth, are created by bombarding the nuclei of stable elements with charged or uncharged subatomic particles. The unnatural radioactive nuclides, called transmutations, are radioactive isotopes of natural elements. Artificial nuclides have been produced from all known natural elements by transmutation. In addition, man-made transuranium elements have been created. All transuranium elements are radioactive transmutations of the element uranium that each contain more than 92 protons in their nuclei. Examples include neptunium, plutonium, americium, and einsteinium. All transuranium elements, like other artificial **radioisotopes**, decay into more stable elements through the emission of radiation.

In contrast to the radiation emitted by artificial nuclides, emanation of natural radiation occurs without human intervention. Natural radiation results from the spontaneous decay of naturally occurring elements like uranium, radium, and polonium. **Radioactive decay** is the disintegration of a **nucleus** of an **atom** into a lighter, more stable nucleus. The type and rate of decay involved in natural radiation depends upon the subatomic nuclear content of the atoms involved. For instance, if a naturally radioactive element contains too many neutrons, it will spontaneously eliminate neutrons or portions of neutrons to stabilize the nucleus. The particles and associated energy released constitute natural radiation.

There are several forms of natural radiation. Alpha emission is the release of an **alpha particle**, two protons and two neutrons bound together. Alpha particles are positively charged and are emitted from very heavy unstable nuclei. For example, the natural radiation of alpha particles from polonium results in stable lead atoms. Beta emission is natural radiation that is the ejection of a negatively charged **electron** or positively charged **positron** from a nucleus that has too many neutrons. **Gamma radiation**, the most energetic electromagnetic radiation, is the emission of high-energy electromagnetic waves from a nucleus as it shifts from an excited state to a lower energy, ground state. Gamma rays have shorter **wavelength** and higher energy than do **x rays**. Cosmic radiation consists of a stream of high-energy charged particles (e.g., nuclei, electrons, positrons, and neutrinos) emanating from stellar processes.

One form of natural radiation, **background radiation**, is ubiquitous. Background radiation, the radiation that is detected by a Geiger counter in the absence of nearby radioactive material, is in part the result of radiation from the **Sun** and other stellar bodies. A constant source of energy, the Sun continually delivers a steady, low-level stream of natural background radiation to Earth. Similarly, slight background radiation may emanate from Earth itself, as tiny concentrations of radioactive elements release natural radiation. One form of natural radiation that poses a threat to human health is the radiation supplied by the decay of radon gas. Radon is an odorless, colorless, radioisotope that can seep into household basements from natural rock deposits. An insidious source of natural radiation, the gas is thought to contribute to the development of lung cancer. In some areas, tests for the presence of radon and special types of substructure venting are required.

See also Alpha radiation; Beta radiation

NATURAL SOUND FREQUENCIES

Natural **sound** frequencies specify the **frequency** attributes of sound **waves** that will efficiently induce vibration in a body or that naturally result from the vibration of that body.

Every object has a unique natural frequency of vibration. Vibration can be induced by the direct forcible disturbance of an object or by the forcible disturbance the medium in contact with an object (e.g. the surrounding air). Once excited, all such vibrators (i.e., vibratory bodies) become **generators** of sound waves. For example, when struck, the prongs of a simple tuning fork undergo sinusoidal **oscillations** and generate a sound wave.

Vibratory bodies can also absorb sound waves. Vibrating bodies can, however, efficiently vibrate only at certain frequencies called the natural frequencies of oscillation. In the case of a tuning fork, if a traveling sinusoidal sound wave has the same frequency as the sound wave naturally produced by the oscillations of the tuning fork, the traveling **pressure** wave can induce vibration of the tuning fork at that particular frequency.

Mechanical resonance occurs with the application of a periodic **force** at the same frequency as the natural vibration frequency. Accordingly, as the pressure fluctuations in a resonant traveling sound wave strike the prongs of the fork, the prongs experience successive forces at appropriate intervals to produce sound generation at the natural vibrational or natural sound frequency. If the resonant traveling wave continues to exert force, the amplitude of oscillation of the tuning fork will increase and the sound wave emanating from the tuning fork will grow stronger. If the frequencies are within the range of human **hearing** the sound will seem to grow louder. Singers are able to break glass by loudly singing a note at the natural vibrational frequency of the glass. Vibrations induced in the glass can become so strong that the glass exceeds its elastic limit and breaks.

All objects have a natural frequency or set of frequencies at which they vibrate. Objects that vibrate and produce sound waves at a single frequency are said to produce pure tones. Objects that vibrate in response to complex sets of frequencies that have a whole number mathematical relationship between them are said to produce rich sound (e.g., flutes produce pure tones, tubas produce a rich tones). The natural sound frequencies of musical wind instruments, for example, is dependent upon the length of the air column of the instrument. The natural sound frequencies of string instruments depends upon the composition, diameter, and length of the string.

The determination of mechanical resonance frequencies and natural sound frequencies is an important consideration in engineering applications, mechanical failure can result if an object is continually subjected to vibration inducing forces.

See also Acoustics; Mechanics; Simple waves

NEBULAE

Nebulae are interstellar clouds of gas and dust. The word nebula is derived from the Latin word for mist or cloud, and historically has been used to refer to almost any diffuse astronomical object. In 1781 French astronomer Charles Messier (1730-1817) published a catalog of 103 nebular objects. Of these, only 11 were true nebulae, the rest were ultimately discovered to be distant galaxies.

There are three general types of nebula: dark, reflection, and emission nebulae, of which the latter may be subdivided into H II regions, planetary nebulae, and supernova remnants. Emission nebulae are also known as gaseous nebulae. The main difference between these are the manner in which they are illuminated, and whether gas or dust is more prominent.

Emission nebulae are visible as a result of **light** emitted by atoms and **ions** which have been excited by nearby hot **stars**. Electrons bound to hydrogen atoms are ejected by energetic ultraviolet photons. When the free electrons recombine with the hydrogen ions, they cascade down through different atomic orbital levels, emitting photons in the process. This mechanism is called **fluorescence**. Thus, the spectrum of an emission nebula exhibits many emission lines, a result of the high **temperature** of the nebular gas (10,000K), and its low **density** (100-1000 atoms per cubic centimeter). At such low densities, electrons will accumulate in metastable levels and remain undisturbed by collisions long enough to cascade down to lower **energy** levels. The photons thus emitted are known as forbidden lines, with the strongest generally arising from twice ionized **oxygen** at 5007 angstroms, causing these nebulae to have a greenish tint. Because photons from forbidden lines can not be reabsorbed by hydrogen or **helium**, they act to cool the nebula. As a result, even though the local stars may have temperatures in excess of 100,000K, nebula rarely reach temperatures over 20,000K. Ratios of the intensities of forbidden lines may be used to determine the temperature and density of the gas.

Many emission nebula emit large amounts of infrared **radiation**. Dust within the nebula absorbs ultraviolet light emitted by the hot stars, and eventually re-emits it as infrared photons. Dust in emission nebulae is much cooler than the gas, with a usual temperature of 300K.

Only very hot stars with temperatures in excess of 25,000K (e.g., spectral type O) emit appreciable numbers of ultraviolet photons with energies sufficient to ionize hydrogen. Such electrons have wavelengths shorter than 912 angstroms. Within the **interstellar gas** surrounding such a star, nearly all the hydrogen within a certain radius will have been ionized. Conversely, very little hydrogen farther from the star will be ionized. Such a region is called an H II region (Stromgen

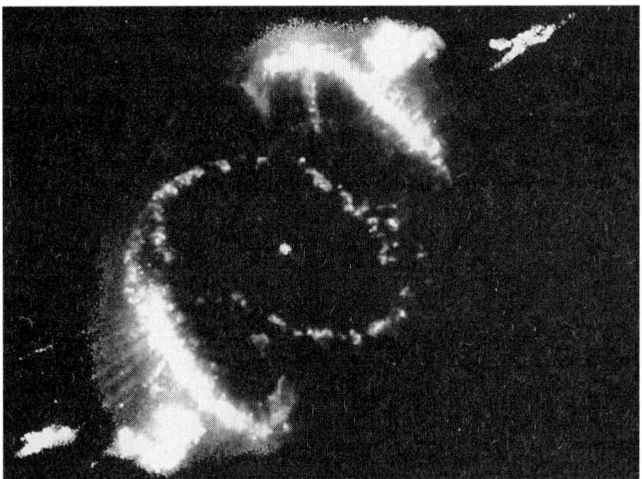

Cat's Eye Nebula (NGC 6543) as seen from the Hubble Telescope. *(Photo courtesy of NASA. Reproduced by permission.)*

sphere). H I denotes neutral hydrogen, while H II denotes H^+. Similarly, O^{++}, noted above, is called O III. The most famous emission nebula, or H II region, is the Great Nebula in the constellation of Orion, where a group of O stars have ionized the surrounding interstellar gas.

Planetary nebulae, are a type of emission nebula in which a red giant star has gently expelled its outer layers into **space**, exposing its superhot carbon-oxygen core. The ultraviolet photons cause the expelled gas to fluoresce. Planetary nebula are short lived, persisting only about 30,000 years before dissipating into the interstellar medium.

Supernovae remnants are similar to planetary nebula, except that the nebula is a result of a violent explosion, and as a result is expanding into space with velocities of up to thousands of kilometers per second. Supernovae remnants also emit large amount of Bremsstralung radiation, emitted by rapidly moving electrons trapped in magnetic fields.

Reflection nebulae result when stars in or near a cloud of gas and dust are not hot enough to ionize the gas. As a result, the light from these stars is not absorbed and re-emitted by the gas, and the nebula is visible only from light reflected by dust particles. About one percent of the **mass** in reflection nebula is contained in dust particles, with diameters ranging from 0.003 mm to 1 nanometer. On average, there is one grain of dust in every 100,000 cubic meters of space. Very small particles such as these scatter blue light much more strongly than red, so light from reflection nebulae tends to be blue. About half the light falling on dust grains is absorbed, with the rest being reflected. Reflection also tends to polarize the light seen from these nebulae, an effect not seen in other nebulae. The Pleiades, located in the constellation Taurus, is composed of about 3000 young stars and provides an excellent example of a reflection nebula plainly visible with a small **telescope**.

Dark nebulae are dense clouds of gas and dust that block light from stars and/or illuminated gas. Several dark nebulae are visible with the naked eye, most notably the Coalsack in the constellation Cygnus, where clouds of cold gas and dust block the light from distant stars of the Milky Way. Another famous dark nebula is the Horsehead Nebulae in IC 434, where dark gas blocks the light from a cloud of hot gas farther away. The convoluted appearance of many dark nebula indicate that the interstellar medium is in complicated **motion**, even far from hot, ionizing stars. The smallest dark nebulae are the Bok globules, which are small dark clouds less than 1 parsec in diameter, containing 10 to 100 solar masses. That at least some Bok globules are contracting to form new stars is indicated by the emission of infrared light, a result of warm centers caused by contraction.

See also Absorbtion spectrum; Astronomy; Evolution of the Universe; Gravitational collapse; Spectral lines; Spectroscopy; Stellar evolution; Stellar life cycle

NEPTUNE

Neptune is the eighth planet in the solar system with respect to its average distance from the **Sun**. It is the outermost of the giant gas **planets** (along with Jupiter, Saturn, and Uranus). The discovery of Neptune was one of the triumphs of mathematical **astronomy**. To understand the perturbations in the orbit of the planet Uranus, astronomers John Couch Adams and Urbain Jean Joseph Leverrier independently calculated the existence and position of the new planet. Then, Johann Gottfried Galle, of the Berlin Observatory, discovered Neptune on September 23, 1846, based on the theoretical mathematical calculations sent to him by Leverrier.

The diameter of Neptune is approximately 3.88 times the size of Earth's diameter, with an equatorial diameter of 30,763 mi (49,528 km) and a polar diameter of 30,236 mi (48,680 km). Its **mass** is about 17.2 times that of Earth's mass, with a value of 7.019×10^{24} slug (1.024×10^{26} kg). Neptune's volume is equivalent to about 60 earth-volumes. Its mean **density** is 3.18 slug/ft^3 (1.64 g/cm^3). Neptune is the coldest planet in the solar system, with an average **temperature** of -318°F (-218°C).

As determined by the **space** probe *Voyager 2*, Neptune's rotation about its axis (or length of day) is 16 hours, 6.7 minutes. The planet's period of revolution about the Sun is 164.79 earth-years. Neptune is about 30 times further from the Sun than Earth. The *Voyager 2* flyby of the planet resulted in the calculation of its minimum distance from the Sun to be 2.7748 billion mi (4.4656 billion km) and its maximum distance from the Sun to be 2.8240 billion mi (4.5461 billion km). Neptune's minimum distance from the Earth is 2.680 billion mi (4.313 billion km) and maximum distance is 2.910 billion mi (4.683 billion km). The tilt of Neptune's equator relative to its orbit is 29.6 degrees, somewhat larger than Earth's 23.4 degrees.

The inside two thirds of Neptune is composed of a mixture of molten rock, water, liquid ammonia, and methane. The outer third is a mixture of heated gases comprised of hydrogen, **helium**, water, and methane. Neptune's magnetic field traps solar wind and galactic cosmic-ray particles in a belt around the planet, similar to Earth's Van Allen **radiation** belts. It is theorized that motions in the interior of Neptune help form its magnetosphere. The magnetosphere of Neptune is smaller than Earth's and highly tilted at 47 degrees from the rotation axis. Scientists believe the extreme orientation may be characteristic of liquid flows in the interior of the planet. The magnetic material of Neptune is thought to be within its icy core. **Heat** generated within Neptune's core helps form the unusual winds of the atmosphere.

Neptune's low mean density (at 1.8 times that of water) and its high proportion of **light** that is reflected off its surface indicate that Neptune has a thick layer of atmosphere. The atmosphere of Neptune is believed to be composed 98% of hydrogen and helium that are nearly invisible. The remaining molecules (mostly in the upper part of the planet's atmosphere) consist of methane gas that absorbs red light. This effect is responsible for Neptune's striking greenish-blue outward appearance. Because the atmosphere of Neptune is so dramatic, changing in a matter of weeks, or even days, a large recurrent disturbance called the great dark spot, seems to disappear and then reappear at frequent intervals. These dark spots are being created by strong upward motions of pockets of methane gas at the higher, colder altitudes of 31-62 mi (50-100 km) of the atmosphere.

The *Voyager 2* probe also confirmed the existence of an orbiting ring system with at least four bands that are thin and very faint. The rings are made up of dust particles thought to have been made by tiny meteorites smashing into Neptune's moons. Mathematical theory suggests that the rings of Neptune affect the **motion** of particles in its magnetosphere.

Neptune has eight known natural satellites. Triton (discovered in 1846) is the largest and brightest of the moons, being only slightly smaller than Earth's **moon**. It has a diameter of 1,700 mi (2,700 km), travels in an unusual retrograde orbit, and has several active geysers of nitrogen ice and gas. Nereid (discovered in 1949) has an estimated diameter of 145-290 mi (235-470 km). The *Voyager 2* space probe encountered Neptune in 1989, and discovered six smaller, dark moons (Naiad, Thalassa, Despina, Galatea, Larissa, and Proteus).

See also Earth; Hubble Space Telescope

NEUTRINO

The neutrino, whose symbol is the Greek letter *nu*, is a neutral massless particle that is created during **beta decay** and travels at the speed of **light**. Its existence was suggested to explain missing **energy** and **angular momentum** when a beta particle was emitted from a radioactive element. **Wolfgang Pauli** suggested in 1930 that beta decay must involve a third particle that would be neutral, have negligible rest **mass** and a spin of one-half. **Enrico Fermi** named this particle the *neutrino*, Italian for little neutral one, and its existence was accepted without further evidence. It was not until 1953 that **Frederick Reines** and **Clyde Cowan** devised an experiment to detect the neutri-

Spherical halo of neutrinos around the Milky Way Galaxy. *(Photo courtesy of NASA. Reproduced by permission.)*

nos that should be produced by the **nuclear reactions** taking place in a nuclear reactor. They succeeded, and Reines was awarded the 1995 Nobel Prize in physics for that **work**.

Discoveries in the field of physics during the second half of the twentieth century have led to the development of a new classification system for atomic and **subatomic particles** that is based on their interactions. Under this system, the neutrino is classified as a **lepton**, the category for light particles (*lepton* is derived from a Greek word meaning light) with a spin of one-half that interact via the weak nuclear **force** but do not respond to the strong force. Because they interact with **matter** only via the weak force, neutrinos are hard to detect. Nevertheless, they have been detected on earth and in **space**. A very important source of neutrinos turned out to be supernovae: studies on SN1987 show that most of the energy of the supernova was emitted as neutrinos. The core of the exploding star was converted to neutrons during the explosion when protons and electrons were forced together in reverse beta decay, producing the neutrinos.

NEUTRON

The discovery of the nuclear **atom** by **Ernest Rutherford** in 1911 was a critical step forward in improving our understanding of the fundamental nature of **matter**. However, it was immediately clear to scientists that Rutherford's atomic model was incomplete. The simplest and most obvious data showed without question that atoms must consist of more than nuclei, which contain protons, and electrons, located outside the **nucleus**.

Those data come from measurements of atomic number and atomic **mass**. An element's atomic number is equal to the total positive charge on—the number of protons in—the nuclei of atoms of which the element is made. The element's atomic mass is equal to the total mass of all particles that make up the atom. Since the mass of a single **proton** on the atomic mass scale is 1, and that of the **electron** is 0.0055, the mass of the atom must be very nearly equal to the mass of the protons present in the atom if protons and electrons are the only particles present in the atom. In such a case, an element's atomic number and atomic mass should be about equal. However, with the exception of hydrogen, this is never the case.

Scientists considered a number of ways in which this dilemma could be resolved. One popular notion was that a nucleus might contain both protons and electrons. The electrons in the nucleus could balance the electrical charge of some of the protons without adding any significant mass to the atom. Unfortunately, there were a number of rather clear reasons that electrons could not exist in the nucleus, and this theory was abandoned.

Rutherford himself suggested that some sort of neutral particle might exist in the nucleus and, during the 1920s, he and a graduate student, **James Chadwick** (1891-1974), searched for such a particle. They were unsuccessful, however, largely because neutrons (the particles for which they were seeking) do not readily ionize atoms and are not, therefore, detected by any standard tools such as cloud chambers or Geiger counters. The clue that led to the resolution of this problem showed up in the research of **Irène Joliot-Curie** and her husband, **Frédéric Joliot-Curie**, in France and of Walther Bothe (1891-1957) in Germany in the early 1930s. The Joliot-Curies and Bothe all observed that the bombardment of **light** elements, such as beryllium and lithium, with alpha particles resulted in the production of a strange, intense form of **radiation**. They were unable, however, to identify that radiation.

Chadwick, in England, repeated the experiments of the Joliot-Curies and Bothe with greater success. He directed the beam of "strange radiation" at a piece of paraffin and observed that protons were ejected from the paraffin. He concluded that the radiation must consist of particles with no charge and a mass about equal to that of the proton. That particle was the neutron.

Chadwick's discovery was a bonanza for nuclear physicists. Unlike the proton, electron, and other charged **subatomic particles**, the neutron is not repelled by either the nucleus or the orbital electrons in an atom. It is a much more efficient "bullet," then, in initiating **nuclear reactions**.

For the next three decades, the neutron was regarded as a fundamental particle that could not be broken down into anything simpler. In the early 1960s, however, the American physicist **Robert Hofstadter** was able to observe internal structure in both neutrons and protons. Hofstadter passed a beam of electrons close to each type of particle and studied the patterns that resulted from the electrons' deflection. He discovered that both protons and neutrons contain a central core of positively charged matter that is surrounded by two shells. In the neutron, one shell is negatively charged, just balancing the positive charge in the particle's core.

NEUTRON ACTIVATION ANALYSIS

Neutron activation analysis is an analytical technique for determining the elements present in a material as well as the amount of each element in the sample. The technique is based on a well-known reaction from nuclear **chemistry**. When an element is bombarded with neutrons, some of the atoms of that element may capture neutrons and incorporate them into their nuclei. Those atoms that do so have the same atomic number (that is, are the same element) as the original target element, but have an atomic **mass** of one higher. Bombarding atoms of sodium-23 with neutrons, for example, converts them to atoms of sodium-24.

In most cases, the heavier atoms formed in this reaction are radioactive. They usually decay with the emission of a beta particle and a **gamma ray**. The **energy** associated with these decay schemes is characteristic of the radioactive **isotope** in question. For example, a gamma ray released in the decay of sodium-24 has an energy of 2.75 or 1.37 MeV.

In practice, nuclear activation analysis is carried out by placing thesample to be examined in a nuclear reactor. The neutrons available in the reactor bring about the n/c (neutron/gamma ray) reactions described above. The radioactive sample is then removed from the reactor and examined with a gamma ray spectrometer. This device measures the type and intensity of **radiation** released by the sample. These data can then be compared to standard tables to determine which elements and the amounts of each are present in the sample.

Neutron activation analysis is valuable as a non-destructive form of analysis. It can be used without fear of damaging or destroying the material being tested. It is also a very precise form of analysis, permitting the detection of very small quantities of an element. One application that illustrates these strengths is in the analysis of archaeological materials that are too fragile or too valuable to expose to other analytical techniques.

NEUTRON STARS

Neutron stars are a class of very compact astrophysical objects which are remnants of massive stars that collapse after exhausting their capacity for thermonuclear burning in their interiors. The typical **mass** of a neutron star is about one and a half times that of the **Sun** (which is about 4.4×10^{30} lb, or 2×10^{30} kg), while its radius is only about 6 mi (10 km)—one hundred thousand times smaller than the solar radius. Such a combination exists through compression of **matter** in the interior of the star to a **density** greater than 2.7×10^{16} lb/in³ (10^{15} g/cm³), several times the density in atomic nuclei. Unlike most naturally occurring systems in **equilibrium**, where matter exists in the form of atoms surrounded by electrons, matter in the cores of neutron stars is believed to be a uniform mixture of nucleons, electrons and possibly other species of elementary particles.

The concept of a star composed of neutrons was first raised by the prominent Russian physicist, L. D. Landau, shortly after the discovery of the neutron in 1932. Landau pointed out that since the neutron has no **electric charge**, a macroscopic number of neutrons could be bound together by **gravity**. Two years later, astronomers F. Badde and W. Zwicky suggested that the unexplained phenomena of Supernovae arise from the collapse of normal star into a neutron star. Both suggestions remained theoretical speculation for several decades, but were revived shortly after the discovery of **pulsars** in 1967. The very accurate pulsation of pulsars is attributed to rotation, and only very compact stars can rotate at the observed rates—with periods as low as millisceonds. Neutron stars are the only viable candidate for such objects, and it is well accepted that pulsars offer conclusive evidence that neutron stars exist. The identification of pulsars in the remnants of known supernovae, such as the Crab and Vela, confirmed Badde and Zwicky's hypothesis about the supernova-neutron star connection as well.

Theoretical models of **stellar evolution** suggest that stars whose initial mass is in the range 10-25 solar masses will end their lives in a supernovae explosion that will leave behind a neutron star. If this is correct, there should be over one hundred million neutron stars in our galaxy, but the vast majority of them are undetectable by any astronomical method. A small minority of the younger neutron stars have sufficiently high surface magnetic fields, typically 10^{12} Gauss (over one million times stronger than the highest manmade magnetic field), and are spinning rapidly enough to emit beamed electromagnetic emission that enables their identification. As of early 2000, over 700 pulsars had been discovered. Several tens of other neutron stars have been identified by other means, mostly as sources of x-ray emission, or by their gravitational effect on an observable binary companion pulsars.

Neutron stars are the only known system in nature where matter at extremely high densities can exist in a stable manner. In this sense, neutron stars are cosmological laboratories for nuclear and hadronic physics, which are dominated by the strong interaction. Moreover, the microscopic physics of matter at supernuclear densities determine the star's main observable macroscopic quantities, such as mass and radii, so that astronomical observation of neutron stars can, in fact, serve as a primary research tool for the study of microscopic physics. A different but closely related topic is matter at densities just below that of atomic nuclei, which includes a mixture of nuclei and free nucleons. Such matter occupies the "crust" of the neutron star (the outer 1/2 mile), which is believed to be responsible for violent quakes that have been detected in several young neutron stars.

Neutron stars also offer laboratories for very strong magnetic field and **gravitational field** physics. In particular, the gravitational filed in a neutron star and its vicinity is large enough that general relativity leads to effects of the order of 10-20% with respect to Newtonian gravity. Observations related to gravitational influence of neutron stars in their close vicinity can indeed serve as probes of general relativity. This applies to surface emission from the star, to inter-stellar material falling on to the star, and to the **motion** of a close binary companion. The "textbook" manifestation of general relativity is the double-neutron star system PSR 1913+16, which undergoes a shifting of its pulsation period. This shift in excellent agreement with predictions general relativity, indicating that gravitational **waves** are emitted from the system (the discovery of this system earned Taylor and Hulse the Nobel prize in 1994).

Many neutron stars are also the sites of strong x-ray sources in the range of a few to a few tens of keV. These sources are neutron stars which accrete material ejected by a binary companion star. The material is compressed as it accretes on to the neutron star, and the very strong gravitational field near the surface causes **temperature** of the accreting material to rise to millions of degrees, hence emitting in the x-ray range. The detailed features of x-ray emission vary considerably between different sources. Some sources are very periodic, probably due to strong emissions forming a hot-spot that rotates along with the neutron star; the "classic" case of her X-1, with a period of 1.24 seconds in its x-ray emission, is such an x-ray pulsar. In other cases, the period of **oscillations** in the x-ray emission seems to vary, implying that more complex phenomena determines the rate of accretion and the emission efficiency. Some sources show strong, transient bursts of **x rays**, which are probably due to thermonuclear ignition of accreted hydrogen on the surface of the star. Since the x-ray emission originates from near or at the surface of the neutron star, it can provide important clues concerning the properties of neutron stars and the physics of strong gravitational fields. Improved space-based detectors of cosmological x-ray sources (and NASA's Chandra x-ray satellite in particular) have great potential in improving our knowledge of these exciting fields in the near future.

See also Supernovae; Relativity, general

NEWTON, ISAAC (1642-1727)

English physicist and mathematician

Born in Woolsthorpe, Lincolnshire, England, Newton was a premature baby who was not expected to live. His father had died three months previous to the birth, and his mother remarried three years later, leaving Newton in the care of his grandparents.

Newton did not distinguish himself in school, and he was removed by his mother in the late 1650s to work on the family farm, but Newton proved a worse farmer than scholar. His uncle, however, encouraged the boy to go to Cambridge in 1660. Five years later Newton graduated, even though he had failed a scholarship exam in 1663 due to his lack of knowledge concerning geometry.

Newton returned to the farm shortly thereafter to escape the Bubonic plague, which at the time was decimating London. While at the farm in 1666, Newton saw an apple fall to the ground, and he began to ponder the **force** that was responsible for the action. While this story has often been considered to be the stuff of legends, Newton confirmed that it did indeed happen.

He first surmised that the apple fell because all **matter** attracts other matter. He then further theorized that the rate of the apple's fall was directly proportional to the attractive force which Earth exerted upon it. In addition, he suggested the inverse square law; force decreases according to the square of the distance from the center of Earth. Then he made a daring

Sir Isaac Newton.

hypothesis, suggesting the force that pulled the apple was also responsible for keeping the **moon** in orbit around Earth.

When Newton made calculations of what the rate of fall for the moon should be, he came up short of what was actually observed and was quite disappointed. The problem was twofold; first, the radius of Earth was not known with precision and the size Newton used was too small. Second, he was not absolutely certain he was correct in making his calculations based on the **gravitational force** at the center of Earth, as opposed to the surface. Because of these issues, he set aside his work on **gravity** for 15 years.

During this same time, Newton began to experiment with **light**, a book written by **Robert Boyle** having prompted his interest. Newton passed a beam of sunlight through a prism of glass and observed it was refracted into a spectrum. He passed the spectrum through a second prism, and the light was recombined into a white spot. For an unexplained reason, Newton did not notice any of the visible dark lines in the spectrum of the **sun**. It was not until over 150 years later that William Hyde Wollaston and **Joseph Fraunhofer** noticed the lines, and Gustav Kirchhoff realized they revealed the composition of heavenly bodies. Newton's experiments led him to conclude that light was comprised of a stream of particles that moved through an **ether**. Although this theory was later disproved, it was kept alive by Newton's influence for nearly a century.

In 1672 Newton was elected to the Royal Society where he reported on his experiments with light and its spectrum. While most members were impressed by his findings, physicist **Robert Hooke**, who had performed similar experiments and drawn different conclusions, dismissed Newton's findings as irrelevant. In response, Newton sent a scathing letter to Hooke, which was later published in the Royal Society's periodical. The feud between Hooke and Newton escalated, though the latter's increasing reputation in the British scientific community forced them to act cordially—at least in public. The two scientists' dislike for one another culminated in 1686 when Newton published his theory of universal gravitation. Hooke, who had outlined a similar theory in a letter seven years earlier, insisted that Newton had plagiarized his idea. In all fairness to Newton, it is possible he was unaware of Hooke's work, which was in any event fundamentally flawed. However, modern scientists now credit Hooke with the ideas that inspired Newton to develop his own, more accurate, theory of gravitation.

Newton became involved in yet another controversy, this time with the German mathematician **Gottfried Leibniz**. Both men had independently developed calculus, although each used different symbols and notations. Leibniz's notation, however, was considered superior and came to be preferred over that of Newton, causing a bitter controversy between the two. Their conflict quickly became a matter of national pride, and English scientists refused to accept Leibniz's version. Unfortunately this stubbornness kept them from significantly contributing to mathematics for nearly a century.

In the 1680s English astronomer **Edmond Halley**, a good friend of Newton, told him of Hooke's claim that he had discovered the laws concerning the **motion** of celestial objects. When Newton replied that he had made the calculations 15 years earlier but had lost interest, Halley convinced him to make the calculations again. With his friend's financial help, Newton published the results of his second effort in *Philosophiae Naturalis Principia Mathematica* (*Mathematical Principles of Natural Philosophy*), which outlined the laws of gravity and motion. Though he now had calculus at his disposal, he apparently thought it more elegant to prove his deductions using only geometry.

In the *Principia Mathematica*, which appeared in 1687, Newton arranged Galileo's findings into three basic laws of motion: a body at rest remains at rest, and a body in motion remains in motion; force is equal to **mass** times **acceleration**; for every action, there is an equal and opposite reaction. These three laws allowed Newton to calculate the gravitational force between Earth and the moon. He asserted these laws acted upon any two objects in the **universe**, establishing the law of universal gravitation, and proceeded to estimate, quite accurately, the masses of Jupiter and Saturn. The *Principia Mathematica* not only explained Johannes **Kepler's laws of planetary motion**, it elegantly showed how celestial bodies and earthly objects followed a single law; the unity of heaven and Earth were at last grasped by reason.

The book was written in only 18 months and apparently exerted an enormous strain on Newton. Coupled with the continual controversy surrounding him, Newton suffered a nervous breakdown in 1692. Although he recovered, he was never the same. He went on to write another book, *Opticks*, which covered his work on light; however, he waited until after Hooke's death before printing it in 1704. In 1705 Queen Anne knighted him.

Newton died on March 20, 1727, at the age of 84, and his vast influence upon science continued, later rivaled only by that of Charles Darwin and **Albert Einstein**. In a letter written to Hooke, Newton stated, "if I have seen further than other men, it is because I stood on the shoulders of giants."

NEWTON'S LAW OF UNIVERSAL GRAVITATION

English physicist Sir Isaac Newton's (1642-1727) 1687 work, *Philosophiae Naturalis Principia Mathematica (Mathematical Principles of Natural Philosophy)* advanced a law of universal gravitation that had a profound impact on the scientific and philosophical world. Newton's law specifies that objects with **mass** attract each other with a **force** that varies directly as the product of their masses and inversely as the square of the distance between the objects. Along with Newton's laws of **motion**, his law of gravitation became one of the key components in the development of physical theory.

Perhaps the most famous contribution of **Isaac Newton** to science was his theory of gravitation. As reported by the French philosopher, Voltaire: "One day in the year 1666, Newton had gone to the country, and seeing the fall of an apple, as his niece told me, let himself be led into a deep meditation on the cause which thus draws every object along a line whose extension could pass almost through the center of the Earth."

In Aristotle' physics, objects simply returned to Earth because everything was thought to return to its natural state, and the natural state of objects was to be on the ground. Newton improved on Aristotle's explanation by providing a mathematical expression for the **gravitational force** between two objects and by identifying the force which causes apples to fall as the same force that keeps **planets** in their orbits. The latter is the more subtle and profound of Newton's contributions because until Newton, people believed that planets were kept in place by divine purpose. Newton's laws of universal gravitation provided a physical explanation for the patterns in planetary motion.

The content of Newton's theory is summarized by the force equation, $F = Gm_1m_2/r^2$. Further equation reveals that the force between two objects is proportional to the product of their masses, m_1 and m_2. The law of universal gravitation also states that the gravitational force between two objects is inversely proportional to the square of the distance r between m_1 and m_2. A gravitational constant, $G = 6.67 \times 10^{-11}$ Nm^2/kg^2

provides a measure value for the gravitational force. Moreover, Newton showed that the attraction between two masses depends only on the distance between their "centers of mass." (When people refer to the "center of gravity" in everyday speech, they mean its center of mass.) In effect, **gravity** acts as if all the mass of an object were concentrated at one point, regardless of the actual shape of the object. The gravitational force is always attractive, masses never repel each other through gravity as can occur between like-charged particles.

Newton's first law of motion states that objects stay in their state of motion unless acted upon by a force. The second law describes how an object's motion can be altered by a force in a way which depends on the object's mass. The mass referred to in the first and second laws is the same, and it is often called the "intertial mass." The mass that appears in the law of gravity is called the "gravitational mass" because it determines the strength of the gravitational force between two objects. In classical **mechanics**, there is no reason why these two masses should be equivalent, but they are. An experiment by Eötvös in 1909 showed that the two quantities differ by less than one part in a billion. An explanation for why this is true follows immediately from Einstein's principle of equivalence and was critically important to the development of the general theory of relativity.

One of the most significant aspects of Newton's theory is the conceptual leap in identifying the force which pulls objects to Earth with the force which keeps planets in their orbits. Newton demonstrated this identity by deriving **Kepler's laws of planetary motion** from his theory of gravity. This triumph encouraged seventeenth- and eighteenth-century European philosophers to think of the **universe** in mechanical terms. God was seen as a "watchmaker" who built the machine of the universe and set it to run according to Newton's laws. Although some argued that explanations of natural phenomena reveled the very nature of God and that God could only be understood within the laws of science. For other scientists, however, the revelation of a mechanistic universe left no place for a divine operator, and they abandoned or modified their religious views.

In the twentieth century, Einstein's general theory of relativity supplanted Newton's theory of gravity as the most accurate theory of gravitation, yet Newton's theory remains the more useful theory for most purposes. The corrections to most calculations arising from relativistic effects are so small that they are not worth the extra effort it takes to use Einstein's much more complicated theory. Accordingly, sending astronauts to the **Moon** required modern technology, but the basis of the calculations used to plot trajectories and orbital paths was centuries old.

See also Grand Unified Theory; Gravitational energy; Gravitational field; Gravitational radiation and waves; Graviton; Inertia; Mass; Relativistic velocity transformations; Relativity, general; Relativity, special

NEWTON'S LAWS OF MOTION

Earthly and heavenly motions were of great interest to Newton. Applying an acute sense for asking the right questions with reasoning, Newton formulated three laws which allowed a complete analysis (mathematical) of **dynamics**, relating all aspects of **motion** to basic causes, **force** and **mass**. So great was Newton's **work** that it is referred to as the first revolution in physics.

Galileo's observation that without **friction** a body would tend to move forever challenged Aristotle's notion that the natural state of motion on Earth was one of rest. **Galileo** deduced that it was a property of **matter** to maintain its state of motion, a property he called **inertia**. Newton, grasping the meaning of inertia and recognizing that Aristotle's reference to what keeps a body in motion (outside influence) really should have been what changes a body's state of motion, set forth a first law of motion which states: A body at rest remains at rest or a body in constant motion remains in constant motion along a straight line unless acted on by an external influence, called force.

(1.) Why use seat belts? Riding in a car you and the car have the same motion. When the brakes are applied, the brakes stop the car. What stops you? Eventually the steering wheel, the dashboard, or the window unless they are replaced by a seat belt, which stops your body. When the accelerator is depressed with the car in gear the motor turns the wheels and the car moves forward. What moves you forward? As the car moves forward the seatback comes forward, contacts, you and pushes you forward.

(2.) While you are riding in the front passenger seat of a car, the driver suddenly turns left. What about you? You continue to move in a straight line until the door to your right, turning left, eventually runs into you. In the car it may appear to you that you slid outward and hit the door.

The first law of motion concentrates on a state of constant motion but adds unless an outside influence, force, acts on it. Force produces a change in the state of motion (**velocity** describes a body's motion); that is, an **acceleration**. Newton found that the greater a body's mass the greater the force required to overcome its inertia and mass is taken as a quantitative measure of a body's inertia. He also found that applying equal force to two different masses, the ratio of their accelerations was inversely proportional to the ratio of their masses. Newton's second law of motion is thus stated: A net force acting on a body produces an acceleration; the acceleration is inversely proportional to its mass and directly proportional to the net force and in the same direction.

This law can be put mathematically $F = ma$ where F is the net force, m is the mass, and a the acceleration. The second law is a cause-effect relationship. The net force acting on a body is determined from all forces acting and the resultant acceleration calculated (assuming a known mass). From the acceleration, velocity and distance traveled can be determined for any time.

(1.) Objects, when released, fall to the ground due to the earth's attraction. Newton's universal law of **gravity** gave the force of attraction between two masses, m and M, as F

$=GmM/R^2$ where G is the gravitational constant and R is the distance between mass centers. This force, **weight**, produces gravitational acceleration g, thus weight $= GmM/R^2 = mg$(2nd Law) giving $g = GM/R^2$. This relationship holds universally. For all objects at the earth's surface, g = 32 ft/sec/sec or 9.0 m/sec^2 downward and on Jupiter 84 ft/sec/sec. Since the dropped object's mass does not appear, g is the same for all objects. Falling objects have their velocity changed downward at the rate of 32 ft/sec each second on earth. Falling from rest, at the end of one second the velocity is 32 ft/sec, after 2 seconds 64 ft/sec, after 3 seconds 96 ft/sec, etc.

For objects thrown upward, gravitational acceleration is still 32 ft/sec/sec downward. A ball thrown upward with an initial velocity of 80 ft/sec has a velocity after one second of 80-32 = 48 ft/sec, after two seconds 48-32 = 16 ft/sec, and after three seconds 16-32 = -16 ft/sec (now downward), etc. At 2.5 seconds the ball had a zero velocity and after another 2.5 seconds it hits the ground with a velocity of 80 ft/sec downward. The up and down motion is symmetrical.

(2.) Friction, a force acting between two bodies in contact, is parallel to the surface and opposite the motion (or tendency to move). By the second law, giving a mass of one kilogram (kg) an acceleration of 1 m/sec/sec requires a force of one Newton (N). However, if friction were 3 N, a force of 4 N must be applied to give the same acceleration. The net force is 4N (applied by someone) minus 3N (friction) or 1N.

Free fall, example (1), assumed no friction. If there were atmospheric friction it would be directed upward since friction always opposes the motion. Air friction is proportional to the velocity; as the velocity increases the friction force (upward) becomes larger. The net force (weight minus friction) and the acceleration are less than due to gravity alone. Therefore, the velocity increases less rapidly, becoming constant when the friction force equals the weight of the falling object (net force=0). This velocity is called the terminal velocity. A greater weight requires a longer time for air friction to equal the weight, resulting in a larger terminal velocity.

(3.) A contemporary and friend of Newton, **Edmond Halley**, observed a comet in 1682 and suspected others had observed it many times before. Using Newton's new **mechanics** (laws of motion and universal law of gravity) Halley calculated that the comet would reappear at Christmas, 1758. Although Halley was dead, the comet reappeared at that time and became known as Halley's comet. This was a great triumph for Newtonian mechanics.

Using Newton's universal law of gravity (see example 1) in the second law results in a general solution (requiring calculus) in which details of the paths of motion (velocity, acceleration, period) are given in terms of G, M, and distance of separation.

While these results agreed with planetary motion known at the time there was now an explanation for differences in motions. These solutions were equally valid for applying to any systems body: Earth's **moon**, Jupiter's moons, galaxies, truly universal.

(4.) Much of Newton's work involved rotational motion, particularly circular motion. The velocity's direction constantly changes, requiring a centripetal acceleration. This centripetal acceleration requires a net force, the centripetal force, acting toward the center of motion. Centripetal acceleration is given by a (central) = v^2/R where v is the velocity's magnitude and R is the radius of the motion. Hence, the centripetal force F (central) = mv^2/R, where m is the mass. These relationships hold for any case of circular motion and furnish the basis for "thrills" experienced on many amusement park rides such as ferris wheels, loop-the-loops, merry-go-rounds, and any other means for changing your direction rather suddenly. Some particular examples follow.

(a.) Newton asked himself why the moon did not fall to the earth like other objects. Falling with the same acceleration of gravity as bodies at the earth's surface, it would have hit the earth. With essentially uniform circular motion about the earth, the moon's centripetal acceleration and force must be due to earth's gravity. With **gravitational force** providing centripetal force, the centripetal acceleration is a (central) = GM/R^2 (acceleration of gravity in example 1 above). Since the moon is about 60 times further from the earth's center than the earth's surface, the acceleration of gravity of the moon is about 0.009 ft/sec/sec. In one second the moon would fall about 0.06 inch but while doing this it is also moving away from the earth with the result that at the end of one second the moon is at the same distance from the earth, R.

(b.) With the gravitational force responsible for centripetal acceleration, equating the two acceleration expressions given above gives the magnitude of the velocity as $v^2 = GM/R$ with the same symbol meanings. For the moon in (a) its velocity would be about 2,250 mph. This relationship can be universally applied.

Long ago it was recognized that this analysis could be applied to artificial "moons" or satellites. If a satellite could be made to encircle the earth at about 200 mi it must be given a tangential velocity of about 18,000 mph and it would encircle the earth every 90 seconds; astronauts have done this many times since 1956, 300 years after Newton gave the means for predicting the necessary velocities.

It was asked: what velocity and height must a satellite have so that it remains stationary above the same point on the earth's surface; that is, have the same rotation period, one day, as the earth. Three such satellites, placed 120 degrees apart around the earth, could make instantaneous communication with all points on the earth's surface possible. From the fact that period squared is proportional to the cube of the radius and the above periods of the moon and satellite and the moon's distance, it is found that the communication satellite would have to be located 26,000 mi from earth's center or 22,000 mi above the surface. Its velocity must be about 6,800 mph. Many such satellites are now in **space** around the earth.

(c.) When a car rounds a curve what keeps it on the road? Going around a curve requires a centripetal force to furnish the centripetal acceleration, changing its direction. If there is not, the car continues in a straight line (first law) moving outward relative to the road. Friction, opposing outward motion, would be inward and the inward acceleration a (cent) = friction/mass = v^2/R. Each radius has a predictable velocity for which the car can make the curve. A caution must be added: when it is raining, friction is reduced and a lower velocity is needed to make the curve safely.

Newton questioned the interacting force an outside agent exerted on another to change its state of motion. He concluded that this interaction was mutual so that when you exert a force on something you get the feeling the other is exerting a force on you. Newton's third law of motion states: When one body exerts a force on a second body, the second body exerts an equal and opposite force on the first body.

In the second law, only forces exerted on a body are important in determining its acceleration. The third law speaks about a pair of forces equal in magnitude and opposite in direction which are exerted on and by two different bodies. This law is useful in determining forces acting on an object by knowing forces it exerts. For example, a book sitting on a table has a net force of zero. Therefore, an upward force equal to its weight must be exerted by the table on the book. According to the third law the book exerts an equal force downward on the table. When two objects are in contact they exert equal and opposite forces on each other and these forces are perpendicular to the contacting surface.

(1.) What enables us to walk? To move forward parallel to the floor we must push backward on the floor with one foot. By the third law, the floor pushes forward, moving us forward. Then the process is repeated with the other foot, etc. This cannot occur unless there is friction between the foot and floor and on a frictionless surface we would not be able to walk.

(2.) How can airplanes fly at high altitudes and space crafts be propelled? High altitude airplanes utilize jet engines; that is, engines burn fuel at high temperatures and expel it backward. In expelling the burnt fuel a force is exerted backward on it and it exerts an equal forward force on the plane. The same analysis applies to space crafts.

(3.) A father takes his eight-year-old daughter to skate. The father and the girl stand at rest facing each other. The daughter pushes the father backwards. What happens? Whatever force the daughter exerts on her father he exerts in the opposite direction equally on her. Since the father has a larger mass his acceleration will be less than the daughter's. With the larger acceleration the daughter will move faster and travel farther in a given time.

Newtonian physics

Newtonian physics are first distinguished from Aristotelian physics and encompass the contributions of **Isaac Newton** (1642-1717), not only as they pertain to the understanding of the physics of **motion** but also as they shaped what is now referred to as the **scientific method**, basically a method which concentrates on a precise, mathematical description of the physical world. With this approach, Newton explained both the motions of heavenly bodies and the motions of objects on or near the surface of Earth by formulating four simple laws: the three laws of motion and the law of universal gravitation. This contribution of Newtonian over pre-Newtonian physics represents one of the most remarkable achievements of mankind's intellectual development. It consisted first of all in radically modifying the two thousand year old prevailing concept of the **universe** which was based on **Aristotle**'s physics. These foundations were first shaken by **Galileo** during the Italian Renaissance and by **Kepler's laws of planetary motion**, but it was Newton who formulated the exact relationship between **force** and motion and revolutionized our view of the universe by showing that the same physical laws apply to all **matter**, whether inert or living. **Aristotle** tried to explain the underlying reasons as to why objects move and he proposed for example that natural motion, such as free-fall, resulted from the objects "wanting" or preferring to lie in their natural state, i.e. in this case, on the ground. He also described another type of motion, "voluntary" motion, such as that shown by a person going from point A to point B because he "wants to." Finally, he believed that a third type of motion could occur, "forced motion," due to an object forcing another to move. Newton formulated the modern concept of force starting with the insight that the only effect which can affect motion are interactions between two objects. Thus, in Newtonian physics, there is only one cause for a change in motion, and it is simply *force*. Forces may be of a different nature, but they all have the same effect, which is to produce changes in the motion of a body and so, we can conclude that two forces are comparable if they produce the same change when applied in the same context. Aristotle also believed that forces could only act on objects which touched each other. Newtonian physics describe such forces as contact forces but also account for noncontact forces, such as **magnetism**.

Newtonian physics is also a term understood to refer to the state of physics before the advent of the theory of relativity in the 1920s. Since that time, Newtonian physics have also been referred to as classical physics or pre-relativistic physics, because its laws describe the physical world in a way which can not account for relativistic gravitational effects. In other words, classical Newtonian laws are valid only insofar as gravitational **potential energy** differences are small compared to mc². But as **velocity** approaches the **speed of light**, they are no longer valid. Thus, the physics of very massive and very fast objects can only be described by relativity. Newtonian physics also fail to describe the physics of very small objects, such as electrons, atoms and elementary particles, which are accounted for by in **quantum theory**, also formulated in the 1920's.

The starting point of Newtonian physics are Newton's three laws of motion. The first one states that all objects have **inertia**; that is, an object will remain at rest or in uniform motion unless acted upon by some outside force. In other words, an object initially at rest will remain at rest if the total force acting on it is zero, and a moving object will remain in motion with the same **velocity** in the same direction. And if the total force on an object is not zero, then the object will accelerate.

Predicting the resulting acceleration is the subject of the second law which states that an object's **acceleration** is in the same direction as the force exerted on it. The force exerted on

that object is equal to the product of its **mass** and acceleration, or to F = ma, and an object's acceleration can thus be predicted given its mass and the force acting on it.

Newton's third law states that for every action there must be an equal and opposite reaction or that forces occur in equal and opposite pairs: whenever object A exerts a force on object B, object B must also be exerting a force on object A. The two forces are equal in magnitude and opposite in direction.

Physics describe motion and the motion of an object relative to a **frame of reference** can defined by its position in a Cartesian x, y, z-coordinate system as a function of time t. The first statement implies that frames of reference are identical to each other and that the laws of **mechanics** apply equally well everywhere. This is because when defining a position, a decision must be made as to where to place x, y and z = 0, and also which respective axis direction will be positive. This involves the selection of the coordinate system and also of a **frame of reference** in which to place the system. The first is concerned with the actual variable x and the second with the observer's point of view. Any frame of reference in which the first law is valid is termed *inertial* and Newtonian physics assume that, if not acted on by an external force, a moving object will then continue its uniform motion relative to the frame of reference. Thus, any transformation from one frame of reference whose velocity is different from the other is then linear and the time coordinate is absolute, i.e., it is the same in any frame of reference. While the second revolution in physics brought about by relativity has clearly shown the limitations of Newton's laws, in that they are only true in frames of reference that are not accelerating, i.e., in inertial frames, and that time is a relative concept, it remains that Newtonian physics, unlike Aristotelian physics, does have a range of validity which accurately describes part of our physical world experience.

See also Force and the laws of motion; Gravity; Lorentz transformations; Magnetism; Momentum and Newton's second law; Newton's laws of motion; Newton's law of universal gravitation; Quantum mechanics; Relativistic velocity transformations; Relativity theory; Space time; Unification of physics

NODES

A node is the point in a periodic system, such as a standing wave, that has no amplitude. The word, node, was derived in the fourteenth century from the Latin *nodus* meaning knot. A node is created when a crest of one wave moving in one direction intersects with a trough from another wave traveling in the opposite direction. The two **waves** cancel each other and create a node that appears to be stationary in a standing wave. There is no **motion** in the standing wave at the node but at a distance of a quarter of a **wavelength** away from each node the wave has a crest, where the amplitude is its largest. These points are called anti-nodes. The distance between successive nodes or anti-nodes is equal to a half of a wavelength. If the standing wave is produced in a string that is clamped at each end to a stationary point then the ends can also be called nodes. These points on the string also remain stationary because they are attached. If the string is loose at one end then this point would be an anti-node because the string would be at maximum amplitude. In order for a **sound** to be produced in a stringed instrument there have to be nodes at each end of the string.

NOETHER, EMMY (1882-1935)

German American mathematician

Emmy Noether was a world-renowned mathematician whose innovative approach to modern abstract algebra inspired colleagues and students who emulated her technique. Dismissed from her university position at the beginning of the Nazi era in Germany—for she was both Jewish and female—Noether emigrated to the United States, where she taught in several universities and colleges. When she died, **Albert Einstein** eulogized her in a letter to *New York Times* as "the most significant creative mathematical genius thus far produced since the higher education of women began."

Noether was born on March 23, 1882, in the small university town of Erlangen in southern Germany. Her first name was Amalie, but she was known by her middle name of Emmy. Her mother, Ida Amalia Kaufmann Noether, came from a wealthy family in Cologne. Her father, Max Noether, a professor at the University of Erlangen, was an accomplished mathematician who worked on the theory of algebraic functions. Two of her three younger brothers became scientists— Fritz was a mathematician and Alfred earned a doctorate in chemistry.

Noether's childhood was unexceptional, going to school, learning domestic skills, and taking piano lessons. Since girls were not eligible to enroll in the gymnasium (college preparatory school), she attended the Städtischen Höheren Töchterschule, where she studied arithmetic and languages. In 1900 she passed the Bavarian state examinations with evaluations of "very good" in French and English (she received only a "satisfactory" evaluation in practical classroom conduct); this certified her to teach foreign languages at female educational institutions.

Begins a Teaching Career

Instead of looking for a language teaching position, Noether decided to undertake university studies. However, since she had not graduated from a gymnasium, she first had to pass an entrance examination for which she obtained permission from her instructors. She audited courses at the University of Erlangen from 1900 to 1902. In 1903 she passed the matriculation exam, and entered the University of Göttingen for a semester, where she encountered such notable mathematicians as **Hermann Minkowski**, Felix Klein, and **David Hilbert**. She enrolled at the University of Erlangen where women were accepted in 1904. At Erlangen, Noether

Amalie Emmy Noether. *(Archive Photos, Inc. Reproduced by permission.)*

studied with Paul Gordan, a mathematics professor who was also a family friend. She completed her dissertation entitled "On Complete Systems of Invariants for Ternary Biquadratic Forms," receiving her Ph.D., summa cum laude, on July 2, 1908.

Noether worked without pay at the Mathematical Institute of Erlangen from 1908 until 1915, where her university duties included research, serving as a dissertation adviser for two students, and occasionally delivering lectures for her ailing father. In addition, Noether began to work with Ernst Otto Fischer, an algebraist who directed her toward the broader theoretical style characteristic of Hilbert. Noether not only published her thesis on ternary biquadratics, but she was also elected to membership in the Circolo Matematico di Palermo in 1908. The following year, Noether was invited to join the German Mathematical Society (Deutsche Mathematiker Vereinigung); she addressed the Society's 1909 meeting in Salzburg and its 1913 meeting in Vienna.

Formulates the Mathematics of Relativity

In 1915, Klein and Hilbert invited Noether to join them at the Mathematical Institute in Göttingen. They were working on the mathematics of the newly announced general theory of relativity, and they believed Noether's expertise would be helpful. Einstein later wrote an article for the 1955 Grolier

Encyclopedia, characterizing the theory of relativity by the basic question, "how must the laws of nature be constituted so that they are valid in the same form relative to arbitrary systems of co-ordinates (postulate of the invariance of the laws of nature relative to an arbitrary transformation of **space** and time)?" It was precisely this type of invariance under transformation on which Noether focused her mathematical research.

In 1918, Noether proved two theorems that formed a cornerstone for general relativity. These theorems validated certain relationships suspected by physicists of the time. One, now known as Noether's Theorem, established the equivalence between an invariance property and a conservation law. The other involved the relationship between an invariance and the existence of certain integrals of the equations of **motion**. The eminent German mathematician Hermann Weyl described Noether's contribution in the July 1935 *Scripta Mathematica* following her death: "For two of the most significant sides of the general theory of **relativity theory** she gave at that time the genuine and universal mathematical formulation."

While Noether was proving these profound and useful results, she was working without pay at Göttingen University, where women were not admitted to the faculty. Hilbert, in particular, tried to obtain a position for her but could not persuade the historians and philosophers on the faculty to vote in a woman's favor. He was able to arrange for her to teach, however, by announcing a class in **mathematical physics** under his name and letting her lecture in his place. By 1919, regulations were eased somewhat, and she was designated a Privatdozent (a licensed lecturer who could receive fees from students but not from the university). In 1922, Noether was given the unofficial title of associate professor, and was hired as an adjunct teacher and paid a modest salary without fringe benefits or tenure.

Noether's enthusiasm for mathematics made her an effective teacher, often conducting classroom discussions in which she and her students would jointly explore some topic. In *Emmy Noether at Byrn Mawr*, Noether's only doctoral student at Bryn Mawr, Ruth McKee, recalls, "Miss Noether urged us on, challenging us to get our nails dirty, to really dig into the underlying relationships, to consider the problems from all possible angles."

Lays the Foundations of Abstract Algebra

Brilliant mathematicians often make their greatest contributions early in their careers; Noether was one of the notable exceptions to that rule. She began producing her most powerful and creative work about the age of 40. Her change in style started with a 1920 paper on noncommutative fields (systems in which an operation such as multiplication yields a different answer for a x b than for b x a). During the years that followed, she developed a very abstract and generalized approach to the axiomatic development of algebra. As Weyl attested, "she originated above all a new and epoch-making style of thinking in algebra."

Noether's 1921 paper on the theory of ideals in rings is considered to contain her most important results. It extended the work of Dedekind on solutions of polynomials—algebraic

expressions consisting of a constant multiplied by variables raised to a positive power—and laid the foundations for modern abstract algebra. Rather than working with specific operations on sets of numbers, this branch of mathematics looks at general properties of operations. Because of its generality, abstract algebra represents a unifying thread connecting such theoretical fields as logic and number theory with applied mathematics useful in chemistry and physics.

During the winter of 1928–29, Noether was a visiting professor at the University of Moscow and the Communist Academy, and in the summer of 1930, she taught at the University of Frankfurt. Recognized for her continuing contributions in the science of mathematics, the International Mathematical Congress of 1928 chose her to be its principle speaker at one of its section meetings in Bologna. In 1932 she was chosen to address the Congress's general session in Zurich.

Noether was a part of the mathematics faculty of Göttingen University in the 1920s when its reputation for mathematical research and teaching was considered the best in the world. Still, even with the help of the esteemed mathematician Hermann Weyl, Noether was unable to secure a proper teaching position there, which was equivalent to her male counterparts. Weyl once commented: "I was ashamed to occupy such a preferred position beside her whom I knew to be my superior as a mathematician in many respects." Nevertheless, in 1932, on Noether's fiftieth birthday, the university's algebraists held a celebration, and her colleague Helmut Hasse dedicated a paper in her honor, which validated one of her ideas on noncommutative algebra. In that same year, she again was honored by those outside her own university, when she was named cowinner of the Alfred Ackermann-Teubner Memorial Prize for the Advancement of Mathematical Knowledge.

Teaches in Exile

The successful and congenial environment of the University of Göttingen ended in 1933, with the advent of the Nazis in Germany. Within months, anti-Semitic policies spread through the country. On April 7, 1933, Noether was formally notified that she could no longer teach at the university. She was a dedicated pacifist, and Weyl later recalled, "her courage, her frankness, her unconcern about her own fate, her conciliatory spirit were, in the midst of all the hatred and meanness, despair and sorrow surrounding us, a moral solace."

For a while, Noether continued to meet informally with students and colleagues, inviting groups to her apartment. But by summer, the Emergency Committee to Aid Displaced German Scholars was entering into an agreement with Bryn Mawr, a women's college in Pennsylvania, which offered Noether a professorship. Her first year's salary was funded by the Emergency Committee and the Rockefeller Foundation.

In the fall of 1933, Noether was supervising four graduate students at Bryn Mawr. Starting in February 1934, she also delivered weekly lectures at the Institute for Advanced Study at Princeton. She bore no malice toward Germany, and maintained friendly ties with her former colleagues. With her characteristic curiosity and good nature, she settled into her new home in America, acquiring enough English to adequately converse and teach, although she occasionally lapsed into German when concentrating on technical material.

During the summer of 1934, Noether visited Göttingen to arrange shipment of her possessions to the United States. When she returned to Bryn Mawr in the early fall, she had received a two-year renewal on her teaching grant. In the spring of 1935, Noether underwent surgery to remove a uterine tumor. The operation was a success, but four days later, she suddenly developed a very high fever and lost consciousness. She died on April 14th, apparently from a post-operative infection. Her ashes were buried near the library on the Bryn Mawr campus.

Over the course of her career, Noether supervised a dozen graduate students, wrote forty-five technical publications, and inspired countless other research results through her habit of suggesting topics of investigation to students and colleagues. After World War II, the University of Erlangen attempted to show her the honor she had deserved during her lifetime. A conference in 1958 commemorated the fiftieth anniversary of her doctorate; in 1982 the university dedicated a memorial plaque to her in its Mathematics Institute. During the same year, the 100th anniversary year of Noether's birth, the Emmy Noether Gymnasium, a coeducational school emphasizing mathematics, the natural sciences, and modern languages, opened in Erlangen.

NON-EUCLIDEAN GEOMETRY

Non-Eucildean geometry is the study of spatial relationships in which the Euclidean parallel postulate is altered or replaced to produce a system in which the traditional properties of **space** are altered.

As a 12-year-old, German-American physicist **Albert Einstein** was given a small book on Euclidean plane geometry. More than half a century later, Einstein referred to it as "that holy geometry booklet." In 1922, in her poem "The Harp Weaver," American poet Edna St. Vincent Millay wrote that "Euclid alone has looked on beauty bare." That beauty, viewed in the third century B.C. by the most prominent mathematician of antiquity, **Euclid** of Alexandria, concerns the relationships between points and lines lying in a two-dimensional space, a plane.

Move a point uniformly, and a straight line is produced. Move that straight line in a direction perpendicular to itself to produce a plane. Euclid demonstrated the properties of figures in a plane (i.e., constructed his geometry) through logical proofs, starting with a number of axioms and postulates. These cannot themselves be proven, but are so intuitive as to be universally accepted. All of Euclid's assumptions, with one exception, seem obviously true. The exception, however, is a crucial one. Euclid's fifth postulate states that, if a straight line falling on two straight lines makes the interior angles on the

same side less than two right angles, the two straight lines if extended indefinitely meet on that side on which the angles are less than two right angles. This postulate, which may be restated as in a given plane, through a given point, there exists not more than one parallel to a given line, constitutes Euclidean geometry. It is needed, for example, to show that the sum of interior angles of a triangle equals 180°. As it happens, the fifth postulate is not required for a consistent geometry to be built.

By 1832, three mathematicians, **Carl Friedrich Gauss**, János Bolyai (1802-1860), and Nikolai Lobachevsky (1792-1856), had independently discovered an infinite two-dimensional space in which all of Euclid's postulates except the fifth are satisfied. Called the space of constant negative curvature, it is most easily visualized as the surface of a saddle, and on it an infinite number of lines can be drawn parallel to a given line. On the space of constant negative curvature, the sum of interior angles of a triangle is less than 180°. On a different surface, that of a sphere (a space of constant positive curvature), no line can be drawn parallel to a given line (since all such lines meet at a pole), and the sum of interior angles of a triangle is greater than 180°. Euclid's plane is flat while these other surfaces are not.

Each of these two-dimensional curved surfaces is said to be non-Euclidean because Euclid's parallel postulate is not obeyed. On each surface, a complete and consistent geometry exists. The properties of figures drawn on these surfaces may be deduced via the remaining postulates. The curvature of these surfaces is easily visualized. A spherical surface, i.e. the surface of a globe, can be seen as it sits in a three-dimensional space, a room. This is termed embedding; the non-Euclidean surfaces are embedded in a higher dimensional Euclidean space.

In 1827, Gauss showed that curved surfaces had both extrinsic properties (those which are due to their embedding in a higher dimensional space), and intrinsic properties (those which are not due to embedding). The surface of a cylinder and a plane have the same intrinsic curvature, since a plane may be rolled up into a cylinder, but a spherical surface and a plane do not. A sphere cannot be flattened into a plane without distortion. Gauss discovered a function, today known as the Gaussian curvature, K, calculable for any surface, which is a measure of the intrinsic curvature.

Gauss dealt only with two-dimensional surfaces. For dimensions greater than two, the description of the geometry becomes more complicated. In 1854, Georg Friedrich Bernhard Riemann (1826-1866) generalized Gauss's work for spaces of arbitrary dimension. It is the Riemannian geometry that was employed by Einstein in 1915 in his general theory of relativity.

In non-Euclidean geometrical applications space is curved and the degree of curvature is related to localized **mass**. Einstein's general theory of relativity advances the concept that the **gravitational force** is a result of an intrinsic curvature of spacetime (a fusion of three dimensional space and time) that can be described as non-Euclidean spacetime. In such desccriptions what appear to be linear displacements are actually displacements along a curved or geodesic path.

See also Relativity, general; Space-time geometry

NONLINEAR DYNAMICS

Nonlinear **dynamics** describes systems in which the forces acting on those systems do not increase linearly with quantities such as position, **mass** or **velocity**. It also describes systems which exhibit a "sensitive dependence on their initial conditions." This expression was first coined in the 1960s by the meteorologist Edward Lorenz, who was working on **weather** prediction models. He realized that a weather parameter curve would diverge from the original input curve if the starting points of the two curves were infinitesimally different. This behavior is now referred to as the butterfly effect, i.e. small changes in initial conditions may lead to qualitatively and quantitatively different long-term behavior. It led Lorenz to conclude that it was impossible to predict the weather accurately because many parameters can cause slight changes to a weather system at one time thus causing the later behavior of the meteorological system to become completely different. This insight also led to the rapid development of chaos theory, a cornerstone of nonlinear dynamics. Mathematically, nonlinearity also implies that a small variation in an input quantity causes a considerable change in the response of the system for which linear algebraic methods are no longer adequate to find solutions to the stated problem.

Nonlinear dynamics are important because no model of a real system is truly linear and as such, nonlinear dynamics apply very concretely to a wide variety of problems in fields such as biology (complex patterns of neuronal growth), **chemistry** (diffusion processes in chemical reactions), economics (the irregular fluctuations of the stock market), engineering (instabilities at rigid interfaces), geology (shock wave patterns in earthquakes), medicine (properties of normal and pathological biological cells), materials science (stress propagation in crystals and glassy materials), **meteorology** (hurricane prediction) and mining (dynamics of fluidized flotation beds), to give but a few examples.

Aside from sensitive dependence on initial conditions, the distinctive behavior of nonlinear systems is also characterized by spatial or temporal symmetry-breaking, which means that the temporal response of the system is not the same as that of the **force** driving the system, or that the spatial patterns will be independent from the boundary conditions. A good example is the case of the nonlinear oscillator. A linear or simple harmonic oscillator, such as an object oscillating on a spring, is characterized by a type of **motion** described by Hooke's law and can be the sum of many contributing harmonic motions if the oscillator is complex. Such an oscillator will always be displaced by a force acting on the mass which increases linearly with its displacement. But the motions involved in the nonlinear oscillator are not the simple linear combinations of the harmonic **oscillations** and

so the nonlinear oscillator will oscillate at several amplitudes unlike the linear oscillator which is characterized by a single resonant maximum amplitude. The contributing factors to the nonlinear behavior of such an oscillator are **friction**, damping forces, or other input elements which vary nonlinearly and which cause the oscillator to be described not by a parabolic potential curve, as is the case with the harmonic oscillator, but by a multitude of non-parabolic potentials. This translates into differential equations with solutions that can only be solved numerically, if at all, while the linear oscillator has analytical solutions.

Nonlinear dynamics describes instabilities, chaos, and pattern formation in systems driven far from thermodynamic **equilibrium**. Such systems are referred to as turbulent systems and they have a special importance in nonlinear dynamics. The major goal of **turbulence** physics, now understood to be a function of dynamical chaos in nonlinear systems, is to predict the effects of turbulence and investigate means of controlling it in applications as diverse as industrial mixers and burners, car engines, **nuclear reactors**, aircraft and ship turbines, and spray nozzles.

Examples of everyday nonlinear phenomena include the erosion pattern created by a stream of water as it flows along a path or a mound of sand generated by a gentle flow of sand piling up on the ground. The stream of water generates repetitive miniature mountains, ponds, and secondary streams as it winds through the soil. This is one of nature's most complex processes and it can be described by highly nonlinear fractal patterns. In the case of the sand, a complex equilibrium keeps the mound stable until it reaches the point of criticality, i.e. the point at which addition of more sand pebbles will cause collapse of the mound. After the collapse, a new mound starts building up if the sand flow is not interrupted. Each new failure cycle relieves **pressure** on the mound and similar patterns of alternating growth and failure are characteristic of many complex systems. But no matter how distinct such complex systems may be, they all reach their respective points of criticality via bifurcations, or dramatic changes of behavior which occur over extremely small parameter ranges. Whereas linear dynamical systems depend on quantities such as **temperature** or pressure which, if changed gradually, produce predictable and gradual variations in the system, nonlinear systems become unpredictable when the point of criticality is abruptly exceeded. Bifurcation theory studies the onset of turbulence in nonlinear systems as some control parameter is continuously varied.

Another important theoretical concept in nonlinear dynamics is that of phase **space**, which consists of the independent dynamical variables used as coordinates to define the space for the system and represents the collection of possible states for that system. An example is the coin toss which has two states in phase space: heads or tails. For a deterministic system, the future state of the system depends on the current state of the system, represented as a point in phase space. The classical pendulum also exists in two-dimensional phase space and according to Newton's laws, knowing its two states (i.e., position and velocity), we can predict its subsequent motion. One or two-dimensional phase space does not exhibit nonlinear behavior, but it becomes possible in three-dimensional phase spaces, which often consist of a classical two-phase state system to which the time dimensionality is added. As the system progresses in time, a trajectory of its state-space evolution is obtained and can be mapped to yield a function of the phase space that will yield the next state of the system. These phase maps have geometrical properties and nonlinear systems are often characterized by chaotic fractal patterns.

Nonlinear dynamics have had a crucial impact on twentieth-century physics. Its contributions are considered as important as those of **quantum mechanics** and relativity in shaping the new physics because manifestations of nonlinearity are an inescapable part of our real physical world.

See also Changes of phase; Chaos and order; Determinism; Dynamics; Fractals; Friction; Game theory; Harmonic motion; Newton's laws; Poincare, Henri; Probability wave; Resonance; Singularity; Statistical physics

NON-RENEWABLE ENERGY RESOURCES

Non-renewable **energy** resources are sources of energy drawn from nature that cannot be replenished. These include wood, coal, petroleum and radioactive material (for **nuclear fission**). The possibility of exhausting these energy resources is an increasingly serious problem in the modern world, as consumption increases with expanding worldwide industrialization and the growing world population. The extraction and refinement of non-renewable energy resources has often caused pollution and interfered with the aesthetics and utility of the environment. Awareness of the scope of the problem has encouraged conservation efforts and the use of alternative fuel sources.

Wood is one of the oldest sources of energy, and for centuries seemed inexhaustible. With the onset of the Industrial Revolution, however, the amount of available wood was not enough to meet the needs of increasing numbers of factories and mills. In addition, other uses of wood (e.g., in furniture and housing) decreases quantities available for use as fuel. Although wood can be replenished as trees are replanted, only recently has the lumber industry invested in large-scale replanting.

Coal became an important source of energy in the Middle Ages. Coal usage gained in prominence as local wood supplies dwindled. Coal was the primary source of **heat** for **steam engines** and was used in furnaces to heat houses and factories. Although a potent fuel, the mining of coal posed dangers to miners exposed to cave-ins and black lung disease (a health hazard developed through breathing coal dust). Strip mining, once a common practice, rendered the surface land above the mine unsightly and useless. In addition to mining problems, refining coal involves the release of dangerous gas that pollute the atmosphere and contribute to acid rain.

Petroleum naturally seeping out of the ground has long been used as a source of heat. The modern reliance on oil began with the first commercial drill built in 1859. Petroleum can be used to produce three fundamental types of petroleum derivative fuels: kerosene, gasoline and heating oil. The distribution of petroleum from its source to its users is hampered by unsure methods of exploring oil fields, the lack of the necessary technology to efficiently excavate the oil reserves, and the logistical problems often involved in the transportation of oil. Although modern technology can more effectively point out the most likely location of the oil field, the only sure way of finding oil is to drill. In many cases exploratory drilling comes up dry. Furthermore, the oil in a reserve must also be able to flow into the well from the surrounding rock. Depending on the position of the drill, two-thirds of the available supply can be expected to be left behind.

Nuclear **power** is seen by some as an alternative to the use of petroleum and other fuels. Nuclear fission as a source of energy, however, still relies on non-renewable energy sources. An important component of nuclear energy is uranium which, like other resources, has to be mined and is in limited supply. Moreover, nuclear fission creates a by-product of spent fuel that cannot be easily used and remains extremely dangerous to handle for thousands of years. The storage of nuclear waste is an important concern for scientists and engineers.

Alternative sources of energy that do not rely on non-renewable fuels generally include utilization of solar electromagnetic **radiation** or a thermal derivative of insolation such as wind power. Various methods of concentrating the heat of the **sun** are being developed to use in houses and commercial buildings. Wind power relies on windmills to turn the turbine of an electric generator.

See also Nuclear fission; Solar-thermal energy

NOVAE AND SUPERNOVAE

Although the names appear similar, novae and supernovae actually represent vastly different physical phenomena from result from quite different physical processes. A nova results from the explosion of a white dwarf star. The explosion results from the increased **mass** of the white dwarf caused by **matter** falling onto its surface from a binary companion (usually a red giant star). Novae **stars** greatly increase from tens to hundreds of thousands of times in brightness (luminosity) before slowly fading back to normal luminosity. A supernova is a much more energetic event that in the course of **stellar evolution** marks the cataclysmic end of star's normal hydrogen fusion burning lifetime. Supernova explosions are among the most energetic events of the **universe** and a supernova may produce more luminosity that all of the other stars on the supernova's galaxy combined.

Stellar observations have provided ample evidence that some stars vary their brightness slowly, in a predictable fashion, whereas others flare up quite suddenly and unexpectedly. On rare occasions, a single star may become so bright that it

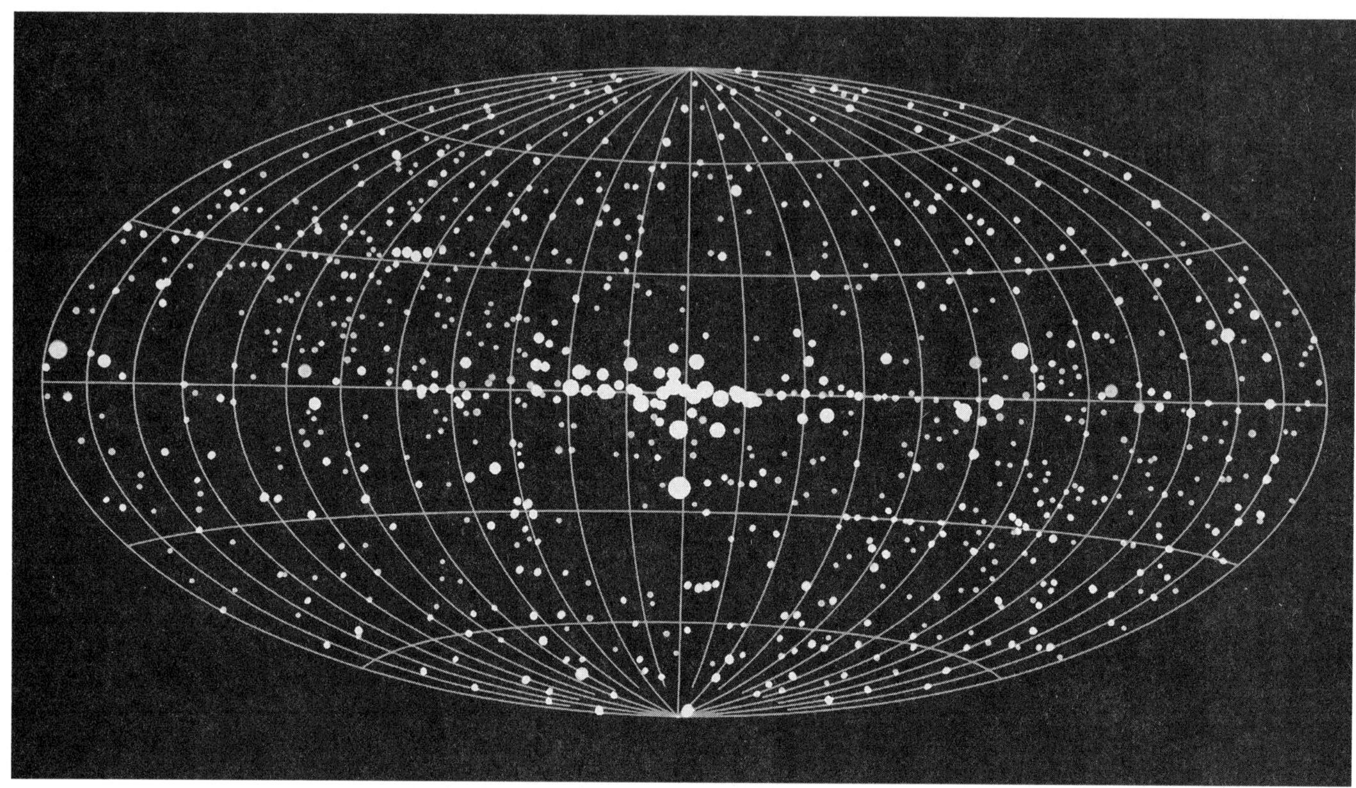

X-ray image of supernova explosion. *(Image courtesy of Naval Research Laboratory. Reproduced by permission.)*

outshines all of the other stars in the sky, and may even become visible even during daytime. These explosive events are commonly known as novae and supernovae, both terms being derived from *nova* (Latin for "new"). This term, is, however, misleading as both nova and supernova processes occur in mature stars.

Classical novae outbursts (referred to in the singular as a nova) originate when a white dwarf (a very compact star made primarily of **carbon**) with a mass comparable to that of our **Sun** in the volume of Earth) accretes hydrogen-rich matter from the envelope of a companion star. The accreting matter forms a thin, rotating disk surrounding the white dwarf, which acts as a kind of reservoir to store matter from the envelope of the companion. Matter then accretes from the disk onto the surface of the white dwarf. Such disks are subject to a wide variety of instabilities, which can cause the accretion rate of the disk onto the star to increase by orders of magnitude over a short time. In such systems, a critical point is reached which allows a sustained nuclear **chain reaction** to begin at the surface of the white dwarf. The reaction then rapidly proceeds to consume the accreted matter in a giant explosion. After some length of time, more mass is accumulated, the process is repeated, and the white dwarf can once again undergoes a novae class explosion. Although the phenomena may vary in intervals as the stars are pushed apart by explosions and attracted by **gravitational force**, this process may repeated many times. According to current theory, the cycle

most likely ends when the stars collide and form a larger unstable star that undergoes carbon detonation.

Swiss astronomer and physicist Fritz Zwicky (1898-1974) was the first astronomer to realize the significance of American astronomer **Edwin Hubble**'s discovery of the extra-galactic length scale for novae studies. Hubble found the distance to many nearby galaxies by identifying individual Cepheid variable stars within them and determining their intrinsic luminosity from the observed period of variability. He discovered that the host galaxies (or **nebulae** as they are now known) which the Cepheids were associated with lay vastly farther away from us than previously expected (i.e., the nebulae he identified were huge distant galaxies much like our own that were very far away). Zwicky realized that this made the intrinsic luminosity of many already-known novae events vastly greater than previously imagined. He termed these much more energetic explosions supernovae and undertook a systematic study of them in the 1930s. Today we understand these events to be orders of magnitude more energetic than classical novae outbursts.

Supernovae are broadly classified into two types: type I supernovae show no hydrogen in their spectra whereas type II supernovae do. There are many subdivisions of these classes, which are indicated by a letter next to the major class. Subclasses based on spectra are indicated by smallcase letters (i.e., type Ia, Ib, Ic, etc.), and subclasses based on **light** curves are indicated by uppercase letters (i.e., type IIL and IIP).

(Zwicky introduced types III, IV, and V, which were motivated by anomalous supernovae that, curiously, all occurred in 1961. They are largely forgotten today.) The kind of supernovae present in a host galaxy is tightly correlated to its type: supernovae types Ib, Ic, and II typically occur in spiral and irregular galaxies, and supernovae type Ia occur in all galaxies. This correlation between host galaxy and supernovae type is a very important piece of evidence, since it is also known that one of the key differences that sets spirals and irregular galaxies apart from other types is that they are sites of active star formation.

The evidence strongly suggests that supernovae types Ib, Ic, and II are the result of explosions of massive stars tha occur at the point where they can no longer sustain the **energy** they need through **nuclear fusion** in their cores. Theoretical models indicate that the imploding core sets off a strong shock wave which launches through the star, and eventually breaks out from the star itself. The resultant explosion is one of the most energetic phenomena in the universe: over a period of a few weeks, the supernova will typically liberate about as much energy as a star like our Sun will over 10 billion years. During that time, the supernova can outshine all of the billions of stars in its host galaxy.

Supernovae type Ia, on the other hand, are believed to be the result of accretion onto a white dwarf star (a very compact star; with a mass comparable to that of our Sun in the volume of Earth) from its companion in a binary star system, much as in the case of classical novae outbursts. However, in this case, the energy of the explosion is far greater, and there is no remnant left behind. It is believed that the accreted matter put the total mass of the white dwarf above the Chandresekhar mass limit (the maximum mass which a white dwarf star can have; about 1.4 times the mass of our Sun), and the white dwarf then completely destroys itself in the resultant explosion.

See also Stellar evolution; Stellar life cycle

NUCLEAR COUNTERTERRORISM

The responsibility for investigating and countering terror threats against U.S. targets has long rested with the Federal Bureau of Investigation. The unique nature of **nuclear weapons** and other weapons of **mass** destruction, however, have necessitated contingency planning that transcend any one agency's jurisdiction, should terrorists ever attempt to employ such weapons against the United States.

The Nuclear Emergency Search Team (NEST), part of the Department of Energy, was formed in 1975 in anticipation of just such a threat. Their responsibilities are threefold; first, to investigate possible cases of threatened nuclear terrorism. Second, if the threat proves credible, to locate the nuclear device. Third, to disarm or contain the device and its nuclear material. At all levels, NEST works with appropriate local, state and federal agencies, including the FBI, CIA, and FEMA (Federal Emergency Management Agency), as circumstances dictate.

In the early years of NEST's existence, the possibility of nuclear terror was seen as remote. Terrorist organizations are by nature isolated and are little trusted, even by their sponsors. The production of even rudimentary atomic weapons is complex enough to be beyond their abilities, and it was felt they had little access to experts or the material needed to build and **atom** bomb. Fissile material in particular, such as plutonium and enriched uranium, require a massive industrial base to produce and have long been carefully controlled by the national powers that have access to them.

All this changed in the early 1990s with the breakup of the Soviet Union and the destabilization of many of the former Soviet republics. The fear has arisen that cash-strapped republics may illicitly sell nuclear weapons or material to terror organizations, or else that unemployed nuclear technicians and scientists may be lured to work for terrorists or rogue states.

So far, NEST has mobilized less than a hundred times in response to threats of nuclear terror. All such threats have turned out to be hoaxes. When NEST mobilizes, its scientists fan out across a city (it is usually assumed that terrorists would attack urban areas, in order to maximize casualties) employing inconspicuous detection gear in an attempt to find a nuclear device prior to detonation. The search, though intense, is usually done in as low-key a manner as possible in order to avoid a panic. The notion of evacuating a major city was long ago ruled out as impossibly impractical and far too time-consuming. Instead, spy planes circle overhead, searching for clues as dozens or hundreds of two-person teams spread out and try to narrow the search with portable detection equipment. Contrary to popular belief, most nuclear devices do not emit large amounts of **gamma radiation**, so the NEST searchers must come within a few feet of a bomb to find it. NEST has practiced its detection procedures in numerous training exercises, and has thus far always managed to located simulated nuclear weapons, although it can take up to several days to do so.

In addition to the threat of atomic detonation, NEST and FEMA recognize the danger of a terrorist attack utilizing a radiological **dispersion** device; that is, a weapon designed to disperse **radioactive fallout** over a wide area. Such a device would be much easier for a terrorist organization to produce, as it requires little more than a quantity of conventional explosive and some type of radioactive material; many of which are easily found at universities and hospitals. Another source of weapons material might actually come from U.S. stockpiles. Although nuclear material in the United States are among the most carefully guarded in the world, the U.S. military admits that at least a dozen nuclear weapons were lost and never found throughout the Cold War years. Although most of these are likely to be irretrievable and inoperable, such as the one that lies thousands of feet deep off New Jersey's Atlantic coast, the possibility remains that at the very least a radiological dispersion device could be fashioned should one of these lost weapons ever be located. The rest of the world's nuclear powers, including Russia, China, France, and the United Kingdom, have never admitted to losing a nuclear weapon, although U.S. experts say it is almost inevitable that such incidents have occurred.

At present it seems that the United States' best protection against nuclear terrorism is preparation. While NEST and FEMA train and plan for nuclear emergencies, the CIA and FBI monitor foreign and domestic threats, and move to counter the credible ones before they get off the ground. The State Department has had great success in identifying and isolating rogue states that traffic in nuclear material. The United States has even been known to pay the salaries of former Soviet scientists to ensure that they do not seek employment elsewhere.

NUCLEAR FISSION

Nuclear fission is a process in which the **nucleus** of an **atom** splits, usually into two pieces. This reaction was discovered when a target of uranium was bombarded by neutrons. Fission fragments were shown to fly apart with a large release of **energy**. The fission reaction was the basis of the **atomic bomb**, which was developed by the United States during World War II. After the war, controlled energy release from fission was applied to the development of **nuclear reactors**. Reactors are utilized for production of **electricity** at nuclear **power** plants, for propulsion of ships and submarines, and for the creation of radioactive isotopes used in medicine and industry.

The fission reaction was discovered in 1938 by two German scientists, **Otto Hahn** and Fritz Strassmann. They had been doing a series of experiments in which they used neutrons to bombard various elements. If they bombarded copper, for example, a radioactive form of copper was produced. Other elements became radioactive in the same way. When uranium was bombarded with neutrons, however, an entirely different reaction seemed to occur. The uranium nucleus apparently underwent a major disruption.

The evidence for this supposed process came from chemical analysis. Hahn and Strassmann published a scientific paper showing that small amounts of barium (element 56) were produced when uranium (element 92) was bombarded with neutrons. It was very puzzling to them how a single **neutron** could transform element 92 into element 56.

Lise Meitner, a long-time colleague of Hahn who had left Germany due to Nazi persecution, suggested a helpful model for such a reaction. One can visualize the uranium nucleus to be like a liquid drop containing protons and neutrons. When an extra neutron enters, the drop begins to vibrate. If the vibration is violent enough, the drop can break into two pieces. Meitner named this process "fission" because it is similar to the process of cell division in biology. It takes only a small amount of hyperlink energy to start the vibration which leads to a major breakup.

Scientists in the United States and elsewhere quickly confirmed the idea of uranium fission, using other experimental procedures. For example, a **cloud chamber** is a device in which vapor trails of moving nuclear particles can be seen and photographed. In one experiment, a thin sheet of uranium was placed inside a cloud chamber. When it was irradiated by neutrons, photographs showed a pair of tracks going in opposite directions from a common starting point in the uranium. Clearly, a nucleus had been photographed in the act of fission.

Another experimental procedure used a Geiger counter, which is a small, cylindrical tube that produces electrical **pulses** when a radioactive particle passes through it. For this experiment, the inside of a modified Geiger tube was lined with a thin layer of uranium. When a neutron source was brought near it, large **voltage** pulses were observed, much larger than from ordinary **radioactivity**. When the neutron source was taken away, the large pulses stopped. A Geiger tube without the uranium lining did not generate large pulses. Evidently, the large pulses were due to uranium fission fragments. The size of the pulses showed that the fragments had a very large amount of energy.

To understand the high energy released in uranium fission, scientists made some theoretical calculations based on Albert Einstein's famous equation $E=mc^2$. The Einstein equation says that **mass** m can be converted into energy E, and the conversion factor is a huge number, c, which is the **velocity** of **light**, squared. One can calculate that the total mass of the fission products remaining at the end of the reaction is slightly less than the mass of the uranium atom plus neutron at the start. This decrease of mass, multiplied by itself, shows numerically why the fission fragments are so energetic.

Through fission, neutrons of low energy can trigger off a very large energy release. With the imminent threat of war in 1939, a number of scientists began to consider the possibility that a new and very powerful "atomic bomb" could be built from uranium. Also, they speculated that uranium perhaps could be harnessed to replace coal or oil as a fuel for industrial power plants.

Nuclear reactions in general are much more powerful than chemical reactions. A chemical change such as burning coal or even exploding TNT affects only the outer electrons of an atom. A nuclear process, on the other hand, causes changes among the protons and neutrons inside the nucleus. The energy of attraction between protons and neutrons is about a million times greater than the chemical **binding energy** between atoms. Therefore, a single fission bomb, using nuclear energy, might destroy a whole city. Alternatively, nuclear electric power plants theoretically could run for a whole year on just a few tons of fuel.

In order to release a substantial amount of energy, many millions of uranium nuclei must split apart. The fission process itself provides a mechanism for creating a so-called "chain reaction." In addition to the two main fragments, each fission event produces two or three extra neutrons. Some of these can enter nearby uranium nuclei and cause them in turn to fission, releasing more neutrons, which cause more fissions, and so forth. In a bomb explosion, neutrons have to increase very rapidly, in a fraction of a second. In a controlled reactor, however, the neutron population has to be kept in a steady state. Excess neutrons must be removed by some type of absorber material.

In 1942, the first nuclear reactor with a self-sustaining **chain reaction** was built in the United States. The principal designer was **Enrico Fermi**, an Italian physicist and the 1938 Nobel Prize winner in physics. He had emigrated to the United States to escape from Benito Mussolini's fascism. Fermi's reactor design had three main components: lumps of uranium (the fuel), blocks of **carbon** (the moderator, which slows down the neutrons), and control rods made of cadmium (an excellent neutron absorber). The reactor was built at the University of Chicago. When the pile of uranium and carbon blocks was about 10 ft (3 m) high and the cadmium control rods were pulled out far enough, Geiger counters showed that a steady-state chain reaction had been successfully accomplished. The power output was only about 200 watts, but it was enough to verify the basic principle of reactor operation. The power level of the chain reaction could be varied by moving the control rods in or out.

General Leslie R. Groves was put in charge of the project to convert the chain reaction experiment into a usable military weapon. Three major laboratories were built under wartime conditions of urgency and secrecy. Oak Ridge, Tennessee, became the site for purifying and separating uranium into bomb-grade material. At Hanford, Washington, four large reactors were built to produce another possible bomb material, plutonium. At Los Alamos, New Mexico, the actual work of bomb design was started in 1943 under the leadership of physicist **J. Robert Oppenheimer.**

The element uranium is a mixture of two isotopes, uranium-235 and uranium-238. Both isotopes have 92 protons in the nucleus, but uranium-238 has three additional neutrons. Both isotopes have 92 orbital electrons to balance the 92 protons, so their chemical properties are identical. When uranium is bombarded with neutrons, the two isotopes have differing nuclear reactions. A high percentage of the uranium-235 nuclei undergo fission, as described previously. The uranium-238, on the other hand, simply absorbs a neutron and is converted to the next heavier **isotope**, uranium-239. It is not possible to build a bomb out of natural uranium. The reason is that the chain reaction would be halted by uranium-238 because it removes neutrons without reproducing any new ones.

The fissionable isotope, uranium-235, constitutes only about 1% of natural uranium, while the non-fissionable neutron absorber, uranium-238, makes up the other 99%. To produce bomb-grade, fissionable uranium-235, it was necessary to build a large isotope separation facility. Since the plant would require much electricity, the site was chosen to be in the region of the Tennessee Valley Authority (TVA). The technology of large-scale isotope separation involved solving many difficult, unprecedented problems. By early 1945, the Oak Ridge Laboratory was able to produce kilogram amounts of uranium-235 purified to better than 95%.

An alternate possible fuel for a fission bomb is plutonium-239. Plutonium does not exist in nature but results from **radioactive decay** of uranium-239. Fermi's chain reaction experiment had shown that uranium-239 can be made in a reactor. However, to produce several hundred kilograms of plutonium required a large increase from the power level of

Fermi's original experiment. Plutonium production reactors were constructed at Hanford, Washington, located near the Columbia river to provide needed cooling water. A difficult technical problem was how to separate plutonium from the highly radioactive fuel rods after irradiation. This was accomplished by means of remote handling apparatus that was manipulated by technicians working behind thick protective glass windows.

With uranium-235 separation started at Oak Ridge and plutonium-239 production under way at Hanford, a third laboratory was set up at Los Alamos, New Mexico, to work on bomb design. In order to create an explosion, many nuclei would have to fission almost simultaneously. The key concept was to bring together several pieces of fissionable material into a so-called "critical mass." In one design, two pieces of uranium-235 were shot toward each other from opposite ends of a cylindrical tube. A second design used a spherical shell of plutonium-239, to be detonated by an "implosion" toward the center of the sphere.

The first atomic bomb was tested at an isolated desert location in New Mexico on July 16, 1945. President Truman then issued an ultimatum to Japan that a powerful new weapon could soon be used against them. On August 8, a single atomic bomb destroyed the city of Hiroshima with over 80,000 casualties. On August 11, a second bomb was dropped on Nagasaki with a similar result. The Japanese leaders surrendered three days later.

The decision to use the atomic bomb has been vigorously debated over the years. It brought a quick end to the war and avoided many casualties that a land invasion of Japan would have cost. However, the civilians who were killed by the bomb and the survivors who developed **radiation** sickness left an unforgettable legacy of fear. The horror of mass annihilation in a nuclear war is made vivid by the images of destruction at Hiroshima. The possibility of a ruthless dictator or a terrorist group getting **nuclear weapons** is a continuing threat to world peace.

The first nuclear reactor designed for producing electricity was put into operation in 1957 at Shippingsport, Pennsylvania. From 1960-1990, more than 100 **nuclear power** plants were built in the United States. These plants now generate about 20% of the nation's electric power. Worldwide, there are over 400 nuclear power stations.

The most common reactor type is the pressurized water reactor (abbreviated PWR). The system operates like a coal-burning power plant, except that the firebox of the coal plant is replaced by a reactor. Nuclear energy from uranium is released in the two fission fragments. The fuel rod becomes very hot because of the cumulative energy of many fissioning nuclei. A typical reactor core contains hundreds of these fuel rods. Water is circulated through the core to remove the **heat.** The hot water is prevented from **boiling** by keeping the system under **pressure.**

The pressurized hot water goes to a heat exchanger where steam is produced. The steam then goes to a turbine, which has a series of fan blades that rotate rapidly when hit by the steam. The turbine is connected to the rotor of an electric generator. Its output goes to cross-country transmission lines that supply the electrical users in the region. The steam that made the turbine rotate is condensed back into water and is recycled to the heat exchanger. This method of generating electricity was developed for coal plants and is known to be very reliable.

Safety features at a nuclear power plant include automatic shutdown of the fission process by insertion of control rods, emergency water cooling for the core in case of pipeline breakage, and a concrete containment shell. It is impossible for a reactor to have a nuclear explosion because the fuel enrichment in a reactor is intentionally limited to about 3% uranium-235, while almost 100% pure uranium-235 is required for a bomb. The worst accident at a PWR would be a steam explosion, which could contaminate the inside of the containment shell. This scenario was played out at the Chernobyl nuclear power plant in the Ukrainian republic of the former USSR on April 26, 1986. Considered the world's worst nuclear power plant disaster, the accident occurred when human errors led to a dangerous heat and steam buildup, triggering two enormous explosions at the Chernobyl.

The fuel in the reactor core consists of several tons of uranium. As the reactor is operated, the uranium content gradually decreases because of fission, and the radioactive waste products (the fission fragments) build up. After about a year of operation, the reactor must be shut down for refueling. The old fuel rods are pulled out and replaced. These fuel rods, which are very radioactive, are stored under water at the power plant site. After five to ten years, much of their radioactivity has decayed. Only those materials with a long radioactive lifetime remain, and eventually they will be stored in a suitable underground depository.

There are vehement arguments for and against nuclear power. The various advantages and problems should be thoroughly aired so that the general public can evaluate for itself whether the benefits outweigh the risks. Additional electric power plants will be required in the future to supply a growing world population that desires a higher standard of living.

All methods of producing electricity have serious environmental impacts. The main objections to nuclear power plants are the fear of possible accidents, the unresolved problem of nuclear waste storage, and the possibility of plutonium diversion for weapons production by a terrorist group. The issue of waste storage becomes particularly emotional because leakage from a waste depository could contaminate ground water. Chemical dump sites have leaked in the past, so there is distrust of all hazardous wastes.

The main advantage of uclear power plants is that they do not cause atmospheric pollution. No smokestacks are needed because nothing is being burned. France initiated a large-scale nuclear program after the Arab oil embargo in 1973 and has been able to reduce its acid rain and carbon dioxide emissions by more than 40%. Nuclear power plants do not contribute to the **global warming** problem. Shipments of fuel are minimal so the hazards of coal transportation and oil spills are avoided.

Environmentalists are divided in their opinions of nuclear power. It is widely viewed as a hazardous technology but there is growing concern about atmospheric pollution making nuclear power more acceptable.

NUCLEAR FUSION

Nuclear fusion is the process by which two light atomic nuclei combine to form one heavier atomic **nucleus**. As an example, a **proton** (the nucleus of a hydrogen **atom**) and a **neutron** will, under the proper circumstances, combine to form a deuteron (the nucleus of an atom of "heavy hydrogen"). In general, the **mass** of the heavier product nucleus is less than the total mass of the two lighter nuclei.

When a proton and neutron combine, for example, the mass of the resulting deuteron is 0.00239 atomic mass unit less than the total mass of the proton and neutron combined. This "loss" of mass is expressed in the form of 2.23 MeV (million electron volts) of **kinetic energy** of the deuteron and other particles and as other forms of energy produced during the reaction. Nuclear fusion reactions are like **nuclear fission** reactions, therefore, in the respect that some quantity of mass is transformed into energy.

The particles most commonly involved in nuclear fusion reactions include the proton, neutron, deuteron, a triton (a proton combined with two neutrons), a helium-3 nucleus (two protons combined with a neutron), and a helium-4 nucleus (two protons combined with two neutrons). Except for the neutron, all of these particles carry at least one positive electrical charge. That means that fusion reactions always require very large amounts of energy in order to overcome the **force** of repulsion between two like charged particles. For example, in order to fuse two protons with each other, enough energy must be provided to overcome the force of repulsion between the two positively charged particles.

As early as the 1930s, a number of physicists had considered the possibility that nuclear fusion reactions might be the mechanism by which energy is generated in the **stars**. Certainly no familiar type of **chemical reaction**, such as oxidation, could possibly explain the vast amounts of energy released by even the smallest star. In 1939, the German-American physicist **Hans Bethe** worked out the mathematics of energy generation in which a proton first fuses with a **carbon** atom to form a nitrogen atom. The reaction then continues through a series of five more steps, the net result of which is that four protons disappear and are replaced by one **helium** atom.

Bethe chose this sequence of reactions because it requires less energy than does the direct fusion of four protons and, thus, is more likely to take place in a star. Bethe was able to show that the total amount of energy released by this sequence of reactions was comparable to that which is actually observed in stars.

The Bethe "carbon cycle" is by no means the only nuclear fusion reaction that one might conceive. A more direct approach, for example, would be one in which two protons

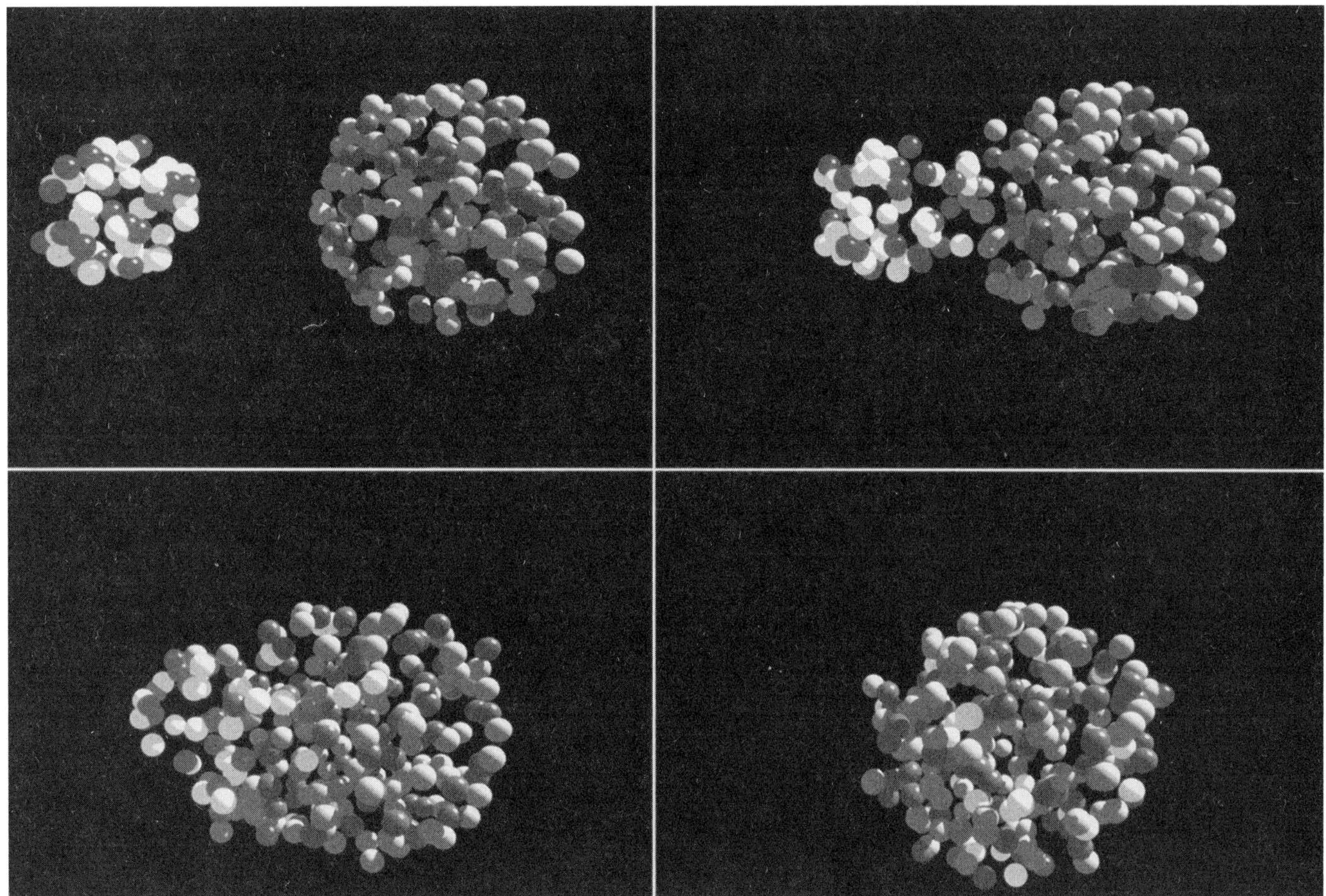

Nuclear fusion, computer simulation sequence of gold and nickel nuclei fusing. *(Image by Jens Konopka and Henning Weber. Photo Researchers, Inc. Reproduced by permission.)*

fuse to form a deuteron. That deuteron could then fuse with a third proton to form a helium-3 nucleus. Finally, the helium-3 nucleus could fuse with a fourth proton to form a helium-4 nucleus. The net result of this sequence of reactions would be the combining of four protons (hydrogen nuclei) to form a single helium-4 nucleus. The only net difference between this reaction and Bethe's carbon cycle is the amount of energy involved in the overall set of reactions.

The term less energy used to describe Bethe's choice of **nuclear reactions** is relative, however, since huge amounts of energy must be provided in order to bring about any kind of fusion reaction. In fact, the reason that fusion reactions can occur in stars is that the temperatures in their interiors are great enough to provide the energy needed to bring about fusion. Since those temperatures generally amount to a few million degrees, fusion reactions are also known as thermonuclear (thermo = **heat**) reactions.

The understanding that fusion reactions might be responsible for energy production in stars brought the accompanying realization that such reactions might be a very useful source of energy for human needs. Imagine that it would be possible to build and operate a small star on the outskirts of your community that operated on nuclear fusion. That **power** plant would be able to supply all of the community's energy needs as far into the future as anyone could see.

The practical problems of building a fusion power plant are incredible, however, and scientists are still a long way from achieving that goal. A much simpler challenge, however, is to construct a fusion power plant that does not need to be controlled; that is, a fusion bomb.

Scientists who worked on the first fission ("atomic") bomb during World War I were aware of the potential for building an even more powerful bomb that operated on fusion principles.

The core of the fusion bomb would consist of an fission bomb, such as the one they were then developing. That core could then be surrounded by a casing filled with isotopes of hydrogen. Isotopes of hydrogen are various forms of hydrogen that all have a single proton in their nucleus, but may have zero, one, or two neutrons. The nuclei of the hydrogen isotopes are the proton, the deuteron, and the triton.

Imagine that a device such as the one described here could be exploded. In the first fraction of a second, the fission bomb would explode, releasing huge amounts of energy. In fact, the **temperature** at the heart of the fission bomb would reach a few millions degrees, the only way that humans know of for producing such high temperatures.

That temperature would not last very long, but in the microseconds that it did exist, it would provide the energy for fusion to begin to occur within the casing surrounding the fission bomb. Protons, deuterons, and tritons would begin fusing with each other, releasing more energy, and initiating other fusion reactions among other hydrogen isotopes. The original explosion of the fission bomb would have ignited a small star-like reaction in the casing surrounding it.

From a military standpoint, the fusion bomb had one powerful advantage over the fission bomb. For technical reasons, there is a limit to the size one can make a fission bomb. But there is no technical limit on the size of a fusion bomb. One simply makes the casing surrounding the fission bomb larger and larger, until there is no longer a way to lift the bomb into the air so that it can be dropped on an enemy.

On August 20, 1953, the Soviet Union announced the detonation of the world's first fusion bomb. It was about 1,000 times more powerful than was the fission bomb that had been dropped on Hiroshima less than a decade earlier. Since that date, both the Soviet Union (now Russia) and the United States have become proficient at manufacturing fusion bombs on an almost assembly-line schedule.

As research on fusion weapons was going on, attempts were also being made to develop peaceful uses for nuclear fusion. The concept of a star power plant just outside the city was never out of sight for a number of nuclear scientists.

The problems to be solved in controlling the nuclear fusion reaction have, however, been enormous. The most obvious challenge is simply to find a way to hold the nuclear fusion reaction in place as it occurs. One can't build a machine made out of metal, plastic, glass, or any other common kind of material. At the temperatures at which fusion occurs, any one of these materials would vaporize instantly. So how does one contain the nuclear fusion reaction?

Traditionally, two general approaches have been developed to solve this problem: magnetic and inertial containment. To understand the first technique, imagine that a mixture of hydrogen isotopes has been heated to a very high temperature. At a sufficiently high temperature, the nature of the mixture begins to change. Atoms totally lose their electrons, and the mixture consists of a swirling mass of positively charged nuclei and electrons. Such a mixture is known as a plasma.

One way to control that plasma is with a magnetic field. One can design such a field so that a swirling hot mass of plasma within it can be held in any kind of a shape one chooses. The best known example of this approach is a doughnut-shaped Russian machine known as a tokamak. In the tokamak, two powerful electromagnets create fields that are so powerful that they can hold hot plasma in place as readily as a person can hold an orange in her hand.

The technique, then, is to heat the hydrogen isotopes to higher and higher temperatures while containing them within a confined space by means of the magnetic fields. At some critical temperatures, nuclear fusion will begin to occur. At that point, the tokamak is producing energy by means of

fusion while the fuel is being held in suspension by the magnetic field.

A second method for creating controlled nuclear fusion makes use of a laser beam or a beam of electrons or atoms. In this approach, hydrogen isotopes are suspended at the middle of the machine in tiny hollow glass spheres known as microballoons. The microballoons are then bombarded by the laser, electron, or atomic beam and caused to implode. During implosion, enough energy is produced to initiate fusion among the hydrogen isotopes within the pellet. The plasma thus produced is then confined and controlled by means of the external beam.

The production of useful nuclear fusion energy by either of these methods depends on three factors: temperature, containment time, and energy release. That is, it is first necessary to raise the temperature of the fuel (the hydrogen isotopes) to a temperature of about 100 million degrees. Then, it is necessary to keep the fuel suspended at that temperature long enough for fusion to begin. Finally, some method must be found for tapping off the energy produced by fusion.

A measure of the success of a machine in producing useful fusion energy is known as the Lawson confinement parameter, the product of the **density** of particle in the plasma and the time the particles are confined. That is, in order for controlled fusion to occur, particles in the plasma must be brought close together and they must be kept together for some critical period of time. All of this must take place, of course, at a temperature at which fusion can occur.

The two nuclear reactions now most commonly used for power production purposes are designated as D-D and D-T reactions. The former stands for deuterium-deuterium and involves the combination of two deuterium nuclei to form a helium-3 nucleus and a free neutron. The second reaction stands for deuterium-tritium and involves the combination of a deuterium nucleus and a tritium nucleus to produce a helium-4 nucleus and a free neutron. The most common form of an inertial confinement machine, for example, uses a fuel that consists of equal parts of deuterium and tritium.

Research on controlled fusion power has now been going on for a half century with somewhat disappointing results. Some experts believe that success may be "just around the corner," but others argue that the problems of an economically feasible fusion power plant may never be solved.

In recent years, scientists have begun to explore approaches to fusion power that depart from the more traditional magnetic and inertial confinement techniques. One such approach is called the PBFA process. In this machine, **electric charge** is allowed to accumulate in capacitors and then discharged in 40-nanosecond micropulses. Lithium **ions** are accelerated by means of these **pulses** and forced to collide with deuterium and tritium targets. Fusion among the lithium and hydrogen nuclei takes place, and energy is released. Thus far, however, the PBFA approach to nuclear fusion has been no more successful than has that of more traditional methods.

The scientific world was astonished in March of 1989 when two electrochemists, Stanley Pons and Martin Fleischmann, reported that they had obtained evidence for the occurrence of nuclear fusion at room temperatures. During the **electrolysis** of **heavy water** (deuterium oxide), it appeared that the fusion of deuterons was made possible by the presence of palladium electrodes used in the reaction. If such an observation could have been confirmed by other scientists, it would have been truly revolutionary. It would have meant that energy could be obtained from fusion reactions at moderate temperatures.

The Pons-Fleischmann discovery was the subject of immediate and intense scrutiny by other scientists around the world. It soon became apparent, however, that evidence for **cold fusion** could not consistently be obtained by other researchers. A number of alternative explanations were developed by scientists for the fusion results that Pons and Fleischmann believed they had obtained. Today, some scientists are still convinced that Pons and Fleischmann had made a real and important breakthrough in the area of fusion research. Most researchers, however, attribute the results they reported to other events that occurred during the electrolysis of the heavy water.

NUCLEAR MAGNETIC RESONANCE

Nuclear magnetic resonance (NMR) is the effect produced when a radiofrequency field is imposed at right angles to a (usually much larger) static magnetic field to perturb the orientation of nuclear magnetic moments generated by spinning electrically charged atomic nuclei. When the perturbed spinning nuclei interact with the very large (10,000-50,000 gauss) static magnetic field, characteristic spectral shifts and fine structure are produced that reflect the molecular or chemical environment seen by the nucleus. Hydrogen nuclei, fluorine, carbon-13, and oxygen-17 all have distinctive magnetic properties that make them suitable for NMR studies.

The sensitivity of nuclear magnetic resonance to **molecular structure** has made it a valuable research tool in organic chemistry, enabling chemists to determine hydrogen locations in crystals, something that cannot be done using x-ray **diffraction**. Nuclear magnetic resonance has also been used to study **electron** densities, chemical bonding, the compositions of mixtures, and to make purity determinations.

The basic requirements for NMR spectroscopy are that the magnetic field be homogeneous over the volume of the sample, that there be a radiofrequency field rotating in a plane perpendicular to the static field, and that there be a means of detecting the interaction of the radiofrequency field with the sample.

Techniques that have been developed for the observation of NMR signals fall into two categories: pulsed and continuous wave. In the case of pulsed methods, an applied rotating (or alternating) magnetic field with a **frequency** at or near the Larmor frequency (i.e., frequency of precession) of the **nucleus** to be studied is directed at a right angle to the static field. If the rotating field is applied at exact resonance, the nuclei precess about that field as though there was no static field. Continuous wave methods are either high resolution or

broadline. Broad line widths are produced by most oriented molecules exhibiting strong magnetic dipolar interactions, so broadline spectroscopy does not permit measurements of chemical shifts and spin-spin coupling. High resolution spectroscopy, on the other hand, has been used to identify molecules, to measure subtle electronic effects, to determine structure, to study reaction intermediates, and to follow the **motion** of molecules or groups of atoms within molecules. For high resolution studies, the magnetic field must be uniform to 1 part in 10^8 for a 100MHz instrument if a resolution of 1 MHz is to be obtained. In the case of broadline studies, 5 parts in 10^6 may be adequate.

Nuclear magnetic resonance has been used to study the physics and chemistry of solids, including metals, **semiconductors**, magnetic solids, and organic materials. Physical phenomena studied include conduction-electron paramagnetism; spin **waves** and magnetic fluctuations in ordered magnetic materials; metal-insulator transitions; charge **density wave phenomena**; spin-freezing in spin glasses; and frequency shift and spin-lattice relaxation effects. At low temperatures, NMR has been used to make **temperature** measurements and to study the superfluid phases of ^3He.

In the fields of organic che mistry and materials science, NMR has been used to study polymers, amorphous systems, and complex molecular solids. In many of these systems, the NMR line widths of the nuclei are dominated by dipolar fields arising from neighboring magnetic moments. These systems exhibit complex NMR spectra due to shifts in nuclear magnetic resonance frequencies.

In the case of complex molecules in liquid environments, the molecules undergo a tumbling motion, producing very sharp NMR spectra. The technique used to study these systems is known as Fourier transform NMR spectroscopy.

Nuclear magnetic resonance has been adapted to medical studies in the form of **magnetic resonance imaging** (MRI), a technique that is capable of producing high quality images of the internal human body without requiring invasive surgery. Begun as a tomographic imaging technique because it produced an image corresponding to a thin slice of the human body, MRI later evolved into a three-dimensional imaging technique.

Like NMR, magnetic resonance imaging is based on the absorption and emission of radiofrequency **energy** under the influence of a magnetic field. On a molecular level, the human body consists largely of hydrogen-rich fat and water molecules, and approximately 63% of the human body contains hydrogen atoms. The hydrogen nuclei emit signals that can be observed by magnetic resonance imaging.

When exposed to a magnetic field, the spin of the protons in the hydrogen nuclei of water, which ordinarily assume random orientations, line up. Although short **pulses** of **radio** waves briefly disturb this spin alignment, the spins promptly realign in the direction of the magnetic field. In the process of realigning, small signals are produced that can be picked up by sensitive scanners. The signals are then compiled by computer and an image is formed.

Because water is the major component of soft tissue, MRI creates excellent images of soft tissue and organs, but poor images of dense structures like bone. MRI thus complements x-ray imaging, which by contrast is fine for dense structures but useless for soft tissue.

NUCLEAR PHYSICS

Nuclear physics is the study of the interaction of particles in a **nucleus**: protons and neutrons. Its goal is to understand the behavior of the nucleus. In early days of nuclear physics, this involved study of **radioactivity** and the nature of the **atom** itself; by the end of the twentieth century, it was primarily concerned with the strong nuclear **force**, which binds the nucleus together despite the protons' **electric charge** repulsion pushing the nucleons apart.

Despite the sophistication of their field of study, nuclear physicists basically do variations on one experiment: the collision. All nuclear physics experiments consist of either a beam of particles hitting a stationary target or of two beams hitting each other. It is often compared to smashing a Swiss watch and trying to deduce its mechanism from watching what parts fly out. This method has had surprising success and his historically used such devices as the electrostatic accelerator, the linear accelerator, the **cyclotron**, the betatron, the **cloud chamber**, and the bubble chamber to control and observe collision experiments.

Although the nuclear model of the atom had not been proposed yet, nuclear physics began with the study of radioactivity at the end of the nineteenth century. In studying elements such as uranium, Antoine Becquerel and the Curies were able to determine that the atoms were emitting other particles. With further study, **Ernest Rutherford** and T. Royds showed that one of the particles emitted, the **alpha particle**, was actually a **helium** nucleus. At this time, the standard model of an atom was Thomson's "plum pudding" model. This model characterized the atom as a sphere of uniform positive charge with small dots of negative charge within it, like raisins in a plum pudding.

Further study led Rutherford to propose in the first nuclear model of the atom in 1911. Collision experiments had shown quite clearly that the Thomson plum pudding model would not work, so Rutherford proposed that the positively charged material was gathered in together in a central nucleus, while the negatively charged electrons were outside it. In 1913, **Niels Bohr** proposed that the electrons orbited the nucleus, somewhat like **planets** around the **sun**. While Bohr's planetary model had some flaws, it produced good approximations of the average **electron** radius in a **hydrogen atom** and provided explanations for mechanisms. When **Erwin Schrödinger** developed his **quantum theory**, including the wave equation, in 1926, more questions about the nature of the atom were answered.

With the basic structure of the nucleus determined, it remained for physicists to probe the internal structure of nuclei. In 1932, **James Chadwick** discovered the **neutron**, which allowed for the known **mass** of the **proton** to be used in calculating atomic masses, without resorting to variations on the plum pudding model with some electrons in the nucleus and some outside. This discovery made it clear that some force independent of charge must be binding the nucleons together, and that this force must have greater strength at short range than the Coulomb repulsion that arises between particles of like charge, such as protons. Several models of nuclear force structure have been proposed, each with varying advantages and points of accuracy or clarity.

One of the earliest models of nuclear structure was proposed in 1936 by Niels Bohr. The compound nucleus theory states that the nucleus can be made up of previous particles or nuclei that have been fused in a collision, and that the later decay of the new nuclei will be independent of their formation. Further postulates were necessary from Bohr and Wheeler in 1939 to explain **nuclear fission**. The "liquid drop model" envisions the nucleus as similar to a drop of water that will reach a mass for which **surface tension** is no longer sufficient to hold it together. Similarly, the nucleus is thought to fission when the strong force is overcome by the Coulomb repulsion. Some of the terms in the formula for calculating the strong force under the liquid drop model are actually the same as for calculating the force on a real liquid drop.

The shell model, proposed in 1949 by **Maria Goeppert-Mayer**, J. Hans, D. Jensen, Haxel, and Suess, proposes that nucleons follow a shell structure somewhat akin to the shell system of electrons. Considerations such as spin were necessary to determine how stable a nucleus would be. This refinement explained why the stability of a nucleus does not depend solely on its size, but also on its place in a nuclear version of a **periodic table**. As with electrons, each atom would have inner nucleons and valence nucleons.

The collective model was developed to address properties of the nucleus as a whole that were not addressed in the shell model, which only dealt with individual electrons. Aage Bohr, Mottelson, and Rainwater developed this model to take into account all of the valence nucleons as a whole, rather than merely the "outermost" valence **nucleon**. The calculations of the magnetic moments of the nucleus were much more accurate from this model than from previous ones. Nonetheless, it must be remembered that all nuclear models are approximations.

In 1964, the quark model of **hadron** interactions was devised by **Murray Gell-Mann** and Zweig. Testing the quark model required bigger, higher **energy** facilities for collisions. More importantly, it was a movement into the next field: **particle physics**. Nucleons were no longer presumed to be the most fundamental units of the atom. While studies of the strong force were still possible and necessary in traditionally nuclear methods and sizes, structure discussions had to include considerations at the quark level.

Nuclear physics arose from a combination of **atomic physics** and quantum **mechanics**; it has in turn given rise to particle physics. While nuclear physics is no longer the field probing the smallest particles known to man, it still has a vital role to play in testing theories about the strong nuclear force.

It has also given rise to technologies ranging from medical imaging methods to nuclear **power**.

See also Atomic structure; European Center for Nuclear Research (CERN); Fermilab; Large Hadron Collider; Nuclear reactions; Nuclear reactors; Particle detectors; Scattering; Strong interactions; Subatomic particles; Synchrotrons

NUCLEAR POWER

Nuclear **power** is any method of doing **work** that makes use of **nuclear fission** or fusion reactions. In its broadest sense, the term refers to both the uncontrolled release of **energy**, as in fission or fusion weapons, and to the controlled release of energy, as in a nuclear power plant. Most commonly, however, the expression nuclear power is reserved for the latter of these two instances.

The world's first exposure to nuclear power came with the detonation of two fission ("atomic") bombs over Hiroshima and Nagasaki, Japan, events that helped bring World War II to a conclusion. A number of scientists and laypersons perceived an optimistic aspect of these terrible events. They hoped that the power of nuclear energy could be harnessed to do much of the work that all human societies face. Those hopes have been realized to only a modest degree, however. Some serious problems associated with the use of nuclear power have never been satisfactorily solved and, after three decades of progress in the development of controlled nuclear power, interest in this energy source has leveled off and, in many nations, declined.

A nuclear power plant is a system in which energy released by fission reactions is captured and used for the generation of **electricity**. Every such plant contains four fundamental elements: the reactor, coolant system, electrical power generating unit, and the safety system.

The source of energy in a nuclear reactor is a fission reaction in which neutrons collide with nuclei of uranium-235 or plutonium-239 (the fuel), causing them to split apart. The products of any fission reaction include not only huge amounts of energy, but also waste products, known as fission products, and additional neutrons. A constant and reliable flow of neutrons is insured in the reactor by means of a moderator, which slows down the speed of neutrons, and control rods, which control the number of neutrons available in the reactor and, hence, the rate at which fission can occur.

Energy produced in the reactor is carried away by means of a coolant, a fluid such as water, liquid sodium, or carbon dioxide gas. The fluid absorbs **heat** from the reactor and then begins to boil itself or to cause water in a secondary system to boil. Steam produced in either of these ways is then piped into the electrical generating unit where it turns the blades of a turbine. The turbine, in turn, powers a generator that produces electrical energy.

The high cost of constructing a modern nuclear power plant reflects in part the enormous range of safety features needed to protect against various possible mishaps. Some of

those features are incorporated into the reactor core itself. For example, all of the fuel in a reactor is sealed in a protective coating made of a zirconium alloy. The protective coating, called a cladding, helps retain heat and **radioactivity** within the fuel, preventing it from escaping into the plant itself.

Every nuclear plant is also required to have an elaborate safety system to protect against the most serious potential problem of all, loss of coolant. If such an accident were to occur, the reactor core might well melt down, releasing radioactive materials to the rest of the plant and, perhaps, to the outside environment. To prevent such an accident from happening, the pipes carrying the coolant are required to be very thick and strong. In addition, back-up supplies of the coolant must be available to replace losses in case of a leak.

On another level, the whole plant itself is required to be encased within a dome-shaped containment structure. The containment structure is designed to prevent the release of radioactive materials in case of an accident within the reactor core.

Another safety feature is a system of high-efficiency filters through which all air leaving the building must pass. These filters are designed to trap microscopic particles of radioactive materials that might otherwise be vented to the atmosphere. Other specialized devices and systems have also been developed for dealing with other kinds of accidents in various parts of the power plant.

Nuclear power plants differ from each other primarily in the methods they use for transferring heat produced in the reactor to the electricity generating unit. Perhaps the simplest design of all is the **boiling** water reactor plant (BWR) in which coolant water surrounding the reactor is allowed to boil and form steam. That steam is then piped directly to turbines, whose spinning drives the electrical generator. A very different type of plant is one that was popular in Great Britain for many years, one that used carbon dioxide as a coolant. In this type of plant, carbon dioxide gas passes through the reactor core, absorbs heat produced by fission reactions, and is piped into a secondary system. There the heated carbon dioxide gas gives up its energy to water, which begins to boil and change to steam. That steam is then used to power the turbine and generator.

In spite of all the systems developed by nuclear engineers, the general public has long had serious concerns about the use of such plants as sources of electrical power. Those concerns vary considerably from nation to nation. In France, for example, more than half of all that country's electrical power now comes from nuclear power plants. The initial enthusiasm for nuclear power in the United States in the 1960s and 1970s soon faded, and no nuclear power plants have been constructed in this country in more than a decade.

One concern about nuclear power plants, of course, is an echo of the world's first exposure to nuclear power, the **atomic bomb** blasts. Many people fear that a nuclear power plant may go out of control and explode like a nuclear weapon. And, in spite of experts' insistence that such an event is impossible, a few major disasters have instigated the fear of nuclear power plants exploding. By far the most serious of those disasters was the explosion that occurred at the Chernobyl Nuclear Power Plant near Kiev in the Ukraine in 1986.

On April 16 of that year, one of the four power-generating units in the Chernobyl complex exploded, blowing the top off the containment building. Hundreds of thousands of nearby residents were exposed to lethal or damaging levels of **radiation** and were removed from the area. Radioactive clouds released by the explosion were detected as far away as western Europe. More than a decade later, the remains of the Chernobyl reactor remain far too radioactive for anyone to spend more than a few minutes in the area.

Critics also worry about the amount of radioactivity released by nuclear power plants on a day-to-day basis. This concern is probably of less importance than is the possibility of a major disaster. Studies have shown that nuclear power plants are so well shielded that the amount of radiation to which nearby residents are exposed is no more than that of a person living many miles away.

In any case, safety concerns in the United States have been serious enough to essentially bring the construction of new plants to a halt in the last decade. Licensing procedures are now so complex and so expensive that few industries are interested in working their way through the bureaucratic maze to construct new plants.

Perhaps the single most troubling issue for the nuclear power industry is waste management. After a period of time, the fuel rods in a reactor are no longer able to sustain a **chain reaction** and must be removed. These rods are still highly radioactive, however, and present a serious threat to human life and the environment. Techniques must be developed for the destruction and/or storage of these wastes.

Nuclear wastes can be classified into two general categories, low-level wastes and high-level wastes. The former consist of materials that release a relatively modest level of radiation and/or that will soon decay to a level where they no longer present a threat to humans and the environment. Storing these materials in underground or underwater reservoirs for a few years or in some other system is usually a satisfactory way of handling these materials.

High-level wastes are a different matter. The materials that make up these wastes are intensely radioactive and are likely to remain so for thousands of years. Short-term methods of storage are unsatisfactory because containers leak and break open long before the wastes are safe.

For more than two decades, the United States government has been attempting to develop a plan for the storage of high-level nuclear wastes. At one time, the plan was to bury the wastes in a salt mine near Lyons, Kansas. Objections from residents of the area and other concerned citizens made that plan infeasible. More recently, the government decided to construct a huge crypt in the middle of Yucca Mountain in Nevada for the burial of high-level wastes. Again, complaints by residents of Nevada and other citizens have delayed placing that plan into operation. The government insists, however, that Yucca Mountain will eventually become the long-term storage site for the nation's high-level radioactive wastes. Until that site is actually put into operation, however, those

wastes are in "temporary" storage at nuclear power sites throughout the United States.

The first nuclear reactor was built during World War II as part of the **Manhattan Project** to build an atomic bomb. The reactor was constructed under the direction of **Enrico Fermi** in a large room beneath the squash courts at the University of Chicago. It was built as the first concrete test of existing theories of nuclear fission.

Until the day on December 2, 1942, when the reactor was first put into operation, scientists had relied entirely on mathematical calculations to determine the effectiveness of nuclear fission as an energy source. It goes without saying that the scientists who constructed the first reactor were taking an extraordinary chance.

That reactor consisted of alternating layers of uranium and uranium oxide with graphite as a moderator.Cadmium control rods were used to control the concentration of neutrons in the reactor. Since the various parts of the reactor were constructed by piling materials on top of each other, the unit was at first known as an atomic "pile." The moment at which Fermi directed the control rods to be withdrawn occurred at 3:45 p.m. on December 2, 1945, and that date can legitimately be regarded as the beginning of the age of controlled nuclear power in human history.

Many scientists believe that the ultimate solution to the world's energy problems may be in the harnessing of **nuclear fusion**. A fusion reaction is one in which two small nuclei combine with each other to form one larger **nucleus**. As an example, two hydrogen nuclei may combine with each other to form the nucleus of an **atom** known as deuterium, or heavy hydrogen.

Many scientists now believe that fusion reactions are responsible for the production of energy in **stars**. They hypothesize that four hydrogen atoms fuse with each other in a series of reactions to form a single **helium** atom. An important byproduct of these fusion reactions is the release of an enormous amount of energy. In fact, gram-for-gram, a fusion reaction releases many times more energy than does a fission reaction.

The world was introduced to the concept of fusion reactions in the 1950s when the Soviet Union and the United States exploded the first fusion ("hydrogen") bombs. The energy released in the explosion of each such bomb was more than 1,000 times greater than the energy released in the explosion of a single fission bomb.

As with fission, scientists and non-scientists alike expressed hope that fusion reactions could someday be harnessed as a source of energy for everyday needs. This line of research has been much less successful, however, than research on fission power plants. In essence, the problem has been to find a way of containing the very high temperatures produced (a few million degrees Celsius) when fusion occurs. Optimistic reports of progress on a fusion power plant appear in the press from time to time, but some authorities now doubt that fusion power will ever be an economic reality.

NUCLEAR REACTIONS

Nuclear decay, **neutron** capture, fission, and fusion are the four major nuclear reactions.

The understanding, use, and control of nuclear reactions is one of the most profound scientific accomplishments. Although chemical reactions of everyday experience generally take place between the electrons surrounding an **atom** there are important and fundamental processes that take place in the **nucleus** of an atom. Nuclear reactions are different than chemical reactions. Nuclear energies are several orders of magnitude larger than the energies involved in chemical reactions.

In 1903, French physicist, **Antoine Henri Becquerel** won the Nobel Prize in physics for his work with spontaneous **radioactivity.** In 1898, French scientists, Marie and **Pierre Curie**, two-time Nobel Laureates, discovered the naturally occurring radioactive elements polonium and radium.

In the 1930s, French physicists Frédéric and Irène Joliot-Curie demonstrated artificial radioactivity by bombarding stable atoms with nuclear particles to create radioactive isotopes (**radioisotopes**). In 1932 British scientists, Sir **John Cockcroft** (1897-1967) and Ernest Walton (1903- 1995), were the first to disintegrate the nucleus by bombarding it with high **energy** projectile-particles.

Before World War II, German chemists **Otto Hahn** and Fritz Strassmann discovered **nuclear fission.** During the war there was a frantic race to develop atomic weapons. The United States assembled a team of leading physicists and chemists directed by **J. Robert Oppenheimer** (1904-1967) at Los Alamos, New Mexico to take part in project code named Trinity. In 1945, they produced the world's first atomic explosion using a fission reaction. In August 1945, atomic bombs were dropped on the Japanese cities of Hiroshima and Nagasaki. As a result the Japanese surrendered and World War II ended.

In elements heavier than hydrogen (hydrogen's nucleus has only a single **proton**), the nucleus is composed of protons and neutrons. The number of protons is described by the atomic number and is unique for each element. Along with the protons are neutrons, particles of **mass** similar to the proton, but without any electrical charge. The sum of t he atomic mass of the protons and neutrons determines the atomic mass or atomic **weight** of the nucleus.

Elements can be found that do not have identical numbers of neutrons, these elements are isotopes of one another and they have nuclei with different atomic mass. Nuclei with differing weights, that is, unique combinations of protons and neutrons are called nuclides.

Not all atomic nuclides are stable. Radioactive nuclides are unstable and undergo various nuclear reactions that transform the particles within the nucleus and simultaneously release energy. It is important to remember that radioactivity is a process not a term to be applied to the energy or substance emitted as a result of the radioactive p rocess.

Nuclear reactions include **alpha decay** (the emission of a **helium** nucleus, He+), **beta decay** (a reaction where a neutron is transformed into a proton and a high energy **electron** is emit-

ted), **positron** decay (a reaction where a nuclear proton converts into a neutron and a high-energy positron is emitted), and decay that produces **gamma radiation** (a high energy **photon** emission).

A common misconception is that **x rays** are a product of nuclear reactions. In fact, x rays are the result of high energy electron transitions (e.g., jumps between electron energy levels or orbits) in heavy elements.

Scientists study nuclear reactions by accelerating atomic particles (e.g., neutrons) that, upon collision with the target nucleus, transfer enough energy to stimulate nuclear reactions. When some elements are bombarded with protons, neutrons, or other accelerated (and therefore highly energetic) atomic particles, their nuclei may be transfor med to create unstable isotopes that are radioactive. These radioactive isotopes are created by nuclear reactions termed neutron capture.

In neutron capture reactions an element's atomic number (Z) remains unchanged but its atomic mass increases by one because a neutron is added to the nucleus.

Two other important nuclear reactions, fusion and fission, also start with the capture of a neutron. In these reactions, however, the energy levels are so high that the transformed nucleus is very unstable. At the lower energy levels involved in simple neutron capture the element simply becomes radioactive, that is, it undergoes decay reactions to form decay products.

The nuclear reactions of fission and fusion have profound scientific and controversial social consequences. Fission involves the splitting of the atomic nucleus. During fission reactions, a nucleus is split into two nuclei. When, under the right circumstances, uranium is bombarded with neutrons it can undergo nuclear fission to produce barium, krypton, and three neutrons (the basis of the **chain reaction**).

The use of fission reactions in **nuclear reactors** remains socially and politically controversial. Fission's most common use is as a trigger in fusion reaction bombs. Other debates usually concern the safe construction and operation of nuclear **power** facilities and the proper disposal of the dangerous radioactive products of termed nuclear waste.

Fusion is the nuclear reaction that fuels the **Sun** and **stars**, the reaction involves the combining of two nuclei into one nucleus. At stellar temperatures of 10-15 million degrees Celsius, hydrogen is converted to helium. The energy from these reactions provides the energy to sustain life on Earth. The basic reaction combines or fuses four hydrogen atoms into one helium atom with the emission of tremendous energy generated by converting mass into energy according to Einstein's equation $E = mc^2$.

The potential benefits to mankind for the proper development of fusion based technology are enormous. However, until methods are developed to control the reaction, its use will be limited to **nuclear weapons.**

As stars age the supply of hydrogen for fusion reactions is used up and stars must use increasingly heavier elements in their fusion reactions. The mass of the star determines what fuels it can use (up to iron). Except for the transformation of elements by nuclear reactions, all of the atoms in the **universe** heavier than helium are the products of the nuclear reactions that take place in dying stars.

NUCLEAR REACTORS

A nuclear reactor is a device by which **energy** is produced as the result of a nuclear reaction, either fission or fusion. At the present time, all commercially available nuclear reactors make use of fission reactions, in which the nuclei of large atoms such as uranium (the fuel) are broken apart into smaller nuclei, with the release of energy. It is theoretically possible to construct reactors that operate on the principle of **nuclear fusion**, in which small nuclei are combined with each other with the release of energy. But after a half century of research on fusion reactors, no practicable device has yet been developed.

When neutrons strike the **nucleus** of a large **atom**, they cause that nucleus to split apart into two roughly equal pieces known as fission products. In that process, additional neutrons and very large amounts of energy are also released. Only three isotopes are known to be fissionable—uranium-235, uranium-233, and plutonium-239. Of these, only the first, uranium-235, occurs naturally. Plutonium-239 is produced synthetically when nuclei of uranium-238 are struck by neutrons and transformed into plutonium. Since uranium-238 always occurs along with uranium-235 in a nuclear reactor, plutonium-239 is produced as a byproduct in all commercial reactors now in operation. As a result, it has become as important in the production of nuclear **power** as uranium-235. Uranium-233 can also be produced synthetically by the bombardment of thorium with neutrons. Thus far, however, this **isotope** has not been put to practical use in nuclear reactors.

Nuclear fission is a promising source of energy for two reasons. First, the amount of energy released during fission is very large compared to that obtained from conventional energy sources. For example, the fissioning of a single uranium-235 nucleus results in the release of about 200 million **electron** volts of energy. In comparison, the oxidation of a single **carbon** atom (as it occurs in the burning of coal or oil) releases about four electron volts of energy. When the different masses of carbon and uranium atoms are taken into consideration, the fission reaction still produces about 2.5 million times more energy than does the oxidation reaction.

Second, the release of neutrons during fission makes it possible for a rapid and continuous repetition of the reaction. Suppose that a single **neutron** strikes a one gram block of uranium-235. The fission of one uranium nucleus in that block releases, on an average, about two to three more neutrons. Each of those neutrons, then, is available for the fission of three more uranium nuclei. In the next stage, about nine neutrons (three from each of three fissioned uranium nuclei) are released. As long as more neutrons are being released, the fission of uranium nuclei can continue.

A reaction of this type that continues on its own once under way is known as a **chain reaction**. During a nuclear chain reaction, many billions of uranium nuclei may fission in less than a second. Enormous amounts of energy are released in a very short time, a fact that becomes visible with the explosion of a nuclear weapon.

Arranging for the uncontrolled, large-scale release of energy produced during nuclear fission is a relatively simple task. Fission (atomic) bombs are essentially devices in which a chain reaction is initiated and then allowed to continue on its own. The problems of designing a system by which fission energy is released at a constant and useable rate, however, are much more difficult.

The heart of any nuclear reactor is the core, which contains the fuel, a moderator, and control rods. The fuel used in some reactors consists of uranium oxide, enriched with about 3-4% of uranium-235. In other reactors, the fuel consists of an alloy made of uranium and plutonium-239. In either case, the amount of fissionable material is actually only a small part of the entire fuel assembly.

The fuel elements in a reactor core consist of cylindrical pellets about 0.6 in (1.5 cm) thick and 0.4 in (1.0 cm) in diameter. These pellets are stacked one on top of another in a hollow cylindrical tube known as the fuel rod and then inserted into the reactor core. Fuel rods tend to be about 12 ft (3.7 m) long and about 0.5 in (1.3 cm) in diameter. They are arranged in a grid pattern containing more than 200 rods each at the center of the reactor. The materials that fuel these pellets must be replaced on a regular basis as the proportion of fissionable nuclei within them decreases.

A nuclear reactor containing only fuel elements would be unusable because a chain reaction could probably not be sustained within it. The reason is that nuclear fission occurs best with neutrons that move at relatively modest speeds, called thermal neutrons. But the neutrons released from fission reactions tend to be moving very rapidly, at about 1/15 the speed of **light**. In order to maintain a chain reaction, therefore, it is necessary to introduce some material that will slow down the neutrons released during fission. Such a material is known as a moderator.

The most common moderators are substances of low atomic **weight** such as **heavy water** (deuterium oxide) or graphite. Hydrides (binary compounds containing hydrogen), hydrocarbons, and beryllium and beryllium oxide have also been used as moderators in certain specialized kinds of reactors.

A chain reaction could easily be sustained in a reactor containing fuel elements and a moderator. In fact, the reaction might occur so quickly that the reactor would explode. In order to prevent such a disaster, the reactor core also contains control rods. Control rods are solid cylinders of metal constructed of some material that has an ability to absorb neutrons. One of the metals most commonly used in the manufacture of control rods is cadmium.

The purpose of control rods is to maintain the ratio of neutrons used up in fission compared to neutrons produced during fission at about 1:1. In such a case, for every one new neutron that is used up in causing a fission reaction, one new neutron becomes available to bring about the next fission reaction.

The problem is that the actual ratio of neutrons produced to neutrons used up in a fission reaction is closer to 2:1 or 3:1. That is, neutrons are produced so rapidly that the chain reaction

Nuclear reactors, interior of containment building at the Trojan power station, Rainier, OR. *(Photo by Mark Marten. Photo Researchers, Inc. Reproduced by permission.)*

goes very quickly and is soon out of control. By correctly positioning control rods in the reactor core, however, many of the excess neutrons produced by fission can be removed from the core and the reaction can be kept under control.

The control rods are, in a sense, the dial by which the rate of fission is maintained within the core. When the rods are inserted completely into the core, most neutrons released during fission are absorbed, and no chain reaction occurs. As the rods are slowly removed from the core, the rate at which fission occurs increases. At some point, the position of the control rods is such that the 1:1 ratio of produced to used up neutrons is achieved. At that point, the chain reaction goes forward, releasing energy, but under precise control of human operators.

In most cases, the purpose of a nuclear reactor is to capture the energy released from fission reactions and put it to some useful service. For example, the **heat** generated by a nuclear reactor in a **nuclear power** plant is used to boil water and make steam, which can then be used to generate **electrici-**

ty. The way that heat is removed from a reactor core is the basis for defining a number of different reactor types.

For example, one of the earliest types of nuclear reactors is the **boiling** water reactor (BWR) in which the reactor core is surrounded by ordinary water. As the reactor operates, the water is heated, begins to boil, and changes to steam. The steam produced is piped out of the reactor vessel and delivered (usually) to a turbine and generator, where electrical power is produced.

Another type of reactor is the pressurized water reactor (PWR). In a PWR, coolant water surrounding the reactor core is kept under high **pressure**, preventing it from boiling. This water is piped out of the reactor vessel into a second building where it is used to heat a secondary set of pipes also containing ordinary water. The water in the secondary system is allowed to boil, and the steam formed is then transferred to a turbine and generator, as in the BWR.

Some efforts have been made to design nuclear reactors in which liquid metals are used as **heat transfer** agents. Liquid sodium is the metal most often suggested. Liquid sodium has many attractive properties as a heat transfer agent, but it has one serious drawback. It reacts violently with water and great care must be taken, therefore, to make sure that the two materials do not come into contact with each other.

At one time, there was also some enthusiasm for the use of gases as heat transfer agents. A group of reactors built in Great Britain, for example, were designed to use carbon dioxide to move heat from the reactor to the power generating station. Gas reactors have, however, not experienced much popularity in other nations.

At the end of World War II, great hopes were expressed for the use of nuclear reactors as a way of providing power for many human energy needs. For example, some optimists envisioned the use of small nuclear reactors as power sources in airplanes, ships, and automobiles. These hopes have been realized to only a limited extent. Nuclear powered submarines, for example, have become a practical reality. But other forms of transportation seldom make use of this source of energy.

Instead, the vast majority of nuclear reactors in use today are employed in nuclear power plants where they supply the energy needed to manufacture electrical energy. In a power reactor, energy released within the reactor core is transferred by a coolant to an external building in which are housed a turbine and generator. Steam obtained from water boiled by reactor heat energy is used to drive the turbine and generator, thereby producing electrical energy.

Reactors with other functions are also in use. For example, a breeder reactor is one in which new reactor fuel is manufactured. By far the most common material in any kind of nuclear reactor is uranium-238. This isotope of uranium does not undergo fission and does not, therefore, make any direct contribution to the production of energy. But the vast numbers of neutrons produced in the reactor core do react with uranium-238 in a different way, producing plutonium-239 as a product. This plutonium-239 can then be removed from the reactor core and used as a fuel in other reactors. Reactors whose primary function it is to generate plutonium-239 are known as breeder reactors.

Research reactors may have one or both of two functions. First, such reactors are often built simply to test new design concepts for the nuclear reactor. When the test of the design element has been completed, the primary purpose of the reactor has been accomplished.

Second, research reactors can also be used to take advantage of the various forms of **radiation** released during fission reactions. These forms of radiation can be used to bombard a variety of materials to study the effects of the radiation on the materials.

NUCLEAR STRUCTURE AND STABILITY

The atomic **nucleus** is consists of a combination of protons, which have a positive electrical charge, and neutrons, which are electrically neutral. Protons and neutrons are both called nucleons. Every atomic nucleus contains at least one **proton**, meaning that the nucleus itself has a positive charge. Except

for hydrogen, all naturally occurring atomic nuclei also contain at least as many neutrons as protons. For elements larger than calcium, which has 20 protons, all naturally occurring nuclei have more neutrons than protons. The excess of neutrons increases with the size of the nucleus. Nuclei with fewer neutrons than these natural limits have been made artificially, but they are all very unstable and decay rapidly. Very large nuclei are also unstable. Unstable nuclei can emit **energy** in the form of gamma rays or particles such as protons, neutrons, alpha particles (two protons and two neutrons), and other nuclei.

In nuclei containing more than one proton—that is, every element other than hydrogen—each proton repels every other proton by the electromagnetic **force**. However, a very powerful force, known as the nuclear, or strong, force also affects the nucleons. This force acts at an extremely short range, about one femtometer (10^{-15} m). The strong force is complicated and only partially understood. At very small distances between nucleons, less than one femtometer, the force is repulsive, pushing the nucleons apart. At distances of about one to two femtometers, roughly the diameter of a **nucleon**, the force is strongly attractive and binds nucleons together by energy millions of times stronger than the electromagnetic repulsion of charged protons. Beyond a few femtometers, the attraction of the strong force drops off rapidly with distance, having a much smaller effect than the nuclear particles than does the electromagnetic force. Because the particles of the nucleus are bound so closely together, the volume of the nucleus is very small compared to that of the **atom** itself. Although most of the atom's **mass** in the nucleus, the nuclear volume is only about one ten-thousandth of the atom.

Since the strong force acts at a short range, each nucleon is attracted only by its immediate neighbors. The amount of attractive energy within the nucleus increases in proportion to the number of nucleons. Because each proton interacts with every other proton in the nucleus, whether they are adjacent or separated by other nucleons, the repulsive electromagnetic energy increases as a square of the number of protons in the nucleus. In large nuclei, this electrical repulsion becomes strong enough to make the nucleus unstable. Fission occurs when the nucleus divides into two smaller nuclei that are then rapidly driven apart by electrical repulsion. This division can occur spontaneously or as the result of added energy of a collision of a rapidly moving particle, such as a **neutron** or another nucleus. A **chain reaction** occurs when the fission process releases energetic particles that strike other nuclei, releasing energy and more particles. All nuclei containing more than 240 nucleons are unstable, as are many smaller nuclei. The energy released during a nuclear reaction is about one million times as strong as that of a **chemical reaction**, due to the relative size of the nucleus to the atom and the magnitude of the strong force at nuclear distances.

The effects on the nucleons of the strong nuclear is also the source of energy release by **nuclear fusion**, the process that fuels the **sun**. Two nuclei of deuterium, hydrogen molecules with one proton and one neutron each, combine to form a **helium** nucleus. A very large amount of energy is released as the nuclear force pulls the pairs of nucleons together. Initially, energy is required to overcome the electrical repulsion of the protons. In the sun, this energy is supplied as **kinetic energy** when the nuclei are heated to temperatures of millions of degrees.

Nuclear **accelerators** are important tools for the study of the nucleus. They create beams of high-energy particles such as electrons, nucleons, alpha particles, or larger nuclei. The particles are accelerated to velocities close to the speed of **light**. At this **velocity**, the particles have sufficient energy to overcome the effects of electromagnetic forces and initiate **nuclear reactions** when they strike target nuclei. The products of these reactions, particles and energy, disclose information about the nature of the nucleus and the forces affecting it. In some cases, the accelerated particle combines with the target nucleus, creating a larger nucleus. This is the source of elements and isotopes that do not occur in nature, such as the transuranium elements.

The strong force can be studied by observing nucleon interactions when a nucleus is bombarded with a beam of protons or neutrons. The effect of the strong force deflects the beam particles passing within a certain distance of the nucleons of the target. By observing these deflections, physicists have calculated the diameter of the nucleons and the intensity and reach of the strong force. Other information of the strong force has been derived from the electromagnetic **radiation** and particles emitted by the nucleus during transitions between **quantum states** of the nucleus. The energy difference between the states is equal to the energy emitted as very high energy, short **wavelength** gamma rays.

According to the **big bang** theory, immediately after the big bang, the **universe** consisted of an unimaginably hot plasma of **quarks** and gluons, the components of nucleons. After approximately one one-millionth of a second, these particles began to combine in groups of three quarks, accompanied by gluons, to form neutrons and protons. In about three minutes, these nucleons began to join together to form atomic nuclei—hydrogen, helium, and a small amount of lithium. Twelve minutes into the life of the universe, it had cooled enough that nucleons no longer had sufficient energy to join together. The larger nuclei did not form until much later, when they were produced in the hot interior of **stars**. Experiments using large accelerators try to duplicate the **density** and **temperature** of the early universe in order to produce conditions that will allow the quarks in individual nucleons to exist separately. These results would provide information the structure of the nucleus and the nature of the strong force as well as insight into the beginning of the universe.

NUCLEAR WEAPONS

The first nuclear explosion on earth occured on July 16, 1945, in Alamagordo, New Mexico, as part of a test for the **Manhattan project** initiated by President Franklin Delano Roosevelt. The first nuclear weapon used in war was deployed less than a month later, when the United States dropped a

Atomic bombs, "Fat Man" and "Little Boy."

Uranium 235 bomb on the Japanese city of Hiroshima on August 6, 1945, near the end of World War II. It killed about 140,000 men women and children. The United States managed to maintain a monopoly on nuclear weapons only for a few years. On August 29, 1949, the Soviet Union exploded its first nuclear bomb, escalating an arms race that would heavily influence international politics for the rest of the century.

This state of world affairs had its origins about 50 years earlier in the seemingly innocent studies of **radioactivity** which was first discovered by **Henri Bequerel** in 1896. In 1898, **Ernst Rutherford** characterized simple radioactivity as consisting of beta rays, later identified as electrons, and alpha rays, later identified as **helium** nuclei made up of two protons and two neutrons. However even as of the 1920s, no one seriously thought that the phenomenon of heavy nuclei ejecting **radiation** could be exploited to harness large amounts of **energy**. It was in 1934 that **Leo Szilard**, an eminent Hungarian physicist, first applied for a patent on the extraction of energy from the **nucleus**. His patent described the notion of a nuclear **chain reaction** which in spirit embodies the mechanism behind the first nuclear weapons. Szilard's ideas soon became appreciated by the physics community. Many physicists, including **Albert Einstein** and **Niels Bohr**, alerted Roosevelt about the possibility of nuclear weapons. These warnings eventually lead to the establishment of the Manhattan Project which in turn culminated in the first successful explosion of an **atomic bomb**.

The physical process underlying the first nuclear explosions is known as fission. The **atom** consists of a cloud of electrons surrounding a relatively tiny, densely packed nucleus consisting of protons and neutrons. Most chemical reactions, including the ones in conventional explosives such as TNT, involve a release of energy due to changes in the configurations of the electrons. Electromagnetic **binding energy** is converted into the **light** and **heat** energy of the explosion. Nuclear weapons employ the same principle, only with the nuclear **force** rather than the electromagnetic force. In a fission reaction a heavy nucleus, such as uranium-235, which has 92 protons and 143 neutrons, is bombarded by a **neutron**, which splits the nucleus into two lighter nuclei plus extra neutrons. The sum of the masses of the fission products is less than the **mass** of the original nucleus, and the extra mass is converted into energy according to Einstein's famous equivalence between mass and energy: $E = mc^2$. Each fission event releases about 200 MeV of energy (1 MeV = 1.6×10^{-13} Joules). This is a microscopic amount of energy but there are typically many fission events in a nuclear explosion. Just 1.99 oz (57 g) of uranium, when fissioned, can convert approximately 0.00199 oz (0.057 g) of mass into energy, yielding an energy release of 1 kiloton.

In order to sustain a large nuclear reaction in a bomb, one needs a **critical mass** of fissionable material. In such a mass the **density** of material is high enough the two or more neutrons released in each fission event have a high probability to induce further fission events which in turn release more neutrons. One then gets an exponential proliferation of fission events which yield the energy of the nuclear explosion. The uranium-235 **isotope** is a good choice for fissionable material because it has a high probability to undergo fission upon capture of a neutron. In nature uranium consists mostly (99.3%) of uranium-238 which absorbs neutrons without fissioning and hence is unsuitable for attaining a critical mass. Through a complicated process of isotopic enrichment one must obtain 90% pure uranium-235. This difficult and expensive process was a main obstacle in the creation of the first bomb. However a baseball sized piece of enriched uranium-235 weighing about 22.1 lb (10 kg) is sufficient for critical mass.

Another technical difficulty in creating an atomic bomb is the triggering mechanism that starts the **nuclear fission** reaction. Fissionable material at critical mass is highly unstable and could explode without warning. Thus atomic bombs must consist of two or more pieces of fissionable material at subcritical mass that are brought together immediately before the bomb is to be detonated. In a gun-type bomb this so called critical assembly is achieved by a gun that fires one subcritical mass of uranium-235 at another. The two masses must be brought together very quickly, before the whole assembly has a chance to fission prematurely. The gun system moves the two masses together at a few millimeters per microsecond which is sufficiently fast. However, in plutonium-239 bombs, the rate of plutonium fission is faster than that of uranium, and the gun assembly is not fast enough to bring the subcritical pieces together. An alternate mechanism can be used known as implosion. In the implosion bomb a subcritical configuration

of plutonium is surround by carefully arranged layers of explosive, which is then symmetrically detonated. The resulting shockwave compresses the plutonium so that it becomes critical and can then undergo a nuclear chain reaction. The bomb that was dropped on Hiroshima was a gun-type uranium-235 bomb, whereas the bomb tested in Los Alamos was an implosion type plutonium bomb.

Besides the fission based bombs there exists a class of more powerful nuclear weapons based upon the principle of **nuclear fusion**, the same process that powers our **sun**. The first fusion bomb, the Mike shot, was detonated on the Eniwetok Atoll in the South Pacific in 1952. The basic principle giving rise to the energy release in a fusion bomb is the fusion of two isotopes of hydrogen, deuterium with two neutrons, and tritium with three neutrons, into a helium-4 nucleus. One neutron is then left over and 17.6 MeV is released. Although 17.6 Mev is much less than the 200 MeV released in a fission event, hydrogen is much lighter than uranium, so many more fusion events can occur than fission events for a given **weight** of explosive material. To obtain an energy release of 1 megaton, 130 lb (57 kg) of uranium-235 or plutonium-239 would have to be fissioned whereas only 31 lb (14 kg) of tritium and deuterium would have to be fused.

The difficult part of the hydrogen bomb is giving the hydrogen nuclei enough energy to overcome the coulumb repulsion of their protons in order to fuse into helium. The phenomenon of quantum mechanical tunnelling through this barrier helps in the process of fusion (as it does in the sun as well) but still considerable heat and **pressure** must be supplied. Usually the source of this heat and pressure is a separate fission bomb called the primary which produces a flood of radiation. The radiation hits the thermonuclear part of the bomb, also known as the secondary which consists of lithium deuteride. A neutron hits the lithium splitting it into tritium and helium-4. The tritium then fuses with the deuterium in the lithium deuteride yielding energy. Lithium deuteride is used in this indirect fashion in order to avoid the cryogenic apparatus that would be required to store the isotopes of hydrogen. In order to enhance the yield of the bomb, it is encased in uranium-238. Excess neutrons hit the uranium-238 which may fission yielding even more neutrons which can subsequently induce further fusion. This combination of fission and fusion forms the mechanism behind the most powerful weapon known to man and can yield up to 100 megatons of energy, compared to the 15 kilotons of energy released by the atomic bomb dropped on Hiroshima.

The effects of a one megaton hydrogen bomb can be severe. Temperatures at the beginning of the nuclear explosion are comparable to the interior of the sun, about 1×10^7 degrees celsius. The surrounding air immediately turns into a fireball which also creates a high pressure shockwave that spreads out from the explosion. The shockwave consists of an overpressure front that exceeds atmospheric pressure by 100 psi 0.5 mi (0.8 km) from the blast. An overpressure of 5 psi (which exists about 4 mi [6.4 km] from the blast) is enough to destroy brick houses. Behind this shockwave comes hurricane force winds. The fireball also rises up into the air and cools off in the process creating the famous mushroom cloud. After about a minute the cloud can be 6 mi (9.6 km) high and still rises at 220 mi (354 km) per hour. It eventually reaches a maximum height of 12 mi (19.3 km). If the bomb is detonated high enough in the air, **radioactive fallout** from the fission events in the casing of the bomb can reach the stratosphere and from there can reach any part of the world, resulting in global fallout.

NUCLEATION EVENT

A nucleation event is the process of condensation or aggregation (gathering) that results in the formation of larger drops or crystals around a material that acts as a structural **nucleus** around which such condensation or aggregation proceeds. Moreover, the introduction of such structural nuclei can often induce the processes of condensation or crystal growth. Accordingly, nucleation is one of the ways that a phase transition can take place in a material. In a phase transition, a material changes from one form to another. For example, ice melts to form liquid water, or a liquid boils to form a gas. Phase transitions occur due to changes in **temperature**. Certain transitions occur smoothly throughout the whole material, while others happen suddenly at different points in the material. When the transitions occur suddenly, a bubble forms at the point where the transition began, with the new phase inside the bubble and the old phase outside. The bubble expands, converting more and more of the material into the new phase. The creation of a bubble is called a nucleation event.

Phase transitions are grouped into two categories, known as first order transitions and second order transitions. Nucleation events happen in first order transitions. In this kind of transition, there is some kind of obstacle to the transition occurring smoothly. A prime example is condensation of water vapor to form liquid water. Because only two or three molecules will not stick together, condensation requires that many water molecules collide and stick together almost simultaneously. This requirement for simultaneous collisions presents a temporary but measurable barrier (temporary because eventually the simultaneous collision of multiple water molecules will happen) to the formation of a bubble of liquid phase. Following formation, the bubble expands as more water molecules strike the surface of the bubble and are absorbed into the liquid phase. Because of the obstacle to the phase transition, a liquid may exist in its gaseous state even though the temperature is well below the **boiling** point. A liquid in this state is said to be supercooled. Accordingly, in order for a liquid to be supercooled, it must be pure, because dust or other impurities act as nucleation centers. If the liquid is very pure, however, it may remain supercooled for a long time. A supercooled state is termed metastable due to its relatively long lifetime.

The other type of phase transition is called second order, and it proceeds simultaneously throughout the whole material. An example of a second order transition is the melting of a solid. As the temperature rises, the magnitude of the thermal vibrations of molecules causes the solid to break apart into a liquid form. As long as the solid is in thermal **equilibrium** and the melting occurs slowly, the transition takes place at the same time everywhere in the solid, rather than taking place through nucleation events at isolated points.

The principles of nucleation were used in cloud seeding **weather** modification experiments where nuclei of inert materials were dispersed into clouds with the hopes of inducing condensation and rainfall.

See also Phase changes and phase equilibrium; Phase diagram; Thermal energy

NUCLEON

Protons and neutrons, the components of the atomic **nucleus**, are both classified as nucleons. Protons carry a positive **electric charge**, while the **neutron** is electrically uncharged, but they display the same behavior in their strong **force** interactions. These particles are classified as nucleons even when they are not inside an **atom**. Except for hydrogen, whose nucleus is composed of a single **proton**, all atomic nuclei consist of a combination both types of nucleons. The nucleons are the most stable hadrons, that is, particles affected by the strong nuclear force. Protons are so stable that they have never been observed to decay and they are believed to be stable for the entire lifetime of the **universe**. Although a free neutron decays to form a proton, an **electron** and an anti-neutrino in about 15 minutes, neutron bound in an atomic nucleus are stabilized and usually do not decay.

The **mass** of the proton is 1,836 times that of the electron and the mass of the neutron is a little more than 1,837 times the electron. Unlike the electrons, which act as though they are point particles, nucleons have a measurable size. By observing the **scattering** effects of particles on a beam of moving electrons, physicists have determined that the positive charge of the proton is distributed in a cloud extending about one femtometer (10^{-15} m) from its center. The neutron is about the same size as the proton. Neutrons are electrically neutral overall, but experiments demonstrate an internal structure with a slight positive charge in the core surrounded by a shell of negative charge. Nucleons cannot penetrate one another and act like hard balls on contact. The distance between nucleons inside a nucleus is always the same. So the **density** of all nuclei is the same.

Nucleons are affected by the three forces of nature that act on all particles, including the electrons—gravitation, **electromagnetism**, and the weak force. However, nucleons also exert a very powerful force on each other—the strong, or nuclear, force. At very small distances between nucleons, less than one femtometer (10^{-15} m), the force is repulsive, pushing the nucleons apart. At distances of about one to two femtometers, roughly the diameter of a nucleon, the force is strongly attractive and binds nucleons together by **energy** millions of times stronger than the electromagnetic repulsion of charged protons. Beyond a few femtometers, the attraction of the strong force drops off rapidly with distance, having a much smaller effect than the nuclear particles than does the electromagnetic force. Because of the short range of the strong force effects, it is not felt outside an atom. Within the nucleus, the strong force overwhelms the electromagnetic repulsion of the charged protons, binding the nucleons together. The strong force interactions between nucleons are the same whether the two particles are protons, neutrons, or a proton and neutron. Because of their electromagnetic repulsion, two protons cannot be bound together without a neutron, so there is no **helium isotope** consisting only of two protons.

NUCLEOSYNTHESIS

Nuclosynthesis refers to the primordial generation of the **light** elements when the **universe** was formed. The most abundant element in our present universe is hydrogen, followed by **helium**. The **big bang** theory owes much of its success to its ability to predict the relative amounts of these elements.

The big bang also holds that the first elements created in the universe were the light elements, namely helium, deuterium and lithium, which were produced in the first few instants. At that point, the **temperature** was so high, all **matter** was fully ionized and dissociated. About three minutes later, the temperature of the universe rapidly cooled from 1,032K to approximately 109K. At this temperature, nucleosynthesis began to occur: protons and neutrons collided to produce deuterium, which consists of one **proton** and one **neutron**. This newly generated deuterium then collided with other protons and neutrons to produce helium as well as a small amount of tritium, i.e., the 3-hydrogen **isotope**. Some trace amounts of 7-lithium from the merging of one tritium and two deuterium nuclei were also produced. The universe was then one second old and its temperature was 1,010K.

The theory initially held that all elements were generated by the big bang, but this was subsequently revised: even at the extremely high temperatures available when helium and lithium nuclei were created, they were still not high enough to smash two helium nuclei together into a heavier **atom**. It is now known, mostly from the work of **George Gamow**, that **stars** had to be created to provide the required temperature ranges before the remaining elements of the **periodic table** could be generated.

Elements heavier than helium originated in the interiors of stars, which formed much later as the universe unfolded. All stars derive their **energy** from the thermonuclear fusion of light elements into heavier elements. For nuclear species to be transformed into other nuclear species by reactions that add or remove protons or neutrons or both, requires very high temperatures to overcome the mutual electrostatic repulsion of the protons in each fusing atomic **nucleus**. For example, the temperature required for the fusion of hydrogen is 5 million degrees and elements with more protons in their nuclei necessitate even higher temperatures, such as **carbon**, which requires a temperature of 1 billion degrees for fusion to proceed. The big bang theory not only successfully accounts for nucleosynthesis but also for the synthesis and cosmic abundance of the elements up to iron, by successive **nuclear fusion**

reactions, and of the elements heavier than iron, by neutron capture. Deuterium and lithium, while consumed in the **nuclear reactions** occurring in stars, are very rarely produced by them. Whatever amounts of these two elements are present in the universe were created some 15 billion years ago, as was most of the helium.

The relative amounts of helium, lithium, deuterium and hydrogen predicted by the big bang theory are confirmed by astronomical observations. For example, the observed hydrogen to helium ratio is of 4:1, which is in close agreement with the predicted ratio. Thus the study and observation of stellar processes not only contributes descriptive knowledge about our physical world but also provides a measure of its history and future evolution.

See also Cosmology; Periodic table: Stellar evolution

NUCLEUS

The word nucleus is a derivative of the Latin word *nux*, meaning nut or kernel. Around 1909, **Ernest Rutherford** gave that name to the dense, central part of the **atom** that he and his colleagues **Hans Geiger** and Ernest Marsden discovered by bombarding thin sheets of gold foil with **alpha particles**. The results of the experiment led Rutherford to propose that all the positive charge of an atom and almost all its **mass** are located in the nucleus. He developed this model based on the behavior of the positively charged alpha particles as they hit the gold foil. Although most of them passed right through, some were deflected at various angles, including angles that directed the alpha particles back toward their source. Since most of the alpha particles passed through undeflected, Rutherford concluded that the positive part of the atom was not evenly distributed throughout the atom. He explained the deflections as the result of the electric repulsion between like charges—a positively charged **alpha particle** and the positive part of the atom. He used measurements of the deflection angles and the number of particles deflected at that angle to calculate the size of the nucleus. Its diameter is about 10^{-15} meters (one femtometer, now known as a fermi) which is ten thousand times smaller than the diameter of the atom itself. About half a century later, **Robert Hofstadter** won the 1961 Nobel Prize in physics for his work verifying that the radius of an atomic nucleus is about one fermi (10^{-15} m). In his experiments, Hofstadter collided electrons with atoms at high energies.

How the positive charges are able to stay in such close proximity in the confines of the nucleus is another question that was raised. The **force**, labeled the strong nuclear force, would have to be attractive between the nuclear particles and act at very small distances. At that time only two forces were recognized. The first, **gravity**, is only significant when large masses are involved. The second, electric force, acts only between charged particles. It is an attractive force between particles of opposite charge, and a repulsive force between particles of same charge. The nuclear force is about one hundred times stronger than this repulsive force between the pro-

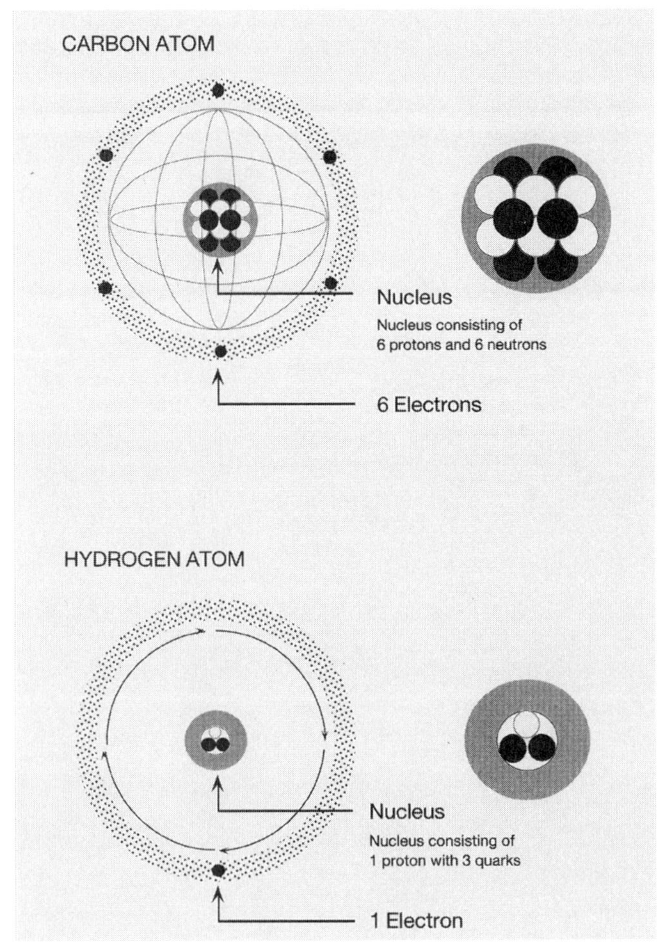

Subatomic particles with nucleus and electron.

tons. Rutherford suggested that there is another particle in the nucleus that has mass but no charge. He called this particle the **neutron**, and postulated that the strong nuclear force acts between the protons and the neutrons to keep the nucleus together. This particle was not discovered until 1932 by English physicist **James Chadwick**. Chadwick won the 1935 Nobel Prize in physics for this discovery. In 1934, **Hideki Yukawa** was the first to attempt to explain the way in which this nuclear force operates. He suggested that the strong nuclear force results from the exchange of a particle between the neutrons and the protons; he named that exchange particle a **meson**. In 1947, English physicist **Cecil Powell** observed Yukawa's mesons, now called pi-mesons or **pions**, in the upper atmosphere, where they were produced by cosmic ray collisions.

Maria Goeppert Mayer shared in the 1963 Nobel Prize in physics for developing the shell model theory of the structure of atomic nuclei. This theory suggests that just like electrons in an atom, protons and neutrons within the nucleus are arranged in quantum shells that fill according to specific magic numbers.

It has been proved that the mass of an atomic nucleus is less than the sum of the masses of the protons and neutrons that it contains. Using Einstein's mass-energy equivalency equation, this mass difference multiplied by the speed of **light** squared gives the **binding energy** of the nucleus. When the sum of the masses of the particles in a nucleus is less than the mass of the nucleus itself, the nucleus will not be stable and that atom will be radioactive (unstable).

O

OHM, GEORG SIMON (1787-1854)

German physicist

A single discovery established Georg Ohm's claim to fame, though a disastrous decision nearly robbed him of it.

Ohm, who was born on March 16, 1789 in Erlangen, Bavaria, was the son of a self-educated locksmith. Ohm's father had a great interest in science, which he conveyed to his son. Ohm went on to study at the University of Erlangen in 1805, but left in the following year to become a teacher in Switzerland. He returned to the university in 1811 and received his Ph.D. Ohm's goal in life was to receive a university appointment. Unfortunately, his mathematical talents were not appreciated by academia, and he ended up teaching at the Erlangen "gymnasium" (secondary school).

Still determined to make an impression and receive an appointment, Ohm decided that producing important research work would better his situation. Knowing of the pioneering work of **Alessandro Volta**, Ohm became interested in studying **electricity**. Because he was unable to afford to purchase equipment with which to experiment, Ohm had to build his own.

Borrowing from the work of Jean-Baptiste-Joseph Fourier, who had learned that the flow of **heat** between two points depended on the **temperature** difference and the material conducting the heat, Ohm applied the concept to the flow of electricity through various electric conductors.

Using wires of differing length and diameter, Ohm discovered that a long, thick wire passed less current than a short, thin wire. This established a relationship between electric resistance, **electric potential (electromotive force)** and current flow.

In 1827 Ohm issued his "law," stating that the amount of current passing through a wire was inversely proportional to the length and directly proportional to the thickness of a wire.

Henry Cavendish had discovered the same relationship nearly fifty years earlier, but neglected to publish the information. Ohm went a step beyond, stating the relationship in mathematical terms: $V = IR$ (the **voltage** of the current is equal to the current in the circuit times the resistance); doubling the voltage doubles the current, and doubling resistance halves the current. On the other hand, conductance (noted with the letter G) is expressed by two equations: $G = I/R$ and $I = GV$. **Ohm's Law** made it possible for scientists to calculate the amount of current, voltage, and resistance in circuitry, establishing the science of electrical engineering. This discovery was Ohm's crowning moment, and he felt certain he would attain his appointment.

Unfortunately, Ohm made a disastrous decision when he published his research: He stressed the mathematical theory behind his work, glossing over the experimental studies. German scientists proceeded to ignore Ohm's equations, believing that mathematics were irrelevant to understanding nature. Ohm received so much ridicule, he not only forfeited an appointment but was also forced to resign the modest teaching position he had at the secondary school.

Ohm spent the next six years in disappointment and poverty. Fortunately his research had made its way to England, where it was fully appreciated. To his delight he began receiving numerous letters of congratulation. He received the Copley Medal in 1841, and was elected to the Royal Society in 1842. In 1843 British scientist Charles Wheatstone credited Ohm for having influenced his own work.

This foreign acclaim began to make an impression on the German scientists, who finally realized the extent of Ohm's discovery. At last, in 1849, Ohm was appointed a professor at the University of Munich, a position he held for the remaining five years of his life. Ohm died at the age of 67 on July 6, 1854.

William Thomson (Lord Kelvin) recognized Ohm's contribution by dubbing the unit of resistance the "ohm" and its

George Simon Ohm.

reciprocal, the unit of conductance, the "mho" (Ohm's name spelled backward).

OHM'S LAW

Ohm's law is a relationship between the **voltage** across an **electric circuit**, the **electrical resistance** in the circuit, and the current in the circuit. This law is named after its discoverer, **Georg Simon Ohm**. Ohm found that for most electric circuits, the voltage across the circuit was equal to the current flowing through the circuit times the electrical resistance of the circuit. For the same voltage, a circuit with a low resistance will have a higher current than a circuit with a higher resistance. The voltage, properly called the potential difference, is measured in volts, and the current in amperes (amps). The resistance is therefore in volts per ampere, which is defined as ohms.

It is important to understand that Ohm's law is not a fundamental law that always applies, such as the law of **gravity**. Rather it is an empirical law that has been found by experiment to work fairly well most of the time. There are times, however, usually in extreme cases, when Ohm's law breaks down. For example, if an extremely high voltage is applied across a circuit, Ohm's law will not predict the correct value for the cur-

rent. Even though Ohm's law does not always apply it works for most everyday situations and is therefore very useful.

For example, why will a short circuit blow a fuse or circuit breaker? When a short circuit occurs, most of the electrical resistance in the circuit is bypassed. In effect, a new circuit with a very low resistance is created. So, according to Ohm's law if the resistance is very low the current must be very high. Fuses and circuit breakers are designed to protect the circuit by blowing when the current becomes too high. Hence, the short circuit will produce a current high enough to blow the fuse. As another application, electronic devices often have resistors placed in the circuit to increase the resistance and therefore limit the current.

OLBERS'S PARADOX

Why is the night sky dark? This apparently simple question, known as Olbers's paradox, actually reveals something deep and fundamental about the **universe**. Although named after the German physician and amateur astronomer Heinrich Olbers (1758-1840), who discussed the problem in 1823, **Johannes Kepler** was the first to consider it in 1610.

Olbers's paradox begins by saying that the apparent brightness of a star on the sky decreases with the square of the distance to the star. But, since the area the star appears to cover also decreases with the square of the distance, this means that the apparent brightness per unit area on the sky does not depend on the distance to the star. If one assumes that the universe is homogeneous, infinitely old and infinitely large, this means that every line of sight should intersect at one of the infinite number of **stars**, and the night sky should be infinitely bright! Hence the paradox.

It is best to note that in order to work, like most paradoxes, Olbers's paradox must make some assumptions, most of which are explicitly stated, while at least one is implicit. These are: (1) the universe is infinitely large; (2) the universe is infinitely old; and (3) the universe is homogeneous.

Since the average **density** of luminous **matter** in the universe is non-zero, assumption (3) combined with (1) leads to the conclusion that the universe has an infinite amount of luminous matter in it. Since it takes **light** a finite time to travel to us from any point in **space**, assumption (2) is necessary in order to see distant stars.

In order to solve the paradox, one or more of these assumptions must be invalidated. Some people might think that dust can blanket the earth, effectively obscuring the **radiation** from the infinite amount of luminous matter in an infinitely old, infinitely large, homogeneous universe. (Indeed, this was Olbers's suggested solution to the paradox.) Assumption (2) must not be true, however, since in an infinitely old universe, even the dust would **heat** up until it reaches **equilibrium** with the radiation by emitting its own thermal emission at the same rate at which it absorbs radiation, and the paradox still applies.

Modern **cosmology** has shown from a variety of observations that the universe is expanding, and therefore has a finite age. Hence, most astronomers today would agree that the resolution to Olbers's paradox is that the universe is not infinitely old.

OORT CLOUD

In 1950, the Dutch Astronomer Jan Hendrik Oort noted several facts about the orbital parameters of the known comets.When Oort examined the population of LPCs (**comets** with orbital periods longer than 200 years), he found that the distribution of orbits was concentrated so that the furthest distance in the orbit was often far beyond the limits of the solar system. Moreover, he noted that the cometary orbits appeared to be random; they approached the solar system from all directions in **space**, and as many orbited with the solar system (prograde) as in the opposite sense (retrograde). Oort hypothesized that the origin of the LPCs was a large spherical cloud of perhaps 10^{12} comets, located at some 50,000 AU (1 AU = distance from Earth to **Sun** = 1.5×10^8 km) distance from the Sun. This hypothetical cloud is termed the *Oort cloud* in his honor.

In this conceptual framework, a gravitational disturbance—for instance, passage of the Sun near another star, passage of the Sun through the plane of the galaxy, or passage of the sun through a dense gaseous molecular cloud—stirs up the Oort cloud, and sends a large number of comets moving in towards the solar system in a vast cometary shower. On their first passage through the solar system, these comets would tend to be scattered by the presence of Jupiter and Saturn, and tend to be captured onto SPC (comets with orbital periods of less than 200 years) orbital trajectories.

Scientists hotly debate whether these comets play a role in episodic extinctions on Earth. Some astronomers have demonstrated that there are definite periods in the flux of cometary bodies; for instance, every 30 millions years, the Sun, executing a slow vertical **motion** in its orbit through the galaxy, passes through the midplane of the disk of the galaxy. During this time, scientists expect a much higher incidence of nearby stellar encounters, which should result in an increase of cometary influx in the inner solar system by about a factor of four or so. Other scientists point to **mass** extinctions in the fossil record which occur on roughly this same timescale, though it is still unclear whether the two are directly related.

OORT, JAN HENDRICK (1900-1992)

Dutch astronomer

Jan Oort received his higher education at the University of Groningen where he obtained a Ph.D. in 1926. He began working at the Leiden Observatory in 1924, and became its director in 1945.

In 1927, Oort had been able to demonstrate that the **Milky Way galaxy** was rotating. Since a galaxy is not a solid object, but comprised of billions of **stars**, it does not rotate as a single solid object. Instead, stars near the center of the

Oort cloud. *(Image by Seth Shostack. Photo Researchers, Inc. Reproduced by permission.)*

galaxy move faster, while those farther move slower. By studying the **motion** of the stars in our vicinity, Oort was able to determine that the center of the galaxy lay in the same direction as the constellation of Sagittarius. That was in agreement with the observations of American astronomer **Harlow Shapley**. Shapley, however, had placed the galactic center at a distance of 50,000 light-years. Oort, taking into account the discovery of dark dust clouds that obscured the **light** of distant stars, thought the galactic center was not that far away. He suggested a distance of 30,000 light-years, a distance that is accepted today.

Karl Jansky had discovered **radio** emissions from outer **space** in the early 1930s, and Grote Reber's 1940s map of their intensity showed a peak toward Sagittarius. Radio **waves** could penetrate the dust that blocked visual light, making radio telescopes ideally suited for the task of examining the galactic center. However, World War II was in progress, and only theoretical work could be carried out. An associate of Oort's, Hendrik van de Hulst, made calculations which indicated that it should be possible to detect radio noise at the 21 cm **wavelength**. This would be produced by the hydrogen gas in **interstellar space**.

Once the war was over, Oort and his group set out to perform the search. They were successful in 1951. Because the hydrogen gas is more concentrated in the arms of the Milky Way galaxy, they were able to use radio telescopes to trace out the spiral structure.

Despite all his work on the structure of the galaxy, Oort is best known for a unique theory to account for the origin of **comets**. He suggested that there was a great shell of cometary material encircling the solar system at a distance of one light-year. Occasionally objects leave the so-called **Oort cloud**, for reasons not entirely understood, and head into the inner solar system. As such objects encounter the Sun's **radiation**, it warms and the gases and ice comprising it develop into the long tails characteristic of a comet.

Jan Oort died on November 12, 1992, his wide-ranging research interests and signal achievements having established him as one of the eminent astronomers of the twentieth century.

OPPENHEIMER, J. ROBERT (1904-1967)

American physicist

Theoretical physicist J. Robert Oppenheimer was a pioneer in the field of **quantum mechanics**, the study of the **energy** of atomic particles. His research on protons and their relation to electrons led directly to the discovery of a new particle, the **positron**. His later work shed **light** on deuterons, the nuclei of heavy hydrogen atoms. He was a charismatic teacher and effective administrator who directed the laboratory at Los Alamos, New Mexico, where the **atomic bomb** was developed during World War II. In the postwar years, however, Oppenheimer staunchly opposed the proliferation of **nuclear weapons**. This stance brought him before Congress during the McCarthy era and cost him his security clearance as a government consultant.

Julius Robert Oppenheimer was born in New York City on April 22, 1904, to a wealthy and cultured family. His father, Julius Oppenheimer, who emigrated from Germany as a young man, had a successful business importing textiles. His mother, the former Ella Friedman, was a painter and a great lover of the arts. Oppenheimer, his parents, and his younger brother, Frank, divided their time between a spacious New York apartment overlooking the Hudson River and a summer house on Long Island.

It became apparent when he was quite young that Oppenheimer had a quick mind, a vast appetite for learning, and a wide range of interests. At age eleven, he was the youngest person ever admitted to the New York Mineralogical Society, and at the age of twelve he presented a paper there. He attended the Ethical Culture School in New York, and after graduating he spent the summer in Europe. Unfortunately, he contracted dysentery there, and needed the following year to recuperate. When he was well again he took his first trip to the West, where the expanse of the Pecos Valley of New Mexico captured his imagination. His family eventually bought a ranch there, returning year after year.

Oppenheimer entered Harvard College in 1922. He studied a broad curriculum, which included several languages as well as chemistry and physics. As an undergraduate he was especially close to physicist **Percy Bridgman**, also a man of many interests, who may have shaped the way Oppenheimer combined physics with philosophy in his later career. Despite his course load, Oppenheimer graduated from Harvard *summa cum laude* in just three years, and in 1925 he left the United States for Europe to study theoretical physics.

It was in Europe during that period that the most important advances were being achieved in the study of the behavior and energy of particles that make up the atom—a discipline known as **quantum mechanics**. Such brilliant theoreti-

cians as **Werner Heisenberg**, **Erwin Schrödinger**, and **Paul Dirac** were formulating their theories about quantifying and predicting the movement and location of atomic particles. Oppenheimer initially went to the Cavendish Laboratory in Cambridge, England, and within just a few months he had submitted his first paper, which used some of the most recent theoretical advances in **nuclear physics** to explain aspects of molecular behavior.

Explains Molecular Activity

In 1926 Oppenheimer left the Cavendish for the University of Göttingen in Germany. He did independent research on **radiation** at Göttingen and also collaborated with the physicist **Max Born** in a further investigation of molecular activity. The two scientists tackled variations in the vibration, rotation, and electronic properties of molecules. Their results led to the so-called "Born-Oppenheimer method," which is in effect a quantum mechanics at the molecular rather than atomic level. After receiving his doctorate in 1927, he left Göttingen for Leiden, Holland, and then went on to Zurich where he worked with another distinguished physicist, **Wolfgang Pauli**. Throughout this period, Oppenheimer consistently demonstrated his ability to synthesize ideas, draw connections between theories, and detect their inherent contradictions.

Oppenheimer's work in Europe had been of such quality that he was able to arrange teaching positions both at the University of California at Berkeley and at the California Institute of Technology. For the next thirteen years he taught and did research at these schools during alternating semesters. During this period, Oppenheimer evolved into an extraordinary, charismatic teacher. Although many students complained that he set an impossible pace in the classroom, he attracted a number of students and even some colleagues to theoretical physics by his quick wit and probing questions. He inspired many of his students, and some even adopted his gestures and way of speaking. Oppenheimer's social life was very much involved with his teaching; he would often discuss subjects such as **astrophysics** and **cosmic rays** or nuclear physics and **electrodynamics** for hours on end.

Some of his best work in **particle physics** was also done during these years. In 1930 he was able to demonstrate that the **proton** is not the **antimatter** equivalent of the **electron** (or antielectron) as had until then been supposed. One of Oppenheimer's students, **Carl Anderson**, used this work in his search for the true antielectron and found the positron. Oppenheimer's earlier work on radiation led to his contribution to the discovery that cosmic ray particles could break down into another generation of particles, a phenomenon commonly called the "cascade process." In 1935, he discovered that it was possible to accelerate deuterons, made up of a proton and a **neutron**, to much higher energies than neutrons alone. Deuterons, as a consequence, could be used to bombard positively charged atomic nuclei at high energies, enabling further research into atomic particles.

Leads the Manhattan Project

For years physicists had been aware of the possibility of manipulating **nuclear fission**. Bombarding the nuclei of certain molecules, they suspected, could result in a **chain reaction** that would release an extremely large amount of energy. In the United States and Nazi Germany, scientists were rushing to work on a weapon that could harness this energy—the atomic bomb. At Berkeley's Radiation Laboratory under the direction of Oppenheimer's colleague, Ernest Orlando Lawrence, researchers selected uranium as the **chemical element** most likely to lend itself to nuclear fission for military purposes. However, no coordinated effort to design and fabricate an actual atomic weapon was made.

According to Oppenheimer's colleague Victor Weisskopf, "many physicists were drawn into this work by fate and destiny rather than enthusiasm," and Oppenheimer was one of them. Early in 1942 he brought together a group of theoretical physicists—many of whom had been working in separate laboratories under the umbrella of the Manhattan Project—and became director of the new research lab at Los Alamos to develop the first nuclear weapon.

The administration of the research at Los Alamos presented its challenges, which Oppenheimer handled effectively. The intense and sustained work that was done on the **Manhattan Project** was due in large part to his evident sense of purpose and his ability to provide a creative, cooperative environment for scientists forced to work in secret. A design for the bomb was ready and a fissionable form of plutonium produced to fuel it by mid–1945. On the fateful morning of July 16, 1945, Oppenheimer stood silently awaiting the detonation of the test bomb nicknamed "Fat Man" at nearby Alamogordo. He wrote in his *Letters and Recollections* that, upon seeing the **power** it unleashed, he thought the play of the light had the "radiance of a thousand suns," but he was also reminded of a dark, foreboding line from the Hindu *Bhagavad-Gita:* "I am become death, the Shatterer of Worlds."

Oppenheimer was one of a panel of four scientists including Ernest Orlando Lawrence, **Enrico Fermi**, and **Arthur Compton** that was asked to formulate an opinion regarding the use of the atomic bomb to end the war against Japan. They were told that there was a choice between a military invasion of Japan, which was certain to cost many American lives, and a nuclear attack on a military target that would also kill many civilians. Confronted with this choice, the panel voted to use the bomb. Oppenheimer later regretted his decision, saying that the intentional slaughter of civilians had been unnecessary and wrong.

Political Controversies

After the war, Oppenheimer became more and more concerned about the devastating potential of atomic weaponry. With this concern foremost in his mind, he cowrote the "Acheson-Lilienthal Report," which opposed the nuclear arms race at its very outset, instead proposing stringent international controls on the development of nuclear arsenals. The report was rewritten and presented to the United Nations as the

J. Robert Oppenheimer.

Baruch Plan, but the Soviet Union vetoed its adoption. In 1946 Oppenheimer became chair of the general advisory committee of the Atomic Energy Commission and continued to advocate controls on the development of **nuclear power**.

In October of 1947, Oppenheimer became director of the Institute for Advanced Study at Princeton University, New Jersey. Under his directorship, the institute became one of the foremost centers of research in theoretical physics, even though Oppenheimer himself did little research from this time forward. He was keenly aware of what others were doing at the institute both inside and outside theoretical physics, but much of his energy was now spent on policy issues rather than science, and many considered his scientific judgment no longer as keen as it had been.

As the 1940s drew to a close, President Truman decided that it was in the country's best interest to develop the hydrogen bomb. Oppenheimer's position was clear; he did not believe in the proliferation of nuclear weapons, and he did not hesitate to express his opinion in public. His position disturbed many of those in power in Washington. In November of 1953 William Borden, former executive director of Congress' Joint Atomic Energy Committee, sent a registered letter to J. Edgar Hoover of the FBI, stating he had considerable evidence to show that Oppenheimer was a Soviet agent. In December Oppenheimer was informed that his security clearance, which he needed to have access to classified information, was revoked on suspicion of unpatriotic activities on his part.

A Congressional hearing giving Oppenheimer the opportunity to clear himself did little to dispel the myth of his lack of patriotism. It was the era of the Cold War and the anti-communist hysteria spearheaded by Senator Joseph McCarthy. Oppenheimer's association with Communists during the 1930s, never a secret nor an obstacle to his receiving clearance when he was director of Los Alamos, was now presented as a blot on his character and a challenge to his patriotic commitment. Both his position on nuclear weapons and his arrogance and ability to argue had made him unpopular in some quarters, and there were those who wanted to see him removed from his public position. After many grueling hours of testimony, the majority of a three-man panel found that Oppenheimer was "a loyal citizen," but still denied him clearance on the basis of "defects of character."

After this very public hearing was finally over, Oppenheimer emerged somewhat aged and wounded. Yet he continued to lecture on science and express his opinions on politics. In 1963 Oppenheimer received the prestigious Enrico Fermi Award, which was something of a public vindication but did not undo all the harm done earlier.

Almost every published photograph showed Oppenheimer with a cigarette or pipe in hand, and he knew for some time that he had throat cancer. He died at home in Princeton on February 18, 1967. After his cremation, his wife Kitty spread his ashes in the sea near their vacation retreat in the Virgin Islands.

OPTICAL INSTRUMENTS

An optical instrument is a device that uses a lens or **lenses** to create an image of an object. A lens is a piece of transparent material that is shaped such that it bends parallel rays of **light**. The shape of a lens is a segment of a sphere or a plane. The bending of light causes the light to cross, forming an image. This image can be larger or smaller than the actual object being viewed through the lens. The bending of light as it passes through a medium such as glass or plastic is called refraction. The early Greeks used lenses in magnifying glasses long before the ideas of refraction were understood. One of the first optical instruments was a pair of eyeglasses, invented in Italy in the late 1200s. The **telescope** was invented in 1608 by the Dutch optician **Hans Lippershey**. Today, lenses are used in many optical instruments, such as cameras, **binoculars**, compound microscopes, and movie projectors.

In order to understand how optical instruments work, one must understand some basic principles about lenses. Lenses can occur in one of two shapes. A lens that is thicker in the middle than at the edges is called a convex or converging lens, and a lens that is thinner in the middle is called a concave or diverging lens. A converging lens causes the parallel light rays that pass through it to converge, or come together to a point, called the focal point, as it exits the lens. A diverging lens causes the light to diverge, or separate, as it exits. A con-

verging lens can produce both a virtual and a real image. A virtual image is one that cannot be focused on a screen. It is perceived only through the human eye. This occurs in a converging lens when the object is between the focal point and the lens. An example of a converging lens producing a virtual image is when a magnifying glass produces a larger image. The object to be magnified, for example, small print in a book, is placed between the focal point and the magnifying glass. The print appears larger as you look at it through the lens. This larger print could not be focused on a screen because the light rays are not focused on a fixed position. If an object is placed far enough away from a converging lens to be past the focal point, the light passing through the lens does converge to a point on the other side of the lens and can be focused on a screen. This is a real image. An example of a converging lens producing a real image is when a camera lens focuses an image on film. A diverging lens can only produce a virtual image, because the light passing through a diverging lens never converges to a point.

A camera uses a lens to focus a real image into a box that is impermeable to light. The image is focused on a light-sensitive film. The lens forms an inverted, or upside-down, image on the film. Sometimes the camera is designed to allow the lens to move back and forth, changing the distance between the lens and the film. The shutter and diaphragm in a camera control the amount of light that reaches the film; the shutter by controlling the length of time light is exposed to the film, and the diaphragm by controlling the size of the opening (or aperture) through which light enters the camera.

A telescope uses a lens to form a real image of an object found far away. An astronomical telescope generally consists of two lenses, the objective and the eyepiece. The objective lens produces a real image, focusing it close to the eyepiece. The eyepiece, then, acts as a magnifying glass, allowing a person to view the image. The image produced by the eyepiece is a virtual image of the image produced by the objective. This arrangement of lenses produces an inverted image. Refracting telescopes are those that use lenses, although many large astronomical telescopes use mirrors instead and are called reflecting telescopes. A terrestrial telescope consists of three lenses, or two lenses and a pair of prisms. The third lens or prisms flip the inverted image, so the viewer sees a right-side up image. This image, although it is right-side up, is much dimmer than that of an astronomical telescope. Two terrestrial telescopes placed side by side are called binoculars.

A **compound microscope** uses two converging lenses to create an enlarged image of a very small object. The two lenses are called the eyepiece and the objective. The objective focuses a real, enlarged image of an object. The eyepiece further magnifies this image to produce a virtual, inverted image. The difference between a compound **microscope** and a telescope is that the objective in a telescope does not enlarge the original image.

A movie or slide projector consists of a light source, a condenser, a slide or movie frame, and a projection lens. The light source shines light through a pair of lenses called the condenser. These lenses direct the light through the slide or movie frame to the projection lens. The projection lens then focuses the image from the slide or movie frame onto a screen. The projection lens is allowed to move back and forth in order to focus the image on the screen.

No lens can create a perfect image. Each lens has imperfections, called aberrations, which cause distortions in the image produced. A camera corrects spherical aberrations, those caused by light passing through the edge of a lens causing the image to be out of focus, by covering the edges of the lens with a diaphragm. Optical instruments use more than one lens in order to further minimize these aberrations.

OPTICS

Optics is the branch of physics that is concerned with **light** and its properties. Physicists who focus on optics study the properties of light. They also apply these properties to phenomena such as **color**, mirrors, and **lenses**.

Ancient Greek philosophers were the first to study light. They theorized that light was made up of tiny particles that could enter the eye, creating vision. The idea of the particulate nature of light was widely accepted even past Sir Isaac Newton's (1642-1727) time, although a few people, such as the Greek philosopher **Empedocles** (490-430 B.C.) and the Dutch scientist **Christiaan Huygens** (1629-1695), believed light was actually a wave. In the nineteenth century, the wave theory of light was accepted. In 1905, **Albert Einstein** (1879-1955) theorized that light behaves both as a particle and a wave. This dual nature of light is the theory scientists agree upon today.

We now know that light is a form of **energy** that travels in a wave with both electric and magnetic behavior. Light, therefore, is called an electromagnetic wave. Other electromagnetic **waves** include **radio** waves, **microwaves**, and **x rays**. Light waves are transverse waves, in other words, they oscillate perpendicular to their direction of travel. A wave that oscillates up and down is vertically polarized, and one that oscillates from side to side is horizontally polarized. Polarized light oscillates in one direction. The light from a common light source, such as a light bulb or the **sun**, is not polarized. Light waves originating from these sources can oscillate at any orientation. When the light passes through a polarizing filter, such as polarized sunglasses, it exits as polarized light. The filter only passes light waves that are oscillating in a certain direction.

Electromagnetic waves occur at different frequencies. The **frequency** of a wave is the number of wave crests that pass by a certain point in a given amount of time, usually expressed as waves per second. The frequency of a given light wave is directly related to color. **Isaac Newton** was the first scientist to study color. He passed sunlight through a prism and found that it could be separated into beams of light of different colors. He showed that visible light actually consists of red, orange, yellow, green, blue, and violet light. Each of these colors corresponds to a particular frequency of light. Newton passed the individual color bands produced by the prism

Fiber optics. *(Digital Stock. Reproduced by permission.)*

through a second prism. This second prism re-combined the individual bands and the light exited the prism as white light. This showed that white light is actually the combination of all of the colors of the spectrum.

The color of an object is due to the frequencies of light absorbed by the object. Most objects absorb the majority of the frequencies of light. Any frequencies that are not absorbed by the object are reflected, giving the object a particular color. If an object absorbs all light except the frequencies found in the red region of the spectrum, the object appears red. Red light is reflected off of the object. White is actually not a color, but a combination of all colors, occurring when all frequencies of light are reflected. Likewise, black is actually the absence of reflected light, occurring when all frequencies of light are absorbed.

Interference patterns can impact light waves two different ways, constructively or destructively. Constructive interference occurs when two or more light waves meet in phase, meaning all of their crests (top of the wave) meet at the same time. This usually results in a more intense or bright resulting light. When the light waves meet out of phase—the crest of one wave meets at the trough (bottom of the wave) of another—destructive interference takes place. Because the two wavelengths are out of phase, they cancel each other and no light is visible.

The concept of interference is important for understanding the phenomena of **diffraction**. Thomas Young's double-slit interference experiment is a classic explanation for diffraction,

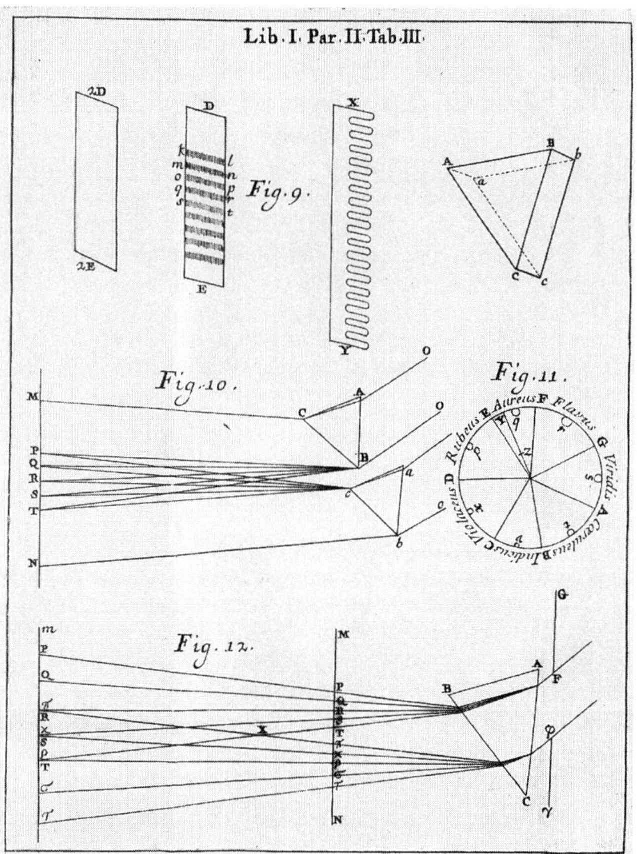

Diagrams from Sir Isaac Newton's "Optics," 1704. (Corbis-Bettmann. Reproduced by permission.)

which is the bending of light as it passes around an object. Young made two small slits relatively close to each other on a dark board. When he shined a light through the slits and observed the light on a screen, he noticed that the light did not pass directly though in two straight lines. Instead, there was a pattern of alternating bright and dark bands of light. This resulted from the light waves fanning out-diffracting-as they passed through the barrier slits, much like water ripples when it passes from a small opening into a larger body of water. Because light waves were passing through two slits, two fans were created that overlapped at certain points. Some of these points experienced destructive interference, while other were constructive, thus leading to the alternating bands of light. The dark bands occurred when light waves canceled each other out.

Other phenomena associated with light are called reflection and refraction. Light is reflected when the light waves bounce off of something and travel in a new direction. A surface that causes light to bounce back is called a reflective surface. A mirror is an example of a reflective surface. The angle an out going light ray makes with a reflective surface will be equal to the angle of the incoming light ray. To an observer, a reflected light ray will appear to come from behind the reflecting surface. For instance, when a person stands in front of a mirror, they will see an image of themselves that appears to be behind the mirror. Because the image appears to

originate from an imaginary point, the image is called a virtual image. A virtual image created by a mirror is the same size as the original object.

Refraction can occur when light travels through one medium into another. The **velocity** of light is different for various materials. For instance, the velocity of light in air is slower than the its velocity in **vacuum** and slower still in glass or plastic. Under the right circumstances, the light ray will be refracted back into the original material. In a sense, the light ray reflects off the boundary. For example, if a waterproof flashlight is held in a bathtub of water at different angles, a particular angle can be found where the beam does not escape the water to shine light through the air above the water surface. The light is refracted at the surface of the water back into the water instead of being passed through the water and into the air. This angle is called the critical angle. Any angle beyond the critical angle will cause total internal reflection.

Optical fibers, also called light pipes, utilize this phenomenon. Light travels through the transparent fibers by a series of total internal reflections, much like a rubber ball would bounce through a pipe. Fiber optics have many different important uses today. Mechanics use optical fibers to shine light deep into engines. Surgeons use them to see inside a patient's body. Optical fibers are also used in communications because they are less bulky and more inexpensive than copper cables. Information in these fibers is carried by light instead of electrical current.

Another form of light that has become indispensable in society is the laser. The light in **lasers** results from photons emitted by highly excited atoms returning to their ground state. The photons are harnessed between two mirrors where they continue to collide until they collectively exit in one direction at a specific **wavelength**. Laser light is a very precise, specific wavelength that can be altered to match the absorption of almost any substance. The laser light will only damage materials whose absorption band matches the lasers wavelength. This controlled intensity makes the laser a handy tool for several applications ranging from surgery to reading compact disks.

See also Reflection, refraction and dispersion

ORBITALS

An orbital describes a probability region for finding an **electron** at a certain **energy** level, distance, and orientation as related to the atomic **nucleus**. Orbitals are described by a set of **quantum numbers** that describe their location shape and electron capacity. In accord with wave-particle duality, the high probability regions for finding electrons also describe the wavelike **motion** of electrons in the three-dimensional region of **space** surrounding an atomic nucleus.

Within each orbital beyond the first orbital there are suborbitals that reflect particular orientations in space (e.g., p orbitals directed along the x, y, or z axis). Each suborbital orbital can hold up to two electrons. Electrons occupy different energy levels in atoms and have different wave properties. The Schrödinger equation relates the energy of a system to its wave properties. This equation employs the three orbital quantum numbers, which describe an orbital, as well as a fourth quantum number, the electron spin quantum number. The shapes of different kinds of orbitals are different because of the interaction of the quantum effects between the different atomic particles.

In 1911, New Zealand-born English physicist **Ernest Rutherford** (1871-1937) developed a picture explaining the structure of the **atom** in terms of a positively charged nucleus surrounded by negatively charged electrons in orbits. Later, in 1913, Danish physicist **Niels Bohr** (1885-1962) developed a theory that explained why electrons can remain in orbits and do not collapse onto the nucleus. **Quantum theory**, developed during the 1920s, explained all phenomena concerning electrons and their role in atoms. The theory showed that all particles have certain associated wave properties and that electrons cannot be pictured as localized particles in space, but rather should be thought of as clouds of negative charge spread out over the entire orbit. These clouds represent orbits, the regions around the nucleus where the electrons are most likely found.

The exact physical manifestation of each orbital depends on many factors that are described by the quantum numbers associated with a particular energy state. The three quantum numbers that determine the shapes are n, the principle quantum number, l, the **angular momentum** quantum number, and m, the magnetic quantum number. The principle quantum number, n, determines the distance, size and general **energy level** of the orbital. It can be any positive integer (1, 2, 3, 4, etc.) with the higher number indicating a larger, less energetic orbital located progressively farther from the nucleus. The angular **momentum** quantum number, l, determines the shape of the orbital and its value takes on all positive integral values from 0 to n-1. Therefore, for any orbital there can be many shapes (i.e., orientation in space) for the associated suborbitals. The first angular momentum quantum numbers are designated as s, p, d, and f orbitals. The third of the quantum numbers, m the magnetic quantum number, determines the orientation with respect to an applied magnetic field. All values of the magnetic quantum number are integral and can range in value from -3 to 3.

The s orbital is spherical. The p orbital is dumb-bell shaped with two parts separated by a nodal plane where the probability of finding the electron approaches zero. Because of the range of possible values of magnetic quantum numbers associated with this orbital, there are three possible orientations of the p orbital, along the x-axis, along the y-axis or along the z-axis. The d orbital has four lobes with nodal planes at the intersection of the four lobes. There are five possible orientations for these orbitals because of the range of possible magnetic quantum numbers associated with this orbital. f orbitals have still more complicated shapes. There are seven possible orientations for these orbitals described by the magnetic quantum numbers. There is a fourth quantum number that describes the spin of the electron in an orbital. This quantum number is called the spin quantum number and can have

a value of either 1/2 or -1/2 since an electron can orient in two ways in an applied magnetic field.

The electronic configuration of atoms is determined by application of the rules for placing electrons in orbitals. The electronic configuration for any atom is governed by the Aufbau principle (electrons occupy the lowest-energy orbitals available, the Hund principle (all sub orbitals receive an electron before electrons are paired) and the **Pauli exclusion principle** (no two electrons can have the same four quantum numbers). To build an electronic configuration of an atom, electrons are placed in the lowest energy orbitals first.

See also Atomic orbitals; Atomic structure; Quantum numbers

ØRSTED, HANS CHRISTIAN (1777-1851)
Danish physicist and philosopher

Danish physicist and philosopher Hans Christian Ørsted is best known for discoverering the connection between **electricity** and **magnetism** known as **electromagnetism**. Ørsted was born in Rudkobing on the Danish island of Lolland. Ørsted became interested in science while working as a young boy for his father, Soren Christian Ørsted, who owned a pharmacy. Educated primarily through self-study at home, both Ørsted and his brother, Anders Sandoe Ørsted, went to Copenhagen in 1793 to take entrance exams for Copenhagen University. The brothers passed and distinguished themselves academically at the University. By 1796, Ørsted receivied honors for his papers in both aesthetics and medicine.

Although he passed his pharmaceutical examination in 1797, Ørsted continued his education and, in 1801 received a travel scholarship and public grant that was to profoundly affect his future. With the awards, Ørsted spent three years traveling in Europe. While in Germany, Ørsted met Johann Ritter (1776-1810), a physicist who was interested in finding a connection between electricity and magnetism. Drawn to the study of physics, Ørsted returned to Copenhagen after three years abroad and accepted a post as a lecturer on physics at Copenhagen University. Largely under Ørsted's guidance, the university went on to develop a comprehensive physics and chemistry program, and to establish new laboratories.

In 1820, which Ørsted described as the happiest year of his life, Ørsted considered a lecture for his students focusing on electricity and magnetism that would involve a new electric battery. During a classroom demonstration, Ørsted saw that a compass needle deflected from magnetic north when the **electric current** from the battery was switched on or off. This deflection interestred Ørsted convincing him that magnetic fields might radiate from all sides of a live wire just as **light** and **heat** do. However, the initial reaction was so slight that Ørsted put off further research for three months until he began more intensive investigations. Shortly afterwards, Ørsted's findings were published, proving that an electric current produces a magnetic field as it flows through a wire. This discovery revealed the fundamental connection between electricity and magnetism, which most scientists thought to be completely unrelated phenomena. The result was an intensive effort throughout the scientific community in **electrodynamics** research, including French physicist **André-Marie Ampère's** (1775-1836) developments of a single mathematical form to represent the magnetic forces between current-carrying conductors. Ørsted's discovery also represented a major step toward a unified concept of **energy**.

Although some accounts of Ørsted's discovery described the event as purely accidental, Ørsted's earlier travels in Europe and wide range of studies had drawn him to a believe in the unity of nature, and that a relationship therefore must exist between most natural phenomena, including electricity and magnetism. If Ørsted had not been looking for such a relationship, he possibly would not have placed the compass near an electrical current to see if there was an effect. Ørsted's discovery, which laid the foundation for the theory of electromagnetism, led to research that created much of the technology in common use today, including **radio**, television, and fiber **optics**. In honor of its discoverer, the physical unit of magnetic field strength was named Oersted in 1932.

Among Ørsted's continued scientific works, he was one of the first scientists to produce pure aluminum and to conduct experiments in the compressibility of **fluids**. Ørsted had a wide range of interests, including education, philosophy, politics, and literary affairs (Ørsted was an ardent supporter of the Dutch fairy tale writer Hans Christian Andersen and introduced the words *brint* and *ilt*, for hydrogen and **oxygen**, respectively, into the Dutch language). In 1824, Ørsted founded the Society for the Dissemination of Natural Science, which he stated was for the purpose of "spreading knowledge about experimental science, the mechanical part as well as the chemical part."

Ørsted, who married Birgitte Ballum and had eight children, spent over 50 years at Copenhagen University and the Technological College. To the Danish people, Ørsted is as much revered for his efforts as a teacher and educator as for his discovery of electromagnetism. His combination of scientific thought and wide-ranging efforts in cultural and political activities helped to transform Danish society. Ørsted believed that Denmark would best progress as a country through a close cooperation between research and industry. To help ensure that all of the small country's human resources were fully utilized, Ørsted emphasized the need for a comprehensive nationwide system of education and training. At the time of his death in 1851, Ørsted was head of the Polytechnical Institute at Copenhagen University. In 1993, nearly a century and a half after his death, the first Dutch satellite was named in Ørsted's honor.

See also Magnetic fields and forces

OSCILLATIONS

An oscillation is a particular kind of **motion** in which an object repeats the same movement over and over. It is easy to see that a child on a swing and the pendulum on a grandfather clock both oscillate when they move back and forth along an arc. A small **weight** hanging from a rubber band or a spring can also oscillate if pulled slightly to start its motion, but this repeated motion is now linear (along a straight line). On a larger scale, you can notice oscillations when bungee jumpers fall to the end of their cords, are pulled back up, fall again, etc. Actually, oscillations are all around us, even in the pages of this book.

Anything, no **matter** how large or small, can oscillate if there is some point where the object is in stable **equilibrium**. Stable equilibrium means that an object always wants to return to that position. Suppose you placed a marble at the exact center inside a very smooth bowl. If you tap the marble slightly to move it a small distance, it rolls back towards the center, overshoots, rolls back, overshoots, etc. The marble is oscillating as it continues to return to center of the bowl, its point of stable equilibrium. If you think of the marble and the bowl as a "unit," you can see that the "unit" stays together even though the marble is oscillating (unless you tap the marble so hard that it flies out of the bowl). This is the reason for using the term stable.

On the other hand, what if you turned the bowl over and tried the same experiment by placing the marble on top at the center? You might succeed in balancing the marble for a short time, but eventually you will touch the table or a breeze will move the marble a small amount and it will fall. When this happens, the "unit" of marble and bowl comes apart and no oscillation can happen. In this case, the center of the bowl would be a point of unstable equilibrium, since you can balance the marble there, but the marble cannot return to that point when disturbed to keep the "unit" from disintegrating.

For the motion of a child on a swing, the bottom of the arc (when the swing hangs straight down) is the point of stable equilibrium. The point of stable equilibrium for a weight on the rubber band is the location at which the weight would hang if it was very slowly lowered. In either case, an oscillation occurs when the object (child or weight) is moved away from stable equilibrium. If we pull the swing back some distance the child will move toward the bottom of the arc. At the instant the swing is at the point of stable equilibrium, the child is moving the fastest since as the swing proceeds up the arc on the other side, it slows down. The higher the swing was when the motion was started, the faster the child moves at the bottom. The swing overshoots stable equilibrium and the child rises to the same distance on this side of the bottom as on the starting side. For a brief instant the swing will stop before the swing begins to retrace its path, traveling in the other direction.

This simple example demonstrates several properties shared by all oscillations: 1) The point of stable equilibrium is the center of the oscillating motion since the object moves the same distance on either side. That distance is called the amplitude of the oscillation. 2) At either end of the motion, the object stops briefly (slowest location) while the fastest location is when the object is just passing through the point of stable equilibrium. 3) The **energy** that an object has when it is oscillating is related to the amplitude. The larger the amplitude, the larger the energy.

Oscillations also have two very specific properties regarding time. Every oscillation takes a certain amount of time before the motion begins to repeat itself. Since the motion repeats, we really only need to worry about what happens in one cycle, or repetition of the oscillation. If we pick any point in the motion and follow the object until it has returned to that same point ready to repeat, then the oscillation has completed one cycle. The amount of time that it took to complete one cycle is called the period, and every cycle will take the same amount of time. Suppose for the child on a swing we pick the point at the bottom of the arc. When the swing moves through that position, we start a timer. The child will swing up, stop, swing back down through the bottom (but traveling in the other direction), swing up, stop, and swing back through our point. Now the child has returned to our starting point and the motion is about to repeat, so we stop our timer. The curious thing is that even when the amplitude is changed, the period stays the same. This is because even though the child moves faster when pulled higher to start the motion, the swing also has farther to travel to complete a cycle.

The other time property is called the **frequency**, which tells how often the motion repeats. This really gives the same information as the period since if it takes 0.5 second for 1 cycle, then the frequency will be (1 cycle)/(0.5 second) = 2 cycles per second. Often a unit called the Hertz (Hz) is used to represent a cycle per second. The cycles per second should **sound** familiar because the magnitude, or amplitude, of the electrical current in most households oscillates at 60 cycles per second. The time properties of an oscillation are very important since they control how best to add energy to the motion.

The child on a swing, weight on a spring, and bungee jumper on an elastic cord are all types of oscillations which we can see with our eyes. However, if you kick a rock (disturb it) and it does not disintegrate, then the atoms within the rock must be in stable equilibrium. The atoms within the rock are therefore capable of oscillating. Small oscillations are actually occurring all the time in every seemingly solid substance, including the rock and this page. We cannot see this motion, but we do feel it. The larger the amplitude of those small oscillations of the atoms, the hotter the object, and that is something we can detect directly.

OXYGEN

Oxygen is the simplest group VIA element and is, under normal atmospheric conditions, usually found as a colorless, odorless and tasteless gas. Oxygen has an atomic number of 8 and an atomic **mass** of 16.0 amu. The liquid and solid forms, which are strongly paramagnetic, are a pale blue **color**. Oxygen has a **boiling** point of -297°F (-182.8°C) and a melting point of -368.7°F (-222.6°C).

Oxygen is the third most abundant element found in the **Sun**, after hydrogen and **helium**, and plays an important role in the carbon-nitrogen cycle. Oxygen composes 21% of Earth's atmosphere by volume and is vital to the existence of carbon-based life forms.

Although English chemist Joseph Priestley (1733-1804) is generally credited with the discovery of oxygen in 1774,

many science historians contend that Swedish chemist Carl Scheele (1742-1786) probably discovered oxygen a few years prior to Priestly. French chemist Antoine Lavoisier's (1743-1794) contributions to the study of the important reactions, combustion and oxidation, were spurred by the discovery of oxygen. Lavoisier noticed that something was absorbed when combustion took place and that it was obtained from the surrounding air. Lavoisier noted that the increase in the **weight** of the substance burned was equal to the decrease in the weight of the air used. His studies lead to Lavoisier's oxidation theory, which eventually superseded the phlogistonists' theory (i.e., that every combustible substance was thought to contain a phlogiston, or inherent principal of fire of fire, liberated through burning, along with a residue) that was widely accepted at that time. Lavoisier eventually named the gas he studied oxygen from the Greek *oxys* meaning acid or sharp, and *geinomial* meaning forming. Lavoisier named the gas oxygen because he noted that the burned materials were converted into acids.

Although oxygen has nine isotopes, natural oxygen is a mixture of only three of these. The most abundant **isotope**, oxygen-18, is stable and available commercially. The most common use for oxygen gas is in enrichment of steel blast furnaces. Large quantities are also used in making synthetic ammonia gas, methanol and ethylene oxide. Oxygen is also consumed in oxy-acetylene welding. Most commercial oxygen is produced in air separation plants. It is estimated that the United States consumes 20 million tons of oxygen in commercial use per year and the demand is expected to increase dramatically.

When oxygen is exposed to ultraviolet **light**, as from the Sun, or an electrical discharge, as from lightening, ozone (O_3) is formed. Although ozone is toxic to breathe, the 0.12 in (3 mm) thick layer of ozone in the Earth's atmosphere provides a shield from harmful ultraviolet rays from the Sun. The **ozone layer** has recently been the subject of intense scientific interest to determine whether, and to what extent, it may be deteriorating, mainly from pollutants in the atmosphere. Unlike pure oxygen gas, ozone has a bluish color and its liquid and solid forms are bluish-black to violet-black.

See also Gases, behavior and properties

OZONE LAYER

The ozone layer is a region of atmosphere 12-30 mi (19-48 km) above Earth's surface that consists of ozone concentrations of as much as 10 parts per million. Ozone is an allotropic form of **oxygen** composed of three atoms of oxygen in each molecule, O_3. The main function of the ozone layer is protection of Earth's surface from harmful ultraviolet rays from the **Sun**. Although this atmospheric layer filters ultraviolet rays, the ozone layer does allow enough ultraviolet rays to pass through to support the activation of vitamin D in humans.

The word ozone is derived from the Greek *oxein* meaning to smell. It is a pale blue gas with a strong odor. Ozone boils at -169.5°F (111.9°C) and melts at -314.5°F (-192.5°C). In the liquid form ozone appears bluish-black, but when frozen it becomes a violet-black **color**. Liquid ozone is also strongly magnetic. Ozone is much more chemically active than ordinary oxygen, and has a variety of uses ranging from bleaching foods to purifying water and sterilizing air. Ozone is harmful to breathe and has been thought to cause certain respiratory problems such as asthma and bronchitis. According to medical studies, exposure to ozone should not exceed 0.2 mg/m^3 on an average basis.

The ozone layer is formed naturally by short ultraviolet rays from the Sun interacting with oxygen molecules. In 1930, British geophysicist Sidney Chapman (1888-1970) formulated the first photochemical theory for the formation and decomposition of ozone in the atmosphere. Oxygen molecules absorb wavelengths of **light** shorter than 240 nm. These excited molecules dissociate to yield oxygen atoms that are very reactive. These atoms combine with other oxygen molecules, forming ozone. Ozone molecules then can absorb light between 240 nm and 320 nm, and undergo a reaction that produces oxygen molecules and a lone oxygen **atom**. The oxygen atom encounters another oxygen atom or molecule to combine into oxygen or ozone. The Dobson unit is the usual scale for measuring the total amount of ozone occupying a column of atmosphere overhead. The ozone layer over the United States has about 300 DU of ozone. If this layer were compressed to 32°F (0°C) and 1 atmosphere, it would be about 0.12 in (3 mm) thick.

Pollutants in the atmosphere may have a detrimental effect on the ozone layer. In the 1970s, scientists discovered that certain chemicals such as chlorofluorocarbons (CFCs) used as refrigerants and propellants in aerosol sprays are broken down in the atmosphere, producing chlorine atoms which then react with and destroy ozone molecules. Up to 100,000 molecules of ozone can be destroyed per molecule of CFC in the atmosphere. Chemicals in fertilizers can also attack ozone molecules and destroy them. A decrease in the ozone layer is expected to cause increases in skin cancers and cataracts as well as damage to crops and certain types of marine plankton. These decreases in plant life are expected to lead to an increase in carbon dioxide levels, which is a component of a potential **global warming** problem. In the early 1980s, scientists discovered a seasonal loss of ozone over Antarctica. There have been several subsequent studies intended to determine whether the overall ozone layer is actually decreasing. By the late 1980s, countries recognizing the problems associated with ozone depletion joined together to phase out the use and production of CFCs and other chemicals known to be detrimental to the ozone layer. In 1995, the Nobel prize in chemistry was awarded to three scientists (Paul Crutzen, Mario Molina, and F. Sherwood Rowland) for their contributions to atmospheric chemistry, particularly their advancements of the understanding of ozone formation and decomposition.

$\mathbf{P}$

PARALLEL CIRCUITS

Electric circuits are essentially paths through which electrical current moves. The two basic types of electric circuits are the direct-current, or DC circuit, in which the current flows only in one direction, and the alternating-current, or AC circuit, in which current alternates at a given **frequency** (usually 60 Hz in North America and 50 Hz in Europe). Circuits consist of localized circuit elements connected by conducting paths (e.g., wires). The three basic circuit elements are resistors, capacitors, and inductors. An example of the simplest circuit possible is provided by resistors connected across an emf source either in series or in parallel. When resistors are connected in series, all the current flowing through one resistor also flows through each of the other resistors and the total resistance of the circuit is the sum of the individual resistances of the resistors. A parallel circuit, however, is one that contains a junction. The current flowing through the circuit divides, as more than one path is available. Accordingly, only part of the current flows through any resistor or circuit element. The higher the resistance of a given path in a parallel circuit, the less current will flow through it. As a result, the **voltage** across each branch of a parallel circuit is the same, but the current may not be the same. This is especially useful in understanding basic circuits, which apply the same voltage across each circuit element (e.g., a device or appliance) whose individual loads draw different amount of current, depending on their **power** requirements.

The behavior of electrical circuits is described by Kirchoff's laws. The current law states that the sum of the currents flowing into a junction is equal to the sum of the currents flowing out. The voltage law states that the sum of the emf's in a closed circuit is equal to the sum of the potential difference of the components. These laws can be illustrated by applying them to a parallel circuit, such as a simple one consisting of two resistors, R_1 and R_2. Since there are two junctions in the circuit, the current is not the same everywhere in the circuit. According to Kirchoff's current law, the current flowing into the junction is equal to the amount of current leaving the junction. From **Ohm's law**, the potential differences across each resistor are equal to $R_1 I_1$ and $R_2 I_2$, respectively. Because the circuit is parallel, there are two possible paths, or two circuits, for the current to flow through. Because there is only one resistor in each circuit, the potential difference across it must be equal to the emf. Kirchoff's voltage law applied to the first circuit is then: emf $= R_1 I_1$, and emf $= R_2 I_2$ for the second one. For the overall circuit, this yields: emf $= R_1 I_1 = R_2 I_2$.

The total resistance of a **series circuit** is equal to the sum of the resistance of each of the circuit elements. This is not the case for a parallel circuit. In a parallel circuit, the total resistance is equal to the potential difference across the parallel segment divided by the current going through it: $R_T = V_T/I_T$.

See also Alternating current circuits; Conductors and insulators; Current and electricity; Electric circuit; Electric current; Electrical resistance; Electricity; Electromotive force; Faraday's laws; Inductors and inductance; Resistance, reactance and impedance; Resistance in series and parallel circuits; Resistors in capacitors and circuits; Resonance; Semiconductors

PARITY

Parity is both an operation and an intrinsic property used to describe particles and their wavefunctions (mathematical representations of one or more particles) in **quantum mechanics** (a branch of physics focusing on particles smaller than an atomic **nucleus**).

The parity operation is a combination of a left-right trade (mirror reflection) with a top-bottom switch. This combination is also called a spatial inversion. How objects behave under a parity operation defines their intrinsic parity. All microscopic particles have an intrinsic parity that helps us tell them apart. An object or group of objects that is the same before and after a parity operation is called parity invariant. A parity invariant object has 'even' or '+1' intrinsic parity. If the parity of an object changes due to a parity operation, it has 'odd' or '-1' intrinsic parity.

Even though people do not obey reflection symmetry (their right and left sides are different), scientists believed that the laws of physics were parity invariant. In 1956 a Chinese-American scientist named Tsung Dao Lee figured out that the idea of parity invariance had not been tested in relation to one of the fundamental forces of physics, the weak **force** (responsible for spontaneous decays of some microscopic particles). This prompted Lee and a colleague, Chen Ning Yang, to think of a clever experiment to test the parity invariance of the weak force. Later in 1956 Dr. **Chien-Shiung Wu** carried out this difficult experiment using a radioactive (spontaneously decaying) element called cobalt. Wu observed the direction of electrons (smallest naturally-occurring charged particles) coming out of the cobalt due to its **radioactive decay**. She found that the electrons did not come out the way she expected. Thus this experiment was not parity invariant. Since it tested the weak force, this meant the weak force was not parity invariant either. This result was so important that Yang and Lee won the 1957 Nobel Prize in physics for it. Now we know when we see parity invariance the weak force is the culprit. All other fundamental forces are parity invariant.

Parity is often studied along with charge conjugation. Charge conjugation changes a particle into its opposite, or **antiparticle**, by changing the sign of its **electric charge**. Even though parity is not conserved by itself, it was thought that the combination of parity (P) and charge conjugation was conserved. In 1964, however, physicists J. H. Christenson, J. W. Cronin, V. L. Fitch, and R. Turlay discovered that CP conservation is not obeyed by studying the decays of particles called Kaons. Scientists know that the laws of nature must obey conservation of the combination of parity, charge conjugation, and time symmetry (T), because of the way they are formulated. This is called CPT symmetry. Since we know CP symmetry is not conserved, it follows that time symmetry must not hold, so that total combination, CPT, can be conserved. Physicists are looking for experimental evidence to disprove CPT conservation. If they find it, we may have to rethink the laws of physics.

PARTICLE ACCELERATORS

Particle **accelerators**, as the name suggests, accelerate particles to ultra-high velocities in order to study high-energy collision processes. Because all **energy** in the collision can be converted to **mass** energy of other particles, new and exotic particles can be created by crashing the original particles together at higher energies. There are a variety of accel-

erator designs based on the size of the energies desired and the amount of space the experimenters have at their disposal. Accelerators may be either fixed-target, where the projectile particle is accelerated and its target is at rest with respect to the ground, or it may be a center-of-mass machine, which accelerates the two particles toward each other at the same time.

The simplest type of accelerator is the linear collider. The linear collider is a very long and straight pipe with an electric field pointed along the pipe. Charged particles, such as electrons and protons, are introduced into the pipe and are accelerated toward the far end. In a fixed-target machine, the target is placed at the far end of the pipe. In a center-of-mass machine, particles of opposite charges are introduced at opposite ends of the pipe and are accelerated toward each other. The main drawback of the linear accelerator is that an extremely long pipe is needed to accelerate the particles to very high energies, but its main advantage is that the particles lose very little energy to **radiation**.

Other types of accelerators make charged particles travel in circular paths, which allows them to be built with a smaller amount of space compared to a linear accelerator of the same energy. The main drawback to circular colliders is that at very high energies, the particles radiate much of their energy in the form of x rays and gamma rays. Therefore, it takes a great deal of energy to keep accelerating the particles past a certain energy. Also, past a certain **energy level**, the advantage of smaller size becomes questionable because very large rings are necessary. The highest energy colliders, such as the Tevatron at Fermilab, are miles across and are built underground. The LEP collider at CERN is so large that it runs under two countries, Switzerland and France.

The simplest kind of circular accelerator is the **cyclotron**. In a cyclotron, a charged particle is introduced into a cavity with a large constant magnetic field directed perpendicular to the particle's **velocity**. The particle will then travel in a circle. The cavity is divided into two halves, called dees, with an electric field between them. As the charged particle crosses from one dee to the other, it is accelerated to a higher velocity. The electric field direction between the dees is quickly reversed, so that when the particle crosses back to the first dee, it is accelerated again. By alternating the electric field direction continuously, the particle can be accelerated to a high energy.

At high enough energies, the particles in a cyclotron become relativistic, so their mass increases. The frequency at which the electric field must be changed depends on the both the mass of the particle and the magnetic field, so the magnetic field must be turned up slowly to counteract the increased mass of the particle. A cyclotron that has this capability is called a synchrocyclotron. At even higher energies, the size of the magnets needed for the synchrocyclotron becomes prohibitive, so a different design called a **synchrotron** is used.

See also Particle physics; Particle detectors

Stanford Linear Accelerator Center, Stanford, CA. *(Photo by David Parler. National Audubon Society Collection/Photo Researchers, Inc. Reproduced by permission.)*

PARTICLE DETECTORS

Particle detectors are instruments designed for the detection and measurement of sub-atomic particles such as those emitted by radioactive materials, produced by particle **accelerators** or observed in **cosmic rays**. They include electrons, protons, neutrons, alpha particles, gamma rays and numerous mesons and baryons. Most detectors utilize in some way the ionization produced when these particles interact with **matter**.

The Geiger counter is one of the oldest and simplest of the many particle detectors. The counter was developed in the early part of the twentieth century by **Hans Geiger** and Wilhelm Muller, shortly after the discovery of **radioactivity**. A schematic diagram of a Geiger counter is shown here. A wire electrode runs along the center line of a cylinder having conducting walls. The tube is usually filled with a monatomic gas such as argon at a **pressure** of about 0.1 atmosphere. A high **voltage**, slightly less than that required to produce a discharge in the gas, is applied between the walls and the central electrode. A rapidly moving charged particle which gets into the tube will ionize some of the gas molecules in the tube, trig-

gering a discharge. The result of each ionizing event is an electrical pulse which can be amplified to activate ear phones or a loud speaker, making the counter useful in searches for radioactive minerals or in surveys to check for radioactive contamination. The counter provides very little information about the particles which trigger it because the signal from it is the same size no matter how it is triggered. However, one can learn quite a bit about the source of **radiation** by inserting various amounts of shielding between source and counter to see how the radiation is attenuated.

Scintillation counters are made from materials which emit **light** when charged particles move through them. To detect these events and to gain information about the radiation, some means of detecting the light must be used. One of the first **scintillation detectors** was a glass screen coated with zinc sulfide. This sort of detector was used by **Ernest Rutherford** in the early versions of his classic experiment in which he discovered the **nucleus** of the **atom** by **scattering** alpha particles from heavy atoms such as gold. The scattered alpha particles hit the scintillating screen and the small flashes produced were observed by experimenters in a darkened room using only the human eye.

The modern scintillation counter usually uses what is called a photo multiplier tube to detect the light. Light incident on the photocathode of such a tube is converted into an electrical signal and amplified millions of times after which it can be sent to appropriate counters. Physicists working at **particle accelerators** often use transparent plastic materials like Lucite or plexiglass to which are added materials to make them scintillate. These plastic scintillators can be cut to convenient shapes, mounted on a photomultipler tube and placed in particle beams to provide a very fast signal when charged particles pass through them.

A very useful scintillation detector, particularly for the measurement of gamma rays, utilizes a transparent crystal of NaI (sodium iodide) mounted on a photomultiplier tube. These crystals are particularly useful because charged particles produce in them an amount of light which is directly proportional to their **energy** over a fairly wide range. A schematic diagram of a **gamma ray** scintillation spectrometer is shown here. Gamma rays have no charge and thus no detector is sensitive to them directly. Fortunately, gamma rays interact with matter and produce charged particles—usually electrons. For the measurement of gamma ray energies, the two most important interactions are the **photoelectric effect** and the **Compton effect**. These two processes can combine to produce energetic electrons in the crystal which scintillates to produce an amount of light directly proportional to the gamma ray energy. These light **pulses** are converted to electrical pulses in the photomultiplier tube. These are amplified and sent to a pulse-height analyzer which sorts out the pulses and displays a pulse height spectrum. A particular gamma ray shows up as a fairly sharp peak in this pulse height distribution.

Similar results with much improved energy resolution, the sharpness of the peaks in the pulse height distribution, can be obtained using solid state detectors made from semi-conducting materials such as silicon or germanium. When properly constructed, the electrical charges released in the material by the passage of charged particles can be collected directly producing a short electrical pulse which can be amplified and analyzed. Germanium detectors made for use with gamma rays can have peaks in the pulse height distribution almost 100 times narrower than the peaks from a sodium iodide detector. To obtain this improved resolution these detectors must be cooled to the **temperature** of liquid nitrogen 77K (-320.8°F; -196°C).

Smaller solid state detectors, usually made from silicon, are also used for measuring the energy of alpha particles, beta rays (electrons) from radioactive materials and **x rays.**

Since neutrons are uncharged, their detection must depend on an interaction with matter which produces energetic charged particles. There are several **nuclear reactions** initiated by neutrons which result in charged particles. One of the most useful for slow neutrons is the reaction in which a **neutron** is incident on a boron nucleus. This reaction produces a lithium nucleus and an **alpha particle**, both of which are rapidly moving. Note that it is the boron **isotope** of **mass** 10, with

a natural abundance of about 20%, that is required for this reaction and that the alpha particle is simply the nucleus of the **helium** atom. The boron is usually incorporated in the gas molecule BF_3 (boron trifluoride) which can be used as the gas in a proportional counter, which is much like a Geiger counter. The difference is simply that the voltages used are lower so that the discharge does not spread disruptively along the whole central electrode with the result that the electrical signal coming from the tube is proportional to the number of **ions** produced. The signals are much smaller than from a Geiger tube and require more amplification but the signal produced by the lithium nucleus and alpha particle, both of which are heavily ionizing, is relatively large and easily distinguishable. For fast neutrons, the probability of this boron reaction becomes very low so that other methods are required. A useful technique is to use a proportional counter filled with hydrogen. Fast neutrons colliding with the protons in hydrogen produce energetic protons which produce a signal from the counter.

When a charged particle moves through a transparent material with a **velocity** v, greater than the velocity of light in that material, it radiates light in the forward direction at an angle whose cosine is equal to c/vn, where n is the index of refraction of the material. This light is called Cerenkov radiation and can be detected with photomultiplier tubes as was the case with scintillation detectors. It is named after the Russian physicist Pavel Cerenkov who discovered it in 1934. The special theory of relativity limits particle velocities to values less than c, the velocity of light in a **vacuum**. Cerenkov detectors can be of two types. A threshold detector merely detects the fact that light is emitted and indicates that the velocity of the particle passing through it is greater than c/n. Other more complicated detectors can actually determine the velocity v by measuring the angle at which light is emitted.

A **cloud chamber** utilizes an enclosed volume of clean air saturated with water vapor. If this volume of air is enclosed in a cylinder with a piston and the volume is suddenly expanded, the temperature of the air falls causing the mixture to become supersaturated. If a charged particle passes through the volume at this time the vapor tends to condense on the ions produced, leaving a trail of water droplets on the path of the charged particle. With proper illumination and timing these trails can be photographed. If a magnetic field is applied, the radius of curvature of these tracks can be measured and this information, combined with the **density** of droplets along the trail can be used to measure the energy of the particle. The cloud chamber was first used by C. T. R. Wilson around the turn of the century and was useful in the early days of **nuclear physics** but suffered from several disadvantages such as the long time required to recycle and the low density of air. In 1932, **Carl D. Anderson** discovered the **positron**, the **antiparticle** of the **electron** while using a cloud chamber to observe cosmic rays.

A rather similar device called the bubble chamber was developed using liquids rather than a gas. Liquefied gases such as hydrogen, xenon, and helium have been used. Pressure is applied to the liquid to keep it a liquid above its normal **boiling** point at atmospheric pressure. If the pressure is suddenly reduced the liquid is superheated but will not boil spontaneously, at least for a short time. In order to boil, the liquid must have small irregularities on which bubbles of vapor form and they can be provided in the bubble chamber by the ions left by charged particles passing through the chamber. Thus tiny bubbles form along the tracks of particles passing through the chamber just after the pressure has been reduced. The bubbles grow very quickly but if the tracks are photographed at just the right time after expansion they are revealed as a thin trail of tiny bubbles. **Bubble chambers** work very well with particle accelerators that are pulsed. The expansion of the chamber can be timed so that particles from the accelerator pass through just after the chamber is expanded. As with the cloud chamber, application of a magnetic field permits measurement of the curvature of the tracks and when this information is combined with the density of bubbles along the track the energy, **momentum**, charge (sign) and mass of the particle can be determined. The bubble chamber was invented in 1953 by the American physicist **Donald Glaser** who used a small glass device containing about 30 cubic centimeters of diethyl **ether**. The use and size of bubble chambers grew during the following decades culminating in the discovery of the omega minus particle in the 80 in (203 cm) bubble chamber at Brookhaven National Laboratory in 1964 and the construction of the 3168 gal (12,000 l) "Gargamelle" chamber at the CERN laboratory in Geneva Switzerland in the early 1970s. In recognition of the great importance of this device to **particle physics** research, Glaser was awarded the Nobel Prize in physics in 1961.

In many nuclear and particle physics experiments, beam lines are constructed along which secondary particles of interest, produced by an accelerator, are maintained in a beam by a series of focusing and bending magnets. Wire chambers are used along these beam lines to actually track individual particles as they move along the beam line. The chambers are similar in a general way to the Geiger counter since they are gas counters. Instead of one wire the chambers have many parallel wires spaced at distances of a few millimeters. The position of charged particles passing through the chamber can be measured with uncertainties even less than the wire spacing, using fast timing circuits. These chamber measurements facilitate identification of the particle and the measurement of its momentum.

The ultimate in particle detectors are probably those being used and constructed at large national and international laboratories such as Fermilab in Illinois and CERN in Geneva, Switzerland. At these locations colliding beam accelerators have been built which produce collisions of fundamental particles, such as electrons and positrons at CERN, and protons and anti-protons at Fermilab. At various points around these large circular accelerators the counter rotating beams cross, and head on collisions can take place making large amounts of energy available for the production of other particles. Huge detectors costing millions of dollars and requiring hundreds of physicists to run them, are constructed surrounding these collision points. At Fermilab, two of these large devices, one called CDF and the other DZero, have recently reconstructed

events, produced in these collisions, which provide strong evidence for the existence of the long sought top quark. To do this the detectors are designed to detect as many of the millions of particles produced in these collisions as possible. At DZero about 400,000 proton-anti-proton collisions occur per second. The detectors, weighing thousands of tons, are constructed in layers and almost completely surround the collision points. They utilize most of the detection techniques discussed above including scintillators, solid state detectors and devices similar to wire chambers which provide much improved performance. These are called silicon microstrip detectors. They are made up of closely spaced strips of silicon detectors which give very fast position measurements of particles accurate to about 0.01 mm. The thousands of individual detectors and detector systems are connected to **computers** which help select the very special events that might involve the top quark from the millions that do not.

PARTICLE PHYSICS

Particle physics is the study of the basic elements of **matter** and the forces which act among these basic elements. The goal of particle physics is to discover the fundamental laws that govern how matter is constructed. From these fundamental laws, particle physics then attempts to expand its theories into the laws that describe the physical **universe**. In fact, it is often the case that the study of the smallest **subatomic particles** gives rise to laws that hold for the largest objects, such as **stars** and galaxies, and provide insight into cosmological processes.

Particle physics is concerned with the search for the fundamental building blocks of matter. Well over 150 particles have thus far been described. Some particles are extraterrestrial, coming through **space** as **cosmic rays**. Others have been manufactured in particle **accelerators**. Particles are catalogued by the Particle Data Group, an international group of particle physicists dedicated to the free exchange of information and standardization of particle physics throughout the globe.

Quantum mechanics is the set of laws that applies to the subatomic world. Such laws can be extended to the macroscopic world as in English physicist and mathematician Sir **Isaac Newton**'s classical laws of mechanics, and Scottish physicist **James Clerk Maxwell**'s equations combining **electricity** and **magnetism**. It is only in the world of the very smallest particles where the old laws of classical physics fail, and **quantum mechanics** is required.

The study of particle physics dates to the fifth century B.C. with argument advanced by the Greek philosopher Democritus (c. 420 B.C.) that matter is made up of very small indivisible particles called atoms. According to Democritus, one could keep dividing matter again and again until reaching the level of atoms that could be divided no further.

Knowledge of particle physics remained stagnate until the 19th century. In the interim, German astronomer and mathematician **Johannes Kepler**, Italian astronomer and physicist **Galileo** Galilei, **Isaac Newton**, **Christiaan Huygens**, **Michael Faraday**, Maxwell, and a host of other scientists had brought physics to point where many felt there was nothing new to be discovered. Only electricity seemed to require further investigation.

The British physicist **William Crookes** (1832-1919) studied the passage of electricity through tubes of gas connected to **vacuum** pumps which controlled the amount of gas present in the tubes. The **electric current** created strange and fascinating glows around the tubes. It was believed that rays called cathode rays caused the gas to glow. In 1897, however, **J. J. Thompson**, professor of physics at Cambridge University, proved that the cathode rays were a stream of particles. The particles were named electrons, and Thompson found their **mass** to be 2,000 times less than the hydrogen **atom**, the lightest known atom. Nothing lighter than a **hydrogen atom** was thought to exist at the time. The first evidence that atoms had a complex inner structure had been found, and a new field of physics, particle physics, was opened for discovery. Thompson's discoveries showed that atoms were not solid like billiard balls but had a complicated internal structure.

In 1895, Wilhelm Roentgen (1845-1923) discovered **x rays**. In 1896, **Henri Becquerel** conducted research on the heavy elements. Becquerel found that uranium gave off rays that darkened photographic plates, and discovered **radiation**. **Marie** and **Pierre Curie** further explored **radioactivity** and radioactive substances, and discovered two new elements, polonium and radium. Rutherford and Frederick Soddy (1877-1956) discovered radioactive transmutations, that is the changing of one element into a different element through the mechanism of **radioactive decay**. Rutherford also came to the conclusion that atoms were like solar systems in miniature, with considerable space between the central **nucleus** and the orbiting electrons. Rutherford also discovered protons by firing alpha particles from radioactive substances at boron, fluorine, sodium, and other elements. As technology improved, new particles were found, and the atom slowly revealed its secrets. **Hans Geiger** (1882-1945) had invented his now famous particle counter ("Geiger counter"), but a better medium than photographic plates or counters was needed to detect particles.

In 1894, **Charles Wilson** invented the **cloud chamber**. With this device the tracks of charged particles could be seen. In 1910, Wilson began to use his new detector. From 1921 to 1924, Wilson allowed alpha particles to bombard nitrogen in the chamber, and captured nuclear transmutation on film. The cloud chamber was an effective detector of subatomic particles.

By the 1920s, the role of the protons carrying the positive charge of the nucleus had become firmly established. The negatively charged electrons were known to orbit the nucleus. A problem lingered, namely that the nucleus contained 99.95% of the atom's mass and the protons should have carried this mass. However, this was not the case. The protons accounted for less than half the nuclear mass. It was **Ernest Rutherford** who proposed the **neutron**, a neutrally charged particle, as making up the remainder of the mass. Based on the work of **Walther Bothe** in Germany, and **Irene Joliot-Curie** and

her husband **Fredrick Joliot** in France, **James Chadwick** discovered the neutron in 1932. The nucleus was ultimately considered to consist of protons and neutrons held together by the strong nuclear **force**. The electromagnetic force attracted the negatively charged electrons to the positively charged nucleus.

Physicists then considered the questions of particle physics, such as what particles make the atom, and what caused some elements to transmutate by radioactive decay. It was known that if a magnetic field were placed across a cloud chamber and a radioactive source placed near the chamber, three types of radiation would be seen. Electrons (**beta radiation**) were bent toward the positively charged side of the chamber, alpha particles (**helium** nuclei) were bent toward the negatively charged side, and gamma rays (found to be similar to Roentgen's x rays) traveled straight through. Occasionally, a particle from outside the system would fire through the chamber. These were cosmic rays, extraterrestrial particles from outer space flying through the atmosphere.

The year 1912 marked the discovery of cosmic rays by Victor Hess (1883-1963). **Robert Millikan** pushed the investigations further using high altitude balloons. His cohort, **Carl Anderson** in 1934 discovered a particle with the exact mass of an **electron** but with a positive charge. Called the **positron**, this particle was the **antimatter** equivalent of the electron. **Hideki Yukawa**, a Japanese theoretical physicist working in Kyoto, predicted a mesotron, later shortened to **meson**, to be a middleweight particle between protons and electrons. Anderson observed such a particle, but it did not fulfill all criteria and was later found to be a **muon**. Yukawa's predicted particle was found a decade later with the discovery of the pi-meson, or pion.

Further investigations by **Cecil Powell** and his team used photographic emulsions on mountain peaks carried to high altitude by polyethylene balloons. Powell's group discovered that primary cosmic rays are either photons or subatomic particles. These primary cosmic rays had enormous energies compared with alpha, beta, and gamma radioactive decay. Apollo astronauts were startled as they slept when a heavy particle cosmic ray would strike their retinas and cause a tiny flash of **light**.

Primary cosmic rays come predominantly from within our galaxy, the Milky Way, although the ultra-high-energy cosmic rays may originate from beyond our galaxy. When these primary rays strike our atmosphere they interact with the nuclei of atoms producing a shower of secondary particles consisting primarily of muons, protons, electrons, **pions** and neutrinos. Some cosmic rays are energetic enough that they come directly through the atmosphere. Before the development of **particle accelerators**, cosmic rays were the only source of high-energy particles.

Particle accelerators were the next accomplishment in particle physics. Rutherford had been the first to "smash the atom" but it was **Ernst Lawrence**'s **cyclotron** in 1931 that lead to today's high-energy accelerators. Lawrence used a magnetic field to bend the path of the particles into a circular orbit so they would run the same accelerating gap many times, a circular accelerator operating in cycles. From this point onward it was only a matter of building larger, more powerful accelerators including the 1.8 mi (3 km) long non-circular Stanford Linear Accelerator at Stanford University and the giant circular rings at Fermilab near Chicago, and CERN near Geneva, which has a 27 km path for particle **acceleration**. CERN (a joint European nation project) has both a Super **Proton** Synchrotron and the Large Electron-Positron collider. All of these accelerators are seeking higher and higher energies at which to bombard or collide particles. They represent earthbound cosmic ray producers that can be controlled.

The Geiger counter and the cloud chamber led to the bubble chamber, a machine which used a superheated liquid to track the trails of particles, similar to the cloud chamber but with much longer lasting and larger tails than a cloud chamber which allowed photographs to be more easily taken.

Contemporary **particle detectors** reflect our most advanced technological innovations. Although occasionally a bubble chamber-like apparatus may still be used, most detectors consist of sophisticated electronic devices and computer systems for the control and analysis of data. As a result, there are now over 200 particles listed by the Particle Data Group. Although the place of many of these particles in the scheme of matter and antimatter remains to be discerned, the standard model has developed into an overall particle-based and dynamic atomic model.

See also Neutrinos; Pions; Quantum theory; Quarks and gluons; Strong interactions

PARTICLE-WAVE DUALITY

Wave-particle duality is a term used to describe the modeling of nature by both wave theory and particle theory. In essence, physical phenomena (e.g. electrons, photons, etc.) can exhibit varying degrees (sometimes determined by the experiment performed) of wave-like behavior and/or particle-like behavior (i.e., the phenomena can best be described by wave related equations or equations that treat phenomena as particles).

In its most usual form, particle-wave duality is a concept generally related to microscopic particles, more specifically their behavior as both particles and **waves**. This seems like a contradiction in terms because waves and particles are mutually exclusive. This seemingly contradictory statement arises because the concepts of particles and waves were developed from observations of macroscopic objects.

In 1905, German-American physicist **Albert Einstein** was concerned with explaining the **photoelectric effect** by extending German physicist **Maxwell Planck**'s concept of **energy** quantization to electromagnetic **radiation**. Planck considered electromagnetic radiation to be a wave. During this time, Einstein proposed that in addition to having wavelike properties, **light** could also be thought to consist of particles, and as such, have properties consistent with particles. Although Einstein's theory of the photoelectric effect agreed with Hungarian physicist and Nobel prize winner **Philipp Lenard**'s work, scientists were reluctant to accept it. Finally, in 1916,

American physicist **Robert Millikan** made definitive tests to prove Einstein's theory. During this time there were also unsuccessful attempts to apply the **Bohr theory** to atoms with more than one **electron**, and to molecules. Eventually it became evident that there was a fundamental error in the Bohr theory. In 1923, French physicist **Louis de Broglie** advanced the idea that just as light shows both wave and particle properties, **matter** also has a particle-wave dual nature. De Broglie proposed that electrons could show particle-like behavior, but could also show wave-like behavior manifesting itself in the quantized energy levels of electrons in atoms and molecules. In 1927, American physicists **Clinton Davisson** and **Lester Germer** confirmed de Broglie's hypothesis by observing **diffraction** effects when an electron beam was reflected from a crystal of nickel. Diffraction effects were also observed when electrons passed through a thin sheet of metal. Since that time several particles have demonstrated diffraction effects showing that de Broglie's hypothesis applies to all particles and not just electrons.

The particle-like nature of an electron can be understood because it has **mass** at rest and repels other objects that are like-charged. The wave-like nature of the electron is more difficult to conceptualize although it is important when calculating the energy levels of electrons in atoms. Erwin Schrödinger formalized an equation relating the energy of a system to its wave properties and employs this idea. On the other hand, the wave-like nature of a **photon** is easy to visualize because of experiences on the macroscopic scale, but its particle-like nature is difficult to fully understand. A photon's rest mass is zero, but photons are never at rest. They are always moving at the **speed of light**. This apparent particle-wave duality of matter and radiation imposes certain limitations on the information obtainable about a microscopic system reflected in the Heisenberg uncertainty principle.

Under certain conditions an electron behaves like a particle, and under other conditions it behaves like a wave. Although it demonstrates these properties, an electron is neither a particle nor a wave, but something that is not fully understand or describable in terms of macroscopic concepts such as particles or waves.

See also Particle physics

PASCAL, BLAISE (1623-1662)
French mathematician, physicist, and philosopher

In his short life, Blaise Pascal was able to make major contributions to many of the fields of science that were rapidly developing in the seventeenth century, including projective geometry probability theory, and hydraulics. He also developed one of the earliest calculating machines for mathematical computation.

Pascal was born in Clermont-Ferrand. His mother died when he was three and the young Pascal was raised by his father, who supervised his education, which focused mainly upon studying ancient languages. Pascal was a weak and sick-

Blaise Pascal.

ly child, but he exhibited a formidable intellect at an early age and showed a keen ability to teach himself the principles of science. His penchant for geometry, which stemmed from his reading of Euclid's Elements at the age of 12, was recognized by his father who encouraged his son in the study of mathematics. By the time he was 16, Pascal was attending meetings of the Academie Parisienne and had become the principal disciple in geometry of the mathematician Girard Desargues (1591-1661). Building on the ideas of his mentor, Pascal published a pamphlet in 1640 on the mathematical attributes of sections of cones. This paper presented the outlines of a much larger essay that would, presumably, come later. Unfortunately, the final treatise was never published, although a manuscript copy was apparently seen by **Gottfried Leibniz**. Even without a complete published work on conics, Pascal's work in this area was enough to contribute greatly to the successful development of projective geometry in the nineteenth century.

Pascal directed his energies to the creation of a computing machine, the *pascaline*, which was capable of adding and subtracting numbers. In 1642, Pascal had worked out the design of his machine but was unable to produce a dependable, and saleable, mechanical version until 1645. The machine never generated the significant profit that Pascal had hoped for. Nevertheless, the pascaline was the predecessor of increasingly sophisticated devices such as adding machines and, ultimately, modern-day calculators and **computers**.

While he was working on the details of manufacturing his calculating machine, Pascal became interested in a scientific controversy that was raging in 1647. The debate centered on whether or not a **vacuum**, the complete absence of **matter**, could actually exist in nature. Eminent scientists such as **René Descartes** maintained that a vacuum was impossible, while others such as Maggiotti, Berti and **Evangelista Torricelli** had performed experiments which seemed to indicate that nature could preserve a vacuum under certain conditions. Torricelli, in particular, had created an instrument called the barometer which he claimed contained a vacuum in its empty portion. This device, said Torricelli, measured the **pressure** of air, or barometric pressure. Pascal began building a number of barometers, and upheld the view that they worked on the principles of air pressure and contained *vacua*. Pascal also arranged a famous experiment in which a barometer was taken up the mountainside of the Puy de Dôme in 1648, showing that, as altitude increased, air pressure decreased and therefore air indeed had **weight**. His defense of the idea of the vacuum led to a more modern understanding of physical forces. It was also part of Pascal's formulation of the hydraulic principle; that pressure exerted on a fluid in a closed vessel is transmitted unchanged throughout the fluid. Today, the applications of this discovery are commonplace in devices such as hydraulic presses, automobile brakes, and aircraft controls.

Pascal became more and more interested in religious matters. In 1646, while involved in his vacuum experiments, Pascal started idenitifying with Jansenism, a Catholic movement strongly influenced by the teachings of St. Augustine. Pascal's acceptance of Jansenism profoundly influenced the direction of the rest of his life. He would spend less time with experiments; and when his father passed away in 1651, Pascal became even more devout. Most of the remainder of his short life would be devoted to meditation and religious writing. (In fact, his spiritual writings, such as the Pensées, would make an impact on religious philosophy nearly equal to the effect his scientific work had upon physics and mathematics.)

Nevertheless, Pascal never completely lost interest in mathematical studies. In 1654, at the age of 31, Pascal entered into a correspondence with Pierre de Fermat in which, together, they studied the frequency of the occurrence of certain combinations in the fall of dice. In pursuing this work, Pascal and Fermat laid the foundations of the calculus of probability, a branch of mathematics dedicated to analyzing how frequently "random" events occur over the long run. In 1658 and 1659, Pascal worked on the problems of calculating areas and volumes of complex shapes using his "theory of indivisibles" which was the forerunner of integral calculus.

Pascal's last important effort, at the age of 39, was the design of a public transit system for the city of Paris which was implemented in 1662, the same year he died. Today, Pascal is remembered for his prodigious contributions to several important branches of science, made more remarkable by his tragically short life.

PAULI EXCLUSION PRINCIPLE

In 1913, **Niels Bohr** proposed a model of the hydrogen **atom** that introduced the concept of quantized **energy** levels. Bohr assumed that the energy which an **electron** in an atom may have is quantized, i.e. it can possess only certain amounts of energy. Bohr assumed that electrons rotate around the **nucleus** in orbits with distances from the nucleus related to a quantum number n. The lowest quantized **energy level** has n = 1; the next highest has n = 2, etc. The properties that are predicted by Bohr's model match the experimentally observed values.

Efforts were begun immediately to extend Bohr's model of the **hydrogen atom**, which has only one electron, to multi-electron atoms.

The electrons of an atom will assume a configuration which has the lowest possible energy, called the ground state. In the hydrogen atom, the single electron is located in the orbit for which the quantum number n = 1. Helium has two electrons. In its ground state, both electrons are located in the lowest energy orbit with n = 1.

The lithium atom has three electrons. Again it might be assumed that in the ground state, the lowest energy configuration would require all three electrons to be in the n = 1 orbit. Experiment indicates, however, that while two electrons are in the n = 1 orbit, the third is in the orbit corresponding to n = 2. Atoms of the element beryllium have four electrons, two with n = 1 and two with n = 2.

For boron, with 5 electrons, two have n = 1, two have n = 2, and we might suppose that the fifth will have n = 3. But for some reason, unexplained by the Bohr model, the energy of this electron is much closer to the n = 2 level than to that of n = 3. A more sophisticated theory of **atomic structure** was needed.

In 1923, **Louis de Broglie** proposed that particles possess wave properties. In 1925, to explain why all electrons in an atom will not be in the lowest n = 1 energy state, **Wolfgang Pauli** proposed that no two electrons can simultaneously have the same wave properties. This proposal evolved into the Pauli exclusion principle.

In 1926, **Erwin Schrödinger** proposed a theory, known as wave **mechanics**, to explain the properties of particles such as electrons. The solutions of the theory's wave equation for hydrogen match exactly the predictions of the **Bohr theory**. Moreover, Schrödinger's wave mechanics has been successfully applied to systems with more than one electron.

Each solution of Schrödinger's wave mechanical equation for atomic systems, called an orbital, has a set of four **quantum numbers** associated with it. The principal quantum number n may have any positive, non-zero integer value. It is the primary determinant of the allowed energy levels of electrons. The **angular momentum** quantum number l may have only integer values from 0 to (n-1). The energy of an orbital depends slightly on the value of l. The magnetic quantum number m_l may have any of the integer values -1 to +1 (including 0); and the spin quantum number s may have only the values +1/2 and -1/2. Each orbital has a different set of these four quantum numbers.

The Pauli exclusion principle may now be stated as requiring that no two electrons in an atom may occupy the same orbital, i.e. may simultaneously possess the same values of the four quantum numbers. Since the spin quantum number may have only the values +1/2 and -1/2, and since the value of spin affects neither the energy nor the rest of the **wave function**, we often speak of each orbital containing two electrons, one with spin +1/2 and the other with spin -1/2.

The quantum numbers corresponding to the possible energy levels of an atom are as follows (with energy increasing as the value of n and l increase):

There is only one orbital with n = 1, because when n = 1, l = 0 and m = 0. This is called the 1s orbital and two electrons may occupy it (one with spin +1/2 and one with spin -1/2).

There are five **orbitals** with 2: a 2s orbital with n = 2, l = 0, m = 0, containing two electrons; three 2p orbitals, of slightly higher energy, for which n = 2, l = 1, and m has the values -1, 0, and +1; each of these p orbitals may contain two electrons; six electrons may thus be accommodated in these p orbitals.

There are nine orbitals with n = 3: a 3s orbital with two electrons; three 3p orbitals of slightly higher energy that may contain a total of six electrons; and five 3d orbitals with n = 3, l = 2, and m has values of -2, -1, 0, +1, and +2. The 3d orbitals have slightly higher energy than the 3p orbitals, and may contain a total of ten electrons.

Similar orbital schemes may be constructed for larger values of the principal quantum number.

Recall that the ground state of an atom is the state of lowest energy, with the electrons of the atom occupying the orbitals of lowest energy.

The lone electron of the hydrogen atom will occupy the 1s orbital in the ground state. The two electrons of the **helium** atom are both in the 1s orbital, one with spin +1/2, the other with spin -1/2. The third electron of lithium must have n = 2 since there are no more unfilled orbitals with n = 1. Both the third and fourth electrons of beryllium reside in the 2s orbital, each with a different spin quantum number. The fifth electron of boron cannot be accommodated in the filled 2s orbital; the next lowest orbital is 2p.

Continuing this process, using the Pauli exclusion principle, the ground state electron configuration of the elements of the **periodic table** can be explained. For instance, a chlorine atom has seventeen electrons. In the ground state, they will be located in the orbitals of lowest possible energy: two electrons in the 1s orbital, two in 2s, six in 2p, two in 3s, and five in 3p. A bromine atom has thirty-five electrons, located as follows: two in 1s, two in 2s, six in 2p, two in 3s, six in 3p, ten in 3d, two in 4s, and five in 4p.

The electrons of molecular bonds can be understood in a similar way. Instead of orbitals located on a single atom, the wave theory treatment of the electrons of molecular bonds results in a set of molecular orbitals which spread over two or more atomic centers. The electrons that are shared by the atoms to form the bond belong to one of these molecular orbitals. The ground state of the molecule is that in which the bonding electrons are in the molecular orbitals of lowest energy. Again, the Pauli exclusion principle holds: only two elec-

Wolfgang Pauli.

trons may be located in each molecular orbital, each with a different spin.

In addition to electrons, the Pauli exclusion principle applies to all sub-atomic particles with half-integral spins, known as **fermions**, such as neutrons and protons. The Pauli exclusion principle does not apply, however, to particles with integral spin, known as bosons, such as photons.

PAULI, WOLFGANG (1900-1958)

Austro-Hungarian Swiss physicist

Wolfgang Pauli 's exclusion, or "Pauli," principle asserted the later-proven existence of the **neutrino**, a chargeless, massless particle. This discovery led to the Nobel Prize-winning theory of a fourth quantum number with only two possible values (a quantum number expresses the distinct state of a quantum system). The exclusion principle limits the number of electrons possible in the first level of **energy** to two, a restriction not previously seen in **quantum physics**.

Wolfgang Ernst Pauli was born on April 25, 1900, in Vienna, in what was then Austria-Hungary. His father was Wolfgang Joseph Pauli, a medical doctor and biochemist who later became a professor at the University of Vienna. His mother was the former Bertha Schültz, an author. R. E. Peierls

reports in *Biographical Memoirs of Fellows of the Royal Society* that Pauli's parents' "background and their acquaintance with the leading authorities in many fields had a profound effect in creating the high standards and the impatience with anything but the best of its kind, which became later an important characteristic of the young Pauli."

Masters Relativity Theory at an Early Age

Pauli was a very bright student who sometimes found Vienna's schools dull and boring. He took to reading advanced treatises on modern physics during his most tedious classes, gaining familiarity with such new and revolutionary concepts as American physicist **Albert Einstein**'s general theory of relativity. In 1918, Pauli graduated from high school in Vienna and entered the University of Munich. He chose Munich because it was the home of German theoretical physicist **Arnold Sommerfeld**, then one of the greatest teachers of theoretical physics alive. Two years into his course of study, Pauli was given a special assignment by Sommerfeld: he was asked to write an entry on the subject of relativity for the forthcoming *Encyclopedia of Mathematical Sciences*. The 250-page article that Pauli produced was not only a summary of all that was known on the subject but also an analysis of the information. In a letter to Einstein, Sommerfeld described Pauli's work as "simply masterful." A year later, after the shortest time allowed by the university, Pauli was awarded his Ph.D. for a thesis on the hydrogen moleculeion.

When Pauli was a graduate student, the field of physics was in a state of disarray, stumbling towards new perspectives on the nature of **matter** and energy. The development of quantum **mechanics** and **relativity theory** was still encumbered by remnants of classical theory. Pauli made important contributions to the clarification of the nature of modern theory, especially with his enunciation of the exclusion principle. In addition, Pauli tackled a then-troublesome problem in which the emission of electrons during **beta decay** (a form of **radiation**) does not appear to obey the law of **conservation of energy**. To resolve this difficulty, Pauli hypothesized the existence of a chargeless, massless particle later given the name neutrino.

After receiving his degree, Pauli was offered a job as assistant in theoretical physics at the University of Göttingen. There he came into contact not only with English physicist **Max Born**, professor of theoretical physics, but also with Danish physicist **Niels Bohr**, who was guest lecturer at Göttingen in 1922. It was through Bohr's lectures that Pauli began to think about some of the fundamental difficulties that still remained in Bohr's **quantum theory** of the **atom**. One of these problems was the existence of various **electron** energy levels within an atom. Nothing in classical physics could explain the fact that electrons are distributed in various energy levels outside the **nucleus**, with each **energy level** having a maximum permitted number of electrons. Bohr had derived such a model from empirical data rather than any theoretical concept, and the fact that the model worked did not detract from questions as to *why* it worked.

Solving the Zeeman Effect Leads to the Exclusion Principle

Over the next three years, Pauli worked on this question. He continued to do so when he went to the University of Copenhagen as Bohr's assistant in 1922 and then to the University of Hamburg as assistant professor of theoretical physics in 1923. At Hamburg, he also began to think about another puzzle in **atomic theory**, the **Zeeman effect**. In 1896, the Dutch physicist **Pieter Zeeman** had found that the presence of a strong magnetic field causes the lines of an atomic spectrum to split. More than twenty years later, no one had yet devised an explanation for this phenomenon, but, early in 1925, Pauli found a possible explanation. He suggested that a fourth quantum number —in addition to the three already known—was needed to describe completely the energy state of an electron. In addition to its principle quantum number (n), its azimuthal quantum number (ℓ), and its magnetic quantum number (m), an electron must also have a fourth quantum number, Pauli said. This fourth number could have one of two (but only two) possible values.

At the time, Pauli had no idea as to how this fourth quantum number could be interpreted in physical terms. That problem was solved three years later when Dutch physicists **Samuel Goudsmit** and George Uhlenbeck discovered electron spin. Goudsmit and Uhlenbeck suggested that an electron could spin in one of two directions, clockwise or counter-clockwise. Their designation of these spins as +1/2 or -1/2 corresponded precisely to the two-valuedness of the fourth quantum number that Pauli had predicted.

Proposes the Exclusion Principle

As significant as the discovery of the fourth quantum numer was, it was perhaps even more important in terms of what it led to: the exclusion principle. Prompted by a 1924 paper by the English physicist E. D. Stoner, Pauli was able to develop an explanation for the fact that all electrons in an atom do not occupy the lowest energy level. The reason, he said, is that no two electrons can have exactly the same set of **quantum numbers**. An electron in the first energy level (n = 1) is restricted to an azimuthal quantum number of zero (because ℓ = n - 1) and a magnetic quantum number of 0 (because m = $\pm\ell$). It can have spin quantum numbers of +1/2 or -1/2. These restrictions mean that electrons in the first energy level can have quantum numbers of 1,0,0,+1/2 or 1,0,0,-1/2, but no others. Therefore, since the exclusion principle says that there can be no more than one electron of each of these two kinds, only two electrons can occupy that first energy level. By a similar argument, it is possible to show how the second energy level can hold no more than eight electrons, the third level no more than eighteen, and so on. In this way, a theoretical basis is provided for the electron orbital capacities originally devised empirically by Bohr more than a decade earlier.

In 1928, Dutch American chemical physicist Peter Debye retired as professor of theoretical physics at the Eidgennössische Technische Hochschule (ETR, or Federal Institute of Technology) in Zürich and Pauli was appointed as his successor. Pauli was to remain at this post until his death, with two brief interruptions. The first came in the academic year 1935–36 when he was visiting professor at the Institute for Advanced Studies in Princeton, New Jersey. The second came five years later when German leader Adolf Hitler's armies had begun to sweep through Europe. Even though he lived in a neutral country, Pauli decided that it would be safer to leave Europe. He traveled back to the Institute for Advanced Studies, where he again held an appointment. Although Pauli returned to Zürich in 1946, he continued to hold a permanent appointment at the Institute and went back for short visits several times.

Pauli Proposes the Existence of the Neutrino

During his first years in Zürich, Pauli turned to another problem troubling physicists: beta decay. According to quantum theory at the time, the loss of a beta particle by a radioactive particle should be accompanied by the loss of a discrete quantity of energy. The spectrum produced by beta decay should, therefore, be characterized by a series of lines. Instead, the spectra associated with beta decay were always continuous spectra. In 1930, Pauli proposed a solution for this dilemma. He suggested that the loss of a beta particle by a nucleus was accompanied by the loss of a second particle. The characteristics of beta decay required that this particle have no electrical charge and no—or almost no—mass. At first, Pauli referred to the particle as a "neutron," a name later given to the chargeless nuclear particle discovered by English physicist **James Chadwick** in 1932. After Chadwick's discovery, American physicist **Enrico Fermi** rechristened Pauli's particle as the "neutrino," or "little neutron." A year later, Fermi had also incorporated the neutrino into an elegant and totally satisfactory mathematical theory of beta decay. Because of the characteristics of this elusive particle, the neutrino itself was not actually discovered for more than two decades after the work of Fermi and Pauli. Shortly before his return to Switzerland in 1945, Pauli was informed of his selection as the winner of the Nobel Prize in physics. After resuming his post in Zürich, he became a Swiss citizen and continued his research on elementary particles.

Pauli was married twice, first to Kate Depner, then to Franciska Bertram on April 4, 1934. There were no children from either marriage. Pauli's work was acknowledged not only by the Nobel Prize but also by the Lorentz Medal of the Royal Dutch Academy of Sciences in 1930, the Franklin Medal of the Franklin Institute in 1952, and the **Max Planck** Medal of the German Physical Society in 1958. Pauli died unexpectedly in Zürich on December 14, 1958.

PENROSE, ROGER (1931-   )
English mathematical physicist

Roger Penrose explored a range of topics in mathematical theory and physics, including **relativity theory**, quantum **mechanics**, **astrophysics**, **cosmology**, possible and impossible geometric shapes, and how the human brain works. With theoretical physicist and professor **Stephen Hawking**, he extended our understanding of **black holes** and the "**big bang**" theory of the origin of the **universe**, and his work with geometric puzzles shed **light** on the nature of quasi-crystals.

Penrose was born August 8, 1931, in Colchester, England. His father, Lionel S. Penrose, was a geneticist, and his mother, Margaret Newman, a doctor; his uncle, Sir Roland Penrose, was a surrealist painter and a biographer of Picasso. Penrose's older brother Oliver became a mathematics professor, while his younger brother Jonathan, a professor of psychology, was British chess champion ten times. As a boy, Penrose shared his father's interest in nature and geometrical puzzles. In school, he showed an aptitude for mathematics, devising geometry problems that challenged his teachers.

As a mathematics student at University College, London, Penrose discovered a theorem concerning eight conics in a plane, for which three well-known theorems turned out to be special cases. He received a Bachelor of Science degree in 1952 and a Ph.D. from Cambridge in 1957, writing his dissertation on algebraic geometry. As a student, Penrose rediscovered and developed mathematician E. H. Moore's generalized inverse matrix, a method of solving equations that involves rectangular arrays of numbers. At St. John's College, Cambridge, Penrose heard lectures by **Paul Dirac** on **quantum mechanics** and by Hermann Bondi on the theory of relativity, and became interested in relating quantum mechanics and **space-time** structure.

Beginning as a research fellow at St. John's College from 1957 to 1960, Penrose pursued a career of research and teaching at major universities in England and the United States. He was a North Atlantic Treaty Organization (NATO) research fellow at Princeton, Syracuse, and Cornell universities from 1959 to 1961. In the 1960s he had visiting appointments at the University of Chicago, Yeshiva University in New York, the University of Texas in Austin, the University of California at Berkeley, and King's College and Bedford College in London. In 1973 Penrose was named the Rouse Ball professor at Oxford University, and in the 1980s he was the Edgar Odell Lovett Professor of Mathematics at Rice University. Penrose's early honors include the 1966 Adams Prize from Cambridge University and the Dannie Heineman Prize for Physics from the American Physical Society and the American Institute of Physics in 1971. He was elected to the Royal Society in 1972.

In 1965 Penrose showed that, as a consequence of Einstein's theory of general relativity, there must inevitably be points in **space** that are infinitely dense and hot. At such points—called singularities—the laws of classical physics do not apply; the **gravitational field** becomes infinite, or other pathological behavior occurs. Penrose's theorem proved that if a star of sufficient **mass** collapsed, a singularity—the core of what later became known as a black hole—would result. Penrose's work first convinced many physicists of the existence of black holes, and inspired the search for them.

A year later, Stephen Hawking of Cambridge University applied Penrose's theorem to cosmology, proving that the universe started in a **singularity**. In 1970, Penrose and Hawking,

working together, succeeded in proving a singularity theorem much more powerful than their earlier efforts. Previous work by others had indicated that our universe did not begin with a "big bang" singularity; Penrose and Hawking's theorem challenged this view, asserting that in any universe with certain fundamental properties, a **big bang** singularity must occur. The two scientists proved their theorem using mathematical techniques from the theory of differential **topology**; using the same techniques, Penrose contributed to Hawking's work on black holes, which showed that the surface area of a black hole must increase as mass is added. In 1975, Penrose and Hawking shared the Royal Astronomical Society's Eddington Medal, and in 1988 they were awarded the Wolf Foundation Prize in Physics.

Penrose's efforts were aided by his invention of the "twistor," a mathematical tool for describing physical objects and space, incorporating **energy**, **momentum**, and spin—the three properties possessed by all objects moving through space-time. A twistor has either six or eight dimensions, each of which involves either movement or change in size.

Another of Penrose's primary areas of study is tiling, which involves completely covering a flat surface with a regular pattern of tiles. He shares this interest with numerous other mathematicians, including **Johannes Kepler** in the seventeenth century. Penrose has said that he inherited his love of puzzles from his father, who used models to understand or explain concepts in genetics. Speaking of his father in an interview for *Omni*, Penrose explained, "With him there was no clear line between making puzzles for his children and his serious work in genetics." While a graduate student in 1954, Penrose saw the drawings of Dutch artist M. C. Escher at a mathematics conference in Amsterdam. Escher's work incorporates geometry and perspective to produce drawings of "impossible" objects. Penrose was fascinated with the illustrations of objects which violate the rules of three-dimensional reality. He proceeded to draw his own such construct—a "tribar" of three beams—which he sent to Escher. The artist later used the tribar figure as the basis for a continuous flow of water in his lithograph, *Waterfall*.

The simplest tiling pattern utilizes identical tiles in the shape of squares, equilateral triangles, or regular hexagons. However, regular pentagons—five-sided tiles—will not tile a surface without leaving gaps. Penrose found that a floor, or any plane surface, can be covered with pentagons plus two other shapes—stars and hat-shaped pieces. The resulting pattern has regularities but does not repeat itself exactly. Some tiling patterns repeat themselves in a certain way: if you placed a sheet of thin paper over the design and traced it, and then moved the paper sideways without rotating it, the tracing would match. Such designs are called "periodic." Penrose's pattern was non-periodic. With some tile designs, such as hexagons, you know from the neighboring tiles where to put down the next one; such patterns are called "local." Penrose's pattern appeared to be non-local; it was necessary to look at the position of pieces some distance away in order to position the next tile correctly.

Penrose worked with pieces of various shapes, trying to find the smallest number of shapes that would force a non-periodic tiling. He discovered several combinations, and finally, in 1974, lowered the number of shapes to two. One of Penrose's most interesting designs used two shapes derived from a rhombus: one piece looked like a kite, and the other resembled a dart or a "stealth" airplane. Following rules about which edges could be fitted together, the two shapes forced a non-periodic tiling. In his book, *Penrose Tiles to Trapdoor Ciphers,* Martin Gardner displays some tiling patterns and explains how to make a set of "Penrose tiles."

Following Penrose's work, others extended the tiling concept to three dimensions, devising solid polyhedrons to fill space without any gaps; these tilings were also non-periodic. In the 1980s, crystallographers became interested in Penrose's findings. For a century, scientists had thought that the atoms in crystals were arranged periodically, and that crystals with five-fold symmetry were impossible. It was generally assumed that the atoms in solid **matter** took one of two forms: either a periodic crystal arrangement or a disordered arrangement in materials such as glass. But in studying an alloy of aluminum and manganese, Dan Schechtman at the National Bureau of Standards saw what appeared to be non-periodic crystals with five-fold symmetry. The complicated sequences in the pattern were only "quasi-periodic." Halfway between crystals and glass-like structures, this new form of matter, quasi-crystals, caused considerable excitement among chemists, physicists, and crystallographers. The similarity between the alloy pattern and the Penrose patterns was recognized by University of Pennsylvania physicist Paul Steinhardt and by crystallographer Alan MacKay of London. Soon scientists found similar non-periodic structures in other alloys. Alloys with five-fold symmetry, as well as seven-fold, nine-fold, and eleven-fold symmetry, proved to be possible.

The apparent non-locality of the structures puzzled scientists, because it was not clear how a growing quasi-crystal would attach new atoms, one at a time, in the right location and sequence. Then in 1988 Steinhardt and George Onoda, an IBM ceramics expert, devised rules for building a Penrose tiling, specifying which vacancy to fill first, which piece to use, and which way the tile should be turned. It was not necessary to pay attention to any distant part of the pattern. These results provided possible clues to the growth of three-dimensional quasi-crystals. Researchers proceeded to study the properties of the new alloys. Because of their intricate structure, it was thought that they might turn out to be harder than crystals and usable as substitutes for industrial diamonds, or as materials in electronic devices.

Much of Penrose's work, including topics such as theoretical mathematics, cosmological singularities, and quasi-crystals, came together in his 1989 book, *The Emperor's New Mind: Concerning Computers, Minds, and the Laws of Physics.* The book was on the *New York Times* best-seller list for nine weeks, attracting interest from people in a wide range of fields, and was the topic of more than twenty book reviews in periodicals in fields ranging from science and artificial intelligence to philosophy, as well as popular newspapers.

In the book Penrose disagrees with the view, held by some researchers in the field of artificial intelligence, that **computers** are capable of mimicking the function of human brains. A computer program, Penrose argues, uses an algorithm, a step-by-step mechanical procedure for working toward an answer from the input data. Certain kinds of human thinking are nonalgorithmic—not formalizable—so, Penrose concludes, the human brain must be making use of nonalgorithmic processes.

Penrose considers the nature of the physics that might underlie conscious thought processes. He sees a significant gap between our physical understanding at the small-scale quantum level (which includes the behavior of molecules, atoms and **subatomic particles**) and at the larger classical level (which includes the behavior of larger objects such as baseballs). Penrose suspects that a greater understanding of the functioning of the human brain may depend on a fundamentally new understanding of physics, to be sought in a radical new theory of **quantum gravity**. He believes that until we understand the borderline between quantum mechanics and classical mechanics, computers will not be able to work as human brains do.

PENZIAS, ARNO (1933-)
German American astrophysicist

Arno Penzias shared the Nobel Prize in physics in 1978 with Robert Wilson for a discovery that supported the **big bang** theory of the **universe**. The two **radio** astronomers at what was then American Telephone & Telegraph's (AT&T) Bell Telephone Laboratories were using a 20-ft (6-m) horn reflector antenna that year to measure the intensity of radio **waves** emitted by the halo of gas surrounding our galaxy. And they were bothered by a persistent noise which they could not explain. At first they pinned it on two pigeons that were nesting in the antenna throat. But even after they evicted the birds, the noise continued. Eventually the scientists were able to conclude that the noise came from cosmic background, or microwave, **radiation**. This came to be widely considered as remnant microwave radiation from the "big bang" in which the universe was created billions of years ago. And the Penzias-Wilson discovery came to be considered a major finding in **astrophysics**.

Arno Allan Penzias was born in Munich, Germany, April 26, 1933, to Jewish parents Karl and Justine (Eisenreich) Penzias. Hitler's campaign to wipe out the Jews of Europe was well underway when the family escaped in 1940. Arriving in New York, Penzias had to acclimate to a new culture and language and suffer through hard times for his family. Naturalized in 1946, he demonstrated scientific acumen at Brooklyn Technical High School and went on to obtain his B.S. at City College in New York in 1954. He married Anne Pearl Barras that same year; the union produced three children. After a two-year stint in the U.S. Army Signal Corps, Penzias obtained both his master's and Ph.D. degrees at Columbia University. He has said he chose to study physics because he asked a professor if he could make a living in the field and was told, "Well, you can do the same things engineers can do and do them better."

Begins Research Career

In 1961 Penzias was hired at Bell Labs in Holmdel, New Jersey. AT&T was a telecommunications monopoly at that time and Bell Labs was its research center, attracting the best and brightest scientific minds. In this context Penzias demonstrated his capabilities early on. Asked to join a committee of older scientists who were trying to devise how to calculate the precise positions of communication satellites by triangulation, young Penzias suggested they use radiostars, which emit characteristic frequencies from fixed positions, as reference points. The distinguished scientists nodded their heads in agreement—and the committee immediately disbanded. For his abilities, Penzias rose through the Bell ranks to become director of the facility's Radio Research Laboratory in 1976, and executive director of the Communications Sciences Research Division in 1979. He also took part in the pioneering Echo and Telstar communications satellite experiments of the 1970s.

Penzias's Cosmic Finding

It was astronomer **George Gamow** who in 1942 first calculated the conditions of **temperature** and **density** that would have been required for a fireball explosion or "big bang" origin of the universe 15 billion years ago. Astronomers Ralph Alpher and Robert Herman later concluded that cosmic radiation would have resulted from this event. This theory was confirmed by Penzias and Wilson. According to the theory, the **background radiation** resulting from the big bang would have lost **energy**; it would have essentially "cooled." Gamow and Alpher calculated in 1948 that the radiation should now be characteristic of a perfectly emitting body—or black body—with a temperature of about 5K, or –268°C. The scientists said this radiation should lie in the microwave region of the spectrum; their calculations were verified by physicists Robert Dicke and P. J. E. Peebles.

Penzias's and Wilson's contribution to the issue began with a 20-foot directional radio antenna, the same kind of radio antenna designed for satellite communication. Investigating an irritating noise emitted by the antenna, the two men realized in May of 1964 that what they heard was not instrumental noise but microwave radiation coming from all directions uniformly. Penzias and Wilson calculated the radiation's temperature as about 3.5K. Dicke and Peebles, who had made the earlier calculations, got reinvolved from nearby Princeton University with a scientific explanation of the Penzias-Wilson discovery. More experiments followed, confirming that the radiation was unchanging when measured from any direction. Even after the duo received the Nobel Prize in 1978 (also awarded that year to **Pyotr Kapitsa** for unrelated work in physics) they continued to collaborate on research into intergalactic hydrogen, galactic radiation and interstellar abundances of the isotopes.

Bringing Bell Labs into a New Era

At the time of the federal lawsuit which led to the breakup of AT&T in 1984, Penzias, who had become vice president of research in 1981, predicted that without the operating companies as a base, Bell Labs would become "a sinking ship." That did not happen. Instead, Penzias in September of 1990 presided over the realignment of Bell Labs into a facility whose research is streamlined and oriented towards the activities of its business units. "We adjusted the food chain," Penzias told *Science* magazine in 1991. "If we'd done everything in the old way we probably would have sunk." Picking up on his former marine metaphor, Penzias added: "But we've fixed the hull; we're back to a healthy operation."

While rearranging Bell Labs, Penzias has kept an eye on the outside world, writing *Ideas and Information: Managing in a High-Tech World* in 1989 and staying involved in the national dialogue regarding the growth of computer technology and competition with the Japanese. "You've got to understand the Japanese are not superhuman," he told *Forbes* magazine in March of 1989. "You go into Sears, the best cordless telephone you can buy is an AT&T phone. It works better. You try it."

In his personal life, Penzias, who is the proud grandfather of three, is also an avid skier, swimmer, and runner with an interest in kinetic sculpture and writing limericks. Penzias is a member of the National Academy of Sciences and the National Academy of Engineering, as well as the vice chairman of the Committee of Concerned Scientists, devoted to political freedom for scientists internationally. He has written over 100 articles and collected 19 honorary degrees. In all this Penzias argues that technology can be liberating. As he wrote in *Fortune* magazine in March of 1990: "Everybody is overstressed.... We've got to stop going to meetings and have them electronically instead.... How far away are we from realizing this dream? My guess is that by the time *Fortune* marks its 100th anniversary, a lot of this will have happened. In fact, long before I retire in 1998, I hope to have at least a multimedia terminal in my office so that I can integrate voice, data, high-definition video, conference video, document access, and shared software... who's going to do all this? I hope it's AT&T. But it could be IBM, Apple—it could be anybody."

PERIODIC TABLE

The Periodic Table is an essential part of the language of **chemistry**. It has much in common with a thesaurus, providing a guide to similarities and differences among the elements. From the way elements are organized in the Periodic Table, we can predict their behavior and write chemical formulas of compounds using just a few general guideli nes. Using such rules is not the same as understanding why elements in certain areas the Periodic Table behave as they do, but the trends that arise from the arrangement of elements in the Periodic Table allows a chemist to remember useful facts about the types of compounds formed from specific elements and their chemical reactions.

In the second half of the nineteenth century, data from laboratories in France, England, Germany and Italy were assembled into a pamphlet by Stanislao Cannizzaro, a teacher

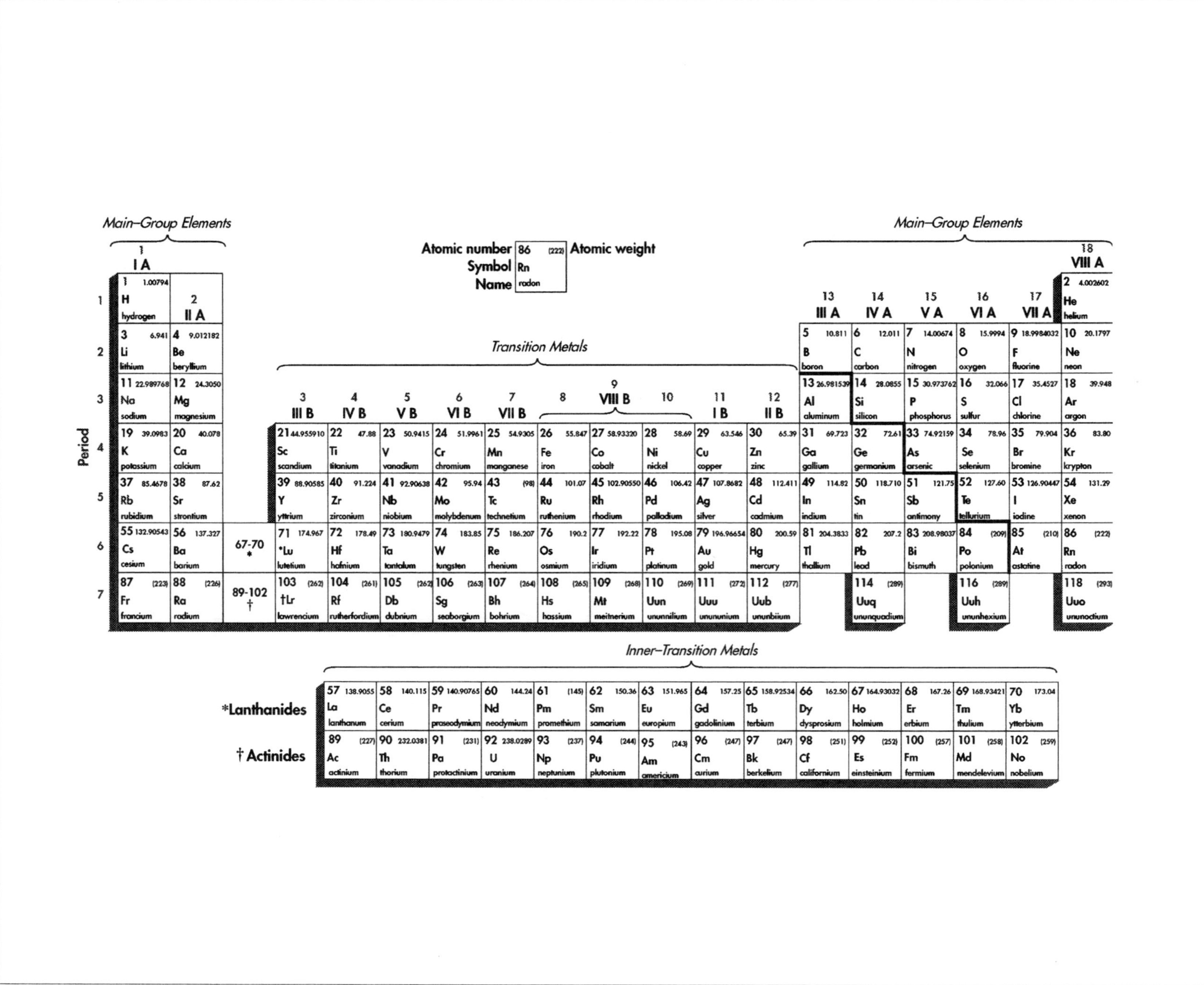

Periodic table of the elements.

in what is now northern Italy. In this pamphlet, Cannizzaro demonstrated a way to determine a consistent set of atomic weights, one **weight** for each of the elements then known. Cannizzaro distributed his pamphlet and explained his ideas at an international meeting held in Karlsruhe, Germany in 1860 organized to discuss new ideas about the theory of atoms. When Dmitri Mendeleyev returned from the meeting to St. Petersburg, Russia, he pondered Cannizzaro's list of atomic weights along with an immense amount of information he had gathered about the properties of elements. Mendeleyev found that when he arranged the elements in order of increasing atomic weight, similar properties were repeated at regular intervals; they displayed periodicity. He used the periodic repetition of chemical and physical properties to construct a chart much like the Periodic Table we currently use.

Early in the twentieth century work initiated by Joseph John Thomson in England led to the discovery of the **electron** and, later, the **proton**. In 1932, **James Chadwick**, also in England, proved the existence of the **neutron** in the atomic **nucleus**. The discovery of these elementary particles and the experimental determination of t heir actual weights led scientists to conclude that different atoms have different weights because they contain different numbers of protons and neutrons. However, it was not yet clear how many **subatomic particles** were present in any but the simplest atoms, such as hydrogen, **helium** and lithium.

In 1913, a third British scientist, **Henry G. J. Moseley**, determined the **frequency** and **wavelength** of x rays emitted by a large number of elements. By this time, the number of protons in the nucleus of some of the lighter elements had been determined. Moseley found the wavelength of the most energetic x ray of an element decreased systematically as the number of protons in the nucleus increased. Moseley then hypothesized the idea could be turned around: he could use the wavelengths of x rays emitted from heavier elements to determine how many protons they had in their nuclei. His work set the stage for a new interpretation of the Periodic Table.

Moseley's results led to the conclusion that the order of elements in the periodic chart was based on some fundamental principle of **atomic structure**. As a result of Moseley's work, scientists were convinced that the periodic nature of the properties of elements is due to differences in the numbers of subatomic particles. As each succeeding element is added across a row on the periodic chart, one proton and one electron are added. The number of neutrons added is unpredictable but can be determined from the total weight of the **atom**.

When Mendeleyev placed elements in his Periodic Table, he had all elements arranged in order of increasing relative atomic weight. However, in the modern Periodic Table, the elements are placed in order of the number of protons in the nucleus. As atomic weight determinations became more precise, discrepancies were found. The first case of a heavier element preceding a lighter one in the modern Periodic Table occurs for cobalt and nickel (58.93 and 58.69, respectively). In Mendeleyev's time, both atomic weights had been determined to be 59. Mendeleyev grouped both elements together, along with iron and copper. From the work of Moseley and others,

the number of protons in the nuclei of the elements cobalt and nickel had been found to be 27 and 28. Therefore, the order of these elements, and the reason for their similar behavior with respect to other members of their chemical families, arises because nickel has one more proton and one more electron than cobalt.

Whereas Mendeleyev based his order of elements on **mass** and chemical and physical properties, the arrangement of the table now arises from the numbers of subatomic particles in the atoms of each element. The stage was now set for examining how the number of subato mic particles affects the chemistry of the elements. The role of the electrons in determining chemical and physical properties was obscure early in the twentieth century, but that would soon change with the pivotal work of Gilbert N. Lewis.

After Moseley's work, the idea that the periodic patterns in chemical reactivity might actually be due to the number of electrons and protons in atoms intrigued many chemists. Among the most notable was G. N. Lewis of the University of California at Berkeley. Lewis explored the relationship between the number of electrons in an atom and its chemical properties, the kinds of substances formed when elements reacted together to form compounds, and the ratios of atoms in the formulas for these compounds. Lewis concluded that chemical properties change gradually from metallic to nonmetallic until a certain "stable" number of electrons is reached.

An atom with this stable set of electrons is a very unreactive species. But if one more electron is added to this stable set of elect rons, the properties and chemical reactivities of this new atom change dramatically: the element is again metallic, with the properties like elements of Group 1. Properties of subsequent elements change gradually until the next stable set of electrons is reached and another very unreactive element completes the row.

A stable number of electrons is defined as the number of electrons found in an unreactive or "noble" gas. Lewis suggested electrons occupied specific areas around the atom, called shells. The noble gas atoms have a complete octet of electrons in the outermost shell.

The observation that each element starting a new row has just one electron in a new shell opens the door to relating chemical properties to the number of electrons in a shell. Mendeleyev put elements together in a family because they had similar reactivities and properties; Lewis proposed that elements have similar properties because they have the same number of electrons in their outer shells.

Many observations of the chemical behavior of elements are consistent with this idea: the number of electrons in the outer shell of an atom (the valence electrons) determines the chemical properties of an element. G. N. Lewis extended his ideas about the importance of the number of valence electrons from the properties of elements to the bonding of atoms together to form compounds. He proposed that atoms bond with each other either by sharing electrons to form covalent bonds or by transferring electrons from one atom to another to form ionic bonds. Each atom forms stable compounds with other atoms when all atoms achieve complete shells. An atom can achieve a complete shell by sharing electrons, by giving them a way completely to another atom, or by accepting electrons from another atom.

Many important compounds are formed from the elements in rows two and three in the Periodic Table. Lewis predicted these elements would form compounds in which the number of electrons about each atom would be a full shell, like the noble gases. The noble gas es of rows two and three, neon or argon, each has eight electrons in the outermost valence shell. Thus, Lewis's rule has become known as the octet rule and simply states that there should be eight electrons in the outer shell of an atom in a compound. An important exception to this is hydrogen for which a full shell consists of only two electrons.

The octet rule is followed in so many compounds it is a useful guide. However, it is not a fundamental law of chemistry. Many exceptions are known, but the octet rule is a good starting point for learning how chemists view compounds and how the periodic chart can be used to make predictions about the likely existence, formulas and reactivities of chemical substances.

Elements in a vertical column of the Periodic Table typically have many properties in common. After all, Mendeleyev used similarities in properties to construct a periodic table in the first place. Because they show common characteristics, elements in a column are known as a family. Sometimes a family had one very important characteristic many chemists knew about: that characteristic became the family name. Four important chemical family names of elements still widely used are the alkali metals, the alkaline earths, the halogens, and the noble gases. The alkali metals are the elements in Group 1, excluding hydrogen, which is a special case. These elements—lithium, sodium, potassium, rubidium, cesium and francium—all react with water to give solutions that change the **color** of a vegetable dye from red to blue. These solutions were said to be highly alkaline or basic; hence the name alkali metals was given to these elements.

The elements of Group 2 are also metals. They combine with **oxygen** to form oxides, formerly called "earths," and these oxides produce alkaline solutions when they are dissolved in water. Hence, the elements are called alkaline earths.

The name for Group 17, the halogens, means salt former because these elements all react with metals to form salts.

The name of Group 18, the noble gases, has changed several times. These elements have been known as the rare gases, but some of them are not especially rare. In fact, argon is the third most prevalent gas in the atmosphere, making up nearly 1% of it. Helium is the second most abundant element in the universe—only hydrogen is more abundant. Another name used for the Group 18 family is the inert gases. But Neil Bartlett, while at the University of British Columbia in Vancouver, Canada, showed over 30 years ago that several of these gases can form well-defined compounds. The members of Group 18 are now known as noble gases—they do not generally react with the common ele ments but do on occasion, especially if the common element is as reactive as fluorine.

Knowing the chemistry of four families of the periodic table—groups 1, 2, 17 and 18, the alkali metals, the alkaline earths, the halogens and the noble gases—enables us to divide the elements in the Periodic Chart into other general categories: metals and nonmetals. We all automatically think of metals as shiny, hard but ductile substances that conduct **electricity**. Groups 1 (excluding hydrogen) and 2 are families of metallic elements. Groups 17 and 18 contain elements with very different properties perhaps best described by what they are not—they are not metals, and hence are called nonmetals. Between Groups 1 and 2 and Groups 17 and 18 is a dividing line between these two types of elements. Most periodic charts have a heavy line cutting between aluminum and silicon and descending downward and to the right in a stair-step fashion. Elements to the left of the line are metallic; those to the right, nonmet allic. The boundary is somewhat fuzzy, however, because the properties of elements change gradually as one moves across and down the chart, and some of the elements touching that border have a blend of characteristics of metals and nonmetals; they are frequently called semi-metals or metalloids.

The elements in the center region of the table, consisting of dozens of metallic elements in Groups 3 - 12, including the lanthanide and actinide elements, are called the transition elements or transition metals. The other elements, Groups 1,2, and 13 - 18, are called the representative elements.

There is a correlation among the representative elements between the number of valence electrons in an atom and the tendency of the element to act as a metal, nonmetal or metalloid. Among the representative elements, the metals are located at the left and have few valence electrons. The nonmetals are at the right and have nearly a full shell of electrons. The metalloids have an intermediate number of valence electrons.

The structure and bonding of a compound determine its chemical and physical properties. Lewis's idea of stable, filled electron shells can be used to predict what atom is bonded to what other atom in a molecule. In many cases, Lewis's octet rule is followed by taking one or more electrons from one atom to form a cation and donating the electron or electrons to another atom to form an anion. Metallic elements on the far left of the Periodic Table can give electrons away and elements on the far right can readily accept electrons. When these elements combine, ionic bonds result. An example of an ionic compound is sodium chloride. The electron transferred from the sodium atom to form the sodium cation, Na^+, is shown with the chloride anion, Cl^-. In covalent bonds, electrons are shared between atoms. Lewis defined a covalent bond as a union between two atoms resulting from the sharing of two electrons. Thus, a covalent bond must be considered a pair of electrons shared by two atoms. Elemental bromine, Br_2, is an example of a covalent compound. Each bromine atom has seven electrons in its outer shell and requires one electron to achieve a noble gas configuration. Each can pick up the needed electron by sharing one with the other bromine atom. Water is a compound formed by the combination of atoms of two nonmetallic elements, hydrogen and oxygen. Each **hydrogen atom** requires just one electron to fill its shell because the first shell (the number of electrons of the noble gas helium) has only two electrons. Oxygen lacks two electrons compared with neon, the nearest noble gas. If each hydrogen can obtain one electron by sharing electrons from the oxygen atom and the oxygen atom can share one electron from each of the two hydrogen atoms, every atom will have a full shell of electrons, and two covalent bonds will be formed as a result of sharing two pairs of electrons.

The tendency for particular elements to form ionic or covalent compounds was known before Mendeleyev had constructed his version of the Periodic Table, but there was no simple way to remember which elements formed which types of compounds until the Periodic Table was formulated. We can now use the table instead of memorizing each element's properties separately. One of the most important properties of an element that can be used to predict bonding characteristics is whether the element is metallic or nonmetallic.

Pure metals are typically shiny and malleable. These properties were evident to people thousands of years ago when elements such as gold, silver, copper, tin, zinc, and lead were recognized as materials of a particular class—the metals. Other properties might require some equipment to determine, for example, electrical conductivity. Chemists have found metals also have common chemical properties. Metals combine in similar ways with other elements and form compounds with common characteristics. Metals combine with nonmetals to form salts. In salts, the metals tend to be cations. Salts conduct electricity well when melted or when dissolved in water or some other solvents but not when they are solid.

Most pure metals, when freshly cut to expose a new surface, are lustrous, but most lose this luster quickly by combining with oxygen, **carbon** dioxide or hydrogen sulfide to form oxides, carbonates or sulfides. Only a few metals such as gold, silver and copper are found pure in nature, uncombined with other elements.

Nonmetals in their elemental form are usually gases or solids. A few are shiny solids, but instead of being metallic grey they are typically black (boron, carbon as graphite), colorless (carbon as diamond) or highly colored (violet iodine, yellow sulfur). At room **temperature**, only one of them is a liquid (bromine).

Nonmetallic elements combine with metallic elements to form salts. In salts, the nonmetallic elements tend to be anions. Non-metals accept electrons in forming anions while metals donate electrons to form cations. This reflects a periodic property o f elements: as you move from left to right across a row on the periodic chart, you move from atoms of metals which tend to give up electrons relatively easily to atoms of nonmetals which do not readily give up electrons in forming chemical bonds. At the start of the next row, the trend is repeated. This periodic property is referred to as electron egativity. The more readily atoms accept electrons in forming a bond, the higher their electronegativity. Metals are characterized by low electronegativities; nonmetals, by high electronegativity. Electronegativity increases across a row on the periodic chart.

Nonmetallic elements combine with each other to form compounds. Although some nonmetallic elements form solutions when mixed with other nonmetallic elements, most react with other nonmetals to form new substances. For example, at the high temperatures and pressures of an internal combustion engine, nitrogen and oxygen gases from the atmosphere react to form nitrogen oxides such as nitric oxide, NO, and nitrogen dioxide, NO_2. Nonmetallic elements form covalent bonds with each other by sharing electron pairs. This tendency to bond by sharing electrons reflects the periodic trend described above: elements on the right side of the periodic chart do not give up electrons easily when forming bonds; their electronegativity is high. They tend to either accept electrons from metals to form salts or share electrons with other nonmetals to form covalent compounds.

Metalloids typically show physical characteristics intermediate between the metals and nonmetals. For example, the electrical conductivity of the metalloid germanium, which is used in semiconductor devices, is 2.2 x 10 4 ohm-1 cm-1 and its thermal conductivity is 0.60 watts cm-1 K-1. By comparison, the electrical conductivity of a metal such as copper is over 106 ohm-1 cm−1 and its thermal conductivity is 4.01 watts cm−1 K-1 thermal conductivity is and the electrical and thermal conductivities of non-metals are orders of magnitude lower than those of germanium. Metalloids typically act more like nonmetals than metals in their chemistry. They more often combine with nonmetals to form covalent compounds rather than salts, but they can do both. This reflects their intermediate position on the Periodic Table. They can, however, also form alloys with metals and with the other metalloids. Semiconductors are typically made from combinations of two metalloids. The minor constituent, for example germanium, is said to be "doped" into the major constituent, which is often silicon.

The boundaries between metals, nonmetals and metalloids are arbitrary. The changes in properties as one moves from element to element on the chart are gradual.

Mendeleyev used several properties when he decided how to arrange the elements in the Periodic Table. He considered metallic and nonmetallic properties, deciding potassium should be placed in the column with sodium, for example. He also considered the ratio of the number of atoms of the metallic element to the number of atoms of the nonmetallic element in salts. On this basis, he put calcium, which forms $CaCl_2$, in the column under magnesium, which forms $MgCl_2$. Another important property he used was the acidic or basic character of oxides formed from the elements.

Because we exist in an atmosphere containing about 20% oxygen and oxygen is quite reactive, most elements can be found in nature as oxides. The alkali metals (Group 1) and alkaline earths (Group 2) were so named because the metallic oxides formed when the metals reacted with oxygen produced basic solutions when dissolved in water. Metallic oxides are known as basic anhydrides (anhydrous, meaning without water), because basic solutions are formed when they are added to water.

Nonmetallic elements combine with oxygen to form oxides, many of which, such as carbon dioxide, sulfur dioxide and nitrogen dioxide, are gases. When oxides of nonmetallic elements are dissolved in water they tend to form acidic solutions or neutral solutions. Nonmetal oxides that form acidic solutions when dissolved in water are called acid anhydrides.

Transition metals react with oxygen to form a wide variety of oxides, some of which are basic and some acidic. A few transition metals are relatively unreactive and may be found in nature as pure elements.

A great deal of chemical information is contained in the Periodic Table. The organization of the modern table gives us insight into the importance of composition of the nucleus and number of electrons in the outer shell of an atom. Several useful theories enable us to use the number of valence electrons to write formulas of compounds and to predict physical properties and chemical reactivities of elements. An element is defined by the number of protons in the nucleus of its atoms, but its chemical reactivity is determined by the number of electrons in its outer shell. Elements in a family of the Periodic Table have the same number of electrons in the outer shell. Because they have the same number of valence electrons, elements in a family often exhibit similar chemical properties.

Metals and nonmetals differ in their tendencies to lose valence electrons, accept valence electrons, or share valence electrons in forming compounds. The octet rule, proposed by G. N. Lewis, is a useful guideline to predict formulas and the likely existence of compounds. Ionic compounds, formed when electrons are transferred from one atom to another to make **ions**, have vastly different properties compared to covalent compounds, which are formed when pairs of electrons are shared between atoms.

Chemical behavior, such as the acidic or basic properties of oxides, is also periodic. Physical properties, such as the ability to conduct electricity or **heat** and malleability or brittleness, may also be predicted from the position of an element on the Periodic Table.

PERPETUAL MOTION

The persuance of perpetual **motion** is to physics and mechanical engineering what alchemy is to modern **chemistry**, which is to say that it is seeking after the unrealizable. The earliest written description of a perpetual motion machine dates to the fifth century A.D. Manuals and scientific treatises abounded throughout the Middle Ages and the Renaissance that described various perpetual motion devices. As late as the early 1800s, a miller in Ohio poured his life savings into building a three-story mill house that was supposed to run perpetually by means of a closed system powered by 500 gal (1,890 L) of water. It didn't work.

Perpetual motion, when defined by its literal meaning, is an impossibility according to Classical physics. A perpetual motion device, when set in motion, never stops and never requires further input of **energy**. Another way of saying that is, it puts out more energy than it consumes. If such a machine could be built on a big enough scale, it might promise limitless kinetic or electrical energy.

The first two laws of **thermodynamics** are what prevent perpetual motion from working. The first law, which address-es the **conservation of energy**, prevents a system from continuing to create its own **power**. The first law negates the possibility of a perfectly efficient system, and requires that some energy always be lost as **heat** or **friction**. Conservation of energy dictates that work and energy seek to reach **equilibrium**, and thus, as energy is dissipated, work slows to a stop. This is why perpetual motion machines that utilize falling water or overbalanced wheels always grind to a halt, even if they power themselves for a startlingly long time.

The **second law of thermodynamics** gives us **entropy**, which says that all closed systems eventually break down. It is this principle that disallows perpetual motion devices that attempt to self-sustain a system based on condensation/evaporation cycles of liquids and gasses in sealed boilers and similar apparatus. Over time, the finite energy within the system is lost as heat, and the system ceases to work.

The laws of thermodynamics are statistical in nature. This is just due to the fact that while it is impossible to measure the motion of every molecule in a system, it is possible to measure the average of these motions. Statistically then, the individual motions are just treated as random fluctuations around the average. Though exceedingly rare, this collective randomness does open up the possibility that the molecules of a system will not precisely follow the predictions of thermodynamics. That means there is a statistical possibility that a pan of water placed on a roaring fire will freeze. It follows that a closed system has a theoretical chance to operate without energy loss. This "loophole" in the physical impediment to the possibility of perpetual motion offers an intriguing, if totally academic twist to this 1,500-year-old quest. Since such a happenstance would be totally unpredictable, it is likewise totally unharnessable. It is also highly unlikely. Chemist Henry Bent, who coined the now famous analogy about a group of monkeys typing out the works of Shakespeare (as a comparison to the likelihood of a local reversal of entropy), also claimed to have calculated the odds of a reversal of entropy resulting in usable energy. He said it would be akin to the monkeys turning out Shakespeare's plays 15 quadrillion times without error.

See also Entropy; Laws of Thermodynamics

PERRIN, JEAN BAPTISTE (1870-1942)
French physicist

The contribution made by French physicist Jean Baptiste Perrin to the study of **atomic physics** was of the most fundamental kind: he helped to prove that atoms and molecules exist. This achievement, which quantitatively extended the original observations of botanist Robert Brown on the movement of pollen grains in water, and put scientific substance into **Albert Einstein**'s exquisite mathematical equations describing the distribution of those particles in solution, won for Perrin the 1926 Nobel Prize in chemistry.

Perrin was born in Lille, France, on September 30, 1870, and raised, along with two sisters, by his widowed mother. His father, an army officer, died of wounds he received during the

Franco-Prussian War. The young Perrin attended local schools and graduated from the Lycée Janson-de-Sailly in Paris. After serving a year of compulsory military service, he entered the Ecole Normale Supérieure in 1891, where his interest in physics flowered and he made his first major discovery.

Between 1894 and 1897 Perrin was an assistant in physics at the Ecole Normale, during which time he studied cathode rays and **x rays**, the basis of his doctoral dissertation. At this time, scientists disagreed over the nature of cathode rays emitted by the negative electrode (cathode) in a **vacuum** tube during an electric discharge. Physicists disagreed among themselves over whether cathode rays were particles—a logical assumption, since they carried a charge—or whether they took the form of **waves**.

Confirms Nature of Cathode Rays

In 1895 Perrin settled the debate simply and decisively using a cathode-ray discharge tube attached to a larger, empty vessel. When the discharge tube generated cathode rays, the rays passed through a narrow opening into the vessel, and produced **fluorescence** on the opposite wall. Nearby, an electrometer, which measures **voltage**, detected a small negative charge. But when Perrin deflected the cathode rays with a magnetic field so they fell on the nearby electrometer, the electrometer recorded a much larger negative charge. This demonstration was enough to prove conclusively that cathode rays carried negative charges and were particles, rather than waves. This work laid the basis of later work by physicist J. J. Thomson, who used Perrin's apparatus to characterize the negatively charged particles, called electrons, which were later theorized to be parts of atoms.

In 1897 Perrin married Henriette Duportal, with whom he had a son and a daughter. He received his doctorate the same year, and began teaching a new course in physical chemistry at the University of Paris (the Sorbonne). He was given a chair in physical chemistry in 1910 and remained at the school until 1940. During his early years at the University of Paris, Perrin continued his study of the **atomic theory**, which held that elements are made up of particles called atoms, and that chemical compounds are made up of molecules, larger particles consisting of two or more atoms. Although the atomic theory was widely accepted by scientists by the end of the nineteenth century, some physicists insisted that atoms and molecules did not actually exist as physical entities, but rather represented mathematical concepts useful for calculating the results of chemical reactions. To them, **matter** was continuous, not made up of discrete particles. Thus, at the dawn of the twentieth century, proving that matter was discontinuous (atomic in nature) was one of the great challenges left in physics. Perrin stood on the side of the "atomists," who believed that these tiny entities existed. In 1901 he even ventured (with no proof) that atoms resembled miniature solar systems. His interest in atomic theory led him to study a variety of related topics, such as osmosis, ion transport, and crystallization. However, it was colloids that led him to study **Brownian motion**, the basis of his Nobel Prize-winning discovery of the atomic nature of matter.

Verifies Einstein's Calculations of Brownian Motion

In 1827 the English botanist Robert Brown reported that pollen grains suspended in water were in violent and irregular **motion**, a phenomenon at first ascribed to differences in **temperature** within the fluid. Before the end of the century, however, scientists generally accepted the notion that the motion might be caused by bombardment of the pollen grains by molecules of the liquid—an apparent triumph for atomic theory. Yet some scientists remained skeptical.

In 1905 Albert Einstein calculated the mathematical basis of Brownian motion, basing his work on the assumption that the motion was due to the action of water molecules bombarding the grains. But Einstein's work, though elegant, lacked laboratory experiments needed to demonstrate the reality of his conclusions. It fell to Perrin to bolster Einstein's calculations with observations. From 1908 to 1913, Perrin, at first unaware of Einstein's published paper on the subject, devoted himself to the extremely tedious but necessary experiments— experiments now considered classics of their kind. He hypothesized that if Brownian movement did result from molecular collisions, the average movements of particles in suspension were related to their size, **density**, and the conditions of the fluid (e.g., **pressure** and density), in accordance with the **gas laws**. Perrin began by assuming that both pollen grains and the molecules of the liquid in which they were suspended behave like gas molecules, despite the much greater size of the grains.

According to Einstein's equations governing Brownian motion, the way the particles maintained their position in suspension against the **force** of **gravity** depended partly on the size of the water molecules. In 1908 Perrin began his painstaking observations of suspensions to determine the approximate size of the water molecules by observing suspensions of particles. He spent several months isolating nearly uniform, 0.1-gram pieces of gamboge—tiny, dense extracts of gum resin, which he suspended in liquid. According to Einstein's molecular theory, not all particles will sink to the bottom of a suspension. The upward **momentum** that some particles achieve by being bombarded by molecules of the fluid will oppose the downward force of gravity. At **equilibrium**, the point at which the reactions balance each other out, the concentrations of particles at different heights will remain unchanged.

Perrin devised an ingenious system to make thousands of observations of just such a system. He counted gamboge particles at various depths in a single drop of liquid only one twelve-hundredth of a millimeter deep. The particle concentration decreased exponentially with height in such close agreement with the mathematical predictions of Einstein's theory that his observations helped to prove that molecules existed.

In essence, his system behaved like the Earth's atmosphere, which becomes increasingly rarified with height, until, at the top of a very tall mountain, people may find it difficult to breathe. Furthermore, it was already known that a change in altitude of five kilometers is required to halve the concentration of **oxygen** molecules in the atmosphere, and that the oxygen **atom** has a **mass** of sixteen. Based on his knowledge of the gas laws, Perrin realized that if, in his tiny system, the height required to halve the concentration of particles was a billion times less than the height it took to halve the concentration of oxygen in the atmosphere, he could, by simple proportion, calculate the mass of a gamboge particle relative to the oxygen molecule.

Einstein had linked to Brownian motion the concept of **Avogadro's number**, the number of molecules in any gas at normal temperature and pressure, now known to be 6.023 x 10^{23}. According to Avogadro's hypothesis, equal volumes of all gases at the same temperature and pressure contain equal numbers of molecules. Furthermore, the total mass of a specific volume of gas is equal to the mass of all the individual molecules multiplied by the total number of these individual molecules. So a gram-molecule of all gases at the same temperature and pressure should contain the same number of molecules. (A gram-molecule, or mole, is a quantity whose mass in grams equals the **molecular weight** of the substance; for example, one gram-molecule of oxygen equals 16 g of oxygen.) Only if this were true would the concept that each individual molecule contributes a minute bit of pressure to the overall pressure hold true, and individual entities called molecules could be said to exist.

Perrin calculated the gram-molecular **weight** of the 0.1-gram particles in the equilibrium system and therefore knew the number of grams in a gram-molecule of the particles. Then he divided the gram-molecular weight by the mass in grams of a single particle. The result, 6.8 x 10^{23}, was extremely close to Avogadro's number. Thus, Perrin had demonstrated that uniform particles in suspension behave like gas molecules, and calculations based on their mass can even be used to calculate Avogadro's number. This demonstrated that Brownian motion is indeed due to bombardment of particles by molecules, and came as close as was possible at the time to detecting atoms without actually seeing them. "In brief," Perrin said during his Nobel Prize acceptance speech, "if molecules and atoms do exist, their relative weights are known to us, and their absolute weights would be known as soon as Avogadro's number is known."

Perrin's work ranged farther afield than equilibrium distribution of particles and Avogadro's number, however. As an officer in the engineering corps of the French army during World War I, he contributed his expertise to the development of acoustic detection of submarines. His commitment to science, however, did not inhibit his social graces. He was a popular figure who took a genuine interest in young people, and held weekly parties for discussion groups in his laboratory. Following the war, Perrin's reputation continued to grow. In 1925, he became one of the first scientists to use an electric generator capable of producing a continuous current of 500,000 volts. At the time, he predicted that someday much larger machines of this type would let physicists bombard atoms, and thus make important discoveries about the structure of these particles.

In 1929 after being appointed director of the newly founded Rothschild Institute for Research in **Biophysics**, he was invited to the United States as a distinguished guest at the opening of Princeton University's new chemical laboratory. In 1936 Perrin replaced Nobel laureate **Irene Joliot-Curie** as French undersecretary of state for scientific research in the government of Premier Léon Blum. The following year, as president of the French Academy of Science, he assumed the chair of the scientific section of an exhibit in the Grand Palais at the 1937 Paris exposition. The project enabled him to help the average person, including children, to appreciate the wonders of science, from astronomy to zoology.

His flourishing reputation was further enhanced in 1938 when he informed the French Academy of Science, of which he had been a member since 1923, and was then president, that his collaborators had discovered the ninety-third **chemical element**, neptunium, a substance heavier than uranium. Four years earlier, **Enrico Fermi** (who was awarded the 1938 Nobel Prize in physics and directed the first controlled nuclear **chain reaction**) had artificially created neptunium, a so-called transuranium element, by bombarding uranium (element 92) with neutrons. Perrin's announcement that neptunium existed in nature excited speculation among physicists that there also might exist even more undiscovered elements, which turned out to be the case.

Condemns Totalitarianism and Warns of Impending War

His blossoming career did not shield the French physicist from concerns over what he considered to be a steady encroachment by totalitarian governments around the world on the freedom of science to express itself. A socialist and outspoken opponent of fascism, Perrin expressed his concerns during a speech delivered at the Royal Opera House in London before the International Peace Conference, reported in the *New York Times*. He asserted that world science stands or falls with democracy, and decried the fact that scientists seemed unable to understand "how financiers and capitalists as a whole cannot see that it is to their interest not to support those powers which, if they are successful, will ruin them." Perrin also criticized what he called "an irrational world that made it difficult to extend higher education or grant more aid to science but relatively easy to raise money for costly armaments." He voiced concern over what he believed was the coming war—World War II—which he feared would cost millions of lives, as well as threaten "the democracy that is the spirit of science." Perrin also warned that the victory of totalitarianism would mean "perhaps a thousand years of ruthless subjugation and standardization of thought, which will destroy the freedom of scientific research and theorizing."

Perrin's fears were realized in September of 1939, when France joined Great Britain in entering World War II against Germany following that country's invasion of Poland. By the end of September, the French government appointed Perrin president of a committee for scientific research to help the war effort. The situation became particularly grim in the summer of 1940, when German troops swept into Paris. Perrin fled the city and took up residence in Lyon as a refugee. In December 1941 he moved to the United States, where he lived with his son, Francis Perrin, a visiting professor of physics and mathematics at Columbia University. While in the United States, Perrin sought American support for the French war effort and helped to establish the French University of New York.

Perrin spoke out against the German occupation and French collaboration with the enemy. He was particularly disturbed when the Germans began operating an armaments industry in the suburbs of Paris using forced labor. Following Allied aerial bombardment of the factories, the *New York Times* reported that Perrin defended the action as "one of the sad necessities" of the war. Speaking before 500 guests at the first dinner of the French American Club in New York City, in March 1942, Perrin asked, "Who does not understand that it was imperative to put an end to this?" A few weeks later, Perrin took ill, and ten days later he died at the age of 71 at Mount Sinai Hospital in New York.

Three years after the defeat of Germany and the end of the war, diplomats and scientists in New York paid homage to Jean Perrin at ceremonies held at the Universal Funeral Chapel. Afterwards, Perrin's ashes were placed aboard the training cruiser Jean d'Arc at Montreal, on which they were transported to France for burial at the Pantheon, a magnificent former eighteenth-century church converted to civic use. Among his many honors in addition to the Nobel Prize, Perrin received the Joule Prize of the Royal Society of London in 1896 and the La Caze Prize of the French Academy of Sciences in 1914. In addition, he held honorary degrees from the universities of Brussels, Liège, Ghent, Calcutta, and Manchester and from New York, Princeton, and Oxford universities. Perrin was also a member of the Royal Society of London and scientific academies in Italy, Czechoslovakia, Belgium, Sweden, Romania, and China.

PERTURBATION THEORY

Perturbation theory is an approximation method useful in dealing with the Schrödinger equation as it relates to systems of interacting particles. In 1924, by using **Albert Einstein**'s special theory of relativity, **Louis de Broglie** showed that particles have **waves** associated with them. It then became obvious that a specific type of partial differential equation should be able to describe their position and that their future behavior could be predicted. In 1926, **Erwin Schrödinger** used partial differential equations and the Hamiltonian function to develop a powerful equation that relates the **energy** of the **electron** to the energy of the electric field in which it is situated. This equation, the Schrödinger equation, relates the energy of a system to its wave properties and allows prediction of the energy of the electron and its future behavior, i.e. the probability of finding the electron in a particular region around the **nucleus**. Although the Schrödinger equation is a powerful equation it is not practical unto itself because it can rarely be solved. To overcome this problem scientists have developed two main methods that allow us to obtain approximations of the energy of a system without solving the Schrödinger equation. These two methods are the variation method and the perturbation theory.

The basic premise behind the perturbation theory is as follows. Assuming that the Hamiltonian of the perturbed system in which we are interested is only slightly different from the Hamiltonian of an unperturbed system for which we are able to solve the Schrödinger equation, the eigenfunctions and eigenvalues (solutions to the Schrödinger equation associated with special values of the electron's energy levels) of the perturbed system should be closely related to those of the unperturbed system. So we can determine the unknown, perturbed system's values from the known, unperturbed system's values. Basically perturbation theory is developed to deal with small corrections to problems which we know how to solve exactly.

The perturbation theory can be used for non-degenerate as well as degenerate systems. So, unlike the variation method that is restricted to the ground state of a system, the perturbation method applies to all the states of an **atom** or molecule. It can be used to predict the energies of electrons in excited atoms or molecules as well as ground state particles. The perturbation method allows one to calculate the energy much more accurately than it allows one to calculate the **wave function**.

PHASE CHANGES AND PHASE EQUILIBRIUM

The two types of material **equilibrium** are phase equilibrium and reaction equilibrium. Phase equilibrium refers to the transport of **matter** between phases of the system without conversion of one species to another. That is, the same chemical species may be present in several different forms in a system. These phases may inter-convert from one to another, but the overall amount of any one particular phase is constant in time if the system is in equilibrium.

The crossing of any two-phase curve in a **phase diagram** is called a transition. Special names are given some particular transitions. Melting is the process of a solid going to a liquid, and freezing is a liquid transitioning to a solid. Vaporization is when a liquid transitions to a gaseous or vapor phase. Condensation is a vapor transitioning to a liquid. **Sublimation** is when a solid transitions to a vapor, and deposition denotes a vapor transitioning to a solid.

Systems in which a net transport of matter from one phase to another are occurring are not in thermodynamic equilibrium. In an isolated system where spontaneous transport of matter between phases is occurring irreversibly, the **entropy** is increasing. This process will continue until the entropy is maximized. Any further process can then only decrease entropy and violate the second law of **thermodynamics**. This yields a critical criterion for equilibrium. In an isolated system the maximization of the system's entropy is the point where the system is in equilibrium. Usually one deals with closed systems (systems that can exchange **heat** and **work** with their surroundings), rather than isolated systems. Under these conditions it is useful to take the system itself plus the surroundings with which it interacts to constitute an isolated system. In this case the condition for phase equilibrium in the system is the maximization of the total entropy of the system plus its surroundings.

The initial application of the laws of thermodynamics to material equilibrium is primarily the work of American mathematician **Josiah W. Gibbs**. Gibbs' work in 1876-78 employed the first and second laws of thermodynamics to deduce the conditions of material equilibrium. Using these principles, Gibbs developed a state function called the Gibbs free **energy**, Gibbs function, or Gibbs energy. Basically, this state function decreases in value during a system's approach to equilibrium at constant **temperature** and **pressure**, and reaches a minimum when the system is at equilibrium. If the volume is increased in a closed container of water, for example, above which is found water vapor, then some of the liquid water will undergo a phase transition to vapor. During this phase transition period the Gibbs free energy is decreasing until enough water has evaporated, and the system is again in equilibrium. Once the system is in equilibrium the Gibbs free energy is at a minimum. A similar state function for a system under constant temperature and volume is called the Helmholtz free energy, Helmholtz energy, or the Helmholtz function. This state function is also found to be continually decreasing during the spontaneous, **irreversible processes** of matter transport between phases under constant temperature and volume until the system reaches equilibrium.

Several interesting phenomena are associated with materials undergoing phase changes. One is that when the temperature of a pure substance is raised through a phase transition, the enthalpy is increased while the temperature remains constant, so long as the two phases coexist. For instance, if a block of ice is in a closed system, at a constant pressure, and heat is added to the system by raising the temperature, the block of ice will start to melt. The temperature of the ice and water will remain constant until all of the ice is transformed into water. Only then will the temperature of the water start to rise. The temperature can rise again only when the whole of the low-temperature form has been converted into the high-temperature form. The enthalpy change at the transition temperature is called the enthalpy of transition, and is a useful thermodynamic quantity.

Another interesting concept associated with phase changes is that of chemical potential. For a pure substance, the chemical potential is simply the molar Gibbs free energy. But although systems can be simple and contain only one substance, they can also be complex and contain mixtures of substances in various phases. In these types of systems the chemical potential is a useful concept in defining equilibrium. In a closed system in thermodynamic equilibrium, the chemical potential of any given component is the same in every phase in which that component is present. For a system that is in thermal and mechanical equilibrium but that has not yet reached phase equilibrium, a substance will go from a phase of higher chemical potential to a phase with lower chemical potential until the chemical potential of that substance has been equalized in all of the phases of the system. The higher the value of the chemical potential of a substance in a particular phase, the greater is the tendency of that substance to leave

A pan of boiling water showing the phase change from liquid to gas. *(Photo by Martin Dohrn, Science Source/Photo Researchers. Reproduced by permission.)*

that phase and flow into a phase with lower chemical potential. This concept yields another useful definition of phase equilibrium. The condition for equilibrium between phases is that for each substance the chemical potential must be the same in every phase of the system in which that substance is present. Chemical potentials are the key properties in chemical thermodynamics, since they determine phase and reaction equilibrium.

Although the overall process of reaching phase equilibrium and the concepts associated with it can be very complex the laws of thermodynamics allow one to understand and predict the behavior of substances in different phases interacting with one another. The concepts associated with phase and chemical equilibrium lie at the heart of chemical thermodynamics.

PHASE DIAGRAM

A phase diagram is a graphical representation of the various states of a substance as they relate to **temperature** and **pressure**. These diagrams can be used to predict the physical behavior of materials. Some phase diagrams are simple with three areas that represent gas, liquid and solid phases, separated by three two-phase lines. Other diagrams can be complex with multiple liquid and or solid phases, but only one gas phase.

The simplest phase diagrams are plotted as temperature versus pressure, and consist of three areas corresponding to the phases of material. In general, at high temperatures and low pressures materials usually exist as gases; at low temperatures materials usually exist as solids. It is possible for two states to exist simultaneously. The two phases are said to coexist in **equilibrium** and can be liquid-vapor, solid-vapor, or solid-liquid. The lines joining the three areas show where material exists in equilibrium between states. Several points on a phase diagram have special significance. The point on a phase diagram that gives a unique temperature and pressure at which solid, liquid, and vapor coexist is called the triple point. The triple point delineates where the three lines joining the different phases meet. Another important point on a phase diagram is the normal **boiling** point. This point is found at the temperature where the vapor pressure of a liquid is equal to the standard atmospheric pressure. The normal melting point is found at the temperature where the vapor pressure of the solid and the vapor pressure of the liquid are equal. The other important point on a phase diagram is called the critical temperature. This is the temperature above which no amount of pressure will liquefy a vapor. Phase diagrams, whether simple or complex, are useful in predicting the behavior of a material under specific conditions. Phase diagrams also delineate regions where the temperature and pressure are such that the solid can go to the gaseous form via the process of **sublimation**.

The phase diagram for water is more complex than a simple diagram. Water has the normal liquid and gaseous phases, but is more complex in that water is polymorphic, that is it has several crystalline forms. The solid phase known as ice I is most familiar, but under high pressures there are another five forms of ice for water. In general, such polymorphism is much more the rule than the exception among substances. Because water can exist in several solid forms, the phase diagram has several triple points in addition to the normal one found at the junction between the solid, liquid and gaseous states. The additional triple points are at the junctions between the different forms of ice.

See also States of matter

PHILOSOPHY OF PHYSICS

Philosophy of physics is a branch of philosophy that deals with the validity of logic and nature of truth (epistemology) in the physical sciences. Modern questions and problems also involve inquiry into the nature of **space** and time, geometry, probability and quantum **mechanics**. The task of scientists and philosophers is to determine what knowledge is, how it is acquired, and on what basis it rests. It is thus the work of philosophers of science, especially those scientists who have sufficient expertise in physics that allow them to accurately and completely evaluate physical theory, to interpret scientific theories, experiments, methodologies and results. Typical issues that arise out of such inquiry include the role of probability in **statistical physics**, the interpretation of measurement in **quantum mechanics**, and the philosophical presuppositions of **space-time** theories.

Before the **Scientific Revolution**, philosophers relied upon logic and reason to explain natural phenomena. For example, the geometry of **Euclid**, published in 300 B.C., allowed tremendous insight into the world using premises (called postulates) derived through reason. Euclid started with simple truths about the world; e.g., that a straight line can be drawn between any two points, or that all right angles are equal to each other. Although such claims seem simple and self-evident, from them, and through the processes of deduction and inference, Euclid and other arrived at more complex truths about the geometric structure of the world.

Until the Scientific Revolution, inquires into the nature and workings of the physical world ultimately led philosophers and theologians to ascribe the "first cause" or the regulation of the natural world to the divine and supernatural. With the Copernican Revolution, the discarding of geocentric theories, the philosophy of science increasing depended upon the interpretation of mechanistic laws and principles.

The idea that pure reason can be used to determine the truth of the world, and that the existence of things depends on how we perceive them is known as idealism. As a reaction to the Enlightenment thinking in the aftermath of Newton's Principia, there was a movement toward idealism, especially among theologians. Despite philosophic and scientific refutations, idealism, and derivatives of idealism still persist in some modern philosophical circles. In the later half of the 18th century there was a synthesis of rationalism and experimentalism in the work of the German philosopher **Immanuel Kant** that profoundly influenced the course of Western thought. Kant acknowledged the limitations on rationalism arising from the interpretation of the material world by the human mind but allowed only a transcendental idealism that did not deny the existence of physical reality outside of human experience. In this regard Kant struck a balance between idealism and an insistence that experimentalism could lead to an absolutely verifiable knowledge of the natural world.

In the nineteenth century, mathematicians discovered non-Euclidean geometries with postulates that conflicted with Euclid's postulates. Each of the geometries gave valid answers over a differing set of conditions. French philosopher Henri Poincaré concluded that our conventions (what we decide as a society) dictate what our measurements should mean. If the angles of a triangle add up to more than 180 degrees (which they do in some non-Euclidean geometries), Poincaré says we can feel justified in claiming that there is a problem with our measuring sticks, and that the angles really do add up to 180 degrees, as stipulated by the conventional Euclidean geometry.

The nature of space and time is an ongoing source of debate in philosophical and scientific circles. Newton argued that space was an absolute structure against which forces, such as **acceleration**, could be observed and measured. More than 200 years later, Einstein's theories of relativity made it clear that there were no absolute **reference frames** except that two observes in different reference frames would measure the same speed of **light**. Einstein's theoretical advances raised interesting philosophical questions regarding simultaneity, i.e., how do we know if an event occurs at the same time for two observers? Out of Einstein's relativity theories emerged the idea that the **universe** is shaped by a fabric of space-time.

Quantum mechanics is another scientific theory rich with philosophical implications. The basic problems in quantum mechanics and **quantum theory** in general relate to the nature of **matter**. Although important only in the description of behavior for relatively small and **subatomic particles**, experiments show that all matter has the properties of both **waves** and particles. The standard interpretation is that a **photon** of light that goes unobserved has a probability of being neither a wave nor a particle; it's a wave of probability. When it is observed, that wave collapses into a point. The wave-particle paradox is illustrated in the **Schrödinger's cat** thought experiment, which, in essence, asks you to imagine that a room contains a cat that is both dead and alive at the same time and that the fate of the cat is determined only when looking in the box.

The philosophy of physics also explores the justification of induction and its methods, the nature of explanation, and the role of falsifiabilty in science.

See also History of Physics

PHOSPHORESCENCE

Phosphorescence is the extended emission of electromagnetic **radiation** (**light**) following excitation or the emission of light without the production of significant **heat** (e.g., in the bioluminescence of fireflies).

Materials that display a persistent afterglow are commonly said to exhibit phosphorescence. The term is derived from *phosphors*, a class of well-known phosphorescent materials. When a luminescent molecule or **atom** absorbs light **quanta** in the ultra-violet or visible region of the **electromagnetic spectrum**, electrons are promoted to a higher **energy** state called an excited state. The radiation absorbed can be emitted either in the form of **fluorescence** or phosphorescence or both. If the ground state is a singlet state, and emission occurs from an excited singlet state, the emission is referred to as fluorescence. But if the emission occurs from an excited electronic state which undergoes a change of spin to a triplet state, the process is called phosphorescence. Singlet to triplet transitions are forbidden under quantum mechanical selection rules and electrons achieve this triplet state via intersystem crossing. This is why phosphorescence occurs on a much slower timescale than fluorescence, ranging between 10^{-4} seconds to hours.

The following Jablonski diagram illustrates the processes involved in phosphorescence.

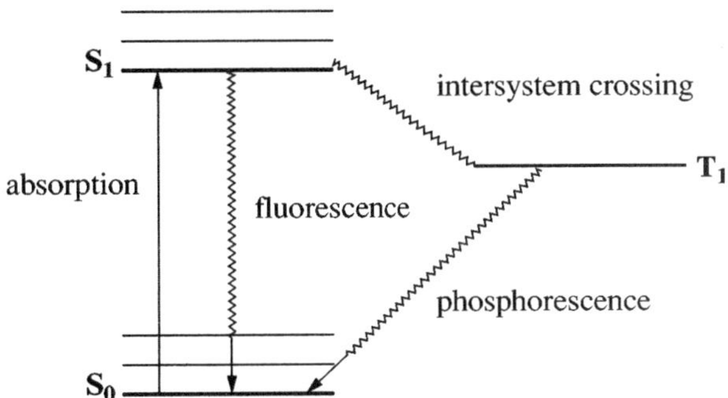

Many aquatic organisms exhibit marine phosphorescence, from bacteria to phytoplankton to jellyfish and deep-sea organisms. Most emit a blue-green phosphorescence in a region of the electromagnetic spectrum which is highly visible in deep and dark waters. Marine phosphorescence is an almost ubiquitous phenomenon in the oceans, it helps marine organisms survive by providing them with a means to search for nutrients and defend themselves against predators. The study of marine luminescent organisms is a very active field of research, providing valuable insights on marine ecology, since it can be used as a tool to determine how these organisms live and how they are affected by factors such as light, pollution and salinity of the ocean.

In contrast, very few land species are phosphorescent, a well-known example is the common firefly, *Photinus pyralis*, which emits a yellow-green light (at a **wavelength** of 562 nm) that scientists assert provides the insect with a means of communication.

See also Absorption spectrum; Luminescence; Molecular orbital theory; Quanta; Quantum numbers; Spectroscopy

PHOTOELECTRIC EFFECT

The process in which visible **light**, **x rays** or gamma rays incident on **matter**, cause an **electron** to be ejected. The ejected electron is called a photoelectron.

The photoelectric effect was discovered by Heinrich Hertz in 1897 while performing experiments that led to the discovery of electromagnetic **waves**. Since this was just about the time that the electron itself was first identified the phenomenon was not really understood. It soon became clear in the next few years that the particles emitted in the photoelectric effect were indeed electrons. The number of electrons emitted depended on the intensity of the light but the **energy** of the photoelectrons did not. No matter how weak the light source was made the maximum **kinetic energy** of these electrons stayed the same. The energy however was found to be directly proportional to the **frequency** of the light. The other perplexing fact was that the photoelectrons seemed to be emitted instantaneously when the light was turned on. These facts were impossible to explain with the then current wave theory of light. If the light were bright enough it seemed reasonable, given enough time, that an electron in an **atom** might acquire enough energy to escape regardless of the frequency. The answer was finally provided in 1905 by **Albert Einstein** who suggested that light, at least sometimes, should be considered to be composed of small bundles of energy or particles called photons. This approach had been used a few years earlier by **Max Planck** in his successful explanation of black body **radiation**. In 1907 Einstein was awarded the Nobel Prize in physics for his explanation of the photoelectric effect.

Einstein's explanation of the photoelectric effect was very simple. He assumed that the kinetic energy of the ejected electron was equal to the energy of the incident **photon** minus the energy required to remove the electron from the material, which is called the **work** function. Thus the photon hits a surface, gives nearly all its energy to an electron and the electron is ejected with that energy less whatever energy is required to get it out of the atom and away from the surface. The energy of a photon is given by $E = hg = hc/l$ where g is the frequency of the photon, l is the **wavelength**, and c is the **velocity** of light. This applies not only to light but also to x rays and gamma rays. Thus the shorter the wavelength the more energetic the photon.

Many of the properties of light such as **interference** and **diffraction** can be explained most naturally by a wave theory while others, like the photoelectric effect, can only be explained by a particle theory. This peculiar fact is often referred to as wave-particle duality and can only be understood using **quantum theory** which must be used to explain what happens on an atomic scale and which provides a unified description of both processes.

The photoelectric effect has many practical applications which include the photocell, photoconductive devices and solar cells. A photocell is usually a **vacuum** tube with two electrodes. One is a photosensitive cathode which emits electrons when exposed to light and the other is an **anode** which is maintained at a positive **voltage** with respect to the cathode. Thus when light shines on the cathode, electrons are attracted to the anode and an electron current flows in the tube from cathode to anode. The current can be used to operate a relay which might turn a motor on to open a door or ring a bell in an alarm system. The system can be made to be responsive to light, as described above, or sensitive to the removal of light as when a beam of light incident on the cathode is interrupted, causing the current to stop. Photocells are also useful as exposure meters for cameras in which case the current in the tube would be measured directly on a sensitive meter.

Closely related to the photoelectric effect is the photoconductive effect which is the increase in electrical conductivity of certain non metallic materials such as cadmium sulfide when exposed to light. This effect can be quite large so that a very small current in a device suddenly becomes quite large when exposed to light. Thus photoconductive devices have many of the same uses as photocells.

Solar cells, usually made from specially prepared silicon, act like a battery when exposed to light. Individual solar cells produce voltages of about 0.6 volts but higher voltages and large currents can be obtained by appropriately connecting many solar cells together. **Electricity** from solar cells is still quite expensive but they are very useful for providing small amounts of electricity in remote locations where other sources are not available. It is likely however that as the cost of producing solar cells is reduced they will begin to be used to produce large amounts of electricity for commercial use.

PHOTOGRAPHY

Photography is the recording of visible images by focusing **light** on light-sensitive materials. English astronomer Sir John Herschel (1792-1871) first used the term photography in 1839. The word comes from the Greek *photos*, meaning light, and *graphos*, meaning writing.

There are many forms of photography. Still photography captures individual self-contained images on single sheets of film. Portraits are an example of still photography. **Motion** photography, also called cinematography, records a series of images illustrating a subject in motion. These photographic images are focused onto a screen using a movie projector. Television, video, even medical **x rays** are based on the principles of photography.

The predecessor to the modern handheld camera was the camera obscura. This giant, cumbersome device consisted of a dark box or room with a tiny hole in one wall. When peered through the tiny opening, an inverted image could be seen on the opposite wall. Leonardo da Vinci described this early form of photography in a manuscript dated 1519. Today such cameras are used by astronomers and physicists to document high **energy** reactions.

Two distinct optical and chemical scientific processes combined to make photography possible. Although scientists knew as early as the 1600s that the chemicals now used for photography darkened when exposed, they believed the reaction was caused by air or **heat**, not by light. Inventors did not discover the importance of light in photography until 200 years later.

The first person to create a photograph in the camera obscura was Joseph-Nicéphore Niepce (1765-1833), a French lithographer. In the summer of 1827, Niepce focused light on an asphalt solution for eight hours to create his picture.

In January 1829, Niepce partnered with the man who eventually was regarded as the inventor of modern photography, Louis-Jacques-Mandé Daguerre (1789-1851). Although Niepce died four years later, Daguerre continued to experiment. Daguerre used a camera consisting of two boxes sliding inside one another. At 50 lb (18 kg), the device was cumbersome, but smaller models soon appeared.

After nearly a decade of trial and error, Daguerre discovered a way of developing photographic plates that significantly reduced Niepce's exposure time from eight hours to just 30 minutes. He also documented that an image could be made permanent or fixed by immersing it in a salt solution. He called his image the Daguerreotype. The Daguerreotype became the first

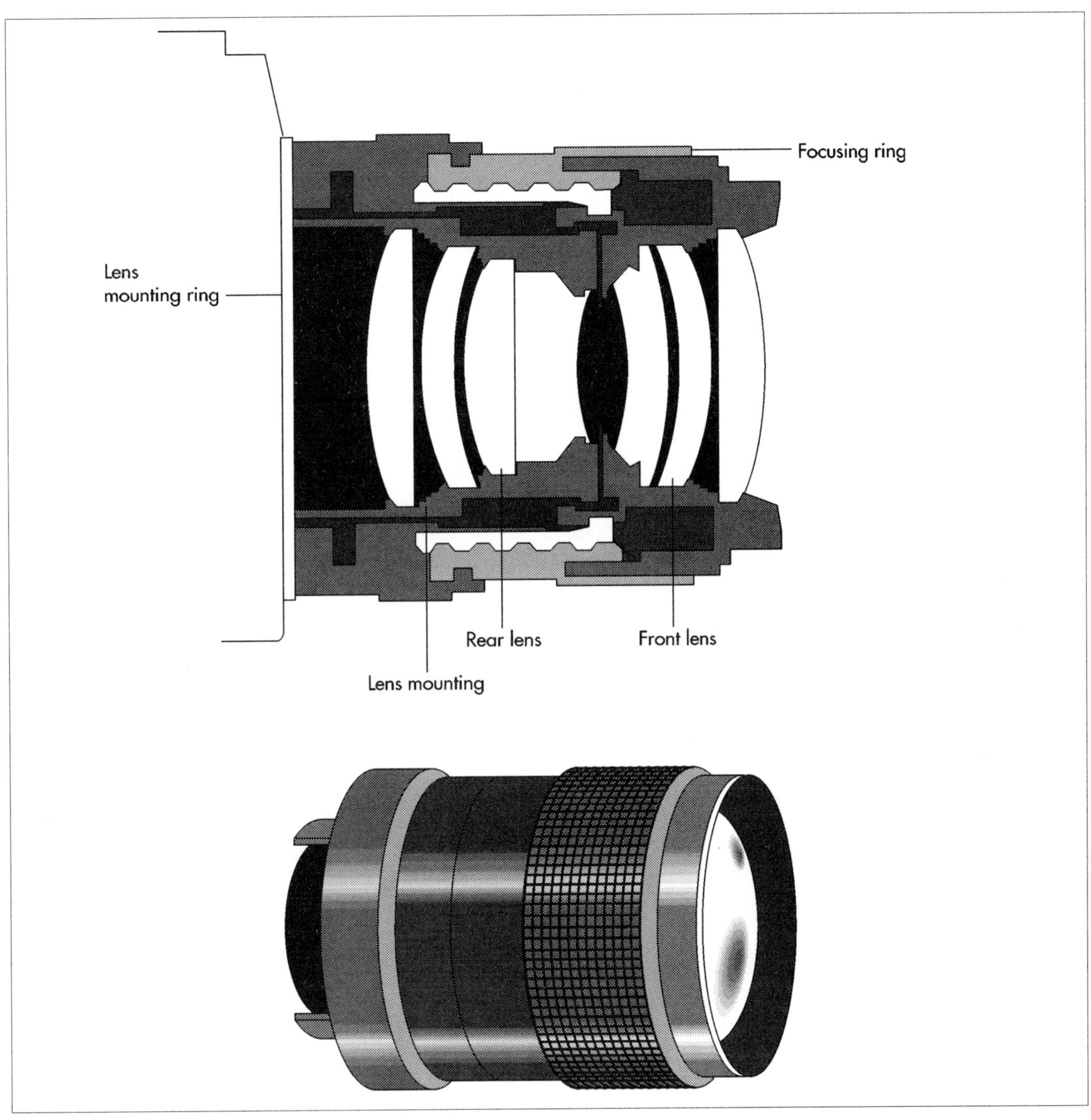

Camera lens.

commercially utilized photographic process. Yet, the daguerreo-type eventually proved unpopular because of its unique positive images, and the fact that it was expensive and could not be mass-produced. By 1851, this novel new art form was abandoned.

The negative-positive photographic process in use today was the idea of William Henry Fox Talbot (1800-1877), a British amateur scientist. His calotype (later called the talbo-type) used paper negatives to yield positive pictures. Inevitably, however, imperfections embedded in the paper were printed along with the image. Compared with Daguerreotypes, the quality of Talbot's images were rough. However, the fact that an unlimited number of positive prints could be created made it an attractive technique. To perfect Talbot's process, many tried to substitute glass for paper. However, the silver solution needed for exposure would not stick to the shiny glass surface.

A solution was offered in 1848, when a nephew of Niepce, Calude-Félix-Abel Niepce de Saint-Victor (1805-1870), coated the glass plate with an egg white solution. The new process, called albumen photography, allowed for very fine detail and much higher photographic quality. Its slow speed, however, made the technique impossible for portraiture use. Saint-Victor also brought the world the first **color** images. Saint-Victor and French physicist Alexandre-Edmond Becquerel produced the colors using treated silver chloride coatings. The team hit a roadblock when they failed to fix the final photograph. Scientists would try again several decades later.

In 1851, English sculptor Frederick Scott Archer (1813-1857) launched a new era in photography. His development of the wet plate allowed for much faster processing speeds. This process used a glass plated coated with a nitrocellulose (collodion) solution containing potassium iodide. The plate was placed in a silver nitrate solution, then exposed inside the camera while still wet. The substance was the first light-sensitive material to allow instantaneous exposure in adequate light.

Archer's wet-collodion plate quickly became the principal photographic process. The first news and feature photographers used the new technique to document the Civil and Crimean wars. But the material's tendency to rapidly evaporate if not immediately developed meant photographers traveled with cumbersome portable darkrooms. Graphic artists, on the other hand, embraced the high-quality, fine-grain negatives and kept the process in use until the middle of the twentieth century. At the same time of Archer's important development, Talbot demonstrated the first high-speed flash photograph using a spark from a battery.

In 1859, Britain's **James Clerk Maxwell** discovered that colors from an image could be reproduced by combining red, green, and blue light in various proportions. Nearly a decade later, Frenchmen L. Ducos du Hauron and C. Clos suggested a new way to form color pictures. The pair's idea to superimpose dye images (subtractive color synthesis) as a means of forming color images provided the basis of assembly colorprint processes from the 1890s onward.

The next major step forward in photographic processing came in 1871, when gelatin was used instead of glass as a basis for the photographic plate. The invention meant photographers were no longer required to coat and sensitize their own processing materials. The new dry plate technique produced images much faster than with any previous technique. The simple fact that dry-plates could be made and stored for months revolutionized photographic development. Readymade emulsions were on sale by 1873.

In 1888, George Eastman (1854-1932) introduced the first box camera, complete with a roll of negative paper, and processing and printing service. His slogan, "You push the button, we do the rest," helped popularize photography. Several other camera designs quickly followed. In 1924 the first commercially successfully miniature camera, the Leica, hit the market. The 24 x 36 mm camera was extremely portable and easy to use, and eventually became the photojournalist's standard.

In the twentieth century, photography emerged into a **mass** hobby. No longer was photography considered the peculiar cousin of painting; instead it was regarded as a unique and evolving art form. By the mid 1900s, consumers could see the photographs they snapped in just seconds using Edwin Herbert Land's instant film and Polaroid camera. In 1950, Eastman Kodak's Kodachrome film was widely marketed for all photographers. The Polaroid Land Company followed suit with the introduction of instant color film in 1963.

As technology advanced, so has the scope of photography. In the 1980s, Sony introduced the first personal camcorder, bringing motion photography into the hands of amateurs. Shortly after, scientists with Kodak created the world's first digital sensor capable of recording more than a million image elements. The feat resulted in the production of 5 x 7 in (12.7 x 17.8 cm) digital prints that looked like photographs. In the 1990s, digital photography escalated as Nikon, Canon, Leaf Systems, Kodak, and others announce new digital cameras for amateurs and professionals alike.

By the beginning of the twenty-first century, photography is considered more than an art form; it is an invaluable means of communication and history preservation. The average person in the United States typically encounters more than 1,000 camera images a day.

PHOTON

One of the fundamental dichotomies in classical physics was that between **energy** and **matter**. By the late nineteenth century, most scientists agreed that matter is composed of tiny, discrete particles with measurable **mass**. At the time, the **atom** was considered to be the ultimate particle of which all matter consists. Energy, on the other hand, was thought to have no material basis, but corresponded to anything that was able to cause a change of position in some form of matter. Most forms of energy were thought to travel through **space** as **waves**. The duality between particles and waves was, therefore, a fundamental tenet of physical theory. By the turn of the century, however, the wave-particle duality began to come apart fairly rapidly. A critical factor in this change was Albert Einstein's analysis of the **photoelectric effect**. Einstein showed that the emission of electrons from a metal that has been exposed to **light** can best be explained by assuming that light consists of tiny "packages" of energy. The energy of each package and its relationship with the **wavelength** of the light is translated in the equation $E = c \div \lambda$, where is **Planck's constant** (6.62617 x 10^{-34} J-sec), is the **speed of light** (2.99792 X 10^8 m sec^{-1}), and λ is the wavelength in meters. The energy is directly proportional to the **frequency** of the photon that is equal to c/λ and expressed in sec^{-1}. The concept of an energy "package" was not originally conceived by Einstein. In 1903, Joseph J. Thomson had found it useful to talk about energy "specks" in describing his **work** on ionization of gases. There is no question, however, that the clearest expression of this concept came from Einstein's work. In the 1920s, Arthur Holly Compton proposed the name *photon* (from the Greek *photos*, for "light") for this "package" of energy. Compton's own work involved the **scattering** of **x rays** by atoms in a crystalline lattice. Compton adopted Einstein's approach to explain the fact that x rays reflected from the crystals have longer wavelengths than the incident x rays. By assigning x rays the property of **momentum**, a property traditionally reserved for particles, Compton was able to calculate exactly the observed properties of the scattered x rays. He later wrote that his experiments probably "were the first to give, at least to physicists in the United States, a conviction of the fundamental validity of the quantum theory." The photon concept is important historically because it marks the first time that language normally used for particles was applied to energy phenomena. Consequently, scientist began to study both matter and energy as two manifestations of a common, or at least related, phenomenon. For example, **Werner Heisenberg** adopted the photon concept in his development of **quantum theory**. In fact, the photon is usually described today as the quantum of light, a particle with zero charge and zero mass that mediates the transfer of the electromagnetic **force**. In modern **particle physics** theory, the photon is described as a boson, a particle with integral spin that acts as carrier of the electromagnetic force.

PHYSICS AND ART

Physics and the arts have always had a complex, interactive relationship. Classical Greek art and philosophy (the forerunner of modern science) shared the same goal: the description of the natural world, the first in concrete terms (through sculpture, painting, or architecture) and the second in contemplative terms (through mathematics and other forms of thought). During the Classical period (c. 475-338 B.C.), sculptors like Phidias and Polyclitus, the architects Mnesicles, Ictinus and Callicrates, and the painters Apollodorus and Zeuxis created works that depicted the natural world as close as possible to visual reality. Zeuxis, for instance, was reported to have painted grapes that appeared so natural that birds tried to eat them. The Athenian Erechtheum (constructed 421-405 B.C.), a temple on the Acropolis, mixes the aesthetics of sculpture with the physics of architecture on its south porch, known as the Porch of the Maidens. Instead of columns, the roof of the Erechtheum was supported by huge female statues, called caryatids. Greek philosophers also explored some of the same ideas as do modern physicists, ranging from the depiction of reality through mathematics pioneered by Pythagoras to the explorations of the nature of the **universe** by Thales and the members of the Ionian school.

Successors to the classical Greeks also investigated the same questions as do modern physicists through art. In his journals, Leonardo da Vinci mixed artistically gifted sketches with experimental notes. In the twentieth century artists and physicists have both been obsessed with the same themes. Salvador Dali did paintings that included hypercubes and warped **space-time**; Rene Magritte included many trains, rulers, clocks, and other favorites of Einsteinian metaphor in his surrealist work.

Some believe that the beginning of the information age will bring about greater cross-fertilization between the fields of physics and art. Others believe that an increase in human knowledge will increase the requirement for specialization, so that physics and art will drift further apart in the years to come.

PHYSICS OF FLIGHT

Any object moving through **space**, above the ground and within our atmosphere can be thought of as being in flight. Birds, planes, helicopters, Frisbees, and baseballs, if moving through the air, are all subject to the physics of flight. The physics of flight, or aeronautics, is similar to other branches of physics in that it is all about the balancing of forces. A body in straight and level flight can be said to be in **equilibrium**.

Four forces act upon a body in flight; two forces aid flight and two oppose it. Forward **momentum** is provided by thrust. Upward movement, away from the ground, is provided by lift. Air resistance is called drag, and the tendency of an object to fall toward Earth is **gravity**.

Thrust is the motive **power** that propels a flying object forward. An airplane's engine, a baseball pitcher's arm, a bird's thoracic muscles; all these are providers of thrust. Thrust can be divided into two categories. Ballistic thrust is the expenditure of **energy** early in the flight, with inertial momentum carrying the object the rest of the way. Bullets, **rockets**, and the thrown baseball all fly ballistically. The other type of thrust is generated inflight, and continuously. Birds generate their thrust with every flap of their wings and airplanes generate it with either the turning of propellers or the combustion of jet fuel.

Drag is the **force** that opposes thrust. Anyone who has ever stuck an arm out of a car window has experienced drag. In its simplest form, drag is **friction**. Drag can be reduced greatly with streamlining, but it can never be eliminated entirely. At very high speeds, drag presents itself as an actual compression of the air layer along the leading edge of the flying body. This form of drag is responsible for, among other things, shock **waves** and sonic booms.

Gravity is probably the most familiar force. **Newtonian physics** tells us that gravity is the attraction between two bodies, proportional to their **mass**. Observation tells us that gravity seeks to pull flying bodies out of the air. Gravity is the aeronautic force that opposes lift.

The most elusive of the aeronautic forces is lift. Since lift opposes gravity, it follows that lift can be quantified based on the mass of the object in flight. The minimum amount of lift needed to get an object off the ground is equal to the object's own **weight**. This concept is easy enough to apply to lighter-than-air flight, but becomes somewhat counterintuitive when one considers the airfoil, or wing.

It is generally agreed that what supplies lift beneath an airfoil is an effect of the Bernoulli principle. **Daniel Bernoulli** discovered that the **pressure** of **fluids**, which in this case includes air, is proportional to **velocity**. Slower moving fluids have higher pressures than do faster moving fluids. Thus, if air passing under a wing is moving slower than the air passing over it, lift is created and the airplane flies. The question is, why is there a differential in velocities of the air over and under the wing?

The most commonly held theory regarding lift involves the shape of the wing. Most aircraft have wings that are thicker toward the leading edge of the wing, and are slightly curved along the top. This shape causes the airflow along the top of the wing to have a further distance to travel than air along the bottom. The air along the bottom, since it has a shorter distance to travel, is moving more slowly, and thus has greater pressure. It is this rise in air pressure below the wings, goes the theory, that produces lift.

One of the problems with this theory is that it does not explain how paper airplanes, balsa wood models, or other flying bodies with flat wings produce lift. In the flat-wing model, the relative velocities of air above and below the wing are presumed to be equal, and their respective pressures are likewise equal. One theory seeks to explain this by suggesting that "trailing vortices" formed at the wing tips cause a pressure differential, resulting in lift. Another supposes that lift is nothing more that the equal-but-opposite reaction to downwash created at the wings' trailing edges.

It is interesting to note that, nearly 100 years after the first manned heavier-than-air flight, physicists are still not in agreement as to what makes it possible. One explanation may be that lift is provided by a combination of the three recognized, and perhaps even other unrecognized phenomena. Regardless, it is clear that powered flight is here to stay, is governed by the laws of physics (whether we fully understand them or not), and needs further investigation.

PIONS

A pion (or pi-meson) is one of the particles that carry the strong nuclear **force** between protons and neutrons (and other hadrons), which binds an atomic **nucleus** together. Pions are denoted π^+, π^-, π^0, depending on charge.

Pions were predicted to exist in 1935 by the Japanese theoretical physicist **Hideki Yukawa** when, in the first paper he ever published, Hideki proposed there was a new force involved in holding the nucleus together. This new force was overcoming the positive charge on each **proton** that was trying to blow the nucleus apart. Eventually this force came to be known as the strong nuclear force. Yukawa predicted that this new force was carried by a massive particle with only a very short range (about 10 cm). Hideki predicted the **mass** of this new particle would be about 300 times that of the **electron**, or about 150 MeV. To account for all the possible interactions between nucleons (protons and neutrons), the particle would have to exist in three charge states: positive, neutral, and negative.

The charged pions should have been observable in cloud chambers. In 1936, American physicist **Carl Anderson**, discoverer of the **positron**, proving the existence of **antimatter**) utilized the **cloud chamber** to discover a new charged particle with mass between that of the electron and the proton.

Anderson called this particle the mesotron, from the Greek, meaning middle. Yukawa claimed this particle as his strong force carrier. As it turned out, the particle was actually a **muon**.

Pions are actually the products of primary **cosmic rays** colliding with Earth's atmosphere where vast numbers of positive, negative, and neutral pions are produced. The charged pions decay within 10 seconds to positive and negative muons, which then decay into positrons or electrons. English physicist **Cecil Powell** and his group finally discovered the charged pions on a mountaintop in 1947. By gaining altitude they had reached a level where the pion had not yet decayed.

Pions create a web of force which holds the nucleons together. They are not classified with the W and Z particles or the **photon**, however, because they are not elementary particles. Pions are made up of **quarks** and transmit the strong force at the larger nuclear level, whereas at the deeper level of quarks the strong force is transmitted by gluons.

In 1949, the neutral pion was the first subatomic particle to be discovered with the aid of an accelerator. A neutral pion decays into two energetic photons gamma rays, each of which then decays into a positron and an electron. The neutral pion and gamma rays, however, left no tracks in the cloud chamber. Using new synchrocyclotron and sophisticated electronic methods, the existence of the neutral pion was later confirmed, and the pion family was complete.

According to contemporary models the pion has three varieties. The pi-zero (zero **electric charge**) has a mass of 135 MeV, and the pi-plus and pi-minus (with one unit of positive and negative charge respectively) have a mass of 140 MeV. The extra mass and **angular momentum** of the pi-plus and pi-minus is accounted for by the giving off of neutrinos during their decay.

Pions are bosons with zero spin. The pi-zero is made up of a quark doublet of an up/antiup quark pair, or a down/antidown quark pair (it can be made up of either combination). The pi-minus is a down/antiup pair, and the pi-plus is an up/antidown quark pair.

See also Particle physics; Strong interactions

PLANCK, MAX (1858-1947)

German physicist

Max Planck is best known as one of the founders of the **quantum theory** of physics. As a result of his research on **heat radiation** he was led to conclude that **energy** can sometimes be described as consisting of discrete units, later given the name *quanta*. This discovery was important because it made possible, for the first time, the use of matter-related concepts in an analysis of phenomena involving energy. Planck also made important contributions in the fields of **thermodynamics**, relativity, and the philosophy of science. He was awarded the 1918 Nobel Prize in physics for his discovery of the quantum effect.

Max Karl Ernst Ludwig Planck was born on April 23, 1858, in Kiel, Germany. His parents were Johann Julius Wilhelm von Planck, originally of Göttingen, and Emma

Patzig, of Griefswald. Johann had previously been married to Mathilde Voigt, of Jena, with whom he had two children. Max was the fourth child of his father's marriage to Emma.

Johann von Planck was descended from a long line of lawyers, clergyman, and public servants and was himself Professor of Civil Law at the University of Kiel. Young Max began school in Kiel, but moved at the age of nine with his family to Münich. There he attended the Königliche Maximillian Gymnasium until his graduation in 1874.

As a child, Planck demonstrated both talent in and enthusiasm for a variety of fields, ranging from mathematics and science to music. He was accomplished at both piano and organ and gave some thought to a career in music. He apparently abandoned that idea, however, when a professional musician told him that he did not seem to have the commitment needed for that field. Planck did, however, maintain a life-long interest in music and its mathematical foundations. Later in life, he held private concerts in his home which featured eminent musicians, such as Joseph Joachim and Maria Scherer, as well as fellow scientists, including **Albert Einstein**, often with Planck himself at the piano.

Planck entered the University of Münich in 1874 with plans to major in mathematics. He soon changed his mind, however, when he realized that he was more interested in practical problems of the natural world than in the abstract concepts of pure mathematics. Although his course work at Münich emphasized the practical and experimental aspects of physics, Planck eventually found himself drawn to the investigation of theoretical problems. It was, biographer Hans Kango points out in *Dictionary of Scientific Biography,* "the only time in [his] life when he carried out experiments."

Planck's tenure at Münich was interrupted by illness in 1875. After a long period of recovery, he transferred to the University of Berlin for two semesters in 1877 and 1878. At Berlin, he studied under a number of notable physicists, including Hermann Helmholtz and Gustav Kirchhoff. By the fall of 1878, Planck was healthy enough to return to Münich and his studies. In October of that year, he passed the state examination for higher level teaching in math and physics. He taught briefly at his alma mater, the Maximillian Gymnasium, before devoting his efforts full time to preparing for his doctoral dissertation. He presented that dissertation on the **second law of thermodynamics** in early 1879 and was granted a Ph.D. by the University of Münich in July of that year.

Planck's earliest field of research involved thermodynamics, an area of physics dealing with heat energy. He was very much influenced by the work of Rudolf Clausius, whose work he studied by himself while in Berlin. He discussed and analyzed some of Clausius's concepts in his own doctoral dissertation. Between 1880 and 1892, Planck carried out a systematic study of thermodynamic principles, especially as they related to chemical phenomena such as osmotic **pressure, boiling** and freezing points of solutions, and the dissociation of gases. He brought together the papers published during this period in his first major book, *Vorlesungen über Thermodynamik,* published in 1897.

Max Planck.

During the early part of this period, Planck held the position of Privat-Dozent at the University of Münich. In 1885, he received his first university appointment as extraordinary professor at the University of Kiel. His annual salary of 2,000 marks was enough to allow him to live comfortably and to marry a childhood sweetheart from Münich, Marie Merck. Marie was eventually to bear Planck three children.

Planck's personal life was beset with tragedy. Both of his twin daughters died while giving birth: Margarete in 1917, and Emma in 1919. His son, Karl, also met an untimely death when he was killed during World War I. Marie had predeceased all her children when she died on October 17, 1909. Planck later married Marga von Hoessli, with whom he had one son.

Planck's research on thermodynamics at Kiel soon earned him recognition within the scientific field. Thus, when Kirchhoff died in 1887, Planck was considered a worthy successor to his former teacher at the University of Berlin. Planck was appointed to the position of assistant professor at Berlin in 1888 and assumed his new post the following spring. In addition to his regular appointment at the university, Planck was also chosen to head the Institute for Theoretical Physics, a facility that had been created especially for him. In 1892, Planck was promoted to the highest professorial rank, ordinary professor, a post he held until 1926.

Once installed at Berlin, Planck turned his attention to an issue that had long interested his predecessor, the problem of black body radiation. A black body is defined as any object that absorbs all frequencies of radiation when heated and then gives off all frequencies as it cools. For more than a decade, physicists had been trying to find a mathematical law that would describe the way in which a black body radiates heat.

The problem was unusually challenging because black bodies do not give off heat in the way that scientists had predicted that they would. Among the many theories that had been proposed to explain this inconsistency was one by the German physicist **Wilhelm Wien** and one by the English physicist John Rayleigh. Wien's explanation worked reasonably well for high **frequency** black body radiation, and Rayleigh's appeared to be satisfactory for low frequency radiation. But no one theory was able to describe black body radiation across the whole spectrum of frequencies. Planck began working on the problem of black body radiation in 1896 and by 1900 had found a solution to the problem. That solution depended on a revolutionary assumption, namely that the energy radiated by a black body is carried away in discrete "packages" that were later given the name *quanta* (from the Latin, *quantum,* for "how much"). The concept was revolutionary because physicists had long believed that energy is always transmitted in some continuous form, such as a wave. The wave, like a line in geometry, was thought to be infinitely divisible.

Planck's suggestion was that the heat energy radiated by a black body be thought of as a stream of "energy bundles," the magnitude of which is a function of the **wavelength** of the radiation. His mathematical expression of that concept is relatively simple: $E = \hbar\upsilon$, where E is the energy of the quantum, υ is the wavelength of the radiation, and $\hbar$ is a constant of proportionality, now known as *Planck's constant*. Planck found that by making this assumption about the nature of radiated energy, he could accurately describe the experimentally observed relationship between wavelength and energy radiated from a black body. The problem had been solved.

The numerical value of **Planck's constant**, hbar 6,50 ·, can be expressed as 6.62×10^{-27} erg second, an expression that is engraved on Planck's headstone in his final resting place at the Stadtfriedhof Cemetery in Göttingen. Today, Planck's constant is considered to be a **fundamental constant** of nature, much like the speed of **light** and the gravitational constant. Although Planck was himself a modest man, he recognized the significance of his discovery. Robert L. Weber in *Pioneers of Science: Nobel Prize Winners in Physics* writes that Planck remarked to his son Erwin during a walk shortly after the discovery of the quantum concept, "Today I have made a discovery which is as important as Newton's discovery." That boast has surely been confirmed. The science of physics today can be subdivided into two great eras, classical physics, involving concepts worked out before Planck's discovery of the quantum, and modern physics, ideas that have been developed since 1900, often as a result of that discovery. In recognition of this accomplishment, Planck was awarded the 1918 Nobel Prize physics.

After completing his study of black body radiation, Planck turned his attention to another new and important field of physics: relativity. Albert Einstein's famous paper on the theory of general relativity, published in 1905, stimulated Planck to look for ways on incorporating his quantum concept into the new concepts proposed by Einstein. He was somewhat successful, especially in extending Einstein's arguments from the field of **electromagnetism** to that of **mechanics**. Planck's work in this respect is somewhat ironic in that it had been Einstein who, in another 1905 paper, had made the first productive use of the quantum concept in his solution of the photoelectric problem.

Throughout his life, Planck was interested in general philosophical issues that extended beyond specific research questions. As early as 1891, he had written about the importance of finding large, general themes in physics that could be used to integrate specific phenomena. His book *Philosophy of Physics,* published in 1959, addressed some of these issues. He also looked beyond science itself to ask how his own discipline might relate to philosophy, religion, and society as a whole. Some of his thoughts on the correlation of science, art, and religion are presented in his 1935 book, *Die Physik im Kampf um die Weltanschauung.*

Planck remained a devout Christian throughout his life, often attempting to integrate his scientific and religious views. Like Einstein, he was never able to accept some of the fundamental concepts of the modern physics that he had helped to create. For example, he clung to the notion of causality in physical phenomena, rejecting the principles of uncertainty proposed by Heisenberg and others. He maintained his belief in God, although his descriptions of the Deity were not anthropomorphic but more akin to natural law itself.

By the time Planck retired from his position at Berlin in 1926, he had become probably the most highly respected scientific figure in Europe, if not the world, except for Einstein. Four years after retirement, he was invited to become president of the Kaiser Wilhelm Society in Berlin, an institution that was then renamed the Max Planck Society in his honor. Planck's own prestige allowed him to speak out against the rise of Nazism in Germany in the 1930s, but his enemies eventually managed to have him removed from his position at the Max Planck Society in 1937. The last years of his life were filled with additional personal tragedies. His son by his second marriage, Erwin, was found guilty of plotting against Hitler and executed in 1944. During an air raid on Berlin in 1945, Planck's home was destroyed with all of his books and papers. During the last two and a half years of his life, Planck lived with his grandniece in Göttingen, where he died on October 4, 1947.

PLANCK'S CONSTANT

Light emitted by a heated solid object is a well-known phenomenon: even before he age of modern science, alchemists noted the correlation between an object's **temperature** and the kind of light it emitted. For example, if a metal object was being heated, a red glow gradually turned to white as the temperature increased.

Max Planck was the first to notice, in 1900, a characteristic regularity in the behavior of heated objects. Assuming that the atoms of a particular heated object will vibrate at a particular **frequency** v, Planck noticed that only certain **energy** levels were possible. These energy levels could only be captured if the value of v was multiplied by a whole number (n and a constant ($\hbar$) whose value was 6.63 10 ^{34} J. The upshot of Planck's discovery was that atoms, unlike objects in the macroscopic world, obeyed energy constraints that could only be explained in terms of a physical constant multiplied by a whole number. As Darrell D. Ebbing and Steven D. Gammon explained, under similar constraints, a car would, for example, increase its speed only by increments of five miles per hour: from zero to five, from five to ten, and so on. In other words, a speed of, say, seven miles per hour would be impossible. The quantum rule may seem unreasonable in our world; in the atomic world, however, energy levels are determined by Planck's constant.

While Planck himself was somewhat uncomfortable with his discovery, younger scientists quickly grasped the universality of Planck's constant. Thus, **Albert Einstein** (1879-1955) posited that Planck's formula $E = \hbar v$ could be applied to light. According to Einstein, Planck's formula accurately expresses the wave-particle duality of light: E represents the energy of a light particle (**photon**), and v represents the frequency of light as a wave. By introducing the concept of light particle, or photon, in 1905, Einstein explained the **photoelectric effect**, or the process whereby an object emits electrons when exposed to light. Einstein theorized that electrons are ejected by photons. According to Einstein, the photoelectric effect clearly manifests the dual nature of light: light approaches an object as a particle (photon) but is absorbed as a wave by the **electron** that is about to be ejected. Following Einstein's insight, scientists later realized that the wave-particle duality applied to **subatomic particles** in general.

PLANETS

Planets are major bodies which orbit **stars**. They are not massive enough to ignite **nuclear reactions** at their cores like stars, yet they are massive enough in some cases to support gaseous atmospheres and complex **chemistry**. Since antiquity, five planets (from the Greek word for "wanderer") were known to move across the background of distant stars: Mercury, Venus, Mars, Jupiter, and Saturn. Early astronomers believed that the **universe** revolved around the Earth, and found it particularly difficult to describe the motions of the planets. Mars and Jupiter exhibited a puzzling behavior known as "retrograde motion" against the dome of the sky in which, on occasion, they seemed to reverse their direction over several days before returning to their original direction of **motion**. The ancient Greek astronomer **Ptolemy** was forced to invent a small circular motion, known as an

"epicycle," to account for the variation of the mean motion of the planet on a circular orbit around Earth. It took the work of **Nicholas Copernicus**, **Galileo**, **Tycho Brahe**, **Johannes Kepler**, and **Isaac Newton** to offer a simpler, more physically motivated model of the solar system with all bodies, including Earth, orbiting the **Sun**.

Three more planets were discovered in our own solar system in modern times. In 1781, the British astronomer Sir **William Herschel** discovered Uranus. In 1846, the French scientist Urbain Leverrier (1811-1877) and the British scientist John Couch Adams (1819-1892) predicted the existence and location of Neptune—the first time such a prediction had ever been made. Lastly, in 1930, the American Clyde Tombaugh (1906-1997) discovered Pluto. While no other planets have been found in our solar system, despite numerous searches, a large number of minor bodies, including **comets** and asteroids, have been discovered over the years.

Most of what is known about the planets in the solar system comes from a variety of unmanned **space** probes launched since the 1960s. A number of American and Russian probes were sent out to our nearest neighbors, the inner rocky planets of Mercury, Venus, and Mars. Notably, the *Viking* missions to Mars in the 1970s could not detect signs of life on the planet today. The outer planets were surveyed with a grand tour of the solar system conducted by *Voyager I* and *Voyager II*, which used gravity-assist trajectories to visit multiple planets. *Voyager I* visited both Jupiter and Saturn, and *Voyager II* visited Jupiter, Saturn, Uranus, and **Neptune**, sending back images and measurements of unprecedented quality from each of these planets.

Our knowledge of planetary bodies had been extremely biased until just a few years ago. All of the known planets in our solar system had formed from the same primordial conditions around the same star. Astronomers had constructed formation hypotheses to explain the origin of the solar system. The prevailing notion was that planets formed from a flattened, spinning disk of dust and gas that formed at the same time as our Sun. The inner planets formed by coagulation of rocky planetismals, and the outer planets, lying in the colder outer portions of the disk, accreted a substantial amount of their **mass** in the form of gas from the surrounding primordial disk.

The scientific world was stunned in 1995 when Michel Mayor and Didier Queloz announced the discovery of the first *extrasolar* planet orbiting a nearby sun-like star, 51 Pegasi, which lies about 40 **light** years from Earth. Since then, a total of 34 extrasolar planets have been discovered, with masses ranging from about one-fifth to ten times that of Jupiter. In a most remarkable detection, Geoff Marcy and Paul Butler announced an eclipsing extrasolar planet orbiting the star HD 209458. From this detection, scientists now know both the mass and radius of the object, which has a mass intermediate between Saturn and Jupiter, and a radius about 60% greater than that of Jupiter. Scientists continue to debate the explanation for these new observations, which have re-invigorated and excited the astronomical community. The outstanding chal-

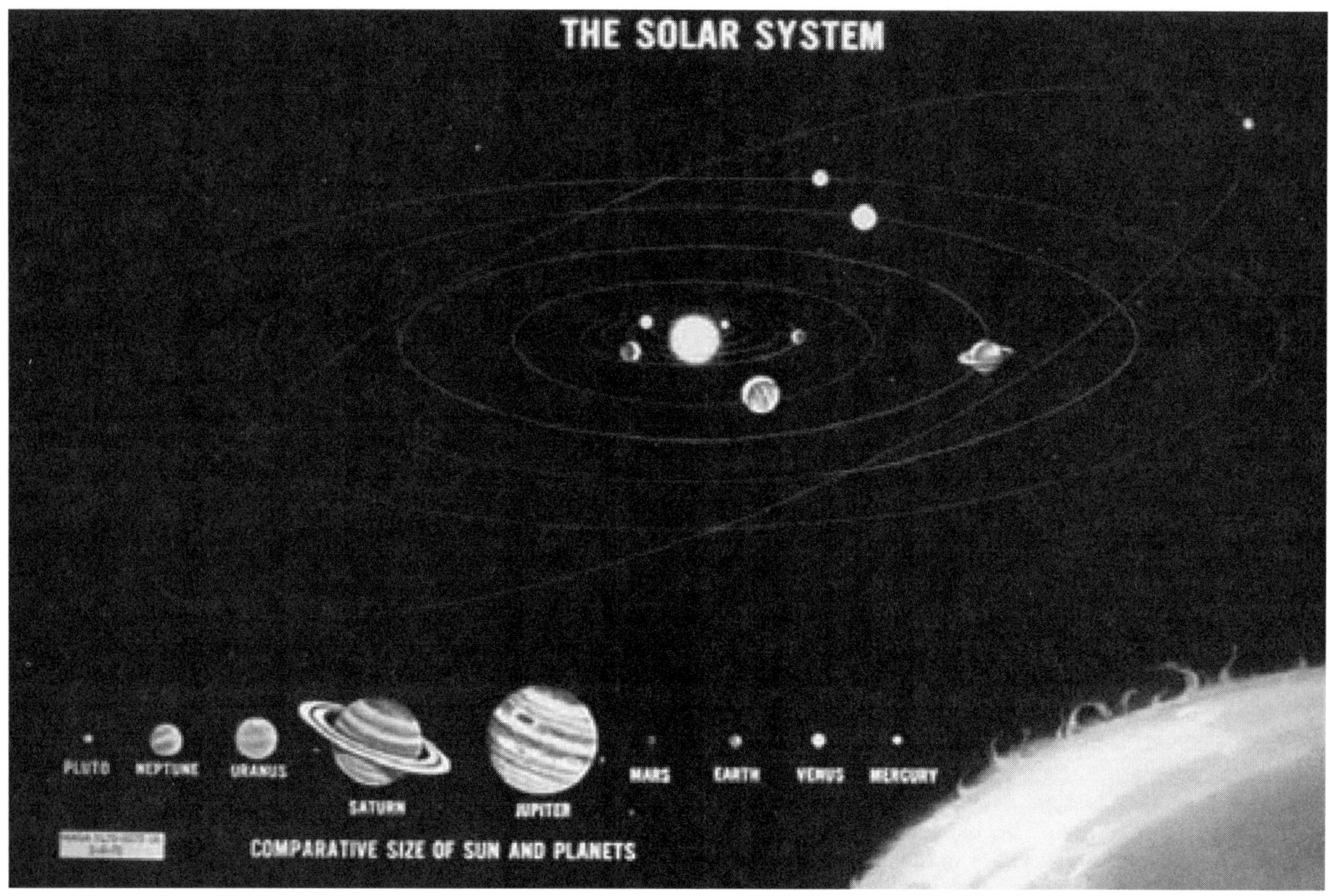

The Solar System. *(Image courtesy of NASA. Reproduced by permission.)*

lenge for the future is to detect a small rocky planet like our own Earth, with a gaseous atmosphere and possibly even life.

PLASMA PHYSICS

Plasma physics is concerned with the study of the physical characteristics, properties, and applications of plasmas. A plasma is usually defined as the "fourth state of matter." A more precise definition is that a plasma is a quasineutral gas in which many of the atoms or molecules are ionized and exhibit collective behavior. Quasineutral refers to the uniform **density** (n) of particles within the plasma and collective behavior means that a plasma can undergo motions which not only depend on local conditions but also on the state of the plasma in more remote regions. To be considered a plasma, a gas must satisfy the three following conditions: $\lambda_D \ll L$; $N_D \gg 1$ and $\omega\tau > 1$. In these expressions, λ_D is the Debye length, or the characteristic distance over which charges are shielded in a plasma, and is equal to $6.9\,(T/n)^{1/2}$cm; N_D is the number of particles contained in a Debye sphere, a sphere around a charged test particle whose radius is equal to the Debye length, and ω is the natural collective oscillation **frequency** of the charged

species (such as electrons, **ions**, etc.) present in the plasma in the absence of a magnetic field and is equal to $9000n^{1/2} \times 2\pi\,\mathrm{sec}^{-1}$; and τ is the average time elapsed between collisions with neutral atoms. Much of our understanding of the physics of plasmas is due to the work of the Swedish physicist H. Alfven (Nobel Prize in physics, 1971) on the theory of magneto-hydrodynamics and its plasma applications.

Plasmas are found first of all in fusion plasmas, and also in much of interplanetary, interstellar, and intergalactic **space**, in the Sun's corona, in solar wind, in Earth's magnetosphere and ionosphere, in parts of the atmosphere around lightning discharges and also in fluorescent light bulbs and other gas-discharge tubes, as well as in very hot flames. Plasmas are accordingly classified as follows: astrophysical, collisionless, cylindrical, electrostatically neutral, inhomogeneous, intergalactic, interstellar, magnetized, nonthermal, partially ionized, relativistic, solid state, strongly coupled, thermal, unmagnetized, and vlasov.

Plasmas are produced in fusion reactions. Fusion combines the nuclei of light elements to form a heavier element. This is a nuclear reaction and results in the release of very large amounts of **energy**. In fusion, the total **mass** of the resultant nuclei is slightly less than the total mass of the original particles. For fusion reactions to occur, temperatures must be high enough to **heat** the particles and the particles must be present in sufficient number and well contained. These conditions are all well met by plasmas. A plasma contains electrons that are stripped from their nuclei and so the plasma consists of charged particles, ions and electrons. Four particle-containment techniques are presently used to confine these hot plasmas: magnetic, inertial, electrostatic and **gravity** confinement. Magnetic confinement makes use of strong magnetic fields, typically 100,000 times greater than the earth's magnetic field, which are arranged in a configuration that prevents the charged particles from leaking out; this is referred to as a magnetic bottle. Inertial confinement uses powerful **lasers** or high energy particle beams to compress the fusion fuel. The plasma is imploded so quickly and the **inertia** of the converging particles is so high that many fuse before they disperse. This is the method used in a hydrogen bomb. Inertial confinement designs for **power** production usually use small pellets of fuel in an attempt to make "miniature" H-bomb type explosions that can be controlled. The electrostatic approach confines the charged particles by means of electric fields, rather than the magnetic fields used in magnetic confinement. Gravity is strong enough to confine plasmas in the **Sun** and other **stars**. An example of a fusion confinement device is the tokamak. The word tokamak is an acronym derived from Russian words meaning "toroidal chamber and magnetic coil." The tokamak has a doughnut-shaped configuration through which a current is introduced up to several million amperes and flows through the plasma, heating it to temperatures exceeding a hundred million degrees centigrade with the use of high-energy particle beams or radio-frequency **waves**.

The present goals of plasma physics research are focused in three main directions: first, achieve the demonstration of a steady-state, high-gain fusion plasma capable of producing reactor-level fusion power; second, improve the present understanding of the underlying physics and advance the state-of-the-art critical enabling technologies. This requires investigating the transport of heat particles from plasma, the contribution of magnetohydrodynamic factors and the effects of large populations of energetic alpha particles. The third goal is to find high strength materials that do not become excessively activated from the fusion neutrons or weakened by the extreme heat for the structural design of functional reactors, as well as improved, high field superconducting magnets to provide the required steady-state confinement of fusion plasmas.

Plasma research has also been successfully applied to the study of the physics of the earth's magnetosphere and its interactions with the solar wind and the ionosphere. Because the important physical processes in the earth's space environment occur on a wide range of spatial and temporal scales, a variety of simulation techniques are being used to investigate them. For example, magnetohydrodynamic (MHD) models are used to study the large scale circulation of plasma in the magnetosphere-ionosphere system which is in many ways similar to the global atmosphere-ocean circulation models used for climate research.

See also Alpha particles; Charges in electric fields; Electric charge; Electric field and forces; Interstellar gas; Lasers; Magnetic confinement fusion; Magnets; Neutrons; Nuclear fusion; Nuclear physics; Nuclear reactions; Particles; States of matter

Plato (CA. 427 B.C.-CA. 347 B.C.)
Greek philosopher

Plato, known as perhaps the greatest of all the ancient Greek philosophers and educators, was more interested in moral rather than natural (or scientific) philosophy. Nevertheless, he made many important contributions to the philosophy of mathematics and physics, primarily through his belief and teachings that mathematics provided the best training for the mind. Born in Athens, Greece, to wealthy and aristocratic parents, Plato was originally named Aristocles. He later was called "Platon," meaning broad, by schoolmates as a nickname because of his broad shoulders. Plato's first inclination was to enter politics, but he grew disillusioned with political life, especially after his teacher, the Greek philosopher Socrates, received a death sentence by the government of Athens. After the death of his mentor, who committed suicide in 399 B.C., Plato left Athens and traveled to Egypt and Italy, where he learned to appreciate mathematics from the disciples of Pythagoras.

When Plato returned to Athens he devoted his life to philosophy and founded the Academy, considered by some to be the first real university. Most of what is known about Plato and his philosophy can be found in Plato's dialogues, a series of writings in which his mentor Socrates discusses issues with students and others. Although Plato never introduces himself into these writings, many of the beliefs and statements are Plato's own. Plato emphasized that ideas represented what is constant and true in life, and was drawn to mathematics because it represented idealized abstractions, and seemed superior to the mundane material world. Plato was enamored with the possibility of developing a type of pure, rather than applied, mathematics to study the **universe**. For example, in his dialogue *Timaeus*, Plato expounds his believe that the heavenly bodies reveal a perfect geometric form. He also describes the existence of what has become known as the five Platonic solids, which he may have learned about from the Pythagoreans. These forms—the tetrahedron, cube, octahedron, icosahedron, and dodecahedron—represented mathematical models of the elements earth, fire, air, water, and, in the case of dodecahedron, the entire universe. German astronomer **Johannes Kepler**, nearly 2,000 years later, developed his own **cosmology**, or theory of the structure of the universe, based on these same shapes.

Plato's emphasis on proofs stemming from a clear hypothesis and accurate definitions, precluded Euclid's later systematic approach to mathematics. In addition, friends or pupils of Plato, including Archytas, conducted much of the most important mathematical work of the fourth century B.C. Through his academy, Plato also advocated that students should progress and learn four fields of mathematical study,

Bust of Plato.

beginning with arithmetic and then progressing to plane geometry, solid geometry, then **astronomy** and **harmonics**, both of which he believed were based on precise mathematical principles. Plato's adherence to the philosophy that mathematics represents precise and definite thinking is illustrated by the inscription over his Academy's main entrance, which read, "Let no one unversed in geometry enter here."

Plato lived to be approximately 80 years of age and is reported to have died peacefully, either in his sleep after attending a student's wedding or while working quietly on his studies. Plato's influence on philosophy and the development of mathematics continued long after his death. His academy remained in existence for about 900 years, until 529 A.D., when the Eastern Roman Emperor, Justinian, ordered it closed because he considered it a pagan establishment in what had become a Christian world.

See also Philosophy of physics

POINCARÉ, JULES HENRI (1854-1912)

French mathematician

Jules Henri Poincaré has been described as the last great universalist—"the last man," E. T. Bell wrote in *Men of*

Mathematics, "to take practically all mathematics, pure and applied, as his province." He made contributions to number theory, theory of functions, differential equations, **topology**, and the foundations of mathematics. In addition, Poincaré was very much interested in astronomy, and some of his best known research is his work on the three-body problem, which concerns the way **planets** act on each other in **space**. He worked in the area of **mathematical physics** and anticipated some fundamental ideas in the theory of relativity. He also participated in the debate about the nature of mathematical thought, and he wrote popular books on the general principles of his field.

Poincaré was born in Nancy, France, on April 29, 1854. The Poincaré family had made Nancy their home for many generations, and they included a number of illustrious scholars. His father was Léon Poincaré, a physician and professor of medicine at the University of Nancy. Poincaré's cousin Raymond Poincaré was later to serve as prime minister of France and as president of the republic during World War I. Poincaré was a frail child with poor coordination; his larynx was temporarily paralyzed when he was five, as a result of a bout with diphtheria. He was also very bright as a child, not necessarily an advantage in dealing with one's peers. All in all he was, according to James Newman in the *World of Mathematics,* "a suitable victim of the brutalities of children his own age."

Poincaré received his early education at home from his mother and then entered the lycée in Nancy. There he began to demonstrate his remarkable mathematical talent and earned a first prize in a national student competition. In 1873, he was admitted to the École Polytechnique, although he scored a zero on the drawing section of the entrance examination. His work was so clearly superior in every other respect that examiners were willing to forgive his perennial inability to produce legible diagrams. He continued to impress his teachers at the École, and he is reputed to have passed all his math courses without reading the textbooks or taking notes in his classes.

After completing his work at the École, Poincaré went on to the École des Mines with the intention of becoming an engineer. He continued his theoretical work in mathematics, however, and three years later he submitted a doctoral thesis. He was awarded his doctorate in 1879 and was then appointed to the faculty at the University of Caen. Two years later, he was offered a post as lecturer in mathematical analysis at the University of Paris, and in 1886 he was promoted to full professor, a post he would hold until his death in 1912.

Discovers Automorphic Functions

One of Poincaré's earliest works dealt with a set of functions to which he gave the name Fuchsian functions, in honor of the German mathematician Lazarus Fuchs. The functions are more commonly known today as automorphic functions, or functions involving sets that correspond to themselves. In this work Poincaré demonstrated that the phenomenon of periodicity, or recurrence, is only a special case of a more general property; in this property, a particular function is restored when its variable is replaced by a number of transformations of itself.

As a result of this work in automorphic functions, Poincaré was elected to the French Academy of Sciences in 1887 at the age of thirty-two.

For all his natural brilliance and formal training, Poincaré was apparently ignorant of much of the literature on mathematics. One consequence of this fact was that each new subject Poincaré heard about drove his interests in yet another new direction. When he learned about the work of Georg Bernhard Riemann and Karl Weierstrass on Abelian functions, for example, he threw himself into that work and stayed with the subject until his death.

Two other fields of mathematics to which Poincaré contributed were topology and probability. With topology, or the geometry of functions, he was working with a subject which had only been treated in bare outlines, and from this he constructed the foundations of modern algebraic topology. In the case of probability, Poincaré not only contributed to the mathematical development of the subject, but he also wrote popular essays about probability which were widely read by the general public. Indeed, he was elected to membership in the literary section of the French Institut in 1908 for the literary quality of these essays.

Contributes to Three-Body Problem

In the field of celestial **mechanics**, Poincaré was especially concerned with two problems, the shape of rotating bodies (such as **stars**) and the three-body problem. In the first of these, Poincaré was able to show that a rotating fluid goes through a series of stages, first taking a spheroidal and then an ellipsoidal shape, before assuming a pear-like form that eventually develops a bulge in it and finally breaks apart into two pieces. The three-body problem involves an analysis of the way in which three bodies, such as three planets, act on each other. The problem is very difficult, but Poincaré made some useful inroads into its solution and also developed methods for the later resolution of the problem. In 1889, his work on this problem won a competition sponsored by King Oscar II of Sweden.

The work Poincaré did in **celestial mechanics** was part of his interest in the application of mathematics to physical phenomena; his title at the University of Paris was actually professor of mathematical physics. Of the roughly 500 papers Poincaré wrote, about seventy deal with topics such as **light**, **electricity**, capillarity, **thermodynamics**, **heat**, **elasticity**, and telegraphy. He also made contributions to the development of **relativity theory**. As early as 1899 he suggested that absolute **motion** did not exist. A year later he also proposed the concept that nothing could travel faster than the **speed of light**. These two propositions are, of course, important parts of Albert Einstein's theory of special relativity, first announced in 1905.

As he grew older, Poincaré devoted more attention to fundamental questions about the nature of mathematics. He wrote a number of papers criticizing the logical and rational philosophies of Bertrand Russell, **David Hilbert**, and Giuseppe Peano, and to some extent his work presaged some of the intuitionist arguments of L. E. J. Brouwer. As E. T. Bell wrote: "Poincaré was a vigorous opponent of the theory that all math-

M. Henri Poincare. *(Photo courtesy of Hulton-Deutsch Collection/Corbis-Bettmann. Reproduced by permission.)*

ematics can be rewritten in terms of the most elementary notions of classical logic; something more than logic, he believed, makes mathematics what it is."

Poincaré died on July 17, 1912, of complications arising from prostate surgery. He was fifty-eight years old. During his lifetime he had received many of the honors then available to a scientist, including election to the Royal Society as a foreign member in 1894. Poincaré was married to Jeanne Louise Marie Poulain D'Andecy, with whom he had four children, one son and three daughters.

POLARIZATION OF LIGHT

Light, like **radio** transmissions and nuclear **radiation**, is emitted in the form of electromagnetic **waves**. Specifically, light travels in transverse waves—that is, while traveling along a generally straight path, light waves will vibrate in many directions perpendicular to that path. However, there are several methods available to filter light, causing it to vibrate in only one plane perpendicular to the path of travel. Such filtered light is called polarized light.

When an **electron** is excited it will vibrate; if excited further, it will sometimes emit a **photon** of light. Since the excited electron vibrates in only one direction, that photon will also vibrate along a single plane and, therefore, be polarized. If all electrons in a material vibrated in the same direction, all the light emitted would be plane-polarized. In reality, though, the material's electrons vibrate in an almost infinite number of directions, producing unpolarized light. In order for ordinary light to be polarized it must either pass through or bounce off a polarizing substance. Depending upon its individual properties, this substance will either align the light along a single plane or will remove the light not vibrating along that plane.

The first person to observe the polarization of light was Etienne-Louis Malus (1775-1812), who found that light passing through a piece of Iceland spar crystal was split into two beams. Thinking that each beam was aligned with some mystical "pole of light" (similar, in theory, to the poles of a magnet) Malus described the two beams as being "polarized." More precise **work** on the subject was conducted by a classmate of Malus', Jean-Baptiste Biot (1774-1862), during the final years of the eighteenth century.

Biot's research was further advanced by Augustin Jean-Fresnel (1788-1827), who used the phenomenon to redesign **lenses** for lighthouses and, more importantly, to support his theory of the transverse nature of light. At this time, debate raging between Europe scientists over whether to subscribe to **Isaac Newton**'s particle theory of light or, like Fresnel, to the much-doubted theory that light acted as a wave.

In 1828 the Scottish physicist William Nicol (1768-1851) used two pieces of Iceland spar to construct a polarizing prism. As light entered the prism it was split into two beams (just as Malus had observed); upon reaching the second crystal one beam would exit the prism while the second, now polarized, would pass through. This was the first reliable instrument for obtaining polarized light, and, through its use, a generation of scientists began to truly understand the nature of light.

Today, there are four basic methods for polarizing light: absorption, reflection, double refraction, and **scattering**. The most popular method, absorption, requires a material whose molecules allow only one plane of light to pass through, absorbing all other planes. Such a material is called a *dichroic*. A common man-made dichroic is *polaroid*, discovered in 1938 by the American entrepreneur E. H. Land. Land figured that, if a large crystal could be used to polarize light, then thousands of tiny crystals should be capable as well. He mixed the crystals into a clear plastic solution which was cooled and stretched into thin sheets; this had the effect of aligning all the crystals and keeping them from drifting, which is a problem in single-crystal systems. Within a few years polaroid replaced nearly all polarizing crystals in scientific and industrial use, and Land built his Polaroid Corporation from its revenues.

Polarization of light by reflection is found more in nature than in industry. When light strikes a flat surface it is polarized to some extent, depending on the angle at which it strikes the surface. If the angle is zero or 90 degrees, the light will not be polarized; if it is less than 90 degrees, some polarization will occur. At one particular angle light will be completely polarized. This phenomenon is called Brewster's law (after its discoverer, Sir David Brewster [1781-1868]) and the polarizing angle is often referred to as Brewster's angle. Many different materials, such as water, snow, and glass, can polarize light through reflection, though the angle at which this occurs is different for each material.

When Nicol perfected his polarizing prism, he was using the method called double refraction. Certain crystals will split incoming light into two separate rays; Iceland spar is one such crystal, as are calcite and quartz. Each ray produced is polarized, although they are aligned perpendicular to each other—if one ray vibrated up-and-down, the other would vibrate side-to-side. Such polarizing crystals, called birefringent crystals, are seldom used for polarization today, for they are far less versatile than man-made dichroics.

The last method for polarizing light is called scattering. It occurs when light enters a system of many particles that can absorb **energy** and re-emit it. Many gases, including air, scatter light. As a photon of light strikes a gas molecule it will excite the molecule, causing it to vibrate. When the molecule rereleases the photon it, too, vibrates along the same plane. If many gas molecules are caused to vibrate along one plane, the scattered light will be, to some degree, polarized. Polarization by scattering is primarily a natural phenomenon and has found almost no use in industry.

While instrumental in the study of light, polarization has many practical uses as well. For example, polarized sunglasses have been developed; in addition to blocking direct sunlight, they also absorb the polarized light that is reflected from ground, water, or snow.

Certain plastic and crystalline materials have the unusual effect of twisting polarized light. With such materials, polarized light enters vibrating along one plane; upon exiting, however, it vibrates along a completely different plane, while still remaining polarized. These materials are said to display optical activity, and are rapidly gaining popularity among engineers—for example, liquid crystal displays (LCDs) rely upon polarization by optically active liquid crystals.

POSITRON

A positron is the **antimatter** equivalent of the negatively charged **electron**. A positron is equal to the electron in **mass**, but has a positive charge. In 1828, English physicist **Paul Dirac** advanced an equation that incorporated both **quantum physics** and the requirements of the theory of special relativity to provide a complete description of the electron. The equation resulted in a particle, however, that could be positively or negatively charged. On this basis, Dirac predicted the existence of the positron.

In 1932, American physicist **Carl Anderson** observed a new kind of particle in his **cloud chamber**. A particle too faint to be a **proton** or **alpha particle** entered the chamber, and then curved toward the negative area of the magnetic field that was around the chamber. It's **velocity** and mass indicated was the same as the electron but it swerved toward the negative pole and therefore had to carry a positive charge. Anderson realized he had discovered Dirac's antimatter particle to the electron, the positron (e+).

The positron has the same rest mass (0.511MeV) and the same spin as the electron, but carries the opposite **electric charge**. The electron has electron number +1 while the positron has –1. Both particles have their corresponding neutrinos; the electron has a **neutrino** with the electron **lepton** number +1, and the positron has its antimatter electron neutrino with electron lepton number –1.

That both particles exist in nature is a classic example of E=mc². Positrons are created when a high **energy** gamma-ray **photon** strikes Earth's atmosphere and decays, or when a high-energy photon itself penetrates the atmosphere and converts its energy into mass creating a positron-electron pair. In addition, some radioactive nuclei spontaneously emit a positron (along with a neutrino).

The E=mc² transformation is reversed when an electron and positron are brought sufficiently close together. The two particles, through a process called annihilation, convert into energy. This energy can then revert to **matter** and antimatter so long as the matter produced exactly matches the antimatter produced. The matter-antimatter does not have to be another electron-positron pair. All energy, however, must be conserved.

As of the year 2000, positrons can be manufactured, stored, and then used in various particle reactions carried out at high energy accelerator facilities. Still, the question remains why electrons and positrons are not constantly annihilating all around us. Dirac's explanation is that there is a sea of positrons residing in energy levels that are full and not accessible to the electron. Periodically, whether spontaneous or manufactured, a positron leaves this sea providing a hole for an electron-positron annihilation to fill when the annihilation leaves behind only a puff of energy. The electron has fallen into the positron hole, releasing energy as it does so, and both the hole and the electron simply disappear from the everyday world, canceling each other out. Or if adequate energy is available, i.e. from an energetic photon, a negative-energy invisible electron may be kicked out of its hole into visibility, creating a positron-electron pair and leaving behind a hole.

In the late 1940s American physicist **Richard Feynman** and others put together a more complete theory called **quantum electrodynamics (QED)** which explains the reaction of photons, electrons, and positrons.

POTENTIAL ENERGY

Matter may store **energy** even when at rest. Energy that is stored and held for future use is referred to as potential energy. If this is confusing, recall that energy is the capacity to do **work**, regardless of whether that energy involves **motion** (**kinetic energy**) or not (potential energy).

The most familiar form of potential energy involves the pull of Earth's **gravity**. The law of **conservation of energy** tells us that a boulder at rest on the edge of a cliff has potential energy in an amount equal to the amount of work it took to raise the boulder to that height from ground level. (Mathematically, the work done in lifting an object is equal to the **weight** of the object times the distance it is raised.)

But Earth's gravity is not required for the existence of potential energy. An example of an object containing nongravitational potential energy is a stretched bow. When one draws the bowstring, the energy required to stretch it is transferred into the bow. **Neptune** has potential energy because it can expend energy by falling into the **Sun** due to the Sun's gravitational pull. Moreover, a nail within the magnetic field of a magnet has potential energy because it can expend energy by moving toward the magnet.

The chemical energy in fossil fuel, for example, is a form of potential energy. This energy is converted to kinetic energy when a chemical change takes place in the material. There is also potential energy in electric batteries, wound alarmclocks, and in the food we eat.

POWELL, CECIL FRANK (1903-1969)
English physicist

Cecil Frank Powell's research into cloud chambers and the detection of **subatomic particles** led to his development of photographic emulsion systems to detect and identify fast-moving particles, especially those found in **cosmic rays**. This enabled him to discover the pi-meson, a particle formed from **nuclear reactions** within cosmic rays. Powell was awarded the 1950 Nobel Prize in physics for his work in this area. He also was a member of the British Atomic Energy Project during World War II, though in his later years he became an advocate for nuclear disarmament.

Powell was born on November 5, 1903, at Tonbridge, Kent, England. His father, Frank Powell, was a gunsmith, and his mother, Elizabeth Caroline Bisacre, came from a family of skilled technicians. Powell developed an interest in science at an early age after becoming captivated by a chemistry book he saw in a store. Inspired by the book to conduct his own chemistry experiments, he eventually convinced his family to let him purchase the makings of a home chemistry set.

Studies at Cambridge under Rutherford and Wilson

In 1914 Powell won a scholarship to the Judd School in Tonbridge. Upon graduation from Judd, he earned two more scholarships that allowed him to attend Sidney Sussex College at Cambridge University. He graduated in 1925 with a degree in physics but turned down a teaching job to continue his graduate work at Cambridge.

At the time **Ernest Rutherford**, the Nobel-Prize winning physicist who had determined the structure of the **atom**, was the director of the Cavendish Laboratories at Cambridge. It was under **C. T. R. Wilson**, the inventor of the **cloud chamber**, that Powell conducted his doctoral research, a study of condensation phenomena in cloud chambers, for which he was awarded his Ph.D. in 1927. Cloud chambers are devices that reveal ionized particles by producing a trail of water droplets from air saturated with water. Powell accepted an appointment as research assistant to A. M. Tyndall at Bristol College. In succession he became lecturer in physics, a reader in physics, the Melville Wills Professor of Physics (1948), the Henry Overton Wills Professor of Physics and director of the H. H. Wills Physics Laboratory (1964), and vice-chancellor of the University at Bristol (1964).

Develops Photographic Detection Devices

Powell's initial work at Bristol involved the study of ion mobility in gases—the way electrically charged atoms behave in gases. By 1938, however, Powell became interested in particle detection devices. That interest, which first developed while he was studying cloud chambers, was rekindled when he learned that photographic emulsions could be used to detect particles in the atmosphere. For a number of years, Wilson's cloud chamber had been the instrument of choice for detecting subatomic particles, such as those produced in radioactive reactions and cosmic rays. However, the cloud chamber possessed one serious disadvantage—it required a brief resting phase each time it was used. In contrast, photographic emulsions were ready at all times to record events.

Like other scientists, Powell had been aware of the potential of photographic emulsions for this purpose, but no one had yet used them successfully. The main problem was that emulsions were not sensitive enough to be used for detection purposes, so Powell decided to find a way to overcome this limitation. His first year of research proved disappointing; he found that neither the emulsions nor the microscopes available were of sufficient quality to obtain the results he wanted. His research was interrupted by World War II, and he became involved with the British Atomic Energy Project for its duration. After the war, he again tackled the technical challenges of using photographic emulsions for detection purposes, this time with much greater success.

In 1946 at Powell's request, Ilford Ltd., a photographic company, developed a new emulsion that could more clearly record particle tracks. Powell and his colleagues used this new detection system to study cosmic **radiation** at altitudes of up to 9,000 ft (27,432 m). These studies resulted in the discovery of a new particle, the pion (or pi-meson), that had been predicted by the Japanese physicist **Hideki Yukawa** in 1935. The pion proved to be a cohesive **force** within the atomic **nucleus**, as was the K-meson, another particle discovered by Powell shortly thereafter. It was partly for these discoveries that Powell was awarded the 1950 Nobel Prize in physics. Over the next decade Powell continued his studies of cosmic radiation. As balloon technology improved, he launched his detectors higher into the atmosphere, in some cases reaching and maintaining altitudes of 90,000 ft (27,432 m) for many hours. A key element in the success of this research program was the collaborative effort among scientists, technicians, and laypersons throughout Europe who collected and monitored his equipment. That experience proved to be especially helpful in the early 1960s, when Powell became involved in organizing the **European Center for Nuclear Research (CERN)** in Geneva, Switzerland. Powell

served as chairman of CERN's Science Policy Committee from 1961 to 1963.

During the 1950s Powell became increasingly concerned about social problems related to scientific and technological development. He served as president of the Association of Scientific Workers from 1952 to 1954, and as president of the World Federation of Scientific Workers from 1956 until his death. A founding member of the Pugwash Movement for Science and World Affairs, he lent his signature to Bertrand Russell's 1955 petition calling for nuclear disarmament.

Powell was married in 1932 to Isobel Therese Artner, with whom he had two daughters. He died on August 9, 1969, at Bellano, Lake Como, Italy, while on vacation to celebrate his retirement from Bristol a few months earlier. Powell's awards in addition to the Nobel Prize included the Hughes Medal in 1949 and the Royal Medal of the Royal Society in 1961, the Rutherford Medal and Prize in 1961, the Lomonosov Gold Medal of the Soviet Academy of Sciences in 1967, and the Guthrie Prize and Medal of the Institute of Physics and Physical Society in 1969.

POWER

The term "power" has many different meanings in the English language. In physics, however, power has a very specific meaning. It is the rate of transfer of **energy**; the measurement of the amount of energy transferred in a unit amount of time.

In physics there is a fundamental principle that says energy can never be used up or destroyed. Energy converts from one form of energy to another. In a car, the chemical energy locked in gasoline is converted to **heat** energy in the explosion within the cylinders. This energy is converted to **kinetic energy** in the pistons and wheels of the car. All this energy is finally converted again to low grade heat energy through the **friction** of the moving parts, tires on the road and the air rushing past the body of the car. What actually takes place is **conservation of energy** as it changes form. The rate at which it does this is the measure of the vehicle's power.

The first attempts at defining this rate of transfer of energy involved the use of the British system of units, namely feet (distance) and pounds (**weight**). A weight being lifted means that energy is being converted. If a one pound weight (1 lb wt) is raised up 1 ft, then the energy converted is 1 ft lb wt. If it was done in one second then the power used was 1 ft lb wt/sec. This is a clumsy set of units, but they are still in limited use today. In the eighteenth century, **James Watt** realized the world needed a bigger unit of power for the new machines he and others were inventing. He developed the unit of one horsepower (1 hp) by using a horse to raise a load from a mine with a pulley and rope. He measured the load and the time the horse took to do the job. He figured that the power of one horse was 550 ft lb wt/sec. This means a horse is supposed to be able to lift a 550 lb weight a distance of 1 ft in 1 sec, or a 1 lb weight a distance of 550 ft in 1 sec, or some other combination. Of course, had Watt used a different horse, a different value for the unit of horsepower would be used today!

Horsepower is still a very common form of power measure in the United States. For example, an outboard motor for boats may be rated as 15 hp, which is the same as 8,250 ft lb wt.

Physicists needed a better and more consistent means of measuring power. The same concepts are still used, however, the fundamental units are now meters and newtons. In this system, **force** is not measured in pounds weight but "newtons". (It takes a force of approximately 10 newtons to lift a **mass** of 1 kg off the earth's surface.) A force of 1 newton raising a load a distance of 1 m is said to be an energy conversion of one joule. If this energy conversion takes place in 1 sec, then the power unit is 1 watt. (Obviously in honor of the pioneering James Watt and his horse!)

One horsepower is approximately equivalent to 746 watts, which obviously makes the watt a small unit. Once again, a larger unit was needed for working with modern equipment. The metric system readily provided the kilowatt, which is one thousand watts. The electric power supplied to homes is measured in kilowatt hours. This unit measures the amount of energy consumed by the home. Power is defined as energy divided by time. This equation can be restated as: energy is power times time. Therefore, the power measured in watts, times the number of hours power was used, results in the energy consumed, watt-hours (wh). One wh=3600 J, but because the numbers are often quite large, energy is more commonly referred to as kilowatt hours (kwh), 1 kwh = 3.6×10^6J. The average yearly electric power consumption for the world rose from 1 trillion (10^{12}) kwh (3.6×10^{18} J) to 11.5 trillion kwh (4.14×10^{19} J) between 1950 and 1990, obviously this larger unit comes in handy.

Pulling a load up a mine shaft seems to bear little relationship to moving an **electric current** through a wire. However, both the load and the current move against opposing forces. The force needed to overcome the opposition in both instances can be measured as power. Most countries use watts and kilowatts to measure the power exerted in both sets of circumstances. The United States, however, still commonly use horsepower to describe engine and motor power and watts when discussing electrical power.

Moving a mass up against a **gravitational field** increases its **potential energy**. Similarly, moving a charge up through an **electric potential** increases its potential energy. The increase in potential energy of a charged particle is defined as the amount of energy converted (joules; J) in moving it. The **voltage** between two points in a circuit is the total amount of energy converted in moving one unit of charged particles between the two points. This can be represented mathematically as:

$$\text{Volts (V)} = \frac{\text{Energy converted (J)}}{\text{Total charges}}$$

The amount of power used is calculated using the following equations:

$$\text{Power (J / s)} = \frac{\text{Energy converted}}{\text{Time (s)}} = \text{Volt (V)} \times \frac{\text{Total Charges}}{\text{Time (s)}}$$

$$\frac{\text{Total charges}}{\text{Time (s)}} = \text{Current (I) (i.e. number of charges passing / sec.)}$$

The above equation can be simplified to its more common format of: $P = V \times I$.

When turned on, a **light** bulb will give out energy in the form of heat and light. If a light bulb produces 100 J of energy each second, it produces 100 W of power. A 100W light bulb will have approximately 1 amp of current or 6.25×10^{18} electronic charges per second flowing through the filament. A 60W light bulb however, only produces 60J of energy per second, which results in a dimmer light.

PRESSURE

Pressure is a physical **force** exerted over a surface. Specifically, it is defined as a force (F) divided by unit area (A). When expressed relative to a pure **vacuum**, this quotient is referred to as absolute pressure. When referred to relative to atmospheric pressure it is called gauge pressure. Pressure is an important concept in many chemical and physical processes. For example, many types of reactor vessels and boilers must be monitored to ensure they do not exceed specific pressure limits, lest they rupture. In atmospheric science, an understanding of air pressure, also known as barometric pressure, is critical for making **weather** predictions.

Pressure is expressed in various units depending on the application. For example, air pressure is usually given in terms of atmospheres (atm), torr, bars, or millimeters of mercury (mm Hg at 0°C). (Although air pressure varies with altitude and weather conditions, on average the air pressure at sea level can support a column of mercury 760 mm in height.) Industrially, pressure is described in terms of pounds per square inch (psi) or kilograms per square meter. In the **International System of Units**, pressure is given as pascals (Pa), which are defined as 1.0 newtons per square meter. The centimeter- gram-second system quantifies pressure in dynes per square centimeter. In terms of equivalent units, 1 atm = 760 mm Hg = 760 torr = 1.01325 bars = 14.6960 psi = 101,325 Pa = 1,013,250 dynes/cm².

Historically, pressure has been measured with gauges that monitor the displacement of a mechanical element. One of the earliest measuring tools was a liquid-filled tube known as a manometer. Manometers were commonly used by scientists in the 1700s and 1800s and are still used today to calibrate other pressure gauges. A manometer consists of a cylindrical glass U-shaped tube that is partially filled with liquid. One end of the tube is exposed to the process that is generating pressure; the other end may be sealed or left open depending on the design of the instrument. The pressure exerts a force on the surface of the liquid and causes it to move a specific distance that is proportional to the magnitude of the force. The pressure differential can be measured by comparing the height of the liquid to calibrated marks on the side of the tube. Another type of manometer is the inverted bell-type that consists of two inverted U-shaped tubes, or bells, mounted on a balance beam.

Each of these bells is located over a pressure inlet tube in such a way that the bell that is subjected to the highest pressure will rise. This type of apparatus is used to measure very small pressure changes such as those found in drying appliances like conveyor dryers and kilns.

Another type of pressure-sensing element is the Bourdon tube, which is a flexible tube made from steel, beryllium, copper, and other special alloys. The tube, which is flattened slightly and sealed on one end, is formed into a coil. When the open end is exposed to pressure, the tube uncoils to a degree that corresponds to the magnitude of the pressure. Positive pressure causes the tube to expand and a vacuum (negative pressure as compared to air pressure) causes it to contract. As the coil moves, it displaces a needle or other indicator that points along a pre-calibrated scale to give the pressure measurement. Bourdon tubes have the advantage of being low cost, simply designed, and useful for measuring both high and low pressure. They also have several disadvantages: they are not extremely accurate at low pressures, the mechanical linkage to the indicator needle may require amplification to give a meaningful reading, and process materials may accumulate inside the tube since one end is sealed.

Other important mechanical pressure sensors include the bellows and diaphragm type elements. The bellows employs a coiled spring instead of a coiled tube as the responsive element. When exposed to pressure, the spring stretches and moves a needle on a measuring gauge. The diaphragm is a flexible disk made of sheet metal with ridges carved into it. When the disk is exposed to pressure it deforms to one side or the other, depending on which side is exposed to the higher pressure. The disks are very sensitive to small changes in pressure and therefore are useful in low pressure measurements as low as 13 Pa (0.1 torr).

Today, mechanical pressure devices have been supplemented with electrical sensors and transmitters. Certain devices, like piezoelectric crystals, may directly register pressure changes. These crystals operate on the principle that certain quartz compounds can be cut and aligned in such a way that they will trigger a small **electric current** when pressure is applied to them. Other electronic sensing devices depend on a mechanical sensor like a Bourdon tube or a diaphragm to register a change in pressure, and then they use electronic components to amplify and transmit the signal. The mechanical element transmits the pressure change to the magnetic coil of a transformer, which in turn translates the pressure to an electronic signal that can be relayed to a variety of output devices. Examples of this type of gauge include strain gauges, which measure the change in **electrical resistance** of a metal wire exposed to pressure via a mechanical sensor such as a Bourdon tube or a bellows, and capacitive pressure transducers that register changes in **capacitance** when an elastic element is displaced by pressure.

PRIMORDIAL BLACK HOLES • See Black holes

PRINCIPLE OF ENTROPY INCREASE

Entropy is a physical quantity that can be interpreted as a measure of the thermodynamic disorder of a physical system. Entropy has the unique property that its global value must always increase or stay the same; this property is reflected in the second law of **thermodynamics**. The fact that entropy must always increase in natural processes introduces the concept of irreversibility, and defines a unique direction for the flow of time.

On a fundamental level, entropy is related to the number of possible physical states of a system, $S = k \log (\text{Gamma})$, where S represents the entropy, k is Boltzmann's constant, and (Gamma) is the number of states of the system. A useful example to illustrate the principle of entropy increase is a closed box containing an ideal gas with a fixed number of molecules. If the enrgy of the box is increased, the number of states increases because there are many ways that the gas molecules reflect the increased **energy** state. One molecule could respresnt the entire increase or two or more molecules could represent the increase.

Although the entropy of a system can be reduced by a reduction in energy (accomplished by doing **work** on the surroundings) there is an increase in the entropy of the system's surroundings. Any process that includes **heat** transfer from one system to another, therefore, increases the total entropy. When two systems are in thermal **equilibrium**, however, the energy is divided equally between them, no **heat transfer** takes place, and the entropy does not change. The entropy of a system, therefore, is greatest when it is in thermal equilibrium with its surroundings.

A process in which the net entropy change is positive is called irreversible, because a process with a negative net entropy change cannot be performed to counteract it. A process which has a zero net entropy change, however, is reversible, because the change can be counteracted by another process with a zero net entropy change. Entropy, therefore, increases in all real processes.

One interpretation of the principle of entropy increase is that it defines a unique direction for the flow of time. If all processes were reversible, then movement forward or backward in time would be impossible to tell; broken glass might spontaneously reassemble itself, for example. Increasing entropy sets the direction of the **arrow of time**.

See also Boltzmann's constant k; Ideal gas law; Irreversible processes; Reversible processes

PROBABILITY AND QUANTUM MECHANICS

Probability is the likelihood that a certain event will occur. Unlike classical physics, **quantum physics** is not completely deterministic. Probability in the context of **quantum mechanics** has to do with the likelihood of finding a particle, such as an **electron**, in a particular region around the **nucleus** at a particular time.

In 1911 **Ernest Rutherford** developed a picture explaining the structure of the **atom** in terms of a positively charged nucleus surrounded by negatively charged electrons in orbits. Later, in 1913, **Niels Bohr** developed a theory that explained why electrons remain in orbits and do not collapse onto the nucleus. **Quantum theory**, developed during the 1920s, explained many of the phenomena concerning electrons and their role in atoms. In 1924, using **Albert Einstein**'s special theory of relativity, **Louis de Broglie** showed that all particles with **mass**, such as electrons, should have wave-like properties, including a **wavelength**. Neils Bohr then took this concept and suggested the quantized orbits of atomic electrons where actually electron standing **waves** around the nucleus. The **Bohr atom** was a simple two dimensional model. Later, the Schrodinger equation was solved in three dimensions for atomic hydrogen. The cloud-like description comes from the spherical symmetry of these solutions. Therefore, electrons cannot be pictured as localized particles in **space** but rather should be thought of as clouds of negative charge spread out over the entire orbit. These clouds represent the regions around the nucleus where the probability of finding an electron is the largest. Electrons occupy different **energy** levels in atoms and have different wave properties.

Because of this idea that electrons behave as particles and waves, it became obvious that a specific type of partial differential equation should be able to describe the position as well as their future behavior. In 1926, **Erwin Schrödinger** used partial differential equations, Werner Heisenberg's uncertainty principle, and the Hamiltonian function to develop a powerful equation that relates the energy of the electron and the energy of the electric field in which it is situated. Atomic electrons, specifically the **hydrogen atom**, provided a good laboratory for developing the theory of **quantum mechanics**. However, quantum mechanics was always intended to be developed generally. The Schrödinger Equation actually derived from a classical mechanics. Schrödinger took a rather standard classical equation of **motion** and quantized it. That is, he applied the Heisenberg Uncertainty Principle to it turning some of the variables into quantum mechanical operators. The success of the Schrödinger Equation with atomic hydrogen was stunning but by no means the only application. Heisenberg had determined that certain pairs of quantum operators cannot be measured with infinite precision at the same time. For instance, **momentum** and position cannot both be measured precisely at the same time. The equation Schrödinger derived relates the energy of a system to its wave properties and allows one to predict the energy of the electron and its future behavior, i.e. the probability of finding the electron in a particular region around the nucleus. Shortly after Schrödinger developed his equation, **Max Born** postulated that the **wave function** could be used to determine the probability of finding a particle in a particular region at a specific time. This probability is written as $-Y(x,t)-2dx$ = probability at time t of finding the particle at a position x at time t. $-Y(x,t)-2$ is the observable probability **density** for finding this particle. If the probability does not change with time then the state is called a stationary state. This does not mean that the particle in a stationary state is not moving but rather that the probability density is stationary.

Because quantum mechanics is basically random in nature, knowing the state does not allow us to predict the result of any measurement with certainty. However, using large numbers of measured instances does allow us to predict the probability of various possible results statistically. Quantum mechanics dver a large region of space as a wave is distributed. Rather that the probability patterns, wave functions, used to describe the electron's motion behave like waves and satisfy wave equations.

Max Born was awarded the Nobel Prize in physics in 1954 for his fundamental research in quantum mechanics, specifically for his statistical interpretation of the wave function. In his collaborations with **Wolfgang Pauli** and **Werner Heisenberg** on quantum theory he produced work of fundamental importance in quantum mechanics. His treatment replaced the original quantum theory, which regarded electrons as particles, with a mathematical description representing their observed behavior more accurately. He suggested that the only observable aspect of the wave function was its square, not the wave function itself, and that its square at a given point in space is proportional to the probability of finding the particle at that point in space.

PROBABILITY WAVE

In 1925, following the suggestions of French physicist **Louis Victor de Broglie** that **matter** particles could behave like **waves**, Austrian physicist **Erwin Schrödinger** advanced a differential wave equation to describe particles such as electrons. Schrödinger's equation is given schematically by H(psi) = E(psi), where E is simply a number describing the **energy** of the **electron**, and H is a mathematical operator known as the Hamiltonian, containing derivative and multiplicative operators. The solution of Schrödinger's equation in simple cases leads to interesting predictions. For example, if electrons are confined to a box, or potential well, the energy states are quantized, meaning the energy may only have certain discrete values. Schrödinger solved his equation for the hydrogen **atom** and discovered that the energies given by its solution matched the observable hydrogen spectra.

Schrödinger and other physicists, however, were unsure of the wavefunctions that were actual solutions of his equation. For photons, whose particle nature was observed later than their wave nature, it was argued that the wavefunctions related to the electric and magnetic fields that make up **light** waves. In contrast some scientists argued that the wavefunctions related to the electron charge or **mass density** and that such an effort to extend the definition to a neutral or massless particle would not make sense.

The problem was eventually solved by German physicist **Max Born**. Born suggested that the wavefunctions were a probability amplitude for the particles. By taking the absolute-value squared of the probability amplitude, the probability density for finding the particle at a specific point in **space** is found. Accordingly, the probability density can also be related to the charge or mass densities of particles, although it exists apart from the charge or mass and is a concept that applies to all particles.

If a certain process can happen more than one way, then the probability amplitudes are added for each way, and finally, the absolute-value squared of the sum is taken to find the probability for the outcome of the process. Because the probability amplitudes are waves, they can interfere with each other. If the amplitudes are in phase, they will add and the probability for finding a particle will increase. If the amplitudes are out of phase, they will cancel and the probability for finding a particle will decrease. For example, in the double-slit experiment, electrons shot at a double slit in a barrier can take one of two possible paths, leading to an **interference** pattern. If a movable detector is placed behind the double slit, the number of electrons will alternate between regions where many electrons are found and regions where few electrons are found. If the electrons are treated classically, no interference pattern will be created. When the experiment is actually performed, an interference pattern is seen, justifying the probability amplitude interpretation of wavefunctions and confirming the wave nature of electrons.

See also Double-slit interference experiment; Hydrogen atom; Matter waves; Photon; Quantum states; Quantum theory; Wave function; Wave interference; Wave superposition

PROTON

A subatomic particle, a proton (*proton* means "first" in Greek) is the fundamental unit of positive **electricity** in the **atom**, with a charge equal to an electron's negative charge. However, a proton's **mass** is 1,836 times the mass of an **electron**. Protons, along with neutrons, constitute an atom's **nucleus**. In **chemistry**, the number of protons in a the atom of a particular element defines that atomic number (Z); an element's atomic number is its identity tag. With one exception (hydrogen), chemical elements in their ideal form (excluding isotopes) have an equal number of protons and neutrons. The sum of protons and neutrons constitutes an element's atomic mass. The atomic number and mass of hydrogen, because it has one proton and no **neutron**, both have the value of one.

Early **atomic theory** assumed that atoms were indivisible. However, when **Michael Faraday** demonstrated, in 1834, that chemical elements were electrical in nature, scientists embarked on a series of experiments which challenged, and eventually demolished, the fundamental assumptions of traditional atomic theory. For example, in 1878, **William Crookes,** who predicted the discovery of isotopes, started experimenting with cathode rays, which occurred when electricity was discharged between two metal plates in a tube (later named a Crookes tube) with gas at an extremely low **pressure**. Crookes believed these rays were negatively charged electrons. In 1897, however, J. J. Thomson (1856-1940) identified them as

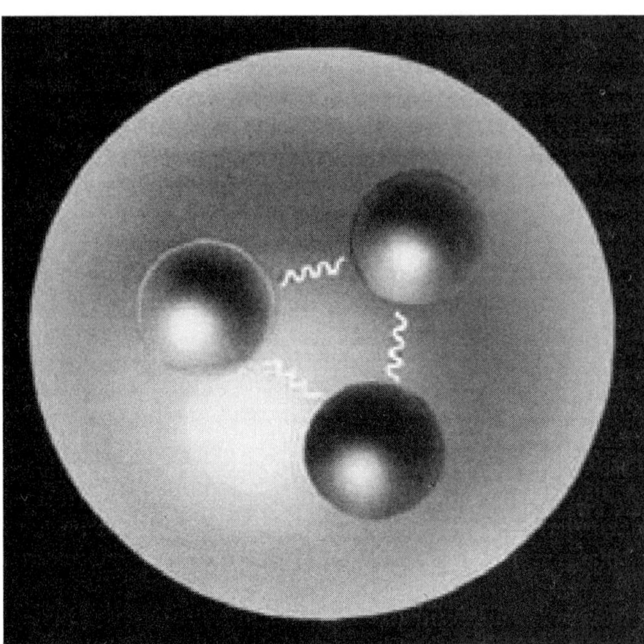

Diagram of the structure of the proton. *(Image by Michael Gilbert. Photo Researchers, Inc. Reproduced by permission.)*

subatomic particles, eventually named electrons. Further experiments led to the discovery of protons. For example, when scientists drilled holes in the positively charged (**anode**) plate of the Crookes tube, they detected rays moving in the opposite direction (from the anode to the negatively charged cathode) of the cathode rays. Named "canal rays," these rays were studied by **Wilhelm Wien**. In 1905, Wien identified some of these rays are hydrogen **ions**. Researchers later established that rays with the least mass were protons.

In 1909, **Ernest Rutherford** instructed his younger colleague **Hans Geiger** and Ernst Marsden (1889-1970), who was still an undergraduate, to perform the gold foil experiment. The experimenters bombarded a thin gold foil with alpha particles (**helium** atoms without electrons). Since the current model of the atom was a positively charged sphere electrons inserted throughout (the plum pudding model), they expected the alpha particles to penetrate the foil without resistance. Indeed, most of the alpha particles did, but a small number were strongly repelled. These results indicated that the gold atom had a positively charged nucleus. In 1919, Rutherford identified the proton as the fundamental unit of positive electrical charge in the atom. The neutron, the major particle constituting the atom's nucleus was discovered in 1932 by **James Chadwick**. Approximately equal to a proton in mass, the neutron has no charge.

PROTON DECAY

If observed, *proton decay* would serve as a clue to new physical theories. It is not predicted by the Standard Model of **particle physics**, and has not yet been observed. The non-observation of **proton** decay implies a lower limit on the proton lifetime of 10^{31} to 10^{32} years, far older than the age of the **universe**.

The proton is made of two up **quarks** and a down quark, so if the proton decayed it would be due to the decay of one of the component quarks. In the Standard Model, decay of an up quark must result in a down quark in the final state, and vice versa. However, both cases are forbidden by **conservation of energy**. If the lone down quark decayed to an up quark, the resulting particle is called a (Delta)++, which has a higher **mass** than the proton; therefore, this decay violates conservation of **energy** and is not allowed. If one of the up quarks decayed to a down quark, the resulting particle would simply be a **neutron**, but the mass of the neutron is slightly greater than that of the proton, so this decay is also not allowed. Therefore, in the Standard Model, the proton is absolutely stable and cannot decay.

Some GUTs, or Grand Unified Theories, predict interactions which allow quarks to decay to purely leptonic final states (that is, states with no quarks). In these theories, the proton is not stable and can decay. The simplest GUT, called SU(5), predicts a maximum lifetime of 10^{30} years, so it is ruled out as a possible theory. More complicated GUTs predict longer lifetimes and are therefore possible.

PTOLEMY (100-170)

Greek astronomer

Very little is known about Ptolemy's early life. Born in Alexandria, Egypt, as Ptolemais Hermii, his name was later latinized as Claudius Ptolemaeus, and later Ptolemy.

Ptolemy's chief contribution to science is a series of books in which he compiled the knowledge of the ancient Greeks, his primary source being Hipparchus (fl. second century B.C.). Because most of Hipparchus' writings have not survived from antiquity, many of the ideas he espoused about the **universe** have become known as the Ptolemaic system.

Ptolemy's system placed Earth directly at the center of the universe. The **Sun, Moon** and **planets** all orbited Earth. However, since such a scheme did not match the observed motions of the planets, Ptolemy added small orbits to the planets called *epicycles*, and introduced other mathematical devices to make a better fit.

Despite its errors and complications, the Ptolemaic system was adequate enough to make predictions of planetary positions, and it influenced thinking for 1,400 years. It was not until 1543 that **Nicholas Copernicus** published his book refuting the Ptolemaic system. After **Tycho Brahe**'s exceptionally accurate measurements of the positions of the planets showed Ptolemy's system was inaccurate, it fell upon **Johannes Kepler** to devise a better explanation of planetary orbits.

Hipparchus had made a catalogue of **stars**, which were grouped into 48 constellations. Ptolemy placed them in his book and gave these patterns the names that are still in use today. He also included Hipparchus' work on trigonometry, his estimate of the distance between Earth and the Moon, which

was fairly accurate, as well as Aristarchus' (fl. third century B.C.) incorrect estimate of Earth's distance from the Sun.

Ptolemy's book was entitled *Mega (mathematike) syntaxis* ("Great [mathematical] compilation") although *Mega* was sometimes replaced by *Megiste* ("Greatest"). When the Arabs adopted the work, they called it Al-majisti ("The Greatest"), which it is known as today. It was translated into Latin in 1175 (as "Almagesti" or "Almagestum") and dominated European thinking for four centuries.

In the field of **optics**, Ptolemy wrote about the reflection and refraction of **light**. He lists tables for the refraction of light as it passes into water at different angles. Another book, *Tetrabiblos*, is a serious treatment of astrology.

Ptolemy also wrote a treatise that dealt with geography and included maps as well as tables of latitude and longitude. It explained how those lines could be mathematically determined, but only a few latitudes were calculated. He had accepted Poseidonius' (c. 135-51 B.C.) erroneously small estimate of the size of Earth, instead of Eratosthenes' (c. 276-194 B.C.) more accurate figure, and Ptolemy unwittingly may have altered the history of the world. After his geography had been translated into Latin, it eventually came to the attention of Christopher Columbus (1451-1506), who accepted the incorrect size and concluded that his search for a short-cut to Asia was possible.

PULSARS

The discovery of pulsars, along with that of **quasars** and the cosmic microwave background, ranks among some of the most important discoveries in **astronomy** in the latter half of the twentieth century. The first pulsar was discovered in 1967 by Jocelyn Bell, Anthony Hewish, and their collaborators at Cambridge University. Bell's group was in the process of observing quasars, or distant, extremely bright galaxies whose source of **energy** to this day remains a bit of a mystery. In the astronomical data, Bell noticed an extremely regular, repetitive signal whose period was one and a third seconds. Its regularity led Bell and Hewish to first conjecture that the source was artificial, but after a careful search, they ruled out such a possibility and concluded that the **pulses** were astronomical in origin.

Currently there are about 1,000 known pulsars, similar to the one discovered by Bell and Hewish. The periods range from approximately 1 ms to almost 5 s. The periods are not entirely regular but rather increase at an extremely slow rate, taking anywhere from a hundred thousand to a thousand trillion years to gain a second. Early researchers realized that such incredibly fast and regular pulses would require a small rotating astronomical object that emitted a highly directional beam of **radiation**. As the object rotated, the beam would rotate with it, sweeping past our earth at regular intervals, much like a lighthouse. The only known objects that rotate at such speeds are **white dwarfs, black holes,** or **neutron stars**. White dwarfs are too large to rotate more than once a second. Black holes, being singularities in spacetime itself, have no physical surface on which radiation emitters can reside. The only remaining theory is that these pulsars are actually rotating **neutron stars**.

Although the exact mechanism by which neutron stars emit the observed pulse like radiation is to this day a **matter** of debate, much is known about the physics of these stars which may in the future shed **light** on the source of their radiation. Neutron stars are extremely dense stellar objects that are produced in supernova explosions. They were first proposed by Baade and Zwicky in 1934 in their pioneering work on supernovae. Their masses are on the order of the **mass** of the **Sun** even though their typical radii are only 10 km. This yields an average interior **density** of 300 trillion kg/c_3, which is greater than the density of a large atomic **nucleus**.

These neutron stars originally come about from supernovae under the gravitational collapse of the stellar core when it reaches a critical size. As the core collapses, the immense gravitational **binding energy** is converted into **kinetic energy**, allowing the core to rotate faster. The mechanism behind this rotation is conservation of **angular momentum** which in this case states that the angular **velocity** of the rotating core is inversely proportional to the square of its radius. This is actually a familiar effect and is entirely analogous to the way a person in a spinning chair spins faster when his or her arms are pulled inwards. Further gravitational collapse is prevented by quantum mechanical Fermi **pressure** of neutrons in the core. This pressure eventually exceeds the binding energy of nuclei at which point protons and electrons interact via the weak interactions of **particle physics**, ejecting a **neutrino** and leaving behind a neutron. The end result is a rapidly spinning, dense condensate of neutron matter that comprises the neutron star.

Neutron stars also carry very strong magnetic fields, on the order of a trillion Tesla. It is these magnetic fields, combined with the rotating nature of the neutron star, that are thought to be responsible for the radiation we observe from pulsars. It is well known from elementary **electromagnetism** that a rotating magnetic dipole (the pulsar) radiates electromagnetic **waves** at the same **frequency** as that of the rotation. In addition this dipole radiation can accelerate charge particles that may be in the vicinity of the pulsar to large, relativistic energies. It is conjectured that these accelerated particles account for the origin of high-energy **cosmic rays**. As the pulsar radiates energy, the rotation frequency of the pulsar also decays, presumably due to torques applied by the magnetic fields themselves. This decay in rotation accounts for the observed increase in the pulse period over time.

There are about 30 known pulsars that derive their source of **power** from an alternate mechanism, that of accretion rather than rotation. Accretion-powered pulsars have slightly longer and less regular pulse periods than rotation-powered pulsars. They are typically found in binary systems with a normal star or a red giant as a companion. Matter from the companion star flows into the neutron star and generates energy in the form of **x rays** or gamma rays.

The study of pulsars, despite their mystery, has nonetheless lead to important results in related fields. Observations of Doppler shifts in radiation as binary pulsars rotate around each other have verified predictions about the nature of gravitational radiation. Similar Doppler shifts lead to the discovery of the first **planets** outside our solar system. Furthermore, pulsars can yield information about extremely dense **states of matter**, often at densities that are beyond the reach of particle physics. Their continued study will yield important clues about the structure of the **universe** at all scales.

PULSES

A pulse is a type of wave that is limited in its extent. A physical example which illustrates pulses is a long taut string. When plucked, a disturbance, consisting of particles moving perpendicularly to the string, travels along the string. This disturbance is localized, which means that particles in the string only move near the center of the wave. Pulses are often constructed so that they are very short; this means that the pulse is highly localized in a very narrow spatial region, so the entire pulse will pass by a given point on the string very quickly. Pulses may be measured in seismic **waves** from an earthquake, or a **light** pulse may be used to cause vibrations in a solid crystal.

PYTHAGORAS OF SAMOS (580 B.C.- 500 B.C.)

Greek philosopher and mathematician

So little is known about the life of Pythagoras that some historians have suggested that such a person did not exist and that the name was for a society of people who studied the mystical properties of numbers. However, the majority of historians, who have based their theories on the works of the later Greek writers **Plato** (ca. 428 B.C.-ca. 348 B.C.) and Herodotus (ca. 484 B.C.-ca. 430 or 420 B.C.), believe that an individual named Pythagoras did indeed exist.

Believed to have been born on the island of Samos about 560 B.C., Pythagoras traveled extensively in Egypt and Babylon, absorbing the mathematical and mystical doctrines of these cultures. Pythagoras settled in the Greek colony of Croton in southern Italy, becoming the leader of a religious, scientific and philosophical brotherhood. Marked by secrecy and mysticism, the Pythagorean society often followed unique practices; for example, they forbade the poking of fire with an iron poker and the eating of beans.

With his goal being the moral reformation of society, Pythagoras created a school of learning, concentrating on mathematics and science. Fascinated by numbers, Pythagoras looked for numerical values and relationships in everything he saw. The numbers 1, 2, 3, and 4 were especially valued because they added up to 10, which was considered a "perfect number" to the Pythagoreans. Despite their advanced knowledge of numbers, they recognized only whole numbers and fractions and dismissed the existence of irrational numbers. Therefore, when Hippasus (ca. 500 B.C.), one of the society members, discovered proof of the existence of irrational numbers, he created quite a scandal. According to legend, Pythagoras threw Hippasus overboard while at sea for the trouble he had caused. Pythagoras attempted to keep the discovery of irrational numbers a secret; members were reportedly banished from the society if they revealed the truth. Possibly due to his possible involvement in a colonial uprising, Pythagoras was driven from Croton, and fled to Metapontum, where he may have been murdered around 500 B.C. Pythagoreanism survived as a cult for centuries following its founder's death.

Many discoveries are owed to the Pythagorean society. Their methods of representing numbers as lines, triangles or squares of pebbles gave us the word calculate, based on the Latin word calculuscalculus, or pebble. Pythagoras's work in geometric principles produced the Pythagorean theorem Pythagorean theorem: the square of the length of the hypotenuse is equal to the sum of the squares of the lengths of the other two sides. More simply stated it allows us to find the length of a side of a right triangle if we know the length of the other two sides. However, historians believe that the theorem was not Pythagoras's discovery, and that he may have learned it from Egyptian mathematicians.

In medicine, the Pythagoreans believed that the blending of the opposites—hot-cold and wet-dry—in the human body was necessary to maintain health. Too much or too little of any one factor would produce sickness. Staunch believers in human reincarnation, they strove to achieve harmony and balance in all things. Their belief that everything was based on numbers was also evident in other aspects of Pythagorean civilization. They reduced astronomy and music to numerical patterns and studied them as mathematical subjects. In the sixth century, the Pythagoreans were the first to suggest that the earth was spherical in shape and proposed

Pythagoras. *(Image courtesy of The Bettmann Archive. Reproduced by permission.)*

the idea that the earth, **moon** and **planets** revolved around a central **Sun**.

Q

QUANTA

Light was once explained as a form of radiant **energy** that was continuous. More recently, however, light is described as existing in discrete units of energy, or infinitesimal packets, that are radiated from **matter**. The smallest quantity, or bundle, of light energy that is emitted is called a quantum of light energy. Quanta, plural for quantum, are light energy packets whose energy is directly proportional to the light **frequency**. Light of long wavelengths has low frequencies and consists of quanta with lower energy than light of small wavelengths and greater frequencies. For example, quanta from ultraviolet light are more energetic than quanta from infrared light.

Quanta are also called photons. When a **photon**, or quantum, collides with an **electron** of an **atom**, it may transfer its energy to the electron. When this occurs, the electron receives all of a photon's energy, and the quantum ceases to exist. The electron, now at a higher energy state, might occupy a more energetic, less stable orbital around the **nucleus** of the atom. If the electron then, returns to its original ground state orbital, a photon, or quantum, of light energy is released from the atom. The light has a frequency proportional to the energy release. This concept explains how neon gas emits light when its atoms' electrons are excited and then return to their original energy states. The branch of physics that examines the nature of quanta is called **quantum mechanics**, and has revolutionized the understanding of the way matter and energy interact.

QUANTUM CHROMODYNAMICS THEORY

Quantum Chromodynamics, or QCD, is the theory describing the **strong interactions** between **quarks** and gluons. These interactions are responsible for binding quarks together into nucleons and mesons, and also for binding the nucleons together inside atomic nuclei.

QCD is quite similar in many respects to **Quantum electrodynamics (QED)**. Instead of the positive and negative electric charges found in QED, QCD has three positive charges called colors. The names given to the colors vary, but a common scheme is red, green and blue. The negative charges usually are called anti-red, anti-green, and anti-blue. QED has one **photon**, a massless force-carrier particle, compared to QCD's eight massless force-carrier particles, the gluons.

The largest difference between QCD and QED lies in the group structure of the two theories. QED has a symmetry under an Abelian group called U(1), similar to the group of two-dimensional rotations. In an Abelian group, symmetry transformations can be performed in any order with the same effect (they commute). QCD has a symmetry under a more general non-Abelian group called SU(3), where the order of the symmetry transformations matters. The upshot of these mathematical notions is found when applying the **renormalization** theory. Forces described by Abelian theories become stronger at higher energies, while the forces described by non-Abelian theories become weaker at higher energies. The converse is also true; non-Abelian forces become stronger at lower energies, but Abelian forces become weaker at lower energies.

The strong **force** described by QCD is extremely strong at low energies, resulting in a property of quarks known as confinement. Quarks are confined in either nucleons or mesons at low energies. They cannot be pulled apart, explaining why lone quarks do not exist. At high energies accessible to particle **accelerators**, however, the QCD force is much weaker. This explains why **scattering** from nucleons can be described accurately by describing the quarks as free particles inside the **nucleon**.

Because of its strength at low energies, low **energy** calculations in QCD theory are very difficult. The usual procedure for accomplishing calculations in **quantum field theory** is called **perturbation theory**, but it requires that the force be weak. Difficult differential equations of QCD, therefore, must be solved explicitly using computer algorithms. One successful technique, lattice QCD, has been developed, where calculations are done on a discrete lattice, and at the end of the calculation, a limit of the answer is taken when the lattice spacing is zero. Because the calculations require much computing **time** and clever algorithms, development of lattice QCD has been very challenging. Although QCD is well verified, it remains to many physicists a frustrating theory because of the difficulty in calculating its low-energy effects.

QUANTUM COMPLEMENTARITY

Quantum complementarity is a notion originated by one of the fathers of **quantum mechanics, Niels Bohr**, in an attempt to explain apparent inconsistencies in the structure of **quantum theory**. The general principle of complementarity states that a complete understanding of nature may require two (or more) mutually exclusive yet equally necessary descriptions of nature. The validity of either description of nature can be tested through specific experiments, but the observation of one particular description requires experiments that preclude the possibility of observing the other mutually exclusive, or complementary, descriptions. The fact that it is experimentally impossible to observe both descriptions at once is crucial in rendering the whole theory consistent.

The idea of wave-particle duality embodies the concept of complementarity, and indeed it was Bohr's attempts to understand wave-particle duality that gave rise to the idea of complementarity in general. **Light**, although thought of classically as a wave phenomenon, was shown by **Albert Einstein** to obey particle like properties in his theory of the **photoelectric effect**. **Matter**, classically thought of as particles, was shown experimentally by **Clinton Davisson** and **Lester Germer** in their **electron diffraction** experiments to act like **waves**. The principle of complementarity asserts both wave and particle descriptions of nature, although seemingly contradictory, are both essential.

Once way to understand this apparent inconsistency is through an analysis of the famous double-slit experiment, in which light is shone through a wall punctured by two slits and the intensity of transmitted light is measured on the other side. If one of the slits is blocked, the intensity pattern will be a hill-shaped distribution centered on the opposite, unblocked slit signifying that the particle like trajectory of light traveled through the unblocked slit. If both slits are open, the resulting intensity profile displays an **interference** pattern that signifies that light, as a wave, traveled through both slits at once.

Now suppose one wishes to observe both the particle and wave aspects of light at once by keeping both slits open but also recording which slit the photons, or particles of light travel through. In order to measure the trajectory of the **photon**, one would have to place a measuring device at one of the slits to detect photons travelling through that slit. This measuring device disturbs the photon travelling through the slit and the final result interrupts the wave-like pattern. Thus the interference pattern characteristic of the wave behavior of light travelling through both slits at once is destroyed in an attempt to observe particle behavior. It is in this sense that light can

exhibit both particle and wave behaviors but not at the same **time**. Certain experiments will exhibit wave behavior, certain experiments exhibit particle behavior, but none can exhibit both.

Heisenberg's uncertainty principle is another manifestation of quantum complementarity. It states that upon measuring both the **momentum** and position of a particle, the product of the uncertainties has a fundamental lower bound proportional to **Planck's constant**. Hence one cannot measure both position and momentum in the same direction to infinite accuracy. They are complementary variables; the measurement of one precludes the measurement of the other. This can be understood relatively simply in terms of experiment. Suppose one wanted to measure the position of a particle to a high degree of precision. One would have to bombard with a short **wavelength** and hence high-energy photon and deduce its position from the reflected photon. But after this measurement, the momentum of the initial particle is completely uncertain. Indeed there is no experimental setup that can measure both quantities to arbitrary accuracy. Thus both in the wave-particle picture and in the position-momentum uncertainty principle, the general principle of quantum complementarity appears not as some sort of mystery but rather takes into account in a consistent way the role of the observer in **quantum mechanics**.

QUANTUM COMPUTERS

The hope to see quantum **computers** materialize in the near future is based on a fundamentally new mode of information processing that is replacing the current theories of computation and information technology with quantum generalizations. They are a direct consequence of concepts derived from quantum **mechanics** among which quantum **interference** and the principles of superposition and entanglement are usually described as the underlying principles of the new science of quantum computing. Used together, they have the potential to solve problems far beyond the capabilities of any present-day computers.

Quantum interference is a principle that states the outcome of a given process depends on its past history and it gives quantum computing the potential of being significantly more powerful than our current computing technology. Superposition is related to the properties of **waves**. A string vibrates in a pattern having evenly spaces **nodes**, or points of zero amplitude displacement, and antinodes, or points of maximum amplitude displacement. The principle of superposition states that when two or more standing waves occur, the displacement at any position will be the sum of the displacements at that position for each individual wave. Since **quantum theory** describes electrons as having both particle and wave properties, they can also adopt several superposition states. Entanglement is defined as the ability of quantum systems to consist of correlated superposed states. In an information processing context, quantum entanglement could allow mathematical operations to be performed on a considerable superposition of numbers in parallel without requiring additional circuits for each part of the superposition.

The possibility of quantum computation started as early as 1982, as **Richard Feynman** was thinking about simulating quantum objects by quantum methods. But the credit is usually given to Paul Benioff of the Argonne National Laboratory who first applied quantum theory to computers in 1981 and to David Deutsch of the University of Oxford who published a theoretical paper in 1985 describing a quantum computer. At the time, the only potential application had to do with solving complex and esoteric mathematical problems. Quantum computers seemed little more than an esoteric specialty of theoretical physics. In 1994 however, the first quantum algorithm was developed by Peter Shor from AT&T's Bell Laboratories in New Jersey and its concrete usefulness for efficient factorization was immediately realized by the computing community. An improved factorization efficiency is required for most common methods of encryption, which rely on the inherent difficulty of factoring large numbers. Among the current most active quantum computing research areas are the following applications: cryptography (development of completely secure communication); searching technology, especially algorithmic searching; rapid factorization of large numbers; and the efficient simulation of quantum-mechanical systems.

The new concept of quantum computation developed over the past few years has the potential of tremendous computational power. To see what would make a quantum computer so much more efficient than a present-day computer, we need to understand the difference between a standard bit and a quantum bit, or qubit. From a physical point of view a bit can be described as a system that can exist in one of two different states represented by two logical values such as yes or no, true or false, or with the binary digits 0 or 1. In a digital computer, the **voltage** between the plates in a capacitor represents the bit of information: when the capacitor is charged, the bit value is 1 and if not, the bit value is 0. But one bit of information could also be encoded using two different electronic states of an **atom** or an ion, because the wavelike property of the atom allows it to exist not only as two distinct electronic states but also as a superposition of the two states. Such "ion" qubits have been made at the National Institute of Standards in Boulder and at the Los Alamos National Laboratory that differ from the standard digital bit in that they can store intermediate values, i.e. between 0 and 1. A standard two-bit register using the base-2 number system can represent any number between 0 and 3, because the two bits, taken together, can have the following values: 00, 01, 10, and 11. These are the base 2 representations for 0, 1, 2, and 3. But a quantum two-qubit register can represent the numbers between 0 and 3 simultaneously, due to its capacity to be put into a superposition of any or all of the states allowed by the standard register. Thus, an eight qubit register represents a superposition of $2^8 = 256$ numbers; similarly, a quantum register of 1,000 qubits could represent approximately 21,000 (~ 10300 numbers) simultaneously. Compared to the standard register of 1,000 bits, which can represent only one at a time any of the integers between 0 and approximately 10300, the qubit-register is considerably more

powerful for generating massive parallel computation. This has led to the realization that, compared to a digital computer, a quantum computer would be much more powerful so as to represent a technology leap significantly greater than the one between an abacus and the world's fastest present-day supercomputer.

Although no usable quantum computer has yet been developed, many of the required mechanisms are being fine-tuned. The real power of quantum information technology lies in the new quantum algorithms recently discovered and allow the solution of problems that cannot be solved on present-day digital computers. The new quantum algorithms can solve such problems because they exploit quantum superposition and thus contain an exponential number of different terms. A new theory of quantum communication and information is emerging that offers much more than faster microprocessors. Potential experimental realizations of quantum computers have been proposed, and quantum gates as well as simple quantum networks have also been demonstrated experimentally. New types of quantum computations offering significant advantages over digital computation are continually being discovered and the new quantum computing science is becoming an integral part of the physics seeking to understand both quantum theory and information processing.

See also Quantum mechanics; Quantum states; Quantum physics; Standing waves and resonance

QUANTUM COSMOLOGY

Physicists believe in a unified description of the world. The complex structures and phenomena we observe today, in every physical system, have at their heart one set of conceptually simple rules, although the mathematical formalism can be staggeringly complex. **Cosmology** is the study of the structure and evolution of one such system, the **Universe**, and the same set of rules must apply; there are no "special systems." And these basic rules, which seem to **work** so extraordinarily well for every other system, are quantum-mechanical in nature.

The gist of quantum cosmology, the description of the entire Universe in terms of **quantum theory**, is simply this: *every* system, including the system we call "the Universe" must be described by a quantum-mechanical state function which encapsulates all the "knowledge" of that system. At the heart of the quantum cosmological program lie two goals. The first is the elucidation of the "correct" state function—that which predicts (or, quantum-mechanically speaking, assigns a high probability to) a late universe with the features that we observe today. The second is gaining an understanding of the mechanism by which that particular initial state is selected from the set of all possible states.

Quantum cosmology is thus the theory of the Universe's initial conditions. And, as the dominant **force** at cosmological scales is **gravity**, the quantum cosmological program requires a quantum treatment of gravity—a quantized version of Einstein's general theory of relativity.

The foundations of the quantum-cosmological program were laid in the 1960s, with the work of Bryce DeWitt, **John Wheeler**, and Charles Misner. (The "equation of motion" in quantum cosmology, analogous to the Schrödinger equation in non-relativistic quantum **mechanics**, is known as the Wheeler-DeWitt equation.) Largely unexplored, the field underwent a rebirth in the early 1980s, when classical cosmology proved unable to explain fully the Universe's very early history.

Why had we been so successful, for so long, in treating the Universe as a classical system? The enormity of cosmological length scales, for all but the first fraction of a second of the Universe's existence, justified that approach. (If the length scale of a system is much larger than the relevant **de Broglie wavelength**, classical laws suffice for its description.) Why, then, did we find that quantum theory must be incorporated in the study of cosmology? There are two reasons, one somewhat philosophical, and one practical. The universe is fundamentally quantum in nature; the observed classical "laws" governing any macroscopic system obtain from the underlying quantum laws in the limit of large **quantum numbers**. This is Bohr's "correspondence principle." In this sense, then, quantum theory explains behavior on *all* scales. And we know there is an epoch in the Universe's history when the classical limit is certainly not justified. At very early times the Universe was very small indeed—big bang cosmology predicts that at the Planck **time**, 10^{-44} s after the **big bang**, the observable universe was forty orders of magnitude smaller than a **proton**. At such length scales a quantum description is mandatory. Today, after 10-20 billion years of expansion, cosmological quantum effects are unimportant—they are too small ever to observe. But, there was an extraordinarily brief (and, as it happens, extraordinarily important) period in the evolution of the Universe in which the classical approximation may not be applied, and quantum effects dominated. Those quantum effects are ultimately responsible for the existence today of large-scale structure—galaxies and galactic clusters. They also provide a physical, rather than metaphysical mechanism for cosmogenesis. The most remarkable prediction of quantum-cosmological models is that the Universe may have originated *ex nihilo* via a process analogous to quantum-mechanical **tunneling**.

QUANTUM ELECTRODYNAMICS (QED)

Quantum **electrodynamics** (QED) is a complex and highly mathematical theory regarding the interaction of electromagnetic **radiation** with **matter**. The development of QED theory was essential in the verification and development of **quantum field theory** and it allows physicists to predict how **subatomic particles** are created or destroyed. QED is a fundamentally important scientific theory that accounts for all observed physical phenomena except those phenomena associated with aspects of general **relativity theory** and **radioactive decay**. QED is compatible with special relativity theory and special relativity equations are incorporated into QED equations.

QED is also termed a gauge-invariant theory because its predictions are not affected by variations in **space** or **time**.

The practical value of QED theory is that it allows physicists to make calculations regarding the absorption and emission of **light** by atoms. In addition, QED provides very accurate predictions regarding the interactions between photons and charged atomic particles such as electrons.

During the first half of the twentieth century physicists struggled to Scottish physicist **James Clerk Maxwell**'s (1831-1879) equations regarding **electromagnetism** with emerging quantum and relativistic theories advanced by German physicist Maxwell Planck, Danish physicist **Niels Bohr**, German-American physicist **Albert Einstein** and others. Prior to WWII, English physicist **Paul Dirac**, German physicist **Werner Heisenberg**, and Austrian-born American physicist **Wolfgang Pauli** made significant independent contributions to the mathematical foundations related to QED. Working with QED theory initially proved difficult, however, because of infinite values in the mathematical calculations (e.g., for emission rates or determinations of **mass**). Early QED predictions often failed to match experimental data. Subsequently, QED calculations were made more reliable by a process termed *renormalization* (allowing positive infinities to cancel out negative infinities) and other advances developed independently by American physicists **Richard Feynman** and **Julian Schwinger**, and Japanese physicist Sin-Itîro Tommaga.

The use of **renormalization** initially allowed QED theorists to use measured values of mass and charge in QED calculations. The result made QED a highly reliable theory with regard to its ability to predict and reflect the observed interactions of electrons and photons. QED theory was, however, revolutionary in theoretical physics because of the nature and methodology of its predictions. QED reflected a growing awareness of limitations on the ability to make predictions regarding behavior of subatomic particles. Instead of making predictions resulting from mechanistic cause-and-effect interactions, QED relies on an understanding of the probabilities associated with the quantum properties and behavior of subatomic particles to allow the calculation of probabilities regarding outcomes of subatomic interactions.

As **quarks**, gluons and other subatomic particles became known, QED became an increasingly important in explaining the structure, properties and reactions of these particles. QED, also known as the **quantum theory** of light, eventually became one of the most precise, accurate, and well tested theories in science. QED predictions of the mass of some subatomic particles, for example, offer results accurate to six significant figures or more.

QED describes the phenomena of light in ways that are counter-intuitive (not typical of everyday experience) because QED treats the quantum properties of light (properties that are conserved and that occur in discrete amounts called **quanta**). According to QED theory, light exists in a particle and wave-like dualities (i.e., the electromagnetic wave has both particle and wave-like properties). Electromagnetism results from the quantum properties of the **photon**, the fundamental particle responsible for the transmission or propagation of electromagnetic radiation. Unlike the particles of everyday experience, photons, can also exist as **virtual particles** that are constantly exchanged between charged particles and the forces of **electricity** and **magnetism** arise from the exchange of these virtual photons between charged particles.

The most accurate and complete definitions of virtual particles (e.g. virtual photons) are mathematical. Most non-mathematical descriptions, however, usually describe virtual photons as wave-like (i.e., exisitng in form like the wave on a surface of water after it is touched). According to QED theory, virtual photons are passed back and forth between the charged particles somewhat like basketball players passing a ball between them as run down the court. Only in their cloaked or hidden state do photons act as mediators of **force** between particles. The force caused by the exchange of virtual photons results from changes charged particles change their **velocity** (speed and/or direction of travel) as they absorb or emit virtual photons.

As virtual particles, photons are cloaked from observation and measurement. Accordingly, as virtual particles, virtual photons can only be detected by their effects. The naked transformation of a virtual particle to a real particle would violate the laws of physics specifying the **conservation of energy** and **momentum**. Photons themselves are electrically neutral and only under special circumstances and as a result of specific interactions do virtual photons become real photons observable as light.

QED theory accounts, for example, for the interactions of electrons, positrons (the positively charged **antiparticle** to the **electron**), and photons. In electron-positron fields, electron-positron pairs come into existence as photons interact with these fields. According to QED theory the process also operates in reverse to allow photons to create a particle and its antiparticle (e.g., an electron and a **positron**).

QED mathematically describes a force similar to **gravity** in that it becomes weaker as the distance between charged particles increases. Like gravity, the force reduces in strength as the inverse square of the distance between charged particles. Moreover, the concept of forces such as electromagnetism arising from the exchange of virtual particles may carry profound implications regarding the advancement of theories relating to the strong, electroweak, and gravitational forces. Some physicists assert that if a unified theory can be found, it will rest on the foundations and methodologies established during the development of QED theory.

See also Feynman diagrams; Quantum chromodynamics theory; Quantum theory; Relativistic electromagnetism

QUANTUM FIELD THEORY

Quantum mechanics was born at the turn of the twentieth century when physicists realized that the behavior of the particles of matter—the atoms composing Max Planck's classic black body, for example—did not obey Isaac Newton's classical

laws. Field theory is more than half a century older. **Michael Faraday** introduced the field concept in 1831.

Unwilling to accept the notion of "action-at-a-distance" to describe the electrostatic interaction between two charges, Faraday postulated the existence of the "electric field" filling all **space** around the particles. **Electric charge** both creates the field—it is the field's "source"—and reacts to the presence of a field. The electric **force** felt by a charge is a measure of the field's strength at that point in space.

By 1926 **Paul Dirac**, **Werner Heisenberg**, **Erwin Schrödinger**, and others had formulated non-relativistic **quantum mechanics**, the quantum analog of classical mechanics. Describing the low-energy interactions between the fundamental particles of **matter**, the theory worked incredibly well. By the mid-1920s, then, theoretical physics handled well the regimes in which most interactions took place. Classical mechanics provided the correct description for interactions between large, slow-moving bodies. Quantum mechanics (which subsumed, rather than replaced classical mechanics) was employed when bodies were small and slow-moving. Relativistic mechanics—which also subsumed classical mechanics—yielded the correct description for large, fast-moving bodies. But the story could not end there. What if one wished to describe interactions between small, fast-moving bodies (the subject matter of what is today called "high-energy physics")? A new model was needed, one which combined both quantum mechanics and special relativity.

Such a theory is required for consistency as well. Particles interact with one another through the mediation of a field, for example, the **electromagnetic field**. Non-relativistic quantum mechanics treats these two entities in fundamentally different ways—the particles are "quantized," but the fields are treated classically. Consequently, non-relativistic quantum mechanics violates the principle of relativity. While **Maxwell's equations** describing the field are Lorentz invariant, the quantum mechanical Schrödinger equation is not.

So, how, exactly, does quantum field theory differ from quantum mechanics? The crucial difference is this: Einstein's mass-energy equivalence implies that **energy** can be converted into **mass** (particles) and vice versa. Thus, the number of particles is not constant during a high-energy interaction. (Or, as a physicist would say, particle number is not conserved.)

Consider an example: in 1911 **Ernest Rutherford** deduced the existence of the atomic **nucleus** by bombarding a thin gold foil with low-energy alpha particles, and observing how the alphas were "scattered" (deflected from their original paths) by their electromagnetic interaction with the gold atoms. The initial state of the system contains two particles, the incoming alpha and the target gold nucleus, and the final state also contains two particles, the gold nucleus and the scattered alpha. There is insufficient energy involved in the collisions to produce (i.e., be converted into) new particles. Schrödinger's equation, which treats the matter particles quantum-mechanically (they are represented by wavefunctions) and the electromagnetic field classically, provides an adequate description of this low-energy interaction.

But high-energy interactions are much more complicated. Neither the number, nor the type of particles need be conserved. An **electron** and a **positron** (the electron's **antiparticle**) can collide and wind up with, say, an electron and a positron, or with two muons, or with an electron and a positron and two muons. Or it finish with no matter particles at all, only photons (the "quanta" of the electromagnetic field). The number of "final states" is limited only by the initial energy of the system.

In order to deal with this, an inherently democratic theory emerges. In the spirit of Niels Bohr's principle of complimentarity, in quantum field theory all entities—matter and fields—are quantized, and thus posses both wavelike and particlelike properties. Formally, matter "particles" and fields alike are represented by "field operators," called such because they perform the operation of creating or destroying the **quanta** of the field. The quanta of the electron field are electrons, the quanta of the positron field are positrons, and the quanta of the electromagnetic field are photons. Interactions are represented by products of these operators, which change the initial state of the interacting system into the final state—which may differ in both the number and type of particles present.

The new synthesis was brought to fruition by a number of physicists, most notably the Americans **Richard Feynman** and **Julian Schwinger**, and the Japanese Shin-Ichiro Tomanaga, who shared the 1965 Nobel Prize in physics for their work. The prototypical quantum field theory, it is called "quantum electrodynamics." The accuracy of the theory is staggering. For example, the measured value of the electron's **magnetic moment** (since an electron is charged, and possess "spin," it acts like a tiny bar magnet), and the theoretical value agree to within one part in one hundred billion. A measurement of the distance between New York and San Francisco, performed to this accuracy, would be off by the thickness of a human hair. This extraordinary level of agreement between theory and experiment provides powerful justification for belief in both the correctness of the principle of relativity, and the description of particle interactions in terms of quantum field theories.

QUANTUM GRAVITY

Relativity and **quantum theory**, the foundations upon which all of modern physics rests, were developed in the first quarter of the twentieth century. Their accuracy in predicting the results of experiment or observation is astounding; the symmetry principles at their core are among the most beautiful and profound concepts in science. The union of these two models of nature in relativistic **quantum field theory** led to the development of the standard model of elementary particles and their interactions. The standard model is incomplete, however, it lacks a quantum explanation of the **gravitational force**.

Gravity is the weakest of the four fundamental forces of nature. Although the effects of gravitational **force** are easily calculated by the classical laws developed by English physicist and mathematician Sir **Isaac Newton** in his *Philosophiae*

Naturalis Principia Mathematica (Mathematical Principles of Natural Philosophy) or by the modern relativity physics developed by German-American physicist **Albert Einstein** (1879-1955), the actual mechanisms of the gravitational force remain elusive.

Almost immediately upon completion of his special theory of relativity in 1905, Einstein realized his work was not yet done. Special relativity allowed only inertial observers, and thus was too restrictive. Any observer, Einstein felt, even an accelerated one, must be free to deduce physical law based on his own observations, and all observers must arrive at the same mathematical structure. By 1915, Einstein published the general theory of relativity. The general theory provides a structure for the geometry of **space-time**, and, because the effects of **acceleration** are indistinguishable from those of gravity, a theory of the gravitational force. In general relativity, gravity is a curvature of space-time.

The predictions of general relativity differ, of course, from those of Newtonian gravitational theory, but those differences become significant only for very strong gravitational fields—those produced by astronomical scale masses. General relativity has long been the purview of those physicists, astrophysicists, and cosmologists who regularly deal with such large scales.

From a purely pragmatic point of view it is not necessary to include gravity in the discussion of elementary particles and their interactions. The gravitational interaction between elementary particles is incredibly weak, twenty-nine orders of magnitude smaller than the weak force. At the elementary particle scale space-time is flat, and special relativity suffices. The (specially) relativistic quantum field theories describing all non-gravitational particle interactions, including the **electroweak theory** of Weinberg and Salam and quantum chromodynamics (the field theory of the strong nuclear force) work extremely well. Regardless, there are two important reasons why a quantum theory of gravity is necessary. The first reason is aesthetic: physicists argue that ours is a quantum **universe**, and all interactions, at the most fundamental level, must therefore have a quantum description. **Quantum physics** itself provides a second reason: Heisenberg's uncertainty relations demand that all fields undergo quantum fluctuations.

Consider a region of **space** in which the **gravitational field** vanishes. Quantum theory tells us that it is really the average value of the field that is zero; for very brief periods of **time**, and over very small regions of space, the field strength undulates, and may become relatively quite large. At the so-called Planck scale (10^{-33} cm), quantum fluctuations in the strength of the gravitational field become comparable to the strength of all other forces, and cannot be ignored. The Planck length is almost unimaginably tiny. Imagine a **proton** the size of the **Milky Way galaxy**. On that scale the Planck length would be about four inches. Accordingly, if we wish to understand the Universe at these scales, a quantum theory of gravity is a necessity.

Undulations in gravitational field strength are undulations in the **space-time geometry**. At the Planck scale, space-time is no longer the smooth, continuous manifold described by general relativity, but rather a frothy, roiling structure, dubbed "space-time foam" by American physicist John A. Wheeler (1911-). The heart of **relativity theory**, a smooth space-time described by Einstein's equations, is incompatible with the heart of quantum theory which allows uncertainty between physical observables. It is for this reason that all attempts to quantize general relativity have failed. All such theories are non-renormalizable: calculations of physically measurable quantities diverge; they yield infinite answers.

If somehow the behavior of space-time at the Planck scale and below never entered the theory, the divergences could be avoided. This is not possible, though, if the theory contains point particles. They are infinitesimally small, and thus require treatment at an extreme space-time scale. But if the fundamental entities had a finite size, space-time behavior at smaller scales would be irrelevant. **String theory**, which replaces the fundamental particles with Planck-sized strings, is renormalizable, and for that reason string theory may offer the best hope of yielding a viable quantum theory of gravity.

See also Bell's theorem; Grand Unified Theory; Gravitational field; Gravitational radiation and waves; Graviton; Gravity; Quantum chromodynamics theory; Quantum cosmology; Quantum electrodynamics (QED); Quantum gravity; Relativity, general; Relativity, special; Renormalization

QUANTUM MECHANICS

Quantum mechanics is the theory used to provide an understanding of the behavior of microscopic particles such as electrons and atoms. However, it is more than just a collection of formulae used by physicists and chemists to calculate, for example, where an **electron** might be. The **quantum theory** also introduced an entirely new way of thinking about very small objects that is strangely different from the way we think about macroscopic objects.

An example of a macroscopic object is a baseball. Whenever we throw a ball into the air it is a good idea to know where it will fall and how fast it will be traveling when it hits something. The most exact way to describe the ball's **motion** is by using *classical mechanics* which predicts the position and **velocity** of the ball at every instant during its flight. This approach fits our everyday experience since we are accustomed to seeing a ball move in a very well-defined path.

The problem comes when we try to apply the classical approach to microscopic objects. If an electron were just an exceptionally small ball, its motion would follow a path predicted by classical mechanics. However, experiments have shown that this is not the case. The best illustration of this is the "double-slit experiment," in which electrons are sent one at a time towards a wall with two small slits or holes. On the other side of the wall is a screen of some sort which detects where electrons hit, perhaps by making a spot. If one slit is covered, electrons can pass through the uncovered hole and strike the screen. The hits on the screen are directly behind the open slit, exactly what we would expect. We get the same

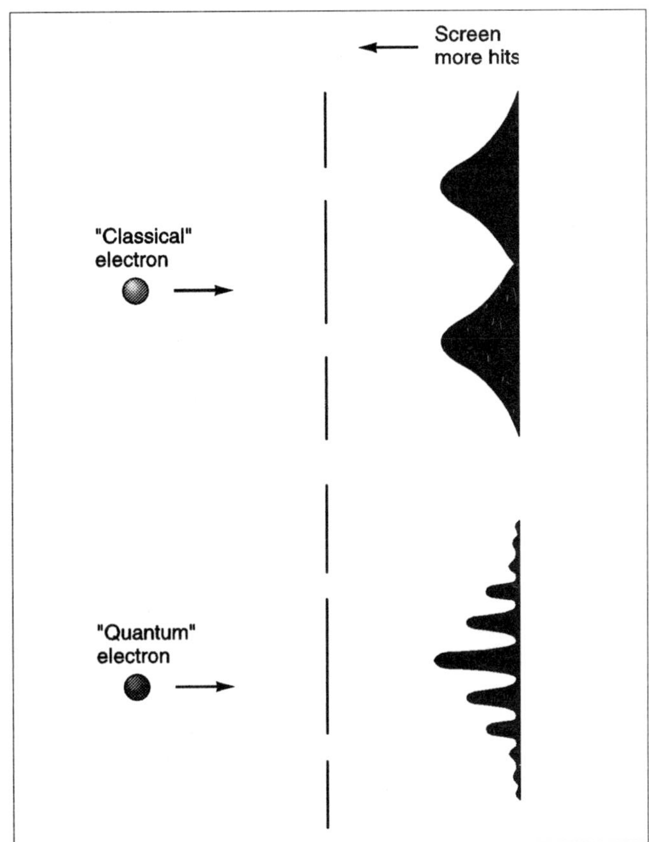

Electrons, clasical and quantum.

result by opening the first slit and covering the second, only now the pattern of spots is behind the first hole.

What if both slits are open? Taking a single baseball and throwing it at the wall, we know it will either pass through one slit or the other. Of course, it might miss the holes and never make it to the screen, but that case is not particularly interesting. Our experience with macroscopic objects makes us think that the electrons will behave the same way so the screen should show electron hits only directly behind the holes. However, this is not what happens. Instead the spots are spread out over the entire screen, even in places that should be blocked by the wall as shown here. The only way for such a pattern to occur is if each electron somehow travels through both open slits! You might think that someone just made a terrible mistake when they conducted the experiment but many scientists have now verified this result (since they didn't want to believe it either after spending years learning classical mechanics). Apparently an entirely new way of thinking must be used for microscopic objects.

The pattern of electron hits that occurs when both slits are open appears very strange for particles. However, similar patterns are commonly produced by *waves* which are disturbances in a medium that carry **energy**. Suppose you filled a bathtub with water and then tapped the surface with your finger after the water had become still. By disturbing the water you create a wave that will move on the surface away from your finger. The water acts as the medium through which the wave moves. If you place a piece of paper in the water, you will notice that when the wave passes through that location the paper will be disturbed. Then the paper becomes still as the wave proceeds on. This is an example of a *traveling wave* which carries the energy of your tap (in the disturbance) from place to place in the medium.

Now put a barrier in the bathtub with two openings in it. What happens if you start a wave moving toward the barrier? You would see that parts of the wave pass through the openings, splitting the original wave into two separate **waves**. As those new waves continue on the other side of the barrier, they recombine (or interfere). At some locations they reinforce and produce an even larger wave (constructive interference) and in others they cancel (destructive interference). You can experience the same effect by listening to the **sound** waves coming from two speakers producing the same musical note. If you move around (some distance away from the speakers), you will find places where the sound is louder. The two speakers are producing separate but identical waves which are interfering. By marking the loudest locations on a map, you would draw an interference pattern similar to that made by the water wave after passing through the two-opening barrier. This is the same pattern that electrons make when both slits are open. The conclusion is that electrons act more like waves than they do like solid objects.

Let's try another wave experiment before talking more about electrons. Suppose instead of one wave, we produce a sequence of waves by tapping the water in a rhythm, or frequency. As the waves move through the water, the piece of paper would bob up and down at the same **frequency**. However, for certain rhythms the traveling waves and their reflections from the walls of the bathtub reinforce in an important way. The paper continues to bob up and down as before, but the disturbance in the water appears to be stationary (the wave does not travel). This is called a standing wave. The certain allowed frequencies that produce standing waves are determined by the dimensions of the bathtub whose walls confine the traveling waves. It is even easier to see a standing wave in the case of a guitar string. Striking the string sends many different traveling waves of different frequencies towards the ends which are held still. Standing waves occur for certain frequencies and those correspond to particular musical notes. Holding the string down at some point confines the waves more, and different standing waves appear on the string producing different notes. The energy of the wave can have any amount, depending on how hard the string was struck, and more energy translates into a louder sound (of an allowed frequency). We encounter these kinds of waves in our macroscopic world and because their motion can be understood using classical mechanics, they might be called classical waves.

Waves produce **interference** patterns like that of electrons in the double-slit experiment and we are accustomed to waves moving in less well-defined paths than solid objects. The new idea of quantum mechanics is to use waves to represent microscopic objects. These waves obey many of the properties of classical waves with an important conceptual difference. The "disturbance" of the new wave is not a physical motion of a medium, like the raised surface of water, it is an increased probability that the object is at a particular location. For an electron in the double-slit experiment, its traveling wave passes through both openings and interferes to produce a pattern of probability on the screen. The most probable locations (where constructive interference occurs) are the places where more electrons will hit and this is how the pattern of probabilities becomes a measurable pattern on the screen. This approach makes sense but we must give up the idea of predicting the path of objects such as electrons. Instead, in quantum mechanics we concentrate on determining the probability of obtaining a certain result when a measurement is made, for example of the position of an electron hit. This makes measurable quantities, called observables, particularly important.

Quantum mechanics requires advanced mathematics to give numerical predictions for the outcome of measurements. However, we can understand many significant results of the theory from the basic properties of the probability waves. An important example is the behavior of electrons within atoms. Since such electrons are confined in some manner, we expect that they must be represented by standing waves that correspond to a set of allowed frequencies. Quantum mechanics states that for this new type of wave, its frequency is proportional to the energy associated with the microscopic particle. Thus, we reach the conclusion that electrons within atoms can only exist in certain states, each of which corresponds to only one possible amount of energy. The energy of an electron in an **atom** is an example of an observable which is quantized, that is it comes in certain allowed amounts, called quanta (like quantities).

When an atom contains more than one electron, quantum mechanics predicts that two of the electrons both exist in the state with the lowest energy, called the ground state. The next eight electrons are in the state of the next highest energy, and so on following a specific relationship. This is the origin of the idea of electron "shells" or "orbits," although these are just convenient ways of talking about the states. The first shell is "filled" by two electrons, the second shell is filled by another eight, etc. This explains why some atoms try to combine with other atoms in chemical reactions. If an atom does not contain enough electrons to fill all of its lowest-energy shells completely, and it comes near another atom which has all its shells filled plus extra electrons, the atoms can become held together. The "bond" between the atoms comes from sharing the extra electrons. Thus quantum mechanics provides an understanding of chemistry by explaining why atoms combine, such as sodium and chlorine in salt.

This idea of electron states also explains why different atoms emit different colors of **light** when they are heated. Heating an object gives extra energy to the atoms inside it and this can transform an electron within an atom from one state to another of higher energy. The atom eventually loses the energy when the electron transforms back to the lower-energy state. Usually the extra energy is carried away in the form of light which we say was produced by the electron making a *transition*, or a change of its state. The difference in energy between the two states of the electron (before and after the transition) is the same for all atoms of the same kind. Thus, those atoms will always give off light of that energy, which corresponds to a specific **color**. Another kind of atom would have electron states with different energies (since the electron is confined differently) and so the same basic process would produce another color. Using this principle, we can determine which hot atoms are present in **stars** by measuring the exact colors in the emitted light.

Although quantum mechanics in its present form is less than 70 years old, the theory has been extremely successful in explaining a wide range of phenomena. Another example would include a description of how electrons move in materials, like those that travel through the chips in a personal computer. Quantum mechanics is also used to understand superconductivity, the decay of nuclei, and how **lasers** work. The list could go on and on. A great number of scientists now use quantum mechanics daily in their efforts to better understand the behavior of microscopic parts of the **universe**. However, the basic ideas of the theory still conflict with our everyday experience and the argument about their meaning continues among the same physicists and chemists who use the quantum theory.

If thinking about electrons and other particles as waves seems unsettling, you're in good company. That idea has been hotly debated since the beginning of quantum mechanics by some of the greatest physicists. The search for a new theory actually began in the late 1800s when several phenomena involving light were discovered which could not be understood using classical mechanics. One of those phenomena is called "blackbody radiation," the behavior of light within a box held at a certain **temperature**. It had been known for many years that light often behaves like a wave so physicists attempted to explain **blackbody radiation** as a case of standing waves within the box. Those standing waves were treated classically, with only certain allowed frequencies but any amount of energy. The results of that approach seemed promising, but did not quite agree with the experiments.

In 1900, **Max Planck** published a paper that explained blackbody **radiation**, but to do so he had to assume that the energy of the standing waves was quantized. Since the frequencies already had only certain allowed values, he made the simplest assumption possible, stating that the energy was equal to the frequency multiplied by a constant. That constant was written as h and it was later named after him. Planck did not know why the energy had to be quantized; he simply had to introduce the idea of energy **quanta** to make the theory agree with the experiments. The quantum idea was used by other physicists including **Albert Einstein** to explain more phenomena involving light which had eluded the classical approach. Einstein thought of light as being a wave "packet" which became known as a *photon.*

The concepts of energy quanta and photons were not accepted immediately, since most physicists believed that light was a classical wave. However, at least the new idea was not totally foreign since **Isaac Newton** had attempted to treat light

as particles about 200 years before. Eventually it was accepted that a light wave could behave like a particle in some circumstances. However, in 1924 **Louis de Broglie** suggested that the opposite might also occur; that a solid object could correspond to what he called a *matter wave.* This was a revolutionary and unsettling idea, but many physicists set about to determine where it might lead. In 1926, **Erwin Schrödinger** developed an equation for those waves so that the theory could be used to make numerical predictions for the behavior of many physical systems. The Schrödinger equation proved to be extremely useful in describing microscopic objects and in fact it remains the cornerstone of quantum mechanics even now. However, the real breakthrough was that energy quanta came naturally from the mathematics instead of having to be assumed as Planck had done. This was because the equation for **matter** waves was similar, but different in a subtle way from the equation for classical waves. Schrödinger called his new approach *wave mechanics* and the name quantum mechanics came later.

The usefulness of quantum mechanics can hardly be argued. However, the basic concepts were a source of debate from the very beginning. That tradition is best exemplified by the historical debates that started in the late 1920s between two giants of physics, Einstein and **Niels Bohr**. In 1927 **Werner Heisenberg** had recognized an important result of treating microscopic particles as waves that described their probability of being in a certain location. Heisenberg derived the *uncertainty relation* that said the uncertainty in a measurement of a particle's position multiplied by the uncertainty in its velocity must always be larger than a specific constant, which unsurprisingly includes **Planck's constant**. For example, if we exactly measure where a particle is at a particular time (uncertainty in position is almost zero), then we do not have any information about the velocity of the particle (uncertainty in velocity is extremely large). That means we have no idea were the particle is going. The uncertainty relation restates that we cannot think of well-defined paths for microscopic particles.

Many different interpretations arose to make sense of this and other facets of quantum mechanics, and Bohr was the leading proponent of the *Copenhagen interpretation* (since that is where he was from). Today this is the accepted viewpoint among most, but not all physicists. The Copenhagen interpretation considers the uncertainty relation to express a fundamental limit to how accurately we can measure properties of microscopic particles. A way to understand this is to think how we detect a macroscopic object such as a car. We actually detect something that has bounced off the car, such as light. Suppose instead of light, we use a baseball (not really recommended) and decide whether or not a car was as at a location at a particular instant based on whether the ball bounces back to us after being thrown. The car is much larger and heavier than the baseball so we can detect where the car is without significantly disturbing the path of the automobile. However, what would happen if we tried to use a baseball to detect another baseball? We could find where the ball to be measured was located, but in the process it would be knocked off its original path into a new direction. If on the other hand we bounce a ping pong ball off a baseball, we have less exact information where the baseball is at a certain instant, but at least we disturb its path less. In every instance we have to interact (disturb) with an object to measure its position and the uncertainty relation simply reflects this. Why don't we notice this everyday? The answer is in the size of Planck's constant. It is so small that for macroscopic objects like a baseball the uncertainties in position and velocity are unmeasurable so the ball moves in a well-defined path. Quantum mechanics is actually around us all the time, but we just don't notice it.

Einstein had used quantum ideas, but he remained dissatisfied with quantum mechanics and particularly with Bohr's interpretation. Beginning in 1927 they began a public debate over the meaning of the new theory that raged for many years in scientific publications and often in person at conferences. Einstein felt that there must be some experiment which permitted exact measurements of position and velocity for microscopic objects. He continually challenged Bohr with suggestions of new experiments. However, in every instance Bohr was able to refute Einstein's experiments with arguments of his own that supported the Copenhagen interpretation. Einstein eventually gave up that approach but still maintained that quantum mechanics was somehow incomplete and that once the missing ideas were found, uncertainty would disappear. This search for a missing piece of the puzzle, if there is one, continues as physicists attempt to devise new experiments to more clearly understand the meaning of quantum mechanics.

QUANTUM NUMBERS

Quantum numbers are of basic physical importance in **quantum physics**. In the early 1900s **energy** in atomic systems was found to exist as **quanta** or discrete, indivisible units. This concept along with the **particle-wave duality** led to the development of quantum physics and hence the Schrödinger equation. Unlike classical **mechanics**, in which energy is continuous and electrons follow fixed orbits described by Bohr's planetary model, **quantum mechanics** describes electrons as effectively taking up entire predefined spaces around the **nucleus** and energy as a quantized phenomenon. Quantum numbers are the four numbers used to describe not only the distribution of electrons in atoms and molecular systems but also the allowable values of certain physical quantities of an electron's behavior. A **wave function** for an **electron** describes the probability of finding that electron at various points in **space**. The first three quantum numbers describe the physical dimensions of space in which that electron can be found. Quantum numbers are used to describe atomic **orbitals** although they are also used to describe hybrid orbitals, orbitals composed of individual atomic orbitals, in molecular systems. The Schrödinger equation cannot be solved exactly for an **atom** containing more than one electron although methods of successive approximations have been used to obtain approximate solutions for systems containing more than one electron. The solutions to the Schrödinger equation reveals the values

of the allowed or permitted energies which can be assumed by every electron in an atom or molecule and are described by the set of quantum numbers for that electron.

The energies of bound electrons are quantized and depend on **angular momentum** and orientation as well as spin of the electron and are derived from the mathematical solution to the Schrödinger or Dirac equations. The Schrödinger equation links the energy and position of electrons in atoms. The locations of the electrons, described by these numbers, are not circular orbits but rather fields of electron **density**. These fields are described by the four quantum numbers: n, l, m_l, and s. Although no two electrons in the same atom or molecule may have an identical set of quantum numbers each quantum number can have values that depend upon the values of the other quantum numbers describing the same electron.

The principle quantum number, n, determines the size of the orbital. It denotes the major shell that contains the electron. When the value of the principle quantum number increases the orbital size increases as well as its distance from the nucleus. It can be any positive integer (1, 2, 3, 4, etc.) with the higher number indicating a larger size. Different values correspond to different energy levels. As n grows larger, the relative spacing of the energy levels become closer together approaching the ionization limit for the atom or molecule.

The angular **momentum** quantum number, l, determines the geometric shape of the orbital and its value can range from zero to n-1. This quantum number is sometimes referred to as the subshell quantum number. So for any orbital there can be many shapes associated with a particular size depending upon what the principle quantum number is. The first five shapes associated with the orbital quantum numbers are referred to as s, p, d, and f. A pure s orbital is spherical. A pure p orbital is dumb-bell shaped with two parts separated by a nodal plane where the probabilit ity of finding the electron is zero. The pure d orbital has four lobes with nodal planes at the intersection of the four lobes. f orbitals have complicated shapes and are too complicated to describe here. Although these are the respective shapes of the orbitals in atomic systems, molecules combine atomic orbitals to make molecule orbitals and so the shapes are hybrid, more complex combinations of the individual atomic orbitals.

The third quantum number is the magnetic quantum number, m_l. It determines the orientation of an orbital with respect to an applied magnetic field and can range in value from $-l$ to l. The magnetic quantum number denotes the energy levels available for occupation within a subshell. Because of the range of possible values of magnetic quantum numbers associated with the $l = 2$ orbital there are three possible orientations of the p orbital. There are five possible orientations for d orbitals because of the range of possible magnetic quantum numbers associated with this orbital and there are seven possible orientations for f orbitals described by the magnetic quantum numbers. The different values of the magnetic quantum number mean little differences in energies of the electrons.

There is a fourth quantum number that describes the spin of the electron in an orbital. This quantum number is called the spin quantum number, s, and can have a value of either 1/2 or -1/2 since an electron can orient in two ways in an applied magnetic field. It is an intrinsic quantum number that is unrelated to the *s*-shaped orbital. Because there are only two values associated with this quantum number each orbital can contain only two electrons and each electron must have a spin that is opposite to the spin of the other electron in that orbital. The electrons in these electron pairs have essentially the same energies.

Quantum numbers describe the distribution of electrons in atoms and molecules. The Pauli principle states that no two electrons may be in the same quantum state, hence no two electrons can have identical sets of quantum numbers. Hund's rule tells us that electrons will enter empty orbitals of equal energy when they are available. Quantum numbers, derived from **quantum theory**, are of basic physical importance in calculating energy as well as describing the physical space occupied by electrons in molecular systems.

QUANTUM PHYSICS

Quantum physics is based on **quantum theory**. The first idea about quantum theory was from **Max Planck** in 1900. He proposed that **energy** was not emitted continuously but rather in **quanta** in black body **radiation**. He was awarded the 1918 Nobel Prize in physics for this **work**. In 1905, **Albert Einstein** extended this idea to the **photoelectric effect**, which treated **light** as a flow of particles called photons. This was further proved by the Compton experiment on **photon scattering**. Einstein was awarded the 1921 Nobel Prize partly because of this work. **Erwin Schrödinger** developed **Schrödinger's equation** and wave **mechanics** at the beginning of the last century. **Werner Heisenberg** developed matrix mechanics and the uncertainty principle at about the same **time**. These two mechanics were proven equivalent later. **Paul Dirac** developed a relativistic version of quantum theory. Their theories became the basis of today's **quantum mechanics**. Schrödinger, Heisenberg, and Dirac were all awarded Nobel Prizes for their work. Many other physicists have continued to contribute to the development of quantum physics. Today quantum physics has become a very broad area.

Some important points of quantum mechanics are wave-particle duality and the uncertainty principle. Any object, whether it is an **electron**, a ball, or a human, can act either as a wave or a particle depending on how it is measured. This is what is meant by wave-particle duality. For instance, light will act like a wave in experiments that measure **diffraction** or refraction and it acts like a particle in experiments that measure reflection. Generally, the wave nature of objects is only evident at the scale of atomic and subatomic physics.

An electron propagates as a quantum wave. This does not mean that the electron is broken into pieces and spreads in **space** as a water wave does. The electron is still an electron. The value of a quantum wave at a certain position determines the probability that it will be observed at that position. Mathematically the quantum wave has complex (real and imaginary) variables, so the square of modular of the quantum

wave value is used to calculate the probability. According to **Louis-Victor de Broglie**, the **wavelength** (λ) of a quantum wave of an object is $\lambda = h/mv$, where h is the Planck constant ($6.63 \times 10^{-34} J \cdot s$), m is the **mass** of the object, and v is its **velocity**. For a flying ball with a mass of 1 lb (0.45 kg) and a speed of 100 ft/s (30 m/s), the wavelength is about 1.6×10^{-34} ft (4.9×10^{-35} m). That is much shorter than a trillion trillionth of one foot. This is so small that its wave nature can be safely ignored.

There is a rule of thumb to determine when quantum physics should be used and when classical physics is still good enough. The Planck constant is used to determine this. Its unit is energy multiplied by time, or **momentum** multiplied by length. We find out the characteristic time and length of the system we are studying, then find the product of energy and time or momentum and length. If the result is much larger than h, then we can use classical mechanics. Otherwise, we must use quantum mechanics. A walking person, for instance, has a mass of about 160 lb (72 kg), a speed of 3 ft/s (0.9 m/s), and a characteristic length (person's height) of 6 ft (1.8 m). The product is 116 joules/s, which is much larger than the Planck constant. Classical mechanics is correct for use in our everyday life.

The uncertainty principle is an important result of quantum mechanics. It says that certain pairs of values, e.g. momentum (or speed) and position, can't be accurately determined simultaneously. The more accurately you can measure one value, the less accurately you measure the other. This seems to completely contradict our everyday experience where both position and speed can be specified simultaneously. Our experience is only approximately correct. The uncertainty principle poses an ultimate limit on how accurately we can measure position and momentum simultaneously. Energy and time is another such pair of variables.

Quantum physics is very important in the microscopic and atomic world. We can see its effects everywhere in the world around us as well. The color of metal, for instance, could not be explained until quantum physics came out because it is a quantum effect. Quantum physics can indeed be used to do calculations in our everyday life although the calculation is usually much more tedious than classical physics. Our everyday experience is only an approximation for the more accurate quantum physics. However, since the quantum wave-like properties of macroscopic objects is imperceptiable to us, classical physics will suffice. It is at the scale of atoms and **subatomic particles** that quatum physics becomes critically important.

See also Quantum states; Matter waves; Schrödinger's cat

QUANTUM STATES

A quantum state is the condition a quantum mechanical system is in. The state is usually described by a **wave function**. It is different from classical states of systems in that the quantum system has separate, discrete levels, whereas the states of a classical system are continuous. This means that a particle can move from one quantum state to another without ever being in a state in between. The discrete steps are known as quantization.

A quantum leap or jump is the transition between quantum states. Although colloquial usage varies, quantum leaps are almost always very small—only a fraction of the system's total **energy**. These quantum state changes are responsible for common phenomena like **light** emission from atoms. They can occur spontaneously (as in **radioactive decay**) or can be stimulated to occur (as in **lasers**).

Each quantum mechanical system has special stationary states known as eigenstates. The eigenstates give mathematical expressions for the quantized states possible for the system. Eigenstates can be finite or infinite in number: for example, electrons can have spin up or spin down, but there are infinitely many positions for a particle to occupy in free **space**. The eigenstates are a basis set: that is, all other states can be expressed in terms of the eigenstates. This expression is sometimes called a superposition of states, in which each state has some percentage of each eigenstate. The most famous superposition experiment is that of **Schrödinger's cat**, where the eigenstates of "alive" and "dead" are superimposed.

When a quantum mechanical state is observed, the system is forced into one or another of the eigenstates. The total state prior to observation could be expressed in terms of the probability a system has of being in each eigenstate. After observation, the system is only in one of them. In the famous Schrodinger's cat example, only "alive" or "dead" are permitted to be the observed eigenstates. The state may be changed and evolve after observation, but it may only be observed in an eigenstate.

There are some state measurements which are incompatible. For example, position and **momentum** cannot simultaneously have their state determined precisely. This principle is known complementarity; the measurements are complementary or conjugate measurements. Complementarity of state measurements derives from Heisenberg's Uncertainty Principle.

The idea of quantum states is useful for expressing the instantaneous behavior of a system, as well as for analyzing potential modes of behavior for it. Each of the eigenstates corresponds to a potential measured reality, and can be seen in the lab.

See also Complementarity principle; Eigenfunction; Energy levels; Quanta; Quantum complementarity; Quantum numbers; Quantum theory; Stationary state; Uncertainty principle; Wave superposition

QUANTUM THEORY

Quantum theory is the basis of the modern understanding of most branches of physics. At the turn of the twentieth century, many physicists assumed that physics was almost totally figured out, with only a few small problems left to be solved. These "small problems" occurred in situations where particles

behaved like **waves** and **light** behaved like particles, and also occurred in situations where **radiation** was important. Exploration of these problems led the way into quantum theory, a true revolution in physics.

The early part of quantum theory was built up out of a series of experiments and their explanations. In 1901, **Max Planck** managed to make one of the problems work out if he assumed that light only occurred in quantized units, somewhat like particles. Then, in 1905, **Albert Einstein** explained the **photoelectric effect** with the same quantization units as Planck had used, postulating that the light actually was quantized.

Atomic models also helped to put quantum theory into practice. In 1913, Bohr's quantum theory of spectra explained the hydrogen **atom** much better than the previous models had. While the Bohr model has since been refined, its major features proved to be both correct and non-classical. **Niels Bohr** assumed that hydrogen exists in discrete **energy** states characterized by the atom's total **angular momentum**, in units of the same constant Planck and Einstein had used. The energy transitions in the **hydrogen atom** were then quantized in units of **Planck's constant** as well.

In 1924, **Wolfgang Pauli** formulated his famous exclusion principle to explain the behavior of electrons around atoms. Pauli hypothesized that some particles (now known to be **fermions**) cannot occupy the same quantum state at the same **time**. The **Pauli exclusion principle** gave **chemistry** its basis in physics, in the difference in behavior for each quantum state of an **electron**.

Louis Victor de Broglie then theorized that if light can behave like a particle, **matter** might well be able to behave like a wave, with a **wavelength** related to its **momentum**. This theory was later tested on electrons by **Clinton Davisson** and **Lester Germer**, completing the wave-particle duality picture. Under quantum mechanical wave-particle duality, energy and matter behave as both waves and particles, depending upon the circumstances of their observation. A **photon** or electron which is observed under "particle" apparatus exhibits particle behavior; apparatus designed to detect waves will do so. In complete contrast to classical systems, the method of observation changes the observation.

One of the fundamental physical underpinnings of **quantum mechanics** is Heisenberg's uncertainty principle. The uncertainty principle states that for conjugate (or complementary) variables, it is impossible to measure both variables with absolute accuracy simultaneously. The measurement of one changes the state of the other and induces error into the previous measurement. Some examples of conjugate variables are position/momentum and energy/time.

Erwin Schrödinger formulated the principles of **quantum mechanics** with his wave equation, which is a quantum mechanical of the **conservation of energy** in the form of a differential equation. It was **Max Born** who suggested that the **wave function** squared represents the probability function. The Schrödinger wave equation is mathematically equivalent to two other ways of expressing quantum mechanical principles: Heisenberg matrices and Feynman sum-over histories. These formulations allowed quantum mechanics to be used as a single, cohesive whole. Quantum mechanical principles have since been applied to the strong **force**, to form quantum chromodynamics, and to the unified electric, magnetic, and weak forces, to yield quantum **electrodynamics** (Q.E.D.). As of the turn of the twentieth-first century, the only force which did not have a quantum mechanical explanation was **gravity**.

The implications of quantum theory were revolutionary: the behavior of a particle under quantum theory was not deterministic. In classical physics, it was assumed that if the state of a particle was known exactly in all parameters, then all future states could be calculated. Quantum theorists discovered that, even if Heisenberg's uncertainty principle did not interfere with knowing the exact values of all properties of a particle, that particle would still only behave probabilistically—that is, one could only determine its likely future state. Albert Einstein had some of the most vociferous objections to this aspect of quantum theory, arguing endlessly with Niels Bohr and other quantum theory advocates about the possibility of further changes to the theory that would allow it to be deterministic. However, injecting **determinism** into quantum mechanics seems to involve "hidden variables"—properties as yet unobserved by physicists—and it has been proven that those cannot exist.

The problem of connecting the microscopic world of quantum mechanics to the everyday observance of classical mechanics then remained. Ehrenfest's principle, which states that classical laws of **motion** hold for **quantum states** on the average, serves to explain why classical behavior is observed at macroscopic levels. On the average, Newton's laws, **Maxwell's equations**, and all other classical laws of physics will also hold in quantum mechanical systems.

Quantum theory has enjoyed enormous success, forming an underpinning for most of modern physics. One of its offshoot theories, quantum electrodynamics, provides the most accurate match of theory and experiment in determining a **fundamental constant** of the **universe**. While many parts of quantum theory seem counterintuitive, it succeeds at that most fundamental aspect of physics: matching theory with experimental facts.

See also Bohr atom; Bohr theory; Complementarity principle; Heisenberg, Werner; Matter waves; Philosophy of physics; Probability and quantum mechanics; Quanta; Quantum complementarity; Quantum field theory; Quantum numbers; Quantum physics; Standard model of particle physics; Uncertainty principle

Quarks

Along with leptons, quarks are basic building blocks of all **matter**. According to the standard model of **particle physics**, **fermions** (quarks and leptons) are the basic constituents of all matter. In the standard model there are six types of quarks and six types of leptons (the **electron**, the **muon**, tau particle and their respective neutrinos). All of these fundamental particles have corresponding antiparticles that have exactly the same

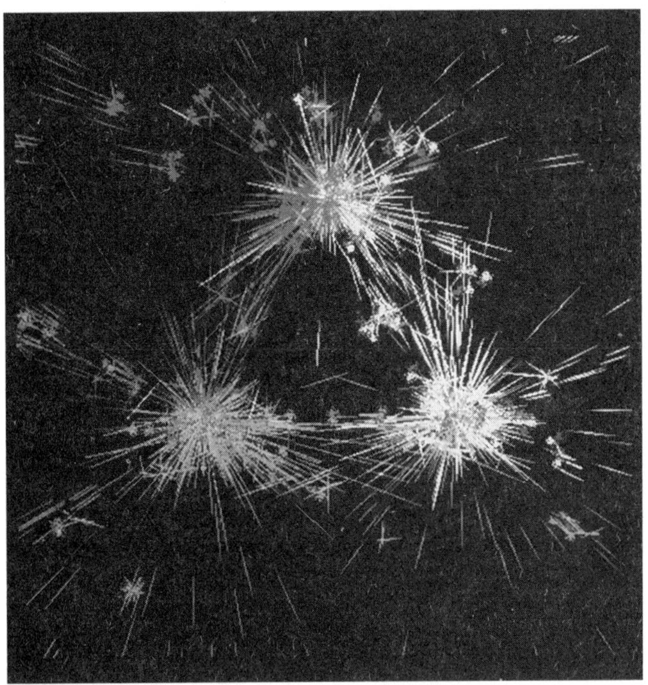

Proton, visualisation of how a proton is produced from three quarks. *(Photo by ArSciMed/Science Photo Library, Photo Researchers Inc. Reproduced by permission.)*

mass but which carry an opposite charge. Fermions exist in three generations (mass and **energy** states). Protons, **neutron** and electrons are in the first generation of particles and the more massive and energetic particles of the second and third generations usually quickly decay into less-massive first generation particles.

Quarks are acted upon by the strong **force** and do not exist in isolation. The confinement of quarks dictates that quarks are found only in groups of three (baryons) or in quark anti-quark pairs (mesons). American particle physicist **Murray Gell-Mann** named the quark based upon his reading of the James Joyce classic, *Finnegan's Wake*. The line "Three quarks for Muster Mark..." inspired Gell-Mann because quarks always occur in sets of three.

In the currently accepted standard model, there are six differing types (flavors) of quarks that together account for the known properties of more than 200 types of mesons and baryons. Quarks are known by their somewhat peculiar names: up, down, strange, charmed, top and bottom quarks. It is important to remember that the names for quarks have no direct relation to their properties (i.e., there is nothing intrinsically "up" about an up quark or "down" about a down quark).

All ordinary matter that we see around us is made of up and down quarks. Quarks also have their associated antiparticles: the anti-up quark, the anti-down, and the anti-strange. All quarks carry a 1/2 spin and a baryon number of 1/3. Accordingly, the articulation of quarks involves their characterization in terms of their relative properties of charge and mass.

Up and down quarks are the quarks that, in sets of three, comprise protons and neutrons. As such they are the most common quarks that comprise most observed or ordinary matter. Both have a mass of 360 MeV. The up (u) quark is assigned a charge of +2/3, while the down (d) quark is assigned a charge of -1/3. The heavier charmed (c) (1500 MeV mass) and strange (s) quarks (540 MeV mass) are respectively assigned charges of +2/3 and -1/3. Strange quarks derive their name from the extraordinary longevity of the **lambda particles** they help comprise. The top (t) quark (sometime called the truth quark) has a mass of 185 GeV and is assigned a +2/3 charge. The bottom (b) quark (sometimes called the beauty quark) has a mass of 5 GeV and is assigned a -1/3 charge. In addition, in accord with quantum chromodynamic theory (QCD) each of the six types of quarks can have three different "colors" (another property unrelated to its name that describes a force acting between quarks that binds them in hadrons).

Because quarks can not be isolated there can be no direct measurement of their mass. Some masses are assigned based upon the mass of the particles they comprise (e.g., Up and down quarks are each assigned a proportional 1/3 of the mass of the **proton** or neutron they comprise). Other masses for the quarks are estimates derived from **scattering** experiments. Experimental analysis actually yields differing masses for quarks depending upon their configuration (i.e., a particular combination of quark such as the up and anti-down quark pairs comprising **pions** or rho vector mesons).

By the 1960s physicists had discovered a large number of new **subatomic particles** that, like the proton and neutron, attract one another through the nuclear force (also called the strong force). Classification of all the particles, such as those known as pions, kaons, and others only seen after the collisions of **cosmic rays**, was reminiscent of the chemist's classification of elements into the **periodic table**. These relationships helped reveal an underlying fundamental structure. Theoretical physicists set out to do something similar with all the new nuclear particles. Gell-Mann, together with the Israeli physicist Yuval Ne'eman and, independently, the American physicist George Zweig, introduced the concept of the quark as a basic building block for all particles under the influence of the nuclear force. Using particle **accelerators**, Maurice Jacob and Peter Lanshoff smashed high-energy electrons into protons in 1980. They found that some electrons bounced off at large angles (or backwards) higher than would be expected if the proton's charge was uniformly spread across its volume. Their results were consistent with the idea that the proton was instead composed of subparticles.

Because quarks have a spin-1/2, it initially seemed as though a group of three would violate the **Pauli exclusion principle** which asserts that two identical spin-1/2 particles cannot be together in the same arrangement. The articulation quark colors (red, green, and blue). QCD gave reality to the idea of **color**, considering it akin to **electric charge**: quarks were attracted to one another through their "color charges." Whereas electrons attract and repel other electric charges by exchanging a **photon**, quarks do so by a new particle called the

gluon. Unlike photons, however, which have no electric charge, the eight differing gluons do have a color charge, or rather, combinations of color charges. The result of color attraction is that quarks do not exist in isolation. Quarks remain in proximity to one another within the diameter of the proton (about 10^{-15} m). When separation of quarks is attempted they emit more and more gluons until everything breaks apart into new combinations of quarks, anti-quarks, and gluons (i.e., more particles, but no individual quarks or gluons). This strange property is called asymptotic freedom. Unlike the force between electric charges, which decreases with distance, the force between color charges increases with distance.

As elementary particle physics progressed through the 1970s and 1980s, physicists found more and more exotic particles, such as the psi-meson in 1974, whose mass is about three times that of the proton. An accurate description of its observed properties required the addition of a fourth quark, the charm quark with an electric charge of +2/3e. In 1977 discovery of the upsilon-meson (ten times the proton mass) required the introduction of the bottom quark, with charge -1/3e. A sixth quark, the top quark, was found in 1994 at the Fermilab National Accelerator in Illinois, with a charge of +2/3e and a mass of about 180 times the proton mass, which is equivalent to that of a gold **atom**. All of these three heavier quarks decay very quickly into other particles (including the other quarks).

The proton comprises two up quarks and a down quark. A neutron is composed of two down quarks and an up quark. Quarks can undergo transformations or decays (e.g., from being an down to an up quark in **beta radiation**) via an exchange of the W bosons of the weak force (weak interaction).

See also Atomic physics; Atomic structure; Dispersion; Electroweak force; Electroweak particles; Electroweak theory; Energy and mass; Higgs boson; K mesons (kaons); Mesons; Nuclear physics; Nuclear reactions; Nucleon; Nucleus; Parity; Proton decay; Radioactive decay; Spin of subatomic particles; Standard model of particle physics; Strange matter; Strange particles; Symmetry and symmetry principles; Symmetry, antimatter and polarity; Tau particles; Vector bosons; Virtual particles; W mesons; Z-boson

QUASARS

At the beginning of the twenty-first century, quasi-stellar **radio** sources (quasars) were the most distant cosmic objects observed by astronomers. Although not visible to the naked eye, quasars were also among the most energetic of cosmic phenomena. Although some quasars may be physically smaller in size than our own solar system some quasars are calculated to be brighter than hundreds of galaxies combined. Quasars and active galaxies appear to be related phenomena each associated with massive rotating **black holes** in their central region. As a type of active galaxy, the enormous **energy** output of quasars can be explained using the theory of general relativity.

The great distance of quasars means that the **light** we observe coming from them was produced when the **universe** was very young. Because of the finite **speed of light**, large cosmic distances translate to looking back in time. The observation of quasars at large distances and of their nearby scarcity argues that quasars were much more common in the early Universe. Correspondingly, quasars may also represent the earliest stages of galactic evolution. This change in the Universe over time (e.g., specifically the rate of quasar formation) contradicted steady state cosmological models that relied on a Universe that was the same in all directions (when averaged over a large span of **space**) and at all times. Along with the discovery of a ubiquitous cosmic **background radiation**, the discovery of quasars and tilted the cosmological argument in favor of **big bang**-based cosmological models.

In 1932, American engineer Karl Janskey (1905-1945) discovered existence of radio **waves** emanating from beyond the solar system. By the mid-1950s an increasing number of astronomers using radio telescopes sought explanations for mysterious radio emissions from optically dim stellar sources.

In 1962 British radio astronomer Cyril Hazard used the **moon** as a occultive shield to discover strong radio emissions traceable to the constellation Virgo. Optical telescopes pinpointed a faint star-like object (subsequently designated quasar 3C273—3rd Cambridge Catalog, 273rd radio source) as the source of the emissions. Of greater interest was an unusual emission spectrum found associated with 3C273. In 1963 American astronomer Marten Schmidt explained the abnormal spectrum from 3C273 as evidence of a highly redshifted spectrum. Redshift describes the Doppler-like shift of spectral emission lines toward longer (hence redder) wavelengths in objects moving away from an observer. Observers measure the light coming from objects moving away from them as redshifted (i.e., at longer wavelengths and at a lower **frequency** when the light was emitted. Conversely, Observers measure the light coming from objects moving toward them as blueshifted (i.e., at shorter wavelengths and at a higher frequency when the light was emitted. Most importantly, the determination of the amount of an object's redshift allows the calculation of a recession **velocity**. Moreover, because the recession rate increases with distance, the recession velocity is a function (known as the Hubble relation) of the distance to the receding object. After 3C273 many other quasars were discovered with similarly redshifted spectra.

Schmidt's calculation of the redshift of the 3C273 spectrum meant that 3C273 was approximately three billion light-years away from Earth It became immediately apparent that, if 3C273 was so distant, it had to be many thousands of times more luminous than a normal galaxy for the light to appear as bright as it did from such a great distance. Refined calculations involving the luminosity of 3C273 indicate that, although dim to optical astronomers, the quasar is actually five trillion times as bright as the **Sun**.

The high redshift of 3C273 also implied a great velocity of recession measuring one-tenth the speed of light.

Astronomers now assert that quasars represent a class of galaxies with extremely energetic centers. Large radio emissions seem most likely associated with large black holes with a large amounts of **matter** available to enter the accretion disk. In fact, prior to more direct observations late in the 20th century, the discovery of quasars provided at least tacit proof of the existence of black holes. Black holes form around a **singularity** (the remnant of a collapsed massive **stars**) with a **gravitational field** so intense that not even light can escape. Located outside the black hole is the accretion disk, an area of intense **radiation** emitted as matter heats and accelerates toward the black hole's **event horizon**. Further, as electrons in the accretion disk are accelerated to near light speed they are influenced by a strong magnetic field to emit quasar-like radio waves in a process termed synchrotron radiation. Electromagnetic waves similar to the electromagnetic waves emanating from quasars are observed on Earth when physicists pass high energy electrons through synchrotron particle **accelerators**. Studies of Quasar 3C273 and other quasars identified jets of radiation blasting tens of thousands of light-years into space.

In addition to radio and visible light emissions, some quasars emit light in other regions of the **electromagnetic spectrum** including ultraviolet, infrared, x-ray, and gamma-ray regions. In 1979, an x-ray quasar was found to have a redshift of 3.2, indicating a recession velocity equaling 97% the speed of light.

Not all quasars or active galaxies are alike. Although they seem optically similar to energetic quasars, at least 90% of active galaxies appear to be radio quiet. Accordingly, Seyfert galaxies or quasi-stellar objects (QSO) may be radio silent or emit electromagnetic radiation at greatly reduced levels. More than 1,500 quasars have now been identified as distant QSO. One hypothesis accounts for these quiet quasars by linking them to smaller black holes or to black holes in regions of space with less matter available for consumption.

The limitations of ground-based telescopes and the need to study quasars was officially cited as one of the principal reasons to build the **Hubble Space Telescope** launched by the United States in 1990. In addition to direct studies of quasars, astronomers use quasars as an electromagnetic backdrop that can be used to study the primitive gas clouds found in the early Universe.

R

RABI, I. I. (1898-1988)

Austrian American physicist

Born in Austria, I. I. Rabi came to the United States with his parents at an early age. He attended Cornell and Columbia Universities, receiving his Ph.D. in physics from the latter in 1927. During a post-doctoral year in Germany Rabi worked with **Otto Stern** and learned about Stern's experiments (conducted with Walther Gerlach) on the analysis of atomic and **molecular structure** by means of atomic and molecular beams. Upon his return to the United States in 1929, Rabi worked on methods for extending and refining the Stern-Gerlach techniques. He eventually made a number of important discoveries regarding the magnetic properties of the **nucleus** and of sub-atomic particles—discoveries that later found application in a number of fields, including **nuclear magnetic resonance**, masers and **lasers**, and time measurement by means of atomic clocks. During World War II, Rabi worked on the development of **radar** devices and **nuclear weapons**. At the war's conclusion, he devoted most of his time and **energy** to the political aspects of scientific and technological development, serving as chairman of the U.S. Atomic Energy Commission from 1952 to 1956. Rabi died in 1988 at the age of 91.

Isidor Isaac Rabi was born on July 29, 1898, in Rymanow (also given as Raymanou or Rymanov), Galicia, then a part of the Austro-Hungarian empire. Rabi's parents were David Rabi and the former Janet (also given as Jennie or Scheindel) Teig. The senior Rabi emigrated to the United States shortly after his son's birth and, in 1899, sent for his family to join him in New York City. David Rabi has been described by various biographers as an unskilled worker, a tailor, and an owner of a grocery store; Rabi himself said that his father started out by doing odd jobs, such as delivering ice, and then "graduated into work in the sweatshop, making women's blouses." Yiddish was the only language spoken in the Rabi household, and young Isidor learned his English on the streets. He was a quick learner, however, and did well in the public schools of New York City. After graduating from Brooklyn's Manual Training High School in 1916, he entered Cornell University with plans to major in electrical engineering. He eventually changed his major to chemistry, though, graduating with a bachelor of science degree in 1919. He then spent three years working as a chemist before returning to Cornell for graduate work. Rabi soon discovered that his real interest was physics, and in 1923 enrolled in a doctoral program in this field at Columbia University. In order to support himself at Columbia, Rabi took a job teaching physics at the City College (now City University) of New York, a post he held until he received his Ph.D. in 1927.

Pursues Postgraduate Studies in Europe

For his postdoctoral studies, Rabi planned a two-year tour of the most important scientific institutions in Europe, including Munich, Copenhagen, Hamburg, Leipzig, and Zürich. While on tour he studied with such leading figures as Arnold Sommerfeld, **Niels Bohr**, **Wolfgang Pauli**, **Werner Heisenberg**, and Otto Stern. The visit with Stern may have been the most significant stop on the tour, because Stern's work at Hamburg closely corresponded to Rabi's own field of interest and the subject of his doctoral thesis, the effects of magnetic fields on **matter**. In 1922, Stern and Walther Gerlach had developed methods for creating beams of atoms or molecules that could be used to study the magnetic properties of the atomic nuclei in these beams. For his discoveries in this field, Stern would go on to win the 1943 Nobel Prize in physics.

In 1929, when Rabi returned to the United States, he began his own research on the use of atomic and molecular beams to study nuclear properties. This work took place at Columbia, where he had been appointed lecturer in physics; over the next decade, he worked his way up the professional ladder, being promoted to assistant professor in 1930, associate professor in 1932, and then full professor of physics in 1937. Throughout this period, Rabi refined his methods of atomic and molecular beam analysis, eventually making a number of important discoveries.

Discovers Atomic Spin Properties

The Stern-Gerlach experiment of 1922 had showed that a molecular beam passing through a magnetic field splits into two parts. The discovery of **electron** spin by George Uhlenbeck and **Samuel Goudsmit** in 1927 explained this phenomenon: they demonstrated that electrons in an **atom** can spin in only one of two directions; hence, electrons spinning in one direction split into one beam, while those spinning in the opposite direction split into another.

As Rabi studied this effect in more detail, he realized that the magnetic properties of an atom are more complex than first suggested by the Stern-Gerlach experiment. In the first place, the nucleus itself spins, creating its own magnetic field. Thus, there will be interactions among the magnetic field of the nucleus, the magnetic fields of the orbital electrons, and any external magnetic field that is applied to the atom.

In his research, Rabi was able to sort out and quantify many of these discrete properties. His most important accomplishment was to determine the **magnetic moment** of the nucleus, an important piece of information essential to the construction of an accurate model of the atom. By 1937, Rabi had made yet another discovery, namely that he could reverse the spin of a nucleus by imposing an external radio-frequency signal on an atomic or molecular beam. That discovery has been used in a number of important applications; one of these, nuclear magnetic resonance (NMR), is now among the most powerful analytical tools available to scientific investigators and medical diagnosticians. Rabi was awarded the 1944 Nobel Prize in physics for his work on "the resonance method for recording the magnetic properties of the atomic nucleus."

During World War II, Rabi took a leave of absence from Columbia to worked on the development of microwave radar devices at the Massachusetts Institute of Technology. Though most of his colleagues in the scientific community were devoting their wartime efforts to the development of atomic weapons, Rabi believed that, of the two projects, radar would be more immediately useful to the U.S. war effort—though he did consult on nuclear weapons projects as part of the **Manhattan Project**. At the war's conclusion, Rabi returned to Columbia as chairman of the physics department. He devoted his time primarily to administrative responsibilities and to the effort by scientists to restrict military control of nuclear technology. "Speaking for the group of men who created these weapons," Rabi once said in *Atlantic Monthly*, "I would say that we are frankly pleased, terrified, and to an even greater degree embarrassed when we contemplate the results of our wartime efforts." In order to monitor the use of atomic energy and weapons, Rabi became a member of the General Advisory Committee of the Atomic Energy Commission in 1945 and then served as chairman of the committee from 1952 to 1956 (after the retirement of J. Robert Oppenheimer). He was an advisor to NATO and the United Nations, and served as a member of the American delegation

to UNESCO, overlooking the **European Center for Nuclear Research (CERN)** in Geneva.

Rabi was married to Helen Newmark in 1926. They had two daughters, Nancy Elizabeth and Margaret Joella. In addition to the Nobel Prize, Rabi won a host of other awards, including the Elliott Cresson Medal of the Franklin Institute in 1942, the U.S. Medal for Merit (the country's highest civilian service award) in 1948, the Niels Bohr International Gold Medal in 1967, the Atoms for Peace Award in 1967, the Franklin Delano Roosevelt Freedom Medal in 1985, and the Public Welfare Medal of the National Academy of Sciences in 1985. Rabi died in New York City on January 11, 1988.

RADAR

In the latter part of the nineteenth century **electricity** and **magnetism** were a prime area of research in Europe. In 1887, Heinrich Hertz in Germany was the first to realize the significance of the behavior of **radio waves**. Hertz found that, like most electromagnetic waves, radio waves could be deflected and reflected by certain materials. Physicists knew that radio waves traveled at a constant 298,051 km/s (186,282 mi/s). If the returning radio waves arrived one second after leaving, then the reflecting object was 149,025 km (93,141 mi) away. The technology that used this principle was developed during the 1930s by the British team led by Sir Robert Watson-Watt, who called it RAdio Detection and Ranging, or RADAR. It was incredibly useful in 1940, when Britain was at war with Germany, as it enabled the British air force to know in advance the position of enemy aircraft even if they were invisible due to cloud cover, darkness or just large distance.

Radar systems consist of four basic parts. A transmitter generates the radio signal. An antenna directs the signal toward the target and captures the reflected signal. A receiver takes the reflected signal from the antenna and processes it. Finally, a display unit takes the processed signal and displays it in such a way that the radar operator can interpret the signal's information.

Radar operates using the physical principles of electromagnetic radiation—energy that moves in wavelike forms at the speed of **light** or less. Most radar systems use **energy** that falls in the longer **wavelength** area of the **electromagnetic spectrum**, in the same general area as microwave and radio **radiation**: beyond the wavelengths of visible light. Radar systems use these wavelengths because they tend not to scatter as easily as shorter waves, such as **x rays** or gamma rays, and a greater percentage of the electromagnetic signal can be received by the unit. Different systems can use different wavelengths, depending on the type of object or phenomena they are trying to detect. Some long-wavelength radar systems can even bounce signals off the earth's ionosphere, providing important information about the atmosphere and the **weather**.

Radar has many uses in the modern world. The same idea is now used in guidance systems for aircraft and **rockets**. Doppler radar is used for used for weather forecasting, since storm clouds reflect certain radio waves. Radar can also be used to measure the **motion** or speed of the detected object. The police have used radar devises to detect the speed of oncoming vehicles for many years. **Computers** can process the echo of a radar signal, removing a lot of the "noise" or "clutter" and revealing the fundamental shape of the object that is reflecting the signal, which can benefit aircraft controllers and anti-ballistic missile defense systems

RADIATION

The term radiation broadly applied to a class of physical phenomena that share a characteristic loss of **energy** in the radiating body or system that results in a propagation of particles and/or **waves** that carry the ability to do **work** and thus change the energy state of an impacted body or system.

Radiation is a catch-all name given to several different types of energetic phenomena. In particular, we think of the entire **electromagnetic spectrum** as being radiation of graduating wavelengths. **Heat** and visible **light**, being parts of the spectrum, were among the first forms of electromagnetic radiation recognized by scientists. Electromagnetic radiation is now understood to encompass a broad spectrum of electromagnetic radiation form gamma rays to **radio** waves that differ only in **wavelength** and **frequency**.

Radiation also aptly describes phenomena specifically related to blackbodies that are described by Planck's law, Kirchhoff's law, Wien's law, and the Stefan-Boltzmann law.

Another form of radiation is the particulate emissions from natural **radioactive decay,** or from induced atomic disruption. In either case, unstable atoms emit protons, neutrons, electrons and neutrinos at varying energy levels. Alpha particles are heavy, positively-charged emissions with low penetrating capabilities; they rarely travel more than a few centimeters from their source and have difficulty penetrating clothes or even skin. Beta particles are high-energy electrons with much greater penetrating **power.**

The primary biological danger of particulate radiation, as well as **gamma radiation**, is their ionizing effect. **Ionizing radiation**, creating **ions** in impacted bodies or systems, that can often result in a break down of **molecular structure**. This breakdown can bring about a variety of physiological effects, depending on the tissues and organs exposed, and the duration and intensity of exposure.

Radiation can propagate in different ways depending on circumstances. Radiation moving through a **vacuum** tends to take the form of waves. When interacting with **matter**, however, radiation (even non-particulate radiation) generally exhibits particle-like behavior. This variance in behavior is called the wave-particle duality and is one of the bases of **quantum theory.**

We are continuously exposed to radiation from solar and cosmic sources. In addition to extraterrestrial sources, there are naturally occurring terrestrial radiation sources (e.g., radioactive elements such as uranium). The decay of uranium deep in the rock strata can also release radon, a radioactive gas that can seep into basements and underground structures.

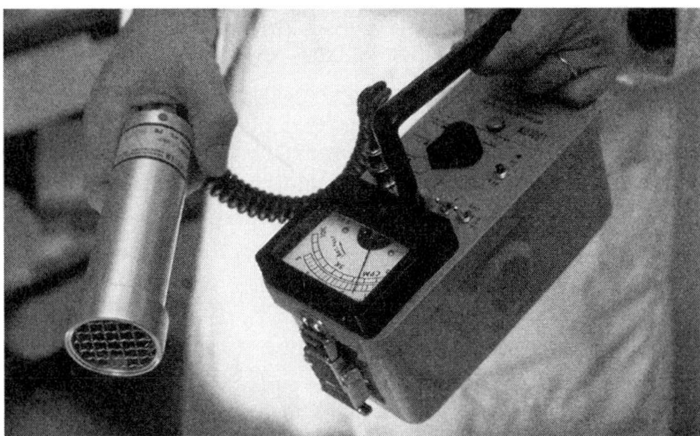

Hand held Geiger counter for measuring radiation intensity. *(Photo by Hank Morgan. National Audubon Society Collection/Photo Researchers, Inc. Reproduced by permission.)*

Radiation can also be generated artificially. A common way of doing this is to bombard a heavy metal such as tungsten with electrons to stimulate **x rays**. Radiation is also produced by the series of high-energy **isotope** conversions that take place during and after **nuclear fission**. The medical dangers of radiation are still being sorted out. One of the reasons for this is that the effects are to some degree governed by probability. The same duration and intensity of radiation exposure can result in widely varying physical effects. Although extreme levels of radiation can be harmful or even fatal, radiation in the form of diagnostic x rays and radiation therapy is routinely used in medical diagnosis and treatment (e.g., in cancer treatment).

See also Beta radiation; Blackbody radiation; Cosmic background radiation; Energy transformations; Gravitational radiation and waves; Kinetic molecular theory; Natural radiation; Quantum electrodynamics (QED); Radioactivity; Thermal radiation

RADIO

Radio **waves** are produced by controlled induction of movement of charges along a transmission antenna to produce a variable electric field that results in the propagation of an electromagnetic wave. An electromagnetic wave is sustained by the induction of magnetic fields by electrical fields and the reciprocal induction of electric fields by magnetic fields. These electromagnetic radio waves impact target antenna and cause corresponding changes in charge distribution along the receiving antenna that can be translated into **sound.**

The culmination of 19th century experimentation with **electricity** and **magnetism** was embodied in the equations of the **electromagnetic field** published by Scottish physicist **James Clerk Maxwell. Maxwell's equations** describing the propagation of electromagnetic waves predicted the existence of a spectrum of such waves that varied only in **frequency** and **wavelength**. In such an **electromagnetic spectrum**, visible **light** comprised but one region.

In 1888, German physicist Henrich Rudolph Hertz demonstrated the existence of the region known as radio waves. Hertz produced radio waves by supplying an **electric charge** to a capacitor and then inducing a short-circuit. The resulting spark radiated electromagnetic waves with a frequency and wavelength now characteristic of the radio region of the electromagnetic spectrum. As predicted by Maxwell's equations, these Hertzian radio waves traveled at the **speed of light**. In separate experiments conducted during the 1890s, British physicist Sir Oliver Lodge (1851-1940) used a device named a *coherer* to detect the presence of radio waves.

In 1895, Italian physicist and inventor **Guglielmo Marconi** devised the first practical apparatus for the purpose of transmitting radio waves. Marconi's apparatus consisted of an aerial, a condenser and a **grounding** connection. In 1898, wireless telegraphic signals were used in British naval maneuvers between ships separated by 50-60 nautical miles. By 1899, limited radio telegraphic communications were possible across the English channel and in 1901 Marconi led a team that made the first intelligible transmission of a telegraphic radio signal across the Atlantic.

Early radios relied on electrolytic and carborundum tubes, but in 1906 a major advance in radio technology came about with American inventor Lee De Forest's (1873-1961) invention of the Audion thermionic grid-triode **vacuum** tube. Along with using the **triode** as a radio wave producing oscillator, the 1913 separate discovery by De Forest and by American engineer Edwin Howard Armstrong (1890-1954) of regenerative feedback spurred the growth of the commercial radio industry. The two inventors subsequently entered into a series of bitter legal feuds over patent rights to several radio related innovations. Armstrong went on to invent the superheterodyne receiver and a frequency modulation (FM) apparatus that were major advances in radio technology and emerging **RADAR** technology. However, it was De Forest who was legally credited with the development of the triode that allowed more sensitive reception and amplification of radio signals.

By 1914 the thermionic triode was used as a radio generator that produced a carrier wave capable of speech modulation. By the 1920s, communication between continents was achieved through the use of powerful shortwave radio transmitters. In the 1920s the rectifying properties of crystals were utilized to permit radio wave detection and improve reception.

During the twentieth century, the development of radio and in the radio industry spurred tremendous technological advances in communications that revolutionized human society. The development of radio allowed people the ability to immediately and spontaneously communicate in their natural language over long distances.

Radio communication is achieved by the controlled modulation of wave amplitude and frequency. Strength of signal is directly related to wave amplitude. Frequency refers to the number of wave crests that pass a given point in a period of time (e.g., cycles per second or hertz). Metric prefixes allow the classification of frequencies in kilohertz and megahertz.

Modulation achieved through amplitude modulation results in a AM signal that is varied between maximum posi-

tive and maximum negative peaks that can eventually be translated into the intended sound. Because there are a great number of natural process (e.g., lightening) that also produce AM signals, AM signals are prone to static **interference**. Also because weak signals are utilized as part of amplitude modulation, AM receivers must be designed to compensate for these signal strength differences. Frequency modulation (FM) was designed to eliminate AM noise. FM signals subtly modulate the frequency of the carrier wave in a band between a maximum and a minimum frequency. Accordingly, FM receivers remove corresponding amplitude variations from the signal and use a discriminator circuit to convert the variable frequencies into the intended sound.

Other types of wave modulations, all variations on AM and FM include single sideband (SSB), double sideband (DSB), and vestigial sideband (VSB), phase shift keying (PSK) and digital modulation. VSB is widely used in broadcast television signals. FM signals also allow the use of subcarrier signals that carry data or allow stereophonic sound. Subcarriers are signals that are modulated like the original carrier wave except at much higher frequencies.

See also Diodes; Electric field and forces

RADIO ASTRONOMY

By measuring regions of the **electromagnetic spectrum** not available to optical astronomers, **radio astronomy** effectively extends the range of astronomical observation. During the twentieth century, radio astronomy advanced to become one of the most productive means of astronomical research. Moreover, the development and success of radio astronomy spurred astronomers to devise new instrumentation designed to investigate other regions of the Cosmic electromagnetic spectrum.

In the nineteenth century Scottish physicist **James Clerk Maxwell** developed a set of equations describing electromagnetic phenomena. Using **Maxwell's equations**, in 1988 German physicist Henrich Rudolph Hertz demonstrated the existence of radio **waves** as a portion of the electromagnetic spectrum. In separate experiments conducted in the 1890s, American inventor **Thomas Edison** and British physicist Sir Oliver Lodge unsuccessfully attempted to detect solar radio activity.

In 1932, while performing an experiment to locate sources of static **interference** vexing radio communication systems, American engineer Karl Jansky (1905-1945) discovered the existence of a unique low-level radio interference. Jansky conducted a search for the source of the radio waves by methodically eliminating various forms of terrestrial interference such as equipment static, thunderstorms, etc. Jansky was also careful to eliminate the **Sun** as a possible source of interference.

During his research, Janksy noticed that the time difference in the daily reception of maximum static shifted by about four minutes each day. Jansky realized that this shift corresponded to the shift in the positions of **stars**. Armed with this evidence, Jansky concluded that the static emanated from a source outside the solar system and Jansky eventually traced the shifting interference pattern to the center of the **Milky Way galaxy**. Jansky attributed the source of the static to hot, charged particles. Although Jansky's findings were initially ignored, Jansky's fundamental contributions to radio astronomy were eventually recognized and the unit of radio-wave emission strength (i.e. the Jansky) was named in his honor.

In 1937, another American radio engineer, Grote Reber (1911-), attempted to extend Jansky's work. Literally in his own backyard, Reber constructed a parabolic reflector dish receiver capable of receiving cosmic radio waves. Reber began a systematic study of the sky at varying radio wavelengths and in 1944 Reber published the first radio **frequency** related celestial maps. Reber's work and writings, published in both scholarly journals and general-interest science magazines, spurred intense interest in radio astronomy.

The development of radio astronomy was, however, inhibited by the secrecy surrounding the development and use of **radar** during World War II. Immediately following the war, however, increasing numbers of scientists turned their attention to the development of radio astronomy. Several large and sophicated radio telescopes were constructed, and—due to the fact that both type of telescopes measure manifestations of electromagnetic radiation—engineers quickly discovered that many of the problems associated with optical telescopes were applicable to the development of radio telescopes. Except for differences in the **wavelength** and frequencies of the electromagnetic **radiation** received, the fundamental physics describing the nature of radio waves is exactly the same as physical laws and formulations describing visible **light** waves. Instead of ground glass lens used in **optical instruments**, radio telescopes use parabolic-shaped metal dishes to focus radio waves at a focal point where they can be amplified and measured.

Radio astronomers quickly pinpointed thousands of sources of extra-terrestrial radio emissions. In 1946, radio astronomers bounced radar signals off of the lunar surface and a number of objects, including meteors not visible to the naked eye, were quickly discovered.

In accord with the principles of optical spectroscopy, light from receding objects becomes shifted toward longer wavelengths (e.g., red-shifted) so what was once visible light appears in different parts of the electromagnetic spectrum, including radio wavelengths and frequencies. Accordingly, radio astronomy allows astronomers to study ancient and distant cosmic objects that have had their energetic light emissions energetically cooled or red-shifted into the radio wave portion of the electromagnetic spectrum as they move away from us in an expanding **universe**. In 1963, American scientists **Arno Penzias** and Robert Wilson, discovered that no matter where in the sky they pointed their antenna, they found radio emissions—including emissions from pats of the sky that were visibly empty. The emissions were attributed to be cosmic **background radiation** left over from the **big bang,** and their discovery was hailed as a major confirmation of big bang-based cosmological models.

National Radio Astronomy Observatory, New Mexico. *(AP/Wide World. Reproduced by permission.)*

Astronomical observations of radio frequencies also allowed the discovery of the origin of very strong radio waves from optically dim stars and other seemingly star-like objects. Eventually known as **quasars** (quasi-stellar radio sources), these enigmatic objects appear to be the most distant—and yet among the most energetic—objects ever observed by astronomers. The discovery of quasars (now thought to be a type of active galaxy containing a black hole) helped confirm theories regarding the formation of **black holes** predicted by general **relativity theory**. Radio astronomers confirmed other significant predictions related to **stellar evolution** with the subsequent identification in the late 1960s of radio **pulsars** (rapidly spinning **neutron stars**).

Radio astronomy has also allowed scientists new insights into the mechanisms operating in solar flares and sunspots, both strong radio sources.

Another source of radio waves are cosmic objects that are cooler than the temperatures required to produce visible light. Accordingly, in addition to allowing astronomers an enhanced ability to detect cooler objects, radio astronomy allowed astronomers to probe obscuring clouds of interstellar dust.

Radio astronomers eventually realized better resolution by electrically linking physically separate telescopes together in a process termed radio **interferometry**. The Very Large Array (VLA) radio **telescope** complex located in New Mexico, for example, utilizes radio interferometry to achieve resolutions exceeding the largest ground-based optical telescopes.

Improved reception also allowed radio astronomers to realistically and systematically listen for evidence of extra-terrestrial life and intelligence in several research projects now commonly referred to as the SETI (Search for Extraterrestrial Intelligence) projects.

RADIOACTIVE DATING

Radioactive dating is a technique which allows for accurate determination of the age of materials. The most common type of radioactive dating is called radiocarbon dating, which can be used to find the age of organic materials. A long-lived **isotope** of **carbon**, ^{14}C, is similar chemically to the stable isotope of carbon, ^{12}C. Living organisms exchange carbon with their environments constantly, so they accumulate ^{14}C. The amount contained in an organism quickly reaches the same abundance that is found on the earth, about one **atom** of ^{14}C for every 7 x 10^{11} atoms of ^{12}C. When the organisms die, they stop accumulating ^{14}C, and the abundance of ^{14}C decreases due to **radioactive decay**. By measuring the **radioactivity** of the sample, the abundance of ^{14}C relative to the amount of ^{12}C can be found, thus determining how long the organism has been dead.

^{14}C is produced in the upper atmosphere from **nuclear reactions** caused by **cosmic rays**. It combines with **oxygen** atoms, just as ^{12}C atoms do, to form carbon dioxide. Eventually, ^{14}C beta decays to a ^{14}N atom, a beta particle (**electron**), and an electron antineutrino. The decay rate for this process in a living organism is about 15 decays per minute per gram of carbon. This process has a **half-life** of approximately 5,730 years. After this amount of time, the amount of ^{14}C and the radioactivity of a given sample is reduced by half.

To determine the age of a sample, the amount of carbon in the sample is found through chemical analysis. The number of decays expected if the sample was a living organism is calculated by multiplying the decay rate by the **mass** of the carbon sample. Next, the actual number of decays is measured using a Geiger counter. After N half-lives, the decay rate decreases by a factor of $(1/2)^{N}$. This is the same of the ratio of the number of decays measured to the number of decays for a living organism. In this way, the number N of half-lives that have elapsed can be found, determining the age of the sample.

One of the most familiar applications of radioactive dating is determining the age of fossilized remains, such as dinosaur bones. Radioactive dating is also used to authenticate the age of rare archaeological artifacts. Because items such as paper documents and cotton garments are produced from plants, they can be dated using radiocarbon dating. Without radioactive dating, a clever forgery might be indistinguishable from a real artifact. There are some limitations, however, to the use of this technique. Samples that were heated or irradiated at some time may yield by radioactive dating an age less than the true age of the object. Because of this limitation, other dating techniques are often used along with radioactive dating to ensure accuracy.

RADIOACTIVE DECAY

Radioactive decay is the process in which an **atom** emits another particle or particles. A radioactive element exhibits fewer decays per unit of time at an exponential rate. The **half-life** is how long it takes the element to decay to half its original concentration. This process obeys the known conservation laws, including **conservation of mass** number at low **energy**.

Radioactive elements can be found naturally (as in radium or uranium) or artificially created in a laboratory. Pierre and **Marie Curie** and **Antoine-Henri Becquerel** were among the first to study naturally occurring **radioactivity**, in substances such as uranium, radium, thorium, and polonium. They jointly won the Nobel Prize in 1903 for studying spontaneous radioactivity. In 1934, Pierre Joliot and Irene Joliot-Curie were the first to induce radioactivity by bombarding a sheet of aluminum to produce a radioactive **isotope** of phosphorus.

There are several kinds of radioactive decay. They are classified by the type of particle emitted. The major decay types are alpha, beta, and **gamma decay**. **Nuclear fission** is sometimes also included in types of decay, but it is usually a large **nucleus** splitting into two roughly equally sized nuclei, rather than one atom emitting a much smaller particle.

Alpha decay occurs when a larger nucleus emits a Helium-4 nucleus, called an alpha-particle. This occurs when the nucleus would be more stable with fewer protons and neutrons. The energy released when an **alpha particle** is emitted is equal to the **binding energy** of that alpha particle-that is, the amount of energy it took for the particle to stay with the rest of the nucleus.

Beta decay occurs when there is an imbalance of protons and neutrons in the nucleus. A **neutron** can decay to a **proton** and an **electron**. In this case, the proton will stay in the nucleus, and the electron (or beta-minus particle) will emerge. Alternately, a proton can decay to a neutron and a **positron** (or beta-plus particle). Here, the positron will be ejected. In many cases, electron capture (pulling an electron into the nucleus, where it joins with a proton to become an electron) is classified as beta decay as well. All of these processes are mediated by the weak nuclear **force**.

Gamma decay puts a nucleus in a lower, more favorable energy state. The gamma-particle is actually a **photon** with about the same amount of energy as is lost by its parent atom. Any atom that can stay together in excited states can exhibit gamma emission.

Radioactive elements may emit many particles before reaching a stable energy. This series of decays can be used to identify an unknown radioactive element, since different nuclei emit at different energies. Radioactive decay can also be used to date a sample of a material by the proportion of radioactive nuclei left and a knowledge of those substances' half-lives.

Radioactive decay has uses in many modern technologies, including medical technologies, nuclear **power**, and **nuclear weapons**. Since the beginning of the twentieth century, great progress has been made in discovering both the physical mechanisms and the applications of radioactive decay.

Chernobyl after meltdown. *(AP/Wide World. Reproduced by permission.)*

RADIOACTIVE FALLOUT

The term "radioactive fallout" refers to the contamination caused by nuclear events that have taken place in the atmosphere. Some of these events have been deliberate, such as the scientific testing of **nuclear weapons**, while others have been accidental, like the nuclear **power** station meltdowns that have occurred in various parts of the world. The radioactive dust and debris from these nuclear events are transported by the wind and contaminate the land, water, and food chain wherever they settle.

Radioactive materials emit several kinds of **radiation**. One type of radiation is the spontaneous emission of sub atomic particles such as protons, neutrons and electrons. Another type is **gamma radiation**, which is similar to X radiation. Both forms of radiation are known to be toxic, and exposure to large doses of such radiation cause death and mutation of offspring due to DNA damage. Smaller doses may cause late onset cancer, sometimes years after the exposure to the radiation.

Background radiation has always been present, originating from outer **space** and minerals in the earth. However, in the 1950s and 1960s it was discovered that the background radiation level was increasing. The increase was caused by the detonation of nuclear bombs in the upper atmosphere as part of the weapons testing programs being conducted by the United States, Union of Soviet Socialist Republics, Britain, France, and China. The winds were scattering the radioactive materials, such as strontium-90, potassium-40, carbon-14, caesium-137,

and iodine-131, all around the world. While heavier particles tended to fall to the earth just downwind of the explosion, the lighter particles stayed in the atmosphere for years.

Studies showed that these radioactive materials become more concentrated at each step of the food chain, and in some cases, food animals were destroyed because of the high level of **radioactivity** contained within their bodies. Radioactive fallout became a worldwide concern and, in an effort to halt the rising levels of background radiation, the nuclear powers negotiated a test ban treaty in 1963. Since that time the background radiation count has fallen.

In 1982, the United States Congress required the National Cancer Institute to investigate the effects of the iodine-131 component of the radioactive fallout from the nuclear bomb testing program of the 1950s and 1960s. The report, released in 1997, indicated that some children in the western states downwind of the Nevada tests had higher than normal doses of the radioactive component from drinking contaminated milk, which has resulted in an increased number of thyroid cancer cases.

With the halt of nuclear bomb testing, accidental release of radioactive fallout by nuclear meltdowns became a growing concern, especially for those living downwind of nuclear facilities. In Britain the accidental releases of radioactive fallout made it necessary to dispose of milk and destroy herds of sheep on several occasions. The Three Mile Island accident, in Pennsylvania, released relatively small amounts of fallout, but nevertheless caused major public concern about the nuclear

energy program in the U.S. The Chernobyl disaster (1986) caused a plume of radioactive fallout to cover northern Europe. Research is being done to measure the effect of this fallout on the reindeer population of northern Sweden.

Radioactive fallout is a controversial issue where extreme positions are often taken. It will continue to be an issue as long as nations continue to test nuclear weapons, build more **nuclear power** stations, and try to dispose of the accumulating nuclear waste.

RADIOACTIVE SERIES

A radioactive series is a chain of radioactive decays which results transforms an unstable **nucleus** into a stable nucleus. Unlike simple alpha, beta, or **gamma decay**, there are multiple decays involved in a radioactive series. The decays of the series occur along the nuclear stability curve, an experimentally determined line on a graph of **neutron** number versus **proton** number (atomic number). For proton numbers less than about 20, the stability curve traces the line of equal numbers of protons and neutrons. Above proton number 20, there normally are more neutrons than protons in stable nuclei.

Each nucleus in the series that results from a previous decay either has excessive neutrons or excessive protons to lie on the stability curve, hence it will also decay. The decay mode is usually determined by whether the nucleus has too many protons or neutrons. If the nucleus has excessive protons, the nucleus will **alpha decay**, reducing the number of protons and neutrons by two. If the nucleus has excessive neutrons, it often beta decays, reducing the number of neutrons by one and increasing the number of protons by one. However, the series is not easy to predict and must be discovered by observations of radioactive samples. The nucleus at the end of the radioactive series lies almost exactly on the stability curve, so it will not decay.

An example of a radioactive series is the thorium series. Thorium-232 is a heavy element that has 90 protons and 142 neutrons. It initially alpha decays to radium-228, which lies on the neutron-heavy side of the stability curve. Radium-228 then beta decays to actinium-228 which beta decays again to thorium-228. Then four consecutive alpha decays follow: thorium-228 to radium-224, to radon-220, to polonium-216, to lead-212. The lead-212 then either beta decays twice to polonium-212, which then alpha decays to lead-208, or it alpha decays to thallium-208 and beta decays to lead-208. In either case, lead-208 is the end of the series, because it contains a stable nucleus.

Knowing the sequence of events in a radioactive series allows scientists to roughly calculate the age of Earth. The nucleus at the top of the decay chain is not produced from other decays, so any amount of that element found in nature must have been present when Earth formed. The age of Earth, therefore, should be less than the **half-life** of the nucleus at the top of the chain. If an element at the top of a decay chain is not found to be naturally occurring and assuming the element should be present on Earth, the age of Earth must be greater than the element's half-life.

See also Beta decay; Nuclear structure and stability; Radioactive dating; Radioactive decay; Radioactivity

RADIOACTIVITY

Radioactivity is the process in which unstable atomic nuclei become more stable by spontaneously emitting highly energetic particles and/or **energy**. Material is said to be radioactive if some of its atomic nuclei are emitting such **radiation**. The radiations emitted by unstable nuclei are capable of ionizing **matter** and disrupting molecules, including DNA; they are therefore a biological hazard in prolonged or intense exposures.

Radioactivity is important to society for two reasons. First, it is produced in large amounts by **nuclear fission** in nuclear **power** plants, and the safe disposal of radioactive waste is a problem. Second, radioactivity is widely used as a diagnostic and therapeutic tool in many important medical applications. It is therefore both a burden and a blessing to society.

Every atomic **nucleus** consists of a certain number of protons, strongly bound to a certain number of neutrons. Among the almost limitless number of kinds of nuclei that can be concocted by combining various numbers of protons with various numbers of neutrons, only certain combinations will be stable. The rest will be unstable or radioactive: they will spontaneously change their proton-neutron composition. Atoms that are radioactive are radionuclides, often called **radioisotopes**. (Nuclide is a generic term meaning "kind of nucleus," just as species is used to denote a specific kind of plant or animal. A radionuclide is a radioactive nuclide.)

Making different nuclides out of protons and neutrons is similar to making different kinds of molecules out of atoms. You can't just make a molecule out of any old combination of atoms and expect it to hold together indefinitely, if at all, because atoms bind together according to certain rules of chemical bonding. For example, two hydrogen atoms and one **oxygen atom** form a perfectly stable molecule, H_2O. But two hydrogen atoms and two oxygen atoms make the unstable molecule H_2O_2, hydrogen peroxide. This compound will slowly break down all by itself into water and oxygen. If you try to make a molecule out of two hydrogen atoms and three oxygen atoms, it won't hold together long enough for you to name it.

Similarly, the nucleus that consists of two protons and two neutrons is absolutely stable; it is a helium-4, the **helium isotope** of **mass** number 4, symbolized ^{4}He. But try to use three neutrons, to make a nucleus of helium-5 (^{5}He), and it'll blow itself apart after only 10^{-21} seconds.

There are certain natural "rules" that govern how many protons and neutrons can bind together to form a stable nucleus. While these rules of nuclear binding aren't understood as deeply as are the rules for molecular bonding, their effects are well known and predictable. Nuclides that are constructed of "rule-breaking" numbers of protons and neutrons will be unstable to varying degrees. They will spontaneously break up, emitting particles and energy in order to change themselves into more stable nuclei. That is, they will be radioactive. This is an example of the general principle that nature always tries to increase the stability of a system, and in order to accomplish that it will use any avenue that is open to it—that is, any process that is energetically possible. In the case of unstable nuclei, there appear to be three options: the nucleus can split apart, it can emit particles, or it can emit pure energetic radiation. The general term for all of these changes is nuclear disintegration. Radioactivity, then, is any process by which unstable nuclei disintegrate in order to become more stable.

When a radioactive nucleus disintegrates by emitting an alpha or beta particle, it changes into a different nucleus—that is, one with a different number of protons and/or neutrons. In those cases in which the number of protons changes, the new nucleus has a different atomic number, and it therefore belongs to a different element. In other words, the radioactive atom has undergone a transmutation from one element to another. When this fact was first proposed at the beginning of the twentieth century, it was very difficult for the scientific world to accept because until that time the notion of changing one element into another had existed solely in the realm of alchemy and magic.

Any observable sample of a radionuclide will contain a huge number of atoms. As the unstable nuclei of these atoms continue to disintegrate one after the other, changing themselves into other kinds of atoms, there will be fewer and fewer of the original kind left. The more unstable that particular radionuclide is, the faster its atoms will be disappearing. Radionuclides are known that are so unstable that their half-lives are only the tiniest fractions of a second that can be measured. Others are known that are so very slightly unstable that their half-lives are trillions and quadrillions of years.

Many nuclides appear to be absolutely stable, and presumably their atoms will last forever. There are no avenues open to them, no spontaneous, energetically possible processes that could change their numbers of protons and neutrons to a more stable combination. Of the roughly 2,000 known nuclides, only 264 are stable. The rest are all radioactive in one way or another. Some of these radionuclides (such as the isotopes of uranium and radium) occur naturally on Earth, but the vast majority of them have been made artificially in **nuclear reactors** and in particle **accelerators**, commonly known as "atom smashing" machines.

In December 1895, the German physicist Wilhelm Roentgen (1845-1923) announced his discovery of mysterious penetrating rays—he called them **x rays**—that could go right through an object such as a human hand, making an image of the bones on a photographic plate placed behind the hand. Because these rays came from fluorescent (glowing) spots in Roentgen's glass **vacuum** tubes, scientists immediately began to test a wide variety of fluorescent materials—materials that glow after being exposed to light—to see if they also emitted x rays.

In early 1896, the French physicist Henri Becquerel (1852-1908), working in Paris, was examining many chemical substances that were known to be fluorescent. He first exposed them to bright sunlight to make them fluoresce and then observed whether they emitted any rays that could go through light-proof paper and expose a photographic plate that was wrapped inside. On one gray February day, he put some sam-

ples away in a drawer until he could expose them on the next sunny day. To his surprise, he found that these samples left an image on a photographic plate that was also stored in the drawer. Apparently, the samples were emitting some kind of penetrating radiation without even having to be exposed to **light**. He soon found that only compounds that contained the element uranium had this ability to emit radiation, and it didn't matter what chemical compounds the uranium atoms were in. Becquerel had discovered that uranium is radioactive.

What was startling about this discovery was that the uranium atoms were apparently a source of energy all by themselves; they didn't have to be "activated" by absorbing light energy or anything else. Where the uranium atoms were getting their energy was the big question. For more than 50 years, scientists had believed in the law of **conservation of energy**: that energy simply cannot come from nowhere. But they also believed strongly that atoms were unchangeable. How could uranium atoms be giving off radiation without changing?

In December 1897, **Marie Curie**, a Polish chemist working on her doctoral thesis at the Sorbonne in Paris, began to try to find out where this mysterious uranium energy was coming from. In the process, she discovered two new elements that are millions and billions of times more radioactive than uranium: radium and polonium. By inventing new chemical techniques for separating radioactive elements, Marie Curie laid the foundation for all of today's applications of radioisotopes in industry and medicine. She was the world's first radiochemist.

Marie Curie's work was an important step toward the understanding of radioactivity. However, it remained for **Ernest Rutherford** and Frederick Soddy (1877-1956) to suggest in 1902 that radioactivity represents an actual disintegration of atoms. Later, Rutherford also discovered the atomic nucleus and the **proton**. In 1931 he was named Baron Rutherford of Nelson (the town in New Zealand where he was born) in recognition of his scientific achievements.

Lord Rutherford found two kinds of rays coming from the radioactive uranium atoms: a slightly penetrating kind that wouldn't even go through paper and a somewhat more penetrating kind that would go through thin sheets of metal. He called them alpha rays and beta rays, respectively. Later, a very penetrating, third kind of radiation, called gamma rays, was discovered. These three types of radiation, symbolized by the Greek letters α, β, and γ, still represent the three major types of radioactivity. We now know that alpha and beta "rays" are actually high-speed **subatomic particles**, while gamma rays are electromagnetic energy **waves** of pure energy.

Nuclear chemists and physicists have found that certain combinations of neutrons and protons seem to make the most stable nuclei. In general, the most stable nuclei will be those that (a) contain nearly equal numbers of protons and neutrons, but with more neutrons than protons, and that (b) have an even (rather than odd) total number of protons plus neutrons. Nuclei that deviate too much from these rules will be unstable to various degrees.

The three kinds of radioactivity, alpha, beta, and gamma, come from nuclei that deviate in three different ways from the stability rules: alpha particles come from nuclei that have too many protons plus neutrons (that is, they are simply too big and heavy); beta particles come from nuclei that have too many protons or too many neutrons; and **gamma radiation** comes from nuclei that simply have too much energy.

Nuclei are known that contain up to 266 neutrons and protons, a very large number. That's a lot of particles to be packed into a volume that has a radius of only 10^{-12} centimeter, especially since almost half of the particles are protons—positively charged particles that, being all of the same charge, are trying hard to repel each other. So when a nucleus is too "big and fat," it tries to reduce by shooting off some of its particles. The most energetically favorable combination of particles that it can shoot off is a tight little package of two protons and two neutrons. Two protons and two neutrons, bound together, constitute a nucleus of helium-4. This nucleus, when shot out at high speed by an unstable nucleus, is called an alpha (α) particle.

Alpha particles, containing two protons, have a charge of +2. So when an **alpha particle** is shot off into the surrounding matter by a radioactive nucleus, it will interact very strongly with the negatively charged electrons in the matter. This has two effects: (1) the alpha particle tears electrons off many atoms as it passes through, that is, it ionizes many atoms, and (2) it slows down and stops very quickly because the ionization process uses up its energy. Alpha particles, therefore, cannot penetrate very far through matter before they slow down completely and stop. Even a sheet of paper will stop most alpha particles.

All nuclides that are heavier than bismuth (atomic number 83, mass number 209) are too heavy to be stable; they are radioactive and emit alpha particles. Among the commonly known alpha emitters are various isotopes of polonium, radium, thorium, uranium, and all of the transuranium elements.

When a nucleus emits an alpha particle, it decreases its mass number—the total number of protons plus neutrons—by four units: it loses two protons and two neutrons. However, losing the two protons also decreases its nuclear charge or atomic number by two units, transforming it into the nucleus of a different element, two spaces to the left in the **periodic table**. For example, when a radium nucleus (atomic number 88) emits an alpha particle, it becomes a nucleus of radon (atomic number 86).

$$^{226}Ra \rightarrow {}^{4}He + {}^{222}Rn + Energy$$

(In these symbols, the superscript is the mass number, the total number of protons and neutrons in the nucleus.) The released energy is mostly in the form of kinetic (movement) energy of the alpha particles, which are emitted at speeds of around one-tenth the **speed of light**, depending on the particular radionuclide that is emitting them.

Alpha particles are both highly energetic and relatively highly charged, so they disrupt many atoms and molecules along their paths before they come to a stop. But they are not much of a hazard to living things because they stop so soon. They can only penetrate about a thousandth of an inch (0.03 mm) of aluminum, for example. Human skin will stop them, so they can't penetrate far enough to disrupt the cells in any vital organs. If alpha-emitting radionuclides are inhaled into

the lungs or ingested into the stomach, however, they can do their damage locally to highly susceptible kinds of tissues, and can therefore be very dangerous. Inhaling radon gas has been blamed for many lung cancers in uranium miners, for example. There is radon gas in uranium mines because the disintegration of uranium leads to radium, which then forms radon as shown in the equation above. The radon is not yet stable, and it emits alpha particles itself.

A second means of disintegration that is open to too-heavy nuclides is spontaneous fission. Most of the transuranium elements have isotopes that disintegrate by fissioning (splitting) in addition to emitting alpha particles.

As previously stated, a nucleus must have roughly equal numbers of protons and neutrons in order to be stable. If a nucleus has too many protons for its number of neutrons, it will be radioactive. Simply shooting out an unwanted proton turns out to require more energy than the nucleus has to give, except in a few very rare cases. (Lutetium-151, which has an extremely large number of protons compared with its number of neutrons, has actually been observed to emit protons from some of its nuclei.)

If a nucleus with too many protons could transform one of them into a **neutron**, however, it would be improving its proportion of protons to neutrons by simultaneously losing a proton and gaining a neutron. And that is what it does, but because the proton has a positive charge and the neutron is neutral, the nucleus somehow has to get rid of a positive charge. It does that by creating and emitting a positron, also known as a positive beta particle. For example (see equation below), potassium-40, an isotope of potassium that constitutes 1.2% of all potassium atoms in nature (including those in our own bodies), is radioactive and emits positrons.

$$^{40}K \rightarrow \beta^{+} + {}^{40}Ar + Energy$$

Positrons are emitted at perhaps nine-tenths the speed of light, depending on the radionuclide that is emitting them. The **positron** is a light particle, identical to an ordinary **electron** except that its charge is +1 instead of -1. Theory predicts that every particle has an opposite, called an antiparticle, and the positron is the **antiparticle** of the electron. Antiparticles don't last long in our world of ordinary particles, however. As soon as a positron (or any antiparticle) meets an ordinary electron (or its ordinary counterpart), the two particles annihilate each other: they both disappear in a puff of energy.

Positrons emitted from radioactive nuclei are light-weight and have only a single unit of charge, so they don't interact as strongly with matter as alpha particles do. They can penetrate farther into matter—about a tenth of an inch of aluminum—before slowing down and annihilating. Therefore, they are also a greater hazard to humans than alpha particles are.

Another kind of radioactivity that accomplishes the same thing as positron emission is electron capture. This is the process in which a proton is converted into a neutron by the nucleus capturing a negative electron from one of the inner orbits of its atom. No particles are emitted, but some x rays are, due to the now-missing atomic electron. Radionuclides that have too many protons often disintegrate by both methods: some of the nuclei by positron emission and some by electron capture.

If a nucleus has too many neutrons in relation to its number of protons, it will try to become more stable by decreasing its number of neutrons. Ejecting a neutron is energetically unfavorable, however, except in one or two rare cases. (Lithium-11 has been observed to emit neutrons.) However, a nucleus with too many neutrons can convert one of them into a proton. To do this, it would have to find an extra positive charge, because the neutron is neutral and the proton is positively charged. But an object can also increase its positive charge by throwing out a negative charge, and that is what the nucleus does: it creates and emits a negative beta particle, which is identical to an ordinary electron except that it comes from a nucleus instead of from the outer parts of an atom.

A common emitter of negative beta particles (see equation below) is carbon-14, the radioactive isotope of **carbon** that is found in all living plants and animals.

$^{14}C \rightarrow \beta^- + ^{14}N + Energy$

Negative beta particles penetrate matter in the same way as positive beta particles do, except that they don't annihilate.

A nucleus can have an unusually large amount of internal energy, just as the electrons in a whole atom can. In both cases, we say that the nucleus or the atom is excited, or in an excited state. A nucleus can find itself in an excited state when, for example, it has just been created through the disintegration of another radioactive nucleus. Just as an excited atom can dispose of its excess energy by emitting x rays, an excited nucleus can emit gamma rays. Gamma rays are electromagnetic radiation just like x rays except that they are generally of higher energy. When a nucleus emits gamma rays, the composition of its protons and neutrons does not change.

Most radionuclides that emit alpha and beta particles also emit gamma rays. This is because the nuclei into which they are converted are often created in excited states; these excited nuclei immediately get rid of their excess energy by emitting gamma rays. This is an important safety consideration because gamma rays are extremely penetrating and can cause biological damage all the way through the body. Almost any radioactive substance must be assumed to be emitting highly penetrating gamma rays, even if the substance is known to be "only" an alpha or beta emitter.

All of the elements heavier than bismuth (atomic number 83) are completely radioactive—that is, they have no stable isotopes. We still find quite a few of the elements in nature, either because they have such long half-lives that they haven't completely died out since the earth was formed some 4.5 billion years ago, or because they are constantly being produced by the disintegration of uranium or thorium, whose half-lives are indeed comparable to the age of the earth. (By an interesting coincidence, uranium-238, the principal isotope of uranium, has a **half-life** of 4.47×10^9 years, just about equal to the age of the earth. Thus, there is almost exactly half as much uranium left on earth today as there was when the earth was formed.)

There are three series of naturally occurring heavy radionuclides. Each one begins with a radionuclide that is long-lived enough to have survived since the earth was formed. By a sequence of alpha and beta disintegrations, it transforms itself into a series of other radionuclides, until it becomes a stable isotope.

One natural **radioactive series** begins with uranium-238 (atomic number 92), which undergoes a long sequence of disintegrations, producing radioactive isotopes of several elements until it reaches the stable isotope, lead-206. A second series of disintegrations begins with uranium-235 (half-life 7.04×10^8 years) and winds up as stable lead-207. The third series begins as thorium-232 (half-life 1.41×10^{10} years) and winds up at stable lead-208.

Along the way, these disintegration series produce radioactive isotopes of protactinium, thorium, actinium, radium, francium, radon, astatine, polonium, bismuth, lead, thallium, and mercury. The first eight of these are those heavier-than-bismuth elements that we find in nature. Depending on the relationship between the half-lives of the radionuclides and the half-lives of their predecessors and successors in the sequences that produce them, various amounts of these elements exist on earth at the present time.

In addition to the heavy radioactive elements, 18 lighter elements have radioisotopes that are found in nature because their half-lives are long compared with the age of the earth. Two others, hydrogen-3 (tritium) and carbon-14, are constantly being produced by special processes. All of the naturally occurring radionuclides, both heavy and light, contribute to a certain amount of radioactivity to which everyone on earth is always being exposed, regardless of humanity's activities in nuclear technology.

Of the roughly 1,700 radioactive nuclides that are known, only about 70 occur in nature. The rest have all been made synthetically: either they were found in the nuclear debris of man-made nuclear fission, or they have been produced in **particle accelerators**. During **nuclear reactions** carried out in accelerators, the numbers of protons and neutrons in atomic nuclei can be changed, thereby transforming them into different, and sometimes new, radionuclides. In fact, the elements with atomic numbers 43, 61, and 85 (technetium, promethium, and astatine, respectively) were unknown on Earth until some of their radioactive isotopes had been produced synthetically. In addition, all of the elements with atomic numbers higher than uranium's (92) were discovered by making them synthetically in particle accelerators.

Both natural and synthetic radionuclides, generally referred to as radioisotopes, are of enormous value in science and industry, and in medical research, diagnosis, and therapy. Applications of radioisotopes are based primarily on two facts: radioactivity can be detected with such astounding sensitivity—the disintegration of single atoms can actually be detected—that extremely tiny amounts of radioactive material can be followed through complex biological and industrial processes by keeping track of where the radiation goes; and the radiations from radioactive materials can be used to destroy living cells, such as harmful microorganisms and human cancer cells.

RADIOISOTOPES

Atomic theory holds that the **nucleus** of an **atom** consists of protons and neutrons and describes the elements of the **periodic table** as atoms consisting of nuclei surrounded by electrons. Identical elements have the same atomic number, Z, which is equal to the number of protons in the nucleus, and the same **mass** number, A, which is equal to the number of protons plus the number of neutrons. Atoms with the same atomic number but a different number of neutrons, or a different mass number A, are called isotopes. For example, hydrogen has an atomic number of Z = 1, so it has one **proton**, but it occurs as three isotopes: 1-hydrogen (Z=1; A=1), deuterium or 2-hydrogen (Z=1, A=2), and tritium or 3-hydrogen (Z=1, A=3) in which the respective nuclei contain zero, one or two neutrons. The isotopes of an element are by convention prefixed by their mass number. The nuclei of elements can undergo **radioactive decay**, during which particles and **radiation** are emitted. The most common types of emitted radioactive particles are the following: alpha particles, which are 4-helium nuclei (two protons and two neutrons)]; beta particles, which are electrons traveling close to the speed of **light** (2.998×10^8 m/s) and which result from the breakdown of a **neutron** into a proton, an **electron** and a **neutrino**; and gamma particles, a form of electromagnetic radiation consisting of high-energy photons of very short **wavelength**. Isotopes are classified as to whether they decay spontaneously or not. The elements whose nuclei do not decay are termed stable isotopes and those isotopes whose nuclei are unstable and disintegrate spontaneously are called radioisotopes.

The first radioisotope experiments were performed in 1896 by the French physicist A.H. Becquerel who studied salts of uranium and observed they produced a latent image on a photographic plate that could be developed in the same way as a latent image produced by visible light. The radiation emitted by the uranium salts could penetrate thin sheets of metal and discharge a charged oscilloscope.

Many radioisotopes occur naturally, such as the heavy atoms near the end of the periodic table. They belong to the uranium, thorium and actinium families. Each group is headed by a long-lived primary **isotope**: the thorium series starts with 232-thorium, the actinium series with 235-uranium, and the uranium-radium series with 238-uranium. The decay of the primary isotope leads to another radioisotope, which decays further. The exhaustion of the decay chain finally leads to formation of a stable isotope. The three families contain altogether some 44 radioisotopes and in each case, the final product is a stable lead isotope. Some lighter radioisotopes also occur. Natural potassium for example, besides stable 39-potassium and 41-potassium, also contains trace quantities (0.012%) of radioactive 40-potassium. Rubidium and samarium also contain radioisotopes. More than a thousand artificial radioisotopes are also known. They are produced by changing the original proton-neutron ratio of a stable nucleus by the initiation of nuclear disintegration processes, for instance by bombarding the nuclei with high-energy particles. Examples of artificially produced radioisotopes include over ten variants of the stable 127-iodine isotope and numerous variants of 22- and 24-sodium.

See also Alpha decay; alpha particle; alpha radiation; Atomic structure; Becquerel, Antoine Henri; Atomic theory; Beta decay; Beta radiation; Gamma decay; Gamma radiation; Nucleon; Photon; Positron; Radioactivity; Radioisotopes in medicine

RADIOISOTOPES IN MEDICINE

Radioisotopes are extensively used in nuclear medicine to allow physicians to explore bodily structures and functions in vivo (in the living body) with a minimum of invasion to the patient. Radioisotopes are also used in radiotherapy (**radiation therapy**) to treat some cancers and other medical conditions that require destruction of harmful cells.

Radioisotopes, containing unstable combinations of protons and neutrons, are created by **neutron** activation involving the capture of a neutron by the **nucleus** of an **atom** resulting in an excess of neutrons (neutron rich). **Proton** rich radioisotopes are manufactured in cyclotrons. During **radioactive decay**, the nucleus of a radioisotope seeks energetic stability by emitting particles (alpha, beta, or **positron**) and photons (including gamma rays).

Although nuclear medicine traces its clinical origins to the 1930s, the invention of the gamma scintillation camera by American engineer Hal Anger in the 1950s, however, brought major advances in nuclear medical imaging and rapidly elevated the use of radioisotpes in medicine. Radioisotopes allow high quality imaging of bones, soft organs (e.g., thyroid, heart, liver, etc.). A number of diagnostic techniques in nuclear medicine use **gamma ray** emitting tracers. The tracers are formed from the bonding of short-lived radioisotopes with chemical compounds that allow the targeting of specific body regions or physiologic processes. Emitted gamma rays (photons) can be detected by gamma cameras and computer enhancement of the resulting images allows quick and relatively non-invasive (compared to surgery) assessments of trauma or physiological impairments.

Technetium-99 (an **isotope** of the artificially-produced element technetium) is a radioisotope widely used in nuclear medical procedures. Technetium-99 decays by an isomeric process which emits gamma rays and low **energy** beta particles (electrons). Technetium is supplied to hospitals from **nuclear reactors** in containment vessels initially containing molybdenum-99 that, with a **half-life** of 66 hours, decays to technetium-99 which is removed by flushing.

American chemist Peter Alfred Wolf's (1923-1998) work with radioisotope utilizing positron emission tomography (PET) led to the clinical diagnostic use of the PET scan. Positron emission tomography utilizes isotopes produced in a **cyclotron**. Positron-emitting radionuclides are injected and allowed to accumulate in the target tissue or organ. As the radionuclide decays it emits a positron that collides with nearby electrons to result in the emission of two identifiable gamma photons. PET scans use rings of detectors that sur-

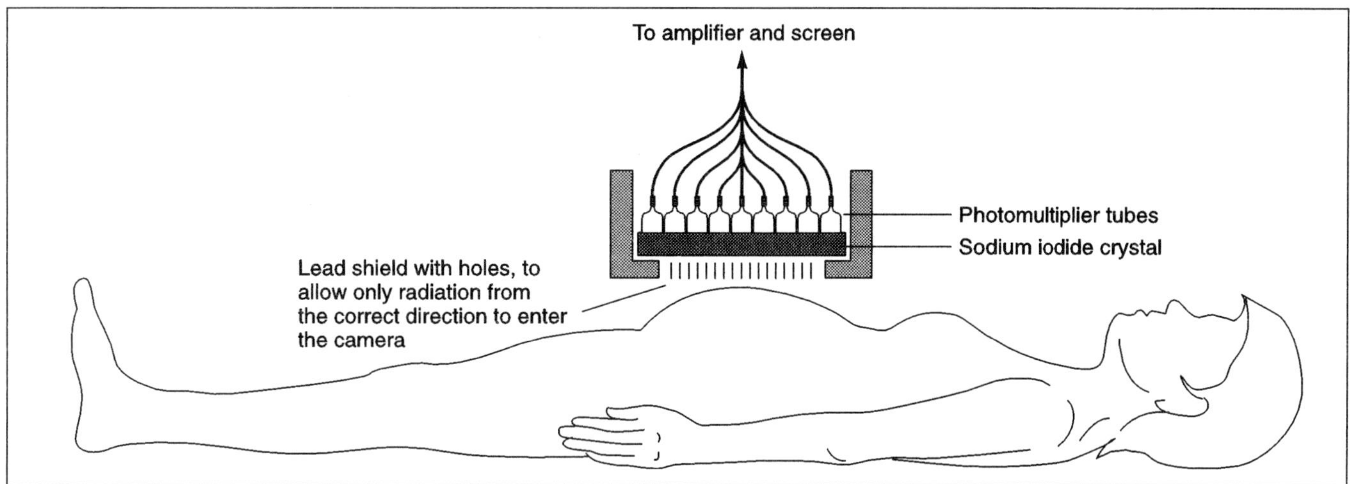

Radioactive tracers (human body covered by lead shield).

round the patient to track the movements and concentrations of radioactive tracers. PET scans have attracted the interest of physicians because of their potential use in research into metabolic changes associated with mental diseases such as schizophrenia and depression. PET scans are used in the diagnosis and characterizations of certain cancers and heart disease, as well as clinical studies of the brain.

Cancer and other rapidly dividing cells are usually sensitive to damage by radiation. Accordingly, some cancerous growths can be restricted or eliminated by radioisotope irradiation. The most common forms of external radiation therapy use gamma and **x rays**. During the last half of the twentieth century the radioisotope cobalt-60 was frequently used source of radiation used in such treatments. More modern methods of irradiation include the production of x rays from linear **accelerators**.

Internal radiotherapy involves the introduction of a radioisotope as a radiation source. Iridium-192 implants emit both gamma and beta rays that destroy surrounding target tissue. In low dose forms, strontium-89 has been used to relieve cancer-induced bone pain. Not all radioisotope techniques involve restricted sites, in the treatment of some diseases requiring bone marrow transplants the malfunctioning marrow is killed with a massive dose of radiation before the introduction of healthy marrow.

Because they can be detected in low doses, radioisotopes can also be used in sophisticated and delicate biochemical assays or analysis. There are many common laboratory tests utilizing radioisotopes to analyze blood, urine, and hormones. Radioisotopes are also finding increasing use in the labeling, identification and study of immunological cells. Shielded radioisotopes are also used to **power** heart pacemakers and sterilize medical instruments.

Iodine-131 and phosphorus-32 are commonly used in radiotherapy. More radical uses of radioisotopes include the use of boron-10 to specifically attack tumor cells. Boron-10 concentrates in tumor cells and is then subjected to neutron beams that result in highly energetic alpha particles that are lethal to the tumor tissue.

The selection of radioisotopes for medical use is governed by several important considerations involving dosage and half-life. Radioisotopes must be administered in sufficient dosages so that emitted radiation is present in sufficient quantity to be measured. Ideally the radioisotope has a short enough half-life that, at the delivered dosage, there is insignificant residual radiation following the desired length of exposure. Regardless, the use of radioisotopes allows increasingly accurate and early diagnosis of serious pathology (e.g. tumors) and earlier diagnosis often results in more favorable outcome for patients.

RAMAN, C. V. (1888-1970)
Indian physicist

Physicist C. V. Raman helped to usher India into the world of twentieth-century science. Raman overcame obstacles of geographical isolation and political oppression to establish himself, and thus India, as a serious contributor to modern Western science. His primary research interests were **acoustics**, musical instruments, and wave **optics**. He was best known for his discovery of the Raman effect (first announced in 1928), a process by which a beam of **light** passing through a solid, liquid, or gas was diffracted and its frequencies (and so its colors) were changed. In recognition of this discovery Raman was knighted in 1929 and awarded the Nobel Prize in physics in 1930.

Prior to Raman's time, India's development had proceeded along the lines of literature, art, and architecture. After the British colonized this land and set up trade, science was imported only to assist in furthering commerce. The British kept native Indians on the periphery of this activity and did not allow them training in modern scientific methods. It was only after the British introduced Western education that native Indians came in direct contact with twentieth-century

Sir Chandrasekhara V. Raman.

European science. Raman, then, had to struggle against his country's delayed interest in science, its geographical isolation, and its oppression by the ruling British. By the end of the nineteenth century when European science had already become mature, science in India was a mere fledgling.

Childhood Education and Early Achievements

Chandrasekhara Venkata Raman, the second of eight children, was born on November 7, 1888, near Trichinopoly on the banks of the Kaveri. His father, Ramanathan Chandrasekaran Iyer, was a lecturer in physics, mathematics, and physical geography. Iyer read avidly, collected books, and played the violin. These pastimes came to have a great influence on his son. Raman's mother, S. Parvati Ammal, was the daughter of Saptarshi Sastri, a great Sanskrit scholar.

As a child, Raman was not strong or athletic, but he excelled at intellectual pursuits. He won many scholarships and prizes at school, and in 1903 at the age of 16 he won a scholarship and entered the prestigious Presidency College as one of its youngest undergraduates. At college Raman pursued his boyhood interest, physics, and also developed a fondness for English. Raman graduated first in his class and won gold medals in physics and English.

After Raman passed his B.A. examination, his teachers encouraged him to go to England to continue his studies. But because of his frail health he was counseled against subjecting himself to the unhealthy English climate. As a result, Raman decided to enroll in the Presidency College's M.A. program in physics. Raman's teacher at this time made few demands on

him and let him explore on his own. Raman, guided by his own interests and goaded by his own motivation, conducted his own experiments in the **diffraction** of light passing through rectangular slits. When he had accumulated a number of findings, he wrote them up and sent a manuscript to the *Philosophical Magazine* in London, which published it as "Unsymmetrical Diffraction Bands Due to a Rectangular Aperture." This accomplishment was remarkable because Raman sent the manuscript on his own, his was the first paper to come out of the Presidency College, and he had done this at the age of 18.

After earning his M.A. in 1907, Raman found that there were no opportunities for a career in research open to Indians in India. So he secured through fierce competition a coveted position in the Indian Civil Service as an accountant. He pursued his scientific research in his spare time at home until 1917 when Calcutta University offered him the Palit Chair for Physics at about half the salary he was receiving from the government. Raman accepted the position without hesitation.

Promoting Indian Science

Because of India's great size and diversity, its centers of research often suffered from their isolation. Raman felt a need for some way of consolidating and promoting his country's interests in science. To that end he worked at inciting interest among Indian scientists in establishing an academy of science. His initial attempts were stymied in Bombay by intramural disputes among various scientific factions, so he independently founded his own academy in Bangalore.

The inaugural meeting of the Indian Academy of Sciences was held in August of 1934 in Bangalore. Raman became its first president, and he retained the office until his death. For this reason some called the organization Raman's Academy. Its stated objectives were to provide a forum for discussing the results of scientific research and to publish the achievements of Indian science. G. Venkataraman, in his *Journey into Light: Life and Science of V.S. Raman,* called the Academy "one of Raman's gifts to India."

The Academy's journal, the *Proceedings of the Indian Academy of Sciences,* appeared in 1934 and featured as its opening paper Raman's work entitled "The Origin of the Colours of the Plumage of Birds." By 1935 the amount of publishable material had grown to such an extent that Raman divided the journal into two parts: physical and mathematical sciences, and biological sciences. Eventually the *Proceedings* was divided into six separate specialty journals in 1977.

Fascination with Light and Sound

Raman's fascination with the phenomenal world seemed to be behind most of his research interests. Raman once admitted to letting his attention wander from his English professor at the Presidency College because the glittering waves of the blue sea, which were visible from his lecture hall, caught his eye. Later, **color** was to become an ever-present aspect of his research on light and optics, which included studies on the color effects of shells, gems, minerals, flowers, and plumage.

Music also influenced Raman's research. His father inspired in his son a love for music, particularly the violin. Raman himself became a competent violinist, and later he approached musical instruments as a physicist. Raman was recognized as the first in this century to rekindle the research into the physics of the violin and other musical instruments. His research in this area included studies on the behavior of bowed strings, the influence of the violin's bridge, and the **frequency** response of the violin (known as the Raman curve). Raman also investigated the sound-producing mechanisms of the piano and of some traditional Indian instruments like the tabla and the tambura.

At one point, Raman was able to combine his interests in **sound** and light in the same research project. The fruit of this project was the Raman-Nath theory, perhaps his greatest achievement during his stay at the Indian Institute of Science. The Raman-Nath theory explained what happens to a beam of light as it passes through a liquid that is agitated by a sound wave. This theory became important later to the research on the propagation of starlight through the atmosphere and the propagation of laser emissions through plasma.

As a student, Raman had learned science in a setting where laboratory resources were modest. As a result he became adept at improvising the equipment he needed for his experiments. In a 1905 experiment on the **surface tension** of liquids, he devised a spark generator that illuminated a suspended drop so its shadow could be photographed. In 1927 while studying the **scattering** of light, Raman needed to keep the sensitivity of an observer's eye at a maximum, so he fashioned a light-tight wooden enclosure for the observer, which came to be called the "black hole of Calcutta." During his research on the physics of the violin, Raman needed a way to study and control the **force** of the bow, so he contrived his experimental apparatus from materials he had on hand including parts from an optical bench and the chain and hubs from a cycle. The result was a mechanical violin player where the violin moved while the bow remained stationary.

Soon after discovering the Raman effect (frequency and color changes with respect to light diffraction), Raman became interested in the diamond. Captivated by its beauty and physical properties, he referred to it as the "prince of solids." He began collecting them at his own expense, and by 1944 he had over 300. Raman even admitted that he had used part of his Nobel Prize money to buy diamonds. When he couldn't afford to buy them, he borrowed diamond rings from wealthy friends. Venkataraman reported that Raman once saw his brother sporting a diamond ring and said to him, "I say, why don't you put that thing on your finger to some use?" So strong was his scientific curiosity for diamonds that at one point every one of his students was studying some aspect of diamonds. Raman studied many optical phenomena in diamonds including light absorption, light scattering, and x-ray diffraction.

Later Interests

Raman did not want to become idle once he retired, so he began planning a new research institute two years before his retirement from the Indian Institute of Science. He conducted his own fund raising, and by 1948, although he had not raised enough money, he had acquired a building. During the first year of its existence, the new Raman Research Institute had no **electricity**, but that did not prevent Raman from conducting some important optical experiments using sunlight. Eventually Raman had the financial support to complete his new facility, which included gardens for the trees and flowers that he loved and a museum for his collections of crystals, gems, minerals, shells, birds, and butterflies. Raman continued his work on optics at the Institute, which included, in conjunction with his nephew, the development of a theory to explain mirages.

Raman was known to be proud and at times even arrogant. Venkataraman reported that in 1924 at the age of 36, Raman was elected Fellow of the Royal Society. When he was congratulated and asked what was next, he replied, "the Nobel Prize, of course." When he actually did receive the prize six years later, it was discovered that he had booked passage to Stockholm, the site of the awards ceremony, four months before he received the official announcement of the award.

In the course of his 66 years as a physicist, Raman published over 450 research papers and inspired almost three times that number of papers among his students. In promoting Indian science, Raman was hailed as one of India's heroes along with Mohandas Gandhi and Motilal Nehru. Raman died at his research institute in Bangalore on November 20, 1970, and his ashes were scattered there among the trees.

RAMSEY, NORMAN FOSTER (1915-)
American physicist

Norman Foster Ramsey is a preeminent physicist whose research has focussed on the properties of molecules, atoms, nuclei, and elementary particles. The numerous awards and honors he won throughout his career culminated in the 1989 Nobel Prize in physics. Although the prize seemed to recognize a lifetime of achievements in the field, the Nobel committee specifically cited his work in developing a method of measuring the differences between atomic **energy** levels. His findings were key to the development of the cesium **atomic clock**, which measures time with an accuracy previously unknown.

Ramsey was born in Washington, D.C., on August 27, 1915. He was named after his father, a graduate of the United States Military Academy who was then a general serving as assistant to the Chief of Ordnance. His mother was Minnie Bauer Ramsey. After several years in Washington, his father was transferred to the Command and General Staff School in Fort Leavenworth, Kansas. There, Norman Jr. attended high school, distinguishing himself as president of his class. He graduated in 1931. Ramsey studied physics at Columbia University in New York City, where he won the Van Aminge and Van Buren prizes in mathematics. He graduated Phi Beta Kappa from Columbia in 1935 and entered the doctoral program there.

As a graduate student, Ramsey studied at Cambridge University, receiving a B.A. in 1937 and an M.A. in 1941. From 1939 to 1940, he was also a Carnegie Fellow at the Carnegie Institution in Washington. In 1940 he received his Ph.D. from

Columbia. His thesis, written on research he had performed with the Nobel laureate **I. I. Rabi**, focused on the rotational magnetic moments of hydrogen molecules. Ramsey and Rabi had found that magnetic moments were dependent on the **weight** of the nuclei, which led to the discovery of a new **force**, called the tensor force, between the **neutron** and the **proton**.

From 1940 to 1942 Ramsey continued his research as an associate at the University of Illinois. As the United States became involved in World War II, he also worked at the Massachusetts Institute of Technology Radiation Laboratory as the head of a group developing the magnetron transmitter for **radar**. The result of this work was the three-centimeter-wavelength radar system, the first of its kind, which was widely used during the war. From 1942 to 1945, Ramsey served as consultant to the Secretary of War, at first advising the Air Force on the use of radar and later consulting with the National Defense Research Committee. In this capacity, he was sent to Los Alamos to study the possibilities of building an **atomic bomb**, and from 1943 to 1945 he was group leader and associate division chief of the Laboratory of the Atomic Energy Project. In 1945 he went with his group to the Tinian Islands bomber base to oversee the first **atom** bombing missions.

During the war, Ramsey remained active in academics. From 1942 to 1945 he was an assistant professor of physics at Columbia, and in 1945 he was promoted to associate professor. After the war he served as executive secretary of the group that founded the Brookhaven National Laboratory in Long Island, New York. He was named head of the physics department there in 1946. In 1947 he took a job as associate professor at Harvard University, becoming a full professor in 1950.

In 1948, Ramsey was named chair of the Harvard Nuclear Physics Committee and director of the Harvard Nuclear Laboratory. In this capacity he was involved with the development of Harvard's first postwar **cyclotron**, built in 1949. The 125,000,000-electron-volt cyclotron was designed to smash atoms in order to study the particles that constitute atomic nuclei and the forces that keep the particles together. By 1956, Harvard and MIT had joined forces to build another cyclotron, the world's best at the time, producing the fastest artificially accelerated particles. Ramsey chaired the committee that oversaw its construction.

Ramsey's research during these years led to several important discoveries. One was that some atoms which were thought to exist near a **temperature** of **absolute zero**, a value based on the **Kelvin temperature scale**, could actually have temperatures that were below absolute zero. It had previously been thought that temperatures of **matter** could never have negative values. Parts of the second law of **thermodynamics** had to be rephrased in order to accomodate this finding.

Discovers Separate Oscillating Fields

Ramsey also challenged the prevailing practice of measuring atomic energy spectra. Beams of atoms had been measured by passing them through an **electromagnetic field** tuned to the difference between the atom's two energy levels. The resulting pattern of **interference** was studied to deduce information regarding the structure and behavior of atoms. However, the accuracy of this technique was limited by the need to maintain a constant magnetic field throughout the process. Ramsey attempted to expose atoms to two separate electromagnetic fields—one as the atoms entered the field and another as they departed. The pattern of interference was much more accurate than the one previously produced by using a homogenous magnetic field. His experiments led to the development of the hydrogen maser (for Microwave Amplification by Stimulated Emission of Radiation).

Improving the techniques of studying atoms led Ramsey to the development of his Nobel-winning achievement: the cesium atomic clock. Announced by Ramsey in 1960, the atomic clock is believed to be 100,000 times more accurate than previous atomic clocks, which used gaseous ammonia molecules. Ramsey's clock uses high-energy atomic hydrogen, which could be measured for the first time due to the method of using separate oscillating fields. It is now the time standard used throughout the world, in which the second is defined as the time in which it takes a cesium atom to make 9,192,631,770 **oscillations**.

Ramsey was awarded half the 1989 Nobel Prize in physics, the other half being divided between Wolfgang Paul and Hans Dehmelt. Ramsey was cited for his work on the hydrogen maser, as well as the atomic clock. Daniel Kleppner, with whom Ramsey often collaborated, told *Science* magazine: "His work was seminal in the theory of chemical shifts, which underlies the use of the **magnetic resonance imaging** units in hospitals."

Kleppner also called Ramsey a "statesman of science," and World War II was not the end of his government service. In 1958, Ramsey was named Scientific Advisor to the North Atlantic Treaty Organization. He took a leave from Harvard to head an advisory committee which oversaw all NATO activities in research and applied science. In the late 1970s he was again asked to be a government adviser, and he became the co-chairman of a federal committee to study the possible practical uses of cold **nuclear fusion**.

In addition to the Nobel Prize, Ramsey also received the Lawrence Award in 1960, and the Davisson-Germer Prize from the American Physical Society in 1974. He was awarded the Karl Compton Prize from the American Institute of Physics, the Rumford Prize, and the National Medal of Science, all in 1985. He was elected to the National Academy of Sciences in 1952.

Ramsey married Elinor Stedman Jameson in 1940; she died in 1983. They had four daughters. In 1985, Ramsey married again, this time to Ellie Welch. He is now Higgins Professor of Physics, Emeritus, at Harvard University.

RANDOM WALKS AND BROWNIAN MOTION

Brownian motion describes how very small solid particles, such as smoke particles or microscopic pollen particles, move when they are bombarded by the molecules surrounding them.

This movement was first observed in 1827 by Scottish botanist Robert Brown (1173-1858). He saw in his **microscope** small pollen particles suspended in water. On closer inspection he noticed the pollen particles appeared to be jiggling about at high speed. It was suggested in 1863 that this **motion** was the result of millions of collisions between the solid pollen particles and the speeding molecules of the water. **Albert Einstein** turned his attention to the nature of this jiggling motion and the important outcome of his work was the calculation of the number of molecules in a standard volume of gas. This value had been unknown until that time.

The first question in considering Brownian motion is the pattern of movement of the particles. Intuitively we can recognize that movement of a smoke particle must be composed of short straight motions punctuated with sudden sharp changes in direction. The motion of a drunken man can serve as a (facetious) example. Suppose our inebriated fellow has absolutely no sense of direction and each step taken bears no relation to the previous one. The motion of the man at any point in time will be completely unpredictable, and the path taken through 10 consecutive steps would bear no relation to the path taken during the next 10 steps. This kind of motion is called a random walk, or sometimes the drunkard's walk. This was the motion analyzed by Einstein in the sober confines of his mind.

The first thing to realize is that although our drunken friend may actually travel quite a long distance the net result of all his efforts will be much reduced because, for some of the time, he will be accidentally reversing his direction. A lot of effort will be expended for very little real progress. If our friend always takes a 2-ft (61-cm) stride then in more sober times 100 steps will produce progress of 200 ft (60 m). However, mathematical analysis of this random walk shows us that, on average, the drunkard will only move the equivalent of 10 steps, or 20 ft (6.1 m). For steps of equal length the distance covered from the start can be calculated by finding the square root of the number of steps and multiplying it by the length of the step. One hundred and forty four steps produces, on average, a real journey of only 12 steps distance from the starting point.

When the length of each step varies, as we might expect in the case of the particles in Brownian motion, then the mathematics is a little more complicated but the result comes out as follows:

For a large number of steps, the distance measured from start to finish is the square root of the number of steps multiplied by the square root of the average squared step size. The square root of the average squared step size is called the root mean square of the steps. (It is calculated by taking the length of each step and squaring it. Then all these squared lengths are added together to get a total and then an average value. The square root of this average value is the root mean square.).

Returning to physics, we can see that every gas molecule will collide over and over with the other gas molecules in much the same way as they collided with the pollen and smoke particles in Brownian motion. The motion of each gas particle, therefore, must mirror the Random Walk described above. So, in

short, we now have a way of connecting the long random erratic journey taken by a molecule with its actual drifting journey through a volume of gas. This turns out to be very useful. It is possible to measure the drift of one gas mixing with another in diffusion experiments. From the kinetic theory of ideal gases we can also calculate the root mean square speed of the gas molecules between collisions. (This requires measuring gas **pressure**, **temperature** etc.) Knowing these two values enables us to use the random walk results to get an estimate for the number of gas molecules that must exist in a given volume of gas. This number, known as **Avogadro's number**, opened the way at the beginning of the twentieth century, to the early understanding the physical dimensions of the molecular world.

See also Gases, behavior and properties; Ideal gas law

RAYLEIGH SCATTERING

Why is the sky blue? Why are sunsets red? The answer involves Rayleigh **scattering**. When **light** strikes small particles, it bounces off in a different direction in a process called scattering. Rayleigh scattering is the scattering that occurs when the particles are smaller than the **wavelength** of the light. Blue light has a wavelength of about 400 nanometers, and red light has a wavelength of about 700 nanometers. Other colors of light are in between. A nanometer is a billionth of a meter. So, for Rayleigh scattering of visible light the particles must be smaller than 400-700 nanometers. Scattering can occur off larger particles, but it will follow a different scattering law.

The Rayleigh scattering law, derived by Lord Rayleigh in 1871, applies to particles smaller than the wavelength of the light being scattered. It states that the percentage of light that will be scattered is inversely proportional to the fourth power of the wavelength. Small particles will scatter a much higher percentage of short wavelength light than long wavelength light. Because the mathematical relationship involves the fourth power of the wavelength even a small wavelength difference can mean a large difference in scattering efficiencies. For example, applying the Rayleigh law to the wavelengths of red and blue light given above shows that small particles will scatter blue light roughly 10 times more efficiently than red light.

What does all this have to do with blue skies and sunsets? Earth's atmosphere contains lots of particles. The dust particles scatter light but are often large enough that the Rayleigh scattering law does not apply. However the nitrogen and **oxygen** molecules in Earth's atmosphere are particles small enough that Rayleigh scattering applies. They scatter blue light about 10 times as much as red light. When the **Sun** is high overhead on a clear day, some of the blue light is scattered. Much of it is scattered more than once before eventually hitting our eyes, so we see blue light coming not directly from the sun but from all over the sky. The sky is then a pretty shade of Carolina blue. In the evening, when there is less blue light coming directly from the Sun it will appear redder than it really is. What about sunsets? When the Sun is low in the sky, the light must travel through much more atmosphere to reach our eyes. Even more of the blue light is scattered, and the Sun appears even redder than when it is overhead. Hence, sunsets and sunrises are red.

RECTIFIERS

Without rectifiers we could not use the **electricity** supplied to our homes to **power** any of our electronic devices such as televisions, **computers**, **microwaves**, battery chargers, etc. With few exceptions, these types of devices require direct current (DC) power to function as a result of their incorporation of integrated circuits in their design, which require uninterrupted DC power. However, the power supplied to our homes is transmitted as an alternating current (AC) signal, and the polarity of this supplied power reverses at 60 hertz (in the United States).

So how is AC turned into DC? A simple circuit device called a rectifier does the trick. The sole purpose of rectifiers is to turn alternating current into direct current. Through the use of **diodes**, which only pass current in one direction, rectifiers are able to take an AC signal as input and then output a DC signal that can be used by electronic devices.

There are two general types of rectifiers: half-wave and full-wave. Half-wave rectifiers simply block one half of the AC signal to the output so that only half the input signal is used. Full-wave rectifiers use the full input signal, but by diverting half of the input the output signal is of a single polarity.

While the rectifier part of any electronic device is relatively small, its importance cannot be over emphasized. Without rectifiers, the power we get from a power station cannot be coupled to the electronic devices upon which we have come to depend.

RED DWARF/RED GIANTS • See Stars

RED SHIFT

Red shift is an apparent increase of the **wavelength** of **light** toward the red end of the **electromagnetic spectrum** caused by relative **motion** between the light-emitting object and an observer. The red shift of light occurs due to relative velocities of source and observer, cosmological expansion, and **gravity**. It is an instance of the **Doppler effect**, which is commonly observed with **sound waves**. Waves emitted by a moving object will be blue shifted (compressed) if approaching an observer and red shifted (elongated) if receding. This occurs with both sound and light, but the Doppler effect in light is different than that of sound because light waves do not need a medium to travel in.

As a source of waves speeds away from an observer, the crest of each wave spread from each other; that is, the waves arrive with longer wavelengths. In light, longer wavelengths are those near the red end of the electromagnetic spectrum. Scientists calculate how fast an object—such as a galaxy—

speeds toward or recedes from the Milky Way by knowing the value of the Doppler shift. The change produced by the Doppler shift is measured by identifying **spectral lines** (bright and dark lines seen in the spectra of **stars** and other luminous objects) of elements—such as hydrogen—that exist in an star. A scientist compares the star's spectral lines with those of the same earthbound element. Such a comparison shows how fast it moves toward or away from Earth. The red shift observed in **astronomy** is equal to the change in **frequency** (the time between each wave crest) of the **radiation** divided by the frequency of the emitted light.

The red shift (and corresponding blue shift) of light was first successfully described in the mid-1800s. At that time, scientists knew the Doppler shift relates to how fast a star or galaxy recedes from Earth. When American astronomer **Edwin Hubble** began investigating galaxies' Doppler shifts in the 1920s, he made some remarkable discoveries and took that notion a step further. Hubble detected variable stars (stars whose brightness changes periodically) that can be used to determine how far away galaxies are from Earth. He compared galaxies' Doppler shifts with their estimated distance from Earth and made a graph of his findings. The result: a majority of the galaxies he observed were receding from us. Not only that, but the farther a galaxy is from Earth, the faster it rushes away. The linear relationship between distance and red shift is called Hubble's Law. This observation was, for many scientists, proof that the **universe** is not static (neither expanding nor contracting) as previously thought. Though some scientists thought that the linear relationship between red shift and distance would change at high red shifts due to the deceleration of the universe, there has not been any reliable observation of such an effect.

Red shift can be caused by gravity, which bends light as predicted by Einstein's theory of general relativity. This effect is very weak, however, and many scientists believe it need not be taken into account when measuring cosmological red shift.

Hubble was the first person to pinpoint the expansion rate of the universe, known as the **Hubble constant** (H_o). He did this by dividing the distance to a galaxy by the speed that it recedes from Earth. Since distance divided by **velocity** (distance over time) is equal to time, the value of the constant tells us the age of the universe. This value, however, is far from certain and has been the subject of much debate among astronomers since its discovery.

Determining the age of the universe also depends on assumptions made about its behavior. Today, cosmologists believe that the universe expands as a result of an event called the **big bang** that occurred probably more than 10 billion years ago. Hubble's law, which says that galaxies' recessional velocities increase with their distance, relies upon an expanding universe model; one in which all galaxies recede from all others in all directions.

REFERENCE FRAMES

A reference frame is a set of coordinate axes that help describe the position or movement of an object. There are two basic types of reference frames: inertial and accelerating. Inertial frames are those where the coordinate axes appear to follow Newton's first law in every direction from the point of view of an observer: things at rest tend to remain at rest unless acted upon by an external **force**. An accelerating frame, on the other hand, is one where the coordinate axis are accelerating from the point of view of an observer, such as a rotating merry go round or an accelerating rocket ship.

Two types of transformation are used when transforming or mapping coordinates from one inertial frame to another. If the constant **velocity** of the frame is zero or small compared to the speed of **light**, a Galilean transformation is appropriate. However, if the velocity is a significant portion of the **speed of light**, a Lorentz transformation is appropriate, since the laws of special relativity apply. The Lorentz transformation equations are actually true at any speed, but Galilean equations are mathematically valid at low speeds.

Galilean Transformation Equations

$$x' = x - vt$$
$$t' = t$$

Lorentz Transformation Equations

$$x' = \gamma(x - v)$$
$$y' = y$$
$$z' = z$$
$$t' = \gamma\left(t - \frac{vx}{c^2}\right)$$

For example, take two carts, one stationary and one moving with a constant velocity. To a stationary observer in each frame, the ball on the stationary cart will fall straight down, but the ball on the moving cart will fall in a parabola. Data is taken on both balls from a non-moving view point in each frame. A Galilean transformation is performed on the stationary cart and checked against the moving cart data.

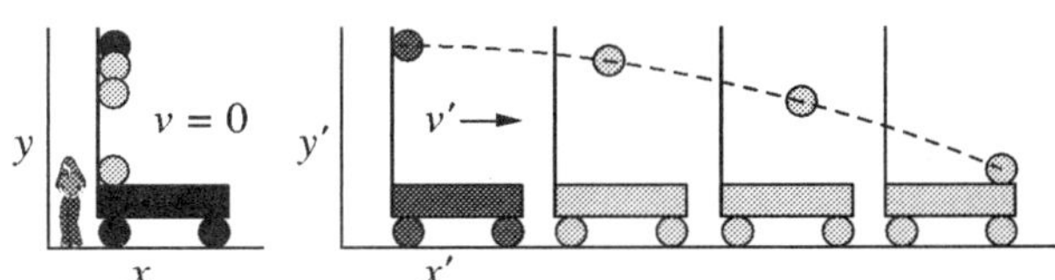

Stationary Cart Data	Moving Cart Data	Galilean Transformation of Moving
$x = 0$	$x' = v't'$	$x = x' - v't' = 0$
$y = \frac{1}{2}gt^2$	$y' = \frac{1}{2}g(t')^2$	$y = \frac{1}{2}gt^2$
$v_x = 0$	$v'_{x'} = v'$	$v_x = \frac{x}{t} = 0$
$v_y = gt$	$v'_{y'} = gt'$	$v_y = gt$
$p_x = 0$	$p'_{x'} = mv'_{x'}$	$p_x = 0$
$p_y = mv_y$	$p'_{y'} = mv'_{y'}$	$p_y = mv_y$

Now, suppose there are two electrons, each in its own **frame of reference**. One is moving with velocity v_1=0.6c and the other with velocity v_2=-0.6c. A Galilean transformation from either frame of reference to the other makes one **electron** moving at v=1.2c while the other is stationary. We know from special relativity that this is not possible; nothing can move faster than c, the speed of light in a **vacuum**. This is where we would have to use a Lorentz transformation on the data to map from one coordinate system to another.

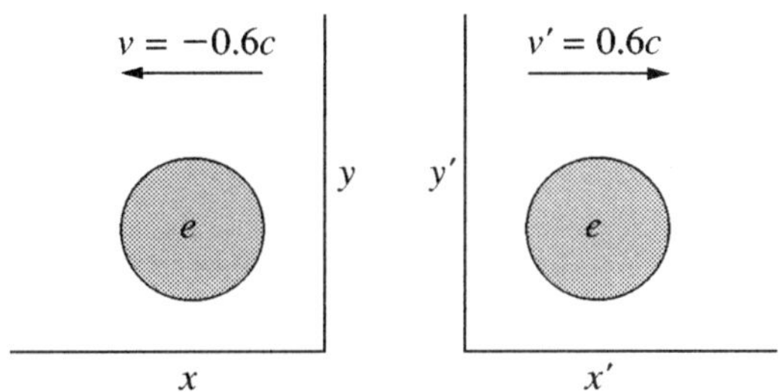

As an example of a Lorentz transformation, let's consider a reference frame, B, moving relative to another frame, A, with a constant velocity, v_b=0.8c along the common direction of their x-axes. All clocks read zero as the origins coincide. An observer in B sees an explosion, E, on his or her x-axis at position x_b=1.5km and at time t_b=0.001s. To find when and where the person in A will observe this event, we must do the transformation.

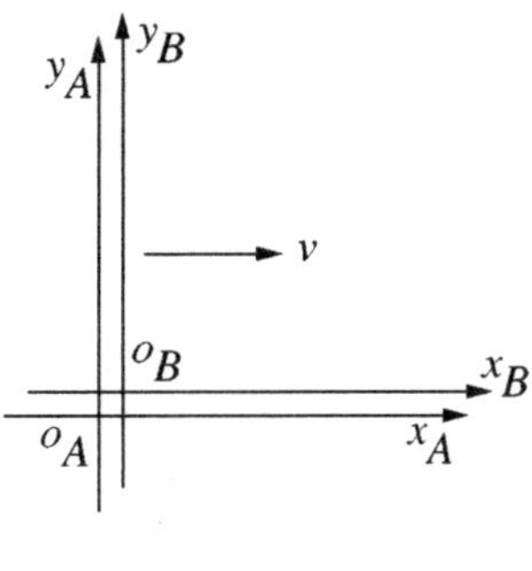

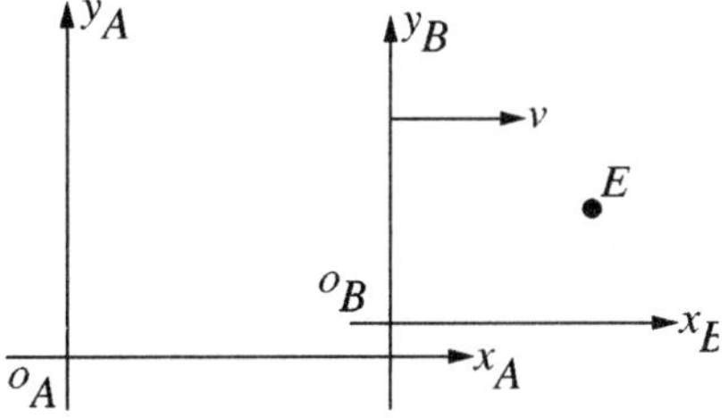

Our givens are: t_b =0.001s, x_b =1.5 x 10³ m, and v_b =0.8c =0.8(3 x 10⁸)m/s.

To find x_b,

$$x_a = \frac{x_b + v_b t_b}{\sqrt{1 - \dfrac{v^2}{c^2}}}$$

$$x_a = 4.025 \times 10^5 \, m$$

To find t_a,

$$t_a = \frac{t_b + x_b \left(\dfrac{v_b}{c_b}\right)}{\sqrt{1 - \dfrac{v^2}{c^2}}}$$

$$t_a = 0.001673 s$$

Notice the time and **length contraction** of the B frame as predicted by special relativity.

As these graphs show, events are relative; they depend on the observer's reference frame. It is very important when working in physics to define correctly the appropriate reference frame.

REFLECTION, REFRACTION, AND DISPERSION

When **light** travels from one medium to another, three phenomena can occur; reflection, refraction, and **dispersion**. Reflection takes place when the light does not pass into the material, but is instead "bounced" off the surface. Refraction occurs when the light passes into the material. It is bent toward or away from the normal according to the index of refraction of the material. The normal is a line perpendicular to the surface of the medium. Dispersion occurs when different wavelengths of light are refracted different amounts, separating the light into its constituent colors. The wave property of light prevents all light from passing directly though a medium. As light passes through different mediums, part of it is reflected and part is transmitted through.

When light travels from one material to another, it is refracted, obeying Snell's Law. This law states that the sine of the angle of incidence for a ray multiplied by the refractive index for the medium, equals the product of the refractive index and the sine of the incidence angle for the next medium. Every translucent material has an index of refraction, n. This is an index that tells how the **speed of light** is effected by the material. Since n=c/v, (where c equals the speed of light in a vaccuum and v equals the speed of light through the medium) the higher the index, the lower the **velocity** of light inside the material. If the light is traveling from a lower index to a higher index, such as going from air (n=1) to glass (n=1.33), the light is bent towards the normal. If traveling from a higher index to a lower one, light is bent

Spoon in a glass of water illustrating light refraction. *(Photo courtesy of Photo Researchers, Inc. Reproduced by permission.)*

away from the normal. The effects of refraction are apparent when viewing fish from a boat. This is because light is being refracted as it passes from air to water. Water has a higher index of refraction so light is bent towards the normal. However, when we view an object, we assume the light travels in a straight path. As a result, we will think the fish is actually closer to the surface of the water. If you aimed a spear at the fish, you'd miss! Your best bet when spear fishing is to jab straight down (perpendicular to the water's surface) since light traveling along the normal is not refracted (the sine of zero degrees is one).

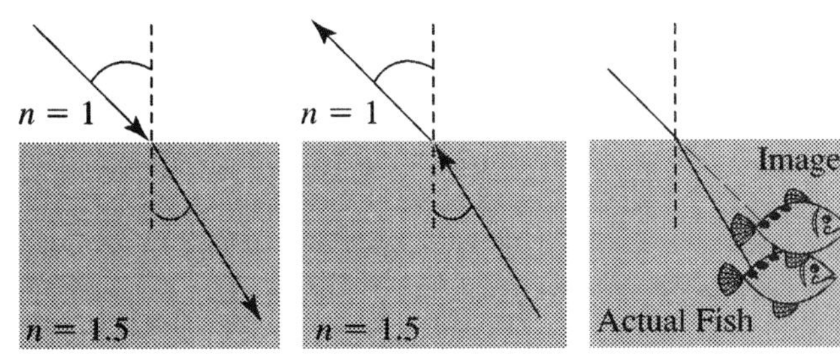

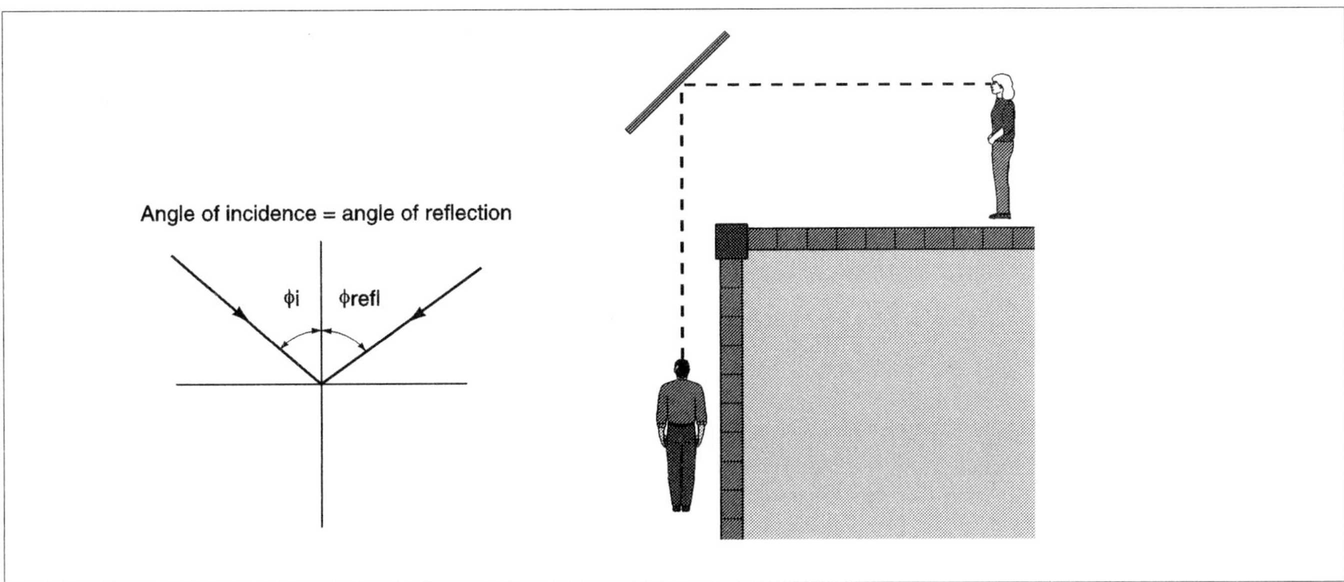

The Law of Reflection.

When light is reflected, the incident angle is equal to the reflection angle, as measured from the normal. When the light is exiting from a medium, total internal reflection can occur if the angle of incidence is greater than the critical angle. The critical angle is the angle that causes the light to travel along the surface of the material when refracted. Total internal reflection is the basic phenomenon behind fiber optic light; once it enters the fiber, is totally internally reflected, keeping it inside the fiber until it exits the other end.

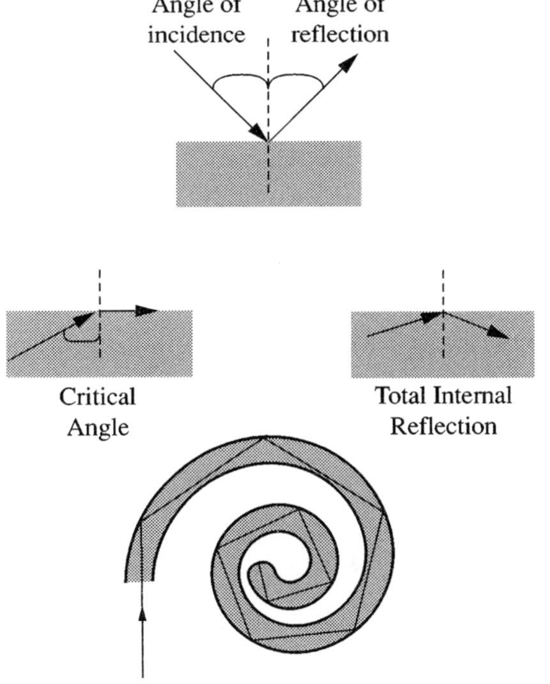

Dispersion is the effect associated with the separation of light into colors by a prism. As the light enters the glass, it is refracted. Each **wavelength** of light is refracted a different amount, with the blue wavelengths being refracted the most. We know that the amount of refraction is dependent on the index, n, which is itself dependent on the velocity of the light in the medium. The velocity is wavelength dependent, v being equal to the **frequency**, f, multiplied by the wavelength, lambda. When light is dispersed by a prism, red light has a higher velocity in the glass and is not refracted as much as the slower, blue light. Any refractive process using non-monochromatic light has an element of dispersion. This leads to effects such as chromatic aberration in **lenses**. Each wavelength of light exiting from a lens that exhibits this kind of aberration has a different focal length; therefore, light is not totally recombined into white light and a slight rainbow effect can be seen in the image.

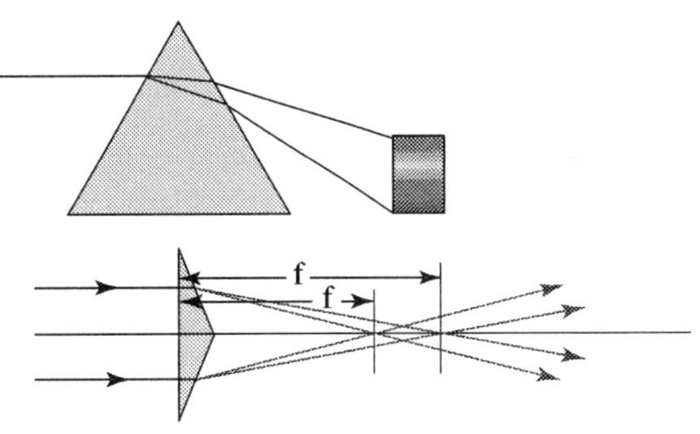

REINES, FREDERICK (1918-1998)

American physicist

Frederick Reines is best known for his discovery of the **neutrino**, a particle that is created during a nuclear reaction. Since that discovery, he continued to concentrate on the search for neutrinos and study their characteristics and behaviors. In later years, Reines studied the possibility of **proton decay**, an experiment whose negative results have had important significance for elementary particle theory.

Reines was born in Paterson, New Jersey, on March 16, 1918. His father was Israel Reines and his mother was Gussie Cohen Reines. After graduation from high school in 1935, Reines enrolled at the Stevens Institute of Technology, from which he earned a B.S. in mechanical engineering in 1939 and an M.A. in science in 1941. He then went to New York University, where he was awarded his Ph.D. in theoretical physics in 1944. His first job was at the Los Alamos Scientific Laboratory, where he was first a staff member and later a group leader responsible for studying the blast effects of **nuclear weapons**. In 1951, he was director of Operation Greenhouse, a group of experiments related to the testing of nuclear weapons at Eniwetok Atoll in the South Pacific.

Discovers the Neutrino

It was during his association with Los Alamos that Reines made the discovery for which he has become most famous, the detection of the neutrino. The neutrino had been predicted by **Wolfgang Pauli** in 1931 as a way of solving a puzzling nuclear phenomenon. When a beta particle is emitted from an unstable **nucleus** during **beta decay** (a radioactive nuclear interaction), the **energy** it carries away is insufficient to account for the energy lost within the nucleus itself. In order to account for this discrepancy, Pauli suggested that a second particle was formed that appropriated the energy discrepancy. **Enrico Fermi** later named this particle the neutrino. The name neutrino ("little neutron") arises from the properties postulated for the particle by Pauli. From the physics of beta decay, it was clear that the particle could have no charge, like the **neutron**, and no or very small **mass**; thus it was dubbed a "little neutron."

These properties guaranteed, however, that finding the neutrino would be very difficult. With no charge and perhaps no mass, it would be able to pass through **matter** (including detection instruments) without undergoing any type of interactions. In the early 1950s, Reines and a colleague, **Clyde Cowan**, undertook a search for the neutrino. The basis for their research was the assumption that although the chance of a neutrino's interacting with matter was very low, it was not zero. The key to success, they hypothesized, was to focus their detectors on a situation in which very large number of neutrinos would be expected to form, thus greatly increasing the chance of observing at least one reaction.

The best source for observing a flood of neutrinos, Reines and Cowan concluded, was a nuclear reactor. Their

first experiments were carried out, therefore, at the Atomic Energy Commission's Hanford Nuclear Laboratory in Washington state. To search for the elusive neutrino, they decided to select one of the many reactions that physicists had hypothesized for it, one in which gamma rays of characteristic energy are generated. They then built a large 80-gallon liquid scintillator with a bank of ninety photomultiplier tubes (a form of **vacuum** tube used for detecting very low levels of **light**) which they placed next to the Hanford reactor. The first evidence for neutrinos began to appear in 1953, but it was not entirely conclusive. Cowan and Reines decided to expand and improve their detection system and to move their experiment to the Savannah River National Laboratory in South Carolina. In 1956, they repeated their experiment there and obtained conclusive evidence for the existence of neutrinos.

In 1959, Reines left Los Alamos and accepted a position as professor of physics and head of the department at Case Institute of Technology (now Case Western Reserve University) in Cleveland. At the same time, he became chair of the Joint Case-Western Reserve High Energy Physics Program. Reines held these positions until 1966, during which time he continued to serve as a consultant at Los Alamos and was also a consultant to the Institute for Defense Analysis (1965–1969) and trustee of the Argonne National Laboratory.

Reines left Case-Western Reserve in 1966 to accept an appointment as professor of physics and the first dean of physical sciences at the University of California at Irvine. Four years later, he was also appointed professor of radiological sciences at Irvine's Medical School. He retired in 1988 and was named Distinguished Professor of Physics, Emeritus at Irvine. Reines died in Orange, California, in 1998.

Searches for Neutrinos in Varied Environments

Following his discovery of the neutrino, Reines continued his research on these elusive particles. He later built huge tanks containing the colorless liquid perchloroethylene in a search for atmospheric neutrinos, that is, neutrinos produced by solar **cosmic rays** in Earth's atmosphere. That search achieved success, although the number of neutrinos detected was significantly less than the number predicted by current theories. The detectors constructed for the search for atmospheric neutrinos were also used to find neutrinos produced by the dramatic eruption of Supernova 1987A in 1987. (A supernova is an exploding star that, at the height of its luminosity, can be brighter than the sun.)

Reines's research has also involved a search for the possible decay of the **proton** in nuclear interactions. Although the pronton has long been considered a stable particle, some current theories of elementary particles suggest that they may have a very long, but not infinite, **half-life**. Reines's research sought to prove that these half-lives could be many billions of years long. Using detection devices that had been successful with neutrino research,

Reines has shown that the minimum proton half-life predicted so far is not possible, although still longer half-lives may be.

Reines married Sylvia Samuels on August 30, 1940. They have two children, Robert and Alisa. Among his numerous honors and awards are the 1981 J. Robert Oppenheimer Memorial Prize, the 1985 National Medal of Science, the Bruno Rossi Prize of the American Astronomical Society in 1989, and the Michelson-Morley Award in 1990. Reines also received the W. K. H. Panofsky Prize and the Franklin Medal of the Benjamin Franklin Institute, both in 1992.

RELATIVISTIC ELECTROMAGNETISM

Relativistic **electromagnetism** is a combination of special relativity and electromagnetism. Many physicists contributed to the development of electromagnetism during the nineteenth century. In 1864, James Maxwell consolidated their **work** and devised a set of equations describing electromagnetism. The equations are called **Maxwell's equations**. In 1905, **Albert Einstein** published the theory of special relativity. According to special relativity, Newtonian **mechanics** are only a low-speed approximation. Maxwell's equations, however, were still correct. They needed to be put into a relativistic form that is much different from the classical Maxwell's equations.

Special relativity tells us about the correct transformation of quantities from one inertial reference frame to another. A set of transformation rules were given by **Hendrik Lorentz**, which were later named **Lorentz transformations**. Quantities obeying these rules are said to be Lorentz covariant. The two variables in electromagnetism, electric field (**E**) and magnetic field (**B**) are not Lorentz covariant. A Lorentz covariant variable, called the field-strength tensor, can be found from **E** and **B**. It is actually a matrix.

$$F = \begin{pmatrix} 0 & -E_1 & -E_2 & -E_3 \\ E_1 & 0 & -B_3 & B_2 \\ E_2 & B_3 & 0 & -B_1 \\ E_3 & -B_2 & B_1 & 0 \end{pmatrix}$$

E and **B** show up in the same Lorentz covariant variable. This means that **E** and **B** are indeed equivalent and are unified in this way. The unification of **E** and **B** implies that what looks like an electrical field can be recognized as also having a magnetic component from another inertial frame. In solving a relativistic electromagnetism problem, the solution is usually calculated in a convenient inertial frame. The results in any other inertial frame can then be calculated using Lorentz transformations from this convenient frame. The field-strength tensor must be used in the transformations.

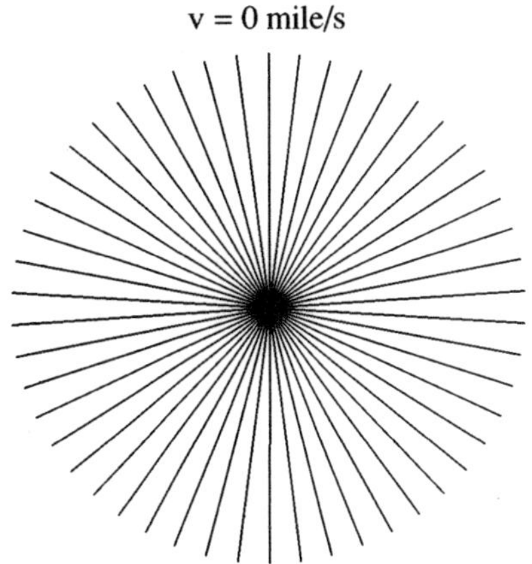

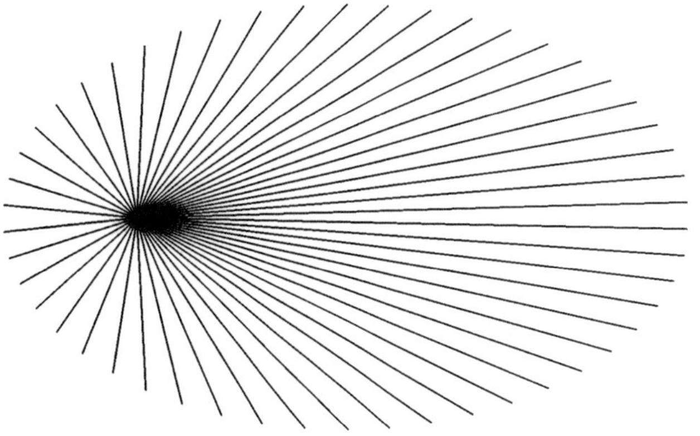

Special relativity has many interesting effects on an object moving at high speed. The above picture shows the **light** emitted by a bulb when it is stationary and when it moves at 135 million mi/s (80% the **speed of light**) with lines representing light. The length of a line is proportional to the **frequency** of the light emitting in that direction. A longer line represents a larger frequency and shorter **wavelength**. The **density** of the lines is proportional to the brightness of the light.

If the bulb is moving toward us, the wavelength will become shorter and the **color** moves to the blue end of the spectrum. If the bulb is moving away from us, the wavelength will become longer and the color moves to the red end of the spectrum. These effects are called relativistic Doppler shift, similar to Doppler shift of **sound**. If the bulb comes toward us, passes us, and goes away from us very quickly, we will see the bulb as much brighter when it is moving toward us than when it is moving away from us. The relativistic Doppler shift also occurs even if the bulb passes in front of us. It is then called transverse Doppler shift.

There is a rule of thumb to determine when relativistic electromagnetism should be used. If the speed of the object is close to the speed of light, the relativistic electromagnetism must be used to obtain a correct result. Otherwise, the relativistic effect can be ignored. This rule of thumb applies to special relativity in general. The speed of light is about one hundred thousand times faster than our typical speed in everyday life. The relativistic effect is consequently very small in most everyday experiences. When you drive towards a red traffic light, from your point of view, the traffic light is moving toward you. You do not need to worry that the light will look green to you because of relativistic Doppler shift (green has a shorter wavelength than red). The relative shift in wavelength is only about 40 billionths if you drive at 30 mph.

Relativistic electromagnetism turns out to be crucially important in two extreme cases. One case is in the **space** around us, the **universe**. Many **stars** are moving fast enough that we must use relativistic electromagnetism to do calculations. The relativistic Doppler shift can be used the other way around to calculate the speed of stars from their shift in wavelength. The speeds of many stars are found this way. Another case is very small, the atomic and subatomic world. Physicists do experiments on small particles like electrons in high **energy** or medium energy laboratories. Fermi National Laboratory (Fermilab) is one of the largest high energy laboratories in the world. In a high energy physics experiment, particles are accelerated to a very high speed. Some can move at more than 99% the speed of light. This extremely fast case is sometimes called the ultra relativistic case. Relativistic electromagnetism is used here to insure the correctness of the results. Relativistic electromagnetism is a base of modern **astronomy** and high energy physics, also called **particle physics**.

RELATIVISTIC MASS AND ENERGY

The relativistic **mass** of an object refers to the measured mass of an object moving relative to the observer. A moving object appears to have a larger mass than the same object at rest. The mass of the object is directly related to its total **energy** according to Albert Einstein's 1905 publication on the theory of special relativity. The observed total energy of the moving object is called its relativistic energy.

The measured mass of a stationary object is called its rest mass. For an object with a rest mass m_0 moving with a speed v, the observed mass of the object is found to be

$$m = m_0 \sqrt{1 - \frac{v^2}{c^2}}$$

where c is the speed of **light** in a **vacuum**, measured to be 186,000 mi/s (300,000 km/s). The total relativistic energy of the object is then given by Einstein's famous equation showing the equivalence of **matter** and energy:

$$E = mc^2 = m_0 c^2 \sqrt{1 - \frac{v^2}{c^2}}$$

As the object's speed increases, both the relativistic mass and energy of the object increase.

The energy associated with the object's **motion** is commonly referred to as **kinetic energy**, and the total relativistic energy of an object is the sum of the object's kinetic energy and its **rest energy**, $E_0 = m_0 c^2$. Therefore, the relativistic kinetic energy is the difference between the total relativistic energy and the rest energy. The deviation of the relativistic kinetic energy from Newton's kinetic energy is noticeable only for particles moving at a considerable fraction of the **speed of light**. For example, a particle moving with a speed equal to c/3 will be observed to have a relativistic mass that is 6% larger than its rest mass. Therefore, there will be a 6% deviation from Newton's kinetic energy formula.

The relativistic mass and energy of a particle must be considered in most modern elementary **particle physics** experiments, such as the experiments taking place at most particle **accelerators** like the one at the Fermi National Accelerator Laboratory in Batavia, Illinois, outside of Chicago.

RELATIVISTIC (MINKOWSKI'S) INTERVAL

The relativistic interval is a quantity used to denote the distance between two events in four-dimensional **space-time**. Similar to the distance between two points in **space**, which is unchanged by relabeling the origin of the coordinate system or rotating it around, the relativistic interval is invariant under **Lorentz transformations**. It is defined by the equation I = $c^2(\Delta)t^2 - (\Delta)x^2$. In this equation, $(\Delta)t$ is the **time** between two events, and $(\Delta)x$ is the distance between two events. Unlike a distance, which is always positive, the interval may be positive, negative, or zero.

Depending on the value of the interval, pairs of events fall into three categories. If the interval is negative, the distance between the two events is larger than the distance **light** can travel in time (Delta)t. In this case, a Lorentz transformation can be done so that an observer sees the two events occur simultaneously, so the events are called "space-like." If the interval is positive, a Lorentz transformation can be done so that the two events appear to occur at the same point in space but at different times, so the events are called "time-like." If the interval is zero, the only way the two events can communicate is by a light signal, so the events are called "light-like."

RELATIVISTIC VELOCITY TRANSFORMATION

Consider two inertial observers. One, Jonathan, sits on a railroad train, which is traveling to the right with a constant **veloc-**

ity u, as measured by Nicholas, who stands by the railway embankment. Jonathan throws a ball towards the front of the train, which, according to him, travels with velocity v'. According to Nicholas, the ball is moving to the right with a speed v, given by the simple (and almost intuitive) Galilean velocity addition rule: $v = v' + u$. This rule follows directly from the Galilean transformations relating **space** and **time** measurements made by our two observers. Einstein showed in 1905, however, that in order for the speed of **light** to have the same value for all inertial observers, the Galilean transformations must be replaced by the **Lorentz transformations**. The requirement that the laws of physics be Lorentz invariant leads to a new, relativistic rule for the addition of velocities: $v = (v' + u)/(1 + (v'u/c^2))$.

This differs from Galileo's result by the presence of the $v'u/c^2$ in the denominator. For all practical purposes, this term is exceedingly small, as either v', u, or both, are much smaller than the **speed of light**, c. The value of v will differ only infinitesimally from the Galilean value. Thus we are free to use the Galilean result in the example above.

But, as is easily checked, its presence in the denominator ensures that the velocity of light is identical for all observers. If Jonathan shines a flashlight toward the front of the train, so that $v' = c$, a simple calculation shows that $v = c$, as well. The relativistic value for v will differ appreciably from the Galilean if either v', u, or both, are appreciable fractions of the speed of light. Then the extra term ensures that v is always less than c.

See also Space-time

RELATIVITY, GENERAL

The general theory of relativity, which **Albert Einstein** completed in 1915, is a theory of **gravity** and **motion**. His special theory of relativity challenged the classical ideas of **space**, **time**, and simultaneity, and the general theory went further to declare that the properties of **space-time** are neither homogeneous (the same at all locations) nor isotropic (the same in every direction regardless of orientation), and do not necessarily remain the same as time passes. Whereas special relativity deals with frames of reference in relative motion at constant **velocity**, general relativity also considers accelerated frames of reference and gravity. Measurements of space and time intervals depend not only on the relative motion between observers, but also on the strength of the local gravitational fields.

The central concept in general relativity is the **equivalence principle**, which states that the effects of gravity and **acceleration** are indistinguishable and equivalent. Using this principle, Einstein described gravity as a geometric effect rather than as a **force** acting at a distance. The presence of **matter** or **energy** curves space-time, redefining the way objects move in the vicinity. This contrasts with special relativity, which assumes space-time to be flat. General relativity predicts that the **universe** as a whole may have curvature. If this

is the case, space, though of a higher dimension, would be analogous to the surface of Earth, which is finite in area, but because of its curvature has no boundaries. On a spherical surface such as Earth's, the shortest route between two points is not a straight line but a great circle or geodesic. The equator is one example of a geodesic. In curved space, geometry differs from that in flat space, and the curvature effects are interpreted as gravitational fields. Like the spherical surface, curved space is in general non-Euclidean, and for example, parallel lines may cross, and the sum of the angles in a triangle may be more than 180°. By picturing space-time and matter as interacting with each other Einstein radically departed from previous theories that had viewed space and time as a fixed stage on which physical objects moved.

Cosmological effects are predicted by the theory. Prior to any observational evidence, Einstein believed the universe to be static, neither expanding nor contracting, and introduced an arbitrary element into his equations, the cosmological term, to force the model to be stable. After expansion of the universe was experimentally confirmed by **Edwin Hubble** in 1929, Einstein withdrew the extra term and called it "the greatest blunder of his life." Extrapolating backwards in time, the expansion suggests that the universe was much smaller in the past, and may have begun with a dense hot explosion, the **big bang**.

Gravitational systems such as **binary stars** are predicted to emit energy in the form of gravitational **radiation**. Such radiation would, like the ordinary effects of gravity, propagate at the speed of **light**, but would be far more feeble than electromagnetic radiation. Gravitational **waves** are thus exceedingly difficult to observe, and despite attempts at detection, convincing evidence of their interaction with an experimental apparatus has yet to be obtained.

Another astronomical prediction of general relativity is the possibility of **black holes**, collapsed **stars** with gravitational fields of such intensity, light cannot escape their surfaces. Stars that can no longer resist gravitational collapse with **nuclear fusion** become **white dwarfs** or **neutron stars**, and if sufficiently massive, the **Pauli exclusion principle** will be unable to prevent further compression, allowing the star to shrink to its Schwarzschild radius and become a black hole. Worm holes, or Einstein-Rosen bridges, are possible in general relativity, though they have not yet been observed. Worm holes are somewhat like tunnels connecting different regions of space, and if real, would act like shortcuts between points that might be separated by long distances in ordinary space.

The differences between the predictions of Newtonian gravity and general relativity tend to be most pronounced with very large masses, and so it is in astronomical phenomena that the results of the two theories diverge. The first confirmation of general relativity came soon after its publication. The point on Mercury's orbit nearest the **Sun**, its perihelion, was known to move slowly around the Sun. Newtonian gravity predicted such a precession but the magnitude was smaller than the measured value. Einstein's theory accurately predicted the measured value.

A novel phenomenon was the bending of light by gravity. Measurements during an eclipse in 1919 showed that the Sun's gravity changed the apparent positions of stars and that the amount of shift was consistent with general relativity. Similar effects are seen in gravitational lensing, which is found in various astronomical contexts. In gravitational lensing light is bent by a concentration of **mass**, such as a cluster of galaxies, so that light from a single object is bent around both sides of the mass concentration, allowing observers on Earth to see multiple images of the object, or arcs of light. The gravity of massive objects such as stars will also extract energy from radiation emitted from them by the process of gravitational **red shift**, which reduces the frequency of photons leaving the star and so lowers their energy.

Einstein and others attempted to broaden the scope of the general theory of relativity to include a description of electromagnetic interactions as manifestations of the geometry of space-time, but a satisfactory unified field theory was not found. Theories unifying the nuclear forces with **electromagnetism** have been produced, but general relativity continues to defy merger with the other forces.

See also Relativity theory; Space-time curvature; Warping of space and time

RELATIVITY, SPECIAL

Einstein's theory of relativity consists of two major portions: the special theory of relativity and the general theory of relativity. Special relativity deals with phenomena that become noticeable when traveling near the speed of **light** and **reference frames** that are moving at a constant **velocity**, inertial reference frames. General relativity deals with reference frames that are accelerating, noninertial reference frames, and with phenomena that occur in strong gravitational fields. General relativity also uses the curvature of **space** to explain **gravity**.

In the seventeenth century, **Isaac Newton** completed a grand synthesis of physics that used three laws of **motion** and the law of gravity to explain motions we observe both on Earth and in the heavens. These laws worked very well, but by the end of the nineteenth century, physicists began to notice experiments that did not **work** quite the way they should according to Newton's classical understanding. These anomalies led to the development of both relativity and quantum **mechanics** in the early part of the twentieth century.

One such experiment was the **Michelson-Morley experiment**. To understand this experiment, imagine a bored brother and sister on a long train ride. (Einstein liked thought experiments using trains.) To pass the time, they get up and start throwing a baseball up and down the aisle of the train. The boy is in the front and the girl in the back. The train is traveling at 60 mph (96.5 km/h), and they can each throw the ball at 30 mph (48 km/h). As seen by an observer standing on the bank outside the train, the ball appears to be traveling 30 mph (60-30) when the boy throws the ball to the girl and 90 mph (145 km/h) (60+30) when the girl throws it back. The Michelson-

Morley experiment was designed to look for similar behavior in light. Earth orbiting the **Sun** takes the place of the train, and the measured **speed of light** (like the baseball's speed) should vary by Earth's orbital speed depending on the direction the light is traveling. The experiment did not work as expected; the speed of light did not vary. Because Einstein took this result as the basic assumption that led to the special theory of relativity, the Michelson-Morley experiment is sometimes referred to as the most significant negative experiment in the history of science.

The orbit of the planet Mercury around the Sun has some peculiarities that can not by explained by Newton's classical laws of physics. The general theory of relativity can explain these peculiarities, so they are described in the article on general relativity.

To understand many concepts in relativity one first needs to understand the concept of a reference frame. A reference frame is a system for locating an object's (or event's) position in both space and time. It consists of both a set of coordinate axes and a clock. An object's position and motion will vary in different reference frames. Go back to the example above of the boy and girl tossing the ball back and forth in a train. The boy and girl are in the reference frame of the train; the observer on the bank is in the reference frame of Earth. The reference frames are moving relative to each other, but there is no absolute reference frame. Either reference frame is as valid as the other.

For his special theory of relativity, published in 1905, Einstein assumed the result of the Michelson-Morley experiment. The speed of light will be the same for any observer in any inertial reference frame, regardless of how fast the observer's reference frame is moving. Einstein also assumed that the laws of physics are the same in all reference frames. In the special theory, Einstein limited himself to the case of nonaccelerating, nonrotating reference frames (moving at a constant velocity), which are called inertial reference frames.

From these assumptions, Einstein was able to find several interesting consequences that are noticeable at speeds close to the speed of light (usually taken as greater than one tenth the speed of light). These consequences may violate our everyday common sense, which is based on the sum total of our experiences. Because we have never traveled close to the speed of light we have never experienced these effects. We can, however, accelerate atomic particles to speeds near the speed of light, and they behave as special relativity predicts.

Special relativity unified our concepts of space and time into the unified concept of spacetime. In essence time is a fourth dimension and must be included with the three space dimensions when we talk about the location of an object or event. As a consequence of this unification of space and time the concept of simultaneous events has no absolute meaning. Whether or not two events occur simultaneously and the order in which different events occurs depends on the reference frame of the observer.

If, for example, you want to meet a friend for lunch, you have to decide both which restaurant to eat at and when to eat lunch. If you get either the time or the restaurant wrong you are not able to have lunch with your friend. You are in essence specifying the space-time coordinates of an event, a shared lunch. Note that both the space and time coordinates are needed, so space and time are unified into the single concept of space-time.

Imagine a rocket ship traveling close to the speed of light. A number of unusual effects occur: Lorentz contraction, **time dilation**, and **mass** increase. These effects are as seen by an outside observer at rest. To the pilot in the reference frame of the rocket ship all appears normal. These effects will occur for objects other than rocket ships and do not depend on there being someone inside the moving object. Additionally, they are not the result of faulty measuring devices (clocks or rulers); they result from the fundamental properties of space-time.

A rocket moving close to the speed of light will appear shorter as seen by an outside observer at rest. All will appear normal to an observer such as the pilot moving close to the speed of light inside the rocket. As the speed gets closer to the speed of light, this effect increases. If the speed of light were attainable the object would appear to have a length of zero to an observer at rest. The length of the rocket (or other moving object) measured by an observer at rest in the reference frame of the rocket, such as the pilot riding in the rocket, is called the proper length. This apparent contraction of a moving object as seen by an outside observer is called the Lorentz contraction.

A similar effect, time dilation, occurs for time. As seen by an outside observer at rest, a clock inside a rocket moving close to the speed of light will move more slowly. The same clock appears normal to the pilot moving along with the rocket. The clock is not defective; the rate at which time flows changes. Observers in different reference frames will measure different time intervals between events. The time interval between events measured both at rest in the reference frame of the events and with the events happening at the same place is called the proper time. This time dilation effect increases as the rocket gets closer to the speed of light. Traveling at the speed of light or faster is not possible according to special relativity, but if it were, time would appear to the outside observer to stop for an object moving at the speed of light and to flow backward for an object moving faster than light. The idea of time dilation is amusingly summarized in a famous limerick: "There was a young lady named Bright, Whose speed was much faster than light. She set out one day, In a relative way, And returned on the previous night." As seen by an outside observer, the mass of the rocket moving close to the speed of light increases. This effect increases as the speed increases so that if the rocket could reach the speed of light it would have an infinite mass. As for the previous two effects to an observer in the rocket, all is normal. The mass of an object measured by an observer in the reference frame in which the object is at rest is called the rest mass of the object.

These three effects are usually thought of in terms of an object, such as a rocket, moving near the speed of light with an outside observer who is at rest. But it is important to remember that according to relativity there is no preferred or absolute reference frame. Therefore the viewpoint of the pilot in the refer-

ence frame of the rocket is equally valid. To the pilot, the rocket is at rest and the outside observer is moving near the speed of light in the opposite direction. The pilot therefore sees these effects for the outside observer. Who is right? Both are.

Think about accelerating the rocket in the above example. To accelerate the rocket (or anything else) an outside **force** must push on it. As the speed increases, the mass appears to increase as seen by outside observers including the one supplying the force (the one doing the pushing). As the mass increases, the force required to accelerate the rocket also increases. (It takes more force to accelerate a refrigerator than a feather.) As the speed approaches the speed of light the mass and hence the force required to accelerate that mass approaches **infinity**. It would take an infinite force to accelerate the object to the speed of light. Because there are no infinite forces no object can travel at the speed of light. An object can be accelerated arbitrarily close to the speed of light, but the speed of light can not be reached. Light can travel at the speed of light only because it has no mass. The speed of light is the ultimate speed limit in the **universe**.

$E=mc^2$. This famous equation means that **matter** and **energy** are interchangeable. Matter can be directly converted to energy, and energy can be converted to matter. The equation, $E=mc^2$, is then a formula for the amount of energy corresponding to a certain amount of matter. E represents the amount of energy, m the mass, and c the speed of light. Because the speed of light is very large a small amount of matter can be converted to a large amount of energy. This change from matter into energy takes place in **nuclear reactions** such as those occurring in the Sun, **nuclear reactors**, and **nuclear weapons**. Nuclear reactions release so much energy and nuclear weapons are so devastating because only a small amout of mass produces a large amount of energy.

A paradox is an apparent contradiction that upon closer examination has a noncontradictory explanation. Several paradoxes arise from the special theory of relativity. The paradoxes are interesting puzzles, but more importantly, help illustrate some of the concepts of special relativity.

Perhaps the most famous is the **twin paradox**. Two twins are initially the same age, as is customary for twins. One of the twins becomes an astronaut and joins the first interstellar expedition, while the other twin stays home. The astronaut travels at nearly the speed of light to another star, stops for a visit, and returns home at nearly the speed of light. From the point of view of the twin who stayed home, the astronaut was traveling at nearly the speed of light. Because of time dilation the homebound twin sees time as moving more slowly for the astronaut, and is therefore much older than the astronaut when they meet after the trip. The exact age difference depends on the distance to the star and the exact speed the astronaut travels. Now think about the astronaut's reference frame. The astronaut is at rest in this frame. Earth moved away and the star approached at nearly the speed of light. Then Earth and star returned to their original position at nearly the speed of light. So, the astronaut expects to be old and reunite with a much younger twin after the trip. The resolution to this paradox lies in the fact that for the twins to reunite, one of them must accelerate by slowing down, turning around, and speeding up. This **acceleration** violates the limitation of special relativity to inertial (nonaccelerating) reference frames. The astronaut, who is in the noninertial frame, is therefore the younger twin when they reunite after the trip. Unlike much science fiction in which star ships go into a fictional warp drive, real interstellar travel will have to deal with the realities of the twin paradox and the speed of light limit.

The garage paradox involves a very fast car and a garage with both a front and back door. When they are both at rest, the car is slightly longer than the garage, so it is not possible to park the car in the garage with both doors closed. Now imagine a reckless driver and a doorman who can open and close both garage doors as fast as he wants but wants only one door open at a time. The driver drives up the driveway at nearly the speed of light. The doorman sees the car as shorter than the garage, opens the front door, allows the car to drive in, closes the front door, opens the back door, allows the car to drive out without crashing, and closes the back door. The driver on the other hand, sees the garage as moving and the car as at rest. Hence, to the driver the garage is shorter than the car. How was it possible, in the driver's reference frame, to drive through the garage without a crash? The driver sees the same events but in a different order. The front door opens, the car drives in, the back door opens, the car drives through, the front door closes, and finally the back door closes. The key lies in the fact that the order in which events appear to occur depends on the reference frame of the observer. (See the section on spacetime.)

Like any scientific theory, the theory of relativity must be confirmed by experiment. So far, relativity has passed all its experimental tests. The special theory predicts unusual behavior for objects traveling near the speed of light. So far no human has traveled near the speed of light. Physicists do, however, regularly accelerate **subatomic particles** with large particle **accelerators** like the recently canceled Superconducting Super Collider (SSC). Physicists also observe **cosmic rays** which are particles traveling near the speed of light coming from space. When these physicists try to predict the behavior of rapidly moving particles using classical **Newtonian physics**, the predictions are wrong. When they use the corrections for Lorentz contraction, time dilation, and mass increase required by special relativity, it works. For example, muons are very short lived subatomic particles with an average lifetime of about 2 millionths of a second. However when they are traveling near the speed of light physicists observe much longer apparent lifetimes for muons. Time dilation is occurring for the muons. As seen by the observer in the lab time moves more slowly for the muons traveling near the speed of light.

Time dilation and other relativistic effects are normally too small to measure at ordinary velocities. But what if we had sufficiently accurate clocks? In 1971 two physicists, J. C. Hafele and R. E. Keating used atomic clocks accurate to about one billionth of a second (1 nanosecond) to measure the small time dilation that occurs while flying in a jet plane. They flew atomic clocks in a jet for 45 hours then compared the clock readings to a clock at rest in the laboratory. To within the accuracy of the clocks they used time dilation occurred for the clocks in the jet as predicted by relativity. Relativistic effects occur at ordinary velocities, but they are too small to measure without very precise instruments.

The formula $E=mc^2$ predicts that matter can be converted directly to energy. Nuclear reactions that occur in the sun, in nuclear reactors, and in nuclear weapons confirm this prediction experimentally.

Albert Einstein's special theory of relativity fundamentally changed the way we understand time and space. So far it has passed all experimental tests. It does not however mean that Newton's law of physics is wrong. Newton's laws are an approximation of relativity. In the approximation of small velocities, special relativity reduces to Newton's laws. If at some time in the future someone does an experiment that does not agree with the theory of relativity, we will have to modify the theory just as relativity modified Newton's classical physics.

RELATIVITY THEORY

Relativity theory, a general term encompassing special and general relativity theories, sets forth a specific set of laws relating **motion** to **mass**, **space**, **time**, and **gravity**. Relativity theory allows calculations of the differences in mass, space and time as measured in different **reference frames**.

At the start of the twentieth century the classical laws of physics contained in Sir Isaac Newton's 1687 work, *Philosophiae Naturalis Principia Mathematica (Mathematical Principles of Natural Philosophy)* adequately described the phenomena of everyday existence. In accord with these laws, more than century of experimental and mathematical work in **electricity** and **magnetism** resulted in Scottish physicist James Clerk Maxwell's four equations describing **light** as an electromagnetic wave. Prior to **Maxwell's equations** it was thought that all **waves** required a medium or **ether** for propagation. Such an ether would also serve as an absolute reference frame against which absolute motion, space and time could be measured. Ironically, although Maxwell's equations established that electromagnetic waves do not require such a medium, Maxwell and others remained unconvinced and the search for an elusive ether continued. For more than three decades, the lack of definable or demonstrable ether was explained away as simply a problem of experimental accuracy. The absence of a need for an ether for the propagation of electromagnetic **radiation** was demonstrated in late nineteenth century experiments conducted by **Albert Michelson** and **Edward Morley**.

In 1904, French mathematician Jules-Henri Poincaré pointed out important problems with concepts of simultaneity by asserting that observers in different reference frames must measure time differently. In 1905, a German-born clerk in the Swiss patent office named **Albert Einstein** published a theory of light that incorporated implications of Maxwell's equations, demonstrated the lack of need for an ether, explained FitzGerald-Lorentz contractions, and explained Poincaré's reservations concerning differential time measurement. Both

Einstein and his special relativity theory went to revolutionize modern physics.

In formulating his special theory of relativity, Einstein assumed that the laws of physics are the same in all inertial (moving) reference frames and that the **speed of light** was measured as a constant regardless of its direction of propagation. Moreover, the measured speed of light was independent of the **velocity** of the observer.

Einstein's special theory of relativity also related mass and **energy**. Einstein published a formula relating mass and energy, $E=mc^2$ (Energy = mass times speed of light squared). Einstein's equation implied that tremendous energies were contained in small masses. Along with advancements in **atomic theory**, Einstein's insights ultimately allowed the development of atomic weapons during World War II and the dawn of the nuclear age.

Special relativity also gave rise to a number of counter-intuitive paradoxes dealing with the passage of time (e.g., the **twin paradox**) and with problems dependent upon an assumptions of simultaneity. According to the postulates of special relativity, under certain conditions it would be impossible to determine when one event happened in relation to another event.

Although Einstein's special theory was limited to special cases dealing with systems in uniform nonaccelerated motion, the theory did away with the need for an absolute **frame of reference**. In addition, the implications of special relativity on the **equivalence of mass and energy** revolutionized classical laws regarding **conservation of mass** and energy. A more complete understanding of the conservation of mass and energy now relies upon mass-energy systems. Einstein's special relativity theory was also important in the development of **quantum theory**. German Physicist **Max Planck** and others who were in the process of developing quantum theory, set out to reconcile (often unsuccessfully) relativity theory with quantum theory.

In 1915, a decade after publishing his special relativity theory, Einstein published his general theory of relativity that soon came to supplant well-understood Newtonian concepts of gravity. Although Newtonian theories of gravity hold valid for most objects, there were small but noticeable errors in calculations regarding the motion of bodies at high velocities or for description of motion in massive gravitational fields. These small errors were completely corrected by the general theory of relativity that described nonuniform, or accelerated, motion.

General relativity's impact on calculations regarding gravity sparked dramatic revisions in **cosmology** that continue today. The conceptual fusion of traditional three-dimensional space with time to create **space-time** also made observers more integral to measurement of phenomena.

In a sophisticated elaboration of **Newton's laws of motion**, in general relativity theory the motion of bodies is explained by the assertion that in the vicinity of mass, space-time curves. The more massive the body the greater the curvature of space-time and, consequently, the greater the **force** of gravitational attraction.

It may be fairly argued that the most stunning philosophical consequence of general relativity was that space-time is a creation of the **universe** itself. Under general relativity, the universe is not simply expanding into preexisting space and time, but rather creating space-time as a consequence of expansion. In this regard, general relativity theory set the stage for the subsequent development of **big bang** theory.

Unlike the esoteric proofs of special relativity, the proofs of general relativity could be measured by conventional experimentation. General relativity's assertion that gravitational fields would bend light was confirmed during a 1919 solar eclipse. Other predictions regarding shifts in the perihelion of Mercury and in redshifted spectra also found confirmation. Using relativity based equations, German physicist Karl Schwarzschild mathematically described the **gravitational field** near massive compact objects. Schwarzschild's work subsequently enabled the predication and discovery of the evolutionary stages in massive **stars** (e.g., **neutron stars**, **black holes**, etc.).

Relativity theory was quickly accepted by the general scientific community, and its implications on general philosophical thought were profound. Although **Newtonian physics** still enjoys widespread utility, along with **quantum physics**, relativity-based theories have replaced Newtonian cosmological concepts.

RENEWABLE ENERGY RESOURCES

A renewable **energy** resource is one that can constantly be tapped because it replenishes itself. The search for viable renewable energy sources is a **matter** of growing urgency as non-renewable sources of energy become more difficult and expensive to discover and exploit. The major sources of renewable energy include solar energy, **wind energy**, bioenergy, **geothermal energy**, hydropower, and ocean energy. However, these sources have limitations that make them more expensive or less efficient in situations worldwide.

Solar energy, or energy derived from the **Sun**, works well in areas that have a good supply of direct sun **light**. Certain types of elements, such as cadmium sulfide or gallium arsenide, separated by layers of crystalline silicon, when exposed to direct sunlight, will produce an **electric current**. Light striking these substances knocks electrons off the element's surface. If the substances are formed into photovoltaic cells, the electrons can produce an electric current as they move from the negative pole of the cell (the cathode) to the positive pole (the **anode**). These photovoltaic cells, placed in arrays, can produce significant amounts of **electricity**. Sometimes it is possible to reduce the number of cells by using **lenses** or concave reflectors to focus an intense beam of light onto the cells. The downside of this approach is the need to keep moving the reflectors and lenses to keep the moving sun focusing on the cells. Unfortunately photovoltaic cells are comparatively inefficient (between 15% and 30%), and they can only produce electricity at a cost of about 30 cents per kilowatt hour, and energy from other sources is currently cheaper to produce.

Wind energy is another growing source of energy in the world. High-tech windmills can be very efficient and currently are beginning to produce increasing amounts of energy in those places where the wind are reasonably strong and consistent. Although modern windmills (sometimes called wind turbines) cause less damage to the environment than some other types of renewable energy resources, they are not a universal solution. In order to be efficient, wind turbines require average wind velocities in excess of 13 mi/hr (21 km/hr) over the course of a year. Other problems associated with these devices include high initial start up costs, noise, spoiling of the visual environment, and bird deaths. Energy storage is also a problem for those times when energy is needed but the wind is not available.

The burning of wood, agricultural waste, and the organic components of domestic and industrial wastes to produce energy is increasingly common. There are several ways that biomass can be converted to energy. Chemicals can turn the biomass into fuel oil or gasoline additives. Ethanol, methanol and biodiesel are three biofuels that can be produced from the biomass. Biodiesel can be produced vegetable oils or animal fat. The generation of electricity from biomass is also possible. Burning the biomass as a supplement with coal can help reduce the use of this non-renewable resource. Biomass can also be turned directly into gas by direct application of **heat** or by natural decay. Some landfills are set up to produce natural gas, which is then tapped and used for heating or to produce electricity—although the process of burning can produce pollution.

Electricity can also be generated by using geothermal heat sources. In certain parts of the world raw steam emerges from vents which can be directed onto turbine blades to generate electricity. In other places very hot water emerges which can then be provided with additional heat to achieve the same result. In those places where no steam or hot water emerges it may be possible to inject a fluid deep into the hot bedrock. This scalding hot fluid can then be passed through pipes that in turn will turn water into steam. Geothermal energy is already used extensively in Iceland, where extensive volcanism makes natural steam sources common.

Mountainous regions offer another renewable energy source in the form of hydropower. In this case a dam is built to form a reservoir. The water is released from the dam to falls under **gravity** onto the blades of turbines that generate electricity. Clearly, the higher fall the greater the energy produced. A different type of hydropower uses pumped water. In this case water is pumped up into a high reservoir at night, when electrical **power** is least used. When electricity is expensive the process can be reserved and water allowed to fall back down onto turbine blades. However, hydroelectric power is of limited use except in areas with major rivers—and the dams used to generate it can cause extensive environmental damage.

Finally, the ocean can also be a source of renewable energy. The **motion** of the tides can be harnessed in the same way as hydroelectric **generators** harness the energy of flowing rivers. Incoming tides flood through hydroelectric dams, turn-ing turbines that generate electric current. The water they bring in is then caught behind the dams and, as the tide recedes, the water can again run through the turbines, again generating electric current. Tidal generators are most efficient where tides are highest. They have been put into place in northwestern France, but such generators are expensive and few others have been built.

RENORMALIZATION

In 1930s, following the initial development quantum electrodynamic theory (QED), American physicist J. Robert Oppenheimer (1904-1967) employed the new theory to calculate the **energy** levels of the hydrogen **atom**. The first such calculation had been performed by **Niels Bohr** in 1913, a full decade before the development of modern **quantum theory**. In the case of hydrogen (and only in the case of hydrogen) the Bohr model works, yielding the same values that are predicted by the full, but non-relativistic quantum **mechanics**. The measured values for other atoms, however, differed from expected values.

QED, like any relativistic **quantum field theory**, incorporates German-American physicist **Albert Einstein**'s mass-energy relation, and thereby explains the creation and absorption of **virtual particles**. A propagating **electron**, for example, can spontaneously emit, and then almost instantaneously reabsorb a **photon**. This process contributes to what is called the electron self-energy, and will minutely shift the **energy level** values obtained with the non-relativistic theory. It was hoped that a full, relativistic treatment would agree with experiment. Oppenheimer set out to calculate this correction. But his calculation showed that rather than being tiny, the energy level shift actually diverged; it was infinite. This was nonsensical prediction, and quantum **electrodynamics**, it seemed, was dead in its infancy.

It is the difference in energy between two levels that is actually measured in the laboratory. By measuring a number of such differences, the values of the levels themselves can be inferred. When the electron in a **hydrogen atom** jumps from a higher level to a lower, it emits a photon. The energy of that photon (which is directly measured) corresponds to the difference in energy between the two levels. Oppenheimer had subtracted two of his values, one **infinity** from another, and found that most of the divergence canceled.

It was left for others, though, to develop general methods for removing the remaining divergence. In particular, English-born American physicist **Freeman Dyson** (1923-) proved in 1949 that all of the divergence that arose in QED could be removed by rescaling or renormalizing the electron's **mass** and charge.

Renormalization is conceptually simple, but calculationally difficult. All of the rules for calculating with a particular quantum field theory are derived (via Hamilton's least action principle) from a mathematical object called the Lagrangian. The Lagrangian contains a number of parameters; in QED there are two, the electron's mass, and charge, that must be measured experimentally before the theory can be used to calculate. The trick is to absorb the divergence by rescaling those parameters. Obviously, this is possible only if the number of types of divergence is finite. If that is the case, a finite number of parameters (a finite number of measurements that must be made before the theory can be used) will suffice to renormalize the theory. QED, for example, appears at first glance to contain three types of divergence, called the vacuum-polarization, the electron self-energy, and the vertex correction. The symmetry inherent in the theory provides a relation between the last two, leaving just two independent divergences. QED will always yield finite results.

In 1971, two Dutch physicists, Gerard't Hooft and **Martinus Veltman**, showed that the **Weinberg-Salam theory** describing the electromagnetic and weak interactions was also renormalizable. In 1999, they were awarded the Nobel Prize in Physics for their work.

Quantum field theories of **gravity**, on the other hand, contain an infinite number of different types of divergence. An infinite number of parameters would be required to absorb them, and thus an infinite number of experiments must be performed (an impossible task) before the theory could be used to calculate. Thus far, **Quantum gravity** is a non-renormalizable theory.

See also Quantum physics

RESISTANCE IN SERIES AND PARALLEL CIRCUITS

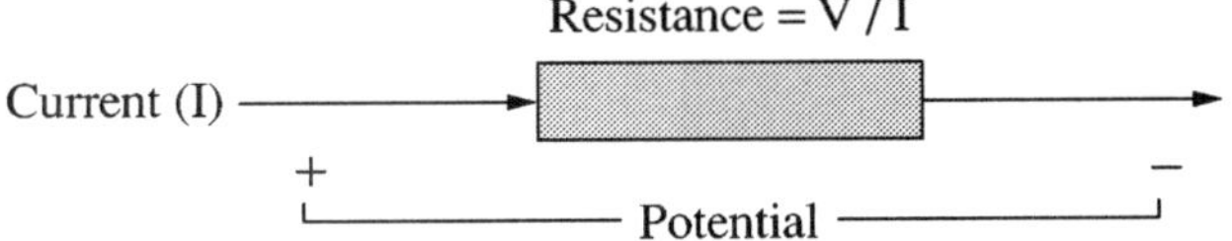

The resistance of any electrical device is defined by the expression, R=V/I, where V (volts) is the potential difference across the device and I (amps) is the electrical current flowing through it.

When a current of one amp flows through a resistor after a potential of one volt is placed across it we say that the resistor has a unit resistance of one ohm. If the resistor is a metallic conductor then, as long as it's **temperature** is constant it's resistance remains constant. In other words, the ratio of potential difference (V) and current flow (I) remains constant. This is **Ohm's Law**. This means that for any given resistor doubling the potential potential across its ends will double the current flowing through it, tripling the potential will triple the current, etc.

The next interesting question is what happens when several resistors are connected end to end as shown in the diagram below. This arrangement is called a Series Connection.

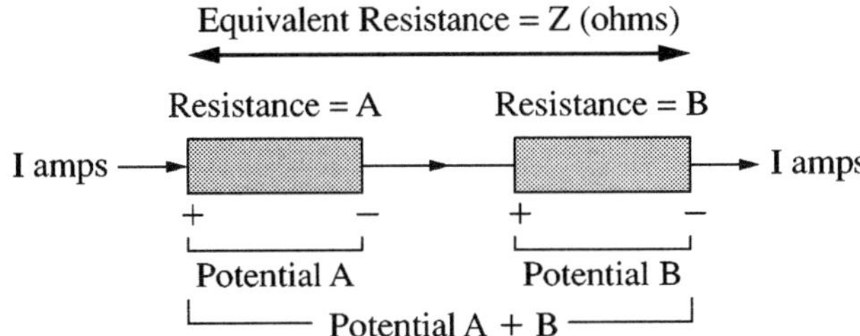

The first thing to be realized is the same current (I amps) flows through both resistances (A and B). The next thing to be understood is that the electrical potential across resistance A plus the potential difference across the resistance B is equal to the total potential difference across both of them.

Knowing these two facts it is now comparatively easy to figure out what would be the effect of connecting two resistors in series as diagramed above. In other words, if we know the resistance of the resistor A and we know the resistance of the resistor B what would be the total resistance (Z) when they are in a series connection.

Our approach to this problem will to use the concept that Potential A + Potential B=Potential (A + B). From Ohm's Law: A ohms = Potential A/I and B ohms = Potential B/I.

Similarly, if we want the value of the total equivalent resistance (Z) of the two resistors connected together in series we can say, again using Ohm's Law:

$$Z \text{ ohms} = \text{Potential } (A + B) / I$$

$$= (\text{Potential A} + \text{Potential B}) / I$$

$$= \text{Potential A} / I + \text{Potential B} / I$$

$$\mathbf{Z = A + B}$$

For resistors connected in series, therefore, the total value for the overall resistance is simply the sum of of all the single resistances that are connected together. For example, if we connect three resistors together in series with values 2 ohms, 3 ohms, and 4 ohms respectively then the resistance of the combination (z) is simply z = 2 ohms + 3 ohms + 4 ohms; z = 10 ohms.

There is another way to connect resistors together. The diagram below shows this other method.

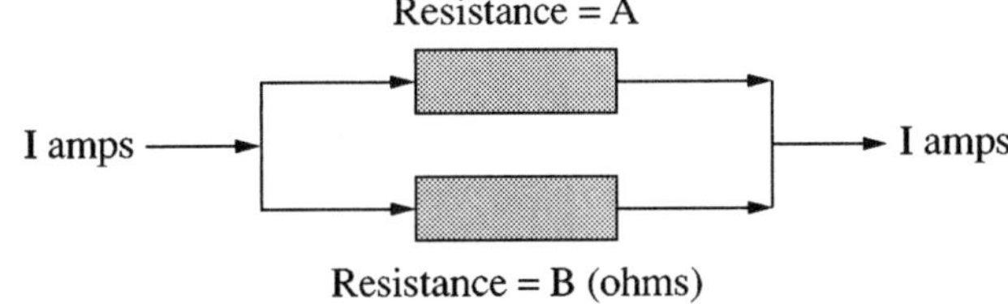

This method of connecting resistors is called resistances in parallel connection.

Once again, we need to know the overall resistance (Z ohms) of this type of combination of resistors. This time it is not as simple as just adding the two resistors A and B together. However, once again, we can use two basic facts about the above circuit. The following diagram will help us.

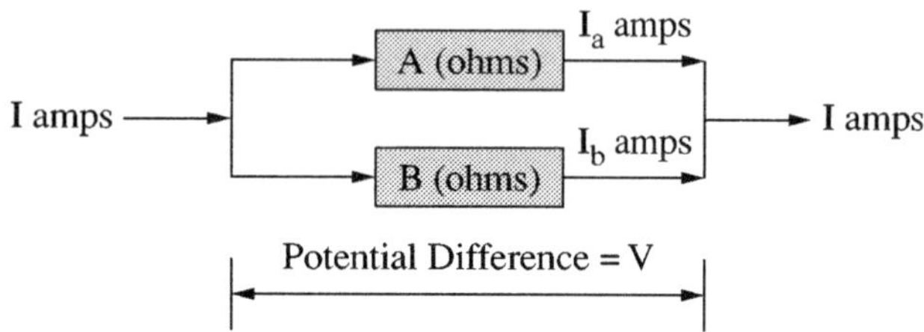

We can make two statements about the above circuit. First, the potential difference (V volts) across the ends of the two resistors A and B must be the same. Second, the input current (I) is divided into two components Ia and Ib such that Ia + Ib = I.

Knowing this we can say the following: from Ohm's Law, A (ohms) = V/Ia or Ia = V/A and, B (ohms) = V/Ib or Ib = V/B and, Z (ohms) = V/I or I = V/Z.

Using what we know about the currents in this circuit, namely Ia + Ib=I, we can go ahead and say V/A + V/B = V/Z. Simplifying this expression we get 1/A + 1/B = 1/Z.

This works for any number of resistors. If four resistors (A, B, C, and D ohms) are all connected in parallel then their equivalent resistance (Z ohms) can be calculated from the following expression: 1/A + 1/B + 1/C + 1/D = 1/Z.

RESISTANCE, REACTANCE, AND IMPEDANCE

Electrical resistance (R) is a measure of the resistance that a material or **electric circuit** presents to the flow of current, and is defined as the ratio of the **voltage** (V) applied to the **electric current** (I) which flows through it. The unit of resistance is the ohm. If the resistance is constant over a wide range of voltage, then **Ohm's law** (I = V/R) can be used to predict the behavior of any material through which a current flows. This definition holds for both DC and AC circuits. Thus, for a given voltage, materials or circuits with a high resistance will allow a small current relative to a material with a low resistance. At the atomic level, currents are defined as the flow of the valence electrons of atoms through a given material. Resistance in this context results from collisions of electrons with other electrons and atoms. Electrical resistance can also be defined in terms of resistivity, a quantity used to describe the resistance of a wire and expressed as the product of its resistance and its cross-sectional area divided by its length. Resistivity and resistance are temperature-dependent and most materials can be classified according to their resistance **temperature** coefficient. For example, the resistance of positive temperature

coefficient (PTC) wires increases strongly with increasing temperature and they are used for current stabilization purposes. In contrast, the resistance of negative temperature coefficient (NTC) wires decreases with increasing temperature and they are used for voltage control purposes.

Electric circuits, the paths followed by electric current, are either DC or AC circuits. The fact that the voltage of AC current varies sinusoidally with time requires the introduction or modification of the quantities required to take this fact into account. Reactance (X), is such a quantity. It is also a measure of the resistance offered by a material or a circuit to the flow of current, but only when the current is alternating. There are two types of reactance, inductive (X_L) and capacitive (X_C). The first is associated with the magnetic field induced around a coil or wire through which AC current flows, also called an inductor. The inductive reactance is a measure of the resistance to the flow of this current and it is proportional to the **frequency** of the alternating current and to inductance (L), an inductor property arising from Faraday's law, and defined in terms of the **electromotive force** (emf) generated to oppose a given change in current. The capacitive reactance is associated with the changing electric field between two parallel conducting plates separated by an insulator. This arrangement defines a capacitor, which tends to oppose any voltage change across its conducting plates. The capacitive reactance is a measure of this resistance and is inversely proportional to the frequency of the alternating current and to **capacitance** (C), a capacitor property equal to the magnitude of the charge stored on each capacitor plate (Q) divided by the voltage applied to the plates.

Because Ohm's law applies only to DC circuits and to resistors in AC circuits, a modified version is used for AC circuits. It is expressed as: I=V/Z where I and V are not maximum values, but rather root-mean-square (rms) or "effective" values of voltage and current. The impedance (Z) is a phase angle, which describes the angle between the current and voltage AC sinusoidal curves.

In a resistor, voltage and current are in phase, so effectively, Z= R. In a capacitor, voltage lags current by 90°, and in an inductor, voltage leads current by 90º. Thus, when an AC circuit includes capacitors or inductors, even though Z is also the ratio of the voltage and current peaks, as is the case for R in the DC formulation of Ohm's law, the I and V peaks are now out-of-phase, that is, they do not occur at the same time. The phase of AC circuits is described by phasor diagrams which represent the circuit phase as a vector in a plane. Zero phase is taken to be the positive x-axis and represents the resistor with in-phase voltage and current. The length of the phasor is directly proportional to the magnitude of the quantity represented, and its angle represents its phase relative to that of the current flowing through the resistor.

Impedance consists of the sum of resistance and reactance contributions. The resistance contribution results from the collisions of the electron-carrying current-with the constituent atoms of the conductor material. The reactance component is an additional resistance to the movement of **electric charge** resulting from the changing magnetic and electric fields occurring in AC circuits. From the modified Ohm's law, the impedance of a circuit is equal to the voltage measured across the circuit, divided by the maximum value of the current through the circuit (Z = V/I). Impedance, like resistance, is expressed in ohms. The reciprocal of the impedance (1/Z), is called the admittance and is expressed in mho.

See also Alternating current circuits; Conductors and insulators; Current and electricity; Electricity; Faraday's laws; Inductors and inductance; Resistors in capacitors and circuits; Resistance in series and parallel circuits; Resonance; Semiconductors

RESISTORS AND CAPACITORS IN CIRCUITS

Both resistors and capacitors are found and used in virtually every electronic device currently made. The uses to which they are put vary from circuit to circuit; however, the functioning of a resistor or capacitor remains the same regardless of its location in any particular circuit.

A resistor impedes the flow of current through the wires that are connected to them. A resistor is a physical embodiment of **Ohm's Law**, which states that the current through any portion of a circuit is directly proportional to the **voltage** applied to that segment. Resistance in any circuit can manifest as a heating of the resistive component. Examples of this range from a resistor in a circuit to the toaster you use in your kitchen. The toaster elements that get red hot do so because they are actually resistors. They impede the current that passes through them, and as they do this they get hot.

While resistors merely impede the flow of current through a section of a circuit, capacitors block it completely. A capacitor device usually consists of two electrical leads separated by an insulator, a nonconducting material. The insulator prevents electrons from traveling between the two leads. Because of the blocking property of capacitors, they are often used to eliminate direct current **electricity** (DC) from entering a circuit where it would be harmful.

More often, however, capacitors are used for their primary feature, the storing of electrical charge. As long as a voltage is applied to the capacitor, the charge continues to build up to some final maximum value. Once the voltage is removed, however, the charge bleeds away. This property makes capacitors valuable in ensuring a continuous, uninterrupted **power** supply. The addition of a capacitor will enable short disruptions in the power supply to be ignored because the capacitor will pick up the slack during the outage. The capacitor can also be used to double the available voltage applied to a circuit for short periods. Once the capacitor is charged, the charge can be released in conjunction with the standard power supply in order to increase the output voltage.

See also Capacitors

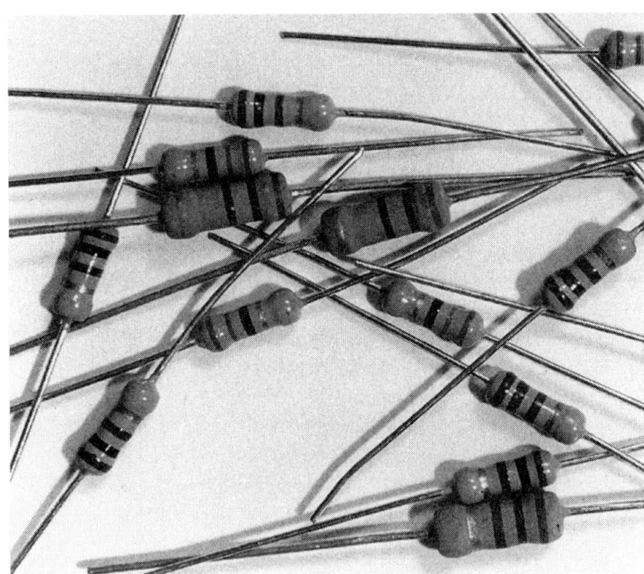

Color-coded resistors. *(Photo by Jim Olenski/Thomas Video. Reproduced by permission.)*

RESONANCE (CIRCUIT)

In **electricity**, the behavior of an alternating circuit (AC) is determined by whether its circuit elements are connected in series or in parallel. Resonance describes a characteristic **frequency** determined by the values of the resistance (in units of ohms), **capacitance** (in units of farads), and inductance (in units of henrys) of the circuit. There are accordingly two types of resonance: series resonance, for which the condition of resonance is minimum impedance and zero phase, and parallel resonance, the most common in electronic applications, for which the condition of resonance is minimum current and a phase change.

A circuit consisting of resistors, inductors, and capacitors connected in series obeys Kirchhoff's **voltage** law which states that the sum of the emf's in a closed circuit is equal to the sum of the potential difference of the components. The reactance of the circuit is equal to: $X = \omega L - 1/\omega C$ where ω is the circuit frequency, L is the inductance and C the capacitance. The reactance is dependent on the frequency of the voltage $f = \omega/(2\pi)$. The impedance (Z) of such a circuit is:

$$Z = \sqrt{R^2 + (\omega L - 1/\omega C)^2}$$

where R is the resistance. The phase angle is: $\Phi_z = \arctan$ î$[\omega L - 1/(\omega C)]/R$ï and resonance occurs in the circuit when the capacitive and inductive reactance increases. The total resistance is then equal to the ohmic resistance R and the current flow is at a maximum. The resonant frequency is given by:

$$f_r = (1/2\pi)(1/\sqrt{LC})$$

At the resonant frequency, current and driving voltage are in phase because the phase angle is zero, the impedance Z is at a minimum, and the **power** dissipated is at a maximum.

A circuit consisting of resistors, inductors and capacitors connected in parallel obeys Kirchhoff's current law which states that the sum of the currents flowing into a junction is equal to the sum of the currents flowing out. In this case, the reactance of the circuit is equal to: $X = \omega C - 1/\omega L$ and it is also dependent on the frequency of the voltage $f = \omega/(2\pi)$. The impedance (Y) of the circuit is:

$$Y = \sqrt{1/R^2 + (\omega C - 1/\omega L)^2}$$

The phase angle is given by: $\Phi_y =$ arctan $[\omega RC - R/(\omega L)]$. Resonance occurs when the circuit frequency ω is such that the inductive and capacitive reactances are fully compensated. The resonant frequency is also equal to:

$$f_r = (1/2\pi)(1/\sqrt{LC})$$

At the resonant frequency of a parallel circuit, the current is at a minimum and the phase angle changes by 180°.

AC circuit resonance is the operating principle for building devices designed to process signals of a given frequency while rejecting others. In a television receiver, for example, resonance occurs when the frequency of one of the incoming signals reaching the circuit is close to the circuit frequency, which can then absorb maximum **energy** from the signal as the current flowing through the circuit surges back and forth in step with the current flowing in the antenna. Another example of electric resonance occurs when tuning a receiver to listen to a broadcast. Tuning modifies the setting of the tuning capacitor of the **radio**. When the radio circuit reaches a frequency resonant with that of the broadcast frequency, the induced voltage is increased and processed into **sound.**

See also Acoustics, Alternating current circuits; Conductors, semiconductors, insulators; Current and electricity; Electric circuit; Electric current; Electrical resistance; Electromotive force; Faraday laws; Inductors and inductance; Ohm's law; Parallel circuits; Resistors in capacitors and circuits; Resistance, reactance and impedance; Resistors in series and parallel circuits

REST ENERGY

The rest **energy** of an object is the amount of energy that would be released were all of its **mass** converted directly into energy according to **Albert Einstein**'s famous equation: $E=mc^2$, where c is the speed of **light** in a **vacuum**, measured to be 186,000 mi/s (300,000 km/s). This relationship states that mass is just another manifestation of energy, and converting mass into energy and vice-versa is possible.

Early in the twentieth century **nuclear fission** reactions demostrated the truth behind Einstein's $E=mc^2$ and his theory of special relativity of 1905. A nuclear fission reaction is when

the **nucleus** of a large **atom**, called the parent nucleus, splits to form many smaller nuclei, called daughter products, that then spread apart. However, the mass of the smaller nuclei does not add up to the mass of the parent. In fact, the total mass of the daughter products is less than the mass of the parent nucleus in the fission nuclear reaction. Einstein's theory explains where the absent mass went. The mass was converted into the **kinetic energy** the daughter products used to spread apart.

The kinetic energy released in the nuclear fission processes is used to produce **heat** and steam to turn electrical **generators** in nuclear **power** plants. In **nuclear fusion** processes, two small nuclei are combined to form a single larger nucleus and other byproducts. In nuclear fusion reactions, mass is also converted into the kinetic energy of the byproducts. In some **nuclear reactions**, the energy released can be violent and tremendous, as is the case for atomic and hydrogen bombs.

REVERSIBLE PROCESSES

A reversible process describes an ideal thermodynamic system where steps in a reaction, or changes in a physical process, can reverse direction such that the resulting **entropy** change for the system and surroundings is zero. If it is possible to restore a thermodynamic system and its surroundings to their original state (including the original entropic state) following a process then such a process is termed a reversible process. All other reactions and processes are irreversible.

Considering both the system and surroundings, all real processes are irreversible. Internally reversible processes demand no irreversibilities in the system and externally reversible process demand no irreversibilities outside the system.

A ideal reversible process describes a reaction or process in which the system its surroundings are in **equilibrium** throughout the process. In considering ideal gas expansions and contractions, for example, the system and surroundings are considered to be in equilibrium if they have the same **temperature** and the same **pressure** throughout the process. If at any time the system or the surrounds diverge from equilibrium then the process becomes irreversible thermodynamically. The definitional (and easily constructed) test of reversibility dictates that if a reaction or process is completely reversible then it can be driven to change direction by an infinitesimal change in some external variable.

To expand an ideal gas by a reversible process demands that outside and inside pressures be kept balanced as the expansion proceeds. The temperatures of system and surroundings must remain the same so that the system and surroundings are in equilibrium at all times. The only way to accomplish this is through a series of thermostatic steps in which the system does a maximum amount **work** on the surroundings (expand) while the surroundings does minimum work on the system.

Although the second law of **thermodynamics** states that all isolated systems move spontaneously toward a state of maximum entropy, as a practical **matter**, nearly reversible process can be designed if reaction or process conditions are not allowed to stray far from equilibrium. Because entropy increases create irreversible process and set a limiting boundary on the approach to reversibility, reversible processes must be designed to reduce entropy increase to a minimum. Physicists studying the thermodynamics of reversible processes or equilibrium **dynamics** use thermostatic techniques to create near reversible processes through a series of slow and gradual sequence steps conducted at, or very near, the equilibrium point of each step. In some cases, thermostatic techniques substitute a series of very slow and gradual changes into the transition from initial to final states.

In many idealized approximations (especially at the high school and undergraduate university level) students encounter idealized problems that are reversible. Very few, if any, real problems are so idealized and entropic and chaotic effects make natural reactions or processes irreversible.

Most reactions are irreversible because **energy** is usually transformed and work is done to cause a change in a system or its surroundings. In additon, in most reactions some energy is lost because of undesirable transformations of energy (e.g., **heat** lost during the use of electrical energy). No matter how well a system is insulated, or designed to reduce such factors as **friction**, there is always energy dissipation and entropy increase. By definition, reversible process demand 100% efficiency in the use and conversion of energy (any loss would preclude a return to the initial state). Because heat (generated by friction, etc.) can never be 100% recovered and reconverted into work, no machine can operate at 100% efficiency. As a consequence, machines can not perform reversible functions.

Owen W. Richardson.

RICHARDSON, OWEN W. (1879-1959)

English physicist

Owen W. Richardson is best known for his work in thermionics, the emission of electrons from a heated surface, and specifically for the law that describes that phenomenon, now called Richardson's law. For his work on thermionics, Richardson was awarded the 1928 Nobel Prize in physics. In addition, Richardson's investigations into gyromagnetics led to his theory that a body's **magnetism** is caused by the movement of its electrons, a phenomenon known as the Richardson-Einstein-de Haas effect.

Owen Willans Richardson was born in Dewsbury, Yorkshire, England, on April 26, 1879. He was the eldest of three children born to Joshua Henry and Charlotte Maria Willans Richardson. Young Owen was a precocious boy and, according to William Wilson in *Biographical Memoirs of Fellows of the Royal Society,* "gained so many scholarships and exhibitions that his education cost his parents nothing at all." Richard entered Trinity College, Cambridge, in 1897. He earned first-class honors in 1900 in botany, chemistry, and physics, and was awarded his B.A. degree. He then stayed on at Trinity for his graduate studies in chemistry and physics and, in 1902, was appointed a fellow at the college. Two years later he was chosen for a Clerk Maxwell scholarship and was awarded his D.Sc. by University College, London.

Begins Thermionic Research at Cavendish

Richardson's earliest research dealt with the properties of the **electron**. As early as 1901 he began the studies for which he was to become most famous, thermionics, a term that Richardson himself coined. The term refers to the emission of electrons from a heated metal. In his first studies, Richardson used platinum metal since it had the highest melting point of any readily available metal, allowing him to heat the material to very high temperatures and observe the emission of electrons. When ductile tungsten (which can be drawn out or hammered down, and whose melting point is much higher than that of platinum) became available in 1913, Richardson was then able to extend his work to even higher temperatures.

The emission of electrons from heated metals was first described as early as 1603 by Sir William Gilbert, but was not studied systematically until the late nineteenth century. Then, a number of scientists, including **Thomas Edison**, described a number of phenomena associated with thermionics. In his own studies with platinum metal, Richardson found that the heating of the metal resulted in the loss of a certain number of electrons for each unit area of surface. In addition, the number of electrons lost from the metal increases exponentially with the **temperature**. He was able to devise a mathematical formula describing these results, a formula now known as Richardson's law. According to that formula, the current released by a metal is a function of certain characteristics of the metal and the temperature. Richardson explained the thermionic phenomenon by assuming that heating a metal provides some electrons on the metal surface with enough **energy** to overcome the attraction of **ions** in the metal These electrons then "evaporate" from the metal surface in much the same way that liquid molecules escape from the surface of a liquid.

Becomes Professor of Physics at Princeton

In 1906 Richardson was offered an appointment as professor of physics at Princeton University. Before leaving England, however, he was married to Lilian Maude Wilson, daughter of a goods manager at the North Eastern Railway Company and sister of Richardson's fellow student at the Cavendish, H. A. Wilson. The couple's three children, two sons and a daughter, were all born during Richardson's tenure at Princeton.

Richardson continued his work on thermionic emission during his Princeton years, but also studied other phenomena, including the **photoelectric effect**, **thermodynamics**, **x rays**, and gyromagnetics. The latter term refers to attempts to explain various gravitational effects in terms of electromagnetic theory. Although this effort failed in its broadest goals, one phenomenon he did discover was eventually called the Richardson-Einstein-de Haas effect. This effect relates an object's magnetism to the movement of the object's electrons.

Richardson was just about to become a naturalized citizen of the United States when, in 1913, he was offered the Wheatstone Professorship of Physics at King's College at the University of London. He accepted the offer and returned to England just before the outbreak of World War I. During the war, Richardson worked on the development of more efficient **vacuum** tubes to be used in military communication systems. At the war's conclusion, Richardson returned to his teaching and research at King's College until 1924, when he was appointed Yarrow Research Professor of the Royal Society. This position relieved him of all teaching responsibilities, and he dedicated the next two decades of his life to research. One of the main foci of this research was an attempt to delineate the connections between chemistry and physics. Even after his retirement from the Royal Society post in 1944, Richardson continued his research activities, publishing his last paper in 1953.

Richardson was awarded the 1928 Nobel Prize in physics for his work on thermionics and was knighted in 1939. He served as president of the Physical Society from 1926 to 1928. After his first wife Lilian died in 1945, Richardson married Henrietta M. G. Rupp, a noted physicist who specialized in electron **diffraction** and **luminescence** in solids. Richardson died in Alton, Hampshire, on February 15, 1959.

RICHTER, BURTON (1931-)

American physicist

Burton Richter is a physicist at Stanford University who was largely responsible for the design of the eight-billion electron-volt accelerator called the Stanford Positron-Electron Accelerating Ring (SPEAR). Designed to cause collisions

Burton Richter. *(Photo by Kevin Fleming/Corbis. Reproduced by permission.)*

between electrons and positrons at almost unprecedented **energy** levels, SPEAR allowed scientists to perform experiments in 1974 that resulted in the discovery of an entirely new and unexpected particle, which Richter named psi (ψ). The discovery had revolutionary implications for **particle physics**; the lifetime of the psi-particle was many thousands of times longer than other particles, which indicated that it possessed a previously undiscovered property of **matter**. At almost the same time Richter made his discovery, Samuel C. C. Ting, working at the Massachusetts Institute of Technology (MIT), discovered the same particle using a different method. The two men were jointly awarded the Nobel Prize in physics in 1976.

Richter was born in Brooklyn, New York, on March 22, 1931. He was the oldest child and only son of Abraham Richter, a textile worker, and Fannie (Pollack) Richter. Burton attended Far Rockaway High School in Queens and Mercersburg Academy in Mercersburg, Pennsylvania. In 1948, he entered MIT, where he hesitated between studying chemistry or physics. His choice was determined by one of his professors, Francis Friedman, who "opened my eyes to the beauty of physics," as Richter is quoted as saying in *Current Biography*. At MIT, Richter worked with both Francis Bitter and Martin Deutsch, writing his senior thesis on the effect of external magnetic fields on the hydrogen spectrum.

After receiving his B.S. in 1952, Richter stayed on at Bitter's laboratory as a research assistant and doctoral student. His first assignment there was to find a way of producing a short-lived **isotope** of mercury, mercury-197, which he accomplished by bombarding gold foil with deuterons from the MIT **cyclotron**. The task accomplished, Richter found that he was interested in further work with particle **accelerators**. Physicist David Frisch provided him with the opportunity to work with one of the world's most powerful accelerators, the cosmotron at the Brookhaven National Laboratory. In 1953, Richter spent six months pursuing his interest in accelerators. After returning to MIT, Richter began his doctoral research on the production of pi-mesons in MIT's own synchrotron accelerator. He received his Ph.D. in 1956.

Richter accepted an appointment as a research assistant at Stanford University in 1956. This decision was based on his interest in pursuing research on **quantum electrodynamics (QED)**, the modern version of electromagnetic theory. In particular, he wanted to find out if QED was valid at very small distances, such as those corresponding to the dimensions of the atomic **nucleus**. The 700-million-electron-volt linear accelerator at Stanford was, Richter later wrote in an edition of the University's *Stanford Linear Accelerator Beam Line,* "the perfect device with which to do an experiment on electron-electron **scattering** which I had been considering." He carried out these studies over the next decade, discovering that QED is indeed valid to distances of at least the diameter of an atomic nucleus, about 10^{-14} centimeter.

Uses Colliding Beams to Discovery the Psi-Particle

Even before the QED experiments had been completed, Richter began thinking about another approach to accelerator design. In all accelerators built at that time, particles were accelerated and caused to collide with a stationary target. There was, however, a major limitation to this approach: a large fraction of the energy of the incident beam was lost in the **kinetic energy** of particles produced in the collision. Richter realized that a more efficient way to study particle collisions would be to have two particle beams collide with each other while travelling in opposite directions. In 1970, Richter obtained funding for the machine he had in mind, the Stanford Positron-Electron Accelerating Ring (SPEAR). Three years later, SPEAR was in operation; with it, Richter was able to make what is considered one of the most important discoveries concerning elementary particles.

In this experiment, two concentric beams of particles, one of electrons and one of positrons, traveled around the circumference of the SPEAR accelerator. At a particular moment, the paths of the two beams were diverted and the particles allowed to collide. A fraction of the enormous energy released in this collision was converted into particles that do not exist at lower energy levels. One of these particles had never been observed before and had dramatically unexpected properties. It was about three times as massive as a **proton** with a lifetime about 5,000 times greater than would have been predicted for such a particle. Richter named the particle psi. Coincidentally, an identical discovery using a totally different approach was made at almost the same time on the other side of the continent. At the Brookhaven National Laboratory, a research team led by Ting discovered the same particle and gave it the name *J*. Richter and Ting made a joint announcement of their discoveries on November 11, 1974, at Stanford. Two years later, the two were awarded the Nobel Prize in physics for the discovery of the particle, now generally called the J/psi.

The primary significance of the J/psi discovery was the questions it raised. Traditional physical theory, which postulated the existence of three kinds of fundamental particles called **quarks**, was unable to explain the J/psi particle's peculiar properties. However, the discovery was consistent with a prediction made by American physicist Sheldon Lee Glashow in 1964; he had theorized that a previously unknown property, which he called charm, was needed to describe fully the properties of at least some elementary particles. The application of Glashow's theory to the J/psi particle was confirmed, and it is now widely accepted that the particle consists of a fourth quark.

Richter has remained at Stanford throughout his academic career. He was promoted to assistant professor in 1959, associate professor in 1963, and full professor in 1967. Since 1980, he has been Paul Pigott Professor of Physical Sciences. In 1975, he received the E. O. Lawrence Memorial Award from the U.S. Energy Research and Development Agency. In addition to his many awards and honors, Richter was Loeb Lecturer at Harvard University in 1974, De Sholit Lecturer at the Weizmann Institute in Israel in 1975, and visiting researcher at the European Center for Nuclear Research in Geneva from 1975 to 1976. Richter married Laurose Becker, an administrative assistant at Stanford, on July 1, 1960; the Richters have a daughter and a son.

RIGID BODIES

In **mechanics**, a rigid body is defined as a system of particles whose relative distances remain constant. This implies that the body does not change its shape during its **motion**, hence the name rigid body. Since all bodies have a certain degree of **elasticity**, a perfect rigid body is only an idealization but a very useful one, as it allows the description of two types of motion of fundamental interest in physics, i.e. translational motion and rotational motion. A rigid body undergoes translation when all of its constituent particles are moving on parallel trajectories and it undergoes rotation when all of its particles are moving on circular trajectories about an axis of rotation. It can also undergo simultaneous translation and rotation. If the rigid body only consist of a point moving in **space**, it has three degrees of freedom, i.e. the number of independent coordinates required to express the value of a given property and in this case, three coordinates are needed to determine its position. But, in three-dimensional space, the motion of the rigid body will require six degrees of freedom since two coordinate systems are used: first, the three coordinates of the **center of mass** with respect to an inertial **frame of reference** and three angles describing the orientation of the rigid body in space.

In rotation about a fixed axis, the rigid body is rotating about a stationary axis and has no translational motion. Every point of the rigid body undergoes circular motion about the axis of rotation. The angular position of the body describes its orientation relative to a reference orientation (usually the x-axis if the z-axis is defined as the rotation axis).

As the body rotates, this angular position is changing and a quantity called the angular displacement (θ) measures how far the body has rotated from its reference orientation. It is measured in degrees or in revolutions, when describing the rotation of a motor. Angular displacement may also be measured in revolutions, as done when describing the rotation of a motor. The angular speed of the body (ω) is then given by the rate of change in angular displacement with time and it is expressed in radians per second or revolutions per minute: $\omega = d\theta/dt$.

A vector quantity (vector ω), the angular **velocity**, can also be defined, its magnitude is the angular speed ω, and its direction is that of the axis of rotation. The sense of rotation is determined using the "right-hand rule." For example, if the axis of rotation of the rigid body is vertical, the fingers of the right hand curl in the direction specifying the sense of the rotation, i.e. either clockwise or counterclockwise, and the extended thumb points in the direction of the axis of rotation, i.e. vertically. When the angular speed varies or when the orientation of the rotation axis changes, the angular velocity also changes.

Whenever the magnitude or direction of the angular velocity changes, the rigid body acquires an angular **acceleration** α, also a vector quantity, which is defined as the rate of change of the angular velocity with time, $d\omega/dt$. It is expressed in radians per second squared.

Angular acceleration about a given point is the result of a **torque** applied at that point. A torque (τ) is a vector quantity and it is defined as the cross product of the moment arm (vector r)—i.e. the perpendicular distance from the center of rotation, or pivot, to the point at which the **force** is applied—and of the component of the force vector that is applied perpendicular to the moment arm: $\tau = \vec{r} \times \vec{F}$.

Thus, the greater the force applied or the longer the moment arm, the larger the resulting torque.

A rotating rigid body always has kinetic **energy**, even when it has no translational motion. Its total rotational **kinetic energy** is equal to half its moment of **inertia** (I) multiplied by the square of its angular speed or $E_{k,rot} = (1/2)I\omega^2$. When the rigid body undergoes both translational and rotational motion, its total kinetic energy is the sum of the translational and rotational kinetic energies. The above concepts of angular displacement, angular speed, angular velocity, angular acceleration, and kinetic energy apply to any type of rigid body either rotating about an internal axis—such as a wheel spinning on its axle—or to bodies revolving around an external point, such as **planets** orbiting the **sun** or satellites Earth. A rigid body can also undergo rotation about a moving axis, as illustrated by a body rotating about itself and falling through space. In this case, the motion can be described as a translation of the rigid body's center of **mass** combined to rotation about an axis that goes through the center of mass. A good example is provided by Earth which rotates about its own central axis of rotation once in 24 hours while orbiting the sun once a year. Another example is the gyroscope, a device that consists of a wheel or disc mounted on a moveable frame so as to be able to tilt and spin about an axis of rotation in any direction. When a gyroscope spins, it has a tendency to retain its initial orientation in space, regardless of external forces applied to it.

ROCKETS

Although rockets are widely viewed as ultra-modern inventions, the principles behind them have been at least generally understood for thousands of years. The first mechanical devices that might be thought of as rockets were developed as steam-driven curiosities in Greece around the fourth century B.C. By the first century A.D. the Chinese had formulated an early type

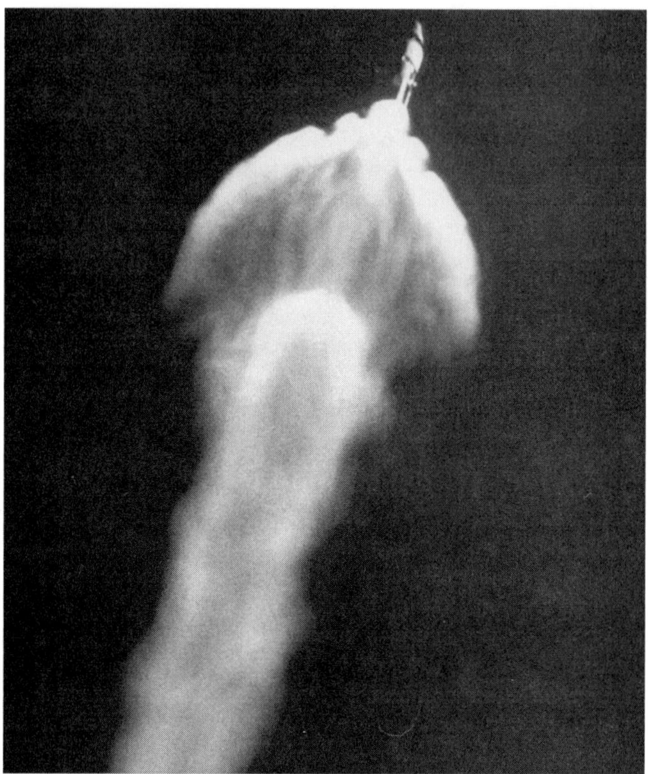

Rocket launching. *(Photo by NASA. Reproduced by permission.)*

of gunpowder and were rolling it into tight paper tubes that could propel an arrow or other missile. The rocket was born.

In theory, the rocket is a fairly simple proposition. It is a vehicle that carries some type of fuel internally. The fuel is caused to expand (usually by way of some **chemical reaction**, such as ignition) and the exhaust from this expansion is forced out through a small opening at the rear. Forward **momentum** is simply the equal-but-opposite reaction to this exhaust. The mighty Saturn V rocket and the humble balloon that is released with being tied closed, both move about by virtue of the same physical forces.

For centuries, rocketry changed little. Most of the innovations that did occur came about from attempts to deploy rockets as offensive weapons. One common type was a crude rocket rigged to explode on impact, and stabilized in flight somewhat by being tethered to a stick. The British Navy employed rockets of this type as shore-bombardment weapons, and it was their use against Fort McHenry during the War of 1812 that inspired Francis Scott Key (1779-1843) to write his famous line about "the rockets' red glare."

Rocketry took a great leap forward in the early twentieth century, largely thanks to the work of the American scientist Robert H. Goddard (1882-1945). Goddard built the first liquid-fuel rocket and was the first to realize that rockets would operate more efficiently in the **vacuum** of **space**.

It was during World War II, though, that modern rocketry developed. The German military had become interested in the possibilities that rockets offered, and sponsored much research and development during the years leading up to the war. The Versailles Treaty that ended World War I forbade Germany from building an air **force**, and rocketry was seen as one way of circumventing the prohibition. After the war began on September 1, 1939, the Reich poured vast resources into the rocket program. Wernher von Braun (1912-1977) was the brilliant young rocket scientist who directed much of the research. By 1944, Germany had developed and were using dozens of different types of offensive rockets; everything from surface-to-air missiles to anti-personnel weapons. Most famous, though, was von Braun's A-4 rocket, more commonly known as the V-2.

The V-2 was deployed as Hitler's great "vengeance weapon" against the people of England. Launched from mobile bases in Western Europe, the rockets could carry a thousand-pound warhead all the way to London in just minutes. The V-2 was a guided missile, carrying gyroscopes coupled to steering vanes that directed it to its target. Since the re-entry phase of the rocket's flight was supersonic, there was no way for the British to stop it. The British people suffered hundreds of V-2 attacks until advancing Allied armies swept German forces out of launching range. Although the V-2 caused much destruction and loss of life, it was not very effective as a strategic weapon. The Germans were never able to deploy them in the numbers needed to impair the Allies' war efforts. Still, as a harbinger of coming technology, the V-2 was an amazing achievement, and may very well have been the first man-made object to travel beyond the stratosphere.

After the war, von Braun and many of his colleagues relocated to the United States to help develop the American rocket program. One of their first successes was the Redstone missile, which was designed as both an offensive weapon and as the space vehicle used in NASA's Mercury launches. Later they built the multi-stage Saturn rocket, still one of the largest ever constructed, that carried American astronauts to the **Moon**. Multi-stage rockets allow fuel (usually a combination of liquid hydrogen and **oxygen**) to be burned quickly during the early stages of the rocket's flight. This helps the rocket develop a very high **velocity** within a few seconds of its launch, and helps it reach escape velocity—a speed high enough to leave the earth's immediate **gravitational field**.

Rocketry today is a very well-understood science, thanks to the work of pioneers like Goddard and von Braun. Modern solid- and liquid-fueled rockets have become reasonably safe and dependable vehicle platforms, used by both science and commerce in applications ranging from powering the space shuttle to placing telecommunication satellites in orbit. Tiny solid-fuel rocket motors even **power** one of the most popular contemporary hobbies: model rocketry. Advances continue to the present day that make rocketry both more accessible and more commonplace.

ROHRER, HEINRICH (1933-)

Swiss physicist

Heinrich Rohrer shared half of the 1986 Nobel Prize in physics with **Gerd Binnig** for their development of an entirely

new type of **microscope** that revealed for the first time the **atomic structure** of the surface of solids. This **scanning tunneling microscope** (STM) has such a vast array of applications in such a wide range of fields that the Royal Swedish Academy of Sciences was prompted to award its prestigious prize even though the device had only been successfully tested for the first time in 1981.

Rohrer was born on June 6, 1933, in Buchs, St. Gallen, Switzerland, the son of Hans Heinrich Rohrer, a distributor of manufactured goods, and Katharina Ganpenbein Rohrer. When he was 16, Rohrer moved with his family from the country to the large city of Zurich. As a student, Rohrer was interested in both physics and chemistry and classical languages, finally settling on the study of physics when he entered the Federal Institute of Technology in Zurich in 1951. He received his diploma in 1955 and his Ph.D. in 1960, both in physics, from the Institute. His doctoral research involved superconductivity.

From 1960-1961 Rohrer was a research assistant at the Institute in Zurich and followed this with two years of postdoctoral research in superconducting at Rutgers University in the United States. On his return to Zurich in 1963, Rohrer joined the staff of the research laboratory of International Business Machines (IBM), eventually becoming manager of the physics department as well as an IBM Fellow. Rohrer has remained at the IBM lab throughout his career, except for an academic year as a visiting scholar at the University of California, Santa Barbara, in 1974-1975, when he studied **nuclear magnetic resonance**. In 1961, Rohrer married Rose-Marie Eggar; the couple had two daughters.

After joining IBM, Rohrer expanded his research in physics beyond superconductivity, investigating magnetic fields and critical phenomena. He became interested in the little-understood and complex atomic structures of the surfaces of materials. While **electron** microscopes had been developed to probe the internal arrangements of atoms in materials, attempts to uncover the very different characteristics of surface atoms had been decidedly unsuccessful. In 1978 Gerd Binnig, a young German who had just received his Ph.D., joined Rohrer's research team in Zurich. Together, Rohrer and Binnig began to explore oxide layers on metal surfaces. They decided to develop a spectroscopic probe and in the process invented an entirely new type of microscope.

Rohrer and Binnig began with the phenomenon called **tunneling**. As revealed through quantum **mechanics**, electrons behave in a wavelike manner that causes them to produce a diffuse cloud as they leak out from the surface of a sample. When electron clouds from two adjacent surfaces overlap, electrons tunnel from one surface or cloud to the other. Tunneling through an insulating layer had been used often to reveal information about the materials on either side of the insulation. Rohrer and Binnig decided to tunnel through a **vacuum** and then use a sharp, needlelike probe within the vacuum to scan the samole's surface. As the scanning tip closely approached the sample, the electron clouds of each overlapped and a tunneling current began to flow. A feedback mechanism used the tunneling current to keep the tip at a constant height above the sample's surface. In this way, the tip followed the contours of the individual atoms of the scanned surface, and a computer processed the tip's **motion** to produce a three-dimensional, high-resolution image of that surface.

From the beginning, Rohrer told *Science* magazine, "We were quite confident. Even at the beginning, we knew it would be a significant development. The surprising thing is that it went so fast." One large problem was the sensitivity of the scanning tip to disturbances from vibration and noise. Here Rohrer's background in superconductors was helpful, because transducers too are extremely sensitive to vibration. Rohrer and Binnig solved the problem by shielding their scanner from disturbances with magnets and a heavy stone table set on inflated rubber tires. They successfully tested their new device in 1981 and then worked to refine it technologically. By the mid-1980s the scanning tunneling microscope could fit in the palm of the hand (except for the vacuum chamber) and could show some details as tiny as 0.1 angstrom (with 1 angstrom being about the diameter of a single **atom**, or 2.5 billionths of an inch). STMs were also developed that worked in water, air, and cryogenic **fluids** as well as vacuums. By 1987 Rohrer's group at IBM had developed an STM the size of a fingertip.

Rohrer may not have been surprised when he and Binnig shared half of the Nobel Prize in 1986 for their STM. After he received the honor, Rohrer told *Business Week* that when he explained to colleagues at the IBM lab what he and Binnig planned to try, "They all said, 'You are completely crazy—but if it works, you'll get the Nobel Prize'." In awarding the prize, the Swedish Academy conceded that the STM was completely new and barely yet developed. Nevertheless, the Academy stated, because of the STM, "It is . . . clear that entirely new fields are opening up for the study of the structure of matter." This study has included living organisms such as viruses, catalysts used to produce chemical reactions in the pharmaceutical and petrochemical industries, and **semiconductors** and metals. (Interestingly, the other recipient of the 1986 Nobel Prize in physics was Ernst Ruska, for his design of the first **electron microscope** in 1931, 50 years before Rohrer and Binnig developed their scanning microscope.)

Rohrer shared other international awards with Binnig for his work on the STM. He is a member of many important scientific societies and has been awarded honorary doctorates by several universities.

Röntgen, Wilhelm Conrad (1845-1923)

German physicist

In 1895, physicist Wilhelm Conrad Röntgen made a unique finding that electrified both the scientific world and the general public. He unintentionally discovered the existence of a mysterious ray that could penetrate various materials. He named the unknown ray the "x ray," and it would rapidly revolutionize both physics and medicine.

Wilhelm Conrad Röntgen.

Röntgen was born in Lennep, Germany, the only child of a cloth manufacturer and merchant. Three years later his family moved to the Netherlands, where Röntgen attended the Utrecht Technical School, which he was expelled unfairly from after being accused of a prank another student had committed. Although Röntgen did not appear to be especially gifted in his schoolwork, he was good at tinkering with and building mechanical objects, a talent that would serve him well later in life as he would build many of his own experimental devices. In 1869, he earned a doctorate in mechanical engineering from the University of Zurich, where he attended lectures by the noted physicist **Rudolf Julius Emmanuel Clausius**. Röntgen published his first work in 1870. His interests ranged from research in the specific heats of gases and thermal conductivity of crystals to the electromagnetic influences that could modify planes of polarized **light**. After several appointments at Strassbourg University, the Academy of Agriculture at Hohenheim in Wurtemberg, and the University of Giessen, Röntgen accepted an appointment as chair of the physics department at the University of Wurzburg, where he worked with Hermann Helmholtz and **Hendrik Lorentz**.

On the evening of November 8, 1895, Röntgen, who was elected Rector of the University of Wurzburg the previous year, was working on an experiment with cathode rays using a Hittorf-Crookes tube. He immediately noticed a fluorescent glow on a nearby screen painted with barium platinocyanide and determined that it was coming from the tube, even though he had surrounded his tube with black paper. A gifted experimentalist, Röntgen conducted a series of subsequent studies of the x ray, which thoroughly outlined the properties of the mysterious ray. In his most famous experiment, he used the hand of his wife and past the **x rays** over a photographic plate to reveal the bone structure beneath her flesh. Within months of revealing his discovery, x rays were heralded as a new diagnostic tool in medicine. So astounded was the world by this new discovery, that several cities named streets after Röntgen. In 1901, he received the first Nobel Prize in Physics.

In 1900, Röntgen had accepted a post at the University of Munich, where he would spend the rest of his career. A quiet, modest, and private man, Röntgen never patented his discovery, which could have made him millions even then. Instead, he wanted his discovery to be available to all for the good of the world. Married to Anna Berta Ludwig, an innkeeper's daughter, Röntgen was content to remain in academia to teach and continue his research, taking time when he could to explore nature and do mountain climbing at his summer home near the Bavarian Alps. Near the end of his career, Röntgen spent more and more time caring for his ailing wife. He retired after she died in 1919. Although world famous and once comfortable financially, Röntgen was nearly bankrupt near the end of his life due to rapid inflation following World War I. He died from cancer of the rectum on February 10, 1923.

ROWLAND, HENRY AUGUSTUS (1848-1901)

American physicist

Henry Augustus Rowland, a founder of modern physics, was recognized primarily for his research into **electromagnetism** and development of **diffraction** grating for spectroscopes. Born in Honesdale, Pennsylvania, Rowland was the son of a long line of Yale-trained ministers. Unlike his forefathers, Rowland was more interested in science than the ministry. As a result, he attended the Rensselaer Polytechnic Institute in Troy, New York, where he earned a degree in civil engineering in 1870. Rowland began his career as a railroad surveyor and as a teacher for the College of Wooster. In 1872, he returned to Rensselaer to teach physics and, in 1876, became chair of the physics department at the newly formed Johns Hopkins University in Balitmore.

Before joining Johns Hopkins, Rowland conducted research in the laboratory of German physicist **Hermann von Helmholtz** in Berlin. It was there that Rowland demonstrated that an **electric charge** on a moving body had the exact same magnetic effect as an **electric current** moving through a conductor. As a result, he was the first to demonstrate conclusively that charged bodies in **motion** produce magnetic effects. Later, Rowland confirmed that the electromagnetic/electrostatic ratio was equal to the speed of the **light**. His

Henry A. Rowland. *(Photo courtesy of Corbis-Bettmann. Reproduced by permission.)*

trailblazing experiments included establishing the value of ohm and the mechanical equivalent of **heat**. Rowland gained the most fame for inventing a new spectral grating machine that automatically focused light. Rowland's spectrograph reduced to a few hours what had previously taken several days to analyze, revolutionizing **optics** and opening the way to modern spectroscopy. Rowland's own use of the spectrograph included re-mapping the solar spectrum and developing highly accurate **wavelength** tables. For his spectrograph, Rowland won a gold medal at the 1890 Paris Exposition.

At the age of 42, Rowland married Henrietta Harrison, and the couple had three children. Although Rowland was an advocate of science for science's sake, not for the purpose of making money (for example, he sold his spectrographs at cost to scientists around the world), he turned his attention to making money after taking a physical for an insurance policy. Diagnosed with diabetes in the days prior to insulin therapy and told that he had only 10-15 years to live, Roland began consulting to the telegraph and hydroelectric **power** businesses to secure his family's financial future. Rowland died in 1901 and according to his wishes, was cremated and buried in the wall of his basement laboratory.

See also Ohm's law

RUBBIA, CARLO (1934-)

Italian physicist

Carlo Rubbia was born in Italy and carried out his first serious scientific experiments as a young boy, using communication equipment abandoned at the end of World War II. Since his postdoctoral year at Columbia University in 1958 and 1959, Rubbia has been particularly interested in the study of elementary particles, and through his affiliations with the **European Center for Nuclear Research (CERN)** he has had some of the most powerful particle **accelerators** in the world available for his research. Since the late 1960s, Rubbia's primary research has involved the search for a trio of particles known as the W+, W-, and Z0 bosons, which were postulated in the 1960s as the **force** particles through which the **electroweak force** exerts its influence. By 1982, Rubbia and his colleagues at CERN had designed the equipment needed to carry out this search and had successfully located the first W particles. For this accomplishment, he and co-worker Simon van der Meer were awarded the 1984 Nobel Prize in physics.

Rubbia was born on March 31, 1934, in the small town of Gorizia, in northern Italy. His father was Silvio R. Rubbia, a telephone worker; his mother, Bice Liceni Rubbia, was a school teacher. When World War II ended in 1945, 11-year-old Carlo "scaveng[ed] **radio** equipment that had been abandoned as various armies marched through on their various advances and retreats," according to Gary Taubes in his book *Nobel Dreams*. He used this equipment to learn everything about radios, becoming something of an "electronics freak," Taubes wrote.

Studies Elementary Particles in Italy and the United States

In 1945, Silvio Rubbia's job brought the family to Pisa, where Rubbia was enrolled at the prestigious Scuola Normale Superiore, a rigorous secondary school affiliated with the University of Pisa. After graduating from the Scuola, Rubbia went on to the university, earning a Ph.D. in physics in 1958 for his dissertation on cosmic **radiation** and particle detection devices. During the academic year 1958–59, Rubbia continued his studies at Columbia University in the United States, where he worked with some of the world's outstanding physicists, including **Tsung-Dao Lee**, Chen Ning Yang, **Chien-Shiung Wu**, Charles H. Townes, **Melvin Schwartz**, Leon Ledeberg, and **Steven Weinberg**, acquiring from them an interest in weak-interaction physics.

Rubbia returned to Italy in 1960 to continue his postdoctoral studies at the University of Rome. A year later, he accepted an appointment at CERN in Geneva. A consortium of more than a dozen European nations, CERN is one of the world's most important centers for the study of elementary particles. Rubbia quickly moved up the hierarchy at CERN and was appointed to the prestigious position of team leader before he was 30 years old.

Rubbia's goal during his first years at CERN was the discovery of three "intermediate **vector bosons**" known as the W+, W-, and Z0 particles. The existence of these particles had been predicted in the 1960s, when **Sheldon Glashow**, **Abdus Salam,** and Steven Weinberg had independently developed an **electroweak theory** proposing that two fundamental forces, the electromagnetic and weak forces, are manifestations of a more fundamental natural force, and predicting the existence of W and Z particles. In 1971, Salam and Weinberg suggested that the neutral Z particle could be detected in "neutral currents" that would be produced by the collision of neutrinos and **matter**. Rubbia's objective was to design and conduct the experiment that would, for the first time, produce this particle.

Improved Accelerator Design Leads to Discovery of W and Z Particles

In 1969, Rubbia joined a project at the **Fermi National Accelerator Laboratory (Fermilab)** near Chicago to search for the W particles. In 1971, however, the team switched their efforts to the attempt to prove the existence of neutral weak currents, thus putting them in direct competition with Rubbia's former colleagues at CERN. In 1973, the CERN team published nearly conclusive evidence that neutral weak currents existed. The Fermilab team immediately rushed to publish their own as yet incomplete results, which also supported the existence of the currents. At about the same time, Rubbia's visa expired and he returned to Euope. Soon thereafter, the Fermi team reconducted their experiments and announced that their original findings had been in error. They then realized, however, that this second set of experiments was flawed, and they retracted their earlier retraction. The confusion temporarily tarnished Rubbia's reputation.

Both teams now turned their attention to the search for the W and Z particles. However, no existing particle accelerator could generate the **energy** needed to produce them. Rubbia proposed a revolutionary new technique in which two particle beams, one composed of protons and one of antiprotons, would be set in **motion** in opposite directions and caused to collide with each other. The amount of energy released in such a collision, Rubbia said, should be sufficient to result in the formation of W and Z particles. Rubbia's idea was ridiculed by a number of physicists, including the director of Fermilab. Managers at CERN were more open-minded, however, and provided the $100 million needed to redesign the center's super **proton** synchrotron to Rubbia's specifications.

By 1982, that work had been completed and the search for the W+, W-, and **Z bosons** began. Within a month's time, the first W particles had been identified and, less than a year later, the first Z's were also discovered. For this accomplishment, Rubbia shared the 1984 Nobel Prize in physics with the Dutch scientist Simon van der Meer, who had devised a means of storing and regulating the erratic antiprotons.

Rubbia has been affiliated with Harvard University since 1970 (1972 according to some biographers), teaching one semester there each year. He has also continued an active program of research at CERN. Rubbia married Marissa Romé, a high school physics teacher, on June 27, 1960. They have two daughters, Laura and Andrea. Rubbia has been described as one of the most controversial figures in modern **particle physics**, a man driven by his love for science and, according to some, his own ego. His reputation at CERN, for example, has been described by Taubes as "very, very good and very, very bad," and an article in the October 25, 1984, issue of *New Scientist* called him "an ebullient yet irascible Italian whom fellow physicists love to hate."

RUMFORD, BENJAMIN THOMPSON (1753-1814)

English American physicist

Benjamin Thompson was something of a soldier-of-fortune: he was a spy for the British during the Revolutionary War; he used his position of influence in the government to take bribes; and he worked for several countries to advance the science of armamentation, through which he first earned international acclaim. However, he was also a shrewd and intuitive scientist and was almost solely responsible for the acceptance of the concept of **heat** as a form of **motion** rather than a fluid. He is better known in the annals of science as Count Rumford.

Benjamin Thompson was born in Woburn, Massachusetts, in 1753. As a teen he worked as an apprentice to a Salem storekeeper. He was apparently a very poor apprentice, viewing himself as destined for greater achievements; in fact, by the age of 17 he had taught himself French, philosophy, and the sport of fencing, all in anticipation of his future position. When he was 19 he became a schoolmaster, moving to nearby Rumford (now Concord, New Hampshire). There he met and married a wealthy older widow.

About this time Thompson was the center of a local controversy. Members of the growing anti-British movement accused him of selling secrets to the British army. Thompson soon fled Rumford, leaving his wife and child behind, and moved to England. It was later discovered that he had indeed been a spy, having always considered himself an Englishman at heart. Thompson continued his career of espionage until the end of the Revolutionary War; upon the defeat of the King's army it became clear that he could never return to the land of his birth.

He began his life of exile in the employ of King George III, eventually holding the titles of Minister of War, Minister of the Interior, and Royal Scientist to the King. In the years shortly after the war Thompson dedicated himself to studying and improving upon the science of weaponry and, in particular, gunpowder. He devised a new type of mortar that could be used to determine the explosive potential of gunpowder. In 1781 he was elected to the Royal Society for his work with explosives; many years later he would establish the gunpowder standard.

The same year Thompson was accepted into the exclusive Royal Society, he was exiled from England for selling naval secrets to the French. However, many of his duties for the crown were top secret; instead of an execution he was given an appointment as a diplomat to Bavaria, where he

would again serve as a spy for Britain. Once he was safely in Germany, though, Thompson severed his ties with King George and entered the service of Elector Karl Theodor of Bavaria.

As an administrator for the Elector, one of Thompson's first duties was to find a solution for Bavaria's large beggar population. Rather than arrest or execute them, Thompson (who was always considered a practical man) established workhouses where they could make army uniforms and winter clothing. During this time he brought to the European continent James Watt's steam engine and the potato. For his accomplishments he earned overnight fame and the title of Count Rumford (a name he chose for the small town he and his wife had lived in).

It was in the year 1798 that Rumford would make his most important contribution to science. While observing the boring of a cannon, he noticed the intense heat that the drill generated. At that time it was believed that heat was a kind of liquid called caloric that existed in all **matter** (this view was no doubt substantiated by the fact that metal expanded when heated—apparently infused with caloric). When two solids such as the cannon and the drill rubbed against each other, a small amount of caloric was released, producing heat. Because the drill ground the metal into shavings, more caloric was released and the **temperature** increased even more. However, Rumford was intrigued by the sheer amount of heat that the boring generated—the metal was far too hot to touch, and the drillers continuously doused the cannon with water just to stay near it. It seemed to Rumford that enough heat had been released by the cannon to melt it down if it were put back in. Also, the drill released more heat when it was blunt than when it successfully cut the metal.

Intrigued, Rumford set out to disprove the existence of caloric. He began by weighing an amount of water in both liquid and solid forms; if caloric did exist, the water should weigh more at the higher temperature. Rumford found no change in the water's **weight**, surmising that caloric must either be suspiciously weightless or nonexistent.

Rumford pronounced that heat was a form of motion, produced (in his cannon example) by the motion of the drill bit against the metal. He even attempted to determine the amount of heat generated by a given amount of mechanical **energy**, but his poorly designed experiments yielded a figure much too high. Many years later, James Joule would announce a much more accurate number for the mechanical equivalent of heat.

In 1799 the Elector of Bavaria died; Rumford, whose abrasive personality had won him few friends, chose this time to return to England. He stayed in London long enough to establish the Royal Institution with Sir Joseph Banks. The institution soon attracted such brilliant young scientists as **Thomas Young** and Humphry Davy as lecturers. They continued Rumford's research into the nature of heat, though it would be nearly 50 years before **James Clerk Maxwell** would finally put the **caloric theory** to rest.

Meanwhile, Rumford himself grew restless in London, and in 1804 he moved to Paris, where he would remain for the rest of his life. There he met and married a wealthy older widow who, coincidentally, had been married to Antoine Lavoisier—the originator of the caloric theory. It was an unhappy marriage that dissolved after four years. Rumford's daughter joined him in Paris in 1811 and cared for him until his death in 1814.

Perhaps to atone for the dishonorable actions of his youth, Rumford made arrangements in his will for his substantial wealth to be used for the advancement of science. He established the Rumford medals at the Royal Society and the Academy of Arts and Science at Boston. The remainder of his fortune he gave to Harvard University for the creation of a Rumford professorship. All of these accolades exist today and are among the most prestigious in their fields.

RUSSELL, HENRY NORRIS (1877-1957)
American astronomer

Henry Russell was born on October 25, 1877, in Oyster Bay, New York. He received his higher education at Princeton University and had earned his Ph.D. by the age of 23. He became the director of Princeton's observatory in 1912.

During the course of his observations, Russell discovered a relationship between a star's **color** (or, analogously, its spectral class), and its brightness. Blue **stars** were hottest, yellow stars were cooler, and red stars were coolest. In general, blue stars should be the brightest and red stars the dimmest, but there were exceptions—some red stars were very bright. For a cool, red star to appear bright, Russell reasoned that it either had to be very close or very large. A star with a large surface area could radiate a significant amount of **light** at low temperatures, so those very bright stars had to be giants. Hence, he called these stars red giant stars.

Armed with observations of a large number of stars, Russell created a diagram on which the stars' spectral classification, roughly equivalent to **temperature**, was plotted along the x-axis, and their luminosity was plotted along the y-axis. On this diagram, most stars fell along a diagonal line running from upper left (the hottest, brightest stars) to lower right (the coolest, faintest stars). The stars on this diagonal line were called main sequence stars. These are healthy, stable stars like the **Sun**. Stars not on the main sequence are typically those in advanced stages of **stellar evolution**, such as the red giants or supergiants.

As it turned out, this same relationship had been discovered by Danish astronomer **Ejnar Hertzsprung** nearly 10 years earlier but had received no publicity. To remedy this oversight, the chart was called the Hertzsprung-Russell diagram, honoring the work of both men. The H-R diagram was used to deduce the path of stellar evolution and it worked fairly well for 25 years. When scientists like **Hans Bethe** began to work out the **mechanics** of the nuclear **energy** that powered the stars, the physical mechanisms that produced the elegant realtionships on the H-R diagram were at last uncovered. By the end of the twentieth century, many variants on the original H-R diagram had been developed, to accommodate the vari-

ous new observational parameters that had become available. Nevertheless, the essential character of the original diagram remained, having served as a useful tool for stellar astronomers for nearly a century.

RUTHERFORD, ERNEST (1871-1937)
New Zealand-English physicist

Ernest Rutherford's explanation of **radioactivity** earned him the 1908 Nobel Prize in chemistry, but his most renowned achievement was his classic demonstration that the **atom** consists of a small, dense **nucleus** surrounded by orbiting electrons. He also demonstrated the transmutation of one element into another by splitting the atom. His direction of laboratories in Canada and Great Britain led to such triumphs as the discovery of the **neutron** and helped to launch high-energy, or particle, physics, which concentrates on the constitution, properties, and interactions of elementary particles of **matter**.

Rutherford was born the fourth of 12 children on August 30, 1871, to James and Martha Thompson Rutherford on the South Island of New Zealand near the village of Spring Grove. Both parents had arrived in New Zealand as children, not long after Great Britain annexed the territory into the Commonwealth in 1840. Rutherford's father, of Scottish descent, logged, cultivated flax, worked in construction, and pursued other endeavors with a mechanical inventiveness inherited from his wheelwright father, George. Martha Rutherford was a schoolteacher of English descent.

Rutherford's early success in school earned him a scholarship to Nelson College, a secondary school in a village on the north end of New Zealand's South Island. He then received a scholarship to Canterbury College at Christchurch, New Zealand, where he earned his bachelor of arts degree in 1892. He continued studying at Canterbury, earning a master of arts degree with honors in mathematics and **mathematical physics** in 1893 and a bachelor of science degree in 1894. In New Zealand, Rutherford met Mary Newton, the woman who would become his wife in 1900. She was the daughter of the woman who provided Rutherford with lodging while he studied at Canterbury. Rutherford and his wife had one daughter, Eileen (1901–1930), who married Ralph Fowler, a laboratory assistant of Rutherford's in the 1920s and 1930s.

While working toward his bachelor of science degree, Rutherford researched the effects of electromagnetic **waves**, produced by rapidly alternating electrical currents, on the magnetization of iron. He observed that, contrary to contemporary expectations, iron did magnetize in high-frequency electromagnetic fields. Conversely, he also showed that electromagnetic waves could demagnetize magnetized iron needles. On the basis of this observation, Rutherford devised a device for picking up electromagnetic waves produced at a distance. Italian physicist **Guglielmo Marconi** would later parlay the same principles into the development of wireless telegraphy, or **radio**.

These experiments earned Rutherford a scholarship in 1895 derived from profits from London's Great Exhibition of

1851. Rutherford attended Trinity College at Cambridge University to work under the direction of English physicist J. J. Thomson at the Cavendish Laboratory as the university's first research student. The laboratory had been established in 1871 for research in experimental physics and was first led by **electromagnetism** pioneer **James Clerk Maxwell**. Rutherford's demonstration of his electromagnetic detector greatly impressed Thomson and other scientists at the Cavendish almost immediately.

Thomson invited Rutherford in 1896 to assist him in studies of the effects of **x rays**, which had been discovered in 1895 by Wilhelm Conrad Röntgen, on the electrical properties of gases. Thomson and Rutherford demonstrated that x rays broke gas molecules into electrically and positively charged **ions**, making the gas electrically conductive. This work brought Rutherford widespread recognition in the British scientific community for the first time.

In 1897, Rutherford took up the study of radioactivity, the phenomenon discovered almost accidentally by French physicist Henri Becquerel in 1896. He began by studying the radioactive emissions of uranium, systematically wrapping uranium in successive layers of aluminum foil to observe the penetrating ability of these emissions. He concluded that uranium emitted two distinct types of **radiation**: a less penetrating type, which he called "alpha," and a more penetrating type, "beta." He also later observed what was described by French physicist Paul Villard as "gamma" radiation, the most penetrating type of all.

Not assured of a professorship at Cambridge, Rutherford applied for and was appointed Second MacDonald Professor of Physics at McGill University in Montreal, Canada, in 1898. McGill University appealed to Rutherford especially because it had perhaps the best-equipped laboratory in North America, if not the world, at the time. At McGill, Rutherford turned from studying uranium to thorium, another radioactive element. Although thorium emits alpha and **beta radiation** as does uranium, emission patterns for thorium substances seemed erratic. Rutherford determined in 1899 that an emanation, or new radioactive substance, was being produced. He also observed that the radioactivity of the emanation gradually decreased geometrically with time, an occurrence now known as the **half-life** of a radioactive substance, which is a measurement of the time it takes for half of a substance to decay. In 1901, Rutherford forged a partnership with Frederick Soddy, an Oxford chemistry demonstrator based at McGill who first encountered Rutherford in a debate on the existence of **subatomic particles**. Rutherford wanted Soddy to help him study the thorium emanations and to explain curious observations of radioactive substances noticed in Europe by Becquerel and by Sir **William Crookes**, who discovered thallium. Both had isolated the active parts of uranium from an apparently inert part. However, Becquerel also observed that the active part soon lost its activity, while the inert remainder regained its activity.

Rutherford and Soddy isolated the active part of radioactive thorium, which they named thorium-X, from the apparently-inert parent thorium. They charted how thorium-X gradually lost its radioactivity while the original thorium regained its

Ernest Rutherford.

activity, illustrating that thorium-X had its own distinctive half-life, which was much shorter than the half-life of thorium. Soddy tried to get thorium-X to interact chemically with other reagents without success. From these observations, Rutherford and Soddy put together the modern understanding of radioactivity in 1903. Thorium-X was a product of the disintegration or decay of thorium. In nature, radioactive elements and their products decay simultaneously. However, when the product is separated out, it continues to decay but is not replenished by decaying thorium atoms, so its radioactivity falls off. Meanwhile, the inert parent thorium eventually regains its radioactivity as it generates new radioactive products. Rutherford's explanation of radioactivity at the atomic level is what caused a sensation in scientific circles. He explained that radioactivity—alpha, beta, and gamma radiation—was the physical manifestation of this disintegration, the pieces of the thorium atom that were released as it decayed. In other words, thorium was steadily being transformed, or transmuted, into a new element that was lower in atomic number. It was this work that earned Rutherford the 1908 Nobel Prize in chemistry.

Rutherford received, and turned down, offers to teach at Yale and Columbia Universities in the United States. He became a Fellow of the Royal Society in 1903 and received the Rumford Medal in 1904. His books on radioactivity became standard textbooks on the subject for years and he was a popular speaker. He attracted a number of talented associates at

McGill, the most famous of whom was **Otto Hahn**, a German physicist who would, with Austrian physicist **Lise Meitner**, demonstrate the fissioning of uranium in 1939.

In 1904, Rutherford was the first to suggest that radioactive elements with extremely long half-lives might provide a source of **energy** for sustaining the **heat** of Earth's interior. This would supply a means for estimating the age of Earth in the billions of years, allowing plenty of time for evolution by natural selection to proceed along lines outlined by the naturalist Charles Darwin in 1859.

In 1906, Sir Arthur Schuster offered Rutherford his chair as professor of physics at Manchester University in Great Britain. Eager to return to what was then the center of the scientific world, Rutherford accepted the position in 1907. Rutherford was again blessed at Manchester with a well-equipped laboratory and talented associates from around the world, such as **Hans Geiger**, who would develop the radioactivity counter; Charles Darwin, grandson of the famous naturalist; and physicists **Niels Bohr**, Ernest Marsden, and H. G. J. Moseley.

Rutherford proceeded with his study of radioactive emissions, particularly the high-energy alpha particles. His research was slowed at first when a sample of radium, his favorite alpha source, was sent by the Radium Institute of the Austrian Academy of Sciences in Vienna to a rival, William Ramsay, discoverer of the noble gases. Rutherford had to request and await another sample from the Institute before he could proceed in earnest with his work.

Rutherford wanted to determine precisely the nature of the alpha particles. In 1903, at McGill, he had succeeded in deflecting alpha particles in electric and magnetic fields, proving they had a positive charge. He was certain that the relatively massive particle must be equivalent to **helium** nuclei, which consist of two protons. At Manchester in 1908, Rutherford and his colleagues proved experimentally through spectroscopic means that the alpha particles were indeed nuclei of helium atoms.

In 1908, Rutherford and Geiger devised a method for counting alpha particles precisely. Alpha particles were fired into a nearly evacuated tube with a strong electric field. The resulting ionizing effect in the gas could be detected by an electrometer, a device that measures **electric charge** in a gas. The alpha particles could then be detected visually as well as they struck a zinc sulfide screen to cause an identifiable flash or scintillation. Geiger would build upon this technique in developing the electric radiation counter that bears his name.

In 1909, Rutherford had instructed Marsden to study the **scattering** of alpha particles at large angles. Marsden observed that when alpha particles were fired at gold foil, a significant number of particles were deflected at unusually large angles; some particles were even reflected backward. Metals with a larger atomic number (such as lead) reflected back even more particles. It was not until late in 1910 that Rutherford postulated from this evidence the modern concept of the atom, which he announced early in 1911. He surmised that the atom did not resemble a "plum pudding" of positively charged

nuclear particles with electrons embedded within like raisins, as suggested by Thomson. Instead, Rutherford suggested that the atom consisted of a very small, dense nucleus surrounded by orbiting electrons. Geiger and Marsden provided the mathematics to support the theory and Bohr linked this concept with **quantum theory** to produce the model of the atom employed today. After World War I broke out in 1914, Rutherford was called upon to serve in the British Navy's Board of Invention and Research. His main area of research for the Board was in devising a method for detecting German U-boats at sea. His work established principles applied later in the development of **sonar**.

The work on **alpha particle** scattering continued during the war. Marsden observed in 1914 that alpha particles fired into hydrogen gas produced anomalous numbers of scintillations. Rutherford first concluded that the scintillations were being caused by hydrogen nuclei. However, Rutherford later observed the same effect when alpha particles were fired into nitrogen. After a long series of experiments to exclude all possible explanations, in 1919 Rutherford determined that the alpha particles were splitting the nitrogen atoms and that the extraneous hydrogen atoms were remnants of that split. Nitrogen was thus transmuted into another element. In 1925, English physicist Paul Maynard Stewart Blackett used the **cloud chamber** apparatus devised by Scottish physicist C. T. R. Wilson to verify Rutherford's observation and to show that the atom split after it had absorbed the alpha particle.

In 1919, Rutherford was persuaded to succeed Thomson as director of the Cavendish Laboratory at Cambridge, a post he would hold until his death. Rutherford directed the Cavendish during its most fruitful research period in its history. English atomic physicist John D. Cockcroft and Irish experimental physicist **Ernest Walton** constructed the world's first particle accelerator in 1932 and demonstrated the transmutation of elements by artificial means. Also in the early 1930s, English physicist **James Chadwick** confirmed the existence of the neutron, which Rutherford had predicted at least a decade earlier. Rutherford, with Chadwick, continued to bombard and split **light** elements with alpha particles. With Marcus Oliphant and Paul Harteck, Rutherford, in 1934, bombarded deuterium with deuterons (deuterium nuclei), achieving the first fusion reaction and production of tritium.

Rutherford was involved significantly in national and international politics during this period, albeit not for himself but for the sake of science. He worked with the civilian Department of Scientific and Industrial Research (DSIR) to obtain grants for his scientific team and served as president of the British Association for the Advancement of Science from 1925 to 1930. Beginning in 1933, Rutherford served as president of the Academic Assistance Council, established to assist refugee Jewish scientists fleeing the advance of Nazi Germany. When the Soviet Union prevented Russian physicist **Pyotr Kapitsa**, a promising Cavendish scientist, from returning to Great Britain from the Soviet Union in 1934, Rutherford launched an ultimately futile effort to convince the Soviets to release him. Rutherford maintained close relations through correspondence with leading scientists in Europe, North America, Australia, and

New Zealand. Despite his preference for experiment, Rutherford corresponded with Bohr, German physicist **Max Planck**, American physicist **Albert Einstein** and other theoretical physicists transforming physics with relativity and quantum **mechanics**. He also remained close to his mother in New Zealand, exchanging letters with her frequently until her death in 1935.

In his long and distinguished career, Rutherford's most prestigious award, aside from his Nobel Prize, may have been the Order of Merit that he received from King George V in 1925. The Order of Merit is Britain's highest civilian honor. Rutherford was knighted in 1914 and made a peer (Baron Rutherford of Nelson or Lord Rutherford) in 1931. He died from complications after surgery on a strangulated hernia on October 19, 1937, in Cambridge. His cremated remains were buried near the graves of **Isaac Newton** and Charles Darwin at Westminster Abbey in London.

RYDBERG CONSTANT

The Rydberg constant, 10,973,731.57 m^{-1}, is the constant used in the Rydberg equation which describes the Rydberg state energies of a Rydberg series. This equation describes many line series of elements very well and the constant is universal to all elements.

In 1885, Swiss mathematician Johann Balmer showed that the **wavelength** of the visible spectrum of the hydrogen **atom** can be described by a formula relating the wavelength to **Planck's constant** times a function of the principle quantum number. Independently, **Johannes Rydberg** analyzed the spectra of many elements. He was the first to relate the principal lines of these spectra to a distinct series. To minimize the number of calculations, he introduced the concept of a wave number, which is the reciprocal of the wavelength and is the number of **waves** per length, more specifically, the number of waves per centimeter. Having made this change he began to see patterns not previously discernable. He found that for a given series when he plotted the difference in wave number versus the principle quantum number he obtained hyperbolic curves that were virtually identical in shape for different series and different elements. From this observation he was able to formulate the Rydberg equation, $E = -R /(n–m)^2$ where E is the **energy** of the line in wave numbers, R is the Rydberg constant, n is the principle quantum number and m is the quantum defect. The quantum defect describes how much the Rydberg series departs from the behavior of the Rydberg states of atomic hydrogen

and is directly related to the interaction of the excited **electron** with the leftover ion core. This equation describes the Rydberg series for the elements. The Rydberg series is a set of bound states of the excited electron for a given set of excited electron **angular momentum quantum numbers** and ion core state.

In 1913 **Niels Bohr** showed that the Rydberg constant can be expressed as a combination of the speed of **light** in a **vacuum** c, Planck's constant h, the electron's charge e, and its **mass** m by $R = me^4/(8eo^2h^3c)$, where eo is the permittivity of vacuum.

RYDBERG, JOHANNES ROBERT (1854-1919)
Swedish mathematician and physicist

Johannes Rydberg was a Swedish physicist noted for his work in spectroscopy. The son of a merchant and ship owner, Rydberg was born in Halmstad. He enrolled in the University of Lund in 1873, receiving his bachelor's degree in mathematics in 1875 and his doctorate in 1879 for his thesis on the construction of conic sections. He then became a lecturer in mathematics at Lund. Soon afterwards, he turned his attention to **mathematical physics** and worked on research with **electricity**.

Rydberg's most important scientific contribution stemmed from his interest in understanding the periodic system of elements. Determined to find some kind of order in the overwhelming data of the atomic spectra, Rydberg discovered a fairly easy mathematical calculation to relate the various lines in the spectra of the elements. To minimize the number of calculations, he introduced the wave number *n*, which expresses the number of wavers per centimeter. Working on the Balmer formula, he developed what is now known as the **Rydberg constant**. Rydberg's work also provided the foundation for the development of the structure theory of the **atom**.

Rydberg married Lydia Eleonora Matilda Carlsson in 1886, and the couple had three children. Despite his important contributions to physics, Rydberg's career was beset with disappointments and setbacks. Although nominated for the Nobel Prize, he never received it. He also had to fight for tenure as an academic, which he received with his appointment as chair of physics at Lund in 1901, at the age of 42. By 1914, his health seriously deteriorated and he took sick leave, never to return. Rydberg died of a brain hemorrhage five years later.

See also Spectroscopy

S

S-MATRIX

The S-matrix is a quantity related to the probability of events occurring in quantum mechanical **scattering** processes. In scattering, two or more particles collide, and the resulting mass-energy can be converted to other particles. The particles that exist before the collision, as well as their energies, momenta, spins, etc., are called the *initial state*, while the particles and their energies, momenta, etc. after the collision are called the *final state*. The S-matrix contains the probability information for all scattering processes.

Mathematically, a matrix is a two dimensional array of numbers arranged in rows and columns. Individual numbers are called matrix elements. Each row of the S-matrix corresponds to a specific initial state, and each column corresponds to a specific final state. The S-matrix element at row i, column j is the probability that initial state i becomes final state j after the scattering occurs. If an S-matrix element is zero, then the scattering process corresponding to its initial and final states cannot occur.

The advantage of the S-matrix formalism is that if the underlying physical theory obeys any symmetry principles, then the S-matrix also obeys those principles. For example, all quantum field theories must be invariant under **Lorentz transformations**. Therefore, the S-matrix is also invariant under Lorentz transformations. It is often easier to evaluate S-matrix elements in a specific reference frame, and then use Lorentz transformations to generalize the answer to any reference frame for use in other situations.

See also Probability and quantum mechanics; Quantum states; Symmetry and symmetry principles

SAGAN, CARL (1934-1996)

American astronomer

Born in 1934 in New York City, Carl Edward Sagan earned his bachelor's and doctoral degrees from the University of Chicago. He conducted research and lectured at the University of California in Berkeley and Harvard University. In 1970, he became the David Duncan Professor of Astronomy and Space Science at Cornell University in Ithaca, New York. He died in 1996 at age 62, having established himself as an insightful researcher, but in particular as arguably the greatest popularizer of astronomy of the twentieth century. For his signal achievements in bringing astronomy into the popular limelight, Sagan was honored with the American Astronomical Society's Annenberg Prize in 1992.

Throughout his career, Sagan's work centered on studying planetary atmospheres, especially the conditions on Earth that gave birth to early life forms. In this area, Sagan made his greatest contributions to science.

Sagan began his research, however, on the surface and atmosphere of Venus. In the early 1960s, it was thought that the surface of the planet was relatively cool, despite conflicting data on **radiation** emissions that put the **temperature** at 600K. Based on **radar** and optical observations, he was able to calculate the height of the clouds on Venus and account for the "greenhouse effect" that raised surface temperature to that of the emissions data.

Next, Sagan turned his energies to simulating the primordial atmosphere of Earth in an effort to duplicate the conditions that supported early life on our planet. Similar to the earlier experiments of Stanley Miller and Harold Urey, Sagan was able to produce amino acids, the building blocks of DNA, by irradiating a methane, ammonia, water, and hydrogen sulfide mixture. He also found five-carbon sugars, which form nucleic acids, glucose, and fructose—essential to life processes—as well as adenosine triphosphate (ATP), which stores **energy** in living cells. From his 1963 discovery, scientists can now conjecture that life on earth arose when complex nucleic acids and proteins formed, using the chemical energy stored in the oceans.

Sagan pushed for planetary exploration that launched the Viking probes that landed on Mars and the Voyager probes that took photographs of Jupiter and Saturn. In 1966, he discovered mountains and cliffs on the surface of Mars using

Carl Sagan. *(UPI/Bettmann. Reproduced by permission.)*

reflecting radar. From information about the planet's dust storms collected in Mariner 9's 1971-72 voyage, he hypothesized that, on Earth, a nuclear war would put clouds of dust and debris in the atmosphere and keep out sunlight, resulting in a nuclear winter. He became a vocal critic of nuclear weaponry, although he did support the use of radioactive sources, such as plutonium, to power interplanetary spacecraft in the far reaches of the solar system. He also found evidence of organic molecules in Jupiter's atmosphere.

A Pulitzer Prize-winning author, Sagan wrote over a dozen books and 400 published scientific articles. He popularized the science of astronomy with several books that deal with **planets** and the evolution of life, strongly supporting the view that life exists on other planets. He served as the editor of the main astronomical journal devoted to planetary astronomy, Icarus, and his 1980 television series "Cosmos" and book of the same name was a runaway hit. In the final years of his life, Sagan's wide-ranging mind turned to the problem of how atmospheric temperature is affected by volcanic activity.

SALAM, ABDUS (1926-1996)

Pakistani physicist

Abdus Salam's major field of interest in the 1950s and 1960s was the relationship between two of the four basic forces gov-

erning nature then known to scientists: the electromagnetic and weak forces. In 1968, Salam published a theory showing how these two forces may be considered as separate and distinct manifestations of a single more fundamental **force**, the **electroweak force**. Experiments conducted at the **European Center for Nuclear Research (CERN)** in 1973 provided the empirical evidence needed to substantiate Salam's theory. For this work, Salam shared the 1979 Nobel Prize in physics with physicists **Sheldon Glashow** and **Steven Weinberg**, who had each independently developed similar theories between 1960 and 1967. Salam's long-time concern for the status of science in Third World nations prompted him in 1964 to push for the establishment of the International Center for Theoretical Physics (ICTP) in Trieste, Italy. The Center provides the kind of instruction for Third World physicists that is generally not available in their own homelands.

Salam was born on January 29, 1926, in the small rural town of Jhang, Pakistan, to Hajira and Muhammed Hussain. Salam's father worked for the local department of education. At the age of 16, Abdus Salam entered the Government College at Punjab University in Lahore, and, in 1946, he was awarded his master's degree in mathematics. Salam then received a scholarship that allowed him to enroll at St. John's College at Cambridge University, where he was awarded a bachelor's degree in mathematics and physics, with highest honors, in 1949.

Attempts to Return to Pakistan

Salam remained at Cambridge as a graduate student for two years, but felt an obligation to return to Pakistan. Accepting a joint appointment as professor of mathematics at the Government College of Lahore and head of the department of mathematics at Punjab University, Salam soon discovered that he had no opportunity to conduct research. "To my dismay," he told Nina Hall for an article in the *New Scientist,* "I learnt that I was the only practicing theoretical physicist in the entire nation. No one cared whether I did any research. Worse, I was expected to look after the college soccer team as my major duty besides teaching undergraduates."

As a result, Salam decided to return to Cambridge, from which he had received a Ph.D. in theoretical physics in 1952. He taught mathematics for two years at Cambridge and, in 1957, was appointed professor of theoretical physics at the Imperial College of Science and Technology in London. He has held that post ever since.

Attacks the Problem of Force Unification

Beginning in the mid-1950s, Salam turned his attention to one of the fundamental questions of modern physics, the unification of forces. Scientists recognize that there are four fundamental forces governing nature—the gravitational, electromagnetic, strong, and weak forces—and that all four may be manifestations of a single basic force. The unity of forces would not actually be observable, they believe, except at **energy** levels much greater than those that exist in the everyday world, energy levels that currently exist only in cosmic **radiation** and in the most powerful of particle **accelerators**.

Attempts to prove unification theories are, to some extent, theoretical exercises involving esoteric mathematical formulations. In the 1960s, three physicists, Salam, Steven Weinberg, and Sheldon Glashow, independently derived a mathematical theory that unifies two of the four basic forces, the electromagnetic and weak forces. A powerful point of confirmation in this work was the fact that essentially the same theory was produced starting from two very different beginning points and following two different lines of reasoning.

One of the predictions arising from the new **electroweak theory** was the existence of previously unknown weak "neutral currents," as anticipated by Salam and Weinberg. These currents were first observed in 1973 during experiments conducted at the CERN in Geneva, and later at the Fermi National Accelerator Laboratory in Batavia, Illinois. A second prediction, the existence of force-carrying particles designated as W^+, W^-, and Z^0 bosons was verified in a later series of experiments also carried out at CERN in 1983. By that time, Salam, Glashow, and Weinberg had been honored for their contributions to the electroweak theory with the 1979 Nobel Prize in physics.

Theoretical physics was only one of Salam's two great passions in life. The other was a concern for the status of theoretical physicists in Third World nations. His own experience in Pakistan had been a lifelong reminder of the need for encouragement, instruction, and assistance for others like himself growing up in developing nations. His concern drove Salam to recommend the establishment of a training center for such individuals. That dream was realized in 1964 with the formation of the ICTP in Trieste, Italy, which invites outstanding theoretical physicists to teach and lecture aspiring students on their own areas of expertise. In addition, the Center acts, according to *New Scientist*'s Nina Hall as a "sort of lonely scientist's club for Brazilians, Nigerians, Sri Lankans, or whoever feels the isolation resulting from lack of resources in their own country." Salam also served as a member of Pakistan's Atomic Energy Commission (1958–1974) and its Science Council (1963–1975), as Chief Scientific Advisor to Pakistan's President (1961–1974) and as chairman of the country's Space and Upper Atmosphere Committee (1962–1963)

Salam, who became director of ICTP when it was founded, took part in a host of other international activities linking scientists to each other and to a variety of governmental agencies. He was a member (1964–1975) and chairman (1971–1972) of the United Nations Advisory Committee on Science and Technology, vice president of the International Union of Pure and Applied Physics (1972–1978), and a member of the Scientific Council of the Stockholm International Peace Research Institute (1970–1997). Salam was awarded more than two dozen honorary doctorates and received more than a dozen major awards, including the Atoms for Peace Award for 1968, the Royal Medal of the Royal Society in 1978, the John Torrence Tate Medal of the American Institute of Physics in 1978, and the Lomonosov Gold Medal of the U.S.S.R. Academy of Sciences in 1983. He died in Oxford, England, in 1996.

Savart, Félix (1791-1841)

French physicist

Félix Savart was an experimental physicist who studied **acoustics**, vibration, and **elasticity**. Savart extended German physicist Ernst Chladni's (1756-1824) studies with vibrating plates of sand, and developed new ways of studying the elasticity of materials. With the French physicist, Jean Baptiste Biot, Savart demonstrated that the magnetic field produced by a current in a wire is inversely proportional to the distance from the wire.

Born in Mézières, France, in 1791, Savart's father, Gérard Savart, was an engineer at the military school in Metz. Savart first studied medicine at the military hospital at Metz, receiving his medical degree from the University of Strasbourg in 1816. Savart's real interests, however, lay in physics. His brother Nicolas, an engineer who had studied at the École Polytechnique, also began to work on the physics of vibration.

Savart moved to Paris where he built his first experimental violin in 1817. Two years later, he delivered a paper on the physics of the violin to the Paris Academy of Sciences. Savart examined how vibrations were transmitted from the strings to the body of the violin. He used Chladni's vibrating sand patterns to study the nodal lines produced by the vibrations of the strings. In an attempt to improve the tone, he built a trapezoid-shaped violin with rectangular holes.

Biot was impressed with Savart's work and found him a position teaching physics in Paris. In 1820, Savart and Biot began measuring the magnetic fields produced by a current. These experiments resulted in the Biot-Savart Law of **Electrodynamics**. Savart also continued his studies on vibrations, building on Chladni's experiments with vibrating plates. He developed methods for studying the vibrations of air, membranes, solids, and various other materials. Savart also studied the vocalizations of animals and humans. He determined the lower **frequency** limits of **hearing**, using a toothed wheel that produced tones of given frequencies. Most of his 27 scientific papers were published in the *Annales de Chimie et de Physique*.

Savart became a member of the Paris Academy in 1827. The following year he was appointed professor of experimental physics at the Collège de France. Savart died in Paris at the age of 49.

Scalar Fields

A field is an object which has a set of numerical values for each point in **space-time**. A scalar field is the simplest example of a field. A scalar field is an object which has a single number as its value at each point in space-time. A familiar example is the **temperature** field in a room. At every point in the room, the temperature field has a single numerical value. The values of the temperature field are higher near sources of **heat**, such as heating vents, and lower near sinks of heat, such as leaky windows.

Scalar fields are the simplest type of field because they have only two numbers associated with them at each point; two numbers, because scalar fields may be complex numbers. If a **quantum theory** of scalar fields is constructed, it is found that scalar fields have no spin. Because of this fact, they cannot describe any of the common particles such as electrons and photons. In order to describe these particles, we must use more complicated fields called spinor and vector fields, respectively, which have four or more components each.

Scalar fields are very important in theoretical **particle physics**, although there is no direct evidence yet for their existence. The **Higgs boson**, which is required for the standard model to be logically consistent, is a scalar particle. Many of the superpartners, particles predicted by supersymmetry and **string theory**, are scalar particles. Detection of scalar particles would therefore give physicists strong clues about the more fundamental theories behind the standard model.

See also Quantum field theory; Bosons; Spin of subatomic particles; Vector bosons; Fermions; Standard model of particle physics; Supersymmetry

SCANNING TUNNELING MICROSCOPE

The latter half of the twentieth century has opened to scientists an entirely new world: the world of atomic and **subatomic particles**. With the inventions of the **electron microscope** and the field ion **microscope**, scientists have been able to observe the microcosm as never before.

Until the early 1980s, however, one mystery that eluded researches was the nature of the surfaces of substances. Since the arrangement of atoms on the surface of a substance differs greatly from that of its bulk, it requires other methods of analysis. Scientists had lacked a mechanism for studying the intricacies of surfaces until 1981, when German physicists **Gerd Binnig** and **Heinrich Rohrer** invented the scanning tunneling microscope (STM).

The word **tunneling** used here describes an effect of quantum **mechanics** theorized upon for years and first verified in the laboratory in 1960. It was known that an **electron** orbits about the **nucleus** of an **atom**, its **motion** random and diffuse. When an atom is placed very close to another, the cloud-orbits of the surface atoms will overlap slightly. The resulting diffusion of electrons is called tunneling.

When Binnig and Rohrer met in 1978 they were both working at IBM research laboratories in different cities, each studying the atomic structures of surfaces. They decided to combine their efforts toward using tunneling to explore these structures. By 1980, they had constructed a prototype STM, and in the spring of 1981, they succeeded in obtaining microscopic images using electron tunneling.

The STM that Binnig and Rohrer had built was actually based upon the field ion microscope invented by Erwin Wilhelm Müller. The field ion microscope uses a tiny sharpened needle placed within a cathode-ray tube; as an electrical field is applied, metal **ions** are emitted from the tip of the nee-

dle, creating an image of the metal's **atomic structure** upon the cathode screen.

In Binnig and Rohrer's device, a similar needle is placed in a **vacuum**, above a specimen to be scanned at a height of less than one nanometer. The sharper the needle, the more precise is the STM reading; the best needle tips are only one or two atoms wide. A very low **voltage** is applied, causing the overlapping clouds of electrons to tunnel from the needle to the specimen. This electron flow is called a tunneling current, and by measuring this flow irregularities in the surface of the specimen can be determined. The scanning tip is swept over the sample, so that the entire surface may be mapped.

Even the very first tests of the STM showed it to be extremely powerful. When scanning crystals of calcium-iridium, Binnig and Rohrer resolved surface hills only one atom high. Maintaining the very small distance between the needle tip and the specimen proved to be difficult, however, since noises as unobtrusive as a footstep would jar the instrument. Binnig and Rohrer used magnets to suspend the microscope over a table equipped with shock absorbers to solve this problem. Other improvements increased the magnification of the STM, and today's tunneling microscopes can resolve features as small as one hundred-billionth of a meter, or about one-tenth the width of a **hydrogen atom**. It has also been discovered that STMs are equally useful in air, water, and cryogenic fluid media.

The incredibly precise three-dimensional images provided by STMs have found varied applications in a number of industries. They are used for quality control in manufacturing digital recording heads as well as in the construction of compact audio disk stampers. In 1991, the STM was used to move and place 35 atoms of xenon in a predetermined pattern; this ability to manipulate **matter** at the level of a single atom may allow scientists to customize molecules, possibly creating ultramicroscopic data storage chips.

Because it is effective in many media and uses a very low voltage, the STM can be used to study the atomic structure of living and biologic matter if that matter readily conducts electrons. Most biologic matter does not, however, and must be coated with a thin layer of a conducting substance. The STM is standard equipment in most atomic research laboratories. It is undeniably the most powerful optical tool yet invented, and for its invention Gerd Binnig and Heinrich Rohrer shared the 1986 Nobel Prize in physics, along with Ernst Ruska, the inventor of the electron microscope.

SCATTERING

Scattering has a very general meaning. Any type of collision can be called a scattering. Another example of scattering is the reflection of **light** on a mirror, in which case the light (more accurately, the photons) is scattered. In the most general sense, any change of state can be considered as a scattering.

Scatterings can be treated using either classical **mechanics** or **quantum mechanics**, depending on the size of the scattering objects. In scattering, the **momentum** of the system is always conserved. This can be derived easily using Newton's third law in classical mechanics. In the microscopic world where we must use quantum mechanics, there are subtle issues because of the uncertainty principle. The momentum of the system long before and long after the scattering is always equal. "Long" means that the period of time is large compared with the scattering time.

The **energy** of the system is not always conserved. If the total energy of the system is conserved, the scattering is called an elastic scattering. If total energy is not conserved, it is called an inelastic scattering. Depending on the actual scattering process, the total energy may increase or decrease. An explosion, for example, is a process where the total energy is conserved. If energy decreases in a collision, it is lost to the environment. For instance, when cars collide some of the energy is lost to deforming the body of the car. Likewise, when energy increases in a collision, it comes from the environment. In the case of explosions, it comes from the stored chemical energy of the explosive material.

An important quantity in scattering is the scattering cross section. It characterizes how easily scattering can occur. The larger the cross section is, the stronger scattering we have. The scattering cross section is generally very difficult to calculate because of the complicated interactions during the scattering. In classical mechanics, depending on the number of objects involved in the scattering (few-body system or many-body system), we can use different approximations. If we have two objects, for instance, we can use classical mechanics to calculate the cross section. For many objects, we can use numerical methods. In the case of quantum mechanics, things become much more complicated because of the interaction of quantum **waves**. One famous approximation tool is Fermi's golden rule.

The molecules in a gas or liquid are constantly colliding, or scattering, with one another. Their scattering is dependent, however, on many variables, including **temperature**, **pressure**, and volume. Some scattering parameters, such as cross section, are directly relates to the thermodynamic properties of the substance. For instance, molecules with a larger cross section should reach thermal **equilibrium** faster than molecules with a small cross section at the same temperature, pressure and volume. In high energy physics, or **particle physics**, scientists accelerate particles and collide them at very high speeds. By calculating the cross section of the scatterings and comparing them to the values measured from experiment, scientists can discover many details about the particles they are studying. Scattering is a basic tool in high energy physics. Almost all **subatomic particles** have been discovered using scattering.

See also Feynman diagrams; Thermodynamics

SCHRÖDINGER, ERWIN (1887-1961)
Austrian physicist

Erwin Schrödinger shared the 1933 Nobel Prize in physics with English physicist **Paul Dirac** in recognition of his devel-

Erwing Schrödinger.

opment of a wave equation describing the behavior of an **electron** in an **atom**. His theory was a consequence of French theoretical physicist Louis Victor de Broglie's hypothesis that particles of **matter** might have properties that can be described by using wave functions. Schrödinger's wave equation provided a **sound** theoretical basis for the existence of electron **orbitals** (**energy** levels), which had been postulated on empirical grounds by Danish physicist **Niels Bohr** in 1913.

Schrödinger was born in Vienna, Austria, on August 12, 1887. His father, Rudolf Schrödinger, enjoyed a wide range of interests, including painting and botany, and owned a successful oil cloth factory. Schrödinger's mother was the daughter of Alexander Bauer, a professor at the Technische Hochschule. For the first 11 years of his life, Schrödinger was taught at home. Though a tutor came on a regular basis, Schrödinger's most important instructor was his father, whom he described as a "friend, teacher, and tireless partner in conversation," as Armin Hermann quoted in *Dictionary of Scientific Biography*. From his father he also developed a wide range of academic interests, including not only mathematics and science but also grammar and poetry. In 1898, he entered the Akademische Gymnasium in Vienna to complete his pre-college studies.

Hasenöhrl Inspires Early Interest in physics

Having graduated from the Gymnasium in 1906, Schrödinger entered the University of Vienna. By all accounts, the most powerful influence on him there was Friedrich Hasenöhrl, a brilliant young physicist who was killed in World War I a decade later. Schrödinger was an avid student of Hasenöhrl's for the full five years he was enrolled at Vienna. He held his teacher in such high esteem that he was later to remark at the 1933 Nobel Prize ceremonies that, if Hasenöhrl had not been killed in the war, it would have been Hasenöhrl, not Schrödinger, being honored in Stockholm.

Schrödinger was awarded his Ph.D. in physics in 1910 and was immediately offered a position at the University's Second Physics Institute, where he carried out research on a number of problems involving, among other topics, **magnetism** and **dielectrics**. He held this post until the outbreak of World War I, at which time he became an artillery officer assigned to the Italian front. As the War drew to a close, Schrödinger looked forward to an appointment as professor of theoretical physics at the University of Czernowitz, located in modern-day Ukraine. However, those plans were foiled with the disintegration of the Austro-Hungarian Empire, and Schrödinger was forced to return to the Second Physics Institute.

During his second tenure at the Institute, on April 6, 1920, Schrödinger married Annemarie Bertel, whom he had met prior to the War. Not long after his marriage, Schrödinger accepted an appointment as assistant to Max Wien in Jena, but remained there only four months. He then moved on to the Technische Hochschule in Stuttgart. Once again, he stayed only briefly—a single semester—before resigning his post and going on to the University of Breslau. He received yet another opportunity to move after being at the University for only a short time: he was offered the chair in theoretical physics at the University of Zürich in late 1921.

Work at Zürich Results in Wave Equation

The six years that Schrödinger spend at Zürich were probably the most productive of his scientific career. At first, his work dealt with fairly traditional topics; one paper of particular practical interest reported his studies on the relationship between red-green and blue-yellow color-blindness. Schrödinger's first interest in the problem of wave **mechanics** did not arise until 1925. A year earlier, de Broglie had announced his hypothesis of the existence of matter **waves**, a concept that few physicists were ready to accept. Schrödinger read about de Broglie's hypothesis in a footnote to a paper by American physicist **Albert Einstein**, one of the few scientists who did believe in de Broglie's ideas.

Schrödinger began to consider the possibility of expressing the movement of an electron in an atom in terms of a wave. He adopted the premise that an electron can travel around the **nucleus** only in a standing wave (that is, in a pattern described by a whole number of wavelengths). He looked for a mathematical equation that would describe the position of such "permitted" orbits. By January of 1926, he was ready to publish the first of four papers describing the results of this

research. He had found a second order partial differential equation that met the conditions of his initial assumptions. The equation specified certain orbitals (energy levels) outside the nucleus where an electron wave with a whole number of wavelengths could be found. These orbitals corresponded precisely to the orbitals that Bohr had proposed on purely empirical grounds 13 years earlier. The wave equation provided a sound theoretical basis for an atomic model that had originally been derived purely on the basis of experimental observations. In addition, the wave equation allowed the theoretical calculation of energy changes that occur when an electron moves from one permitted orbital to a higher or lower one. These energy changes conformed to those actually observed in spectroscopic measurements. The equation also explained why electrons cannot exist in regions between Bohr orbitals since only non-whole number wavelengths (and, therefore, non-permitted waves) can exist there.

After producing unsatisfactory results using relativistic corrections in his computations, Schrödinger decided to work with non-relativistic electron waves in his derivations. The results he obtained in this way agreed with experimental observations and he announced them in his early 1926 papers. The equation he published in these papers became known as "the Schrödinger wave equation" or simply "the wave equation." The wave equation was the second theoretical mechanism proposed for describing electrons in an atom, the first being German physicist Werner Karl Heisenberg's matrix mechanics. For most physicists, Schrödinger's approach was preferable since it lent itself to a physical, rather than strictly mathematical, interpretation. As it turned out, Schrödinger was soon able to show that wave mechanics and matrix mechanics are mathematically identical.

Rise of Nazis Forces Schrödinger to Oxford and Dublin

In 1927, Schrödinger was presented with a difficult career choice. He was offered the prestigious chair of theoretical physics at the University of Berlin left open by German physicist **Max Planck**'s retirement. The position was arguably the most desirable in all of theoretical physics, at least in the German-speaking world; Berlin was the center of the newest and most exciting research in the field. Though Schrödinger disliked the hurried environment of a large city, preferring the peacefulness of his native Austrian Alps, he did accept the position.

Hermann quoted Schrödinger as calling the next six years a "very beautiful teaching and learning period." That period came to an ugly conclusion, however, with the rise of National Socialism in Germany. Having witnessed the dismissal of outstanding colleagues by the new regime, Schrödinger decided to leave Germany and accept an appointment at Magdalene College, Oxford, in England. In the same week he took up his new post he was notified that he had been awarded the 1933 Nobel Prize for physics with Dirac.

Schrödinger's stay at Oxford lasted only three years; then, he decided to take an opportunity to return to his native Austria and accept a position at the University of Graz.

Unfortunately, he was dismissed from the University shortly after German leader Adolf Hitler's invasion of Austria in 1938, but Eamon de Valera, the Prime Minister of Eire and a mathematician, was able to have the University of Dublin establish a new Institute for Advanced Studies and secure an appointment for Schrödinger there.

In September, 1939, Schrödinger left Austria with few belongings and no money and immigrated to Ireland. He remained in Dublin for the next 17 years, during which time he turned to philosophical questions such as the theoretical foundations of physics and the relationship between the physical and biological sciences. During this period, he wrote one of the most influential books in twentieth-century science, *What Is Life?* In this book, Schrödinger argued that the fundamental nature of living organisms could probably be studied and understood in terms of physical principles, particularly those of **quantum mechanics**. The book was later to be read by and become a powerful influence on the thought of the founders of modern molecular biology.

After World War II, Austria attempted to lure Schrödinger home. As long as the nation was under Soviet occupation, however, he resisted offers to return. Finally, in 1956, he accepted a special chair position at the University of Vienna and returned to the city of his birth. He became ill about a year after he settled in Vienna, however, and never fully recovered his health. He died on January 4, 1961, in the Alpine town of Alpbach, Austria, where he is buried.

Schrödinger received a number of honors and awards during his lifetime, including election into the Royal Society, the Prussian Academy of Sciences, the Austrian Academy of Sciences, and the Pontifical Academy of Sciences. He also retained his love for the arts throughout his life, becoming proficient in four modern languages in addition to Greek and Latin. He published a book of poetry and became skilled as a sculptor.

SCHRÖDINGER'S CAT

In 1926, physicist **Erwin Schrödinger** proposed one of the most famous thought experiments of twentieth-century-science. Schrödinger intended to illustrate the difficulties involved in extending the insights of quantum mechanics from the probabilistic world of the vanishingly small to the macroscopic world of everyday life, where classical physics applies. Schrödinger's thought experiment takes its name from the central role played by a hypothetical cat, distinguished by the unusual property of being dead and alive at the same time.

Schrödinger's experiment was simple. Put a cat and a radioactive **atom** with a half life of one hour into a box with a vial of hydrocyanic acid connected to a Geiger counter. In the course of an hour, there is a 50:50 chance the atom will decay. If it does, the detector will register the emission of an **alpha particle** and trigger a hammer that will smash the vial of poison, killing the cat. Seal the box, and leave. At the end of an hour, what can be said about the interior of the box without looking inside?

In the cause-and-effect world of classical physics, the cat is *already* either alive or dead *before* the box is opened. Not so in **quantum mechanics**. The bizarre implications of **quantum theory** seem to require that the cat be *both* alive and dead *until* the box is opened. Only then does one of these possibilities become actualized; in the language of quantum mechanics, only then does the **wave function** describing the cat collapse. Even stranger, it is the seemingly insignificant act of looking in the box that collapses the wave function. The fate of the cat depends on whether or not an observer chooses to look at it. Or does it?

In the quantum world, mutually exclusive possible outcomes, such as whether a radioactive atom has or has not decayed, exist simultaneously. In 1926, Schrödinger discovered the equation, the Schrödinger wave equation, that describes this superposition of possible states (called eigenstates), an achievement for which he shared the 1933 Nobel Prize in physics with Paul M. Dirac (1902-1984). Although the formulation of matrix mechanics by **Werner Heisenberg** has the honor of being the first successful statement of quantum theory, Schrödinger's discovery of wave mechanics was recognized by physicists as a more familiar approach—because it was based on a quantized standard classical equation of motion—and in any case the two were soon found to be mathematically equivalent.

Wave mechanics originated with **Max Planck** and his quantization of the **light**. Einstein then used the concept of wave mechanics to solve the mystery of the **photoelectric effect**. It was Einstein who postulated the wave-particle duality of light. **Louis de Broglie** took the concept a little further by proposing that all **subatomic particles**, including electrons, possess wave-particle duality. De Broglie won the Nobel Prize in Physics in 1929 for his discovery. Neils Bohr (1855-1962) then expanded the idea of **electron waves** to explain the structure of the Rutherford atom and the spectroscopic emission and absorption of light by atoms. Schrödinger's wave equation described this dualistic behavior for all particles in terms of probability. This discovery, together with the Copenhagen interpretation of quantum mechanics developed by Neils Bohr and Heisenberg, shattered the edifice of classical physics. Not only that, the very act of looking for answers changed the nature of the original questions asked, and any answers found, being probabilistic, could not be used to predict with certainty future answers to similar questions. According to Heisenberg's Uncertainty Principle, pairs of related quantities (called observables) are linked. Measuring one observable with infinite accuracy means the other is indeterminate. In practice, you can only measure position with the limited accuracy of your measuring device, which then sets the limit on the accuracy with which you may measure **momentum**. Thus, the two observables of position and momentum are 'fuzzy.' In such a system, it is meaningless to speak of any reality apart from observed reality. Philosophers and physicists have been struggling with the implications ever since.

The Copenhagen interpretation gives rise to what is known as the measurement problem, and it is this subtle but as-yet-unresolved aspect of quantum mechanics that

Schrödinger's cat paradox was designed to illustrate. The fate of the cat, a macroscopic object, is tied to the fate of a radioactive atom, a microscopic object. The fate of the atom is described by a superposition of eigenstates: a wave equation describing the decayed atom and another describing the non-decayed atom. Since it is only when the scientists look inside the box that the wave function of the atom collapses to its decayed or non-decayed state, the wave function of the cat must behave similarly. Reality in the box, then, has come into being through the act of observation, of measurement. This is confusing enough. But what if the box is opened inside a sealed room, and outside the room other scientists are waiting to learn the outcome? When the box is opened, the wave function inside collapses, but the superposition of eigenstates, rather than disappearing, shifts outward. Now it is the *room* that is in a superposition of eigenstates. Until the scientists outside the room open the door, the macro-reality of the experiment has not yet come into being. Where does one draw the line in this infinitely regressing series of realities springing into existence from the froth of collapsing wave functions?

One possible solution to the measurement problem is the so-called many worlds interpretation of quantum mechanics. This states that every possible state of the cat (or any particle) is manifested in parallel universes that come into being the instant the cat (or particle) is observed. If the cat is alive in one **universe**, it is dead in another. Though mathematically consistent with quantum theory, this solution may cause more problems than it solves, not the least of which is the impossibility of experimental verification.

In 1996, physicists at the National Institute of Standards and Technology in Boulder, Colorado, separated the superposition of two eigenstates of a beryllium atom by 80 nanometers, a distance that falls in the mesoscopic scale of sizes between the microscopic and macroscopic, where the properties of quantum mechanics and classical physics are seen to commingle. For a brief time, both versions of the atom were observed in two separate places simultaneously; the "cat" of the beryllium atom was simultaneously "alive" and "dead." While this and similar experiments may not solve the paradox of Schrödinger's Cat, they hold out the promise of major advances in fields such as quantum encryption, quantum computing, and **nanotechnology**.

SCHWARTZ, MELVIN (1932-)

American physicist

Melvin Schwartz's research and experimentation in the weak **force** of the four fundamental forces of nature resulted in the proof of the existence of the **neutrino**, a particle of zero-rest **mass**, and the Nobel Prize-winning discovery and definition of the two existing types of neutrinos, the **electron** neutrino and the **muon** neutrino.

Schwartz was born in New York City on November 2, 1932, to Harry and Hannah Shulman Schwartz. Desperately poor as a result of the Great Depression, his parents "worked extraordinarily hard," Schwartz was later to say, as quoted in

Nobel Prize Winners Supplement, to provide some level of "economic stability" in their lives. He entered the world-famous Bronx High School of Science in 1944, at the age of 12, and graduated five years later. By that time, he had made up his mind to become a theoretical physicist. That decision having been made, his choice for a college education was easy. At the time New York City's own Columbia University had, as Schwartz later characterized it, a physics department that was "unmatched by any in the world."

Schwartz Attacks the Problem of the Weak Force

Schwartz earned his bachelor's degree in mathematics and physics in 1953. That year, Schwartz was married to Marilyn Fenster, with whom he later had two daughters, Diane and Betty Lynne, and a son, David. He began his doctoral studies under the direction of **Jack Steinberger**, whom Schwartz has called "the best experimental physicist I have ever been associated with, and the best teacher," as quoted by Bertram Schwarzschild in *Physics Today*. It was from Steinberger that Schwartz gained his special interest in **particle physics**, an interest that was to dominate much of his research over the next four decades.

Schwartz was awarded his Ph.D. in physics in 1959 and then joined the faculty at Columbia. Within a year, an event was to take place that would dramatically alter Schwartz's future. During an afternoon coffee hour in November 1959, at Columbia's Pupin Laboratory, a group of physicists discussed the problems of studying the weak force, one of the four fundamental forces of nature (the others being gravitation, **electromagnetism** and the strong force). **Tsung-Dao Lee**, a theoretical physicist, challenged his colleagues to find a way to obtain additional empirical evidence on the weak force. The challenge at first seemed an enormous one since at atomic dimensions, the weak force is much weaker—and therefore much harder to observe—than are the electromagnetic and strong forces. According to Schwarzschild, Schwartz later described his feeling about Lee's challenge as one of "hopelessness. There seemed to be no decent way," he said, "of exploring the terribly small cross-sections characteristic of weak interactions." This feeling lasted less than 24 hours, as that evening Schwartz suddenly had the answer. "It was incredibly simple," he decided. "All you had to do was use neutrinos."

Neutrinos turned out to be the perfect tool with which to study the weak force. Because these tiny particles are uncharged and have very small mass, they are essentially unaffected by electromagnetic or strong forces. When a beam of neutrinos passes through **matter**, the only interactions it undergoes are those involving the weak force.

Schwartz Confirms the Existence of Two Kinds of Neutrinos

To work out the details of the neutrino/weak force experiment, Schwartz met with his former doctoral advisor, Steinberger, and Columbia colleague Leon Max Lederman. The three devised a method for generating an intense beam of neutrinos using the Brookhaven National Laboratory's new 30 billion-electron-volt alternating gradient synchrotron (AGS). **Proton** beams from the AGS would be directed at a target of beryllium metal. The collision between beam and target would tear apart beryllium atoms and release an avalanche of **subatomic particles**, neutrinos among them. The neutrinos thus produced would then be directed through a block of steel where at least some would interact with atoms by means of the weak force.

One issue involved in the experiment was the nature of the neutrinos to be used, as relatively little was known about these particles. When the neutrino was discovered, physicists assumed that it existed in only one form, the form now known as the electron neutrino. For various theoretical reasons, however, Columbia theoretical physicist Gerald Feinberg posed the possibility in 1958 that a second neutrino, associated with mu mesons (muons), which are particles of a different **weight**, might also exist. The Schwartz-Steinberger-Lederman experiment was designed to determine also the validity of Feinberg's hypothesis.

By September 1961, the experiment was under way. Eight months later, an estimated 10^{14} neutrinos had been produced, of which 51 interactions with matter were observed. In every one of these cases, the interaction was such that it confirmed the existence of a muon neutrino distinct from the electron neutrino. Feinberg's hypothesis had been confirmed. The 1988 Nobel Prize awarded to Schwartz, Steinberger, and Lederman recognized not only the discovery of the muon neutrino, but also the development of a technique that, the Nobel committee said, promised to provide "entirely new opportunities for research into the innermost structure and **dynamics** of matter."

Schwartz resigned his post at Columbia in 1966 in order to take a position as professor of physics at Stanford University, where the new 2-mi (3.2 km) long linear accelerator was available for his research. He remained at Stanford until 1979 when he decided to establish his own software development business, Digital Pathways, Inc. After dividing his time between Stanford and Digital for four years, he resigned the former post and became a full-time business man. In 1991 Schwartz returned to the academic world by accepting the post of associate director of high **energy** and **nuclear physics** at the Brookhaven National Laboratory. Schwartz is a member of the National Academy of Sciences and is a fellow of the American Physical Society, which awarded him the Hughes Prize in 1964.

SCHWARZCHILD RADIUS

In 1795, French mathematician Pierre Simon de Laplace, armed with a knowledge of Newtonian gravitational theory, asked himself whether a star could be so large that even **light** would be gravitationally bound. He knew that the escape **velocity** (the minimum velocity with which an upwardly moving object would never fall back) is given by $\sqrt{2GM/r}$, where M and r are the **mass** and radius of the planet or star, and G is

the Newtonian gravitational constant. This expression is independent of the projectile's own mass. Thus, by setting the above expression equal to the **speed of light**, c, he arrived at an expression for the radius of a body from which nothing, not even light, could escape: $r=2GM/c^2$. This calculation assumes, of course, that the star's entire mass lies within the radius r.

But Newtonian theory does not describe **gravity** correctly, although it suffices quite well in describing the **gravitational force** between, for example, people and **planets**. In order to describe very strong gravitational fields correctly, Einstein's general theory of relativity must be employed. In 1916, shortly before he was killed in World War I, German astronomer Karl Schwarzchild found the solution for the **space-time** geometry around a spherical mass. Remarkably, Laplace's naive calculation was borne out, and the radius at which a body becomes a black hole now bears Schwarzchild's name. The Schwarzchild radius for bodies of even astronomical mass is quite small; for example Earth's entire mass would have to be compressed into a sphere 3.7 mi (6 km) in diameter in order for light to be unable to escape its pull.

SCHWINGER, JULIAN (1918-1994)
American physicist

American physicist Julian Schwinger worked primarily to develop a **quantum theory** of **radiation**. As a theorist, he produced mathematical frameworks that showed the relationships between charged particles and electromagnetic fields, and his equations eventually united relativity and quantum theory. In recognition of this work, Schwinger received the National Medal of Science in 1964 and the Nobel Prize for physics in 1965, which he shared with American theoretical physicist Richard P. Feynman and Japanese physicist **Sin-Itiro Tomonaga**.

Julian Seymour Schwinger was born on February 12, 1918, in the Jewish Harlem section of New York City. His father was Benjamin Schwinger, a garment manufacturer, and his mother was Belle (Rosenfeld) Schwinger. Julian was the younger of two brothers. As a child, Schwinger had an insatiable appetite for science. He became interested by reading popular scientific magazines. When he entered high school, he had already decided to study physics. He had read all that the *Encyclopedia Britannica* offered in physics, and he scoured the New York Public libraries for all the books he could find on mathematics and physics, starting at the uptown branches and methodically working his way down to the main branch. He also combed used book stores for texts in mathematics and physics.

Schwinger was known to be very shy, but he became a member of the world of adults at a very young age. He skipped three grades in high school and graduated at the age of 14. While he was still in high school, Schwinger studied scientific papers about quantum mechanics by the English physicist **Paul Dirac** as they appeared in the *Proceedings of the Royal Society of London*. Schwinger would later base his mathematical work in **quantum electrodynamics** on Dirac's theories.

Julian Schwinger.

Schwinger started his undergraduate studies at the College of the City of New York, where he began writing papers on theoretical physics. One of these papers was on **quantum mechanics** and was published in the *Physical Review*. It so impressed Isador Issac Rabi, a physicist at Columbia University, that he helped Schwinger to get a scholarship at Columbia. While at Columbia, Schwinger was interested almost solely in science to the neglect of his other classes, particularly English composition. Were it not for Rabi's intervention, Schwinger would have been expelled from Columbia. He excelled in physics, however, and often served as a substitute lecturer in quantum mechanics. Schwinger graduated from Columbia in 1936 at the age of 17. Three years later at Columbia he earned his Ph.D. degree.

Works on Atomic Bomb and Radar

After Columbia, Schwinger continued his research at the University of California under J. Robert Oppenheimer. In 1941 he joined Purdue University as an instructor and left there in 1943 as an assistant professor. In 1943 he went to the Metallurgical Laboratory at the University of Chicago to assist in developing the **atomic bomb**. In the same year he joined the radiation laboratory at the Massachusetts Institute of Technology, where he worked on microwave problems and helped to improve **radar** systems.

During these early years of moving from one position to another, his reputation for maintaining nocturnal working habits and sleeping until the afternoons preceded him. But it was rarely an issue. In 1945 he became an associate professor at Harvard University, and in 1947 he was made one of the youngest full professors in the university's history. In addition to his research in theoretical physics, Schwinger also distinguished himself in his presentations in scientific journals and in lecture halls. His published papers were considered exemplars for the scientific community.

In the lecture hall Schwinger's presentations often drew applause, a rare response from a scientific audience. He gave his lectures extemporaneously, with little preparation, and without the use of notes. Jeremy Bernstein, in an article for the *American Scholar*, said of Schwinger's lecture, "It's like poetry." Schwinger was known for speaking with long measured periods delivered without pause. He took equal care in his accompanying board work; he rarely made mistakes with formulas. In addition, James Gleick in his biography of Richard Feynman, *Genius,* reports that Schwinger was ambidextrous and able to list on the board the steps for two equations simultaneously.

Since his high school days Julian Schwinger had shown a great respect for and faith in the work of Paul Dirac. Dirac's work was a mathematical account of the interaction between electric and magnetic forces inside the **atom**. It was a useful description of these forces, but physicists who studied radiation through observation began to find discrepancies between calculations based on the theory and their measurements in the laboratory. In particular, the theory required the **electron** to behave as though it were a particle with infinite **mass**. The unreal nature of this notion led William Laurence in a 1948 *New York Times* article to call it the "cosmic ghost."

Adjusts Dirac's "Cosmic Ghost" Theory

Because of this discrepancy, quantum physicists were anxious to discard Dirac's theory. Schwinger, however, believed that with some corrections this theory could still be an accurate account of observable phenomena. At a 1948 meeting of the American Physical Society Schwinger showed his colleagues how Dirac's theory could be corrected. Schwinger had found the terms in Dirac's work that produced the ghost of infinite mass and showed how to account for them in a way that both preserved Dirac's theory and kept it congruent with reality. With this work, Schwinger routed the cosmic ghost and eventually earned the Nobel Prize in 1965.

Almost all of Schwinger's writings are abstract and directed to a highly specialized audience of theoretical physicists. One of Schwinger's more recent books, however, *Einstein's Legacy: The Unity of Space and Time,* attempts to reach a general audience. In it, Schwinger explains in lay terms how **Albert Einstein**'s theories of special and general relativity are unified and what consequences they have for experiments.

Julian Schwinger and Clarice Carrol were married in 1947 and had no children. Outside of the lecture hall Schwinger was known for his reserved manner but firm convictions. He was one of the few scientists of his stature open to the possibility of achieving **cold fusion**. He died of cancer in Los Angeles on July 16, 1994.

SCIENTIFIC METHOD

The scientific method is a group of interdependent concepts involving observation and experimentation. In an ideal world, researchers would use the scientific method to conduct experiments under completely controlled conditions, drawing observations based on logic and reason. Perfectly controlled conditions are almost impossible to achieve. However, closely controlled conditions are important because no experiment is valid unless it can be duplicated and verified in other laboratories. This safeguard is built into scientific and medical research to help scientists guard against wasting time and resources on faulty data.

Researchers try to expand on others' findings so new discoveries can be made for everyone's benefit. Therefore, each experiment must serve as a sound building block on which other research can be based. By agreeing on the scientific method as their common framework, researchers give their work international credibility. Scientists who deviate from the scientific method risk being charged with scientific misconduct, a very serious accusation in the research community. Research begins with a hypothesis, or theory, and an experiment is designed to see if the hypothesis is valid. After careful control of experimental methods, analysis, and interpretation, the researcher tries to draw general conclusions that will apply in other situations.

Some of the earliest records of sound research principles began with **Aristotle** in Greece during the fourth century B.C. Aristotle developed a method of observing, classifying, and drawing general conclusions, the foundation for the scientific method. Subsequent scholars admired Aristotle's methods but were unable to expand upon his ideas until the Middle Ages. Medieval scholars copied Aristotle's manuscripts, but it was a time of scholastic conformity, and independent thought was not valued. Scholars apparently lacked the understanding to draw conclusions based on what they observed. By the end of the thirteenth century, an educated person knew no more about the physical **universe** than the ancient Greeks. Parisian scholars broke this impasse in the fourteenth century when they questioned Aristotle's law of **motion**.

The **scientific revolution** of the sixteenth century advanced the scientific method, because scholars were now free to explain physical phenomena on the basis of cause and effect. Previously, educated people had explained most observations as part of God's plan. Historians of science see the Industrial Revolution of the nineteenth century as a further application of the scientific method because for the first time, science was used as a means of designing and improving products for the general public.

The interdependent concepts of the scientific method rest on two primary thought processes: inductive reasoning, where one starts with a particular observation and notices a general pattern; and deductive reasoning, where one goes from the general to the particular.

The experimenter begins with a hypothesis that may spring from past experience or a new idea. It is important to state the hypothesis very precisely, to keep the experiment from being sidetracked by other observations. The experiment's purpose must also be clearly defined. At this point, the researcher is designing the experiment to answer an underlying question and to test the hypothesis. The experimenter will outline the methods and materials he expects to use. After these basics have been addressed, side issues may be explored. The written hypothesis guides the experimenter during the experiment, the interpretation, and any generalizations that might result.

The researcher must choose appropriate methods for obtaining observations and measurements. Does the research design allow for all possible variables? Are the observations and measurements truly representative? Is there a lack of randomness in the design that could lead to accusations of research bias? Before the experiment starts, the researcher must take another look at the methods that will be used to analyze data. How will the record keeping be performed? Experiments that appear to be routine might lend themselves to shortcuts or to reliance on memory rather than on written notes.

After the experiment is carried out and the researcher begins to analyze data, the most difficult step—interpretation—begins. Correct interpretation accomplishes much more than deciding whether the findings confirm or refute the hypothesis. The experimenter has to be able to read experimental data correctly, then place the data in some sort of order. There is a temptation to discard findings that don't fit, but the correct approach is to re-examine the experiment to seek reasons for data that are out of line.

As part of interpreting the data, the researcher must be on guard for four types of variation. Usually the researcher will lack precise control over the variables. There may be variables that are unknown to the experimenter. Perhaps the samples used in the experiment are not representative. These first three forms of variation can be used to determine whether a difference between two samples of data is significant or not. There is also a kind of variation between types of data that can only be seen as the experiment is carried out. In the interpretation, the researcher must also guard against the problem of comparing data when it is not legitimate to do so. For example, data might be organized by the day specimens arrive in the laboratory, but that might actually be irrelevant to the purpose of the experiment. The experimenter must also watch for coincidental relationships between variables and avoid false conclusions of cause and effect. The next step is to compare the actual data with what was expected.

The researcher will present the results in as close to their original form as possible. Summarizing graphs can be used if they help the reader and do not distort the data. The researcher will also evaluate the validity of the hypothesis. The conclusion should help to place the experiment into the body of previous work and should identify areas where further experimentation would bring about greater understanding.

SCIENTIFIC REVOLUTION

The scientific revolution was a fundamental change in the direction of Western thought and scientific practice that may reasonably be said to have begun with the reassertion of heliocentric model of **universe** advocated by Nicolaus Copernicus in 1543 and that culminated with English physicist Sir Isaac Newton's (1642-1727) publication of his profoundly influential 1687 work, *Philosophiae Naturalis Principia Mathematica (Mathematical Principles of Natural Philosophy)*. The scientific revolution was not merely a set of discoveries or technical advances, but brought about a fundamental change in the way the laws of nature were investigated. During this time scientists increasingly supplanted pure deduction with experimentation.

Prior to about 1600, science and philosophy in the West for the most part had made only minor advances since the fall of the Roman Empire in the west (ca. 400). Despite significant technical advances in some fields, during this time no systematic inquiries were made into the workings of nature. Many of the philosophical inquires known to ancient Greek and Roman scientists were suppressed by theological doctrine. When translations of ancient Latin texts once again reawakened interest in the Aristotelian scientific/philosophical systems of the classical world. these were based almost purely upon deductive reasoning. Not until the early 1600s with the research of **Galileo** were scientific theories based upon actual experimentation rather than philosophic ideals.

Astronomers, both ancient and modern, have used the heavens as a laboratory to understand the basic principles of nature. According to the theories of **Aristotle** (384-322 B.C.) the earth occupied the center of the universe, while the heavens were the realm of perfection, in which the **sun** and **planets** orbited the earth in perfectly circular orbits at an unvarying rate of speed. To explain the retrograde **motion** of the planets a complicated system of secondary circular orbits called **epicycles** were superimposed upon the main orbits. This system was perfected by the astronomer **Ptolemy** (ca. A.D. 140). Despite its complexity, it yielded reasonable results in predicting the positions of the planets.

Aristotle further postulated that all physical **matter** was a mixture of four basic elements (air, earth, fire, and water) each of which had certain intrinsic physical properties. For instance, bodies fell not because of **gravity**, but because the earth had an inherent tendency to seek out the lowest possible level. Thus, heavier objects, containing more of the basic element earth, would fall at a faster rate than **light** ones. Other than this 'natural' motion, objects would remain always at rest, unless a **force** was constantly applied to them. Heavenly bodies constantly moved in circular orbits at a uniform rate of speed because this was an inherent property of the perfect material of which they were supposedly made, aptly named the quintessence.

The Aristotelian system based solely upon reasoning rather than experimentation, was deeply flawed as a scientific methodology. Regardless, it advanced the idea that the workings of the universe could be understood using only a small number of basic principles, an underlying principle of scientific inquiry today.

The Polish astronomer Nicolaus Copernicus (1473-1543) realized that planetary motion could be more simply explained by a system in which the planets orbited the sun. However, laboring under the influence of Aristotle, he insisted on perfectly circular orbits and uniform rates of speed as an inherent property of the planets. While Copernicus's heliocentric theory yielded the correct **frame of reference** for further scientific inquiry, it was no more accurate than the Ptolemaic system in predicting the motions of the planets. Perhaps fearing persecution by the religious and scientific establishment of the time, he delayed publishing his theory in his work, *On the Revolution of Planets*, until just prior to his death.

Danish astronomer **Tycho Brahe** (1546-1601), rejecting the both the Copernican and Ptolemaic systems, devised a geocentric model in which planets orbited the sun, which in turn orbited the earth, and amassed a huge data set of planetary positions attempting to prove this theory. After Brahe's death, German astronomer and mathematician **Johannes Kepler** (1571-1630), used Brahe's data to formulate **Kepler's laws of planetary motion**. Kepler mathematically proved that planets orbit the Sun in elliptical orbits and that the square of their orbital periods is proportional to the cube of the semi-major axis of their orbit. In addition, Kepler was able to predict that the orbital speed of the planets was inversely proportional to their distance from the sun-planet **center of mass**. Kepler's work effectively disproved the Aristotelian and Ptolemaic model of planetary motion.

The renowned Italian astronomer and physicist Galileo Galilei (1564-1642) was the first person to use the **telescope** to observe the heavens (1609). His discoveries included the mountains of the **moon**, moons of Jupiter, and the rings of Saturn. Perhaps most importantly he discovered that Venus displays crescent phases similar to the moon's. This last discovery conclusively disproved the validity of geocentric systems.

Galileo was prohibited by Church from advancing heliocentric theory for complex reasons, and forbidden to do more astronomical research. Under house arrest, in his old age he resumed studies of motion. His work included studies on the addition of velocities, the **acceleration** of falling bodies, and most importantly, the principle of motion. Galileo advanced the theory that if no forces (i.e., **friction**) act upon a body set in motion, the body will continue to move forever, effectively overturning the Aristotelian idea that motion of object was an inherent property of it. Most important of Galileo's contributions to science was the idea of experimentation, that the laws of nature can be discovered only through experiments designed to prove or disprove the underlying principle. This idea has become so ingrained in Western civilization that the radicalism of Galileo's approach is often overlooked by historians of science.

Newton's development of the laws of motion and gravitation marked the beginning of what is know known as classical or **Newtonian physics**. In addition to developing calculus, Newton made tremendous advances the understanding of light and **optics**. Without question, Newton's work dominated the Western intellectual landscape for more than two centuries. In particular, Newton made three fundamental realizations regarding planetary motion from which Newtonian physics evolved. The first of these was the idea of a force at a distance, that is, gravity could act upon two bodies even if there was no physical contact between them. Also, the force between the two bodies was equal and opposite. This was a profound and innovative departure from the Aristotelian idea. Newton's second assertion was that the acceleration of a body was equal to the force applied to it divided by its **mass**. Moreover, **Newton's law of universal gravitation** explained and validated the elliptical orbits advanced by Kepler. According to Newton's law of gravity, the force of gravity had to be inversely proportional to the square of the distance between the sun and the planets. To prove these theories, Newton, working independently of German mathematician Gottfried Wilhelm von Leibniz (1646-1716), developed the calculus. The techniques of calculus advanced by newton and Lebniz were immediately applied to a wide variety of problems in physics and **astronomy**.

The advancement of experimentalism during the scientific revolution and the use of calculus resulted in a new scientific methodology that lead to the Enlightenment and to an explosion of scientific inquiry into the natural world.

See also History of physics

Scintillation detectors

Scintillation detectors are used in experiments to determine when ionizing particles have hit a target. When the detector receives that type of **radiation**, it absorbs it and emits **light** in a process called **luminescence**. The intensity of the luminescence from a scintillation detector will be proportional to the **energy** and intensity of the incoming radiation. Scintillation detectors may be used to detect any sort of **ionizing radiation** (radiation that causes atoms to ionize) and not just electromagnetic radiation.

A good scintillation detector must allow the light its own atoms emit to pass through easily. Most often, scintillation detectors are used in combination with photomultiplier tubes, which convert a visible light signal into an electronic pulse. Plastic scintillators are often used in conjunction with clear light guides which funnel the light into photomultiplier tubes. In all cases, the properties of the scintillation detector and the properties of the photomultiplier tube should be well matched to maximize efficiency in detection. Sometimes the entire combination of phosphorescent substance, photomultiplier tube, and associated circuits is referred to as the scintillation detector.

Scintillation detectors were originally developed in the 1950s to solve the problem of low efficiency rates in the gas-

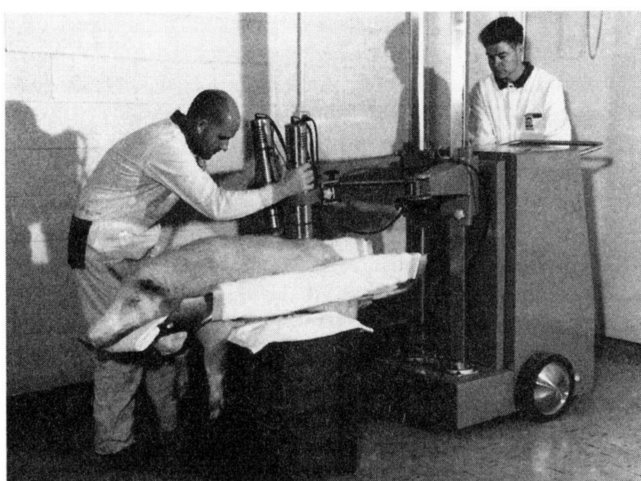

Scintillation detector. *(Corbis. Reproduced by permission.)*

filled detectors in use at the time. They have the added benefit that they are very fast and have very little deadtime, which is the time it takes for a detector to recover from a single measurement and be ready for the next. However, since most of the materials used in scintillation detectors are **semiconductors**, it was not until the 1960s that the industry was developed many of the designs currently in use. Different material choices result in different efficiencies and resolution. In many cases, the efficiency—how many of the photons passing through get detected—has to be balanced with the resolution; in other words, how clearly one can tell the photon's timing and energy. Different factors will be important in different experiments.

Some materials currently used in scintillation detectors are sodium iodide, cesium iodide, bismuth germanate, and barium fluoride. There are also less common organic materials used for detection purposes. In addition, sometimes small impurities are added to make the scintillation detectors more efficient. These impurities are known as activators. Scintillation detectors are used in **nuclear physics**, **astrophysics**, and many other applications. They allow single particles to be counted and tracked.

Second law of thermodynamics

The laws of **thermodynamics** were formulated in the nineteenth century after careful experimentation. Although the **first law of thermodynamics** states that **energy** is conserved in every process, it is the second law of thermodynamics that governs the ordering of events that occur during any process. The second law of thermodynamics can be stated in many ways but probably the most common way is to say that the **entropy** of an isolated system can never decrease. Entropy can be understood as a measure of a system's closeness to **equilibrium** or as a measure of a system's disorder. The second law of thermodynamics can be used to show that **heat** will not pass from a colder region to a hotter one unless **work** is done.

The history of the second law of thermodynamics can be traced back to Nicolas-Leonard-Sadi Carnot in the early 1800s. At that time the **caloric theory** was readily accepted, although Benjamin Thompson had already begun to promote the idea that heat was a form of **motion** instead of a material substance. Sadi Carnot, a French engineer, was the first person to prove that it is not possible for a perfectly efficient engine to exist. He said that because an engine must create exhaust, the limit of efficiency is always less than 100%. This limit of efficiency placed on engines by the second law of thermodynamics is also known as the Carnot cycle. The first of Carnot's works was a paper written in 1822-23 that attempted to find a mathematical expression for the work produced by one kilogram of steam. Later, in 1824, Carnot published his work on **steam engines** and it became well-known after **Benoit Clapeyron** published an analytic reformulation of it in 1834. It was clear from notes he made between 1824 and 1826 that Carnot was moving away from the caloric theory. Carnot's ideas were later incorporated into the thermodynamic theory of **Rudolph Clausius** and **William Thompson** (Lord Kelvin). These two scientists later formulated the second law of thermodynamics into words in 1850-51, some 25 years after Carnot's work on the subject. In 1865 Clausius invented the term entropy and so the second law is usually stated in terms of entropy.

The second law of thermodynamics rules out the possibility of a **perpetual motion** machine's existence because heat cannot be completely converted into work. The entropy of the **universe** increases in the course of a spontaneous process. Entropy can also be thought of as the unavailable energy of a system. In this way, the second law can be stated as in a closed system, available energy can never increase, so its opposite, entropy, can never decrease. In thinking of the second law in terms of available energy it suggests the existence of an absolute **temperature** scale that includes an **absolute zero** temperature, the temperature at which all molecular motion ceases. This leads to the third law of thermodynamics that states that although absolute zero can be approached it can never be reached.

SEEING (ASTRONOMY)

Astronomical seeing refers to the ability to view celestial objects through the obscurations of the earth's atmosphere. These obscurations include opacity, **scattering, turbulence**, atmospheric and thermal emission, and ionization.

Opacity refers to the fact that the earth's atmosphere is transparent only to relatively narrow **wavelength** ranges of **light**. These include visual light, the near infrared, **microwaves** and **radio waves** with wavelengths between about 0.35 millimeter and one meter. The atmosphere is nearly completely opaque to ultraviolet light, x rays, gamma-rays, and radio waves with wavelengths greater than one meter. The need to observe heavenly bodies outside of these narrow wavelength windows, along with the desirability to avoid the degrading

affects of the atmosphere are among the main reasons for the development of space-based telescopes.

Ultraviolet photons are absorbed by **electron** transitions in **oxygen** and ozone atoms in the upper atmosphere. Because the amount of ozone varies greatly with loation and seasonal time of year (e.g., the hole in the **ozone layer** over Antarctica during the winter) so does the amount of ultraviolet light reaching the surface of the earth. Far ultraviolet light is mainly absorbed by molecular nitrogen. Infrared light is absorbed mainly by water vapor and **carbon** dioxide in the earth's atmosphere. Millimeter wavelengths are absorbed by rotational bands of water and molecular oxygen, while long radio waves are absorbed by **ions** high in the earth's atmosphere. Because 50% of the water vapor lies within 1.8 mi (3 km) of the earth's surface, to some degree the IR spectrum may be observed with instruments located on high mountains, or mounted on airplanes or balloons. Wavelength regions at which the atmosphere is transparent are called atmospheric windows.

Scattering of light particles degrades seeing. The mechanism by which molecules scatter visible light is called **Rayleigh scattering** and the degree to which light is scattered is inversely proportional to the fourth power of the wavelength and proportional to the **density** of atmosphere. Thus, blue light is scattered more strongly than red light, accounting for the blue **color** of the sky. Red sunsets are an optical illusion caused by the intense scattering of blue light as the light rays travel through the horizon-level thickest regions of the atmosphere.

Atmospheric turbulence resulting from thermal currents or wind, creates small changes in the density of pockets of air that cause the direction of light rays from point sources such as **stars** to be changed by refraction. In effect, the position of the star seems to shift slightly, and the star appears to twinkle. On a photographic plate, turbulence results in a smearing of the stellar image. Details of planetary features are often highly obscured and degraded by turbulence. The human eye, which processes light nearly instantly, often sees a much sharper image than may be obtained with photographs.

Atmospheric emission of the night sky, also called airglow, is caused by the recombination of electrons with atoms that were ionized during the day by photochemical dissociation. This fluorescent light arises from neutral oxygen atoms and molecules, sodium, hydrogen, and hydroxide molecules, and is emitted about 62 mi (100 km) above the earth's surface. The total airglow over one arcsecond, roughly the apparent size of a star with a large **telescope**, corresponds to a visual magnitude of 22. Thus, stars dimmer than this, or galaxies with a surface area brightness less than this, are difficult to detect.

Thermal emission of the night sky is also a factor in the near infrared part of the spectrum. Any warm object in thermal **equilibrium** will emit black body **radiation** (e.g., iron heated to several hundred degrees will glow red or white). The earth's atmosphere below about 31 mi (50 km) emits a faint light by this mechanism, equivalent to a few Janskys (10^{-26} Watts/m²Hz).

Ionization in the earth's ionosphere degrades radio waves. Turbulence in this layer of the atmosphere causes small fluctuations in the density of free electrons, which alter the direction of radio waves and cause **dispersion** in the frequencies of the radio waves. A radio wave emitted from a star of a particular wavelength or **frequency** will be dispersed into a small range of frequencies by ionization.

See also Reflection; Refraction

SEGRÈ, EMILIO (1905-1989)

Italian American physicist

Emilio Segrè is credited as the co-discoverer of three chemical elements, technetium in 1937, astatine in 1940, and plutonium in 1941. In addition, he was codiscoverer, with his former student Owen Chamberlain, of the antiproton in 1955, an achievement for which he shared the 1959 Nobel Prize in physics. Segrè's early academic career is closely intertwined with that of nuclear physicist **Enrico Fermi**, under whom he received his doctorate at the University of Rome in 1928. Segrè then continued his affiliation with that university for most of the next eight years. In 1936, he was appointed professor of physics at the University of Palermo, but was discharged two years later by the Fascist government of Benito Mussolini. Already in the United States at the time of his dismissal, Segrè accepted an appointment at the University of California at Berkeley, where he remained until his retirement in 1972. He returned to the University of Rome, where a special chair in physics had been created for him by the Italian government.

Emilio Gino Segrè was born on February 1, 1905 in Tivoli, Italy, one of three sons born to Giuseppe Segrè, a manufacturer, and the former Amelia Treves. Segrè attended the local elementary school in Tivoli and graduated in 1922 from the Liceo Mamiani in Rome. He subsequently entered the University of Rome, where he majored in engineering. Segrè eventually switched to physics, however, and completed his Ph.D. in that field in 1928. Biographers have conjectured that Segrè's decision to change majors was strongly influenced by his mentor, Enrico Fermi, on the Rome faculty. In any case, Segrè was Fermi's first doctoral student at Rome.

Discovers Technetium

After receiving his degree, Segrè spent a year in the Italian army and then returned to the University of Rome as an instructor in physics. From 1930 to 1932, he studied at Hamburg under **Otto Stern** and at Amsterdam under **Pieter Zeeman**. He returned to Rome as associate professor of physics, where he again collaborated with Fermi. At Fermi's suggestion, Segrè's early research into atomic spectroscopy, molecular beams, and **x rays** soon gave way to **neutron** physics, Fermi's own field of specialization. As part of Fermi's research team, Segrè was involved in the discovery which showed that slow neutrons are more effective in bringing about **nuclear fission** than are fast neutrons.

In 1936, Segrè was invited to become chairman of the department of physics at the University of Palermo. Shortly after accepting that position, he traveled to the United States to visit Ernest Orlando Lawrence and observe Lawrence's **cyclotron** at the University of California. During his visit, Segrè talked with Lawrence about a possible search for element number 43, one of the two elements known to exist that had not yet been discovered. The discovery of the element had been announced in the 1920s by German chemists Walter Noddack, Ida Tacke, and Otto Berg, but had not been confirmed. Only its missing space in the **periodic table** was evidence that it existed.

Based on his earlier work with Fermi, Segrè reasoned that the bombardment of molybdenum (element 42) with neutrons should result in the production of an element with an atomic number one greater than that of molybdenum—the missing 43. When he left Berkeley to return to Italy, therefore, Segrè obtained from Lawrence a sample of molybdenum that had been bombarded with deuterons (in this case, the equivalent of neutrons). In the following year, in collaboration with a colleague, C. Perrier, Segrè was able to confirm chemically the presence of the anticipated element forty-three in the molybdenum sample. This was the first artificially produced new element in scientific history. Segrè and Perrier suggested the name *technetium* for the element from the Greek word *teknetos,* for "artificial."

Assists in the Discovery of Astatine and Plutonium 239

Segrè returned to the United States again in 1938 to work with Lawrence at Berkeley. While there, he designed an experiment by which he was convinced the last remaining missing element, number 85, could be prepared. Working with Dale R. Corson and K. R. MacKenzie, Segrè bombarded a small sample of polonium (element 84) with deuterons and obtained evidence for the existence of element 85. The team suggested the name *astatine,* from the Greek *astatos,* for "unstable," since the element is a radioactive element without stable isotopes.

Segrè's planned return to Italy in the summer of 1938 was interrupted by an unexpected development. The Fascist government of Italy had instituted a purge of Jews similar to that which was carried out by the Nazis in Germany. Segrè, a Jew, was consequently expelled from his post at Palermo. Segrè's response to this snub, according to Isaac Asimov in *Asimov's Biographical Encyclopedia of Science and Technology,* was that he "shrugged and remained in the United States, becoming a citizen in 1944." Offered a research post at Berkeley, Segrè continued his work on neutron physics and artificial **radioactivity**. Two years later, he was involved in the discovery of yet another new element—plutonium—element number 94. This line of research, along with Segrè's previous work on neutron physics, made him an invaluable asset during work on the development of the **atomic bomb** and, in 1943, he was appointed a group leader at the Los Alamos Scientific Laboratory of the **Manhattan Project**.

Discovers the Antiproton with Chamberlain

After the war, Segrè returned to Berkeley as professor of physics and began his collaboration with Owen Chamberlain on the search for the antiproton. In 1928, the English physicist **Paul Dirac** had predicted the existence of a particle identical to the **electron** in all respects except for its having a positive rather than a negative electrical charge. The discovery of the **positron** in 1932 by Carl David Anderson confirmed Dirac's prediction.

An extension of Dirac's original hypothesis suggested that all **subatomic particles**, not just the electron, should have their "anti-" counterparts. The search for the next **antiparticle**, a negatively-charged **proton**, was hampered by the fact that the particle was known to exist only at very high energies, such as the **energy** of cosmic **radiation**.

The construction of the 6.2 billion-electron-volt bevatron in the early 1950s provided another possible source of antiprotons. The accelerator had the potential to produce levels of energy at which the antiproton might be generated. Working with the bevatron, Segrè and Chamberlain finally discovered the hypothesized particle in 1955. For that discovery and the further confirmation it provided for Dirac's **antimatter** theory, Segrè and Chamberlain were awarded the 1959 Nobel Prize in physics.

Segrè was married to Elfriede Spiro on February 2, 1936. After she died in 1970, he married Rosa Mines. Segrè had three children, Claudio, Amelia, and Fausta, by his first wife. In addition to his Nobel Prize, Segrè was awarded the Hofmann Medal of the German Chemical Society in 1954, the Cannizzaro Medal of the Accademia Nazionale de Lincei in 1956, and honorary doctorates from the University of Palermo, Gustavus Adolphus College, and Tel Aviv University. Segrè died in Lafayette, California, on April 22, 1989, of a heart attack.

SEMICONDUCTORS

Semiconductors are a unique class of materials in which the number of charge carriers, the particles that move **electric charge**, is intermediate between those of insulators and those of conductors. A conductor is a material that contains a large number of free electrons that are able to move freely when a **voltage** is applied. An insulator is a material that has very few free electrons and hence is a poor conductor of **electricity**. There are two main types of semiconductors: intrinsic and extrinsic semiconductors. The conducting **power** of intrinsic semiconductors is greatly influenced by **temperature** and sometimes **light** making them particularly interesting.

Although there are some materials that are semiconductors in their pure state, they are virtually never used as such in electronic devices because the only charge carriers present are those produced by thermal breakdown of the covalent bonds within the material. Two of the most common pure semiconductor materials are germanium and silicon. Each of these materials has atoms with four valence electrons that are shared with the valence electrons of its four nearest neighbors in the solid state. As the temperature is raised the atoms start to vibrate and eventually some electrons have sufficient **energy** to break free of their neighboring atoms and move freely through the material. These free electrons participate in the conduction process if a voltage is applied across the material. For pure semiconductors as the temperature is increased the resistance decreases and so raising the temperature increases the conducting power of these materials. These types of semiconductor materials are called intrinsic semiconductors.

The other type of semiconductor is one in which the charge carriers are created by an impurity. These types of semiconductors are known as extrinsic semiconductors. The addition of impurities, such as arsenic or indium, has a significant influence on the conductive properties of semiconductors. The addition of such impurities is known as doping. Arsenic has five valence electrons and so when it is incorporated into silicon it shares electrons with four of its nearest neighboring silicon atoms but has one **electron** left unshared. This spare electron is loosely bound and so can move freely from the **atom** and participate in conduction. This type of impurity contributes an electron to the overall material and is known as a donor atom. A semiconductor of this type is known as an n-type semiconductor because most of the charge carriers are negative electrons. If the dopant is an atom with three valence electrons, such as indium, the three electrons form bonds with the three neighboring atoms. In this case there is a neighbor with an electron deficiency that is referred to as a hole. Electrons of neighboring bound atoms can jump over to the hole if an electric field is applied resulting in migration of the hole in effect. The impurity atoms added have in effect donated a hole and are called acceptors. The hole will migrate in the direction of the electric field as if it were a positive charge carrier. These types of semiconductor materials are known as p-type semiconductors since the charge carriers are positive holes.

Most devices incorporating semiconductors are fabricated using a combination of p-type and n-type semiconductor materials. The boundary between the two regions—the p-region and n-region—is called the p-n junction. In this region there are no free charge carriers since the electrons and holes that move through it recombine in the area around the junction. In a diode of this type there is a delay in the decay to the normal value of reverse current because of the setup of the semiconductor. This delay is a serious disadvantage for operation at microwave frequencies or in high-speed switching. For high-speed applications a Schottky diode is used. This type of diode uses a metal-semiconductor contact instead of a p-n junction and so gives a superior reverse recovery.

Diodes that are composed of a combination of semiconductor materials, with p-regions and n-regions, are good conductors when a voltage is applied so that a current is driven in one direction but a poor conductor if the voltage is reversed. For example, if a battery is connected to such a device so that the positive end is connected to the p side of the device and the negative end is connected to the n side of the device, the situation is referred to as a forward-biased junction. The charges cross the p-n junction with ease, since there is an accelerating field setup, with electrons driven towards the positive end of the battery and positive holes driven towards the negative end of the battery, that allows them to overcome the opposing electric field. If the battery is connected in the opposite manner, with the polarity reversed, then the junction is said to be reverse-biased. The electric field at the junction is enhanced and very few charges are able to cross the junction because they combine with one another, resulting in a very small current. These types of combination semiconductor devices are very efficient at transferring electrical charge in one direction relative to the opposite direction and are, as such, widely used in commercial applications.

The transistor is a special semiconductor device discovered by **John Bardeen**, Walter Brattain (1902-1987), and William Shockley(1910-1989) in 1948. Their discovery revolutionized the world of **electronics** and the three were awarded the Nobel Prize in physics in 1956. There are two types of **transistors**: p-n-p transistors and n-p-n transistors. The p-n-p transistor consists of a semiconducting material with a very narrow n region sandwiched between two p regions. The n-p-n transistor consists of a p region sandwiched between two n regions. The operation of these two types of transistors is virtually identical. These types of devices are commonly used to amplify small, time dependent signals.

The most common semiconductor materials used today are combinations of group III (aluminum, gallium, indium) and group V (phosphorus, arsenic, antimony) elements. The most popular used in devices are gallium arsenide, indium phosphide, silicon carbide, and gallium nitride. The starting semiconductor materials are grown in boules (pear-shaped synthetically formed masses with the **atomic structure** of a single crystal) that are sliced and polished to form the thin substrates on which the electronic or photonic capabilities are etched or grown. Several novel techniques, such as photolithography, are employed to make the small chips. Eventually the individual devices are separated and packaged for electronic and photonic devices such as cell phones, **lasers**, **computers**, and LEDs. Compound semiconductor devices can perform highly specialized functions and are considered the wave of the future.

SERIES CIRCUIT

In a series circuit each electrical circuit component is joined end-to-end in such a manner that there is only one possible path of current flow.

An **electric circuit** is the path of **electric current**, composed of conductors and conducting devices, including a source of **electromotive force**, such as a battery, that drives the current around the circuit. There are two basic types of connections in electric circuits: parallel and series. A series circuit is a circuit that contains components connected in series. That is, the devices or elements of the circuit are arranged in such a way that the entire current passes through each device without division or branching into **parallel circuits**.

A series circuit is one in which the current flows through the devices one after another. The total resistance in the circuit is equal to the sum of the resistance of each individual device ($R_{total} = R_1 + R_2 + R_3 +...$) and current that flows through each device is the same as the total current in the circuit. The **voltage** drop across the circuit must equal the total voltage supplied by the electromotive force device (i.e., the battery).

In many circuits, fuses are used in series with other devices in the circuit for safety purposes. A fuse has a small metallic strip through which the current passes. If the current is too large, the strip melts and the circuit is opened. Another device, the circuit breaker, is also employed to do the same job. In a circuit breaker the current passes through a bimetallic strip. If the current is too high, the bimetallic strip bends and causes a switch to open the circuit. In this state no current passes until the switch is set and the bimetallic strip has cooled. The circuit breaker system is often employed since it can be reset and does not need to be replaced when the current exceeds the recommended magnitude.

Because there is only one path for the charges to move along in a series circuit, each component terminal end is connected to another component's terminal end. Thus, if the positive terminal end of one component is connected to the negative terminal end of another component, then there is no connection to another component at that same terminal. In this way the charges must move in sequence, from one component to the next. If one component in the circuit is broken, no charge will move through the circuit because it only has one path to travel. In a string of lights connected in series, when one bulb burns out the entire string goes dark because the circuit is broken and there is no other path for the current to follow.

See also Electric conductor

SHAPLEY, HARLOW (1885-1972)

American astronomer

American astronomer Harlow Shapley was known for his accomplishments in astronomy, education, and humanitarian causes. Shapley, whose obituary in the *New York Times* described him as the "dean of American astronomers," was chiefly known for discovering that the Milky Way galaxywas much larger than originally thought and that Earth's solar system was not in fact at the center of the galaxy. For this achievement, he was compared to Nicolaus Copernicus, the Polish astronomer who in 1514 proposed that the earth was not the center of the solar system. As an educator, Shapley helped make the Harvard Observatory the leading center for astronomy in the United States during his time. Finally, through his humanitarian efforts, Shapley helped save Jewish scientists from Nazi Germany; he also founded and supported international causes and organizations such as the United Nations Educational, Scientific and Cultural Organization (UNESCO).

Shapley was born on November 2, 1885, along with his twin brother, Horace, in Nashville, Missouri. The twins were later followed by a younger brother, John. Shapley's father was farmer and teacher Willis Shapley, and his mother was Sarah Stowell, whose ancestors were early settlers in Massachusetts. Shapley's youth was marked by the death of his father and by limited schooling: a mere five grades' worth of education in a local rural school. Shapley received his high school education from the Presbyterian Carthage Collegiate Institute, from which he graduated in two semesters. In addition, during this time Shapley worked as a reporter for the Chanute *Daily Sun* in Kansas; he was the paper's city editor by age twenty.

Shapley attended college beginning at age twenty, enrolling in the University of Missouri. He originally planned to study journalism but switched to astronomy when he found out that the university's journalism school was still a year away from opening. By his junior year at the university, Shapley was working as assistant to the director of the Laws Observatory, Frederick H. Seares. Shapley graduated in 1910 with a B.A. in mathematics and physics; the following year he received his M.A.

From the University of Missouri, Shapley went directly to Princeton Observatory as a recipient of the Thaw fellowship in astronomy. Under the guidance of **Henry Norris Russell**, director of the observatory, Shapley completed his graduate work, receiving his Ph.D. in 1913. In his dissertation, Shapley presented information about the properties of **stars** known as eclipsing binary (or double) stars. He eventually expanded and published his thesis; the resulting work, which analyzed ninety binaries, is considered a significant contribution to the field of astronomy.

In 1913 Shapley interviewed with George Ellery Hale, director of the Mount Wilson Observatory in Pasadena, California, about the possibility of obtaining a position there. Shapley got the post, but before moving to California he spent five months touring Europe with his brother John and then returned to Princeton for several months to complete his work there. In addition, on his way to California in 1914, Shapley married Martha Betz, a former classmate from Missouri. Shapley and his wife eventually had five children. Martha Betz Shapley was herself a mathematician and astronomer, and she wrote several scientific papers with her husband.

The Center of the Galaxy

In his seven years at Mount Wilson, Shapley was to make some of his most significant discoveries. Shapley concentrated his efforts on studying stars known as Cepheid variables and on globular clusters—concentrated areas of stars. Shapley found that the Cepheid variables (named for the first star of this type to be discovered, Delta Cephei) were pulsating stars, whose size and brightness fluctuated regularly. Another Harvard astronomer, **Henrietta Leavitt**, had shown that the rate at which a Cepheid pulsed was directly related to its brightness. Using this information, Shapley could then determine the star's absolute brightness (or luminosity) and by

comparing this to its observed brightness could calculate how far away the star was.

It was this measuring system that allowed Shapley to determine the placement of Earth's solar system in the Milky Way. Shapley charted the distance and location of the globular clusters and found that they were distributed almost symmetrically above and below the plane of the Milky Way. He then used this information as a guide to the general outline of the Milky Way and estimated that the center of the galaxy was located fifty thousand light-years from the **sun**, in the constellation Sagittarius. The Shapley Center, as it came to be known, is actually 33,000 light-years from the sun. Shapley also estimated the diameter of the Milky Way to be about 300,000 light-years, later revising that figure to 150,000 light-years.

At the time, Shapley's conclusions were almost revolutionary: his figures increased the then-estimated size of the Milky Way by a factor of ten and as a result, many astronomers believed he overestimated. In fact, Shapley's conclusions weren't fully accepted until he took part in a now-famous debate at the National Academy of Sciences in Washington in 1920 with Heber D. Curtis of the Lick Observatory. It is generally agreed that at this debate Shapley won his point about the size and structure of the Milky Way.

In addition to his work on the question of the center of the galaxy, Shapley also studied spiral **nebulae** while at Mount Wilson (in that era, the term "nebula" described anything not positively identified as a star). At the time, astronomers were not sure whether the nebulae were part of the Milky Way or were what was known as "island universes"—in modern terms, galaxies. Several new stars had been discovered in the nebulae, and Shapley, using information about the luminosity of the stars, estimated the distance to the Andromeda nebula (now known as the Andromeda galaxy) to be one million light-years. The theories of other scientists at the time caused Shapley to withdraw this estimate; ironically, his figures turned out to be nearly correct.

Shapley also pursued scientific interests that went beyond astronomy at Mount Wilson. He discovered a relationship between **temperature** and the speed at which ants run, for instance, and published five scientific papers on ants. According to Shapley biographers, he was especially proud of his work in this area.

Made Harvard Observatory U.S. Leader

Shapley returned to Cambridge, Massachusetts, in 1921 to become director of the Harvard Observatory and Paine Professor of Practical Astronomy. At Harvard he concentrated his research on stars in the Magellanic Cloud (later revealed as two neighboring galaxies), working on a unique round desk that rotated. A few years into his tenure, in 1924, he was elected to the National Academy of Sciences, and he received that organization's Draper Medal in 1926. Among Shapley's scientific achievements in this period was, in the 1930s, the first discovery of two dwarf galaxies in the constellations of Sculptor and Fornax (in the southern hemisphere). Interestingly, he used the term "galaxy," while his fellow astronomer and rival of the period, **Edwin Hubble**, used the term "extragalactic nebulae." Shapley maintained that galaxies were distributed irregularly, in contrast to Hubble's view that the **universe** was more uniform. Shapley's work at Harvard also included overseeing group projects that cataloged tens of thousands of galaxies in both the northern and southern hemispheres (the Harvard Observatory had observatories in both hemispheres) and that worked on a catalog of the spectral classifications of stars (the **light** from a star, broken into its rainbow or spectrum of colors, reveals its chemical makeup).

Perhaps more significantly, Shapley helped make the Harvard Observatory the leading astronomy education center in the United States in the 1920s. He set up a graduate program in astronomy and attracted scientists from the international community as well as from across America. Among the scientists that Shapley brought to Harvard and the United States were refugees from Nazi Germany: Shapley personally led an effort to bring in Jewish scientists. According to Shapley biographer Bart J. Bok in an article for *Sky and Telescope,* "One of these who came to Harvard was Richard Prager of Berlin Observatory, who told me quietly and seriously that every night at least a thousand Jewish scientists must say a prayer of thanks for Harlow Shapley's humanitarian efforts to help save them and their families." His activities in the early 1940s also included serving as president of the American Academy of Arts and Sciences.

International Efforts

After World War II Shapley continued to focus on international affairs. He helped form the United Nations Educational, Scientific and Cultural Organization, serving as a U.S. representative in writing UNESCO's charter in 1945. Shapley lobbied the U.S. government for the creation of the posts of Secretary of Peace and Welfare, Secretary of Education, and Secretary of Science and Technology after World War II. According to the *New York Times,* Shapley believed that peace should be the first concern of government. Eventually, only the Department of Health, Education, and Welfare was established.

After a visit to the Soviet Union in 1945 as the Harvard representative at the 220th anniversary of the Academy of Sciences in Moscow, Shapley became a supporter of scientific cooperation with the Soviets. He served as chairman of the Independent Citizens Committee of the Arts, Sciences, and Professions, a fund-raising group for liberal congressional candidates. In addition, he was the chairman of several meetings of left-wing groups that dared to invite Russian delegates. Shapley, who believed that the Soviet government's support of science would speed its development, urged the United States to give similar support to research in order to remain competitive. His views were not popular at the time, and in 1946 the House Committee on Un-American Activities subpoenaed Shapley to question him about his views. Shapley's obituary in the *New York Times* reported that Representative John E. Rankin, a Mississippi Democrat who had questioned Shapley as a one-man subcommittee, reported after the meeting, "I

have never seen a witness treat a committee with more contempt." The paper reported that in return Shapley charged the committee with "Gestapo methods" and called for its elimination. The same year, Shapley was elected president of the American Association for the Advancement of Science, a move widely regarded as a show of support by the scientific community. Despite his encounter with reactionary politics, Shapley continued with his efforts against the committee, becoming, for example, acting chairman of the Committee of One Thousand, an organization geared toward protecting First Amendment rights. Physicist **Albert Einstein** and author-lecturer Helen Keller were also members. In March of 1950, Senator Joseph McCarthy "charged that Dr. Shapley had belonged to many Communist-front organizations," according to the *New York Times.* Later that year, a Senate subcommittee cleared him of the charges.

Shapley was director of the Harvard Observatory until 1952; he continued to teach courses there until 1956 and then traveled and lectured through his seventies, speaking on astronomy, religion, philosophy, and evolution. His many appearances during his career included one at the 1939 New York World's Fair, at which he gave the speech "Astronomy in the World of Tomorrow." He also wrote extensively, publishing hundreds of scientific papers as well as newspaper and magazine articles and several popular books, including *Through Rugged Ways to the Stars.* At eighty-five Shapley moved to Boulder, Colorado, where one of his four sons lived; he died there in a nursing home at age eighty-six after a long battle with illness.

SHOCK WAVES

A shock wave is essentially a surface of discontinuity propagating supersonically through a medium. **Density, pressure,** and **temperature** are unequal on the different sides of the wave front, clearly distinguishing between the shocked and unshocked material. This is in sharp contrast to subsonic **sound waves**, where all thermodynamic quantities vary continuously through the medium. In general, the shock compresses and heats the material that it flows through in an irreversible process, again in contrast to sound waves. The fact that the front propagates supersonically with respect to the unshocked material allows for the discontinuity to be maintained as the wave advances, since it impedes information from being transmitted from the region through which the shock has already passed.

Shock waves generally accompany supersonic **motion,** such as a sonic boom or the impact of projectile on a surface (like a meteor hitting the earth). Shock waves are especially associated with events where a large amount of **energy** is deposited in a small volume—an explosion. A special kind of shock wave-induced explosion is a detonation, where energy deposited by the shock is sufficient to drive an exothermic reaction in the material it passes through, as with a gasoline tank or TNT. Shocks are generally transient phenomena: without a constant supply of energy, pressure in the material behind the shock will gradually decrease as its volume increases, until the propagation speed becomes subsonic and the perturbation reduces to a sound wave.

The basic physics of shock waves is usually described by the textbook example of an infinite tube of gas held at one end by a piston at rest. If the piston is accelerated instantaneously to a supersonic speed into the gas, the solution of the flow is a propagating front, where the (shocked) gas behind it settles at uniform density and pressure that are both higher than the unperturbed gas before the front. The properties of the gas on both sides of the discontinuity can be related through the general equations of **conservation of mass**, **momentum**, and energy across the front, known as the Rankine-Hugoniot equations. It can also be shown that the flow settles on a *self-similar* form, meaning that the spatial profile of all quantities in the flow remains constant in time up to a scale, which is the only time-dependent function. Such self-similar behavior is typical of strong shocks.

There exist several sub-classifications of shock waves, according to specific properties. Shocks where the direction of propagation is not normal to the shock front are called oblique shocks, and offer a wide variety of reflection and **diffraction** phenomena. A weak shock, where the pressure in the unshocked material is not negligible with respect to the pressure in the shocked region, also introduces interesting variations with respect to the simpler case of a strong shock. In **astrophysics**, there is great interest in radiative shocks, in which energy can be emitted from the shocked region beyond the shock front (the textbook case, where no energy can overtake the shock front, is called an adiabatic shock).

The origin of shock wave theory can be traced back to B. Riemann's work on the propagation of acoustic disturbances and the experimental study of supersonic flow by E. Mach, both in the second half of the nineteenth century. However, theoretical and experimental investigation of shock phenomena has emerged in earnest as part of the **atomic bomb** project in both the United States and the Soviet Union. The study of shock waves has since become a fundamental aspect of **fluid dynamics** and **plasma physics**, as well as astrophysics, where shocks are very abundant (and also tend to give rise to excessive emission of **radiation**, and hence become observable, even at large distances). Theoretical and experimental studies also concern the response of different types of materials to shock loading and instabilities that arise due to passage of a shock between different materials (Rayleigh-Taylor and Richtmayer-Meshkov instabilities). Physics of the shock front is another subject of interest: the discontinuity is never really infinitely thin, and its width is determined by the chemical processes that distribute the energy and **entropy** which have been deposited. While the width of the front is generally much smaller than any other typical length scale of the flow, its properties are potentially important in determining the stability of shock propagation and the efficiency of detonation fronts.

SHOEMAKER, EUGENE M. (1928-1997)

American geologist

Eugene M. Shoemaker was a geologist whose research interests and expertise ranged far beyond Earth. Shoemaker studied impact craters (craters created by the collision of an asteroid or comet and a planet) on both Earth and the **Moon**, as well as searched for the **comets** and asteroids that could have caused the craters. He was a key figure in guiding the scientific exploration of the lunar surface during the 1960s. Intrigued by the study of craters in general, Shoemaker then turned his focus back to Earth and became a leader in studying the impact history of the planet.

Eugene Merle Shoemaker was born in Los Angeles on April 28, 1928, to George Estel and Muriel May Scott Shoemaker. His mother was a teacher, and his father at different times worked in teaching, business, farming, and for motion picture studios. The family moved to Oregon, New York, Wyoming, and finally back to Los Angeles when Shoemaker was fourteen years old. He had one younger sister. Shoemaker's interest in geology coalesced early in his life. He began collecting rocks, minerals, and fossils when he was seven. He took classes in mineralogy and geology at the science museum in Buffalo, New York, starting when he was nine. By the time he finished high school in Los Angeles he wanted to be a geologist. Shoemaker earned his bachelor's degree in geology from the California Institute of Technology in 1947 and his master's degree the next year.

In July 1948 he moved to Grand Junction, Colorado, to join the United States Geological Survey (USGS) as a geologist. For two years he searched for uranium deposits and investigated salt structures in Colorado and Utah. During the early 1950s he concentrated on the geochemistry, volcanology, and structure of the rocks of the Colorado Plateau. He took a leave from the USGS to study at Princeton University from 1950 to 1951 and 1953 to 1954, receiving his M.A. from Princeton in 1954. During the late 1950s Shoemaker became interested in craters, those created by both meteorite impacts and nuclear explosions. His doctoral thesis was to draw the first detailed geological map of the huge Meteor Crater in Arizona, also known as the Barringer Crater. In 1960, Shoemaker and Edward C. T. Chao discovered the presence of the mineral coesite at Meteor Crater. Since coesite is a high-pressure form of silica that forms when quartz is squeezed at intense pressures, Shoemaker and Chao had essentially proven that the Barringer Crater was indeed formed by a meteorite striking Earth, rather than by volcanic action as some scientists had thought. That year Shoemaker received his Ph.D. from Princeton.

Shoemaker began working on Meteor Crater and other impact-related craters because of his interest in the Moon and its pocketed surface. "I got a bug in my ear in 1948, shortly after I went to work for the U.S. Geological Survey, that during my professional career human beings would go to the Moon. I made up my mind then that I was going to be the first scientist to get to the Moon," he told *Astronomy* in 1993. During the 1960s, the decade that astronauts did indeed walk on the moon, Shoemaker became heavily involved with the science of lunar exploration.

In 1960 Shoemaker established a lunar geological time scale and developed improved ways to map the Moon's surface. He then organized a USGS unit on astrogeology at the Survey's office in Menlo Park, California, where he was working. (The unit later became a full branch and moved to Flagstaff, Arizona, in 1965.) In 1962 and 1963 the USGS loaned Shoemaker to the National Aeronautics and **Space** Administration (NASA) to help develop plans for exploring the Moon. For several months Shoemaker was acting director of NASA's Manned Space Sciences Division, which was charged with formulating the scientific objectives for the first astronauts to land on the moon.

While an administrator at NASA and the USGS, Shoemaker continued to work on terrestrial impact craters in addition to attempting to figure out what the lunar surface was like. He promoted using television systems to investigate the geology of other moons and **planets**. Shoemaker was in charge of the television experiments on Project Ranger, and later, Project Surveyor. (From 1979 to 1990 he was also co-investigator for the television set-up on the two Voyager probes that visited the outer planets of the solar system.) During the mid-1960s Shoemaker studied photographs of the lunar surface sent back by exploratory spacecraft to determine the nature of the surface. With photos from the *Ranger 7* probe, Shoemaker used his method of "photoclinometry," which computed the steepness of craters' edges by the amount of shadow they threw onto the ground. He announced that the surface was quite smooth. With later photographs, Shoemaker determined that the Moon is predominantly gray and that it is covered with shallow, silty soil. (Before the *Surveyor* spacecraft landed on the surface, there was concern that a thick layer of dust might preclude a probe landing.) Shoemaker became the principal scientist for the geological field studies of the Apollo landings from 1965 to 1970.

In 1966 Shoemaker resigned as administrator of the USGS astrogeology branch so that he could return to full-time research as chief scientist in Flagstaff. He began an association with the California Institute of Technology (Caltech), as a visiting professor in 1962 and 1987, and a full-time professor of geology from 1969 to 1985. He chaired Caltech's Division of Geological and Planetary Sciences from 1969 to 1972.

By this time Shoemaker's experience with Moon studies prompted him to consider the objects that create impact craters on Earth as well. His research in the late 1960s shifted to Earth-crossing asteroids, those objects that one day might hit the earth and form another Meteor Crater. He decided to try to catalog the asteroids, a task that would take assiduous sky-searching. At Caltech he hired the young geologist Eleanor Helin to begin an asteroid search program. He and Helin started in 1973 to use the 18-in (45.7 cm) Schmidt **telescope** atop Palomar Mountain in California for the first-ever systematic asteroid search, named the Palomar Planet-Crossing Asteroid Survey. It took them six months to find their first asteroid, and two and a half years for the second one. Eventually they decided to step up their search and spend more nights at the telescope.

Eugene Shoemaker. *(Photo by David Parker, Science Source/Photo Researchers. Reproduced by permission.)*

Shoemaker continued to supervise the hunt for near-Earth asteroids. He calculated that about 1,000 to 1,500 asteroids at least half a mile across are on potential collision courses with Earth. When Helin left the search in 1982, Shoemaker's wife Carolyn soon joined the hunt, renamed the Palomar Asteroid and Comet Survey. The two Shoemakers developed a stereoscopic technique for comparing pairs of film: one eye looks at one photograph, the other eye looks at a snapshot taken 45 minutes to an hour later, and any moving asteroid or comet appears to float above the combined image.

Shoemaker also continued to work on the impact history of Earth. In 1980, his line of research got a shot of excitement when a group led by the physicist **Luis Alvarez** suggested that a cataclysmic impact had struck Earth 65 million years ago, triggering the chain of events that led to the death of the dinosaurs and many other species. Scientists had previously blamed the dinosaurs' demise on many other causes, from climate change to constipation. The Berkeley group offered as proof the thin layer of iridium appearing in 65-million-year-old sediments around the world. Iridium is a rare element on Earth, but extraterrestrial sources like meteorites are relatively rich in iridium. Thus, the Alvarez group suggested, a giant meteorite must have struck Earth, kicking up a cloud of dust that covered Earth, blocking the planet from sunlight and causing a massive extinction.

Soon after the Alvarez hypothesis, the hunt began for the giant scar that must have been left by such an impact. Scientists have identified the Chicxulub crater, buried under water and sediments off the coast of the Yucatan, as the most likely candidate for the impact's calling card, since it is at least 120 mi (193 km) across. However, a smaller crater in Iowa—22 mi (35.4 km) across—has been dated at roughly the same age of 65 million years. Shoemaker suggested that it was too coincidental to have two large craters of the same age. He proposed instead that multiple impacts smashed Earth, one right after another on the scale of geologic time. And since the probability of several huge meteorites coming at the same time is slim, Shoemaker suggested that perhaps a comet had broken up and shed its remains on our planet in a destructive repetition, leaving both the Mexican and the Iowa craters. Most scientists now believe that the Chicxulub crater is the scar of the primary impact that killed off the dinosaurs. The question of multiple impacts has not been resolved.

Shoemaker's career was recognized with numerous honors, including honorary doctorates of science from Arizona State College, Temple University, and the University of Arizona. Among other awards, he received the NASA Medal for Scientific Achievement, the Kuiper Prize of the American Astronomical Society, and the National Medal of Science. Shoemaker died on July 18, 1997 of injuries suffered in a car crash in the Australian outback.

Shoemaker married Carolyn Jean Spellman in 1951. They had two daughters, Christine and Linda, and one son, Patrick. Carolyn Shoemaker, a former teacher, became involved with the Palomar asteroid search after her three children were grown. She has since discovered more comets than any other person in history except for the nineteenth-century observer Jean-Louis Pons. But only twenty-six comets bear Pons's name; that record was broken in 1992, when Carolyn Shoemaker found her twenty-seventh comet. The Shoemakers also discovered more than three-hundred asteroids, some of which are named after their children and grandchildren.

One of the comets found by Carolyn Shoemaker became a once-in-a-lifetime chance for stargazers to witness a rare event. Comet Shoemaker-Levy 9, discovered in March 1993, collided with Jupiter in July 1994. Gene Shoemaker calculated that this sort of celestial collision happens once in a thousand years. "I've been hoping for many decades that I would live long enough to see the impact of a significant object," he told Alexandra Witze in an interview. "I was thinking it might be something small on Earth. I never dreamed we would discover something to impact another planet."

SIEGBAHN, KAI M. (1918-   )

Swedish physicist

Kai M. Siegbahn, the son of a Nobel physics laureate, himself won the Nobel Prize in physics for his development of **electron** spectroscopy for chemical analysis (ESCA). This reliable technique reveals many more details about the atomic and molecular structure of **matter** than was previously possible to determine. Siegbahn's electron spectroscopy soon became widely used around the world in scientific and industrial research labs.

Siegbahn was born to Karl M. G. Siegbahn and Karin Högbom Siegbahn on April 20, 1918, in Lund, Sweden. His father was a lecturer in physics at the University of Lund and the director of the Nobel Institute for Physics of the Royal Swedish Academy of Sciences for nearly thirty years. The elder Siegbahn's discoveries and research in x-ray spectroscopy won him the 1924 Nobel Prize in physics. About growing up with such a role model, Kai Siegbahn was quoted in *Newsweek* as saying, "It's a decided advantage if you start discussing physics every day at the breakfast table."

The younger Siegbahn pursued his interests in mathematics, physics, and chemistry at the University of Uppsala, from which he earned his bachelor of science degree in 1939 and his master of science degree in 1942. He received his doctorate in 1944 from the University of Stockholm and worked as a researcher from 1942 to 1951 at the Nobel Institute of physics. After a few years as a physics professor at the Royal Institute of Technology in Stockholm, Siegbahn left in 1954 to become professor and then head of the physics department at the University of Uppsala, where he remained and conducted his important research.

It was while he was pursuing his graduate degree that Siegbahn's interest in spectroscopy developed. Spectroscopy studies the frequencies, or wavelengths, at which particles of matter emit **light** or **radiation**. Since the frequencies are specific, or characteristic, of the matter being studied, they can yield valuable information about the atomic and molecular structure of that matter. Heinrich Hertz had discovered the **photoelectric effect** in 1883: objects struck by ultraviolet light release electrons, called photoelectrons. The nature of the effect was further examined by **Max Planck** and **Albert Einstein**, and researchers, hoping to advance scientific knowledge about the composition of matter, began attempting to use spectroscopy to analyze the photoelectrons. Electrons collide and scatter as they leave matter, losing unknown amounts of **energy** and creating spectra that are very difficult to read. Existing spectroscopes were inadequate to analyze the particles, but Siegbahn solved this problem.

Adaptations to Instrument Lead to ESCA

When he was in graduate school during the 1940s, Siegbahn studied electrons given off by radioactive nuclei, a process called **beta decay**. If he could measure the energies of these beta-ray electrons, he could learn much more about the nuclei. Existing instruments were of limited use for this purpose, so Siegbahn developed a mushroom-shaped magnet that allowed focusing in two directions. Siegbahn's high-resolution instrument increased the accuracy of measurements of these photoelectrons ten times. By the 1950s Siegbahn had turned his attention to electron spectroscopy—knocking electrons out of nonradioactive atoms with light or x rays—to determine the energies that bind electrons to atoms. Again, he found the spectroscopes he was using unable to produce accurate measurements. He decided to apply the double-focusing method he had developed for **nuclear physics** to the energy spectra of photoelectrons. The outcome was a double-focusing spectrometer with a high resolution that revealed previously unseen, narrow, but well-defined electron lines.

Siegbahn and his colleagues completed the first high-resolution, double-focusing spectrometer in 1954, and again it improved measurement accuracy tenfold. In 1957 the team recorded the first extremely sharp lines that allowed precise measurement of binding energies. In the 1960s the usefulness of the device was greatly extended when Siegbahn and two fellow researchers found that their electron spectrometer also revealed the chemical environment of the atoms that released the photoelectrons being studied, yielding details about chemical bonding as atoms combined into molecules. The technique now became known as electron spectroscopy for chemical analysis, or ESCA. Siegbahn and his team also adapted ESCA for the analysis of gases and liquids as well as solids.

Siegbahn's ESCA technique changed electron spectroscopy from a laboratory concept with a very limited application to a widely-used tool. ESCA provides high-resolution analysis of the atomic, molecular, and chemical characteristics of a nearly unlimited range of scientific, commercial, and industrial materials that includes atmospheric pollutants and surface corrosion. To acknowledge his contribution, Siegbahn was awarded the 1981 Nobel Prize in physics for his role in the development of a technique that provided a better understanding of the nature of matter.

SIEGBAHN, KARL M. G. (1886-1978)
Swedish physicist

Noted for modernizing Swedish physics, Karl M. G. Siegbahn contributed significantly to the field of x-ray spectroscopy begun in Germany and England. His design and application of equipment and techniques in this field vastly improved the accuracy of existing methods by which x-ray wavelengths were distinguished, and led to important discoveries about the nature of **x rays** (electromagnetic **radiation** invisible to the unaided human eye) and **atomic structure**. In recognition of his work, Siegbahn was awarded the 1924 Nobel Prize in physics. He became the first director of the Nobel Institute of Experimental Physics in 1937, and remained director of the institute until his retirement in 1964.

Karl Manne Georg Siegbahn was born in Örebro, Sweden, on December 3, 1886. His father was Nils Reinhold Georg Siegbahn, a station master for the Swedish national railway system, and his mother was the former Emma Sofia Mathilda Zetterberg. After his father's retirement, the family moved to Lund, Sweden, where Siegbahn entered the University of Lund in 1906. He studied astronomy, chemistry, mathematics, and physics there, eventually receiving his bachelor's degree in 1908, and his licentiate degree (comparable to a master's degree) in 1910. Studying **electromagnetism** as a research assistant to **Johannes Rydberg**, Siegbahn was awarded a doctorate in physics in 1911 for his dissertation on the

measurement of magnetic fields. He spent the summer semesters of 1908 and 1909 studying at the universities of Göttingen and Munich respectively, and took up studies in Paris and Berlin during the summer of 1911.

Begins Studies on X rays

Upon receiving his doctorate, Siegbahn remained in his position as assistant to Rydberg, eventually assuming Rydberg's duties as lecturer when the older man was ill for long periods of time. When Rydberg died in 1920, Siegbahn succeeded him as professor of physics. Siegbahn's work under Rydberg had dealt primarily with studies of electrical and magnetic phenomena. By 1914, however, visits to research centers in Paris and Heidelberg turned his attentions toward the study of x-ray spectroscopy, a field to which he would devote the rest of his academic life.

Fundamental research on the properties of x rays had been carried out by the English physicist Charles Barkla in the first decade of the twentieth century. Barkla had found that x rays emitted from certain elements were of two types, "hard" or "K" x rays, and "soft" or "L" x rays, based on an element's atomic **weight**. Barkla's equipment was too primitive, however, to permit more detailed characterization of their wavelengths, or spectrums. In 1912, the German physicist Max von Laue developed a technique for revealing the **spectral lines** of x rays based on the **diffraction** grating technique used for visible **light**. But, because the wavelengths of x rays are much shorter than wavelengths of visible light, von Laue needed a finer grating or spacing device. He discovered that the regular spacing between atoms in a crystal would serve this purpose, so that x rays passed through an analyzing crystal, such as zinc sulfide, would separate into a series of spectral lines similar to those in an optical spectrum. The father-and-son team of physicists **William Henry Bragg** and **William Lawrence Bragg** refined the process and developed the first x-ray spectrometer shortly thereafter.

Perfects the Techniques of X-ray Spectroscopy

By 1914, Siegbahn had embarked on his research program in x-ray spectroscopy, one aspect of which involved the improvement of equipment used in the field. Over a period of more than three decades, Siegbahn designed improvements in the tubes, **vacuum** systems, and spectrometers used in x-ray spectroscopy, all with the objective of attaining more precise measurements of x-ray wavelengths. One of his first developments was the vacuum spectrometer, which enclosed the whole analytical system—x-ray tube, target, and spectroscopic and photographic plates—in a high vacuum, thus reducing the possibility of any extraneous **interference**. Siegbahn also designed more accurate spectrometers capable of reading specific **wavelength** regions.

Each improvement in technology achieved by Siegbahn was soon followed by new discoveries. He was able not only to confirm the existence of Barkla's K and L series of lines, for example, but also to discover two new series, which he named the M and N. He also found that the two K lines discovered by the English physicist **Henry Moseley** were actually doublets and, where Moseley had found four or five L lines, Siegbahn found as many as 28.

These discoveries were especially important because of the information they provided about atomic structure. Spectral lines are emitted when electrons move from one **energy** level to another in an **atom**. **Niels Bohr**'s 1913 model of the atom had offered a simplistic view of **electron** energy levels, one in which electrons were restricted to a small number of **orbitals** around the **nucleus**. With the discovery of elliptical orbitals, magnetic effects within the atom, and electron spin, however, this picture became much more complex. The number and type of electron transitions possible within an atom became much greater and Siegbahn's analysis of x-ray spectra provided invaluable keys to learning more about them. For both his discoveries and contributions to the improvement of x-ray spectroscopic technology, Siegbahn was awarded the 1924 Nobel Prize in physics. He also earned a number of other honors and awards throughout his life, including the Hughes (1934) and Rumford (1940) medals of the Royal Society, and the Duddel Medal of the London Physical Society, as well as honorary degrees from a number of universities throughout the world.

In 1923, Siegbahn was offered the position of professor of physics at the University of Uppsala, a post he held until 1937. He then became the first director of the newly created Nobel Institute of Experimental physics in Stockholm. In addition to his continuing interest in x-ray spectroscopy, Siegbahn established a school of **nuclear physics** at the institute and, by 1938, had overseen the construction of Sweden's first particle accelerator. Siegbahn also served as president of the International Union of Pure and Applied physics from 1938 to 1947.

Siegbahn's influence on the history of science goes beyond his own personal accomplishments. For most of his life, he maintained a strong commitment to the expansion and improvement of scientific facilities and programs of science education in Sweden. At both Uppsala and the Nobel Institute, he took facilities with poor or limited equipment and built them into first-class research institutions. Siegbahn was married to Karin Högbom in 1914. They had two sons, Bo Siegbahn and Kai M. Siegbahn, the latter a Nobel Laureate in physics himself in 1981. Karl M. G. Siegbahn died in Stockholm on September 26, 1978.

SIGMA PARTICLE

In 1947 physicists observed new particles associated with cosmic ray events. The observed particle-streaming from these highly energetic cosmic events came in two types—a neutral particle that decayed into positive and negative particles and another, more massive positively charged particle that decayed into a positively charged heavy **electron** and a **photon**. What was surprising to scientists was the decay time associated with these particles. The decay times due to **strong interactions** are measured in femto seconds (10^{-15} s). The

decay time of the observed **K mesons** was, however, about 10^{-10} s, and was, therefore, more typical of a weak decay. Since the initial discovery both mesonic and baryonic type particles (e.g., **pions**), similar particles have been discovered. Because of their decay times these particles are described as **strange particles**.

The sigma particle was one of the first three strange particles discovered. A particle is considered "strange" if it is produced by the strong **force**, but decays by the weak or the electromagnetic force. Experiments showed that strange particles were always produced in pairs. Their behavior was initially explained by **Murray Gell-Mann** in the United States and, independently, by Kazuhiko Nishijima (1929-) in Japan. If one of the particles was given strangeness (s) s+1 and the other –1 (similar to **electric charge** and the way electrons and positrons are produced), then when an S=+1 and an S=–1 particle were produced by the strong reaction, the total strangeness was zero, and strangeness was conserved. Strangeness is not preserved in the weak or electromagnetic decay reactions. The sigma particle was designated S=–1.

In 1961, Gell-Mann and independently, the Israeli physicist Yuval Ne'eman (1925-) began using some of the conserved characteristics in hadrons and mesons to group them into multiplets of eight. Gell-Mann called his classification "the eightfold way" borrowing the term from Buddhism, which stated the way to nirvana was the eightfold way.

Gell-Mann and others developed a classification scheme similar to Mendeleev's (1834-1907) **periodic table** of the elements. From these schemes Gell-Mann was able to predict nearly exactly the omega minus particle which was found in 1964 at the Brookhaven accelerator laboratory. Gell-Mann believed there was an internal structure to the hadrons. In 1963, he published his paper on **quarks**, the building blocks of hadrons. The sigma particle was known to be a family of three, one with a positive charge, one with a negative charge, and one neutral. They were all strange particles and were baryons, about 10% more massive than a **proton**. They each had ½ unit of spin.

The sigma-plus is made up of two up (u) quarks and one strange (s) quark, the sigma-minus of two down (d) quarks and one strange (s) quark, and the sigma zero of one up (u), one down (d), and one strange (s) quark. The sigma particle played a key role in filling out the classification scheme of Gell-Mann and others, and helped lead the way to an understanding of one of the fundamental elements of **matter**, the quark.

See also Lambda particles; Particle physics; Quantum electrodynamics (QED); Quantum physics

SIMPLE MACHINES

A simple machine is a device that acts to increase or decrease the magnitude of applied **force**. There are six simple machines: inclined plane, lever, wheel and axle, pulley, wedge, and screw. **Energy** is conserved in the operation of simple machines and the total **work** done on a simple machine must equal the total work done by the machine (including work to overcome **friction**, etc.). Simple machines operate by allowing a trade-off between the magnitude of force and the distance over which that force is applied.

In physics, work is defined as the product of force multiplied by the distance over which the force acts. With regard to simple machines, because the total work done on a system must equal the work output (i.e., the work done by the system) the products of force and distance must always be equal for various force and distance configurations. If the magnitude of force applied increases, then the distance over which the force acts must decrease. Correspondingly, if the magnitude of the force decreases, then the distance over which the force acts must increase.

The inverse relationship between magnitude of force and the distance over which the force acts are further described in terms of the mechanical advantage of simple machines. A mechanical advantage states a factor of force increase (i.e., a machine with a mechanical advantage of two exerts twice the force initially applied). Quantitatively, the mechanical advantage is equal to the force exerted by the machine divided by the input or applied force. The actual mechanical advantage of any machine is always less than the maximum mechanical advantage because there is always some undesired energy transformation (e.g. heating due to friction).

Undesired **energy transformations** also reduce the efficiency of machines. The efficiency of a simple machine is based upon the ratio of work output to input (Percent Efficiency = (W_{out}/ W_{in}) x 100).

The inclined plane is the simplest example of the trade-off of force for distance. In building the pyramids, for example, the ancient Egyptians used ramps to move heavy blocks of stone to great heights. The net work done in raising the stones against **gravity** did not depend upon the path taken, because by applying a lower force over a longer distance, the same amount of work was performed on the stone to lift it against gravity. In moving any **mass** against gravity, the work required is the product of mass, gravity, and the height to which the object is raised. By using an inclined plane (a ramp) it is possible to use a lower force over a longer distance to raise the body. This is possible because the inclined plane resolves the **weight** of the body into vertical and horizontal components. A lower lifting force is therefore required to overcome the vertical or gravitational component.

A lever is rigid body designed to rotate about a pivot point. A downward force applied to one end of the lever results in an upward force on the other end. The rotational **torque** at a pivot point is directly proportional to the product of the magnitude of force applied and the length of the lever arm (distance from the force application to the pivot point). For a lever of fixed length, variations in the placement of the pivot point result in varying lengths of lever arms and thus differential rotational torques about the pivot point. Depending on pivot point placement, a lever can multiply either the force or the distance over which the force is applied. **Levers** are divided into three classes, depending on the relative arrangement of

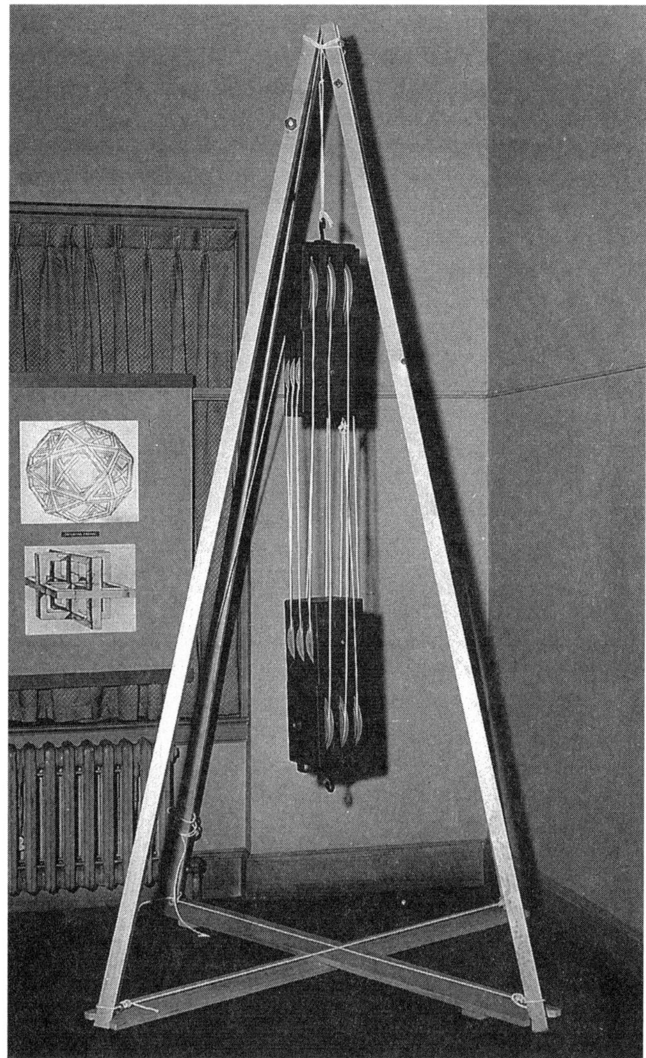

Circular pulley. *(Photo by Grant Smith/Corbis-Bettmann. Reproduced by permission.)*

the pivot point (i.e., the fulcrum), the output force (i.e., the load), and the applied or input force (i.e. the effort). A first class lever (e.g., a crowbar) places the fulcrum between the load and the effort and the load is displaced in the opposite direction to effort. In a second class lever (e.g., a bottle opener or wheelbarrow) the load is placed between the effort and the fulcrum. And in a third class lever (e.g., a fishing rod) the effort is placed between the fulcrum and the load. In both second and third class levers the load is displaced in the same direction as the applied force. Although both first and second class levers amplify applied force, the third class lever acts as a force reducer (i.e., the output force is lower than the applied force.

In a wheel and axle machine, a wheel is fused to an axle so that the angular displacement of one results in the same angular displacement for the other. Because of equal angular displacements, the application of force to the rim of the wheel results in a shorter length but more powerful displacement at the axle. Correspondingly, a force applied to the axle that results in angular displacement turns the wheels through a greater distance (with the same angular displacement). In a fundamental sense a wheel and axle operate on the same principle as the lever. A tangential force acts on the wheel, with the long lever arm equal to the radius of the cylindrical wheel, and the short lever arm equals the radius of the cylindrical axle.

A pulley is a force reversal machine (i.e., pulleys reverse the direction component of force). Pulleys in a system of pulleys (e.g., a block and tackle), however, permit the lifting of heavier loads with the same application of force. The force distance trade-off requires that the rope move a greater distance than the load.

A wedge converts unidirectional force into forces that act at right angles to the wedge. The application of force at right angles to the wedge results in a splitting **motion**.

A screw is a variation of an inclined plane in which the incline is wound around the core or shaft of the screw. Consequently, a screw increases the distance of force application by the application of force along the thread path in such a way that, in turning a screw, rotational force is converted to a linear force that moves the screw body forwards or backwards.

Simple machines form the basic components of more complex machines and all machines, no matter how complex, are composed of combinations of the six different types of simple machines.

See also Center of gravity; Center of mass; Mechanics; Work and potential energy

SIMPLE WAVES • See Waves

SIMULTANEITY AND TIME DILATION

Simultaneity describes the occurrence of two events at the same time. In accord with the postulates of special relativity, simultaneity is a perception governed by the speed of **light** that depends upon the reference frame and the relationship of observers to phenomena. Accordingly, simultaneity is a relative phenomenon rather than an absolute and is affected by **time dilation**.

In 1916, German-American physicist **Albert Einstein** wrote, "There is hardly a simpler law in physics than that according to which light is propagated in empty **space**. Every child at school knows that this propagation takes place in straight lines with a **velocity** $c = 300,000$ km/sec... Who would imagine that this simple law has plunged the conscientiously thoughtful physicist into the greatest intellectual difficulties?" Those intellectual difficulties, as described by Einstein, concern the relativity of simultaneity, and observer-dependent time.

In order to understand the relativity of simultaneity, the concept of simultaneity itself must be defined. There is no problem determining whether two events that occur at the

same place are simultaneous; if they occur at the same time (the same reading of a clock at that point), they are simultaneous. But what of events which are separated in space? Einstein proposed an operational definition. Consider an observer equidistant between the locations of two events (for example, two strokes of lightning). Einstein states, the events are simultaneous if that observer measures or sees the two events as occurring at the same time.

Seeing the two events involves light rays propagating from the two events to the observer's eye. If each stroke is the same distance from the observer, the light signals carrying news of each stroke take the same amount of time to reach the observer. If he sees the strokes simultaneously, they occurred simultaneously. What if there are two different observers, each equidistant from the events, but in relative **motion** with respect to one another? Are two events, which are judged simultaneous by one observer, also judged simultaneous by the other?

To analyze this situation, Einstein proposed his famous thought-experiment involving two observers and a railroad train moving, for example, to the right along a straight embankment. One observer, Jonathan, is on the train, at the center, while the other, Nicholas, stands alongside the embankment. Remembering the principle of relativity, there is no experiment either physicist can perform that can determine which one is moving. All that can be said is that Jonathan is moving uniformly with respect to Nicholas, and Nicholas is moving uniformly (although in the opposite direction) with respect to Jonathan.

At the instant when the positions of Jonathan and Nicholas coincide, two strokes of lightning hit the embankment at the front and rear of the train. Both observers are equidistant from the lightning strokes. The light from each stroke travels with the same speed, c, with respect to each observer (this is the constancy of the **speed of light**). Nicholas, on the embankment, judges the strokes to be simultaneous, as the light from each reaches him at the same time. He sees the flashes simultaneously.

But what does Jonathan see? Since he is moving towards the position of the front stroke, light from it will have less distance to travel by the time it reaches him. Conversely, light from the rear stroke will have a greater distance to traverse. Thus, the light from the front stroke reaches Jonathan before that from the rear. Each observer employs Einstein's intuitive definition of simultaneity, as each was equidistant from the two events. Nicholas judges them to be simultaneous, while Jonathan does not. Simultaneity is thus relative: two events, which are simultaneous for one observer, will not be simultaneous for another in motion with respect to the first.

Considering time interval, the amount of time that elapses between two events, Einstein's thought experiment shows that the time interval between strokes is different for each observer. It is zero as measured by Nicholas's clock, but greater than zero as measured by Jonathan's. According to Jonathan, who, like all inertial observers can consider himself to be at rest, the time interval as measured by a moving clock (Nicholas and his clock are, according to Jonathan, moving to the left) is less than that measured by his. This is a particular example of what is known in general as time dilation, often summed up by the statement that "moving clocks run slow."

According to the postulates of relativity related to time dilation, time in a moving system will be observed by a stationary observer to be running slower by a factor that is the reciprocal of the Lorentz contraction equation. Although the effects are negligible for small velocities, the effects increase asymptotically as velocity approaches the speed of light.

See also Relativity, special

SINGULARITY

A mathematical function is said to be singular at a point if it "blows up" there; that is, if the function evaluated at that point becomes infinite. The point is a *singularity* of that particular function.

In physics, any point at which the mathematical expression for a physical quantity becomes infinite is a singular point. Thus, for example, the origin, $r=0$ is a singularity of the Coulomb expression for the electric field surrounding a point charge, q, given by $E=kq/r^2$.

When discussing the physics of **space-time**, the relevant mathematical function is the curvature, which is represented mathematically by the Riemann curvature tensor. The Riemann tensor has twenty independent components, each of which is a function of the space-time coordinates. Loosely speaking, a space-time point at which any component of the Riemann tensor blows up is called a singular point, or a singularity. There is a curvature singularity in the Schwarzchild **space-time geometry**, again at the origin, $r=0$. As physical quantities can never take on infinite values, the laws of physics break down at a singularity—they lose their power to predict the results of a physical measurement.

This is referred to as "loosely speaking" because it is quite hard (and perhaps impossible) to give meaning to the idea of a singularity as a "place" in space-time if the very concept of space-time breaks down. The rigorous mathematical definition is thus more complicated, but in this, as in many other things, the simple, more intuitive picture certainly suffices as a way of visualization.

SMOOT, GEORGE F. (1945-   )
American physicist

An astrophysicist and cosmologist, George F. Smoot conducted seminal research on the origin of the **universe**, including hard data to support the **big bang** theory of the universe's origin. Born in Yukon, Florida, Smoot gained an early respect for learning, particularly in math and science, from his father, a hydrologist, and his mother, a science teacher. As a youth, he enjoyed reading science fiction and science books.

Although initially interested in medicine when he enrolled at the Massachusetts Institute of Technology (MIT),

George Smoot at a news conference. (AP/Wide World Photos. Reproduced by permission.)

Smoot soon turned his attention to mathematics and physics. He received both his B.S. (1966) and Ph.D. (1970) from MIT, with a focus on **particle physics**. After graduation, Smoot turned his attention to **cosmology**, reasoning that it was a less crowded field where he could make a greater impact while still studying **subatomic particles** as important factors in the universe's origin. In 1970, Smoot joined the University of California at Berkeley, where he has remained throughout his career. At Berkeley, he worked with Nobel Prize-winning physicists Luis W. Alvarez and Emilio Segre, both of whom had studied under **Enrico Fermi**, one of Smoot's heroes since his college days at MIT.

Smoot worked on the high-altitude particle physics experiment (HAPPE) with Alvarez at Berkeley in an effort to find evidence of the big bang theory. First set forth in the late 1920s, the theory gained wide acceptance approximately 40 years later when scientists discovered cosmic **background radiation** as predicted by the theory. Focusing on cosmic **radiation**, which is believed to be a relic of the intense **heat** resulting from the early big bang, Smoot set out to probe the early universe. His efforts included the use of high-altitude balloon studies and high-flying U-2 spy planes to carry specialized probes. In the mid-1970s, Smoot began working with National Aeronautics and **Space** Administration (NASA) to develop the **Cosmic Background Explorer (COBE)**, which would be sent into space to develop a more comprehensive map of **cosmic**

background radiation. Taking more than a decade before it was finally approved, the $160 million COBE satellite was launched on November 18, 1989, from the Vandenberg Air Force base in California to its orbit around the Earth.

Data gathered via COBE enabled Smoot and colleagues to map the radiation intensity from the early big bang. The team also discovered variations of radiation so small that they were believed to be the seeds on which **gravity** worked to develop the galaxies throughout the universe. Cautious because earlier work by other physicists on the big bang theory had proven to be false, Smoot demanded that his team keep their discovery a secret until further analysis was made. In fact, Smoot offered any member of his team who could prove that a mistake had been made a free plane ticket to any destination in the world. Finally, on April 23, 1992, Smoot announced that the group had detected smaller ripples in the cosmic background radiation **temperature** that coincide with the big bang theory, providing the first concrete data for what had previously been speculation.

Smoot is not married and splits his time between homes in Berkeley, California, and Greenbelt, Maryland, where he began working on COBE in the 1970s. Author of *Wrinkles in Time*, a popular book about cosmology, Smoot spends the majority of his time working long hours on the COBE project. His outside interests include philosophy and cross-country skiing. Smoot has served as a research physicist at the Lawrence Berkeley National Laboratory since 1970, and at the University of California Space Sciences Laboratory since 1971.

SNEL, WILLEBRORD (1580-1626)
Dutch physicist, astronomer, and mathematician

Snel is most famous for his research on the refraction of **light**, which yielded Snel's law. He also devoted a considerable amount of time to the calculation of geographic distances through trigonometric triangulation, and he is often credited with founding the modern science of mapmaking.

Snel was born in Leiden, Holland, in 1580, where his father taught mathematics. He was considered something of a prodigy, and at the age of twenty, though he had not yet completed his master's degree, he was allowed to lecture in mathematics at the University of Leiden. Snel spent the next few years traveling across Europe and meeting such influential scientists as **Tycho Brahe** and **Johannes Kepler**. He completed his graduate studies in 1608 and, in 1613, was appointed to his father's chair at the university teaching mathematics, physics, and **optics**.

Shortly after his appointment, Snel began his experiments with triangulation. Very little work in this field had been attempted, and so Snel had to develop most of its techniques along the way. Using a large quadrant, he began to equate distances with the angle by which they were visibly separated. After many repetitions Snel even suggested a figure for the radius of Earth based on his trigonometric find-

Solar collector panel covered with photovoltaic cells. *(Photo Researchers, Inc.)*

ings; this figure has been recently tested and found to be very precise.

Probably Snel's most important contribution to the annals of science was his law of refraction. He noted that a ray of light passing from one transparent medium into another (for example, moving from air into water or glass) would bend slightly at the boundary. This phenomenon had been observed long ago by **Ptolemy**, had never been properly explained. Snel showed that the sine of the two angles (the angle of incidence and the angle of refraction) were related, and that their ratio was a constant. That constant is known as the index of refraction and is determined by the type of material the light is passing through.

There is some controversy over the true authorship of Snel's law. Snel himself never published his findings, and it was not until 1637 that the law of refraction was introduced by **René Descartes**. Descartes's equation differed slightly from Snel's but was blatantly derivative. Some prominent scientists—most notably **Christiaan Huygens**—accused Descartes of plagiarism, particularly after it was found that he had made a number of trips to Leiden after Snel's death.

SOLAR THERMAL ENERGY

The **heat energy** that is supplied to Earth by the **sun** is called solar **thermal energy**. The sun, like all **stars**, is a massive sphere of superheated gas that remains heated by the **nuclear reactions** occurring within its center. Within its core, the fusion of hydrogen atoms creates **helium** atoms. Because the resulting helium atoms have a lower **mass** than hydrogen protons that fused to form them, a "mass defect" is said to occur in which portions of **matter** are converted into energy, some of which is radiated as solar thermal energy. These atomic reactions release so much energy that the **temperature** of the center of the sun is estimated to be approximately 10 to 20 million degrees Celsius. The surface of the sun, however, has a much lower average temperature of about 6,000°C. Once created, solar thermal energy travels a mean distance of 92,960,000 mi (149,591,000 km) to reach Earth where it has many profound effects. For instance, solar thermal energy in conjunction with Earth's rotation creates atmospheric and oceanic currents that affect global **weather** patterns.

Solar thermal energy also has practical applications. Heat from the sun can be used in thermoelectric **generators**. Here, solar thermal energy is used to create or save **electricity** by superheating water or oil for use in generators. Solar energy is also used as a supplemental heat source for households in areas where sunlight is plentiful. For domestic use, solar energy is collected in roof plates, and the heat that is generated can be transferred to water and stored there for distribution throughout the house. Similarly, solar stoves use focused thermal energy from the sun to cook food.

SOLID STATE PHYSICS

Solid state physics is the study of how solids behave, in particular, how the presence of other atoms effect properties of other nearby atoms and the applications of these properties. Band theory, an important aspect of solid state physics, explains how electrons behave in solids. Much like the **energy** levels of a single **atom**, the structure of a solid requires that electrons can only be in specific "bands" of energy. Each band can only hold a specific number of electrons and those bands with the least energy associated with them fill up first. The highest, unfilled band is called the conduction band. The highest filled band is called the valence band. The energy difference between the conduction and valence bands is called the energy gap. If the highest band is only partially filled, the material is a conductor. If the energy gap is large, the material is an insulator. If the energy gap is small, the material is a **semiconductor**.

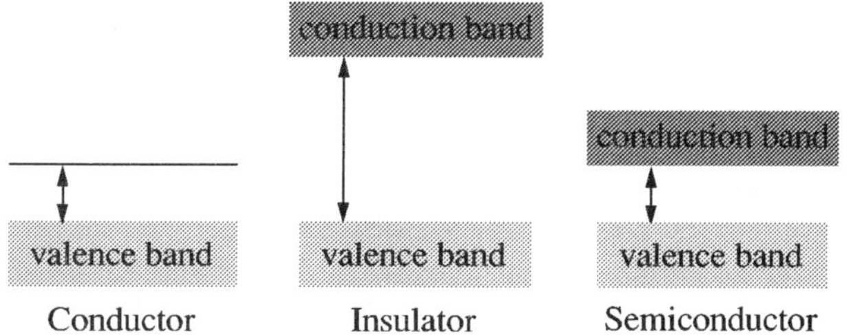

"Doping" a solid is a process that involves adding impurities to a solid, usually to silicon, so that there are either extra electrons or fewer electrons (adding "holes") in the solid. The extra electrons and holes effect the conduction band of the material. Those with extra electrons are called "n-type" solids; those with holes are called "p-type" solids. When these solids are sandwiched together, it creates n-p junction. Combinations

of these junctions are used to create solid state components such as the diode and the transistor.

A diode consists of one n-p junction and acts as a one-way valve for electrical current. When you have an n-p junction, the electrons can drop into the "holes" (lack of **electron**) until there is an area where there are no free charges. This area is called a depletion zone. Applying a potential difference in one way gives the electrons the energy they need to cross the zone; this is called forward biased. When the diode is reverse biased, the potential difference makes the zone harder to cross and so there is no current flow.

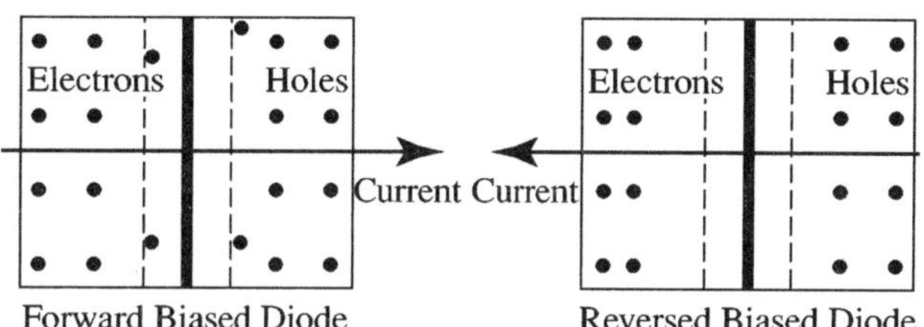

The **Hall effect** is seen when a solid conducting material, such as a strip of copper, has a current running through it while it is in a magnetic field. The charge carriers, usually electrons, are deflected by the magnetic field and a potential difference is set up on the sides of the strip transverse to the current. This allows the sign (either positive or negative) of the charge carriers to be determined.

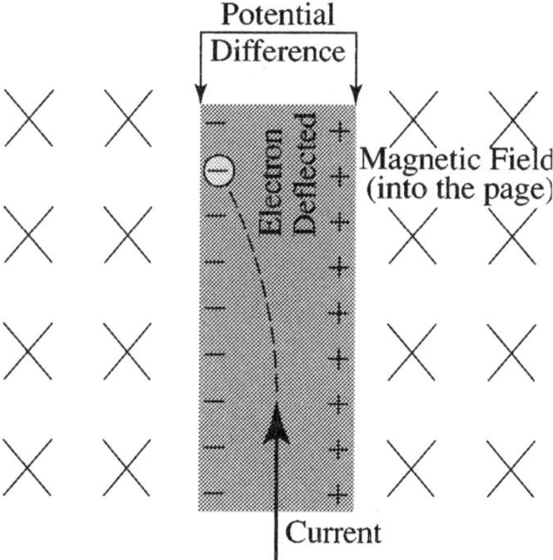

Superconductors are another development from solid state theory. These substances usually show little or no conductivity at room **temperature**. Some, such as ceramics, are actually used as insulators at room temperature. However, when they are cooled to very low temperatures, all the resistivity of the material vanishes, and so the material can superconduct current without resistance. A magnet set spinning over a cooled superconductor will induce an opposing magnetic field inside the material so that the magnet is levitated just above the surface. This is known as the Meissner effect. Superconductors have applications in many areas; they can be used in levitating trains, medical imaging, and high-energy particle research.

SOMMERFELD, ARNOLD JOHANNES WILHELM (1861–1951)

German physicist

Sommerfeld was born in Königsberg, Prussia, on December 5, 1868. He was educated at Königsberg University and then taught at Göttingen, Clausthal, and Aachen. In 1906, he succeeded Boltzmann as professor of theoretical physics at the University of Munich.

The first two decades of Sommerfeld's career were devoted to a wide range of topics, including optical phenomena, the behavior of gyroscopes, and **radio waves**. In 1915, he turned his attention to the subject where he made his greatest contribution: **atomic structure**.

In 1913, **Niels Bohr** had proposed a new model of the hydrogen **atom** in which **electron**s traveled in specific quantized **energy** levels outside the **nucleus**. The distance of each orbit from the atomic nucleus became designated by the principal **quantum number**, n. The movement of electrons between energy levels resulted in the gain or loss of specific quantities of energy that could be observed in experiments as hydrogen's **spectral lines**. Almost immediately, however, it became apparent that the Bohr model was inadequate to explain certain fine structure of spectral lines. That is, upon close examination, some lines that appear to be single in fact consist of two or more lines aligned very close to each other.

Sommerfeld's solution to this problem was to suggest that electrons travel in elliptical, not circular, orbits around the nucleus. This realization allowed Sommerfeld to introduce into Bohr's analysis a second factor, called the azimuthal quantum number, designated by the letter l. This quantum number relates the major and minor axes of the elliptical orbit to each other. He also found that a third quantum number was needed to explain the splitting of spectral lines. This quantum number, m, explains the behavior of an electron in the presence of an external magnetic field.

Sommerfeld continued in his post at Munich until the rise of Adolf Hitler in Germany. He strongly opposed the new Nazi government and was forced into retirement. He survived both Hitler and the war, however, and lived to the age of 83. He died in Munich on April 26, 1951, after being struck by an automobile while out walking.

SONAR

Sonar is a remote sensing technique based on the echolocation of **sound waves** in water. The name Sonar is short for SOund Navigation And Ranging and is closely related to the technology of **RADAR** (RAdio Detection And Ranging). The basic principles behind both technologies are the same. RADAR employs the transmission and reflection of radio waves in air to detect objects in the atmosphere. Similarly, SONAR employs sound waves rather than **radio** waves to detect underwater objects. Sound waves are preferable in underwater applications because radio waves lose too much **energy** when they propagate through the water. Likewise, the propagation of sound waves in air is also inefficient.

SONAR was originally motivated by the desire to detect icebergs after the sinking of the *Titanic* in 1912. World War I subsequently stimulated the development of further SONAR technology in order to detect enemy submarines. However, SONAR now has tremendous civilian, as well as Naval, applications. Such applications include determining water depth, finding fish, mapping the ocean floor, and locating various objects such as pipelines, wellheads, and shipwrecks. It can also be used to study water currents, and determine characteristics of ocean floor sediments. Although a human invention, it can be argued that the principles behind SONAR were discovered far earlier by nature. Some marine mammals use sound waves to navigate and find food in much the same way that the Navy searched for submarines in World War I. Echolocating bats (*Eptesicus fuscus*) are experts at navigating using only waves in the 25–100 kilohertz (kHz) range.

Most SONAR systems fall into two broad categories: active (echolocation) and passive. Passive SONAR is the simplest category and it consists of systems that essentially do nothing but listen for sound vibrations in the water. Active SONAR systems are more complicated and involve the projection of short **pulses** of sound that propagate through water in a narrow beam at about 1500 m/s. These pulses then reflect off possible targets in the beam. A portion of the reflected beam is later detected at the beam's original source. The distance to the target can be calculated as half the time between transmission and detection of the beam times the **velocity** of the beam. The direction of the target can also be determined from the relative orientation of the reflected beam. In passive SONAR the direction of a target can be determined as well from the direction in which sounds waves are detected, but range detection is much harder. However, passive SONAR systems are far preferable in military applications, where it is crucial to detect enemy vessels without divulging one's location.

In active SONAR, the echolocation beam is first transmitted by an electroacoustic transducer that converts an electrical signal of certain duration and **frequency** into a sound signal that radiates into the water. Typically the transducer is reciprocal in the sense that it can also be used to detect the returning echoes. This can be accomplished by using a piezoelectric ceramic such as barium titanate. In the case of transmission, an alternating **voltage** applied to the ceramic causes it to vibrate, which produces sets of sound waves in water. In the case of detection, the alternating **pressure** received from the echo causes the piezoelectric substance to generate a similar electrical signal that, after amplification, yields the required range and direction information about the target.

Most practical transducers are formed from a large number of small piezoelectric elements packed into a transducer array. A large number of elements allows sound waves emitted to constructively and destructively interfere in such a way that the emitted beam has a highly directional nature. Beams emitted from arrays can be as narrow as 0.1 angstroms and can give high resolution sonar images of the target.

NTSB sonar map of EgyptAir flight 990 (which crashed into the Atlantic on October 31, 1999) debris field. *(Image by Reuters Newsmedia Inc./Corbis Corp. Reproduced by permission.)*

There are three basic transducer orientations. In the usual depth sounder, the sound beam is directed downward from a transducer that hanging below the water from the keel of a ship. Such configurations are useful for measuring ocean depth and detecting any fish that may be below. Another configuration is the side scan, where a beam is transmitted on either side of the ship, usually perpendicular to the direction of travel. Such a configuration is useful in scanning large areas and in mapping the ocean floor while moving at constant speed. The third and most popular employs a rotating sound beam that can scan a section of water surrounding a usually stationary sonar platform.

Typical frequencies used in underwater SONAR are dictated by the absorption spectrum of sound energy in water. Roughly, the higher the frequency, the greater the absorption of sound energy, at a rate proportional to the frequency squared. Thus passive SONAR used to detect submarines operate between 3.5-35 kHz, which yields a detection range of about 10 km (6 mi). Active SONAR used to detect much smaller objects must operate at higher frequencies, typically between 100 kHz-1.0 MHz yielding ranges of a few hundred meters or less.

SONIC SPECTRUM

The sonic spectrum is broken into three basic parts: **infrasound**, audible **sound**, and **ultrasound**. Infrasound is sound that is below the human threshold of **hearing**, about 20 Hz.

Ultrasound is sound that is above the human threshold of hearing at about 20,000 Hz.

Acoustics is the science of sound, usually that of the audible range. Acousticians study how sound is produced, absorbed, reflected, and transmitted by objects. They use this information in a variety of ways, including designing better concert halls, testing musical instruments, and studying the development of hearing and speech.

Infra-acoustics studies are mostly of a geophysical nature. Most seismic **waves**, such as those before and after an earthquake, are infrasonic. Scientists have observed that these infrasonic waves can be reflected and transmitted through Earth. Large earthquakes can send **shock waves** to the other side of the globe. Other applications of infra-acoustic studies include animal communication. Animal behavior scientists were long mystified by the sudden reactions of an entire elephant herd to an unobserved event. Recently, it has been shown that elephants communicate with infrasound. These low **frequency**, long **wavelength** sounds are capable of traveling great distances, thus a herd might be warned of danger long before it reaches them.

Ultrasound has many applications, the most prominent of these being medical. It is used as both a diagnostic tool and as a non-surgical method of treatment. As a diagnostic tool, it is used to image internal organs. It is also used to image a fetus while it develops in the mother's womb. The signals are bounced off the tissue and are picked up by a receiver, then translated into an image on a screen. This way, doctors can see inside the body without having to do surgery. As a treatment, ultrasound can be used to disintegrate kidney, gall, and bladder stones so they can be passed by the patient in a normal manner. The ultrasound is set to a resonant frequency for each stone. The stone's response has such a high amplitude that it breaks the stone apart. Each stone has its own resonant frequency based on its size and makeup, so a variable ultrasound source is used. This method also eliminates the need for surgery in most cases. Ultrasound can also be used as a therapeutic treatment. The healing process in lesions and broken bones has been shown to be accelerated by using ultrasound.

SONOLUMINESCENCE

Sonoluminescence is the emission of **light** from bubbles of air trapped in water that contains intense **sound waves**. Hypothesized in 1933 by Reinhardt Mecke of the University of Heidelberg, from the observation that intense sound from military **SONAR** systems could catalyze chemical reactions in water, it was first observed in 1934 by H. Frenzel and H. Schultes at the University of Cologne. Modern experiments show it to be a result of heating in a bubble when the surrounding sound waves compress its volume by approximately a million-fold.

The precise details of light generation within an air bubble are not presently known. Certain general features of the process are well understood, however. Owing to water's high degree of incompressibility, sound travels through it in the form of high speed, high **pressure** waves. Sound waves in air have smaller pressure, since air is highly compressible. The transmitted **power** in a wave is proportional to the product of its pressure and the amplitude of its vibrational **motion**. This means that the **wave motion** is strongly amplified when a sound wave travels from water to air.

Since the **speed of sound** is much greater in water than in air, a small bubble within water-carrying sound waves will be subjected to essentially the same pressure at every point on its surface. Thus, the sound waves within the bubble will be nearly spherical.

Spherical symmetry, along with the large amplitude of displacement of the surface of the bubble, results in extreme compression of the air at the center of the bubble. This compression takes place *adiabatically*, that is, with little loss of **heat**, until the air is at a high enough **temperature** to emit light. Temperatures within a sonoluminescing bubble range between 10,000–100,000K (17,540–179,541°F; 9,727–99,727°C).

Theorists are working on models of sonoluminescing bubbles in which the inward-traveling wave becomes a shock wave near the center of the bubble; this is thought to account for the extremely high temperatures there. Although the maximum temperatures within a sonoluminescing bubble are not known with any certainty, some researchers are investigating the possibility of using these imploding **shock waves** to obtain the million-degree temperatures needed for controlled **nuclear fusion**.

SOUND

Sound is produced when an object vibrates. When something vibrates, it transfers kinetic **energy** to the **molecules** of the medium surrounding it. This **kinetic energy** causes the molecules in the medium to vibrate, transferring kinetic energy to new molecules. This transfer of kinetic energy through a medium causes a series of compressions (areas where the molecules are crowded together) and rarefactions (areas where the molecules are spread out). These compressions and rarefactions move through the medium, away from the original vibration. When a series of compressions and rarefactions move through a medium, it is called a longitudinal wave. Therefore, sound is a longitudinal wave.

When you speak, air moves from your lungs past your vocal cords, causing them to vibrate. Your vocal cords consist of two flaps of tissue that can move closer to or farther away from each other. When they move closer together, they cause a compression in the air between them. When they move farther apart, they cause a rarefaction. The continuous vibration as air moves over your vocal cords and causes a series of compressions and rarefactions, in other words, sound. A stereo speaker operates in the same manner. The speaker vibrates back and forth, causing the air in front of it to vibrate. This vibration is the sound coming from your stereo.

The speed of the sound **waves** depends on several factors. These factors are the **temperature, elasticity,** and **density** of the medium through which the sound waves travel. Sound travels slower at lower temperatures. For example, the **speed of sound** in air at 32 °F (0°C) is 0.2 mi/sec (0.32 km/s), compared to its speed in air at 77°F (25°C), 0.21 mi/sec (0.35 m/s). Sound travels most rapidly in solids and slowest in gases. The molecules in a solid bounce back faster when vibrated, in other words, are more elastic, than liquids or gases. The speed of sound is slowest in a material with a higher density. For example, solid nickel is denser than solid iron. The speed of sound in nickel is 3 mi/sec (4.8 km/s), while the speed of sound in iron is 3.2 mi sec (5.1 km/s). Therefore, the speed of sound would be greatest in a high temperature, low density solid.

Since sound is actually a series of longitudinal waves, sound displays the wave properties of amplitude, **frequency,** and **wavelength**. The amplitude of a wave is the height of a wave crest above its origin. The amplitude determines the loudness of a sound. The frequency is the number of wave crests that travel past a point per second. The wavelength is the distance between successive wave crests. Frequency and wavelength are inversely proportional to each other and they determine the pitch of a sound.

The pitch of a sound is how high or low the sound is. As an object vibrates faster, it will create a sound of higher pitch. In other words, the pitch depends on the frequency of the sound waves. The frequency of sound waves is measured as the number of waves, or cycles, per second, also known as hertz (Hz). One hertz is equal to one cycle per second. The high note of an opera singer may have a frequency of 1,000 Hz, whereas the low sound of **electricity** "humming" has a frequency of around 60 Hz.

If all vibrating objects produce sound, why is there no audible sound when we wave our hand in the air? The human ear can only hear sounds in a relatively narrow range of frequencies, from about 20–20,000 Hz. A sound with a frequency lower than 20 Hz is called infrasonic, and a sound with a frequency higher than 20,000 Hz is called ultrasonic. The sound produced when you wave your hand is infrasonic, or beyond our **hearing** capacity. Elephants can hear and produce infrasonic sounds below 20 Hz. Dogs can hear sounds up to 35,000 Hz and cats can hear sounds up to 65,000 Hz. This is why your dog can hear a dog whistle, which produces an ultrasonic sound, and you cannot.

The amplitude of a sound wave determines how loud a sound is, or its intensity. The larger the amplitude, the more energy is carried by the wave, and the higher the intensity of the waves. High-intensity waves are louder than low-intensity waves. The relative intensity of sounds can be measured using the decibel (dB) scale. A sound with an intensity of 0 dB can barely be heard, while a rock concert produces sound with an intensity of around 120 dB. Any sound over 120 dB can cause pain and hearing loss in humans.

Waves can combine with other waves to produce **interference**. The interference may have an additive effect, called constructive interference, or a deleterious effect, called destructive interference. In sound waves, constructive interference increases its intensity, making the sound louder. Destructive interference decreases the intensity, making the sound softer. Engineers construct auditoriums and concert halls so the sound waves coming from the stage bounce off of the walls, creating constructive interference. These engineers are especially careful to avoid floor plans that cause the sound waves to bounce off of the walls creating destructive interference. Destructive interference can create the absence of sound, or dead spots, in a room. Engineers who design buildings with this phenomenon in mind are called acoustical engineers.

Sound waves can be used for our benefit in ways other than to create pleasing music or soothing sounds such as a babbling brook. Ultrasonic waves are used in Sound Navigation and Ranging, or **SONAR**, systems to explore deep-sea ocean floors. A research vessel can send sound waves down into the water. The sound travels to the ocean floor and bounces back up to the surface. By using the speed of sound in salt water and the time it takes for the sound to travel to the floor and back, researchers can calculate the depth of the floor at that point. Using this procedure, maps of the ocean floors have been created. The auto focus feature on many cameras also uses SONAR to determine how far away the object to be photographed is from the camera lens.

Ultrasonic waves can also be used to clean delicate objects such as fine jewelry. The object is placed in a liquid and ultrasonic waves are sent through the liquid, creating high-speed vibrations that knock dirt and debris from the surface of the object. Physicians can also use ultrasonic waves to create a picture much like an x ray or to send vibrations to damaged muscle, aiding the healing process.

Charles E. Yeager, broke the sound barrier flying a rocket propelled aircraft.

SPACE

Space commonly refers to any region of the **universe** that contains a very low **density** of dust and gas particles, such as **interstellar space**. Space is also defined as the three-dimensional extent of the physical universe. In relativistic terms however, the volume of the universe is associated with a fourth dimension, i.e. that of **time**, thus defining a four-dimensional **space-time**.

Classical representations of space are based on Euclidean geometry, which describes objects in three-dimensional space using the Cartesian coordinate system and following a series of postulates and axioms. The Euclidean model of space is built from systems of parallel lines and objects in space are described as points using equations defining variables representing distances from real or imaginary parallel lines. The Pythagorean theorem provides an illustration of **Euclidean space** by providing a means to calculate the distance between any two points or objects. Some of the Euclidean postulates were shown to be non valid and other types of geometries were developed to represent space, such as hyperbolic geometry, which rejected the Euclidean postulate stating that only one parallel line can be drawn through a point located next to a line, and elliptic geometry, which modified other Euclidean axioms, such as the one stating that two parallel lines are taken to be everywhere equidistant.

It was, however, the work of **Herman Minkowski** that had by far the most significant impact on the classical 3-D concept of space. He realized that the geometry of space had to introduce a time dimension element and he described space as a flat, **four-dimensional space-time** continuum in which the time coordinate of one coordinate system depends on both the time and space coordinates of another system moving relative to the first. This concept was incorporated in the **relativity theory**, formulated by **Albert Einstein.** In the Special theory, space representation uses points to describe events, which are both moments in time and locations in space and straight lines represent particles moving through space as a function of time. In the General theory, the Minkowski concept is modified to introduce space-time curvature and **mass** points, which allow the formulation of the **equivalence principle** stating that gravitational mass is identical to inertial mass.

If we consider a region of interstellar space relatively near some **stars**, and assume the only **force** felt in this region is the gravitational pull of these stars, then all bodies accessing this region will accelerate in precisely the same way. This is equivalent to stating that space has a property responsible for this **acceleration** due to **gravity**. The important conclusion then is that gravity alters the properties of space, which implies that space is a dynamical quantity along with time and they are affected by **matter** and **energy**, i.e. by the bodies present in space. These deformations of space and time in turn determine the subsequent **motion** of the bodies in space-time: matter tells space-time how to curve and space-time tells matter how to move.

See also Euclid; Frame of reference; Hilbert, David; Length contraction; Newtonian physics; Relativity, general; Relativity, special; Space-time geometry; Unification of physics; Warping of space and time

SPACE-TIME

In 1905 **Albert Einstein** published his seminal paper "On the **Electrodynamics** of Moving Bodies." In it he put forth his "Principle of Relativity," which soon gave the work the name by which it has been known since. All inertial (i.e., unaccelerated) observers, he said, regardless of their state of relative **motion**, must find the identical mathematical form for all laws of physics. (This is an example of what is today known as an "invariance principle"; in this case the laws are invariant—unchanging in form—when the **space** and **time** coordinates used by one observer are transformed into those used by another.) The laws known at that time were those of **Isaac Newton**'s **mechanics** and **James Clerk Maxwell**'s electrodynamics. **Newton's laws of motion** obeyed Einstein's principle, but **Maxwell's equations** did not.

The problem, Einstein realized, lay with the set of transformation equations relating position and time measurements made by observers in uniform relative motion. **Galileo** had been the first to consider these, and the relationships between the coordinates used by two such observers were known as the Galilean transformations. Newton's laws, but not those of Maxwell, were "Galilean invariant." A new set of transformations was needed, under which all laws of physics remained unchanged. The Dutch physicist **Hendrik Antoon Lorentz** had discovered the proper set of transformations in 1904 (although for a different reason). Maxwell's equations are Lorentz invariant, while Newton's laws must be modified somewhat to bring them in line. These modifications, which change "classical" mechanics into "relativistic" mechanics, lead to one of the most well known predictions of **relativity theory**, the equivalence of **mass** and **energy**. The mathematical statement of Einstein's Principle of Relativity is simply this: that the laws of physics are Lorentz invariant.

The **Lorentz transformations** do something quite strange; they mix the space and time coordinates in a way that the Galilean transformations do not. Consider two observers, Nicholas at rest, and Jonathan in uniform motion with respect to him. Nicholas measures an object's position. It is no surprise that the position coordinate, as measured by Jonathan, depends upon both the position measured by Nicholas, and the time at which Nicholas made that measurement (since Jonathan is moving). But it is quite surprising that the two observers do not agree upon the *time* at which the position measurement occurred. It had always been assumed that there existed a single "universal time," measured by every observer. Of this unvarying, absolute time Newton had written, "Absolute, true, and mathematical time, of itself and from its own nature, flows equably without relation to anything external." The root of all of the disturbing and counterintuitive results of relativity (e.g., **length contraction** and **time dilation**) is the error of that assumption. Time is observer-dependent.

There is a beautiful symmetry, though, to this mixing of space and time. The structure of the Lorentz transformations is such that the space coordinates and the time behave in an identical manner. Thus, we should not treat the world as three-dimensional, requiring three coordinates for its description, with time a "different thing," a universal parameter, agreed upon by all observers. Rather, the world is four dimensional, time being merely another coordinate needed to localize an event.

In 1908 the mathematician **Hermann Minkowski** showed that the equations of relativity simplify enormously if the three spatial and one time coordinate needed to define an event (a particular place at a particular time) are treated as the four coordinates of a higher dimensional "space"—a four-dimensional "space-time." Speaking of this new view, Minkowski wrote, "Henceforth space by itself, and time by itself, are doomed to fade away into mere shadows, and only a kind of union of the two will preserve an independent reality."

That "independent reality" is made manifest by the fact that it is quantities in space-time, not quantities in space, such as lengths, or time intervals independently, that are Lorentz invariant; i.e., that have the same values for all inertial observers. According to Einstein, "Pure 'space distance' of two events with respect to [one observer] results in 'time-distance' of the same events with respect to [another]." Minkowski showed that the

Lorentz transformations have the mathematical form of rotations in space-time. Just as a three-dimensional vector (for example, the displacement between two points in space) has a length, which is invariant under rotations in space, a "four-vector" (for example, the "interval" between two events in space-time) has a magnitude that is invariant under the Lorentz transformations. The world as we observe it at any particular time, described by physicists in terms of three-dimensional **vectors** such as **momentum**, is simply a three-dimensional "slice" of four-dimensional space-time.

As Einstein wrote in 1916, "The non-mathematician is seized by a mysterious shuddering when he hears of "four-dimensional" things, by a feeling not unlike that awakened by thoughts of the occult. And yet there is no more commonplace statement than that the world in which we live is a four-dimensional space-time.... That we have not been accustomed to regard the world in this sense... is due to the fact, that in physics, before the advent of the theory of relativity, time played a different and more independent role, as compared with the space coordinates.... The four-dimensional mode of consideration of the 'world' is natural in the theory of relativity, since according to this theory time is robbed of its independence."

SPACE-TIME CURVATURE · See Space time geometry

SPACE-TIME GEOMETRY

The Special theory of relativity, which included the concept of **space-time**, dealt only with relations between measurements made by inertial observers. This restriction to non-accelerated observers was a serious one, and German-American physicist **Albert Einstein** almost immediately began the search for a more inclusive theory, one that could relate observations made by any observer, regardless of his state of **motion**.

Any inertial observer perceives a three-dimensional Euclidean **space**, one in which, for example, the distance between two points satisfies the Pythagorean theorem, and the ratio of the circumference of a circle to its radius is π. This is flat space. And, an inertial observer who is at rest with respect to his clocks (no matter where they are located) sees them all run at an identical rate. This is uniform **time**. In the four-dimensional space-time of special relativity, for any inertial observer space is flat, time is uniform, and the interval between two events satisfies the Pythagorean relation $\Delta s^2 = \Delta x^2 + \Delta y^2 + \Delta z^2 - \Delta t^2$. Such Minkowskian space-time is the analog of Euclidean space: Einstein soon realized, however, that Minkowskian space-time could not suffice for a theory that allowed accelerated observers.

Consider, Einstein proposed, an observer (whom we'll call Nicholas), sitting on a uniformly rotating disk. Equipped with a standard meter-stick, this accelerated physicist will, by laying out the stick end to end, measure both the disk's radius and circumference. Previously, while the disk was not rotating,

another physicist (whom we'll call Jonathan), performed the same set of measurements using the same standard meter-stick. When Nicholas measures the circumference, the rotating meter-stick lies parallel to the direction of the disk's motion; when he measures the radius, it lies perpendicular. Thus the stick is relativistically shortened as the circumference is measured, but when measuring the radius, it is not. Nicholas and Jonathan obtain the same value for the radius, but Nicholas obtains a greater value for the circumference than did Jonathan. (He needs to lay out his shortened meter-stick more times than Jonathan did.) Each observer calculates the circumference/radius ratio: non-accelerated Jonathan finds π, while accelerated Nicholas obtains a value greater than π. Accordingly, the accelerated observer perceives a non-Euclidean space.

There are two clocks on the disk, as well. One clock is at the center, and the other on the edge. When the disk was stationary, Jonathan found they ran at the same rate. As the disk rotates, however, Jonathan sees the clock at the edge, which is moving with respect to him, lag behind the other. As Einstein said, "It is obvious that the same effect would be noted by an observer [Nicholas] whom we will imagine sitting alongside his clock at the center of the disk. Thus, on our circular disk, a clock will go more quickly or less quickly, according to the position in which the clock is situated." Time, on the disk, is not uniform.

So, for our accelerated observer, space is not flat, and time is not uniform. Such curved space-time is called Riemannian space-time.

Einstein's crucial insight, made in 1907, which he later called "the happiest thought of my life" argued that the effects of **acceleration** are indistinguishable from those of gravitation. This is the principle of equivalence. The prototypical illustration is an elevator in deep space, free from all gravitational forces, accelerating at a constant rate. There is no experiment a physicist in such an elevator could perform which will determine whether he is indeed in an accelerating elevator, or in an elevator at rest in a uniform **gravitational field**. All objects in the elevator, for example, fall towards the floor with an acceleration independent of their **mass**, the hallmark of Newtonian gravitation. A theory, then, which allows accelerated observers must necessarily be a theory of **gravity**.

The effects of acceleration, we have seen, are a curving of space-time. Those effects are indistinguishable from those of gravity. Thus, gravity is a curvature of space-time. All the gravitational effects we observe, from Newton's apple falling from the tree, to the orbiting of the **Moon** around Earth, are due to the non-flat, Riemannian geometry of space-time.

Gravity, Newton postulated, was an effect of mass. Gravity, Einstein postulated, is a curving of space-time. Thus mass (or its equivalent, **energy**) curves space-time. This is the essence of Einstein's masterwork, the General theory of relativity, completed in 1915.

In the presence of a massive object, a body still traverses the shortest possible path between points, but, since the space-time has been warped, that path is no longer a straight line. It is, rather, a geodesic, and its precise form depends upon the space-time geometry, the mathematical description of the curvature. An easily visualized example is a two-dimensional rubber sheet. When it is flat, its geodesic is a straight line. But add a mass and the sheet distorts, its geodesic will be a curved path. In this model there is no **gravitational force**. Bodies always take the shortest possible path, but the particular mass distribution present will determine what that path will be. Gravity is a manifestation of space-time geometry.

See also Length contraction; Non-Euclidean geometry; Relativity, general; Relativity, special; Relativity theory; Simultaneity and time dilation

SPARK CHAMBER

A spark chamber is a means of detecting particles moving through **matter**. When particles are released through **radioactivity** or **acceleration** and move through matter such as a gas field, they leave a trail of **ions**. The purpose of **particle detectors** is to make the ion trail observable. The spark chamber was developed in the 1960s as a response to the deficiencies of **bubble chambers** and cloud chambers. These conventional particle detectors encountered problems such as **background radiation** interfering with the reaction under study. Particle detectors are important to better understand the nature of **subatomic particles**, the process of nuclear decay, and the effect of external influences (such as magnetic field) on those particles.

The important components of the spark chamber are a field of inert gas and parallel metal plates through which high **voltage** is intermittently passed. When the voltage charges the ions, small, elongated sparks appear and the trail can be detected. The inert gas field usually consists of a neon and **helium** mixture with 90% neon. The metal plates are made up of grids that are synchronized to send out electrical charges as the ions pass. The synchronization of electric charges is coordinated by scintillation counters that detect the presence of the charged particles and trigger the electric discharges in the appropriate place at the appropriate moment.

Although the thin foil parallel plate spark detector is the most common, cylindrical and thick metal plate spark chambers have also been developed for the study of specific phenomena. For example, the thick metal spark chamber can be use to measure the range and energies of ionization loss. All spark chambers maintain a gap of 2–15 mm.

SPECIFIC GRAVITY

When a variety of different materials are compared for their relative weights, some will be "heavier" than others. To make a fair comparison one should take samples that are identical in volume. Physicists compare by using standard volumes of substances such as one cubic centimeter or one cubic meter of the substance.

Physicists know that one cubic centimeter of a pure substance always has the same **mass**. The value of that mass is called its **density**. The density of any substance is the mass of one unit volume of that substance. For example, the density of gold measures is 19.3 grams per cubic centimeter, or 19.3 g/cm^3. In the case of pure gold, the density will always be 19.3 g/cm^3. If it is not 19.3 g/cm^3, then we know that either the gold is impure or the substance is not gold at all.

It is useful to compare the densities of all substances to that of water. Lead has a density 11.3 times greater than water. The ratio of the density of lead to the density of water is 11.3. This ratio is called the specific **gravity** (or more commonly, relative density) of lead.

SPECIFIC HEAT AND STATISTICAL MECHANICS

Current **statistical mechanics** explains why the equipartition of **energy** principle is false. This law states that in a system of thermal **equilibrium**, an equal amount of energy is associated with each independent energy state. Using the equipartition of energy principle, scientists of the late 1800s calculated specific heats of polyatomic gases. However, these calculated values of specific **heat** did not correspond to the values obtained from experiments. The equipartition of energy principle failed because it used the classical mechanical expression for the molecular energy, but molecular vibrations and rotations obey quantum **mechanics**, not classical mechanics. After the formulation of **quantum mechanics** in the early part of the 1900s, the necessary modifications in statistical mechanics were easily made and the correct calculated values of specific heat were determined.

Statistical mechanics, the link between quantum mechanics and **thermodynamics**, provides a method of deducing the macroscopic properties of **matter** from the properties of the molecules composing the system. That is, statistical mechanics deals with both the macroscopic world and the microscopic world. Because there is such a large number of molecules in a macroscopic system, it is useful to employ statistical methods instead of trying to consider the **motion** of each molecule in the system. Statistical mechanics originated in the late 1800s with the work of **James Clerk Maxwell** and **Ludwig Boltzmann** on the kinetic theory of gases. **Josiah Willard Gibbs** made major advances in the theory and methods of practical calculation in *Elementary Principles in Statistical Mechanics* published in 1902. **Albert Einstein** contributed with work he conducted during 1902–1904 to the development of the kinetic theory. Since quantum mechanics had yet to be formulated, these scientists assumed that molecules of a system obeyed classical mechanics. In some cases, this assumption lead to incorrect results, one of which was calculation of the specific heats of some polyatomic gases.

Scientists in the early 1900s attempted to calculate the specific heats of ideal polyatomic gases. Using kinetic theory they were able to show that the **heat capacity** for an ideal monatomic gas was equal to 3/2 of the ideal gas constant, R.

However, **thermal energy** added to a gas of polyatomic molecules can manifest itself in rotational, vibrational and translational energies of the molecules and so the specific heat of a polyatomic gas exceeds that of a monatomic gas. Because of this, these scientists applied the classical mechanical kinetic theory to determine the contributions of molecular vibration and rotation to the specific heat. They first formulated an equation for the average molecular energy of the system: $U = N_A\varepsilon$, where U is the average molecular energy, N_A is Avogadro's constant, and ε is the sum of the energies of the individual molecules. The energy of a single polyatomic molecule, ε, involves several terms and is equal to: $\varepsilon = E_{tr} + E_{ro} + E_v + E_{el}$, where E_{tr} is the translational energy, E_{ro} is the rotational energy, E_v is the vibrational energy, and E_{el} is the electronic energy. Once formulated, they needed to determine the average of each term in the equation for molecular energy. They employed the probability distribution function of the Maxwell kinetic theory, which pertains to gas molecules and the distribution of their velocities. Ludwig Boltzmann enhanced Maxwell's ideas and created a new distribution law that made it applicable to all gas properties, allowed for size determination of gas molecules, and led to the separation of gases via centrifuge. The new Maxwell-Boltzmann distribution law stated that the average value for each term in the equation for molecular energy, ε_E, is equal to $(1/2)kT$, where k is Boltzmann's constant and T is **temperature**. This meant that classical mechanics and the Maxwell-Boltzmann distribution law predict that each quadratic term in the molecular energy equation contributes $(1/2)kT$ to the average energy per molecule. $U = N_A\varepsilon_E = (1/2)N_AkT = (1/2)RT$. So for the overall system, each term in the expression for molecular energy contributes $(1/2)RT$ to the molecular energy of the system. This is the principle of equipartition of energy. Also, since the heat capacity at constant volume, C_V, is equal to $C_V = (dU/dT)_V$, this means that each term contributes $(1/2)R$ to the heat capacity and hence the specific heat of a gas.

When these findings were applied to systems of polyatomic gases, such as **carbon** dioxide gas, the predicted values of specific heat and the experimentally determined values did not agree. In general, the heat capacity increases with temperature and attains the value predicted by the principle of equipartition of energy only at high temperatures. This showed that the equipartition principle was false.

The equipartition principle fails because it employs the classical mechanical expressions for the molecular energy, but molecular vibrations and rotations obey quantum mechanics. The term for molecular vibration, in particular, was not correct because the vibrational levels in a molecule are not closely spaced relative to kT, as rotational and translational levels are. At high temperatures, the harmonic-oscillator motions of the vibrations goes to the classical equipartition value, but at other temperatures modern statistical mechanics that incorporate quantum mechanics need to be employed.

Quantum mechanics was developed in the early 1900s. Quantum mechanics is mainly concerned with the probability of finding a system in a particular state. By using quantum mechanics, it is possible to find the probabilities and quantum mechanical energies of each molecule in a system, and in turn, it is possible to determine the macroscopic property of the average molecular energy of the system. Employing the advances made during the development of quantum mechanics, the problems with statistical mechanics were corrected. The corrected version of statistical mechanics showed that the terms for translation and rotation contributing to heat capacity agree with the classical statistical mechanical equipartition theorem, i.e. that $(1/2)R$ is contributed for each quadratic term in the overall energy equation. The disagreement in the terms for vibration contributing to heat capacity using classical mechanics and statistical mechanics corrected using **quantum theory** was revealed. Using modern statistical mechanics, scientists were able to formulate values for specific heats of polyatomic gases that were in much closer agreement with experimentally obtained values.

For small molecules, the gas-phase thermodynamic properties, such as specific heat, calculated by statistical mechanics are usually more accurate than the values determined experimentally by calorimetric measurements. For liquids and solids, where the **intermolecular forces** are much stronger because of the close proximity of the molecules, the intermolecular forces make the statistical mechanical values less accurate than the experimentally determined ones.

SPECIFIC HEAT CAPACITY

As mankind's knowledge of the nature of **matter** has increased, some of the old concepts about "heat" and "temperature" have changed. The atomic model of matter has atoms in a constant state of **motion**. The motion takes many forms. The atoms may be linked to other atoms to form molecules. This is not a static linkage, however, as the atoms are vibrating as if they were bound with springs. The clusters of atoms (molecules) also may also have a straight-line motion, if the substance is a gas or liquid, speeding along until collisions with other molecules force them to change direction. In addition, the clusters will have shapes that cause them to spin after they collide. In short, the total kinetic **energy** of a substance may be in the form of vibration, rotational and spin energy. There will also be a store of **potential energy** due to the forces of attraction and repulsion that exists between the atoms and molecules. **Heat** energy input changes the total potential and **kinetic energy** of the atoms that make up the substance.

Temperature, on the other hand, is a measure of only part of this total energy. A solid has atoms that vibrate in a bound lattice structure. There is no translational motion or rotational motion. Temperature, in the case of a solid, is a measure of the vibrational kinetic energy of the atoms. In a liquid or a gas, however, the temperature of the substance is defined predominantly by the translational kinetic energy of the molecules and not by the atomic vibrations or the spin of the molecules.

Materials differ in the amount of energy required to increase their temperatures. Lots of factors go into this difference. The size of the atoms and the shape of molecules and how firmly they are aligned with each other are some of the causes for the difference. Physicists have measured how much energy it takes to increase the temperatures of different substances. To do this they defined a standard measure to help them compare substances against each other. This standard measure is the specific **heat capacity**.

The specific heat capacity of a substance is the amount of energy required to raise the temperature of a unit **mass** of a substance by a unit amount. If we use the **International System of units** we would say it is the amount of energy (in Joules) required to raise the temperature of one kilogram of the substance by one degree Celsius. Water is a good example. One kilogram of water will require 4,200 joules of energy to raise its temperature by 1°C if its temperature is around 20°C. Putting 4,200 joules of energy into one kilogram of water by any means (stirring, electrical wires, gas flames etc) will cause the temperature to rise by 1°C. This value of 4,200 joules will change depending on the operating temperature of the water. At close to **absolute zero** the specific heat capacity of all substances tend to drop to zero. This means that it requires very little energy to heat up a substance just 1° from absolute zero. For example, at normal room temperature it requires 390 joules of energy to raise the temperature of 1 kg of copper by 1°C. But at the very low temperature of -250°C the same temperature rise can be achieved with only 140 joules of energy.

Solids generally have lower specific heat capacities than liquids; and liquids have lower specific heat capacities than gases. The reason for this is tied into the various ways in which atoms and molecules can absorb kinetic energy, as was mentioned above. The molecules in a gas have several ways of using kinetic energy. Inputting kinetic energy will cause the molecules to travel faster through **space**, will increase the vibrations of the atoms within the molecule, and will increase the speed of rotation of the molecules. In other words, a gas uses incoming energy in three different ways. This sharing of the kinetic energy results in a lower temperature increase because temperature is predominantly only measuring the translational kinetic energy. In a solid there is little or no translational or rotational energy since the atoms are held tightly in a rigid lattice. Thus all incoming kinetic energy simply gets used to increase the vibrations of the atoms or molecules in the lattice which gives the temperature reading. This means that comparatively small energy input will produce a large increase in temperature. A liquid has some free movement of its molecules and some rotation and vibration but not to the same degree as a gas, so a liquid temperature increase is usually between that of solids and gases for the same energy input.

Accurately measuring the specific heat capacities of various substances has lead to a greater understanding of the atomic and molecular structures of gases, solids and liquids. Early theories have been confirmed or disconfirmed by these measures. The study of phase changes (where gases turn into liquids and liquids into solids) has been helped by these measurements.

SPECTRAL LINES

The spectral lines observed in the absorption and emission spectra of atoms or **molecules** are the result of transitions occurring between different **energy** levels as electromagnetic **radiation** is absorbed or emitted. The **frequency** (ν)and **wavelength** (λ) of this electromagnetic radiation are related through the following equation: $\nu = c/\lambda$, where c is the speed of **light**. Thus, UV light absorbed at 3,000 angstroms (å) corresponds to a frequency of $(2.96 \times 10^{10}$ cm/sec$) / 3 \times 10^{-5}$ cm $= 10^{15}$ sec^{-1} $= 10^{15}$ hertz (Hz).

The variable that can be most accurately measured is λ. However, it is the frequency that is directly related to the energy, so spectroscopists usually locate the position of spectral lines using a unit called the wavenumber, which is inversely proportional to the wavelength and is represented by σ. Thus, if λ is measured in cm, σ is then measured in cm^{-1} and is equal to the number of **waves** in one centimeter. A wavelength of 3,000 å is then equal to: the number of angstroms in one cm/λ $= 10^{8}/3 \times 10^{3} = 33,333$ cm^{-1}.

A typical spectrum consists of a recording of the intensity of spectral lines as a function of wavelength. The position of the observed spectral lines will depend on the energies associated with the transition. Planck's equation, $E = h\nu$, where h is **Planck's constant**, states that electromagnetic radiation of frequency ν occurs in small, discrete units called **quanta**, or photons, and that the amount of energy any one of these quanta contains, $h\nu$, is directly proportional to the frequency. Thus, if electromagnetic radiation is absorbed by a molecule or **atom**, the energy of the quantum absorbed is equal to the difference in energy between two states of the molecule or atom. A wavelength of 3,000 å then corresponds to an energy equal to $E = 6.626 \times 10^{-34}$ joule sec $\times 10^{15}$ sec^{-1} = 6.626×10^{-19} joule. The absorption of electromagnetic radiation at 3,000 å will then result in observation of a spectral line, which is the result of a transition between **energy levels** separated by the above calculated energy and whose width is determined by the Heisenberg **uncertainty principle**. The spectral lines will be observed unless forbidden by the transition selections rules that are for the most part governed by symmetry and formulated within the framework of **molecular orbital theory**.

The spectral lines are always broadened, partly due to the finite resolution of the measuring spectrometer and partly due to Doppler and **pressure** broadening. Doppler broadening is due to the thermal **motion** of the emitting atoms or **ions**. For a Maxwellian **velocity** distribution, the line shape is Gaussian; the full width at half maximum intensity (FWHM) is, in å, equal to: FWHM= $(7.16 \times 10^{-7} \lambda (T/M)^{1/2})$.

Where T is the **temperature** in K, and M the atomic **weight** in atomic **mass** units (amu). Pressure broadening is due to collisions of the emitters with neighboring particles and the lineshapes are often Lorentzian.

The **electromagnetic spectrum** ranges from the higher energy far ultraviolet ($\lambda = 10$ to 200 nm; $\nu= 30,000$ to 1,500 $\times$ 10^{-12} Hz) to the lower energy microwave region ($\lambda = 300,000$ to 10^{6} nm; $\nu = 1$ to 0.0003 Hz). The absorption of radiation from the more energetic region, i.e., the far UV, near UV and visible regions, by molecules or atoms will result in electronic transitions, i.e., the promotion of their electrons to higher energy electronic levels. Absorption of lower energy **photons**, i.e. from the infrared region will result in vibrational or rotational motion of the constituent atoms of a molecule. Thus, the spectral lines observed in **UV visible spectroscopy** are due to **electrons** being promoted to higher energy levels while those recorded by **infrared spectroscopy** correspond to bond bending and stretching vibrations. The spectral lines observed in spectra thus provide information about the electrons and their arrangement in a given atom or molecule, just as the lines recorded in a vibrational spectrum will be indicative of the molecular geometry of the material. As such, they represent one of the most powerful tools available to investigate the fundamental properties of **matter**.

Some specific examples are the atomic spectral lines observed for a given gas which can reveal its elemental composition and concentration, since there is a direct relationship between the intensity of spectral lines and the amount of absorbing or emitting species present. This is how the composition of intergalactic gases has been identified.

See also Absorption spectrum; Microwave background radiation; Quantum numbers; Quantum theory; Spectroscopy; Speed of light; Symmetry and symmetry principles; Transverse and longitudinal waves

SPECTROSCOPE

A spectroscope is an instrument used to observe the atomic spectrum of a given material. Because atoms can absorb or emit **radiation** only at certain specific wavelengths defined by **electron** transitions, the spectrum of each type of **atom** is directly related to its structure. There are two classifications of atomic spectra: absorption and emission.

An absorption spectrum is produced when **light** passes through a cool gas. From **quantum mechanics** we know that the **energy** of light is directly proportional to its **wavelength**. For a given type of atom, a **photon** of light at some specific wavelength can transfer its energy to an electron, moving that electron into a higher **energy level**. The atom is then in an "excited state." The electron absorbs the energy of the photon during this process. Thus, a white light spectrum will show a dark line where light of that energy/wavelength has been absorbed as it passed through the gas. This is called an **absorption spectrum**.

The energy transfer is reversible. Consider the excited state photon in the example above. When that electron relaxes into its normal state, a photon of the same wavelength of light will be emitted. If a gas is heated, rather than bombarded with light, the electrons can be pushed into an excited state and emit photons in much the same way. A spectrum of this emission will show bright lines at specific wavelengths. This is known as an emission spectrum.

Light entering a spectroscope is carrying spectral information. The information is decoded by splitting light into its spectral components. In its simplest form, a spectroscope is a viewing instrument consisting of a slit, a collimator, a dispersing element, and a focusing objective (see figure). Light passes through the slit and enters the collimator. A collimator is a special type of lens that "straightens out" light coming in a various angles so that all the light is travelling the same direction. The wavefront is converted into a planar wavefront; if you wish to think of light as rays, all the light rays are made to travel in parallel.

Next, light enters the dispersing element. A dispersing element spreads light of multiple wavelengths into discrete colors. A prism is an example of a dispersing element. White light entering the prism is separated out into the colors of the spectrum. Another type of dispersing element is a **diffraction** grating. A diffraction grating redirects light at a slightly different angle depending on the wavelength of the light. Diffraction gratings can be either reflection gratings or transmission gratings. A grating is made of a series of fine, closely spaced lines. Light incident on the grating is reflected at an angle that varies as wavelength. Thus, white light will be divided into the spectral colors, and each colorwill appear at a discretely spaced position. A transmission grating works similar to a reflection grating, except that light travels through it and is refracted or bent at different angles depending on wavelength. The focusing objective is just a lens system, such as that on a **telescope**, that magnifies the spectrum and focuses it for viewing by eye.

A spectroscope gives useful information, but it is only temporary. To capture spectroscopic data permanently, the spectrograph was developed. A spectrograph operates on the same principles as a spectroscope, but it contains some means to permanently capture an image of the spectrum. Early spectrographs contained photographic cameras that captured the images on film. Modern spectrographs contain sophisticated charge coupled device (CCD) cameras that convert an optical signal into an electrical signal; they capture the image and transfer it to video or computer for further analysis.

A spectroscopic instrument in great demand today is the spectrometer. A spectrometer can provide information about the amount of **radiation** that a source emits at a certain wavelength. It is similar to the spectroscope described above, except that it has the additional capability to determine the quantity of light detected at a given wavelength.

There are three basic types of spectrometers: monochromators, scanning monochromators, and polychromators. A monochromator selects only one wavelength from the source light, whereas a scanning monochromator is a motorized monochromator that scans an entire wavelength region. A polychromator selects multiple wavelengths from the source.

A spectrophotometer is an instrument for recording absorption spectra. It contains a radiant light source, a sample holder, a dispersive element, and a detector. A sample can be put into the holder in front of the source, and the resulting light is dispersed and captured by a photographic camera, a CCD array, or some other detector.

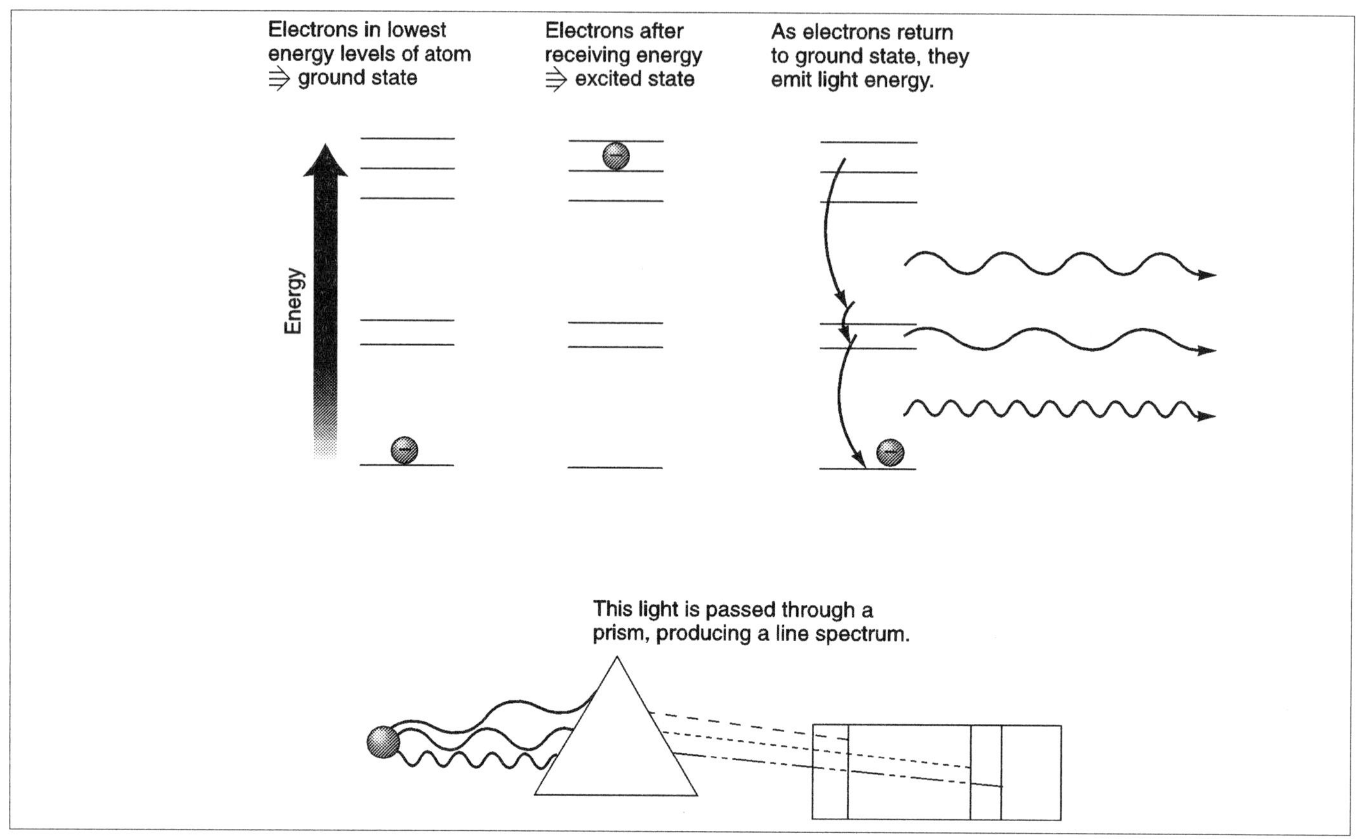

Atomic spectroscopy.

An important class of spectrometer is called an imaging spectrometer. These are remote sensing instruments capable of acquiring images of the Earth's surface from an aircraft or from a satellite in orbit. Quantitative data about the radiant intensity or reflectivity of the scene can be calculated, yielding important diagnostic information about that region. For example, a number of important rock-forming minerals have absorption features in the infrared spectral region. When sunlight hits these rocks and is reflected back, characteristic wavelengths of the light are absorbed for each type of rock. An imaging spectrometer takes a picture of a small region of rocks, splits the light from the image into different wavelengths, and measures how much reflected light is detected at each wavelength. By determining which quantities and wavelengths of light are absorbed by the region being imaged, scientists can determine the composition of the rocks. With similar techniques, imaging spectrometers can be used to map vegetation, track acid rain damage in forests, and track pollutants and effluent in coastal waters.

Another class of spectrometer highly useful to the laser industry is the spectrum analyzer. Although **lasers** are nominally monochromatic sources, there are actually slight variations in the wavelengths of light emitted. Spectrum analyzers provide detailed information about the wavelength and quality of the laser output, critical information for many scientific applications.

SPEED OF LIGHT

The speed of light is the constant speed at which **light** and other electromagnetic **radiation** travel in a **vacuum**. The value of the speed of light in a vacuum is equal to 299,792,458 m/s (186,282.397 mi/s), as recommended by the Committee of Data for Science and Technology on July 1999 for international use in all fields of science and technology. A commonly used approximation for the speed of light is 3.0 x 10⁸ m/s (186,000 mi/s). The electromagnetic constant c denotes the speed of light. Light travels slower in a medium (such as glass or water) than it does in a vacuum. Since **space** is not a perfect vacuum, light travels through space at a very slightly lower speed and decreases somewhat more when the radiation enters, for instance, a planet's atmosphere. Light would take about 1.2 seconds to travel from the **Moon** to Earth, around 8.5 minutes to travel from the **Sun** to Earth, and roughly 100,000 years to travel across the **Milky Way galaxy**.

The first demonstration that light travels at a finite speed was provided by observations from astronomer Ole Röemer in 1675. Röemer observed that the elapsed **time** between eclipses of Jupiter's satellites by Jupiter became shorter as Earth moved closer to Jupiter and became longer as Earth and Jupiter traveled further apart. He correctly concluded that this phenomenon was caused by the time needed for light to cross the increased distance between the two **planets**. Röemer's measurement of this **velocity** was the first reasonable approximation of the currently accepted value for the speed of light.

Physicist Armand-Hippolyte-Louis Fizeau performed the first laboratory measurement of the speed of light in 1849. His procedure sent a beam of light to a mirror by passing it through a gap between two teeth in a wheel. If the wheel were stationary, the mirror would reflect the beam back through the same gap. If the wheel was set in rapid **motion**, by the time the reflected beam reached the wheel a tooth would have moved into the location occupied by the gap when the light first passed the wheel. Knowing the wheel's rotational speed, Fizeau calculated the time required for a tooth to move into the position occupied by the adjacent gap. This was the same time that the light was sent traveling the distance from the wheel to the mirror and back. The result of the experiment was accurate to within four percent of the currently accepted speed of light.

In 1864, following the groundwork laid by Coulomb, Ampére, Gauss, and Faraday, **James Clerk Maxwell** created his classical theory of **electromagnetism**. This theory predicts that **electromagnetic waves** will propagate through a vacuum at a constant speed. Recognizing that this calculated speed was very similar to the speed that had been measured for light, Maxwell suggested that light must be one form of electromagnetic radiation.

There are distinctive features of the speed of light that set it apart from other speeds. One feature is that the speed of light is an absolute barrier, as predicted by **Albert Einstein**'s special theory of relativity. Nothing that imparts **energy** or information can go faster than the speed of light, and no body containing **mass** can travel as fast as the speed of light.

Another feature of the speed of light is that it cannot depend on the motion of the source. Astronomer Willem de Sitter proved this when he found that two **stars** in a double-star system, while revolving about a common **center of mass**, both send light to Earth at a velocity "c" even though one star is sending light while traveling *toward* Earth at a velocity "c + v," while the other star is sending light while traveling *away* from Earth at a velocity "c - v." The speed of light from both stars, de Sitter proved, does not include the speeds of the stars themselves. The motion of the source does not change the speed of light.

A third feature of the speed of light is that its speed is independent of the motion of the observer. A famous experiment that proved this is the **Michelson-Morley experiment** performed in 1887. At that time **Albert Michelson** and **Edward Williams Morley** were attempting to measure the speed of the Earth through "ether" (the so-called medium for transmission of electromagnetic **waves**) by measuring the speed of light in two different directions. However, the pair did not observe any difference in the travel time along two light beams set at right angles to each other. With equipment precise enough to detect the small orbital speed of Earth compared to the speed of light, their result indicated that Earth was stationary (with respect to the expected **ether**). Eventually Michelson and Morley concluded that the speed of light does not depend on the motion of the observer. The theory of special relativity, as laid out by **Albert Einstein** in 1905, assumed as a premise that the speed

of light c is independent of the motion of both source and observer.

See also Relativity, special

SPEED OF SOUND

Sound can be described as a **pressure** disturbance that travels through the constituent particles of a medium: as one particle of the medium is perturbed by the pressure, it exerts a **force** on an adjacent particle, thus perturbing that particle in turn. The speed of sound reflects the speed at which the disturbance is passed from particle to particle, or the distance it travels in **time**. This form of pressure disturbance is propagated through the medium exclusively as a longitudinal wave in the case of gases and liquids, i.e., a type of wave in which the displacement of the medium is parallel to the propagation of the wave. In this case, the speed of sound refers to the distance a compression or a rarefaction point travels per unit time. In solids however, sound may propagate as a transverse wave, i.e., a type of wave in which the displacement of the medium is perpendicular to the direction of propagation of the wave. Unlike **light**, which can travel through a **vacuum**, sound **waves** require a medium to travel though and their propagation speed will not depend on wave properties such as period, **frequency** or amplitude, but on the inertial and elastic properties of the medium through which they travel, such as bulk modulus, **elasticity** and **density**. If the medium is a homogeneous gas, the speed of sound is constant and can be calculated using:

$$v_{sound} = \sqrt{\gamma RT / M}$$

where γ is a constant specific to the gas through which sound is propagating, R is the universal gas constant (8.314 J/mol K), T is the **temperature** (in units of kelvin) and M is the formula **weight** of the gas. In air, the speed of sound is approximated at normal atmospheric pressure by: v_{sound} = 331.5+0.6 T m/s, where T is in Celsius units, and which yields ~340 m/s. The speed of sound propagating through a liquid or a solid can be obtained using:

$$v_{sound} = \sqrt{B / \rho}$$

where B is the bulk modulus of the propagating medium and ρ its density. Generally speaking, the more dense the medium, the faster the speed of sound, and the following expression is usually valid for the speed of sound when other properties are equal: v_{sound} in solids > v_{sound} in liquids > v_{sound} in gases. For example, at standard temperature, the speed of sound in water is ~1485 m/s, compared to 1522 m/s in sea water, ~5790 m/s in bulk stainless steel and ~4700 m/s in bulk brass.

The slow speed of sound in air relative to that of light (i.e. ~340 vs. ~300,000,000 m/s) makes it possible to approximate the distance of thunderstorms by calculating the delay between observing lightning (the light wave) and **hearing** thunder (the sound wave). For example, if thunder is heard five seconds after lightning, then the sound wave has traveled an approximate distance of 340 m/s × 5 s ≡ 1,700 m and the storm is located about a mile away. Echo, or the reflection of a sound wave from a barrier, is another time delay perception used to estimate distances with the speed of sound. The time delay between a voiced cry, shout or holler and its echo corresponds to the time required by the original sound wave to travel the round-trip distance to the barrier and back. It is then possible to calculate the one-way distance to the barrier responsible for the echo. This phenomenon is used by bats to navigate and hunt. They produce short bursts of ultrasonic sound waves that reflect off various bodies and which they can hear to estimate distances in their surroundings. The underlying mechanism is a consequence of the **Doppler effect**, or of the change produced in the frequency of a sound tone due to the relative **motion** between the source of the sound and the observer. Similar to bats, underwater **SONAR** uses the travel time of sound waves to determine distance. This makes the speed of sound in water critical to whales, dolphins, and other marine creatures that use sonar. Likewise it is also critical to submarines and oceanographic researchers who use SONAR to map the ocean floor.

See also Acoustics; Nodes; Oscillations; Reflection, refraction, and dispersion; Standing waves and resonance; Transverse and longitudinal waves; Wave interference; Wavelength

SPIN OF SUBATOMIC PARTICLES

Spin, s, is the rotation of a particle on its axis, as Earth spins on its axis. The spin of a particle is also called intrinsic **angular momentum**. Angular **momentum** is momentum (**mass** times **velocity**) times the perpendicular lever arm (distance between point of rotation and application of **force**). An intrinsic property is one that depends on the essential nature of an object. The total angular momentum of a particle then is the spin combined with the angular momentum from the particle moving around.

The idea of spin has been around for a long time. In 1925, G. E. Uhlenbeck and **Samuel Abraham Goudsmit** proposed that the **electron** has a spin, and the spin of the electron has been proven experimentally. The spin of the electron combined with its **electric charge** gives the electron magnetic qualities because of the electromagnetic force.

The spin of microscopic particles is so small it is measured in special units called "h-bar." Related to **Planck's constant**, h, which is defined as 4.1×10^{-21} MeV seconds, h-bar is defined to be h divided by two and by π (3.14159...).

Quantum **mechanics** is a branch of physics focusing on **subatomic particles**, and dealing in probabilities. One of the rules of **Quantum mechanics** says spin can only have certain values. Another way of saying this is spin must be "quantized." Particles with spin values of one-half h-bar, three-halves h-bar, five halves h-bar, and so on are called **fermions**

and described by a mathematical framework called Fermi-Dirac statistics in quantum mechanics. Particles with spin values of zero h-bar, one h-bar, two h-bar, and so on are called **bosons** and described by a mathematical framework called Bose-Einstein statistics. The quantization of spin means we have to add spins together carefully using special rules for addition of angular momentum in quantum mechanics.

In quantum mechanics, particles can also be represented mathematically using spinors. A spinor is like a vector, but instead of describing something's size and orientation in **space**, it describes the particle in a theoretical space called spin space.

Every particle and every **atom** or molecule (combination of atoms) with a specific **energy** has its own unique spin. Thus spin is a way of classifying particles. Using spin, all particles that make up **matter** are fermions. For example, all **quarks** and **leptons** have spins of one-half h-bar. The particles which mediate, or convey, the fundamental forces are bosons with spins of one h-bar. Baryons are particles made of combinations of three quarks. They have spins of one-half h-bar or three-halves h-bar. **Baryons** include protons (spin one-half h-bar) and neutrons (spin one-half h-bar). **Mesons** are particles made of a quark and an antiquark. They have spins of zero h-bar or one h-bar.

Spin should not be confused with a quantum mechanical idea called isospin, isotopic spin, or isobaric spin. Isospin is the theoretical quality assigned to quarks and their combinations which enables physicists to study the strong force which acts independently of electric charge.

SQUIDS

SQUID is an acronym for Superconducting QUantum Interference Device. **Superconductors** are materials that conduct electric currents with minimum resistance at low temperatures. Examples are common materials such as mercury, lead, and tin, and more exotic compounds that have been developed to superconduct at higher temperatures ranging up to 133K.

SQUIDs operationally rely on quantum mechanical interference similar to that demonstrated in a basic double-slit quantum experiment. In such experiments, a single particle (e.g., an **electron**) is shot at a double slit. Instead of acting as a particle, however, and going through just one of the slits, the particle acts as a wave going through both slits, and an interference pattern appears on a screen behind the double slits.

The SQUID operates in a similar way. A ring made of superconducting material is fabricated so that it has two Josephson junctions on opposite sides of the ring. A **Josephson junction** is a thin layer (a few atomic layers thick) of insulating material sandwiched between two superconducting materials. An **electric current** is run through the SQUID that can take two paths, through either Josephson junction and then recombine on the other side of the ring. In this simple setup, the two currents are mathematically represented by electron-pair wavefunctions. Each wavefunction has the same phase, so the currents constructively interfere yielding the original current.

SQUIDs are highly sensitivity to **magnetic fields**. When a magnetic field is produced that goes through the center of the ring (but not through the superconductor itself), a phase difference is generated in the electron-pair wavefunctions, so that they now interfere destructively instead of constructively. By measuring the current as a function of time, the phase shift can be extracted, and the **magnetic flux** and magnetic field can be determined. In this way, the SQUID operates like the double-slit experiment: the electrons have a choice between two paths, but without trying to detect which path the electrons take, an interference pattern results.

Because of its sensitivity to magnetic fields, the SQUID can be used as a precision magnetometer, a device used for measuring extremely tiny magnetic fields. For example, a ring 5 mm in diameter can detect magnetic fields much less that a millionth the size of Earth's magnetic field. Recently, SQUID "microscopes" have been constructed by various research groups and have been used to map out the magnetic fields of the human heart and the brain.

See also Double-slit interference; Wave function; Wave interference

STANDARD MODEL OF COSMOLOGY

The standard model of **cosmology** is the **big bang** theory. It refers to the event, between 8 and 15 billion years ago, in which the **universe** was born in a cataclysmic explosion. In the aftermath, **planets**, **stars**, and **galaxies** slowly formed as the primordial elements cooled and the universe expanded. According to the big bang theory, the universe had a definite beginning and continues to expand. Neither time nor **space** existed before that event, indeed, there was no "before" the big bang.

The big bang theory first emerged in the 1920s, in the decade after **Albert Einstein** published his theory of general relativity. Though general relativity's complicated equations predict an expanding universe, Einstein refused to believe it. He assumed that the universe is static and eternal, that it has neither beginning nor end.

Russian meteorologist Alexander Friedmann disagreed. He studied Einstein's work and postulated that galaxies and stars are carried along a fabric of space as it expands, like painted spots that move from each other on a balloon as it fills with air. A few years after Friedmann came to his conclusion, astronomer **Edwin Hubble**'s observations of distant galaxies gave empirical evidence to the expanding universe hypothesis. Hubble calculated changes in the wavelengths of **light** from distant galaxies and determined that, for the most part, they are speeding away from us, and each other, at enormous speeds.

In 1950, cosmologist **Fred Hoyle** argued in favor of another model, called the steady-state universe, and jokingly dubbed the expanding universe theory the "big bang." The

steady-state theory claims that the universe expands but new **matter** continuously forms so that the overall **density** of the universe remains constant. The universe, therefore, could have existed forever and will continue to appear basically the same.

Astronomers in the past 50 years have detected **radiation** from distant galaxies that indicate they were much closer together in the past, giving evidence for the big bang theory.

STANDARD MODEL OF PARTICLE PHYSICS

The standard model of **particle physics** is the **quantum theory** that describes all of the known particles and three of the four forces they feel. Although it is called a "model," it is a full-fledged scientific theory that has been extremely well-verified over the past thirty years. Currently, all laboratory and particle collider experiments have agreed with the predictions of the standard model, making it a great success. On the other hand, the only clues pointing to more fundamental physical theories come from disagreements with the standard model, which have been frustratingly hard to find.

The standard model is composed of three parts. The first part is the **matter** particles, which are called **quarks** and leptons. The second part is the forces, or interactions, which include the electromagnetic, weak, and strong forces. (Currently, there is no verified quantum theory of gravity.) The third part is the rules of **quantum mechanics**, which can be used to calculate observable numbers (such as decay rates and cross sections) using the properties of the particles and forces as input.

The matter particles make up the everyday matter we see around us. Protons and neutrons are made of quarks, and electrons combine with protons and neutrons to make atoms. The matter particles are further divided into quarks and leptons, which are spin-1/2 **fermions**. Quarks are particles with fractional **electric charge** and another type of charge referred to as **color**. There are two types of quarks in everyday matter, the up with electric charge +2/3 e and the down with electric charge -1/3 e. Just as negatively charged electrons combine with positively charge protons to make a charge-neutral **atom**, three quarks combine to form a color-neutral **proton** or **neutron**.

The other type of matter particles are called leptons. The **electron** is the most familiar **lepton**, with electric charge -e. The other common lepton is the electron **neutrino**, which is electrically neutral and is massless. (Whether or not neutrinos are actually massless is not known, but in the standard model they are treated as massless particles. In any case, the **mass** is extremely small.)

The forces, or interactions, in the standard model arise due to some symmetry principles. Under a set of global transformations, where the basic objects of the theory are redefined in the same way at each point in spacetime, the theory remains unchanged. However, to make the theory invariant under local transformations, where the basic objects are redefined differently at different points in **space-time**, an extra set of objects are added to the theory to offset the changes. These new objects are referred to as gauge bosons and they are said to mediate the interactions between particles.

In a **quantum field theory**, particles are considered to interact by exchanging gauge bosons. For example, two electrons **scattering** may occur the following way: the first electron emits a **photon** and recoils, and then the second electron absorbs the photon and changes its **motion** appropriately. The benefit of this viewpoint is that there cannot be any instantaneous transmission of **force**, which would violate the principles of relativity. The gauge bosons of the weak force are called the W- and **Z-bosons**, and the gauge bosons of the strong force are referred to as **gluons**.

The symmetry transformations that lead to the existence of a force are called gauge symmetries. **Gauge symmetry** is a very powerful idea, because it predicts that all particles interacting through a given force interact with the same strength. Without the gauge symmetry, it would be perfectly acceptable for the electromagnetic interaction between two protons to be stronger than between a proton and an electron. However, the gauge symmetry requires the same coupling strength to all particles with the same electric charges. This idea is even more useful in the theories of the weak and **strong interactions**, which we have less intuition about.

The standard model also incorporates two heavier copies of each matter particle already mentioned. It does not explain why they exist, but it does describe their interactions and how they can decay into the everyday matter particles. The heavier quarks similar to the down quark are the strange and bottom quarks, those similar to the up quark are called charm and top, and the heavier leptons similar to the electron are the **muon** and the tau. There are also two copies of the electron neutrino, but they also appear to be massless.

The standard model includes one more type of particle, known as the **Higgs boson**. The existence of the Higgs boson is necessary to explain why matter particles and certain gauge bosons have mass. The Higgs boson is the only missing piece of the puzzle, because it has not yet been discovered. However, particle physicists are sure of its existence and expect it to be discovered soon at either the newly refurbished Tevatron collider at Fermilab near Chicago, or at the new Large Hadron Collider at CERN in Geneva, Switzerland. In reality, there may be many Higgs bosons, because there are various models of Higgs physics that explain the existence of mass.

Aside from the problem of detecting the Higgs particle, all other known experimental results can be explained using the standard model. However, physicists desire to go further and explain some of the facts the standard model can only describe, such as why there are three families of particles. In addition, there are some theoretical loopholes in the **renormalization** theory that can be closed by introducing extra particles and symmetries. Because of these and other developments, theoretical physicists are convinced the standard model is not the whole story, but until they receive some signals that disagree with the standard model they will have to wait for verification of any new scenarios.

See also Electroweak force; Strong interactions; Symmetry and symmetry principles; W mesons; Grand Unified Theory

STANDING WAVES AND RESONANCE

One fundamental type of **motion** in the physical world is termed periodic motion. Periodic motion is motion repeats itself in a given cycle and is described by the following quantities: the period (T) of the motion, defined as the **time** required to complete a full cycle (in units of seconds per cycle), the **frequency** (f) which is the reciprocal of the period or 1/T (in units of Hertz), and the amplitude (A) which is the maximum displacement from the **equilibrium** position. A common type of periodic motion is that of the traveling wave, which propagates in the form of a sine wave. Traveling **waves** are observed when the waves are not confined to a specific region of the medium. The most commonly observed traveling wave is a water wave. A traveling wave has a **velocity** of propagation (v) equal to the product of the frequency and the **wavelength** (λ) which is the crest-to-crest distance of the wave, or the distance between two subsequent amplitudes. This fundamental wave relationship is expressed as: $v = f\lambda$. Another parameter is also used to describe traveling waves, the angular wave frequency, which is related to the frequency as follows: $\omega = 2\pi f$. The mechanism by which a traveling wave propagates itself through a medium is described by a wave equation of the type $y(x,t) = A \sin(2\pi/\lambda)(x-vt)$.

The concept of a standing, or stationary, wave can be understood by considering a wave propagating on a string whose ends are held some distance apart. For a string of length L which is fixed at both ends, the solution of the wave equation can take the form of a standing wave: $y(x,t) = A \sin\omega t \sin(n\pi x/L)$. The wave will propagate from one end of the string to the other end, and will then reflect and travel back in the opposite direction. The reflected wave will then interfere with the portion of the wave incoming from the opposite direction. This type of wave **interference** of an incident sine wave with a reflected sine wave produces an irregular and non-repeating wave propagation pattern leading to a likewise irregular and non-repeating motion of the string.

If the string is vibrated at a harmonic frequency, a sine wave pattern can be produced whose amplitude will change over time. Such frequencies can be selected so as to have the interference of the incident and reflected waves occur in such a manner that there are specific points, or **nodes**, along the string which seem to be immobile. Since the observed wave pattern is characterized by nodes which appear to be standing still, such a wave pattern is referred to as a standing wave pattern. Other points along the vibrating string change position over time, but in a regular manner. These points, or antinodes, vibrate back and forth between a large positive and large negative displacement. A standing wave pattern always consists of alternating nodes and antinodes and since antinode displacement occurs at regular time intervals, the motion of the string is also regular and repeating.

The harmonic frequencies required to generate standing waves are also called the natural frequencies of the string, or vibrating object, and they are those vibrations which result in the highest amplitude vibrations with the least input of **energy**. The lowest possible frequency which can generate a standing wave is referred to as the fundamental frequency. Such a wave has no nodes (i.e., it spans from one end of the medium to the other). The first overtone, or harmonic, of a fundamental is a wave with one node, which is exactly at the midpoint of the fundamental. The frequency of each harmonic is proportional to the speed of wave propagation and to the wavelength. The propagation speed is a function of properties such as string tension, diameter, and composition. The wavelength of the harmonic is dependent upon the length of the string and the harmonic number.

An object can be induced into vibrating at one of its harmonic frequencies under the influence of an externally applied oscillatory **force** provided by another object vibrating at one of those harmonic frequencies. This phenomenon is referred to as **resonance** (i.e., when one object vibrating at the same natural frequency of a second object forces that second object into vibrational motion). An example of resonance is the vibration induced in a piano wire of a given pitch when a musical note of the same pitch is played close by. Thus, all objects have resonant frequencies which are determined by the physical properties of the vibrating object. It is easy to induce an object to vibrate at one of its resonant frequencies, and much harder hard to induce it to vibrate at other frequencies. A vibrating object will also select its resonant frequencies from a complex excitation pattern and easily vibrate at those frequencies.

See also Acoustics; Electromagnetic spectrum; Electromagnetic waves; Harmonic motion; Oscillations; Particle-wave duality; Polarization of light; Sound; Transverse and longitudinal waves; Vibrating systems and resonance; Wave interference; Wave phenomena

STARK, JOHANNES (1874–1957)

German physicist

Johannes Stark's life can be divided into two fairly distinct and contrasting halves. During the earlier period, he demonstrated unusual skills as an experimentalist and won acclaim as a brilliant physicist, holding posts at universities throughout Germany. Founder and editor of the prestigious *Jahrbuch der Radioaktivität und Elektronik* (*Yearbook of Radioactivity and Electronics*), he is credited with discovering the **Doppler effect** in canal rays, and the splitting of the **spectral lines** of hydrogen by means of an external electrical field, a phenomenon now known as the Stark effect. For these discoveries, Stark received the 1919 Nobel Prize in physics. After 1913, however, Stark began to withdraw from the scientific community and to ally himself with Adolf Hitler's program of National Socialism. Along with Philipp von Lenard, Stark called for a "purification" of German science, an adoption of a non-Jewish "Aryan science." He failed to receive the recognition he sought in the political arena and eventually found himself ostracized by fellow scientists in Germany and throughout the world.

Stark was born on April 15, 1874, in Schickenhof, Bavaria. Raised on a farm, he attended local schools in Bayreuth and Regensburg. In 1894, he entered the University of Munich as a science major, earning his doctorate in 1897 for a dissertation entitled "Investigations on Lampblack." He then accepted a post as assistant to Eugen Lommel at Munich, a position he held for the next three years. In the spring of 1900, Stark moved to the University of Göttingen as assistant to Eduard Riecke, and was appointed privatdozent in 1903.

Stark Discovers the Doppler Effect in Canal Rays

During his tenure at Göttingen, Stark made the first of his important discoveries, the Doppler effect in canal rays. The Doppler effect is the change in **frequency** that occurs in a wave as its source advances toward or retreats from an observer. **Wavelengths** shorten as they approach, producing a higher pitch or frequency, and lengthen as they recede, producing a lower pitch. The apparent change in pitch of a train whistle as it passes an observer is a familiar example of the Doppler effect. The Doppler effect had been predicted by **Johann Christian Doppler** in 1842, and, by 1900, had been observed by the American astronomer **Edwin Hubble** in the **red shift** of galaxies, though no terrestrial example had yet been described. Stark decided that an appropriate way to observe the Doppler effect in the laboratory was with canal rays, beams of positively charged particles generated in a **vacuum** tube. In 1905, Stark used canal rays of hydrogen atoms to conduct the experiment and observed the predicted Doppler effect in hydrogen spectral lines—as they approached they reached higher frequencies, the violet end of the spectrum, and, like Hubble's galaxies, they shifted to the red, or lower, frequencies as they receded.

In 1906, Stark was appointed lecturer in applied physics and **photography** at the Technical College in Hannover. During his three-year tenure there, Stark was continuously on bad terms with his superior, Julius Brecht. Finally, in 1909, Stark accepted an appointment as professor at the Technical College at Aachen, where he remained for eight years.

Observes the Splitting of Spectral Lines

Since his days at Hannover, Stark had been thinking about a problem originally suggested by the work of the Dutch physicist **Pieter Zeeman**. In 1896, Zeeman had observed that the presence of a magnetic field can cause an element's **spectral lines** to split. This analogy to an electric field was too obvious for physicists to miss, and while a number of them had tried in the first decade of the twentieth century to produce this effect, none were successful. But in 1913, Stark succeeded in splitting spectral lines in an **electric field**. He placed a third electrode a few centimeters from the cathode in a vacuum tube and applied a potential difference of 20,000 volts between the two. When canal rays were generated in the tube, Stark was able to observe the splitting of the spectral lines of hydrogen gas, a phenomenon that is now known as the Stark effect. For his work on the Doppler and Stark effects, Stark was awarded the 1919 Nobel Prize in physics. In addition, he received the Baumgartner Prize of the Vienna Academy of Sciences in 1910, the Vahlbruch Prize of the Göttingen Academy of Sciences in 1914, and the Matteuci Gold Medal of the National Academy of Sciences of Italy.

Shifts from Scientific to Political Activities

After 1913, Stark slowly fell out of the mainstream of scientific research. Scholars have suggested a number of reasons for this change. A major factor seems to have been his inability to get along with other scientists and subsequent failure to receive an appropriate academic appointment. In 1917, he accepted a post as professor of physics at the University of Greifswald. He seems to have been happy in the conservative climate of this university, but decided to leave in 1920 to accept a similar post at the University of Würzburg. Stark was much less comfortable there, as his colleagues and superiors found a number of reasons to object to his presence, including his use of Nobel Prize money to finance the construction of a new ceramics factory. Although he devoted an increasing amount of time and attention to the factory, it eventually failed. Stark's colleagues found his attention to non-academic concerns ethically questionable. In addition, Stark's increasingly conservative political views were not well received in the liberal environment of Würzburg.

By 1922, Stark had become so uncomfortable at Würzburg that he resigned his post. He then became increasingly active politically in opposition to the post-World War I Weimar Republic, and in efforts to establish conservative, anti-governmental scientific organizations. One of these, the Fachgemeinschaft der deutschen Höchschulehrer der Physik (Professional Association of the German Higher Education Teachers of Physics), he established as an attempt to counter-balance the older, more liberal Deutsche Physikalische Gesellschaft (German Physical Society), based in Berlin.

When Stark decided to return to academic life, he found that he had made too many enemies and offended too many colleagues. He was rejected for posts at the universities of Berlin and Tübingen in 1924, Breslau and Marburg in 1926, Heidelberg in 1927, and Munich in 1928. By the early 1930s, Stark had become almost entirely an administrator of science, and then only in posts that his political influence had won for him. During the mid–1930s, he worked diligently to gain control over the direction of German science policy-making, but eventually lost out in that struggle to men who were even more closely allied to Adolf Hitler and the Nazi party.

By the beginning of World War II, Stark had become a self-made loner, having angered and annoyed both his former scientific colleagues and his former political allies. He sat out the war in his estate of Eppenstatt, in Bavaria, where he had constructed a laboratory. Stark had married Louise Uepter and had five children; beyond his professional interests Stark also enjoyed forestry and cultivating fruit trees. In 1947 he was sentenced to four years in a labor camp by a German de-Nazification court; he died in Eppenstatt on June 21, 1957.

STARS

A star is a body of hot gas and dust that shines by its own **light** produced by **nuclear fusion** reactions. The nearest star to Earth is the **Sun**. When compared to other stars in the **universe**, the Sun is of average size, with a diameter of approximately 860,000 mi (1,380,000 km). Over a million Earths could fit inside the volume of the Sun. The Sun is made up of gas (it is about 70% hydrogen and 28% **helium**), and is about 1.5 times as dense as water. The **mass** of the Sun is over 300,000 times that of Earth. Deep inside the Sun, **nuclear reactions** convert hydrogen to helium, changing **matter** to **energy** in a process called fusion. Fusion is responsible for converting over four million tons of the Sun's matter into energy every second.

The fusion reactions that take place within the Sun's, and any star's, core are the same as those that take place in a hydrogen bomb. These nuclear reactions require incredibly high temperatures. About 1% of a hydrogen atom's mass is converted into energy when two hydrogen atoms are converted into a single helium **atom**. This energy is used to maintain such high interior temperatures, sometimes up to millions of degrees Fahrenheit (hundreds of thousands of degrees Celsius). The surface temperatures are slightly lower and vary depending on the type of star. All stars are composed of at least 99% hydrogen and helium, fueling these nuclear reactions. The remaining mass of the star is composed of heavier elements such as nitrogen, **carbon**, or **oxygen**.

Stars can differ from each other in many ways. One difference can be in their densities. Some stars are thousands of times more dense than the sun, while others are thousandths of times less dense than air. Another notable difference between stars is size. Some stars, such as the red supergiants, are thousands of times larger than the Sun, while others, such as **white dwarfs** and **neutron stars**, are smaller than Earth.

The Sun is 93 million mi (150 million km) away from Earth. The next nearest star is 26 million million mi (42 million million km) away. It is difficult to comprehend such large numbers, so a scale other than miles or kilometers is used when talking about stars. Distances between stars are measured in terms of light years. One light year is the distance light travels in one year, approximately 6 million million mi (9,461 trillion km). One light year is the distance you would cover if you circled Earth 118 million times. Stars exist throughout the known universe, up to billions of light years away.

Stars appear brighter or dimmer not only because of their distance from Earth, but also because of a property called luminosity, or brightness. The star's apparent magnitude is how bright a star appears to be from Earth. The brightest stars are called first-magnitude stars, and the dimmest stars (to the naked eye) are sixth-magnitude stars. Each magnitude differs from the next by a factor of 2.5. A first-magnitude star is 2.5 times brighter than a second-magnitude star, which is 2.5 times brighter than a third magnitude star, and so on. Some stars are even brighter than first-magnitude stars. These stars have magnitudes less than one, such as zero-magnitude or even negative-magnitude stars. The brightest star seen from Earth is the Sun, with a magnitude of -27. The second brightest star is Sirius, with an apparent magnitude of -1.43.

The apparent magnitude does not tell how bright a star actually is; it only tells how bright it appears from Earth. The star's luminosity is its true brightness. Luminosity depends on the size and the **temperature** of the star. A star that has a higher temperature than a star of the same size will be more luminous. A star that is larger than another star at the same temperature will also be more luminous. A star's temperature not only determine its brightness, but also its **color**. Stars can be found in many different colors, ranging from red (the coolest stars) to blue-white (the hottest).

The term absolute magnitude is used to describe a star's true luminosity. This is the apparent magnitude a star would have if it were located 32.6 light years from the Sun. The most luminous stars are the red supergiants. These are the largest of all the stars. Red giant stars are less luminous than red supergiants, although their large size also makes them appear very bright. The dwarf stars are less luminous. Stars, then, can be classified according to their luminosity into one of several groups, listed in order of decreasing luminosity: the supergiants, giants, sub-giants, dwarfs, and white dwarfs.

Stars can also be classified according to their spectrum. There are seven spectral types of stars, in order of decreasing temperature: O, B, A, F, G, K, and M. A star's spectrum is related to its temperature because different elements will appear in a spectrum at different temperatures. Type O stars are the hottest stars, with temperatures over 55,000°F (30,000°C). These stars appear blue-white. Type B stars have temperatures around 36,000°F (20,000°C), and are also blue-white. Type A stars are those with temperatures around 20,000°F (11,000°C). These stars appear white. Type F stars are around 13,500°F (7,500°C), and are seen as yellowish white. Type G stars have temperatures around 11,000°F (6,000°C) and are seen as yellow stars, type K stars are around 7,500°F (4,100°C) and are orange stars, and type M stars are around 5,500°F (3,000°C) and are seen as red stars.

Other star types include variable stars and **pulsars**. A variable star, also known as a pulsing star, exhibits variable brightness. These stars appear brighter and dimmer as they expand and contract. This phenomenon usually occurs as a star is forming or dying. Pulsars are **neutron** stars that emit **pulses of radio waves** in a very periodic fashion. These stars produce powerful beams of radio waves due to their intense magnetic fields. As a pulsar rotates, the beam sweeps past Earth, which is measured as a pulse. Stars can be found in formations that, when viewed from Earth, appear to take on various shapes. These formations are called constellations. Stars can also be found in pairs, called binary or double stars, in clusters, and in groups of thousands of stars, called galaxies.

See also Astrophysics; Astronomy; Nuclear reactions; Red dwarf/Red giant; Stellar evolution

STATES OF MATTER

All material must exist in one of the three forms of matter—a solid, a liquid, or a gas. These are different physical states of being and each form has implications for the substance in question.

When we consider chemical substances most can exist in any of the three states. Which state of **matter** is encountered depends upon the physical conditions that they are being studied under. If the conditions are not specified then standard **temperature** and **pressure** is assumed. As a result of this it can be said that sodium chloride (table salt) is a solid. What in fact should be said is that at a pressure of one atmosphere ($1.013 \times 10^5 Nm^2$) and a temperature of 273.15 K (32°F or 0°C) sodium chloride is a solid.

The three states of matter have different ways of responding to changes in temperature and pressure. All three will show an increase in volume (expansion) when the temperature is increased and a decrease in volume (contraction) when the temperature is lowered. This effect is most noticeable with a gas and least noticeable with a solid, with a liquid being intermediate between the two extremes.

The difference between the states of matter are due to the differences between the amounts of **energy** their molecules have. A solid has molecules that are relatively immobile. All the molecules comprising a solid are in close contact with their neighboring molecules. The molecules are not free to move away from each other. This means that a solid has a definite shape and a definite volume, neither of which change much as the conditions of the environment alter. As the temperature increases the molecules are able to vibrate more vigorously. As the temperature decreases the molecules move more slowly and they become more aligned, this makes it easier for the transmission of **electricity**, if the solid is a conductor.

A liquid is very similar to a solid in many respects. The molecules of a liquid are also in close proximity to their neighbors. The liquid molecules are vibrating faster than those of a solid. A liquid has a fixed volume although its shape is not fixed; it will flow to take on the shape of its container. The layers of molecules in a liquid are more capable of moving over each other.

When a gas is considered the situation is very different. Within a gas the molecules have very high energy. The molecules of a gas are not touching any of their neighbors and they are free to act independently. This allows a gas to have neither a fixed volume nor shape. A gas will expand to fill the container into which it is placed. The properties of a gas are described by a series of equations known as the **gas laws**, these are **Boyle's law**, **Charles's law**, and the constant volume law.

The theory by which the physical properties of the three states of matter is explained by reference to the **motion** of the molecules making up the material is know as the kinetic theory of matter.

When a solid changes into a liquid it is by a process called melting. When a liquid changes into a gas it is by **boiling**. A gas changing to a liquid is condensation and a liquid changing to a solid is freezing. Some substances are capable of going directly from a solid to a gas, this is a process known as **sublimation**.

The bonding that is present has a strong influence on the state of matter of a material. Strong **intermolecular forces, van der Waals forces**, are characteristic of a solid. It is these forces that strongly hold together the molecules of a solid. With a liquid the molecules are also held closely together by intermolecular forces, although not as strongly as in a solid. The intermolecular forces in a liquid are not strong enough to keep the molecules from slipping past one another. It is this characteristic that makes the pouring of liquids a practicality. With solids the intermolecular forces virtually lock the molecules in place. The molecules in a solid can take up and retain a regular structure, a lattice. **Kinetic energy** has the tendency to speed up the movement of particles and **force** them apart whereas intermolecular forces tend to draw molecules together and stop them from moving. The particles of a solid and a liquid are fairly close together compared to those of a gas, and solids and liquids are called the condensed states. By altering the kinetic energies, a solid can change to a liquid and then to a gas. The kinetic energy that is applied in a situation such as this has to be sufficient to overcome the various van der Waals forces that are in operation within the molecule.

If pressure is increased on a substance, it forces the molecules closer together, which in turn strengthens the intermolecular forces. If the pressure applied is sufficient, then a state change to a more ordered form can be brought about. Hydrogen chloride in solution provides good indication of the relative strengths of intermolecular bonds compared to molecular bonds. To turn the hydrogen chloride from a liquid into a gas or vapor, only 16 kJ /mol is needed to overcome the intermolecular bonds that are present. Breaking the covalent bond to disassociate the molecule would require 431 kJ/mol of energy. It is due to this difference in energies required to break the different bonds that the majority of substances change from one state to another upon heating rather than disassociating. The majority of these intermolecular forces are due to the polarity of the molecules under consideration, the greater the polarity the stronger the resultant intermolecular forces.

It is possible for all three states of matter to exist in the same place and at the same **time**. This is known as the triple point and it only occurs at one set of temperature and pressure conditions. For example the triple point of water is at a pressure of 4.58 torr and a temperature of 32.0176°F (0.0098°C). At these conditions ice, water and water vapor can all exist together. This relationship can be shown in a **phase diagram**. A phase diagram is a plot of temperature against pressure for a particular substance and it shows the equilibria which exist between different states of matter (phases) under different physical conditions. The equilibria between the two phases are indicated by a line. There is only one point in a phase diagram where all three phases meet, this is the triple point. A phase change is a change from one state of matter to another.

The state of matter of the substances in a chemical equation can be indicated by adding the initial letter of the state—s for solid, l for liquid, and g for gas—within parentheses following the formulas.

STATIC PRESSURE AND DEPTH

Pressure is defined as **force** per unit area. Static pressure is the pressure in or due to a non-moving fluid, such as water. A fluid does not sustain shearing forces, thus a force that compresses the object is the only force transferrable by a liquid. This fact leads to the conclusion that an object submerged in a fluid will feel only a force perpendicular to each surface.

Two kinds of pressure are commonly discussed: absolute pressure and atmospheric pressure. Atmospheric pressure is an arbitrary value based on the atmosphere of the planet, in our case, Earth. We feel a pressure exerted by the atmosphere no **matter** where we are or what we are doing. The average value on Earth is 1.0×10^5 Pascals or 760 mm of Hg (14.7 psi) at sea level, but it changes with altitude and **weather** conditions. Even though everything on the planet feels this pressure, when we use a pressure gauge, it does not register. The pressure shown by the gauge is called "gauge pressure" and it measures how much difference there is above or below atmospheric pressure. Absolute pressure is measured from **absolute zero** and so would always include atmospheric pressure.

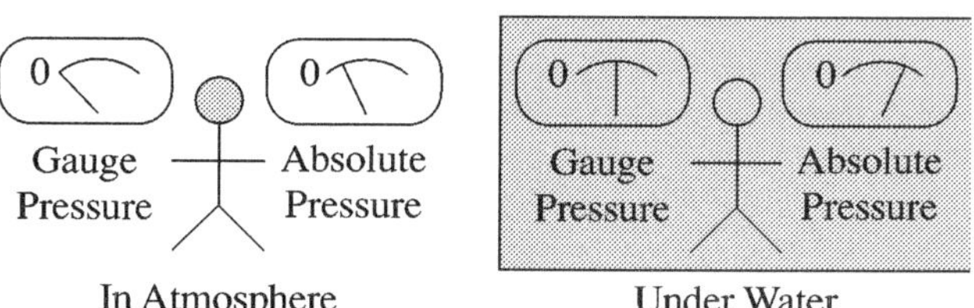

French scientist **Blaise Pascal** was the first to recognize that pressure in a fluid varies with depth. For example, the top of cylinder submerged in a tank of water is at the water's surface, and it has height (h), a cross sectional area (A), and a **mass**, (m). The forces on the cylinder are the **weight** of the cylinder, the atmospheric pressure at the top, and the water pressure at the bottom. All these forces must add up to zero: $P_0A + mg = PA$.

Since we know the mass is dependent on the volume (V) and **density**, ρ, of the object, this equation becomes: $P_0A + \rho Vg = PA$.

The volume of a cylinder is the height (h) multiplied by the cross sectional area, A. Substituting for volume and dividing by area, the equation becomes: $P_0 + \rho gh = P$.

Thus, the pressure, P, at the bottom of the cylinder is greater than the atmospheric pressure, P_0, at the top by an amount equal to ρgh. This is known as Pascal's law. It is only dependent on the height of the object (or depth in the fluid); it is not dependent on the area. Also, the forces perpendicular to the weight of the object are not included; the net force in that direction is always zero for static **fluids**.

Evangelista Torricelli invented a simple device to measure atmospheric pressure; the barometer. A tube with one end open is filled with liquid then turned upside down in a bowl of the same liquid. The liquid flows out of the tube until the atmospheric pressure and **vacuum** in the tube are sufficient to support the weight of the column. Since it is the densest liquid readily available, mercury is usually used in a barometer; a water barometer would need to be several stories tall. Knowing $P=0$ in a vacuum, the density of the liquid, and the height of the column the atmospheric pressure can be measured using Pascal's law.

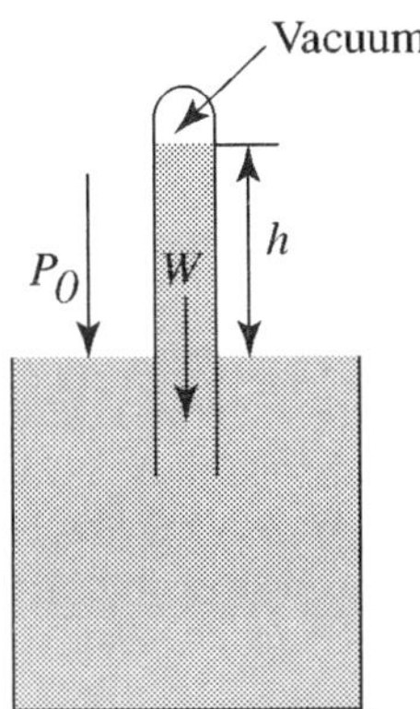

Another simple device to measure pressure also uses Pascal's law. The manometer is a U-shaped tube with two open ends. One end is open to atmospheric pressure, while the other is open to an unknown pressure. The liquid in the tube will come to **equilibrium** with a height difference proportional to the pressure difference. Knowing the value for atmospheric pressure, the height, and the density of the liquid (usually water), the unknown pressure can be calculated.

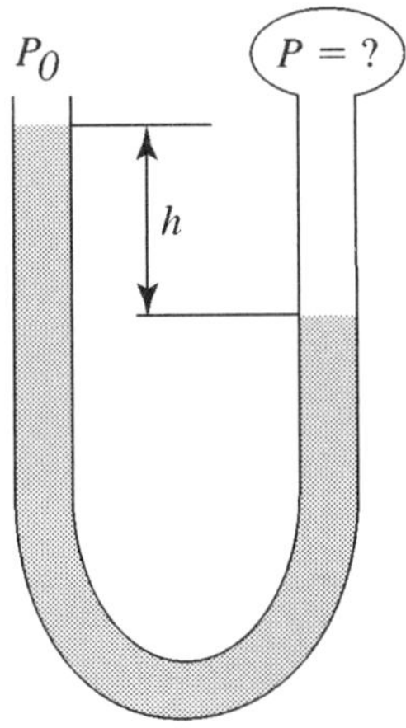

STATISTICAL MECHANICS

Statistical **mechanics** is the study of large groups of particles in **equilibrium**. While the individual behavior of each particle can be described with classical or **quantum mechanics**, this becomes unfeasible with large numbers of particles. Statistical mechanics deals with observable behaviors that can be deduced from the nature of the different particles by using probability. It is the link from the world of single photons and electrons to the world of our everyday lives.

Some elements of statistical mechanics are useful for any type of system. The concept of equipartition of **energy** states that for each degree of freedom, the system will have energy proportional to 1/2kT, where k is **Boltzmanns' constant** and T is the **temperature**. That is, for each way in which the system can be changed in position or **momentum**, the energy will increase according to the above rule. The equipartition of energy is useful for estimating the temperature-related behavior of different substances, an application of **thermodynamics** that crosses the line into statistical mechanics. Ideas such as enthalpy, **entropy**, **pressure**, and temperature, are also useful in both fields.

The basic part of statistical mechanics is the equation of state. This equation links general system properties like volume, pressure, temperature, and number of particles. (The most famous equation of state is the **ideal gas law**: PV = nRT, where P is pressure, V is volume, n is the number of particles, R is a constant, and T is temperature.) There are three different ways to arrive at an equation of state, through what are called different ensembles: the microcanonical ensemble, the canonical ensemble, and the grand canonical ensemble. Each makes a different but equivalent mathematical assumption about equilibrium states, allowing the equation of state to be derived in three equivalent ways.

The microcanonical ensemble assumes that the particles in the system have a total energy that is known. Each set of particle energy states that would sum to give that total energy is equally likely. The microcanonical ensemble is good for dealing with systems that are isolated from the rest of the **universe**, so that no transfer of energy can occur.

The canonical ensemble is not in isolation but has a fixed number of particles. The system's contact with the outside world is limited to transfer of energy into the particles that are already inside the system. This ensemble uses in its derivation the fact that temperatures are constant at equilibrium. The temperature, not the total energy, is assumed to be known.

Finally, the grand canonical ensemble does not require a fixed number of particles, merely a known average number of particles. It, like the canonical ensemble, is in contact with the outside world. The property which is assumed to be constant is the chemical potential, defined as the change in free energy when the number of particles changes but the temperature and volume are known and constant.

The ensembles describe differences in the particle's state: whether it's isolated and what properties are constants. There are also differences in the properties of the particles themselves, and those differences change the behavior significantly. There are three basic ways of expressing particle properties in statistical mechanics situations, and they are reflected in the distributions of particles in different states. They are the Maxwell-Boltzmann distribution, the Bose-Einstein distribution, and the Fermi-Dirac distribution.

The Maxwell-Boltzmann distribution describes essentially classical particles. It is assumed that most possible states of this system are not filled. This distribution is most commonly used for gases but will work for any equilibrium system with gas-like conditions. Low densities and high temperatures will make these statistics work most like a classical gas and will minimize interactions and quantum mechanical behavior.

The Fermi-Dirac distribution applies to quantum mechanical systems of **fermions**, particles that obey the **Pauli exclusion principle**. The largest number of particles possible in a given state is one. Therefore, the statistics derived by **Enrico Fermi** and **Paul Dirac** can be used to describe the behavior of equilibrium systems with this behavior, such as **electron** behavior in metals. This type of statistical mechanics can be used to describe the behavior of electrons in metals as it pertains to temperature, as well as having other applications.

The Bose-Einstein distribution applies to quantum mechanical systems of **bosons**. Since there is no restriction on the number of particles that can be in the same state here, at the lowest temperatures, all particles may be forced into the same ground state. This is known as a Bose-Einstein condensate.

Using these different ensembles and distributions, average behavior can be calculated or predicted for a wide variety of physical systems in equilibrium, and many thermodynamic conclusions can be verified. Statistical mechanics offers insight into such diverse physical systems as a balloon, a Bose-Einstein condensate, and a **neutron** star. The applications of statistical mechanics in the rest of physics are nearly limitless.

STATISTICAL PHYSICS

Many physical processes, for example quantum processes such as **nuclear reactions**, are probabilistic, and can thus only be described statistically. The same consideration applies to many large classical systems that have so many variables, or degrees of freedom, that an exact treatment is not at all possible. Statistical physics can be used to model both of these types of systems. The development of the fundamental theory of statistical physics is chiefly credited to **James Clerk Maxwell**, **Josiah Williams Gibbs**, and **Ludwig Edward Boltzmann** for their contributions in both **thermodynamics** and mathematical formulation.

Statistical physics is divided into two main branches. The first is **statistical mechanics**, which attempts to predict the most probable behavior of a large collection of objects (an ensemble). Ensembles are characterized by specific macroscopic properties, such as volume, **potential energy**, **pressure**, and **temperature**. The individual particles in an ensemble, however, differ from each other as they are distributed over a range of microscopic states. Instead of trying to precisely define the states of the constituent particles of an ensemble, statistical **mechanics** describes their most probable states, attempting to arrive at an acceptable description of their macroscopic properties.

The second division of stastistical physics is statistical thermodynamics. Stastical thermodynamics makes use of the distribution laws of statistical mechanics to calculate and predict the energies and molecular velocities of an ensemble of particles, as well as their most probable energies and velocities. A macroscopic system can be defined by a few properties, such as volume, pressure, **density**, or temperature. From a

microscopic point of view, there are a great number of **quantum states** consistent with the fixed macroscopic properties. In statistical thermodynamics, in order to calculate any property, such as **energy**, the value of that property is calculated in each quantum state. The values are averaged, and it is postulated that this average value corresponds to the macroscopic thermodynamic property. The canonical ensemble is the most basic concept of statistical thermodynamics. A canonical ensemble is an assembly of identical systems, each of which is characterized by a number of components, a given volume, and a temperature. The systems are in thermal contact with each other, enabling energy to circulate from one system to the other. Although the energy of each system varies, the average energy is known. Thus, given an ensemble in which many systems are distributed between macroscopic states (the states of the system being defined by macroscopic quantities such as N, T, and V, i.e., the number of components, the temperature and the pressure, etc.) of a given energy, then the probability of finding a system in a specific state of energy is given by the canonical partition function of the system. This partition function can then be compared with the Boltzmann distribution for an isolated system with a given number of particles which can yield the partition function of the individual particle. The importance of the partition function is that it represents the bridge between the quantum mechanical energy states of a macroscopic state and the overall thermodynamic properties of the system. Statistical thermodynamics also uses other ensembles. For example, the micro-canonical ensemble, in which N, V, and E are fixed, and the isothermal-isobaric ensemble, in which N, T, and P are fixed.

A concrete example describing the use of statistical methods is provided by the determination of the flux of neutrons in a nuclear reactor. The reactor can be simulated by tracking a **neutron** with random initial conditions, i.e. a given starting position and **momentum**, through the entire system. There are different probabilities for absorption, **scattering**, triggering nuclear reactions to produce more neutrons, etc., for the different materials in the reactor. A statistical ensemble of such neutrons then yields the required flux. Statistical methods are applied to a wide variety of other physical processes, such as radioactive processes, quantum processes in semiconducting, conducting and superconducting materials, high energy physics reactions occurring in particle **accelerators** and colliders, determination of the thermal properties of solids, liquids, and gases, or of the phase transitions which they undergo, **heat** conduction phenomena, and studies of the magnetic and electronic properties of metallic alloys.

On a theoretical level, statistical physics represents absolute requirements for understanding the quantum field theories of **electromagnetism** and of the electroweak and **strong interactions**, as well as physics unification theories such as **string theory** and **quantum gravity**. The mathematical approaches used in statistical physics fall into two categories. The first approaches encompass the analytical methods. These involve approximations and provide few exact solutions. They are used to describe the fundamental physics of a given system, but often break down in regions of instabilities, such as phase transitions. The advent of **computers** has significantly contributed to the development and widespread use of the second type of approach, the Monte Carlo methods. These calculations usually involve integrals over a very large number of dimensions or configurations and quantities for any parameter values can be computed. An advantage of these methods is that the computing power allows them to sample more configurations and larger systems, thus decreasing error margins.

See also Electroweak theory; Quantum mechanics; Quantum theory; Probability and quantum mechanics

STEADY-STATE THEORY

Was there a moment of creation for the **universe**, or has the universe always existed? The steady-state theory is a cosmological theory for the origin of the universe that suggests the universe has always existed and did not have a moment of creation. This theory was popular during the 1950s and 1960s, but because of observations made during the 1960s, few, if any, astronomers now think that the steady-state theory is correct. The basic tenet of the steady-state theory is that the universe on a large scale does not change with time (evolve). It has always existed and will always continue to exist looking much as it does now. The universe is, however, known to be expanding. To allow for this expansion in an unchanging universe, the authors of the steady-state theory postulated that hydrogen atoms appeared out of empty **space**. These newly created hydrogen atoms were just enough to fill in the gaps caused by the expansion. Because hydrogen is continuously being created, the steady-state theory is sometimes called the continuous creation theory. This theory achieved great popularity for a couple of decades, but mounting observational evidence caused its decline in the late 1960s. The discovery in 1965 of the **cosmic background radiation** provided one of the most serious blows to the steady-state theory.

The steady-state model is based on a set of four assumptions collectively known as the *perfect cosmological principle*. The first assumption is that physical laws are universal. Any science experiment if performed under identical conditions will have the same result anywhere in the universe because physical laws are the same everywhere in the universe. Second, on a sufficiently large scale the universe is homogeneous. We know there is large scale structure in the universe, such as clusters of galaxies; so, we assume that the universe is homogenous only on scales large enough for even the largest structures to average out. Third, we assume that the universe is *isotropic*, meaning that there is no preferred direction in the universe. Fourth, we assume that over sufficiently long times the universe looks essentially the same at all times.

Collectively the first three of the above assumptions are the *cosmological principle* (not to be confused with the perfect **cosmological principle**). In a nutshell, the cosmological principle states that the universe looks essentially the same at any location in the universe. This principle remains a largely untested assumption because we cannot travel to every location in the universe to perform experiments to test the assumption. Adding the fourth assumption, that the universe does not change on the large scale with time, gives us the perfect cosmological principle. Essentially, the universe looks the same at all times as well as at all locations within the universe. The perfect cosmological principle forms the philosophical foundation for the steady-state theory. With the addition of the fourth assumption, the universe does not evolve in the steady-state theory, so observational evidence that the universe evolves would be evidence against the steady-state theory.

There are a number of observations that astronomers have made to test cosmological theories, including both the steady-state and the **big bang** theory.

When we look at the most distant objects in the universe, we are looking back in time. For example if we observe a quasar that is three billion **light** years away, it has taken the light three billion years to get here, because a light year is the distance light travels in one year. We are therefore seeing the quasar as it looked three billion years ago. **Quasars**, the most distant objects known in the universe, are thought to be very active nuclei of distant **galaxies**. The nearest quasar is about a billion light years away. The fact that we do not see any quasar closer than a billion light years away suggests that quasars disappeared at least a billion years ago. The universe has changed with time. Several billion years ago quasars existed; they no longer do. This observation provides evidence that the perfect cosmological principle is untrue, and therefore that the steady-state theory is incorrect. (Note however that when the steady-state theory and the perfect cosmological principle were first suggested, we had not yet discovered quasars.)

In his work measuring distances to galaxies, **Edwin Hubble**, after whom the **Hubble space telescope** was named, noticed an interesting correlation. The more distant a galaxy is, the faster it is moving away from us. This relationship is called the Hubble law. This relationship can be used to find the distances to additional galaxies, by measuring the speed of recession. More importantly, Hubble deduced the cause of this correlation. The universe is expanding. To visualize this expansion, draw some galaxies on an ordinary balloon and blow it up. Notice how the "galaxies" move farther apart as the balloon expands. Measuring distances between the drawn in galaxies at the rates at which they move apart, would give a relationship similar to Hubble's law.

The expanding universe can be consistent with either the big bang or the steady state theory. However in the steady-state theory, new **matter** must appear to fill in the gaps left by the expansion. Normally as the universe expands, the average distance between galaxies would increase as the **density** of the universe decreases. These evolutionary changes with time would violate the fundamental assumption behind the steady-state theory. Therefore in the steady-state theory, hydrogen atoms appear out of empty space and collect to form new galaxies. With these new galaxies, the average distance between galaxies remains the same even in an expanding universe.

The Hubble plot also provides evidence that the universe changes with time. The slope of the Hubble plot gives us the rate at which the universe is expanding. If the universe is not evolving, this slope should remain the same even for very distant galaxies. The measurements are difficult, but the Hubble plot seems to curve upward for the most distant galaxies. The universe was expanding faster in the distant past, contrary to the prediction of the steady-state theory that the universe is not evolving.

In the mid 1960s, **Arno Penzias** and Robert Wilson were working on a low noise (static) microwave antenna when they made an accidental discovery of cosmic significance. After doing everything possible to eliminate sources of noise, including cleaning out nesting pigeons and their waste, there was still a small noise component left. This weak noise did not vary with direction or with the time of day or year, because it was cosmic in origin. It also corresponded to a **temperature** of 3K (-518°F; -270°C, three degrees above **absolute zero**,). This 3K cosmic **background radiation** turned out to be the leftover **heat** from the initial big bang that had been predicted by proponents of the big bang theory as early as the 1940s.

Because this cosmic background **radiation** was a prior prediction of the big bang theory, it provided strong evidence to support the big bang theory. Proponents of the steady-state theory have been unable to explain in detail how this background radiation could arise in a steady-state universe. The cosmic background radiation therefore gave the steady-state theory its most serious setback. Penzias and Wilson received the 1978 Nobel Prize in physics for their work.

The steady-state theory was inspired at least in part by a 1940s movie entitled *Dead of Night*. The movie had four parts and a circular structure such that at the end the movie was the same as at the beginning. After seeing this movie in 1946, Thomas Gold, Hermann Bondi, and **Fred Hoyle** wondered if the universe might not be constructed the same way. The discussion that followed led ultimately to the steady-state theory.

In 1948, Hermann Bondi and Thomas Gold proposed extending the cosmological principle to the perfect cosmological principle, so that the universe looks the same at all times as well as at all locations. They then proposed the steady-state theory based on the new perfect cosmological principle. Because Hubble had already observed that the universe is expanding, Bondi and Gold proposed the continuous creation of matter. Hydrogen atoms created from nothing combine to form galaxies. In this manner the average density of the universe remains the same as the universe expands. In the steady-state, the rate at which new matter is created must exactly balance the rate at which the universe is expanding. Otherwise the average density of the universe will change and the universe will evolve, violating the perfect cosmological principle. To maintain the steady-state, in a cubic meter of space one hydrogen **atom** must appear out of nothing every five billion years. In a volume of space the size of Earth, the amount of new matter created would amount to roughly a grain of dust in a million years. In the entire observable universe, roughly one new galaxy per year will form from these atoms. Bondi and Gold recognized that a new theory must be developed to

explain how the hydrogen atoms formed out of nothing, but did not suggest a new theory.

In the same year, Fred Hoyle proposed a modification of **Albert Einstein**'s general theory of relativity. Hoyle worked independently of Bondi and Gold, but they did discuss the new theories. Hoyle's modification used a mathematical device to allow the creation of matter from nothing in general relativity. No experiments or observations have been made to justify or contradict this modification of general relativity.

There are a number of problems with the steady-state theory, but at the time the theory was proposed, there were also points in its favor.

The steady-state theory rests on the foundation of the perfect cosmological principle. Hence, any evidence that the universe evolves is evidence against the steady-state theory. The existence of quasars and the change in the expansion rate of the universe a few billion years in the past discussed earlier are evidence against the steady-state. This evidence for the evolution of the universe did not exist in 1948, when the steady-state theory originated. It became part of the cumulative **weight** of evidence that had built up against the steady-state theory by the mid 1960s. In gallant attempts to save the steady-state theory, its proponents, chiefly Hoyle and Jayant Narlikar, have argued that the universe can change over periods of a few billion years without violating the perfect cosmological principle. We must look at even longer time spans to see that these changes with time average out.

The cosmic background radiation is widely considered the final blow to the steady-state theory. Again proponents of the steady-state theory have made gallant efforts to save their theory in the face of what most astronomers consider overwhelming evidence. They argue that the background radiation could be the cumulative radiation of a large number of **radio** sources that are to faint to detect individually. This scheme requires the existence of roughly 100 trillion (about 10,000 times the number of observable galaxies) such sources that about are one millionth as bright as the radio sources we do detect. Few astronomers are willing to go to such great lengths to rescue the steady-state theory.

Another objection raised against the steady-state theory is that it violates one of the fundamental laws of physics as that law is currently understood. The law of conservation of matter and **energy** states that matter and energy are interchangeable and can change between forms, but the total amount of matter and energy in the universe must remain constant. It can be neither created nor destroyed. The steady-state theory requires continuous creation of matter in violation of this law. However, laws of science result from our experimental evidence and are subject to change, not at our whim, but as experimental results dictate. The rate at which matter is created in the steady-state theory is small enough that we would not have noticed. Hence we would not have discovered experimentally any conditions under which matter could be created or any modifications required in this law.

When the steady-state model was first suggested, our best estimate of the age of the universe in the context of the big bang model was about two billion years. However, Earth and solar system are about five billion years old. The oldest **stars** in our galaxy are at least 10-12 billion years old. These age estimates present the obvious problem of a universe younger than the objects it contains. This problem is no longer so severe. Modern estimates for the age of the universe range from about 10 billion years to about 20 billion years. The lower end of this range still has the problem, but the upper end of the range gives a universe old enough to contain the oldest objects we have found so far. In the steady-state theory, the universe has always existed, so there are no problems presented by the ages of objects in the universe.

For some people there are philosophical or esthetic grounds for preferring the steady-state theory over the big bang theory. The big bang theory has a moment of creation, which some people prefer for personal or theological reasons. Those who do not share this preference often favor the steady-state theory. They prefer the grand sweep of a universe that has always existed to a universe that had a moment of creation and may, by inference, also have an end in some far distant future time.

The weight of evidence against the steady-state theory has convinced most modern astronomers that it is incorrect. The steady-state theory does, however, still stand as a major intellectual achievement and as an important part of the history of the development of **cosmology** in the twentieth century.

STEAM ENGINES

The introduction of an efficient steam engine in 1769 by Scottish engineer **James Watt** was one of the starting points of the Industrial Revolution. The steam engine allowed **heat energy** to be converted into mechanical **work**, and eventually, **electric current**. Steam engines propelled ships and trains, and provided energy for factory machinery. With the development of the steam engine, transportation took less time, and goods could be produced more quickly and efficiently.

Although the ancient Greeks are credited with the first use of steam-powered machines, the steam engine as it existed before 1769 was primitive. Prior to Watt's innovations, Englishmen Thomas Savery and Thomas Newcomen built a working steam engine that could be used for demonstrations, however, this engine was inefficient and could not be used for long periods of time. In the Savery-Newcomen engine, a piston in an open cylinder containing water was alternatively heated and cooled. The steam **pressure** forced the piston out, doing work; then the steam was vented, leading to greater pressure outside the cylinder, forcing the piston back in so the cycle could be repeated. Steam was lost in each cycle, however, and the water supply had to be constantly refilled. The efficiency of the engine was poor.

Watt made the engine more efficient with his idea of keeping the piston and cylinder at the same **temperature** as the steam entering the piston. Instead of alternately cooling and heating the water in the cylinder, the boiler producing the steam was constantly heated. Instead of venting the steam to the outside air, the steam was vented into an evacuated con-

denser, surrounded by water, so that the escaping steam would condense back into a liquid form. The water was then pumped back into the boiler to undergo the cycle again. In this way, the boiler did not have to be constantly refilled. The piston was connected to a system of driveshafts and gears to produce the desired mechanical work.

Modern steam engines use turbines to generate mechanical work. Steam from the **boiling** water strikes one side of a turbine, which acts like a fan blade. The turbine blades experience a **torque** and rotate, and the rotation can be connected to a series of gears and driveshafts similar to the piston engine.

Steam engines can be used to generate **electricity**, by using the principle of **electromagnetic induction**. Wire loops called commutators are attached to the rotating shaft of a steam engine. Strong permanent magnets are placed around the commutator, so that there is a changing **magnetic flux** through the wire loops. Changing magnetic flux through a wire loop produces an electric current in the loop (**Faraday**'s **laws**).

See also Heat engines; Thermal Energy; Thermodynamics

STEINBERGER, JACK (1921–)

German American physicist

Jack Steinberger is an experimentalist in high-energy physics who discovered a new type of **neutrino** with the physicists **Leon Max Lederman** and **Melvin Schwartz**. In a groundbreaking experiment, the three scientists were able to create the first laboratory-made beam of neutrinos. Their discovery led to the development of the so-called "standard theory" of **matter**. In recognition of these contributions, Steinberger and his co-investigators were awarded the 1988 Nobel Prize in physics.

Steinberger was born on May 25, 1921, in Bad Kissingen, Germany. His father was Ludwig Lazarus Steinberger, a cantor and leader of the local Jewish community, and his mother was Berta May Steinberger. Jack Steinberger and his brother left Germany in 1934 and immigrated to Chicago, Illinois. There, they lived with the family of Barnard Faroll, a grain broker. Later, Faroll was instrumental in bringing the rest of Steinberger's family to the United States.

Steinberger studied chemical engineering at the Illinois Institute of Technology, but left to enlist in the U.S. army following the attack on Pearl Harbor. The army sent him to the Radiation Laboratory at the Massachusetts Institute of Technology, where he made **radar** bomb sights and studied physics. After the war, he enrolled at the University of Chicago, earning a bachelor of science degree in 1942, and a doctorate in physics in 1948. During his doctoral studies at Chicago, Steinberger had studied **muons** (semi-stable electrical particles) and showed that, when they decay, they yield two neutrons (nuclear particles with the **mass** of **protons** but not the charge) and an **electron** (a negatively charged atomic particle). He proposed at the time that muons could be broken down into muon-neutrinos and electron-neutrinos. (The neutrino is an elusive electrically uncharged basic particle of matter that lacks mass). The neutrino was important because Steinberger and other nuclear physicists believed neutrinos could be harnessed to study the weak nuclear **force**, which is responsible for certain types of **radioactivity**. The **weak force**, like **electromagnetism**, is one of the fundamental interactions between elementary particles.

Creates Neutrino Beam

Steinberger's professional career began at the University of California at Berkeley, with his appointment as professor of physics in 1949. He remained there for a year, and then moved on to Columbia University, also as a professor of physics. It was during a coffee break at the Pupin Physics Building of Columbia University that Melvin Schwartz, then Steinberger's graduate student, broached the idea to Leon Max Lederman and Steinberger of making beams of high-energy neutrinos for use in research. They decided to use the Brookhaven accelerator in Upton, Long Island, New York, to make the beam. They set up the accelerator to produce masses of protons at 15 billion electron volts and shoot them at a beryllium metal target. The protons (elementary positively charged particles) smashed the beryllium atomic nuclei into their component protons and neutrons. The impact also created **pions**, elusive short-lived particles that decay into muons and neutrinos. Seeking to filter out the neutrinos, the researchers constructed a 40-ft (12-m) barrier of steel built from the scrap of the battleship *Missouri*. The massive obstacle filtered out all the particles but the neutrinos, creating the first laboratory made beam of high-intensity neutrinos.

Until then, scientists had only been aware of the existence of neutrinos produced by the type of **radioactive decay** that also creates an electron. But the neutrinos that came out of the steel filter were accompanied by muons. These were muon-neutrinos. This finding led to the development of the "standard model theory," which posits the existence of three generations of matter: a first generation consisting of seven electrons, electron neutrinos and the up and the down **quarks** (hypothetical particles from which fundamental particles are built); a second generation of matter—that revealed by the experiments carried out by Steinberger and his colleagues—consisting of muons, muon-neutrinos and varieties of quarks called charmed and strange quarks; and a third generation, which physicists are exploring now and which includes **tau particles** (particles produced when electrons and anti-electrons are smashed together), tau neutrinos, and the top and the bottom quarks. The discovery of the muon-neutrino was a pioneering step in a road that enabled high-**energy** physicists to use neutrinos in scientific investigations. In astronomy, for instance, neutrino telescopes detect explosions of neutrinos associated with far-off **supernovae** (**stars** that blow up with great brilliance).

Steinberger left Columbia University in 1968, moving on to become director, administrator and researcher of the European Center for Nuclear Research in Geneva. He retired from his administrative duties in 1986, but remained as a staff

physicist. He has continued to work on experiments in elementary particles and conducts research using the Large Electron-Positron Collider. In 1988 Steinberger received extensive public recognition for his discovery of the muon-neutrino and for a lifetime of contributions to physics. He was awarded not only the Nobel Prize in that year, but also received the National Medal of Science from the United States. Steinberger married Joan Beauregard in 1943. The marriage ended in divorce. In 1961 he married Cynthia Eve Alff. Steinberger is the father of three sons, Joseph, Richard, and John, and a daughter, Julia Karen. Steinberger has become a United States citizen.

STEINMETZ, CHARLES P. (1865–1923)
German American electrical engineer and mathematician

Charles P. Steinmetz was a mathematician and electrical engineer whose theories and research fostered the widespread use of electrical **energy**. A scientist of prodigious inventiveness, Steinmetz was granted some two hundred patents. His discovery of the phenomenon known as magnetic hysteresis led to the development of energy-efficient motors. He worked out mathematical theories that made practical the use of alternating current in long-distance **power** transmission. His work in changes in electrical circuits of very short duration was used to develop new cables and improved methods of operating transmission systems. He also built artificial lightning **generators** that led to the development of lightning arrestors to protect electrical apparatus and transmission lines.

Charles Proteus Steinmetz was born in Breslau, Germany (now Wroclaw, Poland), on April 9, 1865. His mother, Caroline Neubert, died when he was one year old; his father, Karl Heinrich, worked for the government-owned railroad. Steinmetz was the third of three sons and had two stepsisters and one half-sister. He changed his name from Karl August Rudolph when he immigrated to America. He was born with an inherited condition that gave him a hunchback. He never married for fear of passing the trait on to his children.

As a young boy, Steinmetz had a difficult first year at school; however, the slow start and concerns soon evaporated as he quickly ascended to the top of the class. At the University of Breslau he studied mathematics and worked toward a Ph.D. He would have graduated in 1888 were it not for his political activities. While at the university, Steinmetz became a Socialist, eventually taking over the publication of a student-run Socialist newspaper. The secret police had a warrant for his arrest when he fled Germany for Switzerland on the eve of his graduation. He immigrated to the United States in 1889.

Steinmetz's first job in the United States was with the Eikemeyer and Osterheld Manufacturing Company in Yonkers, New York, where he set up a small research lab and worked on alternating current (an **electric current** that reverses its direction at regularly recurring intervals) motors. While working at Eikenmeyer's, he discovered how the influence of a changing magnetic field operates in a motor to consume energy. Characterizing the relationship between the fluctuating strength of a magnetic field in a motor and the amount of energy lost—a phenomenon known as magnetic hysteresis—his discovery showed engineers how they could design motors with a minimal loss of energy. Steinmetz published a 178-page paper on the law of hysteresis in 1892 that brought him instant recognition.

Steinmetz then turned his attention to mathematical theories that could make practical the use of alternating current in long-distance power transmission. Alternating current had no constant value or direction, and its large-scale behavior was at that time impossible to predict. Nevertheless, Steinmetz was able to fit the principles of alternating current into predictable mathematical models.

The Eikemeyer plant was bought by General Electric Company (GE) in 1892, when the company was just being organized. Steinmetz was hired by GE and eventually moved to Schenectady, New York, where he lived the rest of his life. It was at GE that he completed his work on the mathematics of alternating current, and published in the late 1890s a three-volume work so complex that most engineers had trouble understanding it at first. For four years, until his theories became well understood, Steinmetz spent much of his time explaining them to engineers throughout the country.

Another of Steinmetz's major achievements was his study of changes in electrical circuits of very short duration, for which he designed a 220,000-volt experimental transformer. His research was used to develop new cables, and new and improved methods of operating transmission systems. He even created artificial lightning by designing generators that could produce **electricity** at very high potential. This was used to study the effect of lightning on the power lines that were just going up around the country, and led to the development of lightning arrestors, devices that protect electrical apparatus and electric transmission lines from damage by lightning. In all, Steinmetz held about two hundred patents.

Steinmetz taught electrical engineering at Union College in Schenectady from 1903 to 1913, heading the new department that was formed owing to his presence. He was known as a tough lecturer, but was popular with the students. Faithful to the philosophy of his youth, he continued to espouse socialism. When George R. Lunn was elected Schenectady's Socialist mayor in 1911, Steinmetz volunteered his services to the administration. He was appointed to the board of education and was later elected president of the board. In 1915 he was elected a member of the common council in Schenectady. As president of the council, he established classes for handicapped children. He ran for state engineer in 1922.

Steinmetz, who never married and had no children, adopted Roy Hayden, his associate at General Electric. When Hayden married, he and his family lived in Steinmetz's house on Wendall Avenue in Schenectady. Steinmetz also legally adopted Hayden's three children as his grandchildren. Steinmetz loved animals and at times kept a full range of wild animals in his backyard. His favorite pets were a pair of crows, John and Mary. He was an avid grower of orchids and cacti.

Steinmetz died in Schenectady on October 26, 1923, of heart failure.

STELLAR EVOLUTION

In **astrophysics** and **cosmology**, stellar evolution refers to the life history of **stars** that is driven by the interplay of internal **pressure** and **gravity**.

Essentially, throughout the life of a star a tension exists between the compressing **force** of the star's own gravity and the expanding pressures generated by the **nuclear reactions** taking place in its core. After cycles of swelling and contraction associated with the burning of progressively heavier nuclear fuels, the star eventually expends its useable nuclear fuel and resumes contraction under the force of its own gravity. There are three possible fates for such a collapsing star. The particular fate for any star determined by the **mass** of the star left after blowing away its outer layers. A star less than 1.44 times the mass of the **Sun** (termed the **Chandrasekhar limit**) collapses until the pressure-compacted **electron** clouds exert enough pressure to balance the compressing force of gravity. Such stars become **white dwarfs** that are contracted to a radius the size of a planet. This is the fate of most stars. If the star retains between 1.4 and roughly three times the mass of the Sun, the pressure of the electron clouds is insufficient to stop the gravitational collapse and stars of this mass continue their collapse to become **neutron stars**. Although **neutron** stars are only a few miles across, they have enormous **density**. Within a neutron star the nuclear forces and the repulsion of the compressed atomic nuclei balance the crushing force of gravity. With still more massive stars, however, there is no known force in the **universe** that can stop the final gravitational collapse and such stars collapse to form a **singularity**, a geometrical point of infinite density. As such a star collapses its **gravitational field** warps **space-time** and a **black hole** forms around the singularity.

Gravitational collapse is the process that provides the **energy** required for star formation, which starts with hydrogen fusion in a protostar at a **heat** of over 15 million K. Gravity, always directed inwards, decreases the radius of **interstellar gas** clouds, causing them to collapse and form a protostar, the immediate precursor of a star. Interstellar gas is initially cold, but it is heated up by the **gravitational energy** released by the cloud contraction process. The radius of the protostar will continue to shrink under the influence of gravity until enough internal **gas pressure**, always directed outwards, builds up to stabilize the collapse. At this stage, the protostar is still too cold for hydrogen fusion to be initiated. Protostars can be detected by **infrared spectroscopy** because the initial warming event releases infrared electromagnetic **radiation**. If the mass of the protostar is less than 0.08 solar masses, the **temperature** of its core never reaches the range required for **nuclear fusion** and the failed star becomes a brown dwarf.

If, however, the mass of the protostar exceeds 0.08 solar masses, hydrogen fusion can proceed and the protostar becomes a main sequence star, with average surface tempera-

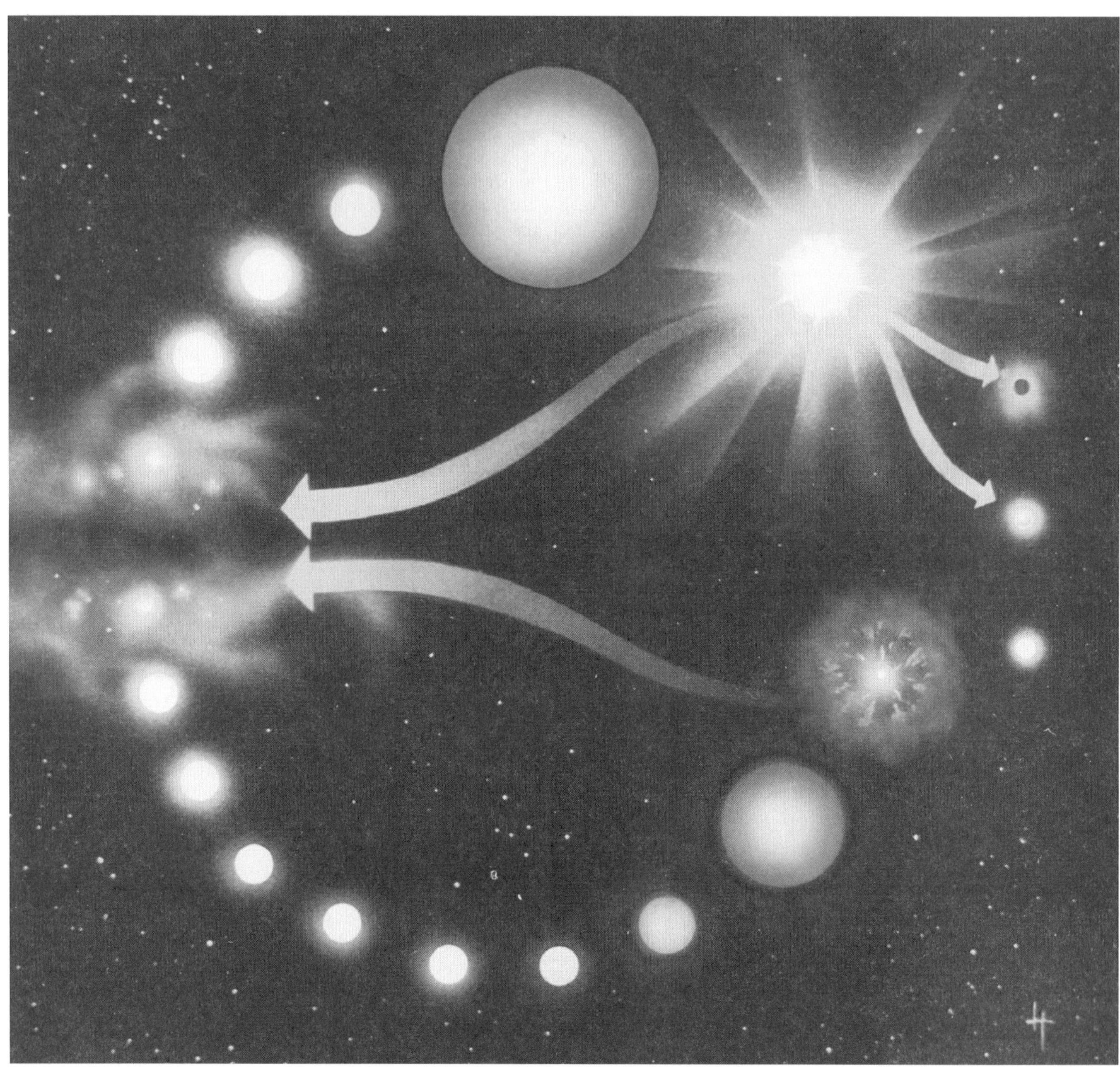

Stellar evolution showing the main stages of star life. *(Image by David A. Hardy. Photo Researchers, Inc. Reproduced by permission.)*

tures of 10,832°F or 6,000°C (the internal and coronal temperatures measure in the millions of degrees). Most stars in the universe are main sequence stars and are found on the diagonal of the **Hertzsprung-Russell diagram**. The main sequence stage of star evolution is the most stable state a medium-sized star can reach, and it can last for billions of years as such stars undergo very gradual and slow changes in luminosity and temperature. This is because pressure and gravitational forces are in **equilibrium** and the core has reached the temperature required for the fusion of hydrogen to **helium** to proceed smoothly. The time spent by a star in the main sequence is a function of its mass. The more massive the star, the less time spent on the main sequence. Although massive stars have large amounts of fuel, hydrogen fusion proceeds so quickly that it is completed within a few hundred thousand years. The fate of such massive stars is to explode violently. Smaller stars fuse their hydrogen at a slower rate. The lightest stars created in the early history of the universe, for example, are still on the main sequence and the Sun is about half way through its main sequence life.

A post-main sequence star has two distinct regions, consisting of a core of helium nuclei and electrons surrounded by an envelope of hydrogen. With two protons in its **nucleus**, helium requires a higher fusion temperature than the one at which hydrogen fusion is proceeding. Without a source of energy to increase its temperature, the core can not counter the effect of gravitational collapse and it starts to collapse, heating up as it does. This heat is transferred to the fusing hydrogen layer, which increases the luminosity of the shell and also causes it to expand. As it expands, the outer layers cool off. At this point, the star is characterized by expansion and cooling of its shell, which causes it to become redder with increased luminosity. This is termed the red giant phase. When the Sun reaches this stage, it will be large enough to include Mercury in its sphere and hot enough to evaporate Earth's oceans. The core temperature of a red giant is on the order of 212 million degrees Fahrenheit (100 million degrees Celsius), which is the threshold temperature for the fusion of helium into **carbon**. A red giant, however, is stable, as pressure and gravity have reached equilibrium.

If helium continues to accumulate in the core as the outer portions of the hydrogen envelope continue to fuse, eventually the helium in the core starts fusing into carbon in a violent event referred to as a helium flash, lasting as little as a few seconds. During this phase, the star gradually blows away its outer atmosphere into an expanding shell of gas known as a planetary nebula.

A star takes thousands of years to go through the red giant phase, after which it evolves into a white dwarf. It is then a small, hot star with a surface temperature as high as 180,032°F (100,000°C) which makes it glow white. Because of its small size, a white dwarf has a very high density. A white dwarf consists of those elements which were created in its previous evolutionary phases. The original hydrogen has been fused into helium which was then totally or partly fused into carbon and there may be some heavier elements fused from the carbon. The temperature of a white dwarf is not high enough to initiate a new cycle of fusion. In time, it eventually becomes a black dwarf as it loses its residual heat over billions of years. The size of a white dwarf is limited by a process called electron degeneracy. Electron degeneracy is the stellar equivalent of the **Pauli exclusion principle**, as is neutron degeneracy. No two electrons can occupy identical states, even under the pressure of a collapsing star of several solar masses. For stellar masses smaller than about 1.44 solar masses, the energy from the gravitational collapse is not sufficient to produce the neutrons of a neutron star, so the collapse is effectively stopped. This maximum mass for a white dwarf is called the Chandrasekhar limit.

When a massive star has fused all of its hydrogen, gravitational collapse is capable of generating sufficient energy so that the core can begin to fuse helium nuclei to form carbon. If the process can go beyond the red giant stage, the star is known as a supergiant. Following fusion and disappearance of the helium, the core can successively burn carbon and other heavier elements until it acquires a core of iron, the last element that can be formed by fusion without the input of energy. Another possible fate of white dwarfs is to evolve into novae or another type of **supernovae**. Novae

occur in binary star systems in which one star is a white dwarf. If the companion star evolves into a red giant, it can expand far enough so that gas from its outer shell can be pulled onto the white dwarf. The white dwarf accumulates the additional gas until it reaches nuclear fusion temperatures, at which point the gas ignites explosively into a nova. Alternatively, a white dwarf may accumulate enough material from its binary star to exceed the Chandrasekhar limit. This results in a sudden and total collapse of the white dwarf, with temperature increases in ranges capable of initiating rapid carbon fusion and subsequent explosion of the white dwarf into a spectacular supernova, which can shine with the brightness of 10 billion suns with a total energy output of ~1,044 joules, equivalent to the total energy output of the sun during its entire lifetime.

See also Astrophysics; Big bang; Binary stars; Chandrasekhar mass; Cosmology; Interstellar space; Primordial black holes; Red dwarf/Red giants; Stellar lifecycle; Supergiants

STELLAR LIFE CYCLE

The stellar life cycle describes the formation and death of **stars**. Stars begin as large clouds of dust and gas. These immense clouds (up to 25 **light** years in diameter) have a low **density** although each cloud contains as many particles as a single star. Hydrogen gas occupies as much as 99% of the cloud; less than one percent is made up of dust particles. The dust particles are tiny, each having a diameter of less than four hundred-thousandths of an inch (one ten-thousandth of a centimeter). These particles contain such compounds as silicon carbide and elements such as **carbon** and nitrogen. These clouds of dust and gas probably begin as a dying star exploding into a nova or a supernova.

The dust particles are gradually pulled close together by the **force** of **gravity**. An outside force, such as a shock wave from a supernova, usually triggers this pulling together of the particles. As the cloud is pulled together, it reaches extremely high temperatures. Eventually, the cloud becomes dense enough to be called a protostellar nebula, Latin for clouds. **Nebulae** can be luminous or dark. Luminous nebulae result when a nebula is located near a bright star, for example, the Great Nebula in the constellation Orion. The light from the star creates a glow in the nebula. Dark nebulae are those that are not located near stars, for example, the Horsehead Nebula, also seen in the constellation Orion. These nebulae block some of the light from distant stars, and appear as dark spots in **space**. Many nebulae are not visible to the unaided eye because they are not near a bright star or they do not block the light of distant stars.

The nebula glows as its **temperature** and density increase. This hot, glowing cloud of gas and dust particles is called a protostar. If the protostar reaches high enough temperatures, fusion reactions begin inside the core. During these reactions, two hydrogen nuclei combine to form one **helium nucleus**. Some of the **mass** from the hydrogen atoms is converted into **energy**. Once the fusion reactions begin, the proto-star has become a star. The energy released from the fusion reactions work against the force of gravity and the cloud of dust and gas stops contracting. When this occurs, the star is said to be in its stable state. Reaching the stable state may take hundreds of thousands to millions of years, depending on the size of the star. Less massive stars reach the stable state faster than stars with great mass.

A star in the stable state remains the same size and emits a steady amount of **radiation** for millions of years. Eventually, the quantity of hydrogen gas driving the fusion reactions diminishes and the star emits less energy. At this point, the energy no longer balances the force of gravity and the star begins to contract. As the star contracts, the temperature of the core increases further, which causes the outer layers of the cloud to expand. This makes the star appear brighter because of its increased size. Simultaneously, fusion reactions begin to take place in the outer layers as the core supply of hydrogen gas is expended. This causes the star to expand, and it is now called a red giant or a supergiant. Sometimes, the star's temperature continues to rise as it expands to a temperature high enough to facilitate another type of fusion reaction in its core. When this occurs, two helium nuclei fuse to form heavier elements such as **oxygen**, carbon, or iron.

Eventually, the hydrogen and helium inside the star are exhausted until no fusion reactions occur. At this point, the inner core cannot support the **weight** of the outside layers, and the giant or **supergiant** collapses, forming a white dwarf. **White dwarfs** are usually smaller than Earth and are formed from stars of equal or less size as the **Sun**. Without the fusion reactions taking place, the dwarf's temperature decreases. Enough **heat** remains from the fusion reactions for the dwarf to glow for a few million years. Occasionally, a white dwarf will become extremely bright when it has collided with another star. The white dwarf then becomes a nova, or new star. The increased luminosity usually lasts only a few years, at which point the nova returns to its original brightness as a white dwarf.

A larger star may form a supernova when it collapses from a red giant. Fusion reactions in the red giant form an iron core. When the red giant cools, the core begins to collapse, causing the temperature and **pressure** inside of the core to increase. The temperature becomes high enough that the iron atoms undergo fusion reactions to produce even heavier elements. When the giant collapses further, a violent explosion occurs and the star's luminosity increases dramatically. A supernova lasts for a few weeks or months at most. After the explosion occurs, only about half of the original star's mass remains. This remaining mass is called a **neutron** star. A neutron star is a highly dense cloud of neutrons. The violent explosion essentially crushed the atoms together, causing electrons to combine with protons to form neutrons. A neutron star may only be several kilometers in diameter, although it is millions of times more dense than the Sun.

The largest red giants, or supergiants, may form a black hole when they collapses. In this scenario, the collapse is so great that all of the atoms are condensed into a very small vol-ume. The force of gravity between the atoms is strong enough that no light can escape. **Black holes** are invisible structures that can only be identified when **matter** from a nearby star is drawn into them.

The stellar life cycle varies from star to star, depending on its original size. The more massive stars have greater temperatures and, therefore, have shorter lives. Smaller stars have longer life cycles.

See also Cosmology; Seeing (astronomy); Stellar evolution

STERN, OTTO (1888–1969)

American physicist

Otto Stern received the 1943 Nobel Prize in physics for his development of molecular beam methods and the use of these methods to determine a number of important physical constants, especially the **magnetic moment** of atoms and nuclei. The molecular beam—the introduction of a gas or vapor into a **vacuum** which results in the formation of a stream of atoms or molecules that is similar to a **light** beam—had been discovered by French physicist Louis Dunoyer. This molecular or atomic beam can be used to study the properties of the gas or vapor of which it is made.

Stern was born in Sohrau, Upper Silesia, Germany (later Zory, Poland), on February 17, 1888. He was the oldest of five children born to Oskar Stern and the former Eugenie Rosenthal. Both parents had come from prosperous grain merchants and flour millers and eagerly encouraged the intellectual development of their children. When Otto was four years old, his family moved to Breslau (later Wroclaw, Poland), where he attended the local Johannes Gymnasium. Most of Stern's training in science came from his home, where books, conversation, and even some simple experimentation were the primary means of instruction.

Earns Doctorate and Joins Einstein

As was common at the time, Stern traveled to a number of universities for his further education after graduating from the gymnasium in 1906. Finally, he undertook a doctoral program in chemistry at Breslau under Otto Sackur. He was awarded his Ph.D. in 1912 for a thesis on the kinetic theory of osmotic **pressure** in concentrated solutions. Through Sackur's influence, Stern was able to take a position as research assistant to **Albert Einstein** from 1912 to 1914, first in Prague and later in Zürich, at the Federal Institute of Technology. Einstein was to have an important influence on Stern's career. It was from the great man, biographer **Emilio Segrè** pointed out in the *Biographical Memoirs of the National Academy of Sciences,* that Stern "learned what were the really important problems of contemporary physics: the nature of the quantum of light with its double aspect of particle and wave, the nature of atoms, and relativity."

With the outbreak of World War I in 1914, Stern was drafted into the German army where he served first as a private and later as a non-commissioned officer. He was assigned with the **Meteorology** Corps on the Russian front. During his four years in the military, Stern had enough spare time to continue his theoretical research, writing two important papers on the application of **quantum theory** to statistical **thermodynamics** during the war years.

Begins Molecular Beam Studies at Frankfurt

As the war drew to a close, Stern was finally able to accept an appointment as Privat dozen at the University of Frankfurt-on-the-Main that he had been offered in 1915, but which the war had prevented his accepting. At Frankfurt, Stern became assistant to **Max Born**, who had just moved there from the University of Berlin. The research topic in which Stern became interested was the use of molecular beams to study atomic and molecular properties. Stern had first learned about molecular beams during the war. The techniques for using such beams had been developed in 1911 by the French physicist Louis Dunoyer. By introducing a gas or vapor into a vacuum, a beam of atoms or molecules is formed. This beam allows the researcher to study **atomic structure** and properties.

Stern spent the greatest part of his professional career developing the molecular beam technology and using it to determine a number of atomic and molecular properties. The first application he studied was the determination of molecular velocities in gases. In the 1850s, **James Clerk Maxwell** had calculated the theoretical distribution of molecular velocities in a gas. Although there was not much doubt about the accuracy of Maxwell's findings, no empirical data supporting his results had ever been obtained. In 1920, Stern completed an experiment using the molecular beam technique to find the actual distribution of molecular velocities in a gas. His results unequivocally substantiated Maxwell's predictions.

The following year, Stern went on to a more significant application of the molecular beam technique. According to quantum theory, the **atom** is an electrically charged particle rotating in **space**. As such, it was thought to have associated with it a magnetic field whose magnetic field could be expressed as its "magnetic moment." Furthermore, the atom's magnetic moment could have only a finite, discrete number of values. This notion was described as the atom's "spatial quantization." (In fact, the same argument applies to the components of an atom, its electrons, **nucleus**, protons, and neutrons, all of which have their own unique magnetic moments.)

In 1922, Stern used molecular beam s to test the theory of spatial quantization. He passed a beam of silver atoms through a nonuniform magnetic field and found that it split into two distinct parts. The now famous experiment was given the name the Stern-Gerlach experiment for Stern and his associate in the work, Walther Gerlach. The experiment not only confirmed the concept of space quantization, but also allowed Stern to calculate the magnetic moment of the silver atom. His result was in agreement with the theoretical values calculated by means of quantum theory. For this experiment, and others like it carried out later, Stern was awarded the 1943 Nobel Prize in physics.

Otto Stern, 1944. *(AP/Wide World Photos. Reproduced by permission.)*

Stern's tenure at Frankfurt had ended in 1921 in the midst of planning for his famous experiment with Gerlach. The experiment was carried out at the University of Rostock, where he had been appointed associate professor of theoretical physics in 1921. Two years later, Stern moved on to the University of Hamburg as professor of physical chemistry and director of the Institute for Physical Chemistry. At Hamburg, Stern continued to explore the use of molecular and atomic beams. He found experimental proof for **Louis de Broglie**'s theory of the wave nature of particles and determined the magnetic moment of the **proton** and deuteron, **subatomic particles**. The value he found for the proton was at least twice that predicted by theory, a discrepancy that has yet to be totally explained.

Adolf Hitler's rise to power in Germany convinced Stern that he had to leave his homeland. Although he was probably not in any personal danger, he decided to resign his post at Hamburg in protest of the new government's anti-Semitic policies. He accepted an appointment as research professor in physics at the Carnegie Institute in Pittsburgh and became a naturalized U.S. citizen on March 8, 1939. During World War II, Stern served as a consultant to the War Department and then, in 1946, resigned his post at Carnegie to move to Berkeley where he could be near his two sisters. He lived largely in isolation for the rest of his life. He died in a Berkeley movie theater on August 17, 1969. Stern never married and was said by biographer Segrè to have "liked good cuisine, excellent cigars, and in general all the refinements of life."

STRAIN ENERGY

The strain **energy** is the **potential energy** stored in a body when that body is *elastically* deformed. It is equal to the amount of **work** required to produce this deformation. A body deforms elastically until the applied stress reaches some maximum value, beyond which the body exhibits permanent deformation. The body need not break at this maximum stress; it will simply not return to its original dimensions if stressed any more. As long as the elastic limit is not exceeded, the strain is directly proportional to the stress. Familiar examples of bodies that behave elastically include rubber bands, diving boards, and springs.

If a uniform rod is stretched slowly and elastically, energy will be stored in the rod as elastic energy. The elastic energy stored per unit volume, or strain energy **density**, is equal to one-half the stress raised to the second power divided, in general, by the Young's modulus. The Young's modulus is defined as the longitudinal stress divided by the longitudinal strain in the elastic region of deformation; strain is defined as the relative change in dimensions or shape in a body as the result of an applied stress, and is a dimensionless quantity; and stress is the magnitude of the applied **force** per unit area, usually measured in units of pounds per square inch or pascals.

STRANGE MATTER

In the late 1940s, a number of never-before-seen **subatomic particles** were discovered which seemed to behave in a most peculiar manner. As was first noticed by physicist Abraham Pais, the new particles, which were unstable, were produced, always in pairs, via the strong interaction (the **force** that holds the atomic **nucleus** together against the mutual electrostatic repulsion of the constituent protons), but that the strong force was not responsible for their decay. There is a Heisenberg uncertainty relation between the interaction timescale and the interaction strength; the stronger the interaction, the shorter the timescale. The particles were produced in about 10^{-23} s, consistent with the strong interaction. They decayed, though, in about 10^{-10} s and thus indicated that a weaker interaction (aptly named the weak interaction) was responsible for their decay. In 1953 **Murray Gell-Mann** and Kazuhiko Nishijima put forth a classification scheme that explained the new phenomena.

A typical production interaction was, for example, $\pi^- + p^+ \to K^+ + \Sigma^-$, in which a negatively charged pion strikes a **proton**, to produce two of the new particles, a positively charged kaon and a negatively charged sigma. A new property, called strangeness, was assigned to the K^+ + and the Σ^-; each possesses a strangeness **quantum number**, s=1, and s=-1, respectively. Notice that the net strangeness on the left-hand side (0) is equal to the net strangeness on the right-hand side (1 + -1 = 0): strangeness is conserved by the strong interaction.

The unstable **strange particles** decay. For example, the sigma might decay via the reaction $\Sigma^- \to n + \pi^-$ becoming a **neutron** and a negative pion. This decay occurs via the weak interaction, which does not conserve strangeness; neither the neutron nor the pion is a strange particle, and we have s=-1 on the left, and s=0 on the right.

By 1960 the number of strongly interacting particles that had been discovered (these are called hadrons) had skyrocketed—in fact there were so many that **Enrico Fermi** once quipped that he should have become a botanist. But in 1964 an enormous simplification occurred. Gell-Mann and George Zweig independently proposed that the hadrons were not fundamental at all. Rather, they were constructed of constituent particles, named **quarks** by Gell-Mann. Just three types (now called flavors) of quark sufficed to construct all the then-known hadrons, the up (u), the down (d), and the strange (s). Strange hadrons contained a strange quark; non-strange hadrons did not.

The moern standard model postulates that there are six flavors of quark, arranged into three pairs, or families, the up and down, the strrange and the charmed, top, and bottom quarks. All stable **matter** is built of up and down quarks. The weak interaction can, for example change a strange into a down. Thus the strange particles are unstable, and can decay into non-strange particles.

It seems that, in nature, quarks come in only two types of packages, either in threes (e.g., a proton contains the quarks uud), or in a quark-antiquark pair (e.g., a K^- is an $s\bar{u}$ pair). But must this be so? Can there exist states with arbitrarily large numbers of quarks? If only the first family of quarks is considered, the answer is no. For example, it can be shown that a deuterium nucleus, a proton (uud) plus a neutron (ddu), is a lower **energy** state than a grouping of the six constituent quarks. Thus deuterium nuclei are abundant, while the six-quark state uuuddd is not observed.

A number of physicists argued that strange quark matter, comprised of roughly equal numbers of up, down, and strange quarks, should be both energetically favorable (under the right conditions), and stable. The reason has to do with the **Pauli exclusion principle**, which forbids two half-integer spin particles from occupying the same quantum state (quarks have spin-1/2.) Configurations of quark matter can exist in which there are no empty down quark states available to receive the down quark that would result from the decay of a strange quark. And strange quark matter, having roughly equal numbers of the three quark types, carries only a small net **electric charge** (up quarks have charge 2/3, while the down and strange quarks each have charge -1/3). With only a tiny internal electrostatic repulsion, strange quark matter could exist in nuggets of almost unlimited size.

Particle physicists propose that this strange quark matter may exist inside **neutron stars**, or that huge amounts of it may have been formed during the first millionth of a second after the **big bang**. This would be in the form of quark nuggets, with diameters between 10^{-7} and 10 cm. These nuggets, comprising between 10^{33} and 10^{42} quarks, and having a **mass** between 10^9 and 10^{18} grams, may make up the **dark matter** that seems to comprise almost 90% of the Universe's content. Experimental searches for strange quark matter, either naturally occurring or produced in **accelerators**, are ongoing.

See also Quarks and gluon; Standard model of particle physics; Particle physics; Quantum theory

STRANGE PARTICLES

In **particle physics**, strange particles are those particles that, in addition to containing up and down **quarks**, contain a strange quark. Strange particles were originally so named because they exhibited greatly increased lifespans in comparison with other particles.

Strange particles are particles that are only produced in pairs in electromagnetic and strong interaction processes. They possess an extra **quantum number** called strangeness, and this quantum number is conserved in all strong and electromagnetic **scattering** and decay processes. Strangeness was later related to the fact that strange particles contain a new quark, known as the strange quark.

Strangeness (and the strange quantum number) are elementary particles that are conserved by the strong **force**. The quantum number reflects the difference between the baryon number and the hypercharge.

By the 1960s, the number of elementary particles was increasing rapidly. Among the strange particles discovered were **K mesons**, K+, K-, K0, and K-bar0, with strangeness +1 or -1; the λ^0, σ^0, σ^+, and σ^- baryons, also with strangeness +1 or -1; the $(\Xi)^0$, $(\Xi)^-$, Y0, Y+, and Y- baryons, with strangeness +2 or -2, and the (Omega) with strangeness -3.

As an example of strangeness conservation, consider the scattering reaction p + $(\pi)^+ \rightarrow \lambda^0 + K^0$. Protons and **pions** are common and have strangeness number 0. This collision occurs via the strong interaction, so strangeness is conserved. Therefore, by conservation of strangeness, the λ^0 and K^0 must have opposite strangeness numbers. By convention, the K^0 has strangeness number +1, requiring the λ^0 to have strangeness number -1.

Strange particles are free to decay via the weak interaction, but are not allowed to decay through the strong interaction. Because of this, strange particles have lifetimes of about 10^{-10} seconds compared with the 10^{-23} second lifetime of an unstable **hadron** with no strangeness.

Once it was understood that hadrons and mesons were made up of quarks, the existence of strange particles required more than just the common up and down quarks. To explain strangeness, the strange quark was postulated. The strange quark is nearly identical to the down quark, but it has a heavier **mass**. In the standard model, strange quarks can only decay via the weak interaction, but strange quark-antiquark pairs can be produced in strong and electromagnetic interactions and thus provide for the conservation of strangeness.

In contrast to the strange quark, the existence of other exotic quarks such as the charm and bottom quark were inferred by looking at resonance in processes. A resonance is a sharp peak in the probability distribution for scattering between two particles, and it occurs when there is the exact amount of **energy** needed to produce a particle-antiparticle pair.

The existence of the strange particles, as well as the existence of the **muon**, were the first clues that a three-family structure of particles exists. Physicists suspect there are three copies of each kind of particle, the only difference being that the masses are larger in the later families. For example, the down, strange, and bottom quarks all interact identically, but the bottom is heavier than the strange which is heavier than the down quark.

See also Electromagnetism; Quarks and gluons; Standard model of particle physics; Strong interactions; Weak force and interactions

STRING THEORY

Sometime in the fifth century B.C., the Greek philosopher Democritus introduced the atomic hypothesis—the idea that all **matter** is composed of extremely tiny, indivisible bits. (Hence the name, **atom**, which comes from the Greek *atomos*, meaning uncuttable.) That reductionist model, in which a few types of fundamental particles suffice to build everything that is, has, remarkably, lasted twenty-five centuries. But the current idea of which particles are actually fundamental has not. Since the turn of the last century it has been discovered that atoms have internal structure, a heavy **nucleus** surrounded by much lighter electrons. That nucleus is constructed of protons and neutrons, which, in their turn, are built of **quarks**.

Until 1967, it was thought that there were just four fundamental forces: the strong nuclear **force**, the weak nuclear force, the electromagnetic force, and **gravity**. But that year **Steven Weinberg** and **Abdus Salam** showed that the weak and electromagnetic forces were but different manifestations of a single "electroweak" interaction. Could it be that all forces derived, in a similar manner, from a single interaction? A number of theorists began the search for such a "Theory of Everything."

Meanwhile it was discovered that the strong nuclear force, which binds quarks together, behaved very strangely indeed. Rather than weakening with distance (like the inverse-square gravitational and electromagnetic forces), it became stronger as the separation between quarks was increased. The three quarks that bind together to form a **proton**, for example, behaved as if they were free—they did not interact with one another "inside" the proton. Yet they were permanently confined—no amount of **energy** could separate them. It was as if they were tied together by elastic bands; the attractive force needed to separate two quarks increased the further one tried to pull them apart.

In 1974 a new model for the strong interaction was introduced, in which the fundamental objects were not point particles, as in all previous models, but were, rather, one-dimensional-they were "strings." Particles were identified as "excitations" of the string (akin to the various frequencies with which, say, a plucked guitar string may vibrate). For a number of reasons, though, the model proved not to be a viable theory of **strong interactions**. In particular, though, there was present a massless, spin-2 excitation. There was no such strongly interacting particle in nature.

But there is a massless spin-2 particle in nature, the **graviton**, the carrier of the **gravitational force**. Gravity had been proving fiendishly difficult to quantize. All proposed models that united both quantum **mechanics** and **Albert Einstein**'s general relativity were beset by a major problem: non-renormalizability. All calculations of physical quantities led to infinite quantities which could not be removed; the models thus had no power to predict. It had not been possible to construct a consistent **quantum field theory** containing a graviton. Here, though, was a new **quantum theory** that contained both half-integral spin excitations (the **fermions** that constitute matter), and integral spin excitations (the bosons that mediate the fundamental interactions), including the spin-2 graviton. Joel Scherk and John Schwarz thus proposed string theory as a possible unified theory of all fundamental forces. The zero-dimensional particles of Democritus were to be replaced by one-dimensional strings.

String theory is "Planck scale physics." The size of the strings is of the order of the Planck length, 10^{-33} cm. A proton is about 10^{-13} cm in diameter. On any "reasonable" scale, then, and certainly on any scale that we are ever likely to be able to directly investigate experimentally, these extended objects "look like" point particles. But it is their extended, rather than point-like nature that allows string theories to be renormalizable.

By the early 1990s, there were five consistent string theories that were acceptable as possible unified models. All were renormalizable and mathematically consistent, and contained the proper particles and forces. All were "supersymmetric"— they were invariant under a set of transformations that could interchange fermions and bosons. And all were wonderful examples of just how abstract fundamental physics had become by the dawn of the twenty-first century.

In all the models the strings live in a ten-dimensional **space-time**. (Ten dimensions are necessary in order for the theories to be anomaly-free.) To recover the four-dimensional space-time in which we live, six space-time dimensions must be compactified; that is, they are curled up very tightly. If we could travel along one of the compact directions, we would return to our starting point after traveling roughly one Planck length. (As an example, visualize a large two-dimensional space—a sheet of paper. Now roll the sheet up into a very long, very narrow soda-straw. The **space** is still two-dimensional, but one of the dimensions has been compactified.) Neither the mechanism by which compactification occurs, nor the geometry taken by the compact six-dimensional space is presently understood-a real problem for string theorists.

Theoretical physicists believe, to paraphrase Einstein, that "God had no choice," that there is only one unique way to construct a **universe**. That five consistent models could be constructed was truly an embarrassment of riches. It now seems, though, that there is an underlying theory of which all five are simply different aspects. It is called "M-theory," which is short for the "Mother of all theories," and is, at this writing, being intensively investigated.

STRONG INTERACTIONS

The strong **force** is a fundamental force that acts across the short range of the atomic **nucleus** to bind it together. The strong force reflects an interact between **quarks** that is mediated by the carrier particle (**boson**) termed the **gluon**. In addition to binding **nucleons** (i.e., neutrons and protons), the strong force controls interactions between mesons and baryons.

At the end of the nineteenth century, two fundamental forces were known, the electromagnetic force and the **gravitational force**. As physicists began to study atoms and subatomic structure, however, they discovered more forces at work. The first new force to be found was the strong force, responsible for binding protons and neutrons into atomic nuclei. Initially, the nucleus presented a puzzle to physicists because, according to electromagnetic theory, nuclei with larger numbers of protons should not be stable. The positive charges would repel each other, and the attraction to neutrons could only be explained by **gravity**. The gravitational force between protons is dwarfed by the electric force, however, and is entirely negligible. The strong force was postulated because there was no other possible explanation for the stability of nuclei.

The strong force in nuclei acts between nucleons. The strong force does not differentiate between whether a **nucleon** is a **proton** or a **neutron**; it simply binds them together. Because of this "flavor-blindness" in the strong force, theorists sought a theory of the strong force based on a symmetry principle where the theory is invariant under exchange of protons and neutrons. In the language of **quantum field theory**, the strong force between nucleons is mediated (carried) by gluons. To describe all the possible interactions between gluons with differing colors (i.e., charges of the strong force) there are eight gluons, each of which generally carries a mixture of a **color** and a differing color anticolor. In connection with the strong force, the term color relates to three possible strong force charges not to the common use of the term color as related to the **electromagnetic spectrum**.

The true nature of the strong force was not discovered until the invention of the quark model. High-energy **scattering** experiments involving protons produced strange effects that could not be explained by assuming that protons were elementary particles. In the quark model, the proton was assumed to be constructed out of three smaller elementary particles, named quarks by American physicist **Murray Gell-Mann**. This model was successful in predicting the observed experimental results. Furthermore, it was found that the quarks possessed a new type of charge, named color. There are three distinct color charges. A nucleon consists of three quarks each of a different color, while a **pion** (the lightest form of **meson**) consists of a quark-antiquark pair of the same color. The force-carrier particle associated with the strong interaction is a massless boson, the gluon. There are eight different gluons necessary for interactions among all combinations of the three color charges. In terms of the quark model, the strong force between nucleons is a result of the residual force "leaking out" of the nucleons through the exchange of particles. This theory of color charges is known as **quantum chromodynamics (QCD)**, and has been verified by further experiments. QCD is the strong force analog of **quantum electrodynamics (QED)**.

See also Nuclear physics; Symmetry and symmetry principles

STRUTT, JOHN WILLIAM, LORD RAYLEIGH, (1842–1919)

English physicist

John William Strutt, who held the title Lord Rayleigh, received almost every honor available to a British scientist, including the Nobel Prize in physics in 1904. Lord Rayleigh never claimed that his methods were original, but he was very versatile in applying them. He specialized in technical physics, now called engineering physics. In his research, Rayleigh alternated between theory and experiment.

During his 50-year scientific career, Rayleigh published nearly 500 technical papers. He was professor at the Cavendish Laboratory at Cambridge University and later served as chancellor at Cambridge. Rayleigh was fellow, secretary and president of the Royal Society of London and received numerous honorary degrees from universities and science academies around the world. Early in his career, Lord Rayleigh started and helped to fund the Rayleigh farms and dairies that paid the family expenses during his academic career.

Rayleigh pursued classical physics in fields such as **mechanics**, **dynamics**, **acoustics**, and **thermodynamics**. Rayleigh looked at the classical nature of **electricity** and many of the advances that later took place in **radio**, **electronics**, and electrical engineering could be traced to Rayleigh's work. Rayleigh's style of research followed the pattern now used in computer analysis. He attacked a complex mathematical expression by breaking the expression into a series of terms and working only with the terms of the lowest order. Rayleigh moved to the next higher order only if the mathematical problem did not fit the observed data. In his calculations, Rayleigh used logarithms but never used a slide rule. It was not unusual for one problem to take as many as 50 handwritten pages. Perhaps because of the lengthy calculations required for physics research, Rayleigh was valued for his ability to go straight to the point and express it simply.

As of 2000, Rayleigh's theory of **scattering** was still taught in physics. The theory of scattering developed when Rayleigh observed that air molecules scatter **light** and this affects the way haze in the atmosphere can be measured.

John William Strutt III, Third Lord Rayleigh.

Rayleigh's theory of **sound**, published in 1877, is useful for the study of acoustics, wave propagation, vibration, and periodic systems. The mathematical formulations used for wave mechanics are adaptations of Rayleigh's work on the theory of sound.

Rayleigh helped lay the groundwork for the quantum revolution. The Rayleigh-Jeans law, a classical formulation in **blackbody radiation**, worked for long wavelengths but not short ones. German physicist, **Wilhelm Wien**, developed an opposite theory that worked for short wavelengths but not long ones. **Max Planck** tried to make sense of the contradictory theories and this led to his discovery of the quantum in action. Rayleigh's research led to further developments in **quantum mechanics** such as the Rayleigh-Ritz approximation and the Schroedinger-Rayleigh method. In 1887, Lord Rayleigh assisted and advised **Albert A. Michelson** in the Michelson-Morley **ether** drift experiments which were a stepping stone to Einstein's **relativity theory**. While Rayleigh accepted the theory of relativity, his own studies were classical and nonrelativistic.

In 1895, Rayleigh isolated the element argon and gave equal credit to Sir William Ramsay.

Rayleigh won the 1904 Nobel Prize for his work on resolution limits. Using a model of the optical system, Rayleigh observed that a shorter **wavelength** and larger aperture result in a finer resolution. The smallest feature that can be resolved by an optical system with a circular aperture is in direct proportion to the wavelength of the light source when divided by the diameter of the lens aperture. Rayleigh's limit is useful to the semiconductor industry because the formula can be applied to determine the smallest **transistors** that can be put into a computer chip.

Rayleigh's six volumes of scientific papers were reprinted in 1964 as physicists continued using them. Rayleigh's continues to be read in the original and cited, whereas the work of most of his contemporaries is not. A memorial tablet in Westminster Abbey honors Rayleigh as "An unerring leader in the advancement of natural knowledge."

See also Rayleigh scattering

SUBATOMIC PARTICLES

Subatomic particles are particles that are smaller than an **atom**. Historically subatomic particles were considered to be electrons, protons, and neutrons. However, the definition of subatomic particles has now been expanded to include elementary particles and all the particles they can be combined to make.

Elementary particles are particles that cannot be divided up into smaller particles. There are two types of elementary particles. One type of elementary particles make up **matter**. Examples of these particles include electrons and **quarks** (which make up protons and neutrons). Baryons and **mesons** are combinations of quarks and are considered subatomic particles. The most famous baryons are protons and neutrons.

The other elementary particles are mediators of the fundamental forces. These mediator particles enable the matter particles to interact with each other. To illustrate this idea, let's say two boys want to play catch. The boys represent the matter particles, and playing catch represents the fundamental **force**. In this case, the ball would represent the mediator particle.

The first subatomic particle to be discovered was the **electron**. While others had deduced the existence of a negatively charged particle in what were called cathode rays, it was **Joseph John Thomson** who in 1897 measured their **velocity** and charge-to-mass ratio. The charge-to-mass ratio was found to be relatively large, and independent of the gas in his experiments, which indicated to him that he had found a true particle. Thomson gave it the name "corpuscle," which was later changed to the **electron.**

The charges of all particles are traditionally measured in terms of the size of the charge of the electron. The electron has a charge, e, of 1.6×10^{-19} Coulombs.

The first mediator particle to be discovered was the **photon.** In 1900, **Max Planck** reported that **light** came in little packages of **energy**, which he called "quanta." In 1905 **Albert Einstein** studied the **photoelectric effect** and proposed that **radiation** is quantized by its nature. A photon (the name was coined by chemist Gilbert Lewis in 1926) is one of these **quanta**, the smallest possible piece of energy in a light wave.

The **proton** was one of the earliest particles known. (The word proton is Greek for "the first one.") In 1906 the first clues to the nature of the proton were seen. Thomson reported detecting positively charged hydrogen "atoms." These were in fact, hydrogen nuclei, or protons, but **atomic structure** was not understood at the time. Conversely, Thomson thought that protons and electrons were randomly scattered throughout the atom, the so-called "plum-pudding model." In 1909–1911 **Ernest Rutherford** and his colleagues **Hans Wilhelm Geiger** and Ernest Marsden, did their famous **scattering** experiments of alpha particles (two protons and two neutrons) onto gold foil. They found that atoms had relatively hard and small centers, thus discovering the atomic **nucleus** and disproving the plum-pudding model.

In 1913, the Bohr model of the atom was introduced. In this model, the **hydrogen atom** consists of an electron orbiting the nucleus (a single proton), much as Earth orbits the **Sun**. The Bohr model also requires that the **angular momentum** (**mass** times velocity times distance) of the electron be limited to certain values (that is, that it be "quantized"), in order that the electron doesn't fall into the nucleus. Rutherford gave the proton its name in 1920.

When the principles of quantum **mechanics** were developed, the Heisenberg **uncertainty principle** meant the **Bohr atom** had to be modified. The Heisenberg uncertainty principle tells us that it is impossible to accurately determine both the position and the **momentum** (mass times velocity) of a subatomic particle at the same time. This means that the electrons in an atom still orbit the nucleus, but their position is scattered throughout a "cloud" instead of in well-defined orbits.

In 1930, scientists started to suspect the existence of another subatomic particle, which came to be known as the **neutrino**. Neutrinos are considered matter particles, but they do not make up Earth matter by themselves. Neutrinos are very common, but we don't observe them because they don't interact much at all with other particles.

In 1930 a problem with a process called nuclear **beta decay** had developed. Nuclear beta decay occurs when an unstable, or radioactive, nucleus decays into a lighter nucleus and an electron. Scientists observed that the energy before the beta decay was greater than the energy after the beta decay. This was puzzling because one of the main laws of physics, the law of **conservation of energy**, states that the amount of energy in any process must remain the same. To keep the idea of energy conservation intact, **Wolfgang Pauli** proposed that another particle carried off the missing energy that was given off in the decay. In 1933 **Enrico Fermi** named this hard-to-detect particle the neutrino, and successfully explained the theory of beta decay.

The neutrino was found in a 1956 experiment. Later, a second type of neutrino, the **muon** neutrino, was found, and a third type, called the tau neutrino, was discovered in the late 1990s. In 1998 physicist discovered that at least some of these three types of neutrinos must have a mass. Though it would have to be very tiny, it must at least be greater than 20-billionths of the mass of the electron—extremely small, but definitely not zero.

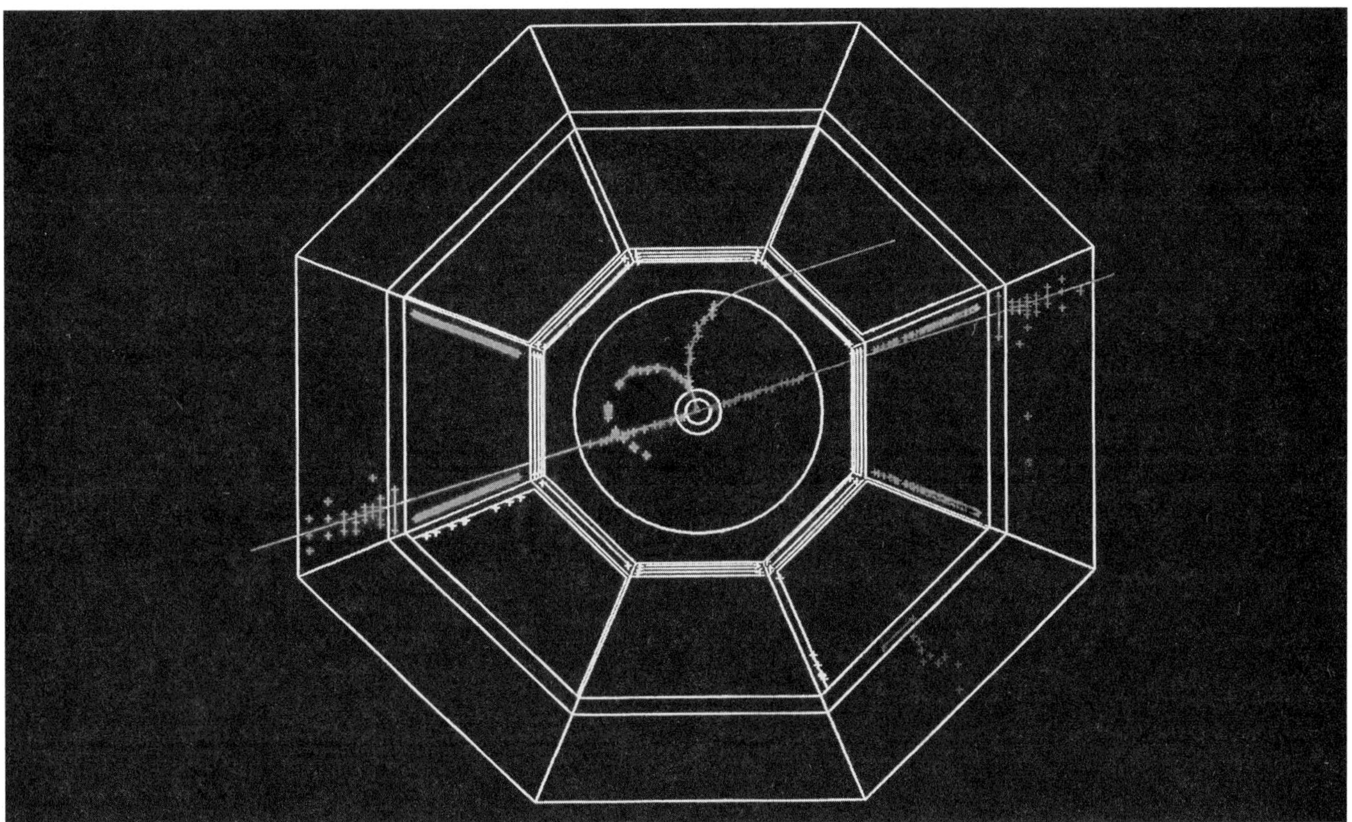

Electronic display of energies from subatomic particles in the CLEO detector. *(Image by Newman Laboratory of Nuclear Science, Cornell University. National Audubon Society Collection/Photo Researchers, Inc. Reproduced by permission.)*

In 1931–1932 **Carl Anderson** experimentally observed the anti-electron, which he called the **positron**, after its positive charge. The positron is an **antiparticle** which had been predicted by **Paul Adrain Maurice Dirac** in 1927–1930. It turns out every particle has an antiparticle partner that has the same properties except for an opposite **electric charge** (and other less obvious quantities used in **quantum mechanics**). Antiparticles make up what is called **antimatter**. As far as we can tell, matter is much more common in our **universe** than antimatter, though it is unknown why this is so.

In 1932 **James Chadwick** discovered another matter particle, the **neutron**. The neutron is very similar to the proton except that it is electrically neutral. Chadwick found it by hitting a chemical called beryllium with alpha particles. When this occurred, highly penetrating radiation was emitted. This "radiation" turned out to be a stream of neutrons. After Chadwick's experiment, **Werner Heisenberg** proposed that the nucleus is made of protons and neutrons, which was later found to be true.

The second mediator particle discovered was the pion. In 1935 **Hideki Yukawa** formulated the idea that protons and neutrons were held together by a nuclear force, which was mediated by a particle called a pion. Yukawa described it in detail. In 1937 the first evidence for it was observed by studying **cosmic rays** (high-energy particles from **space**). By 1947 it became clear that cosmic rays did contain Yukawa's **pions**,

but also contained another particle, a heavy electron, which was given the name muon. In 1947 yet another particle was detected from cosmic rays, the kaon. The kaon is like a heavy pion, and decays into two pions. The kaons are considered **strange particles** because they can be made fairly quickly, but it takes a long time for them to decay. Usually we would expect the time to make a particle and the time for it to decay to be about the same, but this wasn't true for the kaon. The kaon ended up being the first of many new particles to be detected.

In 1980 Maurice Jacob and Peter Lanshoff detected small, hard, objects inside the proton by firing high-energy electrons and protons at it. Most of the high-energy particles seemed to pass right through the proton. However, a few of these high-energy particles were reflected back, as if they had hit something. These and other experiments indicated that the proton contains three small, hard, solid objects. Thus protons are not elementary, but the objects inside them may be. These objects are now called quarks.

Quarks had been postulated much earlier, in 1964, by **Murray Gell-Mann** and independently by George Zweig. The theory describing quarks was called the quark model. In 1964 it was thought that there should be three different quarks. These different quarks each have a unique property called flavor. These first three quarks had flavors that were whimsically named up, down, and strange. Up-flavored quarks have an electric charge of $(2/3)e$, where e is the fundamental quantum of charge such as that of the negatively charged electron. Down- and strange-flavored quarks have an electric charge of $(-1/3)e$. The quark model also says that quarks must remain bound inside their particles—in nature, quarks cannot exist by themselves. This idea is called quark confinement, and is based on the experimental observation that a free quark has never been seen. Since we cannot isolate quarks, it is very difficult to determine their masses.

In 1964 Oscar W. Greenberg suggested each quark has a quality called **color**. The label "color" for this quark property does not refer to the usual definition of color. It is just a way to keep track of quarks. Using this idea of color, the improved quark model says only overall colorless particles can exist in nature. There are only three different kinds of color in the quark model, usually designated red, blue, and green. Color had to be introduced when a particle called the $\Delta++$ (Delta-plus-plus) baryon was discovered to avoid violating the **Pauli exclusion principle**. The Pauli exclusion principle says that each particle in a system of matter particles must have unique properties like electric charge, and mass, and spin. The $\Delta++$ baryon is made of three up quarks. Without color, each of its three up quarks cannot have its own properties. Color has been proven experimentally, and a theory called the standard model of elementary particles has updated the quark model.

There are two kind of elementary (indivisible) matter particles, the quarks and the leptons. The two lowest-mass leptons are the electron (e^-) and its partner the neutrino, usually called the electron-neutrino (v_e). For unknown reasons, this **lepton** pairing is repeated two more times, each time with increasing mass. These leptons are called the muon (μ^-), and muon neutrino (v_μ), and the tau (τ^-), and tau neutrino (v_τ). We say there are three families, or generations, of leptons.

As in the lepton sector, the quark sector has three families. The first family of quarks are the up and down quarks, the second contains the strange and "charmed" quarks, and the third the "botton" and "top" quark. Though all matter we see around us contains only up, down, and strange quarks, physicists have proven the existence of all six flavors of quarks, culminating with the discovery of the top quark in 1995.

Another property of elementary particles is called spin. Spin is akin to the rotation of a particle on its axis, as the earth spins on its axis to give us day and night. (In actuality elementary particles do not rotate; it is that the quantity called spin obeys rules that mathematically are similar to those used to describe certain rotation bodies.) The spin of elementary particles is so small it is measured in special units called h-bar (h-bar is **Planck's constant** divided by 2*pi), and equals 1.1×10^{-34} Joule-seconds. Using the property called spin, all matter particles are **fermions** which have spin one-half h-bar or three-halves h-bar. All quarks and leptons have spins of one-half h-bar. The matter particles and some of their properties are summarized in this table. Masses are given in units of MeV/c^2, where c is the **speed of light** (three-hundred-million meters per second). The quark masses are approximate.

Bosons are particles defined to have spin of zero h-bar, one h-bar, or two h-bar. The elementary mediator particles are bosons with spins of one h-bar. The force we are most familiar with is the electromagnetic force. The electromagnetic force is responsible for keeping electrons and nuclei together to form atoms. The electromagnetic force is mediated by photons, which are massless. The mediators of the strong force are called gluons (g), because they glue quarks together to form mesons and baryons. Like the quarks, the gluons carry the color property, and as a result there are eight different types of gluons.

The weak force is more uncommon—it is responsible for radioactive decays like nuclear beta decay. The mediators of the weak force are the electrically charged W-bosons $(W\pm)$, and the electrically neutral Z bosons (Z^0), both discovered in 1983. Some properties of the mediator particles are given here. Masses are given in MeV/c^2.

One of the main rules of the standard quark model is that combinations of three quarks are called baryons. Protons (p) and neutrons (n) are the most important baryons. Protons are made of two up quarks and one down quark. Neutrons are made of two down quarks and one up quark. Since the quark model requires that naturally-occurring particles be colorless, a baryon must be made of a red, a blue, and a green quark. These combine to make a white, or colorless particle. Spin is also important in classifying baryons. Baryons are fermions and so have spins of one-half h-bar or three-halves h-bar. Table 3 summarizes several kinds of baryons, with masses in MeV/c^2 and spin in terms of h-bar.

The second main idea of the standard quark model is that combinations of one quark and one antiquark are called mesons. Pions (π) and kaons (K) are examples of mesons. Thus now we see Yukawa's nuclear force mediator particle, the pion, is really a matter particle made of a quark and an antiquark. There are several kinds of pions. For example, the positively charged pion, π^+, is made of an up quark and a down antiquark. Similarly there are several kinds of kaons. One kind of kaon, K^+, is made of an up quark and a strange antiquark. The colorless rule requires that mesons must be made of quarks with opposite color, red and anti-red for example. All mesons are bosons and so have spins of zero h-bar or one h-bar.

Subatomic particles are very important in many technologies. Television sets use beams of electrons to create their pictures. A television set picks up the television signal which is then sent to electron guns. Electron guns shoot off beams of electrons which hit the face of the picture tube. When electrons hit the tube, it lights up, creating the picture. A common type of smoke detector that uses subatomic particles is an ionization smoke detector. In an ionization smoke detector alpha particles take away or add electrons ("ionize") to groups of atoms called air molecules. These ionized air molecules cause **electric current** to flow in the detector. If there is a fire, other particles enter the detector and interfere with the flow of the electric current, which makes the alarm go off.

In the 1990s, **particle physics** was particularly exciting, with several important experimental developments. Besides the discovery of the W and **Z bosons** and the top quark, scientists working in Japan in 1998 found evidence that at least some of the three types of neutrinos have a small but nonzero mass. (Their experiment did not allow them to determine the exact value for the mass.) This could be an important breakthrough and may mean that neutrinos play an important part in the ultimate fate of the universe.

The matter that astronomers can observe is called visible matter. There is not enough mass from visible matter in the universe for it to collapse in a big crunch (the opposite of the **big bang**, the explosion that created our universe) due to its own **gravity**. If it does not collapse, the universe will continue to expand forever, and everything, including **stars**, will die. Since neutrinos are not visible matter (to astronomers), they may be considered dark or "missing" matter. If the mass of neutrinos is enough, this neutrino missing matter may be great enough to "close" the universe, allowing it to collapse and be reborn someday. Surprisingly, astronomers found evidence in 1998 that the expansion of the universe is actually accelerating—that there is not enough mass to halt and reverse its current expansion. If true it implies that the universe would probably continue to expand forever.

SUBLIMATION

The process of sublimation is the transformation of a solid substance into a gaseous phase without first traversing through a liquid phase. Typically, when solid **matter** changes its physical state as it is heated, it first becomes a liquid. Then as more **heat energy** is applied, the liquid becomes a gas. When sublimation takes place, however, the solid immediately becomes a gas and no melting occurs. A good example is the sublimation of solid **carbon** dioxide. Carbon dioxide, the gaseous compound found in our atmosphere that is exhaled during respiration, is a solid at very cold temperatures. Solid carbon dioxide is called *dry ice* because, at ordinary pressures, it sublimates into a gas without first passing through a liquid phase. The opaque carbon dioxide vapor that emanates from dry ice is very cold compared to room **temperature** air, and thus flows downward, in a manner opposite of steam. Sublimating solid carbon dioxide is often used to create haunting and mysterious misty effects in theater productions and movies.

Sublimation also occurs with water, the most important inorganic compound for life on Earth. Solid water, or ice, can sublimate very slowly at subfreezing temperatures. For example, thin layers of snow can be seen to disappear slowly from the sides of roadways in the middle of winter when the temperature does not exceed the freezing point of water. The solid water slowly sublimates into water vapor over time, resulting in snow that vanishes, but does not melt. The same process can be observed at an accelerated pace when frost on car windshields sublimates into water vapor before the windshield warms up from the interior and as the car is moving. The increased air flow over the ice crystals of the frost facilitates their sublimation into water vapor. Frost free refrigerator freezers utilize sublimation of ice crystals to prevent frost formation and ice build up.

Sublimation can occur in reverse as well. Sublimation in reverse is sometimes referred to as *deposition*. Water vapor forms solid ice as frost on the windows of airplanes as it sublimates, or deposits, directly from gas to solid. Also, water reverse sublimates from air onto very cold surfaces, like dehumidifier coils. Snowflakes also can be formed by reverse sublimation. Amazingly, substances that do not ordinarily sublimate can be made to sublime by changing temperature and **pressure**. In general, if both temperature and pressure are sufficiently lowered, any solid may sublimate directly into a gaseous state. Under these conditions, the substance cannot exist as a liquid.

SUN

The Sun is one of 2×10^{11} **stars** in the **Milky Way galaxy**. While it may appear quite large, it is neither the largest star nor the smallest. The Sun is, however, the largest object in our solar system. Its **mass** is 334,000 times that of Earth. Scientists classify the Sun as a G2 star, meaning it is the second hottest star of the yellow G class of stars.

The Sun is made almost entirely of hydrogen and **helium**. Sunlight is produced by atomic fusion, the same process that powered the hydrogen bomb. This process occurs across three sections of the Sun: the core, the radiative zone, and the convective zone.

In the Sun's core, temperatures soar to 27 million°F (15 million°C). The **heat**, combined with an internal **pressure** 340 billion times Earth's air pressure at sea level, triggers **thermonuclear reactions** that convert hydrogen gas to helium. The resulting **energy** races through the Sun's middle layer, called the radiative zone, and then boils and bubbles to the surface of the convective zone.

In addition to three zones, the Sun also has several layers: the photosphere, the chromosphere, and the corona. The photosphere is the only layer we can see during normal daylight. This fiery hot layer has a **temperature** of 11,000°F (6,000°C) and a blotchy appearance due to the turbulent eruptions of energy at the surface. The hot gases of this layer produce **light** by a process called incandescence.

The chromosphere and corona are only visible during a solar eclipse. The chromosphere is an irregular layer above the photosphere. When the **Moon** blocks the Sun, the normally transparent chromosphere appears as a red glow. The corona is the Sun's outer atmosphere. During a total solar eclipse, it appears as a pearly white light surrounding the darkened Sun. The most remarkable aspect of the corona is its high temperature. This outermost layer reaches about 1.8 million°F (1 million°C). The source of the corona's heat remains a puzzle to scientists.

Sunlight and heat are not the Sun's only effects on Earth. Sometimes the Sun's corona spits out highly charged particles that zoom through the solar system at a million miles per hour. This is called the solar wind. When the solar wind slams into Earth's atmosphere, it distorts Earth's protective magnetic field. The change in shape can cause current surges in power lines that can cause widespread blackouts. The speeding solar wind can also short-circuit the satellites we use for telecommunications and forecasting. Scientists sometimes refer to this event as a geomagnetic storm or simply, **space weather**.

The solar wind also causes the night sky to brighten around the northern and southern poles. This effect is called the *aurora*.

It is very dangerous to look at the Sun with your naked eye, sunglasses, or even a **telescope**. Astronomers use a special **spectroscope** to observe the structure of the Sun. This instrument reflects the Sun's image onto a white surface or photographic film.

Sunspots are dark patches on the Sun's surface. These relatively cool areas are caused when the Sun's magnetic fields slow down the **radiation** of heat. Sunspot activity peaks every 11 years. The peak is called the solar maximum. During that event, the Sun spits out more charged particles (solar wind) than it does in off-peak years. The solar maximum in 1989 triggered major eruptions from the Sun that knocked out power to six million Canadians.

Scientists expect that in about five billion years, the Sun's fuel will be gone, leaving behind only its collapsed core, a small, dying white dwarf.

SUPERCONDUCTORS AND SUPERCONDUCTIVITY

A superconductor is a material that conducts electrical current with no resistance. The phenomenon of superconductivity was first discovered by H. Kamerlingh Onnes in 1911, when he found that the resistance of mercury dropped to zero at a **temperature** of about 4K. Many other materials, including tin, lead, and niobium, were tested and also found to superconduct at extremely low temperatures, less than 10 K. In 1957, a theory of superconductivity was advanced that explained the phenomenon quite well, and it has been used in further research and development of other superconducting materials.

Superconductivity normally only occurs at extremely low temperatures. Each superconducting material has a different critical temperature; above the critical temperature, the material does not superconduct. As the temperature of the material is lowered, the material undergoes a phase transition from its normal state to the superconducting state. The critical temperature also depends on the external magnetic field present. As the strength of the magnetic field increases, the critical temperature decreases. At a certain field strength, called the critical magnetic field, the material can no longer superconduct at any temperature.

Superconductors have another interesting property: in the presence of a weak external magnetic field, all magnetic field lines will be expelled from the superconductor itself. This effect is called the Meissner Effect. It can be easily observed by trying to place a magnet on top of a piece of superconducting material; the magnet will levitate above the surface of the superconductor. Another interesting phenomenon involving superconductors is quantized **magnetic flux**. If a superconducting ring is placed in a magnetic field above its critical temperature, and then cooled, the magnetic flux will be

Superconductor. *(Photo by Takoshi Takahara. Photo Researchers, Inc. Reproduced by permission.)*

expelled from the superconductor itself, but some flux will be trapped in the hole. It is found that the flux can only exist in quantized units; it turns out that this phenomenon is explained by requiring the wavefunction for electrons in the superconductor to have a single value at each spatial point.

In 1957, **John Bardeen**, Leon N. Cooper, and John Robert Schrieffer published a quantum mechanical theory of superconductors that came to be known as the BCS theory. In this theory, pairs of electrons attract each other by interacting through the crystal lattice of the solid material. These Cooper pairs are the charge carriers of the **electric current**, with charge -2e. The attractive **potential energy** of the electron-lattice interaction sets up an **energy** gap between electrons that interact collectively in pairs, and electrons that interact individually. The energy of the Cooper pairs cannot be dissipated by collisions, because the energy gap between levels is too large compared to the energy that would be lost in a collision. Therefore, there is no **power** loss and hence no resistance.

The possible uses of superconductors are quite exciting. For example, a current set up in a superconducting ring can continue to circulate for an extremely long time without dissipating. Theory predicts a lifetime of at least 100,000 years for a current circulating in a superconducting ring; experimenters have been able to measure the current dissipation in superconducting rings and have found it to be negligible. Power could be transmitted through superconducting power lines, making electrical transmission more efficient and saving energy. Superconducting magnets can be produced that act like permanent magnets. The Meissner effect could be used to produce efficient magnetic-levitation transportation or other technologies.

The obstacle standing in the way of widespread use of superconductors is that superconducting materials have to be cooled to extremely low temperatures. Common superconducting materials, such as mercury and lead, have to be cooled below 10K, around the **boiling** point of liquid **helium**. After much research, materials known as high-T_c superconductors have been created, but they still require temperatures close to the boiling point of nitrogen, 77K. These high-T_c superconductors are exotic compounds containing rare elements; for example, the compound $YBa_2Cu_3O_{6.9}$ has a critical temperature of 90K, but the amount of Yttrium found naturally is probably too small to allow **mass** production of superconducting power lines. It is safe to say that high-T_c superconductor research is one of the largest current research fields, and room-temperature superconductors are one of the most coveted materials in science research. The promising uses for these materials makes their development extremely desirable.

SUPERFLUIDITY

Superfluidity refers to a transition that occurs in the properties of liquid **helium** at a **temperature** of 2.17K, called the lambda point of helium. A discontinuity in **heat capacity** is observed, the **density** of the liquid is decreased, and a fraction of the liquid helium becomes a superfluid, i.e., a zero **viscosity** fluid that exhibits an extraordinarily enhanced fluidity through any medium.

In its elemental state, helium is a gas that can occur as two isotopes, helium-4, which is the most common, and helium-3, which occurs only in very small amounts. Helium-4 has a **nucleus** consisting of two protons and two neutrons surrounded by two electrons, which is the definition of a **boson**, i.e. a species consisting of even numbers of elementary particles. The helium-3 nucleus however, has only one **neutron** and is therefore termed a fermion because it consists of an odd number of spin-1/2 particles in its nucleus. The fundamental distinction between bosons and **fermions** is that the former obey Bose-Einstein statistics, which allows them to condense to the state of lowest possible **energy**, the ground state, as predicted by **Albert Einstein**. In 1924 he proposed that bosons could condense in unlimited numbers into a single ground state because they are not subject to the **Pauli exclusion principle**. A phase transition in which such a phenomenon occurs is referred to as **Bose-Einstein condensation**. Unlike bosons, fermions such as helium-3 obey Fermi-Dirac statistics, which implies that they should not condense in the lowest energy state. Helium-3 does possess characteristics that allow it to become supefluid, however, this requires a much lower temperature than helium-4. These two forms of helium remains unique in their ability to experience superfluidity because other elements will typically freeze at such temperatures even though they may also be spin-zero atoms.

Superfluidity was first discovered in 1937 in the helium-4 **isotope** by the Russian physicist **Pyotr Leonidovich**

Kapitza (Nobel Prize in physics, 1978) and it is considered one of the most remarkable breakthroughs of **low-temperature physics**. Theoretically, it was shown shortly after the discovery by Fritz London and **Lev Landau** (Nobel Prize in physics, 1962) that the effect was due to a two-fluid state of the helium: one fraction remained a normal liquid and the superfluid fraction consisted of those atoms undergoing Bose-Einstein condensation, thus becoming incapable of contributing to the **entropy** or **heat** capacity of the liquid. In 1996, the Nobel prize was conferred to D.M. Lee, D.D. Osheroff, and R.C. Richardson for discovering in the 1970s that the helium-3 isotope can also be made superfluid at a temperature only about two thousandths of a degree above **absolute zero** (-273.15°C). This achievement greatly surprised the physics community since fermions were not expected to undergo Bose-Einstein condensation. The BCS theory for superconductivity in metals, formulated by **John Bardeen**, Leon Cooper, and John Robert Schrieffer (Nobel Prize in physics, 1972) provided the explanation. It was proposed that electrons, being fermions since they consist of one particle, must follow Fermi-Dirac statistics just as the helium-3 isotopes do. However in supercooled metals, two electrons can combine as Cooper pairs, which allows them to exhibit boson-like behavior, such as the capacity to undergo Bose-Einstein condensation.

Superfluidity cannot be understood in terms of classical physics. When helium becomes superfluid, its atoms lose all randomness and move in a coordinated manner, which causes the liquid to totally lack **friction**, thus allowing it to flow out of very small holes, and exhibit other manifestations of non-classical behavior. Understanding the fundamental properties of such a liquid requires an advanced form of **quantum physics**, and for this reason, liquid helium is often referred to as a quantum liquid. Further research developments have shown that helium-3 has at least three different superfluid phases, one of which occurs when the helium is placed in a magnetic field. As a quantum liquid, helium-3 can thus exhibit a considerably more complicated structure than helium-4.

An interesting application of superfluidity in helium-3 is that it is being used to investigate the formation of the cosmic strings required by **string theory**. These enormous hypothetical objects, believed to be involved in the birth of galaxies, could have arisen as a consequence of the rapid phase transitions believed to have taken place a fraction of a second after the **big bang**. Research teams used neutrino-induced **nuclear reactions** to quickly heat superfluid helium-3, followed by rapid re-cooling. This resulted in the production of balls of vortices that were proposed to correspond to cosmic strings. Helium-4 has no notable applications. Observation of the superfluid state of helium-4 results in further analysis of its unusual properties such as its ability to move with zero viscosity, against **gravity**, and not be affected by condensation or evaporation. The fact that it also experiences a tremendous increase in heat conductivity may prevent it from having useful applications.

See also Conduction, heat; Fluids; Gauge symmetry; Isotopes; Neutron stars; Pauli exclusion principle; Superconductors and superconductivity; Viscosity

SUPERGIANTS • See Stars

SUPERGRAVITY THEORY

Supergravity theories merge the notions of supersymmetry and gravitation. Supersymmetry requires that every bosonic particle (integer spin) has a fermionic counterpart (half-integer spin) and vice versa. Spin-0 particles (scalars) and spin-1 particles (**vectors**), together with their identical **mass** spin-½ superpartners, comprise supermultiplets. It is natural to ask whether a gauge transformation can mediate a change from a bosnic degree of freedom to a fermionic one. Such an operation is highly non-trivial. Since spin has units of **angular momentum**, changing the spin of a particle also alters its **space-time** coordinates. In contrast, a gauge transformation within the context of the standard model rephases a field, and this effect is compensated by a shift of the covariant derivative in such a way that the Lagrangian of the theory remains invariant. The key difference here is that the position of the field itself is unchanged.

The general theory of relativity treats the **gravitational force** in terms of the response of **matter** to the curvature of space-time, which in turn results from the presence of matter or **energy**. A gauge transformation in the theory of **gravity** can therefore be thought of as producing a slight displacement in the position of a particle. A theory that allows transitions among particles of different spin must also be consistent with the theory of gravity. In particular, it must include a particle that mediates gravity as a fundamental **force** on an equal footing with the electromagnetic, strong, and weak interactions. That particle is the **graviton**, and it carries spin-2. The supersymmetric partner of the graviton is called a gravitino, and it has spin-3/2.

Different supergravity theories are classified by the types of particles in each supermultiplet. The extensively studied N=8 theory, has supermultiplets with particles of all spins between 0 and 2. Because supergravity treats each of the fundamental forces in a similar way, it was initially viewed as a candidate for the so-called *theory of everything*, which would unify all of physics. Unfortunately, infinities in the theory do not completely cancel because gravity is not renormalizable. However, because supersymmetry and supergravity can be extended to a theory with more than four space-time dimensions, these concepts served as an important step towards constructing superstring theory.

SUPERNOVAE • See Novae and supernovae

SUPERSONIC SPEEDS

The speed of **sound** plays a fundamental role in compressible gas **dynamics**: it is the speed at which a small amplitude disturbance propagates through a gas. Sound is actually a **pres-**sure oscillation, areas of compression and rarefaction passing through a medium (gas, liquid, or solid). When a body is moving faster than the **speed of sound**, it compresses the sound **waves** radiating out from it. These build up into a large pressure front which is the shock wave. As the body passes the observer, the effect of this shock wave is a sudden spike in pressure. In the atmosphere, this is the so-called sonic boom.

Imagine the following thought experiment. Take a platform, and move it at a constant speed through a uniform gaseous medium. Arrange a sound speaker on platform, so that it is continually emitting sound. If the platform is at rest, the sound signal propagates in all directions at the speed of sound. Now, once the body is moving forward with a fixed speed, the signal will propagate relative to the platform. So long as the platform is moving slower than the speed of sound, an observer in front of the platform will always hear the forward propagating signal before the platform itself arrives. However, if the platform moves faster than the speed of sound, an observer in front of the platform will hear absolutely nothing before the platform itself arrives. This thought experiment illustrates how supersonic speeds can play an important role in gas dynamics: they behave quite distinctly from subsonic motions.

A more detailed examination of the problem of a body moving supersonically through a uniform medium reveals that the gas arrives at the observer through a shock wave that propagates supersonically with respect to the background gas. **Shock waves** are like "sudden news" to the observer: because they are propagating supersonically, the observer does not hear them coming. Under ordinary terrestrial conditions, the thickness of the shock front is set by the distance that a free streaming molecule in the advancing shock travels before it impacts another molecule of unshocked gas. This distance is referred to as the collisional mean-free path. Since this distance is so small in comparison to most terrestrial dimensions, pre-shocked gas impacted by the shock is rapidly compressed and heated, and attains the post-shocked state consistent with **conservation of mass**, **momentum**, and **energy**.

Because the speed of sound on Earth is so large, terrestrial supersonic gas dynamics finds relatively few applications: mainly to explosions and projectiles. However, in **astrophysics**, supersonic speeds are very commonplace: they occur everywhere from stellar atmospheres to superonovae explosions in **interstellar space**. For instance, some scientists have argued that the tremendous energy released in supernovae explosions, and transmitted via shock waves to the surrounding interstellar medium, is very effective in heating the interstellar medium gas to temperatures up to millions of degrees Kelvin, where it emits strongly in the x-ray range. The heating of **interstellar gas** via such shock waves plays a very significant role in the overall energy budget of interstellar gas in the galaxy.

SUPERSTRING THEORY • See String theory

SUPERSYMMETRY • See Symmetry

The long legs of a waterstrider (*Gerris paludum*) allow it to distribute its weight across the water and be supported by surface tension. *(Photo by Hermann Eisenbeiss, National Audubon Society Collection/Photo Researchers, Inc. Reproduced by permission.)*

SURFACE TENSION

Surface tension is the result of the cohesive forces that attract water molecules to one another. This surface **force** keeps objects more dense than water (meaning they should not float) from sinking into it. The surface tension of water makes it puddle on the ground and keeps it in a droplet shape when it falls.

If you use a table fork to carefully place a paper clip on the surface of some clean water, you will find that the paper clip, although more dense than water, will remain on the water's surface. If you look closely, you will see that the surface is bent by the **weight** of the paper clip much as your skin bends when you push on it with your finger.

A molecule inside a volume of water is pulled equally in all directions by the other molecules of water that surround it. A molecule on the water's surface, on the other hand, is pulled by the molecules below it and to its sides. The net force on this surface molecule is inward. The result is a surface that behaves as if it were under tension. If a glob of water with an irregular shape is created, the inward forces acting on the molecules at its surface quickly pull it into the smallest possible volume it can have, which is a sphere.

A simple apparatus can be used to measure the forces on a liquid's surface. A force is applied to a wire of known length, forming a circle parallel to the surface of the water. The force balances the surface forces acting on each side of the wire. The surface tension of the liquid, g, is defined as the ratio of the surface force to the length of the wire (the length along which the force acts). For this kind of measurement, g = F/2L. The force, F, applied to the wire is that required to balance the surface forces; L is the length of the wire. The 2 appears in the denominator because there is a surface film on each side of the wire. Thus, surface tension has units of force per length-dynes/cm, newtons/m, oz/in, etc.

Water has a relatively high surface tension that, not surprisingly, decreases with increasing **temperature**. The increased kinetic **energy** of the molecules at higher temperature would oppose the forces of cohesion. In fact, at the **boiling** temperature, the **kinetic energy** of molecules is sufficient to overcome their cohesive forces of attraction and the molecules separate to form a gas.

SYMMETRY AND SYMMETRY PRINCIPLES

The physics definition of symmetry is only slightly more precise than our everyday definition; it is any kind of transformation that leaves the laws of physics that apply to a system unchanged. Another term is often used to describe systems that have symmetry—a system is said to be invariant under a certain transformation if that transformation is a symmetry.

The role of symmetries in physics has changed greatly since the time of **Isaac Newton**. Until the twentieth century, they were mainly invoked to solve the complicated differential equations of classical **mechanics** and **electromagnetism**. However, with the discovery of special relativity by **Albert Einstein** in 1905, a new era of symmetry usage began. Any fundamental theory of physics had to satisfy the constraints placed on it by the symmetries of special relativity. When physicists began to tackle the problems of the **quantum mechanics** of electromagnetism and the nuclear forces, it was discovered that new symmetry principles allowed relatively simple explanations of experiments and also agreed with the experimental results very well. Today, hardly any research in elementary **particle physics** takes place without using symmetry principles, and symmetry ideas have been used in other fields of physics research as well.

Symmetries in physics generally fall into two categories: **space-time** symmetries and internal symmetries. Space-time symmetries are probably the easiest to understand and place stringent constraints on what physical theories can look like. The first class of symmetries are those of special relativity, called the Lorentz group. The Lorentz group includes normal rotations, as well as boosts, where an observer's **velocity** is changed; together they are referred to as **Lorentz transformations**. It is important to note that observers in two different inertial **reference frames** may not agree on certain measurements, as in the well known **time dilation** and **length contraction** effects of special relativity. However, all of the laws of physics such as Maxwell's equations are the same in the two reference frames.

There are three other space-time transformations, called discrete transformations. They are called **parity**, charge conjugation, and time reversal. Parity means a reversal of all position **vectors**, charge conjugation means a reversal of particles and antiparticles, and time reversal means a reversal of the **direction of time**. A theorem of quantum mechanics, known as the CPT theorem, is that the laws of physics are invariant under the combined action of a parity, charge conjugation, and time reversal.

At first, it was also thought that all physical theories should be invariant under parity, charge conjugation, and time reversal independently. The theory of quantum electromagnetism and the strong nuclear **force** satisfy these criteria, but after a number of puzzling experimental results, it was found that the weak nuclear force was not invariant under parity, charge conjugation, or time reversal. The phenomenon of non-invariance under time reversal is currently a topic of intense research in elementary particle physics, where it is usually referred to as "CP-violation."

The second class of symmetries are the internal symmetries. These symmetries do not have to do with changing our point of view, as in space-time symmetries; instead they tell us how free we are to redefine the fundamental quantities in physics theories.

The first internal symmetry principles were actually discovered in the nineteenth century in relation to electromagnetism. In this theory, the electric and magnetic fields are the physically observed quantities, but they can be derived from other quantities called potentials. These potentials are not unique; functions satisfying certain specifications can be added to potentials without changing the observed electric and magnetic fields. Therefore, the potentials can be redefined without changing the laws of physics, so electromagnetism is invariant under these transformations. This symmetry came to be known as a **gauge symmetry**, and the transformations are known as gauge transformations.

Similarly, the theories of the weak and strong nuclear forces have gauge symmetries and are therefore known as gauge theories. In the case of the strong nuclear force, there are three charges (rather than the two charges in **electricity**) called **color**. For example, an up quark can have color charge red, green, or blue (the colors are purely a matter of taste). In this case, a gauge transformation redefines the colors to be three different combinations of red, green, and blue. However, performing this transformation leaves the forces between **quarks** unchanged, so it is a symmetry.

The weak nuclear force is explained by a gauge theory, but it is unlike the strong and electromagnetic theories because it incorporates a broken symmetry. For example, look at the first family of quark particles, called up and down. Similar to the strong nuclear force, the weak theory can be made invariant by redefining the fundamental quark particles to be mixtures of up and down quarks. In fact, this procedure works rather well for making predictions about **scattering** between **subatomic particles**. However, there is a big problem: the up and down quarks have to be massless in order for this symmetry to hold. Experimental observations find that quarks have **mass** and that the up and down masses are not the same. Therefore transformations mixing up and down quarks do not appear to be symmetries.

A way out of this problem was found in the late 1960s and early 1970s. The quarks were stipulated to be massless, but interactions were added between the quarks and a new field called the Higgs field. The Higgs field is a scalar field, similar to the **temperature** distribution in a room. The purpose of adding this field is that interactions with it can generate masses for the quarks, and all other massive particles, without messing up the good features of the massless weak theory. This new theory is said to have a broken symmetry. A prediction of this theory is a new particle called the **Higgs boson**, which is yet to be discovered.

Some of the more recent attempts at constructing particle physics theories are called grand unified theories and incorporate large gauge symmetries. A different sort of symmetry idea, supersymmetry, postulates that each observed particle has a superpartner particle (called sparticles for short) and that the laws of physics should be invariant under mixing normal particles and sparticles. Supersymmetry is also a prediction of string theories. These symmetry ideas are attractive because they could lead to a unification of all the different theories into a single theory of everything.

See also Relativity, special; Group theory; Antimatter; Arrow of time; Standard model of particle physics; Grand Unified Theory; String theory

SYMMETRY, ANTIMATTER, AND POLARITY

The abstract study of symmetry is the object of mathematical **group theory**, which, as its name implies, describes groups defined by the specific symmetry elements they contain and the specific symmetry operations they allow. Groups are also related to each other by symmetry transformations. Group theory provides physics with a powerful means for classifying "everything" in the **universe** and understanding the interdependence of physical processes based on symmetry transformations.

For example, many transformations in the physical world leave the laws of physics invariant, i.e. the laws of physics are the same everywhere in the universe. This is referred to as translational invariance, and it corresponds to the law of conservation of **linear momentum**; similarly, the symmetry that states the laws of physics are the same at all times is equivalent to the law of **conservation of energy**; and the rotational invariance of the laws of physics is equivalent to the law of conservation of **angular momentum**. **Quantum theory** fully describes **atomic structure** in terms of the symmetry properties of atoms, molecules, and **orbitals**. The symmetry of general relativity, much larger than any observed in physics before **Albert Einstein**, combines rotational invariance, translational invariance and Lorentz invariance to form the complete symmetry group of special relativity known as the Poincaré group, a 10-dimensional group (cf. the three-dimensional Euclidean or E(3) group generated by translations in the x direction, translations in the y direction, and rotations in the xy plane).

Particle physics describes elementary particle-antiparticle interactions as well as their associated wave fields entirely on the basis of symmetry considerations, embodied in gauge theory. This theory was formulated in 1918 by H.Weyl, and includes some infinite-dimensional groups. Symmetry or gauge transformations of the wave field variables are the means by which the basic quantum field laws remain valid, or gauge invariant, because they define general restrictions on the way a given field can interact with other fields and elementary particles.

In the 1960s physicists developed quantum field theories to explain the weak and strong nuclear interactions. They realized that **gauge symmetry**, termed the U(1) symmetry, could be generalized to gauge symmetries based on other continuous groups, such as the special orthogonal groups SO(N), the special unitary groups SU(N), the special symplectic groups Sp(N) and the exceptional groups G2, F4, E6, E7, and E8. A unified gauge symmetry could then be made up of combinations of these groups. The groups can combine using a direct product, denoted X, in which both groups are independent subgroups. From tables of particles, physicists were then able to propose, for example, that the strong nuclear interactions used the gauge group SU(3), termed **color**. The weak interaction was experimentally detected, and used SU(2) X U(1) symmetry, broken by a Higgs mechanism and a **Higgs boson**, whose **vacuum** state breaks the symmetry at low energies. With the use of these gauge symmetries, theoretical

physicists were able to construct the complete **standard model of particle physics** by 1972, as well as its particle and corresponding **antiparticle** complement.

Antimatter is defined as any **matter** composed of antiparticles, which are the charge conjugates of ordinary protons, electrons, and neutrons, i.e., they have the same **mass** but opposite electrical charge and **magnetic moment**. The antiparticles also have properties exactly opposite to those of their corresponding particle due to a remarkable symmetry translation. Therefore, every known elementary particle has a counterpart with opposite charge and/or spin, or polarity. The existence of the first antiparticle was predicted by the theory of quantum **mechanics** by **Paul Dirac** in 1928. This was the **positron**, defined as an antielectron with positive charge and **lepton** number -1. It was experimentally observed in **cosmic rays** by **Carl Anderson** in 1932 at the California Institute of Technology. Since then, the full range of antiparticles has either been experimentally observed or predicted. In 1996, scientists at the European Organization for Nuclear Research in Switzerland, discovered the first antiatom, antihydrogen, in a spectacular breakthrough that will significantly contribute to our understanding of matter and antimatter.

Most of the antimatter of the universe is believed to have been canceled out during the initial **big bang** events through a **parity** symmetry violation, yielding an almost antimatter-free universe. Antimatter is now usually produced in particle **accelerators**. Its behavior is well described by gauge theory with the exception of its reaction to **gravity** because it has yet to be produced in sufficient quantities to study these interactions. When matter and antimatter are made to collide in accelerators, both may be annihilated, i.e., they disappear and their kinetic plus rest-mass **energy** is converted into other particles, such as photons and **pions**, according to the equation $E = mc^2$. For example, when positrons and electrons collide they annihilate each other, and their energies are converted into gamma rays. If they are at rest and their spins are oppositely paired, the collision results in the production of two gamma rays, each with an energy of 511,002.7 **electron** volts. Antimatter is also produced during some **radioactive decay** processes. For example, when 14-carbon decays, a **neutron** decays to a **proton** plus an electron and an electron antineutrino. When 19-neon decays, a proton decays to a neutron plus a positron and an electron **neutrino**. The neutrino and electron are leptons while the antineutrino and positron are anti-leptons.

Leptons are defined as point-like particles that can interact with the electromagnetic, weak and gravitational forces, but not with the **strong interactions**. This is a result of their symmetry properties: the only massless particles known are the **photon**, the neutrinos, and the hypothetical **graviton**. These particles are symmetric under an even larger group than the Poincaré group, namely, the conformal group, which is 15-dimensional. **Maxwell's equations** are invariant under the conformal group and so are the Yang-Mills equations, which describe the weak and strong interactions. Thus, in each decay reaction, one lepton and one anti-lepton are produced, and this also illustrates a fundamental symmetry law of physics, i.e., for each new lepton that is produced there is a corresponding new antilepton.

See also Baryons; Baryons and baryonic matter; Cloud chamber; Cosmic ray; Electroweak force; Electroweak particles; Electroweak theory; Energy; Gamma rays; Hadrons; Magnetic moment; Mesons; Particle physics; Quantum electrodynamics; Quantum mechanics; Quantum states; Quarks and gluons; Radioactive decay; Relativity, special; Rest energy; Strong interactions; Supersymmetry; Symmetry and symmetry principles; Tau particles; Weak force and interactions

SYNCHROTRONS

Synchrotrons are a class of particle **accelerators**. The highest **energy** accelerators in use today, such as the Tevatron at Fermilab and the large **electron positron** collider (LEP) at the European Organization for Nuclear Research (CERN). The synchrotron incorporates improvements over the simpler **cyclotron**, which allows it to be used with highly relativistic particles. Instead of one large magnet with a constant magnetic field, as found in a cyclotron, the synchrotron has many smaller magnets placed around a ring. As the particles are accelerated, they become heavier according to special relativity. The path of the particles depends on their **mass** and the magnetic field acting on them; as the particles become more massive, the magnetic field must be turned up in order to keep them traveling around in a circle. Since there are many smaller magnets instead of one large magnet, they can be controlled individually to keep the particles traveling through the accelerator.

A special application is the **radiation** synchrotron. Unlike synchrotrons built to study high-energy physics, which are designed to minimize the amount of radiation produced, radiation synchrotrons are built specifically to produce high-energy **x rays** and gamma rays. They have the advantage of producing very narrow and very bright beams of radiation, which makes them ideal for studying condensed **matter** and biological systems.

SZILARD, LEO (1898–1964)

Hungarian American physicist

A native of Hungary, Leo Szilard was one of the leading contributors to the development of nuclear **energy** and the first atomic weapons. He was also among the earliest and most active campaigners for nuclear arms control. In 1942, with **Enrico Fermi**, he set up the first nuclear **chain reaction**. He later became increasingly interested in the fields of molecular energy and **biophysics**, and helped develop the **electron microscope**.

Leo Szilard was born in Budapest, Hungary, on February 11, 1898, to Louis Szilard, an architect and engineer, and his wife, the former Thekla Vidor. The eldest of three children, Leo was a sickly child, and for a number of years his mother taught him at home.

Educated in Budapest and Berlin

In the fall of 1916, Szilard entered the Budapest Institute of Technology, intending to major in electrical engineering. At the end of his first year at the institute, he was called to service in the Austro-Hungarian army and assigned to officer training school. He became very ill with influenza, however, and had not fully recovered by the time World War I ended in 1918. He returned to the Budapest Institute of Technology for just over a year before transferring first to the Technische Hochschule at Berlin-Charlottenburg.

At Berlin, Szilard's career outlook underwent a significant change. He came into contact with some of the finest physicists in the world, including **Albert Einstein**, **Max Planck**, and **Max von Laue**, the last of whom was to become Szilard's own doctoral advisor. Szilard decided that his real interests lay in the field of physics rather than engineering, and in 1922 he was awarded his Ph.D. in that field. His doctoral thesis, written under the supervision of von Laue, dealt with the statistical implications of the second law of **thermodynamics**. Szilard's work on this topic continued for a number of years, culminating in a paper published in 1929, "On the Decrease of **Entropy** in a Thermodynamic System by the Intervention of Intelligent Beings," in which he investigated the application of thermodynamical laws to information theory. That paper is regarded as an important precursor to modern cybernetic theories.

After receiving his degree, Szilard was appointed first a research assistant and then, in 1925, *Privatdozent* at the Institute of Theoretical Physics at the University of Berlin. Director of the institute at the time was von Laue, his former advisor. In addition to his continuing work on thermodynamics during this period, Szilard also originated a series of studies on x-ray **crystallography**, a field in which von Laue was a world leader. Szilard also worked closely with Albert Einstein on the development of a pump for liquid metals, for which he eventually obtained a patent. In addition, he became interested in the problem of particle **accelerators** and invented a number of devices that were later to be incorporated into early cyclotrons.

Emigrates from Germany to England and the United States

In 1933, the rise of Adolf Hitler convinced Szilard, who was Jewish, that he should leave Germany. Fearing for both his career and his life, he fled first to Vienna and then, six weeks later, to England. There he joined the physics department at St. Bartholomew's Hospital in London. In 1935 he moved to the Clarendon Laboratory at Oxford.

It was during this period that Szilard received news of **Frédéric** and **Irène Joliot-Curie**'s discovery of artificial **radioactivity**. He began to think about the possibility of a nuclear chain reaction in which the nuclear decay of one **atom**, brought about by some type of particle, would result in the production of a new atom with the release of more particles of the kind needed to start the reaction. In such a case, the reaction, once initiated, would be self-sustaining over many, many episodes of decay. The value of such a reaction, Szilard knew, was that energy would be released in each step of the process. After countless repetitions of the reaction, huge amounts of energy—sufficient, for instance, to make a powerful bomb—would be released.

Szilard first explored the possibility of using beryllium in such a chain reaction. He and a colleague, T. A. Chalmers, found that gamma rays directed at a beryllium target would cause the emission of a **neutron** from the beryllium **nucleus**. The two hoped that this reaction could act as the first step in a chain reaction in which beryllium atoms would break apart to form **helium** atoms and more neutrons. The neutrons thus formed would, they hoped, then cause more beryllium atoms to break apart into helium atoms with the release of more neutrons, and so on. More detailed studies showed, however, that such a reaction could not be sustained.

In addition to his research, Szilard continued his efforts to find new jobs for scientists fleeing the Nazi purges on the continent. These efforts were characteristic of Szilard's life-long commitment to helping others. He once said that this humanitarian impulse was largely the result of reading, at the age of ten, Hungarian author Imre Madách's *The Tragedy of Man.*

Toward the end of 1938, Szilard decided to move to the United States. He had no more settled in at his new workplace, Columbia University, than he received startling news from Europe. **Otto Hahn** and **Fritz Strassman** had produced the first fission of an atomic nucleus, an event that was fully understood and explained by **Lise Meitner** in January 1939. Szilard immediately recognized the significance of this discovery. It held the potential for making possible the very kind of nuclear chain reaction on which he had been working in London.

With a colleague, Walter Zinn, Szilard set up a replica of the Hahn-Strassman experiment at Columbia. Their goal was to find out whether the fission of a uranium nucleus would result in the formation of at least one neutron, a condition necessary for the maintenance of a chain reaction. On March 3rd, the experiment was ready. A few flashes of **light** on an oscilloscope gave Szilard and Zinn the answer they sought: neutrons were being released during the fission of uranium. A nuclear chain reaction was possible. Szilard would later say he knew immediately that this discovery would cause the world great sorrow.

Leads Effort to Establish Nuclear Bomb Project

News of the discovery of **nuclear fission** swept through the physics community like wildfire. Few failed to grasp the military potential of the discovery. A group of physicists in the United States who were particularly concerned about this potential became convinced that the U.S. government must take fast and aggressive action to see whether nuclear fission could really be used in the development of weapons. Szilard composed a letter, which Albert Einstein signed, presenting their arguments to President Franklin D. Roosevelt. Roosevelt responded by appointing an Advisory Committee on Uranium to investigate the issue. After some initial hesitation, the com-

Leo Szilard.

mittee produced a favorable recommendation, and the Manhattan Engineering District Project was created to pursue the development of the world's first atomic bombs. The first contract let under the **Manhattan Project** was to a group of scientists at Columbia that included Szilard.

In 1942, Szilard left Columbia to become part of the Manhattan Project's Metallurgical Laboratory at the University of Chicago. Working there with Enrico Fermi, he witnessed the first controlled nuclear reaction on December 2, 1942, when the world's first atomic pile (nuclear reactor) was put into operation. The hopes and dreams—as well as the fears—that Szilard had long held for nuclear chain reactions had become a reality.

Shortly thereafter, Szilard began to argue for a cessation of research on **nuclear weapons**. A number of factors influenced his position. First, he was convinced that the tide of war had turned in favor of the Allies, and he thought the war could soon be ended with conventional weapons. Second, he feared that the successful development of nuclear weapons would lead to an all-out arms race with the Soviet Union after the war. Finally, he recognized the horrible human tragedies that would result from the use of an **atomic bomb**. His suggestions for a demonstration test of nuclear weapons in an uninhabited area to which the Japanese government would be invited fell on deaf ears, however. Instead, on August 6 and 9, 1945, the first atomic bombs were dropped on Hiroshima and Nagasaki in Japan. The war ended a week later.

Turns to Biological Studies

In the post-war years, Szilard spent a major portion of his time working for the control of atomic energy. He joined a large number of his fellow nuclear scientists in forming the Federation of Atomic Scientists, which worked to keep control of atomic energy out of the hands of the military and within a civilian department. He also made efforts to encourage mutual disarmament and the reduction of tensions between the United States and the Soviet Union. To this end, he was active in the formation and planning of the Pugwash Conferences on Science and World Affairs, a series of conferences on nuclear safety that met in the late 1950s and early 1960s. In 1962, he helped found the Council for a Livable World, a Washington, D.C.-based lobby for nuclear arms control.

In the late 1940s, Szilard once again turned to scientific research, but this time in the field of biology. In 1946, he accepted an appointment as professor of biophysics at the University of Chicago. One of his first accomplishments was the development of the chemostat, an instrument that aids in the study of bacteria and viruses by making it possible to regulate various growth factors. Later he became interested in the biology of aging. Another topic in which Szilard became interested, memory and recall, was the subject of his final scientific paper, published after his death.

Szilard became a U.S. citizen in 1943. In 1951 he married Gertrud Weiss, whom he had first met in 1933 in Vienna, where Weiss was a medical student. The couple had no children. Szilard was awarded the Einstein Gold Medal in 1958 and the Atoms for Peace Award in 1959. He died of a heart attack on May 30, 1964, in La Jolla, California, where he had been a resident fellow at the Salk Institute.

T

Tau Particles

In the standard model of **particle physics** the tau particles (e.g., tau leptons and tau neutrinos) are among the fundamental building blocks of **matter**.

The standard model attempts a comprehensive synthesis of fundamental particles and forces. According to the standard model, **fermions** (**quarks** and leptons) comprise all matter. There are twelve particles and twelve antiparticles in this class. In addition to the six known types of quarks (i.e., up, down, charm, strange, top, and bottom quarks) there are three charged leptons (the **electron**, the **muon,** and the tau particle), each with a corresponding neutral particle termed a **neutrino**. All of these fundamental particles have corresponding antiparticles that have exactly the same **mass** but which carry an opposite charge (i.e., the **antiparticle** of the electron is the **positron** and the antiparticle of the tau neutrino is the tau antineutrino). Fermions exist in three generations (mass and **energy** states). Protons, **neutron** and electrons are in the first generation of matter and the more massive and energetic particles of the second and third generations usually quickly decay into less-massive first generation particles.

By the early 1970s, there was a known symmetry between two generations of quarks and leptons. After a study of collisions of electrons and positrons, **Martin Louis Perl** (1927–) at the Stanford University linear **acceleration** (SLAC) accounted for energy enigmatically missing after such collisions with the discovery of a tau particle (τ). In addition to finding a new **lepton**, Perl discovered evidence of a third generation of matter. The tau lepton, like the muon and electron, had a negative charge and had an **antimatter** particle version with a positive charge.

Because the tau particle carries a charge, it reacts via the electromagnetic **force** and is, of course, far less affected by the much weaker **gravitational force**. Interestingly, however, the tau was more massive than expected. The energy for producing the Tau+, Tau- pair was about 3.6 Gev. This energy implied a tau rest mass of approximately 1.8 Gev, (its actual mass is 1.784 Gev), about twice as heavy as a **proton**, 20 times as heavy as the muon, and an elephantine 4,000 times as heavy as an electron. Such a high mass made the tau unstable, with a lifetime of only 3×10^{-13} seconds.

The tau particle decays to produce a tau neutrino, an electron and an anti-electron-neutrino, while the anti-tau decays into an anti-tau-neutrino, an anti-mu and a mu-neutrino.

Neutrinos are the precursor particles to leptons (i.e., tau lepton results from the collision of tau neutrinos). Accordingly, the discovery of the tau lepton in the 1970s by the Direct Observation of the Nu Tau (DONUT) project spurred the quest for an experimental confirmation of the tau neutrino. The quest proved elusive because tau neutrinos can not be observed directly and their existence can only be inferred by the results of their interactions.

According to the standard model the tau neutrino is one of three neutrino particles. The other neutrinos, electron neutrinos and muon neutrinos, were discovered in the late 1950s and early 1960s. Like other neutrinos (Italian for little neutral one), tau neutrinos carry no **electric charge** and have such a small mass that they are able to travel near the speed of **light**. For many years scientists predicted that neutrinos would be massless. In 1998, however, a team of scientists working in Japan made an experimental determination that neutrinos possessed a very small mass (less than one-millionth that of an electron).

In 2000, an international team of scientists working at the Fermi National Accelerator Laboratory finally announced the experimental confirmation of the existence of the tau neutrino. The experimental observations fit perfectly with the predictions of the standard model.

See also Beta radiation; Electroweak particles; Electroweak theory; Nuclear physics; Nuclear structure and stability; Particle accelerators; Particle detectors; Pions; Quantum states; Quantum theory; Quarks and gluons; Radioactive decay; Standard model of particle physics; Weak force and interactions

TAYLOR, JR., JOSEPH H. (1941–)
American astrophysicist

Joseph H. Taylor, Jr. is an astrophysicist who discovered the first **binary pulsar**—two extremely dense, collapsed **stars** in orbit around each other. He made this discovery in 1974 with **Russell A. Huls**e, who was then his graduate student at the University of Massachusetts. The two men used this binary pulsar to verify aspects of **Albert Einstein**'s general theory of relativity, which scientists had not yet had an opportunity to test. Binary **pulsars** became what Taylor and Hulse describe in *Astrophysical Journal Letters* as "a nearly ideal relativity laboratory," and they have made particularly important contributions to the understanding of **gravity**. For their discovery of the binary pulsar and the application of their findings to the theory of relativity, Taylor and Hulse were awarded the 1993 Nobel Prize in physics.

Taylor was born in Philadelphia on March 29, 1941, the son of Joseph and Sylvia Evans Taylor. In 1959, Taylor entered Haverford College where he majored in physics. He graduated in 1963 with a B.A. degree and entered the doctoral program in astronomy at Harvard University. He was awarded his Ph.D. in 1968 and spent the next year as a research fellow and lecturer in astronomy at Harvard. In 1969 he joined the faculty at the University of Massachusetts in Amherst as an assistant professor of astronomy. In 1973, he was named an associate professor and four years later elevated to full professor. In the fall of 1980, Taylor left Massachusetts to become professor of physics at Princeton University; in 1986 he was named James S. McDonnell Distinguished University Professor of Physics.

Begins Search for Pulsars

While at the University of Massachusetts in 1970, Taylor was approached by one of his graduate students, Russell Hulse, in search of a thesis project. The pair agreed on an undertaking involving the use of the 984-ft (300-m) diameter Arecibo **telescope** in Puerto Rico, the world's largest single-element **radio** telescope, to search the skies for the weak radio signals emitted by pulsars. Pulsars were first discovered in 1967 by Susan Jocelyn Bell Burnell and Antony Hewish. They are **neutron stars** whose diameter, in contrast with other stars, is very small, sometimes as small as 6 mi (10 km). Their **mass**, on the other hand, is as great or greater than that of the **Sun**, and the **gravitational field** that surrounds them is extremely high. As a result of their strong gravitational pull, radio **waves** are released from pulsars only at the poles; the beams reach Earth in **pulses** as the star spins in **space**, and on a radio telescope, the waves from a pulsar can resemble the beam from a lighthouse. In analyzing the results of a pulsar detected on July 2, 1974, Taylor and Hulse noticed an unexpected variation in the pulsar's period. The bursts were not perfectly regular like those of known pulsars, and the irregularity

revealed that there were actually two pulsars, orbiting each other.

As a press release from the Royal Swedish Academy observes, "Hulse's and Taylor's discovery in 1974 of the first binary pulsar, called PSR 1913 + 16... brought about a revolution in the field." Binary pulsars orbit at great speeds and at close range—approximately that of the distance from Earth to the Moon—and the discovery of these stars gave scientists an opportunity to study the effects of gravity outside the gravitational field of our solar system. Over a period of almost twenty years, Taylor and Hulse made detailed observations of the behavior of these stars in orbit. They discovered that the path the pulsars followed is changing: their orbit is contracting and the two stars are rotating at greater speeds as they grow closer to each other.

Their examination of the timing of the pulses provided the first evidence for the existence of what *Sky & Telescope* calls the "magnetic aspect of gravity." In 1916, Einstein predicted that two masses in orbit around each other would have certain properties similar to **electromagnetism**. He predicted that bodies would emit what he called gravitational waves and thus lose **energy**. The small changes Taylor and Hulse have detected are consistent with this prediction; even the rate of change very nearly matches the rate Einstein predicted it would follow.

For Taylor, this indirect confirmation of the gravitational waves is only part of the support binary pulsars can offer to the general theory of relativity. As Taylor told the *New York Times*, "Continued study of binary pulsars as they spin off energy over years is essential. We've measured three of the relativity effects on this pulsar to a high accuracy, and two other consequences to somewhat lower accuracy. But there are potentially about a dozen other relativity effects we hope to measure in the future." In 1985, Taylor and a graduate student discovered another binary pulsar. His work has raised the possibility of creating a new branch of astronomy, called gravitational wave astronomy, which would enable astronomers to gather evidence about a number of events in the **universe** that they currently cannot observe.

In addition to the Nobel Prize, Taylor received the Dannie Heineman Prize from the American Astronomical Society and American Institute of Physics in 1980, a MacArthur Fellowship in 1981, the National Academy of Sciences' Henry Draper Medal and Tomalla Foundation Prize in gravitation and **cosmology** in 1985, and the Wolf Prize in Physics in 1992. He is a member of the National Academy of Sciences, the American Philosophical Society, and a fellow of the American Academy of Arts and Sciences. Taylor married Marietta Bisson on January 3, 1976. They have three children.

TELESCOPE

The telescope is a device that intensifies and magnifies the image of distant objects. The telescope enables astronomers, scientists, and amateurs alike, to observe and study **planets**, **stars**, **galaxies**, and other features of the **universe**.

The observation of celestial objects is subjected to the limitations imposed on photons traveling over great distances, in that only a small quantity of photons will actually reach Earth. Consequently, the object appears to be faint, and because of the distance, it will also seem to be small. Therefore, the first function of the optical telescope is to intensify the **light**. This is done by using a lens or a mirror, or both. The larger the diameter of the objective lens, called the aperture, the more photons are captured. The world's largest optical telescope has an aperture of 236 in (6 m). Amateur astronomers will commonly use telescopes with a lens or mirror of 8 in (20 cm) in diameter, thus gathering 512 times more light than the human eye, which has a pupil aperture of less than 1 cm.

The second function of a telescope is to magnify the object. Magnification is achieved by the use of eyepieces of different angular resolution, resolution being defined as the ability to distinguish two close objects or two features on an object. However, magnification is limited by the effects of atmospheric **turbulence** and distortion, which are also inadvertently magnified. Obviously, visual observation using optical telescopes is only possible on clear nights. Two factors that influence the quality of the observation are transparency (the amount of light the atmosphere will transmit), and seeing, which describes the steadiness of the atmosphere. Wind and turbulence in the upper atmosphere distort images and reduce the sharpness of the objects. The geographical location of astronomical observatories is also of great importance in order to reduce the disturbances; thus they are ideally located on mountain tops and far from urban centers. The **Hubble space telescope** with its 96 in (2.4 m) aperture is free from such problems; it was placed in orbit far above any atmospheric **interference**.

Optical telescopes are characterized by the device used to collect the light. The invention of the first refracting telescope in 1609 is credited to Italian astronomer and physicist **Galileo Galilei** (1564–1642). In its simplest form, the light enters the telescope through the objective lens and passes through the eyepiece lens that produces the image. Galileo's largest instrument was about 48 in (120) cm) long and the width of its objective lens was 2 in (5 cm). Today, the world's largest refractor has an aperture of 40 in (100 cm), and was installed in 1897 at the Yerkes Observatory, Wisconsin.

The reflecting telescope developed by English physicist and mathematician Sir **Isaac Newton** (1642–1727) in 1668 relied on mirrors to collect the light. The Newtonian reflector captures light with its primary mirror. The captured light is then deflected onto a diagonal mirror near the front of the instrument which will reflect the image to the side through the eyepiece lens. In 1672, a French astronomer Jacques Cassegrain (1652–1712), developed a variation of the reflector. Cassegrain's telescope had a secondary mirror reflecting the image back toward the primary mirror, which had an opening in its center allowing the image to pass through to the eyepiece lens. The largest existing reflector telescope has a 236-inch (6-m) reflector, and it is located at the Special Astrophysical Observatory in Zelenchukskaya, Russia. Current compound telescopic designs use both **lenses**, usually placed in the front, and mirrors. The most well-known of this type is the Schmidt-

The telescope at Lick Observatory in California.

Cassegrain telescope developed by Estonian born German astronomer Bernhard Voldemar Schmidt 1879–1935). Its 8-in (20-cm) reflector version is widely used by amateur astronomers.

Until the 1930s, astronomical observations were limited to the visible **wavelengths** of the **electromagnetic spectrum**. Since then, technological advances have led to the development of **radio**, x-ray, infrared, and gamma-ray telescopes that allow the exploration of every region of the electromagnetic spectrum. Based upon the discovery of a source of static emanating from outside the solar system by American engineer Karl Janskey (1905–1945), in 1937, another American radio engineer, Grote Reber (1911–   ) constructed a primitive radio telescope using a parabolic reflector dish mounted in his own backyard. The first large radio telescope, which is still operational, was fitted with an 83-yard (76-m) antenna and built in 1957 at Jodrell Bank, England. The largest radio telescope, with its 109-yard (100-m) steerable antenna, is used at the Max Planck Institute for Radio Astronomy at Effelsberg, Germany. In contrast to optical telescopes, radio telescopes can operate at daytime and even in cloudy or rainy conditions. Radio telescopes capture the natural radio signals emitted by objects in **space**. Equipped with spectrometers and **computers** to collect, store, and analyze data, they provide information on **temperature**, **radiation**, and on the molecular and chemical composition of planets, stars, and gaseous envelopes around stars.

See also Astronomy; Celestial mechanics; Cosmology; Milky Way galaxy; Radio astronomy; Reflection, refraction, and dispersion

TELLER, EDWARD (1908–)

Hungarian American physicist

Trained as a theoretical physicist, Edward Teller became a leading authority on **nuclear physics** during the 1930s and was involved with the **Manhattan Project** during World War II. He was an early advocate of thermonuclear weapons, which are many times more powerful than the **atomic bomb**, and he is best known for the leading role he played in the development of the hydrogen bomb between 1949 and 1951. Beginning in the 1940s, Teller figured prominently in policy discussions about America's nuclear arsenal, advising government officials at the highest levels and even testifying against **J. Robert Oppenheimer** at a Congressional **hearing** during the McCarthy era. By the 1980s, a lifetime of advising politicians had given him a network of political contacts that included a friendship with President Ronald Reagan. Teller was instrumental in convincing Reagan that a system could be developed to shoot down incoming ballistic missiles. Many consider him responsible for the president's decision to support the Strategic Defense Initiate (SDI), popularly known as "Star Wars."

Teller was born in Budapest on January 15, 1908. His parents were Jewish but not orthodox, celebrating the sabbath and the high holidays. His father, Max Teller, was a lawyer from Hungarian Moravia. His mother, the former Ilona Deutch, was the daughter of a banker and a cotton-mill owner from a small town in the eastern part of the Austrian-Hungarian Empire. Teller did not speak until he was almost four years old, but once he began he spoke a great deal, and it became clear that he was quite precocious. He delighted in performing mathematical calculations in his head; when he was twelve, his father introduced him to Leopold Klug, a professor of mathematics at the University of Budapest, and Teller started seriously thinking of mathematics as a career. By the age of fourteen, he was reading about **Albert Einstein** and his work on the special and general theories of relativity.

In 1925, when he was eighteen, Teller won first place in a prestigious mathematics contest for all Hungarian high school students. Still, his father worried that mathematics was not a dependable occupation, so Teller agreed to study engineering as well as mathematics in college. He studied first for a short time at the University of Budapest and then at the Institute of Technology in Karlsruhe, Germany. By 1928 he was enrolled in the University of Munich as a physics student.

The year he came to Munich, Teller fell while jumping off a streetcar and lost most of one foot when he slipped beneath its wheels. The amputation was so sudden that he did not even realize what had happened until he saw his boot, with his foot still in it, lying in front of him. What was left of Teller's foot was reconstructed so he could walk without a prosthesis, which he was able to do after the accident, although he usually chose to use an artificial foot as well.

Despite this injury, Teller never allowed the loss of his foot to stand in the way of his career or his life. Late that year he moved again to the University of Leipzig to study with **Werner Heisenberg**. Two years later, in 1930, he was awarded his doctorate; he had written his dissertation on experiments in which he used quantum **mechanics** to calculate **energy** levels in an excited hydrogen molecule.

Teller spent several years as an assistant at the University of Göttingen, and then in 1934, he was awarded a Rockefeller Foundation fellowship. He used it to join **Niels Bohr** at the Copenhagen Institute for Theoretical Physics, where many of the great twentieth century physicists studied. On February 26, 1934, a few weeks after starting his Copenhagen fellowship, Edward married Augusta Maria Harkanyi, who he had known for over a decade. The couple spent a year in England, where Teller was a lecturer at the University of London, and then they moved to the United States, where he took a full professorship in the physics department at George Washington University.

Teller was twenty-six by this time and he had already published almost thirty papers, usually with coauthors. While at George Washington University, he worked closely with **George Gamow**, a Russian exile; together they calculated the rules for one of the major forms of **radioactivity**, which became known as the Gamow-Teller selection rules for **beta decay**. During this early part of Teller's career, he worked on many different problems, including molecular vibrations, magnetic cooling processes, and the absorption of gases on solids. One of his specialties was the behavior of **matter** under unusual conditions, including the interior of **stars**.

Advocates the Development of the Hydrogen Bomb

At the end of 1938, two German chemists, **Otto Hahn** and Fritz Strassmann, succeeded in splitting the **atom**. Within months, **Leo Szilard** and Walter H. Zinn had created a **chain reaction** at Columbia University and it had become clear that atomic weapons were possible. Teller was one of the physicists who lobbied President Franklin D. Roosevelt and his administration to attempt to build such a weapon, and he was involved in the Manhattan Project from the beginning. From 1940 to the end of the war, Teller worked on the atomic bomb, moving between Washington, New York, Chicago, and Los Alamos, New Mexico. In 1941, Teller and his wife became naturalized U.S. citizens; their son was born in 1943 and their daughter in 1946.

One of the tasks Teller was assigned on the Manhattan Project was calculating the possibility of a thermonuclear or hydrogen bomb, also then known among nuclear physicists as the "super." He initially concluded it was impossible, but he redid his figures with Emil Konopinski and realized he had been wrong. Once he understood that a hydrogen weapon was possible, convincing the government to research and build one became a major focus of his energies, although he did continue to offer advice on the atomic bomb. Teller continued to campaign for building the hydrogen bomb even after an atomic bomb was dropped on Hiroshima (an act which Teller opposed) and World War II ended.

There was initially little interest in the hydrogen bomb project, and in 1946 Teller left Los Alamos to teach at the University of Chicago. He focused on issues in theoretical physics, such as the relation of gravitational and electromagnetic forces over time and the origins of the elements and **cosmic rays**. But another arms race began in earnest when the Soviets exploded their atomic bomb in 1949; the development of the H-bomb became a priority for the United States and Teller returned to Los Alamos. It was at this juncture that Teller began to exert a strong influence on nuclear policy. He convinced Paul Nitze, who was then a key State Department official (and would later be President Reagan's top arms control advisor), that the H-bomb could be built. By 1951, Teller, Stanislaw M. Ulam, and Frederic de Hoffman had made a number of crucial breakthroughs, and the path to building a thermonuclear weapon was clear.

J. Robert Oppenheimer, one of the most influential nuclear physicists in the 1930s and the director of the laboratories at Los Alamos, publicly opposed the development of thermonuclear weapons. Over his objections, Teller used his growing influence in government and lobbied for a second **nuclear weapons** laboratory. It was established in July 1952 at Livermore, California, and later named after the famous experimental physicist **Ernest Orlando Lawrence**, who had proposed the site. Teller was associate director of the Lawrence Livermore Laboratory from 1954 to 1958 and director from 1958 to 1960. In 1953, he was also appointed professor at the University of California; he was named University Professor in 1960, a position he would hold until he retired in 1975. At the University of California's Davis campus, he played a major role in establishing the Department of Applied Sciences.

The political battles over the development of the hydrogen bomb and the founding of a second nuclear weapons laboratory in Livermore, California, brought the tensions between Teller and Oppenheimer to a head. There was also a considerable amount of resentment in government circles against Oppenheimer for opposing thermonuclear weapons, and in 1953 the Atomic Energy Commission filed official charges against him, based primarily on his association with communists, including family members. He was denied his security clearance and requested a Congressional hearing. Though many scientists spoke in his favor, Teller appeared to testify against him. According to Daniel J. Kevles in *The Physicists: The History of a Scientific Community in Modern America,* Teller "considered Oppenheimer a Communist and an advocate of Soviet appeasement." At the hearing Teller testified, as quoted by Kevles, that "his actions frankly appeared to me confused and complicated... I would like to see the vital interests of this country in hands which I understand better and therefore trust more... I would feel personally more secure if public matters would rest in other hands." Though the committee did not find Oppenheimer disloyal, they decided not to restore his security clearance.

Teller's testimony alienated most of the nuclear physicists in the country and cost him many of his friendships among other scientists, but it strengthened his ties to military and political leaders. His influence on American strategic policy only increased during the 1950s. In 1956, Teller assured the Navy that Livermore could build a warhead small enough to be fired from a submarine, and four years later the first Polaris submarine was armed with the warheads Teller had enivisoned. He was less successful in fulfilling his promise that Livermore could build a "clean" nuclear weapon that would not produce dangerous radioactivity. It has never been made and many physicists believe it is impossible. During this period, Teller was also active in several other areas of nuclear policy, opposing nuclear test bans and advocating peaceful atomic projects such as **power** reactors and "plowshare" explosions for mining and canal digging. He was the first chairman of the Atomic Energy Commission's Advisory Committee on Reactor Safeguards, which oversaw the production of the first manual on technical aspects of reactor safety. He was especially fascinated by the prospect of **nuclear fusion**, where power is produced by merging atoms together, as opposed to the fission, or splitting, of atoms, and Livermore became a center for fusion research.

Argues for a Nuclear Defense System

All through the sixties and seventies, Teller was an outspoken advocate of **nuclear power,** and as early as 1962, he began advocating an "active defense system" to shoot down attacking enemy missiles. Teller became increasingly convinced that a nuclear shield could remove the threat of a retaliatory nuclear strike. Work at Lawrence Livermore National Laboratories suggested that giant x-ray **lasers** powered by nuclear blasts might possibly be an effective antimissile weapon; this was only a theory, however, and it was based on a number of assumptions which had not been proven and a number of scientific and technical breakthroughs which had not yet taken place.

When Teller retired from the University of California in 1975, he became a senior research fellow at Stanford University's Hoover Institute on War, Revolution, and Peace. He also continued to maintain his political contacts with policymakers both inside and outside of government. When his friend Reagan was elected president in 1980, Teller found himself with the most political influence he had ever had in his career, and he succeeded at last in his campaign to acquire government funding for an active nuclear defense system. Reagan supported the Strategic Defense Initiative (SDI) or "Star Wars," in large part because Teller convinced him it was possible.

Teller's conviction that such a defense system was possible turned out to be premature, if not entirely incorrect. His original plan for nuclear-powered lasers was quickly rejected as infeasible; extensive research was done on burning small holes in the outer sheeting of enemy missiles, thereby causing them to break up in outer **space**. The necessary antimissile satellites for such a project were expected to be cheap but eventually turned out to be extremely costly; their effectiveness also depended upon complex computer technology.

Whatever the feasibility of the Strategic Defense Initiative, the proposal was at least partly designed to push the Soviets into an arms race they could not afford to pursue.

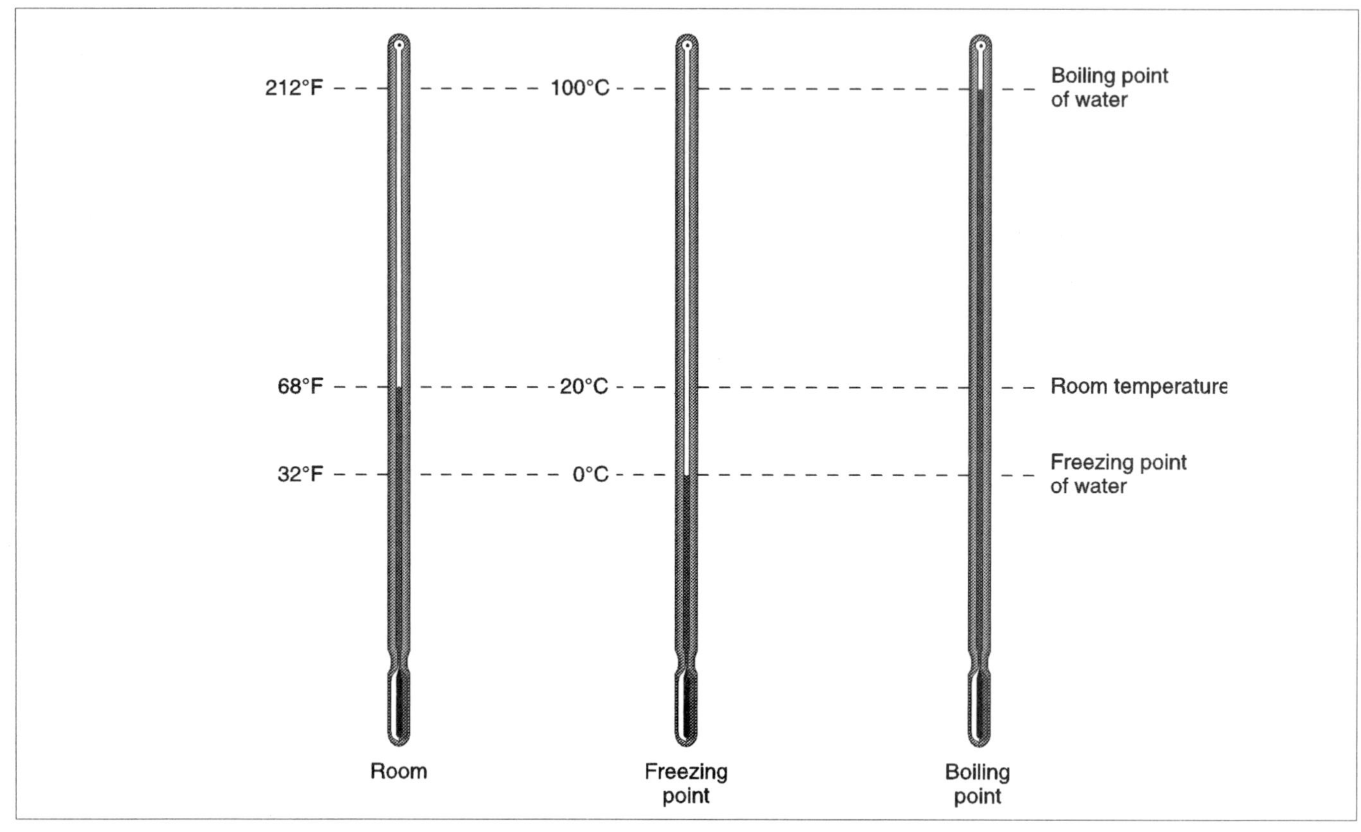

Thermometers showing room temperature, freezing point, and boiling point.

Defense spending was already an enormous share of the Soviet Gross Domestic Product; an expensive antimissile system involving high technology items like lasers and computer-tracking systems was simply beyond their means. Later, commentators in both the West and Russia credited SDI with forcing the societal reforms that ended in the collapse of communism in Eastern Europe and the Soviet Union.

In *Physics Today,* Robert March writes that Teller is "best known to the public, and even to the generation of physicists educated after World War II, as the tireless and single-minded champion of the technological arms race." Teller's political influence has overshadowed many of his scientific contributions, but since World War II, he has done important work in theoretical physics with **Enrico Fermi** and others. In 1962, he was given the Fermi Award, one of the highest honors in physics. In 1983, he was awarded the National Medal of Science for his research on stellar energy, fusion reaction, molecular physics, and nuclear safety. Among other awards Teller has received are the Priestley Memorial Award (1957), the Einstein Award (1959), the General Donovan Memorial Award (1959), the Robins Award (1963), the Leslie R. Groves Gold Medal (1974), the Harvey Prize (1975), the Sylvanus Thayer Award (1986), the Presidential Citizen Medal (1989), and the Order of Banner with Rubies of the Republic of Hungary. Teller is a member of the National Academy of Science and a fellow of the American Nuclear Society, the American Physical Society, the American Academy of Arts and Science, and the American Association for the Advancement of Science.

In the 1990s, Teller remained active in both physics and public policy. "I am amusing myself with a number of problems from **astrophysics** and superconductivity," Teller told contributor Chris Hables Gray. He went on to describe his work organizing a conference among physicists on human health and **radiation** levels. On his political influence, he remarked: "I tried to contribute to the defeat of the Soviets. If I contributed one percent, it is one percent of something enormous." He added that it was a great pleasure, now that Eastern Europe is free, to be able to visit his native Hungary for the first time in fifty-six years.

TEMPERATURE

Temperature is usually defined as the degree of an object's **heat energy**. The higher the **energy level**, the hotter the object. When two objects are placed next to each other, heat energy will always travel from the hotter to the colder object. Temperature plays an important role in **chemistry**, as chemical reactions depend on the temperature of the environment. The rate of reaction generally increases with a temperature increase.

Temperature can have a significant impact on the behavior of substances. For example, at extremely low temperatures, some materials become superfluids, which "escape" out of containers. Other substances, such as mercury, become **superconductors.** Supeconductivity was discovered in 1911, when **Heike Kamerlingh Onnes**, using liquid **helium**, cooled mercury to 4K, and discovered that the metal offered no resistance to **electricity.**

Since our perception of temperature is subjective (terms such as "hot" and "cold" are relative), scientists use **thermometers** to measure temperature. The accepted thermometer scale for scientific use is the Celsius (formerly called centigrade) scale, developed by the Swedish astronomer **Anders Celsius.** Celsius divided his scale into 100 degrees, 100 indicating the melting point of ice and 0 indicating the **boiling** point of water. After Celsius's death, his colleagues at the University of Uppsala reversed the scale, creating the modern form.

While the Celsius scale is widely used by scientists, the International System of Units (SI) uses the Kelvin scale the measure temperature. Suggested by William Thomson, Lord Kelvin (1824–1907) in 1848, this is an absolute scale, based on the concept of **absolute zero**, the lowest possible temperature, when all molecular movement stops. Although absolute zero still remains a purely theoretical concept, scientists working in the field of **low-temperature physics** have managed to create temperatures low enough to be measured in nanodegrees K. The Kelvin scale uses Celsius units, the main difference being that zero on the Kelvin scale is absolute. On the Kelvin scale, ice melts at 273.15K.

In the United States, temperature is also measured by the Fahrenheit scale, according to which ice melts as 32°F and water boils at 212°F. Since the Fahrenheit scale uses 80 units (from 32 to 212) for a temperature range covered by 100 Celsius units, it follows that 9°F equal 5°C.

TEMPERATURE AND MEASUREMENT

There are several equivalent definitions of **temperature**. In **thermodynamics**, temperature is defined as a property of an object that determines the direction of **heat** flow. Heat always flows from a higher temperature region to a lower temperature region. Heat flow stops when the two regions reach an **equilibrium** state, called thermal equilibrium. Two objects are said to be of the same temperature if they are in thermal equilibrium. This is also called the 0th law of thermodynamics. This definition only gives a comparison between the temperatures of two objects. In **statistical physics**, the temperature can be defined without using comparison between two objects. Simply speaking, the temperature is a quantity proportional to the internal **energy** of an object. Internal energy is defined to be the sum of the kinetic and **potential energy** of all the molecules inside the object. Internal energy is difficult to measure so this definition is mainly a theoretical one.

There are two approaches to quantify temperature according to the first definition. The first is to choose a material, often water, and define its temperature arbitrarily in two states, often at melting and **boiling** points. Two temperature scales were originally defined this way, the Fahrenheit scale

and the Celsius scale. The Fahrenheit scale defines the melting point of water to be 32°F and its boiling point to be 212°F. This scale is familiar to people in the United States. The Celsius scale defines the melting point of water to be 0°C and its boiling point to be 100°C. The Celsius scale is used throughout the world and in scientific research and reporting. Once two temperatures are specified, temperatures at any other state can be determined. Another way to define temperature is to use the Carnot engine. Kelvin temperature is defined this way, and is used in scientific research. 0K is defined as the point at which no body can emit heat, and 1K equals 1/273.16 of the temperature of the triple point of water. The triple point is a unique state where water, steam, and ice can exist together. Today the Celsius and Fahrenheit scale are both defined through the Kelvin scale. $°C = K - 273.15$ and $°F = °C(9/5)+32$. Kelvin temperature can be quantified through a Carnot engine. A Carnot engine is a ideal engine. It absorbs a certain amount of heat (Q1) at a hot source (temperature T2), then release a certain amount of heat (Q2) at a cold source (temperature T1) and does some **work**. The process is carried slowly so the thermal equilibrium is always kept. Then we have Q1/T1 = Q2/T2. Kelvin temperature can be determined through this equation. The lowest temperature is 0K. Nothing can reach this temperature in a finite number of steps. This is also called the fourth law of thermodynamics. The third law of thermodynamics is about **entropy**.

There are numerous ways to measure temperature. They can be put into two classes: contact method and non-contact method. The contact method is easier to use. Measuring body temperature in a hospital, for example, is a contact method. Contact methods are based on the 0th law of thermal physics. A thermometer is put in contact with an object, and time passes until they reach thermal equilibrium. The temperature of the object can be read from the thermometer because they have equal temperature in thermal equilibrium. Sometimes a contact method is not feasible or not accurate enough. A non-contact method can be used when we cannot or do not want to touch the object we want to measure or when we want high accuracy. These methods are based on physical laws. An example is the pyrometer. It measures temperature through the **radiation** from an object. Compared with contact methods, it can be used in many more cases. Say, if we want to measure the temperature of a star, a contact thermometer certainly won't do. We can only use a non-contact thermometer. A total-radiation pyrometer is often used in this case. The radiation of the star can be focused by a **telescope** to a blackened foil to which a thermopile is connected. From the E.M.F. (electric potential difference) produced by the thermopile, the total radiation energy and thus the temperature of the star can be calculated.

See also Thermometers; Thermal energy

TERRESTRIAL GRAVITY

Gravity is a universal attractive **force** that is directly related to the **mass** of an object. One mass will always attract another mass, and the size of the attraction is related to the size of the masses involved. Gravity is a comparatively weak force until one or other of the masses gets very large. Earth, being a large mass, exerts an appreciable attraction on any mass close to it. The **gravitational force** of Earth reaches far out into **space**. To reach a spot where Earth's gravitational attraction is one millionth of its value at Earth's surface it would be necessary to travel approximately four million miles into space.

If Earth were a stationary sphere composed of a uniform substance, the gravitational force would be identical at any place on its surface. Earth, however, is none of these things. It is rotating, which produces two notable effects. First, Earth is deformed out of a true spherical shape because of the centripetal force of the spin. Thus it has a greater diameter at the equator than at the poles. This means that people standing at the equator are further from the center of Earth than if they were standing at one or other of the poles. Thus the gravitational force is somewhat less at the equator.

The second effect is that all objects at the equator are tending to fly off the spinning earth. The force throwing them off Earth is, fortunately, much smaller than the attractive force of gravity. However, this ejecting force does reduce the over all effect of gravity to some degree. So bodies at the equator have both of these effects apparently reducing the over all attractive force of the earth. The difference between the apparent force attracting bodies to the surface at the equator and the force at the poles is approximately 0.5%. This means a 200 lb (91 kg) man at the poles appears to weigh only 199 lb (90 kg) at the equator.

Earth has an irregular **gravitational field** for other reasons other than the wider diameter at the equator. Clearly Earth is not a smooth sphere. Mountain ranges and deep valleys will also cause variations in the gravitational attraction. In addition, Earth is not composed of uniform material, or of materials spread uniformly throughout its volume. In other words, Earth has materials of different densities scattered throughout its volume. On Earth, variations in ground water storage, ice and snow cover, oceanic mass distribution, and even the variations in the mass of the atmosphere play a part in deforming the gravitational field. Even the oceanic tides cause an observable fluctuation in Earth's gravity.

These variations are comparatively small but they will effect the orbits of satellites around Earth. Small undulations will occur as a result of this variable gravitational field. In fact, tracking the perturbations of orbiting satellites is one way of mapping Earth's gravitational field. Other methods include measurements taken at Earth's surface and measurements taken from orbiting satellites.

Developing a map of the total gravitational field of Earth has been a major scientific enterprise. The ability to observe changes in this field is important to those studying tidal changes, earthquakes, tectonic shifts, and other terrestrial phenomena.

See also Newton's Law of Universal Gravitation

TESLA, NIKOLA (1856–1943)
American inventor and electrical engineer

The first person to prove and perfect the efficient use of alternating-current **electricity**, Nikola Tesla saw his polyphase system become the standard for **power** transmission throughout the world. He also pioneered research in such areas as artificial lightning, high-frequency and high-tension currents, and **radio** telegraphy. Before his death in 1943, Tesla had acquired more than one hundred patents for high-frequency **generators**, adjustable condensers, thermomagnetic motors, **transformers**, his famous Tesla coil, and other inventions that were to become integral elements in modern technology. Tesla was born on July 10, 1856, the son of Serbian parents in the Croatian village of Smiljan. The settlement was located near the town of Gospić in what was then a part of the Austro-Hungarian empire, an area that later became Yugoslavia. Tesla's father and mother, Milutin Tesla and Djuka Mandić, had expected their son to follow in his father's footsteps as a Greek Orthodox clergyman. However, during his early school years in Smiljan and then in nearby Gospić, where his parents moved when he was six or seven years old, he excelled in math and science. Gradually it became clear that the young and independent-minded Tesla was no candidate for the seminary.

In 1871, when Tesla was fifteen, he attended the higher secondary school at Karlovac, Croatia. After four years, Tesla moved to Graz, Austria, to attend the higher technical school or polytechnic institute in 1875. As before, he excelled in math and science, seemed to have a prodigious memory (he was reputed to have memorized Johann Wolfgang von Goethe's epic drama *Faust*), and showed particular interest in electrical engineering. While attending the technical school in Graz, Tesla commented on the unnecessary (and potentially dangerous) sparks that were emitted by a Gramme dynamo, a direct-current induction motor that was being demonstrated in the classroom. The sparks emerged from where the brushes came into contact with the commutator, and Tesla commented that these sparks could be eliminated by creating a motor without a commutator. The professor was skeptical of the young scientist's theory, and at that time nothing came of the idea. Over the coming years, however, Tesla would continue to work to overcome the problems of direct-current motors.

The details of this period of Tesla's life are unclear, but according to one of his biographers, Margaret Cheney in *Tesla: Man Out of Time,* Tesla's education was interrupted during these years by bouts of malaria and cholera. In any event, Tesla may have attempted to continue his university education at the University of Prague in 1880 (although Cheney indicates that there is no record of this). He was said to have gambled frequently in Prague, wagering for pleasure and in the often vain hope of augmenting his meager income. Tesla appears never to have completed his formal education at Prague, however, possibly because the death of his father forced him to become financially independent. As it was, Tesla may have merely audited classes and used the library without actually enrolling in the university.

Tesla's post-Prague years come into sharper focus. In January 1881 he moved to Budapest where he worked in the Hungarian government's new central telegraph office. During his brief tenure here, Tesla invented a telephone amplifier or loudspeaker, yet for reasons unknown he never patented the device. Tesla also continued to ruminate about the sparks created by the Gramme dynamo in the classroom in Graz, and about rotating magnetic fields, which would later become the basis for all polyphase induction motors. The following year, 1882, Tesla took a position with the Continental Edison Company in Paris.

Tesla's job here was to correct problems in the Edison plants in Germany and France. One of his trips took him to Strasbourg, where he earned local gratitude (but not a promised bonus) for having repaired the railroad station's lighting plant. While in Strasbourg, ever mindful of the sparking problem of direct-current motors, Tesla tried to interest the city's mayor and several of his wealthy colleagues in his design for an alternating-current motor that would eliminate the need for a commutator. In response, the mayor and his friends rewarded Tesla with a few bottles of 1801 St. Estèphe wine but gave no financial support.

Tesla decided to try his luck in the United States where there were interesting developments in electrical engineering and presumably greater opportunities for funding. With a reference from the manager of the Edison company in Paris, Tesla secured a position in Thomas Alva Edison's research laboratory in New York. Tesla embarked for the New World in 1884.

Thomas Edison had already made a reputation for himself as an **electronics** wizard, but he was committed to the use of direct-current electricity. When Tesla explained to Edison his plans for a motor based on alternating current, all he did was create the foundation for a difficult relationship with his unyielding new boss. Edison insisted that Tesla's designs for his new motor were impractical and dangerous. Edison hired Tesla, however, and for a year the new immigrant designed direct-current dynamos and motors for the Edison Machine Works in New York. The experience was limiting and unsatisfying for Tesla, who found that he was unable to overcome the personal and professional differences that separated him from Edison. These factors and a disagreement over compensation that Tesla felt was due to him caused the young Serb to strike out on his own.

In the ensuing year, some entrepreneurs persuaded Tesla to establish an electric company. He established the company's headquarters in Rahway, New Jersey, in 1885. In establishing his own company, Tesla saw an opportunity to apply his ideas on alternating current. His financial supporters, however, seemed mainly interested in providing arc lighting for streets and factories. Again, Tesla faced disappointment and was forced to work for at least part of 1886 as a common laborer. In his spare time, however, Tesla continued to work on his innovations. During this period he managed to acquire seven patents for his work with arc lighting. Growing interest in

Nikola Tesla.

electrical innovations gradually worked to Tesla's advantage, and by 1887, he was able to establish the Tesla Electric Company.

Working within his own organization, Tesla was able to create the first efficient polyphase motor. This was achieved by designing a motor that incorporated several wire-taped blocks that surrounded the rotor. When alternating current is supplied to the wires, with the current to each block being slightly out of phase with the others, a rotating magnetic field is created. The movement of the rotor is achieved as it follows this revolving field. The practical effect of Tesla's invention was that it allowed strong electrical currents to be transmitted over long distances. Edison's direct current, on the other hand, was limited to local use and required many electrical relay stations to distribute the current throughout a given area such as a city. Tesla's invention undermined Edison's assertion that alternating current was impractical, and by 1891, Tesla had acquired forty patents having to do with this technology. His inventions attracted attention, and Tesla began giving lectures in the late 1880s. Perhaps the most notable of these lectures was the one he delivered to the American Institute of Electrical Engineers in May 1888, after which his reputation as a preeminent electrical engineer was firmly established.

George Westinghouse, inventor and manufacturer, bought one of Tesla's patents for the polyphase motor and hired the man to work in his Pittsburgh plant. In 1889 Tesla became an American citizen. He was now famous and his future seemed assured. During the ensuing years, Tesla continued to research and lecture to prestigious organizations across the United States and in Europe. In Britain he addressed the Institution of Electrical Engineers and the Royal Society, and in France, the Society of Electrical Engineers and the French Society of Physics. In these lectures Tesla discussed his work in the transmission of electrical power through radio **waves**. At the Columbian Exposition in Chicago in 1893, the first world's fair to have electricity, Westinghouse provided it using Tesla's system of polyphase alternating current. At the Exposition Tesla also gave lectures and demonstrations of his research.

It was also Tesla's partnership with Westinghouse that allowed Tesla the opportunity to design what may have been the scientist's greatest achievement, the world's first hydroelectric generating plant. The plant, located at Niagara Falls, distributed electrical current to the city of Niagara Falls and to Buffalo, New York, some 23 mi (37 km) away. The Niagara power plant, completed late in 1895, destroyed forever Edison's objections to Tesla's polyphase system of alternating current and established the kind of power system that would eventually be used throughout the United States and the world.

Meanwhile, Tesla had turned his interests to the proposition that radio waves could carry electrical **energy** and in 1897 demonstrated wireless communication over some 25 mi (40 km). Tesla also demonstrated the idea of transmitting electrical energy in 1898 with several radio-controlled model boats that he had constructed. However, the Spanish-American War was underway, distracting the public from this new revelation. Many of Tesla's other inventions would later prove beneficial in a number of applications. His work with high-frequency currents yielded several generating machines that were forerunners to those used in radio communication, and his Tesla coil, a resonant air-core transformer, proved capable of producing currents at a great number of frequencies and magnitudes. In 1898, Tesla moved to the clear, dry air of Colorado Springs, Colorado, where he continued his experiments on electricity, but this time on a grander scale than model boats. As before, his interests focused on transmission of high energy, sending and receiving wireless messages, and related issues pertaining to high **voltage** electricity. The two hundred kilowatt transmitting tower that Tesla built in Colorado Springs could produce lightning bolts that were millions of volts in strength, so powerful they could overload the city's electrical generator. Indeed, during one experiment in creating artificial lightning, Tesla did just that, causing the municipal generator to catch fire and plunging the town into darkness.

Tesla's year of experimentation in Colorado Springs produced no immediate practical results. Tesla's work did provide the basis, however, for research by later scientists. Physicist Robert Golka, for example, modeled his research in **plasma physics** on material he gleaned from Tesla's often cryptic Colorado Springs notes that were housed at the Tesla Museum in Belgrade after World War II. Similarly, Soviet physicist **Pyotr Kapitsa**, who shared the 1978 Nobel Prize for his research on **magnetism**, acknowledged Tesla's work as a model for his own research. Richard Dickinson, a researcher at Cal Tech's Jet Propulsion Laboratory, who was involved in research on the transmission of wireless energy, also invoked Tesla's concepts as a guide to further research.

Although Tesla's work had enduring qualities that inspired the research scientists of later generations, Tesla's influence in the scientific community of his contemporaries began to wane after his year in Colorado Springs. Although he had received royalties from his many patents, that income gradually diminished, due in part to a royalty agreement he had renegotiated with Westinghouse before alternating-current electricity attained prominence. As a result, Tesla realized only a fraction of the fortune that alternating current generated, and he was left with scant resources for his later research. In addition, it appeared at least to some minds that Tesla was beginning to lose his grasp on rigorous scientific inquiry. For example, Tesla had received radio signals while at Colorado Springs that he suggested were from intelligent life on Mars or Venus. Although radio signals from **space** are now a staple of astronomical research, they were not so in the early twentieth century. And to suggest intelligent life as the source of these signals, without the benefit of corroborating evidence, undermined confidence in Tesla's credibility.

During the last four decades of his life, Tesla became reclusive and lived alone in a hotel room in New York City. He continued to perform such experiments as he could with his limited resources, but he never recaptured the glory of his earlier years. Those past accomplishments continued to garner attention, however. Late in 1915 the press rumored that the Nobel Prize committee had listed Tesla and Edison as candidates to share the Nobel Prize in physics. Tesla became indignant because he would have to share the prize with his archrival, but for reasons never made clear, the Nobel Prize committee gave the award in physics to two other candidates. In 1917, a colleague recommended Tesla for the prestigious Edison Medal of the American Institute of Electrical Engineers. Again, because of the award's association with Edison, Tesla at first refused the honor. After he was finally induced to receive the medal and attend the banquet in his honor, he soon drifted from the crowd and was found outside feeding the pigeons.

As Tesla grew older his reclusive and eccentric behavior grew more intense. He was reportedly troubled by phobias—an aversion to pearl earrings and billiard balls, for example—and his ideas seemed ever more bizarre. On his seventy-eighth birthday he told an interviewer that he had plans for an invincible death beam with a potential for 50 million volts that could instantly destroy 10,000 airplanes or one million soldiers. He publicly offered to create such a death beam for the U.S. government, which he said he could create in three months for less than $2 million dollars.

Early in morning of January 8, 1943, the maid at the Hotel New Yorker discovered Tesla's body in his room. He had been ill for the previous two years and had evidently died in his sleep on the evening of January 7 of a coronary thrombosis. He was 86 years old. In death he received much of the adulation that he did not receive during his lifetime. Scores of notable people—Franklin and Eleanor Roosevelt, New York mayor Fiorello H. LaGuardia, political figures from Yugoslavia, Nobel Prize winners, leaders in science—lauded Tesla as a visionary who provided the foundations for modern technology. Indeed, within a year of Tesla's death, the United States Supreme Court ruled that Nikola Tesla, and not **Guglielmo Marconi**, had invented the radio. Yugoslavia made him a national hero and established the Tesla Museum in Belgrade after World War II. In addition to honorary degrees from American and foreign universities (including Columbia and Yale in 1894), and the Edison Medal, Tesla was also recipient during his lifetime of the John Scott Medal. In 1975 Tesla became an inductee into the National Inventors Hall of Fame.

THERMAL ENERGY

Although a body at rest relative to a reference frame has zero kinetic **energy** relative to that reference frame it still maintains an energy of **motion** known as thermal energy.

The essential concepts underlying thermal energy predate **Newtonian physics**. A generation before **Isaac Newton**, the English natural philosopher **Robert Boyle** conjectured that **heat** might be associated with the motion of the individual atoms that constitute **matter**.

A body's thermal energy is that which is due to the random motion of its constituent atoms or molecules. They are constantly moving in all directions or vibrating at varying speeds or frequencies. If, for moving particles, we were to sum up the individual **velocity vectors**, we would, on average, obtain zero; for every particle moving in a particular direction with a specific speed, there is another moving at the same speed, but in the direction opposite. But if we were to square each velocity vector and then sum, so that all the terms were positive, we would, of course, find a nonzero result. Such a sum (with each term multiplied by the individual particle's **mass**) is a measure of the body's internal kinetic or thermal energy.

As a ball falls from a height relative to a particular reference frame (e.g., the ground), its gravitational **potential energy**, which it possesses by virtue of its mass, height, and the **acceleration** due to **gravity**, is converted into **kinetic energy,** which it possesses by virtue of its mass and velocity. Its speed increases as its height decreases. Energy is being conserved. When the ball strikes the ground the ball (assuming it does not bounce) stops, and accordingly, its kinetic energy goes to zero. Ignoring other factors, the KE is transformed into thermal energy. The ball's constituent particles, however, move or oscillate faster than before the collision. The kinetic energy is thus converted into thermal energy, and increases the ball's **temperature**.

Mechanical energy can be converted into thermal energy and thermal energy into mechanical energy. The laws of energy conservation, however, place an important limitation on such transformations. A body containing a certain amount of thermal energy, as measured indirectly by temperature, is able to perform a certain amount of **work** on another object or system. Indeed, these types of transformations are typical in many types of engines. The transformation into work, however, is limited by the second law of **thermodynamics**. An thermal engine will utilize the thermal energy in a hotter body or reservoir and then transform a fraction (called the engine's "efficiency") of that thermal energy into mechanical work and utilizing the remainder to heat the cooler body until both systems are in thermal **equilibrium**. In such processes, a small amount of thermal energy is utilized toward the required increase in **entropy**. As work is done by the engine, the hotter body cools, and the cooler body heats up. When they achieve the same temperature, no more work can be done.

THERMAL EXPANSION

The most easily observed examples of thermal expansion are size changes of materials as they are heated or cooled. Almost all materials (solids, liquids, and gases) expand when they are heated, and contract when they are cooled. Increased **temperature** increases the **frequency** and magnitude of the molecular **motion** of the material and produces more energetic collisions. Increasing the **energy** of the collisions forces the molecules further apart and causes the material to expand.

Different materials expand or contract at different rates. In general, gases expand more than liquids, and liquids expand more than solids. Observation of thermal expansion in a solid object requires careful scrutiny. Several everyday examples are: 1) The sag in outdoor electrical lines is much larger on hot summer days than it is on cold winter days. 2) The rails for trains are installed during warm **weather** and have small gaps between the ends to allow for further expansion during very hot summer days. 3) Because the metal expands more than glass a stuck metal lid on a glass container can be loosened by running hot water over the joint between the lid and the container.

Liquids generally expand by larger amounts than solids. This difference in expansion rate is sometimes observed when the gas tank of a car is filled on a hot day. Gasoline pumped from the underground container is cold and it gradually heats to the temperature of the car as it sits in the gas tank. The gasoline expands in volume faster than the gas tank and overflows onto the ground.

Gases expand even more than liquids when heated. The expansion difference between a gas and a solid can be observed by filling a plastic air mattress in a cool room and then using it on a hot beach. The difference in thermal expansion between the container and the gas could unexpectedly over inflate the mattress and blow a hole in the plastic.

Sometimes man's ingenuity has led him to find practical applications for these differences in thermal expansion

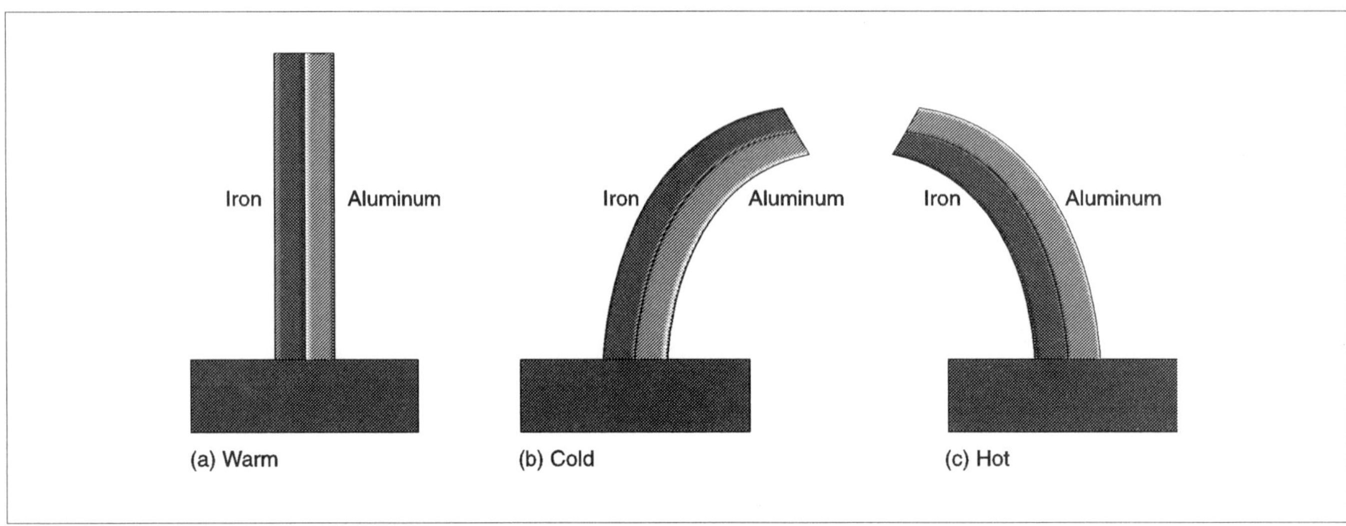

Thermal expansion of iron and aluminum when warm, cold, and hot.

between different materials. In other cases, he has developed technologies or applications that overcome the problems caused by the difference in thermal expansion between different materials.

The thermally induced change in the length of a thin strip of metal differs for each material. For example, when heated, a strip of steel would expand by half as much as an equal length piece of aluminum. Welding together a thin piece of each of these materials produces a bimetallic strip.

The difference in expansion causes the bimetallic strip to bend when the temperature is changed. This movement has many common uses including: thermostats to control temperature, oven **thermometers** to measure temperature, and switches to regulate toasters. Some practical solutions to everyday thermal expansion problems in solids are: 1) The material developed for filling teeth has the same expansion as the natural enamel of the tooth. 2) The steel developed to reinforce concrete has the same expansion as the concrete. 3) Concrete roads are poured with expansion joints between the slabs to allow for thermal expansion (these joints are the cause of the thumping noise commonly experienced when traveling on a concrete highway).

The manufacture of mercury and alcohol thermometers is based upon the expansion difference between solids and liquids. Thermometer fabrication consists of capturing a small amount of liquid (mercury or alcohol) inside an empty tube made of glass or clear plastic.

Because the liquid expands at a faster rate than the tube, it rises as the temperature increases and drops as the temperature decreases. The first step in producing a thermometer scale is to record the height of the liquid at two known temperatures (i.e. the **boiling** point and freezing point of water). The difference in fluid height between these point is divided into equal increments to indicate the temperature at heights between these extremes.

Automobile engine coolant systems provide a practical example of a liquid-thermal expansion problem. If the radiator is filled with coolant when the engine is cold, it will overflow when the engine heats during operation. In older car models, the excess fluid produced by the hot temperatures was released onto the ground. Periodic replacement was required to avoid overheating. Newer cars have an overflow container that collects the released fluid during thermal expansion and returns it to the radiator as the engine cools after operation. This improvement in the coolant system reduces the number of times the coolant fluid level must be checked and avoids the expense of replacing costly antifreeze material mixed with the radiator fluid.

Hot-air balloons are an obvious example of the practical use of the thermal expansion difference between a gas and a solid. Because the hot air inside the balloon bag increases in size faster than the container it stretches so that it expands and displaces the colder (heavier) air outside the bag. The difference between the lower **density** of the air inside the bag compared to the lower density of the air outside the bag causes the balloon to rise. Cooling the air inside the bag causes the balloon to descend.

Water, like most other liquids, expands when heated and contracts when cooled, except in the temperature region between 32°F (0°C) and 39.2°F (4°C). A given **mass** of fresh water decreases in volume until the temperature is decreased to 39.2°F (4°C). Below this temperature, the volume per unit mass increases until the water freezes. This unusual behavior is important to freshwater plants and animals that exist in climates where water freezes in the colder seasons of the year. As the water surface cools to 39.2°F (4°C), it becomes more dense and sinks to the bottom, pushing the warmer water to the surface.

This mixing action continues until all of the water has reached this temperature. The upper layer of water then becomes colder and less compact and stays near the surface where it freezes. When an ice layer forms, it provides an insu-

Bridge expansion joint. *(Image courtesy of JLM Visuals. Reproduced by permission.)*

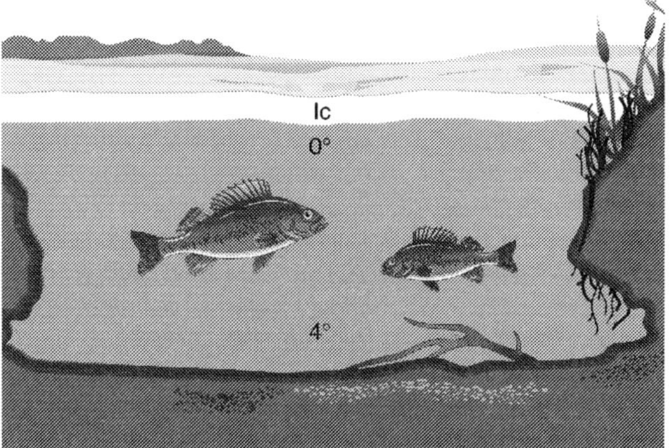

Effects of thermal expansion in a lake.

lation barrier that retards cooling of the remaining water. Without this process, fresh water animal and plant life could not survive the winter.

THERMAL RADIATION

Thermal **radiation** is defined as thermal (**heat**) **energy transformations** caused by the emission of **electromagnetic waves** from an emitting body. Thermal radiation summed over all wavelengths obeys the Stefan-Boltzmann law, which states that the radiation by a blackbody radiator per second per unit area is proportional to the fourth **power** of the absolute **temperature**. Mathematically, it is expressed as: $P/A = \sigma\, T^4$ (in units of J/m^2s), where P is the radiated power, A is the radiative area and σ is Stefan's constant, equal to 5.6703×10^{-8} watt/m $^2K^4$. For radiating bodies other than ideal blackbody radiators, the expression is modified to: $P/A = e\, \sigma\, T^4$ where e is the emissivity of the body; e = 1 for a blackbody. If the temperature of the radiating body is higher than that of its surroundings, the net radiation loss rate takes is given by: $P = e\, \sigma\, A\, (T_r^4 - T_c^4)$ where T_r is the temperature of the radiating body and T_c is that of the surroundings. In accord with the laws of **thermodynamics**, the natural transfer of heat between two systems occurs from a higher to a lower temperature. In the process, the internal **energy** of both systems is changed.

If the temperature of a tungsten filament is gradually increased, it begins to radiate photons and glow a dull red **color** at about 900K. As the temperature is further increased, the filament glows bright red, then orange, then yellow until it finally glows white, at which point the temperature is 2,300K. These types of phenomena were the object of what became one of the fundamental breakthroughs of **quantum theory**, in that classical physics was incapable of accounting for **blackbody radiation**. A perfect absorber, or blackbody, is a material which absorbs all frequencies of electromagnetic radiation and emits none. At the turn of the century, the experimentally observed variation of the radiated power **density**, or radiant excitance, of a blackbody with **wavelength** could not be explained by any model because classical theory assumed that

energy was divided equally between all the vibrations emitting the radiation, with the result that the energy should increase as the wavelength became shorter, as short wavelengths have more vibrational modes. The problem was solved in 1901 German physicist **Maxwell Planck** proposed that energy consisted of discrete units, or **quanta**. From the assumption that the electromagnetic photons were thus quantized in energy (with the quantum of energy equal to the product of **Planck's constant** and the **frequency**) Planck derived a radiation formula stating that the average energy per quantum is the energy of the quantum times the probability that it will be occupied in accordance with the Bose-Einstein distribution function.

See also Electromagnetic spectrum; Heat transfer; Infrared spectroscopy; Power; Quantum theory; Thermal energy; Thermodynamics

THERMODYNAMICS

Thermodynamics is the study of the transformation of **energy**. In **chemistry**, thermodynamics refers to the transformations of energy associated with chemical reactions.

Thermodynamics deals with quantities (e.g., energy, **entropy**) known as state functions. A state function is a property of a system that does not depend on pathways, only on the initial and final states.

The **first law of thermodynamics** states that energy is conserved; it can neither be created nor destroyed. The **second law of thermodynamics** states that, in a isolated system, entropy—a measure of amount of energy in a system unavailable to do work—must increase as time passes. The third law of thermodynamics states that the entropy of a perfect crystal is zero when the **temperature** of the crystal is equal to **absolute zero** (0K). There is a fourth law called the zeroth law that states if two objects are each in thermal **equilibrium** with a third object, then all three bodies are in thermal equilibrium with each other.

Although energy is one of the most fundamental concepts in physics or chemistry— and is used in the fundamental definitions relating to thermodynamics— the term energy is difficult to define. Energy is usually defined as "the ability to do work." The unit of energy in the **International System of Units** (SI), the joule (J), is replacing many older units (e.g., calorie).

Energy is a property of a system—not something that is exchanged between systems. Of course, the energy of a system can increase or decrease, but various mechanisms are required to accomplish these changes.. The mechanisms that accomplish changes in energy are forms of **work**. Heating is one such form of work.

Nobel prize winning physicist and renowned teacher of physics, **Richard Feynman**, once constructed an analogy between the law of **conservation of energy** and a mother's accounting for a child's indestructible blocks. Feynman, with his superb wit and intellect, ended his analogy by stating that accounting for energy was like accounting for the child's blocks except, "there are no blocks." This was Feynman's way of saying that we do not have a precise definition for energy. We can describe what it does and we can account for it (as we do in the laws of thermodynamics) but precisely defining exactly what energy is remains difficult and elusive.

Whatever energy is described to be, the mechanisms of work (e.g., **heat**) are the ways that systems change their state of energy. Energy itself does not flow or move between objects, or between systems and surroundings.

The total internal energy in a system is best described as the sum of the forms of energy of its components. Chemistry is usually concerned with the internal energy (E) of a system where the internal energy is the sum of energy (e.g., potential, kinetic, etc.) of all the particles (atoms, molecules, compounds, or complexes) in the system.

Most chemical reactions are performed under constant **pressure** conditions (at atmospheric pressure). As a result, except for reactions involving gases, most of the changes of energy in a system are accomplished through the transfer of heat. Enthalpy (H) is the thermodynamic property that measures of the heat or heat changes in a system. Enthalpy of formation is defined as the enthalpy change in a reaction that forms a compound from its constituent elements in their naturally occurring forms.

When a **chemical reaction** requires heat in order for it to proceed spontaneously (i.e., proceed without further outside assistance), the reaction is said to be endothermic. When reactions yield heat, they are considered to be exothermic.

Most thermodynamic calculations are carried out at what are termed standard thermodynamic conditions. Standard thermodynamic conditions exist when all gases are at a pressure of one atmosphere, all solids and liquids are pure, and aqueous solutions are concentrations of one mole solute per liter (1 M). Standard enthalpies of formation are almost always reported at a temperature of 298K.

Just as **matter** is conserved, energy is always conserved in chemical reactions. That is, if the masses of reactants used in a reaction are constant then the masses of the products formed and the changes in energy between the reactants and products is also constant.

It is easy to recognize that the first law of thermodynamics is a law of conservation similar to the law of conservation of matter. Considered together, the two laws demonstrate the **equivalence of mass and energy**. Like matter, energy is conserved in all chemical reactions. Further, just as the masses of reactants and products are constant, the amount of energy emitted or absorbed in a specific chemical reaction is a constant. The first law of thermodynamics also demands conservation of energy. As a result, the sum of the energy or a system and its surroundings must remain constant. As the **energy level** of a system drops, the energy level of the system's surroundings must increase by the same amount.

Systems and surroundings are described by variables such as pressure, volume, temperature, specific heat, **density**, compressibility, and the **thermal expansion** coefficient. Regardless, the first law of thermodynamics demands that all of the energy must be accounted for because none is lost or destroyed.

Although, according to the first law of thermodynamics, energy can neither be created or destroyed, it can change its form. There are not, however, an unlimited mechanism to accomplish these transformations. The forms include nuclear energy, **kinetic energy** (i.e., the energy possessed by moving bodies), **potential energy**, heat energy, electrical energy, mechanical energy, **light** energy and chemical energy (i.e., energy stored in chemical bonds).

The second law of thermodynamics implies that the entropy of the **universe** is increasing over time. With regard to isolated system— those that can not do work upon their surroundings nor have work done upon them by the surroundings —entropy must also always increase over time.

When dealing with systems, entropy is a measure of the organization of a system. With regard to chemistry, a change in entropy is reflected in a change in the order or disorder of molecules. Because entropy is actually a measurement of disorder, an increase in entropy means an increase in the disorder of any system. Changes in entropy are measured in joules/Kelvin and is always related to absolute temperature (Kelvin scale).

As a consequence of the second law of thermodynamics, natural process always move toward increasingly disorder or the lowest energy state. The increase in entropy within a system is spontaneous. During many processes—the organization of molecules in living things—entropy in the system may decreases, that is, the system becomes more ordered. This decrease in entropy is never accomplished without work being done on a system. Earth is not an isolated system—liberalizing our definition of energy— the **Sun** is constantly supplying energy to Earth's systems.

According to the second law of thermodynamics, atoms seek the lowest energy level (i.e., ground state). Energy in an excited **atom** is often reduced by conversion into light energy as photons are emitted by the atom. As the energy levels in the excited **hydrogen atom** decrease, the **electron** returns to lower energy **orbitals** closer to the **nucleus**.

One of the earliest statements of the first two laws of thermodynamics was put forth by German physicist **Rudolf Clausius** in a paper published in 1865. Clausius stated that the energy of the universe is constant. and that the entropy of the universe tends to a maximum. Much of Clausius's work was based on earlier writings of French mathematician **Nicolas Léonard Sadi Carnot**. Carnot had meticulously studied and articulated the physical and chemical principles dealing with the operation of **heat engines** (**steam engines**) designed to convert heat into mechanical work.

Carnot's observations laid the intellectual and mathematical groundwork for the formal statement of the first two laws of thermodynamics. An ideal cycle would be performed by a perfectly efficient heat engine, that is, all the heat would be converted to mechanical work. Carnot's calculations dealing with thermodynamic cycles in steam engines proved that an ideal engine—one that was 100% efficient in the transformation of heat energy into mechanical energy could never exist. Because entropy increases some energy in every system is always made useless or unavailable.

There have been many attempts to build perpetual **motion** machines and energy contraptions that would have to violate the laws of thermodynamics if they were to operate as advertised. All have failed or been exposed as frauds either because they were purposefully designed to deceive (somewhat akin to magic tricks) or their proponents were careless in their analysis of the thermodynamics of the system and surroundings. Most of the time, wild claims to have discovered an exception to either of the first two laws of thermodynamics is simply sloppy "accounting for the blocks."

There are no known exceptions to the first two laws of thermodynamics.

The concept of entropy increasing, that is, of a isolated system becoming more random, with increasing amounts of energy unavailable to do work is what gives an **arrow of time** to reactions. We would find it most distressing to have water run uphill or shriveled fruit become ripe and then regress to an embryonic seed. The order of reactions in the natural world is based on the second law of thermodynamics.

Third law of thermodynamics (the entropy of a perfect crystal is zero when the temperature of the crystal is equal to absolute zero [0K]) implies that all molecular motion stops at absolute zero. The atoms in a perfectly pure crystal would be perfectly aligned and would not move. It is important to understand that this means that there is no movement between atoms. Atoms do not "freeze" and there is no cessation of movement of atomic particles within the atom.

The zeroth law of thermodynamics is commonly expressed as heat flowing from hot to cold objects. It would be unnatural to put a cube of ice in a warm drink and not have the ice melt (the increasing temperature allowing the constituent water molecule to move into the liquid phase) or for the drink not to get at least a bit cooler.

Much of the study of thermodynamics involves the use of sophisticated statistical analysis. Indeed, the application of statistical methods to the science is revolutionizing our views of natural processes in much the manner as did the concepts of

relativity and quantum **mechanics**. In fact, an accurate depiction of the universe depends on the understanding and use of all three concepts.

It is a fundamental hypothesis of science, including chemistry, that natural phenomena are governed by laws that can be formulated in such a manner that they can be used to predict the outcome of events and reactions. Nonlinear thermodynamics is part of **chaos** theory and deals with efforts to find structure in systems that are so complex that they are usually termed "unpredictable."

In linear thermodynamics, small changes produce small and predictable changes in systems. A certain measure of heat will always raise the energy level of a system by a certain amount. In nonlinear systems, the result of an action is highly dependent upon the conditions of the system. As a result, the change in energy in a nonlinear system would not only depend on the measure of heat but also on the state of the system.

Although nonlinear theory is more than half a century old, the application of high-speed computing rapidly advancing the application of the fundamental theories. A common feature of nonlinear systems is that their properties can be unexpected, counterintuitive and surprising. Non-linear thermodynamics deals with concepts such as self-organization.

THERMOMETERS

Thermometers are familiar devices used to measure the **temperature** of an object or a system. Thermometers use the principle of thermal **equilibrium**, also known as the zeroth law of **thermodynamics** to find the temperature of an object. Thermometers may be low-cost models that may be found in a medicine cabinet, or high-tech models used in research and engineering. Different types of thermometers must be used over different temperature ranges.

To define a thermometer, the concept of temperature must also be precisely defined. Temperature is an indirect quantitative measure of the average kinetic **energy** in a body or system resulting from the movement or agitation of particles within the body or system. Temperature is usually expressed in units related to a particular scale (e.g., Celsius, Kelvin, etc.) In thermodynamic terms, temperature determines the direction of **heat** flow between systems that are not in thermal equilibrium. Heat "flows" (Note: There is no tangible substance termed heat, it is a property of as body or system) from a region of higher temperature to system or body of lower temperature.

A thermometer measures a body or system by itself becoming in thermal equilibrium with that body or system. The exact mechanism that is used to read the thermometer varies, but all thermometers are calibrated (with another thermometer, for example) so that the temperature can be read directly from the thermometer.

An important consideration when making a thermometer is to minimize the thermometer's ability to change the temperature of the object being measured. Because of this, thermometers should be small enough that they do not absorb much energy from the object being measured.

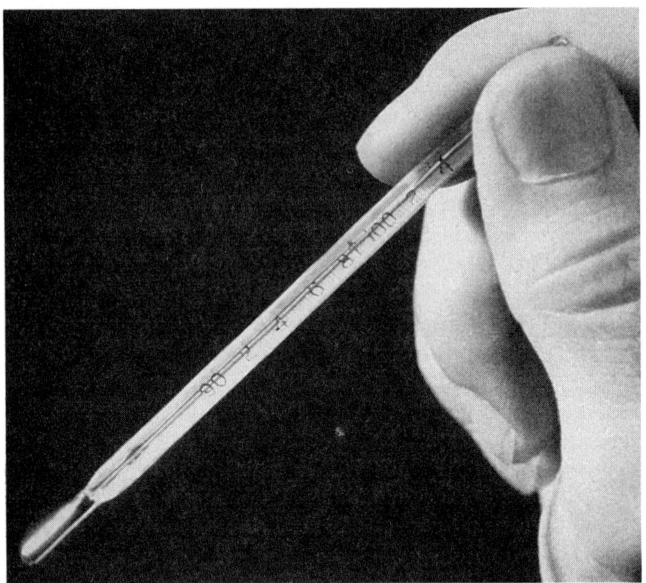

Hand-held thermometer. *(Archive Photos. Reproduced by permission.)*

The simplest kind of thermometer is the common mercury or alcohol thermometer. This thermometer consists of a glass tube with a small amount of fluid inside. When the temperature of the thermometer increases or decreases, the fluid expands or contracts, respectively. Markings on the outside of the thermometer allow measurement of the length of the column of fluid. Essentially mechanical expansion and contraction resulting from **thermal energy** changes is thereby translated into a quantitative measure of temperature.

A different type of thermometer is called a thermocouple. The thermocouple is a resistor, which is sensitive to temperature changes. A resistance meter continuously measures the resistance. Increases or decreases in resistance correspond to increases or decreases in temperature, respectively. The thermocouple is used to measure very low temperatures, such as temperatures of objects submersed in liquid nitrogen (about 77K). A mercury or alcohol thermometer is not useful at this temperature because the fluid inside the thermometer would freeze.

See also Low-temperature physics; Thermal expansion

THERMONUCLEAR REACTIONS

Nuclear reactions, highly energetic processes involving **subatomic particles** found within the nuclei of atoms, exist in two general forms. Fission reactions involve the splitting of atomic nuclei into smaller subatomic particles, in the process releasing relatively large amounts of **energy**. Fission chain reactions are responsible for the massive explosions of the atomic bombs that were employed during World War II. The second variety, fusion reactions, accomplish just the opposite.

Fusion reactions create larger atoms from the nuclei of smaller atoms. Here, when the nuclei of two or more atoms collide with sufficient **force**, they fuse to form a single, larger **nucleus**. During the fusion process, part of the **mass** of the fused nucleus is converted into energy. The energy produced is then released as **heat**, **light**, and various forms of **radiation**. Such nuclear reactions require extremely high temperatures to induce the necessary collisions. Critical temperatures for fusion range from 50 to 400 million degrees Celsius (122 to 752 million degrees Fahrenheit). Because such immense amounts of heat energy are required, fusion reactions are also called thermonuclear reactions. Thermo-nuclear chain reactions in turn release massive amounts of energy once started, and are utilized by thermonuclear bombs (also known as hydrogen bombs). Thermonuclear warheads have destructive potentials many thousands of times greater than their fission counterparts.

As imposing as man-made thermonuclear reactions are, they are merely tiny reproductions of much larger and more significant thermonuclear processes found naturally throughout the **universe**. The **stars** themselves are formed and fueled by thermonuclear reactions. Stars begin their lives as massive, nebulous interstellar clouds of gas, consisting mostly of hydrogen. As **gravity** condenses the hydrogen, the clouds become more and more dense. In the process, the condensed **matter** becomes hotter over time, reaching many millions of degrees. Eventually, the heat energy becomes great enough to initiate thermonuclear reactions, and a new star is born.

Once created, the condensed gaseous matter continues to fuel the thermonuclear reactions of stars. For example, the massive amounts of solar energy released by the **Sun** are produced by thermonuclear fusion of its gaseous contents. In the 1930s, it was discovered that the Sun derives its energy from the fusion of hydrogen atoms, in a process called the proton-proton reaction. This thermonuclear process occurs in relatively cool stars. The proton-proton cycle involves the fusion of four hydrogen nuclei (protons) to form a single **helium** nucleus. Therefore, life as it exists on earth is made possible by only by thermonuclear reactions. A different kind of thermonuclear reaction, called the carbon-nitrogen cycle, creates most of the energy radiated by stars much hotter than the Sun. The carbon-nitrogen cycle involves **isotopes** of the elements hydrogen, **carbon**, nitrogen, and **oxygen** that are formed during a complex chain of fusion events, each releasing energy. As long as the gaseous elemental fuel for their fusion reactions remains plentiful, the stars in the **cosmos** continue to shine with thermonuclear brilliance.

THOMSON, JOSEPH JOHN (1856–1940)
English physicist

Joseph John Thomson, who discovered the **electron** in 1897, won the 1906 Nobel Prize in Physics "in recognition of the great merits of his theoretical and experimental investigations on the conduction of **electricity** by gases." The British scientist discovered that cathode rays consisted of negatively charged particles of subatomic size, which he called corpuscles, and which ultimately became known as electrons. He carried this theory to all **matter**, hypothesizing that matter consisted of negatively charged particles that were surrounded by positively charged particles, and that these charges neutralized each other. He became Sir Joseph in 1908 when he was knighted for his work. As evidence of his amazing mind and great teaching abilities, seven of his research assistants and his son, George, went on to win Nobel Prizes in physics.

Thompson, always known as "J.J.," was born near Manchester, England, the son of a bookseller, who sent him to Owens College until an apprenticeship with a leading engineer opened up. However, his father died before the apprenticeship became available and, by the time it did, his family could not afford his fees. Encouraged by the excellent scientific professors at Owens, Thomson remained in the engineering program after receiving a small scholarship, and worked toward a scholarship in mathematics at Trinity College, Cambridge. He won the small scholarship, and entered Trinity in 1876. In 1880, he placed second in the final honors examinations in mathematics, and the college subsequently awarded him another scholarship in 1881. He studied three different lines of mathematical theory while continuing his own research interests. He was also strongly influenced by physicist **James Maxwell**, a fellow Englishman who proposed the theory of electromagnetics.

In 1882, as a fellow at Trinity, Thomson began designing his first mathematical models while contesting for an academic award called the Adams Prize. The subject matter was "a general investigation of the action upon each other of two closed vortices in a perfect incompressible fluid." He carried his research well beyond the subject matter to the theory of the vortex **atom**. In this theory, atoms of gas in a frictionless fluid are likened to smoke rings floating in the air, except that the atoms are eternal (never degrade). Thompson attempted to analyze the phenomenon mathematically and published his theories in *Treatise*. This work is described in the *Dictionary of Scientific Biography* Vol. III, as a "quantitative, mechanistic, and ultimate account of the physical world...perhaps the most glorious episode in this hopeless struggle."

Much of Thomson's work during this period focused on **atomic structure**, theories of chemical action, and the nature of **light** through "precise calculation and ingenious analogy applied with great virtuosity." He also wrote a dissertation for his fellowship that focused on an idea he had at Owens College—that "potential **energy** in a given system might be replaced by the **kinetic energy** of imaginary masses connected to it in an appropriate way."

In 1884, at the age of just 27, Thomson succeeded Lord Rayleigh as Cavendish Professor of Experimental Physics at Trinity. He remained in that role until 1919. On the advice of Maxwell, he decided to investigate the phenomenon of gas discharge in relation to cathode rays. Thomson suspected these rays to be streams of charged particles, "...matter highly

Sir Joseph J. Thomson.

charged with electricity and moving with great velocities, because the path of the rays could be deflected or curved by a magnetic **force** field." He adhered to his line of thinking, even though German physicists were convinced cathode rays were an "aethere disturbance," or **ether waves**, similar to ultraviolet light, because of their ability to make glass fluoresce (or glow).

To conduct his experiments, Thomson used a glass tube with one end connected to a negative electrode (cathode), a positively charged electrode (**anode**) at the other end, and a tiny paddle wheel in the center. Upon running an **electric current** through a variety of gases inside the tube, he discovered that: 1) The current created different colors in different gases; 2) The anode (positive) end glowed; the negative (cathode) did not; and 3) The paddle wheel spun in the direction of the anode. From this, Thomson deduced that atoms of different gases produced different energies, as shown by the different colors; that, in order to make the paddle spin, the gases must contain particles; and that the current flowed from negative to positive. This latter finding was one of his most important discoveries.

Because **Heinrich Hertz** had discovered that cathode rays could pass through thin gold or aluminum leafing inside a discharge tube, Thomson assumed the rays must consist of very minute components and so pursued the negatively charged particle theory. He began applying the theory of the vortex atom to the gas discharge and discovered an extremely important phenomenon-that gas discharge, like **electrolysis**, occurs through the disruption of chemical bonds. He found that, by placing positive and negative electrodes at the middle of the tube as well, he could "bend" or deflect the cathode ray. This ultimately led him to discover the existence of **ions**, charge carriers much smaller than atoms, and to theorize the structure of the atom. Because he found that negative particles produced in the discharge tube were the same in all residual gases and with different electrodes; and because negative particles could also be produced by heated metals, the action of ultraviolet light, or **x rays** on metals, he predicted that these tiny bodies must be the common component in all atoms.

In 1897, he gave a lecture to the Royal Institution in which he described in detail his apparatus for deflecting cathode rays in both magnetic and electric fields. Thompson called the tiny particles in the cathode rays "corpuscles." It was George Johnstone Stoney (1826–1911) who first proposed they be called "electrines" and eventually "electrons."

Thomson then turned his focus to determining the nature of positively charged ions, or positive electricity. This was to become his final important experimental work. Many years of painstaking research, in collaboration and comparison with other physicists, ultimately allowed him to show that neon gas was composed of two types of ions. Each one had a different charge, different **mass**, or both. He determined this by deflecting the stream of neon gas containing positive ions with both magnetic and electric fields. This produced two invisible streams (or traces), images of which he captured on a photographic plate. This led him to suspect different types of atoms of the same element, with the same atomic number, but with a different mass. Thompson devised many innovative experiments and types of equipment to help analyze these traces. An associate, Francis Ashton (1877–1945), perfected the mass **spectroscope**, for which he also won a Nobel Prize.

Thomson was not only an ingenious scientist but also a gifted teacher. Twenty-seven of his students became fellows of the Royal Society and dozens became physics professors. He personally helped students get papers published and to gain placement in appropriate research institutions. He aimed at improving the level of scientific studies in secondary schools as well as universities; lectured at the Royal Institution where he became Professor of Natural Philosophy in 1905; established the Cavendish Physical Society in 1893, a seminar held every two weeks which reviewed recent works to help his scientists keep abreast of developments; and expanded the size of the Cavendish laboratory twice. His most important papers were published in *Philosophical Magazine*, which he helped become the premier journal on physics in all of England.

Other awards recognizing his tremendous contribution to science included the Order of Merit in 1912; presidency of the Royal Society beginning in 1915 (which placed on him the responsibilities of that Society's contribution to the efforts of World War I); and, in 1918, became master of Trinity, his old college at Cambridge. He continued to fulfill his duties in this capacity until just several months before his death. He was granted the tremendous honor of burial in London's Westminster Abbey.

TIDAL FORCES

Tidal forces are responsible for the ocean tides we observe on Earth. One of the earliest problems considered by physicists was why Earth has two high tides per day. **Galileo** was unable to understand or explain this, but **Isaac Newton** eventually solved the problem using his theory of gravitation. The solution to the problem is that high tides are caused by the differences between the gravitational forces on Earth's water applied by Earth, the **Moon**, and the **Sun**. The water closest to the Moon is pulled toward the Moon, and Earth is also pulled toward the Moon, away from the water furthest from the Moon. This leads to an oval-shaped distribution of water around the earth, rather than a uniform spherical shape, with the thickest parts of the oval along the line between Earth and the Moon. Because Earth is rotating, a point on its surface crosses the line between Earth and the Moon twice a day, where the height of the water is at its maximum. The Sun also exerts tidal forces which are approximately half as large as those due to the Moon. This is the reason for the high and low tides that we observe.

TIME

Time is a measurement to determine the duration of an event, or to determine when an event occurred. Time has different incremental scales (year, day, second, etc.), and it has different ways by which it is reported (Greenwich Mean Time, Universal Time, Ephemeris Time, etc.).

Our time scale is based upon the movement of the **Sun**. Since the Sun appears to move from east to west in the sky, when it is twelve noon in New Jersey, people in Seattle, Washington, say that the Sun has not yet reached its apex, or noontime position. In fact, if you say that the Sun has attained its noontime position, someone who is west of you by one degree of longitude will not see the Sun in its noontime position for approximately four minutes.

In October of 1884, an international agreement divided the planet into 24 time zones (longitudinally into 15° segments), allowing us to standardize time keeping. The starting point, or zero meridian, was the longitude that ran through the Royal Observatory in Greenwich, England. This is the common position from which we measure our time and is generally referred to as Greenwich Mean Time (GMT).

As you move westward from Greenwich, you add one hour for every 15° meridian you cross. As you move eastward from Greenwich, you subtract one hour for every 15° meridian you cross. (On land the time zones are not always straight lines; this is sometimes done to keep countries, states, cities, etc., all within the same time zone.) It should also be noted that these time zones are also listed by letters of the alphabet as well as 15° increments: Greenwich is designated Z for this purpose, the zones to the East are designated A through M, omitting the letter J, and the zones to the West are N through Y.

In North America, most people measure time based upon two 12 hour intervals. To distinguish the first 12 hours from the second, we label the time. The first twelve hours are denoted by the letters A.M. (for the Latin *ante meridiem*, meaning *before noon*—before the Sun has reached its highest point in the sky). The second 12 hours is denoted by P.M. (for *post meridiem*, which means *after noon*).

The International Date Line is near the 180° longitude in the Pacific Ocean. When you cross it going west, you add 24 hours (one day), and when you cross it going east, you subtract 24 hours (one day).

If you ask someone from Europe, or someone in the armed services, what time it is, he or she will answer you using a 24-hour system. Their time starts at midnight of every day, which they call 0000 hours (or 2400 hours of the previous day). Their time then proceeds forward for 24 hours, at which point it starts again. If someone tells you it is 0255 Zulu time, it means that it is 2:55 A.M. in Greenwich, England. This type of time reckoning is extremely important to synchronizing clocks for navigation, etc.

Astronomers found GMT to be awkward to work with in their calculations, because its measurements were based upon the Sun's actual **motion** and period between successive occurrences at its noontime position. Therefore, in 1972 they defined a Universal Time which starts at midnight. They also realized that the solar day, the time from noon to noon, is not necessarily constant because of the irregular motion of Earth. To compensate for this, they have adopted a mean (average) solar day, time from noon to noon, upon which to base their measurements.

Both GMT and Universal Time are solar-based time systems, which means that they are based upon the apparent movement of the Sun. This method is not accurate enough for all scientific measurements because Earth's orbit around the Sun is not circular (it is an ellipse), Earth's rotation rate (how fast it spins on its axis) is not constant, and its rotation axis is nutating (wobbling). Astronomers compensate for these "imperfections" by using a Sidereal Time measurement, a measurement system based upon the repetitive motion of the **stars**. This motion is, in fact, due to Earth's motion. Ephemeris time is a Sidereal Time based on the apparent repetitious motion of the **Moon** and the **planets**.

During World War II, a "war time" was instituted to save **electricity**. Later, this became daylight saving time. However, some states and cities no longer use daylight saving time because they see it as a needless complication.

Albert Einstein's theory of relativity prompted scientists to revise the idea of time as an absolute. For example, the German mathematician **Hermann Minkowski** described time as the fourth dimension of **space**, postulating a **space-time** continuum, which scientists, including Einstein, accepted. Essentially, time cannot be a basis for absolute measurement. Because we are all in motion, in one form or another, our time measurements are dependent upon how we are moving. Relativistic time measurements depend upon who is measuring it, how fast the person is moving, and whether or not they are undergoing an **acceleration**.

We measure time, in general, with a clock. However, not all clocks are based on the same time scale. Measurement systems and types of time differ widely. We have just seen that clocks in different places are calibrated to different schemes, and so are likely to tell different times.

The first clock was natural—the motion of the Sun through the sky. This led to the invention of the sundial to measure time, as well as the evening version using the Moon's position. Later came the hourglass; candles; and clocks that burned wax, or oil, at a specified rate; and water clocks, which allowed water to flow at a specified rate. Early clocks of greater accuracy used a pendulum arrangement invented in the 1656 by the Dutch astronomer **Christiaan Huygens**. Balance wheels then replaced pendulums in some clocks, which allowed clocks to become portable. In recent times, coupled pendulum clocks were used to keep extremely accurate times.

Most of today's clocks and watches use a quartz crystal to keep the time. The specially manufactured quartz crystal, with a specific **frequency voltage** passing through it, will vibrate at a constant (characteristic) frequency. This frequency, when amplified, drives an electric motor which makes the hands of the clock turn, or makes the digital display change accordingly.

Atomic clocks are the most accurate clocks, and are used by the United States Naval Observatory. They vibrate at a constant sustainable frequency. The process is analogous to setting a pendulum in motion, or a quartz crystal vibrating, except the atoms "vibrate" between two different energies. This "vibration" between **energy** levels is associated with a specific frequency, just like the oscillating quartz crystal. This atomic frequency is what drives the clock. The current standard for atomic clocks is one that uses an **isotope** of Cesium, ^{133}Cs. In 1967, the second was redefined to allow for this accuracy. The second is now defined as 9,192,631,770 times one period, the time for one complete oscillation, of ^{133}Cs.

Research continues into ion clocks such as the hydrogen maserclock. This type of clock is more accurate than a ^{133}Cs clock; however, after several days its accuracy drops to, or below, that of a ^{133}Cs clock.

We perceive time as always proceeding forward into the future; there is no way we know of to travel into the past (except through memories). Time reversal refers the attempt to understand whether a process is moving forward or backward in time. For example, if you watch a movie of two isolated billiard balls colliding on a billiard table, can you tell if the movie is being shown forward or backward? If the two balls are exactly the same, and you do not see one of them being hit to start the process, probably not. If we were to film the process of an egg being dropped and hitting the ground, and then show the film, you could definitely determine in which direction the process continued. The ability to distinguish between forward and backward processes, between past and future on all scales, is crucial for scientific research.

To see if there really is an "arrow of time," or preferred direction in which time flows, researchers have been conducting many different types of experiments for more than 40 years. According to theory, we should be able to tell forward time processes from backward time processes. Experimental verifications of this theory delineate how difficult some experiments are to design and perform. The latest are designed to observe atomic particles as they undergo different processes. If the processes are then "run backward," any differences that show up mean that we can tell a forward process from a backward running process.

TIME DILATION

Time dilation is one of the primary phenomena predicted by **Albert Einstein**'s theory of special relativity published in 1905. The theory states that the measured length of time between two events depends upon the **frame of reference** of the observer. In particular, time appears to go slower for objects moving with respect to the observer. Hence, time appears to have dilated, or lengthened.

An event is anything that takes place at a specific point in **space** and at a specific moment in time, like the chiming of a clock. The twelve o'clock chime of a clock occurs at the location of the clock and at the moment when the minute and hour hands of the clock are on the twelve.

The time interval between two events, like the twelve and one o'clock chimes of a clock, depend upon the observer's frame of reference. An observer who is stationary with respect to the clock will measure the proper time interval δt_0, which—assuming the clock is well made—will be 1 hour. A second observer, moving at a speed v with respect to the clock will measure the time interval between the two events to be longer. The second observer measures a time interval of $\delta t = \delta t_0 / sqrt(1 - v^2 / c^2)$, where c is the speed of **light** in **vacuum** measured to be 300,000 km/s (186,000 mi/s).

The phenomenon of time dilation has been tested and observed in experiments since the beginning of the twentieth century. One striking example of time dilation explains why muons are detected at the surface of Earth. Muons are unstable particles that look like electrons with 207 times more **mass**, but they have a mean life of only 0.0000022 seconds before decaying into smaller particles. (This length of time is called the lifetime of the unstable particle.) Muons are produced with velocities close to 99.8% the **speed of light** from **cosmic rays** collisions with atoms in Earth's upper atmosphere, roughly 10 km (6.3 mi) above sea level. Even at these tremendous speeds, the muons would only travel a distance equal to roughly 660 m (2200 ft) before decaying, and therefore they should not be detectable at the surface of Earth.

However, since the **muon** is traveling at 99.8% the speed of light with respect to us on the surface of Earth, the lifetime of the muon appears to be longer by a factor of $1 / \sqrt{1 - (0.998^2)} = 1 / 0.063 = 16$, which means the muon appears to live for 0.000035 seconds. At these speeds, a muon that lives for this length of time can travel approximately 10.4 km (6.5 mi) before it decays. Hence, time dilation explains why we detect muons at sea level.

TING, SAMUEL C. C. (1936–)
Asian American physicist

Samuel C. C. Ting is an American physicist who received the 1976 Nobel Prize for his discovery of the J/psi particle, which led to the detection of many new **subatomic particles**. Ting shared the prize with **Burton Richter**, who had made the same discovery almost simultaneously, using a different experimental technique. Ting is known as a confident, daring theorist, as well as a precise experimenter. He is a consummate practitioner of physics in the era of "big science," when research is conducted by large international teams using costly, complex experimental apparatus.

Ting was born in Ann Arbor, Michigan, on January 27, 1936, while his father, Kuan Hai Ting, was studying engineering at the University of Michigan. He completed his studies when Ting was two months old, and the family returned to mainland China, where his father became an engineering professor. His mother, Tsun-Ying Wang, was a psychology professor. As a child, Ting was cared for mostly by his maternal grandmother while both his parents worked. Although his grandmother emphasized the strong value of education, Ting was not able to begin school until he was twelve years old, because World War II intervened. After the war, the family moved to Taiwan, where Ting's father taught at the National Taiwan University.

In 1956, Ting enrolled at the University of Michigan, studying both mathematics and physics, and in 1959 he earned bachelor's degrees in both subjects. He married Kay Louise Kune, an architect, in 1960, with whom he would have two daughters. Ting received his Ph.D. in physics in 1962, and the next year he went to the **European Center for Nuclear Research (CERN)** in Geneva as a Ford Foundation fellow. He worked with Giuseppe Cocconi on the **proton** synchrotron, a device that accelerates protons (the **nucleus** of an **atom**) for analysis and measurement. In 1965 Ting joined the faculty of Columbia University, where he worked with **Chien-Shiung Wu**.

Ting became interested in the production of **electron** (negatively charged particles of an atom) and **positron** (positively charged particles of an atom) pairs by **photon radiation** after experiments conducted at Harvard raised questions regarding some of the predictions of quantum electrodynamic theory (the theory that deals with the interaction of **matter** with electromagnetic radiation). He took a leave of absence from Columbia and went to Hamburg, Germany, in 1966 to repeat the Harvard experiments at the

German synchrotron facility. There his team built a double-arm spectrometer (an instrument used to analyze and measure particle emissions), which enabled them to measure the **momentum** of two particles simultaneously. It also recorded the angles of their deflection from the radiation beam. The researchers were able to calculate the masses of the particles and their combined **energy**, making identification of the particles easier and clarifying their interrelationships. Results of these experiments confirmed the accuracy of the quantum electrodynamic description of pair production.

Ting's work at Hamburg led him to ponder the nature of heavy photons (particles of radiation). After his return from Germany, he moved to the Massachusetts Institute of Technology (MIT), where he became full professor in 1969. In 1971, while still at MIT, Ting began a project to determine the properties of heavy photons at Brookhaven National Laboratory in Long Island, New York. Rather than the usual method of bombarding a beryllium target with photon beams, he used a proton beam of ten trillion protons per second in hopes of creating a heavy particle that would decay into pairs of electrons and positrons.

Because the search for heavy particles requires such high energy levels, Ting's MIT team redesigned the double-arm spectrometer to detect electron-positron pairs between 1.5 and 5.5 giga-electron volts (a giga equals one billion). The spectrometer also had to be capable of adding precise but small amounts of energy incrementally, as well as detecting their effects on the particle pairs. After several months of searching, the Ting team was rewarded in August 1974 by the appearance of a sharp spike of high-energy electron-positron pairs at 3.1 billion electron volts. This was unexpected. Ting checked his measurements carefully and decided he was looking at evidence of a new particle that had not been predicted, the J/psi particle. It was heavier than known similar particles; it also occupied a very narrow range of energy states, and it lasted a relatively long time.

Ting reported his results to the Frascati Laboratory in Italy, where physicists were able to confirm his observations in only two days. Ting's paper and the results of the Frascati experiment were accepted for publication in *Physical Review Letters*. Just a few days after Ting discussed the paper with the review's editor, he attended a routine scheduling meeting at the Stanford Linear Accelerator Center; here he shared his results with Stanford's Burton Richter. Amazingly, Richter had made the same discovery at virtually the same time by creating collisions between positrons and electrons in an accelerator.

Ting and Richter shared the 1976 Nobel Prize in physics. The two-year period between discovery and award was probably the shortest interval on record and caused considerable comment at the time, because some scientists feared the discovery would not stand the test of time. However, it has since been the basis for a virtual explosion in the detection of many other fundamental particles.

The J/psi particle's lifespan was a thousand times longer than expected for such a heavy particle (three times heavier than a proton). It was believed that most subatomic particles were made up of combinations of even more fundamental particles called **quarks**, of which only three types were thought to exist before the discovery of the J/psi particle. The peculiarities of the J/psi particle (especially its long life) suggested the existence of a fourth type of quark, called charm. The J/psi particle was interpreted to be composed of a charmed quark and an antiquark, creating a property called "charmonium." Charm had been predicted in 1970 and its addition to the family of quarks was thought to unify the electromagnetic and weak forces, further encouraging physicists to believe in the possibility of a grand unifying theory in which the fundamental forces of nature would be shown to be equivalent at very high energies.

There are several stories of how the Ting-Richter particle received its name of J/psi, which is a combination of Ting's name for it (J) and Richter's (psi). Classical particles were traditionally assigned Greek letters for names, while newly discovered particles are labeled with capital letters. One story says Ting called his particle J because he had been working with electromagnetic currents carrying a J label. Another story says the J derives from the physical symbol for **angular momentum**. A third claims Ting chose the Chinese symbol for his name. In any case, the particle has retained the double label. A similar particle, called the psi-prime, was found by Richter's team within ten days of the first discovery.

Ting is a fellow of the American, European, and Italian physical societies as well as several academies of science, including the Academia Sinica. In addition to the Nobel Prize, Ting received the 1976 E. O. Lawrence Award.

TOMONAGA, SIN-ITIRO (1906–1979)
Japanese physicist

Sin-Itiro Tomonaga was a pioneer in the field of quantum **electrodynamics**, a broad theory that uses principles from quantum **mechanics** and special relativity to explain a wide variety of physical phenomena. He developed a theory about **subatomic particles** that was consistent with the **relativity theory** about the same time that **Richard P. Feynman** and **Julian Schwinger** independently reached similar solutions. The three were jointly awarded the 1965 Nobel Prize in physics for their efforts in **quantum theory**.

Sin-Itiro (also transliterated as Sin-Ichiro) Tomonaga was born in Tokyo on March 31, 1906, to Sanjuro and Hide Tomonaga. When Sin-Itiro was a child, his family moved to Kyoto, where his father had been appointed professor of philosophy at the Kyoto Imperial University. Sin-Itiro enrolled at Kyoto's prestigious Third High School where he was a classmate of **Hideki Yukawa**, later to become Japan's first Nobel Prize winner (in the field of physics) in 1949. After graduation, Tomonaga and Yukawa both went on to Kyoto Imperial University, where both majored in physics and earned their

Sin-Itiro Tomanga.

bachelor's degrees in 1929. The two then stayed on as research assistants to Kajuro Tamaki.

In 1932, Tomonaga and Yukawa finally went their separate ways, with Tomonaga accepting a position as research assistant to Yoshio Nishina at the Institute of Physical and Chemical Research in Tokyo. After five years in this post, Tomonaga traveled to the University of Leipzig where he studied under physicist **Werner Heisenberg**. While at Leipzig, Tomonaga wrote a paper on the atomic **nucleus** that earned him a Ph.D. from Kyoto Imperial upon his return to Japan in 1939. In 1941 Tomonaga became professor of physics at Tokyo's Bunrika University (now Tokyo University of Education), a post he held until 1956, when he became president of the university. After leaving Tokyo University in 1962, Tomonaga served as president of the Science Council of Japan and director of the Institute for Optical Research.

Attacks Problems of Quantum Electrodynamics

The topic that dominated Tomonaga's research throughout most of his life was **quantum electrodynamics (QED)**. QED arose during the 1920s when the successes of quantum theory and relativity made it apparent that classical laws of physics were inadequate to explain the behavior of elementary particles. A number of theorists attempted to develop new equations that would take into consideration both **quantum**

mechanics and relativity theory to explain the behavior of particles and their interaction with **energy**.

By the late 1920s, impressive progress had been made in dealing with this problem, especially as a result of the work of the English physicist **Paul Dirac**. Dirac's theory successfully predicted the qualitative properties of atomic particles and the way they interacted with energy. Over time, however, it became apparent that Dirac's theory was quantitatively inadequate. Among the most serious problems of the Dirac theory was the prediction that, under certain circumstances, particles would have infinite **mass** and infinite charge. The physical absurdity of these predictions, commonly known as "divergence difficulties," deeply troubled physicists, many of whom decided that a totally knew approach to QED would be needed.

Success with Quantum Electrodynamics Brings Nobel Prize

Tomonaga, however, took another view. He was convinced that techniques could be found that would allow the retention of Dirac's fundamental approach while resolving the "divergence difficulties" inherent within it. The mathematical technique he used, called **renormalization**, worked. Tomonaga was eventually able to demonstrate that infinite mass and charge are indeed a fundamental part of the Dirac theory, but that they apply to situations that would never be encountered in the real world.

Tomonaga's fundamental paper on QED was published in Japan in 1943. Because of the war, however, news and translation of the paper did not reach the rest of the world until 1947. At about the same time, similar papers dealing with QED written by Schwinger at Harvard and Feynman at the California Institute of Technology were also being published. When the 1965 Nobel Prize in physics was announced, therefore, it was divided among the three physicists for the independent solutions of the "divergence difficulties" problem.

Tomonaga's prize-winning research had been carried out at Tokyo's Bunrika University of Science and Literature, where he had been appointed professor of physics in 1941. He remained in this post, after the university had been incorporated into the Tokyo University of Education, until 1949. He then accepted an invitation to become a visiting scholar the Institute for Advanced Studies at Princeton, New Jersey. He returned to Tokyo in 1951 to become head of the Institute for Scientific Research. Four years later, he was instrumental in the founding of the Institute for Nuclear Studies at the University of Tokyo and then, in 1956, became president of the university. Upon his retirement in 1962, Tomonaga accepted an offer to become president of the Science Council of Japan and director of the Institute for Optical Research, posts he held until 1969.

Tomonaga was married in 1940 to Ryoko Sekiguchi, daughter of the director of the Tokyo Metropolitan Observatory. They had two sons, Atsushi and Makoto, and a daughter, Shigeko. In addition to the Nobel Prize, Tomonaga was awarded the Japan Academy Prize in 1948, the Order of

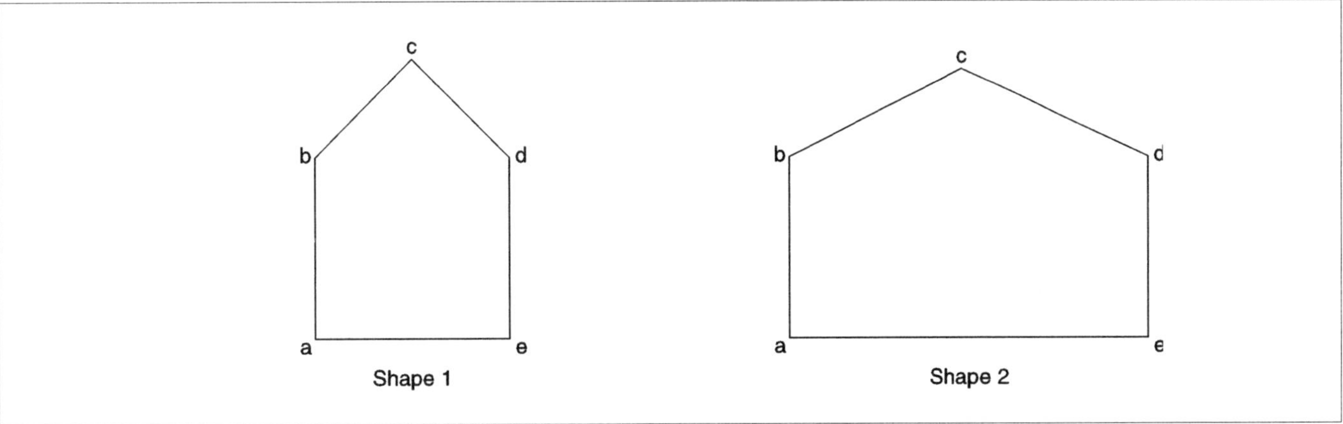

Topology, pentagons.

Culture of Japan in 1952, and the Lomonosov Medal of the Soviet Academy of Sciences in 1964. Tomonaga died in Tokyo on July 8, 1979.

TOPOLOGY

Topology is the study of the properties of spaces that are insensitive to continuous deformations. Physicists recognize several different subdivisions of topology. The simplest is called point set topology, which is needed to define the notion of a topological **space** itself. A topological space is a set of points, possibly continuously infinite in number, which is endowed with a topology, or a collection of open sets that in essence specifies which points in the space are close to each other. This is the minimum mathematical structure needed to define continuous functions from one topological space to another. Continuous functions are simply those functions that take points that are near each other in one space to points that are near each other on the target space. An example of a family of continuous functions is polynomials in one variable which map the real line back to itself.

Two topological spaces are considered equivalent if either can be deformed into the other in a continuous way without making any rips in each space along the way, or without plugging up any holes. For example, a hollow sphere with a hollow handle attached to it is equivalent to a hollow torus or donut, but the sphere without the handle attached is not equivalent. No matter how hard one tries one cannot deform a sphere into a donut. One branch of topology called algebraic topology is especially useful in physics and enables one to characterize when two topological spaces are not equivalent. The basic idea is to attach to each topological space an algebraic object. Two topological spaces will be equivalent only if the corresponding algebraic objects are the same. Thus if one computes the algebraic objects corresponding to two different spaces and finds that they are different,

one has essentially proved that the corresponding spaces are not equivalent.

As an example consider again the sphere and the donut. The critical feature characterizing their topological inequivalence is the existence of non-contractible circles on the donut. Consider for example any circle on the sphere. It can always be contracted to a single point, just like a lasso can slip off any round object. However the hollow donut has two inequivalent circles that cannot be contracted to a point. One circle travels around the equator of the donut. The second circles around one end of the donut, in much the same way one's thumb and index finger would curl around the donut when holding it at one end.

Since the donut has two non-contractible circles, mathematicians say the first homology group of the donut is two-dimensional, whereas the first homology group of the sphere is zero-dimensional. These homology groups are among the simplest algebraic objects that can be associated to topological spaces. This picture can be generalized to higher dimensional objects, and n-dimensional homology groups can be defined, which essentially count the number of inequivalent n-dimensional non-contractible cycles that exist in the topological space. The dimensions of all the homology groups are called the betti numbers. A pivotal result in the field of algebraic topology is that if the betti numbers of two different spaces differ then the two spaces cannot be deformed into each other.

Topology shows up in a crucial way in the modern theories of physics. For example **string theory** presents a picture of the world as a ten-dimensional space consisting of our usual four dimensions of space and time and a tiny six-dimensional space attached to each point. The topology of this six-dimensional space is highly constrained by consistency requirements in physics, and the constraints are naturally expressed in terms of something similar to the betti numbers described earlier. In condensed matter theory as well topology has proved a useful tool in characterizing classical field configurations and is

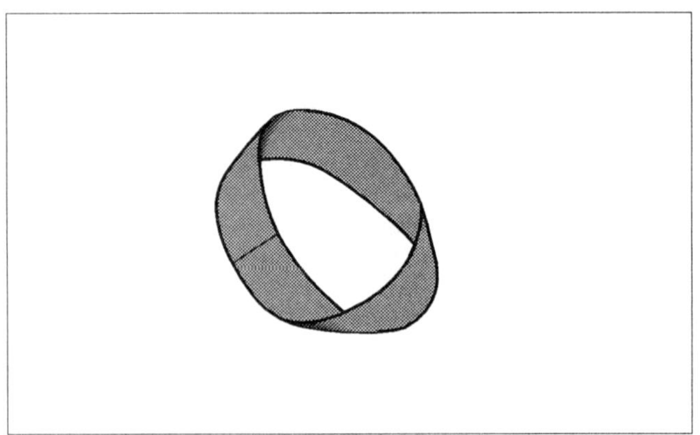

Topology, loop.

indispensable in understanding such notions as defects, vortices, and solitons.

TORQUE

According to **Isaac Newton**, an object at rest will remain at rest, and an object will remain in **motion** unless acted upon by an outside **force**. A force, therefore, is what causes any object to move. Any force that causes an object to rotate, turn, or twist is called a torque. Torque is equal to the amount of force being exerted on the object times the object's rotation point to the location where the force is being applied on the object.

Seesaws are a good example of torque. Many people have had the experience of a large person sitting on one end of the seesaw, and a small person on the other end. If the larger person is sitting closer to the pivot point of the seesaw, the smaller person can lift them with little or no problem. The reason this is possible comes from the difference in torque experienced by each person. Even though the smaller person exerts a smaller force, his distance from the pivot point (lever arm) is longer, hence a large torque. The larger person exerts a larger force; however, because he is closer to the pivot point, his lever arm is shorter, hence a smaller torque.

Wrenches also work by torque. (Some wrenches are even calibrated to display the amount of torque you are applying to a nut; they are called torque wrenches.) The nut (or bolt) is the point of rotation because we either want to tighten or loosen it by having it turn. The force is being exerted by your hand and arm. Since we try to pull or push (exert a force) at right angles on the wrench's handle, the lever arm is then the length of the wrench's handle. To increase the torque on the nut, we must either increase how hard we pull or push on the lever arm; or, increase the length of the lever arm by placing a pipe over the end of the wrench.

TORRICELLI, EVANGELISTA (1608–1647)
Italian physicist

Born near Ravenna, Torricelli was first educated in local Jesuit schools and showed such brilliance that he was sent to Rome to study with Galileo's former student Benedetto Castelli (1578–1643). Through Castelli he first corresponded with and met **Galileo**, finally becoming his secretary and assistant. A few months after Galileo's death in 1642, Torricelli accepted Galileo's old position as court mathematician and philosopher to the Grand Duke of Tuscany, a position he held until his own death, before his fortieth birthday.

As a scientist Torricelli became well known for his study of the **motion** of **fluids** and was declared the father of hydrodynamics by **Ernst Mach**. Torricelli also conducted experiments on what we now call gases, though the term was not then in use. Most notably, Torricelli settled an argument about the nature of gases and the existence of the **vacuum**. **Aristotle** believed that a vacuum could not exist. Though Galileo disagreed, he contended that the action of suction (in a water pump, for example) was produced by a vacuum itself and not by the **pressure** of the air pushing on the liquid being pumped. Despite his argument, Torricelli noticed that water could be pumped only a finite distance through a vertical tube before it ceased to move any further and set out to examine this paradox, inventing the first barometer in the process.

Torricelli first filled a one-ended glass tube with mercury, then immersed this open end in a dish of more mercury, placing the tube in a upright position. He found that about thirty inches of mercury remained in the tube, deducing that a vacuum had been created above the mercury in the tube, and that the mercury was held in place not by the vacuum, but by the pressure of air pushing down the mercury in the dish. Thus he demonstrated the existence of a vacuum, showed why pumps then in use could only move liquids vertically a certain distance (the distance determined by the pressure of the surrounding air), and created an instrument capable of measuring air pressure.

Torricelli's invention of the barometer led to a burst of both theoretical and experimental work in physics and **meteorology**. Torricelli also made a contribution to meteorology with his suggestion that wind was not caused by the "exhalations" of vapors from a damp earth, but by differences in the **density** of air which in turn were caused by differences in the air **temperature**.

Torricelli's investigations in mathematics played an important role in scientific history as well. Based on Francesco Cavalieri's "geometry of indivisibles," Torricelli worked out equations upon curves, solids, and their rotations, helping to bridge the gap between Greek geometry and calculus. Along with the work of Rene Descartes, Pierre de Fermat, Gilles Personne de Roberval, and others, these works enabled **Isaac Newton** and **Gottfried Wilhelm Leibniz** to give calculus its first complete formulation. Though not as great a scientist as his older contemporary, Galileo, Torricelli continued the tradition of Italian scientific pioneering. This tradition was not to last long after his own death in the middle of the seventeenth century, however; by the beginning of the next century the center of scientific progress had shifted to northern Europe.

TOWNES, CHARLES H. (1915-　　)

American physicist

Charles H. Townes was awarded a share of the 1964 Nobel Prize in physics for his discovery in 1951 of the maser, a device that can amplify **microwaves** for practical applications. About six years later, Townes speculated on the possibility of building a maser-like instrument using solid crystals instead of gases. A device of this kind—the laser —was actually constructed two years afterwards by Theodore Maiman.

Charles Hard Townes was born in Greenville, South Carolina, on July 28, 1915. His father was Henry Keith Townes, an attorney, and his mother was the former Ellen Sumter Hard. Townes grew up in an intellectually stimulating environment in which both parents had an avid interest in natural history. He later told Shirley Thomas in a sketch for the book *Men of Space* that "there was always an inclination toward science" in his family. He was convinced that had the opportunity been available, his father "would have become a very excellent scientist."

Townes's genius was obvious early on. His parents allowed him to skip seventh grade, and by age 16 he was ready to enter Furman University in his hometown of Greenville. Although he planned to major in science, he also took a full schedule of language classes. As a result, he was able to graduate in 1935 with a B.S. in physics and a B.A. in modern languages. He continued his mastery of French, German, Italian, Russian, and Spanish throughout his life. In addition to his demanding class work at Furman, Townes was also curator of the college museum and a member of the band, glee club, swimming team, and newspaper staff.

Townes entered Duke University in the fall of 1935 to work on his master's degree. In addition to his thesis research on van der Graaf **generators**, he continued his study of French, Italian, and Russian. Having completed his work at Duke (various biographers credit him with either an M.A. or an M.S., awarded in either 1936 or 1937), Townes headed for the California Institute of Technology (Cal Tech) in Pasadena, California. There he completed his doctoral research on the spin of the carbon–13 **nucleus** in 1939 and was awarded his Ph.D.

War Research Leads to Interest in Microwaves

An offer from Bell Laboratories to pay Townes a salary of $3,016 a year "astonished" him, according to Mary Ann Harrell in *Those Inventive Americans*. He could scarcely believe that a career in physics would be "so highly paid," and he accepted the Bell offer quickly. Townes found an apartment in New York City and began to take full advantage of the city's cultural opportunities. He took evening classes at the Julliard School of Music and, according to Harrell, "changed apartments every three months to explore the city thoroughly."

For most of his time at Bell, Townes worked on projects related to national defense. In particular, he was involved in the development of **radar** systems. At one point he warned the army against using a three-centimeter band radar wave detection system because he knew that water molecules absorb in this range and that the system would, therefore, be ineffective. Having ignored this advice, the Army built 1.2-in (3-cm) radar devices anyway, only to find that they would not work.

As the pressures of war research receded, Townes turned his attention to a problem many scientists and engineers were thinking about: finding a way to amplify microwaves so that they could be used in practical applications, as were **radio** and radar **waves**. **Albert Einstein** had outlined a theoretical method for accomplishing this goal in 1917. He suggested using individual atoms as resonators rather than using macroscopic-sized objects. The problem was that, by 1950, no one had found a way to build a working device based on Einstein's principle.

Townes's breakthrough in this area came while he was sitting on a park bench in Washington D.C.'s Franklin Square in 1951, waiting for a restaurant to open so that he could have breakfast. As Harrell tells the story, "six years of work suddenly blossomed into insight: a way to make ammonia molecules amplify... microwaves by the process Einstein had outlined." In classic storybook fashion, Townes outlined his ideas on the back of an old envelope he found in his pocket.

Producing a device that actually worked according to his theory, however, was no simple matter. Townes worked with two colleagues, James P. Gordon and H. J. Zeiger, for more than two years before they had success. Finally, in late 1953, the three produced a successful model of Townes's Franklin Square idea, a device that amplified an incoming microwave beam while maintaining the signal wave's phase. They gave to their invention the name *maser* (for microwave amplification by stimulated emission of **radiation**). For his discovery of the maser, Townes was awarded a share of the 1964 Nobel Prize in physics. He shared that prize with two Russian scientists, N. G. Basov and Aleksandr Prokhorov, who had independently and somewhat earlier come up with a similar method for using the Einstein principle to build a maser-like device, which they called a "molecular generator."

Outlines Laser Theory

By the time Townes announced the first working maser in 1954, he had already been at Columbia University for six years, having been appointed full professor there in 1954. He continued to teach and do research at Columbia until 1961, when he resigned to become professor of physics and provost at the Massachusetts Institute of Technology (MIT). While still at Columbia, he developed a theory with his brother-in-law Arthur L. Schawlow that described a method by which a maser-like device could be built that would operate with visible **light** instead of microwaves. Theodore Maiman's laser provided proof of Townes and Schawlow's theory some two years later.

In 1966 Townes left MIT to become University Professor of Physics at the University of California at Berkeley, where he remained until his retirement in 1986. Townes was married on May 4, 1941, to the former Francis H. Brown, and they raised four children. Townes has received many honors and awards in addition to his 1964 Nobel Prize in Physics. These include the

Liebmann Memorial Prize of the Institute of Radio Engineers, the Sarnoff Award of the Institute of Electrical Engineers, the Comstock Medal of the National Academy of Sciences, the Ballantine Medal of the Franklin Institute, the Distinguished Public Service Medal of the National Aeronautics and Space Administration, and the National Medal of Science from the National Science Foundation.

TRANSFORMERS

Transformers are responsible for transforming an input **voltage** into a different output voltage. When a portable CD player which runs on 3.5V is plugged into a car which has a 12V battery, it is a transformer that changes the 12V into 3.5V. When virtually any electronic device is plugged into the wall, there is a transformer somewhere in the chassis of the device that turns the 110VAC that is supplied to the home into something that can be used by the equipment.

The physical components of a transformer are simple and have changed little since first introduced. The primary coil of wire and a secondary coil of wire are electrically insulated from each other. They usually share a core of iron or some other material that is easily permeable by a magnetic field, and are often wound one on top of the other so that they are concentric.

The entirety of the physics of transformers is contained in Faraday's Law, which states that a changing magnetic field produces a voltage in a conductor that the magnetic field penetrates. In the case of a transformer, current runs through the primary coil, producing a magnetic field which extends through the secondary coil. Because a changing magnetic field is necessary to produce a voltage, transformers only function when supplied with an AC signal.

Using a different number of turns for the primary and secondary coils, any input voltage can be turned into any output voltage, subject to the restriction that the output voltage is proportional to the input voltage with a proportionality constant equal to the ratio of the number of primary coil turns to the number of secondary coil turns. This may seem incompatible with the law of **conservation of energy**, but is compatible because the **power** transmitted remains constant.

A transformer that takes an input voltage and outputs a higher voltage is called a step-up transformer, while a transformer that reduces the voltage is called a step-down transformer. Using a combination of step-up and step-down transformers, power is transmitted to homes, turned into one voltage level for the entire home, and then transformed again in each device that requires **electricity**.

TRANSIENT ENERGY

Transient **energy** is energy that exists only during a transfer process from one stored energy to another. There are two forms of transient energy: **heat** and **work**. These cannot be stored like other forms of energy, such as **potential energy**; i.e., a hot object does not, by itself, stay hot. Its heat will flow away sponta-

neously. If a book is lifted from the floor to the table, work is being done to move it; but once the book is placed on the table, all of the work has been converted to potential energy.

Work is an energy transfer due to a potential difference other than heat. It can be observed in many different forms, such as mechanical work, expansion and compression work, and work done on charges in a magnetic or electric field. Mechanical work done by a constant **force**, F, acting through a distance, d, is a scalar derived from a dot product. This means that it is the magnitude of the force times the magnitude of the distance times the cosine of the angle in between the force and distance.

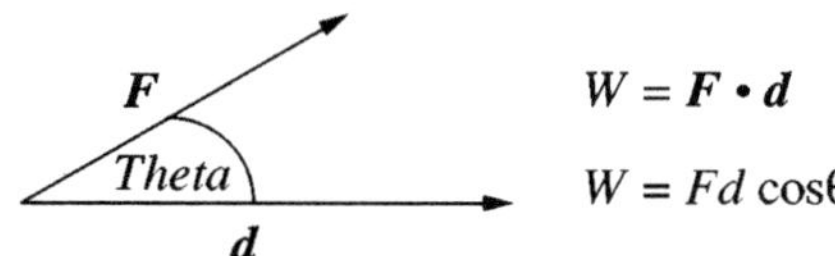

Work done by a variant force is a line integral from x_1 to x_2.

$$W = \int_{x_2}^{x_1} F_x dx$$

Work done during the expansion and compression of a gas is related to the change in **pressure** and volume. It is equal to the area under the curve of a PV diagram, so it is path dependent. This is referred to as isothermal work since the **temperature** is held constant. It is also possible to do isobaric work where the pressure is held constant.

$$W = \int_{V_2}^{V_1} P_V dV$$

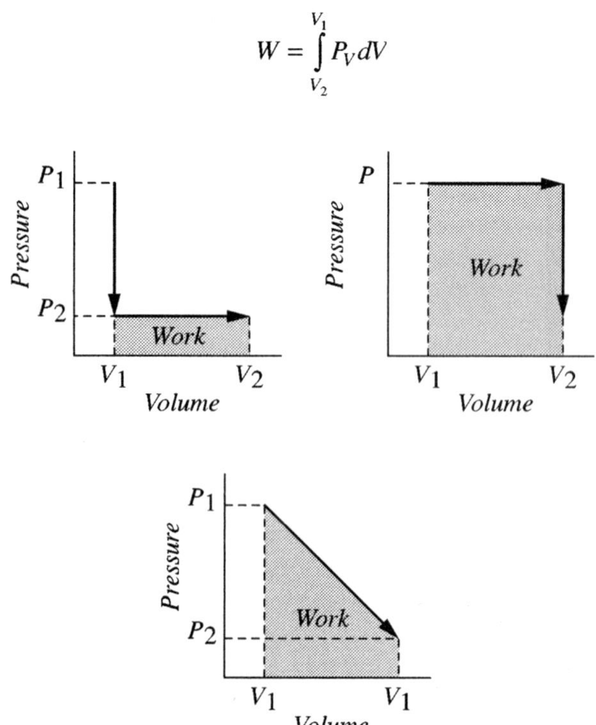

A moving charge in a magnetic field will feel a deflecting force perpendicular to the field and its line of **motion**. This force, F, is dependent on the particle's **velocity** and the magnetic field through which it moves. The work done by the field on the particle is the force carried out through a distance, as in mechanical work. Since an electric field exerts a force on a charged particle, it too can do work.

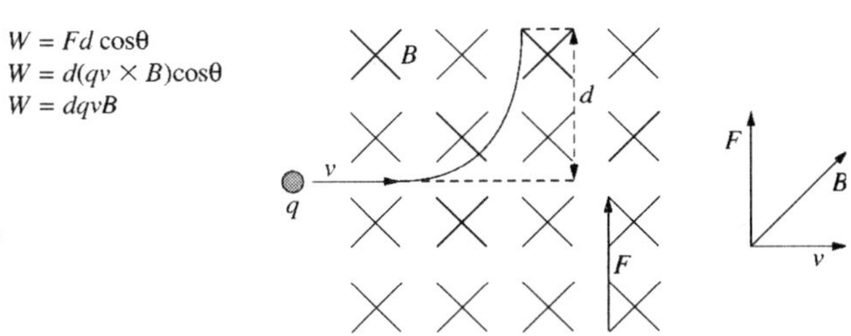

$$W = Fd \cos\theta$$
$$W = d(qv \times B)\cos\theta$$
$$W = dqvB$$

The other form of transient energy is heat, or **thermal energy**. This transient energy allows a system that is governed by temperature differences to produce work. Heat processes are governed by the laws of **thermodynamics**. The **first law of thermodynamics** states that the change in the internal energy of a system is the heat added minus the work done: that heat is equivalent to work and vice versa. The **second law of thermodynamics** states that heat cannot produce an amount of work energy exactly equivalent to the heat energy it absorbed. These two laws describe the **conservation of energy** that is one of the primary defining characteristics of this **universe**.

TRANSISTORS

A transistor, broadly speaking, is a solid-state switch that allows a small signal to control a large signal, such as a current flow. The first transistor was invented by **John Bardeen**, Walter Brattain, and William Shockley in 1945. They were awarded the Nobel Prize for physics in 1956. That transistor was a point-contact transistor that is no longer used today. Since that time, transistors of all types have become crucial for modern **electronics** to function.

A transistor has three terminals. Current flows between two of them while the third one controls the current. In **semiconductors**, there are two charge carriers. One is the **electron**, which carries negative charges. The other is a hole, which carries positive charges. The amount of electrons and holes is often unequal. Consequently, the carrier which has a larger amount is called a majority carrier and the other is called a minority carrier. Depending on what carriers produce the current, transistors can be classified as bipolar or unipolar transistors.

Bipolar Transistor

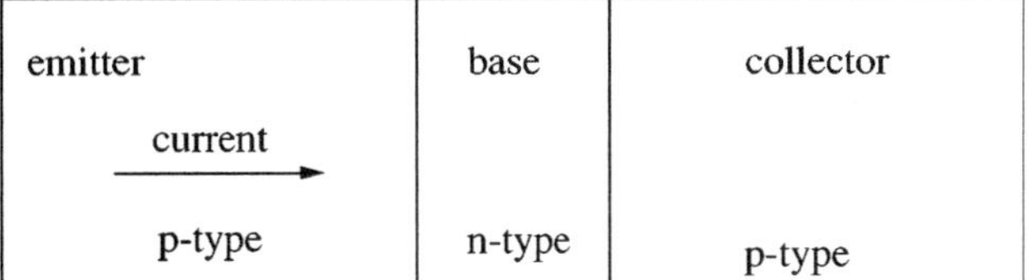

Unipolar Transistor (FET)

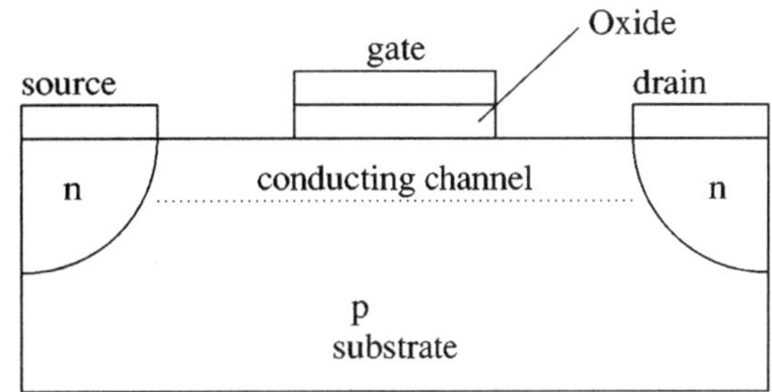

In bipolar transistors, the current is carried by both majority and minority carriers. The three terminals are the emitter, the collector, and the base. Two structures are possible, p-n-p and n-p-n. "p" indicates a type of semiconductor where the majority carriers are holes (positive), while "n" indicates a type of semiconductor where the majority carriers are electrons (negative). In the p-n-p type of transistors, a thin layer of n-type semiconductor is sandwiched between two p-type semiconductor layers. The n-type layer acts as the base. The two p-type layers serve as the emitter and the collector. The n-p-n type of transistor is similar in structure, but reverses the sequence of the layers.

In unipolar transistors, only the majority carriers carry the current. The field-effect transistor (FET) is an example of a unipolar transistor. The three terminals are the source, the drain, and the gate. The region between the source and drain is called a substrate. FETs are used widely in today's electronic industry. Depending on whether there is an insulation layer between the gate and the substrate, there are junction FETs (JFET) or insulated gate FETs (IGFET). Oxide is often used as the insulation material, so IGFET is often called MOSFET (Metal-Oxide-Semiconductor Field Effect Transistor). The MOSFET built using silicon is the heart of today's computer chip.

When building integrated circuits based on millions of transistors, many issues need to be taken into account. One is the **power** consumption. The type of transistors used should not consume too much power or the integrated circuit will become too hot to use. In making integrated circuits, often what matters is the group behavior of many transistors. Some types of transistors behave well when only several are used in a circuit, but not as well when millions of them are put together. These kinds of transistors will not be used because they do not have good scalability. FETs built on silicon turn out to be very good, which is why silicon devices are widely used today.

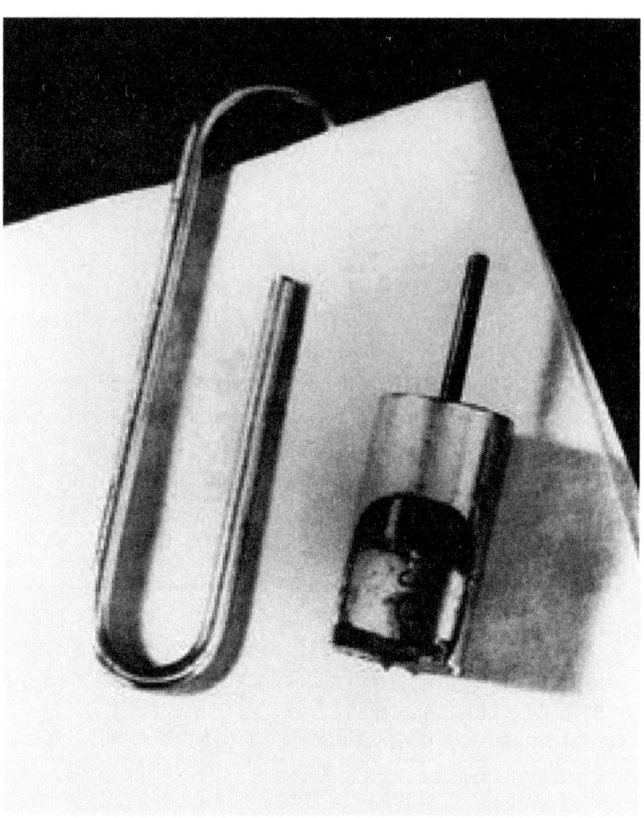

Transistor.

Part of the reason **computers** are becoming faster and smaller today is that the underlying transistors can be made faster and smaller. The fastest CPU for personal computers available in the market today has about 1 GHz **frequency**. There is certainly some limit on how small the semiconductor transistors can be made. An **atom** has a size of about 0.1 nm, which is about one billionth of a foot. The transistor should be at least the size of several layers of atoms, or about one nm. Today's technology labs have reached thicknesses of around 50 nm. As they reach the lower limit, the physics of making transistors will change dramatically.

There are two major complications in making transistors smaller and faster. Nonlinear effects can become important. In a smaller device, electrons will accelerate to high speed and then experience a collision and return to a slower average speed. This effect is called **velocity** overshoot, and it is a nonlinear "hot electron" effect. Another complication is the change of fundamental physics. When the transistor gets even smaller, quantum effects become important. They can manifest in transistors of 50 nm size, which is smaller than what can be seen with the naked eye, but is still much larger than an atom. The nanoscale device is an active research area today as technology enters this region.

See also Nanotechnology

TRANSVERSE AND LONGITUDINAL WAVES

In wave physics, a wave is defined as a periodic disturbance in a medium or in **space**. A single **frequency** traveling wave can take the mathematical form of a sine wave, whose **velocity** of propagation can be expressed by: $v = f/\lambda$ where f is the frequency of the wave and is equal to the reciprocal of its period (1/T), and where λ is the **wavelength**. The period is the time required for a point on the wave to make a complete oscillation through the axis of propagation. The crest of a wave is the highest point that it can reach, while the trough of the wave is the lowest point. These are referred to as the maximum and minimum amplitudes, or displacement, of the wave. **Waves** may be classified on the basis of the direction of movement of the individual particles of the medium relative to the direction in which the waves travel. This leads to distinguishing between transverse and longitudinal waves.

A transverse wave is one in which all particles of the medium oscillate along paths perpendicular to the direction of the wave's advance. Transverse waves require a relatively rigid medium in order to transmit their **energy**. Waves propagating on the surface of liquids such as water, and propagating through solids, such as secondary seismic waves, are examples of transverse waves. Transverse waves are mathematically represented by sine or cosine functions because the amplitude of any point on the curve is a function of the sine or the cosine of an angle. Transverse waves obey the $v = f/\lambda$ relationship. A special case is electromagnetic **radiation** which can be considered as a form of radiant energy propagated as an electromagnetic transverse wave traveling at the same velocity (c) through a **vacuum** (2.99792458×10^8 m/sec). The elctric and magnetic fields vibrate perpendicular to the direction of propagation.

A longitudinal wave is one in which the particles of the medium move in a direction parallel to the direction in which the wave moves. Longitudinal waves can propagate either through liquids or solids, such as primary seismic waves, and they also obey the $v = f/\lambda$ relationship. **Sound** waves in air are also longitudinal waves. Such waves cause a sinusoidal **pressure** variation in the air. The air **motion** resulting from the passage of the sound wave occurs back and forth in the direction of the propagation of the sound, a characteristic of longitudinal waves, in which each particle of **matter** vibrates about its normal rest position and along the axis of propagation. All the other particles of the medium participating in the wave propagation behave in the same way, except that there is a progressive change in the vibration phase. The combined motions result in the advance of alternating regions of compression and rarefaction in the direction of propagation. The crest and trough of a transverse wave correspond respectively to the compression and rarefaction regions. Compression occurs when the particles of the medium are closer together, i.e., when their **density** is greatest, and rarefaction occurs when these particles are further apart than is normal, or when their density is decreased.

See also Acoustics; Diffraction; Electromagnetic spectrum; Electromagnetic waves; Harmonic motion; Nodes; Oscillations; Particle-wave duality; Polarization of light; Reflection, refraction, and dispersion; Standing waves and resonance; Vibrating systems and resonance; Wave interference; Wave phenomena

TRIODE

A triode is an electrical circuit component that uses a small **voltage** to control a relatively large flow of current. A triode contains three electrodes, a cathode, an **anode**, and a control grid. The name triode is derived from the term tri-electrode (i.e., three electrodes).

The triode is actually a variation on the diode, a device that has just two electrodes sealed in **vacuum**. One electrode, the cathode, is attached to a negative **electric potential**, the other electrode, the anode, is attached to a positive potential. This device allows current to pass through it in one direction only.

Under normal operating conditions, the cathode is heated to high **temperature** and the anode remain cooler. Current flow through the device can be increased by a further increase in the temperature of the cathode, by increasing the potential difference between the cathode and anode, or by shortening the physical distance between the cathode and anode.

With a triode, additional control over current flow is achieved by introducing a third electrode between the cathode and anode. Usually this electrode is placed much closer to the cathode than the anode and, instead of being a small metal plate like the anode and cathode, is a grid with holes. The additional control over current flow is caused by the fact that the electrons streaming from the hot cathode must pass through the grid (usually made of a wire mesh) on their journey to the positive anode.

The addition of the grid, hence the creation of a triode, completely revolutionizes the diode's characteristics. The grid is a powerful controller of the **electron** stream flowing from the cathode to the anode. If it is slightly negatively charged it entirely chokes off the flow of electrons since they are sensitive to the repulsive **force** from the grid. Also it only needs a small negative potential to cause this effect. More importantly, it is interesting to consider the effect of a small positive charge on the grid. The cloud of electrons gathered around the hot cathode are repelled by the negative charge of the cathode but they are also powerfully attracted to the positive charge of the grid which is very close to them. They speed towards this positive charge and their **momentum** carries them through the holes in the grid and towards the strong positive potential of the anode. A very small increase in the positive charge on the grid brings about a disproportionately large increase in the electron stream flowing from cathode to anode, and large current changes are accomplished with small voltage changes.

These properties of triodes allowed them to be used in **radio** amplification circuits. A small vibrating electrical signal applied to the grid of a triode will translate into a large signal output from the triode. Accordingly, a small input signal is magnified, or amplified.

A similar effect can be produced in a modern version of the triode, the n-p-n or p-n-p transistor. In either case, a small variation in voltage to the core of the triode causes a large response in potential across the outside electrodes. This mirrors the effect of a small variation on the triode grid producing a large response across cathode and anode. Because of their robustness and the fact they can be produced in extremely small sizes **transistors** have virtually supplanted triodes in amplifier circuit design.

See also Cathode and cathode rays; Diodes; Electric charge; Electric circuit; Electric field and forces; Electricity

TUNNELING

Tunneling, also known as the tunnel effect, is a quantum mechanical phenomenon by which a tiny particle can penetrate a barrier that it could not, by any classical or obvious means, pass. Though seemingly miraculous, the effect does have some intuitive characteristics. For instance, thin barriers allow more particles to tunnel than do thick ones, and low barriers permit more tunneling than do high ones.

Tunneling does not generally show itself in the macroscopic world. It only starts to become a factor for microscopic items. Atoms can tunnel, as can electrons, but things such as tennis balls and grapes, easily seen with the naked eye, will not. For microscopic particles, the barrier heights are described in terms of **energy** instead of distance, but for conceptual purposes there is little difference.

It is important to note that the effect can only be understood with the aid of quantum **mechanics**. Classical mechanics, the system pioneered by **Isaac Newton** that stood unchallenged until the early twentieth century, has no way of explaining tunneling. In the Newtonian philosophy, all particles, even the tiniest of microscopic particles, can be located precisely. Any uncertainty is seen as the result of an imperfect measuring device or a sloppy scientist. In addition, each microscopic particle is considered to be like a tiny pebble and there can be nothing wave-like about it.

The quantum view of the **universe** is fundamentally different from the Newtonian view. Each particle is said to have both corpuscular (pebble-like) *and* wave-like properties. Furthermore, a quantum particle cannot, in general, be located precisely. It has a built-in uncertainty that cannot be taken away by the best of measuring instruments. For these reasons, a particle in **quantum mechanics** is often treated as a **wave function**. This is another way of saying that the particle is akin to a small bundle of **waves**.

Representing a quantum particle as such a bundle has two advantages. For one, it reveals that the particle is, in some sense, blurry and can never be exactly pinned down. It exists over a range of **space**, not a specific point. For another, the wave function format allows particles to exhibit wave-like properties. Tunneling has one of these wave-like properties.

By the more general name of "barrier penetration," it is a well-established characteristic of waves. **Light** waves, for instance, have long been observed to overcome daunting optical barriers.

In a fundamental sense, the quantum mechanical explanation of tunneling can be illustrated by an analogy. If we roll a ball very slowly toward, say, a cement speed bump, we confidently say that it will not surpass the barrier and predict that it will roll back toward us. However, if a very modest ocean swell approaches an offshore sandbar, we are not as sure of the results and rightly so. The character of the wave, even if diminished in size, often makes its way past the sandbar. Similarly, treating a microscopic particle with the mathematical model of a ball clearly tells us tunneling is impossible, while using the mathematical model of a wave just as clearly states that particles will always have a chance to tunnel.

Some **scattering** experiments of the early twentieth century paved the way for tunneling. **Ernest Rutherford**, the pioneer of the scattering method, came across a paradox in a series of uranium experiments in 1910. Scattering involves the probing of a microscopic object by bombarding it with other particles and then keenly observing how the particles are, literally, scattered by the object in question. Specifically, Rutherford tried to pinpoint where the bombarding particles were scattered and how fast they were moving.

An example using tennis balls gives one a simple and accurate picture of a scattering experiment. Say a statue exists in a dark courtyard and we want information about the statue without being able to see it. Mainly, we want to know what its shape is (e.g., a horse or a person) and from what materials it is made (e.g., granite or clay). To learn more about the statue, we stand at the edge of the courtyard and toss tennis balls into the dark area, hoping to hit the statue. After many thousands of tennis balls, we've learned a lot. If only a few of the tennis balls have hit anything, we know the statue is probably quite small. Also, the directions that tennis balls were scattered are important. If most of them come straight back to us, we can ascertain that the statue has a large, flat front like a wall. Moreover, by observing the speed of the scattered tennis balls, we can get an idea of how hard the statue is. Atomic scattering set a similar scene for Rutherford, but he was interested in the nuclei of atoms and he preferred alpha particles to tennis balls.

Through a series of remarkable scattering experiments, Rutherford, working in conjunction with his students **Hans Geiger** and Ernest Marsden, had accurately mapped the insides, or **nucleus**, of the uranium **atom**. Uranium was of great interest at the time because of its radioactive properties. The summation of Rutherford's results for uranium came in the form of a potential diagram. Potential refers to electric **potential energy**, so his diagram mapped out an energy barrier. Any particle that escaped the nucleus (i.e., any particle of nuclear **radiation**) would have to overcome this energy obstacle before it left. Rutherford estimated the top of this barrier to be at least about nine units of energy high. However, when observing the occasional radiated **alpha particle**, he found that it had only four units of energy. The paradox was born. How could a particle with so little energy overcome a barrier of such height? Such a problem can be compared to a man walking next to a huge baseball stadium and suddenly seeing a baseball floating toward him, as if gently tossed. Surely any ball hit out of the stadium would have been moving fast enough to at least sting his palm.

The question lingered for 18 years until **George Gamow**, assisted by his colleagues Edward Uhler Condon and Ronald Gurney, proposed a solution. In 1928, quantum mechanics was gaining credibility, and the three physicists performed a relatively simple calculation, treating the alpha particle as a quantum mechanical wave function. In essence, they analyzed the problem from the viewpoint that an alpha particle was not located precisely at any given spot, but rather that its existence was spread out, like a wave. Their explanation proposed that the alpha particles tunneled out of uranium's energy barrier, and it fit Rutherford's observations perfectly. The acceptance and practical application of tunneling theory had begun.

One of the first applications of tunneling was an **atomic clock** based on the tunneling **frequency** of the nitrogen atom in an ammonia (chemical formula NH_3) molecule. The rate at which the nitrogen atom tunnels back and forth across the energy barrier presented by the hydrogen atoms is so reliable and so easily measured that it was used as the timing mechanism in one of the earliest atomic clocks.

A current and quickly advancing application of tunneling is Scanning Tunneling Microscopy (abbreviated STM). This technique can render high-resolution images, including individual atoms, that accurately map the surface of a material. As with many high-tech tools, its operation is fairly simple in principle, while its actual construction is quite challenging.

The working part of a tunneling **microscope** is an incredibly sharp metal tip. This tip is electrically charged and held near the surface of an object (known as the sample) that is to be imaged. The energy barrier in this case is the gap between the tip and the sample. When the tip gets sufficiently close to the sample surface, the energy barrier becomes thin enough that a noticeable number of electrons begin to tunnel from the tip to the object. Classically, the technique could never work because the electrons would not pass from the tip to the sample until the two actually touched. The number of tunneling electrons, measured by incredibly sensitive equipment, can eventually yield enough information to create a picture of the sample surface.

Another application of tunneling has resulted in the tunnel diode. The tunnel diode is a small electronic switch and, by incorporating **electron** tunneling, it can process electronic signals much faster than any ordinary physical switch. At peak performance, it can switch on and then off again ten billion times in a single second.

TURBULENCE

Turbulence is the formation of eddies in a fluid (liquid or gas). It is produced whenever a fluid—under certain conditions—is in contact with a solid and there is relative **motion** between them; for example: when wind flows past a building or past a mountain; when the ocean flows past an island; when a baseball flies by; when a jet plane moves in the stratosphere; or when a river flows past a bridge pier. In all these cases, eddies form behind the obstacle (i.e., "downstream"), and eventually are carried away by the main body of fluid.

Turbulence has long been observed, but its scientific study began with the work of William John Macquorn Rankine (1820-1872); later, Osborne Reynolds (1842-1912) defined the number bearing his name, and Ludwig Prandtl (1875-1953) put forth the limiting-layer hypothesis.

Today, the study of turbulence, experimental and theoretical, continues; but an agreement between both approaches is still in the future.

There is a number—called Reynolds number—whose values indicate clearly whether the motion of a fluid in a certain region is turbulent or not. It is defined as: Whether an obstacle carries any eddies, and whether these are released into the flow downstream depends upon the speed of the incoming fluid, the size of the obstacle and the internal **friction (viscosity)** of the fluid, according to Karen J. Heywood, writing in *Physics Education*. Just like a solid body, a parcel of liquid or gas has **mass**, and therefore **inertia**. Inertial **force** is the amount of force required to stop a body that is moving along steadily with its own inertia (for example, the force to stop a charging rhino moving towards you would be greater than that needed to stop a hummingbird at the same speed). The inertial force necessary to stop a parcel of water that occupies a unit volume ($1 \ m^3$) is proportional to the square of its speed (v^2) divided by a length typical of the obstacle (d)—say, the diameter of a stone in a river. All **fluids**, as they flow, present friction between their different parts. This property is called viscosity. Liquids are more viscous than gases, and, among liquids, corn syrup is much more viscous than water. The viscous force (internal friction) working on an object of diameter d moving through a fluid at speed v is proportional to $h.d/d\ 2$, where h is the viscosity coefficient of the fluid. Dividing the expression for the inertial force by that for the frictional force, we obtain R as given by the above equation.

If we place an obstacle—e.g., a sphere-at-rest in the midst of a fluid stream—our physical intuition suggests that the part of the fluid which is really in contact with the obstacle must be at rest. However, photographs taken in different laboratories appeared not to support this idea. In order to explain what was happening, Ludwig Prandtl (1875-1953) introduced the hypothesis of the "limiting layer," according to which in the immediate neighborhood of the obstacle there is a very thin layer of fluid whose **velocity** parallel to the surface of the object brows very rapidly, from zero at the surface itself to the velocity of the main body of fluid far from the object. This limiting layer is very thin upstream, but broadens downstream, i.e., behind the obstacle.

Inside the downstream limit layer the fluid begins to move backwards and in circles, until the eddies form when the Reynolds number is $R=5$. As the fluid velocity grows, bringing R to a value of 70, the limit layer broadens still more

downstream, forming what is called a "von Karman vortex street" after Theodor von Karman (born 1881). These eddies finally leave the vicinity of the obstacle and float away with the main fluid current. If R keeps increasing to values of 1000 or 2500, eddies become more frequent and the vortex street broadens still more, finally breaking up and forming a "turbulent wake." At this stage the motion of fluid particles is chaotic and varies in time.

TWIN PARADOX

The twin paradox is a thought experiment that illustrates a consequence of Einstein's theory of special relativity. The theory of special relativity makes two fundamental claims: The speed of **light**, c, is a constant and never changes; and, the laws of physics must be identical for all observers who are either stationary or in constant **velocity motion**. If one sits inside a windowless van that rides along at a constant speed, it is impossible to tell if one is moving or standing still. Any physics experiments that are conducted in the van would have the same results whether the van speeds along or sits in a parking lot.

An effect of a fixed **speed of light** is that measurements of time and **space** vary depending on the motion of the observer. Since most observers generally move much slower than that of light, the effects of special relativity exist but are negligible. As an observer's speed approaches the speed of light, however, time and space change from the position of the observer.

One application of **time dilation** is the twin paradox thought experiment. Twins Larry and Arthur separate. Larry goes for a ride in his rocket ship, while Arthur remains behind on Earth. Larry's spaceship travels very quickly—close to the speed of light. Since Larry is moving at such a high speed, the passage of time on the watch he wears is slower than that of the watch worn by Arthur. When Larry returns to Earth, his brother Arthur has grown very old, while he has aged very little in comparison.

As mentioned above, a postulate of the special relativity is the relativity of uniform motion. In other words, time dilation should be symmetrical because neither Arthur nor Larry is in a privileged **frame of reference**. From Arthur's perspective, Larry travels at a high speed, so his clock must be the one that slows down. But because Larry travels at a constant speed, he can rightfully claim that Arthur is the one moving, so Arthur's clock must slow down. So why is Arthur much older than Larry when Larry returns to Earth?

The reason is that Arthur and Larry were not in constant motion relative to each other. Larry's rocket ship turns around midway through his journey. This introduces a third reference frame into the thought experiment which breaks the symmetry between the **reference frames** of Larry and Arthur. It is true, that from Larry's perspective, Arthur will age more slowly on the outward and return legs of his trip. But there is a gap in Arthur's timeline which Larry doesn't see as he changes direction, and thus changes reference frames, at the midway point.

This gap accounts for Arthur's advanced age when Larry returns to Earth.

See also Space-time; Special relativity

TWO-DIMENSIONAL MOTION

In **mechanics**, **motion** in one dimension is defined as a change of position of an object relative to another object or with respect to a **frame of reference**. Motion in two-dimensional **space** involves the same general concepts and relationships between the position, **velocity** and **acceleration** of a moving object in one dimension simply extended to two dimensions. A classic illustration is provided by free-falling objects.

Italian astronomer and physicist **Galileo** Galilei's (1564-1642) study of free falling objects provided the background for English physicist and mathematician Sir Isaac Newton's (1642-1727) subsequent development of his laws of motion. Free-fall in one dimension occurs in a Cartesian plane along the vertical axis. In a two-dimensional frame of reference, motion along both horizontal and vertical axes are combined into a characteristic parabolic shape and the vertical and horizontal components of the motion are independent from each other. The dependent variables of the object (e.g., position, velocity, acceleration) can be expressed as a function of the independent variable (e.g., **time**). Similarly, the rate of change of a dependent variable can be expressed with respect to the independent variable. An example is displacement: if at time t_1 an object is at position P_1 and if it moves to position P_2 at a time t_2, the displacement (d) in the time interval $t_2 - t_1$ is equal to $P_2 - P_1$. This is usually expressed as the displacement vector Δr $(r_2 - r_1)$ when the displacement occurs in more than one dimension with r being used to represent the position vector. Similarly, the velocity is expressed as $v = \Delta r/\delta t$.

There is an important distinction between the graph of y as a function of an independent variable x, and a curve traced by the motion of an object in two-dimensional space, (i.e., a plot of r *vs* t). In the first graph, the value of y is determined by the value of x. In the second graph, the vector r of the object is a function of time. The fact that the position vector has an x and a y component connects the x and y values together, thus they are not independent from each other. If the vector r is a function of time, then for every value of time t, there is a corresponding value of vector r. But if vector r is expressed in terms of its component **vectors**, the unit vectors are constant and independent of time and it is only the x and y components of vector r which are functions of time.

The variables x and y can each be related to time by plotting them independently as a function of time on separate graphs. Plotting the path of a moving particle in such a way is called a parametric representation of the particle path and the functions relating x to time and y to time are called parametric equations. Parametric equations can be applied, for example, to the case of a particle whose horizontal position changes with time according to a specific equation and whose vertical position changes with time according to a different equation. Similarly, velocity and acceleration can be described in two dimensions. Thus, to say that an object or particle has constant acceleration in two dimensions is equivalent to stating that the acceleration is constant in both magnitude and direction. Important generalizations of **Newtonian physics** which apply to two-dimensional motion are that horizontal motion is independent from vertical motion, and that forces have no perpendicular effects. For example, when a **force** acts on an object, it has no effect on the part of the object's motion that is perpendicular to the force. This is illustrated by the example of a man walking at a fast pace and dropping a ball while continuing to walk. It can be verified that the ball will not fall behind the man, but right at his feet. This leads to the extension of **Newton's laws of motion** to two dimensions, the first being expressed as follows: if the two components of the total force acting on an object are zero, then the object will continue in the same state of motion. Similarly, the second law can be simply stated as: $a_x = F_{x, total}/m$ and $a_y = F_{y, total}/m$, while the third can be expressed as: if two objects interact via forces, then the x and y components of their forces on each other are equal and opposite. In physics, two-dimensional motion is used to describe projectile motion, pendulum motion, circular motion, circular orbits, and to introduce the concepts of **angular momentum** and **torque**.

See also Center of mass; Force and the law of motion; Inertia; Motion; Motion of particles; Reference frames; Rigid bodies; Rotation in three dimensions

Ultrasound and Ultrasonography

Ultrasound is the term used for vibrations at higher frequencies than the human ear can hear. Generally, the upper end of the range of human **hearing** occurs at 20 kHz, or 20,000 vibrations per second. Since the speed of ultrasonic **waves** is the same as the speed of **sound** (1130 ft/s in warm air and standard atmospheric **pressure**), the **wavelength** of these vibrations is extremely short. Ultrasonic waves, like all sound waves, are the pressure waves of a displacement front. However, in the case of ultrasonic waves, the displacements are coming too quickly for the human ear to distinguish them.

Ultrasonic vibrations in nature are usually emitted by animals such as dolphins and bats, although some pure crystal minerals can be mechanically induced to vibrate at ultrasonic frequencies as well. The ultrasonic imaging used by bats and dolphins is also known as **sonar**, and some human machines also use the same echolocation techniques. With sonar, a pulse of ultrasound is emitted. Each wave in the pulse bounces back off of anything it hits and returns to the sender. The reflected waves are recorded and used to create an image in the same way that reflected **light** rays are used by the eye to create an image.

Machines with parts that vibrate at ultrasonic frequencies are often used for medical imaging. These parts are generally working in the range of five to seven MHz. The **frequency** is important because of the time it takes waves to rebound. Ultrasonic imaging is considered desirable because it does not require invasive surgery, is relatively inexpensive to perform, and provides information about the patient right away. It also avoids the use of electromagnetic **radiation**, which can damage tissues. However, ultrasound and x-ray radiation provide complementary medical imaging systems, because **x rays** primarily provide images of firm objects such as bones, whereas ultrasound only works with liquid-based organs. Dense objects like bones absorb too much of the ultrasonic waves to make imaging practical. When a physician per-forms an ultrasound for imaging purposes, he or she is essentially using sonar on the patient. Occasionally, ultrasound will also be used for massage purposes, to relax damaged and misaligned muscles. It may also be used to break up kidney stones, alleviating the need for surgery.

The development of ultrasound technology began in the twentieth century, but it continues to be refined into the twenty-first. Applications in medical imaging and other commercial uses continue to grow, and the development of truly three-dimensional images from ultrasound is on the rise. These images bring human sound technology closer to visible light technology and may prove invaluable for evaluating medical data.

Uncertainty Principle

The Uncertainty principle (also known as the Heisenburg uncertainty principle or Indeterminacy principle) is a fundamental postulate of **quantum theory** used to describe the behavior of **energy** and **matter** on atomic and subatomic scale that states that two complementary properties of a system (i.e., the position and **momentum** of an **electron**) can never both be measured exactly. In this regard the Uncertainty principle established a limit to the accuracy of measurement. Moreover, the Uncertainty principle specifies that the measurement of a system alters the system.

Using matrix mathematics, German physicist **Werner Heisenberg** formulated the first complete and self-consistent theory of **quantum mechanics**. In 1926, Heisenberg put forward his uncertainty principle and the concept quickly became a cornerstone of quantum theory that helped explain the wave-particle nature of **subatomic particles**. The Uncertainty principle mathematically specifies that the position and the momentum of an object cannot both be measured exactly. Moreover, at the quantum level, the concepts of position and momentum have no specified values until a system is measured. For exam-

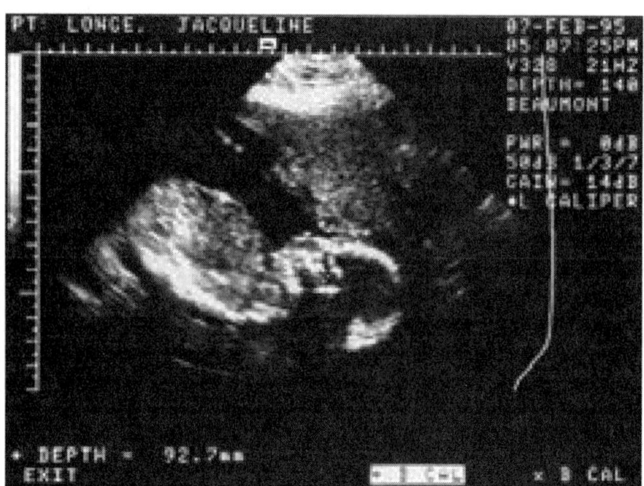

Ultrasound of a twenty-week-old fetus. *(Photo courtesy of Jacqueline Longe.)*

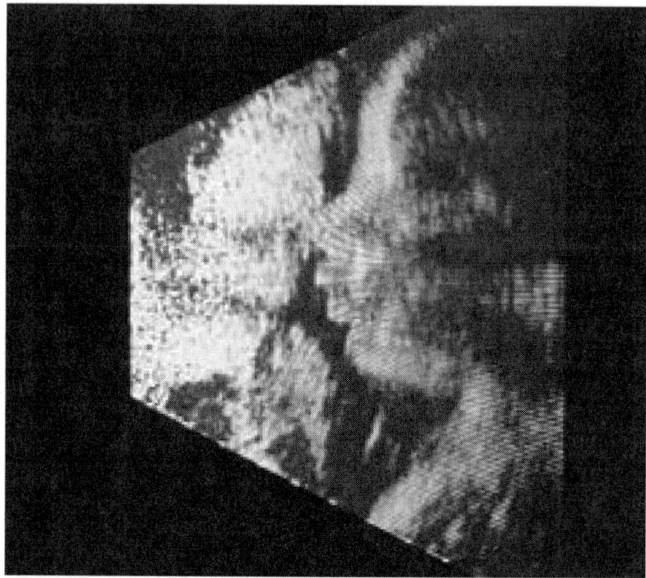

False-color, computer-coded ultrasound image of a human fetus in the womb after seven months of development, showing the face and shoulder. *(Photo courtesy of Jacqueline Longe.)*

ple, Heisenburg asserted that in any measurement of the position or the momentum of an electron, there would be an uncertainty in the measurement due to the fact that the measurement of one of the quantities would necessarily disturb the measurement of the second quantity. For example, the position of an electron can be determined using short **wavelength radiation** which consists of high-energy photons. However, the use of high-energy photons to determine the position of an electron changes the momentum of the electron.

The development of quantum theory, especially the delineation of **Planck's constant** and the articulation of the Heisenburg uncertainty principle carried profound philosophical implications regarding limits on knowledge. Because the product of the uncertainty in the measurement of position and momentum must always be greater than or equal to Planck's constant (h) divided by $2pi$, infinite precision in measurement is not possible.

The uncertainty principle also applies to energy and **time** and explains the broadening of **spectral lines**: if the lifetime of a molecule promoted to an excited state is short, the uncertainty in time will be small and the uncertainty in energy will be large, resulting in a wide range of energy values for the transition and in spectral broadening of the line.

Unification of physics

Many physicists have long argued the conceptual simplicity of the **Universe**, and have sought to explain its workings in a manner consistent with that simplicity. In the late sixteenth century Italian astronomer and physicist **Galileo** Galilei demonstrated that mathematical laws describe Earthly physical phenomena. In the early seventeenth century English physicist and mathematician Sir **Isaac Newton** proved the universality of those physical laws. Ever since, the history of theoretical physics has shown an evolution towards greater simplicity, an evolution towards a description of nature in terms of fewer, and more logically coherent, ideas.

This unification of physical theory is largely due to the realization that a few unifying principles apply to nature as a whole. These principles, which transcend all branches of physics, point to a concise fundamental structure of the world, and can be expressed simply: nature is both economical and symmetrical.

Open any physics text, and one is confronted with scores of equations, English physicist and mathematician Sir Isaac Newton's laws of **motion** and **gravity**, Scottish physicist **James Clerk Maxwell**'s equations, Austrian physicist **Erwin Schrödinger**'s equation, German-American physicist **Albert Einstein**'s relativity equations, and the list seems endless. Yet, every one of them can be derived by invoking a single principle, first enunciated in 1834 by the Irish mathematician Sir William Rowan Hamilton, the principle of least action.

As with most notions in physics with abstract principles, a simple example taken from the realm of **mechanics** can suffice for exposition. A particle travels from point A to point B. Imagine an infinite number of different paths over which the particle may travel on its journey between those two points. How does nature choose the unique path actually taken? For each path, a particular mathematical function, called the action, takes a numerical value. The actual path is the one that yields the minimum value; the action evaluated over any path other than the actual path is larger. In the same manner, consider what will be the electric field surrounding a particular charge distribution, or what will be the **temperature** change in a gas as it expands. In each of these cases, the actual result will be the one that minimizes the action associated with that process. It is a simple idea, and is pervasive in nature.

Various geometrical figures contain a familiar symmetry. The mathematical definition of that symmetry is invariance under a symmetry transformation. The letter A is reflection invariant, since it is identical to its mirror image. Rotate a square by 90 degrees (or any multiple of 90 degrees), and it remains a square; it is rotation invariant under those transformations. The same definition holds for mathematical expressions. When the laws of physics are expressed mathematically they are symmetrical; they are invariant under a large number of symmetry transformations. Those invariances have remarkable consequences. For example, if the laws are translation invariant (which means that the laws here are the laws everywhere), then **momentum** must be conserved. If the laws are **time** translation invariant (which means that the laws now are the laws at any time), then **energy** must be conserved.

But what are the symmetries that physical law obeys? Identifying the proper symmetries leads to uniquely defined unified theories, in which a single mathematical model encompasses what had previously been thought to be unrelated phenomena. For example, try to write down a Lorentz invariant theory of the electric field, and two things happen. First, there is one, and only one set of **Maxwell's equations** that satisfies the symmetry requirement. It must be accepted, or give up the symmetry requirement altogether. Second, that model must, by necessity, describe the magnetic field as well. Thus, if physical law is Lorentz invariant, then Maxwell's theory of **electromagnetism**, which unites **electricity** and **magnetism**, must be the correct description of those phenomena; there is no other choice.

In the 1950s, physicists began exploring the quantum field theories that possessed a symmetry called local gauge invariance. Consider English physicist Paul Dirac's (1902-1984) **quantum theory** of the **electron**. Modify it so that it will be invariant under a particular set of symmetry transformations, and one result is the quantized form of Maxwell's unified theory, **quantum electrodynamics** (QED). In 1967, both **Steven Weinberg** and **Abdus Salam** independently realized that the field theory invariant under a larger group of symmetry transformations unites electromagnetism and the weak nuclear **force** to form what is now understood as the electroweak interaction. The search for the ultimate symmetry group which would unite all four forces in nature (the strong nuclear force, the weak nuclear force, the electromagnetic force, and gravity) was on. Thus far, a more comprehensive unifying theory remains elusive.

See also Electromotive force; Grand unified theory; Gravity; Newtonian physics, Particle physics; Strong interactions; Weak force and interactions

UNITS AND STANDARDS

A unit of measurement is some specific quantity that has been chosen as the standard against which other measurements of the same kind are made. For example, the meter is the unit of measurement for length in the metric system. When an object is said to be four meters long, that means that the object is four times as long as the unit standard (one meter).

The term "standard" refers to the physical object on which the unit of measurement is based. For example, for many years the standard used in measuring length in the metric system was the distance between two scratches on a platinum-iridium bar kept at the Bureau of Standards in Sèvres, France. A standard serves as a model against which other measuring devices of the same kind are made. The meter stick in your classroom or home is thought to be exactly one meter long because it was made from a permanent model kept at the manufacturing plant that was originally copied from the standard meter in France.

All measurements consist of two parts: a scalar (numerical) quantity and the unit designation. In the measurement 8.5 meters, the scalar quantity is 8.5 and the unit designation is meters.

The need for units and standards developed at a point in human history when people needed to know how much of something they were buying, selling, or exchanging. A farmer might want to sell a bushel of wheat, for example, for 10 dollars, but he or she could do so only if the unit "bushel" was known to potential buyers. Furthermore, the unit "bushel" had to have the same meaning for everyone who used the term.

The measuring system that most Americans know best is the British system, with units including the foot, yard, second, pound, and gallon. The British system grew up informally and in a disorganized way over many centuries. The first units of measurement probably came into use shortly after 1215. These units were tied to easily obtained or produced standards. The yard, for example, was defined as the distance from King Henry II's nose to the thumb of his outstretched hand.

The British system of measurement consists of a complex, irrational collection of units whose only advantage is its familiarity. As an example of the problems it poses, the British system has three different units known as the quart. These are the British quart, the United States dry quart, and the United States liquid quart. The exact size of each of these quarts differs.

In addition, a number of different units are in use for specific purposes. Among the units of volume in use in the British system, (in addition to those mentioned above) are the bag, barrel (of which there are three types—British and United States dry, United States liquid, and United States petroleum), bushel, butt, cord, drachm, firkin, gill, hogshead, kilderkin, last, noggin, peck, perch, pint, and quarter.

In an effort to bring some rationality to systems of measurement, the French National Assembly established a committee in 1790 to propose a new system of measurement, with new units and new standards. That system has come to be known as the metric system and is now the only system of measurement used by all scientists and in every country of the world except the United States and the Myanmar Republic. The units of measurement chosen for the metric system were the gram (abbreviated g) for **mass**, the liter (l) for volume, the meter (m) for length, and the second (s) for **time**.

A specific standard was chosen for each of these basic units. The meter was originally defined as one ten-millionth the distance from the north pole to the equator along the prime meridian. As a definition, this standard is perfectly acceptable, but it has one major disadvantage: a person who wants to make a meter stick would have difficulty using that standard to construct a meter stick of his or her own.

As a result, new and more suitable standards were selected over time. One improvement was to construct the platinum-iridium bar standard mentioned above. Manufacturers of measuring devices could ask for copies of the fundamental standard kept in France and then make their own copies from those. As you can imagine, the more copies of copies that had to be made, the less accurate the final measuring device would be.

The most recent standard adopted for the meter solves this problem. In 1983, the international Conference on Weights and Measures defined the meter as the distance that **light** travels in 1/299,792,458 second. The standard is useful because it depends on the most accurate physical measurement known—the second—and because anyone in the world is able, given the proper equipment, to determine the true length of a meter.

In 1960, the metric system was modified somewhat with the adoption of new units of measurement. The modification was given the name of Le Système International d'Unités, or the **International System of Units**—more commonly known as the SI system.

Nine fundamental units make up the SI system. These are the meter (abbreviated m) for length, the kilogram (kg) for mass, the second (s) for time, the ampere (A) for **electric current**, the kelvin (K) for **temperature**, the candela (cd) for light intensity, the mole (mol) for quantity of a substance, the radian (rad) for plane angles, and the steradian (sr) for solid angles.

Many physical phenomena are measured in units that are derived from SI units. As an example, **frequency** is measured in a unit known as the hertz (Hz). The hertz is the number of vibrations made by a wave in a second. It can be expressed in terms of the basic SI unit as s^{-1}. **Pressure** is another derived unit. Pressure is defined as the **force** per unit area. In the metric system, the unit of pressure is the Pascal (Pa) and can be expressed as kilograms per meter per second squared, or $kg/m \, x^2$. Even units that appear to have little or no relationship to the nine fundamental units can, nonetheless, be expressed in these terms. The absorbed dose, for example, indicates that amount of **radiation** received by a person or object. In the metric system, the unit for this measurement is the gray. One gray can be defined in terms of the fundamental units as meters squared per second squared, or $m^2 \, x \, s^2$.

Many other commonly used units can also be expressed in terms of the nine fundamental units. Some of the most familiar are the units for area (square meter: m^2), volume (cubic meter: m^3), **velocity** (meters per second: m/s), concentration (moles per cubic meter: mol/m^3), **density** (kilogram per cubic meter: kg/m^3), luminance (candela per square meter: cd/m^2), and magnetic field strength (amperes per meter: A/m).

A set of prefixes is available that makes it possible to use the fundamental SI units to express larger or smaller amounts of the same quantity. Among the most commonly used prefixes are milli- (m) for one-thousandth, centi- (c) for one-hundredth, micro- (æ) for one-millionth, kilo- (k) for one thousand times, and mega- (M) for one million times. Thus, any volume can be expressed by using some combination of the fundamental unit (liter) and the appropriate prefix. One million liters, using this system, would be a megaliter (ML) and one millionth of a liter, a microliter (æL).

One characteristic of all of the above units is that they have been selected arbitrarily. The committee that established the metric system could, for example, have defined the meter as one one-hundredth the distance between Paris and Sèvres. It was completely free to choose any standard it wished.

Some measurements, however, suggested "natural" units. In the field of **electricity**, for example, the charge carried by a single **electron** would appear to be a natural unit of measurement. That quantity is known as the elementary charge (e) and has the value of $1.6021892 \times 10^{-19}$ coulomb. Other natural units of measurement include the **speed of light** (c: 2.99792458×10^8 m/s), the Planck constant (6.626176×10^{-34} joule per hertz), the mass of an electron (m_e: $0.9109534 \times 10^{-30}$ kg), and the mass of a **proton** (m_p: $1.6726485 \times 10^{-27}$ kg). As you can see, each of these natural units can be expressed in terms of SI units, but they are often used as basic units in specialized fields of science.

For many years, an effort has been made to have the metric system, including SI units, adopted worldwide. As early as 1866, the United States Congress legalized the use of the metric system. More than a hundred years later, in 1976, the Congress adopted the Metric Conversion Act, declaring it the policy of the nation to increase the use of the metric system in the United States.

In fact, little progress has been made in that direction. Indeed, elements of the British system of measurement continue in use for specialized purposes throughout the world. All flight navigation, for example, is expressed in terms of feet, not meters. As a consequence, it is still necessary for an educated person to be able to convert from one system of measurement to the other.

In 1959, English-speaking countries around the world met to adopt standard conversion factors between British and metric systems. To convert from the pound to the kilogram, for example, it is necessary to multiply the given quantity (in pounds) by the factor 0.45359237. A conversion in the reverse direction, from kilograms to pounds, involves multiplying the given quantity (in kilograms) by the factor 2.2046226. Other relevant conversion factors are 1 inch = 2.54 centimeters and 1 yard = 0.9144 meter.

UNIVERSE

Think bigger than home, city, or state. Think bigger than country, continent, or planet. Think even bigger than solar system, galaxy, or supercluster. Only one entity encompasses all of these objects, and this is the universe. What's more, the universe is comprised not only of all that humans can directly observe, but also of everything that we cannot see. Along with the most familiar forms of visible **matter**, elusive entities such as **quarks**, neutrinos, **black holes**, and yet-undiscovered forms of **dark matter** compose the universe.

Simply put, the universe, or **cosmos**, as it is sometimes called, is everything. It is the entire body of objects and phenomena that may be observed or theoretically postulated. As the late **Carl Sagan** revealed with the memorable opening line of his bestselling book *Cosmos*, "The cosmos is all that is or ever was or ever will be."

Scientists continually take on the task of improving our understanding of the universe. They observe the constituents of the universe and the phenomena that pervade its expanse. In so doing, they discover the physical laws that govern the universe. With vast bodies of collected information, scientists then theorize and predict how the universe should behave. Ultimately, subsequent observations serve to test the theories that scientists propose, allowing them to accept those that are consistent, to refine marginal theories, and abandon those that are inconsistent with observation.

The process of bettering human understanding of the universe is cyclic and cumulative. As **Isaac Newton** suggested, when he offered, "If I have seen farther, it is by standing on the shoulders of giants," each generation relies upon the scientific successes of the former.

The evolution of human understanding of the universe has certainly been long and interesting. The ancient Roman atomists viewed the universe as a body of infinite dimension and complexity. A geocentric universe, with Earth occupying the center, eventually replaced this antique model and reigned supreme through the Middle Ages. Later, it was **Nicholas Copernicus** who helped us to understand that the Earth is not the center of the universe, while scientists and thinkers like **Johannes Kepler**, Isaac Newton, and **Immanuel Kant** paved a way for the understanding that galaxies might exist outside of the Milky Way.

Surprisingly enough, it was not until 1923, when Edwin P. Hubble determined conclusively that the Andromeda Nebula was really a galaxy some 2.5 million **light** years away, that our notion of the universe expanded to include innumerable galaxies like our own. Moreover, it was also Hubble who widened our perspective to include a universe which is literally expanding. For it was he who discovered in 1929 that all points in the universe are rushing away from one another.

Through the efforts of cumulative scientific investigation, we now know that the universe is governed by four fundamental physical forces: the strong and weak nuclear forces, the **gravitational force**, and the electrostatic **force**. We know that Earth is not at the center of the universe; that the **Sun** is not at the center of the **Milky Way galaxy**, nor is the Milky Way at the center of the universe; that the universe indeed has no center; that **planets** revolve around several distant **stars**; and that there are at least 100 billion galaxies in the universe, each with 100 billion stars. We know that the universe was likely created some 12-15 billion years ago in an unimaginable event known as the **big bang** and that the universe probably will not end with a big crunch. We have come to know a great deal about the universe in which we live.

With such marked advances and augmentation of our understanding of the universe in the last several thousand years, one can only wonder what will be discovered and learned in the coming millennia. But one thing is certain—the universe is a fascinating place to call home.

UV-VISIBLE SPECTROSCOPY

UV-visible spectroscopy investigates the interactions between ultraviolet or visible electromagnetic **radiation** and **matter**. UV-visible spectroscopic measurements provide precise information about atomic and **molecular structure**. The foundation of UV-visible spectroscopy was laid by English physicist and mathematician Sir **Isaac Newton** who, by diffracting white **light** through a prism, showed that it consisted of light of several colors ranging from violet to red. This is now termed the UV-visible **electromagnetic spectrum**. The red end of the spectrum corresponds to electromagnetic radiation of longer wavelengths (~780 nm) and lower frequencies, while the violet end is associated with shorter wavelengths (~380 nm) and higher frequencies. An understanding of the dual nature of light, however, which consists of particles called photons that also behave like **waves**, and of its interaction with matter, waited for the advent of **quantum theory**. German physicist Maxwell Planck (1858-1947) first postulated that the **energy** of the electromagnetic spectrum was not continuous, but consisted of small, discrete packets which he called **quanta**, and whose energy, E, was directly proportional to the **frequency** of the light ν as expressed in the fundamental equation of spectroscopy, i.e., $E = h\nu$ where h is a proportionality constant called **Planck's constant**. The direct relationship between energy and frequency shows that light of lower frequency will carry fewer energetic photons than light of a higher frequency.

Spectral lines observed in the UV-visible region correspond to the energy difference between two well-defined electronic energy levels of the absorbing **atom** or molecule. When an atom or molecule absorbs energy in that region, its electrons are promoted from a state of lowest energy, i.e., the ground state, to states, or **orbitals**, of higher energy. Quantum theory also explains how these electronic energy levels arise from the electronic configuration of the molecule or atom absorbing the radiation. Energy levels are defined by a set of **quantum numbers**, such as n, the principal quantum number, which specifies the energy of the level and which can take integer values; l, the azimuthal **momentum** quantum number, which designates the **electron** sub-shells and is equal to $n-1$; m_s, the spin quantum number, which can only equal $+\frac{1}{2}$, or $-\frac{1}{2}$; L, the total **angular momentum** quantum number, which is 0 or a positive integer for a single electron; and S, its total spin angular momentum quantum number, which can be 0, a positive integer, or $\frac{1}{2}$. For example, the electronic configuration of hydrogen which has only one electron is $1s^1$ in the ground state, and L=1, n=1, and l=0. In the first excited state, its electronic configuration is $2s^1$, n=2, and l=0 or 1.

UV-visible spectroscopy thus provides direct evidence about electron energy jumps between distinct energy levels.

For example, the emission spectrum of hydrogen consists of a series of sharp spectral lines, each of which results from an electron jumping from one **energy level** to another while emitting a **photon** of energy equal to the energy difference between the two levels. Thus, electron jumps from levels where $n > 3$, down to the $n = 3$ level yield the Paschen series of lines observed in the near infrared region (~780-3,000 nm), and electronic transitions to the $n = 1$ level yield the Lyman series of lines observed in the UV region (~10-380 nm), while transitions to the $n = 2$ level are observed as the Balmer series in the visible region (~380-780 nm). This series is named after the physicist J.J. Balmer, who discovered that the wavelengths of the spectral lines in the visible hydrogen spectrum could fit to an equation yielding the frequency of any line observed in the series: $v = Rc\, (\frac{1}{2}^2 - 1/n_1^2)$ where R is the **Rydberg constant** $(1.097 \times 10^7 \text{ m}^{-1})$.

Associated with the electronic transitions observed in the UV-visible region are changes of vibrational and rotational energy which also influence the shape of the spectral bands. If the material is a gas, the molecular electronic spectrum is sharp and well-resolved, consisting of distinct lines, such as the hydrogen spectrum. If the material is in the liquid state, the spectral bands are blurred and significantly broadened because the vibrational motions can not occur as freely as they could in the gas phase. Not all molecules exhibit pure vibrational-rotational spectra, but all of them can produce electronic spectra because there will always be excited states to which the electrons of molecules can be raised by the absorption of UV-visible radiation. The occurrence of electronic transitions and the intensity of their spectral lines is governed by selection rules which are formulated by **molecular orbital theory**.

UV-visible spectroscopy has been enhanced as an experimental technique by the introduction of **lasers**. This field was developed in the 1950s with the initial work occurring in the microwave region, but it quickly extended to the UV-visible regions. A commonly used laser today is the ruby laser, in which the lasing medium is a rod of sapphire (Al_2O_3) with some **ions** replaced by Cr(III) ions, giving the crystal its bright red **color**. The ruby laser produces light at a **wavelength** of 694.3 nm, which lies at the red end of the visible spectrum. An important characteristic of laser light is that it is monochromatic, i.e., it covers a very narrow range of wavelengths, and this provides laser spectroscopy with distinct advantages compared to conventional spectroscopy.

See also Absorption spectrum; Atomic structure; Fluorescence; Infrared spectroscopy; Light; Microwave background radiation; Particle-wave duality; Phosphorescence; Spectroscopy; Speed of light; Symmetry and symmetry principles; Transverse and longitudinal waves; Uncertainty principle; Zeeman effect

Vacuum

Vacuum is a term that describes conditions where the **pressure** is lower than that of the atmosphere. A sealed container is said to be "under vacuum" in this case whereas it is "pressurized" when the pressure is higher than atmosphere. In a vacuum, it becomes necessary to define pressure microscopically. This means that the pressure, or **force** per unit area, is determined by the number of collisions between the atoms or molecules present and the walls of the container.

The first experiments involving vacuum date back to 1644 when **Evangelista Torricelli** worked with columns of mercury, leading to the first barometer (a device for measuring pressure). The famous experiment of **Otto von Guericke** in 1654 demonstrated the astounding force of vacuum when he evacuated the volume formed by a pair of joined hemispheres and attached each end to a team of horses that were unable to pull the hemispheres apart.

In order to create a vacuum, some kind of pump is needed. Simple mechanical pumps create a pressure difference, or suction force, which can be sufficient to pump water, for example. The most common use of vacuum, the vacuum cleaner, is simply a chamber and hose which are continuously evacuated by a fan. More sophisticated vacuum pumps must be sealed to prevent air from leaking back into the pumping volume too quickly. These pumps increase in complexity as better vacuum is needed. Pumps can generally be grouped into two categories: dynamic pumps, using mechanical or turbo-molecular action, and static pumps, using electrical ionization or low **temperature** (cryogenic) condensation.

Vacuum is important for research and industry, especially for manufacturing. Many industrial processes require vacuum either to be efficient or to be possible at all. Vacuum can be used for the prevention of chemical reactions, such as clotting in blood plasma or the removal of water in the process of freeze drying. Vacuum is also necessary for the prevention of particle collisions with background gas, in a television picture tube for example. For the fabrication of integrated **electronics**, it is very important to avoid impurities on a microscopic scale. It is only with excellent vacuum that such conditions can be obtained.

Vacuum Energy Density

In 1917, just one year after the publication of his general theory of relativity, German American physicist **Albert Einstein** wrote a short paper titled *Cosmological Considerations on the General Theory of Relativity*. For the first time the **universe** was treated as a four-dimensional (three spatial and one temporal) manifold termed **space-time**. The geometry of the model universe was determined by solving the Einstein field equations, which relate the curvature of space-time to **matter** distribution. Modern **cosmology** was born. And so was the physics of the **vacuum**.

Einstein initially argued for a static universe and a dozen years would pass before Milton Humason and American astronomer **Edwin Hubble** would discover the universal expansion of the **cosmos**. Although Einstein's own field equations dictated that the universe could not be static, Einstein attempted to reconcile general relativity with a static universe by adding a new term to his field equations, termed the **cosmological constant**. Using this artificial constant, Einstein was able to model a static, matter-filled universe.

If Einstein's modified equations were applied to an empty universe, the cosmological constant essentially fills the empty universe (a vacuum) with an **energy density** and a (negative) **pressure**. And, of course, it adds to the energy density and pressure in a matter-filled universe. It is this negative vacuum pressure, in fact, that kept Einstein's model universe static.

Tokamak Fusion Test Reactor vacuum vessel. *(AP/Wide World Photos. Reproduced by permission.)*

After Hubble's work demonstrated that the universe continues to expand, Einstein determined there was no need for his ad hoc term in his field equations, and he renounced it. In doing so he also renounced the concept of the vacuum energy that it predicted. Subsequently Einstein would characterize his introduction of the cosmological constant as the greatest blunder of his. But the cosmological constant was not an absolute blunder. Half a century after Einstein's abjuration, **quantum field theory** tells us that there must be (or at least must have been) a large vacuum energy density.

The most current models of the fundamental forces, the Weinberg-Salam **electroweak theory** and quantum chromodynamics, the gauge theory of the strong interaction, are examples of spontaneously broken gauge theories (as would be any **Grand Unified Theory** uniting all three interactions). Models of this type predict the production, at the very high temperatures that existed in the early universe, of regions of space-time possessing very large vacuum energy densities. If the idea of spontaneous symmetry breaking is correct, there were, at a very early time in the universe's evolution, space-time regions with very large effective cosmological constants. The vacuum energy density in a region drives an enormously rapid (though quite brief) period of expansion, known as inflation. One such tiny region can, in this manner, inflate enough to become the entire observable universe known today.

See also Beginning of time; Energy; Euclidean space; Relativistic mass and energy

VACUUM TUBES

A vaccum tube is an evacuated **electron** conducting chamber (tube) designed to eliminate or minimize the influence of atmospheric gas. More commonly, a **vacuum** tube is an electronic mechanism used primarily to convert alternating current into direct current.

The vacuum tube consists of a glass encasement and two or more electrodes between which electrons flow freely. One of these electrodes, called the cathode, consists of a filament heated to the degree that it releases electrons, a process called thermionic emission. The other electrode, called the **anode**, consists of a metal plate attached to a filament. The electrons released by the cathode are attracted to the anode when the anode is positively charged, thus completing an electrical circuit. However, the electrons will not flow to the anode during the phase in an alternating current when the anode is negatively charged. In constructing the vacuum tube, all air and gas is removed from the encasement and the tube is sealed. The flow of electrons inside the tube has no **interference** from the presence of any gas.

The first vacuum tube was invented by British electrical engineer **John Fleming** who was investigating a principle discovered by American scientist **Thomas Edison**. Edison had accidentally discovered that a metal plate prevents electrons from flowing from the anode in an alternating current. Edison treated this discovery as only a scientific oddity. Fleming, who had once been an assistant to Edison, realized the practical applications of the Edison effect. Fleming, while working with Italian physicist **Guglielmo Marconi** on perfecting the **radio**, recognized that the Edison principle provided the means of rectifying an alternating current into a direct current (i.e., the process of rectification). The direct current alleviated some of the interference found in radio transmission and reception. Fleming's vacuum tube on two electrodes is now referred to as a diode. Years later, an American inventor, **Lee De Forest,** developed a **triode**, a vacuum tube with three electrodes. The important difference in De Forest's tube was the introduction of a grid made up of fine wire mesh. This grid, usually surrounding the cathode, can impede or amplify the flow of electrons depending on slight changes in **voltage**. Other types of vacuum tubes, with as many as eight electrodes, have been developed for a variety of uses such as **frequency** converters in radios.

In modern times, the transistor has largely replaced the vacuum tube. The advantages of the transistor over the vacuum tube include durability, compactness and greater utility. Furthermore, the vacuum tube is not as **energy** efficient as the transistor. The vacuum tube requires a greater energy supply and releases much unusable **heat** in thermionic emission.

In spite of the benefits and advancements offered by the transistor, there remain some functions that are better accomplished through a vacuum tube. One of these functions is harden military equipment that may be exposed to the electromagnetic pulse associated with an atmospheric nuclear explosion. Such an explosion produces a voltage pulse that vacuum tubes can withstand but which can destroy **transistors**.

See also Cathode and cathode rays; Diodes

VAN ALLEN, JAMES (1914-)

American physicist

James Van Allen is best known for his discovery of bands of high-level **radiation** surrounding Earth. Popularly known as the Van Allen belts, these belts are part of Earth's magnetosphere. Although Van Allen's discovery of the belts has been the highlight of his career, he has also been associated with other significant scientific research. During World War II Van Allen made his first significant scientific contribution by creating the **radio** proximity fuse, a small tracking device that triggered weapons when they were close to the target. The knowledge and skills Van Allen developed in weapons research and miniaturization later proved useful in his studies of the atmosphere using **rockets** and satellites.

James Alfred Van Allen was born in Mount Pleasant, Iowa, on September 7, 1914. He was the second of four sons of Alfred M. Van Allen, an attorney, and Alma E. (Olney) Van Allen. His interest in science developed early, and when he was twelve, he and a brother built their own electrostatic generator. The boys used their machine to produce bolts of artificial lightning. By the time Van Allen reached high school, physics had become a passion for him. According to biographer D. S. Halacy, Jr., in *They Gave Their Names to Science,* "Teachers sometimes had to forcibly eject him from laboratories after the day was over!"

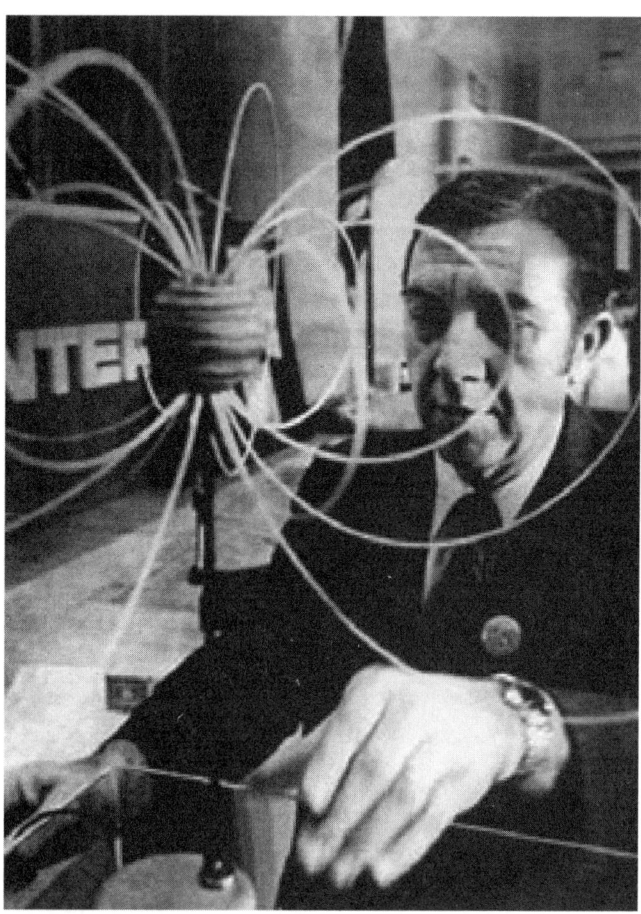

James Van Allen, with model of Jupiter and its radiation belts. *(AP/Wide World Photos. Reproduced by permission.)*

After graduating from high school, Van Allen entered Wesleyan College in his hometown. At Wesleyan he was strongly influenced by Thomas Poulter, later to become director of the Stanford Research Institute at Stanford University. In 1935 Van Allen received his bachelor of science degree from Wesleyan and entered the State University of Iowa at Iowa City for graduate study. He earned his master of science degree in 1936 and his Ph.D. in 1939.

Van Allen's work at the State University of Iowa caught the attention of the Carnegie Institute in Washington, D.C., and he was offered a position as research fellow there in the Department of Terrestrial Magnetism. After three years at Carnegie, he moved on to the Applied Physics Laboratory at Johns Hopkins University in 1942.

War Brings New Opportunities

By 1942, however, World War II had become the dominant factor in Van Allen's life. He left Johns Hopkins to accept a commission in the U.S. Navy, where he served until 1946. Van Allen's first important scientific accomplishment—the radio proximity fuse—came as a result of his war research. The fuse was a device consisting of a radio transmitter and receiver that was attached to weapons. Signals sent out and received by the fuse indicated when the weapon was close to the target, allowing it to explode before actual impact. This greatly increased the efficiency of the missiles, as it eliminated the need for a direct hit.

Van Allen's work on the fuse was to shape his future scientific career in unexpected ways. At the end of World War II, American scientists inherited about one hundred V–2 rockets built by German scientists. Far more advanced than anything produced by Americans at the time, these rockets promised to be valuable for scientists studying Earth's atmosphere. As a result of his research on the fuse, Van Allen had become a leading authority on the miniaturization of instruments, a skill crucial in rocket research. The U.S. Army therefore appointed him to administer and coordinate rocket-based research, and in 1946 Van Allen took charge of the V–2 research program based at the White Sands Proving Ground in New Mexico. His primary responsibilities were to design payloads to be sent into the atmosphere and to select projects to be used in the rocket research. From 1947 to 1958 he also served as chairman of the committee overseeing this research, originally called the V–2 Rocket Panel and, later, the Rocket and Satellite Research Panel.

This change of name reflected a change in actual rocket research itself. With a limited number of V–2 rockets available, scientists soon began to explore alternative instruments. Although American scientists could not produce an equivalent of the V–2, under Van Allen's direction they eventually constructed an acceptable substitute, the Aerobee. The Aerobee carried smaller payloads than the V–2, and although it reached altitudes of only 60 mi (96.6 km), compared to the German rocket's 100 mi (161 km), it became a research workhorse for American scientists over the next decade.

The Return to Iowa

In 1951 Van Allen returned to his native state of Iowa as professor of physics and head of the department of physics and astronomy at the State University of Iowa, positions he held until 1985. Shortly after this move, Van Allen developed a new approach for the study of the atmosphere. For over a century, balloons had been the most effective way to accumulate information about the atmosphere. Instruments were carried aloft in balloons to heights of up to fifteen miles and then parachuted back to Earth with their data.

Van Allen combined ballooning techniques with modern rocket technology to make a "rockoon," which consisted of a balloon carrying a rocket. When the balloon reached its maximum altitude, the rocket was fired off by remote control. It traveled straight upward, through the balloon itself, another 50–70 mi (80.5-112.7 km) into the atmosphere. Readings obtained from such rockoon launchings provided better information about the outer atmosphere than ever before.

Information from two rockoons sent up in 1953 produced data that puzzled Van Allen. He discovered that levels of radiation at an altitude of 30 mi (48.3 km) were much higher than had been expected. These results made Van Allen curious about what he might find at even higher altitudes. At this time scientists were discussing the next stage in rocket devel-

opment: an artificial Earth satellite that would remain in long-term orbit around the planet. As part of the International Geophysical Year planned for 1957–58, the U.S. government had made a commitment to finance such a satellite. The first satellite in that program, *Explorer I,* was launched on January 31, 1958. Results obtained from the satellite were intriguing; the levels of cosmic radiation in the upper atmosphere were much higher than had been anticipated. Instruments on *Explorer II* and *Explorer III* gave similar results.

By May of 1958, Van Allen was ready with an explanation for these results. He hypothesized the existence of two belts of radiation, one at an altitude of 600–3,000 mi (965.6-4,828 km), the other at an altitude of 9,000–15,000 mi (14,484-24,140 km). Later research confirmed this hypothesis and proved that the belts consist of high-velocity protons and electrons spiraling around Earth's magnetic lines of **force**. These are the Van Allen radiation belts. In scientific terms, these belts form part of Earth's magnetosphere, an area around the planet dominated by charged particles that are held there by Earth's magnetic field. For his discovery of the belts and other contributions to science, Van Allen has received many scientific awards and honors, including the Space Flight Award of the American Astronautical Society, the John A. Fleming Award of the American Geophysical Union, and the Elliott Cresson Medal of the Franklin Institute.

Van Allen married Abigail Fithian Halsey on October 13, 1945. They have five children, Cynthia (Schaffner), Margot (Cairns), Sarah (Trimble), Thomas, and Peter.

VAN DE GRAAFF, ROBERT J. (1901-1967)
American physicist

Robert J. Van de Graaff invented a particle accelerator named for him that tremendously advanced research in **nuclear physics**. He devoted his life to the improvement and commercial construction of his Van de Graaff accelerator, which has found widespread application in nuclear research, medicine, and industry around the world. At the time of his death, he was immersed in an effort to further extend the capabilities of his accelerator. Robert Jemison Van de Graaff was born and raised in Tuscaloosa, Alabama. His mother was Minnie Cherokee Hargrove, and his father was Adrian Sebastian Van de Graaff, a jurist. He was educated in Tuscaloosa's public schools, and then attended the University of Alabama, from which he received his B.S. degree in 1922 and his M.S. degree in 1923, both in mechanical engineering. After graduation, Van de Graaff worked for the Alabama Power Company for a year as a research assistant, continuing his studies of the conversion of **heat** into mechanical **energy**.

The young engineer's desire to understand the physical forces that drive natural phenomena—rather than work only with practical methods to harness and use those forces—led him to the study of physics. He began with studies at the Sorbonne in Paris from 1924 to 1925 and while there attended **Marie Curie**'s lectures on **radiation**. In 1925 he went to Oxford University in England as a Rhodes Scholar, where he received another B.S. degree, this time in physics in 1926, followed by his Ph.D. in physics in 1928 for work on ion mobility. He stayed at Oxford for one more year on a fellowship. While at Oxford, Van de Graaff absorbed **Ernest Rutherford**'s 1927 address to the Royal Society expressing that pioneer nuclear experimenter's hope that particles could someday be accelerated to speeds sufficient to disintegrate nuclei. Study of a **nucleus** as it disintegrated and scattered would reveal much about the nature of individual atoms. Only very high-speed particles would have enough **force** to smash apart an atomic nucleus, and nature does not supply enough of these: hence the need for an **acceleration** machine.

Van de Graaff realized the importance of research on atomic nuclei, and, impressed by Rutherford's address, worked out the principle of a device to accelerate elementary particles to high energies, drawing on his mechanical engineering background. He based his design on the oldest kind of **electricity** known: static electricity, the kind that causes small effects like static cling and large effects like lightning. Van de Graaff decided to generate high voltages using a direct-current electrostatic method. A moving belt would carry an **electric charge** inside an insulating tube into an insulated metal sphere, which would store the accumulated electricity on its surface. High-voltage discharges of this electricity into an acceleration tube would provide the required particle acceleration. From his engineering background, Van de Graaff knew that a polished and rounded surface for the electric terminal would greatly reduce the possibility of electric stress and breakdown.

In 1929 Van de Graaff returned to the United States, joining Princeton University's Palmer Physics Laboratory as a National Research Fellow. That fall, he verified his particle acceleration principles by constructing the first working model of his device, which developed 80,000 volts. It was not sophisticated, consisting as it did of a silk ribbon, a tin can, and a small motor. E. Alfred Burrill in *Physics Today* wrote that in order to construct this elementary model, Van de Graaff searched through Princeton's local millinery shops for pure silk, going so far as to set at least one sample on fire—in the shop—to be sure of its purity. More working models followed, and Van de Graaff introduced his invention to fellow physicists in September 1931 at a meeting of the American Physical Society. He followed with a demonstration in November 1931 at the inaugural dinner of the American Institute of Physics, producing over a million volts.

Fellow physicist Karl T. Compton encouraged Van de Graaff in his work at Princeton. When Compton became president of the Massachusetts Institute of Technology (MIT) in 1931, Van de Graaff accepted Compton's invitation to come to MIT as a research associate and further developed his accelerator. In 1932 and 1933, Van de Graaff constructed his first large generator, in an aircraft hangar at Round Hill, on the shore of Buzzard's Bay in South Dartmouth, Massachusetts. The machine was truly enormous. It consisted of two polished aluminum spheres each fifteen feet in diameter mounted on cylindrical insulating columns twenty-five feet high and six feet in diameter. The columns were mobile, mounted on trucks operating on a railway track which boosted the spheres to forty-three feet above ground level. This machine had its debut performance on November 28, 1933, with spectacular effects choreographed for half an hour by Van de Graaff as he directed the switches controlling the generator's current. The spheres, acting as **voltage** terminals, sent out seven-million-volt bolts of blue lightning between each other and between themselves and the hangar's walls and floor. Observers could feel their hair rise from the static force. The *New York Times,* reporting on the demonstration on November 29, 1933, headlined its story "Man Hurls Bolt of 7,000,000 Volts," and went on to describe the "brilliance and the savage fury of [the generator's] unleashed power," the "magic wands which spat out flaming streaks of blue, liquid fire," and the "snakelike tongues of violet, pink and lavender flames [that] lashed out" as the participating scientists calmly monitored their switches. Operating the generator in the hangar caused problems, so the machine was moved to a pressurized enclosure at MIT in 1937. A **vacuum tube** to contain the current was added. By 1940, a modified version was producing voltages of 2.75 MV, and after its productive years as a tool of nuclear and radiography research ended, it became a permanent "atom smasher" exhibit at the Boston Museum of Science.

The Van de Graaff accelerator was an immediate success, as its advantages over existing machines were immediately apparent. John D. Cockcroft and **Ernest Walton** of the Cavendish Laboratory in England had built a successful particle accelerator in 1932, which used voltage-multiplier circuits to produce the required high voltages. This machine was bulky and complicated, however, and maximum voltages were limited. The Van de Graaff device, in contrast, was extremely simple and (ultimately) compact, based as it was on electrostatic generation, employing a simple belt rather than multiple **transformers**. Because its voltage was relatively easy to stabilize, it permitted precisely controllable particle acceleration, which the Cockcroft-Walton accelerator did not, and it achieved significantly higher energies. The simplicity of Van de Graaff's principle is made apparent by the fact that it is possible to produce a small, homemade Van de Graaff generator—and many science students and laboratory assistants do so with, for example, loaf pans (for the base and dome), a plastic fruit-juice mixer (for the insulating tube), a rubber belt, and a toy motor.

Guided by Compton and by Vannevar Bush, MIT's vice president, Van de Graaff toiled through the writing of his patent application and was rewarded with a patent for his invention in 1935. Through the 1930s and 1940s Van de Graaff worked with John G. Trump, a professor of electrical engineering at MIT, and William W. Buechner, a professor in MIT's physics department, to modify and improve his particle accelerator, aiming for higher voltages, vertically mounted and more compact designs, and steadier, more homogeneous particle beams. With Trump, Van de Graaff adapted his design into a machine that could treat cancerous tumors with precisely penetrating **x rays**. The medical Van de Graaff machine was first used clinically in 1937 at Harvard Medical School's Huntington Memorial Hospital.

During World War II, Van de Graaff remained at MIT as director of the High Voltage Radiographic Project, sponsored by the Office of Scientific Research and Development. Working with Buechner, he directed the adaptation of the electrostatic generator to precision radiographic examination of U.S. Navy ordnance. This opened up the possibility of industrial applications for the Van de Graaff accelerator. After the war, a 1945 Rockefeller Foundation grant funded the development of an improved accelerator at MIT, which involved Van de Graaff in continued nuclear research and experimentation. On December 19, 1946, Trump and Van de Graaff formed the High Voltage Engineering Corporation (HVEC) in Burlington, Massachusetts, for the commercial production of particle **accelerators**. Denis M. Robinson, a professor of electrical engineering from England, became president. Trump was technical director. Van de Graaff was chief physicist and a board member; he acted in the capacity of consultant while retaining his post as associate professor of physics at MIT.

Under the direction of these three men, HVEC produced a series of ever more technologically advanced Van de Graaff **particle accelerators**. Soon the company was the leading supplier of electrostatic **generators**, which were used in cancer therapy, in industry for radiography, and in studies of nuclear structure in both chemistry and physics. In 1951 **Luis W. Alvarez** at the University of California in Berkeley rediscovered the tandem principle of particle acceleration first invented by Willard Bennett in 1937. Van de Graaff became very involved in efforts at HVEC to develop a tandem Van de Graaff accelerator, which uses the same high voltage twice to accelerate the particles twice, resulting in particle energies double the machine's voltage. The first of these machines was purchased for use in 1956 by the Chalk River Laboratories of Atomic Energy of Canada and put into use in 1959. HVEC's tandem Van de Graaff accelerators became very successful.

Van de Graaff solved other difficult and limiting problems inherent with the electrostatic generator. His uniform-field electrode configuration and inclined field tubes overcame problems in the insulating tubes that could inhibit acceleration. His insulating-core transformer of the late 1950s generated high-voltage direct current via **magnetic flux** rather than by the electrostatic charging belt, thereby permitting higher dc voltages than previously attainable with electrostatic machines. This invention had important applications in high-voltage electric power utilities and industrial processes, and the inspiration for it came from a magnetic circuit Van de Graaff had observed in an Alabama Power company hydroelectric generator decades earlier. Van de Graaff also devised many methods of controlling particle beams after their acceleration so they could be adapted to precise and individual research requirements. Using Van de Graaff accelerators, experimenters accumulated a vast amount of information on nuclear disintegrations and reactions, which led directly to very sophisticated theories of nuclear structure.

Toward the end of his life, Van de Graaff was absorbed with adapting his insulating-core concept to produce triple tandem accelerators powerful enough to accelerate heavy **ions**, which in turn could smash the nuclei of even the heaviest atoms—in particular, uranium. He envisioned the possibility of creating new elements if a uranium nucleus bombarded by a high-speed **proton** captured the proton rather than disintegrating. He also was excited by the powerful nuclear reaction that would result if two uranium nuclei could be induced to fuse. His goal for particle acceleration was to be able to produce and use precisely controlled charged-particle beams of any element, including heavy ones. In the fall of 1966, shortly before his death, Van de Graaff launched an intensive research program at HVEC bombarding uranium nuclei with uranium ions, thereby producing highly charged uranium atoms and yielding valuable data on heavy particle **motion**.

In 1936, Van de Graaff married Catherine Boyden; they had two sons, John and William. He was an associate professor of physics at MIT from 1934 until 1960, when he resigned to devote himself to his ever-increasing involvement with HVEC. In 1966, he was awarded the Tom W. Bonner Prize by the American Physical Society "for his contribution to and continued development of the electrostatic accelerator, a device that has immeasurably advanced nuclear physics." (The prize was named for a scientist who had used Van de Graaff particle accelerators to achieve the results of his fundamental research.) Van de Graaff also advanced physics in his role as a teacher and mentor, producing many proteges who themselves advanced particle accelerator technology and basic nuclear knowledge. Van de Graaff was courteous and unassuming, but nevertheless was inspirational and effective as a teacher and research leader because he was able to communicate his ideas clearly, visually, and with excitement. He published many articles in scientific journals, often coauthored with colleagues, and was granted many patents, including those for the electrostatic generator and the insulating-core transformer. He received honorary doctorates from several universities and many honors and awards, including the 1947 Duddell Medal of the Physical Society of Great Britain. Van de Graaff died of a heart attack on the morning of January 16, 1967, in Boston at the age of sixty-five. At the time of his death, over five hundred Van de Graaff particle accelerators were in use in more than thirty countries.

VAN DER WAALS, JOHANNES DIDERIK (1837-1923)
Dutch physicist

Johannes Diderik van der Waals received his doctorate in physics from the University of Leiden at the relatively late age of thirty-six. His doctoral dissertation, "On the Continuity of Gaseous and Liquid States," quickly became known among his colleagues and made his reputation almost immediately. The Nobel Prize in physics, awarded him in 1910, recognized the line of work begun in his dissertation, eventually resulting in a famous equation of state relating the **pressure**, volume, and **temperature** of a gas. He also demonstrated why a gas cannot be liquified above its critical temperature. Van der Waals also investigated the weak nonchemical bond forces between molecules that now carry the name *van der Waals forces*.

Dutch physicists Johannes van der Waals (right) and Heike Kammerlingh Onnes photographed at Kammerlingh Onnes's low temperature laboratory in Leiden in about 1919. *(Science Photo Library/Photo Researchers, Inc. Reproduced by permission.)*

Van der Waals was born in Leiden in the Netherlands on November 23, 1837. His parents were Jacobus van der Waals, a carpenter, and the former Elisabeth van den Burg. Van der Waals attended local primary and secondary schools and then took a job teaching elementary school in his hometown. In 1862 he began taking courses at the University of Leiden and, two years later, received the credentials necessary to teach high school physics and mathematics. He then accepted a job teaching physics in the town of Deventer and, a year later in 1866, became headmaster of a secondary school in the Hague.

During his year at Deventer, van der Waals married Anna Magdalena Smit, who bore him three daughters, Anne Madeleine, Jacqueline Elisabeth, and Johanna Diderica, and one son, Johannes Diderik. Biographers note that Anna Magdalena died while the children were still very young; Van der Waals never remarried.

While in the Hague, van der Waals continued to attend the University of Leiden on an informal basis. Since he had never studied Greek and Latin in high school, he was not allowed by federal law to enroll in a doctoral program. When that regulation was abolished in the late 1860s, van der Waals was admitted as a regular graduate student at Leiden. For his dissertation he chose to study the nature and behavior of the particles that make up gases and liquids.

Van der Waals's choice of topics, he later said, was strongly influenced by a paper written by the German physicist **Rudolf Clausius** in 1857. In that paper Clausius had argued that the molecules of a gas can be considered tiny points of **matter** in constant **motion**. From this initial premise, Clausius was able to derive theoretically a law relating **gas pressure** and volume originally stated empirically by **Robert Boyle** in 1662. It occurred to van der Waals that the molecules of both gases and liquids might be considered in the same way, as tiny points of matter. In such a case, according to van der Waals, there might be no fundamental difference between gases and liquids, the latter being only compressed gas at a low temperature.

It was this concept that van der Waals explored in detail in his doctoral thesis, presented to the faculty at Leiden in 1873. He pointed out that two fundamental assumptions of earlier **gas laws** were not valid. In the first place, such laws had assumed that the particles of which a material is made had no effective size. Van der Waals argued that they did have measurable volume and that such volume affected the behavior of a gas. A second assumption of gas laws was that gas particles do not interact with each other. Van der Waals argued instead that particles do indeed exert forces on each other.

Given these modifications in starting assumptions, van der Waals was able to develop an equation that more closely matches the actual behavior of gases. Laws such as those of Robert Boyle and **Jacques César Charles** had been regarded as correct for "ideal" gases, but always failed to some extent when applied to any real gas. Under van der Waals's formulation, the revised gas law applied with remarkable precision to any real gas. Van der Waals's work earned him almost instantaneous fame among his colleagues. His thesis was translated into German, English, and French, and gained him notice in the science world. Van der Waals was elected to the Royal Dutch Academy of Sciences in 1877, and two years later he was appointed professor of physics at the newly created University of Amsterdam. He remained in that post for three decades, retiring in 1907, to be succeeded by his son.

Van der Waals continued to work on the relationship between gases and liquids for the rest of his career. In 1890 he suggested the notion of binary solutions—states in which a substance exists as both a gas and a liquid at the same time. The calculations that van der Waals made on binary solutions later proved crucial in the fledgling field of **cryogenics**, specifying the conditions under which a gas can be converted to a liquid. One of the pioneers of this field, **Heike Kamerlingh-Onnes**, acknowledged his debt to van der Waals in an article in Eduard Farber's book *Great Chemists,* in which he said, "How much I was under the influence of its great importance as much as forty years ago may be best judged by my taking it then as a guide for my own researches."

For many students of science, van der Waals may be best known for the weak **intermolecular forces** that now carry his name. Originally called by him "pseudoassociation," these forces were hypothesized by van der Waals to explain the aggregation of particles in liquid solutions that occurred, for example, during the formation of binary solutions. Today, van der Waals forces are invoked to describe a host of situations in which rapidly shifting **electron** distributions in a molecule result in the formation of weak, but nonzero, transient attractions between molecules.

During the last ten years of his life, van der Waals gradually grew frail; he died in Amsterdam on March 8, 1923. Van der Waals had been elected to membership in the French Academy of Sciences, the British Chemical Society, the U.S. National Academy of Sciences, the Royal Academy of Sciences of Berlin, and the Russian Imperial Society of Naturalists.

VAN DER WAALS FORCE

A Van der Waals **force** between molecules is a relatively weak intermolecular attraction. All neighboring molecules in liquids and solids attract each other. The nature and strength of these interactions depends on the types of **atom** groups or functional groups that comprise the molecules. Some molecules are polar and some have hydrogen bonds. These relatively strong

intermolecular interactions require specific structural features. Polar interactions require a nonsymmetric arrangement of bonds with atoms of different electronegativity—polar bonds. Hydrogen bonding requires that one species have a **hydrogen atom** bonded to a highly electronegative atom such as fluorine, **oxygen**, or nitrogen. The other species must have a highly electronegative atom without a hydrogen atom bonded to it. However, all molecules interact with other molecules through Van der Waals interactions.

Van der Waals forces are the attractive forces of one transient dipole for another. A transient dipole is a temporary imbalance of positive and negative charge. At particular instances, even atoms that are spherical on average, such as those of the noble gases, will have greater **electron density** on one side of the atom than another. At that instant, the atom will possess a temporary dipole with a negative charge concentration on the side of the atom with greater electron density. If this happens in the case of an argon atom in liquid argon, for example, the argon atoms next to the one with temporary dipole would feel the effect of the dipole. An atom near the negative end of the dipole would have its own electrons slightly repelled from the negative concentration of charge, developing a dipole with its positive end near the negative charge of the original atom. An argon atom on the other side of the original temporary dipole would feel its electrons attracted to the positive end of the dipole, developing a dipole with the opposite orientation. In this way, temporary dipoles are propagated through a liquid or solid. The **motion** of the molecules in the liquid or solid soon disrupts the pattern, but similar events take place continually. The larger the size of atoms and the more electrons they possess, the greater the probability of forming substantial transient dipole interactions. Molecules which are non-polar and non-polar functional groups of molecules only experience Van der Waals interactions with other molecules or functional groups.

To understand the differences in properties of larger molecules, the additivity of intermolecular interactions becomes important. In effect, the interaction of each group of atoms of a molecule with a group of atoms of a neighboring molecule can be considered to be independent of the interactions of other groups of atoms of the molecules. The total **energy** required to move two molecules apart is the sum of all the energies of the individual interactions. The more groups and the stronger each individual interaction, the greater the sum of energy of interactions. Among the non-polar linear alkanes, the **boiling** point for a molecule with many -CH_2 groups, such as liquid octane—$CH_3(CH_2)_6CH_3$—is higher than that of gaseous propane—$CH_3CH_2CH_3$—because of the greater number of Van der Waals interactions between the octane molecules. For large molecules such as the higher alkanes (heavy oils and waxes) and polymers such as polyethylene, the total attractive energy due to Van der Waals forces can be greater than the polar interactions or hydrogen bonding interactions of other, smaller molecules. Hence, molecules that have only Van der Waals interactions may still melt or boil at high temperatures.

Van Vleck, John (1899-1980)

American physicist

John Van Vleck was one of the United States' first theoretical physicists, specializing in problems of chemical physics, **magnetism, quantum theory**, and spectroscopy. Some of his work has had important practical applications in such devices as the **atomic clock, lasers** and **transistors**. He shared the 1977 Nobel Prize for Physics for his "fundamental theoretical investigations of the electronic structure of magnetic and disordered systems."

John Hasbrouck Van Vleck was born in Middletown, Connecticut, on March 13, 1899, into a prosperous family with a history of notable intellectual accomplishments. His paternal grandfather had been a professor of astronomy at Connecticut's Wesleyan College, and his father, Edward Burr Van Vleck, was professor of mathematics at Wesleyan at the time of his son's birth. Van Vleck's mother was the former Hester Lawrence (also given as Laurence) Raymond. It has been noted that his parents' overbearing manner may have been responsible for their only child's shyness as a youngster. When Van Vleck was seven years old his father accepted an appointment at the University of Wisconsin. John attended public schools in Madison and then entered the University of Wisconsin, graduating in 1920 with a bachelor's degree in physics. For his graduate work, Van Vleck chose to attend Harvard University, where his father was serving as visiting professor of mathematics.

Chooses a Career in Theoretical Physics

At Harvard, Van Vleck decided to concentrate on theoretical physics, a field with little tradition in American science at the time. In fact, when he received his doctorate in 1922, his thesis was one of the first, if not actually *the* first, in America based on a purely theoretical subject—the ionization **energy** of a particular model of the **helium atom**. Upon completing his degree, Van Vleck was invited to stay on at Harvard as an instructor in physics.

In 1923, based largely on his doctoral work, the University of Minnesota offered Van Vleck a job in its physics department; he accepted and remained at Minnesota until 1928. During his tenure there, Van Vleck worked on problems involving the application of quantum mechanical theory to a variety of physical phenomena. His magnum opus during this period was his first book, *Quantum Principles and Line Spectra,* published in 1926. Though the book came out just as major modifications in quantum theory were being made, much of what he had written remained valid and the book was an unexpected commercial success.

Begins Work on Magnetism

Van Vleck's years at Minnesota were also marked by his first venture into the field for which he is best known, the quantum explanation of magnetic effects. He tried to find a way in which modern developments in quantum theory could be used to explain the various forms of magnetism—efforts that resulted in the publication of *The Theory of Electric and Magnetic Susceptibilities* in 1932. The work on magnetism was by no means the only topic Van Vleck researched at Minnesota, but it would prove to be the most important, earning him both a share of the 1977 Nobel Prize for physics and the title "father of modern magnetism."

Van Vleck was married to Abigail June Pearson on June 10, 1927. The following year he moved to the University of Wisconsin, where he accepted a post as professor of physics. He was attracted to Wisconsin, in part, by the university's visiting scholars program. Each semester, an outstanding authority in some field was invited to be in residence on the Madison campus. Van Vleck knew that the program would be an excellent way for him to stay in touch with developments in modern physics, a field in which American scientists were woefully deficient. In 1934, Van Vleck was offered an opportunity to return to Harvard, one that he accepted. He remained at Harvard until his retirement in 1969, serving the last eighteen years of his tenure there as Hollis Professor of Mathematical and Natural Philosophy, the oldest endowed science chair in North America.

One of Van Vleck's areas of interest at Harvard was crystal field theory. He again used quantum theory to evaluate the relationship between **electron** and ion energy levels in bound systems such as crystals. Understanding these relationships is critical in solid-state theories and in their applications in devices such as lasers and semiconducting devices.

At the beginning of World War II, Van Vleck was asked to serve on a committee evaluating the feasibility of building an **atomic bomb**. That committee's favorable decision eventually led to the creation of the **Manhattan Project**, under which the world's first **nuclear weapons** were designed and built. For the majority of the war years, however, Van Vleck worked on the development of **radar** at the Radio Research Laboratory in Cambridge, Massachusetts.

Van Vleck worked on a host of other problems on his return to Harvard after the war, including **nuclear magnetic resonance**, molecular spectra, and the cohesive energy of metals. In addition to the Nobel Prize, he was awarded the title of Chevalier in the French Legion of Honor, and was the recipient of the Irving Langmuir Award of the General Electric Foundation in 1965, the National Medal of Science in 1966, and the Lorentz Medal of the Royal Netherlands Academy of Science in 1974, among others. Van Vleck died in Cambridge on October 27, 1980.

Variational Principles

In 1696 **Isaac Newton** left Cambridge University to become Warden of the Mint. When he came home one afternoon, he learned of a problem set by Johann Bernoulli as a challenge to the "great mathematicians of the world." Since Newton and **Gottfried Wilhelm Leibnitz** were engaged in a bitter priority dispute over the invention of calculus, it was really a challenge issued by continental mathematicians to Newton. The so-called *brachistochrone* problem was to determine the shape of the wire connecting two nails randomly driven into a wall which would permit a bead that is acted upon only by **gravity** to slide from the top nail to the bottom one in the least amount of time. Newton developed the calculus of variations and solved the problem that very night. Though he published his solution anonymously, upon examining it Bernoulli famously remarked, "I recognize the lion by his claw."

Indeed, the calculus of variations is one of the most powerful tools in all of mathematics. The important results of physics, particularly those describing the **dynamics** of particles and fields, are obtained from it. All of classical **mechanics** is realized as minimizing a quantity known as the action, the integral of the Lagrangian, which is the kinetic **energy** minus the **potential energy** of a system. The general theory of relativity reduces to determining the geodesic, or path, followed by **light** in a curved **space-time**. The path integral approach to **quantum field theory** applies the calculus of variations to the **space** of trajectories followed by a particle.

As with all great ideas, the underlying principles are straightforward. Consider a function f(y, y', x) where f is a known function of the variables y(x), y'(x) = dy/dx, and the parameter x. The functional form of y(x) is not known *a priori*. Let $I = \int_a^b f(y, y', x)\, dx$. The coordinates of the limits of integration, the points a and b in the xy-plane, are known and fixed. The calculus of variations is to determine the function y(x), namely the path of integration, such that the integral I is minimized. The solution to this problem yields a relation for f known as the Euler-Lagrange equation:

$$\frac{\partial f}{\partial y} - \frac{d}{dx}\frac{\partial f}{\partial y'} = 0$$

Suppose we let x be the time t and y(t) be the position of a particle. The first derivative of position with respect to time, $\dot{y}(t)$, is **velocity**. Let f be the Lagrangian of the particle. Solving the Euler-Lagrange equation for y(t) determines how the particle evolves in time.

To see exactly how this idea works in practice, let us consider an explicit example. Attach a **mass** m to the end of a spring with Hooke's constant k. The **kinetic energy** of the mass is written $1/2\, m\dot{y}^2$, and the potential energy is $1/2\, ky^2$. Therefore, the Lagrangian is their difference: $f = 1/2\, m\dot{y}^2 - 1/2\, ky^2$. The Euler-Lagrange equation gives

$$-ky - m\ddot{y} = 0$$

or equivalently,

$$\ddot{y} = -\frac{k}{m}y$$

This is a statement of Newton's second law. The **force** applied by the spring F = –ky is the the product of mass and **acceleration**. The solution to this second-order differential equation in time is y(t) = A cos (ωt) + B sin (ωt) with

$$\omega = \sqrt{k/m}$$

which is indeed the equation of a simple harmonic oscillator. Although this problem is rather simple, the very same principle applies to more complicated systems. Of course, the Euler-Lagrange equation generalizes to problems where the function f depends on more than one variable y. Applying techniques similar in spirit to the previous computation, the differential equations which govern the time evolution of the system can be written immediately from the Lagrangian. Thus, problems which were previously cumbersome to solve (determining the shape of planetary orbits, for example) reduce to solving a system of differential equations.

The shortest distance between any two points on a plane is a straight line. The shortest distance between two points on a sphere lies on a great circle. Fermat's principle, which anticipated variational techniques, argues that the path taken by a beam of light between two points is the one which is an extremum (a maximum, minimum, or saddle point). Such a trajectory is called a geodesic. Einstein's general theory of relativity addresses geodesics in backgrounds that are curved by the presence of **matter**. The determination of geodesics such as the path taken by the Earth in its orbit around the **Sun** is an exercise in the calculus of variations.

Another significant problem which the calculus of variations can address is determining minimal surface areas and volumes. For example, soap bubbles minimize their energy by forming shapes such that the smallest possible surface area encloses a given volume of air. The calculus of variations allows us to deduce that such a minimal area surface is a sphere.

VECTOR BOSONS

Three of the four fundamental forces play a significant role in the interactions of elementary particles. A common feature that **electromagnetism**, the strong **force**, and the weak force share is that the gauge particles that mediate these forces are all **vectors**, which means that they carry spin-1. A **space-time** vector describes the coordinates of a point in three spatial dimensions and time. The Lorentz group (the set of boosts from one inertial reference frame to another together with the set of rotations) acts on a vector in a specified way. A particle is called a vector if it transforms just like a space-time vector does under the action of the Lorentz group.

Consider an **electron** and its anti-particle, the **positron**. In a collision, they annihilate to form a virtual boson. A boson is a particle with an integer spin (1,2,3...). If the interaction is electromagnetic in nature, the virtual particle that is created is a **photon**. If the interaction is mediated through the weak force, the virtual particle is a **Z-boson**. Since electrons do not carry **color**, they do not experience the strong force. **Quarks**, however, can interact through any of the three forces, and the intermediate particle can be a photon, a gluon, a W+, a W-, or a Z. Because **electric charge** is conserved, if the total electric charge of the initial particles is zero as it is in electron-positron annihilation, then either a photon or a neutral "Z" is produced. When the initial particles carry a charge of +1 and interact weakly, only a W+ may be produced. Similarly, a W- results from the weak interaction of initial particles of charge -1.

Let us examine each of the vector bosons. The photon mediates **quantum electrodynamics (QED)**. Because a real photon is massless, it has two polarizations. It is said to be left-handed or right-handed. Heisenberg's uncertainty principle says that the product of the uncertainty in the **energy** and the uncertainty in the propagation time of a particle exceeds a dimensional constant, h/2, so a high energy virtual photon can be produced for a brief interval of time. This virtual photon has a nonzero **mass** and is said to be off-shell; it has three polarizations. The gluon is the gauge boson which mediates the strong force or quantum chromodynamics (QCD). One difference between QED and QCD is that while there is only one electrically neutral photon, there are eight different kinds of gluons. All but one of the gluons carry color, the charge associated to QCD. Photons are electrically neutral so they do not interact with the force they mediate. Gluons, on the other hand, do interact with the strong force. This makes QCD a much more complicated theory than QED.

Unlike the photon and the gluon, the intermediate vector bosons, Z, W+, and W-, are massive. An intermediate vector boson is virtual if its mass is different from its measured mass. The photon, Z, and W arise from a process called spontaneous symmetry breaking. In a model devised independently by **Sheldon Glashow**, **Steven Weinberg**, and **Abdus Salam**, the electroweak symmetry is broken to distinguish between the electromagnetic and weak forces. Symmetry breaking is accomplished through a scalar field called the Higgs. The Higgs particle acquires a **vacuum** energy. Because three of the **generators** of the electroweak symmetry are nonzero when they act on the vacuum energy of the Higgs scalar, three vector bosons (Z, W+, W-) acquire masses. The fourth, the photon, remains massless. The masses of the W and Z are related by the formula $M_W = M_Z cos\theta_W$, where θ_W is the weak mixing angle.

VECTORS

A vector is a quantity that has both magnitude and direction. A scalar, for comparison, has only magnitude. **Velocity** and displacement are everyday examples of vectors. An arrow with a direction is used to draw a vector. An arrow above a letter is used as the notation for a vector. In publications, a bold letter like r is often used instead because it is easier to type.

There are two kinds of vectors. One is called a polar vector; the other is called an axial vector or pseudo-vector. They both obey vector algebra, but they have different behaviors in a symmetric transformation called space inversion. Velocity and displacement are both vectors, while angular velocity, scalars, and infinitesimal angular displacement are examples of pseudo-vectors.

Vectors also have algebraic operations. The ones used most widely are addition, subtraction, and multiplication (dot product and cross product). The set of rules for these operations is called vector algebra. Division of vectors is not defined. There are two methods to do these operations on vec-

tors, geometric method and matrix method. The matrix method is used most frequently in physics, but the geometric method is much more intuitive. The operations below are illustrated using the geometric method.

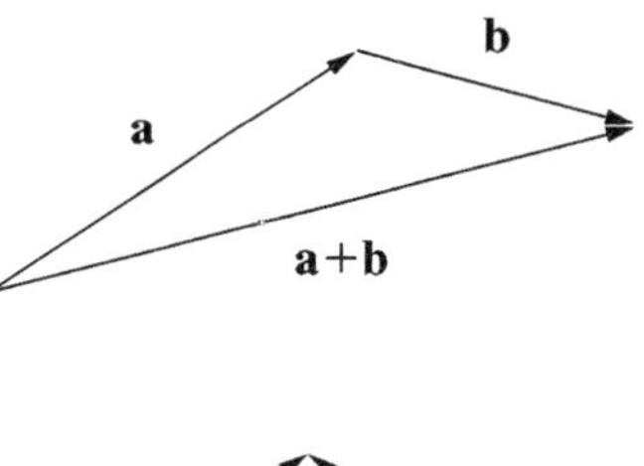

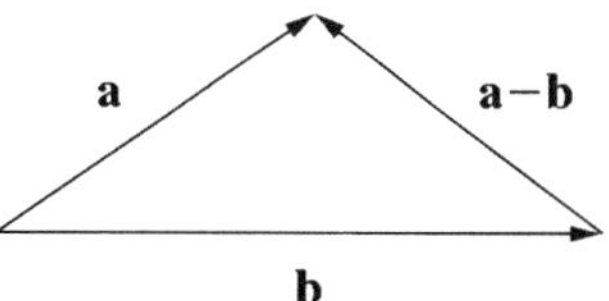

Vectors can be added up in the following way. First draw out the vectors as arrows one by one, putting the tail of the next arrow right at the head of the previous one. After this is done, draw an arrow from the tail of the first vector to the head of the last one. That final arrow is the sum of all the vectors. The order of vectors does not matter. For example, in the addition of velocities, one can suggest that someone walking for three miles north followed by four miles east will end up at the same point as somebody walking five miles in the appropriate north-easterly direction.

A vector can also be subtracted from another vector. To subtract one vector from another, reverse the first vector's direction and subtract it from the other. For example, to calculate $a - b$, draw the two vectors as arrows, making their tails starting at the same point, then draw an arrow from the head of b to the head of a (order is important here). That final arrow is $a - b$. If the order of drawing at the final step is reversed, the result will be $b - a$ instead.

Vector products come in two types. Dot products, also called inner products, give a scalar (an ordinary quality) as a result. For example, 12 km is a scalar product, while 12 km northwest is a vector. One can multiply the scalar of two vectors using the dot product function. Cross product functions, on the other hand, give a third vector as a result.

VECTORS IN SPACE-TIME

The development of relativistic physics requires the use of a geometry adequate to account for the structure of spacetime. Whereas special relativity describes the non-quantum physical world in a nongravitational context in which spacetime is flat, the gravitational effects generated by general relativity must also account for the additional complexity of the curvature of spacetime.

Four-dimensional **vectors** thus take into account the special requirements of space-time geometry, i.e., the fact that space-time has four dimensions, as opposed to the three dimensions of **Euclidean space**. They accordingly have several applications in the context of relativistic physics. These applications include the use of translation vectors, which represent displacements in the four-dimensional continuum space-time model, the more abstract consequences of special relativity such as **time dilation**, **length contraction** or the addition of relativistic velocities, and the more elaborate descriptions of space and time by observers measuring time and space differently because they are moving relative to each other. In this case, the relevant quantities are assembled into 4-vector representations which transform linearly when their values are compared for the observers in different states, say, of uniform **motion**. A 4-vector is then a quantity with four components which changes like spacetime coordinates under a coordinate transformation. 4-vector representations are obtained using the same approach as the one used, for example, to generate a 3-component vector xi from a set of three spatial coordinates:

$$\overline{x}_i = \left(x_1, x_2, x_3\right) \equiv \left(x, y, z\right)$$

Similarly, a 4-vector **space-time** coordinate for an *event* can then be written as:

$$\overline{x}_\mu = \left(x_c, x_1, x_2, x_3\right) \equiv \left(ct, x, y, z\right)$$

By convention, Greek letters are used to specify space-time components and Roman letters are used for spatial components.

An example of the use of four-dimensional vectors in the spacetime context is provided by the calculation of relative speeds. If an observer A measures two objects B and C to be travelling at velocities $u = (u_x, u_y, u_z)$ and $v = (v_x, v_y, v_z)$ respectively, the relative speed between B and C would be given in classical relativity by: $w^2 = (u–v) \cdot (u–v) = (u_x - v_x)^2 + (u_y - v_y)^2 + (u_z - v_z)^2$.

In special relativity the relative speed is instead given by the formula:

$w^2 = [(u–v) \cdot (u–v) - (uXv)^2/c^2] / (1 - (u \cdot v)/c^2)^2$.

where u-$v = (u_x - v_x, u_y - v_y, u_z - v_z)$ is the vector difference of u and v, $u \cdot v = u_x v_x + u_y v_y + u_z v_z$ is the inner product of u and v and uXv is the vector product for which $(uXv)^2 = (u \cdot u)(v \cdot v) - (u \cdot v)^2$. And when $u_y = u_z = v_y = v_z = 0$, the formula reduces to: $w = |u_x - v_x| / (1 - u_x v_x/c^2)$.

See also Energy-momentum, four vector; Event horizon; Four-dimensional space-time; Invariant interval; Lorentz transformations; Michelson-Morley experiment; Minkowski, Herman; Relativistic interval; Relativistic mass and energy; Relativistic velocity transformations; Relativity, special; Space; Space-time geometry

VELOCITY

Velocity is the time rate of change of the position of a body. Mathematically, velocity is a vector quantity having direction as well as magnitude. Speed, on the other hand, is a scalar quantity which has only magnitude. The magnitude of velocity is expressed in units such as miles per hour or meters per second when describing **motion** along a straight or curved path. A body which is rotating about an axis has angular velocity. Angular velocity is also a vector quantity and is expressed as units of angular rotation per unit of time such as revolutions per minute or radians per second.

VELTMAN, MARTINUS J.G. (1931-)
Danish physicist

Over the course of his career, Martinus J.G. Veltman made fundamental contributions to elementary **particle physics** and is particularly renowned for his work on gauge theories and the quantum structure of electroweak interactions, for which he shared the Nobel Prize in physics. Born in the Netherlands, the Dutch physicist received his education at the University of Utrecht, earning a Ph.D. in theoretical physics in 1963. For the next three years he was a fellow at CERN, the European organization for nuclear research. Veltman joined the University of Utrecht as professor of physics in 1966.

While the electroweak interaction theory was developed in the early 1960s, it had reached a standstill in the world of physics, partly because of the complex and time-consuming calculations needed to provide more details about subatomic physical quantities. Undaunted, Veltman started researching the **renormalization** of gauge theories in 1968, incorporating into his work Schoonschip, a computer program that Veltman had developed in 1963 to do complex algebra problems. This program was to prove invaluable in assisting with the tedious calculations and theory testing needed for research of field and gauge theories. In 1969, Veltman was joined in his work by **Gerardus 't Hooft**, a 22-year-old graduate student. Together, the two laid down the foundations for physicists to mathematically predict properties of sub-atomic particles that make up all **matter** in the **universe** and the forces that hold these particles together. As a part of that effort, the duo's work showed that the **electroweak theory** made sense. In 1999, the pair received the Nobel Prize in physics for "elucidating the quantum structure of electroweak interactions in physics," placing particle physics theory on a firmer mathematical foundation. Their work also led to the discovery of the top quark.

Veltman left the Netherlands to join the University of Michigan in 1981 as the John D. and Catherine T. MacArthur Professor of Physics. His other contributions to science and physics include developing a practical method of implementing probability conversion in field theory, known as the Veltman cutting rules. He also made contributions in the **quantum theory** of gravitation and in the theory of radiative corrections. Veltman's many awards include the 1989 Alexander von Humboldt Award, the 1993 High Energy Physics prize from the European Physical Society, and knighthood into the Dutch Order of the Lion.

See also Quarks and gluons

Martinus Veltman. *(AFP/Corbis. Reproduced by permission.)*

VIBRATING SYSTEMS AND RESONANCE

Anyone who has watched a child swinging on a playground swing has witnessed a vibrating system in resonance. The phenomenon of resonance appears often in everyday life; another example is the vibration of the body panels of a car containing a loud car stereo system. In the case of the child on the swing as the "vibrating system," resonance allows for fun and excitement, while in the case of the car stereo the resonance of the body panels is a source of annoyance. Three main concepts underlie vibrating systems and resonance: restoring **force**, natural **frequency**, and frequency matching.

A vibrating system is any system that exhibits a "restoring force." In the case of the child on a swing, the restoring force is the force of **gravity** that constantly acts to restore the swing to its place at the bottom of its arc. The force is called restoring because it always acts to restore the swing back to its vertical resting position. A restoring force results in vibration because any movement of the swing away from vertical will cause the swing to fall back towards vertical. In the case of the vibrating car panels the restoring force arises from the stiffness of the metal panels which forces the panels constantly towards the positions they were molded into by the manufacturer. Air **pressure** from the car stereo deforms the panels temporarily and causes them to vibrate around their molded positions.

Any system exhibiting a restoring force will possess a "natural frequency." In the case of the child on the swing the natural frequency is the number of swings undergone per second. Any swing has a natural frequency with which it tends to vibrate. This is intuitive when one considers that it would be disturbing to arrive at the playground and find the children swinging on the swings at the very low rate of one swing per hour, or the very fast rate of 100 swings per second! It seems the natural frequency of a child on a swing is somewhere near 1 swing per second. In fact, the natural frequency of a swing is largely determined by the length of the swing. Similarly, the body panels of a car have a natural frequency they "prefer" to oscillate at. These frequencies are higher than a swing's frequency, and are high enough that they can be heard as low tones or rattles.

Resonance appears when a vibrating system is forced to vibrate at its natural frequency. The child forces the swing to vibrate at its natural frequency by swinging her legs at the swing's natural frequency. This "frequency matching" causes the vibrations of the swing to grow large. The child exploits this phenomenon for fun and excitement. In the case of the car panels, when the stereo plays a note whose frequency corresponds to the natural frequency of a body panel, the body panel vibration will grow large. One might notice that different body panels will resonate with different played notes because each panel has a different resonant frequency. For the car owner, however, resonance is a source of annoyance and frustration.

VIRTUAL PARTICLES

Virtual particles are quantum mechanical particles that act as short-lived intermediate states during the interactions between real particles. By their nature, they do not live long enough to detect. However, the theory suggesting their existence has brought such success that it is widely accepted by physicists today that they do exist.

Both virtual and real quantum mechanical particles exhibit certain characteristic properties that determine their particle type. For a quantum particle to be considered an **electron**, for example, it must have an electrical charge equal to $-e = -1.6 \times 10^{-19}$ C, a **mass** equal to $m_e = 2.0 \times 10^{-26}$ lb $(9.1 \times 10^{-27}$ kg$)$, and a total intrinsic spin **angular momentum** of h/4pi. (The parameter h is **Planck's constant**, equal to 6.626×10^{-34} Js.) In addition to these characteristic properties, all quantum mechanical particles must obey a set of fundamental conservation laws including the conservation of electrical charge, spin angular **momentum**, **energy**, and the **linear momentum**.

A real quantum mechanical particle is one whose total relativistic energy is related to its rest mass m_0 by the equation $E = m_0 c^2 / \sqrt{(1 - v^2 / c^2)}$, where c is the speed of **light** in a **vacuum**, measured to be 186,000 mi/s (300,000 km/s), and v is the speed of the particle. If this relationship is true for a particle, then we say that such a particle is on the mass shell, or real. A particle whose total energy does not obey this equation is said to be off the mass shell, or virtual. The amount of deviation from this equation, dE, determines how long the virtual particle can live. Heisenberg's uncertainty principle says that such a virtual particle can only exist for an approximate time h/dE.

In electron-positron acceleraators, electrons and **positron** (anti-electrons with all of the same characteristic properties as electrons but having positive electrical charge, +e) collide to form virtual photons. (Photons are the quantum mechanical particles that make up light.) The conservation of electrical charge and spin angular momentum indicates that the intermediate state formed by the electron-positron collision must be a **photon**. However, the **conservation of energy** and linear momentum says that the photon must be virtual. Therefore, the photon lives for a very short time and then can decay into a variety of different final state particles.

VISCOSITY

The viscosity of a fluid is a measure of its resistance to continuous deformation caused by sliding or shearing forces. Imagine a fluid between two flat plates; one plate is stationary and the other is being moved by a **force** at a constant **velocity** parallel to the first plate. The applied force per unit area of the plate is called the shear stress. The applied shear stress keeps the plate in **motion** and, when the plate velocity is steady, this shear stress is in **equilibrium** with the frictional and drag forces within the fluid. The shear stress is proportional to the speed of the plate and inversely proportional to the distance between the plates. The proportionality factor between the shear stress and the velocity difference between the plates is defined as the coefficient of viscosity or simply the viscosity of the fluid. Thick **fluids** such as tar or honey have a high viscosity; thin fluids such as water or alcohol have a low viscosity.

In general, viscosity is a function of **temperature** and **pressure**; however, in some fluids viscosity is dependent on the rate of shear and time. When brushed on (sheared) quickly, fluids such as paint have a low viscosity and flow easily. After paint is applied, only the slow and steady pull of its **weight** causes it to flow; at this slow shear rate the viscosity of paint is high and its resists the tendency to flow or sag. Fluids that behave in this manner are called non-Newtonian fluids. Other examples are liquid plastics and mud. For gases and non-polymeric liquids like water, viscosity is independent of the fluid's shear stress and history. These are called Newtonian fluids. In the case of gases, the viscosity increases with temperature because of the increased molecular activity at higher temperatures. Liquids, conversely, generally show decreasing viscosity with increasing temperature.

The flow of liquids in pipes, the performance of oil-lubricated bearings in engines or oil-filled automotive shock absorbers, and the air resistance on a moving car or airplane are all dependent on the viscosity of the fluids involved.

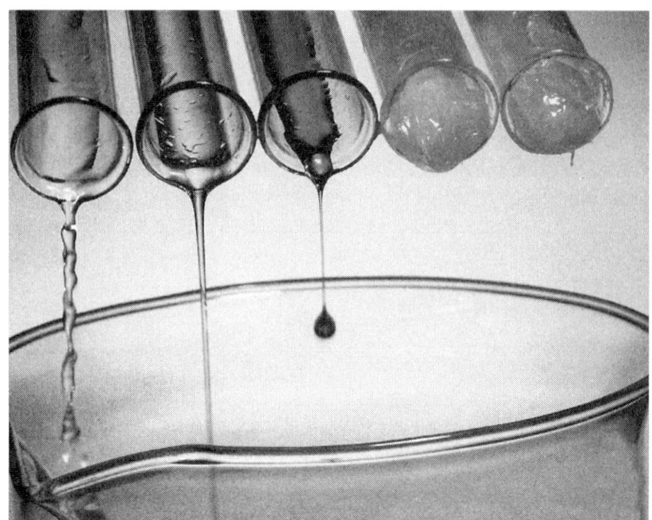

Liquids of different viscosity. (Photo by Yoav Levy/Phototake NYC. Reproduced by permission.)

VOLTA, ALESSANDRO (1745-1827)
Italian physicist

Alessandro Giuseppi Volta became one of the foremost celebrities of his day for his work with electrical currents. Volta was born on February 18, 1745, in Como, Lombardy, Italy. Most of his eight brothers and sisters entered the church, but Alessandro became engrossed in the study of **electricity**. Influenced by a history on the subject written by Joseph Priestley, 14-year-old Volta announced his intention of becoming a physicist, and became educated on the subject.

In 1774 Volta was appointed professor of physics at the high school in Como. There he created one of his most significant inventions; the electrophorous. This device had the ability to store significant electrical charges, and replaced the Leyden jar which, up until that time, had been used for storing smaller charges.

The discovery that led Volta to the invention of the electrophorous actually began with the work of French physicist **Charles Augustin de Coulomb**. Coulomb had discovered that electrical charges were located on the surface of a charged body, and not in its interior. Volta's electrophorous used two metal discs; one was rubbed to produce a negative electrical charge, the second disk was brought close enough to the first to establish a positive charge on the one side, leaving a negative charge on the other. Volta used Coulomb's discovery to draw off the negative charge from one side of his charged disc, leaving just a positive charge on the opposite side. The electrophorous was the predecessor of the modern condenser, which stores electricity in circuits.

In 1776 Volta became involved with an entirely different subject. By studying the components of marsh gas, he was able to discover methane gas. He also exploded hydrogen gas to remove **oxygen** from air and was able to make the first accurate determination of the proportion of oxygen in the air. Later, around 1796, Volta discovered that the vapor **pressure** of a

Count Alessandro Volta

given liquid had nothing to do with the pressure of the surrounding atmosphere; it was solely dependent on **temperature**.

Meanwhile, the electrophorous had established Volta's reputation and, in 1779, he was appointed a professor at the University of Pavia. There he developed an electrometer that allowed him to measure electric currents. Then, in 1791, he was drawn into a controversy that erupted when his compatriot **Luigi Galvani** announced the existence of "animal electricity" that caused muscles in frog's legs to twitch when touched with metal probes of different composition. Volta did not believe such a thing possible and proceeded to experiment himself.

Volta experimented on numerous animals, including himself, and ultimately discovered there was no such thing as "animal electricity," and that it was the action of the two different types of metal probes that was the source of the current. This fact was made painfully certain when Volta placed the two metal probes on his tongue. The result was definitely unpleasant. Volta was surprised to find that his own tongue was a more sensitive detector of electricity than his electrometer.

The controversy raged on until 1800. In that year Volta built a device that produced a large flow of electricity. He filled bowls with a saline solution and "connected" them with strips of different metals. One end of the strip was copper; the other end was tin or zinc. By bending his strips from one bowl into another, Volta was able to create a constant flow of elec-

trical current; the world's first electric battery had been invented. Volta had proven that the metal was the source of the electricity, and *animal electricity* did not exist.

In the interest of making his battery smaller, Volta used round discs of copper, zinc, and cardboard that had been soaked in a saline solution. He stacked his discs one on top of the other. Attaching a wire to the top and bottom of his pile allowed the **electric current** to flow. The invention of this Voltaic pile marked the apex of Volta's career.

The Voltaic pile came to the attention of English chemist William Nicholson (1753-1815), who proceeded to build his own in the same year. Nicholson placed the ends of his wires in water and discovered the flowing current "electrolyzed" the water, breaking it up into hydrogen and oxygen. **Henry Cavendish** had shown those two elements could form water; Nicholson reversed the procedure.

The invention of the Voltaic pile, the earliest form of an electric battery, was the highpoint of Volta's life. He died on March 5, 1827, at the age of 82. In his honor, the unit of **force** that moves electric current was named the volt.

VOLTAGE

Voltage is a difference in electrical potential. Although voltage is sometimes referred to as the **electromotive force**, voltage is actually a potential difference. Usually voltage is expressed in volts (V), a unit of electrical measurement named after Italian physicist **Alessandro Volta**.

Voltage refers to the difference in electrical potential between two points, not the electrical potential itself. The difference in electrical potential between two points A and B is defined as the change in **potential energy** divided by the charge moved. The change in potential **energy** can be thought of as the negative of the **work** done to move a charge in an electric field. When a charge q moves in an electric field E from point A to point B, the work done on the charge by the electric field is qEd, where d is the distance between points A and B. In this case, the change in potential energy is equal to -qEd. When thinking about voltage it is useful to use this concept. For instance, if a 12V battery is connected to two parallel plates, the electric field E between the plates has a magnitude equal to the difference in potential of the two terminals divided by the plate separation.

The voltage of a system is often measured by a voltmeter. The voltmeter measures the difference in electrical potential between two points in a circuit. Two main types of voltmeters are the digital and the analog voltmeter. Analog voltmeters are usually based on the d'Arsonval galvanometer and give voltage readings that can vary over a continuous range. Analog voltmeters usually have a continuous scale and a pointer that indicates the voltage. A digital voltmeter yields voltage readings in groups of digits and has no continuous scale. Digital voltmeters are becoming increasingly popular because of ease of use and reduced costs in production.

A potentiometer is another useful device that is used to continuously vary the voltage of a device. It acts as a variable

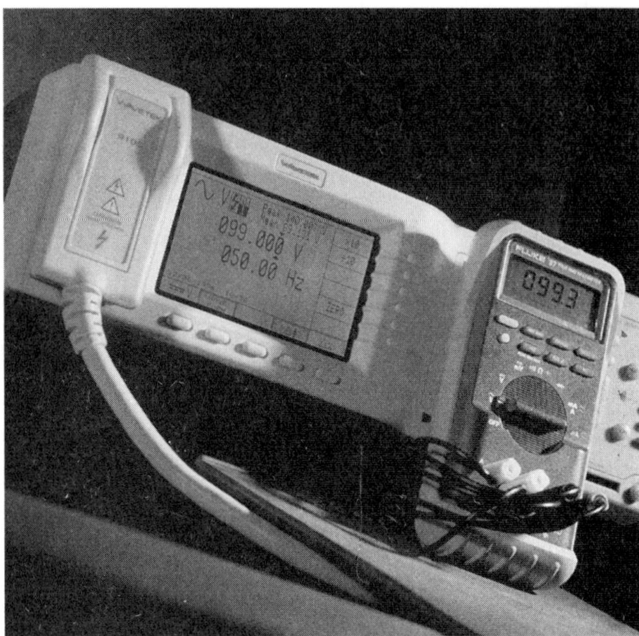

Digital voltmeter measuring electricity supplies (volts, herz, alternating current). (Photo by Steve Worrell. Science Photo Library/Photo Researchers, Inc. Reproduced by permission.)

resistor in a circuit. Potentiometers are usually some form of conductive metal tube or wire that has three terminals. The two resistance terminals are attached to the two input voltage conductors of the battery and the other terminal of the potentiometer, movable with respect to the resistance terminals, is attached to the output voltage conductor. The distance between the two resistance terminals and the third terminal can be varied, hence the percentage of the input voltage applied to the circuit. As well as continuously varying the voltage of a device, the potentiometer can be used to measure the electrical potential difference of a device. It is sometimes more useful to use a potentiometer than a voltmeter because the voltmeter itself often draws significant current.

See also Electric potential

VON KLITZING, KLAUS (1943-)
German physicist

Klaus von Klitzing was awarded the 1985 Nobel Prize in physics for his discovery of the quantized **Hall effect**, a variation on an electrical phenomenon first observed by the American physicist **Edwin Herbert Hall** in about 1880. Von Klitzing's discovery has had a number of profound effects in both theoretical and practical fields of physics, exhibiting one of the first instances in which quantum effects had been observed on a macroscopic scale. In addition, it made possible the establishment of an entirely new international standard for the ohm, the unit of measure of **electrical resistance**.

Von Klitzing was born on June 28, 1943, in Schroda, Germany, close to the Polish border, to Bogislav von Klitzing,

a forester, and Anny Ulbrich. As World War II turned against Germany, the von Klitzing family decided to stay ahead of the advancing Soviet army and moved westward to the town of Lutten. Three years later, in 1948, they moved again to Oldenburg and finally, in 1951, to the northern town of Essen. Von Klitzing eventually completed his secondary education at the Artland Gymnasium in Quakenbrück.

Begins His Studies of Semiconductors

In 1962, von Klitzing entered the Technical University of Braunschweig, intending to major in physics. He was awarded his baccalaureate degree in 1969 for a dissertation on the electrical properties of indium antimonide, a compound of two semiconducting elements, indium and antimony. Von Klitzing then moved to the University of Würzburg for his doctoral studies, planning to continue his work on **semiconductors** there. He also accepted a job at the University teaching physics to premedical students. At Würzburg, von Klitzing became particularly interested in the effects of strong magnetic fields on the conducting properties of semiconductors. In 1971, he published his first scientific paper on this topic, "Resonance Structure in the High Field Magnetoresistance of Tellurium," with G. Landwehr, one of his instructors, and was awarded his Ph.D. the following year for this line of research.

A key feature of von Klitzing's ongoing research was the need for stronger and stronger magnetic fields. He spent the 1975 academic year at Oxford University because of the powerful superconducting magnets being manufactured there and, in 1979, continued his research at the High-Field Magnet Laboratory of the Institute Max von Laue-Paul Langevin in Grenoble, France. It was at Grenoble that von Klitzing made the discovery that earned him the Nobel Prize.

Observes Quantization of the Hall Effect

The Hall effect is a three-dimensional phenomenon in which an electrical current is passed in one direction through a conducting material while a magnetic field is applied at right angles to the current. The accumulation of electrons along one edge of the conductor results in a potential difference along the face of the material that is called the Hall **voltage**. The Hall resistance, then, is the Hall voltage divided by the current in the conductor. Under these conditions, von Klitzing observed a totally unexpected effect. As the magnetic field on the sample was strengthened, the Hall resistance also increased as a linear function over a certain range, and then levelled off. Further increases in the magnetic field had no effect on the Hall resistance over another range, but then the Hall resistance began to increase again. After another period, the Hall resistance again levelled off. The final results of the experiment appear graphically as a series of steps. Under conventional experimental conditions, however, the Hall resistance is a continuous, linear function of the imposed magnetic field.

To analyze his contradictory findings, von Klitzing devised an experiment with a number of conditions. Most important was his use of a very thin sheet of silicon that constrained the movement of electrons to two dimensions rather than the three normally allowed. In addition, the experiment was carried out in a powerful magnetic field at temperatures close to **absolute zero**. In this manner, von Klitzing demonstrated that the Hall effect is quantized; changes in the external magnetic field only induce electrical changes in the silicon in certain steps and not continuously. Von Klitzing found that all the possible quantum steps had a value of a **fundamental constant** divided by an integer number. That constant, 25,813 ohms, is significant because it is the ratio of two fundamental constants of nature, the square of an electron's electrical charge and **Planck's constant**. In fact, an effect of this type had been foreseen in 1975 by three Japanese theoretical physicists, T. Ando, Y. Matsumoto, and Y. Uemura. The theory did not predict, however, the high degree of precision that von Klitzing had found.

Von Klitzing's discovery is a significant one for physics. It is one of the few instances in which quantum effects have been observed directly in the laboratory; such effcts are normally important only at the level of individual particles such as the **electron** or **atom**. In addition, the high precision of von Klitzing's results means that a new and more exact standard for the ohm, the unit of electrical resistance, may be possible.

In 1971, von Klitzing married Renate Falkenberg, with whom he had two sons and a daughter. In addition to the Nobel Prize, von Klitzing was awarded the Walter-Schottley Prize of the German Physical Society in 1981 and the Hewlett Packard Prize of the European Physical Society in 1982.

W

W MESONS

W mesons, or W bosons, are heavy charged particles that mediate the weak nuclear interaction. In standard high-energy units, they have a **mass** of 81 GeV. There are two varieties, the W^+ with charge +e, and its **antiparticle**, the W^-, with charge −e. The W bosons are some of the heaviest elementary particles, with a mass comparable to that of a krypton **nucleus**. They have a very short lifetime on order of 10^{-27} seconds, which requires that they be detected by looking for specific decay signatures in **particle detectors**. They were predicted to exist in the late 1960s and early 1970s by the unified **electroweak theory**, and were eventually detected in the mid-1980s at the large **electron positron** collider, LEP, at the **European Center for Nuclear Research (CERN)** near Geneva, Switzerland.

See also Bosons; Electroweak particles; Particle accelerators; Standard model of particle physics

WALTON, ERNEST (1903-1995)

Irish experimental physicist

Ernest Walton was an Irish experimental physicist who gained renown for achieving, with physicist **John D. Cockcroft**, the first artificial disintegration of an atomic **nucleus**, without the use of radioactive elements. Their breakthrough was accomplished by artificially accelerating a beam of protons (basic particles of the nuclei of atoms that carry a positive charge of **electricity**) and aiming it at a target of lithium, one of the lightest known metals. The resultant emission of alpha particles, that is, positively charged particles given off by certain radioactive substances, indicated not only that some protons had succeeded in penetrating the nuclei of the lithium atoms but also that they had somehow combined with the lithium atoms and had been transformed into something new. Although the process was not an efficient **energy** producer, the work of Walton and Cockcroft stimulated many theoretical and practical developments and influenced the whole course of **nuclear physics**. For their pioneering work, Walton and Cockcroft shared the 1951 Nobel Prize in physics.

Ernest Thomas Sinton Walton was born October 6, 1903, in Dungarven, County Waterford, in the Irish Republic. His father, John Arthur Walton, was a Methodist minister, while his mother, Anna Elizabeth (Sinton) Walton, was from a very old Ulster family, who had lived in the same house in Armagh for over two hundred years. The young Walton was sent to school at Belfast's Methodist College, where he demonstrated an aptitude for science and math. It was no surprise, then, that he decided to enroll in math and experimental science at Dublin's Trinity College in 1922. He graduated in 1926 with a B.A. degree, in 1928 with an M.Sc., and in 1934 with an M.A.

Joins the Cavendish During its Heyday

The following year he headed to Cambridge University, England, on a Clerk Maxwell research scholarship. There, he joined the world-famous Cavendish Laboratory, headed by the great New Zealand-born physicist **Ernest Rutherford**. Walton was assigned cramped laboratory space in a basement room. While his quarters were less than luxurious, he was at least blessed by having roommates with whom he struck up an immediate friendship, physicists T. E. Allibone and John D. Cockcroft. Walton would go on to make scientific history, in collaboration with the latter, for a project that would pave the way for the development of the **atomic bomb**. At the suggestion of Rutherford, Walton began attempting to increase the **velocity** of electrons (the negatively charged particles of the **atom**) by spinning them in the electric field produced by a changing circular magnetic field as a method of nuclear disintegration. Although the method was not successful, he was able to figure out the stability of the orbits of the revolving electrons, and the design and engineering problems of creating an accelerating machine with minimal tools and materials. This early work of Walton's later led to the development of the betatron, that is, a particle accelerator in which electrons are propelled by the inductive action of a rapidly varying magnetic field.

Next, Walton tried to build a high **frequency** linear accelerator. His goal was to produce a stream of alpha particles traveling at high speed which could be used to shed light on various aspects of the atomic nucleus. Rutherford had long been keen to get his hands on such a source of alpha particles but despaired of any short-term breakthrough. As Walton's work progressed, Rutherford's wish was granted sooner than he expected.

What was needed was a fundamentally different way of viewing the problem. Walton and his colleagues at the Cavendish were trying to accelerate electrons to a speed sufficient to enable them to penetrate an atomic nucleus. Such high velocities were necessary, they believed, in order to counteract the repulsive charge of the nuclei. The speeding electrons, they figured, would literally bully their way through. However, achieving such high speeds was easier said than done. It required the application of enormous amounts of electricity, about four million volts, which at that time was impossible to generate in a discharge tube (a tube that contains a gas or metal vapor which conducts an electric discharge in the form of light). A crucial breakthrough came in 1929, when the Russian physicist **George Gamow** visited the Cavendish laboratory. With physicist **Niels Bohr** in Copenhagen, he had worked out a wave-mechanical theory of the penetration of particles, in which they believed particles tunneled through rather than over potential barriers. This meant that particles propelled by about 500,000 volts, as opposed to millions, could possibly permeate the barrier and enter the nucleus if present in sufficiently large numbers. That is, one would need a beam of many thousands of millions of moving particles to produce atomic disintegrations that would be capable of being observed.

Rutherford gave Walton and Cockcroft the go-ahead to test the supposition. It was a measure of his confidence in them—the high **voltage** apparatus they constructed to enable them to accelerate atomic particles cost almost £1,000 (British pounds) to build. It was an enormous sum in those days, and represented almost the entire annual budget for the laboratory.

The machine, the first of its kind ever built, and today on view at the London Science Museum in South Kensington, was built out of an ordinary transformer, enhanced by two stacks of large condensers (or what would today be called capacitors), which could be turned on and off by means of an electronic switch. This arrangement generated up to half a million volts, which were directed at an electrical discharge tube. At the top of the tube protons were produced. The velocity of the protons was increased into a beam which could be used to hit any mark at the bottom of the tubes. Although it would be considered primitive by today's standards, their apparatus was, in fact, an ingenious construction, cobbled together from glass cylinders taken from old fashioned petrol pumps, flat metal sheets, plasticene, and **vacuum** pumps. The current generated by the discharge tube was almost one hundred-thousandth of an ampere, which meant that about 50 million million protons per second were being produced. The availability of such a large and tightly controlled source of particles—compared with that produced by, say, a radioactive source—greatly increased the odds of a nucleus being penetrated by the speeding atomic particles.

Halfway through 1931, while their experiment was still in its early stages, Walton and Cockcroft were forced to vacate their subterranean basement when it was taken over by physical chemists. They were obliged to deconstruct their installation and build it again. As it happened, it turned out to be a lucky break. Their new laboratory was an old lecture theater, whose high ceiling was much more suitable for their purposes. When Walton and Cockcroft went to reassemble their massive apparatus, they used the opportunity to introduce a few modifications. This time around, they incorporated a new voltage multiplying circuit, which Cockcroft had just developed, into their apparatus. It took them until the end of 1931 just to produce a steady stream of five or six hundred volts.

When the accelerator was finally completed, they restarted the laborious process of trying to penetrate an atomic nucleus using a stream of speeded up protons. They positioned a thin lithium target obliquely across from the beam of protons in order to observe the alpha particles on either side of it. In order to detect the alpha particles that they hoped would be produced, they set up a tiny screen made of zinc sulfide, which they observed with a low-power **microscope**, a technique borrowed from Rutherford.

Walton's Scintillating Discovery

The first few months of 1932 were spent in rendering the installation more reliable and measuring the range and speed of the accelerated protons. It was not until April 13, 1932, that they achieved a breakthrough. On that fateful date, Walton first realized that their experiment had been successful. On the tiny screen, he detected flashes, called scintillations. These indicated that not only had the steam of protons succeeded in boring through the atomic nuclei but also that, in the process, a transformation had occurred. The speeding protons had combined with the lithium target to produce a new substance, the alpha particles, which appeared on the screen as scintillations.

Walton and Cockcroft confirmed these observations using a paper recorder with two pens, each operated by a key. Walton worked one key, Cockcroft the other. When either noticed a flash, he pressed his key. As both keys were consistently pressed at the same time, it was clear that the alphas were being emitted in pairs. The implication was that the lithium nucleus, with a **mass** of seven and a charge of three had, on contact with an accelerated **proton**, split into two alpha particles, each of mass four and charge two. In the transformation, a small amount of energy was lost, equivalent to about a quarter of a percent of the mass of lithium.

Walton and Cockcroft's achievement was groundbreaking and historic in many ways. It represented the first time that anyone had produced a change in an atomic nucleus by means totally under human control. They had also discovered a new energy source. Furthermore, they had confirmed George

Gamow's theory that particles could tunnel or burrow their way into a nucleus, despite the repulsion of the electrical charges. And finally, they furnished a valuable confirmation of physicist Albert Einstein's theory that **energy and mass** are interchangeable. The extra energy of the alpha particles, when allowance was made for the energy of the proton, exactly corresponded to the loss of mass.

Walton and Cockcroft's achievement was announced in a letter in *Nature* and later at a meeting of the Royal Society of London on June 15, 1932. By that time, they had succeeded in splitting the nuclei of 15 elements, including beryllium, the lightest, to uranium, the heaviest. All produced alpha particles, although the most spectacular results were obtained from fluorine, lithium, and boron. The news caused a sensation throughout the world. As a result of their discovery, Walton and Cockcroft were the star attractions at the Solvay Conference, an important gathering of international physicists, held in 1933, and at the International Physics Conference, held in London in 1934.

Walton and Cockcroft's particle accelerator spawned many more sophisticated models, including one built by their colleague physicist Marcus Oliphant at the Cavendish. It was capable of producing a more abundant supply of particles; not only protons, but also deuterons (nuclei of heavy hydrogen). With this, many groundbreaking nuclear transformations were carried out. Their invention also inspired the American nuclear physicist **Ernest Orlando Lawrence** to build a **cyclotron**, a cyclical accelerator, capable of reaching tremendous speeds. Although scientists in the close of the twentieth century may regard the equipment Walton and Cockcroft used as primitive, the basic idea behind the particle accelerator has stayed the same.

In 1932, Walton received his Ph.D. from Cambridge and two years later returned to Dublin as a fellow of Trinity College, his reputation preceding him. That same year he married Winifreda Wilson, a former pupil of the Methodist College, Belfast. They had two sons and two daughters, Alan, Marian, Philip, and Jean.

The next few years passed rather uneventfully for Walton. While his erstwhile partner, John D. Cockcroft, went from one high profile position to another, Walton preferred to remain slightly aloof from the mainstream of physics. He concentrated instead on establishing his department's reputation for excellence. His efforts were rewarded in 1946 when he was appointed Erasmus Smith Professor of Natural and Experimental Philosophy.

Shares 1951 Nobel Prize with Cockcroft

In 1951, almost twenty years after achieving the breakthrough that changed the face of nuclear physics, Walton and Cockcroft finally achieved the recognition that many believed was long overdue. The Nobel Prize in physics was awarded to them jointly for their pioneering work on the transmutation of atomic nuclei by artificially accelerated atomic particles. The following year, Walton became chairman of the School of Cosmic Physics of the Dublin Institute for Advanced Studies. He was elected a senior fellow of Trinity College in 1960.

Outside of his scientific work, Ernest Walton was active in committees concerned with the government, the church, research and standards, scientific academies, and the Royal City of Dublin Hospital. He died on June 25, 1995, at the age of 91.

WARPING OF SPACE AND TIME

In relativistic physics, the warping of **space-time** refers to the geometry of the **universe** resulting from the theory of general relativity. This geometry can no longer be described as a flat, three-dimensional Euclidean universe, but rather, as embedded in a type of four-dimensional Minkowski **space**. As formulated by German American physicist **Albert Einstein**, with contributions from other theoreticians such as Dutch physicist **Hendrik Lorentz**, relativity is centered on the space-time interdependence. As it affects space and **time**, the conceptual framework of relativity can be summarized as follows: all observable **motion** in space is relative. The **velocity** of **light** is a universal constant and independent of source or detector velocity. No **energy** can be transmitted at a velocity greater than that of light (c) which travels at 2.9979×10^8 m/s in a **vacuum**. Accordingly, the **speed of light** is often referred to as the velocity limit of the universe. The **mass** of a body in motion is relative and the concept of universal time is not valid. Space and time are interdependent; **matter** results in the warping of a space, or the relativistic Minkowski space changes in a **gravitational field**.

According to **relativity theory**, mass is responsible for the warping of space. Objects traveling in curved space act as through they are deflected as a function of the **gravitational force**. Three types of space curvatures are mathematically possible: the absence of curvature, (zero-curvature), a convex type of curvature (positive curvature, represented by a sphere in three dimensions), and a convex type of curvature, (negative curvature). These different types of warping patterns lead to three possible geometries for the universe, with each geometry related to the amount of mass present in the universe and each predicting a different cosmological evolution. For the universe as a whole, the shape of the curvature then depends on the average **density** of the matter. If the curvature of space is negative, this implies that the amount of mass required to stop the expansion of the universe is not sufficient. This results in an open universe which is boundless and will accordingly, expand forever. If space has zero curvature (i.e., if it is flat) then it also contains just enough mass to stop the expansion of the universe after an infinite amount of time. In this case, the universe is also boundless and is also expanding, but unlike the open universe, the predicted rate of expansion very gradually approaches zero. If the curvature of space is positive, the universe is closed, and there is excess mass which will reach critical density and result in a contraction after expansion has stopped. The type of space curvature of the universe is at present unknown because the exact amount of mass present in the universe has yet to be determined.

Warped space, theory of relativity. *(Image by Tony Craddock. Science Photo Library/Photo Researchers, Inc. Reproduced by permission.)*

A geometrical determination of the curvature of space may be possible in principle. For example, if a triangle could be drawn by reaching far into outer space so as to trace lines connecting three galaxies distant from each other, the curvature of the universe could be evaluated. The angles of such a triangle in an open universe would be greater than 180°, they would be smaller than 180° in a closed universe, and they would add up to precisely 180° in a flat, or Euclidean, universe. Just as the exact mass of the universe can not be determined, no geometric method has yet been successfully applied to the determination of the curvature of three-dimensional space. This would require a uniform distribution of galaxies throughout the universe, which is not the case as shown by astronomical observations, all leading to the conclusion that the galaxies have a finite age and have changed over time.

Visualizing a warped, four-dimensional space may be difficult, especially since the warping occurs on three different scales. The first, microscopic-scale curvature is due to the curvatures of each subatomic particle. An intermediate-scale curvature results from the gravitational pull of interstellar bodies such as galaxies and **black holes**. Finally, large-scale curvature embodies the overall shape of space, the sum total of all the mater and energy present in space. An analogy for the different scales of space warping would be the following consideration: on the large-scale, the surface of Earth is curved into the shape of a sphere. On the intermediate scale, the surface of Earth is covered with curves such as mountains and valleys. And on a smaller scale, the surface of Earth consists of rocks, earth particles and sand pebbles. The warping of space on the large scale is described as hyperbolic space, which has its own specific geometry, called hyperbolic geometry, a **non-Euclidean geometry** that rejects the validity of some Euclidean postulates. Even though all the axioms of Euclidean geometry may hold as long as a very small region of hyperbolic space is being described, this is not the case when the geometry must include the effect of the presence of mass, accounted for by Riemannian representations of space, which yield a curved geometry for space regions containing local concentrations of mass.

The warping of space-time, generated by the mass of the bodies which it contains, thus represents one of the most significant consequences of the theory of relativity because it is the curvature of space which determines the trajectories of the totality of the bodies moving in it. Given the initial positions and velocities of the moving bodies, solutions to the relativity equations can predict their future relative positions and velocities. In the limit where the velocities are significantly below the velocity of light, the predictions of relativity are the same as those obtained using English physicist and mathematician Sir **Isaac Newton**'s laws of motion. At large velocities, however, significant deviations occur and Einstein's theory more accurately describes the physical universe.

See also Astronomy; Cosmological arrow of time; Cosmology; Critical mass; Energy-momentum four-vector; Evolution of the universe; Frame of reference; Length contraction; Lorentz transformations; Newtonian physics; Relativity, special; Schwarzschild radius; Space-time curvature; Space-time geometry; Time dilation; Unification of physics; Vectors in space-time

WATT, JAMES (1736-1819)
Scottish engineer and inventor

The Industrial Revolution in Europe could not have taken place without the work of James Watt, who is commonly credited with inventing the steam engine.

Watt was born in Greenock, Scotland, on January 19, 1736. At an early age he helped his father build ships and was exposed to the various technology of the time. In 1755, he left for London to study the craft of instrument making. He began working with steam in 1764 when Glasgow University, for which he worked, brought him a Newcomen engine for repair.

Watt not only repaired the engine, but he also began improving on it. He noted, for example, that the engine wasted time cooling the piston chamber during every cycle. Within five years Watt built a demonstration model that he patented as "a new method of lessening the consumption of steam and fuel in fire engines." This model introduced a second chamber where the condensing cycle took place, leaving the steam chamber hot and ready for a new batch of steam.

Watt's work with steam caught the attention of English engineer Matthew Boulton. In 1775 Watt entered a partnership with Boulton, whose factory was widely respected for its quality metalwork, plating, and silversmithing. Watt was continually searching for ways to improve existing technology. One of his innovations was the 1772 micrometer. This invention helped John Wilkinson perfect a boring machine that could drill cylinders with unprecedented uniformity, a necessary step towards Watt's next development—introducing steam on both sides of the piston. This innovation forced the piston in both directions rather than allowing atmospheric **pressure** to complete the condensation/vacuum cycle in its own time. These were the engines that he and Boulton began to market in 1775.

In 1781 he devised mechanical attachments to convert the rocking **motion** of the piston to a rotary movement that could more easily be used to **power** machinery. It was this step perhaps more than any other that led the steam engine to the forefront of Europe's Industrial Revolution. It was used to pump bellows for blast furnaces, to power huge hammers for shaping and strengthening forged metals, and to turn machinery at textile mills. For the first time, mills and factories were not limited to locations near streams or windy plains.

In 1788 Watt unveiled another major contribution to the realm of instrumentation and machine control by introducing the centrifugal governor. Since a similar device had been in use in windmills, Watt made no effort to apply for a patent. Still, it became commonly known as the Watt governor. Its importance centered around its automatic control of steam output. The governor consisted of two metal spheres mounted on a vertical rod that was spun by the engine's output of steam. The faster the rod spun, the farther the two spheres were thrown outward by centrifugal **force**. The farther the spheres were thrown, the more they choked off the steam outlet. As steam output decreased, so did intake as well as the engine's power output. As power and steam output decreased, the slower the spheres rotated, re-opening the steam outlet and beginning the reverse of the process. Engine output and speed could be controlled and adjusted between these two limits.

By 1790 Watt's engines had nearly replaced Newcomen's engines which were, for the most part, at least fifty years old. By this time Watt had added many other improvements such as a new type of condenser that used a system of tubes instead of one large chamber, an air pump to maintain a **vacuum** in the condenser, a jacket and "stuffing box" to prevent steam from escaping where the piston rod passed through the cylinder, a *sun and planet gear* that introduced rotary power to a generation of mills in Soho, and the later crankshaft that replaced this gear.

Watt's other ideas include the steam radiator **heat**, an office copying press, and a device that could reproduce sculptured busts.

Watt retired in 1800. He received an honorary doctorate degree from Glasgow University and was elected to the Royal Society. He died near Birmingham, England, on August 19, 1819, a successful and respected inventor whose contributions are used to this day. He is buried beside his former business partner, Matthew Boulton.

James Watt.

WAVE FUNCTION

In quantum **mechanics** the wave function is a term used to describe a quantum mechanical system. The function specifies conditions of **space** and **time** that satisfy **Schrödinger's equation**. In an alternative use, a wave function (i.e., a Dirac function) describes functions of space and time (using space and time coordinates) that can be used to describe particle spin properties.

In 1921, German American physicist **Albert Einstein** was awarded the Nobel Prize for physics. What is surprising, perhaps, is that Einstein was honored not for relativity (still considered in 1921 to be too controversial a subject) but rather for his explanation of the **photoelectric effect**.

Einstein had introduced the concept of light-quanta, or photons, and thus had ascribed particle-like behavior to **light**, which had been understood in terms of wave mechanics since English physicist **Thomas Young**'s famous double-slit experiment in 1801.

A young French aristocrat had become captivated by Einstein's idea, and in his 1924 doctoral thesis in physics, **Louis de Broglie** took it to its logical conclusion: if **waves** behaved as particles, shouldn't particles behave as waves? Shouldn't nature admit only one type of entity, having both wave-like and particle-like properties? "After long reflection in solitude and meditation," de Broglie later explained, "I suddenly had the idea, during the year 1923, that the discovery made by Einstein in 1905 should be generalized by extending it to all material particles, and notably to electrons."

De Broglie associated a fictitious wave to electrons, whose **wavelength** was determined as a function of **momentum**. Combining Einstein's relativistic energy-momentum relation, $E^2 - p^2 = m^2c^4$, and his energy-frequency relation for photons, $E = h\nu$, de Broglie obtained the expression for what is now known as the **de Broglie wavelength**.

The new wave model of the **electron** gave a physical justification for Danish physicist **Niels Bohr**'s model of the hydrogen **atom**. The allowed atomic **energy** levels were those for which a standing wave pattern could be constructed: an integral number of wavelengths must fit into an allowed orbit's circumference. But what did the associated wave represent, and how did it react when acted upon by a **force**? In 1926, the latter question was answered by the Austrian physicist **Erwin Schrödinger**.

Schrödinger introduced a mathematical function, the wave function $\psi(x,t)$, which described the electron wave, and which obeyed a wave equation, today known as the Schrödinger equation. Consider a familiar wave phenomenon, ripples on the surface of a pond. The mathematical function representing the water level at a point on the surface obeys a wave equation; knowing the forces acting on the water allows the wave equation to be solved, and the water level at any point to be determined. In a like manner, knowing the forces acting on an electron allows the Schrödinger equation to be solved, and the associated wave at any point to be determined. The water level, however, is a measurable quantity, and its wave function, the height at any point, is real-valued; it is a real number. Schrödinger realized that the electron's wave-function must be complex-valued; at any point it is a complex number, a number containing the imaginary $\sqrt{-1}$. The wave function cannot represent something directly measurable.

The absolute square of a complex number is a real number. Thus, the absolute square of the electron's wave function can represent something physical and measurable. At first Schrödinger thought it might be the electron's **density** (no longer could electrons be considered point particles). Although it turned out that the absolute square of the electron's wave function could be nonzero in disjoint regions of space, no one had ever seen an electron in pieces. Accordingly, density was excluded. The problem was solved by German physicist **Max Born**, also in 1926, who later explained:

"Again an idea of Einstein's gave me the lead. He had tried to make the duality of particles (light **quanta** or photons) and waves comprehensible by interpreting the square of the [light-wave] amplitudes as probability density for the occurrence of photons. This concept could at once be carried over to the wave function: [In other words, the square of the electron's wave function] ought to represent the probability density for electrons (or other particles)."

Born's probability interpretation is today the standard. The absolute square of the wave function represents the probability that the particle will be found at that point. **Quantum mechanics** does not predict precisely where a particle will be found, as does Newtonian mechanics, but rather specifies a probability region wherein a particle is likely to be found.

See also Particle-wave duality

WAVE INTERFERENCE

The physics of **waves** describes their behavior and **wave phenomena** such as reflection, refraction, **dispersion**, and **interference**. Interference is the phenomenon observed when two traveling wave trains coexist in the same medium. Consider single **frequency** waves described by a sine function and the following quantities: a period T, or the **time** required to complete a full cycle, a frequency f, which is the reciprocal of the period, an amplitude A, which is the maximum displacement from the **equilibrium** position and a **wavelength** λ which is the distance between two wave crests, or positions of maximum amplitude. If the sine crests of two waves traveling in opposite directions through the same medium completely overlap (i.e., become in-phase) their amplitudes add and the interference is referred to as constructive interference. A single wave with double amplitude is observed in the medium. If the two wave crests become half a period or 180° out-of-phase, their amplitudes then subtract and the interference is called destructive interference. The interfering waves have the same maximum displacement, but in opposite directions and thus, their amplitudes cancel out, or subtract if they were not equal before overlap. The occurrence of interference does not modify the properties of the individual waves, nor does it change their path. The shape of the wave resulting from interference is determined by the principle of superposition, which states that when wave interference occurs, the resulting displacement of the medium at any location is the algebraic sum of the displacements of the individual waves at that same location.

An example of interference is provided by the tuning fork, which creates an outward traveling **pressure** wave when it is struck. One wave component has a pressure higher than atmospheric pressure, and another component has a lower pressure. At some specific angles, the high pressure areas of the two waves interfere constructively and a louder **sound** is heard. At other angles, the high pressure part of one wave coincides with the low pressure part of the other wave, resulting in destructive interference and a lower sound. Another example of wave interference is illustrated by the beat phenomenon which occurs when two sound waves of different frequency alternate constructive and destructive interference. This produces a sound which alternates between soft and loud, or beats. The beat frequency is equal to the absolute value of the difference in frequency of the two waves and if it lies in the mid-frequency region, it can be heard by the human ear as a third tone, called a subjective tone. Such tones are used in multiphonics applications. Another use of beat frequencies is in police **radar** speed detectors which bounce microwave **radiation** off of speeding vehicles and detect the reflected waves. These waves are shifted in frequency as a result of the **Doppler effect**, and the beat frequency between the emitted and reflected waves measures the speed of the moving car.

An important consequence of interference in wave physics is that interference is the phenomenon which gives rise to standing waves. Standing waves are the characteristic signatures of a specific vibrational pattern generated within a medium when the vibrational frequency of the source causes reflected waves from one end of the medium to interfere with incident waves from the source in such a way that **nodes** along the medium appear to be standing still.

See also Acoustics; Diffraction; Electromagnetic waves; Frequency; Harmonic motion; Oscillations; Reflection, refraction, and dispersion; Standing waves and resonance; Vibrating systems and resonance; Wave superposition

WAVE MOTION

A wave is nothing more than a disturbance that moves from place to place in some medium, carrying **energy** with it. Since the behavior of **waves** is so closely related to the concept of **oscillations**, that is a good place to start.

There are many examples of simple oscillations, but a very good one is that of an object attached to the end of a spring. We will assume that the other end is held fixed, perhaps by a clamp. Suppose we let the spring hang vertically and slowly lower the object until it becomes stationary. The spring is now stretched enough for its upward pull to balance the **weight** of the object, which at that location is in **equilibrium**. Now we disturb the object by lifting it a short distance above that point and letting go. The object then begins to oscillate vertically as it first falls, until the spring stops it and pulls it back upward to the original position, then it falls again, etc. The energy in the oscillating **motion** came from the original disturbance, which in this case moved the object to a position above the equilibrium point. At that instance, the object was at its maximum displacement whose size is the amplitude. The larger the amplitude, the greater the energy in the motion.

Now, let's try a different experiment. First, let's see what happens if we hang a duplicate object and spring side-by-side with the first one, but without the two being in contact in any way. If we disturb the first object, it oscillates just as before and the second object remains stationary at its equilibrium position. However, the situation becomes different if we connect the two objects with a rubber band and then disturb only the first object. We see that it begins to oscillate as before, but soon the second object starts to oscillate also. The rubber band allows energy to transfer to the second object, which will move with the same **frequency** as the first oscillation. We can make the experiment more complicated if we use a large number of springs hanging in a row, with each object connected to the one before and after it with rubber bands. If only the first

Wave motion, water droplet producing transverse waves.
(Photo by Martin Dohm. Science Photo Library, National Audubon Society Collection/Photo Researchers, Inc. Reproduced by permission.)

object is disturbed, the oscillation will pass its energy through all the springs. After the energy has been transferred to the next few springs, the first spring will become still with its object back at its equilibrium point. This demonstrates exactly how a disturbance can move as a wave through a medium. In this example, the medium is composed of the objects on springs, which act as coupled oscillators.

The way that a disturbance is transferred from one part of a medium to another in the spring model is very similar to the way that waves move in water, air or a guitar string. We can think of the water, perhaps in a bathtub, as being composed of a great number of H_2O molecules lying very close side-by-side. If the water has been undisturbed for a long **time**, its surface will be calm and the molecules will be relatively stationary (of course, they are not completely stationary, but their motion is microscopic). Now suppose we disturb the water by tapping it with a finger. This produces a disturbance as many molecules are forced downward, the opposite of what happened in our spring example. We have all seen a wave move on the surface of the water away from the position of our tap, and this is simply the disturbance being transferred from one molecule to another. Just as before, the larger the original displacement, the greater the energy in the disturbance and the bigger the energy that the wave carries. You might ask how the transfer of energy takes place. Actually, the molecules of water all exert some **force** on their neighbors which are quite close. This holds them together and provides the same effect as the rubber bands.

The wave we just made is an example of a traveling wave since the disturbance moves from place to place in the medium. It can also be classified as a transverse wave because the direction of the disturbance (vertical in this case) is perpendicular to the direction that the wave travels (horizontally). There are also longitudinal waves, like those which carry the energy of **sound** in air, in which the direction of the disturbance is the same as that of the wave motion. This is easy to visualize if you have ever seen a speaker move when the volume is turned up very loud. The sudden movements of the speaker compress the nearby air and that disturbance moves in the same direction toward you and your ears.

A repeated pattern of individual waves, a wave train, often occurs. A wave train can be produced in water by tapping the surface with a specific rhythm, or frequency. A complementary characteristic of a wave train is the **wavelength**. Suppose a friend taps the water with a specific frequency and you take a picture of the resulting waves. In effect, this lets you "freeze" the wave train in time and examine it. You will notice that there is a constant distance between the individual waves; this is the wavelength. The frequency tells how often the wave train repeats itself in time and the inverse of the wavelength tells how often the wave train repeats itself in **space**. Multiplying the numerical values of the frequency and the wavelength gives the speed at which the waves move.

Waves have many interesting properties. They can reflect from surfaces and refract, or change their direction, when they pass from one medium into another. If these properties seem familiar, that is because we are accustomed to light behaving in exactly this way. Obviously **light** reflects, and an example of refraction is the bending of light when it passes from water into air. For this reason, when you look into water, objects appear to be at different locations than their real positions. Light is, therefore, considered in these instances to behave as though it was a wave produced by moving disturbances of electric and magnetic fields.

Waves can also combine, or interfere. For example, two waves can reach a particular point at just the right time for both to disturb the medium in the same way (such as if two water waves both try to lift the surface at the same time). This is constructive **interference**. Likewise, destructive interference happens when the disturbances of different waves cancel. Interference can also lead to standing waves which appear to be stationary—the medium is still disturbed, but the disturbances are oscillating in place. This can only occur within confined regions, like a bathtub or a guitar string (fixed at both ends). For just the right wavelengths, traveling wave trains and their reflections off the boundaries can interfere to produce a wave that appears stationary.

WAVE PHENOMENA

A wave is a uniformly propogating disturbance in a medium. **Waves** can be classified as either mechanical or electromagnetic. Although mechanical waves cannot transmit **energy** through a **vacuum**, **electromagnetic waves** are capable of self propagation and can travel through a vacuum. This distinction, however, does not prevent mechanical and electromagnetic waves from exhibiting similar behavior as they propagate through a medium. When a wave, whatever its type, travels through a medium, it can either reach the end of the medium, encounter a barrier, or transfer to another medium with different inertial (e.g., **density**) or elastic (e.g., bulk modulus) properties. Waves traveling through a medium exhibit wave reflection, refraction, **diffraction**, or **interference**.

Reflection occurs when a wave strikes a reflective surface. If the wave is electromagnetic, it will be partially reflect-

ed and partially transmitted as a refracted ray. In physics, the general law of reflection, which states that the angle of reflection equals the angle of incidence, also applies to waves, in which context it becomes: the angle of incidence of any wave or wavefront onto a straight reflective surface equals the angle of reflection. The angle of incidence can be represented as the angle between the incoming wave and a line perpendicular to the surface of the reflective surface. If the reflective surface is curved, such as a concave parabolic surface, the incoming waves will be reflected back and will cross each other at a point called the focal point, a point at which the waves incident toward the reflective surface and traveling parallel to the principal axis will meet after reflection. The main result of reflection is that the waves abruptly change their direction of travel after striking the reflective surface. Wave reflection is the principle used to design optical fibers. In these materials, electrical signals are converted into optical signals which travel through a glass fiber and are subsequently converted back into electrical signals. They consist of two concentric optical layers, the core and the cladding, which have different properties, and a protective outer coating, the buffer. When **light** is passed into the core of the optical fiber, it bounces on the core-cladding interface and is reflected back into the core. With equal angles of incidence and reflection resulting from the law of reflection, the light wave is able to zigzag down the length of the fiber.

When waves travel from one medium to another, a phenomenon known as refraction occurs, which is always accompanied by a change in the speed and **wavelength** of the wave as well as by a change in the direction of travel. The resulting quantities will depend on the type of wave (i.e., mechanical or electromagnetic), and the properties of the two media. For example, the **speed of light** waves slows down as the waves pass from air to water while the reverse is observed for **sound** waves. Wave refraction proceeds according to the following principles: when a wave travels into a medium where its speed is decreased, its path bends toward the normal to the boundary between the two media, and its wavelength is decreased; when a wave travels into a medium where its speed is increased, its path bends away from the normal, and its wavelength is increased. The extent of bending depends on the respective refraction indices of the two media and is described by Snell's law, which yields a dimensionless unit called the index of refraction, defined as the ratio of the speed of the wave in unrestrained conditions to the speed of the wave in that medium and also yields the angle between the wave and the normal to the boundary between the two media. Examples of refraction include the **dispersion** of light into its constituent wavelengths by prisms and the occurrence of rainbows. Rainbows occur when the Sun's polychromatic (white) light passes into a spherical raindrop. The constituent colors of the light wave are refracted through different angles, reflected off the back of the drop, and then refracted again as they exit the drop. The polychromatic light wave is then dispersed into waves consisting of its component colors, which travel in slightly different directions, thus producing the rainbow.

Diffraction is another wave phenomenon which manifests itself in the apparent bending of waves around small barriers lying in their propagation paths or in the spreading out of waves past small openings. Electromagnetic waves can undergo Frauenhofer diffraction or Fresnel diffraction. Frauenhofer diffraction occurs when the wave approaching the diffracting body is parallel and monochromatic and the image plane is at a large distance relative to the size of the diffracting object. The different slit, or opening, geometries of the diffracting objects will determine the wave diffraction patterns. The general diffraction formulation is referred to as Fresnel diffraction. In diffraction, the wavelength and **frequency** of the waves involved remain constant. The first wave diffraction experiment was carried out in 1801 by a British physicist, **Thomas Young**. This experiment uses two coherent (i.e., with a constant phase difference) light sources placed behind two slits to cause diffraction. The wavefronts, a series of semicircles, overlap at some points, and constructive or destructive interference occurs as a result of superposition, depending on their amplitudes. If a screen is placed a certain distance away, parallel to a line drawn between the two sources, points of constructive and destructive interference can be noted. Points of maximum light intensity are those where constructive interference occurs, and points of minimum intensity are those where destructive interference is taking place. As the wavefronts keep advancing and being superimposed, the screen is eventually reached by fringes of maximum intensity. By calculating the distance between the screen and the slits, the distance between the centers of the two slits, and the average distance between two fringes, the wavelength of the original waves can then be calculated. An application of diffraction is in the design of diffraction gratings, which consist of a large number of parallel and closely spaced slits and which are used in telescopes and spectroscopy instrumentation to separate light of different wavelengths with high resolution, which a prism can not provide.

Interference is a wave phenomenon closely related to diffraction. While interference is the result of the superposition and interaction of waves, diffraction can also be described as the result of the superposition and interaction of waves, but only after the waves pass through a circular opening or go around objects. When two waves moving in opposite directions meet, their displacements at a given point in time are added and the resulting displacement is that of the new wave produced. In interference, unlike diffraction, the original waves continue moving in the same direction, which remains unchanged. When the two waves are in phase, they create a wave of larger amplitude, and the interference is called constructive. If they are out of phase and their amplitudes cancel out, then the interference is called destructive.

See also Electromagnetic spectrum; Harmonic motion; Nodes; Optics; Oscillations; Polarization of light; Reflection, refraction, and dispersion; Sound; Spectroscopy; Standing waves and resonance; Telescope; Transverse and longitudinal waves; Vibrating systems and resonance; Wave interference; Wave motion; Wave superposition

WAVE SUPERPOSITION

When dealing with wave **mechanics** the principle of linear superposition specifies that when two or more **waves** interact they interfere with each other. This **interference** may be constructive or destructive interference. When wave crests match constructive interference takes place, when their form (phases) do not match destructive interference can occur. With identical waves (i.e., waves of the same **wavelength** and amplitude) emitted by the same source, the waves will undergo constructive interference if the distance traveled by one wave is equal to, or differs by an integral number of wavelengths from, the distance traveled by the second wave. Whenever path distances differ by exactly one-half a wavelength, the waves are completely out of phase and undergo complete destructive interference.

The general formulation of the superposition principle is that, when a number of influences act on a system, the total influence on that system is the linear addition of the individual influences. In **quantum mechanics**, the equivalent principle states that the resultant **wave function** due to two or more individual wave functions is the sum of the individual wave functions. In wave physics, the principle of superposition can be applied to waves whenever two or more waves are traveling through the same medium at the same time and they pass through each other without being disturbed. In this case, the net displacement of the medium at any point is simply the sum of the individual wave displacements. This applies to wave **pulses** and to continuous sine waves. For example, the net displacement of two gaussian wave pulses traveling in opposite directions in a given non-dispersive medium and passing through each other without being disturbed is the sum of the two individual displacements. Another example is the case of two sine waves with the same amplitude, **frequency**, and wavelength traveling in the same direction in a given medium. Applying the principle of superposition, the resulting medium displacement may be described by summing the two waves.

The result is a traveling wave whose amplitude depends on the phase angle. When the waves are in-phase, constructive interference occurs with the resulting wave amplitude being twice that of the individual waves. When the waves are out-of-phase, destructive interference is produced and the respective amplitudes cancel each other out. Another example is of two sine waves also with the same amplitude, frequency, and wavelength but that are traveling in opposite directions. Again, applying the principle of superposition, the resulting wave is no longer a traveling wave but a standing wave.

In the case of electromagnetic waves, when two such waves with the same frequency superpose in **space**, the resultant electric and magnetic field strength of any point of space and time is the sum of their respective fields, except when they are in phase or out-of-phase, which leads to interference. In the first case, constructive interference occurs because the resultant field is twice that of each individual wave and the resultant intensity is four times the intensity of each superposed wave because intensity is proportional to the square of the field strength. In the second case, destructive interference occurs since the respective electric field **vectors** are in opposite directions, so the result is zero and no wave results.

See also Dispersion; Electromagnetic waves; Light; Nodes; Oscillations; Quantum computers; Reflection, refraction, and dispersion; Standing waves and resonance; Transverse and longitudinal waves; Vibrating systems and resonance; Wave interference

WAVELENGTH

Wavelength is the distance between two points of the same phase in consecutive cycles on a periodic wave. Wavelength can be expressed in many units of distance, the most logical being the meter. However, because so many kinds of electromagnetic **radiation** are of very short wavelength some smaller units are necessary. Some of the most common units of measure for wavelength are the centimeter, the nanometer ($1x10^{-9}$ m), and the angstrom ($1x10^{-10}$ m). The angstrom is named for the Swedish physicist Anders Ångström who used this unit as a basic unit of measurement for his studies of the solar spectrum in the mid-1800s. Visible **light** (for humans) has wavelengths ranging from roughly 400 nm to about 750 nm.

When light passes from one transparent material to another, with differing refractive indices, it is refracted or bent, the extent of which is determined by the wavelength of light. Violet light, light with a high **frequency**, is bent more than red light, light with a low frequency. The explanation for this phenomenon was put forth by **Isaac Newton**, an English mathematician, in 1666.

A wave will advance the distance of one wavelength in a **time** equal to one period of the vibration. Since all frequencies of electromagnetic radiation travel at the same speed in a **vacuum** it is possible to relate the wavelength to the frequency. The wavelength l is equal to the **speed of light** c divided by the frequency n, the number of cycles passing a given point per unit time, $l = c/n$. The frequency is usually denoted in units of $s-1$. The wave number s, another useful unit of measurement, is related to the wavelength in a reciprocal fashion, $s = 1/l$. The importance of the wave number became known when Rydberg noticed a periodicity when observing the spectra of elements with respect to wave number rather than wavelength. Unlike wavelength, wave number is usually expressed as 1/cm.

There are many distinct regions of the **electromagnetic spectrum**. The regions are divided according to wavelength. **Radio** waves are **waves** whose wavelengths are below 0.1 m. **Microwaves** are waves with length between 0.1 and $1.0x10^{-3}$ m. Infrared light ranges in wavelength from $1.0x10^{-3}$ to $750x10^{-9}$ m. Visible light falls next in the ordering and extends to $400x10^{-9}$ m. Ultraviolet light ranges from $400x10^{-9}$ to about $1.0x10^{-9}$ m. **X rays** range in wavelength from $1.0x10^{-9}$ to about $5.0x10^{-11}$ m. Gamma rays are the highest in **energy** and range from $5.0x10^{-11}$m to shorter wavelengths. The shorter the wavelength, the higher the energy of the photons. The energy-wavelength relationship for electromagnetic radiation is: E =

hc/l, where h is **Planck's constant**, c is the speed of light in a vacuum, and l is the wavelength.

See also Sound waves; Speed of sound; Wave motion

WAVES

A wave is a disturbance that travels either through a medium (mechanical waves, such as **sound** waves) or through **space** (electromagnetic waves, such as **light** waves). As a wave travels, it carries **energy** from one place to another. Waves are used to describe phenomena such as sound, light, and the behavior of particles.

Some waves require a medium in order to propagate. This means that certain waves must actually have physical contact between particles in order to transfer **kinetic energy**, the energy of **motion**. These waves are called mechanical waves. In a mechanical wave, a disturbance causes a portion of the medium to deviate from an **equilibrium** (resting) position. Because it is physically connected to a nearby portion, it causes the nearby portion to deviate from its equilibrium position, but because the influence is by means of a **force**, the new portion accelerates rather than moves instantaneously. The **mass** of the new portion governs how fast it reacts to the force. Thus the influence propagates, not at an infinite speed, but at a speed related to the strength of the connecting force and inversely related to the massiveness of the moving particles. Dimensional analysis then can be used to show that the speed of propagation of a material wave is the square root of a force-like term and a mass-like term.

Some waves do not require a medium. Physical contact between particles is not necessary for the kinetic energy to be transferred. They can be transmitted through a **vacuum**. These waves disturb electric and magnetic fields instead of particles. This type of wave is called an electromagnetic wave. An example of a mechanical wave is a sound wave. Electromagnetic waves include light, **microwaves**, and **radio** waves.

Waves also can include the wave-like behavior of particles such as electrons and photons in a phenomenon called wave-particle duality. Scientists use this duality to describe the idea that particles can act like waves, and waves can act like particles. Wave-particle duality explains **Albert Einstein**'s **photoelectric effect**; that electrons emitted from a metal exposed to high **frequency** light exhibited the properties of frequency and **wavelength**. The wave-particle duality has given rise to equations, called wave functions, which are used to describe the structure of the **atom** and the behavior of **subatomic particles**. The study of wave functions and how they describe atoms is called **quantum mechanics**.

Sound is produced when an object vibrates, causing a transfer of kinetic energy through a medium. This transfer of kinetic energy causes a series of compressions (areas where the surrounding molecules are crowded together) and rarefactions (areas where the molecules are spread out). These compressions and rarefactions move through the medium, away from the original vibration. When a series of compressions and

Electromagnetic wave, San Francisco Exploratorium. *(Photo by Phil Schermeister. Corbis Corp. Reproduced by permission.)*

rarefactions move through a medium, they create a wave disturbance where the movement of the material's particles from equilibrium is parallel to the direction of propagation. These waves are called longitudinal waves. Sound waves in air are longitudinal waves.

Transverse waves oscillate up and down or from side to side. The direction of a transverse wave that travels through a medium is perpendicular to the motion of the medium. The waves that travel on a guitar string when it is plucked are transverse waves.

Other transverse waves oscillate transversely, but do not require a medium. For example, light is energy emitted by electrons in atoms. This energy travels in a wave with both electric and magnetic behavior. Light, therefore, is called an electromagnetic wave. Other electromagnetic waves include infrared waves, ultraviolet waves, and **x rays**. Electromagnetic waves travel through space at the **speed of light**. These waves are the result of combined electric and magnetic fields. Changing an electric field will produce a magnetic field, and changing a magnetic field will produce an electric field. These electric and magnetic fields move through space, combining to produce electromagnetic waves. The electric and magnetic fields move at right angles to each other.

Some waves are neither transverse nor longitudinal, but rather a combination of the two. One such wave is a surface wave. Surface waves occur at the boundary, or surface, of two different mediums. An example of a surface wave would be ocean waves, those occurring at the surface between the ocean water and the air. The water molecules in the ocean do not simply move up and down or back and forth, instead, they move in a circular motion. This circular motion is a combination of transverse and longitudinal **wave motion**.

All waves, with the exception of impulsive waves, share certain characteristics. These characteristics are amplitude, wavelength, and frequency. Impulsive waves do not have a wavelength or a frequency. Any wave also exhibits a phenomenon known as Fourier decomposition. When a wave passes through a medium, the particles of the medium vibrate and are moved from their original position. The amplitude of a wave is the height of a wave crest, or the highest point of a wave. This is the farthest distance a particle moves from its original position. The wavelength is the distance between two successive crests. Wavelength is symbolized by the Greek letter λ.

The frequency of a wave is the number of crests that pass a given point per second. Frequency is measured as waves (or cycles) per second, or hertz (Hz). One hertz is equal to one cycle per second.

The speed of a wave can be described by both its frequency and its wavelength and can be calculated using the equation: speed = frequency x wavelength. The speed of a wave is measured in meters per second when frequency is measured in hertz and wavelength is measured in meters. The speed of a wave is constant in any particular medium. Therefore, if the frequency is increased, the wavelength decreases, and visa versa, as illustrated in the above equation.

The speed of a wave in a particular medium depends on the **elasticity** and **density** of the medium through which the waves travel. Waves travel faster in solids and slower in gases. The molecules in a solid bounce back faster when vibrated than liquids or gases. The particles in a solid are packed more tightly together than a liquid or a gas. As a result, when one particle of a solid is displaced, it displaces its neighboring particles with a greater **acceleration**, resulting in a faster wave. The molecules in a gas are much more spread out than in a liq-

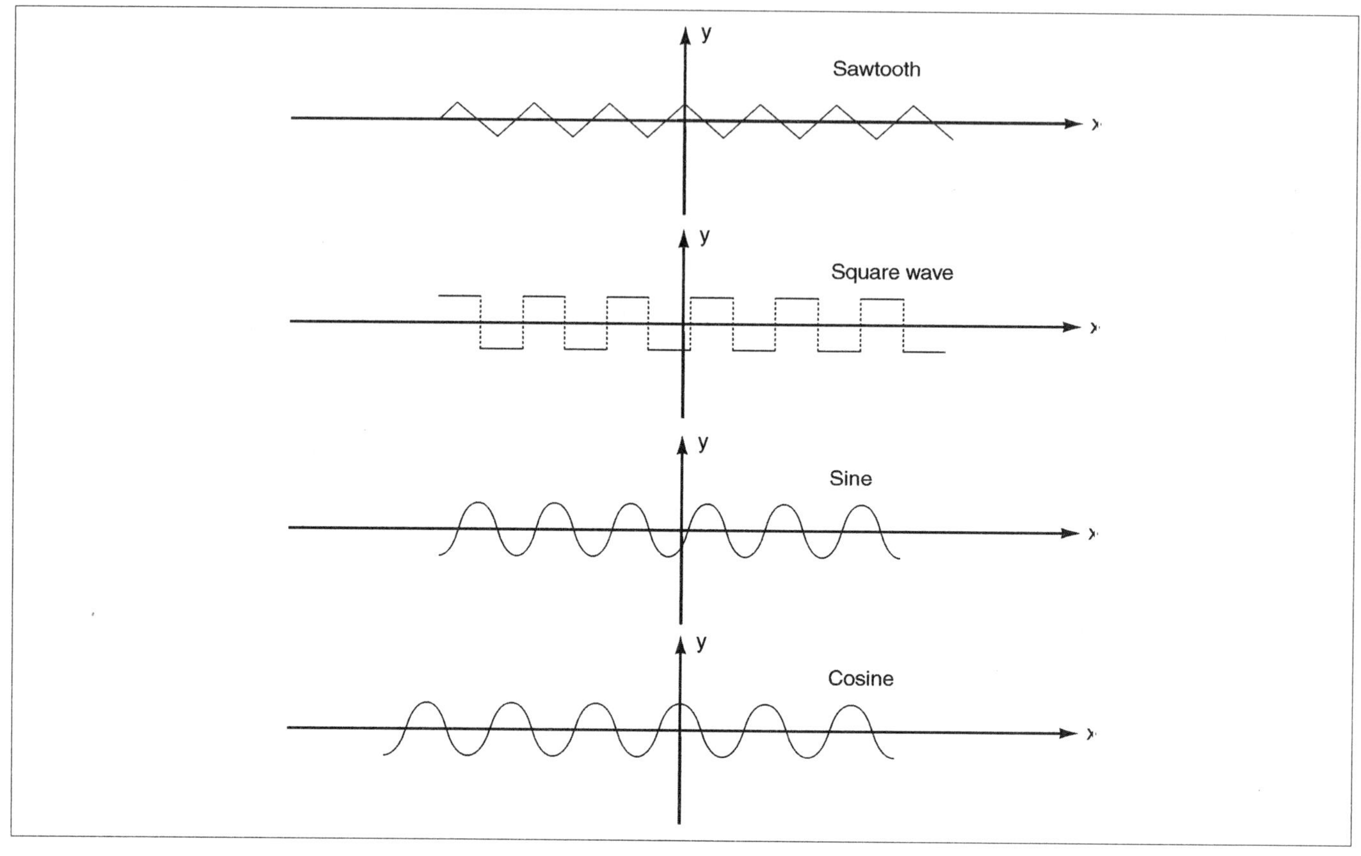

Periodic functions of waves.

uid or a solid. The propagation of a wave in a gas requires the collision of two molecules. As a result, the speed of a wave in a gas is slower than in a liquid or a solid.

Waves can interact with their medium and boundaries in four different ways; through reflection, refraction, **diffraction**, and **interference**. Reflection occurs when a wave bounces back from a barrier or the end of a medium through which it is traveling. Refraction occurs when a wave travels through one medium into another. Refraction is a result of a change in **velocity** of a wave as it passes through different media. Diffraction occurs when waves bend around a barrier in their path. Diffraction explains why we can hear someone's voice when they are around the corner, in a different room. **Christiaan Huygens**, a Dutch scientist, developed a principle to explain diffraction. According to Huygens, the crest of any wave can act as a series of wave sources, equally spaced. As a wave is broken by a barrier, the broken crest can act like a new wave source, prodicing new waves. Waves can combine with other waves to produce interference. The interference may have an additive effect, called constructive interference, or a deleterious effect, called destructive interference. Waves in a linear medium can superimpose. When two or more waves are present at the same point in a medium at the same time, a new wave forms that is a sum of the individual waves.

The study of waves and their properties have provided valuable information for many branches of science. Waves explain the behavior of electrons, light, sound, atoms, and even earthquakes. Physicists, chemists, and physicians have used the study of waves and wave behavior for both theoretical and practical purposes.

WEAK FORCE AND INTERACTIONS

There are four fundamental forces in nature related to physical phenomena. Although **gravity** is, by far, the weakest **force**, the force mediating particle transformations is termed the weak force. The weak force is actually a submicroscopic force working only within the atomic **nucleus**, whereas gravity has an infinite range. When the range of the weak force is considered, it is truly weak compared to the strong nuclear force. Accordingly, the weak force is considered as a weak nuclear force.

The beginning of the discovery of the weak force can be traced back to German physicist **Wilhelm Röntgen** who discovered **x rays**. Röntgen inadvertently left wrapped photographic plates near a Crookes tube, and returned later to find them exposed. Röntgen then wrapped a Crookes tube in black cloth and noted a glow on a piece of paper covered with barium platinocyanide. Röntgen realized invisible rays, which he called x rays, were being given off by the Crookes tube.

French physicist **Antoine-Henri Becquerel** postulated that some substances radiated or gave off after being exposed to sunlight. Becquerel worked with uranium salt. Becquerel put photographic plates in a drawer covered with black cloth and metal plate, on top of which he deposited a golf ball-sized piece of uranium salt. Then he closed the drawer. The day was not sunny and Becquerel wanted to be sure the plates were not exposed. The material lay in the darkened drawer. Three days later when Becquerel opened the drawer, he expected the plates to remain unexposed because the uranium salt had not been exposed to the **Sun**. Instead, Becquerel found a clear image on the plates. The uranium salt had given off invisible rays, even in total darkness. Becquerel continued to experiment with uranium and discovered that it had some very interesting properties. The uranium rays were different from x rays. They did not penetrate materials. The emitted their rays spontaneously, day or night, for seemingly endless periods of time. While x rays required **energy** input into a Crookes tub; uranium gave off its rays spontaneously, and no outside energy source was required. This form of **radiation** is now understood as nuclear or particle **radioactivity**.

At the same time in Paris, French physicists **Marie** and **Pierre Curie** were able to extract two new radioactive elements from uranium ore, the elements polonium and radium. Radium was discovered to emit radiation a million times more intensely than uranium. Every sample of radium pours out energy in torrents. But an unknowing public used it to **light** watch dials, make flashy costumes, and gamble with luminous chips. It was quickly discovered that radium destroyed cancer cells, but also killed healthy tissue, as well as causing cancers and radiation poisoning.

Another powerful discovery was made by New Zealand-born English physicist **Ernest Rutherford**. Using cloud chambers, Geiger counters, and scintillation screens, Rutherford showed that three different types of radiation (alpha, beta, and gamma rays) were given off. Rutherford showed alpha particles were actually **helium** atoms, beta particles were electrons, and gamma rays were similar to x rays. Rutherford also found, when he carefully reviewed all his data, that the **atom** was somewhat like a miniature solar system with negative electrons orbiting a positively charged nucleus.

Danish physicist **Niels Bohr** pushed the idea of a planetary model into the realm of **quantum physics**. Because nucleus was much too massive for the amount of positive charge it was carrying, it could not be made entirely of protons. English physicist **James Chadwick** discovered the **neutron**, a particle with slightly more **mass** than the **proton**, but without an **electric charge**. It became clear that the nucleus was composed of positively charged protons and neutral neutrons around which orbited the negatively charged electrons.

It was beta (**electron**) decay that inspired the notion of the weak nuclear force. Bohr and Austrian physicist **Wolfgang Pauli** studied **beta decay**. It was found that the electron emitted in beta decay had a wide range of **kinetic energy** ranging anywhere from zero up to a maximum value of 156 keV. It was as if the law of the **conservation of energy** was being violated.

Pauli suggested that perhaps an additional new particle was being emitted which was difficult to detect. Such a particle could be carrying off the energy, **momentum**, and **angular momentum** required to maintain the conservation laws. Italian physicist **Enrico Fermi** named such a particle the **neutrino**, meaning little neutral one.

Fermi studied beta decay intensively. The mass of any nucleus was approximately the sum of the masses of the protons and neutrons it contained. The simplest manifestation of beta decay is the decay of a free neutron into an electron, a proton, and an antineutrino. Fermi thought only of a neutrino, until French physicist **Louis de Broglie** introduced the term **antiparticle** in 1934, after which Fermi developed a theory of beta decay that included the antineutrino. Fermi considered the above reaction to be the prototype of weak interactions. Fermi described the wide energy variations of the beta particle (electron) and developed a theory to compensate for such variations. Fermi stated that at a single point in **space-time**, the quantum-mechanical wave function of the neutron is transformed into that of the proton, and that the wave function of the incoming neutrino (which is equivalent to the outgoing antineutrino we actually see) is transformed into that of the electron. Fermi described the reaction as four **fermions** reacting at a single point. In beta decay, the electron emitted is not an orbital electron, but is created within the nucleus itself.

Fermi also postulated the existence of a fourth force in nature which regulates this decay, the weak nuclear force. It was not until 1956 that the existence of the neutrino was confirmed, and Fermi's theory verified. The weak force theory included the possible energy levels of the expelled electron and also explained the range of weak force (less than 10^{-18} meters) and the fact that the fastest beta decays take on the average of about a hundredth of a second, slow compared with the typical time scale of processes caused by the strong nuclear force, which is about a millionth of a second.

There were problems with Fermi's theory. Chief among these was how the weak force depends on the relative orientation of the spins of the participating particles, and why the weak force reaction seemed to violate the conservation of charge conjugative (C) symmetry and **parity** (P) symmetry. These problems have been solved with models using the properties of three observed boson particles that transmit the weak nuclear force, namely, the W^+, W^-, and Z°. The weak force is responsible for beta decay and for interaction of the neutrino with **matter**.

See also Alpha decay; Beta radiation; Beta rays; Electroweak force; Electroweak particles; Electroweak theory; Gamma decay; Nuclear structure and stability; Radioactive decay

WEAKLY INTERACTING MASSIVE PARTICLES

Weakly interacting massive particles, or WIMPs, are a theoretical construct which was proposed to account for apparently missing **mass** in the galaxy. These particles would have to have much more mass than traditional hard-to-detect particles like neutrinos, so that they could account for enough of the mass of the galaxy. However, they would have to interact exclusively with the weak **force**, making it difficult for them to be observed by other, more traditional means. They could not emit **light** or otherwise interact with the electromagnetic strong forces, making them almost impossible to detect directly.

The rate of the universe's expansion can be measured, and by the current measurements, it is much slower than it ought to be for the amount of **matter** detected by current standards. Also, the observed movements of galaxies and galactic clusters is not consistent with the gravitational forces due to the visible matter. That is, there must be a great deal of unseen mass in the **universe** that is affecting the movements of the galaxies. The missing amount of gravitational matter is sometimes called "missing matter," or more commonly "dark matter." It is simply matter that does not emit electromagnetic **radiation** and thus is not visible to our astronomical instruments. The WIMPs would have to have a spherically symmetric distribution throughout the galaxy—that is, they would have to exist in about equal numbers in every direction—in order to explain the observed galactic **dynamics**.

Some physicists think that neutrinos may account for the weakly interacting massive particle mass. However, there is a known upper bound on the **neutrino** mass. Even with the most massive possible neutrino, they could only account for about 10% of the **dark matter**. The missing matter would have to be composed of elementary particles similar to neutrinos, however. Usually WIMP candidates are considered to be non-baryonic, that is, not composed of **quarks** like the **proton** and **neutron**. Baryonic candidates are separated into a different category of matter and are not properly WIMPs since they can undergo strong and electromagnetic interactions.

Many different experimental searches for WIMPS have been carried out. Because WIMPs are massive, they can collect at the center of massive bodies such as the Earth or **Sun**. Some experimental searches have concentrated on looking for WIMP decay products which come from the Earth's core or the direction of the Sun. It is also possible for WIMPs to change the **temperature** profile of the Sun's interior by carrying **energy** out of the core. This, in turn, would affect the neutrino flux coming from the Sun. Other searches have looked for evidence of WIMPs moving through their experimental detectors. To date, no clear evidence of WIMPs has been uncovered.

Given the continuing developments in **astrophysics** detector technology, the search for WIMPs shows promise. However, these particles are still only theoretical, their properties not completely agreed upon. There are even some physicists who think that a whole new view of **gravity** is more appropriate than new exotic particles to explain the behavior of our galaxy. The existence of WIMPs is one of the most interesting problems to combine astrophysical observations and particle physical theory at the beginning of the twenty-first century.

See also Baryons and baryonic matter; Weak force and interactions

WEATHER

Weather refers to the atmospheric conditions at a certain time or over a certain short period in a given area. It is described by a number of meteorological phenomena which include atmospheric **pressure**, wind speed and direction, **temperature**, humidity, sunshine, cloudiness, and precipitation.

The effects of weather are felt by all life forms on Earth, and weather is also a major contributor in shaping Earth's lithosphere (surface crust). The impact of weather is most pronounced during the occurrence of extreme weather situations, such as prolonged periods of **heat**, cold, rain, drought and smog conditions. In addition shorter but intense events such as hurricanes, tornadoes, winter blizzards, freezing rain, and floods also produce often dramatic effects on both the social and geologic landscape. The concern to reduce the impact of weather on public health and property provides an important motivation for the continued efforts by meteorologists and scientists to improve weather forecasting.

The study of meteorological phenomena is an important component in the development of chaos theory. Chaos theories are used to study weather-related complex systems in which, out of seemingly random, disordered processes there arise new processes that are more predictable. In 1956, Edward Lorenz, a professor of **meteorology** at the Massachusetts Institute of Technology was studying the numerical solution to a set of atmospheric models. Lorenz discovered that his set of model atmospheric equations was very sensitive to their initial conditions. Lorenz named this phenomenon the butterfly effect and suggested that, in the mathematical extreme, the flapping of a butterfly's wings in Kansas might be determined to be responsible for a monsoon in India a month later.

Most of the weather elements on which weather forecasting is based cannot be seen directly, they can only be observed by the effects they create. For the most part weather variables are measured and recorded by instruments. For example, air seems to be without substance, yet it subjects everything to considerable pressure. At sea level, the atmosphere exerts approximately 15 lb per square inch (about 1 kg per square cm) of pressure. The standard instrument used to measure atmospheric pressure is the mercury barometer. The physics for the barometer dates to the classic experiments performed for the first time in 1643 by the Italian scientist **Evangelista Torricelli**. A column of mercury is held in a closed glass tube, then inverted and immersed into a mercury dish. The **weight** of the column is thus balanced by the atmospheric pressure and the length of the column affords a measure of that weight. The mean atmospheric pressure at sea level is 760 mmHg or 1,013 millibars. Pressure as well as air **density** decrease with increasing altitude, and barometric pressure will rise or fall as a function of different weather systems. On weather maps, points of equal pressure are represented by isobars.

Wind, by its broadest definition, is any air **mass** in **motion** relative to Earth's surface. It is predominantly a horizontal movement. However, localized vertical air motion, updraft or downdraft, also occurs, for example, as in storms.

Wind is described by two quantities, speed and direction. Wind **velocity** as measured by the anemometer is reported in mi/hr, knots, m/sec or km/hr. The wind direction is given by the compass bearing from which the wind blows, for example, a southerly wind blows from the south. The horizontal air movement near Earth's surface is controlled by four forces: the pressure gradient **force**, the **Coriolis force**, the centrifugal force, and the frictional forces. The existence of barometric differences in the atmosphere sets up the pressure gradient force which causes air to move from a higher to a lower pressure area. The Coriolis force is the apparent deflection of air mass caused by the rotation of Earth, and it was named after the French mathematician **Gaspard-Gustave de Coriolis** who developed the concept in 1835. Because of Earth's rotation, there is an apparent deflection of all **matter** in motion to the right of their path in the northern hemisphere, and to the left in the southern hemisphere. Air from near the equator is thus deflected eastward as it moves toward a low pressure area. This provides the cyclonic counterclockwise rotation of the air around lows in the northern hemisphere. The Coriolis force counteracts the pressure gradient force as it is pulling in the diametrically opposite direction. The perfect balance of these two forces results in air traveling parallel to isobars at a constant speed, and the wind of this idealized scenario is called the geostrophic wind. As for the centrifugal force, it influences wind speed when the motion follows a curved path. The main effect of frictional forces, which exist between the moving air layer and the land and ocean surfaces, is to modify the direction and speed of wind.

Temperature and humidity are crucial in defining the origins and types of air masses. The thermal properties of an air mass are determined by its latitudinal position on the globe, and its moisture content depends on the underlying surface, be it land or water. For example, polar air is cold and dry, whereas tropical air is hot and humid. In essence, it is the convergence of these two types of air masses which is responsible for most global weather activities. The clash of these contrasting air masses leads to the formation of frontal wave depressions moving in an oscillating west-east pattern and steered by the upper-air jet stream. Hot, humid tropical air is also the source material that fuels the devastating force of hurricanes. Across the network of weather stations, readings of temperature and humidity are taken at regular intervals. Standard equipment in an instrumentation shelter consists of a dry and a wet bulb thermometer, and readings from the latter are used to establish the dew point. A pair of special **thermometers** measures the maximum and minimum temperatures occurring during day and nighttime. The hygrometer measures the relative humidity of the air. In fully automated stations, electronic sensors measure and transmit weather information.

In addition to temperature and humidity, daily weather forecasts inform the public about the heat index during summer and about the wind chill index during the winter. These indicators warn about the possible dangers to human health resulting from exposure to summer heat and winter cold. By combining temperature and humidity, the heat index gives a measure of what temperatures actually feel like. In terms of human health, an increased heat index corresponds to physical activity being more exhausting, resulting in possible heat related illnesses, cramps, exhaustion, or heatstroke. By contrast, the wind chill factor relates the risk of cold to exposed skin, which may lead to frostbite and hypothermia. The wind chill factor takes into account the effect of windspeed on temperature. For example, a temperature of 20°F at a windspeed of 20 mph will feel like -10°F. Humidity is the one factor which not only creates weather activity, but also makes life on Earth possible. Water exists in either one of the following three phases: vapor, liquid, or ice. Water vapor, the invisible gaseous form of water, is always present in the atmosphere; it is defined as the partial pressure of the atmosphere and therefore, like air pressure, it is measured in mmHg. Water vapor supplies the moisture for dew and frost, for clouds and fog, and for wet and frozen forms of precipitation.

The visible weather elements are, of course, sunshine, clouds and precipitation. Traditionally, the forecasting of weather was mainly based on the observation of clouds, because their size, shape and location are the visible indicators of air movement and of changes in water going from vapor to liquid or ice. The first important contribution to the classification of clouds was made in 1802-03 by the Englishman Luke Howard. Based on his observations, clouds were grouped according to three basic shapes: cumulus (heaps), stratus (layers), and cirrus (wispy curls). He also attached the term nimbus to clouds associated with precipitation. From this basic scheme has evolved the modern classification system of clouds by which the lower ten miles of the atmosphere are divided into three layers of clouds characterized by their water phase, i.e., low clouds consisting of water droplets, middle clouds containing a mixture of water droplets and ice crystals, and high clouds entirely made up of ice crystals. While some types of clouds are confined to one layer, such as stratus, stratocumulus and smaller type cumuli in the lower layer, altocumulus and altostratus in the middle layer, and cirrus and cirrostratus in the higher layer, other types can occupy two layers, namely, the nimbostratus and the swelling cumulus cloud which can reside in both lower and middle layers, as well as the cirrocumulus found in the middle and higher layers. A third type can expand through all three layers, such as the huge cumulus congestus cloud and, of course, the mighty cumulonimbus with its characteristic anvil.

Warm and cold fronts are also distinct in their cloud cover. The first signs of an approaching warm front are the cirrus and cirrostratus clouds, followed by the obscuring altostratus and the thick nimbostratus with continuous precipitation, and occasionally with the formation of patches of stratus clouds. After the passage of the warm front, precipitation ceases and the cloud cover breaks up. The typical cloud of cold fronts is the cumulonimbus and, depending on the instability of the air, nimbostratus. Precipitation will vary from brief showers to heavy, prolonged downpours with thunder and lightning.

The weather's immediate impact on public health has been demonstrated numerous times by severe events like hurricanes, tornadoes, floods, snow and icestorms, and prolonged periods of extreme heat or cold. In past years, considerable research efforts have been deployed to gain a better understanding of the physics of hurricanes and tornadoes. Better forecasting the path of severe weather systems and broadcasting early warnings have helped decrease the occurrence of weather-related deaths and injuries. Concerns are now increasingly focused on the weather's indirect influence on human health. It has been observed that certain weather situations provide conditions which will, for example, foster the proliferation of insects and consequently the spread of disease. This was the case in 1999 in the eastern regions of the United States, where weeks of drought and heat created the perfect breeding conditions for mosquitoes carrying a type of encephalitis virus. Weather conditions can also heighten the effects of pollution, for example, air pollutants trapped in fog or smog may cause severe respiratory problems. The interrelationship of weather and environmental health issues lends urgency for more meteorology research in order to develop the accurate forecasting capabilities required to lower the impact of adverse weather on public health.

See also Atmospheric refraction; Chaos and order

WEBER (WB)

The Weber (Wb) is the **International System of Units** (or SI) unit for **magnetic flux**, named in honor of German physicist **Wilhelm Eduard Weber**, who in 1851 discussed a standardized system of units which included electromagnetic quantities. Magnetic flux is defined as the area integral of the magnetic field vector dotted with the area element vector. In the SI system, 1 Wb is derived from more fundamental units as 1 V s.

WEBER, WILHELM EDUARD (1804-1891)
German physicist

Wilhelm Eduard Weber was a German physicist interested in **magnetism** and **electricity** and noted for contributions that proved crucial to the development of the electromagnetic theory of **light**. Weber was born in Wittenberg, Germany. His father was a theologian and moved the family to Halle, where Weber attended the University of Halle and received his doctorate in physics in 1826. In 1831, he was appointed professor of physics at Göttingen, where he began a close collaboration and friendship with physicist **Carl Friedrich Gauss**, who was also interested in electricity and magnetism. With Gauss, Weber worked on terrestrial magnetism and developed several magnetic instruments, including an instrument for measuring small electrical currents called the electrodynamometer.

Weber's career was temporarily sidetracked in 1837 when Weber joined six of his colleagues at the university to protest against the abrogation of the German constitution by the King of Hanover. Known as the "Göttingen Seven," all the faculty members were dismissed from their positions. Weber remained in Göttingen without an official appointment until

1843, when he was appointed a physics professor in Leipzig. Three years later, Weber published *Electrodynamical Measurements*, in which he modified the central **force** concepts of physics and established an absolute system of electrical units. After returning to his former position at Göttingen in 1948, Weber worked on the ratio between the electrodynamic and electrostatic units of charge that became fundamental to the later development of the electromagnetic theory of light. Weber continued his work in **electrodynamics** and the electrical structure of **matter**. The magnetic unit, the weber, is named after him.

In addition to his work in magnetism and electricity, Weber co-wrote a treatise on walking called the *Mechanics of the Human Walking Tool* with his brother Eduard Friedrich Weber. With another brother, Ernst Heinrich Weber, he published a book on wave **motion**. Weber died in Göttingen at the age of 86.

See also Electromagetism

WEIGHT

The **force** exerted by a **gravitational field** on a body, such as the **gravity** of Earth, is called the weight of the body. The weight of a body is equal to the product of its **mass** m and the **acceleration** due to gravity g, (w = mg). Weight should not be confused with mass.

Newton's second law states that the net force F acting on an object is equal to the mass of the object m multiplied by its acceleration a (F = ma). Freely falling bodies experience an acceleration (g) due to the gravitational field of Earth. This force of this field is directed towards the center of Earth. By applying Newton's second law to freely falling bodies, with a = g and F = w, the weight of the body is given as w = mg. Because weight depends upon the gravitational field it varies with geographical location. Because g decreases with increasing distance from the center of Earth, bodies weigh less at higher altitudes than at sea level. Because of this, weight, unlike mass, is not an inherent property of a body.

Weight is usually expressed in pounds or grams. Although weight and mass are not synonymous terms they are often used interchangeably. One concept associated with weight is Archimedes's principle that states that a body immersed in a fluid is acted upon by a force equal and opposite in direction to the weight of the displaced fluid. This principle explains the buoyancy of ships, as well as the rise of **helium** filled balloons. It also explains why we feel lighter in water than air. Another interesting concept of weight is associated with **specific gravity**. The specific gravity of a material is the ratio of the weight of a given volume of that substance to the weight of an equal volume of water. For example, because of the salt, the specific gravity of a saltwater solution is greater than one. This high specific gravity gives saltwater its large buoyancy **power** because the weight of the volume displaced by a person in the ocean is larger than the weight of the volume displaced by the same person in a pool. This means that buoyancy is increased (i.e., it easier to float) when swimming in the ocean rather than in fresh water.

The **molecular weight** of a substance is usually expressed in atomic mass units which is exactly 1/12 the mass of a carbon-12 **atom**.

WEINBERG, STEVEN (1933-)

American physicist

Steven Weinberg shared the 1979 Nobel Prize in physics with **Sheldon Glashow** and **Abdus Salam** for his contributions to the development of a theory unifying the electromagnetic and weak forces, two of the four forces governing nature. He predicted that one of the three particles inherent in the **weak force** could be found in "neutral currents." Its subsequent discovery in 1983 may have brought scientists one step closer to a unified theory of the **universe**.

Weinberg was born in New York City on May 3, 1933, to Frederick Weinberg, a court stenographer, and the former Eva Israel. Weinberg's early interests in science were nurtured both at home and at the world-famous Bronx High School of Science, from which he graduated in 1950. Like Glashow, his classmate at the Bronx High School, Weinberg decided to major in physics at Cornell University, graduating with a B.A. in 1954. He then spent a year at the Niels Bohr Institute (formerly the Institute for Theoretical Physics) in Copenhagen, studying under the noted theoretical physicist Gunner Källén. In 1955, Weinberg returned to the United States and Princeton University, where he began his doctoral research under Samuel Treiman; he was awarded a Ph.D. for his thesis on weak interactions in 1957.

Weinberg took teaching positions at Columbia University from 1957 to 1959, the Lawrence Berkeley Laboratories of the University of California from 1959 to 1969, the Massachusetts Institute of Technology (MIT) from 1969 to 1973, and Harvard University in 1973, where he replaced **Julian Schwinger** as Eugene Higgins Professor of Physics. He took a concurrent position beginning in 1973 at the Smithsonian Astrophysical Observatory, where he served as senior scientist. During much of this period, Weinberg was engaged in a relatively wide variety of research, including studies on **muon** physics, **scattering** theory, broken symmetries, and Feynman graphs.

Weinberg Focuses on Problems of the Electroweak Force

For nearly half a century, physicists had recognized the existence of four fundamental forces: **gravity**, the strong nuclear force, the weak nuclear force, and **electromagnetism.** At no time, however, were physicists entirely convinced that these forces acted as separate entities. In fact, there had been a continuing effort to find a theory allying the four forces into a single fundamental force. A model of that nature had already been conceived during the mid-nineteenth century by **James**

Steven Weinberg.

Clerk Maxwell, who showed that two apparently different kinds of forces, the electrical and magnetic forces, were actually two manifestations of a single, more basic electromagnetic force.

By the early 1960s, efforts toward unification were directed specifically at finding ways of relating the electromagnetic and weak forces. Electromagnetism causes such phenomena as sunlight and **radio waves**, and the weak force operates over very short distances within the **nucleus** and is responsible for some forms of **radioactive decay**. The theories put forth all assumed that while the electromagnetic and weak forces are clearly distinct at **energy** levels we encounter in everyday life, they are indistinguishable at much greater energies, such as those encountered in **cosmic rays** and the most powerful particle **accelerators**.

Between 1960 and 1968 a theory unifying the electromagnetic and weak forces was developed independently by Weinberg, Pakistani physicist Abdus Salam, and, to an extent, by Glashow. There were a number of ways in which the premises and conclusions of these theories could be tested, including the detection of "neutral currents." Weinberg in 1971 had indicated that in the collisions of neutrinos and **matter**, a neutral current was produced which contained one of the three weak force carrying particles. In 1973, experiments at the **European Center for Nuclear Research (CERN)** confirmed the existence of these currents, and a decade later, the three types of particles were detected in further experiments carried out at

CERN by a group headed by the Italian and Dutch physicists **Carlo Rubbia** and Simon van der Meer. For their contributions to the development of the **electroweak theory**, Weinberg, Glashow, and Salam were jointly awarded the 1979 Nobel Prize in physics.

Weinberg has continued to work on a wide variety of topics in physics and **cosmology**. In 1974, with Howard Georgi and Helen Quinn, he made the first estimate of the energy at which the strong, weak, and electromagnetic forces would all be unified. In 1982, Weinberg joined the physics and astronomy departments of the University of Texas at Austin as Josey Regental Professor of Science.

Weinberg was married on July 6, 1954, to Louise Goldwasser, a professor of law. The couple has one daughter, Elizabeth. Weinberg's numerous honors include the 1977 Dannie Heineman Prize of the American Physical Society, the 1979 Elliott Cresson Medal of the Franklin Institute, election to the Royal Society of London in 1981, the 1992 National Medal of Science, and many honorary doctoral degrees. Weinberg, who left Harvard in 1983 to become Josey Professor of Science at the University of Texas at Austin, lists his primary leisure time interest as reading history and walking in the Austin Hills.

WEINBERG–SALAM THEORY

Antoine-Henri Becquerel's discovery of **radioactivity** in 1896 was also the indirect discovery of the weak nuclear **force**. Becquerel's photographic plate was fogged by beta particles (electrons) emitted by uranium nuclei in a process known as weak **beta decay**. A **neutron** sitting in a uranium **nucleus** becomes a **proton** while emitting an **electron** and an antineutrino. Physicists now understand that protons and neutrons are not fundamental particles, but rather are each composed of three **quarks**; at the most fundamental level, then, this process transforms a down quark into an up quark: Every type of particle, massive or massless, charged or neutral, quark or **lepton**, interacts via the weak force. In comparison, leptons do not feel the strong force, and neutral particles do not feel the electromagnetic force.

The first theory of weak interactions was devised by **Enrico Fermi** in 1933, and was quite successful in modeling beta decay, a low **energy** process. It was recognized by the 1950s, however, that the theory could not be valid at high energies. The problem with Fermi's theory was that the weak force was not mediated by an intermediate particle.

Quantum electrodynamics (QED), the **quantum field theory** of the electromagnetic interaction, had shown that charged particles interact via the exchange of a **photon**, the quantum of the **electromagnetic field**. This is true for all interactions; **matter** particles interact by exchanging field **quanta**. It is proposed that **gravity** is mediated by the exchange of, as of yet unobserved, gravitons, the strong force by the exchange of gluons, and the weak force by the exchange of intermediate **vector bosons**. A boson is a particle of integer spin; vector refers to spin-1 particles. The properties of the exchange particle can be inferred from the characteristics of the force. The intermediate

vector bosons had to be quite massive, as the weak force has a very short range. And three were needed, the positively charged W⁻, the negatively charged W⁺, and the neutral Z⁰.

It was not difficult to write down the field theory of particles interacting through massive vector bosons, but the new theory had a much more serious and seemingly fatal flaw, it was non-renormalizable. There was no consistent field theory of massive vector bosons.

By the mid 1960s formal developments in quantum field theory placed in the hands of theorists the tools required to construct a viable theory of weak interactions. Gauge theories came first. A field theory of matter particles only, modified to be invariant under a particular set of symmetry transformations (those that form a mathematical group, called the gauge group), yields a field theory with exchange particles as well. The particular gauge group chosen determines both the number of exchange particles and their spin. The theory is renormalizable, but the exchange particles are massless.

In any quantum field theory the state of lowest energy is called the **vacuum** state. It is possible to construct theories in which the vacuum does not share all the symmetries of the theory; this is called spontaneous symmetry breaking. By adding to the model a massless spin-0 particle, which breaks the symmetry of the vacuum in a particular way, some of the exchange particles can become massive, without **mass** being put into the theory. These spontaneously broken gauge theories contain massive vector bosons, and, as was later demonstrated by the Dutch physicist **Gerardus 't Hooft**, are still renormalizable.

By 1967 all the pieces were in place, and were independently exploited by the American **Steven Weinberg** and the Pakistani **Abdus Salam**. They constructed a theory invariant which contains four massless vector bosons. They employed the above-mentioned Higgs mechanism to generate mass (named after its discoverer, the Scottish physicist **Peter Higgs**). After some deft mathematical manipulation, they had a renormalizable theory with three massive exchange particles, the W⁺, the W⁻, and the Z⁰, and one massless boson, the photon. Not only had they produced a theory of the weak interaction, but it incorporated the electromagnetic interaction as well. It was the first unified quantum field theory, and is today called the theory of the electroweak interaction.

Every facet of the Weinberg-Salam model, save one, has been experimentally verified. The three massive bosons were discovered in 1982 at CERN, the European high-energy physics laboratory. They had just the masses predicted by the electroweak model. What has not yet been seen, though, is the **Higgs boson**, the particle required to make the W± and the Z⁰ massive. It is possible that the Higgs is just too massive to be produced by present-day particle **accelerators**. Although the Higgs boson enters the theory as a massless particle, it, too, develops a mass. It is hoped that it will be observed when the next generation of accelerators come on-line early in the twenty-first century.

See also Particle physics; Quantum theory; Quarks and gluons; Renormalization; Spin of subatomic particles; Standard model of particle physics; Virtual particles

WHEELER, JOHN ARCHIBALD (1911-)
American physicist

John Archibald Wheeler's work with **Niels Bohr** and **Edward Teller** helped advance the processes of **nuclear fission** and fusion. Wheeler conducted a variety of military research in association with the **Manhattan Project**, the effort which developed the **atomic bomb** during World War II, and also was instrumental in the creation of the hydrogen bomb. A professor of physics and director of the Center for Theoretical Physics at the University of Texas at Austin, Wheeler added the term "black hole" to astronomy dictionaries. Throughout his career, Wheeler has made fundamental contributions to the studies of nuclear structure, nuclear fission, **scattering** theory, relativity, geometrodynamics, and other subjects. A long-time professor at Princeton University, his students include the Nobel-honored physicist **Richard P. Feynman**. Wheeler was born in Jacksonville, Florida, on July 9, 1911. His parents, Dr. Joseph Lewis Wheeler and Mabel Archibald Wheeler, were both librarians. Dr. Wheeler was later head of the Enoch Pratt Free Library in Baltimore, Maryland, when his son entered Johns Hopkins University as an undergraduate. Wheeler's plans to major in electrical engineering changed after his first year, according to Dennis Overbye's essay on Wheeler in *A Passion To Know,* because of "a frustrating summer spent rewinding electrical motors in a silver mine in New Mexico." Wheeler then changed his major to theoretical physics, a field in which he earned his Ph.D. in 1933.

A postdoctoral fellowship from the National Research Council in 1933 allowed Wheeler to continue his studies first at New York University with Gregory Breit and then at the Institute for Theoretical Physics in Copenhagen. There, he worked closely with Bohr; Wheeler would later tell Overbye that "you can talk about people like Buddha, Jesus, Moses, Confucius, but the thing that really convinced me that such people existed were the conversations with Bohr."

While studying with Bohr, Wheeler began to think about one of the questions that was to occupy his attention for many years: the structure of the atomic **nucleus**. At the time, two models of the nucleus were popular, one which emphasized the properties of the nucleus as a whole and one that emphasized the properties of the nucleons (protons and neutrons) that make up the nucleus. More than a decade later, in 1953, Wheeler and a colleague, D. L. Hill, made one of the first and most successful attempts to combine these two models into a single theory, the "collective model" of the atomic nucleus. Wheeler, in collaboration with Bohr, also devised a theory explaining the process of nuclear fission, predicting the fissility of plutonium produced from the uranium **isotope** ^{238}U. The Wheeler-Bohr discovery was later to become critical in the development of the first atomic weapons.

Throughout his career, Wheeler has pursued some of the most difficult and most fundamental questions in all of physics. Some of his earlier research dealt with the search for a unified field theory, which would show how the fundamental forces of nature (the strong, weak, electromagnetic, and gravitational forces) are related to each other. His 1962 book, *Geometrodynamics,* is a collection of his papers on this subject. He also worked during the 1940s with Feynman on the problem of action at a distance, a line of research for which Feynman shared the 1965 Nobel Prize in physics.

During World War II, Wheeler took a leave of absence from Princeton to consult on various aspects of the Manhattan Project, working at the University of Chicago's Metallurgical Laboratory, where the first atomic pile was constructed; at Washington States' Hanford Engineering Works, where plutonium was manufactured; and at the Los Alamos Scientific Laboratory, where the first atomic bombs were assembled and tested. Wheeler maintained scientific affiliations with the government, serving as a member of the U.S. General Advisory Committee on Arms Control and Disarmament from 1969 to 1976, science advisor to the U.S. Senate Delegation to the 1957 NATO Conference of Parliamentarians, project chairman of the Department of Defense Advance Research Project Agency in 1958, and consultant to the U.S. Atomic Energy Commission in 1958. Wheeler received the Albert Einstein Award of the Strauss Foundation in 1965 for his contributions in the field of nuclear energy.

One of Wheeler's best known and most controversial activities was his participation in the design and construction of the first thermonuclear fusion (hydrogen) bomb. Invited by Teller in 1949 to assist in this project, Wheeler was at first hesitant to become involved, but eventually became convinced that such a weapon was necessary to preserve worldwide peace. Wheeler's part in the project took place primarily between 1951 and 1953 at his Princeton offices under the code name Project Matterhorn.

In 1939, **J. Robert Oppenheimer** described the theoretical effects of the curvature of **space**, when **thermonuclear reactions** cease to function in **stars** and gravitational forces cause their collapse. Wheeler carried out his own investigations into this phenomenon and, in 1967, coined the term "black hole." Expanding upon this concept even further, he rationalized that the whole **universe** might be subject to what he called the Big Crunch. As the universe contracts upon itself to super-dense dimensions, it would cause an explosion similar to that of the **big bang**, creating a totally new universe. As part of this research, Wheeler has developed the concept of "superspace," a highly complex mathematical construct that may be all that remains of the universe after the big crunch. His ideas on the big crunch and superspace have continued to evolve over time, resulting in the better understanding of **black holes** and imaginative theoretical constructs such as "wormholes," which deal with holes in space containing electrical forces.

Wheeler was married to Janette Latourette Zabriskie Hegner in 1935, "three days after [his] return from Copenhagen," according to Overbye. The couple has three children, one son and two daughters. Upon his return, Wheeler accepted a job as assistant professor of physics at the University of North Carolina. He remained there three years before taking a similar post at Princeton University, an affiliation that lasted until 1976. Wheeler took early retirement in order to accept an appointment at the University of Texas at

Austin, as professor of physics and director of the Center for Theoretical Physics and then, in 1979, as Ashbell Smith Professor of Physics. He has also retained his Princeton rank of Joseph Henry Professor Emeritus.

Wheeler has received a host of honors and awards, including the Cressey-Morrison Prize of the New York Academy of Sciences, 1946, the Enrico Fermi Award of the U.S. Energy Research and Development Agency, 1968, the Franklin Medal of the Franklin Institute, 1969, the National Medal of Science, 1971, the Niels Bohr International Gold Medal, 1982, the Oersted Medal, 1983, and the J. Robert Oppenheimer Memorial Prize, 1984.

WHITE DWARFS

White dwarfs are a class of **stars** that are approximately the same size as the Earth but have roughly the same **mass** as our **Sun**. They are about a thousand times dimmer than our Sun, although their surface temperatures are about twice the Sun's. Though they are called "white," the **light** emitted by the stars is actually a bluish **color**. A white dwarf is a stage of evolution in a star's life cycle commonly known as the star's "death," because the star is cooling and will eventually stop radiating light **energy**. Scientists estimate that about 10% of the stars in our galaxy are white dwarfs.

One of the earliest discovered and most famous white dwarfs is Sirius B, which is the binary companion to the bright star Sirius. By observing the orbit of Sirius and using Kepler's laws, the mass of Sirius B was determined to be between 0.75 and 0.95 solar masses, and its brightness was determined to be about 1/360 of that of the Sun. Its **temperature** and radius were also estimated in 1914 by looking at its **radiation** spectrum. However, it was not until 1926, after the invention of **quantum mechanics** and Fermi-Dirac statistics, that scientists began to understand why Sirius B and other white dwarfs did not collapse.

In a "normal" star like our Sun, gravitational forces compress gases (mostly hydrogen) until **nuclear fusion** reactions begin to take place. Visible light and **heat** are emitted in the fusion process, and these exert an outward **pressure**. Eventually, the star settles down into an **equilibrium** state where the pressure due to **gravity** is equal to the radiation pressure, and it stays relatively stable until its nuclear fuel is exhausted.

Once the nuclear fuel is used up, there is no more radiation pressure and the gravitational forces compress the star further. The core of the star starts to collapse and the outer shell is blown off into **space,** forming a planetary nebula. Then a quantum-mechanical effect comes into play, called the **Pauli exclusion principle**, which says that no two spin-1/2 particles (such as electrons) may have the same **quantum numbers** (such as energy, **momentum**, and spin direction). The exclusion principle causes an outward pressure which counteracts the pressure due to gravity, and halting the collapse of the core, thus stabilizing the white dwarf.

As the white dwarf cools, the atoms radiate their remaining **thermal energy** as **blackbody radiation** until their energy becomes close to the Fermi energy. At this point they stop radiating and simply contribute to the stability of the white dwarf. The calculated time scale for a white dwarf to stop radiating completely is 10^9 years, so the process happens very slowly, which explains why we can still observe white dwarfs in the galaxy today.

One of the interesting features of white dwarfs is that they have a maximum mass, first discovered in 1930 by **Subrahmanyan Chandrasekhar** and later called the **Chandrasekhar limit**. In the case where the Fermi energy of the electrons is much greater than their mass-energy, we must treat the problem using special relativity. The Fermi energy only depends on the number of electrons and the size of the star in this case, and as a result, there is a maximum number of electrons that allows the white dwarf to be stable. Multiplying this maximum number by the mass of a **nucleon** and the number of nucleons per **electron** (since the nucleons are so much heavier than the electrons) yields a rough limit of 1.5 solar masses. For the case of a Fermi energy much lower than the electron mass-energy, the mass of the star will always be less than a solar mass. Therefore, all white dwarf stars must have a mass less than the Chandrasekhar limit of about 1.5 solar masses.

See also Fermions; Neutron stars; Spin of subatomic particles; Stellar evolution; Stellar life cycle

WIEN, WILHELM (1864-1928)
German physicist

Wilhelm Wien is best known for his studies of **radiation**. Of the two laws he developed dealing with this topic, one was later confirmed, and is now known as Wien's displacement law. The second law was later shown to be inadequate and was replaced by Max Planck's brilliant theoretical analysis of the quantum nature of **energy** emission. Although Wien never made discoveries later in life of the quality of those from his early career, he eventually became a highly respected leader of German science in the early part of the twentieth century.

Wilhelm Carl Werner Otto Fritz Franz Wien was born on January 13, 1864, on his family's farm at Gaffken, near Fischhausen, in East Prussia. He was the only child of Carl Wien and the former Caroline Gertz, both descended from land-owning Prussian aristocracy. When Wien was two years old, the family moved to a smaller farm at Drachenstein, in the district of Rastenburg. As a young child, Wien received private tutoring and learned to speak French before he could write his native German. He was quite introverted, however, and spent a great deal of time by himself riding and swimming. Wien's mother was a particularly strong influence in her son's life. She was responsible for operating the Drachenstein farm after her husband had become ill, and according to Wien's entry in the *Dictionary of Scientific Biography,* "her excellent knowledge of history and literature stimulated [Wien's] interest in

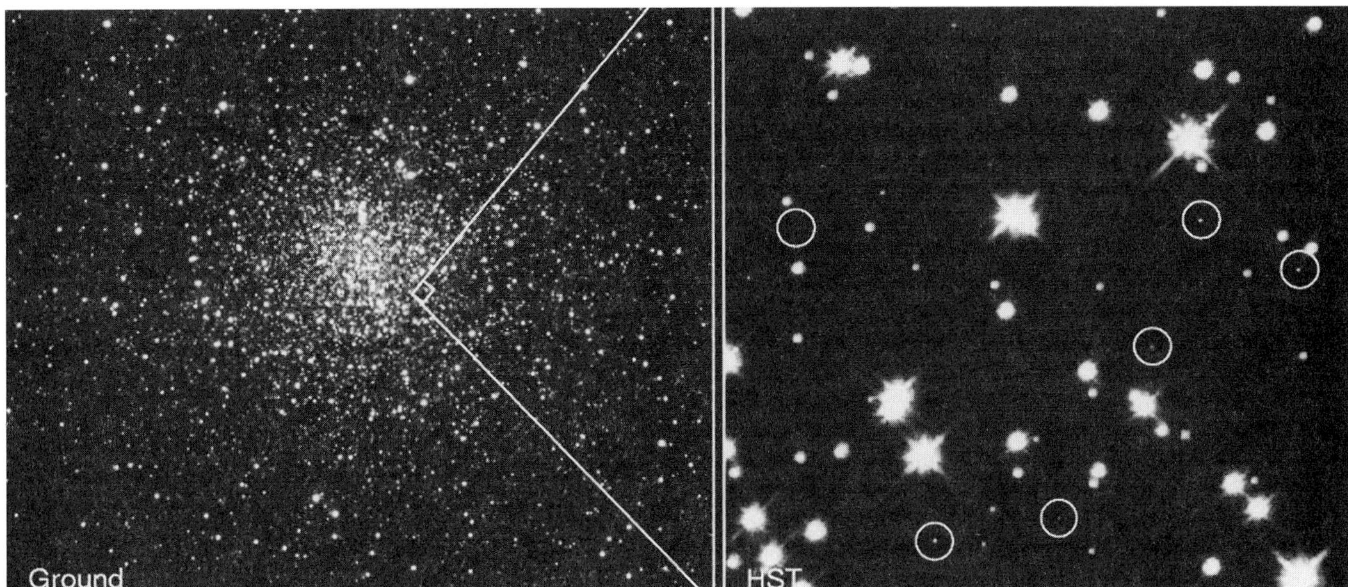

White dwarf stars. *(Images courtesy of NASA.)*

those subjects." Wien was sent to the local Gymnasium at Rastenburg in 1875, but he showed little interest in his classes and was brought home five years later without graduating. For a period of time, he stayed at home learning agriculture and studying with another private tutor. He then returned to formal classes at the Königsberg Altstädtisches Gymnasium, graduating in 1882.

Undecided about Becoming a Farmer

At his mother's urging, Wien then enrolled at the University of Göttingen to study mathematics and natural science. After only one semester, he became bored and left the university, setting off for an extended vacation through the Rhineland and Thüringen. He returned home once again, convinced that as the only child in the family he should take over the farm from his parents. That commitment lasted only a few months, however, and he headed back to school again in the fall of 1883, this time to the University of Berlin.

His academic experience this time was very different. He came under the tutelage of the great German physicist, mathematician, and physiologist **Hermann von Helmholtz** and, according to his own reports, "really came into contact with physics for the first time." He now applied himself vigorously to his studies and in the spring of 1886 received his doctorate. His dissertation dealt with the behavior of **light** diffracted by the sharp edge of a piece of metal.

After receiving his doctorate, Wien returned yet another time to the Drachenstein farm. The occasion for this trip was a disastrous fire that had destroyed some of the farm buildings. For four years Wien remained in a mood of indecision, feeling that he should continue to operate the farm, but still maintaining an interest in physics and continuing to do research on his own.

Joins Helmholtz at Charlottenberg

In 1890 a decision was made for Wien. An extended period of drought forced the Wien family to sell the farm, and Wien decided to take a job as Helmholtz's assistant at the newly established Physikalisch-Technische Reichsanstalt in Charlottenberg, outside Berlin. His parents also moved to Berlin, where his father died less than a year later. In 1892 Wien was promoted to lecturer at Berlin and then, four years later, he was offered a position as professor of physics at the Technical University in Aachen. He remained at Aachen for three years before moving on to the University of Giessen in 1899 and then to the University of Würzburg in 1900.

Wien's most productive period was the decade of the 1890s, when his main area of interest was the nature of **blackbody radiation**. The term *blackbody* refers to a theoretical substance that absorbs all of the radiation that falls on it; the fact that it reflects none of the radiation makes it black. In the 1860s, **Gustav Kirchhoff** had thoroughly studied the thermal properties of blackbodies. He pointed out that they are a perfect tool for studying radiation since when heated they emit radiation of all wavelengths. This fact makes it possible to study in great detail the nature of radiation emitted at different temperatures.

In about 1893 Wien began a theoretical analysis of the characteristics of blackbody radiation beginning with the fundamental laws of **thermodynamics**. He eventually developed two important conclusions. The first of these, now known as Wien's displacement law, says that the **wavelength** of radiation emitted by a blackbody is inversely proportional to the **temperature** of the body. That is, at low temperatures a blackbody will radiate energy with a long wavelength (red light). As the temperature rises, the most abundant wavelength radiated becomes smaller, and the **color** of the emitted light changes to orange, yellow, and then white.

Wien next attempted to find a mathematical formula that would fit the empirical graphical representation of the relationship between the amount of energy radiated at each wavelength for various temperatures. He obtained a complex equation that works fairly well at short wavelengths, but not very well at long wavelengths. He published this result in June 1896. In the meantime **John William Strutt** (Lord Rayleigh) in England had derived a formula that worked well at long wavelengths, but not at short wavelengths. It was not until **Max Planck** introduced the concept of a quantum of energy in 1900 that the problem of blackbody radiation was finally solved.

By 1897 Wien had moved on to a new field of interest, cathode rays. Although he completed some excellent studies in this field, he did not produce any major breakthroughs. His two most notable accomplishments were probably his confirmation of the nature of cathode rays as rapidly moving negatively charged particles (1897–98) and of canal rays as rapidly moving positively charged particles (1905). He also carried out some of the earliest studies on the **diffraction** of **x rays** by crystals, anticipating the discoveries of **Max von Laue** in this area by at least five years.

Wien's tenure at Würzburg lasted for two decades, during which time he was awarded the 1911 Nobel Prize in physics for his work on radiation. He also traveled extensively, including trips to Norway, Spain, Italy, and England (in 1904), to Greece (in 1912), to the United States (where he visited Columbia, Yale, and Harvard in 1913), and to the Baltic states (in 1918). Wien's last academic position was at the University of Munich, where he became professor of physics in 1920. There he supervised construction of a new physics institute and served as rector from 1925 to 1926. He died in Munich on August 30, 1928.

Wien was married to Luise Mehler in 1898. They had two sons, Waltraut and Karl, and two daughters, Gerda and Hildegard. In addition to the Nobel Prize, Wien was honored with membership in the scientific societies of Berlin, Göttingen, Vienna, and Stockholm. He was also a member of the U.S. National Academy of Sciences. From 1906 until his death, he was joint editor with Planck of the prestigious *Annalen der Physik* and later, with F. Harms, of the *Handbuch der Experimental Physik*.

WIGNER, EUGENE PAUL (1902-1995)

American mathematical physicist

Eugene Paul Wigner's enormous contribution to various branches of physics, notably quantum and nuclear, was confirmed by his receipt of the 1963 Nobel Prize in physics (he shared the award with **Maria Goeppert-Mayer** and J. Hans D. Jensen). Recognizing the role of symmetry principles in predicting certain physical processes, Wigner formulated many of the laws governing this theory. Wigner is remembered as being one of the first physicists to call attention to the problems of nuclear **energy**, and also as one of the first scientists to forge links between science and industry around nuclear energy.

Wigner was born in Budapest, Hungary, on November 17, 1902, the son of a businessman. At school, Wigner discovered an interest in physics, but he realized that job opportunities as a physicist in Hungary would be very limited. He therefore decided to study chemical engineering. After receiving a doctorate in chemical engineering from the Technische Hochschule in Berlin in 1925, Wigner returned to Budapest for a year to take up a post in a leather-tanning plant. He left Hungary for the last time in 1926 upon receiving an invitation to return to Berlin to work as an assistant to the well-known physical chemist R. Becker. "The whole of **quantum physics** was being created within my own eyesight," he said of physics in Germany during the 1920s, as quoted in *Pioneers of Science: Nobel Prize Winners in Physics*. Inspired by such inventiveness, Wigner began writing papers of his own; specifically, he was interested in exploring how the mathematical concept known as **group theory** could be used as a tool in the new quantum **mechanics**. On the strength of this work, Wigner was invited in 1927 to join the physics department of the University of Göttingen, as assistant to the mathematician **David Hilbert**.

At Göttingen Wigner developed his law of the conservation of **parity**, which states that no fundamental distinction can be made between left and right in physics. The laws of physics are the same in a right-handed system of coordinates as they are in a left-handed system. Based on Wigner's law of parity conservation, particles emitted during a physical interaction should emanate from the **nucleus** to the right and the left in equal numbers. In practical terms, the law meant that a nuclear process should be indistinguishable from its mirror image, that is, an **electron** emitted from a nucleus will be indifferent as to whether it is ejected to the left or right and will shoot off in equal numbers in both directions along the spin axes of the aligned nucleus. This theory remained steadfast until 1956 when two Chinese American physicists, **Tsung-Dao Lee** and Chen Ning Yang, disproved it by showing experimentally that parity is not conserved in weak interactions. Instead, their experiments revealed that far more electrons were emitted from the south end of the nucleus than from the north. For invalidating the widely held concept of the conservation of parity, they shared the 1957 Nobel Prize in physics.

Wigner returned in 1928 to the Technische Hochschule and continued his work on group theory until 1930, when he moved to the United States to accept a position as lecturer in **mathematical physics** at Princeton University. For eight years he served as a part-time professor at Princeton, until he was elevated to the position of Thomas D. Jones Professor of Mathematical Physics in 1938. The year before, Wigner had become an American citizen.

Develops Short Periodicity of Binding Energies

Wigner's tenure at Princeton afforded him the time and space to do his most important work. As a young physicist, he had become interested in symmetry principles, especially with the patterns found in atomic and molecular spectra. Important discoveries in the 1930s of the binding forces within a nucleus paved the way for Wigner's research. It was found that nuclei containing even numbers of protons (the positively charged particles in the nucleus) and neutrons (the neutral particles) are bound together more strongly than those with an uneven number of protons and neutrons. This is referred to as the short periodicity of binding energies. Longer periods of **binding energy** are also possible, and show especially strong binding when the number of protons or neutrons or both is 2, 8, 20, 28, 40, 50, 82, or 126. The longer binding periods are thought to be caused by the existence in the nucleus of shells or orbits, similar to those that surround the nucleus and contain the electrons, the negatively charged particles. Armed with this data, Wigner forecasted an optical spectra based on the long periodicity model. His findings were published in one of the first papers on the subject. Wigner also contributed significantly to the understanding of short periodicity in his application of mathematical group theory to the energy levels of nuclei up to atomic **weight** 50. His book on group theory has become a classic in the physics canon.

In 1933, the year after **James Chadwick** discovered the **neutron**, Wigner composed a paper which postulated the existence of an energy state of the deuteron which differed from the ground state that had been observed. A deuteron is the nucleus of an **atom** of deuterium, the hydrogen **isotope** that has twice the **mass** of regular hydrogen, and which occurs in water. A deuteron contains one **proton** and one neutron. Wigner's theory provided an explanation for a hitherto unaccounted-for phenomenon: the large deflections of slow neutrons when they pass close to protons. Although Wigner discounted the idea's importance and did not deem it worthy of publication, it proved to be the foundation for numerous other papers.

In 1936, while working with Gregory Breit, Wigner examined the phenomenon of neutron absorption by a compound nucleus. Their Breit-Wigner formula did much to throw **light** on this subject. Continuing his work around atomic nuclei, Wigner postulated in 1937 that protons and neutrons were analogous to isotopes in the **periodic table** of the elements. The manifestation of a particle as a proton or as a neutron could be accounted for by different degrees of spin on the particle, known as isotopic spin or isospin for short.

Turning his attention to **nuclear fission** in 1938, Wigner developed a number of theoretical techniques of reactor calculations, some of which formed the basis of the first controlled **chain reaction** carried out by the Italian physicist **Enrico Fermi**. Together with his fellow Hungarians, **Leo Szilard** and **Edward Teller**, Wigner persuaded **Albert Einstein** to send a letter to President Franklin Roosevelt urging him to beat Hitler in the race to develop an **atomic bomb**. The letter was crucial in convincing the American government to build **nuclear reactors** and was also directly responsible for the establishment of the **Manhattan Project**, on which Wigner played a key role.

In 1936 Wigner married Amelia Z. Frank. She died the following year. In 1941 Wigner married Mary Annette Wheeler, with whom he had two children, David and Martha. After her death, he married Eileen Hamilton and had another daughter, Erika. Despite his many scientific commitments,

Eugene Paul Wigner.

Wigner tried to find time to devote to his family and to pursue his hobbies of bowling and figure skating.

The outbreak of war in Europe caused Wigner to turn his full attention to **nuclear physics**. At the Metallurgical Laboratory at the University of Chicago, he began work on the Manhattan Project as the chief engineer of the water-cooled Hanford plutonium reactors. Wigner's colleagues observed that, for a theorist, he had a remarkably precise knowledge of the engineering design of reactors. Also remarkable was the tremendous speed at which the latest scientific findings in the laboratory were converted into engineered chain reactors. After the war, Wigner accepted a position as director of research and development at the Clinton Laboratories at Oak Ridge from 1946 to 1947.

Career Honored with Copious Awards

Wigner's many contributions to physics have been recognized in a variety of prizes and honorary degrees. He was elected to the National Academy of Sciences in 1945. He was awarded the U.S. Atomic Energy Commission's Enrico Fermi Prize in 1958 and the Atoms for Peace Award in 1960. Most significantly, in 1936, Wigner won the Nobel physics prize for "systemically improving and extending the methods of **quantum mechanics** and applying them widely." Specifically, he was commended for his contribution to the theory of atomic nuclei elementary particles, especially for his discovery and application of fundamental principles of symmetry. This marked an unusual departure for the Nobel Committee, which normally awards the prize for a single discovery or invention.

In 1990 Wigner received the Order of the Banner of the Republic of Hungary with Rubies from his newly democratized birthplace, Hungary. In 1994, he was presented with Hungary's highest recognition, the Order of Merit.

Wigner, who retired from Princeton in 1971, was also active on behalf of other scientists. He was one of thirty-three Nobel Prize winners who sent a telegram to President Podgorny of the former Soviet Union asking that Andrei Sakharov be permitted to receive the Nobel Peace Prize in Stockholm.

Wigner died from pneumonia at the age of 92 on Sunday, January 1, 1995. He is survived by his wife, Eileen Hamilton Wigner, and three children.

WILSON, C. T. R. (1869-1959)
Scottish physicist

A Scottish physicist, C. T. R. Wilson invented the cloud or expansion chamber, which enabled physicists to track the paths of individual atoms and electrons. It was described by the physicist W. B. Lewis as being "to the atomic physicist what the **telescope** is to the astronomer." Lord **Ernest Rutherford** described the **cloud chamber** as "the most original and wonderful [invention] in scientific history." For his invention, Wilson shared the 1927 Nobel Prize for physics with **Arthur Holly Compton**. Wilson is also credited with the discovery of **cosmic rays**, that is, speeding atomic nuclei from outer **space** that enter Earth's atmosphere.

Charles Thomson Rees Wilson was born on the 14th of February, 1869, in Glencorse, near Edinburgh, Scotland, to John Wilson, a sheep farmer, and his second wife, Annie Clark Harper Wilson. When Wilson senior died in 1873, the family moved to Manchester. There Wilson first attended a private school then, when he was fifteen, Ownes College, later renamed the Victoria College of Manchester. Although he had registered as a medical student, he switched to science and graduated with a First Class degree in zoology when he was only eighteen years of age. At the urging of his tutors, he decided to sit for the scholarship exams for Cambridge in practical physics and chemistry. Although he had received no instruction in either subject, he performed so well in the exam that he was awarded a scholarship to Sidney Sussex College.

Begins Career at Cambridge

In 1888, when he was nineteen years old, Wilson entered Cambridge to study physics and chemistry. He found it a thoroughly stimulating environment. He attended physics lectures at its world-renowned Cavendish Laboratory—he was, in fact, the only student in his year taking physics as a main subject—and began to develop what would be a lifelong interest in meteorological physics. After graduating from Cambridge in 1892, Wilson found work as a demonstrator and

a private coach. Although it helped him to keep body and soul together, it left him with precious little time or energy for his own research. At the time, his work was devoted to a comparison of the behavior of different substances in solution. In a bid to secure his future, Wilson somewhat reluctantly accepted the post of science teacher at Bradford Grammar School, but he quickly realized that his career was on the wrong track. In 1894 he decided, therefore, to return to Cambridge, where the university's decision to extend the teaching of physics to medical students had opened up a post as supervisor of the students' practical work.

Returning to Cambridge afforded him more time and space to continue his own work. With the encouragement of the eminent physicist **Joseph John Thomson**, who at the time headed the Cavendish, he began constructing an expansion chamber which would enable him to track the paths of atoms and electrons. In 1895, the first cloud chamber was finished. Wilson intended to use the apparatus to create artificial clouds in the laboratory. His work was inspired by the meteorological phenomena he had witnessed during a visit to Ben Nevis, Britain's highest mountain, in the Scottish Highlands the year before. During a two-week stint as temporary observer to the observatory atop Ben Nevis, he had been captivated by the extraordinary sight of the **Sun** shining on the clouds and the magical optical effects it produced. He was so awestruck that he vowed to create similar clouds to the ones he had witnessed. His theory was that if he expanded moist air in an enclosed chamber, a cloud would thereby be created.

Wilson's hunch was based on results that had been obtained in 1888 by another Scottish physicist, John Aitken. Aitken had discovered that when compressed air is permitted to suddenly expand, a cloud is formed if the air contains dust particles. The latter act as nuclei on which the water vapor can condense—in their absence, the air becomes supersaturated and no clouds are formed. At the beginning of 1895, Wilson began trying to manufacture clouds in the Cavendish Laboratory. Within a short space of time, he succeeded in producing them and in recreating the optical effects he had witnessed on the mountain. In the process, he obtained some interesting results. Using his cloud chamber, Wilson discovered that, contrary to Aitken's theory, a cloud could indeed form even if the air was dust free, so long as the moist air was expanded beyond a certain precise limit. No drops were formed unless the expansion exceeded this limit, in which case a shower of drops was produced.

Encouraged by his preliminary results, Wilson embarked on a series of experiments using more sophisticated equipment. He now discovered that if the air was expanded beyond a second precise limit, equivalent to an approximately eight-fold supersaturation of the vapor, dense clouds were formed in dust-free air. In the process, extraordinary optical effects were also created. Wilson published his results in early 1895. They indicated that two different kinds of clouds could be produced in dust-free air by different degrees of expansion, corresponding to two different types of nuclei. The first kind were less common than the second, which were very numerous, but required more expansion to come into effect.

In 1895, Wilson was appointed one of Cambridge's Clerk Maxwell Students. That same year, he also began investigating electrical fields and thunderstorms inspired, once again, by the **weather**; during another visit to the Scottish Highlands he had witnessed a mighty thunderstorm, which started him thinking about electrical fields in the atmosphere. Over the next five years he developed his basic knowledge of atmospheric **electricity** and put forward the proposition that atmospheric **ions**, which produce electricity, might in fact be produced by sources of **radiation** outside the Earth's atmosphere.

In 1895, **x rays** were discovered. Wilson quickly took advantage of the new technology and used x rays to investigate the behavior of ions as condensation nuclei. He found that x rays produced large numbers of the nuclei in gases that Wilson had been creating in much smaller amounts in his cloud chamber. He found that not only x rays, but also other ionizing agents such as uranium rays and ultraviolet rays, produce condensation nuclei in gases, which are identical with respect to the supersaturation needed to enable water to condense on them and form clouds. These ionizing agents allowed Wilson to successfully test his idea that the two kinds of nuclei he had discovered when he expanded moist air in his dust free cloud chamber were in fact positive and negative ions, of the kind that had been discovered in ionized gases. This work preoccupied him from 1896 to 1898. Between 1898 and 1899, he was engaged full-time in studying condensation on negative and positive ions. Eventually, he was able to render individual ions visible to the eye, and to distinguish between positive and negative ions.

In 1899, Wilson joined the Meteorological Council as a researcher on atmospheric electricity. The following year, he was elected fellow of Sidney Sussex College, Cambridge, and appointed a lecturer in physics. For the next eighteen years, he remained in this position, building a solid advanced physics research team around him and teaching at the Cavendish. Wilson preferred a practical method of instruction. He liked to assign his students minor research problems to carry out in the laboratory themselves rather than having them rely on textbook experiments. His was a close and small coterie, however. He felt comfortable working at a measured pace with just a few, well-chosen pupils to whom he could devote his full attention. He was a dedicated mentor, who took his responsibilities as a teacher extremely seriously. Although he was known as a shy and somewhat plodding lecturer, his insights were considered second to none and inspired subsequent generations of experimental physicists.

Since Wilson had invented his first cloud chamber in 1895, much more had been learned about the phenomena of electricity in ions. But a mystery remained as to where the ions originated. Wilson himself provided the answer in 1900. Using his cloud-expansion chamber, he demonstrated that ions are constantly present in the air, incessantly regenerating. He was able to prove this because, given the right degree of expansion of moist air in the cloud chamber, he was always able to obtain nuclei. Their presence meant that the air should always be conducting. He tested his hypothesis using a gold-leaf electroscope, and discovered that positive and negative ions are con-

stantly being produced in air in equal numbers, at the rate of about 14 per cubic meters of air per second. This discovery provided an explanation as to how an electrical discharge could pass through air, in the laboratory or in the atmosphere, in the form of lightning. Moreover, it represented the first step in the discovery of cosmic rays. These are the streams of atomic nuclei of heterogeneous, extremely penetrating, character that enter Earth's atmosphere from outer space at speeds approaching that of **light**; when they enter Earth's atmosphere, they bombard atmospheric atoms to produce mesons as well as secondary particles possessing some of the original energy. However, it was not until 1912 that Wilson had obtained enough information about the phenomena that he felt confident enough to postulate their existence. His cloud chamber, particularly the more sophisticated versions of it devised by the English physicist Sir Patrick Blackett, was an essential tool in the study of cosmic radiation.

In 1908, at the age of thirty-nine, Wilson married Jessie Fraser Dick. The marriage produced two children. In 1910, he started work on designing a new, much improved cloud chamber. His first two prototypes had been used, respectively, to create clouds and to help in research on ions. The third model was to be used to measure an atom's electrical charge. He proposed to do this by condensing drops of nuclei on atoms and thus rendering them visible. By measuring the total charge and counting the number of drops, the charge per **atom** could be figured. Before Wilson could finish his research, however, another scientist discovered the charge per atom before him.

Wilson's attention turned to the possibility of rendering visible the vapor trail or tracks of positively charged alpha rays. By 1911, he had built an apparatus to carry out the experiment. Although he did not have high hopes of success, he was pleasantly surprised when the tracks were actually revealed. When the cloud chamber was subjected to a magnetic field, the nature of the curved path showed if the charge was positive or negative and what size the particle was.

In 1911, Wilson was appointed observer in meteorological physics at the Solar Physics Observatory. He was promoted reader in electrical **meteorology** in 1918. In 1925, he was made Jacksonian Professor of Natural Philosophy at the University of Cambridge. He retired from Cambridge in 1934. Wilson remained highly active after his retirement. He moved back to Scotland where he was able to indulge his love of mountain climbing until he was well into his eighties. Until the age of eighty-six, he took weather flights over the Outer Scottish Isles as an honorary member of the University of Edinburgh's meteorology department.

Honored with Many Awards

During his long career, Wilson received many honors. He was elected a Fellow of the Royal Society of London in 1900 and awarded its Hughes Medal in 1911. He received the Hopkins Prize of the Cambridge Philosophical Society in 1920, the Royal Society of Edinburgh's Gunning Prize in 1921, and the Franklin Institute's Howard Potts Medal in 1925. He was given the highest honor in 1927 when he was awarded the Nobel Prize for physics, "for his discovery of the vapor condensation method of rendering visible the paths of electrically charged particles." He shared the award with the American theoretical physicist Arthur Holly Compton, whose theory of the **scattering** of x rays upon contact with **matter** was confirmed using Wilson's cloud chamber.

Wilson kept working right up to the end of his long life. In 1956, at the age of eighty-seven, he presented his paper, *A Theory of Thundercloud Electricity,* to the Royal Society. It was the last presentation given by the Society's oldest fellow. Wilson died in 1959 at Carlops, Peebleshire, Scotland.,

WIND ENERGY

Wind is a form of solar **energy**. The irregular heating of Earth's atmosphere by the **Sun** causes the air **mass** to move from regions of high **pressure** to lower pressure areas. The **kinetic energy** of the moving air (wind energy) can be translated directly into mechanical or electrical energy.

Historically, the first use of wind energy was probably to propel boats by use of sails. There is some evidence that wind energy was being used this way as early as 5000 B.C. on the Nile. At some point later, this linear **motion** was translated into rotational motion by means of the windmill. These were in existence in China several centuries before Christ. Windmills were initially used to pump water but over the centuries they have been used to grind grain into flour and to produce **electricity**. They are found in all parts of the world with Holland and Spain historically being most associated with their use.

Today much attention is being paid to wind energy as an alternative to conventional methods of producing electrical energy. The advantages of using the wind as a means of electrical energy production are obvious. In these days of atmospheric pollution it is an attractive proposition to have absolutely clean, non-polluting energy sources such as wind energy. In 1990, California produced enough electrical energy from the wind to offset the emission of 2.5 billion pounds of **carbon** dioxide from conventional power plants.

The basic technology has been in use since the beginning of the twentieth century. The energy of the wind is used to turn windmill blades that rotate a generator to produce electricity. Modern windmills have reached high levels of efficiency by using the latest advances in **electronics**, material technology and aerodynamics. The most recent designs have reached levels of up to 98% efficiency. However, all this comes with a price tag. Roughly 80% of the cost of electrical conversion from wind energy is the price of equipment. Once installed, however, the running costs are low, as no energy sources such as oil and coal are needed.

Obviously, wind energy can only be used if there are fairly predictable winds with enough speed to be efficient. Several areas of the United States (Appalachia, the Plains and the Pacific Northwest) do have very good wind levels. It has been estimated that North Dakota alone could produce 36% of the energy needs of the lower 48 states from the wind it receives. This, of course, would not be a practical proposition. Hawaii has good wind supply and currently boasts some of the largest windmills in the world. The largest have blades up to 100 yd (91 m) in length and stand twenty stories high. Each one can meet the energy requirements of 1,400 homes.

There are certain disadvantages to wind energy. Obviously it can be uncertain and uncontrollable in its delivery. Storage of electricity is difficult but necessary if the wind is going to drop off. Large windmills can be unsightly and noisy. In addition large numbers of bird kills have been recorded over the past few years. Also the price tag is still an issue, although with rising conventional energy cost wind energy is looking a better proposition. It does look as though wind energy has a future as one of the world's energy sources. The U.S. Department of Energy recently announced an initiative to have 5% of the nation's energy being wind generated by 2020.

WIRELESS COMMUNICATION • See Radio

WORK

In the familiar sense, work refers to activities which require exertion. Within the scientific discipline of physics, however, the concept has a more precise definition. Here, work is the transfer of **energy** when a **force** acts on a body over some distance. For example, if a force causes something to move some distance in a straight line, then the work done by the force equals the magnitude of the force multiplied by the distance that the object traveled. When bowling, for instance, the work done by the bowler with a bowling ball is equal to the force of the swing times the distance that the ball travels along the bowling lane. Similarly, the work done by a snow plow is approximately equal to the force exerted by the plow against the snow multiplied by the length of road plowed. In another example, the work performed by the combustion of space shuttle fuel is about equal to the product of the force of the combustion against the gravitational pull on the shuttle and the distance the shuttle travels upward into the atmosphere.

When the work that is accomplished is in the same direction as the displacement, or movement, of the object, the work is said to be positive. If the force is in the opposite direction as the displacement, the work is negative. For example, pushing a door open involves positive work. The work done by the **gravitational force** on the space shuttle in the previous example is an example of negative work since the shuttle proceeds upward. The **time** component of doing work, or the rate of doing work, is called **power**. In order to calculate work, time is not essential, but displacement is. That is, an object must be moved from one location to another. Therefore, while it is a difficult task to hold a textbook stationary at arm's length for a period of time and energy is required, technically no work is done because there is no **motion**. Because the book does not gain or lose any energy from the force of the hand holding it, no energy is transferred from the hand to the book. While machines, like **levers** and pulleys, make tasks easier by reducing the amount of force necessary to move heavy objects, they

Wind turbines, Tehachapi Pass, California. *(Photo courtesy of U.S. Department of Energy, Washington, D.C.)*

do not change the amount of work done because the distance over which the force must be applied also changes.

WORK AND POTENTIAL ENERGY

Work is performed whenever a **force** moves an object. When a person lifts an object, he or she is performing work. More work is done if the object is heavier or if it is lifted higher. Work, then, depends on the force acting on the object as well as the distance it is moved. The amount of work done (W) can be calculated by multiplying the magnitude of the force acting on an object (F) by the distance the object moves (d), using the equation $W = \vec{F} \cdot \vec{d}$. Force and distance are vector quantities, that is, they have both a magnitude and a direction. The direction of the force and distance are important in work calculations, as will be discussed below. Using the English system of measurement, the force is measured in pounds (lb) and the distance is measured in feet (ft). The product of the two, or the work done, is then measured as foot-pounds (ft-lb). With the metric, or SI, system, the unit of force is the Newton (N) and the unit of distance is the meter (m). The resulting unit for work is the Newton-meter (N-m), otherwise known as the joule (J), named after the English physicist **James Prescott**

Joule. One joule is equal to 0.7376 ft-lb. The unit for work, the joule, is the same as the unit for **energy**. When work is performed, energy is transferred from one object to another through mechanical means.

It is important to understand that work is only performed when an object moves. For example, work is not performed when the **velocity** of an object is perpendicular to the direction of the force. An object in circular **motion** about another object can be said to be moved by the other object, and yet no work is being done. Also, if a person holds a 12-lb (53.5 N) box of books while standing still, he is not performing any work. However, if the person lifted the box off the ground, say a distance of 3 ft (1 m), he is performing work. The work performed by the person is the product of the force times the distance, or 12 times 3, or 36 ft-lb (or 53.5 N-m, or 53.5 J). Work is also only done if the force is exerted in the same direction as the movement. If a person holds the 12-lb box of books, he is exerting a force in the upward direction on the box (against the force of **gravity**). If this person is holding the 12-lb box of books while traveling in a car, even though the box is moving, the person does no work on the box because the force he is exerting is not in the same direction as the motion of the box.

A common demonstration of this phenomenon in a physics class is to ask a student to come to the front of the class. The student is instructed to hold a heavy book at arm's length. As the student sweats and strains, the instructor explains that no work is being done. Why then does the student feel as though an effort is being made? The student is experiencing the force of gravity on the book, that is, gravitational **potential energy** (discussed below). Even though no work is being done (no displacement occurs), a force is very much present.

Sometimes a force is exerted on an object at an angle, causing the object to move. An example of this would be a person pushing a stroller. In order to find the work done by the person, you first must know the amount of force exerted on the stroller handle (F) as well as the angle the stroller handle makes with the direction of motion (θ). To calculate the work done in this example, you could draw the force and displacement as vector quantities and determine the resultant vector using vector addition, or you could use a different equation for work: $W = Fd\cos(\theta)$. For example, if a mother pushes a stroller with a force of 5.6 lb (25 N) from her car to the entrance of a grocery store a distance of 60 feet (20 m), and the handle of the stroller makes an angle of 75° with the ground, how much work has she done? Using the equation $W = Fd\cos(\theta)$ and the metric units; 25 N x 20 m x cos(75) = 129 J.

The effect of doing work is the transfer of energy to an object. This energy can take the form of **heat**, as explained in the first law of **thermodynamics**. This law states that the increase of **thermal energy** of a system is the sum of the heat added to it and the work done on it. The energy can take other forms. For example, if the person in the above example lifted the 12-lb box of books 3 feet, placing it on a shelf, the person transferred energy to the box. The box, sitting on the shelf, has the potential to fall, crushing anything sitting on the floor beneath. The box would then do work on the surface of the object it crushes as the surface moves closer to the ground. Energy is the ability to do work, so the box sitting on the shelf has energy.

Energy can take many forms. An object in motion, such as a falling box of books, can do work on another object when a collision occurs, by exerting a force on the object, moving it through a distance. Because an object in motion has the ability to do work, it has energy. This energy of motion, the ability to do work upon collision, is called **kinetic energy**. A box of books on a shelf, even though it is not moving, also has energy. The box has the potential to fall due to the force of gravity. The box, therefore, has what is called gravitational potential energy. The box-Earth system has stored the energy in their combined **gravitational field**. Potential energy is stored energy, giving an object the potential to do work. When the object performs the work, the potential energy is changed into kinetic energy.

Gravitational potential energy (PE) depends on the **mass** of the object (m) as well as its height (h). The height of an object is its distance above a defined zero level, known as the reference level. The reference level is usually the floor of a room or Earth's surface, but it could be the lowest shelf on a bookcase, the top of a table, or the third rung of a ladder. The reference level is arbitrary, but it must be defined. The potential energy of an object is then calculated using its height with

respect to this reference level. The amount of potential energy in an object can be calculated using the equation PE = mgh, where g is equal to the **acceleration** due to gravity, 9.8 m/s². The **weight** of an object is its mass multiplied by the acceleration due to gravity, or mg. Weight is a force. The potential energy equation can then be rewritten as PE = Fh. This is strikingly similar to the equation for work! Using English units, the force, or weight, would be measured in pounds and the height in feet, so the units for potential energy would be foot-pounds. Using the SI system, the unit for force would be kg-m/s2 (equivalent to Newtons) and the unit for height would be meters, so the units for potential energy would be Newton-meters, or joules. The units for energy are the same as the units for work.

In the example of the 12-lb box of books on a shelf, the gravitational potential energy of the box is 12 lb times 3 feet, or 36 ft-lb (53.5 J). Notice that the amount of work done by placing the box on the shelf is the same as the potential energy gained by the box. This example demonstrates that work is simply a method of transferring energy to an object. The amount of work done on the object is equal to the potential energy, or energy of position, gained by the object. This relationship illustrates the law of **conservation of energy**, that the energy in a system is never lost even though it may change forms. Other forms of potential energy include electromagnetic potential energy (for example, the potential energy generated by water turning turbines in a hydroelectric **power** plant), and nuclear potential energy (the potential energy of atoms in **nuclear reactions**, where mass can be converted into energy).

WORMHOLES

General relativity is the theory of **space-time** curvature. It is expressed in the language of differential geometry, the branch of mathematics that deals with the infinitesimal separation between two neighboring space-time points. **Albert Einstein**'s field equations fix the local geometry of space-time, but they say nothing about its global **topology**, how the space-time is connected to itself. Two surfaces (or space-times, but surfaces are easier to visualize) are topologically equivalent if one may be smoothly deformed into the other, without tearing. Thus, a coffee cup and a doughnut are topologically identical, but a sphere and a doughnut are not. It was the introduction of topological considerations into general relativity that led to the concept of "wormholes."

In 1916 (just one year after the publication of Einstein's seminal work), German astrophysicist Karl Schwarzschild solved the field equations to obtain the geometry outside a spherical **mass**, a star for example. He calculated the geometry in a coordinate system in which the star is at rest at the origin. Expressed in those coordinates, the solution exhibited a bizarre property. At a particular distance from the star's center (now called the "Schwarzschild radius," or "horizon") the radial coordinate becomes timelike, and the time coordinate becomes spacelike. Time and **space** exchange places.

The bizarre behavior only occurs around objects whose surface lies within the horizon, since the Schwarzschild solution is only valid outside the mass distribution. This is certainly not the case for a star; the **Schwarzschild radius** of the **Sun**, for example, is 2.9 km. But, if the surface of a dying star, no longer upheld by **radiation pressure**, falls through the horizon, complete gravitational collapse results, producing what mathematicians call a **singularity** of infinite curvature, and what **John Wheeler** called a "black hole."

Now add topology to the mix and things become weirder yet. General relativity arose by requiring the laws of physics to be invariant under general **coordinate transformations**. Any coordinate system can be used to label the space-time points. In 1960 Martin Kruskal at Princeton University expressed the Schwarzschild geometry in a new set of coordinates (which now bear his name), and found a startling result. A collapsing star produces a "bridge," or "wormhole," connecting either two different space-times, or two points in the same space-time (same solution, different topology). It was conjectured that such an object could be used for faster-than-light-speed travel. Send a traveler through the short throat to emerge at a point millions of light-years distant.

But the excitement was short-lived. The wormhole was dynamic, and its time-evolution precluded travel. Initially there are two singularities, at two far-removed space-time points. As time progresses, a non-singular, and hence possibly traversable bridge emerges, joining the two previously singular points. The throat rapidly expands to its maximum radius, but just as rapidly contracts and "pinches off," leaving, once again, two vastly separated singularities. The formation, expansion, and collapse of the bridge occur far too rapidly—any traveler would be caught and crushed as the bridge pinched off.

In 1985 Cal Tech physicist Kip Thorne was contacted by Cornell astrophysicist **Carl Sagan**, who had just finished writing his science fiction novel *Contact*. Could wormholes be used for interstellar travel, as depicted in the novel? Intrigued, Thorne investigated, and found that the wormhole throat could be held open if it was threaded with "exotic" material; that is, the material lining the throat must have an average negative **energy density**. Of course, no such material is known, but by 1994 Thorne had determined that the quantum field fluctuations occurring naturally around the wormhole might have the desired property. Perhaps a suitably advanced civilization could construct a stable wormhole cosmic subway system.

WU, CHIEN-SHIUNG (1912-)
Chinese physicist

Chien-Shiung Wu is perhaps the most respected female physicist in America. She was instrumental in the research that earned two of her colleagues, Dr. **Tsung-Doa Lee** and Dr. Ning Yang, the Nobel Prize in physics. Wu was born in Shanghai in 1915 and was raised in the nearby town of Liu Ho. Her father, Zong-Yee Wu, was the principal of the local elementary school, and he brought to his children a love of books, knowl-

Chien-Shiung Wu, 1964. *(AP/Wide World Photos. Reproduced by permission.)*

edge, and the arts; he also recognized the changes in the political and social landscape of China and prepared his children for them. Chien-Shiung went to high school in Soochow, where she studied English and science and decided to become a physicist. After graduation she enrolled at the National Central University, a government-funded school at Nanking. She received her B.S. in physics in 1936. Once she had graduated from the National Central University, Wu had exhausted China's scientific resources. No graduate program for physics existed in her country so she came to the United States in the fall of 1936 and enrolled at the University of California in Berkeley. This was an exciting time at the Berkeley laboratory: Dr. **Ernest Orlando Lawrence** was the new director and was developing there the first **cyclotron**. Wu was very fortunate in that she was studying under Lawrence as he conducted the research that would ultimately win him a Nobel Prize. In the late 1930s, Lawrence's laboratory was the site of some of the world's finest young physicists. Wu quickly adapted to her new environment. She was given a teaching position in her second semester. Wu also met there a fellow graduate student named Luke Cha-Liou Yuan, whom she eventually married. She was given a position as Dr. Lawrence's research assistant, and in her last year she was elected to Phi Beta Kappa. She received her Ph.D. in 1940. Research continued at the Berkeley laboratory, and Wu remained there after graduation. In 1942, however, Lawrence's laboratory was enlisted to aid the war effort; Wu, who chose to pursue pure research, soon accepted a teaching position at Smith College. At the end of her first year there she received an invitation to teach **nuclear physics** at

Princeton; this was an unprecedented event, as Princeton frowned upon females addressing its male student body—particularly in the sciences. Though she accepted the position, her stay at Princeton ended after just a few months, for in March of 1944 she was asked to join the **Manhattan Project** team at Columbia University. This time she was very eager to contribute to the war effort, and she spent the next two years developing instruments to detect **radiation**. After the war Wu became a research associate at Columbia, where she studied the spectra of **beta decay** (which had been the subject of her doctoral thesis). She was made associate professor in 1952 and began to earn a reputation as an outstanding researcher. In 1956, she was asked to assist doctors Lee and Yang in their experiments. Lee and Yang were concerned with the validity of a concept of nuclear physics known as the principle of **parity**. It states that, on a nuclear level, an object and its mirror image will behave the same way. To understand this principle, imagine standing in front of a mirror while holding a lidded jar. Turn the lid clockwise and it comes off; meanwhile, your mirror image turns the lid counterclockwise, also removing the lid. In nuclear physics, a spinning **nucleus** emits particles as it decays; the principle of parity simply removes the mirror,

asserting that the particles will be emitted no matter which way the nucleus spins. In 1952, a new particle called K-meson was discovered. Strangely, it did not seem to behave in the manner described by the principle of parity. This was unheard of, since the principle of parity had been universally accepted for more than thirty years. To doubt its validity was akin to disbelieving the law of **gravity**. Still, Drs. Lee and Yang announced that the parity principle was flawed. They suggested two types of experiments to disprove the theory and asked Wu to perform them. By cooling a decaying nucleus of cobalt 60 to 0.01 degrees kelvin she was able to manipulate its spin. She then used a scintillation counter to observe the number of electrons emitted. Wu found that many more electrons were emitted as the nucleus spun one direction than in the other—thus disproving the principle of parity. Lee and Yang were awarded the Nobel Prize for their hypothesis, and they acknowledged Wu as being instrumental in its realization. Wu herself has received numerous accolades for her work. She was the seventh woman to join the National Academy of Sciences and the first woman to be granted an honorary doctorate in science from Princeton University in 1958. She is also a member of Academia Sinica, the Academy of Sciences of China.

X

X RAYS

X rays are a form of high-energy electromagnetic **radiation**. Electromagnetic radiation, or **light energy**, occurs as a continuous gradient, or spectrum, of forms. Exhibiting wave-like properties, the **electromagnetic spectrum** consists of light energy of many differing wavelengths and frequencies. **Wavelength** is the distance between the same points on adjacent light **waves**, and **frequency** is defined as the number of waves that pass a given point within a specified **time** period (wave fronts per second, or Hertz). Therefore, smaller wavelengths have greater frequencies, and vice versa. The entire electromagnetic spectrum spans wavelengths smaller than and greater than 100,000 meters. On one end of the spectrum, **radio** waves are very long wavelength light waves (around 100,000 m or more) with low frequency. On the other end, x rays are very short wavelength (10^{-13} meters or less), high frequency waves. X rays specifically have wavelengths that range from 10^{-8} to 10^{-10} meters.

X rays were produced artificially for the first time in 1895 by **Wilhelm Röntgen**. They are produced when electrons, accelerated through high voltages, strike metal. The resulting collision releases x-ray radiation. Also, like other forms of light, x rays are produced naturally. The **Sun**, for example, radiates x rays in addition to the visible spectrum (the spectrum visible to humans). Because x rays are a form of light, their energy is inversely proportional to their wavelength. Thus, since x rays have very short wavelengths, x-ray photons have very high energy. This intense energy makes x rays a form of penetrating radiation. In other words, x rays penetrate, and sometimes pass entirely through, **matter**. X rays can traverse several centimeters of solid material, depending upon its **density**. When x rays pass through living tissue, their energy can damage cells. Damage to cellular DNA is of particular

concern. X rays can cause chromosomal aberrations or mutation that can lead to cancer upon prolonged, or chronic, exposure. Therefore, shielding is used when x rays are employed. The penetrability of x rays is limited, however, and dense materials such as lead effectively block x-ray radiation.

Because of their penetrating energy, x rays have many practical applications. In medicine and industry, x rays are used to visualize interior structures in a manner not possible using visible light wavelengths. X-ray imaging exposes photographic film to x rays that pass through an object (living or inanimate). The darkening of the film is proportional to the degree of x-ray exposure. Thus, lighter areas are denser than darker areas. In this way, cracks in human bone or imperfections in metal castings are detected. A more complex medical application of x rays is Computerized Axial Tomography, or CAT scan, sometimes shortened to CT scan. CAT scans take multiple, rotating x-ray exposures through a subject that are recorded digitally within a computer. As the exposures progress longitudinally along the patient, digital transverse sections are formed that in composite can create a three-dimensional x-ray image of the patient's interior. Because CAT scans have greater sensitivity than x-ray techniques used to assess bone fractures or dental health, they are able to detect some forms of cancerous tumors that would be undetected by other forms of x-ray analysis. X rays are also important to the study of crystals. When x rays pass through crystals, they are scattered by electrons in the atoms that make up the crystal lattice. Because the atoms are close and spaced in a regular fashion, the scattered x-rays will interfere with each other in a process called **diffraction**. The diffraction pattern that is created can be recorded and then used to deduce the structure of the crystal.

See also Crystals and crystallography

Z

Z BOSONS

Z bosons are heavy neutral particles that mediate the weak nuclear interaction. In standard high-energy units, they have a **mass** of 92 GeV. The Z boson has no charge, and like the **photon**, it is its own **antiparticle**. Z bosons interact very much like photons, except that their heavy mass makes the range of the weak **force** much shorter. The Z bosons are some of the heaviest elementary particles, with a mass comparable to that of a niobium **nucleus**. They have a very short lifetime of about 10^{-27} seconds, which requires that they be detected by looking for specific decay signatures in **particle detectors**. They were predicted to exist in the late 1960s and early 1970s by the newly created standard model, and were eventually detected in the mid-1980s at the large **electron positron** collider, LEP, at the **European Center for Nuclear Research (CERN)** near Geneva, Switzerland.

See also Antiparticle; Bosons; Electroweak particles; Particle accelerators; Standard model of particle physics

ZEEMAN, PIETER (1865-1943)

Danish physicist

Publishing his first scientific paper at the age of 18, Pieter Zeeman conducted groundbreaking studies in the physics of optical phenomena that resulted in his winning the Nobel Prize in physics in 1902. Zeeman was born in the small village of Zonnemaire in the Netherlands to Catharinus Forandinus Zeeman, a clergyman, and Wilhelmina Worst. Before he entered college, Zeeman read widely in the sciences, impressing contemporaries such as the young physicist **Heike Kammerlingh-Onnes** (winner of the 1913 Nobel Prize in physics), by his youthful intelligence and early dedication to performing experiments. By the time he entered the university at Leiden in 1885, the prestigious scientific journal *Nature* had already published Zeeman's description and drawings of his observations of the aurora borealis made while he was still in high school.

At the University of Leiden, Zeeman became the pupil and assistant of **Hendrik Lorentz**, who encouraged Zeeman to study the Kerr effect. After receiving his doctorate in 1892, Zeeman became a lecturer at Leiden. In 1896, he began his studies of **magnetism** and its effects on **light**, a natural continuation of his earlier studies with Lorentz of the Kerr effect. In his experiments, Zeeman focused on finding changes in the **spectral lines** of sodium when it was burning in a magnetic field. Zeeman soon discovered that single spectral lines in the light spectrum split into several lines. His observation that spectral lines split when the light source is under the influence of a magnetic field became known as the **Zeeman effect**. This discovery not only confirmed his mentor Lorentz's theory of electromagnetic **radiation**, it provided a means for studying the nature and behavior of **atomic structure**. Zeeman went on to use the Zeeman effect to calculate the charge to **mass** ratio of the sodium atom's vibrating ion and to prove that it was negatively charged. Zeeman shared the Nobel Prize in physics with his former teacher Lorentz in recognition of their "research into the influence of magnetism upon radiation phenomenon."

Zeeman's research also encompassed work on the **Doppler effect** in **optics** and the propagation of light in moving media, and he co-discovered several new isotopes. Zeeman joined the University of Amsterdam in 1897 as a lecturer in physics. However, he was frustrated for many years in pursuing his main interest in magnetism and light. The source of his problems was an inadequate laboratory facility that caused his **spectroscope** to vibrate from passing street traffic and caused anomalies in his data. Finally, in 1923, the university constructed a special laboratory that included a concrete block to serve as a platform for his spectrometer and allowed him to conduct vibration-free experiments.

Zeeman spent the remainder of his career at the University of Amsterdam, where he headed the physics insti-

Pieter Zeeman.

tute, now known as the Zeeman Laboratory of Amsterdam University. His numerous awards include the Rumford Medal of the Royal Society of London, the Franklin Medal, and the Henry Draper Medal of the United States National Academy

of Sciences. Married to Johanna Elizabeth Lebret in 1895, Zeeman had four children, was known as a kind and accomplished teacher, and pursued additional interests in literature and music.

ZEEMAN EFFECT

The Zeeman effect is the splitting of atomic **spectral lines** caused by application of an external magnetic field. **Pieter Zeeman** observed this effect in 1896 and was awarded the Nobel Prize in 1902, along with **Hendrik Lorentz**, for explaining the phenomenon.

All of the atomic states with the same **energy** are said to be degenerate. The Zeeman effect is one of a number of mechanism that lift the degeneracy in energy. That is, there are many mechanisms that give the degenerate states slightly different energies. Each level is composed of states that are degenerate with respect to total **angular momentum** J until an external magnetic field is applied. When an external magnetic field is applied to the system these states are no longer degenerate in energy but are split.

The Normal Zeeman effect was the first attempt to explain the behavior of the spectral lines (energy levels) in a strong magnetic field. It only included the angular **momentum** of the atomic electrons. However, the number of energy levels predicted by this first attempt were different than those that were measured. This result was known as the Anomalous Zeeman effect. Later, the effect of **electron** spin was included into the theory, and the measured levels matched the predicted levels.

See also Atomic structure; Magnetic moment

Books

Abell, G., and D. Morrison. *Explorations of the Universe.* New York: Saunders College Publishing, 1987.

Alic, Margaret. *Hypatia's Heritage: A History of Women in Science from Antiquity through the Nineteenth Century.* Boston: Beacon Press, 1986.

Aller, Lawrence H. *Atoms, Stars, and Nebulae,* 3rd edition. Cambridge: Cambridge University Press, 1991.

Armstead, Christopher H., ed. *Geothermal Energy.* Paris: UNESCO, 1973.

Asimov, Isaac. *The History of Physics.* New York: Walker and Company, 1966.

Barger, V., and M. Olsson. *Classical Mechanics: A Modern Perspective.* New York: McGraw-Hill, 1973.

Barrow, John D. *The Origin of the Universe.* New York: Basic Books, 1994.

Barry, R. G., and R. J. Chorley. *Atmosphere, Weather, and Climate.* London: Methuen and Co., Ltd., 1971.

Beckett, B. *Introduction to Cryptology.* Malden, Massachusetts: Blackwell Scientific, 1988.

Bell, J. S. *Speakable and Unspeakable in Quantum Mechanics.* Cambridge: Cambridge University Press, 1987.

Binney, James, and Michael Merrifield. *Galactic Astronomy.* Princeton, New Jersey: Princeton University Press, 1998.

Blaedel, Niels. *Harmony and Unity.* Translated by Geoffrey French. Madison, Wisconsin: Science Tech, 1988.

Blatt, F. J. *Modern Physics.* New York: McGraw-Hill, 1992.

Bless, R. C. *Introductory Astronomy.* Sausalito, California: University Science Books, 1996.

Boorstin, Daniel J. *The Discoverers.* New York: Random House, 1983.

Bothun, Greg. *Modern Cosmological Observations and Problems.* London: Taylor & Francis Ltd., 1998.

Brown, Julian. *Minds, Machines, and the Multiverse: The Quest for the Quantum Computer.* New York: Simon & Schuster, 2000.

Carroll, Bradley W., and Dale A. Ostlie. *An Introduction to Modern Astrophysics.* Reading, Massachusetts: Addison-Wesley Publishing Company, Inc, 1996.

Chaisson, Eric J. *The Hubble Wars: Astrophysics Meets Astropolitics in the Two-Billion-Dollar Struggle over the Hubble Space Telescope.* London: Harvard University Press, 1994.

Cline, William R. *The Economics of Global Warming.* Washington: Institute for International Economics, 1992.

Connors, Kenneth. *Chemical Kinetics: The Study of Reaction Rates in Solution.* New York: VCH, 1990.

Coughlan, G. D., and J. E. Dodd. *The Ideas of Particle Physics,* 2nd edition. Cambridge: Cambridge University Press, 1991.

Cuttnell, John D., and Kenneth W. Johnson. *Physics,* 3rd edition. New York: John Wiley & Sons, 1995.

Cutnell, John D., and Kenneth W. Johnson *Physics,* 4th edition. New York: John Wiley & Sons, 1998.

Davies, Paul. *The Mind of God.* New York: Simon & Schuster, 1992.

Davies, Paul. *The Last Three Minutes.* New York: Basic Books, 1994.

Duffin, W. J. *The Solid State: From Superconductors to Superalloys.* Translated by André Guinier and Rémi Jullien. Oxford: Oxford University Press, 1989.

Ellis, G. F. R., and R. M. Williams. *Flat and Curved Space-Times.* Oxford: Clarendon Press, 2000.

Elmegreen, Debra Meloy. *Galaxies and Galactic Structure.* Upper Saddle River, New Jersey: Prentice Hall, 1998.

Epstein, Lewis Carroll. *Relativity Visualized.* San Francisco: Insight Press, 1997.

Ferris, Timothy, ed. *The World Treasury of Physics, Astronomy and Mathematics.* Boston: Little, Brown and Company, 1991.

Feynman, Richard P. *The Feynman Lectures on Physics,* Volume 1. Reading, Massachusetts: Addison-Wesley Publishing Company, 1963.

Feynman, Richard P. *The Character of Physical Law.* Cambridge, Massachusetts: MIT Press, 1965.

Feynman, Richard P. *QED: The Strange Theory of Light and Matter.* Princeton, New Jersey: Princeton University Press, 1985.

Fischer, Daniel, and Hilmar Duerbeck. *Hubble: A New Window to the Universe.* Translated by Helmut Jenkner and Douglas Duncan. New York: Copernicus, 1996.

Fisher, David E. *Fire and Ice: The Greenhouse Effect, Ozone Depletion, and Nuclear Winter.* New York: Harper & Row, 1990.

Foster, James, and J. D. Nightingale. *A Short Course in General Relativity.* New York: Springer-Verlag, 1995.

Fraser, P. M. *Ptolemaic Alexandria.* Oxford: Clarendon Press, 1972.

French, A. P. *Special Relativity.* New York: W. W. Norton & Company, 1968.

Galeotti, P., and David N. Schramm, eds. *Dark Matter in the Universe.* Dordrecht, The Netherlands: Kluwer Academic Publishers, 1990.

Gavin, William, Sir. *Ninety Years of Family Farming: The Story of Lord Rayleigh's and Strutt & Parker Farms.* London: Hutchinson, 1967.

Giancoli, Douglas C. *Physics: Principles with Applications,* 5th edition. New Jersey: Prentice Hall, 1998.

Gilmore, Robert. *Alice in Quantumland.* New York: Copernicus Books, 1995.

Gleick, James. *Chaos.* New York: Penguin, USA, 1988.

Goldman, Martin. *The Demon in the Aether, The Story of James Clerk Maxwell.* Edinburgh: Paul Harris Publishing, 1983.

Goldsmith, Donald. *Einstein's Greatest Blunder? The Cosmological Constant and other Fudge Factors in the Physics of the Universe.* Cambridge, Massachusetts: Harvard University Press, 1995.

Green, James A. *Letters on Unified Field Theory: From General Relativity to Unified Field Theory.* Hollywood, California: Greenwood Research, 1999.

Gribbin, John. *Schrodinger's Kittens and the Search for Reality.* London: Weidenfield & Nicholson, 1995.

Gribbin, John. *Q is for Quantum: An Encyclopedia of Particle Physics.* New York: The Free Press, 1998.

Gray, H. J., and Alan Isaacs. *The Penguin Dictionary of Physics.* Middlesex, U.K.: Penguin, 1977.

Greene, Brian. *The Elegant Universe.* New York: W. W. Norton and Company, 1999.

Hamermesh, Morton. *Group Theory and its Application to Physical Problems.* New York: Dover Publications, Inc., 1989.

Hawking, S. *A Brief History of Time.* New York: Bantam Books, 1988.

Hellemans, Alexander, and Bryan Bunch. *The Timetables of Science: A Chronology of the Most Important People and Events in the History of Science.* New York: Simon & Schuster Inc., 1988.

Hewitt, Paul G., John Suchocki, and Leslie A. Hewitt. *Conceptual Physics,* 3rd edition. Menlo Park, California: Scott Foresman Addison-Wesley, 1999.

Hilts, Philip J. *Scientific Temperaments: Three Lives in Contemporary Science.* New York: Simon & Schuster, 1982.

Huang, Kerson. *Statistical Mechanics.* New York: John Wiley & Sons, 1987.

Hughes, R. I. G. *The Structure and Interpretation of Quantum Mechanics.* Cambridge, Massachusetts: Harvard University Press, 1989.

Jones, D. S. *Acoustic and Electromagnetic Waves.* Oxford: Clarendon Press, 1989.

Kamm, L. J. *Understanding Electro-Mechanical Engineering.* Piscataway, New Jersey: IEEE Press, 1995.

Kaplan, D., and L. Glass. *Understanding Nonlinear Dynamics.* Berlin: Springer-Verlag, 1995.

Kaufmann, William. *Black Holes and Warped Spacetime.* New York: W. H. Freeman and Company, 1979.

Kittel, Charles. *Introduction to Solid State Physics.* New York: John Wiley & Sons, 1996.

Kock, Winston. *Lasers and Holography.* Mineola, New York: Dover Books, 1981.

Kondepudi, Dilip, and Ilya Prigogine. *Modern Thermodynamics: From Heat Engines to Dissipative Structures.* New York: John Wiley & Sons, 1998.

Krane, Kenneth. *Introductory Nuclear Physics.* New York: John Wiley & Sons, 1988.

Krapp, Kristine M., ed. *Notable Twentieth Century Scientists.* Detroit: The Gale Group, 1998.

Léna, Pierre. *Observational Astrophysics.* Translated by A. R. King. Berlin: Springer-Verlag, 1988.

Lerner, Adrienne W. "Development of Trigonometry." *Science and Its Times,* Volume 3. Detroit: Gale Group, 2000.

Lerner, B. W. "The Development of High-Tech Medical Diagnostic Tools." *Science and Its Times,* Volume 7. Detroit: Gale Group, 2000.

Lerner, B. W. "Calculators, A Pocket-Sized Revolution." *Science and Its Times,* Volume 7. Detroit: Gale Group, 2000.

Lerner, K. L. *The Development of Quantum Electrodynamics.* London: University Press, 1989.

Lerner, K. L. *The Golden Age of Physics.* Cambridge Press, 1992.

Lerner, K. L. *The Quest: The Search for a Unified Theory of Physics.* Science Research and Policy Institute Press, 1995.

Lerner, K. L. "Bohr Theory." *World of Chemistry.* Detroit: Gale Group, 1999.

Lerner, K. L. "18th Century Astronomers Argue the Existence of God." *Science and Its Times,* Volume 4. Detroit: Gale Group, 2000.

Lerner, K. L. "19th Century Electromagnetic Theory." *Science and Its Times,* Volume 5. Detroit, Michigan: Gale Group, 2000.

Levine, Ira. *Quantum Chemistry,* 4th edition. New Jersey: Prentice Hall, 1991.

Liboff, Richard L. *Introductory Quantum Mechanics.* Reading, Massachusetts: Addison-Wesley Publishing Co., 1997.

Lide, D. R., and H. P. R. Frederikse, eds. *CRC Handbook of Chemistry and Physics 1996-1997: A Ready Reference Book of Chemical and Physical Data,* 77th edition. Boca Raton, Florida: CRC Press, 1996.

Little, James Maxwell. *An Introduction to the Experimental Method: For Students of Biology and the Health Sciences.* Minneapolis: Burgess Publishing Co., 1961.

Lorenz, H. A., A. Einstein, H. Minkowski, and H. Weyl. *The Principle of Relativity.* Translated by W. Perrett and G. B. Jeffery. New York: Dover Publications, Inc., 1952.

Manchester, R. N., and J. H. Taylor. *Pulsars.* San Francisco: W. H. Freeman, 1977.

Mandl, Franz. *Statistical Physics.* New York: John Wiley & Sons, 1988.

Maton, Anthea, et al. *Exploring Physical Science,* 3rd edition. New Jersey: Prentice Hall, 1999.

McKibben, Bill. *The End of Nature.* New York: Random House, 1989.

Menezes, A., P. van Oorschot, and S. Vanstone. *Handbook of Applied Cryptography.* Boca Raton, Florida: CRC Press, 1997.

Menzel, Donald H., ed. *Fundamental Formulas of Physics.* Mineola, New York: Dover Publications, 1960.

Merzbacher, E. *Quantum Mechanics.* New York: John Wiley & Sons, 1997.

Michaels, Patrick J. *Sound and Fury: The Science and Politics of Global Warming.* Washington D. C.: Cato Institute, 1992.

Misner, Charles, Kip Thorne, and John Archibald Wheeler. *Gravitation.* New York: W. H. Freeman and Company, 1973.

Mitton, Jacqueline, and Stephen P. Maran. *Gems of Hubble.* Cambridge: Cambridge University Press, 1996.

Moore, Walter John. *A Life of Erwin Schrodinger.* New York: Cambridge University Press, 1994.

Morrison, F. *The Art of Modeling Dynamic Systems: Forecasting for Chaos, Randomness, and Determinism.* New York: John Wiley & Sons, 1991.

Nilsson, Annika. *Greenhouse Earth.* Chichester: John Wiley & Sons, 1992.

Ohanian, Hans. *Gravitation and Spacetime,* 2nd edition. New York: W. W. Norton & Company, 1994.

Orchin, M., and H. H. Jaffe. *Symmetry, Orbitals, and Spectra.* New York: Wiley-Interscience, 1971.

Pais, Abraham. *Subtle is the Lord?: The Science and the Life of Albert Einstein.* New York: Oxford University Press, 1982.

Peierls, R. E. *Atomic History.* New York: Springer-Verlag, 1997.

Petersen, Carolyn Collins, and John C. Brandt. *Hubble Vision: Further Adventures with the Hubble Space Telescope,* 2nd edition. Cambridge: Cambridge University Press, 1998.

Poole, C. P., Jr. *The Physics Handbook: Fundamentals and Key Equations.* New York: John Wiley & Sons, 1998.

Popper, Karl R. *The Logic of Scientific Discovery.* New York: Basic Books, 1959.

Rayleigh, Lord (Robert John Strutt). *Life of John William Strutt, Third Baron Rayleigh, O. M., F. R. S., by Robert John Strutt, Fourth Baron Rayleigh.* Madison: University of Wisconsin Press, 1968.

Rhodes, Richard. *The Making of the Atomic Bomb.* New York: Touchstone/ Simon & Schuster, 1986.

Rodman, Dorothy, Donald D. Bly, Fred Owens, and Ann-Claire Anderson. *Career Transitions for Chemists.* Salem, Massachusetts: American Chemical Society, 1995.

Ronchi, Vasco. *The Nature of Light: An Historical Survey.* Translated by V. Barocas. Cambridge, Massachusetts: Harvard University Press, 1970.

Rosenblum, Naomi. *A World History of Photography.* New York: Abbeville Press, 1997.

Rosser, W. G. V. *An Introduction to Statistical Physics.* New York: Prentice Hall, 1982.

Rozental, S., ed. *Niels Bohr: His Life and Work as Seen by Friends and Colleagues.* New York: John Wiley & Sons, 1967.

Sagan, Carl. *Cosmos.* New York: Random House, 1980.

Sakurai, J. J. *Advanced Quantum Mechanics.* Reading, Massachusetts: Addison-Wesley Publishing Company, 1967.

Salmon, Wesley C. *Space, Time, and Motion: A Philosophical Introduction.* Minneapolis, Minnesota: University of Minnesota Press, 1980.

Schweber, Silvan S. *QED and the Men Who Made It: Dyson, Feynman, Schwinger, and Tomonaga.* Princeton, New Jersey: Princeton University Press, 1994.

Sciama, D. W., *Modern Cosmology and the Dark Matter Problem.* Cambridge: Cambridge University Press, 1993.

Sears, F. W., and G. L. Salinger. *Thermodynamics, Kinetic Theory and Statistical Thermodynamics.* Reading: Addison-Wesley, 1975.

Seberry, J., and J. Pieprzyk. *Cryptography: An Introduction to Computer Security.* New York: Prentice Hall, 1989.

Serway, Raymond A. *Physics for Scientists and Engineers with Modern Physics,* 3rd edition. Philadelphia: Saunders College, 1990.

Shapiro, Stuart L., and Saul A. Teukolsky. *Black Holes, White Dwarfs, and Neutron Stars.* New York: John Wiley & Sons, 1983.

Shapley, Harlow, and Helen E. Howarth. *A Source Book in Astronomy.* New York: McGraw-Hill, 1929.

Shipman, Harry. *Black Holes, Quasars, and the Universe.* New York: Houghton-Mifflin, 1976.

Shlain, Leonard. *Art and Physics: Parallel Visions in Space, Time, and Light.* New York: Quill, 1993.

Sklar, Lawrence. *Philosophy of Physics.* Boulder, Colorado: Westview Press, 1992.

Starr, Cecie. *Biology: Concepts and Applications.* Belmont, California: Wadsworth Publishing, 1997.

Stroke, H. Henry, ed. *The Physical Review: The First Hundred Years.* Woodbury, New York: AIP Press, 1995.

Taylor, A. W. B. *Superfluidity and Superconductivity,* 2nd edition. Bristol: Adam Hilger, 1986.

Thorne, Kip S. *Black Holes and Time Warps.* New York: Norton, 1994.

Tierney, Brian, and Sidney Painter. *Western Europe in the Middle Ages, 300-1475,* 5th edition. New York: McGraw-Hill, Inc., 1992.

Tipler, Paul A. *Modern Physics.* New York: Worth, 1969.

Tolstoy, Ivan. *James Clerk Maxwell, a Biography.* Chicago: University of Chicago Press, 1981.

Townsend, John S. *A Modern Approach to Quantum Mechanics.* New York: McGraw-Hill, 1992.

Trefil, J. *Space, Time, Infinity.* New York: Pantheon Books, 1985.

Voelkel, James R. *Johannes Kepler and the New Astronomy.* New York: Oxford University Press, 1999.

Wheeler, J. A., and W. H. Zurek, eds. *Quantum Theory and Measurement.* Princeton: Princeton University Press, 1983.

Whittaker, Edmund. *A History of Theories of Aether and Electricity.* New York: Dover Publications, Inc., 1989.

Williams, J. *The Weather Book.* New York: Vintage Books, a Division of Random House, 1992.

Woodhouse, N. M. J. *Special Relativity.* New York: Springer-Verlag, 1992.

Yeomans, Donald K. *Comets: A Chronological History of Observation, Science, Myth, and Folklore.* New York: John Wiley & Sons, 1991.

Zeilick, Michael, and Stephen A. Gregory. *Introductory Astronomy and Astrophysics,* 4th edition. Orlando, Florida, Harcourt Brace & Company, 1998.

Zeldovich, Ya B., and Yu. P. Raizer. *Physics of Shock Waves and High-Temperature Hydrodynamic Phenomena.* San Diego: Academic Press, 1966.

Zitzewitz, Neff, Davids, and Wedding. *Merrill Physics: Principles and Problems.* Westerville, Ohio: Glencoe/McGraw-Hill, 1992.

Journal Articles

Blandford, R., and N. Gehrels. "Revisiting the Black Hole." *Physics Today* (June 1999).

Bohr, N. "Can Quantum-Mechanical Description of Physical Reality Be Considered Complete?" *Physical Review* (October 15, 1935): 696-702.

Browne, Malcolm W. "Physicists Create First Atoms of Antimatter." *The New York Times,* (5 January 1996): sec. A, 1 and 9.

Croswell, K. "The Best Black Hole in the Galaxy." *Astronomy* (March 1992).

Einstein, A., B. Podolsky, and N. Rosen. "Can Quantum-Mechanical Description of Physical Reality Be Considered Complete?" *Physical Review* (May 15, 1935): 777-780.

Feder, Toni. "CERN Council Decides to Build LHC Now—and Pay for It Later." *Physics Today* (February 1997): 58-59.

Hilborn, R. C., and N. B. Tufillaro. "Nonlinear Dynamics." *American Journal of Physics* 65, no. 822 (1997).

Hudson, Richard L. "Supercollider's Post-Mortem Costs May Mount." *The Wall Street Journal.* (26 November 1993): sec. B, 1 and 6.

Monroe, C., D. M. Meekhof, B. E. King, and D. J. Wineland. "A 'Schrodinger Cat' Superposition State of an Atom." *Science* 272 (May 24, 1996).

Spotts, Peter N. "Antimatter Happens, in European Lab." *Christian Science Monitor* (8 January 1996): 4.

Taubes, Gary. "Schizophrenic Atom Doubles as Schrodinger's Cat or Kitten." *Science* 272, no. 5265 (May 24, 1996).

Wearner, R. "The Birth of Radio Astronomy." *Astronomy* (June 1992).

Wilkes, B. "The Emerging Picture of Quasars." *Astronomy* (December 1991).

Wilson, Jim. "Executing Schrodinger's Cat." *Popular Mechanics* 174, no. 10.

Websites

(Editor's Note: As the World Wide Web is constantly expanding, the URLs listed below may be altered or nonexistent as of August 24, 2000.)

Allison, Henry E. "Immanuel Kant." http://www.columbia.edu/~pjs38/biokant.htm (21 August 2000).

American Physical Society. "Henry Augustus Rowland (1848-1901)." 1998. http://www.aps.org/apsnews/1198/119806.html (21 August 2000).

Brock, Henry M. Transcribed by Dennis McCarthy. (copyright: Kevin Knight.) "Gaspard-Gustave de Coriolis." *The Catholic Encyclopedia.* 1999. http://www.newadvent.org/cathen/04370a.htm (21 August 2000).

Brown, Kevin. "The Path to Mass-Energy Equivalence." http://mathpages.com/home/albro/albro8.htm (21 August 2000).

Clyde, John, Johann Schleier-Smith, and Greg Tseng. "Equivalence of Mass and Energy." *Nuclear Physics: Past, Present, and Future.* October 1996. http://library.thinkquest.org/3471/energy_mass_equivalence.html (21 August 2000).

Daniels, Alison. (copyright: The University of Edinburgh.) "Point of Creation: The Higgs Boson." 2000. http://www.cpa.ed.ac.uk/edit/07/articles/03.html (21 August 2000).

Fairbanks, Arthur. Scanned and Proofread by Aaron Gulyas. "Anaxagoras: Fragments and Commentary." 1998. http://history.hanover.edu/texts/presoc/anaxagor.htm (21 August 2000).

Fowler, Michael. "Tycho Brahe." 1995. http://www.phys.virginia.edu/classes/109N/1995/lectures/tychob.html (21 August 2000).

Fox, William. (copyright: Kevin Knight) Transcribed by Thomas J. Bress. "Joseph Von Fraunhofer." *The Catholic Encyclopedia.* 1999. http://www.newadvent.org/cathen/06250a.htm (21 August 2000).

Hamilton, Calvin J. "Neptune." 1997-2000. http://planetscapes.com/solar/eng/neptune.htm (21 August 2000).

Heather, Alan. "Oliver Heaviside." 2000. http://www-groups.dcs.st-and.ac.uk/~history/Miscellaneous/other_links/Heaviside.html (21 August 2000).

Hulse, Russell Alan. (copyright 1999: The Nobel Foundation.) "Autobiography of R. A. Hulse." 1993. http://www.nobel.se/physics/laureates/1993/hulse-autobio.html (21 August 2000).

KDG Wittenberg. "The Weber Family." 2000. Translated by Prof. Timothy Bennett. http://www.wittenberg.de/e/seiten/personen/weber00.html (21 August 2000).

KDG Wittenberg. "Giordano Bruno." 2000. http://www.wittenberg.de/e/seiten/personen/bruno.html (21 August 2000).

Kennedy, D. J. (Copyright: Kevin Knight.) Transcribed by Kevin Cawley. "St. Thomas Aquinas." *The Catholic Encyclopedia.* 1999. http://www.newadvent.org/cathen/14663b.htm (21 August 2000).

Kraan-Korteweg, Renée C., and Ofer Lahav. "Galaxies Behind the Milky Way." *Scientific American.* 2000. http://www.sciam.com/1998/1098issue/1098laham.html (21 August 2000).

Leggat, Robert. "A History of Photography from Its Beginnings till the 1920s." 1999. http://www.kbnet.co.uk/rleggat/photo/ (21 August 2000).

Lienhard, John H. "Engines of Our Ingenuity: Oliver Heaviside." 1988-1997. http://www.uh.edu/engines/epi426.htm (21 August 2000).

Linacre, E., and B. Geerts. "Gustave Coriolis and the Coriolis Effect." January 1999. http://www-das.uwyo.edu/~geerts/cwx/notes/chap11/gustave.html (21 August 2000).

Liss, Tony, and P. L. Tipton. "The Discovery of the Top Quark." 1997. http://www.sciam.com/0997issue/0997tipton.html (21 August 2000).

May, Stephen. "One Summer in Seal Harbor." June 1995. http://www.jhu.edu/~news_info/jhmag/695web/eakins.html (21 August 2000).

Nobel Foundation. "Pieter Zeeman." 2000. http://www.nobel.se/physics/laureates/1902/zeeman-bio.html (21 August 2000).

Nobel Foundation. "Wilhelm Conrad Röntgen." 1999. http://www.nobel.se/physics/laureates/1901/rontgen-bio.html (21 August 2000).

O'Connor, J. J., and E. F. Robertson. "Leonardo da Vinci." 1996. http://turnbull.dcs.st-and.ac.uk/history/Mathematicians/Leonardo.html (21 August 2000).

O'Connor, J. J., and E. F. Robertson. "Jean Baptiste Josephe Fourier." 1997. http://www-history.mcs.st-and.ac.uk/history/Mathematicians/Fourier.html (21 August 2000).

Pallon, Jan. "Did mercury poisoning cause the death of Tycho Brahe?" 1996. http://www.fysik.lu.se/~pixejan/tycho.htm (21 August 2000).

Regents of the University of Michigan. "Biographical Information for Martinus J. G. Veltman." 1998. http://www.physics.lsa.umich.edu/veltmanbiographical.htm (21 August 2000).

Royal Danish Embassy. "H. C." 1998. http://www.denmarkemb.org/oersted.html (21 August 2000).

Schewe, Phillip, and Ben Stein. "Physics News Update: The 1999 Nobel Prize for Physics." 1999. http://www.physics.uq.edu.au/media/Nobel99.html (21 August 2000).

Spergel, David N., Gary Hinshaw, and Charles L. Bennett. "The Milky Way." *Microwave Anisotropy Probe.* May 1, 2000. http://map.gsfc.nasa.gov/html/milky_way.html (21 August 2000).

Stoner, Ron. "Relativistic Dynamics." April 3, 1999. http://fermi.bgsu.edu/~stoner/P202/relative2/sld001.htm (21 August 2000).

Taylor, L. S. "An Anecdotal History of Optics from Aristophanes to Zernike." University of Maryland, Electrical Engineering Department. 1998. http://www.ee.umd.edu/~taylor/optics1.htm (21 August 2000).

Thomas, Dan. "Jean-Baptiste Joseph Fourier." 1996. http://www.chembio.uoguelph.ca/educmat/chm386/rudiment/tourclas/fourier.htm (21 August 2000).

Turner, William. (copyright: Kevin Knight.) "Plato and Platonism." *The Catholic Encyclopedia.* 1999. http://www.newadvent.org/cathen/12159a.htm (21 August 2000).

University of Cambridge. "Our Own Galaxy: The Milky Way." *Cambridge Cosmology.* May 16, 2000. http://www.damtp.cam.ac.uk/user/gr/public/gal_milky.html (21 August 2000).

Van Helden, Albert. "Tycho Brahe." 1995. http://es.rice.edu/ES/humsoc/Galileo/People/tycho_brahe.html (21 August 2000).

Weisstein, Eric W. "Empedocles of Akragas." 1996-2000. http://www.treasure-troves.com/bios/Empedocles.html (21 August 2000).

Weisstein, Eric W. "Weber, Wilhelm." 1996-2000. http://www.treasure-troves.com/bios/WeberWilhelm.html (21 August 2000).

Wills, Jamal. "Coriolis Effect." *Living Universe Foundation.* 1996. http://www.luf.org/bin/view/GIG/CoriolisEffectCoriolisForceTerm (21 August 2000).

Wilson, Prof. Fred L. "Science and Human Values: Plato." 1999. http://www.rit.edu/~flwstv/plato.html (21 August 2000).

c. 900 B.C.

Homer, Greek epic poet, refers to the use of a lodestone for navigation purposes during the siege of Troy.

c. 600 B.C.

Thales of Miletus (624-546 B.C.), Greek philosopher, proposes water as the fundamental substance of the universe. He is also the first to systematically study magnetism.

c. 600 B.C.

Thales of Miletus (624-546 B.C.), Greek philosopher, first notices the electrification of amber (the Greek word for amber is *elektron*) by friction.

c. 550 B.C.

Pythagoras (c. 582-c. 497 B.C.), Greek philosopher, studies sound with experiments on the monochord.

c. 525 B.C.

Anaximenes (c. 570-c. 500 B.C.), Greek philosopher, proposes that air is the fundamental element of the universe. He says that when compressed it can take the form of water and earth.

c. 500 B.C.

Heraclitus (c. 540-c. 475 B.C.), Greek philosopher, states that fire is the fundamental element of the universe. He also states that all things are in constant motion and that nothing is ever lost.

c. 500 B.C.

Parmenides, Greek philosopher, suggests that matter can be neither created nor destroyed.

c. 450 B.C.

Empedocles, Greek philosopher, says that the four elements in the universe—air, earth, fire, and water—display opposing tendencies of integration and disintegration. He also suggests the principle of the conservation of matter.

c. 450 B.C.

Anaxagoras (c. 500-c. 428 B.C.), Greek philosopher, offers one of the first atomic theories, saying that all matter consists of atoms or "seeds of life."

c. 450 B.C.

Leucippus, Greek philosopher, is credited with being the creator of atomism since he is the teacher of Democritus (c. 470-c. 380 B.C.). He is also the first to state the rule of causality, that every natural event has a natural cause.

c. 425 B.C.

Democritus (c. 470-c. 380 B.C.), Greek philosopher, states his atomic theory that all matter consists of infinitesimally tiny particles that are indivisible. These atoms are eternal and unchangeable, although they can differ in their properties. They can also recombine to form new patterns. His intuitive ideas contain much that is found in modern theories of the structure of matter.

c. 380 B.C.

Plato (c. 427-c. 347 B.C.), Greek philosopher, teaches a geometrical theory of matter on which he elaborates in his *Timaeus*.

c. 350 B.C.

Aristotle (384-322 B.C.), Greek philosopher, rejects the atomism of the Greek philosopher Democritus (c. 470-c. 380 B.C.), thus condemning it to oblivion until modern times. He also states that a vacuum does not exist in nature and that sound travels by a succession of impacts on the air. Aristotle argues correctly that sound is not conducted in the absence of air and incorrectly that a body will move only as long as it keeps being pushed. He also says that heavy bodies fall faster than light ones.

c. 300 B.C.
Epicurus (341-270 B.C.), Greek philosopher, elaborates on the atomism of the Greek philosopher Democritus (c. 470-c. 380 B.C.), but substitutes the notion of chance for the determinism of Democritus.

c. 300 B.C.
Strato (c. 340-c. 270 B.C.), Greek physicist, is the first to argue that a body accelerates when falling. He agrees with Aristotle that heavier bodies fall faster than light ones.

c. 300 B.C.
Euclid, Greek mathematician, writes a treatise on optics in which he makes optics a part of geometry by dealing with light rays as though they are straight lines. He also offers a theory of reflection which he treats geometrically.

c. 250 B.C.
Philon, Greek engineer, experiments with air and discovers that it expands when heated.

c. 250 B.C.
Ctesibius, Greek inventor, constructs a water organ in which air is forced through different organ pipes by a falling lead weight. He also improves the ancient Egyptian water clock and builds a clock whose accuracy is not surpassed for eighteen centuries.

c. 220 B.C.
Archimedes (c. 287-c. 212 B.C.), Greek mathematician and engineer, writes his *Treatise on Floating Bodies* in which he tells of his discovery of the principle of buoyancy. Using this principle, he is able to determine if a crown is pure gold by comparing the amount of water it displaces to that displaced by an equal amount of actual gold. He also works out mathematically the principle of the lever and founds the science of statics.

c. 10
Cleomedes, Greek astronomer, discusses the optical properties of water and says that in a similar manner, the Sun may be visible when it has actually gone a bit below the horizon. This is the first consideration of atmospheric refraction until Ptolemy.

c. 50
Hero, Greek engineer, formulates the principle of the motive power of steam, building many steam-powered devices. He also writes of the five simple machines—lever, pulley, wheel, inclined plane, and wedge—and extends and generalizes the law of the lever. He also maintains correctly that air takes up space and is compressible.

c. 100
Nicomachos of Gerasa (c. 50-110), Greek mathematician, writes his *Harmonices* in which his theory

of acoustics and harmonics reaches back to the ideas of the Greek philosopher Pythagoras.

c. 150
Ptolemy, Greek astronomer, writes a treatise on optics in which he considers the refraction of light. He offers an original approach that is both theoretical and experimental.

c. 150
Ptolemy, Greek astronomer, writes a treatise on acoustics in which he synthesizes various traditions and draws on the work of both musicians and mathematicians. His work on acoustics influences medieval thought on the subject.

c. 425
During the reign of the Eastern Roman emperor Theodosius II (401-450), the Greek historian Zosimus is the first to note the electrolytic separation of metals.

c. 500
Anthemius of Tralles, Byzantine mathematician and engineer, writes of the focal property of parabolic mirrors. He also discusses the possibility of making burning mirrors.

517
Johannes Philoponus (c. 490-570), Alexandrian philosopher, also called John the Grammarian, rejects Aristotle's idea that a body will only move as long as it is pushed and offers his own theory of motion. He says that a body will keep moving in the absence of friction or as long as nothing opposes it. He argues that this is why the stars continue to move.

c. 750
With the rise of the Abbasid dynasty and the transfer of its capital to Baghdad, Hellenistic scientific culture with its knowledge of the physical sciences begins to be diffused throughout the Arab world (first in Syriac and then in Arabic) by the Nestorians.

c. 850
Al-Kindi (c. 801-866), Arab physicist, writes a treatise on optics and the reflection of light. He also studies meteorology, the tides, and specific weights and is regarded as the first great Arab author in physics.

980
Moslem sect called the Brethren of Purity is formed. They develop an alternative to the orthodox Moslem physical image of the world and revive the ancient pre-Aristotelian doctrine of atomism. They are maligned but influence the Christian West via Spain.

c. 1000
Alhazen (c. 965-1038), Arabian physicist, rejects the idea that people see because their eyes send out a light which reflects back from an object. He argues correctly that light comes from the Sun or another source and reflects from the object into the eye. He studies all aspects of light, especially reflection and refraction, and also discusses the rainbow.

c. 1025
Al-Biruni (973-1048), Arab physician, astronomer, and mathematician, makes a fairly accurate calculation of the specific weights of eighteen precious

stones and metals. He uses the methods employed by Archimedes.

c. 1100 Ibn Bajja, Arab philosopher also called Avempace, defends the theory of Philoponus and also states that in a void, the speed of a body would be finite and not infinite as Aristotle stated.

1137 Abu Ja'far Alchazin (Al Khazin), Arab mathematician and astronomer, writes a book in which he offers tables of specific densities and a general description of the laws of gravity.

c. 1175 Averroës, Arabian philosopher in Spain, disagrees with Philoponus on the concept of motion and upholds Aristotle's ideas. His commentaries on Aristotle become very influential in the West.

c. 1225 Robert Grosseteste (c. 1175-1253), English scholar, studies optics and experiments with mirrors and lenses. He attempts to explain the rainbow and argues that light is the basic substance of the universe. He is also the teacher of the English scholar Roger Bacon (c.1220-c.1292).

c. 1230 Robert Grosseteste (c. 1175-1253), English scholar, writes his *De Generatione Sonoroum* in which he describes sound as a vibrating motion that passes through the air from its source to the ear.

c. 1250 Gerard of Brussels, Flemish physicist, writes his *Liber de motu* which is the first treatise on the study of motion from a purely kinematic point of view.

1267 Roger Bacon (c. 1220-c. 1292), English scholar, writes his *Opus Majus* which includes sections on optics and physics.

c. 1270 Witelo (c. 1230-c. 1275) of Silesia, also called Vitellio, writes his *Perspectiva* which is a systematic treatment of the optics of the Arabian physicist Alhazen (c. 965-1038). It deals with refraction and reflection as well as the twinkling of stars (caused by motion).

c. 1277 John Pecham (c. 1230-1292), English scholar, writes his *Perspectiva Communis* which becomes the standard textbook on optics for the next three centuries. He states that eyes see by the receipt of rays from an object.

c. 1280 Peter Olivi (1248-1298), French scholar, develops the theory of impetus which states that a moving body may continue in motion even when the propelling force is removed.

c. 1300 Johannes Duns Scotus (c. 1266-1308), Scottish philosopher, discusses the nature of physical laws and distinguishes between causal laws and empirical generalizations. This is an important first step in beginning the scientific method.

c. 1325 William of Ockham (c. 1280-1349), English scholar, argues strongly for the importance of empiricism and lays down the rule called *Ockham's razor*. According to this rule, when two theories equally fit all observed facts, the one requiring the fewest or simplest assumptions is to be accepted as more valid.

c. 1350 Jean Buridan (c. 1300-c. 1385), French philosopher, refutes the Aristotelian notion that an object in motion requires a continuous force, and maintains that only an initial impetus is required. He anticipates Newton's first law of motion by saying that the celestial bodies stay in motion in this manner.

1440 Nicholas of Cusa (1401-1464), German cleric and philosopher, studies hydraulics, and invents the hygrometer for measuring moisture and the bathometer for measuring depths in water.

c. 1500 Leonardo da Vinci (1452-1519), Italian artist and inventor, experiments with hydrostatics and diffraction and offers a version of the principle of inertia (which will not come until the time of Galileo). He displays an ingenious insight into the potential of the five simple machines. He also is the first to describe capillary action (liquid moving up the side of a tube).

1537 Niccolo Tartaglia (1499-1557), Italian mathematician, publishes *La Nouva Scientia* which is the first book written on the theory of projectiles or ballistics. He believes incorrectly that a cannon ball falls straight downward after being propelled forward.

1543 Niccolo Tartaglia (1499-1557), Italian mathematician, publishes the first Latin translation of several previously little-known works by Archimedes such as his famous treatise on floating bodies.

1543 Peter Ramus (1515-1572), French scholar also called Pierre de la Ramee, writes his *Animadversions on Aristotle* in which he attacks Aristotle's physics. He is later forbidden by the Church to teach or write against the great philosopher and is threatened with torture.

1544 Girolamo Cardano (1501-1576), Italian mathematician, publishes his *De subtilitate* in which he discusses motion and mechanics as well as ballistics and magnetism.

1581 Robert Norman, English navigator and instrument maker, publishes a work on the lodestone called *The Newe Attractive*. He discusses the known properties of the magnet and is the first to note that steel does not change its weight when magnetized, as most believe. He also discovers the *magnetic dip* as he suspends a compass needle to allow vertical movement and notes that it points down toward the Earth. This is later used by Gilbert.

1581 Galileo Galilei (1564-1642), Italian astronomer and physicist, discovers the principle of isochronism, or the regularity of the pendulum. Young Galileo's experiments show that a pendulum swings in constant time, irrespective of the width of its swing.

1586 Simon Stevin (1548-1620), Dutch mathematician, publishes a report of his experiment in which he refutes the Aristotelian doctrine that heavy bodies fall faster than light ones. He also founds hydrostatics by demonstrating that the pressure on a liquid varies according to how high above the Earth's surface it is and not upon the shape of the container that holds it. His demonstrations also eliminate many standard arguments in favor of the existence of perpetual motion.

1587 Galileo Galilei (1564-1642), Italian astronomer and physicist, begins experiments that lead to his law of falling bodies. He uses a gently sloping inclined plane and shows that the rate of fall of a body is independent of its weight. He eventually states correctly that all objects will fall at the same rate in a vacuum. He also shows that a body can move under the influence of two forces at one time.

1591 Thomas Harriot (1560-1621), English mathematician, is the first Westerner to note that snowflakes are hexagonal (six-sided). He does not publish his findings. The Chinese, however, knew this from the second century B.C.

1593 Giambattista della Porta (1535-1615) of Italy publishes his *De refractione optices* in which he describes the properties of refracting lenses, studies binocular vision, and compares the pupil of the eye to the lens of a camera obscura.

1600 William Gilbert (1544-1603), English physician and physicist, publishes his *De magnete* which is a full account of his extensive investigations on magnetic bodies and electrical attraction. He suggests that the Earth itself is a great, round magnet, and he is the first to use the terms electric attraction, electrical force, and magnetic pole. He is considered by many to be the father of electrical studies.

1604 Johannes Kepler (1571-1630), German astronomer, publishes his *Ad Vitellionem Paralipomena* in which he discusses the work of Witelo and reestablishes the importance of optics.

1609 Johannes Kepler (1571-1630), German astronomer, theorizes that a force of gravity exists that can exert itself through empty space, and that its strength is related to the size of the bodies involved.

1611 Johannes Kepler (1571-1630), German astronomer, publishes his *Dioptrice* which founds the science of modern optics. He studies the newly invented telescope and works on the manner in which light waves are refracted by lenses, eventually describing the laws of refraction.

1621 Willebrord van Roijen Snell (1580-1626), Dutch mathematician, discovers that when a ray of light passes obliquely from a rarer into a denser medium (such as from air into water), it is bent toward the vertical. He then arrives at a general mathematical relationship to express this refraction of light by relating the degree of the bending of light to the properties of the refractive material. This is a key discovery in optics, but it goes unpublished.

1629 Niccolo Cabeo (1585-1650) of Italy publishes his *Philosophia magnetica* in which he continues Gilbert's experimental work on magnetism. He is the first to describe electrical repulsion.

1632 Galileo Galilei (1564-1642), Italian astronomer and physicist, publishes his famous *Dialogo* and puts many of the words of William Gilbert on magnetism into the mouth of his character "Sagredus."

1635 Henry Gellibrand (1597-1636), English astronomer and mathematician, publishes his findings that offer the first indication that the Earth's magnetic field slowly changes over time. His readings made in London over many years demonstrate this, and to date, there is no satisfactory explanation for this occurrence.

1636 Marin Mersenne (1588-1648), French mathematician and clergyman, publishes his *Harmonie Universelle* in which he studies the physical basis of music. He reports several experiments on instruments that allow him to offer empirically derived rules.

1637 René Descartes (1596-1650), French philosopher and mathematician, publishes his *Discours de la méthode* in which he states that all science should be based on mathematics. While his contribution to experimental science is slight, his thinking exerts an enormous influence on physics.

1637 Pierre de Fermat (1601-1665), French mathematician, produces a mathematical derivation of the law of refraction. It contains the postulate that "nature operates by the simplest and most expeditious ways and means," which becomes known as Fermat's principle.

1638 Galileo Galilei (1564-1642), Italian astronomer and physicist, publishes his *Discorsi e dimostrazioni mathematiche intorno a due nuove scienze* which lays the foundations of modern mechanics. In it he formulates what becomes known as the first law of motion (or the law of inertia), as well as the laws of cohesion and strength of materials, and of the pendulum. It also provides a definition of momentum and details the steps or stages of what becomes known as the experimental method. This work marks the end of Aristotelian physics.

1640 Evangelista Torricelli (1608-1647), Italian physicist, writes his *De motu gravium* in which he applies Galileo's laws of motion to fluids and founds the study of hydrodynamics.

1643 Evangelista Torricelli (1608-1647), Italian physicist, is the first to create a sustained vacuum when he invents the barometer. He fills a four-foot-long glass tube with mercury and inverts it onto a dish. He observes that not all the mercury flows out and that over time, the level remaining in the tube varies. He concludes correctly that these changes are caused by atmospheric pressure.

1644 René Descartes (1596-1650), French philosopher and mathematician, publishes his *Principia philosophiae* which contains the principles of Cartesian physics: the nonexistence of the vacuum, the constancy of the quantity of motion, and the infinite speed of light.

1644 Kenelm Digby (1603-1665), English natural philosopher, observes magnetic and electrical attractions as well as acoustic resonance.

1647 Blaise Pascal (1623-1662), French mathematician, physicist, and philosopher, publishes his *Experiences nouvelles touchant le vide* in which he discusses the problem of a vacuum in a purely experimental way.

1648 Blaise Pascal (1623-1662), French mathematician, physicist, and philosopher, conducts his famous experiment on the Puy-de-Dôme mountain and not only verifies Torricelli's experiments, but goes beyond them to demonstrate that air pressure decreases as altitude increases.

1648 Johannes Marcus Marci von Kronland (1595-1667), Bohemian physician, discovers the diffraction of light, but it does not become a recognized fact until Newton's time.

1650 Otto von Guericke (1602-1686), German physicist, constructs the first air pump with which he is able to create a vacuum. He conducts several experiments using evacuated spheres that show a vacuum's properties. Animals cannot live in a vacuum nor will a candle burn.

1650 Athanasius Kircher (1601-1680), German scholar, conducts experiments with the newly produced vacuum pump and demonstrates that sound cannot be conducted in the absence of air.

1653 Blaise Pascal (1623-1662), French mathematician, physicist, and philosopher, studies fluids and formulates what comes to be known as Pascal's principle—that the pressure at any point in a liquid is the same in all directions. Pascal's principle forms the basis of the hydraulic press which he also describes in theory. This information is not published until a year after his death.

1657 Otto von Guericke (1602-1686), German physicist, conducts one of the more spectacular experiments in the history of physics. He demonstrates the power of a vacuum to Emperor Ferdinand III (1608-1657) by showing that not even two teams of eight straining horses can pull apart two attached copper spheres from which all the air has been evacuated. He puts an end to the ancient notion that "nature abhors a vacuum."

1657 Accademia del Cimento is founded in Florence, Italy. Supported by the Medici family and the Grand Duke of Tuscany nearly a decade before its formal creation, this early scientific society is organized by two of Galileo's pupils, the Italian mathematician Vincenzo Viviani (1622-1703) and the Italian physicist Evangelista Torricelli (1608-1647). They enroll some of the leading men of science to come together for laboratory experimentation and discussion. They eventually publish an account of their work.

1657 Robert Boyle (1627-1691), British physicist and chemist, uses an improved air pump made by English physicist Robert Hooke (1635-1703) and proves that Galileo was correct in stating that all objects fall at the same velocity in a vacuum. He also demonstrates that a clock ticks silently in a vacuum.

1659 Christiaan Huygens (1629-1695), Dutch physicist and astronomer, publishes his *De vi centrifuga* in which he discusses the phenomenon of centrifugal force. He also discovers that a centripetal force is required to maintain a body in circular motion and offers a mathematical expression to govern such a force.

1660 Otto von Guericke (1602-1686), German physicist, invents a frictional electrical device which is the first machine to generate an electrical charge. His hand-rotated globe of sulfur accumulates static electricity, and he is able to conduct several electrical experiments with it, most notably establishing the principle of electrical repulsion.

1660 Vincenzo Viviani (1622-1703), Italian mathematician, and Giovanni Alfonso Borelli (1608-1679), Italian mathematician and physiologist, collaborate on an experiment in which they measure the velocity of sound by using the cannon-flash-and-sound method.

1662 Robert Boyle (1627-1691), British physicist and chemist, formulates the law which states that air is not only compressible but that this compressibility varies with pressure according to a simple inverse relationship. This inverse relationship becomes known as Boyle's law.

1663 One year after the death of the French mathematician, physicist, and philosopher Blaise Pascal, his *Traité de la grande expérience de L'équilibre des liqyuers et de la pesanteur de la masse de l'air* is published. It has much to do with the founding of the study of liquid mechanics.

1665 Robert Hooke (1635-1703), English physicist, publishes his *Micrographia* in which he compares light with waves in water. This is the first serious opposing theory to the classical concept of light as a stream of particles.

1666 Robert Boyle (1627-1691), English physicist and chemist, publishes his *Hydrostatical Paradoxes* in which he details his experiments with fluids and refutes the old doctrine that a light liquid can exert no pressure against a heavier fluid.

1669 Erasmus Bartholin (1625-1698), Danish physician, discovers the optical phenomenon he names double refraction. He notes that images seen through Icelandic feldspar (calcite) are not only doubled but that when the crystal is rotated, one image remains still while the other whirls with the crystal. He calls the stationary light beam the *ordinary beam* and the moving image the *extraordinary beam*. This is not explained until polarized light becomes better understood in the early nineteenth century.

1672 Isaac Newton (1642-1727), English scientist and mathematician, publishes his letter on light in the Royal Society's *Philosophical Transactions*. This letter, which is his first scientific publication, details his prism experiments of 1666 and offers findings that reveal for the first time the nature of light. He recounts how he let a ray of sunlight enter a darkened room through a small hole and then passed the ray through a prism onto a screen. The ray was refracted and a band of consecutive colors in rainbow order appeared. He then passed each separate color through another prism and noted that although the light was refracted, the color did not change. From this, he deduced that sunlight (or white light) consists of a combination of these colors. Later he elaborates further on this ground-breaking experiment.

1672 Jean Richer (1630-1696), French astronomer, finds that the same pendulum clock loses two and one-half minutes in Cayenne, French Guiana as opposed to its former location in Paris. This leads to his correct conclusion that the force of gravity differs in magnitude at different parts of the globe.

1673 Christiaan Huygens (1629-1695), Dutch physicist and astronomer, publishes his *Horologium oscillatorium* in which he details his invention of the pendulum, or grandfather, clock. He employs Galileo's principle of isochronicity and ingeniously adapts it to the inner workings of a clock, beginning the era of accurate timekeeping that is so important to the advancement of physics. This highly original work not only demonstrates great mechanical ability but superior mathematical theorizing as well.

1675 Jean Picard (1620-1682), French astronomer, is the first to observe an electric glow in the vacuum of a barometer.

1676 Edmé Mariotte (1620-1684), French physicist, independently formulates Boyle's law and adds an important qualification to it. Like Boyle, he notes that air expands with rising temperature and contracts with falling temperature, but he adds that the inverse relationship between temperature and pressure only holds if the temperature is kept constant. Because of this, Boyle's law is called Mariotte's law in France.

1676 Gottfried Wilhelm Leibniz (1646-1716), German philosopher and mathematician, criticizes Descartes's ideas of motion and formulates his own theory of dynamics which substitutes kinetic energy for the conservation of movement.

1678 Robert Hooke (1635-1703), English physicist, publishes his *Lectures...Explaining the Power of Springing Bodies* in which he states the law of elasticity, now known as Hooke's law. This says that the force tending to restore a spring to the equilibrium position is proportional to the distance by which the spring is displaced from that equilibrium position.

1678 Francesco Redi (1626-1697), Italian physician, is the first to communicate the fact that the electric shock of the torpedo fish may be transmitted to the fisherman through the line and rod.

1679 Denis Papin (1647-1712), French physicist, develops a steam digester that is the forerunner of the pressure cooker. Work on this device leads him to experiment with steam pushing a piston.

1683 Edmund Halley (1656-1742), English astronomer, states that the Earth's magnetism is caused by four poles of attraction, two of them being in each hemisphere near each pole of the Earth.

1684 Isaac Newton (1642-1727), English scientist and mathematician, provides the first summary exposition of his theory of gravitation in a memoir entitled *De motu corporum*. He expands on it later in his 1687 *Principia*.

1686 Edmund Halley (1656-1742), English astronomer, develops a fairly reliable formula that links the altitudes of various localities with the atmospheric pressure measured there. His altimetric formula is one of the first practical applications of the new barometric discoveries.

1687   Isaac Newton (1642-1727), English scientist and mathematician, publishes his *Philosophiae naturalis principia Mathematica* in which he states his three laws of motion. Regarded as the greatest scientific work ever written, it states that the entire world is subsumed under a single set of laws. They are: (1) a body remains at rest unless it is compelled to change by a force impressed upon it; (2) the change of motion (the change of velocity times the mass of the body) is proportional to the forces impressed; and (3) to every action there is an equal and opposite reaction. From these laws he then deduces his law of universal gravitation.

1687   Denis Papin (1647-1712), French physicist, publishes a work in which he offers details on the use of steam to drive a piston in a cylinder which eventually becomes the basic design for an early steam engine. He never built one of his own design.

1690   Christiaan Huygens (1629-1695), Dutch physicist and astronomer, publishes his *Traité de la lumière* in which he states his wave theory of light. This unpopular theory sees light as a longitudinal wave that undulates in the direction of its motion much as a sound wave does. The worth of this theory is not understood for a full century.

1699   Guillaume Amontons (1663-1705), French physicist, publishes his observations on gases. In his work on different gases he shows that each gas changes in volume by the same amount for a given change in temperature. Implied in this is the notion of absolute zero at which gases can contract no further.

1701   Joseph Sauveur (1653-1716), French mathematician, first introduces the term *acoustics* for the study of sound and examines the relations of the tones of the musical scale.

1704   Isaac Newton (1642-1727), English scientist and mathematician, publishes his *Opticks* which is a comprehensive work containing his main discoveries on the nature of light and color as well as his particle, or corpuscular, theory of light.

1709   Francis Hauksbee (c. 1666-1713), English physicist, publishes his *Physico-Mechanical Experiments on Various Subjects* in which he is the first to study capillary action. He makes the first accurate observations of the effects involving the attractive forces between a liquid and a solid. It is capillary action that causes water to rise within thin tubes and to spread over a flat surface.

1714   Gabriel Daniel Fahrenheit (1686-1736), German-Dutch physicist, invents the mercury thermometer, the first really accurate thermometer. His use of mer-

cury instead of alcohol means that temperatures far above the boiling point of water and well below its freezing point can be measured (since mercury has a higher boiling point than alcohol). He also invents the Fahrenheit temperature scale in which the freezing point of water is 32 degrees and the boiling point is 212 degrees. He arrives at these numbers by adding salt to water to find its lowest freezing point which he calls zero.

1724   Willem Jakob Strom van 's Gravesande (1688-1742), Dutch mathematician, publishes his *De evidentia* in which he defines the difference between physics and mathematics and upholds the autonomy of physics as a separate discipline.

1729   Stephen Gray (c. 1696-1736), English chemist, discovers electrical conduction. He notes that when a long glass tube is electrified by friction, the corks at each end also become charged. He states that electricity, or what he calls electric fluid, has traveled from the glass to the corks. This is now called conduction.

1729   Pieter van Musschenbroek (1692-1761), Dutch physicist, builds the first crude testing machine using a steelyard. This is a form of balance in which the object is suspended from the shorter arm of a lever, and its weight is found by moving a counterpoise along the arm to produce equilibrium. He also uses the term *physics* for the first time in place of *natural philosophy*.

1729   Pierre Bouguer (1698-1758), French physicist, publishes his *Essai d'optique sur la gradation de la lumièere* in which he makes some of the earliest measurements in astronomical photometry. He also investigates the absorption of light in the atmosphere and formulates what comes to be known as Bouguer's law. This concerns the attenuation of a light beam upon passage through a transparent medium.

1730   Stephen Gray (c. 1696-1736), English chemist, demonstrates that the human body can be an electrical conductor and harmlessly electrifies a boy using silk threads.

1730   Pieter van Musschenbroek (1692-1761), Dutch physicist, argues for the autonomy of physics from mathematics, saying its laws are revealed by the senses and that it can do without mathematics altogether.

1730   René Antoine Ferchault de Réaumur (1683-1757), French naturalist and physicist, develops a thermometer independently of Fahrenheit and establishes what comes to be known as the Reaumur temperature scale. This system has zero degrees as the freezing point of water and 80 degrees as the boiling point of water at normal atmospheric pressure.

1732 Pierre Louis Moreau de Maupertuis (1698-1759), French mathematician, publishes his *Discours sur la figure des astres* in which he predicts the shape of the Earth using Newtonian mechanics.

1733 Charles François de Cisternay Du Fay (1698-1739), French physicist, discovers that two electrified objects sometimes attract and sometimes repel each other. He notes that their means of being charged seems to be the difference and states that there are two types of electricity, *resinous* and *vitreous*. He formulates the basic electrical law that "like charges repel and unlike charges attract." Later, American statesman and scientist Benjamin Franklin (1706-1790) calls these two types of electricity *positive* and *negative*.

1733 Johann Heinrich Winckler (1703-1770), German physicist, produces an improved frictional electrical machine and is the first to suggest the use of conductors as a means of protection against lightning.

1736 Leonhard Euler (1707-1783), Swiss mathematician, publishes his *Mechanica: sive motus scientia analytice exposita*, a work of mathematical physics in which he maintains that the principles of mechanics are necessary truths.

1738 Daniel Bernoulli (1700-1782), Swiss mathematician, publishes his *Hydrodynamica* which is a comprehensive work on fluid flow, or hydrodynamics. In it he describes what comes to be called Bernoulli's principle; that as the velocity of fluid increases, its pressure decreases. This marks the beginning of theoretical hydrodynamics.

1742 John Theophile Desaguliers (1683-1744), French-English physicist, publishes *Dissertation Concerning Electricity* in which he is the first to use the word *conductor* to describe those substances that allow electricity to flow. He also names nonconducting materials *insulators*.

1742 Anders Celsius (1701-1744), Swedish astronomer, applies a new scale to his thermometer by dividing the temperature difference between the boiling and freezing points of water into an even 100 degrees (with zero at the boiling point, but eventually this is reversed). His system becomes known as the centigrade scale and eventually is adopted internationally by scientists.

1742 Andrew Gordon (1712-1751), Scottish physicist, abandons the use of glass globes to produce electricity and is the first to use a glass cylinder which he is able to rotate rapidly. He achieves 680 revolutions per minute with his new device, which can generate enough electricity to badly shock a person.

1743 Alexis Claude Clairaut (1713-1765), French mathematician, publishes his *Theorie de la figure de la terre* in which he definitively discusses a rotating body in the shape of the Earth and how it acts under the influence of gravity and centrifugal force.

1743 Jean le Rond D'Alembert (1717-1783), French mathematician, publishes his *Traité de dynamique* in which he expands on Newton's laws of motion. He argues that mechanics is the highest rational science because its principles are necessary truths.

1744 Pierre Louis Moreau de Maupertuis (1698-1759), French mathematician, publishes a paper titled *Accord de différentes lois de la nature qui avaient jusqu'ici paru incompatibles*, in which he first states the principle of least action. This states that physical laws include a rule of economy in which action is a minimum.

1744 Christian Friedrich Ludolf of Germany first demonstrates the ignition of inflammable substances by an electric spark.

1745 Ewald Georg von Kleist (c. 1700-1748), German inventor and clergyman, stores electricity in a bottle, anticipating the Leyden jar.

1745 Christian Gottfried Kratzenstein (1723-1795), German physician, first uses electricity to relieve muscle sprains.

1746 Leonhard Euler (1707-1783), Swiss mathematician, argues against the particle theory of light and suggests correctly that light has a wave form and that color depends on the length of that wave.

1746 William Watson (1715-1787), English physician and botanist, publishes his *Experiments on the Nature of Electricity* in which he is the first to investigate the passage of a current through rarified gas and discover that conductivity increases.

1746 Louis Guillaume Le Monnier (1717-1799), French naturalist, is the first to recognize that the air, or atmosphere, is almost constantly charged.

1746 Gowan Knight (1713-1772), English natural philosopher, is the first to make very powerful, artificial steel magnets.

1746 Pieter van Musschenbroek (1692-1761), Dutch physicist, invents the *Leyden jar*, which is the first really efficient device for storing static energy. While experimenting at the University of Leyden, he finds that a glass vial, or jar, partly filled with water, whose mouth is closed by a cork through which a nail or wire dips into the water, will store static electricity. It releases its charge when the wire is touched. This capture of electricity allows it to be studied at length. The Leyden jar is also a prototype of the modern capacitor.

1748   Mikhail Vasilievich Lomonosov (1711-1765), Russian chemist and writer, publishes several works in which he anticipates Lavoisier and suggests the law of conservation of matter.

1748   Gowan Knight (1713-1772), English natural philosopher, publishes a book in which he discounts the reality of matter and states that the forces of attraction and repulsion account for all observed phenomena.

1750   John Michell (1724-1793), English geologist, publishes *Treatise on Artificial Magnets* in which he anticipates Coulomb and explains his discovery of the inverse-square law for repulsive forces of magnetism. He also invents a torsion balance before Coulomb.

1751   Benjamin Franklin (1706-1790), American statesman and scientist, publishes the first of his two-volume work *Experiments and Observations on Electricity Made at Philadelphia in America*. In this work he formulates a theory of general electrical action, also called his single fluid theory of electricity. His theory is adopted as are several of the terms and concepts he introduces, like positive and negative charge.

1752   Benjamin Franklin (1706-1790), American statesman and scientist, flies a kite carrying a pointed wire in a thunderstorm and attempts to test his theory that atmospheric lightning is an electrical phenomenon similar to the spark produced by an electrical frictional machine. To the kite he attaches a silk thread with a metal key at the end, and as lightning flashes, he puts his hand near the key which sparks, just as a Leyden jar would. He proves his point in this extremely dangerous experiment.

1753   Giovanni Batista Beccaria (1716-1781), Italian physicist, publishes his *Dell' elettricismo artificiale et naturale* in which he summarizes all known electrical experiments in light of Franklin's new unitary theory.

1755   Leonhard Euler (1707-1783), Swiss mathematician, publishes his *Principes généraux de l'état d'équilibre des fluides*. In this work on the equilibrium of fluids, he analyzes the behavior of a fluid under the effects of all types of forces.

1756   John Canton (1718-1772), English physicist, begins three years of careful weather observations and finds that on days when the aurora borealis is very noticeable, a compass needle becomes irregular. This is the first observation of what become known as magnetic storms.

1758   Ruggiero Giuseppe Boscovich (1711-1787), Serbian-Italian astronomer and mathematician, publishes his *Philosophiae naturalis theoria redacta ad unicam legem virium in naturam existentium* in which he rejects the corpuscular theory of matter and proposes a dynamistic theory. He says that matter is made up of perfectly indivisible physical points that exert a force of attraction or repulsion on each other depending on their distances.

1760   Johann Tobias Mayer (1723-1762), German astronomer, is the first to state that magnetic action follows the law of inverse squares.

1760   Ebenezer Kinnersley (1711-1778), English-American experimenter, demonstrates that heat can be produced by electricity.

1762   Johann Georg Sulzer (1720-1779), Swiss physicist, precedes Galvani and notes the bitter taste produced by two different metals connected by a wire.

1766   Horace Benedict de Saussure (1740-1799), Swiss physicist, invents the electrometer, the first device used to measure electronic potential.

1769   Alessandro Giuseppe Antonio Anastasio Volta (1745-1827), Italian physicist, publishes his *De vi attractiva ignis electrica phaenomenis indepentibus*. In what is his first work, Volta introduces the important electrical concept of potential.

1769   Leonhard Euler (1707-1783), Swiss mathematician, publishes his *Dioptrics* in which he demonstrates that refraction and dispersion in lenses are not proportional to each other. The significance of this is that it implies the possibility of building an achromatic objective (one concave and one convex lens, each made from a different type of glass).

1780   Charles Augustin de Coulomb (1736-1806), French physicist, publishes the first of two papers which establish his theory of torsion in thin threads and discusses his invention of the torsion balance. This sensitive instrument can measure the quantity of a force and he later uses it to conduct electrical experiments.

1781   Johan Carl Wilcke (1732-1796), Swedish physicist, introduces the concept of specific heat, which is the ratio of the quantity of heat required to raise the temperature of a body one degree to that required to raise the temperature of an equal mass of water one degree.

1781   Antoine Laurent Lavoisier (1743-1794), French chemist, demonstrates that electricity is produced when solid or fluid bodies pass into a gaseous state.

1783   Horace Benedict de Saussure (1740-1799), Swiss physicist, publishes his *Essais sur l'hygrometrie* in which he details his invention of a hygrometer that uses human hair for measuring humidity. He also discusses the general principles of hygrometry.

1783   Lazare Nicolas Marguérite Carnot (1753-1823), French military engineer and mathematician, publishes his *Essai sur les machines en Géenéral* which is a systematic exposition of mechanics and the concept of force. Carnot offers the earliest demonstration that kinetic energy is lost when bodies collide.

1783   Antoine Laurent Lavoisier (1743-1794), French chemist, and Pierre Simon de Laplace (1749-1827), French astronomer and mathematician, co-author *Mémoire sur la chaleur*. Based on a large amount of experimental data on specific heat, this work helps to lay the foundation of thermochemistry. It also leads Lavoisier to conclude that respiration is a form of combustion.

1785   Charles Augustin de Coulomb (1736-1806), French physicist, publishes the first of seven papers in which he establishes the basic laws of electrostatics and magnetism. Using his newly invented torsion balance, he determines that Newton's law of inverse squares also applies to electrical and magnetic attraction and repulsion. He states that the degree of attraction or repulsion depends on the amount of electric charge or the magnetic pole strength.

1787   Abraham Bennet (1750-1799), English physicist and inventor, is the first to observe the contact electrification of metals. He also invents the gold-leaf electroscope.

1787   Jacques Alexandre Charles (1746-1823), French physicist, repeats the gas-expansion work of his countryman, Amontons, and is the first to offer the rule that the volume of a given quantity of gas is proportional to the absolute temperature if the pressure is held constant.

1787   Ernst Florens Friedrich Chladni (1756-1827), German physicist, publishes his *Theorie des Klanges* in which he is the first to discover the quantitative relationships that rule the transmission of sound. He also creates *Chladni's figures* by spreading sand on thin plates and vibrating them, producing complex patterns from which much is learned about vibrations. He is considered the father of acoustics.

1787   René Just Haüy (1743-1822), French mineralogist, publishes his paper, *Exposition raisonné de la Théeorie de l'électricité et du magnetisme*, in which he makes important contributions to the study of pyroelectricity. He also makes the most extensive and accurate observations on the development of electricity in minerals by friction.

1790   Friedrich Albrecht Carl Gren (1760-1798), German chemist and physicist, founds the *Journal der Physik* as a periodical for the "mathematical and chemical branches of natural science." It is succeeded by the *Neues Journal der Physik* and is known today as the *Annalen der Physik*.

1791   Luigi Galvani (1737-1798), Italian anatomist, publishes his *De viribus electricitatis in motu musculari* in which he investigates what he calls *animal electricity*. During his dissections of frogs, Galvani notes that their legs twitch when touched with a scalpel. He actually discovers a phenomenon of enormous potential—that an electric current can be produced from the contact of two different metals in a moist environment—but he misinterprets its larger meaning and concludes that the electricity is within the animal itself. Although he is partly correct, since bioelectric forces do exist within animal tissue, he overgeneralizes and fails to see the important role of metals. His countryman, Volta, does not make the same mistake.

1791   Pierre Prévost (1751-1839), Swiss physicist, offers his theory of exchanges. He correctly states that all bodies of all temperatures radiate heat, noting that cold does not flow from snow to the hand, but rather it is the loss of heat, and not the gain of cold, that makes the hand feel cold. This is an important and fertile distinction to the later science of thermodynamics.

1793   John Dalton (1766-1844), English chemist, publishes the results of his experiments on electricity in the atmosphere. He shows that the aurora borealis makes a magnetic needle move irregularly.

1796   Pierre Simon de Laplace (1749-1827), French astronomer and mathematician, publishes his *Exposition du système du monde* in which he generalizes Newtonian mechanics. To Laplace, the natural laws are in perfect agreement with all phenomena.

1798   Henry Cavendish (1731-1810), English chemist and physicist, is the first to calculate the Earth's mass. He does this by obtaining what Isaac Newton had not provided—a value for the gravitational constant. He builds a model with light balls and large, heavy ones and uses a sensitive wire to calculate the strength of the attraction between the two. Once he obtains this constant value, he calculates the Earth's mass close to what we know it to be today.

1798   Benjamin Thompson, known as Count Rumford (1753-1814), American-British physicist, bores a cannon with a blunt drill and generates an unlimited amount of heat which makes him question the prevailing *caloric* notion that heat is a fluid or liquid form of matter. He eventually overturns this theory and establishes the beginnings of the modern theory that heat is a form of motion.

1800   William Cumberland Cruikshank (1745-1800),

Scottish chemist and surgeon, performs electrical experiments and observes the liberation of oxygen and hydrogen at different poles of metal in water. He also invents the galvanic trough, which is an improved voltaic pile.

1800   Alessandro Giuseppe Antonio Anastasio Volta (1745-1827), Italian physicist, announces his invention of the voltaic pile which is the first real battery. His work duplicating Galvani's 1791 *animal electricity* experiment leads him to discover that it is the contact of dissimilar metals that causes the electricity. He arranges suitable pairs of metallic plates in a certain order, separates them by pieces of leather soaked in brine, and creates a pile, or battery, that produced a continuous and controllable electric current.

1801   William Hyde Wollaston (1766-1828), English chemist and physicist, establishes that frictional and galvanic electricity are identical. He also discovers, independently of Ritter, the existence of invisible light beyond violet light (ultraviolet light).

1801   Thomas Young (1773-1829), English physicist and physician, reads a paper before the Royal Society in which he demonstrates, via his principle of interference, the wave nature of light. This major breakthrough in understanding light is opposed mainly because it refutes Newton's particle theory.

1801   Johann Wilhelm Ritter (1776-1810), German physicist, discovers the ultraviolet region of the spectrum. Knowing that silver chloride breaks down and turns black in the presence of light, he explores the region beyond the violet and finds that an invisible light also blackens silver chloride. He concludes that radiation must exist that is invisible to the eye.

1801   Nicholas Gautherot (1753-1803), French chemist, discovers that when a current is passed through two plates or wires of the same metal in dilute sulfuric acid, a secondary, reverse (or polarization) current is obtained after disconnecting the battery. This is an important step toward the storage of electricity.

1802   Thomas Young (1773-1829), English physicist and physician, performs his classic experiment on interference in which sunlight is made to pass through two pinholes in an opaque screen. With a wave interpretation of his observations, he is able to provide the first quantitative values for the length of light waves.

1802   Giovanni Domenico Romagnosi (1761-1835), Italian physicist, performs electrical experiments that result in the deflection of a magnetic needle. Although he precedes Ørsted in this, he does not make the discovery of electromagnetism.

1803   Louis Poinsot (1777-1859), French mathematician, publishes his *Éléments de statique* which contains his theory of couples (two parallel forces of equal magnitude but opposite direction form a couple). He also introduces the concept of torque.

1805   Johann Wilhelm Ritter (1776-1810), German physicist, announces that if a current is passed through a solution of copper sulfate, metallic copper can be made to plate out. This marks the beginning of electroplating.

1805   Humphry Davy (1778-1829), English chemist, begins his electrical experiments, producing an electric arc and an extremely strong battery made up of over 250 metallic plates.

1807   Thomas Young (1773-1829), English physicist and physician, first uses the word *energy* in its modern, scientific sense, meaning the property of a system that makes it capable of doing work.

1807   Thomas Young (1773-1829), English physicist and physician, publishes his *Lectures on Natural Philosophy* in which he is the first to treat shear as an elastic strain. This marks the beginning of the concept of practical elasticity in the science of materials.

1808   Etienne Louis Malus (1775-1812), French physicist, accidentally discovers a phenomenon he calls *polarized light* while sending a ray of light through a doubly refracting crystal. Although he believes light to be made of particles and not waves, he does establish polarization as a general property of light.

1808   Dominique François Jean Arago (1786-1853), French physicist, finds the neutral point of reflected light where polarization is absent.

1811   Siméon-Denis Poisson (1781-1840), French mathematician, develops a mathematical theory of heat.

1812   Jean Baptiste Biot (1774-1862), French physicist, experiments with double refraction, or what is now called polarized light, and discovers rotary dispersion. This means that the degree of rotation of the plane of polarization depends on the color of the light. This leads him to another discovery.

1812   Pierre Simon de Laplace (1749-1827), French astronomer and mathematician, suggests that a complete knowledge of the masses, positions, and velocities of all particles at any given moment would allow the calculation of all events, past and future.

1815   Jean Baptiste Biot (1774-1862), French physicist, discovers that polarized light, when passing through an organic substance, might be rotated clockwise or counterclockwise depending upon the optical axis of the substance. He realizes that it is the molecules in the substance that affect the polarized light.

1815   David Brewster (1781-1868), Scottish physicist, discovers that a beam of light can be split into a reflected portion and a refracted portion, at right angles to each other, and that both are then completely polarized. This becomes known as Brewster's law.

1816 Augustin Louis Cauchy (1789-1857), French mathematician, writes a classic paper on turbulence, or the propagation of waves at the surface of a liquid.

1817 Thomas Young (1773-1829), English physicist and physician, proposes correctly that light waves are transverse, or vibrate at right angles to the direction of travel.

1818 Pierre Louis Dulong (1785-1838), French chemist, and Alexis Thérèse Petit (1791-1820), French physicist, collaborate and show that the specific heat of an element is inversely related to its atomic weight.

1818 Augustin Jean Fresnel (1788-1827), French physicist, publishes his *Mémoire sur la diffraction de la lumiére* in which he demonstrates the ability of a transverse wave theory of light to account for such phenomena as reflection, refraction, polarization, interference, and diffraction patterns. His excellent calculations eventually help convert many of his contemporaries to the wave theory.

1820 Hans Christian Ørsted (1777-1851), Danish physicist, experiments with a compass and electricity and demonstrates that a current of electricity creates a magnetic field. He announces his discovery in a short article. This is the first time a real connection can be shown between electricity and magnetism, and it founds the new field of electromagnetism.

1820 André Marie Ampère (1775-1836), French mathematician and physicist, extends Ørsted's work and formulates one of the basic laws of electromagnetism. He discovers that two parallel wires each carrying a current attract each other if the currents are in the same direction, but they repel each other if in the opposite direction. He concludes that magnetism is the result of electricity in motion.

1820 Johann Salomo Schweigger (1779-1857), German physicist, learns of Ørsted's experiment and realizes that the deflection of the compass needle can be used to measure current strength. He then devises the first rudimentary galvanometer.

1821 Thomas Johann Seebeck (1770-1831), Russian-German physicist, discovers thermoelectricity, or the conversion of heat into electricity. He observes that an electric current will flow continuously in a circuit made up of two different metals if they are kept at different temperatures. He does not pursue the implications of what is called much later the *Seebeck effect*.

1822 Jean Baptiste Joseph Fourier (1768-1830), French mathematician, publishes his *Théorie analytique de*

la chaleur which is an influential work of mathematical physics. Studying how heat flows from one point to another through a particular object, he offers what becomes known as Fourier's theorem. It proves highly useful in the study of sound, light, and wave phenomena. It also contains the first statement that a scientific equation must involve a consistent set of units, and thus begins dimensional analysis.

1822 Augustin Louis Cauchy (1789-1857), French mathematician, works on the mechanics of materials and offers what becomes the origin of the theory of stress.

1824 Nicolas Léonard Sadi Carnot (1796-1832), French physicist, publishes his *Réflexions sur la puissance motrice du feu* in which he is the first to consider quantitatively the manner in which heat and work are interconverted. This highly original work also introduces the important concept of cyclic operations and the principle of reversibility. Although this work founds the science of thermodynamics, or the movement of heat, it is neglected for ten years.

1825 William Sturgeon (1783-1850), English physicist, invents and demonstrates the first electromagnet. He begins with Ampère's work and goes further, wrapping a wire scores of times around an iron core that is shaped like a horseshoe. When he sends a current through the wire, each coil reinforces the other since they form parallel lines with the current running in the same direction. The resulting strong magnetic force can lift nine pounds, or twenty times the magnet's weight.

1826 André Marie Ampère (1775-1836), French mathematician and physicist, publishes his *Mémoire sur la théorie mathématique des phénomàmes électrodynamiques* in which he offers the mathematical laws that govern the new field of electricity in motion and founds the study of electrodynamics.

1827 Georg Simon Ohm (1787-1854), German physicist, experiments with electricity using wires of different length and diameter and discovers that a long, thick wire passes less current than a short, thin wire. He states what becomes Ohm's law—that the amount of current passing through a wire is inversely proportional to the thickness of a wire.

1827 Jacques Babinet (1794-1872), French physicist, suggests a new standard of length measurement be adopted. Since all standards to this point are based on uncertain and changeable units, he suggests that something unalterable and true, like the wavelength of some particular type of light ray, be used. It is not until 1960 that technology advances to the point where his suggestion becomes possible.

1827 William Rowan Hamilton (1805-1865), Irish mathematician, publishes his *Theory of Systems of Rays.* This mathematical study of optics contains the correct prediction of conical refraction and helps establish the wave theory of light.

1829   Gaspard Gustave de Coriolis (1792-1843), French physicist, publishes *Du calcul de l'effet des machines* in which he introduces the terms *work* and *kinetic energy* and is the first to give them their modern scientific definitions. The kinetic energy of an object is defined as half its mass times the square of its velocity. Work done upon an object is equal to the force upon it multiplied by the distance it is moved against resistance.

1830   Joseph Henry (1797-1878), American physicist, discovers the principle of electromagnetic induction, showing how an electric current in one coil may set up a current in another through the development of a magnetic field. He puts off further work on this discovery until the following summer, and in the intervening time, the English physicist and chemist Michael Faraday (1791-1867) publishes his discovery first.

1831   Michael Faraday (1791-1867), English physicist and chemist, discovers electromagnetic induction. After laboring for ten years to achieve the opposite of what Ørsted had done—to convert magnetism into electricity—he finally produces for the first time an induction current using a magnet. This is the first electric generator. With such a device, mechanical energy can be converted into electrical energy.

1832   Michael Faraday (1791-1867), English physicist and chemist, announces what becomes known as Faraday's law of electrolysis. He states that the mass of a substance decomposed by an electric current is proportional to the length of time and magnitude of the electric current producing the electrolysis. This establishes the very close connection between electricity and chemistry.

1833   Heinrich Friedrich Emil Lenz (1804-1865), Russian physicist, discovers the manner in which the resistance of a metallic conductor changes with temperature; resistance increases with a temperature increase and decreases with temperature decrease.

1834   Benoit Pierre Clapeyron (1799-1864), French engineer, studies the long-ignored work of Carnot (1834) and publishes his mathematical treatment of that non-mathematical work. His work offers a new presentation of Carnot's work and contains what later becomes the second law of thermodynamics.

1834   Heinrich Friedrich Emil Lenz (1804-1865), Russian physicist, investigates electrical induction and discovers that a current induced by electromagnetic forces always produces effects that oppose those forces. This is the first general description of what is called self-induction, and it is known as Lenz's law.

1834   Jean Charles Athanase Peltier (1785-1845), French physicist, discovers that at the junction of two dissimilar metals, an electric current will produce heat or cold, depending on the direction of current flow. This comes to be known as the Peltier effect and is eventually used in refrigeration.

1835   Jean Baptiste Biot (1774-1862), French physicist, founds polarimetry. He shows with his knowledge of polarized light how the hydrolysis (chemical decomposition in water) of sucrose can be followed by noting the changes in optical rotation.

1835   Gaspard Gustave de Coriolis (1792-1843), French physicist, considers mathematically and experimentally the matter of motion on a spinning surface. He describes what comes to be called Coriolis forces which become concepts of major significance to the study of meteorology, ballistics, and oceanography.

1836   John Frederic Daniell (1790-1845), English chemist and meteorologist, designs an improved electric cell that supplies an even current with continuous operation. The Daniell cell is a major improvement over the voltaic cell and is the first reliable source of electric current.

1840   Alexandre Edmond Becquerel (1820-1891), French physicist, discovers the photovoltaic effect as he demonstrates that when light induces certain chemical reactions it can produce an electric current. This leads him to invent a device that can measure light intensity by determining the intensity of the current produced.

1840   Jean Charles Athanase Peltier (1785-1845), French physicist, introduces the concept of electrostatic induction. This is a method of charging a conductor by closely juxtaposing another charged object to attract all charges of one sign and then grounding the conductor to bleed off the other group of charges, leaving a net charge behind.

1841   James Prescott Joule (1818-1889), English physicist, publishes a paper in which he affirms the existence of a proportionality between the heat manifested in conductors and the square of the strength of the electrical current flowing through them. This becomes known as the Joule effect.

1842   Christian Johann Doppler (1803-1853), Austrian physicist, discovers the effect of motion of the source or observer on the observed frequency of sound waves. This mathematical relationship that relates their pitch to the relative motion of the source or observer is called the Doppler effect.

1842   Julius Robert Mayer (1814-1878), German physicist, produces the first fairly accurate mechanical equivalent of heat. He is one year ahead of Joule (1843) on this and precedes Helmholtz (1847) as well in his statement of the conservation of energy. Circumstances dictate that he gets little credit for either until his later years.

1843 James Prescott Joule (1818-1889), English physicist, publishes his value of the amount of work required to produce a unit of heat, called the mechanical equivalent of heat. The value of the mechanical equivalent of heat is eventually represented by the letter J, and a standard unit of work is now called the joule.

1846 Wilhelm Eduard Weber (1804-1891), German physicist, introduces a logical system of units for electricity and offers a general mathematical law expressing the force between moving charges.

1846 William Robert Grove (1811-1896), British physicist, publishes *On the Correlation of Physical Forces* in which he makes an early statement of the principle of the conservation of energy.

1847 James Prescott Joule (1818-1889), English physicist, offers his complete treatment of the law of conservation of energy, stating that energy can neither be created out of nothing nor destroyed into nothing. He establishes that the various forms of energy—mechanical, electrical, and heat—are basically the same and can be changed from one into another.

1847 Hermann Ludwig Ferdinand von Helmholtz (1821-1894), German physiologist and physicist, publishes his *Über die Erhaltung der Kraft* which is the most detailed and clearest explanation of the concept of conservation of energy.

1848 William Thomson (1824-1907), later known as Lord Kelvin, Scottish mathematician and physicist, explores the concept of absolute zero at which the volume of a gas would be reduced to zero and explains it by stating that it is not that the volume reaches zero but rather that the motion of the gas's molecules stops. He then proposes a new temperature scale with its zero mark at absolute zero and its degrees equal to those on the centigrade scale. It becomes known as the kelvin scale. He also coins the term *thermodynamics*.

1849 Armand Hippolyte Fizeau (1819-1896), French physicist, accurately measures the speed of light in air using a rotating mirror technique. The value he calculates is within five percent of today's value. Previous to this, all such measurements had been done using astronomical methods.

1850 Rudolph Julius Emanuel Clausius (1822-1888), German physicist, publishes a paper which contains what becomes known as the second law of thermodynamics, stating that, "heat cannot, of itself, pass from a colder to a hotter body." He later refines the concept.

1851 Jean Bernard Léon Foucault (1819-1868), French physicist, conducts his spectacular series of experiments associated with the pendulum. He swings a heavy iron ball from a wire more than 200 feet long and demonstrates that the swinging pendulum maintains its plane while the Earth slowly twists under it. The crowd of spectators who witnesses this demonstration come to realize that they are watching the Earth rotate under the pendulum—experimental proof of a moving Earth.

1851 Armand Hippolyte Fizeau (1819-1896), French physicist, measures the speed of light as it flows with a stream of water and as it goes against the stream. He finds that the velocity of light is higher in the former.

1851 William Thomson (1824-1907), later known as Lord Kelvin, Scottish mathematician and physicist, publishes *On the Dynamical Theory of Heat* in which he explores Carnot's work and deduces that all energy tends to run down and dissipate itself as heat. This is another form of the second law of thermodynamics and is advanced further by Clausius at about the same time. Kelvin's work is considered the first nineteenth-century treatise on thermodynamics.

1852 Jean Bernard Léon Foucault (1819-1868), French physicist, learns from his pendulum experiment and invents the gyroscope. He sets a wheel within a heavy rim in rotation and sees that when tipped, it is set right again by the force of gravity.

1852 James Prescott Joule (1818-1889), English physicist, and William Thomson (1824-1907), later known as Lord Kelvin, Scottish mathematician and physicist, jointly establish that as a gas cools it also expands. This becomes known as the Joule-Thomson effect.

1853 Anders Jonas Ångström (1814-1874), Swedish physicist, concludes that an incandescent gas emits spectral lines of the same wavelength as those it absorbs when cold.

1853 Johann Wilhelm Hittorf (1824-1914), German chemist and physicist, offers the notion of the transport number as he suggests that ions in a solution with a current running through it travel at different speeds.

1853 William John Macquorn Rankine (1820-1872), Scottish engineer, introduces into physics the concept of potential energy, also called the energy of position.

1855 Johann Heinrich Wilhelm Geissler (1814-1879), German inventor, devises a mercury pump that produces a much better vacuum than old piston pumps.

Called Geissler tubes, they make possible a more advanced study of electricity and eventually of the atom. It is with Geissler tubes that the English physicist Joseph John Thomson (1856-1940) performs his famous experiments elucidating the nature of electrons.

1858 Julius Plücker (1801-1868), German mathematician and physicist, sends an electric current through a vacuum and describes fluorescent effects in detail. He also discovers that the glow shifts position when placed in the field of an electromagnet. This is the very beginning of an awareness of subatomic particles.

1858 Balfour Stewart (1828-1887), Scottish physicist, experiments with infrared radiation and shows that at each wavelength, the amount of radiation emitted by a body equals the amount it absorbs.

1858 Hermann Ludwig Ferdinand von Helmholtz (1821-1894), German physiologist and physicist, publishes his study on the integrals of hydrodynamic equations that express whirling motion. This becomes the point of departure for new ideas on the structure of matter that eventually replaces the old atomistic concepts.

1859 Gustav Robert Kirchhoff (1824-1887), German physicist, after discovering the relation between emission and absorption spectra, concludes that the ratio of the emissive and absorptive powers of a body at each wavelength is the same for all bodies at the same temperature. This becomes known as Kirchhoff's law.

1860 Gustav Robert Kirchhoff (1824-1887), German physicist, introduces the concepts of black bodies and emissivity. A black body is a surface that absorbs all radiation of any wavelength falling on it. Such a body would emit all wavelengths if it were heated to incandescence. This concept later becomes important to quantum theory.

1860 James Clerk Maxwell (1831-1879), Scottish mathematician and physicist, publishes a paper in which he offers his kinetic theory of gases. He shows that the motion of molecules are responsible for the production of heat, and that when the average velocity increases, so does the temperature.

1861 William Thomson (1824-1907), later known as Lord Kelvin, Scottish mathematician and physicist, publishes his *Physical Considerations Regarding the Possible Age of the Sun's Heat* which contains the theme of the *heat death* of the universe. This is offered in light of the principle of dissipation of energy stated in 1851.

1865 Rudolf Julius Emanuel Clausius (1822-1888), German physicist, refines the second law of thermodynamics and first introduces the term *entropy*, stat-

ing that the energy in a closed system will always eventually run down.

1866 August Adolph Eduard Eberhard Kundt (1839-1894), German physicist, invents a method by which he can make accurate measurements of the speed of sound in the air. He uses a *Kundt's tube* whose inside is dusted with fine powder which is then disturbed by traveling sound waves.

1868 Ludwig Eduard Boltzmann (1844-1906), Austrian physicist, publishes a paper in which he independently applies statistics to the kinetic theory of gases and later shares credit with Scottish mathematician and physicist James Clerk Maxwell (1831-1879) for this theory.

1869 John Tyndall (1820-1893), Irish physicist, discovers that a beam of light passing through pure water is not dispersed or interfered with, but when passed through a colloidal solution (one containing tiny particles), the beam is dispersed. This comes to be called the Tyndall effect.

1869 Johann Wilhelm Hittorf (1824-1914), German chemist and physicist, studies cathode tubes and anticipates some of the discoveries of English physicist William Crookes (1832-1919) as he describes the changes that accompany a lowering of gas pressure. He also establishes that their glow is produced by rays originating at the cathode and traveling in straight lines.

1871 John William Strutt Rayleigh (1842-1919), English physicist, discovers that the degree of scattering of light by very fine particles is a function of the wavelength of light. His equation offers a solution to the question of why the sky is blue.

1871 Ludwig Eduard Boltzmann (1844-1906), Austrian physicist, begins work on his mathematical treatment of the second law of thermodynamics, interpreting it in terms of the statistics of molecular motions. His work lays the foundations of statistical mechanics.

1871 George Johnstone Stoney (1826-1911), Irish physicist, notices that a simple ratio exists between the wavelengths of three lines in the hydrogen spectrum. In this he anticipates the Swiss mathematician and physicist Johann Jakob Balmer (1825-1898), who offers his formula for this phenomenon in 1885.

1871 The Cavendish Laboratory is founded at Cambridge University in England. Its first director is the Scottish mathematician and physicist James Clerk Maxwell (1831-1979). A great deal of the development of modern physics takes place at this laboratory which is endowed by the family of English chemist and physicist Henry Cavendish (1731-1810).

1872 Ernst Mach (1838-1916), Austrian physicist, publishes his *Die Geschichte und die Wurzel des Satzes von der Erhaltung der Arbeit*. In this historical-critical work, he attempts to reduce all natural phenomena to mechanics and argues that all knowledge is but a matter of sensation.

1873 James Clerk Maxwell (1831-1879), Scottish mathematician and physicist, publishes *Treatise on Electricity and Magnetism* in which he identifies light as an electromagnetic phenomenon. He determines this when he finds his mathematical calculations for the transmission speed of both electromagnetic and electrostatic waves are the same as the known speed of light. This landmark work brings together the three main fields of physics—electricity, magnetism, and light.

1873 Johannes Van der Waals (1837-1923), Dutch physicist, offers an equation for the gas laws which contains terms relating to the volumes of the molecules themselves and the attractive forces between them. It becomes known as the Van der Waals equation.

1873 Gabriel Jonas Lippmann (1845-1921), French physicist, invents the capillary electrometer, an instrument for measuring extremely small voltages (as small as one thousandth of a volt). This later becomes useful in some of the early electrocardiograms.

1874 Karl Ferdinand Braun (1850-1918), German physicist, discovers crystal rectifiers. These devices convert alternating current into direct current and eventually are used to improve the range of radio transmitters.

1875 August Adolph Eduard Eberhard Kundt (1839-1894), German physicist, provides experimental confirmation of the kinetic theory of gases.

1876 Josiah Willard Gibbs (1839-1903), American physicist, publishes the first of a series of papers in which he deals with the principles of thermodynamics and applies them to the complex processes involved in chemical reactions. He also offers an elaborate equation that becomes known as the *phase rule*, which governs the several phases of matter. His work is difficult to understand, complicated, and multidisciplinary, and is not fully appreciated until the 1890s.

1876 Henry Augustus Rowland (1848-1901), American physicist, establishes for the first time that a moving electric charge or current is accompanied by electrically charged matter in motion and produces a magnetic field.

1876 Eugen Goldstein (1850-1930), German physicist, first applies the name *cathode rays* to the luminescence produced at the cathode in a vacuum tube.

1876 James Clerk Maxwell (1831-1879), Scottish mathe-

matician and physicist, publishes his *Matter and Motion* in which he considers the categories of space and time and states with great prescience that, "all our knowledge, both of time and place, is essentially relative."

1879 Josef Stefan (1835-1893), Austrian physicist, offers the first quantitative rate for the cooling of bodies. Called the Stefan-Boltzmann law, it states that the amount of heat radiated by a surface is proportional to the fourth power of the absolute temperature.

1879 Edwin Herbert Hall (1855-1938), American physicist, announces his discovery of the phenomenon in which a difference of potential develops across a current-carrying conductor that is placed in magnetic field. This becomes known as the Hall effect.

1880 William Crookes (1832-1919), English physicist, explains the green glow produced when an electric discharge is sent through a vacuum. He proves that the glow caused in his *Crookes tube* is caused by negatively charged particles. However, he cannot explain what the particles are.

1880 Pierre Curie (1859-1906), French chemist, collaborates with his brother, Jacques, and discovers piezoelectricity in crystals. They create ultrasonic sound waves by applying a rapidly changing electric potential to a quartz crystal.

1881 Nikola Tesla (1856-1943), Serbian-American inventor and electrical engineer, discovers the principle of the rotating magnetic field, creating the foundation for the application of alternating-current in industry.

1881 Joseph John Thomson (1856-1940), English physicist, introduces the notion of electromagnetic mass, stating that the mass of an object changes when it is charged with electricity.

1883 Ernst Mach (1838-1916), Austrian physicist, publishes his *Die Mechanik in ihrer Entwickelung historisch-kritisch dargestellt* in which he offers a radical philosophy of science that calls into question the reality of such Newtonian ideas as space, time, and motion. His work influences Einstein and prepares the way for relativity.

1883 Thomas Alva Edison (1847-1931), American inventor, discovers the emission of electrons from hot bodies. He inserts a small metal plate near the filament of a light bulb and finds that the plate draws a current when he connects it to the positive terminal of the light bulb circuit, even though the plate is not touching the filament. Called the Edison effect, this is later a major factor in the invention of the vacuum tube.

1883 George Francis Fitzgerald (1851-1901), Irish physicist, first suggests a method of producing radio waves. From his studies of radiation, he concludes that an oscillating current would produce electromagnetic waves. This is later verified experimentally by Hertz in 1888 and used in the development of wireless telegraphy.

1885 Johann Jakob Balmer (1825-1898), Swiss mathematician and physicist, discovers a simple numerical formula with which the prominent series of lines in the visible region of the hydrogen spectrum can be correlated to wavelength. This is known as the Balmer series.

1886 Roland von Eötvös (1848-1919), Hungarian physicist, first introduces the concept of molecular surface tension. His study of Earth's gravitational field, which leads him to invent a precise torsion balance, results in his proof that inertial mass and gravitational mass are equivalent. This proves to be a major principle in Einstein's general theory of relativity.

1886 Eugen Goldstein (1850-1930), German physicist, discovers what he calls *Kanalstrahlen*, later called channel rays or canal rays. He finds that they travel in the opposite direction of cathode rays. This is the beginning of an awareness of protons.

1887 Ernst Mach (1838-1916), Austrian physicist, is the first to note the sudden change in the nature of the airflow over a moving object that occurs as it approaches the speed of sound. Because of this, the speed of sound in air (under certain temperature conditions) is called Mach 1. Mach 2 is twice that speed and so on.

1887 Albert Abraham Michelson (1852-1931), German-American physicist, collaborates with the American chemist Edward Williams Morley (1838-1923) to test the age-old hypothesis that the Earth moves through luminiferous ether (the supposed light-carrying element that exists outside or above Earth's atmosphere). They use Michelson's interferometer—which splits a beam of light in two, sends each on a different path, and then reunites them—to test out this idea. The failure of this extremely sensitive instrument to detect even the slightest change in the velocity of light proves that the ether does not exist (for it would have had to change slightly if it had gone through a substance). This result overturns all ether-based theories and makes physicists search for some explanation of the invariance of the speed of light.

1888 Heinrich Rudolf Hertz (1857-1894), German physicist, for the first time generates electromagnetic (radio) waves and devises a detector that can measure their wavelength. From this he is able to prove experimentally James Clerk Maxwell's (1831-1879) hypothesis that light is an electromagnetic phenomenon. Hertz's work not only discovers radio waves, but experimentally unites the three main fields of physics—electricity, magnetism, and light.

1889 Johann Philipp Ludwig Julius Elster (1854-1920) and Hans Friedrich Geitel (1855-1923), both German physicists, study the photoelectric effect (when an electric current is created upon the exposure of certain metals to light) and produce the first practical photoelectric cells that can measure the intensity of light.

c. 1890 Oliver Joseph Lodge (1851-1940), English physicist, invents the coherer, a detector of radio waves that, although replaced, makes him one of the pioneers of early radio communication. He also suggests correctly that the Sun emits radio waves.

1890 Hendrik Antoon Lorentz (1853-1928), Dutch physicist, suggests that the atom consists of charged particles whose oscillations can be affected by a magnetic field. This is later confirmed by his pupil, the Dutch physicist Pieter Zeeman (1865-1943), in 1896.

1890 Johannes Robert Rydberg (1854-1919), Swedish physicist, offers an equation containing what becomes the Rydberg constant, which interprets the periodicity of the elements in terms of the structure of the atom.

1891 George Johnstone Stoney (1826-1911), Irish physicist, studies electrolysis and says that it can best be understood by thinking in terms of "units of electricity." He then proposes the name *electron* for this unit of electricity.

1892 George Francis Fitzgerald (1851-1901), Irish physicist, attempts to explain the results of the Michelson-Morley experiment of 1887 and offers what comes to be called the Fitzgerald contraction. He suggests that when in motion, a body is shorter along its line of motion than when it is at rest, and that this shortening may have affected the instruments used in that famous experiment. This idea eventually becomes an integral part of Einstein's special theory of relativity.

1893 Oliver Heaviside (1850-1925), English physicist and electrical engineer, publishes his *Electromagnetic Theory* in which he posits that an electric charge would increase in mass as its velocity increases. This is an anticipation of Einstein's special theory of relativity.

1893 Augusto Righi (1850-1920), Italian physicist, demonstrates that Hertz (radio) waves differ from light only in wavelength and not because of any essential difference in their nature. This helps to establish the existence of the electromagnetic spectrum.

1893 Wilhelm Wien (1864-1928), German physicist, conducts black body research and finds that the wavelengths emitted reach a peak at some intermediate level. He then states his displacement law that this maximum wavelength is inversely proportional to the absolute temperature of the body.

1894 Joseph John Thomson (1856-1940), English physicist, announces that his experiments show the velocity of cathode rays is significantly lower than the velocity of light.

1894 Guglielmo Marconi (1874-1937), Italian electrical engineer, uses Hertz's method of producing radio waves and builds a receiver to detect them. He succeeds in sending his first radio waves 30 feet to ring a bell. The next year, his improved system can send a signal 1.5 miles.

1895 Pierre Curie (1859-1906), French chemist, studies the effect of heat on magnetism and shows that there is a critical temperature point above which magnetic properties will disappear. This comes to be called the Curie point.

1895 Jean Baptiste Perrin (1870-1942), French physicist, demonstrates conclusively that cathode rays must consist of negatively charged material and are therefore particles rather than waves.

1895 Wilhelm Conrad Röntgen (1845-1923), German physicist, discovers x rays, initiating the modern age of physics and revolutionizing medicine. While working on cathode ray tubes and experimenting with luminescence, he notices that a nearby sheet of paper that is coated with a luminescent substance glows whenever the tube is turned on. For seven weeks he continues to experiment, and near the end of the year is able to report the basic properties of the unknown rays he names *x rays*.

1895 Wilhem Conrad Röntgen (1845-1923), German physicist, submits his first paper documenting his discovery of x rays. He tells how this unknown ray, or radiation, can affect photographic plates, and that wood, paper, and aluminum are transparent to it. It also can ionize gases and does not respond to electric or magnetic fields nor exhibit any properties of light. This discovery leads to such a stream of groundbreaking discoveries in physics that it has been called the beginning of the second scientific revolution.

1896 Johann Philipp Ludwig Julius Elster (1854-1920) and Hans Friedrich Geitel (1855-1923), both German physicists, study the newly discovered radioactivity and demonstrate that external effects do not influence the intensity of radiation. They are also the first to characterize radioactivity as being caused by changes that occur within the atom.

1896 Charles Thomson Rees Wilson (1869-1959), Scottish physicist, invents the Wilson cloud chamber. This new instrument proves to be a powerful tool for physicists as it makes visible and allows the photographing of the paths or trajectories of charged particles.

1896 Guglielmo Marconi (1874-1937), Italian electrical engineer, travels to England to apply for and obtain the first patent in the history of radio. By this time, he has sent and received a radio signal over nine miles.

1896 Antoine Henri Becquerel (1852-1908), French physicist, studies fluorescent materials to see if they emit the newly discovered x rays and discovers instead that uranium produces natural radiation that is eventually called *radioactivity* in 1898 by the Polish-French chemist Marie Skłodowska Curie (1867-1934). This is the first observation of natural radioactivity.

1897 Karl Ferdinand Braun (1850-1918), German physicist, modifies a cathode ray tube so that the streaming electrons are affected by the electromagnetic field of a varying current. This results in his invention of the oscillograph, or oscilloscope, whose ability to finely vary an electric current makes it the first step toward television and radar.

1897 Joseph John Thomson (1856-1940), English physicist, performs experiments confirming the existence of the electron. He observes cathode ray experiments and concludes that the rays consist of negatively charged particles that are smaller in mass than atoms. He goes on to propose the first theory of atomic structure stating that these subatomic particles are the building-blocks of all matter. Although an inadequate theory, it leads to remarkable insights into the structure of matter.

1898 James Dewar (1842-1923), Scottish chemist and physicist, achieves the large-scale production of liquid hydrogen. A year later he is able to solidify hydrogen. This enables him to conduct his research in cryogenics, since he can now reach a temperature only 14 degrees above absolute zero.

1898 Marie Skłodowska Curie (1867-1934), Polish-French chemist, and her husband, the French chemist Pierre Curie (1859-1906), refine pitchblende and detect a radioactive substance that they name polonium after her homeland.

1898 Marie Skłodowska Curie (1867-1934), Polish-French chemist, and her husband, the French chemist Pierre Curie (1859-1906), detect only a trace of still another radioactive substance in pitchblende and name it radium. They work for the next four years and eventually refine eight tons of pitchblende to obtain one ounce of radium.

1899   Ernest Rutherford (1871-1937), British physicist, discovers that radioactive substances give off different kinds of rays. He names the positively charged ones alpha rays and the negative ones beta rays.

1899   Joseph John Thomson (1856-1940), English physicist, measures the charge of the electron.

1899   Emile Hilaire Amagat (1841-1915), French mathematician and physicist, publishes *The Laws of Gases*. During his studies on the behavior of gases under various pressures and temperatures, he finds that the more a gas is compressed, the more it deviates from the Boyle-Mariotte law.

1900   Friedrich Ernst Dorn (1848-1916), German physicist, demonstrates that the newly discovered radium gives off a gas as well as produces radioactive radiation. This proves to be the first demonstrable evidence that in the radioactive process, one element is actually transmuted into another.

1900   Max Karl Ernst Ludwig Planck (1858-1947), German physicist, publishes his classic and revolutionary paper on quantum physics. He tells of his discovery that light or energy is not found in nature as a continuous wave or flow but is emitted and absorbed discontinuously in little packets, or *quanta*. Further, each quantum, or packet of energy, is indivisible. Planck's new notion of the quantum seemed to contradict the mechanics of Newton and the electromagnetics of Maxwell, and replace them with new rules. In fact, his quantum physics were new rules in a new type of physics—physics of the very fast and the very small. His theory is soon applied (by Einstein) and incorporated (by Bohr), and becomes the watershed between all physics that comes before it (classical physics) and all that is after (modern physics).

1900   Paul Ulrich Villard (1860-1934), French physicist, discovers what are later called gamma rays. While studying the recently discovered radiation from uranium, he finds that in addition to the alpha rays and beta rays, there are other rays, unaffected by magnets, that are similar to x rays but shorter and more penetrating.

1901   Wilhelm Conrad Röntgen (1845-1923), German physicist, receives the first Nobel Prize in physics for his discovery of x rays.

1901   Antoine Henri Becquerel (1852-1908), French physicist, studies the rays emitted by the natural substance uranium and concludes that the only place they could be coming from is within the atoms of uranium. This marks the first clear understanding of the atom as something more than a featureless sphere. It implies a dynamic reality that might also contain electrons. Becquerel's discovery of radioactivity and his focus on the uranium atom make him the father of modern atomic and nuclear physics.

1901   Jagadis Chandra Bose (1858-1937), Indian physicist and plant physiologist, devises extremely sensitive instruments that are able to demonstrate the minute movements of plants to external stimuli and to measure their rate of growth. He pioneers what becomes biophysics and is also the first scientist from India to become internationally known.

1901   Pyotr Nikolayevich Lebedev (1866-1912), Russian physicist, uses extremely light mirrors in a vacuum and is able to measure the radiation pressure exerted by light for the first time. This also is an experimental confirmation of Maxwell's theory of electromagnetism.

1901   Josiah Willard Gibbs (1839-1903), American physicist, introduces the concept of phase space using only the general laws of mechanics.

1901   Jean Baptiste Perrin (1870-1942), French physicist, proposes the hypothesis that atoms are made up of a central nucleus around which electrons orbit like planets in a solar system.

1901   Guglielmo Marconi (1874-1937), Italian electrical engineer, successfully sends a radio signal from England to Newfoundland. This is the first transatlantic telegraphic radio transmission, and, as such, is considered by most as the day radio is invented.

1902   Oliver Heaviside (1850-1925), English physicist and electrical engineer, and Arthur Edwin Kennelly (1861-1939), British-American electrical engineer, independently and almost simultaneously make the first prediction of the existence of the ionosphere, an electrically conductive layer in the upper atmosphere that reflects radio waves. They theorize correctly that wireless telegraphy works over long distances because a conducting layer of atmosphere exists that allows radio waves to follow the Earth's curvature instead of traveling off into space.

1902   Ernest Rutherford (1871-1937), British physicist, and Frederick Soddy (1877-1956), English chemist, publish a paper in which they offer a theory of disintegration of atoms as an explanation for radioactivity. They suggest that atoms break up and transform into other elements.

1903   William Ramsay (1852-1916), Scottish chemist, and Frederick Soddy (1877-1956), English chemist, discover that helium is continually produced by naturally radioactive substances.

1903   Antoine Henri Becquerel (1852-1908), French physicist, shares the Nobel Prize in physics with the husband-and-wife team of Marie Skłodowska Curie (1867-1934), Polish-French chemist, and Pierre

Curie (1859-1906), French chemist. Becquerel wins for his discovery of natural or spontaneous radioactivity, and the Curies win for their later research on this new phenomenon.

1904   Hendrik Antoon Lorentz (1853-1928), Dutch physicist, extends his idea of local time (different time rates in different locations) and arrives at what are called the Lorentz transformations. These are mathematical formulas describing the increase of mass, shortening of length, and dilation of time that are characteristic of a moving body. These eventually form the basis of Einstein's special theory of relativity.

1904   Hantaro Nagaoka (1865-1950), Japanese physicist, offers his Saturnian model of the atom which says that there is a positively charged object at the atom's center around which electrons revolve (like Saturn's rings).

1904   Bertram Borden Boltwood (1870-1927), American chemist and physicist, discovers that radium is a decay product of, or is descended from, uranium, meaning that radioactive elements are not independent, but that one is descended from another, forming a radioactive series.

1904   Charles Glover Barkla (1877-1944), English physicist, discovers that x rays are transverse waves like those of light rather than longitudinal waves like those of sound.

1904   Ludwig Prandtl (1875-1953), German physicist, discovers that as liquid flows in a tube, a film layer adjacent to the wall forms and does not flow as fast as the rest of the liquid. This is called a viscous boundary layer.

1905   William Weber Coblentz (1873-1962), American physicist, begins his career-long research on radiation beyond the visible spectrum. He demonstrates that compounds can be analyzed by the way they absorb infrared radiation. This is the principle behind the modern infrared spectrophotometer.

1905   Walther Hermann Nernst (1864-1941), German physical chemist, announces his discovery of the third law of thermodynamics. He finds that entropy change approaches zero at a temperature of absolute zero, and deduces from this the impossibility of attaining absolute zero.

1905   Albert Einstein (1879-1955), German-Swiss physicist, uses Planck's theory to develop a quantum theory of light which explains the photoelectric effect. He suggests that light has a dual, wave-particle quality.

1905   Albert Einstein (1879-1955), German-Swiss physicist, submits his first paper on the special theory of

relativity titled *Zur Elektrodynamik bewegter Korpen*. It states that the speed of light is constant for all conditions and that time is relative, or passes at different rates, for objects in constant relative motion. This is a fundamentally new and revolutionary way to look at the universe.

1905   Albert Einstein (1879-1955), German-Swiss physicist, publishes his second paper on relativity in which he includes his famous equation stating the relationship between mass and energy: $E = mc^2$. In this equation, E is energy, m is mass, and c is the velocity of light. This contains the revolutionary concept that mass and energy are simply different aspects of the same phenomenon.

1906   Marie Skłodowska Curie (1867-1934), Polish-French chemist, assumes her husband Pierre's professorship at the Sorbonne after he is killed in a traffic accident. She becomes the first woman ever to teach there.

1906   Charles Glover Barkla (1877-1944), English physicist, discovers that each particular element produces its own kind of x rays. This breakthrough leads eventually to the fulfillment of the notion of atomic numbers.

1906   Reginald Aubrey Fessenden (1866-1932), Canadian-American physicist, successfully transmits music and voice for the first time via radio waves that are picked up by wireless receivers. This becomes possible through his invention of the modulation of radio waves. This system sends out a continuous signal rather than a pulsed one, with the amplitude of the waves varied or modulated. These modulations are then sorted out, amplified, and reconverted into sound at the receiving end.

1907   Pierre Weiss (1865-1940), French physicist, offers his theory explaining the phenomenon of ferromagnetism. He states that iron and other ferromagnetic materials form small *domains* of a certain polarity pointing in various directions. When some external magnetic field forces them to be aligned, they become a single, strong magnetic force. This explanation is still accurate.

1907   Hermann Minkowski (1864-1909), Russian-German mathematician, publishes his *Time and Space* in which he shows that Einstein's relativity theory makes it necessary to regard time as a fourth dimension.

1908   Heike Kamerlingh-Onnes (1853-1926), Dutch physicist, produces liquid helium for the first time. This leads to further discoveries.

1908   Jean Baptiste Perrin (1870-1942), French physicist, uses Einstein's formula for Brownian motion and, by actually counting the distribution of suspended particles of gum resin in a liquid, finds that he can for the first time calculate the approximate size of atoms and molecules by observation. This is the final proof that atoms are real entities and not just convenient fictions.

1908 Ernest Rutherford (1871-1937), British physicist, and Hans Wilhelm Geiger (1882-1945), German physicist, develop an electrical alpha-particle counter. Over the next few years, Geiger continues to improve this device which becomes known as the Geiger counter.

1908 Percy Williams Bridgman (1882-1961), American physicist, begins a lifetime of pioneering work in the field of high pressures. Working eventually in the unexplored field of attaining 400,000 atmospheres, he is forced to invent much of his equipment himself. As he extends the range of pressures, he becomes the founder of the laws of high pressure physics.

1910 William Henry Bragg (1862-1942), English physicist, discovers the role of x rays and gamma rays in causing rarified gases to conduct electricity. This happens because the rays knock the electrons out of the gas molecules. He also notes that in this process, both types of rays behave like particles.

1910 Joseph John Thomson (1856-1940), English physicist, measures deflections of positive rays in a cathode tube and calculates the masses of various atoms. He discovers that neon has two isotopes. This is the first confirmation that isotopes are possible, as predicted by English chemist Frederick Soddy (1877-1956).

1911 Heike Kamerlingh-Onnes (1853-1926), Dutch physicist, first discovers the phenomenon of superconductivity when he studies the properties of certain metals subjected to the low temperatures of liquid helium. He finds that some metals, like mercury and lead, undergo a total loss of electrical resistance. He also discovers that a form of liquid helium called *helium II* is produced which has properties unlike any other substances.

1911 Owen Willans Richardson (1879-1959), English physicist, studies how electrons and ions are given off by heated metals and works out the theory of electron and ion emission. This is called the Edison effect and later makes possible the vacuum tube.

1911 First congress of physicists is sponsored by Belgian chemist Ernest Solvay (1838-1922). It is held in Brussels, Belgium, and radiation theory and quanta are discussed.

1911 Ernest Rutherford (1871-1937), British physicist, presents his theory of the nuclear atom, saying that the atom consists of a very small, positively charged nucleus surrounded by negatively charged electrons.

With this, he lays the foundation for the development of nuclear physics.

1912 Karl Frithiof Sundman (1873-1949), Finnish astronomer, publishes his *Mémoire sur le problème des trois corps*, in which he offers a theoretical solution to the long-unsolved three-body problem.

1912 Dmitry Sergeyevich Rozhdestvenski (1876-1940) details an extremely sensitive method for measuring the intensity of spectral lines. This study of spectra offers a valuable experimental basis for research on the atom.

1912 Max Theodor Felix von Laue (1879-1960), German physicist, obtains a diffraction pattern for x rays through a crystal and offers evidence that x rays are a form of electromagnetic radiation and are waves. This marks the beginning of studies on the physics of solids as an analysis of the periodic and regular disposition of atoms in a crystal.

1912 Walter Friedrich (1883-1968), German physicist, and Paul Knipping make the first experimental observations of the diffusion of x rays by crystalline structures. Their work demonstrates the electromagnetic nature of x rays.

1913 Charles Fabry (1867-1945), French physicist, first demonstrates the presence of ozone in the upper atmosphere. It is found later that ozone functions as a screen, preventing much of the Sun's ultraviolet radiation from reaching Earth's surface. Seventy-five years after the discovery of ozone, in 1985, a *hole* in the ozone layer over Antarctica is discovered via satellite.

1913 Robert Andrews Millikan (1868-1953), American physicist, conducts his famous *oil drop* experiment and determines the value of the electric charge of a single electron. He allows a tiny drop of oil to be charged electrically while dropping under the influence of gravity. By balancing the effects of the electromagnetic attraction and the gravitational attraction downward both before and after, he can calculate the charge of a single electron.

1913 Hans Wilhelm Geiger (1882-1945), German physicist, develops an improved device for detecting energetic subatomic particles called the Geiger counter. In 1928 he collaborates with Walther Müller and produces a more sensitive instrument that has a longer lifetime, known as the Geiger-Müller counter.

1913 Johannes Stark (1874-1957), German physicist, demonstrates that a strong electric field causes a multiplication or splitting of spectral lines. Called the Stark effect, this phenomenon is analogous to the effect of a magnetic field and eventually becomes supportive of the quantum theory.

1913 Niels Henrik David Bohr (1885-1962), Danish physicist, offers an entirely different theoretical model for the hydrogen atom. Searching to understand how an atom radiates light, he incorporates Planck's quantum postulate and states that electrons orbit the nucleus in fixed orbits, giving off quanta when jumping from one orbit to another. This is the first reasonably successful attempt to explain the internal structure of the atom via spectroscopy (as well as explain spectroscopy by that structure).

1913 Henry Gwyn-Jeffreys Moseley (1887-1915), English physicist, discovers that the characteristic wavelength x rays of over fifty elements all decrease regularly as their atomic weight increases. He then discovers that this means that a particular element has the same number of electrons as its atomic number, and that arranging them by atomic number would give scientists, for the first time, foreknowledge of how many elements remain to be discovered.

1913 Frederick Soddy (1877-1956), English chemist, suggests the term *isotope* for atoms of the same element that have different atomic masses.

1914 James Franck (1882-1964), German-American physicist, and the German physicist Gustav Hertz (1887-1975) bombard gases and vapors with electrons of different energies and arrive at an experimental confirmation of the Bohr model of the atom.

1914 Ernest Rutherford (1871-1937), British physicist, discovers a positively charged particle he calls a proton.

1915 William Henry Bragg (1862-1942), English physicist, and his son, the Australian-English physicist William Lawrence Bragg (1890-1971), invent the first x-ray spectrometer with which they are able to measure the wavelength of x rays. For this, the Braggs become the only father-son combination to share a Nobel Prize (for physics in 1915).

1915 Albert Einstein (1879-1955), German-Swiss physicist, completes four years of work on his theory of gravitation, or what becomes known as the general theory of relativity.

1916 Arnold Johannes Wilhelm Sommerfeld (1868-1951), German physicist, modifies the Bohr model of the atom by suggesting that the electron paths are elliptical. By doing this, he applies Einstein's relativity theory to the atom.

1916 Albert Einstein (1879-1955), German-Swiss physicist, publishes a paper stating his general theory of relativity. In this theory about the universe as a whole, he states that light will be deflected by a gravitational field significantly more than would be predicted by Newtonian theory.

1918 Francis William Aston (1877-1945), English chemist and physicist, develops the first mass spectrograph which can separate atoms of different mass and measure their weight with remarkable accuracy. With his new device, he discovers that there are different forms of the same element. Now called isotopes, these differ in their mass but not in their chemical properties.

1919 Ernest Rutherford (1871-1937), British physicist, accomplishes the first form of artificial fission. He succeeds, through atomic bombardment with subatomic particles, in altering an atomic nuclei and transforming one element into another. This can be described as the first man-made nuclear reaction.

1919 Heinrich Barkhausen (1881-1956), German physicist, discovers that a slow, smooth increase of a magnetic field applied to a piece of ferromagnetic material like iron causes it to become magnetized in discontinuous jumps accompanied by loud clicks rather than quietly and continuously. This comes to be called the Barkhausen effect.

1920 Francis William Aston (1877-1945), English chemist and physicist, uses his mass spectrograph to formulate the whole-number rule. This states that with the mass of oxygen defined as exactly 16, all other isotopes have masses that are nearly whole numbers. This notion becomes fundamental to the development of nuclear theory.

1920 William Draper Harkins (1873-1951), American chemist, explores nuclear chemistry and predicts the existence of both the neutron and heavy hydrogen.

1921 Albert Einstein (1879-1955), German-Swiss physicist, wins the Nobel Prize in physics for his discovery of the law of the photoelectric effect.

1923 Louis Victor Pierre Raymond de Broglie (1892-1987), French physicist, introduces the particle-wave hypothesis, stating that an electron or any other subatomic particle can behave either as a particle or as a wave. This idea is confirmed in 1927.

1923 Arthur Holly Compton (1892-1962), American physicist, discovers what is later called the Compton effect. While accurately measuring the wavelengths of scattered x rays, he discovers that some of the rays had lengthened their wavelength in scattering. He accounts for this by stating that a *photon* of light strikes an electron, which recoils and subtracts some energy from the photon, thereby increasing its length. It is Compton who coins the term photon to describe light as a particle.

c. 1923 Franz Eugen Francis Simon (1893-1956), German-British physicist, begins his lifelong work in low temperature physics. He goes on to develop small-scale low-temperature apparatus of ingenious design and eventually provides solid backing for Nernst's third law of thermodynamics.

1924 Karl Manne Georg Siegbahn (1886-1978), Swedish physicist, develops improved techniques to produce x rays whose wavelengths can be accurately measured. His improvements in x-ray spectroscopy are such that he is able to produce an actual x-ray spectra for each element. He also is the first to refract x rays with a prism.

1924 Otto Stern (1888-1969), German-American physicist, studies molecular beams and establishes that atoms and molecules behave like tiny magnets when protons and electrons are charged.

1924 Edward Victor Appleton (1892-1965), English physicist, discovers that radio waves are reflected by an ionized layer of the atmosphere. He studies this layer and measures the distance to it by timing radio signals. It eventually becomes known as the Appleton layer.

1924 Satyendranath Bose (1894-1974), Indian physicist, publishes a paper in which he describes a new statistical method of dealing with Planck's quantum theory. It is adopted by Einstein who uses it to deduce that matter should exhibit wave properties.

1924 Hendrik Anthony Kramers (1894-1952), Dutch physicist, predicts the existence of an inelastic scattering of light that later is called the Raman effect.

1925 Robert Andrews Millikan (1868-1953), American physicist, names the radiation coming from outer space *cosmic rays*.

1925 Werner Karl Heisenberg (1901-1976), German physicist, develops a new system of mathematics called matrix mechanics which employs columns of numbers that describe all possible transitions within an atom. The solutions to the matrix correspond to the wavelengths of the hydrogen spectral lines.

1925 Wolfgang Pauli (1900-1958), Austrian physicist, announces his discovery of the exclusion principle which develops into one of the most powerful basic descriptive tools in physics. It states that two electrons with the same quantum numbers cannot occupy the same atom. This principle serves to explain the chemical properties of the elements.

1925 Patrick Maynard Stuart Blackett (1897-1974), English physicist, takes the first photographs of a nuclear reaction in progress. To achieve this he uses a Wilson cloud chamber and takes over 20,000 photographs of more than 400,000 alpha particle tracks and observes eight actual collisions of an alpha particle and a nitrogen molecule.

1925 George Eugene Uhlenbeck (1900-1988) and Samuel

Abraham Goudsmit (1902-1978), both Dutch-American physicists, consider Pauli's exclusion principle from an experimental standpoint and state that it could also be explained if the electron is considered as spinning on its own axis and able to spin clockwise and counter-clockwise.

1926 John Desmond Bernal (1901-1971), English physicist, advances x-ray crystallography by developing the Bernal chart with which he can deduce crystal structure by analyzing photographs of x-ray diffraction patterns.

1926 Erwin Schrödinger (1887-1961), Austrian physicist, publishes a paper in which he mathematically develops wave mechanics and offers what becomes known as the Schrödinger wave equation. This replaces the electron in the Bohr model of the atom with wave trains. This explanation becomes so satisfactory that it finally places Planck's quantum theory on a firm mathematical basis.

1926 Enrico Fermi (1901-1954), Italian-American physicist, introduces Fermi-Dirac statistical mechanics. This theory explains the behavior of *clouds* of electrons in a substance and also shows that gas particles obey Pauli's exclusion principle.

1926 Llewellyn Hilleth Thomas (1903-1992), English physicist, studies electron distribution in atoms and molecules and completes the development of the concept of electron spin.

1926 Max Born (1882-1970), German-British physicist, publishes his first paper on the probability interpretation of quantum mechanics. In working out the mathematical basis of quantum mechanics, he gives electron waves a probabilistic interpretation as to their behavior.

1926 Gilbert Newton Lewis (1875-1946), American chemist, first introduces the term *photon* to describe the minute, discrete energy packet of electromagnetic radiation that is essential to quantum theory.

1926 Eugene Paul Wigner (1902-1995), Hungarian-American physicist, develops the principles involved in applying group theory to quantum mechanics and evolves the concept of the symmetry in space and time that marks the behavior of subatomic particles.

1927 Clinton Joseph Davisson (1881-1958), American physicist, has a laboratory accident which eventually leads him to discover that electrons and other subatomic particles behave like waves. This confirms de Broglie's 1923 hypothesis.

1927 George Paget Thomson (1892-1975), English physicist, discovers the wave nature of electrons independently of Davisson or Laue. This also confirms de Broglie's 1923 hypothesis.

1927 Fritz Wolfgang London (1900-1954), German-American physicist, and Walter Heinrich Heitler (1904-1981), German physicist, offer a quantum treatment of the hydrogen molecule. Their work is considered a major contribution to the understanding of chemical valence.

1927 Werner Karl Heisenberg (1901-1976), German physicist, postulates his uncertainty principle, also known as the principle of indeterminacy. This states that it is impossible to determine accurately and simultaneously two variables of an electron, such as position and momentum. More generally, it states that when working with atom-sized or subatomic particles, the very act of measuring such particles significantly affects the results obtained. Philosophically, this is a troubling notion, for it calls into question much of our traditional belief in straight-forward cause and effect.

1927 Paul Adrien Maurice Dirac (1902-1984), English physicist, develops equations that unite quantum mechanics and relativity theory.

1927 Niels Henrik David Bohr (1885-1962), Danish physicist, states his principle of complementarity, saying that a phenomenon can be considered in each of two mutually exclusive ways, and each way is valid in its own terms. He applies it specifically to the simultaneous wave and particle behavior of an electron, but it later is used by disciplines besides atomic physics.

1928 Chandrasekhara Venkata Raman (1888-1970), Indian physicist, discovers what becomes known as the Raman effect. While investigating light scattering, he finds that a change in the wavelength of light occurs when a beam of light is deflected by molecules. With Raman scattering, it is possible to determine the identity of the molecules, since the exact wavelengths produced depend on the nature of the molecules causing the scattering.

1928 George Gamow (1904-1968), Russian-American physicist, develops the quantum theory of radioactivity which is the first theory to successfully explain the behavior of radioactive elements, some of which decay in seconds and others after thousands of years.

1928 Hermann Weyl (1885-1955), German mathematician, publishes his *Gruppentheorie und Quatenmechanik* in which he shows that most of the regularities of quantum phenomena on the atomic level can be most simply understood using group theory. His book helps mold modern quantum theory.

1928 Eugene Paul Wigner (1902-1995), Hungarian-American physicist, works out the theory of conservation of parity, stating that an atom can fluctuate between two states but can only be in one at a time.

1929 Walther Wilhelm Georg Franz Bothe (1891-1957), German physicist, invents *coincidence counting* by using two Geiger counters to detect the vertical direction of cosmic rays. This allows the measurement of extremely short time intervals, and he uses this technique to demonstrate that the laws of conservation and momentum are also valid for subatomic particles.

1929 John Douglas Cockcroft (1897-1967), English physicist, and Irish physicist Ernest Thomas Sinton Walton (1903-1995) devise the first particle accelerator which produces proton beam energies up to 600,000 volts, which then bombard a variety of targets.

1929 Paul Adrien Maurice Dirac (1902-1984), English physicist, predicts the existence of antielectrons and antimatter.

1931 Wolfgang Pauli (1900-1958), Austrian-American physicist, posits the existence of an uncharged particle with little or no mass that is carried away from an atomic nucleus when a beta particle is emitted. His prediction is proved correct in 1955 when such a particle is finally detected by Reines and Cowan. The uncharged particle is named *neutrino* in 1932 by the Italian-American physicist Enrico Fermi (1901-1954).

1931 Ernest Orlando Lawrence (1901-1958), American physicist, builds the first cyclotron. He invents a device that accelerates protons, electrons, and other subatomic particles spinning them in spirals and boosting them each time around. This first device is small, but Lawrence builds successively larger ones and gives physicists a major tool with which many advances are made.

1931 Paul Adrien Maurice Dirac (1902-1984), English physicist, proposes the antielectron, or positron, a positively charged electron.

1932 Carl David Anderson (1905-1991), American physicist, discovers the positron. This is the positively charged electron that Dirac's mathematics had earlier suggested, and it is the first form of antimatter to be discovered.

1932 Werner Karl Heisenberg (1901-1976), German physicist, wins the Nobel Prize in physics for the creation of quantum mechanics, whose application has led to the discovery of the allotropic forms of hydrogen.

1932 James Chadwick (1891-1974), English physicist, announces his discovery of the neutral particle (with no charge) he calls the neutron. It is this discovery that offers atomic scientists an efficient "bullet" to initiate nuclear reactions. It works well because it does not have a charge like protons and electrons and so is not repelled by either the nucleus or the orbital electrons in an atom.

1932   John Douglas Cockcroft (1897-1967), English physicist, and Irish physicist Ernest Thomas Sinton Walton (1903-1995) use their new particle accelerator to bombard lithium and produce two alpha particles (having combined lithium and hydrogen to produce helium). This is the first nuclear reaction that has been brought about through the use of artificially accelerated particles and without the use of any form of natural radioactivity. This proves highly significant to the creation of an atomic bomb.

1932   Eugene Paul Wigner (1902-1995), Hungarian-American physicist, introduces the concept of invariance under time reversal into quantum mechanics.

1933   Enrico Fermi (1901-1954), Italian-American physicist, develops a theory of beta decay which uses Pauli's *neutrino* as part of its explanation.

1934   Frédéric Joliot-Curie (1900-1958) and Irène Joliot-Curie (1897-1956), husband-and-wife team of French physicists, discover what they call *artificial radioactivity*. They bombard aluminum to produce a radioactive form of phosphorus. They soon learn that radioactivity is not confined only to heavy elements like uranium, but that any element can become radioactive if the proper isotope is prepared. For producing the first artificial radioactive element they win the Nobel Prize in chemistry the next year.

1934   Leo Szilard (1898-1964), Hungarian-American physicist, first conceives of the idea of a nuclear chain reaction (in which a neutron induces an atomic breakdown, which releases two neutrons which break down two more atoms, and so on). Although his method uses beryllium rather than uranium and would be impractical, it is correct in principle. He keeps it a secret, foreseeing its importance in making nuclear bombs, but it is soon discovered by Hahn and Strassman.

1934   Enrico Fermi (1901-1954), Italian-American physicist, bombards uranium with neutrons and obtains not only a new element, number 93 (neptunium), but also a number of other products that he is unable to identify. What he eventually discovers is that he has not only created the first synthetic element, but he has also produced the first nuclear fission reaction.

1934   Marcus Laurence Elwin Oliphant (b. 1901), Australian physicist, discovers tritium, the radioactive isotope of hydrogen. This eventually leads to hydrogen fusion and the hydrogen bomb.

1934   Pavel Alekseyevich Cherenkov (1904-1990), Russian physicist, discovers that a high-energy particle passing through a medium like water may exceed the speed of light in another medium, and when it does, it throws back a *wake* of light. This becomes known as the Cherenkov effect.

1935   Arthur Jeffrey Dempster (1886-1950), Canadian-American physicist, discovers the rare isotope uranium-235. It is this isotope that will undergo fission and be used to build an atomic bomb.

1935   Patrick Maynard Stuart Blackett (1897-1974), English physicist, demonstrates that when gamma rays pass through lead, they sometimes disappear and give rise to a positron and an electron. This is the first demonstrable case of the conversion of energy into matter, and, as such, is a confirmation of Einstein's famous $E=mc^2$ equation.

1935   Lev Davidovich Landau (1908-1968), Russian physicist, pioneers the mathematical treatment of magnetic domains. Within these small regions in substances like iron, all the atomic magnets are lined up in the same direction, yielding ferromagnetism, the strongest type of magnetism.

1935   Hideki Yukawa (1907-1981), Japanese physicist, posits a theory of nuclear forces that correctly predicts the existence of mesons, transient particles with masses between those of electrons and protons.

1936   Carl David Anderson (1905-1991), American physicist, discovers the *muon*. While studying cosmic radiation, he observes the track of a particle that is more massive than an electron but only a quarter as massive as a proton. He initially calls this new particle, which has a lifetime of only a few millionths of a second, a mesotron, but it later becomes known as a muon to distinguish it from Yukawa's meson.

1936   Eugene Paul Wigner (1902-1995), Hungarian-American physicist, proposes the theory of neutron absorption which comes into play when nuclear reactors are built.

1936   Felix Bloch (1905-1983), Swiss-American physicist, discovers a process by which the magnetic properties of nuclei can be measured. This work leads to the development of a delicate method of chemical analysis called nuclear magnetic resonance.

1937   Igor Yevgenyevich Tamm (1895-1971), Russian physicist, offers a theoretical explanation of Cherenkov radiation, the *wake*, or electromagnetic radiation, caused when electrons accelerate.

1937   Isidor Isaac Rabi (1898-1988), Austrian-American physicist, experiments with molecular beams and finds a method to measure the magnetic properties of atoms and molecules with great accuracy.

1937   Emilio Gino Segrè (1905-1989), Italian-American physicist, searches for and finds the lightest element yet discovered, element number 43. He demonstrates that a sample of molybdenum that has been bombarded with deuterons contains a small amount of the anticipated element 43 that he names technetium.

1939 Leo Szilard (1898-1964), Hungarian-American physicist, and Canadian-American physicist Walter Henry Zinn (b. 1906) confirm that fission reactions (nuclear chain reactions) can be self-sustaining using uranium.

1939 Niels Hendrik David Bohr (1885-1962), Danish physicist, proposes a liquid-drop model of the atomic nucleus and offers his theory of the mechanism of fission. His prediction that it is the uranium-235 isotope that undergoes fission is proved correct when work on an atomic bomb begins in the United States.

1939 John Ray Dunning (1907-1975), American physicist, is the first to experimentally confirm Meitner's theory on uranium fission.

1939 Richard Brooke Roberts (1910-1980), American biophysicist, discovers that uranium fission does not release all the neutrons it produces at one time. This phenomenon of *delayed neutrons* eventually proves to be an important element in the safety of nuclear reactors.

1939 Lise Meitner (1878-1968), Austrian-Swedish physicist, and Otto Robert Frisch (1904-1979), Austrian-British physicist, suggest the theory that uranium breaks into smaller atoms when bombarded. Meitner offers the term *fission* for this process.

1939 Otto Hahn (1870-1968), German physical chemist, and Fritz Strassman (1902-1980), German chemist, publish results in which they observe that fission reactions can be self-sustaining because of the chain reaction that occurs. This discovery eventually makes the construction of an atomic bomb feasible.

1940 Donald William Kerst (1911-1993), American physicist, builds the first betatron. This particle accelerator spins electrons (a cyclotron spins protons) in a circle to study high-energy particles.

1940 Georgii Nikolayevich Flerov (1913-1990), Russian physicist, discovers the spontaneous fission of heavy nuclei. He learns that uranium does not undergo fission only when bombarded with neutrons, but that it also does so naturally, or spontaneously, although extremely slowly.

1941 Peter Leonidovich Kapitza (1894-1984) publishes his research on liquid helium, describing the point at which it becomes a superfluid. He coins the term *superfluidity* to describe its incredible conductivity.

1942 Enrico Fermi (1901-1954), Italian-American physicist, heads a Manhattan Project team at the University of Chicago that produces the first controlled chain reaction in an atomic pile of uranium and graphite. With this first self-sustaining chain reaction, the atomic age begins.

1943 Marcus Laurence Elwin Oliphant (b. 1901), Australian physicist, proposes a design for a new and very powerful accelerator called a proton synchrotron. These eventually are built and are capable of accelerating protons to fantastic energies.

1943 J. Robert Oppenheimer (1904-1967), American physicist, is placed in charge of United States atomic bomb production at Los Alamos, New Mexico. He supervises the work of 4,500 scientists and oversees the successful design construction and explosion of the bomb.

1943 First operational nuclear reactor is activated at the Oak Ridge National Laboratory in Oak Ridge, Tennessee.

1944 Sin-Itiro Tomonaga (1906-1979), Japanese physicist, works out the theoretical basis for quantum electrodynamics independently of American physicists Richard Philips Feynman (1918-1988) and Julian Seymour Schwinger (1918-1994) (who also work independently of one another). This method allows the behavior of electrons to be determined with far greater precision than before. It also leads to the formulation of quantum electrodynamics.

1945 Vladimir Iosifovich Veksler (1907-1966), Ukrainian physicist, suggests a design for a cyclotron that will allow for relativistic changes in the mass of accelerating particles and therefore achieve greater energies. His ideas are later used to build the synchrocyclotron.

1945 First atomic bomb is detonated by the United States near Almagordo, New Mexico. The experimental bomb generates an explosive power equivalent to 15-20 thousand tons of TNT.

1945 United States destroys the Japanese city of Hiroshima with a nuclear fission bomb based on uranium-235 on August 6. Three days later a plutonium-based bomb destroys the city of Nagasaki. Japan surrenders on August 14 and World War II ends. This is the first use of nuclear power as a weapon.

1946 Abraham Pais (b. 1918), Dutch physicist, and Christian Møller (1904-1980), Danish physicist, conduct research on high energy particles and suggest the name *lepton* for lightweight particles not affected by the strong force.

1946 First synchrocyclotron is built at the University of California at Berkeley.

1946 George Dixon Rochester (b. 1908) and Clifford Charles Butler (b. 1922), English physicists, discover *V particles* in a cloud chamber. The role of these new elementary particles is not understood until Gell-Mann's concept of *strangeness* is introduced.

1947 Cecil Frank Powell (1903-1969), English physicist, discovers the pi-meson, or pion, a subatomic particle that decays very quickly but carries a strong force.

1947 Willis Eugene Lamb, Jr. (b. 1913), American physicist, applies a new method to measure the spectrum lines of hydrogen and finds them different from what was predicted. He finds a split in energy levels, called the Lamb shift, that contributes to the development of quantum electrodynamics (QED).

1947 Dennis Gabor (1900-1979), Hungarian-British physicist and electrical engineer, first conceives of the idea of holography (lens-less, three-dimensional photography) and begins to develop its basic techniques. Because available light sources are too weak or too diffuse, he must wait until the 1960 invention of the laser to make his idea commercially feasible.

1948 Maria Goeppert Mayer (1906-1972), German-American physicist, suggests that the atomic nucleus consists of protons and neutrons arranged in shells. This makes it possible to explain why some nuclei are more stable than others, as well as why some elements are rich in isotopes. This concept of nuclear shells is advanced independently at the same time by the German physicist Johannes Hans Daniel Jensen (1907-1973).

1948 Cesare Mansueto Giulio Lattes (b. 1924), Brazilian physicist, and Eugene Gardner, American physicist, confirm the existence of heavy and light mesons and detect the first artificially produced pion.

1950 Alfred Kastler (1902-1984), German-French physicist, develops a system of *optical pumping* in which atoms are excited by light or radio waves allowing him to better study atomic structure. It eventually leads to the development of masers and lasers.

1950 Leo James Rainwater (1917-1986), American physicist, first suggests the idea that the nucleus of an atom is not spherical but rather spheroidal.

1952 Edward Teller (b. 1908), Hungarian-American physicist, develops a device that makes the development of the hydrogen-fusion bomb (H bomb) practical.

1952 Donald Arthur Glaser (b. 1926), American physicist, constructs the first bubble chamber. This new method uses liquid hydrogen and proves far more sensitive than cloud chambers.

1952 First thermo-nuclear device is exploded successfully by the United States at the Eniwetok Atoll in the South Pacific. This hydrogen-fusion bomb (H bomb) is the first such bomb to work by nuclear fusion and is considerably more powerful than the atomic bomb exploded over Hiroshima on August 6, 1945.

1953 Murray Gell-Mann (b. 1929), American physicist, suggests that basic particles contain an intrinsic property known as *strangeness* that can explain a number of new observations being made about them. A similar concept is developed independently by the Japanese physicist Kazuhiko Nishijima (b. 1926).

1954 Luis Walter Alvarez (1911-1988), American physicist, develops the liquid hydrogen bubble chamber in which subatomic particles and their reactions are detected.

1955 Emilio Gino Segrè (1905-1989), Italian-American physicist, and American physicist Owen Chamberlain (b. 1920) are the first to detect antiprotons by bombarding copper atoms with protons.

1955 Frederick Reines (1918-1998) and Clyde Lorrain Cowan (1919-1974), American physicists, detect and observe neutrinos for the first time.

1956 Nicolaas Bloembergen (1920-1981), Dutch-American physicist, designs the first continuous maser. This is a solid-state maser that can steadily produce energy.

1956 Tsung-Dao Lee (b. 1926) and Chen Ning Yang (b. 1922), both Chinese-American physicists, conclude that the conservation of parity rule does not hold for weak interactions. They then work to design an experiment that will prove this conclusively.

1957 Leo Esaki (b. 1925), Japanese physicist, works in solid-state physics and discovers the *tunnel effect*. He finds that there are times when resistance decreases with current intensity instead of increasing as was expected. This is caused by the barrier-crossing ability of electrons.

1957 Chien-Shiung Wu (1912-1997), Chinese physicist, announces that her experiments demonstrate that the principle of parity is flawed. She bases her studies on the theoretical work of two Chinese-American physicists, Chen Ning Yang (b. 1922) and Tsung-Dao Lee (b. 1926). The principle of parity states that, on a nuclear level, an object and its mirror image will behave the same way. Disproving this principle eventually allows new and more correct views about neutrinos to emerge.

1960 Rudolf Ludwig Mössbauer (b. 1929), German physicist, discovers how to make gamma rays with an extremely narrow wavelength which can then be used to measure very accurately any wavelength change. This becomes known as the Mössbauer effect. He uses it to verify Einstein's relativity theory more convincingly than ever.

1960 Theodore Harold Maiman (b. 1927), American physicist, develops the first laser. He uses a ruby cylinder that emits a light that is coherent (all in a single direction) and monochromatic (a single wavelength). He finds that it can travel thousands of miles as a beam without dispersing, and that it can be concentrated into a small, super-hot spot. Laser is an acronym for *Light Amplification by Stimulated Emission of Radiation*.

1961 Murray Gell-Mann (b. 1929), American physicist, and Israeli physicist Yuval Ne'Eman (b. 1925) independently introduce a new way to classify heavy subatomic particles. Gell-Mann names it the *eightfold* way, and this system accomplishes for elementary particles what the periodic table did for the elements.

1961 Robert Hofstadter (1915-1990), American physicist, announces his discovery that protons and neutrons are made up of a central core of positively charged matter, around which are two shells of mesonic material.

1964 James Watson Cronin (b. 1931) and Val Logsdon Fitch (b. 1923), American physicists, show that the conservation rule of charge conjugation does not hold up, and they replace it with a still more general conservation law. This implies that time reversal does not always exactly reverse events on the subatomic level.

1964 Murray Gell-Mann (b. 1929), American physicist, first introduces the concept of *quarks* as he postulates the existence of unusual particles that carry fractional electric charges.

1967 V. M. Lobashov demonstrates that the strong nuclear force violates parity conservation.

1973 H. David Politzer (b. 1949) discovers independently of David Gross and Frank Anthony Wilczek (b. 1951) that quarks ignore each other when touching and have increased attraction to each other the farther apart they are.

1974 Howard M. Georgi (b. 1947) and Sheldon Lee Glashow (b. 1932) develop the first grand unified theory. This accounts for the strong, weak, and electromagnetic forces as parts of a single force that broke apart when the universe cooled after the big bang.

1980 Heinrich Rohrer (b. 1933) and Gerd Karl Binnig (b. 1947) invent the scanning tunneling microscope. This uses electron tunneling from a sharp point that is moved over a surface to detect the distance of the point from that surface. It can provide pictures of individual atoms.

1986 K. Alex Muller (b. 1927) and J. Georg Bednorz (b. 1950) discover an oxide combination that, at 30 degrees kelvin (or 30 degrees above absolute zero), is the highest known temperature for superconductivity yet reached.

1986 Teams of physicists in the United States and Germany independently observe individual quantum jumps in individual atoms for the first time.

1991 A new material called Nitromag is discovered by a team of scientists in Ireland. Belonging to the generation of permanent magnetic alloys, its composition is such that magnets made of this material remain useful at much higher temperatures, allowing them to be used in severe environments.

1993 Researchers state that the results from new observatories indicate that the stars, dust, and other observable matter in space represent less than 10% of all the mass in the universe. These results support a long-held belief that the universe holds a great deal of undetected *dark matter*.

1994 Scientists predict the existence of quark-gluon plasma, a form of matter created when extreme densities or temperatures cause the breakdown of mesons and nucleons into their original constituents.

1994 Scientists discover that photons can travel through a mirror at 1.7 times the speed of light. This phenomenon is explained by the fact that tunneling reshapes the photons' wave packets.

1994 Researchers at Fermilab discover the top quark. Scientists believe that this discovery may provide clues about the genesis of matter.

1995 Scientists create a Bose-Einstein condensate by cooling rubidium-87 to 170 nanokelvins. Predicted in 1924 by Satyendra Nath Bose (1894-1974) and Albert Einstein, a Bose-Einstein condensate is a form of matter in which atoms condense to the point of falling into one quantum state. Essentially, the atoms of a Bose-Einstein condensate behave like one atom.

1996 Quark-gluon plasma is created as a result of collisions between lead nuclei. Quark-gluon plasma is a form of matter which was created in the first few microseconds following the big bang, when the strong nuclear force was not strong enough to hold quarks together in protons. At that early stage of the universe, free quarks existed in a "broth" of gluons, particles which convey the strong force.

1997 Researchers find evidence for the existence of an exotic meson which combines quarks and gluons. A meson consists of a quark-antiquark pair. Quarks are held together by gluons, which convey the strong force. The discovery of this meson confirms the standard particle and force theory.

1998 Scientists publish convincing evidence that neutrinos have mass. If neutrinos have mass, physicists will need to formulate a new, more general, paradigm of particle physics.

1999 Todd Ditmire and co-workers create nuclear fusion at Lawrence Livermore National Laboratory by heating clusters of deuterium atoms. At an extremely high temperature, the atom clusters explode and create helium ions. However, the commercial use of fusion is still several decades away.

2000 Cass Sackett and co-workers at the National Institute of Standards and Technology in Boulder, Colorado, force four beryllium ions into a harmonic state called quantum entanglement. If particles are entangled, changes in one particle will affect the others. Researchers believe that laboratory-induced entanglement may pave the way for the creation of extremely powerful computers based on quantum mechanics.

2000 Scientists discover evidence that some dark matter may be located in intergalactic space. Clouds of hydrogen between galaxies are hard to detect because the hydrogen atoms are stripped of their electrons and cannot radiate light.

2000 Astronomers discover a galaxy that is 18.6 billion light years away from Earth. This is the most remote object ever observed. Scientists speculate that this galaxy was formed when the universe was one-sixteenth its present age.

2000 Physicists make definitive discovery of the tau particle.

A

AB effect, 8
Abbe, Ernst, 125
Abelian forces, 589–590
Absolute magnitude, 699
Absolute zero, **1**
    Bardeen's research on, 42–44
    Bednorz's research on, 48
    Bose-Einstein condensation and, 73–74
    Charles's law and, 106
    Cohen-Tannoudji's research on, 115–116
    condensed matter physics and, 128–129
    conservation of heat and, 131–133
    cosmological heat death and, 140
    electrical conductivity and, 193
    electricity and, 196
    entropy and, 224–225
    Kamerlingh Omnes's research on, 378–379
    Kelvin temperature scale and, 385
    low-temperature physics and, 427–428
    specific heat capacity and, 690
Absorption
    polarization of light and, 578
    spectroscopy and, 691–692
Accelerated reference frames, **1–2,** 623
Acceleration, **2–3**
    ballistics and, 41
    celestial mechanics and, 95–96
    centrifugal/centripetal force and, 97–98
    force and, 251
    mass and force and, **251–253**
    relativity theory and, 186
    special relativity and, 632
    units of, 253
Accelerators, **3–5**
    antiparticle research and, 20
    applications, 5
    circular, 4
    Cockcroft's research on, 115

    cyclotron modifications, 4–5
    linear, 3–4
Accretion disk
    black hole research and, 62
    event horizon and, 231–232
Acoustics, **5–8.** *See also* Sound
    air column vibrations, 6
    application of, 8
    Chladni's and, 111–112
    coherence and, 116–117
    natural sound frequencies, 492–493
    Raman's research on, 617–619
    Savart's research on, 655
    sonic spectrum and, 683–684
    sound production, 5–7
    sound reception, 7–8
    sound transmission, 7
    string vibrations, 5–6
Active galaxies, 271–272
Addition of velocity. *See* Velocity
Additive mixtures, color properties and, 121
Adiabatic demagnetization, 147
AES. *See* Flame analysis
AFM. *See* Atomic force microscopes
Aharonov, Yakir, **8**
Aharonov-Bohm effect, 8
Air column vibrations, 6
Air, fluid dynamics and, 249–250
Airglow, astronomical seeing and, 666
Al Hazan, Ibn Al-Haitham, **9**
    color experiments of, 120
    on light, 419
Alpha decay, **9–10**
    alpha radiation principles and, 11
    nuclear reactions and, 519
    radioactive decay as, 610
Alpha particles, **10–11**
    alpha decay and, 9–10
    atomic structure and, 33
    Bragg's research on, 77

R

S